In memory of my mother, Jeanene, and my dad, Bill.

—AST

To my wife, Carol,

with love and appreciation.

—RDG

About the Authors

Alan S. Tussy

Alan Tussy teaches all levels of developmental mathematics at Citrus College in Glendora, California. He has written nine math books—a paperback series and a hard-cover series. A meticulous, creative, and visionary teacher who maintains a keen focus on his students' greatest challenges, Alan Tussy is an extraordinary author, dedicated to his students' success. Alan received his Bachelor of Science degree in Mathematics from the University of Redlands and his Master of Science degree in Applied Mathematics from California State University, Los Angeles. He has taught up and down the curriculum from prealgebra to differential equations. He is currently focusing on the developmental math courses. Professor Tussy is a member of the American Mathematical Association of Two-Year Colleges.

R. David Gustafson

R. David Gustafson is Professor Emeritus of Mathematics at Rock Valley College in Illinois and coauthor of several best-selling math texts, including Gustafson/Frisk's *Beginning Algebra, Intermediate Algebra, Beginning and Intermediate Algebra: A Combined Approach, College Algebra,* and the Tussy/Gustafson developmental mathematics series. His numerous professional honors include Rock Valley Teacher of the Year and Rockford's Outstanding Educator of the Year. He earned a Master of Arts from Rockford College in Illinois, as well as a Master of Science from Northern Illinois University.

Elementary and Intermediate Algebra

Custom Edition for Lane Community College

Alan S. Tussy | R. David Gustafson

CENGAGE Learning®

Australia • Brazil • Japan • Korea • Mexico • Singapore • Spain • United Kingdom • United States

CENGAGE Learning

Elementary and Intermediate Algebra
Custom Edition for Lane Community College

Senior Project Development Manager:
Linda deStefano

Market Development Manager:
Heather Kramer

Senior Production/Manufacturing Manager:
Donna M. Brown

Production Editorial Manager:
Kim Fry

Sr. Rights Acquisition Account Manager:
Todd Osborne

Printed in the United States of America

Elementary and Intermediate Algebra, Fifth Edition
Alan S. Tussy | R. David Gustafson

Library of Congress Control Number: 2010939897

ISBN-13: 978-1-285-90926-4

ISBN-10: 1-285-90926-7

Cengage Learning
5191 Natorp Boulevard
Mason, Ohio 45040
USA

Cengage Learning is a leading provider of customized learning solutions with office locations around the globe, including Singapore, the United Kingdom, Australia, Mexico, Brazil, and Japan. Locate your local office at: **international.cengage.com/region.**

Cengage Learning products are represented in Canada by Nelson Education, Ltd.
For your lifelong learning solutions, visit **www.cengage.com/custom.**
Visit our corporate website at **www.cengage.com.**

Brief Contents

Contents

CHAPTER 1 ▶ An Introduction to Algebra 1

©Caroline K. Smith, M.D./Shutterstock.com

CHAPTER 2 ▶ Equations, Inequalities, and Problem Solving 101

©barang/Shutterstock.com

CHAPTER 3 Graphing Linear Equations and Inequalities in Two Variables; Functions 185

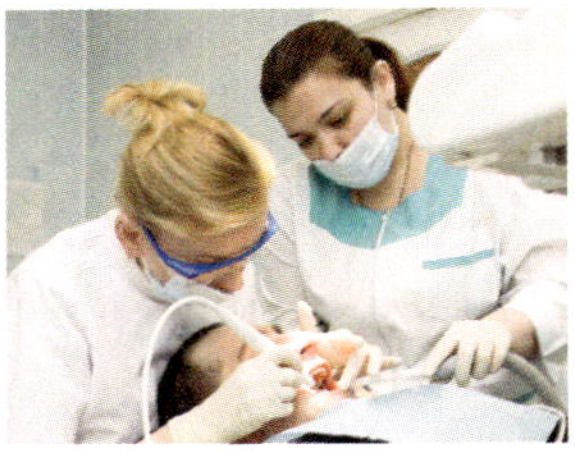

CHAPTER 4 Systems of Linear Equations and Inequalities 283

CHAPTER 5 Exponents and Polynomials 347

CHAPTER 6 Factoring and Quadratic Equations 431

©iStockPhoto.com/Neustockimages

CHAPTER 7 Rational Expressions and Equations 511

©BananaStock/Superstock

CHAPTER 8 Transition to Intermediate Algebra 599

©Ronen/Shutterstock.com

CHAPTER 9 Radical Expressions and Equations 727

CHAPTER 10 Quadratic Equations, Functions, and Inequalities 825

CHAPTER 11 Exponential and Logarithmic Functions 899

PREFACE

We are excited to present the Fifth Edition of *Elementary and Intermediate Algebra*. We believe the revision process has produced an even stronger instructional experience for both students and teachers. First, we have fine-tuned several of the popular features of our series, including the *Strategy* and *Why* example structure, the problem-solving strategy, and the online homework.

Second, we have introduced some new features to further promote student understanding and success. The new *Look Alikes* problems in the *Study Sets* will help students improve their problem recognition. The *Are You Ready?* exercises that begin each section give students the opportunity to review necessary prerequisite skills before they are asked to apply them in the study of new topics. Furthermore, the additional *Campus to Careers* problems inserted in the *Study Sets* expand on the real-life applications of the chapter material.

We want to thank all of you throughout the country who have provided suggestions and input about the previous edition. Your insight has proven invaluable. Throughout this process, our fundamental belief has remained the same: Algebra is a language in its own right. And, as always, the prime objective of this textbook is to teach students how to read, write, speak, and think using the language of algebra.

New to the Fifth Edition

Sections That Begin with Review: *Are You Ready?*

Each section begins with a set of *Are You Ready?* problems. These problems review crucial prerequisite skills that students need to have mastered in order to be successful with the new topics of that section.

ARE YOU READY?

The following problems review some basic skills that are needed to find the slope of a line.

1. Evaluate: $\frac{4-1}{8-3}$

2. Evaluate: $\frac{-10-1}{-4-(-4)}$

3. Multiply: $-\frac{7}{9}\cdot\frac{9}{7}$

4. Simplify: $\frac{15}{18}$

Study Sets with More Problem-Recognition Practice: *Look Alikes*

After a poorer than expected performance on a test, students often tell their instructors, "I could do the homework each night, but when it comes to the test, I get confused." The new *Look Alikes* feature builds students' problem-recognition skills. It requires students to distinguish between similar looking problem types and then to select the correct strategy to solve the problem. Encountering such situations in the homework assignments will better prepare students for quizzes and tests.

Look Alikes . . .

Perform the indicated operations to simplify each expression, if possible.

105. a. $(x-2)+(x^2+2x+4)$ b. $(x-2)(x^2+2x+4)$

106. a. $(a+3)+(a^2-3a+9)$ b. $(a+3)(a^2-3a+9)$

107. a. $(6x^2z^5)-(-3xz^3)$ b. $(6x^2z^5)(-3xz^3)$

108. a. $(-5r^4t^2)-(2r^2t)$ b. $(-5r^4t^2)(2r^2t)$

109. a. $(2x^2-x)-(3x^2-3x)$ b. $(2x^2-x)(3x^2-3x)$

110. a. $(4.9a-b)-(2a+b)$ b. $(4.9a-b)(2a+b)$

And Even More Problem-Recognition Practice: *Try It Yourself*

Designed to promote problem recognition, instructors and reviewers requested more *Try It Yourself* problems for the Fifth Edition. These problem types are thoroughly mixed, giving students an opportunity to practice decision making and strategy selection as they would when taking a test or quiz. With more than 80% more problems added, students will have even more opportunities to practice this essential skill.

TRY IT YOURSELF

Graph each equation. Solve for y first, when necessary.

53. $y = x$
54. $y = 4x$
55. $y = -x - 1$
56. $y = -x + 2$
57. $3y = 12x + 15$
58. $5y = 20x - 30$
59. $y = \frac{3}{8}x - 6$
60. $y = -\frac{3}{2}x + 2$
61. $y = 1.5x - 4$
62. $y = 0.5x + 3$
63. $8x + 4y = 16$
64. $14x + 7y = 28$
65. $y = -\frac{1}{2}x$
66. $y = \frac{3}{4}x$
67. $y = \frac{5}{6}x - 5$
68. $y = \frac{2}{3}x - 2$
69. $-6y = 30x + 12$
70. $-3y = 9x - 15$

Comprehensive Test Preparation: *Chapter Tests*

Instructors often assign an end-of-chapter test for students to use as a means to study for a classroom exam. However, after taking the exam, students often remark that the classroom exam included problem types that weren't in the *Chapter Test.* To address this issue, we have made sure that each *Chapter Test* is a comprehensive collection of problems that covers *all* of the topics discussed in the chapter. As a result, the *Chapter Tests* are lengthy. If students have time to complete a *Chapter Test,* that would be optimal. If, because of time constraints, they are unable to do so, the instructor can assign an appropriate subset of problems that reflects the types of problems that students can see on the exam. This should alleviate the discrepancy between what students practice and what they will see on the test.

2 CHAPTER TEST

1. Fill in the blanks.
   a. To ______ an equation means to find all of the values of the variable that make the equation true.
   b. ______ means parts per one hundred.
   c. The distance around a circle is called its ______.
   d. An ______ is a statement that contains one of the symbols >, ≥, <, or ≤.
   e. The ______ property of ______ says that multiplying both sides of an equation by the same nonzero number does not change its solution.
2. Use a check to determine if 3 is a solution of $5y + 2 = 12$.

Solve each equation.

3. $3h + 2 = 8$
4. $-22 = -x$
5. $\frac{4}{5}t = -4$
6. $\frac{11b - 11}{5} = \frac{3b - 2}{2}$
7. $0.8(x - 1{,}000) + 1.3 = 2.9 + 0.2x$
8. $2(y - 7) - 3y = -(y - 3) - 17$
9. $\frac{m}{2} - \frac{1}{3} = \frac{1}{4} + \frac{m}{6}$
10. $\frac{3}{4}(6n - 2) = 246$
11. $5x = 0$

Solve for the specified variable.

23. $V = \pi r^2 h$ for h
24. $A = P + Prt$ for r
25. $A = \frac{a + b + c + d}{4}$ for c
26. $2x - 3y = 9$ for y
27. **Irons.** Estimate the area of the soleplate of the iron.

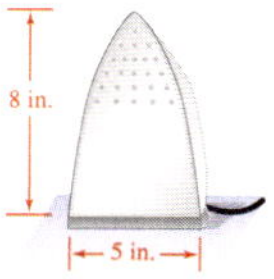

28. **Television.** In a typical 30-minute block of time on TV, the number of programming minutes are 2 less than three times the number of minutes of commercials. How many minutes of programming and how many minutes of commercials are there?

Additional Relevant, Motivating Applications in the *Examples* and *Study Sets*

We have included many new applied examples and problems that involve relevant topics such as our environment and sustainability issues, energy savings, technology and social media, and recycling. For a complete list of topics, see the *Index of Applications* following the *Preface*. To see a sampling of new topics added to each chapter, see *Content Changes by Chapter* in this *Preface.*

EXAMPLE 6 **Water Management.** Lake Mead, on the Nevada-Arizona border, is the largest reservoir in the United States. It is formed by Hoover Dam across the Colorado River. In 2000, the water level in Lake Mead was 89 feet above drought level. By 2010, the water level was 27 feet below drought level. Find the change in the water level over that time span. (Source: Bureau of Reclamation)

©iStockphoto.com/MvH

Strategy We can represent a water level above drought level using a positive number and a water level below drought level using a negative number. To find the change in the water level, we will subtract.

Why In general, *to find the change in a quantity, we subtract the earlier value from the later value.*

Solution $-27 - 89 = -27 + (-89)$ The earlier water level in 2000 (89 ft) is subtracted.

$= -11$

The negative result indic

Self Check 6 Find th normal

Now Try Problem

APPLICATIONS

73. **Social Networks.** Refer to the graph below. In what month did Facebook and MySpace have the same number of unique visitors? Estimate the number.

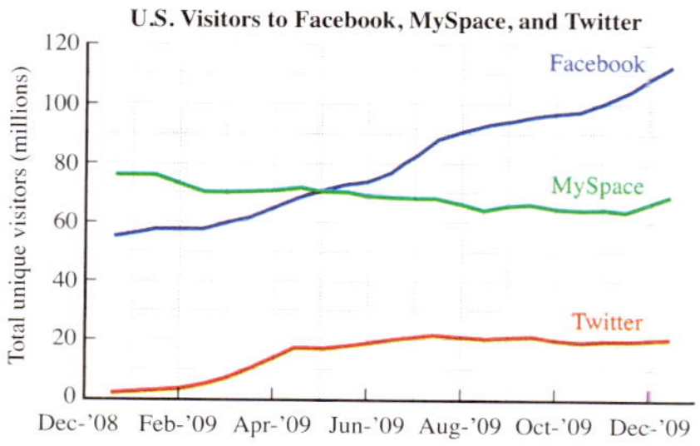

Source: Comscore Media Metrix

A More Precise Problem-Solving Strategy

In an effort to better describe the problem-solving strategy used in this book, we have inserted a new second step in what was formerly a five-step process. This additional step (*Assign a variable to represent an unknown value in the problem.*) better delineates the thought process students should use as they solve application problems. The six steps of the problem-solving strategy are now: *Analyze the problem*, *Assign a variable*, *Form an equation*, *Solve the equation*, *State the conclusion*, and *Check the result*.

EXAMPLE 11

©Christian M/Shutterstock.com

The Language of Algebra

Some phrases that suggest an inequality are:

surpass: >   at least: ≥
not exceed: ≤   at most: ≤
is no less than: ≥
is no greater than: ≤
between: < <

Grades. A student has scores of 72%, 74%, and 78% on three exams. What percent score does he need on the last exam to earn a grade of no less than B (80%)?

Analyze We know three scores. We are to find what the student must score on the last exam to earn a grade of B or higher.

Assign Let x = the score on the fourth (and last) exam.

Form To find the average grade, we add the four scores and divide by 4. To earn a grade of *no less than* B, the student's average must be *greater than or equal to* 80%.

The average of the four grades	must be no less than	80.
$\frac{72+74+78+x}{4}$	$\geq$	80

Solve

$$\frac{224+x}{4} \geq 80$$ Combine like terms in the numerator: 72 + 74 + 78 = 224.

$$4\left(\frac{224+x}{4}\right) \geq 4(80)$$ To clear the inequality of the fraction, multiply both sides by 4.

$$224 + x \geq 320$$ Simplify each side.

$$x \geq 96$$ To isolate x, undo the addition of 224 by subtracting 224 from both sides.

State To earn a B, the student must score 96% or better on the last exam.

Check Pick several exam scores that are 96% or better and verify that the student's average will be 80% or greater. For example, a score of 96% gives the student an average that is exactly 80%.

$$\frac{72+74+78+96}{4} = \frac{320}{4} = 80$$

Self Check 11 **Grades.** A student has scores of 78%, 82%, and 76% on three exams. What percent score does he need on the last test to earn a grade of no less than a B (80%)?

Now Try Problem 103

More Emphasis on "When Will I Use This?"

Each chapter now has three *Campus to Careers* problems that explore the mathematical connections to careers that are presented at the beginning of each chapter.

Systems of Linear Equations and Inequalities

©Andrei Contiu/Shutterstock.com

from Campus to Careers

Photographer

Photographers record our surroundings, the special events in our lives, and people, so that all can be remembered in pictures. Some specialize in weddings and portraits, some photograph landscapes and fashion, while others work on location as photojournalists. Their job responsibilities require a variety of mathematical skills such as: scheduling appointments, keeping financial records, pricing photographs, purchasing supplies, billing customers, and operating digital equipment.

Problem 79 in **Study Set 4.1**, problem 29 in **Study Set 4.4**, and problem 55 in **Study Set 4.5** involve situations that a photographer might encounter on the job. The mathematical concepts discussed in this chapter can be used to solve those problems.

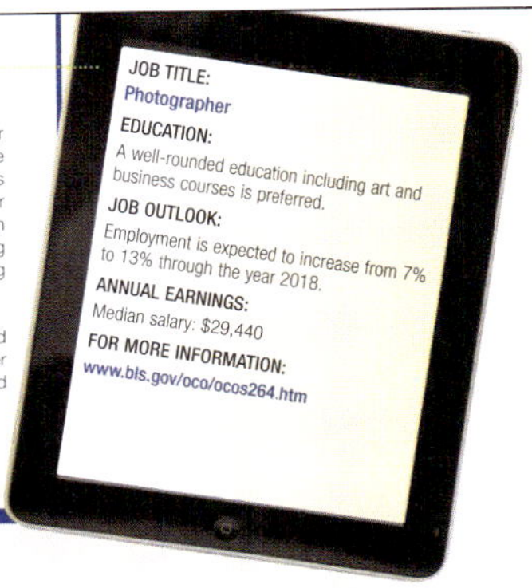

283

79. **from Campus to Careers**

Photographer

© Andrei Contiu/Shutterstock.com

Photographers often use the *rule of thirds* to add more interest to a photo rather than simply centering the subject. They imagine two horizontal and two vertical lines dividing the

55. **from Campus to Careers**

Photographer

© Andrei Contiu/Shutterstock.com

In some cameras, the image that you see in the viewfinder does not exactly match the image that will be recorded through the lens. To understand this difference, graph the solutions of the system

$$\begin{cases} y \leq \frac{1}{4}x + 2 \\ y \geq -\frac{1}{4}x + 2 \end{cases}$$

29. **from Campus to Careers**

Photographer

© Andrei Contiu/Shutterstock.com

Suppose you are a wedding photographer and you sell:

Package 1: one 10 × 14 and ten 8 × 10 color photos for $239.50

Package 2: one 10 × 14 and five 8 × 10 color photos for $134.50

A newlywed couple buys Package 1 and decides that they want one more 10 × 14 and one more 8 × 10 photograph. At the same prices, what should you charge them for each additional photograph?

Reading the Language of Algebra

Students often have difficulty reading the mathematical notation of algebra and, as a consequence, their understanding suffers. To provide assistance in this area, we have inserted notes that explain how to read newly introduced notation. Students will appreciate the ever-present "Read as ..." statements that follow algebraic symbolism that they encounter for the first time.

Product Rule for Exponents	To multiply exponential expressions that have the same base, keep the common base and add the exponents. For any number x and any natural numbers m and n, $x^m \cdot x^n = x^{m+n}$ Read as "x to the mth power times x to the nth power equals x to the m plus nth power."

Trusted Features

- **Examples That Show You How . . . and WHY:** Why? That question is often asked by students as they watch their instructors solve problems in class and as they are working on problems at home. It's not enough to know how a problem is solved. Students gain a deeper understanding of the algebraic concepts if they know why a particular approach was taken. This instructional truth was the motivation for adding a Strategy and Why explanation to each worked example.

- **Examples That Offer Immediate Feedback:** Each worked example includes a Self Check. You can complete these on your own after reading the worked example or your instructor may choose to use them as classroom lecture examples. The Self Check answers can be found in the *Answers to Selected Exercises* in the back of the text.

- **Examples That Ask You to Work Independently:** Each worked example ends with a Now Try problem. These are the final step in the learning process. Each one is linked to a similar problem found within the *Guided Practice* sections of the *Study Sets,* offering you a smooth transition into the homework.

- **Study Sets** found in each section offer a multifaceted approach to practicing and reinforcing the concepts taught in each section. They are designed for students to build their knowledge of the section concepts methodically, from basic recall to increasingly complex problem solving, through reading, writing, and thinking mathematically.

  Vocabulary—Each *Study Set* begins with the important *Vocabulary* discussed in that section. The fill-in-the-blank vocabulary problems emphasize the main concepts taught in the chapter and provide the foundation for learning and communicating the language of algebra.

  Concepts—In *Concepts,* students are asked about the specific subskills and procedures necessary to successfully complete the *Guided Practice* and *Try It Yourself* problems that follow.

  Notation—*Notation* problems review the new symbols introduced in a section. Often, students are asked to fill in steps of a sample solution. This strengthens their ability to read and write mathematics and prepares them for the *Guided Practice* problems by modeling solution formats.

  Guided Practice—The *Guided Practice* section of each *Study Set* consistently provides 1-to-1 linking for each problem type to a single worked example or objective (*i.e.*, See Example 1, See Example 2, See Example 3, and so on). Students will appreciate this 1-to-1 linking as opposed to the all-encompassing linking statements such as See Examples 2–8 or See Example 7–11 that are found in some textbooks.

  Try It Yourself—To promote problem recognition, the *Try It Yourself* problems are thoroughly mixed and are *not* linked to worked examples, giving students an opportunity to practice decision making and strategy selection as they would when taking a test or quiz.

  Applications—The *Applications* provide students the opportunity to apply their newly acquired algebraic skills to relevant and interesting real-life situations.

  Writing—The *Writing* problems help students build mathematical communication skills.

Review—The *Review* problems consist of randomly selected problems from previous chapters. These problems are designed to keep students' successfully mastered skills up-to-date before they move on to the next section.

Challenge Problems—The *Challenge Problems* provide students with an opportunity to stretch themselves and develop their skills beyond the basics. Instructors often find these to be useful as extra-credit problems.

- **Detailed Author Notes** that guide students along in a step-by-step process appear in the solutions to every worked example.
- **The Language of Algebra** boxes draw connections between mathematical terms and everyday references to reinforce the language of algebra thread that runs throughout the text.
- **The Notation, Success Tips, Caution,** and **Calculators** boxes offer helpful tips to reinforce correct mathematical notation, improve students' problem-solving abilities, warn students of potential pitfalls and increase clarity, and offer tips on using scientific calculators.
- **Chapter Tests,** at the end of every chapter, can be used as preparation for the class exam.
- **Cumulative Reviews** follow the end-of-chapter material and keep students' skills current before moving on to the next chapter. Each problem is linked to the associated section from which the problem came for ease of reference. The final *Cumulative Review* often is used by instructors as a final exam review.
- **Using Your Calculator** is an optional feature that is designed for instructors who want to use calculators as part of the instruction in this course. This feature introduces keystrokes and shows how scientific and graphing calculators can be used to solve problems. In the *Study Sets,* icons are used to denote problems that may be solved using a calculator.

Content Changes by Chapter

Based on feedback from colleagues and users of the Fourth Edition, the following changes have been made in an effort to further streamline and update the text.

Chapter 1

- New example and exercise applications include topics such as snowboarding, Lake Mead water levels, melting glaciers, calories burned doing housework, lost luggage, iPhone signal strength, U.S. Federal Budget Deficit/Surplus, and average wait time in airport security lines.
- New instructional features include a screened color 1 that is used when explaining how to build and simplify fractions, additional notes explaining how the notation is read, a worked example demonstrating uses of the commutative and associative properties, a more comprehensive *Chapter Test,* additional cautions, and an upgraded *Group Project.*

Chapter 2

- New example and exercise applications include topics such as Twitter, how tire pressure affects gas mileage, rainforest deforestation, Craigslist ads, *Harry Potter* box office revenue, calculating horsepower, iPhone apps, water usage, consignment shops, signing bonuses, and target heart rates.
- New instructional features include a comparison of linear and nonlinear equations, a new worked example of solving formulas for a specified variable, a new step added to the problem-solving strategy (*Assign a variable.*), additional cautions, additional explanation of consecutive integers, a worked example showing how to clear an inequality of fractions to solve it, and a more comprehensive *Chapter Test.*

Chapter 3

- New example and exercise applications include topics such as the Hollywood sign, René Descartes, endangered species, the U.S. Space Program, Honda Insight gas mileage, renewable energy, dental-assistant programs, firefighting, printing presses, calculating the cost to use an iPad, online games, U.S. credit card debt, the cost of raising a family, managing dental appointment times, and the amount of carbon dioxide in the Earth's atmosphere.
- New instructional features include a comparison of linear and nonlinear equations, tips for constructing a table of solutions for equations of the form $Ax + By = 0$, a comparison of one- and two-dimensional graphs, lines with slopes 1 and -1, additional cautions, a summary table of the forms of linear equations in two variables, and a more comprehensive *Chapter Test*.

Chapter 4

- New example and exercise applications include topics such as social networking, using the *rule of thirds* when taking photographs, the cost of changing CFL light bulbs, the number of women awarded Bachelor's degrees, sources of electricity, newspaper readership, greenhouse gas emissions, and lung cancer statistics.
- New instructional features include additional emphasis on what to write when both variables drop out when using the substitution method, a new step added to the problem-solving strategy (*Assign variables to the unknowns.*), additional cautions, and a comparison of one-variable and two-variable approaches to solving application problems.

Chapter 5

- New example and exercise applications include topics such as threshold hearing, supercomputers, replacing a fan belt, and diabetes diagnoses.
- New instructional features include additional notes explaining how exponential notation is read, additional cautions, a verification that the definitions of zero and negative exponents are consistent with students' previous experience with exponents, a more visible explanation showing how to multiply and divide numbers written in scientific notation, a comparison of polynomials and expressions that are not polynomials, an additional example of polynomial subtraction (vertical form), and more detailed author notes explaining polynomial long division.

Chapter 6

- New example and exercise applications include topics such as making crayons, staining a front door, Grammy nominations, antique shows, and rate of change of Wikipedia entries.
- New instructional features include a worked example in which the terms of a polynomial are rearranged to facilitate factoring by grouping, additional factoring tips, additional insight into factoring using the key number method, more information about how to recognize perfect-square trinomials, alternate factoring approaches, additional *Language of Algebra* boxes, and a more extensive list of types of quadratic equations.

Chapter 7

- New example and exercise applications include topics such as preparing an operating room, number of Tweets per ten seconds, building design, computer hard drives, and exercise equipment depreciation.
- New instructional features include a screened color 1 that is used when explaining how to build and simplify fractions, a worked example in which the common factor $x + 1$ in the numerator and $1 + x$ in the denominator are removed, and increased use of the term *rational expression* in place of the word *fraction*.

Chapter 8

- New example and exercise applications include topics such as websites, packaging, breathing capacity, airline seating, health care, earthquakes, and currency exchange.
- New instructional features include several new *Success Tips,* a more detailed discussion of identities and contradictions, increased emphasis on polynomial functions and their graphs, a more detailed explanation of finding the domain of a function, a more in-depth discussion of rational functions and their graphs, new examples and *Study Set* problems about addition, subtraction, multiplication, and division of functions, a screened color 1 is used when explaining how to build and simplify fractions, and a more comprehensive *Chapter Test.*

Chapter 9

- New example and exercise applications include topics such as matting art and beach pollution.
- New instructional features include some new *Language of Algebra* and *Caution* features, a more in-depth discussion of radical functions and their graphs, additional notes explaining how the notation is read, a screened color 1 is used when explaining how to rationalize expressions, additional worked examples and *Study Set* problems about addition, subtraction, and multiplication of radical expressions, a new worked example and *Study Set* problems about solving equations containing fractional exponents, a more detailed discussion of the complex number system, the midpoint formula has been moved from Section 8.7 to Section 9.6, and a more comprehensive *Chapter Test.*

Chapter 10

- New example and exercise applications include topics such as physics, shopping centers, women in law enforcement, crowd control, hospital emergency departments, comparing job offers, and oceanography.
- New instructional features include a new worked example and *Study Set* problems about quadratic functions, some new *Language of Algebra, Caution,* and *Success Tip* features, a new worked example and *Study Set* problems about quadratic equations with complex-number solutions, a revamped strategy for solving quadratic equations, and a more comprehensive *Chapter Test.*

Chapter 11

- New example and exercise applications include topics such as medication absorption, atmospheric carbon dioxide concentrations, book publishing, U.S. poverty rates, stocking lakes, the Richter scale, social services case loads, buying advertising time, bed bugs, cross-country skiing, weight training progress, gross domestic product of U.S. states, and human growth hormone.
- New instructional features include additional notes explaining how the notation is read, a new worked example and *Study Set* problems about evaluating sum, difference, product, quotient, and composition functions, comparison and contrast of other types of functions to exponential functions, a more detailed discussion of exponential growth and decay models and their graphs, new visuals relating exponential and logarithmic forms, a new visual explaining logarithmic function notation, a new worked example and *Study Set* problems about logarithmic growth, the former Fourth Edition Sections 11.4 and 11.6 were combined to create a new *Section 11.5: Base-e Exponential and Logarithmic Functions,* updated calculator keystrokes instructions, a revamped strategy for solving quadratic equations, more author notes and steps shown in the solutions for worked examples solving logarithmic equations, and a more comprehensive *Chapter Test.*

Chapter 12

- New example and exercise applications include topics such as footwear imports, area codes, investing, interior design, fashion design, physical therapy, hair color treatments, production planning, manufacturing, nutrition, NBA centers, television vs. internet video viewing, student loans, gourmet fruit, weight training, and railroad safety.

- New instructional features include moving the application problems to the end of the chapter in Sections 12.4 and 12.5 so that any of the methods (substitution, elimination, matrices, Cramer's rule) could be used to solve the system of equations, updated graphing calculator keystrokes instructions, a more precise definition of a set of dependent equations, reduced row-echelon form and Gauss-Jordan elimination, more *Notation* and *Language of Algebra* features, additional use of tables for problem solving, and a more comprehensive *Chapter Test.*

Chapter 13

- New example and exercise applications include topics such as civil engineering, landscaping, nuclear power, and advertising.
- New instructional features include several new *Success Tips* about determining h and k, a more in-depth discussion of the role of the center when graphing conics, a discussion of the curvature of the branches of a hyperbola, and a more comprehensive *Chapter Test.*

Chapter 14

- New example and exercise applications include topics such as union membership, manufacturing costs, and flight times.
- New instructional features include several new *Notation* and *Success Tip* features explaining notation, several new *Language of Algebra* features, a discussion of the curvature of the branches of a hyperbola, and a more comprehensive *Chapter Test.*

Student Resources

Print Ancillaries

Complete Course Notebook (1-133-36448-9)
Ann Ostberg
NEW! The *Complete Course Notebook* is your blueprint for success in your algebra course and in all your future studies! This notebook organizes and guides you through your course, offering step-by-step practice related to each section of your textbook. Each chapter contains a chapter readiness assessment, fill-in-the-blank course notes, structured note-taking guides, questions for reflection, activities, and end-of-chapter test prep. The notebook also contains general tip sheets on note-taking, studying, assessing your test scores, and much more.

Student Workbook (1-111-98783-1)
Maria H. Andersen, *Muskegon Community College*
Get a head-start. The *Student Workbook* contains assessments, activities, and worksheets for classroom discussions, in-class activities, and group work.

Student Solutions Manual (1-111-57847-8)
Alexander Lee, *Hinds Community College* and Kristy Hill, *Hinds Community College*
The *Student Solutions Manual* provides worked-out solutions to the odd-numbered problems in the text.

Enhanced WebAssign: Start Smart Guide for Students (0-495-38479-8)
If your instructor has chosen to package *Enhanced WebAssign* with your text, this manual will help you get up and running quickly with the *Enhanced WebAssign* system so you can study smarter and improve your performance in class.

Electronic Ancillaries

Enhanced WebAssign (0-538-73810-3)
Get instant feedback on your homework assignments with *Enhanced WebAssign* (assigned by your instructor). This online homework system is easy to use and includes a multimedia eBook, video examples, and problem-specific tutorials.

Website www.cengagebrain.com or www.cengage.com/math/tussy
Visit us on the web and search by title to access a wealth of learning resources, including *Study Skills Workshop* materials, tutorials, final exams, chapter outlines, chapter reviews, web links, videos, flashcards, study skills handouts, and more!

Printed Access Card for CourseMate with eBook for *Elementary and Intermediate Algebra,* Fifth Edition (1-4282-7475-8)

Instant Access Card for CourseMate with eBook for *Elementary and Intermediate Algebra,* Fifth Edition (1-4282-7474-X)
The more you study, the better the results. Make the most of your study time by accessing everything you need to succeed in one place: read your textbook, take notes, review flashcards, watch videos, and take practice quizzes—online with *CourseMate.*

Acknowledgments

We want to express our gratitude to our accuracy checkers, Diane Koenig and Steve Odrich, as well as many others for their help with this project: Maria H. Andersen, Sheila Pisa, Alexander Lee, Ed Kavanaugh, Karl Hunsicker, Cathy Gong, Dave Ryba, Terry Damron, Marion Hammond, Lin Humphrey, Doug Keebaugh, Robin Carter, Tanja Rinkel, Bob Billups, Jeff Cleveland, Jo Morrison, Sheila White, Jim McClain, Paul Swatzel, Brandon Tussy, Liz Tussy, Dan Davison, Marshall Dean, Dennis Korn, Matt Greenbeck, Joyce Low, Ralph Tippins, Mohamad Trad, and the Citrus College library staff (including Barbara Rugeley) for their help with this project. Your encouragement, suggestions, and insight have been invaluable to us.

We also would like to express our thanks to the Cengage Learning editorial, marketing, production, and design staff for helping us craft this new edition: Charlie Van Wagner, Danielle Derbenti, Gordon Lee, Carrie Jones, Jennifer Cordoba, Heleny Wong, Guanglei Zhang, Maureen Ross, Sam Subity, Jennifer Risden, Vernon Boes, Terri Wright, Ellen Brownstein, Helen Walden, Lori Heckelman, and Graphic World.

Additionally, we would like to say that authoring a textbook is a tremendous undertaking. A revision of this scale would not have been possible without the thoughtful feedback and support from the colleagues listed below. Their contributions to this edition have shaped this revision in countless ways.

Alan S. Tussy
R. David Gustafson

Advisory Reviewers

Ashish Gupta, *William Paterson University*
Katrina Keating, *Diablo Valley College*
Roger Larson, *Anoka Ramsey Community College*
Lorraine Lopez, *San Antonio College*
Paul J. Vroman, *St. Louis Community College at Florissant Valley*

Reviewers of the Fourth and Fifth Editions

Andrea Adlman, *Ventura College*
Khadija Ahmed, *Monroe Community College*
Rodney Alford, *Calhoun Community College*
Maria Andersen, *Muskegon Community College*
Hamid Attarzadeh, *Jefferson Community and Technical College*
Victoria Baker, *University of Houston–Downtown*
Betty Barks, *Lansing Community College*
Scott Barnett, *Henry Ford Community College*
Susan Beane, *University of Houston–Downtown*
David Behrman, *Somerset Community College*
Chad Bemis, *Riverside Community College*
John F. Beyers, *University of Maryland University College*
Barbara Blass, *Oakland Community College*
Candace Blazek, *Anoka Ramsey Community College*
Jennifer Bluth, *Anoka Ramsey Community College*
A. Elena Bogardus, *Camden Community College*
Carilynn Bouie, *Cuyahoga Community College*
Charles A. Bower, *St. Philip's College*
Jeanne Bowman, *University of Cincinnati*
Kim Brown, *Tarrant Community College*
Kirby Bunas, *Santa Rosa Junior College*
Shawna M. Bynum, *Napa Valley College*
Kim Caldwell, *Volunteer State Community College*
Carole Carney, *Brookdale Community College*
Edythe L. Carter, *Amarillo College*

Joe Castillo, *Broward Community College*
Sandra Chandler, *Tidewater Community College*
Carol Cheshire, *Macon State College*
John Close, *Salt Lake Community College*
Chris Copple, *Northwest State Community College*
Tony Craig, *Paradise Valley Community College*
Patrick Cross, *University of Oklahoma*
Mary Deas, *Johnson County Community College*
Suzanne Doviak, *Old Dominion University*
Archie Earl, *Norfolk State University*
Melody Eldred, *State University of New York at Cobleskill*
Peter Embalabala, *Lincoln Land Community College*
Joan Evans, *Texas Southern University*
Mike Everett, *Santa Ana College*
Betsy Farber, *Bucks County Community College*
Rita Fielder, *University of Central Arkansas*
Maggie Flint, *Northeast State*
Anissa Florence, *Jefferson Community and Technical College*
Pat Foard, *South Plains College*
Nancy Forrest, *Grand Rapids Community College*
Tom Fox, *Cleveland State Community College*
Heng Fu, *Thomas Nelson Community College*
Douglas Furman, *SUNY Ulster Community College*
Abel Gage, *Skagit Valley College*
John Garlow, *Tarrant Community College–Southeast Campus*
Vicki Gearhart, *San Antonio College*
Radu Georgescu, *Prince George's Community College*
Rebecca Giles, *Jefferson State Community College*
Alketa Gjikuria, *Cecil College*
Megan Goodwin, *Anoka Ramsey Community College*
Kim Gregor, *Delaware Technical Community College–Wilmington*
Thomas Grogan, *Cincinnati State*
Sally Haas, *Angelina College*
Paula Jean Haigis, *Calhoun Community College*
Haile Kebede Haile, *Minneapolis Community and Technical College*
Kelli Jade Hammer, *Broward Community College*
Mehdi Hakim Hashemi, *Normandale Community College*
Julia Hassett, *Oakton Community College*
Jennifer Hastings, *Northeast Mississippi Community College*
Alan Hayashi, *Oxnard College*
Kristy Hill, *Hinds Community College*
Jim Hodge, *Mountain State University*
Amy Hoherz, *Johnson County Community College*
Laura Hoye, *Trident Technical College*
Becki Huffman, *Tyler Junior College*
Jeffrey Hughes, *Hinds Community College*
Vera Hu-Hyneman, *SUNY–Suffolk Community College*
Angela Jahns, *North Idaho College*
Cassandra Johnson, *Robeson Community College*
Cynthia Johnson, *Heartland Community College*
Leslie Johnson, *John C. Calhoun State Community College*
Pete Johnson, *Eastern Connecticut State University*
Shelbra Jones, *Wake Technical Community College*
Ed Kavanaugh, *Schoolcraft College*
Leonid Khazanov, *Borough of Manhattan Community College*
MC Kim, *Suffolk County Community College*
Lynette King, *Gadsden State Community College*
Mike Kirby, *Tidewater Community College*
Alex Kolesnik, *Ventura College*
Patricia Kopf, *Kellogg Community College*
Elena Kravchuk, *University of Alabama–Birmingham*
Marlene Kutesky, *Virginia Commonwealth University*
Fred Lang, *Art Institute of Washington*
Hoat Le, *San Diego Community College*
Alexander Lee, *Hinds Community College, Rankin Campus*
Wayne (Paul) Lee, *Saint Philip's College*
Richard Leedy, *Polk Community College*
Mary Legner, *Riverside Community College*
Lamar Lider-Manuel, *Seminole Community College*
Daniel Lopez, *Brookdale Community College*
Ann Loving, *J. Sargeant Reynolds Community College*
Yixia Lu, *South Suburban College*
Keith Luoma, *Augusta State University*
Julie L. Mays, *Angelina College*
Mikal McDowell, *Cedar Valley College*
Marcus McGuff, *Austin Community College*
Owen Mertens, *Missouri State University*
Susan Meshulam, *Indiana University/Purdue University Indianapolis*
James Metz, *Kapi'olani Community College*
Trudy Meyer, *El Camino College*
Pam Miller, *Phoenix College*
Molly Misko, *Gadsden State Community College*
Catherine Moushon, *Elgin Community College*
Tania Munding, *Ohlone College*
Charlie Naffziger, *Central Oregon Community College*
Oscar Neal, *Grand Rapids Community College*
Doug Nelson, *Central Oregon Community College*
Elsie Newman, *Owens Community College*
Charlotte Newsom, *Tidewater Community College*
Katrina Nichols, *Delta College*
Randy Nichols, *Delta College*
Stephen Nicoloff, *Paradise Valley Community College*
Megan Nielsen, *St. Cloud State University*
Charles Odion, *Houston Community College*
Jason Pallett, *Longview Community College*
Mary Beth Pattengale, *Sierra College*
Naeemah Payne, *Los Angeles Community College*
Fred Peskoff, *Borough of Manhattan Community College*
Sheila Pisa, *Riverside Community College–Moreno Valley*
Carol Ann Poore, *Hinds Community College*
Jill Rafael, *Sierra College*
Pamela Reed, *North Harris Montgomery Community College*
Pamelyn Reed, *Cy-Fair College*
Nancy Ressler, *Oakton Community College*
Elaine Richards, *Eastern Michigan University*
Harriette Roadman, *New River Community College*
Lilia Ruvalcaba, *Oxnard College*
Jeffrey Saikali, *San Diego Miramar College*
Fary Sami, *Harford Community College*
Emma Sargent, *Tennessee State University*
Ned Schillow, *Lehigh Carbon Community College*
Joe Sedlacek, *Kirkwood Community College*
Wendiann Sethi, *Seton Hall University*

Debra Shafer, *University of North Carolina*
Hazel Shedd, *Hinds Community College*
Patty Sheeran, *McHenry Community College*
Karen Smith, *Nicholls State University*
Christa Solheid, *Santa Ana College*
Donald Solomon, *University of Wisconsin–Milwaukee*
Frankie Solomon, *University of Houston–Downtown*
Jim Spencer, *Santa Rosa Junior College*
John Squires, *Cleveland State Community College*
Michael Stack, *South Suburban College*
Kristen Starkey, *Rose State College*
Robin Steinberg, *Pima Community College*
Kristin Stoley, *Blinn College*
Eleanor Storey, *Front Range Community College–Westminster Campus*
Teresa Sutcliffe, *Los Angeles Valley College*
Eden Thompson, *Utah Valley State College*
Cindy Thore, *Central Piedmont Community College*
Rose Toering, *Kilian Community College*
Fariheh Towfiq, *Palomar College*
James Valade, *Monroe County Community College*
Gowribalan "Ana" Vamadeva, *University of Cincinnati*
Maggie Pasqua Viz, *Brookdale Community College*
Beverly Vredevelt, *Spokane Falls Community College*
Andreana Walker, *Calhoun Community College*
Carol Walker, *Hinds Community College*
Cynthia Wallin, *Central Virginia Community College*
John Ward, *Kentucky Community and Technical College–Jefferson Community College*
Richard Watkins, *Tidewater Community College*
Diane Williams, *Northern Kentucky University*
Antoinette Willis, *St. Philip's College*
Jackie Wing, *Angelina College*
Judith Wood, *Central Florida Community College*
Nazar Wright, *Guilford Technical Community College*
Valerie Wright, *Central Piedmont Community College*
Shishen Xie, *University of Houston–Downtown*
Catalina Yang, *Oxnard College*
Heidi Young, *Bryant and Stratton College*
Mary Young, *Brookdale Community College*
Ghidei Zedingle, *Normandale Community College*
Loris Zucca, *Kingwood College*

Index of Applications

Examples that are applications are shown with **boldface** page numbers.
Exercises that are applications are shown with lightface page numbers.

Business and Industry

Politics, Government, and Military

Recreation and Hobbies

Science and Engineering

Travel and Transportation

An Introduction to Algebra

1

from Campus to Careers

Lead Transportation Security Officer

Since 9/11, Homeland Security is one of the fastest-growing career choices in the United States. A lead transportation security officer works in an airport where he or she searches passengers, screens baggage, reviews tickets, and determines staffing requirements. The job description calls for the ability to perform arithmetic computations correctly and solve practical problems by choosing from a variety of mathematical techniques such as formulas and percentages.

Problem 113 in **Study Set 1.2, Problem 99** in **Study Set 1.5,** and **Problem 117** in **Study Set 1.7** involve situations that a lead transportation security officer might encounter on the job. The mathematical concepts discussed in this chapter can be used to solve those problems.

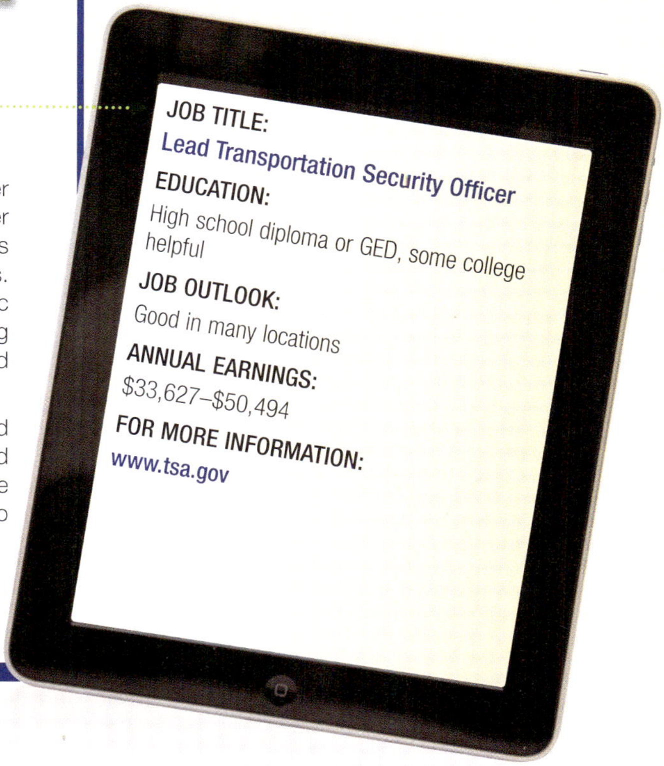

Study Skills Workshop ▶ Committing to the Course

Starting a new course is exciting, but it might also make you a bit nervous. In order to be successful in your algebra class, you need a plan.

MAKE TIME FOR THE COURSE: As a general guideline, 2 hours of independent study time is recommended for every hour in the classroom.

KNOW WHAT IS EXPECTED: Read your instructor's syllabus thoroughly. It lists class policies about attendance, homework, tests, calculators, grading, and so on.

BUILD A SUPPORT SYSTEM: Know where to go for help. Take advantage of your instructor's office hours, your school's tutorial services, the resources that accompany this textbook, and the assistance that you can get from classmates.

Now Try This

Each of the forms referred to below can be found online at: www.cengage.com/math/tussy.

1. To help organize your schedule, fill out the *Weekly Planner Form.*
2. Review the class policies by completing the *Course Information Sheet.*
3. Use the *Support System Worksheet* to build your course support system.

SECTION 1.1 Introducing the Language of Algebra

OBJECTIVES

1 Read tables and graphs.
2 Use the basic vocabulary and notation of algebra.
3 Identify expressions and equations.
4 Use equations to construct tables of data.

ARE YOU READY?

The following problems review some basic arithmetic skills that are needed in this section. Answers to the Are You Ready? *problems are located in Appendix 3 at the back of the book.*

1. Add: $125 + 85$
2. Subtract: $2{,}400 - 650$
3. Multiply: $78 \cdot 14$
4. Divide: $243 \div 27$

Algebra is the result of contributions from many cultures over thousands of years. The word *algebra* comes from the title of the book *Ihm Al-jabr wa'l muqābalah,* written by an Arabian mathematician around A.D. 800. We can think of algebra as a language with its own vocabulary and notation. In this section, we begin to explore the language of algebra by introducing some of its basic components.

1 Read Tables and Graphs.

In algebra, we often use **tables** to show relationships between quantities. For example, the table below lists the number of calories a 160-pound adult burns during 10, 20, 30, and 40 minutes of snowboarding. For a workout of, say, 30 minutes, we locate 30 in the left column and then scan across the table to see that 300 calories are burned.

©Ipatov/Shutterstock.com

Minutes snowboarding	Calories burned
10	100
20	200
30	300
40	400

The information in the table also can be presented in a **bar graph,** as shown on the next page, on the left. The **horizontal axis** of the graph is labeled "Minutes snowboarding," and it

The Language of Algebra

Horizontal is a form of the word *horizon*. Think of the sun setting over the *horizon*. **Vertical** means in an upright position. Professional basketball player LeBron James' *vertical* leap measures more than 40 inches.

is scaled in units of 10 minutes. The **vertical axis,** labeled "Calories burned," is scaled in units of 50 calories. The height of a bar indicates the number of calories burned. For example, the bar over 40 minutes extends upward to 400. This means 400 calories are burned during a 40-minute snowboarding workout.

Another way to present the snowboarding information is with a **line graph.** Instead of using a bar to represent the number of calories burned, we use a dot drawn at the correct height. After drawing the data points for workouts of 10, 20, 30, and 40 minutes, we connect them with line segments to create the graph shown below, on the right.

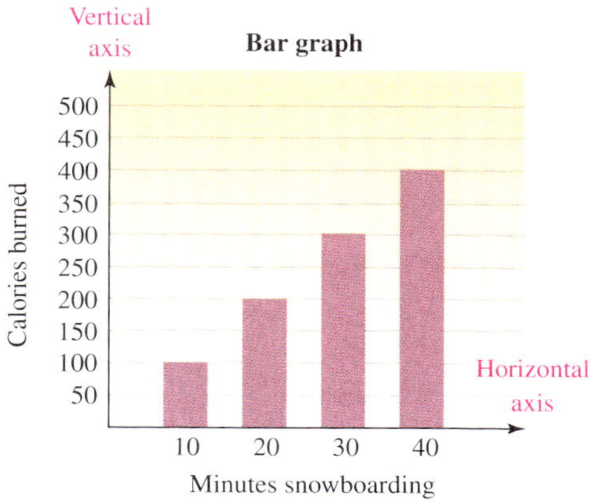

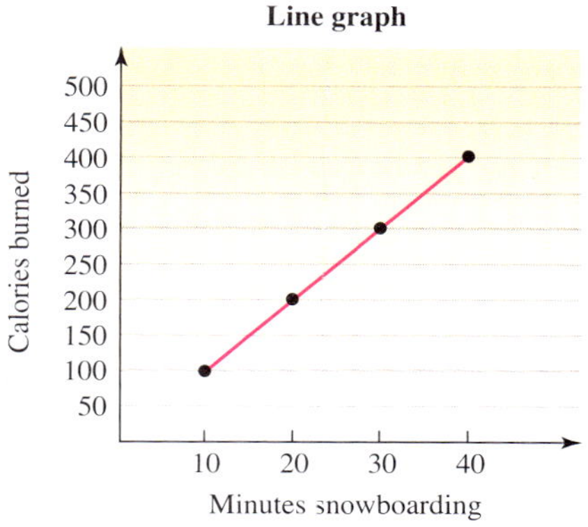

EXAMPLE 1

Fitness. Use the line graph above to find the number of calories burned during a 25-minute snowboarding workout.

Strategy We will start at 25 on the horizontal axis of the graph. Then we will scan up to the red line, and over, to read the number of calories burned on the vertical axis.

Why We start on the horizontal axis because that scale gives the number of minutes of snowboarding. We scan up and over to the vertical axis because that scale gives the number of calories burned.

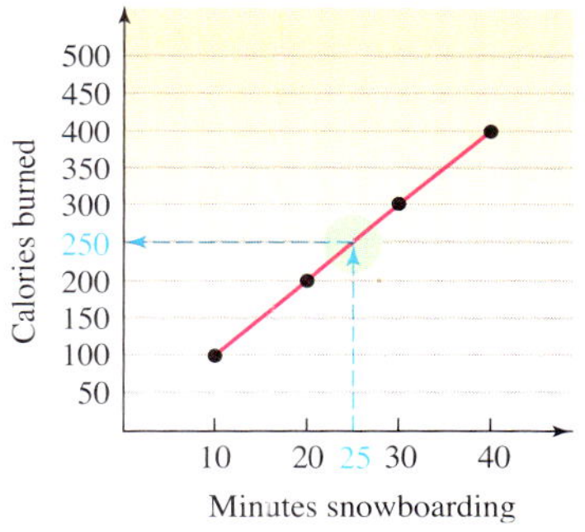

Solution We locate 25 minutes (between 20 and 30 minutes) on the horizontal axis and draw a dashed line upward to intersect the red line. From the point of intersection, we then draw a dashed line to the left that points to the vertical axis at 250. This means that a 25-minute snowboarding workout burns 250 calories.

Success Tip

Answers to the Self Check problems are given in Appendix 3, at the back of the book.

Self Check 1 **Fitness.** Use the graph to find the number of calories burned during a 35-minute snowboarding workout.

Now Try Problem 29

2 Use the Basic Vocabulary and Notation of Algebra.

From the table and graphs, we see that there is a relationship between the number of calories burned and the number of minutes snowboarding. Using words, we can express this relationship as a **verbal model:**

"The number of calories burned is ten times the number of minutes snowboarding."

Since the word **product** indicates the result of a multiplication, we can also write:

"The number of calories burned is the *product* of ten and the number of minutes snowboarding."

The Language of Algebra

The collection of symbols and write-up forms used in this course is called the **notation** of algebra.

To indicate other arithmetic operations, we will use the following words.

- A **sum** is the result of an addition: The sum of 5 and 6 is 11.
- A **difference** is the result of a subtraction: The difference of 3 and 2 is 1.
- A **quotient** is the result of a division: The quotient of 6 and 3 is 2.

Many symbols used in arithmetic are also used in algebra. For example, a + symbol is used to indicate addition, a − symbol is used to indicate subtraction, and an = symbol means *is equal to.*

Since the letter x is often used in algebra and could be confused with the multiplication symbol ×, we usually write multiplication using a **raised dot** or **parentheses.**

Symbols for Multiplication

×	Times symbol	$6 \times 4 = 24$
·	Raised dot	$6 \cdot 4 = 24$
( )	Parentheses	$(6)4 = 24$ or $6(4) = 24$ or $(6)(4) = 24$

In algebra, the symbol most often used to indicate division is the **fraction bar.**

Symbols for Division

÷	Division symbol	$24 \div 4 = 6$
$\overline{)\ \ }$	Long division	$4\overline{)24}$ with quotient 6
—	Fraction bar	$\frac{24}{4} = 6$

EXAMPLE 2 Write each statement in words, using one of the words *sum, product, difference,* or *quotient:*

a. $\frac{22}{11} = 2$ **b.** $22 + 11 = 33$

Strategy We will examine each statement to determine whether addition, subtraction, multiplication, or division is being performed.

Why The word that we should use (*sum, product, difference,* or *quotient*) depends on the arithmetic operation that we have to describe.

Solution

a. Since the fraction bar indicates division, we have: The quotient of 22 and 11 equals 2.

b. The + symbol indicates addition: The sum of 22 and 11 equals 33.

Self Check 2 Write the following statement in words: $22 \cdot 11 = 242$

Now Try Problems 33 and 35

3 Identify Expressions and Equations.

The Language of Algebra

Since the number of minutes snowboarding can *vary,* or change, it is represented using a **variable.**

Another way to describe the relationship between calories burned and snowboarding time uses *variables.* **Variables** are letters that stand for numbers. If we let the letter m represent the number of minutes snowboarding, then the number of calories burned is ten times m, written $10m$. In this notation, the number 10 is an example of a **constant** because it does not change value.

When multiplying a variable by a number, or a variable by another variable, we can omit the symbol for multiplication. For example,

$10m$ means $10 \cdot m$ $\quad$ xy means $x \cdot y$ $\quad$ $8abc$ means $8 \cdot a \cdot b \cdot c$

We call $10m$, xy, and $8abc$ *algebraic expressions.*

Algebraic Expressions Variables and/or numbers can be combined with the operations of addition, subtraction, multiplication, and division to create **algebraic expressions.**

The Language of Algebra

Algebraic expressions are often simply called **expressions.**

Here are some other examples of algebraic expressions.

$4a + 7$ This expression is a combination of the numbers 4 and 7, the variable *a*, and the operations of multiplication and addition.

$\frac{10 - y}{3}$ This expression is a combination of the numbers 10 and 3, the variable *y*, and the operations of subtraction and division.

$15mn(2m)$ This expression is a combination of the numbers 15 and 2, the variables *m* and *n*, and the operation of multiplication.

The Language of Algebra

The equal symbol = can be represented by verbs such as:

is *are* *gives* *yields*

The symbol ≠ is read as *"is not equal to."*

In the snowboarding example, if we let the letter c stand for the number of calories burned, we can translate the verbal model to mathematical symbols.

The number of calories burned	is	ten	times	the number of minutes snowboarding.
c	$=$	10	$\cdot$	m

The statement $c = 10 \cdot m$, or more simply, $c = 10m$, is called an *equation.* An **equation** is a mathematical sentence that contains an = symbol. The = symbol indicates that the expressions on either side of it have the same value. Other examples of equations are

$$3 + 5 = 8 \qquad x + 5 = 20 \qquad 17 - 2r = 14 + 3r \qquad p = 100 - d$$

EXAMPLE 3

©iStockPhoto.com/MvH

Stormy Weather. One way to estimate your distance (in miles) from a lightning strike is to count the number of seconds between the flash of lightning and the sound of thunder and divide by five. Translate this verbal model into an equation.

Strategy We will represent the two unknown quantities using variables and we will use symbols to represent the words *is* and *divided by.*

Why To translate a verbal (word) model into an equation means to write it using mathematical symbols.

Solution Let d = your distance (in miles) from the lightning strike and s = the number of seconds between the lightning and the thunder. Then we have:

Your distance (in miles) from the lightning strike	is	the number of seconds between the lightning and thunder	divided by	five.
d	$=$	s	$\div$	5

If we write the division using a fraction bar, then the verbal model translates to the equation $d = \frac{s}{5}$.

Self Check 3 Translate into an equation: The number of unsold tickets is the difference of 500 and the number of tickets that have been purchased.

Now Try Problems 41 and 45

Success Tip

Throughout this course you will be working with *expressions* and *equations.* It is important to know the difference between them. An equation contains an = symbol. An expression does not.

In the snowboarding example, we have seen that a table, a graph, and an equation can be used to describe the relationship between calories burned and workout time. The equation $c = 10m$ has one major advantage over the other methods. It can be used to accurately determine the number of calories burned during a snowboarding workout of *any* length of time.

EXAMPLE 4 **Fitness.** Use the equation $c = 10m$ to find the number of calories burned during a 36-minute snowboarding workout.

Strategy In $c = 10m$, we will replace m with 36. Then we will multiply 36 by 10 to obtain the value of c.

Why The equation $c = 10m$ indicates that the number of calories burned is found by multiplying the number of minutes snowboarding by 10.

Solution

$c = 10m$ This is the describing equation.

$c = 10(36)$ Replace m, which stands for the number of minutes snowboarding, with 36. Use parentheses to show the multiplication. We also could write $10 \cdot 36$.

$c = 360$ Do the multiplication.

A snowboarding workout of 36 minutes will burn 360 calories.

Self Check 4 **Fitness.** Use the equation $c = 10m$ to find the number of calories burned during a 48-minute snowboarding workout.

Now Try Problem 53

4 Use Equations to Construct Tables of Data.

Equations such as $c = 10m$, which express a relationship between two or more variables, are called **formulas.** Some applications require the repeated use of a formula.

EXAMPLE 5 **Fitness.** Find the number of calories burned during snowboarding workouts of 18 minutes and 65 minutes. Present the results in a table.

Strategy We need to use the formula $c = 10m$ twice.

Why There are two different workouts: one that is 18 minutes long and another that is 65 minutes long.

Solution

Step 1: We construct a two-column table and enter the workout times in the first column, as shown below in red.

The Language of Algebra

To **substitute** means to put or use in place of another, as with a *substitute* teacher. Here, we *substitute* 18 and 65 for m.

$c = 10m$

m	c
18	180
65	650

Since m represents the number of minutes snowboarding, we use it as the heading of the first column.

Since c represents the number of calories burned, we use it as the heading of the second column.

Step 2: We substitute 18 and 65 for m in $c = 10m$ and find each corresponding value of c. The results are entered in the second column of the table, as shown above.

$c = 10m$
$c = 10(18)$
$c = 180$

$c = 10m$
$c = 10(65)$
$c = 650$

Self Check 5 **Fitness.** Find the number of calories burned during snowboarding workouts of 8 minutes and 75 minutes. Present the results in a table.

Now Try Problem 55

Success Tip

Answers to the odd-numbered problems in each Study Set can be found at the back of the book in Appendix 3, beginning on page A-9.

SECTION 1.1 STUDY SET

VOCABULARY

Fill in the blanks.

1. A _____ is the result of an addition. A __________ is the result of a subtraction. A ________ is the result of a multiplication. A ________ is the result of a division.
2. __________ are letters (or symbols) that stand for numbers.
3. A number, such as 8, is called a ________ because it does not change.
4. Variables and numbers can be combined with the operations of addition, subtraction, multiplication, and division to create algebraic __________.
5. An _________ is a mathematical sentence that contains an = symbol. An algebraic __________ does not.
6. An equation such as $c = 10m$, which expresses a relationship between two or more variables, is called a ________.
7. The __________ axis of a graph extends left and right and the vertical axis extends up and down.
8. The word ________ comes from the title of a book written by an Arabian mathematician around A.D. 800.

CONCEPTS

Classify each item as an algebraic expression or an equation.

9. a. $m - 18 = 23$ b. $m + 18$
10. a. $30x$ b. $30x = 600$
11. a. $\frac{c - 7}{5}$ b. $\frac{c - 7}{5} = 7c$
12. a. $r = \frac{2}{3}$ b. $\frac{2}{3}r$
13. What arithmetic operations does the expression $\frac{12 + 9t}{25}$ contain? What variable does it contain?
14. What arithmetic operations does the equation $4y - 14 = 5(6)$ contain? What variable does it contain?
15. Construct a line graph using the data in the following table.

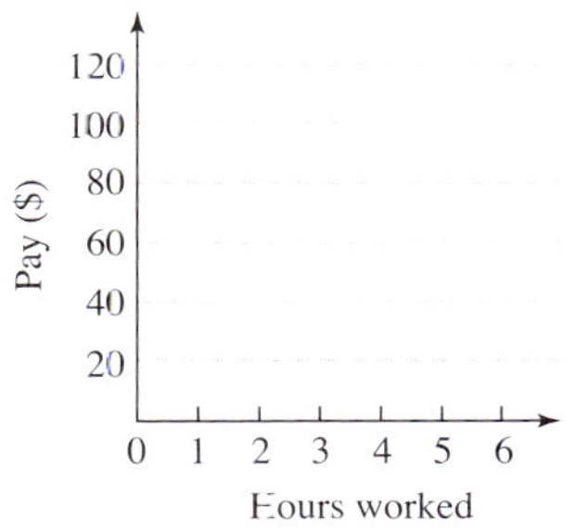

Hours worked	Pay (dollars)
1	20
2	40
3	60
4	80
5	100

16. Use the data in the graph to complete the table.

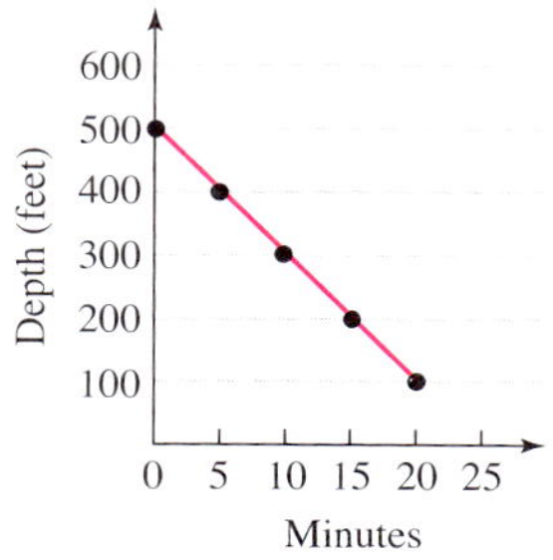

Minutes	Depth (feet)
0	
5	
10	
15	
20	

NOTATION

Fill in the blanks.

17. The symbol $\neq$ means ___ ____ ______ ___.
18. The symbols ( ) are called __________.
19. Write the multiplication 5×6 using a raised dot and then using parentheses.
20. Give four verbs that can be represented by an equal symbol =.

Write each expression without using a multiplication symbol or parentheses.

21. $4 \cdot x$
22. $P \cdot r \cdot t$
23. $2(w)$
24. $(x)(y)$

Write each division using a fraction bar.

25. $32 \div x$
26. $30\overline{)90}$
27. $5\overline{)55}$
28. $h \div 15$

GUIDED PRACTICE

Use the given line graphs to answer the following questions. See Example 1.

29. **Accounting.** Explain what the dashed lines in the graph below help us find.
30. **Accounting.** What is the value of 35-year-old machinery?

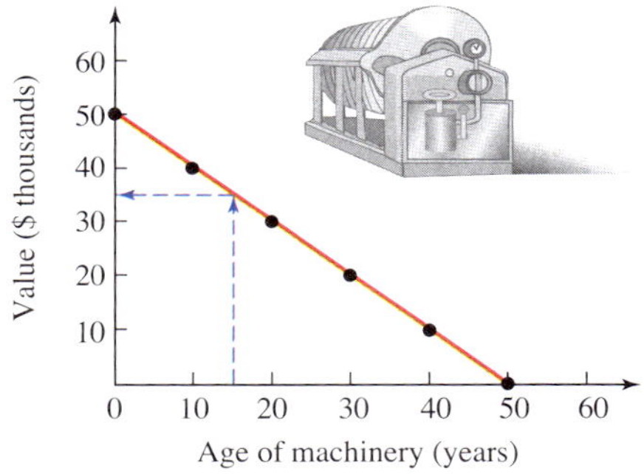

31. **Business.** Refer to the graph below. Find the income received from 30 customers.

32. **Business.** Refer to the graph below. Find the income received from 70 customers.

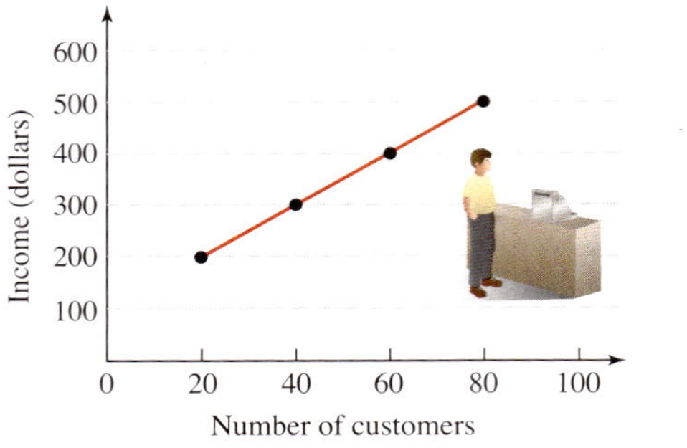

***Express each statement using one of the words* sum, product, difference, *or* quotient. See Example 2.**

33. $8(2) = 16$

34. $45 \cdot 12 = 540$

35. $11 - 9 = 2$

36. $65 + 89 = 154$

37. $x + 2 = 10$

38. $16 - t = 4$

39. $\frac{66}{11} = 6$

40. $12 \div 3 = 4$

***Translate each verbal model into an equation. (Answers may vary, depending on the variables chosen.)* See Example 3.**

41.

The sale price	is	$100	minus	the discount.

42.

The cost of dining out	equals	the cost of the meal	plus	$7 for parking.

43.

7	times	the age of a dog in years	gives	the dog's equivalent human age.

44.

The number of centuries	is	the number of years	divided by	100.

45. The amount of sand that should be used is the product of 3 and the amount of cement used.

46. The number of waiters needed is the quotient of the number of customers and 10.

47. The weight of the truck is the sum of the weight of the engine and 1,200.

48. The number of classes still open is the difference of 150 and the number of classes that are closed.

49. The profit is the difference of the revenue and 600.

50. The distance is the product of the rate and 3.

51. The quotient of the number of laps run and 4 gives the number of miles run.

52. The sum of the tax and 35 gives the total cost.

***Use the formula to complete each table.* See Examples 4 and 5.**

53. $d = 360 + L$

Lunch time (minutes) L	School day (minutes) d
30	
40	
45	

54. $b = 1{,}024k$

Kilobytes k	Bytes b
1	
5	
10	

55. $t = 1{,}500 - d$

Deductions d	Take-home pay t
200	
300	
400	

56. $w = \frac{s}{12}$

Inches of snow s	Inches of water w
12	
24	
72	

Use the data in the table to complete the formula.

57. $d = \frac{e}{}$

Eggs e	Dozens d
24	2
36	3
48	4

58. $p = \quad c$

Canoes c	Paddles p
6	12
7	14
8	16

59. $I = \quad c$

Couples c	Individuals I
20	40
100	200
200	400

60. $t = \frac{p}{}$

Players p	Teams t
5	1
10	2
15	3

APPLICATIONS

61. **Exercise.** The number of calories that a 125-pound adult burns doing general house cleaning chores is three times the number of minutes spent cleaning.

a. Write a verbal model using the word *product* that describes the relationship between calories burned and minutes cleaning.

b. Write a formula using the variables c and m that describes the relationship between calories burned and minutes cleaning.

c. Use your answer to part b to complete the following table.

m	10	20	30	40	50	60
c						

d. Use the data from the table to construct a line graph. Scale the horizontal axis in units of 10 minutes. Scale the vertical axis in units of 30 calories.

62. **Traffic Safety.** As the railroad crossing guard drops, the measure of angle 1 (written $\angle 1$) increases while the measure of $\angle 2$ decreases. At any instant the *sum* of the measures of the two angles is 90°. Complete the table. Then use the data to construct a line graph. Scale each axis in units of 15°.

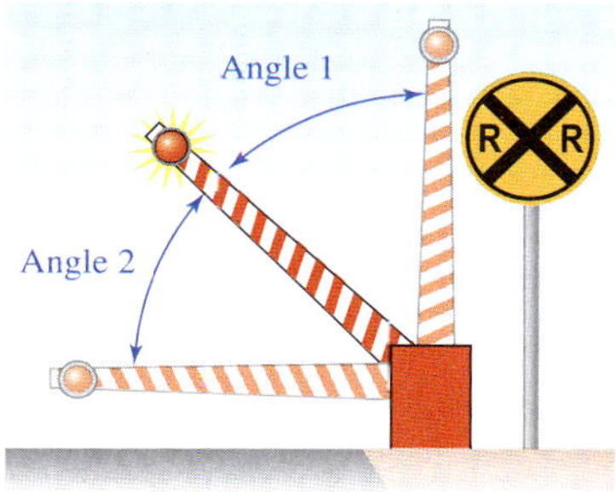

Angle 1 (degrees)	Angle 2 (degrees)
0	
15	
30	
45	
60	
75	
90	

WRITING

63. Many students misuse the word *equation* when discussing mathematics. What is an equation? Give an example.

64. Explain the difference between an algebraic expression and an equation. Give an example of each.

65. In this section, four methods for describing numerical relationships were discussed: tables, verbal models (words), graphs, and equations. Which method do you think is the most useful? Explain why.

66. In your own words, define *horizontal* and *vertical*.

CHALLENGE PROBLEMS

67. Complete the formula. $t = \quad s + \quad$

s	t
18	55
33	100
47	142

68. Suppose $h = 4n$ and $n = 2g$. Complete the following formula: $h = \quad g$

SECTION 1.2

OBJECTIVES

1. Factor and prime factor natural numbers.
2. Recognize special fraction forms.
3. Multiply and divide fractions.
4. Build equivalent fractions.
5. Simplify fractions.
6. Add and subtract fractions.
7. Simplify answers.
8. Compute with mixed numbers.

The Language of Algebra

When we say "factor 8," we are using the word **factor** as a verb. When we say "2 is a *factor* of 8," we are using the word **factor** as a noun.

Fractions

ARE YOU READY?

The following problems review some basic skills that are needed when working with fractions.

1. What is the value of $\frac{8}{8}$?

2. Multiply: $2 \cdot 3 \cdot 5 \cdot 5$

3. Is 42 divisible by 3?

4. Write the fraction $\frac{4}{5}$ in words.

In arithmetic, we add, subtract, multiply, and divide **natural numbers:** 1, 2, 3, 4, 5, and so on. Assuming that you have mastered those skills, we will now review the arithmetic of fractions.

1 Factor and Prime Factor Natural Numbers.

To compute with fractions, we need to know how to *factor* natural numbers. To **factor** a number means to express it as a product of two or more numbers. For example, some ways to factor 8 are

$$1 \cdot 8, \qquad 4 \cdot 2, \qquad \text{and} \qquad 2 \cdot 2 \cdot 2$$

The numbers 1, 2, 4, and 8 that were used to write the products are called *factors* of 8. In general, a **factor** is a number being multiplied.

Sometimes a number has only two factors, itself and 1. We call such numbers *prime numbers*.

Prime Numbers and Composite Numbers

A **prime number** is a natural number greater than 1 that has only itself and 1 as factors. The first ten prime numbers are 2, 3, 5, 7, 11, 13, 17, 19, 23, and 29.

A **composite number** is a natural number, greater than 1, that is not prime. The first ten composite numbers are 4, 6, 8, 9, 10, 12, 14, 15, 16, and 18.

Every composite number can be factored into the product of two or more prime numbers. This product of these prime numbers is called its **prime factorization.**

EXAMPLE 1 Find the prime factorization of 210.

Strategy We will use a series of steps to express 210 as a product of only prime numbers.

Why To *prime factor* a number means to write it as a product of prime numbers.

Solution First, write 210 as the product of two natural numbers other than 1.

$210 = 10 \cdot 21$ The resulting prime factorization will be the same no matter which two factors of 210 you begin with.

The Language of Algebra

Prime factors often are written in **ascending** order. To **ascend** means to move upward.

Neither 10 nor 21 are prime numbers, so we factor each of them.

$210 = 2 \cdot 5 \cdot 3 \cdot 7$ Factor 10 as $2 \cdot 5$ and factor 21 as $3 \cdot 7$.

Writing the factors in order, from least to greatest, the **prime-factored form** of 210 is $2 \cdot 3 \cdot 5 \cdot 7$. Two other methods for prime factoring 210 are shown below.

Work downward. Factor each number as a product of two numbers (other than 1 and itself) until all factors are prime. Circle prime numbers as they appear at the end of a branch.

Factor tree ***Division ladder***

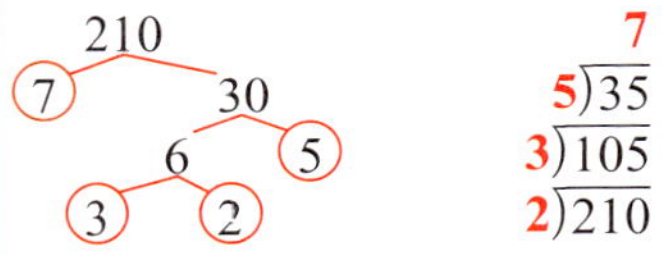

Work upward. Perform repeated division until the final quotient is a prime number. It is helpful to start with the smallest prime, 2, as a trial divisor. Then, in order, try larger primes as divisors: 3, 5, 7, 11, and so on.

Success Tip

The following divisibility rules are helpful when prime factoring.

A whole number is divisible by

- 2 if it ends in 0, 2, 4, 6, or 8
- 3 if the sum of the digits is divisible by 3
- 5 if it ends in 0 or 5
- 10 if it ends in 0

Either way, the factorization is $2 \cdot 3 \cdot 5 \cdot 7$. To check it, multiply the prime factors. The product should be 210.

Self Check 1 Find the prime factorization of 189.

Now Try Problems 15 and 23

2 Recognize Special Fraction Forms.

A **fraction** describes the number of equal parts of a whole. For example, consider the figure below with 5 of the 6 equal parts colored red. We say that $\frac{5}{6}$ (five-sixths) of the figure is shaded.

In a fraction, the number above the **fraction bar** is called the **numerator**, and the number below is called the **denominator.**

The Language of Algebra

The word **fraction** comes from the Latin word *fractio* meaning "breaking in pieces."

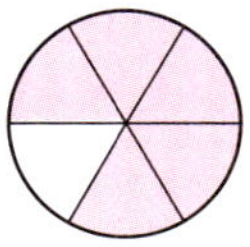

Fraction bar → $\frac{5}{6}$ ← numerator ← denominator

Fractions are also used to indicate division. For example, the fraction bar in $\frac{8}{2}$ indicates that the numerator, 8, is to be divided by the denominator, 2:

$$\frac{8}{2} = 8 \div 2 = 4$$ We know that $\frac{8}{2} = 4$ because of its related multiplication statement: $2 \cdot 4 = 8$.

If the numerator and denominator of a fraction are the same nonzero number, the fraction indicates division of a number by itself, and the result is 1. Each of the following fractions is, therefore, a **form of 1.**

$$1 = \frac{1}{1} = \frac{2}{2} = \frac{3}{3} = \frac{4}{4} = \frac{5}{5} = \frac{6}{6} = \frac{7}{7} = \frac{8}{8} = \frac{9}{9} = \ldots$$

If a denominator is 1, the fraction indicates division by 1, and the result is simply the numerator. For example, $\frac{5}{1} = 5$ and $\frac{24}{1} = 24$.

Special Fraction Forms

For any nonzero number a,

$$\frac{a}{a} = 1 \quad \text{and} \quad \frac{a}{1} = a$$

3 Multiply and Divide Fractions.

The rule for multiplying fractions can be expressed in words and in symbols as follows.

Multiplying Fractions

To multiply two fractions, multiply the numerators and multiply the denominators.

For any two fractions $\frac{a}{b}$ and $\frac{c}{d}$,

$$\frac{a}{b} \cdot \frac{c}{d} = \frac{a \cdot c}{b \cdot d}$$

EXAMPLE 2 Multiply: $\frac{7}{8} \cdot \frac{3}{5}$

Strategy To find the product, we will multiply the numerators, 7 and 3, and multiply the denominators, 8 and 5.

Why This is the rule for multiplying two fractions.

Solution

$$\frac{7}{8} \cdot \frac{3}{5} = \frac{7 \cdot 3}{8 \cdot 5}$$ Multiply the numerators. Multiply the denominators.

$$= \frac{21}{40}$$

Self Check 2 Multiply: $\frac{5}{9} \cdot \frac{2}{3}$

Now Try Problem 27

Success Tip

Every number, except 0, has a reciprocal. Zero has no reciprocal, because the product of 0 and a number cannot be 1.

One number is called the **reciprocal** of another if their product is 1. To find the reciprocal of a fraction, we invert its numerator and denominator.

$\frac{3}{4}$ is the reciprocal of $\frac{4}{3}$, because $\frac{3}{4} \cdot \frac{4}{3} = \frac{12}{12} = 1$.

$\frac{1}{10}$ is the reciprocal of 10, because $\frac{1}{10} \cdot 10 = \frac{10}{10} = 1$.

We use reciprocals to divide fractions.

Dividing Fractions

To divide two fractions, multiply the first fraction by the reciprocal of the second.

For any two fractions $\frac{a}{b}$ and $\frac{c}{d}$, where $c \neq 0$,

$$\frac{a}{b} \div \frac{c}{d} = \frac{a}{b} \cdot \frac{d}{c}$$

EXAMPLE 3 Divide: $\frac{1}{3} \div \frac{4}{5}$

Strategy We will multiply the first fraction, $\frac{1}{3}$, by the reciprocal of the second fraction, $\frac{4}{5}$.

Why This is the rule for dividing two fractions.

Solution

$$\frac{1}{3} \div \frac{4}{5} = \frac{1}{3} \cdot \frac{5}{4}$$ Multiply $\frac{1}{3}$ by the reciprocal of $\frac{4}{5}$, which is $\frac{5}{4}$.

$$= \frac{1 \cdot 5}{3 \cdot 4}$$ Use the rule for multiplying fractions. Multiply the numerators. Multiply the denominators.

$$= \frac{5}{12}$$

Self Check 3 Divide: $\frac{6}{25} \div \frac{1}{2}$

Now Try Problem 31

4 Build Equivalent Fractions.

The two rectangular regions on the right are identical. The first one is divided into 10 equal parts. Since 6 of those parts are red, $\frac{6}{10}$ of the figure is shaded.

The second figure is divided into 5 equal parts. Since 3 of those parts are red, $\frac{3}{5}$ of the figure is shaded. We can conclude that $\frac{6}{10} = \frac{3}{5}$ because $\frac{6}{10}$ and $\frac{3}{5}$ represent the same shaded portion of the figure. We say that $\frac{6}{10}$ and $\frac{3}{5}$ are *equivalent fractions.*

$\frac{6}{10}$ $\frac{3}{5}$

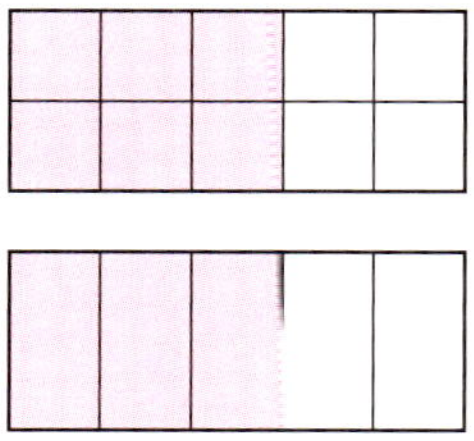

Equivalent Fractions

Two fractions are **equivalent** if they represent the same number. **Equivalent fractions** represent the same portion of a whole.

Writing a fraction as an equivalent fraction with a larger denominator is called **building** the fraction. To build a fraction, we multiply it by a form of 1. Since any number multiplied by 1 remains the same (identical), 1 is called the **multiplicative identity element.**

Multiplication Property of 1

The product of 1 and any number is that number. For any number a,

$$1 \cdot a = a \quad \text{and} \quad a \cdot 1 = a$$

EXAMPLE 4 Write $\frac{3}{5}$ as an equivalent fraction with a denominator of 35.

Strategy We will compare the given denominator to the required denominator and ask, "By what must we multiply 5 to get 35?"

Why The answer to that question helps us determine the form of 1 to be used to build an equivalent fraction.

Solution We need to multiply the denominator of $\frac{3}{5}$ by 7 to obtain a denominator of 35. It follows that $\frac{7}{7}$ should be the form of 1 that is used to build $\frac{3}{5}$. Multiplying $\frac{3}{5}$ by $\frac{7}{7}$ changes its appearance but does not change its value, because we are multiplying it by 1.

$$\frac{3}{5} = \frac{3}{5} \cdot \frac{7}{7} \quad \text{Multiply } \tfrac{3}{5} \text{ by a form of 1: } \tfrac{7}{7} = 1$$

$$= \frac{3 \cdot 7}{5 \cdot 7} \quad \begin{array}{l}\text{Multiply the numerators.}\\ \text{Multiply the denominators.}\end{array}$$

$$= \frac{21}{35}$$

Success Tip

Multiplying $\frac{3}{5}$ by $\frac{7}{7}$ changes its appearance, but does not change its value, because we are multiplying it by a form of 1.

Self Check 4 Write $\frac{5}{8}$ as an equivalent fraction with a denominator of 24.

Now Try Problem 35

Building Fractions

To build a fraction, multiply it by 1 in the form of $\frac{c}{c}$, where c is any nonzero number.

To build an equivalent fraction in Example 4, we multiplied $\frac{3}{5}$ by 1 in the form of $\frac{7}{7}$. As a result of that step, the numerator and the denominator of $\frac{3}{5}$ were multiplied by 7:

$$\frac{3 \cdot 7}{5 \cdot 7} \quad \begin{array}{l}\leftarrow \text{The numerator is multiplied by 7.}\\ \leftarrow \text{The denominator is multiplied by 7.}\end{array}$$

This process illustrates the following property of fractions.

The Fundamental Property of Fractions

If the numerator and denominator of a fraction are multiplied by the same nonzero number, the resulting fraction is equivalent to the original fraction.

Since multiplying the numerator and denominator of a fraction by the same nonzero number produces an equivalent fraction, your instructor may allow you to begin your solution to problems like Example 4 as shown above.

5 Simplify Fractions.

The Language of Algebra

The word **infinitely** is a form of the word **infinite,** which means endless.

Every fraction can be written in infinitely many equivalent forms. For example, some equivalent forms of $\frac{10}{15}$ are:

$$\frac{2}{3} = \frac{4}{6} = \frac{6}{9} = \frac{8}{12} = \frac{10}{15} = \frac{12}{18} = \frac{14}{21} = \frac{16}{24} = \frac{18}{27} = \frac{20}{30} = \ldots$$

Of all of the equivalent forms in which we can write a fraction, we often need to determine the one that is in *simplest form.*

Simplest Form of a Fraction

A fraction is in **simplest form,** or **lowest terms,** when the numerator and denominator have no common factors other than 1.

To **simplify a fraction,** we write it in simplest form by removing a factor equal to 1. For example, to simplify $\frac{10}{15}$, we note that the greatest factor common to the numerator and denominator is 5 and proceed as follows:

$$\frac{10}{15} = \frac{2 \cdot 5}{3 \cdot 5}$$ To prepare to simplify the fraction, factor 10 and 15. Note the form of 1 highlighted in red.

$$= \frac{2}{3} \cdot \frac{5}{5}$$ Use the rule for multiplying fractions in reverse: write $\frac{2 \cdot 5}{3 \cdot 5}$ as the product of two fractions, $\frac{2}{3}$ and $\frac{5}{5}$.

$$= \frac{2}{3} \cdot 1$$ Any nonzero number divided by itself is equal to 1: $\frac{5}{5} = 1$.

$$= \frac{2}{3}$$ Any number multiplied by 1 remains the same.

To simplify $\frac{10}{15}$, we removed a factor equal to 1 in the form of $\frac{5}{5}$. The result, $\frac{2}{3}$, is equivalent to $\frac{10}{15}$.

We can easily identify the greatest common factor of the numerator and the denominator of a fraction if we write them in prime-factored form.

EXAMPLE 5 Simplify each fraction, if possible: **a.** $\frac{63}{42}$ **b.** $\frac{33}{40}$

Strategy We will begin by prime factoring the numerator and denominator of the fraction. Then, to simplify it, we will remove a factor equal to 1.

Why We need to make sure that the numerator and denominator have no common factors other than 1. If that is the case, then the fraction is in *simplest form.*

Solution

a. After prime factoring 63 and 42, we see that the greatest common factor of the numerator and the denominator is $3 \cdot 7 = 21$.

$$\frac{63}{42} = \frac{3 \cdot 3 \cdot 7}{2 \cdot 3 \cdot 7}$$ To prepare to simplify the fraction, write 63 and 42 in prime-factored form.

$$= \frac{3}{2} \cdot \frac{3 \cdot 7}{3 \cdot 7}$$ Write $\frac{3 \cdot 3 \cdot 7}{2 \cdot 3 \cdot 7}$ as the product of two fractions, $\frac{3}{2}$ and $\frac{3 \cdot 7}{3 \cdot 7}$.

$$= \frac{3}{2} \cdot 1$$ Any nonzero number divided by itself is equal to 1: $\frac{3 \cdot 7}{3 \cdot 7} = 1$.

$$= \frac{3}{2}$$ Any number multiplied by 1 remains the same.

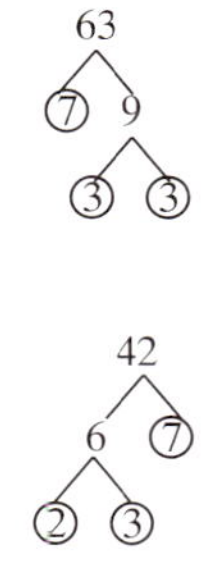

The Language of Algebra

It is best to use the phrase "**simplify** a fraction" as opposed to "reduce a fraction." The word *reduce* suggests getting smaller, and might be confusing. When we simplify a fraction, the result is an equivalent fraction with the same value as the original fraction, not a smaller value.

b. To attempt to simplify the fraction, prime factor 33 and 40.

$$\frac{33}{40} = \frac{3 \cdot 11}{2 \cdot 2 \cdot 2 \cdot 5}$$

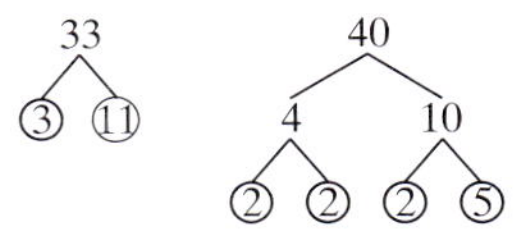

Since the numerator and the denominator have no common factors other than 1, the fraction $\frac{33}{40}$ is in simplest form (lowest terms).

Self Check 5 Simplify each fraction, if possible: **a.** $\frac{24}{56}$ **b.** $\frac{16}{125}$

Now Try Problems 45 and 49

To streamline the simplifying process, we can replace pairs of factors common to the numerator and denominator with the equivalent fraction $\frac{1}{1}$.

EXAMPLE 6 Simplify: $\frac{90}{105}$

Strategy We will begin by prime factoring the numerator, 90, and denominator, 105. Then we will look for any factors common to the numerator and denominator and remove them.

Why When the numerator and/or denominator of a fraction are large numbers, such as 90 and 105, writing their prime factorizations is helpful in identifying any common factors.

Solution

$$\frac{90}{105} = \frac{2 \cdot 3 \cdot 3 \cdot 5}{3 \cdot 5 \cdot 7}$$ To prepare to simplify the fraction, write 90 and 105 in prime-factored form.

$$= \frac{2 \cdot \overset{1}{\cancel{3}} \cdot 3 \cdot \overset{1}{\cancel{5}}}{\underset{1}{\cancel{3}} \cdot \underset{1}{\cancel{5}} \cdot 7}$$ Slashes and 1's are used to show that $\frac{3}{3}$ and $\frac{5}{5}$ are replaced by the equivalent fraction $\frac{1}{1}$. A factor equal to 1 in the form of $\frac{3 \cdot 5}{3 \cdot 5} = \frac{15}{15}$ was removed.

$$= \frac{6}{7}$$ Multiply the remaining factors in the numerator: $2 \cdot 1 \cdot 3 \cdot 1 = 6$. Multiply the remaining factors in the denominator: $1 \cdot 1 \cdot 7 = 7$.

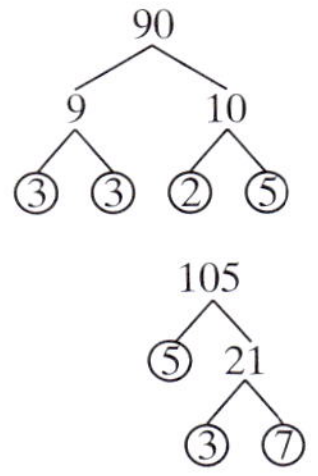

Self Check 6 Simplify: $\frac{126}{70}$

Now Try Problem 53

We can use the following steps to simplify a fraction.

Simplifying Fractions

1. Factor (or prime factor) the numerator and denominator to determine their common factors.
2. Remove factors equal to 1 by replacing each pair of factors common to the numerator and denominator with the equivalent fraction $\frac{1}{1}$.
3. Multiply the remaining factors in the numerator and in the denominator.

6 Add and Subtract Fractions.

In algebra as in everyday life, we can add or subtract only objects that are similar. For example, we can add dollars to dollars, but we cannot add dollars to oranges. This concept is important when adding or subtracting fractions.

Consider the problem $\frac{2}{5} + \frac{1}{5}$. When we write it in words, it is apparent we are adding similar objects.

two-**fifths** + one-**fifth**

Similar objects

Because the denominators of $\frac{2}{5}$ and $\frac{1}{5}$ are the same, we say that they have a **common denominator.**

Adding and Subtracting Fractions That Have the Same Denominator

To add (or subtract) fractions that have the same denominator, add (or subtract) their numerators and write the sum (or difference) over the common denominator. Simplify the result, if possible.

For any fractions $\frac{a}{d}$ and $\frac{b}{d}$,

$$\frac{a}{d} + \frac{b}{d} = \frac{a+b}{d} \quad \text{and} \quad \frac{a}{d} - \frac{b}{d} = \frac{a-b}{d}$$

Caution

We **do not** add fractions by adding the numerators and adding the denominators!

$$\frac{2}{5} + \frac{1}{5} = \frac{2+1}{5+5} = \frac{3}{10}$$

The same caution applies when subtracting fractions.

For example,

$$\frac{2}{5} + \frac{1}{5} = \frac{2+1}{5} = \frac{3}{5} \quad \text{and} \quad \frac{18}{23} - \frac{9}{23} = \frac{18-9}{23} = \frac{9}{23}$$

Now we consider the problem $\frac{2}{5} + \frac{1}{3}$. Since the denominators are not the same, we cannot add these fractions in their present form.

two-**fifths** + one-**third**

Not similar objects

To add (or subtract) fractions with different denominators, we express them as equivalent fractions that have a common denominator. The smallest common denominator, called the **least** or **lowest common denominator,** is always the easiest common denominator to use.

Least Common Denominator (LCD)

The **least** or **lowest common denominator (LCD)** for a set of fractions is the smallest number each denominator will divide exactly (divide with no remainder).

Success Tip

To determine the LCD of two fractions, list the multiples of the larger denominator. The first number in the list that is exactly divisible by the other denominator is their LCD. For $\frac{2}{5}$ and $\frac{1}{3}$, the multiples of the larger denominator, 5, are

5, 10, (15), 20, 25, . . .

Since 15 is the first number in the list that is exactly divisible by the second denominator, 3, the LCD is 15.

The denominators of $\frac{2}{5}$ and $\frac{1}{3}$ are 5 and 3. The numbers 5 and 3 divide many numbers exactly (30, 45, and 60, to name a few), but the smallest number that they divide exactly is 15. Thus, 15 is the LCD for $\frac{2}{5}$ and $\frac{1}{3}$.

To find $\frac{2}{5} + \frac{1}{3}$, we build equivalent fractions that have denominators of 15 and we use the rule for adding fractions.

$$\frac{2}{5} + \frac{1}{3} = \frac{2}{5} \cdot \frac{3}{3} + \frac{1}{3} \cdot \frac{5}{5}$$

Multiply $\frac{2}{5}$ by 1 in the form of $\frac{3}{3}$. Multiply $\frac{1}{3}$ by 1 in the form of $\frac{5}{5}$.

We need to multiply this denominator by 5 to obtain 15. It follows that $\frac{5}{5}$ should be the form of 1 used to build $\frac{1}{3}$.

We need to multiply this denominator by 3 to obtain 15. It follows that $\frac{3}{3}$ should be the form of 1 used to build $\frac{2}{5}$.

$$= \frac{6}{15} + \frac{5}{15}$$

Multiply the numerators and multiply the denominators. Note that the denominators are now the same.

$$= \frac{6 + 5}{15}$$ Add the numerators. Write the sum over the common denominator, 15.

$$= \frac{11}{15}$$ Since 11 and 15 have no common factors other than 1, this fraction is in simplest form.

When adding (or subtracting) fractions with unlike denominators, the least common denominator is not always obvious. Prime factorization is helpful in determining the LCD.

Finding the LCD Using Prime Factorization

1. Prime factor each denominator.
2. The LCD is a product of prime factors, where each factor is used the greatest number of times it appears in any one factorization found in step 1.

EXAMPLE 7 Subtract: $\frac{3}{10} - \frac{5}{28}$

Strategy We will begin by expressing each fraction as an equivalent fraction that has the LCD for its denominator. Then we will use the rule for subtracting fractions with *like* denominators.

Why To add or subtract fractions, the fractions must have like denominators.

Solution To find the LCD, we find the prime factorization of both denominators and use each prime factor the *greatest* number of times it appears in any one factorization.

$$\left.\begin{array}{l} 10 = 2 \cdot 5 \\ 28 = 2 \cdot 2 \cdot 7 \end{array}\right\} \text{LCD} = 2 \cdot 2 \cdot 5 \cdot 7 = 140$$

2 appears twice in the factorization of 28.
5 appears once in the factorization of 10.
7 appears once in the factorization of 28.

Since 140 is the smallest number that 10 and 28 divide exactly, we write $\frac{3}{10}$ and $\frac{5}{28}$ as fractions with the LCD 140.

$$\frac{3}{10} - \frac{5}{28} = \frac{3}{10} \cdot \frac{14}{14} - \frac{5}{28} \cdot \frac{5}{5}$$ We must multiply 10 by 14 to obtain 140. We must multiply 28 by 5 to obtain 140.

$$= \frac{42}{140} - \frac{25}{140}$$ Multiply the numerators and multiply the denominators. Note that the denominators are now the same.

$$= \frac{42 - 25}{140}$$ Subtract the numerators. Write the difference over the common denominator, 140.

$$= \frac{17}{140}$$ Since 17 and 140 have no common factors other than 1, this fraction is in simplest form.

Self Check 7 Subtract: $\frac{11}{48} - \frac{7}{40}$

Now Try Problem 65

We can use the following steps to add or subtract fractions with different denominators.

Adding and Subtracting Fractions That Have Different Denominators

1. Find the LCD.
2. Rewrite each fraction as an equivalent fraction with the LCD as the denominator. To do so, build each fraction using a form of 1 that involves any factors needed to obtain the LCD.
3. Add or subtract the numerators and write the sum or difference over the LCD.
4. Simplify the result, if possible.

7 Simplify Answers.

When adding, subtracting, multiplying, or dividing fractions, remember to express the answer in simplest form.

EXAMPLE 8 Perform the operations and simplify: **a.** $45\left(\frac{4}{9}\right)$ **b.** $\frac{5}{12} + \frac{3}{2} - \frac{1}{4}$

Strategy We will perform the indicated operations and then make sure that the answer is in simplest form.

Why Fractional answers should always be given in simplest form.

Solution

a. $45\left(\frac{4}{9}\right) = \frac{45}{1}\left(\frac{4}{9}\right)$ Write 45 as a fraction: $45 = \frac{45}{1}$.

$= \frac{45 \cdot 4}{1 \cdot 9}$ Multiply the numerators. Multiply the denominators.

$= \frac{5 \cdot \overset{1}{\cancel{9}} \cdot 4}{1 \cdot \underset{1}{\cancel{9}}}$ To simplify the result, factor 45 as $5 \cdot 9$. Then remove the common factor 9 in the numerator and denominator.

$= \frac{20}{1}$ Multiply the remaining factors in the numerator. Multiply the remaining factors in the denominator.

$= 20$ Any number divided by 1 is the number itself.

Caution

Remember that an LCD is **not needed** when multiplying or dividing fractions.

b. Since the smallest number that 12, 2, and 4 divide exactly is 12, the LCD is 12.

$\frac{5}{12} + \frac{3}{2} - \frac{1}{4} = \frac{5}{12} + \frac{3}{2} \cdot \frac{6}{6} - \frac{1}{4} \cdot \frac{3}{3}$ $\frac{5}{12}$ already has a denominator of 12. Build $\frac{3}{2}$ and $\frac{1}{4}$ so that their denominators are 12.

$= \frac{5}{12} + \frac{18}{12} - \frac{3}{12}$ Multiply the numerators and multiply the denominators. The denominators are now the same.

$= \frac{20}{12}$ Add the numerators, 5 and 18, to get 23. From that sum, subtract 3. Write that result, 20, over the common denominator.

$= \frac{\overset{1}{\cancel{4}} \cdot 5}{3 \cdot \underset{1}{\cancel{4}}}$ To simplify $\frac{20}{12}$, factor 20 and 12, using their greatest common factor, 4. Then remove $\frac{4}{4} = 1$.

$= \frac{5}{3}$

The Language of Algebra

Fractions such as $\frac{5}{3}$, with a numerator greater than or equal to the denominator, are called **improper fractions**. In algebra, such fractions are often preferable to their equivalent mixed number form.

Self Check 8 Perform the operations and simplify: **a.** $24\left(\frac{7}{6}\right)$ **b.** $\frac{1}{15} + \frac{31}{30} - \frac{3}{10}$

Now Try Problems 67 and 71

8 Compute with Mixed Numbers.

A **mixed number** represents the sum of a whole number and a fraction. For example, $5\frac{3}{4}$ means $5 + \frac{3}{4}$ and $179\frac{15}{16}$ means $179 + \frac{15}{16}$.

EXAMPLE 9 Divide: $5\frac{3}{4} \div 2$

Strategy We begin by writing the mixed number $5\frac{3}{4}$ and the whole number 2 as fractions. Then we use the rule for dividing two fractions.

Why To multiply (or divide) with mixed numbers, we first write them as improper fractions, and then multiply (or divide) as usual.

Solution

$5\frac{3}{4} \div 2 = \frac{23}{4} \div \frac{2}{1}$ Write $5\frac{3}{4}$ as an improper fraction by multiplying its whole-number part by the denominator: $5 \cdot 4 = 20$. Then add the numerator to that product: $3 + 20 = 23$. Finally, write the result, 23, over the denominator 4. Write 2 as a fraction: $2 = \frac{2}{1}$.

$= \frac{23}{4} \cdot \frac{1}{2}$ Multiply by the reciprocal of $\frac{2}{1}$, which is $\frac{1}{2}$.

$= \frac{23}{8}$ Multiply the numerators. Multiply the denominators.

$= 2\frac{7}{8}$ Since the original problem involves a mixed number, we will express the answer in mixed-number form. Write $\frac{23}{8}$ as a mixed number by dividing the numerator, 23, by the denominator, 8.

$$\begin{array}{r} 2 \\ 8\overline{)23} \\ -16 \\ \hline 7 \end{array}$$

Self Check 9 Multiply: $1\frac{1}{8} \cdot 9$

Now Try Problem 77

EXAMPLE 10 **Freeway Signs.** How far apart are the Downtown San Diego and Sea World Drive exits?

Strategy We can find the distance between exits by finding the difference in the mileages on the freeway sign: $6\frac{1}{2} - 1\frac{3}{4}$.

Why The word *difference* indicates subtraction.

Solution

$$\begin{array}{rcrcrcr} 6\frac{1}{2} & = & 6\frac{2}{4} & = & 5\frac{2}{4} + \frac{4}{4} & = & 5\frac{6}{4} \\ -1\frac{3}{4} & = & -1\frac{3}{4} & = & -1\frac{3}{4} & = & -1\frac{3}{4} \\ \hline & & & & & & 4\frac{3}{4} \end{array}$$

Using vertical form, express $\frac{1}{2}$ as an equivalent fraction with denominator 4. Then, borrow 1 in the form of $\frac{4}{4}$ from 6 to subtract the fractional parts of the mixed numbers.

The Downtown San Diego and Sea World Drive exits are $4\frac{3}{4}$ miles apart.

Self Check 10 Subtract: $9\frac{1}{8} - 2\frac{2}{3}$

Now Try Problem 114

Success Tip

This problem could also be solved by writing the mixed numbers $6\frac{1}{2}$ and $1\frac{3}{4}$ as improper fractions and subtracting them. However, answers to real-world problems are most often given as mixed numbers instead of improper fractions, because mixed numbers are easier to understand.

SECTION 1.2 STUDY SET

VOCABULARY

Fill in the blanks.

1. A factor is a number being ________.
2. Numbers that have only 1 and themselves as factors, such as 23, 37, and 41, are called ______ numbers.
3. When we write 60 as $2 \cdot 2 \cdot 3 \cdot 5$, we say that we have written 60 in ____________ form.
4. The _________ of the fraction $\frac{3}{4}$ is 3, and the __________ is 4.
5. Two fractions that represent the same number, such as $\frac{1}{2}$ and $\frac{2}{4}$, are called _________ fractions.
6. $\frac{2}{3}$ is the _________ of $\frac{3}{2}$, because their product is 1.
7. The ____________ common denominator for a set of fractions is the smallest number each denominator will divide exactly.
8. The ______ number $7\frac{1}{3}$ represents the sum of a whole number and a fraction: $7 + \frac{1}{3}$.

CONCEPTS

Complete each fact about fractions. Assume there are no divisions by 0.

9. a. $\frac{a}{a} =$ b. $\frac{a}{1} =$
   c. $\frac{a}{b} \cdot \frac{c}{d} = \underline{\quad}$ d. $\frac{a}{b} \div \frac{c}{d} = \underline{\quad}$
   e. $\frac{a}{d} + \frac{b}{d} = \underline{\quad}$ f. $\frac{a}{d} - \frac{b}{d} = \underline{\quad}$
10. What two equivalent fractions are shown?

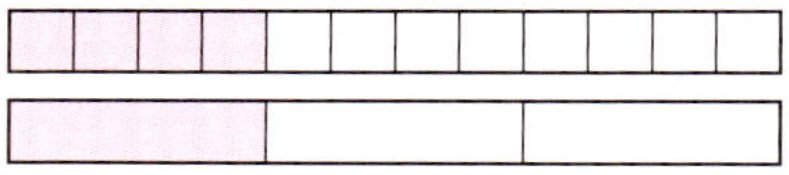

11. Complete each statement.
    a. To simplify a fraction, we remove factors equal to in the form of $\frac{2}{2}$, $\frac{3}{3}$, or $\frac{4}{4}$, and so on.
    b. To build a fraction, we multiply it by in the form of $\frac{2}{2}$, $\frac{3}{3}$, or $\frac{4}{4}$, and so on.
12. What is the LCD for fractions having denominators of:
    a. 3 and 7?
    b. 4 and 6?

NOTATION

Fill in the blanks.

13. a. Multiply $\frac{5}{6}$ by a form of 1 to build an equivalent fraction with denominator 30.

    $\frac{5}{6} \cdot \frac{\quad}{\quad} = \frac{\quad}{\quad}$

    b. Remove common factors to simplify $\frac{12}{42}$.

    $\frac{12}{42} = \frac{2 \cdot \quad \cdot 3}{2 \cdot 3 \cdot \quad} = \frac{\quad}{\quad}$

14. a. Write $2\frac{15}{16}$ as an improper fraction.
    b. Write $\frac{49}{12}$ as a mixed number.

GUIDED PRACTICE

Find the prime factorization of each number. See Example 1.

15. 75
16. 20
17. 28
18. 54
19. 81
20. 125
21. 117
22. 147
23. 220
24. 270
25. 1,254
26. 1,144

Multiply. See Example 2.

27. $\frac{5}{6} \cdot \frac{1}{8}$
28. $\frac{2}{3} \cdot \frac{1}{5}$
29. $\frac{7}{11} \cdot \frac{3}{5}$
30. $\frac{13}{9} \cdot \frac{2}{3}$

Divide. See Example 3.

31. $\frac{3}{4} \div \frac{2}{5}$
32. $\frac{7}{8} \div \frac{6}{13}$
33. $\frac{6}{5} \div \frac{5}{7}$
34. $\frac{4}{3} \div \frac{3}{2}$

Build each fraction or whole number to an equivalent fraction with the indicated denominator. See Example 4.

35. $\frac{1}{3}$, denominator 9
36. $\frac{3}{8}$, denominator 24
37. $\frac{4}{9}$, denominator 54
38. $\frac{9}{16}$, denominator 64
39. 7, denominator 5
40. 12, denominator 3
41. 5, denominator 7
42. 6, denominator 8

Simplify each fraction, if possible. See Examples 5 and 6.

43. $\frac{6}{18}$
44. $\frac{6}{9}$
45. $\frac{24}{28}$
46. $\frac{35}{14}$
47. $\frac{15}{40}$
48. $\frac{22}{77}$
49. $\frac{33}{56}$
50. $\frac{26}{21}$
51. $\frac{26}{39}$
52. $\frac{72}{64}$
53. $\frac{36}{225}$
54. $\frac{175}{490}$

Perform the operations and, if possible, simplify. See Objective 6 and Example 7.

55. $\frac{3}{5} + \frac{3}{5}$
56. $\frac{4}{9} - \frac{1}{9}$
57. $\frac{6}{7} - \frac{2}{7}$
58. $\frac{5}{13} + \frac{6}{13}$
59. $\frac{1}{6} + \frac{1}{24}$
60. $\frac{17}{25} - \frac{2}{5}$
61. $\frac{7}{10} - \frac{1}{14}$
62. $\frac{9}{8} - \frac{5}{6}$

63. $\frac{2}{15} + \frac{7}{9}$

64. $\frac{7}{25} + \frac{3}{10}$

65. $\frac{13}{28} - \frac{1}{21}$

66. $\frac{13}{24} - \frac{3}{40}$

Perform the operations and, if possible, simplify. See Example 8.

67. $16\left(\frac{3}{2}\right)$

68. $30\left(\frac{5}{6}\right)$

69. $18 \cdot \frac{2}{9}$

70. $14 \cdot \frac{3}{7}$

71. $\frac{2}{3} + \frac{5}{18} - \frac{1}{6}$

72. $\frac{3}{5} + \frac{7}{20} - \frac{7}{10}$

73. $\frac{5}{12} + \frac{1}{3} - \frac{2}{5}$

74. $\frac{7}{15} + \frac{1}{5} - \frac{4}{9}$

Perform the operations and, if possible, simplify. See Examples 9 and 10.

75. $4\frac{2}{3} \cdot 7$

76. $7 \cdot 1\frac{3}{28}$

77. $8 \div 3\frac{1}{5}$

78. $15 \div 3\frac{1}{3}$

79. $8\frac{2}{9} - 7\frac{2}{3}$

80. $3\frac{4}{5} - 3\frac{1}{10}$

81. $3\frac{3}{6} + 2\frac{5}{24}$

82. $15\frac{5}{6} + 11\frac{5}{8}$

TRY IT YOURSELF

Perform the operations and, if possible, simplify.

83. $\frac{3}{5} + \frac{2}{3}$

84. $\frac{4}{3} + \frac{7}{2}$

85. $21\left(\frac{10}{3}\right)$

86. $28\left(\frac{4}{7}\right)$

87. $6 \cdot 2\frac{7}{24}$

88. $3\frac{1}{2} \cdot \frac{1}{5}$

89. $\frac{2}{3} - \frac{1}{4} + \frac{1}{12}$

90. $\frac{3}{7} - \frac{2}{5} + \frac{2}{35}$

91. $\frac{21}{35} \div \frac{3}{14}$

92. $\frac{23}{25} \div \frac{46}{5}$

93. $\frac{4}{3}\left(\frac{6}{5}\right)$

94. $\frac{21}{8}\left(\frac{2}{15}\right)$

95. $\frac{4}{63} + \frac{1}{45}$

96. $\frac{5}{18} + \frac{1}{99}$

97. $3 - \frac{3}{4}$

98. $4 - \frac{7}{3}$

99. $\frac{[illegible]}{5} \cdot \frac{3}{5}$

100. $\frac{3}{4} \cdot \frac{5}{7}$

101. $3\frac{1}{3} \div 1\frac{5}{6}$

102. $2\frac{1}{2} \div 1\frac{5}{8}$

103. $\frac{11}{21} - \frac{8}{21}$

104. $\frac{19}{35} - \frac{12}{35}$

105. $\frac{7}{30} - \frac{1}{50} - \frac{19}{75}$

106. $\frac{11}{12} - \frac{7}{15} - \frac{9}{20}$

107. $1\frac{31}{32} \cdot 7\frac{1}{9}$

108. $3\frac{1}{16} \cdot 4\frac{4}{7}$

Look Alikes . . .

109. a. $\frac{4}{9} + \frac{3}{7}$ b. $\frac{4}{9} - \frac{3}{7}$ c. $\frac{4}{9} \cdot \frac{3}{7}$ d. $\frac{4}{9} \div \frac{3}{7}$

110. a. $4\frac{1}{8} + 1\frac{5}{6}$ b. $4\frac{1}{8} - 1\frac{5}{6}$ c. $4\frac{1}{8} \cdot 1\frac{5}{6}$ d. $4\frac{1}{8} \div 1\frac{5}{6}$

APPLICATIONS

111. **Forestry.** A ranger cut down a tree and measured the widths of the outer two growth rings.
   a. What was the growth over this 2-year period?
   b. What is the difference in the widths of the rings?

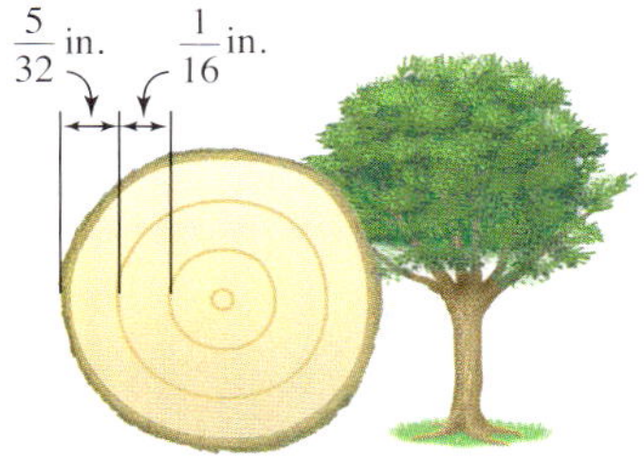

112. **Hardware.** To secure the bracket to the stock, a bolt and a nut are used. How long should the threaded part of the bolt be?

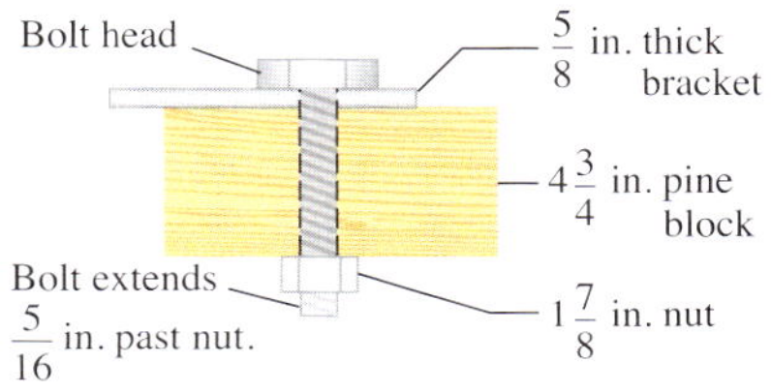

113. **from Campus to Careers**
Lead Transportation Security Officer

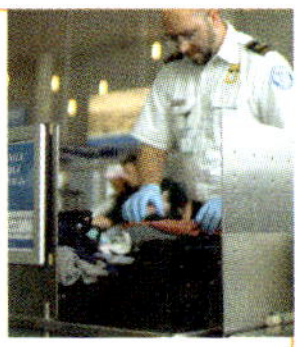

Each year, the Transportation Security Administration (TSA) screens more than 500 million pieces of luggage. On many flights, airlines do not accept luggage whose total dimension (length + width + height) exceeds 62 inches. What is the total dimension figure for the suitcase shown below?

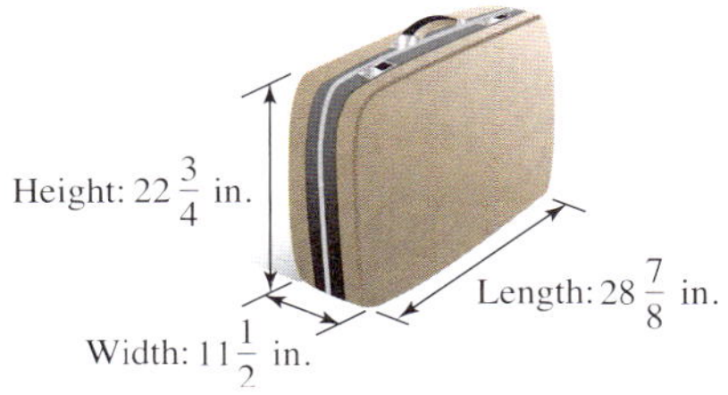

114. **Cooking.** How much butter is left in a $10\frac{1}{2}$-pound tub of butter if $4\frac{3}{4}$ pounds are used to make a wedding cake?

115. Frames. How many inches of molding are needed to make the square picture frame?

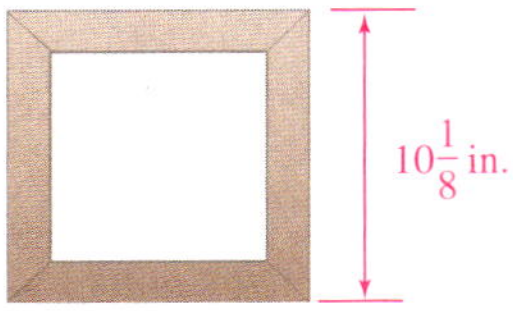

116. Decorating. The materials used to make a pillow are shown below. Examine the inventory list to decide how many pillows can be manufactured in one production run with the materials in stock.

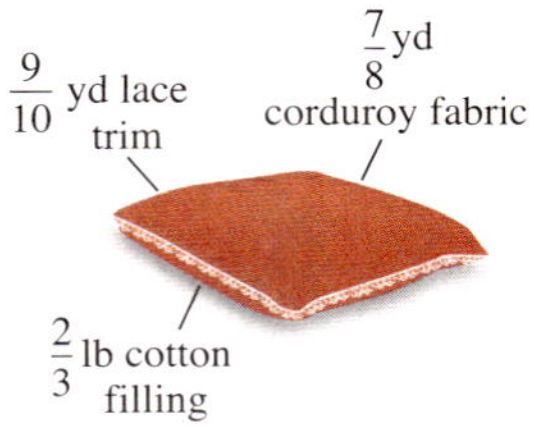

Materials	Amount in stock
Lace trim	135 yd
Corduroy fabric	154 yd
Cotton filling	98 lb

WRITING

117. Explain the error made below in simplifying $\frac{15}{45}$.

$$\frac{15}{45} = \frac{3 \cdot 5}{3 \cdot 3 \cdot 5} = \frac{0}{3} = 0$$

118. To multiply two fractions, must they have like denominators? Explain.

119. What are equivalent fractions? Give an example.

120. Explain the error in the following addition.

$$\frac{4}{3} + \frac{3}{2} = \frac{4 + 3}{3 + 2} = \frac{7}{5}$$

REVIEW

Fill in the blanks.

121. ________ are letters (or symbols) that stand for numbers.

122. A number, such as 10, is called a ________ because it does not change.

CHALLENGE PROBLEMS

123. Which is larger: $\frac{11}{12}$ or $\frac{8}{9}$?

124. If the circle represents a whole, find the missing value.

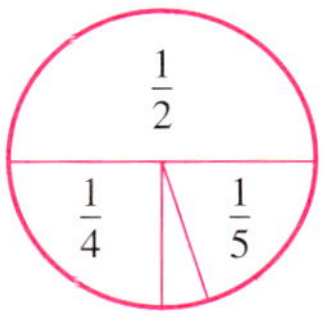

SECTION 1.3

The Real Numbers

OBJECTIVES

1. Define the set of integers.
2. Define the set of rational numbers.
3. Define the set of irrational numbers.
4. Classify real numbers.
5. Graph sets of real numbers on the number line.
6. Find the absolute value of a real number.

ARE YOU READY?

The following problems review several types of numbers that we use in everyday life.

1. Count the number of letters in the word *antidisestablishmentarianism.*
2. What number represents a temperature that is 10 degrees below zero?
3. What type of number is used to express a grade point average (GPA)?
4. Suppose a recipe calls for only part of a full cup of sugar. What type of number is normally used to describe such an amount?

A **set** is a collection of objects, such as a set of golf clubs or a set of dishes. In this section, we will define some important sets of numbers that are used in algebra.

1 Define the Set of Integers.

Natural numbers are the numbers that we use for counting. To write this set, we list its **members** (or **elements**) within **braces** { }.

Natural Numbers	The set of natural numbers is $\{1, 2, 3, 4, 5, \ldots\}$.	Read as "the set containing one, two, three, four, five, and so on."

The natural numbers, together with 0, form the set of **whole numbers.**

Whole Numbers

The set of **whole numbers** is $\{0, 1, 2, 3, 4, 5, \ldots\}$.

Notation

The symbol . . . used in the previous definitions is called an **ellipsis** and it indicates that the established pattern continues forever.

Whole numbers are not adequate for describing many real-life situations. For example, if you write a check for more than what's in your account, the account balance will be less than zero.

We can use the **number line** below to visualize numbers less than zero. A number line is straight and has uniform markings. The arrowheads indicate that it extends forever in both directions. For each natural number on the number line, there is a corresponding number, called its *opposite,* to the left of 0. In the diagram, we see that 3 and -3 (negative three) are opposites, as are -5 (negative five) and 5. Note that 0 is its own opposite.

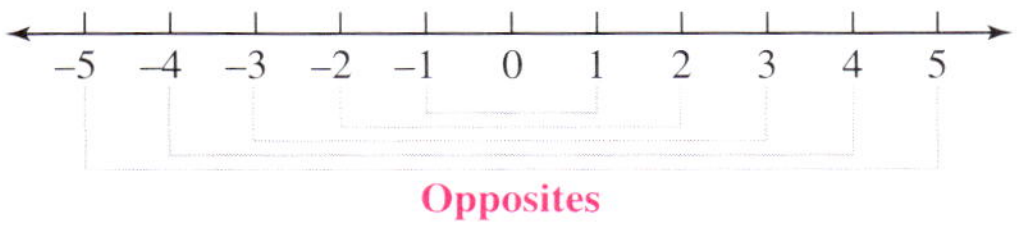

Opposites

Two numbers that are the same distance from 0 on the number line, but on opposite sides of it, are called **opposites.**

The whole numbers, together with their opposites, form the set of **integers.**

Integers

The set of **integers** is $\{\ldots, -4, -3, -2, -1, 0, 1, 2, 3, 4, \ldots\}$.

The Language of Algebra

The **positive integers** are: 1, 2, 3, 4, 5, . . .
The **negative integers** are: $-1, -2, -3, -4, -5, \ldots$
The **nonnegative integers** are: 0, 1, 2, 3, 4, 5, . . .

On the number line, numbers greater than 0 are to the right of 0. They are called **positive numbers.** Positive numbers can be written with or without a **positive sign** $+$. For example, $2 = +2$ (positive two). They are used to describe such quantities as an elevation above sea level (+3,000 ft) or a pay raise ($25).

Numbers less than 0 are to the left of 0 on the number line. They are called **negative numbers.** Negative numbers are always written with a **negative sign** $-$. They are used to describe such quantities as an overdrawn checking account ($-$\$75) or a below-zero temperature ($-12°$).

Positive and negative numbers are called **signed numbers.**

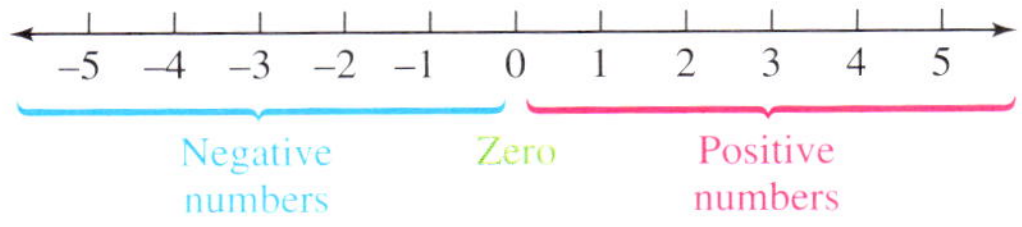

2 Define the Set of Rational Numbers.

We use fractions to describe many situations in daily life. For example, a morning commute might take $\frac{1}{4}$ hour or a recipe might call for $\frac{2}{3}$ cup of sugar. Fractions such as $\frac{1}{4}$ and $\frac{2}{3}$, which are quotients of two integers, are called *rational numbers.*

Rational Numbers

A **rational number** is any number that can be expressed as a fraction (ratio*) with an integer numerator and a nonzero integer denominator.

*Ratios are discussed in more detail in Section 7.8.

The Language of Algebra

Rational numbers are so named because they can be expressed as the *ratio* (quotient) of two integers.

Some other examples of rational numbers are

$$\frac{3}{8}, \quad \frac{41}{100}, \quad \frac{25}{25}, \quad \text{and} \quad \frac{19}{12}$$

To show that negative fractions are rational numbers, we use the following fact.

Negative Fractions

For any numbers a and b where b is not 0,

$$-\frac{a}{b} = \frac{-a}{b} = \frac{a}{-b}$$

To illustrate this rule, we consider $-\frac{11}{16}$ (read as negative eleven-sixteenths). It is a rational number because it can be written as $\frac{-11}{16}$ or as $\frac{11}{-16}$.

Positive and negative mixed numbers are also rational numbers because they can be expressed as fractions. For example,

$$7\frac{5}{8} = \frac{61}{8} \quad \text{and} \quad -6\frac{1}{2} = -\frac{13}{2} = \frac{-13}{2}$$

Any natural number, whole number, or integer can be expressed as a fraction with a denominator of 1. For example, $5 = \frac{5}{1}$, $0 = \frac{0}{1}$, and $-3 = \frac{-3}{1}$. Therefore, every natural number, whole number, and integer is also a rational number.

The Language of Algebra

To **terminate** means to bring to an end. In the movie *The Terminator,* actor Arnold Schwarzenegger plays a heartless machine sent to Earth to bring an end to his enemies.

Many numerical quantities are written in decimal notation. For instance, a candy bar might cost \$0.89, a dragster might travel at 203.156 mph, or a business loss might be −\$4.7 million. These decimals are called **terminating decimals** because their representations terminate (stop). As shown below, terminating decimals can be expressed as fractions. Therefore, terminating decimals are rational numbers.

$$0.89 = \frac{89}{100} \qquad 203.156 = 203\frac{156}{1{,}000} = \frac{203{,}156}{1{,}000} \qquad -4.7 = -4\frac{7}{10} = \frac{-47}{10}$$

Decimals such as 0.3333 . . . and 2.8167167167 . . . , which have a digit (or a block of digits) that repeats, are called **repeating decimals.** Since any repeating decimal can be expressed as a fraction, repeating decimals are rational numbers.

The set of rational numbers cannot be listed in the same way as the natural numbers, the whole numbers, and the integers. Instead, we use **set-builder** notation.

Rational Numbers

The set of rational numbers is

$$\left\{ \frac{a}{b} \,\middle|\, a \text{ and } b \text{ are integers, with } b \neq 0. \right\}$$

Read as "the set of all numbers of the form $\frac{a}{b}$ such that a and b are integers, with $b \neq 0$."

A fraction and its **decimal equivalent** are different forms that represent the same value. To find the decimal equivalent for a fraction, we divide its numerator by its denominator. For example, to write $\frac{1}{4}$ and $\frac{5}{22}$ as decimals, we proceed as follows:

$$\begin{array}{r} 0.25 \\ 4\overline{)1.00} \\ -\ 8 \\ \hline 20 \\ -20 \\ \hline 0 \end{array}$$

Write a decimal point and additional zeros to the right of 1.

The remainder is 0.

$$\begin{array}{r} 0.22727\ldots \\ 22\overline{)5.00000} \\ -4\,4 \\ \hline 60 \\ -44 \\ \hline 160 \\ -154 \\ \hline 60 \\ -44 \\ \hline 160 \end{array}$$

Write a decimal point and additional zeros to the right of 5.

60 and 160 continually appear as remainders. Therefore, 2 and 7 will continually appear in the quotient.

The Language of Algebra

Since every natural number belongs to the set of whole numbers, we say the set of natural numbers is a **subset** of the set of whole numbers. Similarly, the set of whole numbers is a *subset* of the set of integers, and the set of integers is a *subset* of the set of rational numbers.

The decimal equivalent of $\frac{1}{4}$ is 0.25 and the decimal equivalent of $\frac{5}{22}$ is 0.2272727 . . . . We can use an **overbar** to write repeating decimals in more compact form: $0.2272727\ldots = 0.2\overline{27}$. Here are more fractions and their decimal equivalents.

Terminating decimals	*Repeating decimals*
$\frac{1}{2} = 0.5$	$\frac{1}{6} = 0.166666\ldots$ or $0.1\overline{6}$
$\frac{5}{8} = 0.625$	$\frac{1}{3} = 0.333333\ldots$ or $0.\overline{3}$
$\frac{3}{4} = 0.75$	$\frac{5}{11} = 0.454545\ldots$ or $0.\overline{45}$

3 Define the Set of Irrational Numbers.

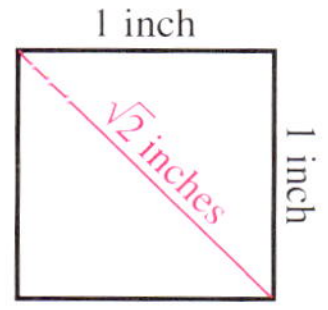

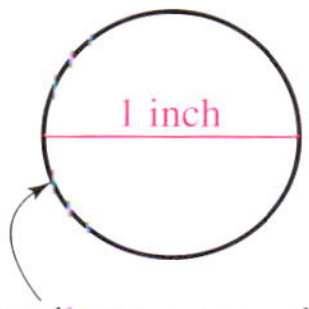

The distance around the circle is π inches.

Not all numbers are rational numbers. One example is the square root of 2, written $\sqrt{2}$. It is the number that, when multiplied by itself, gives 2. That is, $\sqrt{2} \cdot \sqrt{2} = 2$. It can be shown that $\sqrt{2}$ *cannot* be written as a fraction with an integer numerator and an integer denominator. Therefore, it is not rational; it is an *irrational number*. It is interesting to note that a square with sides of length 1 inch has a diagonal that is $\sqrt{2}$ inches long.

The number represented by the Greek letter π (pi) is another example of an irrational number. A circle, with a 1-inch diameter, has a circumference of π inches.

Expressed in decimal form,

$$\sqrt{2} = 1.414213562\ldots \qquad \text{and} \qquad \pi = 3.141592654\ldots$$

These decimals neither terminate nor repeat.

Irrational Numbers

An **irrational number** is a nonterminating, nonrepeating decimal. An irrational number cannot be expressed as a fraction with an integer numerator and an integer denominator.

The Language of Algebra

The prefix **ir** means *not*. For example,

irresponsible ↔ not responsible
irregular ↔ not regular
irrational ↔ not rational

We have seen that $\sqrt{2}$ and π are irrational numbers. Other examples of irrational numbers are:

$$\sqrt{3} = 1.732050808\ldots \qquad -\sqrt{5} = -2.236067977\ldots$$

$$-\pi = -3.141592654\ldots \qquad 3\pi = 9.424777961\ldots$$

3π means $3 \cdot \pi$.

We can use a calculator to approximate the decimal value of an irrational number. To approximate $\sqrt{2}$ using a scientific calculator, we use the square root key $\sqrt{\ }$. To approximate π, we use the *pi* key π.

$$\sqrt{2} \approx 1.414213562 \qquad \text{and} \qquad \pi \approx 3.141592654$$

Read $\approx$ as "is approximately equal to."

Rounded to the nearest thousandth, $\sqrt{2} \approx 1.414$ and $\pi \approx 3.142$.

4 Classify Real Numbers.

The set of **real numbers** is formed by combining the set of rational numbers and the set of irrational numbers. Every real number has a decimal representation. If it is rational, its corresponding decimal terminates or repeats. If it is irrational, its decimal representation is nonterminating and nonrepeating.

The Real Numbers

A **real number** is any number that is a rational number or an irrational number.

The following diagram shows how various sets of numbers are related. Note that a number can belong to more than one set. For example, -6 is an integer, a rational number, and a real number.

The Language of Algebra

The symbol $\mathbb{R}$ is used to represent the set of real numbers.

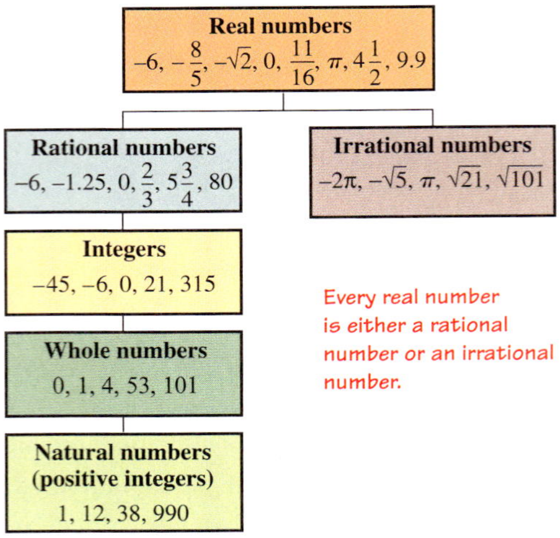

Every real number is either a rational number or an irrational number.

EXAMPLE 1 Which numbers in the following set are natural numbers, whole numbers, integers, rational numbers, irrational numbers, real numbers? $\left\{-3.4,\ \frac{2}{5},\ 0,\ -6,\ 1\frac{3}{4},\ \pi,\ 16\right\}$

Strategy We begin by scanning the given set, looking for any natural numbers. Then we scan it five more times, looking for whole numbers, for integers, for rational numbers, for irrational numbers, and finally, for real numbers.

Why We need to scan the given set of numbers six times, because numbers in the set can belong to more than one classification.

Solution

Natural numbers: 16 — 16 is a member of $\{1, 2, 3, 4, 5, \ldots\}$.

Whole numbers: 0, 16 — 0 and 16 are members of $\{0, 1, 2, 3, 4, 5, \ldots\}$.

Integers: $0, -6, 16$ — $0, -6$, and 16 are members of $\{\ldots, -3, -2, -1, 0, 1, 2, 3, \ldots\}$.

Rational numbers:

$-3.4,\ \frac{2}{5},\ 0,\ -6,\ 1\frac{3}{4},\ 16$ — A rational number can be expressed as a ratio of two integers: $-3.4 = \frac{-34}{10}$, $0 = \frac{0}{1}$, $-6 = \frac{-6}{1}$, $1\frac{3}{4} = \frac{7}{4}$, and $16 = \frac{16}{1}$.

Irrational numbers: π — $\pi = 3.1415\ldots$ is a nonterminating, nonrepeating decimal.

Real numbers:

$-3.4,\ \frac{2}{5},\ 0,\ -6,\ 1\frac{3}{4},\ \pi,\ 16$ — Every natural number, whole number, integer, rational number, and irrational number is a real number.

Self Check 1 Use the instructions for Example 1 with: $\left\{0.1,\ -\frac{2}{7},\ 45,\ -2,\ \frac{13}{4},\ -6\frac{7}{8}\right\}$

Now Try Problem 27

5 Graph Sets of Real Numbers on the Number Line.

Every real number corresponds to a point on the number line, and every point on the number line corresponds to exactly one real number. As we move right on the number line, the values of the numbers increase. As we move left, the values decrease. On the following number line, we see that 5 is greater than −3, because 5 lies to the right of −3. Similarly, −3 is less than 5, because it lies to the left of 5.

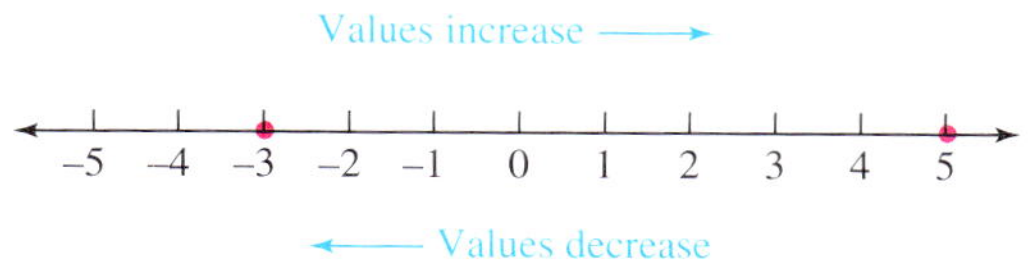

The Language of Algebra

The prefix **in** means *not.* For example:

inaccurate ↔ not accurate
inexpensive ↔ not expensive
inequality ↔ not equal

The **inequality symbol** > means "is greater than." It is used to show that one number is greater than another. The inequality symbol < means "is less than." It is used to show that one number is less than another. For example,

$5 > -3$ Read as "5 is greater than −3."

$-3 < 5$ Read as "−3 is less than 5."

To distinguish between these inequality symbols, remember that each one points to the smaller of the two numbers involved.

$5 > -3$ $\qquad$ $-3 < 5$

Points to the smaller number.

EXAMPLE 2 Use one of the symbols > or < to make each statement true:

a. $-4 \quad 4$ **b.** $-2 \quad -3$ **c.** $4.47 \quad 12.5$ **d.** $\frac{3}{4} \quad \frac{5}{8}$

Strategy To pick the correct inequality symbol to place between a given pair of numbers, we need to determine the position of each number on a number line.

Why For any two numbers on a number line, the number to the *left* is the smaller number and the number to the *right* is the larger number.

Solution

a. Since −4 is to the left of 4 on the number line, we have $-4 < 4$.

b. Since −2 is to the right of −3 on the number line, we have $-2 > -3$.

c. Since 4.47 is to the left of 12.5 on the number line, we have $4.47 < 12.5$.

d. To compare fractions, express them in terms of the same denominator, preferably the LCD. If we write $\frac{3}{4}$ as an equivalent fraction with denominator 8, we see that $\frac{3}{4} = \frac{3}{4} \cdot \frac{2}{2} = \frac{6}{8}$. Therefore, $\frac{3}{4} > \frac{5}{8}$.

To compare the fractions, we also could convert each to its decimal equivalent. Since $\frac{3}{4} = 0.75$ and $\frac{5}{8} = 0.625$, we know that $\frac{3}{4} > \frac{5}{8}$.

Self Check 2 Use one of the symbols < or > to make each statement true:

a. $1 \quad -1$ **b.** $-5 \quad -4$ **c.** $6.7 \quad 4.999$ **d.** $\frac{3}{5} \quad \frac{2}{3}$

Now Try Problems 37 and 42

To **graph a number** means to mark its position on the number line.

EXAMPLE 3 Graph each number in the set: $\left\{-2.43, \sqrt{2}, 1, -0.\overline{3}, 2\frac{5}{6}, -\frac{3}{2}\right\}$

Strategy We locate the position of each number on the number line, draw a bold dot, and label it.

Why To *graph a number* means to make a drawing that represents the number.

Solution It is helpful to approximate the value of a number or to write the number in an equivalent form to determine its location on a number line.

- To locate -2.43, we round it to the nearest tenth: $-2.43 \approx -2.4$.
- To locate $\sqrt{2}$, we use a calculator: $\sqrt{2} \approx 1.4$.
- To locate $-0.\overline{3}$, we recall that $0.\overline{3} = 0.333\ldots = \frac{1}{3}$. Therefore, $-0.\overline{3} = -\frac{1}{3}$.
- In mixed-number form, $-\frac{3}{2} = -1\frac{1}{2}$. This is midway between -1 and -2.

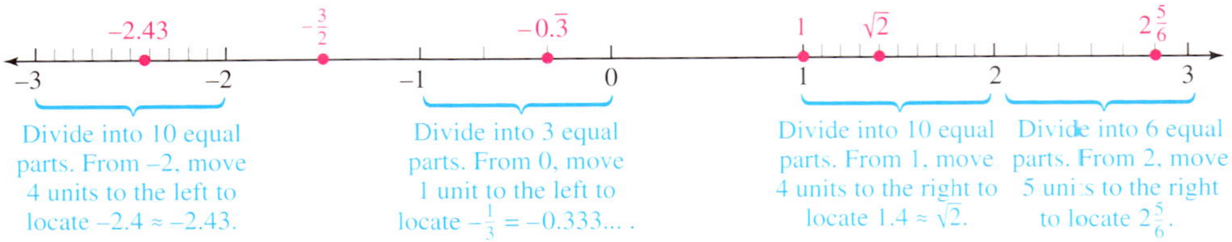

Self Check 3 Graph each number in the set: $\{1.7,\ \pi,\ -1\frac{3}{4},\ 0.\overline{6},\ \frac{5}{2},\ -3\}$

Now Try Problem 57

6 Find the Absolute Value of a Real Number.

A number line can be used to measure the distance from one number to another. For example, on the number line below, we see that the distance from 0 to -4 is 4 units and the distance from 0 to 3 is 3 units.

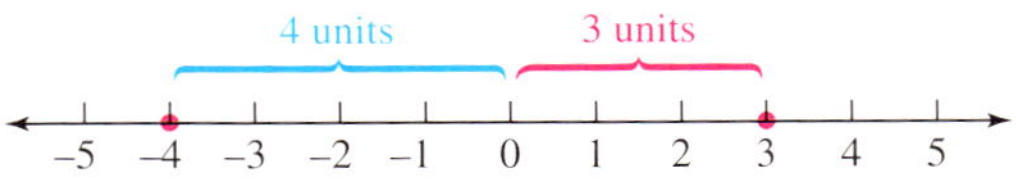

To express the distance that a number is from 0 on a number line, we can use *absolute value*.

Absolute Value The absolute value of a number is its distance from 0 on the number line.

To indicate the absolute value of a number, we write the number between two vertical bars. From the figure above, we see that $|-4| = 4$. This is read as "the absolute value of negative 4 is 4" and it tells us that the distance from 0 to -4 is 4 units. It also follows from the figure that $|3| = 3$.

EXAMPLE 4 Find each absolute value: **a.** $|18|$ **b.** $\left|-\frac{7}{8}\right|$ **c.** $|98.6|$ **d.** $|0|$

Strategy We need to determine the distance that the number within the vertical absolute value bars is from 0.

Why The absolute value of a number is the distance between 0 and the number on a number line.

Solution **a.** Since 18 is a distance of 18 from 0 on the number line, $|18| = 18$.

b. Since $-\frac{7}{8}$ is a distance of $\frac{7}{8}$ from 0 on the number line, $\left|-\frac{7}{8}\right| = \frac{7}{8}$.

Success Tip

Since absolute value expresses distance, the absolute value of a number is always positive or zero, but never negative.

c. Since 98.6 is a distance of 98.6 from 0 on the number line, $|98.6| = 98.6$.

d. Since 0 is a distance of 0 from 0 on the number line, $|0| = 0$.

Self Check 4 Find each absolute value: **a.** $|100|$ **b.** $|-4.7|$ **c.** $|\sqrt{2}|$

Now Try Problems 61 and 67

SECTION 1.3 STUDY SET

VOCABULARY

Fill in the blanks.

1. The set of ______ numbers is $\{0, 1, 2, 3, 4, 5, \ldots\}$.
2. The set of ______ numbers is $\{1, 2, 3, 4, 5, \ldots\}$.
3. The figure 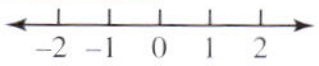 is called a ______ ______.
4. The set of ______ is $\{\ldots, -2, -1, 0, 1, 2, \ldots\}$.
5. Positive and negative numbers are called ______ numbers.
6. 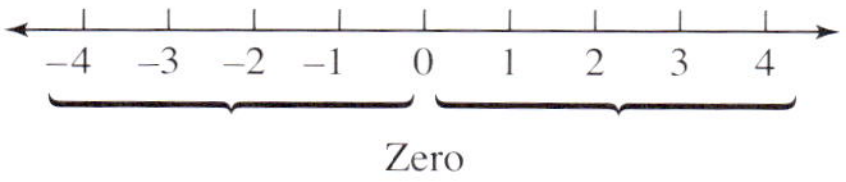

7. The symbols < and > are ______ symbols.
8. A ______ number is any number that can be expressed as a fraction with an integer numerator and a nonzero integer denominator.
9. 0.25 is called a ______ decimal and 0.333 . . . is called a ______ decimal.
10. An ______ number cannot be expressed as a quotient of two integers.
11. An irrational number is a nonterminating, nonrepeating ______.
12. The ______ ______ of a number is the distance on the number line between the number and 0.

CONCEPTS

13. Represent each situation using a signed number.
    a. A loss of \$15 million
    b. A building foundation $\frac{5}{16}$ inch above grade
14. Show that each of the following numbers is a rational number by expressing it as a fraction with an integer numerator and a nonzero integer denominator: $6, -9, -\frac{7}{8}, 3\frac{1}{2}, -0.3, 2.83$.
15. Give the opposite of each number.
    a. 20 b. $-\frac{2}{3}$
16. What two numbers are a distance of 8 away from 5 on the number line?
17. What two numbers are a distance of 5 away from −9 on the number line?
18. Refer to the number line below. Use an inequality symbol, < or >, to make each statement true.
    a. a ___ b b. b ___ a
    c. b ___ 0 and a ___ 0 d. $|a|$ ___ $|b|$

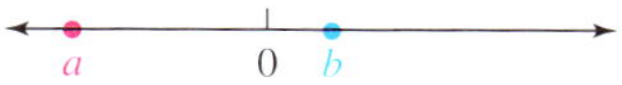

NOTATION

Fill in the blanks.

19. $\sqrt{2}$ is read "the ______ ______ of 2."
20. $|-15|$ is read "the ______ ______ of −15."
21. The symbol ≈ means ______ ______ ______ ______.
22. The symbols { } are called ______.
23. The symbol π is a letter from the ______ alphabet.
24. To find the decimal equivalent for $\frac{2}{3}$, we perform the following long division: $\overline{)\quad}$
25. $-\frac{4}{5} = \frac{\quad}{5} = \frac{4}{\quad}$
26. Write each repeating decimal using an overbar.
    a. 0.666 . . . b. 0.2444 . . .
    c. 0.717171 . . . d. 0.456456456 . . .

GUIDED PRACTICE

Place check marks in the table to show the set or sets to which each number belongs. For example, the check shows that $\sqrt{2}$ is irrational. See Example 1.

27.

	5	0	−3	$\frac{7}{8}$	0.17	$-9\frac{1}{4}$	$\sqrt{2}$	π
Real								
Irrational							✓	
Rational								
Integer								
Whole								
Natural								

28. Which numbers in the following set are natural numbers, whole numbers, integers, rational numbers, irrational numbers, real numbers? $\left\{67, \frac{4}{13}, -5.9, 11\frac{2}{3}, \sqrt{2}, 0, -3, \pi\right\}$

Determine whether each statement is true or false. See Example 1.

29. Every whole number is an integer.
30. Every integer is a natural number.
31. Every integer is a whole number.
32. Every real number is either a rational number or an irrational number.
33. Irrational numbers are real numbers.
34. Every whole number is a rational number.
35. Every rational number can be written as a fraction (ratio) of two integers.
36. Every rational number is a whole number.

Use one of the symbols < or > to make each statement true. See Example 2.

37. $0 \quad -4$
38. $0 \quad 32$
39. $917 \quad 971$
40. $898 \quad 889$
41. $-2 \quad -3$
42. $-5 \quad -4$
43. $-\frac{5}{8} \quad -\frac{3}{8}$
44. $-19\frac{2}{3} \quad -19\frac{1}{3}$
45. $\frac{2}{3} \quad \frac{3}{5}$
46. $\frac{3}{4} \quad \frac{5}{6}$
47. $-6.19 \quad -5.8$
48. $-2.27 \quad -5.25$

Write each fraction as a decimal. If the result is a repeating decimal, use an overbar. See Objective 2.

49. $\frac{5}{8}$
50. $\frac{3}{32}$
51. $\frac{1}{30}$
52. $\frac{7}{9}$
53. $\frac{1}{60}$
54. $\frac{5}{11}$
55. $\frac{21}{50}$
56. $\frac{2}{125}$

Graph each set of numbers on a number line. See Example 3.

57. $\left\{-\pi, 4.25, -1\frac{1}{2}, -0.333\ldots, \sqrt{2}, -\frac{35}{8}, 3\right\}$
58. $\left\{-2\frac{1}{8}, \pi, 2.75, -\sqrt{2}, \frac{17}{4}, 0.666\ldots, -3\right\}$
59. The integers between −5 and 2
60. The whole numbers less than 4

Find each absolute value. See Example 4.

61. $|83|$
62. $|29|$
63. $\left|\frac{4}{3}\right|$
64. $\left|\frac{9}{16}\right|$
65. $|-11|$
66. $|-14|$
67. $|-6.1|$
68. $|-25.3|$

Insert one of the symbols >, <, or = in the blank to make each statement true. See Examples 2 and 4.

69. $|3.4| \quad -3$
70. $0.08 \quad 0.079$
71. $|-1.1| \quad 1.2$
72. $-5.5 \quad -5\frac{1}{2}$
73. $\left|-\frac{15}{2}\right| \quad 7.5$
74. $\left|-2\frac{2}{3}\right| \quad \frac{7}{3}$
75. $\frac{99}{100} \quad 0.99$
76. $|2| \quad |-2|$
77. $0.3 \quad 0.333\ldots$
78. $-0.666\ldots \quad -0.6$
79. $1 \quad \left|-\frac{15}{16}\right|$
80. $\sqrt{2} \quad \pi$

APPLICATIONS

81. **Drafting.** Which dimensions of the aluminum bracket shown below are natural numbers, whole numbers, integers, rational numbers, irrational numbers, and real numbers?

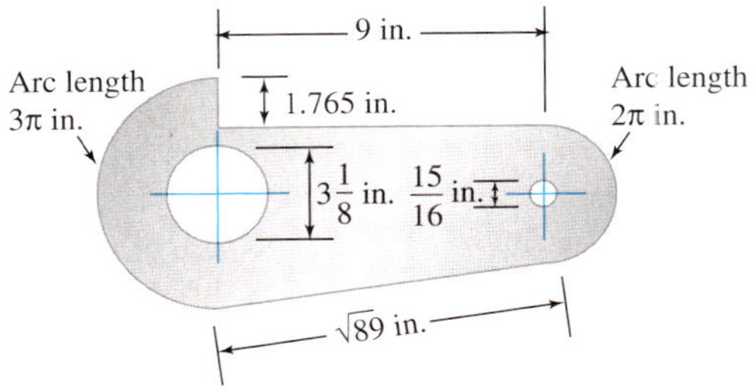

82. **History.** Refer to the time line shown below.
   a. What basic unit was used to scale the time line?
   b. What symbolism is used to represent zero?
   c. Which numbers could be thought of as positive and which as negative?
   d. Express the dates for the Maya civilization using positive and negative numbers.

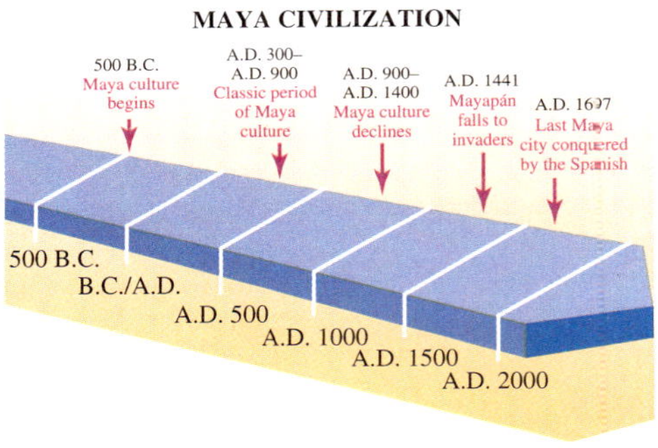

Based on data from *People in Time and Place, Western Hemisphere* (Silver Burdett & Ginn, 1991), p. 129.

83. **iPhones.** You can get a more accurate reading of an iPhone's signal strength by dialing *3001#12345#*. Field test mode is then activated and the standard signal strength bars (in the upper left corner of the display) are replaced by a negative number. The closer the negative number is to zero, the stronger the signal. Which iPhone shown below is receiving the strongest signal?

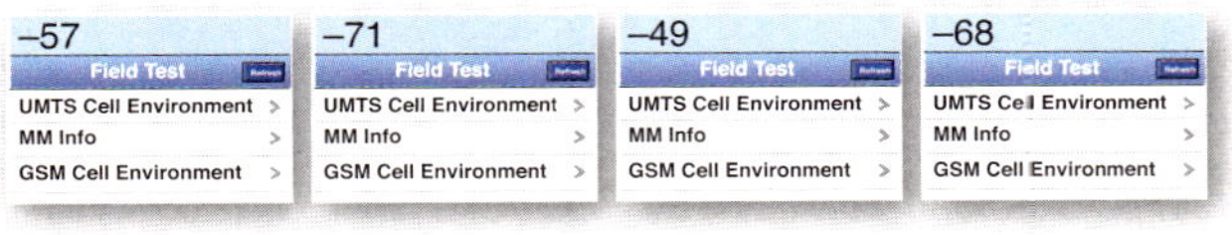

(i) (ii) (iii) (iv)

84. Drafting. On an architect's scale, the edge marked 16 divides each inch into 16 equal parts. Find the decimal form for each fractional part of one inch that is highlighted on the scale.

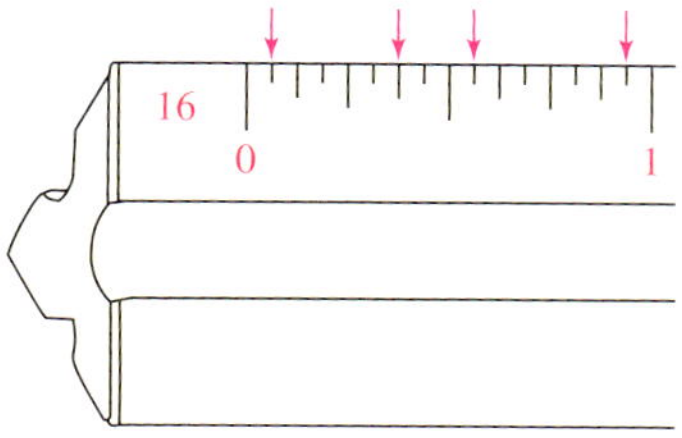

85. Trade. Each year from 1994 through 2009, the United States imported more goods and services from Japan than it exported to Japan. This caused trade *deficits*, which are represented by negative numbers on the following graph.

a. In which year was the deficit the worst? Express that deficit using a signed number.

b. In which year was the deficit the smallest? Express that deficit using a signed number.

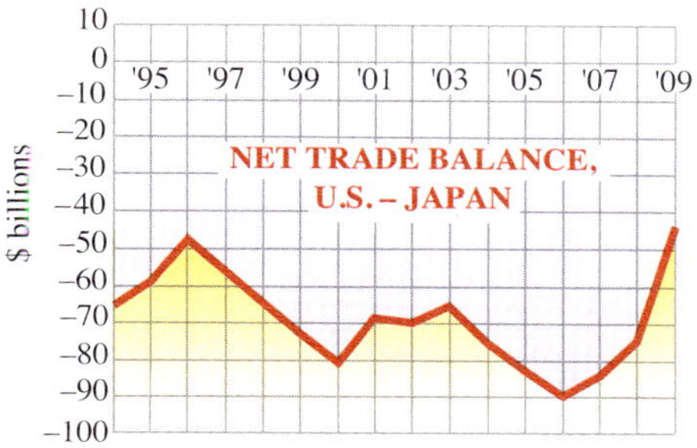

Source: U.S. Bureau of the Census

86. U.S. Budget. A budget *deficit* is a negative number that indicates the government spent more money than it took in that year. A budget *surplus* is a positive number that indicates the government took in more money than it spent that year.

a. Refer to the graph in the next column that shows the U.S. Federal Budget Deficit/Surplus for the years 1980 through 2009. For how many of those years was there a budget surplus?

b. Consider the years in which there was a budget deficit. For how many of those years was it smaller than $300 billion?

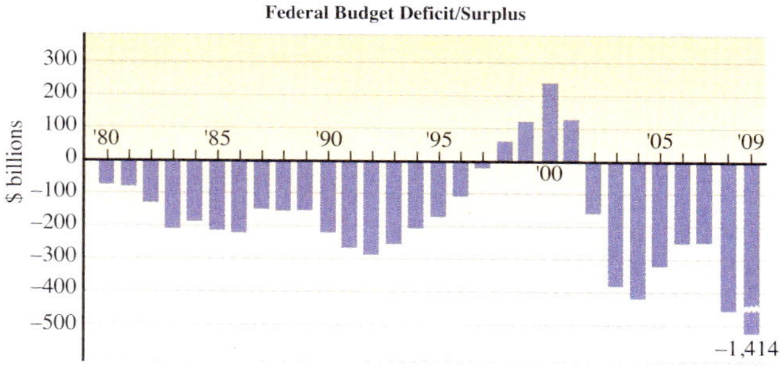

Source: U.S. Bureau of the Census

WRITING

87. Explain the difference between a rational and an irrational number.

88. Can two different numbers have the same absolute value? Explain.

89. Explain how to find the decimal equivalent of a fraction.

90. What is a real number?

91. *Pi Day* (or *Pi Approximation Day*) is an unofficial holiday held to celebrate π. Why do you think Pi Day is observed each year on March 14?

92. Explain why $0.1\overline{333}$ is not the simplest way to represent 0.1333. . . .

REVIEW

93. Simplify: $\frac{24}{54}$

94. Multiply: $\frac{3}{4}\left(\frac{8}{5}\right)$

95. Divide: $5\frac{2}{3} \div 2\frac{5}{9}$

96. Add: $\frac{3}{10} + \frac{2}{15}$

CHALLENGE PROBLEMS

97. How many integers have an absolute value that is less than 1,000?

98. Is 0.10100100010000 . . . a repeating decimal? Explain.

Find a rational number between each pair of numbers.

99. $\frac{1}{8}$ and $\frac{1}{9}$

100. $1.7\overline{1}$ and $1.7\overline{2}$

SECTION 1.4 Adding Real Numbers; Properties of Addition

OBJECTIVES

1. Add two numbers that have the same sign.
2. Add two numbers that have different signs.
3. Use properties of addition.
4. Identify opposites (additive inverses).

ARE YOU READY?

The following problems review some basic concepts that are important when adding positive and negative real numbers.

1. Find $|3|$ and $|-5|$. Which number, 3 or -5, has the larger absolute value?
2. Add: $4.37 + 2.8$
3. Subtract: $710 - 89$
4. Subtract: $\frac{4}{5} - \frac{2}{3}$

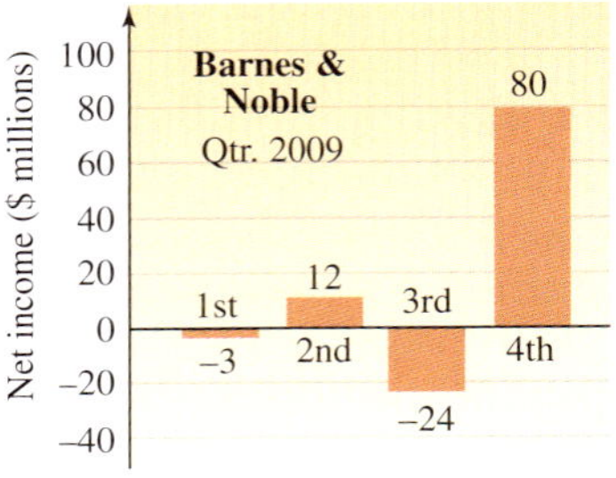

Source: msnbc.msn.com

In the graph to the left, signed numbers are used to show the financial performance of Barnes and Noble Corporation for the year 2009. Positive numbers indicate *profits* and negative numbers indicate *losses*. To find the company's 2009 net income (in millions of dollars), we need to calculate the following sum:

$$\text{Net income} = -3 + 12 + (-24) + 80$$

In this section, we discuss how to perform this addition and others involving signed numbers.

1 Add Two Numbers That Have the Same Sign.

A number line can be used to explain the addition of signed numbers. For example, to compute $5 + 2$, we begin at 0 and draw an arrow five units long that points right. It represents 5. From the tip of that arrow, we draw a second arrow two units long that points right. It represents 2. Since we end up at 7, it follows that $5 + 2 = 7$. The numbers that we added, 5 and 2, are called **addends,** and the result, 7, is called the **sum.**

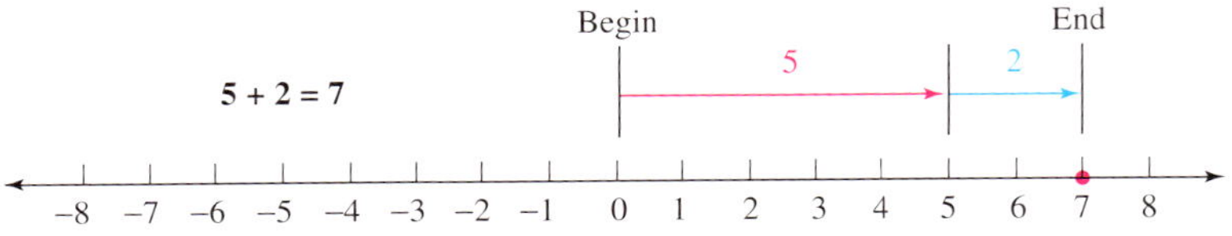

To compute $-5 + (-2)$, we begin at 0 and draw an arrow five units long that points left. It represents -5. From the tip of that arrow, we draw a second arrow two units long that points left. It represents -2. Since we end up at -7, it follows that $-5 + (-2) = -7$.

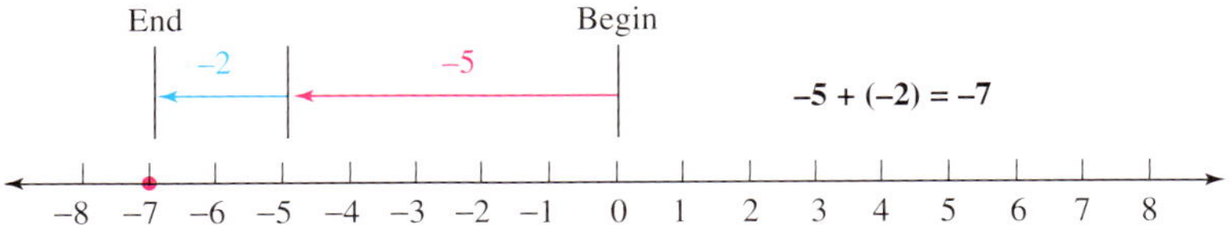

To check this result, think of the problem in terms of money. If you lost \$5 (-5) and then lost another \$2 (-2), you would have lost a total of \$7 (-7).

When we use a number line to add numbers with the same sign, the arrows point in the same direction and they build upon each other. Furthermore, the answer has the same sign as the numbers that we added. These observations suggest the following rules.

Notation

To avoid confusion, we write negative numbers within parentheses to separate the negative sign $-$ from the addition symbol $+$.

$$-5 + (-2)$$

Adding Two Numbers That Have the Same (Like) Signs

1. To add two positive numbers, add them as usual. The final answer is positive.
2. To add two negative numbers, add their absolute values and make the final answer negative.

EXAMPLE 1 Add: **a.** $-20 + (-15)$ **b.** $-7.89 + (-0.6)$ **c.** $-\frac{1}{3} + \left(-\frac{1}{2}\right)$

Strategy We will use the rule for adding two numbers that have the same sign.

Why In each case, we are asked to add two negative numbers.

Solution

a. $-20 + (-15) = -35$ Add their absolute values, 20 and 15, to get 35. Then make the final answer negative.

The Language of Algebra

Two negative numbers, as well as two positive numbers, are said to have **like** signs.

b. To find $-7.89 + (-0.6)$, add their absolute values, 7.89 and 0.6.

$$\begin{array}{r} 7.89 \\ +0.6 \\ \hline 8.49 \end{array}$$

Remember to align the decimal points when adding decimals.

Then make the final answer negative: $-7.89 + (-0.6) = -8.49$.

Success Tip

The sum of two positive numbers is *always* positive. The sum of two negative numbers is *always* negative.

c. To find $-\frac{1}{3} + \left(-\frac{1}{2}\right)$, add their absolute values, $\frac{1}{3}$ and $\frac{1}{2}$.

$$\frac{1}{3} + \frac{1}{2} = \frac{2}{6} + \frac{3}{6}$$ The LCD is 6. Build each fraction: $\frac{1}{3} \cdot \frac{2}{2} = \frac{2}{6}$ and $\frac{1}{2} \cdot \frac{3}{3} = \frac{3}{6}$.

$$= \frac{5}{6}$$ Add the numerators and write the sum over the LCD.

Then make the final answer negative: $-\frac{1}{3} + \left(-\frac{1}{2}\right) = -\frac{5}{6}$.

Self Check 1 Add: **a.** $-51 + (-9)$ **b.** $-12.3 + (-0.88)$ **c.** $-\frac{1}{4} + \left(-\frac{2}{3}\right)$

Now Try Problems 15, 21, and 25

2 Add Two Numbers That Have Different Signs.

To compute $5 + (-2)$, we begin at 0 and draw an arrow five units long that points right. From the tip of that arrow, we draw a second arrow two units long that points left. Since we end up at 3, it follows that $5 + (-2) = 3$. In terms of money, if you won \$5 and then lost \$2, you would have \$3 left.

The Language of Algebra

A positive number and a negative number are said to have **unlike** signs.

To compute $-5 + 2$, we begin at 0 and draw an arrow five units long that points left. From the tip of that arrow, we draw a second arrow two units long that points right. Since we end up at -3, it follows that $-5 + 2 = -3$. In terms of money, if you lost \$5 and then won \$2, you have lost \$3.

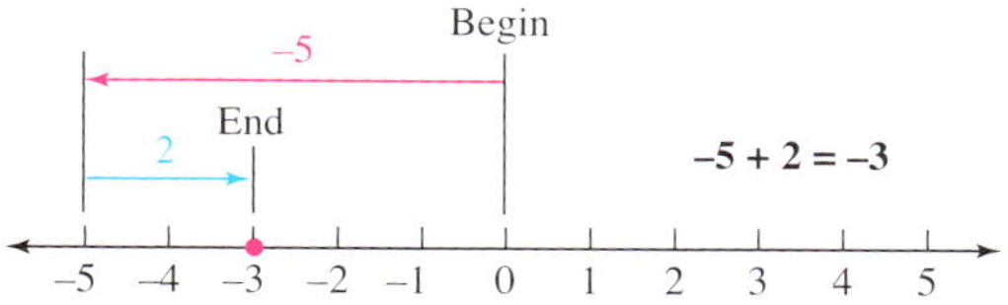

When we use a number line to add numbers with different signs, the arrows point in opposite directions and the longer arrow determines the sign of the answer. If the longer arrow represents a positive number, the sum is positive. If it represents a negative number, the sum is negative. These observations suggest the following rules.

Adding Two Numbers That Have Different (Unlike) Signs

To add a positive number and a negative number, subtract the smaller absolute value from the larger.

1. If the positive number has the larger absolute value, the final answer is positive.
2. If the negative number has the larger absolute value, make the final answer negative.

EXAMPLE 2 Add: **a.** $-20 + 32$ **b.** $5.7 + (-7.4)$ **c.** $-\frac{19}{25} + \frac{2}{5}$

Strategy We will use the rule for adding two numbers that have different (unlike) signs.

Why In each case, we are asked to add a positive number and a negative number.

Solution

a. $-20 + 32 = 12$ Subtract the smaller absolute value from the larger: $32 - 20 = 12$. The positive number, 32, has the larger absolute value, so the final answer is positive.

b. To find $5.7 + (-7.4)$, subtract the smaller absolute value, 5.7, from the larger, 7.4.

$$\begin{array}{r} 7.4 \\ -5.7 \\ \hline 1.7 \end{array}$$

Remember to align the decimal points when subtracting decimals.

Since the negative decimal, -7.4, has the larger absolute value, make the final answer negative: $5.7 + (-7.4) = -1.7$.

c. Since $\frac{2}{5} = \frac{10}{25}$, the fraction $-\frac{19}{25}$ has the larger absolute value. To find $-\frac{19}{25} + \frac{2}{5}$, we subtract the smaller absolute value from the larger:

$$\frac{19}{25} - \frac{2}{5} = \frac{19}{25} - \frac{10}{25}$$ The LCD is 25. Build $\frac{2}{5}$: $\frac{2}{5} \cdot \frac{5}{5} = \frac{10}{25}$.

$$= \frac{9}{25}$$ Subtract the numerators and write the difference over the LCD.

Since the negative fraction $-\frac{19}{25}$ has the larger absolute value, make the final answer negative: $-\frac{19}{25} + \frac{10}{25} = -\frac{9}{25}$.

Success Tip

The sum of two numbers with different signs may be positive or negative. The sign of the sum is the sign of the number with the greater absolute value.

Calculators

Entering negative numbers
We don't do anything special to enter positive numbers on a calculator. To enter a negative number, say -7.4, some calculators require the $-$ sign to be entered before entering 7.4 while others require the $-$ sign to be entered after entering 7.4. Consult your owner's manual to determine the proper keystrokes.

Self Check 2 Add: **a.** $63 + (-87)$ **b.** $-6.27 + 8$ **c.** $-\frac{1}{10} + \frac{1}{2}$

Now Try Problems 29, 33, and 35

EXAMPLE 3 **Accounting.** Find the net income of Barnes and Noble Corporation for the year 2009 using the data in the graph on page 32.

Strategy To find the net income, we will add the quarterly profits and losses (in millions of dollars), performing the additions as they occur from left to right.

Why The phrase *net income* means that we should combine (add) the quarterly profits and losses to determine whether there was an overall profit or loss that year.

Solution

$$-3 + 12 + (-24) + 80 = 9 + (-24) + 80 \quad \text{Add: } -3 + 12 = 9.$$
$$= -15 + 80 \quad \text{Add: } 9 + (-24) = -15.$$
$$= 65$$

In 2009, Barnes and Noble's net income was \$65 million.

Self Check 3 Add: $650 + (-13) + 87 + (-155)$

Now Try Problem 43

The Language of Algebra

Net refers to what remains after all the deductions (losses) have been accounted for. *Net income* is a term used in business that often is referred to as the *bottom line.* Net income indicates what a company has earned (or lost) in a given period of time (usually one year).

3 Use Properties of Addition.

The addition of two numbers can be done in any order and the result is the same. For example, $8 + (-1) = 7$ and $-1 + 8 = 7$. This example illustrates that addition is **commutative.**

The Commutative Property of Addition

Changing the order when adding does not affect the answer.

For any real numbers a and b,

$$a + b = b + a$$

The Language of Algebra

Commutative is a form of the word *commute,* meaning to go back and forth. *Commuter* trains take people to and from work.

In the following example, we add $-3 + 7 + 5$ in two ways. We will use grouping symbols ( ), called **parentheses,** to show this. Standard practice requires that the operation within the parentheses be performed first.

We read $(-3 + 7) + 5$ as "The quantity of –3 plus 7" pause slightly, and then say "plus 5." We read $-3 + (7 + 5)$ as "–3" pause slightly, and then say "plus the quantity of 7 plus 5." The word **quantity** alerts the reader to the parentheses that are used as grouping symbols.

Method 1: Group −3 and 7	*Method 2: Group 7 and 5*
$(-3 + 7) + 5 = 4 + 5$	$-3 + (7 + 5) = -3 + 12$
$= 9$	$= 9$

It doesn't matter how we group the numbers in this addition; the result is 9. This example illustrates that addition is **associative.**

The Associative Property of Addition

Changing the grouping when adding does not affect the answer.

For any real numbers a, b, and c,

$$(a + b) + c = a + (b + c)$$

Sometimes, an application of the associative property can simplify a computation.

EXAMPLE 4 Find the sum: $98 + (2 + 17)$

Strategy We will use the associative property to group 2 with 98. Then, we evaluate the expression by performing the addition within the parentheses first.

Why It is helpful to regroup because 98 and 2 are a pair of numbers that are easily added.

Solution

$$98 + (2 + 17) = (\mathbf{98 + 2}) + 17$$ Use the associative property of addition to regroup. Note that the order of the addends, 98, 2, and 17, is not changed.

$$= \mathbf{100} + 17$$ Do the addition within the parentheses first.

$$= 117$$

The Language of Algebra

Associative is a form of the word *associate,* meaning to join a group. The WNBA (Women's National Basketball Association) is a group of 14 professional basketball teams.

Self Check 4 Find the sum: $(39 + 25) + 75$

Now Try Problem 49

EXAMPLE 5

Game Shows. A contestant on *Jeopardy!* correctly answered the first question to win \$100, missed the second to lose \$200, correctly answered the third to win \$300, and missed the fourth to lose \$400. What is her score after answering four questions?

Strategy We can represent money won by a positive number and money lost by a negative number. Her score is the sum of 100, −200, 300, and −400. Instead of doing the additions from left to right, we will use another approach. Applying the commutative and associative properties, we will add the positives, add the negatives, and then add those results.

Why It is easier to add numbers that have the same sign than numbers that have different signs. This method minimizes the possibility of an error, because we have to add numbers that have different signs only once.

Solution

$$100 + (-200) + 300 + (-400)$$

$$= (\mathbf{100} + \mathbf{300}) + [(\mathbf{-200}) + (\mathbf{-400})]$$ Reorder the numbers. Group the positives together. Group the negatives together using brackets [ ].

$$= \mathbf{400} + (\mathbf{-600})$$ Add the positives. Add the negatives.

$$= -200$$ Add the results.

After four questions, her score was −\$200, which represents a loss of \$200.

Self Check 5 Add: $-6 + 1 + (-4) + (-5) + 9$

Now Try Problem 45

The Language of Algebra

Identity is a form of the word *identical,* meaning the same. You probably have seen *identical* twins.

Whenever we add 0 to a number, the result is the number. Therefore, $8 + 0 = 8$, $2.3 + 0 = 2.3$, and $0 + (-16) = -16$. These examples illustrate the **addition property of 0.** Since any number added to 0 remains the same, 0 is called the **identity element** for addition.

Addition Property of 0 (Identity Property of Addition)

When 0 is added to any real number, the result is the same real number.

For any real number a,

$$a + 0 = a \quad \text{and} \quad 0 + a = a$$

4 Identify Opposites (Additive Inverses).

Recall that two numbers that are the same distance from 0 on a number line, but on opposite sides of it, are called **opposites.** To develop a property for adding opposites, we will find $-4 + 4$ using a number line. We begin at 0 and draw an arrow four units long that points left, to represent −4. From the tip of that arrow, we draw a second arrow, four units long that points right, to represent 4. We end up at 0; therefore, $-4 + 4 = 0$.

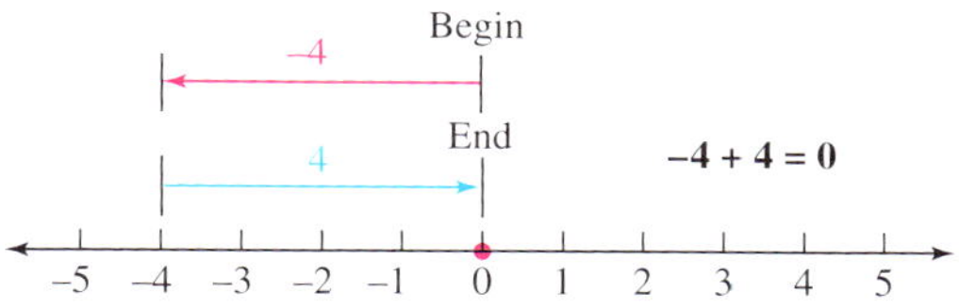

This example illustrates that when we add opposites, the result is 0. Therefore, $1.6 + (-1.6) = 0$ and $-\frac{3}{4} + \frac{3}{4} = 0$. Also, whenever the sum of two numbers is 0, those numbers are opposites. For these reasons, opposites also are called **additive inverses.**

Addition Property of Opposites (Inverse Property of Addition)

The sum of a number and its opposite (additive inverse) is 0.

For any real number a and its opposite or additive inverse $-a$,

$$a + (-a) = 0$$ Read $-a$ as "the opposite of a."

EXAMPLE 6 Add: $12 + (-5) + 6 + 5 + (-12)$

Strategy Instead of working from left to right, we will use the commutative and associative properties of addition to add pairs of opposites.

Why Since the sum of a number and its opposite is 0, it is helpful to identify such pairs in an addition.

Solution

opposites (12 and −12)

$$12 + (-5) + 6 + 5 + (-12) = 0 + 0 + 6$$
$$= 6$$

opposites (−5 and 5)

Self Check 6 Add: $8 + (-1) + 6 + 5 + (-8) + 1$

Now Try Problem 73

SECTION 1.4 STUDY SET

VOCABULARY

Fill in the blanks.

1. In the addition statement $-2 + 5 = 3$, the result, 3, is called the _____.
2. Two numbers that are the same distance from 0 on a number line, but on opposite sides of it, are called _________ or additive _________.
3. The ____________ property of addition states that changing the order when adding does not affect the answer. The ___________ property of addition states that changing the grouping when adding does not affect the answer.
4. Since any number added to 0 remains the same (is identical), the number 0 is called the ________ element for addition.

CONCEPTS

5. For each pair of numbers, which one has the larger absolute value?
   a. 6 or 5
   b. 8.9 or −9.2
6. Determine whether each statement is true or false.
   a. The sum of a number and its opposite is always 0.
   b. The sum of two negative numbers is always negative.
   c. The sum of two numbers with different signs is always negative.
7. For each addition, just determine the sign of the answer.
   a. $39.6 + (-64.9)$
   b. $-18.9 + 19.8$

8. Complete each property of addition. Then give its name.
   a. $a + (-a) =$
   b. $a + 0 =$
   c. $a + b = b +$
   d. $(a + b) + c = a +$

9. Use the commutative property of addition to complete each statement.
   a. $-5 + 1 =$
   b. $15 + (-80.5) =$
   c. $-20 + (4 + 20) = -20 + (\quad)$
   d. $(2.1 + 3) + 6 = (\quad) + 6$

10. Use the associative property of addition to complete each statement.
   a. $(-6 + 2) + 8 =$
   b. $-7 + (7 + 3) =$

11. What properties were used in Step 1 and Step 2 of the solution?

$$\begin{aligned}(99 + 4) + 1 &= (4 + 99) + 1 \quad \text{Step 1}\\ &= 4 + (99 + 1) \quad \text{Step 2}\\ &= 4 + 100\\ &= 104\end{aligned}$$

12. Consider: $-3 + 6 + (-9) + 8 + (-4)$
   a. Add all the positives in the expression.
   b. Add all of the negatives.
   c. Add the results from parts **a** and **b.**

NOTATION

13. a. Express the commutative property of addition using the variables x and y.
   b. Express the associative property of addition using the variables x, y, and z.

14. Fill in the blank: We read $-a$ as "the ________ of a."

GUIDED PRACTICE

Add. See Example 1.

15. $-8 + (-1)$
16. $-3 + (-2)$
17. $-5 + (-12)$
18. $-4 + (-14)$
19. $-29 + (-45)$
20. $-23 + (-31)$
21. $-4.2 + (-6.1)$
22. $-5.1 + (-5.1)$
23. $-\frac{3}{4} + \left(-\frac{2}{3}\right)$
24. $-\frac{1}{5} + \left(-\frac{3}{4}\right)$
25. $-\frac{1}{4} + \left(-\frac{1}{10}\right)$
26. $-\frac{3}{8} + \left(-\frac{1}{3}\right)$

Add. See Example 2.

27. $-7 + 4$
28. $-9 + 7$
29. $50 + (-11)$
30. $27 + (-30)$
31. $15.84 + (-15.84)$
32. $9.19 + (-9.19)$
33. $-6.25 + 8.5$
34. $21.37 + (-12.1)$
35. $-\frac{7}{15} + \frac{3}{15}$
36. $-\frac{8}{11} + \frac{3}{11}$
37. $\frac{1}{2} + \left(-\frac{1}{8}\right)$
38. $\frac{5}{6} + \left(-\frac{1}{4}\right)$

Add. See Examples 3 and 5.

39. $8 + (-5) + 13$
40. $17 + (-12) + (-23)$
41. $21 + (-27) + (-9)$
42. $-32 + 12 + 17$
43. $-27 + (-3) + (-13) + 22$
44. $53 + (-27) + (-32) + (-7)$
45. $-60 + 70 + (-10) + (-10) + 205$
46. $-100 + 200 + (-300) + (-100) + 200$

Apply the associative property of addition to find the sum. See Example 4.

47. $-99 + (99 + 215)$
48. $67 + (-67 + 127)$
49. $(-112 + 56) + (-56)$
50. $(-67 + 5) + (-5)$
51. $\frac{1}{8} + \left(\frac{7}{8} + \frac{2}{3}\right)$
52. $\left(\frac{1}{2} + \frac{9}{16}\right) + \frac{7}{16}$
53. $(12.4 + 1.9) + 1.1$
54. $87.6 + (2.4 + 1.7)$

Add. See Example 6.

55. $-1 + 9 + 1$
56. $5 + 8 + (-5)$
57. $-8 + 11 + (-11) + 8 + 1$
58. $2 + 15 + (-15) + 8 + (-2)$

TRY IT YOURSELF

Add.

59. $-9 + 81 + (-2)$
60. $11 + (-21) + (-13)$
61. $0 + (-6.6)$
62. $0 + (-2.14)$
63. $-\frac{9}{16} + \frac{7}{16}$
64. $-\frac{3}{4} + \frac{1}{4}$
65. $-6 + (-8)$
66. $-4 + (-3)$
67. $-167 + 167$
68. $-25 + 25$
69. $-20 + (-16) + 10$
70. $-13 + (-16) + 4$
71. $19.35 + (-20.21) + 1.53$
72. $33.12 + (-35.7) + 2.98$
73. $-7 + 5 + (-10) + 7$
74. $-3 + 6 + (-9) + (-6)$
75. $19.2 + (-41.3)$
76. $57.93 + (-93.27)$
77. $2{,}345 + (-178)$
78. $-4{,}061 + 5{,}000$
79. $-2.1 + 6.5 + (-8.2) + 2.1$
80. $0.9 + 0.5 + (-0.2) + (-0.9)$
81. $3 + (-6) + (-3) + 74$
82. $4 + (-3) + (-4) + 5$
83. $-\frac{1}{4} + \left(-\frac{2}{7}\right)$
84. $-\frac{3}{32} + \left(-\frac{1}{2}\right)$
85. $-0.2 + (-0.3) + (-0.4)$
86. $-0.9 + (-1.9) + (-2.9)$

Look Alikes . . .

87. a. $12 + 15$ b. $-12 + 15$ c. $-12 + (-15)$ d. $12 + (-15)$
88. a. $432 + 67$ b. $-432 + 67$ c. $-432 + (-67)$ d. $432 + (-67)$
89. a. $\frac{1}{2} + \frac{2}{9}$ b. $-\frac{1}{2} + \frac{2}{9}$ c. $-\frac{1}{2} + \left(-\frac{2}{9}\right)$ d. $\frac{1}{2} + \left(-\frac{2}{9}\right)$
90. a. $0.87 + 0.29$ b. $-0.87 + 0.29$ c. $-0.87 + (-0.29)$ d. $0.87 + (-0.29)$

APPLICATIONS

91. **Military Science.** During a battle, an army retreated 1,500 meters, regrouped, and advanced 2,400 meters. The next day, it advanced another 1,250 meters. Find the army's net gain.

92. **Health.** Find the point total for the six risk factors (in blue) on the medical questionnaire. Then use the table at the bottom of the form to determine the patient's risk of contracting heart disease in the next 10 years.

Age		Total Cholesterol	
Age	Points	Reading	Points
35	−4	280	3
Cholesterol		**Blood Pressure**	
HDL	Points	Systolic/Diastolic	Points
62	−3	124/100	3
Diabetic		**Smoker**	
	Points		Points
Yes	4	Yes	2

10-Year Heart Disease Risk

Total Points	Risk	Total Points	Risk
−2 or less	1%	5	4%
−1 to 1	2%	6	6%
2 to 3	3%	7	6%
4	4%	8	7%

Source: National Heart, Lung, and Blood Institute

93. **Golf.** The leaderboard below shows the top four finishers from the 2009 PGA Championship Golf Tournament. Scores for each round are compared to *par,* the standard number of strokes necessary to complete the course. A score of −2, for example, indicates that the golfer used two strokes less than par to complete the course. A score of 5 indicates five strokes more than par. Determine the tournament total for each golfer.

Leaderboard

	Round 1	2	3	4	Total
Y.E. Yang	+1	−2	−5	−2	
Tiger Woods	−5	−2	−1	+3	
Lee Westwood	−2	0	+1	−2	
Rory McIlroy	−1	+1	−1	−2	

94. **Submarines.** A submarine was cruising at a depth of 1,250 feet. The captain gave the order to climb 550 feet. Compared to sea level, find the new depth of the sub.

95. **Credit Cards.** Refer to the monthly statement. What is the new balance?

Previous Balance	New Purchases, Fees, Advances & Debts	Payments & Credits	New Balance
3,660.66	1,408.78	3,826.58	

Billing Date	Date Payment Due	Credit Line
04/21/11	05/16/11	9,100

96. **Politics.** The following proposal to limit campaign contributions was on the ballot in a state election, and it passed. What will be the net fiscal impact on the state government?

212 Campaign Spending Limits YES ☐ NO ☐

Limits contributions to $200 in state campaigns. Fiscal impact: Costs of $4.5 million for implemen–tation and enforcement. Increases state revenue by $6.7 million by eliminating tax deductions for lobbying.

97. **Movie Losses.** According to the Numbers Box Office Data website, the movie *Stealth,* released in 2005 by Sony Pictures, cost about $176,350,000 to produce, promote, and distribute. It reportedly earned back just $76,700,000 worldwide. Express the dollar loss suffered by Sony as a signed number.

98. **Stocks.** The last entry on the line for June 12 indicates that one share of Walt Disney Co. stock lost $0.81 in value that day. How much did the value of a share of Disney stock rise or fall over the 5-day period from June 12 through June 16?

June 12	43.88	23.38	Disney	.21	0.5	87	−43	40.75	−.81
June 13	43.88	23.38	Disney	.21	0.5	86	−15	40.19	−.56
June 14	43.88	23.38	Disney	.21	0.5	87	−50	41.00	+.81
June 15	43.88	23.38	Disney	.21	0.5	89	−28	41.81	+.81
June 16	43.88	23.38	Disney				−15	41.19	−.63

Based on data from the *Los Angeles Times*

99. **Chemistry.** An atom is composed of protons (with a charge of +1), neutrons (with no charge), and electrons (with a charge of −1). Two simple models of atoms are shown. What is the overall charge of each atom?

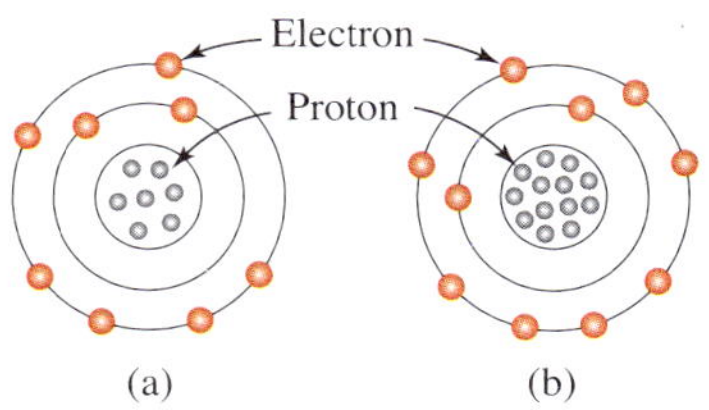

100. **Physics.** In the illustration, arrows show the two forces acting on a lamp hanging from a ceiling. What is the sum of the forces?

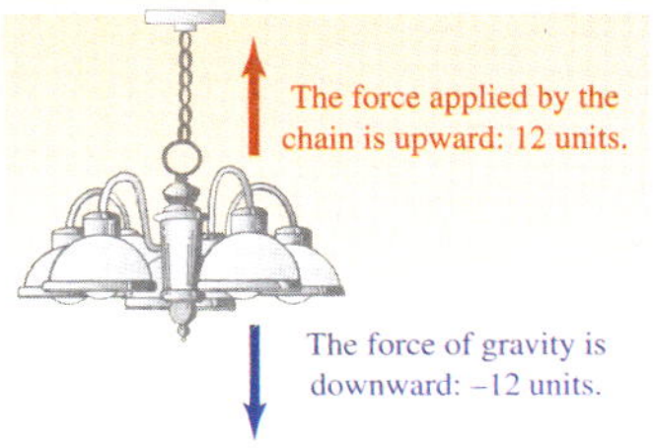

101. **The Big Easy.** The city of New Orleans lies, on average, 6 feet below sea level. What is the elevation of the top of an 85-foot tall building in New Orleans?

102. Electronics. A closed circuit contains two batteries and three resistors. The sum of the voltages in the loop must be 0. Is it?

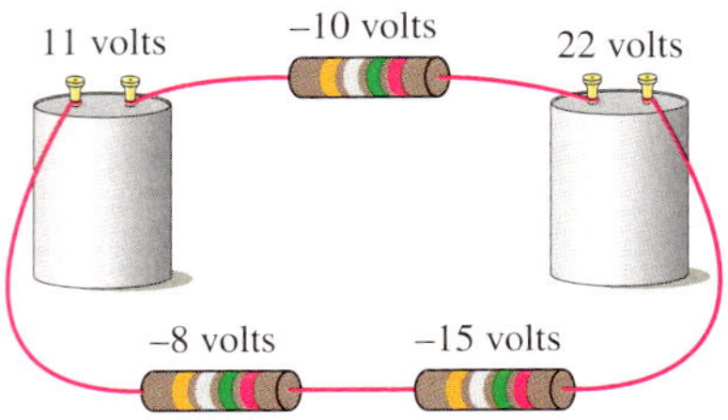

103. Accounting. The 2009 quarterly profits and losses of the Bank of America are shown in the table. Losses are denoted using parentheses. Calculate the company's total net income for 2009.

Quarter	Net income ($ million)
1st	4,247
2nd	3,224
3rd	(1,001)
4th	(194)

Source: www.scottrade.com

104. Politics. Six months before an election, the incumbent trailed the challenger by 18 points. To overtake her opponent, the incumbent decided to use a four-part strategy. Each part of the plan is shown below, with the expected point gain. With these gains, will the incumbent overtake the challenger on election day?

- TV ads +10 pts
- Voter mailing +3 pts
- Union endorsement +2 pts
- Telephone calls +1 pts

WRITING

105. Explain why the sum of two positive numbers is always positive and the sum of two negative numbers is always negative.

106. Explain why the sum of a negative number and a positive number is sometimes positive, sometimes negative, and sometimes zero.

REVIEW

107. True or false: Every real number can be expressed as a decimal.

108. Multiply: $\frac{1}{3} \cdot \frac{1}{3}$

109. What two numbers are a distance of 6 away from -3 on the number line?

110. Graph: $\left\{-2.5, \sqrt{2}, \frac{11}{3}, -0.333\ldots, 0.75\right\}$

CHALLENGE PROBLEMS

111. A set is said to be *closed under addition* if the sum of any two of its members is also a member of the set. Is the set $\{-1, 0, 1\}$ a closed set under addition? Explain.

112. Think of two numbers. First, add the absolute value of the two numbers, and write your answer. Second, add the two numbers, take the absolute value of that sum, and write that answer. Do the two answers agree? Can you find two numbers that produce different answers? When do you get answers that agree, and when don't you?

SECTION 1.5 Subtracting Real Numbers

OBJECTIVES

1. Use the definition of subtraction.
2. Solve application problems using subtraction.

ARE YOU READY?

The following problems review some basic concepts that are important when subtracting positive and negative real numbers.

1. What is the opposite of 6? What is the opposite of -15?
2. Write *twenty-two minus six* in symbols.
3. If 8 is subtracted from 20, what is the result?
4. Add: $-11 + 2$

In this section, we discuss a rule to use when subtracting signed numbers.

1 Use the Definition of Subtraction.

A minus symbol $-$ is used to indicate subtraction. However, this symbol is also used in two other ways, depending on where it appears in an expression.

$5 - 18$	This is read as "five minus eighteen."
-5	This usually is read as "negative five." It also could be read as "the additive inverse of five" or "the opposite of five."
$-(-5)$	This usually is read as "the opposite of negative five." It also could be read as "the additive inverse of negative five."

In $-(-5)$, parentheses are used to write the opposite of a negative number. When such expressions are encountered in computations, we simplify them by finding the opposite of the number within the parentheses.

$-(-5) = 5$ Read as "the opposite of negative five is five."

This observation illustrates the following rule.

Opposite of an Opposite

The opposite of the opposite of a number is that number. For any real number a,

$-(-a) = a$ Read as "the opposite of the opposite of a is a."

EXAMPLE 1 Simplify each expression: **a.** $-(-45)$ **b.** $-(-h)$ **c.** $-|-10|$

Strategy To simplify each expression, we will use the concept of opposite.

Why In each case, the outermost $-$ symbol is read as "the opposite."

Solution

a. The number within the parentheses is -45. Its opposite is 45. Therefore, $-(-45) = 45$.

b. The opposite of the opposite of h is h. Therefore, $-(-h) = h$.

c. The notation $-|-10|$ means "the opposite of the absolute value of negative ten." Since $|-10| = 10$, we have:

$-|\mathbf{-10}| = \mathbf{-10}$ The absolute value bars do not affect the $-$ symbol outside them. Therefore, the result is negative.

Self Check 1 Simplify each expression: **a.** $-(-1)$ **b.** $-(-y)$ **c.** $-|-500|$

Now Try Problems 15, 17, and 19

To develop a rule for subtraction, we consider the following illustration. It represents the subtraction $5 - 2 = 3$.

The Language of Algebra

The names of the parts of a subtraction fact are:

Minuend Subtrahend

$5 - 2 = 3$

Difference

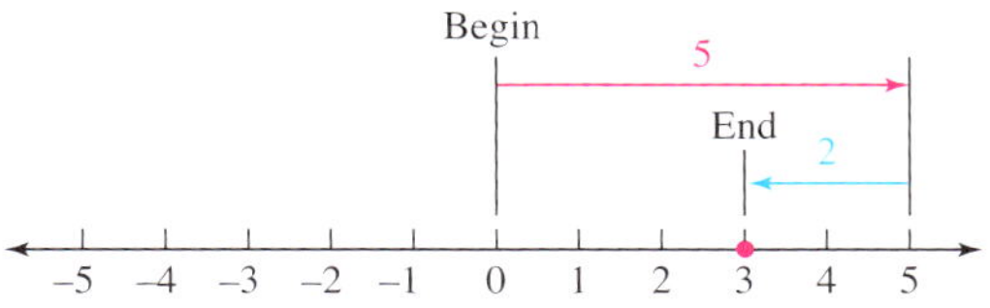

The illustration above also represents the addition $5 + (-2) = 3$. We see that

Subtracting 2 from 5 is the same as adding the opposite of 2 to 5.

$5 - 2 = 3$ $5 + (-2) = 3$

The results are the same.

This observation suggests the following definition.

Subtraction of Real Numbers

To subtract two real numbers, add the first number to the opposite (additive inverse) of the number to be subtracted.

For any real numbers a and b,

$$a - b = a + (-b)$$ Read as "a minus b equals a plus the opposite of b."

EXAMPLE 2 Subtract and check the result:

a. $-13 - 8$ **b.** $-7.6 - (-4.5)$ **c.** $\frac{1}{4} - \left(-\frac{1}{8}\right)$

Strategy To find each difference, we will apply the rule for subtraction: Add the first number to the opposite of the number to be subtracted.

Why It is easy to make an error when subtracting signed numbers. We will probably be more accurate if we write each subtraction as addition of the opposite.

Solution

a. We read $-13 - 8$ as "negative thirteen *minus* eight." Subtracting 8 is the same as adding -8.

Change the subtraction to addition.

$$-13 - 8 = -13 + (-8) = -21$$ Use the rule for adding two numbers with like signs.

No change (on -13). Change the number being subtracted to its opposite.

To check, we add the *difference,* -21, and the *subtrahend,* 8, to obtain the *minuend,* -13.

Check: $-21 + 8 = -13$

b. We read $-7.6 - (-4.5)$ as "negative seven point six *minus* negative four point five." Subtracting -4.5 is the same as adding 4.5.

Add . . .

$$-7.6 - (-4.5) = -7.6 + 4.5 = -3.1$$ Use the rule for adding two numbers with unlike signs.

No change (on -7.6). . . . the opposite

Check: $-3.1 + (-4.5) = -7.6$

c. $\frac{1}{4} - \left(-\frac{1}{8}\right) = \frac{2}{8} - \left(-\frac{1}{8}\right)$ The LCD is 8. Build $\frac{1}{4}$ so that the denominator is 8: $\frac{1}{4} \cdot \frac{2}{2} = \frac{2}{8}$.

$= \frac{2}{8} + \frac{1}{8}$ To subtract, add the opposite. Do not change $\frac{2}{8}$.

$= \frac{3}{8}$

Check: $\frac{3}{8} + \left(-\frac{1}{8}\right) = \frac{2}{8} = \frac{1}{4}$

Self Check 2 Subtract and check the result: **a.** $-32 - 25$ **b.** $1.7 - (-1.2)$ **c.** $-\frac{1}{3} - \left(-\frac{3}{4}\right)$

Now Try Problems 25, 39, and 43

The Language of Algebra

When we change a number to its opposite, we say we have **changed** (or *reversed*) its sign.

The Language of Algebra

The rule for subtracting real numbers is often summarized as: **Subtracting a number is the same as adding its opposite.**

Calculators

The subtraction key
When using a calculator to subtract signed numbers, be careful to distinguish between the *subtraction* key [−] and the keys that are used to enter negative values: [+/−] on a scientific calculator and [(−)] on a graphing calculator.

EXAMPLE 3 **a.** Subtract 0.5 from 4.6 **b.** Subtract 4.6 from 0.5

Strategy We will translate each phrase to mathematical symbols and then perform the subtraction. We must be careful when translating the instruction to subtract one number *from* another number.

Why The order of the numbers in each word phrase must be reversed when we translate it to mathematical symbols.

Solution **a.** The number to be subtracted is 0.5.

Subtract 0.5 from 4.6 — *To translate, reverse the order in which 0.5 and 4.6 appear in the sentence.*

$$4.6 - 0.5 = 4.1$$

b. The number to be subtracted is 4.6.

Subtract 4.6 from 0.5 — *To translate, reverse the order in which 4.6 and 0.5 appear in the sentence. Add the opposite of 4.6.*

$$\begin{aligned} 0.5 - 4.6 &= 0.5 + (-4.6) \\ &= -4.1 \end{aligned}$$

Caution

Notice from parts **a** and **b** that $4.6 - 0.5 \neq 0.5 - 4.6$. This result illustrates an important fact: Subtraction is *not* commutative. When subtracting two numbers, it is important that we write them in the correct order, because, in general, $a - b \neq b - a$.

Self Check 3 **a.** Subtract 2.2 from 4.9 **b.** Subtract 4.9 from 2.2

Now Try Problem 47

EXAMPLE 4 Perform the operations: $-9 - 15 + 20 - (-6)$

Strategy This expression contains addition and subtraction. We will write each subtraction as addition of the opposite and then evaluate the expression.

Why It is easy to make an error when subtracting signed numbers. We probably will be more accurate if we write each subtraction as addition of the opposite.

Solution

$$\begin{aligned} -9 - 15 + 20 - (-6) &= -9 + (-15) + 20 + 6 \\ &= -24 + 26 && \text{Add the negatives. Add the positives.} \\ &= 2 && \text{Add the results.} \end{aligned}$$

Self Check 4 Perform the operations: $-40 - (-10) + 7 - (-15)$

Now Try Problem 51

2 Solve Application Problems Using Subtraction.

Subtraction finds the *difference* between two numbers. When we find the difference between the maximum value and the minimum value of a collection of measurements, we are finding the **range** of the values.

EXAMPLE 5 **U.S. Temperatures.** The record high temperature in the United States of 134°F was set in Death Valley, California, on July 10, 1913. The record low of −80°F was set at Prospect Creek, Alaska, on January 23, 1971. Find the temperature range for these extremes.

Strategy We will subtract the lowest temperature from the highest temperature.

Why The *range* of a collection of data indicates the spread of the data. It is the difference between the largest and smallest values.

Solution

$$134 - (-80) = 134 + 80 \quad \text{134° is the higher temperature and −80° is the lower.}$$
$$= 214$$

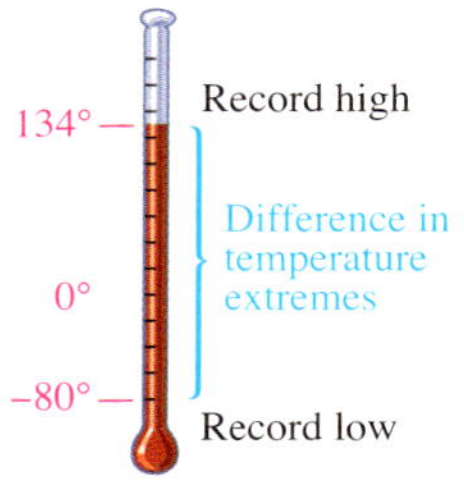

The temperature range for these extremes is 214°F.

Self Check 5 Find the temperature range for a day that had a low of −5°F and a high of 36°F.

Now Try Problem 95

Things are constantly changing in our daily lives. The amount of money we have in the bank, the price of gasoline, and our ages are examples. In mathematics, the operation of subtraction is used to measure change. To find the **change** in a quantity, we subtract the earlier value from the later value.

Change = later value − earlier value

EXAMPLE 6 **Water Management.** Lake Mead, on the Nevada-Arizona border, is the largest reservoir in the United States. It is formed by Hoover Dam across the Colorado River. In 2000, the water level in Lake Mead was 89 feet above drought level. By 2010, the water level was 27 feet below drought level. Find the change in the water level over that time span. (Source: Bureau of Reclamation)

©iStockphoto.com/MvH

Strategy We can represent a water level above drought level using a positive number and a water level below drought level using a negative number. To find the change in the water level, we will subtract.

Why In general, *to find the change in a quantity, we subtract the earlier value from the later value.*

Solution

$$-27 - 89 = -27 + (-89) \quad \text{The earlier water level in 2000 (89 ft) is subtracted from the later water level in 2010 (−27 ft).}$$
$$= -116 \quad \text{Do the addition.}$$

The negative result indicates that the water level *fell* 116 feet in that time span.

Self Check 6 Find the change in water level for a week that started at 4 feet above normal and went to 7 feet below normal level.

Now Try Problem 97

SECTION 1.5 STUDY SET

VOCABULARY

Fill in the blanks.

1. __________ finds the difference between two numbers.
2. In the subtraction $-2 - 5 = -7$, the result of -7 is called the __________.
3. The difference between the maximum and the minimum value of a collection of measurements is called the ______ of the values.
4. To find the ________ in a quantity, subtract the earlier value from the later value.

CONCEPTS

5. Find the opposite (additive inverse) of each number.

   a. 12 b. $-\frac{1}{5}$

   c. 2.71 d. 0

6. Complete each statement.

   a. $a - b = a +$

   To subtract two numbers, add the first number to the ________ of the number to be subtracted.

   b. $-(-a) =$

   The opposite of the opposite of a number is that ________.

7. Apply the rule for subtraction and fill in the blanks.

   $1 - (-9) = 1 \quad \square \quad \square = \square$

8. Use addition to check this subtraction: $15 - (-8) = 7$. Is the result correct?

9. Write each subtraction in the following expression as addition of the opposite.

   $-10 - 8 + (-23) + 5 - (-34)$

10. For each subtraction, just determine the sign of the answer.

    a. $8.76 - 12.91$ b. $8.76 - (-12.91)$

11. Circle any minus signs in each expression.

    a. $-6 - (-4)$ b. $7 + (-3) - 5 - (-2)$

12. In each case, determine what number is being subtracted.

    a. $5 - 8$ b. $-5 - (-8)$

NOTATION

13. Write each phrase using symbols. Then find its value.

    a. One minus negative seven

    b. The opposite of negative two

    c. The opposite of the absolute value of negative three

    d. Subtract 6 from 2

14. Write each expression in words.

    a. $-(-m)$

    b. $-2 - (-3)$

    c. $x - (-y)$

GUIDED PRACTICE

Simplify each expression. See Example 1.

15. $-(-55)$
16. $-(-27.2)$
17. $-(-x)$
18. $-(-t)$
19. $-|-25|$
20. $-|-100|$
21. $-\left|-\frac{3}{16}\right|$
22. $-\left|-\frac{4}{3}\right|$

Subtract. See Example 2.

23. $4 - 7$
24. $1 - 6$
25. $-6 - 4$
26. $-3 - 4$
27. $8 - (-3)$
28. $17 - (-21)$
29. $0 - 6$
30. $0 - 9$
31. $-1 - (-3)$
32. $-1 - (-7)$
33. $20 - (-20)$
34. $30 - (-30)$
35. $-2 - (-7)$
36. $-9 - (-1)$
37. $0 - (-12)$
38. $0 - 12$
39. $-1.4 - 5.5$
40. $-1.3 - 4.7$
41. $-1.5 - 0.81$
42. $-1.57 - (-0.8)$
43. $-\frac{1}{8} - \frac{3}{8}$
44. $-\frac{3}{4} - \frac{1}{4}$
45. $\frac{1}{3} - \frac{3}{4}$
46. $\frac{1}{6} - \frac{5}{8}$

Perform the indicated operation. See Example 3.

47. Subtract -5 from 17.
48. Subtract 45 from -50.
49. Subtract 12 from -13.
50. Subtract -11 from -20.

Perform the operations. See Example 4.

51. $-6 + 8 - (-1) - 10$
52. $-4 + 5 - (-3) - 13$
53. $61 - (-62) + (-64) - 60$
54. $93 - (-92) + (-94) - 95$

TRY IT YOURSELF

Perform the operations.

55. $244 - (-12)$
56. $354 - (-29)$
57. $-20 - (-30) - 50 + 40$
58. $-24 - (-28) - 48 - 44$
59. $-1.2 - 0.9$
60. $-2.52 - 1.72$
61. $\frac{1}{8} - \left(-\frac{5}{7}\right)$
62. $\frac{5}{8} - \left(-\frac{2}{9}\right)$
63. $-62 - 71 - (-37) + 99$
64. $-17 - 32 - (-85) - 51$
65. Subtract 47.5 from 0.
66. Subtract 30.3 from 0.
67. Subtract -137 from 12.
68. Subtract 512 from -47.
69. $-1{,}903 - (-1{,}732)$
70. $-300 - (-11)$
71. $2.83 - (-1.8)$
72. $4.75 - (-1.9)$
73. $-\frac{5}{6} - \frac{3}{4}$
74. $-\frac{3}{7} - \frac{2}{5}$
75. $8 - 9 - 10$
76. $1 - 2 - 3$
77. $-44 - 44$
78. $-33 - 33$
79. $-0.9 - 0.2$
80. $-0.3 - 0.2$
81. $-25 - (-50) - 75$
82. $-33 - (-22) - 44$
83. $6.3 - 9.8$
84. $2.1 - 9.4$
85. $-\frac{9}{16} - \left(-\frac{1}{4}\right)$
86. $-\frac{1}{2} - \left(-\frac{1}{4}\right)$
87. $0 - (-1)$
88. $0 - (-8)$
89. $2 - 15$
90. $3 - 14$

Look Alikes . . .

91. a. $-50 + (-3)$ b. $-50 - (-3)$

92. a. $-\frac{1}{16} + \frac{1}{4}$ b. $-\frac{1}{16} - \frac{1}{4}$

93. a. $-\frac{5}{9} + \left(-\frac{1}{6}\right)$ b. $-\frac{5}{9} - \left(-\frac{1}{6}\right)$

94. a. $2.96 + (-1.78)$ b. $2.96 - (-1.78)$

APPLICATIONS

95. **The Empire State.** New York state's record high temperature of 108°F was set in 1926, and the record low of −52°F was set in 1979. What is the range of these temperature extremes?

96. **Eyesight.** Nearsightedness, the condition where near objects are clear and far objects are blurry, is measured using negative numbers. Farsightedness, the condition where far objects are clear and near objects are blurry, is measured using positive numbers. Find the range in the measurements shown.

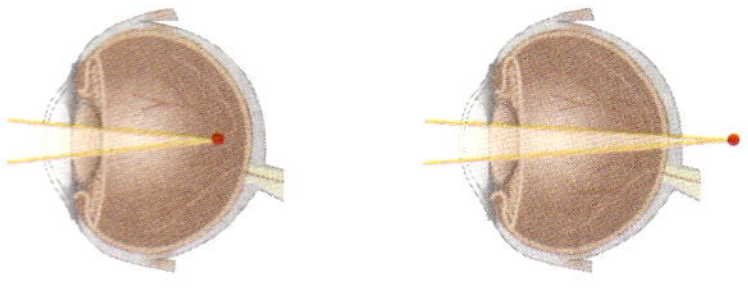

Nearsighted: –2.5 Farsighted: +4.35

97. **Law Enforcement.** A burglar scored −18 on a lie detector test, a score that indicates deception. However, on a second test, he scored +3, a score that is inconclusive. Find the change in the scores.

98. **Racing.** To improve handling, drivers often adjust the angle of the wheels of their car. When the wheel leans out, the degree measure is considered positive. When the wheel leans in, the degree measure is considered negative. Find the change in the position of the wheel shown below.

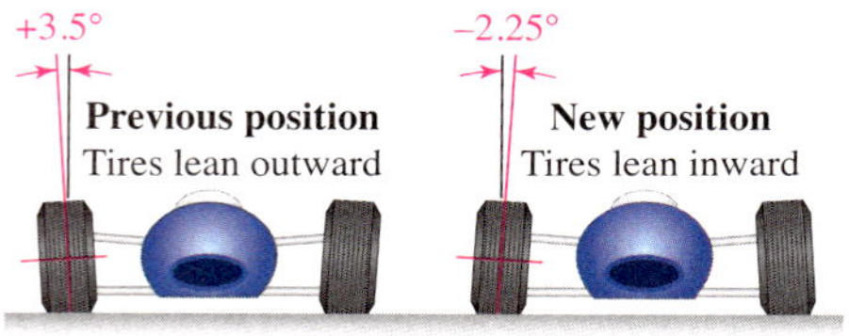

99. **from Campus to Careers**

Lead Transportation Security Officer

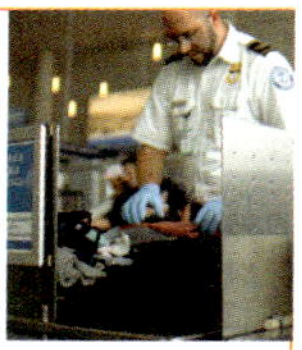

Determine the change in the number of passengers using each airport in 2009 compared with 2008.

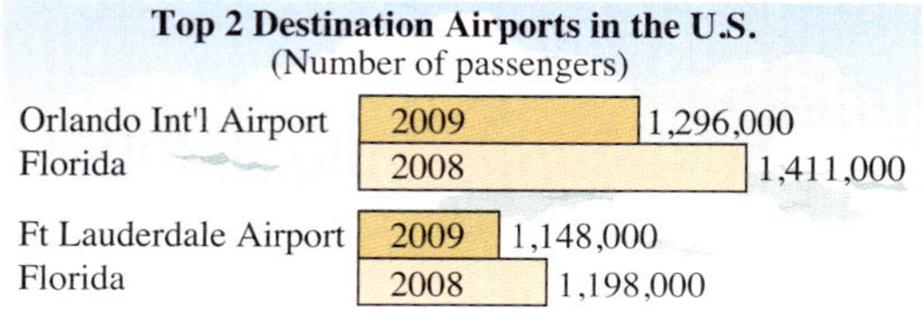

Source: Bureau of Transportation Statistics

100. **U.S. Jobs.** The table lists the three occupations that are predicted to have the largest job declines from 2008 to 2018. Complete the column labeled "Change."

Occupation	Number of jobs 2008	2018	Change
Farmers/ranchers	985,900	906,700	
Sewing machine operators	212,400	140,900	
Order clerks	245,700	181,500	

Source: Bureau of Labor Statistics

101. **Geography.** The elevation of Death Valley, California, is 282 feet below sea level. The elevation of the Dead Sea in Israel is 1,312 feet below sea level. Find the difference in their elevations.

102. **Card Games.** Gonzalo won the second round of a card game and earned 50 points. Matt and Hydecki had to deduct the value of each of the cards left in their hands from their score on the first round. Use this information to update the score sheet below. (Face cards are counted as 10 points, aces as 1 point, and all others have the value of the number printed on the card.)

Running point total	Round 1	Round 2
Matt	+50	
Gonzalo	−15	
Hydecki	−2	

103. **World's Coldest Ice Cream.** Dippin' Dots is an ice cream snack that was invented by Curt Jones in 1987. The tiny multi-colored beads are created by flash freezing ice cream mix in liquid nitrogen at a temperature of −355°F. When they come out of the processor, they are stored at a temperature of −40°F. Find the change in temperature of Dippin' Dots from production to storage. (Source: fundinguniverse.com)

104. **History.** Plato, a famous Greek philosopher, died in 347 B.C. at the age of 81. When was he born?

105. **NASCAR.** Complete the table below to determine how many points the third and fourth place finishers were behind the leader.

2009 Final Driver Standings

Rank	Driver	Points	Points behind leader
1	Jimmie Johnson	6,652	. . .
2	Mark Martin	6,511	−141
3	Jeff Gordon	6,473	
4	Kurt Busch	6,446	

106. Gauges. With the engine off, the ammeter on a car reads 0. If the headlights, which draw a current of 7 amps, and the radio, which draws a current of 6 amps, are both turned on, what will be the new reading?

WRITING

107. Explain what it means when we say that subtraction is *not commutative.*

108. Why is addition of signed numbers taught before subtraction of signed numbers?

109. Explain why we know that the answer to $4 - 10$ is negative without having to do any computation.

110. Is the following statement true or false? Explain.
Having a debt of \$100 forgiven is equivalent to gaining \$100.

REVIEW

111. Find the prime factorization of 30.

112. Write the set of integers.

113. True or false: $-4 > -5$?

114. Use the associative property of addition to simplify the calculation: $-18 + (18 + 89)$

CHALLENGE PROBLEMS

115. Suppose x is positive and y is negative. Determine whether each statement is true or false.

a. $x - y > 0$ **b.** $y - x < 0$
c. $|-x| < 0$ **d.** $-|y| < 0$

116. Find:

$$1 - 2 + 3 - 4 + 5 - 6 + . . . + 99 - 100$$

SECTION 1.6 Multiplying and Dividing Real Numbers; Multiplication and Division Properties

OBJECTIVES

1. Multiply signed numbers.
2. Use properties of multiplication.
3. Divide signed numbers.
4. Use properties of division.

ARE YOU READY?

The following problems review some basic concepts that are important when multiplying positive and negative real numbers.

1. Find $|-14|$ and $|6.75|$.
2. Do the integers -3 and 24 have the same sign or different signs?
3. Multiply: $\frac{7}{8} \cdot \frac{5}{14}$
4. Divide: $2.22 \div 0.6$

In this section, we will develop rules for multiplying and dividing positive and negative numbers.

1 Multiply Signed Numbers.

Multiplication represents repeated addition. For example, 4(3) is equal to the sum of four 3's.

$$\begin{aligned} 4(3) &= 3 + 3 + 3 + 3 \\ &= 12 \end{aligned}$$

The Language of Algebra

The names of the parts of a multiplication fact are:

Factor Factor Product
$4(3) = 12$

This example illustrates that *the product of two positive numbers is positive.*

To develop a rule for multiplying a positive number and a negative number, we will find $4(-3)$, which is equal to the sum of four -3's.

$$\begin{aligned} 4(-3) &= -3 + (-3) + (-3) + (-3) \\ &= -12 \end{aligned}$$

We see that the result is negative. As a check, think in terms of money. If you lose \$3 four times, you have lost a total of \$12, which is written $-\$12$. This example illustrates that *the product of a positive number and a negative number is negative.*

Multiplying Two Numbers That Have Different (Unlike) Signs

To multiply a positive real number and a negative real number, multiply their absolute values. Then make the final answer negative.

EXAMPLE 1 Multiply: **a.** $8(-12)$ **b.** $-151 \cdot 5$ **c.** $(-0.6)(1.2)$ **d.** $\frac{3}{4}\left(-\frac{4}{15}\right)$

Strategy We will use the rule for multiplying two numbers that have different signs.

Why In each case, we are asked to multiply a positive number and a negative number.

Solution

a. $8(-12) = -96$ Multiply the absolute values, 8 and 12, to get 96. Since the signs are unlike, make the final answer negative.

b. $-151 \cdot 5 = -755$ Multiply the absolute values, 151 and 5, to get 755. Since the signs are unlike, make the final answer negative.

c. To find the product of these two decimals with unlike signs, first multiply their absolute values, 0.6 and 1.2.

$$\begin{array}{r} 1.2 \\ \times\ 0.6 \\ \hline 0.72 \end{array}$$

Place the decimal point in the result so that the answer has the same number of decimal places as the sum of the number of decimal places in the factors.

Then make the final answer negative: $(-0.6)(1.2) = -0.72$.

Success Tip

The product of two numbers with unlike signs is *always* negative.

d. $\frac{3}{4}\left(-\frac{4}{15}\right) = -\frac{\overset{1}{\cancel{3}} \cdot \overset{1}{\cancel{4}}}{\underset{1}{\cancel{4}} \cdot \underset{1}{\cancel{3}} \cdot 5}$ Multiply the absolute values $\frac{3}{4}$ and $\frac{4}{15}$. Since the signs are unlike, make the final answer negative.

$= -\frac{1}{5}$ To simplify the fraction, factor 15 as $3 \cdot 5$. Remove the common factors 3 and 4 in the numerator and denominator.

Self Check 1 Multiply: **a.** $20(-3)$ **b.** $-3 \cdot 5$ **c.** $4.3(-2.6)$ **d.** $-\frac{5}{8} \cdot \frac{16}{25}$

Now Try Problems 21, 27, and 29

To develop a rule for multiplying two negative numbers, consider the following list, where we multiply -4 by factors that decrease by 1. We know how to find the first four products. Graphing those results on a number line is helpful in determining the last three products.

This factor decreases by 1 each time — Look for a pattern here

$$\begin{aligned} -4(\mathbf{3}) &= -12 \\ -4(\mathbf{2}) &= -8 \\ -4(\mathbf{1}) &= -4 \\ -4(\mathbf{0}) &= 0 \\ -4(\mathbf{-1}) &= ? \\ -4(\mathbf{-2}) &= ? \\ -4(\mathbf{-3}) &= ? \end{aligned}$$

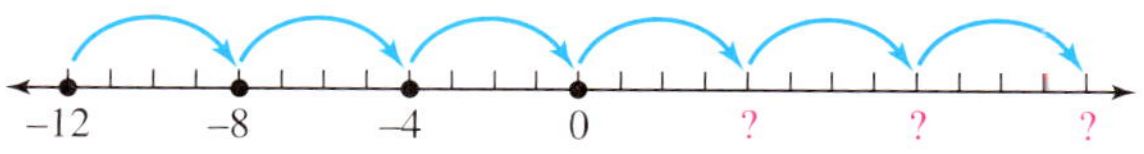

A graph of the products

From the pattern, we see that the product increases by 4 each time. Thus,

$$-4(-1) = 4, \quad -4(-2) = 8, \quad \text{and} \quad -4(-3) = 12$$

These results illustrate that *the product of two negative numbers is positive.* As a check, think of losing four debts of \$3. This is equivalent to gaining \$12. Therefore, $-4(-\$3) = \$12$.

Since the product of two positive numbers is positive, and the product of two negative numbers is also positive, we can summarize the multiplication rule as follows.

Multiplying Two Numbers That Have the Same (Like) Signs

To multiply two real numbers that have the same sign, multiply their absolute values. The final answer is positive.

EXAMPLE 2 Multiply: **a.** $-5(-6)$ **b.** $\left(-\frac{1}{2}\right)\left(-\frac{5}{8}\right)$

Strategy We will use the rule for multiplying two numbers that have the same sign.

Why In each case, we are asked to multiply two negative numbers.

Solution

a. $-5(-6) = 30$ Multiply the absolute values, 5 and 6, to get 30. Since both factors are negative, the final answer is positive.

b. $\left(-\frac{1}{2}\right)\left(-\frac{5}{8}\right) = \frac{5}{16}$ Multiply the absolute values, $\frac{1}{2}$ and $\frac{5}{8}$, to get $\frac{5}{16}$. Since the two factors have the same sign, the final answer is positive.

Success Tip

The product of two numbers with like signs is *always* positive.

Self Check 2 Multiply: **a.** $-15(-8)$ **b.** $-\frac{1}{4}\left(-\frac{1}{3}\right)$

Now Try Problems 33 and 41

2 Use Properties of Multiplication.

The multiplication of two numbers can be done in any order; the result is the same. For example, $-9(4) = -36$ and $4(-9) = -36$. This illustrates that multiplication is **commutative.**

The Commutative Property of Multiplication

Changing the order when multiplying does not affect the answer.

For any real numbers a and b,

$$ab = ba$$

In the following example, we multiply $-3 \cdot 7 \cdot 5$ in two ways. Recall that the operation within the parentheses should be performed first. We read $(-3 \cdot 7)5$ as "the *quantity* of -3 times 7," pause slightly, and then say "times 5." We read $-3(7 \cdot 5)$ as "-3 times the *quantity* of 7 times 5." The word *quantity* alerts the reader to the parentheses that are used as grouping symbols.

Method 1: Group -3 *and* 7	*Method 2: Group* 7 *and* 5
$(-3 \cdot 7)5 = (-21)5$	$-3(7 \cdot 5) = -3(35)$
$= -105$	$= -105$

It doesn't matter how we group the numbers in this multiplication; the result is -105. This example illustrates that multiplication is **associative.**

The Associative Property of Multiplication

Changing the grouping when multiplying does not affect the answer.

For any real numbers a, b, and c,

$$(ab)c = a(bc)$$

EXAMPLE 3 Multiply: **a.** $-5(-37)(-2)$ **b.** $-4(-3)(-2)(-1)$

Strategy First, we will use the commutative and associative properties of multiplication to reorder and regroup the factors. Then we will perform the multiplications.

Why Applying one or both of these properties before multiplying can simplify the computations and lessen the chance of a sign error.

Solution Using the commutative and associative properties of multiplication, we can reorder and regroup the factors to simplify computations.

a. Since it is easy to multiply by 10, we will find $-5(-2)$ first.

$$-5(-37)(-2) = -5(-2)(-37)$$ Use the commutative property of multiplication.

$$= 10(-37)$$

$$= -370$$

b. $-4(-3)(-2)(-1) = 12(2)$ Multiply the first two factors and multiply the last two factors.

$$= 24$$

Self Check 3 Multiply: **a.** $-25(-3)(-4)$ **b.** $-1(-2)(-3)(-3)$

Now Try Problems 43 and 47

In Example 3a, we multiplied three negative numbers. In Example 3b, we multiplied four negative numbers. The results illustrate the following fact.

Multiplying Negative Numbers

The product of an even number of negative numbers is positive. The product of an odd number of negative numbers is negative.

Recall that the product of 0 and any whole number is 0. The same is true for any real number. Therefore, $-6 \cdot 0 = 0$, $\frac{7}{16} \cdot 0 = 0$, and $0(4.51) = 0$.

Multiplication Property of 0

The product of 0 and any real number is 0.

For any real number a,

$$0 \cdot a = 0 \quad \text{and} \quad a \cdot 0 = 0$$

Whenever we multiply a number by 1, the number remains the same. Therefore, $1 \cdot 6 = 6$, $4.57 \cdot 1 = 4.57$, and $1(-9) = -9$. Since any number multiplied by 1 remains the same (is identical), the number 1 is called the **identity element** for multiplication.

Multiplication Property of 1 (Identity Property of Multiplication)

The product of 1 and any number is that number.

For any real number a,

$$1 \cdot a = a \quad \text{and} \quad a \cdot 1 = a$$

Whenever we multiply a number by -1, the result is the opposite of that number. For example, $-1 \cdot 12 = -12$ and $-\frac{5}{8}(-1) = \frac{5}{8}$.

Multiplication Property of −1

The product of -1 and any number is the opposite (or additive inverse) of that number.

For any real number a,

$$-1 \cdot a = -a \quad \text{and} \quad a(-1) = -a$$

The Language of Algebra

Don't confuse the words **opposite** and **reciprocal**. The opposite of 4 is −4 The reciprocal of 4 is $\frac{1}{4}$.

Two numbers whose product is 1 are **reciprocals** or **multiplicative inverses** of each other. For example, 8 is the multiplicative inverse of $\frac{1}{8}$, and $\frac{1}{8}$ is the multiplicative inverse of 8, because $8 \cdot \frac{1}{8} = 1$. Likewise, $-\frac{3}{4}$ and $-\frac{4}{3}$ are multiplicative inverses because $-\frac{3}{4}\left(-\frac{4}{3}\right) = 1$. All real numbers, except 0, have a multiplicative inverse.

Multiplicative Inverses (Inverse Property of Multiplication)

The product of any number and its multiplicative inverse (reciprocal) is 1.

For any nonzero real number a,

$$a\left(\frac{1}{a}\right) = 1$$

EXAMPLE 4 Find the reciprocal of each number: **a.** $\frac{2}{3}$ **b.** $-\frac{2}{3}$ **c.** -11

Strategy To find the reciprocal of a fraction, we invert the numerator and the denominator.

Why We want the product of the given number and its reciprocal to be 1.

Solution

a. The reciprocal of $\frac{2}{3}$ is $\frac{3}{2}$ because $\frac{2}{3}\left(\frac{3}{2}\right) = 1$. *To find the reciprocal of a fraction, invert the numerator and denominator.*

b. The reciprocal of $-\frac{2}{3}$ is $-\frac{3}{2}$ because $-\frac{2}{3}\left(-\frac{3}{2}\right) = 1$.

c. The reciprocal of -11 is $-\frac{1}{11}$ because $-11\left(-\frac{1}{11}\right) = 1$. *Think of −11 as $\frac{-11}{1}$ to find its reciprocal.*

Caution

Do not change the sign of a number when finding its reciprocal.

Self Check 4 Find the reciprocal of each number: **a.** $-\frac{15}{16}$ **b.** $\frac{15}{16}$ **c.** -27

Now Try Problems 51 and 53

3 Divide Signed Numbers.

Every division fact can be written as an equivalent multiplication fact.

Division

For any real numbers a, b, and c, where $b \neq 0$,

$$\frac{a}{b} = c \quad \text{provided that} \quad c \cdot b = a$$

Quotient · divisor = dividend

The Language of Algebra

The names of the parts of a division fact are:

Dividend $\frac{15}{5} = 3$ Quotient

Divisor

We can use this relationship between multiplication and division to develop rules for dividing signed numbers. For example,

$$\frac{15}{5} = 3 \quad \text{because} \quad 3(5) = 15$$

From this example, we see that *the quotient of two positive numbers is positive.*

To determine the quotient of two negative numbers, we consider $\frac{-15}{-5}$.

$$\frac{-15}{-5} = 3 \quad \text{because} \quad 3(-5) = -15$$

From this example, we see that the *quotient of two negative numbers is positive.*

To determine the quotient of a positive number and a negative number, we consider $\frac{15}{-5}$.

$$\frac{15}{-5} = -3 \quad \text{because} \quad -3(-5) = 15$$

From this example, we see that *the quotient of a positive number and a negative number is negative.*

To determine the quotient of a negative number and a positive number, we consider $\frac{-15}{5}$.

$$\frac{-15}{5} = -3 \quad \text{because} \quad -3(5) = -15$$

From this example, we see that *the quotient of a negative number and a positive number is negative.*

We summarize the rules from the previous examples and note that they are similar to the rules for multiplication.

Dividing Two Real Numbers

To divide two real numbers, divide their absolute values.

1. The quotient of two numbers that have the same (*like*) signs is positive.
2. The quotient of two numbers that have different (*unlike*) signs is negative.

EXAMPLE 5 Divide and check the result: **a.** $\frac{-81}{-9}$ **b.** $\frac{45}{-9}$ **c.** $-2.87 \div 0.7$

d. $-\frac{5}{16} \div \left(-\frac{1}{2}\right)$

Strategy We will use the rules for dividing signed numbers. In each case, we need to ask, "Is it a quotient of two numbers with the same sign or different signs?"

Why The signs of the numbers that we are dividing determine the sign of the result.

Solution

a. $\frac{-81}{-9} = 9$ Divide the absolute values, 81 by 9, to get 9. Since the signs are like, the final answer is positive.

Multiply the quotient and the divisor to check the result: $9(-9) = -81$.

b. $\frac{45}{-9} = -5$ Divide the absolute values, 45 by 9, to get 5. Since the signs are unlike, make the final answer negative.

Check: $-5(-9) = 45$ Quotient · divisor = dividend

c. $-2.87 \div 0.7 = -4.1$ Since the signs are unlike, make the final answer negative.

Check: $-4.1(0.7) = -2.87$ Quotient · divisor = dividend

Success Tip

To perform this decimal division, move each decimal point one place to the right.

$0.7\overline{)2.87}$

d. $-\frac{5}{16} \div \left(-\frac{1}{2}\right) = -\frac{5}{16}\left(-\frac{2}{1}\right)$ Multiply the first fraction by the reciprocal of the second fraction. The reciprocal of $-\frac{1}{2}$ is $-\frac{2}{1}$.

$= \frac{5 \cdot 2}{16 \cdot 1}$ Multiply the absolute values $\frac{5}{16}$ and $\frac{2}{1}$. Since the signs are like, the final answer is positive.

$= \frac{5 \cdot \overset{1}{\cancel{2}}}{\underset{1}{\cancel{2}} \cdot 8 \cdot 1}$ To simplify the fraction, factor 16 as $2 \cdot 8$. Then remove the common factor 2.

$= \frac{5}{8}$

Check: $\frac{5}{8}\left(-\frac{1}{2}\right) = -\frac{5}{16}$ Quotient · divisor = dividend

Self Check 5 Divide and check the result: **a.** $\frac{-28}{-4}$ **b.** $\frac{75}{-25}$ **c.** $0.32 \div (-1.6)$ **d.** $\frac{3}{4} \div \left(-\frac{5}{8}\right)$

Now Try Problems 55, 59, 63, and 67

EXAMPLE 6

Depreciation. Over an 8-year period, the value of a $150,000 house fell at a uniform rate to $110,000. Find the amount of depreciation per year.

The Language of Algebra

Depreciation is a form of the word *depreciate,* meaning to lose value. You've probably heard that the minute you drive a new car off the lot, it has *depreciated.*

Strategy The phrase *uniform rate* means that the value of the house fell the same amount each year, for 8 straight years. We can determine the amount it depreciated in one year (per year) by dividing the total change in value of the house by 8.

Why The process of separating a quantity into equal parts (in this case, the change in the value of the house) indicates division.

Solution First, we find the change in the value of the house.

$110{,}000 - 150{,}000 = -40{,}000$ Subtract the previous value from the current value.

The negative result represents a drop in value of $40,000. Since the depreciation occurred over 8 years, we divide $-40{,}000$ by 8.

$\frac{-40{,}000}{8} = -5{,}000$ Divide the absolute values, 40,000 by 8, to get 5,000, and make the quotient negative.

The house depreciated $5,000 per year.

©iStockPhoto.com/EyeMark

Self Check 6 **Depreciation.** Over a 6-year period, the value of a $300,000 house fell at a uniform rate to $286,500. Find the amount of depreciation each year.

Now Try Problem 109

4 Use Properties of Division.

Whenever we divide a number by 1, the quotient is that number. Therefore, $\frac{12}{1} = 12$, $\frac{-80}{1} = -80$, and $7.75 \div 1 = 7.75$. Furthermore, whenever we divide a nonzero number by itself, the quotient is 1. Therefore, $\frac{35}{35} = 1$, $\frac{-4}{-4} = 1$, and $0.9 \div 0.9 = 1$. These observations suggest the following properties of division.

Division Properties

Any number divided by 1 is the number itself. Any number (except 0) divided by itself is 1.

For any real number a,

$$\frac{a}{1} = a \quad \text{and} \quad \frac{a}{a} = 1 \quad (\text{where } a \neq 0)$$

Caution

Division is *not commutative*. For example, $\frac{6}{3} \neq \frac{3}{6}$ and $\frac{-12}{4} \neq \frac{4}{-12}$. In general, $\frac{a}{b} \neq \frac{b}{a}$.

We will now consider division that involves zero. First, we examine division of zero. Let's look at two examples. We know that

$$\frac{0}{2} = 0 \quad \text{because} \quad 0 \cdot 2 = 0 \quad \text{and} \quad \frac{0}{-5} = 0 \quad \text{because} \quad 0(-5) = 0$$

These examples illustrate that *0 divided by a nonzero number is 0.*

To examine division by zero, let's look at $\frac{2}{0}$ and its related multiplication statement.

$$\frac{2}{0} = ? \quad \text{because} \quad ? \cdot 0 = 2$$

The Language of Algebra

When we say a division by 0, such as $\frac{2}{0}$, is **undefined,** we mean that $\frac{2}{0}$ does not represent a real number.

There is no number that can make $0 \cdot ? = 2$ true because any number multiplied by 0 is equal to 0, not 2. Therefore, $\frac{2}{0}$ does not have an answer. We say that such a division is **undefined.**

These results suggest the following division facts.

Division Involving 0

For any nonzero real number a,

$$\frac{0}{a} = 0 \quad \text{and} \quad \frac{a}{0} \text{ is undefined.}$$

EXAMPLE 7

Find each quotient, if possible: **a.** $\frac{0}{8}$ **b.** $\frac{-24}{0}$

Strategy In each case, we need to determine if we have division *of* 0 or division *by* 0.

Why *Division of 0* by a nonzero number is defined, and the result is 0. However, *division by 0* is undefined; there is no result.

Solution

a. $\frac{0}{8} = 0$ because $0 \cdot 8 = 0$. This is division of 0 by 8.

b. $\frac{-24}{0}$ is undefined. This is division of -24 by 0.

The Language of Algebra

Division of 0 by 0, written $\frac{0}{0}$, is called **indeterminate.** This form is studied in advanced mathematics classes.

Self Check 7 Find each quotient, if possible: **a.** $\frac{4}{0}$ **b.** $\frac{0}{17}$

Now Try Problems 71 and 73

SECTION 1.6 STUDY SET

VOCABULARY

Fill in the blanks.

1. The answer to a multiplication problem is called a ________. The answer to a division problem is called a ________.
2. The ____________ property of multiplication states that changing the order when multiplying does not affect the answer.
3. The __________ property of multiplication states that changing the grouping when multiplying does not affect the answer.
4. Division of a nonzero number by 0 is _________.

CONCEPTS

Fill in the blanks.

5. a. The product or quotient of two numbers with like signs is ________.
   b. The product or quotient of two numbers with unlike signs is ________.
6. a. The product of an even number of negative numbers is ________.
   b. The product of an odd number of negative numbers is ________.
7. a. $\frac{-9}{3} = -3$ because $\quad \cdot \quad =$
   b. $\frac{0}{8} = 0$ because $\quad \cdot \quad =$
8. Complete each property of multiplication.
   a. $a \cdot b = b \cdot$
   b. $(ab)c =$
   c. $0 \cdot a =$
   d. $1 \cdot a =$
   e. $a\left(\frac{1}{a}\right) =$
   f. $-1 \cdot a =$
9. Complete each property of division.
   a. $\frac{a}{1} =$
   b. $\frac{a}{a} =$
   c. $\frac{0}{a} =$
   d. $\frac{a}{0}$ is
10. Which property justifies each statement?
   a. $-5(2 \cdot 17) = (-5 \cdot 2)17$
   b. $-5\left(-\frac{1}{5}\right) = 1$
   c. $-5 \cdot 2 = 2(-5)$
   d. $-5(1) = -5$
   e. $-5 \cdot 0 = 0$
11. For each multiplication or division, just determine the sign of the answer
   a. $-\frac{19}{37}\left(-\frac{51}{75}\right)$
   b. $\frac{45.568}{-2.56}$
   c. $-8.2(-4.1)(-6)(-9.3)(-1.5)$
12. Use multiplication to check this division: $\frac{-29.4}{7} = -4.1$. Is the answer correct?
13. Complete each statement using the given property.
   a. $5 \cdot 8 =$ Commutative property of multiplication
   b. $-2(6 \cdot 9) =$ Associative property of multiplication
   c. $5\left(\quad\right) = 1$ Inverse property of multiplication
   d. $\quad(-20) = -20$ Multiplication property of 1
14. Complete the table.

Number	Opposite (additive inverse)	Reciprocal (multiplicative inverse)
2		
$-\frac{4}{5}$		
1.75		

Let POS stand for a positive number and NEG stand for a negative number. Determine the sign of each result, if possible.

15. a. POS · NEG
   b. POS + NEG
   c. POS − NEG
   d. $\frac{\text{POS}}{\text{NEG}}$
16. a. NEG · NEG
   b. NEG + NEG
   c. NEG − NEG
   d. $\frac{\text{NEG}}{\text{NEG}}$

NOTATION

Write each sentence using symbols.

17. The product of negative four and negative five is twenty.
18. The quotient of sixteen and negative eight is negative two.

GUIDED PRACTICE

Multiply. See Example 1.

19. $4(-1)$
20. $6(-1)$
21. $-2 \cdot 8$
22. $-3 \cdot 4$
23. $12(-5)$
24. $(-9)(11)$
25. $3(-22)$
26. $-8 \cdot 9$
27. $1.2(-0.4)$
28. $(-3.6)(0.9)$
29. $\frac{1}{3}\left(-\frac{3}{4}\right)$
30. $\left(-\frac{3}{4}\right)\left(\frac{4}{5}\right)$

Multiply. See Example 2.

31. $(-1)(-7)$
32. $(-2)(-5)$
33. $(-6)(-9)$
34. $(-8)(-7)$

35. $-3(-3)$

36. $-1(-1)$

37. $63(-7)$

38. $43(-6)$

39. $-0.6(-4)$

40. $-0.7(-8)$

41. $\left(-\frac{7}{8}\right)\left(-\frac{2}{21}\right)$

42. $\left(-\frac{5}{6}\right)\left(-\frac{2}{15}\right)$

Multiply. See Example 3.

43. $3.3(-4)(-5)$

44. $(-2.2)(-4)(-5)$

45. $-2(-3)(-4)(-5)(-6)$

46. $-9(-7)(-5)(-3)(-1)$

47. $(-41)(3)(-7)(-1)$

48. $56(-3)(-4)(-1)$

49. $(-6)(-6)(-6)$

50. $(-5)(-5)(-5)$

Find the reciprocal of each number. Then find the product of the given number and its reciprocal. See Example 4.

51. $\frac{7}{9}$

52. $-\frac{8}{9}$

53. -13

54. $\frac{1}{8}$

Divide. See Example 5.

55. $-30 \div (-3)$

56. $-12 \div (-2)$

57. $-6 \div (-2)$

58. $-36 \div (-9)$

59. $\frac{85}{-5}$

60. $\frac{-84}{7}$

61. $\frac{-110}{-110}$

62. $\frac{-200}{-200}$

63. $\frac{-10.8}{1.2}$

64. $\frac{-13.5}{-1.5}$

65. $\frac{0.5}{-100}$

66. $\frac{-1.7}{10}$

67. $-\frac{1}{3} \div \frac{4}{5}$

68. $-\frac{2}{3} \div \frac{7}{8}$

69. $-\frac{9}{16} \div \left(-\frac{3}{20}\right)$

70. $-\frac{4}{5} \div \left(-\frac{8}{25}\right)$

TRY IT YOURSELF

Perform the operations.

71. $\frac{0}{150}$

72. $\frac{0}{-12}$

73. $\frac{-17}{0}$

74. $\frac{225}{0}$

75. $\frac{24}{-6}$

76. $\frac{-78}{6}$

77. $\frac{17}{-17}$

78. $\frac{-24}{24}$

79. $(-2)(-2)(-2)(-2)$

80. $(-3)(-3)(-3)(-3)$

81. $-3(-4)(0)$

82. $15(0)(-22)$

83. $\frac{-23.5}{5}$

84. $\frac{-337.8}{6}$

85. $-5.2 \cdot 100$

86. $-1.17 \cdot 1{,}000$

87. $\frac{1}{2}\left(-\frac{1}{3}\right)\left(-\frac{1}{4}\right)$

88. $\frac{1}{3}\left(-\frac{1}{5}\right)\left(-\frac{1}{7}\right)$

89. $\frac{550}{-50}$

90. $\frac{440}{-20}$

91. $-3\frac{3}{8} \div \left(-2\frac{1}{4}\right)$

92. $-3\frac{4}{15} \div \left(-2\frac{1}{10}\right)$

93. $7.2(-2.1)(-2)$

94. $4.6(-5.4)(-2)$

95. $\frac{1}{2}\left(-\frac{3}{4}\right)$

96. $\frac{1}{3}\left(-\frac{5}{16}\right)$

97. $-\frac{16}{25} \div \frac{64}{15}$

98. $-\frac{15}{16} \div \frac{25}{8}$

99. $\frac{-24.24}{-0.8}$

100. $\frac{-55.02}{-0.7}$

101. $-1\frac{1}{4}\left(-\frac{3}{4}\right)$

102. $-1\frac{1}{8}\left(-\frac{3}{8}\right)$

Look Alikes . . .

103. a. $2.7 + (-0.9)$ b. $2.7 - (-0.9)$

c. $2.7(-0.9)$ d. $\frac{2.7}{-0.9}$

104. a. $-\frac{5}{3} + \left(-\frac{9}{25}\right)$ b. $-\frac{5}{3} - \left(-\frac{9}{25}\right)$

c. $-\frac{5}{3}\left(-\frac{9}{25}\right)$ d. $-\frac{5}{3} \div \left(-\frac{9}{25}\right)$

Use the associative property of multiplication to find each product.

105. $-\frac{1}{2}(2 \cdot 67)$

106. $\left(-\frac{5}{16} \cdot \frac{1}{7}\right)7$

107. $-0.2(-10 \cdot 3)$

108. $-1.5(-100 \cdot 4)$

APPLICATIONS

109. **Real Estate.** Over a 5-year period, the value of a \$200,000 lot fell at a uniform rate to \$160,000. What signed number indicates the amount of depreciation per year?

110. **Tourism.** The ocean liner Queen Mary cost \$22,500,000 to build in 1936. The ship was purchased by the city of Long Beach, California, in 1967 for \$3,450,000. It now serves as a convention center. What signed number indicates the annual average depreciation of the ship over the 31-year period from 1936 to 1967? Round to the nearest dollar.

111. **Fluid Flow.** In a lab, the temperature of a fluid was decreased 6° per hour for 12 hours. What signed number indicates the change in temperature?

112. **Stress on the Job.** A health care provider for a company estimates that 75 hours per week are lost by employees suffering from stress-related illness. In one year, how many hours are lost? Use a signed number to answer.

113. **Weight Loss.** As a result of a diet, Tom has been steadily losing $4\frac{1}{2}$ pounds per month.

a. Which expression below can be used to determine how much heavier Tom was 8 months ago?

i. $-4\frac{1}{2} \cdot 8$ ii. $-4\frac{1}{2}(-8)$

iii. $4\frac{1}{2}(-8)$ iv. $-4\frac{1}{2} - 8$

b. How much heavier was Tom 8 months ago?

114. **Astronomy.** The temperature on Pluto gets as low as −386°F. This is twice as low as the lowest temperature reached on Jupiter. What is the lowest temperature on Jupiter?

115. Car Radiators. The instructions on a container of antifreeze state, "A 50/50 mixture of antifreeze and water protects against freeze-ups down to -34°F, while a 60/40 mix protects against freeze-ups down to one and one-half times that temperature." To what temperature does the 60/40 mixture protect?

116. Accounting. For 2010, the net income of Rite Aid Corporation (the drugstore chain) was about $-\$508$ million. The previous year, the company's net income was even worse, by a factor of about 5.75. What signed number represents Rite Aid's net income in 2009? (Source: moneycentral .msn com)

117. Airlines. In the 2009 income statement for Delta Air Lines below, numbers within parentheses represent a loss. Complete the statement given these facts. The second quarter loss was about 6.4 times the first quarter loss. The fourth quarter loss was about 5 times the second quarter loss. The third quarter loss was about $\frac{5}{16}$ of the fourth quarter loss.

DELTA INCOME STATEMENT				2009
All amounts in millions of dollars	1st Qtr (25)	2nd Qtr (?)	3rd Qtr (?)	4th Qtr (?)

Source: dailyfinance.com

118. Computers. The formula = A1*B1*C1 in cell D1 of the spreadsheet instructs the computer to multiply the values in cells A1, B1, and C1 and to print the result *in place of the formula* in cell D1. (The symbol * represents multiplication.) What value will be printed in the cell D1? What values will be printed in cells D2 and D3?

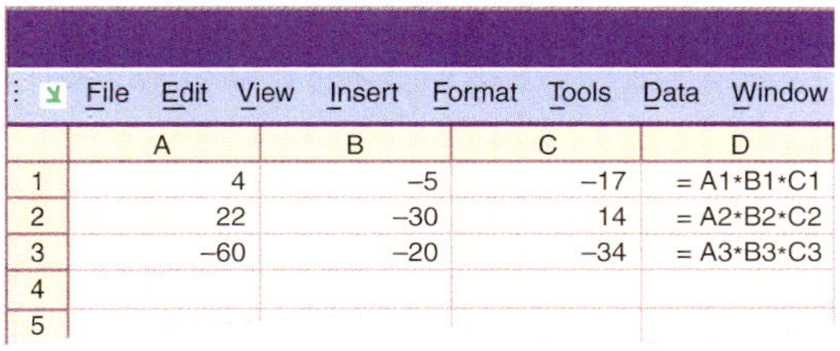

	A	B	C	D
1	4	−5	−17	= A1*B1*C1
2	22	−30	14	= A2*B2*C2
3	−60	−20	−34	= A3*B3*C3
4				
5				

119. Physics. An oscilloscope displays electrical signals as wavy lines on a screen. By switching the magnification dial to ×2, for example, the height of the "peak" and the depth of the "valley" of a graph will be doubled. Use signed numbers to indicate the height and depth of the display for each setting of the dial.

a. normal **b.** ×0.5
c. ×1.5 **d.** ×2

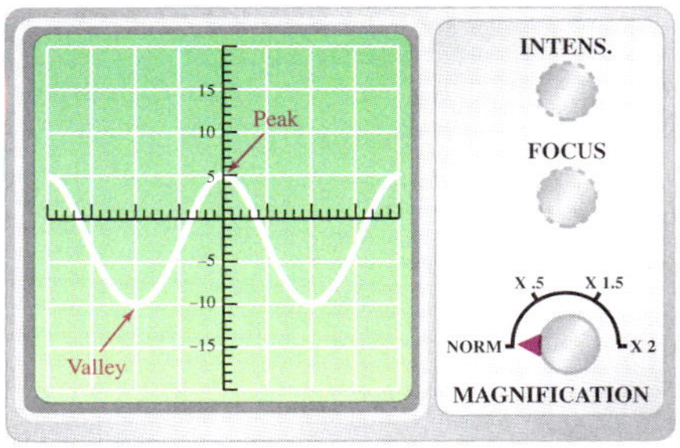

120. Light. Water acts as a selective filter of light. In the illustration, we see that red light waves penetrate water only to a depth of about 5 meters. How many times deeper does

a. yellow light penetrate than red light?
b. green light penetrate than orange light?
c. blue light penetrate than yellow light?

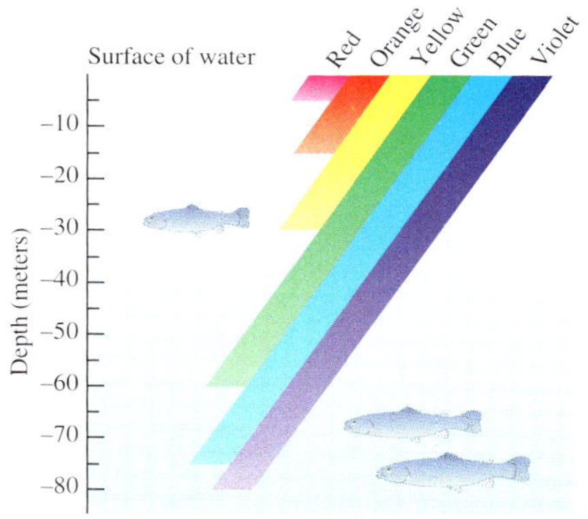

WRITING

121. Explain why $\frac{16}{0}$ is undefined.

122. The commutative property states that changing the order when multiplying does not change the answer. Are the following activities commutative? Explain.

a. Washing a load of clothes; drying a load of clothes
b. Putting on your left sock; putting on your right sock

123. What is wrong with the following statement?

A negative and a positive is a negative.

124. If we multiply two different numbers and the answer is 0, what must be true about one of the numbers? Explain your answer.

REVIEW

125. Add: $-3 + (-4) + (-5) + 4 + 3$

126. Write $-3 - (-5)$ as addition of the opposite.

127. Find $\frac{1}{2} + \frac{1}{4} + \frac{1}{3}$. Answer in decimal form.

128. Which integers have an absolute value equal to 45?

CHALLENGE PROBLEMS

129. If the product of five numbers is negative, how many of them could be negative? Explain.

130. Suppose a is a positive number and b is a negative number. Determine whether the given expression is positive or negative.

a. $-a(-b)$ **b.** $\frac{-a}{b}$

c. $\frac{-a}{a}$ **d.** $\frac{1}{b}$

SECTION 1.7 Exponents and Order of Operations

OBJECTIVES

1. Evaluate exponential expressions.
2. Use the order of operations rule.
3. Evaluate expressions containing grouping symbols.
4. Find the mean (average).

ARE YOU READY?

The following problems review some basic concepts that are important when working with numerical expressions.

1. Name the operations that are involved in the expression: $50 - 2(3)$
2. Name the operations that are involved in the expression: $\dfrac{2 + 6 \cdot 8}{4 + 6}$
3. Multiply: $3 \cdot 3 \cdot 3 \cdot 3$
4. Multiply: $(-5)(-5)(-5)$

In algebra, we often have to find the value of expressions that involve more than one operation. In this section, we introduce an order of operations rule to follow in such cases. But first, we discuss a way to write repeated multiplication using *exponents.*

1 Evaluate Exponential Expressions.

In the expression $3 \cdot 3 \cdot 3 \cdot 3 \cdot 3$, the number 3 repeats as a factor five times. We can use **exponential notation** to write this product in a more compact form.

Exponent and Base	An exponent is used to indicate repeated multiplication. It is how many times the base is used as a factor.

The Language of Algebra

5^2 represents the area of a square with sides 5 units long. 4^3 represents the volume of a cube with sides 4 units long.

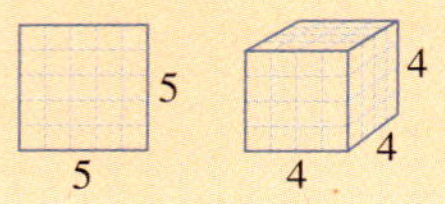

$$\underbrace{3 \cdot 3 \cdot 3 \cdot 3 \cdot 3}_{} = 3^5$$

Five repeated factors of 3. The exponent is 5. The base is 3.

In the **exponential expression** 3^5, the base is 3, and 5 is the exponent. The expression is called a power of 3. Some other examples of exponential expressions are:

5^2 Read as "5 to the second power" or "5 squared."

4^3 Read as "4 to the third power" or "4 cubed."

$(-2)^5$ Read as "-2 to the fifth power."

EXAMPLE 1 Write each product using exponents: **a.** $7 \cdot 7 \cdot 7$ **b.** $(-5)(-5)(-5)(-5)(-5)$ **c.** $8 \cdot 8 \cdot 15 \cdot 15 \cdot 15 \cdot 15$ **d.** $a \cdot a \cdot a \cdot a \cdot a \cdot a$ **e.** $4 \cdot \pi \cdot r \cdot r$

Strategy We need to determine the number of repeated factors in the expression.

Why An exponent can be used to represent repeated multiplication.

Solution

a. The factor 7 is repeated 3 times. We can represent this repeated multiplication with an exponential expression having a base of 7 and an exponent of 3: $7 \cdot 7 \cdot 7 = 7^3$.

b. The factor -5 is repeated five times: $(-5)(-5)(-5)(-5)(-5) = (-5)^5$.

c. $8 \cdot 8 \cdot 15 \cdot 15 \cdot 15 \cdot 15 = 8^2 \cdot 15^4$

d. $a \cdot a \cdot a \cdot a \cdot a \cdot a = a^6$

e. $4 \cdot \pi \cdot r \cdot r = 4\pi r^2$

Self Check 1 Write each product using exponents: **a.** $(12)(12)(12)$ **b.** $2 \cdot 9 \cdot 9 \cdot 9 \cdot 9$ **c.** $(-30)(-30)$ **d.** $y \cdot y \cdot y \cdot y \cdot y \cdot y$ **e.** $8 \cdot b \cdot b \cdot b \cdot c$

Now Try Problems 15 and 21

To **evaluate** (find the value of) an exponential expression, we write the base as a factor the number of times indicated by the exponent. Then we multiply the factors.

EXAMPLE 2 Evaluate each expression: **a.** 5^3 **b.** $\left(-\frac{2}{3}\right)^3$ **c.** 10^1 **d.** $(0.6)^2$ **e.** $(-3)^4$ **f.** $(-3)^5$

Strategy We will rewrite each exponential expression as a product of repeated factors, and then perform the multiplication. This requires that we identify the base and the exponent.

Why The exponent tells the number of times the base is to be written as a factor.

Solution

a. $5^3 = 5 \cdot 5 \cdot 5$ Write the base, 5, as a factor 3 times.

$= 125$ Multiply, working left to right. We say 125 is the *cube of 5*.

Because $5^3 = 125$, we say that 125 is a **power** of 5.

b. $\left(-\frac{2}{3}\right)^3 = \left(-\frac{2}{3}\right)\left(-\frac{2}{3}\right)\left(-\frac{2}{3}\right)$ Since $-\frac{2}{3}$ is the base and 3 is the exponent, we write $-\frac{2}{3}$ as a factor 3 times.

$= \frac{4}{9}\left(-\frac{2}{3}\right)$ Work from left to right: $\left(-\frac{2}{3}\right)\left(-\frac{2}{3}\right) = \frac{4}{9}$.

$= -\frac{8}{27}$

c. $10^1 = 10$ The base is 10. Since the exponent is 1, we write the base once.

d. $(0.6)^2 = (0.6)(0.6)$ Write the base, 0.6, as a factor 2 times.

$= 0.36$ We say 0.36 is the *square of 0.6*.

e. $(-3)^4 = (-3)(-3)(-3)(-3)$ Write the base, -3, as a factor 4 times.

$= 9(-3)(-3)$ Work from left to right.

$= -27(-3)$

$= 81$

f. $(-3)^5 = (-3)(-3)(-3)(-3)(-3)$ Write the base, -3, as a factor 5 times.

$= 9(-3)(-3)(-3)$ Work from left to right.

$= -27(-3)(-3)$

$= 81(-3)$

$= -243$

Caution

Don't make the common mistake of multiplying the base and the exponent:

$5^3 \neq 5 \cdot 3$

Calculators

Finding a power
The squaring key x^2 can be used to find the square of a number. To raise a number to a power, we use the y^x key on a scientific calculator and the ^ key on a graphing calculator.

The Language of Algebra

A number or a variable has an **understood** exponent of 1. For example,

$8 = 8^1$ and $x = x^1$

Self Check 2 Evaluate: **a.** 2^5 **b.** $\left(-\frac{3}{4}\right)^3$ **c.** 9^1 **d.** $(-0.3)^2$ **e.** $(-6)^2$ **f.** $(-5)^3$

Now Try Problems 23, 29, and 33

In Example 2e, we raised -3 to an even power; the result was positive. In part f, we raised -3 to an odd power; the result was negative. These results illustrate the following rule.

Even and Odd Powers of a Negative Number

When a negative number is raised to an even power, the result is positive.

When a negative number is raised to an odd power, the result is negative.

The Language of Algebra

Read $(-4)^2$ as "negative four squared" and -4^2 as "the opposite of the square of four."

Although the expressions $(-4)^2$ and -4^2 look alike, they are not. When we find the value of each expression, it becomes clear that they are not equivalent.

$$(-4)^2 = (-4)(-4) \qquad \text{The base is } -4\text{, the exponent is 2.}$$
$$= 16$$

$$-4^2 = -(4 \cdot 4) \qquad \text{The base is 4, the exponent is 2.}$$
$$= -16$$

Different results

Any real number can be used as a base. However, the base of an exponential expression *does not include* the negative sign unless parentheses are used.

-7^3 Positive base: 7

$(-7)^3$ Negative base: -7

EXAMPLE 3 Evaluate: -2^4

Strategy We will rewrite the expression as a product of repeated factors and then perform the multiplication. We must be careful when identifying the base. It is 2, not -2.

Why Since there are no parentheses around -2, the base is 2.

Solution

$$-2^4 = -(2 \cdot 2 \cdot 2 \cdot 2) \qquad \text{Read as "the opposite of the fourth power of two."}$$
$$= -16 \qquad \text{Do the multiplication within the parentheses to get 16. Then write the opposite of that result.}$$

Self Check 3 Evaluate: -5^4

Now Try Problem 35

2 Use the Order of Operations Rule.

Suppose you have been asked to contact a friend if you see a Rolex watch for sale when you are traveling in Europe. While in Switzerland, you find the watch and send the text message shown on the left. The next day, you get the response shown on the right.

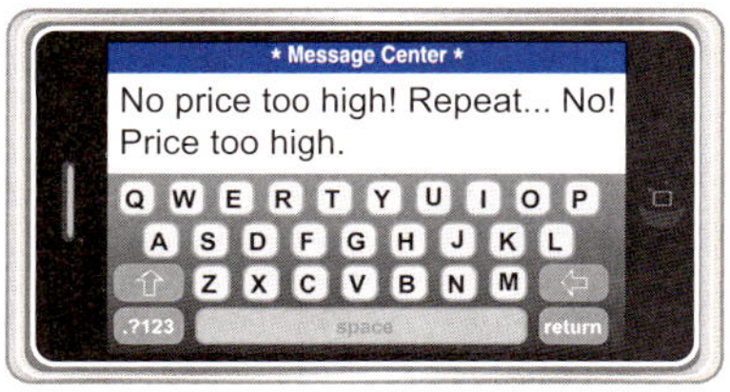

Something is wrong. The first part of the response (No price too high!) says to buy the watch at any price. The second part (No! Price too high.) says not to buy it, because it's too expensive. The placement of the exclamation point makes us read the two parts of the response differently, resulting in different meanings. When reading a mathematical statement, the same kind of confusion is possible. For example, consider the expression

$$2 + 3 \cdot 6$$

We can evaluate this expression in two ways. We can add first, and then multiply. Or we can multiply first, and then add. However, the results are different.

$$\begin{aligned} \mathbf{2 + 3} \cdot 6 &= \mathbf{5} \cdot 6 && \text{Add 2 and 3 first.} \\ &= 30 && \text{Multiply 5 and 6.} \end{aligned} \qquad \begin{aligned} 2 + \mathbf{3 \cdot 6} &= 2 + \mathbf{18} && \text{Multiply 3 and 6 first.} \\ &= 20 && \text{Add 2 and 18.} \end{aligned}$$

Different answers

If we don't establish a uniform order of operations, the expression has two different values. To avoid this possibility, we will always use the following set of priority rules.

Order of Operations

1. Perform all calculations within parentheses and other grouping symbols following the order listed in Steps 2–4 below, working from the innermost pair of grouping symbols to the outermost pair.
2. Evaluate all exponential expressions.
3. Perform all multiplications and divisions as they occur from left to right.
4. Perform all additions and subtractions as they occur from left to right.

When grouping symbols have been removed, repeat Steps 2–4 to complete the calculation.

If a fraction is present, evaluate the expression above and the expression below the bar separately. Then simplify the fraction, if possible.

It isn't necessary to apply all of these steps in every problem. For example, the expression $2 + 3 \cdot 6$ does not contain any parentheses, and there are no exponential expressions. So we look for multiplications and divisions to perform and proceed as follows:

$$\begin{aligned} 2 + \mathbf{3 \cdot 6} &= 2 + \mathbf{18} && \text{Do the multiplication first.} \\ &= 20 && \text{Do the addition.} \end{aligned}$$

EXAMPLE 4 Evaluate: **a.** $3 \cdot 2^3 - 4$ **b.** $-30 - 4 \cdot 5 + 9$ **c.** $24 \div 6 \cdot 2$ **d.** $160 - 4 + 6(-2)(-3)$

The Language of Algebra

Sometimes, for problems like these, the instruction **simplify** is used instead of *evaluate.*

Strategy We will scan the expression to determine what operations need to be performed. Then we will perform those operations, one-at-a-time, following the order of operations rules.

Why If we don't follow the correct order of operations, the expression can have more than one value.

Solution **a.** Three operations need to be performed to evaluate this expression: multiplication, raising to a power, and subtraction. By the order of operations rules, we evaluate 2^3 first.

$$\begin{aligned} 3 \cdot \mathbf{2^3} - 4 &= 3 \cdot \mathbf{8} - 4 && \text{Evaluate the exponential expression: } 2^3 = 8. \\ &= 24 - 4 && \text{Do the multiplication: } 3 \cdot 8 = 24. \\ &= 20 && \text{Do the subtraction.} \end{aligned}$$

b. This expression involves subtraction, multiplication, and addition. The order of operations rule tells us to multiply first.

$$
\begin{aligned}
-30 - 4 \cdot 5 + 9 &= -30 - 20 + 9 && \text{Do the multiplication: } 4 \cdot 5 = 20. \\
&= -50 + 9 && \text{Working from left to right, do the subtraction: } -30 - 20 = -30 + (-20) = -50. \\
&= -41 && \text{Do the addition.}
\end{aligned}
$$

Caution

A common mistake is to forget to work from left to right and incorrectly perform the multiplication before the division. This produces the wrong answer, 2.

$$
\begin{aligned}
24 \div 6 \cdot 2 &= 24 \div 12 \\
&= 2
\end{aligned}
$$

c. Since there are no calculations within parentheses nor are there exponents, we perform the multiplications and divisions as they occur from left to right. The division occurs before the multiplication, so it must be performed first.

$$
\begin{aligned}
24 \div 6 \cdot 2 &= 4 \cdot 2 && \text{Working left to right, do the division: } 24 \div 6 = 4. \\
&= 8 && \text{Do the multiplication.}
\end{aligned}
$$

d. Although this expression contains parentheses, there are no operations to perform within them. Since there are no exponents, we will perform the multiplications as they occur from left to right.

$$
\begin{aligned}
160 - 4 + 6(-2)(-3) &= 160 - 4 + (-12)(-3) && \text{Do the multiplication, working left to right: } 6(-2) = -12. \\
&= 160 - 4 + 36 && \text{Complete the multiplication: } (-12)(-3) = 36. \\
&= 156 + 36 && \text{Working left to right, do the subtraction before the addition.} \\
&= 192 && \text{Do the addition.}
\end{aligned}
$$

Self Check 4 Evaluate: **a.** $2 \cdot 3^2 + 17$ **b.** $-40 - 9 \cdot 4 + 10$ **c.** $18 \div 2 \cdot 3$ **d.** $240 - 8 + 3(-2)(-4)$

Now Try Problems 39, 43, 45, and 49

3 Evaluate Expressions Containing Grouping Symbols.

Grouping symbols serve as mathematical punctuation marks. They help determine the order in which an expression is to be evaluated. Examples of grouping symbols are parentheses ( ), brackets [ ], braces { }, absolute value symbols | |, and the fraction bar —.

EXAMPLE 5 Evaluate each expression: **a.** $(6 - 3)^2$ **b.** $5^3 + 2(-8 - 3 \cdot 2)$

Strategy We will perform the operation(s) within the parentheses first. When there is more than one operation to perform within the parentheses, we follow the order of operations rule.

Why This is the first step of the order of operations rule.

Solution **a.** We read $(6 - 3)^2$ as "the square of the quantity of 6 minus 3."

$$
\begin{aligned}
(6 - 3)^2 &= 3^2 && \text{Do the subtraction within the parentheses: } 6 - 3 = 3. \\
&= 9 && \text{Evaluate the exponential expression.}
\end{aligned}
$$

Notation

Multiplication is indicated when a number is next to a parenthesis or bracket.

$$5^3 + 2(-8 - 3 \cdot 2)$$

b. We begin by performing the operations within the parentheses in the proper order: multiplication first, and then subtraction.

$$5^3 + 2(-8 - \mathbf{3 \cdot 2}) = 5^3 + 2(-8 - \mathbf{6})$$ Do the multiplication: $3 \cdot 2 = 6$.

$$= 5^3 + 2(-14)$$ Do the subtraction: $-8 - 6 = -14$.

$$= 125 + 2(-14)$$ Evaluate 5^3.

$$= 125 + (-28)$$ Do the multiplication: $2(-14) = -28$.

$$= 97$$ Do the addition.

Self Check 5 Evaluate: **a.** $(12 - 6)^3$ **b.** $1^3 + 6(-6 - 3 \cdot 0)$

Now Try Problems 53 and 55

Expressions can contain two or more pairs of grouping symbols. To evaluate the following expression, we begin within the innermost pair of grouping symbols, the parentheses. Then we work within the outermost pair, the brackets.

Innermost pair

$$-4[2 + 3(4 - 8^2)] - 2$$

Outermost pair

EXAMPLE 6 Evaluate: $-4[2 + 3(4 - 8^2)] - 2$

Strategy We will work within the parentheses first and then within the brackets. At each stage, we follow the order of operations rules.

Why By the order of operations, we must work from the *innermost* pair of grouping symbols to the *outermost.*

The Language of Algebra

When one pair of grouping symbols is inside another pair, we say that those grouping symbols are **nested,** or **embedded.**

Solution

$$-4[2 + 3(4 - \mathbf{8^2})] - 2$$

$$= -4[2 + 3(4 - \mathbf{64})] - 2$$ Evaluate the exponential expression within the parentheses: $8^2 = 64$.

$$= -4[2 + 3(-60)] - 2$$ Do the subtraction within the parentheses: $4 - 64 = 4 + (-64) = -60$.

$$= -4[2 + (-180)] - 2$$ Do the multiplication within the brackets: $3(-60) = -180$.

$$= -4[-178] - 2$$ Do the addition within the brackets: $2 + (-180) = -178$.

$$= 712 - 2$$ Do the multiplication: $-4[-178] = 712$.

$$= 710$$ Do the subtraction.

Self Check 6 Evaluate: $-5[4 + 2(5^2 - 15)] - 10$

Now Try Problem 61

EXAMPLE 7 Evaluate: $\dfrac{-3(3 + 2) + 5}{17 - 3(-4)}$

Strategy We will evaluate the expression above and the expression below the fraction bar separately. Then we will simplify the fraction, if possible.

Why Fraction bars are grouping symbols. They group the numerator and denominator. The expression could be written $[-3(3 + 2) + 5] \div [17 - 3(-4)]$.

Solution

$$\frac{-3(3 + 2) + 5}{17 - 3(-4)} = \frac{-3(5) + 5}{17 - (-12)}$$ In the numerator, do the addition within the parentheses. In the denominator, do the multiplication.

$$= \frac{-15 + 5}{17 + 12}$$ In the numerator, do the multiplication. In the denominator, write the subtraction as the addition of the opposite of −12, which is 12.

$$= \frac{-10}{29}$$ Do the additions.

$$= -\frac{10}{29}$$ Write the − sign in front of the fraction: $\frac{-10}{29} = -\frac{10}{29}$. The fraction does not simplify.

Calculators

Order of operations
Calculators have the order of operations built in. A left parenthesis key (and a right parenthesis key) should be used when grouping symbols, including a fraction bar, are needed.

Self Check 7 Evaluate: $\dfrac{-4(-2 + 8) + 6}{8 - 5(-2)}$

Now Try Problem 73

EXAMPLE 8 Evaluate: $10|9 - 15| - 2^5$

Strategy The absolute value bars are grouping symbols. We will perform the calculation within them first.

Why By the order of operations, we must perform all calculations within parentheses and other grouping symbols (such as absolute value bars) first.

Solution

$10|9 - 15| - 2^5 = 10|-6| - 2^5$ Subtract: $9 - 15 = 9 + (-15) = -6$.

$= 10(6) - 2^5$ Find the absolute value: $|-6| = 6$.

$= 10(6) - 32$ Evaluate the exponential expression: $2^5 = 32$.

$= 60 - 32$ Do the multiplication: $10(6) = 60$.

$= 28$ Do the subtraction.

Notation

Multiplication is indicated when a number is next to an absolute value symbol.

$$10|9 - 15| - 2^5$$

Self Check 8 Evaluate: $10^3 + 3|24 - 25|$

Now Try Problem 77

4 Find the Mean (Average).

The **arithmetic mean** (or simply **mean**) of a set of numbers is a value around which the values of the numbers are grouped. The mean is also commonly called the **average**.

Finding an Arithmetic Mean

To find the mean of a set of values, divide the sum of the values by the number of values.

When a value in a set appears more than once, that value has a greater "influence" on the mean than another value that only occurs a single time. To simplify the process of finding a mean, any value that appears more than once can be "weighted" by multiplying it by the number of times it occurs. A mean that is found in this way is called a **weighted mean.**

EXAMPLE 9 **Hotel Reservations.** In an effort to improve customer service, a hotel electronically recorded the number of times the reservation desk telephone rang before it was answered by a receptionist. The results of the week-long survey are shown in the table. Find the average (mean) number of times the phone rang before a receptionist answered.

Number of rings	Number of calls
1	11
2	46
3	45
4	28
5	20

Strategy First, we will determine the total number of times the reservation desk telephone rang during the week. Then we will divide that result by the total number of calls received.

Why To find the *average* value of a set of values, we divide the sum of the values by the number of values.

Solution To find the total number of rings, we multiply each *number of rings* (1, 2, 3, 4, and 5 rings) by the respective number of occurrences and add those subtotals.

$$\text{Total number of rings} = 11(1) + 46(2) + 45(3) + 28(4) + 20(5)$$

The total number of calls received was $11 + 46 + 45 + 28 + 20$. To find the average, we divide the total number of rings by the total number of calls.

$$\text{Average} = \frac{11(1) + 46(2) + 45(3) + 28(4) + 20(5)}{11 + 46 + 45 + 28 + 20}$$

$$= \frac{11 + 92 + 135 + 112 + 100}{150}$$ In the numerator, do the multiplications. In the denominator, do the additions.

$$= \frac{450}{150}$$ Do the addition.

$$= 3$$ Simplify the fraction.

The average number of times the phone rang before it was answered was 3.

Self Check 9 **Evaluations.** On the first question of an instructor's evaluation, 14 students marked 1 for *strongly agree,* 10 students marked 2 for *agree,* 6 students marked 3 for *disagree,* and 4 students marked 4 for *strongly disagree.* What was the average (mean) response for the first question on the evaluation?

Now Try Problem 119

SECTION 1.7 STUDY SET

VOCABULARY

Fill in the blanks.

1. In the exponential expression 7^5, 7 is the _____, and 5 is the _________. 7^5 is the fifth _______ of seven.
2. 10^2 can be read as ten ________, and 10^3 can be read as ten _______.
3. An _________ is used to represent repeated multiplication.
4. To ________ the expression $2(-1 + 4^2)$ means to find its value.
5. The rule for the ______ of operations guarantees that an evaluation of a numerical expression will result in a single answer.
6. To find the arithmetic ______ or average of a set of values, divide the sum of the values by the number of values.

CONCEPTS

7. To evaluate each expression, what operation should be performed first?
   a. $24 - 4 + 2$
   b. $32 \div 8 \cdot 4$
   c. $8 - (3 + 5)^2$
   d. $65 \cdot 3^3$

8. To evaluate $\frac{36 - 4(7)}{2(10 - 8)}$, what operation should be performed first in the numerator? In the denominator?

NOTATION

9. a. Give the name of each grouping symbol: ( ), [ ], { }, | |, and —.
   b. In the expression $-8 + 2[15 - (-6 + 1)]$, which grouping symbols are innermost, and which are outermost?

10. What operation is indicated?

$$2 + 9|5 - (2 + 4)|$$

11. a. In the expression $(-5)^2$, what is the base?
    b. In the expression -5^2, what is the base?

12. Write each expression using symbols. Then evaluate it.
    a. Negative two squared
    b. The opposite of the square of two

Complete the evaluation of each expression.

13. $-19 - 2[(1 + 2)^2 \cdot 3] = -19 - 2[\quad^2 \cdot 3]$
$= -19 - 2[\quad \cdot 3]$
$= -19 - 2[\quad]$
$= -19 -$
$=$

14. $\frac{46 - 2^3}{-3(5) - 4} = \frac{46 - \quad}{\quad - 4}$
$= \frac{\quad}{\quad}$
$= -2$

GUIDED PRACTICE

Write each product using exponents. See Example 1.

15. $8 \cdot 8 \cdot 8$
16. $(-4)(-4)(-4)(-4)$
17. $7 \cdot 7 \cdot 7 \cdot 12 \cdot 12$
18. $5 \cdot 5 \cdot 5 \cdot 5 \cdot 5 \cdot 5 \cdot 7 \cdot 7 \cdot 7$
19. $x \cdot x \cdot x$
20. $b \cdot b \cdot b \cdot b$
21. $r \cdot r \cdot r \cdot r \cdot s \cdot s$
22. $m \cdot m \cdot m \cdot n \cdot n \cdot n \cdot n$

Evaluate each expression. See Example 2.

23. 7^2
24. 9^2
25. 6^3
26. 6^4
27. $(-5)^4$
28. $(-5)^3$
29. $(-0.1)^2$
30. $(-0.8)^2$
31. $\left(-\frac{1}{4}\right)^3$
32. $\left(-\frac{1}{3}\right)^4$
33. $\left(\frac{2}{3}\right)^3$
34. $\left(\frac{3}{4}\right)^3$

Evaluate each expression. See Example 3.

35. $(-6)^2$ and -6^2
36. $(-4)^2$ and -4^2
37. $(-8)^2$ and -8^2
38. $(-9)^2$ and -9^2

Evaluate each expression. See Example 4.

39. $3 - 5 \cdot 4$
40. $-4 \cdot 6 + 5$
41. $32 - 16 \div 4 + 2$
42. $60 - 20 \div 10 + 5$
43. $3^2 \cdot 5 - 6 \div 3$
44. $2^3 \cdot 5 - 4 \div 2$
45. $12 \div 3 \cdot 2$
46. $18 \div 6 \cdot 3$
47. $-22 - 15 + 3$
48. $-33 - 8 + 10$
49. $-2(9) - 2(5)(10)$
50. $-6(7) - 3(-4)(-2)$

Evaluate each expression. See Example 5.

51. $-4(6 + 5)$
52. $-3(5 - 4)$
53. $(9 - 3)(9 - 9)^2$
54. $-(-8 - 6)(6 - 6)^2$
55. $(-1 - 3^2 \cdot 4)2^2$
56. $-1(28 - 5^2 \cdot 2)3^2$
57. $1 + 5(10 + 2 \cdot 5) - 1$
58. $14 + 3(7 - 5 \cdot 3)$

Evaluate each expression. See Example 6.

59. $(-1)^9[-7^2 - (-2)^2]$
60. $[-9^2 - (-8)^2](-1)^{10}$
61. $64 - 6[15 + 2(-3 + 8)]$
62. $4 - 2[26 + 2(5 - 3)]$
63. $-2[2 + 4^2(8 - 9)]^2$
64. $-3[5 + 3^2(4 - 5)]^2$
65. $3 + 2[-1 - (4 - 5)]$
66. $4 + 2[-7 - (3 - 9)]$

Evaluate each expression. See Example 7.

67. $\frac{-2 - 5}{-7 + (-7)}$
68. $\frac{-3 - (-1)}{-2 + (-2)}$
69. $\frac{2 \cdot 2^5 - 60 + (-4)}{5^4 - (-4)(-5)}$
70. $\frac{(6 - 5)^8 - 1}{(-9)(-3) - 4}$
71. $\frac{2(-4 - 2 \cdot 2)}{3(-3)(-2)}$
72. $\frac{3(-3^2 + 2 \cdot 2^2)}{(5 - 8)(7 - 9)}$
73. $\frac{72 - (2 - 2 \cdot 4)}{10^2 - (9 \cdot 10 + 2^2)}$
74. $\frac{13^2 - 5^2}{-3(5 - 3^2)}$

Evaluate each expression. See Example 8.

75. $10 - 2|4 - 8|$
76. $45 - 5|1 - 8|$
77. $-|7 - 2^3(4 - 7)|$
78. $-|9 - 5(1 - 2^3)|$
79. $\frac{(3 + 5)^2 + |-2|}{-2(5 - 8)}$
80. $\frac{|-25| - 8(-5)}{2^4 - 29}$
81. $\frac{|6 - 4| + 2|-4|}{226 - 6^3}$
82. $\frac{4|9 - 7| + |-7|}{6^3 - 211}$
83. $-(2 \cdot 3 - 2^2)^5$
84. $-(3 \cdot 5 - 2 \cdot 6)^2$
85. $2 \cdot 5^2 + 4 \cdot 3^2$
86. $5 \cdot 3^3 - 4 \cdot 2^3$
87. $-2(-1)^2 + 3(-1) - 3$
88. $-4(-3)^2 + 3(-3) - 1$
89. $8 - 3[5^2 - (7 - 3)^2]$
90. $3 - [3^3 + (3 - 1)^3]$

TRY IT YOURSELF

Evaluate each expression.

91. $[6(5) - 5(5)]^3(-4)$
92. $5 - 2 \cdot 3^4 - (-6 + 5)^3$
93. $8 - 6[(130 - 4^3) - 2]$
94. $91 - 5[(150 - 3^3) - 1]$

95. $-2\left(\frac{15}{-5}\right) - \frac{6}{2} + 9$

96. $-6\left(\frac{25}{-5}\right) - \frac{36}{9} + 1$

97. $-5(-2)^3 - |-2 + 1|$

98. $-6(-3)^3 - |-6 + 5|$

99. $\frac{18 - [2 + (1 - 6)]}{16 - (-4)^2}$

100. $\frac{6 - [6(-1) - 88]}{4 - 2^2}$

101. $-|-5 \cdot 7^2| - 30$

102. $2 + |-3 \cdot 2^2 \cdot 8^2 \cdot 1^2|$

103. $(-3)^3\left(\frac{-4}{2}\right)(-1)$

104. $(-2)^3\left(\frac{-6}{2}\right)(-1)$

105. $\frac{1}{2}\left(\frac{1}{8}\right) + \left(-\frac{1}{4}\right)^2$

106. $-\frac{1}{9}\left(\frac{1}{4}\right) + \left(-\frac{1}{6}\right)^2$

107. $\frac{-5^2 \cdot 10 + 5 \cdot 2^5}{-5 - 3 - 1}$

108. $\frac{(-6^2 - 2^4 \cdot 2) + 5}{-4 - 3}$

109. $-\left(\frac{40 - 1^3 - 2^4}{3(2 + 5) + 2}\right)$

110. $-\left(\frac{8^2 - 10}{2(3)(4) - 5(3)}\right)$

Look Alikes . . .

111. **a.** $(-7 - 4)(-2)$ **b.** $(-7 - 4) - 2$

112. **a.** $2 \cdot 3^3$ **b.** $(2 \cdot 3)^3$

113. **a.** $-100 \div 5 \cdot 2$ **b.** $-100 \div (5 \cdot 2)$

114. **a.** $8 + 3[-2 - (6 + 1)]$ **b.** $(8 + 3)[-2 - (6 + 1)]$

APPLICATIONS

115. Light. As light energy passes through the first unit of area, 1 yard away from the bulb, it spreads out. How much area does that light energy cover 2 yards, 3 yards, and 4 yards from the bulb? Express each answer using exponents.

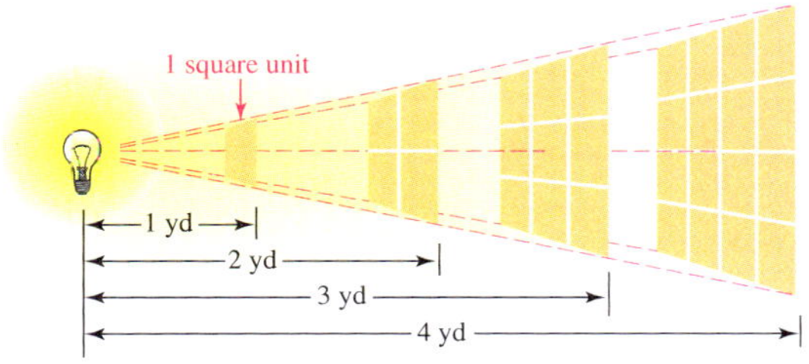

116. Chain Letters. A woman sent two friends a letter with the following request: "Please send a copy of this letter to two of your friends." Assume that all those receiving letters responded and that everyone in the chain received just one letter. Complete the table and then determine how many letters will be circulated in the 10th level.

Level	Number of letters circulated
1st	$2 = 2^1$
2nd	$= 2$
3rd	$= 2$
4th	$= 2$

117 from Campus to Careers

Lead Transportation Security Officer

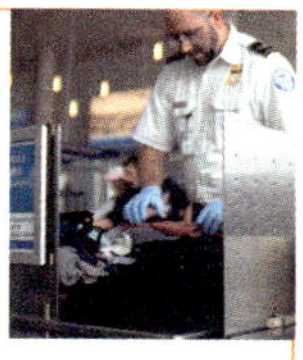

To determine the average afternoon wait time in security lines at an airport, officials monitored four passengers, each at a different gate. The time that each passenger entered a security line and the time the same passenger cleared the checkpoint was recorded, as shown below. Find the average (mean) wait time for these passengers.

	Time entered	Time cleared
Passenger at Gate A	3:05 pm	3:21 pm
Passenger at Gate B	3:03 pm	3:13 pm
Passenger at Gate C	3:01 pm	3:09 pm
Passenger at Gate D	3:02 pm	3:16 pm

118. Energy Usage. Find the average number of therms of natural gas used per month.

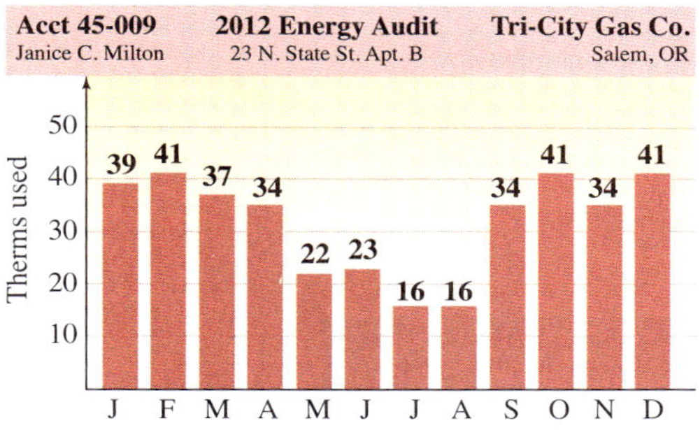

119. Cash Awards. A contest is to be part of a promotional kickoff for a new children's cereal. The prizes to be awarded are shown.

a. How much money will be awarded in the promotion?

b. What is the average cash prize?

YouTube Video Contest

Grand prize: Disney World vacation plus $2,500

Four 1st place prizes of $500
Thirty-five 2nd place prizes of $150
Eighty-five 3rd place prizes of $25

120. Surveys. Some students were asked to rate their college cafeteria food on a scale from 1 to 5. The responses are shown on the tally sheet. Find the average rating.

Poor		Fair		Excellent
1	2	3	4	5
	III	III	𝍸	𝍸 IIII

121. Wrapping Gifts. How much ribbon is needed to wrap the package if 15 inches of ribbon are needed to make the bow?

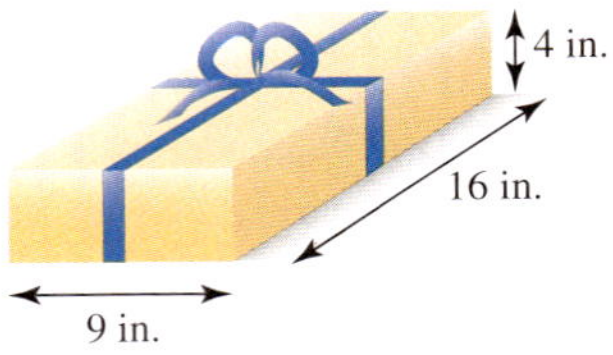

122. Scrabble. Write an expression to determine the number of points received for playing the word QUARTZY and then evaluate it. (The number on each tile gives the point value of the letter.)

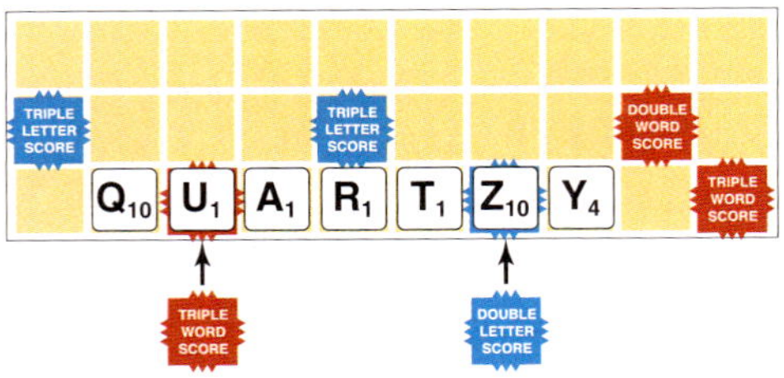

WRITING

123. Explain the difference between 2^3 and 3^2.

124. Why is the order of operations rule necessary?

125. Explain the error. What is the correct answer?

$$40 \div 4 \cdot 2 = 40 \div 8$$
$$= 5$$

126. Explain the error. What is the correct answer?

$$5 + 3(2 - 6) = 5 + 3(-4)$$
$$= 8(-4)$$
$$= -32$$

REVIEW

127. What numbers are a distance of 6 away from -11 on a number line?

128. Fill in the blank with $>$ or $<$: $0.3 \quad \frac{1}{3}$

CHALLENGE PROBLEMS

129. Using each of the numbers 2, 3, and 4 only once, what is the greatest value that the following expression can have?

$$(\square^{\square})^{\square}$$

130. Insert a pair of parentheses into $4 \cdot 3^2 - 4 \cdot 2$ so that it has a value of 40.

Translate the set of instructions to an expression and then evaluate it.

131. Subtract the sum of -9 and 8 from the product of the cube of -3 and the opposite of 4.

132. Increase the square of the reciprocal of -2 by the difference of -0.25 and -1.

SECTION 1.8 Algebraic Expressions

OBJECTIVES

1. Identify terms and coefficients of terms.
2. Translate word phrases to algebraic expressions.
3. Analyze problems to determine hidden operations.
4. Evaluate algebraic expressions.

ARE YOU READY?

The following problems review some basic concepts that are important when working with algebraic expressions.

Write each expression in simpler form.

1. $9 \cdot x$ **2.** $1 \cdot m$ **3.** $-1 \cdot t$ **4.** $\frac{7}{8} \cdot y$

Identify each of the following expressions as either a sum, difference, product, or quotient.

5. $\frac{x}{12}$ **6.** $45 - a$ **7.** $2bc$ **8.** $d + 5$

Since problems in algebra are often presented in words, the ability to interpret what you read is important. In this section, we will introduce several strategies that will help you translate English words into algebraic expressions.

1 Identify Terms and Coefficients of Terms.

Recall that variables and/or numbers can be combined with the operations of arithmetic to create **algebraic expressions.** Addition symbols separate expressions into parts called *terms.* For example, the expression $x + 8$ has two terms.

x	$+$	8
First term		Second term

Since subtraction can be written as addition of the opposite, the expression $a^2 - 3a - 9$ has three terms.

$$a^2 - 3a - 9 = \underset{\text{First term}}{a^2} + \underset{\text{Second term}}{(-3a)} + \underset{\text{Third term}}{(-9)}$$

In general, a **term** is a product or quotient of numbers and/or variables. A single number or variable is also a term. Examples of terms are:

Notation

By the commutative property of multiplication, $r6 = 6r$ and $-15b^2a = -15ab^2$. However, we usually write the numerical factor first and the variable factors in alphabetical order.

$$4, \quad y, \quad 6r, \quad -w^3, \quad 3.7x^5, \quad \frac{3}{n}, \quad -15ab^2$$

The numerical factor of a term is called the **coefficient** of the term. For instance, the term $6r$ has a coefficient of 6 because $6r = 6 \cdot r$. The coefficient of $-15ab^2$ is -15 because $-15ab^2 = -15 \cdot ab^2$. More examples are shown below.

A term such as 4, that consists of a single number, is called a **constant term.**

The Language of Algebra

Terms such as x and y have **implied** coefficients of 1. *Implied* means suggested without being precisely expressed.

Term	Coefficient	
$8y^2$	8	
$-0.9pq$	-0.9	
$\frac{3}{4}b$	$\frac{3}{4}$	This term also could be written as $\frac{3b}{4}$.
$-\frac{x}{6}$	$-\frac{1}{6}$	Because $-\frac{x}{6} = -\frac{1x}{6} = -\frac{1}{6} \cdot x$
x	1	Because $x = 1x$
$-t$	-1	Because $-t = -1t$
27	27	

EXAMPLE 1 Identify the coefficient of each term in the expression: $7x^2 - x + 6$

Strategy We will begin by writing the subtraction as addition of the opposite. Then we will determine the numerical factor of each term.

Why Addition symbols separate expressions into terms.

Solution If we write $7x^2 - x + 6$ as $7x^2 + (-x) + 6$, we see that it has three terms: $7x^2$, $-x$, and 6. The numerical factor of each term is its coefficient.

- The coefficient of $7x^2$ is **7** because $7x^2$ means $\mathbf{7} \cdot x^2$.
- The coefficient of $-x$ is **−1** because $-x$ means $\mathbf{-1} \cdot x$.
- The coefficient of the constant 6 is 6.

Self Check 1 Identify the coefficient of each term in the expression: $p^3 - 12p^2 + 3p - 4$

Now Try Problem 19

It is important to be able to distinguish between the *terms* of an expression and the *factors* of a term.

EXAMPLE 2 Is m used as a *factor* or a *term* in each expression? **a.** $m + 6$ **b.** $8m$

Strategy We will begin by determining whether m is involved in an addition or a multiplication.

Why Addition symbols separate expressions into *terms.* A *factor* is a number being multiplied.

Solution

a. Since m is added to 6, m is a term of $m + 6$.

b. Since m is multiplied by 8, m is a factor of $8m$.

Self Check 2 Is b used as a *factor* or a *term* in each expression?

a. $-27b$ **b.** $5a + b$

Now Try Problems 21 and 23

2 Translate Word Phrases to Algebraic Expressions.

The four tables below show how key phrases can be translated into algebraic expressions.

Caution

Be careful when translating subtraction. Order is important. For example, when a translation involves the phrase *less than*, note how the terms are reversed as we translate from English to mathematical symbols.

18 less than w

$w - 18$

Addition	
the sum of a and 8	$a + 8$
4 plus c	$4 + c$
16 added to m	$m + 16$
4 more than t	$t + 4$
20 greater than F	$F + 20$
T increased by r	$T + r$
exceeds y by 35	$y + 35$

Subtraction	
the difference of 23 and P	$23 - P$
550 minus h	$550 - h$
18 less than w	$w - 18$
7 decreased by j	$7 - j$
M reduced by x	$M - x$
12 subtracted from L	$L - 12$
5 less f	$5 - f$

Multiplication	
the product of 4 and x	$4x$
20 times B	$20B$
twice r	$2r$
double the amount a	$2a$
triple the profit P	$3P$
three-fourths of m	$\frac{3}{4}m$

Division	
the quotient of R and 19	$\frac{R}{19}$
s divided by d	$\frac{s}{d}$
the ratio of c to d	$\frac{c}{d}$
k split into 4 equal parts	$\frac{k}{4}$

Be careful when translating division. As with subtraction, order is important. For example, s divided by d is not written $\frac{d}{s}$.

EXAMPLE 3 Write each phrase as an algebraic expression:

a. one-half of the profit P **b.** 5 less than the capacity c

c. the product of the weight w and 2,000, increased by 300

Strategy We will begin by identifying any key phrases.

Why Key phrases can be translated to mathematical symbols.

Solution

a. Key phrase: *One-half of* **Translation:** multiplication by $\frac{1}{2}$

The algebraic expression is: $\frac{1}{2}P$.

b. Key phrase: *less than* **Translation:** subtraction

Sometimes thinking in terms of specific numbers makes translating easier. Suppose the capacity was 100. Then 5 *less than* 100 would be $100 - 5$. If the capacity is c, then we need to make it 5 less. The algebraic expression is: $c - 5$.

Caution

$5 < c$ is the translation of the statement 5 ***is** less than the capacity c*. It is not the translation of 5 *less than the capacity c*.

The Language of Algebra

The word **of** often indicates multiplication, as in $\frac{1}{2}$ of 8, but it also is used in phrases that indicate other operations, such as the sum *of* 8 and *c*, or the difference *of* *x* and 1.

c. **Key phrase:** *product of* **Translation:** multiplication

Key phrase: *increased by* **Translation:** addition

In the given wording, the comma after 2,000 means w is first multiplied by 2,000; then 300 is added to that product. The algebraic expression is: $2{,}000w + 300$.

If there is no comma, the phrase *the product of the weight w and 2,000 increased by 300* translates to: $w(2{,}000 + 300)$.

Self Check 3 Write each phrase as an algebraic expression:

a. 80 less than the total t **b.** $\frac{2}{3}$ of the time T

c. the difference of twice a and 15, squared

Now Try Problems 25, 31, and 35

To solve application problems, we often let a variable represent an unknown quantity.

EXAMPLE 4 **Swimming.** A pool is to be sectioned into 8 equally wide swimming lanes. Write an algebraic expression that represents the width of each lane.

Strategy We will begin by letting $x =$ the width of the swimming pool in feet. Then we will identify any key phrases.

Why The width of the pool is unknown.

Solution The key phrase, *sectioned into 8 equally wide lanes,* indicates division.

Therefore, the width of each lane is $\frac{x}{8}$ feet.

Self Check 4 **Commuting.** It takes Val m minutes to get to work by bus. If she drives her car, her travel time exceeds this by 15 minutes. How many minutes does it take her to get to work by car?

Now Try Problem 61

EXAMPLE 5 **Painting.** A 10-inch-long paintbrush has two parts: a handle and bristles. Choose a variable to represent the length of one of the parts. Then write an expression to represent the length of the other part.

Strategy There are two approaches. We can let $h =$ the length of the handle or we can let $b =$ the length of the bristles.

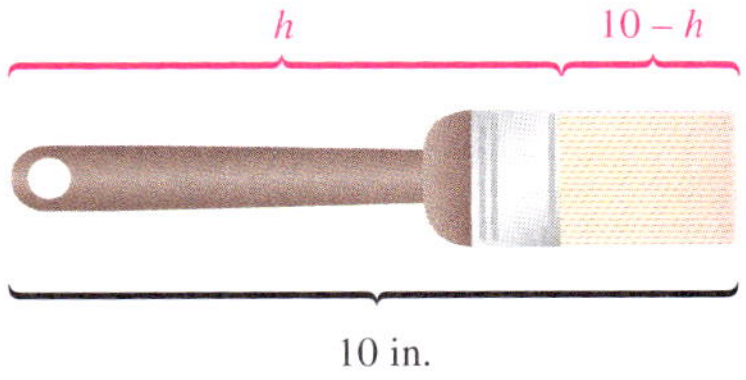

Why Both the length of the handle and the length of the bristles are unknown.

Solution Refer to the drawing on the top. If we let $h =$ the length of the handle (in inches), then the length of the bristles is $10 - h$.

Now refer to the drawing on the bottom. If we let $b =$ the length of the bristles (in inches), then the length of the handle is $10 - b$.

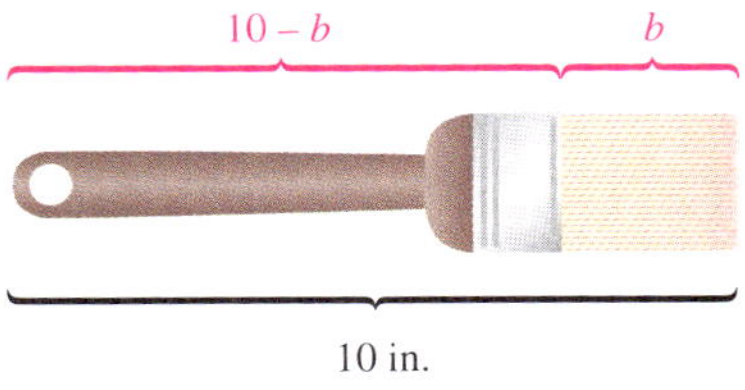

Self Check 5 **Scholarships.** Part of a $900 donation to a college went to the scholarship fund, the rest to the building fund. Choose a variable to represent the amount donated to one of the funds. Then write an expression that represents the amount donated to the other fund.

Now Try Problem 13

EXAMPLE 6 **Enrollments.** Second semester enrollment in a nursing program was 32 more than twice that of the first semester. Let x represent the enrollment for one of the semesters. Write an expression that represents the enrollment for the other semester.

Strategy We will begin by letting x = the enrollment for the first semester.

Why Because the second-semester enrollment is related to the first-semester enrollment.

Solution

Key phrase: *more than* **Translation:** addition

Key phrase: *twice that* **Translation:** multiplication by 2

The second semester enrollment was $2x + 32$.

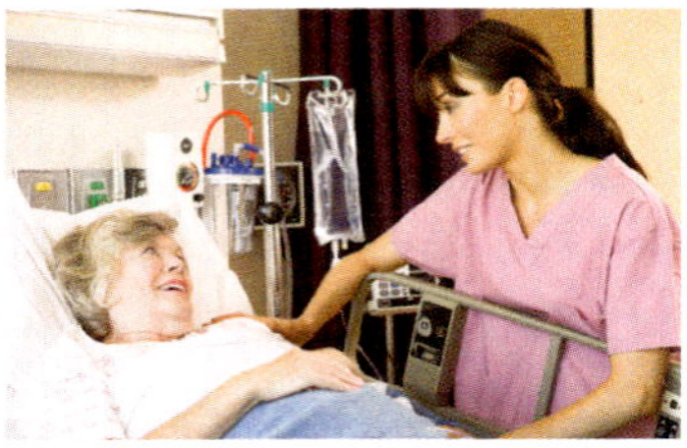

Self Check 6 **Politics.** In an election, the incumbent received 55 fewer votes than three times the challenger's votes. Let x represent the number of votes received by one candidate. Write an expression that represents the number of votes received by the other.

Now Try Problem 109

3 Analyze Problems to Determine Hidden Operations.

Many applied problems require insight and analysis to determine which mathematical operations to use.

EXAMPLE 7 **Vacations.** Disneyland, in California, was in operation 16 years before the opening of Disney World in Florida. Euro Disney, in France, was constructed 21 years after Disney World. Write algebraic expressions to represent the ages (in years) of each Disney attraction.

Strategy We will begin by letting x = the age of Disney World.

Why The ages of Disneyland and Euro Disney are both related to the age of Disney World.

Solution In carefully reading the problem, we see that Disneyland was built 16 years before Disney World. That makes its age 16 years more than that of Disney World. The key phrase *more than* indicates addition.

Attraction	Age
Disneyland	$x + 16$
Disney World	x
Euro Disney	$x - 21$

$x + 16$ = the age of Disneyland

Euro Disney was built 21 years *after* Disney World. That makes its age 21 years less than that of Disney World. The key phrase *less than* indicates subtraction.

$x - 21$ = the age of Euro Disney

Self Check 7 **Tax Forms.** Kayla worked 5 more hours preparing her tax return than she did on her daughter's return. Kayla's son's return took her 2 more hours to prepare than her daughter's. Write expressions to represent the hours she spent on each return.

Now Try Problem 111

EXAMPLE 8 How many months are in x years?

Strategy There are no key phrases so we must analyze the problem carefully. We will begin by considering some specific cases.

Why It's often easier to work with specifics first to get a better understanding of the relationship between the two quantities. Then we can generalize using a variable.

Solution Let's calculate the number of months in 1 year, 2 years, and 3 years. When we write the results in a table, a pattern is apparent.

Number of years	Number of months
1	12
2	24
3	36
x	$12x$

We multiply the number of years by 12 to find the number of months.

The number of months in x years is $12 \cdot x$ or $12x$.

Self Check 8 How many days is h hours?

Now Try Problems 7 and 67

In some problems, we must distinguish between *the number of* and *the value of* the unknown quantity. For example, to find the value of 3 quarters, we multiply the number of quarters by the value (in cents) of one quarter. Therefore, the value of 3 quarters is $3 \cdot 25$ cents $= 75$ cents.

The same distinction must be made if the number is unknown. For example, the value of n nickels is not n cents. The value of n nickels is $n \cdot 5$ cents $= 5n$ cents. For problems of this type, we will use the relationship

Number · value = total value

EXAMPLE 9 Find the total value of: **a.** five dimes **b.** q quarters **c.** $x + 1$ half-dollars

Strategy To find the total value (in cents) of each collection of coins, we multiply the number of coins by the value (in cents) of one coin, as shown in the table.

Why Number · value = total value

Solution

Type of coin	Number	Value	Total value	
Dime	5	10	50	Multiply: $5 \cdot 10 = 50$.
Quarter	q	25	$25q$	Multiply: $q \cdot 25$ can be written $25q$.
Half-dollar	$x + 1$	50	$50(x + 1)$	Multiply: $(x + 1) \cdot 50$ can be written $50(x + 1)$.

Self Check 9 Find the value of: **a.** six \$50 savings bonds
b. t \$100 savings bonds
c. $x - 4$ \$1,000 savings bonds

Now Try Problems 14 and 69

4 Evaluate Algebraic Expressions.

To evaluate an algebraic expression, we substitute given numbers for each variable and perform the necessary calculations in the proper order.

EXAMPLE 10 Evaluate each expression for $x = 3$ and $y = -4$: **a.** $y^3 + y^2$ **b.** $-y - x$
c. $|5xy - 7|$ **d.** $\dfrac{y - 0}{x - (-1)}$

Strategy We will replace each x and y in the expression with the given value of the variable, and evaluate the expression using the order of operation rule.

Why To *evaluate an expression* means to find its numerical value, once we know the value of its variable(s).

Solution

Caution

When replacing a variable with its numerical value, we must often write the replacement number within parentheses to convey the proper meaning.

a. $y^3 + y^2 = (-4)^3 + (-4)^2$ Substitute −4 for each y. We must write −4 within parentheses so that it is the base of each exponential expression.

$= -64 + 16$ Evaluate each exponential expression.

$= -48$ Do the addition.

b. $-y - x = -(-4) - 3$ Substitute −4 for y and 3 for x. Don't forget to write the − sign in front of (−4).

$= 4 - 3$ Simplify: −(−4) = 4.

$= 1$ Do the subtraction.

c. $|5xy - 7| = |5(3)(-4) - 7|$ Substitute 3 for x and −4 for y.

$= |-60 - 7|$ Do the multiplication: 5(3)(−4) = −60.

$= |-67|$ Do the subtraction: −60 − 7 = −60 + (−7) = −67.

$= 67$ Find the absolute value of −67.

d. $\dfrac{y - 0}{x - (-1)} = \dfrac{-4 - 0}{3 - (-1)}$ Substitute 3 for x and −4 for y.

$= \dfrac{-4}{4}$ In the denominator, do the subtraction: 3 − (−1) = 3 + 1 = 4.

$= -1$ Simplify the fraction.

Self Check 10 Evaluate each expression for $a = -2$ and $b = 5$:

a. $|a^3 + b^2|$ **b.** $-a + 2ab$ **c.** $\dfrac{a + 2}{b - 3}$

Now Try Problems 79 and 91

EXAMPLE 11 **Rocketry.** If a toy rocket is shot into the air with an initial velocity of 80 feet per second, its height (in feet) after t seconds in flight is approximated by $-16t^2 + 80t$. How many seconds after the launch will it hit the ground?

Strategy We can substitute positive values for t, the time in flight, until we find the one that gives a height of 0.

Why When the height of the rocket is 0, it is on the ground.

Solution We begin by finding the height after the rocket has been in flight for 1 second ($t = 1$).

$-16t^2 + 80t = -16(1)^2 + 80(1)$ Substitute 1 for t.

$= 64$

As we evaluate $-16t^2 + 80t$ for several more values of t, we record each result in a table. The columns of the table can also be headed with the terms **input** and **output.** The values of t are the inputs into the expression $-16t^2 + 80t$, and the resulting values are the outputs.

Peter Barrett/Shutterstock.com

t	$-16t^2 + 80t$
1	64
2	96
3	96
4	64
5	0

Evaluate for $t = 2$: $-16t^2 + 80t = -16(2)^2 + 80(2) = 96$

Evaluate for $t = 3$: $-16t^2 + 80t = -16(3)^2 + 80(3) = 96$

Evaluate for $t = 4$: $-16t^2 + 80t = -16(4)^2 + 80(4) = 64$

Evaluate for $t = 5$: $-16t^2 + 80t = -16(5)^2 + 80(5) = 0$

Input	Output
1	64
2	96
3	96
4	64
5	0

The height of the rocket is 0 when $t = 5$. The rocket will hit the ground 5 seconds after being launched.

Self Check 11 In Example 11, suppose the height of the rocket is given by $-16t^2 + 112t$. What will be the height of the rocket 6 seconds after launch?

Now Try Problem 97

SECTION 1.8 STUDY SET

VOCABULARY

Fill in the blanks.

1. Variables and/or numbers can be combined with the operations of arithmetic to create algebraic __________.
2. A _____ is a product or quotient of numbers and/or variables. Examples are: $8x$, $\frac{t}{2}$, and $-cd^3$.
3. Addition symbols separate algebraic expressions into parts called ______.
4. A term, such as 27, that consists of a single number is called a ________ term.
5. The __________ of the term $10x$ is 10.
6. To ________ $4x - 3$ for $x = 5$, we substitute 5 for x and perform the necessary calculations.

CONCEPTS

7. Complete the table below on the left to determine the number of days in w weeks.
8. Complete the table below on the right to determine the number of minutes in s seconds.

Number of weeks	Number of days
1	
2	
3	
w	

Number of seconds	Number of minutes
60	
120	
180	
s	

9. The knife shown below is 12 inches long. Write an expression that represents the length (in inches) of the blade.

10. A student inherited \$5,000 and deposits x dollars in American Savings. Write an expression that represents the number of dollars left to deposit in a City Mutual account.

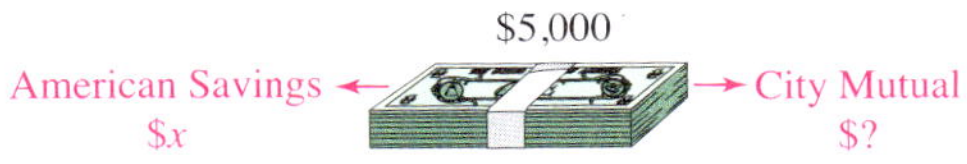

11. Solution 2 is poured into solution 1. Write an expression that represents the number of ounces in the mixture.

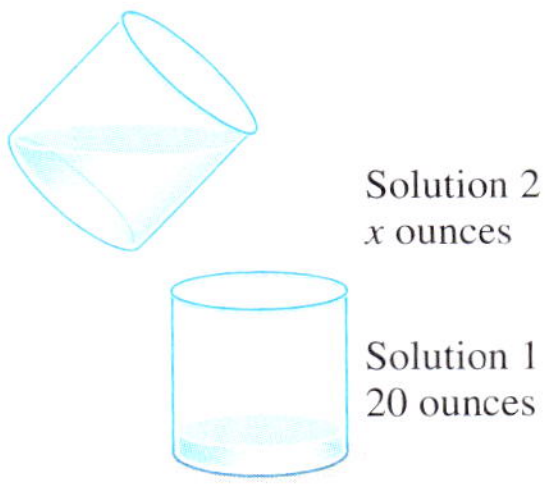

12. Peanuts were mixed with c pounds of cashews to make 100 pounds of a mixture. Write an expression that represents the number of pounds of peanuts that were used.

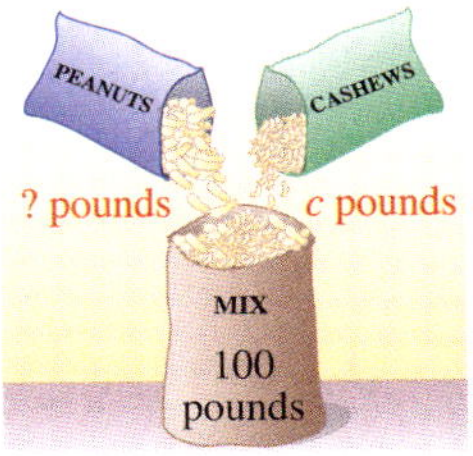

13. a. Let b = the length of the beam shown below (in feet). Write an expression that represents the length of the pipe.
    b. Let p = the length of the pipe (in feet). Write an expression that represents the length of the beam.

14. Complete the table. Give each value in cents.

Coin	Number	· Value	= Total value
Nickel	6		
Dime	d		
Half-dollar	$x + 5$		

NOTATION

Complete each solution. Evaluate each expression for $a = 5$, $x = -2$, and $y = 4$.

15. $9a - a^2 = 9(\quad) - (5)^2$
$= 9(5) - \quad$
$= \quad - 25$
$= 20$

16. $-x + 6y = -(\quad) + 6(\quad)$
$= \quad + 24$
$= 26$

17. Write each term in standard form.
 a. $y8$ **b.** $d2c$
 c. What property of multiplication did you use?

18. Fill in the blanks.
 a. $\frac{w}{2} = \quad w$ **b.** $\frac{2}{3}m = \frac{\quad}{3}$

GUIDED PRACTICE

See Example 1.

19. Consider the expression $3x^3 + 11x^2 - x + 9$.
 a. How many terms does the expression have?
 b. What is the coefficient of each term?

20. Complete the following table.

Term	$6m$	$-75t$	w	$\frac{1}{2}bh$	$\frac{x}{5}$	t
Coefficient						

Determine whether the variable c is used as a factor or as a term. See Example 2.

21. $c + 32$
22. $-24c + 6$
23. $5c$
24. $a + b + c$

Translate each phrase to an algebraic expression. If no variable is given, use x as the variable. See Example 3.

25. The sum of the length l and 15
26. The difference of a number and 10
27. The product of a number and 50
28. Three-fourths of the population p
29. The ratio of the amount won w and lost l
30. The tax t added to c
31. P increased by two-thirds of p
32. 21 less than the total height h
33. The square of k, minus 2,005
34. s subtracted from S
35. 1 less than twice the attendance a
36. J reduced by 500
37. 1,000 split n equal ways
38. Exceeds the cost c by 25,000
39. 90 more than twice the current price p
40. 64 divided by the cube of y
41. 3 times the total of 35, h, and 300
42. Decrease x by -17
43. 680 fewer than the entire population p
44. Triple the number of expected participants
45. The product of d and 4, decreased by 15
46. The quotient of y and 6, cubed
47. Twice the sum of 200 and t
48. The square of the quantity 14 less than x
49. The absolute value of the difference of a and 2
50. The absolute value of a, decreased by 2
51. One-tenth of the distance d
52. Double the difference of x and 18

Translate each algebraic expression into an English phrase. (Answers may vary.) See Example 3.

53. $\frac{3}{4}r$
54. $\frac{2}{3}d$
55. $t - 50$
56. $c + 19$
57. xyz
58. $10ab$
59. $2m + 5$
60. $2s - 8$

Answer with an algebraic expression. See Example 4.

61. A model's skirt is x inches long. The designer then lets the hem down 2 inches. What is the length (in inches) of the altered skirt?

62. A soft drink manufacturer produced c cans of cola during the morning shift. Write an expression for how many six-packs of cola can be assembled from the morning shift's production.

63. The tag on a new pair of 36-inch-long jeans warns that after washing, they will shrink x inches in length. What is the length (in inches) of the jeans after they are washed?

64. A caravan of b cars, each carrying 5 people, traveled to the state capital for a political rally. How many people were in the caravan?

Answer with an algebraic expression. See Example 8.

65. How many minutes are there in h hours?
66. How many feet are in y yards?
67. How many feet are in i inches?
68. How many centuries in y years?

Answer with an algebraic expression. See Example 9.

69. A sales clerk earns \$$x$ an hour; how much does he earn in an 8-hour day?

70. A cashier earns \$$d$ an hour; how much does she earn in a 40-hour week?

71. If a car rental agency charges 49¢ a mile, what is the rental fee if a car is driven x miles?

72. If one egg is worth e cents, find the value (in cents) of one dozen eggs.

73. A ticket to a concert costs \$$t$. What would a pair of concert tickets cost?

74. If one apple is worth a cents, find the value (in cents) of 20 apples.

75. Tickets to a circus cost \$25 each. What will tickets cost for a family of x people if they also pay for two of their neighbors?

76. A certain type of office desk that used to sell for \$$x$ is now on sale for \$50 off. What will a company pay if it purchases 80 of the desks?

Evaluate each expression, for $x = 3$, $y = -2$, and $z = -4$. See Example 10.

77. $-y$
78. $-z$

79. $-z + 3x$

80. $-y - 5x$

81. $3y^2 - 6y - 4$

82. $-z^2 - z - 12$

83. $(3 + x)y$

84. $(4 + z)y$

85. $(x + y)^2 - |z + y|$

86. $[(z - 1)(z + 1)]^2$

87. $-\dfrac{2x + y^3}{y + 2z}$

88. $-\dfrac{2z^2 - x}{2x - y^2}$

Evaluate each expression. See Example 10.

89. $b^2 - 4ac$ for $a = -1$, $b = 5$, and $c = -2$

90. $(x - a)^2 + (y - b)^2$ for $x = -2$, $y = 1$, $a = 5$, and $b = -3$

91. $a^2 + 2ab + b^2$ for $a = -5$ and $b = -1$

92. $\dfrac{a - x}{y - b}$ for $x = -2$, $y = 1$, $a = 5$, and $b = 2$

93. $\dfrac{n}{2}[2a + (n - 1)d]$ for $n = 10$, $a = -4.2$, and $d = 6.6$

94. $\dfrac{a(1 - r^n)}{1 - r}$ for $a = -5$, $r = 2$, and $n = 3$

95. $(27c^2 - 4d^2)^3$ for $c = \frac{1}{3}$ and $d = \frac{1}{2}$

96. $\dfrac{-b^2 + 16a^2 + 1}{2}$ for $a = \frac{1}{4}$ and $b = -10$

Complete each table. See Example 11.

97.

x	$x^3 - 1$
0	
−1	
−3	

98.

g	$g^2 - 7g + 1$
0	
7	
−10	

99.

s	$\frac{5s + 36}{s}$
1	
6	
−12	

100.

a	$2{,}500a + a^3$
2	
4	
−5	

101.

Input x	Output $2x - \frac{x}{2}$
100	
−300	

102.

Input x	Output $\frac{x}{3} + \frac{x}{4}$
12	
−36	

103.

x	$(x + 1)(x + 5)$
−1	
−5	
−6	

104.

x	$\frac{1}{x + 8}$
−7	
−9	
−8	

TRY IT YOURSELF

Look Alikes . . .

Translate each phrase to mathematical symbols. Let x represent the unknown number.

105. a. The sum of a number and 7 squared

b. The sum of a number and 7, squared

106. a. 19 less than a number

b. 19 is less than a number

107. a. 4 times a number increased by 2

b. 4 times a number, increased by 2

108. a. Twice a number decreased by 3

b. Twice a number, decreased by 3

APPLICATIONS

109. **Vehicle Weights.** A Hummer H2 weighs 340 pounds less than twice a Honda Element.

a. Let x represent the weight of one of the vehicles. Write an expression for the weight of the other vehicle.

b. If the weight of the Element is 3,370 pounds, what is the weight of the Hummer?

110. **Sod Farms.** The expression $20{,}000 - 3s$ gives the number of square feet of sod that are left in a field after s strips have been removed. Suppose a city orders 7,000 strips of sod. Evaluate the expression and explain the result.

Strips of sod, cut and ready to be loaded on a truck for delivery

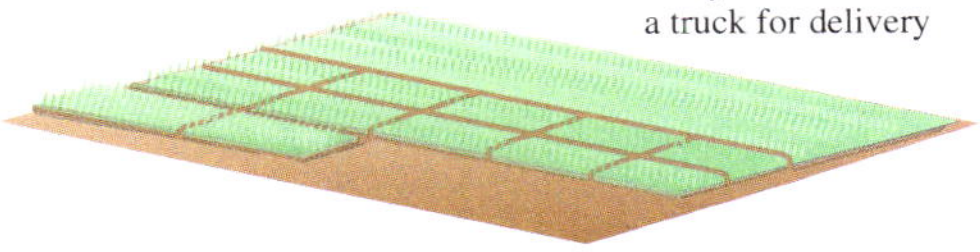

111. **Computer Companies.** IBM was founded 80 years before Apple Computer. Dell Computer Corporation was founded 9 years after Apple.

a. Let x represent the age (in years) of one of the companies. Write expressions to represent the ages (in years) of the other two companies.

b. On April 1, 2008, Apple Computer Company was 32 years old. How old were the other two computer companies then?

112. **Thrill Rides.** The distance in feet that an object will fall in t seconds is given by the expression $16t^2$. Find the distance that riders on "Drop Zone" will fall during the times listed in the table.

©sahua d/Shutterstock

Time (seconds)	Distance (feet)
1	
2	
3	
4	

WRITING

113. What is an algebraic expression? Give some examples.
114. Explain why 2 *less than* x does not translate to $2 < x$.
115. In this section, we substituted a number for a variable. List some other uses of the word *substitute* that you encounter in everyday life.
116. Explain why d dimes are not worth d¢.

REVIEW

117. Find the LCD for $\frac{5}{12}$ and $\frac{1}{15}$.
118. Simplify: $\frac{3 \cdot 3 \cdot 5}{3 \cdot 5 \cdot 5 \cdot 11}$
119. Evaluate: $\left(\frac{2}{3}\right)^3$
120. Find the result when $\frac{7}{8}$ is multiplied by its reciprocal.

CHALLENGE PROBLEMS

121. Evaluate: $(8 - 1)(8 - 2)(8 - 3) \ldots (8 - 49)(8 - 50)$
122. Translate to an expression: The sum of a number decreased by six, and seven more than the quotient of triple the number and five.

SECTION 1.9 Simplifying Algebraic Expressions Using Properties of Real Numbers

OBJECTIVES

1 Use the commutative and associative properties.
2 Simplify products.
3 Use the distributive property.
4 Identify like terms.
5 Combine like terms.

ARE YOU READY?

The following problems review some basic concepts that are important when simplifying expressions using properties of real numbers.

1. How do the expressions $3 + x$ and $x + 3$ differ?
2. How do the expressions $1 + (7 + x)$ and $(1 + 7) + x$ differ?
3. Evaluate $5(2 + 4)$ and $5 \cdot 2 + 5 \cdot 4$ and compare the results.
4. How do the terms $4x$ and $4y$ differ? What do they have in common?

In algebra, we frequently replace one algebraic expression with another that is equivalent and simpler in form. That process, called *simplifying an algebraic expression,* often involves the use of one or more properties of real numbers.

1 Use the Commutative and Associative Properties.

Recall the commutative and associative properties discussed in Sections 1.4 and 1.6.

Commutative Properties

Changing the order when adding or multiplying does not affect the answer.

$$a + b = b + a \quad \text{and} \quad ab = ba$$

Associative Properties

Changing the grouping when adding or multiplying does not affect the answer.

$$(a + b) + c = a + (b + c) \quad \text{and} \quad (ab)c = a(bc)$$

These properties can be applied when working with algebraic expressions that involve addition and multiplication.

EXAMPLE 1 Use the given property to complete each statement.

a. $9 + x =$ ______ *Commutative property of addition*

b. $t \cdot 5 =$ ______ *Commutative property of multiplication*

c. $(a + 92) + 8 =$ ______ *Associative property of addition*

d. $6(10n) =$ ______ *Associative property of multiplication*

Strategy For problems like these, it is important to memorize the properties by name. To fill in each blank, we will determine the way in which the property enables us to rewrite the given expression.

Why We should memorize the properties by name because their names remind us how to use them.

Solution

a. To *commute* means to go back and forth. The commutative property of addition enables us to change the order of the terms, 9 and x.

$9 + x = \underline{x + 9}$ *Commutative property of addition*

b. We change the order of the factors, t and 5.

$t \cdot 5 = \underline{5 \cdot t}$ *Commutative property of multiplication*

Success Tip

In part c, note that the order of the terms does not change. In part d, note that the order of the factors does not change.

c. To *associate* means to group together. The associative property of addition enables us to group the terms in a different way.

$(a + 92) + 8 = \underline{a + (92 + 8)}$ *Associative property of addition*

d. We change the grouping of the factors, 6, 10, and n.

$6(10n) = \underline{(6 \cdot 10)n}$ *Associative property of multiplication*

Self Check 1 Use the given property to complete each statement.

a. $5x + (3x + 1) =$ ______ *Associative property of addition*

b. $a(-15) =$ ______ *Commutative property of multiplication*

Now Try Problems 15 and 17

2 Simplify Products.

The commutative and associative properties of multiplication can be used to simplify certain products. For example, let's simplify $8(4x)$.

$$
\begin{aligned}
8(4x) &= 8 \cdot (4 \cdot x) && \text{Rewrite } 4x \text{ as } 4 \cdot x. \\
&= (8 \cdot 4) \cdot x && \text{Use the associative property of multiplication to group 4 with 8.} \\
&= 32x && \text{Do the multiplication within the parentheses.}
\end{aligned}
$$

We have found that $8(4x) = 32x$. We say that $8(4x)$ and $32x$ are **equivalent expressions** because for each value of x, they represent the same number. For example, both expressions have the value 320 if $x = 10$, and both have the value -96 if $x = -3$.

If $x = 10$		***If*** $x = -3$	
$8(4x) = 8[4(10)]$	$32x = 32(10)$	$8(4x) = 8[4(-3)]$	$32x = 32(-3)$
$= 8(40)$	$= 320$	$= 8(-12)$	$= -96$
$= 320$		$= -96$	

EXAMPLE 2 Multiply: **a.** $-9 \cdot 3b$ **b.** $15a(6)$ **c.** $3(7p)(-5)$ **d.** $\frac{8}{3} \cdot \frac{3}{8}r$ **e.** $35\left(\frac{4}{5}x\right)$

Strategy We will use the commutative and associative properties of multiplication to reorder and regroup the factors in each expression.

Why We want to group all of the numerical factors of an expression together so that we can find their product.

Solution

a. $-9 \cdot 3b = (-9 \cdot 3)b$ Use the associative property of multiplication to group −9 and 3.

$= -27b$ Do the multiplication within the parentheses.

b. $15a(6) = 15(6)a$ Use the commutative property of multiplication to reorder the factors.

$= 90a$ Do the multiplication, working from left to right: 15(6) = 90.

c. $3(7p)(-5) = [3(7)(-5)]p$ Use the commutative and associative properties of multiplication to reorder and regroup the factors.

$= -105p$ Do the multiplication within the brackets.

d. $\frac{8}{3} \cdot \frac{3}{8}r = \left(\frac{8}{3} \cdot \frac{3}{8}\right)r$ Use the associative property of multiplication to regroup the factors.

$= 1r$ Multiply within the parentheses. The product of a number and its reciprocal is 1.

$= r$ The coefficient 1 need not be written.

e. $35\left(\frac{4}{5}x\right) = \left(35 \cdot \frac{4}{5}\right)x$ Use the associative property of multiplication to regroup the factors.

$= \left(\frac{\overset{1}{\cancel{5}} \cdot 7 \cdot 4}{\underset{1}{\cancel{5}}}\right)x$ Factor 35 as 5 · 7 and then remove the common factor 5.

$= 28x$

Success Tip

By the commutative property of multiplication, we can *change* the order of factors. By the associative property of multiplication, we can change the *grouping* of factors.

Self Check 2 Multiply: **a.** $9 \cdot 6s$ **b.** $-4(6u)(-2)$ **c.** $\frac{2}{3} \cdot \frac{3}{2}m$ **d.** $36\left(\frac{2}{9}y\right)$

Now Try Problems 23, 29, 31, and 33

3 Use the Distributive Property.

Another property that is often used to simplify algebraic expressions is the **distributive property.** To introduce it, we will evaluate 4(5 + 3) in two ways.

The Language of Algebra

To **distribute** means to give from one to several. You have probably *distributed* candy to children coming to your door on Halloween.

Use the order of operations:	*Distribute the multiplication:*
$4(5 + 3) = 4(8)$	$4(5 + 3) = 4 \cdot 5 + 4 \cdot 3$
$= 32$	$= 20 + 12$
	$= 32$

Each method gives a result of 32. This observation suggests the following property.

The Distributive Property

For any real numbers a, b, and c,

$a(b + c) = ab + ac$ Read as "a times the quantity of b plus c."

The Language of Algebra

Formally, it is called the **distributive property of multiplication over addition.** When we use it to write a product, such as $5(x + 2)$, as a sum, $5x + 10$, we say that we have **removed** or **cleared** the parentheses.

To illustrate one use of the distributive property, let's consider the expression $5(x + 3)$. *Since we are not given the value of x, we cannot add x and 3 within the parentheses.* However, we can distribute the multiplication by the factor of 5 that is outside the parentheses to x and to 3 and add those products.

$$5(x + 3) = 5 \cdot x + 5 \cdot 3$$ Distribute the multiplication by 5.

$$= 5x + 15$$ Do the multiplication.

EXAMPLE 3 Multiply: **a.** $8(m + 9)$ **b.** $-12(4t + 1)$ **c.** $6\left(\frac{x}{3} + \frac{9}{2}\right)$

Strategy In each case, we will distribute the multiplication by the factor *outside* the parentheses over each term *within* the parentheses.

Why In each case, we cannot simplify the expression within the parentheses. To multiply, we must use the distributive property.

Solution

a. $8(m + 9) = 8 \cdot m + 8 \cdot 9$ Read as "8 times the quantity of *m* plus 9." Distribute the multiplication by 8.

$= 8m + 72$ Do the multiplication.

The Language of Algebra

We read $8(m + 9)$ as "eight times the **quantity** of m plus nine." The word *quantity* alerts us to the grouping symbols in the expression.

b. $-12(4t + 1) = -12 \cdot 4t + -12 \cdot 1$ Distribute the multiplication by −12.

$= -48t + (-12)$ Do the multiplication.

$= -48t - 12$ Write the result in simpler form. Recall that adding −12 is the same as subtracting 12.

c. $6\left(\frac{x}{3} + \frac{9}{2}\right) = 6 \cdot \frac{x}{3} + 6 \cdot \frac{9}{2}$ Distribute the multiplication by 6.

$= \frac{2 \cdot \overset{1}{\cancel{3}} \cdot x}{\underset{1}{\cancel{3}}} + \frac{\overset{1}{\cancel{2}} \cdot 3 \cdot 9}{\underset{1}{\cancel{2}}}$ Factor 6 as $2 \cdot 3$ and then remove the common factors 3 and 2.

$= 2x + 27$

Self Check 3 Multiply: **a.** $7(m + 2)$ **b.** $-80(8x + 3)$ **c.** $24\left(\frac{y}{6} + \frac{3}{8}\right)$

Now Try Problems 35, 37, and 39

Since subtraction is the same as adding the opposite, the distributive property also holds for subtraction.

$$a(b - c) = ab - ac$$

EXAMPLE 4 Multiply: **a.** $3(3b - 4)$ **b.** $-6(-3y - 8)$ **c.** $-1(t - 9)$

Strategy In each case, we will distribute the multiplication by the factor *outside* the parentheses over each term *within* the parentheses.

Why In each case, we cannot simplify the expression within the parentheses. To multiply, we must use the distributive property.

Solution

a. $3(3b - 4) = 3 \cdot 3b - 3 \cdot 4$ Distribute the multiplication by 3.

$= 9b - 12$ Do the multiplication.

Caution

A common mistake is to forget to distribute the multiplication over each of the terms within the parentheses.

$3(3b - 4) = 9b - 4$

b. $-6(-3y - 8) = -6(-3y) - (-6)(8)$ Distribute the multiplication by -6.

$= 18y - (-48)$ Do the multiplication.

$= 18y + 48$ Write the result in simpler form. Add the opposite of -48.

Another approach is to write the subtraction within the parentheses as addition of the opposite. Then we distribute the multiplication by -6 over the addition.

$-6(-3y - 8) = -6[-3y + (-8)]$ Add the opposite of 8.

$= -6(-3y) + (-6)(-8)$ Distribute the multiplication by -6.

$= 18y + 48$ Do the multiplication.

Success Tip

Notice that distributing the multiplication by -1 *changes the sign* of each term within the parentheses.

c. $-1(t - 9) = -1(t) - (-1)(9)$ Distribute the multiplication by -1.

$= -t - (-9)$ Do the multiplication.

$= -t + 9$ Write the result in simpler form. Add the opposite of -9.

Self Check 4 Multiply: **a.** $5(2x - 1)$ **b.** $-9(-y - 4)$ **c.** $-1(c - 22)$

Now Try Problems 43, 47, and 49

Caution

The distributive property applies only to expressions in which multiplication is distributed over addition (or subtraction). For example, to simplify $6(5x)$, do not use the distibutive property.

Correct

$6(5x) = (6 \cdot 5)x = 30x$

Incorrect

$6(5x) = 30 \cdot 6x = 180x$

The distributive property can be extended to several other useful forms. Since multiplication is commutative, we have:

$(b + c)a = ba + ca$ $(b - c)a = ba - ca$

For situations in which there are more than two terms within parentheses, we have:

$a(b + c + d) = ab + ac + ad$ $a(b - c - d) = ab - ac - ad$

EXAMPLE 5 Multiply: **a.** $(6x + 4)\frac{1}{2}$ **b.** $2(a - 3b)8$ **c.** $-0.3(3a - 4b + 7)$

Strategy We will multiply each term within the parentheses by the factor (or factors) outside the parentheses.

Why In each case, we cannot simplify the expression within the parentheses. To multiply, we must use the distributive property.

Solution

a. $(6x + 4)\frac{1}{2} = (6x)\frac{1}{2} + (4)\frac{1}{2}$ Distribute the multiplication by $\frac{1}{2}$.

$= 3x + 2$ Do the multiplication.

b. $2(a - 3b)8 = 2 \cdot 8(a - 3b)$ Use the commutative property of multiplication to reorder the factors.

$= 16(a - 3b)$ Multiply 2 and 8 to get 16.

$= 16a - 48b$ Distribute the multiplication by 16.

c. $-0.3(3a - 4b + 7) = -0.3(3a) - (-0.3)(4b) + (-0.3)(7)$

$= -0.9a + 1.2b - 2.1$ Do each multiplication.

Self Check 5 Multiply: **a.** $(-6x - 24)\frac{1}{3}$ **b.** $6(c - 2d)9$ **c.** $-0.7(2r + 5s - 8)$

Now Try Problems 53, 55, and 57

We can use the distributive property to find the opposite of a sum. For example, to find $-(x + 10)$, we interpret the $-$ symbol as a factor of -1, and proceed as follows:

$-(x + 10) = -1(x + 10)$ Replace the $-$ symbol with -1.

$= -1(x) + (-1)(10)$ Distribute the multiplication by -1.

$= -x - 10$ Do the multiplication.

In general, we have the following property of real numbers.

The Opposite of a Sum

The opposite of a sum is the sum of the opposites.

For any real numbers a and b,

$$-(a + b) = -a + (-b)$$

EXAMPLE 6 Simplify: $-(-9s - 3)$

Strategy We will multiply each term within the parentheses by -1.

Why The $-$ outside the parentheses represents a factor of -1 that is to be distributed.

Solution

$-(-9s - 3) = -1(-9s - 3)$ Replace the $-$ symbol in front of the parentheses with -1.

$= -1(-9s) - (-1)(3)$ Distribute the multiplication by -1.

$= 9s + 3$ Do the multiplication.

Success Tip

Notice that the $-$ symbol in front of the parentheses changes the sign of each term within the parentheses.

Self Check 6 Simplify: $-(-5x + 18)$

Now Try Problem 59

4 Identify Like Terms.

Before we can discuss methods for simplifying algebraic expressions involving addition and subtraction, we need to introduce some new vocabulary.

Like Terms

Like terms are terms containing exactly the same variables raised to exactly the same powers. Any constant terms in an expression are considered to be like terms. Terms that are not like terms are called **unlike terms**.

Success Tip

When looking for like terms, don't look at the coefficients of the terms. Consider only the variable factors of each term. If two terms are like terms, only their coefficients may differ.

Here are several examples.

Like terms		*Unlike terms*	
$4x$ and $7x$	Same variable	$4x$ and $7y$	The variables are not the same.
$-10p^2$ and $25p^2$	Same variable to the same power	$-10p$ and $25p^2$	Same variable, but different powers
$\frac{1}{3}c^3d$ and c^3d	Same variables to the same powers	$\frac{1}{3}c^3d$ and c^3	The variables are not the same.

EXAMPLE 7 List the like terms in each expression: **a.** $7r + 5 + 3r$ **b.** $6x^4 - 6x^2 - 6x$ **c.** $-17m^3 + 3 - 2 + m^3$

Strategy First, we will identify the terms of the expression. Then we will look for terms that contain the same variables raised to exactly the same powers.

Why If two terms contain the same variables raised to the same powers, they are like terms.

Solution

a. $7r + 5 + 3r$ contains the like terms $7r$ and $3r$.

b. Since the exponents on x are different, $6x^4 - 6x^2 - 6x$ contains no like terms.

c. $-17m^3 + 3 - 2 + m^3$ contains two pairs of like terms: $-17m^3$ and m^3 are like terms, and the constant terms, 3 and -2, are like terms.

Self Check 7 List the like terms: **a.** $2x - 2y + 7y$ **b.** $5p^2 - 12 + 17p^2 + 2$

Now Try Problem 65

5 Combine Like Terms.

To add or subtract objects, they must be similar. For example, fractions that are to be added must have a common denominator. When adding decimals, we align columns to be sure to add tenths to tenths, hundredths to hundredths, and so on. The same is true when working with terms of an algebraic expression. They can be added or subtracted only if they are like terms.

This expression can be simplified because it contains like terms.	This expression cannot be simplified because its terms are not like terms.
$3x + 4x$	$3x + 4y$

Recall that the distributive property can be written in the following forms:

$$(b + c)a = ba + ca \qquad (b - c)a = ba - ca$$

The Language of Algebra

Simplifying a sum or difference of like terms is called **combining like terms.**

We can use these forms of the distributive property in reverse to simplify a sum or difference of like terms. For example, we can simplify $3x + 4x$ as follows:

$$3x + 4x = (3 + 4)x \quad \text{Use } ba + ca = (b + c)a.$$
$$= 7x$$

We can simplify $15m^2 - 9m^2$ in a similar way:

$$15m^2 - 9m^2 = (15 - 9)m^2 \quad \text{Use } ba - ca = (b - c)a.$$
$$= 6m^2$$

In each case, we say that we **combined like terms.** These examples suggest the following general rule.

Combining Like Terms

Like terms can be combined by adding or subtracting the coefficients of the terms and keeping the same variables with the same exponents.

EXAMPLE 8 Simplify by combining like terms, if possible: **a.** $2x + 9x$ **b.** $-8p + (-2p) + 4p$ **c.** $0.5s^3 - 0.3s^3$ **d.** $4w + 6$ **e.** $\frac{4}{9}b + \frac{7}{9}b$

Strategy We will use the distributive property in reverse to add (or subtract) the coefficients of the like terms. We will keep the same variables raised to the same powers.

Why To *combine like terms* means to add or subtract the like terms in an expression.

Solution

Success Tip

Just as 2 apples plus 9 apples is 11 apples, $2x + 9x = 11x$.

a. Since $2x$ and $9x$ are like terms with the common variable x, we can combine them.

$2x + 9x = 11x$ Think: $(2 + 9)x = 11x$.

b. $-8p + (-2p) + 4p = -6p$ Think: $[-8 + (-2) + 4]p = -6p$.

c. $0.5s^3 - 0.3s^3 = 0.2s^3$ Think: $(0.5 - 0.3)s^3 = 0.2s^3$.

d. Since $4w$ and 6 are not like terms, they cannot be combined. $4w + 6$ doesn't simplify.

e. $\frac{4}{9}b + \frac{7}{9}b = \frac{11}{9}b$ Think: $\left(\frac{4}{9} + \frac{7}{9}\right)b = \frac{11}{9}b$.

Self Check 8 Simplify, if possible: **a.** $3x + 5x$ **b.** $-6y + (-6y) + 9y$ **c.** $4.4s^4 - 3.9s^4$ **d.** $4a - 2$ **e.** $\frac{10}{7}c - \frac{4}{7}c$

Now Try Problems 67, 69, 71, and 73

EXAMPLE 9 Simplify by combining like terms: **a.** $16t - 15t$ **b.** $16t - t$ **c.** $15t - 16t$ **d.** $16t + t$

Strategy As we combine like terms, we must be careful when working with the terms such as t and $-t$.

Why Coefficients of 1 and -1 are usually not written.

Solution

a. $16t - 15t = t$ Think: $(16 - 15)t = 1t = t$.

b. $16t - t = 15t$ Think: $16t - 1t = (16 - 1)t = 15t$.

c. $15t - 16t = -t$ Think: $(15 - 16)t = -1t = -t$.

d. $16t + t = 17t$ Think: $16t + 1t = (16 + 1)t = 17t$.

Self Check 9 Simplify: **a.** $9h - h$ **b.** $9h + h$ **c.** $9h - 8h$ **d.** $8h - 9h$

Now Try Problems 75 and 77

EXAMPLE 10 Simplify: $6a^2 + 54a - 4a - 36$

Strategy First, we will identify any like terms in the expression. Then we will use the distributive property in reverse to combine them.

Why To *simplify* an expression, we use properties of real numbers to write an equivalent expression in simpler form.

Solution We can combine the like terms that involve the variable a.

$$6a^2 + 54a - 4a - 36 = 6a^2 + 50a - 36 \quad \text{Think: } (54 - 4)a = 50a.$$

Self Check 10 Simplify: $7y^2 + 21y - 2y - 6$

Now Try Problem 85

EXAMPLE 11 Simplify: $4(x + 5) - 5 - (2x - 4)$

Strategy First, we will use the distributive property to remove the parentheses. Then we will identify any like terms and combine them.

Why To *simplify* an expression, we use properties of real numbers, such as the distributive property, to write an equivalent expression in simpler form.

Solution

$$4(x + 5) - 5 - (2x - 4) = 4(x + 5) - 5 - 1(2x - 4)$$ Replace the − symbol in front of (2x − 4) with −1.

$$= 4x + 20 - 5 - 2x + 4$$ Distribute the multiplication by 4 and −1.

$$= 2x + 19$$ Think: $(4 - 2)x = 2x$. Think: $(20 - 5 + 4) = 19$.

Success Tip

Here, the distributive property is used both *forward* (to remove parentheses) and in *reverse* (to combine like terms).

Self Check 11 Simplify: $6(3y - 1) + 2 - (-3y + 4)$

Now Try Problem 87

SECTION 1.9 STUDY SET

VOCABULARY

Fill in the blanks.

1. To ________ the expression $5(6x)$ means to write it in simpler form: $5(6x) = 30x$.
2. $5(6x)$ and $30x$ are __________ expressions because for each value of x, they represent the same number.
3. To perform the multiplication $2(x + 8)$, we use the __________ property.
4. We call $-(c + 9)$ the ________ of a sum.
5. Terms such as $7x^2$ and $5x^2$, which have the same variables raised to exactly the same power, are called _____ terms.
6. When we write $9x + x$ as $10x$, we say we have __________ like terms.

CONCEPTS

7. a. Fill in the blanks to simplify the expression.

   $4(9t) = (\quad \cdot \quad)t = \quad t$

   b. What property did you use in part a?

8. a. Fill in the blanks to simplify the expression.

   $-6y \cdot 2 = \quad \cdot \quad \cdot y = \quad y$

   b. What property did you use in part a?

9. Fill in the blanks.

   a. $2(x + 4) = 2x \quad 8$
   b. $2(x - 4) = 2x \quad 8$
   c. $-2(x + 4) = -2x \quad 8$
   d. $-2(-x - 4) = 2x \quad 8$

10. Fill in the blanks to combine like terms.

   a. $4m + 6m = (\quad)m = \quad m$
   b. $30n^2 - 50n^2 = (\quad)n^2 = \quad n^2$
   c. $12 + 32d + 15 = 32d + \quad$
   d. Like terms can be combined by adding or subtracting the __________ of the terms and keeping the same __________ with the same exponents.

11. Simplify each expression, if possible.

   a. $5(2x)$
   b. $5 + 2x$
   c. $6(-7x)$
   d. $6 - 7x$
   e. $2(3x)(3)$
   f. $2 + 3x + 3$

12. Fill in the blanks: Distributing multiplication by -1 changes the _____ of each term within the parentheses.

$-(x + 10) = \quad (x + 10) = -x \quad 10$

NOTATION

13. Translate to symbols.
    a. Six times the quantity of h minus four.
    b. The opposite of the sum of z and sixteen.
14. Write an equivalent expression for the given expression using fewer symbols.
    a. $1z$ b. $-1d$ c. $0m$
    d. $5x - (-1)$ e. $16t + (-6)$

GUIDED PRACTICE

Use the given property to complete each statement. See Example 1.

15. $8 + (7 + a) =$ __________ Associative property of addition
16. $-2(5b) =$ __________ Associative property of multiplication
17. $y \cdot 11 =$ _____ Commutative property of multiplication
18. $x + x^2 =$ _______ Commutative property of addition
19. $(8a \cdot 2)6 =$ _________ Associative property of multiplication
20. $(-1 + 3a) + 7a =$ _____________ Associative property of addition
21. $9t + (4 + t) = 9t + ($_____$)$ Commutative property of addition
22. $(x - 2)3 =$ _________ Commutative property of multiplication

Simplify each expression. See Example 2.

23. $3 \cdot 4t$
24. $9 \cdot 3s$
25. $5(-7q)$
26. $-7(5t)$
27. $(-5.6x)(-2)$
28. $(-4.4x)(-3)$
29. $5(4c)(3)$
30. $9(2h)(2)$
31. $\frac{5}{3} \cdot \frac{3}{5}g$
32. $\frac{9}{7} \cdot \frac{7}{9}k$
33. $12\left(\frac{5}{12}x\right)$
34. $15\left(\frac{4}{15}w\right)$

Multiply. See Example 3.

35. $5(x + 3)$
36. $4(x + 2)$
37. $-3(4x + 9)$
38. $-5(8x + 9)$
39. $45\left(\frac{x}{5} + \frac{2}{9}\right)$
40. $35\left(\frac{y}{5} + \frac{8}{7}\right)$
41. $0.4(x + 4)$
42. $2.2(2q + 1)$

Multiply. See Example 4.

43. $6(6c - 7)$
44. $9(9d - 3)$
45. $-6(13c - 3)$
46. $-2(10s - 11)$
47. $-15(-2t - 6)$
48. $-20(-4z - 5)$
49. $-1(-4a + 1)$
50. $-1(-2x + 3)$

Multiply. See Example 5.

51. $(3t + 2)8$
52. $(2q + 1)9$
53. $(3w - 6)\frac{2}{3}$
54. $(2y - 8)\frac{1}{2}$
55. $4(7y + 4)2$
56. $8(2a - 3)4$
57. $2.5(2a - 3b + 1)$
58. $5(9s - 12t - 3)$

Multiply. See Example 6.

59. $-(x - 7)$
60. $-(y + 1)$
61. $-(-5.6y + 7)$
62. $-(-4.8a - 3)$

List the like terms in each expression, if any. See Example 7.

63. $3x + 2 - 2x$
64. $3y + 4 - 11y + 6$
65. $-12m^4 - 3m^3 + 2m^2 - m^3$
66. $6x^3 + 3x^2 + 6x$

Simplify by combining like terms. See Example 8.

67. $3x + 7x$
68. $12y - 15y$
69. $-7b^2 + 27b^2$
70. $-2c^3 + 12c^3$

Simplify by combining like terms. See Example 9.

71. $36y + y - 9y$
72. $32a - a + 5a$
73. $\frac{3}{5}t + \frac{1}{5}t$
74. $\frac{3}{16}x - \frac{5}{16}x$
75. $13r - 12r$
76. $25s + s$
77. $43s^3 - 44s^3$
78. $8j^3 - 9j^3$

Simplify by combining like terms. See Example 10.

79. $15y - 10 - y - 20y$
80. $9z - 7 - z - 19z$
81. $3x + 4 - 5x + 1$
82. $4b + 9 - 9b + 9$
83. $9m^2 - 6m + 12m - 4$
84. $6a^2 + 18a - 9a + 5$
85. $4x^2 + 5x - 8x + 9$
86. $10y^2 - 8y + y - 7$

Simplify. See Example 11.

87. $2z + 5(z - 3) - 10$
88. $12(m + 11) - 11 + m$
89. $2(s^2 - 7) - (s^2 - 2)$
90. $4(d^2 - 3) - (d^2 - 1)$

TRY IT YOURSELF

Simplify each expression, if possible.

91. $-\frac{7}{16}x - \frac{3}{16}x$
92. $-\frac{5}{18}x - \frac{7}{18}x$
93. $-9.8c + 6.2c$
94. $-5.7m + 4.3m$
95. $-4(-6)(-4m)$
96. $-5(-9)(-4n)$
97. $-4x + 4x$
98. $-16y + 16y$
99. $-0.2r - (-0.6r)$
100. $-1.1m - (-2.4m)$
101. $8\left(\frac{3}{4}y\right)$
102. $27\left(\frac{2}{3}x\right)$

103. $-9(3r - 9) - 7(2r - 7)$

104. $-6(3t - 6) - 3(11t - 3)$

105. $9(7m)$

106. $12n(8)$

107. $6 - 4(-3c - 7)$

108. $10 - 5(-5g - 1)$

109. $5t \cdot 60$

110. $70a \cdot 10$

111. $36\left(\frac{2}{9}x - \frac{3}{4}\right) + 36\left(\frac{1}{2}\right)$

112. $40\left(\frac{3}{8}y - \frac{1}{4}\right) + 40\left(\frac{4}{5}\right)$

113. $-4r - 7r + 2r - r$

114. $-v - 3v + 6v + 2v$

115. $24\left(-\frac{5}{6}r\right)$

116. $\frac{3}{4} \cdot \frac{1}{2}g$

117. $a + a + a$

118. $t - t - t - t$

119. $60\left(\frac{3}{20}r - \frac{4}{15}\right)$

120. $72\left(\frac{7}{8}f - \frac{8}{9}\right)$

121. $4a + 4b + 4c$

122. $2x + 2y + 2z$

123. $-(c + 7) + 2(c - 3)$

124. $-(z + 2) + 5(3 - z)$

125. $a^3 + 2a^2 + 4a - 2a^2 - 4a - 8$

126. $c^3 - 3c^2 + 9c + 3c^2 - 9c + 27$

Look Alikes . . .

127. **a.** $2(7x)5$ **b.** $2(7x + 5)$

128. **a.** $-3(-4a)(-2)$ **b.** $-3(-4a) - 2$

APPLICATIONS

In Exercises 129 and 130, recall that the perimeter of a figure is equal to the sum of the lengths of its sides.

129. First Aid. Each side of the red cross has length x inches. Write an algebraic expression that represents the perimeter of the cross.

x inches

130. Billiards. Write an algebraic expression that represents the perimeter of the table.

WRITING

131. Explain why the distributive property applies to $2(3 + x)$ but not to $2(3x)$.

132. Explain each error. Then give the correct answer.

a. $9(4b - 2) =$ ~~$36b - 2$~~ **b.** $3(2x) =$ ~~$6 \cdot 3x = 18x$~~

c. $-(23c + 2) =$ ~~$-23c + 2$~~ **d.** $(5n + 1)2 =$ ~~$5n + 2$~~

REVIEW

Evaluate each expression for $x = -3$ and $y = -5$.

133. $\frac{x - y^2}{2y - 1 + x}$

134. $\frac{2y + 1}{x} - x$

CHALLENGE PROBLEMS

135. Fill in the blanks: ___ (___ − ___) $= -187x + 119$

Simplify.

136. $2\{-2[x + 4(2x + 1)] - 5[x + 2(3x + 4)]\} + 106x$

1 Summary & Review

SECTION 1.1 Introducing the Language of Algebra

DEFINITIONS AND CONCEPTS	EXAMPLES
Tables, bar graphs, and **line graphs** are used to describe numerical relationships.	See pages 2 and 3 for examples of tables and graphs.
A **sum** is the result of an addition. A **difference** is the result of a subtraction. A **product** is the result of a multiplication. A **quotient** is the result of a division.	$3 + 15 = \mathbf{18}$ sum $16 - 1 = \mathbf{15}$ difference $7 \cdot 8 = \mathbf{56}$ product $\frac{63}{9} = \mathbf{7}$ quotient

A **variable** is a letter (or symbol) that stands for a number.	Variables: x, a, and y
Algebraic expressions contain variables and numbers combined with the operations of addition, subtraction, multiplication, and division.	Expressions: $5y + 7$, $\frac{12 - x}{5}$, and $8a^2(b - 3)$
An **equation** is a statement that two expressions are equal.	Equations: $3x - 4 = 12$ and $\frac{t}{9} = 12$ Equations contain an = sign. Expressions do not.
Equations that express a relationship between two or more variables are called **formulas.**	$A = lw$ (The formula for the area of a rectangle)

REVIEW EXERCISES

The line graph shows the number of cars in a parking structure from 6 P.M. to 12 midnight on a Saturday.

1. What units are used to scale the horizontal and vertical axes?
2. How many cars were in the parking structure at 11 P.M.?
3. At what time did the parking structure have 500 cars in it?
4. When was the structure empty of cars?

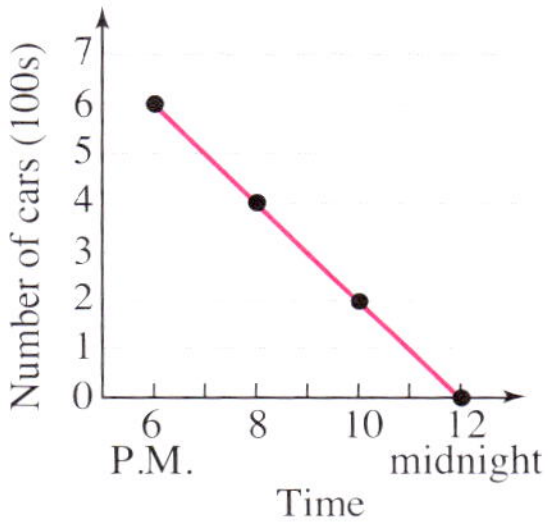

Express each statement in words, using one of these words: sum, difference, product, or quotient.

5. $15 - 3 = 12$
6. $15 + 3 = 18$
7. $15 \div 3 = 5$
8. $15 \cdot 3 = 45$
9. a. Write the multiplication 4×9 with a raised dot and then with parentheses.
   b. Write the division $9 \div 3$ using a fraction bar.
10. Write each multiplication without a multiplication symbol.
    a. $8 \cdot b$ b. $P \cdot r \cdot t$
11. Classify each item as either an expression or an equation.
    a. $5 = 2x + 3$ b. $2x + 3$
12. Use the formula $n = b + 5$ to complete the table.

Brackets (b)	Nails (n)
5	
10	
20	

SECTION 1.2 Fractions

DEFINITIONS AND CONCEPTS	EXAMPLES
A **factor** is a number being multiplied.	$8 \cdot 9 = 72$ (8: Factor, 9: Factor)
A **prime number** is a natural number that is greater than 1 that has only itself and 1 as factors. A **composite number** is a natural number, greater than 1, that is not prime.	Primes: $\{2, 3, 5, 7, 11, 13, 17, 19, 23, . . .\}$ Composites: $\{4, 6, 8, 9, 10, 12, 14, 15, . . .\}$
Any composite number can be factored into the product of two or more prime factors. The product of these prime numbers is called its **prime factorization.**	Find the prime factorization of 98. $98 = 2 \cdot 49 = 2 \cdot 7 \cdot 7$ 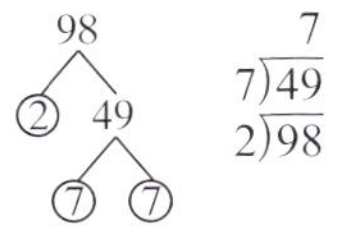
In a fraction, the number above the **fraction bar** is the **numerator** and the number below the fraction bar is called the **denominator.**	$\frac{11}{15}$ (11 ← Numerator, 15 ← Denominator)
Two fractions are **equivalent** if they represent the same number.	Equivalent fractions: $\frac{1}{2} = \frac{2}{4} = \frac{3}{6} = \frac{4}{8} = . . .$
To **multiply two fractions,** multiply their numerators and multiply their denominators.	Multiply: $\frac{5}{8} \cdot \frac{3}{4} = \frac{15}{32}$

One number is the **reciprocal** of another if their product is 1. To **divide two fractions,** multiply the first fraction by the reciprocal of the second fraction.	The reciprocal of $\frac{4}{5}$ is $\frac{5}{4}$ because $\frac{4}{5} \cdot \frac{5}{4} = 1$. Divide: $\frac{4}{7} \div \frac{5}{8} = \frac{4}{7} \cdot \frac{8}{5} = \frac{32}{35}$ The reciprocal of $\frac{5}{8}$ is $\frac{8}{5}$.
Multiplication property of 1: The product of 1 and any number is that number.	$1 \cdot 5 = 5$ and $\frac{7}{8} \cdot 1 = \frac{7}{8}$
To **build a fraction,** multiply it by a form of 1 such as $\frac{2}{2}, \frac{3}{3}, \frac{4}{4}, \ldots$.	Write $\frac{3}{4}$ as an equivalent fraction with a denominator of 20. $\frac{3}{4} = \frac{3}{4} \cdot \frac{5}{5} = \frac{15}{20}$
To **simplify a fraction,** remove pairs of factors common to the numerator and the denominator. A fraction is in **simplest form,** or **lowest terms,** when the numerator and denominator have no common factors other than 1.	Simplify: $\frac{12}{18} = \frac{2 \cdot \overset{1}{\cancel{6}}}{3 \cdot \underset{1}{\cancel{6}}} = \frac{2}{3}$ Factor 12 and 18. Remove the common factor 6 from the numerator and denominator. Slashes and 1's are used to show that $\frac{6}{6}$ is replaced by the equivalent fraction $\frac{1}{1}$. A factor equal to 1 in the form of $\frac{6}{6}$ was removed.
To find the **LCD** of two fractions, prime factor each denominator and find the product of the prime factors, using each factor the greatest number of times it appears in any one factorization.	Find the LCD of $\frac{5}{12}$ and $\frac{7}{8}$. $\left.\begin{array}{l} 12 = 2 \cdot 2 \cdot 3 \\ 8 = 2 \cdot 2 \cdot 2 \end{array}\right\}$ LCD $= 2 \cdot 2 \cdot 2 \cdot 3 = 24$
To **add (or subtract) fractions that have the same denominator,** add (or subtract) the numerators and write the sum (or difference) over the common denominator. Simplify, if possible. To **add (or subtract) fractions that have different denominators,** rewrite each fraction as an equivalent fraction with the LCD as the denominator. Then add (or subtract) as usual. Simplify, if possible.	Add: $\frac{5}{12} + \frac{7}{8} = \frac{5}{12} \cdot \frac{2}{2} + \frac{7}{8} \cdot \frac{3}{3}$ The LCD is 24. Build each fraction. $= \frac{10}{24} + \frac{21}{24}$ The denominators are now the same. $= \frac{31}{24}$ This result does not simplify.
A **mixed number** represents the sum of a whole number and a fraction. In some computations, it is necessary to write mixed numbers as improper fractions.	Mixed numbers: $6\frac{1}{3} = 6 + \frac{1}{3}$ and $1\frac{3}{4} = \frac{7}{4}$

REVIEW EXERCISES

13. a. Write 24 as the product of two factors.

    b. Write 24 as the product of three factors.

    c. List the factors of 24.

14. What do we call fractions, such as $\frac{1}{8}$ and $\frac{2}{16}$, that represent the same number?

Give the prime factorization of each number, if possible.

15. 54
16. 147
17. 385
18. 41

Simplify each fraction.

19. $\frac{20}{35}$
20. $\frac{24}{18}$

Build each number to an equivalent fraction with the indicated denominator.

21. $\frac{5}{8}$, denominator 64
22. 12, denominator 3

What is the LCD for fractions having the following denominators?

23. 10 and 18
24. 21 and 70

Perform each operation and simplify, if possible.

25. $\frac{1}{8} \cdot \frac{7}{8}$
26. $\frac{16}{35} \cdot \frac{25}{48}$
27. $\frac{1}{3} \div \frac{15}{16}$
28. $16\frac{1}{4} \div 5$
29. $\frac{17}{25} - \frac{7}{25}$
30. $\frac{8}{11} - \frac{1}{2}$

31. $\frac{17}{24} + \frac{11}{40}$

32. $4\frac{1}{9} - 3\frac{5}{6}$

33. **The Internet.** A popular website averaged $1\frac{3}{4}$ million hits per day during a 30-day period. How many hits did it receive during that time?

34. **Machine Shops.** How much must be milled off the $\frac{17}{24}$-inch-thick steel rod so that the collar will slip over it?

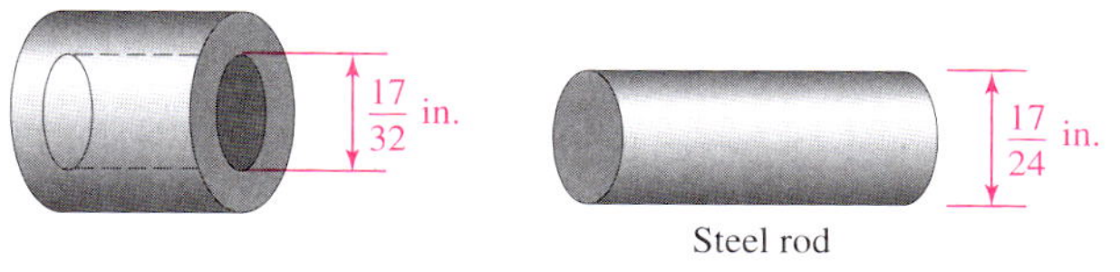

SECTION 1.3 The Real Numbers

DEFINITIONS AND CONCEPTS	EXAMPLES
To write a set, we list its **elements** within **braces** { }.	In the English alphabet, the set of vowels is {a, e, i, o, u}.
The **natural numbers** are the numbers we count with.	Natural numbers: {1, 2, 3, 4, 5, 6, . . .}
The **whole numbers** are the natural numbers together with 0.	Whole numbers: {0, 1, 2, 3, 4, 5, 6, . . .}
Two numbers are called **opposites** if they are the same distance from 0 on the number line but are on opposite sides of it.	The opposite of 3 is -3 and the opposite of -21 is 21.
The **integers** include the whole numbers and their opposites.	Integers: $\{\ldots, -3, -2, -1, 0, 1, 2, 3, \ldots\}$
The **rational numbers** are numbers that can be expressed as fractions with an integer numerator and a nonzero integer denominator.	Rational numbers: $-6,\ -3.1,\ -\frac{1}{2},\ 0,\ \frac{11}{12},\ 9\frac{4}{5},$ and 87 $-6 = \frac{-6}{1},\ -3.1 = \frac{-31}{10},\ 0 = \frac{0}{1},\ 9\frac{4}{5} = \frac{49}{5},\ 87 = \frac{87}{1}$
Terminating and **repeating decimals** can be expressed as fractions and are, therefore, rational numbers.	Terminating decimal: $-0.25 = -\frac{1}{4}$ Repeating decimal: $0.\overline{6} = \frac{2}{3}$
Negative fractions: $-\frac{a}{b} = \frac{-a}{b} = \frac{a}{-b}$	$-\frac{3}{4} = \frac{-3}{4} = \frac{3}{-4}$ and $\frac{-1}{8} = \frac{1}{-8} = -\frac{1}{8}$
An **irrational number** is a nonterminating, nonrepeating decimal. An irrational number cannot be expressed as a fraction with an integer numerator and a nonzero integer denominator. A **real number** is any number that is either a rational or an irrational number. Every real number corresponds to a point on the **number line,** and every point on the number line corresponds to exactly one real number.	Irrational numbers: $\sqrt{2}$, π, and $-\sqrt{7}$ Graph the numbers in the set $\left\{-2, 4, -0.75, 1\frac{3}{4}, \pi, 0, \frac{7}{8}\right\}$ on a number line and classify them. 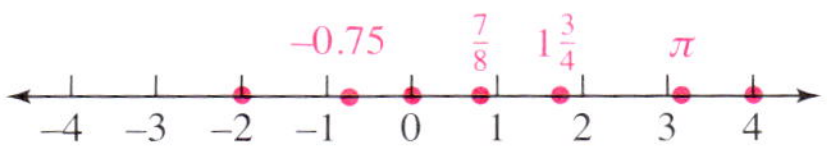 Natural numbers: 4 Whole numbers: 4, 0 Integers: -2, 4, 0 Rational numbers: -2, 4, -0.75, $1\frac{3}{4}$, 0, $\frac{7}{8}$ Irrational numbers: π Real numbers: all
Inequality symbols: $>$ is greater than $<$ is less than	$25 > 15$ and $-2 > -7$ $3.3 < 9.7$ and $-10 < -9$
The **absolute value** of a number is the distance on the number line between the number and 0.	$\|5\| = 5,\ \|-7\| = 7,$ and $-\left\|-\frac{5}{9}\right\| = -\frac{5}{9}$

REVIEW EXERCISES

35. a. Which number is a whole number but not a natural number?
    b. Write the set of integers.

36. Represent 206 feet below sea level with a signed number.

37. Use one of the symbols > or < to make each statement true.
    a. 0 ___ 5
    b. −12 ___ −13

38. Show that each of the following numbers is a rational number by expressing it as a ratio (quotient) of two integers.
    a. 0.7
    b. $4\frac{2}{3}$

Write each fraction as a decimal. Use an overbar if the result is a repeating decimal.

39. $\frac{1}{250}$

40. $\frac{17}{22}$

41. Graph each number on a number line: $\left\{\pi, 0.333\ldots, 3.75, \sqrt{2}, -\frac{17}{4}, \frac{7}{8}, -2\right\}$

42. Determine which numbers in the given set are natural numbers, whole numbers, integers, rational numbers, irrational numbers, and real numbers. $\left\{-\frac{4}{5}, 99.99, 0, \sqrt{2}, -12, 4\frac{1}{2}, 0.666\ldots\ 8\right\}$

Determine whether each statement is true or false.

43. All integers are whole numbers.

44. π is a rational number.

45. The set of real numbers corresponds to all points on the number line.

46. A real number is either rational or irrational.

Insert one of the symbols >, <, or = in the blank to make each statement true.

47. $|-6|$ ___ $|5|$

48. -9 ___ $|-10|$

SECTION 1.4 Adding Real Numbers; Properties of Addition

DEFINITIONS AND CONCEPTS	EXAMPLES
To **add two real numbers with like signs:**	
1. To add two positive numbers, add them as usual. The final answer is positive.	Add: $3 + 5 = 8$
2. To add two negative numbers, add their absolute values and make the final answer negative.	Add: $-5 + (-11) = -16$
To **add two real numbers with unlike signs:**	
1. Subtract their absolute values (the smaller from the larger).	Add: $-8 + 6 = -2$ −8 has the larger absolute value.
2. To that result, attach the sign of the number with the larger absolute value.	Add: $12 + (-5) = 7$ 12 has the larger absolute value.
Properties of Addition	
Commutative property: $a + b = b + a$ *Changing the order when adding does not affect the answer.*	$5 + (-9) = -9 + 5$ Reorder.
Associative property: $(a + b) + c = a + (b + c)$ *Changing the grouping when adding does not affect the answer.*	$(3 + 7) + 5 = 3 + (7 + 5)$ Regroup. Note that the order of the addends, 3, 7, and 5, does not change.
Addition property of 0: $a + 0 = a$ and $0 + a = a$	$-6 + \mathbf{0} = -6$ 0 is the additive identity element.
Addition property of opposites: $a + (-a) = 0$ and $(-a) + a = 0$	$\mathbf{11} + (\mathbf{-11}) = 0$ 11 and −11 are additive inverses.

REVIEW EXERCISES

Add.

49. $-45 + (-37)$

50. $25 + (-13)$

51. $0 + (-7)$

52. $-7 + 7$

53. $12 + (-8) + (-15)$

54. $-9.9 + (-2.4)$

55. $\frac{5}{16} + \left(-\frac{1}{2}\right)$

56. $35 + (-13) + (-17) + 6$

57. Determine what property of addition is shown.
    a. $-2 + 5 = 5 + (-2)$
    b. $(-2 + 5) + 1 = -2 + (5 + 1)$
    c. $80 + (-80) = 0$
    d. $-5.75 + 0 = -5.75$

58. **Temperatures.** Determine Washington State's record high temperature if it is 166° greater than the state's record low temperature of −48°F.

SECTION 1.5 Subtracting Real Numbers

DEFINITIONS AND CONCEPTS	EXAMPLES
The **opposite of the opposite of a number** is that number. For any real number a, $-(-a) = a$.	$-(-13) = 13$ and $-(-x) = x$
To **subtract two real numbers,** add the first to the opposite (additive inverse) of the number to be subtracted. For any real numbers a and b, $a - b = a + (-b)$ To **check** a subtraction, the difference plus the subtrahend should equal the minuend.	Subtract: $4 - 7 = 4 + (-7) = -3$ The opposite of 7 is −7. $6 - (-8) = 6 + 8 = 14$ The opposite of −8 is 8. $-1 - (-2) = -1 + 2 = 1$ The opposite of −2 is 2. To check $-6 - 2 = -8$, verify that $-8 + 2 = -6$.

REVIEW EXERCISES

Write the expression in simpler form.

59. a. The opposite of 10
    b. The additive inverse of -3

60. a. $-\left(-\frac{9}{16}\right)$ b. $-|-4|$

Perform the operations.

61. $45 - 64$

62. Subtract $\frac{1}{3}$ from $-\frac{3}{5}$

63. $-7 - (-12)$

64. $3.6 - (-2.1)$

65. $0 - 10$

66. $-33 + 7 - 5 - (-2)$

67. **Geography.** The tallest peak on Earth is Mount Everest, at 29,028 feet, and the greatest ocean depth is the Mariana Trench, at −36,205 feet. Find the difference in these elevations. Check the result.

68. **History.** Archimedes, a famous Greek mathematician, died in 212 B.C. (−212) at the age of 75. When was he born? Check the result.

SECTION 1.6 Multiplying and Dividing Real Numbers; Multiplication and Division Properties

DEFINITIONS AND CONCEPTS	EXAMPLES
To **multiply two real numbers,** multiply their absolute values. 1. If the numbers have **like signs,** the final answer is positive. 2. If the numbers have **unlike signs,** the final answer is negative.	Multiply: $-5(-7) = 35$ and $14(3) = 42$ Multiply: $6(-6) = -36$ and $-11(5) = -55$
Properties of multiplication	
Commutative property: $ab = ba$ *Changing the order when multiplying does not affect the answer.*	$-8(12) = 12(-8)$ Reorder.
Associative property: $(ab)c = a(bc)$ *Changing the grouping when multiplying does not affect the answer.*	$(-4 \cdot 9) \cdot 7 = -4(9 \cdot 7)$ Regroup. Note that the order of the factors, −4, 9, and 7, does not change.
Multiplication property of 0: $0 \cdot a = 0$ and $a \cdot 0 = 0$	$0 \cdot (-7) = 0$ and $6(5)0 = 0$
Multiplication property of 1: $1 \cdot a = a$ and $a \cdot 1 = a$	$1 \cdot 32 = 32$ 1 is the multiplicative identity.
Multiplication property of −1: $-1 \cdot a = -a$ and $a(-1) = -a$	$-1(8) = -8$, $-1(-32) = 32$, $-x = -1 \cdot x$
Multiplicative inverse property: $a\left(\frac{1}{a}\right) = 1$ and $\frac{1}{a}(a) = 1$	$4\left(\frac{1}{4}\right) = 1$ 4 and $\frac{1}{4}$ are multiplicative inverses.

To **divide two real numbers,** divide their absolute values. **1.** If the numbers have **like signs,** the final answer is positive. **2.** If the numbers have **unlike signs,** the final answer is negative.	Divide: $\frac{16}{8} = 2$ and $\frac{-25}{-5} = 5$ Divide: $\frac{-36}{9} = -4$ and $\frac{56}{-7} = -8$
For any real number, $\frac{a}{1} = a$ and $\frac{a}{a} = 1$, where $a \neq 0$.	$\frac{25}{1} = 25$ and $\frac{-32}{-32} = 1$
Division of zero by a nonzero number is 0. **Division by zero** is undefined.	$\frac{0}{17} = 0$ but $\frac{2}{0}$ is undefined because no number multiplied by 0 gives 2.
To **check** the division $\frac{a}{b} = c$, verify that $c \cdot b = a$.	To check $\frac{6}{-2} = -3$, verify that $-3(-2) = 6$. *Quotient · divisor = dividend.*

REVIEW EXERCISES

Multiply.

69. $-8 \cdot 7$
70. $-9\left(-\frac{1}{9}\right)$
71. $2(-3)(-2)$
72. $(-4)(-1)(-3)$
73. $-1.2(-5.3)$
74. $0.002(-1{,}000)$
75. $-\frac{2}{3}\left(\frac{1}{5}\right)$
76. $-6(-3)(0)(-1)$
77. **Electronics.** The picture on the screen can be magnified by switching a setting on the monitor. What would be the new high and low if every value changed by a factor of 1.5?

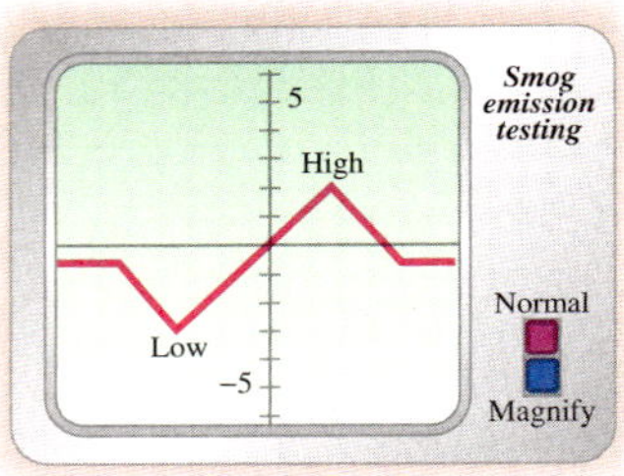

78. Determine what property of multiplication is shown.
    a. $(2 \cdot 3)5 = 2(3 \cdot 5)$
    b. $(-5)(-6) = (-6)(-5)$
    c. $-6 \cdot 1 = -6$
    d. $\frac{1}{2}(2) = 1$

Perform each division, if possible.

79. $\frac{44}{-44}$
80. $\frac{-272}{16}$
81. $\frac{-81}{-27}$
82. $-\frac{3}{5} \div \frac{1}{2}$
83. $\frac{-60}{0}$
84. $\frac{-4.5}{1}$
85. Fill in the blanks: $\frac{0}{18} = 0$ because ___ · ___ = ___.
86. **Gemstones.** A 3-carat yellow sapphire stone valued at $3,000 five years ago is now worth $1,200. What signed number indicates the average annual depreciation of the sapphire?

SECTION 1.7 Exponents and Order of Operations

DEFINITIONS AND CONCEPTS	EXAMPLES
An **exponent** represents repeated multiplication.	$8^5 = 8 \cdot 8 \cdot 8 \cdot 8 \cdot 8$ *The exponent 5 indicates that 8 is to be used as a factor 5 times.*
In a^n, a is the **base** and n is the **exponent.**	In 7^4, the base is 7 and 4 is the exponent.

Order of Operations

1. Perform all calculations within grouping symbols, working from the innermost to the outermost in the following order.
2. Evaluate all exponential expressions.
3. Perform all multiplications and divisions as they occur from left to right.
4. Perform all additions and subtractions as they occur from left to right.

In fractions, evaluate the numerator and denominator separately. Then simplify the fraction.

Grouping Symbols:

Innermost parentheses

$$2[7 + 3(1 - 4)]$$

Outermost brackets

Evaluate:

$$\frac{3(6 - 4^3) - 2^4 + 4}{8 \div 4 \cdot 3} = \frac{3(6 - 64) - 2^4 + 4}{2 \cdot 3}$$ Evaluate: $4^3 = 64$. Divide: $8 \div 4 = 2$.

$$= \frac{3(-58) - 2^4 + 4}{6}$$ Subtract: $6 - 64 = -58$.

$$= \frac{3(-58) - 16 + 4}{6}$$ Evaluate: $2^4 = 16$.

$$= \frac{-174 - 16 + 4}{6}$$ Multiply: $3(-58) = -174$.

$$= \frac{-190 + 4}{6}$$ Subtract: $-174 - 16 = -190$.

$$= \frac{-186}{6}$$ Add: $-190 + 4 = -186$.

$$= -31$$ Divide: -186 by 6.

$$\text{Mean} = \frac{\text{sum of values}}{\text{number of values}}$$

The mean is also called the average.

Find the mean of the test scores of 74, 83, 79, 91, and 73.

$$\text{Mean} = \frac{74 + 83 + 79 + 91 + 73}{5} = 80$$

The mean is a value around which 74, 83, 79, 91, and 73 are grouped.

REVIEW EXERCISES

87. Write each expression using exponents.

a. $8 \cdot 8 \cdot 8 \cdot 8 \cdot 8$ b. $9 \cdot \pi \cdot r \cdot r$

88. Evaluate each expression.

a. 9^2 b. $\left(-\frac{2}{3}\right)^3$

c. 2^5 d. 50^1

Evaluate each expression.

89. $2 + 5 \cdot 3$

90. $-24 \div 2 \cdot 3$

91. $-(16 - 3)^2$

92. $43 + 2(-6 - 2 \cdot 2)$

93. $10 - 5[-3 - 2(5 - 7^2)] - 5$

94. $\dfrac{-4(4 + 2) - 4}{2|-18 - 4(5)|}$

95. $(-3)^3\left(\frac{-8}{2}\right) + 5$

96. $\dfrac{2^4 - (4 - 6)(3 - 6)}{12 + 4[(-1)^8 - 2^2]}$

97. Write each expression in symbols and then evaluate it.

a. Negative nine squared

b. The opposite of the square of nine

98. Walk-A-Thons. Use the data in the table to find the average (mean) donation to a charity walk-a-thon.

Donation	\$5	\$10	\$20	\$50	\$100
Number received	20	65	25	5	10

SECTION 1.8 Algebraic Expressions

DEFINITIONS AND CONCEPTS	EXAMPLES
Addition symbols separate algebraic expressions into **terms.** In a term, the numerical factor is called the **coefficient.** A term that consists of a single number is called a **constant** term.	Since $a^2 + 3a - 5$ can be written as $a^2 + 3a + (-5)$, it has three terms. The coefficient of a^2 is 1, the coefficient of $3a$ is 3, and the coefficient of the constant term -5 is -5.
Key phrases can be translated to algebraic expressions.	*5 more than x* can be expressed as $x + 5$. *25 less than twice y* can be expressed as $2y - 25$. One-half *of c* can be expressed as $\frac{1}{2}c$.
Number · value = total value	The total value (in cents) of n nickels is $n \cdot 5 = 5n$ cents.

To **evaluate algebraic expressions,** we substitute the values of its variables and use the order of operations rule.

Evaluate $\frac{x^2 - y^2}{x + y}$ for $x = 2$ and $y = -3$.

$$\frac{x^2 - y^2}{x + y} = \frac{2^2 - (-3)^2}{2 + (-3)}$$ Substitute 2 for x and −3 for y.

$$= \frac{4 - 9}{-1}$$ Evaluate: $2^2 = 4$ and $(-3)^2 = 9$. Add: $2 + (-3) = -1$.

$$= \frac{-5}{-1}$$ Subtract: $4 - 9 = -5$.

$$= 5$$ Divide −5 by −1.

REVIEW EXERCISES

99. How many terms does each expression have?

a. $3x^2 + 2x - 5$ **b.** $-12xyz$

100. Identify the coefficient of each term of the given expression.

a. $16x^2 - 5x + 25$ **b.** $\frac{x}{2} + y$

Write each phrase as an algebraic expression.

101. 25 more than the height h

102. 15 less than triple the cutoff score s

103. 6 less than one-half of the time

104. The absolute value of the difference of 2 and the square of a

105. Hardware. Let n represent the length of the nail in inches. Write an algebraic expression that represents the length of the bolt (in inches).

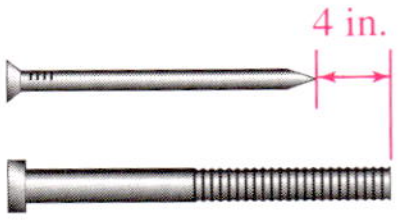

106. Hardware. Let b represent the length of the bolt in inches. Write an algebraic expression that represents the length of the nail (in inches).

107. How many years are in d decades?

108. Five years after a house was constructed, a patio was added. How old, in years, is the patio if the house is x years old?

109. Complete the table below. The units are cents.

Coin	Number	Value	Total value
Nickel	6	5	
Dime	d	10	

110. Complete the table.

x	$20x - x^3$
0	
1	
−4	

Evaluate each algebraic expression for the given values of the variables.

111. $b^2 - 4ac$ for $b = -10$, $a = 3$, and $c = 5$

112. $\frac{x + y}{-x - z}$ for $x = 19$, and $y = 17$, and $z = -18$

SECTION 1.9 Simplifying Algebraic Expressions Using Properties of Real Numbers

DEFINITIONS AND CONCEPTS	EXAMPLES
We often use the *commutative property of multiplication* to reorder factors and the *associative property of multiplication* to regroup factors when **simplifying expressions.**	Simplify: $-5(3y) = (-5 \cdot 3)y = -15y$ −5(3y) and −15y are called equivalent expressions. $-45b\left(\frac{5}{9}\right) = \left(-45 \cdot \frac{5}{9}\right)b = -25b$
The **distributive property** can be used to *remove parentheses:* $a(b + c) = ab + ac$ $a(b - c) = ab - ac$ $a(b + c + d) = ab + ac + ad$	Multiply: $7(x + 3) = 7 \cdot x + 7 \cdot 3 = 7x + 21$ $-0.2(4m - 5n - 7) = -0.2(4m) - (-0.2)(5n) - (-0.2)(7)$ $= -0.8m + n + 1.4$
The **opposite of a sum:** $-(a + b) = -a + (-b)$	$-(3x + 4) = -3x - 4$ and $-(-2y - 12) = 2y + 12$

Like terms are terms with exactly the same variables raised to exactly the same powers.	$3x$ and $-5x$ are like terms. $-4t^3$ and $3t^2$ are unlike terms because the variable t has different exponents. $0.5xyz$ and $3.7xy$ are unlike terms because they have different variables.
Simplifying the sum or difference of like terms is called **combining like terms.** Like terms can be combined by adding or subtracting the coefficients of the terms and keeping the same variables with the same exponents.	Simplify: $4a + 2a = 6a$ Think: $(4 + 2)a = 6a$. $5p^2 + p - p^2 - 9p = 4p^2 - 8p$ Think: $(5 - 1)p^2 = 4p^2$ and $(1 - 9)p = -8p$. $2(k - 1) - 3(k + 2) = 2k - 2 - 3k - 6 = -k - 8$

REVIEW EXERCISES

Use the given property to complete each statement.

113. $a \cdot 150 =$ ______ Commutative property of multiplication

114. $9 + (1 + 7y) =$ __________ Associative property of addition

115. $2.7(10b) =$ ________ Associative property of multiplication

116. $x + 2x^2 =$ ______ Commutative property of addition

Simplify each expression.

117. $-4(7w)$

118. $3(-2x)(-4)$

119. $0.4(5.2f)$

120. $\frac{7}{2} \cdot \frac{2}{7}r$

Use the distributive property to remove parentheses.

121. $5(x + 3)$

122. $-(2x + 3 - y)$

123. $\frac{3}{4}(4c - 8)$

124. $-2(-3c - 7)(2.1)$

Simplify each expression by combining like terms.

125. $8p + 5p - 4p$

126. $-5m + 2 - 2m - 2$

127. $n + n + n + n$

128. $5(p - 2) - 2(3p + 4)$

129. $55.7k^2 - 55.6k^2$

130. $8a^3 + 4a^3 + 2a - 4a^3 - 2a - 1$

131. $\frac{3}{5}w - \left(-\frac{2}{5}w\right)$

132. $36\left(\frac{1}{9}h - \frac{3}{4}\right) + 36\left(\frac{1}{3}\right)$

133. $-(7.6t - 1.9) + (1.4t - 1.2)8 + t$

134. Write an equivalent expression for the given expression using fewer symbols.

a. $1x$

b. $-1x$

c. $4x - (-1)$

d. $4x + (-1)$

1 CHAPTER TEST

1. Fill in the blanks.

a. Two fractions, such as $\frac{1}{2}$ and $\frac{5}{10}$, that represent the same number are called __________ fractions.

b. The result of a multiplication is called a ________.

c. $\frac{8}{7}$ is the __________ of $\frac{7}{8}$ because $\frac{8}{7} \cdot \frac{7}{8} = 1$.

d. $9x^2$ and $7x^2$ are _____ _______ because they have the same variable raised to exactly the same power.

e. For any nonzero real number a, $\frac{a}{0}$ is __________.

2. Security Guards. The graph shows the cost to hire a security guard.

a. What will it cost to hire a security guard for 3 hours?

b. If a school was billed \$40 for hiring a security guard for a dance, for how long did the guard work?

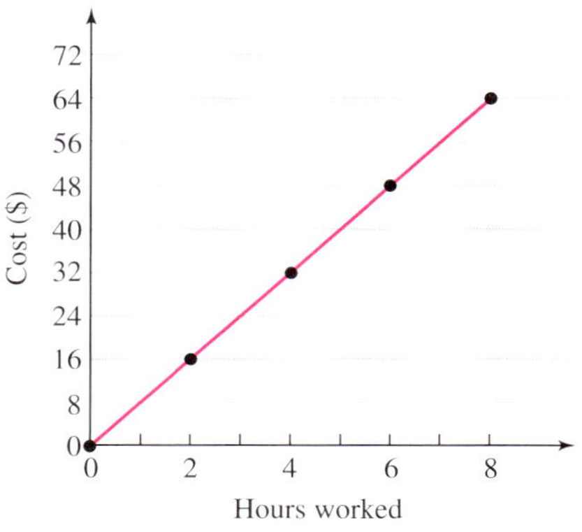

3. Use the formula $f = \frac{a}{5}$ to complete the table.

Square miles (a)	Fire stations (f)
15	
100	
350	

4. Give the prime factorization of 180.

5. Simplify: $\frac{42}{105}$

6. Divide: $\frac{15}{16} \div \frac{5}{8}$

7. Add: $\frac{7}{10} + \frac{1}{14}$

8. Subtract: $8\frac{2}{5} - 1\frac{2}{3}$

9. Shopping.

a. What is the weight of the oranges? Answer using a mixed number.

b. Find the cost of the oranges. Answer in dollars.

10. Write $\frac{5}{6}$ as a decimal.

11. a. Graph the set of numbers on a number line.

$$\left\{-1\tfrac{1}{4},\ 0,\ \sqrt{2},\ -3.75,\ 2,\ \tfrac{7}{2},\ 0.5,\ -3\right\}$$

b. Determine which numbers in the set are natural numbers, whole numbers, integers, rational numbers, irrational numbers, and real numbers.

12. Determine whether each statement is true or false.

a. Every integer is a rational number.

b. Every rational number is an integer.

c. π is an irrational number.

d. 500 is a whole number.

e. A real number is any number that is either a rational or an irrational number.

13. Insert the proper symbol, $>$ or $<$, in the blank.

a. $-2 \quad \square \quad -3$ **b.** $-|-9| \quad \square \quad 8$

c. $|-4| \quad \square \quad -(-5)$ **d.** $\left|-\frac{7}{8}\right| \quad \square \quad 0.5$

14. Television. During "sweeps week," networks try to gain viewers by showing flashy programs. Use the data to determine the average daily gain (or loss) of ratings points by a network for the 7-day "sweeps period."

Day	M	T	W	Th	F	Sa	Su
Point loss/gain	0.6	−0.3	1.7	1.5	−0.2	1.1	−0.2

Perform the operations.

15. $-5.6 + (-2)$

16. $(-6) + 8 + (-4)$

17. $-\frac{1}{2} + \frac{7}{8}$

18. a. $-10 - (-4)$

b. Show a check of the result.

19. a. $\frac{-12.6}{-0.9}$

b. Show a check of the result.

20. $(-2)(-3)(-5)$

21. $-6.1(0.4)$

22. $\frac{0}{-3}$

23. $\left(-\frac{3}{5}\right)^3$

24. $3 + (-3)$

25. $0 - 3$

26. $-30 + 50 - 10 - (-40)$

27. Astronomy. *Magnitude* is a term used in astronomy to describe the brightness of planets and stars. Negative magnitudes are associated with brighter objects. By how many magnitudes do a full moon and the sun differ?

Object	Magnitude
Sun	−26.5
Full moon	−12.5

28. Glaciers. In 2005, the South Cascade Glacier in Washington State gained approximately 87 inches in thickness because of snowfall. That same year it lost approximately 165 inches of thickness due to melting. Was there a net gain or loss in the glacier's thickness that year? How much? (Source: U.S. Geological Survey)

29. Inventory. It was discovered that fifteen MP3 players were missing from the stockroom of an electronics store. If the players cost \$85 each, what signed number represents the store's financial loss?

30. Use the given property to complete each statement.

a. $(-12 + 97) + 3 =$ ______________
Associative property of addition

b. $2(x + 7) =$ ________
Distributive property

c. $-2(m)5 =$ ________
Commutative property of multiplication

d. $\frac{1}{8}(8) =$ ___
Inverse property of multiplication

e. $0 + 15x =$ ____
Identity property of addition

31. Write each product using exponents:

a. $9(9)(9)(9)(9)$ **b.** $3 \cdot x \cdot x \cdot z \cdot z \cdot z$

32. Complete the table.

x	$2x - \frac{30}{x}$
5	
10	
−30	

Evaluate each expression.

33. $8 + 2 \cdot 3^4$

34. $\frac{3(40 - 2^3)}{-2(6 - 4)^2}$

35. $-10^2 - 5 + 6$

36. $9 - 3[45 - 5^2(1^5 - 4)]$

37. $|-50 \div 5 \cdot 2|$

38. Evaluate $3(10x - y) - 5(x + y^2)$ for $x = 2$ and $y = -5$.

39. Translate to an algebraic expression: seven less than twice the width w.

40. a. **Music.** A band recorded x songs for a CD. However, two of the songs were not included in the album because of poor sound quality. Write an algebraic expression that represents the number of songs on the CD.

b. **Money.** Find the value of q quarters in cents.

41. How many terms are in the expression $4x^2 + 5x - 7$?

42. What is the coefficient of each term of $a^3 - 6a^2 - a + 10$?

Simplify each expression.

43. $5(-4x)$

44. $-8(-7t)(4)$

45. $\frac{4}{5}(15a + 5) - 16a$

46. $-1.1d^3 - 3.8d^3 - d^3$

47. $9x + 2(7x - 3) - 9(x - 1)$

48. $m^2 + 4m^2 + 5m - 2m^2 - 3m - 4$

49. Tell whether each statement is true or false.

a. $-\frac{3}{5} = \frac{-3}{5} = \frac{3}{-5}$

b. $-|-8| = -(-8)$

c. $-5^2 = (-5)^2$

d. $-(4a + 7) = -4a + 7$

e. $(-9)^{14}$ is a positive number.

f. The reciprocal of $-\frac{4}{7}$ is $\frac{7}{4}$.

Group Project

WRITING FRACTIONS AS DECIMALS

Overview: This is a good activity to try at the beginning of the course. You can become acquainted with other students in your class while you review the process for finding decimal equivalents of fractions.

Instructions: Form groups of 6 students. Select one person from your group to record the group's responses on the questionnaire. Express the results in fraction form and in decimal form.

What fraction (decimal) of the students in your group . . .	Fraction	Decimal
■ have the letter a in their first names?		
■ have a birthday in January or February?		
■ work full-time or part-time?		
■ have ever been on television?		
■ have downloaded music in the last week?		
■ log into Facebook at least once a day?		
■ send 30 or more text messages a day?		
■ have downloaded at least 10 applications to their cell phones?		

2

Equations, Inequalities, and Problem Solving

©barang/Shutterstock.com

from Campus to Careers

Automotive Service Technician

Anyone whose car has ever broken down appreciates the talents of automotive service technicians. To work on today's high-tech cars and trucks, a person needs strong diagnostic and problem-solving skills. Courses in automotive repair, electronics, physics, chemistry, English, computers, and mathematics provide a good educational background for a career as a service technician.

Problem 47 in **Study Set 2.3, problem 85** in **Study Set 2.4,** and **problem 54** in **Study Set 2.5** involve situations that an automotive service technician might encounter on the job. The mathematical concepts discussed in this chapter can be used to solve those problems.

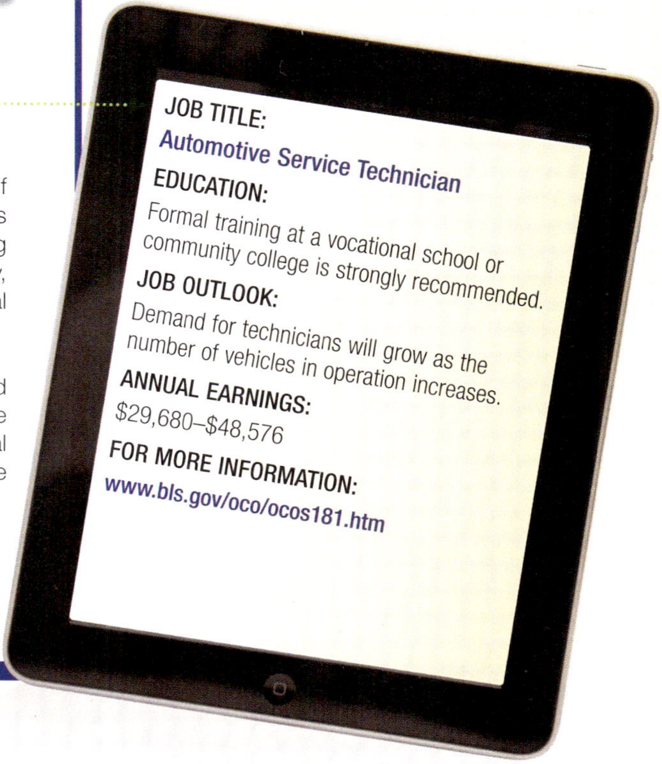

Study Skills Workshop ▶ Preparing to Learn

Many students think that there are two types of people—those who are good at math and those who are not—and that this cannot be changed. This isn't true! Here are some suggestions that can increase your chances for success in algebra.

DISCOVER YOUR LEARNING STYLE: Are you a visual, verbal, or audio learner? Knowing this will help you determine how best to study.

GET THE MOST OUT OF THE TEXTBOOK: This book and the software that comes with it contain many student-support features. Are you taking advantage of them?

TAKE GOOD NOTES: Are your class notes complete so that they are helpful when doing your homework and studying for tests?

Now Try This

1. To determine what type of learner you are, take the *Learning Style Survey* found online at http://www.metamath.com/multiple/multiple_choice_questions.html. Then, write a one-page paper explaining what you learned from the survey results and how you will use the information to help you succeed in the class.
2. To learn more about the student-support features of this book, take the *Textbook Tour* found online at www.cengage.com/math/tussy.
3. Rewrite a set of your class notes to make them more readable and to clarify the concepts and examples covered. If they are not already, write them in outline form. Fill in any information you didn't have time to copy down in class and complete any phrases or sentence fragments.

SECTION 2.1 Solving Equations Using Properties of Equality

OBJECTIVES

1. Determine whether a number is a solution.
2. Use the addition property of equality.
3. Use the subtraction property of equality.
4. Use the multiplication property of equality.
5. Use the division property of equality.

ARE YOU READY?

The following problems review some basic skills that are needed when solving equations. Fill in the blanks.

1. $8 \quad 8 = 0$ **2.** $-1.6 \quad 1.6 = 0$ **3.** $\frac{7}{} = 1$

4. $3 \quad \frac{1}{3} = 1$ **5.** $\frac{2}{5} - \quad = 0$ **6.** $-\frac{9}{8}\left(\quad \right) = 1$

In this section, we introduce four fundamental properties of equality that are used to solve equations.

1 Determine Whether a Number Is a Solution.

An **equation** is a statement indicating that two expressions are equal. An example is $x + 5 = 15$. The equal symbol $=$ separates the equation into two parts: The expression $x + 5$ is the **left side** and 15 is the **right side.** The letter x is the **variable** (or the **unknown**). The sides of an equation can be reversed, so we can write $x + 5 = 15$ or $15 = x + 5$.

- An equation can be true: $6 + 3 = 9$
- An equation can be false: $2 + 4 = 7$
- An equation can be neither true nor false. For example, $x + 5 = 15$ is neither true nor false because we don't know what number x represents.

The Language of Algebra

It is important to know the difference between an **equation** and an **expression**. An equation contains an = symbol and an expression does not.

An equation that contains a variable is made true or false by substituting a number for the variable. If we substitute 10 for x in $x + 5 = 15$, the resulting equation is true: $10 + 5 = 15$. If we substitute 1 for x, the resulting equation is false: $1 + 5 = 15$. A number that makes an equation true when substituted for the variable is called a **solution** and it is said to **satisfy** the equation. Therefore, 10 is a solution of $x + 5 = 15$, and 1 is not. The **solution set** of an equation is the set of all numbers that make the equation true.

EXAMPLE 1 Check to determine whether 9 is a solution of $3y - 1 = 2y + 7$.

Strategy We will substitute 9 for each y in the equation and evaluate the expression on the left side and the expression on the right side separately.

Why If a true statement results, 9 is a solution of the equation. If we obtain a false statement, 9 is not a solution.

Solution

The Language of Algebra

Read $\stackrel{?}{=}$ as "is possibly equal to."

Evaluate the expression on the left side. | Evaluate the expression on the right side.

$$3y - 1 = 2y + 7$$
$$3(9) - 1 \stackrel{?}{=} 2(9) + 7$$
$$27 - 1 \stackrel{?}{=} 18 + 7$$
$$26 = 25$$

Since $26 = 25$ is false, 9 is not a solution of $3y - 1 = 2y + 7$.

Self Check 1 Check to determine whether 25 is a solution of $10 - x = 35 - 2x$.

Now Try Problem 19

2 Use the Addition Property of Equality.

To **solve an equation** means to find all values of the variable that make the equation true. We can develop an understanding of how to solve equations by referring to the scales shown on the right.

The first scale represents the equation $x - 2 = 3$. The scale is in balance because the weights on the left side and right side are equal. To find x, we must add 2 to the left side. To keep the scale in balance, we must also add 2 to the right side. After doing this, we see that x is balanced by 5. Therefore, x must be 5. We say that we have solved the equation $x - 2 = 3$ and that the solution is 5.

In this example, we solved $x - 2 = 3$ by transforming it to a simpler *equivalent equation*, $x = 5$.

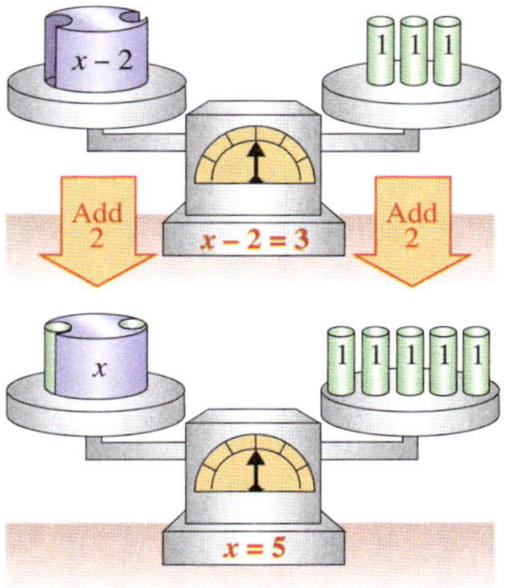

Equivalent Equations

Equations with the same solutions are called **equivalent equations.**

The procedure that we used above suggests the following property of equality.

Addition Property of Equality

Adding the same number to both sides of an equation does not change its solution.

For any real numbers a, b, and c,

if $a = b$, then $a + c = b + c$

When we use this property, the resulting equation is *equivalent to the original one.* We will now show how it is used to solve $x - 2 = 3$ algebraically.

EXAMPLE 2 Solve: $x - 2 = 3$

Strategy We will use a property of equality to isolate the variable on one side of the equation.

Why To solve the original equation, we want to find a simpler equivalent equation of the form **x = a number**, whose solution is obvious.

Solution

$x - 2 = 3$ This is the equation to solve.

$x - 2 + 2 = 3 + 2$ Use the addition property of equality to isolate x on the left side of the equation. Undo the subtraction of 2 by adding 2 to both sides.

$x + 0 = 5$ The sum of a number and its opposite is zero: $-2 + 2 = 0$.

$x = 5$ When 0 is added to a number, the result is the same number.

The Language of Algebra

We solve equations by writing a series of steps that result in an equivalent equation of the form

x = *a number*

or

a number = x

We say the variable is **isolated** on one side of the equation. *Isolated* means alone or by itself.

Since 5 is obviously the solution of the equivalent equation $x = 5$, the solution of the original equation, $x - 2 = 3$, is also 5. To check this result, we substitute 5 for x in the original equation and simplify.

$x - 2 = 3$

$5 - 2 \stackrel{?}{=} 3$ Substitute 5 for x.

$3 = 3$ True

Since the resulting statement is true, 5 is the solution of $x - 2 = 3$. A more formal way to present this result is to write the solution within braces as a solution set: $\{5\}$.

Self Check 2 Solve: $n - 16 = 33$

Now Try Problem 37

EXAMPLE 3 Solve: **a.** $-19 = y - 7$ **b.** $-27 + y = -3$

Strategy We will use a property of equality to isolate the variable on one side of the equation.

Why To solve the original equation, we want to find a simpler equivalent equation of the form **y = a number** or **a number = y**, whose solution is obvious.

Solution

a.

$-19 = y - 7$ This is the equation to solve.

$-19 + 7 = y - 7 + 7$ To isolate y on the right side, use the addition property of equality. Undo the subtraction of 7 by adding 7 to both sides.

$-12 = y$ The sum of a number and its opposite is zero: $-7 + 7 = 0$.

Notation

We may solve an equation so that the variable is isolated on either side of the equation. Note that $-12 = y$ is equivalent to $y = -12$.

Check: $-19 = y - 7$ This is the original equation.

$-19 \stackrel{?}{=} -12 - 7$ Substitute -12 for y.

$-19 = -19$ True

Since the resulting statement is true, the solution is -12. The solution set is $\{-12\}$.

b.

$-27 + y = -3$ The equation to solve.

$-27 + y + 27 = -3 + 27$ To isolate y, use the addition property of equality. Eliminate -27 on the left side by adding its opposite (additive inverse) to both sides.

$y = 24$ The sum of a number and its opposite is zero: $-27 + 27 = 0$.

Caution

After checking a result, be careful when stating your conclusion. Here, it would be incorrect to say:

The solution is -3.

The number we were checking was 24, not -3.

Check:

$$-27 + y = -3 \quad \text{This is the original equation.}$$
$$-27 + 24 \stackrel{?}{=} -3 \quad \text{Substitute 24 for } y.$$
$$-3 = -3 \quad \text{True}$$

The solution is 24. The solution set is $\{24\}$.

Self Check 3 Solve: **a.** $-5 = b - 38$ **b.** $-20 + n = 29$

Now Try Problems 41 and 43

3 Use the Subtraction Property of Equality.

Since any subtraction can be written as an addition by adding the opposite of the number to be subtracted, the following property is an extension of the addition property of equality.

Subtraction Property of Equality

Subtracting the same number from both sides of an equation does not change its solution.

For any real numbers a, b, and c,

if $a = b$, then $a - c = b - c$

When we use this property, the resulting equation is equivalent to the original one.

EXAMPLE 4 Solve: **a.** $x + \frac{1}{8} = \frac{7}{4}$ **b.** $54.9 + x = 45.2$

Strategy We will use a property of equality to isolate the variable on one side of the equation.

Why To solve the original equation, we want to find a simpler equivalent equation of the form x = **a number**, whose solution is obvious.

Solution

a.

$$x + \frac{1}{8} = \frac{7}{4} \quad \text{This is the equation to solve.}$$

$$x + \frac{1}{8} - \frac{1}{8} = \frac{7}{4} - \frac{1}{8}$$ To isolate x, use the subtraction property of equality. Undo the addition of $\frac{1}{8}$ by subtracting $\frac{1}{8}$ from both sides.

$$x = \frac{7}{4} - \frac{1}{8}$$ On the left side, $\frac{1}{8} - \frac{1}{8} = 0$.

$$x = \frac{7}{4} \cdot \frac{2}{2} - \frac{1}{8}$$ To prepare to subtract the fractions, build $\frac{7}{4}$ so that it has a denominator of 8.

$$x = \frac{14}{8} - \frac{1}{8}$$ Multiply the numerators and multiply the denominators.

$$x = \frac{13}{8}$$ Subtract the numerators. Write the result over the common denominator 8. The fraction is in simplest form.

Verify that $\frac{13}{8}$ is the solution by substituting it for x in the original equation and simplifying.

Success Tip

We could also isolate x by adding the additive inverse of $\frac{1}{8}$, which is $-\frac{1}{8}$, to both sides:

$$x + \frac{1}{8} + \left(-\frac{1}{8}\right) = \frac{7}{4} + \left(-\frac{1}{8}\right)$$

Success Tip

It is not necessary to write the solution $\frac{13}{8}$ as a mixed number. In algebra, a solution may be expressed as an improper fraction, as long as it is in simplest form. (The numerator and denominator have no common factors other than 1.)

b.

$54.9 + x = 45.2$	This is the equation to solve.
$54.9 + x - \mathbf{54.9} = 45.2 - \mathbf{54.9}$	To isolate x, use the subtraction property of equality. Undo the addition of 54.9 by subtracting 54.9 from both sides.
$x = -9.7$	On the left side, $54.9 - 54.9 = 0$.

Check:

$54.9 + x = 45.2$	This is the original equation.
$54.9 + (-9.7) \stackrel{?}{=} 45.2$	Substitute -9.7 for x.
$45.2 = 45.2$	True

The solution is -9.7. The solution set is $\{-9.7\}$.

Self Check 4 Solve: **a.** $x + \frac{4}{15} = \frac{11}{5}$ **b.** $0.7 + a = 0.2$

Now Try Problems 45 and 47

4 Use the Multiplication Property of Equality.

To develop another property of equality, consider the first scale shown on the right that represents the equation $\frac{x}{3} = 25$. The scale is in balance because the weights on the left side and right side are equal. To find x, we must triple (multiply by 3) the weight on the left side. To keep the scale in balance, we also must triple the weight on the right side. After doing this, we see in the second illustration that x is balanced by 75. Therefore, x must be 75.

The procedure that we used to keep the scale balanced suggests the following property of equality.

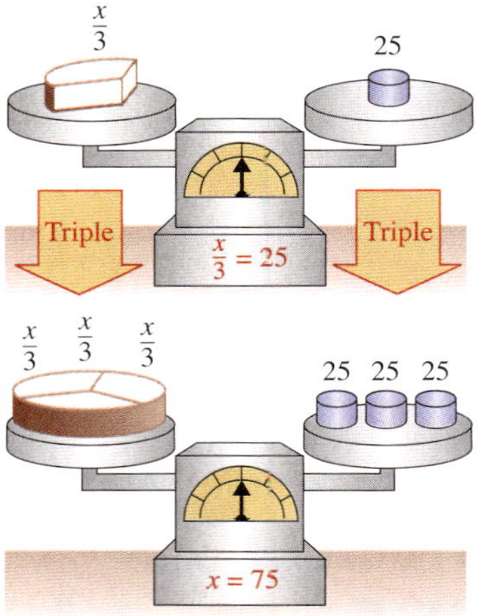

Multiplication Property of Equality	Multiplying both sides of an equation by the same nonzero number does not change its solution. For any real numbers a, b, and c, where c is not 0, if $a = b$, then $ca = cb$

When we use this property, the resulting equation is equivalent to the original one. We will now show how it is used to solve $\frac{x}{3} = 25$ algebraically.

EXAMPLE 5 Solve: $\frac{x}{3} = 25$

Strategy We will use a property of equality to isolate the variable on one side of the equation.

Why To solve the original equation, we want to find a simpler equivalent equation of the form $x = $ **a number**, whose solution is obvious.

Solution

$$\frac{x}{3} = 25$$ This is the equation to solve.

$$3 \cdot \frac{x}{3} = 3 \cdot 25$$ To isolate x, use the multiplication property of equality. Undo the division by 3 by multiplying both sides by 3.

$$\frac{3x}{3} = 75$$ Do the multiplication.

$$1x = 75$$ Simplify $\frac{3x}{3}$ by removing the common factor of 3 in the numerator and denominator: $\frac{3}{3} = 1$.

$$x = 75$$ The coefficient 1 need not be written since $1x = x$.

If we substitute 75 for x in $\frac{x}{3} = 25$, we obtain the true statement $25 = 25$. This verifies that 75 is the solution. The solution set is $\{75\}$.

Self Check 5 Solve: $\frac{b}{24} = 3$

Now Try Problem 49

Since the product of a number and its reciprocal (or multiplicative inverse) is 1, we can solve equations such as $\frac{2}{3}x = 6$, where the coefficient of the variable term is a fraction, as follows.

EXAMPLE 6 Solve: **a.** $\frac{2}{3}x = 6$ **b.** $-\frac{5x}{4} = \frac{3}{16}$

Strategy We will use a property of equality to isolate the variable on one side of the equation.

Why To solve the original equation, we want to find a simpler equivalent equation of the form $x =$ **a number**, whose solution is obvious.

Solution **a.** Since the coefficient of x is $\frac{2}{3}$, we can isolate x by multiplying both sides of the equation by the reciprocal of $\frac{2}{3}$, which is $\frac{3}{2}$.

$$\frac{2}{3}x = 6$$ This is the equation to solve.

$$\frac{3}{2} \cdot \frac{2}{3}x = \frac{3}{2} \cdot 6$$ To undo the multiplication by $\frac{2}{3}$, multiply both sides by the reciprocal of $\frac{2}{3}$.

$$\left(\frac{3}{2} \cdot \frac{2}{3}\right)x = \frac{3}{2} \cdot 6$$ Use the associative property of multiplication to group $\frac{3}{2}$ and $\frac{2}{3}$.

$$1x = 9$$ On the left side, $\frac{3}{2} \cdot \frac{2}{3} = 1$. On the right side, $\frac{3}{2} \cdot 6 = \frac{18}{2} = 9$.

$$x = 9$$ The coefficient 1 need not be written since $1x = x$.

Check:

$$\frac{2}{3}x = 6$$ This is the original equation.

$$\frac{2}{3}(9) \stackrel{?}{=} 6$$ Substitute 9 for x in the original equation.

$$6 = 6$$ On the left side, $\frac{2}{3}(9) = \frac{18}{3} = 6$.

Since the resulting statement is true, 9 is the solution. The solution set is $\{9\}$.

Notation

Variable terms with fractional coefficients can be written in two ways. For example:

$-\frac{5x}{4} = -\frac{5}{4}x$

Caution

If the solution of an equation is a fraction, make sure it is written in simplest form. (The numerator and denominator have no common factors other than 1.)

b. $-\frac{5x}{4} = \frac{3}{16}$ This is the equation to solve.

$-\frac{5}{4}x = \frac{3}{16}$ Write $-\frac{5x}{4}$ as $-\frac{5}{4}x$.

$-\frac{4}{5}\left(-\frac{5}{4}x\right) = -\frac{4}{5}\left(\frac{3}{16}\right)$ To isolate x, undo the multiplication by $-\frac{5}{4}$ by multiplying both sides by the reciprocal of $-\frac{5}{4}$.

$1x = -\left(\frac{4 \cdot 3}{5 \cdot 16}\right)$ On the left side, $-\frac{4}{5}\left(-\frac{5}{4}\right) = 1$. On the right side, multiply the numerators and multiply the denominators. Since the signs of the fractions are unlike, make the final answer negative.

$x = -\frac{\overset{1}{\cancel{4}} \cdot 3}{5 \cdot \underset{1}{\cancel{4}} \cdot 4}$ On the left side, the coefficient 1 need not be written since $1x = x$. On the right side, simplify the fraction by factoring 16 as $4 \cdot 4$ and removing the common factor 4.

$x = -\frac{3}{20}$ Multiply the remaining factors in the numerator and in the denominator.

Verify that $-\frac{3}{20}$ is the solution by checking.

Self Check 6 Solve: **a.** $\frac{7}{2}x = 21$ **b.** $-\frac{3b}{8} = \frac{11}{56}$

Now Try Problems 53 and 55

5 Use the Division Property of Equality.

Since any division can be rewritten as a multiplication by multiplying by the reciprocal, the following property is a natural extension of the multiplication property of equality.

Division Property of Equality

Dividing both sides of an equation by the same nonzero number does not change its solution.

For any real numbers a, b, and c, where c is not 0,

if $a = b$, then $\frac{a}{c} = \frac{b}{c}$

When we use this property, the resulting equation is equivalent to the original one.

EXAMPLE 7 Solve: **a.** $2t = 80$ **b.** $-6.02 = -8.6t$

Strategy We will use a property of equality to isolate the variable on one side of the equation.

Why To solve the original equation, we want to find a simpler equivalent equation of the form **t = a number** or **a number = t**, whose solution is obvious.

Solution

Success Tip

Since division by 2 is the same as multiplication by $\frac{1}{2}$, we also can solve $2t = 80$ using the multiplication property of equality. We can isolate t by multiplying both sides by the *multiplicative inverse* of 2, which is $\frac{1}{2}$:

$$\frac{1}{2} \cdot 2t = \frac{1}{2} \cdot 80$$

a. $2t = 80$ This is the equation to solve.

$\frac{2t}{\mathbf{2}} = \frac{80}{\mathbf{2}}$ To isolate t on the left side, use the division property of equality. Undo the multiplication by 2 by dividing both sides of the equation by 2.

$1t = 40$ Simplify $\frac{2t}{2}$ by removing the common factor of 2 in the numerator and denominator: $\frac{2}{2} = 1$

$t = 40$ The product of 1 and any number is that number: $1t = t$.

If we substitute 40 for t in $2t = 80$, we obtain the true statement $80 = 80$. This verifies that 40 is the solution. The solution set is $\{40\}$.

b. $-6.02 = -8.6t$ This is the equation to solve.

$\frac{-6.02}{\mathbf{-8.6}} = \frac{-8.6t}{\mathbf{-8.6}}$ To isolate t on the right side, use the division property of equality. Undo the multiplication by -8.6 by dividing both sides by -8.6.

$0.7 = t$ Do the division: $8.6\overline{)6.02}$. The quotient of two negative numbers is positive.

The solution is 0.7. Verify that this is correct by checking.

Success Tip

It is usually easier to multiply on each side if the coefficient of the variable term is a *fraction*, and divide on each side if the coefficient is an *integer* or *decimal*.

Self Check 7 Solve: **a.** $16x = 176$ **b.** $10.04 = -0.4r$

Now Try Problems 57 and 59

EXAMPLE 8 Solve: $-x = 3$

Strategy The variable x is not isolated, because there is a $-$ sign in front of it. Since the term $-x$ has an understood coefficient of -1, the equation can be written as $-1x = 3$. We need to select a property of equality and use it to isolate the variable on one side of the equation.

Why To find the solution of the original equation, we want to find a simpler equivalent equation of the form $\boldsymbol{x} = \textbf{a number}$, whose solution is obvious.

Solution To isolate x, we can either multiply or divide both sides by -1.

Multiply both sides by -1:

$-x = 3$ The equation to solve

$-1x = 3$ Write: $-x = -1x$

$(\mathbf{-1})(-1x) = (\mathbf{-1})3$

$1x = -3$

$x = -3$ $1x = x$

Divide both sides by -1:

$-x = 3$ The equation to solve

$-1x = 3$ Write: $-x = -1x$

$\frac{-1x}{\mathbf{-1}} = \frac{3}{\mathbf{-1}}$

$1x = -3$ On the left side, $\frac{-1}{-1} = 1$.

$x = -3$ $1x = x$

Either way, we get the same result, -3.

Check:

$-\boldsymbol{x} = 3$ This is the original equation.

$-(\mathbf{-3}) \stackrel{?}{=} 3$ Substitute -3 for x.

$3 = 3$ On the left side, the opposite of -3 is 3.

Since the resulting statement is true, -3 is the solution. The solution set is $\{-3\}$.

Self Check 8 Solve: $-h = -12$

Now Try Problem 61

SECTION 2.1 STUDY SET

VOCABULARY

Fill in the blanks.

1. A statement indicating that two expressions are equal, such as $x + 1 = 7$, is called an ________.
2. Any number that makes an equation true when substituted for the variable is said to ______ the equation. Such numbers are called ________.
3. To ______ an equation means to find all values of the variable that make the equation true.
4. To solve an equation, we ______ the variable on one side of the equation.
5. Equations with the same solutions are called _________ equations.
6. To ______ the solution of an equation, we substitute the value for the variable in the original equation and determine whether the result is a true statement.

CONCEPTS

7. Given $x + 6 = 12$,
   a. What is the left side of the equation?
   b. Is this equation true, false, or neither?
   c. Is 5 the solution?
   d. Does 6 satisfy the equation?
8. For each equation, determine what operation is performed on the variable. Then explain how to undo that operation to isolate the variable.
   a. $x - 8 = 24$
   b. $x + 8 = 24$
   c. $\frac{x}{8} = 24$
   d. $8x = 24$
9. Complete the following properties of equality. If $a = b$, then
   a. $a + c = b + \square$ and $a - c = b - \square$
   b. $ca = \square b$ and $\frac{a}{c} = \frac{b}{\square}$ (where $c \neq 0$)
10. a. To solve $\frac{h}{10} = 20$, do we multiply both sides of the equation by 10 or 20?
    b. To solve $4k = 16$, do we subtract 4 from both sides of the equation or divide both sides by 4?
11. Simplify each expression.
    a. $x + 7 - 7$
    b. $y - 2 + 2$
    c. $\frac{5t}{5}$
    d. $6 \cdot \frac{h}{6}$
12. a. To solve $-\frac{4}{5}x = 8$, we can multiply both sides by the reciprocal of $-\frac{4}{5}$. What is the reciprocal of $-\frac{4}{5}$?
    b. What is $-\frac{5}{4}\left(-\frac{4}{5}\right)$?

NOTATION

Complete each solution to solve the equation.

13.
$$\begin{aligned} x - 5 &= 45 \\ x - 5 + \square &= 45 + \square \\ x &= \square \end{aligned}$$
Check:
$$\begin{aligned} x - 5 &= 45 \\ \square - 5 &\stackrel{?}{=} 45 \\ \square &= 45 \quad \text{True} \end{aligned}$$
$\square$ is the solution.

14.
$$\begin{aligned} 8x &= 40 \\ \frac{8x}{\square} &= \frac{40}{\square} \\ x &= \square \end{aligned}$$
Check:
$$\begin{aligned} 8x &= 40 \\ 8(\square) &\stackrel{?}{=} 40 \\ \square &= 40 \quad \text{True} \end{aligned}$$
$\square$ is the solution.

15. a. What does the symbol $\stackrel{?}{=}$ mean?
    b. If you solve an equation and obtain $50 = x$, can you write $x = 50$?
16. Fill in the blank: $-x = \square x$

GUIDED PRACTICE

Check to determine whether the number in red is a solution of the equation. See Example 1.

17. **6**, $x + 12 = 28$
18. **110**, $x - 50 = 60$
19. **−8**, $2b + 3 = -15$
20. **−2**, $5t - 4 = -16$
21. **5**, $0.5x = 2.9$
22. **3.5**, $1.2 + x = 4.7$
23. **−6**, $33 - \frac{x}{2} = 30$
24. **−8**, $\frac{x}{4} + 98 = 100$
25. **−2**, $|c - 8| = 10$
26. **−45**, $|30 - r| = 15$
27. **12**, $3x - 2 = 4x - 5$
28. **5**, $5y + 8 = 3y - 2$
29. **−3**, $x^2 - x - 6 = 0$
30. **−2**, $y^2 + 5y - 3 = 0$
31. **1**, $\frac{2}{a + 1} + 5 = \frac{12}{a + 1}$
32. **4**, $\frac{2t}{t - 2} - \frac{4}{t - 2} = 1$
33. $\frac{3}{4}$, $x - \frac{1}{8} = \frac{5}{8}$
34. $\frac{7}{3}$, $-4 = a + \frac{5}{3}$
35. **−3**, $(x - 4)(x + 3) = 0$
36. **5**, $(2x + 1)(x - 5) = 0$

Use a property of equality to solve each equation. Then check the result. See Example 2.

37. $a - 5 = 66$
38. $x - 34 = 19$
39. $9 = p - 9$
40. $3 = j - 88$

Use a property of equality to solve each equation. Then check the result. See Example 3.

41. $-16 = y - 4$
42. $-23 = y - 19$
43. $-3 + a = 0$
44. $-1 + m = 0$

Use a property of equality to solve each equation. Then check the result. See Example 4.

45. $x + \frac{1}{10} = \frac{6}{5}$
46. $x + \frac{1}{14} = \frac{9}{7}$
47. $3.5 + f = 1.2$
48. $9.4 + h = 8.1$

Use a property of equality to solve each equation. Then check the result. See Example 5.

49. $\frac{x}{15} = 3$

50. $\frac{y}{7} = 12$

51. $\frac{d}{8} = -6$

52. $\frac{n}{9} = -2$

Use a property of equality to solve each equation. Then check the result. See Example 6.

53. $\frac{4}{5}t = 16$

54. $\frac{11}{15}y = 22$

55. $-\frac{7r}{2} = \frac{5}{12}$

56. $-\frac{4a}{5} = \frac{11}{15}$

Use a property of equality to solve each equation. Then check the result. See Example 7.

57. $4x = 16$

58. $5y = 45$

59. $-1.7 = -3.4y$

60. $-1.26 = -2.1x$

Use a property of equality to solve each equation. Then check the result. See Example 8.

61. $-x = 18$

62. $-y = 50$

63. $-n = \frac{4}{21}$

64. $-w = \frac{11}{16}$

TRY IT YOURSELF

Solve each equation. Then check the result.

65. $63 = 9c$

66. $40 = 5t$

67. $d - \frac{1}{9} = \frac{7}{9}$

68. $\frac{7}{15} = b - \frac{1}{15}$

69. $0 = \frac{v}{11}$

70. $\frac{d}{49} = 0$

71. $x - 1.6 = -2.5$

72. $y - 1.2 = -1.3$

73. $\frac{2}{3}c = 10$

74. $\frac{9}{7}d = 81$

75. $-100 = -5g$

76. $-80 = -5w$

77. $s + \frac{1}{5} = \frac{4}{25}$

78. $\frac{1}{6} = h + \frac{4}{3}$

79. $\frac{d}{-7} = -3$

80. $\frac{c}{-2} = -11$

81. $8h = 0$

82. $9a = 0$

83. $\frac{y}{0.6} = -4.4$

84. $\frac{y}{0.8} = -2.9$

85. $23b = 23$

86. $16 = 16h$

87. $-\frac{5}{4}h = -5$

88. $-\frac{3}{8}t = -3$

89. $8.9 = -4.1 + t$

90. $7.7 = -3.2 + s$

91. $-2.5 = -m$

92. $-1.8 = -b$

93. $-\frac{9}{8}x = 3$

94. $-\frac{14}{3}c = 7$

95. $\frac{2}{3}n = -\frac{7}{8}$

96. $\frac{4}{9}m = -\frac{3}{5}$

97. $-10 = n - 5$

98. $-8 = t - 2$

99. $\frac{h}{-40} = 5$

100. $\frac{x}{-7} = 12$

101. $-\frac{15}{16}a = -\frac{5}{4}$

102. $-\frac{20}{27}b = -\frac{4}{9}$

103. $-15x = -60$

104. $-14x = -84$

Look Alikes . . .

105. a. $d + \frac{1}{10} = \frac{3}{4}$ b. $d - \frac{1}{10} = \frac{3}{4}$

c. $\frac{1}{10}d = \frac{3}{4}$ d. $10d = \frac{3}{4}$

106. a. $x + 4.2 = -18.9$ b. $x - 4.2 = -18.9$

c. $4.2x = -18.9$ d. $\frac{x}{4.2} = -18.9$

APPLICATIONS

107. **Synthesizers.** To find the unknown angle measure, which is represented by x, solve the equation $x + 115 = 180$.

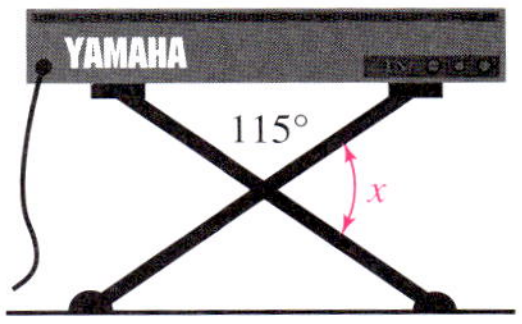

108. **Stop Signs.** To find the degree measure of one angle of the stop sign, which is represented by x, solve the equation $8x = 1{,}080$.

109. **Sharing the Winning Ticket.** When a Florida Lotto Jackpot was won by a group of 16 nurses employed at a Southwest Florida Medical Center, each received \$375,000. To find the amount of the jackpot, which is represented by x, solve the equation $\frac{x}{16} = 375{,}000$.

110. **Social Networks.** In 2009, the annual revenue generated per employee at Twitter was \$142,857. This is \$777,143 less than the annual revenue generated per employee at Facebook. To find the annual revenue generated per employee at Facebook, which is represented by x, solve the equation $x - 777{,}143 = 142{,}857$. (Source: businessinsider.com)

WRITING

111. What does it mean to solve an equation?

112. Explain the error in the following work.

Solve: $-6.4x = 1.1$

$-6.4x + 6.4 = 1.1 + 6.4$

$x = 7.5$

113. Explain the error in the following work.

Solve: $x + 2 = 40$

$x + 2 - 2 = 40$

$x = 40$

114. After solving an equation, how do we check the result?

REVIEW

115. Evaluate $-9 - 3x$ for $x = -3$.

116. Evaluate: $-5^2 + (-5)^2$

117. Translate to symbols: Subtract x from 45

118. Evaluate: $\dfrac{2^3 + 3(5 - 3)}{15 - 4 \cdot 2}$

CHALLENGE PROBLEMS

119. If $a + 81 = 49$, what is $a - 81$?

120. Find two solutions of $\frac{1}{4}|x + 1| = 25$.

SECTION 2.2 More about Solving Equations

OBJECTIVES

1. Use more than one property of equality to solve equations.
2. Simplify expressions to solve equations.
3. Clear equations of fractions and decimals.
4. Identify identities and contradictions.

ARE YOU READY?

The following problems review some basic skills that are needed when solving equations.

1. Simplify: $4x - 12 - 4x$

2. Simplify: $2a + 2 - 2$

3. Simplify: $5m - 3(4m - 6)$

4. Multiply: $5\left(\frac{3}{5}x\right)$

5. Multiply: $18\left(\frac{4}{3}n\right)$

6. Multiply: $100 \cdot 0.08$

We have solved simple equations by using properties of equality. We now will expand our equation-solving skills by considering more complicated equations. We want to develop a general strategy that can be used to solve any kind of *linear equation in one variable.*

Linear Equation in One Variable

A **linear equation in one variable** can be written in the form

$$ax + b = c$$

where a, b, and c are real numbers and $a \neq 0$. Read $\neq$ as "is not equal to 0."

Some examples of linear and nonlinear equations in one variable are shown below.

Linear equations in one variable (x):

$3x + 1 = 4$ Think: $3x^1 + 1 = 4$

$-\frac{5}{3}x - 7 = 0$ Think: $-\frac{5}{3}x^1 - 7 = 0$

Not linear equations in one variable:

$x^2 - x - 6 = 0$ The exponent on x is not 1.

$\frac{1}{2} + \frac{3}{x} = \frac{5}{6}$ x is in the denominator.

1 Use More Than One Property of Equality to Solve Equations.

Sometimes we must use several properties of equality to solve an equation. For example, on the left side of $2x + 6 = 10$, the variable x is multiplied by 2, and then 6 is added to that product. To isolate x, we use the order of operations rule in reverse. First, we undo the addition of 6, and then we undo the multiplication by 2.

$2x + 6 = 10$ This is the equation to solve.

$2x + 6 - 6 = 10 - 6$ To undo the addition of 6, subtract 6 from both sides.

$2x = 4$ Do the subtraction.

$\frac{2x}{2} = \frac{4}{2}$ To undo the multiplication by 2, divide both sides by 2.

$x = 2$ Do the division.

The solution is 2.

Success Tip

Recall that:

- Subtraction undoes addition.
- Addition undoes subtraction.
- Division undoes multiplication.
- Multiplication undoes division.

EXAMPLE 1 Solve: $-12x + 5 = 17$

Strategy First, we will use a property of equality to isolate the *variable term* on one side of the equation. Then we will use a second property of equality to isolate the *variable* itself.

Why To solve the original equation, we want to find a simpler equivalent equation of the form **x = a number**, whose solution is obvious.

Solution On the left side of the equation, x is multiplied by -12, and then 5 is added to that product. To isolate x, we undo the operations in the opposite order.

The Language of Algebra

We subtract 5 from both sides to isolate the **variable term**, $-12x$. Then we divide both sides by -12 to isolate the **variable**, x.

- To isolate the variable term, $-12x$, we subtract 5 from both sides to undo the addition of 5.
- To isolate the variable, x, we divide both sides by -12 to undo the multiplication by -12.

$$-12x + 5 = 17$$ This is the equation to solve.

$$-12x + 5 - 5 = 17 - 5$$ Use the subtraction property of equality: Subtract 5 from both sides to undo the addition and isolate the variable term, $-12x$.

$$-12x = 12$$ Do the subtractions: $5 - 5 = 0$ and $17 - 5 = 12$.

$$\frac{-12x}{-12} = \frac{12}{-12}$$ Use the division property of equality: Divide both sides by -12 to undo the multiplication and isolate the variable x.

$$x = -1$$ Do the division.

Caution

When checking solutions, always use the original equation.

Check:

$$-12x + 5 = 17$$ This is the original equation.

$$-12(-1) + 5 \stackrel{?}{=} 17$$ Substitute -1 for x.

$$12 + 5 \stackrel{?}{=} 17$$ Do the multiplication on the left side.

$$17 = 17$$ True

Since the resulting statement is true, the solution is -1. The solution set is $\{-1\}$.

Self Check 1 Solve: $8x - 13 = 43$

Now Try Problem 13

EXAMPLE 2 Solve: $\frac{5}{8}m - 2 = -12$

Strategy We will use properties of equality to isolate the variable on one side of the equation.

Why To solve the original equation, we want to find a simpler equivalent equation of the form **m = a number**, whose solution is obvious.

Solution We note that the coefficient of m is $\frac{5}{8}$ and proceed as follows.

- To isolate the variable term, $\frac{5}{8}m$, we add 2 to both sides to undo the subtraction of 2.
- To isolate the variable, m, we multiply both sides by $\frac{8}{5}$ to undo the multiplication by $\frac{5}{8}$.

$$\frac{5}{8}m - 2 = -12$$ This is the equation to solve.

$$\frac{5}{8}m - 2 + 2 = -12 + 2$$ Use the addition property of equality: Add 2 to both sides to undo the subtraction and isolate the variable term, $\frac{5}{8}m$.

$$\frac{5}{8}m = -10$$ Do the additions: $-2 + 2 = 0$ and $-12 + 2 = -10$.

$$\frac{8}{5}\left(\frac{5}{8}m\right) = \frac{8}{5}(-10)$$ Use the multiplication property of equality: Multiply both sides by $\frac{8}{5}$ (which is the reciprocal of $\frac{5}{8}$) to isolate the variable, m.

$$m = -16$$ On the left side: $\frac{8}{5}\left(\frac{5}{8}\right) = 1$ and $1m = m$. On the right side, multiply: $\frac{8}{5}(-10) = -\frac{8 \cdot 2 \cdot \cancel{5}}{\cancel{5}} = -16$.

The solution is -16. Verify this by substituting -16 into the original equation. The solution set is $\{-16\}$.

Self Check 2 Solve: $\frac{7}{12}a - 6 = -27$

Now Try Problem 17

EXAMPLE 3 Solve: $-0.2 = -0.8 - y$

Strategy First, we will use a property of equality to isolate the variable term on one side of the equation. Then we will use a second property of equality to isolate the variable itself.

Why To solve the original equation, we want to find a simpler equivalent equation of the form **a number = y**, whose solution is obvious.

Solution

$$-0.2 = -0.8 - y$$ This is the equation to solve.

$$-0.2 + 0.8 = -0.8 - y + 0.8$$ To isolate the variable term $-y$ on the right side, we eliminate -0.8 by adding 0.8 to both sides.

$$0.6 = -y$$ Do the addition.

Since the term $-y$ has an understood coefficient of -1, the equation can be written as $0.6 = -1y$. To isolate y, we can either multiply both sides or divide both sides by -1.

$$0.6 = -1y$$ If it is helpful, write $-y$ as $-1y$.

$$\frac{0.6}{-1} = \frac{-1y}{-1}$$ To isolate y, undo the multiplication by -1 by dividing both sides by -1.

$$-0.6 = y$$

The solution is -0.6. Verify this by substituting -0.6 into the original equation.

Success Tip

We also can multiply both sides of the equation by -1 to isolate y:

$$-1(0.6) = -1(-1y)$$
$$-0.6 = y$$

Self Check 3 Solve: $-6.6 - m = -2.7$

Now Try Problem 23

2 Simplify Expressions to Solve Equations.

When solving equations, we should simplify the expressions that make up the left and right sides before applying any properties of equality. Often, that involves using the distributive property to remove parentheses and/or combining like terms.

EXAMPLE 4 Solve: **a.** $3(k + 1) - 5k = 0$ **b.** $10a - 2(2a - 7) = 68$

Strategy We will use the distributive property along with the process of combining like terms to simplify the left side of each equation.

Why It's best to simplify each side of an equation before using a property of equality.

Solution

a.

$3(k + 1) - 5k = 0$	This is the equation to solve.
$3k + 3 - 5k = 0$	Distribute the multiplication by 3.
$-2k + 3 = 0$	Combine like terms: $3k - 5k = -2k$.
$-2k + 3 - 3 = 0 - 3$	To isolate the variable term $-2k$, undo the addition of 3 by subtracting 3 from both sides.
$-2k = -3$	Do the subtraction: $3 - 3 = 0$ and $0 - 3 = -3$.
$\frac{-2k}{-2} = \frac{-3}{-2}$	To isolate the variable k, undo the multiplication by -2 by dividing both sides by -2.
$k = \frac{3}{2}$	Simplify the fraction: $\frac{-3}{-2} = \frac{3}{2}$.

Success Tip

To check a result, evaluate each side of the equation following the order of operations rule.

Check:

$3(k + 1) - 5k = 0$	This is the original equation.
$3\left(\frac{3}{2} + 1\right) - 5\left(\frac{3}{2}\right) \stackrel{?}{=} 0$	Substitute $\frac{3}{2}$ for k.
$3\left(\frac{5}{2}\right) - 5\left(\frac{3}{2}\right) \stackrel{?}{=} 0$	Do the addition within the parentheses. Think of 1 as $\frac{2}{2}$ and then add: $\frac{3}{2} + \frac{2}{2} = \frac{5}{2}$.
$\frac{15}{2} - \frac{15}{2} \stackrel{?}{=} 0$	Do the multiplication.
$0 = 0$	True

The solution is $\frac{3}{2}$ and the solution set is $\left\{\frac{3}{2}\right\}$.

b.

$10a - 2(2a - 7) = 68$	This is the equation to solve.
$10a - 4a + 14 = 68$	Distribute the multiplication by -2.
$6a + 14 = 68$	Combine like terms: $10a - 4a = 6a$.
$6a + 14 - 14 = 68 - 14$	To isolate the variable term $6a$, undo the addition of 14 by subtracting 14 from both sides.
$6a = 54$	Do the subtraction.
$\frac{6a}{6} = \frac{54}{6}$	To isolate the variable a, undo the multiplication by 6 by dividing both sides by 6.
$a = 9$	Do the division.

Use a check to verify that 9 is the solution.

Self Check 4 Solve: **a.** $4(a + 2) - a = 11$ **b.** $19x - 5(3x - 9) = 1$

Now Try Problems 25 and 27

When solving an equation, if variables appear on both sides, we can use the addition (or subtraction) property of equality to get all variable terms on one side and all constant terms on the other.

EXAMPLE 5 Solve: $3x - 15 = 4x + 36$

Strategy There are variable terms ($3x$ and $4x$) on both sides of the equation. We will eliminate $3x$ from the left side of the equation by subtracting $3x$ from both sides.

Why To solve for x, all the terms containing x must be on the same side of the equation.

Solution

$3x - 15 = 4x + 36$	This is the equation to solve.
$3x - 15 - 3x = 4x + 36 - 3x$	Subtract 3x from both sides to isolate the variable term on the right side.
$-15 = x + 36$	Combine like terms: $3x - 3x = 0$ and $4x - 3x = x$.
$-15 - 36 = x + 36 - 36$	To isolate x, undo the addition of 36 by subtracting 36 from both sides.
$-51 = x$	Do the subtraction.

Check:

$3x - 15 = 4x + 36$	The original equation.
$3(-51) - 15 \stackrel{?}{=} 4(-51) + 36$	Substitute -51 for x.
$-153 - 15 \stackrel{?}{=} -204 + 36$	Do the multiplication.
$-168 = -168$	True

The solution is -51 and the solution set is $\{-51\}$.

Success Tip

We could have eliminated $4x$ from the right side by subtracting $4x$ from both sides. The same solution results.

$$3x - 15 - 4x = 4x + 36 - 4x$$
$$-x - 15 = 36$$

However, it is usually easier to isolate the variable term on the side that will result in a *positive* coefficient.

Self Check 5 Solve: $30 + 6n = 4n - 2$

Now Try Problem 29

3 Clear Equations of Fractions and Decimals.

Equations are usually easier to solve if they don't involve fractions. We can use the multiplication property of equality to clear an equation of fractions by multiplying both sides of the equation by the least common denominator of all the fractions that appear in the equation.

EXAMPLE 6 Solve: $\frac{1}{6}x + \frac{5}{2} = \frac{1}{3}$

Strategy To clear the equation of fractions, we will multiply both sides by the LCD of all the fractions in the equation.

Why It's easier to solve an equation that involves only integers.

Solution

$\frac{1}{6}x + \frac{5}{2} = \frac{1}{3}$	This is the equation to solve.
$6\left(\frac{1}{6}x + \frac{5}{2}\right) = 6\left(\frac{1}{3}\right)$	Multiply both sides by the LCD of $\frac{1}{6}$, $\frac{5}{2}$, and $\frac{1}{3}$, which is 6. Don't forget the parentheses on the left side.
$6\left(\frac{1}{6}x\right) + 6\left(\frac{5}{2}\right) = 6\left(\frac{1}{3}\right)$	On the left side, distribute the multiplication by 6.
$x + 15 = 2$	Do each multiplication: $6\left(\frac{1}{6}\right) = 1$, $6\left(\frac{5}{2}\right) = \frac{30}{2} = 15$, and $6\left(\frac{1}{3}\right) = \frac{6}{3} = 2$. The fractions have been cleared.
$x + 15 - 15 = 2 - 15$	To undo the addition of 15, subtract 15 from both sides.
$x = -13$	

Check the solution by substituting -13 for x in $\frac{1}{6}x + \frac{5}{2} = \frac{1}{3}$.

Success Tip

Some students find it helpful to perform each fraction multiplication as shown below:

$$\overset{1}{\cancel{6}}\left(\frac{1}{\underset{1}{\cancel{6}}}x\right) + \overset{3}{\cancel{6}}\left(\frac{5}{\underset{1}{\cancel{2}}}\right) = \overset{2}{\cancel{6}}\left(\frac{1}{\underset{1}{\cancel{3}}}\right)$$

Self Check 6 Solve: $\frac{1}{4}x + \frac{1}{2} = -\frac{1}{8}$

Now Try Problem 33

If an equation contains decimals, it is often convenient to multiply both sides by a power of 10 to change the decimals in the equation to integers.

EXAMPLE 7 Solve: $0.04(12) + 0.01x = 0.02(12 + x)$

Strategy To clear the equation of decimals, we will multiply both sides by a carefully chosen power of 10.

Why It's easier to solve an equation that involves only integers.

Solution The equation contains the decimals 0.04, 0.01, and 0.02. Since the greatest number of decimal places in any one of these numbers is two, we multiply both sides of the equation by 10^2 or 100. This changes 0.04 to 4, and 0.01 to 1, and 0.02 to 2.

$$0.04(12) + 0.01x = 0.02(12 + x)$$ This is the equation to solve.

$$100[0.04(12) + 0.01x] = 100 \cdot 0.02(12 + x)$$ Multiply both sides by 100. Don't forget the brackets.

$$100 \cdot 0.04(12) + 100 \cdot 0.01x = 100 \cdot 0.02(12 + x)$$ Distribute.

$$4(12) + 1x = 2(12 + x)$$ Multiply each decimal by 100 by moving its decimal point 2 places to the right.

$$48 + x = 24 + 2x$$ Distribute the multiplication by 2.

$$48 + x - 24 - x = 24 + 2x - 24 - x$$ Subtract 24 and x from both sides.

$$24 = x$$ Simplify each side.

$$x = 24$$

The solution is 24. Check by substituting 24 for x in the original equation.

Self Check 7 Solve: $0.08x + 0.07(15{,}000 - x) = 1{,}110$

Now Try Problem 37

Success Tip

Recall that multiplying a decimal by 10 moves the decimal point 1 place to the right, multiplying it by 100 moves it 2 places to the right, and so on.

Success Tip

When we write the decimals in the original equation as fractions, it becomes more apparent why it is helpful to multiply both sides by the LCD, 100.

$$\frac{4}{100}(12) + \frac{1}{100}x = \frac{2}{100}(12 + x)$$

The previous examples suggest the following strategy for solving equations. It is important to note that not every step is needed to solve every equation.

Strategy for Solving Linear Equations in One Variable

1. **Clear the equation of fractions or decimals:** Multiply both sides by the LCD to clear fractions or multiply both sides by a power of 10 to clear decimals.
2. **Simplify each side of the equation:** Use the distributive property to remove parentheses, and then combine like terms on each side.
3. **Isolate the variable term on one side:** Add (or subtract) to get the variable term on one side of the equation and a number on the other using the addition (or subtraction) property of equality.
4. **Isolate the variable:** Multiply (or divide) to isolate the variable using the multiplication (or division) property of equality.
5. **Check the result:** Substitute the possible solution for the variable in the *original* equation to see if a true statement results.

EXAMPLE 8 Solve: $\frac{7m + 5}{5} = -4m + 1$

Strategy We will follow the steps of the equation-solving strategy to solve the equation.

Why This is the most efficient way to solve a linear equation in one variable.

$$\frac{7m + 5}{5} = -4m + 1$$ This is the equation to solve.

Step 1 $$5\left(\frac{7m + 5}{5}\right) = 5(-4m + 1)$$ Clear the equation of the fraction by multiplying both sides by 5.

Step 2 $$7m + 5 = -20m + 5$$ On the left side, remove the common factor 5 in the numerator and denominator. On the right side, distribute the multiplication by 5.

Step 3 $$7m + 5 + 20m = -20m + 5 + 20m$$ To eliminate the term $-20m$ on the right side, add $20m$ to both sides.

$$27m + 5 = 5$$ Combine like terms: $7m + 20m = 27m$ and $-20m + 20m = 0$.

$$27m + 5 - 5 = 5 - 5$$ To isolate the term $27m$, undo the addition of 5 by subtracting 5 from both sides.

$$27m = 0$$ Do the subtraction.

Step 4 $$\frac{27m}{27} = \frac{0}{27}$$ To isolate m, undo the multiplication by 27 by dividing both sides by 27.

$$m = 0$$ 0 divided by any nonzero number is 0.

Step 5 Substitute 0 for m in $\frac{7m + 5}{5} = -4m + 1$ to check that the solution is 0.

Self Check 8 Solve: $6c + 2 = \frac{18 - c}{9}$

Now Try Problem 41

Success Tip

We can remove the common factor 5 from the numerator and the denominator in the following way:

$$\frac{\overset{1}{\cancel{5}}}{1}\left(\frac{7m + 5}{\underset{1}{\cancel{5}}}\right)$$

Caution

Remember that when you multiply one side of an equation by a nonzero number, you must multiply the other side of the equation by the same number.

4 Identify Identities and Contradictions.

Each of the equations that we solved in Examples 1 through 8 had exactly one solution. However, not every linear equation in one variable has a single solution. Some equations are made true by *any* permissible replacement value for the variable. Such equations are called **identities.** An example of an **identity** is

$$x + x = 2x$$ If we substitute -10 for x, we get the true statement $-20 = -20$. If we substitute 7 for x, we get $14 = 14$, and so on.

Since we can replace x with any number and the equation will be true, all real numbers are solutions of $x + x = 2x$. This equation has infinitely many solutions.

Another type of equation, called a **contradiction,** is false for all replacement values for the variable. An example is

$$x = x + 1$$ No number is equal to 1 more than itself.

Since this equation is false for any value of x, it has no solution.

EXAMPLE 9 Solve: $3(x + 8) + 5x = 2(12 + 4x)$

Strategy We will follow the steps of the equation-solving strategy to solve the equation.

Why This is the most efficient way to solve a linear equation in one variable.

Solution

$3(x + 8) + 5x = 2(12 + 4x)$	This is the equation to solve.
$3x + 24 + 5x = 24 + 8x$	Distribute the multiplication by 3 and by 2.
$8x + 24 = 24 + 8x$	Combine like terms: $3x + 5x = 8x$. Note that the sides of the equation are identical.
$8x + 24 - 8x = 24 + 8x - 8x$	To eliminate the term $8x$ on the right side, subtract $8x$ from both sides.
$24 = 24$	Combine like terms on both sides: $8x - 8x = 0$.

In this case, the terms involving x drop out and the result is true. This means that any number substituted for x in the original equation will give a true statement. Therefore, *all real numbers* are solutions and this equation is an identity. Its solution set is written as {all real numbers} or using the symbol $\mathbb{R}$.

Success Tip

At the step $8x - 24 = 24 + 8x$, we know that the equation is an identity because both sides are exactly the same.

Self Check 9 Solve: $3(x + 5) - 4(x + 4) = -x - 1$

Now Try Problem 45

EXAMPLE 10 Solve: $3(d + 7) - d = 2(d + 10)$

Strategy We will follow the steps of the equation-solving strategy to solve the equation.

Why This is the most efficient way to solve a linear equation in one variable.

Solution

$3(d + 7) - d = 2(d + 10)$	This is the equation to solve.
$3d + 21 - d = 2d + 20$	Distribute the multiplication by 3 and by 2.
$2d + 21 = 2d + 20$	Combine like terms: $3d - d = 2d$.
$2d + 21 - 2d = 2d + 20 - 2d$	To eliminate the term $2d$ on the right side, subtract $2d$ from both sides.
$21 = 20$	Combine like terms on both sides: $2d - 2d = 0$.

In this case, the terms involving d drop out and the result is false. This means that any number that is substituted for d in the original equation will give a false statement. Therefore, this equation has *no solution* and it is a contradiction. Its solution set is the **empty set**, which is written as { } or using the symbol $\varnothing$.

The Language of Algebra

Contradiction is a form of the word *contradict*, meaning conflicting ideas. During a trial, evidence might be introduced that *contradicts* the testimony of a witness.

Self Check 10 Solve: $-4(c - 3) + 2c = 2(10 - c)$

Now Try Problem 47

SECTION 2.2 STUDY SET

VOCABULARY

Fill in the blanks.

1. $3x + 8 = 10$ is an example of a linear ________ in one variable.
2. To solve $\frac{s}{3} + \frac{1}{4} = -\frac{1}{2}$, we can ______ the equation of the fractions by multiplying both sides by 12.
3. A linear equation that is true for any permissible replacement value for the variable is called an ________.
4. A linear equation that is false for all replacement values for the variable is called a ____________.

CONCEPTS

Fill in the blanks.

5. a. To solve $3x - 5 = 1$, we first undo the __________ of 5 by adding 5 to both sides. Then we undo the ____________ by 3 by dividing both sides by 3.
   b. To solve $\frac{x}{2} + 3 = 5$, we can undo the ________ of 3 by subtracting 3 from both sides. Then we can undo the ________ by 2 by multiplying both sides by 2.
6. a. Combine like terms on the left side of $6x - 8 - 8x = -24$.
   b. Distribute and then combine like terms on the right side of $-20 = 4(3x - 4) - 9x$.
7. Use a check to determine whether -2 is a solution of $6x + 5 = 7$.
8. Multiply.
   a. $20\left(\frac{3}{5}x\right)$
   b. $100 \cdot 0.02x$
9. a. By what must you multiply both sides of $\frac{2}{3} - \frac{1}{2}b = -\frac{4}{3}$ to clear it of fractions?
   b. By what must you multiply both sides of $0.7x + 0.3(x - 1) = 0.5x$ to clear it of decimals?

Look Alikes . . .

10. a. Simplify: $3x + 5 - x$
    b. Solve: $3x + 5 = 9$
    c. Evaluate $3x + 5 - x$ for $x = 9$
    d. Check: Is -1 a solution of $3x + 5 - x = 9$?

NOTATION

Complete the solution.

11. Solve:

$$2x - 7 = 21$$
$$2x - 7 + \quad = 21 + \quad$$
$$2x = 28$$
$$\frac{2x}{\quad} = \frac{28}{\quad}$$
$$x = 14$$

Check:

$$2x - 7 = 21$$
$$2(\quad) - 7 \quad 21$$
$$\quad - 7 \stackrel{?}{=} 21$$
$$\quad = 21$$

is the solution.

12. A student multiplied both sides of $\frac{3}{4}t + \frac{5}{8} = \frac{1}{2}t$ by 8 to clear it of fractions, as shown below. Explain his error in showing this step.

$$\cancel{8 \cdot \frac{3}{4}t + \frac{5}{8} = 8 \cdot \frac{1}{2}t}$$

GUIDED PRACTICE

Solve each equation and check the result. See Example 1.

13. $-8x + 1 = 73$
14. $-7y + 4 = 60$
15. $-5q - 2 = 23$
16. $-4p + 3 = 43$

Solve each equation and check the result. See Example 2.

17. $\frac{5}{6}k - 5 = 10$
18. $\frac{2}{5}c - 12 = 2$
19. $-\frac{7}{16}h + 28 = 21$
20. $-\frac{5}{8}h + 25 = 15$

Solve each equation and check the result. See Example 3.

21. $-6 - y = -2$
22. $-1 - h = -9$
23. $-1.7 = 1.2 - x$
24. $0.6 = 4.1 - x$

Solve each equation and check the result. See Example 4.

25. $3(2y - 2) - y = 5$
26. $2(-3a + 2) + a = 2$
27. $6a - 3(3a - 4) = 30$
28. $16y - 8(3y - 2) = -24$

Solve each equation and check the result. See Example 5.

29. $7a - 12 = 8a + 9$
30. $10m - 14 = 11m + 13$
31. $60r - 50 = 15r - 5$
32. $100f - 75 = 50f + 75$

Solve each equation and check the result. See Example 6.

33. $\frac{5}{6}x + \frac{2}{9} = \frac{1}{3}$
34. $\frac{2}{3}x + \frac{2}{3} = \frac{3}{4}$
35. $\frac{1}{8}y - \frac{1}{2} = \frac{1}{4}$
36. $\frac{1}{15}x - \frac{4}{5} = \frac{2}{3}$

Solve each equation and check the result. See Example 7.

37. $0.02(62) - 0.08s = 0.06(s + 9)$
38. $0.04(50) + 0.16x = 0.08(x + 50)$
39. $0.09(t + 50) + 0.15t = 52.5$
40. $0.08(x - 100) = 44.5 - 0.07x$

Solve each equation and check the result. See Example 8.

41. $\frac{10 - 5s}{3} = -s + 6$
42. $\frac{40 - 8s}{5} = -2s + 8$
43. $t = \frac{7t - 9}{16}$
44. $-3 = \frac{11r + 68}{3}$

Solve each equation, if possible. See Examples 9 and 10.

45. $8x + 3(2 - x) = 5x + 6$
46. $5(x + 2) = 5x - 2$
47. $-3(s + 2) = -2(s + 4) - s$
48. $21(b - 1) + 3 = 3(7b - 6)$

TRY IT YOURSELF

Solve each equation, if possible. Check the result.

49. $3x - 8 - 4x - 7x = -2 - 8$
50. $-6t - 7t - 5t - 1 = 12 - 3$
51. $\frac{t}{3} + 2 = 6$
52. $\frac{x}{5} - 5 = -12$
53. $4(5b) + 2(6b - 1) = -34$
54. $9(x + 11) + 5(13 - x) = 0$
55. $2x + 5 = 17$
56. $3x - 5 = 13$
57. $\frac{5}{6}(1 - x) = -x + 1$
58. $\frac{3}{8}(14 - u) = -3u + 6$
59. $0.05a + 0.01(90) = 0.02(a + 90)$
60. $0.03x + 0.05(2{,}000 - x) = 99.5$
61. $\frac{7}{2} + \frac{3}{2}d = -9 + 1.5d$
62. $x + 7 = \frac{2x + 6}{2} + 4$
63. $-(19 - 3s) - (8s + 1) = 35$
64. $2(3x) - 5(3x + 1) = 58$
65. $5x = 4x + 7$
66. $3x = 2x + 2$
67. $\frac{3(b + 2)}{2} = \frac{4b - 10}{4}$
68. $\frac{2(5a - 7)}{4} = \frac{9(a - 1)}{3}$
69. $8y - 2 = 4y + 16$
70. $7 + 3w = 4 + 9w$
71. $\frac{1}{6}y + \frac{1}{4}y = -1$
72. $\frac{1}{3}x + \frac{1}{4}x = -2$
73. $0.7 - 4y = 1.7$
74. $0.3 - 2x = -0.9$
75. $-33 = 5t + 2$
76. $-55 = 3w + 5$
77. $-3p + 7 = -3$
78. $-2r + 8 = -1$
79. $2(-3) + 4y = 14$
80. $4(-1) + 3y = 8$
81. $0.06(a + 200) + 0.1a = 172$
82. $0.03x + 0.05(6{,}000 - x) = 280$
83. $8.6y + 3.4 = 4.2y - 9.8$
84. $9.1y + 3.6 = 6.5y - 30.2$
85. $\frac{2}{3}y + 2 = \frac{1}{5} + y$
86. $\frac{2}{5}x + 1 = \frac{1}{3} + x$
87. $0.4b - 0.1(b - 100) = 70$
88. $0.105x + 0.06(20{,}000 - x) = 1{,}740$
89. $\frac{1}{4}(10 - 2y) = 8$
90. $\frac{1}{3}(7 - 7x) = -21$
91. $2 - 3(x - 5) = 4(x - 1)$
92. $2 - (4x + 7) = 3 + 2(x + 2)$
93. $2n - \frac{3}{4}n = \frac{1}{2}n + \frac{13}{3}$
94. $\frac{5}{6}n + 3n = -\frac{1}{3}n - \frac{11}{9}$
95. $10.08 = 4(0.5x + 2.5)$
96. $-3.28 = 8(1.5y - 0.5)$
97. $\frac{3}{4}(d - 8) = \frac{2}{3}(d + 1)$
98. $\frac{3}{2}(c - 2) = \frac{2}{5}(2c + 3)$
99. $2d + 5 = 0$
100. $3c + 8 = 0$
101. $3(A + 2) = 2(A - 7)$
102. $9(T - 1) = 6(T + 2) - T$
103. $4(a - 3) = -2(a - 6) + 6a$
104. $9(t + 2) = -6(t - 3) + 15t$
105. $4(y - 3) - y = 3(y - 4)$
106. $5(x + 3) - 3x = 2(x + 8)$
107. $-(4 - m) = -10$
108. $-(6 - t) = -12$
109. $-\frac{2}{3}z + 4 = 8$
110. $-\frac{7}{5}x + 9 = -5$

Look Alikes . . .

Simplify each expression and solve each equation.

111. a. $-2(9 - 3x) - (5x + 2)$
   b. $-2(9 - 3x) - (5x + 2) = -25$
112. a. $4(x - 5) - 3(12 - x)$
   b. $4(x - 5) - 3(12 - x) = 7$
113. a. $0.6 - 0.2(x + 1)$
   b. $0.6 - 0.2(x + 1) = 0.4$
114. a. $2(6n + 4) + 4(3n + 2) + 2$
   b. $2(6n + 5) = 4(3n + 2) + 2$

WRITING

115. To solve $3x - 4 = 5x + 1$, one student began by subtracting $3x$ from both sides. Another student solved the same equation by first subtracting $5x$ from both sides. Will the students get the same solution? Explain why or why not.
116. What does it mean to clear an equation such as $\frac{1}{4} + \frac{1}{2}x = \frac{3}{8}$ of the fractions?
117. Explain the error in the following solution.

Solve: $2x + 4 = 30$

$$\frac{2x}{2} + 4 = \frac{30}{2}$$
$$x + 4 = 15$$
$$x + 4 - 4 = 15 - 4$$
$$x = 11$$

118. a. Write an equation that is an identity. Explain why every real number is a solution.
   b. Write an equation that is a contradiction. Explain why no real number is a solution.

REVIEW

Name the property that is used.

119. $x \cdot 9 = 9x$
120. $4 \cdot \frac{1}{4} = 1$
121. $(x + 1) + 2 = x + (1 + 2)$
122. $2(30y) = (2 \cdot 30)y$

CHALLENGE PROBLEMS

123. Solve: $\frac{5}{6}\left(-\frac{3}{4}m + 1\right) = -\frac{2}{3}\left(\frac{1}{2}m - 1\right)$
124. In this section, we discussed equations that have no solution, one solution, and an infinite number of solutions. Do you think an equation could have exactly two solutions? If so, give an example.

SECTION 2.3 Applications of Percent

OBJECTIVES

1. Change percents to decimals and decimals to percents.
2. Solve percent problems by direct translation.
3. Solve applied percent problems.
4. Find percent of increase and decrease.
5. Solve discount and commission problems.

ARE YOU READY?

The following problems review some basic skills that are needed to solve percent problems.

1. Multiply: $100 \cdot 0.61$
2. Multiply: $100 \cdot 0.02$
3. Write $\frac{1}{2}$ and $\frac{3}{4}$ as decimals.
4. Divide: $\frac{27}{100}$
5. Multiply: $124 \cdot 0.03$
6. Divide: $\frac{17.82}{0.36}$

In this section, we will use translation skills from Chapter 1 and equation-solving skills from Chapter 2 to solve problems involving percents.

1 Change Percents to Decimals and Decimals to Percents.

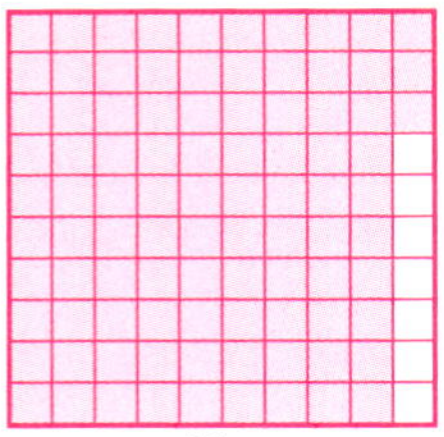

93% or $\frac{93}{100}$ or 0.93 of the figure is shaded.

The word **percent** means parts per one hundred. We can think of the percent symbol % as representing a denominator of 100. Thus, $93\% = \frac{93}{100}$. Since the fraction $\frac{93}{100}$ is equal to the decimal 0.93, it is also true that $93\% = 0.93$. In general, $n\% = \frac{n}{100}$.

When solving percent problems, we must often convert percents to decimals and decimals to percents. To change a percent to a decimal, we drop the % symbol and *divide the given number by 100 by moving the decimal point 2 places to the left.* For example,

$$31\% = 31.0\% = 0.31$$

To change a decimal to a percent, we *multiply the decimal by 100 by moving the decimal point 2 places to the right, and then inserting a % symbol.* For example,

$$0.678 = 67.8\%$$

2 Solve Percent Problems by Direct Translation.

There are three basic types of percent problems. Examples of these are:

Type 1 What number is 8% of 215?

Type 2 102 is 21.3% of what number?

Type 3 31 is what percent of 500?

Every percent problem has three parts: the *amount,* the *percent,* and the *base.* For example, in the question *What number is 8% of 215?,* the words "what number" represent the **amount,** 8% represents the **percent,** and 215 represents the **base.** In these problems, the word "is" means "is equal to," and the word "of" means "multiplication."

What number	is	8%	of	215?
↓	↓	↓	↓	↓
Amount	=	**Percent**	·	**Base**

EXAMPLE 1 What number is 8% of 215?

Strategy We will translate the words of this problem into an equation and then solve the equation.

Why The variable in the translation equation represents the unknown number that we are asked to find.

Solution

The Language of Algebra

Translate the word
- **is** to an equal symbol =
- **of** to multiplication
- **what** to a variable

In this problem, the phrase "what number" represents the amount, 8% is the percent, and 215 is the base.

What number	is	8%	of	215?	
↓	↓	↓	↓	↓	
x	=	0.08	·	215	Change the percent to a decimal: 8% = 0.08.
x	=	17.2			Do the multiplication.

Thus, 8% of 215 is 17.2.

To check, we note that 17.2 out of 215 is $\frac{17.2}{215} = 0.08 = 8\%$.

Self Check 1 What number is 5.6% of 40?

Now Try Problem 13

We will illustrate the other two types of percent problems with application problems.

3 Solve Applied Percent Problems.

One method for solving applied percent problems is to use the given facts to write a **percent sentence** of the form

______ is ______ % of ______ ?

We enter the appropriate numbers in two of the blanks and the words "what number" or "what percent" in the remaining blank. As before, we translate the words into an equation and solve it.

EXAMPLE 2 **Aging Populations.** By the year 2050, a study by the Pew Research Center predicts that about 81 million residents of the U.S. will be age 65 or older. The **circle graph** (or **pie chart**) indicates that age group will make up 18.5% of the population. If this prediction is correct, what will the population of the United States be in 2050? (Round to the nearest million.)

Projection of the 2050 U.S. Population by Age

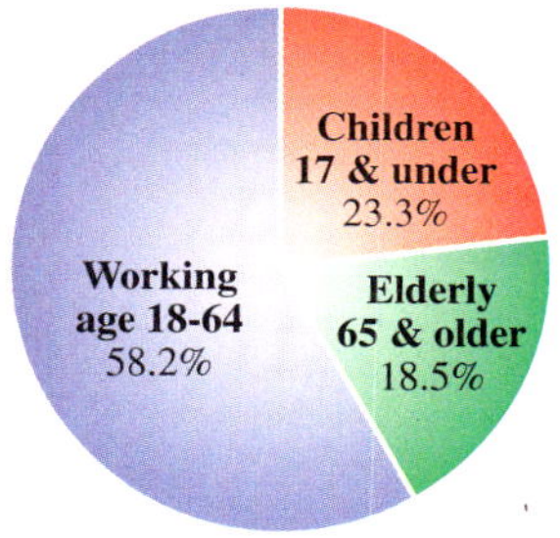

Source: Pew Research Center (2008)

Strategy To find the predicted U.S. population in 2050, we will translate the words of the problem into an equation and then solve the equation.

Why The variable in the translation equation represents the unknown population in 2050 that we are asked to find.

Solution In this problem, 81 is the amount, 18.5% is the percent, and the words "what number" represent the base. The units are millions of people.

Success Tip

Don't forget to change percents to decimal form (or fraction form) before performing any calculations.

81	is	18.5%	of	what number?
↓	↓	↓	↓	↓
81	=	0.185	·	x

$$\frac{81}{\mathbf{0.185}} = \frac{0.185x}{\mathbf{0.185}}$$ To isolate x, undo the multiplication by 0.185 by dividing both sides by 0.185.

$437.8 \approx x$ Do the division.

$438 \approx x$ Round 437.8 million to the nearest million.

The U.S. population is predicted to be about 438 million in the year 2050. We can check using estimation: 81 million out of a population of 438 million is about $\frac{80 \text{ million}}{400 \text{ million}}$, or $\frac{1}{5}$, which is 20%. Since this is close to 18.5%, the answer 438 million seems reasonable.

Self Check 2 **Aging Populations.** By the year 2100, it is predicted that 131 million, or 23%, of the U.S. residents will be age 65 or older. If the prediction is correct, find the population in 2100. (Round to the nearest million.)

Now Try Problem 17

We pay many types of taxes in our daily lives, such as sales tax, gasoline tax, income tax, and Social Security tax. **Tax rates** usually are expressed as percents.

EXAMPLE 3 **Taxes.** A maid makes \$500 a week. One of the deductions from her weekly paycheck is a Social Security tax of \$31. Find her Social Security tax rate.

Strategy To find the tax rate, we will translate the words of the problem into an equation and then solve the equation.

Why The variable in the translation equation represents the unknown tax rate that we are asked to find.

Solution

31	is	what percent	of	500?
↓	↓	↓	↓	↓
31	=	x	·	500

31 is the amount, x is the percent, and 500 is the base.

$$\frac{31}{\mathbf{500}} = \frac{500x}{\mathbf{500}}$$ To undo the multiplication by 500, divide both sides by 500.

$$0.062 = x$$ Do the division.

$$6.2\% = x$$ Change the decimal 0.062 to a percent.

The Social Security tax rate is 6.2%.

We can use estimation to check: \$31 out of \$500 is about $\frac{30}{500}$ or $\frac{6}{100}$, which is 6%. Since this is close to 6.2%, the answer seems reasonable.

Success Tip

One way to check an answer is using *estimation.* If your estimate and your answer are close, you can be reasonably sure that your answer is correct.

Self Check 3 **Medicare Tax.** The maid mentioned in Example 3 also has \$7.25 of Medicare tax deducted from her weekly paycheck. Find her Medicare tax rate.

Now Try Problem 23

4 Find Percent of Increase and Decrease.

Percents often are used to describe how a quantity has changed. For example, a health care provider might increase the cost of medical insurance by 3%, or a police department might decrease the number of officers assigned to street patrols by 10%. To describe such changes, we use **percent of increase** or **percent of decrease.**

EXAMPLE 4 **Identity Theft.** The Federal Trade Commission receives complaints involving the theft of someone's identity information, such as a credit card, Social Security number, or cell phone account. Refer to the data in the table. What was the percent of decrease in the number of complaints from 2008 to 2009? (Round to the nearest percent.)

IDENTITY THEFT
Federal Trade Commission (2010)

Year	2008	2009
Number of Complaints	314,000	278,000

Strategy First, we will subtract to find the *amount of decrease* in the number of complaints. Then we will translate the words of the problem into an equation and solve it.

Why A percent of decrease problem involves finding the *percent of change,* and the change in a quantity is found using subtraction.

Solution To find the *amount of decrease,* we find the difference of 314,000 and 278,000.

$$314{,}000 - 278{,}000 = 36{,}000$$

Caution

The percent of increase (or decrease) is a percent of the *original* number, that is, the number before the change occurred.

Now we translate the percent sentence to an equation, and solve it.

The decrease	is	what percent	of	the number of complaints in 2008?
↓	↓	↓	↓	↓
36,000	=	x	·	314,000

36,000 is the amount, x is the percent, and 314,000 is the base.

$$\frac{36{,}000}{314{,}000} = \frac{314{,}000x}{314{,}000}$$ To undo the multiplication by 314,000, divide both sides by 314,000.

$$0.114649681 \approx x$$ Do the division using a calculator.

$$11.4649681\% \approx x$$ Change the decimal to a percent.

Rounding to the nearest percent, we find that the number of identity theft complaints decreased by about 11% from 2008 to 2009.

A 10% decrease would be 0.10(314,000) or 31,400 fewer complaints. It seems reasonable that 36,000 fewer complaints is an 11% decrease.

Self Check 4 **Identity Theft.** In 2007, there were 259,000 complaints of identity theft. Find the percent increase from 2007 to 2008. (Round to the nearest percent.)

Now Try ▶ Problem 43

5 Solve Discount and Commission Problems.

When the price of an item is reduced, we call the amount of the reduction a **discount.** If a discount is expressed as a percent, it is called the **rate of discount.**

EXAMPLE 5 **Health Club Discounts.** A 30% discount on a 1-year membership for a fitness center amounted to a $90 savings. Find the cost of a 1-year membership before the discount.

Strategy We will translate the words of the problem into an equation and then solve the equation.

Why The variable in the translation equation represents the unknown cost of a 1-year membership before the discount.

Solution We are told that \$90 is 30% of some unknown membership cost.

90	is	30%	of	what number?	
↓	↓	↓	↓	↓	
90	=	0.30	·	x	90 is the amount, 30% is the percent, and x is the base.

$$\frac{90}{\mathbf{0.30}} = \frac{0.30x}{\mathbf{0.30}} \quad \text{To undo the multiplication by 0.30, divide both sides by 0.30.}$$

$$300 = x \quad \text{Do the division.}$$

A one-year membership cost \$300 before the discount.

Self Check 5 **Discounts.** A shopper saved \$6 on a pen that was discounted 5%. Find the original cost.

Now Try Problem 51

Instead of working for a salary or at an hourly rate, many salespeople are paid on **commission.** An employee who is paid a commission is paid a percent of the price of goods or services that he or she sells. We call that percent the **rate of commission.**

EXAMPLE 6 **Commissions.** A real estate agent earned \$14,025 for selling a house. If she received a $5\frac{1}{2}$% commission, what was the selling price?

Strategy We will translate the words of the problem into an equation and then solve the equation.

Why The variable in the translation equation represents the unknown selling price of the house that we are asked to find.

Solution We are told that \$14,025 is $5\frac{1}{2}$% of some unknown selling price of a house.

Success Tip

It is helpful to write percents that involve fractions in an equivalent decimal form. For example,

$$1\tfrac{1}{4}\% = 1.25\%$$
$$5\tfrac{1}{2}\% = 5.5\%$$
$$7\tfrac{3}{4}\% = 7.75\%$$

\$14,025	is	5.5%	of	what number?	Write $5\frac{1}{2}$% as 5.5%.
↓	↓	↓	↓	↓	
14,025	=	0.055	·	x	14,025 is the amount, 5.5% is the percent, and x is the base.

$$\frac{14{,}025}{\mathbf{0.055}} = \frac{0.055x}{\mathbf{0.055}} \quad \text{To undo the multiplication by 0.055, divide both sides by 0.055.}$$

$$255{,}000 = x \quad \text{Do the division.}$$

The selling price of the house was \$255,000.

Self Check 6 **Commissions.** A jewelry store clerk receives a $4\frac{1}{4}$% commission on all sales. What was the price of a gold necklace sold by the clerk if his commission was \$25.50?

Now Try Problem 53

SECTION 2.3 STUDY SET

VOCABULARY

Fill in the blanks.

1. _______ means parts per one hundred.
2. In the statement "10 is 50% of 20," 10 is the ________, 50% is the percent, and 20 is the _____.
3. In percent questions, the word *of* means ____________, and the word ___ means equals.
4. An employee who is paid a ___________ is paid a percent of the cost of goods or services that he or she sells.

CONCEPTS

5. Represent the amount of the figure that is shaded using a fraction, a decimal, and a percent.

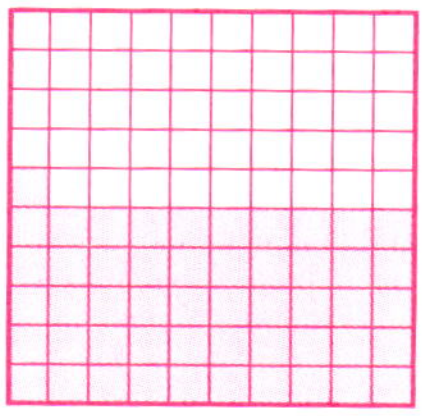

6. Fill in the blanks.
   a. To change a percent to a decimal, drop the % symbol and move the decimal point 2 places to the _____.
   b. To change a decimal to a percent, move the decimal point 2 places to the ______ and insert a % symbol.
7. Fill in the blanks using the words *percent, amount,* and *base.*

   = ·

8. Translate each sentence into an equation. **Do not solve.**
   a. 12 is 40% of what number?
   b. 99 is what percent of 200?
   c. What is 66% of 3?
9.

Number of Earthquakes in the U.S.	
2008	**2009**
3,618	4,257

Source: U.S. Geological Survey

   a. Find the *amount* of increase in the number of earthquakes.
   b. Fill in blanks to find the percent of increase in earthquakes:
   _____ is ______ % of ______?
10. Use estimation to determine if each statement is reasonable.
   a. 18 is 48% of 93.
   b. 47 is 6% of 206.

NOTATION

11. Change each percent to a decimal.
   a. 35%
   b. 8.5%
   c. 150%
   d. $2\frac{3}{4}\%$
   e. 9.25%
   f. $1\frac{1}{2}\%$
12. Change each decimal to a percent.
   a. 0.9
   b. 0.99
   c. 0.999
   d. 9

GUIDED PRACTICE

See Example 1.

13. What number is 48% of 650?
14. What number is 60% of 200?
15. What number is 92.4% of 50?
16. What number is 2.8% of 220?

See Example 2.

17. 75 is 25% of what number?
18. 78 is 6% of what number?
19. 128.1 is 8.75% of what number?
20. 1.12 is 140% of what number?

See Example 3.

21. 78 is what percent of 300?
22. 143 is what percent of 325?
23. 0.42 is what percent of 16.8?
24. 199.92 is what percent of 2,352?

APPLICATIONS

25. **Antiseptics.** Refer to the label on the bottle in figure (a) below. Find the amount of pure hydrogen peroxide in the bottle.

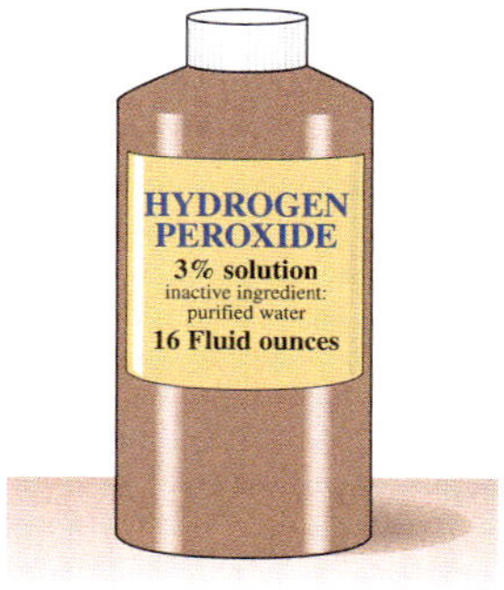

(a)

(b)

26. **Dining Out.** Refer to the sales receipt in figure (b) above. Compute the 15% tip (*rounded up* to the nearest dollar). Then find the total cost of the meal.
27. **U.S. Federal Budget.** The circle graph shows how the government spent $2,980 billion in 2008. How much was spent on
   a. Social Security/Medicare?
   b. Defense/Veterans?

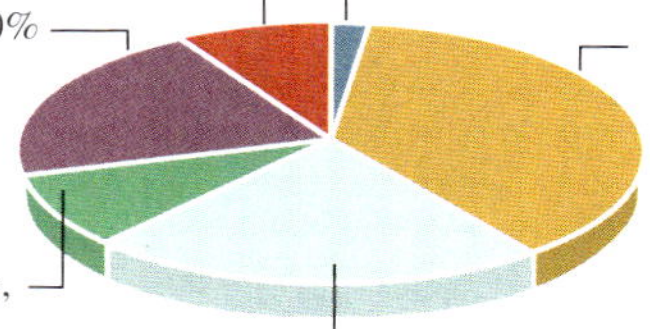

Based on 2009 Federal Income Tax Form 1040

28. Tax Tables. Use the Schedule X table below to compute the amount of federal income tax to be paid on an income of \$39,909.

If your income is over—	But not over—	Your income tax is—	of the amount over—
\$0	\$8,350	.... 10%	\$0
8,350	33,950	\$835.00 + 15%	8,350
33,950	82,250	4,675.00 + 25%	33,950

29. PayPal. Many e-commerce businesses use PayPal to perform payment processing for them. For certain transactions, merchants are charged a fee of 2.9% of the selling price of the item plus \$0.30. What would PayPal charge an online art store to collect payment on a painting selling for \$350?

30. eBay. When a student sold a Play Station 3 on eBay for \$195, she was charged a two-part final value fee: 8% of the first \$50 of the selling price plus 5% of the remainder of the selling price over \$50. Find the fee to sell the Play Station 3 on eBay.

31. Price Guarantees. Home Club offers a "10% Plus" guarantee: If the customer finds the same item selling for less somewhere else, he or she receives the difference in price plus 10% of the difference. A woman bought miniblinds at the Home Club for \$120 but later saw the same blinds on sale for \$98 at another store. How much can she expect to be reimbursed?

32. Room Taxes. A guest at the San Antonio Hilton Airport Hotel paid \$180 for a room plus a 9% city room tax, a $1\frac{3}{4}\%$ county room tax, and a 6% state room tax. Find the total amount of tax that the guest paid on the room.

33. Computer Memory. The *My Computer* screen on a student's computer is shown on the right. What percent of the storage capacity on the hard drive Local Disk (C:) of his computer is used? What percent is free? (GB stands for gigabytes.)

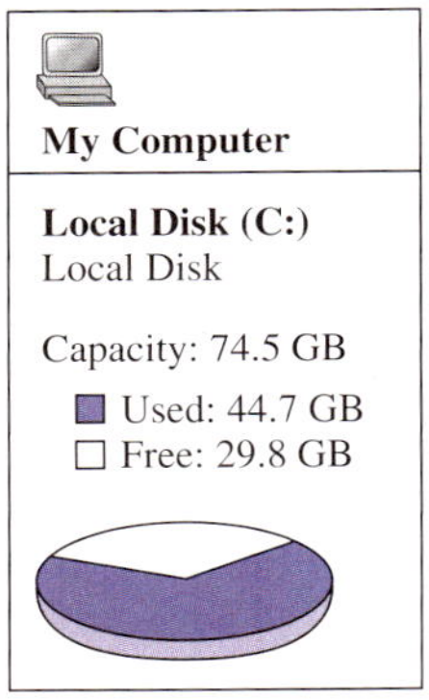

34. Genealogy. Through an extensive computer search, a genealogist determined that worldwide, 180 out of every 10 million people had his last name. What percent is this?

35. Dentistry. Refer to the dental record. What percent of the patient's teeth have fillings? Round to the nearest percent.

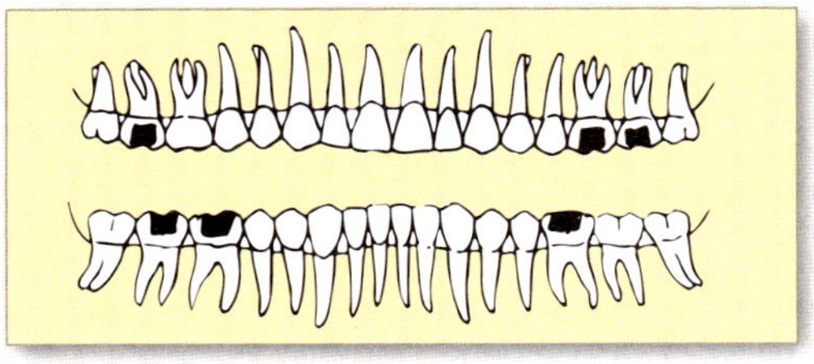

36. Test Scores. The score 175/200 was written by an algebra instructor at the top of a student's test paper. Write the test score as a percent.

37. DMV Written Test. To obtain a learner's permit to drive in Nevada, a score of 80% (or better) on a 50-question multiple-choice test is required. If a teenager answered 33 questions correctly, did he pass the test?

38. iPods. The settings menu screen of an Apple iPod is shown. What percent of the memory capacity is still available? Round to the nearest percent. (GB stands for gigabytes.)

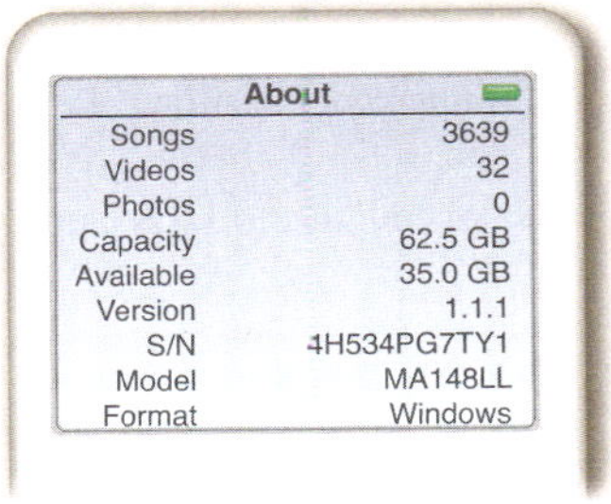

39. Child Care. After the first day of registration, 84 children had been enrolled in a day care center. That represented 70% of the available slots. Find the maximum number of children the center could enroll.

40. Racing Programs. One month before a stock car race, the sale of ads for the official race program was slow. Only 12 pages, or just 30% of the available pages, had been sold. Find the total number of pages devoted to advertising in the program.

41. Nutrition. The Nutrition Facts label from a can of clam chowder is shown.

a. Find the number of grams of saturated fat in one serving. What percent of a person's recommended daily intake is this?

b. Determine the recommended number of grams of saturated fat that a person should consume daily.

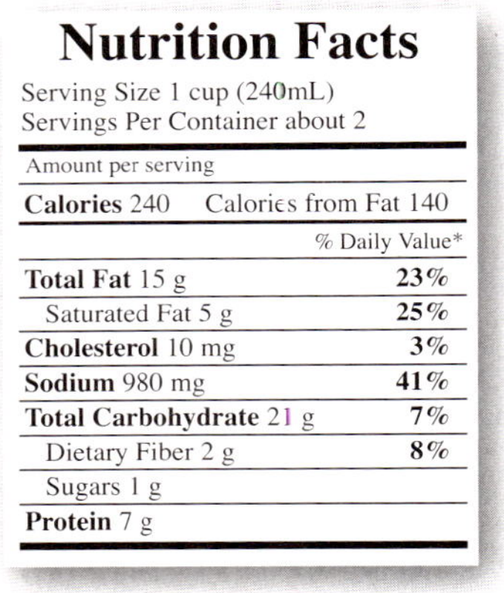

42. Commercials. Jared Fogle credits his tremendous weight loss to exercise and a diet of low-fat Subway sandwiches. His current weight (about 187 pounds) is 44% of his maximum weight (reached in March of 1998). What did he weigh then?

43. Rainforests. Refer to the graph on the next page, which shows the number of square miles of Brazilian Amazon rainforest that has recently been deforested.

a. Find the percent of increase in the number of square miles of the Brazilian Amazon rainforest that was cleared from 2007 to 2008. Round to the nearest percent.

b. Find the percent of decrease in the number of square miles of the Brazilian Amazon rainforest that was cleared from 2008 to 2009. Round to the nearest percent.

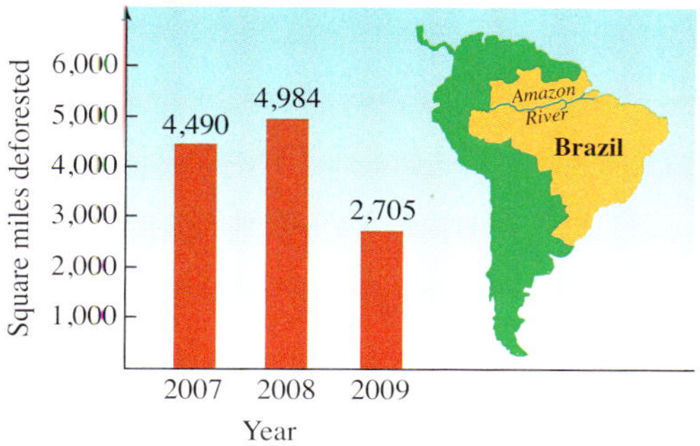

Source: mongabay.com

44. Auctions. A pearl necklace of former First Lady Jacqueline Kennedy Onassis, originally valued at \$700, was sold at auction in 1996 for \$211,500. Find the percent of increase in the value of the necklace. (Round to the nearest percent.)

45. Insurance Costs. A college student's good grades earned her a student discount on her car insurance premium. Find the percent of decrease to the nearest percent if her annual premium was lowered from \$1,050 to \$925.

46. U.S. Life Expectancy. Use the following life expectancy data for 1900 and 2010 to find the percent of increase for males and for females. Round to the nearest percent.

Years of life expected at birth		
	Male	**Female**
1900	46.3	48.3
2010	75.8	80.8

Source: *The World Fact Book*, CIA

47 from Campus To Careers

Automotive Service Technician

© barang/Shutterstock.com

A *single* automobile tire that is underinflated by 5 pounds per square inch (psi) decreases the car's gas mileage by about 1.5%. (Source: auto-buying-tips.com and popularmechanics.com)

a. If all four tires of a car are underinflated by 5 psi, by what percent will its gas mileage be decreased?

b. A 2010 Chevrolet Camaro with properly inflated tires gets 25 miles per gallon on the highway. If each of its tires is underinflated by 5 psi, what will be its highway mileage?

48. Food Labels. To be labeled "Reduced Fat," foods must contain at least 25% less fat per serving than the regular product. One serving of the original Jif peanut butter has 16 grams of fat per serving. The new Jif Reduced Fat product contains 12 grams of fat per serving. Does it meet the labeling requirement?

49. TV Shopping. Jan bought a toy from the QVC home shopping network that was discounted 20%. If she saved \$15, what was the original price of the toy?

50. Discounts. A 12% discount on a watch saved a shopper \$48. Find the price of the watch before the discount.

51. Sales. The price of a certain model patio set was reduced 35% because it was being discontinued. A shopper purchased two of them and saved a total of \$210. Find the price of a patio set before the discount.

52. Craigslist. A post on the classifieds website Craigslist advertised a mountain bike selling for \$800 less than its retail price. If this is a 40% savings, what is the retail price of the bike?

53. Real Estate. The $3\frac{1}{2}\%$ commission paid to a real estate agent on the sale of a condominium earned her \$3,325. Find the selling price of the condo.

54. Consignment. An art gallery agreed to sell an artist's sculpture for a commission of 45%. What must be the selling price of the sculpture if the gallery would like to make \$13,500?

55. Stockbrokers. A stockbroker charges a 2.5% commission to sell shares of a stock for a client. Find the value of stock sold by a broker if the commission was \$640.

56. Agents. A sports agent made one million dollars by charging a 12.5% commission to negotiate a long-term contract for a professional athlete. Find the amount of the contract.

WRITING

57. Explain the error: What is 5% of 8?

$$x = 5 \cdot 8$$
$$x = 40$$

40 is 5% of 8.

58. Write a real-life situation that could be described by "9 is what percent of 20?"

59. Explain why 150% of a number is more than the number.

60. Why is the problem "What is 9% of 100?" easy to solve?

REVIEW

61. Divide: $-\frac{16}{25} \div \left(-\frac{4}{15}\right)$

62. What two numbers are a distance of 8 away from 4 on the number line?

63. Is -34 a solution of $x + 15 = -49$?

64. Evaluate: $2 + 3[24 - 2(2 - 5)]$

CHALLENGE PROBLEMS

65. Soaps. A soap advertises itself as $99\frac{44}{100}\%$ pure. First, determine what percent of the soap is impurities. Then express your answer as a decimal.

66. Express $\frac{1}{20}$ of 1% as a percent using decimal notation.

SECTION 2.4 Formulas

OBJECTIVES

1. Use formulas from business.
2. Use formulas from science.
3. Use formulas from geometry.
4. Solve for a specified variable.

ARE YOU READY?

The following problems review some basic skills that are needed when working with formulas.

1. How many variables does each equation contain?
   a. $4x + 3 = 15$
   b. $P = 2l + 2w$
2. Simplify: $a + d - a$
3. Multiply: $4 \cdot \frac{1}{4}x$
4. Multiply: $c(8 - x)$
5. If $2 = t$, is it also true that $t = 2$?
6. Simplify: $\frac{b}{b}$

A **formula** is an equation that states a mathematical relationship between two or more variables. The variables of a formula represent quantities such as cost, time, and perimeter. Formulas are used in fields such as business, science, and geometry.

1 Use Formulas from Business.

A formula for retail price: To make a profit, a merchant must sell an item for more than he or she paid for it. The price at which the merchant sells the product, called the **retail price,** is the *sum* of what the item cost the merchant plus the **markup.** Using r to represent the retail price, c the cost, and m the markup, we can write this formula as

$r = c + m$ Retail price = cost + markup

A formula for profit: The **profit** a business makes is the *difference* between the **revenue** (the money it takes in) and the cost. Using p to represent the profit, r the revenue, and c the cost, we can write this formula as

$p = r - c$ Profit = revenue − cost

If we are given the values of all but one of the variables in a formula, we can use our equation-solving skills to find the value of the remaining variable.

EXAMPLE 1 **Films.** Estimates are that Warner Brothers made a \$219 million profit on the film *Harry Potter and the Half-Blood Prince.* If the studio received \$469 million in worldwide box office revenue, find the cost to make and distribute the film. (Source: www.thenumbers.com, June 2010)

Strategy To find the cost to make and distribute the film, we will substitute the given values in the formula $p = r - c$ and solve for c.

Why The variable c in the formula represents the unknown cost.

Solution The film made \$219 million (the profit p) and the studio took in \$469 million (the revenue r). To find the cost c, we proceed as follows.

$p = r - c$	This is the formula for profit.
$219 = 469 - c$	Substitute 219 for p and 469 for r.
$219 - 469 = 469 - c - 469$	To eliminate 469 on the right side, subtract 469 from both sides.
$-250 = -c$	Do the subtraction.
$\frac{-250}{-1} = \frac{-c}{-1}$	To solve for c, divide (or multiply) both sides by -1.
$250 = c$	The units are millions of dollars.

It cost \$250 million to make and distribute the film.

Self Check 1 **Fundraisers.** A PTA spaghetti dinner made a profit of \$275.50. If the cost to host the dinner was \$1,235, how much revenue did it generate?

Now Try Problem 13

The Language of Algebra

The word **annual** means occurring once a year. An *annual* interest rate is the interest rate paid per year.

A formula for simple interest: When money is borrowed, the lender expects to be paid back the amount of the loan plus an additional charge for the use of the money, called **interest.** When money is deposited in a bank, the depositor is paid for the use of the money. The money the deposit earns is also called interest.

Interest is calculated in two ways: either as **simple interest** or as **compound interest.** Simple interest is the *product* of the principal (the amount of money that is invested, deposited, or borrowed), the annual interest rate, and the length of time in years. Using I to represent the simple interest, P the principal, r the annual interest rate, and t the time in years, we can write this formula as

$I = Prt$ Interest = principal · rate · time

EXAMPLE 2 **Retirement Income.** One year after investing \$15,000, a retired couple received a check for \$1,125 in interest. Find the interest rate their money earned that year.

Strategy To find the interest rate, we will substitute the given values in the formula $I = Prt$ and solve for r.

Why The variable r in the formula represents the unknown interest rate.

Solution The couple invested \$15,000 (the principal P) for 1 year (the time t) and made \$1,125 (the interest I). To find the annual interest rate r, we proceed as follows.

$I = Prt$ This is the formula for simple interest.

$1{,}125 = 15{,}000r(1)$ Substitute 1,125 for I, 15,000 for P, and 1 for t.

$1{,}125 = 15{,}000r$ Simplify the right side.

$\frac{1{,}125}{15{,}000} = \frac{15{,}000r}{15{,}000}$ To solve for r, undo the multiplication by 15,000 by dividing both sides by 15,000.

$0.075 = r$ Do the division. This is the rate expressed as a decimal.

$7.5\% = r$ To write 0.075 as a percent, multiply 0.075 by 100 by moving the decimal point two places to the right and inserting a % symbol.

Caution

When using the formula $I = Prt$, always write the interest rate r (which is given as a percent) as a decimal (or fraction) before performing any calculations.

The couple received an annual rate of 7.5% that year on their investment. We can display the facts of the problem in a table as shown on the right.

	P ·	r ·	t =	I
Investment	15,000	0.075	1	1,125

Self Check 2 **Home Loans.** A father loaned his daughter \$12,200 at a 2% annual simple interest rate for a down payment on a house. If the interest on the loan amounted to \$610, for how long was the loan?

Now Try Problem 17

2 Use Formulas from Science.

A formula for distance traveled: If we know the average rate (of speed) at which we will be traveling and the time we will be traveling at that rate, we can find the distance traveled. Using d to represent the distance, r the average rate, and t the time, we can write this formula as

$d = rt$ Distance = rate · time

EXAMPLE 3 **Whales.** As they migrate from the Bering Sea to Baja California, gray whales swim for about 20 hours each day, covering a distance of approximately 70 miles. Estimate their average swimming rate in miles per hour. (Source: marinebio.net)

Strategy To find the swimming rate, we will substitute the given values in the formula $d = rt$ and solve for r.

Why The variable r in the formula represents the unknown average swimming rate.

Solution The whales swam 70 miles (the distance d) in 20 hours (the time t). To find their average swimming rate r, we proceed as follows.

$$d = rt \quad \text{This is the formula for distance traveled.}$$

$$70 = r(20) \quad \text{Substitute 70 for } d \text{ and 20 for } t.$$

$$\frac{70}{20} = \frac{20r}{20} \quad \text{To solve for } r\text{, undo the multiplication by 20 by dividing both sides by 20.}$$

$$3.5 = r \quad \text{Do the division.}$$

The whales' average swimming rate is 3.5 miles per hour (mph). The facts of the problem can be displayed in a table, as shown on the right.

	r	$\cdot\ t$	$= d$
Gray whale	3.5	20	70

Caution

When using the formula $d = rt$, make sure the units are consistent. For example, if the rate is given in miles per hour, the time must be expressed in hours.

Self Check 3 **Elevators.** An elevator travels at an average rate of 288 feet per minute. How long will it take the elevator to climb 30 stories, a distance of 360 feet?

Now Try Problem 21

A formula for converting temperatures: In the American system, temperature is measured on the Fahrenheit scale. The Celsius scale is used to measure temperature in the metric system. The formula that relates a Fahrenheit temperature F to a Celsius temperature C is:

$$C = \frac{5}{9}(F - 32)$$

EXAMPLE 4 **Temperature Conversion.** Convert the temperature shown on the City Savings sign to degrees Fahrenheit.

Strategy To find the temperature in degrees Fahrenheit, we will substitute the given Celsius temperature in the formula $C = \frac{5}{9}(F - 32)$ and solve for F.

Why The variable F represents the unknown temperature in degrees Fahrenheit.

Solution The temperature in degrees Celsius is 30°. To find the temperature in degrees Fahrenheit F, we proceed as follows.

$$C = \frac{5}{9}(F - 32) \quad \text{This is the formula for temperature conversion.}$$

$$30 = \frac{5}{9}(F - 32) \quad \text{Substitute 30 for } C\text{, the Celsius temperature.}$$

$$\frac{9}{5} \cdot 30 = \frac{9}{5} \cdot \frac{5}{9}(F - 32) \quad \text{To undo the multiplication by } \tfrac{5}{9}\text{, multiply both sides by the reciprocal of } \tfrac{5}{9}.$$

$$54 = F - 32 \quad \text{Do the multiplication.}$$

$$54 + 32 = F - 32 + 32 \quad \text{To isolate } F\text{, undo the subtraction of 32 by adding 32 to both sides.}$$

$$86 = F \quad \text{Do the addition.}$$

30°C is equivalent to 86°F.

The Language of Algebra

In 1724, Daniel Gabriel **Fahrenheit,** a German scientist, introduced the temperature scale that bears his name. The Celsius scale was invented in 1742 by Swedish astronomer Anders **Celsius**.

Self Check 4 **Planets.** Change $-175°C$, the temperature on Saturn, to degrees Fahrenheit. (Source: universetoday.com)

Now Try Problem 27

3 Use Formulas from Geometry.

The Language of Algebra

When you hear the word **perimeter**, think of the distance around the "rim" of a flat figure.

To find the **perimeter** of a plane (two-dimensional, flat) geometric figure, such as a rectangle or triangle, we find the distance around the figure by computing the sum of the lengths of its sides. Perimeter is measured in American units, such as inches, feet, yards, and in metric units, such as millimeters, meters, and kilometers.

EXAMPLE 5 **Flags.** The largest flag ever flown was an American flag that had a perimeter of 1,520 feet and a length of 505 feet. It was hoisted on cables across Hoover Dam to celebrate the 1996 Olympic Torch Relay. Find the width of the flag.

Perimeter formulas

$P = 2l + 2w$ (rectangle)
$P = 4s$ (square)
$P = a + b + c$ (triangle)

Strategy To find the width of the flag, we will substitute the given values in the formula $P = 2l + 2w$ and solve for w.

Why The variable w in the formula represents the unknown width of the flag.

Solution The perimeter P of the rectangular-shaped flag is 1,520 ft and the length l is 505 ft. To find the width w, we proceed as follows.

©AP Photo/Las Vegas Review Journal/Carol Ferguson

$P = 2l + 2w$ — This is the formula for the perimeter of a rectangle.

$1{,}520 = 2(505) + 2w$ — Substitute 1,520 for P and 505 for l.

$1{,}520 = 1{,}010 + 2w$ — Do the multiplication.

$510 = 2w$ — To undo the addition of 1,010, subtract 1,010 from both sides. This step is done mentally and is not shown.

$255 = w$ — To isolate w, undo the multiplication by 2 by dividing both sides by 2. This step is done mentally and is not shown.

The width of the flag is 255 feet. If its length is 505 feet and its width is 255 feet, its perimeter is $2(505) + 2(255) = 1{,}010 + 510 = 1{,}520$ feet, as given.

Self Check 5 **Flags.** The largest flag that consistently flies is the flag of Brazil in Brasilia, the country's capital. It has a perimeter 1,116 feet and length 328 feet. Find its width.

Now Try Problem 29

Area formulas

$A = lw$ (rectangle)
$A = s^2$ (square)
$A = \frac{1}{2}bh$ (triangle)
$A = \frac{1}{2}h(B + b)$ (trapezoid)

The **area** of a plane (two-dimensional, flat) geometric figure is the amount of surface that it encloses. Area is measured in square units, such as square inches, square feet, square yards, and square meters (written as in.^2, ft^2, yd^2, and m^2, respectively).

EXAMPLE 6 **a.** What is the circumference of a circle with diameter 14 feet? Round to the nearest tenth of a foot. **b.** What is the area of the circle? Round to the nearest tenth of a square foot.

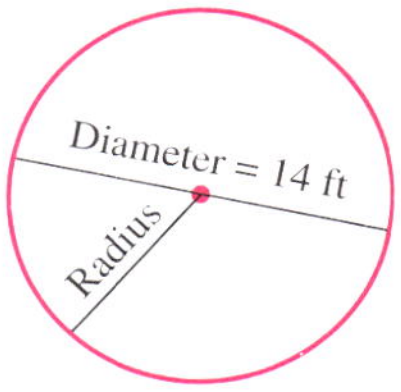

Strategy To find the circumference and area of the circle, we will substitute the proper values into the formulas $C = \pi D$ and $A = \pi r^2$ and find C and A.

Why The variable C in the formula represents the unknown circumference of the circle and A represents the unknown area.

Solution

Circle formulas

$D = 2r$ (diameter)

$r = \frac{1}{2}D$ (radius)

$C = 2\pi r = \pi D$ (circumference)

$A = \pi r^2$ (area)

a. Recall that the circumference of a circle is the distance around it. To find the circumference C of a circle with diameter D equal to 14 ft, we proceed as follows.

$C = \pi D$ — This is the formula for the circumference of a circle. πD means $\pi \cdot D$.

$C = \pi(14)$ — Substitute 14 for D, the diameter of the circle.

$C = 14\pi$ — The exact circumference of the circle is 14π.

$C \approx 43.98229715$ — To use a calculator to approximate the circumference, enter π × 14 =. If you do not have a calculator, use 3.14 as an approximation of π. (Answers may vary slightly depending on which approximation of π is used.)

The circumference is exactly 14π ft. Rounded to the nearest tenth, this is 44.0 ft.

b. The radius r of the circle is one-half the diameter, or 7 feet. To find the area A of the circle, we proceed as follows.

$A = \pi r^2$ — This is the formula for the area of a circle. πr^2 means $\pi \cdot r^2$.

$A = \pi(7)^2$ — Substitute 7 for r, the radius of the circle.

$A = 49\pi$ — Evaluate the exponential expression: $7^2 = 49$. The exact area is 49π ft^2.

$A \approx 153.93804$ — To use a calculator to approximate the area, enter 49 × π =.

The area is exactly 49π ft^2. To the nearest tenth, the area is 153.9 ft^2.

Notation

Since π is an irrational number, its decimal representation has an infinite number of decimal places. When an approximation of π is used in a calculation, it produces an approximate answer. Use an *is approximately equal to* symbol $\approx$ in your solution to show that.

Self Check 6 Find the circumference of a circle with radius 10 inches. Round to the nearest hundredth of an inch.

Now Try Problem 30

The **volume** of a three-dimensional geometric solid is the amount of space it encloses. Volume is measured in cubic units, such as cubic inches, cubic feet, and cubic meters (written as in.^{3}, ft^3, and m^3, respectively).

EXAMPLE 7 Find the volume of the cylinder. Round to the nearest tenth of a cubic centimeter.

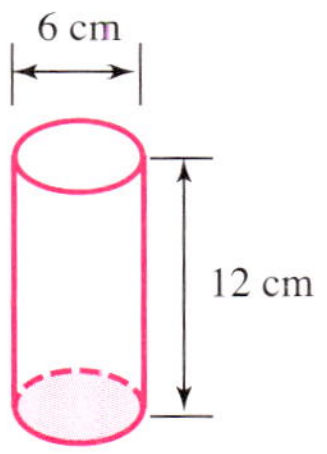

Strategy To find the volume of the cylinder, we will substitute the proper values into the formula $V = \pi r^2 h$ and find V.

Why The variable V in the formula represents the unknown volume.

Success Tip

A complete list of geometric figures and formulas appears on the inside of the back cover of this textbook.

Solution Since the radius of a circle is one-half its diameter, the radius r of the circular base of the cylinder is $\frac{1}{2}(6 \text{ cm}) = 3$ cm. The height h of the cylinder is 12 cm. To find the volume V of the cylinder, we proceed as follows.

$V = \pi r^2 h$ — This is the formula for the volume of a cylinder. $\pi r^2 h$ means $\pi \cdot r^2 \cdot h$.

$V = \pi(3)^2(12)$ — Substitute 3 for r and 12 for h.

$V = \pi(9)(12)$ — Evaluate the exponential expression.

$V = 108\pi$ — Multiply. The exact volume is 108π cm^3.

$V \approx 339.2920066$ — Use a calculator to approximate the volume.

To the nearest tenth, the volume is 339.3 cubic centimeters. This can be written as 339.3 cm^3.

Volume formulas

$V = lwh$ (rectangular solid)

$V = s^3$ (cube)

$V = \frac{4}{3}\pi r^3$ (sphere)

$V = \pi r^2 h$ (cylinder)

$V = \frac{1}{3}\pi r^2 h$ (cone)

Self Check 7 Find the volume of a cone whose base has radius 12 meters and whose height is 9 meters. Round to the nearest tenth of a cubic meter. Use the formula $V = \frac{1}{3}\pi r^2 h$.

Now Try Problem 31

4 Solve for a Specified Variable.

The Language of Algebra

The word **specified** is a form of the word *specify,* which means to select something for a purpose. Here, we select a variable for the purpose of solving for it.

Suppose a shopper wants to calculate the markup m on several items, knowing their retail price r and their cost c to the merchant. It would take a lot of time to substitute values for r and c into the formula for retail price $r = c + m$ and then repeatedly solve for m. A better way is to solve the formula for m first, substitute values for r and c, and then compute m directly.

To **solve a formula for a specified variable** means to isolate that variable on one side of the equation, with all other variables and constants on the opposite side.

EXAMPLE 8 Solve the retail-price formula $r = c + m$ for m.

Strategy To solve for m, we will focus on it as if it is the only variable in the equation. We will use a strategy similar to that used to solve linear equations in one variable to isolate m on one side. (See page 117 if you need to review the strategy.)

Why We can solve the formula as if it were an equation in one variable because all the other variables are treated as if they were numbers (constants).

Solution

We will isolate m on this side of the equation.

$r = c + m$

$r - c = c + m - c$ — To isolate m, undo the addition of c by subtracting c from both sides.

$r - c = m$ — Simplify the right side: $c - c = 0$.

$m = r - c$ — Reverse the sides of the equation so that m is on the left.

The Language of Algebra

We say that the formula is **solved for *m*** because m is alone on one side of the equation and the other side does not contain m.

The resulting formula, $m = r - c$, indicates that the markup on an item is the difference between its retail price and its cost to the merchant.

Self Check 8 Solve the profit formula $p = r - c$, for c.

Now Try Problem 33

EXAMPLE 9 Solve the area of a triangle formula $A = \frac{1}{2}bh$ for b.

Strategy To solve for b, we will treat b as the only variable in the equation and use properties of equality to isolate it on one side. We will treat the other variables as if they were numbers (constants).

Why To solve for a specified variable means to isolate it on one side of the equation.

Solution We use the same steps to solve an equation for a specified variable that we use to solve equations with only one variable.

Success Tip

To solve for b, think of it as the only variable in the equation. Treat A and h as if they were numbers. It is often helpful to first circle the variable in the given formula that you are solving for:

$$A = \frac{1}{2}bh$$

We will isolate b on this side of the equation.

$A = \frac{1}{2}bh$

$2 \cdot A = 2 \cdot \frac{1}{2}bh$ — To clear the equation of the fraction, multiply both sides by 2.

$2A = bh$ — Simplify the right side: $2 \cdot \frac{1}{2} = 1$.

$\frac{2A}{h} = \frac{bh}{h}$ — bh means $b \cdot h$. To isolate b, undo the multiplication by h by dividing both sides by h.

$\frac{2A}{h} = b$ — On the right side, remove the common factor of h: $\frac{b\overset{1}{\cancel{h}}}{\underset{1}{\cancel{h}}} = b$.

$b = \frac{2A}{h}$ — Reverse the sides of the equation so that b is on the left.

Self Check 9 Solve $A = \frac{1}{2}r^2a$ for a.

Now Try Problem 37

EXAMPLE 10 Solve the perimeter of a rectangle formula $P = 2l + 2w$ for l.

Strategy To solve for l, we will treat l as the only variable in the equation and use properties of equality to isolate it on one side. We will treat the other variables as if they were numbers (constants).

Why To solve for a specified variable means to isolate it on one side of the equation.

Solution

We will isolate l on this side of the equation.

$$P = 2l + 2w$$

$$P - 2w = 2l + 2w - 2w \quad \text{To undo the addition of } 2w\text{, subtract } 2w \text{ from both sides.}$$

$$P - 2w = 2l \quad \text{Combine like terms: } 2w - 2w = 0.$$

$$\frac{P - 2w}{2} = \frac{2l}{2} \quad \text{To isolate } l\text{, undo the multiplication by 2 by dividing both sides by 2.}$$

$$\frac{P - 2w}{2} = l \quad \text{Simplify the right side: } \frac{\overset{1}{\cancel{2}}l}{\underset{1}{\cancel{2}}} = l.$$

$$l = \frac{P - 2w}{2} \quad \text{Reverse the sides of the equation so that } l \text{ is on the left.}$$

Caution

Do not try to simplify the result this way:

$$l = \frac{P - \overset{1}{\cancel{2}}w}{\underset{1}{\cancel{2}}}$$

This step is incorrect because 2 is not a factor of the entire numerator.

Self Check 10 Solve $B = 3c + 4d$ for c.

Now Try Problem 41

EXAMPLE 11 In Chapter 3, we will work with equations that involve the variables x and y, such as $3x + 5y = 10$. Solve this equation for y.

Strategy To solve for y, we will treat y as the only variable in the equation and use properties of equality to isolate it on one side.

Why To solve for a specified variable means to isolate it on one side of the equation.

Solution

We will isolate y on this side of the equation.

$$3x + 5y = 10$$

$$3x + 5y - 3x = 10 - 3x \quad \text{To eliminate } 3x \text{ on the left side, subtract } 3x \text{ from both sides.}$$

$$5y = 10 - 3x \quad \text{Combine like terms: } 3x - 3x = 0.$$

$$\frac{5y}{5} = \frac{10 - 3x}{5} \quad \text{To isolate } y\text{, undo the multiplication by 5 by dividing both sides by 5.}$$

$$y = \frac{10}{5} - \frac{3x}{5} \quad \text{Write } \frac{10 - 3x}{5} \text{ as the difference of two fractions with like denominators, } \frac{10}{5} \text{ and } \frac{3x}{5}.$$

$$y = 2 - \frac{3}{5}x \quad \text{Simplify: } \frac{10}{5} = 2. \text{ Write } \frac{3x}{5} \text{ as } \frac{3}{5}x.$$

This result can be written in the following equivalent form:

$$y = -\frac{3}{5}x + 2 \quad \text{On the right side, write the } x\text{-term first.}$$

Success Tip

When solving for a specified variable, there is often more than one way to express the result. Keep this in mind when you are comparing your answers with those in the back of the text.

Self Check 11 Solve $x + 3y = 12$ for y.

Now Try ▶ Problem 45

Sometimes the distributive property is used to solve a formula for a specified variable.

EXAMPLE 12 Solve $S = 2\pi(r^2 + rh)$ for h.

Strategy To solve for h, we will treat it as the only variable in the equation and isolate it on one side.

Why To solve for a specified variable means to isolate it on one side of the equation.

Solution We will isolate h on the right side of the equation. To begin, we use the distributive property to remove the parentheses.

$$S = 2\pi(r^2 + rh)$$ This is the given formula.

$$S = 2\pi r^2 + 2\pi rh$$ Distribute the multiplication by 2π.

$$S - \mathbf{2\pi r^2} = 2\pi r^2 + 2\pi rh - \mathbf{2\pi r^2}$$ To eliminate $2\pi r^2$ on the right side, subtract $2\pi r^2$ from both sides.

$$S - 2\pi r^2 = 2\pi rh$$ On the right side, combine like terms: $2\pi r^2 - 2\pi r^2 = 0$.

$$\frac{S - 2\pi r^2}{\mathbf{2\pi r}} = \frac{2\pi rh}{\mathbf{2\pi r}}$$ $2\pi rh$ means $2 \cdot \pi \cdot r \cdot h$. To isolate h, undo the multiplication by $2\pi r$ by dividing both sides by $2\pi r$.

$$\frac{S - 2\pi r^2}{2\pi r} = h$$ On the right side, remove the common factors of 2, π, and r: $\frac{\overset{1}{\cancel{2}}\overset{1}{\cancel{\pi}}\overset{1}{\cancel{r}}h}{\underset{1}{\cancel{2}}\underset{1}{\cancel{\pi}}\underset{1}{\cancel{r}}} = h$.

$$h = \frac{S - 2\pi r^2}{2\pi r}$$ Reverse the sides of the equation so that h is on the left.

Self Check 12 Solve $A = -xy(3 - s)$ for s.

Now Try ▶ Problem 49

SECTION 2.4 STUDY SET

VOCABULARY

Fill in the blanks.

1. A ________ is an equation that states a mathematical relationship between two or more variables.
2. The distance around a plane geometric figure is called its ________, and the amount of surface that it encloses is called its ______.
3. The ________ of a three-dimensional geometric solid is the amount of space it encloses.
4. The formula $a = P - b - c$ is _______ for a because a is isolated on one side of the equation and the other side does not contain a.

CONCEPTS

5. Use variables to write the formula relating:
   a. Time, distance, rate
   b. Markup, retail price, cost
   c. Costs, revenue, profit
   d. Interest rate, time, interest, principal

6. Complete the table.

	Principal ·	rate ·	time =	interest
Account 1	\$2,500	5%	2 yr	
Account 2	\$15,000	4.8%	1 yr	

7. Complete the table to find how far light and sound travel in 60 seconds. (*Hint:* mi/sec means miles per second.)

	Rate ·	time =	distance
Light	186,282 mi/sec	60 sec	
Sound	1,088 ft/sec	60 sec	

8. Determine which concept (perimeter, area, or volume) should be used to find each of the following. Then determine which unit of measurement, ft, ft^2, or ft^3, would be appropriate.
   a. The amount of storage in a freezer
   b. The amount of ground covered by a sleeping bag lying on the floor
   c. The distance around a dance floor

NOTATION

Complete the solution.

9. Solve $Ax + By = C$ for y.

$$Ax + By = C$$
$$Ax + By - \quad = C - \quad$$
$$By = C - Ax$$
$$\frac{By}{\quad} = \frac{C - Ax}{\quad}$$
$$y = \frac{C - Ax}{\quad}$$

10. Approximate 98π to the nearest hundredth.

11. a. Write $\pi \cdot r^2 \cdot h$ in simpler form.
    b. In the formula $V = \pi r^2 h$, what does r represent? What does h represent?

12. a. What does 45°C mean?
    b. What does 15°F mean?

GUIDED PRACTICE

Use a formula to solve each problem. See Example 1.

13. **Hollywood.** As of 2010, the movie *Titanic* had brought in \$1,842 million worldwide and made a gross profit of \$1,602 million. What did it cost to make the movie? (Source: the numbers.com/movies)

14. **Valentine's Day.** Find the markup on a dozen roses if a florist buys them wholesale for \$12.95 and sells them for \$47.50.

15. **Service Clubs.** After expenses of \$55.15 were paid, a Rotary Club donated \$875.85 in proceeds from a pancake breakfast to a local health clinic. How much did the pancake breakfast gross?

16. **New Cars.** The factory invoice for a minivan shows that the dealer paid \$16,264.55 for the vehicle. If the sticker price of the van is \$18,202, how much over factory invoice is the sticker price?

See Example 2.

17. **Entrepreneurs.** To start a mobile dog-grooming service, a woman borrowed \$2,500. If the loan was for 2 years and the amount of interest was \$175, what simple interest rate was she charged?

18. **Savings.** A man deposited \$5,000 in a credit union paying 6% simple interest. How long will the money have to be left on deposit to earn \$6,000 in interest?

19. **Loans.** A student borrowed some money from his father at 2% simple interest to buy a car. If he paid his father \$360 in interest after 3 years, how much did he borrow?

20. **Banking.** Three years after opening an account that paid simple interest of 6.45% annually, a depositor withdrew the \$3,483 in interest earned. How much money was left in the account?

See Example 3.

21. **Swimming.** In 1930, a man swam down the Mississippi River from Minneapolis to New Orleans, a total of 1,826 miles. He was in the water for 742 hours. To the nearest tenth, what was his average swimming rate?

22. **Parades.** Rose Parade floats travel down the 5.5-mile-long parade route at a rate of 2.5 mph. How long will it take a float to complete the route if there are no delays?

23. **Hot-Air Balloons.** If a hot-air balloon travels at an average of 37 mph, how long will it take to fly 166.5 miles?

24. **Air Travel.** An airplane flew from Chicago to San Francisco in 3.75 hours. If the cities are 1,950 miles apart, what was the average speed of the plane?

See Example 4.

25. **Frying Foods.** One of the most popular cookbooks in U.S. history, *The Joy of Cooking,* recommends frying foods at 365°F for best results. Convert this to degrees Celsius.

26. **Freezing Points.** Saltwater has a much lower freezing point than freshwater does. For saltwater that is saturated as much as it can possibly get (23.3% salt by weight), the freezing point is −5.8°F. Convert this to degrees Celsius.

27. **Biology.** Cryobiologists freeze living matter to preserve it for future use. They can work with temperatures as low as −270°C. Change this to degrees Fahrenheit.

28. **Metallurgy.** Change 2,212°C, the temperature at which silver boils, to degrees Fahrenheit. Round to the nearest degree.

See Examples 5–7. *If you do not have a calculator, use 3.14 as an approximation of π. Answers may vary slightly depending on which approximation of π is used.*

29. **Energy Savings.** One hundred inches of foam weather stripping tape was placed around the perimeter of a rectangular-shaped window. If the length of the window is 30 inches, what is its width?

30. **Rugs.** Find the amount of floor area covered by a circular throw rug that has a radius of 15 inches. Round to the nearest square inch.

31. **Straws.** Find the volume of a 150 millimeter-long drinking straw that has an inside diameter of 4 millimeters. Round to the nearest cubic millimeter.

32. **Rubber Bands.** The world's largest rubber band ball is $5\frac{1}{2}$ ft tall and was made in 2006 by Steve Milton of Eugene, Oregon. Find the volume of the ball. Round to the nearest cubic foot. (*Hint:* The formula for the volume of a sphere is $V = \frac{4}{3}\pi r^3$.) (Source: timesunion.com)

Solve for the specified variable. See Example 8.

33. $r = c + m$ for c
34. $p = r - c$ for r
35. $P = a + b + c$ for b
36. $a + b + c = 180$ for a

Solve for the specified variable. See Example 9.

37. $V = \frac{1}{3}Bh$ for h
38. $C = \frac{1}{7}Rt$ for R
39. $E = IR$ for R
40. $d = rt$ for t

Solve for the specified variable. See Example 10.

41. $T = 2r + 2t$ for r
42. $y = mx + b$ for x
43. $Ax + By = C$ for x
44. $A = P + Prt$ for t

Solve for y. See Example 11.

45. $2x + 7y = 21$
46. $3x + 4y = 20$
47. $9x - 2y = -8$
48. $5x - 6y = -12$

Solve for the specified variable. See Example 12.

49. $T = 4b(a + am)$ for m
50. $f = 7n(d + dz)$ for z
51. $G = g(4r - 1)$ for r
52. $F = f(9n - 1)$ for n

TRY IT YOURSELF

Solve for the specified variable or expression.

53. $A = \dfrac{a + b + c}{3}$ for c
54. $x = \dfrac{a + b}{2}$ for b
55. $3x + y = 9$ for y
56. $-5x + y = 4$ for y
57. $K = \frac{1}{2}mv^2$ for m
58. $V = \frac{1}{3}\pi r^2 h$ for h
59. $C = 2\pi r$ for r
60. $V = \pi r^2 h$ for h
61. $\dfrac{M}{2} - 9.9 = 2.1B$ for M
62. $\dfrac{G}{0.5} + 16r = -8t$ for G
63. $w = \dfrac{s}{f}$ for f
64. $P = \dfrac{ab}{c}$ for c
65. $-x + 3y = 9$ for y
66. $5y - x = 25$ for y
67. $A = \frac{1}{2}h(b + d)$ for b
68. $C = \frac{1}{4}s(t - d)$ for t
69. $c^2 = a^2 + b^2$ for a^2
70. $x^2 + y^2 + z^2 = d^2$ for y^2
71. $\frac{7}{8}c + w = 9$ for c
72. $\frac{3}{4}m - t = 5b$ for m
73. $m = 70 + t(a + b)$ for b
74. $B = 50 + r(x + y)$ for y
75. $V = lwh$ for l
76. $I = Prt$ for r
77. $2E = \dfrac{T - t}{9}$ for t
78. $D = \dfrac{C - s}{n}$ for s
79. $s = 4\pi r^2$ for r^2
80. $E = mc^2$ for c^2

Look Alikes . . .

81. Solve $A = R + ab$
    a. for R
    b. for a
82. Solve $m = (a + d)T$
    a. for T
    b. for d
83. Solve $S = 2(2lw + wh)$
    a. for h
    b. for l
84. Solve $t = -40 + 9(r + az)$
    a. for r
    b. for z

APPLICATIONS

85. **from Campus to Careers**

Automotive Service Technician

One of the formulas that is often used by automotive technicians who service engines is:

$$\text{Torque} = \frac{5{,}252 \cdot \text{Horsepower}}{\text{RPM}}$$

RPM stands for revolutions per minute. Solve the formula for horsepower.

86. Properties of Water. Refer to the illustration below. Use the temperature formula from this section to find the boiling point of water in degrees Fahrenheit and the freezing point of water in degrees Celsius.

87. Avon Products. Complete the financial statement.

Income statement (dollar amounts in millions)	Quarter ending March '09	Quarter ending March '10
Revenue	2,186.9	2,490.4
Cost of goods sold	2,018.5	2,297.6
Operating profit		

Source: Avon Products, Inc.

88. Credit Cards. The finance charge that a student pays on his credit card is 19.8% APR (annual percentage rate). Determine the finance charges (interest) the student would have to pay if the account's average balance for the year was $2,500.

89. Campers. The perimeter of the window of the camper shell is 140 in. Find the length of one of the shorter sides of the window.

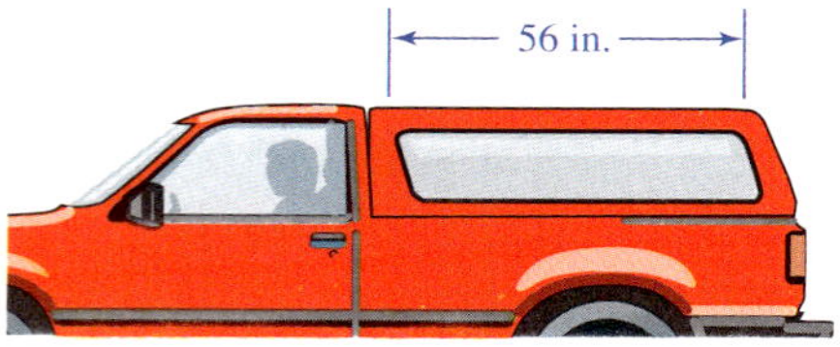

90. Flags. The flag of Eritrea, a country in east Africa, is shown. The perimeter of the flag is 160 inches.

a. What is the width of the flag?

b. What is the area of the red triangular region of the flag?

91. Kites. 650 in.^{2} of nylon cloth were used to make the kite shown. If its height is 26 inches, what is the wingspan?

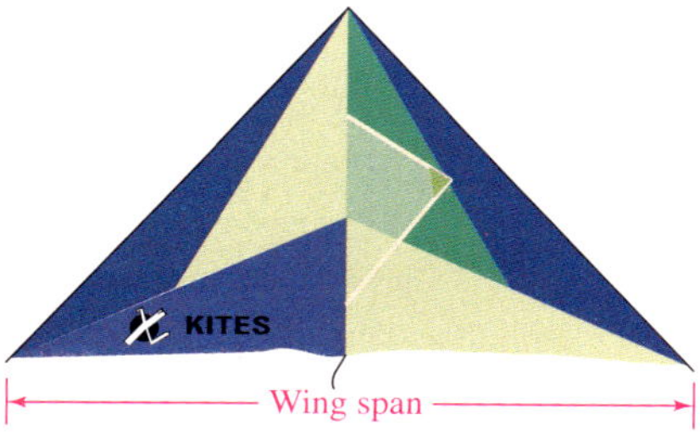

92. Memorials. The Vietnam Veterans Memorial is a black granite wall recognizing the more than 58,000 Americans who lost their lives or remain missing. Find the total area of the two triangular-shaped surfaces on which the names are inscribed.

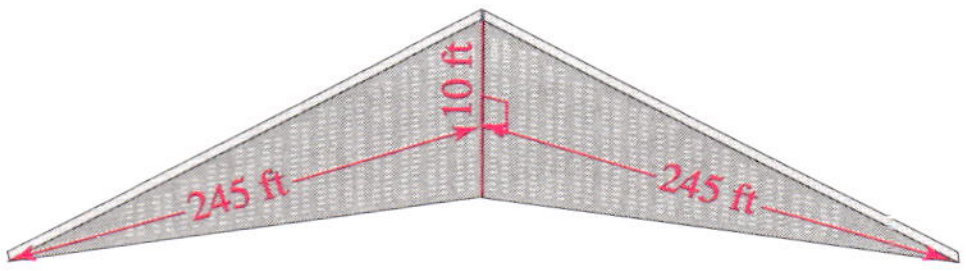

93. Wheelchairs. Find the diameter of the rear wheel and the radius of the front wheel.

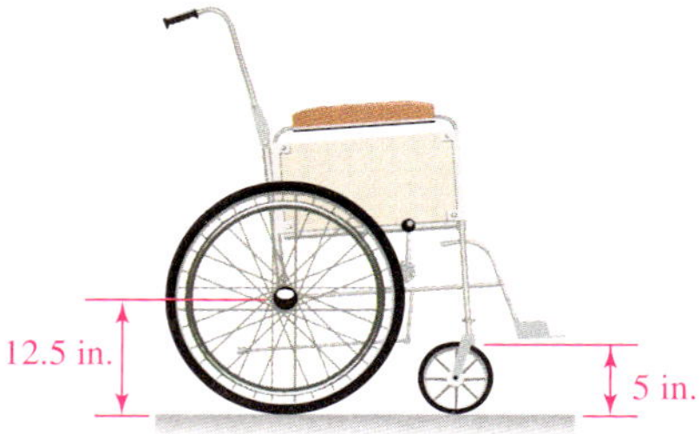

94. Archery. The diameter of a standard archery target used in the Olympics is 48.8 inches. Find the area of the target. Round to the nearest square inch.

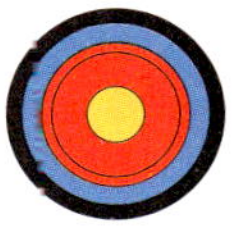

95. Bulls-Eye. See Exercise 94. The diameter of the center yellow ring of a standard archery target is 4.8 inches. What is the area of the bulls-eye? Round to the nearest tenth of a square inch.

96. Geography. The circumference of the Earth is about 25,000 miles. Find its diameter to the nearest mile.

97. Horses. A horse trots in a circle around its trainer at the end of a 28-foot-long rope. Find the area of the circle that is swept out. Round to the nearest square foot.

98. Yo-Yos. How far does a yo-yo travel during one revolution of the "around the world" trick if the length of the string is 21 inches?

99. History. The Inca Empire (1438–1533) was centered in what is now called Peru. A special feature of Inca architecture was the trapezoid-shaped windows and doorways. A standard Inca window was 70 cm high, 50 cm at the base, and 40 cm at the top. Find the area of a window opening.

100. Hamster Habitats. Find the amount of space in the tube.

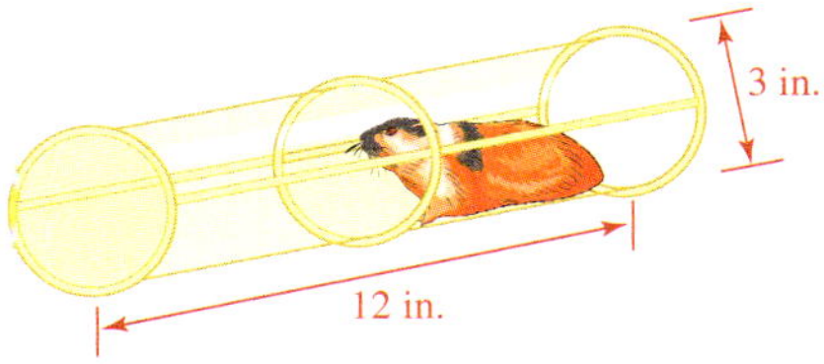

101. Tires. The road surface footprint of a sport truck tire is approximately rectangular. If the area of the footprint is 45 in.², about how wide is the tire?

102. Softball. The strike zone in fast-pitch softball is between the batter's armpit and the top of her knees, as shown. If the area of the strike zone for this batter is 442 in.², what is the width of home plate?

103. Firewood. The cord of wood shown occupies a volume of 128 ft³. How long is the stack?

104. Teepees. The teepees constructed by the Blackfoot Indians were cone-shaped tents about 10 feet high and about 15 feet across at the ground. Estimate the volume of a teepee with these dimensions, to the nearest cubic foot.

105. Igloos. During long journeys, some Canadian Eskimos built winter houses of snow blocks stacked in the dome shape shown. Estimate the volume of an igloo having an interior height of 5.5 feet to the nearest cubic foot.

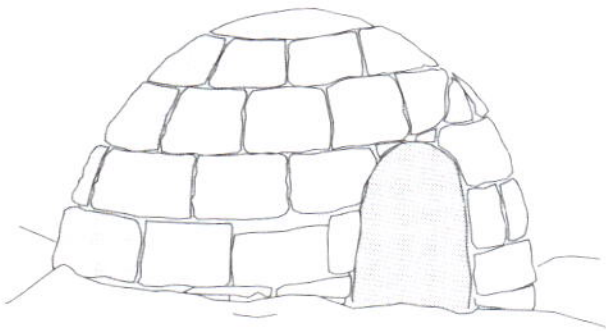

106. Pyramids. The Great Pyramid at Giza in northern Egypt is one of the most famous works of architecture in the world. Find its volume to the nearest cubic foot. (*Hint:* The formula to use is on the inside back cover.)

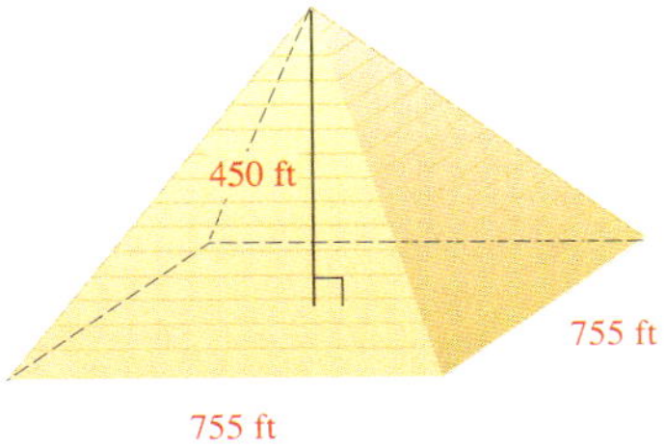

107. Cooking. If the fish shown in the illustration is 18 inches long, what is the area of the grill? Round to the nearest square inch.

108. Skateboarding. A half-pipe ramp is in the shape of a semicircle with a radius of 8 feet. To the nearest tenth of a foot, what is the length of the arc that the rider travels on the ramp?

109. Pulleys. The approximate length L of a belt joining two pulleys of radii r and R feet with centers D feet apart is given by the formula $L = 2D + 3.25(r + R)$. Solve the formula for R.

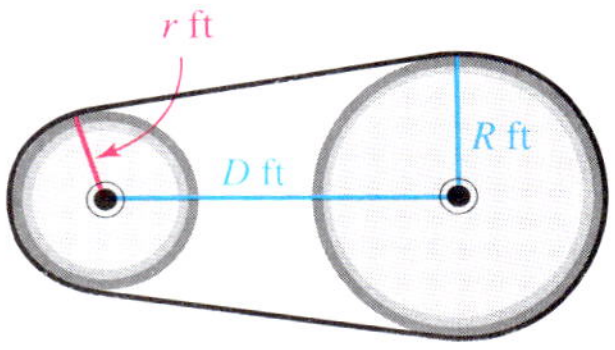

110. Thermodynamics. The Gibbs free-energy formula is given by $G = U - TS + pV$. Solve the formula for the pressure p.

WRITING

111. After solving $A = B + C + D$ for B, a student compared her answer with that at the back of the textbook. Could this problem have two different-looking answers? Explain why or why not.

Student's answer: $B = A - C - D$

Book's answer: $B = A - D - C$

112. A student solved $x + 5c = 3c + a$ for c. His answer was $c = \frac{3c + a - x}{5}$. Explain why the equation is not solved for c.

113. Explain the difference between what perimeter measures and what area measures.

114. Explain the error made below.

$$a = \frac{3b + \overset{1}{\cancel{e}}}{\underset{1}{\cancel{e}}}$$

REVIEW

115. Find 82% of 168.

116. 29.05 is what percent of 415?

117. What percent of 200 is 30?

118. **Shopping.** A woman bought a coat for \$98.95 and some gloves for \$7.95. If the sales tax was 6%, how much did the purchase cost her?

CHALLENGE PROBLEMS

119. In mathematics, letters from the Greek alphabet are often used as variables. Solve the following equation for α (read as "alpha"), the first letter of the Greek alphabet.

$$-7(\alpha - \beta) - (4\alpha - \theta) = \frac{\alpha}{2}$$

120. Solve $B = R - \frac{1}{16}(c - 3D)$ for D.

SECTION 2.5 Problem Solving

OBJECTIVES

1. Apply the steps of a problem-solving strategy.
2. Solve consecutive integer problems.
3. Solve geometry problems.

ARE YOU READY?

The following problems review some basic skills that are needed to solve the application problems in this section.

1. Simplify: $x + x + 1 + 2x + 3$
2. If $x = 8$, find $6x + 1$.
3. If staplers cost \$4.35 each, what is the cost of 9 staplers?
4. Simplify: $x - 0.72x$
5. What is the formula for the perimeter of a rectangle?
6. What is the sum of the measures of the angles of a triangle?
7. Translate to symbols: *8 less than twice a number x*
8. Write 6% as a decimal.

In this section, you will see that algebra is a powerful tool that can be used to solve a wide variety of real-world problems.

1 Apply the Steps of a Problem-Solving Strategy.

To become a good problem solver, you need a plan to follow, such as the following six-step strategy.

Strategy for Problem Solving

1. **Analyze the problem** by reading it carefully to understand the given facts. What information is given? What are you asked to find? What vocabulary is given? Often, a diagram or table will help you understand the facts of the problem.
2. **Assign a variable** to represent an unknown value in the problem. This means, in most cases, to let x = what you are asked to find. If there are other unknown values, represent each of them using an algebraic expression that involves the variable.
3. **Form an equation** by translating the words of the problem into mathematical symbols.
4. **Solve the equation** formed in step 3.
5. **State the conclusion clearly.** Be sure to include the units (such as feet, seconds, or pounds) in your answer.
6. **Check the result** using the original wording of the problem, not the equation that was formed in step 3.

EXAMPLE 1 **California Coastline.** The first part of California's magnificent 17-Mile Drive begins at the Pacific Grove entrance and continues to Seal Rock. It is 1 mile longer than the second part of the drive, which extends from Seal Rock to the Lone Cypress as shown in the map below. The third and final part of the drive winds through Pebble Beach, eventually returning to the entrance. This part of the drive is 1 mile longer than four times the length of the second part. How long is each part of 17-Mile Drive?

Success Tip

When there is more than one unknown value in a problem, let the variable represent the unknown value on which any other unknown values are based.

Analyze The drive is composed of three parts. We need to find the length of each part. We can straighten out the winding 17-Mile Drive and model it with a line segment.

Assign Since the lengths of the first part and of the third part of the drive are related to the length of the second part, we will let x represent the length of the second part. We then express the other lengths in terms of x.

x = the length of the second part of the drive (in miles)

$x + 1$ = the length of the first part of the drive (in miles)

$4x + 1$ = the length of the third part of the drive (in miles)

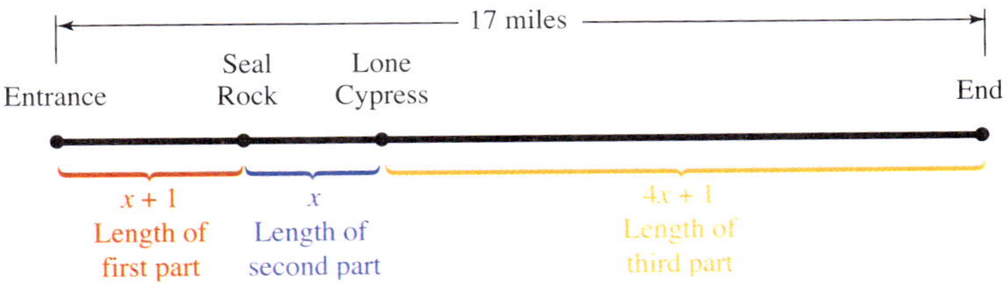

Form Now we translate the words of the problem to an equation.

The length of part 1	plus	the length of part 2	plus	the length of part 3	equals	the total length of the drive.
$x + 1$	$+$	x	$+$	$4x + 1$	$=$	17

Solve

$$x + 1 + x + 4x + 1 = 17$$

$$6x + 2 = 17$$ Combine like terms: $x + x + 4x = 6x$ and $1 + 1 = 2$.

$$6x = 15$$ To undo the addition of 2, subtract 2 from both sides.

$$\frac{6x}{6} = \frac{15}{6}$$ To isolate x, undo the multiplication by 6 by dividing both sides by 6.

$$x = 2.5$$ Do the division.

Recall that x represents the length of the second part of the drive. To find the lengths of the first and third parts, we evaluate $x + 1$ and $4x + 1$ for $x = 2.5$.

Caution

Check the result using the original wording of the problem, not by substituting it into the equation. Why? The equation may have been solved correctly, but the danger is that you may have formed it incorrectly.

First part of drive	*Third part of drive*	
$x + 1 = 2.5 + 1$	$4x + 1 = 4(2.5) + 1$	Substitute 2.5 for *x*.
$= 3.5$	$= 11$	The units are miles.

State The first part of the drive is 3.5 miles long, the second part is 2.5 miles long, and the third part is 11 miles long.

Check Since 3.5 mi + 2.5 mi + 11 mi = 17 mi, the results check.

Self Check 1 **Biking.** The Mountain-Bay State Park Bike Trail in Northeast Wisconsin is 76 miles long. A couple rode the trail in four days. Each day they rode 2 miles more than the previous day. How many miles did they ride each day?

Now Try Problems 13 and 15

EXAMPLE 2 **Computer Logos.** A trucking company had its logo embroidered on the front of baseball caps. It was charged \$8.90 per hat plus a one time set up fee of \$25. If the project cost \$559, how many hats were embroidered?

Analyze

- It cost \$8.90 to have a logo embroidered on a hat.
- The set up charge was \$25.
- The project cost \$559.
- We need to find the number of hats that were embroidered.

Assign Let x = the number of hats that were embroidered.

Success Tip

The *Form* step is often the hardest. To help, write a **verbal model** of the situation (shown here in the boxes) and then translate it into an equation.

Form If x hats are embroidered, at a cost of \$8.90 per hat, the cost to embroider all of the hats is $x \cdot \$8.90$ or $\$8.90x$. Now we translate the words of the problem into an equation.

The cost to embroider one hat	times	the number of hats	plus	the set up charge	equals	the total cost.
8.90	$\cdot$	x	$+$	25	$=$	559

Solve

$$8.90x + 25 = 559$$

$$8.90x = 534 \quad \text{To undo the addition of 25, subtract 25 from both sides.}$$

$$\frac{8.90x}{8.90} = \frac{534}{8.90} \quad \text{To isolate } x\text{, undo the multiplication by 8.90 by dividing both sides by 8.90.}$$

$$x = 60 \quad \text{Do the division.}$$

State The company had 60 hats embroidered.

Check The cost to embroider 60 hats is 60(\$8.90) = \$534. When the \$25 set up charge is added, we get \$534 + \$25 = \$559. The result checks.

Self Check 2 **T-shirts.** A school club had their motto screenprinted on the front of T-shirts. They were charged \$5 per shirt plus a one-time set up fee of \$20. If the project cost \$255, how many T-shirts were printed?

Now Try Problem 23

EXAMPLE 3

Auctions. A classic car owner is going to sell his 1959 Chevy Impala at an auction. He wants to make \$46,000 after paying an 8% commission to the auctioneer. What should be the selling price (called the "hammer price") for the car owner to make this amount of money?

Analyze When the commission is subtracted from the selling price of the car, the owner wants to have \$46,000 left. We need to find the selling price.

Assign Let $x =$ the selling price of the car.

Form The amount of the commission is 8% of x, or $0.08x$. Now we translate the words of the problem to an equation.

The selling price of the car	minus	the auctioneer's commission	should be	\$46,000.
x	$-$	$0.08x$	$=$	46,000

Solve

$$x - 0.08x = 46,000$$

$$0.92x = 46,000$$ Combine like terms: $1.00x - 0.08x = 0.92x$. We could begin with this equation because after the 8% commission is paid, 100%–8% or 92% of the selling price should be \$46,000.

$$\frac{0.92x}{0.92} = \frac{46,000}{0.92}$$ To isolate x, undo the multiplication by 0.92 by dividing both sides by 0.92.

$$x = 50,000$$ Do the division. This is the selling price of the car.

State The owner will make \$46,000 if the car sells for \$50,000.

Check An 8% commission on \$50,000 is $0.08(\$50,000) = \$4,000$. The owner will keep $\$50,000 - \$4,000 = \$46,000$. The result checks.

Self Check 3 **Cattle Auction.** A farmer is going to sell one of his Black Angus cattle at an auction and would like to make \$2,597 after paying a 6% commission to the auctioneer. For what selling price will the farmer make this amount of money?

Now Try Problem 29

The Language of Algebra

Here are some words and phrases that often translate to an equal symbol $=$.

is	are
should be	will be
yields	amounts to
represents	gives
is the same as	was

2 Solve Consecutive Integer Problems.

Integers that follow one another, such as 15 and 16, are called **consecutive integers.** They are 1 unit apart. **Consecutive even integers** are even integers that differ by 2 units, such as 12 and 14. Similarly, **consecutive odd integers** differ by 2 units, such as 9 and 11. When solving consecutive integer problems, if we let $x =$ the first integer, then

- two consecutive integers are x and $x + 1$
- two consecutive even integers are x and $x + 2$
- two consecutive odd integers are x and $x + 2$
- three consecutive integers are x, $x + 1$, and $x + 2$
- three consecutive even integers are x, $x + 2$, and $x + 4$
- three consecutive odd integers are x, $x + 2$, and $x + 4$

EXAMPLE 4

U.S. History. The year George Washington was chosen president and the year the Bill of Rights went into effect are consecutive odd integers whose sum is 3,580. Find the years.

Analyze We need to find two consecutive odd integers whose sum is 3,580. From history, we know that Washington was elected president first and the Bill of Rights went into effect later.

Assign Let $x =$ the first odd integer (the date when Washington was chosen president). The next odd integer is 2 *greater than* x, therefore $x + 2 =$ the next larger odd integer (the date when the Bill of Rights went into effect).

The Language of Algebra

Consecutive means following one after the other in order. Elton John holds the record for the most *consecutive* years with a song on the Top 50 music chart: 31 years (1970 to 2000).

Form

The first odd integer	plus	the second odd integer	is	3,580.
x	$+$	$x + 2$	$=$	3,580

Solve

$$x + x + 2 = 3{,}580$$

$$2x + 2 = 3{,}580 \quad \text{Combine like terms: } x + x = 2x.$$

$$2x = 3{,}578 \quad \text{To undo the addition of 2, subtract 2 from both sides.}$$

$$x = 1{,}789 \quad \text{To isolate } x \text{, undo the multiplication by 2 by dividing both sides by 2.}$$

State George Washington was chosen president in the year 1789. The Bill of Rights went into effect in $1789 + 2 = 1791$.

Check 1789 and 1791 are consecutive odd integers whose sum is $1789 + 1791 = 3{,}580$. The answers check.

Self Check 4 **Dictionaries.** The definitions of the words *little* and *lobby* are on back-to-back pages in a dictionary. If the sum of the page numbers is 1,159, on what page can the definition of *little* be found?

Now Try Problem 37

3 Solve Geometry Problems.

EXAMPLE 5 **Crime Scenes.** Police used 400 feet of yellow tape to fence off a rectangular-shaped lot for an investigation. They used 50 fewer feet of tape for each width than for each length. Find the dimensions of the lot.

Analyze Since the yellow tape surrounded the lot, the concept of perimeter applies. Recall that the formula for the perimeter of a rectangle is $P = 2l + 2w$. We also know that the width of the lot is 50 feet less than the length.

The Language of Algebra

Dimensions are measurements of length and width. We might speak of the *dimensions* of a dance floor or a TV screen.

Assign Since the width of the lot is given in terms of the length, we let l = the length of the lot. Then $l - 50$ = the width.

Form Using the perimeter formula, we have:

2	times	the length	plus	2	times	the width	is	the perimeter.
2	$\cdot$	l	$+$	2	$\cdot$	$(l - 50)$	$=$	400

Solve

$$2l + 2(l - 50) = 400 \quad \text{Write the parentheses so that the entire expression } l - 50 \text{ is multiplied by 2.}$$

$$2l + 2l - 100 = 400 \quad \text{Distribute the multiplication by 2.}$$

$$4l - 100 = 400 \quad \text{Combine like terms: } 2l + 2l = 4l.$$

$$4l = 500 \quad \text{To undo the subtraction of 100, add 100 to both sides.}$$

$$l = 125 \quad \text{To isolate } l \text{, undo the multiplication by 4 by dividing both sides by 4.}$$

Success Tip

When solving geometry problems, a sketch is often helpful.

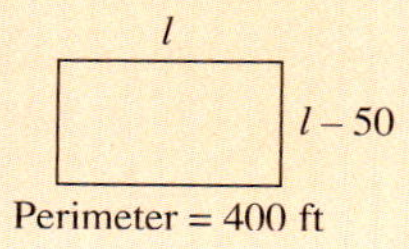

State The length of the lot is 125 feet and width is $125 - 50 = 75$ feet.

Check The width (75 feet) is 50 less than the length (125 feet). The perimeter of the lot is $2(125) + 2(75) = 250 + 150 = 400$ feet. The results check.

Self Check 5 **Counters.** A rectangular counter for the customer service department of a store is 6 feet longer than it is wide. If the perimeter is 32 feet, find the outside dimensions of the counter.

Now Try Problem 43

EXAMPLE 6 **Isosceles Triangles.** If the vertex angle of an isosceles triangle is 56°, find the measure of each base angle.

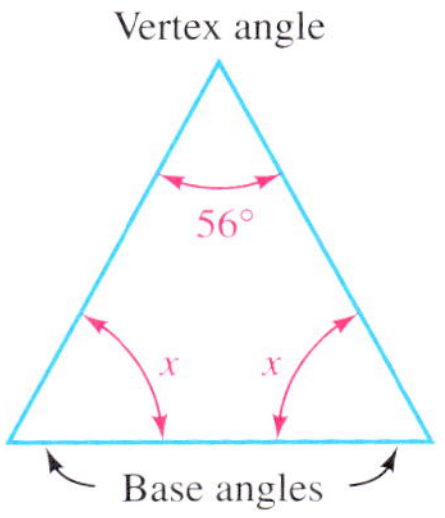

Analyze An **isosceles triangle** has two sides of equal length, which meet to form the **vertex angle.** In this case, the measurement of the vertex angle is 56°. We can sketch the triangle as shown. The **base angles** opposite the equal sides are also equal. We need to find their measure.

Assign If we let $x =$ the measure (in degrees) of one base angle, the measure of the other base angle is also x.

Form Since the sum of the angles of any triangle is 180°, the sum of the base angles and the vertex angle is 180°. We can use this fact to form the equation.

One base angle	plus	the other base angle	plus	the vertex angle	is	180°.
x	$+$	x	$+$	56	$=$	180

Solve

$$x + x + 56 = 180$$

$$2x + 56 = 180 \quad \text{Combine like terms: } x + x = 2x.$$

$$2x = 124 \quad \text{To undo the addition of 56, subtract 56 from both sides.}$$

$$x = 62 \quad \text{To isolate } x\text{, undo the multiplication by 2 by dividing both sides by 2.}$$

State The measure of each base angle is 62°.

Check Since $62° + 62° + 56° = 180°$, the answer checks.

Self Check 6 **Geometry.** The perimeter of an isosceles triangle is 32 cm. If the base is 8 cm, find the length of each remaining side.

Now Try Problem 47

SECTION 2.5 STUDY SET

VOCABULARY

Fill in the blanks.

1. Integers that follow one another, such as 7 and 8, are called ___________ integers.
2. An _________ triangle is a triangle with two sides of the same length.
3. The equal sides of an isosceles triangle meet to form the _______ angle. The angles opposite the equal sides are called _____ angles, and they have equal measures.
4. When asked to find the dimensions of a rectangle, we are to find its _______ and _______.

CONCEPTS

5. A 17-foot pipe is cut into three sections. The longest section is three times as long as the shortest, and the middle-sized section is 2 feet longer than the shortest. Complete the diagram.

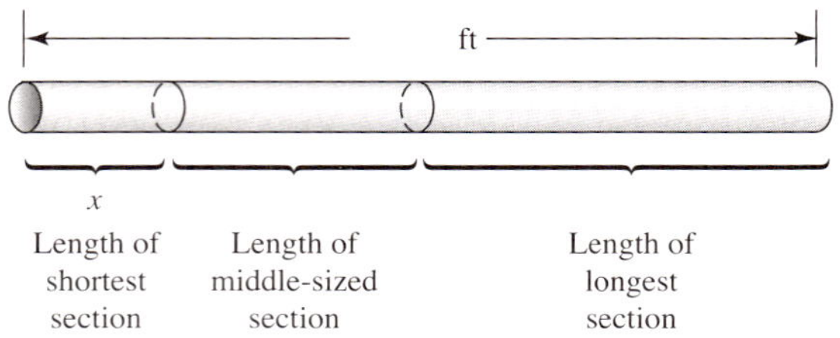

6. It costs $28 per hour to rent a trailer. Write an expression that represents the cost to rent the trailer for x hours.
7. A realtor is paid a 3% commission on the sale of a house. Write an expression that represents the amount of the commission if a house sells for $\$x$.
8. The perimeter of the rectangle below is 15 feet. Fill in the blanks: 2(   ) + $2x$ =

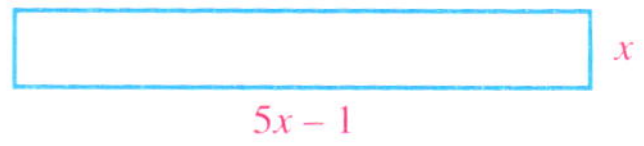

9. What is the sum of the measures of the angles of any triangle?

10. Refer to the isosceles triangle on the right.
    a. Find the missing angle measure.
    b. Find the missing side length.

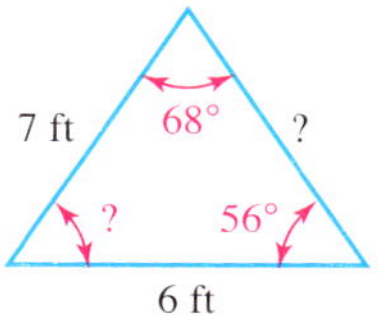

NOTATION

11. a. If x represents an integer, write an expression for the next largest integer.
    b. If x represents an odd integer, write an expression for the next largest odd integer.
    c. If x represents an even integer, write expressions for the next two largest even integers.
12. What does 45° mean?

GUIDED PRACTICE

See Example 1.

13. A 12-foot board has been cut into two sections, one twice as long as the other. How long is each section?

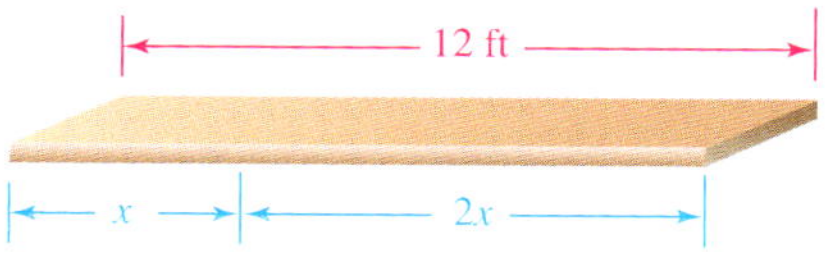

14. The robotic arm will extend a total distance of 18 feet. Find the length of each section.

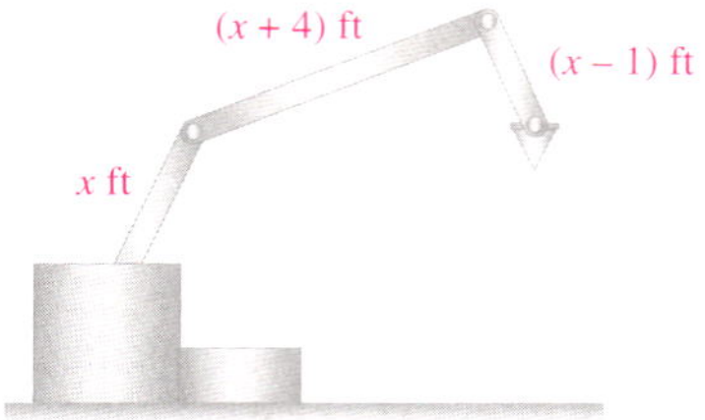

APPLICATIONS

15. **National Parks.** The Natchez Trace Parkway is a historical 444-mile route from Natchez, Mississippi, to Nashville, Tennessee. A couple drove the Trace in four days. Each day they drove 6 miles more than the previous day. How many miles did they drive each day?

16. **Touring.** A rock group plans to travel for a total of 38 weeks, making three concert stops. They will be in Japan for 4 more weeks than they will be in Australia. Their stay in Sweden will be 2 weeks shorter than that in Australia. How many weeks will they be in each country?

17. **Solar Heating.** Two solar panels were installed side-by-side on a roof, as shown below. One panel is 3.4 feet wider than the other. Find the width of each panel.

18. **Accounting.** Determine the 2010 income of Aeropostale Inc. for each quarter from the data in the graph below.

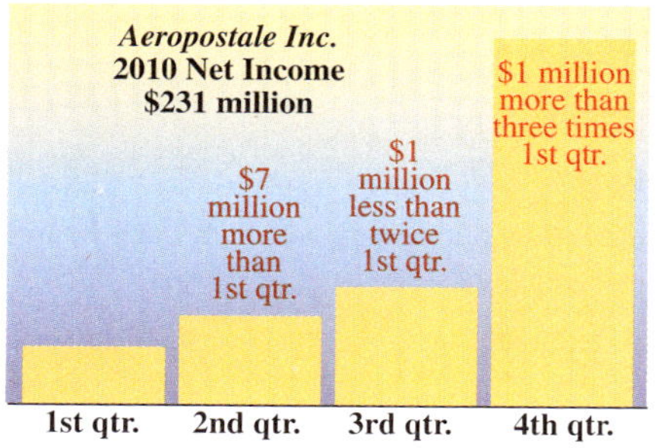

Source: moneycentral.msn.com

19. iPhone Apps. A student spent a total of \$52.97 on the purchase of three applications in the Apps Store on his iPhone. *Call of Duty: World at War: Zombies II* cost \$7 more than *Guitar Hero,* and *Tom Tom USA* cost \$4.11 more than twelve times *Guitar Hero.* Find the cost of each application.

20. Water Usage. It takes about 3.8 times more gallons of water to produce one pound of grain-fed beef than it does to produce one pound of grain-fed chicken. If a combined total of 3,135 gallons of water are needed to produce one pound of each meat, how many gallons does it take to produce one pound of chicken? How many gallons does it take to produce one pound of beef? (Source: The Sierra Club: *The True Cost of Food*)

21. Counting Calories. A slice of pie with a scoop of ice cream has 850 calories. The calories in the pie alone are 100 more than twice the calories in the ice cream alone. How many calories are in each food?

22. Waste Disposal. Two tanks hold a total of 45 gallons of a toxic solvent. One tank holds 6 gallons more than twice the amount in the other. How many gallons does each tank hold?

23. Concerts. The fee to rent a concert hall is \$2,250 plus \$150 per hour to pay for the support staff. For how many hours can an orchestra rent the hall and stay within a budget of \$3,300?

24. Truck Mechanics. An engine repair cost a truck owner \$1,185 in parts and labor. If the parts were \$690 and the mechanic charged \$45 per hour, how many hours did the repair take?

25. Field Trips. It costs a school \$65 a day plus \$0.25 per mile to rent a 15-passenger van. If the van is rented for two days, how many miles can be driven on a \$275 budget?

26. Decorations. A party supply store charges a set-up fee of \$80 plus 35¢ per balloon to make a balloon arch. A business has \$150 to spend on decorations for their grand opening. How many balloons can they have in the arch? (*Hint:* 35¢ = \$0.35.)

27. Tutoring. High school students enrolling in a private tutoring program must first take a placement test (cost \$25) before receiving tutoring (cost \$18.75 per hour). If a family has set aside \$400 to get their child extra help, how many hours of tutoring can they afford?

28. Data Conversion. The *Books2Bytes* service converts old print books to Microsoft Word electronic files for \$20 per book plus \$2.25 per page. If it cost \$1,201.25 to convert a novel, how many pages did the novel have?

29. Cattle Auctions. A cattle rancher is going to sell one of his prize bulls at an auction and would like to make \$45,500 after paying a 9% commission to the auctioneer. For what selling price will the rancher make this amount of money?

30. Listing Price. At what price should a home be listed if the owner wants to make \$567,000 on its sale after paying a 5.5% real estate commission?

31. Selling Used Clothing. A *consignment shop* accepts an item of clothing that no longer fits (or one you have grown tired of) and sells it for you. The shop then charges you an agreed on percent of the selling price as their profit. Suppose the owner of a designer wool coat would like to make \$210 on its sale at a consignment shop. If there is a $12\frac{1}{2}\%$ consignment charge, for what price must the coat be sold?

32. Finder's Fees. A *finder's fee* is an amount of money that is paid to someone who brings people together for business purposes. Suppose the owner of a software company needs to sell it and make \$9,950,000 to pay back creditors. If he expects to pay a finder's fee of $\frac{1}{2}\%$ of the selling price to find a qualified buyer, for what price must the company be sold?

33. Savings Accounts. The balance in a savings account grew by 5% in one year, to \$5,512.50. What was the balance at the beginning of the year?

34. Aluminum Cans. Today's aluminum cans are much thinner and lighter than those of the past. From 1972 to 2010, the number of empty cans produced from one pound of aluminum has increased by about 45%. If 32 cans could be produced from one pound of aluminum in 2010, how many cans could be produced from one pound of aluminum in 1972? Round to the nearest can. (Source: cancentral.com)

Consecutive integer problems

35 Soccer. Ronaldo of Brazil and Gerd Mueller of Germany rank 1 and 2, respectively, with the most goals scored in World Cup play. The number of goals Ronaldo and Mueller have scored are consecutive integers that total 29. Find the number of goals scored by each man. (Source: planetworldcup.com)

36. Dictionaries. The definitions of the words *job* and *join* are on back-to-back pages in a dictionary. If the sum of those page numbers is 1,411, on what page can the definition of *job* be found?

37. TV History. *Friends* and *Leave It to Beaver* are two of the most popular television shows of all time. The number of episodes of each show are consecutive even integers whose sum is 470. If there are more episodes of *Friends,* how many episodes of each were there? (Source: angelfire.com)

38. Time Off. The table shows the average number of days off an employed adult receives for selected countries. Complete the table. The numbers of days are listed in descending order.

Average Number of Days Off per Year*

Country	Days
Brazil	41
Norway	?
South Africa	?
U.S.	25

Consecutive odd integers whose sum is 68.

* Employee has 10 years of service and works 5 days a week
Source: *The Wall Street Journal,* 2009.

39. Celebrity Birthdays. Selena Gomez, Jennifer Lopez, and Sandra Bullock have birthdays (in that order) on consecutive even-numbered days in July. The sum of the calendar dates of their birthdays is 72. Find each birthday.

40. Locks. The three numbers of the combination for a lock are consecutive integers, and their sum is 81. Find the combination.

Geometry problems

41. Tennis. The perimeter of a regulation singles tennis court is 210 feet and the length is 3 feet less than three times the width. What are the dimensions of the court?

42. Swimming Pools. The seawater Orthlieb Pool in Casablanca, Morocco, is the largest swimming pool in the world. With a perimeter of 1,110 meters, this rectangular-shaped pool is 30 meters longer than 6 times its width. Find its dimensions.

43. Art. The *Mona Lisa* was completed by Leonardo da Vinci in 1506. The length of the picture is 11.75 inches shorter than twice the width. If the perimeter of the picture is 102.5 inches, find its dimensions.

44. New York City. Central Park, which lies in the middle of Manhattan, is rectangular-shaped and has a 6-mile perimeter. The length is 5 times the width. What are the dimensions of the park?

45. Engineering. A truss is in the form of an isosceles triangle. Each of the two equal sides is 4 feet shorter than the third side. If the perimeter is 25 feet, find the lengths of the sides.

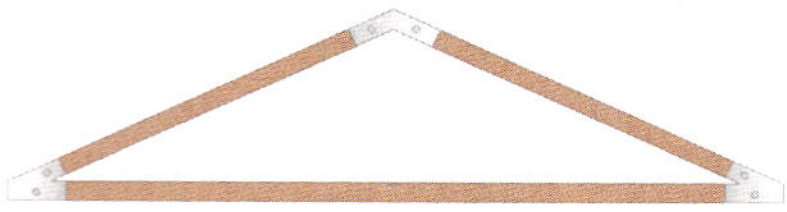

46. First Aid. A sling is in the shape of an isosceles triangle with a perimeter of 144 inches. The longest side of the sling is 18 inches longer than either of the other two sides. Find the lengths of each side.

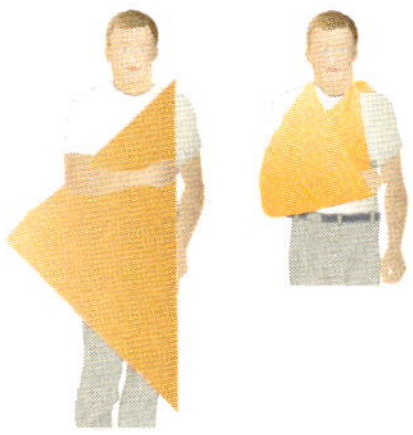

47. TV Towers. A TV tower is supported by several guy wires. Two of the guy wires are attached to the top of the tower to form an isosceles triangle with the ground, as shown on the right. The measure of each of the base angles of the triangle is 4 times the third angle (the vertex angle). Find the measure of the vertex angle.

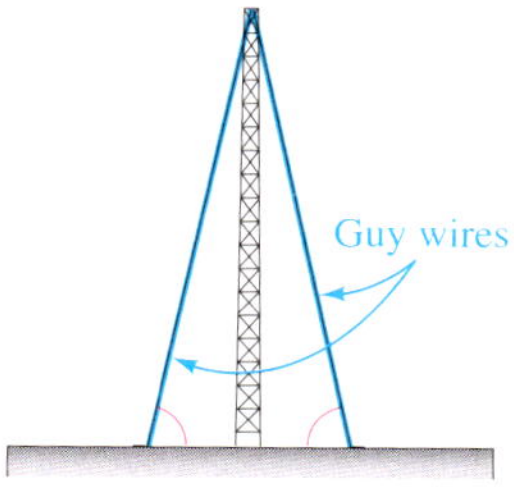

48. Clotheslines. A pair of damp jeans are hung in the middle of a clothesline to dry. Find x, the angle that the clothesline makes with the horizontal.

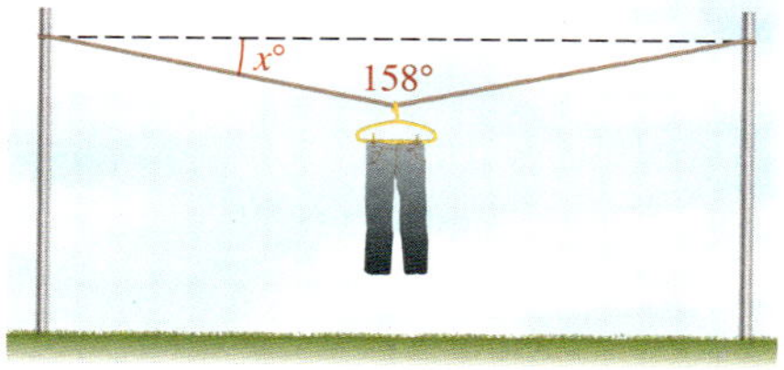

49. Mountain Bicycles. For the bicycle frame shown, the angle that the horizontal crossbar makes with the seat support is 15° less than twice the angle at the steering column. The angle at the pedal gear is 25° more than the angle at the steering column. Find these three angle measures.

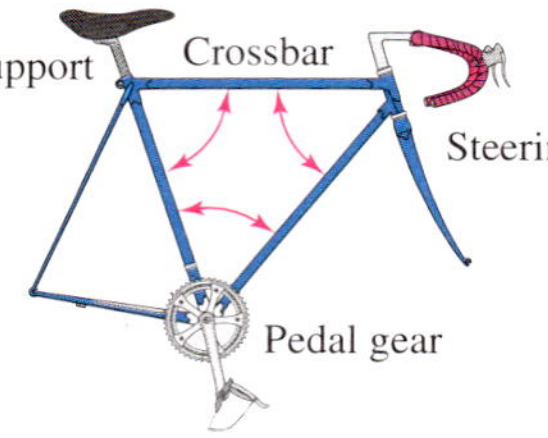

50. Triangles. The measure of $\angle 1$ (read as angle 1) of a triangle is one-half that of $\angle 2$. The measure of $\angle 3$ is equal to the sum of the measures of $\angle 1$ and $\angle 2$. Find each angle measure.

51. Angles. Two angles are called ***complementary angles*** when the sum of their measures is 90°. Refer to the figure on the right. Find x. Then find the measures of the complementary angles.

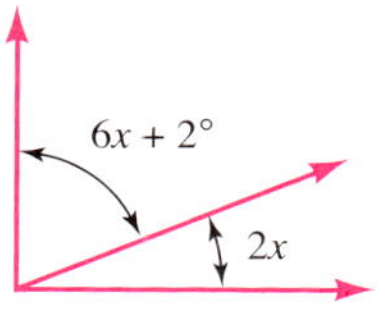

52. Angles. Two angles are called ***supplementary angles*** when the sum of their measures is 180°. Refer to the figure below. Find x. Then find the measures of the supplementary angles.

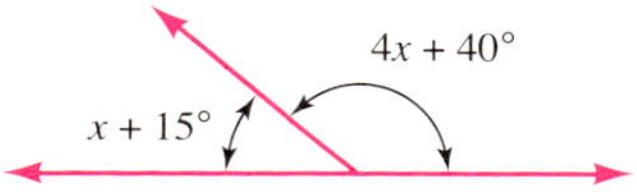

53. "Lightning Bolt." In 2010, Usain Bolt of Jamaica held the world record for the 100 meters and the 200 meters sprints. His *maximum stride angle* shown below is 5° less than 1.5 times its supplement. Find his maximum stride angle. You may need to refer to problem 52 to review the geometry involved. (Source: somaxsports.com)

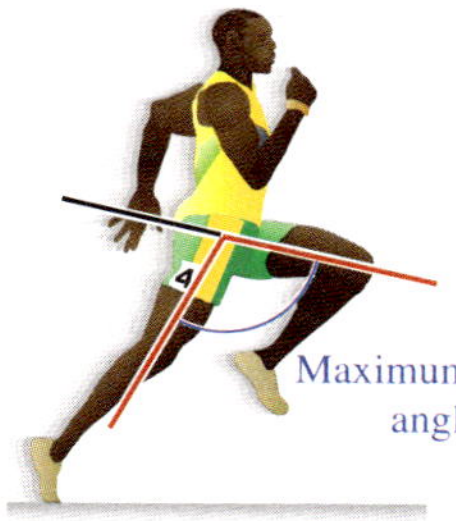

54. **from Campus To Careers**

Automotive Service Technician

© barang/Shutterstock.com

The *sweep angle* of a windshield wiper arm is 115° as shown below. Find x so that the area that is cleared by the wiper is centered on the car's windshield.

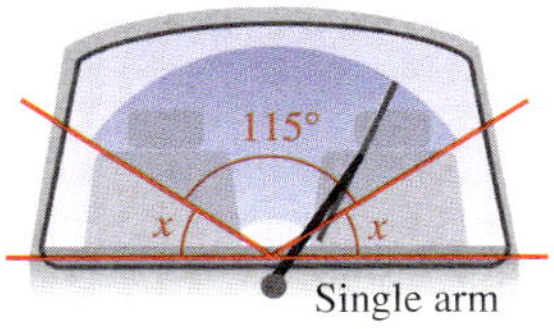

WRITING

55. Create a geometry problem that could be answered by solving the equation $2w + 2(w + 5) = 26$.

56. What information do you need to know to answer the following question?

A business rented a copy machine for \$85 per month plus 4¢ for every copy made. How many copies can be made each month?

57. Make a list of words and phrases that translate to an equal symbol =.

58. Define the word *strategy*.

REVIEW

Solve.

59. $\frac{5}{8}x = -15$

60. $\frac{12x + 24}{13} = 36$

61. $\frac{3}{4}y = \frac{2}{5}y - \frac{3}{2}y - 2$

62. $4.2(y - 4) - 0.6y = -13.2$

CHALLENGE PROBLEMS

63. What concept discussed in this section is illustrated by the following day and time?

Two minutes and three seconds past 1 A.M. on the 5th day of April, 2006

64. **Manufacturing.** A company has two machines that make widgets. The production costs are listed below.

Machine 1: Setup cost \$400 and \$1.70 per widget

Machine 2: Setup cost \$500 and \$1.20 per widget

Find the number of widgets for which the cost to manufacture them on either machine is the same.

SECTION 2.6 More about Problem Solving

OBJECTIVES

1. Solve investment problems.
2. Solve uniform motion problems.
3. Solve liquid mixture problems.
4. Solve dry mixture problems.
5. Solve number-value problems.

ARE YOU READY?

The following problems review basic skills and some formulas that are needed to solve money, motion, and mixture applications.

1. Find the amount of interest earned by \$8,000 invested at a 5% annual simple interest rate for 1 year.
2. At 45 miles per hour, how far will a car travel in 3 hours?
3. A 12-gallon mixture of antifreeze and water is 30% antifreeze. How many gallons of the mixture is antifreeze?
4. At \$2.45 per pound, what is the value of 8 pounds of ground beef?
5. A couple invested \$6,000 of their \$20,000 lottery winnings in bonds. How much do they have left to invest in stocks?
6. Multiply: $100(0.03x)$

In this section, we will solve problems that involve money, motion, and mixtures. Tables are a helpful way to organize the information given in these problems.

1 Solve Investment Problems.

To find the amount of *simple interest* I an investment earns, we use the formula $I = Prt$, where P is the principal (the amount invested), r is the annual interest rate, and t is the time in years.

EXAMPLE 1 **Paying Tuition.** A college student wants to invest the $12,000 inheritance he received and use the annual interest earned to pay his tuition cost of $945. The highest rate offered by a bank is 6% annual simple interest. At this rate, he cannot earn the needed $945, so he decides to invest some of the money in a riskier, but more profitable, investment offering a 9% return. How much should he invest at each rate?

Analyze We know that $12,000 will be invested for 1 year at two rates: 6% and 9%. We are asked to find the amount invested at each rate so that the total return would be $945.

Assign Let x = the amount invested at 6%. Then $12{,}000 - x$ = the amount invested at 9%.

Form To organize the facts of the problem, we enter the principal, rate, time, and interest earned from each account in a table.

Step 1: List each investment in a row of the table.

Bank				
Riskier Investment				

Step 2: Label the columns using $I = Prt$ reversed and also write Total.

	P	$\cdot\ r$	$\cdot\ t =$	I
Bank				
Riskier Investment				
				Total:

Step 3: Enter the rates as decimals, the times, and the total interest.

	P	$\cdot\ r$	$\cdot\ t =$	I
Bank		**0.06**	**1**	
Riskier Investment		**0.09**	**1**	
				Total: **945**

Step 4: Enter each unknown principal.

	P	$\cdot\ r$	$\cdot\ t =$	I
Bank	$\mathbf{x}$	0.06	1	
Riskier Investment	$\mathbf{12{,}000 - x}$	0.09	1	
				Total: 945

Step 5: In the last column, multiply P, r, and t to obtain expressions for the interest earned.

	P	$\cdot\ r$	$\cdot\ t =$	I
Bank	x	0.06	1	$\mathbf{0.06x}$
Riskier Investment	$12{,}000 - x$	0.09	1	$\mathbf{0.09(12{,}000 - x)}$
				Total: 945

← This is $x \cdot 0.06 \cdot 1$.

← This is $(12{,}000 - x) \cdot 0.09 \cdot 1$.

Use the information in this column to form an equation.

The interest earned at 6%	plus	the interest earned at 9%	equals	the total interest.
$0.06x$	$+$	$0.09(12{,}000 - x)$	$=$	945

Solve

$$0.06x + 0.09(12{,}000 - x) = 945$$

$$100[0.06x + 0.09(12{,}000 - x)] = 100(945)$$ Multiply both sides by 100 to clear the equation of decimals.

$$100(0.06x) + 100(0.09)(12{,}000 - x) = 100(945)$$ Distribute the multiplication by 100.

$$6x + 9(12{,}000 - x) = 94{,}500$$ Do the multiplications by 100.

$$6x + 108{,}000 - 9x = 94{,}500$$ Use the distributive property.

$$-3x + 108{,}000 = 94{,}500$$ Combine like terms.

$$-3x = -13{,}500$$ Subtract 108,000 from both sides.

$$x = 4{,}500$$ To isolate x, divide both sides by −3.

Success Tip

We can clear an equation of decimals by multiplying both sides by a power of 10. Here, we multiply 0.06 and 0.09 by 100 to move each decimal point two places to the right:

$100(0.06) = 6 \quad 100(0.09) = 9$

State The student should invest \$4,500 at 6% and \$12,000 − \$4,500 = \$7,500 at 9%.

Check The first investment will earn 0.06(\$4,500), or \$270. The second will earn 0.09(\$7,500), or \$675. Since the total return will be \$270 + \$675 = \$945, the results check.

Self Check 1 **Investments.** A student invested a total of \$4,200 in certificates of deposit, one at 2% and the other at 3%. Find the amount invested at each rate if the first year combined interest income from the two investments was \$102.

Now Try Problem 17

2 Solve Uniform Motion Problems.

If we know the rate r at which we will be traveling and the time t we will be traveling at that rate, we can find the distance d traveled by using the formula $d = rt$.

EXAMPLE 2 **Rescues at Sea.** A cargo ship, heading into port, radios the Coast Guard that it is experiencing engine trouble and that its speed has dropped to 3 knots. (This is 3 sea miles per hour.) Immediately, a Coast Guard cutter leaves port and speeds at a rate of 25 knots directly toward the disabled ship, which is 56 sea miles away. How long will it take the Coast Guard to reach the ship? (Sea miles are also called nautical miles.)

Success Tip

A sketch is helpful when solving uniform motion problems.

Analyze We know the *rate* of each ship (25 knots and 3 knots), and we know that they must close a *distance* of 56 sea miles between them. We don't know the *time* it will take to do this.

Assign Let t = the time it takes the Coast Guard to reach the cargo ship. During the rescue, the ships don't travel at the same rate, but they do travel for the same amount of time. Therefore, t also represents the travel time for the cargo ship.

Form We enter the rates, the variable t for each time, and the total distance traveled by the ships (56 sea miles) in the table. To fill in the last column, we use the formula $r \cdot t = d$ twice to find an expression for each distance traveled: $25 \cdot t = 25t$ and $3 \cdot t = 3t$.

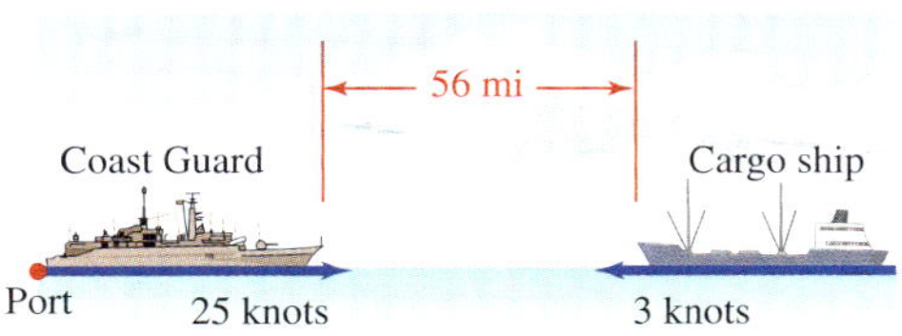

	r	$\cdot\ t$	$= d$
Coast Guard cutter	25	t	$25t$
Cargo ship	3	t	$3t$
			Total: 56

Multiply $r \cdot t$ to obtain an expression for each distance traveled.

Use the information in this column to form an equation.

Caution

A common error is to enter 56 miles as the distance traveled for each ship. However, neither ship traveled 56 miles. Together, they travel 56 miles.

	r	$\cdot\ t$	$= d$
Cutter			56
Cargo			56

The distance the cutter travels	plus	the distance the ship travels	equals	the original distance between the ships.
$25t$	$+$	$3t$	$=$	56

Solve

$$25t + 3t = 56$$

$$28t = 56 \quad \text{Combine like terms: } 25t + 3t = 28t.$$

$$t = \frac{56}{28} \quad \text{To isolate } t\text{, divide both sides by 28.}$$

$$t = 2 \quad \text{Do the division.}$$

State The ships will meet in 2 hours.

Check In 2 hours, the Coast Guard cutter travels $25 \cdot 2 = 50$ sea miles, and the cargo ship travels $3 \cdot 2 = 6$ sea miles. Together, they travel $50 + 6 = 56$ sea miles. Since this is the original distance between the ships, the result checks.

Self Check 2 **Rescues.** Two search-and-rescue teams leave base at the same time looking for a lost boy. The first team, on foot, heads north at 2 mph, and the other, on horseback, heads south at 4 mph. How long will it take them to search a distance of 21 miles between them?

Now Try Problem 29

EXAMPLE 3 **Concert Tours.** While on tour, a country music star travels by bus. Her musical equipment is carried in a truck. How long will it take her bus, traveling 60 mph, to overtake the truck, traveling at 45 mph, if the truck had a $1\frac{1}{2}$-hour head start to her next concert location?

Analyze We know the rate of each vehicle (60 mph and 45 mph) and that the truck began the trip $1\frac{1}{2}$ or 1.5 hours earlier than the bus. We need to determine how long it will take the bus to catch up to the truck.

Assign Let $t =$ the time it takes the bus to overtake the truck. With a 1.5-hour head start, the truck is on the road longer than the bus. Therefore, $t + 1.5 =$ the truck's travel time.

Form We enter each rate and time in the table, and use the formula $r \cdot t = d$ twice to fill in the distance column.

	r ·	t =	d
Bus	60	t	$60t$
Truck	45	$t + 1.5$	$45(t + 1.5)$

Multiply $r \cdot t$ to obtain an expression for each distance traveled.

Enter this information first.

Use the information in this column to form an equation.

When the bus overtakes the truck, they will have traveled the same distance.

The distance traveled by the bus	is the same as	the distance traveled by the truck.
$60t$	=	$45(t + 1.5)$

Solve

$60t = 45(t + 1.5)$

$60t = 45t + 67.5$ Distribute the multiplication by 45: $45(1.5) = 67.5$.

$15t = 67.5$ Subtract $45t$ from both sides: $60t - 45t = 15t$.

$t = 4.5$ To isolate t, divide both sides by 15: $\frac{67.5}{15} = 4.5$.

State The bus will overtake the truck in 4.5 or $4\frac{1}{2}$ hours.

Check In 4.5 hours, the bus travels $60(4.5) = 270$ miles. The truck travels for $1.5 + 4.5 = 6$ hours at 45 mph, which is $45(6) = 270$ miles. Since the distances traveled are the same, the result checks.

Success Tip

We used 1.5 hrs for the head start because it is easier to solve
$60t = 45(t + 1.5)$
than
$60t = 45\left(t + 1\frac{1}{2}\right)$

Self Check 3 **Moving Day.** A moving van, packed with a family's belongings, left their old home for their new home, traveling at 40 miles per hour. Forty-five minutes $\left(\frac{3}{4}\text{ hour}\right)$ later, the family left for their new home, traveling by car at 60 miles per hour. How long did it take the family to overtake the moving van?

Now Try Problem 33

3 Solve Liquid Mixture Problems.

We now discuss how to solve mixture problems. In the first type, a liquid mixture of a desired strength is made from two solutions with different concentrations (strengths).

EXAMPLE 4 **Mixing Solutions.** A chemistry experiment calls for a 30% sulfuric acid solution. If the lab supply room has only 50% and 20% sulfuric acid solutions, how much of each should be mixed to obtain 12 liters of a 30% acid solution?

Success Tip

The strength *(concentration)* of a mixture is always between the strengths of the two solutions used to make it.

Analyze The 50% solution is too strong and the 20% solution is too weak. We must find how much of each should be combined to obtain 12 liters of a 30% solution.

Assign If x = the number of liters of the 50% solution used in the mixture, the remaining $(12 - x)$ liters must be the 20% solution.

Form The amount of pure sulfuric acid in each solution is given by

Amount of solution · strength of the solution = amount of pure sulfuric acid

A table and sketch are helpful in organizing the facts of the problem.

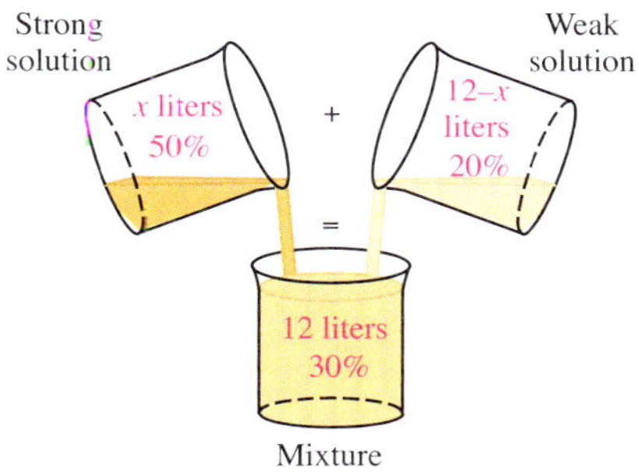

	Amount ·	Strength =	Amount of pure sulfuric acid
Weak	$12 - x$	0.20	$0.20(12 - x)$
Strong	x	0.50	$0.50x$
Mixture	12	0.30	$12(0.30)$

Multiply amount · strength three times to fill in this column.

Enter this information first.

Use the information in this column to form an equation.

The sulfuric acid in the 20% solution	plus	the sulfuric acid in the 50% solution	equals	the sulfuric acid in the mixture.
$0.20(12 - x)$	$+$	$0.50x$	$=$	$12(0.30)$

Solve

$$0.20(12 - x) + 0.50x = 12(0.30)$$

$$2.4 - 0.2x + 0.5x = 3.6 \quad \text{Distribute the multiplication by 0.20.}$$

$$0.3x + 2.4 = 3.6 \quad \text{Combine like terms: } -0.2x + 0.5x = 0.3x.$$

$$0.3x = 1.2 \quad \text{Subtract 2.4 from both sides.}$$

$$x = 4 \quad \text{To isolate } x \text{, undo the multiplication by 0.3 by dividing both sides by 0.3: } \frac{1.2}{0.3} = 4.$$

Success Tip

We could begin by multiplying both sides of the equation by 10 to clear it of the decimals.

State 4 liters of 50% solution and $12 - 4 = 8$ liters of 20% solution should be used.

Check The amount of acid in 4 liters of the 50% solution is $0.50(4) = 2.0$ liters and the amount of acid in 8 liters of the 20% solution is $0.20(8) = 1.6$ liters. Thus, the amount of acid in these two solutions is $2.0 + 1.6 = 3.6$ liters. The amount of acid in 12 liters of the 30% mixture is also $0.30(12) = 3.6$ liters. Since the amounts of acid are equal, the results check.

Self Check 4 **Mixing Solutions.** How many gallons of a 3% salt solution must be mixed with a 7% salt solution to obtain 25 gallons of a 5.4% salt solution?

Now Try Problem 41

4 Solve Dry Mixture Problems.

In another type of mixture problem, a dry mixture of a specified value is created from two differently priced ingredients.

EXAMPLE 5 **Snack Foods.** Because cashews priced at \$9 per pound were not selling, a produce clerk decided to combine them with less expensive peanuts and sell the mixture for \$7 per pound. How many pounds of peanuts, selling at \$6 per pound, should be mixed with 50 pounds of cashews to obtain such a mixture?

Analyze We need to determine how many pounds of peanuts to mix with 50 pounds of cashews to obtain a mixture worth \$7 per pound.

Assign Let x = the number of pounds of peanuts to use in the mixture. Since 50 pounds of cashews will be combined with the peanuts, the mixture will weigh $50 + x$ pounds.

Form The value of the mixture and of each of its ingredients is given by

Amount · the price = the total value

We can organize the facts of the problem in a table.

Caution

To find the number of pounds in the mixture, add the number of pounds of the ingredients:

$$50 + x$$

It would be incorrect to multiply:

$$50 \cdot x$$

	Amount	· Price	= Total value
Peanuts	x	6	$6x$
Cashews	50	9	450
Mixture	$50 + x$	7	$7(50 + x)$

Multiply amount · price three times to fill in this column.

Enter this information first.

Use the information in this column to form an equation.

The value of the peanuts	plus	the value of the cashews	equals	the value of the mixture.
$6x$	$+$	450	$=$	$7(50 + x)$

Solve

$6x + 450 = 7(50 + x)$

$6x + 450 = 350 + 7x$ Distribute the multiplication by 7.

$450 = 350 + x$ To eliminate the term $6x$ on the left side, subtract $6x$ from both sides: $7x - 6x = x$.

$100 = x$ To isolate x, subtract 350 from both sides.

State 100 pounds of peanuts should be used in the mixture.

Check The value of 100 pounds of peanuts, at \$6 per pound, is 100(\$6) = \$600 and the value of 50 pounds of cashews, at \$9 per pound, is 50(\$9) = \$450. Thus, the total value of these two ingredients is \$1,050. Since the value of 150 pounds of the mixture, at \$7 per pound, is also 150(7) = \$1,050, the result checks.

Self Check 5 **Mixing Candy.** Candy worth \$1.90 per pound is to be mixed with 60 lb of a second candy worth \$1.20 per pound. How many pounds of the \$1.90 per pound candy should be used to make a mixture worth \$1.48 per pound?

Now Try Problem 47

5 Solve Number-Value Problems.

When problems deal with collections of different items having different values, we must distinguish between the *number of* and the *value of* the items. For these problems, we will use the fact that

Number · value = total value

EXAMPLE 6 **Dining Area Improvements.** A restaurant owner needs to purchase some tables, chairs, and dinner plates for the dining area of her establishment. She plans to buy four chairs and four plates for each new table. She also plans to buy 20 additional plates in case of breakage. If a table costs \$100, a chair \$50, and a plate \$5, how many of each can she buy if she takes out a loan for \$6,500 to pay for the new items?

Analyze We know the *value* of each item: Tables cost \$100, chairs cost \$50, and plates cost \$5 each. We need to find the *number* of tables, chairs, and plates she can purchase for \$6,500.

Assign The number of chairs and plates she needs depends on the number of tables she buys. So we let t = the number of tables to be purchased. Since every table requires four chairs and four plates, she needs to order $4t$ chairs. Because 20 additional plates are needed, she should order $(4t + 20)$ plates.

Form We can organize the facts of the problem in a table.

	Number	· Value	= Total value
Tables	t	100	$100t$
Chairs	$4t$	50	$50(4t)$
Plates	$4t + 20$	5	$5(4t + 20)$
			Total: 6,500

Multiply number · value three times to fill in this column.

Enter this information first.

Use the information in this column to form an equation.

The value of the tables	plus	the value of the chairs	plus	the value of the plates	equals	the total value of the purchase.
$100t$	+	$50(4t)$	+	$5(4t + 20)$	=	6,500

Solve

$$100t + 50(4t) + 5(4t + 20) = 6{,}500$$

$$100t + 200t + 20t + 100 = 6{,}500 \quad \text{Do the multiplications and distribute.}$$

$$320t + 100 = 6{,}500 \quad \text{Combine like terms: } 100t + 200t + 20t = 320t.$$

$$320t = 6{,}400 \quad \text{Subtract 100 from both sides.}$$

$$t = 20 \quad \text{To isolate } t\text{, divide both sides by 320.}$$

To find the number of chairs and plates to buy, we evaluate $4t$ and $4t + 20$ for $t = 20$.

Chairs: $4t = 4(20)$
$= 80$

Plates: $4t + 20 = 4(20) + 20$
$= 100$

Substitute 20 for *t*.

State The owner needs to buy 20 tables, 80 chairs, and 100 plates.

Check The total value of 20 tables is 20($100) = $2,000, the total value of 80 chairs is 80($50) = $4,000, and the total value of 100 plates is 100($5) = $500. Because the total purchase is $2,000 + $4,000 + $500 = $6,500, the results check.

Self Check 6 **Electronics.** A small electronics store buys iPods for $189, iPod skins for $32, and iTunes cards for $15. If they place an order for three times as many iPods as skins and 20 more iTunes cards than skins, how many of each did they order if the items totaled $2,756?

Now Try Problem 55

SECTION 2.6 STUDY SET

VOCABULARY

Fill in the blanks.

1. Problems that involve depositing money are called __________ problems, and problems that involve moving vehicles are called uniform ________ problems.
2. Problems that involve combining ingredients are called ________ problems, and problems that involve collections of different items having different values are called __________ problems.

CONCEPTS

3. Complete only the *principal column* given that part of $30,000 is invested in stocks and the rest in art.

	P ·	r ·	t =	I
Stocks	x			
Art	?			

4. A man made two investments that earned a combined annual simple interest of $280. Complete the table and then form an equation for this investment problem.

	P ·	r ·	t =	I
Bank	x	0.04	1	
Stocks	$6{,}000 - x$	0.06	1	
			Total:	

5. Complete only the *rate column* given that the east-bound plane flew 150 mph slower than the west-bound plane.

	r ·	t =	d
West	r		
East	?		

6. a. Complete only the *time column* given that a runner wants to overtake a walker and the walker had a $\frac{1}{2}$-hour head start.

	r ·	t =	d
Runner		t	
Walker		?	

b. Complete only the *time column* given that part of a 6-hour drive was in fog and the other part was in clear conditions.

	r ·	t =	d
Foggy		t	
Clear		?	

7. A husband and wife drive in opposite directions to work. Their drives last the same amount of time and their workplaces are 80 miles apart. Complete the table and then form an equation for this distance problem.

	r ·	t =	d
Husband	35	t	
Wife	45		
		Total:	

8. a. How many gallons of acetic acid are there in barrel 2?
b. Suppose the contents of the two barrels are poured into an empty third barrel. How many gallons of liquid will the third barrel contain?
c. Estimate the strength of the solution in the third barrel: 15%, 35%, or 60% acid?

9. **a.** Two antifreeze solutions are combined to form a mixture. Complete the table and then form an equation for this mixture problem.

	Amount ·	Strength =	Pure antifreeze
Weak	x	0.25	
Strong	6	0.50	
Mixture		0.30	

b. Two oil-and-vinegar salad dressings are combined to make a new mixture. Complete the table and then form an equation for this mixture problem.

	Amount ·	Strength =	Pure vinegar
Weak		0.03	
Strong	x	0.06	
Mixture	10	0.05	

10. The value of all the nylon brushes that a paint store carries is \$670. Complete the table and then form an equation for this number-value problem.

	Number ·	Value =	Total value
1-inch	$2x$	4	
2-inch	x	5	
3-inch	$x + 10$	7	
		Total:	

NOTATION

11. Write 6% and 15.2% in decimal form.

12. By what power of 10 should each decimal be multiplied to make it a whole number?

a. 0.08 **b.** 0.162

GUIDED PRACTICE

Solve each equation. See Example 1.

13. $0.18x + 0.45(12 - x) = 0.36(12)$

14. $0.12x + 0.20(4 - x) = 0.6$

15. $0.08x + 0.07(15{,}000 - x) = 1{,}110$

16. $0.108x + 0.07(16{,}000 - x) = 1{,}500$

APPLICATIONS

Investment problems. See Example 1.

17. **Corporate Investments.** The financial board of a corporation invested \$25,000 overseas, part at 4% and part at 7% annual simple interest. Find the amount invested at each rate if the first-year combined income from the two investments was \$1,300.

18. **Loans.** A credit union loaned out \$50,000, part at an annual simple rate of 5% and the rest at an annual simple rate of 8%. They collected combined interest of \$3,400 from the loans that year. How much was loaned out at each rate?

19. **Old Coins.** A salesperson used her \$3,500 year-end bonus to purchase some old gold and silver coins. She earned 15% annual simple interest on the gold coins and 12% annual simple interest on the silver coins. If she saw a return on her investment of \$480 the first year, how much did she invest in each type of coin?

20. **High-Risk Companies.** An investment club used funds totaling \$200,000 to invest in a bio-tech company and in an ethanol plant, with hopes of earning 11% and 14% annual simple interest, respectively. Their hunch paid off. The club made a total of \$24,250 interest the first year. How much was invested at each rate?

21. **Retirement.** A professor wants to supplement her pension with investment interest. If she invests \$28,000 at 6% annual simple interest, how much would she have to invest at 7% annual simple interest to achieve a goal of \$3,500 per year in supplemental income?

22. **Extra Income.** An investor wants to receive \$1,000 annually from two investments. He has put \$4,500 in a money market account paying 4% annual simple interest. How much should he invest in a stock fund that pays 10% annual simple interest to achieve his goal?

23. **1099 Forms.** The form below shows the interest income Terrell Washington earned in 2011 from two savings accounts. He deposited a total of \$15,000 at the first of that year, and made no further deposits or withdrawals. How much money did he deposit in account 822 and in account 721?

RECIPIENT'S name TERRELL WASHINGTON	USA HOME SAVINGS This is important tax information and is being furnished to the Internal Revenue Service.	2011
Account Number	Annual Percent Yield	Interest earned
822	5%	?
721	4.5%	?
FORM 1099	Total Interest Income \$720.00	

24. **Investment Plans.** A financial planner recommends a plan for a client who has \$65,000 to invest. (See the chart at the right.) At the end of the presentation, the client asks, "How much will be invested at each rate?" Answer this question using the given information.

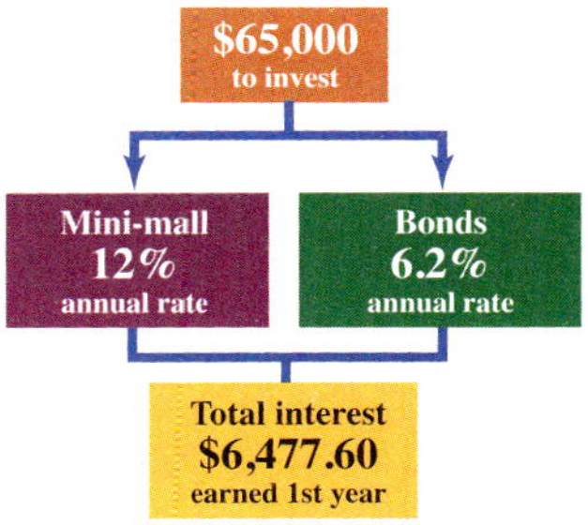

25. **Investments.** Equal amounts are invested in each of three accounts paying 7%, 8%, and 10.5% annual simple interest. If one year's combined interest income is \$1,249.50, how much is invested in each account?

26. **Personal Loans.** Maggy lent her brother some money at 2% annual simple interest. She lent her sister twice as much money at half of the interest rate. In one year, Maggy collected combined interest of \$200 from her brother and sister. How much did she lend each of them?

27. Bad Investments. A newly hired MBA graduate received an \$18,000 signing bonus from her employer. She invested part of it in a credit union account earning 3% annual simple interest and the rest in utility stocks that suffered a 7% loss of value. The net income from both investments for the first year was only \$90. How much of her bonus was originally placed in each investment?

28. Losses. A financial planner invested a portion of his client's \$190,000 in a high-yield mutual fund that earned 11% annual simple interest. The remainder of the money was invested in a mini-mall development. Unfortunately, that investment lost 25% of its value the first year. Find the amount originally made in each investment if the first-year net income was \$6,500.

Uniform motion problems. See Example 2.

29. Tornadoes. During a storm, two teams of scientists leave a university at the same time in vans to search for tornadoes. The first team travels east at 20 mph and the second travels west at 25 mph. If their radios have a range of up to 90 miles, how long will it be before they lose radio contact?

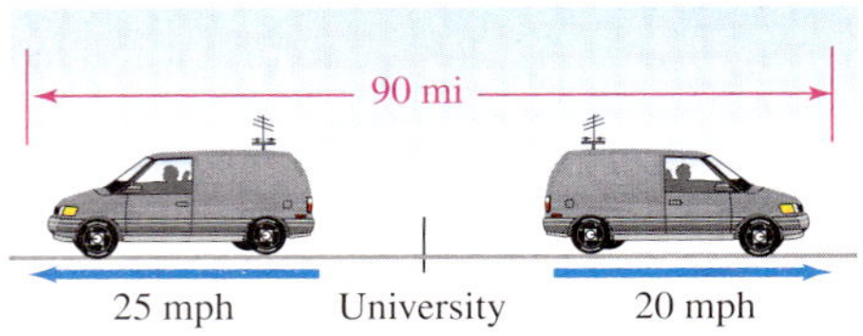

30. Unmanned Aircraft. Two remotely controlled unmanned aircraft are launched in opposite directions. One flies east at 78 mph and the other west at 82 mph. How long will it take the aircraft to fly a combined distance of 560 miles?

31. Hello/Goodbye. A husband and wife work different shifts at the same plant. When the husband leaves from work to make the 20-mile trip home, the wife leaves their home and drives to work. They travel on the same road. The husband's driving rate is 45 mph and the wife's is 35 mph. How long into their drives can they wave at each other when passing on the road?

32. Air Traffic Control. An airliner leaves Berlin, Germany, headed for Montreal, Canada, flying at an average speed of 450 mph. At the same time, an airliner leaves Montreal headed for Berlin, averaging 500 mph. If the airports are 3,800 miles apart, when will the air traffic controllers have to make the pilots aware that the planes are passing each other?

33. Cycling. A cyclist leaves his training base for a morning workout, riding at the rate of 18 mph. One and one-half hours later, his support staff leaves the base in a car going 45 mph in the same direction. How long will it take the support staff to catch up with the cyclist?

34. Parenting. How long will it take a mother, running at 4 feet per second, to catch up with her toddler, running down the sidewalk at 2 feet per second, if the child had a 5-second head start?

35. Road Trips. A car averaged 40 mph for part of a trip and 50 mph for the remainder. If the 5-hour trip covered 210 miles, for how long did the car average 40 mph?

36. Cross-Training. An athlete runs up a set of stadium stairs at a rate of 2 stairs per second, immediately turns around, and then descends the same stairs at a rate of 3 stairs per second. If the workout takes 90 seconds, how long does it take him to run up the stairs?

37. Winter Driving. A trucker drove for 4 hours before he encountered icy road conditions. He reduced his speed by 20 mph and continued driving for 3 more hours. Find his average speed during the first part of the trip if the entire trip was 325 miles.

38. Speed of Trains. Two trains are 330 miles apart, and their speeds differ by 20 mph. Find the speed of each train if they are traveling toward each other and will meet in 3 hours.

Liquid mixture problems. See Example 3.

39. Salt Solutions. How many gallons of a 3% salt solution must be mixed with 50 gallons of a 7% solution to obtain a 5% solution?

40. Photography. A photographer wishes to mix 2 liters of a 5% acetic acid solution with a 10% solution to get a 7% solution. How many liters of 10% solution must be added?

41. Making Cheese. To make low-fat cottage cheese, milk containing 4% butterfat is mixed with milk containing 1% butterfat to obtain 15 gallons of a mixture containing 2% butterfat. How many gallons of each milk must be used?

42. Antifreeze. How many quarts of a 10% antifreeze solution must be mixed with 16 quarts of a 40% antifreeze solution to make a 30% solution?

43. Printing. A printer has ink that is 8% cobalt blue color and ink that is 22% cobalt blue color. How many ounces of each ink are needed to make one-half gallon (64 ounces) of ink that is 15% cobalt blue color?

44. Flood Damage. One website recommends a 6% chlorine bleach-water solution to remove mildew. A chemical lab has 3% and 15% chlorine bleach-water solutions in stock. How many gallons of each should be mixed to obtain 100 gallons of the mildew spray?

45. Interior Decorating. The colors on the paint chip card below are created by adding different amounts of orange tint to a white latex base. How many gallons of Desert Sunrise should be mixed with 1 gallon of Bright Pumpkin to obtain Cool Cantaloupe?

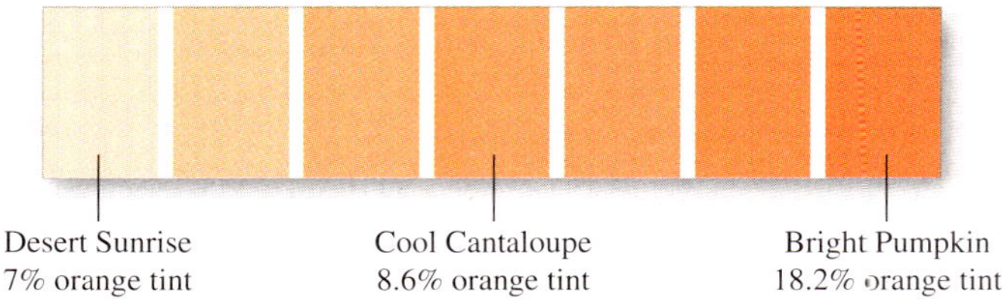

46. Antiseptics. A nurse wants to add water to 30 ounces of a 10% solution of benzalkonium chloride to dilute it to an 8% solution. How much water must she add? (*Hint:* Water is 0% benzalkonium chloride.)

Dry mixture problems. See Example 4.

47. **Lawn Seed.** A store sells bluegrass seed for $6 per pound and ryegrass seed for $3 per pound. How much ryegrass must be mixed with 100 pounds of bluegrass to obtain a blend that will sell for $5 per pound?

48. **Coffee Blends.** A store sells regular coffee for $8 a pound and gourmet coffee for $14 a pound. To get rid of 40 pounds of the gourmet coffee, a shopkeeper makes a blend to put on sale for $10 a pound. How many pounds of regular coffee should he use?

49. **Raisins.** How many scoops of natural seedless raisins costing $3.45 per scoop must be mixed with 20 scoops of golden seedless raisins costing $2.55 per scoop to obtain a mixture costing $3 per scoop?

50. **Fertilizer.** Fertilizer with weed control costing $38 per 50-pound bag is to be mixed with a less expensive fertilizer costing $6 per 50-pound bag to make 16 bags of fertilizer that can be sold for $28 per bag. How many bags of cheaper fertilizer should be used?

51. **Packaged Salad.** How many 10-ounce bags of Romaine lettuce must be mixed with fifty 10-ounce bags of Iceberg lettuce to obtain a blend that sells for $2.50 per ten-ounce bag?

52. **Mixing Candy.** Lemon drops worth $3.80 per pound are to be mixed with jelly beans that cost $2.40 per pound to make 300 pounds of a mixture worth $2.96 per pound. How many pounds of each candy should be used?

53. **Bronze.** A pound of tin is worth $1 more than a pound of copper. Four pounds of tin are mixed with 6 pounds of copper to make bronze that sells for $3.65 per pound. How much is a pound of tin worth?

54. **Snack Foods.** A bag of peanuts is worth $0.30 less than a bag of cashews. Equal amounts of peanuts and cashews are used to make 40 bags of a mixture that sells for $1.05 per bag. How much is a bag of cashews worth?

Number-value problems. See Example 5.

55. **Rentals.** The owners of an apartment building rent equal numbers of 1-, 2-, and 3-bedroom units. The monthly rent for a 1-bedroom is $550, a 2-bedroom is $700, and a 3-bedroom is $900. If the total monthly income is $36,550, how many of each type of unit are there?

56. **Warehousing.** A store warehouses 40 more portables than big-screen TV sets, and 15 more consoles than big-screen sets. The monthly storage cost for a portable is $1.50, a console is $4.00, and a big-screen is $7.50. If storage for all the televisions costs $276 per month, how many big-screen sets are in stock?

57. **Software.** Three software applications are priced as shown. Spreadsheet and database programs sold in equal numbers, but 15 more word processing applications were sold than the other two combined. If the three applications generated sales of $72,000, how many spreadsheets were sold?

Software	Price
Spreadsheet	$150
Database	$195
Word processing	$210

58. **Inventories.** With summer approaching, the number of air conditioners sold is expected to be double that of stoves and refrigerators combined. Stoves sell for $350, refrigerators for $450, and air conditioners for $500, and sales of $56,000 are expected. If stoves and refrigerators sell in equal numbers, how many of each appliance should be stocked?

59. **Piggy Banks.** When a child emptied his coin bank, he had a collection of pennies, nickels, and dimes. There were 20 more pennies than dimes and the number of nickels was triple the number of dimes. If the coins had a value of $5.40, how many of each type coin were in the bank?

60. **Wishing Wells.** A scuba diver, hired by an amusement park, collected $121 in nickels, dimes, and quarters at the bottom of a wishing well. There were 500 nickels, and 90 more quarters than dimes. How many quarters and dimes were thrown into the wishing well?

61. **Basketball.** Epiphanny Prince, of New York, scored 113 points in a high school game on February 1, 2006, breaking a national prep record that was held by Cheryl Miller. Prince made 46 more 2-point baskets than 3-point baskets, and only 1 free throw. How many 2-point and 3-point baskets did she make?

62. **Museum Tours.** The ticket prices for the Coca-Cola Museum in Atlanta are shown. A family purchased 3 more children's tickets than adult tickets, and 1 less senior ticket than adult tickets. The total cost of the tickets was $131. How many of each type did they purchase?

WRITING

63. Create a mixture problem of your own, and solve it.

64. Write an investment problem to fit the following equation, and then solve it.

$$0.08x + 0.07(10{,}000 - x) = 770$$

65. Explain the error in each statement.

a. If 3 pounds of Mocha coffee are mixed with x pounds of Java coffee, there will be $3x$ pounds of the Mocha-Java blend.

b. A financial manager has a total of \$5,000 to invest in two accounts. If \$$x$ is invested in the first account, then \$$(x - 5{,}000)$ is left to be invested in the second account.

66. Is it possible to mix a 10% sugar solution with a 20% sugar solution to get a 30% sugar solution? Explain.

REVIEW

Multiply.

67. $-12(3a + 4b - 32)$

68. $\frac{1}{2}(4b - 8)$

69. $3(5t + 1)2$

70. $2.9(4c - 12)$

CHALLENGE PROBLEMS

71. Evaporation. How much water must be boiled away to increase the concentration of 300 milliliters of a 2% salt solution to a 3% salt solution?

72. Diluting Solutions. How much water should be added to 20 ounces of a 15% solution of alcohol to dilute it to a 10% alcohol solution?

73. Financial Planning. A plumber has a choice of two investment plans:

- An insured fund that pays 11% interest
- A risky investment that pays a 13% return

If the same amount invested at the higher rate would generate an extra \$150 per year, how much does the plumber have to invest?

74. Investments. The amount of annual interest earned by \$8,000 invested at a certain rate is \$200 less than \$12,000 would earn at a rate 1% lower. At what rate is the \$8,000 invested?

SECTION 2.7 Solving Inequalities

OBJECTIVES

1. Determine whether a number is a solution of an inequality.
2. Graph solution sets and use interval notation.
3. Solve linear inequalities.
4. Solve compound inequalities.
5. Solve inequality applications.

ARE YOU READY?

The following problems review some basic skills that are needed to solve inequalities.

1. Fill in the blanks: The symbol $<$ means "____ ____ ____."

2. Is $-5 > -6$ a true or false statement?

3. Graph each number in the set $\left\{-4, -1.7, 2, \frac{13}{4}\right\}$ on a number line.

4. Express the fact that $10 > 0$ using an $<$ symbol.

In our daily lives, we often speak of one value being *greater than* or *less than* another. For example, a sick child might have a temperature *greater than* 98.6°F or a granola bar might contain *less than* 2 grams of fat. In mathematics, we use *inequalities* to show that one expression is greater than or is less than another expression.

1 Determine Whether a Number Is a Solution of an Inequality.

An **inequality** is a statement that contains one or more of the following symbols.

Inequality Symbols			
	$<$ is less than	$>$ is greater than	$\neq$ is not equal to
	$\leq$ is less than or equal to	$\geq$ is greater than or equal to	

An inequality can be true, false, or neither true nor false. For example,

- $9 \geq 9$ is true because $9 = 9$.
- $37 < 24$ is false.
- $x + 1 > 5$ is neither true nor false because we don't know what number x represents.

The Language of Algebra

Because $<$ requires one number to be strictly less than another number and $>$ requires one number to be strictly greater than another number, $<$ and $>$ are called **strict inequalities**.

An inequality that contains a variable can be made true or false depending on the number that is substituted for the variable. If we substitute 10 for x in $x + 1 > 5$, the resulting inequality is true: $10 + 1 > 5$. If we substitute 1 for x, the resulting inequality is false: $1 + 1 > 5$. A number that makes an inequality true is called a **solution** of the inequality, and we say that the number *satisfies* the inequality. Thus, 10 is a solution of $x + 1 > 5$ and 1 is not.

In this section, we will find the solutions of *linear inequalities in one variable.*

Linear Inequality in One Variable

A **linear inequality in one variable** can be written in one of the following forms where a, b, and c are real numbers and $a \neq 0$.

$$ax + b > c \qquad ax + b \geq c \qquad ax + b < c \qquad ax + b \leq c$$

EXAMPLE 1 Is 9 a solution of $2x + 4 \leq 21$?

Strategy We will substitute 9 for x and evaluate the expression on the left side.

Why If a true statement results, 9 is a solution of the inequality. If we obtain a false statement, 9 is not a solution.

Solution

$2x + 4 \leq 21$

$2(9) + 4 \overset{?}{\leq} 21$ Substitute 9 for x. Read $\overset{?}{\leq}$ as "is possibly less than or equal to."

$18 + 4 \overset{?}{\leq} 21$ Do the multiplication.

$22 \leq 21$ This inequality is false.

The statement $22 \leq 21$ is false because neither $22 < 21$ nor $22 = 21$ is true. Therefore, 9 is not a solution of $2x + 4 \leq 21$.

The Language of Algebra

A **linear inequality** in one variable is similar to a linear equation in one variable except that the equal symbol is replaced with an inequality symbol.

Equation	*Inequality*
$2x + 1 = 9$	$2x + 1 > 9$

Self Check 1 Is 2 a solution of $3x - 1 \geq 0$?

Now Try Problem 15

2 Graph Solution Sets and Use Interval Notation.

The **solution set** of an inequality is the set of all numbers that make the inequality true. Some solution sets are easy to find. For example, if we replace the variable in $x > -3$ with a number greater than -3, the resulting inequality will be true. Because there are infinitely many real numbers greater than -3, it follows that $x > -3$ has infinitely many solutions. Since there are too many solutions to list, we use **set-builder notation** to describe the solutions set.

$$\{x \mid x > -3\}$$

Read as "the set of all x such that x is greater than −3."

We can illustrate the solution set by **graphing the inequality** on a number line. To graph $x > -3$, a **parenthesis** or **open circle** is drawn on the endpoint -3 to indicate that -3 is not part of the graph. Then we shade all of the points on the number line to the right of -3. The right arrowhead is also shaded to show that the solutions continue forever to the right.

Notation

The parenthesis (opens in the direction of the shading and indicates that an endpoint is not included in the shaded interval.

All real numbers greater than −3

Notation

The positive infinity symbol ∞ does not represent a real number. It indicates that an interval extends to the right without end.

The graph of $x > -3$ is an example of an **interval** on the number line. We can write intervals in a compact form called **interval notation.**

The interval notation that represents the graph of $x > -3$ is $(-3, \infty)$. As on the number line, a left parenthesis is written next to -3 to indicate that -3 is not included in the interval. The **positive infinity symbol** ∞ that follows indicates that the interval continues without end to the right. With this notation, *a parenthesis is always used next to an infinity symbol.*

The illustration below shows the relationship between the symbols used to graph an interval and the corresponding interval notation. If we begin at -3 and move to the right, the shaded arrowhead on the graph indicates that the interval approaches positive infinity ∞.

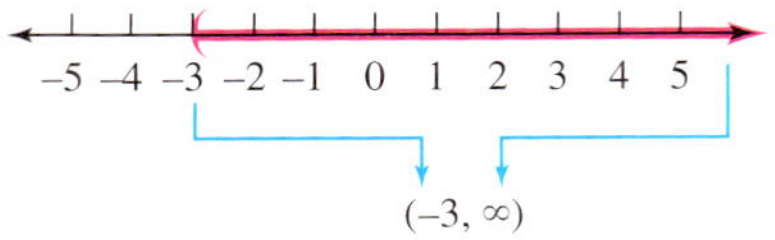

Success Tip

Think of interval notation as a way to tell someone how to draw the graph, from left to right, giving them only a "start" and a "stop" instruction.

We now have three ways to describe the solution set of an inequality.

Set-builder notation	*Number line graph*	*Interval notation*
$\{x \mid x > -3\}$	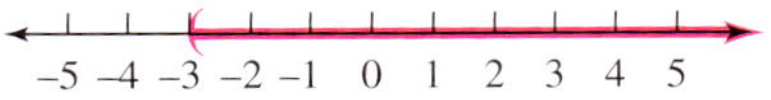	$(-3, \infty)$

EXAMPLE 2 Graph: $x \le 2$

Strategy We need to determine which real numbers, when substituted for x, would make $x \le 2$ a true statement.

Why To graph $x \le 2$ means to draw a "picture" of all of the values of x that make the inequality true.

Solution If we replace x with a number less than or equal to 2, the resulting inequality will be true. To graph the solution set, a **bracket** or a **closed circle** is drawn at the endpoint 2 to indicate that 2 is part of the graph. Then we shade all of the points on the number line to the left of 2 as well as the left arrowhead.

Notation

The *bracket*] opens in the direction of the shading and indicates that an endpoint is included in the shaded interval.

Method 1: bracket

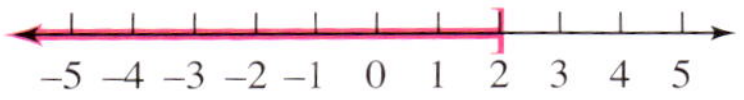

Method 2: closed circle

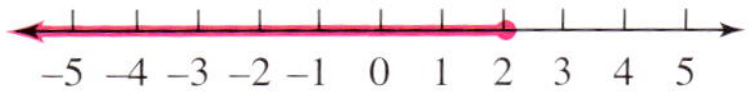

All real numbers less than or equal to 2

The interval is written as $(-\infty, 2]$. The right bracket indicates that 2 is included in the interval. The **negative infinity symbol** $-\infty$ shows that the interval continues forever to the left. The illustration below shows the relationship between the symbols used to graph the interval and the corresponding interval notation.

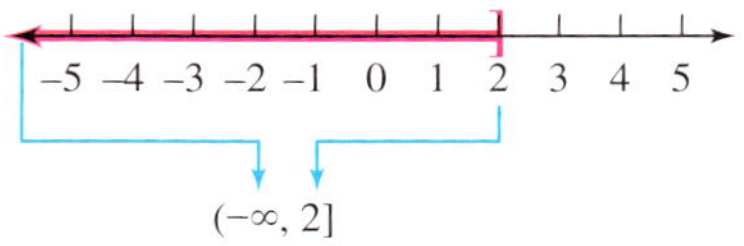

Self Check 2 Graph: $x \ge 0$

Now Try ▶ Problem 19

3 Solve Linear Inequalities.

To **solve an inequality** means to find all values of the variable that make the inequality true. As with equations, there are properties that we can use to solve inequalities.

Addition and Subtraction Properties of Inequality

Adding the same number to, or subtracting the same number from, both sides of an inequality does not change its solutions.

For any real numbers a, b, and c,

If $a < b$, then $a + c < b + c$.   If $a < b$, then $a - c < b - c$.

Similar statements can be made for the symbols $\leq$, $>$, and $\geq$.

After applying one of these properties, the resulting inequality is equivalent to the original one. **Equivalent inequalities** have the same solution set.

Like equations, inequalities are solved by isolating the variable on one side.

EXAMPLE 3 Solve $x + 3 > 2$. Write the solution set in interval notation and graph it.

Strategy We will use a property of inequality to isolate the variable on one side.

Why To solve the original inequality, we want to find a simpler equivalent inequality of the form **$x >$ a number** or **$x <$ a number**, whose solution is obvious.

Solution We will use the subtraction property of inequality to isolate x on the left side of the inequality. We can undo the addition of 3 by subtracting 3 from both sides.

$$x + 3 > 2 \quad \text{This is the inequality to solve.}$$
$$x + 3 - 3 > 2 - 3 \quad \text{Subtract 3 from both sides.}$$
$$x > -1 \quad \text{Do the subtraction: } 3 - 3 = 0 \text{ and } 2 - 3 = -1.$$

Success Tip

We solve linear inequalities by writing a series of steps that result in an equivalent inequality of the form

$x >$ a number

or

$x <$ a number

Similar statements apply to linear inequalities containing $\leq$ and $\geq$.

All real numbers greater than -1 are solutions of $x + 3 > 2$. The solution set can be written in set-builder notation as $\{x \mid x > -1\}$ and in interval notation as $(-1, \infty)$. The graph of the solution set is shown below.

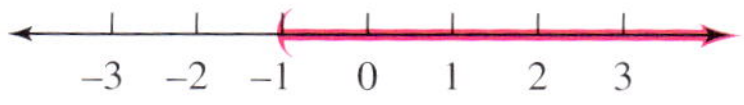

Since there are infinitely many solutions, we cannot check all of them. As an informal check, we can pick some numbers in the graph, say 0 and 30, substitute each number for x in the original inequality, and see whether true statements result.

Check:

$$x + 3 > 2 \qquad\qquad x + 3 > 2$$
$$0 + 3 \overset{?}{>} 2 \quad \text{Substitute 0 for } x. \qquad 30 + 3 \overset{?}{>} 2 \quad \text{Substitute 30 for } x.$$
$$3 > 2 \quad \text{True} \qquad\qquad 33 > 2 \quad \text{True}$$

The solution set appears to be correct.

Self Check 3 Solve $x - 3 < -2$. Write the solution set in interval notation and graph it.

Now Try Problem 23

As with equations, there are properties for multiplying and dividing both sides of an inequality by the same number. To develop what is called *the multiplication property of inequality,* we consider the true statement $2 < 5$. If both sides are multiplied by a positive number, such as 3, another true inequality results.

$$2 < 5 \quad \text{This inequality is true.}$$
$$3 \cdot 2 < 3 \cdot 5 \quad \text{Multiply both sides by 3.}$$
$$6 < 15 \quad \text{This inequality is true.}$$

Caution

If the inequality symbol is not reversed when both sides of a true inequality are multiplied by a negative number, the result is a false inequality. For example,

$$3 < 6 \quad \text{True}$$
$$-3 \cdot 3 < -3 \cdot 6$$
$$-9 < -18 \quad \text{False}$$

The same is true if we divide both sides of a true inequality by a negative number. Divide both sides of $3 < 6$ by -3 to see for yourself.

However, if we multiply both sides of $2 < 5$ by a negative number, such as -3, the direction of the inequality symbol must be reversed to produce another true inequality.

$$2 < 5 \qquad \text{This inequality is true.}$$
$$-3 \cdot 2 > -3 \cdot 5 \qquad \text{Multiply both sides by } -3 \text{ and reverse the direction of the inequality.}$$
$$-6 > -15 \qquad \text{This inequality is true.}$$

The inequality $-6 > -15$ is true because -6 is to the right of -15 on the number line.

Dividing both sides of an inequality by the same negative number also requires that the direction of the inequality symbol be reversed.

$$-4 < 6 \qquad \text{This inequality is true.}$$
$$\frac{-4}{-2} > \frac{6}{-2} \qquad \text{Divide both sides by } -2 \text{ and change } < \text{ to } >.$$
$$2 > -3 \qquad \text{This inequality is true.}$$

These examples illustrate the **multiplication and division properties of inequality.**

Multiplication and Division Properties of Inequality

Multiplying or dividing both sides of an inequality by the same **positive number** does not change its solutions.

For any real numbers a, b, and c, where c is **positive**,

$$\text{If } a < b, \quad \text{then } ac < bc. \qquad \text{If } a < b, \quad \text{then } \frac{a}{c} < \frac{b}{c}.$$

If we multiply or divide both sides of an inequality by the same **negative number**, the direction of the inequality symbol must be reversed for the inequalities to have the same solutions.

For any real numbers a, b, and c, where c is **negative,**

$$\text{If } a < b, \quad \text{then } ac > bc. \qquad \text{If } a < b, \quad \text{then } \frac{a}{c} > \frac{b}{c}.$$

Similar statements can be made for the symbols $\leq$, $>$, and $\geq$.

EXAMPLE 4 Solve each inequality. Write the solution set in interval notation and graph it.

a. $-\frac{3}{2}t \geq -12$ **b.** $-5t < 55$

Strategy We will use a property of inequality to isolate the variable on one side.

Why To solve the original inequality, we want to find a simpler equivalent inequality, whose solution is obvious.

Solution **a.** To undo the multiplication by $-\frac{3}{2}$, we multiply both sides by the reciprocal, which is $-\frac{2}{3}$.

$$-\frac{3}{2}t \geq -12 \qquad \text{This is the inequality to solve.}$$
$$-\frac{2}{3}\left(-\frac{3}{2}t\right) \leq -\frac{2}{3}(-12) \qquad \text{Multiply both sides by } -\tfrac{2}{3}\text{. Since we are multiplying both sides by a negative number, reverse the direction of the } \geq \text{ symbol.}$$
$$t \leq 8 \qquad \text{Do the multiplication: } -\tfrac{2}{3}\left(-\tfrac{3}{2}\right) = 1 \text{ and } -\tfrac{2}{3}(-12) = 8.$$

The solution set is $(-\infty, 8]$ and it is graphed as shown.

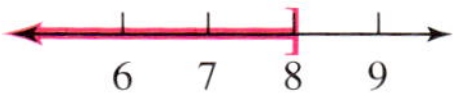

Success Tip

You may find it easier to graph the solution set first and then use the symbols that occur in that form of the result to write the interval notation.

b. To undo the multiplication by -5, we divide both sides by -5.

$$-5t < 55 \quad \text{This is the inequality to solve.}$$

$$\frac{-5t}{-5} > \frac{55}{-5} \quad \text{To isolate } t\text{, undo the multiplication by } -5 \text{ by dividing both sides by } -5. \text{ Since we are dividing both sides by a negative number, reverse the direction of the } < \text{ symbol.}$$

$$t > -11 \quad \text{Do the division.}$$

The solution set is $(-11, \infty)$ and it is graphed as shown.

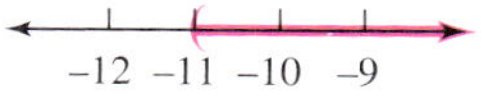

Self Check 4 Solve each inequality. Write the solution set in interval notation and graph it. **a.** $-\frac{h}{20} \le 10$ **b.** $-12a > -144$

Now Try Problems 27 and 31

EXAMPLE 5 Solve $-5 > 3x + 7$. Write the solution set in interval notation and graph it.

Strategy First we will use a property of inequality to isolate the *variable term* on one side. Then we will use a second property of inequality to isolate the *variable* itself.

Why To solve the original inequality, we want to find a simpler equivalent inequality of the form **$x >$ a number** or **$x <$ a number**, whose solution is obvious.

Solution

$$-5 > 3x + 7 \quad \text{This is the inequality to solve.}$$

$$-5 - 7 > 3x + 7 - 7 \quad \text{To isolate the variable term } 3x \text{ on the right side, undo the addition of 7 by subtracting 7 from both sides.}$$

$$-12 > 3x \quad \text{Do the subtraction: } -5 - 7 = -12 \text{ and } 7 - 7 = 0.$$

$$\frac{-12}{3} > \frac{3x}{3} \quad \text{To isolate } x\text{, undo the multiplication by 3 by dividing both sides by 3.}$$

$$-4 > x \quad \text{Do the division.}$$

Success Tip

Don't be confused by the negative number on the left side. We didn't reverse the $>$ symbol because we divided both sides by *positive* 3.

$$\frac{-12}{3} > \frac{3x}{3}$$

To determine the solution set, it is useful to rewrite the inequality $-4 > x$ in an equivalent form with the variable on the left side.

$$x < -4 \quad \text{If } -4 \text{ is greater than } x\text{, it follows that } x \text{ must be less than } -4.$$

The solution set is $(-\infty, -4)$ and it is graphed as shown.

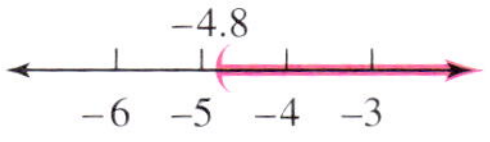

Self Check 5 Solve $-13 < 2r - 7$. Write the solution set in interval notation and graph it.

Now Try Problem 35

EXAMPLE 6 Solve $5.1 - 3k < 19.5$. Write the solution set in interval notation and graph it.

Strategy We will use properties of inequality to isolate the variable on one side.

Why To solve the original inequality, we want to find a simpler equivalent inequality of the form **$k >$ a number** or **$k <$ a number**, whose solution is obvious.

Solution

$5.1 - 3k < 19.5$	This is the inequality to solve.
$5.1 - 3k - 5.1 < 19.5 - 5.1$	To isolate $-3k$ on the left side, subtract 5.1 from both sides.
$-3k < 14.4$	Do the subtraction.
$\frac{-3k}{-3} > \frac{14.4}{-3}$	To isolate k, undo the multiplication by -3 by dividing both sides by -3 and reverse the direction of the $<$ symbol.
$k > -4.8$	Do the division.

The solution set is $(-4.8, \infty)$, and it is graphed as shown.

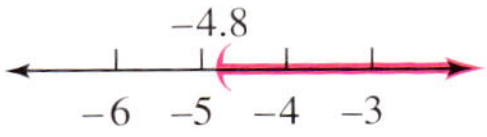

Self Check 6 Solve $-9n + 1.8 > -17.1$. Write the solution set in interval notation and graph it.

Now Try Problem 39

The equation-solving strategy in Section 2.2 can be applied to inequalities. However, when solving inequalities, we must remember to *reverse the direction of the inequality symbol when multiplying or dividing both sides by a negative number.*

EXAMPLE 7 Solve $8(y + 1) \geq 2(y - 4) + y$. Write the solution set in interval notation and graph it.

Strategy We will follow the steps of the equation-solving strategy (adapted to inequalities) to solve the inequality.

Why This is the most efficient way to solve a linear inequality in one variable.

Solution

$8(y + 1) \geq 2(y - 4) + y$	This is the inequality to solve.
$8y + 8 \geq 2y - 8 + y$	Distribute the multiplication by 8 and by 2.
$8y + 8 \geq 3y - 8$	Combine like terms: $2y + y = 3y$.
$8y + 8 - 3y \geq 3y - 8 - 3y$	To eliminate $3y$ from the right side, subtract $3y$ from both sides.
$5y + 8 \geq -8$	Combine like terms: $8y - 3y = 5y$ and $3y - 3y = 0$.
$5y + 8 - 8 \geq -8 - 8$	To isolate the variable term $5y$ on the left side, undo the addition of 8 by subtracting 8 from both sides.
$5y \geq -16$	Do the subtraction: $8 - 8 = 0$ and $-8 - 8 = -16$.
$\frac{5y}{5} \geq \frac{-16}{5}$	To isolate y, undo the multiplication by 5 by dividing both sides by 5. Do not reverse the direction of the $\geq$ symbol.
$y \geq -\frac{16}{5}$	

Success Tip

As an informal check, substitute a number on the graph that is shaded, such as -3, into $8(y + 1) \geq 2(y - 4) + y$. A true statement should result. Then substitute a number on the graph that is not shaded, such as -4, into the inequality. A false statement should result.

The solution set is $\left[-\frac{16}{5}, \infty\right)$ and it is graphed as shown.

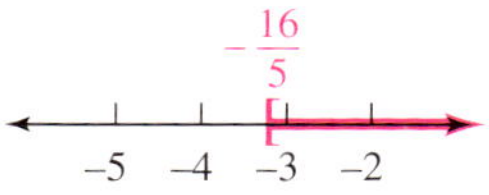

A solution involving an improper fraction is perfectly acceptable. Just make sure the fraction is in simplified form. To locate the endpoint of the interval on the graph, it is helpful to note that $-\frac{16}{5} = -3\frac{1}{5}$.

Self Check 7 Solve $5(b - 2) \geq -(b - 3) + 2b$. Write the solution set in interval notation and graph it.

Now Try Problem 45

EXAMPLE 8 Solve $\frac{3}{4} + \frac{x}{2} > \frac{6}{7}$. Write the solution set in interval notation and graph it.

Strategy The first step of the equation-solving strategy (adapted to inequalities) is to clear the inequality of fractions by multiplying both sides by the LCD.

Why It's easier to solve an inequality that involves only integers.

Solution

$$\frac{3}{4} + \frac{x}{2} > \frac{6}{7}$$ This is the inequality to solve.

$$28\left(\frac{3}{4} + \frac{x}{2}\right) > 28\left(\frac{6}{7}\right)$$ Clear the inequality of fractions by multiplying both sides by the LCD of $\frac{3}{4}$, $\frac{x}{2}$, and $\frac{6}{7}$, which is 28.

$$28\left(\frac{3}{4}\right) + 28\left(\frac{x}{2}\right) > 28\left(\frac{6}{7}\right)$$ On the left side, distribute the multiplication by 28.

$$21 + 14x > 24$$ Multiply: $\overset{7}{28}\left(\frac{3}{\underset{1}{4}}\right) = 21$, $\overset{14}{28}\left(\frac{x}{\underset{1}{2}}\right) = 14x$, and $\overset{4}{28}\left(\frac{6}{\underset{1}{7}}\right) = 24$.

$$21 + 14x - 21 > 24 - 21$$ To isolate the variable term 14x on the left side, undo the addition of 21 by subtracting 21 from both sides.

$$14x > 3$$ Do the subtraction: $21 - 21 = 0$ and $24 - 21 = 3$.

$$\frac{14x}{14} > \frac{3}{14}$$ To isolate x, undo the multiplication by 14 by dividing both sides by 14.

$$x > \frac{3}{14}$$

The solution set is $\left(\frac{3}{14}, \infty\right)$ and it is graphed as shown.

Self Check 8 Solve $\frac{1}{6} + \frac{2a}{9} < \frac{5}{2}$. Write the solution set in interval notation and graph it.

Now Try ▶ Problem 47

4 Solve Compound Inequalities.

The Language of Algebra

The word **compound** means made up of two or more parts. For example, a *compound* inequality has three parts. Other examples are: a *compound* sentence, a *compound* fracture, and a chemical *compound*.

Two inequalities can be combined into a **compound inequality** to show that an expression lies between two fixed values. For example, $-2 < x < 3$ is a combination of

$-2 < x$ and $x < 3$

It indicates that x is greater than -2 and that x is also less than 3. The solution set of $-2 < x < 3$ consists of all numbers that lie between -2 and 3, and we write it as the interval $(-2, 3)$. The graph of the compound inequality is shown below.

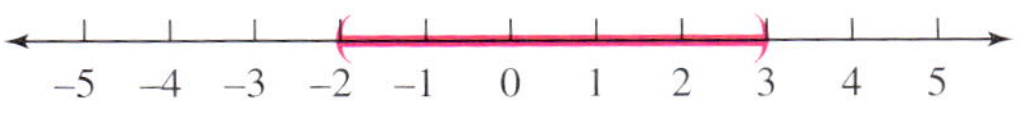

EXAMPLE 9 Graph $-4 \le x < 0$ and write the solution set in interval notation.

Strategy We need to determine which real numbers, when substituted for x, would make $-4 \le x < 0$ a true statement.

Why To graph $-4 \le x < 0$ means to draw a "picture" of all of the values of x that make the compound inequality true.

Solution If we replace the variable in $-4 \le x < 0$ with a number between -4 and 0, including -4, the resulting compound inequality will be true. Therefore, the solution set is the interval $[-4, 0)$.

Notation

Note that the two inequality symbols in $-4 \le x < 0$ point in the same direction and both point to the smaller number.

To graph the interval, we draw a bracket at -4, a parenthesis at 0, and shade in between, as shown.

To check, we pick a number in the graph, such as -2, and see whether it satisfies the inequality. Since $-4 \le -2 < 0$ is true, the answer appears to be correct.

Self Check 9 Graph $-2 \le x < 1$ and write the solution set in interval notation.

Now Try Problem 51

To solve these types of compound inequalities, we isolate the variable in the middle part of the inequality. To do this, we apply the properties of inequality to all *three* parts of the inequality: the left, the middle, and the right.

EXAMPLE 10 Solve $-4 < 2(x - 1) \le 4$. Write the solution set in interval notation and graph it.

Strategy We will use properties of inequality to isolate the variable by itself as the middle part of the inequality.

Why To solve the original inequality, we want to find a simpler equivalent inequality of the form **a number $< x \le$ a number**, whose solution is obvious.

Solution

$$-4 < 2(x - 1) \le 4$$ This is the compound inequality to solve.

$$-4 < 2x - 2 \le 4$$ In the middle, distribute the multiplication by 2.

$$-4 + 2 < 2x - 2 + 2 \le 4 + 2$$ To isolate the variable term $2x$, undo the subtraction of 2 by adding 2 to all three parts.

$$-2 < 2x \le 6$$ Do the addition: $-4 + 2 = -2$, $-2 + 2 = 0$, and $4 + 2 = 6$.

$$\frac{-2}{2} < \frac{2x}{2} \le \frac{6}{2}$$ To isolate x, we undo the multiplication by 2 by dividing all three parts by 2.

$$-1 < x \le 3$$ Do the division.

The solution set is $(-1, 3]$ and it is graphed as shown.

Self Check 10 Solve $-6 \le 3(t + 2) \le 6$. Write the solution set in interval notation and graph it.

Now Try Problem 59

5 Solve Inequality Applications.

When solving application problems, phrases such as "not more than," or "should exceed" suggest that the problem involves an inequality rather than an equation.

EXAMPLE 11

The Language of Algebra

Some phrases that suggest an inequality are:

surpass: $>$
not exceed: $\leq$
is no less than: $\geq$
is no greater than: $\leq$
at least: $\geq$
at most: $\leq$
between: $\square < \square < \square$

Grades. A student has scores of 72%, 74%, and 78% on three exams. What percent score does he need on the last exam to earn a grade of no less than B (80%)?

Analyze We know three scores. We are to find what the student must score on the last exam to earn a grade of B or higher.

Assign Let $x =$ the score on the fourth (and last) exam.

Form To find the average grade, we add the four scores and divide by 4. To earn a grade of *no less than* B, the student's average must be *greater than or equal to* 80%.

The average of the four grades	must be no less than	80.
$\frac{72 + 74 + 78 + x}{4}$	$\geq$	80

Solve

$$\frac{224 + x}{4} \geq 80$$ Combine like terms in the numerator: $72 + 74 + 78 = 224$.

$$4\left(\frac{224 + x}{4}\right) \geq 4(80)$$ To clear the inequality of the fraction, multiply both sides by 4.

$$224 + x \geq 320$$ Simplify each side.

$$x \geq 96$$ To isolate x, undo the addition of 224 by subtracting 224 from both sides.

State To earn a B, the student must score 96% or better on the last exam.

Check Pick several exam scores that are 96% or better and verify that the student's average will be 80% or greater. For example, a score of 96% gives the student an average that is exactly 80%.

$$\frac{72 + 74 + 78 + 96}{4} = \frac{320}{4} = 80$$

Self Check 11 **Grades.** A student has scores of 78%, 82%, and 76% on three exams. What percent score does he need on the last test to earn a grade of no less than a B (80%)?

Now Try Problem 103

SECTION 2.7 STUDY SET

VOCABULARY

Fill in the blanks.

1. An __________ is a statement that contains one of the symbols: $>$, $\geq$, $<$, or $\leq$. An equation is a statement that contains an __________ symbol.
2. To ______ an inequality means to find all the values of the variable that make the inequality true.
3. The solution set of $x > 2$ can be expressed in ________ notation as $(2, \infty)$.
4. The inequality $-4 < x \leq 10$ is an example of a __________ inequality.

CONCEPTS

Fill in the blanks.

5. a. Adding the same number to _____ sides of an inequality does not change the solutions.
   b. Multiplying or dividing both sides of an inequality by the same ________ number does not change the solutions.
   c. If we multiply or divide both sides of an inequality by a ________ number, the direction of the inequality symbol must be reversed for the inequalities to have the same solutions.
6. To solve $-4 \leq 2x + 1 < 3$, properties of inequality are applied to all _____ parts of the inequality: left, middle, and right.

7. Rewrite the inequality $32 < x$ in an equivalent form with the variable on the left side.

8. The solution set of an inequality is graphed below. Which of the four numbers, 3, −3, 2, and 4.5, when substituted for the variable in that inequality, would make it true?

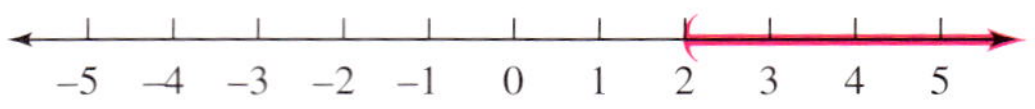

Write the inequality that is represented by each graph. Then describe the graph using interval notation.

9. a. 

$x <$; (, −1)

b. 

$x \geq$; [2,)

10. a. 

$< x \leq$; (,]

b. 

4 x 6; 4, 6

NOTATION

11. Write each symbol.
   a. is less than or equal to
   b. infinity
   c. bracket
   d. is greater than

12. Consider the graph of the interval [4, 8).
   a. Is 4 included in the graph?
   b. Is 8 included in the graph?

Complete the solution to solve each inequality.

13. $4x - 5 \geq 7$

$4x - 5 + \quad \geq 7 +$

$4x \geq$

$\frac{4x}{\quad} \geq \frac{12}{\quad}$

$x \geq 3$ Solution set: [, ∞)

14. $-6x > 12$

$\frac{-6x}{\quad} \quad \frac{12}{-6}$

$x <$ Solution set: (, −2)

GUIDED PRACTICE

See Example 1.

15. Determine whether each number is a solution of $3x - 2 > 5$.
   a. 5
   b. −4

16. Determine whether each number is a solution of $3x + 7 < 4x - 2$.
   a. 12
   b. 9

17. Determine whether each number is a solution of $-5(x - 1) \geq 2x + 12$.
   a. 1
   b. −1

18. Determine whether each number is a solution of $\frac{4}{5}a \geq -2$.
   a. $-\frac{5}{4}$
   b. −15

Graph each inequality and describe the graph using interval notation. See Example 2.

19. $x < 5$
20. $x \geq -2$
21. $-3 < x \leq 1$
22. $-4 \leq x \leq 2$

Solve each inequality. Write the solution set in interval notation and graph it. See Example 3.

23. $x + 2 > 5$
24. $x + 5 \geq 2$
25. $g - 30 \geq -20$
26. $h - 18 \leq -3$

Solve each inequality. Write the solution set in interval notation and graph it. See Example 4.

27. $-\frac{3}{16}x \geq -9$
28. $-\frac{7}{8}x \leq 21$
29. $\frac{2}{3}x \geq 2$
30. $\frac{3}{4}x < 3$
31. $-3y \leq -6$
32. $-6y \geq -6$
33. $8h < 48$
34. $2t > 22$

Solve each inequality. Write the solution set in interval notation and graph it. See Example 5.

35. $64 < 9x + 1$
36. $32 > 4x + 8$
37. $-20 \geq 3m - 5$
38. $-29 \leq 7t - 1$

Solve each inequality. Write the solution set in interval notation and graph it. See Example 6.

39. $1.3 - 2x \geq 0.5$
40. $1.04 - 7x > 0.2$
41. $24.9 - 12a < -3.9$
42. $37.5 - 16t \leq 99.9$

Solve each inequality. Write the solution set in interval notation and graph it. See Example 7.

43. $9a + 4 > 5a - 16$
44. $8t + 1 < 4t - 19$
45. $8(2n + 1) \leq 4(6n + 7) + 4n$
46. $5(2 - d) \leq 3(d - 5) + 3d$

Solve each inequality. Write the solution set in interval notation and graph it. See Example 8.

47. $\frac{1}{2} + \frac{n}{5} > \frac{3}{4}$
48. $\frac{1}{3} + \frac{c}{5} > -\frac{3}{2}$
49. $\frac{1}{2} - \frac{x}{24} \geq -\frac{1}{8}$
50. $\frac{4}{3} - \frac{x}{5} \geq \frac{4}{15}$

Graph each compound inequality and describe the graph using interval notation.

51. $-2 \leq x < 3$
52. $-1 < x \leq 4$
53. $-\frac{7}{4} < x < 2$
54. $0 \leq x \leq \frac{11}{3}$

Solve each compound inequality. Write the solution set in interval notation and graph it. See Example 10.

55. $2 < x - 5 < 5$

56. $-8 < t - 8 < 8$

57. $0 \le x + 10 \le 10$

58. $-9 \le x + 8 < 1$

59. $3 \le 2x - 1 < 5$

60. $4 < 3x - 5 \le 7$

61. $-9 < 6x + 9 \le 45$

62. $-30 \le 10d + 20 < 90$

TRY IT YOURSELF

Solve each inequality or compound inequality. Write the solution set in interval notation and graph it.

63. $\frac{6x + 1}{4} \le x + 1$

64. $\frac{3x - 10}{5} \le x + 4$

65. $17(3 - x) \ge 3 - 13x$

66. $7x + 6 \ge -(x - 6)$

67. $0 < 5(x + 2) \le 15$

68. $-18 \le 9(x - 5) < 27$

69. $0.4x \le 0.1x + 0.45$

70. $0.9s \le 0.3s + 0.54$

71. $-\frac{2}{3} \ge \frac{2y}{3} - \frac{3}{4}$

72. $-\frac{2}{9} \ge \frac{5x}{6} - \frac{1}{3}$

73. $\frac{m}{-42} - 1 > -1$

74. $\frac{a}{-25} + 3 < 3$

75. $6 - x \le 3(x - 1)$

76. $3(3 - x) \ge 6 + x$

77. $6 < -2(x - 1) < 12$

78. $4 \le -4(x - 2) < 20$

79. $-1 \le -\frac{1}{2}n$

80. $-3 \ge -\frac{1}{3}t$

81. $-m - 12 > 15$

82. $-5x + 7 \le 12$

83. $y - \frac{1}{7} \le \frac{2}{3}$

84. $m - \frac{1}{9} \ge \frac{4}{5}$

85. $9x + 13 \ge 2x + 6x$

86. $7x - 16 < 2x + 4x$

87. $7 < \frac{5}{3}a + (-3)$

88. $5 < \frac{7}{2}a + (-9)$

89. $-8 \le \frac{y}{8} - 4 \le 2$

90. $-12 < \frac{b}{3} < 0$

91. $0.04x + 1.04 \le 0.01x + 1.085$

92. $0.005 + 2.08x \le 2.05x - 0.07$

93. $\frac{5}{3}(x + 1) \ge -x + \frac{2}{3}$

94. $\frac{5}{2}(7x - 15) \ge \frac{11}{2}x - \frac{3}{2}$

95. $\frac{4}{5}x < \frac{2}{5}$

96. $\frac{11}{9}x > \frac{5}{9}$

97. $2x + 3(2x + 3) \le 7(x + 1) + 1$

98. $3(3x + 3) - 2 \le 2(2x - 1) + 6x$

Look Alikes . . .

Solve each equation and inequality. Write the solution set of each inequality in interval notation and graph it.

99. a. $\frac{3}{8} + \frac{b}{3} > \frac{5}{12}$ b. $\frac{3}{8} + \frac{b}{3} = \frac{5}{12}$

100. a. $7(a - 3) < 2(5a - 8)$ b. $7(a - 3) = 2(5a - 8)$

101. a. $4 \le 2x - 6$ b. $4 \le 2x - 6 < 18$

102. a. $-16 < 4(x + 8) \le 8$ b. $-16 = 4(x + 8) + 8$

APPLICATIONS

103. **Grades.** A student has test scores of 68%, 75%, and 79% in a government class. What must she score on the last exam to earn a B (80% or better) in the course?

104. **Occupational Testing.** An employment agency requires applicants average at least 70% on a battery of four job skills tests. If an applicant scored 70%, 74%, and 84% on the first three exams, what must he score on the fourth test to maintain a 70% or better average?

105. **Gas Mileage.** A car manufacturer produces three models in equal quantities. One model has an economy rating of 17 miles per gallon, and the second model is rated for 19 mpg. If government regulations require the manufacturer to have a fleet average that exceeds 21 mpg, what economy rating is required for the third model?

106. **Service Charges.** When the average daily balance of a customer's checking account falls below $500 in any week, the bank assesses a $5 service charge. The table shows the daily balances of one customer. What must Friday's balance be to avoid the service charge?

Day	Balance
Monday	$540.00
Tuesday	$435.50
Wednesday	$345.30
Thursday	$310.00

107. **Geometry.** The perimeter of an equilateral triangle is at most 57 feet. What could the length of a side be? (*Hint:* All three sides of an equilateral triangle are equal.)

108. **Geometry.** The perimeter of a square is no less than 68 centimeters. How long can a side be?

109. Counter Space. A rectangular counter is being built for the customer service department of a store. Designers have determined that the outside perimeter of the counter (shown in red) needs to exceed 30 feet. Determine the acceptable values for x.

110. Number Puzzles. What numbers satisfy the condition: Four more than three times the number is at most 10?

111. Graduations. It costs a student \$18 to rent a cap and gown and 80 cents for each graduation announcement that she orders. If she doesn't want her spending on these graduation costs to exceed \$50, how many announcements can she order?

112. Telephones. A cellular telephone company has currently enrolled 36,000 customers in a new calling plan. If an average of 1,200 people are signing up for the plan each day, in how many days will the company surpass their goal of having 150,000 customers enrolled?

113. Windows. An architect needs to design a triangular-shaped bathroom window that has an area no greater than 100 in.^{2}. If the base of the window must be 16 inches long, what window heights will meet this condition?

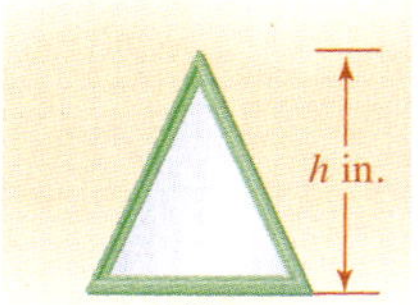

114. Room Temperatures. To hold the temperature of a room between 19° and 22° Celsius, what Fahrenheit temperatures must be maintained? *Hint:* Use the formula $C = \frac{5}{9}(F - 32)$.

115. Number Puzzles. What *whole* numbers satisfy the condition: Twice the number decreased by 1 is between 50 and 60?

116. Exercise. The graph in the next column shows the target heart beat range for different ages and exercise intensity levels. If we let b represent the number of beats per minute, then the compound inequality that estimates the heart beat rate range for a 30-year-old involved in a high-intensity workout is about $168 \le b \le 198$. Use a compound inequality to estimate the heart beat rate range for the following ages and zones.

a. 45-year-old, fat-burning zone
b. 70-year-old, high-intensity zone
c. 25-year-old, intermediate zone

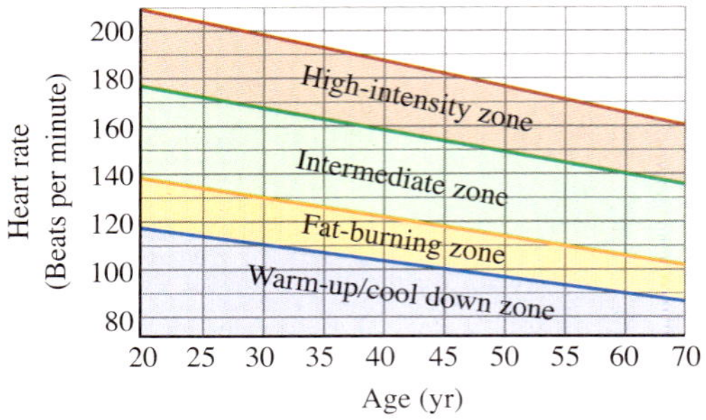

Source: elitefitness.co.nz

WRITING

117. Explain why multiplying both sides of an inequality by a negative number reverses the direction of the inequality.

118. a. What number is a solution of $3x - 26 \ge 4$ but is not a solution of $3x - 26 > 4$? Explain your reasoning.

b. What numbers are solutions of $3x - 26 \ge 4$ but not solutions of $3x - 26 = 4$? Explain your reasoning.

REVIEW

Complete each table.

119.

x	$x^2 - 3$
-2	
0	
3	

120.

x	$\frac{x}{3} + 2$
-6	
0	
12	

CHALLENGE PROBLEMS

Solve each inequality. Write the solution set in interval notation and graph it.

121. $3 - x < 5 < 7 - x$

122. $\frac{1}{x} > 1$. (*Hint:* Use a guess-and-check approach.)

123. $2(5x - 6) > 4x - 15 + 6x$

124. $\frac{3a - 4}{-5} < \frac{3a + 15}{-5}$

2 Summary & Review

SECTION 2.1 Solving Equations Using Properties of Equality

DEFINITIONS AND CONCEPTS	EXAMPLES
An **equation** is a statement indicating that two expressions are equal. The equal symbol = separates an equation into two parts: the *left side* and the *right side*.	Equations: $2x + 4 = 10 \quad -5(a + 4) = -11a \quad \frac{3}{2}t + 6 = t - \frac{1}{3}$
A number that makes an equation a true statement when substituted for the variable is called a **solution** of the equation.	Use a check to determine whether 2 is a solution of $x + 4 = 3x$. ***Check:*** $x + 4 = 3x$ $2 + 4 \stackrel{?}{=} 3(2)$ Substitute 2 for each x. $6 = 6$ True Since the resulting statement $6 = 6$ is true, 2 is a solution of $x + 4 = 3x$.
Equivalent equations have the same solutions.	$x - 2 = 6$ and $x = 8$ are equivalent equations because they have the same solution, 8.
To **solve an equation** isolate the variable on one side of the equation by undoing the operations performed on it using properties of equality. **Addition (Subtraction) property of equality:** If the same number is added to (or subtracted from) both sides of an equation, the result is an equivalent equation.	Solve: $x - 5 = 7$ $x - 5 + 5 = 7 + 5$ $x = 12$ The solution is 12. The solution set is $\{12\}$. Solve: $c + 9 = 16$ $c + 9 - 9 = 16 - 9$ $c = 7$ The solution is 7. The solution set is $\{7\}$.
Multiplication (Division) property of equality: If both sides of an equation are multiplied (or divided) by the same nonzero number, the result is an equivalent equation.	Solve: $\frac{1}{3}m = 2$ $3\left(\frac{1}{3}m\right) = 3(2)$ $m = 6$ The solution is 6. Solve: $10y = 50$ $\frac{10y}{10} = \frac{50}{10}$ $y = 5$ The solution is 5.

REVIEW EXERCISES

Use a check to determine whether the given number is a solution of the equation.

1. 84, $x - 34 = 50$
2. 3, $5y + 2 = 12$
3. -30, $\frac{x}{5} = 6$
4. 2, $|a^2 - a - 1| = 0$
5. -3, $5b - 2 = 3b - 8$
6. 1, $\frac{2}{y + 1} = \frac{12}{y + 1} - 5$

Fill in the blanks.

7. An ________ is a statement indicating that two expressions are equal.
8. To solve $x - 8 = 10$ means to find all the values of the variable that make the equation a _____ statement.

Solve each equation and check the result.

9. $x - 9 = 12$
10. $-y = -32$
11. $a + 3.7 = -16.9$
12. $100 = -7 + r$
13. $120 = 5c$
14. $t - \frac{1}{3} = \frac{3}{7}$
15. $\frac{4}{3}t = -12$
16. $3 = \frac{q}{-2.6}$
17. $6b = 0$
18. $\frac{15}{16}s = -3$

SECTION 2.2 More about Solving Equations

DEFINITIONS AND CONCEPTS | **EXAMPLES**

A five-step **strategy for solving linear equations:**

1. *Clear* the equation of fractions or decimals.
2. *Simplify* each side. Use the distributive property and combine like terms when necessary.
3. *Isolate the variable term.* Use the addition and subtraction properties of equality.
4. *Isolate the variable.* Use the multiplication and division properties of equality.
5. *Check* the result in the original equation.

Solve: $2(y + 2) + 4y = 11 - y$

$2y + 4 + 4y = 11 - y$ Distribute the multiplication by 2.

$6y + 4 = 11 - y$ Combine like terms: $2y + 4y = 6y$.

$6y + 4 + y = 11 - y + y$ To eliminate $-y$ on the right, add y to both sides.

$7y + 4 = 11$ Combine like terms: $6y + y = 7y$ and $-y + y = 0$.

$7y + 4 - 4 = 11 - 4$ To isolate the variable term $7y$, undo the addition of 4 by subtracting 4 from both sides.

$7y = 7$ Simplify each side of the equation.

$\frac{7y}{7} = \frac{7}{7}$ To isolate y, undo the multiplication by 7 by dividing both sides by 7.

$y = 1$ The solution is 1.

It is easier to solve an equation that involves only integers.

To clear an equation of fractions, multiply both sides of an equation by the LCD.

To clear an equation of decimals, multiply both sides by a power of 10 to change the decimals in the equation to integers.

To solve $\frac{1}{2} + \frac{x}{3} = \frac{3}{4}$, first clear the fractions by multiplying both sides by 12:

$12\left(\frac{1}{2} + \frac{x}{3}\right) = 12\left(\frac{3}{4}\right)$ The LCD of $\frac{1}{2}$, $\frac{x}{3}$, and $\frac{3}{4}$ is 12.

To solve $0.5(x - 4) = 0.1x + 0.2$, first clear the decimals by multiplying both sides by 10:

$10[0.5(x - 4)] = 10(0.1x + 0.2)$

Not every equation in one variable has a single solution. Some equations are made true by *any* permissible replacement value for the variable. Such equations are called **identities.**

An equation that is not true for any value of its variable is called a **contradiction.**

When we solve $x + 5 + x = 2x + 5$, the variable drops out and we obtain a true statement $5 = 5$. All real numbers are solutions.

When we solve $y + 2 = y$, the variable drops out and we obtain a false statement $2 = 0$. The equation has no solutions.

REVIEW EXERCISES

Solve each equation. Check the result.

19. $5x + 4 = 14$
20. $98.6 - t = 129.2$
21. $\frac{n}{5} + (-2) = 4$
22. $\frac{b - 5}{4} = -6$
23. $5(2x - 4) - 5x = 0$
24. $-2(x - 5) = 5(-3x + 4) + 3$
25. $\frac{3}{4} = \frac{1}{2} + \frac{d}{5}$
26. $\frac{5(7 - x)}{4} = 2x - 3$
27. $\frac{3(2 - c)}{2} = \frac{-2(2c + 3)}{5}$
28. $\frac{b}{3} + \frac{11}{9} + 3b = -\frac{5}{6}b$
29. $0.15(x + 2) + 0.3 = 0.35x - 0.4$
30. $0.5 - 0.02(y - 2) = 0.16 + 0.36y$
31. $3(a + 8) = 6(a + 4) - 3a$
32. $2(y + 10) + y = 3(y + 8)$

SECTION 2.3 Applications of Percent

DEFINITIONS AND CONCEPTS

To solve **percent problems,** use the facts of the problem to write a sentence of the form:

☐ is ☐% of ☐?

Translate the sentence to mathematical symbols: *is* translates to an = symbol and *of* means multiply. Then solve the equation.

EXAMPLES

648 is 30% of what number?

$648 = 30\% \cdot x$ Translate.

$648 = 0.30x$ Change 30% to a decimal: 30% = 0.30.

$\frac{648}{0.30} = \frac{0.30x}{0.30}$ To isolate *x*, undo the multiplication by 0.30 by dividing both sides by 0.30.

$2{,}160 = x$ Do the division.

Thus, 648 is 30% of 2,160.

DEFINITIONS AND CONCEPTS

To find the **percent of increase** or **the percent of decrease,** find what percent the increase or decrease is of the original amount.

EXAMPLES

Sales Prices. To find the percent of decrease when ground beef prices are reduced from \$4.89 to \$4.25 per pound, we first find the amount of decrease: $4.89 - 4.25 = 0.64$. Then we determine what percent 0.64 is of 4.89 (the original price).

0.64 is what % of 4.89?

$0.64 = x \cdot 4.89$ Translate.

$0.64 = 4.89x$

$\frac{0.64}{4.89} = \frac{4.89x}{4.89}$ To isolate *x*, undo the multiplication by 4.89 by dividing both sides by 4.89.

$0.130879346 \approx x$ Do the division.

$013.0879346\% \approx x$ Write the decimal as a percent.

To the nearest tenth of a percent, the percent of decrease is 13.1%.

REVIEW EXERCISES

33. Fill in the blanks.
    a. ________ means parts per one hundred.
    b. When the price of an item is reduced, we call the amount of the reduction a ________.
    c. An employee who is paid a __________ is paid a percent of the goods or services that he or she sells.
34. 4.81 is 2.5% of what number?
35. What number is 15% of 950?
36. What percent of 410 is 49.2?
37. **Internet Users.** The circle graph below shows the percent of the U.S. population that used the Internet in 2010.
    a. What percent of the population did not use the Internet?
    b. If the U.S. population in 2010 was about 310 million, how many people used the Internet that year? Round to the nearest million.

U.S. Internet Usage,* 2010

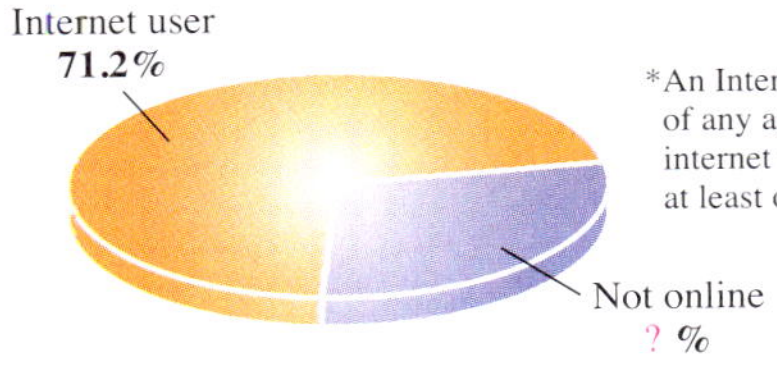

*An Internet user is a person of any age who uses the internet from any location at least once per month.

Source: emarketer

38. **Cost of Living.** A retired trucker receives a monthly Social Security check of \$764. If she is to receive a 3.5% cost-of-living increase soon, how much larger will her check be?
39. **Family Budgets.** It is recommended that a family pay no more than 30% of its monthly income (after taxes) on housing. If a family has an after-tax income of \$1,890 per month and pays \$625 in housing costs each month, are they within the recommended range?
40. **Discounts.** A shopper saved \$148.50 on a food processor that was discounted 33%. What did it originally cost?
41. **Tupperware.** The hostess of a Tupperware party is paid a 25% commission on her in-home party's sales. What would the hostess earn if sales totaled \$600?
42. **Collectibles.** A collector of football trading cards paid \$6 for a 1984 Dan Marino rookie card several years ago. If the card is now worth \$100, what is the percent of increase in the card's value? (Round to the nearest percent.)

SECTION 2.4 Formulas

DEFINITIONS AND CONCEPTS	EXAMPLES
A **formula** is an equation that states a relationship between two or more variables.	Retail price: $r = c + m$ Profit: $p = r - c$ Simple Interest: $I = Prt$ Distance: $d = rt$ Temperature: $C = \frac{5}{9}(F - 32)$
The **perimeter** of a plane geometric figure is the distance around it. The **area** of a plane geometric figure is the amount of surface that it encloses. The **volume** of a three-dimensional geometric solid is the amount of space it encloses.	Rectangle: $P = 2l + 2w$ Circle: $C = \pi D = 2\pi r$ $A = lw$ $A = \pi r^2$ Rectangular solid: $V = lwh$ Cylinder: $V = \pi r^2 h$ *See inside the back cover of the text for more geometric formulas.
If we are given the values of all but one of the variables in a formula, we can use our equation-solving skills to find the value of the remaining variable.	**Bedding.** The area of a standard queen-size bed sheet is 9,180 in.^{2}. If the length is 102 inches, what is the width? $A = lw$ This is the formula for the area of a rectangle. $9{,}180 = 102w$ Substitute 9,180 for the area A and 102 for the length l. $\frac{9{,}180}{102} = \frac{102w}{102}$ To isolate w, undo the multiplication by 102 by dividing both sides by 102. $90 = w$ Do the division. The width of a standard queen-size bed sheet is 90 inches.
To solve a formula for a specific variable means to isolate that variable on one side of the equation, with all other variables and constants on the opposite side. Treat the specified variable as if it is the only variable in the equation. Treat the other variables as if they were numbers (constants).	Solve the formula for the volume of a cone for h. $V = \frac{1}{3}\pi r^2 h$ This is the formula for the volume of a cone. $3(V) = 3\left(\frac{1}{3}\pi r^2 h\right)$ To clear the equation of the fraction, multiply both sides by 3. $3V = \pi r^2 h$ Simplify. $\frac{3V}{\pi r^2} = \frac{\pi r^2 h}{\pi r^2}$ To isolate h, undo the multiplication by πr^2 by dividing both sides by πr^2. $\frac{3V}{\pi r^2} = h$ or $h = \frac{3V}{\pi r^2}$

REVIEW EXERCISES

43. **Shopping.** Find the markup on a CD player whose wholesale cost is \$219 and whose retail price is \$395.

44. **Restaurants.** One month, a restaurant had sales of \$13,500 and made a profit of \$1,700. Find the expenses for the month.

45. **Snails.** A typical garden snail travels at an average rate of 2.5 feet per minute. How long would it take a snail to cross a 20-foot-long flower bed?

46. **Certificates of Deposit.** A \$26,000 investment in a CD earned \$1,170 in interest the first year. What was the annual interest rate?

47. **Jewelry.** Gold melts at about 1,065°C. Change this temperature to degrees Fahrenheit.

48. **Camping.**
   a. Find the perimeter of the air mattress.
   b. Find the amount of sleeping area on the top surface of the air mattress.
   c. Find the volume of the air mattress if it is 3 inches thick.

49. Find the area of a triangle with a base 17 meters long and a height of 9 meters.

50. Find the area of a trapezoid with bases 11 inches and 13 inches long and a height of 12 inches.

In Problems 51–53, the answers may vary slightly depending on which approximation of π is used.

51. a. Find the circumference of a circle with a radius of 8 centimeters. Round to the nearest hundredth.
    b. Find the area of the circle. Round to the nearest square centimeter.

52. Find the volume of a 12-foot tall cylinder whose circular base has a radius of 0.5 feet. Give the result to the nearest tenth.

53. **Halloween.** After being cleaned out, a spherical-shaped pumpkin has an inside diameter of 9 inches. To the nearest hundredth, what is its volume?

54. Find the volume of a pyramid that has a square base, measuring 6 feet on a side, and a height of 10 feet.

Solve each formula for the specified variable.

55. $A = 2\pi rh$ for h

56. $A - BC = \frac{G - K}{3}$ for G

57. $C = \frac{1}{4}s(t - d)$ for t

58. $4y - 3x = 16$ for y

SECTION 2.5 Problem Solving

DEFINITIONS AND CONCEPTS	EXAMPLES
To solve application problems, use the six-step problem-solving strategy: 1. **Analyze** the problem. 2. **Assign** a variable. 3. **Form** an equation. 4. **Solve** the equation. 5. **State** the conclusion. 6. **Check** the result. (See page 142 for the more detailed list of steps.) In this section, we solved application problems involving: ■ More than one unknown (Example 1) ■ Set-up fees (Example 2) ■ Commissions (Example 3) ■ Consecutive integers (Example 4) ■ Perimeter (Example 5) ■ Isosceles triangles (Example 6)	**Income Taxes.** After taxes, an author kept \$85,340 of her total annual earnings. If her earnings were taxed at a 15% rate, how much did she earn that year? **Analyze** The author earned some unknown amount of money. On that amount, she paid 15% in taxes. The difference between her total earnings and the taxes paid was \$85,340. **Assign** If we let x = the author's total earnings, the amount of taxes that she paid was 15% of x or $0.15x$. **Form** We can use the words of the problem to form an equation. Her total earnings / minus / the taxes that she paid / equals / the money that she kept. x − $0.15x$ = 85,340 **Solve** $x - 0.15x = 85{,}340$ $0.85x = 85{,}340$ — Combine like terms: $1x - 0.15x = 0.85x$. $x = 100{,}400$ — To isolate x, divide both sides by 0.85. **State** The author earned \$100,400 that year. **Check** The taxes were 15% of \$100,400 or \$15,060. If we subtract the taxes from her total earnings, we get \$100,400 − \$15,060 = \$85,340. The answer checks.

REVIEW EXERCISES

59. **Sound Systems.** A 45-foot-long speaker wire is to be cut into three pieces. One piece is to be 15 feet long. Of the remaining pieces, one must be 2 feet less than 3 times the length of the other. Find the length of the shorter piece.

60. **Signing Petitions.** A professional signature collector is paid \$50 a day plus \$2.25 for each verified signature he gets from a registered voter. How many signatures are needed to earn \$500 a day?

61. **Lottery Winnings.** After taxes, a lottery winner was left with a lump sum of \$1,800,000. If 28% of the original prize was withheld to pay federal income taxes, what was the original cash prize?

62. **NASCAR.** The car numbers of drivers Bobby Labonte and Kyle Petty are consecutive odd integers whose sum is 88. If Labonte's number is the smaller, find the numbers of each car.

63. Art History. *American Gothic* was painted in 1930 by Grant Wood. The length of the rectangular painting is 5 inches more than the width. Find the dimensions of the painting if it has a perimeter of 109.5 inches.

64. Geometry. Find the missing angle measures of the triangle.

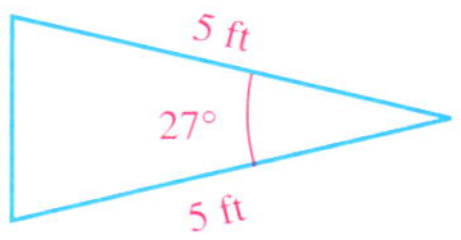

SECTION 2.6 More on Problem Solving

DEFINITIONS AND CONCEPTS

To solve application problems, use the six-step problem-solving strategy.

1. **Analyze** the problem.
2. **Assign** a variable.
3. **Form** an equation.
4. **Solve** the equation.
5. **State** the conclusion.
6. **Check** the result.

(See page 142 for the more detailed list of steps.)

Tables are a helpful way to organize the facts of a problem.

In this section, we solved application problems involving:

- Money/investments (Example 1)
- Motion (Example 2 and Example 3)
- Liquid mixture (Example 4)
- Dry mixture (Example 5)
- Number-value (Example 6)

EXAMPLES

Trucking. Two trucks leave from the same place at the same time traveling in opposite directions. One travels at a rate of 60 mph and the other at 50 mph. How long will it take them to be 165 miles apart?

Analyze We know that one truck travels at 60 mph and the other at 50 mph. Together, the trucks will travel a distance of 165 miles.

Assign Let t = the number of hours until the distance between the trucks is 165 miles.

Form We enter each rate in the table under the heading r. Since the trucks travel for the same length of time, say t hours, we enter t for each truck under the heading t. We enter the distances traveled under the heading d in the table.

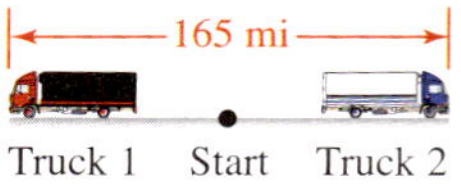

	r	$\cdot\ t$	$= d$
Truck 1	60	t	$60t$
Truck 2	50	t	$50t$
			Total: 165

Multiply $r \cdot t$ to obtain an expression for the distance traveled by each truck.

Use the information in this column to form an equation.

The distance the first truck travels	plus	the distance the second truck travels	is	165 miles.
$60t$	$+$	$50t$	$=$	165

Solve $60t + 50t = 165$

$110t = 165$ Combine like terms: $60t + 50t = 110t$.

$\frac{110t}{\mathbf{110}} = \frac{165}{\mathbf{110}}$ To isolate t, undo the multiplication by 110 by dividing both sides by 110.

$t = 1.5$ Do the division.

State The trucks will be 165 miles apart in 1.5 hours.

Check If the first truck travels 60 mph for 1.5 hours, it will go $60(1.5) = 90$ miles. If the second truck travels 50 mph for 1.5 hours, it will go $50(1.5) = 75$ miles. Since 90 miles + 75 miles = 165 miles, the result checks.

REVIEW EXERCISES

65. Investment Income. A woman has \$27,000. Part is invested for 1 year in a certificate of deposit paying 7% interest, and the remaining amount in a cash management fund paying 9%. After 1 year, the total interest on the two investments is \$2,110. How much is invested at each rate?

66. Walking and Bicycling. A bicycle path is 5 miles long. A man walks from one end at the rate of 3 mph. At the same time, a friend bicycles from the other end, traveling at 12 mph. In how many minutes will they meet?

67. Airplanes. How long will it take a jet plane, flying at 450 mph, to overtake a propeller plane, flying at 180 mph, if the propeller plane had a $2\frac{1}{2}$-hour head start?

68. Autographs. Kesha collected the autographs of 8 more television celebrities than she has of movie stars. Each TV celebrity autograph is worth \$75 and each movie star autograph is worth \$250. If her collection is valued at \$1,900, how many of each type of autograph does she have?

69. Mixtures. A store manager mixes candy worth 90¢ per pound with gumdrops worth \$1.50 per pound to make 20 pounds of a mixture worth \$1.20 per pound. How many pounds of each kind of candy does he use?

70. Eliminating Mildew. How many gallons of a 2% fungicide solution must be mixed with 4 gallons of a 5% fungicide solution to get a 4% fungicide solution?

SECTION 2.7 Solving Inequalities

DEFINITIONS AND CONCEPTS	EXAMPLES
An **inequality** is a mathematical statement that contains an $>$, $<$, $\geq$, or $\leq$ symbol.	Inequalities: $3x < 8$ $\frac{1}{2}y - 4 \geq 12$ $2z + 4 \leq z - 5$
A **solution of an inequality** is any number that makes the inequality true.	Determine whether 3 is a solution of $2x - 7 < 5$. ***Check:*** $2x - 7 < 5$ $2(3) - 7 \overset{?}{<} 5$ Substitute 3 for x. $-1 < 5$ True Since the resulting statement $-1 < 5$ is true, 3 is a solution of $2x - 7 < 5$.
We **solve inequalities** as we solve equations. However, if we **multiply or divide both sides by a negative number,** we must *reverse* the inequality symbol. **Interval notation** can be used to describe the solution set of an inequality. A **parenthesis** indicates that a number is not included in the solution set of an inequality. A **bracket** indicates that a number is included in the solution set.	Solve: $-3(z - 1) \geq -6$ $-3z + 3 \geq -6$ Distribute the multiplication by -3. $-3z \geq -9$ To isolate the variable term $-3z$, undo the addition of 3 by subtracting 3 from both sides. $\frac{-3z}{-3} \leq \frac{-9}{-3}$ To isolate z, undo the multiplication by dividing both sides by -3. Reverse the $\geq$ symbol. $z \leq 3$ Do the division. The solution set is $(-\infty, 3]$ and it is graphed as shown. 

REVIEW EXERCISES

Solve each inequality. Write the solution set in interval notation and graph it.

71. $3x - 2 < 5$

72. $-\frac{3}{4}x \geq -9$

73. $\frac{3}{4} < \frac{d}{5} + \frac{1}{2}$

74. $5(3 - x) \leq 3(x - 3)$

75. $\frac{t}{-5} - (-1.8) \geq -6.2$

76. $a + 5 - 2(10 - a) > 6$

77. $24 < 3(x + 2) < 39$

78. $0 \leq 3 - 2x < 10$

79. Sports Equipment. The acceptable weight w of Ping-Pong balls used in competition can range from 2.40 to 2.53 grams. Express this range using a compound inequality.

80. Signs. A large office complex has a strict policy about signs. Any sign to be posted in the building must be rectangular in shape, its width must be 18 inches, and its perimeter is not to exceed 132 inches. What possible sign lengths meet these specifications?

2 CHAPTER TEST

1. Fill in the blanks.
   a. To ______ an equation means to find all of the values of the variable that make the equation true.
   b. _______ means parts per one hundred.
   c. The distance around a circle is called its ___________.
   d. An _________ is a statement that contains one of the symbols $>$, $\geq$, $<$, or $\leq$.
   e. The ___________ property of _______ says that multiplying both sides of an equation by the same nonzero number does not change its solution.
2. Use a check to determine if 3 is a solution of $5y + 2 = 12$.

Solve each equation.

3. $3h + 2 = 8$
4. $-22 = -x$
5. $\frac{4}{5}t = -4$
6. $\frac{11b - 11}{5} = \frac{3b - 2}{2}$
7. $0.8(x - 1{,}000) + 1.3 = 2.9 + 0.2x$
8. $2(y - 7) - 3y = -(y - 3) - 17$
9. $\frac{m}{2} - \frac{1}{3} = \frac{1}{4} + \frac{m}{6}$
10. $\frac{3}{4}(6n - 2) = 246$
11. $5x = 0$
12. $6a + (-7) = 3a - 7 + 2a$
13. $9 - 5(2x + 10) = -1$
14. $24t = -6(8 - 4t)$
15. What is 15.2% of 80?
16. **Down Payments.** To buy a house, a woman was required to make a down payment of \$11,400. What did the house sell for if this was 15% of the purchase price?
17. **Body Temperatures.** Suppose a person's body temperature rises from 98.6°F to a dangerous 105°F. What is the percent increase? Round to the nearest percent.
18. **Commissions.** An appliance store salesperson receives a commission of 5% of the price of every item that she sells. What will she make if she sells a \$599.99 refrigerator?
19. **Grand Openings.** On its first night of business, a pizza parlor brought in \$445. The owner estimated his profits that night to be \$150. What were the costs?
20. Find the Celsius temperature reading if the Fahrenheit reading is 14°.
21. **Sound.** The speed of sound at sea level is about 1,108 feet per second. How far will sound travel in 1 minute?
22. **Pets.** The spherical fishbowl is three-quarters full of water. To the nearest cubic inch, find the volume of water in the bowl. (*Hint:* The volume of a sphere is given by $V = \frac{4}{3}\pi r^3$.)

Solve for the specified variable.

23. $V = \pi r^2 h$ for h
24. $A = P + Prt$ for r
25. $A = \frac{a + b + c + d}{4}$ for c
26. $2x - 3y = 9$ for y
27. **Irons.** Estimate the area of the soleplate of the iron.

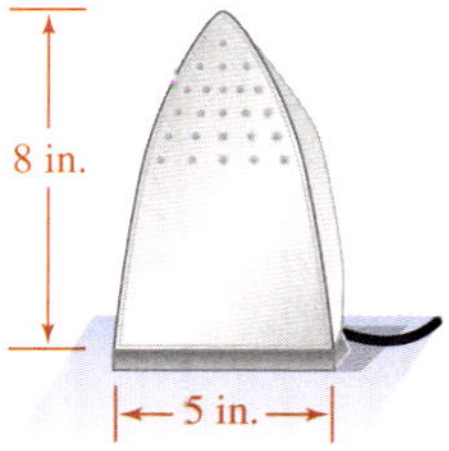

28. **Television.** In a typical 30-minute block of time on TV, the number of programming minutes are 2 less than three times the number of minutes of commercials. How many minutes of programming and how many minutes of commercials are there?
29. **Plumbing Bills.** A part of the invoice for plumbing work shown below is torn away. What is the cost per hour for labor?

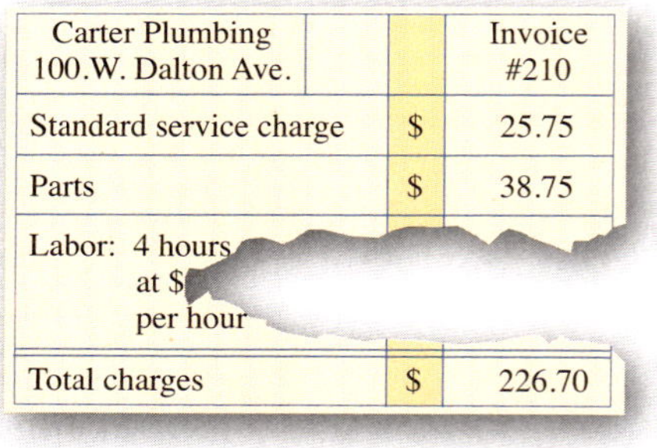

Carter Plumbing 100.W. Dalton Ave.		Invoice #210
Standard service charge	\$	25.75
Parts	\$	38.75
Labor: 4 hours at \$ per hour		
Total charges	\$	226.70

30. **Concert Seating.** Two types of seating were sold for a concert. Floor seats cost \$12.50 a ticket and balcony seats cost \$20.50. Ten times as many floor seats were sold as balcony seats. If the total receipts from the sale of both types of tickets were \$11,640, how many of each type of ticket were sold?
31. **Home Sales.** A condominium owner cleared \$114,600 on the sale of his condo, after paying a 4.5% real estate commission. What was the selling price?
32. **Colorado.** The state of Colorado is approximately rectangular-shaped with perimeter 1,320 miles. Find the length (east to west) and width (north to south), if the length is 100 miles longer than the width.
33. **Tea.** How many pounds of green tea, worth \$40 a pound, should be mixed with herbal tea, worth \$50 a pound, to produce 20 pounds of a blend worth \$42 a pound?

34. Reading. A bookmark is inserted between two page numbers whose sum is 825. What are the page numbers?

35. Travel Times. A car leaves Rockford, Illinois, at the rate of 65 mph, bound for Madison, Wisconsin. At the same time, a truck leaves Madison at the rate of 55 mph, bound for Rockford. If the cities are 72 miles apart, how long will it take for the car and the truck to meet?

36. Pickles. To make pickles, fresh cucumbers are soaked in a salt water solution called *brine*. How many liters of a 2% brine solution must be added to 30 liters of a 10% brine solution to dilute it to an 8% solution?

37. Exercise. How long will it take a bicyclist, traveling at 20 mph, to catch up with a jogger, traveling at 8 mph, if the jogger had a half-hour head start?

38. Geometry. If the vertex angle of an isosceles triangle is 44°, find the measure of each base angle.

39. Investments. Part of \$13,750 is invested at 9% annual interest, and the rest is invested at 8%. After one year, the accounts paid \$1,185 in interest. How much was invested at the lower rate?

40. Use a check to determine whether -3 is a solution of $4 - 9w < -4w + 19$.

Solve each inequality. Write the solution set in interval notation and graph it.

41. $-8x - 20 \le 4$

42. $-8.1 > \frac{t}{2} + (-11.3)$

43. $-12 \le 2(x + 1) < 10$

44. $\frac{1}{3}(a - 5) > \frac{1}{2}(a + 1)$

45. $-9(h - 3) + 2h \le 8(4 - h)$

46. Awards. A city honors its citizen of the year with a framed certificate. An artist charges \$15 for the frame and 75 cents per word for writing out the proclamation. If a city regulation does not allow gifts in excess of \$150, what is the maximum number of words that can be written on the certificate?

Group Project

TRANSLATING KEY WORDS AND PHRASES

Overview: Students often say that the most challenging step of the six-step problem-solving strategy is forming an equation. This activity is designed to make that step easier by improving your translating skills.

Instructions: Form groups of 3 or 4 students. Select one person from your group to record the group's responses. Determine whether addition, subtraction, multiplication, or division is suggested by each of the following words or phrases. Then use the word or phrase in a sentence to illustrate its meaning. (If a word stumps everyone in your group, a dictionary can be helpful.)

deflate	recede	partition	evaporate	amplify
bisect	augment	hike	erode	boost
annexed	diminish	plummet	upsurge	wane
quadruple	corrode	taper off	trisect	broaden

COMPUTER SPREADSHEETS

Overview: In this activity, you will get some experience working with a spreadsheet.

Instructions: Form groups of 3 or 4 students. Examine the following spreadsheet, which consists of cells named by column and row. For example, 7 is entered in cell B3. In any cell you may enter data or a formula. For each formula in cells D1–D4 and E1–E4, the computer performs a calculation using values entered in other cells and prints the result in place of the formula. Find the value that will be printed in each formula cell. The symbol * means multiply, / means divide, and ^ means raise to a power.

	A	B	C	D	E
1	−8	20	−6	= 2*B1 − 3*C1 + 4	= B1 − 3*A1^2
2	39	2	−1	= A2/(B2 − C2)	= B3*B2*C2*2
3	50	7	3	= A3/5 + C3^3	= 65 − 2*(B3 − 5)^5
4	6.8	−2.8	−0.5	= 100*A4 + B4*C4	= A4/10 + A3/2*5

Graphing Linear Equations and Inequalities in Two Variables; Functions

3

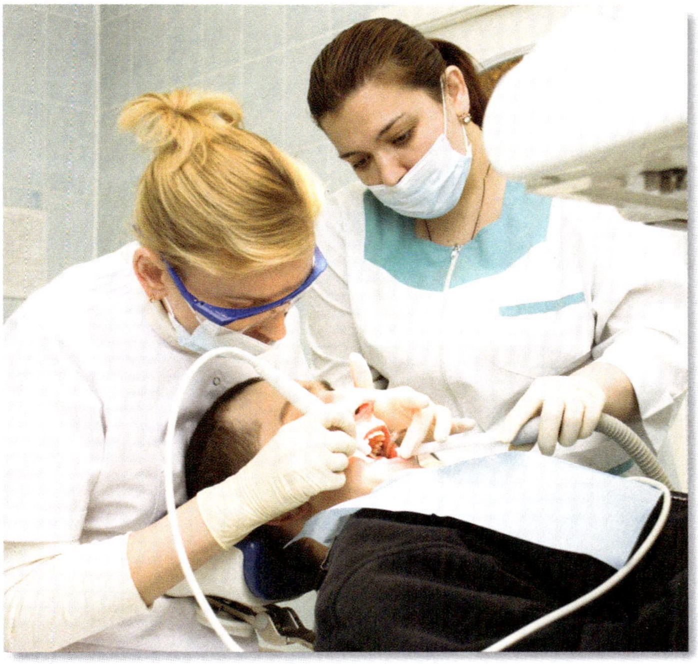

from Campus to Careers

Dental Assistant

A dental assistant is a valuable member of the dental health team who prepares patients for treatment, takes x-rays, sterilizes instruments, and keeps records. Part of the training of a dental assistant includes learning about a *coordinate system* that is used to identify the location of teeth in the mouth. This coordinate system is much like one used in algebra to graph points, lines, and curves.

Problem 33 in **Study Set 3.1**, **problem 105** in **Study Set 3.4**, and **problem 77** in **Study Set 3.7** involve situations that a dental assistant might encounter on the job. The mathematical concepts discussed in this chapter can be used to solve those problems.

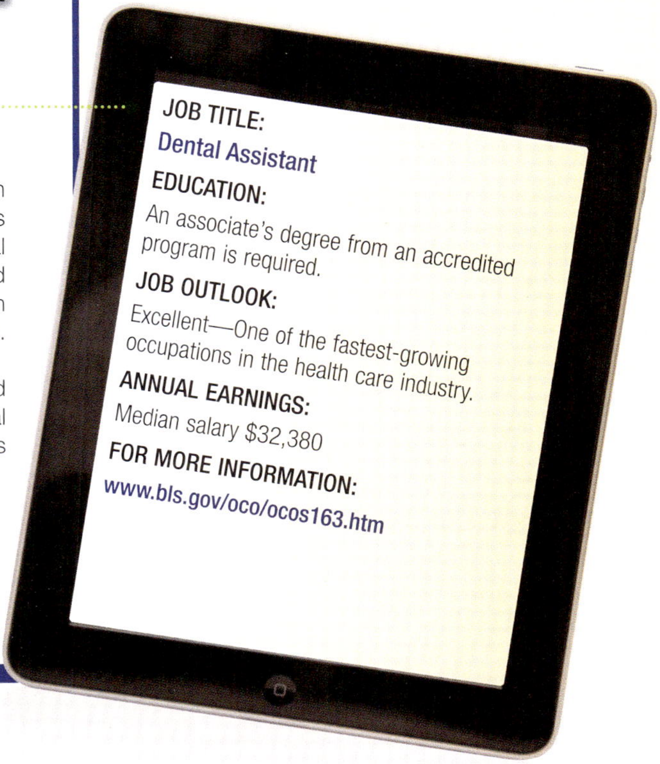

Taking a math test doesn't have to be an unpleasant experience. Here are some suggestions that can make it more enjoyable and also improve your score.

PREPARING FOR THE TEST: Begin studying several days before the test rather than cramming your studying into one marathon session the night before.

TAKING THE TEST: Follow a test-taking strategy so you can maximize your score by using the testing time wisely.

EVALUATING YOUR PERFORMANCE: After your graded test is returned, classify the types of errors that you made on the test so that you do not make them again.

Now Try This

1. Write a study session plan that explains how you will prepare on each of the 4 days before the test, as well as on test day. For some suggestions, see *Preparing for a Test.**
2. Develop your own test-taking strategy by answering the survey questions found in *How to Take a Math Test.**
3. Use the outline found in *Analyzing Your Test Results** to classify the errors that you made on your most recent test.

*Found online at: www.cengage.com/math/tussy

SECTION 3.1 Graphing Using the Rectangular Coordinate System

OBJECTIVES

1. Construct a rectangular coordinate system.
2. Plot ordered pairs and determine the coordinates of a point.
3. Graph paired data.
4. Read line graphs.

ARE YOU READY?

The following problems review some basic skills that are needed when graphing ordered pairs.

1. Graph each number in the set $\left\{\frac{7}{3}, -3, 0, 4, -1.5\right\}$ on a number line.
2. **a.** What number is 8 units to the right of 0 on a number line?
   b. What number is 3.5 units to the left of 0 on a number line?
3. List the first four Roman numerals.
4. Write $\frac{9}{2}$ and $-\frac{11}{3}$ in mixed-number form.

It is often said, "A picture is worth a thousand words." This is certainly true in algebra, where we often use mathematical pictures called *rectangular coordinate graphs* to illustrate numerical relationships.

The Language of Algebra

A rectangular coordinate system is a **grid**—a network of uniformly spaced perpendicular lines. At times, some U.S. cities have such horrible traffic congestion that vehicles can barely move, if at all. The condition is called *gridlock*.

1 Construct a Rectangular Coordinate System.

When designing the Gateway Arch in St. Louis, architects created a mathematical model called a **rectangular coordinate graph.** This graph, shown on the next page, is drawn on a grid called a **rectangular coordinate system.** This coordinate system also is called a **Cartesian coordinate system,** after the 17th-century French mathematician René Descartes.

HHsu/Shutterstock.com

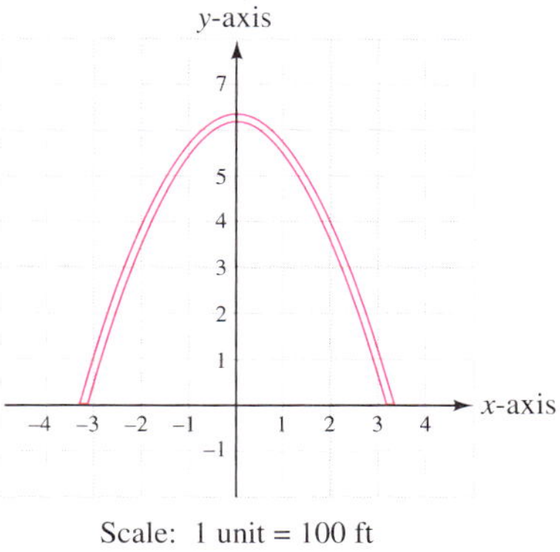

> **The Language of Algebra**
>
> The word **axis** is used in mathematics and science. For example, Earth rotates on its *axis* once every 24 hours. The plural of *axis* is **axes,** which is pronounced ak-seas.

A rectangular coordinate system is formed by two perpendicular number lines. The horizontal number line is usually called the **x-axis,** and the vertical number line is usually called the **y-axis.** On the x-axis, the positive direction is to the right. On the y-axis, the positive direction is upward. Each axis should be scaled to fit the data. For example, the axes of the graph of the arch are scaled in units of 100 feet.

The point where the axes intersect is called the **origin.** This is the zero point on each axis. The axes form a **coordinate plane,** and they divide it into four regions called **quadrants,** which are numbered counterclockwise using Roman numerals.

> **The Language of Algebra**
>
> A **coordinate plane** can be thought of as a perfectly flat surface extending infinitely far in every direction.

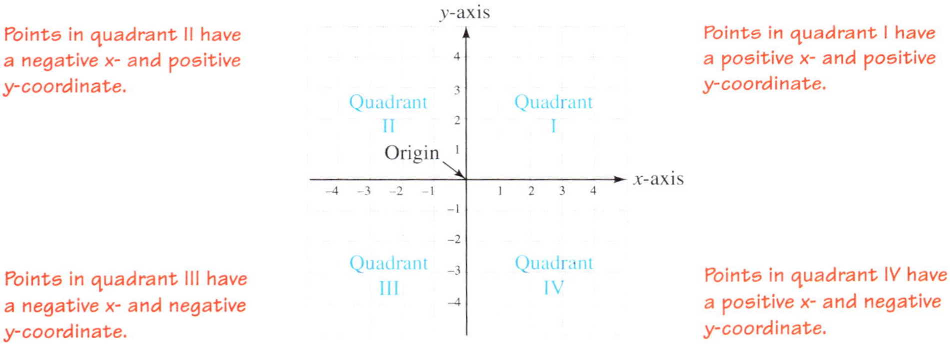

> **Notation**
>
> Don't be confused by this new use of parentheses. $(3, -4)$ represents a point on the coordinate plane, whereas $3(-4)$ indicates multiplication. Also, don't confuse the ordered pair with interval notation.

Each point in a coordinate plane can be identified by an **ordered pair** of real numbers x and y written in the form (x, y). The first number, x, in the pair is called the **x-coordinate,** and the second number, y, is called the **y-coordinate.** Some examples of such pairs are $(3, -4)$, $\left(-1, -\frac{3}{2}\right)$, and $(0, 2.5)$.

$(3, -4)$ Read as "the point three, negative four" or as "the ordered pair three, negative four."

↑ The x-coordinate is listed first.

↑ The y-coordinate is listed second.

2 Plot Ordered Pairs and Determine the Coordinates of a Point.

The process of locating a point in the coordinate plane is called **graphing** or **plotting** the point. On the next page, we use blue arrows to show how to graph the point with coordinates $(3, -4)$. Since the x-coordinate, 3, is positive, we start at the origin and move 3 units to the *right* along the x-axis. Since the y-coordinate, -4, is negative, we then move *down* 4 units and draw a dot. This locates the point $(3, -4)$.

In the figure, red arrows are used to show how to plot the point $(-4, 3)$. We start at the origin, move 4 units to the *left* along the x-axis, then move *up* 3 units and draw a dot. This locates the point $(-4, 3)$.

The Language of Algebra

Note that the points $(3, -4)$ and $(-4, 3)$ have different locations. Since the order of the coordinates of a point is important, we call them **ordered pairs.**

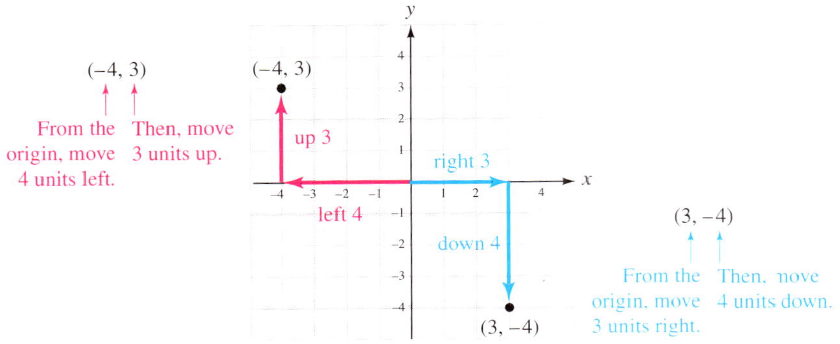

EXAMPLE 1 Plot each point. Then state the quadrant in which it lies or the axis on which it lies.

a. $(4, 4)$ **b.** $\left(-1, -\frac{7}{2}\right)$ **c.** $(0, 2.5)$ **d.** $(-3, 0)$ **e.** $(0, 0)$

Strategy After identifying the x- and y-coordinates of the ordered pair, we will move the corresponding number of units left, right, up, or down to locate the point.

Why The coordinates of a point determine its location on the coordinate plane.

Solution

a. Since the x-coordinate, 4, is positive, we start at the origin and move 4 units to the *right* along the x-axis. Since the y-coordinate, 4, is positive, we then move *up* 4 units and draw a dot. This locates the point $(4, 4)$. The point lies in quadrant I.

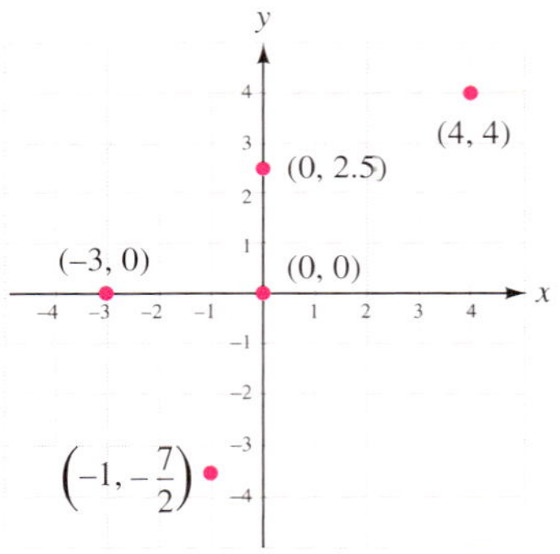

b. To plot $\left(-1, -\frac{7}{2}\right)$, we begin at the origin and move 1 unit to the *left,* because the x-coordinate is -1. Then, since the y-coordinate is negative, we move $\frac{7}{2}$ units, or $3\frac{1}{2}$ units, *down.* The point lies in quadrant III.

c. To plot $(0, 2.5)$, we begin at the origin and do not move right or left, because the x-coordinate is 0. Since the y-coordinate is positive, we move 2.5 units *up.* The point lies on the y-axis.

d. To plot $(-3, 0)$, we begin at the origin and move 3 units to the *left,* because the x-coordinate is -3. Since the y-coordinate is 0, we do not move up or down. The point lies on the x-axis.

e. To plot $(0, 0)$, we begin at the origin, and we remain there because both coordinates are 0. The point with coordinates $(0, 0)$ is the origin.

Caution

When drawing a rectangular coordinate system, always label each axis with the appropriate letter (or title) and **scale** each axis. Here, the scaling shows that each square in the grid is 1 unit long and 1 unit wide.

Success Tip

Points with an x-coordinate that is 0 lie on the y-axis. Points with a y-coordinate that is 0 lie on the x-axis. Points that lie on an axis are not considered to be in any quadrant.

Self Check 1 Plot each point: $(2, -2)$, $(-4, 0)$, $\left(1.5, \frac{5}{2}\right)$, and $(0, 5)$

Now Try Problem 17

EXAMPLE 2 Find the coordinates of points A, B, C, D, E, and F plotted in figure (a) below.

Notation

Points are often labeled with capital letters. For example, the notation $A(2, 3)$ indicates that point A has coordinates $(2, 3)$.

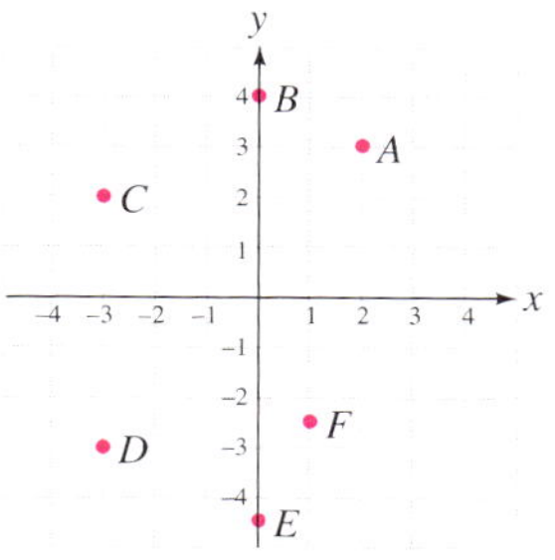

(a)

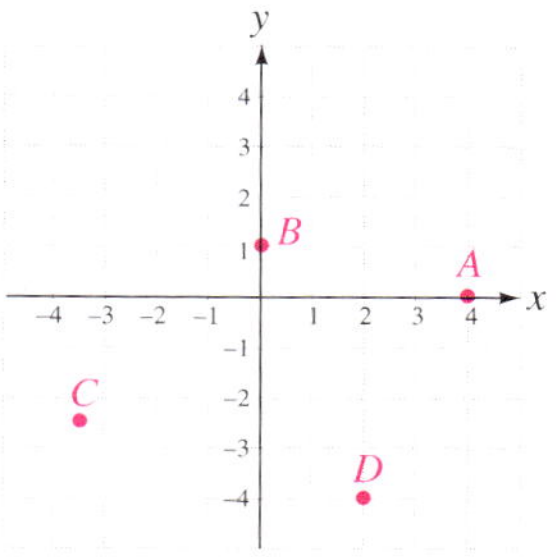

(b)

Strategy We will start at the origin and count to the left or right on the x-axis, and then up or down to reach each point.

Why The movement left or right gives the x-coordinate of the ordered pair and the movement up or down gives the y-coordinate.

Solution To locate point A, we start at the origin, move 2 units to the right on the x-axis, and then 3 units up. Its coordinates are $(2, 3)$. The coordinates of the other points are found in the same manner.

$B(0, 4)$ $\quad$ $C(-3, 2)$ $\quad$ $D(-3, -3)$ $\quad$ $E(0, -4.5)$ $\quad$ $F(1, -2.5)$

The coordinates of points E and F could also be given as $E\left(0, -4\frac{1}{2}\right)$ and $F\left(1, -2\frac{1}{2}\right)$.

Self Check 2 Find the coordinates of each point in figure (b) above.

Now Try Problem 20

3 Graph Paired Data.

Every day, we deal with quantities that are related:

- The time it takes to cook a roast depends on the weight of the roast.
- The money we earn depends on the number of hours we work.
- The sales tax that we pay depends on the price of the item purchased.

We can use graphs to visualize such relationships. For example, suppose a tub is filling with water, as shown below. Obviously, the amount of water in the tub depends on how long the water has been running. To graph this relationship, we can use the measurements that were taken as the tub began to fill.

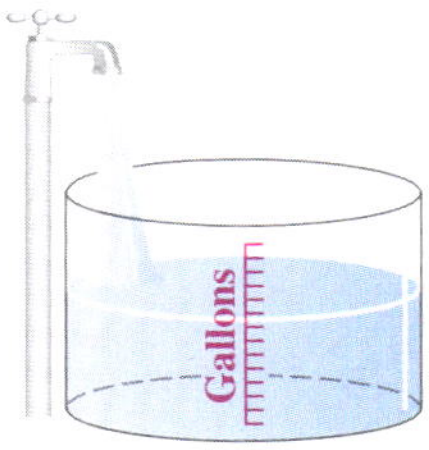

Time (min)	Water in tub (gal)	
0	0	← (0, 0)
1	8	← (1, 8)
3	24	← (3, 24)
4	32	← (4, 32)

↑ x-coordinate ↑ y-coordinate

The data in the table can be expressed as ordered pairs (x, y).

The data in each row of the table can be written as an ordered pair and plotted on a rectangular coordinate system. Since the first coordinate of each ordered pair is a time, we

The Language of Algebra

Data that can be represented as an ordered pair are called **paired data.**

label the x-axis *Time (min)*. The second coordinate is an amount of water, so we label the y-axis *Amount of water (gal)*. The y-axis is scaled in larger units (multiples of 4 gallons) because the size of the data ranges from 0 to 32 gallons.

After plotting the ordered pairs, we use a straightedge to draw a line through the points. As expected, the completed graph shows that the amount of water in the tub increases steadily as the water is allowed to run.

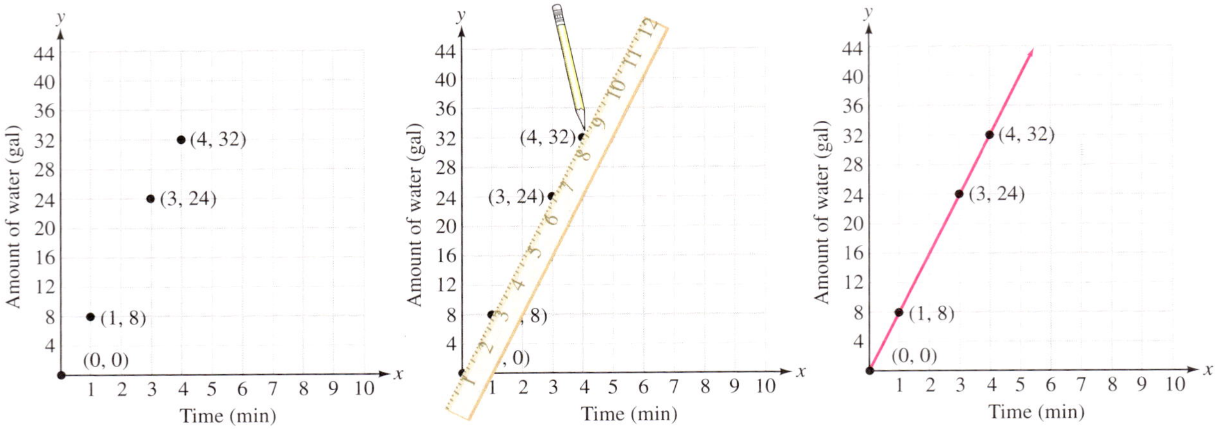

We can use the graph to determine the amount of water in the tub at various times. For example, the green dashed line on the graph at the right shows that in 2 minutes, the tub will contain 16 gallons of water. This process, called **interpolation,** uses known information to predict values that are not known but are *within* the range of the data. The blue dashed line on the graph shows that in 5 minutes, the tub will contain 40 gallons of water. This process, called **extrapolation,** uses known information to predict values that are not known and are *outside* the range of the data.

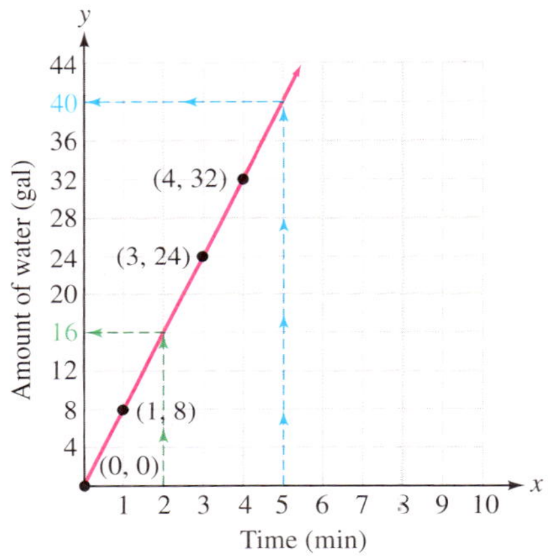

4 Read Line Graphs.

Since graphs are a popular way to present information, the ability to read and interpret them is very important.

EXAMPLE 3 **TV Shows.** The following graph shows the number of people in an audience before, during, and after the taping of a television show. Use the graph to answer the following questions.

a. How many people were in the audience when the taping began?

b. At what times were there exactly 100 people in the audience?

c. How long did it take the audience to leave after the taping ended?

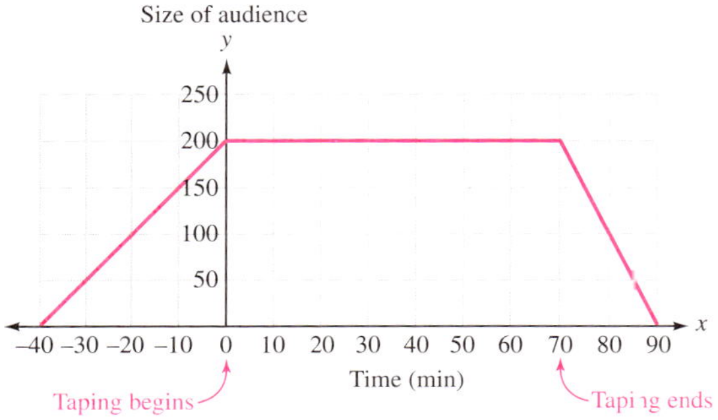

Strategy We will use an ordered pair of the form *(time, size of audience)* to describe each situation mentioned in parts (a), (b), and (c).

Why The coordinates of specific points on the graph can be used to answer each of these questions.

Solution

a. The time when the taping began is represented by 0 on the x-axis. The point on the graph directly above 0 is (0, 200). The y-coordinate indicates that 200 people were in the audience when the taping began. This result is shown in the first row of the table in the margin.

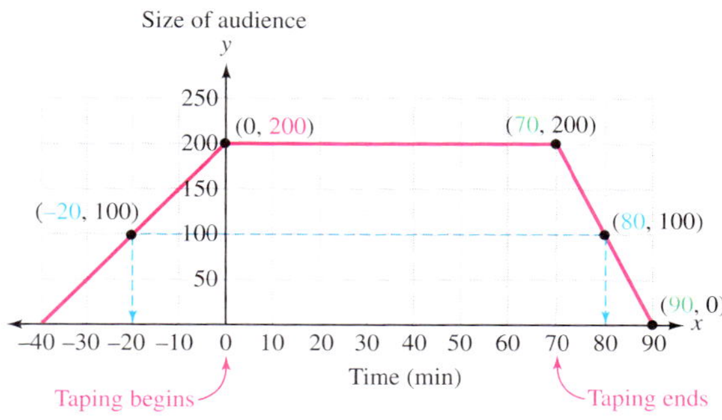

Time	Audience
0	200
−20	100
80	100
70	200
90	0

b. We can draw a horizontal line passing through 100 on the y-axis. Since the line intersects the graph twice, at (**−20**, 100) and at (**80**, 100), there are two times when 100 people were in the audience. These results are shown in the second and third rows of the table. The x-coordinates of the points tell us those times: 20 minutes before the taping began, and 80 minutes after.

c. The x-coordinate of the point (**70**, 200) tells us when the audience began to leave. The x-coordinate of (**90**, 0) tells when the exiting was completed. Subtracting the x-coordinates, we see that it took $90 - 70 = 20$ minutes for the audience to leave.

Self Check 3 Use the graph in Example 3 to answer the following questions.

a. At what times were there exactly 50 people in the audience?

b. How many people were in the audience when the taping took place?

c. When were the first audience members allowed into the taping session?

Now Try Problems 21 and 23

SECTION 3.1 STUDY SET

VOCABULARY

Fill in the blanks.

1. (7, 1) is called an ________ pair.
2. In the ordered pair (2, −5), the ___ -coordinate is 2 and the y-__________ is −5.
3. A rectangular coordinate system is formed by two perpendicular number lines called the x-_____ and the y-_____. The point where the axes cross is called the _______.
4. The x- and y-axes divide the coordinate plane into four regions called __________.
5. The point with coordinates (4, 2) can be graphed on a __________ coordinate system.
6. The process of locating the position of a point on a coordinate plane is called _______________ the point.

CONCEPTS

Fill in the blanks.

7. **a.** To plot (−5, 4), we start at the _______ and move 5 units to the _____ and then move 4 units _____.
   b. To plot $\left(6, -\frac{3}{2}\right)$, we start at the _______ and move 6 units to the ______ and then move $\frac{3}{2}$ units _______.
8. In which quadrant is each point located?
   a. (−2, 7)
   b. $\left(\frac{1}{2}, \frac{15}{16}\right)$
   c. (−1, −2.75)
   d. (50, −16)
9. **a.** In which quadrants are the second coordinates of points positive?
   b. In which quadrants are the first coordinates of points negative?
   c. In which quadrant do points with a positive x-coordinate and a negative y-coordinate lie?
   d. On what axis are the first coordinates of points zero?

10. Farming. The number of bushels of wheat produced per acre depends on the amount of water it receives. Plot the data in the table as ordered pairs and draw a straight line through the points. Use the graph to determine how many bushels per acre will be produced if

a. 6 inches of rain fall. **b.** 10 inches of rain fall.

Inches of rain	Bushels per acre
2	20
4	30
8	50

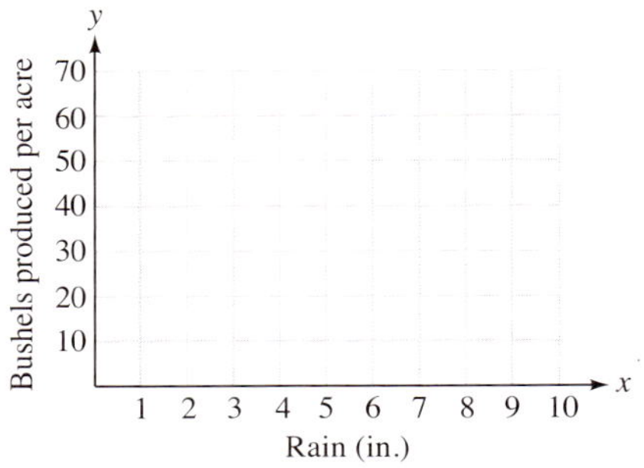

NOTATION

11. Explain the difference between (3, 5) and 3(5).

12. In a paired-data table, does the *first* or *second* column contain values associated with the vertical axis of a graph?

13. Do these ordered pairs name the same point?

$$\left(2.5, -\tfrac{7}{2}\right), \left(2\tfrac{1}{2}, -3.5\right), \left(2.5, -3\tfrac{1}{2}\right)$$

14. Do (3, 2) and (2, 3) represent the same point?

15. In the ordered pair (4, 5), is the number 4 associated with the horizontal or the vertical axis?

16. Fill in the blank: In the notation $P(4, 5)$, the capital letter P is used to name a ______.

GUIDED PRACTICE

See Examples 1 and 2.

17. Plot each point:
$(-3, 4), (4, 3.5), \left(-2, -\tfrac{5}{2}\right), (0, -4), \left(\tfrac{3}{2}, 0\right), (2.7, -4.1)$

18. Plot each point:
$(4, 4), (0.5, -3), (-3.9, -3.2), (0, -1), (0, 0), (0, 3), (-2, 0)$

19. Complete the coordinates for each point in figure (a) below.

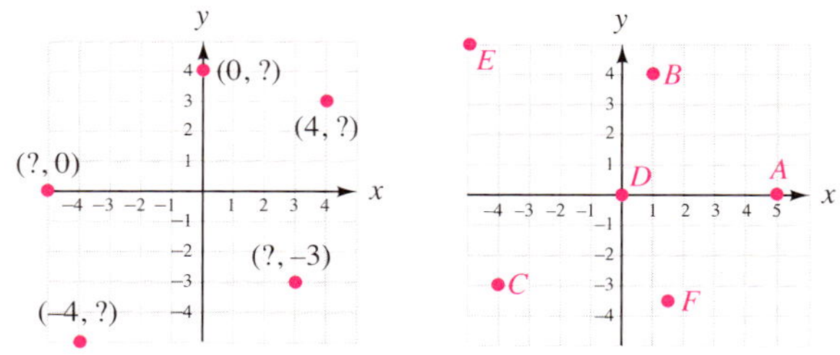

(a) (b)

20. Find the coordinates of points A, B, C, D, E, and F in figure (b) above.

***The following graph gives the heart rate of a woman before, during, and after an aerobic workout. Use it to answer Problems 21–24.* See Example 3.**

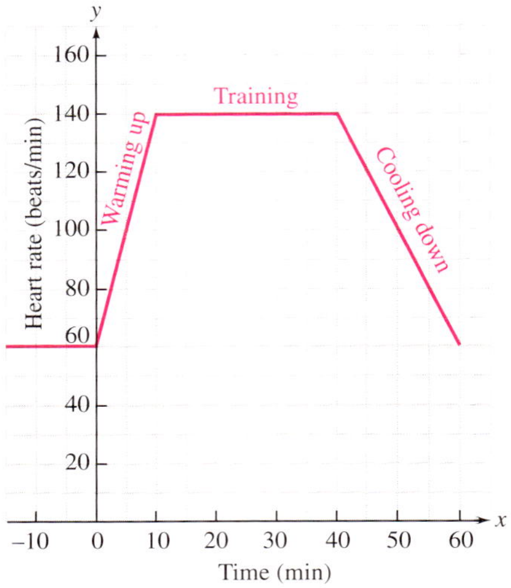

21. a. What was her heart rate before beginning the workout?

b. After beginning her workout, how long did it take the woman to reach her training-zone heart rate?

22. a. What was the woman's heart rate half an hour after beginning the workout?

b. For how long did the woman work out at her training zone?

23. a. At what time was her heart rate 100 beats per minute?

b. How long was her cool-down period?

24. a. What was the difference in the woman's heart rate before the workout and after the cool-down period?

b. What was her approximate heart rate 8 minutes after beginning?

***The following graph shows the depths of a submarine at certain times after it leaves port. Use the graph to answer Problems 25–28.* See Example 3.**

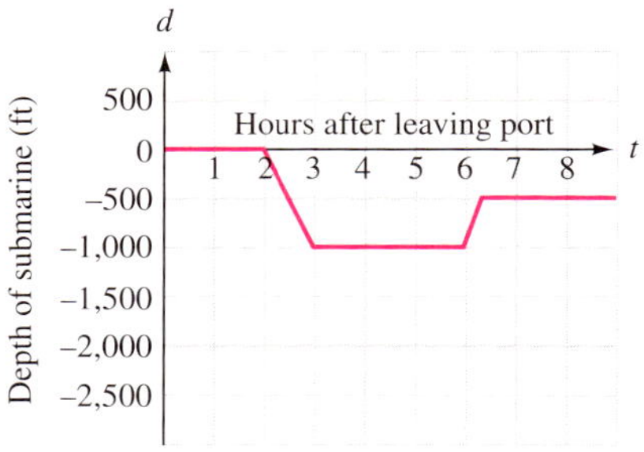

25. a. For how long does the sub travel at sea level?

b. What is the depth of the sub 5 hours after leaving port?

26. a. Once the sub begins to dive, how long does it take to reach −1,000 feet in depth?

b. For how long does the sub travel at a depth of 1,000 feet?

27. a. Explain what happens 6 hours after the sub leaves port.
    b. What is the depth of the sub 8 hours after leaving port?

28. a. How long does it take the sub to first reach −500 feet in depth?
    b. Approximate the time when the sub reaches −500 feet in depth for the second time.

APPLICATIONS

29. **Bridge Construction.** Find the coordinates of each rivet, weld, and anchor.

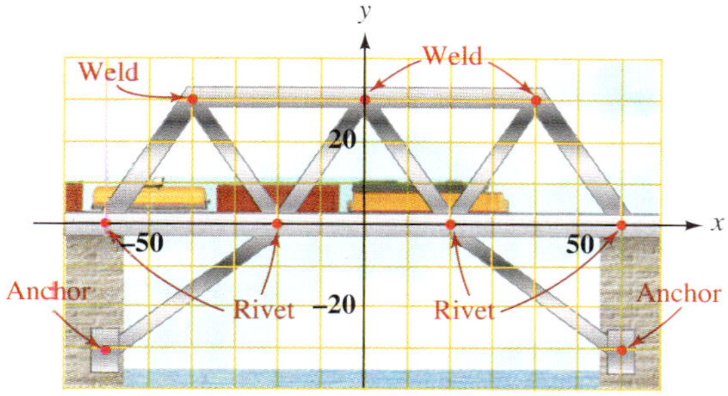

30. **Golf.** A golfer is videotaped and then has her swing displayed on a computer monitor so that it can be analyzed. Give the coordinates of the three highlighted points in red.

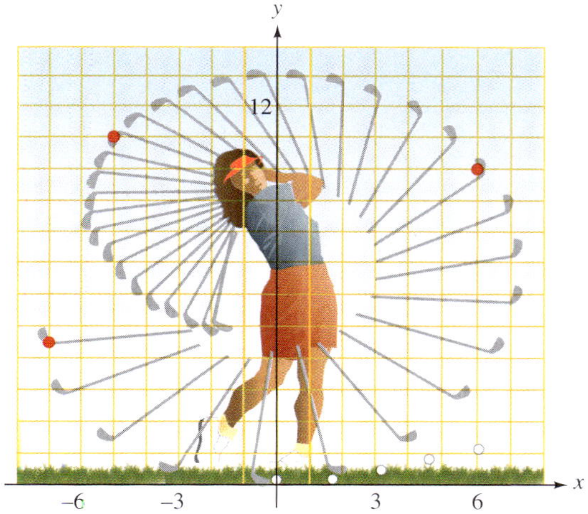

31. **Games.** In the game *Battleship,* players use coordinates to drop depth charges from their ships to hit submarines. What coordinates should be used to make three hits on the submarine seen here? Express each answer in the form (letter, number).

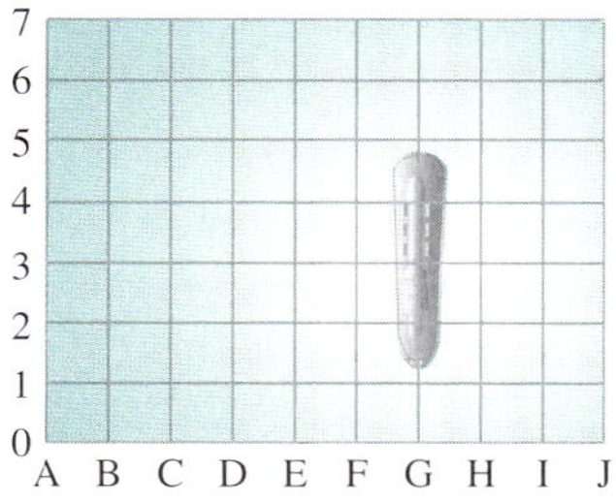

32. **Maps.** Use coordinates of the form (number, letter) to locate each of the following on the map: Tempe, Glendale, Paradise Valley, Sky Harbor Airport, and the intersection of Camelback Road and 7th Avenue.

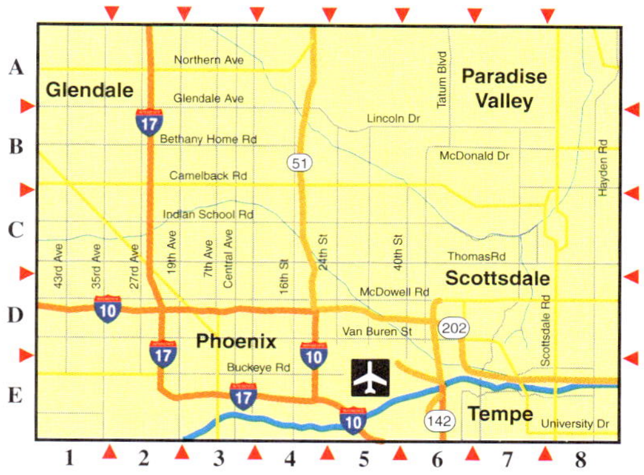

33. **from Campus to Careers**

Dental Assistant

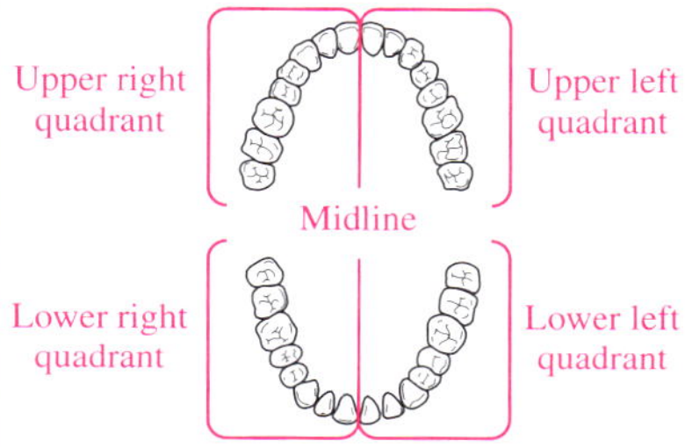

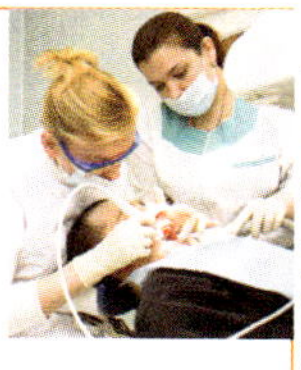

Dentists describe teeth as being located in one of four *quadrants* as shown above.

a. How many teeth are in each quadrant?

b. Why would the upper left quadrant appear on the right in the illustration?

34. **Geography.** A coordinate system that describes the location of any place on the surface of the Earth uses a series of *latitude* and *longitude* lines, as shown below. Estimate the location of the Deep Water Horizon (the oil drilling rig in the Gulf of Mexico that exploded in 2010) using an ordered pair of the form (latitude, longitude). (Source: sailwx.info)

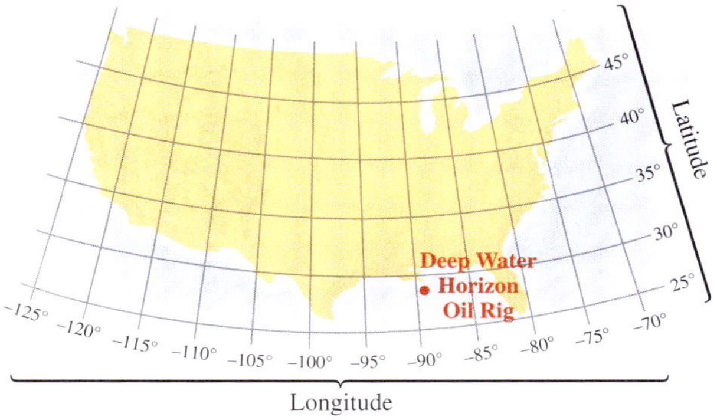

35. Water Pressure. The graphs show how the path of a stream of water changes when the hose is held at two different angles.

a. At which angle does the stream of water shoot up higher? How much higher?

b. At which angle does the stream of water shoot out farther? How much farther?

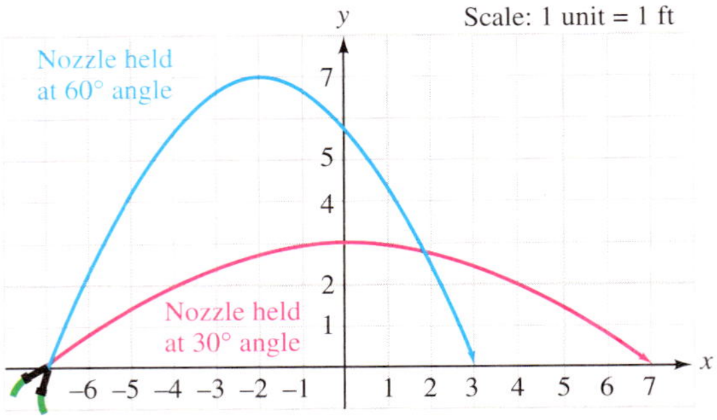

36. Area. Three vertices (corners) of a rectangle are (2, 1), (6, 1), and (6, 4). Find the coordinates of the fourth vertex. Then find the area of the rectangle.

37. Area. Three vertices (corners) of a right triangle are $(-1, -7)$, $(-5, -7)$, and $(-5, -2)$. Find the area of the triangle.

38. Landmarks. A scale model of the block letter H in the Hollywood sign can be drawn by plotting the following points and connecting them: (0, 0), (13, 0), (13, 16), (26, 16), (26, 0), (39, 0), (39, 45), (26, 45), (26, 29), (13, 29), (13, 45), and (0, 45). The scale is 1 unit on the graph is equal to 1 foot on the actual sign. If a gallon of paint covers 350 square feet, how many gallons are needed to paint the front side of the letter H? Round to the nearest gallon.

©Cary Kalscheuer/Shutterstock.com

39. Trucks. The table below shows the number of miles that an 18-wheel truck can be driven on a given number of gallons of diesel fuel. Plot the data in the table as ordered pairs. Then draw a straight line through the points.

a. How far can the truck go on 4 gallons of fuel?

b. How many gallons of fuel are needed to travel a distance of 30 miles?

c. How far can the truck go on 7 gallons of fuel?

Fuel (gal)	Distance (mi)
2	10
3	15
5	25

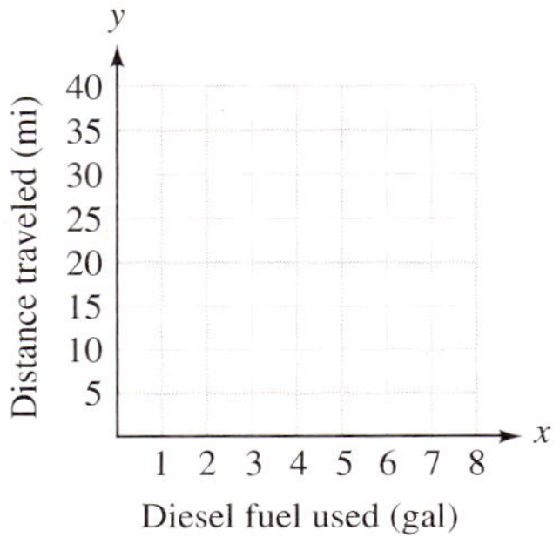

40. Boating. The table below shows the cost to rent a sailboat for a given number of hours. Plot the data in the table as ordered pairs. Then draw a straight line through the points.

a. What does it cost to rent the boat for 3 hours?

b. For how long can the boat be rented for $60?

c. What does it cost to rent the boat for 9 hours?

Rental time (hr)	Cost ($)
2	20
4	30
6	40

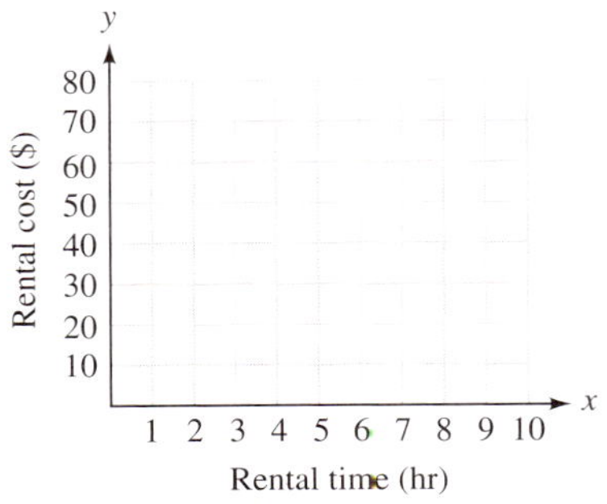

41. Depreciation. The table below shows the value (in thousands of dollars) of a color copier at various lengths of time after its purchase. Plot the data in the table as ordered pairs. Then draw a straight line passing through the points.

a. What does the point (3, 7) on the graph tell you?

b. Find the value of the copier when it is 7 years old.

c. After how many years will the copier be worth $2,500?

Age (yr)	Value ($1,000)
3	7
4	5.5
5	4

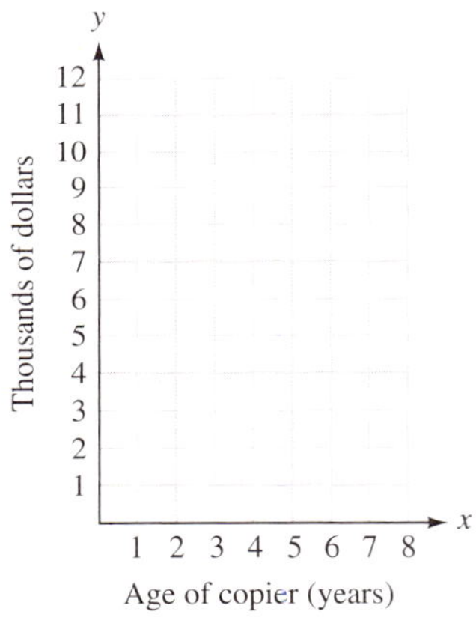

42. Swimming. The table below shows the number of people at a public swimming pool at various times during the day. (0 represents noon, 1 represents 1 PM, and so on.) Plot the data in the table as ordered pairs. Then draw a straight line passing through the points.

a. How many people will be at the pool at 6 P.M.?

b. At what time will there be 250 people at the pool?

c. At what time will the number of people at the pool be half of what it was at noon?

Time	Number of people
0	350
3	200
5	100

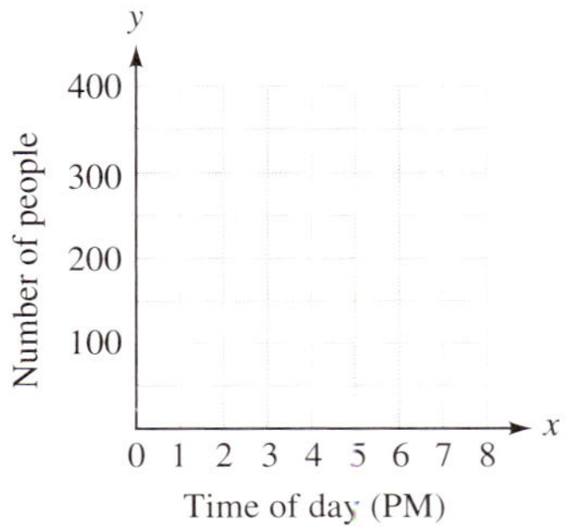

WRITING

43. Explain why the point $(-3, 3)$ is not the same as the point $(3, -3)$.
44. Explain how to plot the point $(-2, 5)$.
45. Use the Internet to perform a search of the name René Descartes. After reading about him, explain how a fly on his bedroom ceiling provided the inspiration for the concept of a rectangular coordinate system.
46. Explain this diagram.

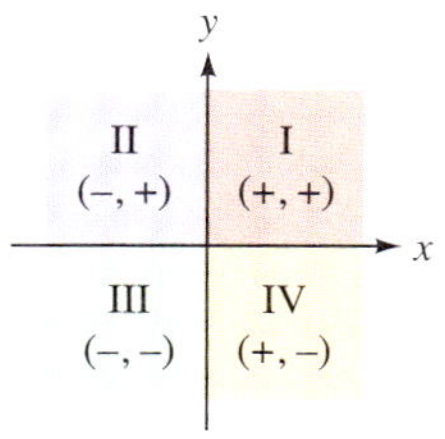

REVIEW

47. Solve $AC = \frac{2}{3}h - T$ for h.
48. Solve $5(x + 1) \le 2(x - 3)$. Write the solution set in interval notation and graph it.
49. Evaluate: $\dfrac{-4(4 + 2) - 2^3}{|-12 - 4(5)|}$
50. Simplify: $\dfrac{24}{54}$

CHALLENGE PROBLEMS

51. In what quadrant does a point lie if the *sum* of its coordinates is negative and the *product* of its coordinates is positive?
52. Draw line segment $\overline{AB}$ with endpoints $A(6, 5)$ and $B(-4, 5)$. Suppose that the x-coordinate of a point C is the average of the x-coordinates of points A and B, and the y-coordinate of point C is the average of the y-coordinates of points A and B. Find the coordinates of point C. Why is C called the midpoint of $\overline{AB}$?

SECTION 3.2

Graphing Linear Equations

OBJECTIVES

1. Determine whether an ordered pair is a solution of an equation.
2. Complete ordered-pair solutions of equations.
3. Construct a table of solutions.
4. Graph linear equations by plotting points.
5. Use graphs of linear equations to solve applied problems.

ARE YOU READY?

The following problems review some basic skills that are needed when graphing linear equations.

1. Is $2(4) + 3(-1) = 4$ a true or false statement?
2. Evaluate $-6x - 7$ for $x = -2$.
3. Solve: $4(2) + 3y = 14$
4. Solve for y: $4x + 5y = -15$
5. Multiply: $\frac{3}{2}(2)$
6. Multiply: $-\frac{5}{9}(-9)$

In this section, we will discuss equations that contain two variables. Such equations are often used to describe algebraic relationships between two quantities. To see a mathematical picture of these relationships, we will construct graphs of their equations.

1 Determine Whether an Ordered Pair Is a Solution of an Equation.

We have previously solved **equations in one variable.** For example, $x + 3 = 9$ is an equation in x. If we subtract 3 from both sides, we see that 6 is the solution. To verify this, we replace x with 6 and note that the result is a true statement: $9 = 9$.

In this chapter, we extend our equation-solving skills to find solutions of **equations in two variables.** To begin, let's consider $y = x - 1$, an equation in x and y.

Notation

Equations in two variables often involve the variables x and y. However, other letters can be used. For example, $a - 3b = 5$ and $n = 4m + 6$ are equations in two variables.

A solution of $y = x - 1$ is a pair of values, one for x and one for y, that make the equation true. To illustrate, suppose x is 5 and y is 4. Then we have:

$y = x - 1$ This is the given equation.

$4 \stackrel{?}{=} 5 - 1$ Substitute 5 for x and 4 for y.

$4 = 4$ True

Since the result is a true statement, $x = 5$ and $y = 4$ is a solution of $y = x - 1$. We write the solution as the ordered pair (5, 4), with the value of x listed first. We say that (5, 4) **satisfies** the equation.

In general, a **solution of an equation in two variables** is an ordered pair of numbers that makes the equation a true statement.

EXAMPLE 1 Is $(-1, -3)$ a solution of $y = x - 1$?

Strategy We will substitute -1 for x and -3 for y and see whether the resulting equation is true.

Why An ordered pair is a solution of $y = x - 1$ if replacing the variables with the values of the ordered pair results in a true statement.

Solution

$y = x - 1$ This is the given equation.

$-3 \stackrel{?}{=} -1 - 1$ Substitute −1 for x and −3 for y.

$-3 = -2$ False

Since $-3 = -2$ is false, $(-1, -3)$ is not a solution of $y = x - 1$.

Self Check 1 Is (9, 8) a solution of $y = x - 1$?

Now Try Problem 17

2 Complete Ordered-Pair Solutions of Equations.

If only one of the values of an ordered-pair solution is known, we can substitute it into the equation to determine the other value.

EXAMPLE 2 Complete the solution $(-5, \quad)$ of the equation $y = -2x + 3$.

Strategy We will substitute the known x-coordinate of the solution into the given equation.

Why We can use the resulting equation in one variable to find the unknown y-coordinate of the solution.

Solution In the ordered pair $(-5, \quad)$, the x-value is -5; the y-value is not known. To find y, we substitute -5 for x in the equation and evaluate the right side.

$y = -2x + 3$ This is the given equation.

$y = -2(-5) + 3$ Substitute −5 for x.

$y = 10 + 3$ Do the multiplication.

$y = 13$ This is the missing y-coordinate of the solution.

The completed ordered pair is $(-5, 13)$.

Self Check 2 Complete the solution $(-2, \quad)$ of the equation $y = 4x - 2$.

Now Try Problem 29

Solutions of equations in two variables are often listed in a **table of solutions** (or **table of values**).

EXAMPLE 3 Complete the table of solutions for $3x + 2y = 5$.

x	y	(x, y)
7		(7,)
	4	(, 4)

Strategy In each case we will substitute the known coordinate of the solution into the given equation.

Why We can solve the resulting equation in one variable to find the unknown coordinate of the solution.

Solution In the first row, we are given an x-value of 7. To find the corresponding y-value, we substitute 7 for x and solve for y.

Caution

When solving equations, we will no longer show all of the steps involving the use of a property of equality, as we did in Chapter 2. For example, on the right, we do not show the subtraction of 21 from both sides nor both sides being divided by 2. These calculations are done mentally.

$3x + 2y = 5$ This is the given equation.

$3(7) + 2y = 5$ Substitute 7 for x.

$21 + 2y = 5$ Do the multiplication.

$2y = -16$ To isolate the variable term 2y, subtract 21 from both sides.

$y = -8$ To isolate y, divide both sides by 2. This is the missing y-coordinate of the solution.

A solution of $3x + 2y = 5$ is $(7, -8)$. It is entered in the table on the left.

x	y	(x, y)
7	−8	(7, −8)

In the second row, we are given a y-value of 4. To find the corresponding x-value, we substitute 4 for y and solve for x.

$3x + 2y = 5$ This is the given equation.

$3x + 2(4) = 5$ Substitute 4 for y.

$3x + 8 = 5$ Do the multiplication.

$3x = -3$ To isolate the variable term 3x, subtract 8 from both sides.

$x = -1$ To isolate x, divide both sides by 3. This is the missing x-coordinate of the solution.

Another solution is $(-1, 4)$. It is entered in the table on the left.

x	y	(x, y)
7	−8	(7, −8)
−1	4	(−1, 4)

Self Check 3 Complete the table of solutions for $3x + 2y = 5$.

x	y	(x, y)
	−2	(, −2)
5		(5,)

Now Try Problem 37

3 Construct a Table of Solutions.

To find a solution of an equation in two variables, we can select a number, substitute it for one of the variables, and find the corresponding value of the other variable. For example, to find a solution of $y = x - 1$, we can select a value for x, say, -4, substitute -4 for x in the equation, and find y.

$y = x - 1$

$y = -4 - 1$ Substitute −4 for x.

$y = -5$ Do the subtraction.

The ordered pair $(-4, -5)$ is a solution. We list it in the table on the left.

x	y	(x, y)
−4	−5	(−4, −5)

To find another solution of $y = x - 1$, we select another value for x, say, -2, and find the corresponding y-value.

x	y	(x, y)
−4	−5	(−4, −5)
−2	**−3**	**(−2, −3)**

$y = x - 1$

$y = -2 - 1$ Substitute −2 for x.

$y = -3$ Do the subtraction.

A second solution is $(-2, -3)$, and we list it in the table of solutions.

If we let $x = 0$, we can find a third ordered pair that satisfies $y = x - 1$.

x	y	(x, y)
−4	−5	(−4, −5)
−2	−3	(−2, −3)
0	**−1**	**(0, −1)**

$y = x - 1$

$y = 0 - 1$ Substitute 0 for x.

$y = -1$ Do the subtraction.

A third solution is $(0, -1)$, which we also add to the table of solutions.

We can find a fourth solution by letting $x = 2$, and a fifth solution by letting $x = 4$.

x	y	(x, y)
−4	−5	(−4, −5)
−2	−3	(−2, −3)
0	−1	(0, −1)
2	**1**	**(2, 1)**
4	**3**	**(4, 3)**

$y = x - 1$ | $y = x - 1$

$y = 2 - 1$ Substitute 2 for x. | $y = 4 - 1$ Substitute 4 for x.

$y = 1$ Do the subtraction. | $y = 3$ Do the subtraction.

A fourth solution is $(2, 1)$ and a fifth solution is $(4, 3)$. We add them to the table.

Since we can choose any real number for x, and since any choice of x will give a corresponding value of y, it is apparent that the equation $y = x - 1$ has *infinitely many solutions*. We have found five of them: $(-4, -5)$, $(-2, -3)$, $(0, -1)$, $(2, 1)$, and $(4, 3)$.

4 Graph Linear Equations by Plotting Points.

It is impossible to list the infinitely many solutions of the equation $y = x - 1$. However, to show all of its solutions, we can draw a mathematical "picture" of them. We call this picture the *graph of the equation*.

Notation

The graph only shows a part of the line. The arrowheads indicate that it extends indefinitely in both directions.

To graph $y = x - 1$, we plot the ordered pairs shown in the table on a rectangular coordinate system. Then we draw a straight line through the points, because the graph of any solution of $y = x - 1$ will lie on this line. We also draw arrowheads on either end of the line to indicate that the solutions continue indefinitely in both directions, beyond what we can see on the coordinate grid. We call the line the **graph of the equation.** It represents all of the solutions of $y = x - 1$.

$y = x - 1$

x	y	(x, y)
−4	−5	(−4, −5)
−2	−3	(−2, −3)
0	−1	(0, −1)
2	1	(2, 1)
4	3	(4, 3)

↑ Select x ↑ Find y ↑ Plot (x, y)

Construct a table of solutions.

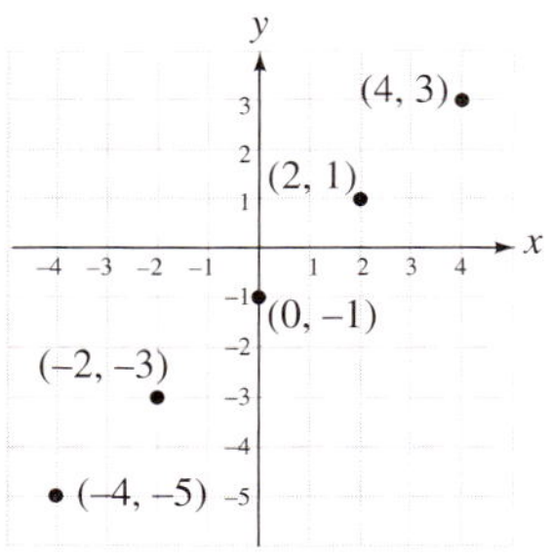

Plot the ordered pairs.

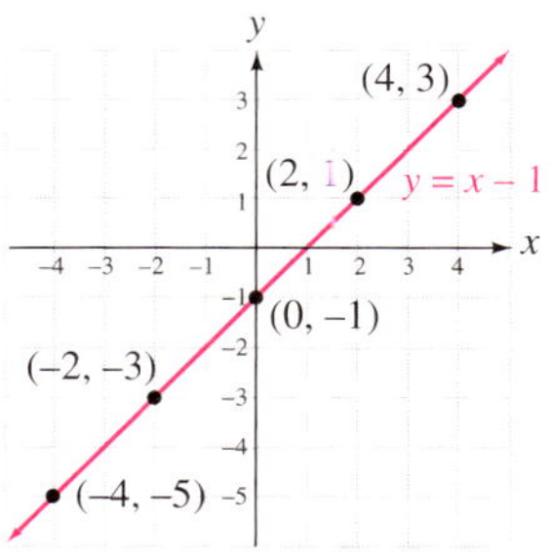

Draw a straight line through the points. This is the *graph of the equation*.

The equation $y = x - 1$ is said to be *linear* and its graph is a line. By definition, a linear equation in two variables is any equation that can be written in the following form, where the variable terms appear on one side of an equal symbol and a constant appears on the other.

Linear Equations

A linear equation in two variables is an equation that can be written in the form

$$Ax + By = C$$

where A, B, and C are real numbers and A and B are not both 0. This form is called **standard form.***

*In some textbooks, the definition of the standard form of a linear equation in two variables contains additional requirements, such as: A, B, and C are integers, $A > 0$, and the greatest common factor of A, B, and C is 1.

Every linear equation in two variables has an infinite number of ordered-pair solutions. The graph of a linear equation in two variables is a straight line. Every point on the line corresponds to a solution.

Some more examples of linear equations are

$$y = 2x + 4, \qquad 2x + 3y = 12, \qquad \text{and} \qquad 3x = 5y$$

The exponent on each variable of a linear equation in two variables is an understood 1. For example, $y = 2x + 4$ can be thought of as $y^1 = 2x^1 + 4$ and $2x + 3y = 12$ can be thought of as $2x^1 + 3y^1 = 12$.

Some examples of equations in two variables that are *not* linear are shown below. You will see later in this course (and in more advanced courses) that the graphs of these equations are not straight lines.

$$y = x^2 + 3, \qquad y = \sqrt{x}, \qquad y = 4^x, \qquad x^2 + y^2 = 25, \qquad \text{and} \qquad y = \frac{1}{x}$$

Linear equations can be graphed in several ways. Generally, the form in which an equation is written determines the method that we use to graph it. To graph linear equations solved for y, such as $y = 2x + 4$, we can use the following **point-plotting method.**

Graphing Linear Equations Solved for y by Plotting Points

1. Find three ordered pairs that are solutions of the equation by selecting three values for x and calculating the corresponding values of y.
2. Plot the solutions on a rectangular coordinate system.
3. Draw a straight line passing through the points. If the points do not lie on a line, check your calculations.

EXAMPLE 4 Graph: $y = 2x + 4$

Strategy We will find three solutions of the equation, plot them on a rectangular coordinate system, and then draw a straight line passing through the points.

Why To *graph* a linear equation in two variables means to make a drawing that represents all of its solutions.

Solution To find three solutions of this linear equation, we select three values for x that will make the calculations easy. Then we find each corresponding value of y.

Success Tip

When selecting x-values for a table of solutions, a rule of thumb is to choose a negative number, a positive number, and 0. When $x = 0$, the calculations to find y are usually quite simple.

***If* $x = -2$:**

$y = 2x + 4$
$y = 2(-2) + 4$
$y = -4 + 4$
$y = 0$

$(-2, 0)$ is a solution.

***If* $x = 0$:**

$y = 2x + 4$
$y = 2(0) + 4$
$y = 0 + 4$
$y = 4$

$(0, 4)$ is a solution.

***If* $x = 2$:**

$y = 2x + 4$
$y = 2(2) + 4$
$y = 4 + 4$
$y = 8$

$(2, 8)$ is a solution.

Success Tip

Since two points determine a line, only two points are needed to graph a linear equation. However, we should plot a third point as a check. If the three points do not lie on a straight line, then at least one of them is in error.

We enter the results in a table of solutions and plot the points. Then we draw a straight line through the points and label it $y = 2x + 4$.

$y = 2x + 4$

x	y	(x, y)
−2	0	(−2, 0)
0	4	(0, 4)
2	8	(2, 8)

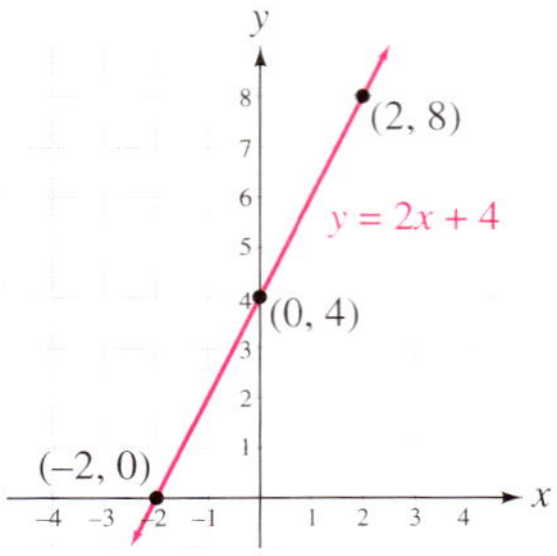

As a check, we can pick two points that the line appears to pass through, such as (1, 6) and (−1, 2). When we substitute their coordinates into the given equation, the two true statements that result indicate that (1, 6) and (−1, 2) are solutions and that the graph of the line is correctly drawn.

***Check* (1, 6):** $y = 2x + 4$

$6 \stackrel{?}{=} 2(1) + 4$

$6 \stackrel{?}{=} 2 + 4$

$6 = 6$ True

***Check* (−1, 2):** $y = 2x + 4$

$2 \stackrel{?}{=} 2(-1) + 4$

$2 \stackrel{?}{=} -2 + 4$

$2 = 2$ True

Self Check 4 Graph: $y = 2x - 2$

Now Try Problem 41

EXAMPLE 5 Graph: $y = -3x$

Strategy We will find three solutions of the equation, plot them on a rectangular coordinate system, and then draw a straight line passing through the points.

Why To *graph* a linear equation in two variables means to make a drawing that represents all of its solutions.

Solution To find three solutions, we begin by selecting three x-values: −1, 0, and 1. Then we find the corresponding values of y. If $x = -1$, we have

$y = -3x$ This is the equation to graph.

$y = -3(-1)$ Substitute −1 for x.

$y = 3$ Do the multiplication.

(−1, 3) is a solution.

Notation

Tables of solutions can be written vertically, as shown in Examples 3–6, or they can be written horizontally, as shown below.

x	−1	0	1
y	3	0	−3

In a similar manner, we find the y-values for x-values of 0 and 1, and record the results in a table of solutions. After plotting the ordered pairs, we draw a straight line through the points and label it $y = -3x$.

$y = -3x$

x	y	(x, y)
−1	3	(−1, 3)
0	0	(0, 0)
1	−3	(1, −3)

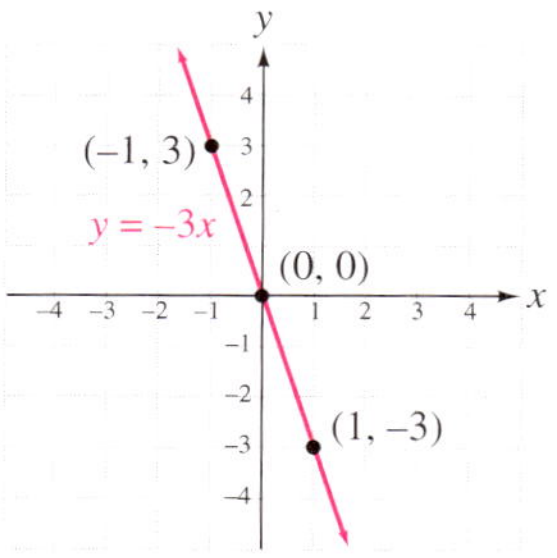

Self Check 5 Graph: $y = -4x$

Now Try Problem 45

To graph linear equations in x and y using the method discussed in this section, the variable y must be isolated on one side of the equation.

EXAMPLE 6 Graph $2x + 3y = -12$ by first solving for y.

Strategy We will use properties of equality to solve the given equation for y. Then we will use the point-plotting method of this section to graph the resulting equivalent equation.

Why The calculations to find several solutions of a linear equation in two variables are usually easier when the equation is solved for y.

Solution To solve for y, we proceed as follows.

$$2x + 3y = -12$$ This is the given equation.

$$2x + 3y - 2x = -2x - 12$$ To isolate the variable term 3y on the left side, subtract 2x from both sides. When solving for y, it is common practice to write the subtraction (or addition) of a variable term before the constant term.

$$3y = -2x - 12$$ On the left side, combine like terms: 2x − 2x = 0.

$$\frac{3y}{3} = \frac{-2x}{3} - \frac{12}{3}$$ To isolate the variable y, undo the multiplication by 3 by dividing both sides, term-by-term, by 3.

$$y = -\frac{2}{3}x - 4$$ Write $\frac{-2x}{3}$ as $-\frac{2}{3}x$. Simplify: $\frac{12}{3} = 4$.

Success Tip

The division by 3 on the right side of the equation is done term-by-term instead of with a single fraction bar. We write:

$$\frac{-2x}{3} - \frac{12}{3} \quad \text{not} \quad \frac{-2x - 12}{3}$$

Since $y = -\frac{2}{3}x - 4$ is equivalent to $2x + 3y = -12$, we can use it to draw the graph of $2x + 3y = -12$.

To find solutions of $y = -\frac{2}{3}x - 4$, each value of x must be multiplied by $-\frac{2}{3}$. This calculation is made easier if we select x-values that are *multiples of the denominator 3*, such as −3, 0, and 6. For example, if $x = -3$, we have

$$y = -\frac{2}{3}x - 4$$ This is the equation to graph.

$$y = -\frac{2}{3}(-3) - 4$$ Substitute −3 for x.

$$y = 2 - 4$$ Multiply: $-\frac{2}{3}(-3) = 2$.

$$y = -2$$ Do the subtraction.

Thus, $(-3, -2)$ is a solution.

Success Tip

When we chose x-values that are multiples of the denominator 3, the corresponding y-values are integers, and not difficult-to-plot fractions.

Two more solutions, one for $x = 0$ and one for $x = 6$, can be found in a similar way, and entered in a table. We plot the ordered pairs, draw a straight line through the points, and label the line as $y = -\frac{2}{3}x - 4$ or as $2x + 3y = -12$.

$2x + 3y = -12$
or
$y = -\frac{2}{3}x - 4$

x	y	(x, y)
−3	−2	(−3, −2)
0	−4	(0, −4)
6	−8	(6, −8)

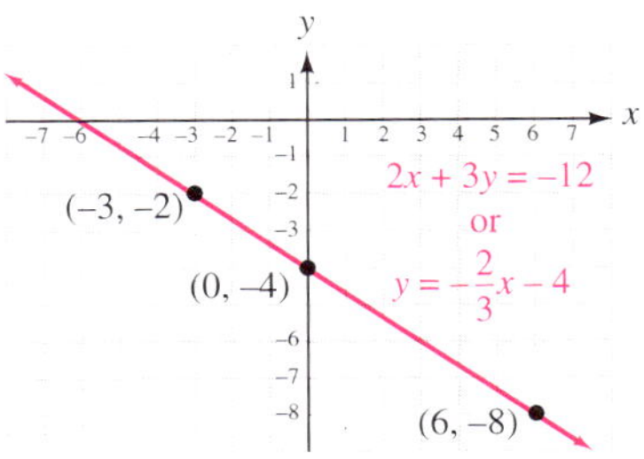

Self Check 6 Graph $5x - 2y = -2$ by first solving for y.

Now Try Problem 49

5 Use Graphs of Linear Equations to Solve Applied Problems.

When linear equations are used to model real-life situations, they are often written in variables other than x and y. In such cases, we must make the appropriate changes when labeling the table of solutions and the graph of the equation.

EXAMPLE 7 **Cleaning Windows.** The linear equation $A = -0.03n + 32$ estimates the amount A of glass-cleaning solution (in ounces) that is left in the bottle after the sprayer trigger has been pulled a total of n times. Graph the equation and use the graph to estimate the amount of solution that is left after 500 sprays.

Strategy We will find three solutions of the equation, plot them on a rectangular coordinate system, and then draw a straight line passing through the points.

Why We can use the graph to estimate the amount of solution left after any number of sprays.

Solution Since A depends on n in the equation $A = -0.03n + 32$, solutions will have the form (n, A). To find three solutions, we begin by selecting three values of n. Because the number of sprays cannot be negative, and the calculations to find A involve decimal multiplication, we select 0, 100, and 1,000. For example, if $n = 100$, we have

$A = -0.03\mathbf{n} + 32$ This is the equation to graph.

$A = -0.03(\mathbf{100}) + 32$ Substitute 100 for n.

$A = -3 + 32$ Multiply by moving the decimal point in -0.03 two places to the right: $-0.03(100) = -3$.

$A = 29$ Do the addition.

Thus, (100, 29) is a solution. It indicates that after 100 sprays, 29 ounces of cleaner will be left in the bottle.

In the same way, solutions are found for $n = 0$ and $n = 1{,}000$ and listed in the table. Then the ordered pairs are plotted and a straight line is drawn through the points.

To graphically estimate the amount of solution that is left after 500 sprays, we draw the dashed blue lines, as shown. Reading on the vertical A-axis, we see that after 500 sprays, about 17 ounces of glass cleaning solution would be left.

Success Tip

It is often helpful, especially with applications, to scale the axes differently. Since we selected large n-values such as 100 and 1,000, the horizontal n-axis was scaled in units of 100. Since the corresponding A-values range from 2 to 32, the vertical A-axis was scaled in units of 4.

$A = -0.03n + 32$

n	A	(n, A)
0	32	(0, 32)
100	29	(100, 29)
1,000	2	(1,000, 2)

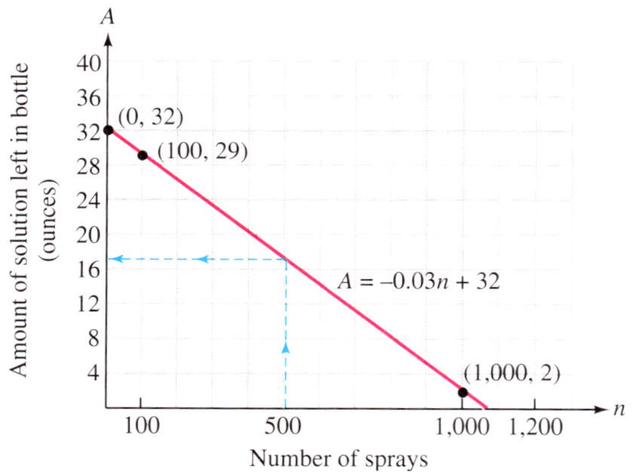

Self Check 7 **Parties.** A laser tag business offers a party package that includes invitations, a party room, and 2 rounds of laser tag. The cost is $15 plus $10 per child. Write a linear equation that will give the cost for a party of any size, and then graph the equation.

Now Try Problems 87 and 89

SECTION 3.2 STUDY SET

VOCABULARY

Fill in the blanks.

1. $y = 9x + 5$ is an equation in _____ variables, x and y.
2. A ________ of an equation in two variables is an ordered pair of numbers that makes the equation a true statement.
3. Solutions of equations in two variables are often listed in a _____ of solutions.
4. The line that represents all of the solutions of a linear equation is called the _______ of the equation.
5. $y = 3x + 8$ is a _______ equation and its graph is a line.
6. The _________ form of a linear equation in two variables is $Ax + By = C$.

CONCEPTS

7. Consider: $y = -3x + 6$
   a. How many variables does the equation contain?
   b. Does $(4, -6)$ satisfy the equation?
   c. Is $(-2, 0)$ a solution of the equation?
   d. How many solutions does this equation have?
8. To graph a linear equation, three solutions were found, they were plotted (in black), and a straight line was drawn through them, as shown below.
   a. Looking at the graph, complete the table of solutions.
   b. From the graph, determine three other solutions of the equation.

x	y	(x, y)
-4		(,)
-1		(,)
1		(,)

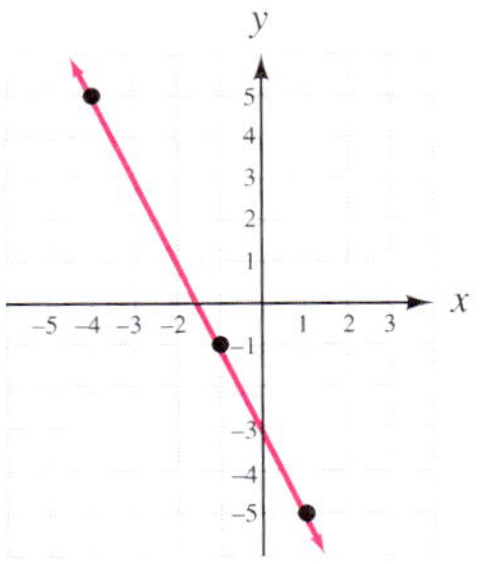

9. The graph of $y = -2x - 3$ is shown in Problem 8. Fill in the blanks: Every point on the graph represents an ordered-pair _________ of $y = -2x - 3$ and every ordered-pair solution is a ______ on the graph.
10. The graph of a linear equation is shown.
   a. If the coordinates of point M are substituted into the equation, will the result be true or false?
   b. If the coordinates of point N are substituted into the equation, will the result be true or false?

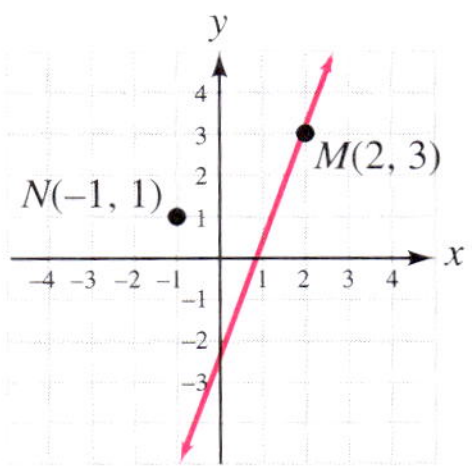

11. Suppose you are making a table of solutions for each given equation. What three x-values would you select to make the calculations for finding the corresponding y-values the easiest?
   a. $y = \frac{4}{5}x + 2$
   b. $y = 0.6x + 500$
12. A table of solutions for a linear equation is shown below. When constructing the graph of the equation, how would you scale the x-axis and the y-axis?

x	y	(x, y)
-20	600	$(-20, 600)$
5	100	$(5, 100)$
35	-500	$(35, -500)$

NOTATION

Complete each solution.

13. Verify that $(-2, 6)$ is a solution of $y = -x + 4$.

$$y = -x + 4$$
$$\stackrel{?}{=} -(\quad) + 4$$
$$6 \stackrel{?}{=} \quad + 4$$
$$6 =$$

14. Solve $5x + 3y = 15$ for y.

$$5x + 3y - 5x = \quad + 15$$
$$= -5x + 15$$
$$\frac{3y}{\quad} = \frac{-5x}{\quad} + \frac{15}{\quad}$$
$$y = \quad x + \quad$$

15. a. In the linear equation $y = \frac{1}{2}x + 7$ what are the understood exponents on the variables?
    b. Explain why $y = x^2 + 2$ and $y = x^3 - 4$ are not linear equations.

16. Complete the labeling of the table of solutions and the axes of the graph of $c = -a + 4$.

		(,)
−1	5	(−1, 5)
0	4	(0, 4)
2	2	(2, 2)

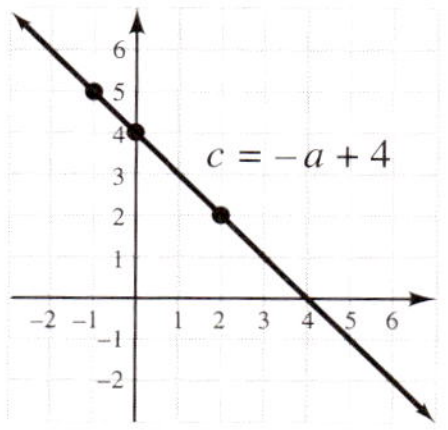

GUIDED PRACTICE

Determine whether each equation has the given ordered pair as a solution. See Example 1.

17. $y = 5x - 4$; $(1, 1)$
18. $y = -2x + 3$; $(2, -1)$
19. $7x - 2y = 3$; $(2, 6)$
20. $10x - y = 10$; $(0, 0)$
21. $x + 12y = -12$; $(0, -1)$
22. $-2x + 3y = 0$; $(-3, -2)$
23. $3x - 6y = 12$; $(-3.6, -3.8)$
24. $8x + 4y = 10$; $(-0.5, 3.5)$
25. $y - 6x = 12$; $\left(\frac{5}{6}, 7\right)$
26. $y + 8x = 4$; $\left(\frac{3}{4}, 2\right)$
27. $y = -\frac{3}{4}x + 8$; $(-8, 12)$
28. $y = \frac{1}{6}x - 2$; $(-12, 4)$

For each equation, complete the solution. See Example 2.

29. $y = -5x - 4$; $(-3, \quad)$
30. $y = 8x + 30$; $(-6, \quad)$
31. $4x - 5y = -4$; $(\quad, 4)$
32. $7x + y = -12$; $(\quad, 2)$
33. $y = \frac{x}{4} + 9$; $(16, \quad)$
34. $y = \frac{x}{6} - 8$; $(48, \quad)$
35. $7x = 4y$; $\left(\quad, -2\right)$
36. $11x = 16y$; $\left(\quad, -3\right)$

Complete each table of solutions. See Example 3.

37. $y = 2x - 4$

x	y	(x, y)
8		
	8	

38. $y = 3x + 1$

x	y	(x, y)
−3		
	−2	

39. $3x - y = -2$

x	y	(x, y)
−5		
	−1	

40. $5x - 2y = -15$

x	y	(x, y)
5		
	0	

Construct a table of solutions and then graph each equation. See Example 4.

41. $y = 2x - 3$
42. $y = 3x + 1$
43. $y = 5x - 4$
44. $y = 6x - 3$

Construct a table of solutions and then graph each equation. See Example 5.

45. $y = -6x$
46. $y = -2x$
47. $y = -7x$
48. $y = -8x$

Solve each equation for y and then graph it. See Example 6.

49. $2x + 3y = -3$
50. $2x + 3y = 9$
51. $5y - x = 20$
52. $4y - x = 8$

TRY IT YOURSELF

Graph each equation. Solve for y first, when necessary.

53. $y = x$
54. $y = 4x$
55. $y = -x - 1$
56. $y = -x + 2$
57. $3y = 12x + 15$
58. $5y = 20x - 30$
59. $y = \frac{3}{8}x - 6$
60. $y = -\frac{3}{2}x + 2$
61. $y = 1.5x - 4$
62. $y = 0.5x + 3$
63. $8x + 4y = 16$
64. $14x + 7y = 28$
65. $y = -\frac{1}{2}x$
66. $y = \frac{3}{4}x$
67. $y = \frac{5}{6}x - 5$
68. $y = \frac{2}{3}x - 2$
69. $-6y = 30x + 12$
70. $-3y = 9x - 15$
71. $y = \frac{x}{3}$
72. $y = -\frac{x}{3} - 1$
73. $y = -2x + 1$
74. $y = -3x + 2$
75. $7x - y = 1$
76. $2x - y = -3$
77. $7y = -2x$
78. $6y = -4x$
79. $y = -2.5x + 5$
80. $y = -3.5x + 4$

APPLICATIONS

81. **Billiards.** The path traveled by the black 8-ball is described by the equations $y = 2x - 4$ and $y = -2x + 12$. Construct a table of solutions for $y = 2x - 4$ using the x-values 1, 2, and 4. Do the same for $y = -2x + 12$, using the x-values 4, 6, and 8. Then graph the path of the 8-ball.

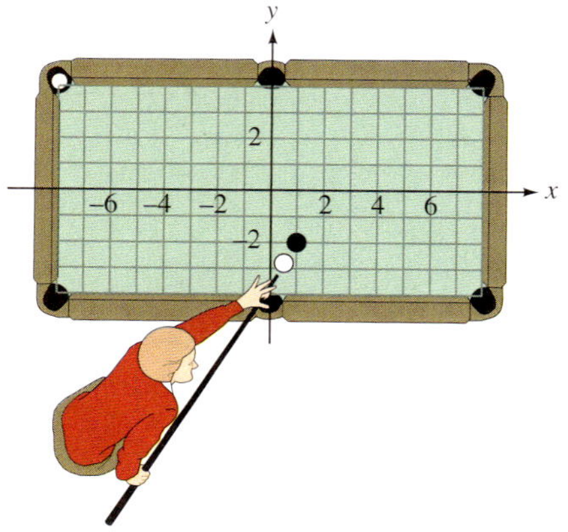

82. **Ping-Pong.** The path traveled by a Ping-Pong ball is described by the equations $y = \frac{1}{2}x + \frac{3}{2}$ and $y = -\frac{1}{2}x - \frac{3}{2}$. Construct a table of solutions for $y = \frac{1}{2}x + \frac{3}{2}$ using the x-values 7, 3, and -3. Do the same for $y = -\frac{1}{2}x - \frac{3}{2}$, using the x-values -3, -5, and -7. Then graph the path of the ball.

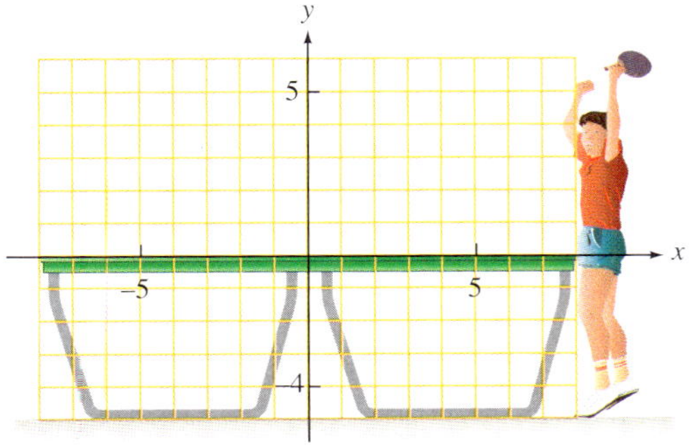

83. **Defrosting Poultry.** The number of hours h needed to defrost a turkey weighing p pounds in the refrigerator can be estimated by $h = 5p$. Graph the equation and use the graph to estimate the time needed to defrost a 25-pound turkey. (Source: helpwithcooking.com.)

84. **Owning a Car.** In 2010, the average cost c (in dollars) to own and operate a car was estimated by $c = 0.57m$, where m represents the number of miles driven. Graph the equation and use the graph to estimate the cost in 2010 of operating a car that is driven 25,000 miles. (Source: AAA Auto Club)

85. **Housekeeping.** The linear equation $A = -0.02n + 16$ estimates the amount A of furniture polish (in ounces) that is left in the bottle after the sprayer trigger has been pulled a total of n times. Graph the equation and use the graph to estimate the amount of polish that is left after 650 sprays.

86. **Sharpening Pencils.** The linear equation $L = -0.04t + 8$ estimates the length L (in inches) of a pencil after it has been inserted into a sharpener and the handle turned a total of t times. Graph the equation and use the graph to estimate the length of the pencil after 75 turns of the handle.

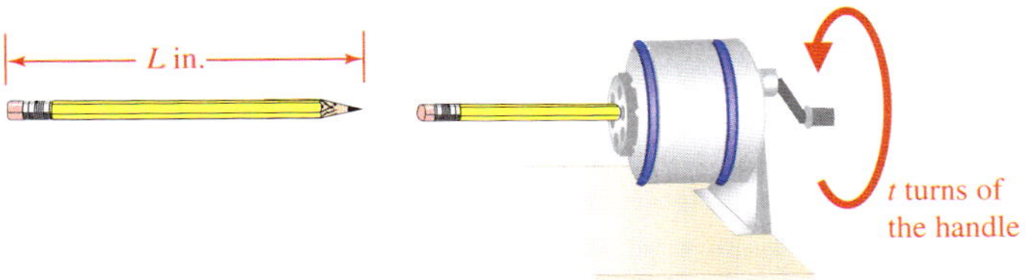

87. **NFL Tickets.** The average ticket price p to a National Football League game during the years 1990–2009 is approximated by $p = 2.7t + 20$, where t is the number of years after 1990. Graph this equation and use the graph to predict the average ticket price in 2020. (Source: Team Marketing Report, NFL.)

88. **U.S. Automobile Accidents.** The number n of lives saved by seat belts during the years 2000–2009 is estimated by $n = 170t + 13{,}800$, where t is the number of years after 2000. Graph this equation and use the graph to predict the number of lives that will be saved by seat belts in 2015. (Source: NHTSA National Center for Statistics.)

89. **Raffles.** A private school is going to sell raffle tickets as a fund raiser. Suppose the number n of raffle tickets that will be sold is predicted by the equation $n = -20p + 300$, where p is the price of a raffle ticket in dollars. Graph the equation and use the graph to predict the number of raffle tickets that will be sold at a price of \$6.

90. **Endangered Species.** The number n of endangered plant and animal species in the U.S. during the years 2000–2010 is estimated by $n = 9t + 960$, where t is the number of years after 2000. Graph this equation and use the graph to predict the number of endangered species in the U.S. in 2022. (Source: *The World Almanac and Book of Facts, 2010*).

91. **U.S. Space Program.** Since 1980, the Gallup Poll organization has surveyed Americans to see whether they think the space program has brought enough benefits to the country to justify its cost. The percent p responding "yes" is estimated by $p = \frac{3}{5}t + 40$, where t is the number of years after 1980. Graph the equation. If the polling trend continues, when will the percent that respond "yes" reach 70%? (Source: galluppoll.com)

92. **Gas Mileage.** The mileage for a Honda Insight traveling between 55 mph and 75 mph is estimated by the equation $m = -\frac{3}{4}s + 95$, where s is the speed of the car (in mph) and m is the mileage (in miles per gallon). Graph the equation for s between 55 and 75. Estimate the speed at which the mileage of the car drops below 40 miles per gallon. (Source: *Consumer Reports* 9/10/2009)

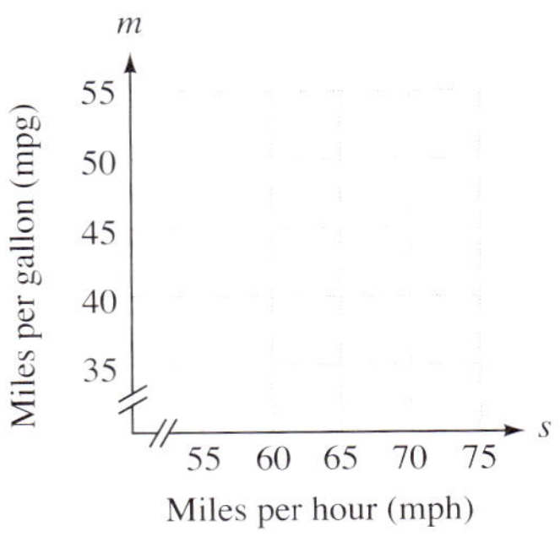

The symbol ⧣ is used to show the break in the scale on an axis. Such a break enables us to omit large portions of empty space on a graph.

WRITING

93. When we say that $(-2, -6)$ is a solution of $y = x - 4$, what do we mean?

94. What is a table of solutions?

95. What does it mean when we say that a linear equation in two variables has infinitely many solutions?

96. A linear equation and a graph are two ways of describing a relationship between two quantities. Which do you think is more informative and why?

97. From geometry, we know that two points determine a line. Why is it a good practice when graphing linear equations to find and plot three solutions instead of just two?

98. A student found three solutions of a linear equation in two variables and plotted them as shown. What conclusion can be made about the location of the points?

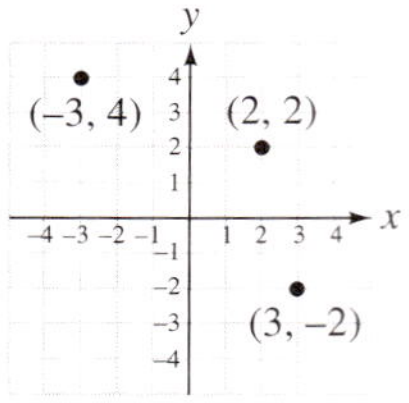

99. To graph $y = 3x - 1$, one student made the table of solutions on the left. Another student made the table on the right. The tables are different. Could they both be correct? Explain.

x	y	(x, y)
0	−1	(0, −1)
2	5	(2, 5)
3	8	(3, 8)

x	y	(x, y)
−2	−7	(−2, −7)
−1	−4	(−1, −4)
1	2	(1, 2)

100. Both graphs below are of the same linear equation $y = 10x$. Why do the graphs have a different appearance?

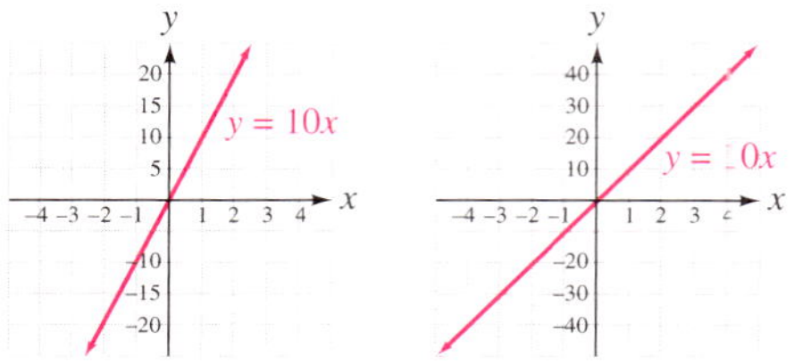

REVIEW

101. Simplify: $-(-5 - 4c)$

102. Write the set of integers.

103. Find the volume, to the nearest tenth of a cubic foot, of a sphere with radius 6 feet.

104. Solve: $-2(a + 3) = 3(a - 5)$

CHALLENGE PROBLEMS

Graph each of the following nonlinear *equations in two variables by constructing a table of solutions consisting of seven ordered pairs. These equations are called nonlinear, because their graphs are not straight lines.*

105. $y = x^2 + 1$

106. $y = x^3 - 2$

107. $y = |x| - 2$

108. $y = (x + 2)^2$

SECTION 3.3 Intercepts

OBJECTIVES

1. Identify intercepts of a graph.
2. Graph linear equations by finding intercepts.
3. Identify and graph horizontal and vertical lines.
4. Obtain information from intercepts.

ARE YOU READY?

The following problems review some basic skills that are needed when finding the intercepts of lines.

1. Graph the points (2, 0), (−4, 0), (0, 1) and (0, −3).
2. What is the x-coordinate of any point that lies on the y-axis?
3. What point lies on both the x-axis and the y-axis?
4. Solve: $3(0) + 2y = 10$

In this section, we will graph linear equations by determining the points where their graphs intersect the x-axis and the y-axis. These points are called the *intercepts* of the graph.

The Language of Algebra

Note the difference in spelling. The point where a line **intersects** the x- or y-axis is called an **intercept.**

1 Identify Intercepts of a Graph.

The graph of $y = 2x - 4$ is shown on the next page. We see that the graph intersects (crosses) the y-axis at the point $(0, -4)$; this point is called the **y-intercept** of the graph. The graph intersects (crosses) the x-axis at the point $(2, 0)$; this point is called the **x-intercept** of the graph.

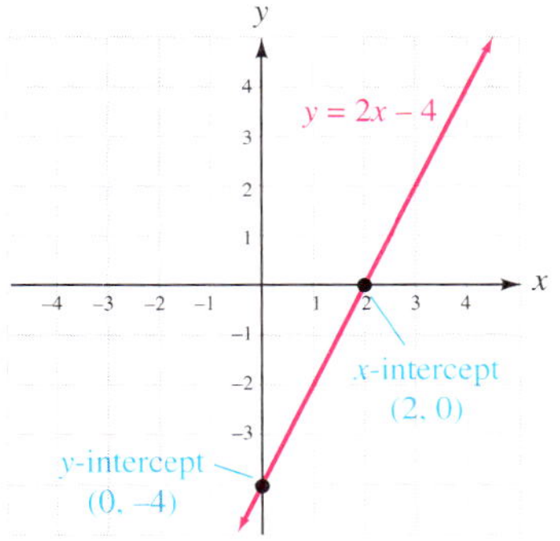

Recall that the x-coordinate of any point on the y-axis is 0.

Recall that the y-coordinate of any point on the x-axis is 0.

EXAMPLE 1 For the graphs in figures (a) and (b), identify the x- and y-intercepts.

a.

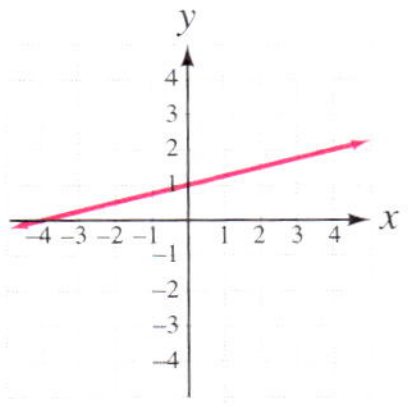

b.

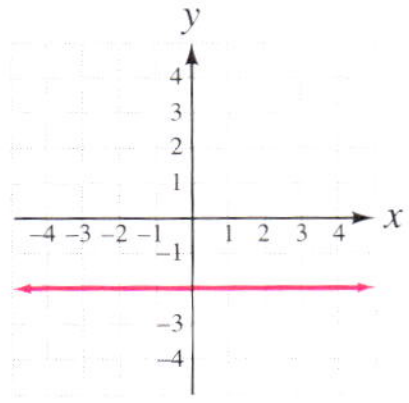

c.

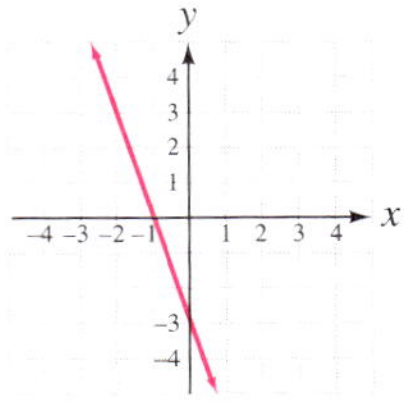

The Language of Algebra

Intercepts are ordered pairs. However, other language is sometimes used to describe them. When referring to an x-intercept of $(-4, 0)$, for example, some textbooks use a single number and say the x-intercept is at -4 on the x-axis.

Strategy We will determine where each graph (shown in red) crosses the x-axis and the y-axis.

Why The point at which a graph crosses the x-axis is the x-intercept and the point at which a graph crosses the y-axis is the y-intercept.

Solution

a. In figure (a), the graph crosses the x-axis at $(-4, 0)$. This is the x-intercept. The graph crosses the y-axis at $(0, 1)$. This is the y-intercept.

b. In figure (b), the horizontal line does not cross the x-axis; there is no x-intercept. The graph crosses the y-axis at $(0, -2)$. This is the y-intercept.

Self Check 1 Identify the x- and y-intercepts of the graph in figure (c).

Now Try Problem 11

From the previous examples, we see that a y-intercept has an x-coordinate of 0, and an x-intercept has a y-coordinate of 0. These observations suggest the following procedures for finding the intercepts of a graph from its equation.

Finding Intercepts

To find the y-intercept, substitute 0 for x in the given equation and solve for y.

To find the x-intercept, substitute 0 for y in the given equation and solve for x.

2 Graph Linear Equations by Finding Intercepts.

Plotting the x- and y-intercepts of a graph and drawing a line through them is called the **intercept method of graphing a line.** This method is useful when graphing linear equations written in the standard form $Ax + By = C$.

EXAMPLE 2 Graph $x - 3y = 6$ by finding the y- and x-intercepts.

Strategy We will let $x = 0$ to find the y-intercept of the graph. We will then let $y = 0$ to find the x-intercept.

Why Since two points determine a line, the y-intercept and x-intercept are enough information to graph this linear equation.

Solution

***y*-intercept: Let $x = 0$**

$x - 3y = 6$

$0 - 3y = 6$ Substitute 0 for x.

$-3y = 6$ Subtract.

$y = -2$ Divide both sides by −3.

The y-intercept is $(0, -2)$.

***x*-intercept: Let $y = 0$**

$x - 3y = 6$

$x - 3(0) = 6$ Substitute 0 for y.

$x - 0 = 6$ Multiply.

$x = 6$ Subtract.

The x-intercept is $(6, 0)$.

Since each intercept of the graph is a solution of the equation, we enter the intercepts in the table of solutions below.

As a check, we find one more point on the line. We select a convenient value for x, say, 3, and find the corresponding value of y.

Success Tip

The check point should lie on the same line as the x- and y-intercepts. If it does not, check your work to find the incorrect coordinate or coordinates.

$x - 3y = 6$ This is the equation to graph.

$3 - 3y = 6$ Substitute 3 for x.

$-3y = 3$ To isolate the variable term −3y, subtract 3 from both sides.

$y = -1$ To isolate y, divide both sides by −3.

Therefore, $(3, -1)$ is a solution. It is also entered in the table.

We plot the intercepts and the check point, draw a straight line through them, and label the line as $x - 3y = 6$.

The Language of Algebra

Points that lie on the same line are said to be **collinear.**

$x - 3y = 6$

x	y	(x, y)	
0	−2	(0, −2)	← y-intercept
6	0	(6, 0)	← x-intercept
3	−1	(3, −1)	← Check point

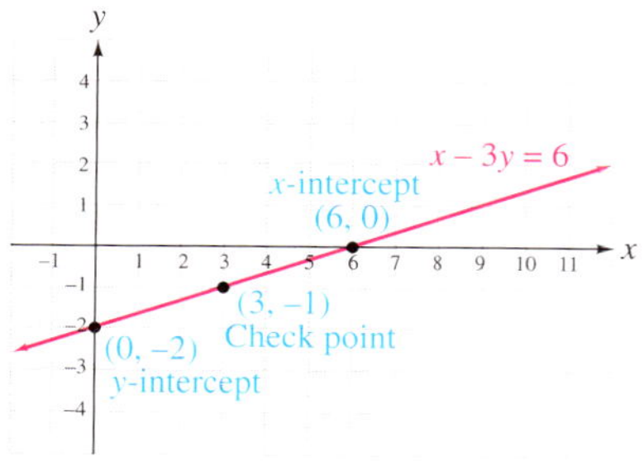

Self Check 2 Graph $x - 2y = 2$ by finding the intercepts.

Now Try Problem 27

The calculations for finding intercepts can be simplified if we realize what occurs when we substitute 0 for y or 0 for x in an equation written in the form $Ax + By = C$.

EXAMPLE 3 Graph $40x + 3y = -120$ by finding the y- and x-intercepts.

Strategy We will let $x = 0$ to find the y-intercept of the graph. We will then let $y = 0$ to find the x-intercept.

Why Since two points determine a line, the y-intercept and x-intercept are enough information to graph this linear equation.

The Language of Algebra

This method to find the intercepts of the graph of a linear equation is commonly referred to as the **cover-over method.**

Caution

When using the cover-over method to find the y-intercept, be careful not to cover the sign in front of the y-term.

Success Tip

To fit y-values of 40 and -40 on the graph, the y-axis was scaled in units of 10.

Solution When we substitute 0 for x, it follows that the term $40x$ will be equal to 0. Therefore, to find the y-intercept, we can cover over the $40x$ and solve the remaining equation for y.

$$40x + 3y = -120 \qquad \text{If } x = 0, \text{ then } 40x = 40(0) = 0. \text{ Cover over the } 40x \text{ term.}$$
$$y = -40 \qquad \text{To solve } 3y = -120, \text{ divide both sides by 3.}$$

The y-intercept is $(0, -40)$. This is entered in the table below.

When we substitute 0 for y, it follows that the term $3y$ will be equal to 0. Therefore, to find the x-intercept, we can cover over the $3y$ and solve the remaining equation for x.

$$40x + 3y = -120 \qquad \text{If } y = 0, \text{ then } 3y = 3(0) = 0. \text{ Cover over the } 3y \text{ term.}$$
$$x = -3 \qquad \text{To solve } 40x = -120, \text{ divide both sides by 40.}$$

The x-intercept is $(-3, 0)$. This is entered in the table below.

We can find a third solution by selecting a convenient value for x and finding the corresponding value for y. If we choose $x = -6$, we find that $y = 40$. The solution $(-6, 40)$ is entered in the table, and the equation is graphed as shown.

$40x + 3y = -120$

x	y	(x, y)	
0	-40	$(0, -40)$	← y-intercept
-3	0	$(-3, 0)$	← x-intercept
-6	40	$(-6, 40)$	← Check point

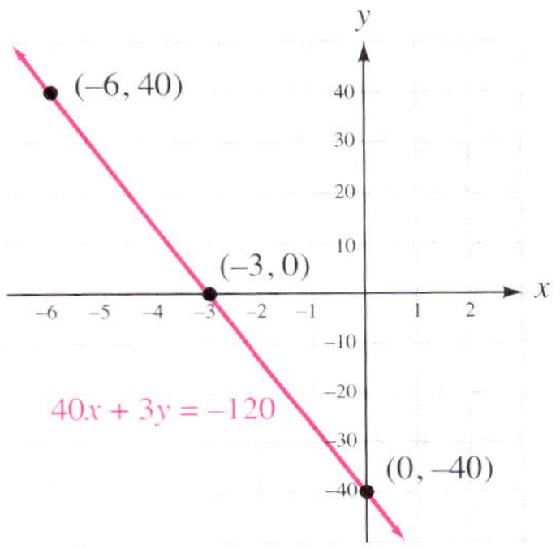

If a different scale is used on the y-axis, the same equation $(40x + 30y = -120)$ will produce a graph with a somewhat different look.

Self Check 3 Graph $32x + 5y = -160$ by finding the intercepts.

Now Try Problem 35

EXAMPLE 4 Graph $3x = -5y + 8$ by finding the intercepts.

Strategy We will let $x = 0$ to find the y-intercept of the graph. We will then let $y = 0$ to find the x-intercept.

Why Since two points determine a line, the y-intercept and x-intercept are enough information to graph this linear equation.

Solution We find the intercepts and select $x = 1$ to find a check point.

***y*-intercept: Let $x = 0$**

$$3x = -5y + 8$$
$$3(0) = -5y + 8$$
$$0 = -5y + 8$$
$$-8 = -5y$$
$$\frac{8}{5} = y$$
$$1\tfrac{3}{5} = y$$

The y-intercept is $\left(0, 1\frac{3}{5}\right)$.

***x*-intercept: Let $y = 0$**

$$3x = -5y + 8$$
$$3x = -5(0) + 8$$
$$3x = 8$$
$$x = \frac{8}{3}$$
$$x = 2\tfrac{2}{3}$$

The x-intercept is $\left(2\frac{2}{3}, 0\right)$.

***Check point*: Let $x = 1$**

$$3x = -5y + 8$$
$$3(1) = -5y + 8$$
$$3 = -5y + 8$$
$$-5 = -5y$$
$$1 = y$$

A check point is $(1, 1)$.

Success Tip

When graphing, it is often helpful to write any coordinates that are improper fractions as terminating decimals or mixed numbers. For example:

$$\left(0, \frac{8}{5}\right) = \left(0, 1\frac{3}{5}\right) = (0, 1.6)$$

and

$$\left(\frac{8}{3}, 0\right) = \left(2\frac{2}{3}, 0\right)$$

The ordered pairs are plotted as shown, and a straight line is then drawn through them.

$3x = -5y + 8$

x	y	(x, y)	
0	$\frac{8}{5} = 1\frac{3}{5}$	$\left(0, 1\frac{3}{5}\right)$	← y-intercept
$\frac{8}{3} = 2\frac{2}{3}$	0	$\left(2\frac{2}{3}, 0\right)$	← x-intercept
1	1	(1, 1)	← Check point

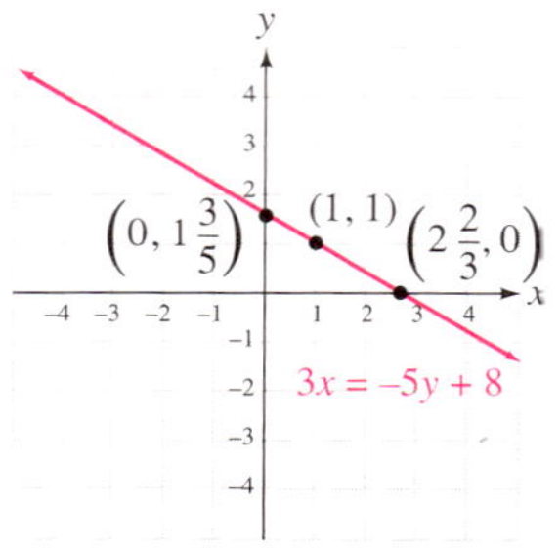

Self Check 4 Graph $8x = -4y + 15$ by finding the intercepts.

Now Try Problem 39

EXAMPLE 5 Graph $2x + 3y = 0$ by finding the intercepts.

Strategy We will let $x = 0$ to find the y-intercept of the graph. We will then let $y = 0$ to find the x-intercept.

Why Since two points determine a line, the y-intercept and x-intercept are enough information to graph this linear equation.

Solution When we find the y- and x-intercepts (shown below), we see that they are both (0, 0). In this case, the line passes through the origin. Since we are using two points and a check point to graph lines, we need to find two more ordered-pair solutions.

If $x = 3$, we see that $(3, -2)$ is a solution. And if $x = -3$, we see that $(-3, 2)$ is also a solution. These two solutions and the origin are plotted and a straight line is drawn through them to give the graph of $2x + 3y = 0$.

***y*-intercept: Let $x = 0$**

$$2x + 3y = 0$$
$$2(0) + 3y = 0$$
$$3y = 0$$
$$y = 0$$

The y-intercept is (0, 0).

***x*-intercept: Let $y = 0$**

$$2x + 3y = 0$$
$$2x + 3(0) = 0$$
$$2x = 0$$
$$x = 0$$

The x-intercept is (0, 0).

The intercepts are the same.

Let $x = 3$

$$2x + 3y = 0$$
$$2(3) + 3y = 0$$
$$6 + 3y = 0$$
$$3y = -6$$
$$y = -2$$

$(3, -2)$ is a solution.

Let $x = -3$

$$2x + 3y = 0$$
$$2(-3) + 3y = 0$$
$$-6 + 3y = 0$$
$$3y = 6$$
$$y = 2$$

$(-3, 2)$ is a solution.

$2x + 3y = 0$

x	y	(x, y)	
0	0	(0, 0)	← The x-intercept and y-intercept.
3	−2	(3, −2)	← A solution.
−3	2	(−3, 2)	← This solution serves as a check point.

Equations that can be written in the form $Ax + By = 0$, such as $2x + 3y = 0$, have graphs that pass through the origin. To find another point on such a line that has integer coordinates, select an x-value equal to the coefficient of y or the opposite of the coefficient of y. Then substitute that x-value into the equation and solve for y.

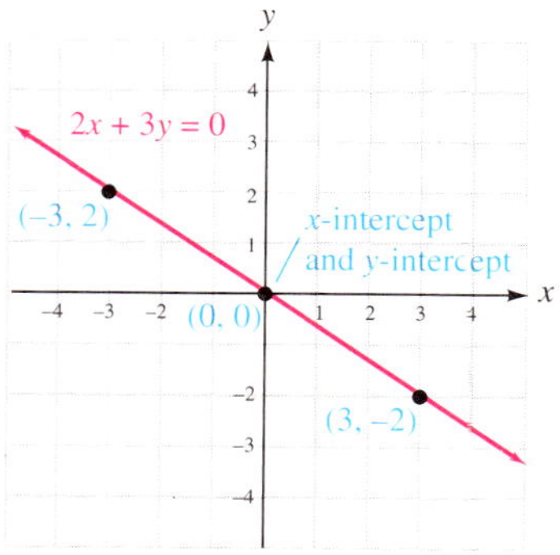

Self Check 5 Graph $5x - 2y = 0$ by finding the intercepts.

Now Try Problem 47

3 Identify and Graph Horizontal and Vertical Lines.

Equations such as $y = 4$ and $x = -3$ are linear equations in two variables, because they can be written in the standard form $Ax + By = C$. For example, $y = 4$ is equivalent to $0x + 1y = 4$ and $x = -3$ is equivalent to $1x + 0y = -3$. We can graph these types of equations using point-plotting.

EXAMPLE 6 Graph: $y = 4$

Strategy To find three ordered-pair solutions of this equation to plot, we will select three values for x and use 4 for y each time.

Why The equation requires that $y = 4$.

Solution We can write the equation in standard form as $0x + y = 4$. Since the coefficient of x is 0, the numbers selected for x have no effect on y. The value of y is always 4. For example, if $x = 2$, we have

$$0x + y = 4$$ This is the given equation, y = 4, written in standard form: Ax + By = C.

$$0(2) + y = 4$$ Substitute 2 for x.

$$y = 4$$ Simplify the left side.

One solution is (2, 4). To find two more solutions, we select $x = 0$ and $x = -3$. For any x-value, the y-value is always 4, so we enter (0, 4) and (−3, 4) in the table. If we plot the ordered pairs and draw a straight line through the points, the result is a horizontal line. The y-intercept is (0, 4) and there is no x-intercept.

Success Tip

An equation with only the variable y is a horizontal line that intersects the y-axis.

$y = 4$

x	y	(x, y)
2	4	(2, 4)
0	4	(0, 4)
−3	4	(−3, 4)

Select any number for x. Each value of y must be 4.

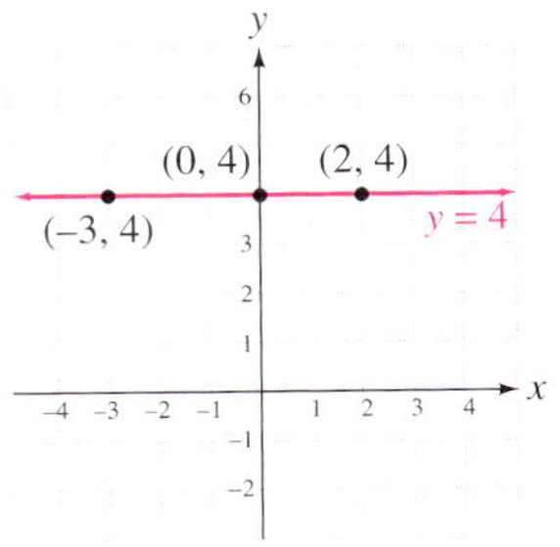

Self Check 6 Graph: $y = -2$

Now Try Problem 51

EXAMPLE 7 Graph: $x = -3$

Strategy To find three ordered-pair solutions of this equation to plot, we must select -3 for x each time.

Why The equation requires that $x = -3$.

Success Tip

If we graph $x = -3$ in one dimension (as in Chapter 2), the graph is a single point on a number line.

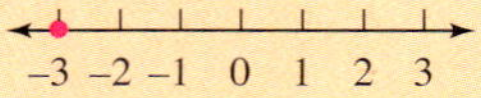

If we graph $x = -3$ in two dimensions (as in Chapter 3), the result is the line shown in Example 7 at the right.

Success Tip

An equation with only the variable x is a vertical line that intersects the x-axis.

Solution We can write the equation in standard form as $x + 0y = -3$. Since the coefficient of y is 0, the value of y has no effect on x. The value of x is always -3. For example, if $y = -2$, we have

$$x + 0y = -3 \quad \text{This is the given equation, } x = -3\text{, written in standard form: } Ax + By = C.$$

$$x + 0(-2) = -3 \quad \text{Substitute } -2 \text{ for } y.$$

$$x = -3 \quad \text{Simplify the left side.}$$

One solution is $(-3, -2)$. To find more solutions, we must again select -3 for x. Any number can be used for y. If $y = 0$, then a second solution is $(-3, 0)$. If $y = 3$, a third solution is $(-3, 3)$. The three solutions are entered in the table below. When we plot the ordered pairs and draw a straight line through the points, the result is a vertical line. The x-intercept is $(-3, 0)$ and there is no y-intercept.

$x = -3$

x	y	(x, y)
-3	-2	$(-3, -2)$
-3	0	$(-3, 0)$
-3	3	$(-3, 3)$

Each value of x we select must be -3. Any number can be used for y.

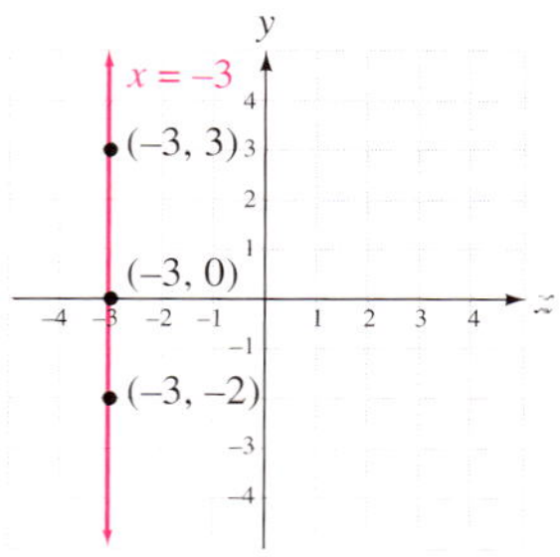

Self Check 7 Graph: $x = 4$

Now Try Problem 55

From the results of Examples 6 and 7, we have the following facts.

Equations of Horizontal and Vertical Lines

The graph of $y = b$ is a horizontal line with y-intercept $(0, b)$.

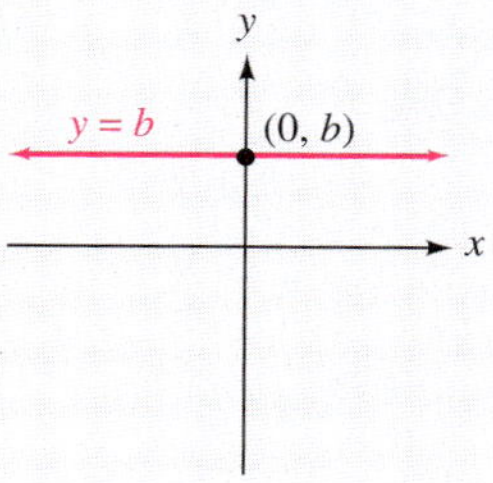

The graph of $x = a$ is a vertical line with x-intercept $(a, 0)$.

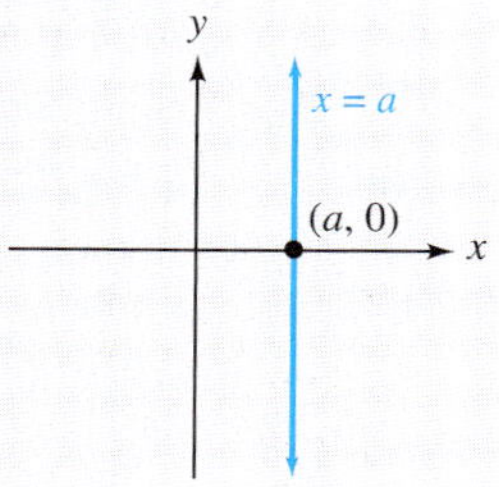

The graph of $y = 0$ is the x-axis. The graph of $x = 0$ is the y-axis.

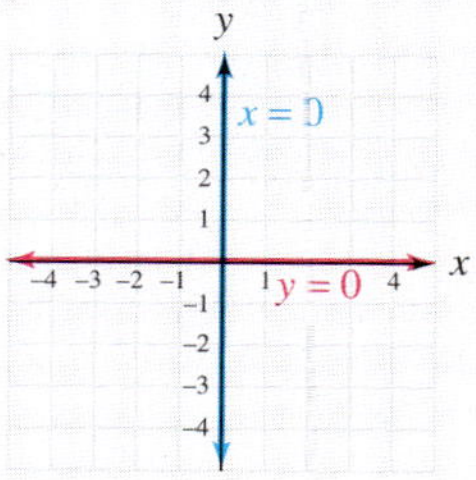

4 Obtain Information from Intercepts.

The ability to read and interpret graphs is a valuable skill. When analyzing a graph, we should locate and examine the intercepts. As the following example illustrates, the coordinates of the intercepts can give useful information.

EXAMPLE 8 **Hybrid Mileage.** Figure (a) shows mileage data for a 2010 Toyota Prius Hybrid. What information do the intercepts give about the car?

©Michael Doolittle/Alamy

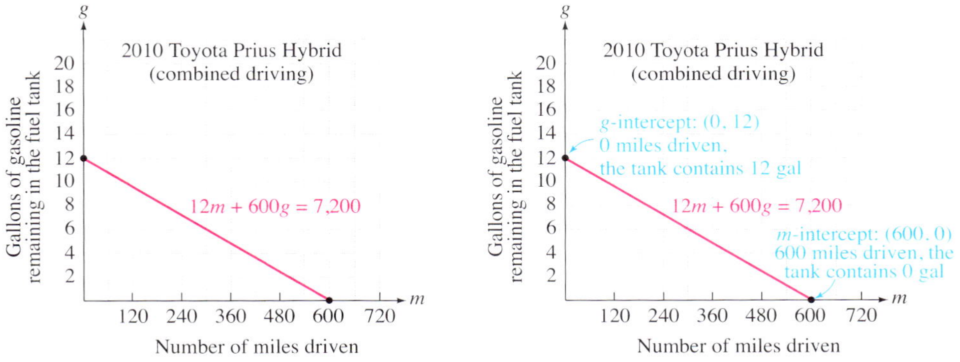

(a) (b)

Strategy We will determine where the graph (the line in red) intersects the g-axis and where it intersects the m-axis.

Why Once we know the intercepts, we can interpret their meaning.

Solution See figure (b). The g-intercept (0, 12) indicates that when the car has been driven 0 miles, the fuel tank contains 12 gallons of gasoline. That is, the Prius has a 12-gallon fuel tank.

The m-intercept (600, 0) indicates that after 600 miles of combined driving, the fuel tank contains 0 gallons of gasoline. Thus, 600 miles of combined driving can be done on 1 tank of gas in a Prius.

Now Try Problems 79 and 81

Using Your Calculator ▶ Use a Calculator to Graph Linear Equations. (Optional)

Courtesy of Texas Instruments Incorporated

So far, we have graphed linear equations by making tables of solutions and plotting points. A graphing calculator can make the task of graphing much easier. However, a graphing calculator does not take the place of a working knowledge of the topics discussed in this chapter. It should serve as an aid to enhance your study of algebra.

The Viewing Window The screen on which a graph is displayed is called the **viewing window.** The **standard window** has settings of

$$\text{Xmin} = -10, \quad \text{Xmax} = 10, \quad \text{Ymin} = -10, \quad \text{and} \quad \text{Ymax} = 10$$

which indicate that the minimum x- and y-coordinates used in the graph will be -10, and that the maximum x- and y-coordinates will be 10.

Graphing an Equation To graph $y = x - 1$ using a graphing calculator, we press the **Y =** key and enter $x - 1$ after the symbol Y_1. Then we press the **GRAPH** key to see the graph.

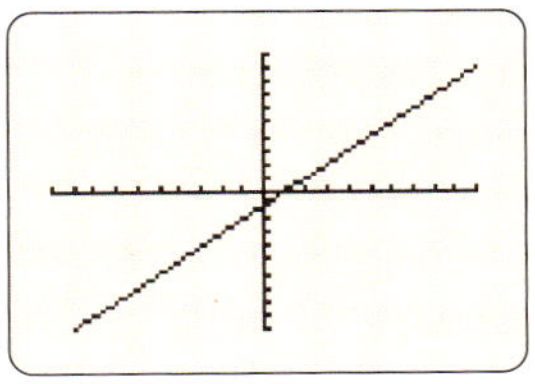

Change the Viewing Window We can change the viewing window by pressing the **WINDOW** key and entering -4 for the minimum x- and y-coordinates and 4 for the maximum x- and y-coordinates. Then we press the **GRAPH** key to see the graph of $y = x - 1$ in more detail.

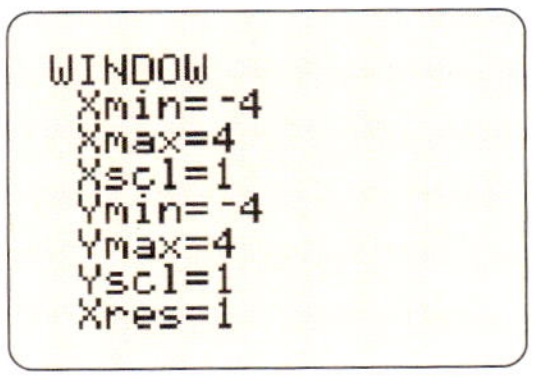

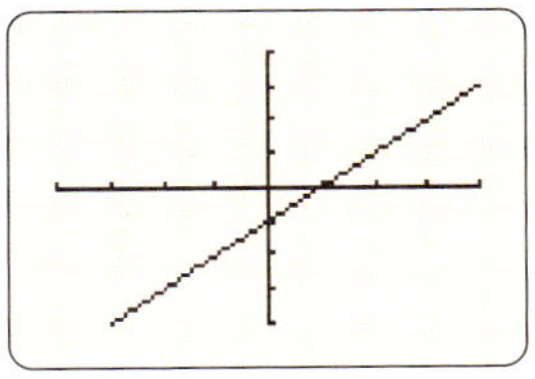

Solving an Equation for *y* To graph $3x + 2y = 12$, we must first solve the equation for y.

$$3x + 2y = 12$$
$$2y = -3x + 12 \quad \text{Subtract } 3x \text{ from both sides.}$$
$$y = -\frac{3}{2}x + 6 \quad \text{Divide both sides by 2.}$$

Next, we press the **WINDOW** key to reenter the standard window settings, press **Y =** and enter $-\frac{3}{2}x + 6$, as shown below. Then press **GRAPH** to see the graph.

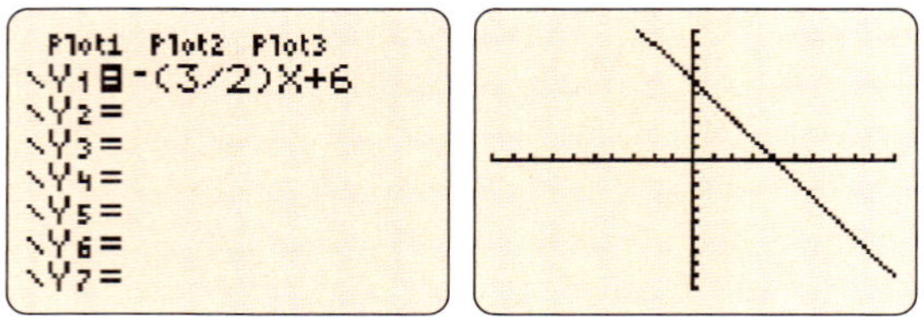

SECTION 3.3 STUDY SET

VOCABULARY

Fill in the blanks.

1. The __________ of a line is the point where the line intersects/crosses the x-axis.
2. The y-intercept of a line is the point where the line ______________ the y-axis.
3. The graph of $y = 4$ is a __________ line and the graph of $x = 6$ is a ________ line.
4. The intercept method is useful when graphing linear equations written in the ________ form $Ax + By = C$.

CONCEPTS

5. Fill in the blanks.
   a. To find the y-intercept of the graph of a line, substitute for x in the equation and solve for .
   b. To find the x-intercept of the graph of a line, substitute for y in the equation and solve for .
6. Complete the table of solutions and fill in the blanks.

$3x + 2y = 6$

x	y	(x, y)	
0			← -intercept
	0		← -intercept
−2			← point

7. a. Refer to the graph. Which intercept tells the purchase price of the machinery? What was that price?
   b. Which intercept indicates when the machinery will have lost all of its value? When is that?

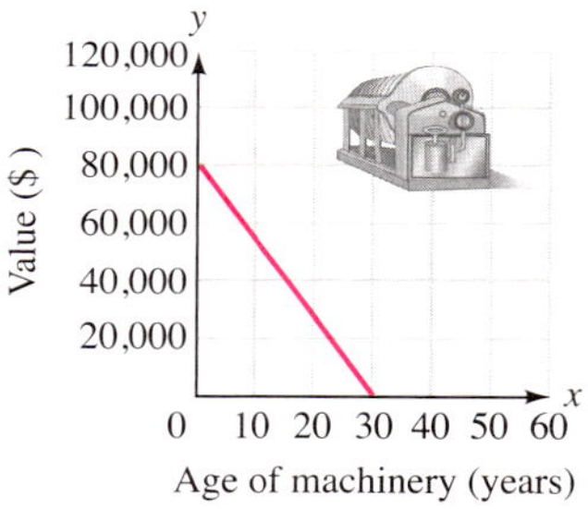

8. Match each graph with its equation.
   a. $x = 2$ b. $y = 2$ c. $y = 2x$
   d. $2x - y = 2$ e. $y = 2x + 2$ f. $y = -2x$

i.

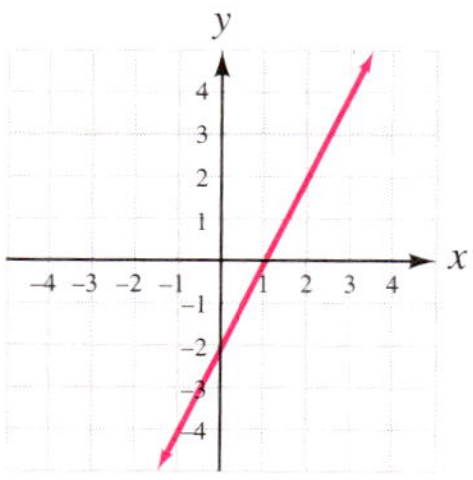

ii.

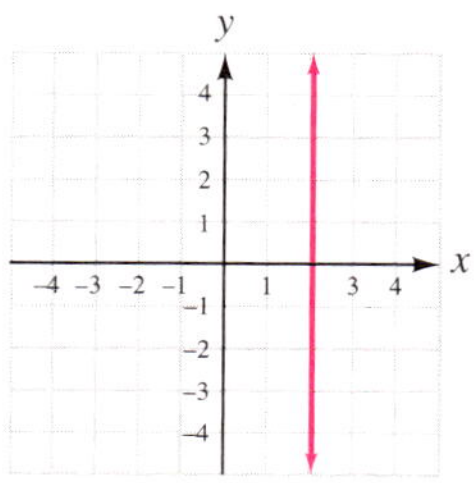

iii.

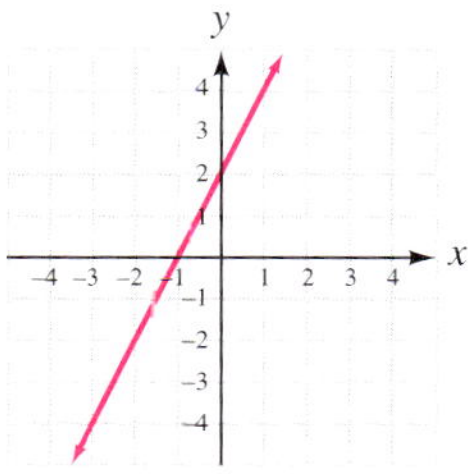

iv.

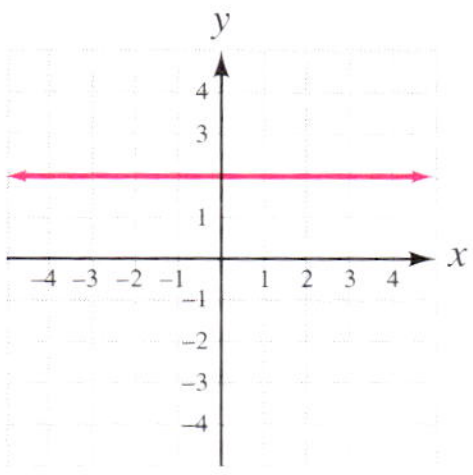

v.

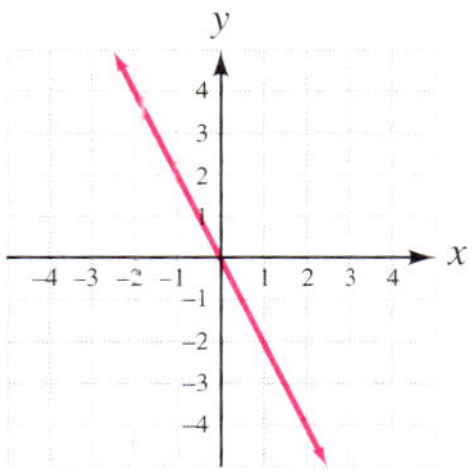

vi.

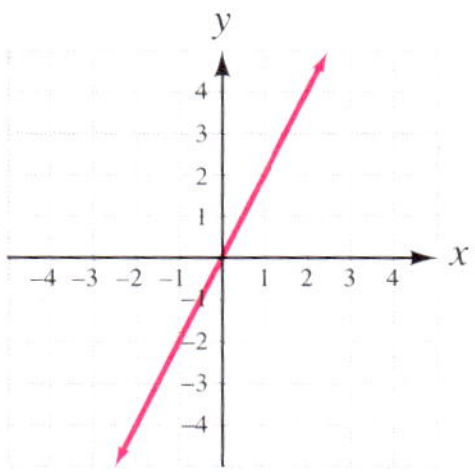

NOTATION

9. What is the equation of the x-axis? What is the equation of the y-axis?

10. Write the coordinates that are improper fractions as mixed numbers.

a. $\left(\frac{7}{2}, 0\right)$ b. $\left(0, -\frac{17}{3}\right)$

GUIDED PRACTICE

Give the coordinates of the intercepts of each graph. See Example 1.

11.

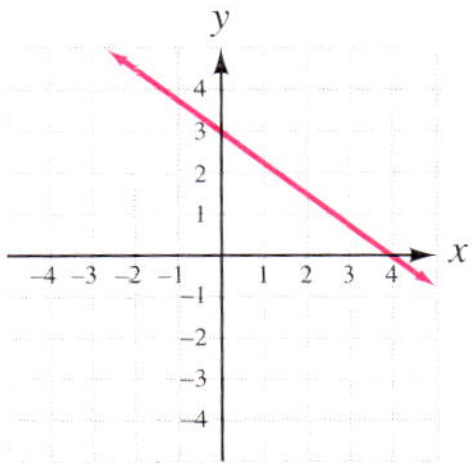

12.

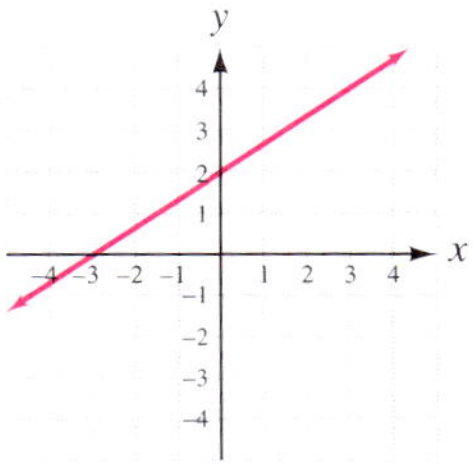

13.

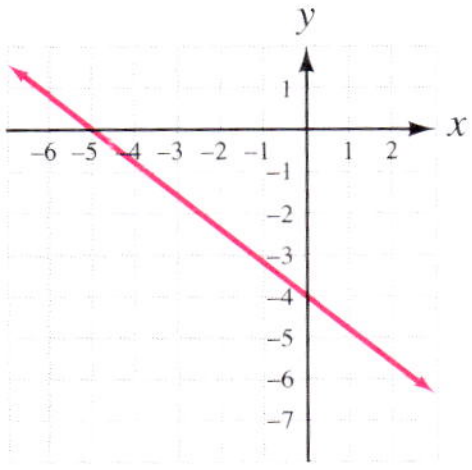

14.

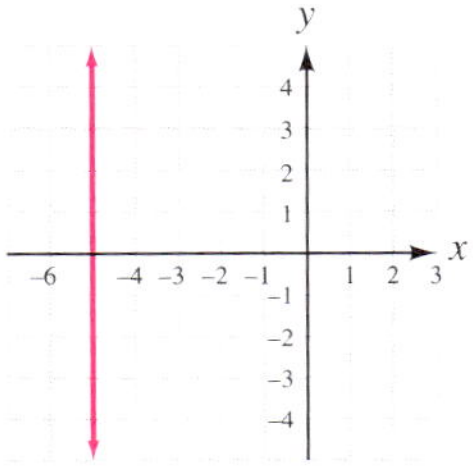

15.

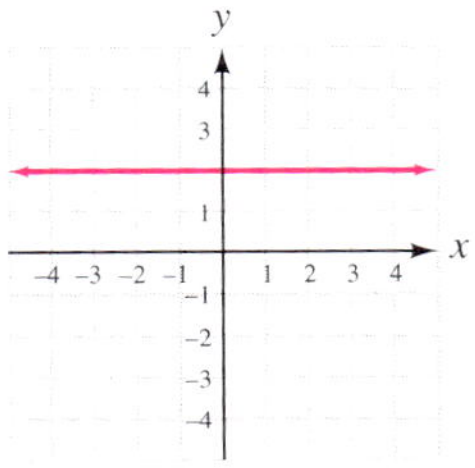

16.

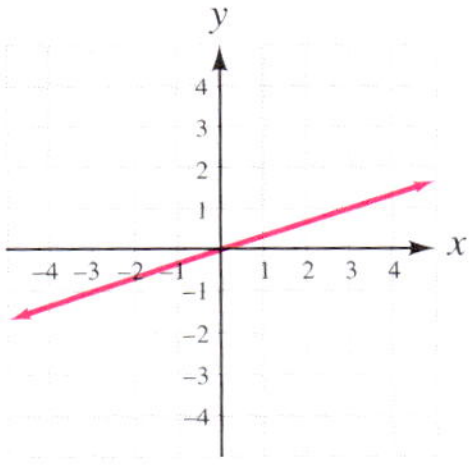

Estimate the coordinates of the intercepts of each graph. (Some are not integers.) See Example 1. *(Answers may vary.)*

17.

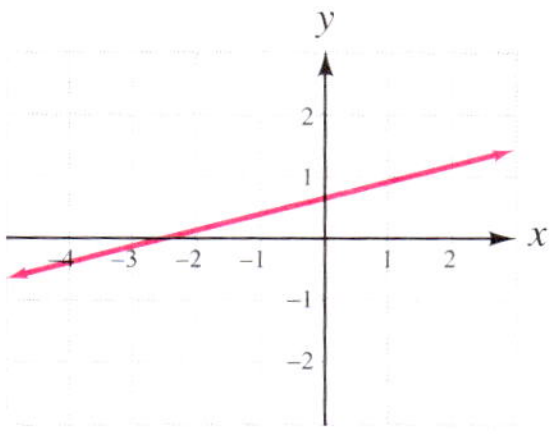

18.

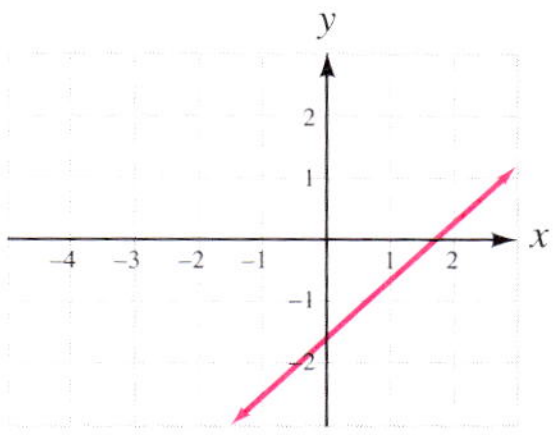

Find the x- and y-intercepts of the graph of each equation. Do not graph the line. See Example 2.

19. $8x + 3y = 24$

20. $5x + 6y = 30$

21. $7x - 2y = 28$

22. $2x - 9y = 36$

23. $-5x - 3y = 10$

24. $-9x - 5y = 25$

25. $6x + y = 9$

26. $x + 8y = 14$

Use the intercept method to graph each equation. See Example 2.

27. $4x + 5y = 20$

28. $3x + 4y = 12$

29. $5x + 15y = -15$

30. $8x + 4y = -24$

31. $x - y = -3$

32. $x - y = 3$

33. $x + 2y = -2$

34. $x + 2y = -4$

Use the intercept method to graph each equation. See Example 3.

35. $30x + y = -30$

36. $20x - y = -20$

37. $4x - 20y = 60$

38. $6x - 30y = 30$

Use the intercept method to graph each equation. See Example 4.

39. $3x + 4y = 8$

40. $2x + 3y = 9$

41. $-9x + 4y = 9$

42. $-5x + 4y = 15$

43. $3x - 4y = 11$

44. $5x - 4y = 13$

45. $9x + 3y = 10$

46. $4x + 4y = 5$

Use the intercept method to graph each equation. See Example 5.

47. $3x + 5y = 0$

48. $4x + 3y = 0$

49. $2x - 7y = 0$

50. $6x - 5y = 0$

Graph each equation. See Examples 6 and 7.

51. $y = 5$

52. $y = -3$

53. $y = 0$

54. $x = 0$

55. $x = -2$

56. $x = 5$

57. $x = \dfrac{4}{3}$

58. $y = -\dfrac{1}{2}$

59. $y - 2 = 0$ (*Hint:* Solve for y first.)

60. $x + 1 = 0$ (*Hint:* Solve for x first.)

61. $5x = 7.5$ (*Hint:* Solve for x first.)

62. $3y = 4.5$ (*Hint:* Solve for y first.)

TRY IT YOURSELF

Graph each equation.

63. $7x = 4y - 12$
64. $7x = 5y - 15$
65. $4x - 3y = 12$
66. $5x - 10y = 20$
67. $x = -\frac{5}{3}$
68. $y = \frac{5}{2}$
69. $y - 3x = -\frac{4}{3}$
70. $y - 2x = -\frac{9}{8}$
71. $7x + 3y = 0$
72. $4x - 5y = 0$
73. $-4x = 8 - 2y$
74. $-5x = 10 + 5y$
75. $3x = -150 - 5y$
76. $x = 50 - 5y$
77. $-3y = 3$
78. $-2x = 8$

APPLICATIONS

79. **Chemistry.** The relationship between the temperature T and volume V of a gas kept in a sealed container at a constant pressure is graphed below. The T-intercept of this graph is a very important scientific fact. It represents the lowest possible temperature, called **absolute zero.**

a. Estimate absolute zero.

b. What is the volume of the gas when the temperature is absolute zero?

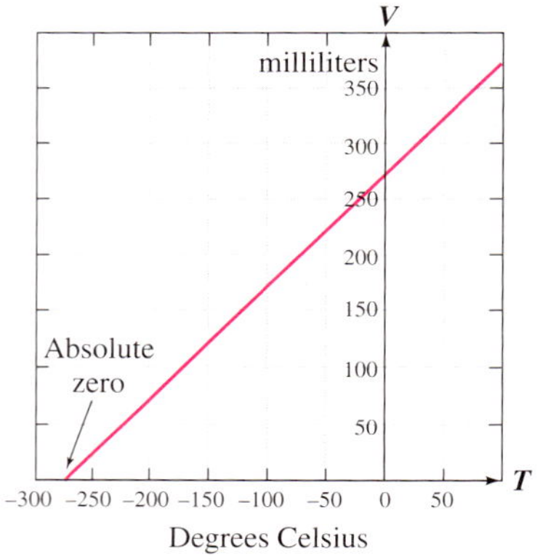

80. **Physics.** The graph shows the length L of a stretched spring (in inches) as different weights w (in pounds) are attached to it. What information about the spring does the L-intercept give?

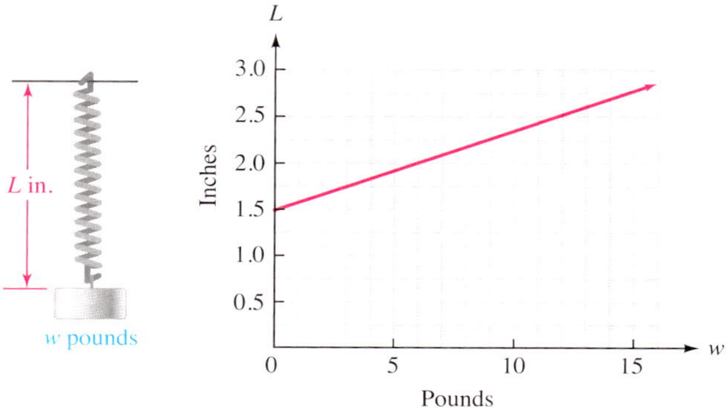

81. **Bottled Water Dispenser.** The graph below shows the number of gallons g of water remaining in a bottle after c six-ounce cups have been served from it. Find the intercepts of the graph. What information do they give?

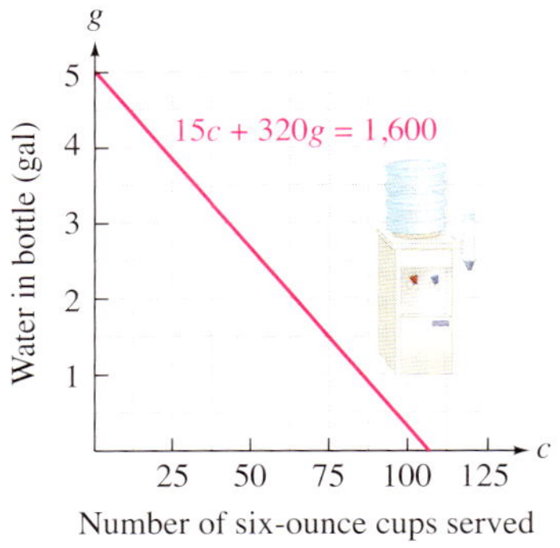

82. **Renewable Energy.** The equation $p = 50s - 300$ estimates the power output in watts from a propeller blade turbine driven by a wind of speed s miles per hour. What information does the s-intercept of the graph of the equation give? (Source: otherpower.com)

83. **Landscaping.** A developer is going to purchase x trees and y shrubs to landscape a new office complex. The trees cost \$50 each and the shrubs cost \$25 each. His budget is \$5,000. This situation is modeled by the equation $50x + 25y = 5{,}000$. Use the intercept method to graph it.

a. What information is given by the y-intercept?

b. What information is given by the x-intercept?

84. **Eggs.** The number of eggs eaten by an average American in one year has remained almost constant since the year 2000. See the graph below. Draw a horizontal line that passes through, or near, the data points. What is the equation of the line?

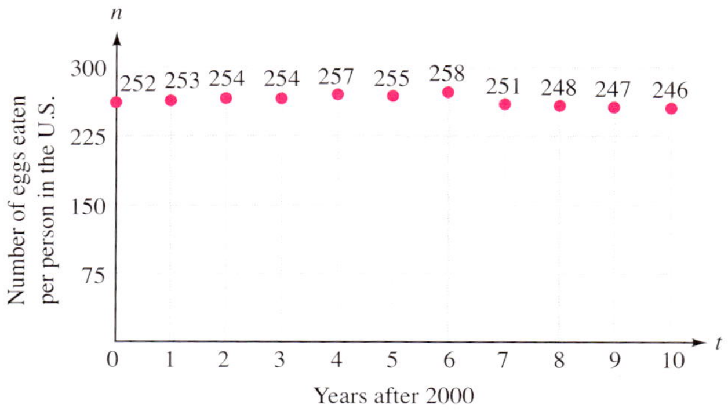

Source: United Egg

WRITING

85. To graph $3x + 2y = 12$, a student found the intercepts and a check point, and graphed them, as shown in figure (a). Instead of drawing a crooked line through the points, what should he have done?

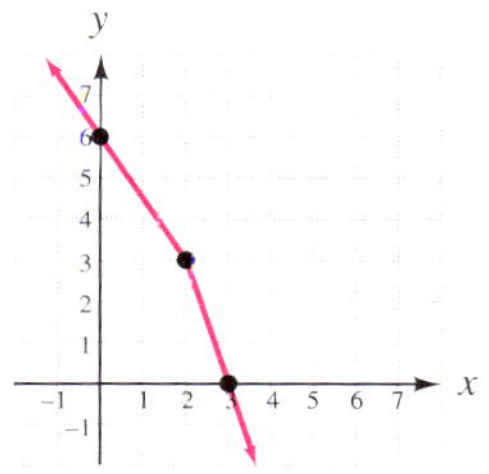

(a)

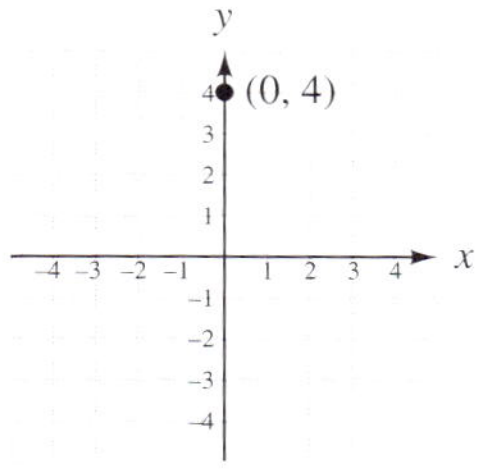

(b)

86. A student graphed the linear equation $y = 4$, as shown above in figure (b). Explain her error.

87. How do we find the intercepts of the graph of an equation without having to graph the equation?

88. In Section 3.2, we discussed a method to graph $y = 2x - 3$. In Section 3.3, we discussed a method to graph $2x + 3y = 6$. Briefly explain the steps involved in each method.

REVIEW

89. Simplify: $\frac{3 \cdot 5 \cdot 5}{3 \cdot 5 \cdot 5 \cdot 5}$

90. Simplify: $4\left(\frac{d}{2} - 3\right) - 5\left(\frac{2}{5}d - 1\right)$

91. Translate: Six less than twice x

92. Is -5 a solution of $2(3x + 10) = 5x + 6$?

CHALLENGE PROBLEMS

93. Where will the line $y = b$ intersect the line $x = a$?

94. Write an equation of the line that has an x-intercept of $(4, 0)$ and a y-intercept of $(0, 3)$.

95. What is the least number of intercepts a line can have? What is the greatest number a line can have?

96. On a rectangular coordinate system, draw a circle that has exactly two intercepts.

SECTION 3.4 Slope and Rate of Change

OBJECTIVES

1. Find the slope of a line from its graph.
2. Find the slope of a line given two points.
3. Find slopes of horizontal and vertical lines.
4. Solve applications of slope.
5. Calculate rates of change.
6. Determine whether lines are parallel or perpendicular using slope.

ARE YOU READY?

The following problems review some basic skills that are needed to find the slope of a line.

1. Evaluate: $\frac{4 - 1}{8 - 3}$
2. Evaluate: $\frac{-10 - 1}{-4 - (-4)}$
3. Multiply: $-\frac{7}{9} \cdot \frac{9}{7}$
4. Simplify: $\frac{15}{18}$

In this section, we introduce a method to measure the steepness (or slant) of a line. We call this measure the *slope of the line,* and it can be found in several ways.

1 Find the Slope of a Line from Its Graph.

The **slope of a line** is a ratio that compares the vertical change with the corresponding horizontal change as we move along the line from one point to another.

As an example, let's find the slope of the line graphed on the right. To begin, we select two points on the line and call them P and Q. One way to move from P to Q is to start at point P and count upward 5 grid squares. Then, moving to the right, we count 6 grid squares to reach point Q. The vertical change in this movement is called the **rise.** The horizontal change is called the **run.** Notice that a right triangle, called a **slope triangle,** is created by this process.

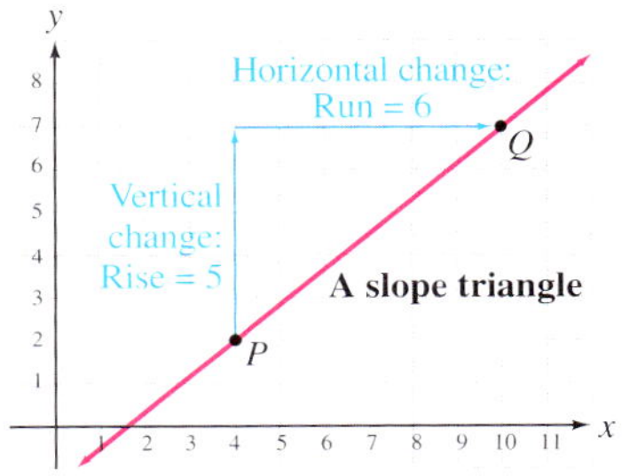

The Language of Algebra

A **ratio** is a comparison of two numbers using a quotient. Ratios are used in many settings. Mechanics speak of gear ratios. Colleges like to advertise their low student-to-teacher ratios.

The Language of Algebra

Many historians believe that m was chosen to represent the slope of a line because it is the first letter of the French word **monter**, meaning to ascend or climb.

The slope of a line is defined to be *the ratio of the vertical change to the horizontal change.* By tradition, the letter m is used to represent slope. For the line graphed on the previous page, we have

$$m = \text{slope} = \frac{\text{vertical change}}{\text{horizontal change}} = \frac{\text{rise}}{\text{run}} = \frac{5}{6}$$

This ratio is a comparison of the rise and the run using a quotient.

The slope of the line is $\frac{5}{6}$. This indicates that there is a rise (vertical change) of 5 units for each run (horizontal change) of 6 units.

EXAMPLE 1 Find the slope of the line graphed in figure (a) below.

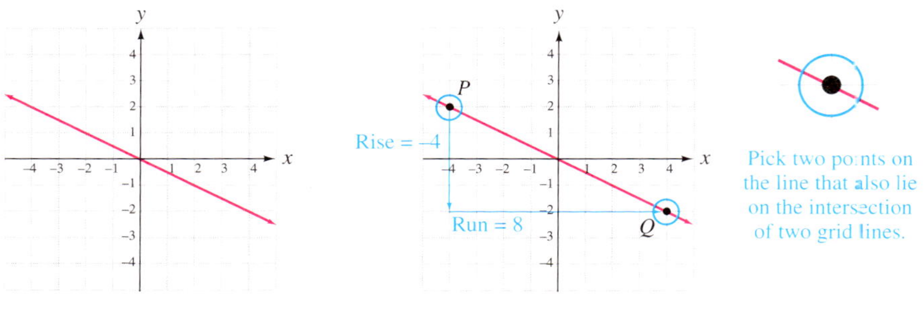

(a) (b)

Strategy We will pick two points on the line, construct a slope triangle, and find the rise and run. Then we will write the ratio of the rise to the run.

Why The slope of a line is the ratio of the rise to the run.

Solution We begin by choosing two points on the line, P and Q, as shown in figure (b). One way to move from P to Q is to start at point P and count *downward* 4 grid squares. Because this movement is downward, the rise is -4. Then, moving right, we count 8 grid squares to reach Q. This indicates that the run is 8.

To find the slope of the line, we write a ratio of the rise to the run in simplified form.

$$m = \frac{\text{rise}}{\text{run}} = \frac{-4}{8} = -\frac{1}{2}$$

Always simplify slope fractions.

The slope of the line is $-\frac{1}{2}$.

Success Tip

When drawing a slope triangle, movement upward or to the right is positive. Movement downward or to the left is negative.

- Negative rise: downward
- Negative run: to the left

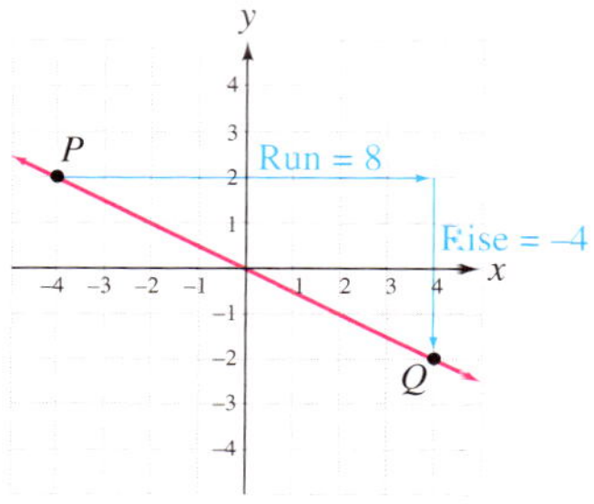

Arrowheads are used to show horizontal movement (left or right) and vertical movement (up or down).

The movement from P to Q can be reversed. Starting at P, we can move to the right, a run of 8; and then downward, a rise of -4, to reach Q. With this approach, the slope triangle is above the line. When we form the ratio to find the slope, we get the *same result* as before:

$$m = \frac{\text{rise}}{\text{run}} = \frac{-4}{8} = -\frac{1}{2}$$

Self Check 1 Find the slope of the line shown above using two points different from those used in the solution of Example 1.

Now Try Problem 21

The identical answers from Example 1 and its Self Check illustrate an important fact: *For any line, the same value will be obtained no matter which two points on the line are used to find the slope.*

2 Find the Slope of a Line Given Two Points.

The Language of Algebra

The prefix **sub** means below or beneath, as in submarine or subway. In x_1, x_2, y_1, and y_2, the **subscripts** 1 and 2 are written lower than the variable. They are not exponents, and they do not represent a calculation.

Success Tip

Recall that subtraction is used to measure change.

We can generalize the graphic method for finding slope to develop a slope formula. To begin, we select points P and Q on the line shown in the figure on the right. To distinguish between the coordinates of these two points, we use **subscript notation.** Point P has coordinates (x_1, y_1), which are read as "x sub 1 and y sub 1." Point Q has coordinates (x_2, y_2), which are read as "x sub 2 and y sub 2."

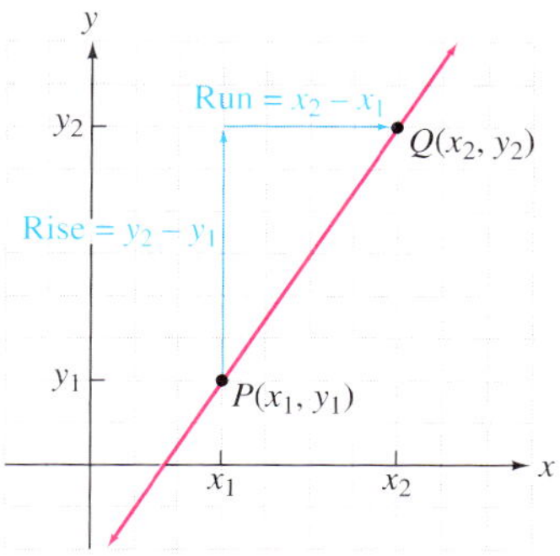

As we move from point P to point Q, the rise is the difference of the y-coordinates: $y_2 - y_1$. We call this difference the **change in *y*.** The run is the difference of the x-coordinates: $x_2 - x_1$. This difference is called the **change in *x*.** Since the slope is the ratio $\frac{\text{rise}}{\text{run}}$, we have the following formula for calculating slope.

Slope of a Line

The slope m of a line that passes through points (x_1, y_1) and (x_2, y_2) is

$$m = \frac{\text{vertical change}}{\text{horizontal change}} = \frac{\text{rise}}{\text{run}} = \frac{\text{change in } y}{\text{change in } x} = \frac{y_2 - y_1}{x_2 - x_1} \quad \text{if } x_2 \neq x_1$$

EXAMPLE 2 Find the slope of the line that passes through (1, 2) and (3, 8).

Strategy We will use the slope formula to find the slope of the line.

Why We know the coordinates of two points on the line.

Solution When using the slope formula, it makes no difference which point you call (x_1, y_1) and which point you call (x_2, y_2). If we let (x_1, y_1) be (1, 2) and (x_2, y_2) be (3, 8), then

Success Tip

When using the slope formula, you may find it helpful to begin your solution by labeling the coordinates of the points in this way:

$$\begin{matrix} (x_1, y_1) & (x_2, y_2) \\ \downarrow \ \downarrow & \downarrow \ \downarrow \\ (1, 2) & (3, 8) \end{matrix}$$

$m = \frac{y_2 - y_1}{x_2 - x_1}$ This is the slope formula.

$m = \frac{8 - 2}{3 - 1}$ Substitute 8 for y_2, 2 for y_1, 3 for x_2, and 1 for x_1.

$m = \frac{6}{2}$ Do the subtraction.

$m = 3$ Simplify. Think of this as a $\frac{3}{1}$ rise-to-run ratio.

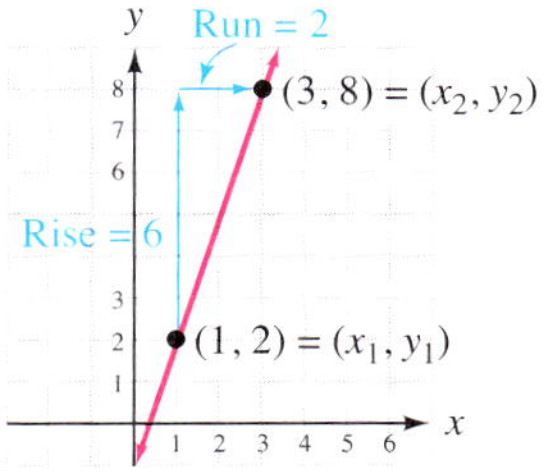

The slope of the line is 3. Note that we obtain the same value for the slope if we let $(x_1, y_1) = (3, 8)$ and $(x_2, y_2) = (1, 2)$.

$$m = \frac{y_2 - y_1}{x_2 - x_1} = \frac{2 - 8}{1 - 3} = \frac{-6}{-2} = 3$$

Success Tip

The slope formula is a valuable tool because it allows us to calculate the slope of a line without having to view its graph.

Although it is not necessary, the line passing through (1, 2) and (3, 8) has been graphed on the right. The graph of the line, including the slope triangle, verifies that the rise is 6 and the run is 2, and that $m = \frac{6}{2} = 3$.

CAUTION When using the slope formula, always subtract the y-coordinates and their corresponding x-coordinates in the same order. Otherwise, your answer will have the wrong sign. If we let (x_1, y_1) be (1, 2) and (x_2, y_2) be (3, 8):

$$\cancel{m = \frac{y_2 - y_1}{x_1 - x_2}} = \frac{8 - 2}{1 - 3} = \frac{6}{-2} = -3 \quad \text{and} \quad \cancel{m = \frac{y_1 - y_2}{x_2 - x_1}} = \frac{2 - 8}{3 - 1} = \frac{-6}{2} = -3$$

Self Check 2 Find the slope of the line that passes through (2, 1) and (4, 11).

Now Try ▶ Problem 33

EXAMPLE 3 Find the slope of the line that passes through $(-2, 4)$ and $(5, -6)$.

Strategy We will use the slope formula to find the slope of the line.

Why We know the coordinates of two points on the line.

Solution If we let (x_1, y_1) be $(-2, 4)$ and (x_2, y_2) be $(5, -6)$, then

$$m = \frac{y_2 - y_1}{x_2 - x_1}$$ This is the slope formula.

$$m = \frac{-6 - 4}{5 - (-2)}$$ Substitute -6 for y_2, 4 for y_1, 5 for x_2, and -2 for x_1.

$$m = -\frac{10}{7}$$ Do the subtraction. We can write the result as $\frac{-10}{7}$ or $-\frac{10}{7}$.

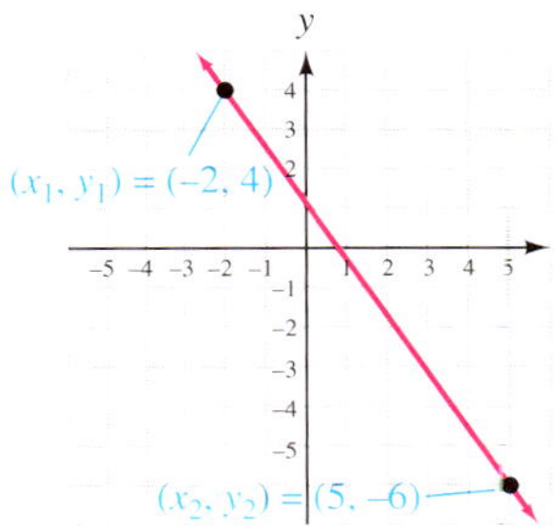

The slope of the line is $-\frac{10}{7}$.

If we graph the line by plotting the two points, we see that the line falls from left to right—a fact indicated by its negative slope.

Caution

Slopes normally are written as fractions, sometimes as decimals, but never as mixed numbers.

As with any fractional answer, always express slope in simplified form.

Self Check 3 Find the slope of the line that passes through $(-1, -2)$ and $(1, -7)$.

Now Try ▶ Problem 39

In Example 2, the slope of the line was positive. In Examples 1 and 3, the slopes of the lines were negative. In general, lines that rise from left to right have a positive slope. Lines that fall from left to right have a negative slope.

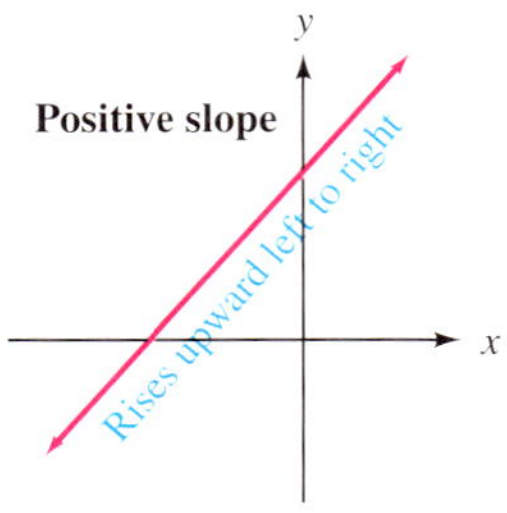

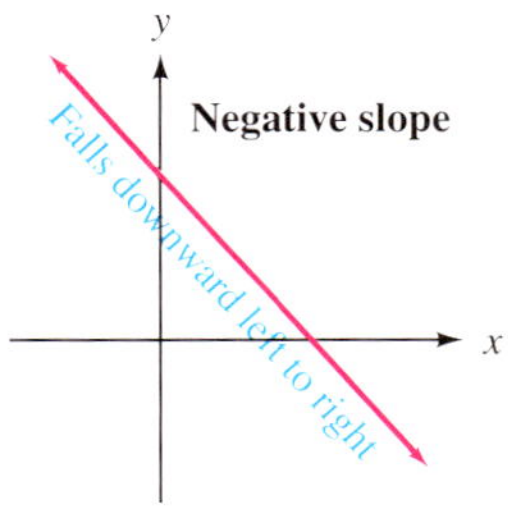

Success Tip

To classify the slope of a line as positive or negative, follow it from left to right, as you would read a sentence in a book.

In figure (a) on the next page, we see that a line with slope 3 is steeper than a line with slope $\frac{5}{6}$, and a line with slope $\frac{5}{6}$ is steeper than a line with slope $\frac{1}{4}$. In general, *the larger the absolute value of the slope, the steeper the line.*

Lines with slopes of 1 and -1 are graphed in figure (b) on the next page. In each case, there is a special relationship between the rise and the run. When $m = 1$, the rise and run are, of course, the same number. When $m = -1$ the rise and run are opposites. Note that both lines create a 45° angle with the horizontal x-axis.

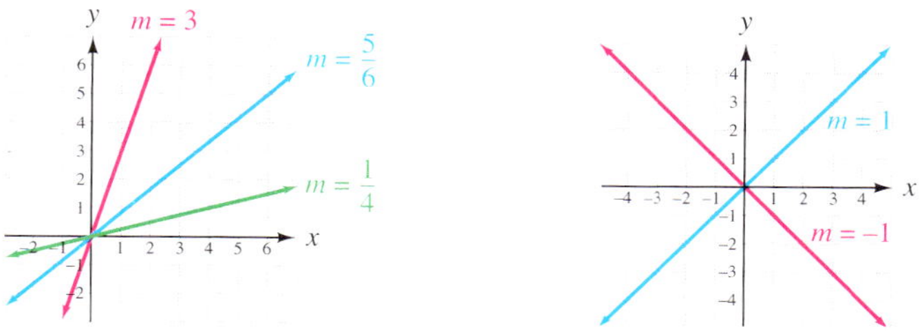

(a) (b)

3 Find Slopes of Horizontal and Vertical Lines.

In the next two examples, we calculate the slope of a horizontal line and we show that a vertical line has no defined slope.

EXAMPLE 4 Find the slope of the line $y = 3$.

Strategy We will find the coordinates of two points on the line.

Why We can then use the slope formula to find the slope of the line.

Solution The graph of $y = 3$ is a horizontal line. To find its slope, we select two points on the line: $(-2, 3)$ and $(3, 3)$. If (x_1, y_1) is $(-2, 3)$ and (x_2, y_2) is $(3, 3)$, we have

$$m = \frac{y_2 - y_1}{x_2 - x_1}$$ This is the slope formula.

$$m = \frac{3 - 3}{3 - (-2)}$$ Substitute 3 for y_2, 3 for y_1, 3 for x_2, and -2 for x_1.

$$m = \frac{0}{5}$$ Simplify the numerator and the denominator.

$$m = 0$$ 0 divided by any nonzero number is equal to 0.

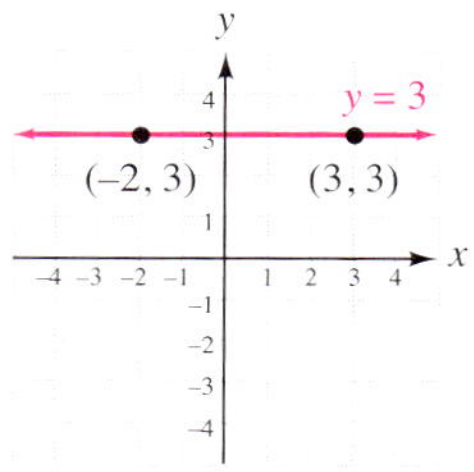

The rise = 0 for these two points.

The slope of the line $y = 3$ is 0.

Self Check 4 Find the slope of the line $y = -10$.

Now Try ▶ Problem 61

The y-coordinates of any two points on a horizontal line will be the same, and the x-coordinates will be different. Thus, the numerator of $\frac{y_2 - y_1}{x_2 - x_1}$ will always be zero, and the denominator will always be nonzero. Therefore, the slope of a horizontal line is 0.

EXAMPLE 5 Find the slope of the line $x = -2$.

Strategy We will find the coordinates of two points on the line.

Why We can then use the slope formula to find the slope of the line.

Solution

Notation

This example explains why the definition of slope includes the restriction that $x_1 \neq x_2$.

The graph of $x = -2$ is a vertical line. To find its slope, we select two points on the line: $(-2, 3)$ and $(-2, -1)$. If (x_1, y_1) is $(-2, -1)$ and (x_2, y_2) is $(-2, 3)$, we have

$$m = \frac{y_2 - y_1}{x_2 - x_1} \quad \text{This is the slope formula.}$$

$$m = \frac{3 - (-1)}{-2 - (-2)} \quad \text{Substitute 3 for } y_2\text{, } -1 \text{ for } y_1\text{, } -2 \text{ for } x_2\text{, and } -2 \text{ for } x_1\text{. Note that } x_1 = x_2.$$

$$m = \frac{4}{0} \quad \text{Simplify the numerator and the denominator.}$$

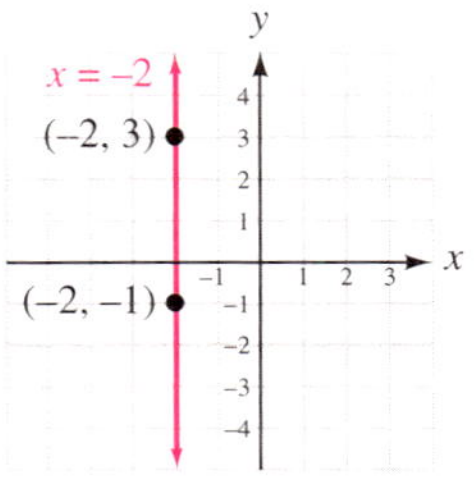

The run = 0 for these two points.

Since division by zero is undefined, $\frac{4}{0}$ has no meaning. The slope of the line $x = -2$ is undefined.

Self Check 5 Find the slope of the line $x = 12$.

Now Try Problem 67

The Language of Algebra

Undefined and **0** do not mean the same thing. A horizontal line has a defined slope; it is 0. A vertical line does not have a defined slope; we say its slope is *undefined.*

The y-coordinates of any two points on a vertical line will be different, and the x-coordinates will be the same. Thus, the numerator of $\frac{y_2 - y_1}{x_2 - x_1}$ will always be nonzero, and the denominator will always be 0. Therefore, the slope of a vertical line is undefined.

We now summarize the results from Examples 4 and 5.

Slopes of Horizontal and Vertical Lines

Horizontal lines (lines with equations of the form $y = b$) have **slope 0.**

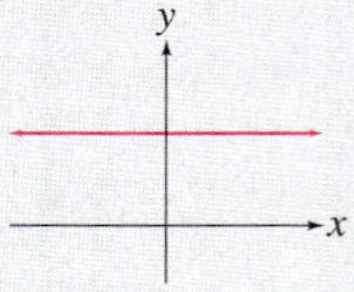

Vertical lines (lines with equations of the form $x = a$) have **undefined slope.**

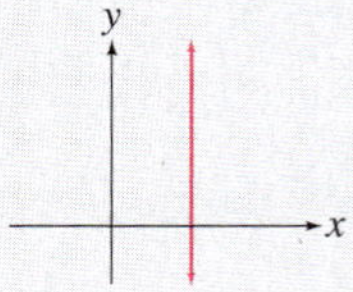

4 Solve Applications of Slope.

The concept of slope has many applications. For example, architects use slope when designing ramps and roofs. Truckers must be aware of the slope, or **grade,** of the roads they travel. Mountain bikers ride up rocky trails and snow skiers speed down steep slopes.

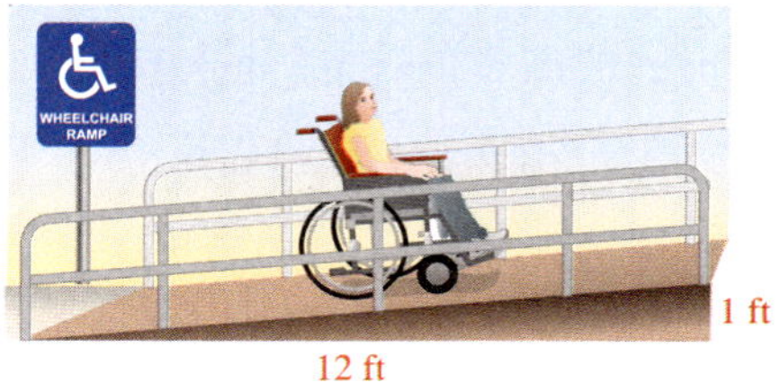

The *Americans with Disabilities Act* provides a guideline for the steepness of a ramp. The maximum slope for a wheelchair ramp is 1 foot of rise for every 12 feet of run: $m = \frac{1}{12}$.

The grade of an incline is its slope expressed as a percent: A 15% grade means a rise of 15 feet for every run of 100 feet: $m = \frac{15}{100}$, which simplifies to $\frac{3}{20}$. A grade is always expressed as a positive percent.

EXAMPLE 6 **Architecture.** **Pitch** is the incline of a roof written as a ratio of the vertical rise to the horizontal run. It is always expressed as a positive number. Find the pitch of the roof shown in the illustration.

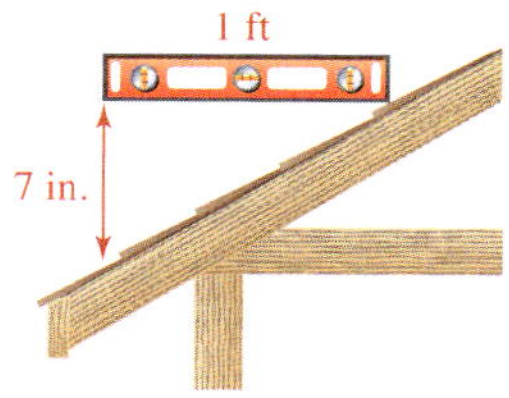

Strategy We will determine the rise and the run of the roof from the illustration. Then we will write the ratio of the rise to the run.

Why The pitch of a roof is its slope, and the slope of a line is the ratio of the rise to the run.

Solution A level is used to create a slope triangle. The rise of the slope triangle is given as 7 inches. Since a ratio is a quotient of two quantities with the *same* units, we will express the length of the one-foot-long level as 12 inches. Therefore, the run of the slope triangle is 12 inches.

$$m = \frac{\text{rise}}{\text{run}} = \frac{7}{12}$$ The roof has a $\frac{7}{12}$ pitch. This means that the roof rises 7 units for every 12 units in the horizontal direction.

Self Check 6 **Roofing.** Find the pitch of the roof.

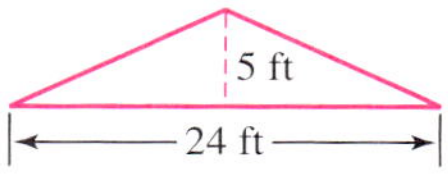

Now Try Problem 99

5 Calculate Rates of Change.

We have seen that the slope of a line compares the change in y to the change in x. This is called the **rate of change** of y with respect to x. In our daily lives, we often make many such comparisons of the change in one quantity with respect to another. For example, we might speak of snow melting at the rate of 6 inches per day or a tourist exchanging money at the rate of 12 pesos per dollar.

When finding rates of change in application problems, we attach units to the numerator and denominator in the slope calculation.

EXAMPLE 7 **Banking.** A bank offers a business account with a fixed monthly fee, plus a service charge for each check written. The relationship between the monthly cost y and the number x of checks written is graphed below. At what rate does the monthly cost change?

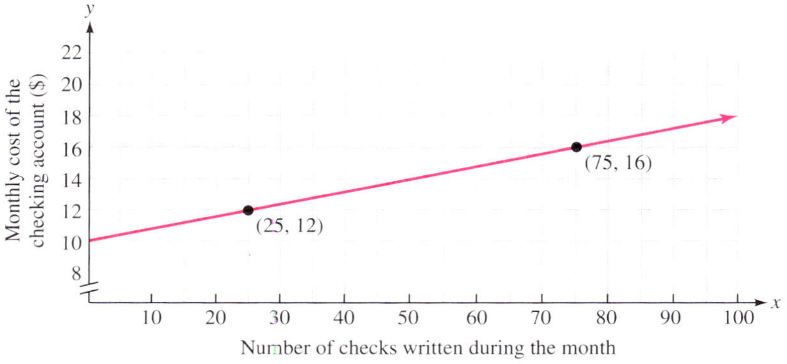

Notation

In the graph, the symbol ≠ indicates a break in the labeling of the vertical axis. The break enables us to omit a large portion of the grid that would not be used.

Strategy We will use the slope formula to calculate the slope of the line and attach the proper units to the numerator and denominator.

Why We know the coordinates of two points on the line.

Solution Two points on the line are (25, 12) and (75, 16). We will let $(x_1, y_1) = (25, 12)$ and $(x_2, y_2) = (75, 16)$, so we have

$$\text{Rate of change} = \frac{(y_2 - y_1)\text{ dollars}}{(x_2 - x_1)\text{ checks}} = \frac{(16 - 12)\text{ dollars}}{(75 - 25)\text{ checks}} = \frac{4\text{ dollars}}{50\text{ checks}} = \frac{2\text{ dollars}}{25\text{ checks}}$$

Since the y-axis is scaled in dollars, we attach the units of "dollars" to the calculation of $y_2 - y_1$ in the numerator. Since x-axis indicates the number of checks written, we attach the units of "checks" to the calculation of $x_2 - x_1$ in the denominator. The monthly cost of the checking account increases $2 for every 25 checks written.

We can express $\frac{2}{25}$ in decimal form by dividing the numerator by the denominator. Then we can write the rate of change in two other ways, using the word *per,* which indicates division.

Rate of change = $0.08 per check or Rate of change = 8¢ per check

The Language of Algebra

The preposition **per** means for each, or for every. When we say the rate of change is 8¢ *per* check, we mean 8¢ for each check.

Self Check 7 **Ethanol.** In 2005, the U.S. produced about 1,600 million bushels of corn for ethanol. By 2009, production had risen to 4,200 million bushels. Find the rate of change in the number of bushels of corn produced for ethanol over that time span. (Source: USDA, ERS Feed Outlook)

Now Try Problem 103

6 Determine Whether Lines Are Parallel or Perpendicular Using Slope.

Two lines that lie in the same plane but do not intersect are called **parallel lines.** Parallel lines have the same slope and different y-intercepts. For example, the lines graphed in figure (a) are parallel because they both have slope $-\frac{2}{3}$.

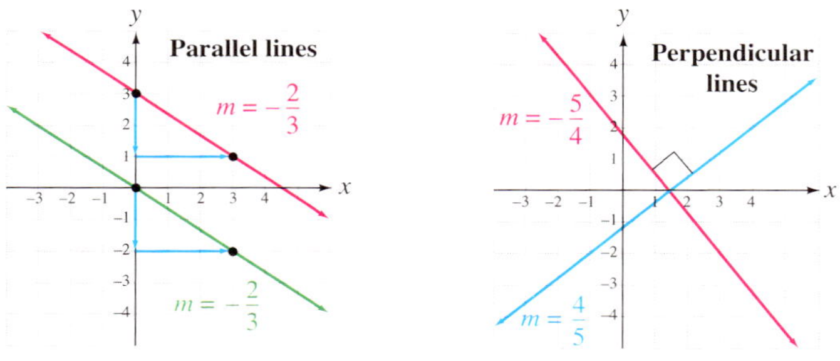

(a) (b)

The Language of Algebra

The words **parallel** and **perpendicular** are used in many settings. For example, the gymnast on the *parallel* bars is in a position that is *perpendicular* to the floor.

©iStockPhoto.com/Galina Barskaya

Lines that intersect to form four right angles (angles with measure 90°) are called **perpendicular lines.** If the product of the slopes of two lines is −1, the lines are perpendicular. This means that the slopes are **negative** (or **opposite**) **reciprocals.** In figure (b), we know that the lines with slopes $\frac{4}{5}$ and $-\frac{5}{4}$ are perpendicular because

$$\frac{4}{5}\left(-\frac{5}{4}\right) = -\frac{20}{20} = -1$$ $\frac{4}{5}$ and $-\frac{5}{4}$ are negative reciprocals.

Slopes of Parallel and Perpendicular Lines

1. Two lines with the same slope are parallel.
2. Two lines are perpendicular if the product of their slopes is −1; that is, if their slopes are negative reciprocals.
3. A horizontal line is perpendicular to any vertical line, and vice versa.

EXAMPLE 8 Determine whether the line that passes through (7, −9) and (10, 2) and the line that passes through (0, 1) and (3, 12) are parallel, perpendicular, or neither.

Strategy We will use the slope formula to find the slope of each line.

Why If the slopes are equal, the lines are parallel. If the slopes are negative reciprocals, the lines are perpendicular. Otherwise, the lines are neither parallel nor perpendicular.

Solution To calculate the slope of each line, we use the slope formula.

Success Tip

An example of two lines that are neither parallel nor perpendicular would be lines with slopes of $\frac{5}{2}$ and −2. When graphed, the lines intersect, but they do not form right angles.

The line through **(7, −9)** *and* **(10, 2):**

$$m = \frac{y_2 - y_1}{x_2 - x_1} = \frac{2 - (-9)}{10 - 7} = \frac{11}{3}$$

The line through **(0, 1)** *and* **(3, 12):**

$$m = \frac{y_2 - y_1}{x_2 - x_1} = \frac{12 - 1}{3 - 0} = \frac{11}{3}$$

Since the slopes are the same, the lines are parallel.

Self Check 8 Determine whether the line that passes through (2, 1) and (6, 8) and the line that passes through (−1, 0) and (−5, 7) are parallel, perpendicular, or neither.

Now Try Problems 73 and 75

EXAMPLE 9 Find the slope of a line perpendicular to the line that passes through (1, −4) and (8, 4).

Strategy We will use the slope formula to find the slope of the line passing through (1, −4) and (8, 4).

Why We can then form the negative reciprocal of the result to produce the slope of a line perpendicular to the given line.

Solution The slope of the line that passes through (1, −4) and (8, 4) is

$$m = \frac{y_2 - y_1}{x_2 - x_1} = \frac{4 - (-4)}{8 - 1} = \frac{8}{7}$$

We let $(x_1, y_1) = (1, -4)$ and $(x_2, y_2) = (8, 4)$.

The slope of a line perpendicular to the given line has slope that is the negative (or opposite) reciprocal of $\frac{8}{7}$, which is $-\frac{7}{8}$.

Self Check 9 Find the slope of a line perpendicular to the line that passes through (−4, 1) and (9, 5).

Now Try Problem 85

SECTION 3.4 STUDY SET

VOCABULARY

Fill in the blanks.

1. The ______ of a line is a measure of the line's steepness. It is the ______ of the vertical change to the horizontal change.

2. $m = \frac{\text{\_\_\_\_\_\_}}{\text{horizontal change}} = \frac{\text{rise}}{\text{\_\_\_\_}} = \frac{\text{change in } y}{\text{\_\_\_\_\_\_}}$

3. The rate of ________ of a linear relationship can be found by finding the slope of the graph of the line and attaching the proper units.

4. ________ lines do not intersect. ______________ lines intersect to form four right angles.

CONCEPTS

5. Which line graphed has
   a. a positive slope?
   b. a negative slope?
   c. zero slope?
   d. undefined slope?

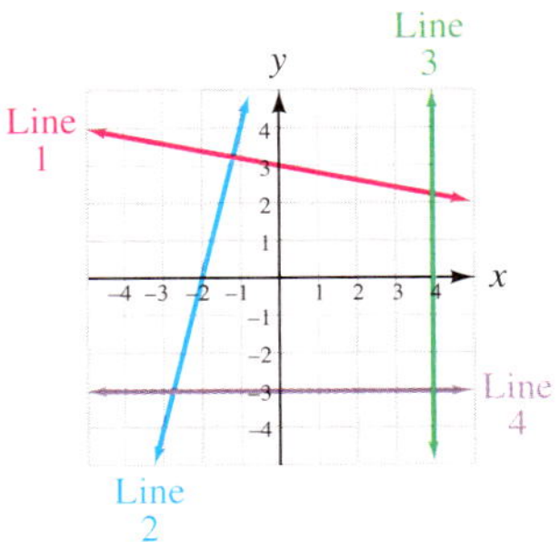

6. Consider each graph of a line and the slope triangle. What is the rise? What is the run? What is the slope of the line?

a.

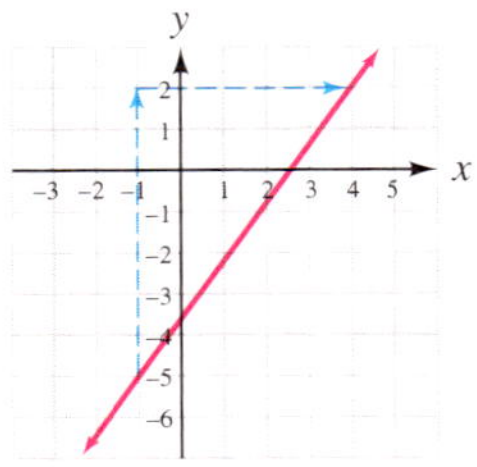

b.

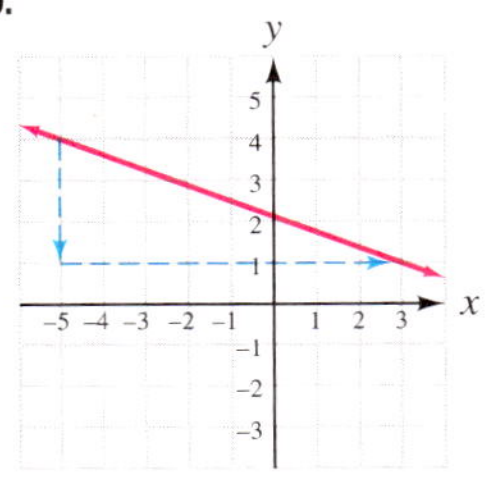

7. For each graph, determine which line has the greater slope.

a.

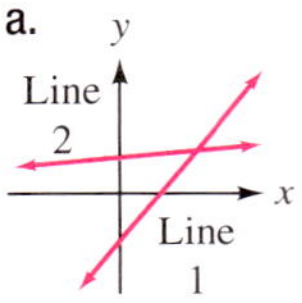

b.

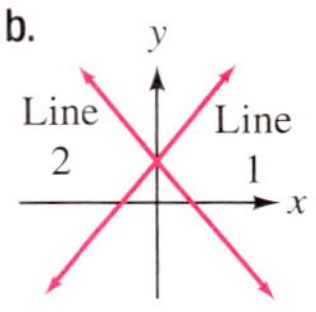

c.

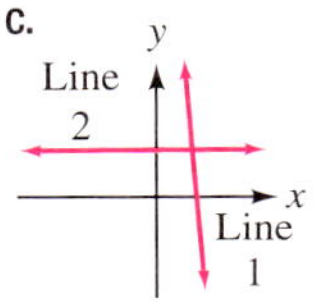

8. Which two labeled points should be used to find the slope of the line?

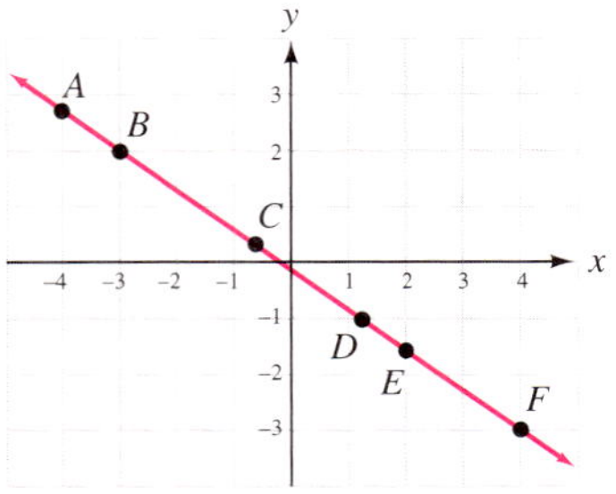

9. a. Find the slope of the line below using the points in black.
   b. Find the slope of the line using the points in green.
   c. Fill in the blank: When finding the slope of a line, the ______ value will be obtained no matter which two points on the line are used.

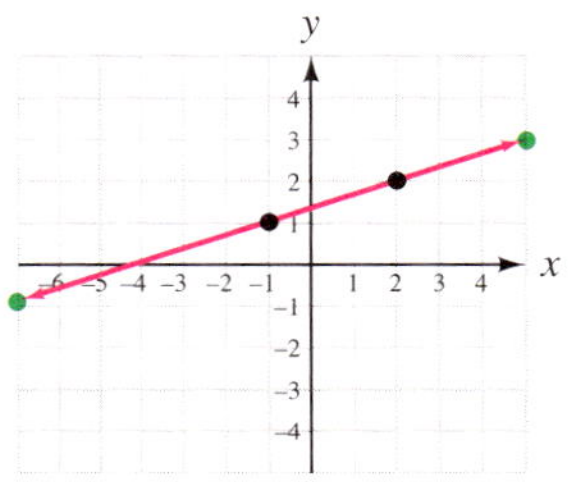

10. Evaluate each expression.
    a. $\frac{10 - 4}{6 - 5}$
    b. $\frac{-1 - 1}{-2 - (-7)}$

11. Write each slope in a better way.
    a. $m = \frac{0}{6}$
    b. $m = \frac{8}{0}$
    c. $m = \frac{3}{12}$
    d. $m = \frac{-10}{-5}$

12. Fill in the blanks: __________ lines have a slope of 0. Vertical lines have __________ slope.

13. The *grade* of an incline is its slope expressed as a percent. Express the slope $\frac{2}{5}$ as a grade.

14. **Growth Rates.** The graph shows how a child's height increased from ages 2 through 5. Fill in the correct units to find the rate of change in the child's height.

$$\text{Rate of change} = \frac{(40 - 31)}{(5 - 2)}$$

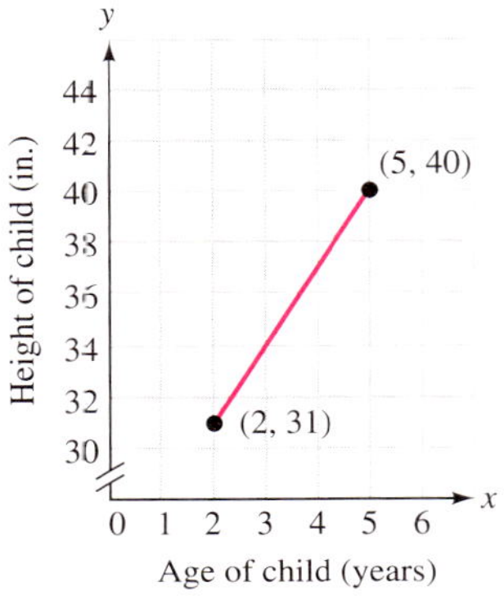

15. Find the negative reciprocal of each number.
    a. 6
    b. $-\frac{7}{8}$
    c. -1

16. Fill in the blanks.
    a. Two different lines with the same slope are ________.
    b. If the slopes of two lines are negative reciprocals, the lines are ____________.
    c. The product of the slopes of perpendicular lines is ____.

NOTATION

17. a. What is the formula used to find the slope of a line passing through (x_1, y_1) and (x_2, y_2)?
    b. Fill in the blanks to state the slope formula in words: m equals y _____ two minus y _____ one ________________ x sub _____ minus x sub _____.

18. Explain the difference between y^2 and y_2.

19. Consider the points (7, 2) and (−4, 1). If we let $x_1 = 7$, then what is y_2?

20. The symbol ≑ is used when graphing to indicate a ______ in the labeling of an axis.

GUIDED PRACTICE

Find the slope of each line. See Example 1.

21.

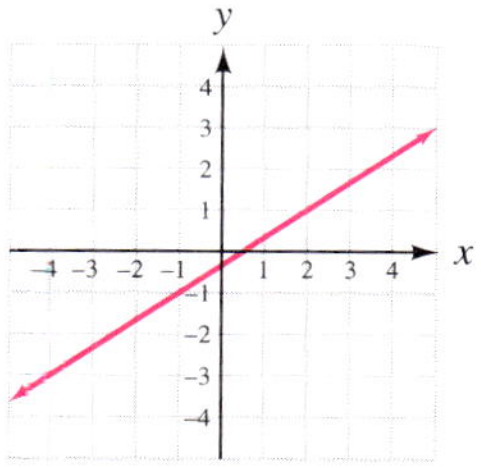

22.

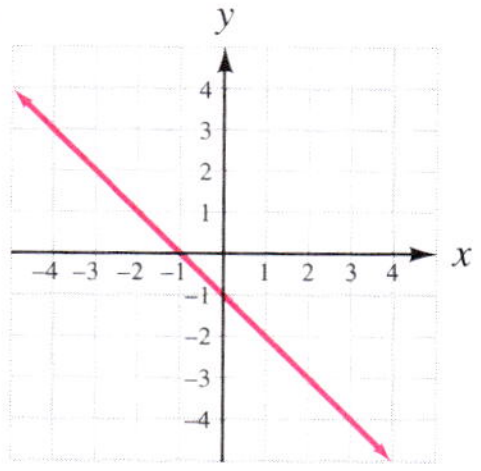

23.

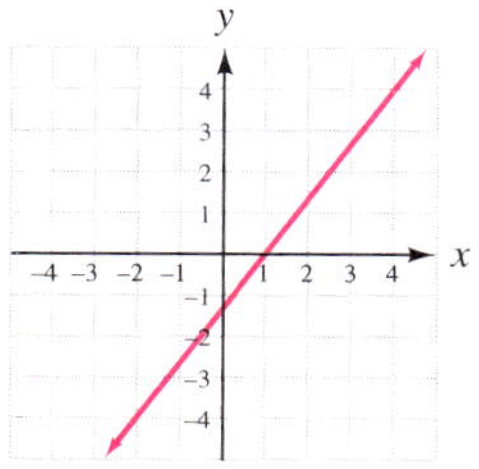

24.

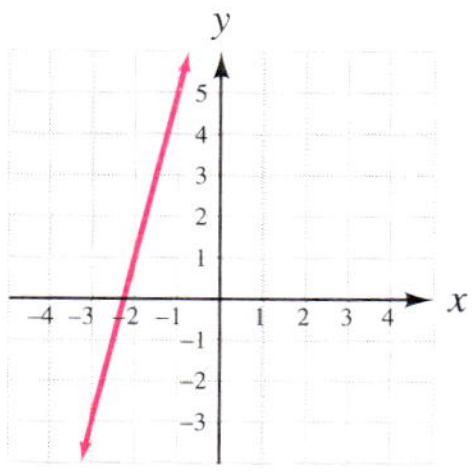

25.

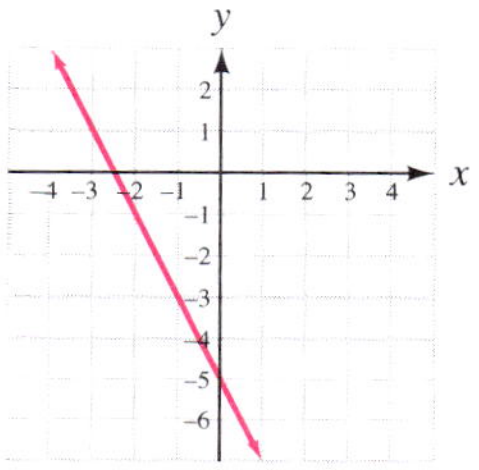

26.

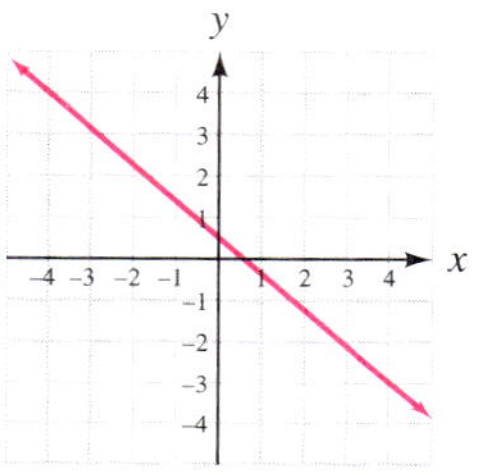

27.

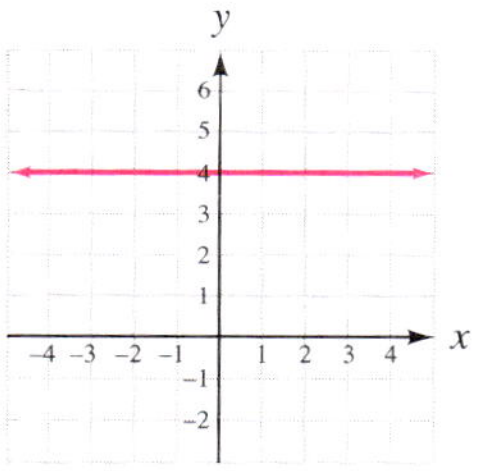

28.

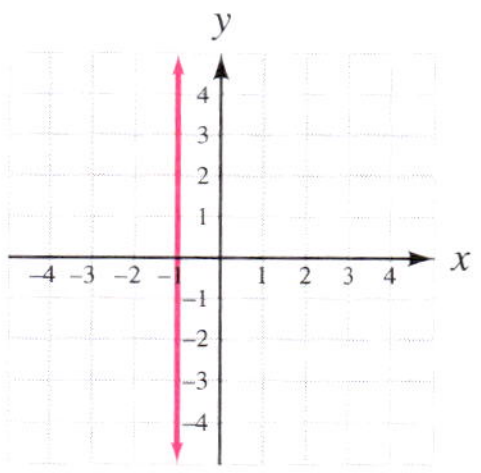

29.

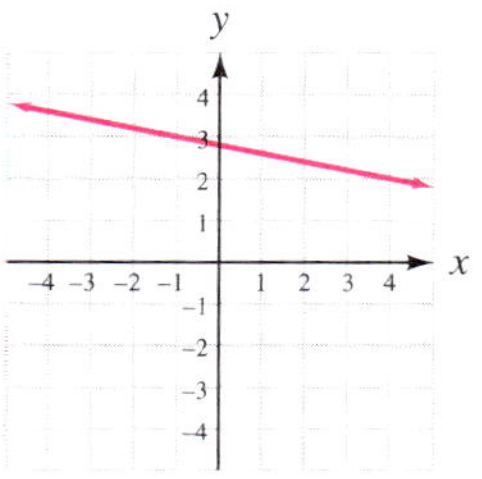

30.

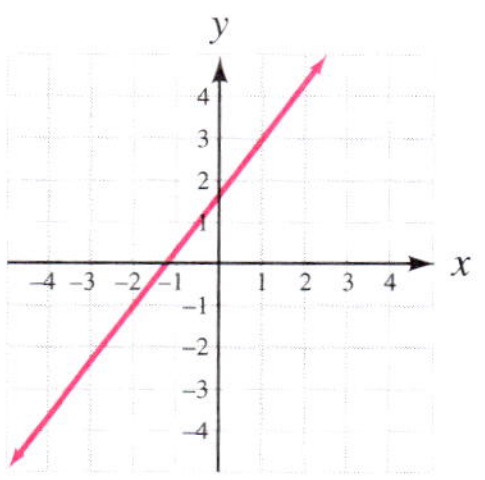

31.

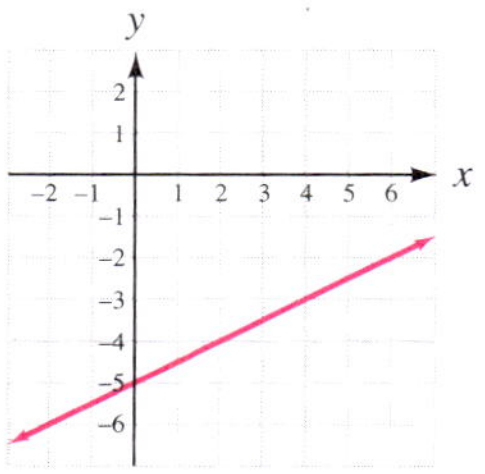

32.

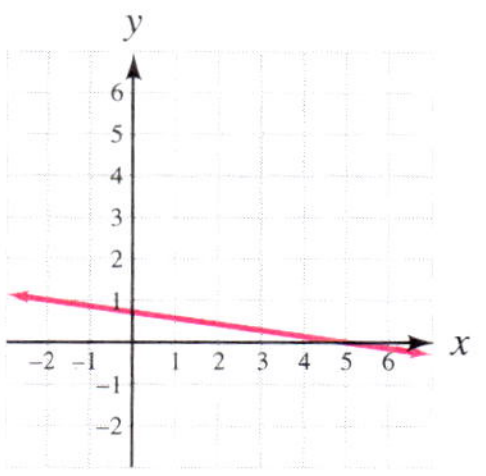

Find the slope of the line passing through the given points. See Examples 2 and 3.

33. $(1, 3)$ and $(2, 4)$
34. $(1, 3)$ and $(2, 5)$
35. $(3, 4)$ and $(2, 7)$
36. $(3, 6)$ and $(5, 2)$
37. $(0, 0)$ and $(4, 5)$
38. $(4, 3)$ and $(7, 8)$
39. $(-3, 5)$ and $(-5, 6)$
40. $(6, -2)$ and $(-3, 2)$
41. $(-2, -2)$ and $(-12, -8)$
42. $(-1, -2)$ and $(-10, -5)$
43. $(5, 7)$ and $(-4, 7)$
44. $(-1, -12)$ and $(6, -12)$
45. $(8, -4)$ and $(8, -3)$
46. $(-2, 8)$ and $(-2, 15)$
47. $(-6, 0)$ and $(0, -4)$
48. $(0, -9)$ and $(-6, 0)$
49. $(-2.5, 1.75)$ and $(-0.5, -7.75)$
50. $(6.4, -7.2)$ and $(-8.8, 4.2)$
51. $(-2.2, 18.6)$ and $(-1.7, 18.6)$
52. $(4.6, 3.2)$ and $(4.6, -4.8)$
53. $\left(-\frac{4}{7}, -\frac{1}{5}\right)$ and $\left(\frac{3}{7}, \frac{6}{5}\right)$
54. $\left(-\frac{4}{9}, -\frac{1}{8}\right)$ and $\left(\frac{5}{9}, \frac{3}{8}\right)$
55. $\left(-\frac{3}{4}, \frac{2}{3}\right)$ and $\left(\frac{4}{3}, -\frac{1}{6}\right)$
56. $\left(\frac{1}{2}, \frac{3}{4}\right)$ and $\left(-\frac{11}{16}, -\frac{1}{2}\right)$

Determine the slope of the graph of the line that has the given table of solutions. See Examples 2 and 3.

57.

x	y	(x, y)
-3	-1	$(-3, -1)$
1	2	$(1, 2)$

58.

x	y	(x, y)
-3	6	$(-3, 6)$
0	2	$(0, 2)$

59.

x	y	(x, y)
-3	6	$(-3, 6)$
0	6	$(0, 6)$

60.

x	y	(x, y)
4	-5	$(4, -5)$
4	0	$(4, 0)$

Find the slope of each line. See Examples 4 and 5.

61. $y = -11$
62. $y = -2$
63. $y = 0$
64. $x = 0$
65. $x = 6$
66. $x = 4$
67. $x = -10$
68. $y = 8$
69. $y - 9 = 0$
70. $x + 14 = 0$
71. $3x = -12$
72. $2y + 2 = -6$

Determine whether the lines through each pair of points are parallel, perpendicular, or neither. See Example 8.

73. $(5, 3)$ and $(1, 4)$
$(-3, -4)$ and $(1, -5)$

74. $(2, 4)$ and $(-1, -1)$
$(8, 0)$ and $(11, 5)$

75. $(-4, -2)$ and $(2, -3)$
$(7, 1)$ and $(8, 7)$

76. $(-2, 4)$ and $(6, -7)$
$(-6, 4)$ and $(5, 12)$

77. $(2, 2)$ and $(4, -3)$
$(-3, 4)$ and $(-1, 9)$

78. $(-1, -3)$ and $(2, 4)$
$(5, 2)$ and $(8, -5)$

79. $(-1, 8)$ and $(-6, 8)$
$(3, 3)$ and $(3, 7)$

80. $(11, 0)$ and $(11, -5)$
$(14, 6)$ and $(25, 6)$

81. $(6, 4)$ and $(2, 5)$
$(-2, -3)$ and $(2, -4)$

82. $(-3, -1)$ and $(3, -2)$
$(8, 2)$ and $(9, 8)$

83. $(4, 2)$ and $(5, -3)$
$(-5, 3)$ and $(-2, 9)$

84. $(8, -3)$ and $(8, -8)$
$(11, 3)$ and $(22, 3)$

Find the slope of a line perpendicular to the line passing through the given two points. See Example 9.

85. $(0, 0)$ and $(5, -9)$

86. $(0, 0)$ and $(5, 12)$

87. $(-1, 7)$ and $(1, 10)$

88. $(-7, 6)$ and $(0, 4)$

89. $\left(-2, \frac{1}{2}\right)$ and $\left(-1, \frac{3}{2}\right)$

90. $\left(\frac{1}{3}, -1\right)$ and $\left(\frac{4}{3}, -2\right)$

91. $(-1, 2)$ and $(-3, 6)$

92. $(5, -4)$ and $(-1, -7)$

APPLICATIONS

93. Pools. Find the slope of the bottom of the swimming pool as it drops off from the shallow end to the deep end.

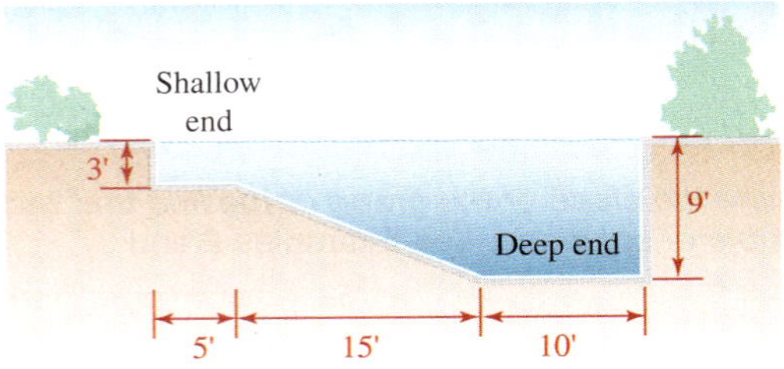

94. Drainage. Find the slope of the concrete patio slab using the 1-foot ruler, level, and 10-foot-long board shown in the illustration. (*Hint:* 10 feet = 120 in.)

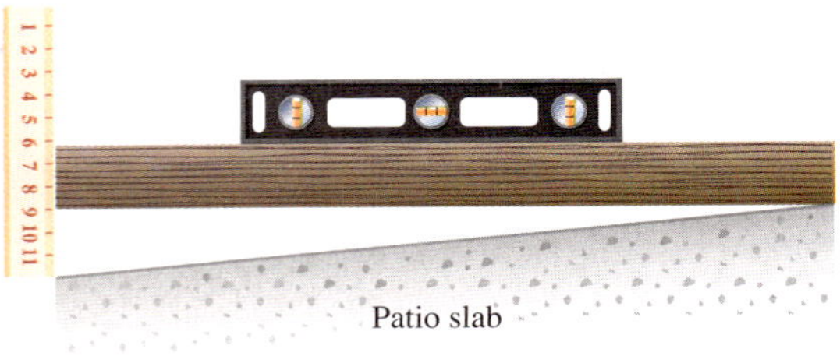

95. Grade of a Road. Refer to the illustration below. Find the slope of the decline and use that information to find the grade of the road.

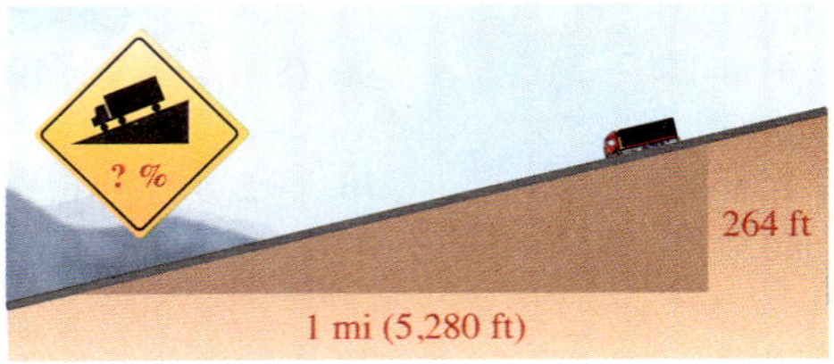

96. Streets. One of the steepest streets in the United States is Eldred Street in Highland Park, California (near Los Angeles). It rises approximately 220 feet over a horizontal distance of 665 feet. What is the grade of the street?

Vince Compagnone/©LAT

97. Treadmills. Find the slope of the jogging surface of the treadmill for a height setting of 6 inches. Then express the incline as a percent.

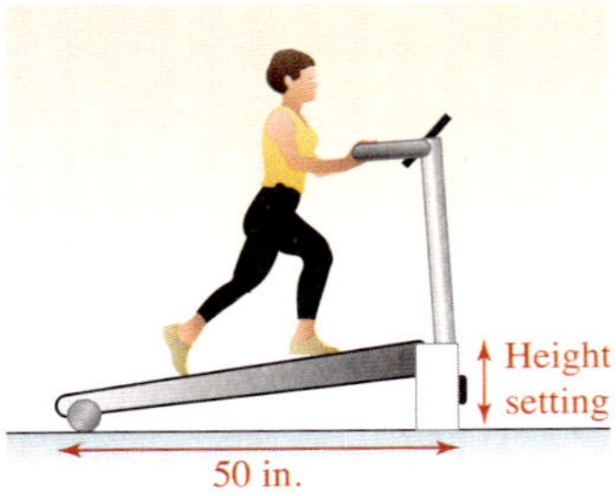

98. Architecture. Locate the coordinates of the peak of the roof if it is to have a pitch of $\frac{2}{5}$ and the roof line is to pass through the two given points in black.

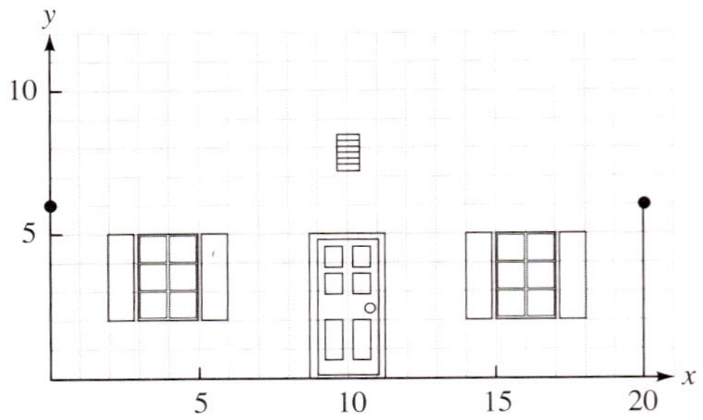

99. Carpentry. Find the pitch of each roof.

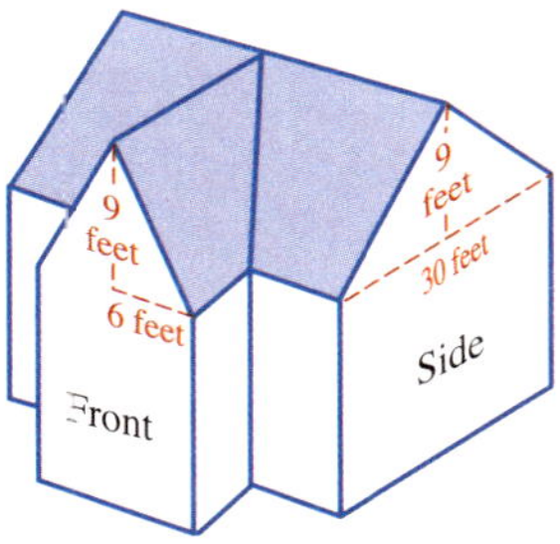

100. Doll Houses. Find x so that the pitch of the roof of the doll house is $\frac{4}{3}$.

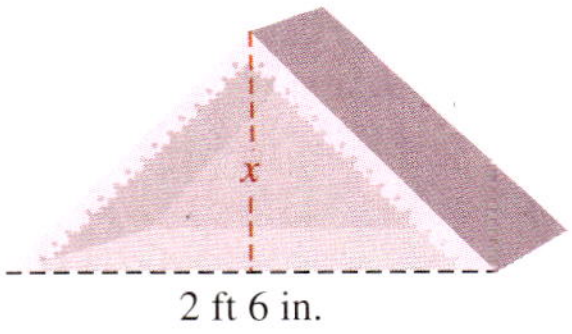

101. Irrigation. The graph on the next page shows the number of gallons of water remaining in a reservoir as water is used from it to irrigate a field. Find the rate of change in the number of gallons of water in the reservoir.

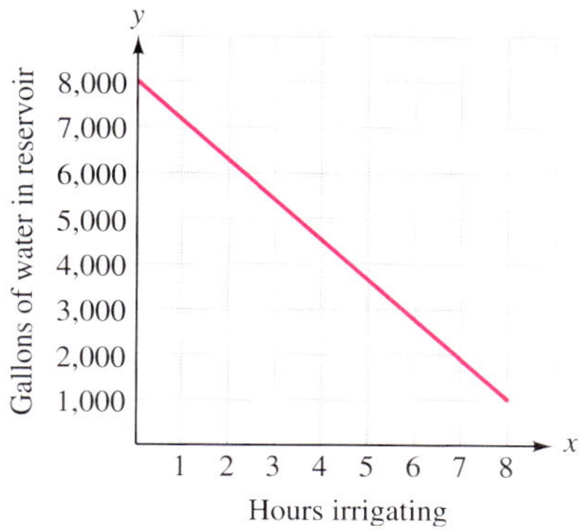

102. Commercial Jets. Examine the graph and consider trips of more than 7,000 miles by a Boeing 777. Use a rate of change to estimate how the maximum payload decreases as the distance traveled increases.

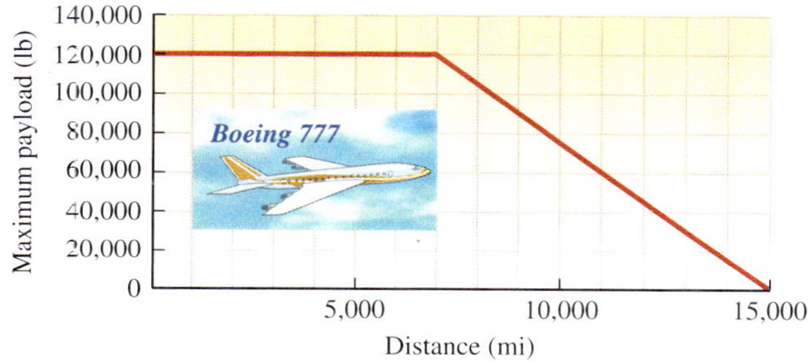

Based on data from Lawrence Livermore National Laboratory and *Los Angeles Times* (October 22, 1998).

103. Milk Production. The following graph approximates the amount of milk produced per cow in the United States for the years 1996–2009. Find the rate of change.

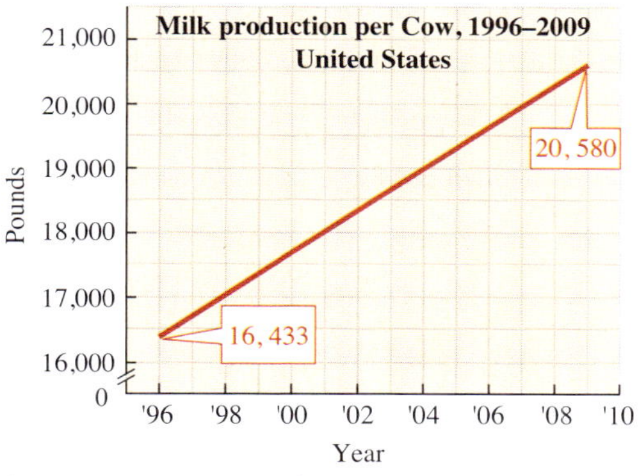

Source: USDA; Agricultural and Applied Economics, UW Madison

104. Wal-Mart. The graph in the next column approximates the sales revenue of Wal-Mart for the years 1991–2009. Find the rate of change in revenue for the years

a. 1991–1998

b. 1998–2009

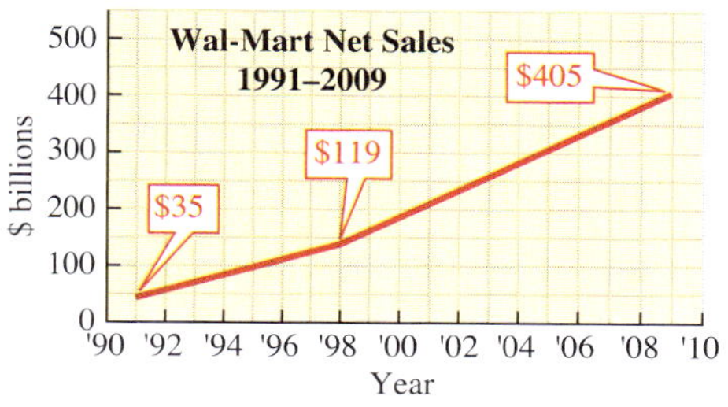

Source: wikinvest.com

105. from **Campus to Careers**

Dental Assistant

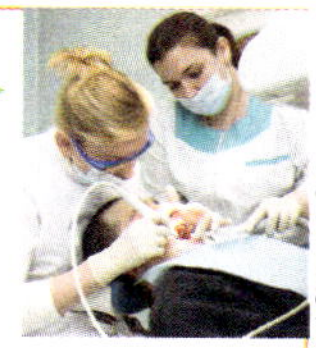

©Yuri Shirokov/Shutterstock.com

In 2000, there were approximately 6,600 students enrolled in dental assisting programs in the U.S. By 2008, that number had steadily increased to about 9,200 students. Find the rate of change in the number of students studying to be dental assistants from 2000 to 2008. (Source: American Dental Education Association)

106. Firefighting. When flames are tilted due to effects of wind, firefighters measure what is called the **slope percent** of the flames. Calculate the slope percent of the flame shown below by expressing its "slope" as a percent. (Source math.fire.org)

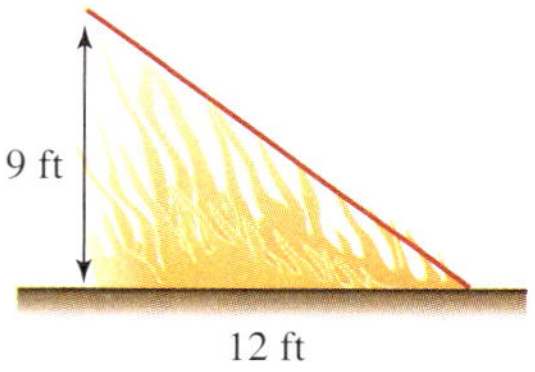

WRITING

107. Explain why the slope of a vertical line is undefined.

108. How do we distinguish between a line with positive slope and a line with negative slope?

109. Explain the error in the following solution: *Find the slope of the line that passes through* (6, 4) *and* (3, 1).

$$m = \frac{1-4}{6-3} = \frac{-3}{3} = -1$$

110. Explain the difference between a rate of change that is positive and one that is negative. Give an example of each.

REVIEW

111. Halloween Candy. A candy maker wants to make a 60-pound mixture of two candies to sell for $2 per pound. If black licorice bits sell for $1.90 per pound and orange gumdrops sell for $2.20 per pound, how many pounds of each should be used?

112. Medications. A doctor prescribes an ointment that is 2% hydrocortisone. A pharmacist has 1% and 5% concentrations in stock. How many ounces of each should the pharmacist use to make a 1-ounce tube?

CHALLENGE PROBLEMS

113. Use the concept of slope to determine whether $A(-50, -10)$, $B(20, 0)$, and $C(34, 2)$ all lie on the same straight line.

114. A line having slope $\frac{2}{3}$ passes through the point $(10, -12)$. What is the y-coordinate of another point on the line whose x-coordinate is 16?

115. Subscripts are used in other disciplines besides mathematics. In what disciplines are the following symbols used?
   a. H_2O and CO_2
   b. C_7 and G_7
   c. B_6 and B_{12}

116. Evaluate $2a_2^2 + 3a_3^3 + 4a_4^4$ for $a_2 = 2$, $a_3 = 3$, and $a_4 = 4$.

SECTION 3.5 Slope–Intercept Form

OBJECTIVES

1. Use slope–intercept form to identify the slope and y-intercept of a line.
2. Write a linear equation in slope–intercept form.
3. Write an equation of a line given its slope and y-intercept.
4. Use the slope and y-intercept to graph a linear equation.
5. Recognize parallel and perpendicular lines.
6. Use slope–intercept form to write an equation to model data.

ARE YOU READY?

The following problems review some basic skills that are needed when working with equations of lines in slope–intercept form.

1. a. Identify each term in the expression $3x - 6$.
   b. What is the coefficient of the first term?
2. Solve for y: $2x + 5y = 15$
3. True or false: $\frac{x}{4} = \frac{1}{4}x$
4. Write 3 as a fraction.
5. On what axis does the point $(0, 6)$ lie?
6. True or false: $-\frac{7}{8} = \frac{-7}{8} = \frac{7}{-8}$

Of all of the ways in which a linear equation can be written, one form, called *slope–intercept form,* is probably the most useful. When an equation is written in this form, two important features of its graph are evident.

1 Use Slope–Intercept Form to Identify the Slope and y-Intercept of a Line.

To explore the relationship between a linear equation and its graph, let's consider $y = 2x + 1$. To graph this equation, three values of x were selected (-1, 0, and 1), the corresponding values of y were found, and the results were entered in the table. Then the ordered pairs were plotted and a straight line was drawn through them, as shown below.

$y = 2x + 1$

x	y	(x, y)
-1	-1	$(-1, -1)$
0	1	$(0, 1)$
1	3	$(1, 3)$

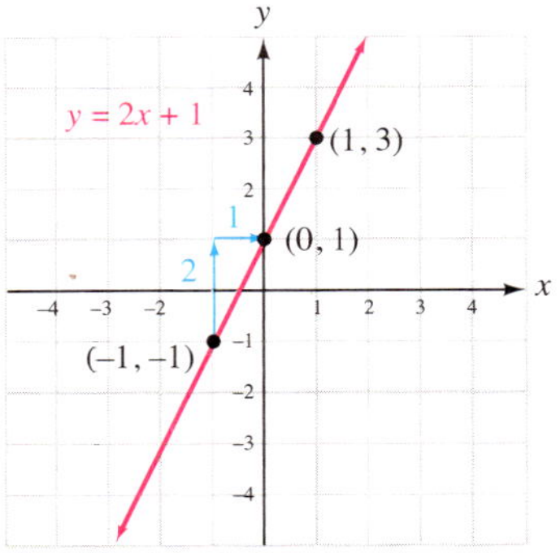

To find the slope of the line, we pick two points on the line, $(-1, -1)$ and $(0, 1)$, and draw a slope triangle and count grid squares:

$$\text{Slope} = \frac{\text{rise}}{\text{run}} = \frac{2}{1} = 2$$

From the equation and the graph, we can make two observations:

- The graph crosses the y-axis at 1. This is the same as the constant term in $y = 2x + 1$.
- The slope of the line is 2. This is the same as the coefficient of x in $y = 2x + 1$.

The Language of Algebra

Recall that a term that contains only a number is called a **constant term.**

This illustrates that the slope and y-intercept of the graph of $y = 2x + 1$ can be determined from the equation.

$$y = 2x + 1$$

The slope of the line is 2. The y-intercept is (0, 1).

These observations suggest the following form of an equation of a line.

Slope–Intercept Form of the Equation of a Line

If a linear equation is written in the form

$$y = mx + b$$

the graph of the equation is a line with slope m and y-intercept $(0, b)$.

When an equation of a line is written in slope–intercept form, the coefficient of the x-term is the line's slope and the constant term gives the y-coordinate of the y-intercept.

$$y = mx + b$$

Slope $\quad$ y-intercept: $(0, b)$

Caution

For equations in $y = mx + b$ form, the slope of the line is the *coefficient* of x, not the x-term. For the graph of $y = 6x - 2$:

$m = 6$
not
$m = 6x$

Linear equation	*Equation written in slope–intercept form*	*Slope*	*y-intercept*
$y = 6x - 2$	$y = 6x + (-2)$	6	$(0, -2)$
$y = -\frac{5}{4}x$	$y = -\frac{5}{4}x + 0$	$-\frac{5}{4}$	$(0, 0)$
$y = \frac{x}{2} + 3$	$y = \frac{1}{2}x + 3$	$\frac{1}{2}$	$(0, 3)$
$y = -\frac{7}{8} - x$	$y = -x + \left(-\frac{7}{8}\right)$	-1	$\left(0, -\frac{7}{8}\right)$

2 Write a Linear Equation in Slope–Intercept Form.

The equation of any nonvertical line can be written in slope–intercept form.

Slope-Intercept form

To write a linear equation in two variables in slope–intercept form, solve the equation for y.

EXAMPLE 1

Find the slope and y-intercept of the line with the given equation.

a. $8x + y = 9$ $\quad$ **b.** $x + 4y = 16$ $\quad$ **c.** $-9x - 3y = 11$

Strategy We will write each equation in slope–intercept form by solving for y.

Why When the equations are written in slope–intercept form, the slope and y-intercept of their graphs become apparent.

Solution

The Language of Algebra

When working with the slope–intercept form, use precise language: b is not the y-intercept of the graph of the equation $8x + y = 9$. b is the *y-coordinate of the y-intercept.*

a. The slope and y-intercept of the graph of $8x + y = 9$ are not obvious because the equation is not in slope–intercept form. To write it in $y = mx + b$ form, we isolate y.

$8x + y = 9$ — This is the given equation.

$8x + y - 8x = -8x + 9$ — To isolate y on the left side, subtract $8x$ from both sides. Since we want the right side of the equation to have the form $mx + b$, we show the subtraction from that side as $-8x + 9$ rather than $9 - 8x$.

$y = -8x + 9$ — On the left side, combine like terms: $8x - 8x = 0$.

Since $m = -8$ and $b = 9$, the slope is -8 and the y-intercept is $(0, 9)$.

b. To write the equation in slope–intercept form, we solve for y.

$x + 4y = 16$ — This is the given equation.

$x + 4y - x = -x + 16$ — To isolate the term $4y$ on the left side, subtract x from both sides. Write the subtraction before the constant term 16.

$4y = -x + 16$ — On the left side. combine like terms: $x - x = 0$.

$y = \frac{-x}{4} + \frac{16}{4}$ — To isolate y, undo the multiplication by 4 by dividing both sides by 4, term-by-term.

$y = -\frac{1}{4}x + 4$ — Write $\frac{-x}{4}$ as $-\frac{1}{4}x$. Simplify: $\frac{16}{4} = 4$.

Since $m = -\frac{1}{4}$ and $b = 4$, the slope is $-\frac{1}{4}$ and the y-intercept is $(0, 4)$.

Success Tip

Since we want the right side of the equation to have the form $mx + b$, we show the division by 4 as

$-\frac{x}{4} + \frac{16}{4}$ not $\frac{-x + 16}{4}$

c. To write the equation in $y = mx + b$ form, we isolate y on the left side.

$-9x - 3y = 11$ — This is the given equation.

$-3y = 9x + 11$ — To isolate the term $-3y$ on the left side, add $9x$ to both sides. Write the addition before the constant term 11.

$\frac{-3y}{-3} = \frac{9x}{-3} + \frac{11}{-3}$ — To isolate y, undo the multiplication by -3 by dividing both sides by -3, term-by-term.

$y = -3x - \frac{11}{3}$ — Simplify.

Since $m = -3$ and $b = -\frac{11}{3}$, the slope is -3 and the y-intercept is $\left(0, -\frac{11}{3}\right)$.

Self Check 1 Find the slope and y-intercept of the line with the given equation. **a.** $9x + y = 4$ **b.** $-x + 11y = -22$ **c.** $-10x - 2y = 7$

Now Try Problems 11, 31, and 39

3 Write an Equation of a Line Given Its Slope and *y*-Intercept.

If we are given the slope and y-intercept of a line, we can write an equation of the line by substituting for m and b in the slope–intercept form.

EXAMPLE 2 Write an equation of the line with slope -1 and y-intercept $(0, 9)$.

Strategy We will use the slope–intercept form, $y = mx + b$, to write an equation of the line.

Why We know the slope of the line and its y-intercept.

Solution If the slope is -1 and the y-intercept is $(0, 9)$, then $m = -1$ and $b = 9$.

$y = mx + b$ This is the slope–intercept form.

$y = -1x + 9$ Substitute -1 for m and 9 for b.

$y = -x + 9$ Simplify: $-1x = -x$.

The equation of the line with slope -1 and y-intercept $(0, 9)$ is $y = -x + 9$.

Self Check 2 Write an equation of the line with slope 1 and y-intercept $(0, -12)$.

Now Try Problem 43

EXAMPLE 3 Write an equation of the line graphed in figure (a).

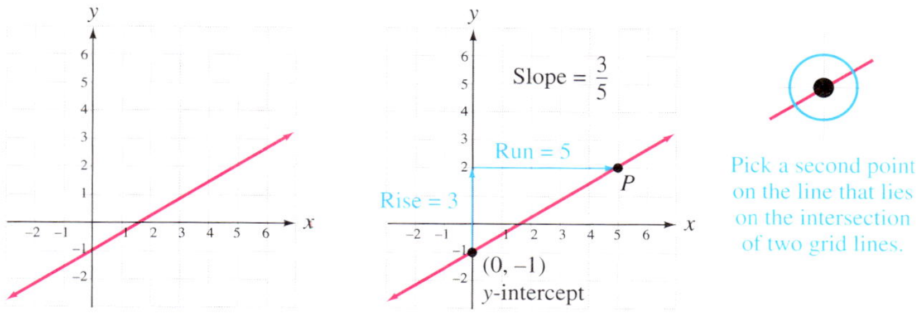

(a) (b)

Strategy We will use the slope–intercept form, $y = mx + b$, to write an equation of the line.

Why We can determine the slope and y-intercept of the line from the given graph.

Solution In figure (b), we highlight and label the y-intercept of the line, $(0, -1)$. Then we pick a convenient second point on the line and label it point P. Moving from the y-intercept to point P, we draw a slope triangle to find that the slope of the line is $\frac{3}{5}$. When we substitute $\frac{3}{5}$ for m and -1 for b into the slope–intercept form $y = mx + b$, we obtain an equation of the line: $y = \frac{3}{5}x - 1$.

Self Check 3 Write an equation of the line graphed here.

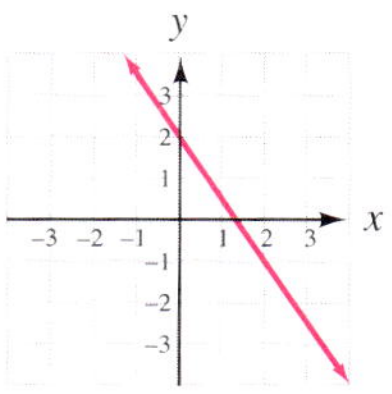

Now Try Problem 55

4 Use the Slope and y-Intercept to Graph a Linear Equation.

If we know the slope and y-intercept of a line, we can graph the line.

EXAMPLE 4 Use the slope and y-intercept to graph $y = 5x - 4$.

Strategy We will examine the equation to identify the slope and the y-intercept of the line to be graphed. Then we will plot the y-intercept and use the slope to determine a second point on the line.

Why Once we locate two points on the line, we can draw the graph of the line.

Solution Since $y = 5x - 4$ is written in $y = mx + b$ form, we know that its graph is a line with a slope of 5 and a y-intercept of $(0, -4)$. To draw the graph, we begin by plotting the y-intercept. The slope can be used to find another point on the line.

If we write the slope as the fraction $\frac{5}{1}$, the rise is 5 and the run is 1. From $(0, -4)$, we move 5 units *upward* (because the numerator, 5, is positive) and 1 unit to the right (because the denominator, 1, is positive). This locates a second point on the line, $(1, 1)$. The line through $(0, -4)$ and $(1, 1)$ is the graph of $y = 5x - 4$.

Notation

Slopes that are integers can be written in $\frac{\text{rise}}{\text{run}}$ form with the run equal to 1. For example, $m = 5 = \frac{5}{1}$.

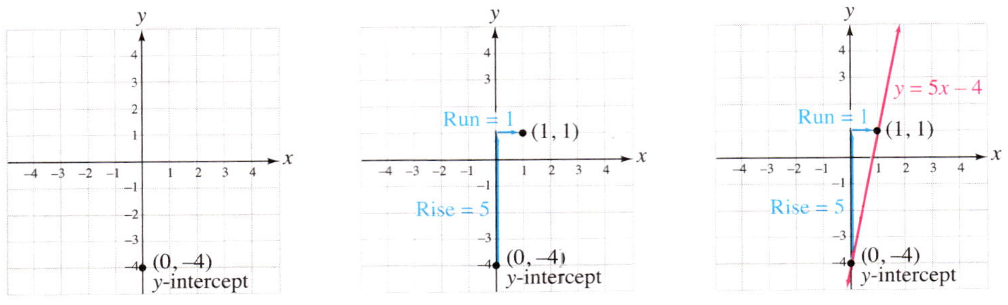

Plot the y-intercept.

From $(0, -4)$, draw the rise and run parts of the slope triangle for $m = \frac{5}{1}$ to find another point on the line.

Use a straightedge to draw a line through the two points.

Caution

When using the y-intercept and the slope to graph a line, remember to draw the slope triangle from the y-intercept, **not the origin.**

Since $\frac{5}{1} = \frac{-5}{-1}$, an alternate way to find another point on the line is to write the slope as $\frac{-5}{-1}$. Like before, we begin at the y-intercept $(0, -4)$. Since the rise is negative, we move 5 units *downward,* and since the run is negative, we then move 1 unit to the *left.* We arrive at $(-1, -9)$, another point on the graph of $y = 5x - 4$.

Self Check 4 Use the slope and y-intercept to graph $y = 2x - 3$.

Now Try Problem 63

EXAMPLE 5 Use the slope and y-intercept to graph $4x + 3y = 6$.

Strategy We will write the equation of the line in slope–intercept form, $y = mx + b$. Then we will identify the slope and y-intercept of its graph.

Why We can use that information to plot two points that the line passes through.

Solution To write $4x + 3y = 6$ in slope–intercept form, we isolate y on the left side.

$$4x + 3y = 6 \quad \text{This is the given equation.}$$

$$3y = -4x + 6 \quad \text{To isolate the term } 3y \text{ on the left side, subtract } 4x \text{ from both sides. Write the subtraction before the constant term 6.}$$

$$\frac{3y}{3} = \frac{-4x}{3} + \frac{6}{3} \quad \text{To isolate } y\text{, undo the multiplication by 3 by dividing both sides by 3, term-by-term.}$$

$$y = -\frac{4}{3}x + 2 \quad m = -\frac{4}{3} \text{ and } b = 2.$$

The slope of the line is $-\frac{4}{3}$ and the y-intercept is $(0, 2)$. To draw the graph, we begin by plotting the y-intercept. If we write the slope as $\frac{-4}{3}$, the rise is -4 and the run is 3. From $(0, 2)$, we then move 4 units *downward* (because the numerator is negative) and 3 units to the *right* (because the denominator is positive). This locates a second point on the line, $(3, -2)$.

Notation

The negative sign in $-\frac{4}{3}$ can be attached to either the numerator or the denominator, since $-\frac{4}{3} = \frac{-4}{3} = \frac{4}{-3}$.

Since $\frac{-4}{3} = \frac{4}{-3}$, we can find another point on the graph by writing the slope as $\frac{4}{-3}$. In this case, the rise is 4 and the run is -3. Again, we begin at the y-intercept (0, 2), but this time, we move 4 units *upward* because the rise is positive. Then we move 3 units to the *left,* because the run is negative, and arrive at the point $(-3, 6)$. The line that passes through (0, 2), $(3, -2)$, and $(-3, 6)$ is the graph of $4x + 3y = 6$.

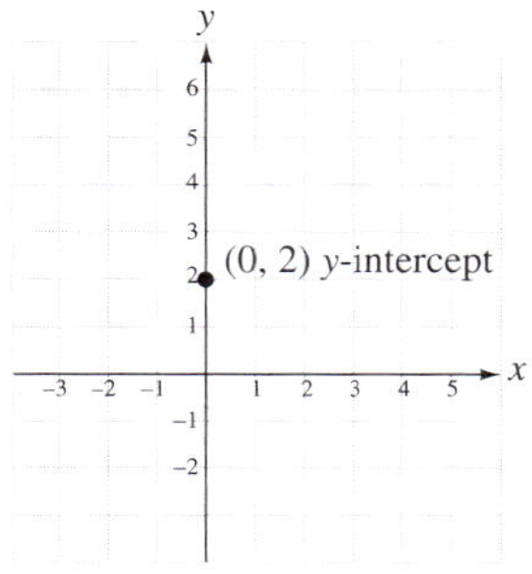

Plot the y-intercept.

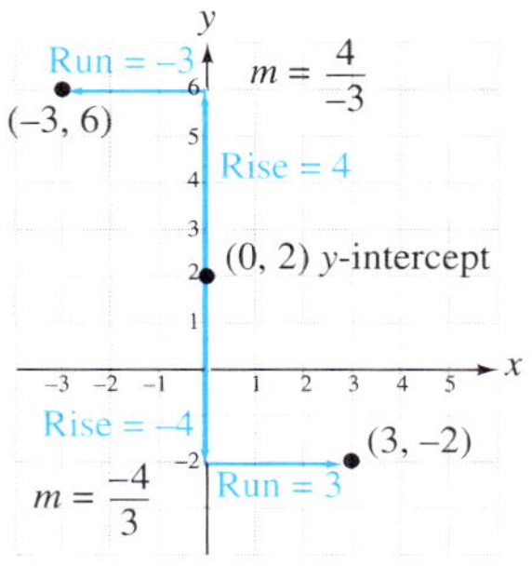

From (0, 2), draw the rise and run parts of the slope triangle for $m = \frac{-4}{3}$ $\left(\text{or } m = \frac{4}{-3}\right)$ to find another point on the line.

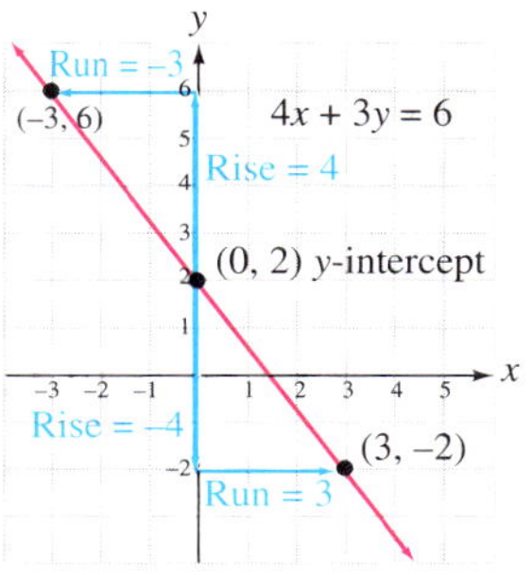

Use a straightedge to draw a line through the points.

To check the graph, verify that $(3, -2)$ and $(-3, 6)$ satisfy $4x + 3y = 6$.

Self Check 5 Use the slope and y-intercept to graph $5x + 6y = 12$.

Now Try Problem 71

5 Recognize Parallel and Perpendicular Lines.

The slope–intercept form enables us to quickly identify parallel and perpendicular lines.

EXAMPLE 6 Are the graphs of $y = -5x + 6$ and $x - 5y = -10$ parallel, perpendicular, or neither?

Strategy We will find the slope of each line and then compare the slopes.

Why If the slopes are equal, the lines are parallel. If the slopes are negative reciprocals, the lines are perpendicular. Otherwise, the lines are neither parallel nor perpendicular.

Solution The graph of $y = -5x + 6$ is a line with slope -5. To find the slope of the graph of $x - 5y = -10$, we will write the equation in slope–intercept form.

Success Tip

Graphs are not necessary to determine if two lines are parallel, perpendicular, or neither. We simply examine the slopes of the lines.

$$x - 5y = -10 \quad \text{This is the second given equation.}$$

$$-5y = -x - 10 \quad \text{To isolate the term } -5y \text{ on the left side, subtract } x \text{ from both sides.}$$

$$\frac{-5y}{-5} = \frac{-x}{-5} - \frac{10}{-5} \quad \text{To isolate } y\text{, undo the multiplication by } -5 \text{ by dividing both sides by } -5 \text{ term-by-term.}$$

$$y = \frac{x}{5} + 2 \quad m = \tfrac{1}{5} \text{ because } \tfrac{x}{5} = \tfrac{1}{5}x.$$

The graph of $y = \frac{x}{5} + 2$ is a line with slope $\frac{1}{5}$. Since the slopes -5 and $\frac{1}{5}$ are negative reciprocals, the lines are perpendicular.

Self Check 6 Determine whether the graphs of $y = 4x + 6$ and $x - 4y = -8$ are parallel, perpendicular, or neither.

Now Try Problem 81

6 Use Slope–Intercept Form to Write an Equation to Model Data.

The concepts that we have studied in the first four sections of this chapter can be used to write equations that mathematically describe, or **model,** many real-world situations. To make the equation more descriptive of the given situation, we can replace the variables x and y in $y = mx + b$ with other letters.

EXAMPLE 7

Group Discounts. To promote group sales for an Alaskan cruise, a travel agency reduces the regular ticket price of \$4,500 by \$5 for each person traveling in the group.

a. Write a linear equation that gives the per-person cost c of the cruise, if p people travel together.

b. Use the equation to determine the per-person cost if 55 teachers travel together.

Strategy We will determine the slope and the y-intercept of the graph of the equation from the given facts about the cruise.

Why If we know the slope and y-intercept, we can use the slope–intercept form, $y = mx + b$, to write an equation to model the situation.

Solution

a. We will let p represent the number of people traveling in the group and c represent the per-person cost of the cruise. Since the cost depends on the number of people in the group, the linear equation that models this situation is

$c = mp + b$ This is the slope-intercept form $y = mx + b$ with the variable c in place of y and the variable p in place of x.

Since the per-person cost of the cruise steadily decreases as the number of people in the group increases, the rate of change of $-\$5$ per person is the slope of the graph of the equation. Thus, m is -5.

If 0 people take the cruise, there will be no discount and the per-person cost of the cruise will be \$4,500. Written as an ordered pair of the form (p, c), we have $(0, 4,500)$. When graphed, this point would be the c-intercept. Thus, b is 4,500.

Substituting for m and b in the slope–intercept form $c = mp + b$, we obtain the linear equation that models the pricing arrangement.

A graph of the equation for groups of up to 100 $(c \leq 100)$ is shown on the right.

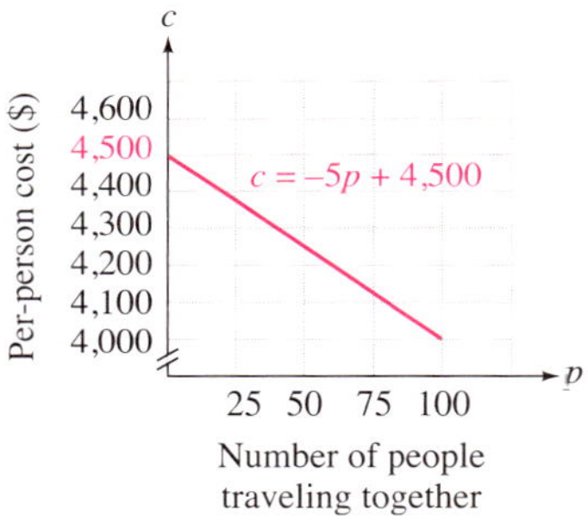

$c = -5p + 4,500$ $m = -5$ and $b = 4,500$.

The Language of Algebra

To determine the slope m for modeling problems like this, look for a phrase that describes a **rate of change,** such as:

- \$5 for each person
- \$20 per unit
- 5° every minute
- 12 feet a year

b. To find the per-person cost of the cruise for a group of 55 people, we substitute 55 for p and evaluate the right side of the equation.

$c = -5p + 4,500$

$c = -5(55) + 4,500$ Substitute 55 for p.

$c = -275 + 4,500$ Do the multiplication.

$c = 4,225$ Do the addition.

If a group of 55 people travel together, the cruise will cost each person \$4,225.

Self Check 7 **Group Discounts.** Write a linear equation in slope–intercept form that gives the cost of the cruise of Example 7 if a \$10.50-per-person discount is offered for groups.

Now Try Problem 91

SECTION 3.5 STUDY SET

VOCABULARY

Fill in the blanks.

1. The equation $y = mx + b$ is called the ______________ form of the equation of a line.
2. The graph of the linear equation $y = mx + b$ has __________ $(0, b)$ and ______ m.

CONCEPTS

3. Determine whether each equation is in slope–intercept form.
   a. $7x + 4y = 2$ b. $5y = 2x - 3$
   c. $y = 6x + 1$ d. $x = 4y - 8$
4. a. Fill in the blank: To write a linear equation in two variables in slope–intercept form, solve the equation for ___ .
   b. Solve $4x + y = 9$ for y.
5. Simplify the right side of each equation.
   a. $y = \frac{4x}{2} + \frac{16}{2}$ b. $y = \frac{15x}{-3} + \frac{9}{-3}$
   c. $y = \frac{2x}{6} - \frac{6}{6}$ d. $y = \frac{-9x}{-5} - \frac{20}{-5}$
6. Find the slope and y-intercept of each line graphed below. Then use that information to write an equation for that line.

a.

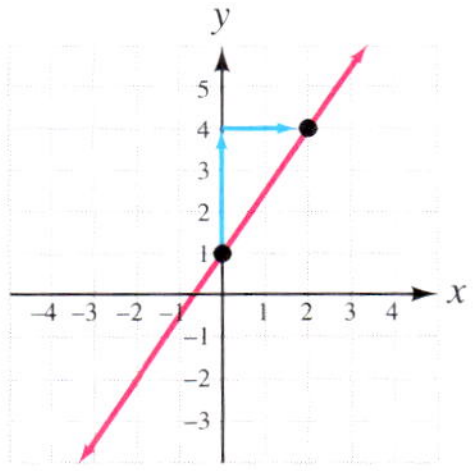

b.

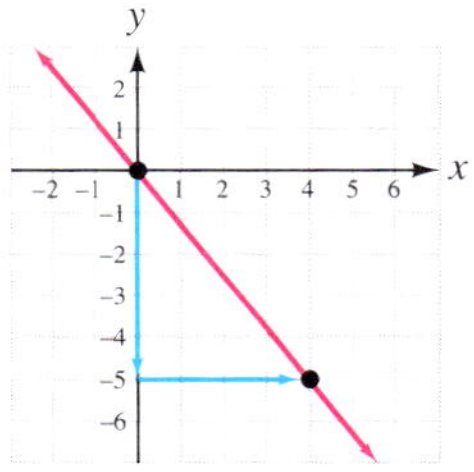

NOTATION

Complete the solution by solving the equation for y. Then find the slope and the y-intercept of its graph.

7. $2x + 5y = 15$

$2x + 5y - 2x = \underline{\quad} + 15$

$\underline{\quad} = -2x + 15$

$\frac{5y}{\underline{\quad}} = \frac{-2x}{\underline{\quad}} + \frac{15}{\underline{\quad}}$

$y = \underline{\quad}x + \underline{\quad}$

The slope is ___ and the y-intercept is (___ , ___).

8. What is the slope–intercept form of the equation of a line?
9. Fill in the blanks: $-\frac{3}{2} = \frac{3}{\underline{\quad}} = \frac{\underline{\quad}}{2}$

10. Determine whether each statement is true or false.
    a. $\frac{x}{6} = \frac{1}{6}x$ b. $\frac{5}{3}x = \frac{5x}{3}$

GUIDED PRACTICE

Find the slope and the y-intercept of the line with the given equation. **See Example 1.**

11. $y = 4x + 2$
12. $y = 7x + 3$
13. $y = -5x - 8$
14. $y = -4x - 2$
15. $y = 25x - 9$
16. $y = 6x - 1$
17. $y = 11 - x$
18. $y = 12 - 4x$
19. $y = \frac{1}{2}x + 6$
20. $y = \frac{4}{5}x - 9$
21. $y = \frac{x}{4} - \frac{1}{2}$
22. $y = \frac{x}{15} - \frac{3}{4}$
23. $y = -5x$
24. $y = 14x$
25. $y = x$
26. $y = -x$
27. $y = -2$
28. $y = 30$
29. $-5y - 2 = 0$
30. $3y - 13 = 0$
31. $x + y = 8$
32. $x - y = -30$
33. $6y = x - 6$
34. $2y = x + 20$
35. $-4y = 6x - 4$
36. $-6y = 8x + 6$
37. $2x + 3y = 6$
38. $4x + 5y = 25$
39. $3x - 5y = 15$
40. $x - 6y = 6$
41. $-6x + 6y = -11$
42. $-4x + 4y = -9$

Write an equation of the line with the given slope and y-intercept and graph it. **See Example 2.**

43. Slope 5, y-intercept $(0, -3)$
44. Slope -2, y-intercept $(0, 1)$
45. Slope -3, y-intercept $(0, 6)$
46. Slope 4, y-intercept $(0, -1)$
47. Slope $\frac{1}{4}$, y-intercept $(0, -2)$
48. Slope $\frac{1}{3}$, y-intercept $(0, -5)$
49. Slope $-\frac{8}{3}$, y-intercept $(0, 5)$
50. Slope $-\frac{7}{6}$, y-intercept $(0, 2)$
51. Slope $\frac{6}{5}$, y-intercept $(0, 0)$
52. Slope $\frac{5}{4}$, y-intercept $(0, 0)$
53. Slope -2, y-intercept $\left(0, \frac{1}{2}\right)$
54. Slope -3, y-intercept $\left(0, -\frac{1}{2}\right)$

Write an equation for each line shown below. **See Example 3.**

55.

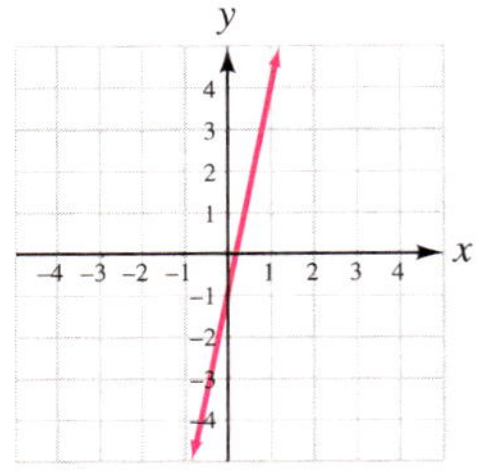

56.

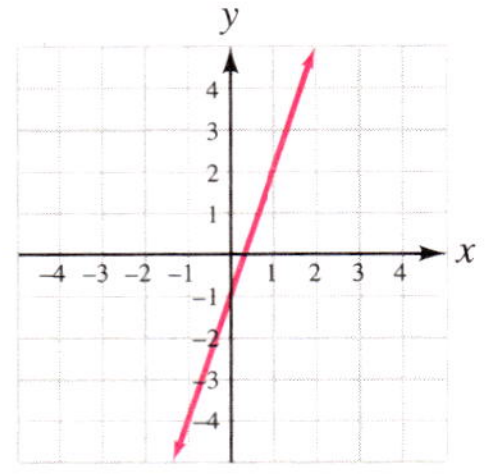

57.

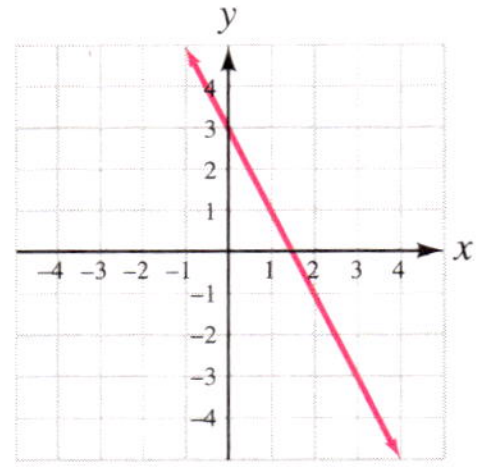

58.

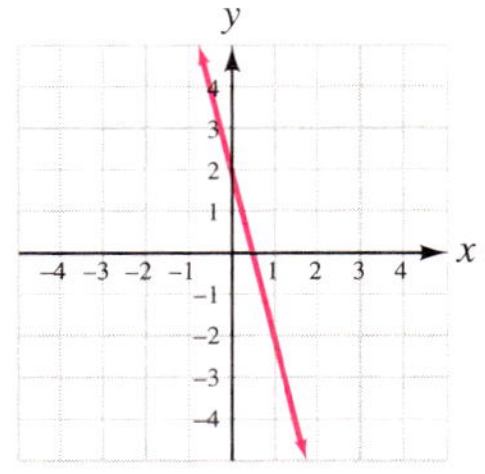

59.

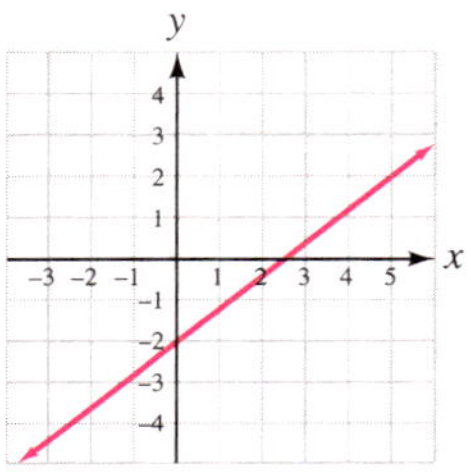

60.

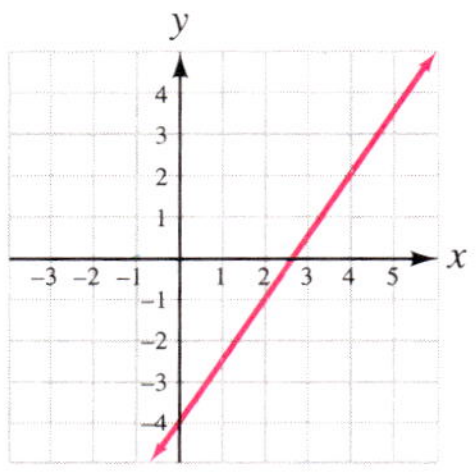

61.

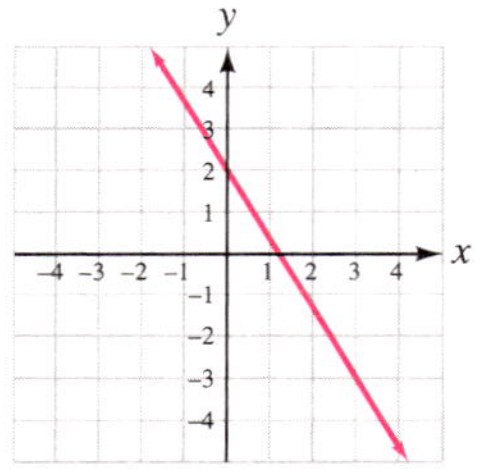

62.

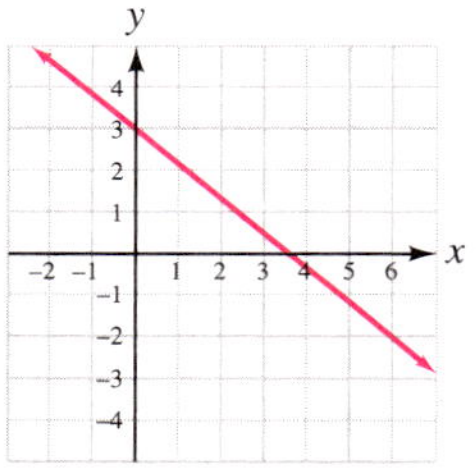

Find the slope and the y-intercept of the graph of each equation and graph it. **See Examples 4 and 5.**

63. $y = 3x + 3$

64. $y = -3x + 5$

65. $y = \frac{1}{2}x + 2$

66. $y = \frac{x}{3}$

67. $y = -3x$

68. $y = -4x$

69. $4x + y = -4$

70. $2x + y = -6$

71. $3x + 4y = 16$

72. $2x + 3y = 9$

73. $10x - 5y = 5$

74. $4x - 2y = 6$

For each pair of equations, determine whether their graphs are parallel, perpendicular, or neither. **See Example 6.**

75. $y = 6x + 8$
$y = 6x$

76. $y = -9x - 3$
$y = -9x$

77. $y = x$
$y = -x$

78. $y = 3x$
$y = 4x$

79. $y = \frac{1}{2}x - \frac{4}{5}$
$y = 0.5x + 3$

80. $y = 3x - 15$
$y = -\frac{1}{3}x + 4$

81. $y = -2x - 9$
$2x - y = 9$

82. $y = \frac{3}{4}x + 1$
$4x - 3y = 15$

83. $3x = 5y - 10$
$5x = 1 - 3y$

84. $-2y = 2 - x$
$2x - 3 = 4y$

85. $x - y = 12$
$-2x + 2y = -23$

86. $y = -3x + 1$
$3y = x - 5$

87. $x = 9$
$y = 8$

88. $-x + 4y = 10$
$2y + 16 = -8x$

89. $-4x + 3y = -12$
$8x + 6y = 54$

90. $-5x + 2y = -8$
$15x + 6y = -48$

APPLICATIONS

91. Production Costs. A television production company charges a basic fee of \$5,000 and then \$2,000 an hour when filming a commercial.

a. Write a linear equation that describes the relationship between the total production costs c and the hours h of filming.

b. Use your answer to part a to find the production costs if a commercial required 8 hours of filming.

92. College Fees. Each semester, students enrolling at a community college must pay tuition costs of \$20 per unit as well as a \$40 student services fee.

a. Write a linear equation that gives the total fees t to be paid by a student enrolling at the college and taking x units.

b. Use your answer to part a to find the enrollment cost for a student taking 12 units.

93. Chemistry. A portion of a student's chemistry lab manual is shown on the right. Use the information to write a linear equation relating the temperature F in degrees Fahrenheit of the compound to the time t (in minutes) elapsed during the lab procedure.

> Chem. Lab #1 Aug. 13
> **Step 1:** Removed freezer compound from freezer @ –10°F.
>
> **Step 2:** Used heating unit to raise temperature of compound 5° F every minute.

94. Rentals. Use the information in the newspaper advertisement to write a linear equation that gives the amount of income A (in dollars) the apartment owner will receive when the unit is rented for m months.

> **APARTMENT FOR RENT**
> 1 bedroom/1 bath, with garage
> \$500 per month +
> \$250 nonrefundable one-time security fee.

95. Employment Services. A policy statement of LIZCO, Inc., is shown below. Suppose a secretary had to pay an employment service \$500 to get placed in a new job at LIZCO. Write a linear equation that tells the secretary the actual cost c of the employment service to her m months after being hired.

Policy no. 23452—A new hire will be reimbursed by LIZCO for any employment service fees paid by the employee at the rate of \$20 per month.

96. Printing Presses. Every three minutes, 100 feet of paper is used off of an 8,000 foot-roll to print the pages of a magazine. Write a linear equation that relates the number of feet of paper that remain on the roll and the number of minutes the printing press has been operating.

97. Sewing Costs. A tailor charges a basic fee of \$20 plus \$5 per letter to sew an athlete's name on the back of a jacket. Write a linear equation that will find the cost to have a name containing x letters sewn on the back of a jacket.

98. Salad Bars. For lunch, a delicatessen offers a "Salad and Soda" special where customers serve themselves at a well-stocked salad bar. The cost is \$2.00 for the drink and 42¢ an ounce for the salad. Write a linear equation that will find the cost of a "Salad and Soda" lunch when a salad weighing x ounces is purchased.

99. iPads. When a student purchased an Apple iPad with Wi-Fi + 3G for \$629.99, he also enrolled in a 250 MB data plan that cost \$14.95 per month.

a. Write a linear equation that gives the cost for him to purchase and use the iPad for m months.

b. Use your answer to part a to find the cost to purchase and use the iPad for 2 years.

100. Online Games. A new Playstation 3 costs \$310.50 and membership in an online videogame multiplayer network cost \$18.49 per month.

a. Write a linear equation that gives the cost for someone to buy the machine and belong to the online network for m months.

b. Use your answer to part a to find the cost to buy the machine and belong to the network for 3 years.

101. Navigation. The graph below shows the recommended speed at which a ship should proceed into head waves of various heights.

a. What information does the y-intercept of the line give?

b. What is the rate of change in the recommended speed of the ship as the wave height increases?

c. Write an equation of the line.

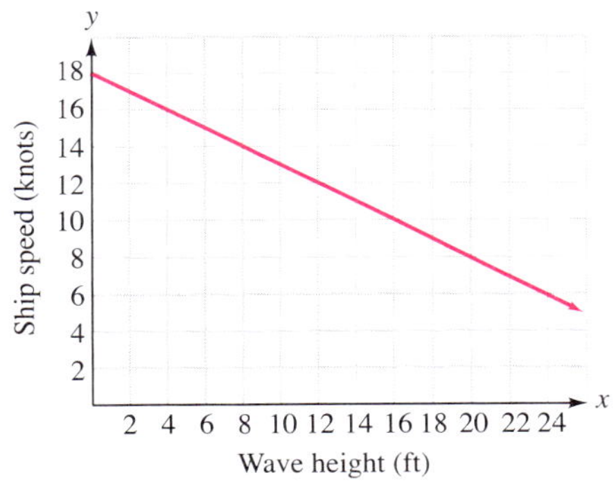

102. Debt. The graph below estimates the amount of credit card debt per U.S. household for the years 1990–2010.

a. What information does the d-intercept of the line give?

b. What was the rate of change in the amount of debt per household?

c. Write an equation of the line, where d is the approximate credit card debt and t is the number of years since 1990.

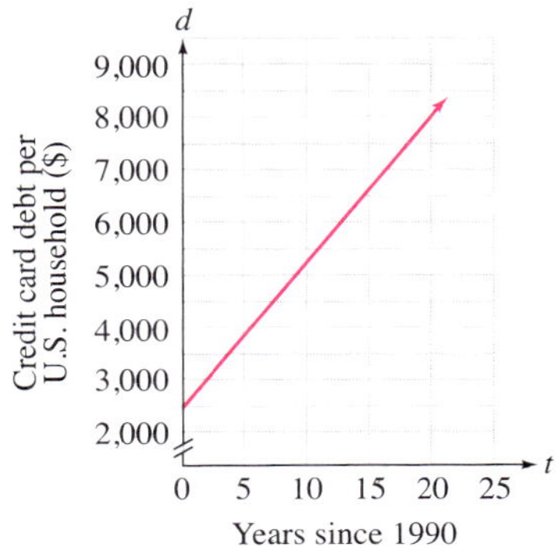

Source: mybudget 360.com

WRITING

103. Why is $y = mx + b$ called the slope–intercept form of the equation of a line?

104. On a quiz, a student was asked to find the slope of the graph of $y = 2x + 3$. She answered: $m = 2x$. Her instructor marked it wrong. Explain why the answer is incorrect.

REVIEW

105. Cable TV. A 186-foot television cable is to be cut into four pieces. Find the length of each piece if each successive piece is 3 feet longer than the previous one.

106. Investments. Joni received \$25,000 as part of a settlement in a class action lawsuit. She invested some money at 10% and the rest at a 9% simple interest rate. If her total annual income from these two investments was \$2,430, how much did she invest at each rate?

CHALLENGE PROBLEMS

107. If the graph of $y = mx + b$ passes through quadrants I, II, and IV, what can be known about the constants m and b?

108. Which of the following equations has the steeper graph, $103x - 200y = -400$ or $17x - 33y = -66$?

SECTION 3.6

Point–Slope Form

OBJECTIVES

1. Use point–slope form to write an equation of a line.
2. Write an equation of a line given two points on the line.
3. Write equations of horizontal and vertical lines.
4. Use a point and the slope to graph a line.
5. Write linear equations that model data.

ARE YOU READY?

The following problems review some basic skills that are needed when working with equations of lines in point–slope form.

1. Find the slope of the line that passes through $(-2, 0)$ and $(-12, -8)$.
2. Simplify: $x - (-5)$
3. Solve $y + 2 = 6(x - 7)$ for y.
4. Add: $-\frac{3}{4} + 8$

If we know the slope of a line and its y-intercept, we can use the slope–intercept form to write the equation of the line. The question that now arises is, can *any* point on the line be used in combination with its slope to write its equation? In this section, we answer this question.

1 Use Point–Slope Form to Write an Equation of a Line.

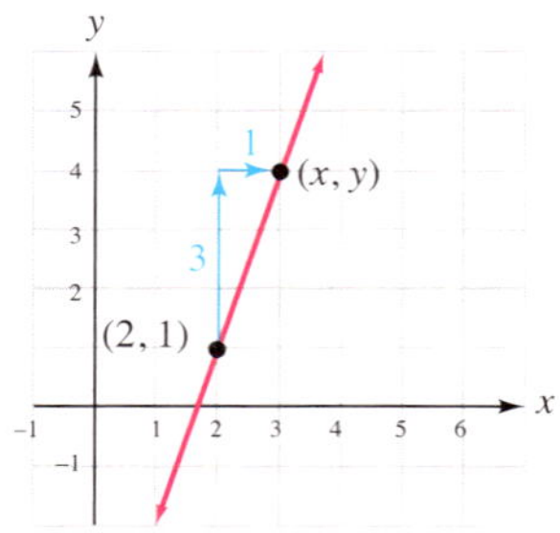

Refer to the line graphed on the left, with slope 3 and passing through the point (2,1). To develop a new form for the equation of a line, we will find the slope of this line in another way.

If we pick another point on the line with coordinates (x, y), we can find the slope of the line by substituting the coordinates of the points (x, y) and $(2, 1)$ into the slope formula.

$$\frac{y_2 - y_1}{x_2 - x_1} = m$$

$$\frac{y - 1}{x - 2} = m \quad \text{Let } (x_1, y_1) \text{ be } (2, 1) \text{ and } (x_2, y_2) \text{ be } (x, y).$$

Substitute y for y_2, 1 for y_1, x for x_2, and 2 for x_1.

Since the slope of the line is 3, we can substitute 3 for m in the previous equation.

$$\frac{y - 1}{x - 2} = 3$$

We then multiply both sides by $x - 2$ to clear the equation of the fraction.

$$\frac{y - 1}{x - 2}(x - 2) = 3(x - 2)$$

$$y - 1 = 3(x - 2) \quad \text{Simplify the left side. Remove the common factor } x - 2$$

in the numerator and denominator: $\frac{y-1}{\cancel{x-2}_1} \cdot \frac{\cancel{x-2}^1}{1}$.

The resulting equation displays the slope of the line and the coordinates of one point on the line:

Slope of the line

$$y - 1 = 3(x - 2)$$

y-coordinate of the point (1); x-coordinate of the point (2)

In general, suppose we know that the slope of a line is m and that the line passes through the point (x_1, y_1). Then if (x, y) is any other point on the line, we can use the definition of slope to write

$$\frac{y - y_1}{x - x_1} = m$$

If we multiply both sides by $x - x_1$ to clear the equation of the fraction, we have

$$y - y_1 = m(x - x_1)$$

This form of a linear equation is called **point–slope form.** It can be used to write the equation of a line when the slope and one point on the line are known.

Point–Slope Form of the Equation of a Line

If a line with slope m passes through the point (x_1, y_1), the equation of the line is

$$y - y_1 = m(x - x_1)$$ Read as "y minus y sub 1 equals m times the quantity x minus x sub 1."

EXAMPLE 1 Find an equation of a line that has slope -8 and passes through $(-1, 5)$. Write the answer in slope–intercept form.

Strategy Although the problem asks for an answer in slope–intercept form, we will begin with the point–slope form to write an equation of the line.

Why We know the slope of the line and the coordinates of a point that it passes through.

Solution The given point is $(-1, 5)$, so $x_1 = -1$ and $y_1 = 5$. The given slope is -8, so $m = -8$. We will subsitute these values into the point–slope form and simplify the right side of the equation.

Caution

When using the point–slope form, never substitute values for x or y.

$$y - y_1 = m(x - x_1)$$

Only substitute values for x_1, y_1, and m.

$y - y_1 = m(x - x_1)$ This is the point–slope form.

$y - 5 = -8[x - (-1)]$ Substitute -8 for m, -1 for x_1, and 5 for y_1. Brackets are used to enclose $x - (-1)$.

$y - 5 = -8(x + 1)$ Simplify the expression within the brackets.

To write an equivalent equation in slope–intercept form, we solve for y.

$y - 5 = -8(x + 1)$ This is the simplified point–slope form.

$y - 5 = -8x - 8$ Distribute the multiplication by -8.

$y - 5 + 5 = -8x - 8 + 5$ To isolate y, undo the subtraction of 5 by adding 5 to both sides.

$y = -8x - 3$ This is the requested slope–intercept form.

In slope–intercept form, the equation is $y = -8x - 3$.

To verify this result, we note that $m = -8$. Therefore, the slope of the line is -8, as required. To see whether the line passes through $(-1, 5)$, we substitute -1 for x and 5 for y in the equation. If this point is on the line, a true statement should result.

$$y = -8x - 3$$
$$5 \stackrel{?}{=} -8(-1) - 3$$
$$5 \stackrel{?}{=} 8 - 3$$
$$5 = 5 \quad \text{True}$$

Self Check 1 Find an equation of the line that has slope -2 and passes through $(4, -3)$. Write the answer in slope–intercept form.

Now Try Problems 13 and 19

2 Write an Equation of a Line Given Two Points on the Line.

In the next example, we show that it is possible to write the equation of a line when we know the coordinates of two points on the line.

EXAMPLE 2 Find an equation of the line that passes through $(-2, 6)$ and $(4, 7)$. Write the equation in slope–intercept form.

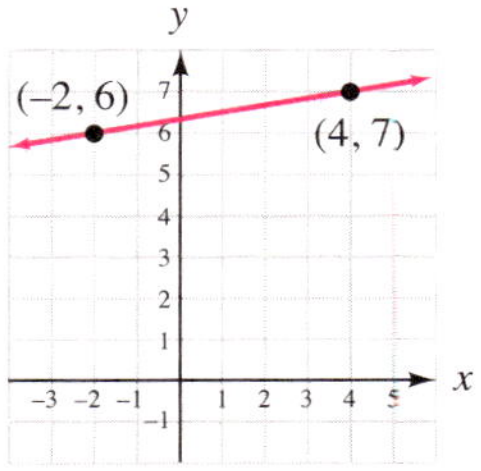

Strategy We will use the point–slope form, $y - y_1 = m(x - x_1)$, to write an equation of the line.

Why We know the coordinates of a point that the line passes through and we can calculate the slope of the line using the slope formula.

Solution To find the slope of the line, we use the slope formula.

$$m = \frac{y_2 - y_1}{x_2 - x_1} = \frac{7 - 6}{4 - (-2)} = \frac{1}{6}$$ Substitute 7 for y_2, 6 for y_1, 4 for x_2, and -2 for x_1.

Either point on the line can serve as (x_1, y_1). If we use $(4, 7)$, we have

$$y - y_1 = m(x - x_1)$$ This is the point–slope form.

$$y - 7 = \frac{1}{6}(x - 4)$$ Substitute $\frac{1}{6}$ for m, 7 for y_1, and 4 for x_1.

To write an equivalent equation in slope–intercept form, we solve for y.

$$y - 7 = \frac{1}{6}x - \frac{2}{3}$$ Distribute the multiplication by $\frac{1}{6}$: $\frac{1}{6}(-4) = -\frac{4}{6} = -\frac{2}{3}$.

$$y - 7 + 7 = \frac{1}{6}x - \frac{2}{3} + 7$$ To isolate y, add 7 to both sides.

$$y = \frac{1}{6}x - \frac{2}{3} + \frac{21}{3}$$ Simplify the left side. On the right side, express 7 as $\frac{21}{3}$ to prepare to add the fractions with the common denominator 3.

$$y = \frac{1}{6}x + \frac{19}{3}$$ Add the fractions: $-\frac{2}{3} + \frac{21}{3} = \frac{19}{3}$. This is slope–intercept form.

An equation of the line that passes through $(-2, 6)$ and $(4, 7)$ is $y = \frac{1}{6}x + \frac{19}{3}$.

Success Tip

In Example 2, either of the given points can be used as (x_1, y_1) when writing the point–slope equation. The results will be the same.

Looking ahead, we usually choose the point whose coordinates will make the calculations the easiest.

Success Tip

To check this result, verify that $(-2, 6)$ and $(4, 7)$ satisfy $y = \frac{1}{6}x + \frac{19}{3}$ using substitution.

Self Check 2 Find an equation of the line that passes through $(-5, 4)$ and $(8, -6)$. Write the equation in slope–intercept form.

Now Try Problem 29

3 Write Equations of Horizontal and Vertical Lines.

We have previously graphed horizontal and vertical lines. We will now discuss how to write their equations.

EXAMPLE 3 Write an equation of each line and graph it. **a.** A horizontal line that passes through $(-2, -4)$ **b.** A vertical line that passes through $(1, 3)$

Strategy We will use the appropriate form, either $y = b$ or $x = a$, to write an equation of each line.

Why These are the standard forms for the equations of a horizontal and a vertical line.

Solution

a. The equation of a horizontal line can be written in the form $y = b$. Since the y-coordinate of $(-2, -4)$ is -4, the equation of the line is $y = -4$. The graph is shown in the figure.

b. The equation of a vertical line can be written in the form $x = a$. Since the x-coordinate of $(1, 3)$ is 1, the equation of the line is $x = 1$. The graph is shown in the figure.

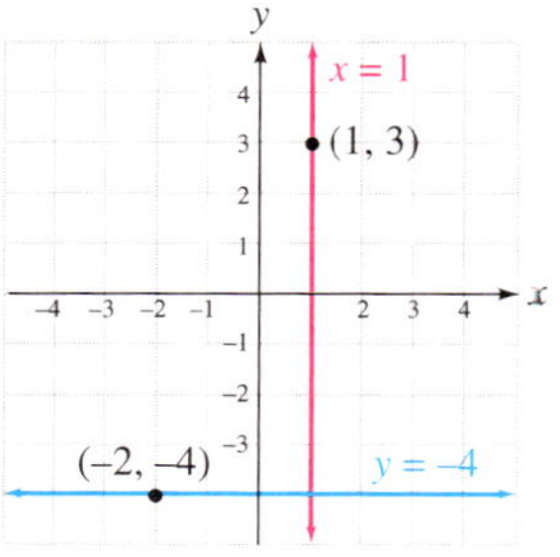

Self Check 3 Write an equation of each line. **a.** A horizontal line that passes through (3, 2) **b.** A vertical line passing through (−1, −3)

Now Try Problems 41 and 43

4 Use a Point and the Slope to Graph a Line.

If we know the coordinates of a point on a line, and if we know the slope of the line, we can use the slope to determine a second point on the line.

EXAMPLE 4 Graph the line with slope $\frac{2}{5}$ that passes through $(-1, -3)$.

Strategy First, we will plot the given point $(-1, -3)$. Then we will use the slope to find a second point that the line passes through.

Why Once we determine two points that the line passes through, we can draw the graph of the line.

Solution To draw the graph, we begin by plotting the point $(-1, -3)$. From there, we move 2 units up (rise) and then 5 units to the right (run), since the slope is $\frac{2}{5}$. This locates a second point on the line, $(4, -1)$. We then draw a straight line through the two points.

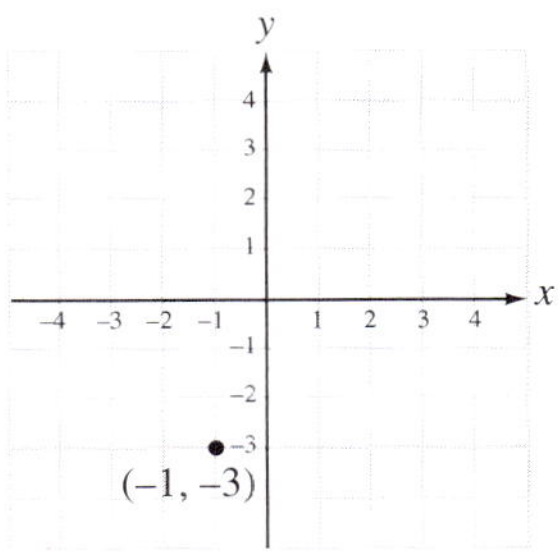

Plot the given point (–1, –3).

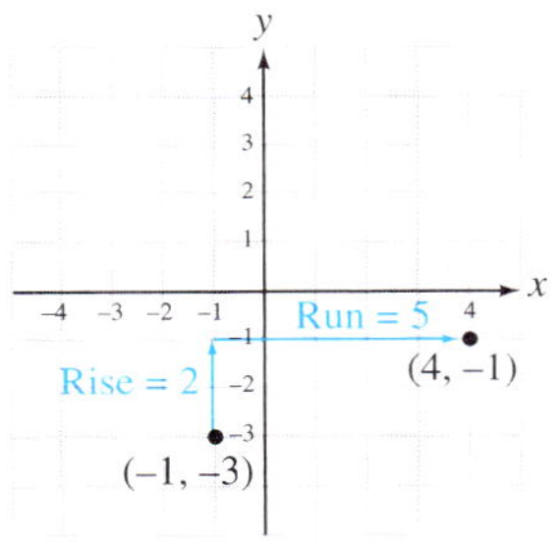

From (–1, –3) draw the rise and run parts of the slope triangle for $m = \frac{2}{5}$ to find another point on the line.

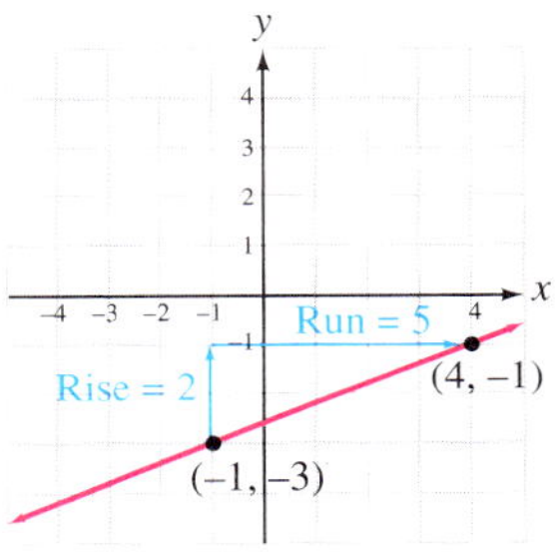

Use a straightedge to draw a line through the points.

Self Check 4 Graph the line with slope −4 that passes through (−4, 2).

Now Try Problem 45

The following table summarizes what you should know about each form of an equation of a line.

Form	Example	Comments
Standard form $Ax + By = C$	$2x + 5y = 9$	■ To graph, find the x- and y-intercepts by letting $y = 0$ and finding x, and letting $x = 0$ and finding y. Also find a checkpoint.
Slope–intercept form $y = mx + b$	$y = \frac{5}{3}x + 4$	■ To graph, plot the y-intercept $(0, b)$. From there, draw a slope triangle using the rise and run to locate another point. ■ Use this form to write a line's equation if you know its slope and y-intercept.
Point–slope form $y - y_1 = m(x - x_1)$	$y - 1 = 6(x - 8)$	■ To graph, plot the point (x_1, y_1). From there, draw a slope triangle using the rise and run of the slope m to locate another point. ■ Use this form to write a line's equation if you know a point on the line and the slope. If two points on the line are known, find the slope and use it with one of the points to write the equation.
Horizontal line $y = b$	$y = 7$	■ To graph, draw a horizontal line $(m = 0)$ with y-intercept $(0, b)$.
Vertical line $x = a$	$x = -1$	■ To graph, draw a vertical line (undefined slope) with x-intercept $(a, 0)$.

5 Write Linear Equations That Model Data.

In the next two examples, we will see how the point–slope form can be used to write linear equations that model certain real–world situations.

EXAMPLE 5 **Men's Shoe Sizes.** The length (in inches) of a man's foot is not his shoe size. For example, the smallest adult men's shoe size is 5, and it fits a 9-inch-long foot. There is, however, a linear relationship between the two. It can be stated this way: Shoe size increases by 3 sizes for each 1-inch increase in foot length.

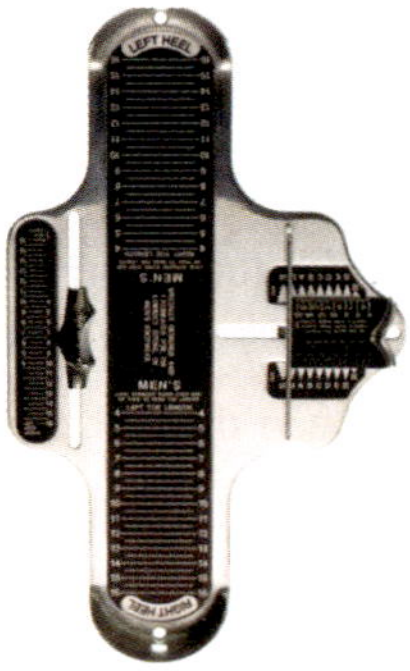

a. Write a linear equation that relates shoe size s to foot length L.

b. Shaquille O'Neal, a famous basketball player, has a foot that is about 14.6 inches long. Find his shoe size.

Strategy We will first find the slope of the line that describes the linear relationship between shoe size and the length of a foot. Then we will determine the coordinates of a point on that line.

Why Once we know the slope and the coordinates of one point on the line, we can use the point–slope form to write the equation of the line.

Solution **a.** Since shoe size s depends on the length L of the foot, ordered pairs have the form (L, s). Because the relationship is linear, the graph of the desired equation is a line.

- The line's slope is the rate of change: $\frac{3 \text{ sizes}}{1 \text{ inch}}$. Therefore, $m = 3$.
- A 9-inch-long foot wears size 5, so the line passes through $(9, 5)$.

We substitute 3 for m and the coordinates of the point into the point–slope form and solve for s.

$s - s_1 = m(L - L_1)$ This is the point–slope form using the variables L and s.

$s - 5 = 3(L - 9)$ Substitute 3 for m, 9 for L_1, and 5 for s_1.

$s - 5 = 3L - 27$ Distribute the multiplication by 3.

$s = 3L - 22$ To isolate s, add 5 to both sides: $-27 + 5 = -22$.

The equation relating men's shoe size and foot length is $s = 3L - 22$.

b. To find Shaquille's shoe size, we substitute 14.6 inches for L in the equation.

$s = 3L - 22$

$s = 3(14.6) - 22$

$s = 43.8 - 22$ Do the multiplication.

$s = 21.8$ Do the subtraction.

Since men's shoes come in only full- and half-sizes, we round 21.8 up to 22. Shaquille O'Neal wears size 22 shoes.

Self Check 5 **Comparing Temperature Scales.** Celsius and Fahrenheit measures of temperature are not the same. There is, however, a linear relationship between the two. Degrees Fahrenheit increase by 9° for each 5° increase in Celsius. If a 212° Fahrenheit temperature measure is the same as 100° Celsius, write a linear equation that relates Fahrenheit measure to Celsius measure.

Now Try Problem 77

EXAMPLE 6

The Language of Algebra

The term *scatter diagram* is somewhat misleading. Often, the data points are not scattered loosely about. In this case, they fall, more or less, along an imaginary straight line, indicating a linear relationship.

Studying Learning. In a series of 40 trials, a rat was released in a maze to search for food. Researchers recorded the trial number and the time that it took the rat to complete the maze as ordered pairs on a **scatter diagram** shown below. All of the points fell on or near the line drawn in red. Write an equation of the line in slope–intercept form.

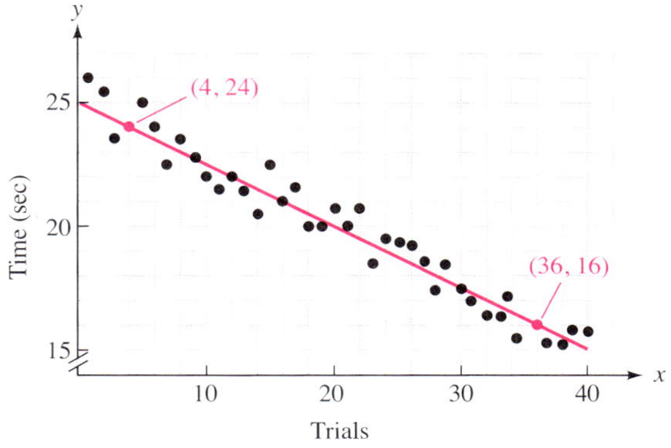

Strategy From the graph, we will determine the coordinates of two points on the line.

Why We can write an equation of a line when we know the coordinates of two points on the line. (See Example 2.)

Solution To write a point–slope equation, we need to know the slope of the line. The line passes through several points; we will use (4, 24) and (36, 16) to find its slope.

$$m = \frac{y_2 - y_1}{x_2 - x_1} = \frac{16 - 24}{36 - 4} = \frac{-8}{32} = -\frac{1}{4}$$

Any point on the line can serve as (x_1, y_1). We will use (4, 24).

$$y - y_1 = m(x - x_1) \quad \text{This is the point–slope form.}$$

$$y - 24 = -\frac{1}{4}(x - 4) \quad \text{Substitute } -\tfrac{1}{4} \text{ for } m\text{, 4 for } x_1\text{, and 24 for } y_1.$$

To write this equation in slope–intercept form, solve for y.

$$y - 24 = -\frac{1}{4}x + 1 \quad \text{Distribute the multiplication by } -\tfrac{1}{4}\text{: } -\tfrac{1}{4}(-4) = 1.$$

$$y = -\frac{1}{4}x + 25 \quad \text{To isolate } y\text{, add 24 to both sides: } 1 + 24 = 25.$$

A linear equation that models the rat's performance on the maze is $y = -\frac{1}{4}x + 25$, where x is the number of the trial and y is the time it took, in seconds.

Self Check 6 **Awards.** Orders for awards to be given to math team members were placed on two separate occasions. The first order of 32 awards cost \$172 and the second order of 5 awards cost \$37. Write a linear equation that gives the cost for an order of any number of awards.

Now Try Problem 85

SECTION 3.6 STUDY SET

VOCABULARY

Fill in the blanks.

1. $y - y_1 = m(x - x_1)$ is called the __________ form of the equation of a line. In words, we read this as y minus y _____ one equals m ______ the quantity of x _______ x sub _____.
2. $y = mx + b$ is called the _______________ form of the equation of a line.

CONCEPTS

3. Determine in what form each equation is written.
   a. $y - 4 = 2(x - 5)$
   b. $y = 2x + 15$
4. What point does the graph of each equation pass through, and what is the line's slope?
   a. $y - 2 = 6(x - 7)$
   b. $y + 3 = -8(x + 1)$
5. Refer to the following graph of a line.
   a. What highlighted point does the line pass through?
   b. What is the slope of the line?
   c. Write an equation of the line in point–slope form.

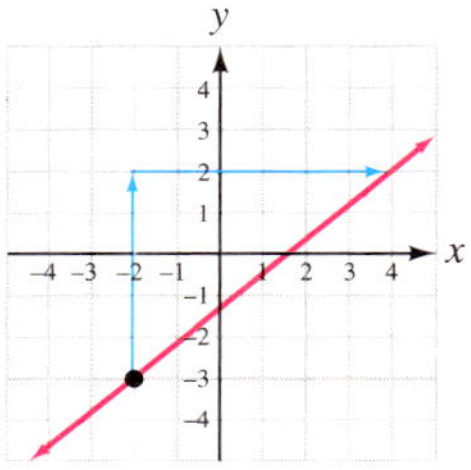

6. On a quiz, a student was asked to write the equation of a line with slope 4 that passes through $(-1, 3)$. Explain how the student can check her answer, $y = 4x + 7$.
7. Suppose you are asked to write an equation of the line in the scatter diagram below. What two points would you use to write the point–slope equation?

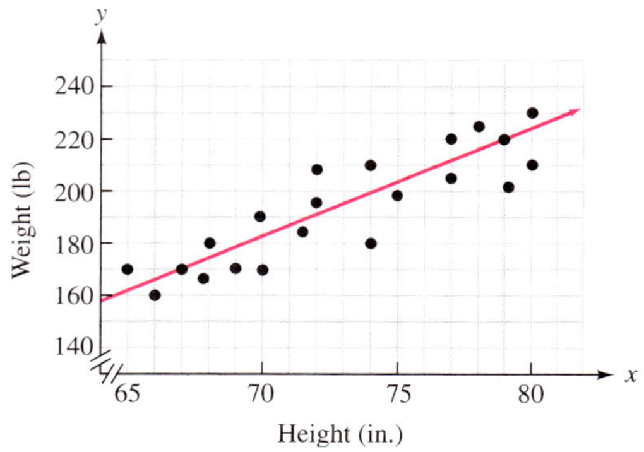

8. In each case, a linear relationship between two quantities is described. If the relationship were graphed, what would be the slope of the line?
   a. The sales of new cars increased by 15 every 2 months.
   b. There were 35 fewer robberies for each dozen police officers added to the force.
   c. One acre of forest is being destroyed every 30 seconds.

NOTATION

Complete the solution.

9. Find an equation of the line with slope -2 that passes through the point $(-1, 5)$. Write the answer in slope–intercept form.

$$y - y_1 = m(x - x_1)$$
$$y - \square = -2[x - (\square)]$$
$$y - 5 = -2[x \;\square\; 1]$$
$$y - 5 = -2x - \square$$
$$y = -2x + \square$$

10. What is the point–slope form of the equation of a line?
11. Consider the steps below and then fill in the blanks:

$$y - 3 = 2(x + 1)$$
$$y - 3 = 2x + 2$$
$$y = 2x + 5$$

The original equation was in __________ form. After solving for y, we obtain an equation in ____________ form.

12. Fill in the blanks: The equation of a horizontal line has the form $\square = b$ and the equation of a vertical line has the form $\square = a$.

GUIDED PRACTICE

Use the point–slope form to write an equation of the line with the given slope and point. Leave the equation in that form. **See Example 1.**

13. Slope 3, passes through (2, 1)
14. Slope 2, passes through (4, 3)
15. Slope $\frac{4}{5}$, passes through $(-5, -1)$
16. Slope $\frac{7}{8}$, passes through $(-2, -9)$

Use the point–slope form to find an equation of the line with the given slope and point. Then write the equation in slope–intercept form. **See Example 1.**

17. Slope 2, passes through (3, 5)
18. Slope 8, passes through (2, 6)
19. Slope -5, passes through $(-9, 8)$
20. Slope -4, passes through $(-2, 10)$
21. Slope -3, passes through the origin
22. Slope -1, passes through the origin
23. Slope $\frac{1}{5}$, passes through (10, 1)
24. Slope $\frac{1}{4}$, passes through (8, 1)
25. Slope $-\frac{4}{3}$,

x	y
6	-4

26. Slope $-\frac{3}{2}$,

x	y
-2	1

27. Slope $-\frac{11}{6}$, passes through $(2, -6)$

28. Slope $-\frac{5}{4}$, passes through $(2, 0)$

Find an equation of the line that passes through the two given points. Write the equation in slope–intercept form, if possible. See Example 2.

29. Passes through $(1, 7)$ and $(-2, 1)$

30. Passes through $(-2, 2)$ and $(2, -8)$

31.

x	y
-4	3
2	0

32.

x	y
-1	-4
1	-2

33. Passes through $(5, 5)$ and $(7, 5)$

34. Passes through $(-2, 1)$ and $(-2, 15)$

35. Passes through $(5, 1)$ and $(-5, 0)$

36. Passes through $(-3, 0)$ and $(3, 1)$

37. Passes through $(-8, 2)$ and $(-8, 17)$

38. Passes through $\left(\frac{2}{3}, 2\right)$ and $(0, 2)$

39. Passes through $\left(\frac{2}{3}, \frac{1}{3}\right)$ and $(0, 0)$

40. Passes through $\left(\frac{1}{2}, \frac{3}{4}\right)$ and $(0, 0)$

Write an equation of each line. See Example 3.

41. Vertical, passes through $(4, 5)$

42. Vertical, passes through $(-2, -5)$

43. Horizontal, passes through $(4, 5)$

44. Horizontal, passes through $(-2, -5)$

Graph the line that passes through the given point and has the given slope. See Example 4.

45. $(1, -2)$, $m = -1$

46. $(-4, 1)$, $m = -3$

47. $(5, -3)$, $m = \frac{3}{4}$

48. $(2, -4)$, $m = \frac{2}{3}$

49. $(-2, -3)$, slope 2

50. $(-3, -3)$, slope 4

51. $(4, -3)$, slope $-\frac{7}{8}$

52. $(4, 2)$, slope $-\frac{1}{5}$

TRY IT YOURSELF

Use either the slope–intercept form (from Section 3.5) or the point–slope form (from Section 3.6) to find an equation of each line. Write each result in slope–intercept form, if possible.

53. Passes through $(5, 0)$ and $(-11, -4)$

54. Passes through $(7, -3)$ and $(-5, 1)$

55. Horizontal, passes through $(-8, 12)$

56. Horizontal, passes through $(9, -32)$

57. Slope $-\frac{1}{4}$, y-intercept $\left(0, \frac{7}{8}\right)$

58. Slope $-\frac{9}{5}$, y-intercept $\left(0, \frac{11}{3}\right)$

59. Slope $-\frac{2}{3}$, passes through $(3, 0)$

60. Slope $-\frac{2}{5}$, passes through $(15, 0)$

61. Slope 8, passes through $(2, 20)$

62. Slope 6, passes through $(1, -2)$

63. Vertical, passes through $(-3, 7)$

64. Vertical, passes through $(12, -23)$

65. Slope 7 and y-intercept $(0, -11)$

66. Slope 3 and y-intercept $(0, 4)$

67. Passes through $(-2, -1)$ and $(-1, -5)$

68. Passes through $(-3, 6)$ and $(-1, -4)$

69. x-intercept $(7, 0)$ and y-intercept $(0, -2)$

70. x-intercept $(-3, 0)$ and y-intercept $(0, 7)$

71. Slope $\frac{1}{10}$, passes through the origin

72. Slope $\frac{9}{8}$, passes through the origin

73. Undefined slope, passes through $\left(-\frac{1}{8}, 12\right)$

74. Undefined slope, passes through $\left(\frac{2}{5}, -\frac{5}{6}\right)$

75. Slope 1.7, y-intercept $(0, -2.8)$

76. Slope 9.5, y-intercept $(0, -14.3)$

APPLICATIONS

77. Anatomy. There is a linear relationship between a woman's height and the length of her radius bone. It can be stated this way: Height increases by 3.9 inches for each 1-inch increase in the length of the radius. Suppose a 64-inch-tall woman has a 9-inch-long radius bone. Use this information to find a linear equation that relates height h to the length r of the radius. Write the equation in slope–intercept form.

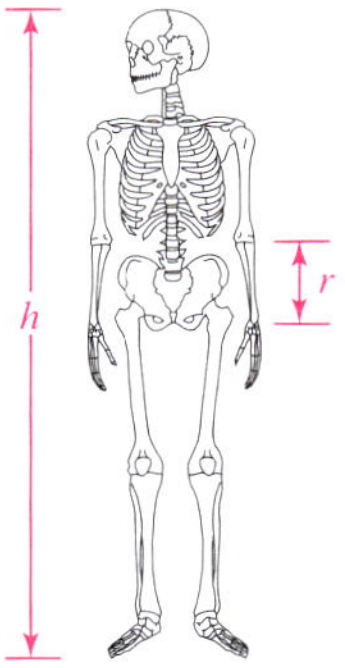

78. Automation. An automated production line uses distilled water at a rate of 300 gallons every 2 hours to make shampoo. After the line had run for 7 hours, planners noted that 2,500 gallons of distilled water remained in the storage tank. Find a linear equation relating the time t in hours since the production line began and the number g of gallons of distilled water in the storage tank. Write the equation in slope–intercept form.

79. Pole Vaulting. Find the equations of the lines that describe the positions of the pole for parts 1, 3, and 4 of the jump. Write the equations in slope–intercept form, if possible.

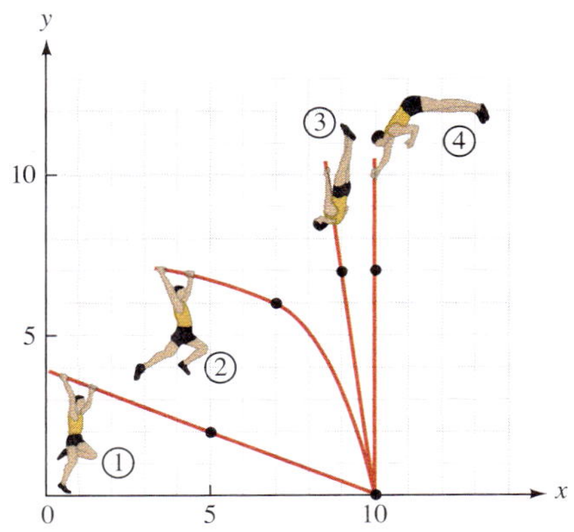

80. Freeway Design. The graph below shows the route of a proposed freeway.

a. Give the coordinates of the points where the proposed Freeway 133 will join Interstate 25 and Highway 40.

b. Find an equation of the line that describes the route of the proposed freeway. Write the equation in slope–intercept form.

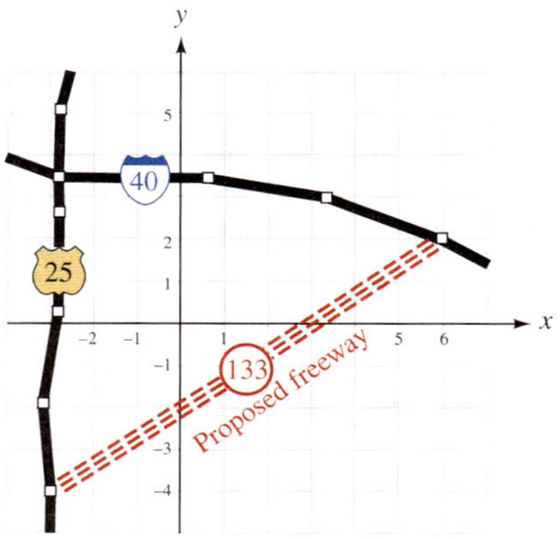

81. Toxic Cleanup. Three months after cleanup began at a dump site, 800 cubic yards of toxic waste had yet to be removed. Two months later, that number had been lowered to 720 cubic yards.

a. Find an equation that describes the linear relationship between the length of time m (in months) the cleanup crew has been working and the number of cubic yards y of toxic waste remaining. Write the equation in slope–intercept form.

b. Use your answer to part (a) to predict the number of cubic yards of waste that will still be on the site one year after the cleanup project began.

82. Depreciation. To lower its corporate income tax, accountants of a company depreciated a word processing system over several years using a linear model, as shown in the worksheet.

a. Find a linear equation relating the years since the system was purchased, x, and its value, y, in dollars. Write the equation in slope–intercept form.

b. Find the purchase price of the system.

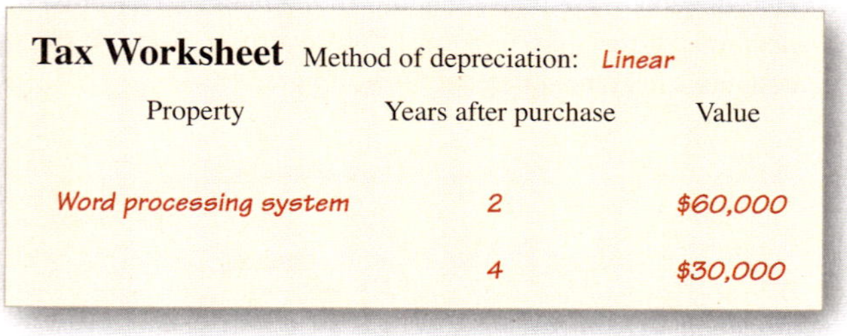
Tax Worksheet Method of depreciation: *Linear*

Property	Years after purchase	Value
Word processing system	*2*	*$60,000*
	4	*$30,000*

83. Trampolines. There is a linear relationship between the length of the protective pad that wraps around a trampoline and the radius of the trampoline. Use the data in the table to find an equation that gives the length l of pad needed for any trampoline with radius r. Write the equation in slope–intercept form. Use units of feet for both l and r.

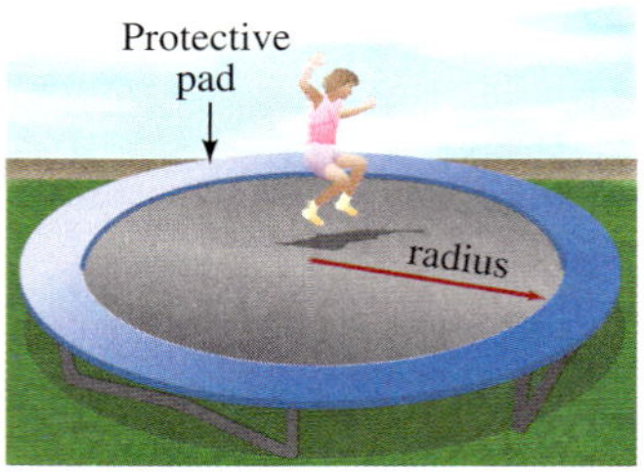

Radius	Pad length
3 ft	19 ft
7 ft	44 ft

84. Raising a Family. In the report *"Expenditures on Children by Families,"* the U.S. Department of Agriculture projected the yearly child-rearing expenditures on children from birth through age 17. For a child born in 2010 to a two-parent middle-income family, the report estimated annual expenditures of $10,808 when the child is 6 years old, and $14,570 when the child is 15 years old.

a. Write two ordered pairs of the form (child's age, annual expenditure).

b. Assume the relationship between the child's age a and the annual expenditures E is linear. Use your answers to part (a) to write an equation in slope–intercept form that models this relationship.

c. What are the projected child-rearing expenses when the child is 17 years old?

85. Got Milk? The scatter diagram shows the amount of milk that an average American drank in one year for the years 1980–2008. A straight line can be used to model the data.

a. Use the two highlighted points on the line to find its equation. Write the equation in slope–intercept form.

b. Use your answer to part (a) to predict the amount of milk that an average American will drink in 2020.

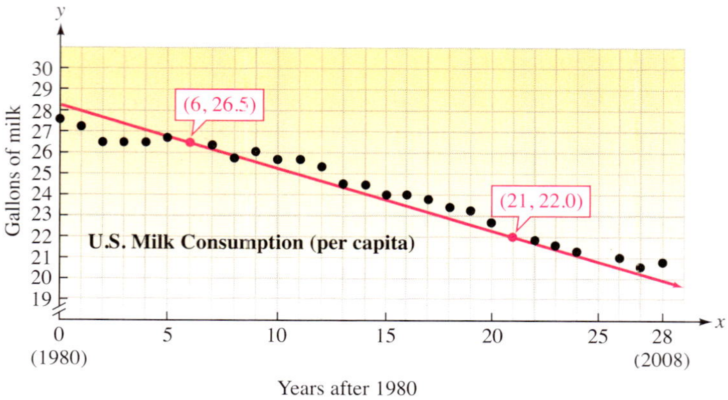

Source: United States Department of Agriculture.

86. **Engine Output.** The horsepower produced by an automobile engine was recorded for various engine speeds in the range of 2,400 to 4,800 revolutions per minute (rpm). The data were recorded on the following scatter diagram. Find an equation of the line that models the relationship between engine speed s and horsepower h. Write the equation in slope–intercept form.

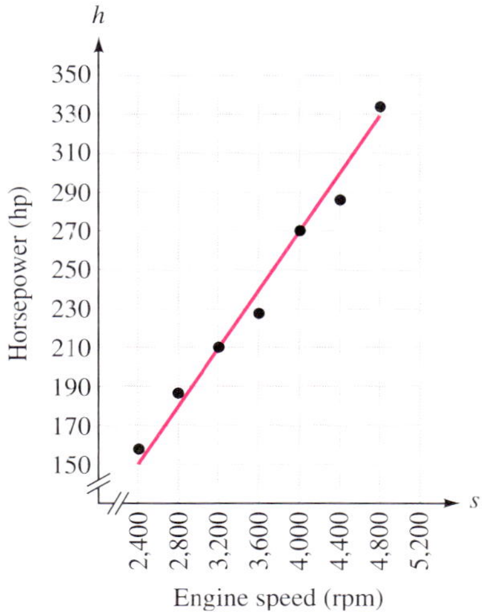

WRITING

87. Why is $y - y_1 = m(x - x_1)$ called the point–slope form of the equation of a line?
88. If we know two points that a line passes through, we can write its equation. Explain how this is done.
89. Explain the steps involved in writing $y - 6 = 4(x - 1)$ in slope–intercept form.
90. Think of the points on the graph of the horizontal line $y = 4$. What do the points have in common? How do they differ?

REVIEW

91. **Frames.** The length of a rectangular picture is 5 inches greater than twice the width. If the perimeter is 112 inches, find the dimensions of the frame.
92. **Speed of an Airplane.** Two planes are 6,000 miles apart, and their speeds differ by 200 mph. They travel toward each other and meet in 5 hours. Find the speed of the slower plane.

CHALLENGE PROBLEMS

93. Find an equation of the line that passes through (2, 5) and is parallel to the line $y = 4x - 7$. Write the equation in slope–intercept form.
94. Find an equation of the line that passes through $(-6, 3)$ and is perpendicular to the line $y = -3x - 12$. Write the equation in slope–intercept form.

SECTION 3.7 Graphing Linear Inequalities

OBJECTIVES

1. Determine whether an ordered pair is a solution of an inequality.
2. Graph a linear inequality in two variables.
3. Graph inequalities with a boundary through the origin.
4. Solve applied problems involving linear inequalities in two variables.

ARE YOU READY?

The following problems review some basic skills that are needed when graphing linear inequalities.

1. True or false: $2(-3) + 1 > -4$
2. True or false: $8 \le 8$
3. Graph: $2x - 3y = 6$
4. Determine whether each of the following points lies *above, below,* or *on* the line graphed in problem 3.

   a. $(2, -4)$ b. $(-3, -4)$ c. $(0, 0)$

Solution set for: $x + 6 < 8$

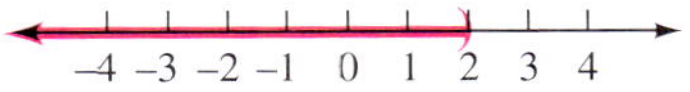

Solution set for: $5x + 3 \ge 4x$

Recall that an **inequality** is a statement that contains one of the symbols $<$, $\le$, $>$, or $\ge$. Inequalities in one variable, such as $x + 6 < 8$ and $5x + 3 \ge 4x$, were solved in Section 2.7. Because they have an infinite number of solutions, we represented their solution sets graphically, by shading intervals on a number line. Two examples of this type of shading are shown in the left margin.

We now extend that concept to linear inequalities *in two variables,* as we introduce a procedure that is used to graph their solution sets.

1 Determine Whether an Ordered Pair Is a Solution of an Inequality.

If the = symbol in a linear equation in two variables is replaced with an inequality symbol, we have a **linear inequality in two variables.**

Linear Inequalities

A linear inequality in two variables is an inequality that can be written in one of the forms

$$Ax + By > C, \quad Ax + By < C, \quad Ax + By \geq C, \quad \text{or} \quad Ax + By \leq C$$

where A, B, and C are real numbers and A and B are not both 0.

Some examples of linear inequalities in two variables are

$$x - y \leq 5, \quad 4x + 3y < -6, \quad y > 2x \quad \text{and} \quad x < -3$$

As with linear equations, an ordered pair (x, y) is a **solution of an inequality** in x and y if a true statement results when the values of the variables are substituted into the inequality.

EXAMPLE 1 Determine whether each ordered pair is a solution of $x - y \leq 5$. Then graph each solution:
a. $(4, 2)$ **b.** $(0, -6)$ **c.** $(1, -4)$

Strategy We will substitute each ordered pair of coordinates into the inequality.

Why If the resulting statement is true, the ordered pair is a solution.

Solution

Notation

The symbol $\stackrel{?}{\leq}$ is read as "is possibly less than or equal to."

a. For $(4, 2)$:

$x - y \leq 5$ This is the given inequality.

$4 - 2 \stackrel{?}{\leq} 5$ Substitute 4 for x and 2 for y.

$2 \leq 5$ True

Because $2 \leq 5$ is true, $(4, 2)$ is a solution of $x - y \leq 5$. We say that $(4, 2)$ *satisfies* the inequality. This solution is graphed as shown on the right.

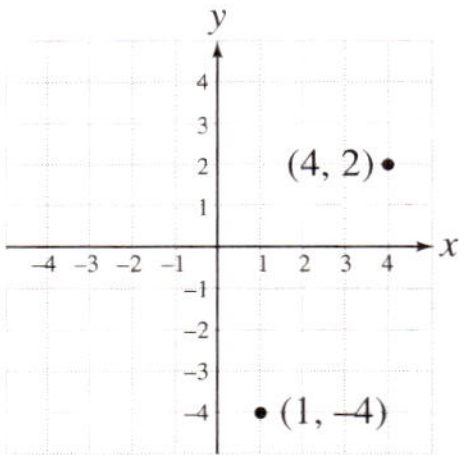

Two solutions of $x - y \leq 5$.

b. For $(0, -6)$:

$x - y \leq 5$ This is the given inequality.

$0 - (-6) \stackrel{?}{\leq} 5$ Substitute 0 for x and −6 for y.

$6 \leq 5$ False

Because $6 \leq 5$ is false, $(0, -6)$ is not a solution.

c. For $(1, -4)$:

$x - y \leq 5$ This is the given inequality.

$1 - (-4) \stackrel{?}{\leq} 5$ Substitute 1 for x and −4 for y.

$5 \leq 5$ True

Because $5 \leq 5$ is true, $(1, -4)$ is a solution, and we graph it as shown.

Self Check 1 Using the inequality in Example 1, determine whether each ordered pair is a solution: **a.** $(8, 2)$ **b.** $(4, -1)$ **c.** $(-2, 4)$ **d.** $(-3, -5)$

Now Try Problem 19

In Example 1, we graphed two of the solutions of $x - y \leq 5$. Since there are infinitely more ordered pairs (x, y) that make the inequality true, it would not be reasonable to plot all of them. Fortunately, there is an easier way to show all of the solutions.

2 Graph a Linear Inequality in Two Variables.

The graph of a linear inequality is a picture that represents the set of all points whose coordinates satisfy the inequality. In general, such graphs are regions bounded by a line. We call those regions **half-planes,** and we use a two-step procedure to find them.

EXAMPLE 2 Graph: $x - y \leq 5$

Strategy We will graph the **related equation** $x - y = 5$ to establish a boundary line between two regions of the coordinate plane. Then we will determine which region contains points whose coordinates satisfy the given inequality.

Why The graph of a linear inequality in two variables is a region of the coordinate plane on one side of a boundary line.

Solution Since the inequality symbol $\leq$ includes an equal symbol, the graph of $x - y \leq 5$ includes the graph of $x - y = 5$.

Notation

The inequality $x - y \leq 5$ means

$x - y = 5$ or $x - y < 5$

Step 1: To graph $x - y = 5$, we use the intercept method, as shown in part (a) of the illustration below. The resulting line, called a **boundary line,** divides the coordinate plane into two half-planes. To show that the points on the boundary line are solutions of $x - y \leq 5$, we draw it as a solid line.

A table of solutions to graph the boundary

$x - y = 5$

x	y	(x, y)
0	-5	$(0, -5)$
5	0	$(5, 0)$
6	1	$(6, 1)$

Let $x = 0$ and find y.
Let $y = 0$ and find x.
As a check, let $x = 6$ and find y.

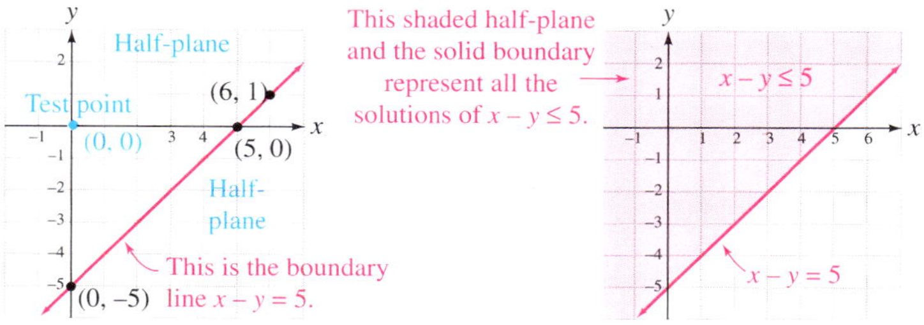

(a) (b)

Step 2: Since the inequality $x - y \leq 5$ also allows $x - y$ to be less than 5, other ordered pairs, besides those on the boundary, satisfy the inequality. For example, consider the origin, with coordinates (0, 0). If we substitute 0 for x and 0 for y in the given inequality, we have

$x - y \leq 5$ This is the given inequality.

$0 - 0 \overset{?}{\leq} 5$ Substitute.

$0 \leq 5$ True

Because $0 \leq 5$, the coordinates of the origin satisfy $x - y \leq 5$. In fact, the coordinates of every point on the same side of the boundary as the origin satisfy the inequality. To indicate this, we shade in red the half-plane that contains the test point (0, 0), as shown in part (b). Every point in the shaded half-plane and every point on the boundary line satisfies $x - y \leq 5$. On the other hand, the points in the unshaded half-plane *do not* satisfy $x - y \leq 5$.

Success Tip

All the points in the unshaded region below the boundary line have coordinates that satisfy $x - y > 5$.

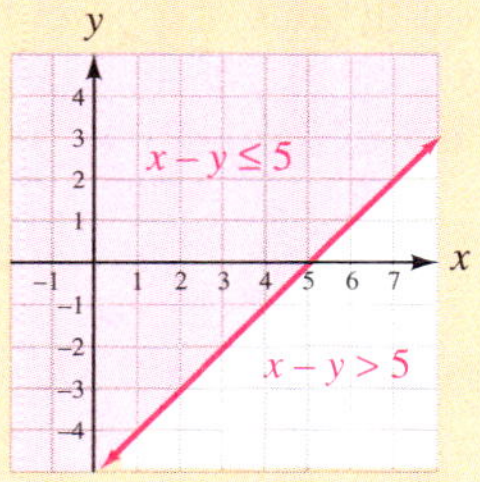

As an informal check, we can pick an ordered pair that lies in the shaded region and one that does not lie in the shaded region. When we substitute their coordinates into the inequality, we should obtain a true statement and then a false statement.

For* (3, 1)*, in the shaded region:

$x - y \leq 5$

$3 - 1 \overset{?}{\leq} 5$ Substitute.

$2 \leq 5$ True

For* (5, −4)*, not in the shaded region:

$x - y \leq 5$

$5 - (-4) \overset{?}{\leq} 5$ Substitute.

$9 \leq 5$ False

Self Check 2 Graph: $x - y \leq 2$

Now Try Problem 35

The previous example suggests the following **test-point method** to graph linear inequalities in two variables.

Graphing Linear Inequalities in Two Variables

1. Replace the inequality symbol with an equal symbol = and graph the boundary line of the region. If the original inequality allows the possibility of equality (the symbol is either $\leq$ or $\geq$), draw the boundary line as a solid line. If equality is not allowed ($<$ or $>$), draw the boundary line as a dashed line.
2. Pick a test point that is on one side of the boundary line. (Use the origin if possible.) Replace x and y in the inequality with the coordinates of that point. If a true statement results, shade the side that contains that point. If a false statement results, shade the other side of the boundary.

EXAMPLE 3 Graph: $4x + 3y < -6$

Strategy We will graph the related equation $4x + 3y = -6$ to establish the boundary line between two regions of the coordinate plane. Then we will determine which region contains points that satisfy the given inequality.

Why The graph of a linear inequality in two variables is a region of the coordinate plane on one side of a boundary line.

Solution To find the boundary line, we replace the inequality symbol with an equal symbol = and graph $4x + 3y = -6$ using the intercept method. Since the inequality symbol $<$ does not include an equal symbol, the points on the graph of $4x + 3y = -6$ will not be part of the graph of $4x + 3y < -6$. To show this, we draw the boundary line as a dashed line. See part (a) of the illustration below.

A table of solutions to graph the boundary

$4x + 3y = -6$

x	y	(x, y)
0	-2	$(0, -2)$
$-\frac{3}{2}$	0	$\left(-\frac{3}{2}, 0\right)$
-3	2	$(-3, 2)$

Let $x = 0$ and find y.
Let $y = 0$ and find x.
As a check, let $x = -3$ and find y.

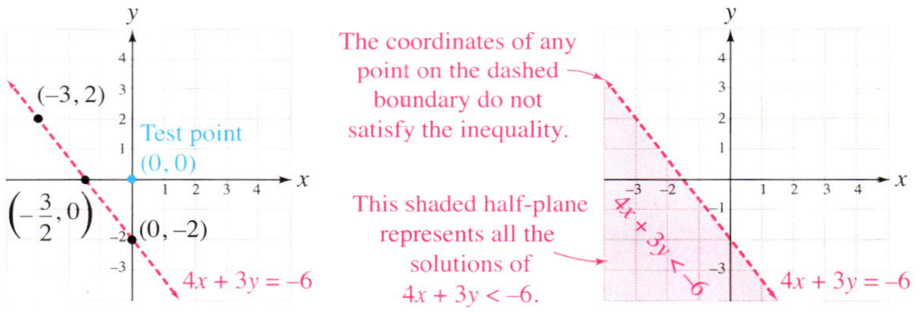

(a) (b)

To determine which half-plane to shade, we substitute the coordinates of a point that lies on one side of the boundary line into $4x + 3y < -6$. We choose the origin (0, 0) as the test point because the calculations are easy when they involve 0. We substitute 0 for x and 0 for y in the inequality.

Caution

When using a test point to determine which half-plane to shade, remember to substitute the coordinates into the given inequality, not the equation for the boundary.

~~$4(0) + 3(0) = -6$~~

$4x + 3y < -6$ This is the given inequality.

$4(0) + 3(0) \overset{?}{<} -6$ The symbol $\overset{?}{<}$ is read as "is possibly less than."

$0 + 0 \overset{?}{<} -6$

$0 < -6$ False

Since $0 < -6$ is a false statement, the point (0, 0) does not satisfy the inequality. In fact, no point in the half-plane containing (0, 0) is a solution. Therefore, we shade the other side of the boundary line—the half-plane that does not contain (0, 0). The graph of the solution set of $4x + 3y < -6$ is the half-plane below the dashed line, as shown in part (b).

Self Check 3 Graph: $5x + 6y < -15$

Now Try Problem 37

3 Graph Inequalities with a Boundary through the Origin.

In the next example, the boundary line passes through the origin. In such cases, the ordered pair (0, 0) should not be used as a test point to determine which half-plane to shade.

EXAMPLE 4 Graph: $y > 2x$

Strategy We will graph the related equation $y = 2x$ to establish the boundary line between two regions of the coordinate plane. Then we will determine which region contains points that satisfy the given inequality.

Why The graph of a linear inequality in two variables is a region of the coordinate plane on one side of a boundary line.

Success Tip

Draw a solid boundary line if the inequality has ≤ or ≥. Draw a dashed line if the inequality has < or >.

Solution To find the boundary line, we graph $y = 2x$. Since the symbol > does *not* include an equal symbol, the points on the graph of $y = 2x$ are not part of the graph of $y > 2x$. Therefore, the boundary line should be dashed, as shown in part (a) of the illustration below.

A table of solutions to graph the boundary

$y = 2x$

x	y	(x, y)
0	0	(0, 0)
−1	−2	(−1, −2)
1	2	(1, 2)

Select three values for x and find the corresponding values of y.

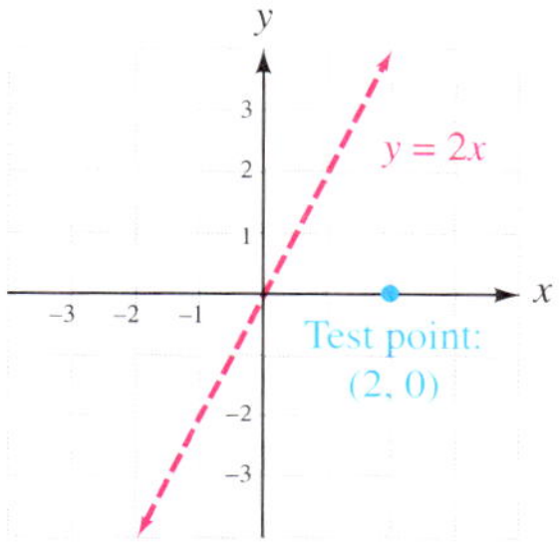

(a)

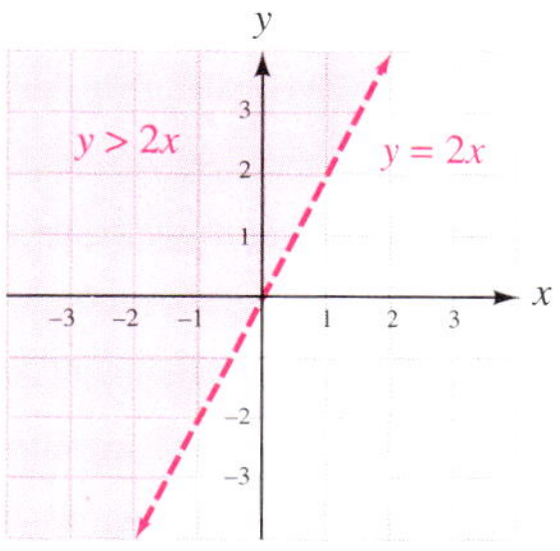

(b)

Success Tip

The origin (0, 0) is a smart choice for a test point because calculations involving 0 are usually easy. If the origin is on the boundary, choose a test point not on the boundary that has one coordinate that is 0, such as (0, 1) or (2, 0).

To determine which half-plane to shade, we substitute the coordinates of a point that lies on one side of the boundary line into $y > 2x$. Since the origin is on the boundary, it cannot serve as a test point. One of the many possible choices for a test point is (2, 0), because it does not lie on the boundary line. To see whether it satisfies $y > 2x$, we substitute 2 for x and 0 for y in the inequality.

$y > 2x$ This is the given inequality.

$0 \overset{?}{>} 2(2)$ The symbol $\overset{?}{>}$ is read as "is possibly greater than."

$0 > 4$ False

Since $0 > 4$ is a false statement, the point (2, 0) does not satisfy the inequality. We shade the half-plane that does not contain (2, 0), as shown in part (b).

Self Check 4 Graph: $y < 3x$

Now Try Problem 55

The graphs of some linear inequalities in two variables have boundary lines that are horizontal or vertical.

EXAMPLE 5 Graph each linear inequality: **a.** $x < -3$ **b.** $y \geq 0$

Strategy We will use the procedure for graphing linear inequalities in two variables.

Why Since the inequalities can be written as $x + 0y < -3$ and $0x + y \geq 0$, they are linear inequalities in two variables.

Solution **a.** Because $x < -3$ contains an $<$ symbol, we draw the boundary, $x = -3$, as a dashed vertical line. See figure (a) below. We can use (0, 0) as the test point.

$x < -3$ This is the given inequality.

$0 < -3$ Substitute 0 for x. The y-coordinate of the test point (0, 0) is not used.

Since the result is false, we shade the half-plane that does not contain (0, 0), as shown in figure (b) below. Note that the solution consists of all points that have an x-coordinate that is less than -3.

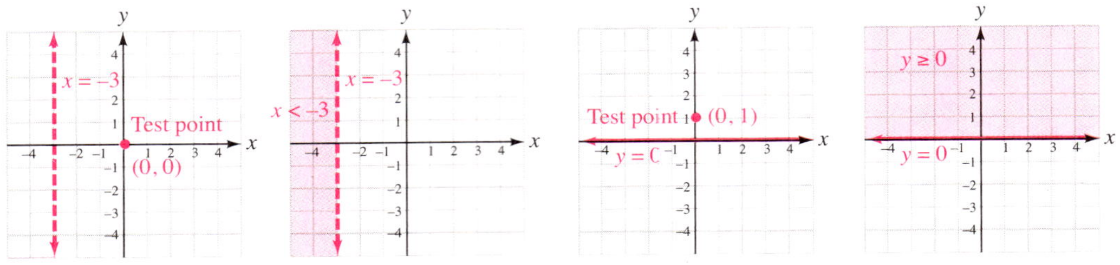

(a) (b) (c) (d)

b. Because $y \geq 0$ contains an $\geq$ symbol, we draw the boundary, $y = 0$, as a solid horizontal line. (Recall that the graph of $y = 0$ is the x-axis. See figure (c) above.) Next, we choose a test point not on the boundary. The point (0, 1) is a convenient choice. See figure (c) above.

$y \geq 0$ This is the given inequality.

$1 \geq 0$ Substitute 1 for y. The x-coordinate of the test point (0, 1) is not used.

Since the result is true, we shade the half-plane that contains (0, 1), as shown in part (d) above. Note that the solution consists of all points that have a y-coordinate that is greater than or equal to 0.

Self Check 5 Graph each linear inequality: **a.** $x \geq 2$ **b.** $y < 4$

Now Try Problems 63 and 65

4 Solve Applied Problems Involving Linear Inequalities in Two Variables.

When solving applied problems, phrases such as *at least, at most,* and *should not exceed* indicate that an inequality should be used.

EXAMPLE 6 **Working Two Jobs.** Carlos has two part-time jobs, one paying \$10 per hour and another paying \$12 per hour. If x represents the number of hours he works on the first job, and y represents the number of hours he works on the second, the graph of $10x + 12y \geq 240$ shows the possible ways he can schedule his time to earn at least \$240 per week to pay his college expenses. Find four possible combinations of hours he can work to achieve his financial goal.

Strategy We will graph the inequality and find four points whose coordinates satisfy the inequality.

Why The coordinates of these points will give four possible combinations.

Solution The graph of the inequality is shown below in part (a) of the illustration. Any point in the shaded region represents a possible way Carlos can schedule his time and earn \$240 or more per week. If each shift is a whole number of hours long, the red highlighted points in part (b) represent four of the many acceptable combinations.

Success Tip

Since Carlos cannot work a negative number of hours on job 1 or on job 2, we need to graph only the inequality in quadrant I.

(6, 24): 6 hours on the first job, 24 hours on the second job
(12, 12): 12 hours on the first job, 12 hours on the second job
(22, 4): 22 hours on the first job, 4 hours on the second job
(26, 20): 26 hours on the first job, 20 hours on the second job

To verify one combination, suppose Carlos works 22 hours on the first job and 4 hours on the second job. He will earn

$$\$10(22) + \$12(4) = \$220 + \$48$$
$$= \$268$$ This is at least \$240 per week.

A table of solutions to graph the boundary
$10x + 12y = 240$

x	y	(x, y)
0	20	(0, 20)
24	0	(24, 0)

Let x = 0 and find y.
Let y = 0 and find x.

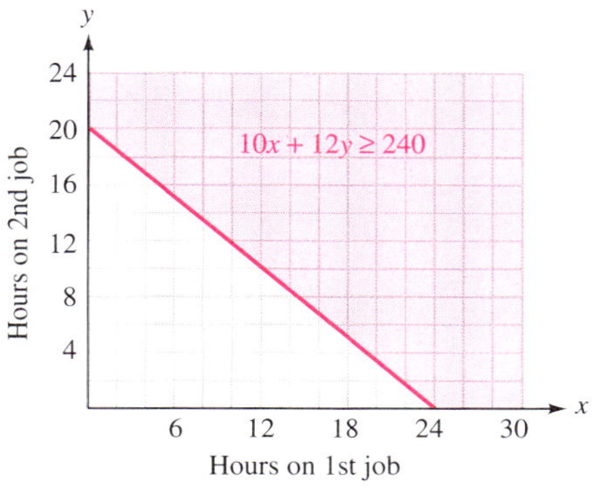

(a)

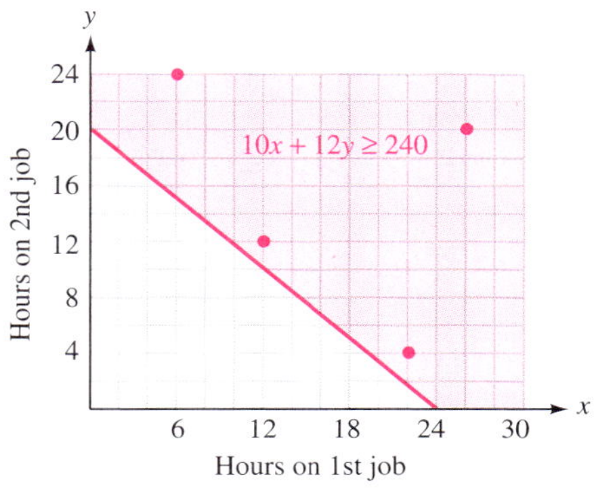

(b)

Self Check 6 **iTunes.** Brianna and Ashley pool their money to purchase some songs and movies on iTunes. If songs cost \$1 and movies cost \$15, write an inequality to represent the number of songs and movies they can buy if they want to spend \$150 or less.

Now Try Problem 79

SECTION 3.7 STUDY SET

VOCABULARY

Fill in the blanks.

1. $2x - y \le 4$ is a linear __________ in two variables.
2. An ordered pair (x, y) is a _________ of a linear inequality in two variables if a true statement results when the values of the variables are substituted into the inequality.
3. (7, 2) is a solution of $x - y > 1$. We say that (7, 2) _________ the inequality.
4. In the graph, the line $2x - y = 4$ is the __________ line.
5. In the graph, the line $2x - y = 4$ divides the coordinate plane into two ___________.
6. When graphing a linear inequality, we determine which half-plane to shade by substituting the coordinates of a test ______ into the inequality.

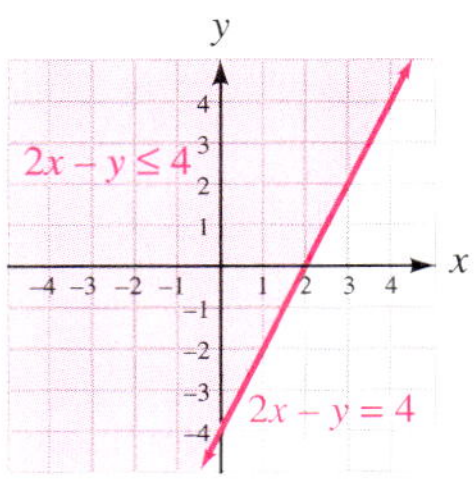

CONCEPTS

7. Determine whether $(-3, -5)$ is a solution of $5x - 3y \ge 0$.
8. Determine whether $(3, -1)$ is a solution of $x + 4y < -1$.

9. Fill in the blanks: A ________ line indicates that points on the boundary are not solutions and a ______ line indicates that points on the boundary are solutions.

10. The boundary for the graph of a linear inequality is shown. Why can't the origin be used as a test point to decide which side to shade?

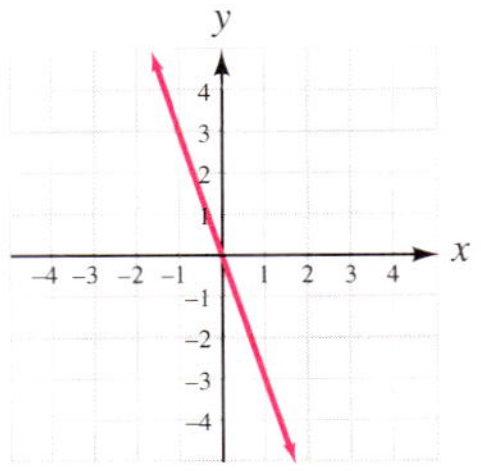

11. If a false statement results when the coordinates of a test point are substituted into a linear inequality, which half-plane should be shaded to represent the solution of the inequality?

12. A linear inequality has been graphed. Determine whether each point satisfies the inequality.

a. $(1, -3)$
b. $(-2, -1)$
c. $(2, 3)$
d. $(3, -4)$

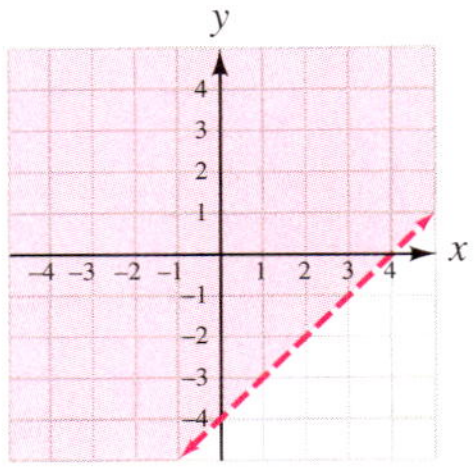

13. A linear inequality has been graphed. Determine whether each point satisfies the inequality.

a. $(2, 1)$
b. $(-2, -4)$
c. $(4, -2)$
d. $(-3, 4)$

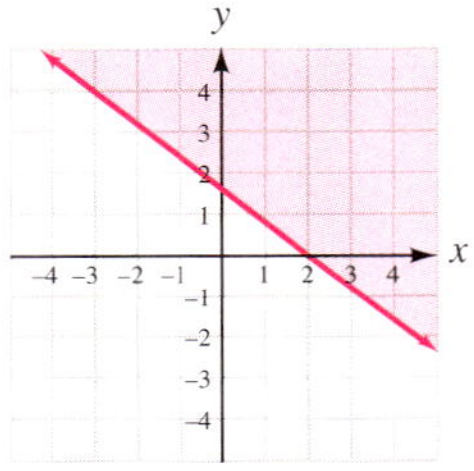

14. To graph linear inequalities, we must be able to graph boundary lines. Complete the table of solutions for each given boundary line.

a. $5x - 3y = 15$

x	y	(x, y)
0		
	0	
1		

b. $y = 3x - 2$

x	y	(x, y)
−1		
0		
2		

NOTATION

15. Write the meaning of each symbol in words.

a. $<$
b. $\geq$
c. $\leq$
d. $\stackrel{?}{>}$

16. a. When graphing linear inequalities, which inequality symbols are associated with a dashed boundary line?

b. When graphing linear inequalities, which inequality symbols are associated with a solid boundary line?

17. Fill in the blanks: The inequality $4x + 2y \leq 9$ means $4x + 2y$ ☐ 9 or $4x + 2y$ ☐ 9.

18. Fill in the blanks: The inequality $-x + 8y \geq 1$ means $-x + 8y$ ☐ 1 or $-x + 8y$ ☐ 1.

GUIDED PRACTICE

Determine whether each ordered pair is a solution of the given inequality. See Example 1.

19. $2x + y > 6; (3, 2)$

20. $4x - 2y \geq -6; (-2, 1)$

21. $-5x - 8y < 8; (-8, 4)$

22. $x + 3y > 14; (-3, 8)$

23. $4x - y \leq 0; \left(\frac{1}{2}, 1\right)$

24. $9x - y \leq 2; \left(\frac{1}{3}, 1\right)$

25. $-5x + 2y > -4; (0.8, 0.6)$

26. $6x - 2y < -7; (-0.2, 1.5)$

Complete the graph by shading the correct side of the boundary. See Example 2.

27. $x - y \geq -2$

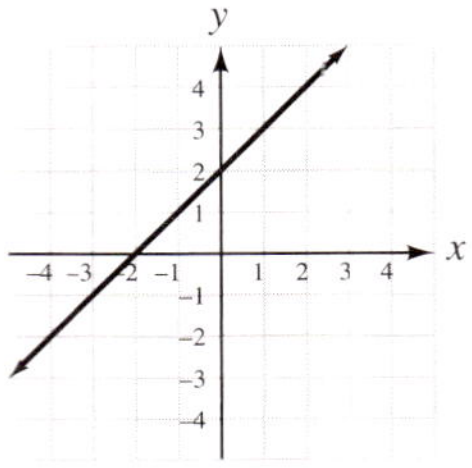

28. $x - y < 3$

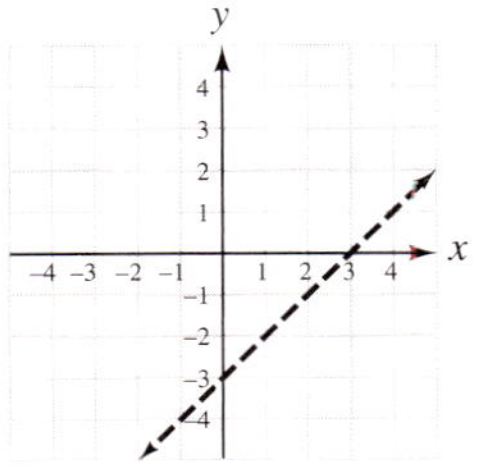

29. $y > 2x - 4$

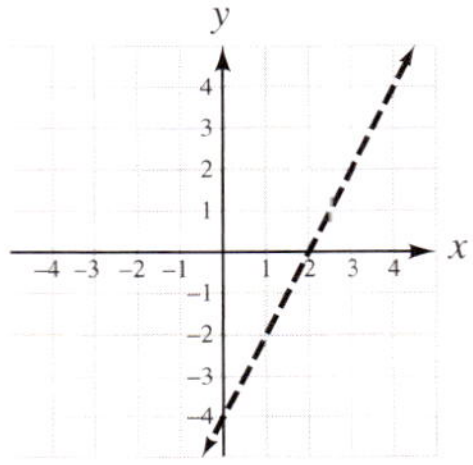

30. $y \leq -x + 1$

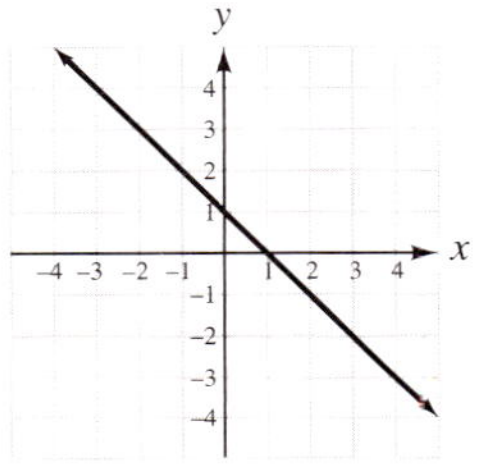

31. $x - 2y \geq 4$

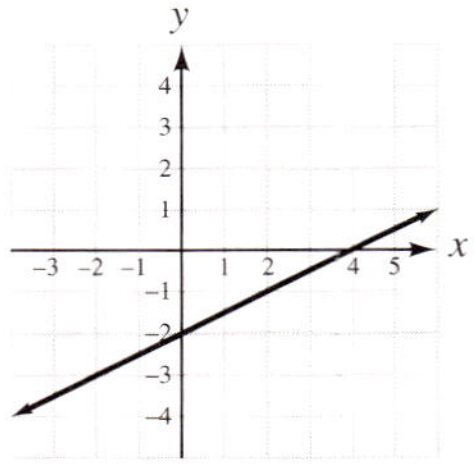

32. $3x + 2y > 12$

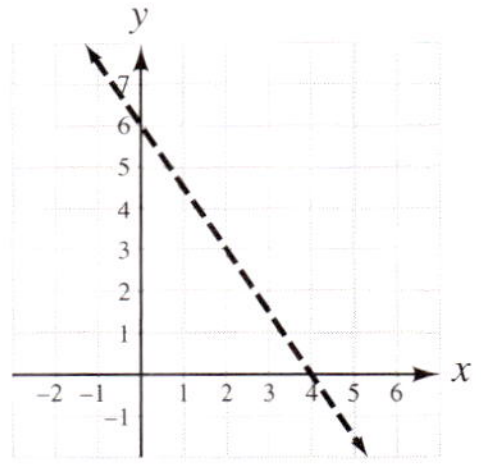

33. $y \leq 4x$

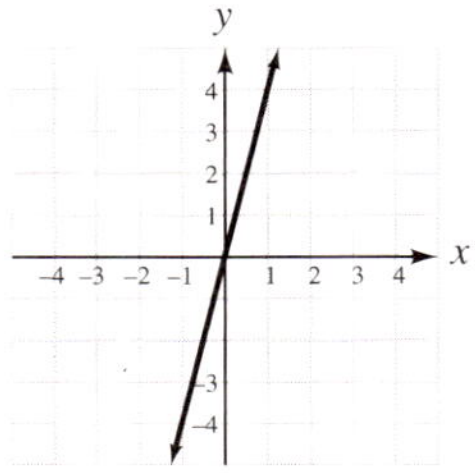

34. $y + 2x < 0$

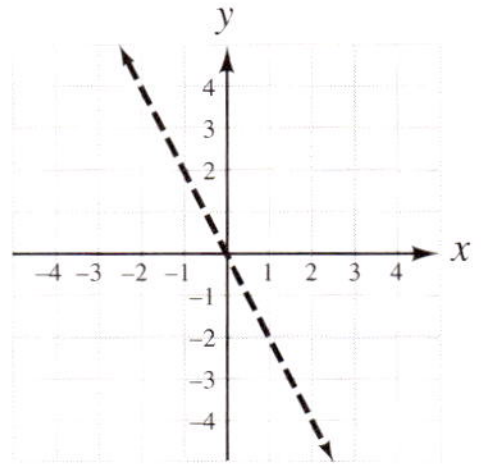

Graph each inequality. See Examples 2 and 3.

35. $x + y \geq 3$

36. $x + y < 2$

37. $3x - 4y > 12$

38. $5x + 4y \geq 20$

39. $2x + 3y \leq -12$

40. $3x - 2y > 6$

41. $y < 2x - 1$

42. $y > x + 1$

43. $y < -3x + 2$

44. $y \geq -2x + 5$

45. $y \geq -\frac{3}{2}x + 1$
46. $y < \frac{x}{3} - 1$
47. $x - 2y \geq 4$
48. $4x + y \geq -4$
49. $2y - x < 8$
50. $y + 9x \geq 3$
51. $7x - 2y < 21$
52. $3x - 3y \geq -10$
53. $2x - 3y \geq 4$
54. $4x + 3y < 6$

Graph each inequality. See Example 4.

55. $y \geq 2x$
56. $y < 3x$
57. $y < -\frac{x}{2}$
58. $y \geq x$
59. $y + x < 0$
60. $y - x < 0$
61. $5x + 3y < 0$
62. $2x + 5y > 0$

Graph each inequality. See Example 5.

63. $x < 2$
64. $y > -3$
65. $y \leq 1$
66. $x \geq -4$
67. $y - 2.5 > 0$
68. $x - 1.5 \leq 0$
69. $x \leq 0$
70. $y < 0$

TRY IT YOURSELF

Look Alikes . . .

Graph the given inequality in part a. Then use your answer to part a to help you quickly graph the associated inequality in part b. (Hint: *If you spot the relationship between the inequalities, the graph in part b can be completed without having to use the test-point method.)*

71. a. $5x - 3y \geq -15$ b. $5x - 3y < -15$
72. a. $y > -\frac{2}{3}x + 2$ b. $y \leq -\frac{2}{3}x + 2$
73. a. $y + 2x < 0$ b. $y + 2x \geq 0$
74. a. $y \leq \frac{1}{4}x$ b. $y > \frac{1}{4}x$

APPLICATIONS

75. **Deliveries.** To decide the number x of pallets and the number y of barrels that a truck can hold, a driver refers to the graph below. Can a truck make a delivery of 4 pallets and 10 barrels in one trip?

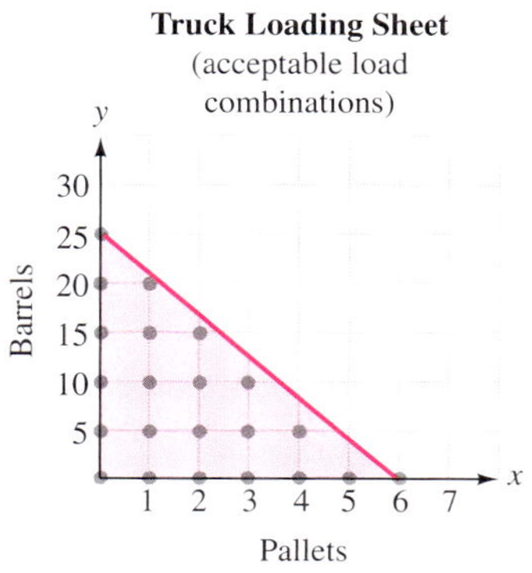

76. **Zoos.** To determine the allowable number of juvenile chimpanzees x and adult chimpanzees y that can live in an enclosure, a zookeeper refers to the graph. Can 6 juvenile and 4 adult chimps be kept in the enclosure?

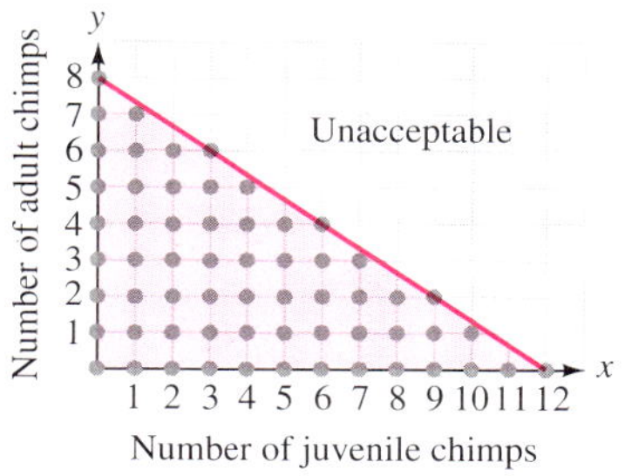

77. **from Campus to Careers**

Dental Assistant

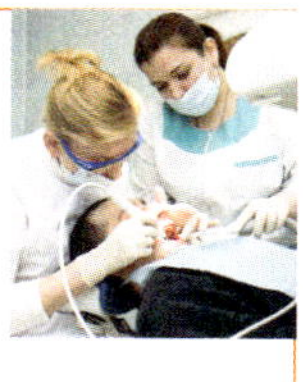
©Yuri Shirokov/Shutterstock.com

A dentist's office schedules 1-hour long appointments for adults and $\frac{3}{4}$-hour long appointments for children. The appointment times do not overlap. Let c represent the number of appointments scheduled for children and a represent the number of appointments scheduled for adults. The graph of $\frac{3}{4}c + a \leq 9$ shows the possible ways the time for seeing patients can be scheduled so that it does not exceed 9 hours per day. Graph the inequality. Label the horizontal axis c and the vertical axis a. Then find three possible combinations of children/adult appointments.

78. **Rolling Dice.** The points on the graph represent all of the possible outcomes when two fair dice are rolled a single time. For example, (5, 2), shown in red, represents a 5 on the first die and a 2 on the second. Which of the following sentences best describes the outcomes that lie in the shaded area?

(i) Their sum is at most 6.
(ii) Their sum exceeds 6.
(iii) Their sum does not exceed 6.
(iv) Their sum is at least 6.

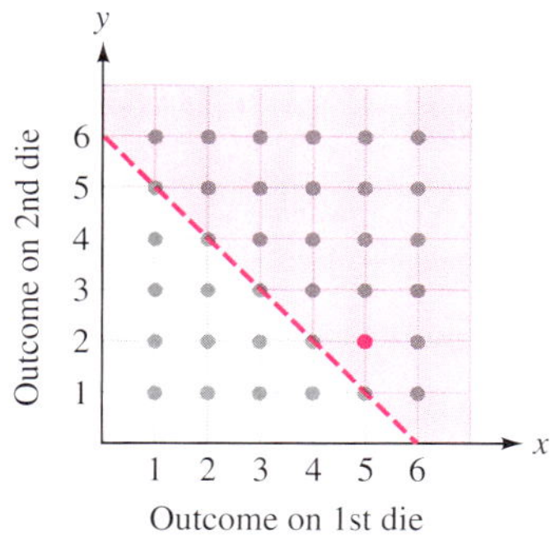

1st die

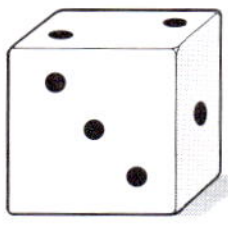
2nd die

79. **Production Planning.** It costs a bakery \$3 to make a cake and \$4 to make a pie. If x represents the number of cakes made, and y represents the number of pies made, the graph of $3x + 4y \leq 120$ shows the possible combinations of cakes and pies that can be produced so that costs do not exceed \$120 per day. Graph the inequality. Then find three possible combinations of cakes and pies that can be made so that the daily costs are not exceeded.

80. Hiring Babysitters. Mrs. Cansino has a choice of two babysitters. Sitter 1 charges \$6 per hour, and Sitter 2 charges \$7 per hour. If x represents the number of hours she uses Sitter 1 and y represents the number of hours she uses Sitter 2, the graph of $6x + 7y \le 42$ shows the possible ways she can hire the sitters and not spend more than \$42 per week. Graph the inequality. Then find three possible ways she can hire the babysitters so that her weekly budget for babysitting is not exceeded.

81. Inventories. A clothing store advertises that it maintains an inventory of at least \$4,400 worth of men's jackets at all times. At the store, leather jackets cost \$100 and nylon jackets cost \$88. If x represents the number of leather jackets in stock and y represents the number of nylon jackets in stock, the graph of $100x + 88y \ge 4{,}400$ shows the possible ways the jackets can be stocked. Graph the inequality. Then find three possible combinations of leather and nylon jackets so that the store lives up to its advertising claim.

82. Making Sporting Goods. A sporting goods manufacturer allocates at least 2,400 units of production time per day to make baseballs and footballs. It takes 20 units of time to make a baseball and 30 units of time to make a football. If x represents the number of baseballs made and y represents the number of footballs made, the graph of $20x + 30y \ge 2{,}400$ shows the possible ways to schedule the production time. Graph the inequality. Then find three possible combinations of production time for the company to make baseballs and footballs.

WRITING

83. Explain how to decide which side of the boundary line to shade when graphing a linear inequality in two variables.
84. Why is the origin usually a good test point to choose when graphing a linear inequality?
85. Why is (0, 0) not an acceptable choice for a test point when graphing a linear inequality whose boundary passes through the origin?
86. Explain the difference between the graph of the solution set of $x + 1 > 8$, an inequality in one variable, and the graph of $x + y > 8$, an inequality in two variables.

REVIEW

87. Solve $A = P + Prt$ for t.
88. What is the sum of the measures of the three angles of any triangle?
89. Simplify: $40\left(\frac{3}{8}x - \frac{1}{4}\right) + 40\left(\frac{4}{5}\right)$
90. Evaluate: $-4 + 5 - (-3) - 13$

CHALLENGE PROBLEMS

91. Find a linear inequality that has the graph shown.

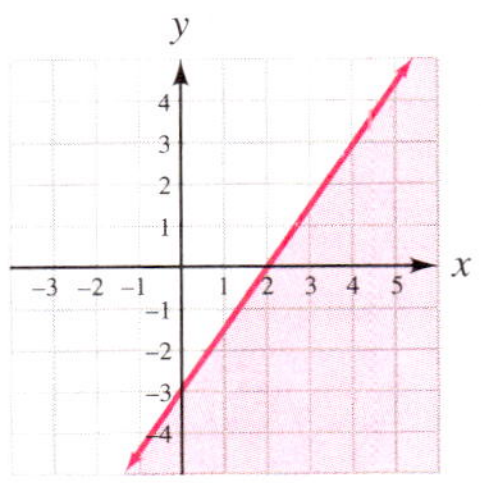

92. Graph the inequality: $4x - 3(x + 2y - 1) \ge -6\left(y - \frac{1}{2}\right)$

SECTION 3.8 An Introduction to Functions

OBJECTIVES

1. Find the domain and range of a relation.
2. Identify functions and their domains and ranges.
3. Use function notation.
4. Graph functions.
5. Use the vertical line test.
6. Solve applications involving functions.

ARE YOU READY?

The following problems review some basic skills that are needed when working with functions.

1. If $y = 3x - 1$, find the value for y when $x = 6$.
2. Which of the following ordered pairs have the same x-coordinate?
   (3, 5), (2, 9), (3, 0), (−1, 5)
3. Which of the following ordered pairs have the same y-coordinate?
   (8, 4), (−7, 6), (0, 0), (−3, 4)
4. Find each absolute value:
   a. $|-5|$ b. $|2|$

In this section, we will discuss *relations* and *functions*. These two concepts are included in our study of graphing because they involve ordered pairs.

1 Find the Domain and Range of a Relation.

The following table shows the number of medals won by American athletes at several recent Winter Olympics.

USA Winter Olympic Medal Count

Year	1988	1992	1994*	1998	2002	2006	2010
Medals	6	11	13	13	34	25	37
	Calgary CAN	Albertville FRA	Lillehammer NOR	Nagano JPN	Salt Lake City USA	Turin ITA	Vancouver CAN

* The Winter Olympics were moved ahead two years so that the winter and summer games would alternate every two years.

We can display the data in the table as a set of ordered pairs, where the **first component** represents the year and the **second component** represents the number of medals won by American athletes:

$$\{(1988, 6), (1992, 11), (1994, 13), (1998, 13), (2002, 34), (2006, 25), (2010, 37)\}$$

A set of ordered pairs, such as this, is called a **relation.** The set of all first components is called the **domain** of the relation and the set of all second components is called the **range** of the relation.

EXAMPLE 1 Find the domain and range of the relation $\{(1, 7), (4, -6), (-3, 1), (2, 7)\}$.

Strategy We will examine the first and second components of the ordered pairs.

Why The set of first components is the domain and the set of second components is the range.

Solution The relation $\{(1, 7), (4, -6), (-3, 1), (2, 7)\}$ has the domain $\{-3, 1, 2, 4\}$ and the range is $\{-6, 1, 7\}$. The elements of the domain and range are usually listed in increasing order, and if a value is repeated, it is listed only once.

Self Check 1 Find the domain and range of the relation $\{(8, 2), (-1, 10), (6, 2), (-5, -5)\}$.

Now Try Problem 15

2 Identify Functions and Their Domains and Ranges.

An **arrow** or **mapping diagram** can be used to define a relation. The data from the Winter Olympics example are shown on the right in that form. Relations are also often defined using **two-column tables.**

Notice that for each year, there corresponds exactly one medal count. That is, this relation assigns to each member of the domain exactly one member of the range. Relations that have this characteristic are called *functions.*

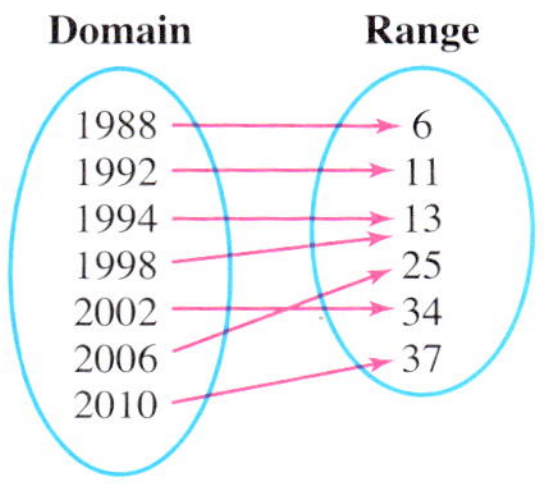

Function A function is a set of ordered pairs (a relation) in which to each first component there corresponds exactly one second component.

We may also think of a function as a rule that assigns to each value of one variable exactly one value of another variable. Since we often worked with sets of ordered pairs of the form (x, y), it is helpful to define a function in an alternate way using the variables x and y.

y Is a Function of x

Given a relation in x and y, if to each value of x in the domain there is assigned exactly one value of y in the range, then y is said to be a function of x.

EXAMPLE 2 Determine whether each relation defines y to be a function of x. If a function is defined, give its domain and range.

a.

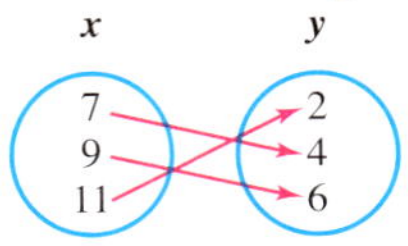

b.

x	y
2	3
5	7
2	1
6	5

c. $\{(0, 8), (3, 8), (4, 8), (9, 8)\}$

d.

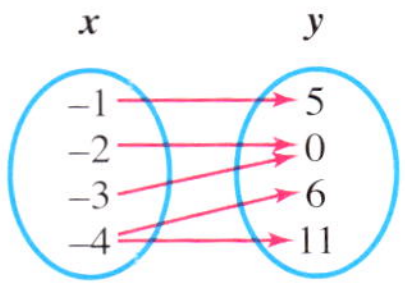

Strategy We will check to see whether each value of x is assigned exactly one value of y.

Why If this is true, then y is a function of x.

Solution

a. The arrow diagram defines a function because to each value of x there is assigned exactly one value of y: $7 \rightarrow 4$, $9 \rightarrow 6$, and $11 \rightarrow 2$.

The domain of the function is $\{7, 9, 11\}$ and the range is $\{2, 4, 6\}$.

b. The table does not define a function, because to the x value 2 there is assigned more than one value of y: $2 \rightarrow 3$ and $2 \rightarrow 1$.

c. Since to each number x exactly one value y is assigned, the set of ordered pairs defines y to be a function of x. It also illustrates an important fact about functions: *The same value of y can be assigned to different values of x.* In this case, each number x is assigned the y-value 8.

The domain of the function is $\{0, 3, 4, 9\}$ and the range is $\{8\}$.

d. The arrow diagram does not define a function, because to the x value -4 there is assigned more than one value of y: $-4 \rightarrow 6$ and $-4 \rightarrow 11$.

Success Tip

Every function is a relation, but not every relation is a function.

Self Check 2 Determine whether each relation defines y to be a function of x. If a function is defined, give its domain and range.

a.

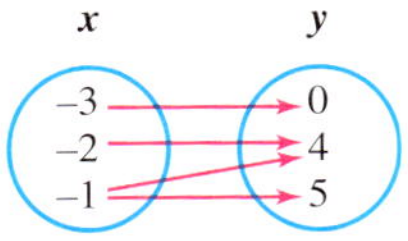

b.

x	y
−6	−6
5	5
4	8

c. $\{(1, 4), (4, 9), (9, 4), (3, 9)\}$

Now Try Problems 19, 25, and 31

3 Use Function Notation.

A function can be defined by an equation. For example, $y = 2x - 3$ is a rule that assigns to each value of x exactly one value of y. To find the y-value that is assigned to the x-value 4, we substitute 4 for x and evaluate the right side of the equation.

$$y = 2x - 3$$

$$y = 2(4) - 3 \quad \text{Substitute 4 for } x.$$

$$= 8 - 3 \quad \text{Evaluate the right side.}$$

$$= 5$$

The function $y = 2x - 3$ assigns the y-value 5 to an x-value of 4. When making such calculations, the value of x is called an **input** and its corresponding value of y is called an **output.**

A special notation is used to name functions that are defined by equations.

Function Notation

The notation $y = f(x)$ indicates that the variable y is a function of x.

Since $y = f(x)$, the equations $y = 2x - 3$ and $f(x) = 2x - 3$ are equivalent. We read $f(x) = 2x - 3$ as "f of x is equal to $2x$ minus 3."

Notation

If y is a function of x, then y and $f(x)$ are interchangeable.

This is the variable used to represent the input value.

$$f(x) = 2x - 3$$

This is the name of the function.

This expression shows how to obtain an output from a given input.

Function notation provides a compact way of representing the value that is assigned to some number x. For example, if $f(x) = 2x - 3$, the value that is assigned to an x-value 5 is represented by $f(5)$.

$f(x) = 2x - 3$

$f(5) = 2(5) - 3$ Substitute the input 5 for each x.

$= 10 - 3$ Evaluate the right side.

$= 7$ The output is 7.

Caution

The symbol $f(x)$ **does not** mean $f \cdot x$. We read $f(x)$ as "f of x" or as "the value of f at x."

Thus, $f(5) = 7$. We read this as "f of 5 is 7." The output 7 is called a **function value.**

To see why function notation is helpful, consider these two sentences, which ask you to do the same thing:

1. If $y = 2x - 3$, find the value of y when x is 5.
2. If $f(x) = 2x - 3$, find $f(5)$.

Sentence 2, which uses $f(x)$ notation, is much more compact.

EXAMPLE 3

For $f(x) = 5x + 7$, find each of the following function values:

a. $f(2)$ **b.** $f(-4)$ **c.** $f(0)$

Strategy We will substitute 2, -4, and 0 for x in the expression $5x + 7$ and then evaluate it.

Why The notation $f(x) = 5x + 7$ indicates that we are to multiply each input (each number written within the parentheses) by 5 and then add 7 to that product.

Solution

a. To find $f(2)$, we substitute the number within the parentheses, 2, for each x in $f(x) = 5x + 7$, and evaluate the right side of the equation.

The Language of Algebra

Another way to read $f(2) = 17$ is to say "the value of f at 2 is 17."

$f(x) = 5x + 7$

$f(2) = 5(2) + 7$ Substitute the input 2 for each x.

$= 10 + 7$ Evaluate the right side.

$= 17$ The output is 17.

Thus, $f(2) = 17$.

b. $f(x) = 5x + 7$

$f(-4) = 5(-4) + 7$ Substitute the input −4 for each x.

$= -20 + 7$ Evaluate the right side.

$= -13$ The output is −13.

Thus, $f(-4) = -13$.

c. $f(x) = 5x + 7$

$f(0) = 5(0) + 7$ Substitute the input 0 for each x.

$= 0 + 7$ Evaluate the right side.

$= 7$ The output is 7.

Thus, $f(0) = 7$.

Self Check 3 For $f(x) = -2x + 3$, find each of the following function values:
a. $f(4)$ **b.** $f(-1)$ **c.** $f(0)$

Now Try Problem 35

We can think of a function as a machine that takes some input x and turns it into some output $f(x)$, as shown in part (a) of the figure below. In part (b), the function machine for $f(x) = x^2 + 2x$ turns the input 4 into the output $4^2 + 2(4) = 24$, and we have $f(4) = 24$.

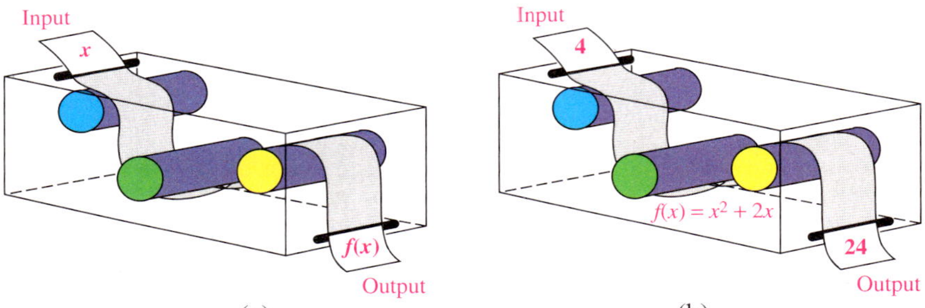

(a) (b)

The letter f used in the notation $y = f(x)$ represents the word *function*. However, other letters, such as g and h, can also be used to name functions.

EXAMPLE 4 For $g(x) = 3 - 2x$ and $h(x) = x^3 + x^2 - 1$, find: **a.** $g(3)$ **b.** $h(-2)$

Strategy We will substitute 3 for x in $3 - 2x$ and substitute -2 for x in $x^3 + x^2 - 1$, and then evaluate each expression.

Why The numbers 3 and -2, which are within the parentheses, are inputs that should be substituted for the variable x.

Solution **a.** To find $g(3)$, we use the function rule $g(x) = 3 - 2x$ and replace x with 3.

$g(x) = 3 - 2x$ Read g(x) as "g of x."

$g(3) = 3 - 2(3)$ Substitute the input 3 for each x.

$= 3 - 6$ Evaluate the right side.

$= -3$ The output is −3.

Thus, $g(3) = -3$.

b. To find $h(-2)$, we use the function rule $h(x) = x^3 + x^2 - 1$ and replace x with -2.

$$h(x) = x^3 + x^2 - 1 \quad \text{Read } h(x) \text{ as "h of x."}$$
$$h(-2) = (-2)^3 + (-2)^2 - 1 \quad \text{Substitute the input } -2 \text{ for each } x.$$
$$= -8 + 4 - 1 \quad \text{Evaluate the right side.}$$
$$= -5 \quad \text{This is the output.}$$

Thus, $h(-2) = -5$.

Self Check 4 Find $g(0)$ and $h(4)$ for the functions in Example 4.

Now Try Problem 41

4 Graph Functions.

We have seen that a function such as $f(x) = 4x + 1$ assigns to each value of x a single value $f(x)$. The input-output pairs generated by a function can be written in the form $(x, f(x))$. These ordered pairs can be plotted on a rectangular coordinate system to give the **graph of the function.**

EXAMPLE 5 Graph: $f(x) = 4x + 1$

Strategy We can graph the function by creating a table of function values and plotting the corresponding ordered pairs.

Why After drawing a line through the plotted points, we will have the graph of the function.

Solution To make a table, we choose several values for x and find the corresponding values of $f(x)$. If x is -1, we have

$$f(x) = 4x + 1 \quad \text{This is the function to graph.}$$
$$f(-1) = 4(-1) + 1 \quad \text{Substitute the input } -1 \text{ for each } x.$$
$$= -4 + 1 \quad \text{Evaluate the right side.}$$
$$= -3 \quad \text{This is the output.}$$

Thus, $f(-1) = -3$. This means when x is -1, $f(x)$ or y is -3, and that the ordered pair $(-1, -3)$ lies on the graph of $f(x)$.

Function notation	***Ordered-pair notation***
$f(-1) = -3$	$(-1, -3)$

Similarly, we find the corresponding values of $f(x)$ for x-values of 0 and 1. Then we plot the resulting ordered pairs and draw a straight line through them to get the graph of $f(x) = 4x + 1$. Since $y = f(x)$, the graph of $f(x) = 4x + 1$ is the same as the graph of the equation $y = 4x + 1$.

Notation

A table of function values is similar to a table of solutions, except that the second column is usually labeled $f(x)$ instead of y.

x	$f(x)$

x	y

$f(x) = 4x + 1$

x	$f(x)$	
-1	-3	→ $(-1, -3)$
0	1	→ $(0, 1)$
1	5	→ $(1, 5)$

The function generates these ordered pairs.

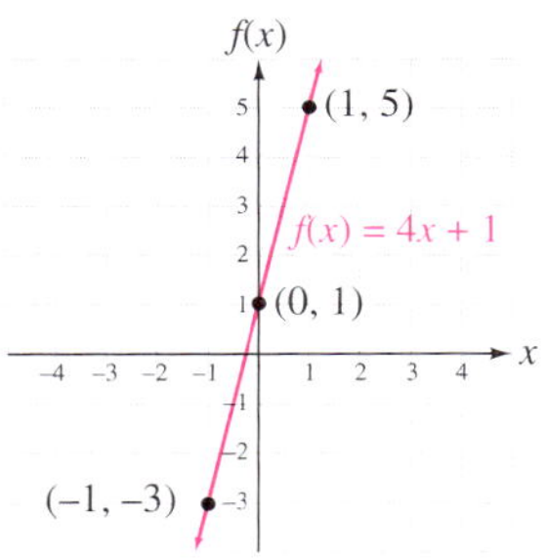

The vertical axis can be labeled y or $f(x)$.

Self Check 5 Graph: $f(x) = -3x - 2$

Now Try Problem 47

We call $f(x) = 4x + 1$ from Example 5 a **linear function** because its graph is a nonvertical line. Any linear equation, except those of the form $x = a$, can be written using function notation by writing it in slope–intercept form ($y = mx + b$) and then replacing y with $f(x)$.

The graphs of some functions are not straight lines.

EXAMPLE 6 Graph: $f(x) = |x|$

Strategy We can graph the function by creating a table of function values and plotting the corresponding ordered pairs.

Why After drawing a "V" shape through the plotted points, we will have the graph of the function.

Solution To create a table of function values, we choose values for x and find the corresponding values of $f(x)$. For $x = -4$ and $x = 3$, we have

$$f(x) = |x| \qquad\qquad f(x) = |x|$$
$$f(-4) = |-4| \qquad\qquad f(3) = |3|$$
$$= 4 \qquad\qquad = 3$$

The results $f(-4) = 4$ and $f(3) = 3$ produce the ordered pairs $(-4, 4)$ and $(3, 3)$.

Similarly, we find the corresponding values of $f(x)$ for several other x-values. When we plot the resulting ordered pairs, we see that they lie in a "V" shape. We join the points to complete the graph as shown. We call $f(x) = |x|$ an **absolute value function.**

$f(x) = |x|$

x	$f(x)$	
−4	4	→ (−4, 4)
−3	3	→ (−3, 3)
−2	2	→ (−2, 2)
−1	1	→ (−1, 1)
0	0	→ (0, 0)
1	1	→ (1, 1)
2	2	→ (2, 2)
3	3	→ (3, 3)
4	4	→ (4, 4)

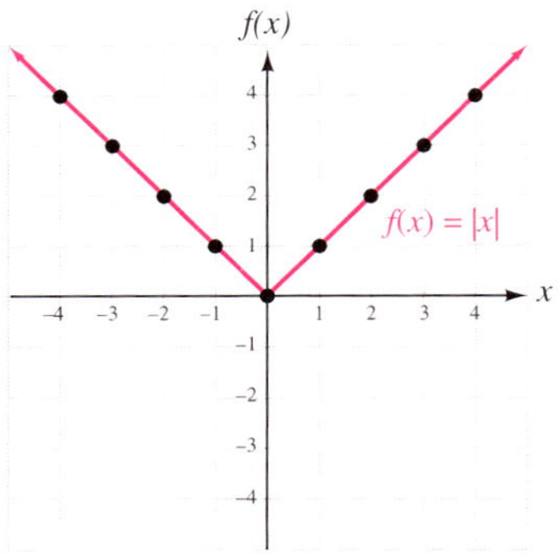

Self Check 6 Graph: $f(x) = |x| + 2$

Now Try Problem 49

5 Use the Vertical Line Test.

If any vertical line intersects a graph more than once, the graph cannot represent a function, because to one value of x there would correspond more than one value of y.

The Vertical Line Test If a vertical line intersects a graph in more than one point, the graph is not the graph of a function.

The graph shown on the right in red does not represent a function, because a vertical line intersects the graph at more than one point. The points of intersection indicate that the x-value -1 corresponds to two different y-values, 3 and -1.

When the coordinates of the two points of intersection are listed in a table, it is easy to see that the x-value of -1 is assigned two different y-values. Thus, this is not the graph of a function.

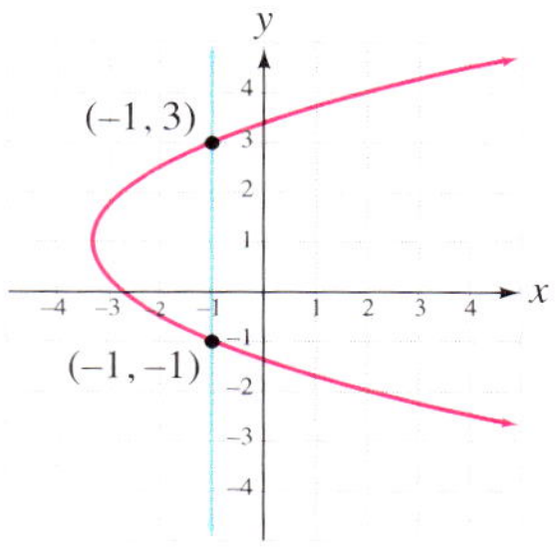

Not a function

x	y
-1	3
-1	-1

EXAMPLE 7 Determine whether each of the graphs shown in red is the graph of a function.

a.

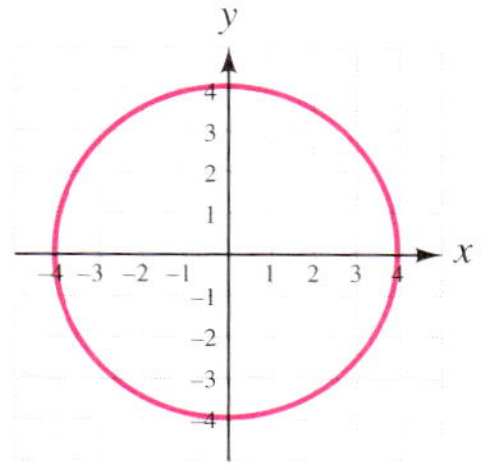

b.

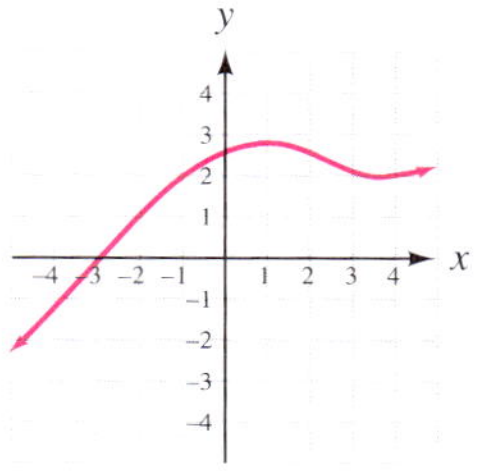

c.

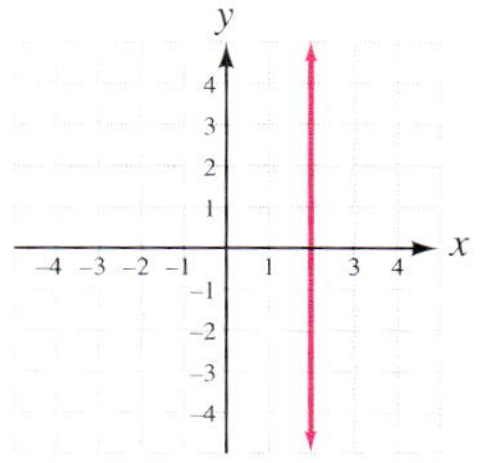

Strategy We will check to see whether any vertical line intersects the graph more than once.

Why If any vertical line does intersect the graph more than once, the graph is not a function.

Solution

a. Refer to figure (a) below. The graph shown in red is not the graph of a function, because a vertical line can be drawn that intersects the graph at more than one point. The points of intersection of the graph and the line reveal that the x-value 3 is assigned two different y-values, 2.5 and -2.5.

b. Refer to figure (b) below. The graph shown in red is a graph of a function, because no vertical line can be drawn that intersects the graph at more than one point. Several vertical lines are drawn in blue to illustrate this.

c. Refer to figure (c) below. The graph shown in red is not the graph of a function, because a vertical line can be drawn that intersects the graph at more than one point. In fact, it intersects it at infinitely many points. From this example, we can conclude that any vertical line will fail the vertical line test. Thus, vertical lines are not functions.

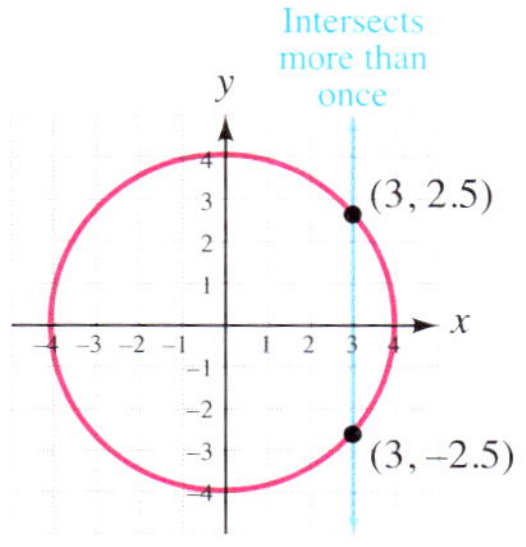

Not a function
(a)

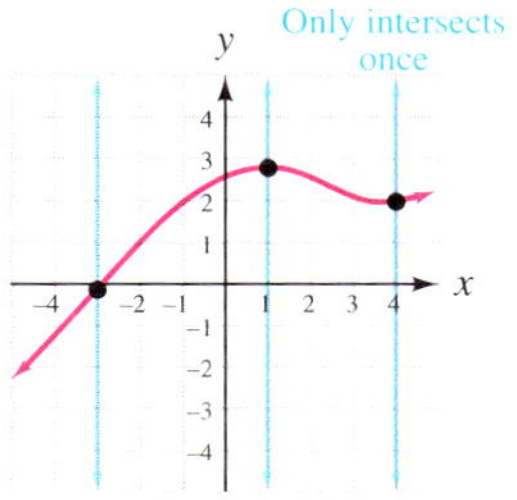

A function
(b)

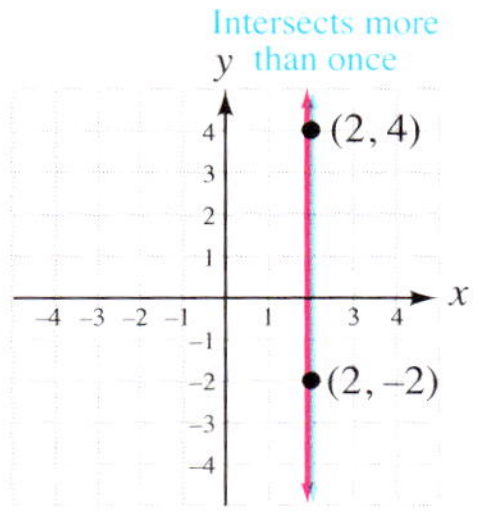

Not a function
(c)

Self Check 7 Determine whether each of the following is the graph of a function.

a.

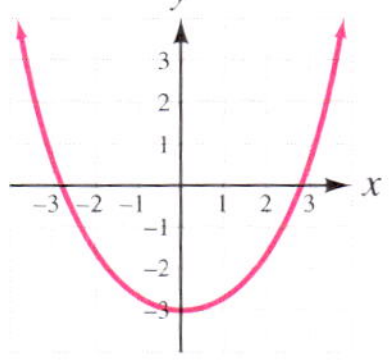

b.

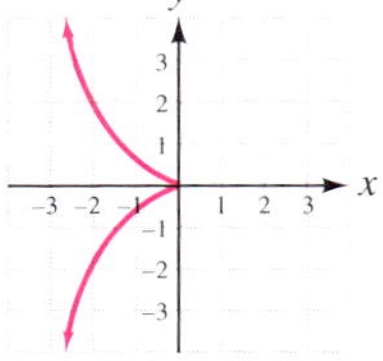

Now Try Problems 55 and 57

6 Solve Applications Involving Functions.

Functions are used to describe certain relationships where one quantity depends on another. Letters other than f and x are often chosen to more clearly describe these situations.

EXAMPLE 8 **Bounce Houses.** The function $C(h) = 80 + 15(h - 4)$ gives the cost in dollars to rent an inflatable jumper for h hours. (The terms of the rental agreement require a 4-hour minimum.) Find the cost of renting the jumper for 10 hours.

Strategy To find the cost to rent the jumper for 10 hours, we will substitute 10 for each h in $C(h) = 80 + 15(h - 4)$ and evaluate the right side.

Why In $C(h) = 80 + 15(h - 4)$, the variable h represents the number of hours that the jumper is rented. We need to find $C(10)$.

Solution For this application involving hours and cost, the notation $C(h)$ is used. The input variable is h and the name of the function is C. If the jumper is rented for 10 hours, then h is 10 and we must find $C(10)$.

$C(h) = 80 + 15(h - 4)$ Read $C(h)$ as "C of h."

$C(10) = 80 + 15(10 - 4)$ Substitute the input 10 for each h.

$= 80 + 15(6)$ Evaluate the right side.

$= 80 + 90$ Do the multiplication.

$= 170$ This is the output.

It costs \$170 to rent the jumper for 10 hours.

Self Check 8 **Bounce Houses.** Find the cost of renting the jumper for 8 hours.

Now Try Problem 65

SECTION 3.8 STUDY SET

VOCABULARY

Fill in the blanks.

1. A set of ordered pairs is called a ________.
2. A ________ is a rule that assigns to each x-value exactly one y-value.
3. The set of all input values for a function is called the ________, and the set of all output values is called the _______.
4. We can think of a function as a machine that takes some ______ x and turns it into some output _____.
5. If $f(2) = -3$, we call -3 a function _______.
6. The graph of a ______ function is a straight line and the graph of an ________ value function is V-shaped.

CONCEPTS

7. **Federal Minimum Hourly Wage.** The following table is an example of a function. Use an arrow diagram to illustrate this.

Year	1992	1994	1996	1998	2000	2002	2004	2006	2008	2010
Minimum wage (\$)	4.25	4.25	4.75	5.15	5.15	5.15	5.15	5.15	6.55	7.25

Source: infoplease.com

8. The arrow diagram describes a function. What is the domain and what is the range of the function?

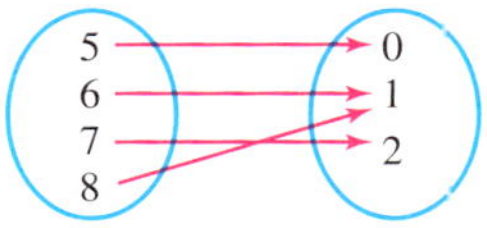

9. For the given input, what value will the function machine output?

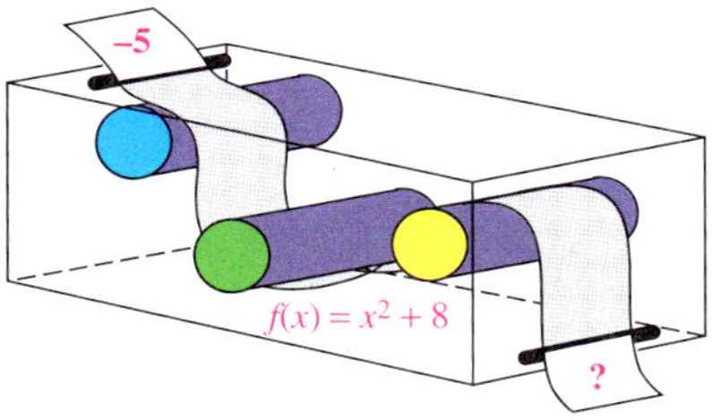

10. a. Fill in the blank: If a ________ line intersects a graph in more than one point, the graph is not the graph of a function.

b. Give the coordinates of the points where the given vertical line intersects the graph.

c. Is this the graph of a function?

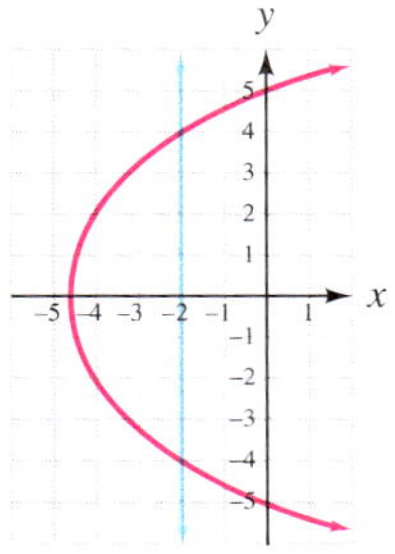

NOTATION

Fill in the blanks.

11. We read $f(x) = 5x - 6$ as "f ____ x is $5x$ minus 6."

12. Since $y =$ _____, the following two equations are equivalent:

$$y = 3x + 2 \quad \text{and} \quad f(x) = 3x + 2$$

13. The notation $f(4) = 5$ indicates that when the x-value ____ is input into a function rule, the output is ____. This fact can be shown graphically by plotting the ordered pair (___ , ___).

14. When graphing the function $f(x) = -x + 5$, the vertical axis of the coordinate system can be labeled ____ or ____.

GUIDED PRACTICE

Find the domain and range of each relation. See Example 1.

15. $\{(6, -1), (-1, -10), (-6, 2), (8, -5)\}$

16. $\{(11, -3), (0, 0), (4, 5), (-3, -7)\}$

17. $\{(0, 9), (-8, 50), (6, 9)\}$

18. $\{(1, -12), (-6, 8), (5, 8)\}$

Determine whether the relation defines y to be a function of x. If a function is defined, give its domain and range. If it does not define a function, find two ordered pairs that show a value of x that is assigned more than one value of y. See Example 2.

19.

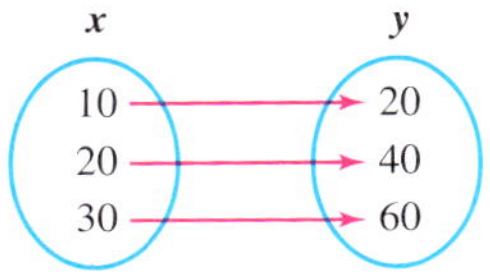

20.

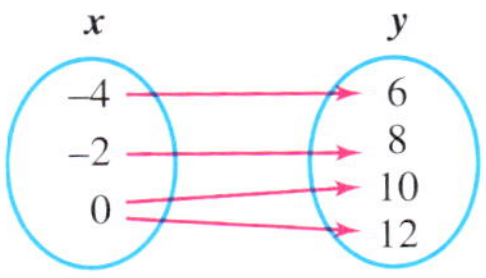

21.

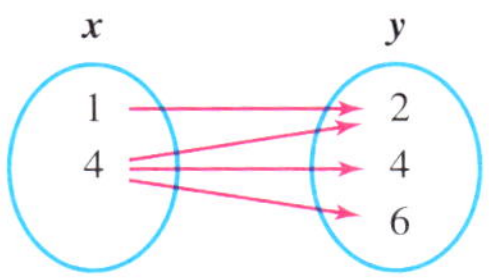

22.

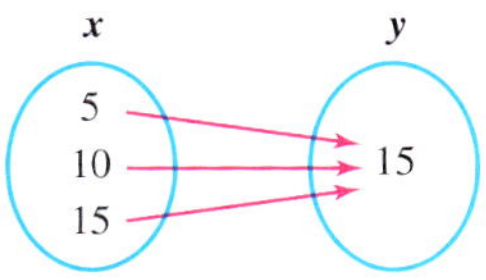

23.

x	y
1	7
2	15
3	23
4	16
5	8

24.

x	y
30	2
30	4
30	6
30	8
30	10

25.

x	y
−4	6
−1	0
0	−3
2	4
−1	2

26.

x	y
1	1
2	2
3	3
4	4

27.

x	y
3	4
3	−4
4	3
4	−3

28.

x	y
−1	1
−3	1
−5	1
−7	1
−9	1

29.

x	y
6	0
−3	−8
1	9
5	4

30.

x	y
1.6	0
−3	−1
2.5	20
−7	0.1
1.6	19

31. $\{(3, 4), (3, -4), (4, 3), (4, -3)\}$

32. $\{(-1, 1), (-3, 1), (-5, 1), (-7, 1), (-9, 1)\}$

33. $\{(-2, 7), (-1, 10), (0, 13), (1, 16)\}$

34. $\{(-2, 4), (-3, 8), (-3, 12), (-4, 16)\}$

Find each function value. See Examples 3 and 4.

35. $f(x) = 4x - 1$

a. $f(1)$ b. $f(-2)$ c. $f\left(\frac{1}{4}\right)$ d. $f(50)$

36. $f(x) = 1 - 5x$

a. $f(0)$ b. $f(-75)$ c. $f(0.2)$ d. $f\left(-\frac{4}{5}\right)$

37. $f(x) = 2x^2$

a. $f(0.4)$ b. $f(-3)$ c. $f(1{,}000)$ d. $f\left(\frac{1}{8}\right)$

38. $g(x) = 6 - x^2$

a. $g(30)$ b. $g(6)$ c. $g(-1)$ d. $g(0.5)$

39. $h(x) = |x - 7|$

a. $h(0)$ b. $h(-7)$ c. $h(7)$ d. $h(8)$

40. $f(x) = |2 + x|$

a. $f(0)$ b. $f(2)$ c. $f(-2)$ d. $f(-99)$

41. $g(x) = x^3 - x$

a. $g(1)$ b. $g(10)$ c. $g(-3)$ d. $g(6)$

42. $g(x) = x^4 + x$

a. $g(1)$ b. $g(-2)$ c. $g(0)$ d. $g(10)$

43. $s(x) = (x + 3)^2$

a. $s(3)$ b. $s(-3)$ c. $s(0)$ d. $s(-5)$

44. $s(x) = (x - 8)^2$

a. $s(8)$ b. $s(-8)$ c. $s(1)$ d. $s(12)$

45. If $f(x) = 3.4x^2 - 1.2x + 0.5$, find $f(-0.3)$.

46. If $g(x) = x^4 - x^3 + x^2 - x$, find $g(-12)$.

Complete each table of function values and then graph each function. **See Examples 5 and 6.**

47. $f(x) = -3x - 2$

x	$f(x)$
−2	
−1	
0	
1	

48. $f(x) = -2x + 8$

x	$f(x)$
−1	
0	
1	
2	

49. $h(x) = |1 - x|$

x	$h(x)$
−2	
−1	
0	
1	
2	
3	
4	

50. $h(x) = |x + 2|$

x	$h(x)$
−5	
−4	
−3	
−2	
−1	
0	
1	

Graph each function. **See Examples 5 and 6.**

51. $f(x) = \frac{1}{2}x - 2$

52. $f(x) = -\frac{2}{3}x + 3$

53. $h(x) = -|x|$

54. $g(x) = |x| - 2$

Determine whether each graph is the graph of a function. If it is not, find ordered pairs that show a value of x that is assigned more than one value of y. **See Example 7.**

55.
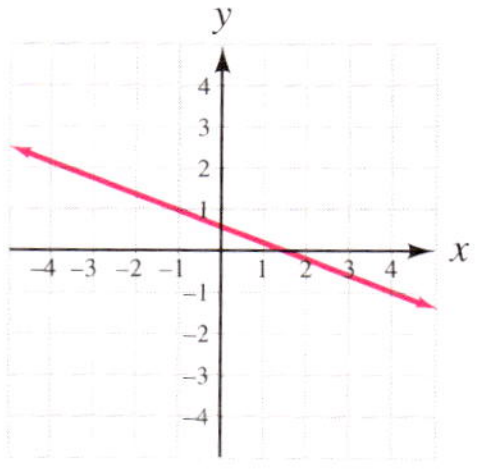

56.
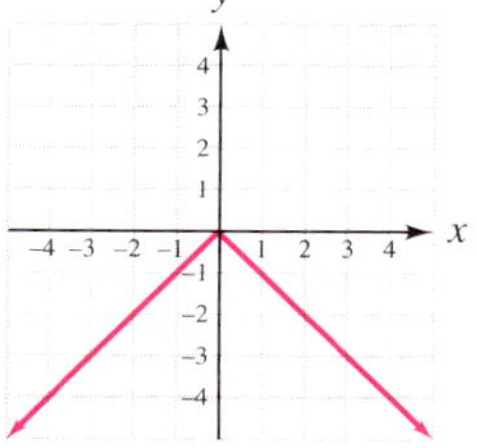

57.
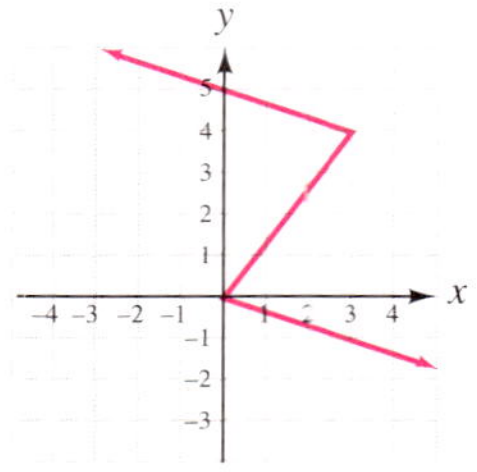

58.
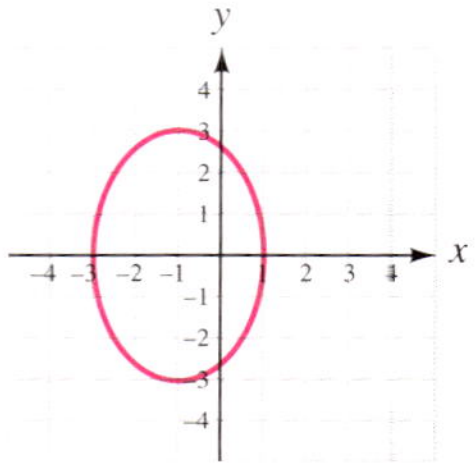

59.
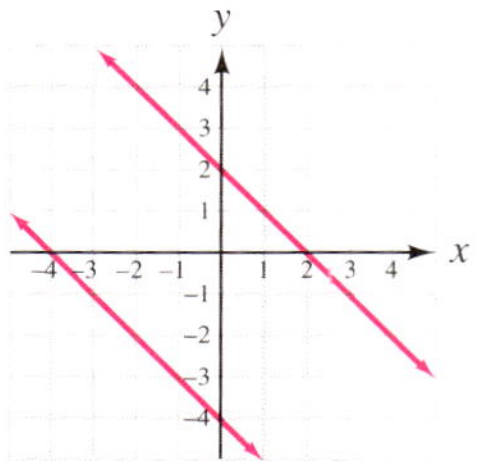

60.
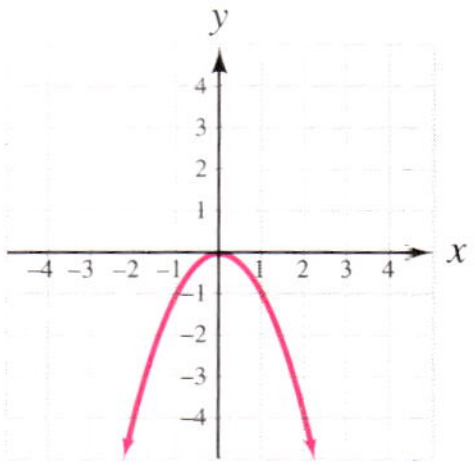

61.
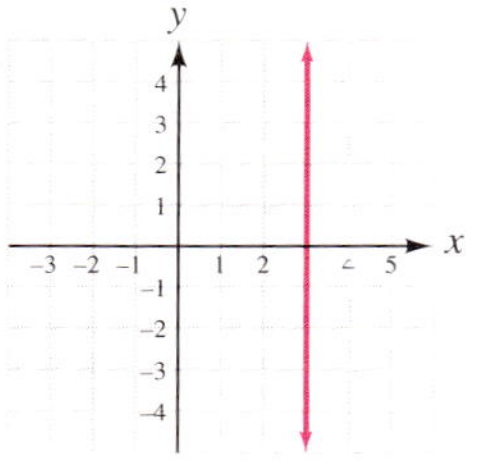

62.
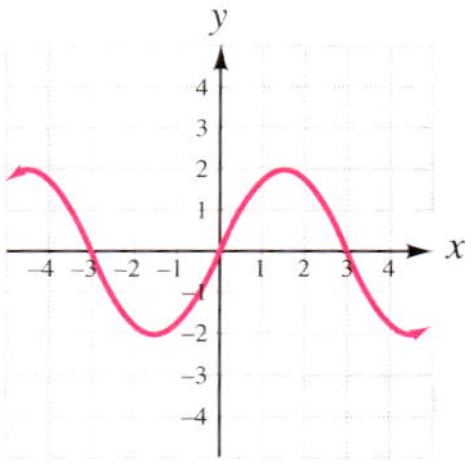

APPLICATIONS

63. **Reflections.** When a beam of light hits a mirror, it is reflected off the mirror at the same angle that the incoming beam struck the mirror. What type of function could serve as a mathematical model for the path of the light beam shown here?

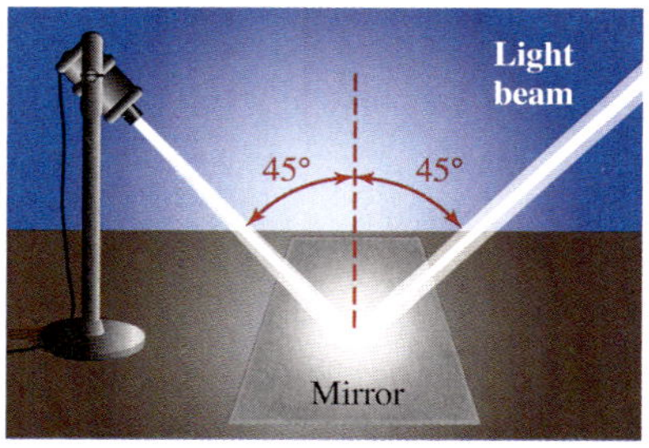

64. **Lightning.** The function $D(t) = \frac{t}{5}$ gives the approximate distance in miles that you are from a lightning strike, where t is the number of seconds between seeing the lightning and hearing the thunder. Find $D(5)$ and explain what it means.

65. Vacationing. The function $C(d) = 500 + 100(d - 3)$ gives the cost in dollars to rent an RV motor home for d days. (The terms of the rental agreement require a 3-day minimum.) Find the cost of renting the RV for a vacation that will last 7 days.

66. Structural Engineering. The maximum safe load in pounds of the rectangular beam shown in the figure is given by the function $S(t) = \dfrac{1{,}875t^2}{8}$, where t is the thickness of the beam, in inches. Find the maximum safe load if the beam is 4 inches thick.

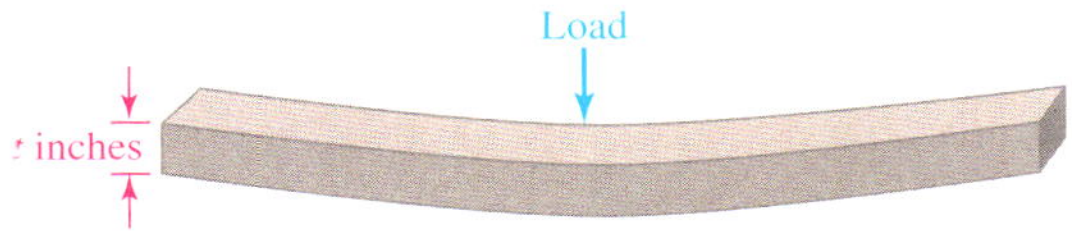

67. Lawn Sprinklers. The function $A(r) = \pi r^2$ can be used to determine the area in square feet that will be watered by a rotating sprinkler that sprays out a stream of water. Find $A(5)$ and $A(20)$. Round to the nearest tenth.

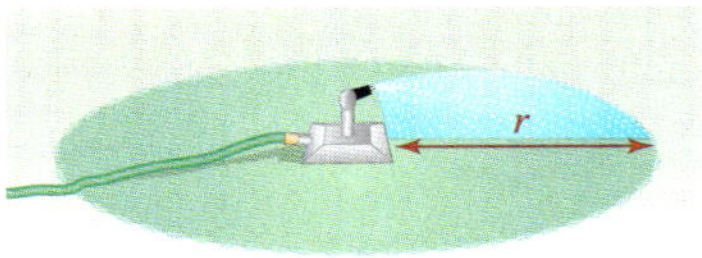

68. Parts Lists. The function $f(r) = 2.30 + 3.25(r + 0.40)$ approximates the length (in feet) of the belt that joins the two pulleys, where r is the radius (in feet) of the smaller pulley. Find the belt length needed for each pulley in the parts list.

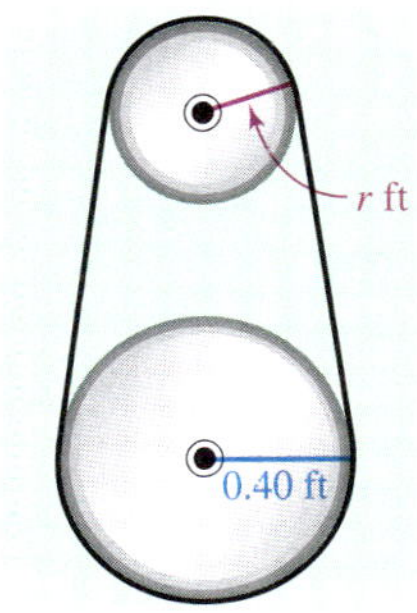

Parts list		
Pulley	r	**Belt length**
P-45M	0.32	
P-08D	0.24	

69. Postage. The **step graph** below shows how the cost of a first class U.S. postage stamp increased from 1990 through 2010. An open circle at the end of a line segment means the endpoint of the segment is not included. Is this the graph of a function?

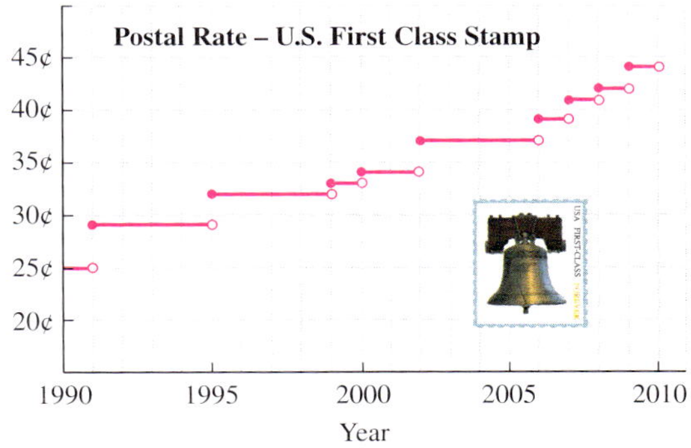

Source: U.S. Postal Service

70. Sound. We cannot see sound waves, but certain scientific instruments are used to draw mathematical models of them. Is the graph of a sound wave shown below a function?

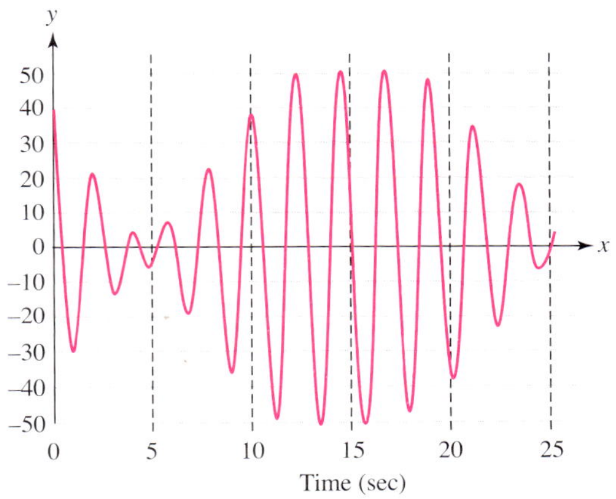

WRITING

71. In the function $y = -5x + 2$, why do you think the value of x is called the *input* and the corresponding value of y the *output*?

72. Explain what a politician meant when she said, "The speed at which the downtown area will be redeveloped is a function of the number of low-interest loans made available to the property owners."

73. Explain the following diagram:

$f(4) = 11$ $(4, 11)$

74. Explain the error in the following solution.

If $f(x) = x^2 + 7x + 1$, find $f(10)$.

$$\begin{aligned} f(10) &= 10^2 + 7x + 1 \\ &= 100 + 7x + 1 \\ &= 101 + 7x \end{aligned}$$

75. A student was asked to determine whether the graph on the right is the graph of a function. What is wrong with the following reasoning?

When I draw a vertical line through the graph, it intersects the graph only once. By the vertical line test, this is the graph of a function.

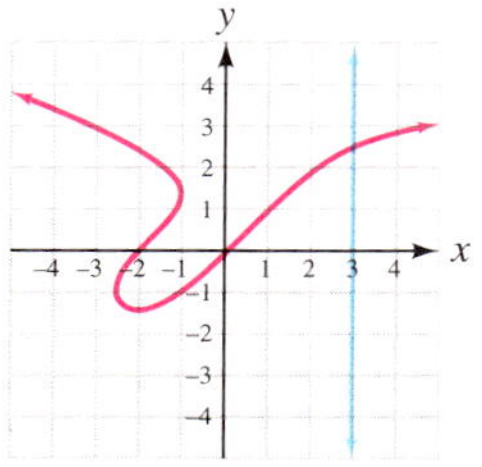

76. In your own words, what is a function?

REVIEW

77. Coffee Blends. A store sells regular coffee for \$4 a pound and gourmet coffee for \$7 a pound. To get rid of 40 pounds of the gourmet coffee, the shopkeeper plans to make a gourmet blend that he will put on sale for \$5 a pound. How many pounds of regular coffee should be used?

78. Photographic Chemicals. A photographer wishes to mix 2 liters of a 5% acetic acid solution with a 10% acetic solution to get a 7% acetic solution. How many liters of 10% acetic solution must be added?

CHALLENGE PROBLEMS

79. Is the graph of $y \geq 3 - x$ a function? Explain.

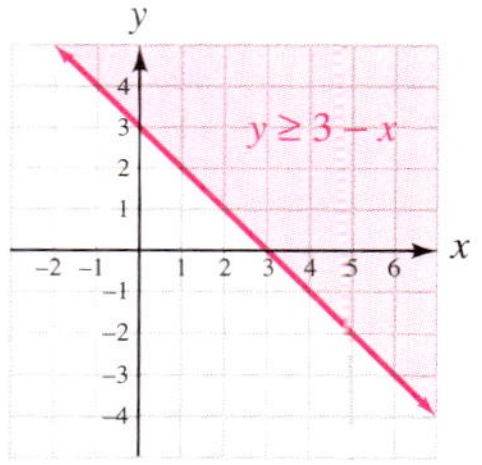

80. If $f(x) = x^2 + x$, find: $f\left(\frac{4}{5}r\right)$

81. Let $f(x) = -2x + 5$. For what value of x is $f(x) = -7$?

82. Let $f(x) = x - 2$ and $g(x) = 3x$. Find $f(g(6))$.

3 Summary & Review

SECTION 3.1 Graphing Using the Rectangular Coordinate System

DEFINITIONS AND CONCEPTS	EXAMPLES
A **rectangular coordinate system** is composed of a horizontal number line called the ***x*-axis** and a vertical number line called the ***y*-axis.** The two axes intersect at the **origin.** To **plot** or **graph** ordered pairs means to locate their position on a rectangular coordinate system. The x- and y-axes divide the **coordinate plane** into four regions called **quadrants.**	Plot the points: $(2, 3), (-4, 2), (-3, -1), (0, -2.5), (4, -2)$ To graph each point, start at the origin and count the appropriate number of units in the x-direction and then the appropriate number of units in the y-direction. 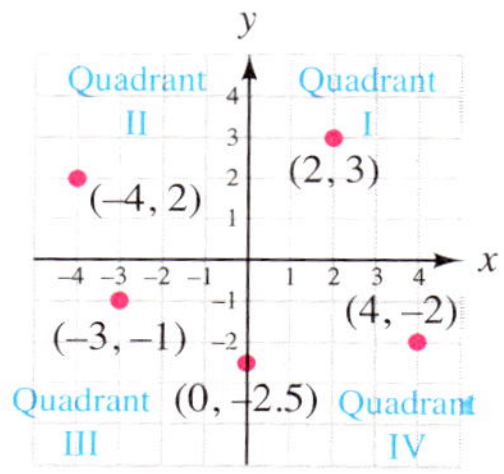

REVIEW EXERCISES

1. Graph the points with coordinates $(-1, 3)$, $(0, 1.5)$, $(-4, -4)$, $\left(2, \frac{7}{2}\right)$, and $(4, 0)$.

2. Hawaii. Estimate the coordinates of Oahu using an ordered pair of the form (longitude, latitude).

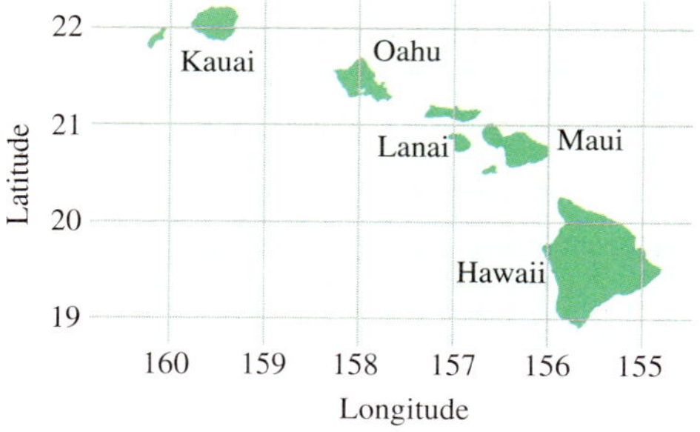

3. In what quadrant does the point $(-3, -4)$ lie?

4. What are the coordinates of the origin?

5. Geometry. Three vertices (corners) of a square are $(-5, 4)$, $(-5, -2)$, and $(1, -2)$. Find the coordinates of the fourth vertex and then find the area of the square.

6. College Enrollments. The graph gives the number of students enrolled at a college for the period from 4 weeks before to 5 weeks after the semester began.

a. What was the maximum enrollment and when did it occur?

b. How many students were enrolled 2 weeks before the semester began?

c. When was the enrollment 2,250?

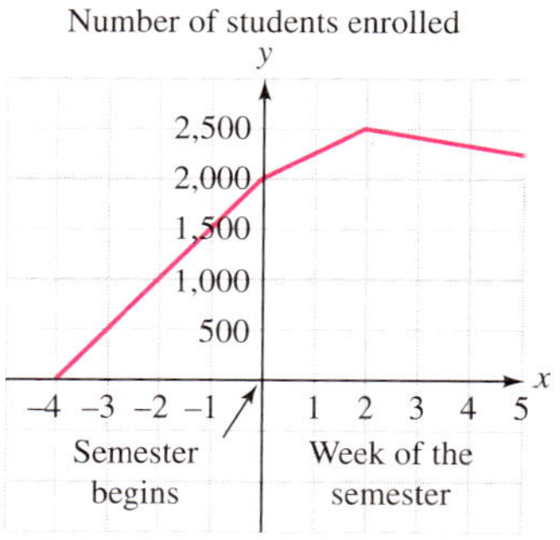

SECTION 3.2 Graphing Linear Equations

DEFINITIONS AND CONCEPTS | **EXAMPLES**

A **solution of an equation in two variables** is an ordered pair of numbers that makes the equation a true statement when the numbers are substituted for the variables.

The **standard form** of a linear equation in two variables is $Ax + By = C$, where A, B, and C are real numbers and A and B are not both zero.

Determine whether $(2, -3)$ is a solution of $2x - y = 7$.

$2x - y = 7$	This is the given equation.
$2(2) - (-3) \stackrel{?}{=} 7$	Substitute 2 for x and −3 for y.
$4 + 3 \stackrel{?}{=} 7$	Evaluate the left side.
$7 = 7$	True

Since the result $7 = 7$ is true, $(2, -3)$ is a solution of $2x - y = 7$.

If only one coordinate of an ordered-pair solution is known:

1. Substitute it into the equation for the appropriate variable.
2. Solve the resulting equation to find the unknown coordinate.

To complete the solution (, 8) for $3x + y = -1$, we substitute 8 for y and solve the resulting equation for x.

$3x + y = -1$	This is the given equation.
$3x + 8 = -1$	Substitute 8 for y.
$3x = -9$	Subtract 8 from both sides.
$x = -3$	To isolate x, divide both sides by 3.

The solution is $(-3, 8)$.

To **graph a linear equation** solved for y:

1. Find three solutions by **selecting three values** of x and finding the corresponding values of y.
2. **Plot** each ordered-pair solution.
3. **Draw** a line through the points.

Graph: $y = -2x + 1$

$y = -2x + 1$

x	y	(x, y)
−1	3	(−1, 3)
0	1	(0, 1)
2	−3	(2, −3)

We construct a table of solutions, plot the points, and draw the line.

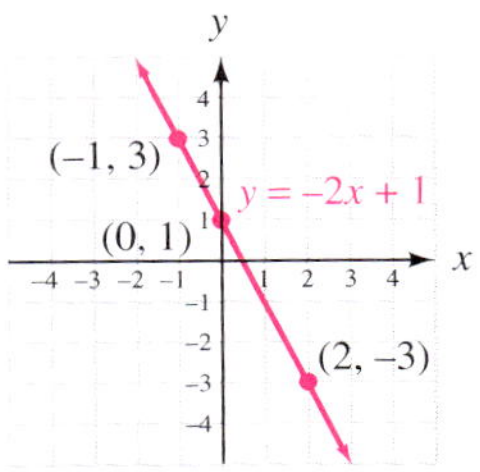

REVIEW EXERCISES

7. Is $(-3, -2)$ a solution of $y = 2x + 4$?
8. Complete the table of solutions.

$3x + 2y = -18$

x	y	(x, y)
−2		
	3	

9. Which of the following equations are not linear equations?

$8x - 2y = 6 \qquad y = x^2 + 1 \qquad y = x \qquad y - x^3 = 0$

10. The graph of a linear equation is shown.
    a. When the coordinates of point A are substituted into the equation, will a true or false statement result?
    b. When the coordinates of point B are substituted into the equation, will a true or false statement result?

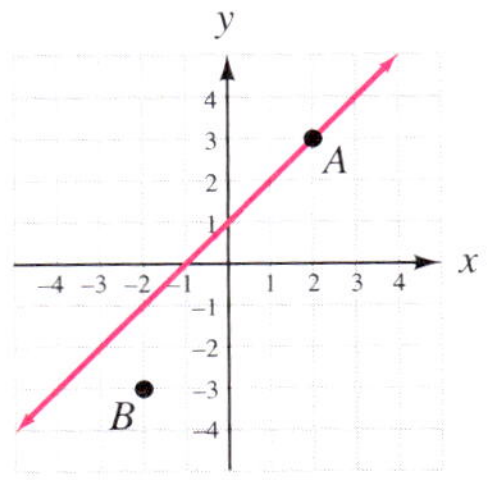

Graph each equation by constructing a table of solutions.

11. $y = 4x - 2$
12. $y = \frac{3}{4}x$
13. $5y = -5x + 15$ (*Hint:* Solve for y first.)
14. $6y = -4x$ (*Hint:* Solve for y first.)
15. **Birthday Parties.** A restaurant offers a party package for children that includes everything: food, drinks, cake, and favors. The cost c, in dollars, is given by the equation $c = 8n + 50$, where n is the number of children attending the party. Graph the equation and use the graph to estimate the cost of a party if 18 children attend.
16. Determine whether each statement is true or false.
    a. It takes three or more points to determine a line.
    b. A linear equation in two variables has infinitely many solutions.

SECTION 3.3 Intercepts

DEFINITIONS AND CONCEPTS	EXAMPLES
The point where a line intersects the x-axis is called the **x-intercept.** The point where a line intersects the y-axis is called the **y-intercept.** To **find the y-intercept,** substitute 0 for x in the given equation and solve for y. To **find the x-intercept,** substitute 0 for y and solve for x. Plotting the x- and y-intercepts of a graph and drawing a line through them is called the **intercept method for graphing a line.**	Use the y- and x-intercepts to graph $3x + 4y = -6$. ***y*-intercept: $x = 0$** $3x + 4y = -6$ $3(0) + 4y = -6$ $4y = -6$ $y = -\frac{3}{2}$ ***x*-intercept: $y = 0$** $3x + 4y = -6$ $3x + 4(0) = -6$ $3x = -6$ $x = -2$ 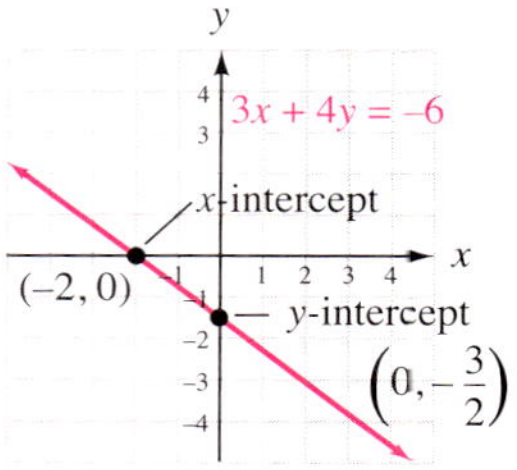 The y-intercept is $\left(0, -\frac{3}{2}\right)$ and the x-intercept is $(-2, 0)$.
The equation $y = b$ represents the **horizontal line** that intersects the y-axis at $(0, b)$. The equation $x = a$ represents the **vertical line** that intersects the x-axis at $(a, 0)$.	Graph: $y = 2$ and $x = -1$ on the same rectangular coordinate system. 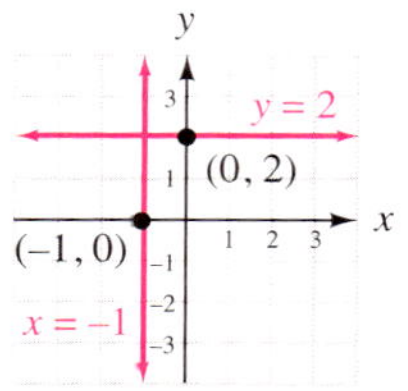

REVIEW EXERCISES

17. Identify the x- and y-intercepts of the graph shown in figure (a) below.

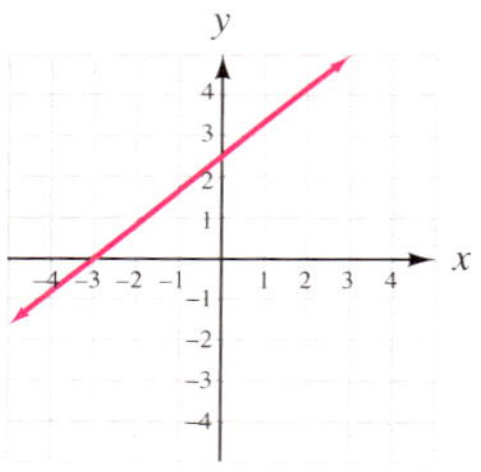

(a)

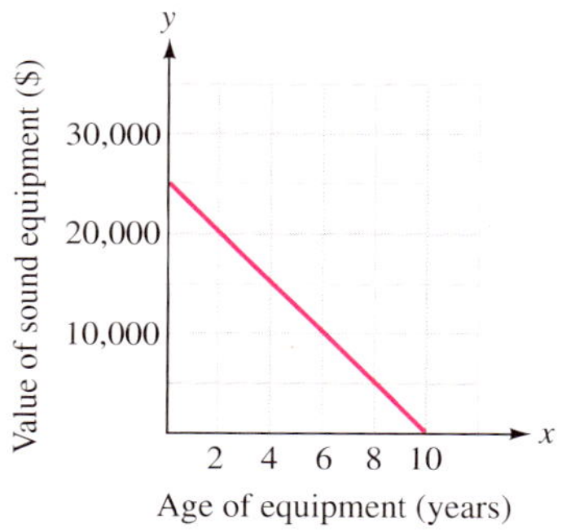

(b)

18. Depreciation. The graph in figure (b) above shows how the value of some sound equipment decreased over the years. Find the intercepts of the graph. What information do the intercepts give about the equipment?

Use the intercept method to graph each equation.

19. $-4x + 2y = 8$

20. $5x - 4y = 13$

21. Graph: $y = 4$

22. Graph: $x = -1$

SECTION 3.4 Slope and Rate of Change

DEFINITIONS AND CONCEPTS	EXAMPLES
The **slope** m of a line is a ratio that compares the vertical and horizontal change as we move along the line from one point to another. We can find the slope of a line graphically using the ratio $m = \frac{\text{rise}}{\text{run}}$.	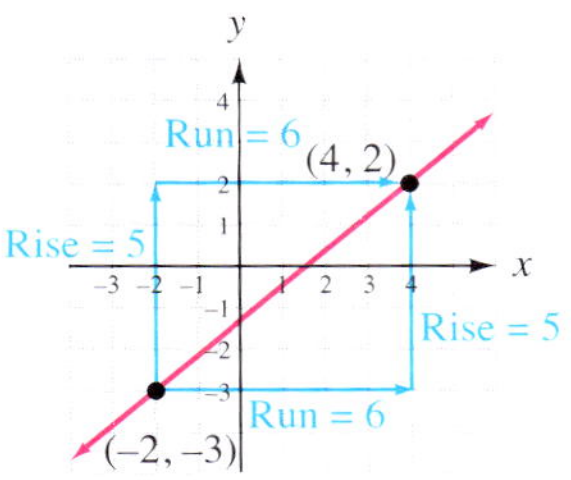 To find the slope of the line graphed on the left, we can use the slope triangle method: $m = \frac{\text{rise}}{\text{run}} = \frac{5}{6}$
We also can find the slope of a line using the **slope formula:** $m = \frac{y_2 - y_1}{x_2 - x_1}$ if $x_1 \neq x_2$	To find the slope of the line that passes through the points $(-2, -3)$ and $(4, 2)$, we substitute into the **slope formula:** $m = \frac{y_2 - y_1}{x_2 - x_1} = \frac{2 - (-3)}{4 - (-2)} = \frac{5}{6}$
Lines that rise from left to right have a **positive slope,** and lines that fall from left to right have a **negative slope.** Horizontal lines have **zero slope** and vertical lines have **undefined slope.**	
When units are attached to a slope, the slope is called a **rate of change.** The **grade** of an incline is its slope expressed as a percent.	An example of a rate of change is: $\frac{300 \text{ pounds}}{1 \text{ year}}$ Read as "300 pounds per year."
Parallel lines have the same slope. The slopes of **perpendicular lines** are negative reciprocals. The product of their slopes is -1.	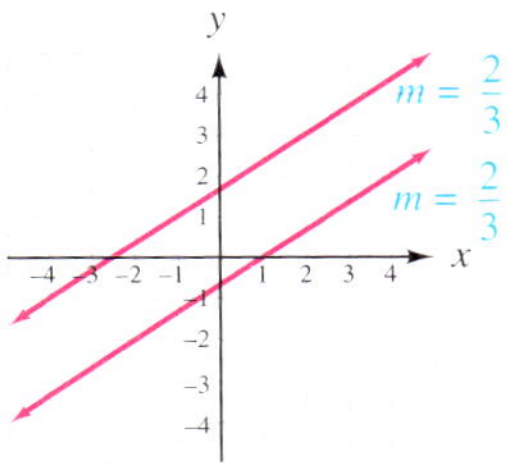 Parallel lines 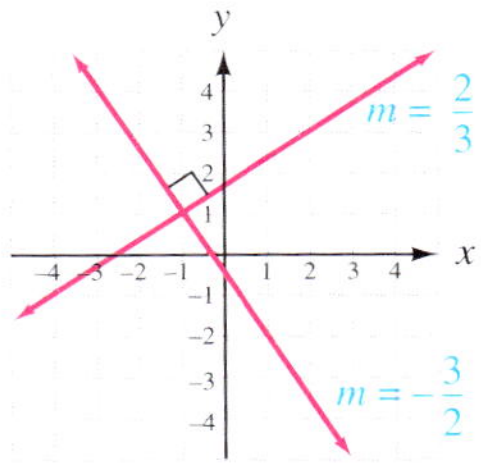 Perpendicular lines

REVIEW EXERCISES

In each case, find the slope of the line.

23.

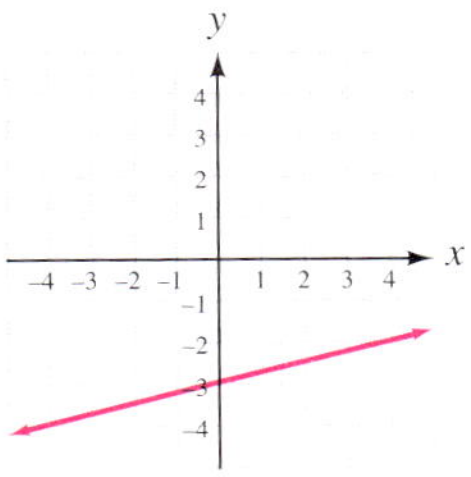

24.

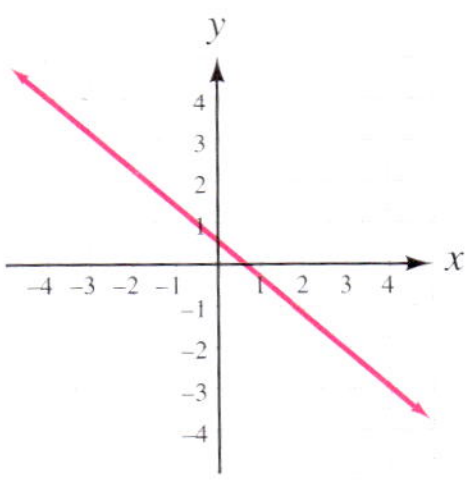

25. The line with this table of solutions:

x	y	(x, y)
2	-3	$(2, -3)$
4	-17	$(4, -17)$

26. The line passing through the points $(1, -4)$ and $(3, -7)$

27. Draw a line having a slope that is

a. Positive b. Negative c. 0 d. Undefined

28. **Carpentry.** If a truss like the one shown below is used to build the roof of a shed, find the slope (pitch) of the roof.

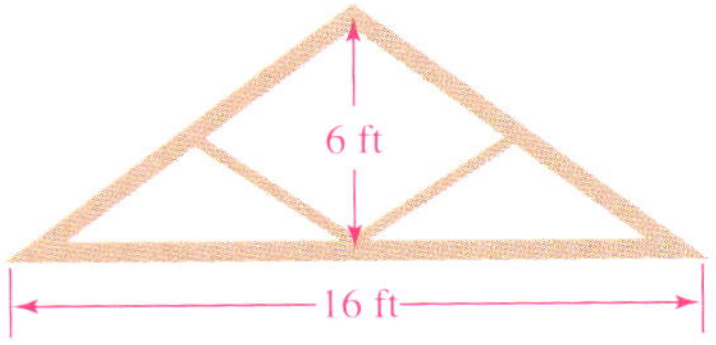

29. **Ramps.** Find the grade of the ramp shown below. Round to the nearest tenth of a percent.

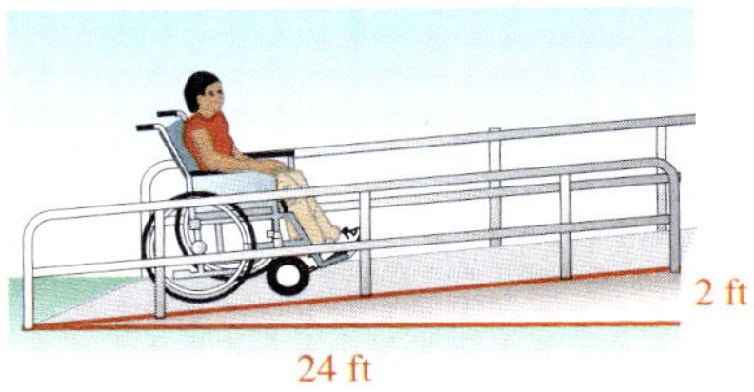

30. Bottled Water. Refer to the graph below. Find the rate of change in the amount of bottled water consumed per person in the U.S. from 2000 through 2008.

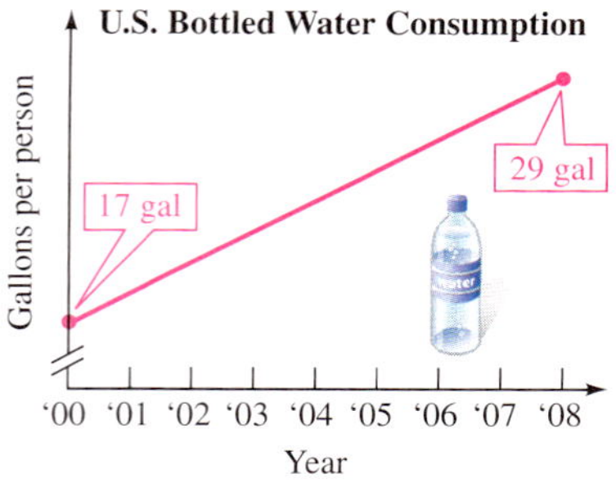

Source: Beverage Marketing Corporation

31. Without graphing, determine whether the line that passes through (6, 6) and (4, 2) and the line that passes through (2, −10) and (−2, −2) are parallel, perpendicular, or neither.

32. Find the slope of a line perpendicular to the line passing through (−1, 9) and (−8, 4).

SECTION 3.5 Slope–Intercept Form

DEFINITIONS AND CONCEPTS	EXAMPLES
If a linear equation is written in **slope–intercept form** $y = mx + b$ the graph of the equation is a line with slope m and y-intercept $(0, b)$. To write a linear equation in two variables in slope–intercept form, solve the equation for y.	Find the slope and y-intercept of the line whose equation is $5x + 3y = 3$. To find the slope and y-intercept, we solve the equation for y. $5x + 3y = 3$ $3y = -5x + 3$ Subtract 5x from both sides. $y = -\frac{5}{3}x + 1$ To isolate y, divide both sides by 3. $m = -\frac{5}{3}$ and $b = 1$. The slope of the line is $-\frac{5}{3}$ and the y-intercept is (0, 1).
To **graph a line in slope–intercept form,** plot the y-intercept and use the slope to determine a second point on the line.	Graph: $y = -\frac{5}{3}x + 1$ $y = -\frac{5}{3}x + 1$ $m = \frac{\text{rise}}{\text{run}} = \frac{-5}{3}$ $b = 1$ y-intercept: (0, 1) 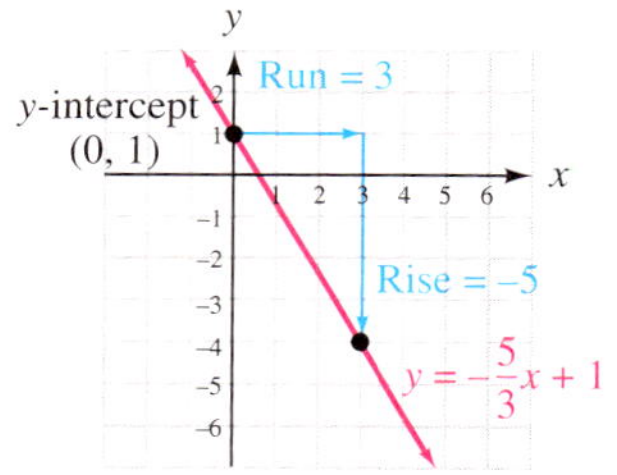
If we know the slope of a line and its y-intercept, we can write its equation.	The equation of a line with slope $\frac{1}{8}$ and y-intercept $(0, -5)$ is $y = \frac{1}{8}x - 5$.
Two different lines with the same slope are **parallel**. If the slopes of two lines are negative reciprocals, the product of their slopes is −1 and the lines are **perpendicular.**	Lines with equations $y = 3x + 4$ and $y = 3x - 12$ are parallel because each line has slope 3. Lines with equations $y = 3x + 4$ and $y = -\frac{1}{3}x - 12$ are perpendicular because their slopes, 3 and $-\frac{1}{3}$, are negative reciprocals.

REVIEW EXERCISES

Find the slope and the y-intercept of each line.

33. $y = \frac{3}{4}x - 2$

34. $y = -4x$

35. $y = \frac{x}{8} + 10$

36. $7x + 5y = -21$

37. Graph the line with slope -4 and y-intercept $(0, -1)$. Write an equation of the line.

38. Write an equation for the line shown here.

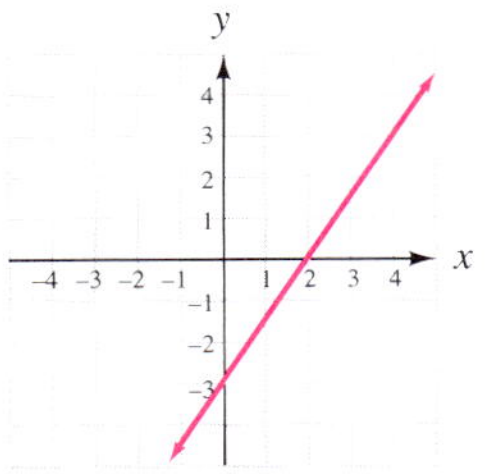

39. Find the slope and the y-intercept of the line whose equation is $9x - 3y = 15$. Then graph it.

40. **Copiers.** A business buys a used copy machine that has already produced 75,000 copies.
   a. If the business plans to run 300 copies a week, write a linear equation that would find the number of copies c the machine has made in its lifetime after the business has used it for w weeks.
   b. Use your result in part a to predict the total number of copies that will have been made on the machine 1 year, or 52 weeks, after being purchased by the business.

Without graphing, determine whether graphs of the given pairs of lines are parallel, perpendicular, or neither.

41. $y = -\frac{2}{3}x + 6$
    $y = -\frac{2}{3}x - 6$

42. $x + 5y = -10$
    $y - 5x = 0$

SECTION 3.6 Point–Slope Form

DEFINITIONS AND CONCEPTS	EXAMPLES
If a line with slope m passes through the point with coordinates (x_1, y_1), the equation of the line in **point–slope form** is $y - y_1 = m(x - x_1)$	Find an equation of the line with slope -3 that passes through $(-2, 4)$. Write the equation in slope–intercept form. We substitute the slope and the coordinates of the point into the point–slope form. $y - y_1 = m(x - x_1)$ This is point–slope form. $y - 4 = -3[x - (-2)]$ Substitute. $y - 4 = -3(x + 2)$ Simplify within the brackets. $y - 4 = -3x - 6$ Distribute. $y = -3x - 2$ To isolate y, add 4 to both sides. This is slope–intercept form.
If we know **two points that a line passes through,** we can write its equation.	Find an equation of the line that passes through (2, 5) and (3, 7). Write the equation in slope–intercept form. The slope of the line is: $m = \frac{y_2 - y_1}{x_2 - x_1} = \frac{7 - 5}{3 - 2} = 2$ Either point on the line can serve as (x_1, y_1). If we use (2, 5), we have: $y - y_1 = m(x - x_1)$ This is point–slope form. $y - 5 = 2(x - 2)$ Substitute: $x_1 = 2$, $y_1 = 5$, and $m = 2$. $y - 5 = 2x - 4$ Distribute. $y = 2x + 1$ To isolate y, add 5 to both sides. This is the slope–intercept form.

REVIEW EXERCISES

Find an equation of the line with the given slope that passes through the given point. Write the equation in slope–intercept form and graph the equation.

43. $m = 3, (1, 5)$

44. $m = -\frac{1}{2}, (-4, -1)$

Find an equation of the line with the following characteristics. Write the equation in slope–intercept form.

45. Passing through $(3, 7)$ and $(-6, 1)$

46. Horizontal, passing through $(6, -8)$

47. Car Registration. When it was 2 years old, the annual registration fee for a Dodge Caravan was \$380. When it was 4 years old, the registration fee dropped to \$310. If the relationship is linear, write an equation that gives the registration fee f in dollars for the van when it is x years old.

48. The Atmosphere. The scatter diagram below shows the amount of carbon dioxide in the Earth's atmosphere as measured at Hawaii's Mauna Loa Observatory from 1960 through 2010. A straight line can be used to model the data.

a. Use the two highlighted points in red to write the equation of the line. Write the answer in slope–intercept form.

b. Use your answer to part a to predict the amount of carbon dioxide in the atmosphere in 2020.

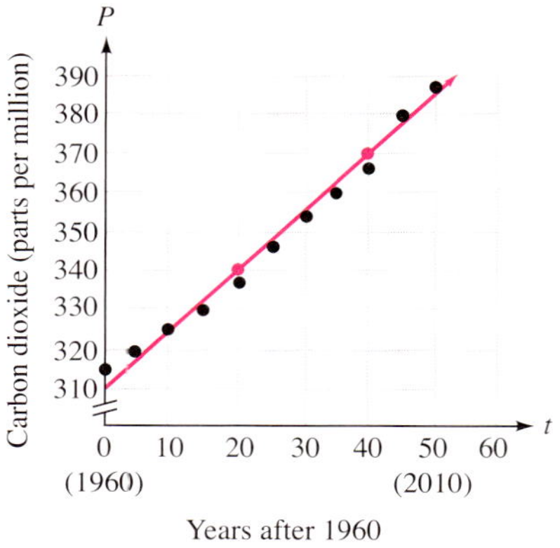

Source: National Oceanic and Atmospheric Administration

SECTION 3.7 Graphing Linear Inequalities

DEFINITIONS AND CONCEPTS | **EXAMPLES**

An ordered pair (x, y) is a **solution of an inequality** in x and y if a true statement results when the variables are replaced by the coordinates of the ordered pair.

Determine whether $(-2, 5)$ is a solution of $x + 3y > -6$.

We substitute the coordinates into the inequality and see if a true statement results.

$x + 3y > -6$ This is the given inequality.

$-2 + 3(5) \overset{?}{>} -6$ Substitute.

$13 > -6$ True

Since the result $13 > -6$ is true, $(-2, 5)$ is a solution.

To graph a linear inequality:

1. Replace the inequality symbol with an $=$ symbol and graph the **boundary line.** Draw a solid line if the inequality contains $\le$ or $\ge$ and a dashed line if it contains $<$ or $>$.
2. Pick a **test point** not on the boundary. Substitute its coordinates into the inequality. If the inequality is satisfied, shade the half-plane that contains the test point. If the inequality is not satisfied, shade the other half-plane.

Graph: $2x - y \le 4$

1. Graph the boundary line $2x - y = 4$ and draw it as a solid line because the inequality symbol is $\le$.
2. Test the point $(0, 0)$:

$2x - y \le 4$ This is the given inequality.

$2(0) - 0 \overset{?}{\le} 4$ Substitute.

$0 \le 4$ True

Since the coordinates of the test point satisfy the inequality, we shade the side of the boundary line that contains $(0, 0)$.

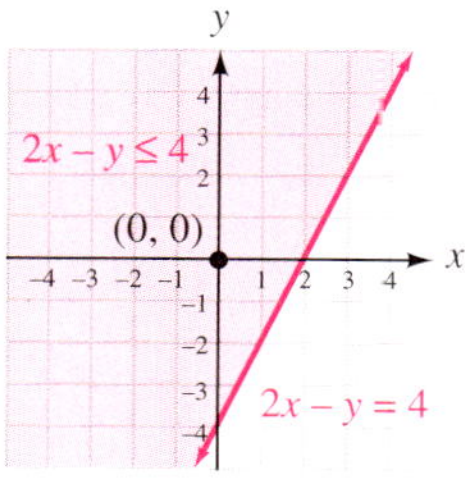

REVIEW EXERCISES

49. Determine whether each ordered pair is a solution of $2x - y \le -4$.
   a. $(0, 5)$ b. $(2, 8)$
   c. $(-3, -2)$ d. $\left(\frac{1}{2}, -5\right)$

50. Fill in the blanks: $2x - 3y \ge 6$ means $2x - 3y$ ___ 6 or $2x - 3y$ ___ 6.

Graph each inequality.

51. $x - y < 5$
52. $2x - 3y \ge 6$
53. $y \le -2x$
54. $y < -4$
55. The graph of a linear inequality is shown in the next column. Would a true or a false statement result if the coordinates of
   a. point A were substituted into the inequality?
   b. point B were substituted into the inequality?
   c. point C were substituted into the inequality?

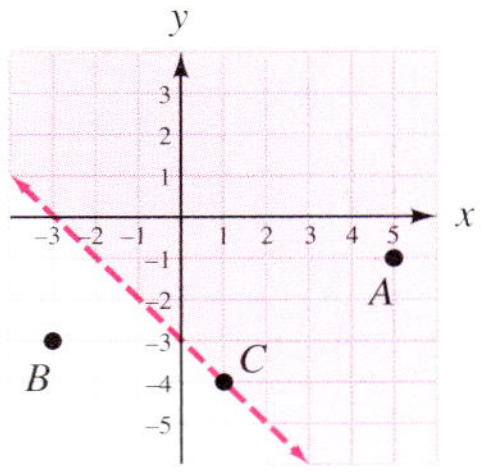

56. **Work Schedules.** A student told her employer that during the school year, she would be available for up to 30 hours a week, working either 3- or 5-hour shifts. If x represents the number of 3-hour shifts she works and y represents the number of 5-hour shifts she works, the inequality $3x + 5y \le 30$ shows the possible combinations of shifts she can work. Graph the inequality and find three possible combinations.

SECTION 3.8 An Introduction to Functions

DEFINITIONS AND CONCEPTS | **EXAMPLES**

A **relation** is a set of ordered pairs. The set of all **first components** is called the **domain** of the relation and the set of all **second components** is called the **range** of a relation.

The relation $\{(4, 7), (0, -3), (-3, 8), (1, 7)\}$ has the domain $\{-3, 0, 1, 4\}$ and the range is $\{-3, 7, 8\}$.

A **function** is a set of ordered pairs (a relation) in which to each first component there corresponds exactly one second component.

If to each value of x in the domain there is assigned exactly one value of y in the range, then **y is a function of x.**

x	y
1	2
5	3
6	8
5	9

y is not a function of x: $5 \to 3$ and $5 \to 9$

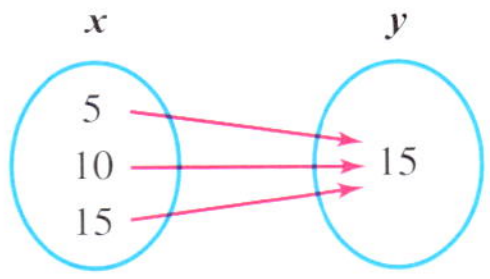

y is a function of x

$\{(4, 7), (7, 4), (4, 1)\}$

Not a function: $4 \to 7$ and $4 \to 1$

A function can be defined by an equation. The notation $y = f(x)$ indicates that the variable y is a function of x. It is read as "f of x."

We can think of a function as a machine that takes some **input** x and turns it into some **output** $f(x)$, called a **function value.**

For the function $f(x) = 8x + 5$, $f(-2)$ is the value of $f(x)$ when $x = -2$.

$$f(x) = 8x + 5$$
$$f(-2) = 8(-2) + 5 \quad \text{Substitute the input } -2 \text{ for each } x.$$
$$= -16 + 5 \quad \text{Evaluate the right side.}$$
$$= -11$$

Thus, $f(-2) = -11$.

The **vertical line test:** If a vertical line intersects a graph in more than one point, the graph is not the graph of a function.

Vertical lines intersect only once

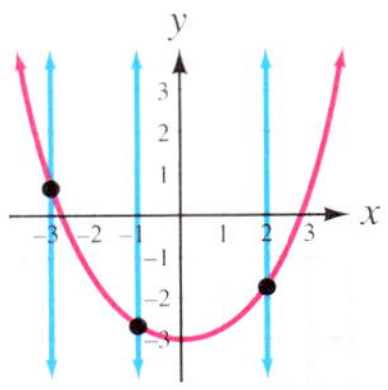

A function

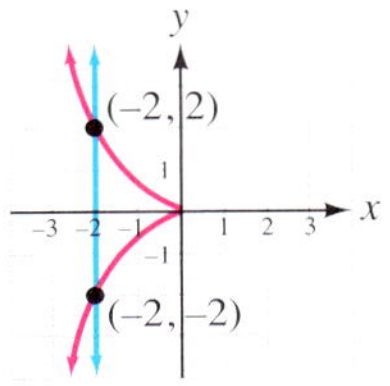

Not a function

A vertical line intersects more than once

x	y
−2	2
−2	2

The input-output pairs that a function generates can be written as ordered pairs and plotted on a rectangular coordinate system to give the **graph of a function.**

Function notation	***Ordered-pair notation***
$f(-3) = 5$	$(-3, 5)$

Graph the function: $f(x) = -\frac{2}{3}x + 3$

We make a table of function values, plot the points, and draw the graph.

$f(x) = -\frac{2}{3}x + 3$

x	$f(x)$	
-3	5	→ $(-3, 5)$
0	3	→ $(0, 3)$
3	1	→ $(3, 1)$

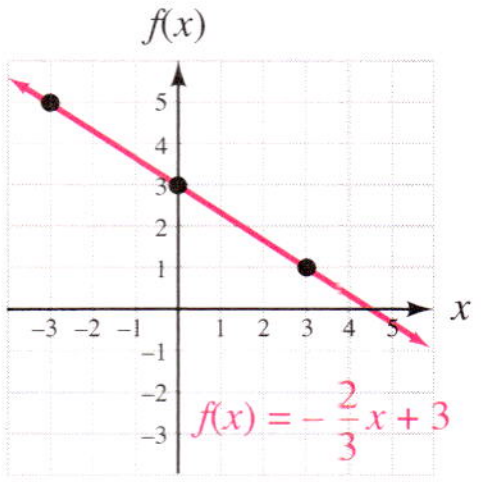

REVIEW EXERCISES

Find the domain and range of each relation.

57. $\{(7, -3), (-5, 9), (4, 4), (0, -11)\}$

58. $\{(2, -2), (15, -8), (-6, 9), (1, -8)\}$

Determine whether each relation defines y to be function of x. If a function is defined, give its domain and range. If it does not define a function, find ordered pairs that show a value of x that corresponds to more than one value of y.

59.

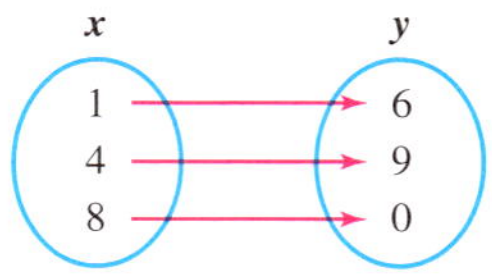

60.

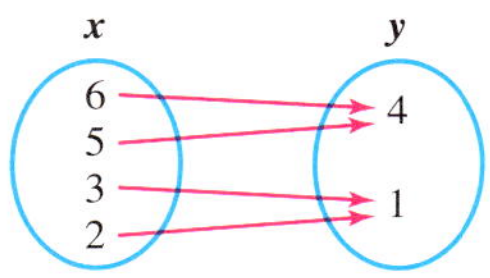

61.

x	y
9	81
7	49
5	25
3	9

62.

x	y
-1	2
0	3
-1	4
1	5

63. $\{(-1, 6), (0, 6), (1, 6), (2, 6)\}$

64. $\{(4, 4), (6, 4), (4, 6)\}$

Fill in the blanks.

65. The set of all input values for a function is called the ________, and the set of all output values is called the ________.

66. Fill in the blank: Since $y =$ ______, the equations $y = 2x - 8$ and $f(x) = 2x - 8$ are equivalent.

For $f(x) = x^2 - 4x$, find each of the following function values.

67. $f(1)$ 68. $f(0)$ 69. $f(-3)$ 70. $f\left(\frac{1}{2}\right)$

For $g(x) = 1 - 6x$, find each of the following function values.

71. $g(1)$ 72. $g(-6)$ 73. $g(0.5)$ 74. $g\left(\frac{3}{2}\right)$

Determine whether each graph is the graph of a function. If it is not, find two ordered pairs that show a value of x that corresponds to more than one value of y.

75.

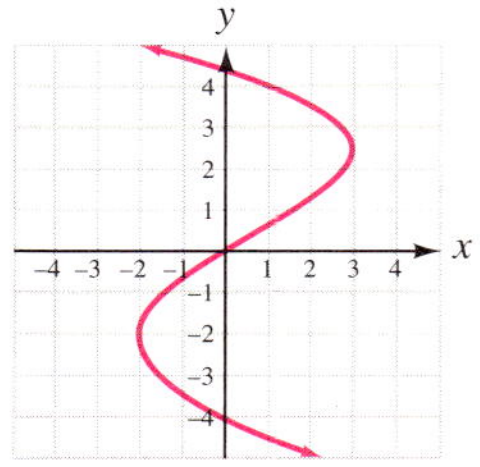

76.

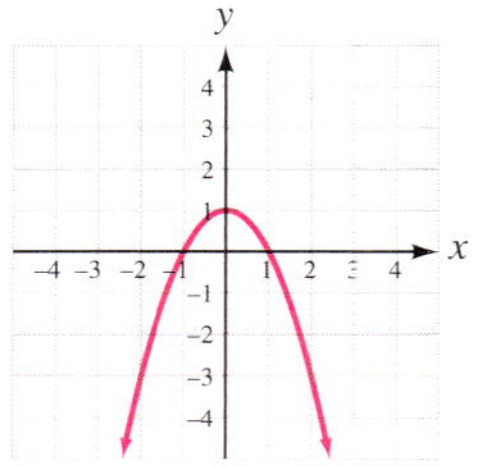

77. Complete the table of function values for $f(x) = 1 - |x|$. Then graph the function.

x	$f(x)$
0	
1	
3	
-1	
-3	

78. **Aluminum Cans.** The function $V(r) = 15.7r^2$ estimates the volume in cubic inches of a can 5 inches tall with a radius of r inches. Find the volume of the can shown in the illustration.

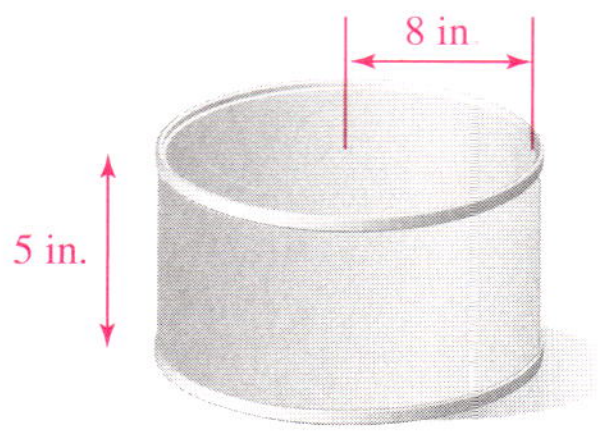

3 CHAPTER TEST

1. Fill in the blanks.
   a. A rectangular coordinate system is formed by two perpendicular number lines called the x-______ and the y-______.
   b. A ________ of an equation in two variables is an ordered pair of numbers that makes the equation a true statement.
   c. $3x + y = 10$ is a ______ equation in two variables and its graph is a line.
   d. The ______ of a line is a measure of steepness.
   e. A ________ is a set of ordered pairs in which to each first component there corresponds exactly one second component.

The graph shows the number of dogs being boarded in a kennel over a 3-day holiday weekend.

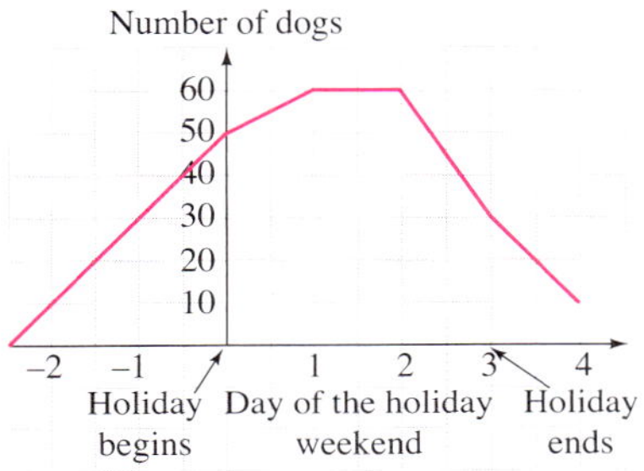

2. How many dogs were in the kennel 2 days before the holiday?
3. What is the maximum number of dogs that were boarded on the holiday weekend at any one time?
4. When were there 30 dogs in the kennel?
5. What information does the y-intercept of the graph give?
6. Plot each point on a rectangular coordinate system: $(1, 3)$, $(-2, 4)$, $(-3, -2)$, $(3, -2)$, $(-1, 0)$, $(0, -1)$, and $\left(-\frac{1}{2}, \frac{7}{2}\right)$.
7. Find the coordinates of each point shown in the graph.

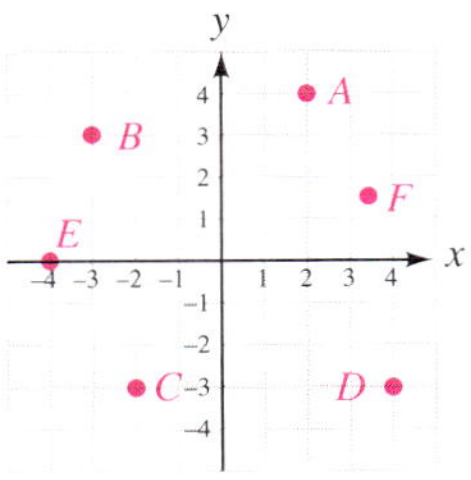

8. In which quadrant is each point located?
   a. $(-1, -5)$
   b. $\left(6, -2\frac{3}{4}\right)$
9. Is $(-3, -4)$ a solution of $3x - 4y = 7$?
10. Complete the table of solutions for the linear equation $x + 4y = 6$.

x	y	(x, y)
2		
	3	

11. The graph of a linear equation is shown.
    a. If the coordinates of point C are substituted into the equation, will the result be true or false?
    b. If the coordinates of point D are substituted into the equation, will the result be true or false?

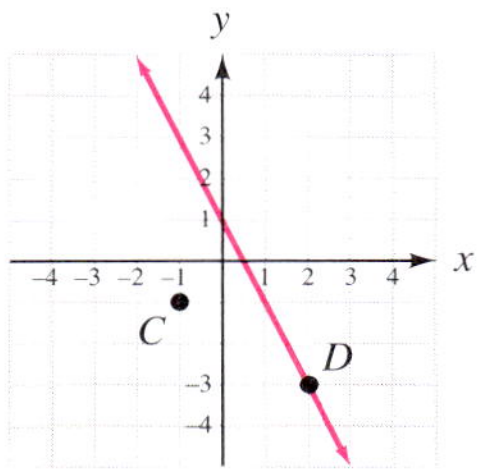

12. Graph: $y = \frac{x}{3}$
13. What are the x- and y-intercepts of the graph of $2x - 3y = 6$?
14. Graph: $8x + 4y = -24$
15. Find the slope of the line.

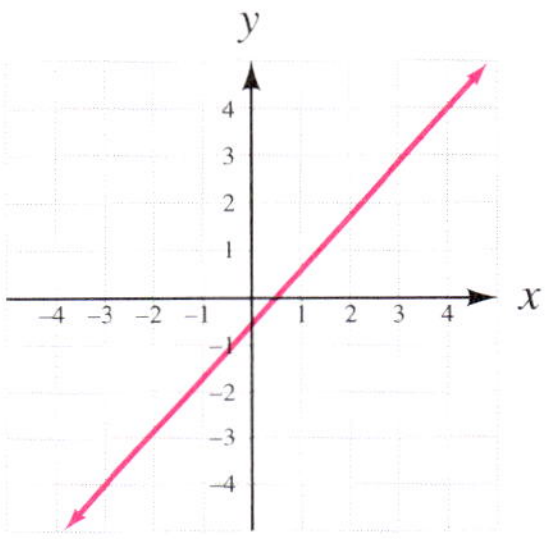

16. Find the slope of the line passing through $(-1, 3)$ and $(3, -1)$.
17. What is the slope of a horizontal line?
18. **Ramps.** Find the grade of a ramp that rises 2 feet over a horizontal distance of 20 feet.
19. One line passes through $(9, 2)$ and $(6, 4)$. Another line passes through $(0, 7)$ and $(2, 10)$. Without graphing, determine whether the lines are parallel, perpendicular, or neither.
20. When graphed, are the lines $y = 2x + 6$ and $2x - y = 0$ parallel, perpendicular, or neither?

In Problems 21 and 22, refer to the illustration that shows the elevation changes for part of a 26-mile marathon course.

21. Find the rate of change of the decline on which the woman is running.
22. Find the rate of change of the incline on which the man is running.

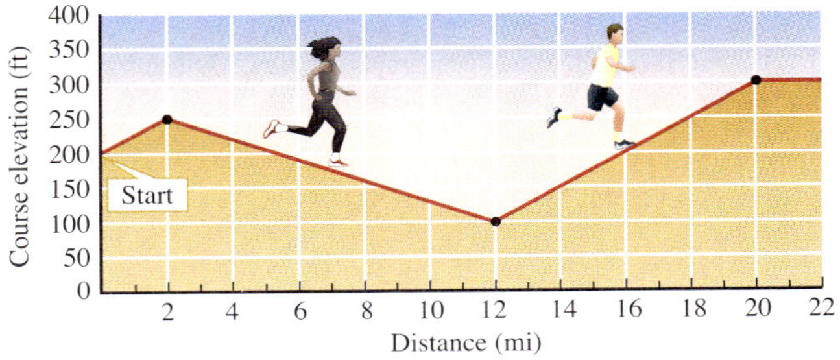

23. Graph: $x = -4$

24. Graph the line passing through $(-2, -4)$ having slope $\frac{2}{3}$.

25. Find the slope and the y-intercept of the graph of $x + 2y = 8$.

26. Find an equation of the line passing through $(-2, 5)$ with slope 7. Write the equation in slope–intercept form.

27. Find an equation for the line shown. Write the equation in slope–intercept form.

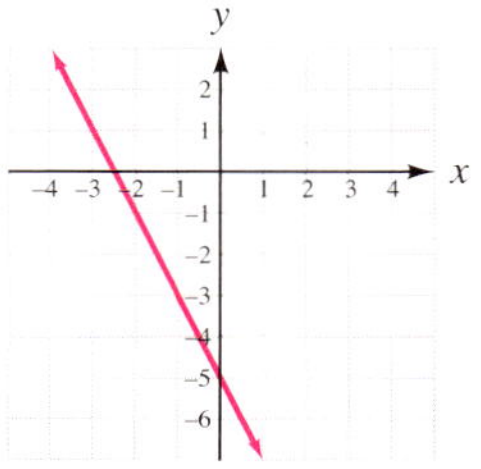

28. **Depreciation.** After it is purchased, a \$15,000 computer loses \$1,500 in resale value every year.

a. Write a linear equation that gives the resale value v of the computer x years after being purchased.

b. Use your answer to part (a) to predict the value of the computer 8 years after it is purchased.

29. Check to determine whether $(6, 1)$ is a solution of $2x - 4y \geq 8$.

30. **Water Heaters.** The scatter diagram shows how excessively high temperatures affect the life of a water heater. Write an equation of the line that models the data for water temperatures between 140° and 180°. Let T represent the temperature of the water in degrees Fahrenheit and y represent the expected life of the heater in years. Give the answer in slope–intercept form.

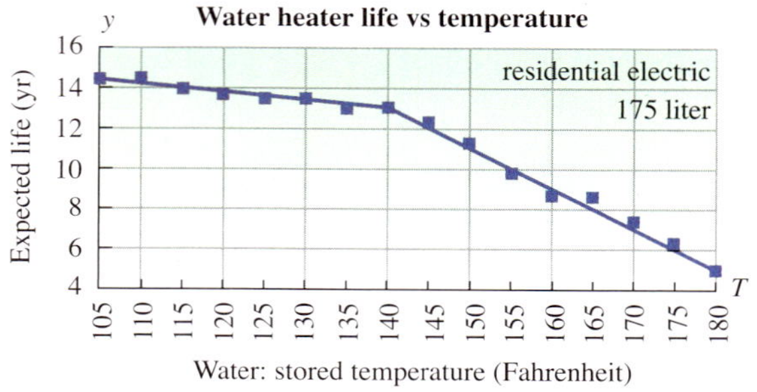

Source: www.uniongas.com/WaterHeating

31. A linear inequality has been graphed below. Determine whether each point satisfies the inequality.

a. $(-2, 3)$
b. $(3, -4)$
c. $(0, 0)$

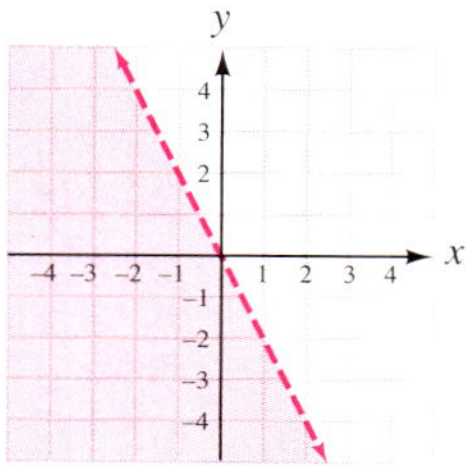

32. Is $(-20, -2)$ a solution of $\frac{1}{2}x - 3y \geq -4$?

33. Graph the inequality: $2x - 5y \leq -10$

34. Find the domain and range of the relation: $\{(5, 3), (1, 12), (-4, 3), (0, -8)\}$

Determine whether the relation defines y to be function of x. If a function is defined, give its domain and range. If it does not define a function, find ordered pairs that show a value of x that corresponds to more than one value of y.

35.

x	y
1	4
2	3
3	2
4	1

36.

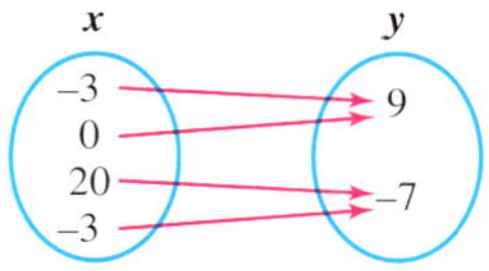

37. $\{(6, 5), (7, 5), (8, 5), (9, 5), (10, 5)\}$

38. $\{(-9, 41), (2, 6), (4, -9), (2, 2)\}$

39.

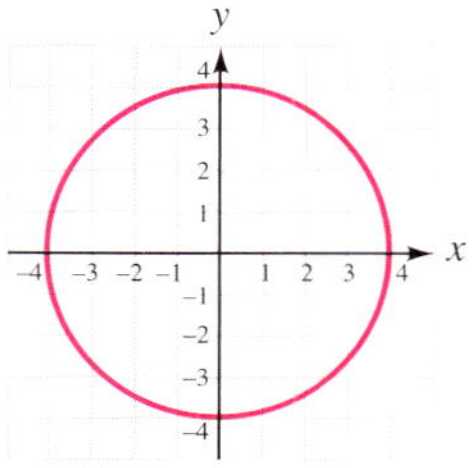

40.

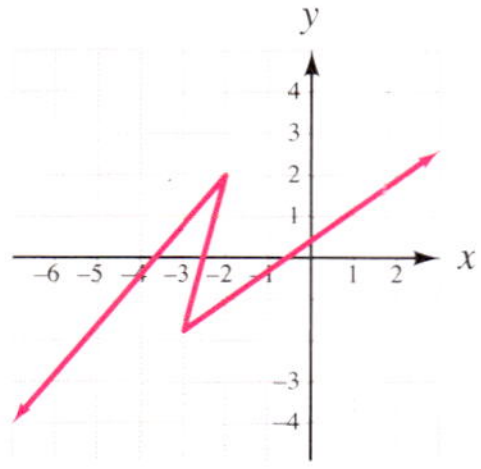

41. If $f(x) = 2x - 7$, find: $f(-3)$

42. If $g(x) = 3.5x^3$ find: $g(6)$

43. **Telephone Calls.** The function $C(n) = 0.30n + 15$ gives the cost C per month in dollars for making n phone calls. Find $C(45)$ and explain what it means.

44. Graph: $f(x) = |x| - 1$

Group Project

Overview: In this activity, you will explore the relationship between a person's height and arm span. Arm span is defined to be the distance between the tips of a person's fingers when his or her arms are held out to the side.

Instructions: Form groups of 5 or 6 students. Measure the height and arm span of each person in your group, and record the results in a table like the one shown below.

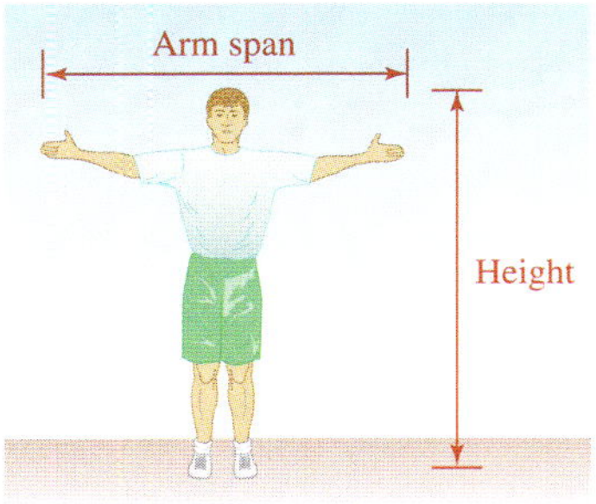

Name	Height (in.)	Arm span (in.)
1.		
2.		
3.		
4.		
5.		
6.		

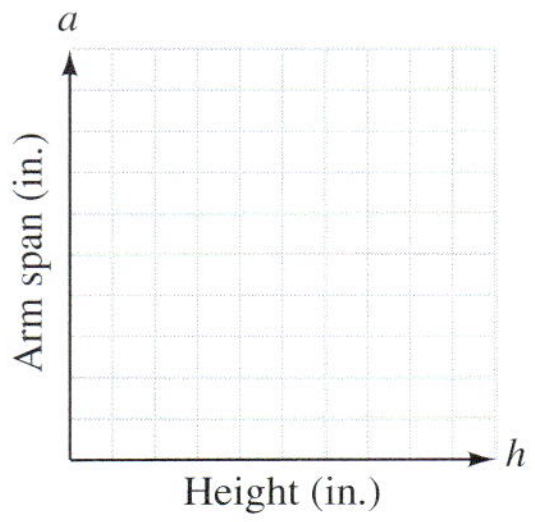

Plot the data in the table as ordered pairs of the form (height, arm span) on a graph like the one shown above. Then draw a straight-line model that best fits the data points.

Pick two convenient points on the line and find its slope. Use the point–slope form $a - a_1 = m(h - h_1)$ to find an equation of the line. Then, write the equation in slope–intercept form.

Ask a person from another group for his or her height measurement. Substitute that value into your linear model to predict that person's arm span. How close is your prediction to the person's actual arm span?

(From *Activities for Beginning and Intermediate Algebra* by Debbie Garrison, Judy Jones, and Jolene Rhodes)

CUMULATIVE REVIEW Chapters 1–3

1. Find the prime factorization of 108. [Section 1.2]
2. Write $\frac{1}{250}$ as a decimal. [Section 1.3]
3. Determine whether each statement is true or false. [Section 1.3]
   a. Every whole number is an integer.
   b. Every integer is a real number.
   c. 0 is a whole number, an integer, and a rational number.

Perform the operations.

4. $-27 + 21 + (-9)$ [Section 1.4]
5. $-1.57 - (-0.8)$ [Section 1.5]
6. $-9(-7)(5)(-3)$ [Section 1.6]
7. $\dfrac{-180}{-6}$ [Section 1.6]
8. Evaluate: $\left|\dfrac{(6-5)^4 - (-21)}{-27 + 4^2}\right|$ [Section 1.7]
9. Evaluate $b^2 - 4ac$ for $a = 2$, $b = -8$, and $c = 4$. [Section 1.8]
10. Suppose x sheets from a 500-sheet ream of paper have been used. How many sheets are left? [Section 1.8]
11. How many terms does the algebraic expression $3x^2 - 2x + 1$ have? What is the coefficient of the second term? [Section 1.8]
12. Multiply. [Section 1.9]
    a. $2(x + 4)$
    b. $-2(x - 4)$

Simplify each expression. [Section 1.9]

13. $5a + 10 - a$
14. $-7(9t)$
15. $-2b^2 + 6b^2$
16. $5(-17)(0)(2)$
17. $(a + 2) - (a - 2)$
18. $-4(-5)(-8a)$
19. $-y - y - y$
20. $\dfrac{3}{2}(4x - 8) + x$

Solve each equation. [Sections 2.1 and 2.2]

21. $3x - 5 = 13$
22. $1.2 - x = -1.7$
23. $\dfrac{2x}{3} - 2 = 4$
24. $\dfrac{y - 2}{7} = -3$
25. $-3(2y - 2) - y = 5$
26. $9y - 3 = 6y$
27. $\dfrac{1}{3} + \dfrac{c}{5} = -\dfrac{3}{2}$
28. $5(x + 2) = 5x - 2$
29. $-x = 99$
30. $3c - 2 = \dfrac{11(c - 1)}{5}$

31. **Pennies.** A 2010 telephone survey of adults asked whether the penny should be discontinued from the national currency. The results are shown in the circle graph. If 869 people favored keeping the penny, how many took part in the survey? [Section 2.3]

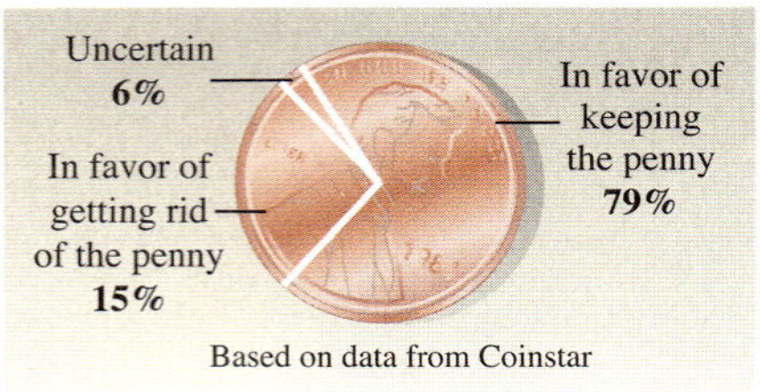

32. Solve for h: $S = 2\pi rh + 2\pi r^2$ [Section 2.4]

33. **Band Aids.** Find the perimeter and the area of the gauze pad of the bandage. [Section 2.4]

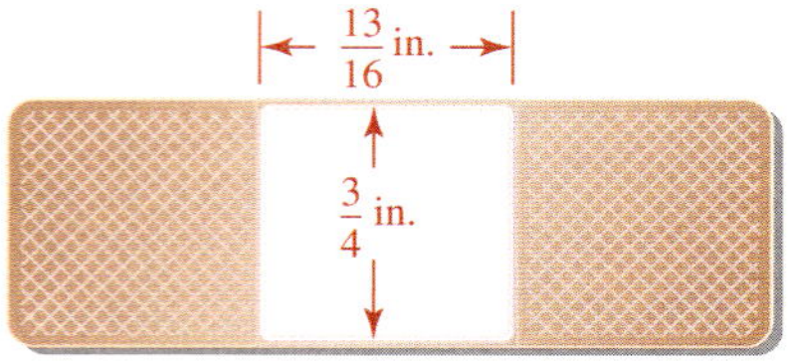

34. **High Heels.** Find the unknown angle measure represented by x. [Section 2.5]

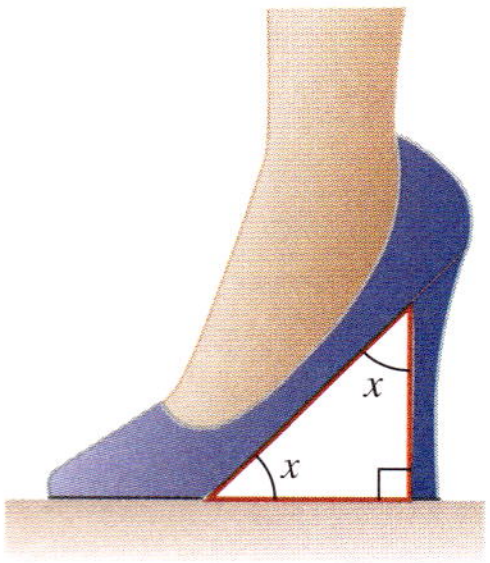

35. Complete the table. [Section 2.6]

	% acid	Liters	Amount of acid
50% solution	0.50	x	
25% solution	0.25	$13 - x$	
30% mixture	0.30	13	

36. **Road Trips.** A bus, carrying the members of a marching band, and a truck, carrying their instruments, leave a high school at the same time. The bus travels at 60 mph and the truck at 50 mph. In how many hours will they be 75 miles apart? [Section 2.6]

37. **Mixing Candy.** Candy corn worth \$2.85 per pound is to be mixed with black gumdrops that cost \$1.80 per pound to make 200 pounds of a mixture worth \$2.22 per pound. How many pounds of each candy should be used? [Section 2.6]

Solve each inequality. Write the solution set in interval notation and graph it. [Section 2.7]

38. $-\frac{3}{16}x \geq -9$

39. $8x + 4 > 3x + 4$

40. In which quadrants are the second coordinates of ordered pairs positive? [Section 3.1]

41. Is $(-2, 4)$ a solution of $y = 2x - 8$? [Section 3.2]

Graph each equation.

42. $y = x$ [Section 3.2]

43. $2x + 4y = -8$ [Section 3.3]

44. What is the slope of the graph of the line $y = 5$? [Section 3.4]

45. What is the slope of the line passing through $(-2, 4)$ and $(5, -6)$? [Section 3.4]

46. **Roofing.** Find the pitch of the roof. [Section 3.4]

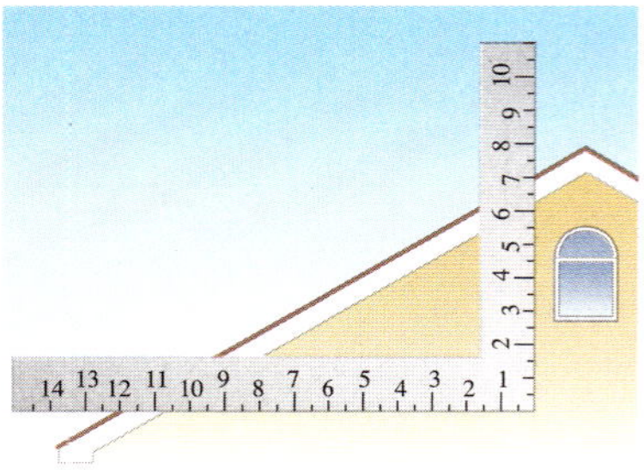

47. Find the slope and the y-intercept of the graph of the line described by $4x - 6y = -12$. [Section 3.5]

48. Write an equation of the line that has slope -2 and y-intercept $(0, 1)$. [Section 3.5]

49. Find an equation of the line that has slope $-\frac{7}{8}$ and passes through $(2, -9)$. Write the equation in point–slope form and in slope–intercept form. [Section 3.6]

50. Is $(-2, -4)$ a solution of $x + y \leq -6$? [Section 3.7]

51. Graph: $y \geq x + 1$ [Section 3.7]

52. Graph $x < 4$ on a rectangular coordinate system. [Section 3.7]

53. If $f(x) = x^4 + x$, find: $f(-3)$ [Section 3.8]

54. Is this the graph of a function? [Section 3.8]

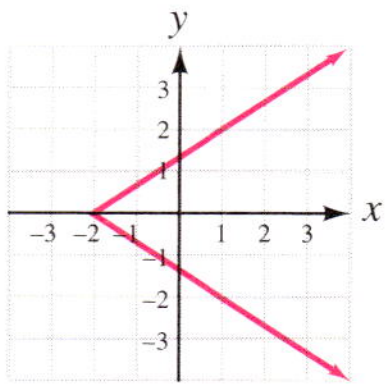

Systems of Linear Equations and Inequalities

4

from Campus to Careers

Photographer

Photographers record our surroundings, the special events in our lives, and people, so that all can be remembered in pictures. Some specialize in weddings and portraits, some photograph landscapes and fashion, while others work on location as photojournalists. Their job responsibilities require a variety of mathematical skills such as: scheduling appointments, keeping financial records, pricing photographs, purchasing supplies, billing customers, and operating digital equipment.

Problem 79 in **Study Set 4.1, problem 29** in **Study Set 4.4,** and **problem 55** in **Study Set 4.5** involve situations that a photographer might encounter on the job. The mathematical concepts discussed in this chapter can be used to solve those problems.

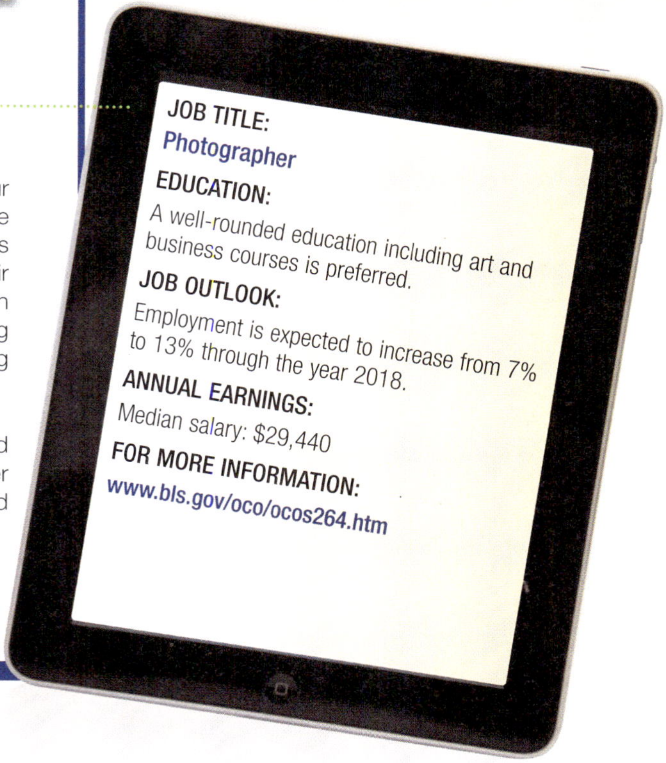

Study Skills Workshop ▶ Making Homework a Priority

Attending class and taking notes are important, but they are not enough. The only way that you are really going to learn algebra is by doing your homework.

WHEN TO DO YOUR HOMEWORK: Homework should be started on the day it is assigned, when the material is fresh in your mind. It's best to break your homework sessions into 30-minute periods, allowing for short breaks in between.

HOW TO BEGIN YOUR HOMEWORK: Review your notes and the examples in your text before starting your homework assignment.

GETTING HELP WITH YOUR HOMEWORK: It's normal to have some questions when doing homework. Talk to a tutor, a classmate, or your instructor to get those questions answered.

Now Try This ▶

1. Write a one-page paper that describes *when, where,* and *how* you go about completing your algebra homework assignments.
2. For each problem on your next homework assignment, find an example in this book that is similar. Write the example number next to the problem.
3. Make a list of questions that you have while doing your next assignment. Then decide whom you are going to ask to get those questions answered.

SECTION 4.1 Solving Systems of Equations by Graphing

OBJECTIVES

1. Determine whether a given ordered pair is a solution of a system.
2. Solve systems of linear equations by graphing.
3. Use graphing to identify inconsistent systems and dependent equations.
4. Identify the number of solutions of a linear system without graphing.
5. Use a graphing calculator to solve a linear system (optional).

ARE YOU READY?

The following problems review some basic skills that are needed when solving systems of equations by graphing.

1. Is $(-2, -6)$ a solution of $y = 3x - 1$?
2. Use the slope and the y-intercept to graph $y = -4x + 2$.
3. Graph $3x + 4y = 12$ by finding the x- and y-intercepts.
4. Without graphing, determine whether the graphs of $y = \frac{1}{2}x - 3$ and $x - 2y = 2$ are parallel, perpendicular, or neither.

The following illustration shows the average amounts of chicken and beef eaten per person each year in the United States from 1985 to 2010. Plotting both graphs on the same coordinate system makes it easy to compare recent trends. The point of intersection of the graphs indicates that Americans ate equal amounts of chicken and beef in 1992—about 66 pounds of each, per person.

In this section, we will use a similar graphical approach to solve systems of equations.

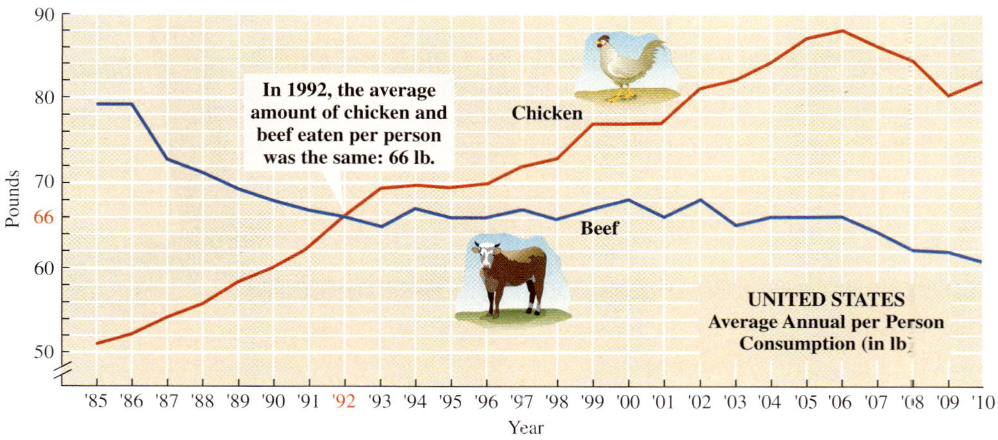

Source: U.S. Department of Agriculture

1 Determine Whether a Given Ordered Pair Is a Solution of a System.

We previously have discussed equations in two variables, such as $x + y = 3$. Because there are infinitely many pairs of numbers whose sum is 3, there are infinitely many pairs (x, y) that satisfy this equation. Some of these pairs are listed in table (a).

Now consider the equation $x - y = 1$. Because there are infinitely many pairs of numbers whose difference is 1, there are infinitely many pairs (x, y) that satisfy $x - y = 1$. Some of these pairs are listed in table (b).

The Language of Algebra

We say that (2, 1) **satisfies** $x + y = 3$, because the x-coordinate, 2, and the y-coordinate, 1, make the equation true when substituted for x and y: $2 + 1 = 3$. To *satisfy* means to make content, as in *satisfy* your thirst or a *satisfied* customer.

$x + y = 3$

x	y	(x, y)
0	3	(0, 3)
1	2	(1, 2)
2	**1**	**(2, 1)**
3	0	(3, 0)

(a)

$x - y = 1$

x	y	(x, y)
0	−1	(0, −1)
1	0	(1, 0)
2	**1**	**(2, 1)**
3	2	(3, 2)

(b)

From the two tables, we see that (2, 1) satisfies both equations.

When two equations with the same variables are considered simultaneously (at the same time), we say that they form a **system of equations.** Using a left brace { , we can write the equations from the previous example as a system:

$$\begin{cases} x + y = 3 \\ x - y = 1 \end{cases}$$ Read as "the system of equations $x + y = 3$ and $x - y = 1$."

Because the ordered pair (2, 1) satisfies both of these equations, it is called a **solution of the system.** In general, a system of linear equations can have exactly one solution, no solution, or infinitely many solutions.

EXAMPLE 1 Determine whether (−2, 5) is a solution of each system of equations.

a. $\begin{cases} 3x + 2y = 4 \\ x - y = -7 \end{cases}$ **b.** $\begin{cases} 4y = 18 - x \\ y = 2x \end{cases}$

Strategy We will substitute the x- and y-coordinates of (−2, 5) for the corresponding variables in both equations of the system.

Why If both equations are satisfied (made true) by the x- and y-coordinates, then the ordered pair is a solution of the system.

Solution

The Language of Algebra

A system of equations is two (or more) equations that we consider **simultaneously**—at the same time. On June 3, 2007, in Kansas City, more than 1,680 guitarists set a world record for the most people playing the same song *simultaneously.* The song was Deep Purple's *Smoke on the Water.*

a. Recall that in an ordered pair, the first number is the x-coordinate and the second number is the y-coordinate. To determine whether (−2, 5) is a solution, we substitute −2 for x and 5 for y in each equation.

Check:

$3x + 2y = 4$ The first equation.
$3(-2) + 2(5) \stackrel{?}{=} 4$
$-6 + 10 \stackrel{?}{=} 4$
$4 = 4$ True

$x - y = -7$ The second equation.
$-2 - 5 \stackrel{?}{=} -7$
$-7 = -7$ True

Since (−2, 5) satisfies both equations, it is a solution of the system.

b. Again, we substitute −2 for x and 5 for y in each equation.

Check:

$4y = 18 - x$ The first equation.
$4(5) \stackrel{?}{=} 18 - (-2)$
$20 \stackrel{?}{=} 18 + 2$
$20 = 20$ True

$y = 2x$ The second equation.
$5 \stackrel{?}{=} 2(-2)$
$5 = -4$ False

Although (−2, 5) satisfies the first equation, it does not satisfy the second. Because it does not satisfy both equations, (−2, 5) is not a solution of the system.

Self Check 1 Determine whether $(4, -1)$ is a solution of: $\begin{cases} x - 2y = 6 \\ y = 3x - 11 \end{cases}$

Now Try Problem 15

2 Solve Systems of Linear Equations by Graphing.

To **solve a system of equations** means to find all of the solutions of the system. One way to solve a system of linear equations is to graph the equations on the same rectangular coordinate system.

EXAMPLE 2 Solve the system of equations by graphing: $\begin{cases} 2x + 3y = 2 \\ 3x - 2y = 16 \end{cases}$

Strategy We will graph both equations on the same coordinate system.

Why Recall that the graph of a linear equation is a "picture" of its solutions. If both equations are graphed on the same coordinate system, we can see whether they have any common solutions.

Solution The intercept-method is a convenient way to graph equations such as $2x + 3y = 2$ and $3x - 2y = 16$, because they are in standard $Ax + By = C$ form.

Success Tip

Accuracy is crucial when using the graphing method to solve a system. Here are some suggestions for improving your accuracy:

- Use graph paper.
- Use a sharp pencil.
- Use a ruler or straightedge.

$2x + 3y = 2$

x	y	(x, y)
0	$\frac{2}{3}$	$(0, \frac{2}{3})$
1	0	$(1, 0)$
-2	2	$(-2, 2)$

$3x - 2y = 16$

x	y	(x, y)
0	-8	$(0, -8)$
$\frac{16}{3}$	0	$(\frac{16}{3}, 0)$
2	-5	$(2, -5)$

To find the y-intercept, let $x = 0$ and solve for y.
To find the x-intercept, let $y = 0$ and solve for x.
As a check, pick another x-value, such as -2 or 2, and find y.

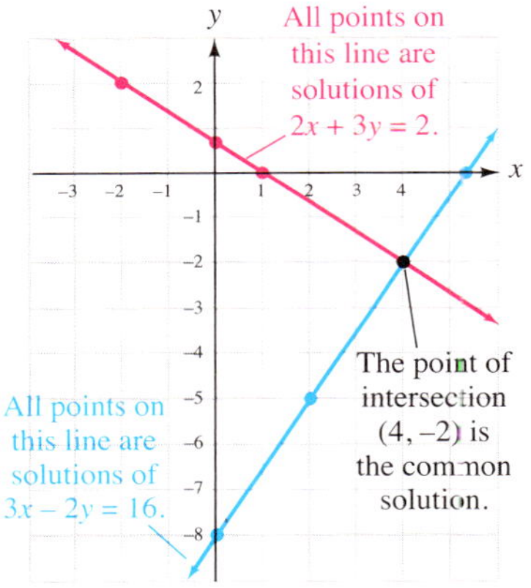

Success Tip

When determining the coordinates of a point of intersection from a graph, realize that they are simply estimates. Only after algebraically checking a proposed solution can we be sure that it is an actual solution.

The coordinates of each point on the line graphed in red satisfy $2x + 3y = 2$ and the coordinates of each point on the line graphed in blue satisfy $3x - 2y = 16$. Because the point of intersection is on both graphs, its coordinates satisfy both equations.

It appears that the graphs intersect at the point $(4, -2)$. To verify that it is the solution of the system, we substitute 4 for x and -2 for y in each equation.

Check:

$$\begin{aligned} 2x + 3y &= 2 && \text{The first equation.} \\ 2(4) + 3(-2) &\stackrel{?}{=} 2 \\ 8 + (-6) &\stackrel{?}{=} 2 \\ 2 &= 2 && \text{True} \end{aligned}$$

$$\begin{aligned} 3x - 2y &= 16 && \text{The second equation.} \\ 3(4) - 2(-2) &\stackrel{?}{=} 16 \\ 12 - (-4) &\stackrel{?}{=} 16 \\ 16 &= 16 && \text{True} \end{aligned}$$

Since $(4, -2)$ makes both equations true, it is the solution of the system. The solution set is written as $\{(4, -2)\}$.

Self Check 2 Solve the system of equations by graphing: $\begin{cases} 2x - y = -5 \\ x + y = -1 \end{cases}$

Now Try Problem 25

To solve a system of linear equations in two variables by graphing, follow these steps.

The Graphing Method

1. Carefully graph each equation on the same rectangular coordinate system.
2. If the lines intersect, determine the coordinates of the point of intersection of the graphs. That ordered pair is the solution of the system.
3. Check the proposed solution in each equation of the original system.

3 Use Graphing to Identify Inconsistent Systems and Dependent Equations.

A system of equations that has at least one solution, like that in Example 2, is called a **consistent system.** A system with no solution is called an **inconsistent system.**

EXAMPLE 3 Solve the system of equations by graphing: $\begin{cases} y = -2x - 6 \\ 4x + 2y = 8 \end{cases}$

Strategy We will graph both equations on the same coordinate system.

Why If both equations are graphed on the same coordinate system, we can see whether they have any common solutions.

Solution Since $y = -2x - 6$ is written in slope–intercept form, we can graph it by plotting the y-intercept $(0, -6)$ and then drawing a slope triangle whose rise is -2 and whose run is 1. We can graph $4x + 2y = 8$ using the intercept method.

$y = -2x - 6$

$m = -2 \quad b = -6$

Slope: $\frac{\text{Rise}}{\text{Run}} = \frac{-2}{1}$

y-intercept: $(0, -6)$

$4x + 2y = 8$

x	y	(x, y)
0	4	(0, 4)
2	0	(2, 0)
1	2	(1, 2)

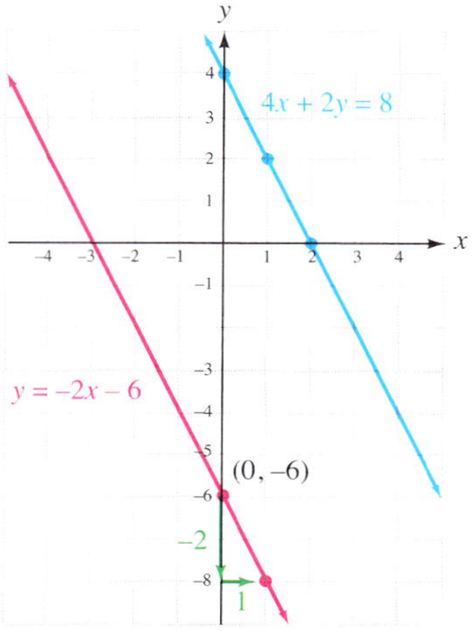

The lines in the graph appear to be parallel. We can verify this by writing the second equation in slope–intercept form and observing that the lines have the same slope, -2, and different y-intercepts, $(0, -6)$ and $(0, 4)$.

$y = -2x - 6 \qquad 4x + 2y = 8$

$2y = -4x + 8$ Subtract $4x$ from both sides.

$y = -2x + 4$ To isolate y, divide both sides by 2.

Different y-intercepts

Same slope

Because the lines are parallel, there is no point of intersection. Such a system has *no solution*. The solution set is the empty set, which is written as $\{ \ \}$ or as $\varnothing$.

Caution

A common error is to graph the parallel lines, but forget to answer with the words *no solution*.

Self Check 3 Solve the system of equations by graphing: $\begin{cases} y = \frac{3}{2}x \\ 3x - 2y = 6 \end{cases}$

Now Try Problem 29

Some systems of equations have infinitely many solutions.

EXAMPLE 4 Solve the system of equations by graphing: $\begin{cases} y = 2x + 4 \\ 4x + 8 = 2y \end{cases}$

Strategy We will graph both equations on the same coordinate system.

Why If both equations are graphed on the same coordinate system, we will be able to see if they have any solutions in common.

Solution To graph $y = 2x + 4$, we use the slope and y-intercept, and to graph $4x + 8 = 2y$, we use the intercept method.

The Language of Algebra

The graphs of these lines **coincide.** That is, they occupy the same location. To illustrate this concept, think of a clock. At noon and at midnight, the hands of the clock coincide.

$y = 2x + 4$

$m = 2 \quad b = 4$

Slope: $\frac{\text{Rise}}{\text{Run}} = \frac{2}{1}$

y-intercept: $(0, 4)$

$4x + 8 = 2y$

x	y	(x, y)
0	4	$(0, 4)$
-2	0	$(-2, 0)$
-3	-2	$(-3, -2)$

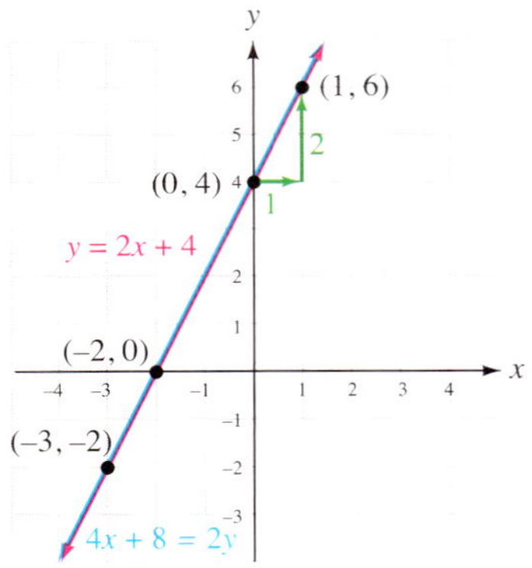

The graphs appear to be identical. We can verify this by writing the second equation in slope–intercept form and observing that it is the same as the first equation.

$y = 2x + 4$ The first equation.

$4x + 8 = 2y$ The second equation.

$2y = 4x + 8$ Reverse the sides.

$\frac{2y}{2} = \frac{4x}{2} + \frac{8}{2}$ To isolate y, divide both sides by 2.

$y = 2x + 4$

This confirms that $y = 2x + 4$ and $4x + 8 = 2y$ are different forms of the same equation. Thus, the equations of this system are equivalent and their graphs are indeed identical.

Since the graphs are the same line, they have infinitely many points in common. All of the points that lie on the common line are solutions because the coordinates of each of those points satisfy both equations of the system. In cases like this, we say that there are *infinitely many solutions.*

From the graph, it appears that four of the infinitely many solutions are $(-3, -2)$, $(-2, 0)$, $(0, 4)$, and $(1, 6)$. Checks for two of these ordered pairs follow.

Caution

A common error is to graph the identical lines, but forget to answer with the words *infinitely many solutions.*

Check $(-3, -2)$:

$$4x + 8 = 2y \qquad\qquad y = 2x + 4$$
$$4(-3) + 8 \stackrel{?}{=} 2(-2) \qquad -2 \stackrel{?}{=} 2(-3) - 4$$
$$-12 + 8 \stackrel{?}{=} -4 \qquad -2 \stackrel{?}{=} -6 + 4$$
$$-4 = -4 \qquad -2 = -2$$

Check $(0, 4)$:

$$4x + 8 = 2y \qquad\qquad y = 2x + 4$$
$$4(0) + 8 \stackrel{?}{=} 2(4) \qquad 4 \stackrel{?}{=} 2(0) + 4$$
$$0 + 8 \stackrel{?}{=} 8 \qquad 4 \stackrel{?}{=} 0 + 4$$
$$8 = 8 \qquad 4 = 4$$

Self Check 4 Solve the system of equations by graphing: $\begin{cases} 6x - 4 = 2y \\ y = 3x - 2 \end{cases}$

Now Try Problem 33

In Examples 2 and 3, the graphs of the equations of the system were different lines. We call equations with different graphs **independent equations.** The equations in Example 4 have the same graph and are equivalent. Because they are different forms of the same equation, they are called **dependent equations.**

There are three possible outcomes when we solve a system of two linear equations using the graphing method.

Solving Systems by Graphing

The two lines intersect at one point.	The two lines are parallel.	The two lines are identical.
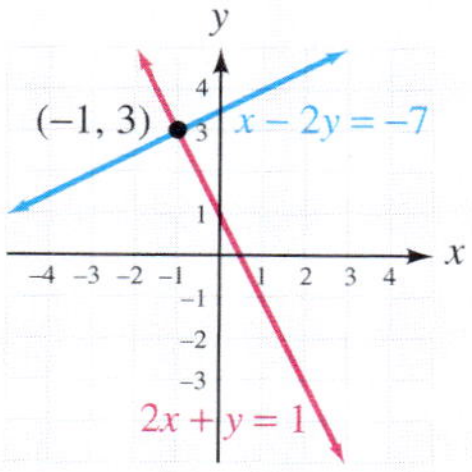	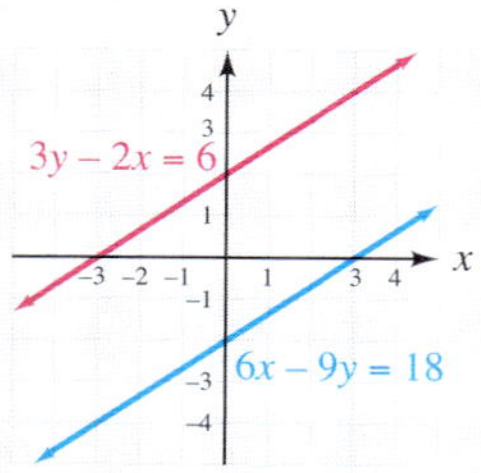	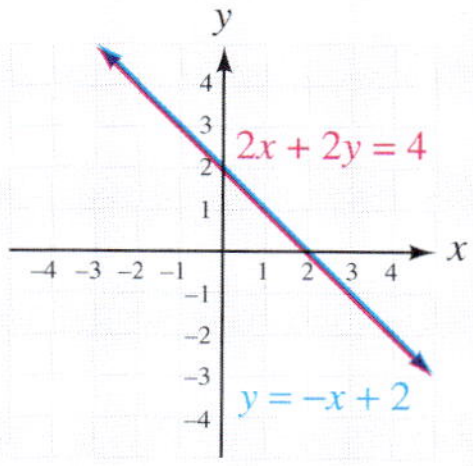
Exactly one solution (the point of intersection)	No solution	Infinitely many solutions (any point on the line is a solution)
Consistent system *Independent equations*	*Inconsistent system* *Independent equations*	*Consistent system* *Dependent equations*

4 Identify the Number of Solutions of a Linear System Without Graphing.

We can determine the number of solutions that a system of two linear equations has by writing each equation in slope–intercept form.

- If the lines have different slopes, they intersect, and the system has one solution. (See Example 2.)
- If the lines have the same slope and different y-intercepts, they are parallel, and the system has no solution. (See Example 3.)
- If the lines have the same slope and same y-intercept, they are the same line, and the system has infinitely many solutions. (See Example 4.)

EXAMPLE 5 Without graphing, determine the number of solutions of: $\begin{cases} 5x + y = 5 \\ 3x + 2y = 8 \end{cases}$

Strategy We will write both equations in slope–intercept form.

Why We can determine the number of solutions of a linear system by comparing the slopes and y-intercepts of the graphs of the equations.

Solution To write each equation in slope–intercept form, we solve for y.

$$5x + y = 5 \quad \text{The first equation.}$$
$$y = -5x + 5$$

$$3x + 2y = 8 \quad \text{The second equation.}$$
$$2y = -3x + 8$$
$$y = -\frac{3}{2}x + 4$$

Different slopes

Since the slopes are different, the lines are neither parallel nor identical. Therefore, they will intersect at one point and the system has one solution.

Self Check 5 Without graphing, determine the number of solutions of:
$\begin{cases} 3x + 6y = 1 \\ 2x + 4y = 0 \end{cases}$

Now Try Problem 41

5 Use a Graphing Calculator to Solve a Linear System (Optional).

A graphing calculator can be used to solve systems of equations, such as

$$\begin{cases} 2x + y = 12 \\ 2x - y = -2 \end{cases}$$

Before we can enter the equations into the calculator, we must solve them for y.

$$\begin{aligned} 2x + y &= 12 \quad \text{The first equation.} \\ y &= -2x + 12 \end{aligned} \qquad \begin{aligned} 2x - y &= -2 \quad \text{The second equation.} \\ -y &= -2x - 2 \\ y &= 2x + 2 \end{aligned}$$

We enter the resulting equations as Y_1 and Y_2 and graph them on the same axes. If we use the standard window setting, their graphs will look like figure (a).

To find the solution of the system, we can use the INTERSECT feature found on most graphing calculators. With this feature, after pushing enter three times to identify each graph and a guess for the point, the cursor automatically moves to the point of intersection of the graphs and displays the coordinates of that point. In figure (b), we see that the solution is (2.5, 7).

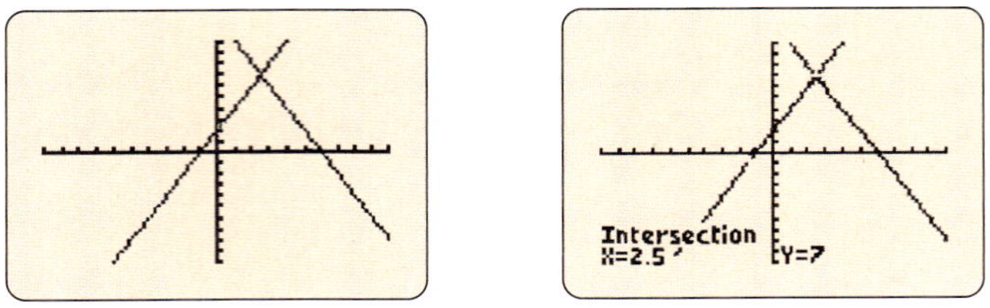

(a) (b)

SECTION 4.1 STUDY SET

VOCABULARY

Fill in the blanks.

1. The pair of equations $\begin{cases} x - y = -1 \\ 2x - y = 1 \end{cases}$ is called a _______ of linear equations.
2. Because the ordered pair (2, 3) satisfies both equations in Problem 1, it is called a ________ of the system of equations.
3. The point of ___________ of the lines graphed in part (a) below is (1, 2).

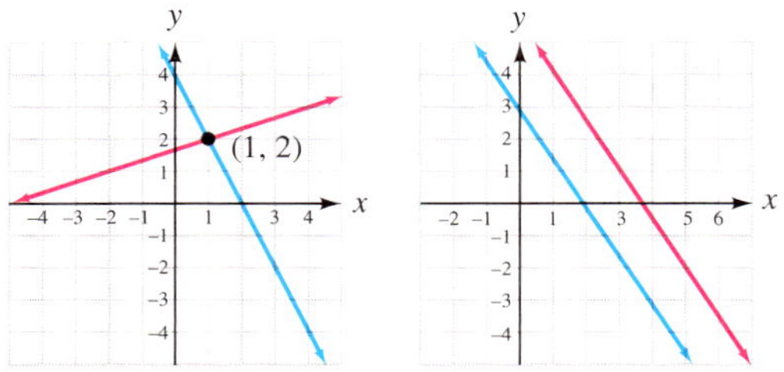

(a) (b)

4. The lines graphed in part (b) above do not intersect. They are ________ lines.
5. A system of equations that has at least one solution is called a __________ system. A system with no solution is called an ____________ system.
6. We call equations with different graphs ____________ equations. Because __________ equations are different forms of the same equation, they have the same graph.

CONCEPTS

7. Refer to the illustration below.
   a. If the coordinates of point A are substituted into the equation for Line 1, will the result be true or false?
   b. If the coordinates of point C are substituted into the equation for Line 1, will the result be true or false?
8. Refer to the illustration on the right.
   a. If the coordinates of point C are substituted into the equation for Line 2, will the result be true or false?
   b. If the coordinates of point B are substituted into the equation for Line 1, will the result be true or false?

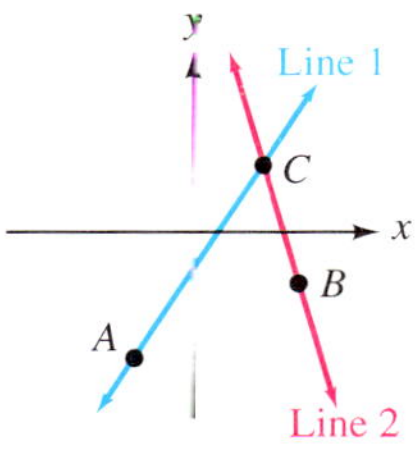

9. a. To graph $5x - 2y = 10$, we can use the intercept method. Complete the table.

x	y
0	
	0

b. To graph $y = 3x - 2$, we can use the slope and y-intercept. Fill in the blanks.

Slope: $\quad = \frac{}{1}$ $\quad$ y-intercept:

10. What is the apparent solution of the system graphed in figure (a) below? Is the system consistent or inconsistent?

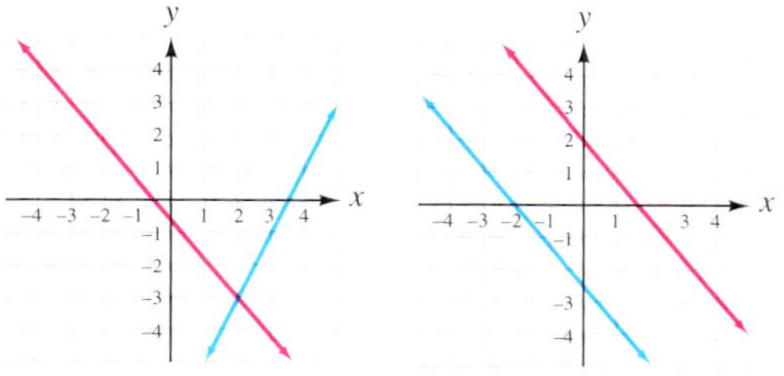

(a) (b)

11. How many solutions does the system graphed in figure (b) above have? Are the equations dependent or independent?

12. How many solutions does the system graphed on the right have? Give three of the solutions. Is the system consistent or inconsistent?

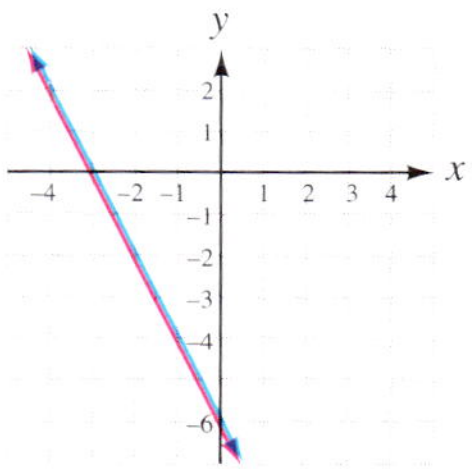

GUIDED PRACTICE

Determine whether the ordered pair is a solution of the given system of equations. See Example 1.

13. $(1, 1)$, $\begin{cases} x + y = 2 \\ 2x - y = 1 \end{cases}$

14. $(1, 3)$, $\begin{cases} 2x + y = 5 \\ 3x - y = 0 \end{cases}$

15. $(3, -2)$, $\begin{cases} 2x + y = 4 \\ y = 1 - x \end{cases}$

16. $(-2, 4)$, $\begin{cases} 2x + 2y = 4 \\ 3y = 10 - x \end{cases}$

17. $(12, 0)$, $\begin{cases} x - 9y = 12 \\ y = 10 - x \end{cases}$

18. $(15, 0)$, $\begin{cases} x - 2y = 15 \\ y = 16 - x \end{cases}$

19. $(-2, -4)$, $\begin{cases} 4x + 5y = -23 \\ -3x + 2y = 0 \end{cases}$

20. $(-5, 2)$, $\begin{cases} -2x + 7y = 17 \\ 3x - 4y = -19 \end{cases}$

21. $\left(\frac{1}{2}, 3\right)$, $\begin{cases} 2x + y = 4 \\ 4x - 11 = 3y \end{cases}$

22. $\left(2, \frac{1}{3}\right)$, $\begin{cases} x - 3y = 1 \\ -2x + 6 = -6y \end{cases}$

23. $(2.5, 3.5)$, $\begin{cases} 4x - 3 = 2y \\ 4y + 1 = 6x \end{cases}$

24. $(0.2, 0.3)$, $\begin{cases} 20x + 10y = 7 \\ 20y = 15x + 3 \end{cases}$

Solve each system of equations by graphing. See Example 2.

25. $\begin{cases} 2x + 3y = 12 \\ 2x - y = 4 \end{cases}$

26. $\begin{cases} 5x + y = 5 \\ 5x + 3y = 15 \end{cases}$

27. $\begin{cases} x + y = 4 \\ x - y = -6 \end{cases}$

28. $\begin{cases} x + y = 4 \\ x - y = -2 \end{cases}$

Solve each system by graphing. If a system has no solution or infinitely many solutions, so state. See Example 3.

29. $\begin{cases} y = -\frac{1}{3}x - 4 \\ x + 3y = 6 \end{cases}$

30. $\begin{cases} y = -\frac{1}{2}x - 3 \\ x + 2y = 2 \end{cases}$

31. $\begin{cases} y = 3x \\ y - 3x = -3 \end{cases}$

32. $\begin{cases} y = -2x \\ 2x + y = -2 \end{cases}$

Solve each system by graphing. If a system has no solution or infinitely many solutions, so state. See Example 4.

33. $\begin{cases} y = x - 1 \\ 3x - 3y = 3 \end{cases}$

34. $\begin{cases} y = -x + 1 \\ 4x + 4y = 4 \end{cases}$

35. $\begin{cases} 4x + 6y = 12 \\ 2x + 3y = 6 \end{cases}$

36. $\begin{cases} 2x - y = 0 \\ 2y - 4x = 0 \end{cases}$

Find the slope and the y-intercept of the graph of each line in the system of equations. Then, use that information to determine the number of solutions of the system. See Example 5.

37. $\begin{cases} y = 6x - 7 \\ y = -2x + 1 \end{cases}$

38. $\begin{cases} y = \frac{1}{2}x + 8 \\ y = 4x - 10 \end{cases}$

39. $\begin{cases} 3x - y = -3 \\ y - 3x = 3 \end{cases}$

40. $\begin{cases} x + 4y = 4 \\ 12y = 12 - 3x \end{cases}$

41. $\begin{cases} x + y = 6 \\ x + y = 8 \end{cases}$

42. $\begin{cases} 5x + y = 0 \\ 5x + y = 6 \end{cases}$

43. $\begin{cases} 6x + y = 0 \\ 2x + 2y = 0 \end{cases}$

44. $\begin{cases} x + y = 1 \\ 2x - 2y = 5 \end{cases}$

Use a graphing calculator to solve each system. See Objective 5.

45. $\begin{cases} y = 4 - x \\ y = 2 + x \end{cases}$

46. $\begin{cases} 3x - 6y = 4 \\ 2x + y = 1 \end{cases}$

47. $\begin{cases} 6x - 2y = 5 \\ 3x = y + 10 \end{cases}$

48. $\begin{cases} x - 3y = -2 \\ 5x + y = 10 \end{cases}$

TRY IT YOURSELF

Solve each system of equations by graphing.

49. $\begin{cases} y = 3x + 6 \\ y = -2x - 4 \end{cases}$

50. $\begin{cases} y = x + 3 \\ y = -2x - 3 \end{cases}$

51. $\begin{cases} 2y = 3x + 2 \\ 3x - 2y = 6 \end{cases}$

52. $\begin{cases} 3x - 6y = 18 \\ x = 2y + 3 \end{cases}$

53. $\begin{cases} x + y = 2 \\ y = x - 4 \end{cases}$

54. $\begin{cases} x + y = 1 \\ y = x + 5 \end{cases}$

55. $\begin{cases} x = 3 \\ 3y = 6 - 2x \end{cases}$

56. $\begin{cases} x = 4 \\ 2y = 12 - 4x \end{cases}$

57. $\begin{cases} y = \frac{3}{4}x + 3 \\ y = -\frac{x}{4} - 1 \end{cases}$

58. $\begin{cases} y = \frac{2}{3}x + 4 \\ y = -\frac{x}{3} + 7 \end{cases}$

59. $\begin{cases} y = -x - 2 \\ y = -3x + 6 \end{cases}$

60. $\begin{cases} y = 2x - 4 \\ y = -5x + 3 \end{cases}$

61. $\begin{cases} -x + 3y = -11 \\ 3x - y = 17 \end{cases}$

62. $\begin{cases} 2x - 3y = -18 \\ 3x + 2y = -1 \end{cases}$

63. $\begin{cases} x + y = 2 \\ y = x \end{cases}$

64. $\begin{cases} x + y = 4 \\ y = x \end{cases}$

65. $\begin{cases} 4x - 2y = 8 \\ y = 2x - 4 \end{cases}$

66. $\begin{cases} 2y = -6x - 12 \\ 3x + y = -6 \end{cases}$

67. $\begin{cases} x + 4y = -2 \\ y = -x - 5 \end{cases}$

68. $\begin{cases} 3x + 2y = -8 \\ 2x - 3y = -1 \end{cases}$

69. $\begin{cases} y = -3 \\ -x + 2y = -4 \end{cases}$

70. $\begin{cases} y = -4 \\ -2x - y = 8 \end{cases}$

71. $\begin{cases} x + 2y = -4 \\ x - \frac{1}{2}y = 6 \end{cases}$

72. $\begin{cases} \frac{2}{3}x - y = -3 \\ 3x + y = 3 \end{cases}$

APPLICATIONS

73. **Social Networks.** Refer to the graph below. In what month did Facebook and MySpace have the same number of unique visitors? Estimate the number.

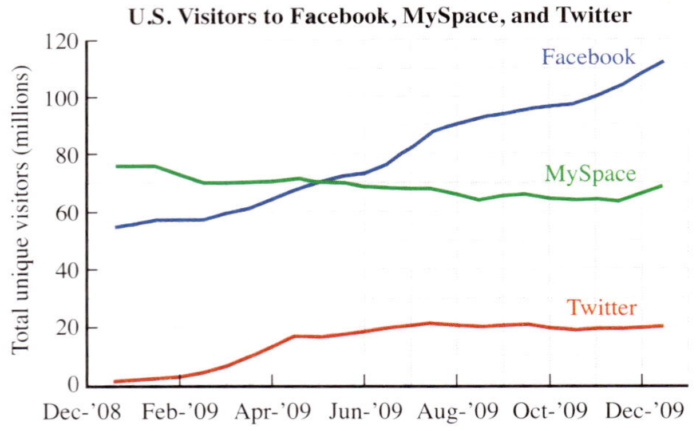

Source: Comscore Media Metrix

74. **Beverages.** Refer to the graph below. In what year was average number of gallons of milk and carbonated soft drinks consumed per person the same? Estimate the number of gallons.

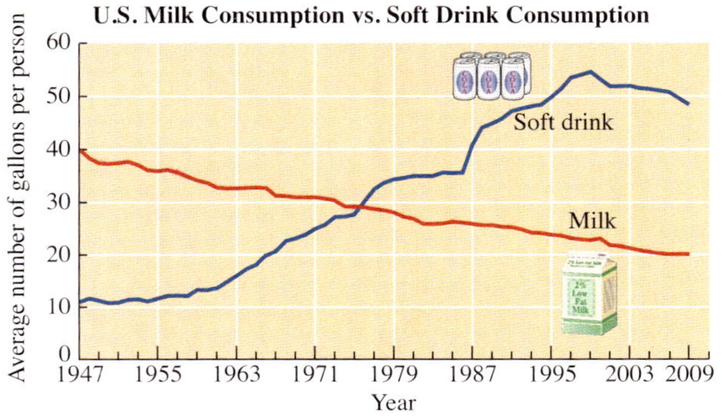

Source: USDA. Economic Research Service

75. **Latitude and Longitude.** Refer to the following map.

a. Name three American cities that lie on a latitude line of 30°.

b. Name three American cities that lie on a longitude line of −90°.

c. What city lies on both lines?

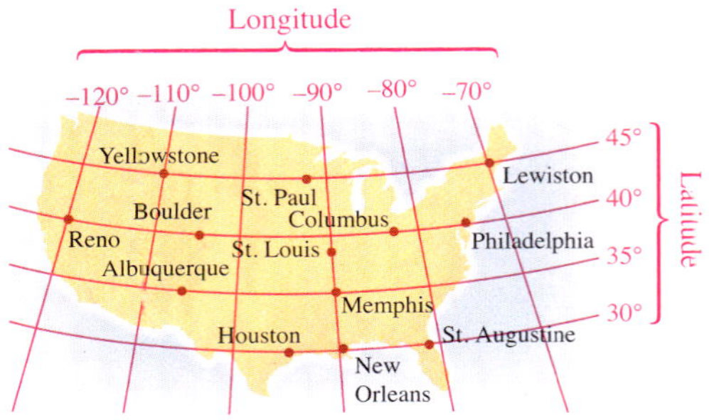

76. **Economics.** The following graph illustrates the law of supply and demand.

a. Complete each sentence with the word *increases* or *decreases*. As the price of an item increases, the supply of the item __________. As the price of an item increases, the demand for the item __________.

b. For what price will the supply equal the demand? How many items will be supplied for this price?

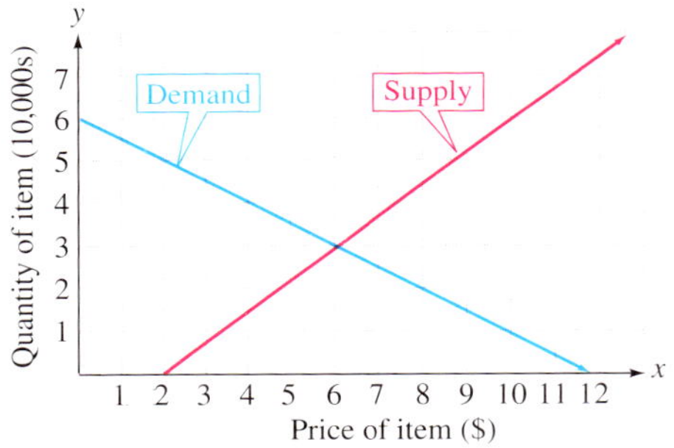

77. **Daily Tracking Polls.** Refer to the graph below.

a. Which political candidate was ahead on October 28 and by how much?

b. On what day did the challenger pull even with the incumbent?

c. If the election was held November 4, who did the poll predict would win, and by how many percentage points?

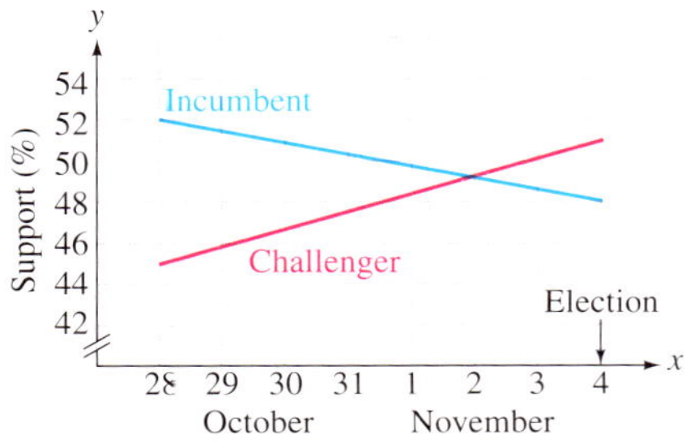

78. Air Traffic Control. The equations describing the paths of two airplanes are $y = -\frac{1}{2}x + 3$ and $3y = 2x + 2$. Graph each equation on the radar screen shown. Is there a possibility of a midair collision? If so, where?

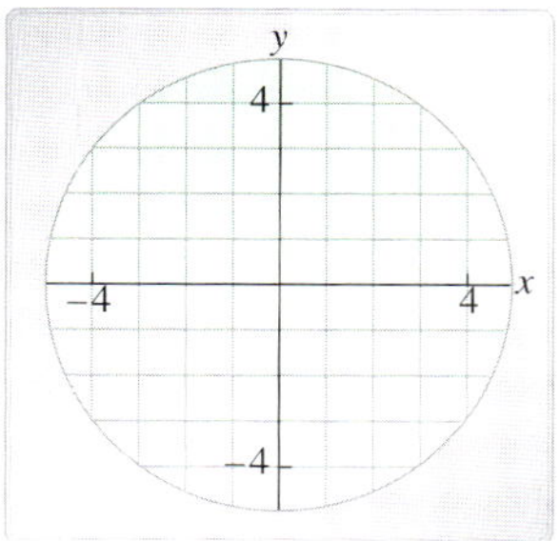

79. from Campus to Careers

Photographer

© Andrei Contiu/Shutterstock.com

Photographers often use the *rule of thirds* to add more interest to a photo rather than simply centering the subject. They imagine two horizontal and two vertical lines dividing the image into a grid of 9 equal parts (like a tic-tac-toe game). The four points where the lines intersect are possible "sweet spots" for placing the subject of the picture. Give the coordinates of the four sweet spots for the image below. Is the sea gull properly placed?

80. TV Coverage. A television camera is located at $(-2, 0)$ and will follow the launch of a space shuttle, as shown here. (Each unit in the illustration is 1 mile.) As the shuttle rises vertically on a path described by $x = 2$, the farthest the camera can tilt back is a line of sight given by $y = \frac{5}{2}x + 5$. For how many miles of the shuttle's flight will it be in view of the camera?

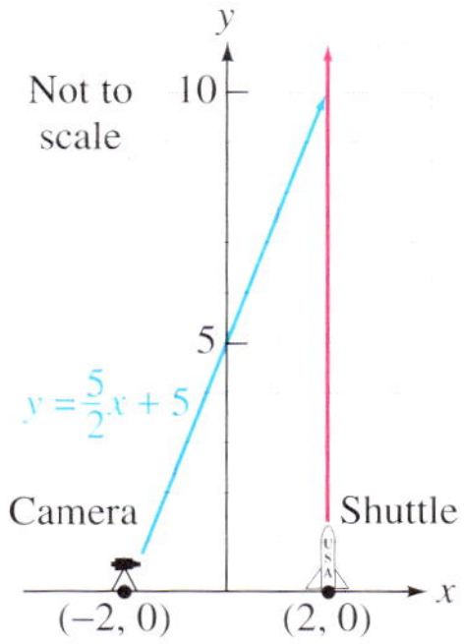

WRITING

81. Explain why it is difficult to determine the solution of the system in the graph.

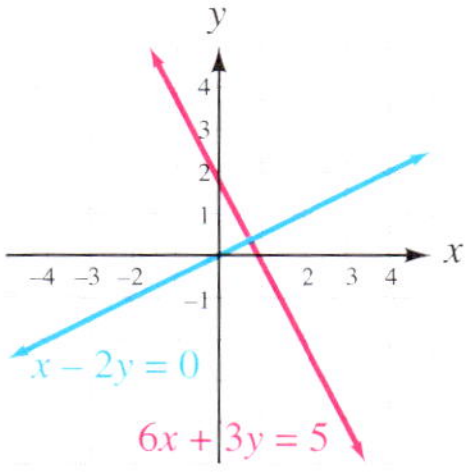

82. Without graphing, how can you tell that the graphs of $y = 2x + 1$ and $y = 3x + 2$ intersect?

83. Could a system of two linear equations have exactly two solutions? Explain why or why not.

84. What is an inconsistent system? What are dependent equations?

85. Suppose the graphs of the two linear equations of a system are the same line. What is wrong with the following statement? *The system has infinitely many solutions. Any ordered pair is a solution of the system.*

86. The Swine Flu. The graph below is for the month of January, 2010. It shows the total number of doses of the H1N1 Swine Flu vaccine that had been produced to date. It also shows the total number of doses that U.S. health officials had ordered to date. Did the total number of doses ordered ever equal or surpass the total number of doses produced? How can you tell?

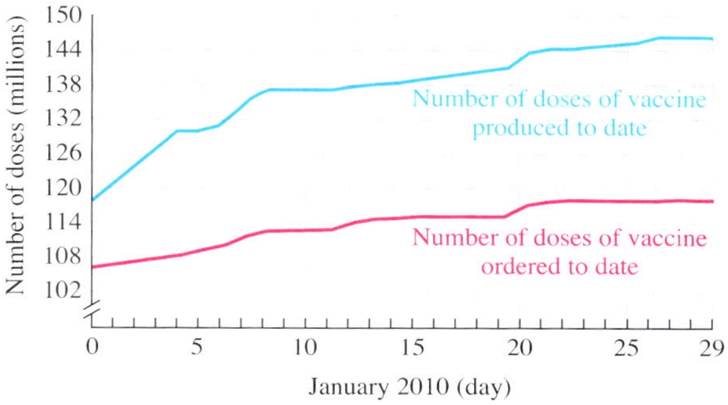

REVIEW

Solve each inequality. Write the solution set in interval notation and graph it.

87. $-4(2y + 2) \le 4y + 28$

88. $-5 < 3t + 4 \le 13$

89. $-1 \le -\frac{1}{2}n$

90. $\frac{1}{3} + \frac{c}{5} > -\frac{3}{2}$

CHALLENGE PROBLEMS

91. Can a system of two linear equations in two variables be inconsistent but have dependent equations? Explain.

92. Construct a system of two linear equations that has a solution of $(-2, 6)$.

93. Write a system of two linear equations such that $(2, 3)$ is a solution of the first equation but is not a solution of the second equation.

94. Solve by graphing: $\begin{cases} \frac{1}{3}x - \frac{1}{2}y = \frac{1}{6} \\ \frac{2x}{5} + \frac{y}{2} = \frac{13}{10} \end{cases}$

SECTION 4.2 Solving Systems of Equations by Substitution

OBJECTIVES

1. Solve systems of linear equations by substitution.
2. Find a substitution equation.
3. Solve systems of linear equations that contain fractions.
4. Use substitution to identify inconsistent systems and dependent equations.

ARE YOU READY?

The following problems review some basic skills that are needed when solving systems of equations by substitution.

1. In $8x + y = 2$, what is the coefficient of y?
2. Solve $2x - y = -3$ for y.
3. What is the LCD for the fractions in the equation $\frac{x}{6} + \frac{y}{9} = \frac{3}{2}$?
4. Substitute 4 for x in $y = 5x - 3$ and find y.
5. Multiply: $8\left(\frac{3x}{8}\right)$

When solving a system of equations by graphing, it is often difficult to determine the coordinates of the intersection point. For example, a solution of $\left(\frac{7}{8}, \frac{3}{5}\right)$ would be almost impossible to identify accurately. In this section, we will discuss a second, more precise method for solving systems that does not involve graphing.

1 Solve Systems of Linear Equations by Substitution.

One algebraic method for solving a system of equations is the **substitution method.** It is introduced in the following example.

EXAMPLE 1 Solve the system: $\begin{cases} y = 3x - 2 \\ 2x + y = 8 \end{cases}$

Strategy Note that the first equation is solved for y. Because y and $3x - 2$ are equal (represent the same value), we will substitute $3x - 2$ for y in the second equation.

Why The objective is to obtain one equation containing only one unknown. When $3x - 2$ is substituted for y in the second equation, the result will be just that—an equation in one variable, x.

Solution Since the right side of $y = 3x - 2$ is used to make a substitution, $y = 3x - 2$ is called the **substitution equation.**

Success Tip

Throughout the course, we have been substituting numbers for variables. With this method, we substitute a *variable expression for a variable.*

Caution

When using the substitution method, a common error is to find the value of one of the variables, say x, and forget to find the value of the other. Remember that a solution of a linear system of two equations is an ordered pair (x, y).

$$\begin{cases} y = 3x - 2 \\ 2x + y = 8 \end{cases}$$

To find the solution of the system, we proceed as follows:

$2x + y = 8$ This is the second equation of the system. It has two variables.

$2x + 3x - 2 = 8$ Substitute $3x - 2$ for y. This equation has only one variable.

The resulting equation can be solved for x.

$2x + 3x - 2 = 8$

$5x - 2 = 8$ Combine like terms: $2x + 3x = 5x$.

$5x = 10$ To isolate the variable term, $5x$, add 2 to both sides.

$x = 2$ Divide both sides by 5. This is the x-value of the solution.

We can find the y-value of the solution by substituting 2 for x in either equation of the original system. We will use the substitution equation because it is already solved for y.

$y = 3x - 2$ — This is the substitution equation.

$y = 3(2) - 2$ — Substitute 2 for x.

$y = 6 - 2$ — Do the multiplication.

$y = 4$ — This is the y-value of the solution. We would have obtained the same result if we had substituted 2 for x in $2x + y = 8$ and solved for y.

The ordered pair (2, 4) appears to be the solution of the system. To check, we substitute 2 for x and 4 for y in each equation.

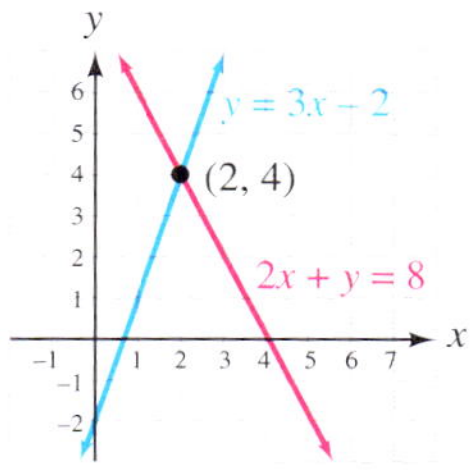

Check:

$y = 3x - 2$	The first equation.	$2x + y = 8$	The second equation.
$4 \stackrel{?}{=} 3(2) - 2$		$2(2) + 4 \stackrel{?}{=} 8$	
$4 \stackrel{?}{=} 6 - 2$		$4 + 4 \stackrel{?}{=} 8$	
$4 = 4$	True	$8 = 8$	True

Since (2, 4) satisfies both equations, it is the solution. The solution set is written as $\{(2, 4)\}$. A graph of the equations of the system shows an intersection point of (2, 4). This illustrates an important fact: *The solution found using the substitution method will be the same as the solution found using the graphing method.*

Self Check 1 Solve the system: $\begin{cases} x + 4y = 7 \\ x = 6y - 3 \end{cases}$

Now Try Problem 15

The substitution method works well for solving systems where one equation is solved, or can be easily solved, for one of the variables. To solve a system of equations in x and y by the substitution method, follow these steps.

The Substitution Method

1. Solve one of the equations for either x or y. If this is already done, go to step 2. (We call this equation the **substitution equation.**)
2. Substitute the expression for x or for y obtained in step 1 into the other equation and solve that equation.
3. Substitute the value of the variable found in step 2 into the substitution equation to find the value of the remaining variable.
4. Check the proposed solution in each equation of the original system. Write the solution as an ordered pair.

EXAMPLE 2 Solve the system: $\begin{cases} 4x + 27 = 7y \\ x = -5y \end{cases}$

Strategy We will use the substitution method to solve this system.

Why The substitution method works well when one of the equations of the system (in this case, $x = -5y$) is solved for a variable.

Solution *Step 1:* Because x and $-5y$ represent the same value, we can substitute $-5y$ for x in the first equation.

Success Tip

The basic objective of this method is to use an appropriate substitution to obtain one equation in one variable.

$$\begin{cases} 4x + 27 = 7y \\ x = -5y \end{cases}$$ This is the substitution equation.

Step 2: When we substitute $-5y$ for x in the first equation, the resulting equation contains only one variable, and it can be solved for y.

$4x + 27 = 7y$	This is the first equation of the system. It has two variables.
$4(-5y) + 27 = 7y$	Substitute $-5y$ for x. Don't forget the parentheses. This equation has only one variable.
$-20y + 27 = 7y$	Do the multiplication.
$27 = 27y$	To eliminate $-20y$ on the left side, add $20y$ to both sides.
$1 = y$	Divide both sides by 27. This is the y-value of the solution.

Caution

We don't have to find the values of the variables in alphabetical order. Here, we found y first.

Step 3: To find x, substitute 1 for y in the equation $x = -5y$.

$x = -5y$	This is the substitution equation.
$x = -5(1)$	Substitute 1 for y.
$x = -5$	This is the x-value of the solution.

Step 4: The following check verifies that the solution is $(-5, 1)$.

Check:

$$\begin{aligned} 4x + 27 &= 7y \quad \text{The first equation.} \\ 4(-5) + 27 &\stackrel{?}{=} 7(1) \\ -20 + 27 &\stackrel{?}{=} 7 \\ 7 &= 7 \quad \text{True} \end{aligned}$$

$$\begin{aligned} x &= -5y \quad \text{The second equation.} \\ -5 &\stackrel{?}{=} -5(1) \\ -5 &= -5 \quad \text{True} \end{aligned}$$

Self Check 2 Solve the system: $\begin{cases} 3x + 40 = 8y \\ x = -4y \end{cases}$

Now Try Problem 17

2 Find a Substitution Equation.

Sometimes neither equation of a system is solved for a variable. In such cases, we can find a substitution equation by solving one of the equations for one of its variables.

EXAMPLE 3 Solve the system: $\begin{cases} 4x + y = 3 \\ 3x + 5y = 15 \end{cases}$

Strategy Since the system does not contain an equation solved for x or y, we must choose an equation and solve it for x or y. We will solve for y in the first equation, because y has a coefficient of 1. Then we will use the substitution method to solve the system.

Why Solving $4x + y = 3$ for x, or solving $3x + 5y = 15$ for x or y, would involve working with cumbersome fractions.

Solution

Success Tip

To find a substitution equation, solve one of the equations of the system for one of its variables. If possible, solve for a variable whose coefficient is 1 or -1 to avoid working with fractions.

Step 1: To find a substitution equation, we proceed as follows:

$4x + y = 3$	This is the first equation of the system. Think: $4x + 1y = 3$.
$y = 3 - 4x$	To isolate y, subtract $4x$ from both sides. This is the substitution equation.

Because y and $3 - 4x$ are equal, we can substitute $3 - 4x$ for y in the second equation of the system.

$$\begin{cases} 4x + y = 3 \quad \rightarrow y = 3 - 4x \\ 3x + 5y = 15 \end{cases}$$

Caution

Here, use parentheses when substituting $3 - 4x$ for y so that the multiplication by 5 is distributed over both terms of $3 - 4x$

$$3x + 5(3 - 4x) = 15$$

Step 2: When we substitute for y in the second equation, the resulting equation contains only one variable and can be solved for x.

$3x + 5y = 15$ — This is the second equation of the system. It has two variables.

$3x + 5(3 - 4x) = 15$ — Substitute $3 - 4x$ for y. Don't forget the parentheses.

$3x + 15 - 20x = 15$ — Distribute the multiplication by 5.

$15 - 17x = 15$ — Combine like terms.

$-17x = 0$ — To isolate the variable term, $-17x$, subtract 15 from both sides.

$x = 0$ — Divide both sides by -17. This is the x-value of the solution.

Step 3: To find y, substitute 0 for x in the equation $y = 3 - 4x$.

$y = 3 - 4x$ — This is the substitution equation.

$y = 3 - 4(0)$ — Substitute 0 for x.

$y = 3 - 0$

$y = 3$ — This is the y-value of the solution.

Step 4: The solution appears to be $(0, 3)$. Check it in the original equations.

Caution

Always use the original equations when checking a solution. Do not use a substitution equation or an equivalent equation that you found algebraically. If an error was made, a proposed solution that would not satisfy the original system might appear to be correct.

Check:

$4x + y = 3$ — The first equation.

$4(0) + 3 \stackrel{?}{=} 3$

$0 + 3 \stackrel{?}{=} 3$

$3 = 3$ — True

$3x + 5y = 15$ — The second equation.

$3(0) + 5(3) \stackrel{?}{=} 15$

$0 + 15 \stackrel{?}{=} 15$

$15 = 15$ — True

Self Check 3 Solve the system: $\begin{cases} 2x - 3y = 10 \\ 3x + y = 15 \end{cases}$

Now Try Problem 21

EXAMPLE 4 Solve the system: $\begin{cases} 3a - 3b = 5 \\ 3 - a = -2b \end{cases}$

Strategy Since the coefficient of a in the second equation is -1, we will solve that equation for a. Then we will use the substitution method to solve the system.

Why If we solve for the variable with a numerical coefficient of -1, we can avoid having to work with fractions.

Solution

Step 1: To find a substitution equation, we proceed as follows:

$3 - a = -2b$ — This is the second equation of the system. Think: $3 - 1a = -2b$.

$-a = -2b - 3$ — To isolate the variable term, $-a$, subtract 3 from both sides.

To obtain a on the left side, multiply both sides of the equation by -1.

$-1(-a) = -1(-2b - 3)$ — Don't forget the parentheses.

$a = 2b + 3$ — Do the multiplication. This is the substitution equation.

Caution

If you solve for a variable in the first equation, be sure to substitute into the second equation. If you solve for a variable in the second equation, be sure to substitute into the first equation.

Because a and $2b + 3$ represent the same value, we can substitute $2b + 3$ for a in the first equation.

$$\begin{cases} 3a - 3b = 5 \\ 3 - a = -2b \rightarrow a = 2b + 3 \end{cases}$$

Step 2: Substitute $2b + 3$ for a in the first equation and solve for b.

$3a - 3b = 5$ This is the first equation of the system. It has two variables.

$3(\mathbf{2b + 3}) - 3b = 5$ Substitute $2b + 3$ for a. Don't forget the parentheses.

$6b + 9 - 3b = 5$ Distribute the multiplication by 3.

$3b + 9 = 5$ Combine like terms: $6b - 3b = 3b$.

$3b = -4$ To isolate the variable term, $3b$, subtract 9 from both sides.

$b = -\frac{4}{3}$ Divide both sides by 3. This is the b-value of the solution.

Step 3: To find a, substitute $-\frac{4}{3}$ for b in the equation $a = 2b + 3$.

Notation

Unless told otherwise, list the values of the variables of a solution in *alphabetical* order. For example, if the equations of a system involve the variables a and b, write the solution in the form (a, b).

$a = 2\mathbf{b} + 3$ This is the substitution equation.

$a = 2\left(-\frac{4}{3}\right) + 3$ Substitute $-\frac{4}{3}$ for b.

$a = -\frac{8}{3} + \frac{9}{3}$ Do the multiplication. To add, we must write 3 as a fraction with a denominator of 3: $3 = \frac{9}{3}$.

$a = \frac{1}{3}$ Add the fractions. This is the a-value of the solution.

Step 4: The solution is $\left(\frac{1}{3}, -\frac{4}{3}\right)$. Check it in the original equations.

Self Check 4 Solve the system: $\begin{cases} 2s - t = 4 \\ 3s - 5t = 2 \end{cases}$

Now Try Problem 25

3 Solve Systems of Linear Equations that Contain Fractions.

It is usually helpful to clear any equations of fractions and combine any like terms before performing a substitution.

EXAMPLE 5 Solve the system: $\begin{cases} \frac{y}{4} = -\frac{x}{2} - \frac{3}{4} \\ 2x - y = -1 + y - x \end{cases}$

Strategy We will use properties of equality to write each equation of the system in simpler, equivalent form. Then we will use the substitution method to solve the resulting system.

Why The first equation will be easier to work with if we clear it of fractions. The second equation will be easier to work with if we eliminate the variable terms on the right side.

Solution We can clear the first equation of fractions by multiplying both sides by the LCD.

Notation

We number the equations (1) and (2) to help describe how the system is solved using the substitution method.

$\frac{y}{4} = -\frac{x}{2} - \frac{3}{4}$ This is the first equation of the system.

$\mathbf{4}\left(\frac{y}{4}\right) = \mathbf{4}\left(-\frac{x}{2} - \frac{3}{4}\right)$ Multiply both sides by the LCD, 4. Don't forget the parentheses.

$\mathbf{4}\left(\frac{y}{4}\right) = \mathbf{4}\left(-\frac{x}{2}\right) - \mathbf{4}\left(\frac{3}{4}\right)$ Distribute the multiplication by 4.

(1) $y = -2x - 3$ Simplify. Call this equation 1.

Notation

When a linear equation is written in standard $Ax + By = C$ form, the *variable terms* are on the left side of the equal symbol and the *constant* term is on the right side.

We can write the second equation of the system in standard $Ax + By = C$ form by adding x and subtracting y from both sides.

$$2x - y = -1 + y - x \quad \text{This is the second equation of the system.}$$
$$2x - y + x - y = -1 + y - x + x - y$$
$$\textbf{(2)} \quad 3x - 2y = -1 \quad \text{Combine like terms. Call this equation 2.}$$

Step 1: Equations 1 and 2 form an equivalent system, which has the same solution as the original one. To find x, we proceed as follows:

$$\begin{aligned} &\textbf{(1)} \quad \begin{cases} y = -2x - 3 \\ 3x - 2y = -1 \end{cases} \quad \text{This is the substitution equation.} \\ &\textbf{(2)} \end{aligned}$$

Step 2: To find x, substitute $-2x - 3$ for y in equation 2 and proceed as follows:

$$\begin{aligned} 3x - 2y &= -1 && \text{This is equation 2. It has two variables.} \\ 3x - 2(-2x - 3) &= -1 && \text{Substitute } -2x - 3 \text{ for } y\text{. Don't forget the parentheses.} \\ 3x + 4x + 6 &= -1 && \text{Distribute the multiplication by } -2. \\ 7x + 6 &= -1 && \text{Combine like terms: } 3x + 4x = 7x. \\ 7x &= -7 && \text{To isolate the variable term, } 7x\text{, subtract 6 from both sides.} \\ x &= -1 && \text{Divide both sides by 7. This is the } x\text{-value of the solution.} \end{aligned}$$

Step 3: To find y, we substitute -1 for x in equation 1.

$$\begin{aligned} y &= -2x - 3 && \text{This is equation 1.} \\ y &= -2(-1) - 3 && \text{Substitute } -1 \text{ for } x. \\ y &= 2 - 3 && \text{Do the multiplication.} \\ y &= -1 && \text{Subtract. This is the } y\text{-value of the solution.} \end{aligned}$$

Step 4: The solution is $(-1, -1)$. Check it in the original system.

Self Check 5 Solve the system: $\begin{cases} \frac{y}{6} = \frac{x}{3} + \frac{1}{2} \\ 2x - y = -3 + y - x \end{cases}$

Now Try Problem 29

4 Use Substitution to Identify Inconsistent Systems and Dependent Equations.

In the previous section, we solved inconsistent systems and systems of dependent equations graphically. We also can solve these systems using the substitution method.

EXAMPLE 6 Solve the system: $\begin{cases} 4y - 12 = x \\ y = \frac{1}{4}x \end{cases}$

Strategy We will use the substitution method to solve this system.

Why The substitution method works well when one of the equations of the system $\left(\text{in this case, } y = \frac{1}{4}x\right)$ is solved for a variable.

Solution To try to solve this system, substitute $\frac{1}{4}x$ for y in the first equation and solve for x.

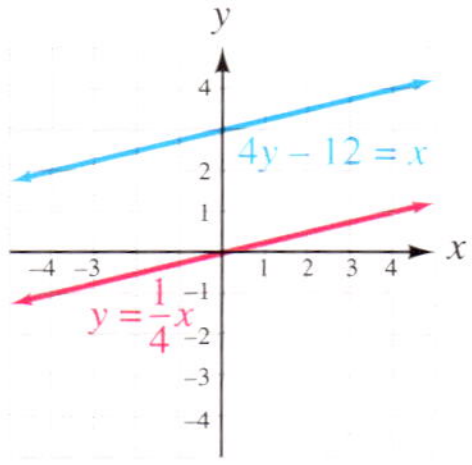

$$4y - 12 = x$$ This is the first equation of the system.

$$4\left(\frac{1}{4}x\right) - 12 = x$$ Substitute $\frac{1}{4}x$ for y.

$$x - 12 = x$$ Do the multiplication: $4\left(\frac{1}{4}\right) = 1$.

$$x - 12 - x = x - x$$ To eliminate x on the right side, subtract x from both sides.

$$-12 = 0$$ False

Here, the terms involving x drop out, and we get $-12 = 0$. This false statement indicates that the system has **no solution** and is inconsistent. The solution set is the empty set, $\varnothing$. The graphs of the equations of the system help to verify this; they are parallel lines.

Self Check 6 Solve the system: $\begin{cases} x - 4 = y \\ -2y = 4 - 2x \end{cases}$

Now Try Problem 33

EXAMPLE 7 Solve the system: $\begin{cases} x = -3y + 6 \\ 2x + 6y = 12 \end{cases}$

Strategy We will use the substitution method to solve this system.

Why The substitution method works well when one of the equations of the system (in this case, $x = -3y + 6$) is solved for a variable.

Solution To solve this system, substitute $-3y + 6$ for x in the second equation and solve for y.

$$2x + 6y = 12$$ This is the second equation of the system.

$$2(-3y + 6) + 6y = 12$$ Substitute $-3y + 6$ for x. Don't forget the parentheses.

$$-6y + 12 + 6y = 12$$ Distribute the multiplication by 2.

$$12 = 12$$ True

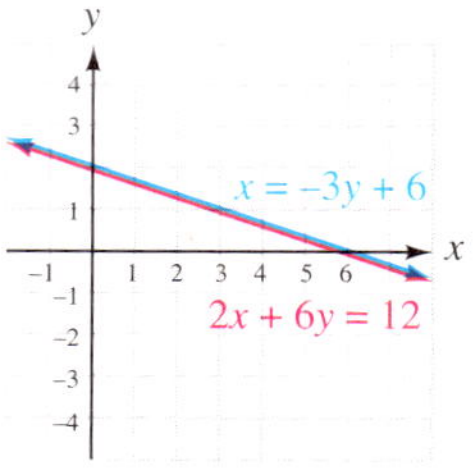

Here, the terms involving y drop out, and we get $12 = 12$. This true statement indicates that the two equations of the system are equivalent. Therefore, they are dependent equations and the system has **infinitely many solutions**. The graphs of the equations help to verify this; they are the same line.

Any ordered pair that satisfies one equation of this system also satisfies the other. To find several of the infinitely many solutions, we can substitute some values of x, say 0, 3, and 6, in either equation and solve for y. The results are: (0, 2), (3, 1), and (6, 0).

Self Check 7 Solve the system: $\begin{cases} y = 2 - x \\ 3x + 3y = 6 \end{cases}$

Now Try Problem 37

Now we summarize the results from Examples 6 and 7. If both variables are eliminated in the substitution process, there are two possible outcomes.

- If the resulting equation is false, write "*no solution*" as the answer. The system is inconsistent. (*Caution:* Do not answer with "false.")
- If the resulting equation is true, write "*infinitely many solutions*" as the answer. The system's equations are dependent. (*Caution:* Do not answer with "true.")

SECTION 4.2 STUDY SET

VOCABULARY

Fill in the blanks.

1. To solve the system $\begin{cases} x = y + 1 \\ 3x + 2y = 8 \end{cases}$ using the method discussed in this section, we begin by __________ $y + 1$ for x in the second equation.
2. We say that the equation $y = 2x + 4$ is solved for ___.

CONCEPTS

3. If the substitution method is used to solve $\begin{cases} 5x + y = 2 \\ y = -3x \end{cases}$, which equation should be used as the substitution equation?
4. Suppose the substitution method will be used to solve $\begin{cases} x - 2y = 2 \\ 2x + 3y = 11 \end{cases}$. Find a substitution equation by solving one of the equations for one of the variables.
5. Suppose $x - 4$ is substituted for y in the equation $x + 3y = 8$. Insert parentheses in $x + 3x - 4 = 8$ to show the substitution.
6. Fill in the blank. With the substitution method, the objective is to use an appropriate substitution to obtain one equation in _____ variable.
7. A student uses the substitution method to solve the system $\begin{cases} 4a + 5b = 2 \\ b = 3a - 11 \end{cases}$ and finds that $a = 3$. What is the easiest way for her to determine the value of b?
8. a. Clear $\frac{x}{5} + \frac{2y}{3} = 1$ of fractions.
   b. Write $2x + y = x - 5y + 3$ in the form $Ax + By = C$.
9. Suppose $-2 = 1$ is obtained when a system is solved by the substitution method.
   a. Does the system have a solution?
   b. Which of the following is a possible graph of the system?

i.
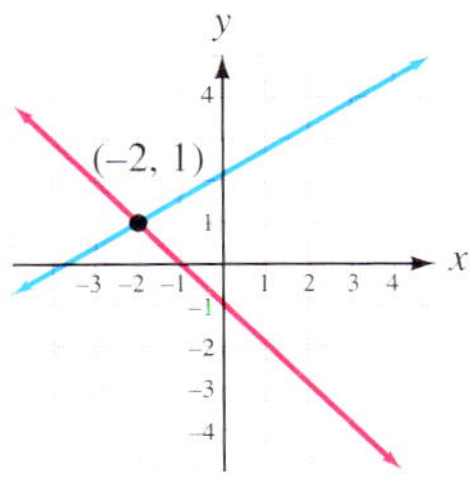

ii.
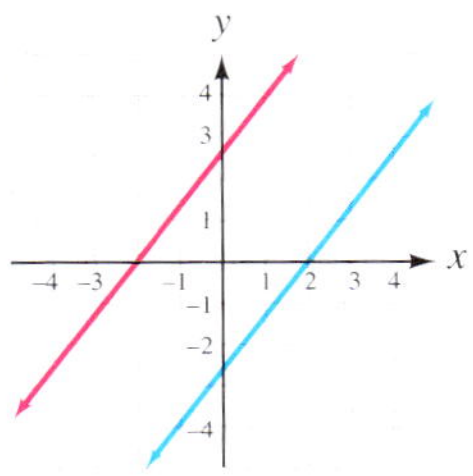

10. Suppose $2 = 2$ is obtained when a system is solved by the substitution method.
    a. Does the system have a solution?
    b. Which graph below is a possible graph of the system?

i.
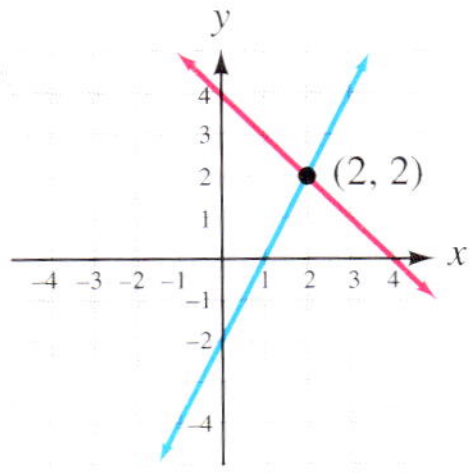

ii.
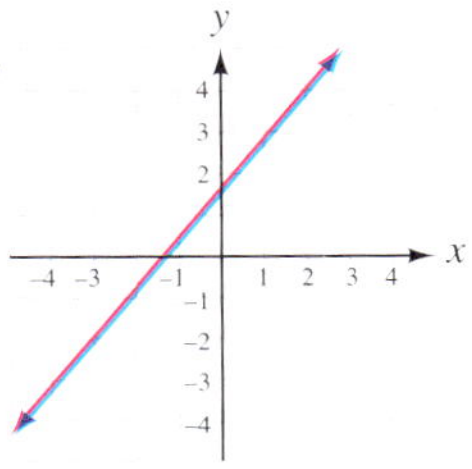

NOTATION

Complete the solution to solve the system.

11. Solve: $\begin{cases} y = 3x \\ x - y = 4 \end{cases}$

$x - y = 4$ This is the second equation.

$x - (\quad) = 4$

$-2x =$

$x =$ This is the x-value of the solution.

$y = 3x$ This is the first equation.

$y = 3(\quad)$

$y =$ This is the y-value of the solution.

The solution is $(\quad, \quad)$.

12. The system $\begin{cases} a = 3b + 2 \\ a + 3b = 8 \end{cases}$ was solved, and it was found that $b = 1$ and $a = 5$. Write the solution as an ordered pair.

GUIDED PRACTICE

Solve each system by substitution. See Example 1.

13. $\begin{cases} y = 2x \\ x + y = 6 \end{cases}$
14. $\begin{cases} y = 3x \\ x + y = 4 \end{cases}$
15. $\begin{cases} y = 2x - 6 \\ 2x + y = 6 \end{cases}$
16. $\begin{cases} 2x + y = -7 \\ y = 3x + 8 \end{cases}$

Solve each system by substitution. See Example 2.

17. $\begin{cases} x + 3y = -4 \\ x = -5y \end{cases}$
18. $\begin{cases} x + 5y = -3 \\ x = -4y \end{cases}$
19. $\begin{cases} y = -5x \\ x + 3y = -28 \end{cases}$
20. $\begin{cases} y = -2x \\ 3x + 2y = -1 \end{cases}$

Solve each system by substitution. See Example 3.

21. $\begin{cases} r + 3s = 9 \\ 3r + 2s = 13 \end{cases}$
22. $\begin{cases} x - 2y = 2 \\ 2x + 3y = 11 \end{cases}$
23. $\begin{cases} 4x + y = -15 \\ 2x + 3y = 5 \end{cases}$
24. $\begin{cases} 4x + y = -5 \\ 2x - 3y = -13 \end{cases}$

Solve each system by substitution. See Example 4.

25. $\begin{cases} 8x - 6y = 4 \\ 2x - y = -2 \end{cases}$
26. $\begin{cases} 5x + 4y = 0 \\ 2x - y = 0 \end{cases}$
27. $\begin{cases} 4x + 5y = 2 \\ 3x - y = 11 \end{cases}$
28. $\begin{cases} 5u + 3v = 5 \\ 4u - v = 4 \end{cases}$

Solve each system by substitution. See Example 5.

29. $\begin{cases} \frac{x}{4} + \frac{y}{4} = -\frac{1}{2} \\ x - 2y = y - 24 - x \end{cases}$
30. $\begin{cases} 5a + 2b = 15 - 5a - b \\ \frac{a}{3} - \frac{2}{3}b = \frac{13}{3} \end{cases}$

31. $\begin{cases} x - 5y = 20 - 4x - y \\ \dfrac{y}{3} = \dfrac{x}{2} - \dfrac{5}{2} \end{cases}$

32. $\begin{cases} x - 6y = 9 - 5x - 2y \\ \dfrac{x}{2} - \dfrac{3}{4} = 2y \end{cases}$

Solve each system by substitution. See Example 6.

33. $\begin{cases} 2a + 4b = -24 \\ a = 20 - 2b \end{cases}$

34. $\begin{cases} 8y = 15 - 4x \\ x + 2y = 4 \end{cases}$

35. $\begin{cases} 6 - y = 4x \\ 2y = -8x - 20 \end{cases}$

36. $\begin{cases} 2x - y = x + y \\ -2x + 4y = 6 \end{cases}$

Solve each system by substitution. See Example 7.

37. $\begin{cases} y - 3x = -5 \\ 21x = 7y + 35 \end{cases}$

38. $\begin{cases} 3a + 6b = -15 \\ a = -2b - 5 \end{cases}$

39. $\begin{cases} x = -3y + 6 \\ 2x + 4y = 6 + x + y \end{cases}$

40. $\begin{cases} 9x = 3y + 12 \\ 4 = 3x - y \end{cases}$

TRY IT YOURSELF

Solve each system by substitution. If a system has no solution or infinitely many solutions, so state.

41. $\begin{cases} -y = 11 - 3x \\ 2x + 5y = -4 \end{cases}$

42. $\begin{cases} -x = 10 - 3y \\ 2x + 8y = -6 \end{cases}$

43. $\begin{cases} \dfrac{x}{2} + \dfrac{y}{6} = \dfrac{2}{3} \\ \dfrac{x}{3} - \dfrac{y}{4} = \dfrac{1}{12} \end{cases}$

44. $\begin{cases} \dfrac{c}{2} + \dfrac{d}{14} = 1 \\ \dfrac{c}{5} - \dfrac{d}{2} = -\dfrac{33}{10} \end{cases}$

45. $\begin{cases} y + x = 2x + 2 \\ 6x - 4y = 21 - y \end{cases}$

46. $\begin{cases} y - x = 3x \\ 2x + 2y = 14 - y \end{cases}$

47. $\begin{cases} y - 4 = 2x \\ y = 2x + 2 \end{cases}$

48. $\begin{cases} x + 3y = 6 \\ x = -3y + 6 \end{cases}$

49. $\begin{cases} a + b = 1 \\ a - 2b = -1 \end{cases}$

50. $\begin{cases} 2b - a = -1 \\ 3a + 10b = -1 \end{cases}$

51. $\begin{cases} 5x = \dfrac{1}{2}y - 1 \\ \dfrac{1}{4}y = 10x - 1 \end{cases}$

52. $\begin{cases} \dfrac{x}{4} + y = \dfrac{1}{4} \\ \dfrac{y}{2} + \dfrac{11}{20} = \dfrac{x}{10} \end{cases}$

53. $\begin{cases} x + 2y = -6 \\ x = y \end{cases}$

54. $\begin{cases} y = 2x - 9 \\ x + 3y = 8 \end{cases}$

55. $\begin{cases} b = \dfrac{2}{3}a \\ 8a - 3b = 3 \end{cases}$

56. $\begin{cases} a = \dfrac{2}{3}b \\ 9a + 4b = 5 \end{cases}$

57. $\begin{cases} 6x - 3y = 5 \\ x + 2y = 0 \end{cases}$

58. $\begin{cases} 5s + 10t = 3 \\ 2s + t = 0 \end{cases}$

59. $\begin{cases} 2x + 3 = -4y \\ x - 6 = -8y \end{cases}$

60. $\begin{cases} 5y + 2 = -4x \\ x + 2y = -2 \end{cases}$

61. $\begin{cases} 2x + 5y = -2 \\ y = -\dfrac{x}{2} \end{cases}$

62. $\begin{cases} y = -\dfrac{x}{2} \\ 2x - 3y = -7 \end{cases}$

63. $\begin{cases} 3x + 4y = -19 \\ 2y - x = 3 \end{cases}$

64. $\begin{cases} 5x - 2y = -7 \\ 5 - y = -3x \end{cases}$

65. $\begin{cases} 3(x - 1) + 3 = 8 + 2y \\ 2(x + 1) = 8 + y \end{cases}$

66. $\begin{cases} 4(x - 2) = 19 - 5y \\ 3(x - 2) - 2y = -y \end{cases}$

67. $\begin{cases} x = \dfrac{1}{3}y - 1 \\ x = y + 5 \end{cases}$

68. $\begin{cases} x = \dfrac{1}{2}y + 2 \\ x = y - 6 \end{cases}$

69. $\begin{cases} 2a - 3b = -13 \\ -b = -2a - 7 \end{cases}$

70. $\begin{cases} a - 3b = -1 \\ -b = -2a - 2 \end{cases}$

71. $\begin{cases} x = 7y - 10 \\ 2x - 14y + 20 = 0 \end{cases}$

72. $\begin{cases} y - 1 = 5x \\ 10x - 2y = 2 \end{cases}$

73. $\begin{cases} 4x + 1 = 2x + 5 + y \\ 2x + 2y = 5x + y + 6 \end{cases}$

74. $\begin{cases} 6x = 2(y + 20) + 5x \\ 5(x - 1) = 3y + 4(x + 10) \end{cases}$

75. $\begin{cases} 2a + 3b = 7 \\ 6a - b = 1 \end{cases}$

76. $\begin{cases} 3a + 5b = -6 \\ 5b - a = -3 \end{cases}$

77. $\begin{cases} 2x - 3y = -4 \\ x = -\dfrac{3}{2}y \end{cases}$

78. $\begin{cases} x = -\dfrac{3}{8}y \\ 8x - 3y = 4 \end{cases}$

79. $\begin{cases} \dfrac{9x}{7} - \dfrac{3y}{7} = \dfrac{12}{7} \\ y - 3x = -4 \end{cases}$

80. $\begin{cases} a = 9 - 2b \\ 2(a + b) = 13 + a \end{cases}$

81. $\begin{cases} 4x + 5y + 1 = -8 + 3x \\ x - 3y + 2 = -3 - x \end{cases}$

82. $\begin{cases} 6x + y = -8 + 3x - y \\ 3x + y = 2y + x - 3 \end{cases}$

APPLICATIONS

83. **Offroading.** The *angle of approach* indicates how steep of an incline a vehicle can drive up without damaging the front bumper. The *angle of departure* indicates a vehicle's ability to exit an incline without damaging the rear bumper. The angle of approach a and the departure angle d for an H3 Hummer are described by the system $\begin{cases} a + d = 77 \\ a = d + 3 \end{cases}$. Use substitution to solve the system. (Each angle is measured in degrees.)

84. High School Sports. The equations shown in the following graph model the number of boys and girls taking part in high school soccer programs. In both models, x is the number of years after 2000, and y is the number of participants. If the trends continue, the graphs will intersect. Use the substitution method to predict the year when the number of boys and girls participating in high school soccer will be the same.

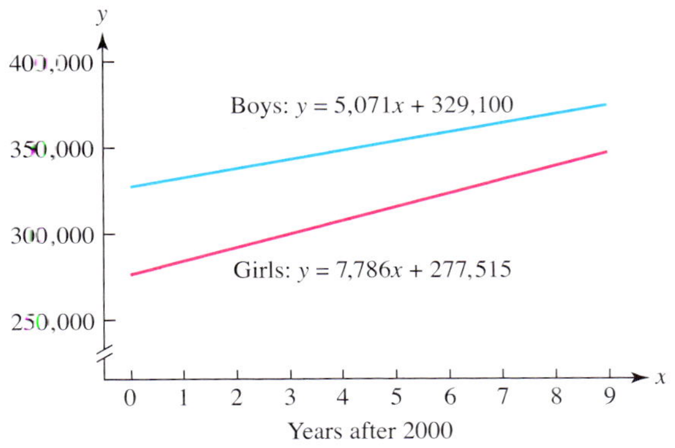

Source: National Federation of State High School Associations

85. Geometry. In the illustration, x and y represent the degree measures of angles. If $x + y = 90$ and $y = 3x$, find x and y.

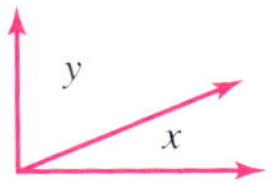

86. Geometry. In the illustration, x and y represent the degree measures of angles. If $x + y = 180$ and $x = 4y$, find x and y.

WRITING

87. What concept does this diagram illustrate?

$$\begin{cases} 6x + 5y = 11 \\ y = 3x - 2 \end{cases}$$

88. Explain the error.

Solve: $\begin{cases} 3a + 4b = 1 \\ a + 2b = 9 \end{cases}$

$a = 9 - 2b$	Solve for a in the second equation.
$9 - 2b + 2b = 9$	Substitute $9 - 2b$ for a.
$9 = 9$	The b-terms drop out. True

The system has infinitely many solutions.

89. When using the substitution method, how can you tell whether
   a. a system of linear equations has no solution?
   b. a system of linear equations has infinitely many solutions?

90. When solving a system, what advantages are there with the substitution method compared with the graphing method?

91. Consider the equation $5x + y = 12$. Explain why it is easier to solve for y than it is for x.

92. Could the substitution method be used to solve the following system? Explain why or why not. If not, what method could be used?

$$\begin{cases} y = -2 \\ x = 5 \end{cases}$$

REVIEW

93. Find the prime factorization of 189.

94. Complete each statement. For any nonzero number a,
   a. $\frac{0}{a} =$
   b. $\frac{a}{0}$ is

95. Add: $\frac{5}{12} + \frac{1}{4}$

96. Divide: $\frac{1}{3} \div \frac{4}{5}$

CHALLENGE PROBLEMS

Use the substitution method to solve each system.

97. $\begin{cases} \frac{6x - 1}{3} - \frac{5}{3} = \frac{3y + 1}{2} \\ \frac{1 + 5y}{4} + \frac{x + 3}{4} = \frac{17}{2} \end{cases}$

98. $\begin{cases} 0.5x + 0.5y = 6 \\ 0.001x - 0.001y = -0.004 \end{cases}$

99. $\begin{cases} 2(2x + 3y) = 5 \\ 8x = 3(1 + 3y) \end{cases}$

100. The system $\begin{cases} \frac{1}{2}x = y + 3 \\ x - 2y = 6 \end{cases}$ has infinitely many solutions. Find three of them.

SECTION 4.3

Solving Systems of Equations by Elimination (Addition)

OBJECTIVES

1. Solve systems of linear equations by the elimination method.
2. Use multiplication to eliminate a variable.
3. Use the elimination method twice to solve a system.
4. Use elimination to identify inconsistent systems and dependent equations.
5. Determine the most efficient method to use to solve a linear system.

ARE YOU READY?

The following problems review some basic skills that are needed when solving systems of equations by elimination (addition).

1. In $4x - 6y = 7$, what is the *opposite* of the coefficient of y?
2. Add: $8x + (-8x)$
3. Substitute 2 for y in $5x + 6y = 7$ and find x.
4. Multiply both sides of the equation $3x + 9y = 1$ by -4.

We have seen that graphing can be an inaccurate method for solving a system of equations because we must estimate the coordinates of the point of intersection. In Section 4.2, we learned that the substitution method is more precise. However, it too has its drawbacks. Recall that in the first step of the substitution method, we solve one of the equations for one of the variables. At times, this can be difficult, especially if none of the variables has a coefficient of 1 or -1. This is the case for the system

$$\begin{cases} 2x + 5y = 11 \\ 7x - 5y = 16 \end{cases}$$

Solving either equation for x or y involves working with cumbersome fractions. For example, if we solve the first equation for x, the resulting substitution equation is $x = \frac{11 - 5y}{2}$. Fortunately, we can solve systems like this one using an easier algebraic method called the **elimination** or the **addition method.**

1 Solve Systems of Linear Equations by the Elimination Method.

The elimination method for solving a system is based on the **addition property of equality:** *When equal quantities are added to both sides of an equation, the results are equal.* In symbols, if $A = B$ and $C = D$, then adding the left sides and the right sides of these equations, we have $A + C = B + D$. This procedure is called *adding the equations.*

Add the terms on the left sides.

$$\begin{aligned} A &= B \\ C &= D \\ \hline A + C &= B + D \end{aligned}$$

Add the terms on the right sides.

EXAMPLE 1 Solve the system: $\begin{cases} 2x + 5y = 11 \\ 6x - 5y = 13 \end{cases}$

Strategy Since the coefficients of the y-terms are opposites (5 and -5), we will add the left sides and the right sides of the given equations.

Why When we add the equations in this way, the result will be an equation that contains only one variable, x.

Solution Since $6x - 5y$ and 13 are equal quantities, we can add $6x - 5y$ to the left side and 13 to the right side of the first equation, $2x + 5y = 11$.

$$\begin{aligned} 2x + 5y &= 11 \\ 6x - 5y &= 13 \\ \hline 8x \quad\quad &= 24 \end{aligned}$$

To add the equations, add the like terms, column by column.

$11 + 13 = 24$

$5y + (-5y) = 0$

$2x + 6x = 8x$

The Language of Algebra

The **elimination** method, or **addition** method as it is also known, is so named because one of the variables is eliminated using addition.

Because the sum of the terms $5y$ and $-5y$ is 0, we say that the variable y has been eliminated. Since the resulting equation has only one variable, we can solve it for x.

$$8x = 24$$
$$x = 3 \quad \text{Divide both sides by 8. This is the x-value of the solution.}$$

To find the y-value of the solution, substitute 3 for x in *either* equation of the original system.

$2x + 5y = 11$ — This is the first equation of the system.
$2(3) + 5y = 11$ — Substitute 3 for x.
$6 + 5y = 11$ — Multiply.
$5y = 5$ — Subtract 6 from both sides.
$y = 1$ — Divide both sides by 5. This is the y-value of the solution.

Now we check the proposed solution (3, 1) in the equations of the original system.

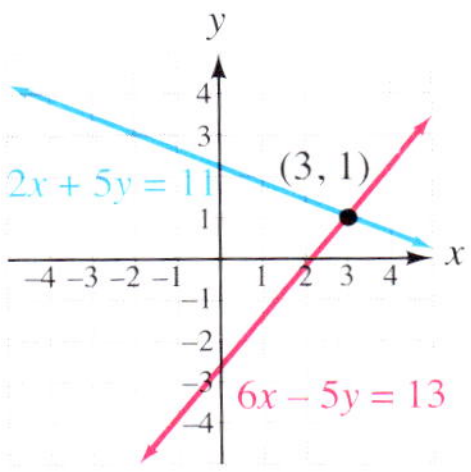

Check:

The first equation.	The second equation.
$2x + 5y = 11$	$6x - 5y = 13$
$2(3) + 5(1) \stackrel{?}{=} 11$	$6(3) - 5(1) \stackrel{?}{=} 13$
$6 + 5 \stackrel{?}{=} 11$	$18 - 5 \stackrel{?}{=} 13$
$11 = 11$ True	$13 = 13$ True

Since (3, 1) satisfies both equations, it is the solution. The graph on the left helps to verify this and it illustrates an important fact: *The solution found using the elimination (addition) method will be the same as the solution found using the graphing method.* The solution set is written $\{(3, 1)\}$.

Self Check 1 Solve the system: $\begin{cases} -4x + 3y = 4 \\ 4x + 5y = 28 \end{cases}$

Now Try Problem 17

To solve a system of equations in x and y by the elimination (addition) method, follow these steps.

The Elimination (Addition) Method

1. Write both equations of the system in standard $Ax + By = C$ form.
2. If necessary, multiply one or both of the equations by a nonzero number chosen to make the coefficients of x (or the coefficients of y) opposites.
3. Add the equations to eliminate the terms involving x (or y).
4. Solve the equation resulting from step 3.
5. Find the value of the remaining variable by substituting the solution found in step 4 into any equation containing both variables. Or, repeat steps 2–4 to eliminate the other variable.
6. Check the proposed solution in each equation of the original system. Write the solution as an ordered pair.

2 Use Multiplication to Eliminate a Variable.

In Example 1, the coefficients of the terms $5y$ in the first equation and $-5y$ in the second equation were opposites. When we added the equations, the variable y was eliminated. For many systems, however, we are not able to eliminate a variable immediately by adding. In such cases, we use the multiplication property of equality to create coefficients of x or y that are opposites.

EXAMPLE 2 Solve the system: $\begin{cases} 2x + 7y = -18 \\ 2x + 3y = -10 \end{cases}$

Strategy We will use the elimination method to solve this system.

Why Since none of the variables has a coefficient of 1 or -1, it would be difficult to solve this system using substitution.

Solution *Step 1:* Both equations are in standard $Ax + By = C$ form. We see that neither the coefficients of x nor the coefficients of y are opposites. Adding these equations as written does not eliminate a variable.

Step 2: To eliminate x, we can multiply both sides of the second equation by -1. This creates the term $-2x$, whose coefficient is opposite that of the $2x$ term in the first equation.

$$\begin{cases} 2x + 7y = -18 \\ 2x + 3y = -10 \end{cases} \xrightarrow[\text{Multiply by } -1]{\text{Unchanged}} \begin{aligned} 2x + 7y &= -18 \\ -1(2x + 3y) &= -1(-10) \end{aligned} \xrightarrow[\text{Simplify}]{\text{Unchanged}} \begin{cases} 2x + 7y = -18 \\ -2x - 3y = 10 \end{cases}$$

Success Tip

It doesn't matter which variable is eliminated first. We don't have to find the values of the variables in alphabetical order. Choose the one that is the easier to eliminate. The objective of this method is to obtain two equations whose sum will be one equation in one variable.

Step 3: When the equations are added, x is eliminated.

$$\begin{aligned} 2x + 7y &= -18 \\ -2x - 3y &= 10 \\ \hline 4y &= -8 \end{aligned}$$

In the left column: $2x + (-2x) = 0$.

This equation has only one variable.

Step 4: Solve the resulting equation to find y.

$$4y = -8$$
$$y = -2 \quad \text{Divide both sides by 4. This is the } y\text{-value of the solution.}$$

Step 5: To find x, we can substitute -2 for y in either of the equations of the original system, or in $-2x - 3y = 10$. It appears the calculations will be simplest if we use $2x + 3y = -10$.

$$\begin{aligned} 2x + 3y &= -10 && \text{This is the second equation of the original system.} \\ 2x + 3(-2) &= -10 && \text{Substitute } -2 \text{ for } y. \\ 2x - 6 &= -10 && \text{Multiply.} \\ 2x &= -4 && \text{Add 6 to both sides.} \\ x &= -2 && \text{Divide both sides by 2. This is the } x\text{-value of the solution.} \end{aligned}$$

The Language of Algebra

Recall that $2x$ and $-2x$ are called **additive inverses** because their sum is equal to 0:

$$2x + (-2x) = 0$$

Step 6: The solution is $(-2, -2)$. Check this result in the original equations.

Self Check 2 Solve the system: $\begin{cases} 2x + 7y = -27 \\ 3x + 7y = -30 \end{cases}$

Now Try Problem 21

EXAMPLE 3 Solve the system: $\begin{cases} 7x + 2y - 14 = 0 \\ 9x = 4y - 28 \end{cases}$

Strategy We will use the elimination method to solve this system.

Why Since none of the variables has coefficient 1 or -1, it would be difficult to solve this system using substitution.

Solution *Step 1:* To compare coefficients, write each equation in the standard $Ax + By = C$ form. Since each of the original equations will be written in an equivalent form, the resulting system will have the same solution as the original system.

$$\begin{cases} 7x + 2y = 14 & \text{Add 14 to both sides of } 7x + 2y - 14 = 0. \\ 9x - 4y = -28 & \text{Subtract } 4y \text{ from both sides of } 9x = 4y - 28. \end{cases}$$

Success Tip

We choose to eliminate y because the coefficient -4 is a *multiple* of the coefficient 2. The same cannot be said for 7 and 9, the coefficients of x.

Caution

When using the elimination method, don't forget to multiply both sides of an equation by the appropriate number.

Multiply both sides by 2.
$2(7x + 2y) = 2(14)$

Step 2: Neither the coefficients of x nor the coefficients of y are opposites. To eliminate y, we can multiply both sides of the first equation by 2. This creates the term $4y$, whose coefficient is opposite that of the $-4y$ term in the second equation.

$$\begin{cases} 7x + 2y = 14 \\ 9x - 4y = -28 \end{cases} \xrightarrow[\text{Unchanged}]{\text{Multiply by 2}} \begin{matrix} 2(7x + 2y) = 2(14) \\ 9x - 4y = -28 \end{matrix} \xrightarrow[\text{Unchanged}]{\text{Simplify}} \begin{cases} 14x + 4y = 28 \\ 9x - 4y = -28 \end{cases}$$

Step 3: When the equations are added, y is eliminated.

$$\begin{aligned} 14x + 4y &= 28 \\ 9x - 4y &= -28 \\ \hline 23x \quad &= 0 \end{aligned}$$

In the middle column: $4y + (-4y) = 0$.

This equation has only one variable.

Step 4: Solve the resulting equation to find x.

$$23x = 0$$
$$x = 0 \quad \text{Divide both sides by 23. This is the x-value of the solution.}$$

Step 5: To find y, we can substitute 0 for x in any equation that contains both variables. It appears the computations will be simplest if we use $7x + 2y = 14$.

$$\begin{aligned} 7x + 2y &= 14 && \text{This is the first equation of the original system.} \\ 7(0) + 2y &= 14 && \text{Substitute 0 for x.} \\ 0 + 2y &= 14 && \text{Multiply.} \\ 2y &= 14 && \text{Simplify the left side.} \\ y &= 7 && \text{Divide both sides by 2. This is the y-value of the solution.} \end{aligned}$$

Step 6: The solution is (0, 7). Check this result in the original equations.

Self Check 3 Solve the system: $\begin{cases} 3x = 10 - 2y \\ 5x - 6y + 30 = 0 \end{cases}$

Now Try Problem 25

Sometimes we must apply the multiplication property of equality to both equations to create coefficients of one variable that are opposites.

EXAMPLE 4 Solve the system: $\begin{cases} 4a + 7b = -8 \\ 5a + 6b = 1 \end{cases}$

Strategy We will use the elimination method to solve this system.

Why Since none of the variables has coefficient 1 or -1, it would be difficult to solve this system using substitution.

Solution ***Step 1:*** Both equations are written in standard $Ax + By = C$ form.

Step 2: In this example, we must write *both* equations in equivalent forms to obtain like terms that are opposites. To eliminate a, we can multiply the first equation by 5 to create the term $20a$, and we can multiply the second equation by -4 to create the term $-20a$.

$$\begin{cases} 4a + 7b = -8 \\ 5a + 6b = 1 \end{cases} \xrightarrow[\text{Multiply by } -4]{\text{Multiply by 5}} \begin{matrix} 5(4a + 7b) = 5(-8) \\ -4(5a + 6b) = -4(1) \end{matrix} \xrightarrow[\text{Simplify}]{\text{Simplify}} \begin{cases} 20a + 35b = -40 \\ -20a - 24b = -4 \end{cases}$$

Success Tip

We create the term $20a$ from $4a$ and the term $-20a$ from $5a$. Note that the *least common multiple* of 4 and 5 is 20:

4, 8, 12, 16, **20**, 24, 28, . . .
5, 10, 15, **20**, 25, 30, . . .

Step 3: When we add the resulting equations, a is eliminated.

$$\begin{aligned} 20a + 35b &= -40 \\ -20a - 24b &= -4 \\ \hline 11b &= -44 \end{aligned}$$

In the left column: $20a + (-20a) = 0$.
This equation has only one variable.

Step 4: Solve the resulting equation to find b.

$$\begin{aligned} 11b &= -44 \\ b &= -4 \end{aligned}$$

Divide both sides by 11. This is the b-value of the solution.

Success Tip

With this method, it doesn't matter which variable is eliminated. We could have created terms of $42b$ and $-42b$ to eliminate b. We will get the same solution, $(5, -4)$.

Step 5: To find a, we can substitute -4 for b in any equation that contains both variables. It appears the calculations will be simplest if we use $5a + 6b = 1$.

$5a + 6\mathbf{b} = 1$	This is the second equation of the original system.
$5a + 6(\mathbf{-4}) = 1$	Substitute -4 for b.
$5a - 24 = 1$	Multiply.
$5a = 25$	Add 24 to both sides.
$a = 5$	Divide both sides by 5. This is the a-value of the solution.

Step 6: Written in (a, b) form, the solution is $(5, -4)$. Check it in the original equations.

Self Check 4 Solve the system: $\begin{cases} 5a + 3b = -7 \\ 3a + 4b = 9 \end{cases}$

Now Try Problem 29

3 Use the Elimination Method Twice to Solve a System.

Sometimes it is easier to find the value of the second variable of a solution by using elimination a second time.

EXAMPLE 5 Solve the system: $\begin{cases} \frac{1}{6}x + \frac{1}{2}y = \frac{1}{3} \\ -\frac{x}{9} + y = \frac{5}{9} \end{cases}$

Strategy We will begin by clearing each equation of fractions. Then we will use the elimination method to solve the resulting equivalent system.

Why It is easier to create a pair of terms that are opposites if their coefficients are integers rather than fractions.

Solution ***Step 1:*** To clear the equations of the fractions, multiply both sides of the first equation by 6 and both sides of the second equation by 9.

$$\begin{cases} \frac{1}{6}x + \frac{1}{2}y = \frac{1}{3} \xrightarrow{\text{Multiply by 6}} \mathbf{6}\left(\frac{1}{6}x + \frac{1}{2}y\right) = \mathbf{6}\left(\frac{1}{3}\right) \xrightarrow{\text{Simplify}} \\ -\frac{x}{9} + y = \frac{5}{9} \xrightarrow{\text{Multiply by 9}} \mathbf{9}\left(-\frac{x}{9} + y\right) = \mathbf{9}\left(\frac{5}{9}\right) \xrightarrow{\text{Simplify}} \end{cases} \begin{cases} x + 3y = 2 \\ -x + 9y = 5 \end{cases}$$

Step 2: The coefficients of x are opposites.

Step 3: The variable x is eliminated when we add the resulting equations.

$$\begin{aligned} x + 3y &= 2 \\ -x + 9y &= 5 \\ \hline 12y &= 7 \end{aligned}$$

In the left column: $x + (-x) = 0$.
This equation has only one variable.

Success Tip

Some students find it helpful to cross out the terms that are eliminated:

$$\begin{aligned} \cancel{x} + 3y &= 2 \\ \cancel{-x} + 9y &= 5 \\ \hline 12y &= 7 \end{aligned}$$

Step 4: Solve the resulting equation to find y.

$$12y = 7$$

$$y = \frac{7}{12} \quad \text{Divide both sides by 12. This is the y-value of the solution.}$$

Step 5: We can find x by substituting $\frac{7}{12}$ for y in any equation containing both variables. However, that calculation could be complicated, because $\frac{7}{12}$ is a fraction. Instead, we can begin again with the system that is cleared of fractions, but this time, eliminate y. If we multiply both sides of the first equation by -3, this creates the term $-9y$, whose coefficient is opposite that of the $9y$ term in the second equation.

$$\begin{cases} x + 3y = 2 \\ -x + 9y = 5 \end{cases} \xrightarrow[\text{Unchanged}]{\text{Multiply by } -3} \begin{matrix} -3(x + 3y) = -3(2) \\ -x + 9y = 5 \end{matrix} \xrightarrow[\text{Unchanged}]{\text{Simplify}} \begin{cases} -3x - 9y = -6 \\ -x + 9y = 5 \end{cases}$$

When we add the resulting equations, y is eliminated.

$$\begin{aligned} -3x - 9y &= -6 \quad &&\text{In the middle column: } 9y + (-9y) = 0. \\ -x + 9y &= 5 \\ \hline -4x \quad &= -1 \quad &&\text{This equation has only one variable.} \end{aligned}$$

Now we solve the resulting equation to find x.

$$-4x = -1$$

$$x = \frac{1}{4} \quad \text{Divide both sides by } -4\text{. This is the x-value of the solution.}$$

Step 6: The solution is $\left(\frac{1}{4}, \frac{7}{12}\right)$. To verify this, check it in the original equations.

Self Check 5 Solve the system: $\begin{cases} -\frac{1}{5}x + y = \frac{8}{5} \\ \frac{x}{8} + \frac{y}{2} = \frac{1}{4} \end{cases}$

Now Try Problem 33

4 Use Elimination to Identify Inconsistent Systems and Dependent Equations.

We have solved inconsistent systems and systems of dependent equations by substitution and by graphing. We also can solve these systems using the elimination method.

EXAMPLE 6 Solve the system: $\begin{cases} 3x - 2y = 2 \\ -3x + 2y = -12 \end{cases}$

Strategy We will use the elimination method to solve this system.

Why The terms $3x$ and $-3x$ are immediately eliminated.

Solution

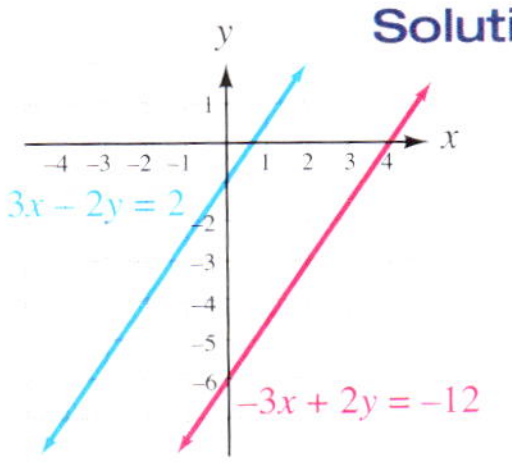

$$\begin{aligned} 3x - 2y &= 2 \quad &&\text{In the left column: } 3x + (-3x) = 0. \\ -3x + 2y &= -12 \quad &&\text{In the middle column: } -2y + 2y = 0. \\ \hline 0 &= -10 \quad &&\text{In the right column: } 2 + (-12) = -10. \end{aligned}$$

In eliminating x, the variable y is eliminated as well. The resulting false statement, $0 = -10$, indicates that the system has no solution and is inconsistent. The graphs of the equations help to verify this; they are parallel lines.

Self Check 6 Solve the system: $\begin{cases} 2x - 7y = 5 \\ -2x + 7y = 3 \end{cases}$

Now Try Problem 37

EXAMPLE 7 Solve the system: $\begin{cases} \dfrac{2x - 5y}{15} = \dfrac{8}{15} \\ -0.2x + 0.5y = -0.8 \end{cases}$

Strategy We will begin by clearing the equations of fractions and decimals. Then we will use the elimination method to solve the resulting equivalent system.

Why In this form, the equations do not contain terms with coefficients that are opposites.

Solution We can multiply both sides of the first equation by **15** to clear it of fractions and both sides of the second equation by **10** to clear it of decimals.

$$\begin{cases} \dfrac{2x - 5y}{15} = \dfrac{8}{15} & \rightarrow & 15\left(\dfrac{2x - 5y}{15}\right) = 15\left(\dfrac{8}{15}\right) & \rightarrow & 2x - 5y = 8 \\ -0.2x + 0.5y = -0.8 & \rightarrow & 10(-0.2x + 0.5y) = 10(-0.8) & \rightarrow & -2x + 5y = -8 \end{cases}$$

We add the resulting equations to get

$$\begin{array}{rcrl} 2x - 5y & = & 8 & \text{In the left column: } 2x + (-2x) = 0. \\ -2x + 5y & = & -8 & \text{In the middle column: } -5y + 5y = 0. \\ \hline 0 & = & 0 & \text{In the right column: } 8 + (-8) = 0. \end{array}$$

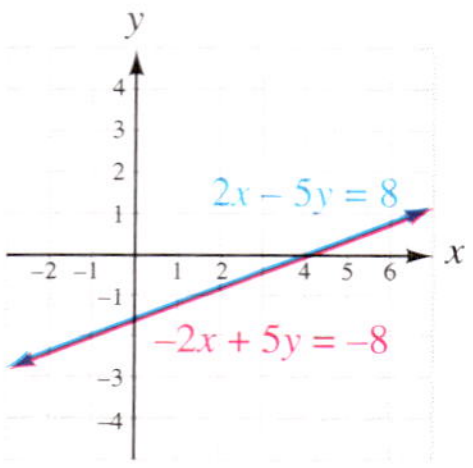

As in Example 6, both variables are eliminated. However, this time a true statement, $0 = 0$, is obtained. It indicates that the equations are dependent and that the system has infinitely many solutions. The graphs of the equations help to verify this; they are identical.

To find several of the infinitely many solutions, we can substitute some values of x, say -1, 4, and 9, in either equation and solve for y. The results are: $(-1, -2)$, $(4, 0)$, and $(9, 2)$.

Self Check 7 Solve the system: $\begin{cases} \dfrac{3x + y}{6} = \dfrac{1}{3} \\ -0.3x - 0.1y = -0.2 \end{cases}$

Now Try Problem 41

5 Determine the Most Efficient Method to Use to Solve a Linear System.

If no method is specified for solving a particular system of two linear equations, the following guidelines can be helpful in determining whether to use graphing, substitution, or elimination.

1. If you want to show trends and see the point that the two graphs have in common, then use the **graphing method.** However, this method is not exact and can be lengthy.
2. If one of the equations is solved for one of the variables, or easily solved for one of the variables, use the **substitution method.**
3. If both equations are in standard $Ax + By = C$ form, and no variable has a coefficient of 1 or -1, use the **elimination method.**
4. If the coefficient of one of the variables is 1 or -1, you have a choice. You can write each equation in standard $(Ax + By = C)$ form and use elimination, or you can solve for the variable with coefficient 1 or -1 and use substitution.

Here are some examples of suggested approaches:

$$\begin{cases} 2x + 3y = 1 \\ y = 4x - 3 \end{cases}$$
Substitution

$$\begin{cases} 5x + 3y = 9 \\ 8x + 4y = 3 \end{cases}$$
Elimination

$$\begin{cases} 4x - y = -6 \\ 3x + 2y = 1 \end{cases}$$
Elimination

$$\begin{cases} x - 23 = 6y \\ 7x - 9y = -3 \end{cases}$$
Substitution

Each method that we use to solve systems of equations has advantages and disadvantages.

Method	Advantages	Disadvantages
Graphing	■ You see the solution(s). ■ The graphs allow you to observe trends.	■ Inaccurate when the solutions are not integers or are large numbers off the graph
Substitution	■ Always gives the exact solutions ■ Works well if one of the equations is solved for one of the variables, or if it is easy to solve for one of the variables	■ You do not see the solution. ■ If no variable has a coefficient of 1 or −1, solving for one of the variables often involves fractions.
Elimination	■ Always gives the exact solutions ■ Works well if no variable has a coefficient of 1 or −1	■ You do not see the solution. ■ The equations must be written in the form $Ax + By = C$.

SECTION 4.3 STUDY SET

VOCABULARY

Fill in the blanks.

1. The coefficients of $3x$ and $-3x$ are __________.

2. When the given equations are added, the variable y will be __________.
$$\begin{array}{r} 5x - 6y = 10 \\ \underline{-3x + 6y = 24} \end{array}$$

CONCEPTS

3. In the given system, which terms have coefficients that are opposites?
$$\begin{cases} 3x + 7y = -25 \\ 4x - 7y = 12 \end{cases}$$

4. Fill in the blank. The objective of the elimination method is to obtain two equations whose sum will be one equation in one __________.

5. Add each pair of equations.
   a. $\begin{array}{r} 2a + 2b = -6 \\ \underline{3a - 2b = 2} \end{array}$
   b. $\begin{array}{r} x - 3y = 15 \\ \underline{-x - y = -14} \end{array}$

6. a. Multiply both sides of $4x + y = 2$ by 3.
   b. Multiply both sides of $x - 3y = 4$ by -2.

7. If the elimination method is used to solve
$$\begin{cases} 3x + 12y = 4 \\ 6x - 4y = 7 \end{cases}$$
   a. By what would we multiply the first equation to eliminate the variable x?
   b. By what would we multiply the second equation to eliminate the variable y?

8. Suppose the following system is solved using the elimination method and it is found that x is 2. Find the value of y.
$$\begin{cases} 4x + 3y = 11 \\ 3x - 2y = 4 \end{cases}$$

9. What algebraic step should be performed to
   a. Clear $\frac{2}{3}x + 4y = -\frac{4}{5}$ of fractions?
   b. Clear $0.2x - 0.9y = 6.4$ of decimals?

10. a. Suppose $0 = 0$ is obtained when a system is solved by the elimination method. Does the system have a solution? Which of the following is a possible graph of the system?
    b. Suppose $0 = 2$ is obtained when a system is solved by the elimination method. Does the system have a solution? Which of the following is a possible graph of the system?

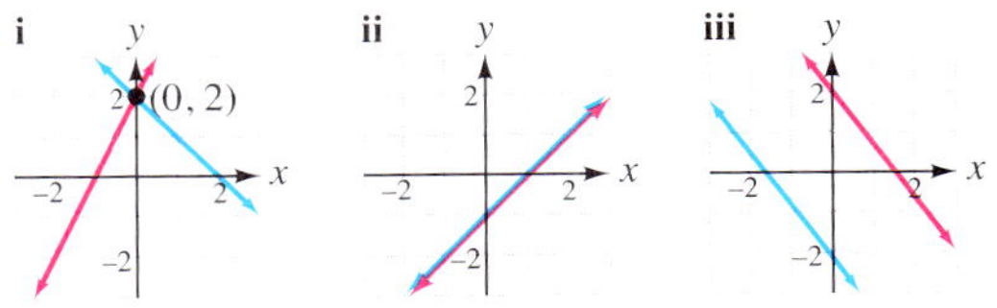

NOTATION

Complete the solution to solve the system.

11. Solve: $\begin{cases} x + y = 5 \\ x - y = -3 \end{cases}$

$$\begin{aligned} x + y &= 5 \quad \text{Add the equations.} \\ x - y &= -3 \\ \hline &= 2 \\ x &= \end{aligned}$$

$$\begin{aligned} x + y &= 5 \quad \text{This is the first equation.} \\ + y &= 5 \\ y &= \end{aligned}$$

The solution is (,).

12. Write each equation in $Ax + By = C$ form:

$\begin{cases} 7x + y + 3 = 0 \\ 8x + 4 = -y \end{cases} \rightarrow \begin{cases} \\ \end{cases}$

GUIDED PRACTICE

Use the elimination method to solve each system. See Example 1.

13. $\begin{cases} x + y = 5 \\ x - y = 1 \end{cases}$

14. $\begin{cases} x - y = 4 \\ x + y = 8 \end{cases}$

15. $\begin{cases} x + y = -5 \\ -x + y = -1 \end{cases}$

16. $\begin{cases} -x + y = -3 \\ x + y = 1 \end{cases}$

17. $\begin{cases} 4x + 3y = 24 \\ 4x - 3y = -24 \end{cases}$

18. $\begin{cases} -9x + 5y = -9 \\ -9x - 5y = -9 \end{cases}$

19. $\begin{cases} 2s + t = -2 \\ -2s - 3t = -6 \end{cases}$

20. $\begin{cases} -2x + 4y = 12 \\ 2x + 4y = 28 \end{cases}$

Use the elimination method to solve each system. See Example 2.

21. $\begin{cases} x + 3y = -9 \\ x + 8y = -4 \end{cases}$

22. $\begin{cases} x + 7y = -22 \\ x + 9y = -24 \end{cases}$

23. $\begin{cases} 7x - y = 10 \\ 8x - y = 13 \end{cases}$

24. $\begin{cases} 6x - y = 4 \\ 9x - y = 10 \end{cases}$

Use the elimination method to solve each system. See Example 3.

25. $\begin{cases} 7x + 4y - 14 = 0 \\ 3x = 2y - 20 \end{cases}$

26. $\begin{cases} 5x - 14y - 32 = 0 \\ -x = 6y + 20 \end{cases}$

27. $\begin{cases} 7x - 50y + 43 = 0 \\ x = 4 - 3y \end{cases}$

28. $\begin{cases} x - 2y + 1 = 0 \\ 12x = 23 - 11y \end{cases}$

Use the elimination method to solve each system. See Example 4.

29. $\begin{cases} 4x + 3y = 7 \\ 3x - 2y = -16 \end{cases}$

30. $\begin{cases} 3x - 2y = 20 \\ 2x + 7y = 5 \end{cases}$

31. $\begin{cases} 5a + 8b = 2 \\ 11a - 3b = 25 \end{cases}$

32. $\begin{cases} 7a - 5b = 24 \\ 12a + 8b = 8 \end{cases}$

Use the elimination method to solve each system. See Example 5.

33. $\begin{cases} \frac{1}{8}x - \frac{1}{8}y = \frac{3}{8} \\ \frac{x}{4} + \frac{y}{4} = \frac{1}{2} \end{cases}$

34. $\begin{cases} \frac{1}{8}x + \frac{1}{4}y = \frac{1}{4} \\ \frac{x}{2} + \frac{y}{4} = \frac{1}{2} \end{cases}$

35. $\begin{cases} \frac{3}{4}x - \frac{5}{8}y = \frac{1}{24} \\ \frac{5x}{6} - y = \frac{1}{4} \end{cases}$

36. $\begin{cases} \frac{1}{3}x + \frac{1}{2}y = \frac{5}{3} \\ \frac{x}{7} - \frac{y}{7} = -\frac{1}{7} \end{cases}$

Use the elimination method to solve each system. If there is no solution, or infinitely many solutions, so state. See Example 6.

37. $\begin{cases} 3x - 5y = -29 \\ 3x - 5y = 15 \end{cases}$

38. $\begin{cases} 2a - 3b = -6 \\ 2a - 3b = 8 \end{cases}$

39. $\begin{cases} 3x - 16 = 5y \\ -3x + 5y - 33 = 0 \end{cases}$

40. $\begin{cases} \frac{-18x + y}{2} = \frac{7}{2} \\ 18x = y \end{cases}$

Use elimination to solve each system. If there is no solution, or infinitely many solutions, so state. See Example 7.

41. $\begin{cases} 0.4x - 0.7y = -1.9 \\ -x + \frac{7y}{4} = \frac{19}{4} \end{cases}$

42. $\begin{cases} 0.1x + 2y + 0.2 = 0 \\ -\frac{x}{4} - 5y = \frac{1}{2} \end{cases}$

43. $\begin{cases} \frac{x - 6y}{2} = 7 \\ -x + 6y + 14 = 0 \end{cases}$

44. $\begin{cases} 2x + 5y - 13 = 0 \\ -2x + 13 = 5y \end{cases}$

TRY IT YOURSELF

Solve the system by either the substitution or the elimination method.

45. $\begin{cases} y = -3x + 9 \\ y = x + 1 \end{cases}$

46. $\begin{cases} x = 5y - 4 \\ x = 9y - 8 \end{cases}$

47. $\begin{cases} 4x + 6y = 5 \\ 8x - 9y = 3 \end{cases}$

48. $\begin{cases} 3a + 4b = 36 \\ 6a - 2b = -21 \end{cases}$

49. $\begin{cases} 6x - 3y = -7 \\ y + 9x = 6 \end{cases}$

50. $\begin{cases} 9x + 4y = 31 \\ y - 5 = 6x \end{cases}$

51. $\begin{cases} x + y = 1 \\ x - y = 5 \end{cases}$

52. $\begin{cases} x - y = -5 \\ x + y = 1 \end{cases}$

53. $\begin{cases} 4(x - 2y) = 36 \\ 3x - 6y = 27 \end{cases}$

54. $\begin{cases} 2(x + 2y) = 15 \\ 3x = 8 - 6y \end{cases}$

55. $\begin{cases} x = y \\ 0.1x + 0.2y = 1.0 \end{cases}$

56. $\begin{cases} x = y \\ 0.4x - 0.8y = -0.5 \end{cases}$

57. $\begin{cases} 2x + 11y = -10 \\ 5x + 4y = 22 \end{cases}$

58. $\begin{cases} 3x + 4y = 12 \\ 4x + 5y = 17 \end{cases}$

59. $\begin{cases} 7x = 21 - 6y \\ 4x + 5y = 12 \end{cases}$

60. $\begin{cases} -4x = -3y - 13 \\ -6x + 8y = -16 \end{cases}$

61. $\begin{cases} 9x - 10y = 0 \\ \frac{9x - 3y}{63} = 1 \end{cases}$

62. $\begin{cases} 8x - 9y = 0 \\ \frac{2x - 3y}{6} = -1 \end{cases}$

63. $\begin{cases} \frac{m}{4} + \frac{n}{3} = -\frac{1}{12} \\ \frac{m}{2} - \frac{5}{4}n = \frac{7}{4} \end{cases}$

64. $\begin{cases} \frac{x}{2} - \frac{y}{3} = -2 \\ \frac{x}{3} + \frac{2}{3}y = \frac{4}{3} \end{cases}$

65. $\begin{cases} x - \frac{4}{3}y = \frac{1}{3} \\ 2x + \frac{3}{2}y = \frac{1}{2} \end{cases}$

66. $\begin{cases} x + y = -\frac{1}{4} \\ x - \frac{y}{2} = -\frac{3}{2} \end{cases}$

67. $\begin{cases} 4x - 7y + 32 = 0 \\ 5x = 4y - 2 \end{cases}$

68. $\begin{cases} 6x = -3y \\ 5x + 15 = 5y \end{cases}$

69. $\begin{cases} 3(x + 4y) = -12 \\ x = 3y + 10 \end{cases}$

70. $\begin{cases} 3x + 2y = 3 \\ y = 2(x - 8) \end{cases}$

71. $\begin{cases} 4a + 7b = 2 \\ 9a - 3b = 1 \end{cases}$

72. $\begin{cases} 5a - 7b = 6 \\ 7a - 6b = 8 \end{cases}$

73. $\begin{cases} 3a - b = 12.3 \\ 4a - b = 14.9 \end{cases}$

74. $\begin{cases} -7x - y = 8.5 \\ 4x - y = -12.4 \end{cases}$

75. $\begin{cases} 5x - 4y = 8 \\ -5x - 4y = 8 \end{cases}$

76. $\begin{cases} 2r + s = -8 \\ -2r + 4s = 28 \end{cases}$

77. $\begin{cases} 9a + 16b = -36 \\ 7a + 4b = 48 \end{cases}$

78. $\begin{cases} 4a + 7b = -24 \\ 9a + b = 64 \end{cases}$

79. $\begin{cases} 8x + 12y = -22 \\ 3x - 2y = 8 \end{cases}$

80. $\begin{cases} 3x + 2y = 45 \\ 5x - 4y = 20 \end{cases}$

81. $\begin{cases} 6x + 5y + 29 = 0 \\ 0.02x = 0.03y - 0.05 \end{cases}$

82. $\begin{cases} 3x = 20y + 1 \\ 0.04x + 0.05y - 0.33 = 0 \end{cases}$

83. $\begin{cases} c = d - 9 \\ 5c = 3d - 35 \end{cases}$

84. $\begin{cases} a = b + 7 \\ 3a - 15 = 5b \end{cases}$

85. $\begin{cases} 0.9x + 2.1 = 0.3y \\ 0.4x = 0.7y + 1.9 \end{cases}$

86. $\begin{cases} 0.7x + 1.1 = 0.4y \\ 0.4x = 0.7y + 2.2 \end{cases}$

87. $\begin{cases} 5c + 2d = -5 \\ 6c + 2d = -10 \end{cases}$

88. $\begin{cases} 11c + 3d = -68 \\ 10c + 3d = -64 \end{cases}$

89. $\begin{cases} \frac{2}{15}x - \frac{1}{5}y = \frac{1}{3} \\ \frac{2}{15}x - \frac{1}{5}y = \frac{1}{10} \end{cases}$

90. $\begin{cases} \frac{1}{5}x + \frac{3}{5}y = \frac{4}{5} \\ \frac{1}{6}x + \frac{1}{2}y = \frac{2}{3} \end{cases}$

APPLICATIONS

91. **Education.** The graph shows educational trends during the years 1980–2009 for persons 25 years or older in the United States. The equation $9x + 11y = 352$ approximates the percent y that had less than high school completion. The equation $5x - 11y = -198$ approximates the percent y that had a Bachelor's or higher degree. In each case, x is the number of years since 1980. Use the elimination method to determine in what year the percents were equal.

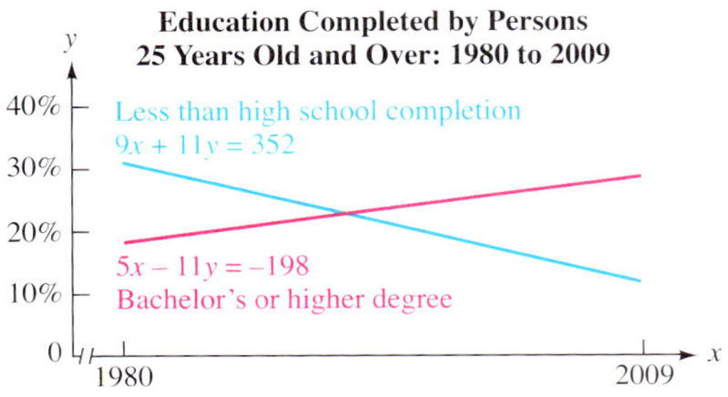

Source: U.S. Department of Commerce, Census Bureau

92. **Newspapers.** The graph shows the trends in the newspaper publishing industry during the years 1990–2008 in the United States. The equation $37x - 2y = -1{,}128$ models the number y of morning newspapers published and $31x + y = 1{,}059$ models the number y of evening newspapers published. In each case, x is the number of years since 1990. Use the elimination method to determine in what year there were an equal number of morning and evening newspapers being published.

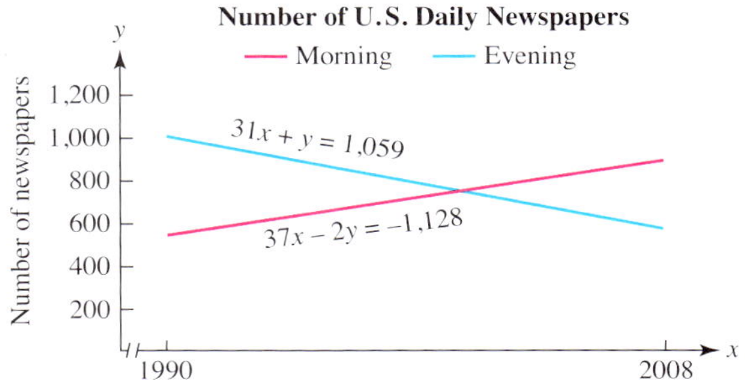

Source: Editor and Publisher Yearbook data

93. **CFL Bulbs.** The graph below shows how a more expensive, but more energy-efficient, compact fluorescent light bulb eventually costs less to use than an incandescent light bulb. The equation $60c - d = 96$ approximates the cost c (in dollars) to purchase and use a CFL bulb 8 hours a day for d days. The equation $15c - d = 6$ does the same for an incandescent bulb. Use the elimination method to determine after how many days the upgrade to a CFL bulb begins to save money.

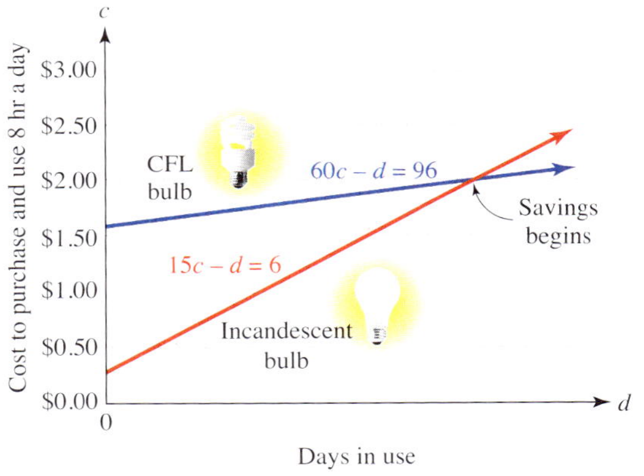

Source: Whitesites.com

94. **The Human Skeleton.** The equation $h + f = 53$ models the fact that the number of bones in the hand and foot totals 53. The equation $h - f = 1$ models the fact that the difference between the number of bones in the hand and foot is just 1. Use the elimination method to find h and f.

WRITING

95. Why is the method for solving systems that is discussed in this section called the *elimination method?* Why is it also referred to as the *addition method?*

96. If the elimination method is to be used to solve this system, what is wrong with the form in which it is written?

$$\begin{cases} 2x - 5y = -3 \\ -2y - 10 = -5x \end{cases}$$

97. Can the system $\begin{cases} 2x + 3y = 13 \\ 7x - 3y = -5 \end{cases}$ be solved more easily using the elimination method or the substitution method? Explain.

98. Explain the error in the following work.

Solve: $\begin{cases} x + y = 1 \\ x - y = 5 \end{cases}$

$$\begin{array}{r} x + y = 1 \\ +x - y = 5 \\ \hline 2x \quad\; = 6 \\ \frac{2x}{2} = \frac{6}{2} \\ \boxed{x = 3} \end{array}$$

The solution is 3.

REVIEW

99. Find an equation of the line with slope $-\frac{11}{6}$ that passes through $(2, -6)$. Write the equation in slope–intercept form.

100. Solve $S = 2\pi rh + 2\pi r^2$ for h.

101. Evaluate: $-10(18 - 4^2)^3$

102. Evaluate: -5^2

CHALLENGE PROBLEMS

Use the elimination method to solve each system.

103. $\begin{cases} \dfrac{x-3}{2} = \dfrac{11}{6} - \dfrac{y+5}{3} \\ \dfrac{x+3}{3} - \dfrac{y+3}{4} = \dfrac{5}{12} \end{cases}$

104. $\begin{cases} \dfrac{4(x+1)}{34} = \dfrac{1}{2} - \dfrac{3(y-1)}{34} \\ 0.2(x + 0.2) + 0.3(y - 0.3) = 0.75 \end{cases}$

SECTION 4.4 Problem Solving Using Systems of Equations

OBJECTIVES

1. Assign variables to two unknowns.
2. Use systems to solve geometry problems.
3. Use systems to solve number-value problems.
4. Use systems to solve interest, uniform motion, and mixture problems.

ARE YOU READY?

The following problems review some basic skills that are needed when solving application problems using systems of equations.

1. What is the formula for the perimeter of a rectangle?
2. At \$12.95 per pound, what is the value of 3 pounds of crab legs?
3. Find the amount of interest earned by \$55,000 invested at a 4% annual simple interest rate for 1 year.
4. At 35 miles per hour, how far will a Canada goose travel in 12 hours?
5. A 12-ounce mixture of window cleaner is 8% alcohol. How many ounces of the mixture is alcohol?

In previous chapters, many applied problems were modeled and solved with an equation in one variable. In this section, the application problems involve two unknowns. It is often easier to solve such problems using a two-variable approach.

1 Assign Variables to Two Unknowns.

The following steps are helpful when solving problems involving two unknown quantities.

Problem-Solving Strategy

1. **Analyze the problem** by reading it carefully to understand the given facts. What information is given? What are you asked to find? What vocabulary is given? Often a diagram or table will help you understand the facts of the problem.
2. **Assign variables** to represent unknown values in the problem. This means, in most cases, to let x = one of the unknowns that you are asked to find, and y = the other unknown.
3. **Form a system of equations** by translating the words of the problem into mathematical symbols.
4. **Solve the system** of equations using graphing, substitution, or elimination.
5. **State the conclusion clearly.** Be sure to include the units.
6. **Check the results** using the words of the problem, not the equations that were formed in step 3.

EXAMPLE 1 **Motion Pictures.** Each year, Academy Award winners are presented with Oscars. The 13.5-inch statuette has a base on which a gold-plated figure stands. The figure itself is 7.5 inches taller than its base. Find the height of the figure and the height of the base.

Analyze

- The statuette is a total of 13.5 inches tall.
- The figure is 7.5 inches taller than the base.
- Find the height of the figure and the height of the base.

Assign Let x = the height of the figure, in inches, and y = the height of the base, in inches.

Form We can translate the words of the problem into two equations, each involving x and y.

The height of the figure	plus	the height of the base	is	13.5 inches.
x	$+$	y	$=$	13.5

The height of the figure	is	the height of the base	plus	7.5 inches.
x	$=$	y	$+$	7.5

The resulting system is: $\begin{cases} x + y = 13.5 \\ x = y + 7.5 \end{cases}$

Solve Since the second equation is solved for x, we will use substitution to solve the system.

$$\begin{cases} x + y = 13.5 \\ x = y + 7.5 \end{cases}$$

$x + y = 13.5$ This is the first equation of the system.

$y + 7.5 + y = 13.5$ Substitute $y + 7.5$ for x.

$2y + 7.5 = 13.5$ Combine like terms: $y + y = 2y$.

$2y = 6$ Subtract 7.5 from both sides.

$y = 3$ Divide both sides by 2. This is the height of the base.

To find x, substitute 3 for y in the second equation of the system.

$x = y + 7.5$ This is the substitution equation.

$x = 3 + 7.5$ Substitute 3 for y.

$x = 10.5$ This is the height of the figure.

State The height of the figure is 10.5 inches and the height of the base is 3 inches.

Check The sum of 10.5 inches and 3 inches is 13.5 inches, and the 10.5-inch figure is 7.5 inches taller than the 3-inch base. The results check.

Caution

If two variables are used to represent two unknown quantities, we must form a system of two equations to find the unknowns.

Caution

In this problem we are to find two unknowns, the height of the figure and height of the base. Remember to give both in the *State* step of the solution.

Self Check 1 **Woodworking.** A carpenter wants to cut a 12-foot board into two pieces. The longer piece is to be twice as long as the shorter piece. Find the length of each piece.

Now Try Problem 17

2 Use Systems to Solve Geometry Problems.

Two angles are said to be **complementary** if the sum of their measures is 90°. Two angles are said to be **supplementary** if the sum of their measures is 180°.

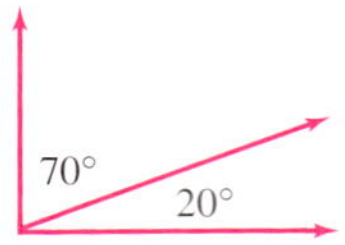

Complementary angles
$20° + 70° = 90°$

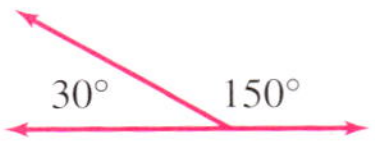

Supplementary angles
$150° + 30° = 180°$

EXAMPLE 2

Angles. The difference of the measures of two complementary angles is 6°. Find the measure of each angle.

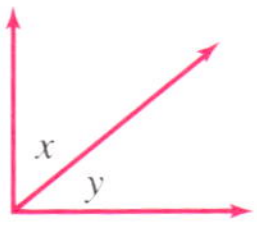

Analyze

- Since the angles are complementary, the sum of their measures is 90°.
- The word *difference* indicates subtraction. If the measure of the smaller angle is subtracted from the measure of the larger angle, the result will be 6°.
- Find the measure of the larger angle and the measure of the smaller angle.

Assign Let x = the measure of the larger angle and y = the measure of the smaller angle.

Form We can translate the words of the problem into two equations, each involving x and y.

The measure of the larger angle	plus	the measure of the smaller angle	is	90°.
x	$+$	y	$=$	90

The measure of the larger angle	minus	the measure of the smaller angle	is	6°.
x	$-$	y	$=$	6

The resulting system is: $\begin{cases} x + y = 90 \\ x - y = 6 \end{cases}$

Solve Since the coefficients of y are opposites, we will use elimination to solve the system.

$$x + y = 90$$
$$\underline{x - y = 6} \quad \text{Add the equations to eliminate } y.$$
$$2x \quad = 96$$
$$x = 48 \quad \text{Divide both sides by 2. This is the measure of the larger angle.}$$

To find y, substitute 48 for x in the first equation of the system.

$$x + y = 90$$
$$48 + y = 90 \quad \text{Substitute 48 for x.}$$
$$y = 42 \quad \text{Subtract 48 from both sides. This is the measure of the smaller angle.}$$

State The measure of the larger angle is 48° and the measure of the smaller angle is 42°.

Check The sum of 48° and 42° is 90°, and the difference is 6°. The results check.

Self Check 2 **Angles.** The difference of the measures of two supplementary angles is 22°. Find the measure of each angle.

Now Try Problem 13

Success Tip

The substitution method also could be used to solve this system. If we solve $x + y = 90$ for x, we obtain the substitution equation $x = 90 - y$.

Success Tip

When solving problems using the two-variable approach, be sure to check the results in two ways.

EXAMPLE 3 **History.** In 1917, James Montgomery Flagg created the classic *I Want You* poster to help recruiting for World War I. The perimeter of the poster is 114 inches, and its length is 9 inches less than twice its width. Find the length and the width of the poster.

Analyze

- The perimeter of the rectangular poster is 114 inches.
- The length is 9 inches less than twice the width.
- Find the length and the width of the poster.

Assign Let l = the length of the poster, in inches, and w = the width of the poster, in inches.

Form The perimeter of a rectangle is the sum of two lengths and two widths, as given by the formula $P = 2l + 2w$, so we have

2	times	the length of the poster	plus	2	times	the width of the poster	is	114 inches.
2	$\cdot$	l	$+$	2	$\cdot$	w	$=$	114

If the length of the poster is 9 inches less than twice the width, we have

The length of the poster	is	2	times	the width of the poster	minus	9 inches.
l	$=$	2	$\cdot$	w	$-$	9

The resulting system is: $\begin{cases} 2l + 2w = 114 \\ l = 2w - 9 \end{cases}$

Solve Since the second equation is solved for l, we will use substitution to solve the system.

$$2l + 2w = 114$$ This is the first equation of the system.

$$2(2w - 9) + 2w = 114$$ Substitute $2w - 9$ for l. Don't forget the parentheses.

$$4w - 18 + 2w = 114$$ Distribute the multiplication by 2.

$$6w - 18 = 114$$ Combine like terms: $4w + 2w = 6w$.

$$6w = 132$$ Add 18 to both sides.

$$w = 22$$ Divide both sides by 6. This is the width of the poster.

To find l, substitute 22 for w in the second equation of the system.

$$l = 2w - 9$$
$$l = 2(22) - 9$$
$$l = 44 - 9$$
$$l = 35$$ This is the length of the poster.

State The length of the poster is 35 inches and the width is 22 inches.

Check The perimeter is $2(35) + 2(22) = 70 + 44 = 114$ inches, and 35 inches is 9 inches less than twice 22 inches. The results check.

Self Check 3 **Gardening.** Tom has 150 feet of fencing to enclose a rectangular garden. If the garden's length is to be 5 feet less than 3 times its width, find the length and width of the garden.

Now Try Problem 23

Success Tip

In this section, to solve the application problems that involve two unknowns, we:

- Assign **two** variables.
- Form **two** equations.
- Solve a system of **two** equations.
- State **two** answers.
- Perform **two** checks.

3 Use Systems to Solve Number-Value Problems.

EXAMPLE 4 **Photography.** At a school, two picture packages are available, as shown in the illustration. Find the cost of a class picture and the cost of an individual wallet-size picture.

Analyze

- Package 1 contains 1 class picture and 10 wallet-size pictures.
- Package 2 contains 2 class pictures and 15 wallet-size pictures.
- Find the cost of a class picture and the cost of a wallet-size picture.

Assign Let c = the cost of one class picture and w = the cost of one wallet-size picture.

Form We can use the fact that **number · value = total value** to write an equation that models the first package. We note that (in dollars) the cost of 1 class picture is $1 \cdot c = c$ and the cost of 10 wallet-size pictures is $10 \cdot w = 10w$.

The cost of 1 class picture	plus	the cost of 10 wallet-size pictures	is	$19.
c	$+$	$10w$	$=$	19

To write an equation that models the second package, we note that (in dollars) the cost of 2 class pictures is $2 \cdot c = 2c$, and the cost of 15 wallet-size pictures is $15 \cdot w = 15w$.

The cost of 2 class pictures	plus	the cost of 15 wallet-size pictures	is	$31.
$2c$	$+$	$15w$	$=$	31

The resulting system is: $\begin{cases} c + 10w = 19 \\ 2c + 15w = 31 \end{cases}$

Solve We can use substitution or elimination to solve this system. If we use elimination, we can eliminate c as follows.

$$-2c - 20w = -38 \quad \text{Multiply both sides of } c + 10w = 19 \text{ by } -2.$$
$$2c + 15w = 31$$
$$-5w = -7 \quad \text{Add the equations to eliminate } c.$$
$$w = 1.4 \quad \text{Divide both sides by } -5. \text{ This is the cost of a wallet-size picture.}$$

To find c, substitute 1.4 for w in the first equation of the original system.

$$c + 10w = 19$$
$$c + 10(1.4) = 19 \quad \text{Substitute 1.4 for } w.$$
$$c + 14 = 19 \quad \text{Multiply.}$$
$$c = 5 \quad \text{Subtract 14 from both sides. This is the cost of a class picture.}$$

State A class picture costs $5 and a wallet-size picture costs $1.40.

Check Package 1 has 1 class picture and 10 wallets: $5 + 10($1.40) = $5 + $14 = $19. Package 2 has 2 class pictures and 15 wallets: 2($5) + 15($1.40) = $10 + $21 = $31. The results check.

Self Check 4 **Sporting Goods.** Getting ready for soccer season, a coach purchased 6 practice cones and a portable goal for $23. On another trip to the sporting goods store, the coach purchased 4 more cones and 2 more goals, spending $42. Find the cost of a practice cone and a portable goal.

Now Try Problem 27

4 Use Systems to Solve Interest, Uniform Motion, and Mixture Problems.

We can solve investment problems like those in Section 2.6 using a two-variable approach.

EXAMPLE 5 **White-Collar Crime.** Investigators discovered that a small business secretly moved \$150,000 out of the country to avoid paying income tax. Some of the money was invested in a Swiss bank account that paid 8% annual simple interest. The remainder was deposited in a Cayman Islands account, paying 7% annual simple interest. The investigation also revealed that the combined interest earned the first year was \$11,500. How much money was invested in each account?

Caution

It is incorrect to let

x = the amount invested in each account

This implies that *equal amounts* were invested in the Swiss and Cayman Island accounts. We do not know that.

Analyze We are told that an unknown part of the \$150,000 was invested at an annual rate of 8% and the rest at 7%. Together, the accounts earned \$11,500 in interest.

Assign Let x = the amount invested in the Swiss account and y = the amount invested in the Cayman Islands account.

Form Because the total investment was \$150,000, we have

The amount invested in the Swiss account	plus	the amount invested in the Cayman Islands account	is	\$150,000.
x	$+$	y	$=$	150,000

We can use the formula $I = Prt$ to determine that x dollars invested for 1 year at 8% earns $x \cdot 0.08 \cdot 1 = 0.08x$ dollars. Similarly, y dollars invested for 1 year at 7% earns $y \cdot 0.07 \cdot 1 = 0.07y$ dollars. If the total combined interest earned was \$11,500, we have

The income on the 8% investment	plus	the income on the 7% investment	is	\$11,500.
$0.08x$	$+$	$0.07y$	$=$	11,500

The resulting system is: $\begin{cases} x + y = 150{,}000 \\ 0.08x + 0.07y = 11{,}500 \end{cases}$

Solve First, clear the second equation of decimals. Then we can use substitution or elimination to solve the system. If we use elimination, we can eliminate x as follows.

$$-8x - 8y = -1{,}200{,}000 \quad \text{Multiply both sides of } x + y = 150{,}000 \text{ by } -8.$$
$$8x + 7y = 1{,}150{,}000 \quad \text{Multiply both sides of } 0.08x + 0.07y = 11{,}500 \text{ by } 100.$$
$$-y = -50{,}000$$
$$y = 50{,}000 \quad \text{Multiply both sides by } -1.$$

To find x, substitute 50,000 for y in the first equation of the original system.

$$x + y = 150{,}000$$
$$x + \mathbf{50{,}000} = 150{,}000 \quad \text{Substitute 50,000 for } y.$$
$$x = 100{,}000 \quad \text{Subtract 50,000 from both sides.}$$

State \$100,000 was invested in the Swiss bank account, and \$50,000 was invested in the Cayman Islands account.

Check

$$\$100{,}000 + \$50{,}000 = \$150{,}000 \quad \text{The two investments total \$150,000.}$$
$$0.08(\$100{,}000) = \$8{,}000 \quad \text{The Swiss bank account earned \$8,000.}$$
$$0.07(\$50{,}000) = \$3{,}500 \quad \text{The Cayman Islands account earned \$3,500.}$$

The combined interest is \$8,000 + \$3,500 = \$11,500. The results check.

Self Check 5 **Investments.** A woman invested \$10,000, some at 9% and the remainder at 10% annual simple interest. The annual income from these two investments was \$975. How much was invested at each rate?

Now Try Problem 35

EXAMPLE 6 **Boating.** A boat traveled 30 miles downstream in 3 hours and made the return trip in 5 hours. Find the speed of the boat in still water and the speed of the current.

Analyze Traveling downstream, the speed of the boat will be faster than it would be in still water. Traveling upstream, the speed of the boat will be slower than it would be in still water.

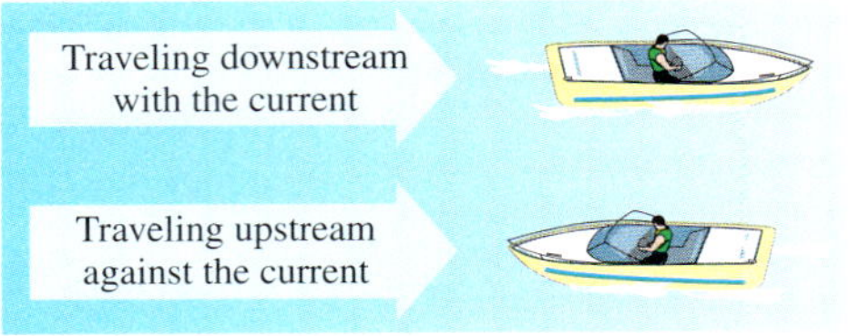

Assign Let s = the speed of the boat in still water and c = the speed of the current.

Form The speed of the boat going downstream is $s + c$ and the speed of the boat going upstream is $s - c$. Using the formula $d = rt$, we find that $3(s + c)$ represents the distance traveled downstream and $5(s - c)$ represents the distance traveled upstream. We can organize the facts of the problem in a table.

	Rate	· Time	= Distance
Downstream	$s + c$	3	$3(s + c)$
Upstream	$s - c$	5	$5(s - c)$

Multiply $r \cdot t$ to obtain an expression for each distance traveled.

Enter this information first.

Set each of these expressions for distance traveled equal to 30 to form the system of equations.

Since each trip is 30 miles long, the Distance column of the table helps us to write two equations in two variables. To write each equation in standard form, use the distributive property.

$$\begin{cases} 3(s + c) = 30 \\ 5(s - c) = 30 \end{cases} \xrightarrow{\text{Distribute}} \begin{cases} 3s + 3c = 30 \\ 5s - 5c = 30 \end{cases}$$

Success Tip

At this stage, you also could divide both sides of $3(s + c) = 30$ by 3 and divide both sides of $5(s - c) = 30$ by 5.

Solve To eliminate c, we proceed as follows.

$$15s + 15c = 150 \quad \text{Multiply both sides of } 3s + 3c = 30 \text{ by 5.}$$
$$15s - 15c = 90 \quad \text{Multiply both sides of } 5s - 5c = 30 \text{ by 3.}$$
$$30s = 240$$
$$s = 8 \quad \text{Divide both sides by 30. This is the speed of the boat in still water.}$$

To find c, it appears that the calculations will be easiest if we use $3s + 3c = 30$.

$$3s + 3c = 30$$
$$3(8) + 3c = 30 \quad \text{Substitute 8 for } s.$$
$$24 + 3c = 30 \quad \text{Multiply.}$$
$$3c = 6 \quad \text{Subtract 24 from both sides.}$$
$$c = 2 \quad \text{Divide both sides by 3. This is the speed of the current.}$$

State The speed of the boat in still water is 8 mph and the speed of the current is 2 mph.

Check With a 2-mph current, the boat's downstream speed will be $8 + 2 = 10$ mph. In 3 hours, it will travel $10 \cdot 3 = 30$ miles. With a 2-mph current, the boat's upstream speed will be $8 - 2 = 6$ mph. In 5 hours, it will cover $6 \cdot 5 = 30$ miles. The results check.

Self Check 6 **Boating.** A boat traveled 24 miles downstream in 2 hours and made the return trip in 3 hours. Find the speed of the boat in still water and the speed of the current.

Now Try Problem 41

The liquid and dry mixture problems that we studied in Section 2.6 can be solved with a two-variable approach.

EXAMPLE 7 **Medical Technology.** A laboratory technician has one batch of antiseptic that is 40% alcohol and a second batch that is 60% alcohol. She would like to make 8 fluid ounces of solution that is 55% alcohol. How many fluid ounces of each batch should she use?

Analyze Some 60% solution must be added to some 40% solution to make a 55% solution.

Assign Let $x =$ the number of ounces to be used from batch 1 and $y =$ the number of ounces to be used from batch 2.

Form The amount of alcohol in each solution is given by

Amount of solution · **strength of solution** = **amount of alcohol**

We can organize the facts of the problem in a table.

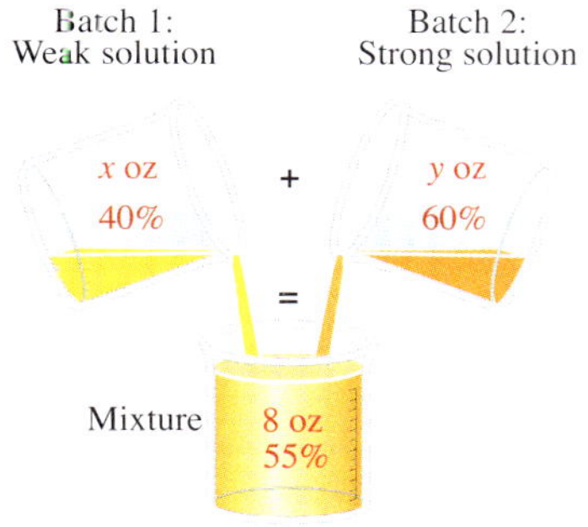

	Amount	· Strength	= Amount of alcohol
Batch 1 (too weak)	x	0.40	$0.40x$
Batch 2 (too strong)	y	0.60	$0.60y$
Mixture	8	0.55	$0.55(8)$

One equation comes from information in this column. (Amount column)

40%, 60%, and 55% have been expressed as decimals. (Strength column)

Another equation comes from information in this column. (Amount of alcohol column)

The information in the table provides two equations.

$$\begin{cases} x + y = 8 \\ 0.40x + 0.60y = 0.55(8) \end{cases}$$

The number of ounces of batch 1 plus the number of ounces of batch 2 equals the total number of ounces in the mixture.

The amount of alcohol in batch 1 plus the amount of alcohol in batch 2 equals the total amount of alcohol in the mixture.

Solve We can solve this system by elimination. To eliminate x, we proceed as follows.

$$-40x - 40y = -320 \quad \text{Multiply both sides of the first equation by } -40.$$
$$40x + 60y = 440 \quad \text{Multiply both sides of the second equation by 100.}$$
$$20y = 120$$
$$y = 6 \quad \text{Divide both sides by 20. This is the number of ounces of batch 2 needed.}$$

To find x, we substitute 6 for y in the first equation of the original system.

$x + y = 8$

$x + 6 = 8$ Substitute.

$x = 2$ Subtract 6 from both sides. This is the number of ounces of batch 1 needed.

State The technician should use 2 fluid ounces of the 40% solution and 6 fluid ounces of the 60% solution.

Check Note that 2 ounces + 6 ounces = 8 ounces, the required number. Also, the amount of alcohol in the two solutions is equal to the amount of alcohol in the mixture.

Alcohol in batch 1: $0.40x = 0.40(2) = 0.8$ ounces
Alcohol in batch 2: $0.60y = 0.60(6) = 3.6$ ounces
Total: 4.4 ounces
Alcohol in the mixture: $0.55(8) = 4.4$ ounces

The results check.

Self Check 7 **Dairy Products.** How much 1% milk and how much 4% milk must be combined to obtain 60 liters of 2% milk?

Now Try Problem 45

EXAMPLE 8 **Breakfast Cereal.** One ounce of raisins (by weight) sells for 22¢ and one ounce of bran flakes (by weight) sells for 12¢. How many ounces of each should be used to create a 20-ounce box of raisin bran cereal that can be sold for 15¢ an ounce?

Analyze We will use a two-variable approach to solve this dry mixture problem.

Assign Let x = the number of ounces of raisins and y = the number of ounces of bran flakes that should be mixed.

Form The value of the mixture and the value of each of its components is given by

Amount · price = total value

Thus, the value of x ounces of raisins is $x \cdot 22$¢ or $22x$¢ and the value of y ounces of bran flakes is $y \cdot 12$¢ or $12y$¢. The sum of these values is also equal to the total value of the final mixture, that is $20 \cdot 15$¢ or 300¢. This information is shown in the table.

	Amount	· Price	= Total value
Raisins	x	22	$22x$
Bran flakes	y	12	$12y$
Mixture	20	15	$20(15)$

Multiply amount · price three times to fill in this column.

One equation comes from this column. (Amount)

One equation comes from this column. (Total value)

The facts of the problem give the following two equations:

The number of ounces of raisins	plus	the number of ounces of bran flakes	is	20.
x	$+$	y	$=$	20

The value of the raisins	plus	the value of the bran flakes	is	the value of the mixture.
$22x$	$+$	$12y$	$=$	$20(15)$

Solve To find out how many ounces of raisins and bran flakes are needed we solve the following system:

$$\begin{cases} x + y = 20 \\ 22x + 12y = 300 \end{cases}$$ Multiply: 20(15) = 300.

To solve this system by substitution, we can solve the first equation for x:

$$x + y = 20$$
$$x = 20 - y$$ This is the substitution equation.

Then we substitute $20 - y$ for x in the second equation of the system and solve for y.

$$22x + 12y = 300$$
$$22(20 - y) + 12y = 300$$ Substitute $20 - y$ for x.
$$440 - 22y + 12y = 300$$ Distribute the multiplication by 22.
$$440 - 10y = 300$$ Combine like terms: $-22y + 12y = -10y$.
$$-10y = -140$$ Subtract 440 from both sides.
$$y = 14$$ Divide both sides by -10. This is the number of ounces of bran flakes needed.

To find x, we substitute 14 for y in the substitution equation and simplify the right side.

$$x = 20 - y$$
$$= 20 - 14$$ Substitute 14 for y.
$$= 6$$ This is the number of ounces of raisins needed.

State To obtain 20 ounces of raisin bran cereal, 6 ounces of raisins and 14 ounces of bran flakes should be combined.

Check When 6 ounces of raisins and 14 ounces of bran flakes are combined, the result is 20 ounces of raisin bran cereal. The 6 ounces of raisins are valued at $6 \cdot 22¢ = 132¢$ and the 14 ounces of bran flakes are valued at $14 \cdot 12¢ = 168¢$. The sum of those values, $132¢ + 168¢ = 300¢$, is the same as the value of the mixture, $20 \cdot 15¢ = 300¢$. The results check.

Self Check 8 **Gardening.** How much planting mix, worth \$36 per cubic yard, and how much topsoil, worth \$24 per cubic yard, should be combined to make 200 cubic yards of a mixture worth \$31.20 per cubic yard?

Now Try Problem 49

SECTION 4.4 STUDY SET

VOCABULARY

Fill in the blanks.

1. Two angles are said to be ______________ if the sum of their measures is 90°. Two angles are said to be ______________ if the sum of their measures is 180°.
2. Problems that involve moving vehicles are called uniform ________ problems. Problems that involve combining ingredients are called ________ problems. Problems that involve collections of different items having different values are called number-______ problems.

CONCEPTS

3. A length of pipe is to be cut into two pieces. The longer piece is to be 1 foot less than twice the shorter piece. Write two equations that model the situation.

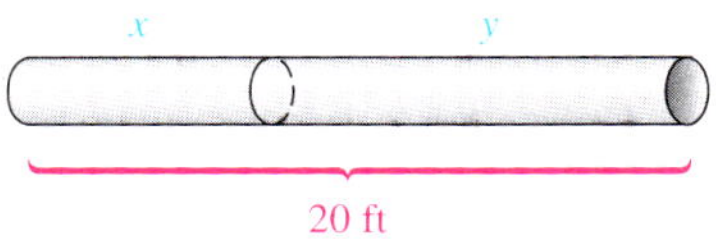

4. Two angles are complementary. The measure of the larger angle is four times the measure of the smaller angle. Write two equations that model the situation.

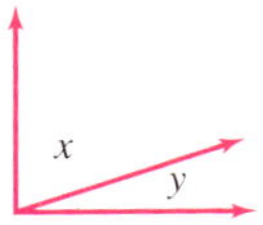

5. Two angles are supplementary. The measure of the smaller angle is 25° less than the measure of the larger angle. Write two equations that model the situation.

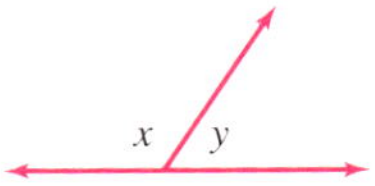

6. The perimeter of the following Ping-Pong table is 28 feet. The length is 4 feet more than the width. Write two equations that model the situation.

7. Let x = the cost of a chicken taco, in dollars, and y = the cost of a beef taco, in dollars. Write an equation that models the offer shown in the advertisement.

8. a. Complete the following table.

	Principal	· Rate	· Time	= Interest
City Bank	x	5%	1 yr	
USA Savings	y	11%	1 yr	

b. A total of $50,000 was deposited in the two accounts. Use that information to write an equation about the principal.

c. A total of $4,300 was earned by the two accounts. Use that information to write an equation about the interest.

9. For each case on the right, write an algebraic expression that represents the speed of the canoe in miles per hour if its speed in still water is x mph.

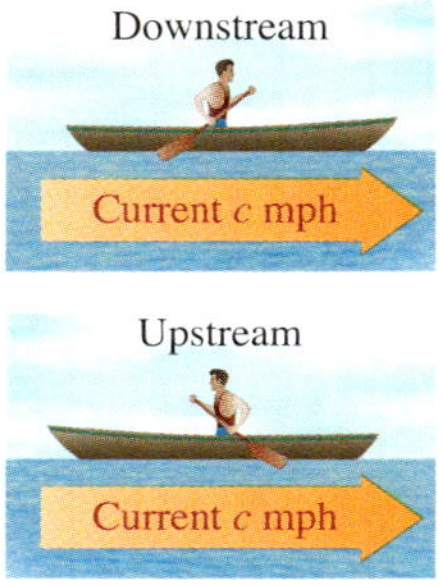

10. Complete the table, which contains information about an airplane flying in windy conditions.

	Rate	· Time	= Distance
With wind	$x + y$	3	
Against wind	$x - y$	5	

11. a. If the contents of the two test tubes are poured into a third tube, how much solution will the third tube contain? (mL stands for milliliter. A milliliter is about 15 drops from an eyedropper.)

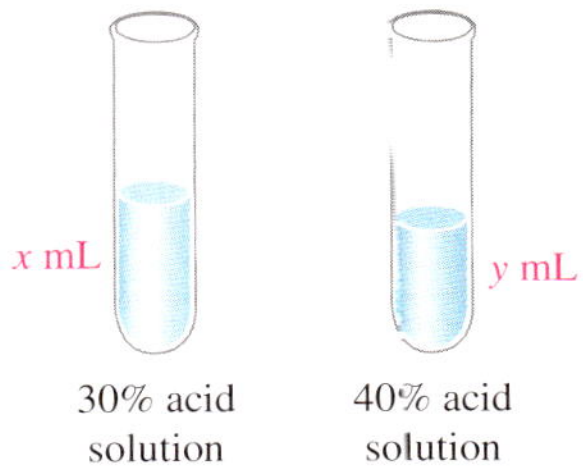

b. Which of the following strengths could the mixture possibly be: 27%, 33%, or 44% acid solution?

12. a. Complete the table, which contains information about mixing two salt solutions to get 12 gallons of a 3% salt solution.

	Amount	· Strength	= Amount of salt
Weak	x	0.01	
Strong	y	0.06	
Mix			

b. Use the information from the Amount column to write an equation.

c. Use the information from the Amount of salt column to write an equation.

GUIDED PRACTICE

See Example 2.

13. **Complementary Angles.** Two angles are complementary. The measure of one angle is 10° more than three times the measure of the other. Find the measure of each angle.

14. **Supplementary Angles.** Two angles are supplementary. The measure of one angle is 20° less than 19 times the measure of the other. Find the measure of each angle.

15. **Supplementary Angles.** The difference of the measures of two supplementary angles is 80°. Find the measure of each angle.

16. **Complementary Angles.** Two angles are complementary. The measure of one angle is 15° more than one-half of the measure of the other. Find the measure of each angle.

APPLICATIONS

Write a system of two equations in two variables to solve each problem.

17. **Tree Trimming.** When fully extended, the arm on a tree service truck is 51 feet long. If the upper part of the arm is 7 feet shorter than the lower part, how long is each part of the arm?

18. Alaska. Most of the 1,422-mile-long Alaskan Highway is actually in Canada. Find the length of the highway that is in Alaska and the length of the highway that is in Canada if it is known that the difference in the lengths is 1,020 miles.

19. Government. The salaries of the president and vice president of the United States total \$627,300 a year. If the president makes \$172,700 more than the vice president, find each of their salaries.

20. Causes of Death. According to the *National Vital Statistics Reports,* in 2007, the number of Americans who died from motor vehicle accidents was about twice the number who died from falls. If the total number of deaths from these two causes was approximately 66,000, how many Americans died from each cause in 2007?

21. Monuments. The Marine Corps War Memorial in Arlington, Virginia, portrays the raising of the U.S. flag on Iwo Jima during World War II. Find the measures of the two angles shown if the measure of $\angle 1$ is 15° less than twice the measure of $\angle 2$.

22. Physical Therapy. To rehabilitate her knee, an athlete does leg extensions. Her goal is to regain a full 90° range of motion in this exercise. Use the information in the illustration to determine her current range of motion in degrees and the number of degrees of improvement she still needs to make.

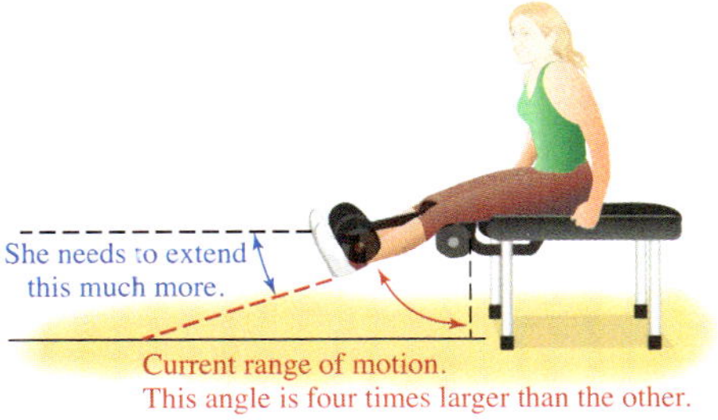

23. Theater Screens. At an IMAX theater, the giant rectangular movie screen has a width 26 feet less than its length. If its perimeter is 332 feet, find the length and the width of the screen.

24. Art. In 1770, Thomas Gainsborough painted *The Blue Boy.* The sum of the length and width of the painting is 118 inches. The difference of the length and width is 22 inches. Find the length and width.

© Huntington Library/Superstock

25. Geometry. A 50-meter path surrounds a rectangular garden. The width of the garden is two-thirds its length. Find the length and width.

26. Ballroom Dancing. A rectangular-shaped dance floor has a perimeter of 200 feet. If the floor were 20 feet wider, its width would equal its length. Find the length and width of the dance floor.

27. Empty Cartridges. A bank recycles its empty printer and copier cartridges. In January, the bank received \$40 for recycling 5 printer and 2 copier cartridges. In February, the bank received \$57 for recycling 6 printer and 3 copier cartridges. How much is the bank paid for an empty printer cartridge and for an empty copier cartridge?

28. Thanksgiving Dinner. There are a total of 510 calories in 6 ounces of turkey and one slice of pumpkin pie. There are a total of 580 calories in 4 ounces of turkey and two slices of pumpkin pie. How many calories are there in 1 ounce of turkey and in one slice of pumpkin pie?

29. from **Campus to Careers**

Photographer

© Andrei Contiu/Shutterstock.com

Suppose you are a wedding photographer and you sell:

Package 1: one 10×14 and ten 8×10 color photos for \$239.50

Package 2: one 10×14 and five 8×10 color photos for \$134.50

A newlywed couple buys Package 1 and decides that they want one more 10×14 and one more 8×10 photograph. At the same prices, what should you charge them for each additional photograph?

30. Buying Painting Supplies. Two partial receipts for paint supplies are shown. (Assume no sales tax was charged.) Find the cost of one gallon of latex paint and the cost of one paint brush.

31. Collecting Stamps. Determine the price of an Elvis Presley stamp and a Statue of Liberty stamp given the following information.

© United States Post Office

© EPE/USPS

- One Elvis stamp and one Liberty stamp cost a total of 63¢.
- A sheet of 40 Elvis stamps and a sheet of 20 Liberty stamps cost a total of \$18.40. (*Hint:* \$18.40 = 1,840¢.)

32. Recycling. A boy scout troop earned \$24 by recycling a total of 330 beverage containers. The recycling rates are shown below. How many of the small capacity containers and how many of the large capacity containers did they recycle? (*Hint:* 5¢ = \$.05 and 10¢ = \$0.10)

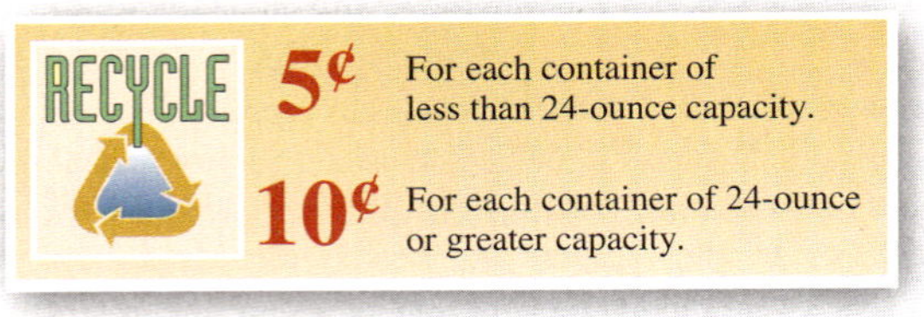

33. Selling Ice Cream. At a store, ice cream cones cost \$1.80 and sundaes cost \$3.30. One day the receipts for a total of 148 cones and sundaes were \$360.90. How many cones were sold? How many sundaes?

34. Buying Tickets. The ticket prices for a movie are shown. Receipts for one showing were \$1,740 for an audience of 190 people. How many general admission tickets and how many senior citizen tickets were sold?

35. Student Loans. A college used a \$5,000 gift from an alumnus to make two student loans. The first was at 5% annual simple interest to a nursing student. The second was at 7% annual simple interest to a business major. If the college collected \$310 in interest the first year, how much was loaned to each student?

36. Financial Planning. In investing \$6,000 of a couple's money, a financial planner put some of it into a savings account paying 6% annual simple interest. The rest was invested in a riskier mini-mall development plan paying 12% annual simple interest. The combined interest earned for the first year was \$540. How much money was invested at each rate?

37. Investing a Bonus. A businessman invested part of his \$40,000 end-of-the-year bonus in an international fund that paid an annual yield of 8%. The rest of the bonus was invested in an offshore bank that paid an annual yield of 9%. Find the amount of each investment if he made a total of \$3,415 in interest from them the first year.

38. Pension Funds. A state employees' pension fund invested a total of one million dollars in two accounts that earned 3.5% and 4.5% annual simple interest. At the end of the year, the total interest earned from the two investments was \$39,000. How much was invested at each rate?

39. Losses. A CEO deposited part of \$22,000 in an account paying 4% annual simple interest. The rest of the money was invested in a biotech company that, after only one year, caused him to lose 3% of his initial investment in it. Find the amount of each investment if the net interest he earned the first year was only \$110.

40. Lottery Winnings. After winning \$60,000 in the lottery, a retired teacher gave \$10,000 of it to her grandchildren. She invested part of the remainder in a growth fund that earned 4.4% annually and the rest in certificates of deposit paying a 5.8% annual percentage yield. The interest that she received on these two investments totaled \$2,732 at the end of the first year. Find the amount of each investment.

41. The Gulf Stream. The gulf stream is a warm ocean current of the North Atlantic Ocean that flows northward, as shown below. Heading north with the gulf stream, a cruise ship traveled 300 miles in 10 hours. Against the current, it took 15 hours to make the return trip. Find the speed of the ship in still water and the speed of the current.

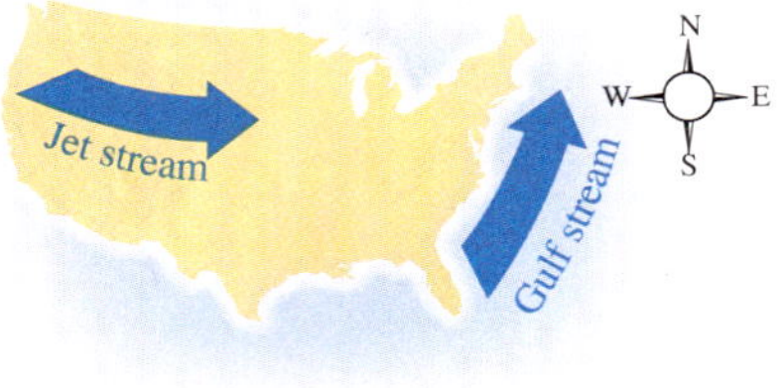

42. The Jet Stream. The jet stream is a strong wind current that flows across the United States, as shown above. Flying with the jet stream, a plane flew 3,000 miles in 5 hours. Against the same wind, the trip took 6 hours. Find the speed of the plane in still air and the speed of the wind current.

43. Aviation. An airplane can fly with the wind a distance of 800 miles in 4 hours. However, the return trip against the wind takes 5 hours. Find the speed of the plane in still air and the speed of the wind.

44. Boating. A boat can travel 24 miles downstream in 2 hours and can make the return trip in 3 hours. Find the speed of the boat in still water and the speed of the current.

45. **Marine Biology.** A marine biologist wants to set up an aquarium containing 3% salt water. He has two tanks on hand that contain 6% and 2% salt water. How much water from each tank must he use to fill a 32-gallon aquarium with a 3% saltwater mixture?

46. **Commemorative Coins.** A foundry has been commissioned to make souvenir coins. The coins are to be made from an alloy that is 40% silver. The foundry has on hand two alloys, one with 50% silver content and one with a 25% silver content. How many kilograms of each alloy should be used to make 20 kilograms of the 40% silver alloy?

47. **Cleaning Floors.** A custodian is going to mix a 4% ammonia solution and a 12% ammonia solution to get 1 gallon (128 fluid ounces) of a 9% ammonia solution. How many fluid ounces of the 4% solution and the 12% solution should be used?

48. **Mouthwash.** A pharmacist has a mouthwash solution that is 6% ethanol alcohol and another that is 18% ethanol alcohol. How many milliliters of each must be mixed to make 750 milliliters of a mouthwash that is 10% ethanol alcohol?

49. **Coffee Sales.** A coffee supply store waits until the orders for its special blend reach 100 pounds before making up a batch. Columbian coffee selling for $8.75 a pound is blended with Brazilian coffee selling for $3.75 a pound to make a product that sells for $6.35 a pound. How much of each type of coffee should be used to make the blend that will fill the orders?

50. **Mixing Nuts.** A merchant wants to mix peanuts with cashews, as shown in the illustration, to get 48 pounds of mixed nuts that will be sold at $6 per pound. How many pounds of each should the merchant use?

51. **Gourmet Foods.** A New York delicatessen sells marinated mushrooms for $12 a pint and stuffed Kalamata olives for $9 a pint. How many pints of each should be used to get 20 pints of a mixture that will sell for $10 a pint?

52. **Herbs.** Ginger root powder sells for $6.50 a pound and ginkgo leaf powder sells for $9.50 a pound. How many pounds of each should be used to make 15 pounds of a mixture that sells for $7 a pound?

WRITING

53. Explain why a table is helpful in solving uniform motion and mixture problems.

54. A man paid $89 for two shirts and four pairs of socks. If we let x = the cost of a shirt, in dollars, and y = the cost of a pair of socks, in dollars, an equation modeling the purchase is $2x + 4y = 89$. Explain why there is not enough information to determine the cost of a shirt or the cost of a pair of socks.

REVIEW

Graph each inequality. Then describe the graph using interval notation.

55. $x < 4$

56. $x \geq -3$

57. $-1 < x \leq 2$

58. $-2 \leq x \leq 0$

CHALLENGE PROBLEMS

59. Three types of hardware items, nails, bolts, and nuts, are placed on a scale as shown below. On the last scale, how many nails will it take to balance 1 nut?

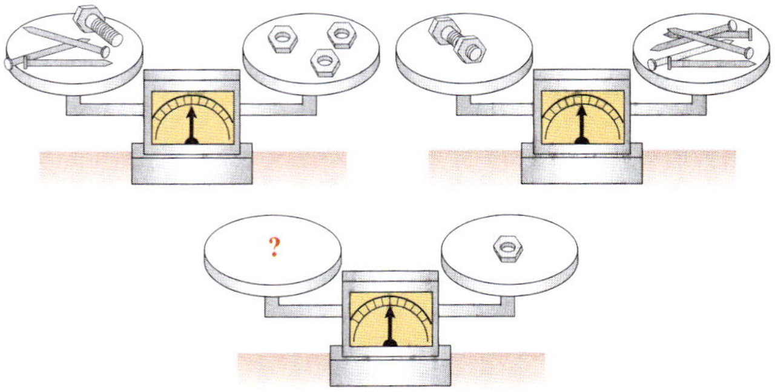

60. **Farming.** In a pen of goats and chickens, there are 40 heads and 130 feet. How many goats and chickens are in the pen?

SECTION 4.5

OBJECTIVES

1 Solve a system of linear inequalities by graphing.

2 Solve application problems involving systems of linear inequalities.

Solving Systems of Linear Inequalities

ARE YOU READY?

The following problems review some basic skills that are needed when solving systems of linear inequalities.

1. True or false: $3(-4) + 7 > -9$
2. Graph: $4x - 3y = -12$
3. Determine whether each of the following points lies *above*, *below*, or *on* the line graphed in problem 2.
   a. $(-2, 3)$ **b.** $(-4, -5)$ **c.** $(-3, 0)$
4. Is the boundary of the graph of $y < 3x - 1$ solid or dashed?

In Section 4.1, we solved systems of linear *equations* graphically by finding the point of intersection of two lines. Now we consider **systems of linear inequalities**, such as

$$\begin{cases} x + y \geq -1 \\ x - y \geq 1 \end{cases}$$

To solve systems of linear inequalities, we again find the points of intersection of graphs. In this case, however, we are not looking for an intersection of two lines, but an intersection of two regions.

1 Solve a System of Linear Inequalities by Graphing.

A **solution of a system of linear inequalities** is an ordered pair that makes each inequality true. *To solve a system of linear inequalities* means to find all of its solutions. This can be done by graphing each inequality on the same set of axes and finding the points that are common to every graph in the system.

EXAMPLE 1 Graph the solutions of the system: $\begin{cases} x + y \geq -1 \\ x - y \geq 1 \end{cases}$

Strategy We will graph the solutions of $x + y \geq -1$ in one color and the solutions of $x - y \geq 1$ in another color on the same coordinate system.

Why We need to see where the graphs of the two inequalities intersect (overlap).

Solution To graph $x + y \geq -1$, we begin by graphing the boundary line $x + y = -1$. Since the inequality contains an $\geq$ symbol, the boundary is a solid line. Because the coordinates of the test point (0, 0) satisfy $x + y \geq -1$, we shade (in red) the side of the boundary that contains (0, 0). See part (a) of the figure on the next page.

Graph the boundary: The intercept method

$x + y = -1$

x	y	(x, y)
0	-1	$(0, -1)$
-1	0	$(-1, 0)$

***Shading: Check the test point* (0, 0)**

$x + y \geq -1$

$0 + 0 \stackrel{?}{\geq} -1$ Substitute.

$0 \geq -1$ True

(0, 0) is a solution of $x + y \geq -1$.

The Language of Algebra

To solve a system of linear inequalities, we **superimpose** the graphs of the inequalities. That is, we place one graph over the other. Most video camcorders can *superimpose* the date and time over the picture being recorded.

In part (b) of the figure, we superimpose the graph of $x - y \geq 1$ on the graph of $x + y \geq -1$ so that we can determine the points that the graphs have in common. To graph $x - y \geq 1$, we graph the boundary $x - y = 1$ as a solid line. Since the test point (0, 0) does not satisfy $x - y \geq 1$, we shade (in blue) the half-plane that does not contain (0, 0).

Graph the boundary: The intercept method	*Shading: Check the test point* **(0, 0)**
$x - y = 1$	$x - y \geq 1$ $0 - 0 \overset{?}{\geq} 1$ Substitute. $0 \geq 1$ False (0, 0) is not a solution of $x - y \geq 1$.

x	y	(x, y)
0	−1	(0, −1)
1	0	(1, 0)

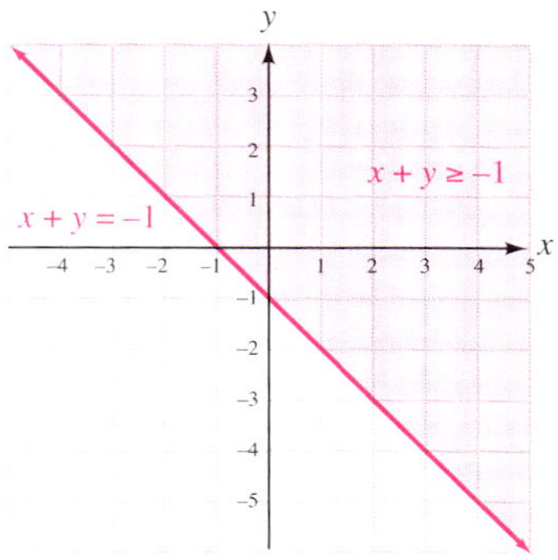

The graph of $x + y \geq -1$ is shaded in red.

(a)

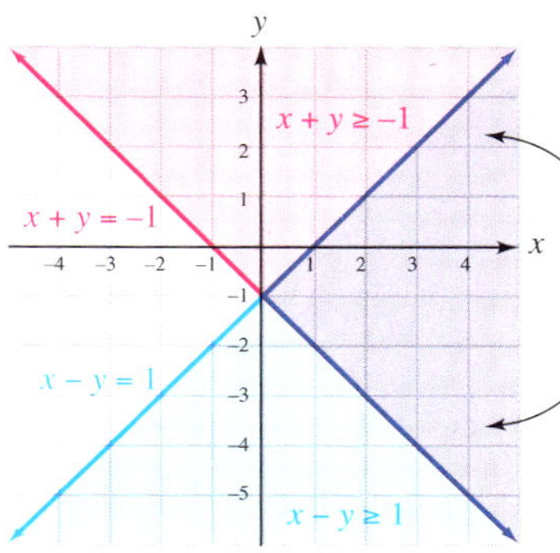

The solutions of the system are shaded in purple. The purple region is the intersection or overlap of the red and blue shaded regions. It includes portions of each boundary.

The graph of $x - y \geq 1$ is shaded in blue. It is drawn over the graph of $x + y \geq -1$.

(b)

Success Tip

Colored pencils often are used to graph systems of inequalities. A standard pencil also can be used. Just draw different patterns of lines instead of shading.

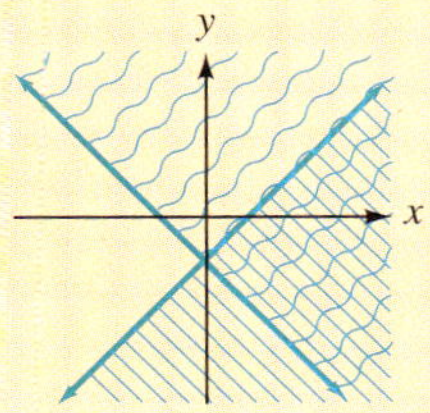

In part (b) of the figure, the area that is shaded twice represents the solutions of the given system. Any point in the **doubly shaded region** in purple (including the purple portions of each boundary) has coordinates that satisfy *both* inequalities.

Since there are infinitely many solutions, we cannot check each of them. However, as an informal check, we can select one ordered pair, say (4, 1), that lies in the doubly shaded region and show that its coordinates satisfy both inequalities of the system.

Check:

$x + y \geq -1$	The first inequality.	$x - y \geq 1$	The second inequality.
$4 + 1 \overset{?}{\geq} -1$		$4 - 1 \overset{?}{\geq} 1$	
$5 \geq -1$	True	$3 \geq 1$	True

The resulting true statements verify that (4, 1) is a solution of the system. If we pick a point that is not in the doubly shaded region, such as (1, 3), (−2, −2), or (0, −4), the coordinates of that point will fail to satisfy one or both of the inequalities.

Self Check 1 Graph the solutions of the system: $\begin{cases} x - y \leq 2 \\ x + y \geq -1 \end{cases}$

Now Try Problem 15

In general, to solve systems of linear inequalities, we will follow these steps.

Solving Systems of Linear Inequalities

1. Graph each inequality on the same rectangular coordinate system.
2. Use shading to highlight the intersection of the graphs (the region where the graphs overlap). The points in this region are the solutions of the system.
3. As an informal check, pick a point from the region where the graphs intersect and verify that its coordinates satisfy each inequality of the original system.

EXAMPLE 2 Graph the solutions of the system: $\begin{cases} y > 3x \\ 2x + y < 4 \end{cases}$

Strategy We will graph the solutions of $y > 3x$ in one color and the solutions of $2x + y < 4$ in another color on the same coordinate system to see where the graphs intersect.

Why The solution set of the system is the set of all points in the intersection of the two graphs.

Solution To graph $y > 3x$, we begin by graphing the boundary line $y = 3x$. Since the inequality contains an $>$ symbol, the boundary is a dashed line. Because the boundary passes through (0, 0), we use (2, 0) as the test point instead. Since (2, 0) does not satisfy $y > 3x$, we shade (in red) the half-plane that does not contain (2, 0). See part (a) of the following figure.

Success Tip

You do not have to use the same test point when graphing the two inequalities.

Graph the boundary: Slope and y-intercept

$y = 3x + 0$

$m = 3 \quad b = 0$

Slope: $\frac{\text{Rise}}{\text{Run}} = \frac{3}{1}$ y-intercept: (0, 0)

***Shading: Check the test point* (2, 0)**

$y > 3x$

$0 \overset{?}{>} 3(2)$ Substitute.

$0 > 6$ False

Since $0 > 6$ is false, (2, 0) is not a solution of $y > 3x$.

In part (b) of the figure, we superimpose the graph of $2x + y < 4$ on the graph of $y > 3x$ to determine the points that the graphs have in common. To graph $2x + y < 4$, we graph the boundary $2x + y = 4$ as a dashed line. Then we shade (in blue) the half-plane that contains (0, 0), because the coordinates of the test point satisfy $2x + y < 4$.

Success Tip

The ordered pairs that lie on a dashed boundary line are never part of the solution of a system of linear inequalities.

Graph the boundary: The intercept method

$2x + y = 4$

x	y	(x, y)
0	4	(0, 4)
2	0	(2, 0)

***Shading: Use the test point* (0, 0)**

$2x + y < 4$

$2(0) + 0 \overset{?}{<} 4$ Substitute.

$0 < 4$ True

Since $0 < 4$ is true, (0, 0) is a solution of $2x + y < 4$.

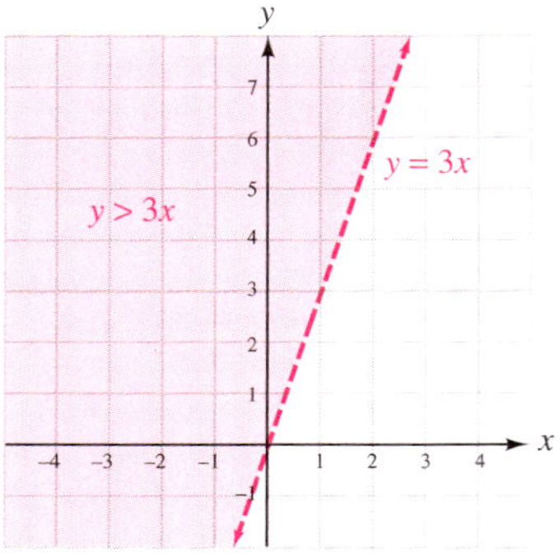

The graph of $y > 3x$ is shaded in red.

(a)

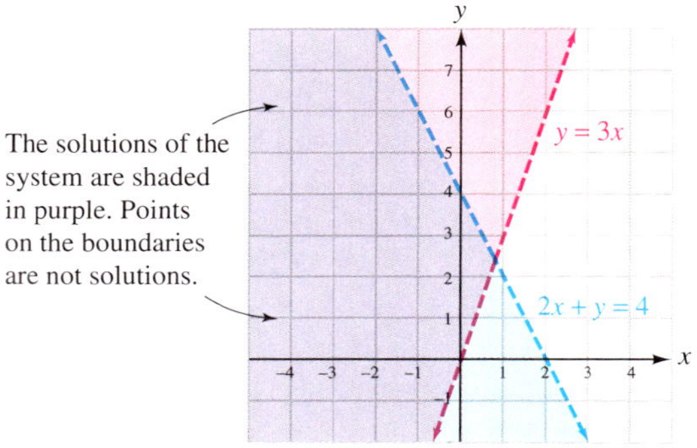

The graph of $2x + y < 4$ is shaded in blue. It is drawn over the graph of $y > 3x$.

(b)

In part (b) of the figure, the area that is shaded twice represents the solutions of the given system. Any point in the doubly shaded region in purple has coordinates that satisfy both inequalities. Pick a point in the region and show that this is true. Note that the region does not include either boundary; points on the boundaries are not solutions of the system.

Self Check 2 Graph the solutions of the system: $\begin{cases} x + 3y < 3 \\ y > \frac{1}{3}x \end{cases}$

Now Try Problem 19

EXAMPLE 3 Graph the solutions of the system: $\begin{cases} x \leq 2 \\ y > 3 \end{cases}$

Strategy We will graph the solutions of $x \leq 2$ in one color and the solutions of $y > 3$ in another color on the same coordinate system to see where the graphs of the two inequalities intersect.

Why The solution set of the system is the set of all points in the intersection of the two graphs.

Solution The boundary of the graph of $x \leq 2$ is the line $x = 2$. Since the inequality contains the symbol $\leq$, we draw the boundary as a solid line. The test point (0, 0) makes $x \leq 2$ true, so we shade the side of the boundary that contains (0, 0). See part (a) of the figure below.

Graph the boundary: A table of solutions

$x = 2$

x	y	(x, y)
2	0	(2, 0)
2	2	(2, 2)
2	4	(2, 4)

***Shading: Check the test point* (0, 0)**

$x \leq 2$

$0 \leq 2$ True

Since $0 \leq 2$ is true, (0, 0) is a solution of $x \leq 2$.

In part (b) of the figure, the graph of $y > 3$ is superimposed over the graph of $x \leq 2$. The boundary of the graph of $y > 3$ is the line $y = 3$. Since the inequality contains the symbol $>$, we draw the boundary as a dashed line. The test point (0, 0) makes $y > 3$ false, so we shade the side of the boundary that does not contain (0, 0).

Graph the boundary: A table of solutions

$y = 3$

x	y	(x, y)
0	3	(0, 3)
1	3	(1, 3)
4	3	(4, 3)

***Shading: Check the test point* (0, 0)**

$y > 3$

$0 > 3$ False

Since $0 > 3$ is false, (0, 0) is not a solution of $y > 3$.

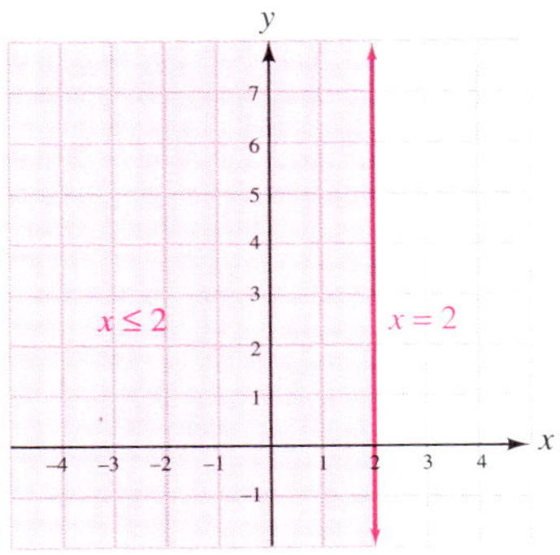

The graph of $x \leq 2$ is shaded in red.

(a)

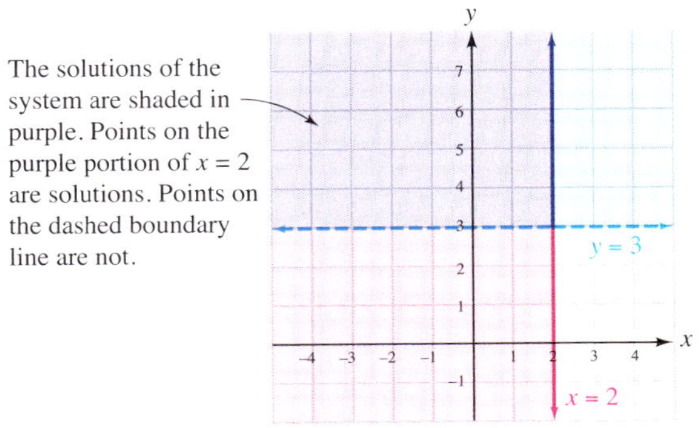

The graph of $y > 3$ is shaded in blue. It is drawn over the graph of $x \leq 2$.

(b)

The area that is shaded twice represents the solutions of the system of inequalities. Any point in the doubly shaded region in purple has coordinates that satisfy both inequalities, including the purple portion of the $x = 2$ boundary. Pick a point in the region and show that this is true.

Self Check 3 Graph the solutions of the system: $\begin{cases} y \le 1 \\ x > 2 \end{cases}$

Now Try Problem 23

EXAMPLE 4 Graph the solutions of the system: $\begin{cases} x \ge 0 \\ y \ge 0 \\ x + 2y \le 6 \end{cases}$

Strategy We will graph the solutions of $x \ge 0$, $y \ge 0$, and $x + 2y \le 6$ on the same coordinate system to see where all three graphs intersect (overlap).

Why The solution set of the system is the set of all points in the intersection of the three graphs.

Solution This is a system of three linear inequalities. If shading is used to graph them on the same set of axes, it can become difficult to interpret the results. Instead, we can draw directional arrows attached to each boundary line in place of the shading.

- The graph of $x \ge 0$ has the boundary $x = 0$ and includes all points on the y-axis and to the right.
- The graph of $y \ge 0$ has the boundary $y = 0$ and includes all points on the x-axis and above.
- The graph of $x + 2y \le 6$ has the boundary $x + 2y = 6$. Because the coordinates of the origin satisfy $x + 2y \le 6$, the graph includes all points on and below the boundary.

The solutions of the system are the points that lie on triangle OPQ and the shaded triangular region that it encloses.

$x + 2y = 6$

x	y	(x, y)
0	3	(0, 3)
6	0	(6, 0)

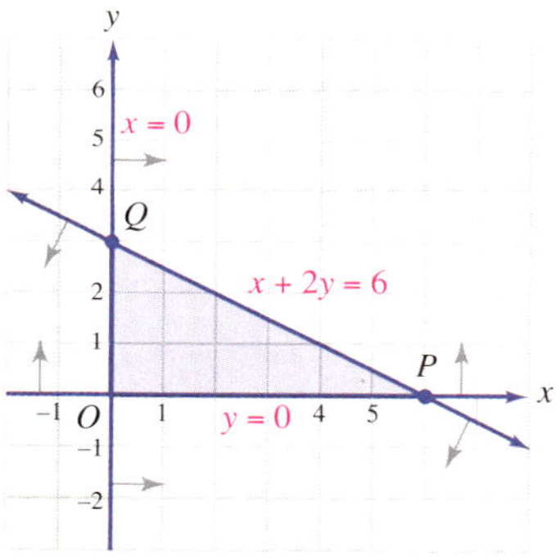

The solutions of the system are shaded purple.

Self Check 4 Graph the solutions of the system: $\begin{cases} x \le 1 \\ y \le 2 \\ 2x - y \le 4 \end{cases}$

Now Try Problem 27

2 Solve Application Problems Involving Systems of Linear Inequalities.

EXAMPLE 5

Landscaping. A homeowner budgets from \$300 to \$600 for trees and bushes to landscape his yard. After shopping around, he finds that good trees cost \$150 each and mature bushes cost \$75 each. What combinations of trees and bushes can he buy?

Analyze

- At least \$300 but not more than \$600 is to be spent for trees and bushes.
- Trees cost \$150 each and bushes cost \$75 each.
- What combination of trees and bushes can he buy?

Assign Let x = the number of trees purchased and y = the number of bushes purchased.

Form We can form the following system of inequalities:

The cost of a tree	times	the number of trees purchased	plus	the cost of a bush	times	the number of bushes purchased	should at least be	\$300.
\$150	$\cdot$	x	$+$	\$75	$\cdot$	y	$\geq$	\$300

The cost of a tree	times	the number of trees purchased	plus	the cost of a bush	times	the number of bushes purchased	should not be more than	\$600.
\$150	$\cdot$	x	$+$	\$75	$\cdot$	y	$\leq$	\$600

Solve To solve the following system of linear inequalities

$$\begin{cases} 150x + 75y \geq 300 \\ 150x + 75y \leq 600 \end{cases}$$

we use the graphing methods discussed in this section. Neither a negative number of trees nor a negative number of bushes can be purchased, so we restrict the graph to Quadrant I.

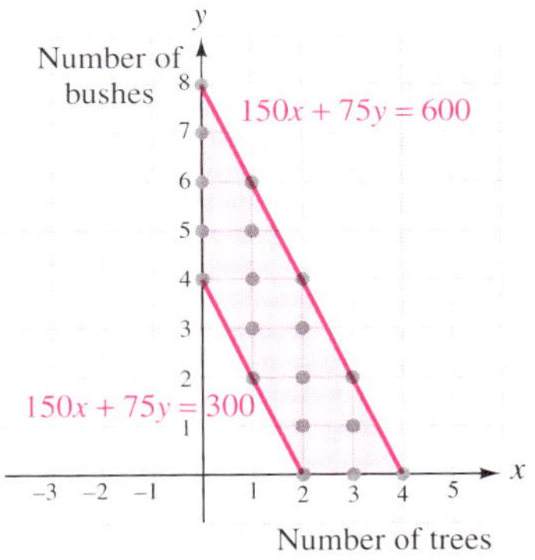

State The coordinates of each point highlighted in the graph give a possible combination of the number of trees, x, and the number of bushes, y, that can be purchased. Written as ordered pairs, these possibilities are

(0, 4), (0, 5), (0, 6), (0, 7), (0, 8),
(1, 2), (1, 3), (1, 4), (1, 5), (1, 6),
(2, 0), (2, 1), (2, 2), (2, 3), (2, 4),
(3, 0), (3, 1), (3, 2), (4, 0)

Check Suppose the homeowner picks the combination of 3 trees and 2 bushes, as represented by (3, 2). Show that this point satisfies both inequalities of the system.

The Language of Algebra

Phrases such as **must be at least** and **cannot go below** can be represented by the inequality symbol $\geq$. Phrases such as **is not more than** and **should not surpass** can be represented by the inequality symbol $\leq$.

Self Check 5 **Buying CDs and DVDs.** An electronics store sells CDs for \$10 and DVDs for \$20. Donna wants to spend at least \$100 but no more than \$200 on (x) CDs and (y) DVDs. What combinations of CDs and DVDs can she afford to buy?

Now Try Problem 51

SECTION 4.5 STUDY SET

VOCABULARY

Fill in the blanks.

1. $\begin{cases} x + y > 2 \\ x + y < 4 \end{cases}$ is a system of linear __________.
2. To graph the linear inequality $x + y > 2$, first graph the __________ $x + y = 2$. Then pick the test _____ (0, 0) to determine which half-plane to shade.
3. To find the solutions of a system of two linear inequalities graphically, look for the __________, or overlap, of the two shaded regions.
4. The phrase *should not surpass* can be represented by the inequality symbol ____ and the phrase *must be at least* can be represented by the inequality symbol ____.

CONCEPTS

5. a. What is the equation of the boundary line of the graph of $3x - y < 5$?
   b. Is the boundary a solid or dashed line?
6. a. What is the equation of the boundary line of the graph of $y \geq 4x$?
   b. Is the boundary a solid or dashed line?
   c. Why can't (0, 0) be used as a test point to determine what to shade?
7. Find the slope and the y-intercept of the line whose equation is $y = 4x - 3$.
8. Complete the table to find the x- and y-intercepts of the line whose equation is $8x - 3y = -24$.

x	y
0	
	0

9. The boundary of the graph of $2x + y > 4$ is shown.
   a. Does the point (0, 0) make the inequality true?
   b. Should the region above or below the boundary be shaded?

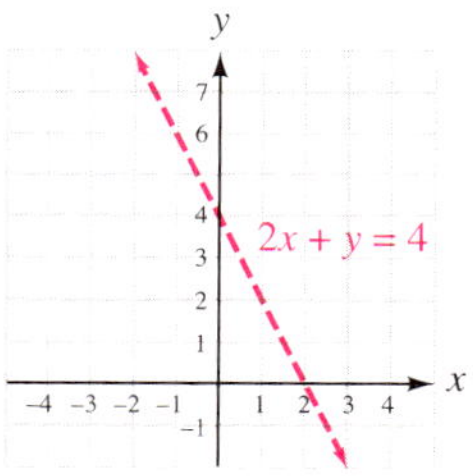

10. Linear inequality 1 is graphed in red below and linear inequality 2 is graphed in blue. Determine whether a true or false statement results when
   a. The coordinates of point A are substituted into inequality 1
   b. The coordinates of point A are substituted into inequality 2
   c. The coordinates of point B are substituted into inequality 1
   d. The coordinates of point B are substituted into inequality 2
   e. The coordinates of point C are substituted into inequality 1
   f. The coordinates of point C are substituted into inequality 2

Inequality 1 solutions are in red

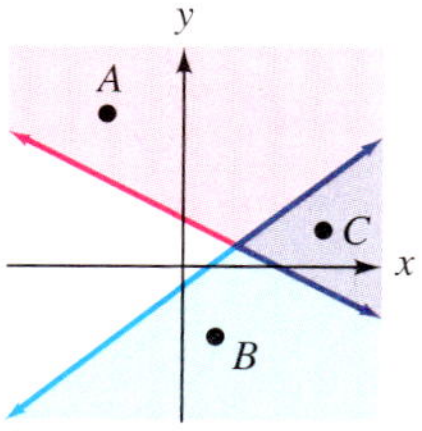

Inequality 2 solutions are in blue

11. The graph of a system of two linear inequalities is shown. Determine whether each point is a solution of the system.
   a. (4, −2)
   b. (1, 3)
   c. the origin

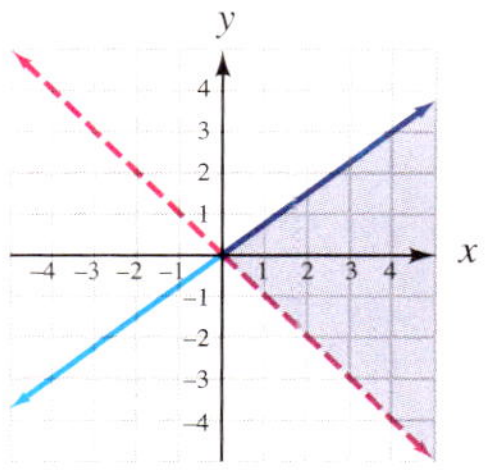

12. Use a check to determine whether each ordered pair is a solution of the system.

$$\begin{cases} x + 2y \geq -1 \\ x - y < 2 \end{cases}$$

   a. (1, 4) b. (−2, 0)

13. Match each equation, inequality, or system with the graph of its solution.
   a. $x + y = 2$
   b. $x + y \geq 2$
   c. $\begin{cases} x + y = 2 \\ x - y = 2 \end{cases}$
   d. $\begin{cases} x + y \geq 2 \\ x - y \leq 2 \end{cases}$

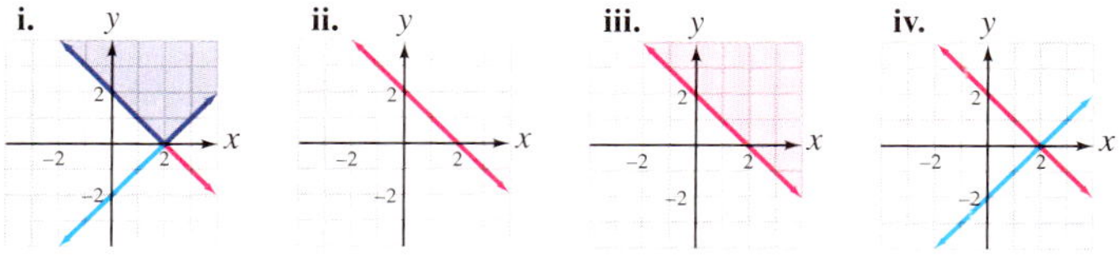

14. Match the system of inequalities with the correct graph.
   a. $\begin{cases} x \geq 2 \\ y < 1 \end{cases}$
   b. $\begin{cases} x > 2 \\ y \leq 1 \end{cases}$
   c. $\begin{cases} x \geq 2 \\ y \geq 1 \end{cases}$
   d. $\begin{cases} x > 2 \\ y > -1 \end{cases}$

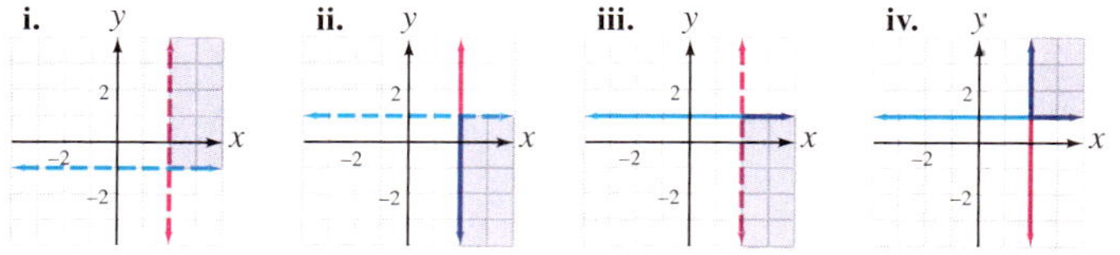

GUIDED PRACTICE

Graph the solutions of each system. See Example 1.

15. $\begin{cases} x + 2y \leq 3 \\ 2x - y \geq 1 \end{cases}$
16. $\begin{cases} 2x + y \geq 3 \\ x - 2y \leq -1 \end{cases}$
17. $\begin{cases} x + y < -1 \\ x - y > -1 \end{cases}$
18. $\begin{cases} x + y > 2 \\ x - y < -2 \end{cases}$

Graph the solutions of each system. See Example 2.

19. $\begin{cases} y > 2x \\ x + 2y < 6 \end{cases}$
20. $\begin{cases} y \leq 2x \\ x + y < 4 \end{cases}$
21. $\begin{cases} y \geq x \\ y \leq \frac{1}{3}x + 1 \end{cases}$
22. $\begin{cases} y > 3x \\ y \leq -x - 1 \end{cases}$

Graph the solutions of each system. See Example 3.

23. $\begin{cases} x \geq 2 \\ y \leq 3 \end{cases}$

24. $\begin{cases} x \geq -1 \\ y > -2 \end{cases}$

25. $\begin{cases} x > 0 \\ y > 0 \end{cases}$

26. $\begin{cases} x \leq 0 \\ y < 0 \end{cases}$

Graph the solutions of each system. See Example 4.

27. $\begin{cases} x \geq 0 \\ y \geq 0 \\ x + y \leq 3 \end{cases}$

28. $\begin{cases} x - y \leq 6 \\ x + 2y \leq 6 \\ x \geq 0 \end{cases}$

29. $\begin{cases} x - y < 4 \\ y \leq 0 \\ x \geq 0 \end{cases}$

30. $\begin{cases} 2x + y \leq 2 \\ y > x \\ x \geq 0 \end{cases}$

TRY IT YOURSELF

Graph the solutions of each system.

31. $\begin{cases} 2x - 3y \leq 0 \\ y \geq x - 1 \end{cases}$

32. $\begin{cases} y > 2x - 4 \\ y \geq -x - 1 \end{cases}$

33. $\begin{cases} x + y < 2 \\ x + y \leq 1 \end{cases}$

34. $\begin{cases} y > -x + 2 \\ y < -x + 4 \end{cases}$

35. $\begin{cases} 3x + 4y \geq -7 \\ 2x - 3y \geq 1 \end{cases}$

36. $\begin{cases} 3x + y \leq 1 \\ 4x - y \geq -8 \end{cases}$

37. $\begin{cases} 2x + y < 7 \\ y > 2 - 2x \end{cases}$

38. $\begin{cases} 2x + y \geq 6 \\ y \leq 2(2x - 3) \end{cases}$

39. $\begin{cases} 2(x - 2y) > -6 \\ 3x + y \geq 5 \end{cases}$

40. $\begin{cases} 2x - 3y < 0 \\ 2x + 3y \geq 12 \end{cases}$

41. $\begin{cases} 3x - y + 4 \leq 0 \\ 3y > -2x - 10 \end{cases}$

42. $\begin{cases} 3x + 2y - 12 \geq 0 \\ x < -2 + y \end{cases}$

43. $\begin{cases} x \geq -1 \\ y \leq -x \\ x - y \leq 3 \end{cases}$

44. $\begin{cases} y > -2.5 \\ 2x - y \geq 2 \\ x \leq 2 \end{cases}$

45. $\begin{cases} x + y > 0 \\ y - x < -2 \end{cases}$

46. $\begin{cases} y + 2x \leq 0 \\ y \leq \frac{1}{2}x + 2 \end{cases}$

Look Alikes . . .

In part a, graph the solution of each system. Use your answer to part a to determine the solution of the system of equations in part b. (No new work is needed.)

47. a. $\begin{cases} x + y > -1 \\ y \geq x - 3 \end{cases}$ b. $\begin{cases} x + y = -1 \\ y = x - 3 \end{cases}$

48. a. $\begin{cases} x - 2y \geq 6 \\ y < -\frac{1}{2}x + 1 \end{cases}$ b. $\begin{cases} x - 2y = 6 \\ y = -\frac{1}{2}x + 1 \end{cases}$

APPLICATIONS

49. **Birds of Prey.** Parts (a) and (b) of the illustration show the fields of vision for each eye of an owl. In part (c), shade the area where the fields of vision overlap—that is, the area that is seen by both eyes.

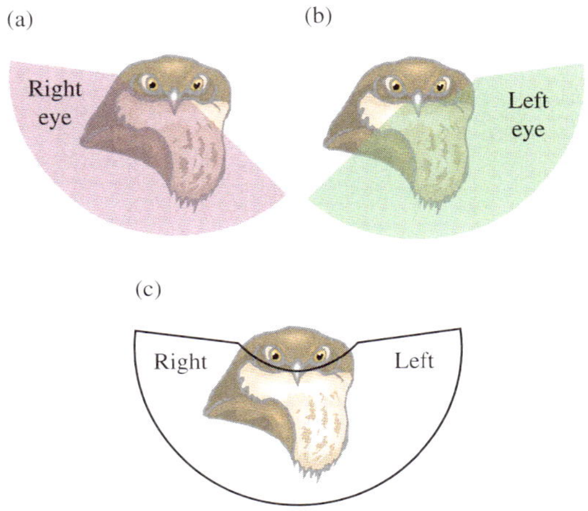

50. **Earth Science.** Shade the area of the earth's surface that is north of the Tropic of Capricorn and south of the Tropic of Cancer.

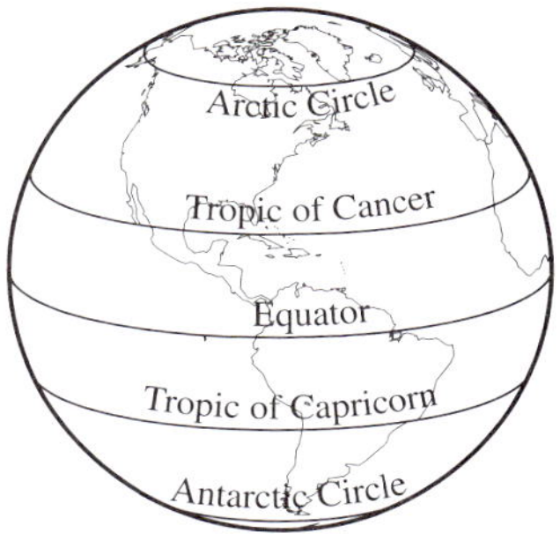

In Problems 51–54, graph each system of inequalities and give two possible solutions. See Example 5.

51. **Buying Compact Discs.** Melodic Music has compact discs on sale for either \$10 or \$15. If a customer wants to spend at least \$30 but no more than \$60 on CDs, graph a system of inequalities showing the possible combinations of \$10 CDs (x) and \$15 CDs (y) that the customer can buy.

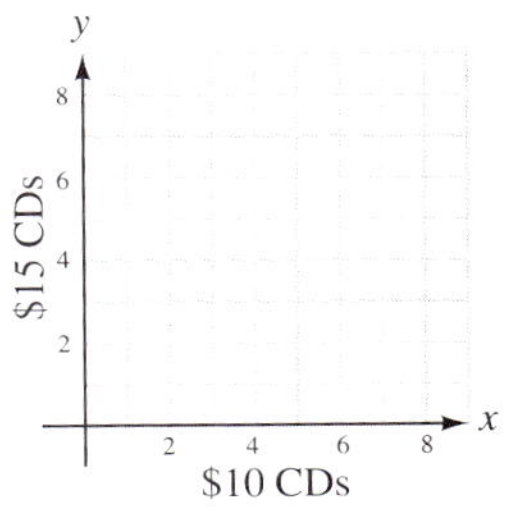

52. **Boats.** Boatworks wholesales aluminum boats for \$800 and fiberglass boats for \$600. Northland Marina wants to make a purchase totaling at least \$2,400 but no more than \$4,800. Graph a system of inequalities showing the possible combinations of aluminum boats (x) and fiberglass boats (y) that can be ordered.

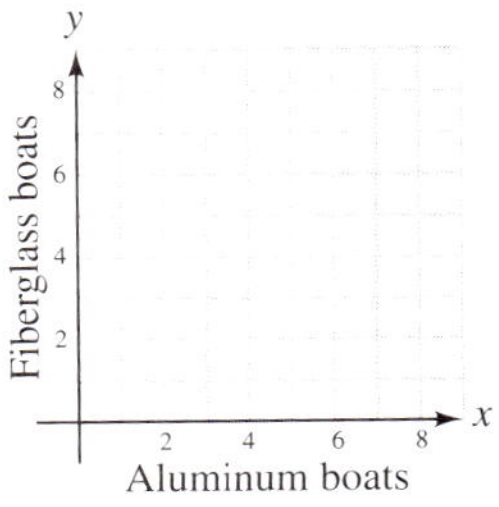

53. Furniture. A distributor wholesales desk chairs for \$150 and side chairs for \$100. Best Furniture wants its order to total no more than \$900; Best also wants to order more side chairs than desk chairs. Graph a system of inequalities showing the possible combinations of desk chairs (x) and side chairs (y) that can be ordered.

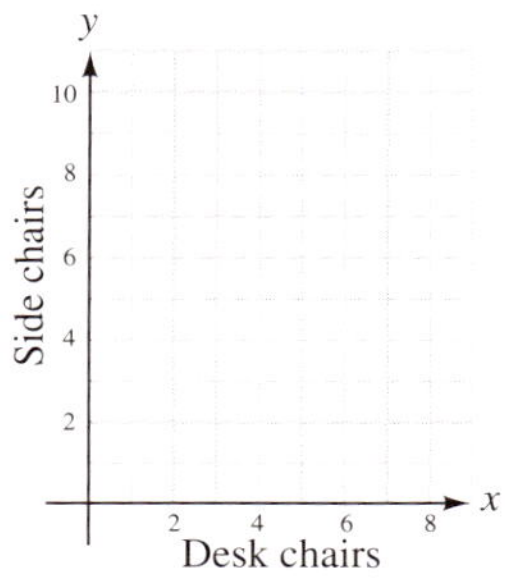

54. Air Quality. J. Bolden Heating Company wants to order no more than \$2,000 worth of electronic air cleaners and humidifiers from a wholesaler that charges \$500 for air cleaners and \$200 for humidifiers. If Bolden wants more humidifiers than air cleaners, graph a system of inequalities showing the possible combinations of air cleaners (x) and humidifiers (y) that can be ordered.

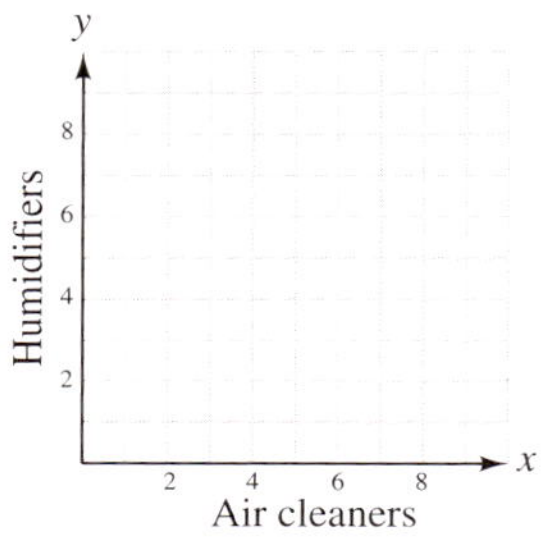

55. from Campus to Careers

Photographer

In some cameras, the image that you see in the viewfinder does not exactly match the image that will be recorded through the lens. To understand this difference, graph the solutions of the system

$$\begin{cases} y \le \frac{1}{4}x + 2 \\ y \ge -\frac{1}{4}x + 2 \end{cases}$$

on the grid below. The shaded solution shows how the viewfinder image is slightly different from the lens image for extreme close-ups.

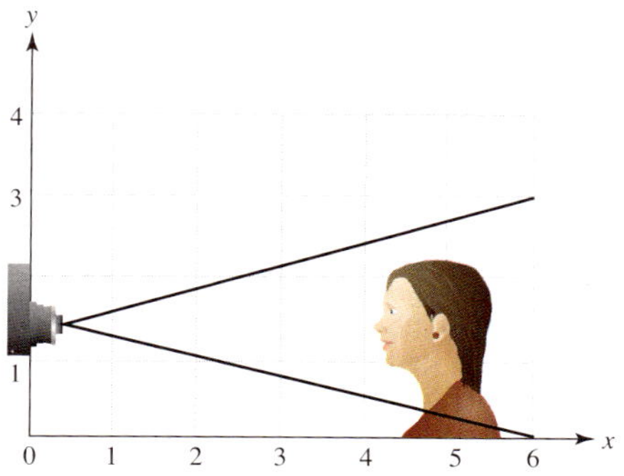

56. Pesticides. To get rid of fruit flies, helicopters sprayed an area of a city that can be described by $y \ge -2x + 1$ (within the city limits). Two weeks later, more spraying was ordered over the area described by $y \ge \frac{1}{4}x - 4$ (within the city limits). Show the part of the city that was sprayed twice.

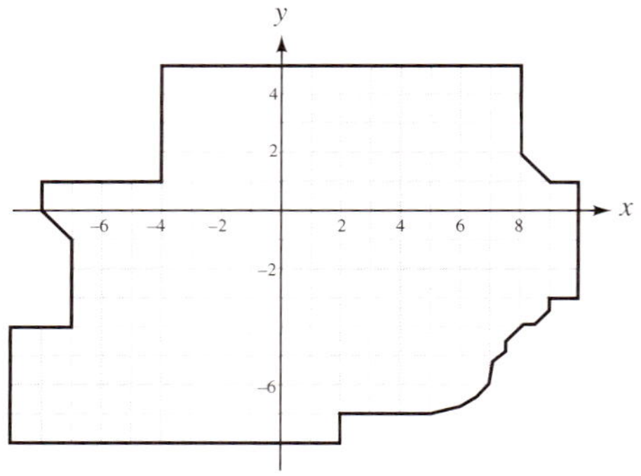

WRITING

57. Explain the error.

Graph: $\begin{cases} y > x \\ x + y < 1 \end{cases}$

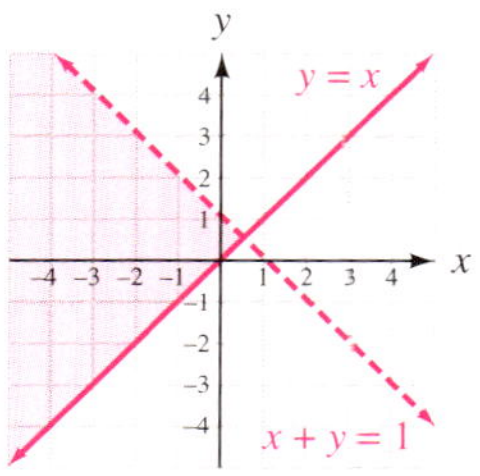

58. When a solution of a system of linear inequalities is graphed, what does the shading represent?

59. Describe how the graphs of the solutions of these systems are similar and how they differ.

$$\begin{cases} x + y = 4 \\ x - y = 4 \end{cases} \quad \text{and} \quad \begin{cases} x + y \ge 4 \\ x - y \ge 4 \end{cases}$$

60. Explain when a system of inequalities will have no solutions.

REVIEW

Simplify each expression.

61. $8\left(\frac{3}{4}t\right)$

62. $-\frac{2}{3}(3w - 6)$

63. $-\frac{7}{16}x - \frac{3}{16}x$

64. $60\left(\frac{3}{20}r - \frac{4}{15}\right)$

CHALLENGE PROBLEMS

Graph the solutions of each system.

65. $\begin{cases} \frac{x}{3} - \frac{y}{2} < -3 \\ \frac{x}{3} + \frac{y}{2} > -1 \end{cases}$

66. $\begin{cases} 3x + y < -2 \\ y > 3(1 - x) \end{cases}$

67. $\begin{cases} 2x + 3y \le 6 \\ 3x + y \le 1 \\ x \le 0 \end{cases}$

68. $\begin{cases} x \ge 0 \\ y \ge 0 \\ 9x + 3y \le 18 \\ 3x + 6y \le 18 \end{cases}$

4 Summary & Review

SECTION 4.1 Solving Systems of Equations by Graphing

DEFINITIONS AND CONCEPTS

When two equations are considered at the same time, we say that they form a **system of equations.**

A **solution of a system** of equations in two variables is an ordered pair that satisfies both equations of the system.

EXAMPLES

Is (4, 3) a solution of the system $\begin{cases} x + y = 7 \\ x - y = 5 \end{cases}$?

To answer this question, we substitute 4 for x and 3 for y in each equation.

$x + y = 7$	The first equation.	$x - y = 5$	The second equation.
$4 + 3 \stackrel{?}{=} 7$	Substitute.	$4 - 3 \stackrel{?}{=} 5$	Substitute.
$7 = 7$	True	$1 = 5$	False

Although (4, 3) satisfies the first equation, it does not satisfy the second. Because it does not satisfy both equations, (4, 3) is not a solution of the system.

DEFINITIONS AND CONCEPTS

To **solve a system graphically:**

1. Graph each equation on the same coordinate system.
2. Determine the coordinates of the **point of intersection** of the graphs. That ordered pair is the solution.
3. Check the solution in each equation of the original system.

EXAMPLES

Use graphing to solve the system: $\begin{cases} y = -2x + 3 \\ x - 2y = 4 \end{cases}$

Step 1: Graph each equation.

$y = -2x + 3$

$m = \frac{\text{Rise}}{\text{Run}} = \frac{-2}{1}$

$b = 3$

y-intercept: (0, 3)

$x - 2y = 4$

x	y
0	−2
4	0

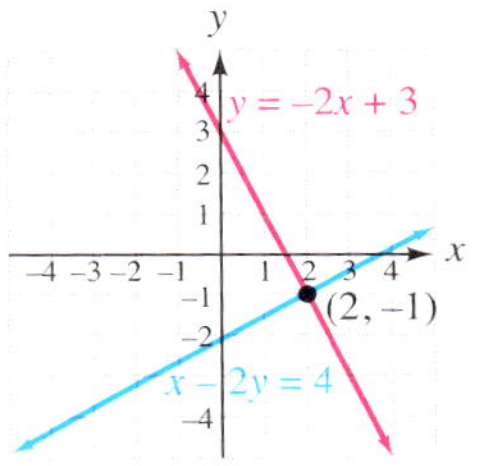

Step 2: It appears that the graphs intersect at the point (2, −1). To verify that this is the solution of the system, substitute 2 for x and −1 for y in each equation.

Step 3: Check

$y = -2x + 3$	The first equation.	$x - 2y = 4$	The second equation.
$-1 \stackrel{?}{=} -2(2) + 3$	Substitute.	$2 - 2(-1) \stackrel{?}{=} 4$	Substitute.
$-1 = -1$	True	$4 = 4$	True

Since (2, −1) makes both equations true, it is the solution of the system.

DEFINITIONS AND CONCEPTS

A system of equations that has at least one solution is called a **consistent system.** If the graphs of the equations of the system are parallel lines, the system has no solution and is called an **inconsistent system.**

Equations with different graphs are called **independent equations.** If the graphs of the equations in a system are the same line, the system has infinitely many solutions. The equations are called **dependent equations.**

We can determine the **number of solutions** that a system of two linear equations has by writing each equation in slope-intercept form, $y = mx + b$, and comparing the slopes and y-intercepts.

EXAMPLES

There are three possible outcomes when solving a system by graphing.

Consistent system
Independent equations

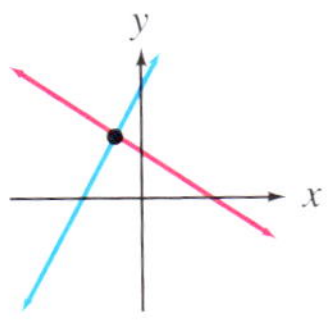

- Exactly one solution
- The lines have different slopes.

Inconsistent system
Independent equations

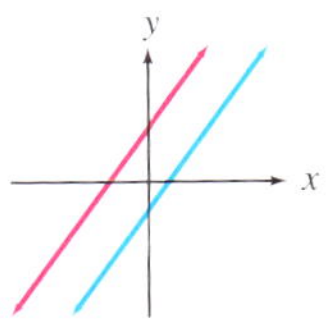

- No solution
- The lines have the same slope but different y-intercepts.

Consistent system
Dependent equations

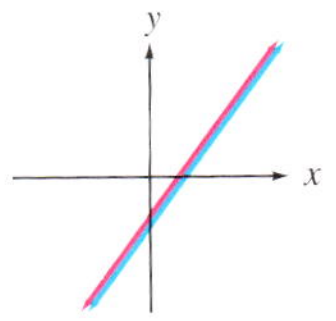

- Infinitely many solutions
- The lines have the same slope and same y-intercept.

REVIEW EXERCISES

Check to determine whether the ordered pair is a solution of the system.

1. $(2, -3)$, $\begin{cases} 3x - 2y = 12 \\ 2x + 3y = -5 \end{cases}$

2. $\left(\frac{7}{2}, -\frac{2}{3}\right)$, $\begin{cases} 3y = 2x - 9 \\ 2x + 3y = 6 \end{cases}$

Use the graphing method to solve each system.

3. $\begin{cases} x + y = 7 \\ 2x - y = 5 \end{cases}$

4. $\begin{cases} 2x + y = 5 \\ y = -\dfrac{x}{3} \end{cases}$

5. $\begin{cases} 3x + 6y = 6 \\ x + 2y - 2 = 0 \end{cases}$

6. $\begin{cases} 6x + 3y = 12 \\ y = -2x + 2 \end{cases}$

7. Find the slope and the y-intercept of the graph of each line in the system $\begin{cases} y = -2x + 1 \\ 8x + 4y = 3 \end{cases}$. Then, use that information to determine the number of solutions of the system.

8. **Bachelor's Degrees.** Estimate the point of intersection of the graphs below. Explain its significance.

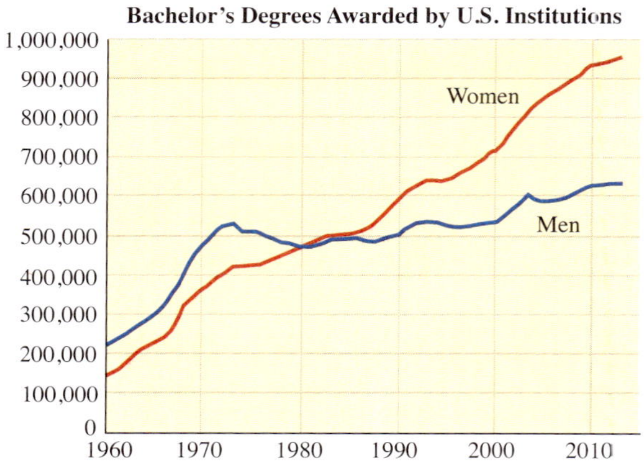

SECTION 4.2 Solving Systems of Equations by Substitution

DEFINITIONS AND CONCEPTS	EXAMPLES
To solve a system of equations in x and y by **substitution:** 1. Solve one of the equations for either x or y. If this is already done, go to step 2. (We call this equation the **substitution equation.**) 2. Substitute the expression for x (or for y) obtained in step 1 into the other equation and solve the equation. 3. Substitute the value of the variable found in step 2 into the substitution equation to find the value of the remaining variable. 4. Check the proposed solution in the equations of the original system. With the substitution method, the objective is to use an appropriate substitution to obtain *one equation in one variable.* If in step 2 the variable drops out and a false statement results, the system has **no solution.** If a true statement results, the system has **infinitely many solutions.**	Use substitution to solve the system: $\begin{cases} y = x + 7 \\ x + 2y = 5 \end{cases}$ **Step 1:** The first equation is already solved for y. **Step 2:** Substitute $x + 7$ for y in the second equation. $x + 2y = 5$ $x + 2(x + 7) = 5$ Substitute $x + 7$ for y. $x + 2x + 14 = 5$ Distribute the multiplication by 2. $3x + 14 = 5$ Combine like terms. $3x = -9$ Subtract 14 from both sides. $x = -3$ Divide both sides by 3. This is the x-value of the solution. **Step 3:** $y = x + 7$ This is the substitution equation. $y = -3 + 7$ Substitute -3 for x. $y = 4$ This is the y-value of the solution. **Step 4:** The following check verifies that the solution is $(-3, 4)$. $y = x + 7$ The first equation. $x + 2y = 5$ The second equation. $4 \stackrel{?}{=} -3 + 7$ Substitute. $-3 + 2(4) \stackrel{?}{=} 5$ Substitute. $4 = 4$ True $5 = 5$ True

REVIEW EXERCISES

Use the substitution method to solve each system.

9. $\begin{cases} y = 15 - 3x \\ 7y + 3x = 15 \end{cases}$

10. $\begin{cases} x = y \\ 5x - 4y = 3 \end{cases}$

11. $\begin{cases} 6x + 2y = 8 - y + x \\ 3x = 2 - y \end{cases}$

12. $\begin{cases} r = 3s + 7 \\ r = 2s + 5 \end{cases}$

13. $\begin{cases} 9x + 3y - 5 = 0 \\ 3x + y = \dfrac{5}{3} \end{cases}$

14. $\begin{cases} \dfrac{x}{2} + \dfrac{y}{2} = 11 \\ \dfrac{5x}{16} - \dfrac{3y}{16} = \dfrac{15}{8} \end{cases}$

15. When solving a system using the substitution method, suppose you obtain the result $8 = 9$.
    a. How many solutions does the system have?
    b. Describe the graph of the system.
    c. What term is used to describe the system?

16. Fill in the blank. With the substitution method, the objective is to use an appropriate substitution to obtain one equation in _____ variable.

SECTION 4.3 Solving Systems of Equations by Elimination (Addition)

DEFINITIONS AND CONCEPTS

To solve a system of equations in x and y using **elimination (addition):**

1. Write each equation in the standard $Ax + By = C$ form.
2. Multiply one (or both) equations by nonzero quantities to make the coefficients of x (or y) opposites.
3. Add the equations to eliminate the terms involving x (or y).
4. Solve the equation obtained in step 3.
5. Find the value of the other variable by substituting the value of the variable found in step 4 into any equation containing both variables.
6. Check the solution in the equations of the original system.

With the elimination method, the basic objective is to obtain two equations whose sum will be one equation in one variable.

If in step 3 both variables drop out and a false statement results, the system has **no solution.** If a true statement results, the system has **infinitely many solutions.**

EXAMPLES

Use elimination to solve: $\begin{cases} 2x - 3y = 4 \\ 3x + y = -5 \end{cases}$

Step 1: Both equations are written in $Ax + By = C$ form.

Step 2: Multiply the second equation by 3 so that the coefficients of y are opposites, -3 and 3.

Step 3:

$$\begin{aligned} 2x - 3y &= 4 \\ 9x + 3y &= -15 \\ 11x \quad &= -11 \end{aligned}$$

In the middle column, $-3y + 3y = 0$.
Add the like terms, column by column.

Step 4: Solve for x.

$11x = -11$

$x = -1$ Divide both sides by 11. This is the x-value of the solution.

Step 5: Find y.

$3x + y = -5$ This is the second equation.

$3(-1) + y = -5$ Substitute -1 for x.

$y = -2$ This is the y-value of the solution.

Step 6: The following check verifies that the solution is $(-1, -2)$.

$2x - 3y = 4$ First equation.

$2(-1) - 3(-2) \stackrel{?}{=} 4$ Substitute.

$-2 + 6 \stackrel{?}{=} 4$

$4 = 4$ True

$3x + y = -5$ Second equation.

$3(-1) + (-2) \stackrel{?}{=} -5$ Substitute.

$-3 - 2 \stackrel{?}{=} -5$

$-5 = -5$ True

REVIEW EXERCISES

17. Write each equation of the system $\begin{cases} 4x + 2y - 7 = 0 \\ 3y = 5x + 6 \end{cases}$ in standard $Ax + By = C$ form.

18. Fill in the blank. With the elimination method, the basic objective is to obtain two equations whose sum will be one equation in _____ variable.

Solve each system using the elimination (addition) method.

19. $\begin{cases} 2x + y = 1 \\ 5x - y = 20 \end{cases}$

20. $\begin{cases} x + 8y = 7 \\ x - 4y = 1 \end{cases}$

21. $\begin{cases} 5a + b = 2 \\ 3a + 2b = 11 \end{cases}$

22. $\begin{cases} 11x + 3y = 27 \\ 8x + 4y = 36 \end{cases}$

23. $\begin{cases} 9x + 3y = 15 \\ 3x = 5 - y \end{cases}$

24. $\begin{cases} 0.02x + 0.05y = 0 \\ 0.3x - 0.2y = -1.9 \end{cases}$

25. $\begin{cases} \dfrac{a}{4} - \dfrac{b}{3} = \dfrac{1}{12} \\ \dfrac{a}{2} - \dfrac{5b}{4} = \dfrac{7}{4} \end{cases}$

26. $\begin{cases} -\dfrac{1}{4}x = 1 - \dfrac{2}{3}y \\ 6x - 18y = 5 - 2y \end{cases}$

For each system, determine which method, substitution or elimination (addition), would be easier to use to solve the system and explain why.

27. $\begin{cases} 6x + 2y = 5 \\ 3x - 3y = -4 \end{cases}$

28. $\begin{cases} x = 5 - 7y \\ 3x - 3y = -4 \end{cases}$

SECTION 4.4 Problem Solving Using Systems of Equations

DEFINITIONS AND CONCEPTS

We can solve many types of problems using a system of two linear equations in two variables:

- Geometry problems
- Number-value problems
- Interest problems
- Uniform motion problems
- Liquid and dry mixture problems

To solve problems involving two unknown quantities:

1. **Analyze** the facts of the problem. Make a table or a diagram if it is helpful.
2. **Assign** different variables to represent the two unknown quantities.
3. Translate the words of the problem to **form two equations** involving those variables.
4. **Solve** the system of equations using graphing, substitution, or elimination.
5. **State** the conclusion.
6. **Check** the results.

Two angles are said to be **complementary** if the sum of their measures is 90°. Two angles are said to be **supplementary** if the sum of their measures is 180°.

EXAMPLES

The difference of the measures of two supplementary angles is 40°. Find the measure of each angle.

Analyze Since the angles are supplementary, the sum of their measures is 180°. If we subtract the smaller angle from the larger, the result should be 40°.

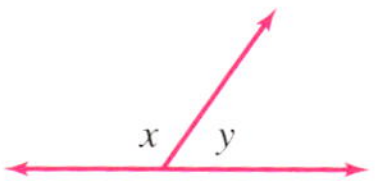

Assign Let x = the measure (in degrees) of the larger angle and y = the measure (in degrees) of the smaller angle.

Form $\begin{cases} x + y = 180 \\ x - y = 40 \end{cases}$ Their sum is 180°. Their difference is 40°.

Solve If we add the equations, we get

$$\begin{aligned} x + y &= 180 \\ x - y &= 40 \\ \hline 2x \quad &= 220 \\ x &= 110 \end{aligned}$$

Divide both sides by 2. This is the measure of the larger angle.

We can use the first equation of the system to find y.

$$x + y = 180$$

$$110 + y = 180 \quad \text{Substitute 110 for } x.$$

$$y = 70 \quad \text{Subtract 110 from both sides.}$$

This is the measure of the smaller angle.

State The angles measure 110° and 70°.

Check Angles with measures of 110° and 70° are supplementary (their sum is 180°) and their difference is 40°. The results check.

REVIEW EXERCISES

Write a system of two equations in two variables to solve each problem.

29. Elevations. The elevation of Las Vegas, Nevada, is 20 times greater than that of Baltimore, Maryland. The sum of their elevations is 2,100 feet. Find the elevation of each city.

30. Painting Equipment. When fully extended, a ladder is 35 feet in length. If the extension is 7 feet shorter than the base, how long is each part of the ladder?

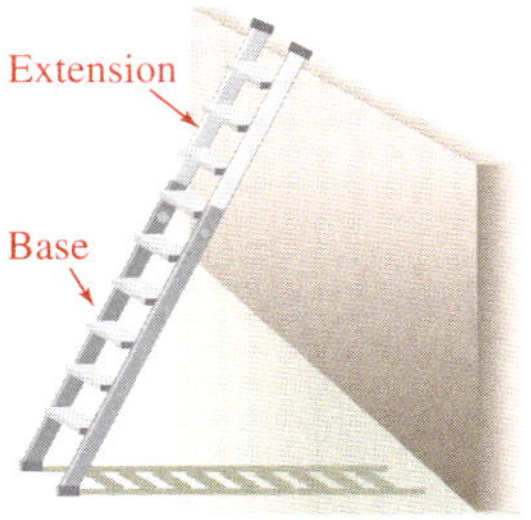

31. Geometry. Two angles are complementary. The measure of one is 15° more than twice the measure of the other. Find the measure of each angle.

32. Crash Investigation. In an effort to protect evidence, investigators used 420 yards of yellow "Police Line—Do Not Cross" tape to seal off a large rectangular-shaped area around an airplane crash site. How much area will the investigators have to search if the width of the rectangle is three-fourths of the length?

33. Complete each table.

a.

	Amount ·	Strength =	Amount of pesticide
Weak	x	0.02	
Strong	y	0.09	
Mixture	100	0.08	

b.

	Rate ·	Time =	Distance
With the wind	$s + w$	5	
Against the wind	$s - w$	7	

c.

	P ·	r ·	t =	I
Mack Financial	x	0.11	1	
Union Savings	y	0.06	1	

d.

	Amount ·	Price =	Total value
Caramel corn	x	4	
Peanuts	y	8	
Mixture	10	5	

34. Candy Store. A merchant wants to mix gummy worms worth \$6 per pound and gummy bears worth \$3 per pound to make 30 pounds of a mixture worth \$4.20 per pound. How many pounds of each type of candy should he use?

35. Boating. It takes a motorboat 4 hours to travel 56 miles down a river, and 3 hours longer to make the return trip. Find the speed of the current.

36. Shopping. Packages containing two bottles of contact lens cleaner and three bottles of soaking solution cost \$63.40, and packages containing three bottles of cleaner and two bottles of soaking solution cost \$69.60. Find the cost of a bottle of cleaner and a bottle of soaking solution.

37. Investing. Carlos invested part of \$3,000 in a 10% certificate account and the rest in a 6% passbook account. The total annual interest from both accounts is \$270. How much did he invest at 6%?

38. Antifreeze. How much of a 40% antifreeze solution must a mechanic mix with a 70% antifreeze solution if she needs 20 gallons of a 50% antifreeze solution?

SECTION 4.5 Solving Systems of Linear Inequalities

DEFINITIONS AND CONCEPTS	EXAMPLES
A solution of a **system of linear inequalities** is an ordered pair that satisfies each inequality.	Graph the solutions of the system: $\begin{cases} y \le x + 1 \\ y > -1 \end{cases}$
To **solve a system of linear inequalities:**	**Step 1:** Graph each inequality on the same coordinate system as shown.
1. Graph each inequality on the same coordinate system.	**Step 2:** Use shading to highlight where the graphs intersect.
2. Use shading to highlight the intersection of the graphs. The points in this region are the solutions of the system.	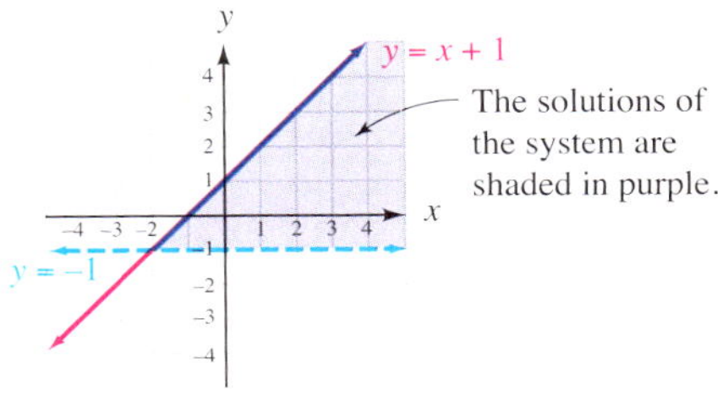
3. As an informal check, pick a point from the region and verify that its coordinates satisfy each inequality of the original system.	**Step 3:** Pick a point from the solution region such as (1, 0) and verify that it satisfies both inequalities.

REVIEW EXERCISES

Solve each system of inequalities.

39. $\begin{cases} 5x + 3y < 15 \\ 3x - y > 3 \end{cases}$

40. $\begin{cases} 3y \le x \\ y > 3x \end{cases}$

41. $\begin{cases} x \le 0 \\ y < 0 \end{cases}$

42. $\begin{cases} y \ge x \\ y \le \frac{1}{3}x + 1 \\ x > -3 \end{cases}$

43. Use a check to determine whether each ordered pair is a solution of the system: $\begin{cases} x + 2y \le 3 \\ 2x - y > 1 \end{cases}$

a. $(5, -4)$ b. $(-1, -3)$

44. Gift Shopping. A grandmother wants to spend at least \$40 but no more than \$60 on school clothes for her grandson. If T-shirts sell for \$10 each and pants sell for \$20 each, write a system of inequalities that describes the possible numbers of T-shirts x and pairs of pants y that she can buy. Graph the system and give two possible solutions.

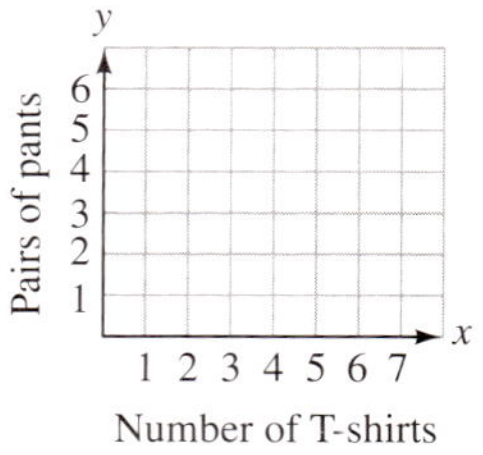

4 CHAPTER TEST

Use a check to determine whether the ordered pair is a solution of the system.

1. $(5, 3)$, $\begin{cases} 3x + 2y = 21 \\ x + y = 8 \end{cases}$

2. $(-2, -1)$, $\begin{cases} 4x + y = -9 \\ 2x - 3y = -7 \end{cases}$

3. Fill in the blanks.

a. A ________ of a system of linear equations is an ordered pair that satisfies each equation.

b. A system of equations that has at least one solution is called a __________ system.

c. A system of equations that has no solution is called an ____________ system.

d. Equations with different graphs are called ___________ equations.

e. A system of __________ equations has an infinite number of solutions.

f. Two angles are said to be _____________ if the sum of their measures is 180°.

4. Energy. The graphs below show U.S. electricity generation from natural gas and nuclear sources from 1995 through 2009. What is the intersection point of the graphs? Explain its significance.

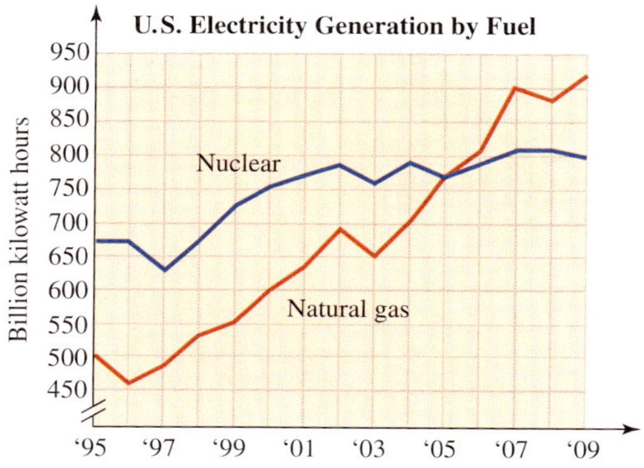

Solve each system by graphing.

5. $\begin{cases} y = 2x - 1 \\ x - 2y = -4 \end{cases}$

6. $\begin{cases} x + y = 5 \\ y = -x \end{cases}$

7. Find the slope and the y-intercept of the graph of each line in the system $\begin{cases} y = 4x - 10 \\ x - 2y = -16 \end{cases}$. Then, use that information to determine the *number of solutions* of the system. **Do not solve the system.**

8. How many solutions does the system of two linear equations graphed on the right have? Give three of the solutions.

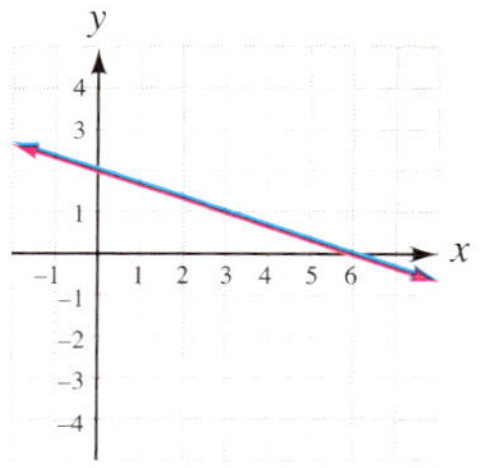

Solve each system by substitution.

9. $\begin{cases} y = x - 1 \\ 2x + y = -7 \end{cases}$

10. $\begin{cases} 3x + 6y = -15 \\ x + 2y = -5 \end{cases}$

Solve each system using elimination.

11. $\begin{cases} 3x - y = 2 \\ 2x + y = 8 \end{cases}$

12. $\begin{cases} 4x + 3y = -3 \\ -3x = -4y + 21 \end{cases}$

Solve each system using substitution or elimination.

13. $\begin{cases} 3x - 5y - 16 = 0 \\ \frac{x}{2} - \frac{5}{6}y = \frac{1}{3} \end{cases}$

14. $\begin{cases} 3a + 4b = -7 \\ 2b - a = -1 \end{cases}$

15. $\begin{cases} y = 3x - 1 \\ y = 2x + 4 \end{cases}$

16. $\begin{cases} 0.6c + 0.5d = 0 \\ 0.02c + 0.09d = 0 \end{cases}$

17. $\begin{cases} a - 1 = 2b \\ 3a + 1 = -10b \end{cases}$

18. $\begin{cases} \dfrac{x+2}{6} = \dfrac{y+3}{4} \\ \dfrac{x}{5} = \dfrac{3y-3}{6} \end{cases}$

19. $\begin{cases} 4(a-2) + 5y = 19 \\ 3(a-2) - y = 0 \end{cases}$

20. $\begin{cases} 3x + 1 = 2x - 4y + 2 \\ 7x - y - 1 = 9y + 5x + 10 \end{cases}$

Write a system of two equations in two variables to solve each problem.

21. **Child Care.** On a mother's 22-mile commute to work, she drops her daughter off at a child care center. The first part of the trip is 6 miles less than the second part. How long is each part of her morning commute?

22. **Vacationing.** It cost a family of 7 a total of \$219 for general admission tickets to the San Diego Zoo. How many adult tickets and how many child tickets were purchased?

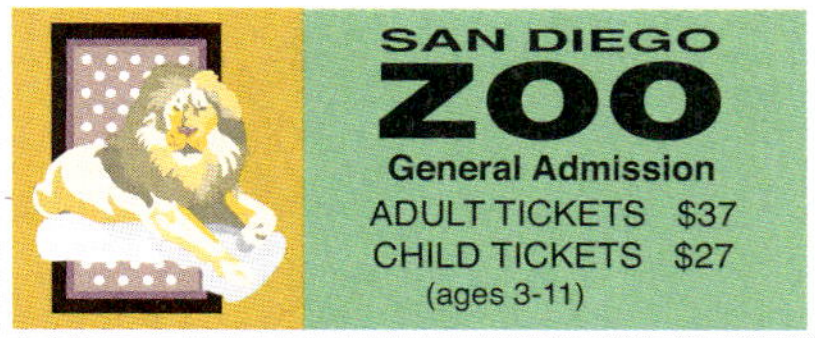

23. **Financial Planning.** A woman invested some money at 8% and some at 9% annual simple interest. The interest for 1 year on the combined investment of \$10,000 was \$840. How much was invested at each rate?

24. **Tailwinds/Headwinds.** Flying with a tailwind, a pilot flew an airplane 450 miles in 2.5 hours. Flying into a headwind, the return trip took 3 hours. Find the speed of the plane in calm air and the speed of the wind.

25. **Tether Ball.** The angles shown in the illustration are complementary. The measure of the larger angle is 10° more than three times the measure of the smaller angle. Find the measure of each angle.

26. **Antifreeze.** How many pints of a 5% antifreeze solution and how many pints of a 20% antifreeze solution must be mixed to obtain 12 pints of a 15% solution?

27. **Sunscreen.** A sunscreen selling for \$1.50 per ounce is to be combined with another sunscreen selling for \$0.80 per ounce. How many ounces of each are needed to make 10 ounces of a sunscreen mix that sells for \$1.01 per ounce?

28. Use a check to determine whether (3, 1) is a solution of the system: $\begin{cases} y \le 2x - 1 \\ x + 3y > 6 \end{cases}$

Solve each system by graphing.

29. $\begin{cases} 3x + 2y < 6 \\ y \ge x + 1 \end{cases}$

30. $\begin{cases} x - y < 3 \\ y \le 0 \\ x \ge 0 \end{cases}$

31. **Clothes Shopping.** This system of inequalities describes the number of \$20 shirts, x, and \$40 pairs of pants, y, a person can buy if he or she plans to spend not less than \$80 but not more than \$120. Graph the system. Then give three solutions.

$$\begin{cases} 20x + 40y \ge 80 \\ 20x + 40y \le 120 \end{cases}$$

32. Match each equation, inequality, or system with the graph of its solution.

a. $2x + y = 2$

b. $2x + y \ge 2$

c. $\begin{cases} 2x + y = 2 \\ 2x - y = 2 \end{cases}$

d. $\begin{cases} 2x + y \ge 2 \\ 2x - y \le 2 \end{cases}$

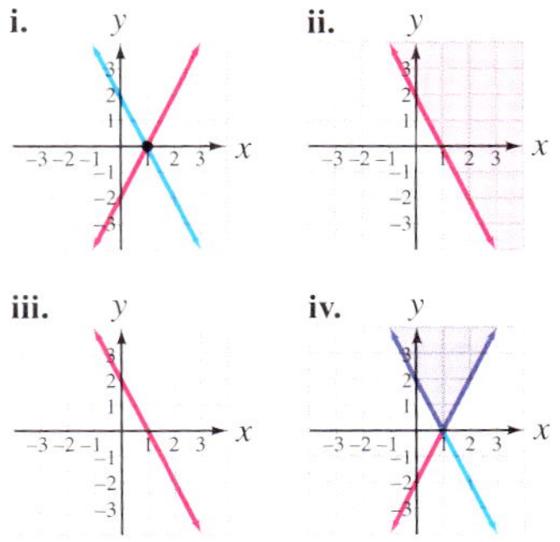

Group Project

WRITING APPLICATION PROBLEMS

Overview: In Section 4.4, you solved application problems by translating the words of the problem into a system of two equations. In this activity, you will reverse these steps.

Instructions: Form groups of 2 or 3 students. For each type of application, write a problem that could be solved using the given equations. If you need help getting started, refer to the specific problem types in the text. When finished writing the five applications, pick one problem and solve it completely.

A rectangle problem:

$$\begin{cases} 2l + 2w = 320 \\ l = w + 40 \end{cases}$$

A number-value problem:

$$\begin{cases} 5x + 2y = 23 \\ 3x + 7y = 37 \end{cases}$$

An interest problem:

$$\begin{cases} x + y = 75{,}000 \\ 0.03x + 0.05y = 2{,}750 \end{cases}$$

A with–against the wind problem:

$$\begin{cases} 2(x + y) = 600 \\ 3(x - y) = 600 \end{cases}$$

A liquid mixture problem:

$$\begin{cases} x + y = 36 \\ 0.50x + 0.20y = 0.30(36) \end{cases}$$

CUMULATIVE REVIEW Chapters 1–4

1. Fill in the blanks. The answer to an addition problem is called a _____. The answer to a subtraction problem is called a _________. The answer to a multiplication problem is called a ________. The answer to a division problem is called a ________. [Section 1.1]
2. Give the prime factorization of 100. [Section 1.2]
3. Divide: $\frac{3}{4} \div \frac{6}{5}$ [Section 1.2]
4. Subtract: $\frac{7}{10} - \frac{1}{14}$ [Section 1.2]
5. Is π a rational or irrational number? [Section 1.3]
6. Graph each member of the set on the number line. [Section 1.3]

   $$\left\{-2\frac{1}{4}, \sqrt{2}, -1.75, \frac{7}{2}, 0.5\right\}$$

7. Write $\frac{2}{3}$ as a decimal. [Section 1.3]
8. What property of real numbers is illustrated? [Section 1.6]

   $$3(2x) = (3 \cdot 2)x$$

Evaluate each expression.

9. $-3^2 + |4^2 - 5^2|$ [Section 1.7]
10. $(4 - 5)^{20}$ [Section 1.7]
11. $\dfrac{-3 - (-7)}{2^2 - 3}$ [Section 1.7]
12. $12 - 2[1 - (-8 + 2)]$ [Section 1.7]
13. **Racing.** Suppose a driver has completed x laps of a 250-lap race. Write an expression for how many more laps he must make to finish the race. [Section 1.8]
14. What is the value of d dimes in cents? [Section 1.8]

Simplify each expression.

15. $13r - 12r$ [Section 1.9]
16. $27\left(\frac{2}{3}x\right)$ [Section 1.9]
17. $4(d - 3) - (d - 1)$ [Section 1.9]
18. $(13c - 3)(-6)$ [Section 1.9]

Solve each equation. Check each result.

19. $3(x - 5) + 2 = 2x$ [Section 2.2]
20. $\dfrac{x - 5}{3} - 5 = 7$ [Section 2.2]
21. $\frac{2}{5}x + 1 = \frac{1}{3} + x$ [Section 2.2]
22. $-\frac{5}{8}h = 15$ [Section 2.2]
23. **Gymnastics.** After the first day of registration, 119 children had been enrolled in a Gymboree class. That represented 85% of the available slots. Find the maximum number of children the center could enroll. [Section 2.3]

24. Greenhouse Gases. In 2008, the total U.S. greenhouse gas emissions were approximately 6,957 teragrams of carbon dioxide equivalent. Use the graph below to determine the greenhouse emissions that come from transportation sources. Round to the nearest teragram. [Section 2.3]

U.S. Greenhouse Gas Emissions, by source

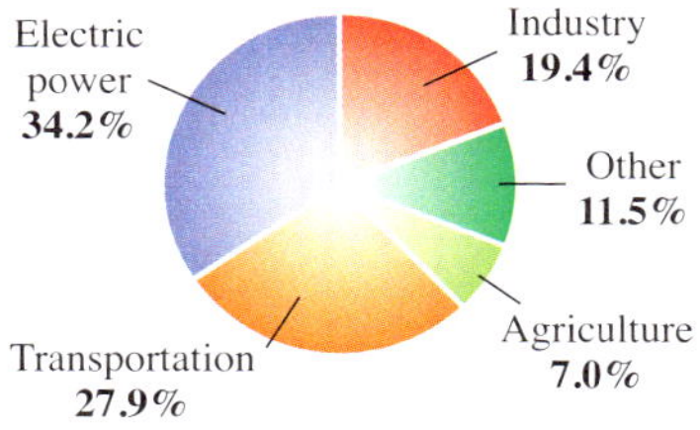

Source: U.S. Environmental Protection Agency

25. Solve $A = \frac{1}{2}h(b + B)$ for h. [Section 2.4]

26. Cancer. According to the National Lung Cancer Partnership, approximately 219,000 people are diagnosed with lung cancer in the U.S. each year, with 13,000 more of them being men than women. Approximately how many men and how many women are diagnosed with lung cancer each year? [Section 2.5]

27. Mixing Candy. The owner of a candy store wants to make a 30-pound mixture of two candies to sell for \$4 per pound. If red licorice bits sell for \$3.80 per pound and lemon gumdrops sell for \$4.40 per pound, how many pounds of each should be used? [Section 2.6]

28. Solve $8(4 + x) > 10(6 + x)$. Write the solution set in interval notation and graph it. [Section 2.7]

29. New York. In Manhattan, avenues run north and south, and streets run east and west to form a grid, as shown below. Give the location of the Chipotle Mexican Grill as an ordered pair of the form (avenue, street). [Section 3.1]

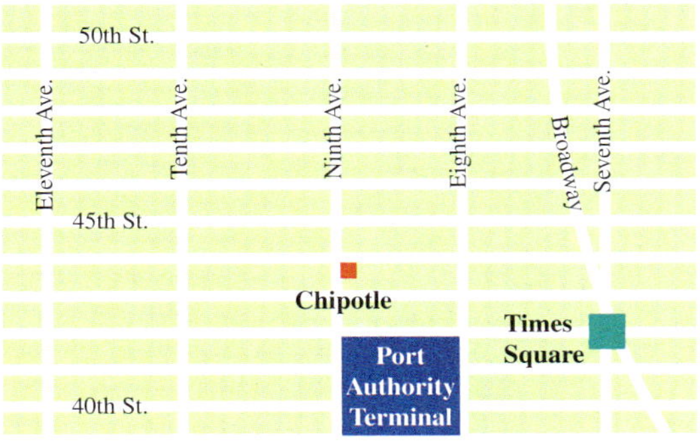

30. Perimeter. The eight vertices (corners) of a figure are $(-4, 6)$, $(4, 6)$, $(4, -3)$, $(2, -3)$, $(2, -5)$, $(-2, -5)$, $(-2, -3)$, and $(-4, -3)$. Find the perimeter of the figure. [Section 3.1]

31. In what quadrant does $(-3.5, 6)$ lie? [Section 3.1]

32. Is $(-2, 8)$ a solution of $y = -2x + 3$? [Section 3.2]

Graph each equation.

33. $x = 4$ [Section 3.3]

34. $4x - 3y = 12$ [Section 3.3]

Find the slope of the line with the given properties.

35. Passing through $(-2, 4)$ and $(6, 8)$ [Section 3.4]

36. A line that is horizontal [Section 3.4]

37. An equation of $2x - 3y = 12$ [Section 3.5]

38. Are the graphs of the lines whose equations are given below parallel or perpendicular? [Section 3.5]

$$y = -\frac{3}{4}x + \frac{15}{4} \qquad 4x - 3y = 25$$

39. Find the slope and the x- and y-intercepts of the line graphed on the right. [Sections 3.3 and 3.4]

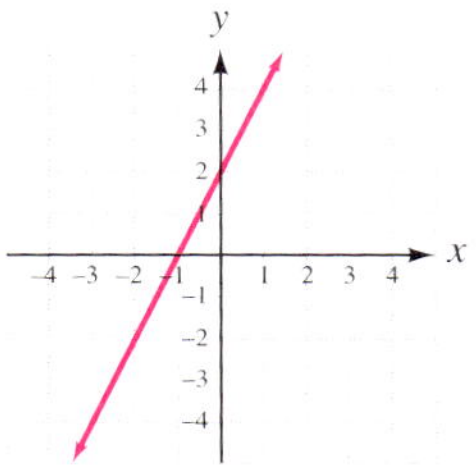

40. Newspapers. The line graph below approximates the percent of adults ages 25–34 who read a newspaper on a regular basis. Find the rate of change in newspaper readership for that age group. [Section 3.4]

Percent of those ages 25–34 who read a newspaper regularly

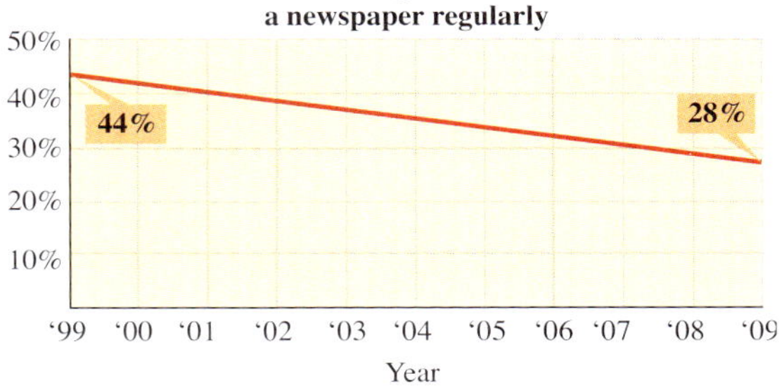

Source: The State of the News Media, 2010

Find an equation of the line with the following properties. Write the equation in slope–intercept form.

41. Slope $= \frac{2}{3}$, y-intercept $= (0, 5)$ [Section 3.5]

42. Passing through $(-2, 4)$ and $(6, 10)$ [Section 3.6]

43. A horizontal line passing through $(2, 4)$ [Section 3.6]

44. Graph: $y < \frac{x}{3} - 1$ [Section 3.7]

45. If $f(x) = -2x^2 - 3x^3$, find $f(-1)$. [Section 3.8]

46. Refer to the graph on the right. Is this the graph of a function? If it is, give the domain and range. If it is not, find ordered pairs that show a value of x that is assigned more than one value of y. [Section 3.8]

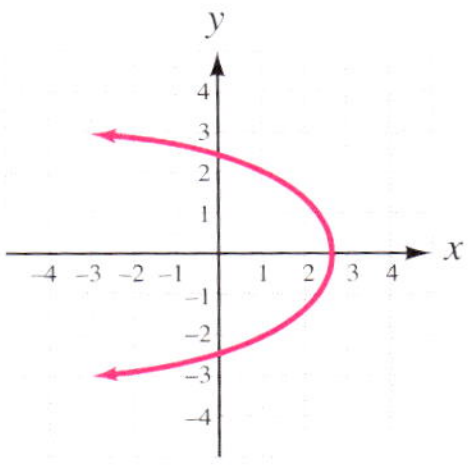

Solve each system by graphing.

47. $\begin{cases} x + 4y = -2 \\ y = -x - 5 \end{cases}$ [Section 4.1]

48. $\begin{cases} 2x - 3y < 0 \\ y > x - 1 \end{cases}$ [Section 4.5]

49. Solve $\begin{cases} x - 2y = 2 \\ 2x + 3y = 11 \end{cases}$ by substitution. [Section 4.2]

50. Solve $\begin{cases} \frac{3}{2}x - \frac{2}{3}y = 0 \\ \frac{3}{4}x + \frac{4}{3}y = \frac{5}{2} \end{cases}$ by elimination. [Section 4.3]

Write a system of two equations in two variables to solve each problem.

51. **Nutrition.** The table shows per serving nutritional information for egg noodles and rice pilaf. How many servings of each food should be eaten to consume exactly 22 grams of protein and 21 grams of fat? [Section 4.4]

	Protein (g)	Fat (g)
Egg noodles	5	3
Rice pilaf	4	5

52. **Investment Clubs.** Part of \$8,000 was invested by an investment club at 10% interest and the rest at 12%. If the annual income from these investments is \$900, how much was invested at each rate? [Section 4.4]

Recall that like terms are terms with exactly the same variables raised to exactly the same powers. To add or subtract exponential expressions, they must be like terms. To multiply or divide exponential expressions, only the bases need to be the same.

$x^5 + x^2$ — These are not like terms; the exponents are different. We cannot add.

$x^2 + x^2 = 2x^2$ — These are like terms; we can add. Recall that $x^2 = 1x^2$.

$x^5 \cdot x^2 = x^7$ — The bases are the same; we can multiply.

$\frac{x^5}{x^2} = x^3$ — The bases are the same; we can divide.

4 Raise Exponential Expressions to a Power.

To develop another rule for exponents, we consider $(5^3)^4$. Here, an exponential expression, 5^3, is raised to a power. Since 5^3 is the base and 4 is the exponent, $(5^3)^4$ can be written as $5^3 \cdot 5^3 \cdot 5^3 \cdot 5^3$. Because each of the four factors of 5^3 contains three factors of 5, there are $4 \cdot 3$ or 12 factors of 5.

The Language of Algebra

An exponential expression raised to a power, such as $(5^3)^4$, is also called a **power of a power.**

$$(5^3)^4 = 5^3 \cdot 5^3 \cdot 5^3 \cdot 5^3 = \overbrace{\underbrace{5 \cdot 5 \cdot 5}_{5^3} \cdot \underbrace{5 \cdot 5 \cdot 5}_{5^3} \cdot \underbrace{5 \cdot 5 \cdot 5}_{5^3} \cdot \underbrace{5 \cdot 5 \cdot 5}_{5^3}}^{\text{12 factors of x}} = 5^{12}$$

We can find this result quickly if we keep the common base of 5 and multiply the exponents.

$$(5^3)^4 = 5^{3 \cdot 4} = 5^{12}$$

This example illustrates the following rule for exponents.

Power Rule for Exponents

To raise an exponential expression to a power, keep the base and multiply the exponents.

For any number x and any natural numbers m and n,

$$(x^m)^n = x^{m \cdot n} = x^{mn}$$

Read as "x to the mth power raised to the nth power equals x to the mnth power."

EXAMPLE 7 Simplify: **a.** $(2^3)^7$ **b.** $[(-6)^2]^5$ **c.** $(z^8)^8$

Strategy In each case, we want to write an equivalent expression using one base and one exponent. We will use the power rule for exponents to do this.

Why Each expression is a power of a power.

Solution

a. $(2^3)^7 = 2^{3 \cdot 7} = 2^{21}$ — Read as "2 cubed raised to the seventh power." Keep the base, 2, and multiply the exponents.

b. $[(-6)^2]^5 = (-6)^{2 \cdot 5} = (-6)^{10}$ — Read as "negative six squared raised to the fifth power." Keep the base, −6, and multiply the exponents. Since $(-6)^{10}$ is a very large number, we will leave the answer in this form.

c. $(z^8)^8 = z^{8 \cdot 8} = z^{64}$ — Keep the base, z, and multiply the exponents.

Caution

Do not confuse the power and product rules for exponents:

$(2^3)^7 = 2^{3 \cdot 7} = 2^{21}$ — Multiply exponents.

$2^3 \cdot 2^7 = 2^{3+7} = 2^{10}$ — Add exponents.

Self Check 7 Simplify: **a.** $(4^6)^5$ **b.** $(y^5)^2$

Now Try Problems 53 and 55

EXAMPLE 8 Simplify: **a.** $(x^2x^5)^8$ **b.** $(z^2)^4(z^3)^3$

Strategy In each case, we want to write an equivalent expression using one base and one exponent. We will use the product and power rules for exponents to do this.

Why The expressions involve multiplication of exponential expressions that have the same base and they involve powers of powers.

Solution

a. $(x^2x^5)^8 = (x^7)^8$ Read as "the quantity of x squared times x to the fifth power, raised to the eighth power." Within the parentheses, keep the common base, x, and add the exponents.

$= x^{56}$ Keep the base, x, and multiply the exponents.

b. $(z^2)^4(z^3)^3 = z^8z^9$ For each power of z raised to a power, keep the base and multiply the exponents.

$= z^{17}$ Keep the common base, z, and add the exponents.

Self Check 8 Simplify: **a.** $(a^4a^3)^3$ **b.** $(a^3)^3(a^4)^2$

Now Try Problems 59 and 63

5 Find Powers of Products and Quotients.

To develop more rules for exponents, we consider the expression $(2x)^3$, which is a *power of the product* of 2 and x, and the expression $\left(\frac{2}{x}\right)^3$, which is a *power of the quotient* of 2 and x.

$$(2x)^3 = 2x \cdot 2x \cdot 2x = (2 \cdot 2 \cdot 2)(x \cdot x \cdot x) = 2^3x^3 = 8x^3$$

$$\left(\frac{2}{x}\right)^3 = \frac{2}{x} \cdot \frac{2}{x} \cdot \frac{2}{x}$$ Assume $x \neq 0$.

$$= \frac{2 \cdot 2 \cdot 2}{x \cdot x \cdot x}$$ Multiply the numerators. Multiply the denominators.

$$= \frac{2^3}{x^3}$$

$$= \frac{8}{x^3}$$ Evaluate: $2^3 = 8$.

These examples illustrate the following rules for exponents.

Powers of a Product and a Quotient

To raise a product to a power, raise each factor of the product to that power. To raise a quotient to a power, raise the numerator and the denominator to that power.

For any numbers x and y, and any natural number n,

$$(xy)^n = x^ny^n \quad \text{and} \quad \left(\frac{x}{y}\right)^n = \frac{x^n}{y^n}, \quad \text{where } y \neq 0$$

EXAMPLE 9 Simplify: **a.** $(3c)^4$ **b.** $(x^2y^3)^5$ **c.** $\left(-\frac{1}{4}a^3b\right)^2$

Strategy In each case, we want to write the expression in an equivalent form in which each base is raised to a single power. We will use the power of a product rule for exponents to do this.

Why Within each set of parentheses is a product, and each of those products is raised to a power.

Caution

Never multiply an exponent and a base, or an exponent and a factor of the base!

$$(3c)^4 = 12c^4$$

When simplifying expressions, *exponents interact only with other exponents* using the operations of addition, subtraction, or multiplication.

Solution

a. $(3c)^4 = 3^4c^4$ Raise each factor of the product 3c to the 4th power.

$= 81c^4$ Evaluate: $3^4 = 81$.

b. $(x^2y^3)^5 = (x^2)^5(y^3)^5$ Raise each factor of the product x^2y^3 to the 5th power.

$= x^{10}y^{15}$ For each power of a power, keep each base, x and y, and multiply the exponents.

c. $\left(-\frac{1}{4}a^3b\right)^2 = \left(-\frac{1}{4}\right)^2(a^3)^2b^2$ Raise each factor of the product $-\frac{1}{4}a^3b$ to the 2nd power.

$= \frac{1}{16}a^6b^2$ Evaluate: $\left(-\frac{1}{4}\right)^2 = \frac{1}{16}$. Keep the base a and multiply the exponents.

Self Check 9 Simplify: **a.** $(2t)^4$ **b.** $(c^3d^4)^6$ **c.** $\left(-\frac{1}{3}ab^5\right)^3$

Now Try Problems 67 and 73

EXAMPLE 10

Simplify: $\frac{(a^3b^4)^2}{ab^5}$

Strategy We want to write the expression in an equivalent form in which each base is raised to a single power. We will use the power of a product rule and the quotient rule for exponents to do this.

Why The expression involves a power of a product and it is the quotient of exponential expressions that have the same base.

Solution

$\frac{(a^3b^4)^2}{ab^5} = \frac{(a^3)^2(b^4)^2}{a^1b^5}$ In the numerator, raise each factor within the parentheses to the second power. In the denominator, it is helpful to write a as a^1.

$= \frac{a^6b^8}{a^1b^5}$ In the numerator, for each power of a power, keep each base, a and b, and multiply the exponents.

$= a^{6-1}b^{8-5}$ Keep each of the bases, a and b, and subtract the exponents.

$= a^5b^3$

Self Check 10 Simplify: $\frac{(c^4d^5)^3}{c^2d^3}$

Now Try Problem 75

EXAMPLE 11

Simplify: $\frac{(5b)^9}{(5b)^6}$

Strategy We want to write the expression in an equivalent form in which the base is raised to a single power. We will begin by using the quotient rule for exponents.

Why The expression involves division of exponential expressions that have the same base, 5b.

Solution

$\frac{(5b)^9}{(5b)^6} = (5b)^{9-6}$ Keep the common base, 5b, and subtract the exponents.

$= (5b)^3$

$= 5^3b^3$ Use the power of a product rule: Raise each factor within the parentheses to the 3rd power. Do not multiply the exponent 3 and the base 5.

$= 125b^3$ Evaluate 5^3.

Self Check 11 Simplify: $\dfrac{(-2h)^{20}}{(-2h)^{14}}$

Now Try Problem 79

EXAMPLE 12 Simplify: **a.** $\left(\dfrac{4}{k}\right)^3$ **b.** $\left(\dfrac{3x^2}{2y^3}\right)^5$

Strategy We want to write each expression in an equivalent form using each base raised to a single power. We will use the power of a quotient rule for exponents to do this.

Why Within each set of parentheses is a quotient, and each of those quotients is raised to a power.

Solution **a.** Since $\frac{4}{k}$ is the quotient of 4 and k, the expression $\left(\frac{4}{k}\right)^3$ is a power of a quotient.

$$\left(\frac{4}{k}\right)^3 = \frac{4^3}{k^3} = \frac{64}{k^3}$$

Raise the numerator and denominator to the 3rd power. Then evaluate: $4^3 = 64$. Do not multiply the exponent 3 and the base 4.

b. $\left(\dfrac{3x^2}{2y^3}\right)^5 = \dfrac{(3x^2)^5}{(2y^3)^5}$ Raise the numerator and the denominator to the 5th power.

$= \dfrac{3^5(x^2)^5}{2^5(y^3)^5}$ Raise each factor within parentheses to the 5th power. Do not multiply the base 3 and the exponent 5 or the base 2 and the exponent 5.

$= \dfrac{243x^{10}}{32y^{15}}$ Evaluate: $3^5 = 243$ and $2^5 = 32$. For each power of a power, keep the base and multiply the exponents.

Self Check 12 Simplify: **a.** $\left(\dfrac{x}{7}\right)^3$ **b.** $\left(\dfrac{2x^3}{3y^2}\right)^4$

Now Try Problems 83 and 85

The following rules for exponents are used so often in this course, you need to memorize them.

Rules for Exponents

If m and n represent natural numbers and there are no divisions by zero, then

Exponent of 1	Product rule	Power rule
$x^1 = x$	$x^m x^n = x^{m+n}$	$(x^m)^n = x^{mn}$
Quotient rule	Power of a product	Power of a quotient
$\dfrac{x^m}{x^n} = x^{m-n}$	$(xy)^n = x^n y^n$	$\left(\dfrac{x}{y}\right)^n = \dfrac{x^n}{y^n}$

SECTION 5.1 STUDY SET

VOCABULARY

Fill in the blank.

1. Expressions such as x^4, 10^3, and $(5t)^2$ are called ________ expressions.

2. Match each expression below with the proper description on the next page.

$\dfrac{a^8}{a^2}$ $\quad (a^4b^2)^5$ $\quad \left(\dfrac{a^6}{a}\right)^3$ $\quad (a^8)^4$ $\quad a^5 \cdot a^3$

a. Product of exponential expressions with the same base
b. Quotient of exponential expressions with the same base
c. Power of an exponential expression
d. Power of a product
e. Power of a quotient

CONCEPTS

Fill in the blanks.

3. a. $(3x)^4 =$ ____ · ____ · ____ · ____
   b. $(-5y)(-5y)(-5y) =$ ____
4. a. $x = x$ ____   b. $x^m x^n =$ ____
   c. $(xy)^n =$ ____   d. $(a^b)^c =$ ____
   e. $\frac{x^m}{x^n} =$ ____   f. $\left(\frac{a}{b}\right)^n =$ ____
5. To simplify each expression, determine whether you add, subtract, multiply, or divide the exponents.
   a. $\frac{x^8}{x^2}$ b. $b^6 \cdot b^9$
   c. $(n^8)^4$ d. $(a^4b^2)^5$
6. a. To simplify $(2y^3z^2)^4$, what factors within the parentheses must be raised to the fourth power?
   b. To simplify $\left(\frac{y^3}{z^2}\right)^4$, what two expressions must be raised to the fourth power?

Simplify each expression, if possible.

7. a. $x^2 + x^2$ b. $x^2 \cdot x^2$
   c. $x^2 + x$ d. $x^2 \cdot x$
8. a. $x^3 - x^2$ b. $\frac{x^3}{x^2}$
   c. $4^2 \cdot 2^4$ d. $\frac{x^3}{y^2}$

NOTATION

Complete each solution to simplify each expression.

9. $(x^4x^2)^3 = (\quad)^3 = x$
10. $\frac{a^3a^4}{a^2} = \frac{\quad}{a^2} = a^{\quad -2} = a$

Fill in the blanks.

11. a. We read 9^4 as "nine to the fourth ____."
    b. We read $(a^2b^6)(a^4b^5)$ as "the ____ of a^2b^6 times the ____ of a^4b^5."
    c. We read $(3^6)^9$ as "3 to the ____ power, raised to the ____ power."
12. a. We read n^2n^3n as "n ____ times n ____ times n."
    b. We read $\frac{x^7}{x^5}$ as "x to the seventh power ____ by x to the ____ power."
    c. We read $(b + 5)^6(b + 5)^8$ as "the ____ of $b + 5$, ____ to the sixth power, times the ____ of $b + 5$, raised to the ____ power."

GUIDED PRACTICE

Identify the base and the exponent in each expression. See Example 1.

Look Alikes . . .

13. a. 4^3 b. -4^3 c. $(-4)^3$
14. a. x^5 b. $-x^5$ c. $(-x)^5$
15. a. $(-3x)^2$ b. $-3x^2$ c. $-(-3x)^2$
16. a. $-\frac{1}{3}x^6$ b. $\left(-\frac{1}{3}x\right)^6$ c. $-\left(-\frac{1}{3}x\right)^6$
17. a. $9m^{12}$ b. $(9m)^{12}$ c. $-9m^{12}$
18. a. $(y + 9)^4$ b. $y + 9^4$ c. $8(y + 9)^4$

Write each expression in an equivalent form using an exponent. See Example 2.

19. $4t \cdot 4t \cdot 4t \cdot 4t$
20. $-5u(-5u)(-5u)(-5u)(-5u)$
21. $-4 \cdot t \cdot t \cdot t \cdot t \cdot t$
22. $-5 \cdot u \cdot u \cdot u$
23. $\frac{t}{2} \cdot \frac{t}{2} \cdot \frac{t}{2}$
24. $\frac{x}{c} \cdot \frac{x}{c} \cdot \frac{x}{c} \cdot \frac{x}{c}$
25. $(x - y)(x - y)$
26. $(m + 4)(m + 4)$

Use the product rule for exponents to simplify each expression. Write the results using exponents. See Example 3.

27. $5^3 \cdot 5^4$
28. $3^4 \cdot 3^6$
29. bb^2b^3
30. aa^3a^5
31. $(y - 2)^5(y - 2)^2$
32. $(t + 1)^5(t + 1)^3$
33. $(a^2b^3)(a^3b^3)$
34. $(u^3v^5)(u^4v^5)$

Find an expression that represents the area or volume of each figure. Recall that the formula for the volume of a rectangular solid is V = length · width · height. See Example 4.

35.
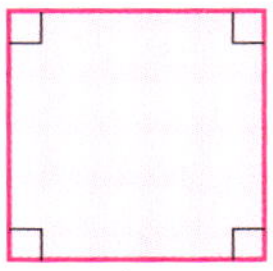

36.

37.

38.
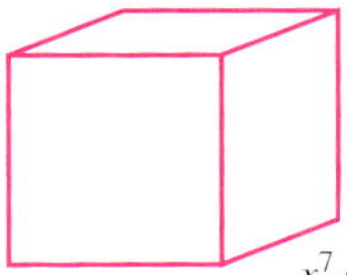

Use the quotient rule for exponents to simplify each expression. Write the results using exponents. See Example 5.

39. $\frac{8^{12}}{8^4}$
40. $\frac{10^4}{10^2}$
41. $\frac{x^{15}}{x^3}$
42. $\frac{y^6}{y^3}$
43. $\frac{(3.7p)^7}{(3.7p)^2}$
44. $\frac{(0.25y)^9}{(0.25y)^3}$
45. $\frac{c^3d^7}{cd}$
46. $\frac{r^8s^9}{rs}$

Use the product and quotient rules for exponents to simplify each expression. **See Example 6.**

47. $\dfrac{y^3y^4}{yy^2}$
48. $\dfrac{b^4b^5}{b^2b^3}$
49. $\dfrac{a^2a^3a^4}{a^8}$
50. $\dfrac{h^3h^6h}{h^9}$

Use the power rule for exponents to simplify each expression. Write the results using exponents. **See Example 7.**

51. $(3^2)^4$
52. $(4^3)^3$
53. $[(-4.3)^3]^8$
54. $[(-1.7)^9]^8$
55. $(m^{50})^{10}$
56. $(n^{25})^4$
57. $(y^5)^3$
58. $(b^3)^6$

Use the product and power rules for exponents to simplify each expression. **See Example 8.**

59. $(x^2x^3)^5$
60. $(y^3y^4)^4$
61. $(p^2p^3)^5$
62. $(r^3r^4)^2$
63. $(t^3)^4(t^2)^3$
64. $(b^2)^5(b^3)^2$
65. $(u^4)^2(u^3)^2$
66. $(v^5)^2(v^3)^4$

Use the power of a product rule for exponents to simplify each expression. **See Example 9.**

67. $(6a)^2$
68. $(3b)^3$
69. $(5y)^4$
70. $(4t)^4$
71. $(-2r^2s^3)^3$
72. $(-2x^2y^4)^5$
73. $\left(-\dfrac{1}{3}y^2z^4\right)^5$
74. $\left(-\dfrac{1}{4}t^3u^8\right)^2$

Use rules for exponents to simplify each expression. **See Example 10.**

75. $\dfrac{(ab^2)^3}{a^2b^2}$
76. $\dfrac{(m^3n^4)^3}{m^3n^6}$
77. $\dfrac{(r^4s^3)^4}{r^3s^9}$
78. $\dfrac{(x^2y^5)^5}{x^6y^2}$

Use rules for exponents to simplify each expression. **See Example 11.**

79. $\dfrac{(6k)^7}{(6k)^4}$
80. $\dfrac{(-3a)^{12}}{(-3a)^{10}}$
81. $\dfrac{(3q)^5}{(3q)^3}$
82. $\dfrac{(ab)^8}{(ab)^4}$

Use the power of a quotient rule for exponents to simplify each expression. **See Example 12.**

83. $\left(\dfrac{a}{b}\right)^3$
84. $\left(\dfrac{r}{s}\right)^4$
85. $\left(\dfrac{8a^2}{11b^5}\right)^2$
86. $\left(\dfrac{7g^4}{6h^3}\right)^2$

TRY IT YOURSELF

Simplify each expression, if possible.

87. $\left(\dfrac{x^2}{y^3}\right)^5$
88. $\left(\dfrac{u^4}{v^2}\right)^6$
89. $y^3y^2y^4$
90. y^4yy^6
91. $\dfrac{15^9}{15^6}$
92. $\dfrac{25^{13}}{25^7}$
93. $\dfrac{t^5t^6t}{t^2t^3}$
94. $\dfrac{m^5m^{12}m}{m^7m^4}$
95. $\dfrac{(k-2)^{15}}{(k-2)}$
96. $\dfrac{(m+8)^{20}}{(m+8)}$
97. $cd^4 \cdot cd$
98. $ab^3 \cdot ab^4$
99. $\left(\dfrac{y^3y^5}{yy^2}\right)^3$
100. $\left(\dfrac{s^5s^6}{s^2s^2}\right)^4$
101. $\dfrac{s^2s^2s^2}{s^3s}$
102. $\dfrac{w^4w^4w^4}{w^2w}$
103. $(-6a^3b^2)^3$
104. $(-10r^3s^2)^2$
105. $\left(\dfrac{3m^4}{2n^5}\right)^5$
106. $\left(\dfrac{2s^2}{3t^5}\right)^5$
107. $\dfrac{(a^2b^2)^{15}}{(ab)^9}$
108. $\dfrac{(s^3t^3)^4}{(st)^2}$
109. $(n^4n)^3(n^3)^6$
110. $(y^3y)^2(y^2)^2$
111. $\dfrac{(6h)^8}{(6h)^6}$
112. $\dfrac{(-7r)^{10}}{(-7r)^8}$
113. $\dfrac{x^4y^7}{xy^3}$
114. $\dfrac{p^7q^{10}}{p^2q^7}$
115. $\left(\dfrac{m}{3}\right)^4$
116. $\left(\dfrac{n}{5}\right)^3$

Look Alikes . . .

117. a. $a^3 \cdot a^3$ b. $(a^3)^3$ c. $a^3 + a^3$
118. a. $(m^5)^7$ b. $m^5 \cdot m^7$ c. $m^5 - m^7$
119. a. $b^3b^2b^4$ b. $(b^3b^2)^4$ c. $\dfrac{b^3b^2}{b^4}$
120. a. $(2n^4n)^5$ b. $2n^4n$ c. $\left(\dfrac{2n^4}{n}\right)^5$

APPLICATIONS

121. **Art History.** Leonardo da Vinci's drawing relating a human figure to a square and a circle is shown. Find an expression for the following:

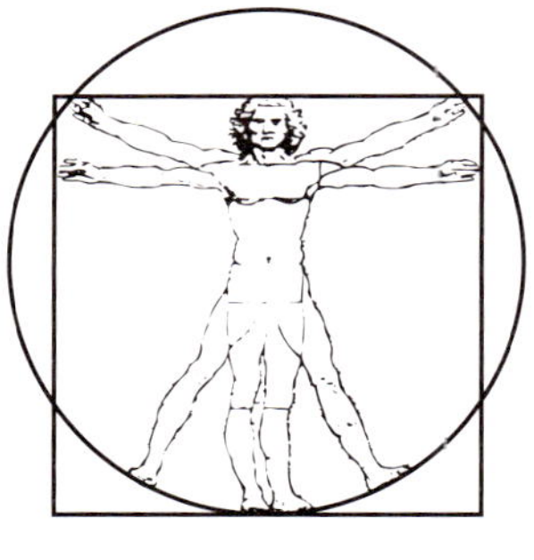

a. The area of the square if the man's height is $5x$ feet
b. The area of the circle if the waist-to-feet distance is $3a$ feet. Leave π in your answer.

122. **Packaging.** A bowling ball fits tightly against all sides of a cardboard box that it is packaged in. Find expressions for the volume of the ball and box. Leave π in your answer.

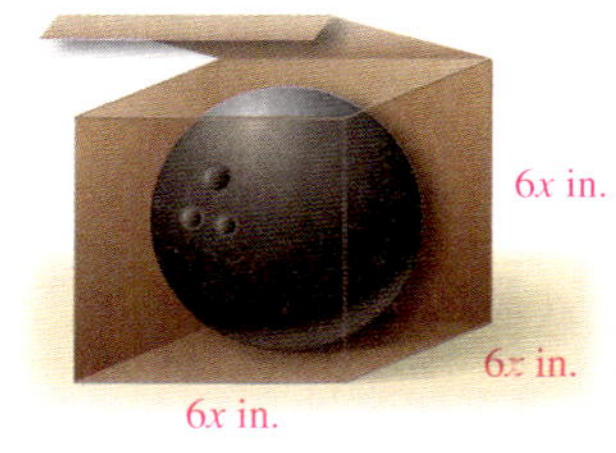

123. Childbirth. Mr. and Mrs. Emory Harrison, of Johnson City, Tennessee, had 13 sons in a row during the 1940s and 1950s. The **probability** of a family of 13 children all being male is $\left(\frac{1}{2}\right)^{13}$. Evaluate this expression.

© Bettmann/CORBIS

124. Toys. A Super Ball is dropped from a height of 1 foot and always rebounds to four-fifths of its previous height. The rebound height of the ball after the third bounce is $\left(\frac{4}{5}\right)^3$ feet. Evaluate this expression. Is the third bounce more or less than $\frac{1}{2}$ foot high?

WRITING

125. Explain the mistake in the following work.

a. $2^3 \cdot 2^2 = 4^5 = 1{,}024$ b. $(5d^2)^3 = 15d^6$

126. Explain why we can simplify $x^4 \cdot x^5$, but cannot simplify $x^4 + x^5$.

REVIEW

Match each equation with its graph below.

127. $y = 2x - 1$ **128.** $y = 3x - 1$

129. $y = 3$ **130.** $x = 3$

a.

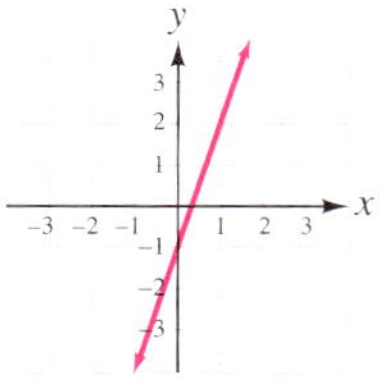

b.

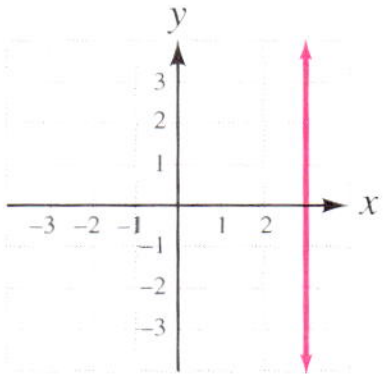

c.

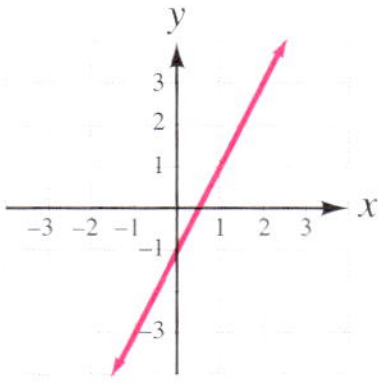

d.

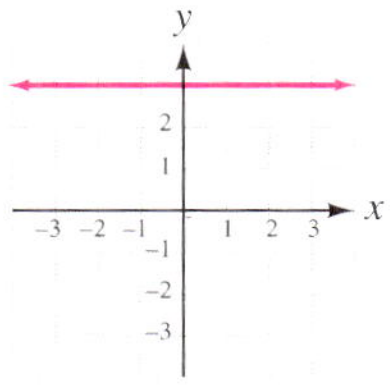

CHALLENGE PROBLEMS

131. Simplify each expression. The variables represent natural numbers.

a. $x^{2m}x^{3m}$ b. $(y^{5c})^4$

c. $\dfrac{m^{8x}}{m^{4x}}$ d. $(2a^{6y})^4$

132. Evaluate the following expression without using a calculator:

$$\frac{(108{,}642)^4}{(54{,}321)^4}$$

SECTION 5.2 Zero and Negative Exponents

OBJECTIVES

1. Use the zero exponent rule.
2. Use the negative integer exponent rule.
3. Use exponent rules to change negative exponents in fractions to positive exponents.
4. Use all exponent rules to simplify expressions.

ARE YOU READY?

The following problems review some basic skills that are needed when working with zero and negative exponents.

1. Simplify: $\dfrac{2 \cdot 2}{2 \cdot 2 \cdot 2}$

2. Simplify: $\dfrac{x \cdot x}{x \cdot x \cdot x \cdot x}$

3. Find the reciprocal: **a.** $\dfrac{1}{5}$ **b.** w

4. Identify the base and exponent: **a.** m^0 **b.** 7^{-2}

We now extend the discussion of natural-number exponents to include exponents that are zero and exponents that are negative integers.

1 Use the Zero Exponent Rule.

To develop the definition of a zero exponent, we will simplify the expression $\frac{5^3}{5^3}$ in two ways and compare the results. First, we apply the quotient rule for exponents, where we subtract the equal exponents in the numerator and denominator. The result is 5^0, which is read as "5 to the zero power." In the second approach, we write 5^3 as $5 \cdot 5 \cdot 5$ and remove the common factors of 5 in the numerator and denominator. The result is 1.

$$\frac{5^3}{5^3} = 5^{3-3} = 5^0 \qquad \frac{5^3}{5^3} = \frac{\overset{1}{\cancel{5}} \cdot \overset{1}{\cancel{5}} \cdot \overset{1}{\cancel{5}}}{\underset{1}{\cancel{5}} \cdot \underset{1}{\cancel{5}} \cdot \underset{1}{\cancel{5}}} = 1$$

These results must be equal.

Since $\frac{5^3}{5^3} = 5^0$ and $\frac{5^3}{5^3} = 1$, we conclude that $5^0 = 1$. This observation illustrates the following definition.

Zero Exponents

Any nonzero base raised to the 0 power is 1.

For any nonzero real number x,

$x^0 = 1$ Read as "x to the zero power equals 1."

EXAMPLE 1 Simplify. Assume $a \neq 0$: **a.** $(-8)^0$ **b.** $\left(\frac{14}{15}\right)^0$ **c.** $(3a)^0$ **d.** $3a^0$

Strategy We note that each exponent is 0. To simplify the expressions, we will identify the base and use the zero-exponent rule.

Why If an expression contains a nonzero base raised to the 0 power, we can replace it with 1.

Solution

a. $(-8)^0 = 1$ Because the base is −8 and the exponent is 0.

b. $\left(\frac{14}{15}\right)^0 = 1$ Because the base is $\frac{14}{15}$ and the exponent is 0.

c. $(3a)^0 = 1$ Because of the parentheses, the base is 3a. The exponent is 0.

d. $3a^0 = 3 \cdot a^0$ Read as "3 times a to the zero power." Since there are no parentheses, the base is a, not 3a. The exponent is 0.

$= 3 \cdot 1$ Simplify: $a^0 = 1$.

$= 3$

The Language of Algebra

The zero exponent definition does not define 0^0. This expression is called an **indeterminate form,** which is beyond the scope of this book.

Self Check 1 Simplify each expression: **a.** $(0.75)^0$ **b.** $-5c^0d$ **c.** $(5c)^0$

Now Try Problems 13 and 17

2 Use the Negative Integer Exponent Rule.

The Language of Algebra

The **negative integers** are: $-1, -2, -3, -4, -5, \ldots$

To develop the definition of a negative exponent, we will simplify $\frac{6^2}{6^5}$ in two ways and compare the results.

If we apply the quotient rule for exponents, where we subtract the greater exponent in the denominator from the lesser exponent in the numerator, we get 6^{-3}. In the second approach, we remove the two common factors of 6 to get $\frac{1}{6^3}$.

$$\frac{6^2}{6^5} = 6^{2-5} = 6^{-3} \qquad \frac{6^2}{6^5} = \frac{\overset{1}{\cancel{6}} \cdot \overset{1}{\cancel{6}}}{\underset{1}{\cancel{6}} \cdot \underset{1}{\cancel{6}} \cdot 6 \cdot 6 \cdot 6} = \frac{1}{6^3}$$

These must be equal.

Since $\frac{6^2}{6^5} = 6^{-3}$ and $\frac{6^2}{6^5} = \frac{1}{6^3}$, we conclude that $6^{-3} = \frac{1}{6^3}$. Note that 6^{-3} is equal to the reciprocal of 6^3. This observation illustrates the following definition.

Negative Exponents

For any nonzero real number x and any integer n,

$$x^{-n} = \frac{1}{x^n}$$ Read as "x to the negative nth power equals 1 over x to the nth power."

In words, x^{-n} is the reciprocal of x^n.

From the definition, we see that another way to write x^{-n} is to write its reciprocal and change the sign of the exponent. For example,

$$5^{-4} = \frac{1}{5^4}$$ Think of the reciprocal of 5^{-4}. Then change the sign of the exponent.

The definitions of a zero exponent and a negative integer exponent have been written in such a way as to remain consistent with the definition for a natural-number exponent that we learned earlier. This can be seen in the following list.

$2^4 = 16$
Divide by 2.
$2^3 = 8$
Divide by 2.
$2^2 = 4$
Divide by 2.
$2^1 = 2$

For natural number exponents, each time that we decrease the exponent by 1, the value of the exponential expression is divided by 2.

$2^0 = 1$ For the pattern to continue, we define $2^0 = 1$.

$2^{-1} = \frac{1}{2}$ For the pattern to continue, we define $2^{-1} = \frac{1}{2^1} = \frac{1}{2}$.

$2^{-2} = \frac{1}{4}$ For the pattern to continue, we define $2^{-2} = \frac{1}{2^2} = \frac{1}{4}$.

$2^{-3} = \frac{1}{8}$ For the pattern to continue, we define $2^{-3} = \frac{1}{2^3} = \frac{1}{8}$.

EXAMPLE 2 Express using positive exponents and simplify, if possible: **a.** 3^{-2} **b.** y^{-1} **c.** $(-2)^{-3}$ **d.** $5^{-2} - 10^{-2}$

Strategy Since each exponent is a negative number, we will use the negative exponent rule.

Why This rule enables us to write an exponential expression that has a negative exponent in an equivalent form using a positive exponent.

Solution

a. $3^{-2} = \frac{1}{3^2} = \frac{1}{9}$ Read as "3 to the negative second power." Write the reciprocal of 3^{-2} and change the sign of the exponent.

b. $y^{-1} = \frac{1}{y^1} = \frac{1}{y}$ Read as "y to the negative first power." Write the reciprocal of y^{-1} and change the sign of the exponent.

c. $(-2)^{-3} = \frac{1}{(-2)^3}$ Read as "−2 to the negative third power." Because of the parentheses, the base is −2. Write the reciprocal of $(-2)^{-3}$ and change the exponent.

$= -\frac{1}{8}$ Evaluate: $(-2)^3 = -8$. Write the − symbol in front of the fraction.

Caution

A negative exponent **does not,** itself, make the simplified expression negative. It indicates a reciprocal. Avoid these common mistakes:

~~$3^{-2} = -9$~~ ~~$3^{-2} = -\frac{1}{9}$~~

d. $5^{-2} - 10^{-2} = \frac{1}{5^2} - \frac{1}{10^2}$ Write the reciprocal of 5^{-2} and 10^{-2} and change the sign of each exponent.

$= \frac{1}{25} - \frac{1}{100}$ Evaluate: $5^2 = 25$ and $10^2 = 100$.

$= \frac{4}{100} - \frac{1}{100}$ Build $\frac{1}{25}$ to have a denominator of 100 so that the fractions can be subtracted: $\frac{1}{25} \cdot \frac{4}{4} = \frac{4}{100}$.

$= \frac{3}{100}$

Self Check 2 Express using positive exponents and simplify, if possible:
a. 8^{-2} **b.** x^{-5} **c.** $(-3)^{-3}$ **d.** $2^{-3} - 4^{-2}$

Now Try Problems 21, 25, and 29

EXAMPLE 3 Simplify. Do not use negative exponents in the answer. **a.** $9m^{-3}$ **b.** -5^{-2}

Strategy We note that each exponent is a negative number. We will identify the base for each negative exponent and then use the negative exponent rule.

Why This rule enables us to write an exponential expression that has a negative exponent in an equivalent form using a positive exponent.

Solution

a. $9m^{-3} = 9 \cdot m^{-3}$ Read as "9 times *m* to the negative third power." The base is *m*. The exponent is −3.

$= 9 \cdot \frac{1}{m^3}$ Write the reciprocal of m^{-3} and change the sign of the exponent. Since 9 is not part of the base, it is not part of the reciprocal.

$= \frac{9}{m^3}$ Multiply.

b. $-5^{-2} = -1 \cdot 5^{-2}$ Read as "the opposite of 5 to the negative second power." The base is 5. The exponent is −2.

$= -1 \cdot \frac{1}{5^2}$ Write the reciprocal of 5^{-2} and change the sign of the exponent. Since −1 is not part of the base, it is not part of the reciprocal.

$= -\frac{1}{25}$ Evaluate 5^2 and multiply. The result is negative because of the − sign in front of -5^{-2}.

Self Check 3 Simplify. Do not use negative exponents in the answer. **a.** $12h^{-9}$
b. -2^{-4}

Now Try Problems 33 and 37

3 Use Exponent Rules to Change Negative Exponents in Fractions to Positive Exponents.

Negative exponents can appear in the numerator and/or the denominator of a fraction. To develop rules for such situations, we consider the following example.

$$\frac{a^{-4}}{b^{-3}} = \frac{\frac{1}{a^4}}{\frac{1}{b^3}} = \frac{1}{a^4} \div \frac{1}{b^3} = \frac{1}{a^4} \cdot \frac{b^3}{1} = \frac{b^3}{a^4}$$

Read $\frac{a^{-4}}{b^{-3}}$ as "*a* to the negative fourth power over (or divided by) *b* to the negative third power."

We can obtain this result in a simpler way. In $\frac{a^{-4}}{b^{-3}}$, we can move a^{-4} from the numerator to the denominator and change the sign of the exponent, and we can move b^{-3} from the denominator to the numerator and change the sign of the exponent.

The Language of Algebra

Factors of a numerator or denominator may be moved **across the fraction bar** if we change the sign of their exponent.

$$\frac{a^{-4}}{b^{-3}} = \frac{b^3}{a^4}$$

This example illustrates the following rules.

Changing from Negative to Positive Exponents

A factor can be moved from the denominator to the numerator or from the numerator to the denominator of a fraction if the sign of its exponent is changed.

For any nonzero real numbers x and y, and any integers m and n,

$$\frac{1}{x^{-n}} = x^n \quad \text{and} \quad \frac{x^{-m}}{y^{-n}} = \frac{y^n}{x^m}$$

These rules streamline the process when simplifying fractions involving negative exponents.

EXAMPLE 4 Simplify. Do not use negative exponents in the answer. **a.** $\frac{1}{d^{-10}}$ **b.** $\frac{2^{-3}}{3^{-4}}$ **c.** $\frac{-6s^{-2}}{t^{-9}}$

Strategy We will move any factors in the numerator that have a negative exponent to the denominator. Then we will move any factors in the denominator that have a negative exponent to the numerator.

Why In this process, the sign of a negative exponent changes to positive.

Solution

a. $\frac{1}{d^{-10}} = d^{10}$ Read as "1 over d to the negative tenth power." Move d^{-10} to the numerator and change the sign of the exponent.

b. $\frac{2^{-3}}{3^{-4}} = \frac{3^4}{2^3}$ Move 2^{-3} to the denominator and change the sign of the exponent. Move 3^{-4} to the numerator and change the sign of the exponent.

$= \frac{81}{8}$ Evaluate: $3^4 = 81$ and $2^3 = 8$.

c. $\frac{-6s^{-2}}{t^{-9}} = \frac{-6t^9}{s^2}$ Since $-6s^{-2}$ has no parentheses, s is the base. Move only s^{-2} to the denominator and change the sign of the exponent. Do not move -6. Move t^{-9} to the numerator and change the sign of the exponent.

Caution

A common error is to mistake the $-$ sign in -6 for a negative exponent. *It is not an exponent.* The -6 should not be moved to the denominator and have its sign changed.

$$\frac{-6s^{-2}}{t^{-9}} \neq \frac{t^9}{6s^2}$$

Self Check 4 Simplify. Do not use negative exponents in the answer.

a. $\frac{1}{w^{-5}}$ **b.** $\frac{5^{-2}}{4^{-3}}$ **c.** $\frac{-8h^{-6}}{a^{-7}}$

Now Try Problems 41, 45, and 47

When a fraction is raised to a negative power, we can use rules for exponents to change the sign of the exponent. For example, we see that

$$\left(\frac{a}{2}\right)^{-3} = \frac{a^{-3}}{2^{-3}} = \frac{2^3}{a^3} = \left(\frac{2}{a}\right)^3$$

The exponent is the opposite of -3.

The base is the reciprocal of $\frac{a}{2}$.

This process can be streamlined using the following rule.

Negative Exponents and Reciprocals

A fraction raised to a power is equal to the reciprocal of the fraction raised to the opposite power.

For any nonzero real numbers x and y, and any integer n,

$$\left(\frac{x}{y}\right)^{-n} = \left(\frac{y}{x}\right)^{n}$$

EXAMPLE 5 Simplify: $\left(\frac{4}{m}\right)^{-2}$

Strategy We want to write this fraction that is raised to a negative power in an equivalent form that involves a positive power. We will use the negative exponent and reciprocal rules to do this.

Why It is usually easier to simplify exponential expressions if the exponents are positive.

Solution

$$\left(\frac{4}{m}\right)^{-2} = \left(\frac{m}{4}\right)^{2}$$ Read as "the quantity of 4 over m, raised to the negative second power." The base is the fraction $\frac{4}{m}$ and the exponent is -2. Write the reciprocal of the base and change the sign of the exponent.

$$= \frac{m^2}{4^2}$$ Use the power of a quotient rule: Raise the numerator m and denominator 4 to the second power.

$$= \frac{m^2}{16}$$ Evaluate: $4^2 = 16$.

Self Check 5 Simplify: $\left(\frac{c}{9}\right)^{-2}$

Now Try Problem 53

4 Use All Exponent Rules to Simplify Expressions.

The rules for exponents involving products, powers, and quotients are also true for zero and negative exponents.

Summary of Exponent Rules

If m and n represent integers and there are no divisions by zero, then

Product rule

$x^m \cdot x^n = x^{m+n}$

Power rule

$(x^m)^n = x^{mn}$

Power of a product

$(xy)^n = x^n y^n$

Quotient rule

$\frac{x^m}{x^n} = x^{m-n}$

Power of a quotient

$\left(\frac{x}{y}\right)^n = \frac{x^n}{y^n}$

Exponents of 0 and 1

$x^0 = 1$ and $x^1 = x$

Negative exponent

$x^{-n} = \frac{1}{x^n}$

Negative exponents appearing in fractions

$\frac{1}{x^{-n}} = x^n$ $\quad \frac{x^{-m}}{y^{-n}} = \frac{y^n}{x^m}$ $\quad \left(\frac{x}{y}\right)^{-n} = \left(\frac{y}{x}\right)^n$

The rules for exponents are used to simplify expressions involving products, quotients, and powers. In general, an expression involving exponents is simplified when

- Each base occurs only once
- No powers are raised to powers
- There are no parentheses
- There are no negative or zero exponents

EXAMPLE 6 Simplify. Do not use negative exponents in the answer. **a.** $x^5 \cdot x^{-3}$ **b.** $\frac{x^3}{x^7}$ **c.** $(x^3)^{-2}$ **d.** $(2a^3b^{-5})^3$ **e.** $\left(\frac{3}{b^5}\right)^{-4}$

Strategy In each case, we want to write an equivalent expression such that each base is used only once and is raised to a positive power. We will use rules for exponents to do this.

Why These expressions are not in simplest form. In parts a and b, the base x occurs more than once. In parts a, c, d, and e, there is a negative exponent.

Solution

a. $x^5 \cdot x^{-3} = x^{5+(-3)} = x^2$ Use the product rule: Keep the base, x, and add exponents.

b. $\frac{x^3}{x^7} = x^{3-7}$ Use the quotient rule: Keep the base, x, and subtract the exponents.

$= x^{-4}$ Do the subtraction: $3 - 7 = -4$.

$= \frac{1}{x^4}$ Write the reciprocal of x^{-4} and change the sign of the exponent.

c. $(x^3)^{-2} = x^{-6}$ Use the power rule: Keep the base, x, and multiply exponents.

$= \frac{1}{x^6}$ Write the reciprocal of x^{-6} and change the sign of the exponent.

Success Tip

We can use the negative exponent rule to simplify $(x^3)^{-2}$ in an alternate way:

$$(x^3)^{-2} = \frac{1}{(x^3)^2} = \frac{1}{x^6}$$

d. $(2a^3b^{-5})^3 = 2^3(a^3)^3(b^{-5})^3$ Raise each factor of the product $2a^3b^{-5}$ to the 3rd power.

$= 8a^9b^{-15}$ Use the power rule: Multiply exponents.

$= \frac{8a^9}{b^{15}}$ Move b^{-15} to the denominator and change the sign of the exponent.

e. $\left(\frac{3}{b^5}\right)^{-4} = \left(\frac{b^5}{3}\right)^4$ The base is the fraction $\frac{3}{b^5}$ and its exponent is -4. Write the reciprocal of the base and change the sign of the exponent.

$= \frac{(b^5)^4}{(3)^4}$ Use the power of a quotient rule: Raise the numerator and denominator to the 4th power.

$= \frac{b^{20}}{81}$ Use the power rule: Keep the base, b, and multiply the exponents 5 and 4. Evaluate: $3^4 = 81$.

Self Check 6 Simplify. Do not use negative exponents in the answer.

a. $t^8 \cdot t^{-4}$ **b.** $\frac{a^3}{a^8}$ **c.** $(n^4)^{-5}$

d. $(4c^2d^{-1})^3$ **e.** $\left(\frac{c^4}{2}\right)^{-3}$

Now Try Problems 57, 61, 65, 69, and 71

EXAMPLE 7 Simplify. Do not use negative exponents in the answer.

a. $\dfrac{y^{-4}y^{-3}}{y^{-20}}$ **b.** $\dfrac{7^{-1}a^3b^4}{6^{-2}a^5b^2}$ **c.** $\left(\dfrac{x^{-3}y^2}{xy^{-3}}\right)^2$

Strategy In each case, we want to write an equivalent expression such that each base is used only once and is raised to a positive power.

Why These expressions are not in simplest form. The bases occur more than once and the expressions contain negative exponents.

Solution

a. $\dfrac{y^{-4}y^{-3}}{y^{-20}} = \dfrac{y^{-7}}{y^{-20}}$ In the numerator, use the product rule: Keep the common base, y, and add exponents: $-4 + (-3) = -7$.

$= y^{-7-(-20)}$ Use the quotient rule: Keep the common base, y, and subtract exponents.

$= y^{13}$ Do the subtraction: $-7 - (-20) = -7 + 20 = 13$.

Alternate solution: To avoid working with negative numbers, we could move each factor across the fraction bar and change the sign of its exponent.

$$\frac{y^{-4}y^{-3}}{y^{-20}} = \frac{y^{20}}{y^4y^3} = \frac{y^{20}}{y^7} = y^{13}$$

Caution

We cannot use the alternate approach if the negative exponents occur in a sum or difference of terms. For example:

$$\frac{y^{-4}+y^{-3}}{y^{-20}} \neq \frac{y^{20}}{y^4+y^3}$$

You will study this situation in more detail in your next algebra course.

b. $\dfrac{7^{-1}a^3b^4}{6^{-2}a^5b^2} = \dfrac{6^2a^3b^4}{7^1a^5b^2}$ Move 7^{-1} to the denominator. Change the sign of the exponent. Move 6^{-2} to the numerator. Change the sign of the exponent.

$= \dfrac{36a^{3-5}b^{4-2}}{7}$ Use the quotient rule twice: Keep each base, a and b, and subtract exponents.

$= \dfrac{36a^{-2}b^2}{7}$ Do the subtractions: $3 - 5 = -2$ and $4 - 2 = 2$.

$= \dfrac{36b^2}{7a^2}$ Move a^{-2} to the denominator and change the sign of the exponent.

c. $\left(\dfrac{x^{-3}y^2}{xy^{-3}}\right)^2 = [x^{-3-1}y^{2-(-3)}]^2$ Within the parentheses, use the quotient rule twice: Keep each base, x and y, and subtract exponents.

$= (x^{-4}y^5)^2$ Do the subtractions: $-3 - 1 = -4$ and $2 - (-3) = 5$.

$= x^{-8}y^{10}$ Raise each factor within the parentheses to the second power.

$= \dfrac{y^{10}}{x^8}$ Move x^{-8} to the denominator and change the sign of its exponent. y^{10} does not move.

Alternate solution: To simplify the expression, we can begin on the "outside" by using the power of a quotient rule first.

$$\left(\frac{x^{-3}y^2}{xy^{-3}}\right)^2 = \frac{(x^{-3})^2(y^2)^2}{x^2(y^{-3})^2} = \frac{x^{-6}y^4}{x^2y^{-6}} = \frac{y^4y^6}{x^6x^2} = \frac{y^{10}}{x^8}$$

Success Tip

Since the rules for exponents can be applied in different orders, there are several equally valid ways to simplify these expressions.

Self Check 7 Simplify. Do not use negative exponents in the answer.

a. $\dfrac{a^{-4}a^{-5}}{a^{-3}}$ **b.** $\dfrac{1^{-4}x^5y^3}{9^{-2}x^3y^6}$ **c.** $\left(\dfrac{c^{-2}d^2}{c^4d^{-3}}\right)^3$

Now Try Problems 73, 77, and 79

SECTION 5.2 STUDY SET

VOCABULARY

Fill in the blanks.

1. In the expression 5^{-1}, the exponent is a ________ integer.
2. x^{-n} is the __________ of x^{n}.
3. We read a^0 as "a to the _____ power."
4. We read 3^{-4} as "3 to the _________ _______ power."

CONCEPTS

5. Complete the table.

Expression	Base	Exponent
4^{-2}		
$6x^{-5}$		
$\left(\frac{3}{y}\right)^{-8}$		
-7^{-1}		
$(-2)^{-3}$		
$10a^0$		

6. Complete each rule for exponents.
   a. $x^m \cdot x^n =$
   b. $x^0 =$
   c. $(x^m)^n =$
   d. $(xy)^n =$
   e. $\left(\frac{x}{y}\right)^n =$
   f. $x^{-n} =$
   g. $\frac{1}{x^{-n}} =$
   h. $\frac{x^{-m}}{y^{-n}} =$
   i. $\frac{x^m}{x^n} =$
   j. $\left(\frac{x}{y}\right)^{-n} =$

Complete each table.

7.

x	3^x
2	
1	
0	
−1	
−2	

8.

x	$(-9)^x$
2	
1	
0	
−1	
−2	

9. Fill in the blanks.
   a. $2^{-3} = \frac{1}{2}$
   b. $\frac{1}{t^{-6}} = t$
   c. A factor can be moved from the denominator to the numerator or from the numerator to the denominator of a fraction if the _____ of its exponent is changed.
   $\frac{5^{-2}}{6^{-3}} = \frac{6}{5}$
   d. A fraction raised to a power is equal to the __________ of the fraction raised to the opposite power.
   $\left(\frac{3}{d}\right)^{-2} = \left(\frac{d}{3}\right)$

10. Determine whether each statement is true or false.
    a. $6^{-2} = -36$
    b. $6^{-2} = \frac{1}{36}$
    c. $\frac{x^3}{y^{-2}} = \frac{y^2}{x^3}$
    d. $\frac{-6x^{-5}}{y^{-6}} = \frac{y^6}{6x^5}$

NOTATION

Complete each solution to simplify each expression.

11. $(y^5y^3)^{-5} = (\quad)^{-5} = y \quad = \frac{1}{y}$

12. $\left(\frac{a^2b}{a^{-3}b^3}\right)^3 = \left(a^{2-} \quad b^{1-} \quad\right)^3$
    $= (a \quad b \quad)^3$
    $= a \quad b$
    $= \frac{a^{15}}{b}$

GUIDED PRACTICE

Simplify each expression. See Example 1.

13. 7^0
14. 9^0
15. $\left(\frac{1}{4}\right)^0$
16. $\left(\frac{3}{8}\right)^0$
17. $2x^0$
18. $8t^0$
19. $\frac{5}{2x^0}$
20. $\frac{4}{3a^0}$

Express using positive exponents and simplify, if possible. See Example 2.

21. 2^{-2}
22. 7^{-2}
23. 6^{-1}
24. 5^{-1}
25. b^{-5}
26. c^{-4}
27. $(-5)^{-1}$
28. $(-8)^{-1}$
29. $2^{-2} + 4^{-1}$
30. $-9^{-1} + 9^{-2}$
31. $9^0 - 9^{-1}$
32. $7^{-1} - 7^0$

Simplify. Do not use negative exponents in the answer. See Example 3.

33. $15g^{-6}$
34. $16t^{-3}$
35. $5x^{-3}$
36. $27m^{-3}$
37. -3^{-3}
38. -6^{-3}
39. -8^{-2}
40. -4^{-2}

Simplify. Do not use negative exponents in the answer. See Example 4.

41. $\frac{1}{5^{-3}}$
42. $\frac{1}{3^{-3}}$
43. $\frac{8}{s^{-1}}$
44. $\frac{6}{k^{-2}}$

45. $\dfrac{2^{-4}}{3^{-1}}$

46. $\dfrac{7^{-2}}{2^{-3}}$

47. $\dfrac{-4d^{-1}}{p^{-10}}$

48. $\dfrac{-9m^{-1}}{n^{-30}}$

Simplify. See Example 5.

49. $\left(\dfrac{1}{6}\right)^{-2}$

50. $\left(\dfrac{1}{7}\right)^{-2}$

51. $\left(\dfrac{1}{2}\right)^{-3}$

52. $\left(\dfrac{1}{5}\right)^{-3}$

53. $\left(\dfrac{c}{d}\right)^{-8}$

54. $\left(\dfrac{a}{x}\right)^{-10}$

55. $\left(\dfrac{3}{m}\right)^{-4}$

56. $\left(\dfrac{2}{t}\right)^{-4}$

Simplify. Do not use negative exponents in the answer. See Example 6.

57. $y^{8} \cdot y^{-2}$

58. $m^{10} \cdot m^{-6}$

59. $b^{-7} \cdot b^{14}$

60. $c^{-9} \cdot c^{14}$

61. $\dfrac{y^4}{y^5}$

62. $\dfrac{t^7}{t^{10}}$

63. $\dfrac{h^{-5}}{h^2}$

64. $\dfrac{y^{-3}}{y^4}$

65. $(x^4)^{-3}$

66. $(y^{-3})^2$

67. $(b^2)^{-4}$

68. $(n^3)^{-5}$

69. $(6s^4t^{-7})^2$

70. $(11r^{10}s^{-3})^2$

71. $\left(\dfrac{4}{x^3}\right)^{-3}$

72. $\left(\dfrac{2}{b^5}\right)^{-2}$

Simplify. Do not use negative exponents in the answer. See Example 7.

73. $\dfrac{y^{-3}}{y^{-4}y^{-2}}$

74. $\dfrac{x^{-12}}{x^{-3}x^{-4}}$

75. $\dfrac{a^{-5}a^{-9}}{a^{-8}}$

76. $\dfrac{b^{-2}b^{-3}}{b^{-9}}$

77. $\dfrac{2^{-1}a^4b^2}{3^{-2}a^2b^4}$

78. $\dfrac{6^{-2}b^9c^3}{5^{-3}b^4c^8}$

79. $\left(\dfrac{y^3z^{-2}}{y^{-4}z^3}\right)^2$

80. $\left(\dfrac{xy^3}{x^{-1}y^{-1}}\right)^3$

TRY IT YOURSELF

Simplify. Do not use negative exponents in the answer. Assume that no variables are 0.

81. $\left(\dfrac{a^4}{2b}\right)^{-3}$

82. $\left(\dfrac{n^8}{9m}\right)^{-2}$

83. $\dfrac{r^{-50}}{r^{-70}}$

84. $\dfrac{m^{-30}}{m^{-40}}$

85. $\dfrac{a^{-5}}{b^{-2}}$

86. $\dfrac{r^{-6}}{s^{-1}}$

87. $(-10)^{-3}$

88. $(-9)^{-2}$

89. $\left(\dfrac{a^2b^3}{ab^4}\right)^0$

90. $\left(\dfrac{xyz}{x^2y}\right)^0$

91. $\dfrac{9^{-2}s^6t}{4^{-3}s^4t^5}$

92. $\dfrac{2^{-5}m^{10}n^6}{5^{-2}m^6n^{10}}$

93. $(2u^{-2}v^5)^5$

94. $(3w^{-8}x^3)^4$

95. $\left(\dfrac{y^4}{3}\right)^{-2}$

96. $\left(\dfrac{p^3}{2}\right)^{-2}$

97. $-15x^0y$

98. $24g^0h^2$

99. $\left(\dfrac{4}{h^{10}}\right)^{-2}$

100. $\left(\dfrac{x^4}{3}\right)^{-4}$

101. $x^{-3} \cdot x^{-3}$

102. $y^{-2} \cdot y^{-2}$

103. $\left(\dfrac{c^3d^{-4}}{c^{-1}d^5}\right)^3$

104. $\left(\dfrac{s^2t^{-8}}{s^{-9}t^2}\right)^4$

105. $15(-6x)^0$

106. $4(-12y)^0$

107. $\dfrac{2^{-2}g^{-2}h^{-3}}{9^{-1}h^{-3}}$

108. $\dfrac{5^{-1}x^{-2}y^{-3}}{8^{-2}x^{-11}}$

109. $(5d^{-2})^3$

110. $(9s^{-6})^2$

111. $\left(\dfrac{x^2y^{-2}}{x^{-5}y^3}\right)^4$

112. $\left(\dfrac{r^4s^{-3}}{r^{-3}s^7}\right)^3$

113. $(2x^3y^{-2})^5$

114. $(3u^{-2}v^3)^3$

115. $\dfrac{t(t^{-2})^{-2}}{t^{-5}}$

116. $\dfrac{d(d^{-3})^{-3}}{d^{-7}}$

117. $\dfrac{-4s^{-5}}{t^{-2}}$

118. $\dfrac{-9k^{-8}}{m^{-2}}$

119. $(x^{-4}x^3)^3$

120. $(y^{-2}y)^3$

Look Alikes . . .

121. a. 8^{-1} b. -8^{-1} c. $(-8)^{-1}$ d. $-(-8)^{-1}$

122. a. 9^{-2} b. -9^{-2} c. $(-9)^{-2}$ d. $-(-9)^{-2}$

123. a. $4xy^{-2}$ b. $(4xy)^{-2}$ c. $4x^{-2}y$ d. $4^{-2}xy$

124. a. $\left(\dfrac{r}{s}\right)^{-1}$ b. $\dfrac{r^{-1}}{s}$ c. $\dfrac{r}{s^{-1}}$ d. $\dfrac{r^{-1}}{s^{-1}}$

APPLICATIONS

125. from **Campus to Careers**

Sound Engineering Technician

©iStockphoto.com/beetle8

The faintest sound that the typical human ear can detect (called the *threshold of hearing*) has an intensity of 10^{-12} units. A sound intensity of 10^4 units will cause an instant rupture of the eardrum. Consider the types of sounds in the table below. In the intensity column, write the most appropriate exponential expression from the following list. Each expression is used only once. (Source: physicsclassroom.com)

10^{-10} 10^{-6} 10^{-4} 10^{-1} 10^{2}

Lowest intensity **Greatest intensity**

Type of sound	Intensity
Front row rock concert	
Normal conversation	
Vacuum cleaner	
Military jet takeoff	
Whisper	

126. **Electronics.** The total resistance R of a certain circuit is given by

$$R = \left(\frac{1}{R_1} + \frac{1}{R_2}\right)^{-1} + R_3$$

Find R if $R_1 = 4$, $R_2 = 2$, and $R_3 = 1$.

WRITING

127. Explain how you would help a friend understand that 2^{-3} is not equal to -8.

128. Explain each error.

a. $\frac{-5x^{-2}}{y^{-2}} = \frac{y^2}{5x^2}$

b. $4^{-2} = -\frac{1}{16}$

REVIEW

Find the slope of the line that passes through the given points.

129. $(1, -4)$ and $(3, -7)$

130. $(1, 3)$ and $(3, -1)$

131. Write an equation of the line having slope $\frac{3}{4}$ and y-intercept -5.

132. Find an equation of the line that passes through $(4, 4)$ and $(-6, -6)$. Write the answer in slope–intercept form.

CHALLENGE PROBLEMS

133. Simplify each expression. Do not use negative exponents in the answer. The variable m represents a positive integer.

a. $r^{5m}r^{-6m}$

b. $\frac{x^{3m}}{x^{6m}}$

134. Write an expression equivalent to $\left(\frac{2x^3y^7}{3z^5}\right)^9$ that involves only *negative* exponents.

SECTION 5.3 Scientific Notation

OBJECTIVES

1. Convert from scientific to standard notation.
2. Write numbers in scientific notation.
3. Perform calculations with scientific notation.

ARE YOU READY?

The following problems review some basic skills that are needed when working with scientific notation.

1. Evaluate: 10^2
2. Multiply: $1{,}000 \cdot 4.528$
3. Evaluate: 10^{-1}
4. Multiply: $0.01 \cdot 6.22$

Scientists often deal with extremely large and small numbers. For example, the distance from the Earth to the sun is approximately 150,000,000 kilometers. The influenza virus, which causes flu symptoms of cough, sore throat, and headache, has a diameter of 0.00000256 inch.

©Sam D Cruz/Shutterstock.com

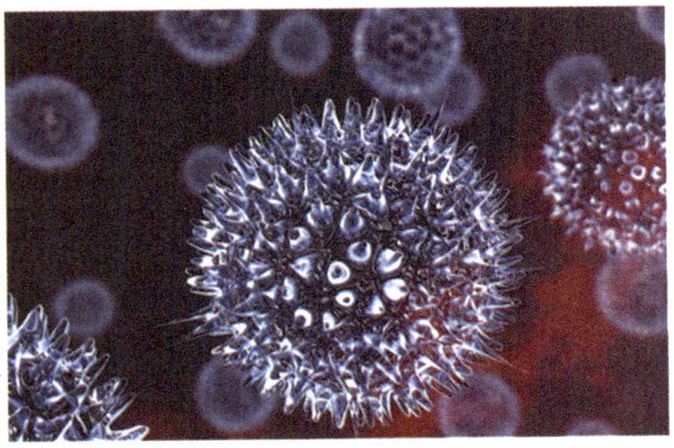

©iStockphoto.com/Björn Meyer

The numbers 150,000,000 and 0.00000256 are written in **standard notation,** which is also called **decimal notation.** Because they contain many zeros, they are difficult to read and cumbersome to work with in calculations. In this section, we will discuss a more convenient form in which we can write such numbers.

1 Convert from Scientific to Standard Notation.

Scientific notation provides a compact way of writing very large or very small numbers.

Scientific Notation

A positive number is written in **scientific notation** when it is written in the form $N \times 10^n$, where $1 \le N < 10$ and n is an integer.

To write numbers in scientific notation, you need to be familiar with **powers of 10,** like those listed in the table below.

Power of 10	10^4	10^3	10^2	10^1	10^0	10^{-1}	10^{-2}	10^{-3}	10^{-4}
Value	10,000	1,000	100	10	1	$\frac{1}{10} = 0.1$	$\frac{1}{100} = 0.01$	$\frac{1}{1,000} = 0.001$	$\frac{1}{10,000} = 0.0001$

Three examples of numbers written in scientific notation are shown below. Note that each of them is the product of a decimal number (between 1 and 10) and a power of 10.

An integer exponent

$3.67 \times 10^2 \qquad 2.158 \times 10^{-3} \qquad 4.0 \times 10^{57}$

A decimal that is at least 1, but less than 10

Notation

The × sign for multiplication is usually used to write scientific notation. However, some books use a raised dot · instead.

A number written in scientific notation can be converted to standard notation by performing the indicated multiplication. For example, to convert 3.67×10^2, we recall that multiplying a decimal by 100 moves the decimal point 2 places to the right.

$$3.67 \times \mathbf{10^2} = 3.67 \times \mathbf{100} = 367$$

To convert 2.158×10^{-3} to standard notation, we recall that dividing a decimal by 1,000 moves the decimal point 3 places to the left.

$$2.158 \times \mathbf{10^{-3}} = 2.158 \times \frac{\mathbf{1}}{\mathbf{10^3}} = 2.158 \times \frac{\mathbf{1}}{\mathbf{1,000}} = \frac{2.158}{\mathbf{1,000}} = 0.002158$$

In 3.67×10^2 and 2.158×10^{-3}, the exponent gives the number of decimal places that the decimal point moves, and the sign of the exponent indicates the direction in which it moves. Applying this observation to several other examples, we have

$5.32 \times 10^6 = 5320000.$ Move the decimal point 6 places to the right.

$1.95 \times 10^{-5} = 0.0000195$ Move the decimal point $|-5| = 5$ places to the left.

$9.7 \times 10^0 = 9.7$ There is no movement of the decimal point.

Success Tip

Since $10^0 = 1$, scientific notation involving 10^0 is easily simplified. For example,

$$9.7 \times \mathbf{10^0} = 9.7 \times \mathbf{1} = 9.7$$

The following procedure summarizes our observations.

Converting from Scientific to Standard Notation

1. If the exponent is positive, move the decimal point the same number of places to the right as the exponent.
2. If the exponent is negative, move the decimal point the same number of places to the left as the absolute value of the exponent.

EXAMPLE 1 Convert to standard notation: **a.** 3.467×10^5 **b.** 8.9×10^{-4}

Strategy In each case, we need to identify the exponent on the power of 10 and consider its sign.

Why The exponent gives the number of decimal places that we should move the decimal point. The sign of the exponent indicates whether it should be moved to the right or the left.

Solution

The Language of Algebra

Standard notation is also called **decimal notation.**

a. Since the exponent in 10^5 is 5, the decimal point moves 5 places to the right.

3 4 6 7 0 0. To move 5 places to the right, two placeholder zeros must be written.

Thus, $3.467 \times 10^5 = 346{,}700$.

b. Since the exponent in 10^{-4} is -4, the decimal point moves 4 places to the left.

0.0 0 0 8 9 To move 4 places to the left, three placeholder zeros must be written.

Thus, $8.9 \times 10^{-4} = 0.00089$.

Self Check 1 Convert to standard notation: **a.** 4.88×10^6 **b.** 9.8×10^{-3}

Now Try Problems 13 and 17

2 Write Numbers in Scientific Notation.

To write a number in scientific notation $(N \times 10^n)$ we first determine N and then n.

EXAMPLE 2 Write each number in scientific notation: **a.** 150,000,000 **b.** 0.00000256 **c.** 432×10^5

Strategy We will write each number as the product of a number between 1 and 10 and a power of 10.

Why Numbers written in scientific notation have the form $N \times 10^n$.

Solution

Caution

At first glance, the following numbers might look like they are written in scientific notation, but they are not. In each case, the first factor is not a number between 1 and 10.

16.38×10^{24} 0.39×10^{-11}

Caution

Don't apply the negative exponent rule when writing numbers in scientific notation.

2.56×10^{-6} ~~$2.56 \times \frac{1}{10^6}$~~

a. We must write 150,000,000 (the distance in kilometers from the Earth to the sun) as the product of a number between 1 and 10 and a power of 10. We note that 1.5 lies between 1 and 10. To obtain 150,000,000, we must move the decimal point in 1.5 exactly 8 places to the right.

1.5 0 0 0 0 0 0 0

This will happen if we multiply 1.5 by 10^8. Therefore,

$150{,}000{,}000 = 1.5 \times 10^8$ This is the distance (in kilometers) from the Earth to the sun.

b. We must write 0.00000256 (the diameter in inches of a flu virus) as the product of a number between 1 and 10 and a power of 10. We note that 2.56 lies between 1 and 10. To obtain 0.00000256, the decimal point in 2.56 must be moved 6 places to the left.

0 0 0 0 0 0 2.56

This will happen if we multiply 2.56 by 10^{-6}. Therefore,

$0.00000256 = 2.56 \times 10^{-6}$ This is the diameter (in inches) of a flu virus.

Calculators

Scientific notation

Calculators automatically change into scientific notation when a result is too large or too small to fit the answer display.

$453.46^5 = 1.917321395 \times 10^{13}$

1.917321395 13

$0.0005^{12} = 2.44140625 \times 10^{-40}$

2.44140625 -40

c. The number 432×10^5 is not written in scientific notation because 432 is not a number between 1 and 10. To write this number in scientific notation, we proceed as follows:

$$432 \times 10^5 = 4.32 \times 10^2 \times 10^5$$ Write 432 in scientific notation.

$$= 4.32 \times 10^7$$ Use the product rule to find $10^2 \times 10^5$. Keep the base, 10, and add the exponents.

Written in scientific notation, 432×10^5 is 4.32×10^7.

Self Check 2 Write each number in scientific notation: **a.** 93,000,000 **b.** 0.00009055 **c.** 85×10^{-3}

Now Try Problems 31, 35, and 55

The results from Example 2 illustrate the following forms to use when converting numbers from standard to scientific notation.

For real numbers between 0 and 1: $\square \times 10^{\text{negative integer}}$

For real numbers at least 1, but less than 10: $\square \times 10^0$

For real numbers greater than or equal to 10: $\square \times 10^{\text{positive integer}}$

3 Perform Calculations with Scientific Notation.

Another advantage of scientific notation becomes apparent when we evaluate products or quotients that involve very large or small numbers. If we express those numbers in scientific notation, we can use rules for exponents to make the calculations easier.

To multiply two numbers written in scientific notation, use the following rule:

$$(a \times 10^m)(b \times 10^n) = (a \cdot b) \times 10^{m+n}$$

EXAMPLE 3 **Astronomy.** Except for the sun, the nearest star visible to the naked eye from most parts of the United States is Sirius. Light from Sirius reaches Earth in about 70,000 hours. If light travels at approximately 670,000,000 mph, how far from Earth is Sirius?

Strategy We can use the formula $d = rt$ to find the distance from Sirius to Earth.

Why We know the *rate* at which light travels and the *time* it takes to travel from Sirius to the Earth.

Solution The rate at which light travels is 670,000,000 mph and the time it takes the light to travel from Sirius to Earth is 70,000 hr. To find the distance from Sirius to Earth, we proceed as follows:

NASA

$d = rt$ This is the formula for distance traveled.

$d = 670{,}000{,}000(70{,}000)$ Substitute 670,000,000 for r and 70,000 for t.

$= (6.7 \times 10^8)(7.0 \times 10^4)$ Write each number in scientific notation.

$= (6.7 \cdot 7.0) \times (10^8 \times 10^4)$ Group the decimals together and the powers of 10 together.

$= (6.7 \cdot 7.0) \times 10^{8+4}$ Use the product rule to find $10^8 \times 10^4$. Keep the base, 10, and add exponents.

$= 46.9 \times 10^{12}$ Do the multiplication and the addition.

We note that 46.9 is not between 0 and 1, so 46.9×10^{12} is not written in scientific notation. To answer in scientific notation, we proceed as follows.

$= \mathbf{4.69 \times 10^1} \times 10^{12}$ Write 46.9 in scientific notation as 4.69×10^1.

$= 4.69 \times 10^{13}$ Use the product rule to find $10^1 \times 10^{12}$. Keep the base, 10, and add the exponents.

Sirius is approximately 4.69×10^{13} or 46,900,000,000,000 miles from Earth.

Self Check 3 Use scientific notation to evaluate: (2,540,000,000,000)(0.00041)

Now Try Problem 61

To divide two numbers written in scientific notation, use the following rule:

$$\frac{a \times 10^m}{b \times 10^n} = \frac{a}{b} \times 10^{m-n}$$

EXAMPLE 4 **Atoms.** As an example of how scientific notation is used in chemistry, we can approximate the weight (in grams) of one atom of the element uranium by evaluating the following expression.

$$\frac{2.4 \times 10^2}{6 \times 10^{23}}$$

Strategy To simplify, we will divide the numbers and powers of 10 separately.

Why We can then use the quotient rule for exponents to simplify the calculations.

Solution

$\frac{2.4 \times 10^2}{6 \times 10^{23}} = \frac{2.4}{6} \times \frac{10^2}{10^{23}}$ Divide the decimals and the powers of 10 separately.

$= \frac{2.4}{6} \times 10^{2-23}$ For the powers of 10, use the quotient rule. Keep the base, 10, and subtract the exponents.

$= \mathbf{0.4} \times 10^{-21}$ Divide the decimals. Subtract the exponents. The result is not in scientific notation form.

$= \mathbf{4 \times 10^{-1}} \times 10^{-21}$ Write 0.4 in scientific notation as 4×10^{-1}.

$= 4 \times 10^{-22}$ Use the product rule to find $10^{-1} \times 10^{-21}$. Keep the base, 10, and add the exponents.

Calculators

Entering scientific notation

To evaluate the expression in Example 4 on a scientific calculator, we enter the numbers using the EE key:

2.4 EE 2 ÷ 6 EE 23 =

One atom of uranium weighs 4×10^{-22} gram or 0.0000000000000000000004 g.

Self Check 4 Find the approximate weight (in grams) of one atom of gold by evaluating: $\frac{1.98 \times 10^2}{6 \times 10^{23}}$

Now Try Problem 65

SECTION 5.3 STUDY SET

VOCABULARY

Fill in the blanks.

1. 4.84×10^5 is written in ________ notation. 484,000 is written in ________ notation.
2. 10^3, 10^{50}, and 10^{-4} are ________ of 10.

CONCEPTS

Fill in the blanks.

3. When we multiply a decimal by 10^5, the decimal point moves 5 places to the _____. When we multiply a decimal by 10^{-7}, the decimal point moves 7 places to the _____.
4. Describe the procedure for converting a number from scientific notation to standard form.
   a. If the exponent on the base of 10 is positive, move the decimal point the same number of places to the _____ as the exponent.
   b. If the exponent on the base of 10 is negative, move the decimal point the same number of places to the ____ as the absolute value of the exponent.
5. a. When a real number greater than or equal to 10 is written in scientific notation, the exponent on 10 is a ________ integer.
   b. When a real number between 0 and 1 is written in scientific notation, the exponent on 10 is a ________ integer.
6. The arrows show the movement of a decimal point. By what power of 10 was each decimal multiplied?
   a. 0.0 0 0 0 0 0 5 5 6 b. 8, 0 4 1, 0 0 0, 0 0 0 .

Fill in the blanks to write each number in scientific notation.

7. a. $7{,}700 = \_\_\_\_ \times 10^3$ b. $500{,}000 = \_\_\_\_ \times 10^5$
   c. $114{,}000{,}000 = 1.14 \times 10^{\_\_}$
8. a. $0.0082 = \_\_\_\_ \times 10^{-3}$
   b. $0.0000001 = \_\_\_\_ \times 10^{-7}$
   c. $0.00003457 = 3.457 \times 10^{\_\_}$
9. Write each expression so that the decimal numbers are grouped together and the powers of ten are grouped together.
   a. $(5.1 \times 10^9)(1.5 \times 10^{22})$
   b. $\dfrac{8.8 \times 10^{30}}{2.2 \times 10^{19}}$
10. Simplify each expression.
   a. $10^{24} \times 10^{33}$ b. $\dfrac{10^{50}}{10^{36}}$ c. $\dfrac{10^{15} \times 10^{27}}{10^{40}}$

NOTATION

11. Fill in the blanks. A positive number is written in scientific notation when it is written in the form $N \times 10^n$, where $\_\_ \le N < \_\_$ and n is an ________.
12. Express each power of 10 in fraction form and decimal form.
   a. 10^{-3} b. 10^{-6}

GUIDED PRACTICE

Convert each number to standard notation. See Example 1.

13. 2.3×10^2
14. 3.75×10^4
15. 8.12×10^5
16. 1.2×10^3
17. 1.15×10^{-3}
18. 4.9×10^{-2}
19. 9.76×10^{-4}
20. 7.63×10^{-5}
21. 6.001×10^6
22. 9.998×10^5
23. 2.718×10^0
24. 3.14×10^0
25. 6.789×10^{-2}
26. 4.321×10^{-1}
27. 2.0×10^{-5}
28. 7.0×10^{-6}

Write each number in scientific notation. See Example 2.

29. 23,000
30. 4,750
31. 1,700,000
32. 290,000
33. 0.062
34. 0.00073
35. 0.0000051
36. 0.04
37. 5,000,000,000
38. 7,000,000
39. 0.0000003
40. 0.0001
41. 909,000,000
42. 7,007,000,000
43. 0.0345
44. 0.000000567
45. 9
46. 2
47. 11
48. 55
49. 1,718,000,000,000,000,000
50. 44,180,000,000,000,000,000
51. 0.0000000000000123
52. 0.0000000000000000555
53. 73×10^4
54. 99×10^5
55. 201.8×10^{15}
56. 154.3×10^{17}
57. 0.073×10^{-3}
58. 0.0017×10^{-4}
59. 36.02×10^{-20}
60. 56.29×10^{-30}

Use scientific notation to perform the calculations. Give all answers in scientific notation and standard notation. See Examples 3 and 4.

61. $(3.4 \times 10^2)(2.1 \times 10^3)$
62. $(4.1 \times 10^{-3})(3.4 \times 10^4)$
63. $(8.4 \times 10^{-13})(4.8 \times 10^9)$
64. $(5.5 \times 10^{-15})(2.2 \times 10^{13})$
65. $\dfrac{2.24 \times 10^4}{5.6 \times 10^7}$
66. $\dfrac{2.47 \times 10^5}{3.8 \times 10^{-5}}$
67. $\dfrac{9.3 \times 10^2}{3.1 \times 10^{-2}}$
68. $\dfrac{7.2 \times 10^6}{1.2 \times 10^8}$
69. $\dfrac{0.00000129}{0.0003}$
70. $\dfrac{169{,}000{,}000{,}000}{26{,}000{,}000}$
71. $(0.0000000056)(5{,}500{,}000)$
72. $(0.000000061)(3{,}500{,}000{,}000)$

73. $\dfrac{96{,}000}{(12{,}000)(0.00004)}$

74. $\dfrac{(0.48)(14{,}400{,}000)}{96{,}000{,}000}$

75. $\dfrac{2{,}475}{(132{,}000{,}000{,}000{,}000)(0.25)}$

76. $\dfrac{147{,}000{,}000{,}000{,}000}{(0.000049)(25)}$

Find each power.

77. $(456.4)^6$

78. $(0.009)^{-6}$

79. 225^{-5}

80. $\left(\frac{1}{3}\right)^{-55}$

APPLICATIONS

81. **Astronomy.** The distance from Earth to Alpha Centauri (the nearest star outside our solar system) is about 25,700,000,000,000 miles. Write this number in scientific notation.

82. **Water.** According to the U.S. Geological Survey, the total water supply of the world is 366,000,000,000,000,000,000 gallons. Write this number in scientific notation.

83. **Earth, Sun, Moon.** The surface area of Earth is 1.97×10^8 square miles, the surface area of the sun is 1.09×10^{17} square miles, and the surface area of the moon is 1.46×10^7 square miles. Convert each number to standard notation.

84. **Atoms.** The number of atoms in 1 gram of iron is approximately 1.08×10^{22}. Convert this number to standard notation.

85. **Sand.** The mass of one grain of beach sand is approximately 0.00000000045 ounce. Write this number in scientific notation.

86. **Molecules.** The mass of a water molecule is approximately 0.0000000000000000000000001056 ounce. Write this number in scientific notation.

87. **Wavelengths.** Examples of the most common types of electromagnetic waves are given in the table in the next column. List the wavelengths in order from shortest to longest.

This distance between the two crests of the wave is called the wavelength.

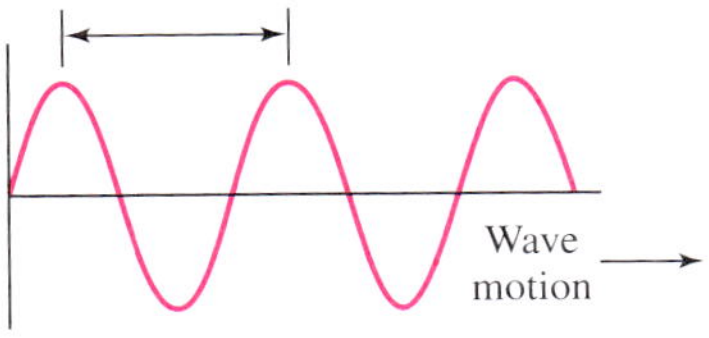

Type	Use	Wavelength (in meters)
visible light	lighting	9.3×10^{-6}
infrared	photography	3.7×10^{-5}
x-ray	medical	2.3×10^{-11}
radio wave	communication	3.0×10^{2}
gamma ray	treating cancer	8.9×10^{-14}
microwave	cooking	1.1×10^{-2}
ultraviolet	sun lamp	6.1×10^{-8}

88. **Exploration.** On July 4, 1997, the *Pathfinder,* carrying the rover vehicle called Sojourner, landed on Mars. The distance from Mars to Earth is approximately 3.5×10^7 miles. Use scientific notation to express this distance in feet. (*Hint:* 5,280 feet = 1 mile.)

89. **from Campus to Careers**

Sound Engineering Technician

The speed of sound in air is approximately 3.3×10^4 centimeters per second. Use scientific notation to express this speed in kilometers per second. (*Hint:* 100 centimeters = 1 meter and 1,000 meters = 1 kilometer.)

90. **Protons.** The mass of one proton is approximately 1.7×10^{-24} gram. Use scientific notation to express the mass of 1 million protons.

91. **Light Years.** One light year is about 5.87×10^{12} miles. Use scientific notation to express this distance in feet. (*Hint:* 5,280 feet = 1 mile.)

92. **Oil.** As of 2009, Saudi Arabia was believed to have crude oil reserves of about 2.65×10^{11} barrels. A barrel contains 42 gallons of oil. Use scientific notation to express Saudi Arabia's oil reserves in gallons. (Source: British Petroleum)

93. **Insured Deposits.** As of June 2009, the total insured deposits in U.S. banks and savings and loans was approximately 7.6×10^{12} dollars. If this money was invested at a rate of 4% simple annual interest, how much would it earn in 1 year? Use scientific notation to express the answer. (Source: Federal Deposit Insurance Corporation.)

94. **Currency.** As of December 2009, the number of \$20 bills in circulation was approximately 6.4×10^9. What was the total value of the currency? Express the answer in scientific notation and standard notation. (Source: The Federal Reserve.)

95. **Powers of 10.** In the United States, we use Latin prefixes in front of "illion" to name extremely large numbers. Write each number in scientific notation.

One million: 1,000,000

One billion: 1,000,000,000

One trillion: 1,000,000,000,000

One quadrillion: 1,000,000,000,000,000

One quintillion: 1,000,000,000,000,000,000

96. **Supercomputers.** As of June 2009, the world's fastest computer was the Cray Jaguar, owned by the Oak Ridge National Laboratory in Tennessee. If it could make 1.75×10^{15} calculations in one second, how many could it make in one minute? Answer in scientific notation. (Source: datacenterknowledge.com)

WRITING

97. In what situations would scientific notation be more convenient than standard notation?

98. To multiply a number by a power of 10, we move the decimal point. Which way, and how far? Explain.

99. 2.3×10^{-3} contains a negative sign but represents a positive number. Explain.

100. Explain why 237.8×10^{8} is not written in scientific notation.

REVIEW

101. If $y = -1$, find the value of $-5y^{55}$.

102. What is the y-intercept of the graph of $y = -3x - 5$?

103. **Counseling.** At the end of her first year of practice, a family counselor had 75 clients. At the end of her second year, she had 105 clients. If a linear trend continues, write an equation that gives the number of clients c the counselor will have at the end of t years.

104. Is $(0, -5)$ a solution of $2x + 3y \geq -14$?

CHALLENGE PROBLEMS

105. Consider 2.5×10^{-4}. Answer the following questions in scientific notation form.

a. What is its opposite?

b. What is its reciprocal?

106. a. Write the numbers one million and one millionth in scientific notation.

b. By what number must we multiply one millionth to get one million?

SECTION 5.4 Polynomials

OBJECTIVES

1. Know the vocabulary for polynomials.
2. Evaluate polynomials.
3. Graph equations defined by polynomials.

ARE YOU READY?

The following problems review some basic skills that are needed when working with polynomials.

1. How many terms does the expression $2x^2 - 5x + 8$ have?
2. What is the coefficient of the term $6a^3$?
3. In $4b^3 + 5b^4 - 16b + b^2$, which term has the largest exponent?
4. For $y = x^3 + 1$, what is the value of y when $x = -3$?

In this section, we will discuss a special type of algebraic expression called a *polynomial.*

1 Know the Vocabulary for Polynomials.

Recall from Chapter 1 that a *term* is a product or quotient of numbers and/or variables. A single number or variable is also a term. Some examples of terms are:

$$14, \quad x, \quad -6y^3, \quad 9cd^2, \quad \text{and} \quad \frac{5}{y}$$

Polynomials

A **polynomial** is a single term or a sum of terms in which all variables have whole-number exponents and no variable appears in a denominator.

The Language of Algebra

The prefix **poly** means many. Some other words that begin with this prefix are *polygon*, *polyester*, and *polyunsaturated*.

Here are some examples of polynomials. Note that polynomials are expressions, not equations.

$$3x + 2, \quad 4y^2 - 2y - 3, \quad a^3 + 3a^2b + 3ab^2 + b^3, \quad \text{and} \quad -8xy^2z$$

Some examples of expressions that are *not polynomials* are:

$6x^3 + 4x^{-2}$ The variable x has an exponent that is not a whole number.

$y^2 + \frac{5}{y} + 1$ The variable y appears in the denominator.

The polynomial $3x + 2$ is the sum of two terms, $3x$ and 2, and we say it is a **polynomial in one variable, *x*.** A single number is called a **constant,** and so its last term, 2, is called the **constant term.**

A polynomial is defined as a single term or the sum of several terms. Since $4y^2 - 2y - 3$ can be written as the sum $4y^2 + (-2y) + (-3)$, it has three terms, $4y^2$, $-2y$, and -3. It is written in **descending powers** of y, because the exponents on y decrease from left to right. When a polynomial is written in descending powers, the first term, in this case $4y^2$, is called the **leading term.** The coefficient of the leading term, in this case 4, is called the **leading coefficient.**

A polynomial can have more than one variable. For example, $a^3 + 3a^2b + 3ab^3 + b^3$ is a **polynomial in two variables,** a and b. It has four terms and is written in descending powers of a and **ascending powers** of b. The polynomial $-8xy^2z$ is a polynomial in three variables, x, y, and z, and has only one term.

The Language of Algebra

The prefix **mono** means one; Jay Leno begins the *Tonight Show* with a monologue. The prefix **bi** means two, as in bicycle or binoculars. The prefix **tri** means three, as in triangle or the *Lord of the Rings Trilogy.*

Polynomials are classified according to the number of terms they have. A polynomial with exactly one term is called a **monomial;** exactly two terms, a **binomial;** and exactly three terms, a **trinomial.** Polynomials with four or more terms have no special names.

Polynomials			
Monomials	**Binomials**	**Trinomials**	**No special name**
$-6x$	$9u - 4$	$5t^2 + 4t + 3$	$a^4 + 2b^2 - 3b - 9$
$5.5x^3y^2$	$-29z^4 - z^2$	$27x^3 - 6x^2 - 2x$	$2.1 - 6.2t + 5.9t^2 + t^3$
11	$18a^2b + 4ab$	$\frac{1}{2}a^2 + 2ab + b^2$	$mn^2 - mn + m - n$

Polynomials and their terms can be described by the exponents on their variables.

Degree of a Term of a Polynomial

The **degree of a term** of a polynomial in one variable is the value of the exponent on the variable. If a polynomial is in more than one variable, the **degree of a term** is the sum of the exponents on the variables in that term. The **degree of a nonzero constant** is 0.

Think of the *degree of a term* as the number of variable factors in that term.

- $9x^6$ has degree **6**. x^6 represents 6 variable factors: $x \cdot x \cdot x \cdot x \cdot x \cdot x$.
- $-2a^4$ has degree **4**. a^4 represents 4 variable factors: $a \cdot a \cdot a \cdot a$.
- $47x^2y^{11}$ has degree **13**. Because $2 + 11 = 13$.
- 8 has degree **0** since it can be written as $8x^0$. There is no variable factor.

We determine the *degree of a polynomial* by considering the degrees of each of its terms.

Degree of a Polynomial

The **degree of a polynomial** is the same as the highest degree of any term of the polynomial.

EXAMPLE 1 Use the vocabulary of this section to describe each polynomial: **a.** $d^4 + 9d^2 - 16$

b. $\frac{1}{2}x^2 - x$ **c.** $-6y^{14} - 1.5y^9z^9 + 2.5y^8z^{10} + yz^{11}$

Strategy First, we will identify the variable(s) in the polynomial and determine whether it is written in ascending or descending powers. Then we will count the number of terms in the polynomial and determine the degree of each term.

Why The number of terms determines the type of polynomial. The highest degree of any term of the polynomial determines its degree.

Solution

a. $d^4 + 9d^2 - 16$ is a polynomial in one variable that is written in descending powers of d. If we write the subtraction as addition of the opposite, we see that it has 3 terms, d^4, $9d^2$, and -16, and is therefore a trinomial. The leading term is d^4 and the leading coefficient is 1. The highest degree of any of its terms is 4, so it is of degree 4.

$$d^4 + 9d^2 - 16 = \underset{\text{1st Term}}{d^4} + \underset{\text{2nd Term}}{9d^2} + \underset{\text{3rd Term}}{(-16)}$$

Term	Coefficient	Degree
d^4	1	4
$9d^2$	9	2
-16	-16	0

Degree of the polynomial: 4

The Language of Algebra

To **descend** means to move from higher to lower. When we write $d^4 + 9d^2 - 16$ as $d^4 + 0d^3 + 9d^2 + 0d^1 - 16d^0$, we more clearly see the *descending* powers of d.

b. $\frac{1}{2}x^2 - x$ is a polynomial in one variable. It is written in descending powers of x. Since it has two terms, it is a binomial. The leading term is $\frac{1}{2}x^2$ and the leading coefficient is $\frac{1}{2}$. The highest degree of any of its terms is 2, so it is of degree 2.

Term	Coefficient	Degree
$\frac{1}{2}x^2$	$\frac{1}{2}$	2
$-x$	-1	1

Degree of the polynomial: 2

Success Tip

Recall that *terms* are separated by + symbols and that the *coefficient* of a term is the numerical factor of the term.

c. $-6y^{14} - 1.5y^9z^9 + 2.5y^8z^{10} + yz^{11}$ is a polynomial in two variables, y and z. It is written in descending powers of y and ascending powers of z. It has 4 terms, and therefore has no special name. The leading term is $-6y^{14}$ and the leading coefficient is -6. The highest degree of any term is 18, so it is of degree 18.

Term	Coefficient	Degree
$-6y^{14}$	-6	14
$-1.5y^9z^9$	-1.5	18
$2.5y^8z^{10}$	2.5	18
yz^{11}	1	12

Degree of the polynomial: 18

Self Check 1 Describe each polynomial: **a.** $x^2 + 4x - 16$ **b.** $-14s^5t + s^4t^2$

Now Try Problems 17 and 35

2 Evaluate Polynomials.

A polynomial can have different values depending on the number that is substituted for its variable (or variables).

EXAMPLE 2 Evaluate $3x^2 + 4x - 5$ for $x = 0$ and $x = -2$.

Strategy We will substitute the given value for each x in the polynomial and follow the rules for the order of operations.

Why To *evaluate a polynomial* means to find its numerical value, once we know the value of its variable.

Solution

For $x = 0$:

$$3x^2 + 4x - 5 = 3(0)^2 + 4(0) - 5$$
$$= 3(0) + 4(0) - 5$$
$$= 0 + 0 - 5$$
$$= -5$$

For $x = -2$:

$$3x^2 + 4x - 5 = 3(-2)^2 + 4(-2) - 5$$
$$= 3(4) + 4(-2) - 5$$
$$= 12 + (-8) - 5$$
$$= -1$$

Caution

Recall that to evaluate $3(-2)^2$, the rules for the order of operations require that we find $(-2)^2$ first, and then multiply that result by 3.

Self Check 2 Evaluate $-x^3 + x - 2x + 3$ for $x = -3$.

Now Try Problem 55

EXAMPLE 3 **Supermarket Displays.** The polynomial $\frac{1}{3}c^3 + \frac{1}{2}c^2 + \frac{1}{6}c$ gives the number of cans used in a display shaped like a square pyramid, having a square base formed by c cans per side. Find the number of cans used in the display.

Strategy We will evaluate the polynomial for $c = 4$.

Why From the illustration, we see that each side of the square base is formed by 4 cans.

Solution

$$\frac{1}{3}c^3 + \frac{1}{2}c^2 + \frac{1}{6}c = \frac{1}{3}(4)^3 + \frac{1}{2}(4)^2 + \frac{1}{6}(4) \quad \text{Substitute 4 for } c.$$
$$= \frac{1}{3}(64) + \frac{1}{2}(16) + \frac{1}{6}(4) \quad \text{Evaluate the exponential expressions first.}$$
$$= \frac{64}{3} + 8 + \frac{2}{3} \quad \text{Do the multiplication, and then simplify: } \frac{4}{6} = \frac{2}{3}.$$
$$= 30 \quad \text{Add the fractions: } \frac{64}{3} + \frac{2}{3} = \frac{66}{3} = 22.$$

There are 30 cans of soup in the display.

Self Check 3 Find the number of cans used in a display having a square base formed by 5 cans per side.

Now Try Problem 81

In the following example, we evaluate a polynomial in two variables.

EXAMPLE 4 Evaluate $3p^2q - 4pq^2$ for $p = 2$ and $q = -3$.

Strategy We will substitute the given values for each p and q in the polynomial and follow the rules for the order of operations.

Why To evaluate a polynomial means to find its numerical value, once we know the value of its variables.

Solution

$$3p^2q - 4pq^2 = 3(2)^2(-3) - 4(2)(-3)^2 \quad \text{Substitute 2 for } p \text{ and } -3 \text{ for } q.$$
$$= 3(4)(-3) - 4(2)(9) \quad \text{Find the powers.}$$
$$= -36 - 72 \quad \text{Do the multiplication.}$$
$$= -108 \quad \text{Do the subtraction.}$$

Self Check 4 Evaluate $3a^3b^2 + 2a^2b$ for $a = 2$ and $b = -1$.

Now Try Problem 61

3 Graph Equations Defined by Polynomials.

In Chapter 3, we graphed equations in two variables such as $y = x$ and $y = 2x - 3$ using the point-plotting method. Recall that these equations are called *linear equations* and that their graphs are straight lines. Note that the right side of the first two equations shown below is a polynomial of degree 1.

$y = x$ $\quad y = 2x - 3$ — The degree of each polynomial is 1.

$y = x^2$ — The degree of this polynomial is 2.

$y = x^3 + 1$ — The degree of this polynomial is 3.

We also can graph equations defined by polynomials with degrees greater than 1 such as $y = x^2$ and $y = x^3 + 1$ by plotting points.

EXAMPLE 5 Graph: $y = x^2$

Strategy We will find several solutions of the equation, plot them on a rectangular coordinate system, and then draw a smooth curve passing through the points.

Why To *graph* an equation in two variables means to make a drawing that represents all of its solutions.

Solution To find some solutions of this equation, we select several values of x that will make the calculations easy. Then we find each corresponding value of y. If $x = -3$, we substitute -3 for x in $y = x^2$ and find y.

$$y = x^2 = (-3)^2 = 9$$

Thus, $(-3, 9)$ is a solution. In a similar manner, we find the corresponding y-values for x-values of $-2, -1, 0, 1, 2$, and 3. If we plot the ordered pairs listed in the table and join the points with a smooth curve, we get the graph shown below, which is called a **parabola.**

Success Tip

When constructing a table of solutions, it is wise to select some positive and negative integer-values for x, as well as 0.

$y = x^2$

x	y	(x, y)
-3	9	$(-3, 9)$
-2	4	$(-2, 4)$
-1	1	$(-1, 1)$
0	0	$(0, 0)$
1	1	$(1, 1)$
2	4	$(2, 4)$
3	9	$(3, 9)$

Select x. Find y. Plot (x, y).

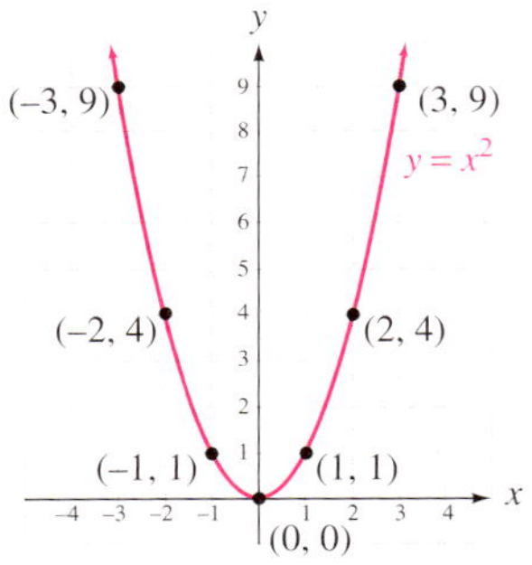

The Language of Algebra

The cup-like shape of a **parabola** has many real-life applications. The word *parabolic* (pronounced par • a • bol • ic) means having the form of a parabola. Did you know that a satellite TV dish is more formally known in the electronics industry as a *parabolic* dish?

Self Check 5 Graph: $y = x^2 - 2$

Now Try Problem 69

Because the variable x is squared, $y = x^2$ is not a linear equation in two variables. We call it a **nonlinear equation.** In Examples 6 and 7, we will graph other nonlinear equations.

EXAMPLE 6 Graph: $y = -x^2 + 2$

Strategy We will find several solutions of the equation, plot them on a rectangular coordinate system, and then draw a smooth curve passing through the points.

Why To *graph* an equation in two variables means to make a drawing that represents all of its solutions.

Solution To make a table of solutions, we select x-values of $-3, -2, -1, 0, 1, 2,$ and 3 and find each corresponding y-value. For example, if $x = -3$, we have

Success Tip

To graph equations of lines, we must find at least two points on the line to draw its graph. Because the graphs of these equations are more complicated, more work is required. We must find a sufficient number of points on the graph so that its entire shape is revealed.

$y = -x^2 + 2$ This is the equation to graph.

$y = -(-3)^2 + 2$ Substitute -3 for x.

$y = -(9) + 2$ Evaluate the exponential expression first: $(-3)^2 = 9$.

$y = -7$ Do the addition: $-9 + 2 = -7$.

The ordered pair $(-3, -7)$ is a solution. Six other solutions appear in the table. After plotting each pair, we join the points with a smooth curve to obtain the graph, a parabola opening downward.

$y = -x^2 + 2$

x	y	(x, y)
−3	−7	(−3, −7)
−2	−2	(−2, −2)
−1	1	(−1, 1)
0	2	(0, 2)
1	1	(1, 1)
2	−2	(2, −2)
3	−7	(3, −7)

Select x. Find y. Plot (x, y).

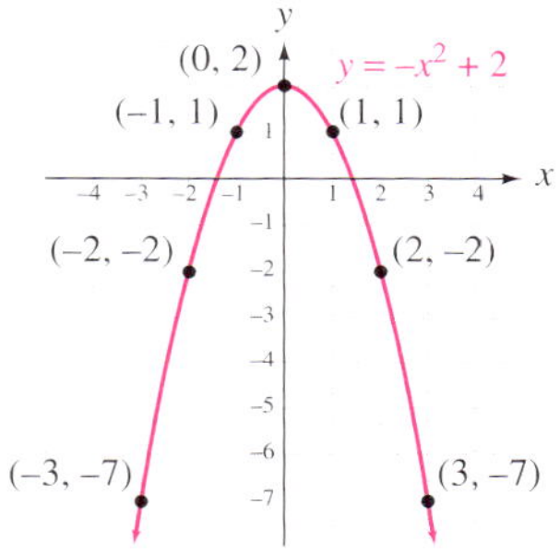

Self Check 6 Graph: $y = -x^2$

Now Try Problem 71

EXAMPLE 7 Graph: $y = x^3 + 1$

Strategy We will find several solutions of the equation, plot them on a rectangular coordinate system, and then draw a smooth curve passing through the points.

Why To *graph* an equation in two variables means to make a drawing that represents all of its solutions.

Solution If we let $x = -2$, we have

$y = x^3 + 1$ This is the equation to graph.

$y = (-2)^3 + 1$ Substitute -2 for x.

$y = -8 + 1$ Evaluate the exponential expression first: $(-2)^3 = -8$.

$y = -7$ Do the addition.

The ordered pair $(-2, -7)$ is a solution. This pair and others that satisfy the equation are listed in the table. Plotting the ordered pairs and joining the points with a smooth curve gives us the graph.

$y = x^3 + 1$

x	y	(x, y)
-2	-7	$(-2, -7)$
-1	0	$(-1, 0)$
0	1	$(0, 1)$
1	2	$(1, 2)$
2	9	$(2, 9)$

Select x. Find y. Plot (x, y).

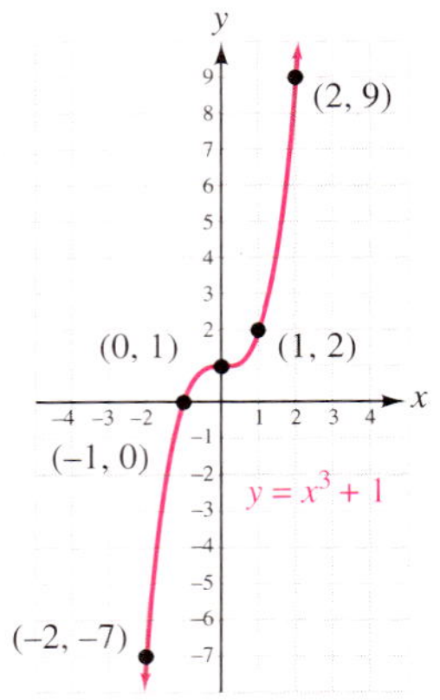

Self Check 7 Graph: $y = x^3 - 1$

Now Try Problem 75

SECTION 5.4 STUDY SET

VOCABULARY

Fill in the blanks.

1. A __________ is a term or a sum of terms in which all variables have whole-number exponents and no variable appears in a denominator.
2. The ______ of a polynomial are separated by + symbols.
3. $x^3 - 6x^2 + 9x - 2$ is a polynomial in ____ variable, and is written in __________ powers of x, and $c^3 + 2c^2d - d^2$ is a polynomial in ____ variables and is written in _________ powers of d.
4. For the polynomial $6x^2 + 3x - 1$, the ________ term is $6x^2$, and the leading __________ is 6. The ________ term is -1.
5. A __________ is a polynomial with exactly one term. A _________ is a polynomial with exactly two terms. A _________ is a polynomial with exactly three terms.
6. The _______ of the term $3x^7$ is 7 because x appears as a factor 7 times: $3 \cdot x \cdot x \cdot x \cdot x \cdot x \cdot x \cdot x$.
7. To ________ the polynomial $x^2 - 2x + 1$ for $x = 6$, we substitute 6 for x and follow the rules for the order of operations.
8. The graph of $y = x^2$ is a cup-shaped curve called a ________.

CONCEPTS

Determine whether each expression is a polynomial.

9. a. $x^3 - 5x^2 - 2$ b. $x^{-4} - 5x$
   c. $x^2 - \dfrac{1}{2x} + 3$ d. $x^3 - 1$
   e. $x^2 - y^2$ f. $a^4 + a^3 + a^2 + a$

10. Fill in the blank so that the term has degree 5.
    a. $9x$ b. $-\dfrac{2}{3}xy$

Make a term-coefficient-degree table like that shown in Example 1 for each polynomial.

11. $8x^2 + x - 7$

Term	Coefficient	Degree

Degree of the polynomial:

12. $y^4 - y^3 + 16y^2 + 3y$

Term	Coefficient	Degree

Degree of the polynomial:

13. $8a^6b^3 - 27ab$

Term	Coefficient	Degree

Degree of the polynomial:

14. $-1.2c^4 + 2.4c^2d^2 - 3.6d^4$

Term	Coefficient	Degree

Degree of the polynomial:

NOTATION

15. a. Write $x - 9 + 3x^2 + 5x^3$ in descending powers of x.

    b. Write $-2xy + y^2 + x^2$ in ascending powers of y.

16. Complete the solution. Evaluate $-2x^2 + 3x - 1$ for $x = -2$.

$$-2x^2 + 3x - 1 = -2(\quad)^2 + 3(\quad) - 1$$
$$= -2(\quad) + 3(\quad) - 1$$
$$= \quad + (-6) - 1$$
$$=$$

GUIDED PRACTICE

Classify each polynomial as a monomial, a binomial, a trinomial, or none of these. See Example 1.

17. $3x + 7$
18. $3y - 5$
19. $y^2 + 4y + 3$
20. $9xy$
21. $\frac{3}{2}z^2$
22. $\frac{3}{5}x^4 - \frac{2}{5}x^3 + \frac{3}{5}x - 1$
23. $t - 32$
24. $12z^4$
25. $s^2 - 23s + 31$
26. $2x^3 - 5x^2 + 6x - 3$
27. $6x^5 - x^4 - 3x^3 + 7$
28. x^3
29. $3m^3n - 4m^2n^2 + mn - 1$
30. $4p^3q^2 + 7p^2q^3 + pq^4 - q^5$
31. $2a^2 - 3ab + b^2$
32. $a^3b - ab^3$

Find the degree of each polynomial. See Example 1.

33. $3x^4$
34. $3x^5$
35. $-2x^2 + 3x + 1$
36. $-5x^4 + 3x^2 - 3x$
37. $\frac{1}{3}x - 5$
38. $\frac{1}{2}y^3 + 4y^2$
39. $-5r^2s^2 - r^3s + 3$
40. $4r^2s^3 - 5r^2s^8$
41. $x^{12} + 3x^2y^3$
42. $17ab^5 - 12a^3b$
43. 38
44. -24
45. $\frac{3}{2}m^7 - \frac{3}{4}m^{18}$
46. $\frac{7}{8}t^{10} - \frac{1}{8}t^{16}$
47. $5.5tw - 6.5t^2w - 7.5t^3$
48. $0.4h + 0.6h^4c + 0.6h^5$

Evaluate each expression. See Example 2 and 3.

49. $x^2 - x + 1$ for
    a. $x = 2$
    b. $x = -3$
50. $x^2 - x + 7$ for
    a. $x = 6$
    b. $x = -2$
51. $4t^2 + 2t - 8$ for
    a. $t = -1$
    b. $t = 0$
52. $3s^2 - 2s + 8$ for
    a. $s = 1$
    b. $s = 0$
53. $\frac{1}{2}a^2 - \frac{1}{4}a$ for
    a. $a = 4$
    b. $a = -8$
54. $\frac{1}{3}b^2 - \frac{1}{9}b$ for
    a. $b = 9$
    b. $b = -9$
55. $-9.2x^2 + x - 1.4$ for
    a. $x = -1$
    b. $x = -2$
56. $-10.3x^2 - x + 6.5$ for
    a. $x = -1$
    b. $x = -2$
57. $x^3 + 3x^2 + 2x + 4$ for
    a. $x = 2$
    b. $x = -2$
58. $x^3 - 3x^2 - x + 9$ for
    a. $x = 3$
    b. $x = -3$
59. $y^4 - y^3 + y^2 + 2y - 1$ for
    a. $y = 1$
    b. $y = -1$
60. $-y^4 + y^3 + y^2 + y + 1$ for
    a. $y = 1$
    b. $y = -1$

Evaluate each polynomial for $a = -2$ and $b = 3$. See Example 4.

61. $6a^2b$
62. $4ab^2$
63. $a^3 + b^3$
64. $a^3 - b^3$
65. $a^2 + 5ab - b^2$
66. $a^3 - 2ab + b^3$
67. $5ab^3 - ab - b + 10$
68. $-a^3b + ab - a - 21$

Construct a table of solutions and then graph the equation. See Examples 5–7.

69. $y = x^2 + 1$
70. $y = x^2 - 4$
71. $y = -x^2 - 2$
72. $y = -x^2 + 1$
73. $y = 2x^2 - 3$
74. $y = -2x^2 + 2$
75. $y = x^3 + 2$
76. $y = x^3 + 4$
77. $y = x^3 - 3$
78. $y = x^3 - 2$
79. $y = -x^3 - 1$
80. $y = -x^3$

APPLICATIONS

81. **Supermarkets.** A grocer plans to set up a pyramid-shaped display of cantaloupes like that shown in Example 3. If each side of the square base of the display is made of six cantaloupes, how many will be used in the display?

82. **Packaging.** The polynomial $4x^3 - 44x^2 + 120x$ gives the volume (in cubic inches) of the resulting box when a square with sides x inches long is cut from each corner of a 10 in. × 12 in. piece of cardboard. Find the volume of a box if 3-inch squares are cut out.

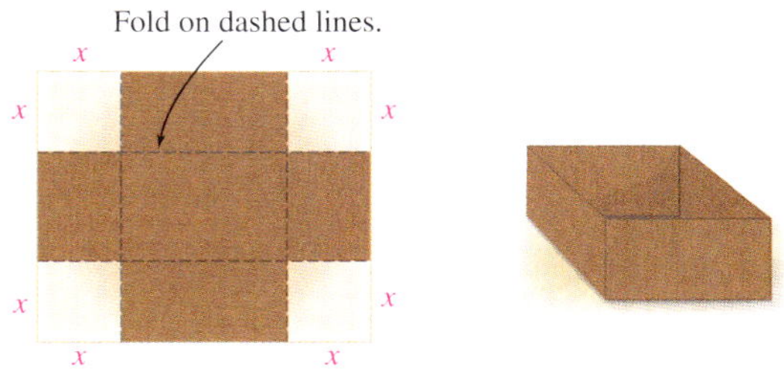

83. **Stopping Distance.** The number of feet that a car travels before stopping depends on the driver's reaction time and the braking distance, as shown in the illustration. For one driver, the stopping distance is given by the polynomial $0.04v^2 + 0.9v$ where v is the velocity of the car. Find the stopping distance when the driver is traveling at 30 mph.

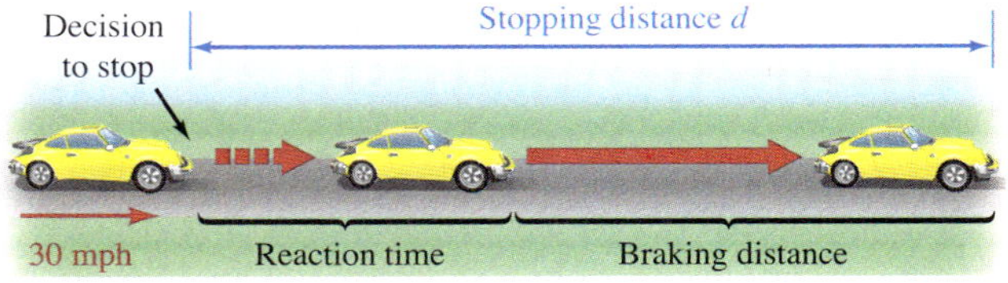

84. **Suspension Bridges.** The following polynomial $-0.0000001s^4 + 0.0066667s^2 + 400$ approximates the length of the cable between the two vertical towers of a bridge, where s is the sag in the cable (in feet). Estimate the length of the cable if the sag is 24.6 feet.

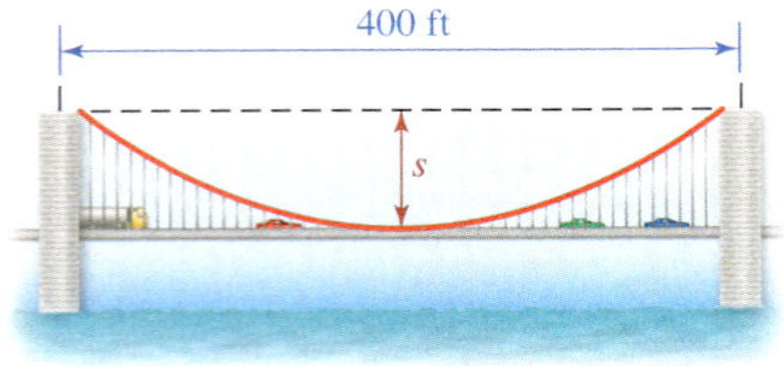

85. from **Campus to Careers**

Sound Engineering Technician

© iStockphoto.com/beetle8

Many people in the recording industry have been impressed by the success of Apple's iTunes Music Store. The polynomial $0.32x^2 - 0.36x + 0.21$ approximates the number of song downloads from iTunes. When $x = 0$, the polynomial estimates the number of downloads (in billions) as of January 2004. When $x = 1$, it estimates the number of downloads (in billions) as of January 2005, and so on. Suppose the trend continues. Use the polynomial to estimate the number of iTunes downloads as of January 2014. (Source: eagereyes.org)

86. **Twitter.** When $x = 1$, the polynomial $4.4x^2 + 36.2x + 42.5$ approximates the number of Tweets (in millions) on the social network Twitter for January, 2009. When $x = 2$, it approximates the number of Tweets (in millions) for the month of February, 2009, and so on. Use the polynomial to find the number of Tweets (in millions) for the month of October, 2009. (Source: pingdom.com)

87. **Science History.** The Italian scientist Galileo Galilei (1564–1642) built an incline plane like that shown to study falling objects. As the ball rolled down, he measured the time it took the ball to travel different distances. Graph the data and then connect the points with a smooth curve.

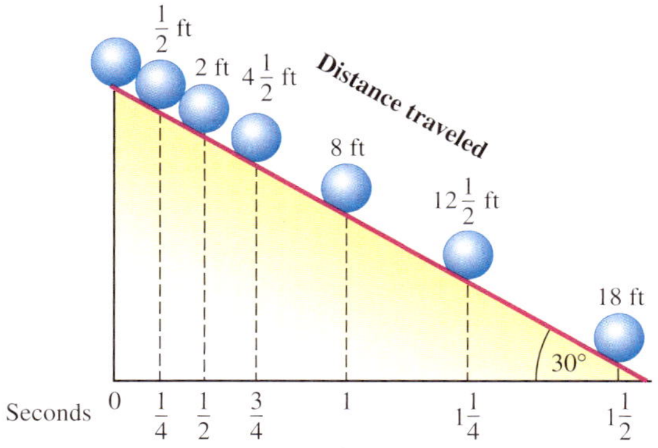

88. **Dolphins.** At a marine park, three trained dolphins jump in unison over an arching stream of water whose path can be described by the equation $y = -0.05x^2 + 2x$. Given the takeoff points for each dolphin, how high must each jump to clear the stream of water?

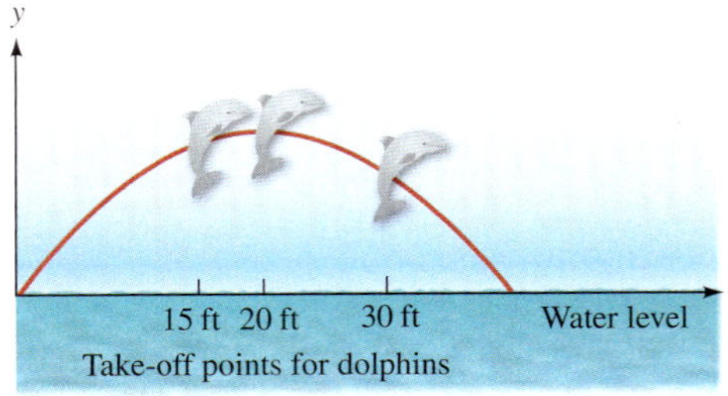

WRITING

89. Describe how to determine the degree of a polynomial.

90. List some words that contain the prefixes *mono, bi,* or *tri.*

91. To graph $y = x^2 - 4$, a table of solutions is constructed and a graph is drawn. Explain the error.

$y = x^2 - 4$

x	y	(x, y)
0	−4	(0, −4)
2	0	(2, 0)

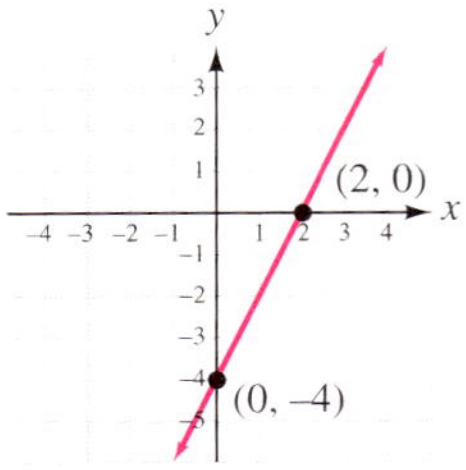

92. The expression $x + y$ is a binomial. Is xy also a binomial? Explain.

REVIEW

Solve each inequality. Write the solution set in interval notation and graph it.

93. $-4(3y + 2) \le 28$

94. $-5 < 3t + 4 \le 13$

Simplify each expression. Do not use negative exponents in the answer.

95. $(x^2x^4)^3$

96. $(a^2)^3(a^3)^2$

97. $\left(\frac{y^2y^5}{y^4}\right)^3$

98. $\left(\frac{2t^3}{t}\right)^{-4}$

CHALLENGE PROBLEMS

99. Find a three-term polynomial of degree 2 whose value will be 1 when it is evaluated for $x = 2$.

100. Graph: $y = 2x^3 - 3x^2 - 11x + 6$

SECTION 5.5

Adding and Subtracting Polynomials

OBJECTIVES

1 Simplify polynomials by combining like terms.
2 Add polynomials.
3 Subtract polynomials.

ARE YOU READY?

The following problems review some basic skills that are needed when adding and subtracting polynomials.

Combine like terms, if possible.

1. $8x + 9x$

2. $2.5a^3 - 1.3a^3$

3. $2t^2u - (-3t^2u)$

4. $8b^2 + 7b$

5. Simplify: $-(4b^2 - 9b + 1)$

6. Subtract 10 from 2.

If we are to add (or subtract) objects, they must be similar. For example, we can add dollars to dollars and inches to inches, but we can't add dollars to inches. If you keep this concept in mind, then adding and subtracting polynomials will be easy. It simply involves combining like terms.

1 Simplify Polynomials by Combining Like Terms.

Recall that **like terms** have the same variables with the same exponents. Only the coefficients may differ.

Like terms	*Unlike terms*	
$-7x$ and $15x$	$-7x$ and $15a$	Different variables
$9.4y^3$ and $1.6y^3$	$9.4y^3$ and $1.6y^2$	Different exponents on the same variable
$\frac{1}{2}x^5y^2$ and $-\frac{1}{3}x^5y^2$	$\frac{1}{2}x^5y^2$ and $-\frac{1}{3}x^2y^5$	Different exponents on different variables

Also recall that to **combine like terms,** we combine their coefficients and keep the same variables with the same exponents. For example,

$$4y + 5y = (4 + 5)y = 9y \qquad 8x^2 - x^2 = (8 - 1)x^2 = 7x^2$$

Polynomials with like terms can be simplified by combining like terms.

The Language of Algebra

Simplifying the sum or difference of like terms is called **combining like terms.**

EXAMPLE 1

Simplify each polynomial by combining like terms:

a. $4x^4 + 81x^4$ **b.** $-0.3r - 0.4r + 0.6r$

c. $17x^2y^2 + 2x^2y - 6x^2y^2$ **d.** $\frac{3}{4}p^2 + \frac{1}{2}q^2 - 7 + \frac{1}{3}p^2 - \frac{5}{4}q^2 + 4$

Strategy We will use the distributive property in reverse to add (or subtract) the coefficients of the like terms. We will keep the same variables raised to the same powers.

Why To *combine like terms* means to add or subtract the like terms in an expression.

Solution

a. $4x^4 + 81x^4 = 85x^4$ Think: $(4 + 81)x^4 = 85x^4$.

b. $-0.3r - 0.4r + 0.6r = -0.1r$ Think: $(-0.3 - 0.4 + 0.6)r = -0.1r$.

c. The first and third terms are like terms.

$$17x^2y^2 + 2x^2y - 6x^2y^2 = 11x^2y^2 + 2x^2y$$ Think: $(17 - 6)x^2y^2 = 11x^2y^2$.

Caution

When combining like terms, simply combine their coefficients. Don't incorrectly add the exponents. The exponents on the variables *stay the same.*

d. $\frac{3}{4}p^2 + \frac{1}{2}q^2 - 7 + \frac{1}{3}p^2 - \frac{5}{4}q^2 + 4$

$= \left(\frac{3}{4} + \frac{1}{3}\right)p^2 + \left(\frac{1}{2} - \frac{5}{4}\right)q^2 - 7 + 4$ Combine like terms.

$= \left(\frac{9}{12} + \frac{4}{12}\right)p^2 + \left(\frac{2}{4} - \frac{5}{4}\right)q^2 - 7 + 4$ Build equivalent fractions: $\frac{3}{4}\cdot\frac{3}{3} = \frac{9}{12}, \frac{1}{3}\cdot\frac{4}{4} = \frac{4}{12}$, and $\frac{1}{2}\cdot\frac{2}{2} = \frac{2}{4}$

$= \frac{13}{12}p^2 - \frac{3}{4}q^2 - 3$ Do the addition and the subtraction.

CAUTION Do not try to clear this expression of fractions by multiplying it by the LCD 12. That strategy works only when we multiply *both sides of an equation* by the LCD.

$$12\left(\frac{3}{4}p^2 + \frac{1}{2}q^2 - 7 + \frac{1}{3}p^2 - \frac{5}{4}q^2 + 4\right)$$

Self Check 1 Simplify each polynomial: **a.** $6m^4 + 3m^4$
b. $-19x + 21x - x$ **c.** $1.7s^3t + 0.3s^2t - 0.6s^3t$
d. $\frac{1}{8}c^5 + \frac{1}{3}d^5 - 9 + \frac{5}{4}c^5 - \frac{3}{5}d^5 + 1$

Now Try Problems 13, 23, and 27

Notice that the terms of each solution in Example 1 are written in **descending powers** of one variable. When working with polynomials, answers are almost always presented in this form because it makes them easy to compare.

2 Add Polynomials.

When adding polynomials horizontally, each polynomial is usually enclosed within parentheses. For example,

$$(3x^2 + 6x + 7) + (2x - 5)$$

is the sum of a trinomial and a binomial. To find the sum, we reorder and regroup the terms using the commutative and associative properties of addition so that like terms are together.

$(3x^2 + 6x + 7) + (2x - 5) = 3x^2 + (6x + 2x) + (7 - 5)$ The x-terms are together. The constant terms are together.

$= 3x^2 + 8x + 2$ Combine like terms.

This example illustrates the following rule.

Adding Polynomials To add polynomials, combine their like terms.

EXAMPLE 2 Add the polynomials: **a.** $(-6a^3 + 5a^2 - 7a + 9) + (4a^3 - 5a^2 - a - 8)$
b. $\left(\frac{1}{2}m^2 + \frac{2}{3}m + 1\right) + \left(\frac{3}{4}m^2 - \frac{7}{9}m - 4\right)$ **c.** $(16g^2 - h^2) + (4g^2 + 2gh + 10h^2)$

Strategy We will reorder and regroup to get the like terms together. Then we will combine like terms.

Why To add polynomials means to combine their like terms.

Solution

Success Tip

Combine the like terms in order: a^3-terms first, a^2-terms second, a-terms third, and constants last. Then the answer will be in descending powers of a.

a. $(-6a^3 + 5a^2 - 7a + 9) + (4a^3 - 5a^2 - a - 8)$

$= (-6a^3 + 4a^3) + (5a^2 - 5a^2) + (-7a - a) + (9 - 8)$ Group like terms together.

$= -2a^3 + 0a^2 + (-8a) + 1$ Combine like terms.

$= -2a^3 - 8a + 1$ It is not necessary to write $0a^2$.

b. $\left(\frac{1}{2}m^2 + \frac{2}{3}m + 1\right) + \left(\frac{3}{4}m^2 - \frac{7}{9}m - 4\right)$

$= \left(\frac{1}{2}m^2 + \frac{3}{4}m^2\right) + \left(\frac{2}{3}m - \frac{7}{9}m\right) + (1 - 4)$ Group like terms together.

$= \left(\frac{2}{4}m^2 + \frac{3}{4}m^2\right) + \left(\frac{6}{9}m - \frac{7}{9}m\right) + (1 - 4)$ Build equivalent fractions: $\frac{1}{2} \cdot \frac{2}{2} = \frac{2}{4}$ and $\frac{2}{3} \cdot \frac{3}{3} = \frac{6}{9}$.

$= \frac{5}{4}m^2 - \frac{1}{9}m - 3$ Combine like terms.

c. $(16g^2 - h^2) + (4g^2 + 2gh + 10h^2)$

$= (16g^2 + 4g^2) + 2gh + (-h^2 + 10h^2)$ Group like terms together.

$= 20g^2 + 2gh + 9h^2$ Combine like terms.

Self Check 2 Add the polynomials:

a. $(2a^2 - a + 4) + (5a^2 + 6a - 5)$

b. $\left(\frac{3}{2}b^3 + \frac{4}{5}b + 7\right) + \left(\frac{3}{4}b^3 - \frac{11}{10}b - 10\right)$

c. $(7x^2 - 2xy - y^2) + (4x^2 - y^2)$

Now Try Problems 29, 31, and 35

EXAMPLE 3 Find a polynomial that represents the perimeter of the trapezoid.

Strategy We will add the polynomials that represent the lengths of the sides of the trapezoid.

Why To find the perimeter of a figure, we find the distance around the figure by finding the sum of the lengths of its sides.

Solution To add the four polynomials that represent the lengths of the sides of the trapezoid, we combine their like terms.

The Language of Algebra

A **trapezoid** is a four-sided figure with exactly two sides parallel.

$(2a^2 - 5) + (a^2 + a) + (4a^2 - 2a + 1) + 3a^2$

$= (2a^2 + a^2 + 4a^2 + 3a^2) + (a - 2a) + (-5 + 1)$ Reorder and regroup terms.

$= 10a^2 - a - 4$ Combine like terms. Think: $2 + 1 + 4 + 3 = 10$, $1 - 2 = -1$, and $-5 + 1 = -4$.

The perimeter of the trapezoid is $(10a^2 - a - 4)$ ft.

Self Check 3 Find a polynomial that represents the perimeter of a triangle with sides of length $(19h^2 + 11h - 2)$ in., $(4h^2 - 4h + 4)$ in., and $(6h^2 - 22h - 2)$ in.

Now Try Problem 39

Polynomials also can be added vertically by aligning like terms in columns.

EXAMPLE 4 Add $4x^2 - 3$ and $3x^2 - 8x + 8$ using vertical form.

Strategy First, we will write one polynomial underneath the other and draw a horizontal line beneath them. Then we will add the like terms, column by column, and write each result under the line.

Why *Vertical form* means to use an approach similar to that used in arithmetic to add two numbers.

Solution When performing vertical addition, any missing term may be written with a coefficient of 0. Since the first polynomial does not have an x-term, we insert a **placeholder** term $0x$ in the second column so that the constant terms line up in the third column.

Success Tip

In arithmetic, we use vertical form so that we add digits in like place-value columns. In polynomial addition, we combine the like terms in each column.

Hundreds, Tens, Ones

$$\begin{array}{r} 403 \\ +\underline{388} \end{array}$$

x^2-terms	x-terms	Constants	
$4x^2$	$+ 0x$	$- 3$	In the x^2-column, find $4x^2 + 3x^2$.
$3x^2$	$- 8x$	$+ 8$	In the x-column, find $0x + (-8x)$.
$7x^2$	$-8x$	$+5$	In the constant column, find $-3 + 8$.

The sum is $7x^2 - 8x + 5$.

Self Check 4 Add $4q^2 - 7$ and $2q^2 - 8q + 9$ using vertical form.

Now Try Problem 41

3 Subtract Polynomials.

Recall from Chapter 1 that we can use the distributive property to find the opposite of several terms enclosed within parentheses. For example, consider $-(2a^2 - a + 9)$.

$$-(2a^2 - a + 9) = -1(2a^2 - a + 9)$$ Replace the $-$ symbol in front of the parentheses with -1.

$$= -2a^2 + a - 9$$ Use the distributive property to remove parentheses.

This example illustrates the following method of subtracting polynomials.

Subtracting Polynomials To subtract two polynomials, change the signs of the terms of the polynomial being subtracted, drop the parentheses, and combine like terms.

EXAMPLE 5 Subtract the polynomials: **a.** $(3a^2 - 4a - 6) - (2a^2 - a + 9)$ **b.** $(-t^3u + 2t^2u - u + 1) - (-3t^2u - u + 8)$

Strategy In each case, we will change the signs of the terms of the polynomial being subtracted, drop the parentheses, and combine like terms.

Why This is the method for subtracting two polynomials.

Solution **a.** $(3a^2 - 4a - 6) - (2a^2 - a + 9)$

$$= 3a^2 - 4a - 6 - 2a^2 + a - 9$$ Change the sign of each term of $2a^2 - a + 9$ and drop the parentheses.

$$= a^2 - 3a - 15$$ Combine like terms.

Success Tip

After some practice, you will be able to reorder and regroup the terms of the polynomials in your head to combine them.

b. $(-t^3u + 2t^2u - u + 1) - (-3t^2u - u + 8)$

$= -t^3u + 2t^2u - u + 1 + 3t^2u + u - 8$ Change the sign of each term of $-3t^2u - u + 8$ and drop the parentheses.

$= -t^3u + 5t^2u - 7$ Combine like terms.

Self Check 5 Subtract the polynomials:

a. $(8a^3 - 5a^2 + 5) - (a^3 - a^2 - 7)$

b. $(x^2y - 2x + y - 2) - (6x + 9y - 2)$

Now Try Problems 49 and 55

Polynomials can also be subtracted vertically by aligning like terms in columns.

EXAMPLE 6

a. Subtract $3x^2 - 2x + 3$ from $2x^2 + 4x - 1$ using vertical form.

b. Subtract $4x^3 - 6x^2 + x$ from $7x^3 - 2x$ using vertical form.

Strategy Since the first polynomial is to be subtracted from the second, we will write the first underneath the second, change the sign of each of its terms and add, column-by-column.

Why *Vertical form* means to arrange the like terms in columns.

Solution

a.

$$\begin{array}{r} 2x^2 + 4x - 1 \\ -(3x^2 - 2x + 3) \\ \hline \end{array} \xrightarrow{\text{Change signs and add}} \begin{array}{r} 2x^2 + 4x - 1 \\ -3x^2 + 2x - 3 \\ \hline -x^2 + 6x - 4 \end{array}$$

In the x^2-column, find $2x^2 + (-3x^2)$.
In the x-column, find $4x + 2x$.
In the constant column, find $-1 + (-3)$.

The difference is $-x^2 + 6x - 4$.

The Language of Algebra

Recall that we must be careful of the order when translating English phrases containing the words **subtract from.** For example, to subtract x from y, we write $y - x$.

b. Since $7x^3 - 2x$ is missing an x^2-term, we will insert the placeholder term $0x^2$ to ensure that like terms are in the same column.

$$\begin{array}{r} 7x^3 + 0x^2 - 2x \\ -(4x^3 - 6x^2 + x) \\ \hline \end{array} \xrightarrow{\text{Change signs and add}} \begin{array}{r} 7x^3 + 0x^2 - 2x \\ -4x^3 + 6x^2 - x \\ \hline 3x^3 + 6x^2 - 3x \end{array}$$

In the x^3-column, find $7x^3 + (-4x^3)$.
In the x^2-column, find $0x^2 + 6x^2$.
In the x-column, find $-2x + (-x)$.

The difference is $3x^3 + 6x^2 - 3x$.

Self Check 6

a. Subtract $2p^2 + 2p - 8$ from $5p^2 - 6p + 7$ using vertical form.

b. Subtract $4m^3 - 6m^2 + 7m$ from $-8m^3 + 16m$.

Now Try Problem 57

EXAMPLE 7 Subtract $1.2a^4 - 0.7a$ from the sum of $0.6a^4 + 1.5a$ and $0.4a^4 - 1.1a$.

Strategy First, we will translate the words of the problem into mathematical symbols. Then we will perform the indicated operations.

Why The words of the problem contain the key phrases *subtract from* and *sum.*

Solution Since $1.2a^4 - 0.7a$ is to be subtracted from the sum, the order must be reversed when we translate to mathematical symbols.

Subtract $1.2a^4 - 0.7a$ from the sum of $0.6a^4 + 1.5a$ and $0.4a^4 - 1.1a$.

$[(0.6a^4 + 1.5a) + (0.4a^4 - 1.1a)] - (1.2a^4 - 0.7a)$ Use brackets [ ] to enclose the sum.

Caution

A common error is to change the sign of only the first term of the polynomial being subtracted and forget to change the signs of its other terms.

Next, we change the sign of each term within $(1.2a^4 - 0.7a)$ and drop the parentheses.

$$= 0.6a^4 + 1.5a + 0.4a^4 - 1.1a - 1.2a^4 + 0.7a$$

$$= -0.2a^4 + 1.1a \quad \text{Combine like terms.}$$

Self Check 7 Subtract $-0.2q^2 - 0.2q$ from the sum of $0.1q^2 - 0.6q$ and $0.3q^2 + 0.1q$.

Now Try Problem 65

EXAMPLE 8

Fireworks. Two firework shells are fired upward at the same time from different platforms. The height, after t seconds, of the first shell is $(-16t^2 + 160t + 3)$ feet. The height, after t seconds, of a higher-flying second shell is $(-16t^2 + 200t + 1)$ feet.

a. Find a polynomial that represents the difference in the heights of the shells.

b. In 5 seconds, the first shell reaches its peak and explodes. How much higher is the second shell at that time?

Strategy To find the difference in their heights, we will subtract the height of the first shell from the height of the higher-flying second shell.

Why The key word *difference* indicates that we should subtract the polynomials.

Solution

a. Since the height of the higher flying second shell is represented by $-16t^2 + 200t + 1$ and the height of the lower flying shell is represented by $-16t^2 + 160t + 3$, we can find their difference by performing the following subtraction.

$$(-16t^2 + 200t + 1) - (-16t^2 + 160t + 3)$$

$$= -16t^2 + 200t + 1 + 16t^2 - 160t - 3 \quad \text{Change the sign of each term of } -16t^2 + 160t + 3 \text{ and drop the parentheses.}$$

$$= 40t - 2 \quad \text{Combine like terms.}$$

The difference in the heights of the shells t seconds after being fired is $(40t - 2)$ feet.

b. To find the difference in their heights after 5 seconds, we will evaluate the polynomial found in part (a) at a value of 5 seconds. If we substitute 5 for t, we have

$$40t - 2 = 40(5) - 2 = 200 - 2 = 198$$

When the first shell explodes, the second shell will be 198 feet higher than the first shell.

Self Check 8 **Property values.** A real estate investor purchased two houses on the same day. The value of the first house, x years after its purchase, is given by $\$(2{,}500x + 95{,}000)$. The value of the second house, x years after its purchase, is given by $\$(4{,}500x + 125{,}000)$. Find a polynomial that represents the total value of the houses after x years.

Now Try Problem 109

SECTION 5.5 STUDY SET

VOCABULARY

Fill in the blanks.

1. $(b^3 - b^2 - 9b + 1) + (b^3 - b^2 - 9b + 1)$ is the sum of two ____________.
2. $(b^2 - 9b + 11) - (4b^2 - 14b)$ is the __________ of a trinomial and a binomial.
3. _____ terms have the same variables with the same exponents.
4. The polynomial $2t^4 + 3t^3 - 4t^2 + 5t - 6$ is written in __________ powers of t.

CONCEPTS

Fill in the blanks.

5. To add polynomials, ________ their like terms.
6. To subtract polynomials, _______ the signs of the terms of the polynomial being subtracted, drop parentheses, and combine like terms.
7. Simplify each polynomial, if possible.
   a. $2x^2 + 3x^2$ b. $15m^3 - m^3$
   c. $8a^3b - a^3b$ d. $6cd + 4c^2d$
8. What is the result of the addition in the x-column?

$$\begin{array}{r} 4x^2 + x - 12 \\ \underline{5x^2 - 8x + 23} \end{array}$$

9. Write without parentheses.
   a. $-(5x^2 - 8x + 23)$ b. $-(-5y^4 + 3y^2 - 7)$
10. What is the result of the subtraction in the x-column?

$$\begin{array}{r} 8x^2 - 7x - 1 \\ \underline{-(4x^2 + 6x - 9)} \end{array} \longrightarrow \begin{array}{r} 8x^2 - 7x - 1 \\ \underline{-4x^2 - 6x + 9} \end{array}$$

NOTATION

Fill in the blanks to add (subtract) the polynomials.

11. $(6x^2 + 2x + 3) + (4x^2 - 7x + 1)$
    $= (6x^2 + \quad) + (\quad - 7x) + (3 + \quad)$
    $= \quad - 5x + \quad$
12. $(6x^2 + 2x + 3) - (4x^2 - 7x + 1)$
    $= 6x^2 + 2x + 3 \quad 4x^2 \quad 7x - 1$
    $= \quad + 9x + \quad$

GUIDED PRACTICE

Simplify each polynomial and write it in descending powers of one variable. See Example 1.

13. $8t^2 + 4t^2$
14. $15x^2 + 10x^2$
15. $18x^2 - 19x + 2x^2$
16. $17y^2 - 22y - y^2$
17. $10x^2 - 8x + 9x - 9x^2$
18. $-3y^2 - y - 6y^2 + 7y$
19. $\frac{1}{5}x^2 - \frac{3}{8}x + \frac{2}{3}x^2 + \frac{1}{4}x$
20. $\frac{6}{7}y^2 + \frac{1}{2}y - \frac{2}{3}y^2 + \frac{1}{5}y$
21. $0.6x^3 + 0.8x^4 + 0.7x^3 + (-0.8x^4)$
22. $1.9m^4 - 2.4m^6 - 3.7m^4 + 2.8m^6$
23. $\frac{1}{2}st + \frac{3}{2}st$
24. $\frac{2}{5}at + \frac{1}{5}at$
25. $-4ab + 4ab - ab$
26. $xy - 4xy - 2xy$
27. $4x^2y + 5 - 6x^3y - 3x^2y + 2x^3y$
28. $5b - 9ab^2 + 10a^3b - 8ab^2 - 9a^3b$

Add the polynomials. See Example 2.

29. $(3q^2 - 5q + 7) + (2q^2 + q - 12)$
30. $(2t^2 + 11t - 15) + (-5t^2 - 13t + 10)$
31. $\left(\frac{2}{3}y^3 + \frac{3}{4}y^2 + \frac{1}{2}\right) + \left(\frac{1}{3}y^3 + \frac{1}{5}y^2 - \frac{1}{6}\right)$
32. $\left(\frac{1}{16}r^6 + \frac{1}{2}r^3 - \frac{11}{12}\right) + \left(\frac{9}{16}r^6 + \frac{9}{4}r^3 + \frac{1}{12}\right)$
33. $(0.3p + 2.1q) + (0.4p - 3q)$
34. $(-0.3r - 5.2s) + (0.8r - 5.2s)$
35. $(2x^2 + xy + 3y^2) + (5x^2 - y^2)$
36. $(-4a^2 - ab + 15b^2) + (5a^2 - b^2)$

Find a polynomial that represents the perimeter of the figure. See Example 3.

37.
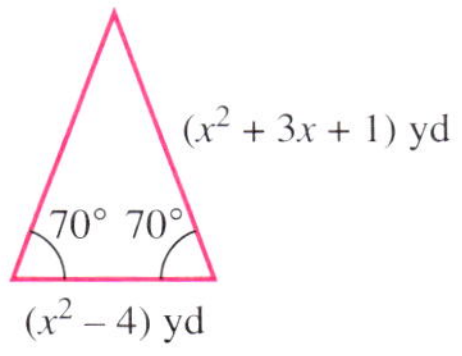

38.
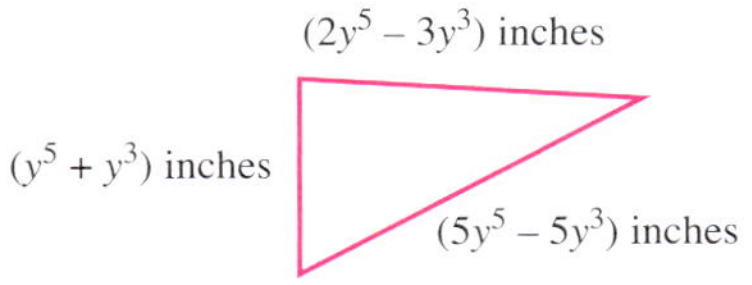

39.
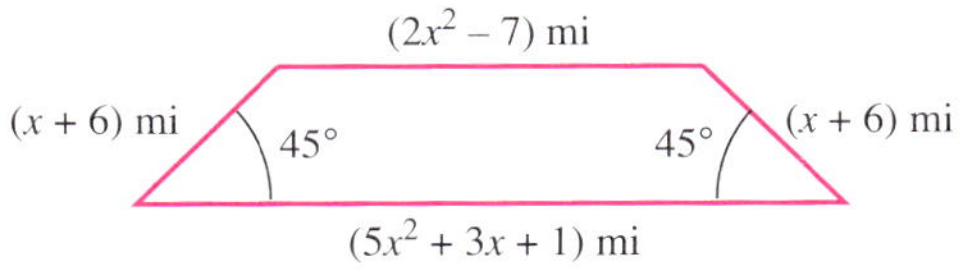

40.
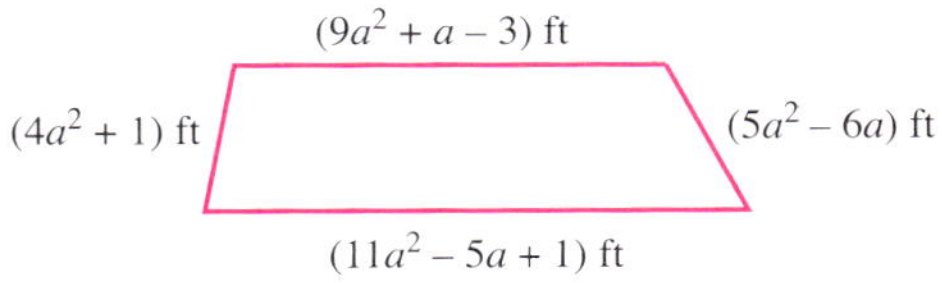

Use vertical form to add the polynomials. See Example 4.

41.
$$\begin{array}{r} 3x^2 + 4x + 5 \\ \underline{2x^2 - 3x + 6} \end{array}$$

42.
$$\begin{array}{r} 6x^3 - 4x^2 + 7 \\ \underline{7x^3 + 9x^2 + 12} \end{array}$$

43. $\begin{array}{r} 6a^2 + 7a + 9 \\ \underline{-9a^2 \qquad - 2} \end{array}$

44. $\begin{array}{r} -2c^2 - 3c - 5 \\ \underline{14c^2 \qquad - 1} \end{array}$

45. $\begin{array}{r} z^3 + 6z^2 - 7z + 16 \\ \underline{9z^3 - 6z^2 + 8z - 18} \end{array}$

46. $\begin{array}{r} 3x^3 + 4x^2 - 3x + 5 \\ \underline{3x^3 - 4x^2 - x - 7} \end{array}$

47. $\begin{array}{r} -3x^3y^2 + 4x^2y - 4x + 9 \\ \underline{2x^3y^2 \qquad\quad + 9x - 3} \end{array}$

48. $\begin{array}{r} 3x^2y^2 + 4xy + 25 \\ \underline{5x^2y^2 \qquad - 12} \end{array}$

Subtract the polynomials. See Example 5.

49. $(3a^2 - 2a + 4) - (a^2 - 3a + 7)$
50. $(2b^2 + 3b - 5) - (2b^2 - 4b - 9)$
51. $(-4h^3 + 5h^2 + 15) - (h^3 - 15)$
52. $(-c^5 + 5c^4 - 12) - (2c^5 - c^4)$
53. $\left(\frac{3}{8}s^8 - \frac{3}{4}s^7\right) - \left(\frac{1}{3}s^8 + \frac{1}{5}s^7\right)$
54. $\left(\frac{5}{6}q^9 - \frac{4}{5}q^8\right) - \left(\frac{1}{4}q^9 + \frac{3}{8}q^8\right)$
55. $(5ab + 2b^2) - (2 + ab + b^2)$
56. $(mn + 8n^2) - (6 - 5mn + n^2)$

Use vertical form to subtract the polynomials. See Example 6.

57. Subtract $2x^2 - 2x + 3$ from $3x^2 + 4x + 5$.

58. Subtract $3y^2 - 6y + 7$ from $6y^2 + 4y + 13$.

59. Subtract $(s^2 + 4s + 2)$ from $(5s^2 + 9)$.

60. Subtract $(4p^2 - 4p - 40)$ from $(10p^2 - 30)$.

61. Subtract $8a^3 + 8a^2 - 3a + 1$ from $17a^3 + 25a - 10$.

62. Subtract $m^3 + 20m^2 - 15m + 39$ from $-4m^3 - m + 22$.

63. $\begin{array}{r} 0.8x^3 \qquad\qquad - 2.3x + 0.6 \\ \underline{-(0.2x^3 - 1.2x^2 - 3.6x + 0.9)} \end{array}$

64. $\begin{array}{r} 9.7y^3 \qquad\qquad + \quad y + 1.1 \\ \underline{-(6.3y^3 - 4.4y^2 + 2.7y + 8.8)} \end{array}$

Perform the operations. See Example 7.

65. Subtract $(3x^2 + 4x - 7)$ from the sum of $(-2x^2 - 7x + 1)$ and $(-4x^2 + 8x - 7)$.

66. Subtract $(32x^2 - 17x + 45)$ from the sum of $(23x^2 - 12x - 7)$ and $(-11x^2 + 12x + 7)$.

67. Subtract $t^3 - 2t^2 + 2$ from the sum of $3t^3 + t^2$ and $-t^3 + 6t - 3$.

68. Subtract $-3z^3 - 4z + 7$ from the sum of $2z^2 + 3z - 7$ and $-4z^3 - 2z - 3$.

TRY IT YOURSELF

Perform the operations.

69. $(9a^2 + 3a) - (2a - 4a^2)$
70. $(4b^2 + 3b) - (7b - b^2)$
71. Subtract $(-y^5 + 5y^4 - 1.2)$ from $(2y^5 - y^4)$.

72. Subtract $(-4w^3 + 5w^2 + 7.6)$ from $(w^3 - 15w^2)$.

73. $3r^4 - 4r + 7r^4$

74. $-2b^4 + 7b - 3b^4$

75. $(0.03f^2 + 0.25f + 0.91) - (0.17f^2 - 1.18)$

76. $(0.05r^2 - 0.33r) - (0.48r^2 + 0.15r + 2.14)$

77. $\left(\frac{7}{8}r^4 + \frac{5}{9}r^2 - \frac{9}{4}\right) - \left(-\frac{3}{8}r^4 - \frac{2}{3}r^2 - \frac{1}{4}\right)$

78. $\left(\frac{4}{5}t^4 - \frac{1}{3}t^2 + \frac{1}{2}\right) - \left(-\frac{1}{2}t^4 + \frac{3}{8}t^2 - \frac{1}{16}\right)$

79. $\begin{array}{r} 8c^2 - 4c - 5 \\ \underline{-(-c^2 + 2c + 9)} \end{array}$

80. $\begin{array}{r} 3t^3 - 4t^2 - 3t + 5 \\ \underline{+11t^3 \qquad\quad - 8t - 2} \end{array}$

81. $(12.1h^3 + 9.9h^2) + (7.3h^3 + 1.1h^2)$
82. $(5.7n^3 - 2.1n) + (-6.2n^3 - 3.9n)$
83. $(20 - 4rt - 5r^2t) + (10 - 5rt)$
84. $(5m^2 - 8m + 8) - (-20m^2 + m)$
85. $(3x^2 - 3x - 2) + (3x^2 + 4x - 3)$
86. $(4c^2 + 3c - 2) + (3c^2 + 4c + 2)$
87. $\frac{2}{3}d^2 - \frac{1}{4}c^2 + \frac{5}{6}c^2 - \frac{1}{2}cd + \frac{1}{3}d^2$
88. $\frac{3}{5}s^2 - \frac{2}{5}t^2 - \frac{1}{2}s^2 - \frac{7}{10}st - \frac{3}{10}st$
89. $(3x + 7) + (4x - 3)$
90. $(2y - 3) + (4y + 7)$
91. Subtract $1.7t^2 - 1.1t$ from the sum of $-2.7t^2 + 2.1t - 1.7$ and $3.1t^2 - 2.5t + 2.3$.

92. Subtract $1.07x^2 - 2.07x$ from the sum of $1.04x^2 - 5.01$ and $1.33x - 1.9x^2 + 5.02$.

93. $-32u^3 - 16u^3$

94. $-25x^3 - 7x^3$

95. $(9d^2 + 6d) + (8d - 4d^2)$
96. $(2c^2 - 4c) + (8c - c^2)$

97. $\begin{array}{r} 3x^3y^2 + 4x^2y + 7x + 12 \\ \underline{-(-4x^3y^2 + 6x^2y + 9x - 3)} \end{array}$

98. $\begin{array}{r} -2x^2y^2 \qquad\quad + 12y^2 \\ \underline{-(10x^2y^2 + 9xy - 24y^2)} \end{array}$

99. $(2x^2 - 3x + 1) - (4x^2 - 3x + 2) + (2x^2 + 3x + 2)$

100. $(-3z^2 - 4z + 7) + (2z^2 + 2z - 1) - (2z^2 - 3z + 7)$

101. $\begin{array}{r} 4x^3 + 4x^2 - 3x + 10 \\ +(5x^3 - 2x^2 - 4x - 4) \\ \hline \end{array}$

102. $\begin{array}{r} 7m^5 + m^3 + 9m^2 - m \\ -(8m^5 - 2m^3 + m^2 + m) \\ \hline \end{array}$

Look Alikes . . .

103. **a.** $(-8x^2 - 3x) + (-11x^2 + 6x + 10)$
b. $(-8x^2 - 3x) - (-11x^2 + 6x + 10)$

104. **a.** $(10 - 2st - 3s^2t) + (4 - 6st)$
b. $(10 - 2st - 3s^2t) - (4 - 6st)$

APPLICATIONS

105. Greek Architecture.

a. Find a polynomial that represents the difference in the heights of the columns.

b. If the columns were stacked one atop the other, to what height would they reach?

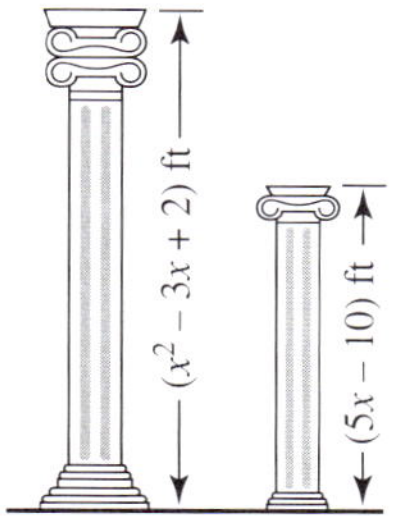

106. Jets. Find a polynomial that represents the length of the larger jet.

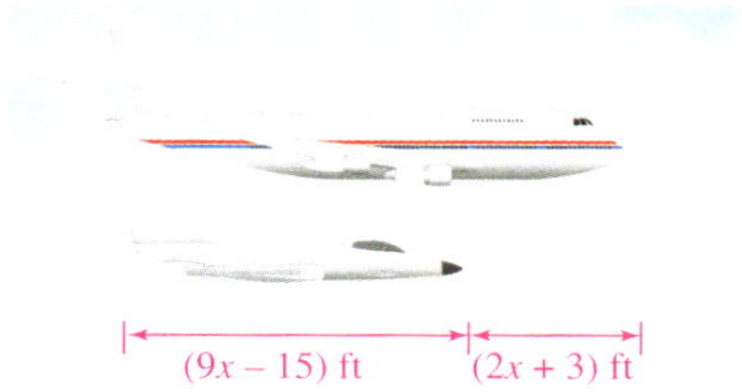

107. Piñatas. Find a polynomial that represents the length of the rope used to hold up the piñata.

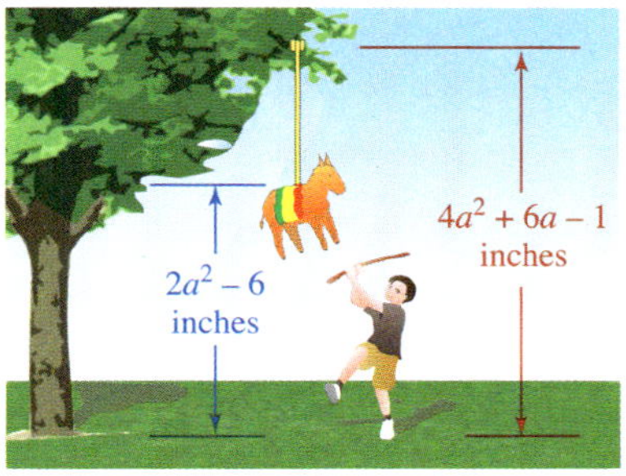

108. Reading Blueprints. Find a polynomial that represents

a. the difference in the length and width of the one-bedroom apartment.

b. the perimeter of the apartment.

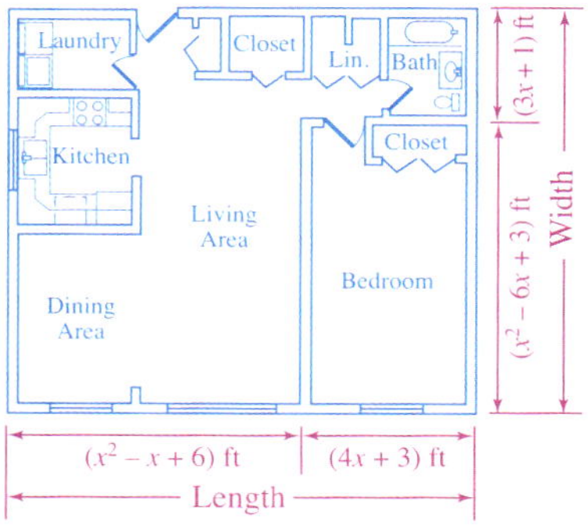

109. Naval Operations. Two warning flares are fired upward at the same time from different parts of a ship. The height of the first flare is $(-16t^2 + 128t + 20)$ feet and the height of the higher-traveling second flare is $(-16t^2 + 150t + 40)$ feet, after t seconds.

a. Find a polynomial that represents the difference in the heights of the flares.

b. In 4 seconds, the first flare reaches its peak, explodes, and lights up the sky. How much higher is the second flare at that time?

110. Auto Mechanics. The length of a fan belt that wraps around three pulleys is $(3x^2 + 11x + 4.5\pi)$ in. Find a polynomial that represents the unknown length of a part of the belt shown in the illustration below.

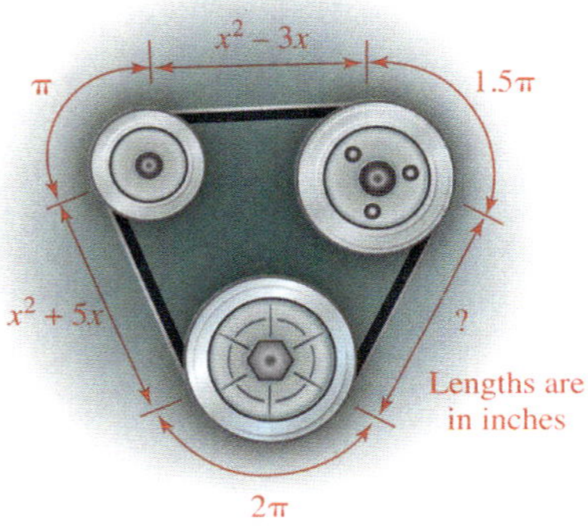

WRITING

111. How do you recognize like terms?

112. Explain why the vertical form used in algebra to add $2x^2 + 4x + 3$ and $5x^2 + 3x + 6$ is similar to the vertical form used in arithmetic to add 243 and 536.

113. Explain the error below.

$$7x^2y + 6x^2y = 13x^4y^2$$

114. Explain the error below.

$$(12x^2 - 4) - (3x^2 - 1) = 12x^2 - 4 - 3x^2 - 1$$
$$= 9x^2 - 5$$

115. A student was asked to simplify $\frac{1}{6}x^2 - 3 + \frac{2}{3}x^2$. Explain the error below:

$$6\left(\frac{1}{6}x^2 - 3 + \frac{2}{3}x^2\right) = x^2 - 18 + 4x^2$$
$$= 5x^2 - 18$$

116. Explain the error below.

Subtract $(2d^2 - d - 3)$ from $(d^2 - 9)$:

$$(2d^2 - d - 3) - (d^2 - 9) = d^2 - d + 6$$

REVIEW

117. What is the sum of the measures of the angles of a triangle?

118. What are the formulas for the area of a circle and the area of a triangle?

119. Graph: $y = -\frac{1}{2}x + 2$

120. Graph: $2x + 3y = 9$

CHALLENGE PROBLEMS

121. What polynomial must be added to $2x^2 - x + 3$ so that the sum is $6x^2 - 7x - 8$?

122. Is the sum of two trinomials always a trinomial? Explain why or why not.

SECTION 5.6 Multiplying Polynomials

OBJECTIVES

1. Multiply monomials.
2. Multiply a polynomial by a monomial.
3. Multiply binomials.
4. Multiply polynomials.

ARE YOU READY?

The following problems review some basic skills that are needed when multiplying polynomials.

1. Multiply: $5 \cdot 10a$
2. Multiply: $5(2x - 3)$
3. Multiply: $(4y + 8)7$
4. Multiply: $x^8 \cdot x^6$
5. Multiply: $-2(3)(5)$
6. Simplify by combining like terms: $9x^2 + 6x - 3x + 2$

We now discuss multiplying polynomials. We will begin with the simplest case—finding the product of two monomials.

1 Multiply Monomials.

To find the product of two monomials, such as $8x^2$ and $3x^4$, we use the commutative and associative properties of multiplication to reorder and regroup the factors.

$$(8x^2)(3x^4) = (8 \cdot 3)(x^2 \cdot x^4)$$ Group the coefficients together and the variables together.

$$= 24x^6$$ Multiply: $8 \cdot 3 = 24$. Simplify: $x^2 \cdot x^4 = x^{2+4} = x^6$.

This example illustrates the following rule.

Multiplying Two Monomials

To multiply two monomials, multiply the numerical factors (the coefficients) and then multiply the variable factors.

EXAMPLE 1 Multiply: **a.** $6r \cdot r$ **b.** $3t^4(-2t^5)$ **c.** $\left(\frac{1}{3}a^2b^3\right)(21ab^2)$ **d.** $-4y^5z^2(2y^3z^3)(3yz)$

Strategy We will multiply the numerical factors and then multiply the variable factors.

Why The commutative and associative properties of multiplication enable us to reorder and regroup the factors.

Solution In each case, we will *multiply* the coefficients and *add* the exponents of the like bases.

a. $6r \cdot r = 6r^2$ Recall that $r = 1r$. Think: $6 \cdot 1 = 6$ and $r \cdot r = r^2$.

b. $(3t^4)(-2t^5) = -6t^9$ Think: $3(-2) = -6$ and $t^4 \cdot t^5 = t^{4+5} = t^9$.

c. $\left(\frac{1}{3}a^2b^3\right)(21ab^2) = 7a^3b^5$ Think: $\frac{1}{3} \cdot 21 = \frac{21}{3} = 7$, $a^2 \cdot a = a^3$, and $b^3 \cdot b^2 = b^5$.

d. $-4y^5z^2(2y^3z^3)(3yz) = -24y^9z^6$ Think: $-4(2)(3) = -24$, $y^5 \cdot y^3 \cdot y = y^9$, and $z^2 \cdot z^3 \cdot z = z^6$.

Caution

Only like terms can be added or subtracted.

Like terms:

$6r + r = 7r$

Unlike terms:

$3t^4 - 2t^5$ Does not simplify

However, like and unlike terms may be multiplied.

$6r \cdot r = 6r^2$

$(3t^4)(-2t^5) = -6t^9$

Self Check 1 Multiply: **a.** $18t \cdot t$ **b.** $-10d^8(-6d^3)$ **c.** $(16y^{12})\left(\frac{1}{4}y^2\right)$ **d.** $(5a^3b^3)(-6a^3b^4)(ab)$

Now Try Problems 13, 19, and 21

2 Multiply a Polynomial by a Monomial.

We can use the distributive property to find the product of a monomial and a binomial such as $5x$ and $2x + 4$:

$$5x(2x + 4) = 5x(2x) + 5x(4)$$ Read as "5x times the quantity of 2x plus 4." Distribute the multiplication by 5x.

$$= 10x^2 + 20x$$ Multiply the monomials.

This example suggests the following rule.

Multiplying Polynomials by Monomials

To multiply a monomial and a polynomial, multiply each term of the polynomial by the monomial.

EXAMPLE 2 Multiply: **a.** $3n^6(16n^{15} + n^{10})$ **b.** $3a^2(3a^2 - 5a + 2)$ **c.** $-2xz^3(6x^3z + x^2z^2 - xz^3 + 7z^4)$ **d.** $(-m^4 - 2.5)(4.1m^3)$

Strategy To find each product, we will multiply each term of the polynomial by the monomial.

Why We use the distributive property to multiply a monomial and a polynomial.

Solution

a. Multiply each term of the binomial $16n^{15} + n^{10}$ by the monomial $3n^6$.

$$3n^6(16n^{15} + n^{10}) = 3n^6(16n^{15}) + 3n^6(n^{10})$$ Distribute the multiplication by $3n^6$.

$$= 48n^{21} + 3n^{16}$$ Multiply the monomials.

b. Multiply each term of the trinomial $3a^2 - 5a + 2$ by the monomial $3a^2$.

$$3a^2(3a^2 - 5a + 2)$$

$$= 3a^2(3a^2) + 3a^2(-5a) + 3a^2(2)$$ Distribute the multiplication by $3a^2$.

$$= 9a^4 - 15a^3 + 6a^2$$ Multiply the monomials.

Success Tip

The rectangle below can be used to picture polynomial multiplication. The total area is $x(x + 2)$ and the sum of the two smaller areas is $x^2 + 2x$. Thus,

$$x(x + 2) = x^2 + 2x$$

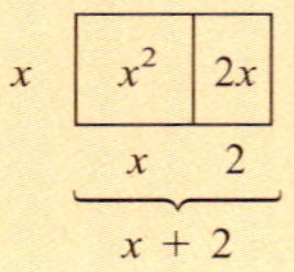

c. Multiply each term of $6x^3z + x^2z^2 - xz^3 + 7z^4$ by the monomial $-2xz^3$.

$$-2xz^3(6x^3z + x^2z^2 - xz^3 + 7z^4)$$
$$= -2xz^3(6x^3z) - 2xz^3(x^2z^2) - 2xz^3(-xz^3) - 2xz^3(7z^4)$$
$$= -12x^4z^4 - 2x^3z^5 + 2x^2z^6 - 14xz^7 \quad \text{Multiply the monomials.}$$

d. Multiply each term of the binomial $-m^4 - 2.5$ by the monomial $4.1m^3$.

$$(-m^4 - 2.5)(4.1m^3) = -m^4(4.1m^3) - 2.5(4.1m^3) \quad \text{Distribute the multiplication by } 4.1m^3.$$
$$= -4.1m^7 - 10.25m^3 \quad \text{Multiply the monomials.}$$

Self Check 2 Multiply: **a.** $22x(10x^5 + x^4)$
b. $5c^2(4c^3 - 9c - 8)$
c. $-s^2t^2(-s^4t^2 + s^3t^3 - s^2t^4 + 7s)$
d. $(w^7 - 2w)6w^5$

Now Try Problems 27, 33, and 35

EXAMPLE 3 Find a polynomial that represents the area of the parallelogram.

Strategy We will multiply the length of the base of the parallelogram by its height.

Why The area of a parallelogram is equal to the product of the length of its base and its height.

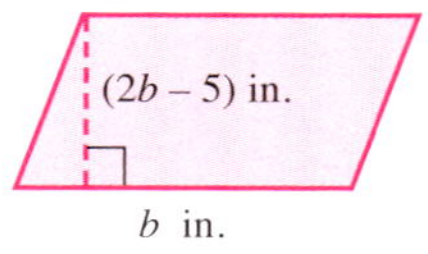

Solution

$$\text{Area} = \text{base} \cdot \text{height} \quad \text{This is the formula for the area of a parallelogram.}$$
$$= b(2b - 5) \quad b \text{ is the length of the base. Substitute } 2b - 5 \text{ for the height.}$$
$$= 2b^2 - 5b \quad \text{Distribute the multiplication by } b.$$

The area of the parallelogram is $(2b^2 - 5b)$ square inches, which can be written as $(2b^2 - 5b)$ in.^{2}.

The Language of Algebra

A **parallelogram** is a four-sided figure whose opposite sides are parallel.

Self Check 3 Find a polynomial that represents the area of a rectangle with length n^3 meters and width $(3n^2 + 2n - 9)$ meters.

Now Try Problem 37

3 Multiply Binomials.

The distributive property also can be used to multiply binomials. For example, to multiply $2a + 4$ and $3a + 5$, we think of $2a + 4$ as a single quantity and distribute it over each term of $3a + 5$.

$$(2a + 4)(3a + 5) = (2a + 4)3a + (2a + 4)5 \quad \text{Read as “the quantity of } 2a + 4 \text{ times the quantity of } 3a + 5\text{.”}$$
$$= (2a + 4)3a + (2a + 4)5$$
$$= (2a)3a + (4)3a + (2a)5 + (4)5 \quad \text{Distribute the multiplication by } 3a \text{ and by } 5.$$
$$= 6a^2 + 12a + 10a + 20 \quad \text{Multiply the monomials.}$$
$$= 6a^2 + 22a + 20 \quad \text{Combine like terms.}$$

In the third line of the solution, notice that each term of $3a + 5$ has been multiplied by each term of $2a + 4$. This example suggests the following rule.

Multiplying Two Binomials

To multiply two binomials, multiply each term of one binomial by each term of the other binomial, and then combine like terms.

EXAMPLE 4 Multiply: $(5x - 8)(x + 1)$

Strategy To find the product, we will multiply $x + 1$ by $5x$ and by -8.

Why To multiply two binomials, each term of one binomial must be multiplied by each term of the other binomial.

Solution

Notation

We will write the answers to polynomial multiplication in descending powers.

$(5x - 8)(x + 1) = 5x(x + 1) - 8(x + 1)$ Read as "the quantity of $5x - 8$ times the quantity of $x + 1$." Multiply $x + 1$ by $5x$ and multiply $x + 1$ by -8.

$= 5x^2 + 5x - 8x - 8$ Distribute the multiplication by $5x$. Distribute the multiplication by -8.

$= 5x^2 - 3x - 8$ Combine like terms.

Self Check 4 Multiply: $(9y + 3)(y - 4)$

Now Try Problem 45

The Language of Algebra

An **acronym** is an abbreviation of several words in such a way that the abbreviation itself forms a word. The *acronym* FOIL helps us remember the order to follow when multiplying two binomials: First, Outer, Inner, Last.

We can use a shortcut method, called the **FOIL method,** to multiply binomials. FOIL is an acronym for **F**irst terms, **O**uter terms, **I**nner terms, **L**ast terms. The FOIL method is a form of the distributive property. To use the FOIL method to multiply $2a + 4$ by $3a + 5$, we

1. Multiply the **F**irst terms $2a$ and $3a$ to obtain $6a^2$,
2. Multiply the **O**uter terms $2a$ and 5 to obtain $10a$,
3. Multiply the **I**nner terms 4 and $3a$ to obtain $12a$, and
4. Multiply the **L**ast terms 4 and 5 to obtain 20.

Then we simplify the resulting polynomial by combining like terms, if possible.

Outer, First, Inner, Last

$(2a + 4)(3a + 5) = \overset{F}{2a(3a)} + \overset{O}{2a(5)} + \overset{I}{4(3a)} + \overset{L}{4(5)}$ Distribute $2a$ over $3a + 5$. Distribute 4 over $3a + 5$.

$= 6a^2 + 10a + 12a + 20$ Multiply the monomials.

$= 6a^2 + 22a + 20$ Combine like terms: $10a + 12a = 22a$.

EXAMPLE 5 Multiply: **a.** $(x + 5)(x + 7)$ **b.** $(3x + 4)(2x - 3)$

c. $\left(2r - \frac{1}{2}\right)\left(2r + \frac{5}{2}\right)$ **d.** $(3a^2 - 7b)(a^2 - b)$

Strategy We will use the FOIL method.

Why In each case we are to find the product of two binomials, and the FOIL method is a shortcut for multiplying two binomials.

Solution

Success Tip

The area of the large rectangle is given by $(x + 5)(x + 7)$. The sum of the areas of the smaller rectangles is $x^2 + 7x + 5x + 35$ or $x^2 + 12x + 35$. Thus,

$$(x + 5)(x + 7) = x^2 + 12x + 35$$

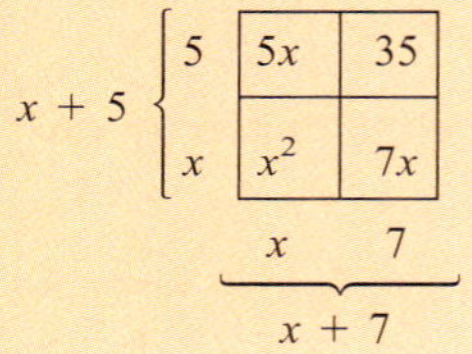

a.

$$(x + 5)(x + 7) = \overset{F}{x(x)} + \overset{O}{x(7)} + \overset{I}{5(x)} + \overset{L}{5(7)}$$

$$= x^2 + 7x + 5x + 35 \quad \text{Multiply the monomials.}$$

$$= x^2 + 12x + 35 \quad \text{Combine like terms.}$$

b.

$$(3x + 4)(2x - 3) = \overset{F}{3x(2x)} + \overset{O}{3x(-3)} + \overset{I}{4(2x)} + \overset{L}{4(-3)}$$

$$= 6x^2 - 9x + 8x - 12 \quad \text{Multiply the monomials.}$$

$$= 6x^2 - x - 12 \quad \text{Combine like terms.}$$

c.

$$\left(2r - \frac{1}{2}\right)\left(2r + \frac{5}{2}\right) = \overset{F}{2r(2r)} + \overset{O}{2r\left(\frac{5}{2}\right)} - \overset{I}{\frac{1}{2}(2r)} - \overset{L}{\frac{1}{2}\left(\frac{5}{2}\right)}$$

$$= 4r^2 + 5r - r - \frac{5}{4} \quad \text{Multiply the monomials.}$$

$$= 4r^2 + 4r - \frac{5}{4}$$

Combine like terms. We cannot clear the fraction because we began with an expression, not an equation.

d.

$$(3a^2 - 7b)(a^2 - b) = \overset{F}{3a^2(a^2)} + \overset{O}{3a^2(-b)} - \overset{I}{7b(a^2)} - \overset{L}{7b(-b)}$$

$$= 3a^4 - 3a^2b - 7a^2b + 7b^2 \quad \text{Multiply the monomials.}$$

$$= 3a^4 - 10a^2b + 7b^2 \quad \text{Combine like terms.}$$

Self Check 5 Multiply: **a.** $(y + 3)(y + 1)$
b. $(2a - 1)(3a + 2)$
c. $\left(4x - \frac{1}{2}\right)\left(4x + \frac{3}{4}\right)$
d. $(5y^3 - 2b)(2y^3 - 7b)$

Now Try Problems 47, 51, and 55

4 Multiply Polynomials.

To develop a general rule for multiplying any two polynomials, we will find the product of $2x + 3$ and $3x^2 + 3x + 5$. In the solution, the distributive property is used four times.

$$(2x + 3)(3x^2 + 3x + 5) = (2x + 3)3x^2 + (2x + 3)3x + (2x + 3)5 \quad \text{Distribute } (2x + 3).$$

$$= (2x + 3)3x^2 + (2x + 3)3x + (2x + 3)5$$

$$= (2x)3x^2 + (3)3x^2 + (2x)3x + (3)3x + (2x)5 + (3)5 \quad \text{Distribute.}$$

$$= 6x^3 + 9x^2 + 6x^2 + 9x + 10x + 15 \quad \text{Multiply the monomials.}$$

$$= 6x^3 + 15x^2 + 19x + 15 \quad \text{Combine like terms.}$$

Success Tip

In this section, you will see that every polynomial multiplication is basically a series of monomial multiplications. Then, if necessary, we combine like terms.

In the third line of the solution, note that each term of $3x^2 + 3x + 5$ has been multiplied by each term of $2x + 3$. This example suggests the following rule.

Multiplying Two Polynomials

To multiply two polynomials, multiply each term of one polynomial by each term of the other polynomial, and then combine like terms.

EXAMPLE 6 Multiply: $(7y + 3)(6y^2 - 8y + 1)$

Strategy We will multiply each term of the trinomial, $6y^2 - 8y + 1$, by each term of the binomial, $7y + 3$.

Why To multiply two polynomials, we must multiply each term of one polynomial by each term of the other polynomial.

Solution

Success Tip

The FOIL method cannot be applied here—only to products of two binomials.

$(7y + 3)(6y^2 - 8y + 1)$ Read as "the quantity of $7y + 3$ times the quantity of $6y^2 - 8y + 1$."

$= 7y(6y^2) + 7y(-8y) + 7y(1) + 3(6y^2) + 3(-8y) + 3(1)$

$= 42y^3 - 56y^2 + 7y + 18y^2 - 24y + 3$ Multiply the monomials.

$= 42y^3 - 38y^2 - 17y + 3$ Combine like terms.

Self Check 6 Multiply: $(3a^2 - 1)(2a^4 - a^2 - a)$

Now Try Problem 59

It is often convenient to multiply polynomials using a vertical form similar to that used to multiply whole numbers.

EXAMPLE 7 Multiply using vertical form: **a.** $(3a^2 - 4a + 7)(2a + 5)$ **b.** $(6y^3 - 5y + 4)(-4y^2 - 3)$

Strategy First, we will write one polynomial underneath the other and draw a horizontal line beneath them. Then, we will multiply each term of the upper polynomial by each term of the lower polynomial.

Why *Vertical form* means to use an approach similar to that used in arithmetic to multiply two numbers.

Solution

Success Tip

Multiplying two polynomials in vertical form is much like multiplying two numbers in arithmetic.

$$\begin{array}{r} 347 \\ \times\ 25 \\ \hline 1{,}735 \\ +694 \\ \hline 8{,}675 \end{array}$$

a. Multiply:

$$\begin{array}{rl} 3a^2 - 4a + 7 & \\ 2a + 5 & \\ \hline 15a^2 - 20a + 35 & \text{Multiply } 3a^2 - 4a + 7 \text{ by } 5. \\ 6a^3 - 8a^2 + 14a & \text{Multiply } 3a^2 - 4a + 7 \text{ by } 2a. \text{ Line up like terms.} \\ \hline 6a^3 + 7a^2 - 6a + 35 & \text{In each column, combine like terms.} \end{array}$$

b. With this method, it is often necessary to leave a space for a missing term to align like terms vertically.

Multiply:

$$\begin{array}{rl} 6y^3 - 5y + 4 & \\ -4y^2 - 3 & \\ \hline -18y^3 \qquad + 15y - 12 & \text{Multiply } 6y^3 - 5y + 4 \text{ by } -3. \\ -24y^5 + 20y^3 - 16y^2 \qquad\qquad & \text{Multiply } 6y^3 - 5y + 4 \text{ by } -4y^2. \\ \hline -24y^5 + 2y^3 - 16y^2 + 15y - 12 & \text{Leave a space for any missing powers of } y. \\ & \text{In each column, combine like terms.} \end{array}$$

The Language of Algebra

The pair of polynomials written below the first horizontal line, $-18y^3 + 15y - 12$ and $-24y^5 + 20y^3 - 16y^2$, are called partial products.

Self Check 7 Multiply using vertical form:

a. $(3x + 2)(2x^2 - 4x + 5)$

b. $(-2x^2 + 3)(2x^2 - 4x - 1)$

Now Try Problem 65

Multiplying Three Polynomials

To multiply three polynomials, multiply any two of them, and then multiply that result by the third polynomial.

EXAMPLE 8 Multiply: $-3a(4a + 1)(a - 7)$

Strategy We will find the product of $4a + 1$ and $a - 7$ and then multiply that result by $-3a$.

Why It is wise to perform the most difficult multiplication first. (In this case, that would be the product of the two binomials). Save the simpler multiplication by $-3a$ for last.

Solution

$$-3a(4a + 1)(a - 7) = -3a(4a^2 - 28a + a - 7)$$ Multiply the two binomials.

$$= -3a(4a^2 - 27a - 7)$$ Combine like terms within the parentheses: $-28a + a = -27a$.

$$= -12a^3 + 81a^2 + 21a$$ Distribute the multiplication by $-3a$.

Self Check 8 Multiply: $-2y(y + 3)(3y - 2)$

Now Try Problem 69

SECTION 5.6 STUDY SET

VOCABULARY

Fill in the blanks.

1. $(2x^3)(3x^4)$ is the product of two ________ and $(2a - 4)(3a + 5)$ is the product of two ________.
2. We read $(x + 7)(2x - 3)$ as "the ________ of $x + 7$ ______ the quantity of $2x - 3$."
3. In the acronym FOIL, F stands for _____ terms, O for ______ terms, I for ______ terms, and L for _____ terms.
4. $(2a - 4)(3a^2 + 5a - 1)$ is the product of a ________ and a ________.

CONCEPTS

Fill in the blanks.

5. a. To multiply two polynomials, multiply ______ term of one polynomial by ______ term of the other polynomial, and then combine like terms.
   b. When multiplying three polynomials, we begin by multiplying _____ two of them, and then we multiply that result by the ______ polynomial.
6. Label each arrow using one of the letters F, O, I, or L. Then fill in the blanks.

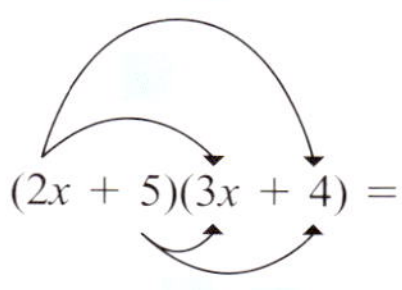

First Outer Inner Last

$(2x + 5)(3x + 4) = \quad + \quad + \quad + \quad$

7. Simplify each polynomial by combining like terms.
   a. $6x^2 - 8x + 9x - 12$
   b. $5x^4 + 3ax^2 + 5ax^2 + 3a^2$
8. $(3a)(2a^2)$ can be classified as a monomial · monomial. Classify the following products by identifying the types of polynomial factors.
   a. $6x(x - 7)$
   b. $(9a + 1)(5a - 3)$
   c. $(c - d)(c^2 - c + d)$
   d. $6m(m^2 + 1)(m^2 - 1)$

NOTATION

Complete each solution.

9. $(9n^3)(8n^2) = (9 \cdot \quad)(\quad \cdot n^2) =$

10. $7x(3x^2 - 2x + 5) = \quad (3x^2) - \quad (2x) + \quad (5)$
$= \quad - 14x^2 + 35x$

11. $(2x - 5)(3x - 2) = 2x(3x) - \quad (2) + \quad (3x) - \quad (2)$
$= 6x^2 - \quad + \quad - 10$
$= 6x^2 + \quad - 10$

12.
$$\begin{array}{r} 3x^2 + 4x - 2 \\ 2x + 3 \\ \hline \quad + 12x - 6 \\ 6x^3 + 8x^2 - 4x \quad \\ \hline \quad + 17x^2 + \quad - 6 \end{array}$$

GUIDED PRACTICE

Multiply. See Example 1.

13. $5m \cdot m$
14. $4s \cdot s$
15. $(3x^2)(4x^3)$
16. $(-2a^3)(11a^2)$
17. $(1.2c^3)(5c^3)$
18. $(2.5h^4)(2h^4)$
19. $(3b^2)(-2b)(4b^3)$
20. $(3y)(7y^2)(-y^4)$
21. $(2x^2y^3)(4x^3y^2)$
22. $(-5x^3y^6)(2x^2y^2)$
23. $(8a^5)\left(-\frac{1}{4}a^6\right)$
24. $\left(-\frac{2}{3}x^6\right)(9x^3)$

Multiply. See Example 2.

25. $3x(x + 4)$
26. $3a(a + 2)$
27. $-4t(t^2 - 7)$
28. $-6s(s^2 - 3)$
29. $-2x^3(3x^2 - x + 1)$
30. $-4b^3(2b^2 - 2b + 2)$
31. $\frac{5}{8}t^2(t^6 + 8t^2)$
32. $\frac{4}{9}a^2(9a^3 + a^2)$
33. $-4x^2z(3x^2 + z^2 + xz - 1)$
34. $-3x^2y(x^2 + y^2 + xy - 1)$
35. $(x^2 - 12x)(6x^{12})$
36. $(w^9 - 11w)(2w^7)$

Find a polynomial that represents the area of the parallelogram or rectangle. See Example 3.

37.
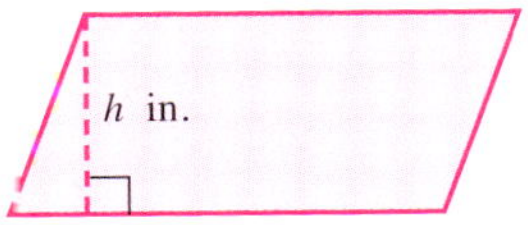

38.
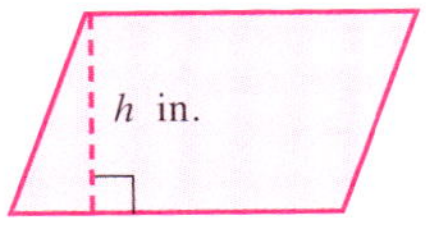

39.
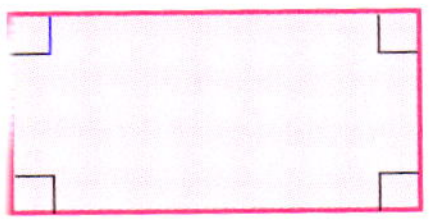

40.
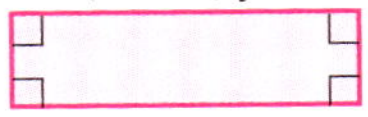

Multiply. See Examples 4 and 5.

41. $(y + 3)(y + 5)$
42. $(a + 4)(a + 5)$
43. $(m + 6)(m - 9)$
44. $(n + 8)(n - 10)$
45. $(4y - 5)(y + 7)$
46. $(3x - 4)(x + 5)$
47. $(2x - 3)(6x - 5)$
48. $(5x - 3)(2x - 3)$
49. $(3.8y - 1)(2y - 1)$
50. $(2.6x - 3)(2x - 1)$
51. $\left(6m - \frac{2}{3}\right)\left(3m - \frac{4}{3}\right)$
52. $\left(8t - \frac{1}{2}\right)\left(4t - \frac{5}{2}\right)$
53. $(t^2 - 3)(t^2 - 4)$
54. $(s^3 - 6)(s^3 - 8)$
55. $(3a - 2b)(4a + b)$
56. $(2t + 3s)(3t - s)$

Multiply. See Example 6.

57. $(x + 2)(x^2 - 2x + 3)$
58. $(x - 5)(x^2 + 2x - 3)$
59. $(4t + 3)(t^2 + 2t + 3)$
60. $(3x + 1)(2x^2 - 3x + 1)$
61. $(x^2 + 6x + 7)(2x - 5)$
62. $(y^2 - 2y + 1)(4y + 8)$
63. $(r^2 - r + 3)(r^2 - 4r - 5)$
64. $(w^2 + w - 9)(w^2 - w + 3)$

Multiply using vertical form. See Example 7.

65. $\begin{array}{r} x^2 - 2x + 1 \\ x + 2 \\ \hline \end{array}$

66. $\begin{array}{r} 5r^2 + r + 6 \\ 2r - 1 \\ \hline \end{array}$

67. $\begin{array}{r} 4x^2 + 3x - 4 \\ 3x + 2 \\ \hline \end{array}$

68. $\begin{array}{r} x^2 - x + 1 \\ x + 1 \\ \hline \end{array}$

Multiply. See Example 8.

69. $4x(2x + 1)(x - 2)$
70. $5a(3a - 2)(2a + 3)$
71. $-3a(a + b)(a - b)$
72. $-2r(r + s)(r + s)$
73. $(-2a^2)(-3a^3)(3a - 2)$
74. $(3x)(-2x^2)(x + 4)$
75. $(x - 4)(x + 1)(x - 3)$
76. $(x + 6)(x - 2)(x - 4)$

TRY IT YOURSELF

Multiply.

77. $(5x - 2)(6x - 1)$
78. $(8x - 1)(3x - 7)$
79. $(3x^2 + 4x - 7)(2x^2)$
80. $(2y^2 - 7y - 8)(3y^3)$
81. $2(t + 4)(t - 3)$
82. $4(x + 7)(x - 6)$

83. $\begin{array}{r} 2a^2 + 3a + 1 \\ \underline{3a^2 - 2a + 4} \end{array}$

84. $\begin{array}{r} 3y^2 + 2y - 4 \\ \underline{2y^2 - 4y + 3} \end{array}$

85. $(t + 2s)(9t - 3s)$

86. $(4t - u)(3t + u)$

87. $\left(\frac{1}{2}a\right)(4a^4)(a^5)$

88. $(12b)\left(\frac{7}{6}b\right)(b^4)$

89. $\left(4a - \frac{5}{4}r\right)\left(4a + \frac{3}{4}r\right)$

90. $\left(5c - \frac{2}{5}t\right)\left(10c + \frac{1}{5}t\right)$

91. $(a + b)(a + b)$

92. $(m + n)(m + n)$

93. $(x + 6)(x^3 + 5x^2 - 4x - 4)$

94. $(x - 8)(x^3 - 4x^2 - 2x - 2)$

95. $9x^2(x^2 - 2x + 6)$

96. $4y^2(y^2 + 5y - 10)$

97. $4y(y + 3)(y + 7)$

98. $2t(t + 8)(t + 10)$

99. $0.3p^5(0.4p^4 - 6p^2)$

100. $0.5u^5(0.4u^6 - 0.5u^3)$

101. $8.2pq(2pq - 3p + 5q)$

102. $5.3ab(2ab + 6a - 3b)$

103. $(-3x + y)(x^2 - 8xy + 16y^2)$

104. $(3x - y)(x^2 + 3xy - y^2)$

Look Alikes . . .

Perform the indicated operations to simplify each expression, if possible.

105. a. $(x - 2) + (x^2 + 2x + 4)$ **b.** $(x - 2)(x^2 + 2x + 4)$

106. a. $(a + 3) + (a^2 - 3a + 9)$ **b.** $(a + 3)(a^2 - 3a + 9)$

107. a. $(6x^2z^5) - (-3xz^3)$ **b.** $(6x^2z^5)(-3xz^3)$

108. a. $(-5r^4t^2) - (2r^2t)$ **b.** $(-5r^4t^2)(2r^2t)$

109. a. $(2x^2 - x) - (3x^2 - 3x)$ **b.** $(2x^2 - x)(3x^2 - 3x)$

110. a. $(4.9a - b) - (2a + b)$ **b.** $(4.9a - b)(2a + b)$

111. a. $3a + (4a - 1) + (6a + 2)$

b. $3a(4a - 1)(6a + 2)$

112. a. $\left(\frac{1}{2}c - 3d\right) + \left(\frac{3}{4}c + d\right)$ **b.** $\left(\frac{1}{2}c - 3d\right)\left(\frac{3}{4}c + d\right)$

APPLICATIONS

113. Stamps. Find a polynomial that represents the area of the stamp.

114. Parking. Find a polynomial that represents the total area of the van-accessible parking space and its access aisle.

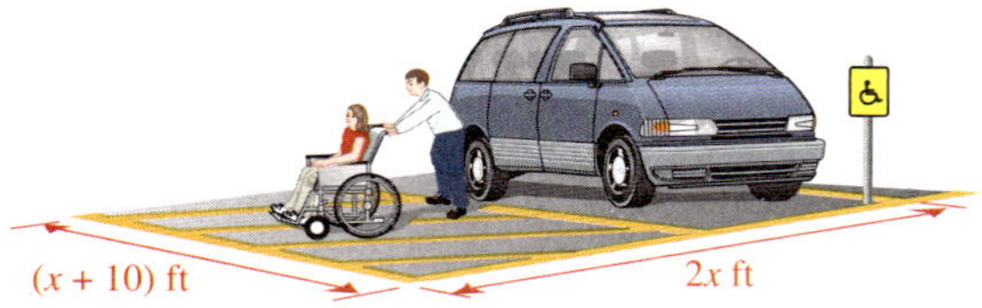

115. Sunglasses. An ellipse is an oval-shaped curve. The area of an ellipse is approximately $0.785lw$, where l is its length and w is its width. Find a polynomial that represents the approximate area of one of the elliptical-shaped lenses of the sunglasses.

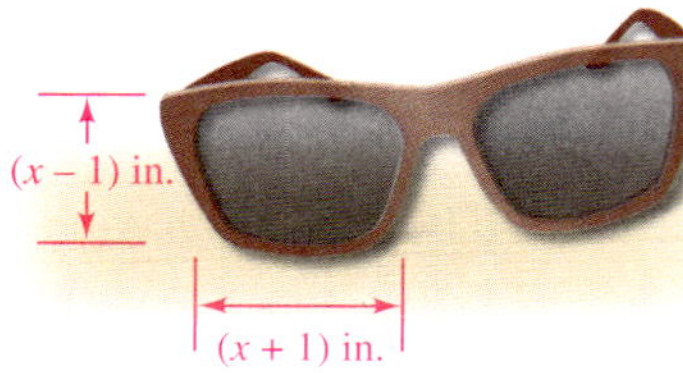

116. Gardening. Refer to the illustration below.

a. Find the area of the region planted with corn, tomatoes, beans, and carrots. Add your answers to find the total area of the garden.

b. Find the length and width of the garden. Multiply your answers to find its area.

c. How do the answers from parts (a) and (b) for the area of the garden compare?

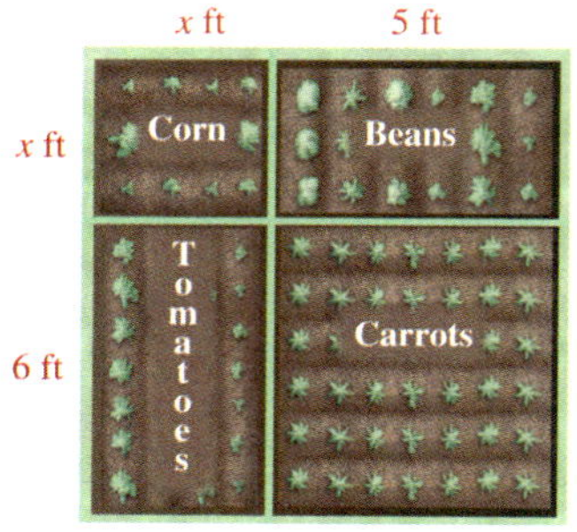

117. Luggage. Find a polynomial that represents the volume of the garment bag. (Recall that the formula for the volume of a rectangular solid is $V = lwh$.)

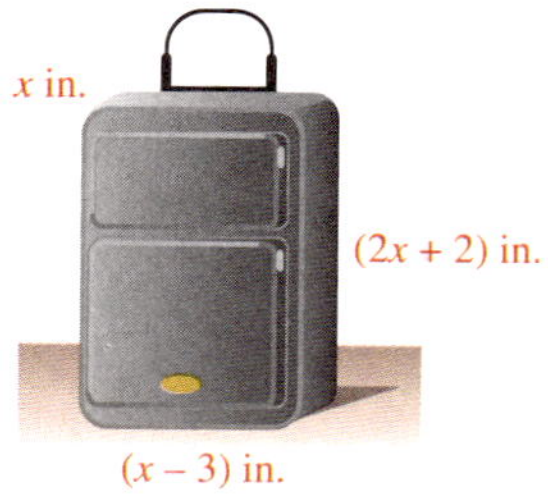

118. **Baseball.** Find a polynomial that represents the volume within the batting cage.

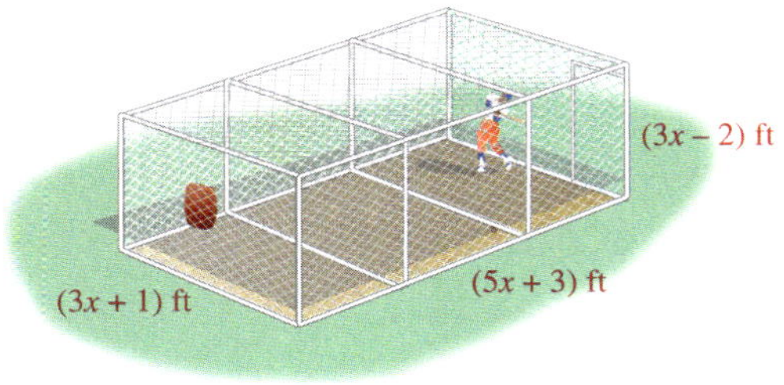

WRITING

119. Is the product of a monomial and a monomial always a monomial? Explain.

120. Explain this diagram.

$$(5x + 6)(7x - 1)$$

121. Explain why the FOIL method cannot be used to find $(3x + 2)(4x^2 - x + 10)$.

122. Explain the error: $(x + 3)(x - 2) = x^2 - 6$

123. Explain why the vertical form used in algebra to multiply $2x^2 + 3x + 1$ and $3x + 2$ is similar to the vertical form used in arithmetic to multiply 231 and 32.

124. Would the OLIF method give the same result as the FOIL method when multiplying two binomials? Explain why or why not.

REVIEW

125. What is the slope of
   a. Line 1?
   b. Line 2?
   c. Line 3?
   d. the x-axis?

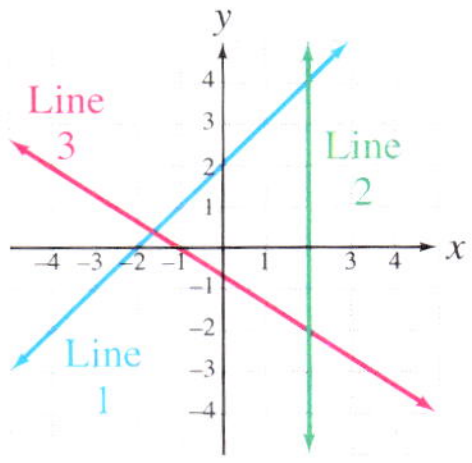

126. a. What is the y-intercept of Line 1?
   b. What is the x-intercept of Line 1?

CHALLENGE PROBLEMS

127. a. Find each of the following products.
   i. $(x - 1)(x + 1)$
   ii. $(x - 1)(x^2 + x + 1)$
   iii. $(x - 1)(x^3 + x^2 + x + 1)$
   b. Write a product of two polynomials such that the result is $x^5 - 1$.

128. Solve: $(y - 1)(y + 6) = (y - 3)(y - 2) + 8$

SECTION 5.7 Special Products

OBJECTIVES

1. Square a binomial.
2. Multiply the sum and difference of the same two terms.
3. Find higher powers of binomials.
4. Simplify expressions containing polynomial multiplication.

ARE YOU READY?

The following problems review some basic skills that are needed when finding special products.

1. Simplify by combining like terms: $x^2 + 2x + 2x + 4$
2. In each expression, identify the base and the exponent:
   a. $(x + 4)^2$
   b. $(x - 1)^3$
3. Multiply: $(a + 5)(a + 5)$
4. Multiply: $(a + 5)(a - 5)$

Certain products of binomials, called **special products,** occur so often that it is worthwhile to learn their forms.

1 Square a Binomial.

To develop a rule to find the *square of a binomial sum,* we consider $(x + y)^2$. We can use the definition of exponent and the procedure for multiplying two binomials to find the product.

$$\begin{aligned}(x + y)^2 &= (x + y)(x + y) && \text{In } (x + y)^2\text{, the base is } (x + y) \text{ and the exponent is 2.}\\ &= x^2 + \mathbf{xy} + \mathbf{xy} + y^2 && \text{Multiply the binomials.}\\ &= x^2 + \mathbf{2xy} + y^2 && \text{Combine like terms: } xy + xy = 1xy + 1xy = 2xy.\end{aligned}$$

Success Tip

The illustration can be used to visualize a special product. The area of the large square is $(x + y)(x + y) = (x + y)^2$. The sum of the four smaller areas is $x^2 + xy + xy + y^2$ or $x^2 + 2xy + y^2$. Thus,

$$(x + y)^2 = x^2 + 2xy + y^2$$

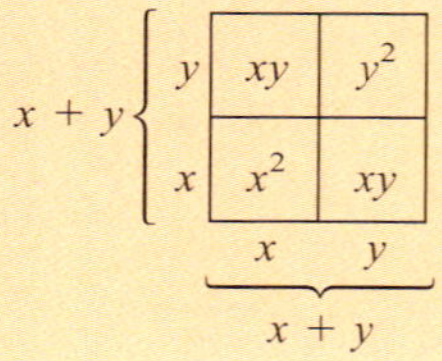

Note that the terms of the resulting trinomial are related to the terms of the binomial that was squared.

$$(x + y)^2 = x^2 + 2xy + y^2$$

- y^2: The square of the second term, y.
- $2xy$: Twice the product of the first and second terms, x and y.
- x^2: The square of the first term, x.

To develop a rule to find the *square of a binomial difference,* we consider $(x - y)^2$.

$(x - y)^2 = (x - y)(x - y)$ In $(x - y)^2$, the base is $(x - y)$ and the exponent is 2.

$= x^2 - xy - xy + y^2$ Multiply the binomials.

$= x^2 - 2xy + y^2$ Combine like terms: $-xy - xy = -1xy - 1xy = -2xy$.

Again, the terms of the resulting trinomial are related to the terms of the binomial that was squared.

$$(x - y)^2 = x^2 - 2xy + y^2$$

- y^2: The square of the second term, $-y$.
- $-2xy$: Twice the product of the first and second terms, x and $-y$.
- x^2: The square of the first term, x.

The observations from these two examples illustrate the following **special-product rules.**

Squaring a Binomial

The square of a binomial is a trinomial, such that:

- Its first term is the square of the first term of the binomial.
- Its last term is the square of the second term of the binomial.
- Its middle term is twice the product of both terms of the binomial.

$$(A + B)^2 = A^2 + 2AB + B^2 \qquad (A - B)^2 = A^2 - 2AB + B^2$$

EXAMPLE 1

Find each square: **a.** $(t + 9)^2$ **b.** $(8a - 5)^2$ **c.** $(d + 0.5)^2$ **d.** $\left(c^3 - \frac{7}{2}a\right)^2$

Strategy To find each square of a binomial, we will use one of the special-product rules.

Why This approach is faster than using the FOIL method.

Solution

a. $(t + 9)^2$ is the square of a binomial sum. The first term is t and the second term is 9.

$$(t + 9)^2 = \underbrace{t^2}_{\text{The square of the first term, } t.} + \underbrace{2(t)(9)}_{\text{Twice the product of both terms.}} + \underbrace{9^2}_{\text{The square of the second term, } 9.}$$

$$= t^2 + 18t + 81$$

The Language of Algebra

When squaring a binomial, the result is called a *perfect-square trinomial.*

$(t + 9)^2 = t^2 + 18t + 81$

Perfect-square trinomial

b. $(8a - 5)^2$ is the square of a binomial difference. The first term is $8a$ and the second term is -5.

$$(8a - 5)^2 = \underbrace{(8a)^2}_{\text{The square of the first term, } 8a.} + \underbrace{2(8a)(-5)}_{\text{Twice the product of both terms.}} + \underbrace{(-5)^2}_{\text{The square of the second term, } -5.}$$

$= 64a^2 - 80a + 25$ Use the power of a product rule: $(8a)^2 = 8^2a^2 = 64a^2$.

Caution

The square of a binomial is a *trinomial.* A common error when squaring a binomial is to forget the middle term of the product.

$$(8a - 5)^2 \neq 64a^2 + 25$$

Missing $-80a$

c. $(d + 0.5)^2$ is the square of a binomial sum. The first term is d and the second term is 0.5.

$$(d + 0.5)^2 = \underbrace{(d)^2}_{\text{The square of the first term, } d.} + \underbrace{2(d)(0.5)}_{\text{Twice the product of both terms.}} + \underbrace{(0.5)^2}_{\text{The square of the second term, 0.5.}}$$

$$= d^2 + d + 0.25$$

d. $\left(c^3 - \frac{7}{2}d\right)^2$ is the square of a binomial difference. The first term is c^3 and the second term is $-\frac{7}{2}d$.

$$\left(c^3 - \frac{7}{2}d\right)^2 = \underbrace{(c^3)^2}_{\text{The square of the first term, } c^3.} + \underbrace{2(c^3)\left(-\frac{7}{2}d\right)}_{\text{Twice the product of both terms.}} + \underbrace{\left(-\frac{7}{2}d\right)^2}_{\text{The square of the second term, } -\frac{7}{2}d.}$$

$$= c^6 - 7c^3d + \frac{49}{4}d^2 \quad \text{Use rules for exponents to find } (c^3)^2 \text{ and } \left(-\tfrac{7}{2}d\right)^2.$$

Self Check 1 Find each square: **a.** $(r + 6)^2$ **b.** $(7g - 2)^2$ **c.** $(v + 0.8)^2$ **d.** $\left(w^4 - \frac{3}{2}y\right)^2$

Now Try Problems 9, 15, and 23

2 Multiply the Sum and Difference of the Same Two Terms.

A final special product that occurs often has the form $(A + B)(A - B)$. In these products, one binomial is the sum of two terms and the other binomial is the difference of the same two terms. To develop a rule to find such products, consider the following multiplication:

$$(x + y)(x - y) = x^2 - xy + xy - y^2 \quad \text{Multiply the binomials.}$$
$$= x^2 - y^2 \quad \text{Combine like terms: } -xy + xy = 0.$$

Note that when we combined like terms, we added opposites. This will always be the case for products of this type; the sum of the outer and inner products will be 0. The first and last products will be squares.

$$(x + y)(x - y) = x^2 - y^2$$

The square of the second term, y.
The square of the first term, x.

These observations suggest a third **special-product rule.**

Success Tip

We can use the FOIL method to find each of the special products discussed in this section. However, these forms occur so often, it is worthwhile to learn the special-product rules.

Multiplying the Sum and Difference of Two Terms

The product of the sum of two terms and difference of the same two terms is the square of the first term minus the square of the second term.

$$(A + B)(A - B) = A^2 - B^2$$

EXAMPLE 2 Multiply: **a.** $(m + 2)(m - 2)$ **b.** $(3y + 4)(3y - 4)$ **c.** $\left(b - \frac{2}{3}\right)\left(b + \frac{2}{3}\right)$ **d.** $(t^4 - 6u)(t^4 + 6u)$

Strategy To find the product of each pair of binomials, we will use the special-product rule for the sum and difference of the same two terms.

Why This approach is faster than using the FOIL method.

Solution

The Language of Algebra

When multiplying the sum and difference of two terms, the result is called a *difference of two squares.*

$$(m + 2)(m - 2) = m^2 - 4$$

Difference of two squares

a. $(m + 2)$ and $(m - 2)$ are the sum and difference of the same two terms, m and 2.

$$(m + 2)(m - 2) = \underbrace{m^2}_{\text{The square of the first term, } m.} - \underbrace{2^2}_{\text{The square of the second term, 2.}}$$

$$= m^2 - 4$$

b. $(3y + 4)$ and $(3y - 4)$ are the sum and difference of the same two terms, $3y$ and 4.

$$(3y + 4)(3y - 4) = \underbrace{(3y)^2}_{\text{The square of the first term, } 3y.} - \underbrace{4^2}_{\text{The square of the second term, 4.}}$$

$$= 9y^2 - 16$$

c. By the commutative property of multiplication, the special-product rule can be written with the factor containing the $-$ symbol first: $(A - B)(A + B) = A^2 - B^2$. Since $\left(b - \frac{2}{3}\right)$ and $\left(b + \frac{2}{3}\right)$ are the difference and sum of the same two terms, b and $\frac{2}{3}$, we have

$$\left(b - \frac{2}{3}\right)\left(b + \frac{2}{3}\right) = \underbrace{b^2}_{\text{The square of the first term, } b.} - \underbrace{\left(\frac{2}{3}\right)^2}_{\text{The square of the second term, } \frac{2}{3}.}$$

$$= b^2 - \frac{4}{9}$$

d. $(t^4 - 6u)$ and $(t^4 + 6u)$ are the difference and sum of the same two terms, t^4 and $6u$.

$$(t^4 - 6u)(t^4 + 6u) = \underbrace{(t^4)^2}_{\text{The square of the first term, } t^4.} - \underbrace{(6u)^2}_{\text{The square of the second term, } 6u.}$$

$$= t^8 - 36u^2$$

Self Check 2 Multiply: **a.** $(b + 4)(b - 4)$ **b.** $(5m + 9)(5m - 9)$ **c.** $\left(s - \frac{3}{4}\right)\left(s + \frac{3}{4}\right)$ **d.** $(c^3 + 2d)(c^3 - 2d)$

Now Try Problems 25, 27, and 31

3 Find Higher Powers of Binomials.

When we find the third, fourth, or even higher powers of a binomial, we say that we are **expanding the binomial.** The special-product rules can be used in such cases. The result is an expression that has more terms than the original binomial.

EXAMPLE 3 Expand: $(x + 1)^3$

Strategy We will use a special-product rule to find the third power of $x + 1$.

Why Since $(x + 1)^3$ can be written as $(x + 1)(x + 1)^2$, we can use a special-product rule to find $(x + 1)^2$ quickly.

Solution

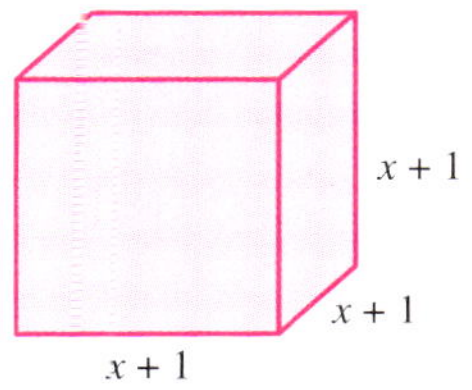

$$(x+1)^3 = (x+1)(x+1)^2$$ Read as "the quantity of x + 1, cubed."

$$= (x+1)(x^2+2x+1)$$ Find $(x+1)^2$ using the rule for the square of a sum.

$$= (x+1)(x^2+2x+1)$$ Multiply the binomial and the trinomial.

$$= x(x^2) + x(2x) + x(1) + 1(x^2) + 1(2x) + 1(1)$$ Multiply each term of $x^2 + 2x + 1$ by each term of $x + 1$.

$$= x^3 + 2x^2 + x + x^2 + 2x + 1$$ Multiply the monomials.

$$= x^3 + 3x^2 + 3x + 1$$ Combine like terms.

Notation

$(x+1)^3$ represents the volume of a cube with sides of length $x + 1$ units.

Self Check 3 Expand: $(n-3)^3$

Now Try Problem 37

4 Simplify Expressions Containing Polynomial Multiplication.

We can use a modified version of the order of operations rule to simplify expressions that involve polynomial addition, subtraction, multiplication, and raising to a power.

Order of Operations with Polynomials

1. If possible, simplify any polynomials within parentheses by combining like terms.
2. Square (or expand) all polynomials raised to powers using the FOIL method or a special-product rule.
3. Perform all polynomial multiplications using the distributive property, the FOIL method, or a special-product rule.
4. Perform all polynomial additions and subtractions by combining like terms.

EXAMPLE 4 Simplify each expression:

a. $-8(y^2 - 2y + 3) - 4(2y^2 + y - 6)$
b. $(x+1)(x-2) + 3x(x+3)$
c. $(3y-2)^2 - (y-5)(y+5)$

Strategy We will follow the rules for the order of operations to simplify each expression.

Why If we don't follow the correct order of operations, we can obtain different results that are not equivalent.

Solution **a.** The two polynomials within the parentheses do not simplify further and no polynomials are raised to a power. To perform the multiplication, we will use the distributive property twice. Then we will combine like terms.

$$-8(y^2 - 2y + 3) - 4(2y^2 + y - 6) = -8y^2 + 16y - 24 - 8y^2 - 4y + 24$$ Distribute.

$$= -16y^2 + 12y$$ Add and subtract to combine like terms.

b. The three polynomials within parentheses do not simplify further and no polynomials are raised to a power. To perform the multiplications, we use the FOIL method and the distributive property. Then we will combine like terms.

$(x + 1)(x - 2) + 3x(x + 3)$

$= x^2 - x - 2 + 3x^2 + 9x$ Use the FOIL method to find $(x + 1)(x - 2)$. Distribute the multiplication by $3x$.

$= 4x^2 + 8x - 2$ Combine like terms.

c. The three polynomials within parentheses do not simplify further. To square $3y - 2$, we use a special-product rule. To find the product of $(y - 5)(y + 5)$, we will use the special-product rule for the sum and difference of the same two terms. Then we will combine like terms.

$(3y - 2)^2 - (y - 5)(y + 5)$

$= 9y^2 - 12y + 4 - (y^2 - 25)$ Write $y^2 - 25$ within parentheses so that both terms are subtracted.

$= 9y^2 - 12y + 4 - y^2 + 25$ Change the sign of each term within $(y^2 - 25)$ and drop the parentheses.

$= 8y^2 - 12y + 29$ Combine like terms.

Self Check 4 Simplify each expression: **a.** $2(a^2 - 3a) + 5(a^2 + 2a)$
b. $(x - 4)(x + 6) + 5x(2x - 1)$
c. $(a + 9)(a - 9) - (2a - 4)^2$

Now Try Problems 45, 47, and 48

EXAMPLE 5 Find a polynomial that represents the area of the triangle.

Strategy We will multiply one-half, the length of the base, and the height of the triangle.

Why The area of a triangle is equal to one-half the product of the length of its base and its height.

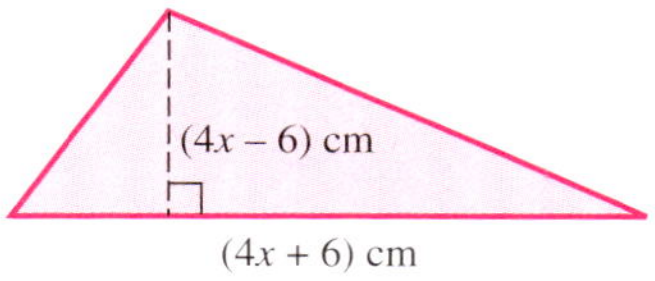

Solution We begin by substituting $4x + 6$ for the length of the base and $4x - 6$ for the height in the formula for the area of a triangle. It is wise to find $(4x + 6)(4x - 6)$ first, using a special-product rule, and then to multiply that result by $\frac{1}{2}$.

Success Tip

Recall that to multiply three polynomials, begin by multiplying any two of them. Then multiply that result by the third polynomial.

$\text{Area} = \frac{1}{2} \cdot \textbf{base} \cdot \textbf{height}$ This is the formula for the area of a triangle.

$= \frac{1}{2}(4x + 6)(4x - 6)$ Substitute.

$= \frac{1}{2}(16x^2 - 36)$ Use the special product rule for a sum and difference of two terms.

$= 8x^2 - 18$ Distribute the multiplication by $\frac{1}{2}$.

The area of the triangle is $(8x^2 - 18)$ square centimeters, which can be written as $(8x^2 - 18)\text{ cm}^2$.

Self Check 5 Find a polynomial that represents the area of a triangle with height $(12a - 2)$ ft and the length of the base $(12a + 2)$ ft.

Now Try Problem 53

SECTION 5.7 STUDY SET

VOCABULARY

Fill in the blanks.

1. Expressions of the form $(x + y)^2$, $(x - y)^2$, and $(x + y)(x - y)$ occur so frequently in algebra that they are called special ________.
2. $(2x + 3)^2$ is the _______ of a binomial and $(a + 6)(a - 6)$ is the product of the sum and difference of the ______ two terms.

CONCEPTS

3. Fill in the blanks to describe each special product.
   a. $(x + y)^2 = x^2 + 2xy + y^2$
      - y^2: The _______ of the second term
      - $2xy$: _______ the product of the first and second terms
      - x^2: The square of the _____ term
   b. $(x + y)(x - y) = x^2 - y^2$
      - y^2: The square of the _______ term
      - x^2: The _______ of the first term
4. Consider the binomial $5x + 4$.
   a. What is the square of its first term?
   b. What is twice the product of its two terms?
   c. What is the square of its second term?

NOTATION

Complete each solution to find the product.

5. $(x - 4)^2 = \;\;^2 + 2(x)(\;\;) + \;\;^2$
   $= x^2 + \;\;\; + 16$
6. $(6r - 1)^2 = (\;\;)^2 \;\; 2(6r)(1) + (-1)^2$
   $= \;\;\; - \;\;\; + 1$
7. $(s + 5)(s - 5) = \;\;^2 - \;\;^2$
   $= s^2 -$
8. True or false: $(t + 7)(t - 7) = (t - 7)(t + 7)$?

GUIDED PRACTICE

Find each product. See Example 1.

9. $(x + 1)^2$
10. $(y + 7)^2$
11. $(m - 6)^2$
12. $(b - 1)^2$
13. $(4x + 5)^2$
14. $(6y + 3)^2$
15. $(7m - 2)^2$
16. $(9b - 2)^2$
17. $(1 - 3y)^2$
18. $(1 - 4a)^2$
19. $(y + 0.9)^2$
20. $(d + 0.2)^2$
21. $(a^2 + b^2)^2$
22. $(c^2 + d^2)^2$
23. $\left(s + \frac{3}{4}\right)^2$
24. $\left(y - \frac{5}{3}\right)^2$

Find each product. See Example 2.

25. $(x + 3)(x - 3)$
26. $(y + 6)(y - 6)$
27. $(2p + 7)(2p - 7)$
28. $(5t + 4)(5t - 4)$
29. $(3n + 1)(3n - 1)$
30. $(5a + 4)(5a - 4)$
31. $\left(c + \frac{3}{4}\right)\left(c - \frac{3}{4}\right)$
32. $\left(m + \frac{4}{5}\right)\left(m - \frac{4}{5}\right)$
33. $(0.4 - 9m^2)(0.4 + 9m^2)$
34. $(0.3 - 2c^2)(0.3 + 2c^2)$
35. $(5 - 6g)(5 + 6g)$
36. $(6 - c^2)(6 + c^2)$

Expand each binomial. See Example 3.

37. $(x + 4)^3$
38. $(y + 2)^3$
39. $(n - 6)^3$
40. $(m - 5)^3$
41. $(2g - 3)^3$
42. $(3x - 2)^3$
43. $(a + b)^3$
44. $(c - d)^3$

Perform the operations. See Example 4.

45. $2(x^2 + 7x - 1) - 3(x^2 - 2x + 2)$
46. $2t(t + 2) + (t - 1)(t + 9)$
47. $(3x + 4)(2x - 2) - (2x + 1)(x + 3)$
48. $(5a - 1)^2 - (a - 8)(a + 8)$
49. $-5d(4d - 1)^2$
50. $-2h(7h - 2)^2$
51. $4d(d^2 + g^3)(d^2 - g^3)$
52. $8y(x^2 + y^2)(x^2 - y^2)$

Find a polynomial that represents the area of the figure. Leave π in your answer. See Example 5.

53.
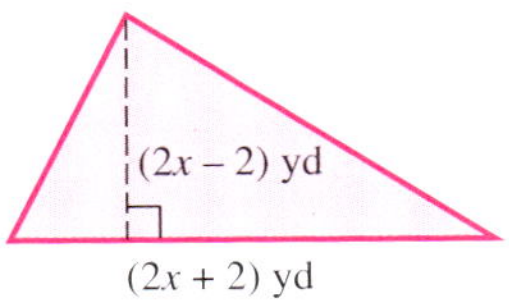

54.
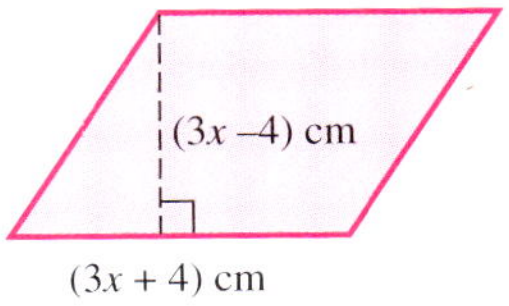

55.
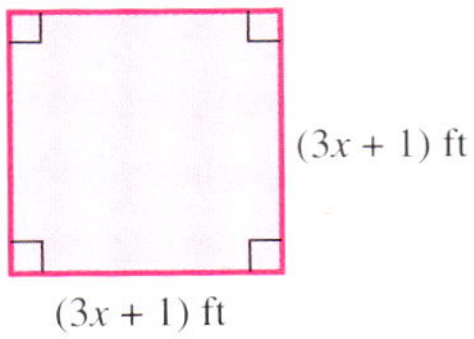

56.
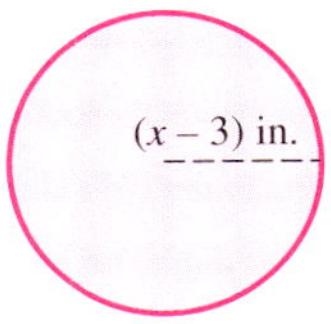

TRY IT YOURSELF

Perform the operations.

57. $(2v^3 - 8)^2$
58. $(8x^4 - 3)^2$
59. $3x(2x + 3)(2x + 3)$
60. $4y(3y + 4)(3y + 4)$
61. $(4f + 0.4)(4f - 0.4)$
62. $(4t + 0.6)(4t - 0.6)$
63. $(r^2 + 10s)^2$
64. $(m^2 + 8n)^2$
65. $2(x + 3) + 4(x - 2)$
66. $3(y - 4) - 5(y + 3)$
67. $\left(d^4 + \frac{1}{4}\right)^2$
68. $\left(q^6 + \frac{1}{3}\right)^2$
69. $(d + 7)(d - 7)$
70. $(t + 2)(t - 2)$
71. $(2a - 3b)^2$
72. $(2x + 5y)^2$
73. $(n + 6)(n - 6)$
74. $(a + 12)(a - 12)$
75. $(m + 10)^2 - (m - 8)^2$
76. $(5y - 1)^2 - (y + 7)(y - 7)$
77. $(2m + n)^3$
78. $(p - 2q)^3$
79. $\left(5m - \frac{6}{5}\right)^2$
80. $\left(6m - \frac{7}{6}\right)^2$
81. $(r^2 - s^2)^2$
82. $(t^2 - u^2)^2$
83. $(x - 2)^2$
84. $(a + 2)^2$
85. $(r + 2)^2$
86. $(n + 10)^2$
87. $(n - 2)^4$
88. $(c + d)^4$
89. $5(y^2 - 2y - 6) + 6(2y^2 + 2y - 5)$
90. $(4b + 1)^2 - (b - 7)(b + 7)$
91. $(3x - 2)^2 + (2x + 1)^2$
92. $(4a - 3)^2 + (a + 6)^2$
93. $(f - 8)^2$
94. $(w - 9)^2$
95. $\left(6b + \frac{1}{2}\right)\left(6b - \frac{1}{2}\right)$
96. $\left(4h + \frac{2}{3}\right)\left(4h - \frac{2}{3}\right)$
97. $3y(y + 2) + (y + 1)(y - 1)$
98. $(x + y)(x - y) + x(x + y)$
99. $(6 - 2d^3)^2$
100. $(6 - 5p^2)^2$
101. $(2e + 1)^3$
102. $(3m - 2n)^3$
103. $(8x + 3)^2$
104. $(4b - 8)^2$

Look Alikes . . .

Perform the indicated operations.

105. a. $(xy)^2$ b. $(x + y)^2$
106. a. $(cd)^2$ b. $(c - d)^2$
107. a. $(2b^2d)^2$ b. $(2b^2 + d)^2$
108. a. $(mn)^3$ b. $(m + n)^3$

APPLICATIONS

109. **Playpens.** Find a polynomial that represents the area of the floor of the playpen.

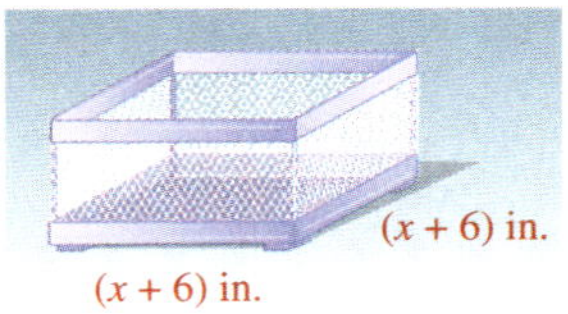

110. **Storage.** Find a polynomial that represents the volume of the cubicle.

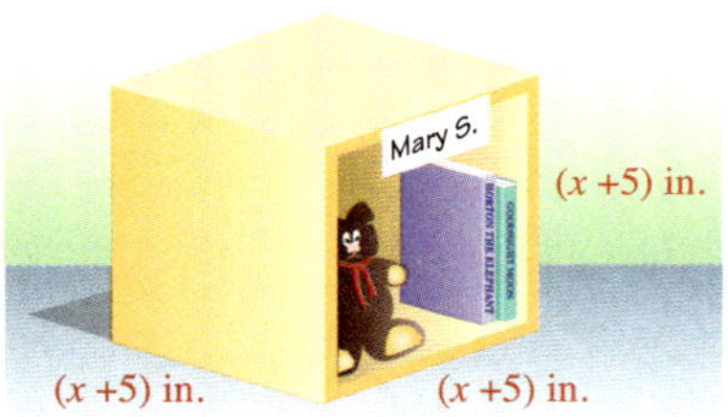

111. **Paper Towels.** The amount of space (volume) occupied by the paper on the roll of paper towels is given by the expression $\pi h(R + r)(R - r)$, where R is the outer radius and r is the inner radius. Perform the indicated multiplication.

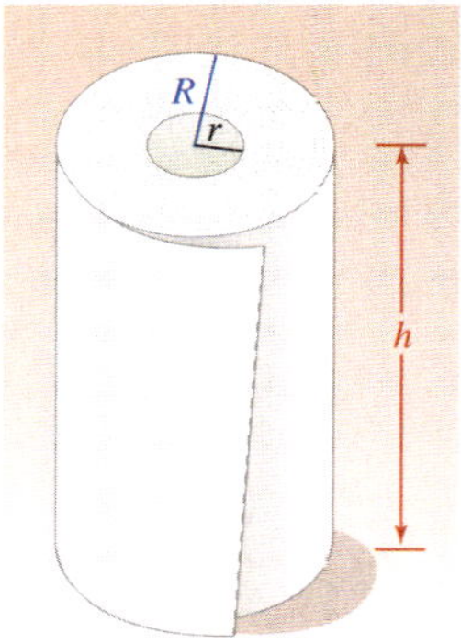

112. Signal Flags. Refer to the illustration below. Find a polynomial that represents the area in blue of the maritime signal flag for the letter p. The dimensions are given in centimeters.

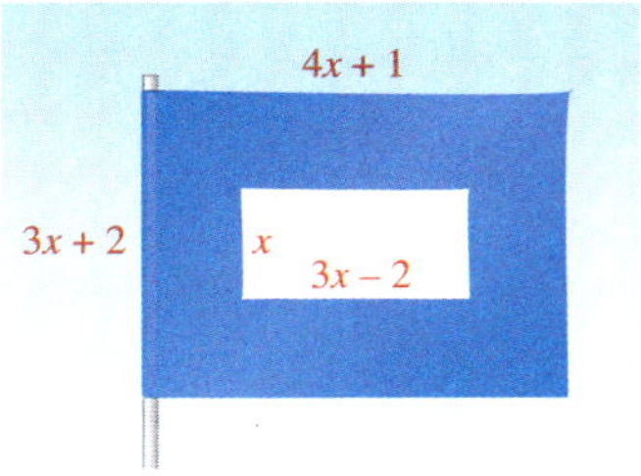

WRITING

113. What is a binomial? Explain how to square it.

114. Writing $(x + y)^2$ as $x^2 + y^2$ illustrates a common error. Explain.

115. We can find $(2x + 3)^2$ and $(5y - 6)^2$ using the FOIL method or using special product rules. Explain why the special product rules are faster.

116. a. Fill in the blanks: $(xy)^2$ is the _______ of a product and $(x + y)^2$ is the _______ of a sum.

b. Explain why $(xy)^2 \neq (x + y)^2$.

REVIEW

117. Simplify: $\frac{30}{36}$

118. Add: $\frac{5}{12} + \frac{1}{4}$

119. Multiply: $\frac{7}{8} \cdot \frac{3}{5}$

120. Divide: $\frac{1}{3} \div \frac{4}{5}$

CHALLENGE PROBLEMS

121. a. Find two binomials whose product is a binomial.

b. Find two binomials whose product is a trinomial.

c. Find two binomials whose product is a four-term polynomial.

122. A special-product rule can be used to find $31 \cdot 29$.

$$\begin{aligned} 31 \cdot 29 &= (30 + 1)(30 - 1) \\ &= 30^2 - 1^2 \\ &= 900 - 1 \\ &= 899 \end{aligned}$$

Use this method to find $52 \cdot 48$.

SECTION 5.8 Dividing Polynomials

OBJECTIVES

1. Divide a monomial by a monomial.
2. Divide a polynomial by a monomial.
3. Divide a polynomial by a polynomial.

ARE YOU READY?

The following problems review some basic skills that are needed when dividing polynomials.

1. Simplify: **a.** $\frac{6}{15}$ **b.** $\frac{a^6}{a^4}$

2. Add: $\frac{a}{d} + \frac{b}{d}$

3. Divide: $24\overline{)864}$

4. Subtract: $\begin{array}{r} 2x^3 + 8x^2 \\ -(2x^3 + 6x^2) \\ \hline \end{array}$

In this section, we will conclude our study of operations with polynomials by discussing division of polynomials. To begin, we consider the simplest case, the quotient of two monomials.

1 Divide a Monomial by a Monomial.

To divide monomials, we can use the method for simplifying fractions or a method that involves a rule for exponents.

EXAMPLE 1 Divide the monomials: **a.** $\frac{21x^5}{7x^2}$ **b.** $\frac{10r^6s}{6rs^3}$

Strategy We can use the rules for simplifying fractions and/or the quotient rule for exponents.

Why We need to make sure that the numerator and denominator have no common factors other than 1. If that is the case, then the fraction is in *simplest form.*

Solution

Success Tip

In this section, you will see that regardless of the number of terms involved, every polynomial division is a series of monomial divisions.

By simplifying fractions

a. $$\frac{21x^5}{7x^2} = \frac{3 \cdot \overset{1}{\cancel{7}} \cdot \overset{1}{\cancel{x}} \cdot \overset{1}{\cancel{x}} \cdot x \cdot x \cdot x}{\underset{1}{\cancel{7}} \cdot \underset{1}{\cancel{x}} \cdot \underset{1}{\cancel{x}}}$$

$$= 3x^3$$

b. $$\frac{10r^6s}{6rs^3} = \frac{\overset{1}{\cancel{2}} \cdot 5 \cdot \overset{1}{\cancel{r}} \cdot r \cdot r \cdot r \cdot r \cdot r \cdot \overset{1}{\cancel{s}}}{\underset{1}{\cancel{2}} \cdot 3 \cdot \underset{1}{\cancel{r}} \cdot \underset{1}{\cancel{s}} \cdot s \cdot s}$$

$$= \frac{5r^5}{3s^2}$$

Using the rules for exponents

$$\frac{21x^5}{7x^2} = 3x^{5-2}$$ Divide the coefficients.

$$= 3x^3$$ Subtract the exponents.

$$\frac{10r^6s}{6rs^3} = \frac{5}{3}r^{6-1}s^{1-3}$$ Simplify $\frac{10}{6}$.

$$= \frac{5}{3}r^5s^{-2}$$ Subtract exponents.

$$= \frac{5r^5}{3s^2}$$ Move s^{-2} to the denominator and change the sign of the exponent.

Self Check 1 Divide the monomials: **a.** $\frac{30y^4}{5y^2}$ **b.** $\frac{8c^2d^6}{32c^5d^2}$

Now Try Problems 15 and 21

2 Divide a Polynomial by a Monomial.

Recall that to add two fractions with the same denominator, we add their numerators and keep their common denominator.

$$\frac{a}{d} + \frac{b}{d} = \frac{a+b}{d}$$

We can use this rule in reverse to divide polynomials by monomials.

Dividing a Polynomial by a Monomial

To divide a polynomial by a monomial, divide each term of the polynomial by the monomial.

If A, B, and D represent monomials, where $D \neq 0$, then

$$\frac{A+B}{D} = \frac{A}{D} + \frac{B}{D}$$

EXAMPLE 2 Divide: **a.** $\frac{9x^2 + 3x}{3x}$ **b.** $\frac{12a^4b^3 - 18a^3b^2 + 2a^2}{6a^2b^2}$

Strategy We will divide each term of the polynomial in the numerator by the monomial in the denominator.

Why A fraction bar indicates division of the numerator by the denominator.

Solution **a.** Here, we have a binomial divided by a monomial.

$$\frac{9x^2 + 3x}{3x} = \frac{9x^2}{3x} + \frac{3x}{3x}$$ Divide each term of the numerator, $9x^2 + 3x$, by the denominator, $3x$.

$$= 3x^{2-1} + 1x^{1-1}$$ Do each monomial division. Divide the coefficients. Keep each base and subtract the exponents.

$$= 3x^1 + 1x^0$$

$$= 3x + 1$$ Recall that $x^0 = 1$.

Check: We multiply the divisor, $3x$, and the quotient, $3x + 1$. The result should be the dividend, $9x^2 + 3x$.

$$3x(3x + 1) = 9x^2 + 3x$$ The answer checks.

b. Here, we have a trinomial divided by a monomial.

$$\frac{12a^4b^3 - 18a^3b^2 + 2a^2}{6a^2b^2} = \frac{12a^4b^3}{6a^2b^2} - \frac{18a^3b^2}{6a^2b^2} + \frac{2a^2}{6a^2b^2}$$ Divide each term of the numerator by the denominator, $6a^2b^2$.

$$= 2a^{4-2}b^{3-2} - 3a^{3-2}b^{2-2} + \frac{a^{2-2}}{3b^2}$$ Do each monomial division. Simplify: $\frac{2}{6} = \frac{1}{3}$.

$$= 2a^2b - 3a + \frac{1}{3b^2}$$ $b^{2-2} = b^0 = 1$ and $a^{2-2} = a^0 = 1$.

Recall that the variables in a polynomial must have whole-number exponents. Therefore, the result, $2a^2b - 3a + \frac{1}{3b^2}$, is not a polynomial because the last term has a variable in the denominator.

Check:

$$6a^2b^2\left(2a^2b - 3a + \frac{1}{3b^2}\right) = 12a^4b^3 - 18a^3b^2 + 2a^2$$ The answer checks.

Caution

A common error is to "remove" the $3x$'s in the expression incorrectly. Recall that only *factors* common to the numerator and denominator can be removed. Here, $3x$ is not a factor of the numerator—it is a *term*.

$$\frac{9x^2 + \cancel{3x}^{\,1}}{\cancel{3x}_{\,1}}$$

The Language of Algebra

The names of the parts of a division statement are

$$\underbrace{\frac{9x^2 + 3x}{3x}}_{\text{Dividend over Divisor}} = \underbrace{3x + 1}_{\text{Quotient}}$$

Dividend $9x^2 + 3x$, *Divisor* $3x$, *Quotient* $3x + 1$

Self Check 2 Divide: **a.** $\frac{50h^3 + 5h^2}{5h^2}$ **b.** $\frac{22s^5t^2 - s^4t^3 + 44s^2t^4}{11s^2t^2}$

Now Try Problems 25, 29, and 33

3 Divide a Polynomial by a Polynomial.

To divide a polynomial by a polynomial (other than a monomial), we use a method similar to long division in arithmetic.

EXAMPLE 3 Divide $x^2 + 5x + 6$ by $x + 2$.

Strategy We will use the long division method. The dividend is $x^2 + 5x + 6$ and the divisor is $x + 2$.

Why Since the divisor has more than one term, we must use the long division method to divide the polynomials.

Solution We write the division using a long division symbol $\overline{)\quad}$ and proceed as follows:

Step 1:

$$x + 2\overline{)\,x^2 + 5x + 6}$$ with quotient x written above.

Divide the first term of the dividend by the first term of the divisor: $\frac{x^2}{x} = x$. Write the result, x, above the long division symbol.

The Language of Algebra

Notice how the instruction translates:

Divide $x^2 + 5x + 6$ by $x + 2$

$$x + 2\overline{)x^2 + 5x + 6}$$

Success Tip

Notice that this method is much like that used for division of whole numbers.

$$\begin{array}{r} 13 \\ 12\overline{)156} \\ -12\downarrow \\ \hline 036 \\ -36 \\ \hline 0 \end{array}$$

Hundreds
Tens
Ones

Success Tip

The long division method aligns like terms vertically.

$$\begin{array}{r} x + 3 \\ x + 2\overline{)x^2 + 5x + 6} \\ -(x^2 + 2x) \\ \hline 3x + 6 \\ -(3x + 6) \\ \hline 0 \end{array}$$

x^2-terms
x-terms
Constants

Step 2:

$$\begin{array}{r} x \\ x + 2\overline{)x^2 + 5x + 6} \\ \underline{x^2 + 2x} \end{array}$$

Multiply each term of the divisor by x. Write the result, $x^2 + 2x$, under $x^2 + 5x$, and draw a line. Be sure to align the like terms.

Step 3:

$$\begin{array}{r} x \\ x + 2\overline{)x^2 + 5x + 6} \\ -(x^2 + 2x)\downarrow \\ \hline 3x + 6 \end{array}$$

Write parentheses around $x^2 + 2x$ so that both of its terms are subtracted. Subtract $x^2 + 2x$ from $x^2 + 5x$. Work column by column: $x^2 - x^2 = 0$ and $5x - 2x = 3x$.
Bring down the next term, 6.

Step 4:

$$\begin{array}{r} x + 3 \\ x + 2\overline{)x^2 + 5x + 6} \\ -(x^2 + 2x) \\ \hline 3x + 6 \end{array}$$

Divide the first term of $3x + 6$ by the first term of the divisor: $\frac{3x}{x} = 3$. Write $+ 3$ above the long division symbol to form the second term of the quotient.

Step 5:

$$\begin{array}{r} x + 3 \\ x + 2\overline{)x^2 + 5x + 6} \\ -(x^2 + 2x) \\ \hline 3x + 6 \\ \underline{3x + 6} \end{array}$$

Multiply each term of the divisor by 3. Write the result, $3x + 6$, under $3x + 6$ and draw a line. Be sure to align the like terms.

Step 6:

$$\begin{array}{r} x + 3 \\ x + 2\overline{)x^2 + 5x + 6} \\ -(x^2 + 2x) \\ \hline 3x + 6 \\ -(3x + 6) \\ \hline 0 \end{array}$$

Write parentheses around $3x + 6$ so that both of its terms are subtracted. Subtract $3x + 6$ from $3x + 6$. Work vertically: $3x - 3x = 0$ and $6 - 6 = 0$. There are no more terms to bring down.

This is the remainder.

The quotient is $x + 3$ and the remainder is 0.

Step 7: Check the result by verifying that $(x + 2)(x + 3)$ is $x^2 + 5x + 6$.

$$\begin{aligned}(x + 2)(x + 3) &= x^2 + 3x + 2x + 6 \\ &= x^2 + 5x + 6\end{aligned}$$

Divisor · quotient = dividend (if 0 remainder)

The result checks.

Self Check 3 Divide $x^2 + 7x + 12$ by $x + 3$.

Now Try Problem 37

Dividing a Polynomial by a Polynomial

To divide a polynomial by a polynomial (other than a monomial) use long division. If there is a remainder, write the result in the form: quotient $+ \frac{\text{remainder}}{\text{divisor}}$.

The long division method used in algebra can have a remainder just as long division in arithmetic often does.

EXAMPLE 4 Divide: $(6x^2 - 7x - 2) \div (2x - 1)$

Strategy We will use the long division method. The dividend is $6x^2 - 7x - 2$ and the divisor is $2x - 1$.

Why Since the divisor has more than one term, we must use the long division method to divide the polynomials.

Solution

Caution

The long-division method is a series of four steps:
- Divide
- Multiply
- Subtract
- Bring down the next term

Use parentheses to avoid sign errors when subtracting.

Step 1: $2x - 1\overline{)6x^2 - 7x - 2}$ with quotient $3x$

Divide the first term of the dividend by the first term of the divisor: $\frac{6x^2}{2x} = 3x$. Write the result, $3x$, above the long division symbol.

Step 2: $2x - 1\overline{)6x^2 - 7x - 2}$ with quotient $3x$; below: $6x^2 - 3x$

Multiply each term of the divisor by $3x$. Write the result, $6x^2 - 3x$, under $6x^2 - 7x$, and draw a line.

Step 3:

$$\begin{array}{r} 3x \phantom{{}-2} \\ 2x - 1\overline{)6x^2 - 7x - 2} \\ -(6x^2 - 3x)\phantom{{}-2} \\ \hline -4x - 2 \end{array}$$

Write parentheses around $6x^2 - 3x$ so that both of its terms are subtracted. Subtract $6x^2 - 3x$ from $6x^2 - 7x$. Work vertically: $6x^2 - 6x^2 = 0$ and $-7x - (-3x) = -7x + 3x = -4x$. Bring down the next term, -2.

Step 4:

$$\begin{array}{r} 3x - 2 \\ 2x - 1\overline{)6x^2 - 7x - 2} \\ -(6x^2 - 3x)\phantom{{}-2} \\ \hline -4x - 2 \end{array}$$

Divide the first term of $-4x - 2$ by the first term of the divisor: $\frac{-4x}{2x} = -2$. Write -2 above the long division symbol to form the second term of the quotient.

Step 5:

$$\begin{array}{r} 3x - 2 \\ 2x - 1\overline{)6x^2 - 7x - 2} \\ -(6x^2 - 3x)\phantom{{}-2} \\ \hline -4x - 2 \\ -4x + 2 \\ \hline \end{array}$$

Multiply each term of the divisor by -2. Write the result, $-4x + 2$, under $-4x - 2$, and draw a line.

Step 6:

$$\begin{array}{r} 3x - 2 \\ 2x - 1\overline{)6x^2 - 7x - 2} \\ -(6x^2 - 3x)\phantom{{}-2} \\ \hline -4x - 2 \\ -(-4x + 2) \\ \hline -4 \end{array}$$

Write $-4x + 2$ in parentheses so that both of its terms are subtracted. Subtract $-4x + 2$ from $-4x - 2$. Work vertically: $-4x - (-4x) = -4x + 4x = 0$ and $-2 - 2 = -4$. There are no more terms to bring down.

This is the remainder.

Success Tip

The long division method for polynomials continues until the degree of the remainder is less than the degree of the divisor. Here, the remainder, -4, has degree 0. The divisor, $2x - 1$, has degree 1. Therefore, the division ends.

The quotient is $3x - 2$ and the remainder is -4. It is common to write the answer in $\textit{Quotient} + \frac{\textit{remainder}}{\textit{divisor}}$ form as either

$$3x - 2 + \frac{-4}{2x - 1} \quad \text{or} \quad 3x - 2 - \frac{4}{2x - 1}$$

Step 7: We can check the result using the fact that for any division:

$$\begin{aligned} \text{Divisor} \cdot \text{quotient} + \text{remainder} &= \text{dividend} \\ (2x - 1)(3x - 2) + (-4) &= 6x^2 - 4x - 3x + 2 + (-4) \\ &= 6x^2 - 7x - 2 \end{aligned}$$

The result checks.

Self Check 4 Divide: $(8x^2 + 6x - 3) \div (2x + 3)$. Check the result.

Now Try Problem 43

The division method works best when the terms of the divisor and the dividend are written in descending powers of the variable. If the powers in the dividend or divisor are not in descending order, we use the commutative property of addition to write them that way.

EXAMPLE 5 Divide $4x^2 + 2x^3 + 12 - 2x$ by $x + 3$.

Strategy We will write the dividend in descending powers of x and use the long division method.

Why It is easier to align like terms in columns when the powers of the variable are written in descending order.

Solution

$$\begin{array}{r} 2x^2 - 2x + 4 \\ x + 3 \overline{)2x^3 + 4x^2 - 2x + 12} \\ -(2x^3 + 6x^2) \\ \hline -2x^2 - 2x \\ -(-2x^2 - 6x) \\ \hline 4x + 12 \\ -(4x + 12) \\ \hline 0 \end{array}$$

The first division: $\frac{2x^3}{x} = 2x^2$.

The second division: $\frac{-2x^2}{x} = -2x$.

The third division: $\frac{4x}{x} = 4$.

Check: $(x + 3)(2x^2 - 2x + 4) = 2x^3 - 2x^2 + 4x + 6x^2 - 6x + 12$
$= 2x^3 + 4x^2 - 2x + 12$ The result checks.

Self Check 5 Divide $x^2 - 10x + 6x^3 + 4$ by $2x - 1$.

Now Try Problem 45

When we write the terms of a dividend in descending powers, we must determine whether some powers of the variable are missing. If any are missing, insert **placeholder terms** with a coefficient of 0 or leave blank spaces for them. This keeps like terms in the same column, which is necessary when performing the subtraction in vertical form.

EXAMPLE 6 Divide: $\frac{27x^3 + 1}{3x + 1}$

Strategy The divisor is $3x + 1$. The dividend, $27x^3 + 1$, does not have an x^2-term or an x-term. We will insert a $0x^2$ term and a $0x$ term as placeholders, and use the long division method.

Why We insert placeholder terms so that like terms will be aligned in the same column when we subtract.

Solution

$$\begin{array}{r} 9x^2 - 3x + 1 \\ 3x + 1 \overline{)27x^3 + 0x^2 + 0x + 1} \\ -(27x^3 + 9x^2) \\ \hline -9x^2 + 0x \\ -(-9x^2 - 3x) \\ \hline 3x + 1 \\ -(3x + 1) \\ \hline 0 \end{array}$$

The first division: $\frac{27x^3}{3x} = 9x^2$.

The second division: $\frac{-9x^2}{3x} = -3x$.

The third division: $\frac{3x}{3x} = 1$.

Check: $(3x + 1)(9x^2 - 3x + 1) = 27x^3 - 9x^2 + 3x + 9x^2 - 3x + 1$
$= 27x^3 + 1$ The result checks.

Self Check 6 Divide: $\frac{x^2 - 9}{x - 3}$. Check the result.

Now Try Problem 49

EXAMPLE 7 **Toys.** The area of an Etch A Sketch screen is represented by the polynomial $(35x^2 + 43x + 12)$ in.^{2}. If the width of the screen is $(5x + 4)$ inches, what polynomial represents its length?

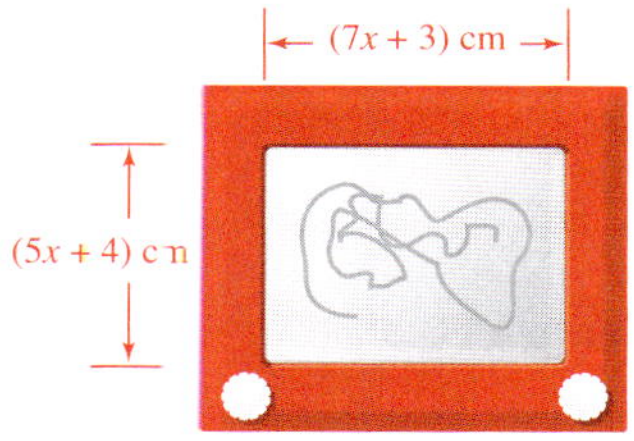

Strategy We will find the length of the screen by dividing its area, $(35x^2 + 43x + 12)$ in.^{2}, by its width, $(5x + 4)$ inches.

Why Recall that the area of a rectangle is given by the formula $A = lw$. If we divide both sides of the formula by w, we see that $l = \frac{A}{w}$.

Solution

$$\text{Length} = \frac{\text{area}}{\text{width}}$$ This is the formula for the length of a rectangle.

$$= \frac{35x^2 + 43x + 12}{5x + 4}$$ Substitute $35x^2 + 43x + 12$ for the area and $5x + 4$ for the width.

To divide $35x^2 + 43x + 12$ by $5x + 4$, we use long division.

$$\begin{array}{r} 7x + 3 \\ 5x + 4\overline{)35x^2 + 43x + 12} \\ -(35x^2 + 28x) \downarrow \\ \hline 15x + 12 \\ -(15x + 12) \\ \hline 0 \end{array}$$

The first division: $\frac{35x^2}{5x} = 7x$.

The second division: $\frac{15x}{5x} = 3$.

The length of the Etch A Sketch screen is $(7x + 3)$ inches.

Self Check 7 **Televisions.** The area of the rectangular screen of a plasma TV is represented by the polynomial $(36x^2 - 51x - 21)$ in^2. If the width of the screen is $(4x - 7)$ inches, what polynomial represents its length?

Now Try Problem 103

SECTION 5.8 STUDY SET

VOCABULARY

Fill in the blanks.

1. The expression $\frac{18x^7}{9x^4}$ is a monomial divided by a ________ .

2. The expression $\frac{6x^3y - 4x^2y^2 + 8xy^3 - 2y^4}{2x^4}$ is a ________ divided by a monomial.

3. The expression $\frac{x^2 - 8x + 12}{x - 6}$ is a trinomial divided by a ________.

4.
$$\begin{array}{r} \downarrow \qquad x - 2 \leftarrow \\ x - 6\overline{)x^2 - 8x - 4} \leftarrow \\ -(x^2 - 6x) \\ \hline -2x - 4 \\ -(-2x + 12) \\ \hline -16 \leftarrow \end{array}$$

CONCEPTS

5. The long division method is a series of four steps that are repeated. Put them in the correct order:

subtract multiply bring down divide

6. In the following long divisions, find the answer to the subtraction that must be performed at this stage.

a. $$\begin{array}{r} 3x \\ 2x + 1\overline{)6x^2 + 9x - 10} \\ -(6x^2 + 3x) \\ \hline \end{array}$$

b. $$\begin{array}{r} x \\ x - 7\overline{)x^2 + 10x + 21} \\ -(x^2 - 7x) \\ \hline \end{array}$$

c. $$\begin{array}{r} x \\ x - 7\overline{)x^2 - 9x - 6} \\ -(x^2 - 7x) \\ \hline \end{array}$$

d. $$\begin{array}{r} x \\ x + 4\overline{)x^2 + 0x - 16} \\ -(x^2 + 4x) \\ \hline \end{array}$$

7. Fill in the blanks: To check an answer of a long division, we use the fact that

Divisor · ______ + remainder = ______

8. Check to see whether the following result of a long division is correct.

$$\frac{x^2 + 4x - 20}{x - 3} = x + 7 + \frac{1}{x - 3}$$

NOTATION

Complete each solution.

9. $\frac{28x^5 - x^3 + 5x^2}{7x^2} = \frac{28x^5}{\square} - \frac{\square}{7x^2} + \frac{5x^2}{\square}$

$= 4x^{\square} - \frac{x^{3-2}}{\square} + \frac{5x^{\square}}{7}$

$= \square - \frac{x}{7} + \frac{\square}{\square}$

10.
$$\begin{array}{r} \square + 2 \\ x + 2\overline{)x^2 + 4x + 5} \\ \underline{-(x^2 + \square)} \\ \square + 5 \\ \underline{-(2x + 4)} \\ \square \end{array}$$

11. Write the polynomial $2x^2 - 1 + 5x^4$ in descending powers of x and insert placeholders for each missing term.

12. True or false: $6x + 4 + \frac{-3}{x + 2} = 6x + 4 - \frac{3}{x + 2}$

GUIDED PRACTICE

Divide the monomials. See Example 1.

13. $\frac{x^5}{x^2}$

14. $\frac{a^{12}}{a^8}$

15. $\frac{12h^8}{9h^6}$

16. $\frac{22b^9}{6b^6}$

17. $\frac{-3d^4}{15d^8}$

18. $\frac{-4x^3}{16x^5}$

19. $\frac{10s^2}{s^3}$

20. $\frac{16y^3}{y^4}$

21. $\frac{8x^3y^2}{40xy^6}$

22. $\frac{3y^3z}{18yz^6}$

23. $\frac{-16r^3y^2}{-4r^2y^7}$

24. $\frac{-35xz^6}{-7x^8z^2}$

Divide the polynomial by the monomial. See Example 2.

25. $\frac{6x + 3}{3}$

26. $\frac{8x + 4}{4}$

27. $\frac{a - a^3 + a^4}{a^4}$

28. $\frac{b^2 + b^3 - b^4}{b^4}$

29. $\frac{6h^{12} + 48h^9}{24h^{10}}$

30. $\frac{4x^{14} - 36x^8}{36x^{12}}$

31. $\frac{9s^8 - 18s^5 + 12s^4}{3s^3}$

32. $\frac{16b^{10} + 4b^6 - 20b^4}{4b^2}$

33. $\frac{7c^5 + 21c^4 - 14c^3 - 35c}{7c^2}$

34. $\frac{12r^{15} - 48r^{12} + r^{10} - 18r^8}{6r^{10}}$

35. $\frac{-25x^2y^3 + 30xy^2 - 5xy}{-5x^2y^2}$

36. $\frac{-30a^4b^4 - 15a^3b - 10a^2b^2}{-10a^2b^3}$

Perform each division. See Examples 3 and 4.

37. Divide $x^2 + 8x + 12$ by $x + 2$.

38. Divide $x^2 + 5x + 6$ by $x + 2$.

39. Divide $x^2 - 5x + 6$ by $x - 3$.

40. Divide $x^2 - 12x + 32$ by $x - 4$.

41. $\frac{2x^2 + 5x + 2}{2x + 3}$

42. $\frac{3x^2 - 8x + 3}{3x - 2}$

43. $\frac{6x^2 - 11x + 2}{3x - 1}$

44. $\frac{4x^2 + 6x - 1}{2x + 1}$

Perform each division. See Example 5.

45. $x + 2\overline{)3x + 2x^2 - 2}$

46. $x + 3\overline{)-x + 2x^2 - 21}$

47. $(3 + 11x + 10x^2) \div (5x + 3)$

48. $(6x + 1 + 9x^2) \div (3x + 1)$

Perform each division. See Example 6.

49. $(a^2 - 25) \div (a + 5)$

50. $(b^2 - 36) \div (b + 6)$

51. $(x^2 - 1) \div (x - 1)$

52. $(x^2 - 9) \div (x + 3)$

53. $\frac{4x^2 - 9}{2x + 3}$

54. $\frac{25x^2 - 16}{5x - 4}$

55. $\frac{81b^2 - 49}{9b - 7}$

56. $\frac{16t^2 - 121}{4t + 11}$

TRY IT YOURSELF

Perform each division.

57. Divide $y^2 + 13y + 13$ by $y + 1$.

58. Divide $z^2 + 7z + 14$ by $z + 3$.

59. $\frac{15a^8b^2 - 10a^2b^5}{5a^3b^2}$

60. $\frac{9a^4b^3 - 16a^3b^4}{12a^2b}$

61. $3x + 2\overline{)2 + 7x + 6x^3 + 10x^2}$

62. $3x - 2\overline{)4x - 4 + 6x^3 - x^2}$

63. $\frac{8x^9 - 32x^6}{4x^4}$

64. $\frac{30y^8 + 40y^7}{10y^6}$

65. $\frac{6a^2 + 5a - 6}{2a + 3}$

66. $\frac{3b^2 - 5b + 2}{3b - 2}$

67. $\frac{45m^{10}}{9m^5}$

68. $\frac{24n^{12}}{8n^4}$

69. $\frac{3b^2 + 11b + 6}{3b + 2}$

70. $\frac{8a^2 + 2a - 3}{2a - 1}$

71. $2x - 7\overline{)-x - 21 + 2x^2}$

72. $2x - 1\overline{)x - 2 + 6x^2}$

73. $\dfrac{x^3 + 1}{x + 1}$

74. $\dfrac{x^3 - 8}{x - 2}$

75. $\dfrac{-65rs^2}{15r^2s^5}$

76. $\dfrac{112uz^4}{-42u^3z^8}$

77. $\dfrac{-18w^6 - 9}{9w^4}$

78. $\dfrac{-40f^4 + 16}{8f^3}$

79. $\dfrac{9m - 6}{m}$

80. $\dfrac{10n - 6}{n}$

81. $\dfrac{y^3 - y}{y - 2}$

82. $\dfrac{a^3 + a}{a + 3}$

83. $\dfrac{5x^4 - 10x}{25x^3}$

84. $\dfrac{24x^7 - 32x^2}{16x^3}$

85. $3 + 4x\overline{)3 - 5x^2 - 2x + 4x^3}$

86. $2x + 3\overline{)7x^2 - 3 + 4x + 2x^3}$

87. $(x^2 + 6x + 15) \div (x + 5)$

88. $(x^2 + 10x + 30) \div (x + 6)$

89. $\dfrac{12x^3y^2 - 8x^2y - 4x}{4xy}$

90. $\dfrac{12a^2b^2 - 8a^2b - 4ab}{4ab}$

91. $a - 5\overline{)a^2 - 17a + 64}$

92. $b - 2\overline{)b^2 - 4b + 6}$

93. $\dfrac{a^3 - 1}{a - 1}$

94. $\dfrac{y^3 + 8}{y + 2}$

95. $\dfrac{6x^3 + x^2 + 2x + 1}{3x - 1}$

96. $\dfrac{3y^3 - 4y^2 + 2y + 3}{y + 3}$

97. $\dfrac{8x^{17}y^{20}}{15x^{15}y^{30}}$

98. $\dfrac{21a^{30}b^{15}}{14a^{40}b^{12}}$

99. $(6m^2 - m - 40) \div (2m + 5)$

100. $(12d^2 - 20d + 3) \div (6d - 1)$

Look Alikes . . .

Perform the indicated operations.

101. a. $\dfrac{16x^2 - 16x - 5}{4x}$ b. $\dfrac{16x^2 - 16x - 5}{4x + 1}$

102. a. $\dfrac{9x^3 + 3x^2 + 4x + 4}{3x}$ b. $\dfrac{9x^3 + 3x^2 + 4x + 4}{3x + 2}$

APPLICATIONS

103. **Furnace Filters.** The area of the rectangular-shaped furnace filter is $(x^2 - 2x - 24)$ square inches. What expression represents its height?

104. **Mini-Blinds.** The area covered by the mini-blinds is $(3x^3 - 6x)$ square feet. What expression represents the height of the blinds?

105. **Pool.** The rack shown in the illustration is used to set up the balls for a game of pool. If the perimeter of the rack, in inches, is given by the polynomial $6x^2 - 3x + 9$, what expression represents the approximate length of one side?

106. **Communications.** Telephone poles were installed every $(2x - 3)$ feet along a stretch of railroad track $(8x^3 - 6x^2 + 5x - 21)$ feet long. What expression represents the number of poles that were used?

WRITING

107. Explain how to check the following long division.

$$\begin{array}{r} x + 5 \\ 3x + 5\overline{)3x^2 + 20x - 5} \\ -(3x^2 + 5x) \\ \hline 15x - 5 \\ -(15x + 25) \\ \hline -30 \end{array}$$

108. Explain the error: $\dfrac{18x^2 + \overset{1}{\cancel{6x}}}{\underset{1}{\cancel{6x}}} = 18x^2 + 1$

109. How do you know when to stop the long division method when dividing polynomials?

110. When dividing $x^3 + 1$ by $x + 1$, why is it helpful to write $x^3 + 1$ as $x^3 + 0x^2 + 0x + 1$?

REVIEW

111. Write an equation of the line with slope $-\frac{11}{6}$ that passes through $(2, -6)$. Write the answer in slope–intercept form.

112. Solve $S = 2\pi rh + 2\pi r^2$ for h.

CHALLENGE PROBLEMS

Perform each division.

113. $\dfrac{6a^3 - 17a^2b + 14ab^2 - 3b^3}{2a - 3b}$

114. $(2x^4 + 3x^3 + 3x^2 - 5x - 3) \div (2x^2 - x - 1)$

115. $(x^6 + 2x^4 - 6x^2 - 9) \div (x^2 + 3)$

116. $\dfrac{6x^{6m}y^{6n} + 15x^{4m}y^{7n} - 24x^{2m}y^{8n}}{3x^{2m}y^n}$

117. $\dfrac{a^8 + a^6 - 4a^4 + 5a^2 - 3}{a^4 + 2a^2 - 3}$

118. $\dfrac{-17x^2 + 5x + x^4 + 2}{4x + x^2 - 1}$

5 Summary & Review

SECTION 5.1 Rules for Exponents

DEFINITIONS AND CONCEPTS	EXAMPLES
An **exponent** indicates repeated multiplication. It tells how many times the **base** is to be used as a factor. Exponent → $x^n = \underbrace{x \cdot x \cdot x \cdot \cdots \cdot x}_{n \text{ factors of } x}$ ← Base	Identify the base and the exponent in each expression. $2^6 = 2 \cdot 2 \cdot 2 \cdot 2 \cdot 2 \cdot 2$ 2 is the base and 6 is the exponent. $(-xy)^3 = (-xy)(-xy)(-xy)$ $-xy$ is the base and 3 is the exponent. $5t^4 = 5 \cdot t \cdot t \cdot t \cdot t$ t is the base and 4 is the exponent. $8^1 = 8$ 8 is the base and 1 is the exponent.
Rules for Exponents: If m and n represent integers and there are no divisions by 0, then **Product rule:** $x^m x^n = x^{m+n}$ **Quotient rule:** $\dfrac{x^m}{x^n} = x^{m-n}$ **Power rule:** $(x^m)^n = x^{m \cdot n} = x^{mn}$ **Power of a product rule:** $(xy)^m = x^m y^m$ **Power of a quotient rule:** $\left(\dfrac{x}{y}\right)^n = \dfrac{x^n}{y^n}$	Simplify each expression: $5^2 5^7 = 5^{2+7} = 5^9$ $\dfrac{t^7}{t^3} = t^{7-3} = t^4$ $(6^3)^7 = 6^{3 \cdot 7} = 6^{21}$ $(2p)^5 = 2^5 p^5 = 32p^5$ $\left(\dfrac{s}{4}\right)^4 = \dfrac{s^4}{4^4} = \dfrac{s^4}{256}$

REVIEW EXERCISES

1. Identify the base and the exponent in each expression.
   a. n^{12}
   b. $(2x)^6$
   c. $3r^4$
   d. $(y - 7)^3$
2. Write each expression in an equivalent form using an exponent.
   a. $m \cdot m \cdot m \cdot m \cdot m$
   b. $-3 \cdot x \cdot x \cdot x \cdot x$
   c. $(x + 8)(x + 8)$
   d. $\left(\frac{1}{2}pq\right)\left(\frac{1}{2}pq\right)\left(\frac{1}{2}pq\right)$

Simplify each expression. Assume there are no divisions by 0.

3. $7^4 \cdot 7^8$
4. $mmnn$
5. $(y^7)^3$
6. $(3x)^4$
7. $\dfrac{b^{12}}{b^3}$
8. $-b^3b^4b^5$
9. $(-16s^3)^2s^4$
10. $(2.1x^2y)^2$
11. $[(-9)^3]^5$
12. $(a^5)^3(a^2)^4$
13. $\left(\frac{1}{2}x^2x^3\right)^3$
14. $\left(\dfrac{x^7}{3xy}\right)^2$
15. $\dfrac{(m - 25)^{16}}{(m - 25)^4}$
16. $\dfrac{(5y^2z^3)^3}{(yz)^5}$
17. $\dfrac{a^5a^4a^5}{a^2a}$
18. $\dfrac{(cd)^9}{(cd)^4}$

Find an expression that represents the area or the volume of each figure, whichever is appropriate.

19.

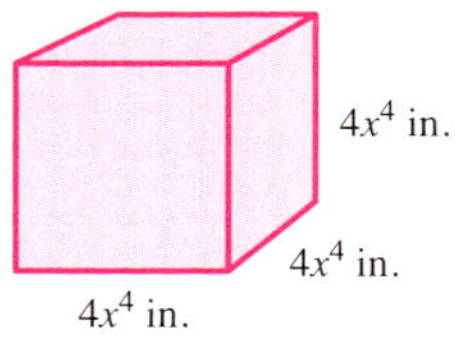

20.

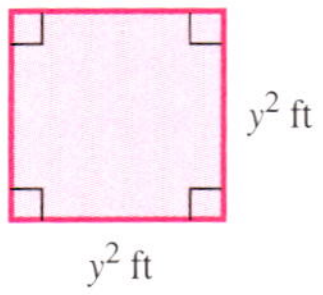

SECTION 5.2 Zero and Negative Exponents

DEFINITIONS AND CONCEPTS	EXAMPLES
Rules for exponents: For any nonzero real numbers x and y and any integers m and n,	Simplify each expression. Do not use negative exponents in the answer.
Zero exponent: $x^0 = 1$	$5^0 = 1$
Negative exponents: $x^{-n} = \frac{1}{x^n}$	$4^{-2} = \frac{1}{4^2} = \frac{1}{16}$ and $7c^{-6} = \frac{7}{c^6}$
Negative to positive rules: $\frac{1}{x^{-n}} = x^n$ $\quad \frac{x^{-m}}{y^{-n}} = \frac{y^n}{x^m}$	$\frac{1}{t^{-8}} = t^8$ and $\frac{2^{-4}}{x^{-6}} = \frac{x^6}{2^4} = \frac{x^6}{16}$
Negative exponents and reciprocals: $\left(\frac{x}{y}\right)^{-n} = \left(\frac{y}{x}\right)^n$	$\left(\frac{x}{10}\right)^{-3} = \left(\frac{10}{x}\right)^3 = \frac{10^3}{x^3} = \frac{1,000}{x^3}$

REVIEW EXERCISES

Simplify each expression. Do not use negative exponents in the answer.

21. x^0
22. $(3x^2y^2)^0$
23. $3x^0$
24. 10^{-3}
25. -5^{-2}
26. $\frac{t^4}{t^{10}}$
27. $\frac{8}{x^{-5}}$
28. $-6y^4y^{-5}$
29. $\frac{7^{-2}}{2^{-3}}$
30. $(x^{-3}x^{-4})^{-2}$
31. $\left(\frac{-3r^4r^{-3}}{r^{-3}r^7}\right)^3$
32. $\left(\frac{4z^4}{z^3}\right)^{-2}$
33. $\frac{3^{-2}c^3d^3}{2^{-3}c^2d^8}$
34. $\frac{t^{-30}}{t^{-60}}$
35. $\frac{w(w^{-3})^{-4}}{w^{-9}}$
36. $\left(\frac{4}{f^4}\right)^{-10}$

SECTION 5.3 Scientific Notation

DEFINITIONS AND CONCEPTS	EXAMPLES
A positive number is written in **scientific notation** when it is written in the form $N \times 10^n$, where $1 \le N < 10$ and n is an integer.	Write each number in scientific notation. $32{,}500 = 3.25 \times 10^4$ (4 decimal places) and $0.0025 = 2.5 \times 10^{-3}$ (3 decimal places) Write each number in standard notation. $1.91 \times 10^5 = 191{,}000$ (5 decimal places) and $4.7 \times 10^{-6} = 0.0000047$ (6 decimal places)
Scientific notation provides an easier way to perform computations involving very large or very small numbers.	Use scientific notation to perform the calculation: $\frac{684{,}000{,}000}{456{,}000} = \frac{6.84 \times 10^8}{4.56 \times 10^5} = \frac{6.84}{4.56} \times \frac{10^8}{10^5} = 1.5 \times 10^3$ or $1{,}500$

REVIEW EXERCISES

Write each number in scientific notation.

37. 720,000,000

38. 9,370,000,000,000,000

39. 0.00000000942

40. 0.00013

41. 0.018×10^{-2}

42. 853×10^{3}

Write each number in standard notation.

43. 1.26×10^{5}

44. 3.919×10^{-8}

45. 2.68×10^{0}

46. 5.76×10^{1}

Evaluate each expression by first writing each number in scientific notation. Express the result in scientific notation and standard notation.

47. $\dfrac{(0.000012)(0.000004)}{0.00000016}$

48. $\dfrac{(4{,}800{,}000)(20{,}000{,}000)}{600{,}000}$

49. World Population. As of January 2007, the world's population was estimated to be 6.57 billion. Write this number in standard notation and in scientific notation.

50. Atoms. The illustration shows a cross section of an atom. How many nuclei (plural for nucleus), placed end to end, would it take to stretch across the atom?

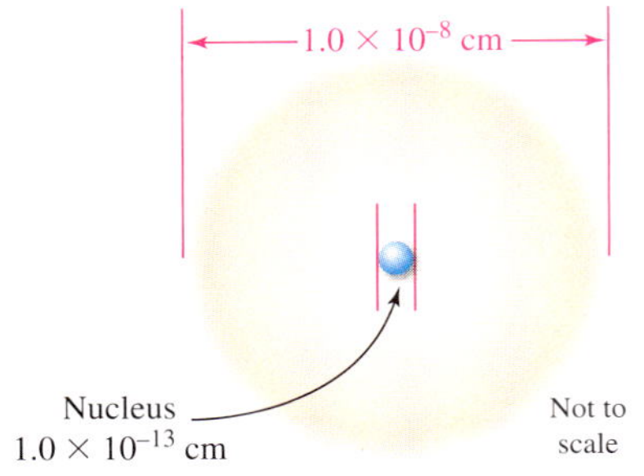

SECTION 5.4 Polynomials

DEFINITIONS AND CONCEPTS	EXAMPLES
A **polynomial** is a single term or a sum of terms in which all variables have whole-number exponents and no variable appears in a denominator.	Polynomials: 32, $-5x^2y^3$, $7p^3 - 14q^3$, $4m^2 + 5m - 12$ Not Polynomials: $y^2 - y^{-5}$, $4x^3 - \dfrac{7}{x} + 3x$
A polynomial with exactly one term is called a **monomial.** A polynomial with exactly two terms is called a **binomial.** A polynomial with exactly three terms is called a **trinomial.**	***Monomials***: $3x^2$, $-12m^3n^2$ ***Binomials***: $2y^3 + 3y$, $87t - 25$ ***Trinomials***: $3p^2 - 7p + 12$, $4p^2q^3 - 8p^2q^2 + 12p^2q$
The **coefficient of a term** is its numerical factor. The **degree of a term** of a polynomial in one variable is the value of the exponent on the variable. If a polynomial has more than one variable, the **degree of a term** is the sum of the exponents on the variables. The **degree of a nonzero constant** is 0.	***Term*** / ***Coefficient*** / ***Degree of the term*** $6a^7$ / 6 / 7 $-7.3x^5y^4$ / -7.3 / $5 + 4 = 9$ 32 / 32 / 0
The **degree of a polynomial** is equal to the highest degree of any term of the polynomial.	***Polynomial*** / ***Degree of the polynomial*** $7m^3 - 4m^2 + 5m - 12$ / 3 $\frac{1}{2}a^4b + \frac{3}{4}a^3b^2 - \frac{2}{3}a^2b^4$ / $2 + 4 = 6$
To **evaluate a polynomial** for a given value, substitute the value for the variable and follow the rules for the order of operations.	Evaluate $3x^2 - 4x + 2$ for $x = 2$. $3x^2 - 4x + 2 = 3(2)^2 - 4(2) + 2$ Substitute 2 for each x. $= 3(4) - 8 + 2$ Evaluate $(2)^2$ first. $= 6$

We can **graph equations defined by polynomials** such as $y = x^2 - 2$, $y = -x^2$, and $y = x^3 + 1$.

The graph of $y = x^2 - 2$ is a cup-shaped curve called a **parabola.**

Graph: $y = x^2 - 2$

Find several solutions of the equation, plot them on a rectangular coordinate system, and then draw a smooth curve passing through the points.

$y = x^2 - 2$

x	y	(x, y)
-2	2	$(-2, 2)$
-1	-1	$(-1, -1)$
0	-2	$(0, -2)$
1	-1	$(1, -1)$
2	2	$(2, 2)$

Select x-values. Find y-values. Plot points.

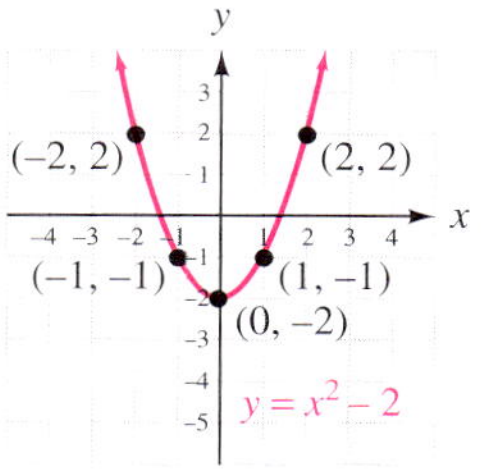

REVIEW EXERCISES

51. Consider the polynomial $3x^3 - x^2 + x + 10$.

a. How many terms does the polynomial have?

b. What is the lead term?

c. What is the coefficient of each term?

d. What is the constant term?

52. Find the degree of each polynomial and classify it as a monomial, binomial, trinomial, or none of these.

a. $13x^7$

b. $-16a^2b$

c. $5^3x + x^2$

d. $-3x^5 + x - 1$

e. $9xy^2 + 21x^3y^3$

f. $4s^4 - 3s^2 + 5s + 4$

53. Evaluate $-x^5 - 3x^4 + 3$ for $x = 0$ and $x = -2$.

54. Diving. The number of inches that the woman deflects the diving board is given by the polynomial $0.1875x^2 - 0.0078125x^3$ where x is the number of feet that she stands from the front anchor point of the board. Find the amount of deflection if she stands on the end of the diving board, 8 feet from the anchor point.

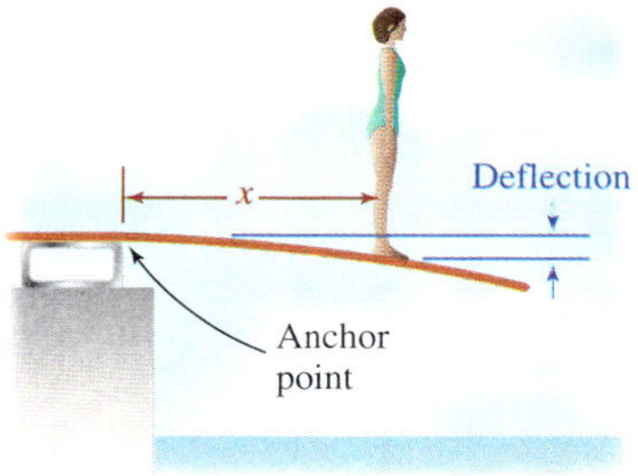

Construct a table of solutions like the one shown here and then graph the equation.

55. $y = x^2$

x	-2	-1	0	1	2
y					

56. $y = x^3 + 1$

x	-2	-1	0	1	2
y					

SECTION 5.5 Adding and Subtracting Polynomials

DEFINITIONS AND CONCEPTS	EXAMPLES
To **simplify a polynomial,** combine like terms.	Simplify: $3r^4 - 4r^3 + 7r^4 + 8r^2$ $= 10r^4 - 4r^3 + 8r^2$ Combine like terms. Think: $(3 + 7)r^4 = 10r^4$.
To **add polynomials,** combine their like terms.	Add: $(4x^2 + 9x + 4) + (3x^2 - 5x - 1)$ $= (4x^2 + 3x^2) + (9x - 5x) + (4 - 1)$ Group like terms. $= 7x^2 + 4x + 3$ Combine like terms.
To **subtract two polynomials,** change the signs of the terms of the polynomial being subtracted, drop the parentheses, and combine like terms.	Subtract: $(8a^3b - 4ab^2) - (-3a^3b + 9ab^2)$ $= 8a^3b - 4ab^2 + 3a^3b - 9ab^2$ Change the sign of each term of $-3a^3b + 9ab^2$ and drop the parentheses. $= 11a^3b - 13ab^2$ Combine like terms.

REVIEW EXERCISES

Simplify each polynomial and write the result in descending powers of one variable.

57. $6y^3 + 8y^4 + 7y^3 + (-8y^4)$
58. $4a^2b + 5 - 6a^3b - 3a^2b + 2a^3b + 1$
59. $\frac{5}{6}x^2 + \frac{1}{3}y^2 - \frac{1}{4}x^2 - \frac{3}{4}xy + \frac{2}{3}y^2$
60. $7.6c^5 - 2.1c^3 - 0.9c^5 + 8.1c^4$

Perform the operations.

61. $(2r^6 + 14r^3) + (23r^6 - 5r^3 + 5r)$
62. $(7.1a^2 + 2.2a - 5.8) - (3.4a^2 - 3.9a + 11.8)$
63. $(3r^3s + r^2s^2 - 3rs^3 - 3s^4) + (r^3s - 8r^2s^2 - 4rs^3 + s^4)$
64. $\left(\frac{7}{8}m^4 - \frac{1}{5}m^3\right) - \left(\frac{1}{4}m^4 + \frac{1}{5}m^3\right) - \frac{3}{5}m^3$
65. Find the difference when $(-3z^3 - 4z + 7)$ is subtracted from the sum of $(2z^2 + 3z - 7)$ and $(-4z^3 - 2z - 3)$.
66. **Gardening.** Find a polynomial that represents the length of the wooden handle of the shovel.

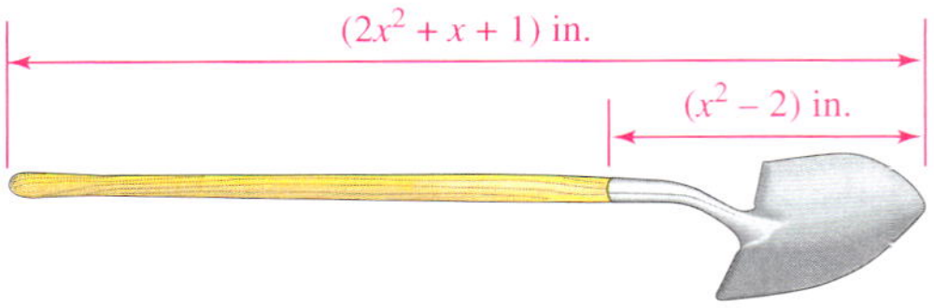

67. Add:

$$\begin{array}{r} 3x^2 + 5x + 2 \\ \underline{x^2 - 3x + 6} \end{array}$$

68. Subtract:

$$\begin{array}{r} 20x^3 \qquad\quad + 12x \\ \underline{-(12x^3 + 7x^2 - \;\; 7x)} \end{array}$$

SECTION 5.6 Multiplying Polynomials

DEFINITIONS AND CONCEPTS	EXAMPLES
To **multiply two monomials,** multiply the numerical factors (the coefficients) and then multiply the variable factors.	Multiply: $(5p^6)(2p^5) = (5 \cdot 2)(p^6 \cdot p^5)$ Group the coefficients together and the variables together. $= 10p^{11}$ Think: $5 \cdot 2 = 10$ and $p^6 \cdot p^5 = p^{6+5} = p^{11}$.
To **multiply a monomial and a polynomial,** multiply each term of the polynomial by the monomial.	Multiply: $3r^2(2r^4 + 7r^2 - 4)$ $= 3r^2(2r^4) + 3r^2(7r^2) + 3r^2(-4)$ Distribute the multiplication by $3r^2$. $= 6r^6 + 21r^4 - 12r^2$ Multiply the monomials.
To **multiply two binomials,** use the *FOIL method:* F: First O: Outer I: Inner L: Last	Multiply: $(3m + 4)(2m - 5) = 3m(2m) + 3m(-5) + 4(2m) + 4(-5)$ (F O I L) $= 6m^2 - 15m + 8m - 20$ Multiply the monomials. $= 6m^2 - 7m - 20$ Combine like terms.
To **multiply two polynomials,** multiply each term of one polynomial by each term of the other polynomial and then combine like terms.	Multiply: $(a - b)(6a^2 - 4ab + b^2)$ $= a(6a^2) + a(-4ab) + a(b^2) - b(6a^2) - b(-4ab) - b(b^2)$ $= 6a^3 - 4a^2b + ab^2 - 6a^2b + 4ab^2 - b^3$ Multiply the monomials. $= 6a^3 - 10a^2b + 5ab^2 - b^3$ Combine like terms.

When finding the **product of three polynomials,** begin by multiplying any two of them, and then multiply that result by the third polynomial.

Multiply: $-9x^4(x-1)(x-7) = -9x^4(x^2 - 7x - x + 7)$ Multiply the two binomials.

$= -9x^4(x^2 - 8x + 7)$ Combine like terms within the parentheses.

$= -9x^6 + 72x^5 - 63x^4$ Distribute the multiplication by $-9x^4$.

REVIEW EXERCISES

Multiply.

69. $(2x^2)(5x)$
70. $(-6x^4z^3)(x^6z^2)$
71. $5b^3 \cdot 6b^2 \cdot 4b^6$
72. $\frac{2}{3}h^5(3h^9 + 12h^6)$
73. $3n^2(3n^2 - 5n + 2)$
74. $x^2y(y^2 - xy)$
75. $2x(3x^4)(x + 2)$
76. $-a^2b^2(-a^4b^2 + a^3b^3 - ab^4 + 7a)$
77. $(x + 3)(x + 2)$
78. $(2x + 1)(x - 1)$
79. $(3t - 3)(2t + 2)$
80. $(3n^4 - 5n^2)(2n^4 - n^2)$
81. $-a^5(a^2 - b)(5a^2 + b)$
82. $6.6(a - 1)(a + 1)$
83. $\left(3t - \frac{1}{3}\right)\left(6t + \frac{5}{3}\right)$
84. $(5.5 - 6b)(2 - 4b)$
85. $(2a - 3)(4a^2 + 6a + 9)$
86. $(8x^2 + x - 2)(7x^2 + x - 1)$
87. Multiply using vertical form:
$$\begin{array}{r} 4x^2 - 2x + 1 \\ 2x + 1 \\ \hline \end{array}$$
88. Refer to the illustration below. Find a polynomial that represents
  a. the perimeter of the base of the dishwasher.
  b. the area of the base of the dishwasher.
  c. the volume occupied by the dishwasher.

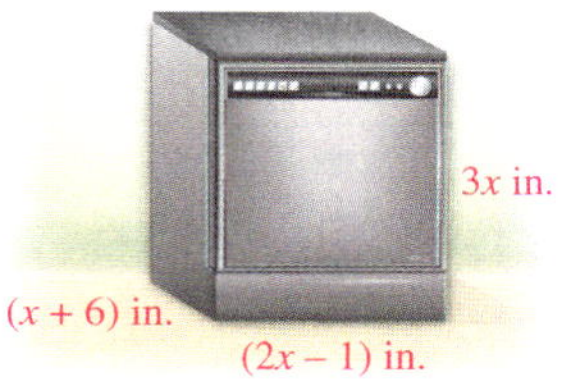

SECTION 5.7 Special Products

DEFINITIONS AND CONCEPTS

The following **special products** occur so often that it is worthwhile to learn their forms.

Square of a binomial:

$(A + B)^2 = A^2 + 2AB + B^2$

This is the square of a binomial sum.

$(A - B)^2 = A^2 - 2AB + B^2$

This is the square of a binomial difference.

Multiplying the Sum and Difference of the Same Two Terms:

$(A - B)(A - B) = A^2 - B^2$

EXAMPLES

Multiply: $(n + 4)^2 = n^2 + 2(n)(4) + 4^2$

The square of the first term, n. Twice the product of both terms. The square of the second term, 4.

$= n^2 + 8n + 16$

Multiply: $(5a - 1)^2 = (5a)^2 + 2(5a)(-1) + (-1)^2$

The square of the first term, $5a$. Twice the product of both terms. The square of the second term, -1.

$= 25a^2 - 10a + 1$

Multiply: $(x + 8)(x - 8) = x^2 - 8^2$

The square of the first term, x. The square of the second term, 8.

$= x^2 - 64$

REVIEW EXERCISES

Find each product.

89. $(a - 3)^2$
90. $(m + 2)^3$
91. $(x + 7)(x - 7)$
92. $(2x - 0.9)(2x + 0.9)$
93. $(2y + 1)^2$
94. $(y^2 + 1)(y^2 - 1)$
95. $(6r^2 + 10s)^2$
96. $-(8a - 3c)^2$
97. $80s(r^2 + s^2)(r^2 - s^2)$
98. $4b(3b - 4)^2$
99. $\left(t - \frac{3}{4}\right)^2$
100. $\left(x + \frac{4}{3}\right)^2$

Perform the operations.

101. $3(9x^2 + 3x + 7) - 2(11x^2 - 5x + 9)$
102. $(5c - 1)^2 - (c + 6)(c - 6)$

103. **Graphic Arts.** A Dr. Martin Luther King poster has his picture with a $\frac{1}{2}$-inch wide blue border around it. The length of the poster is $(x + 3)$ inches and the width is $(x - 1)$ inches. Find a polynomial that represents the area of the *picture* of Dr. King.

National Archives

104. Find a polynomial that represents the area of the triangle.

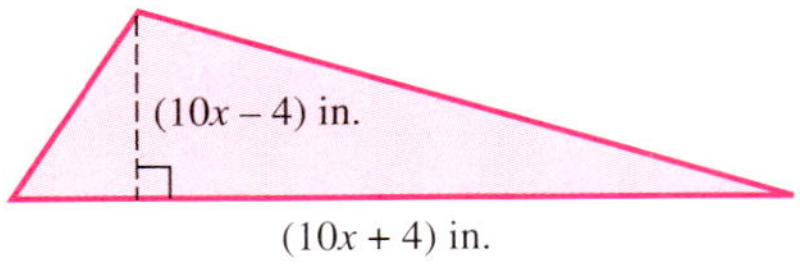

SECTION 5.8 Dividing Polynomials

DEFINITIONS AND CONCEPTS	EXAMPLES
To **divide monomials,** use the method for simplifying fractions and/or the quotient rule for exponents.	Divide the monomials: $\frac{8p^2q}{4pq^3} = \frac{2 \cdot \cancel{4} \cdot \cancel{p} \cdot p \cdot \cancel{q}}{\cancel{4} \cdot \cancel{p} \cdot \cancel{q} \cdot q \cdot q} = \frac{2p}{q^2}$ or $\frac{8p^2q}{4pq^3} = \frac{8}{4}p^{2-1}q^{1-3}$ Keep each base and subtract the exponents. $= 2p^1q^{-2}$ $= \frac{2p}{q^2}$ Move q^{-2} to the denominator and change the sign of the exponent.
To **divide a polynomial by a monomial,** divide each term of the numerator by the denominator.	Divide: $\frac{9a^2b^4 - 12a^3b^6}{\mathbf{18ab^5}} = \frac{9a^2b^4}{\mathbf{18ab^5}} - \frac{12a^3b^6}{\mathbf{18ab^5}}$ $= \frac{a}{2b} - \frac{2a^2b}{3}$ Perform each monomial division.
Long division can be used to **divide a polynomial by a polynomial** (other than a monomial). The long division method is a series of four steps that are repeated: Divide, multiply, subtract, and bring down the next term. When the division has a remainder, write the answer in the form: Quotient $+ \frac{\text{remainder}}{\text{divisor}}$.	Divide $4x^2 - 4x + 5$ by $2x + 1$. $\begin{array}{r} 2x - 3 \\ 2x + 1\overline{)4x^2 - 4x + 5} \\ -(4x^2 + 2x) \downarrow \\ -6x + 5 \\ -(-6x - 3) \\ 8 \end{array}$ The first division: $\frac{4x^2}{2x} = 2x$. The second division: $\frac{-6x}{2x} = -3$. The remainder is 8. The result is: $2x - 3 + \frac{8}{2x + 1}$

The long division method works best when the terms of the divisor and the dividend are written in **descending powers of the variable.**

When the dividend has **missing terms,** insert such terms with a coefficient of 0, or leave a blank space.

Set up each long division.

$\frac{5x + x^3 + 3 + 3x^2}{x + 1}$

The terms of the dividend are not in descending powers of x.

$x + 1\overline{)x^3 + 3x^2 + 5x + 3}$

$\frac{x^2 - 9}{x - 3}$

The dividend is missing a term.

$x - 3\overline{)x^2 + 0x - 9}$

REVIEW EXERCISES

Divide. Do not use negative exponents in the answer.

105. $\frac{16n^8}{8n^5}$

106. $\frac{-14x^2y}{21xy^3}$

107. $\frac{a^{15} - 24a^8}{6a^{12}}$

108. $\frac{15a^5b + ab^2 - 25b}{5a^2b}$

109. $x - 1\overline{)x^2 - 6x + 5}$

110. $\frac{2x^2 + 3 + 7x}{x + 3}$

111. $(15x^2 - 8x - 8) \div (3x + 2)$

112. Divide $25y^2 - 9$ by $5y + 3$.

113. $3x + 1\overline{)-13x - 4 + 9x^3}$

114. $2x - 1\overline{)6x^3 + x^2 + 1}$

115. Use multiplication to show that $(3y^2 + 11y + 6) \div (y + 3)$ is $3y + 2$.

116. **Bedding.** The area of a rectangular-shaped bed sheet is represented by the polynomial $(4x^3 + 12x^2 + x - 12)$ in.^{2}. If the width of the sheet is $(2x + 3)$ inches, find a polynomial that represents its length.

5 CHAPTER TEST

1. Fill in the blanks.
   a. In the expression y^{10}, the _____ is y and the _________ is 10.
   b. We call a polynomial with exactly one term a _________, with exactly two terms a _________, and with exactly three terms a _________.
   c. The _______ of a term of a polynomial in one variable is the value of the exponent on the variable.
   d. $(x + y)^2$, $(x - y)^2$, and $(x + y)(x - y)$ are called _______ products.
2. Use exponents to rewrite $2xxxyyyy$.

Simplify each expression. Do not use negative exponents in the answer.

3. $y^2(yy^3)$

4. $\left(\frac{1}{2}x^3\right)^5 (x^2)^3$

5. $3.5x^0$

6. $2y^{-5}y^2$

7. 5^{-3}

8. $\frac{(x + 1)^{15}}{(x + 1)^6}$

9. $\frac{(y^{-5})^{-4}}{yy^{-2}}$

10. $\left(\frac{a^2b^{-1}}{4a^3b^{-2}}\right)^3$

11. $\left(\frac{8}{m^6}\right)^{-2}$

12. $\frac{-6a}{b^{-9}}$

13. Find an expression that represents the volume of a cube that has sides of length $10y^4$ inches.

14. **Electricity.** One ampere (amp) corresponds to the flow of 6,250,000,000,000,000,000 electrons per second past any point in a direct current (DC) circuit. Write this number in scientific notation.

15. Write 9.3×10^{-5} in standard notation.

16. Evaluate $(2.3 \times 10^{18})(4.0 \times 10^{-15})$. Write the answer in scientific notation and standard notation.

17. Identify $x^4 + 8x^2 - 12$ as a monomial, binomial, or trinomial. Then complete the table.

Term	Coefficient	Degree

Degree of the polynomial:

18. Find the degree of the polynomial $3x^3y + 2x^2y^3 - 5xy^2 - 6y$.

19. Complete the table of solutions for $y = x^2 + 2$ and then graph the equation.

x	-2	-1	0	1	2
y					

20. Free Fall. A visitor standing on the rim of the Grand Canyon drops a rock over the side. The distance (in feet) that the rock is from the canyon floor t seconds after being dropped is given by the polynomial $-16t^2 + 5{,}184$. Find the position of the rock 18 seconds after being dropped. Explain your answer.

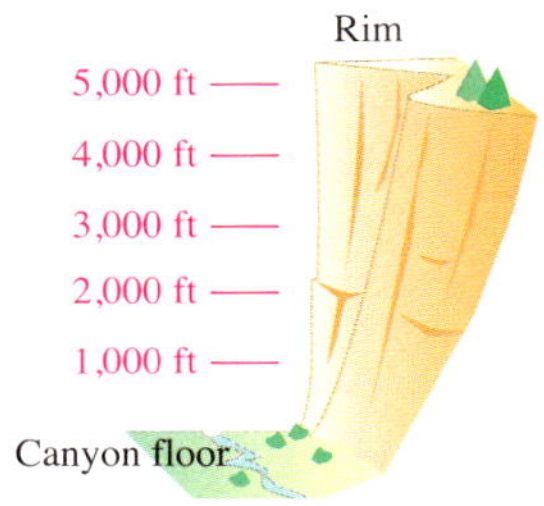

Simplify each polynomial.

21. $\frac{3}{5}x^2 + 6 + \frac{1}{4}x - 8 - \frac{1}{2}x^2 + \frac{1}{3}x$

22. $4a^2b + 5 - 6a^3b - 3a^2b + 2a^3b$

Perform the operations.

23. $(12.1h^3 - 9.9h^2 + 9.5) + (7.3h^3 - 1.2h^2 - 10.1)$

24. Subtract $b^3c - 3bc + 12$ from the sum of $6b^3c - 3bc$ and $b^3c - 2bc$.

25. Subtract:

$$\begin{array}{r} -5y^3 + 4y^2 \qquad + 3 \\ -(-2y^3 - 14y^2 + 17y - 32) \\ \hline \end{array}$$

26. Find a polynomial that represents the perimeter of the rectangle.

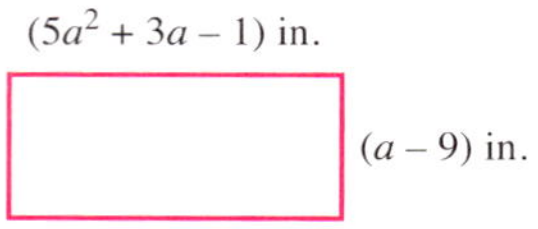

Multiply.

27. $(2x^3y^3)(5x^2y^8)$

28. $9b^3(8b^4)(-b)$

29. $3y^2(y^2 - 2y + 3)$

30. $0.6p^5(0.4p^6 - 0.9p^3)$

31. $\frac{3}{4}s^3t^9(s^4t^8 + 16st)$

32. $(x - 5)(3x + 4)$

33. $\left(6t + \frac{1}{2}\right)\left(2t - \frac{3}{2}\right)$

34. $(3.8m - 1)(2m - 1)$

35. $(a^3 - 6)(a^3 + 7)$

36. $(2x - 3)(x^2 - 2x + 4)$

37. $(1 + 10c)(1 - 10c)$

38. $(7b^3 - 3t)^2$

39. $(2.2a)(a + 5)(a - 3)$

40. Perform the operations: $(x + y)(x - y) + (x + y)^2$

Divide.

41. $\dfrac{6a^2 - 12b^2}{24ab}$

42. $\dfrac{x^2 + x - 6}{x + 3}$

43. $2x - 1\overline{)1 + x^2 + 6x^3}$

44. Find a polynomial that represents the width of a rectangle if its area is represented by the polynomial $(x^2 - 6x + 5)$ ft^2 and the length is $(x - 1)$ ft.

45. Use a check to determine whether $(5m^2 - 29m - 6) \div (5m + 1) = m - 6$.

46. Is $(a + b)^2 = a^2 + b^2$? Show why or why not.

Group Project

BINOMIAL MULTIPLICATION AND THE AREA OF RECTANGLES

Overview: In this activity, rectangles are used to illustrate binomial multiplication.

Instructions: Form groups of 2 or 3 students. Study the figure. The area of the largest rectangle (outlined in blue) is given by $(x + 2)(x + 3)$. Its area is also the sum of the areas of the four smaller rectangles: $x^2 + 3x + 2x + 6$. Thus,

$$(x + 2)(x + 3) = x^2 + 3x + 2x + 6$$
$$= x^2 + 5x + 6$$

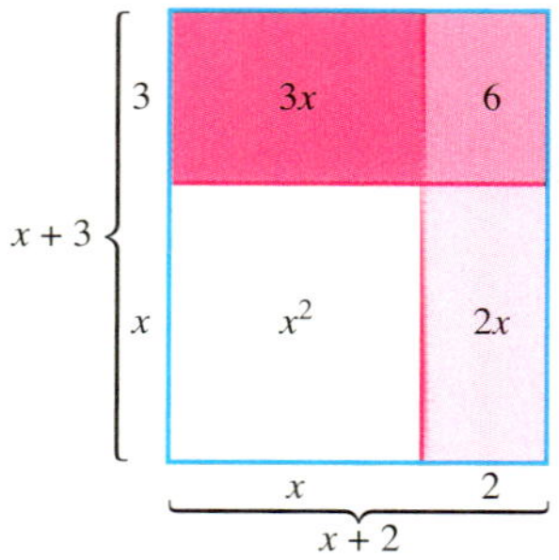

Draw three similar models to represent the following products.

1. $(x + 4)(x + 5)$ 2. $(x + 8)^2$ 3. $x(x + 6)$ 4. $(2x + 1)^2$

Determine the missing number so that each rectangle below has the given area.

5. Area: $x^2 + 9x + 14$ 6. Area: $x^2 + 16x + 55$

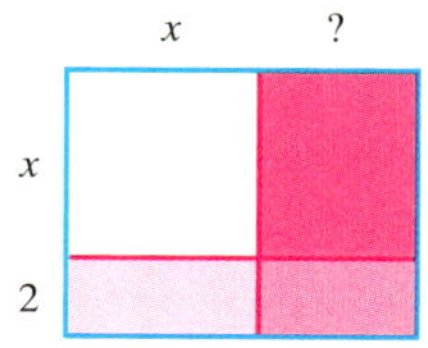

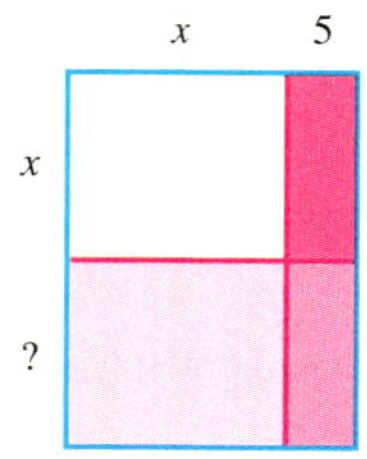

7. Draw a model that represents a rectangle with area $x^2 + 10x + 24$. Label it completely.

CUMULATIVE REVIEW Chapters 1–5

1. Use exponents to write the prime factorization of 270. [Section 1.2]

2. a. Use the variables a and b to state the commutative property of addition. [Section 1.4]

   b. Use the variables x, y, and z to state the associative property of multiplication. [Section 1.6]

Evaluate each expression.

3. $3 - 4[-10 - 4(-5)]$ [Section 1.7]

4. $\dfrac{|-45| - 2(-5) + 1^5}{2 \cdot 9 - 2^4}$ [Section 1.7]

Simplify each expression.

5. $27\left(\dfrac{2}{3}x\right)$ [Section 1.9]

6. $3x^2 + 2x^2 - 5x^2$ [Section 1.9]

Solve each equation.

7. $2 - (4x + 7) = 3 + 2(x + 2)$ [Section 2.2]

8. $\dfrac{2}{5}y + 3 = 9$ [Section 2.2]

9. **Candy Sales.** The circle graph shows how \$6.5 billion in seasonal candy sales for 2009 was spent. Find the candy sales for Halloween. [Section 2.3]

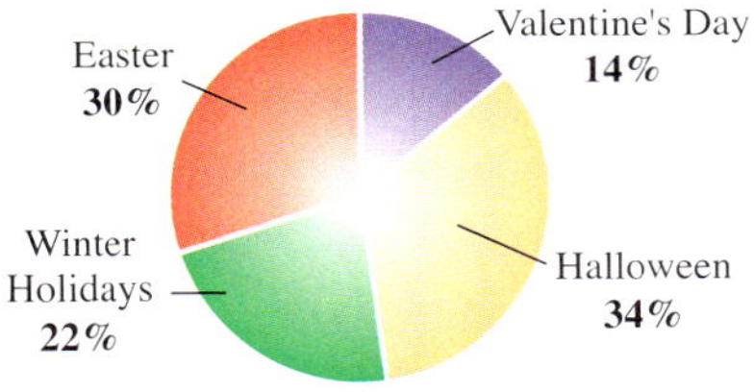

Source: National Confectioners Association

10. **Air Conditioning.** Find the volume of air contained in the duct. Round to the nearest tenth of a cubic foot. [Section 2.4]

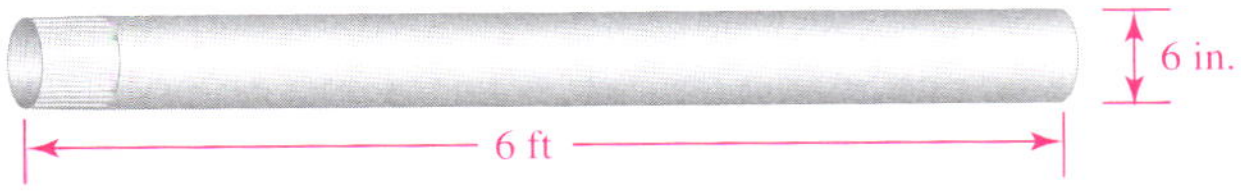

11. **Angle of Elevation.** Find x. [Section 2.5]

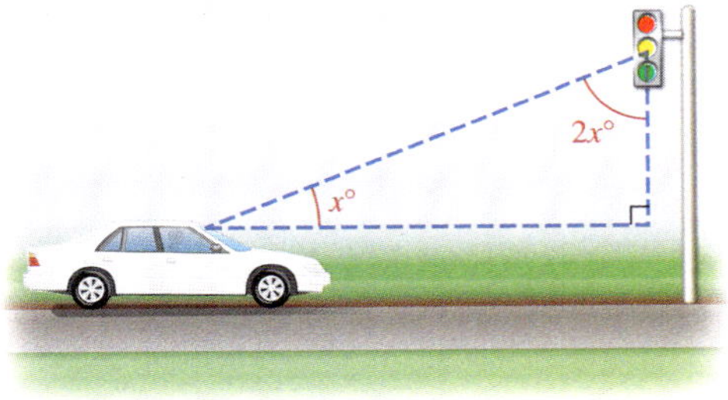

12. **Livestock Auction.** A farmer is going to sell one of her prize hogs at an auction and would like to make \$6,000 after paying a 4% commission to the auctioneer. For what selling price will the farmer make this amount of money? [Section 2.5]

13. **Stock Market.** An investment club invested part of \$45,000 in a high-yield mutual fund that earned 12% annual simple interest. The remainder of the money was invested in Treasury bonds that earned 6.5% simple annual interest. The two investments earned \$4,300 in one year. How much was invested in each account? [Section 2.6]

14. Solve $-4x + 6 > 17$ and graph the solution set. Then describe the graph using interval notation. [Section 2.7]

Graph each equation.

15. $y = 3x$ [Section 3.2]

16. $x = -2$ [Section 3.3]

17. Find the slope of the line passing through $(6, -2)$ and $(-3, 2)$. [Section 3.4]

18. **Diabetes.** The graph below shows the number of people in the United States diagnosed with diabetes from 1996 to 2008. Find the rate of change in the number over this time span. [Section 3.4]

Number of People Diagnosed with Diabetes in U.S.

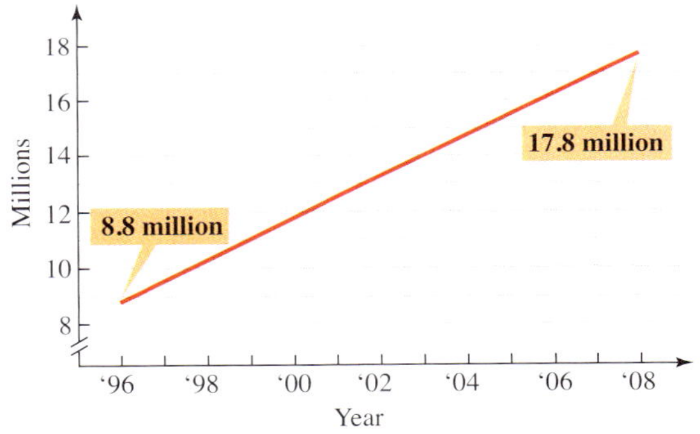

Source: U.S. Centers for Disease Control

19. Find the slope and y-intercept of the line. Then write the equation of the line. [Section 3.5]

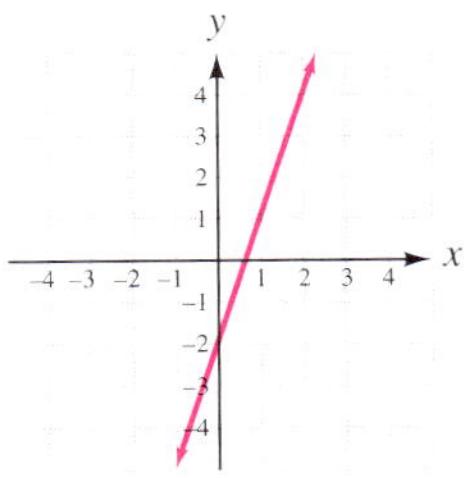

20. Without graphing, determine whether the graphs of $y = \frac{3}{2}x - 1$ and $2x + 3y = 10$ are parallel, perpendicular, or neither. [Section 3.5]

21. Write the equation of the line that passes through $(-2, 10)$ with slope -4. Write the result in slope–intercept form. [Section 3.6]

22. Is $(-2, 1)$ a solution of $2x - 3y \geq -6$? [Section 3.7]

23. If $f(x) = 2x^2 + 3x - 9$, find $f(-5)$. [Section 3.8]

24. Determine whether the graph below is the graph of a function. If it is not, find ordered pairs that show a value of x that is assigned more than one value of y. [Section 3.8]

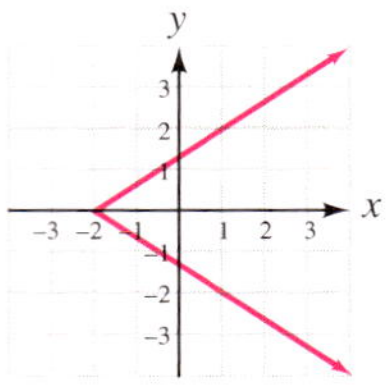

25. Is $\left(\frac{2}{3}, -1\right)$ a solution of the system $\begin{cases} y = -3x + 1 \\ 3x + 3y = -2 \end{cases}$? [Section 4.1]

26. Solve the system $\begin{cases} 3x + 2y = 14 \\ y = \frac{1}{4}x \end{cases}$ by graphing. [Section 4.1]

27. Solve the system $\begin{cases} 2b - 3a = 18 \\ a + 3b = 5 \end{cases}$ by substitution. [Section 4.2]

28. Solve the system $\begin{cases} 8s + 10t = 24 \\ 11s - 3t = -34 \end{cases}$ by elimination (addition). [Section 4.3]

29. **Vacations.** One-day passes to Universal Studios Hollywood cost a family of 5 (2 adults and 3 children) \$275. A family of 6 (3 adults and 3 children) paid \$336 for their one-day passes. Find the cost of an adult one-day pass and a child's one-day pass to Universal Studios. [Section 4.4]

30. Graph: $\begin{cases} y \leq 2x - 1 \\ x + 3y > 6 \end{cases}$ [Section 4.5]

Simplify. Do not use negative exponents in the answer.

31. $(-3x^2y^4)^2$ [Section 5.1]

32. $(v^5)^2(v^3)^4$ [Section 5.1]

33. $ab^3c^4 \cdot ab^4c^2$ [Section 5.1]

34. $\left(\frac{4t^3t^4t^5}{3t^2t^6}\right)^3$ [Section 5.1]

35. $(2y)^{-4}$ [Section 5.2]

36. $\frac{a^4b^0}{a^{-3}}$ [Section 5.2]

37. -5^{-2} [Section 5.2]

38. $\left(\frac{a}{x}\right)^{-10}$ [Section 5.2]

Write each number in scientific notation.

39. 615,000 [Section 5.3]

40. 0.0000013 [Section 5.3]

41. Graph: $y = x^2$ [Section 5.4]

42. **Musical Instruments.** The amount of deflection of the horizontal beam (in inches) is given by the polynomial $0.01875x^4 - 0.15x^3 + 1.2x$, where x is the distance (in feet) that the gong is hung from one end of the beam. Find the deflection if the gong is hung in the middle of the support. [Section 5.4]

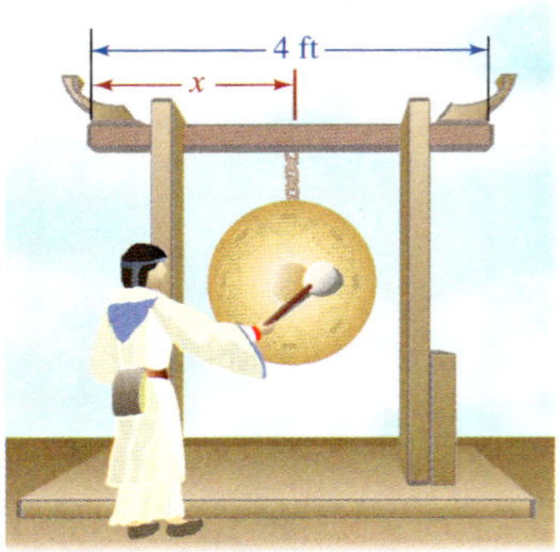

Perform the indicated operations.

43. $(4c^2 + 3c - 2) + (3c^2 + 4c + 2)$ [Section 5.5]

44. Subtract: $\begin{array}{r} 17x^4 - 3x^2 - 65x - 12 \\ \underline{-(23x^4 + 14x^2 + 3x - 23)} \end{array}$ [Section 5.5]

45. $(2t + 3s)(3t - s)$ [Section 5.6]

46. $3x(2x + 3)^2$ [Section 5.7]

47. $5x + 3\overline{)11x + 10x^2 + 3}$ [Section 5.8]

48. $\frac{2x - 32}{16x}$ [Section 5.8]

Factoring and Quadratic Equations

6

©iStockphoto.com/Neustockimages

from Campus to Careers

Elementary School Teacher

It has been said that a teacher takes a hand, opens a mind, and touches a heart. That is certainly true for the thousands of dedicated elementary school teachers across the country. Elementary school teachers use their training in mathematics in many ways. Besides teaching math on a daily basis, they calculate student grades, analyze test results, and order instructional materials and supplies. They use measurement and geometry for designing bulletin board displays and they construct detailed schedules so that the classroom time is used wisely.

Problem 129 in **Study Set 6.1, problem 103** in **Study Set 6.3,** and **problem 17** in **Study Set 6.8** involve situations that an elementary school teacher might encounter on the job. The mathematical concepts discussed in this chapter can be used to solve those problems.

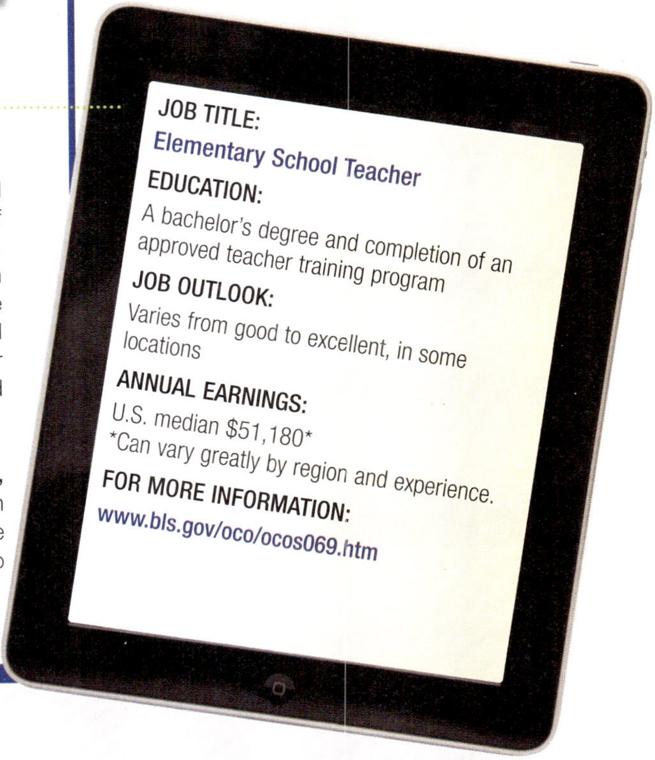

Study Skills Workshop ▶ Reading the Textbook

Reading an algebra textbook is different from reading a newspaper or a novel. Here are two ways that you should be reading this textbook.

SKIMMING FOR AN OVERVIEW: This is a quick way to look at material just before it is covered in class. It helps you become familiar with the new vocabulary and notation that will be used by your instructor in the lecture. It lays a foundation.

READING FOR UNDERSTANDING: This in-depth type of reading is done more slowly, with a pencil and paper at hand. Don't skip anything—every word counts! You should do just as much writing as you do reading. Highlight the important points and work each example. If you become confused, stop and reread the material until you understand it.

Now Try This

Choose a section from this chapter and . . .

1. . . . quickly skim it. Write down any terms in bold face type and the titles of any properties, definitions, or strategies that are given in the colored boxes.
2. . . . work each *Self Check* problem. Your solutions should look like those in the *Examples*. Be sure to include your own "author notes" (the sentences in red to the right of each step of a solution).

SECTION 6.1 The Greatest Common Factor; Factoring by Grouping

OBJECTIVES

1. Find the greatest common factor of a list of terms.
2. Factor out the greatest common factor.
3. Factor by grouping.

ARE YOU READY?

The following problems review some basic skills that are needed when factoring expressions.

1. Find the prime factorization of 54.
2. Write $a \cdot a \cdot a \cdot a \cdot a$ in an equivalent form using exponents.
3. Simplify: $-(x - 7)$
4. How many terms does the expression $3x^3 + x^2 + 4x + 8$ have?
5. Multiply: $4d(d^3 + 2d)$
6. Multiply and simplify: $x^2(x + 1) - 3(x + 1)$

In Chapter 5, we learned how to multiply polynomials. For example, to multiply $3x + 5$ by $4x$, we use the distributive property, as shown below.

$$4x(3x + 5) = 4x \cdot 3x + 4x \cdot 5$$
$$= 12x^2 + 20x$$

In this section, we reverse the previous steps and determine what factors were multiplied to obtain $12x^2 + 20x$. We call that process *factoring the polynomial*.

Multiplication: Given the factors, we find a polynomial.

$$4x(3x + 5) = 12x^2 + 20x$$

Factoring: Given a polynomial, we find the factors.

To **factor a polynomial** means to express it as a product of two (or more) polynomials. The first step when factoring a polynomial is to determine whether its terms have any common factors.

Success Tip

On the game show *Jeopardy!*, answers are revealed and contestants respond with the appropriate questions. Factoring is similar. Answers to multiplications are given. You are to respond by telling what factors were multiplied.

1 Find the Greatest Common Factor of a List of Terms.

To determine whether two or more integers have common factors, it is helpful to write them as products of prime numbers. For example, the prime factorizations of 42 and 90 are given below.

$$42 = 2 \cdot 3 \cdot 7 \qquad 90 = 2 \cdot 3 \cdot 3 \cdot 5$$

The highlighting shows that 42 and 90 have one factor of 2 and one factor of 3 in common. To find their *greatest common factor* (*GCF*), we multiply the common factors: $2 \cdot 3 = 6$. Thus, the GCF of 42 and 90 is 6.

The Greatest Common Factor (GCF)

The **greatest common factor (GCF)** of a list of integers is the largest common factor of those integers. To find the greatest common factor of two (or more) integers:

1. Prime factor each number.
2. Identify the common prime factors.
3. The GCF is a product of all the common prime factors found in Step 2. If there are no common prime factors, the GCF is 1.

Recall from arithmetic that the factors of a number divide the number exactly, leaving no remainder. Therefore, the greatest common factor of two or more integers is the largest integer that divides each of them exactly.

EXAMPLE 1 Find the GCF of each list of numbers: **a.** 21 and 140 **b.** 24, 60, and 96 **c.** 9, 10, and 30

Strategy We will prime factor each number in the list. Then we will identify the common prime factors and find their product.

Why The product of the common prime factors is the GCF of the numbers in the list.

Solution

a. The prime factorization of each number is shown:

$$21 = 3 \cdot 7$$

$$140 = 2 \cdot 2 \cdot 5 \cdot 7 \quad \text{This can be written as } 2^2 \cdot 5 \cdot 7.$$

$3\,\underline{|21}$, 7 $\qquad$ $2\,\underline{|140}$, $2\,\underline{|70}$, $5\,\underline{|35}$, 7

Since the only prime factor common to 21 and 140 is 7, the GCF of 21 and 140 is 7.

Success Tip

Note that the GCF, 7, divides 21 and 140 exactly:

$$\frac{21}{7} = 3 \qquad \frac{140}{7} = 20$$

b. To find the GCF of three numbers, we proceed in a similar way by first finding the prime factorization of each number in the list.

$$24 = 2 \cdot 2 \cdot 2 \cdot 3 \quad \text{This can be written as } 2^3 \cdot 3.$$

$$60 = 2 \cdot 2 \cdot 3 \cdot 5 \quad \text{This can be written as } 2^2 \cdot 3 \cdot 5.$$

$$96 = 2 \cdot 2 \cdot 2 \cdot 2 \cdot 2 \cdot 3 \quad \text{This can be written as } 2^5 \cdot 3.$$

$2\,\underline{|24}$, $2\,\underline{|12}$, $2\,\underline{|6}$, 3 $\qquad$ $2\,\underline{|60}$, $2\,\underline{|30}$, $3\,\underline{|15}$, 5 $\qquad$ $2\,\underline{|96}$, $2\,\underline{|48}$, $2\,\underline{|24}$, $2\,\underline{|12}$, $2\,\underline{|6}$, 3

The highlighting shows that 24, 60, and 96 have two factors of 2 and one factor of 3 in common. The GCF of 24, 60, and 96 is the product of their common prime factors.

$$\text{GCF} = 2 \cdot 2 \cdot 3 = 2^2 \cdot 3^1 = 12$$

Success Tip

The exponent on any factor in a GCF is the *smallest* exponent that appears on that factor in all of the numbers under consideration.

c. Since there are no prime factors common to 9, 10, and 30, their GCF is 1.

$$9 = 3 \cdot 3$$

$$10 = 2 \cdot 5$$

$$30 = 2 \cdot 3 \cdot 5$$

$3\,\underline{|9}$, 3 $\qquad$ $2\,\underline{|10}$, 5 $\qquad$ $2\,\underline{|30}$, $3\,\underline{|15}$, 5

Self Check 1 Find the GCF of each list of numbers: **a.** 24 and 70 **b.** 22, 25, and 98 **c.** 45, 60, and 75

Now Try Problems 17 and 21

To find the greatest common factor of a list of terms, we can use the following approach.

Strategy for Finding the GCF

1. Write each coefficient as a product of prime factors.
2. Identify the numerical and variable factors common to each term.
3. Multiply the common numerical and variable factors identified in Step 2 to obtain the GCF. If there are no common factors, the GCF is 1.

EXAMPLE 2 Find the GCF of each list of terms: **a.** $12x^2$ and $20x$ **b.** $9a^5b^2$, $15a^4b^2$, and $90a^3b^3$

Strategy We will prime factor each coefficient of each term in the list. Then we will identify the numerical and variable factors common to each term and find their product.

Why The product of the common factors is the GCF of the terms in the list.

Solution

Success Tip

One way to identify common factors is to circle them:

$$12x^2 = 2 \cdot 2 \cdot 3 \cdot x \cdot x$$
$$20x = 2 \cdot 2 \cdot 5 \cdot x$$
$$\text{GCF} = 2 \cdot 2 \cdot x = 4x$$

a. ***Step 1:*** We write each coefficient, 12 and 20, as a product of prime factors. Recall that an exponent, as in x^2, indicates repeated multiplication.

$$12x^2 = 2 \cdot 2 \cdot 3 \cdot x \cdot x \quad \text{This can be written as } 2^2 \cdot 3 \cdot x^2.$$
$$20x = 2 \cdot 2 \cdot 5 \cdot x \quad \text{This can be written as } 2^2 \cdot 5 \cdot x.$$

Step 2: There are two common factors of 2 and one common factor of x.

Step 3: We multiply the common factors, 2, 2, and x, to obtain the GCF.

$$\text{GCF} = 2 \cdot 2 \cdot x = 2^2 \cdot x = 4x$$

Success Tip

The exponent on any variable in a GCF is the *smallest* exponent that appears on that variable in all of the terms under consideration.

b. ***Step 1:*** We write the coefficients, 9, 15, and 90, as products of primes. The exponents on the variables represent repeated multiplication.

$$9a^5b^2 = 3 \cdot 3 \cdot a \cdot a \cdot a \cdot a \cdot a \cdot b \cdot b \quad \text{This can be written as } 3^2 \cdot a^5 \cdot b^2.$$
$$15a^4b^2 = 3 \cdot 5 \cdot a \cdot a \cdot a \cdot a \cdot b \cdot b \quad \text{This can be written as } 3 \cdot 5 \cdot a^4 \cdot b^2.$$
$$90a^3b^3 = 2 \cdot 3 \cdot 3 \cdot 5 \cdot a \cdot a \cdot a \cdot b \cdot b \cdot b \quad \text{This can be written as } 2 \cdot 3^2 \cdot 5 \cdot a^3 \cdot b^3.$$

Step 2: The highlighting shows one common factor of 3, three common factors of a, and two common factors of b.

Step 3: $\text{GCF} = 3 \cdot a \cdot a \cdot a \cdot b \cdot b = 3a^3b^2$

Self Check 2 Find the GCF of each list of terms: **a.** $33c$ and $22c^4$ **b.** $42s^3t^2$, $63s^2t^4$, and $21s^3t^3$

Now Try Problems 29 and 35

2 Factor Out the Greatest Common Factor.

The concept of greatest common factor is used to factor polynomials. For example, to factor $12x^2 + 20x$, we note that there are two terms, $12x^2$ and $20x$. We previously determined that the GCF of $12x^2$ and $20x$ is $4x$. With this in mind, we write each term of $12x^2 + 20x$ as a product of the GCF and one other factor. Then we apply the distributive property in reverse: $ab + ac = a(b + c)$.

The Language of Algebra

We have expressed a **sum of terms** as a **product of factors**.

$$\underbrace{12x^2 + 20x}_{\text{Sum of terms}} = \underbrace{4x}_{\text{Factor}}\underbrace{(3x + 5)}_{\text{Factor}}$$

$12x^2 + 20x = \mathbf{4x} \cdot 3x + \mathbf{4x} \cdot 5$ — Write $12x^2$ and $20x$ as the product of the GCF, $4x$, and one other factor.

$= \mathbf{4x}(3x + 5)$ — Write an expression so that the multiplication by $4x$ distributes over the sum of the terms $3x$ and 5.

We have found that the factored form of $12x^2 + 20x$ is $4x(3x + 5)$. This process is called **factoring out the greatest common factor.**

EXAMPLE 3 Factor: **a.** $8m + 24$ **b.** $35a^3b^2 - 14a^2b^3$ **c.** $3x^4 - 5x^3 + x^2$

Strategy First, we will determine the GCF of the terms of the polynomial. Then we will write each term of the polynomial as the product of the GCF and one other factor.

Why We can then use the distributive property to factor out the GCF.

Solution

a. Since the GCF of $8m$ and 24 is 8, we write $8m$ and 24 as the product of 8 and one other factor.

$8m + 24 = \mathbf{8} \cdot m + \mathbf{8} \cdot 3$ — This step can be done mentally.

$= \mathbf{8}(m + 3)$ — Factor out the GCF, 8.

Success Tip

Always verify a factorization by performing the indicated multiplication. The result should be the original polynomial.

To check, we multiply: $8(m + 3) = 8 \cdot m + 8 \cdot 3 = 8m + 24$. Since we obtain the original polynomial, $8m + 24$, the factorization is correct.

CAUTION Remember to factor out the greatest common factor, not just a common factor. If we factored out 4 in the previous example, we would get

$$8m + 24 = 4(\mathbf{2m} + \mathbf{6})$$

However, the terms in red within parentheses have a common factor of 2, indicating that the factoring is not complete.

b. First, find the GCF of $35a^3b^2$ and $14a^2b^3$.

$35a^3b^2 = 5 \cdot 7 \cdot a \cdot a \cdot a \cdot b \cdot b$

$14a^2b^3 = 2 \cdot 7 \cdot a \cdot a \cdot b \cdot b \cdot b$

The GCF is $7a^2b^2$.

$5 \underline{|35}$ $\quad 2\underline{|14}$
$\quad 7 \qquad\quad 7$

Now, we write $35a^3b^2$ and $14a^2b^3$ as the product of the GCF, $7a^2b^2$, and one other factor.

$35a^3b^2 - 14a^2b^3 = \mathbf{7a^2b^2} \cdot 5a - \mathbf{7a^2b^2} \cdot 2b$ — $\frac{35a^3b^2}{7a^2b^2} = 5a$ and $\frac{14a^2b^3}{7a^2b^2} = 2b$.

$= \mathbf{7a^2b^2}(5a - 2b)$ — Factor out the GCF, $7a^2b^2$.

Success Tip

With practice, you will learn how to factor out the GCF and write the correct terms within the parentheses in one step.

We check by multiplying: $7a^2b^2(5a - 2b) = 35a^3b^2 - 14a^2b^3$.

c. We factor out the GCF of the three terms, which is x^2.

$3x^4 - 5x^3 + x^2 = \mathbf{x^2}(3x^2) - \mathbf{x^2}(5x) + \mathbf{x^2}(\mathbf{1})$ — Write the last term, x^2, as $x^2(1)$.

$= \mathbf{x^2}(3x^2 - 5x + \mathbf{1})$ — Factor out the GCF, x^2.

Caution

If the GCF of the terms of a polynomial is the same as one of the terms, leave a 1 in place of that term when factoring out the GCF.

We check by multiplying: $x^2(3x^2 - 5x + 1) = 3x^4 - 5x^3 + x^2$.

Self Check 3 Factor: **a.** $6f + 36$ **b.** $24s^2t^2 - 42s^3t$ **c.** $y^6 - 10y^4 - y^3$

Now Try Problems 41, 49, and 51

When asked to factor a polynomial whose leading coefficient is negative, factor out the *opposite of the GCF.*

EXAMPLE 4 Factor -1 from each polynomial: **a.** $-a^3 + 2a^2 - 4$ **b.** $6 - x$

Strategy We will write each term of the polynomial as the product of -1 and one other factor.

Why We can then use the distributive property to factor out the -1.

Solution

Success Tip

This result suggests a quick way to factor out -1. Simply change the sign of each term of $-a^3 + 2a^2 - 4$ and write a $-$ symbol in front of the parentheses.

a. $-a^3 + 2a^2 - 4 = (-1)a^3 + (-1)(-2a^2) + (-1)4$ This step can be done mentally.

$= -1(a^3 - 2a^2 + 4)$ Factor out -1.

$= -(a^3 - 2a^2 + 4)$ The 1 need not be written.

We check by multiplying: $-(a^3 - 2a^2 + 4) = -a^3 + 2a^2 - 4$.

b. First, we will write the terms of $6 - x$ in descending order to determine whether the leading coefficient is negative.

$6 - x = -x + 6$ Write in descending powers of x.

$= -1(x) + (-1)(-6)$ This step can be done mentally.

$= -1(x - 6)$ Factor out -1.

$= -(x - 6)$ The 1 need not be written.

Self Check 4 Factor -1 from each polynomial: **a.** $-b^4 - 3b^2 + 2$ **b.** $9 - t$

Now Try Problems 57 and 61

EXAMPLE 5 Factor out the opposite of the GCF in $-20m + 30$.

Strategy First, we will determine the GCF of the terms of the polynomial. Then we will write each term of the polynomial as the product of the opposite of the GCF and one other factor.

Why We can then use the distributive property to factor out the opposite of the GCF.

Solution Since the GCF is 10, the opposite of the GCF is -10. We write each term of the polynomial as the product of -10 and another factor. Then we factor out -10.

Success Tip

It is standard practice to factor in such a way that the leading coefficient of the polynomial within the parentheses is positive.

$-20m + 30 = (-10)(2m) + (-10)(-3)$ This step can be done mentally.

$= -10(2m - 3)$ Note that the leading coefficient of the polynomial within the parentheses is positive.

We check by multiplying: $-10(2m - 3) = -20m + 30$.

Self Check 5 Factor out the opposite of the GCF in $-44c + 55$.

Now Try Problem 67

EXAMPLE 6 Factor: $x(x + 4) + 3(x + 4)$

Strategy We will identify the terms of the expression and find their GCF.

Why We can then use the distributive property to factor out the GCF.

Solution The expression has two terms: $\underbrace{x(x + 4)}_{\text{The first term}} + \underbrace{3(x + 4)}_{\text{The second term}}$

The Language of Algebra

We say that the terms have the common binomial factor $x + 4$.

The GCF of the two terms is the binomial $x + 4$, which can be factored out.

$$x(x + 4) + 3(x + 4) = (x + 4)x + (x + 4)3$$ Write each term as the product of $(x + 4)$ and one other factor.

$$= (x + 4)(x + 3)$$ Factor out the common factor, $(x + 4)$. Caution: Don't write $(x + 4)^2$.

Self Check 6 Factor: $2y(y - 1) + 7(y - 1)$

Now Try Problem 73

3 Factor by Grouping.

Although the terms of many polynomials don't have a common factor, other than 1, it is possible to factor some of them by arranging their terms in convenient groups. This method is called **factoring by grouping.**

EXAMPLE 7 Factor by grouping: **a.** $2x^3 + x^2 + 12x + 6$ **b.** $5c - 5d + cd - d^2$

Strategy We will factor out a common factor from the first two terms and a common factor from the last two terms.

Why This often produces a common binomial factor that can then be factored out.

Solution **a.** Except for 1, there is no factor that is common to all four terms. However, the first two terms, $2x^3$ and x^2, have a common factor, x^2, and the last two terms, $12x$ and 6, have a common factor, 6.

$$2x^3 + x^2 \quad + \quad 12x + 6$$
$$\downarrow \qquad\qquad \downarrow$$
$$x^2(2x + 1) \quad + \quad 6(2x + 1)$$

We now see that $2x^3 + x^2$ and $12x + 6$ have a common binomial factor, $2x + 1$, which can be factored out.

$$2x^3 + x^2 + 12x + 6 = x^2(2x + 1) + 6(2x + 1)$$ Factor $2x^3 + x^2$ and $12x + 6$. Don't forget the blue + sign.

$$= (2x + 1)(x^2 + 6)$$ Factor out $2x + 1$. Caution: Don't write $(2x + 1)^2$.

Caution

Factoring by grouping can be attempted on any polynomial with four or more terms. However, not every such polynomial can be factored in this way.

We can check the factorization by multiplying. The result should be the original polynomial.

$$(2x + 1)(x^2 + 6) = 2x^3 + 12x + x^2 + 6$$
$$= 2x^3 + x^2 + 12x + 6$$ Rearrange the terms to get the original polynomial.

Caution

Don't think that $5(c - d) + d(c - d)$ is in factored form. It is a sum of two terms. To be in factored form, the result must be a product.

b. The first two terms have a common factor, 5, and the last two terms have a common factor, d.

$$5c - 5d + cd - d^2 = 5(c - d) + d(c - d)$$ Factor out 5 from $5c - 5d$ and d from $cd - d^2$. Don't forget the blue + sign.

$$= (c - d)(5 + d)$$ Factor out the common binomial factor, $c - d$. Caution: Don't write $(c - d)^2$.

We can check by multiplying:

$$(c - d)(5 + d) = 5c + cd - 5d - d^2$$
$$= 5c - 5d + cd - d^2$$

Self Check 7 Factor by grouping: **a.** $3n^3 + 2n^2 + 9n + 6$ **b.** $7x - 7y + xy - y^2$

Now Try Problems 77 and 79

Factoring a Four-Termed Polynomial by Grouping

1. Group the terms of the polynomial so that the first two terms have a common factor and the last two terms have a common factor.
2. Factor out the common factor from each group.
3. Factor out the resulting common binomial factor. If there is no common binomial factor, regroup the terms of the polynomial and repeat steps 2 and 3.

By the multiplication property of 1, we know that 1 is a factor of every term. We can use this observation to factor certain polynomials by grouping.

EXAMPLE 8 Factor: **a.** $x^3 + 6x^2 + x + 6$ **b.** $x^2 - ax - x + a$

Strategy We will follow the steps for factoring a four-termed polynomial.

Why Since the terms of the polynomials do not have a common factor (other than 1), the only option is to attempt to factor these polynomials by grouping.

Solution **a.** The first two terms, x^3 and $6x^2$, have a common factor of x^2. The only common factor of the last two terms, x and 6, is 1.

$$x^3 + 6x^2 + x + 6 = x^2(x + 6) + 1(x + 6)$$ Factor out x^2 from $x^3 + 6x^2$. Factor out 1 from $x + 6$. Don't forget the blue + sign.

$$= (x + 6)(x^2 + 1)$$ Factor out the common binomial factor, $x + 6$. Caution: Don't write $(x + 6)^2$.

Check the factorization by multiplying.

Success Tip

When we factor out -1 from the last two terms,

$$x^2 - ax \quad -x + a$$
$$= x(x - a) - 1(x - a)$$

the signs of those terms change within the parentheses. The binomials within both sets of parentheses are then identical.

b. Since x is a common factor of the first two terms, we can factor it out.

$$x^2 - ax \quad -x + a = x(x - a) - x + a \qquad \text{Factor out } x \text{ from } x^2 - ax.$$

When factoring four terms by grouping, if the coefficient of the 3rd term is negative, we often factor out a negative coefficient from the last two terms. If we factor -1 from $-x + a$, a common binomial factor $x - a$ appears within the second set of parentheses, which we can factor out.

$$x^2 - ax \quad -x + a = x(x - a) - 1(x - a) \qquad \text{To factor out } -1\text{, change the sign of } -x \text{ and } a\text{, and write } -1 \text{ in front of the parentheses.}$$
$$= (x - a)(x - 1) \qquad \text{Factor out the common factor, } x - a. \text{ Caution: Don't write } (x - a)^2.$$

Check by multiplying.

Self Check 8 Factor: **a.** $a^5 + 11a^4 + a + 11$
b. $b^2 - bc - b + c$

Now Try Problems 85 and 87

EXAMPLE 9 Factor: $5x^3 - 8 + 10x^2 - 4x$

Strategy We will follow the steps for factoring by grouping.

Why Since the four terms of the polynomial do not have a common factor (other than 1), we will attempt to factor it by grouping.

Solution We note that the four terms of the polynomial do not have a common factor.

$$5x^3 - 8 \quad + \quad 10x^2 - 4x \qquad \text{We cannot factor the polynomial in its current form.}$$

No common factor (other than 1) — GCF = $2x$

We will write the polynomial in descending powers of x and attempt to factor by grouping again.

$$5x^3 - 8 + 10x^2 - 4x = 5x^3 + 10x^2 \quad -4x - 8$$
$$= 5x^2(x + 2) - 4(x + 2) \qquad \text{Factor } 5x^2 \text{ from } 5x^3 + 10x^2 \text{ and } -4 \text{ from } -4x - 8.$$
$$= (x + 2)(5x^2 - 4) \qquad \text{Factor out the GCF, } x + 2.$$

Success Tip

An equivalent factorization, $(5x^2 - 4)(x + 2)$, results if the terms are arranged as $5x^3 - 4x + 10x^2 - 8$ or as $5x^3 - 4x - 8 + 10x^2$.

Self Check 9 Factor: $y^3 - 6 + 3y^2 - 2y$

Now Try Problem 93

The next example illustrates that when factoring a polynomial, we should *always look for a common factor first.*

EXAMPLE 10 Factor: $10k + 10m - 2km - 2m^2$

Strategy Since all four terms have a common factor of 2, we factor it out first. Then we will factor the resulting polynomial by grouping.

Why Factoring out the GCF first makes factoring by any method easier.

Solution After factoring out 2 from all four terms, notice that within the parentheses, the first two terms have a common factor of 5, and the last two terms have a common factor of $-m$.

Success Tip

Here is a factoring guideline: If a new pair of parentheses is produced in the factoring process, always see if the resulting expression within those parentheses can be factored further.

$10k + 10m - 2km - 2m^2$

$= 2(5k + 5m - km - m^2)$ Factor out the GCF, 2.

$= 2[5(k + m) - m(k + m)]$ Factor out 5 from $5k + 5m$. Factor out $-m$ from $-km - m^2$. This causes the signs of $-km$ and $-m^2$ to change within the second set of parentheses. Enclose the factoring by grouping process within brackets [ ].

$= 2[(k + m)(5 - m)]$ Factor out the common binomial factor, $k + m$.

$= 2(k + m)(5 - m)$ Drop the unnecessary brackets.

Check by multiplying.

Self Check 10 Factor: $4t + 4s + 4tz + 4sz$

Now Try Problem 97

SECTION 6.1 STUDY SET

VOCABULARY

Fill in the blanks.

1. To ______ a polynomial means to express it as a product of two (or more) polynomials.
2. GCF stands for ________ ________ ______. When we write $2x + 4$ as $2(x + 2)$, we say that we have ________ out the GCF, 2.
3. To factor $m^3 + 3m^2 + 4m + 12$ by ________, we begin by writing $m^2(m + 3) + 4(m + 3)$.
4. The terms $x(x - 1)$ and $4(x - 1)$ have the common ________ factor $x - 1$.

CONCEPTS

5. Complete each factorization.
   a. $6x = 2 \cdot \quad \cdot x$
   b. $35h^2 = 5 \cdot \quad \cdot h \cdot$
   c. $18y^3z = 2 \cdot \quad \cdot 3 \cdot \quad \cdot y \cdot \quad \cdot z$
6. a. Find the GCF of $30x^2$ and $105x^3$.

   $30x^2 = 2 \cdot 3 \cdot 5 \cdot x \cdot x$

   $105x^3 = 3 \cdot 5 \cdot 7 \cdot x \cdot x \cdot x$

   b. Find the GCF of $12a^2b^2$, $15a^3b$, and $75a^4b^2$.

   $12a^2b^2 = 2 \cdot 2 \cdot 3 \cdot a \cdot a \cdot b \cdot b$

   $15a^3b = 3 \cdot 5 \cdot a \cdot a \cdot a \cdot b$

   $75a^4b^2 = 3 \cdot 5 \cdot 5 \cdot a \cdot a \cdot a \cdot a \cdot b \cdot b$
7. a. Write a binomial such that the GCF of its terms is 2.

   b. Write a trinomial such that the GCF of its terms is x.
8. Check to determine whether each factorization is correct.
   a. $9y^3 + 5y^2 - 15y = 3y(3y^2 + 2y - 5)$
   b. $3s^3 + 2s^2 + 6s + 4 = (3s + 2)(s^2 + 2)$

Fill in the blanks to complete each factorization.

9. $2x^2 + 6x = \mathbf{2x} \cdot x + \mathbf{2x} \cdot 3$

   $= \quad (\quad)$
10. $3t^3 - t^2 + 15t - 5 = t^2\mathbf{(3t - 1)} + 5\mathbf{(3t - 1)}$

    $= (\quad)(\quad)$
11. Consider the polynomial $2k - 8 + hk - 4h$.
    a. How many terms does the polynomial have?
    b. Is there a common factor of all the terms, other than 1?
    c. What is the GCF of the first two terms and what is the GCF of the last two terms?
12. What is the first step in factoring $8y^2 - 16yz - 6y + 12z$?

NOTATION

Complete each factorization.

13. $8m^2 - 32m + 16 = \quad (m^2 - 4m + 2)$
14. $10a^4 - 15a^3 = 5a \quad (2a - 3)$
15. $b^3 - 6b^2 + 2b - 12 = \quad (b - 6) + \quad (b - 6)$

    $= (\quad)(b^2 + 2)$
16. $12 + 8n - 3m - 2mn = 4(3 + 2n) \quad m(3 + 2n)$

    $= (3 + 2n)(4 \quad m)$

GUIDED PRACTICE

Find the GCF of each list of numbers. See Example 1.

17. 6, 10
18. 10, 15
19. 18, 24
20. 60, 72
21. 14, 21, 42
22. 16, 24, 48
23. 40, 32, 24
24. 28, 35, 21

Find the GCF of each list of terms. See Example 2.

25. m^4, m^3
26. c^2, c^7
27. $15x, 25$
28. $9a, 21$
29. $20c^2, 12c$
30. $18r, 27r^3$
31. $18a^2, 9a^3, 27a^3$
32. $33m^5, 22m^6, 11m^5$
33. $24a^2, 16a^3b, 40ab$
34. $12r^2, 15rs, 9r^2s^2$
35. $6m^4n, 12m^3n^2, 9m^3n^3$
36. $15c^2d^4, 10c^2d, 40c^3d^3$
37. $4(x + 7), 9(x + 7)$
38. $2(y - 1), 5(y - 1)$
39. $4(p - t), p(p - t)$
40. $a(b + c), 3(b + c)$

Factor out the GCF. See Example 3.

41. $3x - 6$
42. $18x + 24$
43. $18m - 9$
44. $15s - 35$
45. $d^2 - 7d$
46. $a^2 + 9a$
47. $15c^2 + 25$
48. $33h^4 - 22$
49. $24a - 16a^2$
50. $18r - 30r^2$
51. $14x^2 - 7x - 7$
52. $27a^2 - 9a + 9$
53. $t^4 + t^3 + 2t^2$
54. $b^4 - b^3 - 3b^2$
55. $21x^2y^3 + 3xy^2$
56. $3x^2y^3 - 9x^4y^3$

Factor out −1 from each polynomial. See Example 4.

57. $-a - b$
58. $-x - 2y$
59. $-x^2 - x + 16$
60. $-t^2 - 9t + 1$
61. $5 - x$
62. $10 - m$
63. $9 - 4a$
64. $7 - 8b$

Factor each polynomial by factoring out the opposite of the GCF. See Example 5.

65. $-3x^2 - 6x$
66. $-4a^2 - 6a$
67. $-4a^2b + 12a^3$
68. $-25x^4 + 30x^2$
69. $-24x^4 - 48x^3 + 36x^2$
70. $-28a^5 - 42a^4 + 14a^3$
71. $-4a^3b^2 + 14a^2b^2 - 10ab^2$
72. $-30x^4y^3 + 24x^3y^2 - 60x^2y$

Factor. See Example 6.

73. $y(x + 2) + 3(x + 2)$
74. $r(t + v) + 3(t + v)$
75. $m(p - q) - 5(p - q)$
76. $ab(c - 7) - 12(c - 7)$

Factor by grouping. See Example 7.

77. $2x + 2y + ax + ay$
78. $bx + bz + 5x + 5z$
79. $rs - ru + 8sw - 8uw$
80. $12ab - 4ac + 3db - dc$
81. $7m^3 - 2m^2 + 14m - 4$
82. $9s^3 - 2s^2 + 36s - 8$
83. $5x^3 - x^2 + 10x - 2$
84. $6a^3 - a^2 + 18a - 3$

Factor by grouping. See Example 8.

85. $ab + ac + b + c$
86. $xy + 3y^2 + x + 3y$
87. $rs + 4s^2 - r - 4s$
88. $tx + tz - x - z$
89. $2ax + 2bx - 3a - 3b$
90. $rx + sx - ry - sy$
91. $mp - np - mq + nq$
92. $9p - 9q - mp + mq$

Factor by grouping. See Example 9.

93. $5m^3 + 6 + 5m^2 + 6m$
94. $4t^3 + 14 + 28t^2 + 2t$
95. $y^3 - 12 + 3y - 4y^2$
96. $h^3 - 8 + h - 8h^2$

Factor by grouping. Remember to factor out the GCF first. See Example 10.

97. $ax^3 - 2ax^2 + 5ax - 10a$
98. $x^3y^2 - 2x^2y^2 + 3xy^2 - 6y^2$
99. $6x^3 - 6x^2 + 12x - 12$
100. $3x^3 - 6x^2 + 15x - 30$

TRY IT YOURSELF

Factor.

101. $h^2(14 + r) + 5(14 + r)$
102. $x(y + 9) - 21(y + 9)$
103. $22a^3 - 33a^2$
104. $39r^3 + 26r^2$
105. $ax + bx - a - b$
106. $2xy + y^2 - 2x - y$
107. $15r^8 - 18r^6 - 30r^5$
108. $24cm - 12cn + 16c$
109. $27mp + 9mq - 9np - 3nq$
110. $4abc + 4ac^2 - 2bc - 2c^2$
111. $-60p^2t^2 - 80pt^3$
112. $-25x^5y^7 + 75x^3y^2$
113. $-2x + 5$
114. $-3x + 8$
115. $6x^2 - 2xy - 15x + 5y$
116. $m^3 + 5m^2 + m + 5$
117. $2x^3z - 4x^2z + 32xz - 64z$
118. $4a^2b + 12a^2 - 8ab - 24a$
119. $12uvw^3 - 54uv^2w^2$
120. $14xyz - 16x^2y^2z$

121. $x^3 + x^2 + x + 1$
122. $m^4 + m^3 + 2m + 2$
123. $-3r + 2s - 3$
124. $-6yz + 12xz + 5xy$

Look Alikes . . .

125. a. $5t^3 + 6t^2 + 15t + 18$ b. $3t^3 + 6t^2 + 15t + 18$

126. a. $x^2 + xy + x + y$ b. $2x^2 + 2xy + 2x + 2y$

APPLICATIONS

127. **Geometry.** The dimensions of the rectangle shown below can be found by factoring the polynomial that represents its area. Find the polynomials that represent the length and the width of the rectangle.

Area = $(x^3 + 4x^2 + 5x + 20)$ ft^2

128. **Front Doors.** Find a polynomial that represents the amount of surface on the front face of the door shown on the right that needs to be stained. Then factor the polynomial.

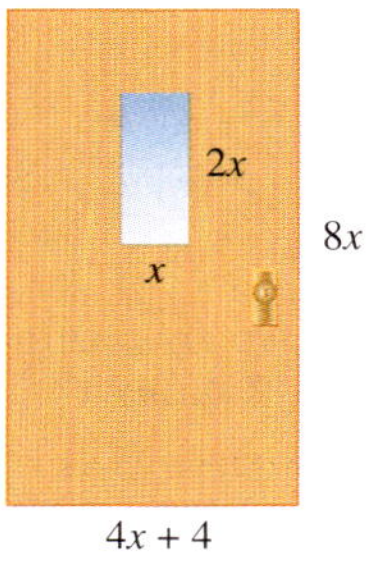

129. **from Campus to Careers**
Elementary School Teacher

©iStockphoto.com/Neustockimages

A teacher is going to purchase a block of wax so that the children in her classroom can melt it down, add coloring, pour it into a mold, and make their own crayons. The amount of wax to make one crayon is given by the volume formula:

$$V = \pi r^2 h_1 + \frac{1}{3}\pi r^2 h_2$$

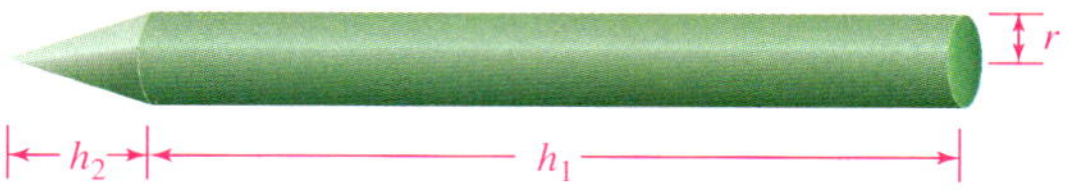

a. Rewrite the formula by factoring the expression on the right side.
b. Use the formula to estimate what size block of wax (in cubic inches) she needs to purchase if there are 20 students in her class, and each student will get 5 crayons. The dimensions for the crayon molds they will use are: $h_1 = 1\frac{5}{6}$ in., $h_2 = \frac{1}{2}$ in., and $r = \frac{1}{4}$ in. Round up to the nearest cubic inch.

130. **Interior Decorating.** The expression $\pi rs + \pi Rs$ can be used to find the amount of material needed to make the lamp shade shown. Factor the expression.

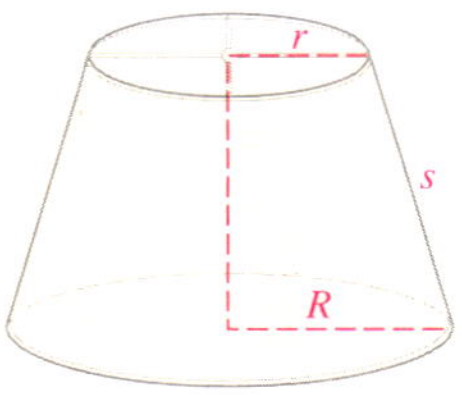

WRITING

131. Explain how to find the GCF of $32a^3$ and $16a^2$.
132. Explain this diagram.

Multiplication ⟶

$$3x^2(5x^2 - 6x + 4) = 15x^4 - 18x^3 + 12x^2$$

⟵ Factoring

133. Explain how factorizations of polynomials are checked. Give an example.
134. Explain the error.

Factor out the GCF: $30a^3 - 12a^2 = 6a(5a^2 - 2a)$

REVIEW

135. **Insurance Costs.** A college student's good grades earned her a student discount on her car insurance premium. What was the percent of decrease, to the nearest percent, if her annual premium was lowered from \$1,050 to \$925?
136. **Calculating Grades.** A student has test scores of 68%, 75%, and 79% in a government class. What must she score on the last exam to earn a B (80% or better) in the course?

CHALLENGE PROBLEMS

137. Factor: $6x^{4m}y^n + 21x^{3m}y^{2n} - 15x^{2m}y^{3n}$

138. Factor $ab - b^2 - bc + ac - bc - c^2$ by grouping.

SECTION 6.2

OBJECTIVES

1. Factor trinomials of the form $x^2 + bx + c$.
2. Factor trinomials of the form $x^2 + bx + c$ after factoring out the GCF.
3. Factor trinomials of the form $x^2 + bx + c$ using the grouping method.

Factoring Trinomials of the Form $x^2 + bx + c$

ARE YOU READY?

The following problems review some basic skills that are needed when factoring trinomials.

1. In $x^2 - 4x + 10$, what is the coefficient of the leading term?
2. Multiply: $(x + 8)(x - 1)$
3. What is the GCF of the terms of $2x^2 - 10x + 4$?
4. $x^2 - 4xy - 5y^2$ is a polynomial in how many variables?
5. Find two integers whose product is 8 and whose sum is 6.
6. Find two integers whose product is -15 and whose sum is 2.

In Chapter 5, we learned how to multiply binomials. For example, to multiply $x + 1$ and $x + 2$, we proceed as follows.

$$(x + 1)(x + 2) = x^2 + 2x + x + 2$$
$$= x^2 + 3x + 2$$

To *factor the trinomial* $x^2 + 3x + 2$, we will reverse the multiplication process and determine what factors were multiplied to obtain this result. Since the product of two binomials is often a trinomial, many trinomials factor into two binomials.

Multiplication: Given the binomial factors, we find a trinomial.

$$(x + 1)(x + 2) = x^2 + 3x + 2$$

Factoring: Given a trinomial, we find the binomial factors.

The Language of Algebra

Recall that when a polynomial in one variable is written in descending powers of that variable, the coefficient of the first term is called the **leading coefficient.**

To begin the discussion of trinomial factoring, we consider trinomials of the form $x^2 + bx + c$, such as

$$x^2 + 8x + 15, \qquad y^2 - 13y + 12, \qquad a^2 + a - 20, \qquad \text{and} \qquad z^2 - 20z - 21$$

In each case, the **leading coefficient**—the coefficient of the squared variable—is 1.

1 Factor Trinomials of the Form $x^2 + bx + c$.

To develop a method for factoring trinomials, we will find the product of $x + 6$ and $x + 4$ and make some observations about the result.

$$\begin{aligned}(x + 6)(x + 4) &= \overset{\mathbf{F}}{x \cdot x} + \overset{\mathbf{O}}{x \cdot 4} + \overset{\mathbf{I}}{6 \cdot x} + \overset{\mathbf{L}}{6 \cdot 4} \quad \text{Use the FOIL method.}\\ &= x^2 + 4x + 6x + 24\\ &= x^2 + 10x + 24\end{aligned}$$

First term (x^2) Middle term ($10x$) Last term (24)

The Language of Algebra

If a term of a trinomial is a number only, it is called a **constant term.**

$x^2 + 10x + 24$

Constant term (24)

The result is a trinomial, where

- the first term, x^2, is the product of x and x
- the last term, 24, is the product of 6 and 4
- the coefficient of the middle term, 10, is the sum of 6 and 4

These observations suggest a strategy to use to factor trinomials that have 1 as the leading coefficient.

EXAMPLE 1 Factor: $x^2 + 8x + 15$

Strategy We will assume that $x^2 + 8x + 15$ is the product of two binomials and we will use a systematic method to find their terms.

Why Since the terms of $x^2 + 8x + 15$ do not have a common factor (other than 1), the only option available is to try to factor it as the product of two binomials.

Solution We represent the binomials using two sets of parentheses. Since the first term of the trinomial is x^2, we enter x and x as the first terms of its binomial factors.

$$x^2 + 8x + 15 = (x\ \square)(x\ \square)$$ Because $x \cdot x$ will give x^2.

The second terms of the binomials must be two integers whose product is 15 and whose sum is 8. Since the integers must have a positive product and a positive sum, we consider only pairs of positive integer factors of 15. The only such pairs, $1 \cdot 15$ and $3 \cdot 5$, are listed in the table. Then we find the sum of each pair and enter each result in the table.

Positive factors of 15	Sum of the positive factors of 15
$1 \cdot 15 = 15$	$1 + 15 = 16$
$3 \cdot 5 = 15$	$3 + 5 = 8$

List all of the pairs of positive integers that multiply to give 15.

Add each pair of factors.

Notation

By the commutative property of multiplication, the order of the binomial factors in a factorization does not matter. Thus, we can also write:

$x^2 + 8x + 15 = (x + 5)(x + 3)$

The second row of the table contains the correct pair of integers 3 and 5, whose product is 15 and whose sum is 8. To complete the factorization, we enter 3 and 5 as the second terms of the binomial factors.

$$x^2 + 8x + 15 = (x + 3)(x + 5)$$

We can check the factorization by multiplying:

$$(x + 3)(x + 5) = x^2 + 5x + 3x + 15$$
$$= x^2 + 8x + 15$$ This is the original trinomial.

Self Check 1 Factor: $y^2 + 7y + 10$

Now Try Problem 15

EXAMPLE 2 Factor: $y^2 - 13y + 12$

Strategy We will assume that $y^2 - 13y + 12$ is the product of two binomials and we will use a systematic method to find their terms.

Why Since the terms of $y^2 - 13y + 12$ do not have a common factor (other than 1), the only option available is to try to factor it as the product of two binomials.

Solution We represent the binomials using two sets of parentheses. Since the first term of the trinomial is y^2, the first term of each binomial factor must be y.

$$y^2 - 13y + 12 = (y\ \square)(y\ \square)$$ Because $y \cdot y$ will give y^2.

The second terms of the binomials must be two integers whose product is 12 and whose sum is -13. Since the integers must have a positive product and a negative sum, we only consider pairs of negative integer factors of 12. The possible pairs are listed in the table.

Negative factors of 12	Sum of the negative factors of 12
$-1(-12) = 12$	$-1 + (-12) = -13$
$-2(-6) = 12$	$-2 + (-6) = -8$
$-3(-4) = 12$	$-3 + (-4) = -7$

You can stop listing the factors after finding the correct combination.

The Language of Algebra

Make sure you understand this vocabulary: **Many trinomials factor as the product of two binomials.**

Trinomial — Product of two binomials

$y^2 - 13y + 12 = (y - 1)(y - 12)$

The first row of the table contains the correct pair of integers -1 and -12, whose product is 12 and whose sum is -13. To complete the factorization, we enter -1 and -12 as the second terms of the binomial factors.

$$y^2 - 13y + 12 = (y - 1)(y - 12)$$

We check the factorization by multiplying:

$$(y - 1)(y - 12) = y^2 - 12y - y + 12$$
$$= y^2 - 13y + 12 \quad \text{This is the original trinomial.}$$

Self Check 2 Factor: $p^2 - 6p + 8$

Now Try Problem 19

EXAMPLE 3 Factor: $x^2 + x - 20$

Strategy We will assume that $x^2 + x - 20$ is the product of two binomials and we will use a systematic method to find their terms.

Why Since the terms of $x^2 + x - 20$ do not have a common factor (other than 1), the only option available is to try to factor it as the product of two binomials.

Solution We represent the binomials using two sets of parentheses. Since the first term of the trinomial is x^2, the first term of each binomial factor must be x.

$$x^2 + x - 20 = (x \square)(x \square) \quad \text{Because } x \cdot x \text{ will give } x^2.$$

To determine the second terms of the binomials, we must find two integers whose product is -20 and whose sum is 1. Because the integers must have a negative product, their signs must be different. The possible pairs are listed in the table.

It is wise to follow an order when listing the factors in the table so that you don't skip the correct combination. Here, the first factors 1, 2, 4, 5, 10, and 20 are listed from least to greatest.

Factors of -20	Sum of the factors of -20
$1(-20) = -20$	$1 + (-20) = -19$
$2(-10) = -20$	$2 + (-10) = -8$
$4(-5) = -20$	$4 + (-5) = -1$
$5(-4) = -20$	$5 + (-4) = 1$
$10(-2) = -20$	$10 + (-2) = 8$
$20(-1) = -20$	$20 + (-1) = 19$

The fourth row of the table contains the correct pair of integers 5 and -4, whose product is -20 and whose sum is 1. To complete the factorization, we enter 5 and -4 as the second terms of the binomial factors.

$$x^2 + x - 20 = (x + 5)(x - 4)$$

Check the factorization by multiplying.

Self Check 3 Factor: $m^2 + m - 42$

Now Try Problem 23

EXAMPLE 4 Factor: $z^2 - 4z - 21$

Strategy We will assume that $z^2 - 4z - 21$ is the product of two binomials and we will use a systematic method to find their terms.

Why Since the terms of $z^2 - 4z - 21$ do not have a common factor (other than 1), the only option available is to try to factor it as the product of two binomials.

Solution We represent the binomials using two sets of parentheses. Since the first term of the trinomial is z^2, the first term of each binomial factor must be z.

$$z^2 - 4z - 21 = (z \square)(z \square) \quad \text{Because } z \cdot z \text{ will give } z^2.$$

To determine the second terms of the binomials, we must find two integers whose product is -21 and whose sum is -4. Because the integers must have a negative product, their signs must be different. The possible pairs are listed in the table.

Success Tip

Show your students how to fix a factorization that is close, but not quite correct, quickly. For example, $(z + 7)(z - 3)$ does not produce the proper sign for the middle term: $(z + 7)(z - 3) = z^2 + 4z - 21$. However, a simple change of signs within the binomials does: $(z - 7)(z + 3) = z^2 - 4z - 21$.

Factors of -21	Sum of the factors of -21
$1(-21) = -21$	$1 + (-21) = -20$
$3(-7) = -21$	$3 + (-7) = -4$
$7(-3) = -21$	$7 + (-3) = 4$
$21(-1) = -21$	$21 + (-1) = 20$

The second row of the table contains the correct pair of integers 3 and -7, whose product is -21 and whose sum is -4. To complete the factorization, we enter 3 and -7 as the second terms of the binomial factors.

$$z^2 - 4z - 21 = (z + 3)(z - 7)$$

Check by multiplying.

Self Check 4 Factor: $q^2 - 2q - 24$

Now Try Problem 29

The following guidelines are helpful when factoring trinomials.

Factoring Trinomials Whose Leading Coefficient Is 1

To factor a trinomial of the form $x^2 + bx + c$, find two numbers whose product is c and whose sum is b.

1. If c is positive, the numbers have the same sign.
2. If c is negative, the numbers have different signs.

Then write the trinomial as a product of two binomials. You can check by multiplying.

$$x^2 + bx + c = (x \square)(x \square)$$

The product of these numbers must be c and their sum must be b.

EXAMPLE 5 Factor: $-h^2 + 2h + 63$

Strategy We will factor out -1 and then factor the resulting trinomial.

Why It is easier to factor trinomials that have a positive leading coefficient.

Solution After factoring out -1, we factor the trinomial within the parentheses.

$$-h^2 + 2h + 63 = -1(h^2 - 2h - 63) \quad \text{Factor out } -1.$$
$$= -(h^2 - 2h - 63) \quad \text{The 1 need not be written.}$$
$$= -(h + 7)(h - 9) \quad \text{Factor } h^2 - 2h - 63.$$

Check:

$$-(h + 7)(h - 9) = -(h^2 - 9h + 7h - 63) \quad \text{Multiply the binomials first.}$$
$$= -(h^2 - 2h - 63) \quad \text{Combine like terms.}$$
$$= -h^2 + 2h + 63 \quad \text{Drop the } - \text{ sign and change the sign of every term within the parentheses.}$$

The result is the original trinomial.

Self Check 5 Factor: $-x^2 + 11x - 28$

Now Try Problem 33

We factor trinomials in two variables in a similar way.

EXAMPLE 6 Factor: $x^2 - 4xy - 5y^2$

Strategy We will assume that $x^2 - 4xy - 5y^2$ is the product of two binomials and we will use a systematic method to find their terms.

Why Since the terms of $x^2 - 4xy - 5y^2$ do not have a common factor (other than 1), the only option available is to try to factor it as the product of two binomials.

Solution We represent the binomials using two sets of parentheses. Since the first term of the trinomial is x^2, the first term of each binomial factor must be x. Since the third term contains y^2, the last term of each binomial factor must contain y. To complete the factorization, we need to determine the coefficient of each y-term.

$$x^2 - 4xy - 5y^2 = (x \square y)(x \square y) \quad \text{Because } x \cdot x \text{ will give } x^2 \text{ and } y \cdot y \text{ will give } y^2.$$

The Language of Algebra

The trinomial $x^2 - 4xy - 5y^2$ is in two variables, x and y. It is written in **descending** powers of x and **ascending** powers of y.

The coefficients of y must be two integers whose product is -5 and whose sum is -4. Such a pair is 1 and -5. Instead of writing the first factor as $(x + 1y)$, we write it as $(x + y)$, because $1y = y$.

$$x^2 - 4xy - 5y^2 = (x + y)(x - 5y)$$

Check: $(x + y)(x - 5y) = x^2 - 5xy + xy - 5y^2$

$= x^2 - 4xy - 5y^2$ This is the original trinomial.

Self Check 6 Factor: $s^2 + 6st - 7t^2$

Now Try Problem 41

2 Factor Trinomials of the Form $x^2 + bx + c$ After Factoring Out the GCF.

If the terms of a trinomial have a common factor, it should be factored out first. The instruction "Factor" means for you to *factor the given expression completely.* A trinomial is **factored completely** when no factor can be factored further. Each factor of a completely factored expression will be prime.

EXAMPLE 7 Factor: $2x^4 + 26x^3 + 80x^2$

Strategy We will factor out the GCF, $2x^2$, first. Then we will factor the resulting trinomial.

Why The first step in factoring any polynomial is to factor out the GCF. Factoring out the GCF first makes factoring by any method easier.

Solution We begin by factoring out the GCF, $2x^2$, from $2x^4 + 26x^3 + 80x^2$.

$$2x^4 + 26x^3 + 80x^2 = 2x^2(x^2 + 13x + 40)$$

Next, we factor $x^2 + 13x + 40$. The integers 8 and 5 have a product of 40 and a sum of 13, so the completely factored form of the given trinomial is

$$2x^4 + 26x^3 + 80x^2 = 2x^2(x + 8)(x + 5)$$ The complete factorization must include $2x^2$.

Check: $2x^2(x + 8)(x + 5) = 2x^2(x^2 + 13x + 40)$

$= 2x^4 + 26x^3 + 80x^2$ This is the original trinomial.

Caution

When prime factoring 30, for example, you wouldn't stop here because 6 can be factored:

30
5 6

If a new pair of parentheses is produced in the factoring process, always see if the resulting expression within the parentheses can be factored further.

Self Check 7 Factor: $4m^5 + 8m^4 - 32m^3$

Now Try Problem 47

EXAMPLE 8 Factor: $-13g^2 + 36g + g^3$

Strategy We will write the terms of the trinomial in descending powers of g.

Why It is easier to factor a trinomial if its terms are written in descending powers of one variable.

Solution

$-13g^2 + 36g + g^3 = g^3 - 13g^2 + 36g$ Rearrange the terms.

$= g(g^2 - 13g + 36)$ Factor out the GCF, g.

$= g(g - 9)(g - 4)$ Factor the trinomial.

Check the factorization by multiplying.

Caution

For multistep factorizations, don't forget to write the GCF in the final factored form.

Self Check 8 Factor: $-12t + t^3 + 4t^2$

Now Try Problem 59

If a trinomial with integer coefficients cannot be factored using only integers, it is called a **prime trinomial.**

EXAMPLE 9 Factor: $x^2 + 2x + 3$

Strategy We will assume that $x^2 + 2x + 3$ is the product of two binomials and we will use a systematic method to find their terms.

Why Since the terms of $x^2 + 2x + 3$ do not have a common factor (other than 1), the only option available is to try to factor it as the product of two binomials.

Solution To factor the trinomial, we must find two integers whose product is 3 and whose sum is 2. The possible factorizations are shown in the table.

The Language of Algebra

When a trinomial is not factorable using only integers, we say it is **prime** and that it does not factor **over** the integers.

Factors of 3	Sum of the factors of 3
$1(3) = 3$	$1 + 3 = 4$
$-1(-3) = 3$	$-1 + (-3) = -4$

Since there are no two integers whose product is 3 and whose sum is 2, the trinomial $x^2 + 2x + 3$ cannot be factored and is a *prime trinomial.*

Self Check 9 Factor: $x^2 - 4x + 6$

Now Try Problem 63

3 Factor Trinomials of the Form $x^2 + bx + c$ Using the Grouping Method.

Another way to factor trinomials is to write them as equivalent four-termed polynomials and factor by grouping. To factor $x^2 + 8x + 15$ using this method, we proceed as follows.

1. First, identify a as the coefficient of the x^2 term, b as the coefficient of the x-term, and c as the last (constant) term. In this section, we are only factoring trinomials that have a leading coefficient of 1. Thus, a will always be equal to 1.

$$\left.\begin{array}{c} ax^2 + bx + c \\ \downarrow \quad\; \downarrow \quad\; \downarrow \\ 1x^2 + 8x + 15 \end{array}\right\} \quad a = 1,\ b = 8, \text{ and } c = 15$$

Then, find the product ac, called the **key number:** $ac = 1(15) = 15$.

2. Next, find two integers whose product is the key number, 15, and whose sum is $b = 8$. Since the integers must have a positive product and a positive sum, we consider only positive factors of 15.

Key number = 15 $\qquad$ $b = 8$

Positive factors of 15	Sum of the positive factors of 15
$1 \cdot 15 = 15$	$1 + 15 = 16$
$3 \cdot 5 = 15$	$3 + 5 = 8$

The second row of the table contains the correct pair of integers 3 and 5, whose product is the key number 15 and whose sum is $b = 8$.

3. Express the middle term, $8x$, of the trinomial as the *sum of two terms,* using the integers 3 and 5 found in step 2 as coefficients of the two terms.

$$x^2 + 8x + 15 = x^2 + 3x + 5x + 15$$ Express $8x$ as $3x + 5x$.

4. Factor the equivalent four-term polynomial by grouping:

$$x^2 + 3x + 5x + 15 = x(x + 3) + 5(x + 3)$$ Factor x out of $x^2 + 3x$ and 5 out of $5x + 15$.

$$= (x + 3)(x + 5)$$ Factor out $x + 3$.

Check the factorization by multiplying.

The grouping method is an alternative to the method for factoring trinomials discussed earlier in this section. It is especially useful when the constant term, c, has many factors.

Factoring Trinomials of the Form $x^2 + bx + c$ Using Grouping

To factor a trinomial of the form $ax^2 + bx + c$, where $a = 1$:

1. Identify a, b, and c, and the key number, ac.
2. Find two integers whose product is the key number and whose sum is b.
3. Express the middle term, bx, as the sum (or difference) of two terms. Enter the two numbers found in step 2 as coefficients of x in the form shown below. Then factor the equivalent four-term polynomial by grouping.

$$x^2 + \square x + \square x + c$$

The product of these numbers must be ac, and their sum must be b.

4. Check the factorization using multiplication.

EXAMPLE 10 Factor by grouping: $x^2 + x - 20$

Strategy We will express the middle term, x, of the trinomial as the difference of two carefully chosen terms.

Why We want to produce an equivalent four-term polynomial that can be factored by grouping.

Solution Since $x^2 + x - 20 = 1x^2 + 1x - 20$, we identify a as 1, b as 1, c as -20, and the key number ac as $1(-20) = -20$. We must find two integers whose product is -20 and whose sum is $b = 1$. Since the integers must have a negative product, their signs must be different.

Key number $= -20$ $\qquad b = 1$

Factors of -20	Sum of the factors of -20
$1(-20) = -20$	$1 + (-20) = -19$
$2(-10) = -20$	$2 + (-10) = -8$
$4(-5) = -20$	$4 + (-5) = -1$
$5(-4) = -20$	$5 + (-4) = 1$
$10(-2) = -20$	$10 + (-2) = 8$
$20(-1) = -20$	$20 + (-1) = 19$

Success Tip

We could also express the middle term as $-4x + 5x$. We obtain the same binomial factors, but in reverse order.

$$x^2 - 4x + 5x - 20$$
$$= x(x - 4) + 5(x - 4)$$
$$= (x - 4)(x + 5)$$

The fourth row of the table contains the correct pair of integers 5 and -4, whose product is -20 and whose sum is 1. They serve as the coefficients of $5x$ and $-4x$, the two terms that we use to represent the middle term, x, of the trinomial.

$$x^2 + x - 20 = x^2 + 5x - 4x - 20 \quad \text{Express the middle term, } x\text{, as } 5x - 4x.$$
$$= x(x + 5) - 4(x + 5) \quad \text{Factor } x \text{ out of } x^2 + 5x \text{ and } -4 \text{ out of } -4x - 20.$$
$$= (x + 5)(x - 4) \quad \text{Factor out } x + 5.$$

Check the factorization by multiplying.

Self Check 10 Factor by grouping: $m^2 + m - 42$

Now Try Problem 23

EXAMPLE 11 Factor by grouping: $x^2 - 4xy - 5y^2$

Strategy We will express the middle term, $-4xy$, of the trinomial as the sum of two carefully chosen terms.

Why We want to produce an equivalent four-term polynomial that can be factored by grouping.

Solution In $1x^2 - 4xy - 5y^2$, we identify a as 1, b as -4, c as -5, and the key number ac as $1(-5) = -5$. We must find two integers whose product is -5 and whose sum is $b = -4$. Such a pair is -5 and 1. They serve as the coefficients of $-5xy$ and $1xy$, the two terms that we use to represent the middle term, $-4xy$, of the trinomial.

Key number = -5 $b = -4$

Factors	Sum
$-5(1) = -5$	$-5 + 1 = -4$

$$x^2 - 4xy - 5y^2 = x^2 - 5xy + 1xy - 5y^2 \quad \text{Express the middle term, } -4xy\text{, as } -5xy + 1xy. \text{ (} 1xy - 5xy \text{ could also be used.)}$$
$$= x(x - 5y) + y(x - 5y) \quad \text{Factor } x \text{ out of } x^2 - 5xy \text{ and } y \text{ out of } 1xy - 5y^2.$$
$$= (x - 5y)(x + y) \quad \text{Factor out } x - 5y.$$

Check the factorization by multiplying.

Self Check 11 Factor by grouping: $q^2 - 2qt - 24t^2$

Now Try Problem 41

EXAMPLE 12 Factor: $2x^3 - 20x^2 + 18x$

Strategy We will factor out the GCF, $2x$, first. Then we will factor the resulting trinomial using the grouping method.

Why The first step in factoring any polynomial is to factor out the GCF.

Solution We begin by factoring out the GCF, $2x$, from $2x^3 - 20x^2 + 18x$.

$$2x^3 - 20x^2 + 18x = 2x(x^2 - 10x + 9)$$

To factor $x^2 - 10x + 9$ by grouping we identify a as 1, b as -10, and c as 9. We must find two integers whose product is the key number $ac = 1(9) = 9$ and whose sum is $b = -10$. Such a pair is -9 and -1.

Key number = 9 $\quad b = -10$

Factors	Sum
$-9(-1) = 9$	$-9 + (-1) = -10$

$$x^2 - 10x + 9 = x^2 - 9x - 1x + 9$$ Express $-10x$ as $-9x - 1x$. ($-1x - 9x$ could also be used.)

$$= x(x - 9) - 1(x - 9)$$ Factor x out of $x^2 - 9x$ and -1 out of $-1x + 9$.

$$= (x - 9)(x - 1)$$ Factor out $x - 9$.

The complete factorization of the original trinomial is

$$2x^3 - 20x^2 + 18x = 2x(x - 9)(x - 1)$$ Don't forget to write the GCF, $2x$.

Check the factorization by multiplying.

Self Check 12 Factor: $3m^3 - 27m^2 + 24m$

Now Try Problem 47

SECTION 6.2 STUDY SET

VOCABULARY

Fill in the blanks.

1. The trinomial $x^2 - x - 12$ ________ as the product of two binomials: $(x - 4)(x + 3)$.
2. A ________ trinomial cannot be factored by using only integers.
3. The ________ coefficient of $x^2 - 3x + 2$ is 1.
4. A trinomial is factored ____________ when no factor can be factored further.

CONCEPTS

Fill in the blanks.

5. a. Before attempting to factor a trinomial, be sure that it is written in ____________ powers of a variable.
   b. Before attempting to factor a trinomial into two binomials, always factor out any __________ factors first.
6. $x^2 + x - 56 = (x \boxed{})(x \boxed{})$
   The product of these numbers must be ___ , and their sum must be ___ .
7. $x^2 + 5x + 3$ cannot be factored because we cannot find two integers whose product is ___ and whose sum is ___ .
8. Complete the following table.

Factors of 8	Sum of the factors of 8
1(8)	
2( )	
−1(−8)	
(−4)	

9. Check to determine whether each factorization is correct.
   a. $x^2 - x - 20 = (x + 5)(x - 4)$
   b. $4a^2 + 12a - 16 = 4(a - 1)(a + 4)$
10. Find two integers whose
    a. product is 10 and whose sum is 7.
    b. product is 8 and whose sum is -6.
    c. product is -6 and whose sum is 1.
    d. product is -9 and whose sum is -8.
11. Consider a trinomial of the form $x^2 + bx + c$.
    a. If c is positive, what can be said about the two integers that should be chosen for the factorization?
    b. If c is negative, what can be said about the two integers that should be chosen for the factorization?
12. Fill in each blank to explain how to factor $x^2 + 7x + 10$ by grouping.

    We express the middle term, $7x$, as the sum of ____ terms:
    $x^2 + 7x + 10 = x^2 + \boxed{}x + \boxed{}x + 10$
    The product of these numbers must be ___ , and their sum must be ___ .

NOTATION

13. To factor a trinomial, a student made a table and circled the correct pair of integers, as shown. Complete the factorization of the trinomial.

    $(x \quad)(x \quad)$

Factors	Sum
1(−6)	−5
2(−3)	−1
(3(−2))	(1)
6(−1)	5

14. To factor a trinomial by grouping, a student made a table and circled the correct pair of integers, as shown. Enter the correct coefficients for the first stage in the factorization process.

$x^2 + \quad x + \quad x + 16$

Key number = 16

Factors	Sum
1 · 16	17
(2 · 8)	(10)
4 · 4	8

GUIDED PRACTICE

Factor. See Example 1 or Objective 1.

15. $x^2 + 3x + 2$

16. $y^2 + 4y + 3$

17. $z^2 + 7z + 12$

18. $x^2 + 7x + 10$

Factor each trinomial. See Example 2 or Example 10.

19. $m^2 - 5m + 6$

20. $n^2 - 7n + 10$

21. $t^2 - 11t + 28$

22. $c^2 - 9c + 8$

Factor. See Example 3 or Example 10.

23. $x^2 + 5x - 24$

24. $u^2 + u - 42$

25. $t^2 + 13t - 48$

26. $m^2 + 2m - 48$

Factor each trinomial. See Example 4 or Example 10.

27. $a^2 - 6a - 16$

28. $a^2 - 10a - 39$

29. $b^2 - 9b - 36$

30. $x^2 - 3x - 40$

Factor. See Example 5.

31. $-x^2 - 7x - 10$

32. $-x^2 + 9x - 20$

33. $-t^2 - t + 30$

34. $-t^2 - 15t + 34$

35. $-r^2 - 3r + 54$

36. $-d^2 - 2d + 63$

37. $-m^2 + 18m - 77$

38. $-n^2 + 14n - 33$

Factor. See Example 6 or Example 11.

39. $a^2 + 4ab + 3b^2$

40. $a^2 + 6ab + 5b^2$

41. $x^2 - 6xy - 7y^2$

42. $x^2 + 10xy - 11y^2$

43. $r^2 + rs - 2s^2$

44. $m^2 + mn - 6n^2$

45. $a^2 - 5ab + 6b^2$

46. $p^2 - 7pq + 10q^2$

Factor. See Example 7 or Example 12.

47. $2x^2 + 10x + 12$

48. $3y^2 - 21y + 18$

49. $6a^2 - 30a + 24$

50. $4b^2 + 12b - 16$

51. $5a^2 - 25a + 30$

52. $2b^2 - 20b + 18$

53. $-z^3 + 29z^2 - 100z$

54. $-m^3 + m^2 + 56m$

Write each trinomial in descending powers of one variable and factor. See Example 8.

55. $80 - 24x + x^2$

56. $y^2 + 100 + 25y$

57. $10y + 9 + y^2$

58. $x^2 - 13 - 12x$

59. $r^3 - 16r + 6r^2$

60. $u^3 - 12u - u^2$

61. $4r^2x + r^3 + 3rx^2$

62. $a^3 + 5ab^2 + 6a^2b$

Factor. See Example 9.

63. $u^2 + 10u + 15$

64. $v^2 + 9v + 15$

65. $r^2 + 2r - 4$

66. $r^2 - 9r - 12$

TRY IT YOURSELF

Choose the correct method from Section 6.1 or Section 6.2 to factor each of the following.

67. $5x + 15 + xy + 3y$

68. $ab + b + 2a + 2$

69. $26n^2 - 8n$

70. $40c^2 - 12c$

71. $a^2 - 4a - 5$

72. $t^2 - 5t - 50$

73. $-x^2 + 21x + 22$

74. $-r^2 + 14r - 45$

75. $4xy - 4x + 28y - 28$

76. $3xy - 3x + 15y - 15$

77. $12b^4 - 48b^3 - 36b^2$

78. $14n^5 - 42n^4 - 28n^3$

79. $r^2 - 9r + 18$

80. $y^2 - 17y + 72$

81. $-n^4 + 28n^3 + 60n^2$

82. $-c^5 + 16c^4 + 80c^3$

83. $x^2 + 4xy + 4y^2$

84. $m^2 - 8mn + 16n^2$

85. $a^2 - 4ab - 12b^2$

86. $p^2 + pq - 6q^2$

87. $4x^4 + 16x^3 + 16x^2$

88. $3a^4 + 30a^3 + 75a^2$

89. $a^2 - 46a + 45$

90. $r^2 - 37r + 36$

91. $r^2 - 2r + 4$

92. $m^2 + 3m - 20$

93. $t(x + 2) + 7(x + 2)$

94. $r(t - v) + 10(t - v)$

95. $s^4 + 11s^3 - 26s^2$

96. $x^4 + 14x^3 + 45x^2$

97. $15s^3 + 75$

98. $33g^4 - 99$

99. $-13y + y^2 - 14$

100. $-3a + a^2 + 2$

101. $2x^2 - 12x + 16$

102. $6t^2 - 18t - 24$

Look Alikes . . .

103. **a.** $x^2 - 10x + 24$ **b.** $x^2 - 10x - 24$

104. **a.** $x^2 - 5x + 6$ **b.** $x^2 - 5x - 6$

APPLICATIONS

105. Pets. The cage shown on the right is used for transporting dogs. Its volume is $(x^3 + 12x^2 + 27x)$ in.^{3}. The dimensions of the cage can be found by factoring. If the cage is longer than it is tall and taller than it is wide, find the polynomials that represent its length, width, and height.

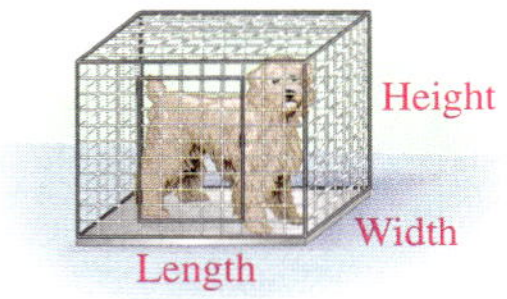

106. Photography. A picture cube is a clever way to display 6 photographs in a small amount of space. Suppose the surface area of the entire cube is given by the polynomial $(6s^2 + 12s + 6)$ in.^{2}. Find the polynomial that represents the length of an edge of the cube.

WRITING

107. Explain what it means when we say that a trinomial is the product of two binomials. Give an example.

108. Are $2x^2 - 12x + 16$ and $x^2 - 6x + 8$ factored in the same way? Explain.

109. When factoring $x^2 - 2x - 3$, one student got $(x - 3)(x + 1)$, and another got $(x + 1)(x - 3)$. Are both answers acceptable? Explain.

110. In the partial solution shown below, a student began to factor the trinomial. Write a note to the student explaining his mistake.

Factor: $x^2 - 2x - 63$

$(x - \quad)(x - \quad)$

111. Explain the error in the following factorization.

$$x^3 + 8x^2 + 15x = x(x^2 + 8x + 15)$$
$$= (x + 3)(x + 5)$$

112. Explain why the factorization is not complete.

$$2y^2 - 12y + 16 = 2(y^2 - 6y + 8)$$

REVIEW

Simplify each expression. Write each answer without negative exponents.

113. $\dfrac{x^{12}x^{-7}}{x^3x^4}$

114. $\dfrac{a^4a^{-2}}{a^2a^0}$

115. $(x^{-3}x^{-2})^2$

116. $\left(\dfrac{18a^2b^3c^{-4}}{3a^{-1}b^2c}\right)^{-3}$

CHALLENGE PROBLEMS

Factor completely.

117. $x^2 - \dfrac{6}{5}x + \dfrac{9}{25}$

118. $x^2 - 0.5x + 0.06$

119. $x^{2m} - 12x^m - 45$

120. $x^2(y + 1) - 3x(y + 1) - 70(y + 1)$

121. Find all positive integer values of c that make $n^2 + 6n + c$ factorable.

122. Find all integer values of b that make $x^2 + bx - 44$ factorable.

SECTION 6.3

Factoring Trinomials of the Form $ax^2 + bx + c$

OBJECTIVES

1. Factor trinomials using the trial-and-check method.
2. Factor trinomials after factoring out the GCF.
3. Factor trinomials using the grouping method.

ARE YOU READY?

The following problems review some basic skills that are needed when factoring trinomials.

1. In $3x^2 - x + 8$, what is the coefficient of the leading term?
2. Multiply: $5y \cdot y$
3. Multiply: $(2x - 5)(3x + 1)$
4. Find two integers whose product is -3. (There are two possible answers.)
5. Identify the coefficients of each term of $6x^2 - 3x + 7$.
6. Factor by grouping: $10b^2 + 15b - 2b - 3$

In this section, we will factor trinomials with leading coefficients other than 1, such as

$$2x^2 + 5x + 3, \qquad 6a^2 - 17a + 5, \qquad \text{and} \qquad 4b^2 + 8bc - 45c^2$$

We can use two methods to factor these trinomials. With the first method, we make educated guesses and then check them using multiplication. The correct factorization is determined through a process of elimination. The second method is an extension of factoring by grouping.

1 Factor Trinomials Using the Trial-and-Check Method.

EXAMPLE 1 Factor: $2x^2 + 5x + 3$

Strategy We will assume that $2x^2 + 5x + 3$ is the product of two binomials and we will use a systematic method to find their terms.

Why Since the terms of $2x^2 + 5x + 3$ do not have a common factor (other than 1), the only option available is to try to factor it as the product of two binomials.

Solution We represent the binomials using two sets of parentheses. Since the first term of the trinomial is $2x^2$, we enter $2x$ and x as the first terms of the binomial factors.

$$(2x \quad \square)(x \quad \square)$$ Because $2x \cdot x$ will give $2x^2$.

The Language of Algebra

To **interchange** means to put each in the place of the other. We create all of the possible factorizations by *interchanging* the second terms of the binomials.

$(2x + 1)(x + 3)$

$(2x + 3)(x + 1)$

The second terms of the binomials must be two integers whose product is 3. Since the coefficients of the terms of $2x^2 + 5x + 3$ are positive, we consider only pairs of positive integer factors of 3. Since there is just one such pair, $1 \cdot 3$, we can enter 1 and 3 as the second terms of the binomials, or we can reverse the order and enter 3 and 1.

$$(2x + 1)(x + 3) \quad \text{or} \quad (2x + 3)(x + 1)$$

The first possibility is incorrect, because when we find the outer and inner products and combine like terms, we obtain an incorrect middle term of $7x$.

Outer: $6x$

$$(2x + 1)(x + 3)$$ Multiply and add to find the middle term: $6x + x = 7x$. $7x$ is not the middle term we want.

Inner: x

The second possibility is correct, because it gives a middle term of $5x$.

Outer: $2x$

$$(2x + 3)(x + 1)$$ Multiply and add to find the middle term: $2x + 3x = 5x$. $5x$ is the middle term we want.

Inner: $3x$

Thus, the factorization is:

$$2x^2 + 5x + 3 = (2x + 3)(x + 1)$$

Check the factorization by multiplying:

$$(2x + 3)(x + 1) = 2x^2 + 2x + 3x + 3$$
$$= 2x^2 + 5x + 3$$ This is the original trinomial.

Self Check 1 Factor: $2x^2 + 5x + 2$

Now Try Problem 19

EXAMPLE 2 Factor: $6a^2 - 17a + 5$

Strategy We will assume that $6a^2 - 17a + 5$ is the product of two binomials and we will use a systematic method to find their terms.

Why Since the terms of $6a^2 - 17a + 5$ do not have a common factor (other than 1), the only option available is to try to factor it as the product of two binomials.

Solution We represent the binomials using two sets of parentheses. Since the first term is $6a^2$, the first terms of the factors must be $6a$ and a or $3a$ and $2a$.

$$(6a \quad \square)(a \quad \square) \quad \text{or} \quad (3a \quad \square)(2a \quad \square)$$ Because $6a \cdot a$ or $3a \cdot 2a$ will give $6a^2$.

The second terms of the binomials must be two integers whose product is 5. Since the last term of $6a^2 - 17a + 5$ is positive and the coefficient of the middle term is negative, we consider only negative integer factors of the last term. Since there is just one such pair, $-1(-5)$, we can enter -1 and -5, or we can reverse the order and enter -5 and -1 as second terms of the binomials.

$(6a - 1)(a - 5)$ (outer: $-30a$; inner: $-a$) $-30a - a = -31a$ Incorrect middle term

$(6a - 5)(a - 1)$ (outer: $-6a$; inner: $-5a$) $-6a - 5a = -11a$ Incorrect middle term

$(3a - 1)(2a - 5)$ (outer: $-15a$; inner: $-2a$) $-15a - 2a = -17a$ Correct middle term

$(3a - 5)(2a - 1)$ (outer: $-3a$; inner: $-10a$) $-3a - 10a = -13a$ Incorrect middle term

Only the possibility shown in blue gives the correct middle term of $-17a$. Thus,

$$6a^2 - 17a + 5 = (3a - 1)(2a - 5)$$

We check by multiplying: $(3a - 1)(2a - 5) = 6a^2 - 17a + 5$.

Self Check 2 Factor: $6b^2 - 19b + 3$

Now Try Problem 27

EXAMPLE 3 Factor: $3y^2 - 7y - 6$

Strategy We will assume that $3y^2 - 7y - 6$ is the product of two binomials and we will use a systematic method to find their terms.

Why Since the terms of $3y^2 - 7y - 6$ do not have a common factor (other than 1), the only option available is to try to factor it as the product of two binomials.

Solution Since the first term is $3y^2$, the first terms of the binomial factors must be $3y$ and y.

$$(3y \quad \square)(y \quad \square)$$ Because $3y \cdot y$ will give $3y^2$.

The second terms of the binomials must be two integers whose product is -6. There are four such pairs: $1(-6)$, $-1(6)$, $2(-3)$, and $-2(3)$. When these pairs are entered, and then reversed as second terms of the binomials, there are eight possibilities to consider. Four of them can be discarded because they include a binomial whose terms have a common factor. If the terms of $3y^2 - 7y - 6$ do not have a common factor (other than 1), neither can any of its binomial factors.

Success Tip

If the terms of a trinomial do not have a common factor other than 1, the terms of each of its binomial factors will not have a common factor other than 1.

***For* 1 *and* −6:** $(3y + 1)(y - 6)$ (outer: $-18y$; inner: y) $-18y + y = -17y$ Incorrect middle term

or ~~$(3y - 6)(y + 1)$~~ See the Success Tip on the left. $3y - 6$ has a common factor of 3. This can't be the factorization.

For −1 and 6: $(3y - 1)(y + 6)$ or ~~$(3y + 6)(y - 1)$~~

$18y$, $-y$

$18y - y = 17y$
Incorrect middle term

$3y + 6$ has a common factor of 3.
This can't be the factorization.

For 2 and −3: $(3y + 2)(y - 3)$ or ~~$(3y - 3)(y + 2)$~~

$-9y$, $2y$

$-9y + 2y = -7y$
Correct middle term

$3y - 3$ has a common factor of 3.
This can't be the factorization.

For −2 and 3: $(3y - 2)(y + 3)$ or ~~$(3y + 3)(y - 2)$~~

$9y$, $-2y$

$9y - 2y = 7y$
Incorrect middle term

$3y + 3$ has a common factor of 3.
This can't be the factorization.

Success Tip

Reversing the signs within the binomial factors reverses the sign of the middle term. For example, the factors 2 and −3 give the middle term $-7y$, while −2 and 3 give the middle term $7y$.

Success Tip

All eight possible factorizations are listed. In practice, you will often find the correct factorization without having to examine the entire list of possibilities.

Only the possibility shown in green gives the correct middle term of $-7y$. Thus,

$$3y^2 - 7y - 6 = (3y + 2)(y - 3)$$

Check the factorization by multiplying.

Self Check 3 Factor: $5t^2 - 23t - 10$

Now Try Problem 35

EXAMPLE 4 Factor: $4b^2 + 8bc - 45c^2$

Strategy We will assume that $4b^2 + 8bc - 45c^2$ is the product of two binomials and we will use a systematic method to find their terms.

Why Since the terms of $4b^2 + 8bc - 45c^2$ do not have a common factor (other than 1), the only option available is to try to factor it as the product of two binomials.

Solution Since the first term is $4b^2$, the first terms of the binomial factors must be $4b$ and b or $2b$ and $2b$. Since the last term contains c^2, the second terms of the binomial factors must contain c.

Notation

The trinomial is in two variables, b and c. It is written in descending powers of b and ascending powers of c.

$$(4b \ \square \ c)(b \ \square \ c) \quad \text{or} \quad (2b \ \square \ c)(2b \ \square \ c)$$

Because $4b \cdot b$ or $2b \cdot 2b$ gives $4b^2$, and because $c \cdot c$ gives c^2.

The coefficients of c must be two integers whose product is −45. Since the coefficient of the last term is negative, the signs of the integers must be different. If we pick factors of $4b$ and b for the first terms, and −1 and 45 for the coefficients of c, the multiplication gives an incorrect middle term of $179bc$.

$$(4b - c)(b + 45c)$$

$180bc$, $-bc$

$180bc - bc = 179bc$
Incorrect middle term

Success Tip

When using the trial-and-check method to factor $ax^2 + bx + c$, here is one suggestion: If b is relatively small, test the factorizations that use the smaller factors of a and c first.

If we pick factors of $4b$ and b for the first terms, and 15 and -3 for the coefficients of c, the multiplication gives an incorrect middle term of $3bc$.

$-12bc$

$$(4b + 15c)(b - 3c)$$

$15bc$

$-12bc + 15bc = 3bc$ Incorrect middle term

If we pick factors of $2b$ and $2b$ for the first terms, and -5 and 9 for the coefficients of c, we have

$18bc$

$$(2b - 5c)(2b + 9c)$$

$-10bc$

$18bc - 10bc = 8bc$ Correct middle term

which gives the correct middle term of $8bc$. Thus,

$$4b^2 + 8bc - 45c^2 = (2b - 5c)(2b + 9c)$$

Check the factorization by multiplying.

Self Check 4 Factor: $4x^2 + 4xy - 3y^2$

Now Try Problem 43

Because guesswork is often necessary, it is difficult to give specific rules for factoring trinomials with leading coefficients other than 1. However, the following hints are helpful when using the **trial-and-check method.**

Factoring Trinomials with Leading Coefficients Other Than 1

To factor trinomials with leading coefficients other than 1:

1. Factor out any GCF (including -1 if that is necessary to make a positive in a trinomial of the form $ax^2 + bx + c$).
2. Write the trinomial as a product of two binomials. The coefficients of the first terms of each binomial factor must be factors of a, and the last terms must be factors of c.

The product of these numbers must be a.

$$ax^2 + bx + c = (\square x\ \square)(\square x\ \square)$$

The product of these numbers must be c.

3. If c is positive, the signs within the binomial factors match the sign of b. If c is negative, the signs within the binomial factors are opposites.
4. Try combinations of the first terms and the second terms of the binomial factors until you find the one that gives the proper middle term in the trinomial. If no combination works, the trinomial is prime.
5. Check by multiplying.

2 Factor Trinomials After Factoring Out the GCF.

If the terms of a trinomial have a common factor, *the GCF (or the opposite of the GCF) should always be factored out first.*

EXAMPLE 5 Factor: $2x^2 - 8x^3 + 3x$

Strategy We will write the trinomial in descending powers of x and factor out the common factor , $-x$.

Why It is easier to factor trinomials that have a positive leading coefficient.

Solution Write the trinomial in descending powers of x: $-8x^3 + 2x^2 + 3x$.

$$-8x^3 + 2x^2 + 3x = -x(8x^2 - 2x - 3)$$ Factor out the opposite of the GCF, $-x$.

We now factor $8x^2 - 2x - 3$. Its factorization has the form

$$(8x \quad \square)(x \quad \square) \quad \text{or} \quad (2x \quad \square)(4x \quad \square)$$ Because $8x \cdot x$ or $4x \cdot 2x$ gives $8x^2$.

The second terms of the binomials must be two integers whose product is -3. There are two such pairs: $1(-3)$ and $-1(3)$. Since the coefficient of the middle term of the trinomial, $-2x$, is small, we pick the smaller factors of $8x^2$, which are $2x$ and $4x$, for the first terms and 1 and -3 for the second terms.

$$(2x + 1)(4x - 3)$$ (outer: $-6x$; inner: $4x$) $-6x + 4x = -2x$

This factorization gives the correct middle term of $-2x$. Thus,

$$8x^2 - 2x - 3 = (2x + 1)(4x - 3)$$

We can now give the complete factorization of the original trinomial.

$$-8x^3 + 2x^2 + 3x = -x(8x^2 - 2x - 3)$$
$$= -x(2x + 1)(4x - 3)$$ Don't forget this factor.

Check the factorization by multiplying.

Self Check 5 Factor: $12y - 14y^3 + 22y^2$

Now Try Problem 53

The Language of Algebra

When asked to *factor* a polynomial, that means we should *factor completely.*

Caution

For multistep factorization, don't forget to write the GCF (or its opposite) in the final factored form

3 Factor Trinomials Using the Grouping Method.

Another way to factor a trinomial of the form $ax^2 + bx + c$ is to write it as an equivalent four-termed polynomial and factor it by grouping. For example, to factor $2x^2 + 5x + 3$, we proceed as follows.

1. Identify the values of a, b, and c.

$$\left.\begin{matrix} ax^2 + bx + c \\ \downarrow \quad \downarrow \quad \downarrow \\ 2x^2 + 5x + 3 \end{matrix}\right\} a = 2,\ b = 5, \text{ and } c = 3$$

Then, find the product ac, called the **key number:** $ac = 2(3) = 6$.

2. Next, find two integers whose product is $ac = 6$ and whose sum is $b = 5$. Since the integers must have a positive product and a positive sum, we consider only positive factors of 6.

Key number = 6 $b = 5$

Positive factors of 6	Sum of the positive factors of 6
$1 \cdot 6 = 6$	$1 + 6 = 7$
$2 \cdot 3 = 6$	$2 + 3 = 5$

The second row of the table contains the correct pair of integers 2 and 3, whose product is 6 and whose sum is 5.

3. Express the middle term, $5x$, of the trinomial as the *sum of two terms,* using the integers 2 and 3 found in step 2 as coefficients of the two terms.

$$2x^2 + 5x + 3 = 2x^2 + 2x + 3x + 3 \quad \text{Express } 5x \text{ as } 2x + 3x.$$

4. Factor the equivalent four-term polynomial by grouping:

$$2x^2 + 2x + 3x + 3 = 2x(x + 1) + 3(x + 1) \quad \text{Factor } 2x \text{ out of } 2x^2 + 2x \text{ and 3 out of } 3x + 3.$$
$$= (x + 1)(2x + 3) \quad \text{Factor out } x + 1.$$

Check by multiplying.

Factoring by grouping is especially useful when the leading coefficient, a, and the constant term, c, have many factors.

Factoring Trinomials by Grouping

To factor a trinomial by grouping:

1. Factor out any GCF (including -1 if that is necessary to make a positive in a trinomial of the form $ax^2 + bx + c$).
2. Identify a, b, and c, and find the key number ac.
3. Find two integers whose product is the key number and whose sum is b.
4. Express the middle term, bx, as the sum (or difference) of two terms. Enter the two numbers found in step 3 as coefficients of x in the form shown below. Then factor the equivalent four-term polynomial by grouping.

$$ax^2 + \square x + \square x + c$$

The product of these numbers must be ac and their sum must be b.

5. Check the factorization by multiplying.

EXAMPLE 6 Factor by grouping: $10x^2 + 13x - 3$

Strategy We will express the middle term, $13x$, of the trinomial as the sum of two carefully chosen terms.

Why We want to produce an equivalent four-term polynomial that can be factored by grouping.

Solution In $10x^2 + 13x - 3$, we have $a = 10$, $b = 13$, and $c = -3$. The key number is $ac = 10(-3) = -30$. We must find a factorization of -30 such that the sum of the factors is $b = 13$. The possible factor pairs are listed in the table. Since the factors must have a negative product, their signs must be different.

It is wise to follow an order when listing the factors in the table so that you don't skip the correct combination. Here, the first factors 1, 2, 3, 5, 6, 10, 15, and 30 are listed from least to greatest.

Key number $= -30$ $b = 13$

Factors of -30	Sum of the factors of -30
$1(-30) = -30$	$1 + (-30) = -29$
$2(-15) = -30$	$2 + (-15) = -13$
$3(-10) = -30$	$3 + (-10) = -7$
$5(-6) = -30$	$5 + (-6) = -1$
$6(-5) = -30$	$6 + (-5) = 1$
$10(-3) = -30$	$10 + (-3) = 7$
$15(-2) = -30$	$15 + (-2) = 13$
$30(-1) = -30$	$30 + (-1) = 29$

Notation

The middle term, $13x$, may be expressed as $15x - 2x$ or as $-2x + 15x$ when using factoring by grouping. The resulting factorizations will be equivalent.

The seventh row contains the correct pair of numbers 15 and -2, whose product is -30 and whose sum is 13. They serve as the coefficients of $15x$ and $-2x$, the two terms that we use to represent the middle term, $13x$, of the trinomial.

$$10x^2 + 13x - 3 = 10x^2 + 15x - 2x - 3 \quad \text{Express } 13x \text{ as } 15x - 2x.$$

Finally, we factor by grouping.

$$10x^2 + 15x - 2x - 3 = 5x(2x + 3) - 1(2x + 3) \quad \text{Factor out } 5x \text{ from } 10x^2 + 15x. \text{ Factor out } -1 \text{ from } -2x - 3.$$

$$= (2x + 3)(5x - 1) \quad \text{Factor out } 2x + 3.$$

So $10x^2 + 13x - 3 = (2x + 3)(5x - 1)$. Check the factorization by multiplying.

Self Check 6 Factor by grouping: $15a^2 + 17a - 4$

Now Try Problems 19, 27, and 35

EXAMPLE 7 Factor: $12x^5 - 17x^4 + 6x^3$

Strategy We will factor out the GCF, x^3, first. Then we will factor the resulting trinomial using the grouping method.

Why The first step in factoring any polynomial is to factor out the GCF.

Solution The GCF of the three terms of the trinomial is x^3.

$$12x^5 - 17x^4 + 6x^3 = x^3(12x^2 - 17x + 6)$$

To factor $12x^2 - 17x + 6$, we must find two integers whose product is $12(6) = 72$ and whose sum is -17. Two such numbers are -8 and -9. They serve as the coefficients of $-8x$ and $-9x$, the two terms that we use to represent the middle term, $-17x$, of the trinomial.

Key number $= 72$ $b = -17$

Factors	Sum
$-8(-9) = 72$	$-8 + (-9) = -17$

$$12x^2 - 17x + 6 = 12x^2 - 8x - 9x + 6$$ Express $-17x$ as $-8x - 9x$. ($-9x - 8x$ could also be used.)

$$= 4x(3x - 2) - 3(3x - 2)$$ Factor out $4x$ and factor out -3.

$$= (3x - 2)(4x - 3)$$ Factor out $3x - 2$.

The complete factorization of the original trinomial is

$$12x^5 - 17x^4 + 6x^3 = x^3(3x - 2)(4x - 3)$$ Don't forget to write the GCF, x^3.

Check the factorization by multiplying.

Self Check 7 Factor: $21a^4 - 13a^3 + 2a^2$

Now Try Problem 53

SECTION 6.3 STUDY SET

VOCABULARY

Fill in the blanks.

1. The ________ coefficient of $3x^2 - x - 12$ is 3.
2. Given $5y^2 + 16y + 3 = (5y + 1)(y + 3)$. We say that $5y^2 + 16y + 3$ factors as the product of two __________.
3. The first terms of the binomial factors $(5y + 1)(y + 3)$ are ____ and ___. The second terms of the binomial factors are ___ and ___.
4. To factor $2m^2 + 11m + 12$ by _________, we write it as $2m^2 + 8m + 3m + 12$.

CONCEPTS

5. If $10x^2 - 27x + 5$ is to be factored as the product of two binomials, what are the possible *first terms* of the binomial factors?
6. Complete each sentence.

The product of these numbers must be .

$$5x^2 + 6x - 8 = (\square x \ \square)(\square x \ \square)$$

The product of these numbers must be .

7. a. Fill in the blanks. When factoring a trinomial, we write it in ___________ powers of the variable. Then we factor out any ______ (including -1 if that is necessary to make the leading ___________ positive).
   b. What is the GCF of the terms of $6s^4 + 33s^3 + 36s^2$?
   c. Factor out -1 from $-2d^2 + 19d - 8$.
8. Check to determine whether $(3t - 1)(5t - 6)$ is the correct factorization of $15t^2 - 19t + 6$.

A trinomial has been partially factored. Complete each statement that describes the type of integers we should consider for the blanks.

9. $5y^2 - 13y + 6 = (5y \ \square)(y \ \square)$
   Since the last term of the trinomial is positive and the middle term is negative, the integers must be _________ factors of 6.
10. $5y^2 + 13y + 6 = (5y \ \square)(y \ \square)$
    Since the last term of the trinomial is positive and the middle term is positive, the integers must be _________ factors of 6.
11. $5y^2 - 7y - 6 = (5y \ \square)(y \ \square)$
    Since the last term of the trinomial is negative, the signs of the integers will be _________.
12. $5y^2 + 7y - 6 = (5y \ \square)(y \ \square)$
    Since the last term of the trinomial is negative, the signs of the integers will be _________.
13. Complete the key number table.

Negative factors of 12	Sum of the negative factors of 12
$-1(-12)$	
$-2(\quad)$	
(-4)	

14. Complete the sentence to explain how to factor $3x^2 + 16x + 5$ by grouping.

$$3x^2 + 16x + 5 = 3x^2 + \square x + \square x + 5$$

The product of these numbers must be and their sum must be .

NOTATION

15. a. Suppose we wish to factor $12b^2 + 20b - 9$ by grouping. Identify a, b, and c.
    b. What is the key number, ac?

16. To factor $6x^2 + 13x + 6$ by grouping, a student made a table and circled the correct pair of integers, as shown. Enter the correct coefficients for the first stage in the factorization process.

$6x^2 + \quad x + \quad x + 6$

$ac = 36 \quad b = 13$

Factors	Sum
1 · 36	37
2 · 18	20
3 · 12	15
(4 · 9)	(13)
6 · 6	12

Complete each step of the factorization of the trinomial by grouping.

17. $12t^2 + 17t + 6 = 12t^2 + 9t + 8t + 6$
$= \quad (4t + 3) + \quad (4t + 3)$
$= (\quad\quad)(3t + 2)$

18. $35t^2 - 11t - 6 = 35t^2 + 10t - 21t - 6$
$= 5t(7t + 2) \quad 3(7t + 2)$
$= (\quad\quad)(5t - 3)$

GUIDED PRACTICE

Factor. See Example 1 or Example 6.

19. $2x^2 + 3x + 1$
20. $3x^2 + 4x + 1$
21. $3a^2 + 10a + 3$
22. $2b^2 + 7b + 3$
23. $5x^2 + 7x + 2$
24. $7t^2 + 12t + 5$
25. $7x^2 + 18x + 11$
26. $5n^2 + 12n + 7$

Factor. See Example 2 or Example 6.

27. $4x^2 - 8x + 3$
28. $4z^2 - 13z + 3$
29. $8x^2 - 22x + 5$
30. $15a^2 - 28a + 5$
31. $15t^2 - 26t + 7$
32. $10x^2 - 9x + 2$
33. $6y^2 - 13y + 2$
34. $6y^2 - 43y + 7$

Factor. See Example 3 or Example 6.

35. $3x^2 - 2x - 21$
36. $3u^2 - 44u - 15$
37. $5m^2 - 7m - 6$
38. $5y^2 - 18y - 8$
39. $7y^2 + 55y - 8$
40. $7x^2 + 33x - 10$
41. $11y^2 + 7y - 4$
42. $13y^2 + 9y - 4$

Factor. See Example 4.

43. $6r^2 - rs - 2s^2$
44. $3m^2 + 5mn + 2n^2$
45. $4x^2 - 8xy + 3y^2$
46. $4b^2 + 15bc - 4c^2$
47. $8m^2 + 91mn + 33n^2$
48. $2m^2 + 17mn - 9n^2$
49. $15x^2 - xy - 6y^2$
50. $4a^2 - 15ab + 9b^2$

Factor. See Example 5 or Example 7.

51. $-26x + 6x^2 - 20$
52. $-28 + 6a^2 - 2a$
53. $15a + 8a^3 - 26a^2$
54. $16r - 40r^2 + 25r^3$
55. $2u^2 - 6v^2 - uv$
56. $6a^2 + 6b^2 - 13ab$
57. $36y^2 - 88y + 32$
58. $70a^2 - 95a + 30$

TRY IT YOURSELF

Factor. If an expression is prime, so indicate.

59. $6t^2 - 7t - 20$
60. $6w^2 + 13w + 5$
61. $15p^2 - 2pq - q^2$
62. $8c^2 - 10cd + 3d^2$
63. $4t^2 - 16t + 7$
64. $9x^2 - 32x + 15$
65. $130r^2 + 20r - 110$
66. $170h^2 - 210h - 260$
67. $8y^2 - 2y - 1$
68. $14y^2 + 11y + 2$
69. $18x^2 + 31x - 10$
70. $20y^2 - 93y - 35$
71. $-y^3 - 13y^2 - 12y$
72. $-2xy^2 - 8xy + 24x$
73. $10u^2 - 13u - 6$
74. $8m^2 + 5m - 10$
75. $-6x^4 + 15x^3 + 9x^2$
76. $-9y^4 - 3y^3 + 6y^2$
77. $6p^2 + pq - q^2$
78. $12m^2 - 11mn + 2n^2$
79. $30r^5 + 63r^4 - 30r^3$
80. $6s^5 - 26s^4 - 20s^3$
81. $16m^3n + 20m^2n^2 + 6mn^3$
82. $-28u^3v^3 + 26u^2v^4 - 6uv^5$
83. $3x^2 + x + 6$
84. $2u^2 + 3u + 25$
85. $-12y^2 - 12 + 25y$
86. $-10t^2 + 1 + 3t$

Choose the correct method from Sections 6.1, 6.2, or 6.3 to factor each of the following.

87. $m^2 + 3m - 28$
88. $-b^2 - 5b + 24$
89. $6a^3 + 15a^2$
90. $9x^4 + 27x^6$
91. $x^3 - 2x^2 + 5x - 10$
92. $x^3 - x^2 + 2x - 2$
93. $5y^2 + 3 - 8y$
94. $3t^2 + 7 - 10t$
95. $-2x^2 - 10x - 12$
96. $4y^2 + 36y + 72$

97. $12x^3y^3 - 18x^2y^3 + 15x^2y^2$

98. $15c^2d^3 - 25c^3d^2 - 10c^4d^4$

99. $a^2 - 7ab + 10b^2$

100. $x^2 - 13xy + 12y^2$

101. $9u^6 - 71u^5 - 8u^4$

102. $25n^8 - 49n^7 - 2n^6$

APPLICATIONS

103. from **Campus to Careers**

Elementary School Teacher

The area of a teacher's desktop is represented by the trinomial $(4x^2 + 20x - 11)$ in.^{2}. Factor it to find the polynomials that represent its length and width.

©iStockphoto.com/Neustockimages

104. **Storage.** The volume of an 8-foot-wide portable storage container is represented by the trinomial $(72x^2 + 120x - 400)$ ft^3. Its dimensions can be determined by factoring the trinomial. Find the polynomials that represent the height and the length of the container.

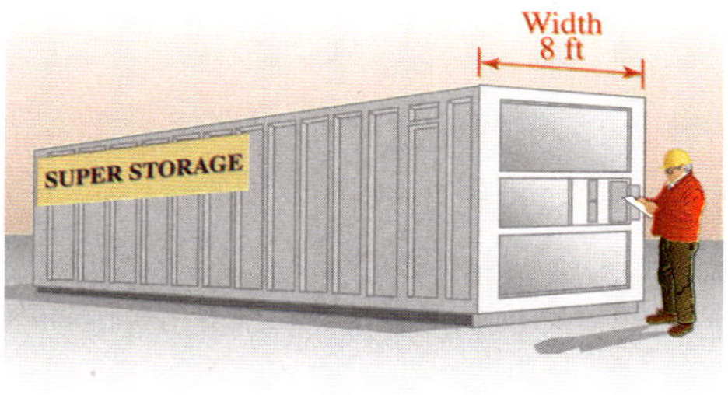

WRITING

105. Two students factor $2x^2 + 20x + 42$ and get two different answers: $(2x + 6)(x + 7)$ and $(x + 3)(2x + 14)$.

Do both answers check? Why don't they agree? Is either answer completely correct? Explain.

106. Why is the process of factoring $6x^2 - 5x - 6$ more complicated than the process of factoring $x^2 - 5x - 6$?

107. Suppose a factorization check of $(3x - 9)(5x + 7)$ gives a middle term $-24x$, but a middle term of $24x$ is actually needed. Explain how to quickly obtain the correct factorization.

108. Suppose we want to factor $2x^2 + 7x - 72$. Explain why $(2x - 1)(x + 72)$ is not a wise choice to try first.

REVIEW

Evaluate each expression.

109. -7^2

110. $(-7)^2$

111. 7^0

112. 7^{-2}

113. $\frac{1}{7^{-2}}$

114. $2 \cdot 7^2$

CHALLENGE PROBLEMS

Factor.

115. $6a^{10} + 5a^5 - 21$

116. $3x^4y^2 - 29x^2y + 56$

117. $8x^2(c^2 + c - 2) - 2x(c^2 + c - 2) - (c^2 + c - 2)$

118. Find all integer values of b that make $2x^2 + bx - 5$ factorable.

SECTION 6.4

Factoring Perfect-Square Trinomials and the Difference of Two Squares

OBJECTIVES

1. Recognize perfect-square trinomials.
2. Factor perfect-square trinomials.
3. Factor the difference of two squares.

ARE YOU READY?

The following problems review some basic skills that are needed when factoring binomials and trinomials.

1. Multiply: $(3y + 2)(3y + 2)$
2. Multiply: $(m + 9)(m - 9)$
3. Simplify: $(8d)^2$
4. Translate to symbols: the sum of x^2 and 25

In this section, we will discuss a method that can be used to factor two types of trinomials, called *perfect-square trinomials.* We also develop techniques for factoring a type of binomial called the *difference of two squares.*

1 Recognize Perfect-Square Trinomials.

We have seen that the square of a binomial is a trinomial. We also have seen that the special-product rules shown below can be used to find the square of a sum and the square of a difference quickly. The terms of the resulting trinomial are related to the terms of the binomial that was squared.

Success Tip

To prepare for this section, it would be helpful to review Section 5.7 Special Products.

$$(A + B)^2 = A^2 + 2AB + B^2$$

This is the square of the first term of the binomial. This is twice the product of A and B, or its opposite. This is the square of the last term of the binomial.

$$(A - B)^2 = A^2 - 2AB + B^2$$

Trinomials that are squares of a binomial are called **perfect-square trinomials.** Some examples are

$y^2 + 6y + 9$ Because it is the square of $(y + 3)$: $(y + 3)^2 = y^2 + 6y + 9$

$t^2 - 14t + 49$ Because it is the square of $(t - 7)$: $(t - 7)^2 = t^2 - 14t + 49$

$4m^2 - 20m + 25$ Because it is the square of $(2m - 5)$: $(2m - 5)^2 = 4m^2 - 20m + 25$

EXAMPLE 1

Determine whether the following are perfect-square trinomials: **a.** $x^2 + 10x + 25$ **b.** $c^2 - 12c - 36$ **c.** $25y^2 - 30y + 9$ **d.** $4t^2 + 18t + 81$

Strategy We will compare each trinomial, term-by-term, to one of the special-product forms discussed in Section 5.7.

Why If a trinomial matches one of these forms, it is a perfect-square trinomial.

Solution

a. To determine whether this is a perfect-square trinomial, we note that

$$x^2 + 10x + 25$$

The first term is the square of x. The middle term is twice the product of x and 5: $2 \cdot x \cdot 5 = 10x$. The last term is the square of 5.

Thus, $x^2 + 10x + 25$ is a perfect-square trinomial.

b. To determine whether this is a perfect-square trinomial, we note that

$c^2 - 12c - 36$ The last term, -36, is not the square of a real number.

Since the last term is negative, $c^2 - 12c - 36$ is not a perfect-square trinomial.

The Language of Algebra

The expressions $25y^2$ and 9 are called *perfect squares* because $25y^2 = (5y)^2$ and $9 = 3^2$.

c. To determine whether this is a perfect-square trinomial, we note that

$$25y^2 - 30y + 9$$

The first term is the square of 5y. The middle term is the opposite of twice the product of 5y and 3: $-2(5y)(3) = -30y$. The last term is the square of 3.

Thus, $25y^2 - 30y + 9$ is a perfect-square trinomial.

d. To determine whether this is a perfect-square trinomial, we note that

$$4t^2 + 18t + 81$$

The first term is the square of 2t. The middle term is not twice the product of 2t and 9, because $2(2t)(9) = 36t$. The last term is the square of 9.

Thus, $4t^2 + 18t + 81$ is not a perfect-square trinomial.

Self Check 1 Determine whether the following are perfect-square trinomials:

a. $y^2 + 4y + 4$ **b.** $b^2 - 6b - 9$

c. $4z^2 + 4z + 4$ **d.** $49x^2 - 28x + 16$

Now Try Problems 13 and 17

Recognizing Perfect-Square Trinomials

1. The first and last terms are squares of integers or monomials.
2. The middle term is twice the product of the expressions that are squared to produce the first and last terms, or its opposite.

2 Factor Perfect-Square Trinomials.

We can factor perfect-square trinomials using the methods previously discussed in Sections 6.2 and 6.3. However, in many cases, we can factor them more quickly by inspecting their terms and applying the special-product rules in reverse.

Factoring Perfect-Square Trinomials

Perfect-square trinomial		Square of a binomial
$A^2 + 2AB + B^2$	=	$(A + B)^2$
$A^2 - 2AB + B^2$	=	$(A - B)^2$

Each of these trinomials factors as the square of a binomial.

When factoring perfect-square trinomials, it is helpful to know the following perfect squares printed in red. The number 400, for example, is a perfect square, because $400 = 20^2$.

$1 = 1^2$	$25 = 5^2$	$81 = 9^2$	$169 = 13^2$	$289 = 17^2$
$4 = 2^2$	$36 = 6^2$	$100 = 10^2$	$196 = 14^2$	$324 = 18^2$
$9 = 3^2$	$49 = 7^2$	$121 = 11^2$	$225 = 15^2$	$361 = 19^2$
$16 = 4^2$	$64 = 8^2$	$144 = 12^2$	$256 = 16^2$	$400 = 20^2$

EXAMPLE 2 Factor: **a.** $x^2 + 20x + 100$ **b.** $9x^2 - 30xy + 25y^2$

Strategy The terms of each trinomial do not have a common factor (other than 1). We will determine whether each is a perfect-square trinomial.

Why If it is, we can factor it using a special-product rule in reverse.

Solution **a.** $x^2 + 20x + 100$ is a perfect-square trinomial, because:

- The first term x^2 is the square of x.
- The last term 100 is the square of **10**.
- The middle term is twice the product of x and **10**: $2(x)(10) = 20x$.

Success Tip

The sign of the middle term of a perfect-square trinomial is the same as the sign of the second term of the squared binomial.

Same sign

$A^2 + 2AB + B^2 = (A + B)^2$

$A^2 - 2AB + B^2 = (A - B)^2$

Same sign

To find the factorization, we match $x^2 + 20x + 100$ to the proper rule for factoring a perfect-square trinomial.

$$A^2 + 2\ A\ \ B + B^2 = (A + B)^2$$

$$x^2 + 20x + 10^2 = x^2 + 2 \cdot x \cdot 10 + 10^2 = (x + 10)^2$$

Therefore, $x^2 + 20x + 10^2 = (x + 10)^2$. Check by finding $(x + 10)^2$.

b. $9x^2 - 30xy + 25y^2$ is a perfect-square trinomial, because:

- The first term $9x^2$ is the square of $3x$: $(3x)^2 = 9x^2$.
- The last term $25y^2$ is the square of $5y$: $(5y)^2 = 25y^2$.
- The middle term is the opposite of twice the product of $3x$ and $5y$: $-2(3x)(5y) = -30xy$.

We can use these observations to write the trinomial in one of the special-product forms that then leads to its factorization.

$$\begin{aligned} 9x^2 - 30xy + 25y^2 &= (3x)^2 - 2(3x)(5y) + (5y)^2 \\ &= (3x - 5y)^2 \end{aligned}$$

Therefore, $9x^2 - 30xy + 25y^2 = (3x - 5y)^2$. Check by finding $(3x - 5y)^2$.

Self Check 2 Factor: **a.** $x^2 + 18x + 81$ **b.** $16x^2 - 8xy + y^2$

Now Try Problems 21 and 29

EXAMPLE 3 Factor: $4a^3 - 4a^2 + a$

Strategy We will factor out the GCF, a, first. Then we will factor the resulting perfect-square trinomial using a special-product rule in reverse.

Why The first step in factoring any polynomial is to factor out the GCF.

Solution The terms of $4a^3 - 4a^2 + a$ have the common factor a, which should be factored out first. Within the parentheses, we recognize $4a^2 - 4a + 1$ as a perfect-square trinomial of the form $A^2 - 2AB + B^2$, and factor it as such.

$$\begin{aligned} 4a^3 - 4a^2 + a &= a(4a^2 - 4a + 1) && \text{Factor out the GCF, } a. \\ &= a(2a - 1)^2 && \text{Since } 4a^2 = (2a)^2, 1 = (1)^2, \text{ and } -4a = -2(2a)(1), \\ & && 4a^2 - 4a + 1 \text{ can be factored as a perfect-square trinomial.} \end{aligned}$$

Success Tip

Perfect-square trinomials, such as $4a^2 - 4a + 1$, also can be factored using the trial-and-check method or factoring by grouping. However, if you recognize that a trinomial is a perfect-square trinomial, the method of this section is often easier and faster.

Self Check 3 Factor: $49x^3 - 14x^2 + x$

Now Try Problem 33

3 Factor the Difference of Two Squares.

Recall the special-product rule for multiplying the sum and difference of the same two terms:

$$(A + B)(A - B) = A^2 - B^2$$

The binomial $A^2 - B^2$ is called a **difference of two squares,** because A^2 is the square of A and B^2 is the square of B. If we reverse this rule, we obtain a method for factoring a difference of two squares.

The Language of Algebra

The expression $A^2 - B^2$ is a **difference of two squares,** whereas $(A - B)^2$ is the **square of a difference.** They are not equivalent because $(A - B)^2 \neq A^2 - B^2$.

Factoring →

$$A^2 - B^2 = (A + B)(A - B)$$

This pattern is easy to remember if we think of a difference of two squares as the square of a **F**irst quantity minus the square of a **L**ast quantity.

Factoring a Difference of Two Squares

To factor the square of a First quantity minus the square of a Last quantity, multiply the First plus the Last by the First minus the Last.

$$F^2 - L^2 = (F + L)(F - L)$$

To factor differences of two squares, it will be helpful if you recognize the perfect squares listed in red on page 466.

EXAMPLE 4 Factor: **a.** $x^2 - 9$ **b.** $16 - b^2$ **c.** $n^2 - 45$ **d.** $a^2 + 81$

Strategy The terms of each binomial do not have a common factor (other than 1). The only option available is to attempt to factor each as a difference of two squares.

Why If a binomial is a difference of two squares, we can factor it using a special-product rule in reverse.

Solution

a. $x^2 - 9$ is the difference of two squares because it can be written as $x^2 - 3^2$. We can match it to the rule for factoring a difference of two squares to find the factorization.

$$F^2 - L^2 = (F + L)(F - L)$$
$$\downarrow \quad \downarrow \qquad \downarrow \quad \downarrow \quad \downarrow \quad \downarrow$$
$$x^2 - 3^2 = (x + 3)(x - 3)$$ 9 is a perfect square: $9 = 3^2$.

Therefore, $x^2 - 9 = (x + 3)(x - 3)$.

Check by multiplying: $(x + 3)(x - 3) = x^2 - 9$.

Notation

By the commutative property of multiplication, the factors of a difference of two squares can be written in either order. For example, we can write:

$$x^2 - 9 = (x - 3)(x + 3)$$

b. $16 - b^2$ is the difference of two squares because $16 - b^2 = 4^2 - b^2$. Therefore,

$$16 - b^2 = (4 + b)(4 - b)$$ 16 is a perfect square: $16 = 4^2$.

Check by multiplying.

c. Since 45 is not a perfect square, $n^2 - 45$ cannot be factored using integers. It is *prime*.

d. $a^2 + 81$ can be written $a^2 + 9^2$, and is, therefore, the **sum of two squares.** We might attempt to factor $a^2 + 81$ as $(a + 9)(a + 9)$ or $(a - 9)(a - 9)$. However, the following checks show that neither product is $a^2 + 81$.

$$(a + 9)(a + 9) = a^2 + 18a + 81 \qquad (a - 9)(a - 9) = a^2 - 18a + 81$$

In general, **the sum of two squares (with no common factor other than 1) cannot be factored using real numbers.** Thus, $a^2 + 81$ is *prime*.

The Language of Algebra

An expression of the form $A^2 + B^2$ is called the **sum of two squares,** whereas $(A + B)^2$ is the **square of a sum.** They are not equivalent because $(A + B)^2 \neq A^2 + B^2$.

Self Check 4 Factor: **a.** $c^2 - 4$ **b.** $121 - t^2$ **c.** $x^2 - 24$ **d.** $s^2 + 36$

Now Try Problems 37 and 45

Terms containing variables such as a^4, $25x^2$, and $4y^4$ are perfect squares, because they can be written as the square of a quantity. For example:

$$a^4 = (a^2)^2, \qquad 25x^2 = (5x)^2, \qquad \text{and} \qquad 4y^4 = (2y^2)^2$$

EXAMPLE 5 Factor: **a.** $25x^2 - 49$ **b.** $-121z^2 + 4y^4$

Strategy In each case, the terms of the binomial do not have a common factor (other than 1). To factor them, we will write each binomial in a form that clearly shows it is a difference of two squares.

Why We can then use a special-product rule in reverse to factor it.

Solution **a.** We can write $25x^2 - 49$ in the form $(5x)^2 - 7^2$ and match it to the rule for factoring the difference of two squares:

$$\mathbf{F}^2 - \mathbf{L}^2 = (\mathbf{F} + \mathbf{L})(\mathbf{F} - \mathbf{L})$$

$$(\mathbf{5x})^2 - \mathbf{7}^2 = (\mathbf{5x} + \mathbf{7})(\mathbf{5x} - \mathbf{7})$$

Therefore, $25x^2 - 49 = (5x + 7)(5x - 7)$. Check by multiplying.

Success Tip

Remember that a *difference of two squares* is a binomial. Each term is a square and the terms have different signs. The powers of the variables in the terms must be even.

b. If we reorder the terms, the resulting binomial, $4y^4 - 121z^2$ is obviously a difference of two squares. We can write $4y^4 - 121z^2$ in the form $(2y^2)^2 - (11z)^2$ and match it to the rule for factoring the difference of two squares:

$$\mathbf{F}^2 - \mathbf{L}^2 = (\mathbf{F} + \mathbf{L})(\mathbf{F} - \mathbf{L})$$

$$(\mathbf{2y^2})^2 - (\mathbf{11z})^2 = (\mathbf{2y^2} + \mathbf{11z})(\mathbf{2y^2} - \mathbf{11z})$$

Therefore, $-121z^2 + 4y^4 = 4y^4 - 121z^2 = (2y^2 + 11z)(2y^2 - 11z)$. Check by multiplying.

An alternate approach is first to factor out -1: $-(121z^2 - 4y^4)$. Then factor the difference of two squares within the parentheses to get an equivalent result: $-(11z + 2y^2)(11z - 2y^2)$.

Self Check 5 Factor: **a.** $16y^2 - 9$
b. $9m^2 - 64n^4$ **c.** $-a^4 + 100$

Now Try Problems 49 and 53

EXAMPLE 6 Factor: $8x^2 - 8$

Strategy We will factor out the GCF, 8, first. Then we will factor the resulting difference of two squares.

Why The first step in factoring any polynomial is to factor out the GCF.

Solution

$$8x^2 - 8 = 8(x^2 - 1) \quad \text{The GCF is 8.}$$

$$= 8(x + 1)(x - 1) \quad \text{Think of } x^2 - 1 \text{ as } x^2 - 1^2 \text{ and factor the difference of two squares.}$$

Check:

$$8(x + 1)(x - 1) = 8(x^2 - 1) \quad \text{Multiply the binomials first.}$$

$$= 8x^2 - 8 \quad \text{Distribute the multiplication by 8.}$$

Self Check 6 Factor: $2p^2 - 200$

Now Try Problem 57

Sometimes we must factor a difference of two squares more than once to factor a polynomial completely.

EXAMPLE 7 Factor: $x^4 - 16$

Strategy The terms of $x^4 - 16$ do not have a common factor (other than 1). To factor this binomial, we will write it in a form that clearly shows it is a difference of two squares.

Why We can then use a special-product rule in reverse to factor it.

Solution

$x^4 - 16 = (x^2)^2 - 4^2$ Write x^4 as $(x^2)^2$ and 16 as 4^2.

$= (x^2 + 4)(x^2 - 4)$ Factor the difference of two squares.

$= (x^2 + 4)(x + 2)(x - 2)$ Factor $x^2 - 4$, which is itself a difference of two squares. The binomial $x^2 + 4$ is a sum of two squares and does not factor further.

Success Tip

Factoring a polynomial is complete when no factor can be factored further.

Self Check 7 Factor: $a^4 - 81$

Now Try Problem 63

SECTION 6.4 STUDY SET

VOCABULARY

Fill in the blanks.

1. $x^2 + 6x + 9$ is a ________-square trinomial because it is the square of the binomial $x + 3$.
2. The binomial $x^2 - 25$ is called a __________ of two squares and it factors as $(x + 5)(x - 5)$. The binomial $x^2 + 25$ is a _____ of two squares and since it does not factor using integers, it is _______.

CONCEPTS

Fill in the blanks.

3. Consider $25x^2 + 30x + 9$.
   a. The first term is the square of ___.
   b. The last term is the square of ___.
   c. The middle term is twice the product of ___ and ___.
4. Consider $49x^2 - 28xy + 4y^2$.
   a. The first term is the square of ___.
   b. The last term is the square of ___.
   c. The middle term is the opposite of twice the product of ___ and ___.
5. a. $x^2 + 2xy + y^2 = (\quad + \quad)^2$
   b. $x^2 - 2xy + y^2 = (x \quad y)^2$
   c. $x^2 - y^2 = (x \quad y)(\quad - \quad)$
6. a. $36x^2 = (\quad)^2$ b. $100x^4 = (\quad)^2$
   c. $4x^2 - 9 = (\quad)^2 - (\quad)^2$
7. List the squares of the integers from 1 through 20.
8. Use multiplication to determine if each factorization is correct.
   a. $9y^2 - 12y + 4 = (3y - 2)^2$
   b. $n^2 - 16 = (n + 8)(n - 8)$

NOTATION

Complete each factorization.

9. $x^2 + 10x + 25 = (x + 5)$
10. $9b^2 - 12b + 4 = (3b \quad 2)^2$
11. $x^2 - 64 = (x \quad 8)(x \quad 8)$
12. $16t^2 - 49 = (4t + \quad)(4t - \quad)$

GUIDED PRACTICE

Determine whether each of the following is a perfect-square trinomial. See Example 1.

13. $x^2 + 18x + 81$
14. $x^2 + 14x + 49$
15. $y^2 + 2y + 4$
16. $y^2 + 4y - 16$
17. $9n^2 - 30n - 25$
18. $9a^2 - 48a - 64$
19. $4y^2 - 12y + 9$
20. $9y^2 - 30y + 25$

Factor. See Example 2.

21. $x^2 + 6x + 9$
22. $x^2 + 10x + 25$
23. $b^2 + 2b + 1$
24. $m^2 + 12m + 36$
25. $c^2 - 12c + 36$
26. $d^2 - 10d + 25$
27. $9 + 4x^2 + 12x$
28. $121 + 4x^2 + 44x$
29. $36m^2 + 60mn + 25n^2$
30. $25a^2 + 30ab + 9b^2$

31. $81x^2 - 72xy + 16y^2$

32. $9x^2 - 48xy + 64y^2$

Factor. See Example 3.

33. $3u^2 - 18u + 27$

34. $3v^2 - 42v + 147$

35. $36x^3 + 12x^2 + x$

36. $4x^4 - 20x^3 + 25x^2$

Factor. If a polynomial can't be factored, write "prime." See Example 4.

37. $x^2 - 4$

38. $x^2 - 9$

39. $x^2 - 16$

40. $x^2 - 25$

41. $36 - y^2$

42. $49 - w^2$

43. $t^2 - 25$

44. $h^2 - 144$

45. $a^2 + b^2$

46. $121a^2 + b^2$

47. $y^2 - 63$

48. $x^2 - 27$

Factor. See Example 5.

49. $25t^2 - 64$

50. $49d^2 - 16$

51. $81y^2 - 1$

52. $400z^2 - 1$

53. $9x^4 - y^2$

54. $4x^2 - z^4$

55. $-49d^4 + 16c^2$

56. $-121b^4 + 36a^2$

Factor. See Example 6.

57. $8x^2 - 32y^2$

58. $2a^2 - 200b^2$

59. $63a^2 - 7$

60. $20x^2 - 5$

Factor. See Example 7.

61. $81 - s^4$

62. $y^4 - 625$

63. $b^4 - 256$

64. $m^4n^4 - 16$

TRY IT YOURSELF

Factor.

65. $a^4 - 144b^2$

66. $81y^4 - 100z^2$

67. $9x^2y^2 + 30xy + 25$

68. $s^2t^2 - 20st + 100$

69. $16t^4 - 16s^4$

70. $2p^4 - 32q^4$

71. $t^2 - 20t + 100$

72. $r^2 + 24r + 144$

73. $9y^2 - 24y + 16$

74. $49z^2 - 14z + 1$

75. $z^2 - 64$

76. $25 + B^2$

77. $25m^4 - 25$

78. $9 - 9n^4$

79. $18a^5 + 84a^4b + 98a^3b^2$

80. $32b^6 + 80b^5c + 50b^4c^2$

81. $x^3 - 144x$

82. $g^3 - 121g$

83. $49t^2 - 28ts + 4s^2$

84. $81p^2 - 36pq + 4q^2$

85. $3m^4 - 3n^4$

86. $5a^4 - 80b^4$

87. $25m^2 + 70m + 49$

88. $25x^2 + 20x + 4$

89. $-100t^2 + 20t - 1$

90. $-81r^2 - 18r - 1$

91. $6x^4 - 6x^2y^2$

92. $4b^2y - 16c^2y$

93. $100a^2 + 81$

94. $25y^2 + 16$

95. $-169 + 25x^2$

96. $-196 + 49f^2$

Choose the correct method from Section 6.1, Section 6.2, Section 6.3, or Section 6.4 to factor each of the following:

97. $x^2 + x - 42$

98. $rx - sx + r - s$

99. $x^2 - 9$

100. $3a^2 - 4a - 4$

101. $24a^3b - 16a^2b$

102. $20ns^2 - 60nu + 100n$

103. $-2r^2 + 28r - 80$

104. $10s - 39 + s^2$

105. $x^3 + 3x^2 + 4x + 12$

106. $2y^2 - 128z^2$

107. $4b^2 - 20b + 25$

108. $a^2 - 4ab - 12b^2$

APPLICATIONS

109. **Genetics.** The Hardy–Weinberg equation, one of the fundamental concepts in population genetics, is $p^2 + 2pq + q^2 = 1$, where p represents the frequency of a certain dominant gene and q represents the frequency of a certain recessive gene. Factor the left side of the equation.

110. **Signal Flags.** The maritime signal flag for the letter X is shown. Find the polynomial that represents the area of the shaded region and express it in factored form.

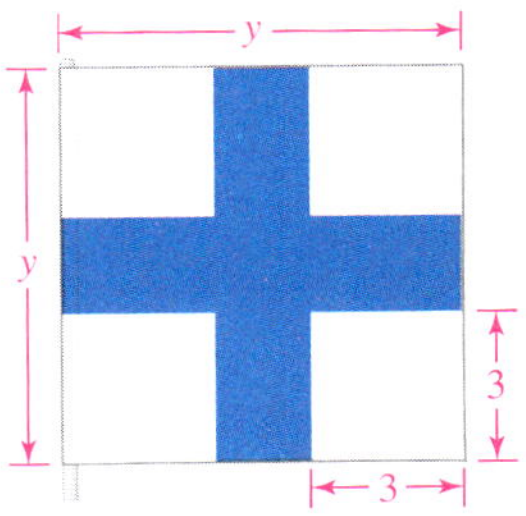

111. Physics. The illustration shows a time-sequence picture of a falling apple. Factor the expression, which gives the difference in the distance fallen by the apple during the time interval from t_1 to t_2 seconds.

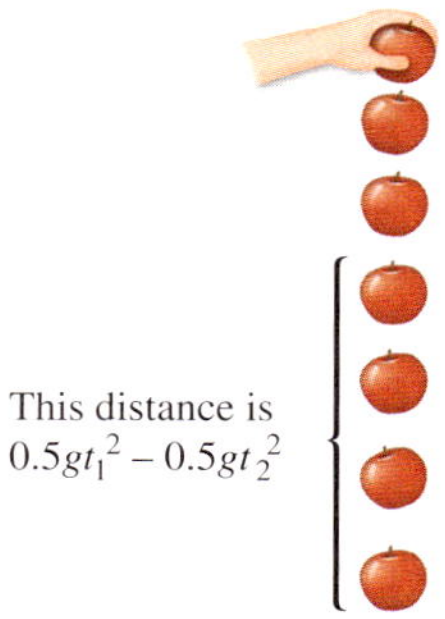

112. Darts. A circular dart board has a series of rings around a solid center, called the bullseye. To find the area of the outer black ring, we can use the formula $A = \pi R^2 - \pi r^2$. Factor the expression on the right side of the equation.

WRITING

113. When asked to factor $x^2 - 25$, one student wrote $(x + 5)(x - 5)$, and another student wrote $(x - 5)(x + 5)$. Are both answers correct? Explain.

114. Explain the error that was made in the following factorization:

$$x^2 - 100 = (x + 50)(x - 50)$$

115. Explain why the following factorization isn't complete.

$$x^4 - 625 = (x^2 + 25)(x^2 - 25)$$

116. Explain why $a^2 + 2a + 1$ is a perfect-square trinomial and why $a^2 + 4a + 1$ isn't a perfect-square trinomial.

REVIEW

Perform each division.

117. $\dfrac{-30c^2d^2 - 15c^2d - 10cd^2}{-10cd}$

118. $2a - 1\overline{)a - 2 + 6a^2}$

CHALLENGE PROBLEMS

119. For what value of c does $80x^2 - c$ factor as $5(4x + 3)(4x - 3)$?

120. Find all values of b so that $0.16x^2 + bxy + 0.25y^2$ is a perfect-square trinomial.

Factor completely.

121. $81x^6 + 36x^3y^2 + 4y^4$

122. $p^2 + p + \dfrac{1}{4}$

123. $c^2 + 1.6c + 0.64$

124. $x^{2n} - y^{4n}$

125. $(x + 5)^2 - y^2$

126. $\dfrac{1}{2} - 2a^2$

127. $c^2 - \dfrac{1}{16}$

128. $t^2 - \dfrac{9}{25}$

SECTION 6.5 Factoring the Sum and Difference of Two Cubes

OBJECTIVES

1 Factor the sum and difference of two cubes.

ARE YOU READY?

The following problems review some basic skills that are needed when factoring binomials and trinomials.

1. Multiply: $(x + 4)(x^2 - 4x + 16)$

2. Multiply: $(2h - 1)(4h^2 + 2h + 1)$

3. Evaluate: **a.** 3^3 **b.** 5^3

4. Explain why $x^2 - 2x + 4$ is prime.

In this section, we will discuss how to factor two types of binomials, called the *sum* and the *difference of two cubes*.

1 Factor the Sum and Difference of Two Cubes.

We have seen that the sum of two squares, such as $x^2 + 4$ or $25a^2 + 9b^2$, cannot be factored. However, the sum of two cubes and the difference of two cubes can be factored.

The sum of two cubes		*The difference of two cubes*	
$x^3 + 8$		$a^3 - 64b^3$	
This is x cubed.	This is 2 cubed: $2^3 = 8$.	This is a cubed.	This is $4b$ cubed: $(4b)^3 = 64b^3$.

The Language of Algebra

The expression $x^3 + y^3$ is a **sum of two cubes,** whereas $(x + y)^3$ is the **cube of a sum.** If you expand $(x + y)^3$, you will see that $(x + y)^3 \neq x^3 + y^3$.

To find rules for factoring the sum of two cubes and the difference of two cubes, we need to find the products shown below. Note that each term of the trinomial is multiplied by each term of the binomial.

$$(x + y)(x^2 - xy + y^2) = x^3 - x^2y + xy^2 + x^2y - xy^2 + y^3$$
$$= x^3 + y^3 \quad \text{Combine like terms: } -x^2y + x^2y = 0 \text{ and } xy^2 - xy^2 = 0.$$

$$(x - y)(x^2 + xy + y^2) = x^3 + x^2y + xy^2 - x^2y - xy^2 - y^3$$
$$= x^3 - y^3 \quad \text{Combine like terms.}$$

These results justify the rules for factoring the **sum** or **difference of two cubes.** They are easier to remember if we think of a sum (or a difference) of two cubes as the cube of a **F**irst quantity plus (or minus) the cube of the **L**ast quantity.

Factoring the Sum and Difference of Two Cubes

To factor the cube of a First quantity plus the cube of a Last quantity, multiply the First plus the Last by the First squared, minus the First times the Last, plus the Last squared.

$$F^3 + L^3 = (F + L)(F^2 - FL + L^2)$$

To factor the cube of a First quantity minus the cube of a Last quantity, multiply the First minus the Last by the First squared, plus the First times the Last, plus the Last squared.

$$F^3 - L^3 = (F - L)(F^2 + FL + L^2)$$

To factor the sum or difference of two cubes, it's helpful to know the cubes of integers from 1 to 10 shown in red below. The number 216, for example, is a **perfect cube,** because $216 = 6^3$.

1 $= 1^3$	**27** $= 3^3$	**125** $= 5^3$	**343** $= 7^3$	**729** $= 9^3$
8 $= 2^3$	**64** $= 4^3$	**216** $= 6^3$	**512** $= 8^3$	**1,000** $= 10^3$

EXAMPLE 1 Factor: $x^3 + 8$

Strategy We will write $x^3 + 8$ in a form that clearly shows it is the sum of two cubes.

Why We can then use the rule for factoring the sum of two cubes.

Solution $x^3 + 8$ is the sum of two cubes because it can be written as $x^3 + 2^3$. We can match it to the rule for factoring the sum of two cubes to find its factorization.

$$F^3 + L^3 = (F + L)(F^2 - F\ L + L^2)$$
$$x^3 + 2^3 = (x + 2)(x^2 - x \cdot 2 + 2^2)$$
$$= (x + 2)(x^2 - 2x + 4)$$

To write the trinomial factor:
- Square the first term of the binomial factor.
- Multiply the terms of the binomial factor.
- Square the last term of the binomial factor.

$x^2 - 2x + 4$ does not factor. It is prime.

Therefore, $x^3 + 8 = (x + 2)(x^2 - 2x + 4)$. We can check by multiplying.

$$(x + 2)(x^2 - 2x + 4) = x^3 - 2x^2 + 4x + 2x^2 - 4x + 8$$
$$= x^3 + 8 \quad \text{This is the original binomial.}$$

Caution

A common error is to try to factor $x^2 - 2x + 4$. It is not a perfect square trinomial, because the middle term needs to be $-4x$. Furthermore, it cannot be factored by the methods of Section 6.2. It is prime.

Self Check 1 Factor: $h^3 + 27$

Now Try Problem 17

Terms containing variables such as a^6, $64b^3$, and $27m^6$ are also perfect cubes, because they can be written as the cube of a quantity:

$$a^6 = (a^2)^3, \qquad 64b^3 = (4b)^3, \qquad \text{and} \qquad 27m^6 = (3m^2)^3$$

EXAMPLE 2 Factor: $a^3 - 64b^3$

Strategy We will write $a^3 - 64b^3$ in a form that clearly shows it is the difference of two cubes.

Why We can then use the rule for factoring the difference of two cubes.

Solution $a^3 - 64b^3$ is the difference of two cubes because it can be written as $a^3 - (4b)^3$. We can match it to the rule for factoring the difference of two cubes to find its factorization.

The Language of Algebra

An expression of the form $a^3 - 64b^3$ is called the **difference of two cubes,** whereas $(a - 64b)^3$ is the **cube of a difference.** They are not equivalent because $(a - 64b)^3 \neq a^3 - 64b^3$.

$$F^3 - L^3 = (F - L)(F^2 + F\,L + L^2)$$
$$a^3 - (4b)^3 = (a - 4b)[a^2 + a \cdot 4b + (4b)^2]$$
$$= (a - 4b)(a^2 + 4ab + 16b^2)$$

To write the trinomial factor:
- Square the first term of the binomial factor.
- Multiply the terms of the binomial factor.
- Square the last term of the binomial factor.

$a^2 + 4ab + 16b^2$ does not factor.

Therefore, $a^3 - 64b^3 = (a - 4b)(a^2 + 4ab + 16b^2)$. Check by multiplying.

Self Check 2 Factor: $8c^3 - 1$

Now Try Problem 37

You should memorize the rules for factoring the sum and the difference of two cubes. Note that the right side of each rule has the form

(a binomial)(a trinomial)

and that there is a relationship between the signs that appear in these forms.

The Sum of Two Cubes

$$F^3 + L^3 = (F + L)(F^2 - FL + L^2)$$

The same sign (between $F^3 + L^3$ and $F + L$); Opposite signs (between $F + L$ and $-FL$); Always plus (before L^2)

The Difference of Two Cubes

$$F^3 - L^3 = (F - L)(F^2 + FL + L^2)$$

The same sign (between $F^3 - L^3$ and $F - L$); Opposite signs (between $F - L$ and $+FL$); Always plus (before L^2)

Success Tip

An easy way to remember the sign patterns within the binomial and trinomial of these factoring forms is with the word "SOAP."

Same, **O**pposite, **A**lways **P**lus

If the terms of a binomial have a common factor, the GCF (or the opposite of the GCF) should always be factored out first.

EXAMPLE 3 Factor: $-2t^5 + 250t^2$

Strategy We will factor out the common factor, $-2t^2$. We can then factor the resulting binomial as a difference of two cubes.

Why The first step in factoring any polynomial is to factor out the GCF, or its opposite.

Solution

$$-2t^5 + 250t^2 = -2t^2(t^3 - 125) \quad \text{Factor out the opposite of the GCF, } -2t^2.$$
$$= -2t^2(t - 5)(t^2 + 5t + 25) \quad \text{Factor } t^3 - 125.$$

Therefore, $-2t^5 + 250t^2 = -2t^2(t - 5)(t^2 + 5t + 25)$. Check by multiplying.

Self Check 3 Factor: $4c^3 + 4d^3$

Now Try Problem 43

SECTION 6.5 STUDY SET

VOCABULARY

Fill in the blanks.

1. $x^3 + 27$ is the _____ of two cubes and $a^3 - 125$ is the difference of two _______.
2. The factorization of $x^3 + 8$ is $(x + 2)(x^2 - 2x + 4)$. The first factor is a binomial and the second is a _________.

CONCEPTS

Fill in the blanks.

3. a. $F^3 + L^3 = (\quad + \quad)(F^2 - FL + L^2)$
   b. $F^3 - L^3 = (F \quad L)(\quad + FL + \quad)$
4. $m^3 + 64$
   This is cubed. This is cubed.
5. $216n^3 - 125$
   This is cubed. This is cubed.
6. a. $x^3 + 64y^3 = (\quad)^3 + (\quad)^3$
   b. $8x^3 - 27 = (\quad)^3 - (\quad)^3$
7. List the first ten positive integer cubes.
8. $(x - 2)(x^2 + 2x + 4)$ is the factorization of what binomial?
9. Use multiplication to determine if the factorization is correct.
   $$b^3 + 27 = (b + 3)(b^2 + 3b + 9)$$
10. The factorization of $y^3 + 27$ is $(y + 3)(y^2 - 3y + 9)$. Is this factored completely, or does $y^2 - 3y + 9$ factor further?

NOTATION

Complete each factorization.

11. $a^3 + 8 = (a + 2)(a^2 - \quad + 4)$
12. $x^3 - 1 = (x - 1)(x^2 + \quad + 1)$
13. $b^3 + 27 = (\quad)(b^2 - 3b + 9)$
14. $z^3 - 125 = (z - 5)(\quad + 5z + \quad)$

Give an example of each type of expression.

15. a. the sum of two cubes
    b. the cube of a sum
16. a. the difference of two cubes
    b. the cube of a difference

GUIDED PRACTICE

Factor. See Example 1.

17. $y^3 + 125$
18. $b^3 + 216$
19. $a^3 + 64$
20. $n^3 + 1$
21. $n^3 + 512$
22. $t^3 + 729$
23. $8 + t^3$
24. $27 + y^3$
25. $a^3 + 1{,}000b^3$
26. $8u^3 + w^3$
27. $125c^3 + 27d^3$
28. $64m^3 + 343n^3$

Factor. See Example 2.

29. $a^3 - 27$
30. $r^3 - 8$
31. $m^3 - 343$
32. $y^3 - 216$
33. $216 - v^3$
34. $125 - t^3$
35. $8s^3 - t^3$
36. $27a^3 - b^3$
37. $1{,}000a^3 - w^3$
38. $s^3 - 64t^3$
39. $64x^3 - 27y^3$
40. $27x^3 - 1{,}000y^3$

Factor. See Example 3.

41. $2x^3 + 2$
42. $8y^3 + 8$
43. $3d^3 + 81$
44. $2x^3 + 54$
45. $x^4 - 216x$
46. $x^5 - 125x^2$
47. $64m^3x - 8n^3x$
48. $16r^4 - 128rs^3$

TRY IT YOURSELF

Choose the correct method from Section 6.1 through Section 6.5 and factor completely.

49. $x^2 + 8x + 16$
50. $64p^3 - 27$
51. $9r^2 - 16s^2$
52. $-63 - 13x + 6x^2$
53. $xy - ty + sx - st$
54. $12p^2 + 14p - 6$
55. $4p^3 + 32q^3$
56. $56a^4 - 15a^3 + a^2$
57. $16c^3t^2 + 20c^2t^3 + 6ct^4$
58. $-t^2 - 9t + 1$
59. $36e^4 - 36$
60. $3(z + 4) - a(z + 4)$
61. $35a^3b^2 - 14a^2b^3 + 14a^3b^3$
62. $-y^2 - 15y + 34$
63. $36r^2 + 60rs + 25s^2$
64. $16u^2 - 16$

Look Alikes . . .

65. a. $x^2 - 1$ b. $x^3 - 1$
66. a. $x^2 - 64$ b. $x^3 - 64$
67. a. $x^2 + 2x$ b. $x^2 + 2x + 1$
68. a. $x^2 - 4x$ b. $x^2 - 4x + 4$

APPLICATIONS

69. **Mailing Breakables.** Write a polynomial that describes the amount of space in the larger box that must be filled with styrofoam chips if the smaller box containing a glass tea cup is to be placed within the larger box for mailing. Then factor the polynomial.

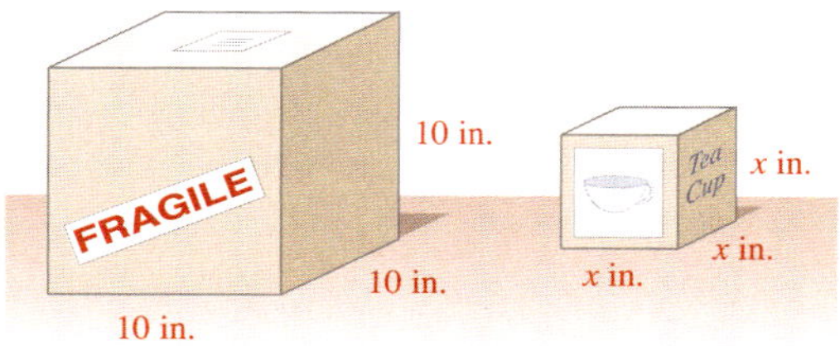

70. **Melting Ice.** In one hour, the block of ice shown below had melted to the size shown on the right. Write a polynomial that describes the volume of ice that melted away. Then factor the polynomial.

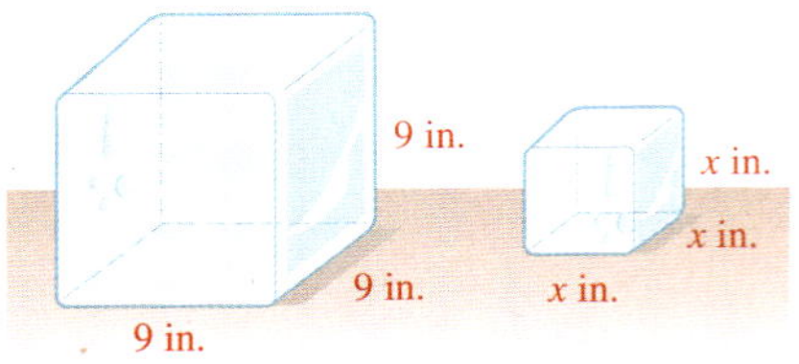

WRITING

71. Explain why $x^3 - 25$ is not a difference of two cubes.
72. Explain this diagram. Then draw a similar diagram for the difference of two cubes.

The same

$$F^3 + L^3 = (F + L)(F^2 - FL + L^2)$$

Opposite Always plus

REVIEW

73. When expressed as a decimal, is $\frac{7}{9}$ a terminating or a repeating decimal?
74. Solve: $x + 20 = 4x - 1 + 2x$
75. Solve: $2x + 2 = \frac{2}{3}x - 2$
76. Check to determine whether 4 is a solution of $3(m - 8) + 2m = 4 - (m + 2)$.

CHALLENGE PROBLEMS

77. Consider: $x^6 - 1$. Write the binomial as a difference of two squares. Then factor.
78. What binomial multiplied by $(a^2b^2 + 7ab + 49)$ produces a difference of two cubes?

Factor.

79. $x^6 - y^9$
80. $\frac{125}{8}s^3 + \frac{1}{27}t^3$
81. $64x^{12} + y^{15}z^{18}$
82. $x^{3m} - y^{3n}$

SECTION 6.6 A Factoring Strategy

OBJECTIVE

1 Use a general strategy for factoring polynomials.

ARE YOU READY?

The following problems review some basic skills that are needed when factoring polynomials.

1. How many terms does each expression have?
   a. $x^4 - 5x^3 + x - 5$
   b. $16x - 5x^2 + 6x^3$
2. Do any of the factors in $(m^2 + 9)(m + 3)(m - 3)$ factor further?
3. Multiply: $3n(n + 5)(n - 14)$
4. What is the greatest common factor of the terms of $15c^3d^2 - 10cd^3$?

The factoring methods discussed so far will be used in the remaining chapters to simplify expressions and solve equations. In such cases, we must determine the factoring method—it will not be specified. This section will give you practice in selecting the appropriate factoring method to use given a randomly chosen polynomial.

1 Use a General Strategy for Factoring Polynomials.

The following strategy is helpful when factoring polynomials.

Steps for Factoring a Polynomial

1. Is there a common factor? If so, factor out the GCF, or the opposite of the GCF so that the leading coefficient is positive. Remember to include it in your final answer.
2. How many terms does the polynomial have?

   If it has **two terms,** look for the following problem types:
   a. The difference of two squares
   b. The sum of two cubes
   c. The difference of two cubes

   If it has **three terms,** look for the following problem types:
   a. A perfect-square trinomial
   b. If the trinomial is not a perfect square, use the trial-and-check-method or the grouping method.

   If it has **four or more terms,** try to factor by grouping.
3. Can any factors be factored further? If so, factor them completely.
4. Does the factorization check? Check by multiplying.

EXAMPLE 1 Factor: $2x^4 - 162$

Strategy We will answer the four questions listed in the *Steps for Factoring a Polynomial.*

Why The answers to these questions help us determine which factoring techniques to use.

Solution ***Is there a common factor?*** Yes. Factor out the GCF, which is 2.

$$2x^4 - 162 = 2(x^4 - 81)$$

The Language of Algebra

Recall that **to factor a polynomial** means to express it as a product of two (or more) polynomials.

How many terms does it have? The polynomial within the parentheses, $x^4 - 81$, has two terms. It is a difference of two squares.

$$2x^4 - 162 = 2(x^4 - 81) \quad \text{Think of } x^4 - 81 \text{ as } (x^2)^2 - 9^2.$$
$$= 2(x^2 + 9)(x^2 - 9) \quad \text{Factor the difference of two squares.}$$

Is it factored completely? No. $x^2 - 9$ is also the difference of two squares and can be factored.

$$2x^4 - 162 = 2(x^4 - 81)$$
$$= 2(x^2 + 9)(x^2 - 9) \quad \text{Think of } x^2 - 9 \text{ as } x^2 - 3^2.$$
$$= 2(x^2 + 9)(x + 3)(x - 3) \quad x^2 + 9 \text{ is a sum of two squares and does not factor.}$$

The Language of Algebra

Remember that the instruction to **factor** means to *factor completely.* A polynomial is *factored completely* when no factor can be factored further.

Therefore, $2x^4 - 162 = 2(x^2 + 9)(x + 3)(x - 3)$.

Does it check? Yes.

$$2(x^2 + 9)(x + 3)(x - 3) = 2(x^2 + 9)(x^2 - 9) \quad \text{Multiply } (x + 3)(x - 3) \text{ first.}$$
$$= 2(x^4 - 81) \quad \text{Multiply } (x^2 + 9)(x^2 - 9).$$
$$= 2x^4 - 162 \quad \text{This is the original polynomial.}$$

Self Check 1 Factor: $11a^6 - 11a^2$

Now Try Problem 21

EXAMPLE 2 Factor: $-4c^5d^2 - 12c^4d^3 - 9c^3d^4$

Strategy We will answer the four questions listed in the *Steps for Factoring a Polynomial.*

Why The answers to these questions help us determine which factoring techniques to use.

Solution ***Is there a common factor?*** Yes. Factor out the opposite of the GCF, $-c^3d^2$, so that the leading coefficient is positive.

$$-4c^5d^2 - 12c^4d^3 - 9c^3d^4 = -c^3d^2(4c^2 + 12cd + 9d^2)$$

How many terms does it have? The polynomial within the parentheses has three terms. It is a perfect-square trinomial because $4c^2 = (2c)^2$, $9d^2 = (3d)^2$, and $12cd = 2 \cdot 2c \cdot 3d$.

$$-4c^5d^2 - 12c^4d^3 - 9c^3d^4 = -c^3d^2(4c^2 + 12cd + 9d^2)$$
$$= -c^3d^2(2c + 3d)^2$$

Success Tip

Recall that if a pair of parentheses is produced in the factoring process, always see if the resulting expression within the parentheses can be factored further.

Is it factored completely? Yes. The binomial $2c + 3d$ does not factor further.

Therefore, $-4c^5d^2 - 12c^4d^3 - 9c^3d^4 = -c^3d^2(2c + 3d)^2$.

Does it check? Yes.

$$-c^3d^2(2c + 3d)^2 = -c^3d^2(4c^2 + 12cd + 9d^2) \quad \text{Use a special-product rule.}$$
$$= -4c^5d^2 - 12c^4d^3 - 9c^3d^4 \quad \text{This is the original polynomial.}$$

Self Check 2 Factor: $-32h^4 - 80h^3 - 50h^2$

Now Try Problem 33

EXAMPLE 3 Factor: $y^4 - 3y^3 + y - 3$

Strategy We will answer the four questions listed in the *Steps for Factoring a Polynomial.*

Why The answers to these questions help us determine which factoring techniques to use.

Solution ***Is there a common factor?*** No. There is no common factor (other than 1).

How many terms does it have? Since the polynomial has four terms, we will try factoring by grouping.

$$y^4 - 3y^3 + y - 3 = y^3(y - 3) + 1(y - 3)$$ Factor out y^3 from $y^4 - 3y^3$. Factor out 1 from $y - 3$.

$$= (y - 3)(y^3 + 1)$$

Is it factored completely? No. We can factor $y^3 + 1$ as a sum of two cubes.

$$y^4 - 3y^3 + y - 3 = y^3(y - 3) + 1(y - 3)$$

$$= (y - 3)(y^3 + 1)$$ Think of $y^3 + 1$ as $y^3 + 1^3$.

$$= (y - 3)(y + 1)(y^2 - y + 1)$$ $y^2 - y + 1$ does not factor further.

Therefore, $y^4 - 3y^3 + y - 3 = (y - 3)(y + 1)(y^2 - y + 1)$.

Does it check? Yes.

$$(y - 3)(y + 1)(y^2 - y + 1) = (y - 3)(y^3 + 1)$$ Multiply the last two factors.

$$= y^4 + y - 3y^3 - 3$$ Use the FOIL method.

$$= y^4 - 3y^3 + y - 3$$ This is the original polynomial.

Self Check 3 Factor: $b^4 + b^3 + 8b + 8$

Now Try Problem 37

Success Tip

Something as simple as counting the number of terms that a polynomial has is very important when determining how to factor it.

EXAMPLE 4 Factor: $32n - 4n^2 + 4n^3$

Strategy We will answer the four questions listed in the *Steps for Factoring a Polynomial.*

Why The answers to these questions help us determine which factoring techniques to use.

Solution ***Is there a common factor?*** Yes. When we write the terms in descending powers of n, we see that the GCF is $4n$.

$$4n^3 - 4n^2 + 32n = 4n(n^2 - n + 8)$$

How many terms does it have? The polynomial within the parentheses has three terms. It is not a perfect-square trinomial because the last term, 8, is not a perfect square.

Negative factors of 8	Sum of the negative factors of 8
$-1(-8) = 8$	$-1 + (-8) = -9$
$-2(-4) = 8$	$-2 + (-8) = -10$

To factor the trinomial $n^2 - n + 8$, we must find two integers whose product is 8 and whose sum is -1. As we see in the table, there are no such integers. Thus, $n^2 - n + 8$ is prime.

Is it factored completely? Yes.

Therefore, $4n^3 - 4n^2 + 32n = 4n(n^2 - n + 8)$. Remember to write the GCF, $4n$, from the first step.

Does it check? Yes.

$$4n(n^2 - n + 8) = 4n^3 - 4n^2 + 32n$$ This is equivalent to the original polynomial.

Self Check 4 Factor: $6m^2 - 54m + 6m^3$

Now Try Problem 45

EXAMPLE 5 Factor: $3y^3 - 4y^2 - 4y$

Strategy We will answer the four questions listed in the *Steps for Factoring a Polynomial.*

Why The answers to these questions help us determine which factoring techniques to use.

Solution ***Is there a common factor?*** Yes. The GCF is y.

$$3y^3 - 4y^2 - 4y = y(3y^2 - 4y - 4)$$

How many terms does it have? The polynomial within the parentheses has three terms. It is not a perfect-square trinomial because the first term, $3y^2$, is not a perfect square.

If we use grouping to factor $3y^2 - 4y - 4$, the key number is $ac = 3(-4) = -12$. We must find two integers whose product is -12 and whose sum is $b = -4$.

The Language of Algebra

It is important to learn the names of all the factoring methods we have studied. Educational research has found that words play an important role in concept formation.

Key number $= -12$ $\qquad$ $b = -4$

Factors of -12	Sum of the factors of -12
$2(-6) = -12$	$2 + (-6) = -4$

From the table, the correct pair is 2 and -6. They serve as the coefficients of $2y$ and $-6y$, the two terms that we use to represent the middle term, $-4y$, of the trinomial.

$$3y^2 - 4y - 4 = 3y^2 + 2y - 6y - 4$$ Express $-4y$ as $2y - 6y$.

$$= y(3y + 2) - 2(3y + 2)$$ Factor y from the first two terms and factor -2 from the last two terms.

$$= (3y + 2)(y - 2)$$ Factor out $3y + 2$.

The trinomial $3y^2 - 4y - 4$ factors as $(3y + 2)(y - 2)$.

Is it factored completely? Yes. Because $3y + 2$ and $y - 2$ do not factor.

Therefore, $3y^3 - 4y^2 - 4y = y(3y + 2)(y - 2)$. Remember to write the GCF, y, from the first step.

Does it check? Yes.

$$y(3y + 2)(y - 2) = y(3y^2 - 4y - 4)$$ Multiply the binomials.

$$= 3y^3 - 4y^2 - 4y$$ This is the original polynomial.

Self Check 5 Factor: $6y^3 + 21y^2 - 12y$

Now Try Problem 67

SECTION 6.6 STUDY SET

VOCABULARY

Fill in the blanks.

1. To factor a polynomial means to express it as a ________ of two (or more) polynomials.
2. A polynomial is factored ________ when each factor is prime.

CONCEPTS

For each of the following polynomials, which factoring method would you use first?

3. $2x^5y - 4x^3y$
4. $9b^2 + 12y - 5$
5. $x^2 + 18x + 81$
6. $ax + ay - x - y$
7. $x^3 + 27$
8. $y^3 - 64$
9. $m^2 - 3mn + 2n^2$
10. $16 - 25z^2$
11. What is the first question that should be asked when using the strategy of this section to factor a polynomial?
12. Use multiplication to determine whether the factorization is correct.

$$5c^3d^2 - 40c^2d^3 + 35cd^4 = 5cd^2(c - 7d)(c - d)$$

NOTATION

Complete each factorization.

13. $6m^3 - 28m^2 + 16m = 2m(3m^2 - \quad + 8)$
$= 2m(3m - 2)(\quad - 4)$
14. $2a^3 + 3a^2 - 2a - 3$
$= \quad (2a + 3) - 1(\quad + 3)$
$= (\quad)(a^2 - 1)$
$= (2a + 3)(a + 1)(\quad)$

TRY IT YOURSELF

The following is a list of random factoring problems. Factor each expression. If an expression is not factorable, write "prime." See Examples 1–5.

15. $2b^2 + 8b - 24$
16. $32 - 2t^4$
17. $8p^3q^7 + 4p^2q^3$
18. $8m^2n^3 - 24mn^4$
19. $2 + 24y + 40y^2$
20. $6r^2 + 3rs - 18s^2$
21. $8x^4 - 8$
22. $t - 90 + t^2$
23. $14c - 147 + c^2$
24. $ab^2 - 4a + 3b^2 - 12$
25. $x^2 + 7x + 1$
26. $3a^3 + 24b^3$
27. $-2x^5 + 128x^2$
28. $16 - 40z + 25z^2$
29. $a^2c + a^2d^2 + bc + bd^2$
30. $6t^4 + 14t^3 - 40t^2$
31. $-9x^2 + 6x - 1$
32. $x^2y^2 - 2x^2 - y^2 + 2$
33. $-20m^3 - 100m^2 - 125m$
34. $5x^3y^3z^4 + 25x^2y^4z^2 - 35x^3y^2z^5$
35. $2c^2 - 5cd - 3d^2$
36. $125p^3 - 64y^3$
37. $p^4 - 2p^3 - 8p + 16$
38. $a^2 + 8a + 3$
39. $a^2(x - a) - b^2(x - a)$
40. $70p^4q^3 - 35p^4q^2 + 49p^5q^2$
41. $a^2b^2 - 144$
42. $-16x^4y^2z + 24x^5y^3z^4 - 15x^2y^3z^7$
43. $2x^3 + 10x^2 + x + 5$
44. $u^2 - 18u + 81$
45. $8v^2 - 14v^3 + v^4$
46. $28 - 3m - m^2$
47. $18a^2 - 6ab + 42ac - 14bc$
48. $81r^4 - 256$
49. $8a^2x^3 - 2b^2x$
50. $12x^2 + 14x - 6$
51. $6x^2 - 14x + 8$
52. $12x^2 - 12$
53. $4x^2y^2 + 4xy^2 + y^2$
54. $81r^4s^2 - 24rs^5$
55. $4m^5 + 500m^2$
56. $ae + bf + af + be$
57. $a^3 - 24 - 4a + 6a^2$
58. $6x^2 - x - 16$
59. $4x^2 + 9y^2$
60. $x^4y + 216xy^4$
61. $16a^5 - 54a^2$
62. $25x^2 - 16y^2$
63. $27x - 27y - 27z$
64. $12x^2 + 52x + 35$
65. $xy - ty + xs - ts$
66. $bc + b + cd + d$
67. $35x^8 - 2x^7 - x^6$
68. $x^3 - 25$
69. $5(x - 2) + 10y(x - 2)$
70. $16x^2 - 40x^3 + 25x^4$
71. $49p^2 + 28pq + 4q^2$
72. $16d^2 - 56dz + 49z^2$
73. $4t^2 + 36$
74. $r^5 + 3r^3 + 2r^2 + 6$
75. $m^2n^2 - 9m^2 + 3n^2 - 27$
76. $z^2 + 6yz^2 + 9y^2z^2$

WRITING

77. Which factoring method do you find the most difficult? Why?
78. What four questions make up the factoring strategy for polynomials discussed in this section?
79. What does it mean to factor a polynomial?
80. How is a factorization checked?

REVIEW

81. Graph the real numbers $-3, 0, 2,$ and $-\frac{3}{2}$ on a number line.
82. Graph the interval $(-2, 3]$ on a number line.
83. Graph: $y = \frac{1}{2}x + 1$
84. Graph: $y < 2 - 3x$

CHALLENGE PROBLEMS

Factor using rational numbers.

85. $x^6 - 4x^3 - 12$
86. $x(x - y) - y(y - x)$
87. $24 - x^3 + 8x^2 - 3x$
88. $25b^2 + 14b + \frac{49}{25}$
89. $x^9 + y^6$
90. $\frac{1}{4} - \frac{u^2}{81}$
91. $x^4 - 13x^2 + 36$
92. $x^4 - 2x^2 - 8$
93. $x^2y^2 - 6xy - 16$
94. $5x + 4y + 25x^2 - 16y^2$

SECTION 6.7

Solving Quadratic Equations by Factoring

OBJECTIVES

1. Define quadratic equations.
2. Solve quadratic equations using the zero-factor property.
3. Solve third-degree equations by factoring.

ARE YOU READY?

The following problems review some basic skills that are needed when solving quadratic equations.

1. Evaluate: $0 \cdot 5$
2. Fill in the blank: $8 \cdot \quad = 0$
3. Solve: $x + 4 = 0$
4. Solve: $8x = 0$
5. Factor: $x^2 - x - 6$
6. Factor: $3n^2 - n - 2$

The factoring methods that we have discussed have many applications in algebra. In this section, we will use factoring to solve *quadratic equations.* These equations are different from those that we solved in Chapter 2. They contain a term in which the variable is raised to the second power, such as x^2 or t^2.

1 Define Quadratic Equations.

In a linear, or first degree equation, such as $2x + 3 = 8$, the exponent on the variable is an unwritten 1. A quadratic, or second degree equation, has a term in which the exponent on the variable is 2, and has no other terms of higher degree.

Quadratic Equations

A **quadratic equation** is an equation that can be written in the **standard form**

$$ax^2 + bx + c = 0$$

where a, b, and c represent real numbers, and $a \neq 0$.

The Language of Algebra

Quadratic equations involve the square of a variable, not the 4th power as *quad* might suggest. Why is this? Because the origin of the word **quadratic** is the Latin word *quadratus,* meaning square.

Some examples of quadratic equations are

$$x^2 - 2x - 63 = 0, \quad x^2 - 25 = 0, \quad 6x^2 - 12x = 0, \quad \text{and} \quad 2x^2 + 3x = 2$$

The first three equations are in standard form. To write the fourth equation in standard form, we subtract 2 from both sides to get $2x^2 + 3x - 2 = 0$.

Even though it does not have an x-term, the equation $x^2 - 25 = 0$ is a quadratic equation because the definition allows b, the coefficient of x, to equal 0. And even though the equation $6x^2 - 12x = 0$ does not have a constant term, it is also a quadratic equation, because the definition allows c, the constant term, to equal 0. However, a quadratic equation must have a variable-squared term, because the definition requires that $a \neq 0$.

2 Solve Quadratic Equations Using the Zero-Factor Property.

To **solve a quadratic equation,** we find all values of the variable that make the equation true. The methods that we used to solve linear equations in Chapter 2 cannot be used to solve a quadratic equation, because we cannot isolate the variable on one side of the equation. However, we often can solve quadratic equations using factoring and the following property of real numbers.

The Zero-Factor Property

When the product of two real numbers is 0, at least one of them is 0.

If a and b represent real numbers, and

if $ab = 0$, then $a = 0$ or $b = 0$

EXAMPLE 1 Solve: $(4x - 1)(x + 6) = 0$

Strategy We will set $4x - 1$ equal to 0 and $x + 6$ equal to 0 and solve each equation.

Why If the product of $4x - 1$ and $x + 6$ is 0, then, by the zero-factor property, $4x - 1$ must equal 0, or $x + 6$ must equal 0.

Solution If $(4x - 1)(x + 6) = 0$ is to be a true statement, then either

$$4x - 1 = 0 \quad \text{or} \quad x + 6 = 0$$

Now we solve each of these linear equations using the methods from Chapter 2.

$4x - 1 = 0$ or $x + 6 = 0$

$4x = 1$ Add 1 to both sides. | $x = -6$ Subtract 6 from both sides.

$x = \frac{1}{4}$ Divide both sides by 4.

The results must be checked separately to see whether each of them produces a true statement. We substitute $\frac{1}{4}$ and then -6 for x in the original equation and evaluate the left side.

Check $\frac{1}{4}$:

$$(4x - 1)(x + 6) = 0$$

$$\left[4\left(\frac{1}{4}\right) - 1\right]\left(\frac{1}{4} + 6\right) \stackrel{?}{=} 0$$

$$(1 - 1)\left(\frac{25}{4}\right) \stackrel{?}{=} 0$$

$$0\left(\frac{25}{4}\right) \stackrel{?}{=} 0$$ The factor $4x - 1$ is 0 when x is $\frac{1}{4}$.

$$0 = 0$$ True

Check -6:

$$(4x - 1)(x + 6) = 0$$

$$[4(-6) - 1](-6 + 6) \stackrel{?}{=} 0$$

$$(-24 - 1)(0) \stackrel{?}{=} 0$$ The factor $x + 6$ is 0 when x is -6.

$$-25(0) \stackrel{?}{=} 0$$

$$0 = 0$$ True

The resulting true statements indicate that $(4x - 1)(x + 6) = 0$ has two solutions: $\frac{1}{4}$ and -6. Recall from Chapter 2 that the *solution set* of an equation is the set of all numbers that make the equation true. Thus, the solution set is $\left\{-6, \frac{1}{4}\right\}$.

Self Check 1 Solve: $(x - 12)(5x + 6) = 0$

Now Try Problem 15

Caution

It would not be helpful to multiply $(4x - 1)$ and $(x + 6)$. We want the left side of the equation to be in factored form so that we can use the zero-factor property.

The Language of Algebra

In the zero-factor property, the word **or** means one or the other or both. If the product of two numbers is 0, then one factor is 0, or the other factor is 0, or both factors can be 0.

In Example 1, the left side of $(4x - 1)(x + 6) = 0$ is in factored form and the right side is 0, so we can immediately use the zero-factor property. However, to solve many quadratic equations, we must factor before using the zero-factor property.

EXAMPLE 2 Solve: $x^2 - 2x - 63 = 0$

Strategy We will factor the trinomial on the left side of the equation and use the zero-factor property.

Why To use the zero-factor property, we need one side of the equation to be factored completely and the other side to be 0.

Solution

Success Tip

When you see the word **solve** in this example, you probably think of steps from Chapter 2 such as combining like terms, distributing, or doing something to both sides. However, to solve this quadratic equation, we begin by factoring $x^2 - 2x - 63$.

$x^2 - 2x - 63 = 0$ This is the equation to solve.

$(x + 7)(x - 9) = 0$ Factor the trinomial, $x^2 - 2x - 63$.

$x + 7 = 0$ or $x - 9 = 0$ Set each factor equal to 0.

$x = -7$ | $x = 9$ Solve each equation using the methods from Chapter 2.

To check the results, we substitute -7 and then 9 for x in the original equation and evaluate the left side.

Check −7:

$$x^2 - 2x - 63 = 0$$
$$(-7)^2 - 2(-7) - 63 \stackrel{?}{=} 0$$
$$49 - (-14) - 63 \stackrel{?}{=} 0$$
$$63 - 63 \stackrel{?}{=} 0$$
$$0 = 0 \quad \text{True}$$

Check 9:

$$x^2 - 2x - 63 = 0$$
$$(9)^2 - 2(9) - 63 \stackrel{?}{=} 0$$
$$81 - 18 - 63 \stackrel{?}{=} 0$$
$$63 - 63 \stackrel{?}{=} 0$$
$$0 = 0 \quad \text{True}$$

The solutions of $x^2 - 2x - 63 = 0$ are -7 and 9, and the solution set is $\{-7, 9\}$.

Self Check 2 Solve: $x^2 + 5x + 6 = 0$

Now Try Problem 27

The previous examples suggest the following strategy to solve quadratic equations by factoring.

The Factoring Method for Solving a Quadratic Equation

1. Write the equation in standard form: $ax^2 + bx + c = 0$ or $0 = ax^2 + bx + c$.
2. Factor completely.
3. Use the zero-factor property to set each factor equal to 0.
4. Solve each resulting equation.
5. Check the results in the original equation.

With this method, we factor *expressions* to solve *equations*.

EXAMPLE 3 Solve: $x^2 - 25 = 0$

Strategy We will factor the binomial on the left side of the equation and use the zero-factor property.

Why To use the zero-factor property, we need one side of the equation to be factored completely and the other side to be 0.

Solution We factor the difference of two squares on the left side of the equation and proceed as follows.

Notation

Although $x^2 - 25 = 0$ is missing a term involving x, it is a quadratic equation in standard $ax^2 + bx + c = 0$ form, where $a = 1$, $b = 0$, and $c = -25$.

$$x^2 - 25 = 0 \quad \text{This is the equation to solve.}$$

$$(x + 5)(x - 5) = 0 \quad \text{Factor the difference of two squares, } x^2 - 25.$$

$$x + 5 = 0 \quad \text{or} \quad x - 5 = 0 \quad \text{Set each factor equal to 0.}$$

$$x = -5 \quad \Big| \quad x = 5 \quad \text{Solve each equation.}$$

Check each result by substituting it into the original equation.

***Check* −5:**	***Check* 5:**
$x^2 - 25 = 0$	$x^2 - 25 = 0$
$(-5)^2 - 25 \stackrel{?}{=} 0$	$5^2 - 25 \stackrel{?}{=} 0$
$25 - 25 \stackrel{?}{=} 0$	$25 - 25 \stackrel{?}{=} 0$
$0 = 0$ True	$0 = 0$ True

The solutions of $x^2 - 25 = 0$ are -5 and 5, and the solution set is $\{-5, 5\}$.

Self Check 3 Solve: $x^2 - 49 = 0$

Now Try Problem 35

EXAMPLE 4 Solve: $6x^2 = 12x$

Strategy We will subtract $12x$ from both sides of the equation to get 0 on the right side. Then we will factor the resulting binomial and use the zero-factor property.

Why To use the zero-factor property, we need one side of the equation to be factored completely and the other side to be 0.

Solution The equation is not in standard form, $ax^2 + bx + c = 0$. To get 0 on the right side, we proceed as follows.

Caution

A creative, but incorrect, approach to solve $6x^2 = 12x$ is to divide both sides by $6x$.

$$\frac{6x^2}{6x} = \frac{12x}{6x}$$

You will obtain $x = 2$; however, you will lose the second solution, 0.

$$6x^2 = 12x \quad \text{This is the equation to solve.}$$

$$6x^2 - 12x = 12x - 12x \quad \text{Use the subtraction property of equality to get 0 on the right side: Subtract } 12x \text{ from both sides.}$$

$$6x^2 - 12x = 0 \quad \text{Combine like terms: } 12x - 12x = 0. \text{ Although it is missing a constant term, this is a quadratic equation in standard } ax^2 + bx + c = 0 \text{ form, where } a = 6, b = -12, \text{ and } c = 0.$$

To solve this equation, we factor the left side and proceed as follows.

$$6x(x - 2) = 0 \quad \text{Factor out the GCF, } 6x.$$

$$6x = 0 \quad \text{or} \quad x - 2 = 0 \quad \text{Set each factor equal to 0.}$$

$$x = \frac{0}{6} \quad \Big| \quad x = 2 \quad \text{Solve each equation using the methods from Chapter 2.}$$

$$x = 0$$

The solutions are 0 and 2 and the solution set is $\{0, 2\}$. Check each solution in the original equation, $6x^2 = 12x$.

Self Check 4 Solve: $5x^2 = 25x$

Now Try Problem 47

EXAMPLE 5 Solve: $2x^2 - 2 = -3x$

Strategy We will add $3x$ to both sides of the equation to get 0 on the right side. Then we will factor the resulting trinomial and use the zero-factor property.

Why To use the zero-factor property, we need one side of the equation to be factored completely and the other side to be 0.

Solution The equation is not in standard form, $ax^2 + bx + c = 0$. To get 0 on the right side, we proceed as follows.

$2x^2 - 2 = -3x$ — This is the equation to solve.

$2x^2 + 3x - 2 = -3x + 3x$ — Use the addition property of equality to get 0 on the right side: Add $3x$ to both sides.

$2x^2 + 3x - 2 = 0$ — Combine like terms: $-3x + 3x = 0$. This equation is in standard form.

$(2x - 1)(x + 2) = 0$ — Factor the trinomial.

$2x - 1 = 0$ or $x + 2 = 0$ — Set each factor equal to 0.

$2x = 1$ | $x = -2$ — Solve each equation using the methods from Chapter 2.

$x = \frac{1}{2}$

The solutions are $\frac{1}{2}$ and -2 and the solution set is $\left\{-2, \frac{1}{2}\right\}$. Check each solution in the original equation, $2x^2 - 2 = -3x$.

Self Check 5 Solve: $3x^2 - 8 = -10x$

Now Try Problem 51

Unlike linear equations, quadratic equations have two solutions. In some cases, however, the two solutions are the same number.

EXAMPLE 6 Solve: $x(9x - 12) = -4$

Strategy To write the equation in standard form, we will distribute the multiplication by x and add 4 to both sides. Then we will factor the resulting trinomial and use the zero-factor property.

Why To use the zero-factor property, we need one side of the equation to be factored completely and the other side to be 0.

Solution

$x(9x - 12) = -4$ — This is the equation to solve.

$9x^2 - 12x = -4$ — Distribute the multiplication by x.

$9x^2 - 12x + 4 = -4 + 4$ — To get 0 on the right side, add 4 to both sides.

$9x^2 - 12x + 4 = 0$ — Combine like terms: $-4 + 4 = 0$. This equation is in standard form.

$(3x - 2)(3x - 2) = 0$ — Factor the trinomial, $9x^2 - 12x + 4$.

$3x - 2 = 0$ or $3x - 2 = 0$ — Set each factor equal to 0.

$3x = 2$ | $3x = 2$ — Solve each equation using the methods from Chapter 2.

$x = \frac{2}{3}$ | $x = \frac{2}{3}$

After solving both equations, we see that $\frac{2}{3}$ is a **repeated solution.** Thus, the solution set is $\left\{\frac{2}{3}\right\}$. Check this result by substituting it into the original equation.

Caution

To use the zero-factor property, one side of the equation must be 0. In this example, it would be incorrect to set each factor equal to -4.

~~$x = -4$ or $9x - 12 = -4$~~

If the product of two numbers is -4, one of them does not have to be -4. For example, $2(-2) = -4$.

Self Check 6 Solve: $x(4x + 12) = -9$

Now Try Problem 59

3 Solve Third-Degree Equations by Factoring.

Some equations involving polynomials with degrees higher than 2 also can be solved by using the factoring method. In such cases, we use an extension of the zero-factor property: When the product of two *or more* real numbers is 0, at least one of them is 0.

EXAMPLE 7 Solve: $6x^3 + 12x = 17x^2$

Strategy This equation is not quadratic, because it contains a term involving x^3. However, we can solve it by using factoring. First we get 0 on the right side by subtracting $17x^2$ from both sides. Then we factor the polynomial on the left side and use an extension of the zero-factor property.

Why To use the zero-factor property, we need one side of the equation to be factored completely and the other side to be 0.

Solution

$$6x^3 + 12x = 17x^2$$ This is the equation to solve.

$$6x^3 - 17x^2 + 12x = 17x^2 - 17x^2$$ To get 0 on the right side, subtract $17x^2$ from both sides.

$$6x^3 - 17x^2 + 12x = 0$$ Combine like terms: $17x^2 - 17x^2 = 0$.

$$x(6x^2 - 17x + 12) = 0$$ Factor out the GCF, x.

$$x(2x - 3)(3x - 4) = 0$$ Factor the trinomial, $6x^2 - 17x + 12$.

If $x(2x - 3)(3x - 4) = 0$, then at least one of the factors is equal to 0.

$$x = 0 \quad \text{or} \quad 2x - 3 = 0 \quad \text{or} \quad 3x - 4 = 0$$ Set each factor equal to 0.

$$2x = 3 \qquad 3x = 4$$ Solve each equation.

$$x = \frac{3}{2} \qquad x = \frac{4}{3}$$

The solutions are 0, $\frac{3}{2}$, and $\frac{4}{3}$ and the solution set is $\left\{0, \frac{4}{3}, \frac{3}{2}\right\}$. Check each solution in the original equation, $6x^3 + 12x = 17x^2$.

The Language of Algebra

Since the highest degree of any term in $6x^3 + 12x = 17x^2$ is 3, it is called a **third-degree** equation. Note that it has three solutions.

Self Check 7 Solve: $10x^3 + x^2 = 2x$

Now Try Problem 63

SECTION 6.7 STUDY SET

VOCABULARY

Fill in the blanks.

1. $2x^2 + 3x - 1 = 0$ and $x^2 - 36 = 0$ are examples of ________ equations.
2. $ax^2 + bx + c = 0$ is called the ________ form of a quadratic equation.
3. The ________ property states that if the product of two numbers is 0, at least one of them is 0: If $ab = 0$, then $a =$ or $b =$.
4. Since the highest degree of any term in $x^3 - 5x^2 - 6x = 0$ is 3, it is called a ______-degree equation.

CONCEPTS

5. Which of the following are quadratic equations?
   a. $x^2 + 2x - 10 = 0$
   b. $2x - 10 = 0$
   c. $x^2 = 15x$
   d. $x^3 + x^2 + 2x = 0$
6. Write each equation in the standard form $ax^2 + bx + c = 0$.
   a. $x^2 + 2x = 6$
   b. $x^2 = 5x$
   c. $3x(x - 8) = -9$
   d. $4x^2 = 25$
7. Set $5x + 4$ equal to 0 and solve for x.
8. What step should be performed first to solve $x^2 - 6x - 16 = 0$?
9. What step (or steps) should be performed first before factoring is used to solve each equation?
   a. $x^2 + 7x = -6$
   b. $x(x + 7) = 3$
10. Check to determine whether the given number is a solution of the given quadratic equation.
    a. $x^2 - 4x = 0$; 4
    b. $x^2 - 2x - 7 = 0$; -2

NOTATION

Complete each solution to solve the equation.

11. $(x - 1)(x + 7) = 0$

$x - 1 = \quad$ or $\quad = 0$

$x = 1 \quad | \quad x =$

12. $7y^2 + 14y = 0$

$(y + 2) = 0$

$7y = 0 \quad | \quad y + 2 = 0$

$y = \quad | \quad y = -2$

13. $p^2 - p - 6 = 0$

$(\quad - 3)(p + 2) = 0$

$= 0$ or $p + 2 =$

$p = \quad | \quad p =$

14. $4y^2 - 25 = 0$

$(2y + \quad)(2y - \quad) = 0$

$2y + 5 = \quad$ or $\quad = 0$

$2y = \quad | \quad 2y = 5$

$y = -\frac{5}{2} \quad | \quad y =$

GUIDED PRACTICE

Solve each equation. See Example 1.

15. $(x - 3)(x - 2) = 0$
16. $(x + 2)(x + 3) = 0$
17. $(x + 7)(x - 7) = 0$
18. $(x - 8)(x + 8) = 0$
19. $6x(2x - 5) = 0$
20. $5x(5x + 7) = 0$
21. $-7a(3a + 10) = 0$
22. $-6t(2t - 9) = 0$
23. $t(t - 6)(t + 8) = 0$
24. $n(n + 1)(n - 6) = 0$
25. $(x - 1)(x + 2)(x - 3) = 0$
26. $(x + 2)(x + 3)(x - 4) = 0$

Solve each equation. See Example 2.

27. $x^2 - 13x + 12 = 0$
28. $x^2 + 7x + 6 = 0$
29. $x^2 - 4x - 21 = 0$
30. $x^2 + 2x - 15 = 0$
31. $x^2 - 9x + 8 = 0$
32. $x^2 - 14x + 45 = 0$
33. $a^2 + 8a + 15 = 0$
34. $a^2 - 17a + 60 = 0$

Solve each equation. See Example 3.

35. $x^2 - 81 = 0$
36. $x^2 - 36 = 0$
37. $t^2 - 25 = 0$
38. $m^2 - 49 = 0$
39. $4x^2 - 1 = 0$
40. $9y^2 - 1 = 0$
41. $9y^2 - 49 = 0$
42. $16z^2 - 25 = 0$

Solve each equation. See Example 4.

43. $w^2 = 7w$
44. $x^2 = 5x$
45. $s^2 = 16s$
46. $p^2 = 20p$
47. $4y^2 = 12y$
48. $5m^2 = 15m$
49. $3x^2 = -8x$
50. $3s^2 = -4s$

Solve each equation. See Example 5.

51. $3x^2 + 5x = 2$
52. $3x^2 + 14x = -8$
53. $2x^2 + x = 3$
54. $2x^2 - 5x = -2$
55. $5x^2 + 1 = 6x$
56. $6x^2 + 1 = 5x$
57. $2x^2 - 3x = 20$
58. $2x^2 - 3x = 14$

Solve each equation. See Example 6.

59. $4r(r + 7) = -49$
60. $5m(5m + 8) = -16$
61. $9a(a - 3) = 3a - 25$
62. $3x(3x + 10) = 6x - 16$

Solve each equation. See Example 7.

63. $x^3 + 3x^2 + 2x = 0$
64. $x^3 - 7x^2 + 10x = 0$
65. $k^3 - 27k - 6k^2 = 0$
66. $j^3 - 22j - 9j^2 = 0$

TRY IT YOURSELF

Solve each equation.

67. $4x^2 = 81$
68. $9y^2 = 64$

69. $x^2 - 16x + 64 = 0$

70. $h^2 + 2h + 1 = 0$

71. $(2s - 5)(s + 6) = 0$

72. $h(3h - 4)(h + 1) = 0$

73. $3b^2 - 30b = 6b - 60$

74. $2m^2 - 8m = 2m - 12$

75. $k^3 + k^2 - 20k = 0$

76. $n^3 - 6n^2 + 8n = 0$

77. $x^2 - 100 = 0$

78. $z^2 - 25 = 0$

79. $z(z - 7) = -12$

80. $p(p + 1) = 6$

81. $3y^2 - 14y - 5 = 0$

82. $4y^2 - 11y - 3 = 0$

83. $(x - 2)(x^2 - 8x + 7) = 0$

84. $(x - 1)(x^2 + 5x + 6) = 0$

85. $(n + 8)(n - 3) = -30$

86. $(2s + 5)(s + 1) = -1$

87. $x^3 - 6x^2 = -9x$

88. $m^3 - 8m^2 + 16m = 0$

89. $4a^2 + 1 = 8a + 1$

90. $3b^2 - 6 = 12b - 6$

91. $2b(6b + 13) = -12$

92. $5f(5f - 16) = -15$

93. $3a^3 + 4a^2 + a = 0$

94. $10b^3 - 15b^2 - 25b = 0$

95. $2x^3 = 2x(x + 2)$

96. $x^3 + 7x^2 = x^2 - 9x$

97. $-15x^2 + 2 + 7x = 0$

98. $-8x^2 + 3 - 10x = 0$

99. $4p^2 - 121 = 0$

100. $q^2 - \frac{1}{4} = 0$

101. $d(8d - 9) = -1$

102. $6n^3 - 6n = 0$

Look Alikes . . .

Factor the expression in part a and solve the equation in part b.

103. a. $x^2 + 4x - 21$ b. $x^2 + 4x - 21 = 0$

104. a. $4a^2 - 8a$ b. $4a^2 - 8a = 0$

105. a. $12n^2 - 5n - 2$ b. $12n^2 - 5n - 2 = 0$

106. a. $x^2 - 36$ b. $x^2 - 36 = 0$

WRITING

107. Explain the zero-factor property.

108. Find the error in the following solution.

$$x(x + 1) = 6$$

$$x = 6 \quad \text{or} \quad x + 1 = 6$$

$$x = 5$$

The solutions are 6 and 5.

109. A student solved $x^2 - 5x + 6 = 0$ and obtained two solutions: 2 and 3. Explain the error in his check.

Check:

$$x^2 - 5x + 6 = 0$$

$$2^2 - 5(3) + 6 \stackrel{?}{=} 0$$

$$4 - 15 + 6 \stackrel{?}{=} 0$$

$$-5 = 0 \quad \text{False}$$

2 is not a solution. 3 is not a solution.

110. In this section, we solved quadratic equations by factoring. Did we always obtain two different solutions? Explain.

111. What is wrong with the step used to solve $x^2 = 2x$ shown below?

$$x^2 = 2x$$

$$\frac{x^2}{x} = \frac{2x}{x}$$

$$x = 2$$

The solution is 2.

112. Explain the error in the following solution.

Factor: $x^2 - 5x + 6$

$$(x - 2)(x - 3) = 0$$

$$x - 2 = 0 \quad \text{or} \quad x - 3 = 0$$

$$x = 2 \qquad\qquad x = 3$$

The solutions are 2 and 3.

REVIEW

113. **Exercise.** A doctor advises a patient to exercise at least 15 minutes but less than 30 minutes per day. Use a compound inequality to express the range of these times t in minutes.

114. **Snacks.** A bag of peanuts is worth \$0.30 less than the same size bag of cashews. Equal amounts of peanuts and cashews are used to make 40 bags of a mixture that is worth \$1.05 per bag. How much is a bag of cashews worth?

CHALLENGE PROBLEMS

Solve each equation.

115. $x^4 - 625 = 0$

116. $2a^3 + a^2 - 32a - 16 = 0$

117. $(x - 3)^2 = 2x + 9$

118. $(x + 3)^2 = (2x - 1)^2$

SECTION 6.8

OBJECTIVES

1. Solve problems involving geometric figures.
2. Solve problems involving consecutive integers.
3. Solve problems using the Pythagorean theorem.
4. Solve problems given the quadratic equation model.

Applications of Quadratic Equations

ARE YOU READY?

The following problems review some basic skills that are needed when solving application problems involving quadratic equations.

1. What is the formula for the area of a rectangle?
2. What is the formula for the area of a triangle?
3. Let x = a number. Write an algebraic expression that represents 1 more than the number.
4. How many sides does a triangle have?

In Chapter 2, we solved mixture, investment, and uniform motion problems. To model those situations, we used *linear equations* in one variable. We will now consider situations that are modeled by *quadratic equations.*

1 Solve Problems Involving Geometric Figures.

We can use the six-step problem-solving strategy and the factoring method for solving quadratic equations to find the dimensions of certain figures, given their area.

EXAMPLE 1

Painting. In 2002, the pop art painting *The American Sweetheart,* by artist Robert Indiana, sold for \$614,500. The area of the rectangular painting is 32 square feet. Find the dimensions of the painting if it is twice as long as it is wide.

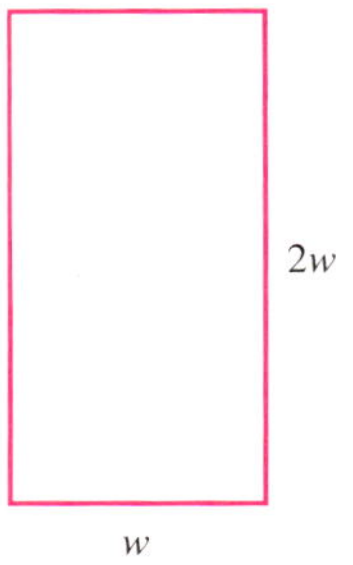

Analyze

- The area of the painting is 32 ft^2.
- The length is twice as long as the width.
- Find the length and width (the dimensions).

Assign Since the length is related to the width, let w = the width of the painting in feet. Then $2w$ = the length of the painting.

Form To form an equation, we use the formula for the area of a rectangle, $A = lw$, where $A = 32$.

The area of the rectangle	equals	the length	times	the width.
32	=	$2w$	$\cdot$	w

Solve

$32 = 2w \cdot w$ — This is the equation to solve.

$32 = 2w^2$ — Multiply $2w$ and w. This is a quadratic equation but it is not in standard form.

$0 = 2w^2 - 32$ — To get 0 on the left side, subtract 32 from both sides.

$0 = 2(w^2 - 16)$ — Factor out the GCF, 2.

$0 = 2(w + 4)(w - 4)$ — Factor the difference of two squares, $w^2 - 16$.

$w + 4 = 0$ or $w - 4 = 0$ — Since 2 cannot equal 0, discard that possibility. Set each factor that contains a variable equal to 0.

~~$w = -4$~~ | $w = 4$ — Solve each equation.

The Language of Algebra

When solving real-world application problems, we **discard** or **reject** any solutions of equations that do not make sense, such as a negative width of a painting.

State The solutions of the equation are -4 and 4. Since w represents the width of the picture, and the width cannot be negative, we discard -4. Thus, the width of the picture is 4 feet and the length is $2 \cdot 4 = 8$ feet.

Check A rectangle with dimensions 4 feet by 8 feet has an area of 32 ft^2, and the length is twice the width. The answers check.

Self Check 1 **Geometry.** A rectangle has an area of 55 square meters. Its length is 1 meter more than twice its width. Find the dimensions of the rectangle.

Now Try ▶ Problem 13

EXAMPLE 2

Windmills. The height of a triangular canvas sail of a windmill is 1 foot less than twice the length of its base. If the sail has an area of 22.5 ft^2, find the length of the base and the height.

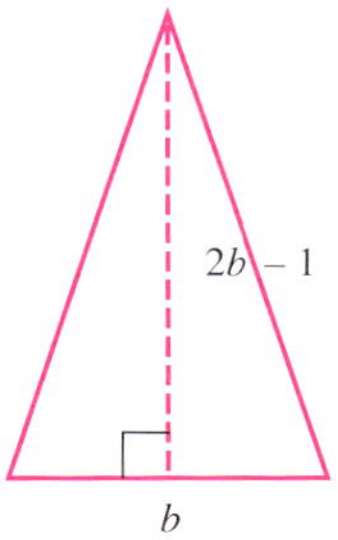

©Manamana/Shutterstock.com

Analyze

- The height is 1 ft less than twice the length of the base.
- The area is 22.5 ft^2.
- Find the length of the base and the height.

Assign Since the height is related to the length of the base, we let b = the length of the base of the sail in feet. Then $2b - 1$ = the height of the sail in feet.

Form To form an equation, we use the formula for the area of a triangle: $A = \frac{1}{2}bh$, where $A = 22.5$.

The area of the triangle	equals	one-half	times	the length of the base	times	the height.
22.5	=	$\frac{1}{2}$	$\cdot$	b	$\cdot$	$(2b - 1)$

Solve

$$22.5 = \frac{1}{2}b(2b - 1)$$ This is the equation to solve.

$$2 \cdot 22.5 = 2 \cdot \frac{1}{2}b(2b - 1)$$ To clear the equation of the fraction, multiply both sides by 2.

$$45 = b(2b - 1)$$ Multiply: $2 \cdot 22.5 = 45$ and $2 \cdot \frac{1}{2} = 1$.

$$45 = 2b^2 - b$$ Distribute the multiplication by b. This is a quadratic equation but it is not in standard form.

$$0 = 2b^2 - b - 45$$ To get 0 on the left side, subtract 45 from both sides.

$$0 = (2b + 9)(b - 5)$$ Factor the trinomial.

$$2b + 9 = 0 \quad \text{or} \quad b - 5 = 0$$ Set each factor equal to 0.

$$2b = -9 \qquad b = 5$$ Solve each equation.

$$b = -\frac{9}{2}$$ (crossed out)

Caution

A common error is to "distribute" the 2 incorrectly on the right side of the equation.

$$2 \cdot 22.5 = 2 \cdot \frac{1}{2}b(2b - 1)$$ (crossed out)

State The solutions of the equation are $-\frac{9}{2}$ and 5. Since b represents the length of the base of the sail, and it cannot be negative, we discard $-\frac{9}{2}$. The length of the base is then 5 feet, and the height is $2(5) - 1 = 9$ feet.

Check A triangle with height 9 feet and base 5 feet has area $\frac{1}{2}(9)(5) = 22.5$ ft^2, and the height is 1 foot less than twice the base. The answers check.

Self Check 2 **Sailing.** The height of a triangular sail is 4 yards more than twice the length of the base. If the sail has an area of 15 yd^2, find the base and the height.

Now Try Problem 19

2 Solve Problems Involving Consecutive Integers.

Consecutive integers are integers that follow one another, such as 15 and 16. When solving consecutive integer problems, if we let $x =$ the first integer, then:

- two consecutive integers are x and $x + 1$
- two consecutive even integers are x and $x + 2$
- two consecutive odd integers are x and $x + 2$

EXAMPLE 3 **Women's Tennis.** In the 1998 Australian Open, sisters Venus and Serena Williams played against each other for the first time as professionals. Venus was victorious over her younger sister. At that time, their ages were consecutive integers whose product was 272. How old were Venus and Serena when they met in this match?

Analyze

- Venus is older than Serena.
- Their ages were consecutive integers.
- The product of their ages was 272.
- Find Venus' and Serena's ages when they played this match.

Assign Let $x =$ Serena's age when she played in the 1998 Australian Open. Since their ages were consecutive integers, and since Venus is older, we let $x + 1 =$ Venus' age.

Form The word *product* indicates multiplication.

Serena's age	times	Venus' age	was	272.
x	$\cdot$	$(x + 1)$	$=$	272

Solve

$$x(x + 1) = 272$$

$$x^2 + x = 272$$ Distribute the multiplication by x. This is a quadratic equation but it is not in standard form.

$$x^2 + x - 272 = 0$$ Subtract 272 from both sides to make the right side 0.

$$(x + 17)(x - 16) = 0$$ Factor $x^2 + x - 272$. Two numbers whose product is -272 and whose sum is 1 are 17 and -16.

$x + 17 = 0$ or $x - 16 = 0$ Set each factor equal to 0.

~~$x = -17$~~ | $x = 16$ Solve each equation.

State The solutions of the equation are -17 and 16. Since x represents Serena's age, and it cannot be negative, we discard -17. Thus, Serena Williams was 16 years old and Venus Williams was $16 + 1 = 17$ years old when they played against each other for the first time as professionals.

Success Tip

The prime factorization of 272 is helpful in determining that $272 = 17 \cdot 16$.

$$16\left\{\begin{array}{r} 2\,\underline{|272} \\ 2\,\underline{|136} \\ 2\,\underline{|68} \\ 2\,\underline{|34} \end{array}\right. \\ 17$$

Check Since 16 and 17 are consecutive integers, and since $16 \cdot 17 = 272$, the answers check.

Self Check 3 The product of two consecutive integers is 552. Find the integers.

Now Try Problem 23

3 Solve Problems Using the Pythagorean Theorem.

Caution

It doesn't matter which leg of the right triangle is labeled a and which is labeled b. However, the hypotenuse must be labeled c.

A **right triangle** is a triangle that has a 90° (right) angle. The longest side of a right triangle is the **hypotenuse,** which is the side opposite the right angle. The remaining two sides are the **legs** of the triangle. The **Pythagorean theorem** provides a formula relating the lengths of the three sides of a right triangle.

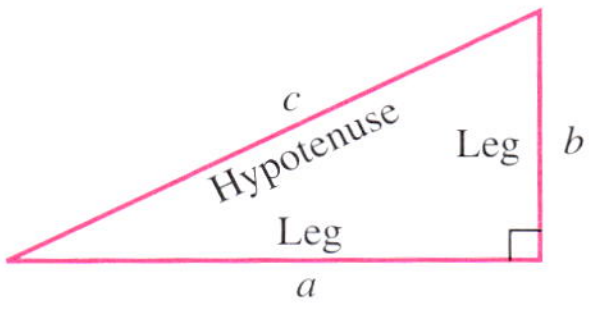

The Pythagorean Theorem

If a and b are the lengths of the legs of a right triangle and c is the length of the hypotenuse, then

$$a^2 + b^2 = c^2$$

In a right triangle, the sum of the squares of the lengths of the two legs is equal to the square of the length of the hypotenuse.

EXAMPLE 4

The Language of Algebra

A **theorem** is a mathematical statement that can be proved. The *Pythagorean theorem* is named after *Pythagoras,* a Greek mathematician who lived about 2,500 years ago. He is thought to have been the first to prove the theorem.

Right Triangles. The longer leg of a right triangle is 3 units longer than the shorter leg. If the hypotenuse is 6 units longer than the shorter leg, find the lengths of the sides of the triangle.

Analyze We begin by drawing a right triangle and labeling the legs and the hypotenuse.

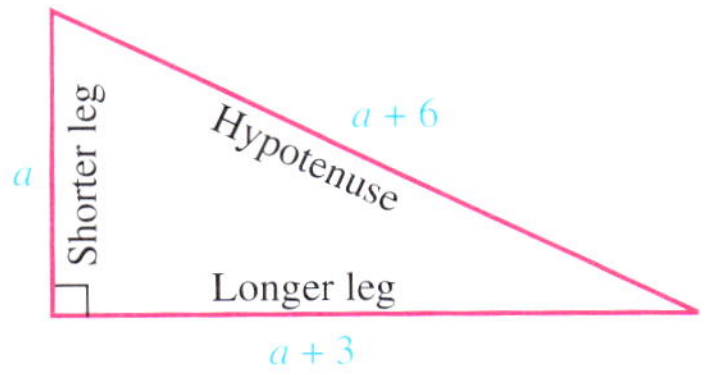

Assign Let a = the length of the shorter leg. Then the length of the hypotenuse is $a + 6$ and the length of the longer leg is $a + 3$.

Form By the Pythagorean theorem, we have

(The length of the shorter leg)^2	plus	(the length of the longer leg)^2	equals	(the length of the hypotenuse)^2
a^2	$+$	$(a + 3)^2$	$=$	$(a + 6)^2$

Solve

$$a^2 + (a + 3)^2 = (a + 6)^2$$

$$a^2 + a^2 + 6a + 9 = a^2 + 12a + 36$$ Find $(a + 3)^2$ and $(a + 6)^2$. Don't forget the middle terms.

$$2a^2 + 6a + 9 = a^2 + 12a + 36$$ On the left side: $a^2 + a^2 = 2a^2$. This is a quadratic equation but it is not in standard form.

$$a^2 - 6a - 27 = 0$$ To get 0 on the right side, subtract a^2, $12a$, and 36 from both sides. This is a quadratic equation.

$$(a - 9)(a + 3) = 0$$ Factor the trinomial.

$$a - 9 = 0 \quad \text{or} \quad a + 3 = 0$$ Set each factor equal to 0.

$$a = 9 \quad \bigg| \quad \cancel{a = -3}$$ Solve each equation.

State Since a side cannot have a negative length, we discard the solution -3. Thus, the shorter leg is 9 units long, the hypotenuse is $9 + 6 = 15$ units long, and the longer leg is $9 + 3 = 12$ units long.

Check The longer leg, 12, is 3 units longer than the shorter leg, 9. The hypotenuse, 15, is 6 units longer than the shorter leg, 9, and the side lengths satisfy the Pythagorean theorem. So the results check.

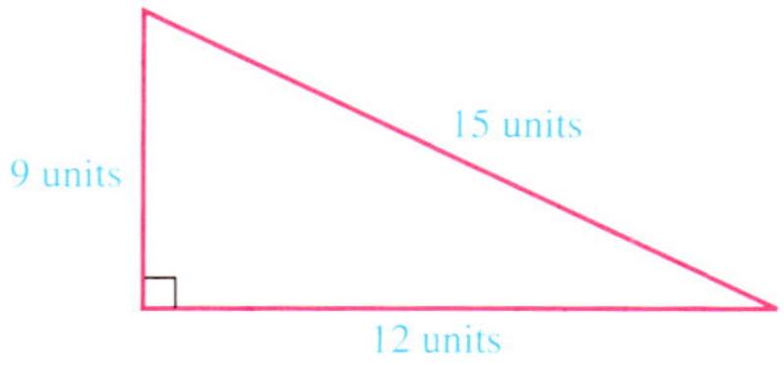

$$9^2 + 12^2 \stackrel{?}{=} 15^2$$
$$81 + 144 \stackrel{?}{=} 225$$
$$225 = 225$$

Self Check 4 **Right Triangles.** The longer leg of a right triangle is 7 inches longer than the shorter leg. If the hypotenuse is 9 inches longer than the shorter leg, find the lengths of the sides of the triangle.

Now Try Problem 31

4 Solve Problems Given the Quadratic Equation Model.

A quadratic equation can be used to describe the height of an object that is projected upward, such as a ball thrown into the air or an arrow shot into the sky.

EXAMPLE 5 **College Pranks.** A student uses rubber tubing to launch a water balloon from the roof of his dormitory. The height h (in feet) of the balloon, t seconds after being launched, is approximated by the formula $h = -16t^2 + 48t + 64$. After how many seconds will the balloon hit the ground?

Analyze When the water balloon hits the ground, its height will be 0 feet.

Assign To find the time that it takes for the balloon to hit the ground, we set h equal to 0, and solve the quadratic equation for t, the time.

Form

$$h = -16t^2 + 48t + 64$$
$$0 = -16t^2 + 48t + 64$$ Substitute 0 for the height, h. This is a quadratic equation.

Solve

$0 = -16t^2 + 48t + 64$ This is the equation to solve.

$0 = -16(t^2 - 3t - 4)$ Factor out the opposite of the GCF, -16.

$0 = -16(t + 1)(t - 4)$ Factor the trinomial.

$t + 1 = 0$ or $t - 4 = 0$ Since -16 cannot equal 0, discard that possibility. Set each factor that contains a variable equal to 0.

~~$t = -1$~~ | $t = 4$ Solve each equation.

State The equation has two solutions, -1 and 4. Since t represents time, and, in this case, time cannot be negative, we discard -1. The second solution, 4, indicates that the balloon hits the ground 4 seconds after being launched.

Check Check this result by substituting 4 for t in $h = -16t^2 + 48t + 64$. You should get $h = 0$.

Success Tip

Note that the common factor, -16, divides -16, 48, and 64 exactly:

$$\frac{-16}{-16} = 1 \quad \frac{48}{-16} = -3$$
$$\frac{64}{-16} = -4$$

Self Check 5 **Archery.** An arrow is shot into the air from a balcony. The height h (in feet) of the tip of the arrow, t seconds after being shot, is approximated by $h = -16t^2 + 77t + 15$. After how many seconds will the arrow hit the ground?

Now Try Problem 37

SECTION 6.8 STUDY SET

VOCABULARY

Fill in the blanks.

1. Integers that follow one another, such as 6 and 7, are called ___________ integers.
2. A _____ triangle is a triangle that has a 90° angle.
3. The longest side of a right triangle is the ___________. The remaining two sides are the _____ of the triangle.
4. The ___________ theorem is a formula that relates the lengths of the three sides of a right triangle.

CONCEPTS

5. A rectangle has an area of 40 in.^{2}. The length is 3 inches longer than the width. Which rectangle below meets these conditions?

i. 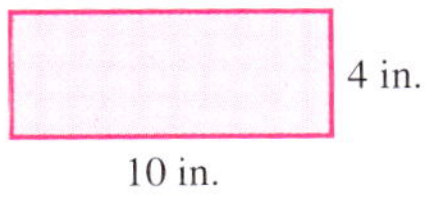

ii. 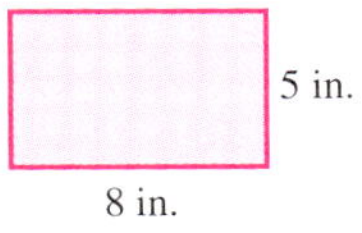

6. A triangle has an area of 15 ft^2. The height is 7 feet less than twice the length of the base. Which triangle below meets these conditions?

i. 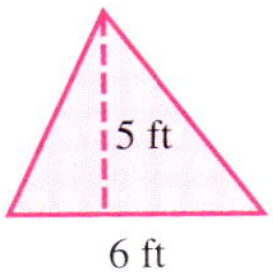

ii. 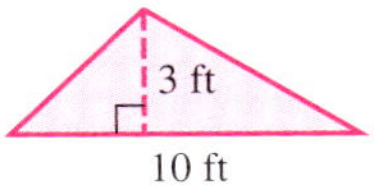

7. Multiply both sides of the equation by 2. ***Do not solve.***

$$10 = \frac{1}{2}b(b + 5)$$

8. Fill in the blanks.
   a. If the length of the hypotenuse of a right triangle is c and the lengths of the other two legs are a and b, then $\qquad = c^2$.
   b. In a right triangle, the sum of the ________ of the lengths of the two legs is equal to the square of the length of the ________.

9. a. What kind of triangle is shown?
   b. What are the lengths of the legs of the triangle?
   c. How much longer is the hypotenuse than the shorter leg?

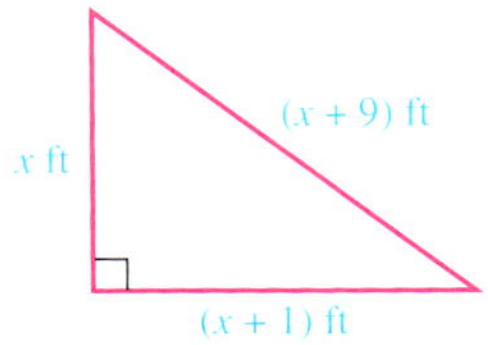

10. A ball is thrown into the air. Its height h in feet, t seconds after being released, is given by the formula $h = -16t^2 + 24t + 6$. When the ball hits the ground, what is the value of h?

NOTATION

Complete the solution to solve the equation.

11. $0 = -16t^2 + 32t + 48$

$0 = \qquad (t^2 - 2t - 3)$

$0 = -16(t - 3)(t + \quad)$

$t - 3 = \qquad$ or $t + 1 = \qquad$

$t = \qquad$ | $t = \qquad$

12. Fill in the blanks.
    a. Consecutive integers can be represented by x and ______.
    b. Consecutive odd integers can be represented by x and ______.
    c. Consecutive even integers can be represented by x and ______.

APPLICATIONS

Geometry Problems

13. **Flags.** The length of the flag of Australia is twice as long as the width. Find the dimensions of an Australian flag if its area is 18 ft^2.

14. **Billiards.** Pool tables are rectangular, and their length is twice the width. Find the dimensions of a pool table if it occupies 50 ft^2 of floor space.

15. X-Rays. A rectangular-shaped x-ray film has an area of 80 square inches. The length is 2 inches longer than the width. Find its width and length.

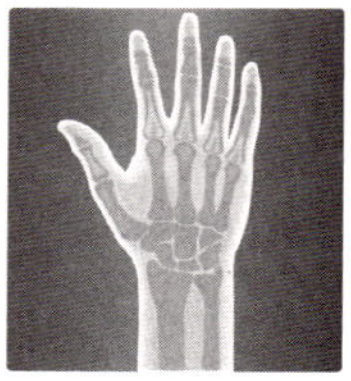

16. Insulation. The area of the rectangular slab of foam insulation in the illustration is 36 square meters. Find the dimensions of the slab.

17. from **Campus to Careers**

Elementary School Teacher

Suppose you are an elementary school teacher. You want to order a rectangular bulletin board to mount on a classroom wall that has an area of 90 square feet. Fire code requirements allow for no more than 30% of a classroom wall to be covered by a bulletin board. If the length of the board is to be three times as long as the width, what are the dimensions of the largest bulletin board that meets fire code?

18. Tubing. Refer to the diagram below. A piece of cardboard in the shape of a parallelogram is twisted to form the tube. The parallelogram has an area of 60 square inches. If its height h is 7 inches more than the length of the base b, what is the length of the base? (*Hint:* The formula for the area of a parallelogram is $A = bh$.)

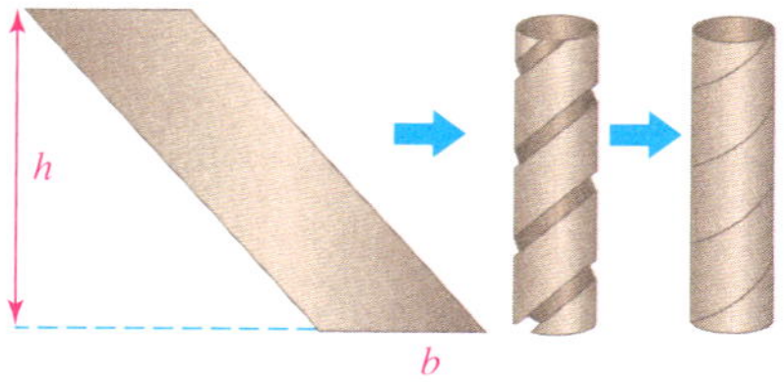

19. Jeans. The height of the triangular-shaped logo on a pair of jeans is 1 centimeter less than the length of its base. If the area of the logo is 15 square centimeters, find the length of the base and the height.

20. Shuffleboard. The area of the numbered triangle on a shuffleboard court is 27 ft^2. Its height is 3 feet more than the length of the base. Find the length of the base and the height.

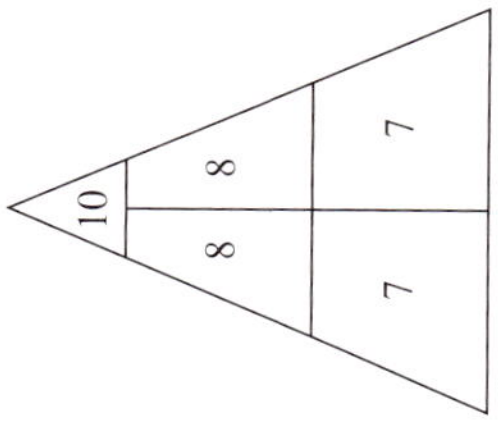

21. Sailboats. Refer to the diagram of a sail shown here. The length of the *luff* is 3 times longer than the length of the *foot* of the sail. Find the length of the foot and the length of the luff.

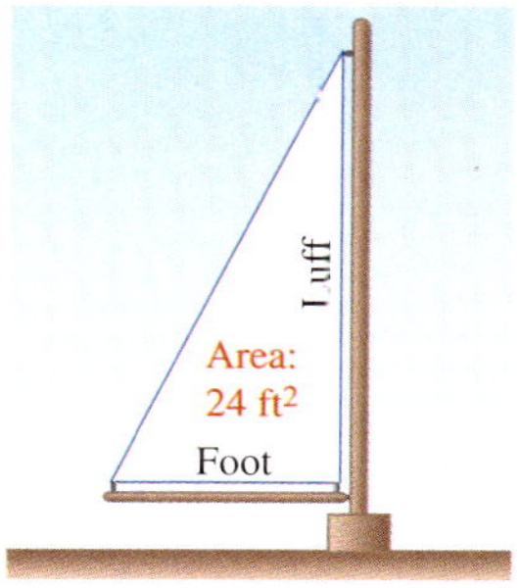

22. Designing Tents. The length of the base of the triangular sheet of canvas above the door of a tent is 2 feet more than twice its height. The area is 30 square feet. Find the height and the length of the base of the triangle.

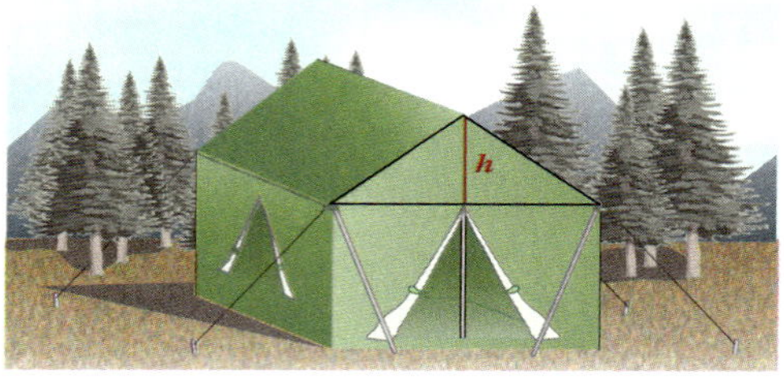

Consecutive Integer Problems

23. NASCAR. The car numbers of drivers Kasey Kahne and Scott Riggs are consecutive positive integers whose product is 90. If Kahne's car number is the smaller, what is the number of each car?

24. Baseball. Catcher Thurman Munson and pitcher Whitey Ford are two of the sixteen New York Yankees who have had their uniform numbers retired. Their uniform numbers are consecutive integers whose product is 240. If Munson's was the smaller number, determine the uniform number of each player.

25. Customer Service. At a pharmacy, customers take a ticket to reserve their turn for service. If the product of the ticket number now being served and the next ticket number to be served is 156, what number is now being served?

26. History. Delaware was the first state to enter the Union and Hawaii was the 50th. If we order the positions of entry for the rest of the states, we find that Kentucky entered the Union right after Vermont, and the product of their order-of-entry numbers is 210. Use the given information to complete these statements:

Kentucky was the ___ th state to enter the Union.

Vermont was the ___ th state to enter the Union.

27. **Plotting Points.** The x-coordinate and y-coordinate of a point in quadrant I are consecutive odd integers whose product is 143. The x-coordinate is less than the y-coordinate. Find the coordinates of the point.

28. **Presidents.** George Washington was born on 2-22-1732 (February 22, 1732). He died in 1799 at the age of 67. The month in which he died and the day of the month on which he died are consecutive even integers whose product is 168. When did Washington die?

Pythagorean Theorem Problems

29. **High-Ropes Adventures Courses.** A builder of a high-ropes adventure course wants to secure a pole by attaching a support cable from the anchor stake 8 yards from its base to a point 6 yards up the pole. How long should the cable be?

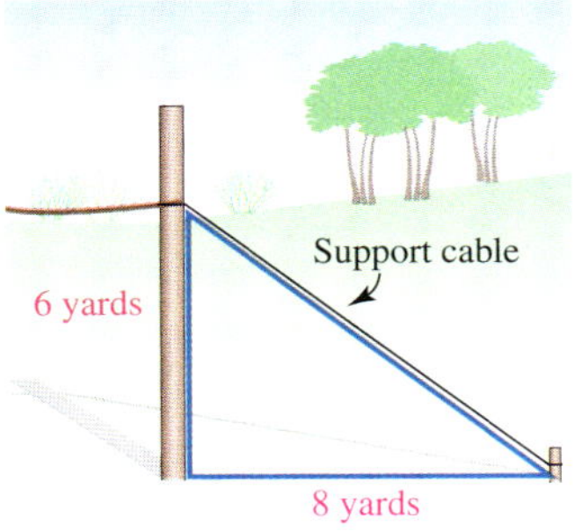

30. **Wind Damage.** A tree was blown over in a wind storm. Find x. Then find the height of the tree when it was standing upright.

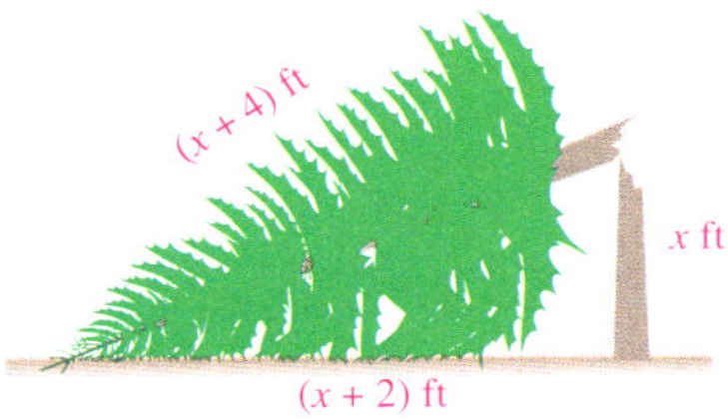

31. **Moto X.** Find x, the height of the landing ramp.

32. **Gardening Tools.** The dimensions (in millimeters) of the teeth of a pruning saw blade are given in the illustration. Find each length.

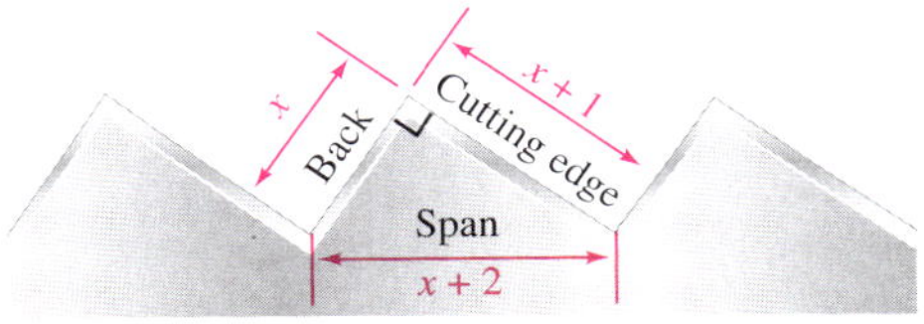

33. **Boating.** The inclined ramp of the boat launch shown in the next column is 8 meters longer than the rise of the ramp. The run is 7 meters longer than the rise. How long are the three sides of the ramp?

34. **Car Repairs.** To create some space to work under the front end of a car, a mechanic drives it up steel ramps. A ramp is 1 foot longer than the back, and the base is 2 feet longer than the back of the ramp. Find the length of each side of the ramp.

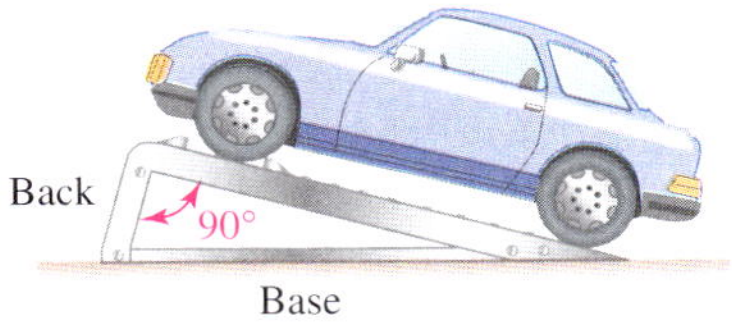

Quadratic Equation Model Problems

35. **Thrill Rides.** At the peak of a roller coaster ride, a rider's sunglasses fly off his head. The height h (in feet) of the glasses, t seconds after he loses them, is given by $h = -16t^2 + 64t + 80$. After how many seconds will the glasses hit the ground? (*Hint:* Factor out -16.)

36. **Parades.** A celebrity on the top of a parade float is tossing pieces of candy to the people on the street below. The height h (in feet) of a piece of candy, t seconds after being thrown, is given by $h = -16t^2 + 16t + 32$. After how many seconds will the candy hit the ground? (*Hint:* Factor out -16.)

37. **Softball.** A pitcher can throw a fastball underhand at 63 feet per second (about 45 mph). If she throws a ball into the air with that velocity, its height h in feet, t seconds after being released, is given by $h = -16t^2 + 63t + 4$. After the ball is thrown, in how many seconds will it hit the ground? (*Hint:* Factor out -1.)

38. **Officiating.** Before a football game, a coin toss is used to determine which team will kick off. The height h (in feet) of a coin above the ground t seconds after being flipped up into the air is given by $h = -16t^2 + 22t + 3$. How long does a team captain have to call heads or tails if it must be done while the coin is in the air? (*Hint:* Factor out -1.)

39. Dolphins. Refer to the illustration. The height h in feet reached by a dolphin t seconds after breaking the surface of the water is given by $h = -16t^2 + 32t$. How long will it take the dolphin to jump out of the water and touch the trainer's hand?

40. Exhibition Diving. In Acapulco, Mexico, men diving from a cliff to the water 64 feet below are quite a tourist attraction. A diver's height h above the water (in feet), t seconds after diving, is given by $h = -16t^2 + 64$. How long does a dive last?

41. Choreography. For the finale of a musical, 36 dancers are to assemble in a triangular-shaped series of rows, where each row has one more dancer than the previous row. The illustration shows the beginning of such a formation. The relationship between the number of rows r and the number of dancers d is given by $d = \frac{1}{2}r(r + 1)$. Determine the number of rows in the formation.

42. Crafts. The illustration shows how a wall hanging can be created by stretching yarn from peg to peg across a wooden ring. The relationship between the number of pegs p placed evenly around the ring and the number of yarn segments s that criss-cross the ring is given by the formula $s = \frac{p(p - 3)}{2}$. How many pegs are needed if the designer wants 27 segments to criss-cross the ring? (*Hint:* Multiply both sides of the equation by 2.)

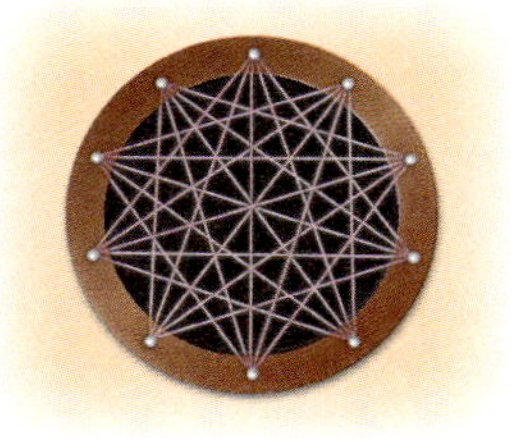

WRITING

43. A student was asked to solve the following problem: *The length of a rectangular room is 2 feet more than twice the width. If the area of the room is 60 square feet, find its dimensions.* Here is the student's solution:

Since $10 \cdot 6 = 60$, the length of the room is 10 feet and the width is 6 feet.

Explain why his solution is incorrect.

44. Suppose that to find the length of the base of a triangle, you write a quadratic equation and solve it to find $b = 6$ or $b = -8$. Explain why one solution should be discarded.

45. What error is apparent in the following illustration?

46. When naming the legs of a right triangle, explain why it doesn't matter which leg you label a and which leg you label b.

REVIEW

Find each special product.

47. $(5b - 2)^2$

48. $(2a + 3)^2$

49. $(s^2 + 4)^2$

50. $(m^2 - 1)^2$

51. $(9x + 6)(9x - 6)$

52. $(5b + 2)(5b - 2)$

CHALLENGE PROBLEMS

53. Pool Borders. The owners of a 10-meter-wide by 25-meter-long rectangular swimming pool want to surround the pool with a crushed-stone border of uniform width. They have enough stone to cover 74 square meters. How wide should they make the border?

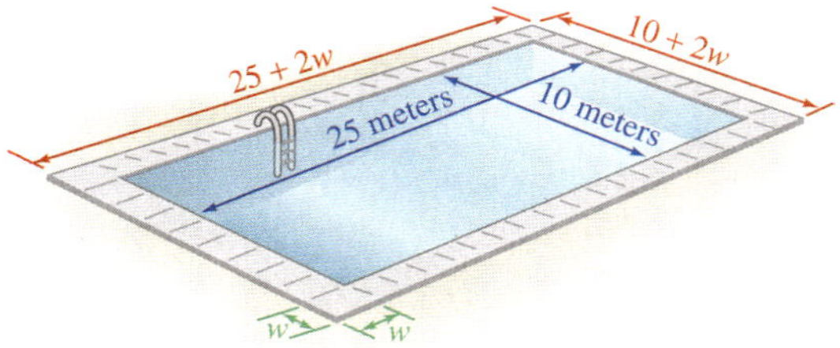

54. Find h.

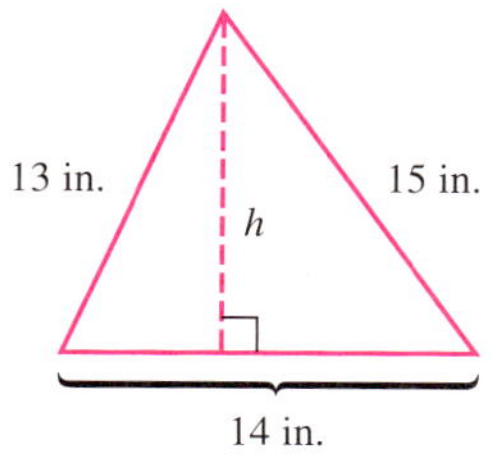

6 Summary & Review

SECTION 6.1 The Greatest Common Factor; Factoring by Grouping

DEFINITIONS AND CONCEPTS	EXAMPLES
Factoring is multiplication reversed. To **factor a polynomial** means to express it as a product of two (or more) polynomials.	Multiplication: Given the factors, we find a polynomial. → $2x(5x + 3) = 10x^2 + 6x$ ← Factoring: Given a polynomial, we find the factors.
A natural number is in **prime-factored form** when it is written as the product of prime numbers.	The prime-factored form of 28 is $2 \cdot 2 \cdot 7 = 2^2 \cdot 7$.
To find the **greatest common factor, GCF,** of a list of terms **1.** Write each coefficient as a product of prime factors. **2.** Identify the numerical and variable factors common to each term. **3.** Multiply the common numerical and variable factors identified in step 2 to obtain the GCF. If there are no common factors, the GCF is 1.	Find the GCF of $35x^4$, $63x^3$, and $42x^2$. $35x^4 = 5 \cdot 7 \cdot x \cdot x \cdot x \cdot x$ $63x^3 = 3 \cdot 3 \cdot 7 \cdot x \cdot x \cdot x$ $42x^2 = 2 \cdot 3 \cdot 7 \cdot x \cdot x$ GCF $= 7 \cdot x \cdot x = 7x^2$
The first step of factoring a polynomial is to see whether the terms of the polynomial have a common factor. If they do, **factor out the GCF.**	Factor: $35x^4 + 63x^3 - 42x^2$ $= 7x^2(5x^2 + 9x - 6)$ Factor out the GCF, $7x^2$. Use multiplication to check the factorization: $7x^2(5x^2 + 9x - 6) = 35x^4 + 63x^3 - 42x^2$ This is the original polynomial.
If a polynomial has four terms, try **factoring by grouping.** **1.** Group the terms of the polynomial so that the first two terms have a common factor and the last two terms have a common factor. **2.** Factor out the common factor from each group. **3.** Factor out the resulting common binomial factor. If there is no common binomial factor, regroup the terms of the polynomial and repeat steps 2 and 3.	Factor: $ax - bx + ay - by$ $= x(a - b) + y(a - b)$ Factor out x from $ax - bx$ and y from $ay - by$. $= (a - b)(x + y)$ Factor out the common binomial factor, $(a - b)$.

REVIEW EXERCISES

Find the prime-factorization of each number.

1. 35
2. 96

Find the GCF of each list.

3. 28 and 35
4. $36a^4$, $54a^3$, and $126a^6$

Factor.

5. $3x + 9y$
6. $5ax^2 + 15a$
7. $7s^5 + 14s^3$
8. $\pi ab - \pi ac$
9. $24x^3 + 60x^2 - 48x$
10. $x^5y^3z^2 + xy^5z^3 - xy^3z^2$
11. $-5ab^2 + 10a^2b - 15ab$
12. $4(x - 2) - x(x - 2)$

Factor out −1.

13. $-a - 7$
14. $-4t^2 + 3t - 1$

Factor.

15. $2c + 2d + ac + ad$
16. $3xy + 18x - 5y - 30$
17. $2a^3 + 2a^2 - a - 1$
18. $4m^2n + 12m^2 - 8mn - 24m$

SECTION 6.2 Factoring Trinomials of the Form $x^2 + bx + c$

DEFINITIONS AND CONCEPTS	EXAMPLES
Many trinomials factor as the product of two binomials. To **factor a trinomial** of the form $x^2 + bx + c$, whose **leading coefficient is 1,** find two integers whose product is c and whose sum is b. $x^2 + bx + c = (x \square)(x \square)$ The product of these numbers must be c and their sum must be b. Use the FOIL method to check the factorization.	Factor: $p^2 + 7p + 12$ $= (p \square)(p \square)$ $= (p + 3)(p + 4)$ Positive factors of 12 / Sum of positive factors of 12: $1 \cdot 12 = 12$, $1 + 12 = 13$; $2 \cdot 6 = 12$, $2 + 6 = 8$; $3 \cdot 4 = 12$, $3 + 4 = 7$ **Check:** $(p + 3)(p + 4) = p^2 + 4p + 3p + 12$ $= p^2 + 7p + 12$
Before factoring a trinomial, write it in **descending powers** of one variable. Also, factor out -1 if that is necessary to make the **leading coefficient positive.**	Factor: $7q - q^2 - 6$ $= -q^2 + 7q - 6$ Write the terms in descending powers of q. $= -(q^2 - 7q + 6)$ Factor out -1. $= -(q - 1)(q - 6)$ Factor the trinomial.
If a trinomial cannot be factored using only integers, it is called a **prime trinomial.**	$t^2 + 2t - 5$ is a prime trinomial because there are no two integers whose product is -5 and whose sum is 2.
The GCF should always be factored out first. A trinomial is **factored completely** when no factor can be factored further. Use multiplication to check the factorization.	Factor: $3m^3 - 6m^2 - 24m$ $= 3m(m^2 - 2m - 8)$ Factor out the GCF, $3m$, first. $= 3m(m - 4)(m + 2)$ Factor the trinomial.
To factor a trinomial of the form $ax^2 + bx + c$ (where $a = 1$) by **grouping,** write it as a equivalent four-term polynomial: $x^2 + \square x + \square x + c$ The product of these numbers must be ac, and their sum must be b. Then factor the four-term polynomial by grouping. Use the FOIL method to check the factorization.	Factor by grouping: $p^2 + 7p + 12$ We must find two numbers whose product is $ac = 1(12) = 12$ and whose sum is $b = 7$. Two such numbers are 4 and 3. They serve as the coefficients of $4p$ and $3p$, the two terms that we use to represent the middle term, $7p$, of the trinomial. $p^2 + 7p + 12 = p^2 + 4p + 3p + 12$ Express $7p$ as $4p + 3p$. $= p(p + 4) + 3(p + 4)$ Factor p out of $p^2 + 4p$ and 3 out of $3p + 12$. $= (p + 4)(p + 3)$ Factor out $(p + 4)$.

REVIEW EXERCISES

19. What is the leading coefficient of $x^2 + 8x - 9$?
20. Complete the table.

Factors of 6	Sum of the factors of 6
1(6)	
2(3)	
−1(−6)	
−2(−3)	

Factor.

21. $x^2 + 2x - 24$
22. $x^2 - 18x - 40$
23. $x^2 - 14x + 45$
24. $t^2 + 10t + 15$
25. $-y^2 + 15y - 56$
26. $10y + 9 + y^2$
27. $c^2 + 3cd - 10d^2$
28. $-3mn + m^2 + 2n^2$
29. Explain how we can check to determine whether $(x - 4)(x + 5)$ is the factorization of $x^2 + x - 20$.
30. Explain why $x^2 + 7x + 11$ is prime.

Factor.

31. $5a^5 + 45a^4 - 50a^3$
32. $-4x^2y - 4x^3 + 24xy^2$

SECTION 6.3 Factoring Trinomials of the Form $ax^2 + bx + c$

DEFINITIONS AND CONCEPTS

We can use the **trial-and-check method** to factor trinomials with **leading coefficients other than 1.** Write the trinomial as the product of two binomials and determine four integers.

$$ax^2 + bx + c = (\square x\ \square)(\square x\ \square)$$

The product of these numbers must be *a*. (the two leading boxes)

The product of these numbers must be *c*. (the two trailing boxes)

Use the FOIL method to check the factorization to obtain the correct middle term.

EXAMPLES

Factor: $2x^2 - 5x - 12$

Since the first term is $2x^2$, the first terms of the binomial factors must be $2x$ and x.

$$(2x\ \square)(x\ \square)$$ Because $2x \cdot x$ will give $2x^2$

The second terms of the binomials must be two integers whose product is -12. There are six such pairs:

$12(-1)$, $6(-2)$, $4(-3)$, $3(-4)$, $2(-6)$, and $1(-12)$

The pair in blue gives the correct middle term when we use the FOIL method to check:

$(2x + 3)(x - 4)$ — Outer: $-8x$; Inner: $3x$

Combine like terms: $-8x + 3x = -5x$. Correct middle term.

Thus, $2x^2 - 5x - 12 = (2x + 3)(x - 4)$.

DEFINITIONS AND CONCEPTS

To factor $ax^2 + bx + c$ by **grouping,** write it as an equivalent four-term polynomial:

$$ax^2 + \square x + \square x + c$$

The product of these numbers must be *ac*, and their sum must be *b*.

Then factor the four-term polynomial by grouping.

Use the FOIL method to check your work.

EXAMPLES

Factor by grouping: $2x^2 - 5x - 12$

We must find two numbers whose product is $ac = 2(-12) = -24$ and whose sum is $b = -5$. Two such numbers are -8 and 3. They serve as the coefficients of $-8x$ and $3x$, the two terms that we use to represent the middle term, $-5x$, of the trinomial.

$$\begin{aligned} 2x^2 - 5x - 12 &= 2x^2 - 8x + 3x - 12 && \text{Express } -5x \text{ as } -8x + 3x. \\ &= 2x(x - 4) + 3(x - 4) && \text{Factor out } 2x \text{ and } 3. \\ &= (x - 4)(2x + 3) && \text{Factor out } (x - 4). \end{aligned}$$

REVIEW EXERCISES

Factor.

33. $2x^2 - 5x - 3$

34. $35y^2 + 11y - 10$

35. $-3x^2 + 13x + 30$

36. $-33p^2 - 6p + 18p^3$

37. $4b^2 - 17bc + 4c^2$

38. $7y^2 + 7y - 18$

39. Entertaining. The rectangular area occupied by a table setting is $(12x^2 - x - 1)$ square inches. Factor the polynomial to find the binomials that represent the length and width of the table setting.

40. In the following work, a student began to factor $5x^2 - 8x + 3$. Explain his mistake.

$$(5x - \quad)(x + \quad)$$

SECTION 6.4 Factoring Perfect-Square Trinomials and the Difference of Two Squares

DEFINITIONS AND CONCEPTS	EXAMPLES
Trinomials that are squares of a binomial are called **perfect-square trinomials.** We can factor perfect-square trinomials by applying the special-product rules in reverse. $A^2 + 2AB + B^2 = (A + B)^2$ $A^2 - 2AB + B^2 = (A - B)^2$	Factor: $g^2 + 8g + 16$ and $m^2 - 18mn + 81n^2$ We match each trinomial to a special-product form shown in the left column. $g^2 + 8g + 16 = g^2 + 2 \cdot 4 \cdot g + 4^2 = (g + 4)^2$ $m^2 - 18mn + 81n^2 = m^2 - 2 \cdot m \cdot 9n + (9n)^2 = (m - 9n)^2$
To factor the **difference of two squares,** use the rule $F^2 - L^2 = (F + L)(F - L)$ It will be helpful to review the table of **squares of integers** shown on page 466.	Factor: $25b^2 - 36$ $= (5b)^2 - 6^2$ This is a difference of two squares. $= (5b + 6)(5b - 6)$
In general, the **sum of two squares** (with no common factor other than 1) cannot be factored using real numbers.	$x^2 + 100$ and $36y^2 + 49$ are prime polynomials.

REVIEW EXERCISES

Factor.

41. $x^2 + 10x + 25$
42. $9y^2 + 16 - 24y$
43. $-z^2 + 2z - 1$
44. $25a^2 + 20ab + 4b^2$
45. $x^2 - 9$
46. $49t^2 - 121y^2$
47. $x^2y^2 - 400$
48. $8at^2 - 32a$
49. $c^4 - 256$
50. $h^2 + 36$

SECTION 6.5 Factoring the Sum and Difference of Two Cubes

DEFINITIONS AND CONCEPTS	EXAMPLES
To factor the **sum** and **difference of two cubes,** use the following rules. $F^3 + L^3 = (F + L)(F^2 - FL + L^2)$ $F^3 - L^3 = (F - L)(F^2 + FL + L^2)$ It will be helpful to review the table of **cubes of integers** shown on page 473.	Factor: $p^3 + 64$ and $125a^3 - 27b^3$ We match each binomial to a factoring rule shown in the left column. $p^3 + 64 = p^3 + 4^3$ This is a sum of two cubes. $= (p + 4)(p^2 - p \cdot 4 + 4^2)$ $= (p + 4)(p^2 - 4p + 16)$ $125a^3 - 27b^3 = (5a)^3 - (3b)^3$ This is a difference of two cubes. $= (5a - 3b)[(5a)^2 + 5a \cdot 3b + (3b)^2]$ $= (5a - 3b)(25a^2 + 15ab + 9b^2)$

REVIEW EXERCISES

Factor.

51. $b^3 + 1$
52. $x^3 - 216$
53. $p^3 + 125q^3$
54. $16x^5 - 54x^2y^3$

SECTION 6.6 A Factoring Strategy

DEFINITIONS AND CONCEPTS	EXAMPLES
To factor a random polynomial, use the **factoring strategy** discussed in Section 6.6 on page 477. Remember that the instruction to factor means to **factor completely.** A polynomial is factored completely when no factor can be factored further.	Factor: $a^5 + 8a^2 + 4a^3 + 32$ ***Is there a common factor?*** No. There is no common factor (other than 1). ***How many terms does it have?*** Since the polynomial has four terms, try factoring by grouping. $a^5 + 8a^2 + 4a^3 + 32 = a^2(a^3 + 8) + 4(a^3 + 8)$ Factor a^2 from $a^5 + 8a^2$ and 4 from $4a^3 + 32$. $= (a^3 + 8)(a^2 + 4)$ Factor out $a^3 + 8$. ***Is it factored completely?*** No. We can factor $a^3 + 8$ as a sum of two cubes. $a^5 + 8a^2 + 4a^3 + 32$ $= (a^3 + 8)(a^2 + 4)$ $a^2 + 4$ is prime. $= (a + 2)(a^2 - 2a + 4)(a^2 + 4)$ $a^2 - 2a + 4$ is prime. ***Does it check?*** Use multiplication to check.

REVIEW EXERCISES

Factor

55. $14y^3 + 6y^4 - 40y^2$
56. $5s^2t + 5s^2u^2 + 5tv + 5u^2v$
57. $j^4 - 16$
58. $-3j^3 - 24$
59. $400x + 400 - m^2x - m^2$
60. $12w^4 - 36w^3 + 27w^2$
61. $2t^3 + 10$
62. $121p^2 + 36q^2$
63. $x^2z + 64y^2z + 16xyz$
64. $18c^3d^2 - 12c^3d - 24c^2d$

SECTION 6.7 Solving Quadratic Equations by Factoring

DEFINITIONS AND CONCEPTS	EXAMPLES
A **quadratic equation** is an equation that can be written in the **standard form** $ax^2 + bx + c = 0$, where a, b, and c are real numbers and $a \neq 0$.	Examples of quadratic equations are: $5x^2 + 25x = 0$, $4a^2 - 9 = 0$, and $y^2 - 13y = 6$
The Zero-Factor Property If the product of two (or more) numbers is 0, then at least one of the numbers is 0.	If $(x + 2)(x - 3) = 0$ then, $x + 2 = 0$ or $x - 3 = 0$.
To use the **factoring method to solve a quadratic equation:** **1.** Write the equation in standard form: $ax^2 + bx + c = 0$ or $0 = ax^2 + bx + c$ **2.** Factor completely. **3.** Use the *zero-factor property* to set each factor equal to 0. **4.** Solve each resulting equation. **5.** Check each result in the original equation.	Solve: $5x^2 + 25x = 0$ $5x(x + 5) = 0$ $5x = 0$ or $x + 5 = 0$ $x = 0$ | $x = -5$ The solutions are 0 and −5. The solution set is $\{0, -5\}$. Check each result in the original equation. Solve: $4a^2 - 9 = 0$ $(2a + 3)(2a - 3) = 0$ $2a + 3 = 0$ or $2a - 3 = 0$ $2a = -3$ | $2a = 3$ $a = -\frac{3}{2}$ | $a = \frac{3}{2}$ The solutions are $-\frac{3}{2}$ and $\frac{3}{2}$. The solution set is $\left\{-\frac{3}{2}, \frac{3}{2}\right\}$.

To use the zero-factor property to solve a quadratic equation, we need one side of the equation to be factored completely and the other side to be 0.	Solve: $5y^2 - 13y = 6$ — This equation is not in standard form. $5y^2 - 13y - 6 = 6 - 6$ — To get 0 on the right side, subtract 6 from both sides. $5y^2 - 13y - 6 = 0$ — Do the subtraction. $(5y + 2)(y - 3) = 0$ — Factor the trinomial. $5y + 2 = 0$ or $y - 3 = 0$ — Set each factor equal to 0. $5y = -2$ | $y = 3$ — Solve each equation. $y = -\frac{2}{5}$ The solutions are $-\frac{2}{5}$ and 3. Check each result in the original equation.

REVIEW EXERCISES

Solve each equation by factoring.

65. $8x(x - 6) = 0$
66. $(4x - 7)(x + 1) = 0$
67. $x^2 + 2x = 0$
68. $x^2 - 9 = 0$
69. $144x^2 - 25 = 0$
70. $a^2 - 7a + 12 = 0$
71. $2t^2 + 28t + 98 = 0$
72. $2x - x^2 = -24$
73. $5a^2 - 6a + 1 = 0$
74. $2p^3 = 2p(p + 2)$

SECTION 6.8 Applications of Quadratic Equations

DEFINITIONS AND CONCEPTS	EXAMPLES
To solve application problems, use the **six-step problem-solving strategy:** **1.** Analyze the problem **2.** Assign a variable **3.** Form an equation **4.** Solve the equation **5.** State the conclusion **6.** Check the result	Find two consecutive positive integers whose product is 72. **Analyze** *Consecutive integers* are integers that follow each other. The word *product* indicates multiplication. **Assign** Let x = the smaller positive integer. Then $x + 1$ = the larger integer. **Form** The smaller integer times the larger integer equals 72. $x \cdot (x + 1) = 72$ **Solve** $x(x + 1) = 72$ — This is the equation to solve. $x^2 + x = 72$ — Distribute the multiplication by x. This is a quadratic equation but it is not in standard form. $x^2 + x - 72 = 0$ — To get 0 on the right side, subtract 72 from both sides. $(x + 9)(x - 8) = 0$ — Factor the trinomial. $x + 9 = 0$ or $x - 8 = 0$ — Set each factor equal to 0. $x = -9$ | $x = 8$ — Solve each linear equation. **State** Since we are looking for positive integers, the solution −9 must be discarded. Thus, the smaller integer is 8 and the larger integer is $x + 1 = 9$ **Check** The integers 8 and 9 are consecutive positive integers and their product is 72.

The Pythagorean Theorem: If a and b are the lengths of the legs of a right triangle and c is the length of the hypotenuse, then $a^2 + b^2 = c^2$	To show that a triangle with sides of 5, 12, and 13 units is a right triangle, we verify that $5^2 + 12^2 = 13^2$. $5^2 + 12^2 \stackrel{?}{=} 13^2$; $25 + 144 \stackrel{?}{=} 169$; $169 = 169$ True 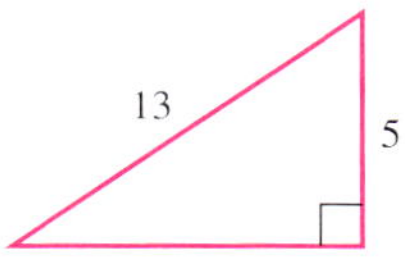

REVIEW EXERCISES

75. Sandpaper. A standard piece of sandpaper is 2 inches longer than it is wide. Find the dimensions of a piece of sandpaper if it has an area of 99 in².

76. Construction. The face of the triangular concrete panel has an area of 45 square meters, and its base is 3 meters longer than twice its height. Find the length of its base.

77. Fill in the blanks. If we let x = the first integer, then:

- two consecutive integers are x and
- two consecutive even integers are x and
- two consecutive odd integers are x and

78 Music Awards. The record for the most Grammy nominations in one year is held by Michael Jackson. Kanye West is currently in second place. The number of times Jackson and West were nominated are consecutive even integers whose product is 120. How many times was each artist nominated? (Source: *Wikipedia*)

79. Tightrope Walkers. A circus performer intends to walk up a taut cable shown in the illustration to a platform at the top of a pole. How high above the ground is the platform?

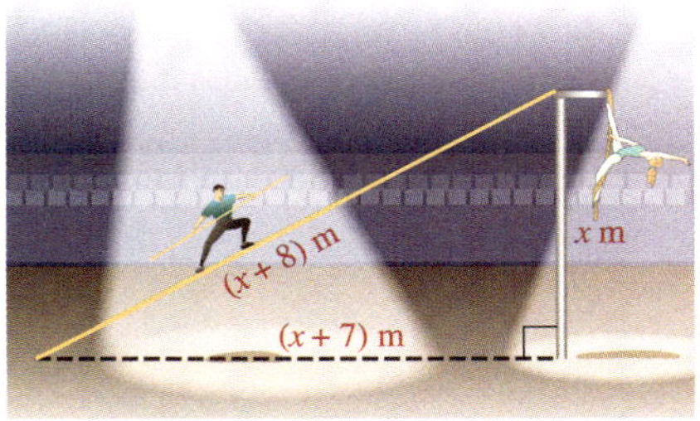

80. Ballooning. A hot-air balloonist accidentally dropped his camera overboard while traveling at a height of 1,600 ft. The height h in feet of the camera t seconds after being dropped is given by $h = -16t^2 + 1{,}600$. In how many seconds will the camera hit the ground?

6 CHAPTER TEST

1. Fill in the blanks.
   a. The letters GCF stand for ________ ________ ________.
   b. To factor a polynomial means to express it as a ________ of two (or more) polynomials.
   c. The ________ theorem provides a formula relating the lengths of the three sides of a right triangle.
   d. $y^2 - 25$ is a ________ of two squares.
   e. The trinomial $x^2 + x - 6$ factors as the product of two ________: $(x + 3)(x - 2)$.
2. a. Find the prime factorizations of 45 and 30.
   b. Find the greatest common factor of $45x^4$ and $30x^3$.

Factor. If an expression cannot be factored, write "prime."

3. $4x + 16$
4. $q^2 - 81$
5. $30a^2b^3 - 20a^3b^2 + 5ab$
6. $x^2 + 9$
7. $2x(x + 1) + 3(x + 1)$
8. $x^2 + 4x + 3$
9. $-x^2 + 9x + 22$
10. $60x^2 - 32x^3 + x^4$
11. $9a - 9b + ax - bx$
12. $2a^2 + 5a - 12$
13. $18x^2 + 60xy + 50y^2$
14. $x^3 + 8$

15. $60m^8 - 45m^6$

16. $3a^3 - 81$

17. $16x^4 - 81$

18. $a^3 + 5a^2 + a + 5$

19. $a^4 - 24 - 4a + 6a^3$

20. $3d - 4 + 10d^2$

21. $8m^2 - 800$

22. $36n^2 - 84n + 49$

23. $8r^2 - 14r + 3$

24. $t^2 - 6t + 10$

25. Checkers. The area of a square checkerboard is represented by $(25x^2 - 40x + 16)$ in.^{2}. Find the polynomial that represents the length of a side of the checkerboard.

©iStockphoto.com/Philip Bonneau

26. Factor $x^2 - 3x - 54$. Show a check of your answer.

Solve each equation.

27. $(x + 3)(x - 2) = 0$

28. $x^2 - 25 = 0$

29. $36x^2 = 6x$

30. $x^2 + 6x = -9$

31. $6x^2 + x - 1 = 0$

32. $a(a - 7) = 18$

33. $x^3 + 7x^2 = -6x$

34. Driving Safety. All cars have a blind spot where it is difficult for the driver to see a car behind and to the right. The area of the rectangular blind spot shown is 54 ft^2. Its length is 3 feet longer than its width. Find its dimensions.

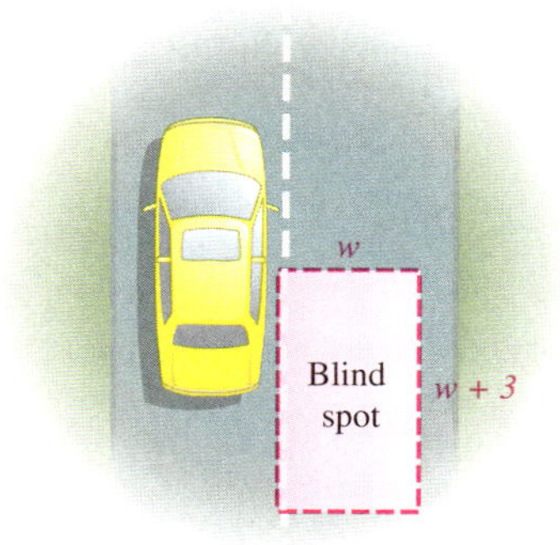

35. Rocketry. The height h, in feet, of a toy rocket t seconds after being launched is given by $h = -16t^2 + 80t$. After how many seconds will the rocket hit the ground?

36. ATV's. The area of a triangular-shaped safety flag on an all-terrain vehicle is 33 in.^{2}. Its height is 1 inch less than twice the length of the base. Find the length of the base and the height of the flag.

37. Find two consecutive positive integers whose product is 156.

38. Find the length of the hypotenuse of the right triangle shown.

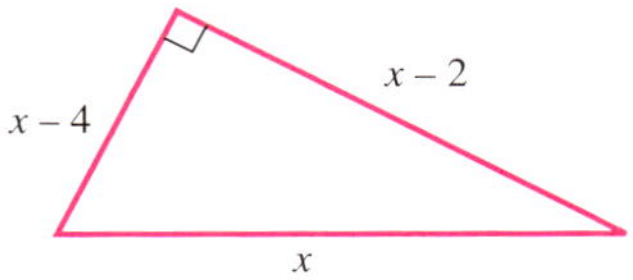

39. What is a quadratic equation? Give an example.

40. If the product of two numbers is 0, what conclusion can be drawn about the numbers?

Group Project

FACTORING MODELS

Overview In this activity, you will construct geometric models to find factorizations of several trinomials.

Instructions Form groups of 2 or 3 students.

1. Copy and cut out each of the following figures. On each figure, write its area.

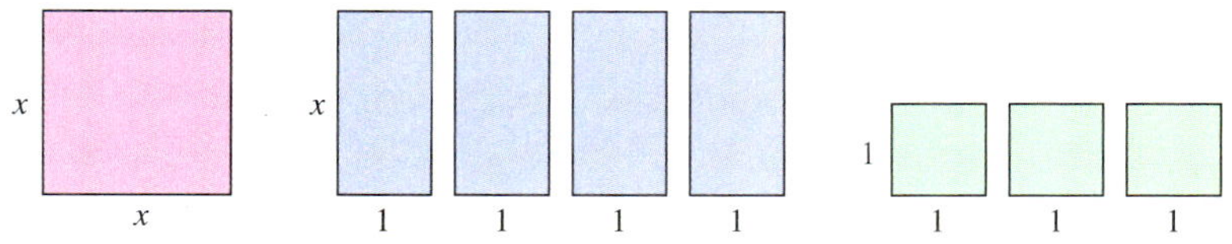

Write a trinomial that represents the *sum* of the areas of the eight figures by combining any like terms: _____ + _____ + _____

2. Now assemble the eight figures to form the large rectangle shown below.

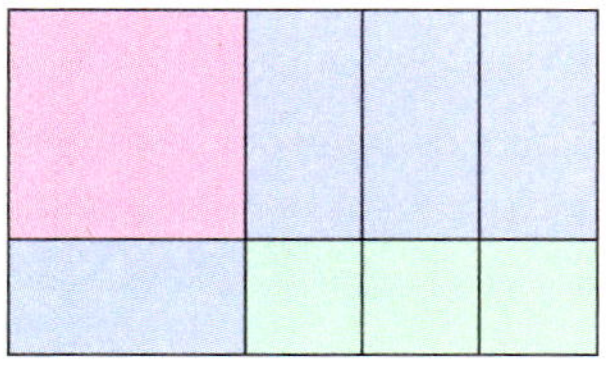

Write an expression that represents the *length* of the rectangle: _____ + _____

Write an expression that represents the *width* of the rectangle: _____ + _____

Express the area of the rectangle as the product of its length and width: (________)(________)

3. The set of figures used in step 1 and the set of figures used in step 2 are the same. Therefore, the expressions for the areas must be equal. Set your answers from steps 1 and 2 equal to find the factorization of the trinomial $x^2 + 4x + 3$.

Answer from step 1	=	Answer from step 2

4. Make a new model to find the factorization of $x^2 + 5x + 4$. (*Hint:* You will need to make one more 1-by-x figure and one more 1-by-1 figure.)

________________ = ________________

5. Make a new model to find the factorization of $2x^2 + 5x + 2$. (*Hint:* You will need to make one more x-by-x figure.)

________________ = ________________

CUMULATIVE REVIEW Chapters 1–6

1. **Heart Rates.** Refer to the graph. Determine the difference in the maximum heart beat rate for a 70-year-old as compared to someone half that age. [Section 1.1]

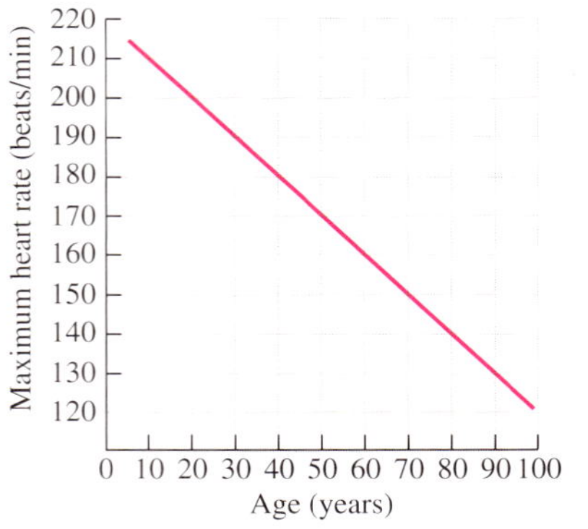

Based on data from *Cardiopulmonary Anatomy and Physiology: Essentials for Respiratory Care*, 2nd ed.

2. Find the prime factorization of 250. [Section 1.2]
3. Find the quotient: $\frac{16}{5} \div \frac{10}{3}$ [Section 1.2]
4. Write $\frac{124}{125}$ as a decimal. [Section 1.3]
5. Determine whether each statement is true or false. [Section 1.2]
   a. Every integer is a whole number.
   b. Every integer is a rational number.
   c. π is a real number.
6. Which division is undefined, $\frac{0}{5}$ or $\frac{5}{0}$? [Section 1.6]

Evaluate each expression.

7. $3 + 2[-1 - 4(5)]$ [Section 1.7]
8. $\dfrac{|-25| - 2(-5)}{9 - 2^4}$ [Section 1.7]
9. What is -3 cubed? [Section 1.7]
10. What is the value of x twenty-dollar bills? [Section 1.8]
11. Evaluate $\frac{-x - a}{y - b}$ for $x = -2$, $y = 1$, $a = 5$, and $b = 2$. [Section 1.8]
12. Identify the coefficient of each term in the expression $8x^2 - x + 9$. [Section 1.8]

Simplify each expression.

13. $-8y^2 - 5y^2 + 6$ [Section 1.9]
14. $3z + 2(y - z) + y$ [Section 1.9]

Solve each equation.

15. $-(3a + 1) + a = 2$ [Section 2.2]
16. $2 - (4x + 7) = 3 + 2(x + 2)$ [Section 2.2]
17. $\dfrac{3t - 21}{2} = t - 6$ [Section 2.2]
18. $-\dfrac{1}{3} - \dfrac{x}{5} = \dfrac{3}{2}$ [Section 2.2]
19. **Watermelons.** The heaviest watermelon on record weighed 270 pounds. If watermelon is 92% water by weight, what was its water weight? Round to the nearest pound. (Source: *Guinness World Records*) [Section 2.3]
20. Find the distance traveled by a truck traveling for $5\frac{1}{2}$ hours at a rate of 60 miles per hour. [Section 2.4]
21. What is the formula for simple interest? [Section 2.4]
22. **Geometry Tools.** A compass is adjusted so that the distance between the pointed ends is 2 inches. Then a circle is drawn. What will the area of the circle be? Round to the nearest tenth of a square inch. [Section 2.4]

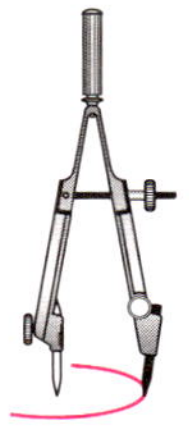

23. Solve $A = P + Prt$ for t. [Section 2.4]
24. **History.** George Washington was the first president of the United States. John Adams was the second, and Thomas Jefferson was the third, and so on. Grover Cleveland was president two *different* times, as shown in the illustration. The sum of the numbers of Cleveland's presidencies is 46. Find these two numbers. [Section 2.5]

25. **Antique Shows.** A traveling antique show will be on the road for 17 weeks, visiting three cities. They will be in Los Angeles for 2 weeks longer than they will be in Las Vegas. Their stay in Dallas will be 1 week less than twice that in Las Vegas. How many weeks will they be in each city? [Section 2.5]
26. **Photographic Chemicals.** A photographer wishes to mix 6 liters of a 5% acetic acid solution with a 10% solution to get a 7% solution. How many liters of 10% solution must be added? [Section 2.6]
27. **Dried Fruits.** Dried apple slices cost \$4.60 per pound, and dried banana chips sell for \$3.40 per pound. How many pounds of each should be used to create a 10-pound mixture that sells for \$4 per pound? [Section 2.6]
28. Solve: $-\frac{x}{2} + 4 > 5$. Write the solution set in interval notation and graph it. [Section 2.8]
29. Is $(-2, 5)$ a solution of $3x + 2y = 4$? [Section 3.2]
30. Graph: $y = 2x - 3$ [Section 3.2]

31. Is the graph of $x = 3$ a vertical or horizontal line? [Section 3.3]

32. If two lines are parallel, what can be said about their slopes? [Section 3.4]

33. **Encyclopedias.** The graph below approximates the total number of articles on the English-language edition of *Wikipedia* for the years 2005 through 2010. Find the rate of change in the number of articles over that time span. [Section 3.4] (Source: Wikipedia)

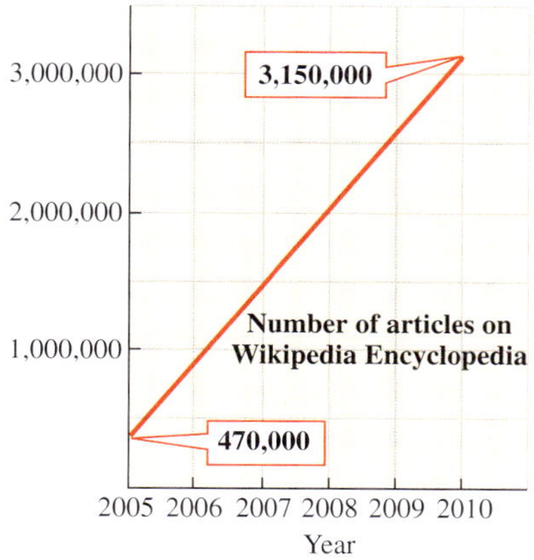

34. Find the slope and the y-intercept of the graph of $3x - 3y = 6$. [Section 3.5]

35. Find an equation of the line passing through $(-2, 5)$ and $(-3, -2)$. Write the equation in slope–intercept form. [Section 3.5]

36. Graph the line passing through $(-4, 1)$ that has slope -3. [Section 3.6]

37. Graph: $8x + 4y \geq -24$ [Section 3.7]

38. If $f(x) = 3x^2 - 2x + 1$, find $f(-2)$. [Section 3.8]

39. Is $\left(\frac{1}{2}, 1\right)$ a solution of the system $\begin{cases} 4x - y = 1 \\ 2x + y = 2 \end{cases}$? [Section 4.1]

40. Solve the system $\begin{cases} 3x - 2y = 6 \\ x - y = 1 \end{cases}$ by graphing. [Section 4.1]

41. Solve the system $\begin{cases} y = -4x + 1 \\ 4x - y = 5 \end{cases}$ by substitution. [Section 4.2]

42. Solve the system $\begin{cases} 5a + 3b = -8 \\ 2a + 9b = 2 \end{cases}$ by elimination (addition). [Section 4.3]

43. **Fundraising.** A Rotary Club held a city-wide recycling drive. They collected a total of 14 tons of newspaper and cardboard that earned them \$356. They were paid \$31 per ton for the newspaper and \$18 per ton for the cardboard. How many tons of each did they collect? [Section 4.4]

44. Graph: $\begin{cases} 4x + 3y \geq 12 \\ y < 4 \end{cases}$ [Section 4.5]

Simplify each expression. Write each answer without negative exponents.

45. $-y^2(4y^3)$ [Section 5.1]

46. $\dfrac{(x^2y^5)^5}{(x^3y)^2}$ [Section 5.1]

47. $\dfrac{b^5}{b^{-2}}$ [Section 5.2]

48. $2x^0$ [Section 5.1]

49. Write 0.00009011 in scientific notation. [Section 5.3]

50. Write 1,700,000 in scientific notation. [Section 5.3]

51. Find the degree of $7y^3 + 4y^2 + y + 3$. [Section 5.4]

52. Graph: $y = x^3 + 2$ [Section 5.4]

Perform the operations.

53. $(x^2 - 3x + 8) - (3x^2 + x + 3)$ [Section 5.5]

54. $4b^3(2b^2 - 2b)$ [Section 5.6]

55. $(3x - 2)(x + 4)$ [Section 5.6]

56. $(y - 6)^2$ [Section 5.7]

57. $\dfrac{12a^2b^2 - 8a^2b - 4ab}{4ab}$ [Section 5.8]

58. $x - 3\overline{)2x^2 - 5x - 3}$ [Section 5.8]

59. **Playpens.** Find an expression that represents the

a. perimeter of the playpen. [Section 5.5]

b. area of the floor of the playpen. [Section 5.6]

c. volume of the playpen. [Section 5.6]

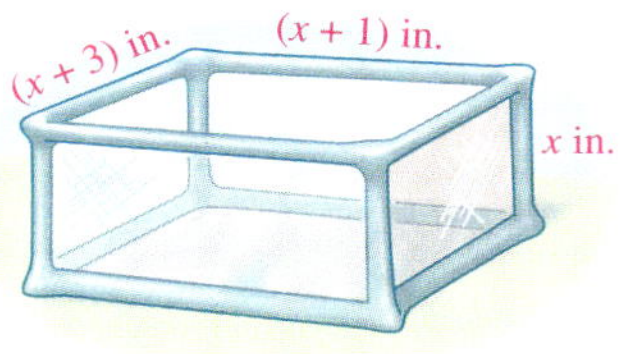

60. Find the GCF of $24x^5y^8$ and $54x^6y$. [Section 6.1]

Factor.

61. $9b^3 - 27b^2$ [Section 6.1]

62. $ax + bx + ay + by$ [Section 6.1]

63. $u^2 - 3 + 2u$ [Section 6.2]

64. $10x^2 + x - 2$ [Section 6.3]

65. $4a^2 - 12a + 9$ [Section 6.4]

66. $9z^2 - 1$ [Section 6.4]

67. $t^3 - 8$ [Section 6.5]

68. $3a^2b^2 - 6a^2 - 3b^2 + 6$ [Section 6.6]

Solve each equation.

69. $15s^2 - 20s = 0$ [Section 6.7]

70. $2x^2 - 5x = -2$ [Section 6.7]

71. $x^3 + 3x^2 + 2x = 0$ [Section 6.7]

72. **Camping.** The rectangular-shaped cooking surface of a small camping stove is 108 in.^{2}. If its length is 3 inches longer than its width, what are its dimensions? [Section 6.8]

Rational Expressions and Equations

7

©BananaStock/SuperStock

from Campus to Careers

Recreation Director

People of all ages enjoy participating in activities, such as arts and crafts, camping, sports, and the performing arts. Recreation directors plan, organize, and oversee these activities in local playgrounds, camps, community centers, religious organizations, theme parks, and tourist attractions. The job of recreation director requires mathematical skills such as budgeting, scheduling, and forecasting trends.

Problem 33 in **Study Set 7.7** and **problem 77** in **Study Set 7.8** involve situations that a recreation director might encounter on the job. The mathematical concepts discussed in this chapter can be used to solve those problems.

JOB TITLE:
Recreation Director
EDUCATION:
An associate's or bachelor's degree in parks and recreation is preferred.
JOB OUTLOOK:
Employment is expected to increase 15% through the year 2018.
ANNUAL EARNINGS:
From \$31,000 up to \$60,000
FOR MORE INFORMATION:
http://stats.bls.gov/oco/ocos058.htm

Study Skills Workshop ▸ Study Groups

Study groups give students an opportunity to ask their classmates questions, share ideas, compare lecture notes, and review for tests. If something like this interests you, here are some suggestions.

GROUP SIZE: A study group should be small—from 3 to 6 people is best.

TIME AND PLACE: You should meet regularly in a place where you can spread out and talk without disturbing others.

GROUND RULES: The study group will be more effective if, early on, you agree on some rules to follow.

Now Try This ▸

Would you like to begin a study group? If so, you need to answer the following questions.

Who will be in your group? Where will your group meet? How often will it meet? For how long will each session last? Will you have a group leader? What will be the leader's responsibilities? What will you try to accomplish each session? How will the members prepare for each meeting? Will you follow a set agenda each session? How will the members share contact information? When will you discuss ways to improve the study sessions?

SECTION 7.1 Simplifying Rational Expressions

OBJECTIVES

1. Evaluate rational expressions.
2. Find numbers that cause a rational expression to be undefined.
3. Simplify rational expressions.
4. Simplify rational expressions that have factors that are opposites.

ARE YOU READY?

The following problems review some basic skills that are needed when simplifying rational expressions.

1. Evaluate: **a.** $\frac{0}{7}$ **b.** $\frac{7}{0}$

2. Simplify: **a.** $\frac{18}{24}$ **b.** $\frac{5}{45}$

3. Factor: $12x - 8$

4. Factor: $a^2 - 16$

5. Factor: $5x^2 - 23x - 10$

6. Evaluate: $\frac{6}{-6}$

Fractions that are the quotient of two integers are *rational numbers.* Examples are $\frac{1}{2}$ and $\frac{9}{5}$. Fractions such as

$$\frac{3}{2y}, \qquad \frac{x}{x+2}, \qquad \text{and} \qquad \frac{2a^2 - 8a}{a^2 - 6a + 8}$$

that are the quotient of two polynomials are called **rational expressions.**

Rational Expressions

A **rational expression** is an expression of the form $\frac{A}{B}$, where A and B are polynomials and B does not equal 0.

1 Evaluate Rational Expressions.

Rational expressions can have different values depending on the number that is substituted for the variable.

EXAMPLE 1 Evaluate $\frac{2x - 1}{x^2 + 1}$ for $x = -3$ and for $x = 0$.

Strategy We will replace each x in the rational expression with the given value of the variable. Then we will evaluate the numerator and denominator separately, and simplify, if possible.

Why Recall from Chapter 1 that to *evaluate an expression* means to find its numerical value, once we know the value of its variable.

Solution

For $x = -3$:

$$\frac{2x - 1}{x^2 + 1} = \frac{2(-3) - 1}{(-3)^2 + 1} = \frac{-6 - 1}{9 + 1} = -\frac{7}{10}$$

For $x = 0$:

$$\frac{2x - 1}{x^2 + 1} = \frac{2(0) - 1}{(0)^2 + 1} = \frac{0 - 1}{0 + 1} = -1$$

Self Check 1 Evaluate $\frac{2x - 1}{x^2 + 1}$ for $x = 7$ and for $x = -2$.

Now Try Problems 13 and 21

2 Find Numbers That Cause a Rational Expression to Be Undefined.

The fraction bar in a rational expression indicates division. Since division by 0 is undefined, we must make sure that the denominator of a rational expression is not equal to 0.

EXAMPLE 2

Find all real numbers for which the expression is undefined: **a.** $\frac{7x}{x - 5}$ **b.** $\frac{3x - 2}{x^2 - x - 6}$ **c.** $\frac{8}{x^2 + 1}$ **d.** $\frac{x + 9.4}{12}$

Strategy To find the real numbers for which each expression is undefined, we will find the values of the variable that make the *denominator* 0.

Why We don't need to examine the numerator of the rational expression; it can be any value, including 0. It's a denominator of 0 that makes a rational expression undefined, because a denominator of 0 indicates division by 0.

Solution

a. The denominator of $\frac{7x}{x - 5}$ will be 0 if we replace x with 5.

$$\frac{7x}{x - 5} = \frac{7(5)}{5 - 5} = \frac{35}{0}$$

Since $\frac{35}{0}$ is undefined, the expression $\frac{7x}{x - 5}$ is undefined for $x = 5$.

b. $\frac{3x - 2}{x^2 - x - 6}$ will be undefined for values of x that make the denominator 0. To find these values, we set $x^2 - x - 6$ equal to 0, and solve for x.

$x^2 - x - 6 = 0$ Set the denominator of $\frac{3x - 2}{x^2 - x - 6}$ equal to 0.

$(x - 3)(x + 2) = 0$ To solve the quadratic equation, factor the trinomial.

$x - 3 = 0$ or $x + 2 = 0$ Set each factor equal to 0.

$x = 3$ | $x = -2$ Solve each equation.

The Language of Algebra

Another way that Example 2 could be phrased is: State the **restrictions** on the variable. For $\frac{3x - 2}{x^2 - x - 6}$, we can state the restrictions by writing $x \neq 3$ and $x \neq -2$.

Since 3 and -2 make the denominator 0, the expression $\frac{3x - 2}{x^2 - x - 6}$ is undefined for $x = 3$ and $x = -2$. To check the results, we proceed as follows.

For $x = 3$:

$$\frac{3x - 2}{x^2 - x - 6} = \frac{3(3) - 2}{3^2 - 3 - 6} = \frac{9 - 2}{9 - 3 - 6} = \frac{7}{0}$$

This expression is undefined.

For $x = -2$:

$$\frac{3x - 2}{x^2 - x - 6} = \frac{3(-2) - 2}{(-2)^2 - (-2) - 6} = \frac{-6 - 2}{4 + 2 - 6} = \frac{-8}{0}$$

This expression is undefined.

c. No matter what real number is substituted for x, the denominator, $x^2 + 1$, will not be equal to 0. (A number squared plus 1 cannot equal 0.) Thus, no real numbers make $\frac{8}{x^2 + 1}$ undefined.

d. Since the denominator of $\frac{x + 9.4}{12}$ does not contain a variable, the denominator can never be equal to 0. Thus, no real numbers make the expression undefined.

Self Check 2 Find all real numbers for which the expression is undefined:

a. $\frac{x}{x + 9}$ **b.** $\frac{9x + 7}{x^2 - 25}$ **c.** $\frac{4 - x}{x^2 + 64}$ **d.** $\frac{x + 3.8}{100}$

Now Try Problems 23, 27, 31, and 33

3 Simplify Rational Expressions.

In Section 1.2, we simplified fractions by removing factors equal to 1. For example, to simplify $\frac{6}{15}$, we factor 6 and 15, and then remove the factor $\frac{3}{3}$.

$$\frac{6}{15} = \frac{2 \cdot 3}{5 \cdot 3} = \frac{2}{5} \cdot \frac{3}{3} = \frac{2}{5} \cdot 1 = \frac{2}{5}$$

To streamline this process, we can replace $\frac{3}{3}$ in $\frac{2 \cdot 3}{5 \cdot 3}$ with the equivalent fraction $\frac{1}{1}$.

$$\frac{6}{15} = \frac{2 \cdot 3}{5 \cdot 3} = \frac{2 \cdot \overset{1}{\cancel{3}}}{5 \cdot \underset{1}{\cancel{3}}} = \frac{2}{5}$$ We are removing $\frac{3}{3} = 1$.

We can simplify rational expressions in a similar manner using a procedure that is based on the following property.

The Fundamental Property of Rational Expressions

If A, B, and C are polynomials, and B and C are not 0,

$$\frac{AC}{BC} = \frac{A}{B}$$

A rational expression is **simplified** if its numerator and denominator have no common factors other than 1. To simplify a rational expression, follow these steps.

Simplifying Rational Expressions

1. Factor the numerator and denominator completely to determine their common factors.
2. Remove factors equal to 1 by replacing each pair of factors common to the numerator and denominator with the equivalent fraction $\frac{1}{1}$.
3. Multiply the remaining factors in the numerator and in the denominator.

EXAMPLE 3 Simplify: $\frac{21x^3}{14x^2}$

Strategy We will write the numerator and denominator in factored form and then remove pairs of factors that are equal to 1.

Why The rational expression is simplified when the numerator and denominator have no common factor other than 1.

Solution

$$\frac{21x^3}{14x^2} = \frac{3 \cdot 7 \cdot x \cdot x \cdot x}{2 \cdot 7 \cdot x \cdot x}$$ To prepare to simplify the rational expression, factor the numerator and the denominator.

$$= \frac{3 \cdot \overset{1}{\cancel{7}} \cdot \overset{1}{\cancel{x}} \cdot \overset{1}{\cancel{x}} \cdot x}{2 \cdot \underset{1}{\cancel{7}} \cdot \underset{1}{\cancel{x}} \cdot \underset{1}{\cancel{x}}}$$ Simplify by replacing $\frac{7}{7}$ and $\frac{x}{x}$ with the equivalent fraction $\frac{1}{1}$. This removes the factor $\frac{7 \cdot x \cdot x}{7 \cdot x \cdot x}$, which is equal to 1.

$$= \frac{3x}{2}$$ Multiply the remaining factors in the numerator: $3 \cdot 1 \cdot 1 \cdot 1 \cdot x = 3x$.
Multiply the remaining factors in the denominator: $2 \cdot 1 \cdot 1 \cdot 1 = 2$.

We say that $\frac{21x^3}{14x^2}$ simplifies to $\frac{3x}{2}$. Since $\frac{21x^3}{14x^2}$ is undefined for $x = 0$, the expressions $\frac{21x^3}{14x^2}$ and $\frac{3x}{2}$ are equal only if $x \neq 0$. That is,

$$\frac{21x^3}{14x^2} = \frac{3x}{2} \quad \text{provided } x \neq 0$$

Caution

Unless otherwise stated, we will now assume that the variables in rational expressions represent real numbers for which the denominator is not zero. Therefore, as we simplify rational expressions, warnings like the "provided $x \neq 0$" statement on the right are not necessary.

An alternate approach for Example 3 is to use rules for exponents to simplify rational expressions that are the quotient of two monomials.

$$\frac{21x^3}{14x^2} = \frac{3 \cdot \overset{1}{\cancel{7}} \cdot x^{3-2}}{2 \cdot \underset{1}{\cancel{7}}} = \frac{3x^1}{2} = \frac{3x}{2}$$ To divide exponential expressions with the same base, keep the base and subtract the exponents.

Self Check 3 Simplify: $\frac{32a^3}{24a}$

Now Try Problem 41

To simplify rational expressions, we often make use of the factoring methods discussed in Chapter 6.

EXAMPLE 4 Simplify: **a.** $\frac{30t - 6}{36}$ **b.** $\frac{x^2 + 13x + 12}{x^2 + 12x}$ **c.** $\frac{x^3 + x^2}{1 + x}$

Strategy We will begin by factoring the numerator and denominator. Then we will remove any factors common to the numerator and denominator.

Why We need to make sure that the numerator and denominator have no common factor other than 1. When this is the case, the rational expression is simplified.

Solution

a. $$\frac{30t - 6}{36} = \frac{6(5t - 1)}{6 \cdot 6}$$ To prepare to simplify the rational expression, factor the numerator: The GCF is 6. Factor the denominator.

$$= \frac{\overset{1}{\cancel{6}}(5t - 1)}{\underset{1}{\cancel{6}} \cdot 6}$$ Simplify by removing a factor equal to 1. Replace $\frac{6}{6}$ with $\frac{1}{1}$.

$$= \frac{5t - 1}{6}$$ Multiply the remaining factors in the numerator: $1 \cdot (5t - 1) = 5t - 1$. Multiply the remaining factors in the denominator: $1 \cdot 6 = 6$.

The Language of Algebra

Sometimes, a common factor of the numerator and denominator has two or more terms. Here, we remove the common **binomial factor**, $x + 12$.

b. $$\frac{x^2 + 13x + 12}{x^2 + 12x} = \frac{(x + 1)(x + 12)}{x(x + 12)}$$ To prepare to simplify the rational expression, factor the numerator. Factor the denominator: The GCF is x.

$$= \frac{(x + 1)\overset{1}{\cancel{(x + 12)}}}{x\underset{1}{\cancel{(x + 12)}}}$$ Simplify by replacing $\frac{x + 12}{x + 12}$ with the equivalent fraction $\frac{1}{1}$. This removes the factor $\frac{x + 12}{x + 12} = 1$.

$$= \frac{x + 1}{x}$$ This rational expression cannot be simplified further.

c. $\frac{x^3 + x^2}{1 + x} = \frac{x^2(x + 1)}{1 + x}$ To prepare to simplify the rational expression, factor the numerator. The GCF is x^2.

$= \frac{x^2\overset{1}{\cancel{(x + 1)}}}{\underset{1}{\cancel{1 + x}}}$ By the commutative property of addition, $x + 1 = 1 + x$. Simplify by removing a factor equal to 1. Replace $\frac{x+1}{1+x}$ with $\frac{1}{1}$.

$= \frac{x^2}{1}$ Multiply in the the numerator: $x^2 \cdot 1 = x^2$.

$= x^2$ Any number divided by 1 is itself.

Self Check 4 Simplify: **a.** $\frac{4t - 20}{12}$ **b.** $\frac{x^2 - x - 6}{x^2 - 3x}$ **c.** $\frac{2x^4 + 4x^3}{2 + x}$

Now Try Problems 43 and 47

When a rational expression is simplified, the result is an **equivalent expression.** In Example 4b, for instance, this means that $\frac{x^2 + 13x + 12}{x^2 + 12x}$ and $\frac{x + 1}{x}$ have the same value for **all** values of x, except those that make either denominator 0. We can use that fact to perform an informal check of our work. If we let $x = 1$, for example, we see that the original rational expression and the simplified expression have the same value, 2.

The original expression

$$\frac{x^2 + 13x + 12}{x^2 + 12x} = \frac{(1)^2 + 13(1) + 12}{(1)^2 + 12(1)} = \frac{1 + 13 + 12}{1 + 12} = \frac{26}{13} = 2$$

The simplified expression

$$\frac{x + 1}{x} = \frac{1 + 1}{1} = \frac{2}{1} = 2$$

Same result

If the results are different, an error has been made, and the problem should be reworked.

EXAMPLE 5 Simplify: **a.** $\frac{3x^2 - 8x - 3}{2x^5 - 18x^3}$ **b.** $\frac{(x - y)^4}{x^2 - 2xy + y^2}$

Strategy We will begin by factoring the numerator and denominator using the methods discussed in Chapter 6. Then we will remove any factors common to the numerator and denominator.

Why We need to make sure that the numerator and denominator have no common factor other than 1. When this is the case, then the rational expression is simplified.

Solution **a.** $\frac{3x^2 - 8x - 3}{2x^5 - 18x^3} = \frac{(3x + 1)(x - 3)}{2x^3(x^2 - 9)}$ To prepare to simplify, factor the trinomial in the numerator. Factor the denominator: The GCF is $2x^3$.

$= \frac{(3x + 1)(x - 3)}{2x^3(x + 3)(x - 3)}$ In the denominator, factor the difference of two squares, $x^2 - 9$.

$= \frac{(3x + 1)\overset{1}{\cancel{(x - 3)}}}{2x^3(x + 3)\underset{1}{\cancel{(x - 3)}}}$ Simplify by replacing $\frac{x-3}{x-3}$ with the equivalent fraction $\frac{1}{1}$. This removes the factor $\frac{x-3}{x-3} = 1$.

$= \frac{3x + 1}{2x^3(x + 3)}$ It is not necessary to perform the multiplication $2x^3(x + 3)$ in the result. It is usually more convenient to leave the denominator in factored form.

b. $\dfrac{(x-y)^4}{x^2-2xy+y^2} = \dfrac{(x-y)^4}{(x-y)^2}$ To prepare to simplify, factor the perfect-square trinomial $x^2 - 2xy + y^2$ in the denominator.

$= \dfrac{(x-y)(x-y)(x-y)(x-y)}{(x-y)(x-y)}$ Write the repeated multiplication indicated by each exponent.

$= \dfrac{\overset{1}{\cancel{(x-y)}}\overset{1}{\cancel{(x-y)}}(x-y)(x-y)}{\underset{1}{\cancel{(x-y)}}\underset{1}{\cancel{(x-y)}}}$ Simplify by replacing each $\frac{x-y}{x-y}$ with $\frac{1}{1}$.

$= (x-y)^2$ Use an exponent to write the repeated multiplication in the numerator. It is not necessary to find $(x-y)^2$. The result can be presented in factored form.

Self Check 5 Simplify: **a.** $\dfrac{4x^2-4x-15}{8x^3-50x}$ **b.** $\dfrac{(a+3b)^5}{a^2+6ab+9b^2}$

Now Try Problems 51 and 53

CAUTION When simplifying rational expressions, we can remove only factors common to the entire numerator and denominator. *It is incorrect to remove any terms common to the numerator and denominator.*

$\dfrac{\overset{1}{\cancel{x}}+1}{\underset{1}{\cancel{x}}}$ x is a term of $x+1$.

$\dfrac{a^2-3a+\overset{1}{\cancel{2}}}{a+\underset{1}{\cancel{2}}}$ 2 is a term of $a^2 - 3a + 2$ and a term of $a + 2$.

$\dfrac{\overset{1}{\cancel{y^2}}-36}{\underset{1}{\cancel{y^2}}-y-7}$ y^2 is a term of $y^2 - 36$ and a term of $y^2 - y - 7$.

EXAMPLE 6 Simplify: $\dfrac{5(x+3)-5}{7(x+3)-7}$

Strategy We will begin by simplifying the numerator, $5(x+3)-5$, and the denominator, $7(x+3)-7$, separately. Then we will factor each result and remove any common factors.

Why We cannot remove $x+3$ immediately because it is not a factor of the *entire* numerator and the *entire* denominator.

Solution

$\dfrac{5(x+3)-5}{7(x+3)-7} = \dfrac{5x+15-5}{7x+21-7}$ Use the distributive property in the numerator and in the denominator.

$= \dfrac{5x+10}{7x+14}$ Combine like terms: $15 - 5 = 10$ and $21 - 7 = 14$.

$= \dfrac{5(x+2)}{7(x+2)}$ To prepare to simplify, factor the numerator: The GCF is 5. Factor the denominator: The GCF is 7.

$= \dfrac{5\overset{1}{\cancel{(x+2)}}}{7\underset{1}{\cancel{(x+2)}}}$ Simplify by replacing $\frac{x+2}{x+2}$ with the equivalent fraction $\frac{1}{1}$. This removes the factor $\frac{x+2}{x+2} = 1$.

$= \dfrac{5}{7}$

The Language of Algebra

Some rational expressions cannot be simplified. For example, to attempt to simplify the following rational expression, we factor its numerator and denominator. Since there are no common factors, we say it **does not simplify** or that it is in **simplest form.**

$$\frac{x^2+x-2}{x^2+x} = \frac{(x+2)(x-1)}{x(x+1)}$$

Self Check 6 Simplify: $\dfrac{4(x-2)+4}{3(x-2)+3}$

Now Try Problem 55

4 Simplify Rational Expressions That Have Factors That Are Opposites.

If the terms of two polynomials are the same, except that they are opposite in sign, the polynomials are **opposites.** For example, the following pairs of polynomials are opposites.

$2a - 1$ and $1 - 2a$ — Compare terms: $2a$ and $-2a$; -1 and 1.

$-3x^2 - x + 5$ and $3x^2 + x - 5$ — Compare terms: $-3x^2$ and $3x^2$; $-x$ and x; 5 and -5.

Success Tip

When a difference is reversed, the original binomial and the resulting binomial are opposites. Here are some pairs of opposites:

$b - 11$ and $11 - b$

$x^2 - 4$ and $4 - x^2$

Notice that the sum of a polynomial and its opposite is 0.

$$(2a - 1) + (1 - 2a) = 0 \qquad (-3x^2 - x + 5) + (3x^2 + x - 5) = 0$$

We have seen that the quotient of two real numbers that are opposites is always -1:

$$\frac{2}{-2} = -1 \qquad \frac{-78}{78} = -1 \qquad \frac{3.5}{-3.5} = -1$$

Likewise, the quotient of two polynomials that are opposites is always -1.

EXAMPLE 7 Simplify: $\frac{2a - 1}{1 - 2a}$

Strategy We will rearrange the terms of the numerator, $2a - 1$, and factor out -1.

Why This step is useful when the numerator and denominator contain factors that are opposites, such as $2a - 1$ and $1 - 2a$. It produces a common factor that can be removed.

Solution

$$\frac{2a - 1}{1 - 2a} = \frac{-1 + 2a}{1 - 2a}$$ Think of the numerator, $2a - 1$, as $2a + (-1)$. Then change the order of the terms: $2a + (-1) = -1 + 2a$.

$$= \frac{-1(1 - 2a)}{1 - 2a}$$ To prepare to simplify: factor out -1 from the two terms of the numerator.

$$= \frac{-1\overset{1}{\cancel{(1 - 2a)}}}{\underset{1}{\cancel{1 - 2a}}}$$ Simplify by replacing $\frac{1 - 2a}{1 - 2a}$ with the equivalent fraction $\frac{1}{1}$. This removes the factor $\frac{1 - 2a}{1 - 2a} = 1$.

$$= \frac{-1}{1}$$ Multiply the remaining factors in the numerator.

$$= -1$$ Any number divided by 1 is itself.

Success Tip

The result would have been the same if, instead, we had factored out -1 from the denominator, $1 - 2a$.

Self Check 7 Simplify: $\frac{3p - 2}{2 - 3p}$

Now Try Problem 59

In general, we have this fact.

The Quotient of Opposites The quotient of any nonzero polynomial and its opposite is -1.

Caution

Don't incorrectly apply the rule for opposites to a rational expression such as $\frac{x + 6}{6 + x}$. This is the quotient of a binomial and itself. The result is 1, not -1.

$$\frac{x + 6}{6 + x} = \frac{\overset{1}{\cancel{x + 6}}}{\underset{1}{\cancel{x + 6}}} = 1$$

For each of the following rational expressions, the numerator and denominator are opposites. Thus, each expression is equal to -1.

$$\frac{x - 6}{6 - x} = -1 \qquad \frac{2a - 9b}{9b - 2a} = -1 \qquad \frac{-3x^2 - x + 5}{3x^2 + x - 5} = -1$$

This fact can be used to simplify certain rational expressions by removing a factor equal to -1. If a factor of the numerator is the opposite of a factor of the denominator, we can replace them with the equivalent fraction $\frac{-1}{1}$, as shown in the following example.

EXAMPLE 8 Simplify: **a.** $\frac{y^2 - 1}{3 - 3y}$ **b.** $\frac{t + 8}{t - 8}$

Strategy We will begin by factoring the numerator and denominator. Then we look for common factors, or factors that are opposites, and remove them.

Why We need to make sure that the numerator and denominator have no common factor (or opposite factors) other than 1. When this is the case, then the rational expression is simplified.

Solution **a.**

$$\frac{y^2 - 1}{3 - 3y} = \frac{(y + 1)(y - 1)}{3(1 - y)}$$

To prepare to simplify, factor the numerator, and factor the denominator.

$$= \frac{(y + 1)\overset{-1}{\cancel{(y - 1)}}}{3\underset{1}{\cancel{(1 - y)}}}$$

Since $y - 1$ and $1 - y$ are opposites, simplify by replacing $\frac{y - 1}{1 - y}$ with the equivalent fraction $\frac{-1}{1}$. This removes the factor $\frac{y - 1}{1 - y} = -1$.

$$= \frac{-(y + 1)}{3}$$

In the numerator, $-1 \cdot (y + 1)$ can be written as $-(y + 1)$. In the denominator, $3 \cdot 1 = 3$.

This result may be written in several other equivalent forms.

$$\frac{-(y + 1)}{3} = -\frac{y + 1}{3}$$

The $-$ symbol in $-(y + 1)$ can be written in the front of the fraction, and the parentheses can be dropped.

$$\frac{-(y + 1)}{3} = \frac{-y - 1}{3}$$

The $-$ symbol in $-(y + 1)$ represents a factor of -1. Distribute the multiplication by -1 in the numerator.

$$\frac{-(y + 1)}{3} = \frac{y + 1}{-3}$$

The $-$ symbol in $-(y + 1)$ can be applied to the denominator. However, we don't usually use this form.

Caution

A $-$ symbol in front of a fraction may be applied to the numerator or to the denominator, but not to both:

$$-\frac{y + 1}{3} \neq \frac{-(y + 1)}{-3}$$

b. The binomials $t + 8$ and $t - 8$ are not opposites because their first terms do not have opposite signs. Thus, $\frac{t + 8}{t - 8}$ does not simplify.

Self Check 8 Simplify: **a.** $\frac{m^2 - 100}{10m - m^2}$ **b.** $\frac{2x - 3}{2x + 3}$

Now Try ▶ Problem 63

SECTION 7.1 STUDY SET

VOCABULARY

Fill in the blanks.

1. A quotient of two polynomials, such as $\frac{x^2 + x}{x^2 - 3x}$, is called a ________ expression.
2. To simplify a rational expression, we remove common _______ of the numerator and denominator.
3. Because of the division by 0, the expression $\frac{8}{0}$ is _________.
4. The binomials $x - 15$ and $15 - x$ are called _________, because their terms are the same, except that they are opposite in sign.

CONCEPTS

5. When we simplify $\frac{x^2 + 5x}{4x + 20}$, the result is $\frac{x}{4}$. These equivalent expressions have the same value for all real numbers, except $x = -5$. Show that they have the same value for $x = 1$.
6. Determine whether each pair of polynomials are opposites. Write *yes* or *no*.
   - **a.** $y + 7$ and $y - 7$
   - **b.** $b - 20$ and $20 - b$
   - **c.** $x^2 + 2x - 1$ and $-x^2 - 2x - 1$

7. Simplify each expression, if possible.

a. $\dfrac{x-8}{x-8}$ b. $\dfrac{x-8}{8-x}$

c. $\dfrac{x+8}{8+x}$ d. $\dfrac{x+8}{x}$

8. Simplify each expression.

a. $\dfrac{(x+2)(x-2)}{(x+1)(x+2)}$ b. $\dfrac{y(y-2)}{9(2-y)}$

c. $\dfrac{(2m+7)(m-5)}{(2m+7)}$ d. $\dfrac{x \cdot x}{x \cdot x(x-30)}$

NOTATION

Complete the solution to simplify the rational expression.

9. $\dfrac{x^2+2x+1}{x^2+4x+3} = \dfrac{(x+1)(\quad+1)}{(x+3)(x+\quad)}$

$= \dfrac{(x+1)\overset{1}{\cancel{(x+1)}}}{(x+3)\underset{1}{}}$

$= \dfrac{x+1}{\quad}$

10. In the following table, a student's answers to three homework problems are compared with the answers in the back of the book. Are the answers equivalent?

Student's Answer	Book's answer	Equivalent?
$\dfrac{-3}{x+3}$	$-\dfrac{3}{x+3}$	
$\dfrac{-(x-4)}{6x+1}$	$\dfrac{-x+4}{6x+1}$	
$\dfrac{x+7}{(x-4)(x+2)}$	$\dfrac{x+7}{(x+2)(x-4)}$	
$\dfrac{x^2+6}{-x}$	$-\dfrac{x^2+6}{x}$	

GUIDED PRACTICE

Evaluate each expression for $x = 6$. See Example 1.

11. $\dfrac{x-2}{x-5}$ 12. $\dfrac{3x-2}{x-2}$

13. $\dfrac{x^2-4x-12}{x^2+x-2}$ 14. $\dfrac{x^2-36}{x^3-1}$

15. $\dfrac{-x+1}{x^2-5x-6}$ 16. $\dfrac{-2x^2-3}{x-6}$

Evaluate each expression for $y = -3$. See Example 1.

17. $\dfrac{y+5}{3y-2}$ 18. $\dfrac{2y+9}{y^2+25}$

19. $-\dfrac{y}{y^2-y+6}$ 20. $-\dfrac{y^3}{3y^2+1}$

21. $\dfrac{y^2+9}{9-y^2}$ 22. $\dfrac{-y-11}{y^2+2y-3}$

Find all real numbers for which the rational expression is undefined. See Example 2.

23. $\dfrac{15}{x-2}$ 24. $\dfrac{5x}{x+5}$

25. $\dfrac{x+5}{8x}$ 26. $\dfrac{4x-1}{6x}$

27. $\dfrac{15x+2}{x^2+6}$ 28. $\dfrac{x^2-4x}{x^2+4}$

29. $\dfrac{x+1}{2x-1}$ 30. $\dfrac{-6x}{3x-1}$

31. $\dfrac{x^2-6x}{9}$ 32. $\dfrac{x^3-x^2}{15}$

33. $\dfrac{30x}{x^2-36}$ 34. $\dfrac{2x-15}{x^2-49}$

35. $\dfrac{15}{x^2+x-2}$ 36. $\dfrac{x-20}{x^2+2x-8}$

37. $\dfrac{16}{20-x}$ 38. $\dfrac{44}{57-x}$

Simplify. See Example 3.

39. $\dfrac{45}{9a}$ 40. $\dfrac{48}{16y}$

41. $\dfrac{6x^4}{4x^2}$ 42. $\dfrac{9x^3}{6x}$

Simplify. See Example 4.

43. $\dfrac{6x+3}{9}$ 44. $\dfrac{4x+12}{16}$

45. $\dfrac{x+3}{3x+9}$ 46. $\dfrac{2x-14}{x-7}$

47. $\dfrac{x^2-4}{x^2-6x+8}$ 48. $\dfrac{y^2-25}{y^2-3y-10}$

49. $\dfrac{x^2+5x+4}{x^2+4x}$ 50. $\dfrac{x^2-10x+21}{x^2-3x}$

Simplify. See Example 5.

51. $\dfrac{m^2-2mn+n^2}{7m^2-7n^2}$ 52. $\dfrac{11c^2-11d^2}{c^2-2cd+d^2}$

53. $\dfrac{4b^2+4b+1}{(2b+1)^3}$ 54. $\dfrac{9y^2-12y+4}{(3y-2)^3}$

Simplify. See Example 6.

55. $\dfrac{10(c-3)+10}{3(c-3)+3}$ 56. $\dfrac{6(d+3)-6}{7(d+3)-7}$

57. $\dfrac{6(x+3)-18}{3x-18}$ 58. $\dfrac{4(t-1)+4}{4t+4}$

Simplify. See Example 7.

59. $\dfrac{2x-7}{7-2x}$ 60. $\dfrac{18-d}{d-18}$

61. $\dfrac{3-4t}{8t-6}$ 62. $\dfrac{5t-1}{3-15t}$

Simplify. See Example 8.

63. $\dfrac{2 - a}{a^2 - a - 2}$

64. $\dfrac{4 - b}{b^2 - 5b + 4}$

65. $\dfrac{25 - 5m}{m^2 - 25}$

66. $\dfrac{36 - 6h}{h^2 - 36}$

TRY IT YOURSELF

Simplify. If an expression cannot be simplified, write "Does not simplify."

67. $\dfrac{a^3 - a^2}{a^4 - a^3}$

68. $\dfrac{2c^4 + 2c^3}{4c^5 + 4c^4}$

69. $\dfrac{4 - x^2}{x^2 - x - 2}$

70. $\dfrac{81 - y^2}{y^2 + 10y + 9}$

71. $\dfrac{6x - 30}{5 - x}$

72. $\dfrac{6t - 42}{7 - t}$

73. $\dfrac{x^2 + 3x + 2}{x^2 + x - 2}$

74. $\dfrac{x^2 + x - 6}{x^2 - x - 2}$

75. $\dfrac{15x^2y}{5xy^2}$

76. $\dfrac{12xz}{4xz^2}$

77. $\dfrac{x^8 + 9x^7}{9 + x}$

78. $\dfrac{x^9 + 50x^8}{50 + x}$

79. $\dfrac{x(x - 8) + 16}{16 - x^2}$

80. $\dfrac{x^2 - 3(2x - 3)}{9 - x^2}$

81. $\dfrac{4c + 4d}{d + c}$

82. $\dfrac{a + b}{5b + 5a}$

83. $\dfrac{3x^2 - 27}{2x^2 - 5x - 3}$

84. $\dfrac{2x^2 - 8}{3x^2 - 5x - 2}$

85. $\dfrac{-3x^2 + 10x + 77}{x^2 - 4x - 21}$

86. $\dfrac{-2x^2 + 5x + 3}{x^2 + 2x - 15}$

87. $\dfrac{42c^3d}{18cd^3}$

88. $\dfrac{49m^4n^5}{35mn^6}$

89. $\dfrac{16a^2 - 1}{4a + 4}$

90. $\dfrac{25m^2 - 1}{5m + 5}$

91. $\dfrac{8u^2 - 2u - 15}{4u^4 + 5u^3}$

92. $\dfrac{6n^2 - 7n + 2}{3n^3 - 2n^2}$

93. $\dfrac{(2x + 3)^4}{4x^2 + 12x + 9}$

94. $\dfrac{(3y - 2)^5}{9y^2 - 12y + 4}$

95. $\dfrac{6c + 3(a + 2) + 12}{a + 2}$

96. $\dfrac{2y + 4(y - 1) - 2}{y - 1}$

97. $\dfrac{15x - 3x^2}{25y - 5xy}$

98. $\dfrac{18c - 2c^2}{81d - 9cd}$

99. $\dfrac{2x^2}{x + 2}$

100. $\dfrac{5y^2}{y + 5}$

101. $\dfrac{18 + 2x}{x^2 - 81}$

102. $\dfrac{12 + 6x}{x^2 - 4}$

APPLICATIONS

103. **Organ Pipes.** The number of vibrations n per second of an organ pipe is given by the formula $n = \dfrac{512}{L}$ where L is the length of the pipe in feet. How many times per second will a 6-foot pipe vibrate?

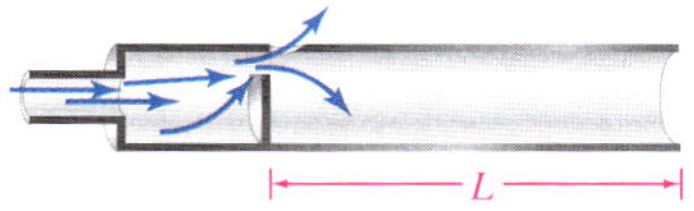

104. **Raising Turkeys.** The formula $T = \frac{2{,}000m}{m + 1}$ gives the number T of turkeys on a poultry farm m months after the beginning of the year. How many turkeys will there be on the farm by the end of July?

105. **Medical Dosages.** The formula $c = \frac{4t}{t^2 + 1}$ gives the concentration c (in milligrams per liter) of a certain dosage of medication in a patient's bloodstream t hours after the medication is administered. Suppose the patient received the medication at noon. Find the concentration of medication in his blood at the following times later that afternoon.

106. **Manufacturing.** If a company produces x child car seats, the average cost c (in dollars) to produce one car seat is given by the formula $c = \frac{50x + 50{,}000}{x}$. Find the company's average production cost if 1,000 are produced.

WRITING

107. Explain why $\frac{x - 7}{7 - x} = -1$.

108. Explain why $\frac{x - 3}{x + 4}$ is undefined for $x = -4$ but defined for $x = 3$.

109. Explain the error in the following work:

Simplify: $\dfrac{x}{x + 2} = \dfrac{\overset{1}{\cancel{x}}}{\underset{1}{\cancel{x}} + 2}$

$= \dfrac{1}{3}$

110. Explain the error in the following work:

Simplify: $\dfrac{30t - 6}{36} = \dfrac{\overset{1}{\cancel{2}} \cdot \overset{1}{\cancel{3}} \cdot 5 \cdot t - \overset{1}{\cancel{2}} \cdot \overset{1}{\cancel{3}}}{\underset{1}{\cancel{2}} \cdot \underset{1}{\cancel{2}} \cdot \underset{1}{\cancel{3}} \cdot \underset{1}{\cancel{3}}}$

$= 5t - 1$

111. Explain why there are no values for x for which $\frac{x - 7}{x^2 + 49}$ is undefined.

112. Write a rational expression that is not defined for $x = 5$. Then explain why that is so.

REVIEW

State each property using the variables a, b, and when necessary, c.

113. a. The associative property of addition

b. The commutative property of multiplication

114. a. The distributive property

b. The zero-factor property

CHALLENGE PROBLEMS

Simplify.

115. $\dfrac{(x^2 + 2x + 1)(x^2 - 2x + 1)}{(x^2 - 1)^2}$

116. $\dfrac{2x^2 + 2x - 12}{x^3 + 3x^2 - 4x - 12}$

117. $\dfrac{x^3 - 27}{x^3 - 9x}$

118. $\dfrac{b^3 + a^3}{a^2 - ab + b^2}$

119. $\dfrac{m^3 + 64}{m^3 + 4m^2 + 3m + 12}$

120. $\dfrac{s^3 + s^2 - 6s - 6}{s^3 + 1}$

SECTION 7.2 Multiplying and Dividing Rational Expressions

OBJECTIVES

1 Multiply rational expressions.

2 Divide rational expressions.

3 Convert units of measurement.

ARE YOU READY?

The following problems review some basic skills that are needed when multiplying and dividing rational expressions.

1. Multiply: $\dfrac{3}{4} \cdot \dfrac{1}{5}$

2. What is the reciprocal of $\dfrac{8}{9}$?

3. Divide: $\dfrac{11}{16} \div \dfrac{7}{8}$

4. Factor: $x - x^2$

In this section, we will extend the rules for multiplying and dividing fractions to problems involving multiplication and division of rational expressions.

1 Multiply Rational Expressions.

Recall that to multiply fractions, we multiply their numerators and multiply their denominators. For example,

$$\frac{4}{7} \cdot \frac{3}{5} = \frac{4 \cdot 3}{7 \cdot 5}$$ Multiply the numerators and multiply the denominators.

$$= \frac{12}{35}$$

We use the same procedure to multiply rational expressions.

Multiplying Rational Expressions

To multiply two rational expressions, multiply their numerators and their denominators. Then, if possible, factor and simplify.

For any two rational expressions, $\frac{A}{B}$ and $\frac{C}{D}$,

$$\frac{A}{B} \cdot \frac{C}{D} = \frac{AC}{BD}$$

EXAMPLE 1 Multiply: **a.** $\frac{x+1}{x} \cdot \frac{9}{4x^2}$ **b.** $\frac{35x^3}{17y} \cdot \frac{y}{5x}$

Strategy To find the product, we will use the rule for multiplying rational expressions. In the process, we must be prepared to factor the numerators and denominators so that any common factors can be removed.

Why We want to give the result in simplified form, which requires that the numerator and denominator have no common factors other than 1.

Solution **a.** $\frac{x+1}{x} \cdot \frac{9}{4x^2} = \frac{9(x+1)}{4x^3}$ Multiply the numerators. Multiply the denominators.

Since the numerator and denominator do not share any common factors, $\frac{9(x+1)}{4x^3}$ cannot be simplified. We can leave the numerator in factored form, or we can distribute the multiplication by 9 and write the result as $\frac{9x+9}{4x^3}$.

b. $\frac{35x^3}{17y} \cdot \frac{y}{5x} = \frac{35x^3 \cdot y}{17y \cdot 5x}$ Multiply the numerators. Multiply the denominators.

It is obvious that the numerator and denominator of $\frac{35x^3 \cdot y}{17y \cdot 5x}$ have several common factors, such as 5, x, and y. These common factors become more apparent when we factor the numerator and denominator completely.

$$\frac{35x^3 \cdot y}{17y \cdot 5x} = \frac{5 \cdot 7 \cdot x \cdot x \cdot x \cdot y}{17 \cdot y \cdot 5 \cdot x}$$ To prepare to simplify, factor $35x^3$.

$$= \frac{\overset{1}{\cancel{5}} \cdot 7 \cdot \overset{1}{\cancel{x}} \cdot x \cdot x \cdot \overset{1}{\cancel{y}}}{17 \cdot \underset{1}{\cancel{y}} \cdot \underset{1}{\cancel{5}} \cdot \underset{1}{\cancel{x}}}$$ Simplify by replacing $\frac{5}{5}$, $\frac{x}{x}$, and $\frac{y}{y}$ with the equivalent fraction $\frac{1}{1}$. This removes the factor $\frac{5 \cdot x \cdot y}{5 \cdot x \cdot y} = 1$.

$$= \frac{7x^2}{17}$$ Multiply the remaining factors in the numerator. Multiply the remaining factors in the denominator.

Caution

When multiplying rational expressions, always write the result in simplest form by removing any factors common to the numerator and denominator.

Self Check 1 Multiply: **a.** $\frac{a+7}{a} \cdot \frac{6}{5a^3}$ **b.** $\frac{a^4}{8b} \cdot \frac{24b}{11a^3}$

Now Try Problems 13 and 17

EXAMPLE 2 Multiply: **a.** $\frac{x+3}{2x+4} \cdot \frac{6}{x^2-9}$ **b.** $\frac{8x^2-8x}{x^2+x-56} \cdot \frac{3x^2-22x+7}{x-x^2}$

Strategy To find the product, we will use the rule for multiplying rational expressions. In the process, we need to factor the monomials, binomials, or trinomials that are not prime, so that any common factors can be removed.

Why We want to give the result in simplified form, which requires that the numerator and denominator have no common factor other than 1.

Solution **a.** $\frac{x+3}{2x+4} \cdot \frac{6}{x^2-9} = \frac{(x+3)6}{(2x+4)(x^2-9)}$ Multiply the numerators and multiply the denominators.

$$= \frac{(x+3) \cdot 3 \cdot 2}{2(x+2)(x+3)(x-3)}$$ Factor 6. Factor out the GCF, 2, from $2x + 4$. Factor the difference of two squares, $x^2 - 9$.

$$= \frac{\overset{1}{\cancel{(x+3)}} \cdot 3 \cdot \overset{1}{\cancel{2}}}{\underset{1}{\cancel{2}}(x+2)\underset{1}{\cancel{(x+3)}}(x-3)}$$ Simplify by replacing $\frac{x+3}{x+3}$ and $\frac{2}{2}$ with $\frac{1}{1}$. This removes the factor $\frac{2 \cdot (x+3)}{2 \cdot (x+3)} = 1$.

$$= \frac{3}{(x+2)(x-3)}$$ Multiply the remaining factors in the numerator. Multiply the remaining factors in the denominator.

Notation

It is not necessary to multiply $(x+2)(x-3)$ in the denominator. When we add and subtract rational expressions in the next section, it is usually more convenient to leave the denominator in factored form as shown here.

b. $$\frac{8x^2 - 8x}{x^2 + x - 56} \cdot \frac{3x^2 - 22x + 7}{x - x^2}$$

$$= \frac{(8x^2 - 8x)(3x^2 - 22x + 7)}{(x^2 + x - 56)(x - x^2)}$$ Multiply the numerators and multiply the denominators.

$$= \frac{8x(x - 1)(3x - 1)(x - 7)}{(x + 8)(x - 7)x(1 - x)}$$ To prepare to simplify, factor all four polynomials.

$$= \frac{\overset{1}{8\cancel{x}}\overset{-1}{\cancel{(x - 1)}}(3x - 1)\overset{1}{\cancel{(x - 7)}}}{(x + 8)\underset{1}{\cancel{(x - 7)}}\underset{1}{\cancel{x}}\underset{1}{\cancel{(1 - x)}}}$$ Simplify. Since $x - 1$ and $1 - x$ are opposites, replace $\frac{x - 1}{1 - x}$ with $\frac{-1}{1}$. This removes the factor $\frac{x - 1}{1 - x} = -1$.

$$= \frac{-8(3x - 1)}{x + 8}$$ Multiply the remaining factors in the numerator. Multiply the remaining factors in the denominator.

The result can also be written as $-\frac{8(3x - 1)}{x + 8}$ or $-\frac{24x - 8}{x + 8}$.

Notation

We could distribute in the numerator and write the result as $\frac{-24x + 8}{x + 8}$. Check with your instructor to see which form of the result he or she prefers.

Self Check 2 Multiply: **a.** $\frac{3n - 9}{3n + 2} \cdot \frac{9n^2 - 4}{6}$ **b.** $\frac{m^2 - 4m - 5}{2m - m^2} \cdot \frac{2m^2 - 4m}{3m^2 - 14m - 5}$

Now Try Problems 21 and 29

EXAMPLE 3 Multiply: **a.** $63x\left(\frac{1}{7x}\right)$ **b.** $5a\left(\frac{3a - 1}{a}\right)$

Strategy We will write each of the monomials, $63x$ and $5a$, as rational expressions with denominator 1. (Remember, any number divided by 1 remains unchanged.) Then we will use the rule for multiplying rational expressions.

Why Writing $63x$ and $5a$ over 1 is helpful during the multiplication process when we multiply numerators and multiply denominators.

Solution **a.** $$63x\left(\frac{1}{7x}\right) = \frac{63x}{1}\left(\frac{1}{7x}\right)$$ Write $63x$ as a fraction: $63x = \frac{63x}{1}$.

$$= \frac{63x \cdot 1}{1 \cdot 7 \cdot x}$$ Multiply the numerators and multiply the denominators.

$$= \frac{9 \cdot \overset{1}{\cancel{7}} \cdot \overset{1}{\cancel{x}} \cdot 1}{1 \cdot \underset{1}{\cancel{7}} \cdot \underset{1}{\cancel{x}}}$$ Write $63x$ in factored form as $9 \cdot 7 \cdot x$. Then simplify by removing a factor equal to 1: $\frac{7x}{7x}$.

$$= 9$$ Because $\frac{9}{1} = 9$

b. $$5a\left(\frac{3a - 1}{a}\right) = \frac{5a}{1}\left(\frac{3a - 1}{a}\right)$$ Write $5a$ as a fraction: $5a = \frac{5a}{1}$.

$$= \frac{5\overset{1}{\cancel{a}}(3a - 1)}{1 \cdot \underset{1}{\cancel{a}}}$$ Multiply the numerators and multiply the denominators. Then simplify by removing a factor equal to 1: $\frac{a}{a}$.

$$= 5(3a - 1)$$

$$= 15a - 5$$ Distribute the multiplication by 5.

Self Check 3 Multiply: **a.** $36b\left(\frac{1}{6b}\right)$ **b.** $4x\left(\frac{x + 3}{x}\right)$

Now Try Problems 31 and 37

2 Divide Rational Expressions.

Recall that one number is the **reciprocal** of another if their product is 1. To find the reciprocal of a fraction, we invert its numerator and denominator. We have seen that to divide fractions, we multiply the first fraction by the reciprocal of the second fraction.

$$\frac{4}{7} \div \frac{3}{5} = \frac{4}{7} \cdot \frac{5}{3}$$ Invert $\frac{3}{5}$ and change the division to a multiplication.

$$= \frac{20}{21}$$ Multiply the numerators and multiply the denominators.

We use the same procedure to divide rational expressions.

Dividing Rational Expressions

To divide two rational expressions, multiply the first by the reciprocal of the second. Then, if possible, factor and simplify.

For any two rational expressions, $\frac{A}{B}$ and $\frac{C}{D}$, where $\frac{C}{D} \neq 0$,

$$\frac{A}{B} \div \frac{C}{D} = \frac{A}{B} \cdot \frac{D}{C} = \frac{AD}{BC}$$

EXAMPLE 4 Divide: **a.** $\frac{a}{13} \div \frac{17}{26}$ **b.** $\frac{9x}{35y} \div \frac{15x^2}{14}$

Strategy We will use the rule for dividing rational expressions. After multiplying by the reciprocal, we will factor the monomials that are not prime, and remove any common factors of the numerator and denominator.

Why We want to give the result in simplified form, which requires that the numerator and denominator have no common factor other than 1.

Solution

a. $\frac{a}{13} \div \frac{17}{26} = \frac{a}{13} \cdot \frac{26}{17}$ Multiply by the reciprocal of $\frac{17}{26}$.

$$= \frac{a \cdot 2 \cdot 13}{13 \cdot 17}$$ Multiply the numerators and multiply the denominators. Then, to prepare to simplify, factor 26 as $2 \cdot 13$.

$$= \frac{a \cdot 2 \cdot \overset{1}{\cancel{13}}}{\underset{1}{\cancel{13}} \cdot 17}$$ Simplify by removing common factors of the numerator and denominator.

$$= \frac{2a}{17}$$ Multiply the remaining factors in the numerator. Multiply the remaining factors in the denominator.

Caution

When dividing rational expressions, always write the result in simplest form, by removing any factors common to the numerator and denominator.

b. $\frac{9x}{35y} \div \frac{15x^2}{14} = \frac{9x}{35y} \cdot \frac{14}{15x^2}$ Multiply by the reciprocal of $\frac{15x^2}{14}$.

$$= \frac{3 \cdot 3 \cdot x \cdot 2 \cdot 7}{5 \cdot 7 \cdot y \cdot 3 \cdot 5 \cdot x \cdot x}$$ Multiply the numerators and multiply the denominators. Then, to prepare to simplify, factor 9, 35, 14, and $15x^2$.

$$= \frac{3 \cdot \overset{1}{\cancel{3}} \cdot \overset{1}{\cancel{x}} \cdot 2 \cdot \overset{1}{\cancel{7}}}{5 \cdot \underset{1}{\cancel{7}} \cdot y \cdot \underset{1}{\cancel{3}} \cdot 5 \cdot \underset{1}{\cancel{x}} \cdot x}$$ Simplify by removing factors equal to 1.

$$= \frac{6}{25xy}$$ Multiply the remaining factors in the numerator. Multiply the remaining factors in the denominator.

Self Check 4 Divide: $\frac{8a}{3b} \div \frac{16a^2}{9b^2}$

Now Try Problems 41 and 45

EXAMPLE 5 Divide: $\frac{x^2 + x}{3x - 15} \div \frac{(x + 1)^2}{6x - 30}$

Strategy To find the quotient, we will use the rule for dividing rational expressions. After multiplying by the reciprocal, we will factor the binomials that are not prime, and remove any common factors of the numerator and denominator.

Why We want to give the result in simplified form, which requires that the numerator and denominator have no common factor other than 1.

Solution

$$\frac{x^2 + x}{3x - 15} \div \frac{(x + 1)^2}{6x - 30}$$

$$= \frac{x^2 + x}{3x - 15} \cdot \frac{6x - 30}{(x + 1)^2}$$ Multiply by the reciprocal of $\frac{(x + 1)^2}{6x - 30}$.

$$= \frac{x(x + 1) \cdot 2 \cdot 3(x - 5)}{3(x - 5)(x + 1)(x + 1)}$$ Multiply the numerators and multiply the denominators. Then, to prepare to simplify, factor the binomials. Write $(x + 1)^2$ as repeated multiplication.

$$= \frac{x\overset{1}{\cancel{(x + 1)}} \cdot 2 \cdot \overset{1}{\cancel{3}}\overset{1}{\cancel{(x - 5)}}}{\underset{1}{\cancel{3}}\underset{1}{\cancel{(x - 5)}}\underset{1}{\cancel{(x + 1)}}(x + 1)}$$ Simplify by removing common factors of the numerator and denominator.

$$= \frac{2x}{x + 1}$$ Multiply the remaining factors in the numerator. Multiply the remaining factors in the denominator.

The Language of Algebra

To find the reciprocal of $\frac{(x + 1)^2}{6x - 30}$, we invert it. To **invert** means to turn upside down: $\frac{6x - 30}{(x + 1)^2}$. Some amusement park thrill rides have giant loops where the riders become inverted.

Self Check 5 Divide: $\frac{z^2 - 9}{z^2 + 4z + 3} \div \frac{z^2 - 3z}{(z + 1)^2}$

Now Try Problem 51

EXAMPLE 6 Divide: $\frac{2x^2 - 3xy - 2y^2}{2x + y} \div (4y^2 - x^2)$

Strategy We begin by writing $4y^2 - x^2$ as a rational expression by inserting a denominator 1. Then we will use the rule for dividing rational expressions.

Why Writing $4y^2 - x^2$ over 1 is helpful when we invert its numerator and denominator to find its reciprocal.

Solution

$$\frac{2x^2 - 3xy - 2y^2}{2x + y} \div (4y^2 - x^2)$$

$$= \frac{2x^2 - 3xy - 2y^2}{2x + y} \div \frac{4y^2 - x^2}{1}$$ Write $4y^2 - x^2$ as a fraction with a denominator of 1.

$$= \frac{2x^2 - 3xy - 2y^2}{2x + y} \cdot \frac{1}{4y^2 - x^2}$$ Multiply by the reciprocal of $\frac{4y^2 - x^2}{1}$.

$$= \frac{(2x + y)(x - 2y) \cdot 1}{(2x + y)(2y + x)(2y - x)}$$ Multiply the numerators and denominators. Then, to simplify, factor $2x^2 - 3xy - 2y^2$ and $4y^2 - x^2$.

$$= \frac{\overset{1}{\cancel{(2x+y)}}\overset{-1}{\cancel{(x-2y)}} \cdot 1}{\underset{1}{\cancel{(2x+y)}}(2y+x)\underset{1}{\cancel{(2y-x)}}}$$
Since $x - 2y$ and $2y - x$ are opposites, simplify by replacing $\frac{x-2y}{2y-x}$ with $\frac{-1}{1}$.

$$= \frac{-1}{2y+x}$$
Multiply the remaining factors in the numerator. Multiply the remaining factors in the denominator.

$$= -\frac{1}{2y+x}$$
Write the − sign in front of the fraction.

Self Check 6 Divide : $(b-a) \div \frac{a^2-b^2}{a^2+ab}$

Now Try Problem 63

3 Convert Units of Measurement.

We can use the concepts discussed in this section to make conversions from one unit of measure to another. *Unit conversion factors* play an important role in this process. A **unit conversion factor** is a fraction that has a value of 1. For example, we can use the fact that 1 square yard = 9 square feet to form two unit conversion factors:

$$\frac{1 \text{ yd}^2}{9 \text{ ft}^2} = 1$$ Read as "1 square yard per 9 square feet."

$$\frac{9 \text{ ft}^2}{1 \text{ yd}^2} = 1$$ Read as "9 square feet per 1 square yard."

Since a unit conversion factor is equal to 1, multiplying a measurement by a unit conversion factor does not change the measurement, it only changes the units of measure.

Success Tip

Remember that unit conversion factors are equal to 1. Some examples are:

$$\frac{12 \text{ in}}{1 \text{ ft}} = 1 \qquad \frac{60 \text{ min}}{1 \text{ hr}} = 1$$

EXAMPLE 7

Carpeting. A roll of carpeting is 12 feet wide and 150 feet long. Find the number of square yards of carpeting on the roll.

Strategy We will begin by determining the number of square feet of carpeting on the roll. Then we will multiply that result by a unit conversion factor.

Why A properly chosen unit conversion factor can convert the number of square feet of carpeting on the roll to the number of square yards on the roll.

The Language of Algebra

The method we are using to convert from one unit of measurement to another is often referred to as **dimensional analysis.**

Solution

When unrolled, the carpeting forms a rectangular shape with an area of $12 \cdot 150 = 1{,}800$ square feet. We will multiply $1{,}800 \text{ ft}^2$ by a unit conversion factor such that the units of ft^2 are removed and the units of yd^2 are introduced. Since $1 \text{ yd}^2 = 9 \text{ ft}^2$, we will use $\frac{1\text{yd}^2}{9\text{ft}^2}$.

©iStockphoto.com/thumb

$$\frac{1{,}800 \text{ ft}^2}{1 \text{ roll}} = \frac{1{,}800 \text{ ft}^2}{1 \text{ roll}} \cdot \frac{\mathbf{1\ yd^2}}{\mathbf{9\ ft^2}}$$
Multiply by a unit conversion factor that relates yd^2 to ft^2.

$$= \frac{1{,}800 \cancel{\text{ft}^2}}{1 \text{ roll}} \cdot \frac{1 \text{ yd}^2}{9 \cancel{\text{ft}^2}}$$
Remove the units of ft^2 that are common to the numerator and denominator.

$$= \frac{200 \text{ yd}^2}{1 \text{ roll}}$$
Divide 1,800 by 9 to get 200.

There are 200 yd^2 of carpeting on the roll.

Self Check 7 Convert $5{,}400 \text{ ft}^2$ to square yards.

Now Try Problem 67

EXAMPLE 8 **The Speed of Light.** The speed with which light moves through space is about 186,000 miles per second. Express this speed in miles per minute.

Strategy The speed of light can be expressed as $\frac{186{,}000 \text{ mi}}{1 \text{ sec}}$. We will multiply that fraction by a unit conversion factor.

Why A properly chosen unit conversion factor can convert the number of miles traveled per second to the number of miles traveled per minute.

Solution We will multiply $\frac{186{,}000 \text{ mi}}{1 \text{ sec}}$ by a unit conversion factor such that the units of seconds are removed and the units of minutes are introduced. Since 60 seconds = 1 minute, we will use $\frac{60 \text{ sec}}{1 \text{ min}}$.

$$\frac{186{,}000 \text{ mi}}{1 \text{ sec}} = \frac{186{,}000 \text{ mi}}{1 \text{ sec}} \cdot \frac{60 \text{ sec}}{1 \text{ min}}$$ Multiply by a unit conversion factor that relates seconds to minutes.

$$= \frac{186{,}000 \text{ mi}}{1 \not{\text{sec}}} \cdot \frac{60 \not{\text{sec}}}{1 \text{ min}}$$ Remove the units of seconds that are common to the numerator and denominator.

$$= \frac{11{,}160{,}000 \text{ mi}}{1 \text{ min}}$$ Multiply 186,000 and 60 to get 11,160,000.

The speed of light is about 11,160,000 miles per minute.

Success Tip

We can remove common units just as we remove factors that are common to the numerator and denominator when multiplying rational expressions.

Self Check 8 **Insects.** A mosquito beats it wings about 600 times per second. How many times is that per minute?

Now Try Problem 71

SECTION 7.2 STUDY SET

VOCABULARY

Fill in the blanks.

1. The __________ of $\frac{x^2 + 6x + 1}{10x}$ is $\frac{10x}{x^2 + 6x + 1}$.

2. A _____ conversion factor is a fraction that is equal to 1, such as $\frac{3 \text{ ft}}{1 \text{ yd}}$.

CONCEPTS

Fill in the blanks.

3. a. To multiply rational expressions, multiply their __________ and multiply their __________. To divide two rational expressions, multiply the first by the __________ of the second. In symbols,

   b. $\frac{A}{B} \cdot \frac{C}{D} = \underline{\quad}$ and $\frac{A}{B} \div \frac{C}{D} = \frac{A}{B} \cdot \underline{\quad}$

Simplify each expression.

4. $\frac{(x + 7) \cdot 2 \cdot 5}{5(x + 1)(x + 7)(x - 9)}$

5. $\frac{y \cdot y \cdot y(15 - y)}{y(y - 15)(y + 1)}$

6. a. Write $3x + 5$ in fractional form.

   b. What is the reciprocal of $18x$?

7. Find the product of the rational expression and its reciprocal.

$$\frac{3}{x + 2} \cdot \frac{x + 2}{3}$$

8. Use the fact that 1 tablespoon = 3 teaspoons to write two unit conversion factors.

NOTATION

9. What units are common to the numerator and denominator of the following product?

$$\frac{45 \text{ ft}}{1} \cdot \frac{1 \text{ yd}}{3 \text{ ft}}$$

10. a. What fact is indicated by the unit conversion factor $\frac{1 \text{ day}}{24 \text{ hours}}$?

    b. Fill in the blank: $\frac{1 \text{ day}}{24 \text{ hours}} = \quad$.

GUIDED PRACTICE

Multiply, and then simplify, if possible. See Example 1.

11. $\frac{3}{7} \cdot \frac{y}{2}$

12. $\frac{2}{7} \cdot \frac{z}{3}$

13. $\frac{y + 2}{y} \cdot \frac{3}{y^2}$

14. $\frac{4}{a + 1} \cdot \frac{a}{7}$

15. $\frac{35n}{12} \cdot \frac{16}{7n^2}$

16. $\frac{11m}{21} \cdot \frac{14}{55m^3}$

17. $\frac{2x^2y}{3xy} \cdot \frac{3xy^2}{2}$

18. $\frac{2x^2z}{z} \cdot \frac{5x}{z}$

Multiply, and then simplify, if possible. See Example 2.

19. $\frac{x+5}{5} \cdot \frac{x}{x+5}$

20. $\frac{a-9}{9} \cdot \frac{8a}{a-9}$

21. $\frac{2x+6}{x+3} \cdot \frac{3}{4x}$

22. $\frac{3y-9}{y-3} \cdot \frac{y}{3y^2}$

23. $\frac{(x-1)^2}{x+2} \cdot \frac{x+2}{x+1}$

24. $\frac{(y-3)^2}{y-5} \cdot \frac{y-5}{y-3}$

25. $\frac{x^2-x}{x} \cdot \frac{3x-6}{3-3x}$

26. $\frac{5z-10}{z+2} \cdot \frac{3}{6-3z}$

27. $\frac{x^2+x-6}{5x} \cdot \frac{5x-10}{x+3}$

28. $\frac{z^2+4z-5}{5z-5} \cdot \frac{5z}{z+5}$

29. $\frac{m^2-2m-3}{2m+4} \cdot \frac{m^2-4}{m^2+3m+2}$

30. $\frac{p^2-p-6}{3p-9} \cdot \frac{2p^2-5p-3}{p^2-3p}$

Multiply, and then simplify, if possible. See Example 3.

31. $7m\left(\frac{5}{m}\right)$

32. $9p\left(\frac{10}{p}\right)$

33. $15x\left(\frac{x+1}{5x}\right)$

34. $30t\left(\frac{t-7}{10t}\right)$

35. $12y\left(\frac{5y-8}{6y}\right)$

36. $16x\left(\frac{3x+8}{4x}\right)$

37. $24\left(\frac{3a-5}{2a}\right)$

38. $28\left(\frac{8-3t}{4t}\right)$

Divide, and then simplify, if possible. See Example 4.

39. $\frac{2}{y} \div \frac{4}{3}$

40. $\frac{3}{a} \div \frac{9}{5}$

41. $\frac{3a}{25} \div \frac{1}{5}$

42. $\frac{3y}{8} \div \frac{3}{2}$

43. $\frac{x^3}{18y} \div \frac{x}{6y}$

44. $\frac{21x}{z^2} \div \frac{7x^3}{z^5}$

45. $\frac{27p^4}{35q} \div \frac{9p}{21q}$

46. $\frac{12}{25s^5} \div \frac{10}{15s^2}$

Divide, and then simplify, if possible. See Example 5.

47. $\frac{9a-18}{28} \div \frac{9a^3}{35}$

48. $\frac{3x+6}{40} \div \frac{3x^2}{24}$

49. $\frac{x^2-4}{3x+6} \div \frac{2-x}{x+2}$

50. $\frac{x^2-9}{5x+15} \div \frac{3-x}{x+3}$

51. $\frac{x^2+7x}{5x-10} \div \frac{(x+7)^2}{15x-30}$

52. $\frac{x^2-10x}{7x+7} \div \frac{(x-10)^2}{35x+35}$

53. $\frac{m^2+m-20}{m} \div \frac{4-m}{m}$

54. $\frac{n^2+4n-21}{n} \div \frac{3-n}{n}$

55. $\frac{t^2+5t-14}{t} \div \frac{t-2}{t}$

56. $\frac{r^2+12r+11}{r} \div \frac{r+11}{r}$

57. $\frac{x^2-2x-35}{3x^2+27x} \div \frac{3x^2+17x+10}{18x^2+12x}$

58. $\frac{x^2-x-6}{2x^2+9x+10} \div \frac{x^2-25}{2x^2+15x+25}$

Divide, and then simplify, if possible. See Example 6.

59. $\frac{x^2-1}{3x-3} \div (x+1)$

60. $\frac{x^2-16}{x-4} \div (3x+12)$

61. $\frac{n^2-10n+9}{n-9} \div (n-1)$

62. $\frac{r^2-11r+18}{r-9} \div (r-2)$

63. $\frac{2r-3s}{12} \div (4r^2-12rs+9s^2)$

64. $\frac{3m+n}{18} \div (9m^2+6mn+n^2)$

65. $24n^2 \div \frac{18n^3}{n-1}$

66. $12m \div \frac{16m^2}{m+4}$

Complete each unit conversion. See Examples 7 and 8.

67. $\frac{150 \text{ yards}}{1} \cdot \frac{3 \text{ feet}}{1 \text{ yard}} = ?$

68. $\frac{60 \text{ inches}}{1} \cdot \frac{1 \text{ feet}}{12 \text{ inches}} = ?$

69. $\frac{6 \text{ pints}}{1} \cdot \frac{1 \text{gallon}}{8 \text{ pints}} = ?$

70. $\frac{4 \text{ cups}}{1} \cdot \frac{1 \text{gallon}}{16 \text{ cups}} = ?$

71. $\frac{30 \text{ miles}}{1 \text{ hour}} \cdot \frac{1 \text{ hour}}{60 \text{ minutes}} = ?$

72. $\frac{300 \text{ meters}}{3 \text{ months}} \cdot \frac{12 \text{ months}}{1 \text{ year}} = ?$

73. $\frac{30 \text{ meters}}{1 \text{ seconds}} \cdot \frac{60 \text{ seconds}}{1 \text{ minute}} = ?$

74. $\frac{288 \text{ inches}^2}{1 \text{ year}} \cdot \frac{1 \text{ feet}^2}{144 \text{ inches}^2} = ?$

TRY IT YOURSELF

Perform the operations and simplify, if possible.

75. $\frac{b^2-5b+6}{b^2-10b+16} \div \frac{b^2+2b}{b^2-6b-16}$

76. $\frac{m^2+m-6}{m^2-6m+9} \div \frac{m^2-4}{m^2-9}$

77. $\frac{5x+5}{25} \cdot \frac{5}{(x+1)^3}$

78. $\frac{7t-7}{28} \cdot \frac{4}{(t-1)^4}$

79. $\frac{6a^2}{a^2+6a+9} \cdot \frac{(a+3)^4}{4a^5}$

80. $\frac{9b^3}{b^2-8b+16} \cdot \frac{(b-4)^4}{15b^8}$

81. $\frac{36c^2-49d^2}{3d^3} \div \frac{12c+14d}{d^4}$

82. $\frac{25y^2-16z^2}{2yz} \div \frac{10y-8z}{y^2}$

83. $10h\left(\frac{5h-3}{2h}\right)$

84. $33r\left(\frac{5r+4}{11r}\right)$

85. $\dfrac{n^2-9}{n^2-3n} \div \dfrac{n+3}{n^2-n}$

86. $\dfrac{b^2-b}{b+2} \div \dfrac{b^2-2b}{b^2-4}$

87. $\dfrac{10r^2s}{6rs^2} \cdot \dfrac{3r^3}{2rs}$

88. $\dfrac{3a^3b}{25cd^3} \cdot \dfrac{5cd^2}{6ab}$

89. $\dfrac{7}{3p^3} \cdot \dfrac{p+2}{p}$

90. $\dfrac{5t^2}{11} \cdot \dfrac{2t}{t-5}$

91. $\dfrac{5x^2+13x-6}{x+3} \div \dfrac{5x^2-17x+6}{x-2}$

92. $\dfrac{3p^2+5p-2}{p^3+2p^2} \div \dfrac{6p^2+13p-5}{2p^3+5p^2}$

93. $\dfrac{4x^2-12xy+9y^2}{x^3y^2} \cdot \dfrac{x^3y}{4x^2-9y^2}$

94. $\dfrac{ab^4}{25a^2-16b^2} \cdot \dfrac{25a^2-40ab+16b^2}{a^2b^4}$

95. $\dfrac{x-2}{x} \cdot \dfrac{2x}{2-x}$

96. $\dfrac{y-3}{y} \cdot \dfrac{3y}{3-y}$

Look Alikes . . .

97. a. $\dfrac{3x+6}{4} \cdot \dfrac{4x+8}{3}$ b. $\dfrac{3x+6}{4} \div \dfrac{4x+8}{3}$

98. a. $\dfrac{4a-8}{5} \cdot \dfrac{5a-10}{4}$ b. $\dfrac{4a-8}{5} \div \dfrac{5a-10}{4}$

99. a. $\dfrac{x^2-5x+6}{2x-4} \cdot \dfrac{2x-6}{x-2}$ b. $\dfrac{x^2-5x+6}{2x-4} \div \dfrac{2x-6}{x-2}$

100. a. $\dfrac{x^2+9x+20}{9x+36} \cdot \dfrac{9x+45}{x+4}$ b. $\dfrac{x^2+9x+20}{9x+36} \div \dfrac{9x+45}{x+4}$

APPLICATIONS

101. **Geometry.** Find the area of the rectangle.

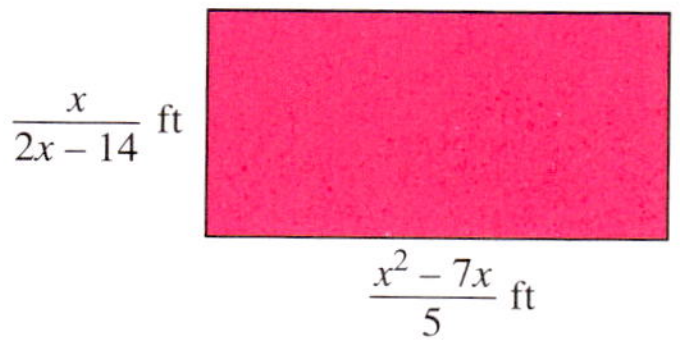

102. **Motion.** The table contains algebraic expressions for the rate an object travels and the time traveled at that rate. Complete the table.

Rate (mph)	Time (hr)	Distance (mi)
$\dfrac{k^2+k-6}{k-3}$	$\dfrac{k^2-9}{k^2-4}$	

103. **Talking.** According to the *Sacramento Bee* newspaper, the number of words an average man speaks a day is about 12,000. How many words does an average man speak in 1 year? (*Hint:* 365 days = 1 year.)

104. **Classroom Space.** The recommended size of an elementary school classroom in the United States is approximately 900 square feet. Convert this to square yards.

105. **Natural Light.** According to the University of Georgia School Design and Planning Laboratory, the basic classroom should have at least 72 square feet of windows for natural light. Convert this to square yards.

106. **Trucking.** A cement truck holds 9 cubic yards of concrete. How many cubic feet of concrete does it hold? (*Hint:* 27 cubic feet = 1 cubic yard.)

107. **Bears.** The maximum speed a grizzly bear can run is about 30 miles per hour. What is its maximum speed in miles per minute?

108. **Fuel Economy.** Use the information that follows to determine the miles per fluid ounce of gasoline for city and for highway driving for the Ford Ranger. (*Hint:* 1 gallon = 128 fluid ounces.)

2010 Ford Ranger

Fuel Economy

Fuel Type	Regular
MPG (city)	**16**
MPG (highway)	**20**

109. **TV Trivia.** On the comedy television series *Green Acres* (1965–1971), New York socialites Oliver Wendell Douglas (played by Eddie Albert) and his wife, Lisa Douglas (played by Eva Gabor), move from New York to purchase a 160-acre farm in Hooterville. Convert this to square miles. (*Hint:* 1 square mile = 640 acres.)

110. **Camping.** The capacity of backpacks is usually given in cubic inches. Convert a backpack capacity of 5,400 cubic inches to cubic feet. (*Hint:* 1 cubic foot = 1,728 cubic inches.)

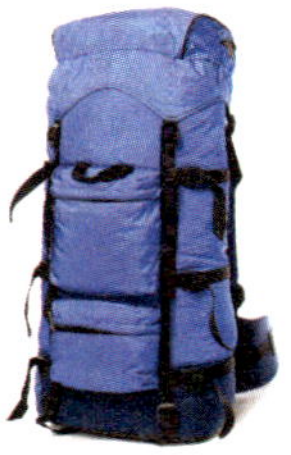

WRITING

111. Explain how to multiply rational expressions.

112. To divide rational expressions, you must first know how to multiply rational expressions. Explain why.

113. Explain why 60 miles per hour and 1 mile per minute are the same speed.

114. Explain why the unit conversion factor $\frac{1 \text{ ft}}{12 \text{ in.}}$ is equal to 1.

REVIEW

115. Hardware. A brace has a length that is 2 inches less than twice the width of the shelf that it supports. The brace is anchored to the wall 8 inches below the shelf. Find the width of the shelf and the length of the brace.

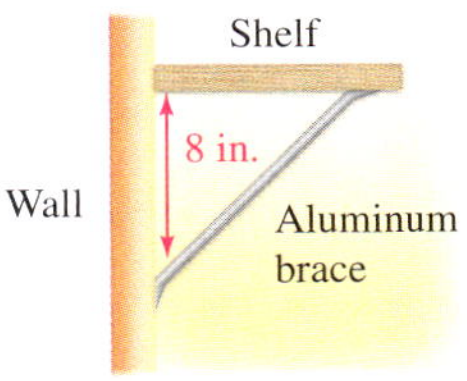

116. Solve $A = \frac{1}{2}h(b + d)$ for b.

CHALLENGE PROBLEMS

Perform the operations. Simplify, if possible.

117. $\frac{c^3 - 2c^2 + 5c - 10}{c^2 - c - 2} \cdot \frac{c^3 + c^2 - 5c - 5}{c^4 - 25}$

118. $\frac{x^3 - y^3}{x^3 + y^3} \div \frac{x^3 + x^2y + xy^2}{x^2y - xy^2 + y^3}$

119. $\frac{-x^3 + x^2 + 6x}{3x^3 + 21x^2} \div \left(\frac{2x + 4}{3x^2} \div \frac{2x + 14}{x^2 - 3x}\right)$

120. $\frac{x^2 - y^2}{2x^4 - 2x^3} \div \left(\frac{x - y}{2x^2} \div \frac{x + y}{x^2 + 2xy + y^2}\right)$

SECTION 7.3 Adding and Subtracting with Like Denominators; Least Common Denominators

OBJECTIVES

1. Add and subtract rational expressions that have the same denominator.
2. Find the least common denominator.
3. Build rational expressions into equivalent expressions.

ARE YOU READY?

The following problems review some basic skills that are needed when adding and subtracting rational expressions with like denominators.

1. Add: $\frac{5}{11} + \frac{3}{11}$

2. Subtract: $\frac{17}{21} - \frac{10}{21}$

3. Simplify: $x^2 + 8x - (3x - 1)$

4. Simplify: $\frac{x + 2}{x^2 - 5x - 14}$

5. Factor completely: $24x^2$

6. Factor: $5x + 5$

In this section, we extend the rules for adding and subtracting fractions to problems involving addition and subtraction of rational expressions.

1 Add and Subtract Rational Expressions That Have the Same Denominator.

Recall from Chapter 1 that to add (or subtract) fractions that have the same denominator, we add (or subtract) their numerators and write the sum (or difference) over the common denominator. For example,

$$\frac{3}{7} + \frac{2}{7} = \frac{3 + 2}{7} = \frac{5}{7} \qquad \text{and} \qquad \frac{18}{25} - \frac{9}{25} = \frac{18 - 9}{25} = \frac{9}{25}$$

The Language of Algebra

We can describe $\frac{3}{7}$ and $\frac{2}{7}$ as having the **same** denominator, **common** denominators, or **like** denominators.

We use the same procedure to add and subtract rational expressions with like denominators.

Adding and Subtracting Rational Expressions That Have the Same Denominator

To add (or subtract) rational expressions that have same denominator, add (or subtract) their numerators and write the sum (or difference) over the common denominator. Then, if possible, factor and simplify.

If $\frac{A}{D}$ and $\frac{B}{D}$ are rational expressions,

$$\frac{A}{D} + \frac{B}{D} = \frac{A + B}{D} \qquad \text{and} \qquad \frac{A}{D} - \frac{B}{D} = \frac{A - B}{D}$$

EXAMPLE 1 Add: **a.** $\frac{x}{8} + \frac{3x}{8}$ **b.** $\frac{4s - 9}{9t} + \frac{7}{9t}$

Strategy We will add the numerators and write the sum over the common denominator. Then, if possible, we will factor and simplify.

Why This is the rule for adding rational expressions, such as these, that have the same denominator.

Solution

a. The given rational expressions have the same denominator, 8.

$$\frac{x}{8} + \frac{3x}{8} = \frac{x + 3x}{8}$$

$$= \frac{4x}{8}$$ Combine like terms in the numerator: $x + 3x = 4x$. This result can be simplified.

$$= \frac{\overset{1}{\cancel{4}} \cdot x}{2 \cdot \underset{1}{\cancel{4}}}$$ Factor 8 as $4 \cdot 2$. Then simplify by removing a factor equal to 1.

$$= \frac{x}{2}$$

Caution

We *do not* add rational expressions by adding numerators and adding denominators!

$$\cancel{\frac{x}{8} + \frac{3x}{8} = \frac{4x}{16}}$$

The same caution applies when subtracting rational expressions.

b. The given rational expressions have the same denominator, $9t$.

$$\frac{4s - 9}{9t} + \frac{7}{9t} = \frac{4s - 9 + 7}{9t}$$ Add the numerators. Write the sum over the common denominator, $9t$.

$$= \frac{4s - 2}{9t}$$ Combine like terms in the numerator: $-9 + 7 = -2$.

To attempt to simplify the result, we factor the numerator to get $\frac{2(2s - 1)}{9t}$. Since the numerator and denominator do not have any common factors, $\frac{4s - 2}{9t}$ cannot be simplified. Thus,

$$\frac{4s - 9}{9t} + \frac{7}{9t} = \frac{4s - 2}{9t}$$

Notation

The numerator of the result may be written two ways:

Not factored	Factored
$\frac{4s - 2}{9t}$	$\frac{2(2s - 1)}{9t}$

Check with your instructor to see which form he or she prefers.

Self Check 1 Add: **a.** $\frac{2x}{15} + \frac{4x}{15}$ **b.** $\frac{3m - 8}{23n} + \frac{2}{23n}$

Now Try Problems 17 and 19

EXAMPLE 2 Add: **a.** $\frac{3x + 21}{5x + 10} + \frac{8x + 1}{5x + 10}$ **b.** $\frac{x^2 + 9x - 7}{2x(x - 6)} + \frac{x^2 - 9x}{(x - 6)2x}$

Strategy We will add the numerators and write the sum over the common denominator. Then, if possible, we will factor and simplify.

Why This is the rule for adding rational expressions that have the same denominator.

Solution

a. $$\frac{3x + 21}{5x + 10} + \frac{8x + 1}{5x + 10} = \frac{3x + 21 + 8x + 1}{5x + 10}$$ Add the numerators. Write the sum over the common denominator, $5x + 10$.

$$= \frac{11x + 22}{5x + 10}$$ Combine like terms in the numerator: $3x + 8x = 11x$ and $21 + 1 = 22$.

$$= \frac{11\overset{1}{\cancel{(x + 2)}}}{5\underset{1}{\cancel{(x + 2)}}}$$ Factor the numerator: The GCF is 11. Factor the denominator: The GCF is 5. Then simplify by removing a factor equal to 1.

$$= \frac{11}{5}$$

Caution

When adding or subtracting rational expressions, always write the result in simplest form by removing any factors common to the numerator and denominator.

Caution

Don't make the mistake of removing the 2's in the result. This is incorrect because 2 is not a factor of the *entire* numerator.

$$\frac{\overset{1}{\cancel{2}}x^2 - 7}{\underset{1}{\cancel{2}}x(x - 6)}$$

b. By the commutative property of multiplication, $2x(x - 6) = (x - 6)2x$. Therefore, the denominators are the same. We add the numerators and write the sum over the common denominator.

$$\frac{x^2 + 9x - 7}{2x(x - 6)} + \frac{x^2 - 9x}{(x - 6)2x} = \frac{x^2 + 9x - 7 + x^2 - 9x}{2x(x - 6)}$$

$$= \frac{2x^2 - 7}{2x(x - 6)}$$ Combine like terms in the numerator: $x^2 + x^2 = 2x^2$ and $9x - 9x = 0$.

Since the numerator, $2x^2 - 7$, does not factor, $\frac{2x^2 - 7}{2x(x - 6)}$ is in simplest form.

Self Check 2 Add: **a.** $\frac{m + 3}{3m - 9} + \frac{m - 9}{3m - 9}$ **b.** $\frac{c^2 - c}{(c - 1)(c + 2)} + \frac{c^2 - 10c}{(c + 2)(c - 1)}$

Now Try Problems 21 and 23

The method used to subtract rational expressions with like denominators is similar to the method used for adding such expressions.

EXAMPLE 3 Subtract: $\frac{x + 6}{x^2 + 4x - 5} - \frac{1}{x^2 + 4x - 5}$

Strategy We will subtract the numerators and write the sum over the common denominator. Then, if possible, we will factor and simplify.

Why This is the rule for subtracting rational expressions that have the same denominator.

Solution

$$\frac{x + 6}{x^2 + 4x - 5} - \frac{1}{x^2 + 4x - 5} = \frac{x + 6 - 1}{x^2 + 4x - 5}$$ Subtract the numerators. Write the difference over the common denominator, $x^2 + 4x - 5$.

$$= \frac{x + 5}{x^2 + 4x - 5}$$ Combine like terms in the numerator: $6 - 1 = 5$.

$$= \frac{\overset{1}{\cancel{x + 5}}}{\underset{1}{\cancel{(x + 5)}}(x - 1)}$$ Factor the denominator. Then simplify by removing a factor equal to 1.

$$= \frac{1}{x - 1}$$

Self Check 3 Subtract: $\frac{n - 3}{n^2 - 16} - \frac{1}{n^2 - 16}$

Now Try Problem 29

Be very careful when subtracting rational expressions. There is great potential for making sign errors.

EXAMPLE 4 Subtract: **a.** $\frac{x^2 + 10x}{x + 3} - \frac{4x - 9}{x + 3}$ **b.** $\frac{x^2}{(x + 7)(x - 8)} - \frac{-x^2 + 14x}{(x + 7)(x - 8)}$

Strategy We will use the rule for subtracting rational expressions that have the same denominators. In both cases, it is important to note that the numerator of the second fraction has *two* terms.

Why We must make sure that the entire numerator (not just the first term) of the second fraction is subtracted.

Solution

a. To subtract the numerators, each term of $4x - 9$ must be subtracted from $x^2 + 10x$.

This − symbol applies to the entire numerator $4x - 9$.

This numerator is written within parentheses to make sure that we subtract both of its terms.

$$\frac{x^2 + 10x}{x + 3} - \frac{4x - 9}{x + 3} = \frac{x^2 + 10x - (4x - 9)}{x + 3}$$ Subtract the numerators. Write the difference over the common denominator.

$$= \frac{x^2 + 10x - 4x + 9}{x + 3}$$ In the numerator, use the distributive property: $-(4x - 9) = -1(4x - 9) = -4x + 9$.

$$= \frac{x^2 + 6x + 9}{x + 3}$$ Combine like terms in the numerator: $10x - 4x = 6x$.

$$= \frac{(x + 3)(x + 3)}{x + 3}$$ To see if the result simplifies, factor the numerator.

$$= \frac{\overset{1}{\cancel{(x + 3)}}(x + 3)}{\underset{1}{\cancel{x + 3}}}$$ Simplify by removing a factor equal to 1.

$$= x + 3$$

b. We subtract the numerators and write the difference over the common denominator.

$$\frac{x^2}{(x + 7)(x - 8)} - \frac{-x^2 + 14x}{(x + 7)(x - 8)} = \frac{x^2 - (-x^2 + 14x)}{(x + 7)(x - 8)}$$ Write the second numerator within parentheses.

$$= \frac{x^2 + x^2 - 14x}{(x + 7)(x - 8)}$$ Use the distributive property: $-(-x^2 + 14x) = x^2 - 14x$.

$$= \frac{2x^2 - 14x}{(x + 7)(x - 8)}$$ In the numerator, combine like terms: $x^2 + x^2 = 2x^2$.

In an attempt to simplify, we can factor $2x^2 - 14x$ as $2x(x - 7)$. However, the numerator and denominator have no common factors. The result is in simplest form.

Caution

Don't make this common error by forgetting to write $4x - 9$ within parentheses.

$$\frac{x^2 + 10x - 4x - 9}{x + 3}$$

Notation

A fraction bar is a grouping symbol. Think of parentheses around the terms of the numerator and the denominator.

$$\frac{4x - 9}{x + 3} = \frac{(4x - 9)}{(x + 3)}$$

Success Tip

$-(-x^2 + 14x)$ means $-1(-x^2 + 14x)$. The multiplication by -1 changes the sign of each term within the parentheses:

$$-(-x^2 + 14x) = x^2 - 14x$$

Self Check 4 Subtract: **a.** $\frac{x^2 + 3x}{x - 1} - \frac{5x - 1}{x - 1}$

b. $\frac{3y^2}{(y + 3)(y - 3)} - \frac{-3y^2 + y}{(y + 3)(y - 3)}$

Now Try Problems 33 and 39

2 Find the Least Common Denominator.

We will now discuss two skills that are needed for adding and subtracting rational expressions that have unlike denominators. To begin, let's consider

$$\frac{11}{8x} + \frac{7}{18x^2}$$

To add these expressions, we must express them as equivalent expressions with a common denominator. The **least common denominator (LCD)** is usually the easiest one to use.

Finding the LCD

To find the LCD of a set of rational expressions:

1. Factor each denominator completely.
2. The LCD is a product that uses each different factor obtained in step 1 the greatest number of times it appears in any one factorization.

EXAMPLE 5 Find the LCD of each pair of rational expressions: **a.** $\frac{11}{8x}$ and $\frac{7}{18x^2}$ **b.** $\frac{20}{x}$ and $\frac{4x}{x-9}$

Strategy We will begin by factoring completely the denominator of each rational expression. Then we will form a product using each factor the greatest number of times it appears in any one factorization.

Why Since the LCD must contain the factors of each denominator, we need to write each denominator in factored form.

Solution

a. $8x = 2 \cdot 2 \cdot 2 \cdot x$ Prime factor 8.

$18x^2 = 2 \cdot 3 \cdot 3 \cdot x \cdot x$ Prime factor 18. Factor x^2.

$$\begin{array}{r|l} 2 & 8 \\ \hline 2 & 4 \\ \hline & 2 \end{array} \qquad \begin{array}{r|l} 2 & 18 \\ \hline 3 & 9 \\ \hline & 3 \end{array}$$

Success Tip

The factorizations can be written:

$$8x = 2^3 \cdot x$$
$$18x^2 = 2 \cdot 3^2 \cdot x^2$$

Note that the highest power of each factor is used to form the LCD.

$$\text{LCD} = 2^3 \cdot 3^2 \cdot x^2 = 72x^2$$

The factorizations of $8x$ and $18x^2$ contain the factors 2, 3, and x. The LCD of $\frac{11}{8x}$ and $\frac{7}{18x^2}$ should contain each factor of $8x$ and $18x^2$ the greatest number of times it appears in any one factorization.

The greatest number of times the factor 2 appears is three times.
The greatest number of times the factor 3 appears is twice.
The greatest number of times the factor x appears is twice.

$$\text{LCD} = 2 \cdot 2 \cdot 2 \cdot 3 \cdot 3 \cdot x \cdot x$$
$$= 72x^2$$

The LCD for $\frac{11}{8x}$ and $\frac{7}{18x^2}$ is $72x^2$.

b. Since the denominators of $\frac{20}{x}$ and $\frac{4x}{x-9}$ are completely factored, the factor x appears once and the factor $x - 9$ appears once. Thus, the LCD is $x(x - 9)$.

Self Check 5 Find the LCD of each pair of rational expressions:
a. $\frac{y+7}{6y^3}$ and $\frac{7}{75y}$ **b.** $\frac{a-3}{a+3}$ and $\frac{21}{a}$

Now Try Problems 41 and 47

EXAMPLE 6 Find the LCD of each pair of rational expressions:
a. $\frac{x}{7x+7}$ and $\frac{x-2}{5x+5}$ **b.** $\frac{6-x}{x^2+8x+16}$ and $\frac{15x}{x^2-16}$

Strategy We will begin by factoring completely each binomial and trinomial in the denominators of the rational expressions. Then we will form a product using each factor the greatest number of times it appears in any one factorization.

Why Since the LCD must contain the factors of each denominator, we need to write each denominator in factored form.

Solution

a. Factor each denominator completely.

$7x + 7 = 7(x + 1)$ The GCF is 7.

$5x + 5 = 5(x + 1)$ The GCF is 5.

Notation

Rather than performing the multiplication, it is often better to leave an LCD in factored form:

$$\text{LCD} = 35(x + 1)$$

The factorizations of $7x + 7$ and $5x + 5$ contain the factors 7, 5, and $x + 1$. The LCD of $\frac{x}{7x+7}$ and $\frac{x-2}{5x+5}$ should contain each factor of $7x + 7$ and $5x + 5$ the greatest number of times it appears in any one factorization.

The greatest number of times the factor 7 appears is once.
The greatest number of times the factor 5 appears is once.
The greatest number of times the factor $x + 1$ appears is once.

$$\text{LCD} = 7 \cdot 5 \cdot (x + 1) = 35(x + 1)$$

b. Factor each denominator completely.

$$x^2 + 8x + 16 = (x + 4)(x + 4) \quad \text{Factor the trinomial.}$$
$$x^2 - 16 = (x + 4)(x - 4) \quad \text{Factor the difference of two squares.}$$

The factorizations of $x^2 + 8x + 16$ and $x^2 - 16$ contain the factors $x + 4$ and $x - 4$.

The greatest number of times the factor $x + 4$ appears is twice.
The greatest number of times the factor $x - 4$ appears is once.

$$\text{LCD} = (x + 4)(x + 4)(x - 4) = (x + 4)^2(x - 4)$$

Self Check 6 Find the LCD: **a.** $\frac{x^3}{x^2 - 6x}$ and $\frac{25x}{2x - 12}$
b. $\frac{m + 1}{m^2 - 9}$ and $\frac{6m^2}{m^2 - 6m + 9}$

Now Try Problems 51 and 57

3 Build Rational Expressions into Equivalent Expressions.

Recall from Chapter 1 that writing a fraction as an equivalent fraction with a larger denominator is called **building the fraction.** For example, to write $\frac{3}{5}$ as an equivalent fraction with a denominator of 35, we multiply it by 1 in the form of $\frac{7}{7}$:

$$\frac{3}{5} = \frac{3}{5} \cdot \frac{7}{7} = \frac{21}{35} \quad \begin{array}{l}\text{Multiply the numerators.}\\ \text{Multiply the denominators.}\end{array}$$

It is important to note that multiplying $\frac{3}{5}$ by $\frac{7}{7}$ changes its appearance but not its value, because we are multiplying it by 1.

To add and subtract rational expressions with different denominators, we must write them as equivalent expressions having a common denominator. To do so, we build rational expressions.

Building Rational Expressions

To build a rational expression, multiply it by 1 in the form of $\frac{c}{c}$, where c is any nonzero number or expression.

EXAMPLE 7 Write each rational expression as an equivalent expression with the indicated denominator:

a. $\frac{7}{15n}$, denominator $30n^3$ **b.** $\frac{6x}{x + 4}$, denominator $(x + 4)(x - 4)$

Strategy We will begin by asking, "By what must we multiply the given denominator to get the required denominator?"

Why The answer to that question helps us determine the form of 1 to be used to build an equivalent rational expression.

Solution

a. We need to multiply the denominator of $\frac{7}{15n}$ by $2n^2$ to obtain a denominator of $30n^3$. It follows that $\frac{2n^2}{2n^2}$ is the form of 1 that should be used to build an equivalent expression.

$$\frac{7}{15n} = \frac{7}{15n} \cdot \frac{2n^2}{2n^2} \quad \text{Multiply the given rational expression by 1, in the form of } \frac{2n^2}{2n^2}.$$
$$= \frac{14n^2}{30n^3} \quad \begin{array}{l}\text{Multiply the numerators.}\\ \text{Multiply the denominators.}\end{array}$$

The Language of Algebra

We say that $\frac{7}{15n}$ and $\frac{14n^2}{30n^3}$ are **equivalent expressions** because they have the same value for all values of n, except those that make either denominator 0.

b. We need to multiply the denominator of $\frac{6x}{x + 4}$ by $x - 4$ to obtain a denominator of $(x + 4)(x - 4)$. It follows that $\frac{x - 4}{x - 4}$ is the form of 1 that should be used to build an equivalent expression.

$$\frac{6x}{x + 4} = \frac{6x}{x + 4} \cdot \frac{x - 4}{x - 4}$$ Multiply the given rational expression by 1, in the form of $\frac{x - 4}{x - 4}$.

$$= \frac{6x(x - 4)}{(x + 4)(x - 4)}$$ Multiply the numerators. Multiply the denominators.

$$= \frac{6x^2 - 24x}{(x + 4)(x - 4)}$$ In the numerator, distribute the multiplication by $6x$. Leave the denominator in factored form.

To get this answer, we multiplied the factors in the numerator to obtain a polynomial in unfactored form: $6x^2 - 24x$. However, we left the denominator in factored form. This approach is beneficial in the next section when we add and subtract rational expressions with unlike denominators.

Self Check 7 Write each rational expression as an equivalent expression with the indicated denominator: **a.** $\frac{7}{20m^2}$, denominator $60m^3$ **b.** $\frac{2c}{c + 1}$, denominator $(c + 1)(c + 3)$

Now Try Problems 65 and 67

EXAMPLE 8 Write $\frac{x + 1}{x^2 + 6x}$ as an equivalent expression with a denominator of $x(x + 6)(x + 2)$.

Strategy We will begin by factoring the denominator of $\frac{x + 1}{x^2 + 6x}$. Then we will compare the factors of $x^2 + 6x$ to those of $x(x + 6)(x + 2)$.

Why This comparison will enable us to answer the question, "By what must we multiply $x^2 + 6x$ to obtain $x(x + 6)(x + 2)$?"

Solution We factor the denominator to determine what factors are missing.

$$\frac{x + 1}{x^2 + 6x} = \frac{x + 1}{x(x + 6)}$$ Factor out the GCF, x, from $x^2 + 6x$.

It is now apparent that we need to multiply the denominator by $x + 2$ to obtain a denominator of $x(x + 6)(x + 2)$. It follows that $\frac{x + 2}{x + 2}$ is the form of 1 that should be used to build an equivalent expression.

$$\frac{x + 1}{x^2 + 6x} = \frac{x + 1}{x(x + 6)} \cdot \frac{x + 2}{x + 2}$$ Multiply the given rational expression by 1, in the form of $\frac{x + 2}{x + 2}$.

$$= \frac{(x + 1)(x + 2)}{x(x + 6)(x + 2)}$$ Multiply the numerators. Multiply the denominators.

$$= \frac{x^2 + 3x + 2}{x(x + 6)(x + 2)}$$ In the numerator, use the FOIL method to multiply $(x + 1)(x + 2)$. Leave the denominator in factored form.

Success Tip

When building rational expressions, write the numerator of the result as a polynomial in unfactored form. Write the denominator in factored form.

Self Check 8 Write $\frac{x - 3}{x^2 - 4x}$ as an equivalent expression with a denominator of $x(x - 4)(x + 8)$.

Now Try Problem 71

SECTION 7.3 STUDY SET

VOCABULARY

Fill in the blanks.

1. The rational expressions $\frac{7}{6n}$ and $\frac{n+1}{6n}$ have the common ___________ $6n$.
2. The _____ common denominator of $\frac{x-8}{x+6}$ and $\frac{6-5x}{x}$ is $x(x+6)$.
3. To ______ a rational expression, we multiply it by a form of 1. For example: $\frac{2}{n^2} \cdot \frac{8n}{8n} = \frac{16n}{8n^3}$
4. $\frac{16n}{8n^3}$ and $\frac{2}{n^2}$ are __________ expressions. They have the same value for all values of n, except for $n = 0$.

CONCEPTS

Fill in the blanks.

5. To add or subtract rational expressions that have the same denominator, add or subtract the __________, and write the sum or difference over the common ___________. In symbols, $\frac{A}{D} + \frac{B}{D} = \_\_\_\_\_$ and $\frac{A}{D} - \frac{B}{D} = \_\_\_\_\_$.
6. When adding or subtracting rational expressions, always write the result in ________ form by removing any factors common to the numerator and denominator.
7. The sum of two rational expressions is $\frac{4x+4}{5(x+1)}$. Factor the numerator and then simplify the result.
8. Factor each denominator completely.
   a. $\frac{17}{40x^2}$
   b. $\frac{x+25}{2x^2-6x}$
9. Consider the following factorizations.

$$18x - 36 = 2 \cdot 3 \cdot 3 \cdot (x-2)$$
$$3x - 6 = 3(x-2)$$

   a. What is the greatest number of times the factor 3 appears in any one factorization?
   b. What is the greatest number of times the factor $x - 2$ appears in any one factorization?
10. Fill in the blanks. To write $\frac{x}{x-9}$ as an equivalent rational expression with a denominator of $3x(x-9)$, we need to multiply the denominator by ___. It follows that — is the form of 1 that should be used to build $\frac{x}{x-9}$.

NOTATION

Complete the solution.

11. $$\frac{6a-1}{4a+1} + \frac{2a+3}{4a+1} = \frac{6a - 1 + (2a + \quad)}{4a+1}$$
$$= \frac{8a + \quad}{4a+1}$$
$$= \frac{2(\quad\quad)}{4a+1}$$
$$=$$

12. The type of multiplication that is used to build rational expressions is shown below. Fill in the blanks.
   a. $\frac{4x}{5} \cdot \frac{2}{2} = \frac{\quad}{10}$
   b. $\frac{3}{t} \cdot \frac{t-2}{t-2} = \frac{\quad}{t(t-2)}$
   c. $\frac{m+1}{m-3} \cdot \frac{m-5}{m-5} = \frac{\quad}{(m-3)(m-5)}$

GUIDED PRACTICE

Add and simplify the result, if possible. See Example 1.

13. $\frac{9}{x} + \frac{2}{x}$
14. $\frac{4}{s} + \frac{4}{s}$
15. $\frac{x}{18} + \frac{5}{18}$
16. $\frac{7}{10} + \frac{3y}{10}$
17. $\frac{a-5}{3a^3} + \frac{5}{3a^3}$
18. $\frac{b^3-8}{10b^4} + \frac{8}{10b^4}$
19. $\frac{x+3}{2y} + \frac{x+5}{2y}$
20. $\frac{y+2}{10z} + \frac{y+4}{10z}$

Add and simplify the result, if possible. See Example 2.

21. $\frac{2}{r^2-3r-10} + \frac{r}{r^2-3r-10}$
22. $\frac{1}{h^2-4h-5} + \frac{h}{h^2-4h-5}$
23. $\frac{3x-5}{x-2} + \frac{6x-13}{x-2}$
24. $\frac{8x-7}{x+3} + \frac{2x+37}{x+3}$

Subtract and simplify the result, if possible. See Example 3.

25. $\frac{2x}{25} - \frac{x}{25}$
26. $\frac{16c}{11} - \frac{4c}{11}$
27. $\frac{m-1}{6m^2} - \frac{5}{6m^2}$
28. $\frac{c+7}{4c^4} - \frac{3}{4c^4}$
29. $\frac{t}{t^2+t-2} - \frac{1}{t^2+t-2}$
30. $\frac{r}{r^2-2r-3} - \frac{3}{r^2-2r-3}$
31. $\frac{11w+6}{3w(w-9)} - \frac{11w}{3w(w-9)}$
32. $\frac{y+8}{2y(y-14)} - \frac{y}{2y(y-14)}$

Subtract and simplify the result, if possible. See Example 4.

33. $\frac{3y-2}{2y+6} - \frac{2y-5}{2y+6}$
34. $\frac{5x+8}{3x+15} - \frac{3x-2}{3x+15}$
35. $\frac{6x^2}{3x+2} - \frac{11x+10}{3x+2}$
36. $\frac{8a^2}{2a+5} - \frac{4a^2+25}{2a+5}$
37. $\frac{6x-5}{3xy} - \frac{3x-5}{3xy}$
38. $\frac{7x+7}{5y} - \frac{2x+7}{5y}$
39. $\frac{2-p}{p^2-p} - \frac{-p+2}{p^2-p}$
40. $\frac{2-7n}{n^2+5} - \frac{-7n+2}{n^2+5}$

Find the LCD of each pair of rational expressions. See Example 5.

41. $\frac{1}{2x}, \frac{9}{6x}$

42. $\frac{4}{9y}, \frac{11}{3y}$

43. $\frac{33}{15a^3}, \frac{9}{10a}$

44. $\frac{m-21}{12m^4}, \frac{m+1}{18m}$

45. $\frac{35}{3a^2b}, \frac{23}{a^2b^3}$

46. $\frac{27}{c^3d}, \frac{17}{2c^2d^3}$

47. $\frac{8}{c}, \frac{8-c}{c+2}$

48. $\frac{d^2-5}{d+9}, \frac{d-3}{d}$

Find the LCD of each pair of rational expressions. See Example 6.

49. $\frac{3x+1}{3x-3}, \frac{3x}{4x-4}$

50. $\frac{b+1}{5b-10}, \frac{b}{6b-12}$

51. $\frac{b-9}{4b+8}, \frac{b}{6}$

52. $\frac{b^2-b}{10b-15}, \frac{11b}{10}$

53. $\frac{6-k}{2k+4}, \frac{11}{8k}$

54. $\frac{5m+6}{4m+12}, \frac{7}{6m}$

55. $\frac{-2x}{x^2-1}, \frac{5x}{x+1}$

56. $\frac{7-y^2}{y^2-4}, \frac{y-49}{y+2}$

57. $\frac{4x-5}{x^2-4x-5}, \frac{3x+1}{x^2-25}$

58. $\frac{44}{s^2-9}, \frac{s+9}{s^2-s-6}$

59. $\frac{5n^2-16}{2n^2+13n+20}, \frac{3n^2}{n^2+8n+16}$

60. $\frac{4y+25}{y^2+10y+25}, \frac{y^2-7}{2y^2+17y+35}$

Build each rational expression into an equivalent expression with the given denominator. See Example 7.

61. $\frac{5}{r}$; $10r$

62. $\frac{4}{y}$; $7y$

63. $\frac{8}{x}$; x^2y

64. $\frac{7}{y}$; xy^2

65. $\frac{9}{4b}$; $12b^2$

66. $\frac{7}{6c}$; $30c^2$

67. $\frac{3x}{x+1}$; $(x+1)^2$

68. $\frac{5y}{y-2}$; $(y-2)^2$

Build each rational expression into an equivalent expression with the given denominator. See Example 8.

69. $\frac{x+9}{x^2+5x}$; $x^2(x+5)$

70. $\frac{a+11}{a^2+9a}$; $a^2(a+9)$

71. $\frac{t+5}{4t+8}$; $4(t+2)(t+9)$

72. $\frac{x+7}{3x-15}$; $3(x-5)(x+10)$

73. $\frac{y+3}{y^2-5y+6}$; $4y(y-2)(y-3)$

74. $\frac{3x-4}{x^2+3x+2}$; $8x(x+1)(x+2)$

75. $\frac{12-h}{h^2-81}$; $3(h+9)(h-9)$

76. $\frac{m^2}{m^2-100}$; $9(m+10)(m-10)$

TRY IT YOURSELF

Perform the operations. Then simplify, if possible.

77. $\frac{3t}{t^2-8t+7} - \frac{3}{t^2-8t+7}$

78. $\frac{10x}{x^2-2x+1} - \frac{10}{x^2-2x+1}$

79. $\frac{c}{c^2-d^2} - \frac{d}{c^2-d^2}$

80. $\frac{b}{b^2-4} - \frac{2}{b^2-4}$

81. $\frac{a^2+a}{4a^2-8a} + \frac{2a^2-7a}{4a^2-8a}$

82. $\frac{3b^2+16b}{6b^2+9b} + \frac{7b^2-b}{6b^2+9b}$

83. $\frac{17a}{2a+4} - \frac{7a}{2a+4}$

84. $\frac{10b}{3b-18} - \frac{4b}{3b-18}$

85. $\frac{8}{9-3x^2} - \frac{-6x+8}{9-3x^2}$

86. $\frac{5}{10-5t^2} - \frac{-15t+5}{10-5t^2}$

87. $\frac{11n}{(n+4)(n-2)} - \frac{4n-1}{(n-2)(n+4)}$

88. $\frac{1}{(t-1)(t+1)} - \frac{6-t}{(t+1)(t-1)}$

89. $\frac{5r-27}{3r^2-9r} + \frac{4r}{3r^2-9r}$

90. $\frac{9a}{5a^2+25a} + \frac{a+50}{5a^2+25a}$

91. $\frac{11}{36y} + \frac{9}{36y}$

92. $\frac{13}{24w} + \frac{17}{24w}$

93. $\frac{-4x}{3x^2-7x+2} - \frac{-3x-2}{3x^2-7x+2}$

94. $\frac{-3c}{5c^2-16c+3} - \frac{-2c-3}{5c^2-16c+3}$

95. $\frac{3x^2}{x+1} - \frac{-x+2}{x+1}$

96. $\frac{8b^2}{3b-2} - \frac{-b^2+4}{3b-2}$

Look Alikes . . .

97. a. $\frac{t}{12} + \frac{5t}{12}$ b. $\frac{t}{12} \cdot \frac{5t}{12}$ c. $\frac{t}{12} \div \frac{5t}{12}$

98. a. $\frac{x}{9} + \frac{2x}{9}$ b. $\frac{x}{9} \cdot \frac{2x}{9}$ c. $\frac{x}{9} \div \frac{2x}{9}$

99. a. $\frac{m+6}{5} - \frac{m+2}{5}$ b. $\frac{m+6}{5} \cdot \frac{m+2}{5}$ c. $\frac{m+6}{5} \div \frac{m+2}{5}$

100. **a.** $\dfrac{2r+9}{2r} - \dfrac{1}{2r}$ **b.** $\dfrac{2r+9}{2r} \cdot \dfrac{1}{2r}$

c. $\dfrac{2r+9}{2r} \div \dfrac{1}{2r}$

APPLICATIONS

101. Geometry. What is the difference of the length and width of the rectangle?

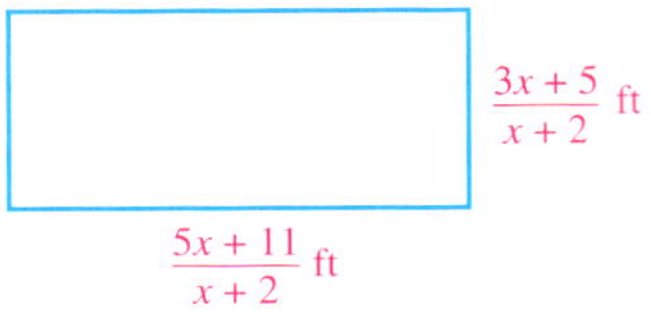

102. Geometry. What is the perimeter of the rectangle in Problem 101?

WRITING

103. Explain how to add fractions with the same denominator.

104. Explain how to find a least common denominator.

105. Explain the error in the following solution:

$$\frac{2x+3}{x+5} - \frac{x+2}{x+5} = \frac{2x+3-x+2}{x+5} = \frac{x+5}{x+5} = 1$$

106. Explain the error in the following solution:

$$\frac{y+4}{y} - \frac{1}{y} = \frac{y+4-1}{y+y} = \frac{y+3}{2y}$$

107. **a.** Explain why the LCD of $\frac{5}{h^2}$ and $\frac{3}{h}$ is h^2 and not h^3.

b. Explain why the LCD of $\frac{1}{x}$ and $\frac{1}{x-9}$ is $x(x-9)$ and not $x-9$.

108. Explain how multiplication by 1 is used to build a rational expression. Give an example.

REVIEW

Give the formula for . . .

109. **a.** simple interest

b. the area of a triangle

c. the perimeter of a rectangle

110. **a.** the slope of a line

b. distance traveled

c. the area of a circle

CHALLENGE PROBLEMS

Perform the operations. Simplify the result, if possible.

111. $\dfrac{3xy}{x-y} - \dfrac{x(3y-x)}{x-y} - \dfrac{x(x-y)}{x-y}$

112. $\dfrac{9t^3-12t^2}{27t^3-64} - \dfrac{-3t+4}{27t^3-64}$

113. $\dfrac{2a^2+2}{a^3+8} + \dfrac{a^3+a}{a^3+8}$

114. Find the LCD of

$$\frac{2}{a^3+8},\ \frac{a}{a^2-4},\ \text{and}\ \frac{2a+5}{a^3-8}.$$

SECTION 7.4 Adding and Subtracting with Unlike Denominators

OBJECTIVES

1 Add and subtract rational expressions that have unlike denominators.

2 Add and subtract rational expressions that have denominators that are opposites.

ARE YOU READY?

The following problems review some basic skills that are needed when adding and subtracting rational expressions with unlike denominators.

1. Write $\dfrac{5}{4a}$ as an equivalent expression with a denominator of $16a^2$.

2. Factor completely: $36x^3$

3. Add: $\dfrac{6x}{9} + \dfrac{5x}{9}$

4. What is the opposite of $8 - t$?

We have discussed a method for finding the least common denominator (LCD) of two rational expressions. We have also built rational expressions into equivalent expressions having a given denominator. We will now use these skills to add and subtract rational expressions with unlike denominators.

1 Add and Subtract Rational Expressions That Have Unlike Denominators.

The following steps summarize how to add or subtract rational expressions that have different denominators.

Adding and Subtracting Rational Expressions That Have Unlike Denominators

1. Find the LCD.
2. Rewrite each rational expression as an equivalent expression with the LCD as the denominator. To do so, build each fraction using a form of 1 that involves any factor(s) needed to obtain the LCD.
3. Add or subtract the numerators and write the sum or difference over the LCD.
4. Simplify the result, if possible.

EXAMPLE 1 Add: $\frac{9x}{7} + \frac{3x}{5}$

Strategy We will use the procedure for adding rational expressions that have unlike denominators. The first step is to determine the LCD.

Why If we are to add (or subtract) rational expressions, their denominators must be the same. Since the denominators of these rational expressions are different, we cannot add them in their present form.

$$\text{sevenths} \quad \frac{9x}{7} + \frac{3x}{5} \quad \text{fifths}$$

Not the same number

Solution *Step 1:* The denominators are 7 and 5. The LCD is $7 \cdot 5 = 35$.

Step 2: We need to multiply the denominator of $\frac{9x}{7}$ by 5 and we need to multiply the denominator of $\frac{3x}{5}$ by 7 to obtain the LCD, 35. It follows that $\frac{5}{5}$ and $\frac{7}{7}$ are the forms of 1 that should be used to write the equivalent rational expressions.

$$\frac{9x}{7} + \frac{3x}{5} = \frac{9x}{7} \cdot \frac{5}{5} + \frac{3x}{5} \cdot \frac{7}{7}$$ Build the rational expressions so that each has a denominator of 35.

$$= \frac{45x}{35} + \frac{21x}{35}$$ Multiply the numerators: $9x \cdot 5 = 45x$ and $3x \cdot 7 = 21x$. Multiply the denominators. Now the denominators are like.

Step 3: $$= \frac{45x + 21x}{35}$$ Add the numerators. Write the sum over the common denominator, 35.

$$= \frac{66x}{35}$$ Combine like terms in the numerator: $45x + 21x = 66x$.

Step 4: Since 66 and 35 have no common factor other than 1, the result cannot be simplified.

Caution

In Step 2, don't simplify $\frac{45x}{35}$ and $\frac{21x}{35}$, because that will take you back to the original rational expressions and you will lose the common denominator.

Self Check 1 Add: $\frac{y}{2} + \frac{6y}{7}$

Now Try Problem 13

EXAMPLE 2 Subtract: $\frac{13}{18b^2} - \frac{1}{24b}$

Strategy We will use the procedure for subtracting rational expressions that have unlike denominators. The first step is to determine the LCD.

Why If we are to subtract rational expressions, their denominators must be the same. Since the denominators of these rational expressions are different, we cannot subtract them in their present form.

Solution *Step 1:* To find the LCD, we form a product that uses each different factor of $18b^2$ and $24b$ the greatest number of times it appears in any one factorization.

$$\left.\begin{aligned} 18b^2 &= 2 \cdot 3 \cdot 3 \cdot b \cdot b \\ 24b &= 2 \cdot 2 \cdot 2 \cdot 3 \cdot b \end{aligned}\right\} \text{LCD} = 2 \cdot 2 \cdot 2 \cdot 3 \cdot 3 \cdot b \cdot b = 72b^2$$

Notation

When checking your answers with those in the back of the book, remember that the results can often be presented in several equivalent forms. For example, this result can also be expressed as

$$\frac{-3b + 52}{72b^2}$$

Step 2: We need to multiply $18b^2$ by 4 to obtain $72b^2$, and $24b$ by $3b$ to obtain $72b^2$. It follows that we should use $\frac{4}{4}$ and $\frac{3b}{3b}$ to build the equivalent rational expressions.

$$\frac{13}{18b^2} - \frac{1}{24b} = \frac{13}{18b^2} \cdot \frac{4}{4} - \frac{1}{24b} \cdot \frac{3b}{3b}$$ Build the rational expressions so that each has a denominator of $72b^2$.

$$= \frac{52}{72b^2} - \frac{3b}{72b^2}$$ Multiply the numerators. Multiply the denominators. Now the denominators are like.

Step 3: $$= \frac{52 - 3b}{72b^2}$$ Subtract the numerators. Write the difference over the common denominator, $72b^2$.

Step 4: Since $52 - 3b$ does not factor, the result cannot be simplified.

Self Check 2 Subtract: $\frac{5}{21z^2} - \frac{3}{28z}$

Now Try Problem 21

EXAMPLE 3 Add: $\frac{3}{2x + 18} + \frac{27}{x^2 - 81}$

Strategy We use the procedure for adding rational expressions when the denominators are binomials. The first step is to find the LCD.

Why Since the denominators are different, we cannot add these rational expressions in their present form.

Solution After factoring the denominators, we see that the greatest number of times each of the factors 2, $x + 9$, and $x - 9$ appear in any one of the factorizations is once.

$$\left.\begin{aligned} 2x + 18 &= 2(x + 9) \\ x^2 - 81 &= (x + 9)(x - 9) \end{aligned}\right\} \text{LCD} = 2(x + 9)(x - 9)$$

Since we need to multiply $2(x + 9)$ by $x - 9$ to obtain the LCD and $(x + 9)(x - 9)$ by 2 to obtain the LCD, $\frac{x - 9}{x - 9}$ and $\frac{2}{2}$ are the forms of 1 to use to build the equivalent rational expressions.

$$\frac{3}{2x+18} + \frac{27}{x^2-81} = \frac{3}{2(x+9)} + \frac{27}{(x+9)(x-9)}$$

Write each denominator in factored form.

$$= \frac{3}{2(x+9)} \cdot \frac{x-9}{x-9} + \frac{27}{(x+9)(x-9)} \cdot \frac{2}{2}$$

Build the expressions so that each has a denominator of $2(x + 9)(x - 9)$.

Multiply the numerators to prepare to combine like terms later.

Don't multiply the denominators. Leave them in factored form to possibly simplify the result later.

$$= \frac{3x-27}{2(x+9)(x-9)} + \frac{54}{2(x+9)(x-9)}$$

Multiply: $3(x - 9) = 3x - 27$ and $27 \cdot 2 = 54$. Now the denominators are like.

Although it is not required, the factors of each denominator are written in the same order.

$$= \frac{3x-27+54}{2(x+9)(x-9)}$$

Add the numerators. Write the sum over the common denominator, $2(x + 9)(x - 9)$.

$$= \frac{3x+27}{2(x+9)(x-9)}$$

Combine like terms in the numerator: $-27 + 54 = 27$.

Caution

Always write the result in simplest form by removing any factors common to the numerator and denominator.

$$= \frac{3\overset{1}{\cancel{(x+9)}}}{2\underset{1}{\cancel{(x+9)}}(x-9)}$$

Factor the numerator. Then simplify the expression by removing a factor equal to 1.

$$= \frac{3}{2(x-9)}$$

This is the result in simplest form.

Self Check 3 Add: $\dfrac{2}{5x+25} + \dfrac{4}{x^2-25}$

Now Try Problem 29

EXAMPLE 4 Subtract: $\dfrac{x}{x-1} - \dfrac{x-6}{x-4}$

Strategy We use the same procedure for subtracting rational expressions when the denominators are binomials. The first step is to find the LCD.

Why Since the denominators are different, we cannot subtract these rational expressions in their present form.

Solution The denominators of $\frac{x}{x-1}$ and $\frac{x-6}{x-4}$ are completely factored. The factor $x - 1$ appears once and the factor $x - 4$ appears once. Thus, the LCD $= (x - 1)(x - 4)$.

We need to multiply the first denominator by $x - 4$ to obtain the LCD and the second denominator by $x - 1$ to obtain the LCD. It follows that $\frac{x-4}{x-4}$ and $\frac{x-1}{x-1}$ are the forms of 1 to use to build the equivalent rational expressions.

$$\frac{x}{x-1} - \frac{x-6}{x-4} = \frac{x}{x-1} \cdot \frac{x-4}{x-4} - \frac{x-6}{x-4} \cdot \frac{x-1}{x-1}$$

Build the rational expressions so that each has a denominator of $(x - 1)(x - 4)$.

Multiply the numerators to prepare to combine like terms later.

Don't multiply the denominators. Leave them in factored form to possibly simplify later.

$$= \frac{x^2-4x}{(x-1)(x-4)} - \frac{x^2-7x+6}{(x-4)(x-1)}$$

Multiply: $x(x - 4) = x^2 - 4x$.
Multiply: $(x - 6)(x - 1) = x^2 - 7x + 6$.
Now the denominators are like.

By the commutative property of multiplication, these are like denominators.

$$= \frac{x^2 - 4x - (x^2 - 7x + 6)}{(x - 1)(x - 4)}$$ Subtract the numerators. Remember the parentheses. Write the difference over the common denominator, $(x - 1)(x - 4)$.

$$= \frac{x^2 - 4x - x^2 + 7x - 6}{(x - 1)(x - 4)}$$ In the numerator, use the distributive property: $-(x^2 - 7x + 6) = -1(x^2 - 7x + 6) = -x^2 + 7x - 6$.

$$= \frac{3x - 6}{(x - 1)(x - 4)}$$ Combine like terms in the numerator: $x^2 - x^2 = 0$ and $-4x + 7x = 3x$.

Although, the numerator factors as $3(x - 2)$, the numerator and denominator do not have a common factor. Therefore, the result is in simplest form.

Self Check 4 Subtract: $\frac{x}{x + 9} - \frac{x - 7}{x + 8}$

Now Try Problem 37

EXAMPLE 5

Subtract: $\frac{m}{m^2 + 5m + 6} - \frac{2}{m^2 + 3m + 2}$

Strategy We use the same procedure for subtracting rational expressions when the denominators are trinomials. The first step is to find the LCD.

Why Since the denominators are different, we cannot subtract these rational expressions in their present form.

Solution Factor each denominator and form the LCD.

$$\left.\begin{aligned} m^2 + 5m + 6 &= (m + 2)(m + 3) \\ m^2 + 3m + 2 &= (m + 2)(m + 1) \end{aligned}\right\} \quad \text{LCD} = (m + 2)(m + 3)(m + 1)$$

Examining the factored forms, we see that the first denominator must be multiplied by $m + 1$, and the second must be multiplied by $m + 3$ to obtain the LCD. To build the expressions, we will use $\frac{m + 1}{m + 1}$ and $\frac{m + 3}{m + 3}$.

$$\frac{m}{m^2 + 5m + 6} - \frac{2}{m^2 + 3m + 2}$$

$$= \frac{m}{(m + 2)(m + 3)} - \frac{2}{(m + 2)(m + 1)}$$ Write each denominator in factored form.

$$= \frac{m}{(m + 2)(m + 3)} \cdot \frac{m + 1}{m + 1} - \frac{2}{(m + 2)(m + 1)} \cdot \frac{m + 3}{m + 3}$$ Build each expression so that it has a denominator of $(m + 2)(m + 3)(m + 1)$.

Multiply the numerators. → Leave the denominators in factored form. →

$$= \frac{m^2 + m}{(m + 2)(m + 3)(m + 1)} - \frac{2m + 6}{(m + 2)(m + 1)(m + 3)}$$ Multiply: $m(m + 1) = m^2 + m$ and $2(m + 3) = 2m + 6$. Now the denominators are like.

By the commutative property of multiplication, these are like denominators.

$$= \frac{m^2 + m - (2m + 6)}{(m + 2)(m + 3)(m + 1)}$$ Subtract the numerators. Remember the parentheses. Write the difference over the common denominator, $(m + 2)(m + 3)(m + 1)$.

$$= \frac{m^2 + m - 2m - 6}{(m + 2)(m + 3)(m + 1)}$$ Use the distributive property: $-(2m + 6) = -1(2m + 6) = -2m - 6$.

$$= \frac{m^2 - m - 6}{(m + 2)(m + 3)(m + 1)}$$ Combine like terms in the numerator: $m - 2m = -m$.

$$= \frac{(m - 3)\overset{1}{\cancel{(m + 2)}}}{\underset{1}{\cancel{(m + 2)}}(m + 3)(m + 1)}$$ Factor the numerator and simplify the expression by removing a factor equal to 1.

$$= \frac{m - 3}{(m + 3)(m + 1)}$$

Self Check 5 Subtract: $\frac{b}{b^2 - 2b - 8} - \frac{6}{b^2 + b - 20}$

Now Try Problem 45

EXAMPLE 6 Add: $\frac{4b}{a - 5} + b$

Strategy We will begin by writing the second addend, b, as $\frac{b}{1}$ and then find the LCD.

Why To add b to the rational expression, $\frac{4b}{a - 5}$, we must rewrite b as a rational expression.

Solution The LCD of $\frac{4b}{a - 5}$ and $\frac{b}{1}$ is $1(a - 5)$, or simply $a - 5$. Since we must multiply the denominator of $\frac{b}{1}$ by $a - 5$ to obtain the LCD, we will use $\frac{a - 5}{a - 5}$ to write an equivalent rational expression.

Success Tip

Since the denominator of $\frac{4b}{a - 5}$ is the LCD, we *do not* need to build it by multiplying it by a form of 1. The step below is unnecessary.

$$\cancel{\frac{4b}{a - 5} \cdot \frac{1}{1}}$$

$$\frac{4b}{a - 5} + b = \frac{4b}{a - 5} + \frac{b}{1} \cdot \frac{a - 5}{a - 5}$$ Build $\frac{b}{1}$ so that it has a denominator of $a - 5$.

$$= \frac{4b}{a - 5} + \frac{ab - 5b}{a - 5}$$ Multiply numerators: $b(a - 5) = ab - 5b$. Multiply denominators: $1(a - 5) = a - 5$. Now the denominators are like.

$$= \frac{4b + ab - 5b}{a - 5}$$ Add the numerators. Write the sum over the common denominator.

$$= \frac{ab - b}{a - 5}$$ Combine like terms in the numerator: $4b - 5b = -b$.

Although the numerator factors as $b(a - 1)$, the numerator and denominator do not have a common factor. Therefore, the result is in simplest form.

Self Check 6 Add: $\frac{10y}{n + 4} + y$

Now Try Problem 47

2 Add and Subtract Rational Expressions That Have Denominators That Are Opposites.

Recall that two polynomials are **opposites** if their terms are the same but they are opposite in sign. For example, $x - 4$ and $4 - x$ are opposites. If we multiply one of these binomials by -1, the subtraction is reversed, and the result is the other binomial.

$$-1(x - 4) = -x + 4$$
$$= 4 - x$$ Write the expression with 4 first.

$$-1(4 - x) = -4 + x$$
$$= x - 4$$ Write the expression with x first.

These results suggest that when a polynomial is multiplied by -1, the result is its opposite. This fact can be used when adding or subtracting rational expressions whose denominators are opposites.

Rational Expressions with Opposite Denominators

When adding or subtracting two rational expressions whose denominators are opposites, multiply either expression by 1 in the form of $\frac{-1}{-1}$ to obtain a common denominator.

EXAMPLE 7 Subtract: $\frac{x}{x-7} - \frac{1}{7-x}$

Strategy We note that the denominators are opposites. Either can serve as the LCD; we will choose $x - 7$. To obtain a common denominator, we will multiply $\frac{1}{7-x}$ by $\frac{-1}{-1}$.

Why When $7 - x$ is multiplied by -1, the subtraction is reversed, and the result is $x - 7$.

Solution We must multiply the denominator of $\frac{1}{7-x}$ by -1 to obtain the LCD. It follows that $\frac{-1}{-1}$ should be the form of 1 that is used to write an equivalent rational expression.

$$\frac{x}{x-7} - \frac{1}{7-x} = \frac{x}{x-7} - \frac{1}{7-x} \cdot \frac{-1}{-1}$$
Build $\frac{1}{7-x}$ so that it has a denominator of $x - 7$.

$$= \frac{x}{x-7} - \frac{-1}{-7+x}$$
Multiply the numerators: $1(-1) = -1$. Multiply the denominators.

$$= \frac{x}{x-7} - \frac{-1}{x-7}$$
Rewrite the second denominator: $-7 + x = x - 7$. Now the denominators are like.

$$= \frac{x-(-1)}{x-7}$$
Subtract the numerators. Remember the parentheses. Write the difference over the common denominator, $x - 7$.

$$= \frac{x+1}{x-7}$$
Do the subtraction in the numerator: $x - (-1) = x + 1$.

The result does not simplify.

Success Tip

Either denominator can serve as the LCD. However, it is common to have a result whose denominator is written in descending powers of the variable. Therefore, we chose $x - 7$, as opposed to $7 - x$, as the LCD.

Self Check 7 Add: $\frac{n}{n-8} + \frac{12}{8-n}$

Now Try Problem 51

SECTION 7.4 STUDY SET

VOCABULARY

Fill in the blanks.

1. $\frac{x}{x-7}$ and $\frac{1}{x-7}$ have like denominators. $\frac{x+5}{x-7}$ and $\frac{4x}{x+7}$ have _______ denominators.
2. The polynomials $x - 3$ and $3 - x$ are _________.

CONCEPTS

3. Write each denominator in factored form.
   a. $\frac{x+1}{20x^2}$
   b. $\frac{3x^2-4}{x^2+4x-12}$
4. The factorizations of the denominators of two rational expressions are given. Find the LCD.
   a. $12a = 2 \cdot 2 \cdot 3 \cdot a$
      $18a^2 = 2 \cdot 3 \cdot 3 \cdot a \cdot a$
   b. $x^2 - 36 = (x+6)(x-6)$
      $3x - 18 = 3(x-6)$

5. What is the LCD for $\frac{x-1}{x+6}$ and $\frac{1}{x+3}$?

6. The LCD for $\frac{1}{9n^2}$ and $\frac{37}{15n^3}$ is $3 \cdot 3 \cdot 5 \cdot n \cdot n \cdot n = 45n^3$. If we want to add these rational expressions, what form of 1 should be used

a. to build $\frac{1}{9n^2}$?

b. to build $\frac{37}{15n^3}$?

Fill in the blanks.

7. To build $\frac{x}{x+2}$ so that it has a denominator of $5(x+2)$, we multiply it by 1 in the form of $\frac{\ }{\ }$.

8. To build $\frac{8x}{2-x}$ so that it has a denominator of $x-2$, we multiply it by 1 in the form of $\frac{\ }{\ }$.

NOTATION

Complete the solution.

9. $$\frac{2}{5}+\frac{7}{3x}=\frac{2}{5}\cdot\frac{\ }{3x}+\frac{7}{3x}\cdot\frac{\ }{5}$$
$$=\frac{6x}{\ }+\frac{35}{\ }$$
$$=\frac{6x+\ }{15x}$$

10. Are the student's answers and the book's answers equivalent?

Student's answer	Book's answer	Equivalent?
$\frac{m^2+2m}{(m-1)(m-4)}$	$\frac{m^2+2m}{(m-4)(m-1)}$	
$\frac{-5x^2-7}{4x(x+3)}$	$-\frac{5x^2-7}{4x(x+3)}$	
$\frac{-2x}{x-y}$	$-\frac{2x}{x-y}$	

GUIDED PRACTICE

Perform the operations. Simplify, if possible.
See Example 1.

11. $\frac{x}{3}-\frac{2x}{7}$

12. $\frac{y}{4}+\frac{3y}{5}$

13. $\frac{7a}{8}+\frac{4a}{5}$

14. $\frac{5t}{6}+\frac{4t}{7}$

Perform the operations. Simplify, if possible.
See Example 2.

15. $\frac{7}{m^2}-\frac{2}{m}$

16. $\frac{6}{n^2}-\frac{2}{n}$

17. $\frac{3}{5p^2}-\frac{5}{10p}$

18. $\frac{15}{16a}-\frac{3}{4a^2}$

19. $\frac{1}{6t}-\frac{11}{8t^3}$

20. $\frac{3}{10a}-\frac{13}{15a^3}$

21. $\frac{1}{6c^4}-\frac{8}{9c^2}$

22. $\frac{7}{8b^2}-\frac{5}{6b^3}$

Perform the operations. Simplify, if possible.
See Example 3.

23. $\frac{1}{2a-4}+\frac{5}{a^2-4}$

24. $\frac{5}{p^2-9}+\frac{2}{3p+9}$

25. $\frac{2}{3a-2}+\frac{5}{9a^2-4}$

26. $\frac{2}{5b-3}+\frac{5}{25b^2-9}$

27. $\frac{4}{a+2}-\frac{7}{a^2+4a+4}$

28. $\frac{9}{b^2-2b+1}-\frac{2}{b-1}$

29. $\frac{6}{5m^2-5m}-\frac{3}{5m-5}$

30. $\frac{9}{2c^2-2c}-\frac{5}{2c-2}$

Perform the operations. Simplify, if possible.
See Example 4.

31. $\frac{9}{t+3}+\frac{8}{t+2}$

32. $\frac{2}{m-3}+\frac{7}{m-4}$

33. $\frac{3x}{2x-1}-\frac{2x}{2x+3}$

34. $\frac{2y}{5y-1}-\frac{2y}{3y+2}$

35. $\frac{s+7}{s+3}-\frac{s-3}{s+7}$

36. $\frac{t+5}{t-5}-\frac{t-5}{t+5}$

37. $\frac{3m}{m-2}-\frac{m-3}{m+5}$

38. $\frac{2x}{x+2}-\frac{x+1}{x-3}$

Perform the operations. Simplify, if possible.
See Example 5.

39. $\frac{4}{s^2+5s+4}+\frac{s}{s^2+2s+1}$

40. $\frac{d}{d^2+6d+5}-\frac{3}{d^2+5d+4}$

41. $\frac{5}{x^2-9x+8}-\frac{3}{x^2-6x-16}$

42. $\frac{3}{t^2+t-6}+\frac{1}{t^2+3t-10}$

43. $\frac{2}{a^2+4a+3}+\frac{1}{a+3}$

44. $\frac{1}{c+6}+\frac{4}{c^2+8c+12}$

45. $\frac{8}{y^2-16}-\frac{7}{y^2-y-12}$

46. $\frac{6}{s^2-9}-\frac{5}{s^2-s-6}$

Perform the operations. Simplify, if possible.
See Example 6.

47. $\frac{9y}{x-4}+y$

48. $\frac{9n}{m+4}+n$

49. $\frac{8}{x}+z$

50. $\frac{2}{y}+z$

Perform the operations. Simplify, if possible.
See Example 7.

51. $\frac{7}{a-4}+\frac{5}{4-a}$

52. $\frac{4}{b-6}+\frac{b}{6-b}$

53. $\dfrac{c}{7c-d} - \dfrac{d}{d-7c}$

54. $\dfrac{a}{5a-3b} - \dfrac{b}{3b-5a}$

TRY IT YOURSELF

Perform the operations and simplify, if possible.

55. $\dfrac{x-7}{x^2+4x-5} - \dfrac{x-9}{x^2+3x-10}$

56. $\dfrac{r}{r^2+5r+6} - \dfrac{2}{r^2+3r+2}$

57. $\dfrac{3d-3}{d-9} - \dfrac{3d}{9-d}$

58. $\dfrac{2x+2}{x-2} - \dfrac{2x}{2-x}$

59. $\dfrac{10}{x-1} + y$

60. $\dfrac{3}{s-8} + t$

61. $\dfrac{b}{b+1} - \dfrac{b-1}{b+2}$

62. $\dfrac{x}{x-2} - \dfrac{x+2}{x+3}$

63. $\dfrac{g}{g^2-4} + \dfrac{2}{4-g^2}$

64. $\dfrac{h}{h^2-49} + \dfrac{7}{49-h^2}$

65. $\dfrac{5y}{6} + \dfrac{5y}{3}$

66. $\dfrac{4x}{3} + \dfrac{x}{6}$

67. $\dfrac{1}{5x} + \dfrac{7x}{x+5}$

68. $\dfrac{10h}{h-3} + \dfrac{7}{9h}$

69. $\dfrac{11}{5x} - \dfrac{5}{6x}$

70. $\dfrac{5}{9y} - \dfrac{1}{4y}$

71. $\dfrac{x}{x+1} + \dfrac{x-1}{x}$

72. $\dfrac{t-2}{t} + \dfrac{t}{t+3}$

73. $\dfrac{y}{y-1} - \dfrac{4}{1-y}$

74. $\dfrac{1}{t-7} - \dfrac{t}{7-t}$

75. $\dfrac{n}{5} - \dfrac{n-2}{15}$

76. $\dfrac{m}{9} - \dfrac{m+1}{27}$

77. $\dfrac{y+2}{5y^2} + \dfrac{y+4}{15y}$

78. $\dfrac{x+3}{x^2} + \dfrac{x+5}{2x}$

79. $\dfrac{x}{x-2} + \dfrac{4+2x}{x^2-4}$

80. $\dfrac{y}{y+3} - \dfrac{2y-6}{y^2-9}$

81. $b - \dfrac{3}{a^2}$

82. $c - \dfrac{5}{3b}$

83. $\dfrac{7}{3a} + \dfrac{1}{a-2}$

84. $\dfrac{5}{9x} + \dfrac{4}{x+6}$

85. $\dfrac{3}{x^2} + \dfrac{17}{x}$

86. $\dfrac{7}{c} + \dfrac{14}{c^2}$

87. $\dfrac{x+2}{x+1} - 5$

88. $\dfrac{y+8}{y-8} - 4$

89. $\dfrac{4b}{3} - \dfrac{5b}{12}$

90. $\dfrac{21y}{12} - \dfrac{7y}{6}$

Look Alikes . . .

91. a. $\dfrac{5}{2x} + \dfrac{4x}{15}$ b. $\dfrac{5}{2x} \cdot \dfrac{4x}{15}$

92. a. $\dfrac{2a+4}{3} - \dfrac{9}{a+2}$ b. $\dfrac{2a+4}{3} \cdot \dfrac{9}{a+2}$

93. a. $\dfrac{t}{t-5} - \dfrac{t}{t^2-25}$ b. $\dfrac{t}{t-5} \div \dfrac{t}{t^2-25}$

94. a. $\dfrac{1}{m+2} - \dfrac{2}{m^2+4m+4}$ b. $\dfrac{1}{m+2} \div \dfrac{2}{m^2+4m+4}$

APPLICATIONS

95. Find the total height of the funnel.

96. **Funnels.** Refer to the illustration on the right. What is the difference between the diameter of the opening at the top of the funnel and the diameter of its spout?

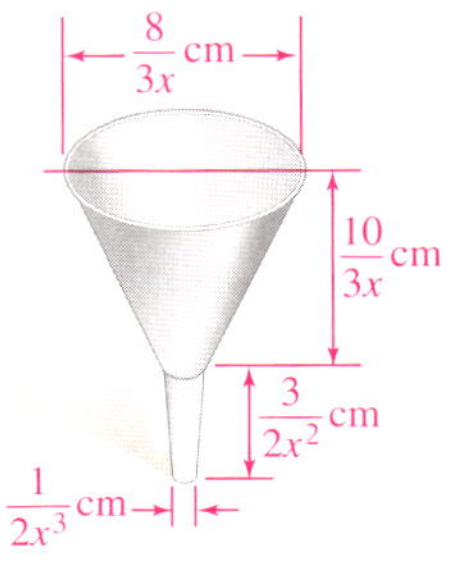

WRITING

97. Explain the error:

a. $\dfrac{3}{x} + \dfrac{8}{y} = \dfrac{3+8}{x+y}$
$= \dfrac{11}{x+y}$

b. $\dfrac{3}{x} + \dfrac{x}{3} = \dfrac{\overset{1}{\cancel{3}}}{\underset{1}{\cancel{x}}} + \dfrac{\overset{1}{\cancel{x}}}{\underset{1}{\cancel{3}}}$
$= 1 + 1$
$= 2$

98. Explain how to add two rational expressions with unlike denominators.

99. When will the LCD of two rational expressions be the product of the denominators of those rational expressions? Give an example.

100. Explain how multiplication by $\frac{-1}{-1}$ is used in this section.

REVIEW

101. Find the slope and y-intercept of the graph of $y = 8x + 2$.

102. Find the slope and y-intercept of the graph of $3x + 4y = -36$.

103. What is the slope of the graph of $y = 2$?

104. Is the graph of the equation $x = 0$ the x-axis or the y-axis?

CHALLENGE PROBLEMS

Perform the operations and simplify the result, if possible.

105. $\dfrac{a}{a-1} - \dfrac{2}{a+2} + \dfrac{3(a-2)}{a^2+a-2}$

106. $\dfrac{2x}{x^2-3x+2} + \dfrac{2x}{x-1} - \dfrac{x}{x-2}$

107. $\dfrac{1}{a+1} + \dfrac{a^2-7a+10}{2a^2-2a-4} \cdot \dfrac{2a^2-50}{a^2+10a+25}$

108. $1 - \dfrac{(x-2)^2}{(x+2)^2}$

SECTION 7.5

Simplifying Complex Fractions

OBJECTIVES

1. Simplify complex fractions using division.
2. Simplify complex fractions using the LCD.

ARE YOU READY?

The following problems review some basic skills that are needed when simplifying complex fractions.

1. What operation is indicated by the fraction bar in $\frac{56}{7}$?

2. Divide: $\frac{2}{9} \div \frac{4}{27}$

3. What is the LCD of $\frac{5}{2}$ and $\frac{3}{x}$?

4. Multiply: $10a\left(\frac{4}{5} - \frac{1}{a}\right)$

A **complex rational expression,** also called a **complex fraction,** is a rational expression whose numerator and/or denominator contains one or more rational expressions. The expression above the main fraction bar of a complex fraction is the numerator, and the expression below the main fraction bar is the denominator. Two examples of complex fractions are:

$$\frac{\frac{5x}{3}}{\frac{2x}{9}} \qquad \frac{\frac{1}{2} - \frac{1}{x}}{\frac{x}{3} + \frac{1}{5}}$$

← Numerator of complex fraction →
← Main fraction bar →
← Denominator of complex fraction →

In this section, we will discuss two methods for simplifying complex fractions. To **simplify a complex fraction** means to write it in the form $\frac{A}{B}$, where A and B are polynomials that have no common factors.

1 Simplify Complex Fractions Using Division.

One method for simplifying complex fractions uses the fact that the main fraction bar indicates division.

Simplifying Complex Fractions Method 1: Using Division

1. Add or subtract in the numerator and/or denominator so that the numerator is a single rational expression and the denominator is a single rational expression.
2. Perform the indicated division by multiplying the numerator of the complex fraction by the reciprocal of the denominator.
3. Simplify the result, if possible.

EXAMPLE 1

Simplify: $\frac{\frac{5x^2}{3}}{\frac{2x^3}{9}}$

Strategy We will perform the division indicated by the main fraction bar using the procedure for dividing rational expressions from Section 7.2.

Why We can skip the first step of method 1 and immediately divide because the numerator and the denominator of the complex fraction are already single rational expressions.

Solution

$$\frac{\frac{5x^2}{3}}{\frac{2x^3}{9}} = \frac{5x^2}{3} \div \frac{2x^3}{9}$$ Write the division indicated by the main fraction bar using a ÷ symbol.

$$= \frac{5x^2}{3} \cdot \frac{9}{2x^3}$$ To divide rational expressions, multiply the first by the reciprocal of the second.

$$= \frac{5x^2 \cdot 9}{3 \cdot 2x^3}$$ Multiply the numerators. Multiply the denominators.

$$= \frac{5 \cdot \overset{1}{\cancel{x}} \cdot \overset{1}{\cancel{x}} \cdot \overset{1}{\cancel{3}} \cdot 3}{\underset{1}{\cancel{3}} \cdot 2 \cdot \underset{1}{\cancel{x}} \cdot \underset{1}{\cancel{x}} \cdot x}$$ Factor 9 as 3 · 3. Then simplify by removing factors equal to 1.

$$= \frac{15}{2x}$$ Multiply the remaining factors in the numerator. Multiply the remaining factors in the denominator.

The Language of Algebra

The second step of this method could also be phrased: Perform the division by **inverting the denominator of the complex fraction and multiplying.**

Self Check 1 Simplify: $\frac{\frac{7y^3}{8}}{\frac{21y^2}{20}}$

Now Try Problem 17

In the next example, we must simplify the numerator and denominator of the complex fraction separately before the indicated division can be performed.

EXAMPLE 2 Simplify: $\frac{\frac{1}{2} - \frac{1}{x}}{\frac{x}{3} + \frac{1}{5}}$

Strategy We will simplify the expressions above and below the main fraction bar separately to write $\frac{1}{2} - \frac{1}{x}$ and $\frac{x}{3} + \frac{1}{5}$ as single rational expressions. Then we will perform the indicated division.

Why The numerator and the denominator of the complex fraction must be written as single rational expressions before dividing.

Solution To write the numerator as a single rational expression, we build $\frac{1}{2}$ and $\frac{1}{x}$ to have an LCD of $2x$, and then subtract. To write the denominator as a single rational expression, we build $\frac{x}{3}$ and $\frac{1}{5}$ to have an LCD of 15, and then add.

Success Tip

Notice that with this method, separate LCD's are found for the numerator and the denominator of the complex fraction.

$$\frac{\frac{1}{2} - \frac{1}{x}}{\frac{x}{3} + \frac{1}{5}} = \frac{\frac{1}{2} \cdot \frac{x}{x} - \frac{1}{x} \cdot \frac{2}{2}}{\frac{x}{3} \cdot \frac{5}{5} + \frac{1}{5} \cdot \frac{3}{3}}$$ ← The LCD for the numerator is 2x. Build each fraction so that each has a denominator of 2x. ← The LCD for the denominator is 15. Build each fraction so that each has a denominator of 15.

$$= \frac{\frac{x}{2x} - \frac{2}{2x}}{\frac{5x}{15} + \frac{3}{15}}$$ Perform each of the four multiplications shown in color.

$$= \frac{\frac{x - 2}{2x}}{\frac{5x + 3}{15}}$$ Subtract in the numerator and add in the denominator of the complex fraction.

Now that the numerator and the denominator of the complex fraction are single rational expressions, we perform the indicated division.

$$\frac{\frac{x-2}{2x}}{\frac{5x+3}{15}} = \frac{x-2}{2x} \div \frac{5x+3}{15}$$ Write the division indicated by the main fraction bar using a ÷ symbol.

$$= \frac{x-2}{2x} \cdot \frac{15}{5x+3}$$ Multiply by the reciprocal of $\frac{5x+3}{15}$.

$$= \frac{15(x-2)}{2x(5x+3)}$$ Multiply the numerators. Multiply the denominators. Since the numerator and denominator have no common factor, the result does not simplify.

$$= \frac{15x-30}{10x^2+6x}$$ Distribute 15 and distribute 2x.

Self Check 2 Simplify: $\dfrac{\frac{1}{3}+\frac{1}{x}}{\frac{x}{5}-\frac{1}{2}}$

Now Try Problem 23

EXAMPLE 3 Simplify: $\dfrac{\frac{6}{x}+y}{\frac{6}{y}+x}$

Strategy We will simplify the expressions above and below the main fraction bar separately to write $\frac{6}{x}+y$ and $\frac{6}{y}+x$ as single rational expressions. Then we will perform the indicated division.

Why The numerator and the denominator of the complex fraction must be written as single rational expressions before dividing.

Solution To write $\frac{6}{x}+y$ as a single rational expression, we build y into a fraction with a denominator of x and add. To write $\frac{6}{y}+x$ as a single rational expression, we build x into a fraction with a denominator of y and add.

$$\frac{\frac{6}{x}+y}{\frac{6}{y}+x} = \frac{\frac{6}{x}+\frac{y}{1}\cdot\frac{x}{x}}{\frac{6}{y}+\frac{x}{1}\cdot\frac{y}{y}}$$

Write y as $\frac{y}{1}$. The LCD for the numerator is x. Build $\frac{y}{1}$ so that it has a denominator of x.

Write x as $\frac{x}{1}$. The LCD for the denominator is y. Build $\frac{x}{1}$ so that it has a denominator of y.

$$= \frac{\frac{6}{x}+\frac{xy}{x}}{\frac{6}{y}+\frac{xy}{y}}$$ Perform the two multiplications shown in color.

$$= \frac{\frac{6+xy}{x}}{\frac{6+xy}{y}}$$ Add in the numerator and in the denominator of the complex fraction.

Now that the numerator and the denominator of the complex fraction are single rational expressions, we can perform the division.

Success Tip

Simplifying using division (method 1) works well when a complex fraction is written, or can be easily written, as a quotient of two single rational expressions.

$$\frac{\frac{6+xy}{x}}{\frac{6+xy}{y}} = \frac{6+xy}{x} \div \frac{6+xy}{y}$$ Write the division indicated by the main fraction bar using a ÷ symbol.

$$= \frac{6+xy}{x} \cdot \frac{y}{6+xy}$$ Multiply by the reciprocal of $\frac{6+xy}{y}$.

$$= \frac{y(6+xy)}{x(6+xy)}$$ Multiply the numerators. Multiply the denominators.

$$= \frac{y\overset{1}{\cancel{(6+xy)}}}{x\underset{1}{\cancel{(6+xy)}}}$$ Simplify the result by removing a factor equal to 1.

$$= \frac{y}{x}$$

Self Check 3 Simplify: $\dfrac{\frac{2}{a}-b}{\frac{2}{b}-a}$

Now Try Problem 31

2 Simplify Complex Fractions Using the LCD.

A second method for simplifying complex fractions uses the concepts of LCD and multiplication by a form of 1. The multiplication by 1 produces a simpler, equivalent expression, which will not contain rational expressions in its numerator or denominator.

Simplifying Complex Fractions Method 2: Multiplying by the LCD

1. Find the LCD of all rational expressions within the complex fraction.
2. Multiply the complex fraction by 1 in the form $\frac{\text{LCD}}{\text{LCD}}$.
3. Perform the operations in the numerator and denominator. No rational expressions should remain within the complex fraction.
4. Simplify the result, if possible.

We will use method 2 to rework Example 2.

EXAMPLE 4 Simplify: $\dfrac{\frac{1}{2}-\frac{1}{x}}{\frac{x}{3}+\frac{1}{5}}$

Strategy Using method 1 to simplify this complex fraction, we worked with $\frac{1}{2}-\frac{1}{x}$ and $\frac{x}{3}+\frac{1}{5}$ separately. With method 2, we will use the LCD of *all four* rational expressions within the complex fraction.

Why Multiplying a complex fraction by 1 in the form of $\frac{\text{LCD}}{\text{LCD}}$ clears its numerator and denominator of fractions.

Solution The denominators of all the rational expressions within the complex fraction are 2, x, 3, and 5. Thus, their LCD is $2 \cdot x \cdot 3 \cdot 5 = 30x$.

We now multiply the complex fraction by a factor equal to 1, using the LCD: $\frac{30x}{30x} = 1$.

$$\frac{\frac{1}{2} - \frac{1}{x}}{\frac{x}{3} + \frac{1}{5}} = \frac{\frac{1}{2} - \frac{1}{x}}{\frac{x}{3} + \frac{1}{5}} \cdot \frac{30x}{30x}$$

$$= \frac{\left(\frac{1}{2} - \frac{1}{x}\right)30x}{\left(\frac{x}{3} + \frac{1}{5}\right)30x}$$

← Multiply the numerators.

← Multiply the denominators.

$$= \frac{\frac{1}{2}(30x) - \frac{1}{x}(30x)}{\frac{x}{3}(30x) + \frac{1}{5}(30x)}$$

← In the numerator, distribute the multiplication by $30x$.

← In the denominator, distribute the multiplication by $30x$.

$$= \frac{15x - 30}{10x^2 + 6x}$$

Perform each of the four multiplications by $30x$. Notice that no fractional expressions remain within the complex fraction.

To attempt to simplify the result, factor the numerator and denominator. Since they do not have a common factor, the result is in simplest form. This is the same result obtained in Example 2, which used method 1.

Success Tip

With method 2, each term of the numerator and each term of the denominator of the complex fraction is multiplied by the LCD. Arrows can be helpful in showing this.

$$\frac{\left(\frac{1}{2} - \frac{1}{x}\right)}{\left(\frac{x}{3} + \frac{1}{5}\right)} \cdot \frac{30x}{30x}$$

Success Tip

When simplifying a complex fraction, the same result will be obtained regardless of the method used. See Example 2.

Self Check 4 Use method 2 to simplify: $\frac{\frac{1}{4} - \frac{1}{x}}{\frac{x}{5} + \frac{1}{3}}$

Now Try Problem 37

EXAMPLE 5

Simplify: $\frac{\frac{1}{8} - \frac{1}{y}}{\frac{8 - y}{4y^2}}$

Strategy Using method 1, we would work with $\frac{1}{8} - \frac{1}{y}$ and $\frac{8 - y}{4y^2}$ separately. With method 2, we use the LCD of all three rational expressions within the complex fraction.

Why Multiplying a complex fraction by 1 in the form of $\frac{\text{LCD}}{\text{LCD}}$ clears its numerator and denominator of fractions.

Solution The denominators of all the rational expressions within the complex fraction are 8, y, and $4y^2$. Therefore, the LCD is $8y^2$ and we multiply the complex fraction by a factor equal to 1, using the LCD: $\frac{8y^2}{8y^2} = 1$.

$$\frac{\frac{1}{8} - \frac{1}{y}}{\frac{8 - y}{4y^2}} = \frac{\frac{1}{8} - \frac{1}{y}}{\frac{8 - y}{4y^2}} \cdot \frac{8y^2}{8y^2}$$

Success Tip

Notice that with this method, all the rational expressions in the complex fraction are considered to find one universal LCD.

$$= \frac{\left(\frac{1}{8} - \frac{1}{y}\right)8y^2}{\left(\frac{8 - y}{4y^2}\right)8y^2}$$

← Multiply the numerators.

← Multiply the denominators.

$$= \frac{\frac{1}{8}(8y^2) - \frac{1}{y}(8y^2)}{\left(\frac{8 - y}{4y^2}\right)(8y^2)}$$

In the numerator, distribute the multiplication by $8y^2$.

$$= \frac{y^2 - 8y}{(8 - y)2}$$

Perform each of the three multiplications by $8y^2$.

$$= \frac{y\overset{-1}{\cancel{(y - 8)}}}{\underset{1}{\cancel{(8 - y)}}2}$$

In the numerator, factor out the GCF, y. Since $y - 8$ and $8 - y$ are opposites, simplify by replacing $\frac{y - 8}{8 - y}$ with $\frac{-1}{1}$.

$$= -\frac{y}{2}$$

The Language of Algebra

After multiplying a complex fraction by $\frac{\text{LCD}}{\text{LCD}}$ and performing the multiplications, the numerator and denominator of the complex fraction will be **cleared of fractions.**

Self Check 5 Simplify: $\dfrac{\frac{10 - n}{5n^2}}{\frac{1}{10} - \frac{1}{n}}$

Now Try Problem 41

EXAMPLE 6 Simplify: $\dfrac{1}{1 + \frac{1}{x + 1}}$

Strategy Although either method can be used, we will use method 2 to simplify this complex fraction.

Why Method 2 is often easier when the complex fraction contains a sum or difference.

Solution The only rational expression within the complex fraction has the denominator $x + 1$. Therefore, the LCD is $x + 1$. We multiply the complex fraction by a factor equal to 1, using the LCD: $\frac{x + 1}{x + 1} = 1$.

$$\frac{1}{1 + \frac{1}{x + 1}} = \frac{1}{1 + \frac{1}{x + 1}} \cdot \frac{x + 1}{x + 1}$$

$$= \frac{1(x + 1)}{\left(1 + \frac{1}{x + 1}\right)(x + 1)}$$

Multiply the numerators.
Multiply the denominators.

$$= \frac{1(x + 1)}{1(x + 1) + \frac{1}{x + 1}(x + 1)}$$

In the denominator, distribute the multiplication by $x + 1$.

$$= \frac{x + 1}{x + 1 + 1}$$

Perform each of the three multiplications by $x + 1$.

$$= \frac{x + 1}{x + 2}$$

Combine like terms in the denominator.

The result does not simplify.

Success Tip

Simplifying using the LCD (method 2) works well when the complex fraction has sums and/or differences in the numerator and/or denominator.

Self Check 6 Simplify: $\dfrac{2}{\frac{1}{x+2}+2}$

Now Try Problem 47

SECTION 7.5 STUDY SET

VOCABULARY

Fill in the blanks.

1. The expression $\dfrac{\frac{2}{3}-\frac{1}{x}}{\frac{x-3}{4}}$ is called a ________ rational expression or a ________ fraction.
2. In a complex fraction, the numerator is above the ______ fraction bar and the __________ is below it.

CONCEPTS

Fill in the blanks.

3. Method 1: To simplify a complex fraction, write its numerator and denominator as ______ rational expressions. Then perform the indicated ________ by multiplying the numerator of the complex fraction by the __________ of the denominator.
4. Method 2: To simplify a complex fraction, find the LCD of ____ the rational expressions within the complex fraction. Multiply the complex fraction by 1 in the form .
5. Consider: $\dfrac{\frac{x-3}{4}}{\frac{1}{12}-\frac{x}{6}}$
   a. What is the numerator of the complex fraction? Is it a single rational expression?
   b. What is the denominator of the complex fraction? Is it a single rational expression?
6. Consider the complex fraction: $\dfrac{\frac{1}{y}-\frac{1}{3}}{\frac{5}{6}+\frac{1}{y}}$
   a. What is the LCD of all the rational expressions in the complex fraction?
   b. To simplify the complex fraction using method 2, it should be multiplied by what form of 1?

NOTATION

Fill in the blanks.

7. $\dfrac{\frac{12}{y^2}}{\frac{4}{y^3}}$ means $\dfrac{12}{y^2} \quad \dfrac{4}{y^3}$

8. $\dfrac{\left(\frac{1}{5}-\frac{1}{a}\right)}{\left(\frac{a}{4}+\frac{2}{a}\right)} \cdot \dfrac{20a}{20a} = \dfrac{\frac{1}{5}(\quad) - \frac{1}{a}(\quad)}{\frac{a}{4}(\quad) + \frac{2}{a}(\quad)}$

   $= \dfrac{4a - \quad}{\quad + \quad}$

GUIDED PRACTICE

Simplify each complex fraction. See Example 1.

9. $\dfrac{\frac{2}{3}}{\frac{3}{4}}$
10. $\dfrac{\frac{3}{5}}{\frac{2}{7}}$
11. $\dfrac{\frac{x}{2}}{\frac{6}{5}}$
12. $\dfrac{\frac{9}{4}}{\frac{7}{x}}$
13. $\dfrac{\frac{x}{y}}{\frac{1}{x}}$
14. $\dfrac{\frac{y}{x}}{\frac{x}{xy}}$
15. $\dfrac{\frac{n}{8}}{\frac{1}{n^2}}$
16. $\dfrac{\frac{1}{m}}{\frac{m^3}{15}}$
17. $\dfrac{\frac{4a}{11}}{\frac{6a^4}{55}}$
18. $\dfrac{\frac{14}{15m}}{\frac{21}{25m^6}}$
19. $\dfrac{-\frac{x^4}{30}}{\frac{7x^2}{15}}$
20. $\dfrac{-\frac{5x^2}{24}}{\frac{x^5}{56}}$

Simplify each complex fraction. See Examples 2 or 4.

21. $\dfrac{\frac{1}{2}+\frac{3}{4}}{\frac{3}{2}+\frac{1}{4}}$
22. $\dfrac{\frac{2}{3}-\frac{5}{2}}{\frac{2}{3}-\frac{3}{2}}$
23. $\dfrac{\frac{1}{4}+\frac{1}{y}}{\frac{y}{3}-\frac{1}{2}}$
24. $\dfrac{\frac{2}{x}-\frac{1}{3}}{\frac{2}{3}+\frac{x}{5}}$
25. $\dfrac{\frac{1}{y}-\frac{5}{2}}{\frac{3}{y}}$
26. $\dfrac{\frac{1}{6}-\frac{5}{s}}{\frac{2}{s}}$
27. $\dfrac{\frac{4}{c}-\frac{c}{6}}{\frac{2}{c}}$
28. $\dfrac{\frac{10}{n}-\frac{n}{4}}{\frac{8}{n}}$

Simplify each complex fraction. See Examples 3 or 5.

29. $\dfrac{\frac{2}{3}+1}{\frac{1}{3}+1}$
30. $\dfrac{\frac{3}{5}-2}{\frac{2}{5}-2}$
31. $\dfrac{\frac{1}{x}-3}{\frac{5}{x}+2}$
32. $\dfrac{\frac{1}{y}+3}{\frac{3}{y}-2}$

33. $\dfrac{\frac{2}{x}+2}{\frac{4}{x}+2}$

34. $\dfrac{\frac{3}{x}-3}{\frac{9}{x}-3}$

35. $\dfrac{\frac{3y}{x}-y}{y-\frac{y}{x}}$

36. $\dfrac{\frac{y}{x}+3y}{y+\frac{2y}{x}}$

Simplify each complex fraction. See Example 4.

37. $\dfrac{\frac{1}{6}-\frac{2}{x}}{\frac{1}{6}+\frac{1}{x}}$

38. $\dfrac{\frac{3}{4}+\frac{1}{y}}{\frac{5}{6}-\frac{1}{y}}$

39. $\dfrac{\frac{a}{7}-\frac{7}{a}}{\frac{1}{a}+\frac{1}{7}}$

40. $\dfrac{\frac{t}{9}-\frac{9}{t}}{\frac{1}{t}+\frac{1}{9}}$

Simplify each complex fraction. See Example 5.

41. $\dfrac{\frac{d^2}{4}+\frac{4d}{5}}{\frac{d+1}{2}}$

42. $\dfrac{\frac{d+2}{2}}{\frac{d}{3}-\frac{d}{4}}$

43. $\dfrac{\frac{2}{x}}{\frac{2}{y}-\frac{4}{x}}$

44. $\dfrac{\frac{2y}{3}}{\frac{2y}{3}-\frac{8}{y}}$

Simplify each complex fraction. See Example 6.

45. $\dfrac{\frac{1}{x+1}}{1+\frac{1}{x+1}}$

46. $\dfrac{\frac{1}{x-1}}{1-\frac{1}{x-1}}$

47. $\dfrac{\frac{x}{x+2}}{\frac{x}{x+2}+x}$

48. $\dfrac{\frac{2}{x-2}}{\frac{2}{x-2}-1}$

TRY IT YOURSELF

Simplify each complex fraction.

49. $\dfrac{\frac{1}{p}+\frac{1}{q}}{\frac{1}{p}}$

50. $\dfrac{\frac{m}{n}+1}{1-\frac{m}{n}}$

51. $\dfrac{40x^2}{\frac{20x}{9}}$

52. $\dfrac{18n^2}{\frac{6n}{13}}$

53. $\dfrac{\frac{1}{c}+\frac{1}{2}}{\frac{1}{c^2}-\frac{1}{4}}$

54. $\dfrac{\frac{1}{m}-\frac{1}{n}}{\frac{m}{n}-\frac{n}{m}}$

55. $\dfrac{\frac{1}{r+1}+1}{\frac{3}{r-1}+1}$

56. $\dfrac{5+\frac{1}{n+7}}{4-\frac{2}{n+7}}$

57. $\dfrac{\frac{b^2-81}{18a^2}}{\frac{4b-36}{9a}}$

58. $\dfrac{\frac{8x-64}{y}}{\frac{x^2-64}{y^2}}$

59. $\dfrac{\frac{10x}{x-3}}{\frac{6}{x-3}}$

60. $\dfrac{\frac{18a}{a-4}}{\frac{12}{a-4}}$

61. $\dfrac{4-\frac{1}{8h}}{12+\frac{3}{4h}}$

62. $\dfrac{12+\frac{1}{3b}}{12-\frac{1}{b^2}}$

63. $\dfrac{\frac{m}{n}+\frac{n}{m}}{\frac{m}{n}-\frac{n}{m}}$

64. $\dfrac{\frac{2a}{b}-\frac{b}{a}}{\frac{2a}{b}+\frac{b}{a}}$

65. $\dfrac{\frac{2}{c^2}}{\frac{1}{c}+\frac{5}{4}}$

66. $\dfrac{\frac{7}{s^2}}{\frac{1}{s}+\frac{10}{3}}$

67. $\dfrac{\frac{4t-8}{t^2}}{\frac{8t-16}{t^5}}$

68. $\dfrac{\frac{9m-27}{m^6}}{\frac{2m-6}{m^8}}$

69. $\dfrac{\frac{2}{s}-\frac{2}{s^2}}{\frac{4}{s^3}+\frac{4}{s^2}}$

70. $\dfrac{\frac{2}{x^3}-\frac{2}{x}}{\frac{4}{x}+\frac{8}{x^2}}$

71. $\dfrac{1+\frac{6}{t}+\frac{8}{t^2}}{1+\frac{1}{t}-\frac{12}{t^2}}$

72. $\dfrac{1-p+\frac{2}{p}}{\frac{6}{p^2}+\frac{1}{p}-1}$

73. $\dfrac{1}{\frac{1}{x}+\frac{1}{y}}$

74. $\dfrac{1}{\frac{b}{a}-\frac{a}{b}}$

75. $\dfrac{-\frac{25}{16x^2}}{\frac{15}{32x^5}}$

76. $\dfrac{\frac{21}{8g^3}}{-\frac{35}{16g^8}}$

77. $\dfrac{3+\frac{3}{x-1}}{3-\frac{3}{x-1}}$

78. $\dfrac{2-\frac{2}{x+1}}{2+\frac{2}{x+1}}$

79. $\dfrac{1-\frac{9}{d^2}}{2+\frac{6}{d}}$

80. $\dfrac{1-\frac{16}{a^2}}{\frac{12}{a}+3}$

81. $\dfrac{\frac{1}{a^2b}-\frac{5}{ab}}{\frac{3}{ab}-\frac{7}{ab^2}}$

82. $\dfrac{\frac{3}{ab^2}+\frac{6}{a^2b}}{\frac{6}{a}-\frac{9}{b^2}}$

83. $\dfrac{m - \dfrac{1}{2m + 1}}{1 - \dfrac{m}{2m + 1}}$

84. $\dfrac{1 - \dfrac{r}{2r + 1}}{r - \dfrac{1}{2r + 1}}$

APPLICATIONS

85. **Slope.** We can use the slope formula shown below to find the slope of a line that passes through $\left(\frac{1}{2}, \frac{1}{3}\right)$ and $\left(\frac{3}{4}, \frac{5}{8}\right)$. Simplify the complex fraction to find m.

$$m = \frac{\frac{5}{8} - \frac{1}{3}}{\frac{3}{4} - \frac{1}{2}}$$

86. **Pitching.** The earned run average (ERA) is a statistic that gives the average number of earned runs a pitcher allows. For a softball pitcher, this is based on a six-inning game. The formula for ERA is shown below. Simplify the complex fraction on the right side of the formula.

$$\text{ERA} = \frac{\text{earned runs}}{\frac{\text{innings pitched}}{6}}$$

87. **Electronics.** In electronic circuits, resistors are tiny components that limit the flow of an electric current. An important formula about two resistors in a circuit is shown below. Simplify the complex fraction on the right side of the formula.

$$\text{Total resistance} = \frac{1}{\frac{1}{R_1} + \frac{1}{R_2}}$$

(Recall that R_1 is read as R sub one.)

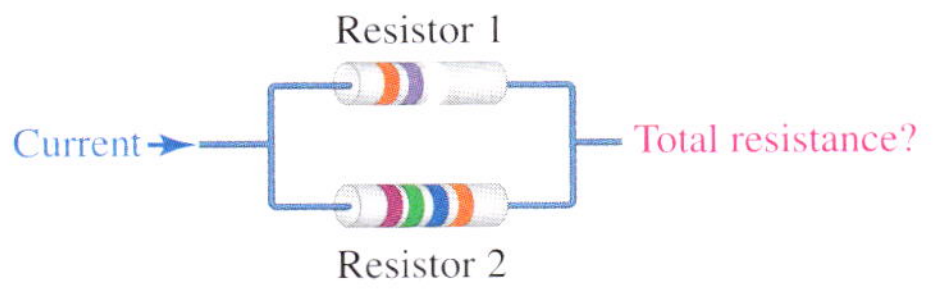

88. **Data Analysis.** Use the data in the table to find the average measurement for the three-trial experiment.

	Trial 1	Trial 2	Trial 3
Measurement	$\frac{k}{2}$	$\frac{k}{3}$	$\frac{k}{2}$

WRITING

89. What is a complex fraction? Give several examples.

90. Explain how to use method 1 to simplify: $\dfrac{1 + \frac{1}{x}}{3 - \frac{1}{x}}$

91. Explain how to use method 2 to simplify the expression in Problem 90.

92. a. List an advantage and a disadvantage of using method 1 to simplify a complex fraction.
    b. List an advantage and a disadvantage of using method 2 to simplify a complex fraction.

REVIEW

Simplify each expression. Write each answer without negative exponents.

93. $(8x)^0$

94. $\left(-\dfrac{3r}{4r^3}\right)^4$

95. $\left(\dfrac{4x^3}{5x^{-3}}\right)^{-2}$

96. $\left(\dfrac{12xy^{-3}}{3x^{-2}y^2}\right)^{-2}$

CHALLENGE PROBLEMS

Simplify.

97. $\dfrac{\dfrac{h}{h^2 + 3h + 2}}{\dfrac{4}{h + 2} - \dfrac{4}{h + 1}}$

98. $\dfrac{\dfrac{2}{b^2 - 1} - \dfrac{3}{ab - a}}{\dfrac{3}{ab - a} - \dfrac{2}{b^2 - 1}}$

99. $a + \dfrac{a}{1 + \dfrac{a}{a + 1}}$

100. $\dfrac{y^{-2} + 1}{y^{-2} - 1}$

SECTION 7.6 Solving Rational Equations

OBJECTIVES

1 Solve rational equations.

2 Solve for a specified variable in a formula.

ARE YOU READY?

The following problems review some basic skills that are needed when solving rational equations.

1. What is the LCD of $\frac{7}{2x}$, $\frac{1}{9x}$, and $\frac{17}{6x}$?

2. Multiply: $5(x - 1)\left(\dfrac{x}{x - 1}\right)$

3. Solve: $x^2 - x - 56 = 0$

4. Find all real numbers for which the rational expression $\dfrac{x + 1}{x + 8}$ is undefined.

In Chapter 2, we solved equations such as $\frac{1}{6}x + \frac{5}{2} = \frac{1}{3}$ by multiplying both sides by the LCD. With this approach, the equation that results is equivalent to the original equation, but easier to solve because it is cleared of fractions.

In this section, we will extend the fraction-clearing strategy to solve another type of equation, called a *rational equation.*

Rational Equations

A **rational equation** is an equation that contains one or more rational expressions.

Rational equations often have a variable in a denominator. Some examples of rational equations are:

$$\frac{2x}{3} = \frac{x}{6} + \frac{3}{2} \qquad \frac{2}{x} + \frac{1}{4} = \frac{5}{2x} \qquad \frac{11x}{x-5} = 6 + \frac{55}{x-5}$$

1 Solve Rational Equations.

To **solve a rational equation,** we find all the values of the variable that make the equation true. Any value of the variable that makes a denominator in a rational equation equal to 0 cannot be a solution of the equation. Such a number must be rejected, because division by 0 is undefined.

The goal when solving a rational equation is to use the multiplication property of equality to find an equivalent equation that we already know how to solve, such as a linear equation in one variable or a quadratic equation.

Strategy for Solving Rational Equations

1. Determine which numbers cannot be solutions of the equation.
2. Multiply both sides of the equation by the LCD of all rational expressions in the equation. This clears the equation of fractions.
3. Solve the resulting equation.
4. Check all possible solutions in the original equation.

EXAMPLE 1 Solve: $\frac{2x}{3} = \frac{x}{6} + \frac{3}{2}$

Strategy We will use the multiplication property of equality to clear this rational equation of fractions by multiplying both sides by the LCD.

Why Equations that contain only integers are usually easier to solve than equations that contain fractions.

Solution There are no restrictions on x, because no value of x ever makes a denominator 0. Since the denominators are 3, 6, and 2, we multiply both sides of the equation by the LCD, 6.

$\frac{2x}{3} = \frac{x}{6} + \frac{3}{2}$ This is the equation to solve.

$6\left(\frac{2x}{3}\right) = 6\left(\frac{x}{6} + \frac{3}{2}\right)$ Multiply both sides of the equation by the LCD of $\frac{2x}{3}$, $\frac{x}{6}$, and $\frac{3}{2}$, which is 6.

$6\left(\frac{2x}{3}\right) = 6\left(\frac{x}{6}\right) + 6\left(\frac{3}{2}\right)$ Distribute the multiplication by 6.

$2 \cdot \overset{1}{\cancel{3}}\left(\frac{2x}{\underset{1}{\cancel{3}}}\right) = \overset{1}{\cancel{6}}\left(\frac{x}{\underset{1}{\cancel{6}}}\right) + \overset{1}{\cancel{2}} \cdot 3\left(\frac{3}{\underset{1}{\cancel{2}}}\right)$ Perform the three multiplications by 6 by first removing common factors of the numerator and denominator. Try to do this step in your head.

Caution

Always enclose the left and right sides of an equation within parentheses when multiplying both sides by the LCD.

Notation

Here is an alternate way to remove common factors when multiplying by the LCD:

$\overset{2}{\cancel{6}}\left(\frac{2x}{\underset{1}{\cancel{3}}}\right)$ and $\overset{3}{\cancel{6}}\left(\frac{3}{\underset{1}{\cancel{2}}}\right)$

$$4x = x + 9$$ Simplify. Note that the fractions have been cleared. The result is a linear equation in one variable.

$$3x = 9$$ To eliminate x on the right side, subtract x from both sides.

$$x = 3$$ To undo the multiplication by 3, divide both sides by 3.

To check, we replace each x with 3 in the *original* equation.

$$\frac{2x}{3} = \frac{x}{6} + \frac{3}{2}$$

$$\frac{2(3)}{3} \stackrel{?}{=} \frac{3}{6} + \frac{3}{2}$$ Substitute 3 for x.

$$2 \stackrel{?}{=} \frac{1}{2} + \frac{3}{2}$$ Simplify: $\frac{2(\cancel{3})}{\cancel{3}} = 2$ and $\frac{3}{6} = \frac{1}{2}$.

$$2 = 2$$ Add: $\frac{1}{2} + \frac{3}{2} = \frac{4}{2} = 2$.

Since we obtain a true statement, 3 is the solution of $\frac{2x}{3} = \frac{x}{6} + \frac{3}{2}$. The solution set is $\{3\}$.

Self Check 1 Solve: $\frac{3x}{5} = \frac{x}{2} + \frac{1}{10}$

Now Try Problem 15

EXAMPLE 2 Solve: $\frac{2}{x} + \frac{1}{4} = \frac{5}{2x}$

Strategy This equation contains two rational expressions that have a variable in their denominators. We begin by asking, "What value(s) of x make either denominator 0?" Then we will clear the equation of fractions by multiplying both sides by the LCD.

Why If a number makes the denominator of a rational expression 0, that number cannot be a solution of the equation because division by 0 is undefined.

Solution If x is 0, the denominators of $\frac{2}{x}$ and $\frac{5}{2x}$ are 0 and the rational expressions would be undefined. Therefore, 0 cannot be a solution.

Since the denominators are x, 4, and $2x$, we multiply both sides of the equation by the LCD, $4x$, to clear the equation of fractions.

$$\frac{2}{x} + \frac{1}{4} = \frac{5}{2x}$$ This is the equation to solve.

$$4x\left(\frac{2}{x} + \frac{1}{4}\right) = 4x\left(\frac{5}{2x}\right)$$ Write each side of the equation within parentheses, and then multiply both sides by $4x$.

$$4x\left(\frac{2}{x}\right) + 4x\left(\frac{1}{4}\right) = 4x\left(\frac{5}{2x}\right)$$ On the left side, distribute the multiplication by $4x$.

$$4\overset{1}{\cancel{x}}\left(\frac{2}{\underset{1}{\cancel{x}}}\right) + \overset{1}{\cancel{4}}x\left(\frac{1}{\underset{1}{\cancel{4}}}\right) = 2 \cdot \overset{1}{\cancel{2}} \cdot \overset{1}{\cancel{x}}\left(\frac{5}{\underset{1}{\cancel{2}} \cdot \underset{1}{\cancel{x}}}\right)$$ On the right side, factor $4x$ as $2 \cdot 2 \cdot x$. Perform the three multiplications by $4x$ by first removing common factors of each numerator and denominator. Try to do this step in your head.

$$8 + x = 10$$ Simplify. Note that the fractions have been cleared. The result is a linear equation in one variable.

$$x = 2$$ To undo the addition of 8, subtract 8 from both sides.

The solution of $\frac{2}{x} + \frac{1}{4} = \frac{5}{2x}$ is 2. The solution set is $\{2\}$. Check by substituting 2 for each x in the *original* equation.

Caution

After multiplying both sides by the LCD and simplifying, the equation should not contain any fractions. If it does, check for an algebraic error, or perhaps your LCD is incorrect.

Success Tip

Don't confuse procedures. To *simplify the expression* $\frac{2}{x} + \frac{1}{4}$, we build each fraction and keep the LCD $4x$, add the numerators, and write the sum over the LCD. To *solve the equation* $\frac{2}{x} + \frac{1}{4} = \frac{5}{2x}$, we multiply both sides by the LCD $4x$ to eliminate the denominators.

Self Check 2 Solve: $\frac{1}{6} + \frac{4}{3x} = \frac{5}{x}$

Now Try Problem 25

EXAMPLE 3 Solve: $y - \frac{12}{y} = 4$

Strategy Since the only denominator is y, we will multiply both sides of the equation by y.

Why Multiplying both sides by y will clear the equation of the fraction, $\frac{12}{y}$.

Solution If y is 0, the denominator of $\frac{12}{y}$ is 0 and the fraction would be undefined. Therefore, 0 cannot be a solution.

Caution

By the multiplication property of equality, each term on both sides of the equation must be multiplied by the LCD. Here, it would be incorrect to multiply only the second term, $\frac{12}{y}$, by the LCD.

$$y - y \cdot \frac{12}{y} = 4$$

$y - \frac{12}{y} = 4$ — This is the equation to solve.

$y\left(y - \frac{12}{y}\right) = y(4)$ — Write each side of the equation within parentheses and then multiply both sides by the LCD, y.

$y(y) - y\left(\frac{12}{y}\right) = y(4)$ — Distribute the multiplication by y.

$y^2 - 12 = 4y$ — Simplify: $\overset{1}{\cancel{y}}\left(\frac{12}{\underset{1}{\cancel{y}}}\right) = 12$. Note that the fraction has been cleared.

Note the y^2-term. The result is a quadratic equation.

We can solve the resulting quadratic equation using the factoring method.

$y^2 - 4y - 12 = 0$ — Subtract $4y$ from both sides to get 0 on the right side.

$(y - 6)(y + 2) = 0$ — Factor the trinomial.

$y - 6 = 0$ or $y + 2 = 0$ — Set each factor equal to 0.

$y = 6$ | $y = -2$ — Solve each equation.

There are two possible solutions, 6 and -2, to check in the original equation.

Check $y = 6$:	**Check $y = -2$:**	
$y - \frac{12}{y} = 4$	$y - \frac{12}{y} = 4$	This is the original equation.
$6 - \frac{12}{6} \stackrel{?}{=} 4$	$-2 - \frac{12}{-2} \stackrel{?}{=} 4$	
$6 - 2 \stackrel{?}{=} 4$	$-2 - (-6) \stackrel{?}{=} 4$	
$4 = 4$ True	$4 = 4$ True	

Thus, the solutions of $y - \frac{12}{y} = 4$ are 6 and -2. The solution set is $\{-2, 6\}$.

Self Check 3 Solve: $x - \frac{24}{x} = -5$

Now Try Problem 31

EXAMPLE 4 Solve: $\dfrac{11x}{x-5} = 6 + \dfrac{55}{x-5}$

Strategy Since both denominators are $x - 5$, we multiply both sides by the LCD, $x - 5$.

Why This will clear the equation of fractions.

Solution If x is 5, the denominators of $\frac{11x}{x-5}$ and $\frac{55}{x-5}$ are 0, and the rational expressions are undefined. Therefore, 5 cannot be a solution of the equation.

Caution

Even if you do not make an arithmetic or algebraic error when solving a rational equation, a possible solution may not check.

$$\frac{11x}{x-5} = 6 + \frac{55}{x-5}$$ This is the equation to solve.

$$(x-5)\left(\frac{11x}{x-5}\right) = (x-5)\left(6 + \frac{55}{x-5}\right)$$ Write each side of the equation within parentheses and then multiply both sides by $x - 5$.

$$\overset{1}{\cancel{(x-5)}}\left(\frac{11x}{\underset{1}{\cancel{x-5}}}\right) = (x-5)6 + \overset{1}{\cancel{(x-5)}}\left(\frac{55}{\underset{1}{\cancel{x-5}}}\right)$$ Distribute the multiplication by $x - 5$. Remove the common binomial factor $(x - 5)$ of the numerator and denominator.

$$11x = (x-5)6 + 55$$ Simplify. Note that the fractions have been cleared. The result is a linear equation in one variable.

$$11x = 6x - 30 + 55$$ To solve the resulting equation, distribute the 6.

$$11x = 6x + 25$$ Combine like terms: $-30 + 55 = 25$.

$$5x = 25$$ To eliminate $6x$ on the right side, subtract $6x$ from both sides.

$$x = 5$$ To undo the multiplication by 5, divide both sides by 5.

Earlier, we determined that 5 makes both denominators in the original equation 0. Therefore, 5 cannot be a solution. Since 5 is the only possible solution, and it must be rejected, it follows that $\frac{11x}{x-5} = 6 + \frac{55}{x-5}$ has *no solution*. The solution set is written as { } or $\varnothing$.

When solving an equation, a possible solution that does not satisfy the original equation is called an **extraneous solution.** In this example, 5 is an extraneous solution.

The Language of Algebra

Extraneous means not a vital part. Mathematicians speak of *extraneous* solutions. Rock groups don't want *extraneous* sounds (like feedback) coming from their amplifiers. Artists erase *extraneous* marks on their sketches.

Self Check 4 Solve $\frac{9x}{x-6} = 3 + \frac{54}{x-6}$, if possible.

Now Try Problem 41

EXAMPLE 5 Solve: $\dfrac{x+5}{x+3} + \dfrac{1}{x^2 + 2x - 3} = 1$

Strategy We will multiply both sides by the LCD of the two rational expressions in the equation. But first, we must factor the second denominator.

Why To determine the restrictions on the variable and to find the LCD, we need to write $x^2 + 2x - 3$ in factored form.

Solution Since the trinomial $x^2 + 2x - 3$ factors as $(x + 3)(x - 1)$, we can write the given equation as:

$$\frac{x+5}{x+3} + \frac{1}{(x+3)(x-1)} = 1$$ If x is -3, the first denominator is 0. If x is -3 or 1, the second denominator is 0.

We see that -3 and 1 cannot be solutions of the equation, because they make rational expressions in the equation undefined.

Since the denominators are $x + 3$ and $(x + 3)(x - 1)$, we multiply both sides of the equation by the LCD, $(x + 3)(x - 1)$, to clear the fractions.

$$(x + 3)(x - 1)\left[\frac{x + 5}{x + 3} + \frac{1}{(x + 3)(x - 1)}\right] = (x + 3)(x - 1)[1]$$ Write each side within brackets [ ].

$$\overset{1}{\cancel{(x + 3)}}(x - 1)\frac{x + 5}{\underset{1}{\cancel{x + 3}}} + \overset{1}{\cancel{(x + 3)}}\overset{1}{\cancel{(x - 1)}}\frac{1}{\underset{1}{\cancel{(x + 3)}}\underset{1}{\cancel{(x - 1)}}} = (x + 3)(x - 1)[1]$$ Distribute the multiplication by $(x + 3)(x - 1)$ and remove common factors.

$$(x - 1)(x + 5) + 1 = (x + 3)(x - 1)$$ Simplify. The fractions are cleared.

The Language of Algebra

We say that the LCD is a **multiplier** that clears a rational equation of fractions.

To solve the resulting equation, we multiply the binomials on the left side and the right side, and proceed as follows.

$$x^2 + 4x - 5 + 1 = x^2 + 2x - 3$$ Find $(x - 1)(x + 5)$ and $(x + 3)(x - 1)$.

$$x^2 + 4x - 4 = x^2 + 2x - 3$$ Combine like terms: $-5 + 1 = -4$.

$$4x - 4 = 2x - 3$$ Subtract x^2 from both sides. The x^2-terms are eliminated. The result is a linear equation in one variable.

$$2x - 4 = -3$$ To eliminate $2x$ on the right side, subtract $2x$ from both sides.

$$2x = 1$$ To undo the subtraction of 4, add 4 to both sides.

$$x = \frac{1}{2}$$ To undo the multiplication by 2, divide both sides by 2.

A check will show that $\frac{1}{2}$ is the solution of the original equation.

Self Check 5 Solve: $\frac{1}{x + 3} + \frac{1}{x - 3} = \frac{5}{x^2 - 9}$

Now Try Problem 45

2 Solve for a Specified Variable in a Formula.

Many formulas are expressed as rational equations. To solve such formulas for a specified variable, we use the same steps, in the same order, as we do when solving rational equations having only one variable.

EXAMPLE 6 **Determining a Child's Dosage.** The formula $C = \frac{AD}{A + 12}$ is called **Young's rule.** It is a way to find the approximate child's dose C of a medication, where A is the age of the child in years and D is the recommended dosage for an adult. Solve the formula for D.

Strategy As we have done in the previous examples, we will begin by multiplying both sides of the equation by the LCD to clear it of the fraction.

Why To isolate D on the right side of the equation, we must first isolate the term AD on that side. That calls for clearing the right side of the denominator $A + 12$.

Solution

$$C = \frac{AD}{A + 12}$$ This is Young's rule.

$$(A + 12)(C) = (A + 12)\left(\frac{AD}{A + 12}\right)$$ Write each side of the formula within parentheses, and then multiply both sides by the LCD, $A + 12$.

Success Tip

It is often helpful to circle the variable in the formula that you are solving for:

$$C = \frac{A\textcircled{D}}{A + 12}$$

$$(A + 12)C = AD$$ Simplify the right side: $(\overset{1}{\cancel{A + 12}})\left(\frac{AD}{\underset{1}{\cancel{A + 12}}}\right)$.

$$AC + 12C = AD$$ Distribute the multiplication by C.

$$\frac{AC + 12C}{A} = \frac{AD}{A}$$ To undo the multiplication by A on the right side and isolate D, divide both sides by A.

$$\frac{AC + 12C}{A} = D$$ Simplify the right side: $\frac{\overset{1}{\cancel{A}}D}{\underset{1}{\cancel{A}}} = D$.

Solving Young's rule for D, we have $D = \frac{AC + 12C}{A}$.

Self Check 6 Solve $R = \frac{eS}{T - 10}$ for S.

Now Try Problem 49

EXAMPLE 7

Photography. The design of a camera lens uses the formula $\frac{1}{f} = \frac{1}{p} + \frac{1}{q}$, where f is the focal length of the lens, p is the distance from the lens to the object, and q is the distance from the lens to the image. Solve the formula for q.

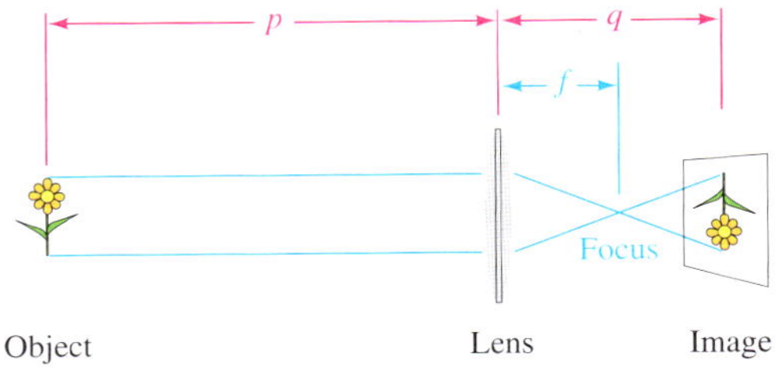

Strategy We will begin by multiplying both sides of the equation by the LCD.

Why It will be easier to isolate q if there are no fractions.

Solution

$$\frac{1}{f} = \frac{1}{p} + \frac{1}{q}$$ This is the given formula.

$$fpq\left(\frac{1}{f}\right) = fpq\left(\frac{1}{p} + \frac{1}{q}\right)$$ Write each side of the formula within parentheses and then multiply both sides by the LCD, fpq.

$$\overset{1}{\cancel{f}}pq\left(\frac{1}{\underset{1}{\cancel{f}}}\right) = f\overset{1}{\cancel{p}}q\left(\frac{1}{\underset{1}{\cancel{p}}}\right) + fp\overset{1}{\cancel{q}}\left(\frac{1}{\underset{1}{\cancel{q}}}\right)$$ Distribute the multiplication by fpq and then remove the common factors of each numerator and denominator.

$$pq = fq + fp$$ Simplify. Note that q is on both sides of the resulting equation.

Caution

A common error is to divide both sides of $pq = fq + fp$ by p to solve for q:

$$\cancel{q = \frac{fq + fp}{p}}$$

The formula is not solved for q because q appears on *both sides* of the equation.

If we subtract fq from both sides, all terms that contain q will be on the left side.

$$pq - fq = fp$$ Subtract fq from both sides.

$$q(p - f) = fp$$ Factor out the GCF, q, from the two terms on the left side.

$$\frac{q(p - f)}{p - f} = \frac{fp}{p - f}$$ To undo the multiplication by $(p - f)$ and isolate q, divide both sides by $p - f$.

$$q = \frac{fp}{p - f}$$ Simplify the left side: $\frac{q\overset{1}{\cancel{(p - f)}}}{\underset{1}{\cancel{p - f}}} = q$.

Solving the formula for q, we have $q = \frac{fp}{p - f}$.

Self Check 7 Solve the formula in Example 7 for p.

Now Try Problem 51

SECTION 7.6 STUDY SET

VOCABULARY

Fill in the blanks.

1. Equations that contain one or more rational expressions, such as $\frac{x}{x+2} = 4 + \frac{10}{x+2}$, are called ________ equations.
2. To ______ a rational equation we find all the values of the variable that make the equation true.
3. To ______ a rational equation of fractions, multiply both sides by the LCD of all rational expressions in the equation.
4. When solving a rational equation, if we obtain a number that does not satisfy the original equation, the number is called an __________ solution.

CONCEPTS

5. Is 5 a solution of the given rational equation?
   a. $\frac{1}{x-1} = 1 - \frac{3}{x-1}$
   b. $\frac{x}{x-5} = 3 + \frac{5}{x-5}$
6. A student was asked to solve a rational equation. The first step of his solution is as follows:
   $$12x\left(\frac{5}{x} + \frac{2}{3}\right) = 12x\left(\frac{7}{4x}\right)$$
   a. What equation was he asked to solve?
   b. What LCD is used to clear the equation of fractions?
7. Consider the rational equation $\frac{x}{x-3} = \frac{1}{x} + \frac{2}{x-3}$.
   a. What values of x make a denominator 0?
   b. What values of x make a rational expression undefined?
   c. What numbers can't be solutions of the equation?
8. A student solved a rational equation and found 8 to be a possible solution. When she checked 8, she obtained $\frac{3}{0} = \frac{1}{0} + \frac{2}{3}$. What conclusion can be drawn?

By what should both sides of the equation be multiplied to clear it of fractions?

9. a. $\frac{1}{y} = 20 - \frac{5}{y}$ b. $\frac{x}{x^2 - 4} = \frac{4}{x-2}$
10. a. $\frac{x}{5} = \frac{3x}{10} + \frac{7}{2x}$ b. $\frac{2x}{x-6} = 4 + \frac{1}{x-6}$
11. Perform each multiplication.
    a. $4x\left(\frac{3}{4x}\right)$ b. $(x+6)(x-2)\left(\frac{3}{x-2}\right)$
12. Fill in the blanks.
    $$\underbrace{8x\left(\frac{3}{4x}\right)} = \underbrace{8x\left(\frac{1}{8x}\right)} + \underbrace{8x\left(\frac{5}{4}\right)}$$
    $$\quad = \quad +$$

NOTATION

Complete the solution to solve the equation.

13. $$\frac{2}{a} + \frac{1}{2} = \frac{7}{2a}$$
    $$\left(\frac{2}{a} + \frac{1}{2}\right) = \left(\frac{7}{2a}\right)$$
    $$\left(\frac{2}{a}\right) + \left(\frac{1}{2}\right) = \left(\frac{7}{2a}\right)$$
    $$+ a =$$
    $$4 + a - 4 = 7 -$$
    $$a =$$
14. Can $5x\left(\frac{2}{x} + \frac{4}{5}\right)$ be written as $5x \cdot \frac{2}{x} + \frac{4}{5}$? Explain.

GUIDED PRACTICE

Solve each equation and check the result. If an equation has no solution, so indicate. See Example 1.

15. $\frac{2}{3} = \frac{1}{2} + \frac{x}{6}$
16. $\frac{7}{4} = \frac{x}{8} + \frac{5}{2}$
17. $\frac{s}{12} - \frac{s}{2} = \frac{5s}{4}$
18. $\frac{n}{18} - \frac{n}{6} = \frac{4n}{3}$
19. $\frac{x}{18} = \frac{1}{3} - \frac{x}{2}$
20. $\frac{x}{4} = \frac{1}{2} - \frac{3x}{20}$
21. $\frac{b}{4} + \frac{1}{2} = \frac{b}{3} - \frac{1}{4}$
22. $\frac{n}{6} + \frac{2}{3} = \frac{n}{3} - \frac{1}{36}$

Solve each equation and check the result. If an equation has no solution, so indicate. See Example 2.

23. $\frac{5}{3k} + \frac{1}{k} = -2$
24. $\frac{3}{4h} + \frac{2}{h} = 1$
25. $\frac{1}{4} - \frac{5}{6} = \frac{1}{a}$
26. $\frac{5}{9} - \frac{1}{3} = \frac{1}{b}$
27. $\frac{1}{8} + \frac{2}{b} - \frac{1}{12} = 0$
28. $\frac{1}{14} + \frac{2}{n} - \frac{2}{21} = 0$
29. $\frac{4}{5} - \frac{1}{10x} = \frac{7}{15}$
30. $\frac{5}{14} - \frac{1}{2x} = \frac{3}{7}$

Solve each equation and check the result. If an equation has no solution, so indicate. See Example 3.

31. $x + \frac{8}{x} = 6$
32. $z - \frac{16}{z} = 6$
33. $\frac{10}{t} - t = 3$
34. $\frac{7}{p} - p = -6$
35. $\frac{20}{c} + c = -9$
36. $d = 4 + \frac{21}{d}$
37. $4 + \frac{15}{p} = 3p$
38. $2x = 6 + \frac{8}{x}$

Solve each equation and check the result. If an equation has no solution, so indicate. See Example 4.

39. $\frac{x}{x-5} = 3 + \frac{5}{x-5}$

40. $\frac{3}{y-2} = \frac{3}{y-2} + 1$

41. $\frac{a^2}{a+2} - a = \frac{4}{a+2}$

42. $\frac{z^2}{z+1} + 2 = \frac{1}{z+1}$

Solve each equation and check the result. If an equation has no solution, so indicate. See Example 5.

43. $\frac{x+6}{x+4} + \frac{1}{x^2+x-12} = 1$

44. $\frac{x+7}{x+2} + \frac{1}{x^2-3x-10} = 1$

45. $\frac{2x}{x^2+x-2} + \frac{2}{x+2} = 1$

46. $\frac{4x}{x^2+2x-3} + \frac{3}{x+3} = 1$

Solve each formula for the specified variable. See Example 6.

47. $h = \frac{2A}{b+d}$ for A

48. $T = \frac{3R}{M-n}$ for R

49. $I = \frac{E}{R+r}$ for r

50. $\frac{S}{k+h} = E$ for k

51. $\frac{5}{x} - \frac{4}{y} = \frac{5}{z}$ for x

52. $\frac{2}{c} + \frac{2}{d} = \frac{1}{h}$ for c

53. $\frac{1}{r} + \frac{1}{s} = \frac{1}{t}$ for r

54. $\frac{1}{x} - \frac{1}{y} = \frac{1}{z}$ for x

Solve each formula for the specified variable. See Example 7.

55. $\frac{P}{n} = rt$ for P

56. $\frac{F}{m} = a$ for F

57. $\frac{a}{b} = \frac{c}{d}$ for d

58. $\frac{pc}{s} = \frac{t}{r}$ for c

59. $\frac{1}{a} + \frac{1}{b} = 1$ for a

60. $\frac{1}{a} - \frac{1}{b} = 1$ for b

61. $F = \frac{L^2}{6d} + \frac{d}{2}$ for L^2

62. $H = \frac{J^3}{cd} - \frac{K^3}{d}$ for J^3

TRY IT YOURSELF

Solve each equation and check the result. If an equation has no solution, so indicate.

63. $\frac{1}{2} + \frac{2}{x-3} = 1$

64. $\frac{3}{5} + \frac{7}{x+2} = 2$

65. $\frac{7}{q^2-q-2} + \frac{1}{q+1} = \frac{3}{q-2}$

66. $\frac{3}{x-1} - \frac{1}{x+9} = \frac{18}{x^2+8x-9}$

67. $\frac{2}{3-t} = \frac{-t}{t+3}$

68. $\frac{n}{n+1} = \frac{6}{n+7}$

69. $\frac{1}{8} + \frac{2}{y} = \frac{1}{y} + \frac{1}{10}$

70. $\frac{7}{10} + \frac{4}{c} = \frac{1}{c} + \frac{11}{15}$

71. $4 - \frac{8}{x+1} = \frac{8x}{x+1}$

72. $\frac{x}{x-2} = \frac{2}{x-2} + 2$

73. $\frac{5a}{a+1} - 4 = \frac{3}{a+1}$

74. $\frac{4}{b-3} = \frac{b+5}{b-3} - 5$

75. $\frac{2}{y+1} + 5 = \frac{12}{y+1}$

76. $\frac{3}{p+6} - 2 = \frac{7}{p+6}$

77. $\frac{3}{x+1} = \frac{x-2}{x+1} + \frac{x-2}{2}$

78. $\frac{2}{x-1} + \frac{x-2}{3} = \frac{4}{x-1}$

79. $\frac{z-4}{z-3} = \frac{z+2}{z+1}$

80. $\frac{a+2}{a+8} = \frac{a-3}{a-2}$

81. $\frac{3}{x} + 2 = 3$

82. $\frac{2}{x} + 9 = 11$

83. $\frac{4}{y^2-4} = \frac{1}{y-2} + \frac{1}{y+2}$

84. $\frac{2w}{w^2-9} = \frac{1}{w+3} - \frac{4}{w-3}$

85. $\frac{3}{5d} + \frac{4}{3} = \frac{9}{10d}$

86. $\frac{2}{3d} + \frac{1}{4} = \frac{11}{6d}$

87. $\frac{n}{n^2-9} + \frac{n+8}{n+3} = \frac{n-8}{n-3}$

88. $\frac{7}{x-5} = \frac{40}{x^2-25} + \frac{3}{x+5}$

89. $\frac{3}{x-2} + \frac{1}{x} = \frac{6x+4}{x^2-2x}$

90. $\frac{x}{x-1} - \frac{12}{x^2-x} = \frac{-1}{x-1}$

91. $y + \frac{2}{3} = \frac{2y-12}{3y-9}$

92. $1 - \frac{3}{b} = \frac{-8b}{b^2+3b}$

93. $\frac{a-1}{7} - \frac{a-2}{14} = \frac{1}{2}$

94. $\frac{3x-1}{6} - \frac{x+3}{2} = \frac{3x+4}{3}$

Look Alikes . . .

For each expression, perform the indicated operations and then simplify, if possible. Solve each equation and check the result.

95. a. $\frac{a}{3} + \frac{3}{5} + \frac{a}{15}$ b. $\frac{a}{3} + \frac{3}{5} = \frac{a}{15}$

96. a. $\frac{1}{6x} - \frac{2}{x-6}$ b. $\frac{1}{6x} = \frac{2}{x-6}$

97. a. $\frac{x}{x-2} - \frac{1}{x-3}$ b. $\frac{x}{x-2} - \frac{1}{x-3} = 1$

98. a. $\dfrac{u^2+1}{u^2-u} - \dfrac{u}{u-1}$ b. $\dfrac{u^2+1}{u^2-u} - \dfrac{u}{u-1} = \dfrac{1}{u}$

APPLICATIONS

99. **Medicine.** Radioactive tracers are used for diagnostic work in nuclear medicine. The ***effective half-life H*** of a radioactive material in an organism is given by the formula $H = \frac{RB}{R+B}$ where R is the radioactive half-life and B is the biological half-life of the tracer. Solve the formula for R.

100. **Chemistry.** Charles's law describes the relationship between the volume and temperature of a gas that is kept at a constant pressure. It can be expressed as $\frac{V_1}{V_2} = \frac{T_1}{T_2}$ where V_1 and V_2 are variables representing two different volumes, and T_1 and T_2 are variables representing two different temperatures. (Recall that the notation V_1 is read as *V sub one.*) Solve for V_2.

101. **Electronics.** Most electronic circuits require resistors to make them work properly. Resistors are components that limit current. An important formula about resistors in a circuit is $\frac{1}{r} = \frac{1}{r_1} + \frac{1}{r_2}$. Solve for r.

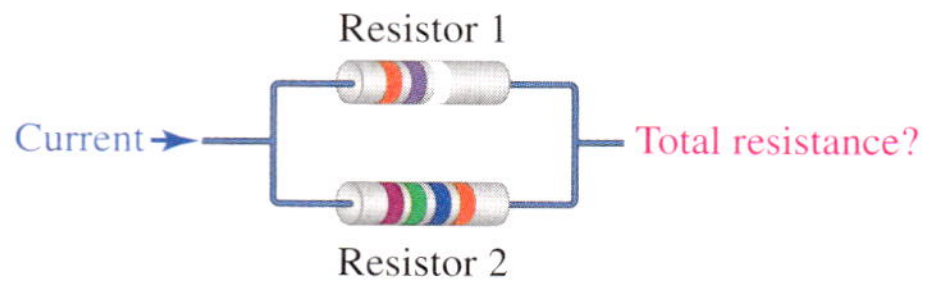

102. **Mathematical Formulas.** To quickly find the sum $\frac{1}{2} + \frac{1}{4} + \frac{1}{8} + \frac{1}{16} + \frac{1}{32} + \frac{1}{64} + \frac{1}{128}$, mathematicians use the formula $S = \dfrac{a(1-r^n)}{1-r}$. Solve the formula for a.

WRITING

103. Explain how the multiplication property of equality is used to solve rational equations. Give an example.

104. When solving rational equations, how do you know whether a solution is extraneous?

105. What is meant by clearing a rational equation of fractions? Give an example.

106. Explain the difference between the procedure used to simplify $\frac{1}{x} + \frac{1}{3}$ and the procedure used to solve $\frac{1}{x} + \frac{1}{3} = \frac{1}{2}$.

REVIEW

107. **Uniforms.** A cheerleading squad had their school mascot embroidered on the front of their uniform sweaters. They were charged $18.50 per sweater plus a one time setup fee of $75. If the project cost $445, how many sweaters were embroidered?

108. **Geometry.** The vertex angle of an isosceles triangle is 46°. Find the measure of each base angle.

CHALLENGE PROBLEMS

Solve each equation and check the result. If an equation has no solution, so indicate.

109. $\dfrac{x-4}{x-3} + \dfrac{x-2}{x-3} = x - 3$

110. $\dfrac{3}{x} = \dfrac{1 - \frac{1}{x}}{3 - \frac{7}{x}}$

111. $x^{-2} + 2x^{-1} + 1 = 0$

112. **Engines.** A formula that is used in the design and testing of diesel engines is $E = 1 - \dfrac{T_4 - T_1}{a(T_3 - T_2)}$. Solve the formula for T_1.

SECTION 7.7 Problem Solving Using Rational Equations

OBJECTIVES

1. Solve number problems.
2. Solve uniform motion problems.
3. Solve shared-work problems.
4. Solve investment problems.

ARE YOU READY?

The following problems review some basic skills that are needed when using rational equations to solve application problems.

1. Solve the uniform motion formula $d = rt$ for t.
2. Multiply: $\dfrac{1}{5} \cdot x$
3. What is the simple interest formula?
4. What is the LCD for the fractions in the rational equation $\dfrac{x}{9} + \dfrac{x}{7} = 1$?

We will now use the six-step problem-solving strategy to solve application problems from a variety of areas, including banking, petroleum engineering, sports, and travel. In each case, we will use a rational equation to model the situation. We begin with an example in which we find an unknown number.

1 Solve Number Problems.

EXAMPLE 1 **Number Problem.** If the same number is added to both the numerator and the denominator of the fraction $\frac{3}{5}$, the result is $\frac{4}{5}$. Find the number.

Analyze

- Begin with the fraction $\frac{3}{5}$.
- Add the same number to the numerator and to the denominator.
- The result is $\frac{4}{5}$.
- Find the number.

Assign Let n= the unknown number.

Form To form an equation, add the unknown number to the numerator and to the denominator of $\frac{3}{5}$. Then set the result equal to $\frac{4}{5}$.

$$\frac{3+n}{5+n}=\frac{4}{5}$$

Solve To solve this rational equation, we begin by clearing it of fractions.

$$\frac{3+n}{5+n}=\frac{4}{5}$$

$$5(5+n)\left(\frac{3+n}{5+n}\right)=5(5+n)\left(\frac{4}{5}\right)$$ Multiply both sides by the LCD, $5(5+n)$. Then remove common factors of the numerator and denominator.

$$5(3+n)=(5+n)4$$ Simplify. The fractions have been cleared.

$$15+5n=20+4n$$ Distribute the multiplication by 5 and by 4.

$$15+n=20$$ To isolate the variable term on the left side, subtract $4n$ from both sides.

$$n=5$$ To undo the addition of 15, subtract 15 from both sides.

State The number is 5.

Check When we add 5 to both the numerator and denominator of $\frac{3}{5}$, we get

$$\frac{3+5}{5+5}=\frac{8}{10}=\frac{4}{5}$$

The result checks.

Self Check 1 **Number Problem.** If the same number is added to both the numerator and denominator of the fraction $\frac{7}{9}$, the result is $\frac{8}{9}$. Find the number.

Now Try Problem 13

2 Solve Uniform Motion Problems.

The Language of Algebra

In uniform motion problems, the word **speed** is often used in place of the word **rate.** For example, we can say a car travels at a *rate* of 50 mph or its *speed* is 50 mph.

Recall that we use the distance formula $d = rt$ to solve motion problems. The relationship between distance, rate, and time can be expressed in another way by solving for t.

$$d = rt$$ Distance = rate · time.

$$\frac{d}{r}=\frac{rt}{r}$$ To undo the multiplication by r and isolate t, divide both sides by r.

$$\frac{d}{r}=t$$ Simplify the right side: $\frac{rt}{r}=t$.

$$t=\frac{d}{r}$$

This result suggests an alternate form of the distance formula, time $= \frac{\text{distance}}{\text{rate}}$, that is used to solve the next example.

EXAMPLE 2 **Runners.** A coach can run 10 miles in the same amount of time as his best student-athlete can run 12 miles. If the student runs 1 mile per hour (mph) faster than the coach, find the running speeds of the coach and the student.

©Exactostock/SuperStock

Analyze

- The coach runs 10 miles in the same time that the student runs 12 miles.
- The student runs 1 mph faster than the coach.
- Find the speed that each runs.

Assign Since the student's speed is 1 mph faster than the coach's, let r = the speed that the coach can run. Then, $r + 1$ = the speed that the student can run.

Form The expressions for the rates are entered in the Rate column of the table. The distances run by the coach and by the student are entered in the Distance column of the table.

Using $t = \frac{d}{r}$, we find that the time it takes the coach to run 10 miles, at a rate of r mph, is $\frac{10}{r}$ hours. Similarly, we find that the time it takes the student to run 12 miles, at a rate of $(r + 1)$ mph, is $\frac{12}{r + 1}$ hours. These expressions are entered in the Time column of the table.

Success Tip

We could have let r = the speed that the student can run. Then, since the coach runs slower than the student, $r - 1$ = the speed of the coach.

	Rate ·	Time =	Distance
Coach	r	$\frac{10}{r}$	10
Student	$r + 1$	$\frac{12}{r+1}$	12

Enter the information in these two columns first.

To get these entries, divide the distance by the rate to obtain an expression for the time: $t = \frac{d}{r}$.

The time it takes the coach to run 10 miles	is the same as	the time it takes the student to run 12 miles.
$\frac{10}{r}$	$=$	$\frac{12}{r + 1}$

Solve To solve this rational equation, we begin by clearing it of fractions.

$$\frac{10}{r} = \frac{12}{r + 1}$$

$$r(r + 1)\left(\frac{10}{r}\right) = r(r + 1)\left(\frac{12}{r + 1}\right)$$ Multiply both sides by the LCD, $r(r + 1)$. Then remove common factors of the numerator and denominator.

$$(r + 1)10 = 12r$$ Simplify. The fractions have been cleared.

$$10r + 10 = 12r$$ On the left side, distribute the multiplication by 10.

$$10 = 2r$$ To isolate the variable term on the right, subtract $10r$ from both sides.

$$5 = r$$ To undo the multiplication by 2, divide both sides by 2.

If $r = 5$, then $r + 1 = 6$.

State The coach's running speed is 5 mph and the student's running speed is 6 mph.

Check The coach will run 10 miles in $\frac{10 \text{ miles}}{5 \text{ mph}} = 2$ hours. The student will run 12 miles in $\frac{12 \text{ miles}}{6 \text{ mph}} = 2$ hours. The times are the same; the results check.

Caution

Check the result *in the words of the problem,* not by substituting it into the equation. Why? The equation may have been solved correctly, but the danger is that you may have formed it incorrectly.

Self Check 2 **Cycling.** A cyclist can ride 24 miles in the same amount of time that her friend can walk 8 miles. If the cyclist travels 8 mph faster than the walker, find the speed of the walker.

Now Try Problem 23

3 Solve Shared-Work Problems.

Problems in which two or more people (or machines) work together to complete a job are called *shared-work problems*. To solve such problems, we must determine the **rate of work** for each person (or machine) involved. For example, suppose it takes you 4 hours to clean your house. Your rate of work can be expressed as $\frac{1}{4}$ of the job is completed per hour. If someone else takes 5 hours to clean the same house, they complete $\frac{1}{5}$ of the job per hour. In general, a rate of work can be determined in the following way.

Rate of Work

If a job can be completed in t units of time, the rate of work can be expressed as:

$\frac{1}{t}$ of the job is completed per unit of time.

To solve shared-work problems, we also must determine what fractional part of a job is completed. To do this, we use the formula

Work completed = rate of work · time worked or $W = rt$

EXAMPLE 3 **Payroll.** At the end of a pay period, it takes the president of a company 15 minutes to sign all of her employees' payroll checks. What fractional part of the job is completed if the president signs checks for 10 minutes?

Strategy We will begin by finding the president's check-signing rate. Then we can use the formula $W = rt$ to find the part of the job that is completed.

Why We know the time worked is 10 minutes. To use the work formula to find what part of the job is completed, we also need to know the president's work rate.

Solution If all of the checks can be signed in 15 minutes, the president's work rate is $\frac{1}{15}$ job per minute. Substituting into the work formula, we have

$$W = rt$$

$$= \frac{1}{15} \cdot 10 \quad \text{Substitute } \tfrac{1}{15} \text{ for } r\text{, the work rate, and 10 for } t\text{, the time worked.}$$

$$= \frac{10}{15} \quad \text{Do the multiplication. Think: } \tfrac{1}{15} \cdot \tfrac{10}{1} = \tfrac{10}{15}.$$

$$= \frac{2}{3} \quad \text{Simplify by removing the common factor of 5: } \tfrac{2 \cdot \cancel{5}}{3 \cdot \cancel{5}} = \tfrac{2}{3}.$$

In 10 minutes, the president will complete $\frac{2}{3}$ of the job of signing the payroll checks.

Self Check 3 **Farming.** It takes a farmer 8 days to harvest a wheat crop. What part of the job is completed in 6 days?

Now Try Problem 5

EXAMPLE 4 **Filling a Tank.** An inlet pipe can fill an oil storage tank in 7 days, and a second inlet pipe can fill the same tank in 9 days. If both pipes are used, how long will it take to fill the tank?

Analyze

- The first pipe can fill the tank in 7 days.
- The second pipe can fill the tank in 9 days.
- How long will it take the two pipes, working together, to fill the tank?

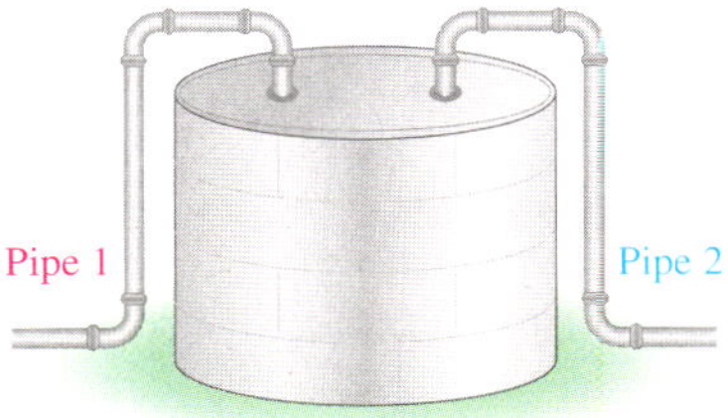

Assign Let x = the number of days it will take to fill the tank if both pipes are used.

Form It is helpful to organize the facts of the problem in a table. Since the pipes will be open for the same amount of time as they fill the tank, enter x as the time worked for each pipe.

The first pipe can fill the tank in 7 days; its rate working alone is $\frac{1}{7}$ of the job per day. The second pipe can fill the tank in 9 days; its rate working alone is $\frac{1}{9}$ of the job per day. To determine the work completed by each pipe, multiply the rate by the time.

Success Tip

The tables and the formulas used to solve uniform motion problems and shared-work problems are similar:

Motion: $d = rt$ Work: $W = rt$

	Rate ·	Time =	Work completed
1st pipe	$\frac{1}{7}$	x	$\frac{x}{7}$
2nd pipe	$\frac{1}{9}$	x	$\frac{x}{9}$

Think: $\frac{1}{7} \cdot \frac{x}{1} = \frac{x}{7}$.

Think: $\frac{1}{9} \cdot \frac{x}{1} = \frac{x}{9}$.

Enter this information first. Multiply to get each of these entries: $W = rt$.

In shared-work problems, the number 1 represents one whole job completed. So we have

The part of job done by 1st pipe	plus	part of job done by 2nd pipe	equals	1 job completed.
$\frac{x}{7}$	$+$	$\frac{x}{9}$	$=$	1

Solve

$$\frac{x}{7} + \frac{x}{9} = 1$$ This is a rational equation.

$$63\left(\frac{x}{7} + \frac{x}{9}\right) = 63(1)$$ Clear the equation of fractions by multiplying both sides by the LCD, 63.

$$63\left(\frac{x}{7}\right) + 63\left(\frac{x}{9}\right) = 63$$ On the left side, distribute the multiplication by 63.

$$9x + 7x = 63$$ Simplify the left side: $\overset{1}{\cancel{7}} \cdot 9\left(\frac{x}{\underset{1}{\cancel{7}}}\right) = 9x$ and $7 \cdot \overset{1}{\cancel{9}}\left(\frac{x}{\underset{1}{\cancel{9}}}\right) = 7x$. The fractions have been cleared.

$$16x = 63$$ Combine like terms.

$$x = \frac{63}{16}$$ To undo the multiplication by 16 and isolate x, divide both sides by 16.

State If both pipes are used, it will take $\frac{63}{16}$ or $3\frac{15}{16}$ days to fill the tank.

Check To check, we use the work formula and multiply each rate by the time. In $\frac{63}{16}$ days, the first pipe fills $\frac{1}{7} \cdot \frac{63}{16} = \frac{9}{16}$ of the tank and the second pipe fills $\frac{1}{9} \cdot \frac{63}{16} = \frac{7}{16}$ of the tank. The sum of these efforts, $\frac{9}{16} + \frac{7}{16}$, is $\frac{16}{16}$ or 1 full tank. The result checks.

Notation

For this application problem, the mixed-number form of the result, $3\frac{15}{16}$ days, is preferable to the improper fraction form, $\frac{63}{16}$ days.

Self Check 4 **Mailing Flyers.** A school secretary can prepare a mass mailing of an informational flyer in 6 hours. A student worker would take 8 hours to prepare the mailing. How long will it take to prepare the mailing if they work together?

Now Try Problem 37

Strategy for Solving Shared-Work Problems

Equations that model shared-work problems involving two people (or machines) have the form

$$\frac{x}{a} + \frac{x}{b} = 1$$

where x represents the time they work together on the job, and a and b represent the respective times each worker needs to complete the job alone.

Example 4 can be solved in a different way by considering the amount of work done by each pipe in 1 day. As before, if we let $x =$ the number of days it will take to fill the tank if both inlet pipes are used, then together, in 1 day, they will complete $\frac{1}{x}$ of the job. If we add what the first pipe can do in 1 day to what the second pipe can do in 1 day, the sum is what they can do together in 1 day.

What the first inlet pipe can do in 1 day	plus	what the second inlet pipe can do in 1 day	equals	what they can do together in 1 day.
$\frac{1}{7}$	$+$	$\frac{1}{9}$	$=$	$\frac{1}{x}$

To solve the equation, begin by clearing it of fractions.

$$\frac{1}{7} + \frac{1}{9} = \frac{1}{x}$$

$$63x\left(\frac{1}{7} + \frac{1}{9}\right) = 63x\left(\frac{1}{x}\right) \quad \text{Multiply both sides by the LCD, } 63x.$$

$$9x + 7x = 63 \quad \text{Distribute the multiplication by } 63x \text{ and simplify.}$$

$$16x = 63 \quad \text{Combine like terms.}$$

$$x = \frac{63}{16} \quad \text{To isolate } x\text{, divide both sides by 16.}$$

This is the same answer as the one obtained in Example 4.

4 Solve Investment Problems.

We have used the interest formula $I = Prt$ to solve investment problems. The relationships among interest, principal, rate, and time can be expressed in another way, by solving for P.

$$I = Prt \quad \text{Interest} = \text{principal} \cdot \text{rate} \cdot \text{time.}$$

$$\frac{I}{rt} = \frac{Prt}{rt} \quad \text{To undo the multiplication by } rt \text{ and isolate } P\text{, divide both sides by } rt.$$

$$\frac{I}{rt} = P \quad \text{Simplify the right side: } \frac{\overset{1\,1}{P\cancel{r}\cancel{t}}}{\underset{1\,1}{\cancel{r}\cancel{t}}} = P.$$

$$P = \frac{I}{rt} \quad \text{Reverse the sides of the equation so that } P \text{ is on the left.}$$

This alternate form of the interest formula, Principal $= \frac{\text{Interest}}{\text{rate} \cdot \text{time}}$, is used to solve the next example.

EXAMPLE 5 **Comparing Investments.** An amount of money invested for one year in bonds will earn \$120. At a bank, that same amount of money will only earn \$75 interest, because the interest rate paid by the bank is 3% less than that paid by the bonds. Find the rate of interest paid by each investment.

Analyze

- The investment in bonds earns \$120 in one year.
- The same amount of money, invested in a bank, earns \$75 in one year.
- The interest rate paid by the bank is 3% less than that paid by the bonds.
- Find the bond's rate of interest and the bank's rate of interest.

Caution

A common mistake is to forget to express 3% as the decimal 0.03 and incorrectly let $r - 3 =$ the bank's rate of interest.

Assign Since the interest rate paid by the bank is 3% less than that paid by the bonds, let $r =$ the bond's rate of interest, and $r - 0.03 =$ the bank's interest rate. (Recall that $3\% = 0.03$.)

Form If an investment earns \$120 interest in 1 year at some rate r, we can use $P = \frac{I}{rt}$ to find that the principal invested was $\frac{120}{r}$ dollars. Similarly, if another investment earns \$75 interest in 1 year at some rate $r - 0.03$, the principal invested was $\frac{75}{r - 0.03}$ dollars. We can organize the facts of the problem in a table.

	Principal ·	Rate ·	Time =	Interest
Bonds	$\frac{120}{r}$	r	1	120
Bank	$\frac{75}{r - 0.03}$	$r - 0.03$	1	75

Divide to get each of these entries: $P = \frac{I}{rt}$. Enter this information first.

The amount invested in the bonds	equals	the amount invested in the bank.
$\frac{120}{r}$	$=$	$\frac{75}{r - 0.03}$

Solve

$$\frac{120}{r} = \frac{75}{r - 0.03}$$ This is a rational equation.

$$\overset{1}{\cancel{r}}(r - 0.03)\left(\frac{120}{\underset{1}{\cancel{r}}}\right) = r\overset{1}{\cancel{(r - 0.03)}}\left(\frac{75}{\underset{1}{\cancel{r - 0.03}}}\right)$$ Multiply both sides by the LCD, $r(r - 0.03)$. Then remove common factors of the numerator and denominator.

$$(r - 0.03)120 = 75r$$ Simplify. The fractions have been cleared.

$$120r - 3.6 = 75r$$ On the left side, distribute the multiplication by 120.

$$45r - 3.6 = 0$$ To isolate the variable term on the left side, subtract $75r$ from both sides.

$$45r = 3.6$$ To undo the subtraction of 3.6, add 3.6 to both sides.

$$r = 0.08$$ To undo the multiplication by 45 and isolate r, divide both sides by 45.

If $r = 0.08$, then the bank's interest rate is given by $r - 0.03 = 0.05$.

State The bonds pay 0.08, or 8%, interest. The bank's interest rate is 5%.

Check The amount invested at 8% that will earn \$120 interest in 1 year is $\frac{120}{(0.08)1} = \$1{,}500$. The amount invested at 5% that will earn \$75 interest in 1 year is $\frac{75}{(0.05)1} = \$1{,}500$. The amounts invested in the bonds and the bank are the same. The results check.

Self Check 5 **Comparing Investments.** An amount of money invested for one year in a certificate of deposit will earn \$210. The same amount of money in a savings account will earn \$70. If the certificate of deposit's interest rate is 2% more than the savings account's rate, find the interest rate of the savings account.

Now Try Problem 41

SECTION 7.7 STUDY SET

VOCABULARY

Fill in the blanks.

1. In this section, problems that involve:
   - moving vehicles are called uniform ________ problems.
   - depositing money are called ___________ problems.
   - people completing jobs are called shared-______ problems.
2. In the formula $W = rt$, the variable W stands for the ______ completed, r is the _____, and t is the _____.

CONCEPTS

3. Choose the equation that can be used to solve the following problem: *If the same number is added to the numerator and the denominator of the fraction $\frac{5}{8}$, the result is $\frac{2}{3}$. Find the number.*

   (i) $\frac{5}{8} + x = \frac{2}{3}$ (ii) $\frac{5 + x}{8} = \frac{2}{3}$

   (iii) $\frac{5 + x}{8 + x} = \frac{2}{3}$ (iv) $\frac{5}{8} = \frac{2 + x}{3 + x}$

4. Fill in the blank: If a job can be completed in t hours, then the rate of work can be expressed as $\frac{1}{}$ of the job is completed per hour.
5. **a.** It takes a night security officer 45 minutes to check each of the doors in an office building to make sure they are locked. What is the officer's rate of work?

   b. It takes an elementary school teacher 4 hours to make out the semester report cards. What part of the job does she complete in x hours?
6. **Hospitals.** An experienced employee can sterilize an operating room in 3 hours. It takes a new employee 5 hours to sterilize the same room. Select the best estimate below of the time it will take them to sterilize the room if they work together.
   - Less than 3 hours
   - Between 3 and 5 hours
   - More than 5 hours
7. **a.** Solve $d = rt$ for t.

   b. Solve $I = Prt$ for P.
8. Complete the table.

	r ·	t =	d
Snowmobile	r		4
4 × 4 truck	$r - 5$		3

9. Complete the table.

	Rate ·	Time =	Work completed
1st printer	$\frac{1}{15}$	x	
2nd printer	$\frac{1}{8}$	x	

10. Complete the table.

	P ·	r ·	t =	I
City savings bank		r	1	50
Credit union		$r - 0.02$	1	75

NOTATION

11. Write $\frac{55}{9}$ days using a mixed number.
12. **a.** Write 9% as a decimal.

    b. Write 0.035 as a percent.

GUIDED PRACTICE

Solve each of these number problems. See Example 1.

13. If the same number is added to both the numerator and the denominator of $\frac{2}{5}$, the result is $\frac{2}{3}$. Find the number.
14. If the same number is subtracted from both the numerator and the denominator of $\frac{11}{13}$, the result is $\frac{3}{4}$. Find the number.
15. If the denominator of $\frac{3}{4}$ is increased by a number, and the numerator is doubled, the result is 1. Find the number.
16. If a number is added to the numerator of $\frac{7}{8}$, and the same number is subtracted from the denominator, the result is 2. Find the number.

17. If a number is added to the numerator of $\frac{3}{4}$, and twice as much is added to the denominator, the result is $\frac{4}{7}$. Find the number.

18. If a number is added to the numerator of $\frac{5}{7}$, and twice as much is subtracted from the denominator, the result is 8. Find the number.

19. The sum of a number and its reciprocal is $\frac{13}{6}$. Find the number.

20. The sum of the reciprocals of two consecutive even integers is $\frac{7}{24}$. Find each integer.

APPLICATIONS

21. **Cooking.** If the same number is added to both the numerator and the denominator of the amount of butter used in the following recipe for toffee, the result is the amount of brown sugar to be used. Find the number.

Toffee

$\frac{2}{3}$ cup pecans, chopped

$\frac{3}{4}$ cup brown sugar (packed)

$\frac{1}{2}$ package (6-oz size) semisweet chocolate

$\frac{1}{4}$ cup butter

22. **Tape Measures.** If the same number is added to both the numerator and the denominator of the first measurement, the result is the second measurement. Find the number.

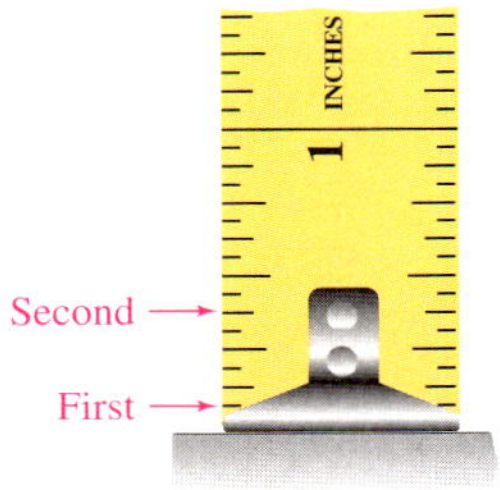

23. **Tour De France.** Maurice Garin of France won the first Tour de France bicycle road race in 1903. In 2005, American Lance Armstrong won his seventh consecutive Tour de France. Armstrong's average speed in 2005 was 10 mph faster than Garin's in 1903. In the time it took Garin to ride 80 miles, Armstrong could have ridden 130 miles. Find each cyclist's average speed.

24. **Physical Fitness.** A woman can bicycle 28 miles in the same time as it takes her to walk 8 miles. She can ride 10 mph faster than she can walk. How fast can she walk?

25. **Packaging Fruit.** The diagram below shows how apples are processed for market. Although the second conveyor belt is shorter, an apple spends the same amount of time on each belt because the second conveyor moves 1 foot per second slower than the first. Determine the speed of each conveyor belt.

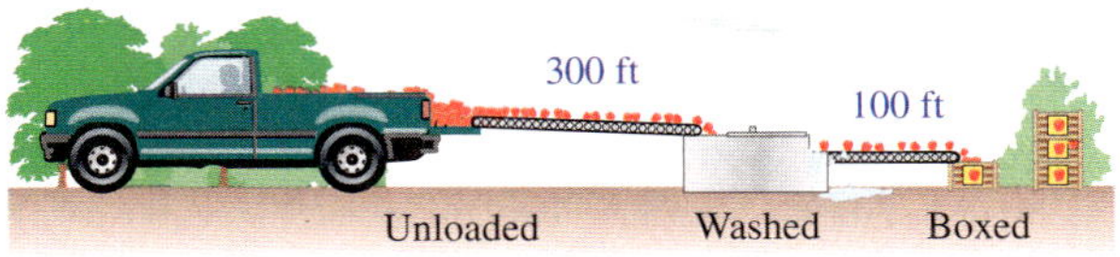

26. **Comparing Travel.** A plane can fly 300 miles in the same time as it takes a car to go 120 miles. If the car travels 90 mph slower than the plane, find the speed of the plane.

27. **Birds in Flight.** Although flight speed is dependent upon the weather and the wind, in general, a Canada goose can fly about 10 mph faster than a great blue heron. In the same time that a Canada goose travels 120 miles, a great blue heron travels 80 miles. Find their flying speeds.

28. **Fast Cars.** The top speed of a Dodge Charger SRT8 is 33 mph less than the top speed of a Chevrolet Corvette Z06. At their top speeds, a Corvette can travel 6 miles in the same time that a Charger can travel 5 miles. Find the top speed of each car.

©Jim West Alamy

29. **Wind Speed.** When a plane flies downwind, the wind pushes the plane so that its speed is the *sum* of the speed of the plane in still air and the speed of the wind. Traveling upwind, the wind pushes against the plane so that its speed is the *difference* of the speed of the plane in still air and the speed of the wind. Suppose a plane that travels 255 mph in still air can travel 300 miles downwind in the same time as it takes to travel 210 miles upwind. Complete the following table and find the speed of the wind, represented by x.

	Rate ·	Time =	Distance
Downwind	$255 + x$		300
Upwind	$255 - x$		210

30. **Boating.** A boat that travels 18 mph in still water can travel 22 miles downstream in the same time as it takes to travel 14 miles upstream. Find the speed of the current in the river. (See problem 29.)

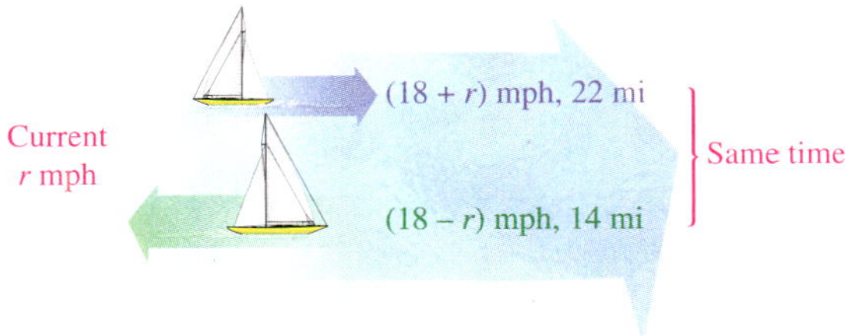

31. **Roofing Houses.** A homeowner estimates that it will take her 7 days to roof her house. A professional roofer estimates that he could roof the house in 4 days. How long will it take if the homeowner helps the roofer?

32. **Holiday Decorating.** One crew can put up holiday decorations in the mall in 8 hours. A second crew can put up the decorations in 10 hours. How long will it take if both crews work together to decorate the mall?

33. from Campus to Careers

Recreation Director

©BananaStock/SuperStock

Suppose you are a recreation director at a summer camp. The water in the camp swimming pool was drained out for the winter and it is now time to refill the pool. One pipe can fill the empty pool in 12 hours and another can fill the empty pool in 18 hours. Suppose both pipes are opened at 8:00 A.M. and you have scheduled a swimming activity for 2:00 P.M. that day. Will the pool be filled by then?

34. Groundskeeping. It takes a groundskeeper 45 minutes to prepare a softball field for a game. It takes his assistant 55 minutes to prepare the same field. How long will it take if they work together to prepare the field?

35. Filling a Pool. One inlet pipe can fill an empty pool in 4 hours, and a drain can empty the pool in 8 hours. How long will it take the pipe to fill the pool if the drain is left open?

36. Sewage Treatment. A sludge pool is filled by two inlet pipes. One pipe can fill the pool in 15 days, and the other can fill it in 21 days. However, if no sewage is added, continuous waste removal will empty the pool in 36 days. How long will it take the two inlet pipes to fill an empty sludge pool?

37. Grading Papers. On average, it takes a teacher 30 minutes to grade a set of quizzes. It takes her teacher's aide twice as long to do the same grading. How long will it take if they work together to grade a set of quizzes?

38. Dog Kennels. It takes the owner/operator of a dog kennel 6 hours to clean all of the cages. It takes his assistant 2 hours more than that to clean the same cages. How long will it take if they work together?

39. Printers. It takes a printer 6 hours to print the class schedules for all of the students enrolled in a community college. A faster printer can print the schedules in 4 hours. How long will it take the two printers working together to print $\frac{3}{4}$ of the class schedules?

40. Office Work. In 5 hours, a secretary can address 100 envelopes. Another secretary can address 100 envelopes in 6 hours. How long would it take the secretaries, working together, to address 300 envelopes? (*Hint:* Think of addressing 300 envelopes as three 100-envelope jobs.)

41. Comparing Investments. An amount of money invested for 1 year in tax-free bonds will earn \$300. In a certain credit union account, that same amount of money will only earn \$200 interest in a year, because the interest paid is 2% less than that paid by the bonds. Find the rate of interest paid by each investment.

42. Comparing Investments. An amount of money invested for 1 year in a savings account will earn \$1,500. That same amount of money invested in a mini-mall development will earn \$6,500 interest in a year, because the interest paid is 10% more than that paid by the savings account. Find the rate of interest paid by each investment.

43. Comparing Investments. Two certificates of deposit (CDs) pay interest at rates that differ by 1%. Money invested for 1 year in the first CD earns \$175 interest. The same principal invested in the second CD earns \$200. Find the two rates of interest.

44. Comparing Interest Rates. Two bond funds pay interest at rates that differ by 2%. Money invested for 1 year in the first fund earns \$315 interest. The same amount invested in the second fund earns \$385. Find the lower rate of interest.

WRITING

45. In Example 4, one inlet pipe could fill an oil tank in 7 days, and another could fill the same tank in 9 days. We were asked to find how long it would take if both pipes were used. Explain why each of the following approaches is incorrect.

The time it would take to fill the tank

- is the *sum* of the lengths of time it takes each pipe to fill the tank: 7 days + 9 days = 16 days.
- is the *difference* in the lengths of time it takes each pipe to fill the tank: 9 days − 7 days = 2 days.
- is the *average* of the lengths of time it takes each pipe to fill the tank:

$$\frac{7\text{ days} + 9\text{ days}}{2} = \frac{16\text{ days}}{2} = 8\text{ days.}$$

46. Write a shared-work problem that can be modeled by the equation:

$$\frac{x}{3} + \frac{x}{4} = 1$$

REVIEW

47. Solve using substitution: $\begin{cases} x + y = 4 \\ y = 3x \end{cases}$

48. Solve using elimination (addition): $\begin{cases} 5x - 4y = 19 \\ 3x + 2y = 7 \end{cases}$

49. Use a check to determine whether $\frac{21}{5}$ is a solution of: $x + 20 = 4x - 1 + 2x$

50. Solve: $4x^2 + 8x = 0$

CHALLENGE PROBLEMS

51. River Tours. A river boat tour begins by going 60 miles upstream against a 5-mph current. There, the boat turns around and returns with the current. What still-water speed should the captain use to complete the tour in 5 hours?

52. Travel Time. A company president flew 680 miles one way in the corporate jet, but returned in a smaller plane that could fly only half as fast. If the total travel time was 6 hours, find the speeds of the planes.

53. Sales. A dealer bought some radios for a total of \$1,200. She gave away 6 radios as gifts, sold the rest for \$10 more than she paid for each radio, and broke even. How many radios did she buy?

54. Furnace Repairs. A repairman purchased several furnace-blower motors for a total cost of \$210. If his cost per motor had been \$5 less, he could have purchased one additional motor. How many motors did he buy at the regular rate?

SECTION 7.8 Proportions and Similar Triangles

OBJECTIVES

1. Write ratios and rates in simplest form.
2. Solve proportions.
3. Use proportions to solve problems.
4. Use proportions to solve problems involving similar triangles.

ARE YOU READY?

The following problems review some basic skills that are needed when working with ratios and proportions.

1. Simplify: $\frac{42}{54}$

2. Multiply: **a.** $bd \cdot \frac{a}{d}$ **b.** $bd \cdot \frac{c}{d}$

3. Solve: $16x = 136$

4. Solve: $x^2 - 7x = 18$

In this section, we will discuss a problem-solving tool called a *proportion.* A proportion is a type of rational equation that involves two *ratios* or two *rates.*

1 Write Ratios and Rates in Simplest Form.

Ratios are used to compare two numbers or two quantities measured in the same units. Here are some examples.

- To prepare fuel for a lawnmower, gasoline is mixed with oil in a 50-to-1 ratio.
- In the stock market, winning stocks might outnumber losers by a ratio of 7 to 4.
- Gold is combined with other metals in the ratio of 14 to 10 to make 14-karat jewelry.

Ratios

A ratio is the quotient of two numbers or the quotient of two quantities that have the same units.

There are three ways to write a ratio: as a fraction, using the word *to,* or with a colon. For example, the comparison of the number of winning stocks to the number of losing stocks mentioned earlier can be written as

$\frac{7}{4}$, 7 to 4, or 7 : 4   Each of these forms can be read as "the ratio of 7 to 4."

EXAMPLE 1 Translate each phrase into a ratio written in fractional form: **a.** The ratio of 5 to 9 **b.** 12 ounces to 2 pounds

Strategy To translate, we need to identify the number (or quantity) before the word *to* and the number (or quantity) after it.

Why The number before the word *to* is the numerator of the ratio and the number after it is the denominator.

Solution

a. The ratio of 5 ***to*** 9 is written $\frac{5}{9}$. (5: numerator; 9: denominator)

b. To write a ratio of two quantities with the same units, we must express 2 pounds in terms of ounces. Since 1 pound = 16 ounces, 2 pounds = 32 ounces. The ratio of 12 ounces to 32 ounces can be simplified so that no units appear in the final form.

$$\frac{12 \text{ ounces}}{32 \text{ ounces}} = \frac{3 \cdot \overset{1}{\cancel{4}} \cancel{\text{ ounces}}}{\underset{1}{\cancel{4}} \cdot 8 \cancel{\text{ ounces}}} = \frac{3}{8}$$

Caution

A ratio that is the quotient of two quantities having the same units should be simplified so that no units appear in the final answer.

Self Check 1 Translate each phrase into a ratio written in fractional form: **a.** The ratio of 15 to 2 **b.** 12 hours to 2 days

Now Try Problem 25

A quotient that compares quantities with different units is called a **rate.** For example, if the 495-mile drive from New Orleans to Dallas takes 9 hours, the average rate of speed is the quotient of the miles driven and the length of time the trip takes.

$$\text{Average rate of speed} = \frac{495 \text{ miles}}{9 \text{ hours}} = \frac{\overset{1}{\cancel{9}} \cdot 55 \text{ miles}}{\underset{1}{\cancel{9}} \cdot 1 \text{ hours}} = \frac{55 \text{ miles}}{1 \text{ hour}}$$

Rates A rate is a quotient of two quantities that have different units.

2 Solve Proportions.

If two ratios or two rates are equal, we say that they are *in proportion.*

Proportion A proportion is a mathematical statement that two ratios or two rates are equal.

The Language of Algebra

The word **proportion** implies a comparative relationship in size. For a picture to appear realistic, the artist must draw the shapes in the proper *proportion.* Remember the Y2K scare? The massive computer failures predicted by some experts were blown way out of *proportion.*

Some examples of proportions are:

$$\frac{1}{2} = \frac{3}{6} \qquad \frac{3 \text{ waiters}}{7 \text{ tables}} = \frac{9 \text{ waiters}}{21 \text{ tables}} \qquad \frac{a}{b} = \frac{c}{d}$$

- The proportion $\frac{1}{2} = \frac{3}{6}$ can be read as "1 is to 2 as 3 is to 6."
- The proportion $\frac{3 \text{ waiters}}{7 \text{ tables}} = \frac{9 \text{ waiters}}{21 \text{ tables}}$ can be read as "3 waiters is to 7 tables as 9 waiters is to 21 tables."
- The proportion $\frac{a}{b} = \frac{c}{d}$ can be read as "a is to b as c is to d."

Each of the four numbers in a proportion is called a **term.** The first and fourth terms are called the **extremes,** and the second and third terms are called the **means.**

First term → a, Second term → b, Third term → c, Fourth term → d: $\frac{a}{b} = \frac{c}{d}$

a and *d* are the extremes. *b* and *c* are the means.

For the proportion $\frac{a}{b} = \frac{c}{d}$, we can show that the product of the extremes, ad, is equal to the product of the means, bc, by multiplying both sides of the proportion by bd, and observing that $ad = bc$.

$$\frac{a}{b} = \frac{c}{d}$$

$$\overset{1}{\cancel{b}}d \cdot \frac{a}{\underset{1}{\cancel{b}}} = b\overset{1}{\cancel{d}} \cdot \frac{c}{\underset{1}{\cancel{d}}}$$ To clear the fractions, multiply both sides by the LCD, bd. Remove common factors of the numerator and denominator.

$$ad = bc$$ Simplify: $\frac{b}{b} = 1$ and $\frac{d}{d} = 1$.

Since $ad = bc$, the product of the extremes equals the product of the means.

The same products ad and bc can be found by multiplying diagonally in the proportion $\frac{a}{b} = \frac{c}{d}$. We call ad and bc **cross products.**

ad bc

$$\frac{a}{b} = \frac{c}{d}$$

The Fundamental Property of Proportions

In a proportion, the product of the extremes is equal to the product of the means.

If $\frac{a}{b} = \frac{c}{d}$, then $ad = bc$ and if $ad = bc$, then $\frac{a}{b} = \frac{c}{d}$.

EXAMPLE 2 Determine whether each equation is a proportion: **a.** $\frac{3}{7} = \frac{9}{21}$ **b.** $\frac{8}{3} = \frac{13}{5}$

Strategy We will check to see whether the product of the extremes is equal to the product of the means.

Why If the product of the extremes equals the product of the means, the equation is a proportion. If the cross products are not equal, the equation is not a proportion.

Solution

a. The product of the extremes is $3 \cdot 21 = 63$. The product of the means is $7 \cdot 9 = 63$. Since the cross products are equal, $\frac{3}{7} = \frac{9}{21}$ is a proportion.

$3 \cdot 21 = 63$ $7 \cdot 9 = 63$

$$\frac{3}{7} = \frac{9}{21}$$ Each cross product is 63.

b. The product of the extremes is $8 \cdot 5 = 40$. The product of the means is $3 \cdot 13 = 39$. Since the cross products are not equal, the equation is not a proportion: $\frac{8}{3} \neq \frac{13}{5}$.

$8 \cdot 5 = 40$ $3 \cdot 13 = 39$

$$\frac{8}{3} = \frac{13}{5}$$ One cross product is 40 and the other is 39.

Caution

We cannot remove common factors "across" an = symbol.

$$\frac{\overset{1}{\cancel{3}}}{7} = \frac{9}{\underset{7}{\cancel{21}}}$$

When this is done, the original proportion, $\frac{3}{7} = \frac{9}{21}$, which we found to be true, produces the false statement: $\frac{1}{7} = \frac{9}{7}$.

Self Check 2 Determine whether the equation $\frac{6}{13} = \frac{24}{53}$ is a proportion.

Now Try Problem 29

We have seen that a proportion contains four terms. If we know only three of the four terms of a proportion, we can use the fundamental property of proportions to find the value of the fourth term. This process is called **solving the proportion.**

EXAMPLE 3 Solve: $\frac{3}{2} = \frac{9}{x}$

Strategy To solve for x, we will set the cross products equal.

Why This equation is a proportion, and in a proportion the product of the extremes equals the product of the means.

Solution If $x = 0$, the denominator of $\frac{9}{x}$ is 0 and the fraction would be undefined. Therefore, 0 cannot be a solution.

Caution

Remember that a cross product is the product of the means or extremes of a proportion. It would be incorrect to try to calculate cross products to solve $\frac{12}{18} = \frac{4}{x} + \frac{1}{2}$ because there is more than one term on the right side of the equation. It is *not* a proportion.

$\frac{3}{2} = \frac{9}{x}$ This is the given proportion. Since it is a type of rational equation, we can solve it by multiplying both sides by the LCD, 2x. However, it is often easier to solve a proportion using the cross products.

$3 \cdot x = 2 \cdot 9$ Find each cross product and set them equal.

$3x = 18$ Do the multiplication.

$\frac{3x}{3} = \frac{18}{3}$ To isolate x, divide both sides by 3.

$x = 6$ Do the division.

Check: To check the result, we substitute 6 for x in $\frac{3}{2} = \frac{9}{x}$ and find the cross products.

$3 \cdot 6 = 18$ $2 \cdot 9 = 18$

$\frac{3}{2} \stackrel{?}{=} \frac{9}{6}$ Each cross product is 18.

Since the cross products are equal, the solution of $\frac{3}{2} = \frac{9}{x}$ is 6. The solution set is $\{6\}$.

Self Check 3 Solve: $\frac{15}{x} = \frac{25}{40}$

Now Try Problem 35

EXAMPLE 4 Solve: $\frac{a}{2} = \frac{4}{a - 2}$

Strategy To solve for a, we will set the cross products equal.

Why Since this equation is a proportion, the product of the means equals the product of the extremes.

Solution If $a = 2$, the denominator of $\frac{4}{a - 2}$ is 0 and the fraction would be undefined. Therefore, 2 cannot be a solution.

Success Tip

Since proportions are rational equations, they can also be solved by multiplying both sides by the LCD. Here, an alternate approach is to multiply both sides by $2(a - 2)$:

$\frac{a}{2} = \frac{4}{a - 2}$ This is the given proportion. Since it is a type of rational equation, we can solve it by multiplying both sides by the LCD, 2(a − 2), as shown in the margin. However, it is often easier to solve a proportion using the cross products.

$a(a - 2) = 2 \cdot 4$ Find each cross product and set them equal. Don't forget to write the parentheses.

$a^2 - 2a = 8$ On the left side, distribute the multiplication by a. This is a quadratic equation. On the right side, multiply.

$a^2 - 2a - 8 = 0$ To get 0 on the right side of the equation, subtract 8 from both sides.

$(a + 2)(a - 4) = 0$ Factor $a^2 - 2a - 8$.

$a + 2 = 0$ or $a - 4 = 0$ Set each factor equal to 0.

$a = -2$ | $a = 4$ Solve each equation.

The solutions are -2 and 4. Verify this using a check.

Self Check 4 Solve: $\frac{6}{c} = \frac{c - 1}{5}$

Now Try Problem 47

3 Use Proportions to Solve Problems.

We can use proportions to solve many problems. If we are given a ratio (or rate) comparing two quantities, the words of the problem can be translated into a proportion, and we can solve it to find the unknown.

EXAMPLE 5

iStockPhoto.com/Graeme Gilmour

Grocery Shopping. If 6 apples cost \$1.38, how much will 16 apples cost?

Analyze We know the cost of 6 apples; we are to find the cost of 16 apples.

Assign Let c = the cost of 16 apples.

Form If we compare the number of apples to their cost, the two ratios must be equal.

6 apples is ***to*** *\$1.38* *as* *16 apples is* ***to*** *\$c.*

Number of apples → 6, Cost → 1.38; 16 ← Number of apples, c ← Cost

$$\frac{6}{1.38} = \frac{16}{c}$$

The units can be written outside the proportion.

Solve We drop the units, find each cross product, set them equal, and then solve the resulting equation for c.

$$6 \cdot c = 1.38(16)$$ In a proportion, the product of the extremes equals the product of the means.

$$6c = 22.08$$ Multiply: $1.38(16) = 22.08$.

$$\frac{6c}{6} = \frac{22.08}{6}$$ To undo the multiplication by 6 and isolate c, divide both sides by 6.

$$c = 3.68$$ Recall that c represents the cost of 16 apples.

State Sixteen apples will cost \$3.68.

Check We can use estimation to check the result. 16 apples are about 3 times as many as 6 apples, which cost \$1.38. If we multiply \$1.38 by 3, we get an estimate of the cost of 16 apples: $\$1.38 \cdot 3 = \$4.14$. The result, \$3.68, seems reasonable.

The Language of Algebra

Remember that the word **to** separates the numerator and denominator of a ratio. If the units are written outside the ratio, we can write 6 apples is *to* \$1.38 in fraction form as:

$$\frac{6}{1.38}$$

Self Check 5 **Concert Tickets.** If 9 tickets to a concert cost \$112.50, how much will 15 tickets cost?

Now Try Problem 73

Be careful when solving problems using proportions. We must make sure that the units of both numerators are the same and the units of both denominators are the same. In Example 5, it would be incorrect to write

Cost of 6 apples → 1.38, 6 apples → 6; 16 ← 16 apples, c ← Cost of 16 apples

$$\frac{1.38}{6} = \frac{16}{c}$$

EXAMPLE 6

Carousel ratio
1 inch:160 inches

Miniatures. A **scale** is a ratio (or rate) that compares the size of a model, drawing, or map with the size of an actual object. The scale indicates that 1 inch on the model carousel is equivalent to 160 inches on the actual carousel. How wide should the model be if the actual carousel is 35 feet wide?

Analyze We are asked to determine the width of the miniature carousel if a ratio of 1 inch to 160 inches is used. We would like the width of the model to be given in inches, not feet, so we will express the 35-foot width of the actual carousel as $35 \cdot 12 = 420$ inches.

Assign Let w = the width of the model.

The Language of Algebra

Architects, interior decorators, landscapers, and automotive engineers are a few of the professionals who construct **scale** drawings or *scale* models of the projects they are designing.

Form The ratios of the dimensions of the model to the corresponding dimensions of the actual carousel are equal.

*1 inch is **to** 160 inches as w inches is **to** 420 inches.*

$$\frac{1}{160} = \frac{w}{420}$$

Model → 1, Actual size → 160; Model ← w, Actual size ← 420

Solve We drop the units, find each cross product, set them equal, and then solve the resulting equation for w.

$420 = 160w$ — In a proportion, the product of the extremes is equal to the product of the means.

$\frac{420}{\mathbf{160}} = \frac{160w}{\mathbf{160}}$ — To undo the multiplication by 160 and isolate w, divide both sides by 160.

$2.625 = w$ — Recall that w represents the width of the model.

State The width of the miniature carousel should be 2.625 in., or $2\frac{5}{8}$ in.

Check A width of $2\frac{5}{8}$ in. is approximately 3 in. When we write the ratio of the model's approximate width to the width of the actual carousel, we get $\frac{3}{420} = \frac{1}{140}$, which is about $\frac{1}{160}$. The answer seems reasonable.

Self Check 6 **Blueprints.** The scale for a blueprint indicates that $\frac{1}{4}$ inch on the print is equivalent to 1 foot for the actual building. If the width of the building on the print is 7.5 inches, what is the width of the actual building?

Now Try Problem 89

When shopping, *unit prices* can be used to compare costs of different sizes of the same brand to determine the best buy. The **unit price** gives the cost per unit, such as cost per ounce, cost per pound, or cost per sheet. We can find the unit price of an item using a proportion.

EXAMPLE 7 **Comparison Shopping.** Which size of toothpaste is the better buy?

$2.19

$2.79

Strategy We will find the unit price for each tube of toothpaste. Then we will identify which tube has the lower unit price.

Why The better buy is the tube that has the lower unit price.

Solution

The Language of Algebra

A **unit price** indicates the cost of 1 unit of an item, such as 1 ounce of bottled water or 1 pound of hamburger. In advanced mathematics, we study unit circles—circles that have a radius of 1 unit.

To find the unit price for each tube, we let x = the price of 1 ounce of toothpaste. Then we set up and solve the following proportions.

For the 4-ounce tube:

Price → 2.19, Ounces → 4; Price ← x, Ounce ← 1

$$\frac{2.19}{4} = \frac{x}{1}$$

$$2.19 = 4x$$

$$\frac{2.19}{4} = x$$

$0.55 \approx x$ — The unit price is approximately \$0.55.

For the 6-ounce tube:

Price → 2.79, Ounces → 6; Price ← x, Ounce ← 1

$$\frac{2.79}{6} = \frac{x}{1}$$

$$2.79 = 6x$$

$$\frac{2.79}{6} = x$$

$0.47 \approx x$ — The unit price is approximately \$0.47.

The price of 1 ounce of toothpaste from the 4-ounce tube is about 55¢. The price for 1 ounce of toothpaste from the 6-ounce tube is about 47¢. Since the 6-ounce tube has the lower unit price, it is the better buy.

Self Check 7 **Comparison Shopping.** Which is the better buy: 3 pounds of hamburger for \$6.89 or 5 pounds for \$12.49?

Now Try Problem 93

4 Use Proportions to Solve Problems Involving Similar Triangles.

If two angles of one triangle have the same measures as two angles of a second triangle, the triangles have the same shape. Triangles with the same shape, but not necessarily the same size, are called **similar triangles.** In the following figure, $\Delta ABC \sim \Delta DEF$. (Read the symbol $\sim$ as "is similar to.")

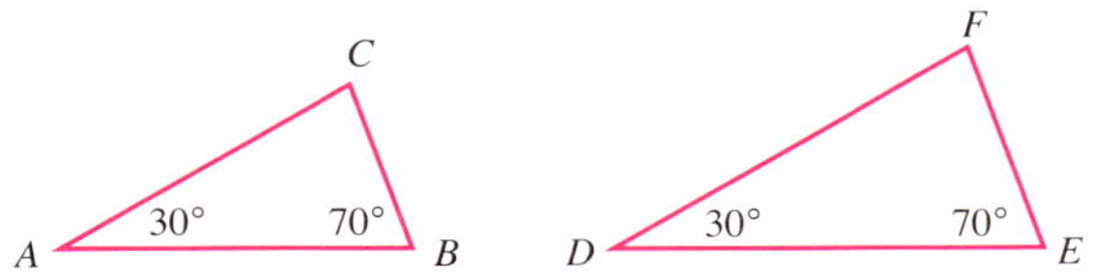

Property of Similar Triangles

If two triangles are **similar,** all pairs of corresponding sides are in proportion.

For the similar triangles previously shown, the following proportions are true.

$$\frac{AB}{DE} = \frac{BC}{EF}, \quad \frac{BC}{EF} = \frac{CA}{FD}, \quad \text{and} \quad \frac{CA}{FD} = \frac{AB}{DE}$$

Read AB as "the length of segment AB."

EXAMPLE 8 **Finding the Height of a Tree.** A tree casts a shadow 18 feet long at the same time as a woman 5 feet tall casts a shadow 1.5 feet long. Find the height of the tree.

Analyze The figure shows the similar triangles determined by the tree and its shadow and the woman and her shadow. Since the triangles are similar, the lengths of their corresponding sides are in proportion. We can use this fact to find the height of the tree.

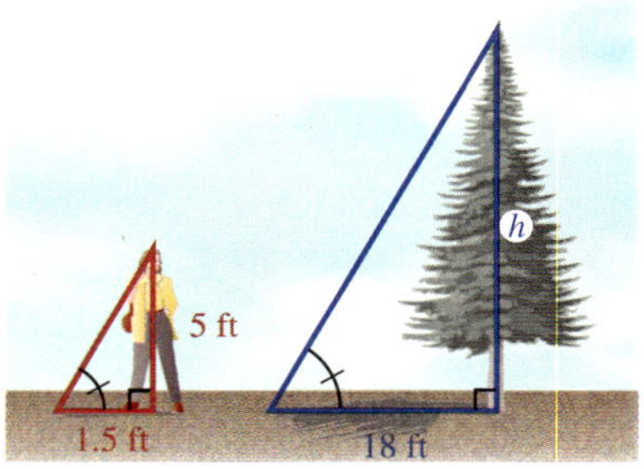

Each triangle has a right angle. Since the sun's rays strike the ground at the same angle, the angles highlighted with a tick mark have the same measure. Therefore, two angles of the smaller triangle have the same measures as two angles of the larger triangle; the triangles are similar.

Assign Let h = the height of the tree.

Form We can find h by solving the following proportion.

$$\frac{h}{5} = \frac{18}{1.5}$$

$\frac{\text{Height of the tree}}{\text{Height of the woman}} = \frac{\text{Length of shadow of the tree}}{\text{Length of shadow of the woman}}$

Success Tip

Similar triangles do not have to be positioned the same. When they are placed differently, be careful to match their corresponding letters correctly. Here, $\Delta RST \sim \Delta MNO$.

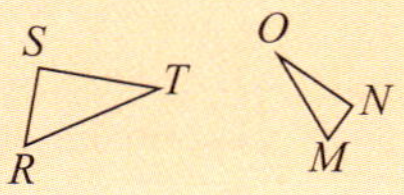

Success Tip

The quickest way to determine if two triangles are similar is to see if the measures of two angles of one triangle have the same measures as two angles of the other triangle.

Solve

$$1.5h = 5(18)$$ In a proportion, the product of the extremes equals the product of the means.

$$1.5h = 90$$ Multiply: $5(18) = 90$.

$$\frac{1.5h}{\mathbf{1.5}} = \frac{90}{\mathbf{1.5}}$$ To undo the multiplication by 1.5 and isolate h, divide both sides by 1.5.

$$h = 60$$ Do the decimal division, $1.5\overline{)90}$, to get 60.

State The tree is 60 feet tall.

Check $\frac{18}{1.5} = 12$ and $\frac{60}{5} = 12$. Since the ratios are the same, the result checks.

Self Check 8 **Shadows.** Find the height of the tree in Example 8 if the woman is 5 feet 6 inches tall and her shadow is 1.5 feet long.

Now Try Problems 55 and 101

SECTION 7.8 STUDY SET

VOCABULARY

Fill in the blanks.

1. A ______ is the quotient of two numbers or the quotient of two quantities with the same units. A _____ is a quotient of two quantities that have different units.
2. A __________ is a mathematical statement that two ratios or two rates are equal.
3. In $\frac{50}{3} = \frac{x}{9}$, the terms 50 and 9 are called the _________ and the terms 3 and x are called the _______ of the proportion.
4. The ______ products for the proportion $\frac{5}{2} = \frac{6}{x}$ are $5x$ and 12.
5. Examples of _____ prices are \$1.65 per gallon, 17¢ per day, and \$50 per foot.
6. Two triangles with the same shape, but not necessarily the same size, are called _______ triangles.

CONCEPTS

7. Fill in the blanks: In a proportion, the product of the extremes is ______ to the product of the means. In symbols,

$$\text{If } \frac{a}{b} = \frac{c}{d}, \text{ then } \quad = \quad .$$

8. Is 45 a solution of $\frac{5}{3} = \frac{75}{x}$?
9. **Snacks.** In a sample of 25 bags of potato chips, 2 were found to be underweight. Complete the following proportion that could be used to find the number of underweight bags that would be expected in a shipment of 1,000 bags of potato chips.

Number of bags → $\frac{25}{\ } = \frac{\ }{\ }$ ← Number of bags
Number underweight → ← Number underweight

10. **Miniatures.** A model of the Seattle Space Needle is to be made using a scale of 2 inches to 35 feet. Complete the following proportion to determine the height h of the model.

$$\frac{2}{35} = \frac{\ }{\ }$$

11. **Kleenex.** Complete the following proportion that can be used to find the unit price of facial tissue if a box of 85 tissues sells for \$2.19.

Price → $\frac{\ }{85} = \frac{x}{\ }$ ← Price
Number of sheets → ← Number of sheets

12. The two triangles shown in the following illustration are similar. Complete the proportion.

$$\frac{x}{\ } = \frac{\ }{10}$$

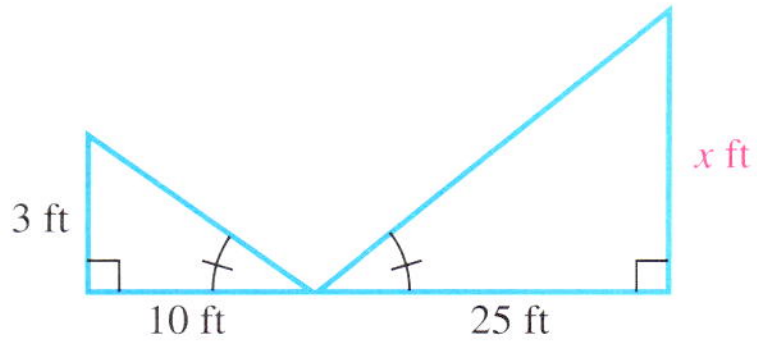

NOTATION

Complete the solution.

13. Solve for x: $\frac{12}{18} = \frac{x}{24}$

$$12 \cdot 24 = 18 \cdot \_\_\_$$
$$\_\_\_ = 18x$$
$$\frac{288}{\_\_\_} = \frac{18x}{\_\_\_}$$
$$\_\_\_ = x$$

14. Write the ratio of 25 to 4 in two other forms.

15. Fill in the blanks: The proportion $\frac{20}{1.6} = \frac{100}{8}$ can be read: 20 is to 1.6 ____ 100 is ____ 8.

16. Fill in the blank: We read $\Delta XYZ \sim \Delta MNO$ as: triangle XYZ is ________ to triangle MNO.

GUIDED PRACTICE

Translate each ratio into a fraction in simplest form. **See Example 1.**

17. 4 boxes to 15 boxes
18. 2 miles to 9 miles
19. 18 watts to 24 watts
20. 11 cans to 121 cans
21. 30 days to 24 days
22. 45 people to 30 people
23. 90 minutes to 3 hours
24. 20 inches to 2 feet
25. 8 quarts to 4 gallons
26. 6 feet to 12 yards
27. 6,000 feet to 1 mile (*Hint:* 1 mi = 5,280 ft)
28. 5 tons to 4,000 pounds (*Hint:* 1 ton = 2,000 lb)

Determine whether each equation is a true proportion. **See Example 2.**

29. $\frac{7}{3} = \frac{14}{6}$
30. $\frac{7}{16} = \frac{3}{7}$
31. $\frac{5}{8} = \frac{12}{19.4}$
32. $\frac{9}{32} = \frac{4.5}{16}$

Solve each proportion. **See Example 3.**

33. $\frac{2}{3} = \frac{x}{6}$
34. $\frac{3}{6} = \frac{x}{8}$
35. $\frac{63}{g} = \frac{9}{2}$
36. $\frac{27}{x} = \frac{9}{4}$
37. $\frac{x+1}{5} = \frac{3}{15}$
38. $\frac{x-1}{7} = \frac{2}{21}$
39. $\frac{5-x}{17} = \frac{13}{34}$
40. $\frac{4-x}{13} = \frac{11}{26}$
41. $\frac{15}{7b+5} = \frac{5}{2b+1}$
42. $\frac{8}{3n+6} = \frac{16}{3n-3}$
43. $\frac{8x}{3} = \frac{11x+9}{4}$
44. $\frac{3x}{16} = \frac{x+2}{5}$

Solve each proportion. **See Example 4.**

45. $\frac{2}{3x} = \frac{x}{6}$
46. $\frac{y}{4} = \frac{4}{y}$
47. $\frac{b-5}{3} = \frac{2}{b}$
48. $\frac{2}{q} = \frac{q-3}{2}$
49. $\frac{a-4}{a} = \frac{15}{a+4}$
50. $\frac{s}{s-5} = \frac{s+5}{24}$
51. $\frac{t+3}{t+5} = \frac{-1}{2t}$
52. $\frac{5h}{14h+3} = \frac{1}{h}$

Each pair of triangles is similar. Find the missing side length. **See Example 8.**

53.
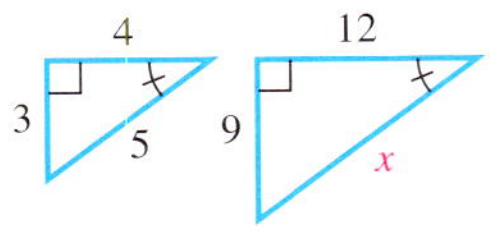

54.
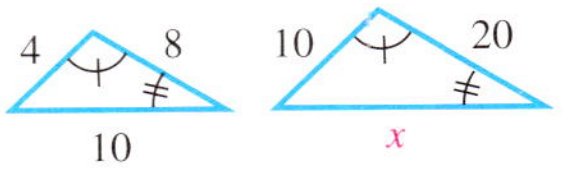

55.
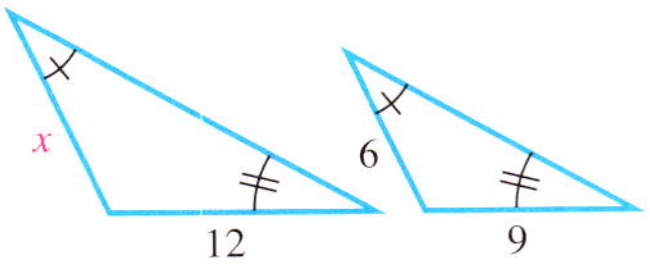

56.
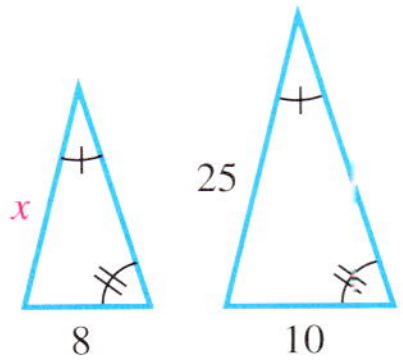

TRY IT YOURSELF

Solve each proportion.

57. $\frac{x-1}{x+1} = \frac{2}{3x}$
58. $\frac{2}{x+6} = \frac{-2x}{5}$
59. $\frac{x+1}{4} = \frac{3x}{8}$
60. $\frac{x-1}{9} = \frac{2x}{3}$
61. $\frac{y-4}{y+1} = \frac{y+3}{y+6}$
62. $\frac{r-6}{r-8} = \frac{r+1}{r-4}$
63. $\frac{c}{10} = \frac{10}{c}$
64. $\frac{-6}{r} = \frac{r}{-6}$
65. $\frac{m}{3} = \frac{4}{m+1}$
66. $\frac{n}{2} = \frac{5}{n+3}$
67. $\frac{3}{3b+4} = \frac{2}{5b-6}$
68. $\frac{2}{4d-1} = \frac{3}{2d+1}$

Look Alikes . . .

Solve each equation.

69. a. $-\frac{2}{5} = \frac{3}{4x}$ b. $\frac{4}{x} - \frac{2}{5} = \frac{3}{4x}$
70. a. $\frac{1}{4} = \frac{2}{3a}$ b. $\frac{5}{6a} + \frac{1}{4} = \frac{2}{3a}$
71. a. $\frac{3}{a-1} = \frac{8}{a}$ b. $\frac{3}{a-1} + \frac{8}{a} = 3$
72. a. $\frac{4}{3x} = \frac{1}{3}$ b. $\frac{4}{3x} - \frac{1}{3} = x$

APPLICATIONS

73. **Shopping for Clothes.** If shirts are on sale at two for \$25, how much do five shirts cost?

74. **Mixing Perfume.** A perfume is to be mixed in the ratio of 3 drops of pure essence to 7 drops of alcohol. How many drops of pure essence should be mixed with 56 drops of alcohol?

75. **CPR.** A first aid handbook states that when performing cardiopulmonary resuscitation on an adult, the ratio of chest compressions to breaths should be 30:2. If 210 compressions were administered to an adult patient, how many breaths should have been given?

76. **Cooking.** A recipe for wild rice soup follows. Find the amounts of chicken broth, rice, and flour needed to make 15 servings.

Wild Rice Soup

A sumptuous side dish with a nutty flavor

3 cups chicken broth	1 cup light cream
$\frac{2}{3}$ cup uncooked rice	2 tablespoons flour
$\frac{1}{4}$ cup sliced onions	$\frac{1}{8}$ teaspoon pepper
$\frac{1}{2}$ cup shredded carrots	Serves: 6

77. **from Campus to Careers**

Recreation Director

©BananaStock/SuperStock

A total of 966 boys and girls are members of a community recreation center.

a. If 504 are boys, how many members are girls?

b. Find the ratio of girls to boys who are members of the recreation center.

78. **Gear Ratios.** Write each ratio in two ways: as a fraction in simplest form and using a colon.

a. The number of teeth of the larger gear to the number of teeth of the smaller gear

b. The number of teeth of the smaller gear to the number of teeth of the larger gear

79. **Computing a Paycheck.** Billie earns $412 for a 40-hour week. If she missed 10 hours of work last week, how much did she get paid?

80. **Waves.** If the peak height to wavelength ratio is greater than 1:7, a wave becomes unstable and it breaks forward. (See the figure below.) What is the maximum height a wave with wavelength 637 feet can have before it breaks forward?

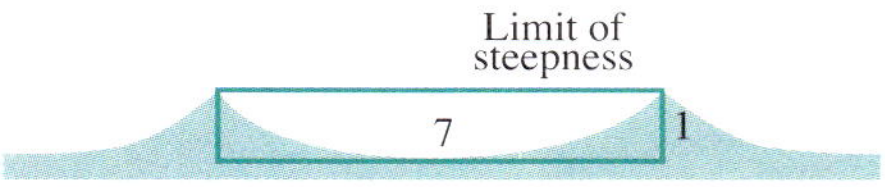

81. **Twitter.** According to a June, 2010 article in the Blog Herald, approximately 7,500 tweets are sent every 10 seconds. At this rate, about how many tweets are sent in one minute?

82. **Engineering.** A portion of a bridge is shown. Use the fact that $\frac{AB}{BC}$ is in proportion to $\frac{FE}{ED}$ to find FE.

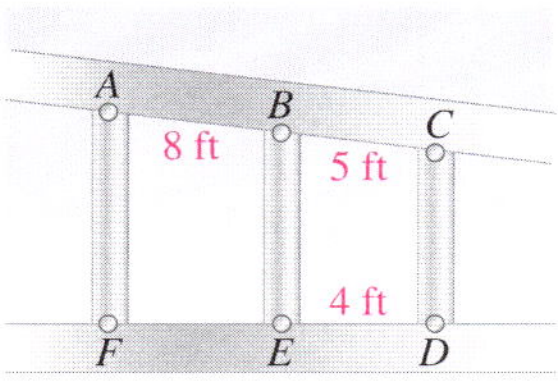

83. **Nutrition.** The table shows the nutritional facts about a 10-oz chocolate milkshake sold by a fast-food restaurant. Use the information to complete the table for the 16-oz shake. Round to the nearest unit when an answer is not exact.

	Calories	Fat (gm)	Protein (gm)
10-oz chocolate milkshake	355	8	9
16-oz chocolate milkshake			

84. **Photo Enlargements.** The 3-by-5 photo is to be blown up to the larger size. Find x.

85. **Mixing Fuel.** The instructions on a can of oil intended to be added to lawnmower gasoline are shown below. Are these instructions correct? (*Hint:* There are 128 ounces in 1 gallon.)

Recommended	Gasoline	Oil
50 to 1	6 gal	16 oz

86. **Driver's Licenses.** Of the 50 states, Alabama has one of the highest ratios of licensed drivers to residents. If the ratio is 399:500 and the population of Alabama is about 4,500,000, how many residents of that state have a driver's license?

87. **Capture–Release Method.** To estimate the ground squirrel population on his acreage, a farmer trapped, tagged, and then released a dozen squirrels. Two weeks later, the farmer trapped 35 squirrels and noted that 3 were tagged. Use this information to estimate the number of ground squirrels on his acreage.

88. **Concrete.** A 2:3 concrete mix means that for every two parts of sand, three parts of gravel are used. How much sand should be used in a mix composed of 25 cubic feet of gravel?

89. **Model Railroads.** An HO scale model railroad engine is 6 inches long. If the HO scale is 1 to 87, how long is a real engine, in inches? In feet?

90. **Model Railroads.** An N scale model railroad caboose is 4.5 inches long. If the N scale is 1 to 160, how long is a real caboose, in inches? In feet?

91. Blueprints. The scale for the drawing shown means that a $\frac{1}{4}$-inch length $\left(\frac{1}{4}''\right)$ on the drawing corresponds to an actual size of 1 foot (1′-0″). Suppose the length of the kitchen is $2\frac{1}{2}$ inches on the drawing. How long is the actual kitchen?

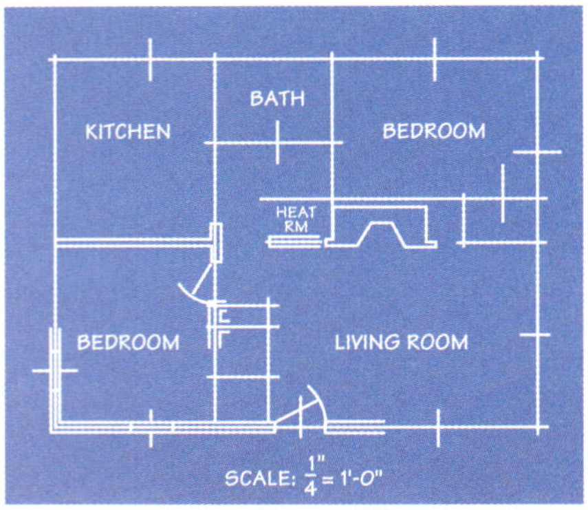

92. The Titanic. A 1:144 scale model of the *Titanic* is to be built. If the ship was 882 feet long, find the length of the model.

***For each of the following purchases, determine the better buy.* See Example 7.**

93. Trumpet lessons: 45 minutes for \$25 or 60 minutes for \$35

94. Memory for a computer: 128 megabytes for \$26 or 512 megabytes for \$110

95. Business cards: 100 for \$9.99 or 150 for \$12.99

96. Dog food: 20 pounds for \$7.49 or 44 pounds for \$14.99

97. Soft drinks: 6-pack for \$1.50 or a case (24 cans) for \$6.25

98. Donuts: A dozen for \$6.24 or a baker's dozen (13) for \$6.65

99.

100.

101. Height of a Tree. A tree casts a shadow of 26 feet at the same time as a 6-foot man casts a shadow of 4 feet. Find the height of the tree.

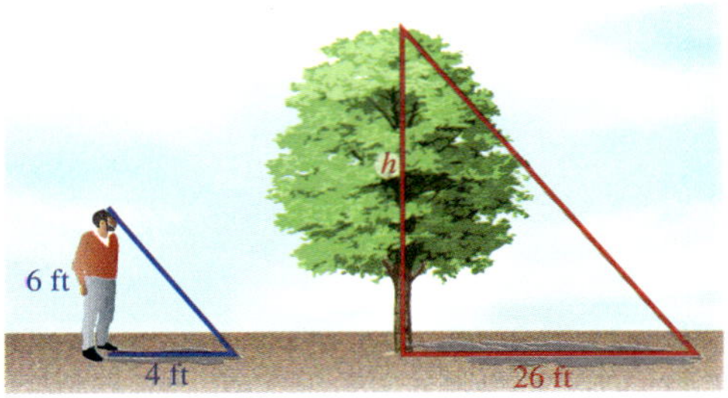

102. Height of a Building. A man places a mirror on the ground and sees the reflection of the top of a building, as shown. The two triangles in the illustration are similar. Find the height, h, of the building.

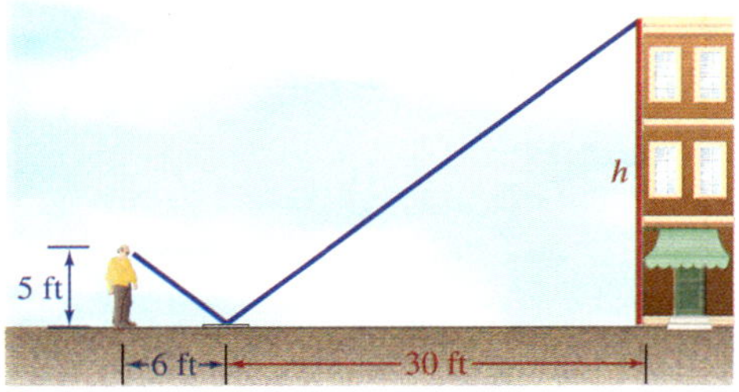

103. Surveying. To find the width of a river, a surveyor laid out the following similar triangles. Find w.

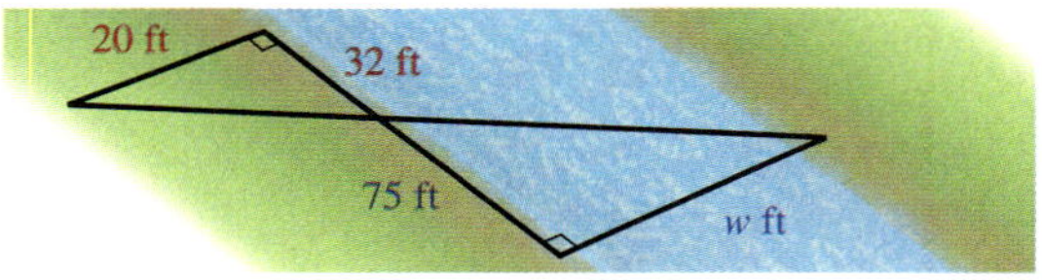

104. Flight Paths. An airplane ascends 100 feet as it flies a horizontal distance of 1,000 feet. How much altitude will it gain as it flies a horizontal distance of 1 mile? (*Hint:* 5,280 feet = 1 mile.)

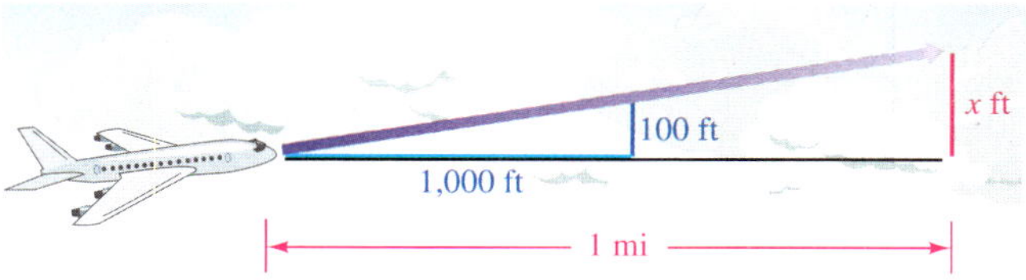

105. Slope. Find the unknown rise of the smaller slope triangle in the figure below.

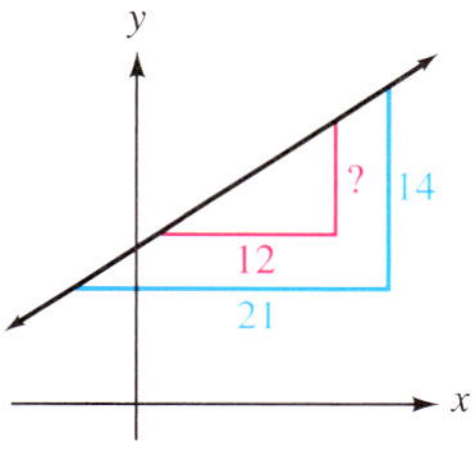

106. Washington, D.C. The Washington Monument casts a shadow of $166\frac{1}{2}$ feet at the same time as a 5-foot-tall tourist casts a shadow of $1\frac{1}{2}$ feet. Find the height of the monument.

WRITING

107. Explain how to solve the equation $\frac{7}{6} = \frac{2}{x}$ and how to simplify the expression $\frac{7}{6} \cdot \frac{2}{x}$.

108. Explain why the concept of cross products cannot immediately be used to solve the equation:

$$\frac{x}{3} - \frac{3x}{4} = \frac{1}{12}$$

109. What are similar triangles?

110. What is a unit price? Give an example.

REVIEW

111. Change $\frac{9}{10}$ to a percent.

112. Change $33\frac{1}{3}\%$ to a fraction.

113. Find 30% of 1,600.

114. Shopping. Maria bought a dress for 25% off the original price of \$98. How much did the dress cost?

CHALLENGE PROBLEMS

115. Suppose $\frac{a}{b} = \frac{c}{d}$. Write three other proportions using a, b, c, and d.

116. a. Verify that $\frac{3}{5} = \frac{12}{20} = \frac{3 + 12}{5 + 20}$.

b. Is the following rule always true? Explain.

$$\frac{a}{b} = \frac{c}{d} = \frac{a + c}{b + d}$$

7 Summary & Review

SECTION 7.1 Simplifying Rational Expressions

DEFINITIONS AND CONCEPTS	EXAMPLES
A **rational expression** is an expression of the form $\frac{A}{B}$, where A and B are polynomials and B does not equal 0.	Rational expressions: $\frac{8}{7t}$, $\frac{a}{a - 3}$, and $\frac{4x^2 - 16x}{x^2 - 6x + 8}$
To **evaluate a rational expression,** we substitute the values of its variables and simplify.	Evaluate $\frac{3x + 1}{x - 2}$ for $x = 5$. $\frac{3x + 1}{x - 2} = \frac{3(5) + 1}{5 - 2} = \frac{16}{3}$ Substitute 5 for x.
To find the real numbers for which a **rational expression is undefined,** find the values of the variable that make the denominator 0.	For which real numbers is $\frac{11}{2x - 3}$ undefined? $2x - 3 = 0$ Set the denominator equal to 0 and solve for x. $2x = 3$ Add 3 to both sides. $x = \frac{3}{2}$ The expression is undefined for $x = \frac{3}{2}$.
To **simplify a rational expression:** **1.** Factor the numerator and the denominator completely. **2.** Remove factors equal to 1. **3.** Multiply the remaining factors in the numerator and denominator.	Simplify: $\frac{x^2 - 4}{x^2 - 7x + 10} = \frac{(x + 2)\overset{1}{\cancel{(x - 2)}}}{(x - 5)\underset{1}{\cancel{(x - 2)}}}$ Factor and simplify. $= \frac{x + 2}{x - 5}$ Multiply the remaining factors in the numerator and the denominator.
The quotient of any nonzero expression and its **opposite** is -1.	$\frac{2t - 3}{3 - 2t} = -1$ Because $2t - 3$ and $3 - 2t$ are opposites.

REVIEW EXERCISES

1. Find the values of x for which the rational expression $\frac{x-1}{x^2-16}$ is undefined.

2. Evaluate $\frac{x^2-1}{x-5}$ for $x = -2$.

Simplify each rational expression, if possible. Assume that no denominators are zero.

3. $\frac{3x^2}{6x^3}$

4. $\frac{5xy^2}{2x^2y^2}$

5. $\frac{x^2}{x^2+x}$

6. $\frac{a^2-4}{a+2}$

7. $\frac{3p-2}{2-3p}$

8. $\frac{8-x}{x^2-5x-24}$

9. $\frac{2x^2-16x}{2x^2-18x+16}$

10. $\frac{x^2+x-2}{x^2-x-2}$

11. $\frac{x^2-2xy+y^2}{(x-y)^3}$

12. $\frac{4(t+3)+8}{3(t+3)+6}$

13. Explain the error in the following work: $\frac{x+1}{x} = \frac{\overset{1}{\cancel{x}}+1}{\underset{1}{\cancel{x}}} = \frac{2}{1} = 2$.

14. **Dosages.** Cowling's rule is a formula that can be used to determine the dosage of a prescription medication for children. If C is the proper child's dosage, D is an adult dosage, and A is the child's age in years, then $C = \frac{D(A+1)}{24}$. Find the daily dosage of an antibiotic for an 11-year-old child if the adult daily dosage is 300 milligrams.

SECTION 7.2 Multiplying and Dividing Rational Expressions

DEFINITIONS AND CONCEPTS	EXAMPLES
To **multiply rational expressions,** multiply their numerators and multiply their denominators. $\frac{A}{B} \cdot \frac{C}{D} = \frac{AC}{BD}$ Then simplify, if possible.	Multiply: $\frac{4b}{b+2} \cdot \frac{7}{b} = \frac{4b \cdot 7}{(b+2)b}$ Multiply the numerators. Multiply the denominators. $= \frac{4\overset{1}{\cancel{b}} \cdot 7}{(b+2)\underset{1}{\cancel{b}}}$ Simplify. $= \frac{28}{b+2}$ Multiply the remaining factors in the numerator and the denominator.
To find the **reciprocal** of a rational expression, invert its numerator and denominator.	The reciprocal of $\frac{c}{c+7}$ is $\frac{c+7}{c}$.
To **divide rational expressions,** multiply the first expression by the reciprocal of the second. $\frac{A}{B} \div \frac{C}{D} = \frac{A}{B} \cdot \frac{D}{C} = \frac{AD}{BC}$ Then simplify, if possible.	Divide: $\frac{t}{t+1} \div \frac{8}{t^2+t} = \frac{t}{t+1} \cdot \frac{t^2+t}{8}$ Multiply by the reciprocal. $= \frac{t \cdot t\overset{1}{\cancel{(t+1)}}}{\underset{1}{\cancel{(t+1)}}8}$ Factor and simplify. $= \frac{t^2}{8}$ Multiply the remaining factors in the numerator and the denominator.
A **unit conversion factor** is a fraction that has a value of 1.	$\frac{1\text{ yd}^2}{9\text{ ft}^2} = 1$ and $\frac{1\text{ mi}}{5{,}280\text{ ft}} = 1$

REVIEW EXERCISES

Multiply and simplify, if possible.

15. $\dfrac{3xy}{2x}\cdot\dfrac{4x}{2y^2}$

16. $56x\left(\dfrac{12}{7x}\right)$

17. $\dfrac{x^2-1}{x^2+2x}\cdot\dfrac{x}{x+1}$

18. $\dfrac{x^2+x}{3x-15}\cdot\dfrac{6x-30}{x^2+2x+1}$

Divide and simplify, if possible.

19. $\dfrac{3x^2}{5x^2y}\div\dfrac{6x}{15xy^2}$

20. $\dfrac{x^2-x-6}{1-2x}\div\dfrac{x^2-2x-3}{2x^2+x-1}$

21. Determine whether the given fraction is a unit conversion factor.

a. $\dfrac{1\text{ ft}}{12\text{ in.}}$ b. $\dfrac{60\text{ min}}{1\text{ day}}$

c. $\dfrac{2{,}000\text{ lb}}{1\text{ ton}}$ d. $\dfrac{1\text{ gal}}{4\text{ qt}}$

22. **Traffic Signs.** Convert the speed limit on the sign from miles per hour to miles per minute.

SECTION 7.3 Adding and Subtracting with Like Denominators; Least Common Denominators

DEFINITIONS AND CONCEPTS / EXAMPLES

To **add (or subtract) rational expressions** that have the same denominator, add (or subtract) their numerators and write the sum (or difference) over their common denominator.

$$\frac{A}{D}+\frac{B}{D}=\frac{A+B}{D}\qquad \frac{A}{D}-\frac{B}{D}=\frac{A-B}{D}$$

Then simplify, if possible.

Add: $\dfrac{2b}{3b-9}+\dfrac{b}{3b-9}=\dfrac{2b+b}{3b-9}$ Add the numerators and write the sum over the LCD, $3b-9$.

$=\dfrac{\overset{1}{\cancel{3}}b}{\underset{1}{\cancel{3}}(b-3)}$ Factor and simplify.

$=\dfrac{b}{b-3}$

Subtract: $\dfrac{x+1}{x}-\dfrac{x-1}{x}=\dfrac{x+1-(x-1)}{x}$ Don't forget the parentheses.

$=\dfrac{x+1-x+1}{x}$

$=\dfrac{2}{x}$ Combine like terms.

To find the **LCD** of several rational expressions, factor each denominator completely. Form a product using each different factor the greatest number of times it appears in any one factorization.

Find the LCD of $\dfrac{3}{x^3-x^2}$ and $\dfrac{x}{x^2-1}$.

$$\left.\begin{aligned}x^3-x^2&=x\cdot x\cdot(x-1)\\ x^2-1&=(x+1)(x-1)\end{aligned}\right\}\ \text{LCD}=x\cdot x\cdot(x-1)(x+1)$$

To **build an equivalent rational expression,** multiply the given expression by 1 in the form of $\frac{c}{c}$ where $c\neq 0$.

$\dfrac{7}{4t}=\dfrac{7}{4t}\cdot\dfrac{3t}{3t}=\dfrac{21t}{12t^2}$

and

$\dfrac{x+1}{x-7}=\dfrac{x+1}{x-7}\cdot\dfrac{x-1}{x-1}=\dfrac{(x+1)(x-1)}{(x-7)(x-1)}=\dfrac{x^2-1}{x^2-8x+7}$

REVIEW EXERCISES

Add or subtract and simplify, if possible.

23. $\dfrac{13}{15d} - \dfrac{8}{15d}$

24. $\dfrac{x}{x+y} + \dfrac{y}{x+y}$

25. $\dfrac{3x}{x-7} - \dfrac{x-2}{x-7}$

26. $\dfrac{a}{a^2-2a-8} + \dfrac{2}{a^2-2a-8}$

Find the LCD of each pair of rational expressions.

27. $\dfrac{12}{x}, \dfrac{1}{9}$

28. $\dfrac{1}{2x^3}, \dfrac{5}{8x}$

29. $\dfrac{7}{m}, \dfrac{m+2}{m-8}$

30. $\dfrac{x}{5x+1}, \dfrac{5x}{5x-1}$

31. $\dfrac{6-a}{a^2-25}, \dfrac{a^2}{a-5}$

32. $\dfrac{4t+25}{t^2+10t+25}, \dfrac{t^2-7}{2t^2+17t+35}$

Build each rational expression into an equivalent fraction having the denominator shown in red.

33. $\dfrac{9}{a}$, $7a$

34. $\dfrac{2y+1}{x-9}$, $x(x-9)$

35. $\dfrac{b+7}{3b-15}$, $6(b-5)$

36. $\dfrac{9r}{r^2+6r+5}$, $(r+1)(r-4)(r+5)$

SECTION 7.4 Adding and Subtracting with Unlike Denominators

DEFINITIONS AND CONCEPTS	EXAMPLES
To **add (or subtract) rational expressions** with unlike denominators: 1. Find the LCD. 2. Write each rational expression as an equivalent expression whose denominator is the LCD. 3. Add (or subtract) the numerators and write the sum (or difference) over the LCD. 4. Simplify the resulting rational expression if possible.	Add: $\dfrac{4x}{x} + \dfrac{2}{x-1} = \dfrac{4x}{x} \cdot \dfrac{x-1}{x-1} + \dfrac{2}{x-1} \cdot \dfrac{x}{x}$ The LCD is $x(x-1)$. Multiply → / Don't multiply → $= \dfrac{4x(x-1)}{x(x-1)} + \dfrac{2x}{x(x-1)}$ Build so that each expression has a denominator of $x(x-1)$. $= \dfrac{4x^2-4x+2x}{x(x-1)}$ Distribute the multiplication by 4x. $= \dfrac{4x^2-2x}{x(x-1)}$ Combine like terms. $= \dfrac{2\overset{1}{\cancel{x}}(2x-1)}{\underset{1}{\cancel{x}}(x-1)}$ Factor and simplify. $= \dfrac{2(2x-1)}{x-1}$
When a polynomial is multiplied by -1, the result is its opposite. This fact is used when adding or subtracting rational expressions whose **denominators are opposites.**	Add: $\dfrac{c}{c-4} + \dfrac{1}{4-c} = \dfrac{c}{c-4} + \dfrac{1}{4-c} \cdot \dfrac{-1}{-1}$ $= \dfrac{c}{c-4} + \dfrac{-1}{c-4}$ $-1(4-c) = c-4$ $= \dfrac{c-1}{c-4}$ Add the numerators. Write the sum over the LCD, $c-4$.

REVIEW EXERCISES

Add or subtract and simplify, if possible.

37. $\frac{1}{7} - \frac{1}{a}$

38. $\frac{x}{x - 1} + \frac{1}{x}$

39. $\frac{2t + 2}{t^2 + 2t + 1} - \frac{1}{t + 1}$

40. $\frac{x + 2}{2x} - \frac{2 - x}{x^2}$

41. $\frac{6}{b - 1} - \frac{b}{1 - b}$

42. $\frac{8}{c} + 6$

43. $\frac{n - 7}{n + 3} - \frac{n - 3}{n + 7}$

44. $\frac{4}{t + 2} - \frac{7}{(t + 2)^2}$

45. $\frac{6}{a^2 - 9} - \frac{5}{a^2 - a - 6}$

46. $\frac{2}{3y - 6} + \frac{3}{4y + 8}$

47. Working on a homework assignment, a student added two rational expressions and obtained $\frac{-5n^3 - 7}{3n(n + 6)}$. The answer given in the back of the book was $-\frac{5n^3 + 7}{3n(n + 6)}$. Are the answers equivalent?

48. Digital Video Cameras. Find the perimeter and the area of the LED screen of the camera.

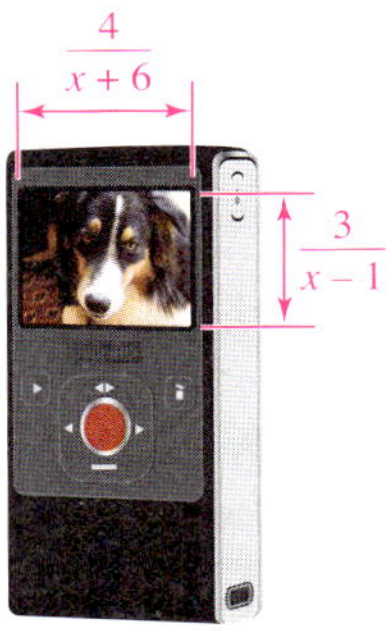

SECTION 7.5 Simplifying Complex Fractions

DEFINITIONS AND CONCEPTS	EXAMPLES
Complex fractions contain fractions in their numerators and/or their denominators.	Complex fractions: $\dfrac{\frac{2}{t}}{\frac{5}{4t}}$ and $\dfrac{\frac{3}{m} + \frac{m}{4}}{\frac{m}{2}}$
To **simplify a complex fraction:** **Method 1** Write the numerator and the denominator as single rational expressions and perform the indicated division.	Simplify: $\dfrac{\frac{3}{m} + \frac{m}{2}}{\frac{m}{4}} = \dfrac{\frac{3}{m} \cdot \frac{2}{2} + \frac{m}{2} \cdot \frac{m}{m}}{\frac{m}{4}}$ In the numerator, build to have an LCD of 2*m*. $= \dfrac{\frac{6}{2m} + \frac{m^2}{2m}}{\frac{m}{4}}$ In the numerator, multiply the fractions. $= \dfrac{\frac{6 + m^2}{2m}}{\frac{m}{4}}$ Add the fractions in the numerator. $= \frac{6 + m^2}{2m} \div \frac{m}{4}$ The main fraction bar indicates division. $= \frac{(6 + m^2) \cdot \overset{1}{\cancel{2}} \cdot 2}{\underset{1}{\cancel{2}}m \cdot m}$ Multiply by the reciprocal of $\frac{m}{4}$. Factor 4 and simplify. $= \frac{12 + 2m^2}{m^2}$ Distribute the multiplication by 2 in the numerator.

Method 2

Determine the LCD of all the rational expressions in the complex fraction and multiply the complex fraction by 1, written in the form $\frac{\text{LCD}}{\text{LCD}}$.

Simplify: $\dfrac{\frac{3}{m}+\frac{m}{2}}{\frac{m}{4}} = \dfrac{\frac{3}{m}+\frac{m}{2}}{\frac{m}{4}} \cdot \dfrac{4m}{4m}$ — The LCD for all the rational expressions is $4m$.

$= \dfrac{\frac{3}{m}\cdot 4m + \frac{m}{2}\cdot 4m}{\frac{m}{4}\cdot 4m}$ — In the numerator, distribute the multiplication by $4m$.

$= \dfrac{12+2m^2}{m^2}$ — Perform each multiplication by $4m$.

REVIEW EXERCISES

Simplify each complex fraction.

49. $\dfrac{\frac{n^4}{30}}{\frac{7n}{15}}$

50. $\dfrac{\frac{r^2-81}{18s^2}}{\frac{4r-36}{9s}}$

51. $\dfrac{\frac{1}{y}+1}{\frac{1}{y}-1}$

52. $\dfrac{\frac{7}{a^2}}{\frac{1}{a}+\frac{10}{3}}$

53. $\dfrac{\frac{2}{x-1}+\frac{x-1}{x+1}}{\frac{1}{x^2-1}}$

54. $\dfrac{\frac{1}{x^2y}-\frac{5}{xy}}{\frac{3}{xy}-\frac{7}{xy^2}}$

SECTION 7.6 Solving Rational Equations

DEFINITIONS AND CONCEPTS	EXAMPLES

To **solve a rational equation** we use the multiplication property of equality to clear the equation of fractions. Use these steps:

1. Determine which numbers cannot be solutions.
2. Multiply both sides of the equation by the LCD of the rational expressions contained in the equation.
3. Solve the resulting equation.
4. Check all possible solutions in the *original* equation. A possible solution that does not satisfy the original equation is called an **extraneous solution.**

Solve: $\dfrac{y}{y-2} - 1 = \dfrac{1}{y}$ — Since no denominators can be 0, $y \neq 2$ and $y \neq 0$.

$y(y-2)\left(\dfrac{y}{y-2} - 1\right) = y(y-2)\left(\dfrac{1}{y}\right)$ — The LCD is $y(y-2)$.

$\overset{1}{y(\cancel{y-2})}\left(\dfrac{y}{\underset{1}{\cancel{y-2}}}\right) - y(y-2)1 = \overset{1}{\cancel{y}}(y-2)\left(\dfrac{1}{\underset{1}{\cancel{y}}}\right)$ — Simplify.

$y \cdot y - y(y-2) = (y-2)\cdot 1$

$y^2 - y^2 + 2y = y - 2$ — Multiply.

$2y = y - 2$ — Combine like terms.

$y = -2$ — Solve for y.

REVIEW EXERCISES

Solve each equation and check the result. If an equation has no solution, so indicate.

55. $\dfrac{3}{x} = \dfrac{2}{x-1}$

56. $\dfrac{a}{a-5} = 3 + \dfrac{5}{a-5}$

57. $\dfrac{2}{3t} + \dfrac{1}{t} = \dfrac{5}{9}$

58. $a = \dfrac{3a-50}{4a-24} - \dfrac{3}{4}$

59. $\dfrac{4}{x+2} - \dfrac{3}{x+3} = \dfrac{6}{x^2+5x+6}$

60. $\dfrac{3}{x+1} - \dfrac{x-2}{2} = \dfrac{x-2}{x+1}$

61. **Engineering.** The efficiency E of a Carnot engine is given by the following formula. Solve it for T_1.

$$E = 1 - \frac{T_2}{T_1}$$

62. Solve for y: $\dfrac{1}{x} = \dfrac{1}{y} + \dfrac{1}{z}$

SECTION 7.7 Problem Solving Using Rational Equations

DEFINITIONS AND CONCEPTS	EXAMPLES
To solve application problems, follow these steps: **1.** Analyze the problem. **2.** Assign a variable. **3.** Form an equation. **4.** Solve the equation. **5.** State the conclusion. **6.** Check the result. **Rate of Work:** If a job can be completed in t units of time, the rate of work can be expressed as $\frac{1}{t}$ of the job is completed per unit of time. **Shared-work problems:** Work completed = rate of work · time worked	**Washing Cars.** Working alone, Carlos can wash the family SUV in 30 minutes. Victor, his brother, can wash the same SUV in 20 minutes working alone. How long will it take them if they wash the SUV together? **Analyze** It takes Carlos 30 minutes and it takes Victor 20 minutes. How long will it take working together? **Assign** Let x = the number of minutes it will take Carlos and Victor, working together, to wash the SUV. **Form** Enter the data in a table. (see table below) The part of the job done by Carlos plus the part of the job done by Victor equals 1 job completed. $\frac{x}{30} + \frac{x}{20} = 1$ **Solve** $60\left(\frac{x}{30} + \frac{x}{20}\right) = 60(1)$ Multiply both sides by the LCD, 60. $60\left(\frac{x}{30}\right) + 60\left(\frac{x}{20}\right) = 60(1)$ On the left side, distribute the multiplication by 60. $2x + 3x = 60$ Perform each multiplication by 60. $5x = 60$ Combine like terms. $x = \frac{60}{5}$ Divide both sides by 5. $x = 12$ **State** Working together, it will take Carlos and Victor 12 minutes to wash the family SUV. **Check** In 12 minutes, Carlos will do $\frac{12}{30} = \frac{24}{60}$ of the job and Victor will do $\frac{12}{20} = \frac{36}{60}$ of the job. Together they will do $\frac{24}{60} + \frac{36}{60} = \frac{60}{60}$ or 1 whole job. The result checks.
Uniform motion problems: $\text{Time} = \frac{\text{distance}}{\text{rate}}$	See Example 2 in Section 7.7.
Investment problems: $\text{Principal} = \frac{\text{Interest}}{\text{rate} \cdot \text{time}}$	See Example 5 in Section 7.7.

	Rate ·	Time =	Work Completed
Carlos	$\frac{1}{30}$	x	$\frac{x}{30}$
Victor	$\frac{1}{20}$	x	$\frac{x}{20}$

REVIEW EXERCISES

63. Number Problems. If a number is subtracted from the denominator of $\frac{4}{5}$ and twice as much is added to the numerator, the result is 5. Find the number.

64. Exercise. A woman can bicycle 30 miles in the same time that it takes her to jog 10 miles. If she can ride 10 mph faster than she can jog, how fast can she jog?

65. House Cleaning. A maid can clean a house in 4 hours. What is her rate of work?

66. House Painting. If a homeowner can paint a house in 14 days and a professional painter can paint it in 10 days, how long will it take if they work together?

67. Investments. In 1 year, a student earned $100 interest on money she deposited at a savings and loan. She later learned that the money would have earned $120 if she had deposited it at a credit union, because the credit union paid 1% more interest at the time. Find the rate she received from the savings and loan.

68. Wind Speed. A plane flies 400 miles downwind in the same amount of time as it takes to travel 320 miles upwind. If the plane can fly at 360 mph in still air, find the velocity of the wind.

SECTION 7.8 Proportions and Similar Triangles

DEFINITIONS AND CONCEPTS	EXAMPLES
A **ratio** is the quotient of two numbers or two quantities with the same units.	Ratios: $\frac{2}{3}$, $\frac{1}{50}$, and 2:3
A **rate** is the quotient of two quantities with different units.	Rates: $\frac{4\text{ oz}}{6\text{ lb}}$, $\frac{525\text{ mi}}{15\text{ hr}}$, and $\frac{\$1.95}{2\text{ lb}}$
A **proportion** is a statement that two ratios or two rates are equal. In the proportion $\frac{a}{b} = \frac{c}{d}$, a and d are the **extremes** and b and c are the **means.**	A proportion: $\frac{4}{9} = \frac{28}{63}$ Extremes: 4 and 63 Means: 9 and 28
In any proportion, the product of the extremes is equal to the product of the means. (The **cross products** are equal.)	A proportion: $\frac{4}{9} = \frac{28}{63}$ Cross product: $4 \cdot 63 = 252$ Cross product: $9 \cdot 28 = 252$
To **solve a proportion,** set the product of the extremes equal to the product of the means and solve the resulting equation.	Solve the proportion: $\frac{3}{2} = \frac{x}{10}$ $3 \cdot 10 = 2 \cdot x$ Set the cross products equal. $30 = 2x$ $15 = x$ Solve for x.
Triangles with the same shape but not necessarily the same size are called **similar triangles.** The lengths of the corresponding sides of two similar triangles are in proportion.	In these similar triangles: $\frac{a}{d} = \frac{b}{e} = \frac{c}{f}$ 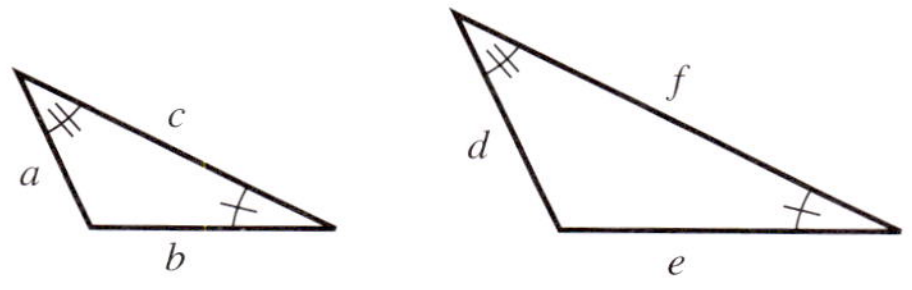
A **scale** is a ratio (or rate) that compares the size of a model to the size of an actual object.	See Example 6 in Section 7.8.
Unit prices can be used to compare costs of different sizes of the same brand to determine the best buy.	For the same item, a cost of $\frac{\$1.95}{1\text{ lb}}$ is a better buy than a cost of $\frac{\$1.99}{1\text{ lb}}$. See Example 7 in Section 7.8.

REVIEW EXERCISES

Determine whether each equation is a proportion.

69. $\frac{4}{7} = \frac{20}{34}$

70. $\frac{5}{7} = \frac{30}{42}$

Solve each proportion.

71. $\frac{3}{x} = \frac{6}{9}$

72. $\frac{x}{3} = \frac{x}{5}$

73. $\frac{x-2}{5} = \frac{x}{7}$

74. $\frac{2x}{x+4} = \frac{3}{x-1}$

75. **Dentistry.** The diagram in the next column was displayed in a dentist's office. According to the diagram, if the dentist has 340 adult patients, how many will develop gum disease?

3 out of 4 adults will develop gum disease.

76. **Utility Poles.** A telephone pole casts a shadow 12 feet long at the same time that a man 6 feet tall casts a shadow of 3.6 feet. How tall is the pole?

77. **Porcelain Figurines.** A model of a flutist, standing and playing at a music stand, was made using a 1/12th scale. If the scale model is 5.5 inches tall, how tall is the flutist?

78. **Comparison Shopping.** Which is the better buy for recordable compact discs: 150 for \$60 or 250 for \$98?

7 CHAPTER TEST

1. Fill in the blanks.
   a. A quotient of two polynomials, such as $\frac{x + 7}{x^2 + 2x}$, is called a ________ expression.
   b. Two triangles with the same shape, but not necessarily the same size, are called ________ triangles.
   c. A __________ is a mathematical statement that two ratios or two rates are equal.
   d. To ______ a rational expression, we multiply it by a form of 1. For example, $\frac{2}{5x} \cdot \frac{8}{8} = \frac{16}{40x}$.
   e. To simplify $\frac{x - 3}{(x + 3)(x - 3)}$, we remove common ________ of the numerator and denominator.

2. **Memory.** The formula $n = \frac{35 + 5d}{d}$ approximates the number of words n that a certain person can recall d days after memorizing a list of 50 words. How many words will the person remember in 1 week?

For what real numbers is each rational expression undefined?

3. $\dfrac{6x - 9}{5x}$

4. $\dfrac{x}{x^2 + x - 6}$

5. **The Internet.** A dial-up modem transmits up to 56K bits per second (K is an abbreviation for one thousand). Convert this to bits per minute.

6. Explain the error: $\dfrac{x + 5}{5} = \dfrac{x + \overset{1}{\cancel{5}}}{\underset{1}{\cancel{5}}} = x + 1$

Simplify each rational expression.

7. $\dfrac{48x^2y}{54xy^2}$

8. $\dfrac{7m - 49}{7 - m}$

9. $\dfrac{2x^2 - x - 3}{4x^2 - 9}$

10. $\dfrac{3(x + 2) - 3}{6x + 5 - (3x + 2)}$

Find the LCD of each pair of rational expressions.

11. $\dfrac{19}{3c^2d}, \dfrac{6}{c^2d^3}$

12. $\dfrac{4n + 25}{n^2 - 4n - 5}, \dfrac{6n}{n^2 - 25}$

Perform the operations. Simplify, if possible.

13. $\dfrac{12x^2y}{15xy} \cdot \dfrac{25y^2}{16x}$

14. $\dfrac{x^2 + 3x + 2}{3x + 9} \cdot \dfrac{x + 3}{x^2 - 4}$

15. $\dfrac{x - x^2}{3x^2 + 6x} \div \dfrac{3x - 3}{3x^3 + 6x^2}$

16. $\dfrac{a^2 - 16}{a - 4} \div (6a + 24)$

17. $\dfrac{3y + 7}{2y + 3} - \dfrac{-3y - 2}{2y + 3}$

18. $\dfrac{2n}{5m} - \dfrac{n}{2}$

19. $\dfrac{x + 1}{x} + \dfrac{x - 1}{x + 1}$

20. $\dfrac{a + 3}{a - 1} - \dfrac{a + 4}{1 - a}$

21. $\dfrac{9}{c - 4} + c$

22. $\dfrac{6}{t^2 - 9} - \dfrac{5}{t^2 - t - 6}$

Simplify each complex fraction.

23. $\dfrac{\frac{3m - 9}{8m}}{\frac{5m - 15}{32}}$

24. $\dfrac{\frac{3}{as^2} + \frac{6}{a^2s}}{\frac{6}{a} - \frac{9}{s^2}}$

Solve each equation. If an equation has no solution, so indicate.

25. $\dfrac{1}{3} + \dfrac{4}{3y} = \dfrac{5}{y}$

26. $\dfrac{9n}{n - 6} = 3 + \dfrac{54}{n - 6}$

27. $\dfrac{7}{q^2 - q - 2} + \dfrac{1}{q + 1} = \dfrac{3}{q - 2}$

28. $\dfrac{2}{3} = \dfrac{2c - 12}{3c - 9} - c$

29. $\dfrac{y}{y - 1} = \dfrac{y - 2}{y}$

30. $\dfrac{a}{a - 3} + \dfrac{4}{a + 3} = \dfrac{18}{a^2 - 9}$

31. Solve for B: $H = \dfrac{RB}{R + B}$

32. Solve for s: $\dfrac{1}{r} + \dfrac{1}{s} = \dfrac{1}{t}$

33. **Health Risks.** A medical newsletter states that a "healthy" waist-to-hip ratio for men is 19:20 or less. Does the patient shown in the illustration fall within the "healthy" range?

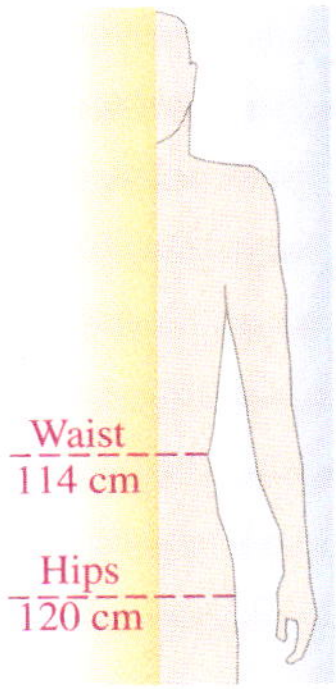

34. **Currency Exchange Rates.** Preparing for a visit to London, a New York resident exchanged 3,500 U.S. dollars for British pounds. (A pound is the basic monetary unit of Great Britain.) If the exchange rate was 100 U.S. dollars for 51 British pounds, how many British pounds did the traveler receive?

35. **TV Towers.** A television tower casts a shadow 114 feet long at the same time that a 6-foot-tall television reporter casts a shadow of 4 feet. Find the height of the tower.

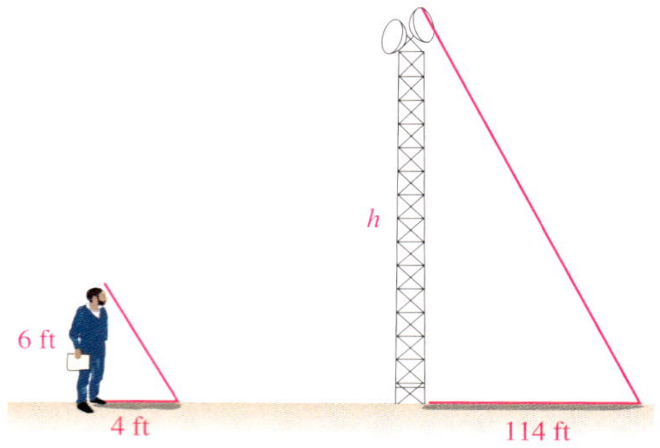

36. **Comparison Shopping.** Which is the better buy for fabric softener: 80 sheets for \$3.89 or 120 sheets for \$6.19?

37. **Cleaning Highways.** One highway worker can pick up all the trash on a strip of highway in 7 hours, and his helper can pick up the trash in 9 hours. How long will it take them if they work together?

38. **Physical Fitness.** A man roller-blades at a rate 6 miles per hour faster than he jogs. In the same time it takes him to roller-blade 5 miles he can jog 2 miles. How fast does he jog?

39. **Number Problem.** If a number is subtracted from the numerator of $\frac{5}{8}$ and twice as much is added to the denominator, the result is $\frac{1}{4}$. Find the number.

40. Explain the difference between the procedure used to simplify $\frac{1}{x} + \frac{1}{4}$ and the procedure used to solve $\frac{1}{x} + \frac{1}{4} = \frac{1}{2}$.

Group Project

WHAT IS π?

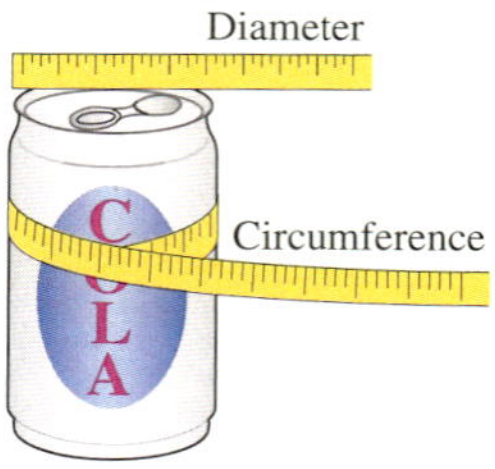

Overview In this activity, you will discover an important fact about the ratio of the circumference to the diameter of a circle.

Instructions Form groups of 2 or 3 students. With a piece of string or a cloth tape measure, find the circumference and the diameter of objects that are circular in shape. You can measure anything that is round: for example, a coin, the top of a can, a tire, or a wastepaper basket. Enter your results in a table, as shown below. Convert each measurement to a decimal, and then use a calculator to determine a decimal approximation of the ratio of the circumference C to diameter d.

Object	Circumference	Diameter	$\frac{C}{d}$ (approx.)
A quarter	$2\frac{15}{16}$ in. = 2.9375 in.	$\frac{15}{16}$ in. = 0.9375 in.	3.13333

Since early history, mathematicians have known that the ratio of the circumference to the diameter of a circle is the same for any size circle, approximately 3. Today, following centuries of study, we know that this ratio is exactly 3.141592653589. . . .

$$\frac{C}{d} = 3.141592653589\ldots$$

The Greek letter π (pi) is used to represent the ratio of circumference to diameter:

$$\pi = \frac{C}{d}, \quad \text{where } \pi = 3.141592653589\ldots$$

Are the ratios in your table numerically close to π? Give some reasons why they aren't exactly 3.141592653589 . . . in each case.

CUMULATIVE REVIEW Chapters 1–7

1. Determine whether each statement is true or false. [Section 1.3]
   a. Every integer is a whole number.
   b. 0 is not a rational number.
   c. π is an irrational number.
   d. The set of integers is the set of whole numbers and their opposites.

2. Insert the proper symbol, $<$ or $>$, in the blank to make a true statement.

   $|2 - 4|$ ___ $-(-6)$ [Section 1.5]

3. Evaluate: $9^2 - 3[45 - 3(6 + 4)]$ [Section 1.7]

4. Find the average (mean) test score of a student in a history class with scores of 80, 73, 61, 73, and 98. [Section 1.7]

5. Simplify: $8(c + 7) - 2(c - 3)$ [Section 1.9]

6. Solve: $\frac{4}{5}d = -4$ [Section 2.1]

7. Solve: $2 - 3(x - 5) = 4(x - 1)$ [Section 2.2]

8. **Grand King Size Beds.** Because Americans are taller compared to 100 years ago, bed manufacturers are making larger models. Find the percent of increase in sleeping area of the new grand king size bed compared to the standard king size. [Section 2.3]

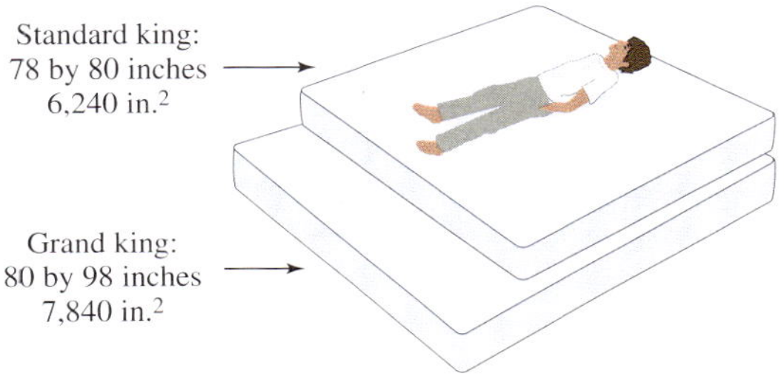

9. Solve $A - c = 2B + r$ for B. [Section 2.4]

10. Change 40°C to degrees Fahrenheit. [Section 2.4]

11. Find the volume of a pyramid that has a square base, measuring 6 feet on a side, and whose height is 20 feet. [Section 2.4]

12. **Blending Tea.** One grade of tea (worth \$6.40 per pound) is to be mixed with another grade (worth \$4 per pound) to make 20 pounds of a mixture that will be worth \$5.44 per pound. How much of each grade of tea must be used? [Section 2.6]

13. **Speed of a Plane.** Two planes are 6,000 miles apart and their speeds differ by 200 mph. If they travel toward each other and meet in 5 hours, find the speed of the slower plane. [Section 2.6]

14. Solve $7x + 2 \geq 4x - 1$. Write the solution set in interval notation and graph it. [Section 2.7]

15. Graph: $y = 2x - 3$ [Section 3.2]

16. Find the slope of the line that passes through $(-1, 3)$ and $(3, -1)$. [Section 3.4]

17. **Cutting Steel.** The graph shows the amount of wear (in millimeters) on a cutting blade for a given length of a cut (in meters). Find the rate of change in the length of the cutting blade. [Section 3.4]

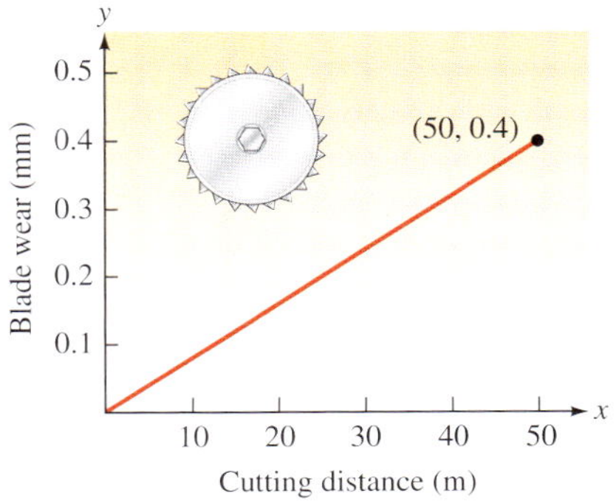

18. What is the slope of a line perpendicular to the line $y = -\frac{7}{8}x - 6$? [Section 3.5]

19. Write an equation of the line that has slope 3 and passes through the point (1, 5). Write the answer in slope–intercept form. [Section 3.6]

20. Graph: $3x - 2y \leq 6$ [Section 3.7]

21. If $f(x) = -3x^2 - 6x$, find $f(-2)$. [Section 3.8]

22. Fill in the blanks. The set of all possible input values for a function is called the ________ and the set of all output values is called the ________. [Section 3.8]

23. Solve the system $\begin{cases} x + y = 1 \\ y = x + 5 \end{cases}$ by graphing. [Section 4.1]

24. Solve the system: $\begin{cases} x = 3y - 1 \\ 2x - 3y = 4 \end{cases}$ [Section 4.2]

25. Solve the system: $\begin{cases} 2x + 3y = -1 \\ 3x + 5y = -2 \end{cases}$ [Section 4.3]

26. **Poker.** After a night of cards, a poker player finished with some red chips (worth \$5 each) and some blue chips (worth \$10 each). He received \$190 when he cashed in the 23 chips. How many of each colored chip did he finish with? [Section 4.4]

Simplify each expression. Write each answer without using negative exponents.

27. x^4x^3 [Section 5.1]

28. $(x^2x^3)^5$ [Section 5.1]

29. $\left(\frac{y^3y}{2yy^2}\right)^3$ [Section 5.1]

30. $\left(\frac{-2a}{b}\right)^5$ [Section 5.1]

31. $(a^{-2}b^3)^{-4}$ [Section 5.2]

32. $\frac{9b^0b^3}{3b^{-3}b^4}$ [Section 5.2]

33. Write 290,000 in scientific notation. [Section 5.3]

34. What is the degree of the polynomial $5x^3 - 4x + 16$? [Section 5.4]

35. Graph: $y = -x^3$ [Section 5.4]

36. **Concentric Circles.** The area of the ring between the two concentric circles of radius r and R is given by the formula

$$A = \pi(R + r)(R - r)$$

Do the multiplication on the right side of the equation. [Section 5.7]

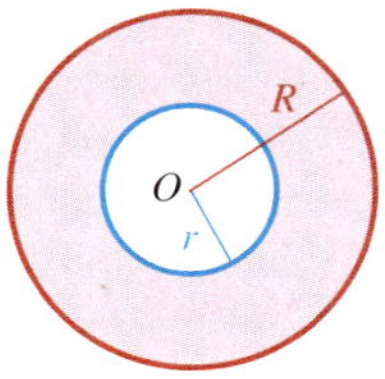

Perform the operations.

37. $(3x^2 - 3x - 2) + (3x^2 + 4x - 3)$ [Section 5.5]

38. $\left(\frac{1}{16}t^3 + \frac{1}{2}t^2 - \frac{1}{6}t\right) - \left(\frac{9}{16}t^3 + \frac{9}{4}t^2 - \frac{1}{12}t\right)$ [Section 5.5]

39. $(2x^2y^3)(3x^2y^2)$ [Section 5.6]

40. $(2y - 5)(3y + 7)$ [Section 5.6]

41. $-4x^2z(3x^2 - z)$ [Section 5.6]

42. $(3a - 4)^2$ [Section 5.7]

43. $\frac{6x + 9}{3}$ [Section 5.8]

44. $2x + 3\overline{)2x^3 + 7x^2 + 4x - 3}$ [Section 5.8]

Factor each polynomial completely.

45. $k^3t - 3k^2t$ [Section 6.1]

46. $2ab + 2ac + 3b + 3c$ [Section 6.1]

47. $u^2 - 18u + 81$ [Section 6.2]

48. $-r^2 + 2 + r$ [Section 6.2]

49. $u^2 + 10u + 15$ [Section 6.2]

50. $6x^2 - 63 - 13x$ [Section 6.3]

51. $2a^2 - 200b^2$ [Section 6.4]

52. $b^3 + 125$ [Section 6.5]

Solve each equation by factoring.

53. $5x^2 + x = 0$ [Section 6.7]

54. $6x^2 - 5x = -1$ [Section 6.7]

55. **Cooking.** The electric griddle shown has a cooking surface of 160 square inches. Find the length and the width of the griddle. [Section 6.7]

56. For what values of x is the rational expression $\frac{3x^2}{x^2 - 25}$ undefined? [Section 7.1]

Perform the operations. Simplify, if possible.

57. $\frac{2x^2 - 8x}{x^2 - 6x + 8}$ [Section 7.1]

58. $\frac{x^2 - 16}{4 - x} \div \frac{3x + 12}{x^3}$ [Section 7.2]

59. $\frac{8m^2}{2m + 5} - \frac{4m^2 + 25}{2m + 5}$ [Section 7.3]

60. $\frac{4}{x - 3} + \frac{5}{3 - x}$ [Section 7.4]

61. $\frac{m}{m^2 + 5m + 6} - \frac{2}{m^2 + 3m + 2}$ [Section 7.4]

62. Simplify: $\dfrac{2 - \frac{2}{x + 1}}{2 + \frac{2}{x}}$ [Section 7.5]

Solve each equation.

63. $\frac{7}{5x} - \frac{1}{2} = \frac{5}{6x} + \frac{1}{3}$ [Section 7.6]

64. $\frac{u}{u - 1} + \frac{1}{u} = \frac{u^2 + 1}{u^2 - u}$ [Section 7.6]

65. **Draining a Tank.** If one outlet pipe can drain a tank in 24 hours, and another pipe can drain the tank in 36 hours, how long will it take for both pipes to drain the tank? [Section 7.7]

66. **Height of a Tree.** A tree casts a shadow of 29 feet at the same time as a vertical yardstick casts a shadow of 2.5 feet. Find the height of the tree. [Section 7.8]

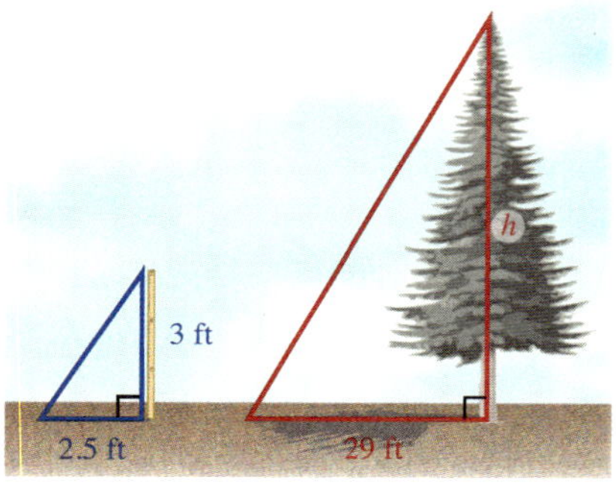

Transition to Intermediate Algebra

8

from Campus to Careers

Webmaster

If you use the Internet, then you have seen firsthand what webmasters do. They design and maintain websites for individuals and companies on the World Wide Web. The job of webmaster requires excellent computer and technical skills. A background in business, art, and design is also helpful. Since webmasters are often called on to be troubleshooters when technical difficulties arise, those considering entering the field are encouraged to study mathematics to strengthen their problem-solving abilities.

Problem 113 in Study Set 8.1, **problem 107** in Study Set 8.2, and **problem 33** in Study Set 8.9 involve situations that a webmaster might encounter on the job. The mathematical concepts discussed in this chapter can be used to solve those problems.

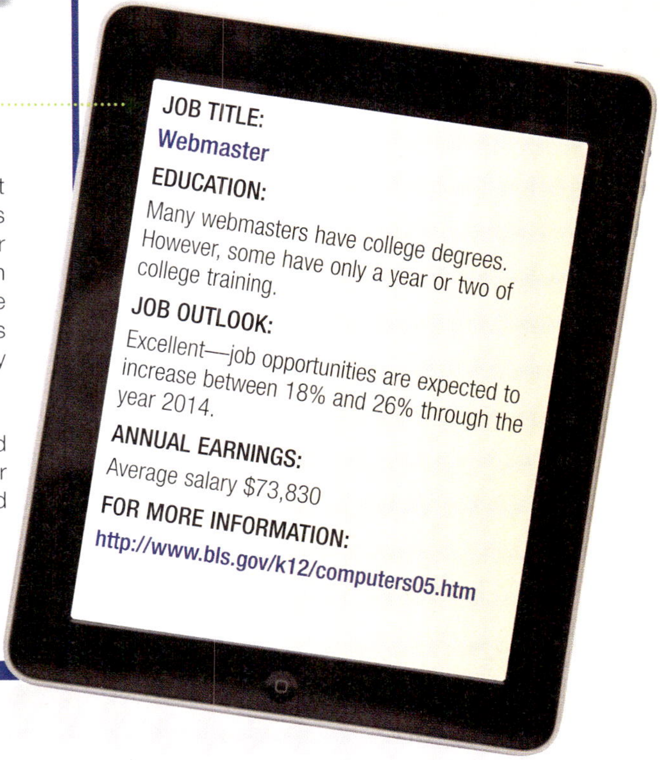

Study Skills Workshop ▶ Maintaining a Positive Attitude

Since a student's level of effort is significantly influenced by his or her attitude, you shoulc strive to maintain a positive mental outlook for the entire term. Here are three suggestions to help you do that:

PERSONAL REMINDERS: From time to time, remind yourself of the ways in which you w ll benefit by passing the course.

DON'T DWELL ON THE NEGATIVE: Counterproductive feelings of stress and math anxiety often can be overcome with extra preparation, support services, and even relaxation techniques.

ACCOMPLISH GOALS AND EARN REWARDS: Reward yourself after studying, learning a difficult concept, or completing a homework assignment. The reward can be small, like listening to music, reading a novel, playing a sport, or spending some time with friends.

Now Try This

1. List six ways in which you will benefit by passing this course. For example, it will get you one step closer to a college degree or it will improve your problem-solving abilities.
2. List three ways in which you can to respond to feelings of stress or math anxiety, should they arise during the term.
3. List some simple ways that you can reward yourself when you complete one of the class goals that you set for yourself.

SECTION 8.1

OBJECTIVES

1. Use properties of equality to solve linear equations.
2. Identify identities and contradictions.
3. Solve formulas for a specified variable.
4. Solve linear inequalities.
5. Use equations and inequalities to solve application problems.

Review of Solving Linear Equations, Formulas, and Linear Inequalities; Applications

ARE YOU READY?

The following problems review some basic skills that are needed when solving equations and inequalities.

1. Simplify: $5m - 2(8m - 4)$
2. Multiply: $21\left(\frac{6}{7}n\right)$
3. Multiply: $100 \cdot 0.27$
4. True or false: $-15 \leq -15$
5. Graph the set of real numbers greater than or equal to -2 on a number line.
6. Express the fact that $6 > x$ using an $<$ symbol.

One of the most useful concepts in algebra is the equation. Writing and then solving an equation is a powerful problem-solving strategy.

The Language of Algebra

It is important to know the difference between an **equation** and an **expression**. An equation contains an = symbol; an expression does not.

1 Use Properties of Equality to Solve Linear Equations.

Recall that an **equation** is a statement indicating that two expressions are equal. A number that makes an equation true when substituted for the variable is called a **solution,** and it is said to *satisfy* the equation. The **solution set** of an equation is the set of all numbers that make the equation true.

In this section, we will solve *linear equations in one variable.*

Linear Equations in One Variable

A **linear equation in one variable** can be written in the form $ax + b = c$, where a, b, and c are real numbers, and $a \neq 0$.

Some examples of linear equations in one variable are

$$5a + 2 = 8, \qquad \frac{7}{3}y = -28, \qquad \text{and} \qquad 3(2x - 1) = 8x + 9 - 6x$$

Success Tip

You may want to review the more detailed explanations of the properties of equality on pages 103, 105, 106, and 108.

We can solve a linear equation in one variable by using the following **properties of equality** to replace it with simpler **equivalent equations** that have the same solution set. We continue this process until the variable is isolated on one side of the = symbol.

Properties of Equality

1. Adding the same number to, or subtracting the same number from, both sides of an equation does not change its solution.
2. Multiplying or dividing both sides of an equation by the same nonzero number does not change its solution.

When solving equations, we should simplify the expressions that make up the left and right sides before applying any properties of equality. Often that involves using the **distributive property** to remove parentheses and/or **combining like terms.**

EXAMPLE 1 Solve: $3(2x - 1) = 8x + 9 - 6x$

Strategy We will use the distributive property on the left side of the equation and combine like terms on the right side.

Why It's best to simplify each side of an equation before using a property of equality. This makes it easier to determine the steps needed to get all the terms containing x on the same side of the equation.

Solution

$3(2x - 1) = 8x + 9 - 6x$	This is the equation to solve.
$6x - 3 = 2x + 9$	On the left side, distribute the multiplication by 3. On the right side, combine like terms: $8x - 6x = 2x$.
$6x - 3 - 2x = 2x + 9 - 2x$	To eliminate 2x from the right side, subtract 2x from both sides.
$4x - 3 = 9$	Combine like terms: $6x - 2x = 4x$ and $2x - 2x = 0$.
$4x - 3 + 3 = 9 + 3$	To isolate the variable term 4x, undo the subtraction of 3 on the left side by adding 3 to both sides.
$4x = 12$	Combine like terms: $-3 + 3 = 0$.
$x = 3$	To isolate the variable x, undo the multiplication by 4 by dividing both sides by 4.

Success Tip

Since division by 4 is the same as multiplication by $\frac{1}{4}$, we can solve $4x = 12$ using the multiplication property of equality by multiplying both sides by the *multiplicative inverse* of 4, which is $\frac{1}{4}$.

$$\frac{1}{4} \cdot 4x = \frac{1}{4} \cdot 12$$
$$x = 3$$

Check: We substitute 3 for x in the original equation to see whether it satisfies the equation.

Evaluate the expression on the left side.

$$3(2x - 1) = 8x + 9 - 6x$$
$$3(2 \cdot 3 - 1) \stackrel{?}{=} 8 \cdot 3 + 9 - 6 \cdot 3$$
$$3(5) \stackrel{?}{=} 24 + 9 - 18$$
$$15 = 15 \quad \text{True}$$

Evaluate the expression on the right side.

The Language of Algebra

Read $\stackrel{?}{=}$ as "is possibly equal to."

Since the resulting statement 15 = 15 is true, 3 satisfies the equation, and we say that 3 is a solution of $3(2x - 1) = 8x + 9 - 6x$. The solution set is {3}.

Self Check 1 Solve: $2(3x - 2) = 10x - 13 - 7x$

Now Try Problem 21

In general, we will follow these steps to solve linear equations in one variable. Not every step is needed to solve every equation.

Strategy for Solving Linear Equations in One Variable

1. **Clear the equation of fractions or decimals:** Multiply both sides by the LCD to clear fractions or multiply both sides by a power of 10 to clear decimals.
2. **Simplify each side of the equation:** Use the distributive property to remove parentheses and combine like terms on each side.
3. **Isolate the variable term on one side:** Add (or subtract) to get the variable term on one side of the equation and a number on the other using the addition (or subtraction) property of equality.
4. **Isolate the variable:** Multiply (or divide) to isolate the variable using the multiplication (or division) property of equality.
5. **Check the result:** Substitute the possible solution for the variable in the *original* equation to see if a true statement results.

EXAMPLE 2 Solve: $\frac{1}{3}(6x + 17) = \frac{3}{2}(x + 2) - 2$

Strategy We will follow the strategy for solving equations.

Why This is the most efficient way to solve a linear equation in one variable.

Solution *Step 1:* We can clear the equation of fractions by multiplying both sides by the least common denominator (LCD) of $\frac{1}{3}$ and $\frac{3}{2}$. The LCD of these fractions is the smallest number that can be divided by both 2 and 3 exactly. That number is 6.

$$\frac{1}{3}(6x + 17) = \frac{3}{2}(x + 2) - 2$$ This is the equation to solve.

$$6\left[\frac{1}{3}(6x + 17)\right] = 6\left[\frac{3}{2}(x + 2) - 2\right]$$ To eliminate the fractions, multiply both sides by the LCD, 6.

$$2(6x + 17) = 6 \cdot \frac{3}{2}(x + 2) - 6 \cdot 2$$ On the left side, perform the multiplication: $6 \cdot \frac{1}{3} = 2$. On the right side, distribute the multiplication by 6.

$$2(6x + 17) = 9(x + 2) - 12$$ Multiply on the right side: $6 \cdot \frac{3}{2} = 9$ and $6 \cdot 2 = 12$. Don't forget the parentheses shown in blue.

Success Tip

Before multiplying both sides of an equation by the LCD, frame the left side and frame the right side with parentheses or brackets.

Step 2: Simplify the expressions on each side of the equation.

$$12x + 34 = 9x + 18 - 12$$ Distribute the multiplication by 2 and the multiplication by 9.

$$12x + 34 = 9x + 6$$ Combine like terms: $18 - 12 = 6$.

Step 3: To get the variable term on the left side of the equation and the constant on the right side, subtract $9x$ and 34 from both sides.

$$12x + 34 - 9x - 34 = 9x + 6 - 9x - 34$$

$$3x = -28$$ On each side, combine like terms.

Step 4: To isolate x, we undo the multiplication by 3 by dividing both sides by 3.

$$\frac{3x}{3} = \frac{-28}{3}$$

$$x = -\frac{28}{3}$$

Step 5: Verify that $-\frac{28}{3}$ is the solution by substituting it into the original equation and evaluating each side.

Self Check 2 Solve: $\frac{1}{3}(2x - 5) = \frac{3}{4}(5x + 1) + 2$

Now Try Problem 29

2 Identify Identities and Contradictions.

The equations discussed so far are called **conditional equations.** For these equations, some real numbers satisfy the equation and others do not. Other equations are made true by *any* permissible replacement value for the variable. Such an equation is called an **identity.** Still other equations are false for *all* replacement values for the variable. We call such an equation a **contradiction.**

EXAMPLE 3 Solve: **a.** $-2(x - 1) - 4 = -4(1 + x) + 2x + 2$
b. $-6.2(-x - 1) - 4 = 4.2x - (-2x)$

Strategy In each case, we will follow the strategy for solving equations.

Why This is the most efficient way to solve a linear equation in one variable.

Solution

a. Since there are no fractions to clear, we will begin by using the distributive property to remove the parentheses on the left and right sides of the equation.

$-2(x - 1) - 4 = -4(1 + x) + 2x + 2$	This is the equation to solve.
$-2x + 2 - 4 = -4 - 4x + 2x + 2$	On each side, use the distributive property.
$-2x - 2 = -2x - 2$	On each side, combine like terms.
$-2x - 2 + 2x = -2x - 2 + 2x$	To attempt to isolate the variable on one side of the equation, add 2x to both sides.
$-2 = -2$	True

In the solution process, the terms involving x drop out. The resulting true statement indicates that the original equation is true for every permissible value of x. Therefore, *all real numbers* are solutions and this equation is an identity. Its solution set is written as {all real numbers} or using the symbol $\mathbb{R}$.

b.

$-6.2(-x - 1) - 4 = 4.2x - (-2x)$	This is the equation to solve.
$-6.2(-x - 1) - 4 = 4.2x + 2x$	Simplify: $-(-2x) = 2x$.
$10[-6.2(-x - 1) - 4] = 10(4.2x + 2x)$	To clear the decimals, multiply both sides by 10.
$10(-6.2)(-x - 1) - 10 \cdot 4 = 10 \cdot 4.2x + 10 \cdot 2x$	On each side, distribute the multiplication by 10.
$-62(-x - 1) - 40 = 42x + 20x$	Perform each multiplication by 10.
$62x + 62 - 40 = 42x + 20x$	On the left side, distribute -62.
$62x + 22 = 62x$	On each side, combine like terms.
$62x + 22 - 62x = 62x - 62x$	To attempt to isolate the variable on one side of the equation, subtract 62x from both sides.
$22 = 0$	False

In the solution process, the terms involving x drop out. The resulting false statement indicates that no value for x makes the original equation true. Therefore, this equation has *no solution* and it is a contradiction. Its solution set is the empty set, which is written as { } or using the symbol $\varnothing$.

Success Tip

We know the given equation is an identity because we see in Step 3 that it s equivalent to the equation $-2x - 2 = -2x - 2$, which is true for all values of x.

The Language of Algebra

Contradiction is a form of the word *contradict,* meaning conflicting ideas. During a trial, evidence might be introduced that contradicts the testimony of a witness.

Self Check 3 Solve: **a.** $3(a + 1) - (20 + a) = 5(a - 1) - 3(a + 4)$

b. $0.3(a + 4) + 0.2 = 0.2(a - 1) + 0.1a + 1.9$

Now Try Problems 31 and 35

3 Solve Formulas for a Specified Variable.

Real-world applications sometimes call for a formula solved for one variable to be solved for a different variable. **To solve a formula for a specified variable** means to isolate that variable on one side of the equal symbol, with all other variables and constants on the other side.

EXAMPLE 4 **Banking.** The formula $A = P + Prt$ gives the amount of money in an account at the end of a specific time t. A represents the amount, P the principal, and r the simple rate of interest. Solve the formula for t.

Strategy To solve for t, we treat it as if it were the only variable in the equation. To isolate t, we will use the same strategy that we used to solve linear equations in one variable.

Why We can solve this formula as if it were an equation in one variable because all the other variables, A, P, and r, are treated as if they were numbers (constants).

Solution

The Language of Algebra

We say the formula is **solved for** t because t is alone on one side of the equation and the other side does not contain t.

$A = P + Prt$ — To solve for t, we will isolate it on this side of the equation.

$A - P = Prt$ — To isolate the term involving t, subtract P from both sides. This step is done mentally.

$\frac{A - P}{Pr} = \frac{Prt}{Pr}$ — To isolate t, divide both sides by Pr (or multiply both sides by $\frac{1}{Pr}$).

$\frac{A - P}{Pr} = t$ — On the right side, remove the common factors: $\frac{\overset{1}{\cancel{P}}\overset{1}{\cancel{r}}t}{\underset{1}{\cancel{P}}\underset{1}{\cancel{r}}}$.

$t = \frac{A - P}{Pr}$ — It is common practice to write the equation so that the specified variable, in this case t, is on the left side.

Self Check 4 Solve $A = P + Prt$ for r.

Now Try Problem 39

EXAMPLE 5 **Geometry.** The formula for the area of a trapezoid is $A = \frac{1}{2}h(b_1 + b_2)$. Solve the formula for b_1.

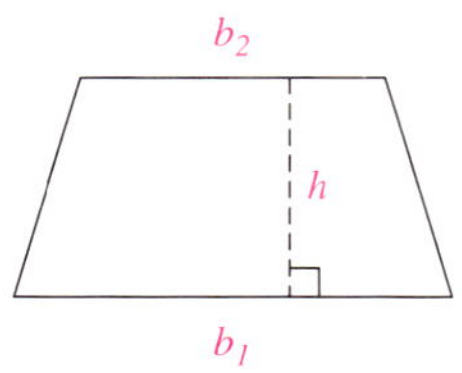

Trapezoid

Strategy To solve for b_1, we will treat it as if it were the only variable in the equation. To isolate b_1, we will use the same strategy that we used to solve linear equations in one variable.

Why We can solve the formula as if it were an equation in one variable because all the other variables, A, h, and b_2, are treated as if they were numbers (constants).

Solution

To solve for b_1, we will isolate b_1 on this side of the equation.

$$A = \frac{1}{2}h(b_1 + b_2)$$ Read b_1 as "b-sub-one" and b_2 as "b-sub-two."

$$2 \cdot A = 2 \cdot \frac{1}{2}h(b_1 + b_2)$$ Multiply both sides by 2 to clear the equation of the fraction.

$$2A = h(b_1 + b_2)$$ Simplify each side of the equation.

$$2A = hb_1 + hb_2$$ Distribute the multiplication by h.

$$2A - hb_2 = hb_1$$ Subtract hb_2 from both sides to isolate the variable term hb_1 on the right side. This step is done mentally.

$$\frac{2A - hb_2}{h} = \frac{hb_1}{h}$$ To isolate b_1, undo the multiplication by h by dividing both sides by h.

$$\frac{2A - b_2h}{h} = b_1$$ On the right side, remove the common factor of h: $\frac{\overset{1}{\cancel{h}}b_1}{\underset{1}{\cancel{h}}}$.

$$b_1 = \frac{2A - hb_2}{h}$$ Reverse the sides of the equation so that b_1 is on the left.

Caution

Do not try to simplify the result in the following way. It is incorrect because h is not a factor of the entire numerator.

$$b_1 = \frac{2A - \overset{1}{\cancel{h}}b_2}{\underset{1}{\cancel{h}}}$$

When solving formulas for a specified variable, there is often more than one way to express the result. In this case, we could perform the division by h on the right side term-by-term: $b_1 = \frac{2A}{h} - \frac{hb_2}{h}$. After removing the common factor of h in the numerator and denominator of the second fraction, we obtain the following equivalent form of the result: $b_1 = \frac{2A}{h} - b_2$.

Self Check 5 Solve $S = \frac{180(t - 2)}{7}$ for t.

Now Try Problem 47

4 Solve Linear Inequalities.

Inequalities are statements indicating that two quantities are unequal and they contain one or more of the following symbols.

Inequality Symbols

$<$ is less than	$>$ is greater than	$\neq$ is not equal to
$\leq$ is less than or equal to	$\geq$ is greater than or equal to	

In this section, we will solve *linear inequalities in one variable*. A linear *inequality* in one variable is similar to a linear *equation* in one variable except that the equal symbol is replaced with an inequality symbol.

Linear Inequalities

A linear inequality in one variable (say, x) is any inequality that can be expressed in one of the following forms, where a, b, and c represent real numbers and $a \neq 0$.

$$ax + b < c \qquad ax + b \leq c \qquad ax + b > c \qquad ax + b \geq c$$

Some examples of linear inequalities are

$$3(2x - 9) < 9, \qquad -11x - 8.2 \leq x + 32.2, \qquad \text{and} \qquad 3x > 0$$

Success Tip

You may want to review the more detailed explanations of properties of inequality on pages 165 and 166.

To **solve a linear inequality** means to find all values that, when substituted for the variable, make the inequality true. The set of all solutions of an inequality is called its **solution set.** Most inequalities that we will solve have infinitely many solutions. To represent such solutions, we can use the graph of an **interval** on the number line and two special types of notation: **interval notation** and **set-builder notation.**

We use the following properties to solve inequalities.

Properties of Inequality

1. Adding the same number to, or subtracting the same number from, both sides of an inequality does not change the solutions.
2. Multiplying or dividing both sides of an inequality by the same positive number does not change the solutions.
3. If we multiply or divide both sides of an inequality by a negative number, the direction of the inequality symbol must be *reversed* for the inequalities to have the same solutions.

After applying one of these properties, the resulting inequality is equivalent to the original one. Like equivalent equations, **equivalent inequalities** have the same solution set.

EXAMPLE 6 Solve $3(2x - 9) < 9$. Graph the solution set and write it using interval notation.

Strategy We will use the properties of inequality and the same strategy for solving equations to isolate x on one side of the inequality.

Why Once we have obtained an equivalent inequality, with the variable isolated on one side, the solution set will be obvious.

Solution

$3(2x - 9) < 9$	This is the inequality to solve.
$6x - 27 < 9$	Distribute the multiplication by 3.
$6x < 36$	To isolate the variable term 6x, undo the subtraction of 27 by adding 27 to both sides: 9 + 27 = 36.
$x < 6$	To isolate x, undo the multiplication by 6 by dividing both sides by 6.

The solution set is the interval $(-\infty, 6)$, whose graph is shown. We also can write the solution set using set-builder notation: $\{x \mid x < 6\}$. We read this notation as "the set of all real numbers x such that x is less than 6."

Notation

When graphing solution sets on a number line, parentheses are used to exclude endpoints and brackets are used to include endpoints.

Since the solution set contains infinitely many real numbers, we cannot check all of them to see whether they satisfy the original inequality. However, as an informal check, we can pick one number in the graph, near the endpoint, such as 5, and see whether it satisfies the inequality. We also can pick one number not in the graph, but near the endpoint, such as 7, and see whether it fails to satisfy the inequality.

Check a value in the graph: $x = 5$

$$3(2x - 9) < 9$$
$$3[2(5) - 9] \overset{?}{<} 9$$
$$3(10 - 9) \overset{?}{<} 9$$
$$3(1) \overset{?}{<} 9$$
$$3 < 9 \quad \text{True}$$

Check a value not in the graph: $x = 7$

$$3(2x - 9) < 9$$
$$3[2(7) - 9] \overset{?}{<} 9$$
$$3(14 - 9) \overset{?}{<} 9$$
$$3(5) \overset{?}{<} 9$$
$$15 < 9 \quad \text{False}$$

Since 5 satisfies $3(2x - 9) < 9$ and 7 does not, the solution set appears to be correct.

Self Check 6 Solve $2(3x + 2) > -44$. Graph the solution set and write it using interval notation.

Now Try Problem 53

EXAMPLE 7 Solve $-11x + 8.2 \leq x + 32.2$. Graph the solution set and write it using interval notation.

Strategy We will use the properties of inequality and the same strategy for solving equations to isolate x on one side of the inequality.

Why Once we have obtained an equivalent inequality, with the variable isolated on one side, the solution set will be obvious.

Solution

Success Tip

We must remember to reverse the inequality symbol every time we multiply or divide both sides by a negative number.

$-11x + 8.2 \leq x + 32.2$ This is the inequality to solve.

$-12x + 8.2 \leq 32.2$ To eliminate x from the right side, subtract x from both sides.

$-12x \leq 24$ To isolate the variable term $-12x$, undo the addition of 8.2 by subtracting 8.2 from both sides: $32.2 - 8.2 = 24$.

$\frac{-12x}{-12} \geq \frac{24}{-12}$ To isolate the x, undo the multiplication by -12 by dividing both sides by -12. Because we are dividing by a negative number, reverse the $\leq$ symbol.

$x \geq -2$ Do the division.

The solution set is $\{x | x \geq -2\}$ or the interval $[-2, \infty)$, whose graph is shown.

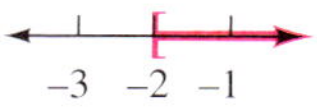

Self Check 7 Solve: $-5x + 6 \leq x$. Graph the solution set and write it using interval notation.

Now Try Problem 59

EXAMPLE 8 Solve $3a - 4 < 3(a + 5)$. Graph the solution set and write it using interval notation.

Strategy We will use the properties of inequality and the same strategy for solving equations to isolate a on one side of the inequality.

Why Once we have obtained an equivalent inequality, with the variable isolated on one side, the solution set is obvious.

Solution

Success Tip

If the terms involving the variable drop out when solving an inequality and the result is false, the solution set contains no elements and is denoted by ∅.

$3a - 4 < 3(a + 5)$ This is the inequality to solve.

$3a - 4 < 3a + 15$ Distribute the multiplication by 3.

$3a - 4 - 3a < 3a + 15 - 3a$ To eliminate 3a on the right side, subtract 3a from both sides.

$-4 < 15$ True

In the solution process, the terms involving a drop out. The resulting true statement indicates that the original inequality is true for every permissible value of a. Therefore, *all real numbers* are solutions and this inequality is an identity. Its solution set is the set of real numbers, written as $(-\infty, \infty)$ or $\mathbb{R}$, and its graph is as shown.

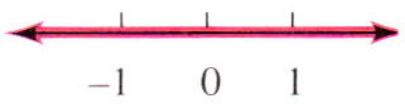

Self Check 8 Solve $-8n + 10 \geq 1 - 2(4n - 2)$. Graph the solution set and write it using interval notation.

Now Try Problem 67

5 Use Equations and Inequalities to Solve Application Problems.

To become a good problem solver, you need a plan to follow such as the following six-step problem-solving strategy: **Analyze** the problem, **Assign** a variable, **Form** an equation (or inequality), **Solve** the equation (or inequality), **State** the conclusion, **Check** the result.

EXAMPLE 9

Travel Promotions. The price of a 7-day Alaskan cruise, normally \$2,752 per person, is reduced by \$1.75 per person for large groups traveling together. How large a group is needed for the price to be \$2,500 per person?

Analyze It is often helpful to consider some specific situations before attempting to form an equation. For a group of, say, 20 people, the price of the cruise would be reduced by 20(\$1.75), and each person would pay \$2,752 − 20(\$1.75). For a group of 30 people, the price of the cruise would be reduced by 30(\$1.75), and each person would pay \$2,752 − 30(\$1.75).

Assign We let x = the group size necessary for the price of the cruise to be \$2,500 per person.

Form We can form the following equation:

The price of the cruise	is	\$2,752	minus	the number of people in the group	times	\$1.75.
2,500	=	2,752	−	x	·	1.75

Solve

$$2{,}500 = 2{,}752 - 1.75x$$

$$2{,}500 - \mathbf{2{,}752} = 2{,}752 - 1.75x - \mathbf{2{,}752}$$ To isolate the variable term −1.75x, subtract 2,752 from both sides.

$$-252 = -1.75x$$ Do the subtraction.

$$\frac{-252}{\mathbf{-1.75}} = \frac{-1.75x}{\mathbf{-1.75}}$$ To isolate x, divide both sides by −1.75.

$$144 = x$$ Use a calculator to do the division. This is the required size of the group.

Success Tip

You may want to review the six steps of the problem-solving strategy in detail. They are listed on page 142.

State If 144 people travel together, the price will be \$2,500 per person.

Check For 144 people, the cruise cost of \$2,752 will be reduced by 144(\$1.75) = \$252. If we subtract, \$2,752 − \$252 = \$2,500. The result checks.

Self Check 9 **Broadway Shows.** Tickets to a Broadway show normally sell for \$90 per person. This price is reduced by \$0.50 per person for large groups. How large a group is needed for the price to be \$75 per person?

Now Try Problem 107

To determine whether to use an equation or an inequality to solve a problem, look for key words and phrases. For example, phrases like *does not exceed, is no more than,* and *is at least* translate into inequalities.

EXAMPLE 10

Communication. A satellite phone is a mobile phone that sends and receives calls using satellites instead of cellular broadcasting towers. Satellite phone calls can be made from anywhere, such as the Sahara desert or the top of Mount Everest. If a satellite telephone company charges callers \$5.50 for the first three minutes and 88¢ for each additional minute, for how many minutes can a call last if the cost is not to exceed \$20?

Analyze We are given the rate at which a call is billed. Since the cost of a call is not to exceed \$20, the cost must be *less than or equal to* \$20. This phrase indicates that we should write an inequality to find how long a call can last.

Assign We will let x = the total number of minutes that a call can last.

Form If the call lasts x minutes, then the cost of a call will be \$5.50 for the first three minutes plus 88¢ times the number of additional minutes, where the number of *additional* minutes is $x - 3$ (the total number of minutes minus the first 3 minutes). With this information, we can form an inequality.

The cost of the first three minutes	plus	the cost of the additional minutes	is not to exceed	\$20.
5.50	$+$	$0.88(x - 3)$	$\le$	20

Success Tip

You may want to review the phrases that translate to an inequality. They are listed on page 171.

Solve First, we clear the inequality of decimals.

$5.50 + 0.88(x - 3) \le 20$ — Write 88¢ as \$0.88.

$550 + 88(x - 3) \le 2{,}000$ — To eliminate the decimals, multiply both sides by 100.

$550 + 88x - 264 \le 2{,}000$ — Distribute the multiplication by 88.

$88x + 286 \le 2{,}000$ — Combine like terms: $550 - 264 = 286$.

$88x \le 1{,}714$ — To isolate $88x$, subtract 286 from both sides.

$x \le 19.4772727\ldots$ — To isolate x, divide both sides by 88. Use a calculator.

State Since the phone company doesn't bill for part of a minute, the longest time a call can last is 19 minutes. If a call lasts for $x = 19.4772727\ldots$ minutes, it will be charged as a 20-minute call, and the cost will be $\$5.50 + \$0.88(17) = \$20.46$.

Check If the call lasts 19 minutes, the cost will be $\$5.50 + \$0.88(16) = \$19.58$. This is less than \$20. The result checks.

Self Check 10 **Renting Cars.** Great Value Car Rental charges \$12 a day and \$0.15 per mile to rent a Ford Fusion. Be Thrifty Car Rental's daily charge for the same car is \$15 and \$0.12 per mile. If a businessman wants to rent the car for one day, for what range of miles driven is the Be Thrifty rental plan better?

Now Try Problem 115

SECTION 8.1 STUDY SET

VOCABULARY

Fill in the blanks.

1. An ________ is a statement indicating that two expressions are equal.
2. $2x + 1 = 4$ is an example of a ______ equation in one variable.
3. A number that makes an equation true when substituted for the variable is called a ________.
4. If two equations have the same solution set, they are called __________ equations.
5. An equation that is made true by any permissible replacement value for the variable is called an ________.
6. An equation that is false for all replacement values for the variable is called a ____________.
7. $<$, $>$, $\le$, and $\ge$ are ___________ symbols.
8. To ______ an inequality means to find all values of the variable that make the inequality true.

CONCEPTS

Fill in the blanks.

9. a. Adding the ______ number to, or subtracting the same number from, both sides of an equation does not change its solution.
   b. Multiplying or dividing both sides of an equation by the ______ nonzero number does not change its solution.
10. If we multiply both sides of an inequality by a negative number, the direction of the inequality must be _________ for the inequalities to have the same solutions.
11. Use a check to determine whether -5 is a solution of the following equation and inequality.
    a. $5(2x + 7) = 2x - 4$
    b. $3x + 6 \le -9$

12. The solution set of a linear inequality in x is graphed on the right. Determine whether a true or false statement results when

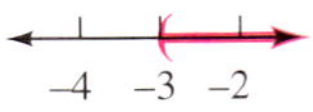

a. -4 is substituted for x.
b. -3 is substituted for x.
c. 0 is substituted for x.

NOTATION

13. Match each interval with its graph.

a. $(-\infty, -1]$ b. $(-\infty, 1)$ c. $[-1, \infty)$

i. 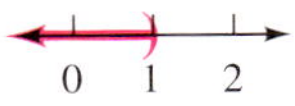

ii. 

iii. 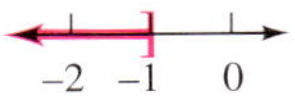

14. a. Suppose that when solving a linear equation, the variable drops out, and the result is $7 = -1$. What is the solution set?

b. Suppose that when solving a linear inequality, the variable drops out, and the result is $6 \leq 10$. Write the solution set in interval notation and graph it.

GUIDED PRACTICE

Solve each equation. Check the result. See Example 1.

15. $4x + 1 = 13$
16. $4x - 8 = 16$
17. $3(x + 1) = 15$
18. $-2(x + 5) = 30$
19. $2x + 6(2x + 3) = -10$
20. $3(2y - 4) - 6 = 3y$
21. $7(a + 2) = 11a + 17 - 7a$
22. $5(5 - a) = 4a + 37 - 6a$

Solve each equation. Check the result. See Example 2.

23. $\frac{1}{2}x - 4 = -1 + 2x$
24. $2x + 3 = \frac{2}{3}x - 1$
25. $\frac{x}{2} - \frac{x}{3} = 4$
26. $\frac{x}{2} + \frac{x}{3} = 10$
27. $\frac{1}{6}(x + 12) + 1 = \frac{x}{3}$
28. $\frac{3}{2}(y + 4) = \frac{20 - y}{2}$
29. $\frac{1}{5}(x + 6) = \frac{5}{8}(2x - 1) + 2$
30. $\frac{2}{3}(x - 3) = \frac{1}{6}(7x + 29) + 3$

Solve each equation. If an equation is an identity or a contradiction, so indicate. See Example 3.

31. $2x - 6 = -2x + 4(x - 2)$
32. $-3x = -2x + 1 - (5 + x)$
33. $2y + 1 = 5(0.2y + 1) - (4 - y)$
34. $4(2 - 3t) + 6t = -6t + 8$
35. $\frac{7}{2}(y - 1) + \frac{1}{2} = \frac{1}{2}(7y - 6)$
36. $2(x - 3) = \frac{3}{2}(x - 4) + \frac{x}{2}$
37. $0.3(x - 4) + 0.6 = -0.2(x + 4) + 0.5x$
38. $0.5(y + 2) + 0.7 - 0.3y = 0.2(y + 9)$

Solve each formula for the specified variable. See Examples 4 and 5.

39. $P = 2l + 2w$ for w
40. $P = 2l + 2w$ for l
41. $V = \frac{1}{3}Bh$ for B
42. $A = \frac{1}{2}bh$ for b
43. $T - W = ma$ for W
44. $G = U - TS + PV$ for S
45. $z = \frac{x - \mu}{\sigma}$ for x
46. $P = L + \frac{s}{f}i$ for s
47. $S = \frac{n(a + l)}{2}$ for l
48. $h = 48t + \frac{1}{2}at^2$ for a
49. $l = a + (n - 1)d$ for d
50. $P = 2(w + h + l)$ for h

Solve each inequality. Graph the solution set and write it in interval notation. See Example 6.

51. $5x - 3 > 7$
52. $7x - 9 < 5$
53. $9a + 11 \leq 29$
54. $3b - 26 \geq 4$
55. $3(z - 2) \leq 2(z + 7)$
56. $5(3 + z) > -3(z + 3)$
57. $2x + 4 + 6x > 2 - 3x + 2$
58. $5x + 6 + 2x \geq 2 - x + 4$

Solve each inequality. Write the solution set in interval notation and then graph it. See Example 7.

59. $-3x - 1 \leq 5$
60. $-2x + 6 \geq 16$
61. $-5t + 3 \leq 5$
62. $-9t + 6 \geq 16$
63. $-7y + 5 > -5y - 1$
64. $8 - 9y \geq -y$
65. $t + 1 - 3t \geq t - 20$
66. $a + 4 - 10a > a - 16$

Solve each inequality. Graph the solution set and write it in interval notation. See Example 8 and the Success Tip in the margin.

67. $2(5x - 6) > 4x - 15 + 6x$
68. $3(4x - 2) > 14x - 7 - 2x$
69. $\frac{3b + 7}{3} \leq \frac{2b - 9}{2}$
70. $-\frac{5x}{4} > \frac{3 - 5x}{4}$

TRY IT YOURSELF

Solve each equation. If an equation is an identity or a contradiction, so indicate.

71. $2r - 5 = 1 - r$
72. $5s - 13 = s - 1$
73. $0.2(a - 5) - 0.1(3a + 1) = 0$
74. $0.8(3a - 5) - 0.4(2a + 3) = 1.2$

75. $-\frac{4}{5}s = 2$

76. $-\frac{9}{8}s = 3$

77. $\frac{1}{2}(3y + 2) - \frac{5}{8} = \frac{3}{4}y$

78. $-\frac{3}{4}(4c - 3) + \frac{7}{8}c = \frac{19}{16}$

79. $8x + 3(2 - x) = 5x + 6$

80. $2(2a + 1) - 1 = 4a + 1$

81. $12 + 3(x - 4) - 21 = 5[5 - 4(4 - x)]$

82. $1 + 3[-2 + 6(4 - 2x)] = -(x + 3)$

83. $\frac{3 + p}{3} - 4p = 1 - \frac{p + 7}{2}$

84. $\frac{4 - t}{2} - \frac{3t}{5} = 2 + \frac{t + 1}{3}$

85. $5x + 10 = 5x$

86. $4(t - 2) - t = -(9 - 3t)$

87. $0.06(a + 200) + 0.1a = 172$

88. $0.03x + 0.05(6{,}000 - x) = 280$

89. $-4[p - (3 - p)] = 3(6p - 2)$

90. $2[5(4 - a) + 2(a - 1)] = 3 - a$

Solve each inequality. Graph the solution set and write it in interval notation.

91. $-3(a + 2) > 2(a + 1)$

92. $-4(y - 1) < y + 8$

93. $\frac{x - 7}{2} - \frac{x - 1}{5} \le -\frac{x}{4}$

94. $\frac{3a + 1}{3} - \frac{4 - 3a}{5} \le -\frac{1}{15}$

95. $5(2n + 2) - n > 3n - 3(1 - 2n)$

96. $-1 + 4(y - 1) + 2y \le \frac{1}{2}(12y - 30) + 15$

97. $0.4x + 0.4 \le 0.1x + 0.85$

98. $0.05 + 0.8x \le 0.5x - 0.7$

99. $\frac{1}{2}y + 2 \ge \frac{1}{3}y - 4$

100. $\frac{1}{4}x - \frac{1}{3} \le x + 2$

101. $7 < \frac{5}{3}a - 3$

102. $5 > \frac{7}{2}a - 9$

Look Alikes . . .

Simplify each expression and solve each equation.

103. a. $\frac{1}{2}(6x + 8) - 10 - \frac{2}{3}(6x - 9)$

b. $\frac{1}{2}(6x + 8) - 10 = -\frac{2}{3}(6x - 9)$

104. a. $6.31w + 9.22 + 5(7.21w - 1.13)$

b. $6.31w + 9.22 = 5(7.21w - 1.13)$

Solve the inequality in part a. Graph the solution set and write it in interval notation. Then use your answer to part a to determine the solution set for the inequality in part b. (No new work is necessary!) Graph the solution set and write it in interval notation.

105 a. $12x - 33.16 \le 5.84$

b. $12x - 33.16 > 5.84$

106. a. $-\frac{3}{4}x > -\frac{21}{32}$

b. $-\frac{3}{4}x \le -\frac{21}{32}$

APPLICATIONS

107. **Spring Tours.** A group of junior high students will be touring Washington, D.C. Their chaperons will have the \$1,810 cost of the tour reduced by \$15.50 for each student they supervise. How many students will a chaperon have to supervise so that his or her cost to take the tour will be \$1,500?

108. **Machining.** Each pass through a lumber plane shaves off 0.015 inch of thickness from a board. How many times must a board, originally 0.875 inch thick, be run through the planer if a board of thickness 0.74 inch is desired?

109. **Moving Expenses.** To help move his furniture, a man rents a truck for \$41.50 per day plus 35¢ per mile. If he has budgeted \$150 for transportation expenses, how many miles will he be able to drive the truck if the move takes 1 day?

110. **Computing Salaries.** A student working for a delivery company earns \$57.50 per day plus \$4.75 for each package she delivers. How many deliveries must she make each day to earn \$200 a day?

111. **Fencing Pens.** A man has 150 feet of fencing to build the two-part pen shown in the illustration. If one end is a square and the other a rectangle, find the outside dimensions of the pen.

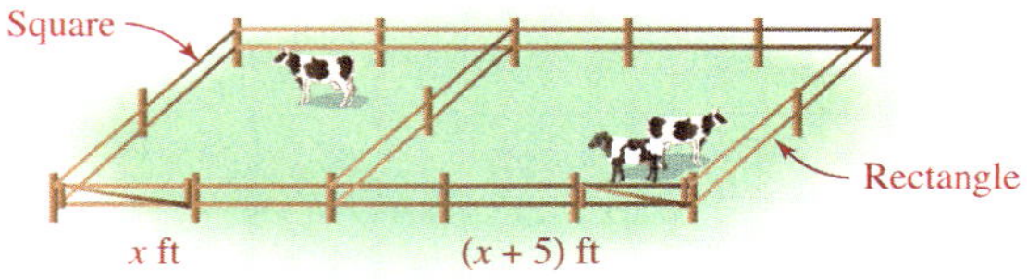

112. **Fencing Pastures.** A farmer has 624 feet of fencing to enclose a pasture. Because a river runs along one side, fencing will be needed on only three sides. Find the dimensions of the pasture if its length is double its width.

113. **from Campus to Careers**

Webmaster

One of the most important duties that a webmaster has is to monitor the web traffic of the site that he or she oversees. Suppose a counter for a news organization website has recorded 650,568,999 *views* for April 1 through April 29. How many views does the site need on the last day of the month (April 30) to average 22,000,000 daily visitors?

114. Averaging Grades. A student has scores of 70, 77, and 85 on three government exams. What score does she need on a fourth exam to give her an average of 80 or better?

115. Fundraising. A school PTA wants to rent a dunking tank for its annual school fundraising carnival. The cost is \$85.00 for the first 3 hours and then \$19.50 for each additional hour or part thereof. How long can the tank be rented if up to \$185 is budgeted for this expense?

116. Work Schedules. A student works two part-time jobs. He earns \$8 an hour for working at the college library and \$15 an hour for construction work. To save time for study, he limits his work to 25 hours a week. If he enjoys the work at the library more, how many hours can he work at the library and still earn at least \$300 a week?

117. Scheduling Equipment. An excavating company charges \$300 an hour for the use of a backhoe and \$500 an hour for the use of a bulldozer. (Part of an hour counts as a full hour.) The company employs one operator for 40 hours per week to operate the machinery. If the company wants to bring in at least \$18,500 each week from equipment rental, how many hours per week can it schedule the operator to use a backhoe?

118. Video Game Systems. A student who can afford to spend up to \$1,000 sees the ad shown in the illustration. If she decides to buy the video game system, find the greatest number of video games that she also can purchase. (Disregard sales tax.)

WRITING

119. Explain why the equation $x = x + 1$ doesn't have a real-number solution.

120. Explain what is wrong with the following statement:

When solving inequalities involving negative numbers, the direction of the inequality symbol must be reversed.

REVIEW

Simplify each expression. Write answers using only positive exponents.

121. $\left(\dfrac{t^3t^5t^{-6}}{t^2t^{-4}}\right)^{-3}$

122. $\left(\dfrac{a^{-2}b^3a^5b^{-2}}{a^6b^{-5}}\right)^{-4}$

CHALLENGE PROBLEMS

123. Find the value of k that makes 4 a solution of the following linear equation in x.

$$k + 3x - 6 = 3kx - k + 16$$

124. Solve: $0.75(x - 5) - \dfrac{4}{5} = \dfrac{1}{6}(3x + 1) + 3.2$

125. Consider the following "solution" of the inequality $\frac{1}{3} > \frac{1}{x}$ where it appears that the solution set is the interval $(3, \infty)$.

$$\frac{1}{3} > \frac{1}{x}$$

$$3x\left(\frac{1}{3}\right) > 3x\left(\frac{1}{x}\right)$$

$$x > 3$$

a. Show that $x = -1$ makes the original inequality true.

b. If $x = -1$ makes the original inequality true, there must be an error in the "solution." Where is it?

126. Medical Plans. A college provides its employees with a choice of the two medical plans shown in the following table. For what size hospital bills is Plan 2 better for the employee than Plan 1? (*Hint:* The cost to the employee includes both the deductible payment and the employee's coinsurance payment.)

Plan 1	Plan 2
Employee pays \$100	Employee pays \$200
Plan pays 70% of the rest	Plan pays 80% of the rest

SECTION 8.2

OBJECTIVES

1. Define relation, domain, and range.
2. Identify functions.
3. Use function notation.
4. Find the domain of a function.
5. Graph linear functions.
6. Write equations of linear functions.
7. Evaluate polynomial functions.

Functions

ARE YOU READY?

The following problems review some basic skills that are needed when working with functions.

1. Which of the ordered pairs in the following set have the same x-coordinate: $\{(3, 5), (2, 9), (6, -7), (-1, 5), (3, 0), (1, 9)\}$?
2. Substitute 8 for x in $y = \frac{1}{2}x + 3$ and find y.
3. What are the slope and the y-intercept of the line described by the equation $y = 3x - 8$?
4. What is the slope of a line perpendicular to the graph of the line that is described by the equation $y = \frac{2}{3}x + 1$?

The concept of a *function* is one of the most important ideas in all of mathematics. To introduce this topic, we will begin with a table that might be seen on television or printed in a newspaper.

1 Define Relation, Domain, and Range.

The following table shows the number of women serving in the U.S. House of Representatives for several recent sessions of Congress.

©Chris Howey/Shutterstock.com

Women in the U.S. House of Representatives							
Session of Congress	105th	106th	107th	108th	109th	110th	111th
Number of Women Representatives	54	56	59	59	68	71	78

Source: womenincongress.house.gov

We can display the data in the table as a set of ordered pairs, where the **first component** represents the session of Congress and the **second component** represents the number of women representatives serving during that session:

$$\{(105, 54), \quad (106, 56), \quad (107, 59), \quad (108, 59), \quad (109, 68), \quad (110, 71), \quad (111, 78)\}$$

Sets of ordered pairs like this are called **relations.** The set of all first components is called the **domain of the relation,** and the set of all second components is called the **range of the relation.** A relation may consist of a finite (countable) number of ordered pairs or an infinite (unlimited) number of ordered pairs.

EXAMPLE 1 Find the domain and range of the relation: $\{(3, 2), (5, -7), (-8, 2), (9, 0)\}$

Strategy We will identify the first components and the second components of the ordered pairs.

Why The set of all first components is the domain of the relation, and the set of all second components is the range.

Solution The first components of the ordered pairs are highlighted in red, and the second components are highlighted in blue: $\{(3, 2), (5, -7), (-8, 2), (9, 0)\}$. When listing the elements of the domain and range, they are usually written in increasing order, and if a value is repeated, it is listed only once.

The domain of the relation is $\{-8, 3, 5, 9\}$ and the range of the relation is $\{-7, 0, 2\}$.

Self Check 1 Find the domain and range of the relation:
$\{(5, 6), (-12, 4), (8, 6), (-6, -6), (5, 4)\}$

Now Try ▶ Problem 19

2 Identify Functions.

The relation in Example 1 was defined by a set of ordered pairs. Relations also can be defined using an **arrow** or **mapping diagram.** The data from the U.S. House of Representatives example is presented on the right in that form. Relations are also often defined using **two-column tables.**

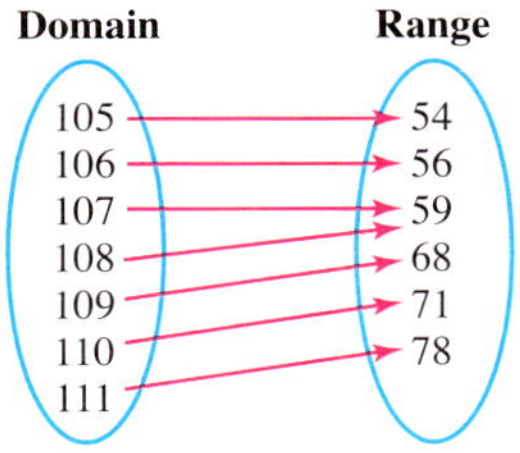

Notice that to each session of Congress, there corresponds exactly one number of women representatives. That is, to each member of the domain, there corresponds exactly one member of the range. Relations that have this characteristic are called *functions.*

Function, Domain, Range

A function is a set of ordered pairs (a relation) in which to each first component there corresponds exactly one second component. The set of first components is called the **domain of the function,** and the set of second components is called the **range of the function.**

Since we often will work with sets of ordered pairs of the form (x, y), it is helpful to define a function using the variables x and y.

y* Is a Function of *x

Given a relation in x and y, if to each value of x in the domain there corresponds exactly one value of y in the range, then y is said to be a function of x.

In the previous definition, since y depends on x, we call x the **independent variable** and y the **dependent variable.** The set of all possible values that can be used for the independent variable is the **domain** of the function, and the set of all values of the dependent variable is the **range** of the function.

EXAMPLE 2 Determine whether the relation defines y to be a function of x.

a.

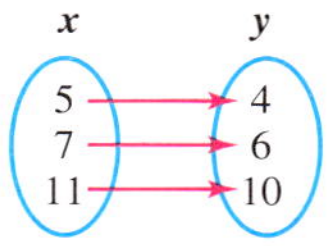

b.

x	y
8	2
1	4
8	3
9	9

c. $\{(-2, 3), (-1, 3), (0, 3), (1, 3)\}$

Strategy We will determine whether there is more than one value of y that corresponds to a single value of x.

Why If to any x-value there corresponds more than one y-value, then y is not a function of x.

Solution

a. The arrow diagram defines a function because to each value of x there corresponds exactly one value of y.

- $5 \rightarrow 4$ To the x-value 5, there corresponds exactly one y-value, 4.
- $7 \rightarrow 6$ To the x-value 7, there corresponds exactly one y-value, 6.
- $11 \rightarrow 10$ To the x-value 11, there corresponds exactly one y-value, 10.

Success Tip

Every function is, by definition, a relation. However, not every relation is a function, as we see in part (b).

b. The table does not define a function, because to the x-value 8 there corresponds more than one y-value.

- In the first row, to the x-value 8, there corresponds the y-value 2.
- In the third row, to the same x-value 8, there corresponds a different y-value, 3.

When the correspondence in the table is written as a set of ordered pairs, it is apparent that the relation does not define a function:

The same x-value

$\{(8, 2), (1, 4), (8, 3), (9, 9)\}$ This is not a function.

Different y-values

c. Since to each value of x, there corresponds exactly one value of y, the set of ordered pairs defines y to be a function of x.

- $(-2, 3)$ To the x-value -2, there corresponds exactly one y-value, 3.
- $(-1, 3)$ To the x-value -1, there corresponds exactly one y-value, 3.
- $(0, 3)$ To the x-value 0, there corresponds exactly one y-value, 3.
- $(1, 3)$ To the x-value 1, there corresponds exactly one y-value, 3.

In this case, the same y-value, 3, corresponds to each x-value.

The results from parts (b) and (c) illustrate an important fact: *Two different ordered pairs of a function can have the same y-value, but they cannot have the same x-value.*

Self Check 2 Determine whether the relation defines y to be a function of x.

a.

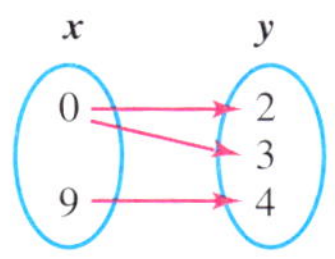

b.

x	y
-1	-60
0	55
3	0

c. $\{(4, -1), (9, 2), (16, 15), (4, 4)\}$

d. $\{(9,5), (10,5), (11,5), (12,5)\}$

Now Try Problems 23, 27, and 31

A function also can be defined by an equation. For example, $y = \frac{1}{2}x + 3$ sets up a rule in which to each value of x there corresponds exactly one value of y. To find the y-value (called an **output**) that corresponds to the x-value 4 (called an **input**), we substitute 4 for x and evaluate the right side of the equation.

The Language of Algebra

We also can think of a function as a rule or correspondence that assigns exactly one range value to each domain value.

$$y = \frac{1}{2}x + 3$$

$$y = \frac{1}{2}(4) + 3 \quad \text{Substitute 4 for x. The input is 4.}$$

$$y = 2 + 3 \quad \text{Do the multiplication: } \frac{1}{2}(4) = 2.$$

$$y = 5 \quad \text{Do the addition. This is the output.}$$

For the function defined by $y = \frac{1}{2}x + 3$, a y-value of 5 corresponds to an x-value of 4.

Not all equations define functions, as we will see in the next example.

EXAMPLE 3 Determine whether each equation defines y to be a function of x.

a. $y = 2x - 5$ **b.** $y^2 = x$

Strategy In each case, we will determine whether there is more than one value of y that corresponds to a single value of x.

Why If to any x-value there corresponds more than one y-value, then y is not a function of x.

Solution

a. To find the output value y that corresponds to an input value x, we multiply x by 2 and subtract 5. Since this arithmetic gives one result, to each value of x there corresponds exactly one value of y. Thus, $y = 2x - 5$ defines y to be a function of x.

b. The equation $y^2 = x$ does not define y to be a function of x, because more than one value of y corresponds to a single value of x. For example, if x is 16, then the equation becomes $y^2 = 16$ and y can be either 4 or -4. This is because $4^2 = 16$ and $(-4)^2 = 16$.

x	y
16	4
16	-4

Self Check 3 Determine whether each equation defines y to be a function of x.

a. $y = -x + 1$ **b.** $\left|\frac{1}{2}y\right| = x$

Now Try Problems 35 and 39

3 Use Function Notation.

A special notation is used to name functions that are defined by equations.

Function Notation The notation $y = f(x)$ indicates that y is a function of x.

If y is a function of x, the symbols y and $f(x)$ are interchangeable. In Example 3a, we saw that $y = 2x - 5$ defines y to be a function of x. To write this equation using function notation, we replace y with $f(x)$ to get $f(x) = 2x - 5$. This is read as "f of x is equal to $2x$ minus 5." In this notation, the parentheses do not indicate multiplication.

Caution

The symbol $f(x)$ denotes a function. It does not mean f times x.

This variable represents the input.

$$f(x) = 2x - 5$$

This is the name of the function.

This expression shows how to obtain an output from a given input.

Function notation provides a compact way of representing the output value that corresponds to some input value. For example, if $f(x) = 2x - 5$, the value that corresponds to $x = 6$ is represented by $f(6)$.

$f(x) = 2x - 5$

$f(6) = 2(6) - 5$ Substitute 6 for each x. (The input is 6.)

$= 12 - 5$ Do the multiplication.

$= 7$ Do the subtraction. This is the output.

Thus, $f(6) = 7$.

We read this result as "f of 6 equals 7." It means that when $x = 6$, $f(x)$ or y is 7. The output 7 is called a **function value.** The input, 6, and output, 7, also can be written as an ordered pair of the form (6, 7).

The Language of Algebra

Another way to read $f(6) = 7$ is to say "the value of f at 6 is 7."

To see why function notation is useful, we consider two sentences that ask you to do the same thing:

1. If $y = 2x - 5$, find the value of y when x is 6.
2. If $f(x) = 2x - 5$, find $f(6)$.

Statement 2, which uses $f(x)$ notation, is more compact.

The letter f used in the notation $f(x)$ represents the word *function.* However, other letters are often used to name functions. For example, in $g(x) = -5x + 15$, the name of the function is g and in $h(x) = x^3 - 7x + 9$, the name of the function is h.

Letters other than x can be used to represent the input of a function. Examples are $f(a) = 4a$, $g(t) = t^2 - 2t$, and $h(n) = -\frac{n^2 + 2}{2}$.

EXAMPLE 4 Let $f(x) = 4x + 3$ and $g(t) = t^2 - 2t$. Find: **a.** $f(3)$ **b.** $f(-1)$ **c.** $f(r + 1)$ **d.** $g(-2.4)$

Strategy We will substitute 3, -1, and $r + 1$ for each x in $f(x) = 4x + 3$ and evaluate the right side. We will substitute -2.4 for each t in $g(t) = t^2 - 2t$ and evaluate the right side.

Why Whatever appears within the parentheses in $f(\ )$ is to be substituted for each x in $f(x) = 4x + 3$. Whatever appears within the parentheses in $g(\ )$ is to be substituted for each t in $g(t) = t^2 - 2t$.

Solution

a. To find $f(3)$, we replace each x with 3.

$$\begin{aligned} f(x) &= 4x + 3 \\ f(3) &= 4(3) + 3 \\ &= 12 + 3 \\ &= 15 \end{aligned}$$

Thus, $f(3) = 15$ and the corresponding ordered pair is (3, 15).

b. To find $f(-1)$, we replace each x with -1:

$$\begin{aligned} f(x) &= 4x + 3 \\ f(-1) &= 4(-1) + 3 \\ &= -4 + 3 \\ &= -1 \end{aligned}$$

Thus, $f(-1) = -1$ and the corresponding ordered pair is $(-1, -1)$.

Notation

Note how function notation and ordered pair notation are related:

$$f(3) = 15$$

$x \uparrow\downarrow$ $y \uparrow\downarrow$

(3, 15)

c. To find $f(r + 1)$, we replace each x with $r + 1$:

$$\begin{aligned} f(x) &= 4x + 3 \\ f(r + 1) &= 4(r + 1) + 3 \\ &= 4r + 4 + 3 \\ &= 4r + 7 \end{aligned}$$

d. To find $g(-2.4)$, we replace each t with -2.4:

$$\begin{aligned} g(t) &= t^2 - 2t \\ g(-2.4) &= (-2.4)^2 - 2(-2.4) \\ &= 5.76 + 4.8 \\ &= 10.56 \end{aligned}$$

Read as "g of t."

The Language of Algebra

In part (d), if we use $g(t) = t^2 - 2t$ or $g(x) = x^2 - 2x$ or $g(c) = a^2 - 2a$, the results are the same: $g(-2.4) = 10.56$. The independent variable is really just a place holder, and for this reason, the letter that is used is often referred to as a **dummy variable**.

Self Check 4 If $f(x) = -2x - 1$ and $h(n) = -\frac{n^2 + 2n}{2}$, find: **a.** $f(2)$ **b.** $f(-3)$ **c.** $f(-t)$ **d.** $h(-0.6)$

Now Try Problems 47, 67, and 71

In the next example, we are asked to find the input of a function when we are given the corresponding output.

EXAMPLE 5 Let $f(x) = 5x + 4$. For what value of x is $f(x) = -26$?

Strategy We will substitute -26 for $f(x)$ and solve for x.

Why In the equation, there are two unknowns, x and $f(x)$. If we replace $f(x)$ with -26, we can use our equation-solving skills to find x.

Solution

$$f(x) = 5x + 4 \quad \text{This is the given function.}$$
$$-26 = 5x + 4 \quad \text{Substitute } -26 \text{ for } f(x).$$
$$-30 = 5x \quad \text{To isolate the variable term } 5x\text{, subtract 4 from both sides.}$$
$$-6 = x \quad \text{To isolate } x\text{, divide both sides by 5.}$$

We have found that $f(x) = -26$ when $x = -6$. To check this result, we can substitute -6 for x and verify that $f(-6) = -26$.

$$f(x) = 5x + 4 \quad \text{This is the given function.}$$
$$f(-6) = 5(-6) + 4 \quad \text{Substitute } -6 \text{ for } x.$$
$$= -30 + 4 \quad \text{Do the multiplication.}$$
$$= -26 \quad \text{This is the desired output.}$$

Self Check 5 Let $f(x) = 7x + 200$. For what value of x is $f(x) = 11$?

Now Try Problem 75

4 Find the Domain of a Function.

We can think of a function as a machine that takes some input x and turns it into some output $f(x)$, as shown in figure (a). The machine shown in figure (b) turns the input -6 into the output -11. The set of numbers that we put into the machine is the domain of the function, and the set of numbers that comes out is the range.

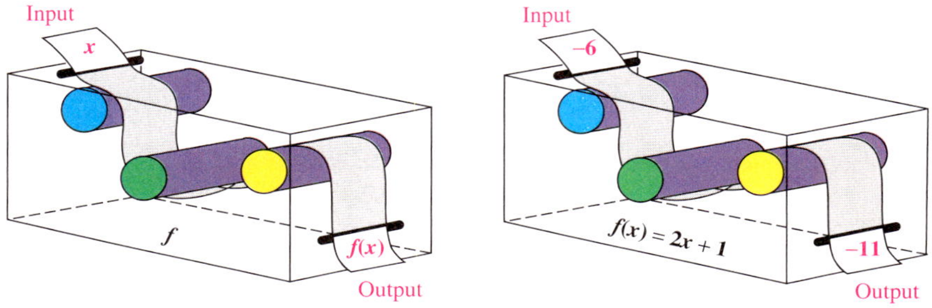

(a) (b)

Success Tip

To find the domain of a function defined by an equation, we must identify all the values of the input variable that produce a real-number output.

EXAMPLE 6 Find the domain of each function: **a.** $f(x) = 3x + 1$ **b.** $f(x) = \dfrac{x}{3x - 6}$

Strategy We will ask, "What values of x are permissible replacements for x in $3x + 1$ and $\dfrac{x}{3x - 6}$?"

Why These values of x form the domain of the function.

Solution

a. We will be able to evaluate $3x + 1$ for any value of x that is a real number. Thus, the domain of the function is *the set of real numbers,* which can be represented by the symbol $\mathbb{R}$.

b. Since division by 0 is undefined, we will not be able to evaluate $\frac{x}{3x - 6}$ for any number x that makes the denominator equal to 0. To find such x-values, we set the denominator equal to 0 and solve for x.

$$3x - 6 = 0$$
$$3x = 6 \quad \text{To isolate the variable term } 3x\text{, add 6 to both sides.}$$
$$x = 2 \quad \text{To isolate the variable, divide both sides by 3.}$$

We have found that 2 must be excluded from the domain of the function because it makes the denominator 0. However, all other real numbers are permissible replacements for x. Thus, the domain of the function *is the set of all real numbers except* 2.

The Language of Algebra

The last two letters in the word *domain* help us remember that it is the set of all *inputs* of a function.

Success Tip

Notice that to determine the domain of $f(x) = \dfrac{x}{3x - 6}$, we don't have to examine the numerator. It can be any value, including 0.

Self Check 6 Find the domain of each function: **a.** $f(x) = 2x - 6$

b. $f(x) = \dfrac{2}{-4x - 12}$

Now Try Problem 79

5 Graph Linear Functions.

In this course, we will study several important "families" of functions. We will begin here with the most basic family, *linear functions.*

Linear Functions

A linear function is a function that can be defined by an equation of the form $f(x) = mx + b$, where m and b are real numbers. The graph of a linear function is a straight line with slope m and y-intercept $(0, b)$.

Some examples of linear functions are:

$$f(x) = \frac{1}{2}x + 3, \qquad f(x) = -6x - 10, \qquad \text{and} \qquad f(x) = x$$

The input-output pairs that a linear function such as $f(x) = \frac{1}{2}x + 3$ generates can be plotted on a rectangular coordinate system to get the **graph of the function.** Since the symbols y and $f(x)$ are interchangeable, we can graph $f(x) = \frac{1}{2}x + 3$ as we would $y = \frac{1}{2}x + 3$, using the methods of Sections 3.2, 3.3, and 3.5.

To use the **point-plotting method** to graph $f(x) = \frac{1}{2}x + 3$, we begin by constructing a table of function values. To make the table, we select several values for x, and find the corresponding values of $f(x)$. Then we plot the ordered pairs and draw a straight line through the points to get the graph of the function, as shown on the next page.

To use the **slope–intercept method** to graph $f(x) = \frac{1}{2}x + 3$, we begin by identifying m and b in the equation. Then we plot the y-intercept and use the slope to determine a second point on the line. Finally, we draw a line through the points to obtain a graph like that shown on the next page.

$$f(x) = \frac{1}{2}x + 3$$

Slope (the $\frac{1}{2}$) — y-intercept: $(0, 3)$

To use the **intercept method** to graph $f(x) = \frac{1}{2}x + 3$, we begin by identifying b in the equation to determine the y-intercept. Then we let y, which in this case is $f(x)$, equal 0 and solve for x to find the x-intercept.

Find the y-intercept: Identify b

$$f(x) = \frac{1}{2}x + 3$$

The y-intercept is $(0, 3)$.

Find the x-intercept: Let f(x) = 0

$$f(x) = \frac{1}{2}x + 3$$

$$0 = \frac{1}{2}x + 3$$

$$-3 = \frac{1}{2}x$$

$$-6 = x$$

The x-intercept is $(-6, 0)$.

When we plot the intercepts $(0, 3)$ and $(-6, 0)$ and draw a straight line through them, we obtain a graph like that on the next page.

Notation

A table of function values is similar to a table of solutions, except that the second column is usually labeled $f(x)$ instead of y.

x	$f(x)$

x	y

$f(x) = \frac{1}{2}x + 3$

x	$f(x)$	
-2	2	$\rightarrow (-2, 2)$
0	3	$\rightarrow (0, 3)$
2	4	$\rightarrow (2, 4)$

Select x. Find f(x). Plot.

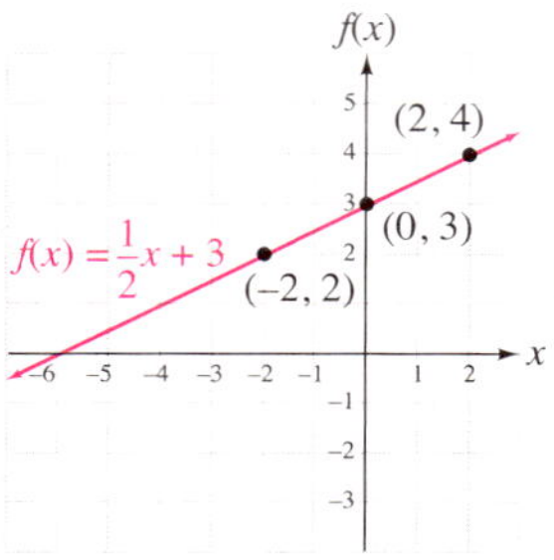

The vertical axis can be labeled f(x) or y.

The most basic linear function is $f(x) = x$. It is called the **identity function** because it assigns each real number to itself. The graph of the identity function is a line with slope 1 and y-intercept (0, 0), as shown below in figure (a).

A linear function defined by $f(x) = b$ is called a **constant function,** because for any input x, the output is the constant b. The graph of a constant function is a horizontal line. The graph of $f(x) = 2$ is shown below in figure (b).

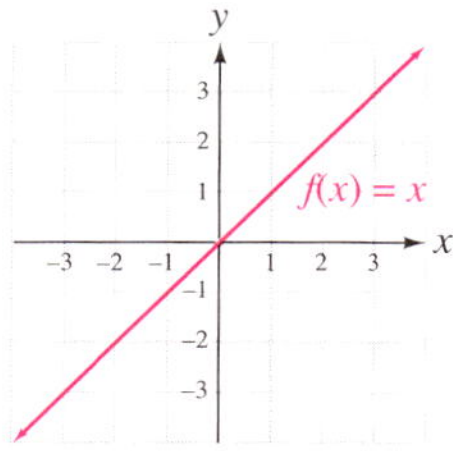

The identity function

(a)

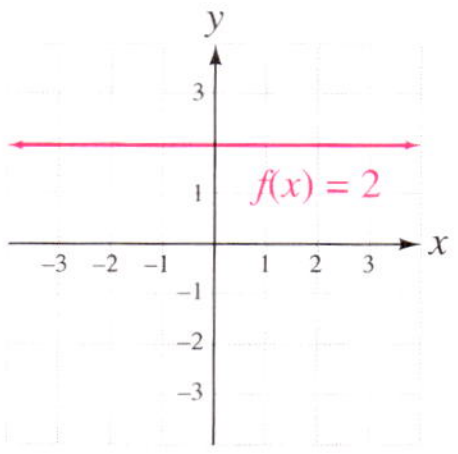

A constant function

(b)

6 Write Equations of Linear Functions.

The slope–intercept and point–slope forms for the equation of a line that we studied in Sections 3.5 and 3.6 can be adapted to write equations of linear functions.

EXAMPLE 7

a. Write an equation for the linear function whose graph has slope $-\frac{3}{4}$ and y-intercept (0, 2).

b. Write an equation for the linear function whose graph passes through (6, 4) with slope 5.

c. Write an equation for the linear function whose graph passes through $(-16, -12)$ and is perpendicular to the graph of $g(x) = -8x - 1$.

Strategy We will use either the slope–intercept or point–slope form to write each equation.

Why Writing equations of linear functions is similar to writing equations of linear equations in two variables, except that we replace y with $f(x)$.

Solution

a. If the slope is $-\frac{3}{4}$ and the y-intercept is (0, 2), then $m = -\frac{3}{4}$ and $b = 2$.

$f(x) = mx + b$ This is the slope–intercept form in which y is replaced with f(x).

$f(x) = -\frac{3}{4}x + 2$ Substitute $-\frac{3}{4}$ for m and 2 for b.

b. If the slope is 5 and the graph passes through (6, 4), then $m = 5$ and $(x_1, y_1) = (6, 4)$.

$f(x) - y_1 = m(x - x_1)$ This is the point–slope form in which y is replaced with f(x).

$f(x) - 4 = 5(x - 6)$ Substitute 4 for y_1, 5 for m, and 6 for x_1.

$f(x) - 4 = 5x - 30$ Distribute the multiplication by 5.

$f(x) = 5x - 26$ To isolate f(x), add 4 to both sides.

c. The slope of the line represented by $g(x) = -8x - 1$ is the coefficient of x: -8. Since the desired function is to have a graph that is perpendicular to the graph of $g(x) = -8x - 1$, its slope must be the negative reciprocal of -8, which is $\frac{1}{8}$.

$$f(x) - y_1 = m(x - x_1)$$ This is the point-slope form in which y is replaced with $f(x)$.

$$f(x) - (-12) = \frac{1}{8}[x - (-16)]$$ Substitute -12 for y_1, $\frac{1}{8}$ for m, and -16 for x_1.

$$f(x) + 12 = \frac{1}{8}(x + 16)$$ Simplify each side.

$$f(x) + 12 = \frac{1}{8}x + 2$$ Distribute the multiplication by $\frac{1}{8}$: $\frac{1}{8} \cdot 16 = 2$.

$$f(x) = \frac{1}{8}x - 10$$ To isolate $f(x)$, subtract 12 from both sides.

Self Check 7 **a.** Write an equation for the linear function whose graph has slope $-\frac{1}{5}$ and y-intercept $(0, 9)$. **b.** Write an equation for the linear function whose graph passes through $(7, 1)$ with slope 2. **c.** Write an equation for the linear function whose graph passes through $(-20, -4)$ and is perpendicular to the graph of $g(x) = -10x$.

Now Try Problems 95, 97, and 103

EXAMPLE 8

The Language of Algebra

The phrase **constant rate** indicates that a linear function should be used to model the situation.

Alzheimer's Disease. The graph on the right is from a recent Alzheimer's Association report. It shows how the number of people in the United States with Alzheimer's disease is expected to increase steadily at a constant rate. The report estimated there were 4,800,000 people with Alzheimer's in 2005 and 5,100,000 with the disease in 2010. (Source: Medill Reports)
a. Write the linear function that models this situation. **b.** Use the function to estimate the number of people who will have Alzheimer's disease in 2025, assuming the trend continues.

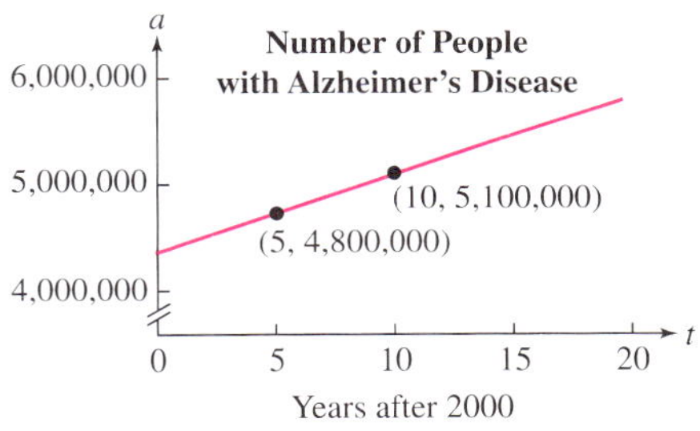

Strategy We will use the point–slope form to write a linear function.

Why We are given two points that lie on the graph of the function.

Solution Let a represent the approximate number of people in the U.S. with Alzheimer's disease and t represent the number of years after 2000. Two ordered pairs of the form (t, a) that lie on the graph of the function are:

- (5, 4,800,000) Since the year 2005 is 5 years after 2000, $t = 5$.
- (10, 5,100,00) Since the year 2010 is 10 years after 2000, $t = 10$.

Since we know two points that lie on the graph, we can write its equation. First, we find the slope of the line. If we write the slope formula using the variables t and a, and let $(t_1, a_1) = (5, 4{,}800{,}000)$ and $(t_2, a_2) = (10, 5{,}100{,}000)$, we have:

$$m = \frac{a_2 - a_1}{t_2 - t_1} = \frac{5{,}100{,}000 - 4{,}800{,}000}{10 - 5} = \frac{300{,}000}{5} = 60{,}000$$

The result indicates the number of people with Alzheimer's disease will increase at a rate of 60,000 per year.

a. To find the linear function, we substitute 60,000 for m, 5 for t_1, and 4,800,000 for a_1 in the point–slope form and simplify.

$a(t) - a_1 = m(t - t_1)$	This is the point–slope form using the variable t and replacing a with $a(t)$.
$a(t) - 4{,}800{,}000 = 60{,}000(t - 5)$	Substitute for m, t_1, and a_1.
$a(t) - 4{,}800{,}000 = 60{,}000t - 300{,}000$	Distribute the multiplication by 60,000.
$a(t) = 60{,}000t + 4{,}500{,}000$	To isolate $a(t)$, add 4,800,000 to both sides.

The approximate number of people with Alzheimer's disease t years after 2000 is given by $a(t) = 60{,}000t + 4{,}500{,}000$.

b. To estimate the number of people who will have Alzheimer's disease in 2025, which is 25 years after 2000, we find $a(25)$.

$a(t) = 60{,}000t + 4{,}500{,}000$	This is the linear function model.
$a(25) = 60{,}000(25) + 4{,}500{,}000$	Substitute 25 for t.
$= 1{,}500{,}000 + 4{,}500{,}000$	Do the multiplication.
$= 6{,}000{,}000$	Do the addition.

In 2025, there will be approximately 6,000,000 people in the U.S. with Alzheimer's disease.

Self Check 8 **Energy.** The world's annual energy consumption can be modeled by a linear function. In 2004, the world consumed about 430 quadrillion Btu. By the year 2006, that number had increased to about 446 quadrillion Btu. (Source: Energy Information Administration)
a. Let t be the number of years after 2000 and E be the amount of energy (in quadrillion Btu). Write a linear function $E(t)$ to model the situation. **b.** Predict the world's energy consumption in 2030, if the trend continues.

Now Try ▶ Problem 111

Using Your Calculator ▶ Evaluating Functions

We can use a graphing calculator to find function values. For example, suppose the linear function $f(c) = 12c - 18$ gives the daily income earned by a cosmetologist from serving c customers.

To find the income she earns for different numbers of customers, we first graph the income function $f(c) = 12c - 18$ as $y = 12x - 18$, using window settings of [0, 10] for x and [0, 100] for y to obtain figure (a). To find her income when she serves seven customers, we trace and move the cursor until the x-coordinate on the screen is nearly 7, as in figure (b). From the screen, we see that her income is about \$66.26.

To find her income when she serves nine customers, we trace and move the cursor until the x-coordinate is nearly 9, as in figure (c). From the screen, we see that her income is about \$90.51.

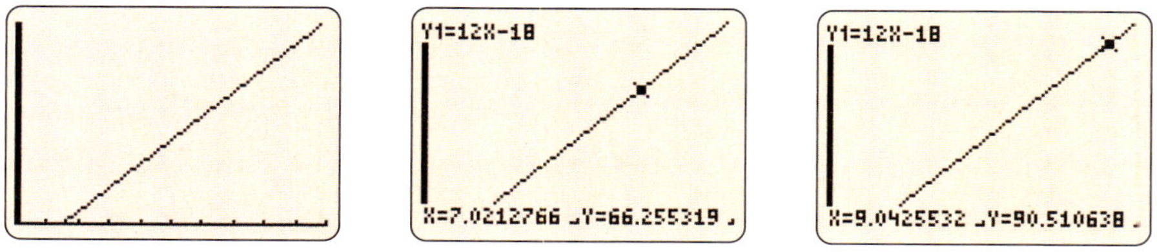

(a) (b) (c)

With some graphing calculator models, we can evaluate a function by entering function notation. To find $f(15)$, the income earned by the cosmetologist if she serves 15 customers, we use the following steps on a TI-84 Plus calculator.

With $f(c) = 12c - 18$ entered as $Y_1 = 12x - 18$, we call up the home screen by pressing 2nd QUIT. Then we enter VARS ▶ 1 ENTER. The symbolism Y_1 will be displayed. See figure (a). Next, we enter the input value 15 within parentheses, as shown in figure (b), and press ENTER. In figure (c) we see that $Y_1(15) = 162$. That is, $f(15) = 162$. The cosmetologist will earn \$162 if she serves 15 customers in one day.

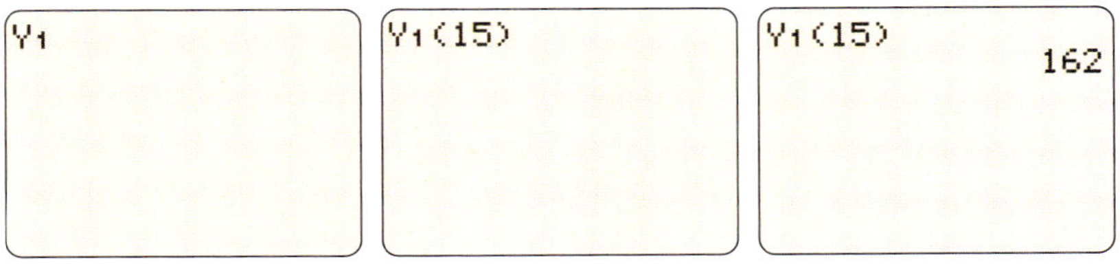

(a) (b) (c)

7 Evaluate Polynomial Functions.

We have seen that linear functions are defined by equations of the form $f(x) = mx + b$. Some examples of linear functions are

$$f(x) = 8x - 10, \qquad g(x) = -\frac{1}{2}x + 1, \qquad \text{and} \qquad h(x) = 5x$$

In each case, the right side of the equation is a polynomial. For this reason, linear functions are members of a larger class of functions known as *polynomial functions.*

Polynomial Functions

A polynomial function is a function whose equation is defined by a polynomial in one variable.

Another example of a polynomial function is $f(x) = x^2 + 6x - 8$. This is a second-degree polynomial function, called a **quadratic function.** Quadratic functions are of the form $f(x) = ax^2 + bx + c$, where $a \neq 0$.

An example of a third-degree polynomial function is $f(x) = x^3 - 3x^2 - 9x + 2$. Third-degree polynomial functions, also called **cubic functions,** are of the form $f(x) = ax^3 + bx^2 + cx + d$, where $a \neq 0$.

Polynomial functions can be used to model many real-life situations. If we are given a polynomial function model, we can learn more about the situation by evaluating the function at specific values. To **evaluate a polynomial function** at a specific value, we replace the variable in the defining equation with that value, called the *input.* Then we simplify the resulting expression to find the *output.*

Success Tip

Recall that a polynomial is a single term or the sum of terms in which all variables have whole-number exponents. No variable appears in the denominator.

EXAMPLE 9 **Packaging.** To make boxes, a manufacturer cuts equal-sized squares from each corner of 10 in. × 12 in. pieces of cardboard and then folds up the sides. The polynomial function $f(x) = 4x^3 - 44x^2 + 120x$ gives the volume (in cubic inches) of the resulting box when a square with sides x inches long is cut from each corner. Find the volume of a box if 3-inch squares are cut out.

Strategy We will find $f(3)$.

Why The notation $f(3)$ represents the volume of the box when 3-inch squares are cut out of the corners of the piece of cardboard.

Solution

$f(x) = 4x^3 - 44x^2 + 120x$ *This is the given function.*

$f(3) = 4(3)^3 - 44(3)^2 + 120(3)$ *Substitute 3 for each x. (The input is 3.)*

$= 4(27) - 44(9) + 120(3)$ *Evaluate the exponential expressions.*

$= 108 - 396 + 360$ *Do the multiplication.*

$= 72$ *The output is 72.*

If 3-inch squares are cut out of the corners, the resulting box has a volume of 72 in.^{3}.

Self Check 9 In Example 9, find the volume of a box if 2-inch squares are cut out.

Now Try Problem 119

SECTION 8.2 STUDY SET

VOCABULARY

Fill in the blanks.

1. A set of ordered pairs is called a ________. The set of all first components of the ordered pairs is called the ________ and the set of all second components is called the ______.
2. A ________ is a set of ordered pairs (a relation) in which to each first component there corresponds exactly one second component.
3. Given a relation in x and y, if to each value of x in the domain there corresponds exactly one value of y in the range, y is said to be a ________ of x. We call x the independent ________ and y the ________ variable.
4. For a function, the set of all possible values that can be used for the independent variable is called the ________. The set of all values of the dependent variable is called the ______.
5. A ______ function is a function that can be defined by an equation of the form $f(x) = mx + b$. A polynomial function is a function whose equation is defined by a polynomial in _____ variable.
6. We call $f(x) = x$ the ________ function because it assigns each real number to itself. We call $f(x) = 2$ a ________ function, because for any input x, the output is always 2.

CONCEPTS

7. **U.S. Recycling.** The following table gives the approximate number of aluminum cans (in billions) collected each year for the years 2000–2006.
   a. Display the data in the table as a relation, that is, as a set of ordered pairs.
   b. Find the domain and range of the relation.
   c. Use an arrow diagram to show how members of the range correspond to members of the domain.

Year	2000	2001	2002	2003	2004	2005	2006
Billions of aluminum cans	63	56	54	50	52	51	51

Source: Aluminum Association of America, U.S. Dept. of Commerce

8. For the given input, what value will the function machine output?

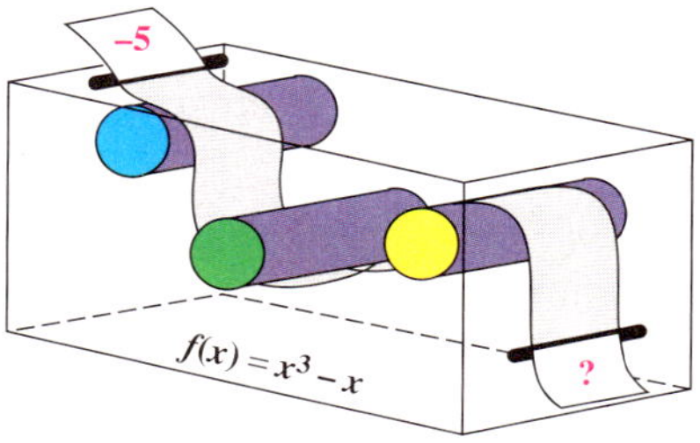

9. Explain why -4 is not in the domain of $f(x) = \frac{1}{x + 4}$.

10. Consider the linear function $y = -\frac{4}{5}x + 3$.
    a. What is the slope of its graph?
    b. What is the y-intercept of its graph?

11. Consider the linear function $f(x) = -6x - 4$.
    a. What is the y-intercept of its graph?
    b. What is the x-intercept of its graph?

12. The graphing calculator display shows a table of values for a function f. Fill in the blanks.

    $f(-1) =$ $f(3) =$

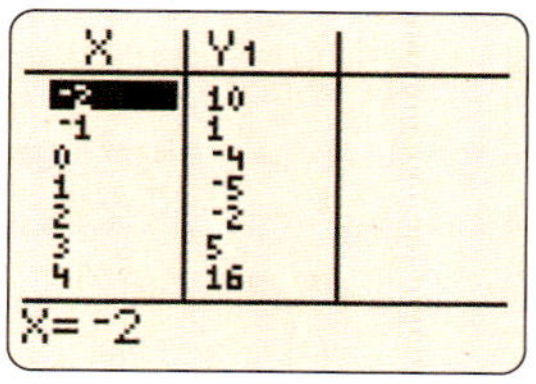

NOTATION

Fill in the blanks.

13. a. We read $f(x) = 5x - 6$ as "f x is equal to $5x$ minus 6."
    b. We read $g(t) = t + 9$ as "g t is equal to t plus 9."

14. This variable represents the ______.

$$f(x) = 2x - 5$$

This is the ______ of the function. Use this expression to find the ______.

15. Fill in the blank so that the following statements are equivalent:
- If $y = 5x + 1$, find the value of y when $x = 8$.
- If $f(x) = 5x + 1$, find ______.

16. If $f(2) = 7$, the input 2 and the output 7 can be written as the ordered pair (,).

17. When graphing $f(x) = -x + 5$, the vertical axis of the rectangular coordinate system can be labeled ______ or ______.

18. To write the slope–intercept form $y = mx + b$ and the point–slope form $y - y_1 = m(x - x_1)$ using function notation, we simply replace y with ______.

GUIDED PRACTICE

Find the domain and range of each relation. See Example 1.

19. $\{(7, -1), (-1, -11), (-5, 3), (8, -6)\}$

20. $\{(15, -3), (0, 0), (4, 6), (-3, -8)\}$

21. $\{(0, 1), (-23, 35), (7, 1)\}$

22. $\{(1, -12), (-6, 8), (5, 8), (0, 0), (1, 4)\}$

Determine whether the relation defines y to be a function of x. If it does not, find two ordered pairs where more than one value of y corresponds to a single value of x. See Example 2.

23.

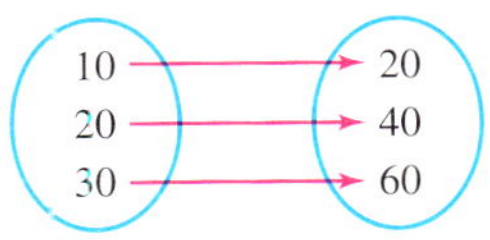

24.

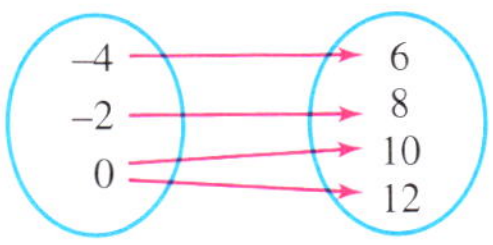

25.

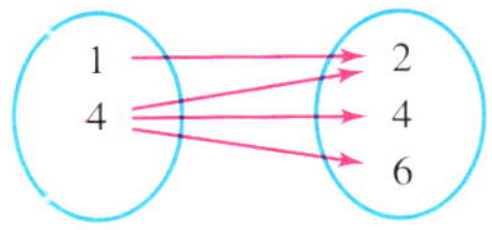

26.

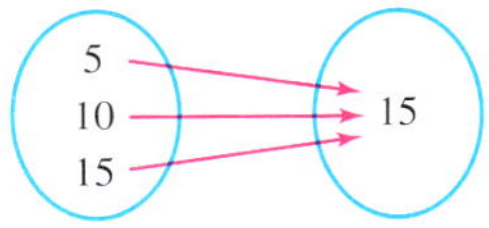

27. $\{(3, 4), (3, -4), (4, 3), (4, -3)\}$

28. $\{(-1, 1), (-3, 1), (-5, 1), (-7, 1), (-9, 1)\}$

29. $\{(-2, 7), (-1, 10), (0, 13), (1, 16)\}$

30. $\{(-2, 4), (-3, 8), (-3, 12), (-4, 16)\}$

31.

x	y
1	7
2	15
3	23
4	16
5	8

32.

x	y
30	2
30	4
30	6
30	8
30	10

33.

x	y
−4	6
−1	0
0	−3
2	4
−1	2

34.

x	y
1	1
2	1
3	1
4	1

Determine whether each equation defines y to be a function of x. If it does not, find two ordered pairs where more than one value of y corresponds to a single value of x. See Example 3.

35. $y = 2x + 3$

36. $y = 4x - 1$

37. $y = 4x^2$

38. $y^2 = x$

39. $y^4 = x$

40. $y = \frac{1}{x}$

41. $xy = 9$

42. $y = |x|$

43. $y = \frac{1}{x^2}$

44. $x + 1 = |y|$

45. $x = |y|$

46. $xy = -4$

Find f(3) and f(−1). See Example 4.

47. $f(x) = 3x$

48. $f(x) = -4x$

49. $f(x) = 2x - 3$

50. $f(x) = 3x - 5$

Find g(2) and g(3). See Example 4.

51. $g(x) = x^2 - 10$

52. $g(x) = x^2 - 2$

53. $g(x) = -x^3 + x$

54. $g(x) = x^3 - x$

55. $g(x) = (x + 1)^2$

56. $g(x) = (x - 3)^2$

57. $g(x) = 2x^2 - x + 1$

58. $g(x) = 5x^2 + 2x + 2$

Find h(5) and h(−2). See Example 4.

59. $h(x) = |x| + 2$

60. $h(x) = |x| - 5$

61. $h(x) = \frac{1}{x + 3}$

62. $h(x) = \frac{3}{x - 4}$

63. $h(x) = \frac{x}{x - 3}$

64. $h(x) = \frac{x}{x^2 + 2}$

65. $h(x) = \frac{x^2 + 2x - 35}{x^2 + 5x + 6}$

66. $h(x) = \frac{x^2 + x - 2}{x^2 - 5x}$

Complete each table. See Example 4.

67. $f(t) = |t - 2|$

t	$f(t)$
−1.7	
0.9	
5.4	

68. $f(r) = -2r^2 + 1$

Input	Output
−1.7	
0.9	
5.4	

69. $g(a) = a^3$

Input	Output
$-\frac{3}{4}$	
$\frac{1}{6}$	
$\frac{5}{2}$	

70. $g(b) = 2\left(-b - \frac{1}{4}\right)$

b	$g(b)$
$-\frac{3}{4}$	
$\frac{1}{6}$	
$\frac{5}{2}$	

Find g(w) and g(w + 1). See Example 4.

71. $g(x) = 2x$
72. $g(x) = -3x$
73. $g(x) = 3x - 5$
74. $g(x) = 2x - 7$

Let $f(x) = -2x + 5$. For what value of x does function f have the given value? See Example 5.

75. $f(x) = 5$
76. $f(x) = -7$

Let $f(x) = \frac{3}{2}x - 2$. For what value of x does function f have the given value? See Example 5.

77. $f(x) = -\frac{1}{2}$
78. $f(x) = \frac{2}{3}$

Find the domain of each function. See Example 6.

79. a. $h(x) = 3x + 6$
    b. $f(x) = \frac{1}{x - 4}$
80. a. $g(x) = |x - 7|$
    b. $f(x) = \frac{5}{x + 1}$
81. a. $f(x) = x^2$
    b. $s(x) = \frac{9}{2x + 1}$
82. a. $h(x) = x^3$
    b. $t(x) = \frac{15}{1 - 3x}$

Graph each function. See Objective 5.

83. $f(x) = 2x - 1$
84. $f(x) = -x + 2$
85. $f(x) = -\frac{3}{2}x - 3$
86. $f(x) = \frac{2}{3}x - 2$
87. $f(x) = x$
88. $f(x) = -x$
89. $f(x) = -4$
90. $f(x) = 2$
91. $g(x) = 0.75x$
92. $g(x) = -0.25x$
93. $s(x) = \frac{7}{8}x + 2$
94. $s(x) = -\frac{4}{5}x + 3$

Write an equation for a linear function whose graph has the given characteristics. See Example 7.

95. Slope 5, y-intercept $(0, -3)$
96. Slope 2, y-intercept $(0, 11)$
97. Slope $\frac{1}{5}$, passes through $(10, 1)$
98. Slope $\frac{1}{4}$, passes through $(8, 1)$
99. Passes through $(1, 7)$ and $(-2, 1)$
100. Passes through $(-2, 2)$ and $(2, -8)$
101. Passes through $(3, 0)$, parallel to the graph of $g(x) = -\frac{2}{3}x - 4$
102. Passes through $(2, 20)$, parallel to the graph of $g(x) = 8x + 1$
103. Passes through $(1, 2)$, perpendicular to the graph of $g(x) = -\frac{x}{6} + 1$
104. Passes through $(54, 0)$, perpendicular to the graph of $g(x) = 9x + 5$
105. Horizontal, passes through $(-8, 12)$
106. Horizontal, passes through $(9, -32)$

APPLICATIONS

107. from **Campus to Careers**

Webmaster

© Ronen/Shutterstock.com

According to data from *Netcraft Web Server Survey,* the total number of active websites available over the Internet (in millions) is approximated by the function $w(t) = 12.67t + 29.15$, where t is the number of years after 2005. Approximately how many active websites were there in 2011?

108. **Decongestants.** The temperature in degrees Celsius that is equivalent to a temperature in degrees Fahrenheit is given by the linear function $C(F) = \frac{5}{9}(F - 32)$. Refer to the label from a bottle of decongestant shown below. Use this function to find the low and high temperature extremes, in degrees Celsius, in which the bottle should be stored.

> **DIRECTIONS:** Adults and children 12 years of age and over: Two teaspoons every 4 hours. DO NOT EXCEED 6 DOSES IN A 24-HOUR PERIOD. Store at a controlled room temperature between 68°F and 77°F.

109. **Concessionaires.** A baseball club pays a vendor \$125 per game for selling bags of peanuts for \$4.75 each.
    a. Write a linear function that describes the profit the vendor makes for the baseball club during a game if she sells b bags of peanuts.
    b. Find the profit the baseball club will make if the vendor sells 110 bags of peanuts during a game.

110. **Home Construction.** In a proposal to some clients, a housing contractor listed the following costs:

Fees, permits, miscellaneous	\$12,000
Construction, per square foot	\$95

    a. Write a linear function that the clients could use to determine the cost of building a home having f square feet.
    b. Find the cost to build a home having 1,950 square feet.

111. **Nurses.** The demand for full-time registered nurses in the United States can be modeled by a linear function. In 2005, approximately 2,175,500 nurses were needed. By the year 2015, that number is expected to increase to about 2,586,500. (Source: National Center for Health Workforce Analysis)
    a. Let t be the number of years after 2000 and N be the number of full-time registered nurses needed in the U.S. Write a linear function $N(t)$ to model the demand for nurses.
    b. Use your answer to part a to predict the number of full-time registered nurses that will be needed in 2025, if the trend continues.

112. Wood Production. The total world wood production can be modeled by a linear function. In 1960, approximately 2,400 million cubic feet of wood were produced. Since then, the amount of increase has been approximately 25.5 million cubic feet per year. (Source: Earth Policy Institute)

a. Let t be the number of years after 1960 and W be the number of million cubic feet of wood produced. Write a linear function $W(t)$ to model the production of wood.

b. Use your answer to part a to estimate how many million cubic feet of wood the world produced in 2010.

113. Breathing Capacity. When fitness instructors prescribe exercise workouts for elderly patients, they must take into account age-related loss of lung function. Studies show that the percent of remaining breathing capacity for someone over 30 years old can be modeled by a linear function. (Source: alsearsmd.com)

a. At 35 years of age, approximately 90% of maximal breathing capacity remains and at 55 years of age, approximately 66% of maximal breathing capacity remains. Let a be the age of a patient and L be the percent of her maximal breathing capacity that remains. Write a linear function $L(a)$ to model this situation.

b. Use your answer to part a to estimate the percent of maximal breathing capacity that remains in an 80-year-old.

114. Chemical Reactions. When students mixed solutions of acetone and chloroform, they found that heat was generated. However, as time passed, the mixture cooled down. The graph shows data points of the form (time, temperature) taken by the students.

a. The linear function $T(t) = -\frac{t}{240} + 30$ models the relationship between the elapsed time t since the solutions were combined and the temperature $T(t)$ of the mixture. Graph the function.

b. Predict the temperature of the mixture immediately after the two solutions are combined.

c. Is $T(180)$ more or less than the temperature recorded by the students for $t = 300$?

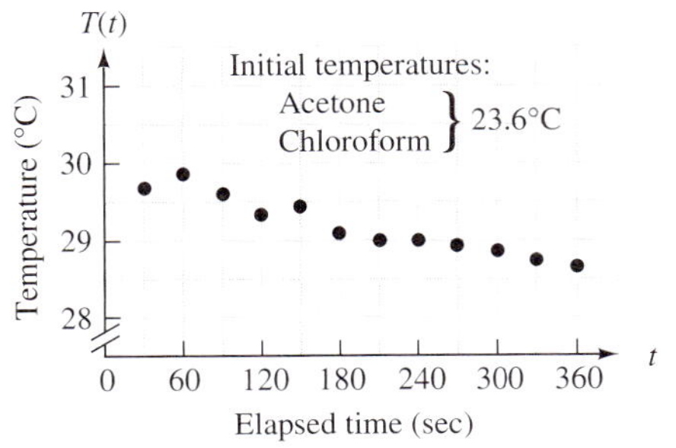

115. Taxes. The function

$$T(a) = 837.50 + 0.15(a - 8{,}375)$$

(where a is adjusted gross income) is a model of the instructions given on the first line of the following tax rate Schedule X.

a. Find $T(25{,}000)$ and interpret the result.

b. Write a function that models the second line on Schedule X.

Schedule X–Use if your filing status is Single | 2010

If your adjusted gross income is: Over —	But not over —	Your tax is	of the amount over —
$ 8,375	$34,000	**$ 837.50 + 15%**	**$ 8,375**
$34,000	$82,400	**$4,681.25 + 25%**	**$34,000**

116. Storage Tanks. The volume $V(r)$ of the gasoline storage tank, in cubic feet, is given by the polynomial function $V(r) = 4.2r^3 + 37.7r^2$, where r is the radius in feet of the cylindrical part of the tank. What is the capacity of the tank if its radius is 4 feet?

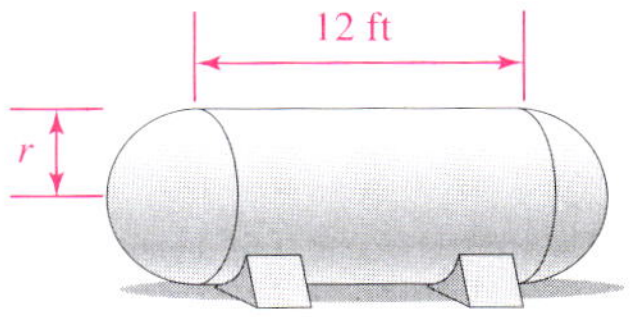

117. Roller Coasters. The polynomial function $f(x) = 0.001x^3 - 0.12x^2 + 3.6x + 10$ models the path of a portion of the track of a roller coaster. Use the function equation to find the height of the track for $x = 0$, 20, 40, and 60.

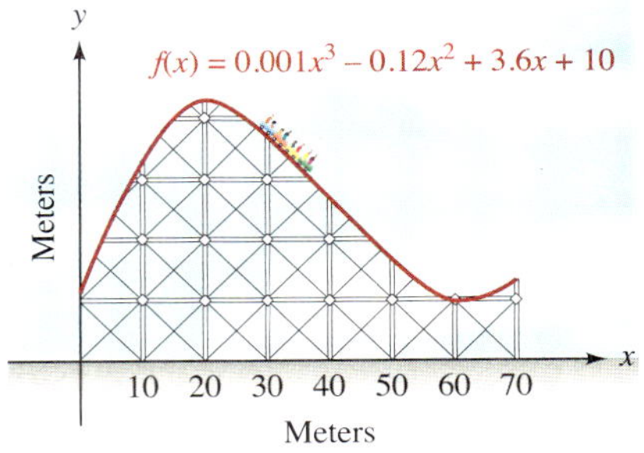

118. Customer Service. A software service hotline has found that on Mondays, the polynomial function $C(t) = -0.0625t^4 + t^3 - 6t^2 + 16t$ approximates the number of callers to the hotline at any one time. Here, t represents the time, in hours, since the hotline opened at 8:00 A.M. How many service technicians should be on duty on Mondays at noon if the company doesn't want any callers to the hotline waiting to be helped by a technician?

119. Rain Gutters. A rectangular sheet of metal will be used to make a rain gutter by bending up its sides, as shown. If the ends are covered, the capacity $f(x)$ of the gutter is a polynomial function of x: $f(x) = -240x^2 + 1{,}440x$. Find the capacity of the gutter if x is 3 inches.

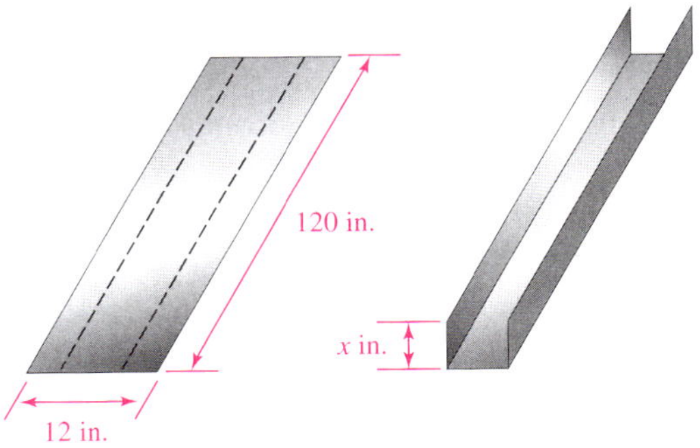

120. **Stopping Distances.** The number of feet that a car travels before stopping depends on the driver's reaction time and the braking distance. For one driver, the stopping distance $d(v)$, in feet, is given by the polynomial function $d(v) = 0.04v^2 + 0.9v$, where v is the velocity of the car in mph. Find the stopping distance at 60 mph.

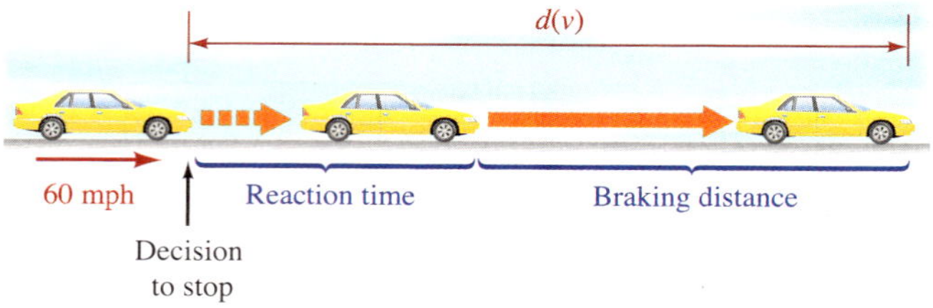

WRITING

121. Explain why 8 is not in the domain of the function $f(x) = \dfrac{5x - 7}{x - 8}$.

122. Explain why we can think of a function as a machine.

123. Consider the function defined by $y = 6x + 4$. Why do you think x is called the *independent* variable and y the *dependent* variable?

124. A website selling nutritional supplements contains the following sentence: "Health is a *function* of proper nutrition." Explain what this statement means.

REVIEW

Solve each equation. If the equation is an identity or a contradiction, so indicate.

125. $-2(t + 4) + 5t + 1 = 3(t - 4) + 7$

126. $\dfrac{3}{2}(a - 4) = 2(a - 3) - \dfrac{a}{2}$

CHALLENGE PROBLEMS

127. Let $f(x) = 4x + 6$, function g be defined by $\{(4, 6), (6, 8), (8, 4)\}$, and function h be defined by the arrow diagram below. Find $\dfrac{f(8) + g(8)}{h(8)}$.

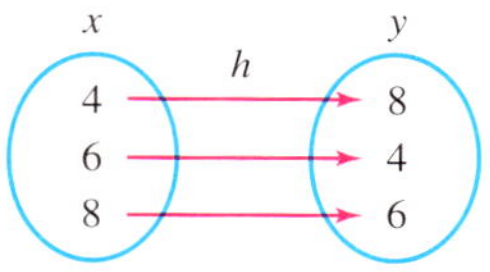

128. Find the domain of $f(x) = \dfrac{1}{5[9(x - 2) - 6(x - 3) + 3]}$.

SECTION 8.3 Graphs of Functions

OBJECTIVES

1. Find function values graphically.
2. Find the domain and range of a function graphically.
3. Graph nonlinear functions.
4. Translate graphs of functions.
5. Reflect graphs of functions.
6. Find function values and the domain and range of polynomial functions graphically.
7. Use the vertical line test.

ARE YOU READY?

The following problems review some basic skills that are needed when graphing functions.

1. Fill in the blanks: $f(3) = 9$ corresponds to the ordered pair (,).

2. If $f(x) = x^2$, find $f(-1)$ and $f(1)$.

3. If $f(x) = x^3$, find $f(-2)$ and $f(2)$.

4. If $f(x) = |x|$, find $f(-4)$ and $f(4)$.

Since a graph is often the best way to describe a function, we need to know how to construct and interpret their graphs.

1 Find Function Values Graphically.

From the graph of a function, we can determine function values. In general, the value of $f(a)$ is given by the y-coordinate of a point on the graph of function f with x-coordinate a.

EXAMPLE 1 Refer to the graph of function f in figure (a) on the next page.

a. Find $f(-3)$ **b.** Find the value of x for which $f(x) = -2$.

Strategy In each case, we will use the information provided by the function notation to locate a specific point on the graph and determine its x- and y-coordinates.

Why Once we locate the specific point, one of its coordinates will equal the value that we are asked to find.

Solution **a.** To find $f(-3)$, we need to find the y-coordinate of the point on the graph of f whose x-coordinate is -3. If we draw a vertical line through -3 on the x-axis, as shown in figure (b), the line intersects the graph of f at $(-3, 5)$. Therefore, 5 corresponds to -3, and it follows that $f(-3) = 5$.

b. To find the input value x that has an output value $f(x) = -2$, we draw a horizontal line through -2 on the y-axis, as shown in figure (c) and note that it intersects the graph of f at $(4, -2)$. Since -2 corresponds to 4, it follows that $f(x) = -2$ if $x = 4$.

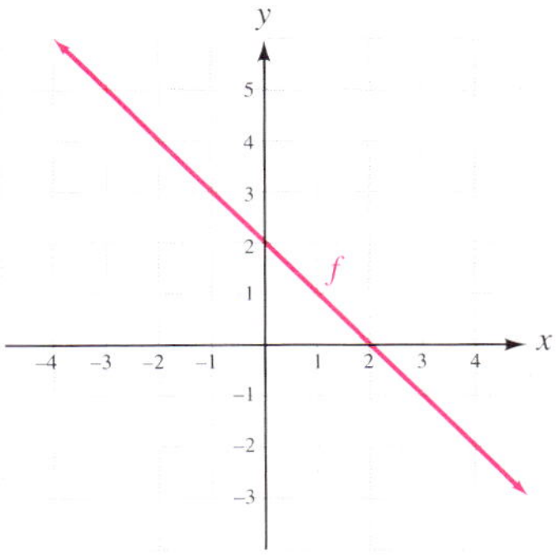

(a)

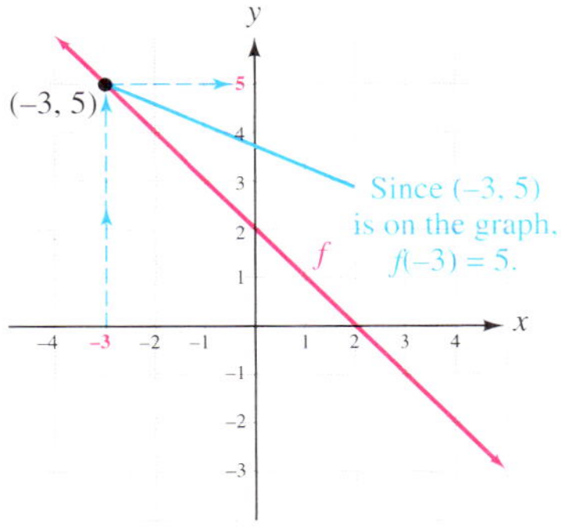

(b)

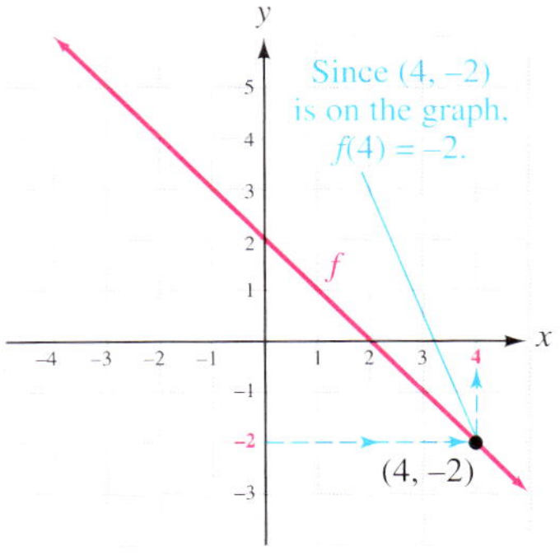

(c)

Self Check 1 Refer to the graph of function g on the right. **a.** Find $g(-3)$. **b.** Find the x-value for which $g(x) = 2$.

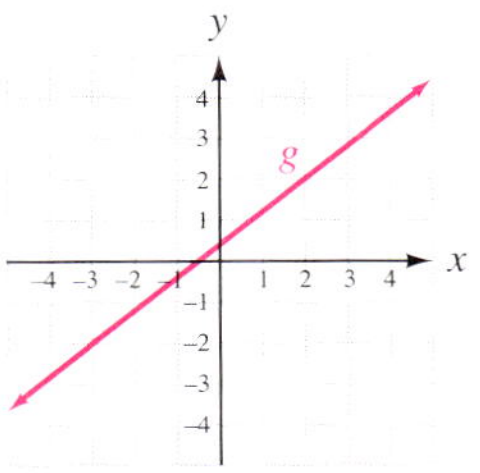

Now Try Problem 17

2 Find the Domain and Range of a Function Graphically.

We can find the domain and range of a function from its graph. For example, to find the domain of the linear function graphed in figure (a), we *project* the graph onto the x-axis. Because the graph of the function extends indefinitely to the left and to the right, the projection includes all the real numbers. Therefore, the domain of the function is the set of real numbers.

To find the range of the same linear function, we project the graph onto the y-axis, as shown in figure (b). Because the graph of the function extends indefinitely upward and downward, the projection includes all the real numbers. Therefore, the range of the function is the set of real numbers.

The Language of Algebra

Think of the **projection** of a graph on an axis as the "shadow" that the graph makes on the axis.

Project the graph onto the x-axis.

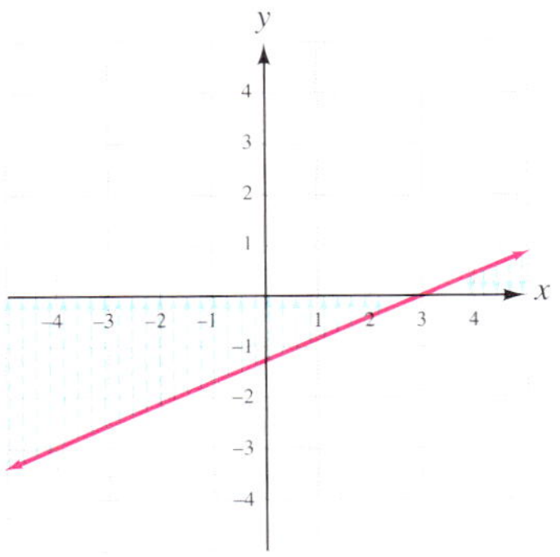

Domain: all real numbers

(a)

Project the graph onto the y-axis.

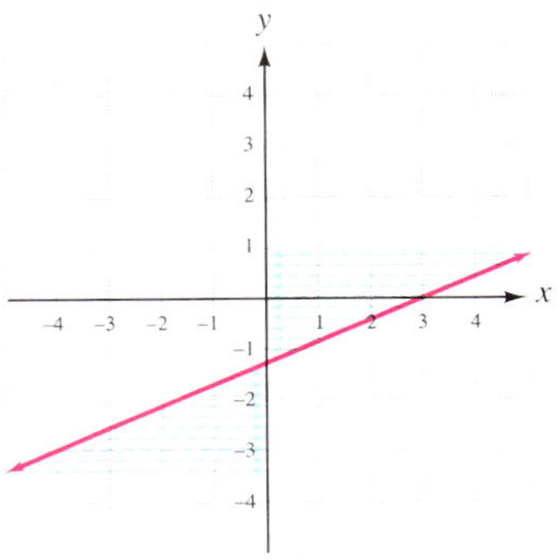

Range: all real numbers

(b)

3 Graph Nonlinear Functions.

We have seen that the graph of a linear function is a line. We will now consider several examples of **nonlinear functions** whose graphs are not lines. We will begin with $f(x) = x^2$, called the **squaring function.** We can graph this function using the **point-plotting method.**

EXAMPLE 2 Graph $f(x) = x^2$ and find its domain and range.

Strategy We will graph the function by creating a table of function values and plotting the corresponding ordered pairs.

Why After drawing a smooth curve through the plotted points, we will have the graph.

Solution To graph the function, we select several x-values and find the corresponding values of $f(x)$. For example, if we select -3 for x, we have

$$f(\mathbf{x}) = \mathbf{x}^2 \quad \text{This is the function to graph.}$$
$$f(\mathbf{-3}) = (\mathbf{-3})^2 \quad \text{Substitute } -3 \text{ for each } x.$$
$$= 9$$

Since $f(-3) = 9$, the ordered pair $(-3, 9)$ lies on the graph of f. In a similar manner, we find the corresponding values of $f(x)$ for six other x-values and list the ordered pairs in the table of values. Then we plot the points and draw a smooth curve through them to get the graph, called a **parabola.**

The Language of Algebra

The cuplike shape of a **parabola** has many real-life applications. For example, a satellite TV dish is called a **parabolic** dish.

$f(x) = x^2$

x	$f(x)$	
-3	9	$\rightarrow (-3, 9)$
-2	4	$\rightarrow (-2, 4)$
-1	1	$\rightarrow (-1, 1)$
0	0	$\rightarrow (0, 0)$
1	1	$\rightarrow (1, 1)$
2	4	$\rightarrow (2, 4)$
3	9	$\rightarrow (3, 9)$

↑ Select x. ↑ Find f(x). ↑ Plot the point.

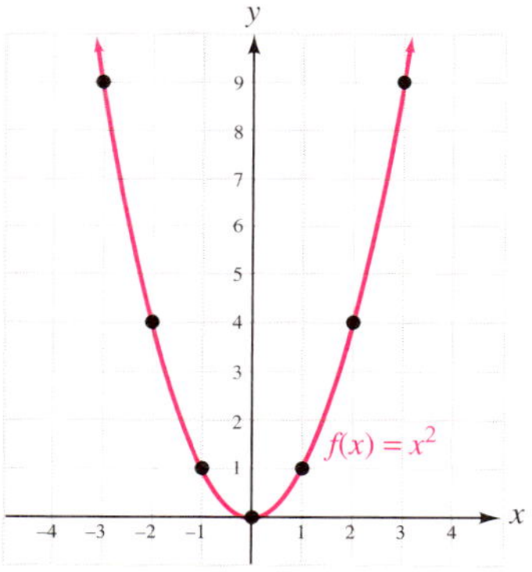

Because the graph extends indefinitely to the left and to the right, the projection of the graph onto the x-axis includes all the real numbers. See figure (a). This means that the domain of the squaring function is the set of real numbers.

The Language of Algebra

The set of **nonnegative real numbers** is the set of real numbers greater than or equal to 0.

Project the graph onto the x-axis.

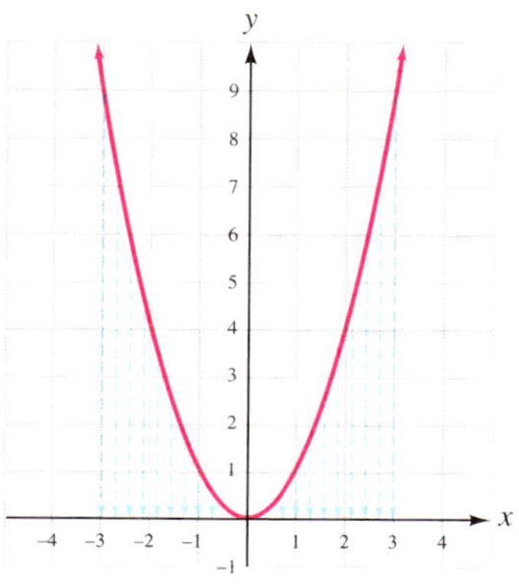

Domain: all real numbers

(a)

Project the graph onto the y-axis.

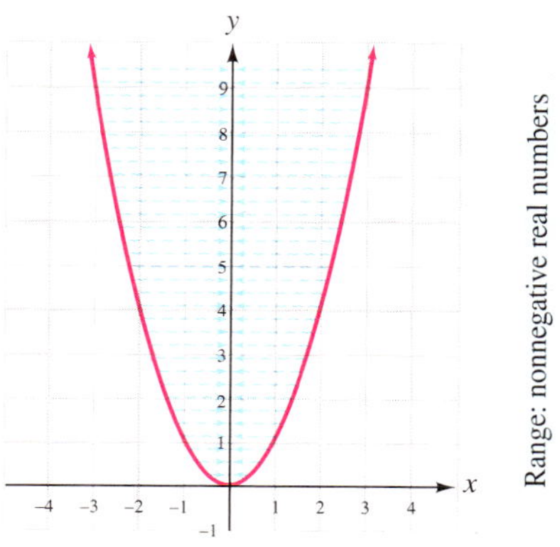

(b)

Because the graph extends upward indefinitely from the point (0, 0), the projection of the graph on the y-axis includes only positive real numbers and 0. See figure (b) on the previous page. This means that the range of the squaring function is the set of nonnegative real numbers.

Self Check 2 Graph $g(x) = x^2 - 2$ by plotting points. Then find its domain and range. Compare the graph to the graph of $f(x) = x^2$.

Now Try Problem 29

Another important nonlinear function is $f(x) = x^3$, called the **cubing function.**

EXAMPLE 3 Graph $f(x) = x^3$ and find its domain and range.

Strategy We will graph the function by creating a table of function values and plotting the corresponding ordered pairs.

Why After drawing a smooth curve through the plotted points, we will have the graph.

Solution To graph the function, we select several values for x and find the corresponding values of $f(x)$. For example, if we select -2 for x, we have

$$f(x) = x^3 \quad \text{This is the function to graph.}$$
$$f(-2) = (-2)^3 \quad \text{Substitute } -2 \text{ for each } x.$$
$$= -8$$

Since $f(-2) = -8$, the ordered pair $(-2, -8)$ lies on the graph of f. In a similar manner, we find the corresponding values of $f(x)$ for four other x-values and list the ordered pairs in the table. Then we plot the points and draw a smooth curve through them to get the graph.

Success Tip

To graph a linear function, it is recommended that you find three points on the line to draw its graph. Because the graphs of nonlinear functions are more complicated, more work is required. You need to find a sufficient number of points on the graph so that its entire shape is revealed. At least five well-chosen input values are recommended.

$f(x) = x^3$

x	$f(x)$	
-2	-8	$\rightarrow (-2, -8)$
-1	-1	$\rightarrow (-1, -1)$
0	0	$\rightarrow (0, 0)$
1	1	$\rightarrow (1, 1)$
2	8	$\rightarrow (2, 8)$

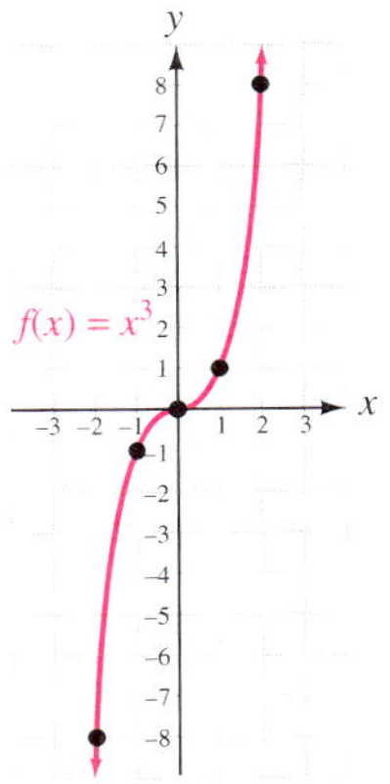

Because the graph of the function extends indefinitely to the left and to the right, the projection includes all the real numbers. Therefore, the domain of the cubing function is the set of real numbers.

Because the graph of the function extends indefinitely upward and downward, the projection includes all the real numbers. Therefore, the range of the cubing function is the set of real numbers.

Self Check 3 Graph $g(x) = x^3 + 1$ by plotting points. Then find its domain and range. Compare the graph to the graph of $f(x) = x^3$.

Now Try Problem 31

A third nonlinear function is $f(x) = |x|$, called the **absolute value function.**

EXAMPLE 4 Graph $f(x) = |x|$ and find its domain and range.

Strategy We will graph the function by creating a table of function values and plotting the corresponding ordered pairs.

Why After drawing straight lines through the plotted points, we will have the graph.

Solution To graph the function, we select several x-values and find the corresponding values for $f(x)$. For example, if we choose -3 for x, we have

$$f(x) = |x| \quad \text{This is the function to graph.}$$
$$f(-3) = |-3| \quad \text{Substitute } -3 \text{ for each x.}$$
$$= 3$$

Since $f(-3) = 3$, the ordered pair $(-3, 3)$ lies on the graph of f. In a similar manner, we find the corresponding values of $f(x)$ for six other x-values and list the ordered pairs in the table. Then we plot the points and connect them to get the following V-shaped graph.

Success Tip

To determine the entire shape of the graph, several positive and negative values, along with 0, were selected as x-values when constructing the table of values.

$f(x) = |x|$

x	$f(x)$	
-3	3	$\rightarrow (-3, 3)$
-2	2	$\rightarrow (-2, 2)$
-1	1	$\rightarrow (-1, 1)$
0	0	$\rightarrow (0, 0)$
1	1	$\rightarrow (1, 1)$
2	2	$\rightarrow (2, 2)$
3	3	$\rightarrow (3, 3)$

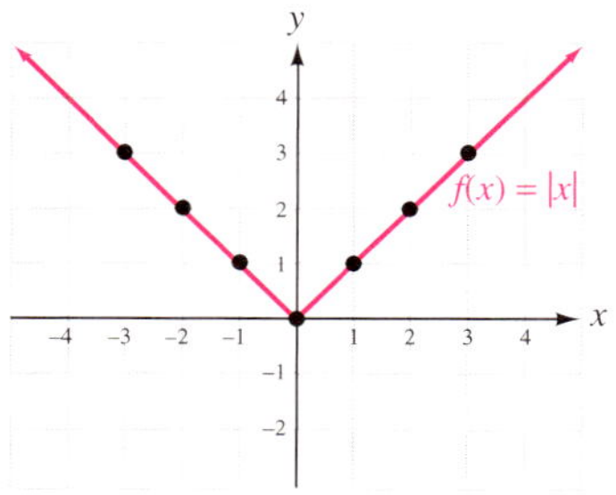

Because the graph extends indefinitely to the left and to the right, the projection of the graph onto the x-axis includes all the real numbers. Thus, the domain of the absolute value function is the set of real numbers.

Because the graph extends upward indefinitely from the point (0, 0), the projection of the graph on the y-axis includes only positive real numbers and 0. Thus, the range of the absolute value function is the set of nonnegative real numbers.

Self Check 4 Graph $g(x) = |x - 2|$ by plotting points. Then find its domain and range. Compare the graph to the graph of $f(x) = |x|$.

Now Try Problem 33

Using Your Calculator ▶ Graphing Functions

We can graph nonlinear functions with a graphing calculator. For example, to graph $f(x) = x^2$ in a standard window of $[-10, 10]$ for x and $[-10, 10]$ for y, we first press Y = . Then we enter the function by typing x ^ 2 (or x followed by x^2), and press the GRAPH key. We will obtain the graph shown in figure (a).

To graph $f(x) = x^3$, we enter the function by typing x ^ 3 and then press the GRAPH key to obtain the graph in figure (b). To graph $f(x) = |x|$, we enter the function by selecting abs(from the NUM option within the MATH menu, typing x, and pressing the GRAPH key to obtain the graph in figure (c).

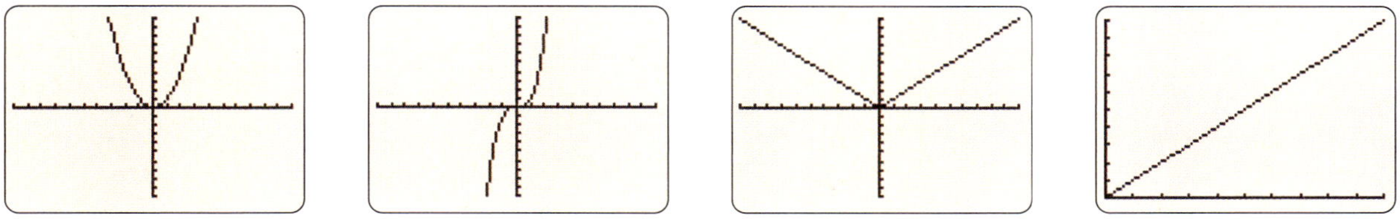

(a) (b) (c) (d)

When using a graphing calculator, we must be sure that the viewing window does not show a misleading graph. For example, if we graph $f(x) = |x|$ in the window [0, 10] for x and [0, 10] for y, we will obtain a misleading graph that looks like a line. See figure (d). This is not correct. The proper graph is the V-shaped graph shown in figure (c). One of the challenges of using graphing calculators is finding an appropriate viewing window.

4 Translate Graphs of Functions.

Examples 2, 3, and 4 and their Self Checks suggest that the graphs of different functions may be identical except for their positions in the coordinate plane. For example, the figure on the right shows the graph of $f(x) = x^2 + k$ for three different values of k. If $k = 0$, we get the graph of $f(x) = x^2$, shown in red. If $k = 3$, we get the graph of $f(x) = x^2 + 3$, shown in blue. Note that it is identical to the graph of $f(x) = x^2$ except that it is shifted 3 units upward. If $k = -4$, we get the graph of $f(x) = x^2 - 4$, shown in green. It is identical to the graph of $f(x) = x^2$ except that it is shifted 4 units downward. These shifts are called **vertical translations.**

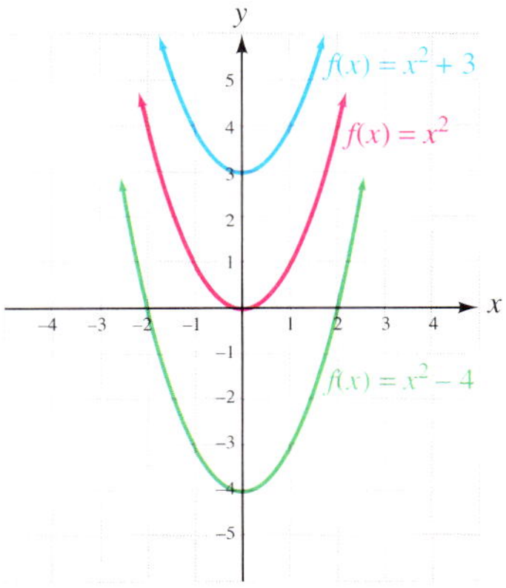

In general, we can make these observations.

Vertical Translations

If f is a function and k represents a positive number, then

- The graph of $y = f(x) + k$ is identical to the graph of $y = f(x)$ except that it is translated k units upward.
- The graph of $y = f(x) - k$ is identical to the graph of $y = f(x)$ except that it is translated k units downward.

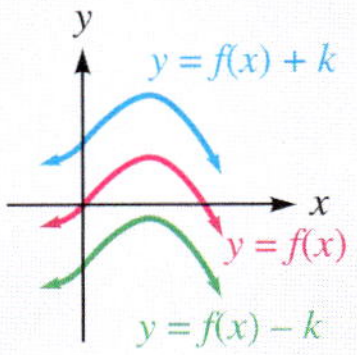

EXAMPLE 5 Graph: $g(x) = |x| + 2$

Strategy We will graph $g(x) = |x| + 2$ by translating (shifting) the graph of $f(x) = |x|$ upward 2 units.

Why The addition of 2 in $g(x) = |x| + 2$ causes a vertical shift of the graph of the absolute value function 2 units upward.

Solution Each point used to graph $f(x) = |x|$, which is shown in gray, is shifted 2 units upward to obtain the graph of $g(x) = |x| + 2$, which is shown in red. This is a vertical translation.

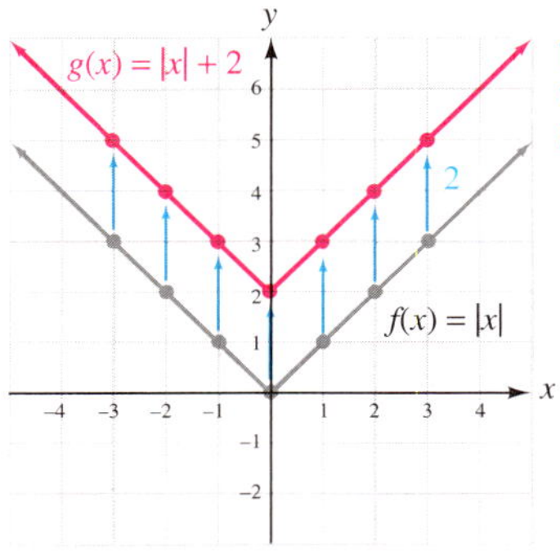

To graph $g(x) = |x| + 2$, translate each point on the graph of $f(x) = |x|$ up 2 units.

Self Check 5 Graph: $g(x) = |x| - 3$

Now Try Problem 37

The figure on the right shows the graph of $f(x) = (x + h)^2$ for three different values of h. If $h = 0$, we get the graph of $f(x) = x^2$, shown in red. The graph of $f(x) = (x - 3)^2$ shown in green is identical to the graph of $f(x) = x^2$ except that it is shifted 3 units to the right. The graph of $f(x) = (x + 2)^2$ shown in blue is identical to the graph of $f(x) = x^2$ except that it is shifted 2 units to the left. These shifts are called **horizontal translations.**

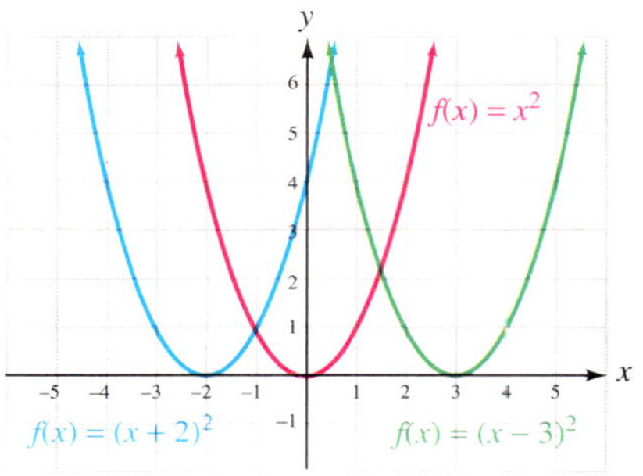

In general, we can make these observations.

Horizontal Translations

If f is a function and h is a positive number, then

- The graph of $y = f(x - h)$ is identical to the graph of $y = f(x)$ except that it is translated h units to the right.
- The graph of $y = f(x + h)$ is identical to the graph of $y = f(x)$ except that it is translated h units to the left.

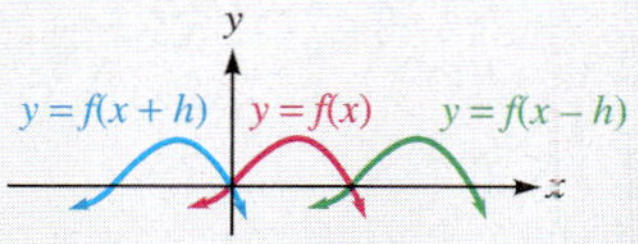

EXAMPLE 6 Graph: $g(x) = (x + 3)^3$

Success Tip

To determine the direction of the horizontal translation, find the value of x that makes the expression within the parentheses, in this case, $x + 3$, equal to 0. Since -3 makes $x + 3 = 0$, the translation is 3 units to the left.

Strategy We will graph $g(x) = (x + 3)^3$ by translating (shifting) the graph of $f(x) = x^3$ to the left 3 units.

Why The addition of 3 to x in $g(x) = (x + 3)^3$ causes a horizontal shift of the graph of the cubing function 3 units to the left.

To graph $g(x) = (x + 3)^3$, translate each point on the graph of $f(x) = x^3$ to the left 3 units.

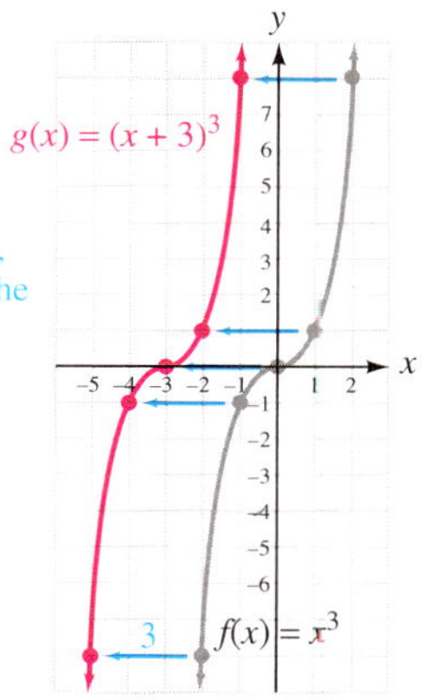

Solution Each point used to graph $f(x) = x^3$, which is shown in gray, is shifted 3 units to the left to obtain the graph of $g(x) = (x + 3)^3$, which is shown in red. This is a horizontal translation.

Self Check 6 Graph: $g(x) = (x - 2)^2$

Now Try ▶ Problem 47

The graphs of some functions involve horizontal and vertical translations.

EXAMPLE 7 Graph: $g(x) = (x - 5)^2 - 2$

Strategy To graph $g(x) = (x - 5)^2 - 2$, we will perform two translations by shifting the graph of $f(x) = x^2$ to the right 5 units and then 2 units downward.

Why The subtraction of 5 from x in $g(x) = (x - 5)^2 - 2$ causes a horizontal shift of the graph of the squaring function 5 units to the right. The subtraction of 2 causes a vertical shift of the graph 2 units downward.

Solution Each point used to graph $f(x) = x^2$, which is shown in gray, is shifted 5 units to the right and 2 units downward to obtain the graph of $g(x) = (x - 5)^2 - 2$, which is shown in red. This is a horizontal and vertical translation.

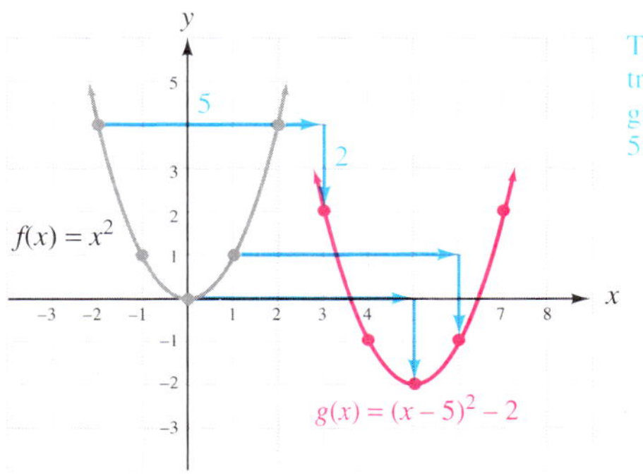

To graph $g(x) = (x - 5)^2 - 2$, translate each point on the graph of $f(x) = x^2$ to the right 5 units and then 2 units downward.

Self Check 7 Graph: $g(x) = |x + 2| - 3$

Now Try ▶ Problem 53

5 Reflect Graphs of Functions.

The following figure shows a table of values for $f(x) = x^2$ and for $g(x) = -x^2$. We note that for a given value of x, the corresponding y-value in the tables are opposites. When graphed, we see that the $-$ sign in $g(x) = -x^2$ has the effect of flipping the graph of $f(x) = x^2$ over the x-axis so that the parabola opens downward. We say that the graph of $g(x) = -x^2$ is a **reflection** of the graph of $f(x) = x^2$ about the x-axis.

$f(x) = x^2$

x	$f(x)$	
-2	4	⟶ $(-2, 4)$
-1	1	⟶ $(-1, 1)$
0	0	⟶ $(0, 0)$
1	1	⟶ $(1, 1)$
2	4	⟶ $(2, 4)$

$g(x) = -x^2$

x	$g(x)$	
-2	-4	⟶ $(-2, -4)$
-1	-1	⟶ $(-1, -1)$
0	0	⟶ $(0, 0)$
1	-1	⟶ $(1, -1)$
2	-4	⟶ $(2, -4)$

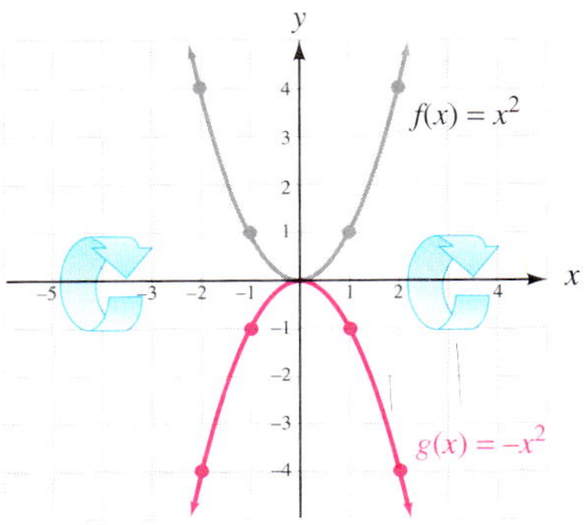

Reflection of a Graph The graph of $y = -f(x)$ is the graph of $y = f(x)$ reflected about the x-axis.

EXAMPLE 8 Graph: $g(x) = -x^3$

Strategy We will graph $g(x) = -x^3$ by reflecting the graph of $f(x) = x^3$ about the x-axis.

Why Because of the $-$ sign in $g(x) = -x^3$, the y-coordinate of each point on the graph of function g is the opposite of the y-coordinate of the corresponding point on the graph $f(x) = x^3$.

Solution To graph $g(x) = -x^3$, we use the graph of $f(x) = x^3$ from Example 3. First, we reflect the portion of the graph of $f(x) = x^3$ in quadrant I to quadrant IV, as shown. Then we reflect the portion of the graph of $f(x) = x^3$ in quadrant III to quadrant II.

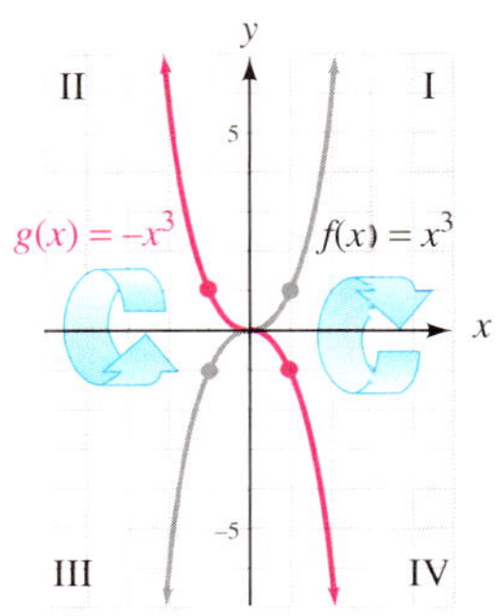

Self Check 8 Graph: $g(x) = -|x|$

Now Try Problem 63

6 Find Function Values and the Domain and Range of Polynomial Functions Graphically.

We have seen that the graphs of polynomial functions of degree 1, such as $f(x) = 4x - 1$ or $f(x) = \frac{1}{2}x + 3$, are straight lines and the graphs of polynomial functions of degree 2, such as $f(x) = x^2$ or $f(x) = x^2 - 2$, are parabolas. The graphs of polynomial functions of degree 3 or higher are often more complicated. Such graphs are always *smooth* and *continuous*. That is, they consist of only rounded curves with no sharp corners, and there are no breaks.

We can obtain important information about a polynomial function from its graph.

EXAMPLE 9 Refer to the graph of polynomial function $f(x) = x^3 - 3x^2 - 9x + 2$ in figure (a) on the next page. **a.** Find $f(2)$. **b.** Find any values of x for which $f(x) = -25$. **c.** Find the domain and range of f.

Strategy For parts a and b, we will use the information provided by the function notation to locate a specific point on the graph and determine its x- and y-coordinates. For part c, we will project the graph onto each axis.

Why Once we locate the specific point, one of its coordinates will equal the value that we are asked to find. Projecting the graph onto the x-axis gives the domain and projecting it onto the y-axis gives the range.

Solution

a. To find $f(2)$, we need to find the y-coordinate of the point on the graph of f whose x-coordinate is 2. If we draw a vertical line downward, from 2 on the x-axis, as shown in figure (b) on the next page, the line intersects the graph of f at $(2, -20)$. Therefore, -20 corresponds to 2, and it follows that $f(2) = -20$.

b. To find any input value x that has an output value $f(x) = -25$, we draw a horizontal line through -25 on the y-axis, as shown in figure (c) on the next page, and note that it intersects the graph of f at $(-3, -25)$ and $(3, -25)$. Therefore, $f(-3) = -25$ and $f(3) = -25$. It follows that the values of x for which $f(x) = -25$ are -3 and 3.

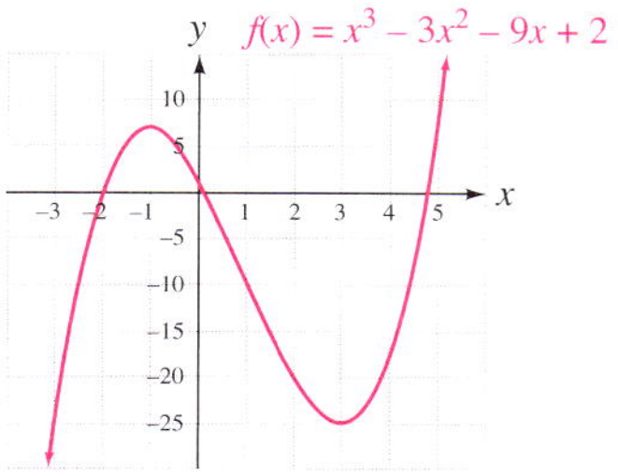

(a)

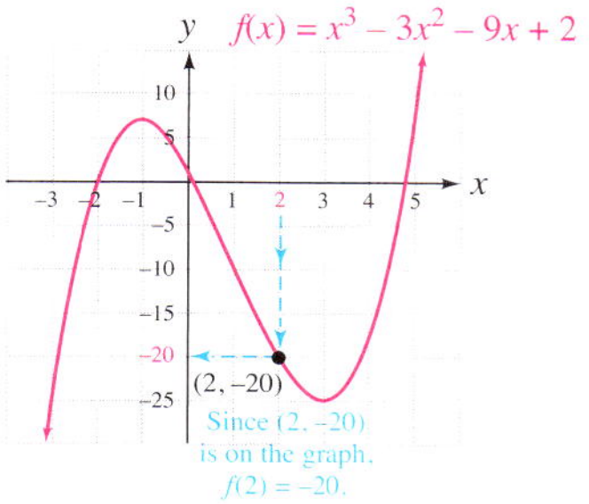

(b)

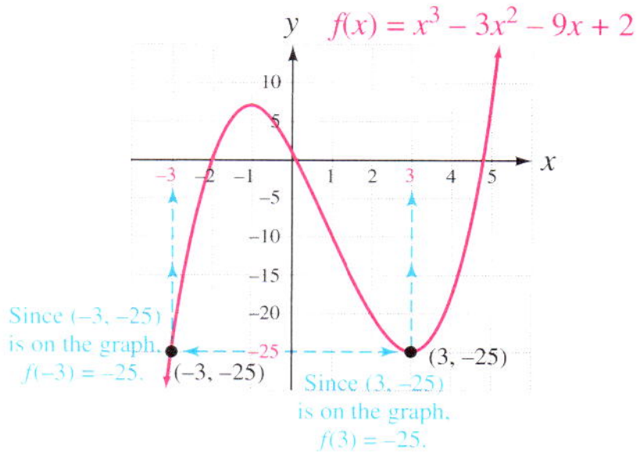

(c)

c. To find the domain of $f(x) = x^3 - 3x^2 - 9x + 2$, we project its graph onto the x-axis as shown in figure (d) below. Because the graph extends indefinitely to the left and right, the projection includes all real numbers. Therefore, the domain of the function is the set of real numbers, which can be written in interval notation as $(-\infty, \infty)$.

To determine the range of the same polynomial function, we project the graph onto the y-axis, as shown in figure (e) below. Because the graph of the function extends indefinitely upward and downward, the projection includes all real numbers. Therefore the range of the function is the set of real numbers, written $(-\infty, \infty)$.

Success Tip

The graphs of many polynomial functions of degree 3 and higher have "peaks" and "valleys" as shown here.

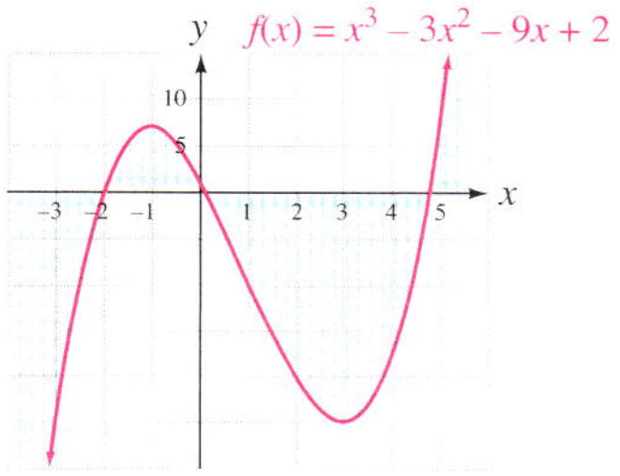

Domain: $(-\infty, \infty)$

(d)

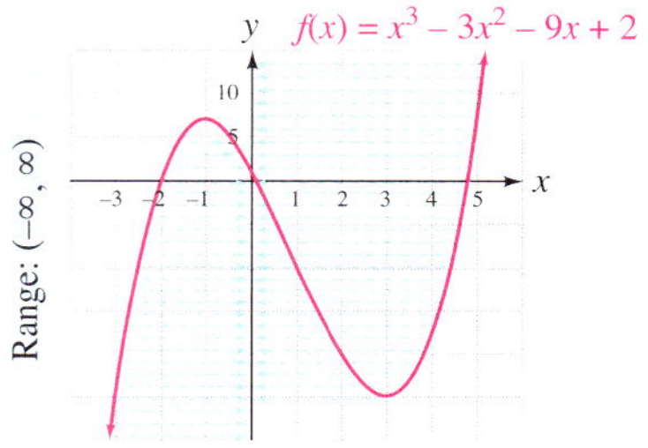

(e)

Self Check 9 Refer to the graph of function f in Example 9. Estimate each of the following: **a.** $f(1)$ **b.** Any x-values for which $f(x) = 0$

Now Try Problem 65

7 Use the Vertical Line Test.

Some graphs define functions and some do not. If a vertical line intersects a graph more than once, the graph does not represent a function, because to one value of x there would correspond more than one value of y.

The Vertical Line Test

If a vertical line intersects a graph in more than one point, the graph is not the graph of a function.

EXAMPLE 10 Determine whether the graph in figure (a) and figure (c) is the graph of a function.

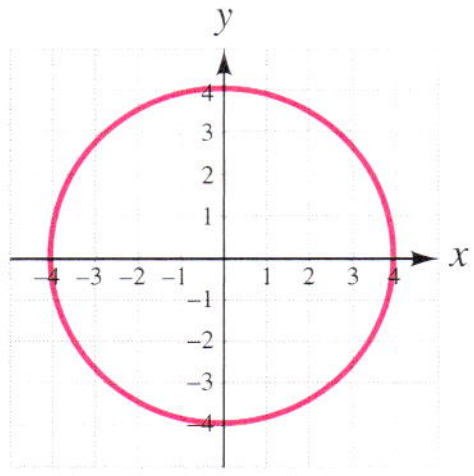

(a)

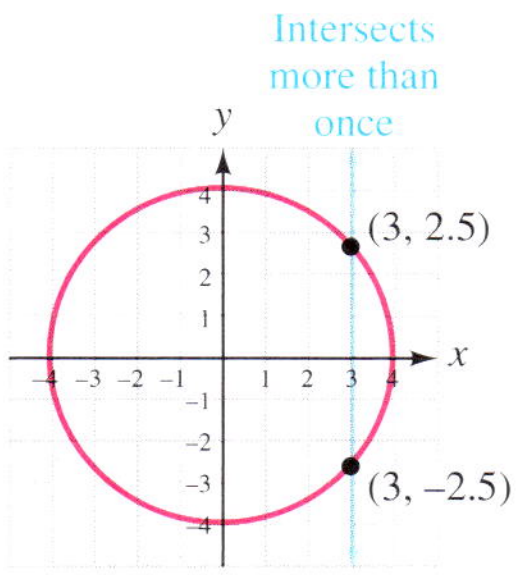

(b)

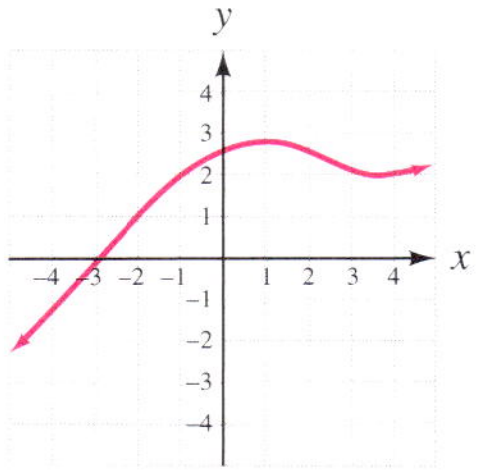

(c)

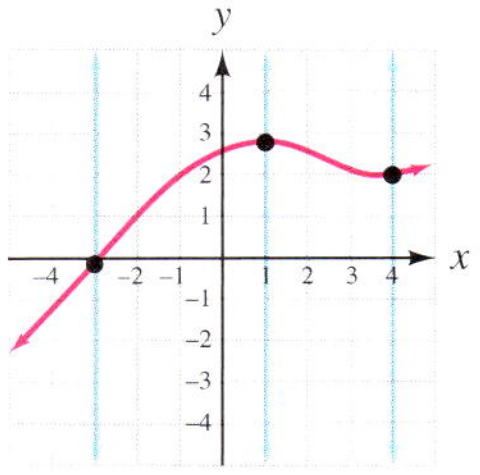

(d)

Strategy We will check to see whether any vertical lines intersect the graph more than once.

Why If any vertical line intersects the graph more than once, it is not the graph of a function.

Solution

a. Refer to figure (b) on the previous page. The graph shown in red is not the graph of a function because a vertical line intersects the graph more than once. The points of intersection of the graph and the vertical line indicate that two values of y (2.5 and -2.5) correspond to the x-value 3.

x	y
3	2.5
3	-2.5

b. Refer to figure (d) on the previous page. The graph shown in red is the graph of a function, because no vertical line intersects the graph more than once.

Self Check 10 Determine whether the following graph is the graph of a function.

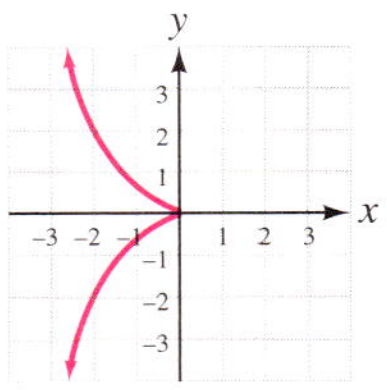

Now Try Problems 69 and 71

SECTION 8.3 STUDY SET

VOCABULARY

Fill in the blanks.

1. Functions whose graphs are not lines are called __________ functions.
2. The graph of $f(x) = x^2$ is a cuplike shape called a __________.
3. The set of __________ real numbers is the set of real numbers greater than or equal to 0.
4. A shift of the graph of a function upward or downward is called a vertical __________.

CONCEPTS

5. Graph each basic function by plotting points and give its name.
   a. $f(x) = x^2$
   b. $f(x) = x^3$
   c. $f(x) = |x|$
6. Complete each sentence about finding function values graphically.
   a. To find $f(-3)$, we find the y-coordinate of the point on the graph whose x-coordinate is ____.
   b. To find the value of x for which $f(x) = -2$, we find the x-coordinate of the point(s) on the graph whose y-coordinate is ____.
   c. Suppose for a function f that $f(5) = 9$. The corresponding ordered pair that will be on the graph of the function is (___ , ___).
7. Fill in the blank. The graph of $g(x) = -x^2$ is the __________ of the graph of $f(x) = x^2$ about the x-axis.
8. Fill in the blanks. The illustration shows the projection of the graph of function f on the ______. We see that the ______ of f is the set of real numbers less than or equal to 0.

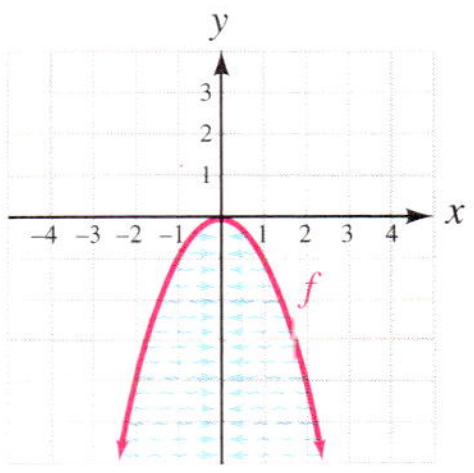

9. Consider the graph of the function f.
   a. Label each arrow in the illustration with the appropriate term: *domain* or *range*.
   b. Give the domain and range of f.

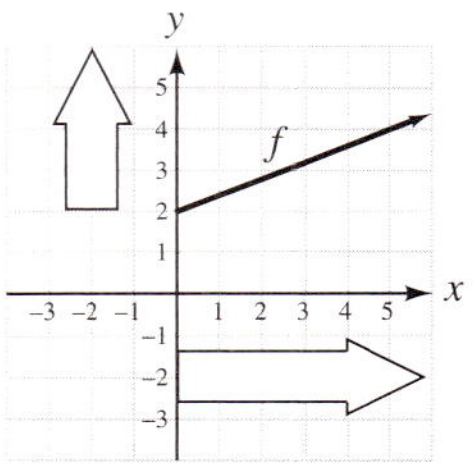

10. The graph of $f(x) = x^2 + k$ for three values of k is shown on the right. Find the value of k for
    a. the blue graph
    b. the red graph
    c. the green graph

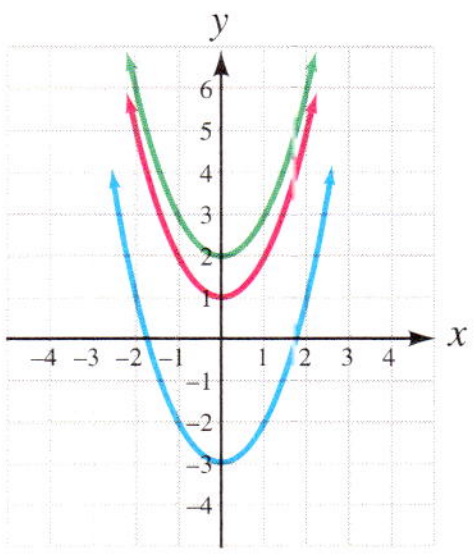

11. The graph of $f(x) = |x + h|$ for three values of h is shown. Find the value of h for
    a. the blue graph
    b. the red graph
    c. the green graph

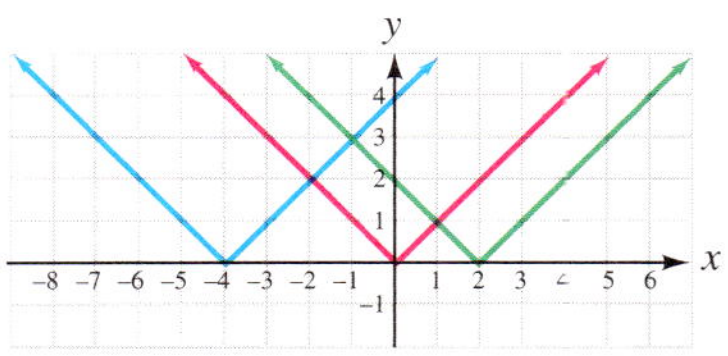

12. **a.** Translate each point plotted on the graph below to the left 5 units and then up 1 unit.

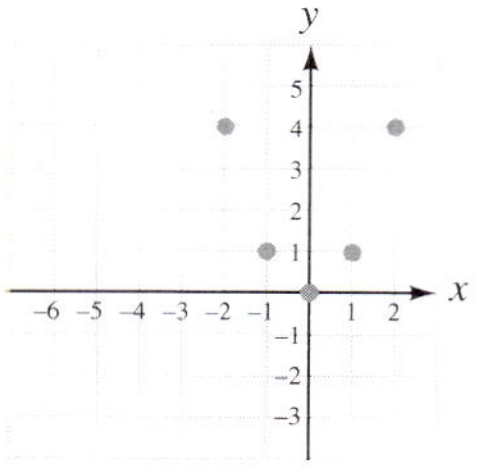

b. Translate each point plotted on the graph below to the right 4 units and then down 3 units.

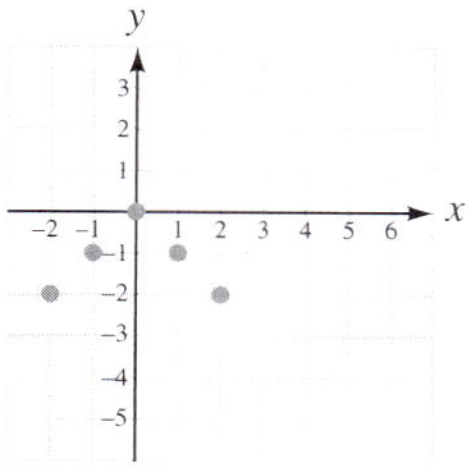

13. **a.** Give the coordinates of the points where the given vertical line intersects the graph in red.

b. Is this the graph of a function? Explain.

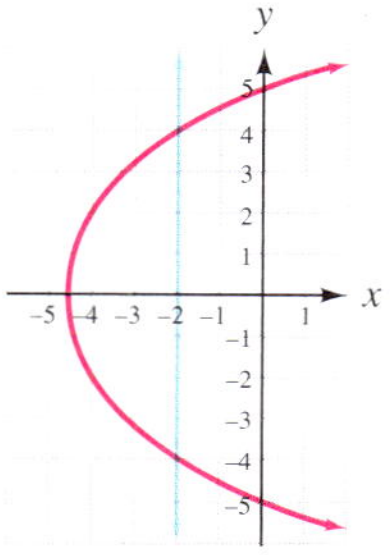

NOTATION

Fill in the blanks.

14. Fill in the blanks. The ________ line test: If a vertical line intersects a graph in more than one point, the graph is not the graph of a ________.

15. **a.** The graph of $f(x) = (x + 4)^3$ is the same as the graph of $f(x) = x^3$ except that it is shifted __ units to the _____.

b. The graph of $f(x) = x^3 + 4$ is the same as the graph of $f(x) = x^3$ except that it is shifted __ units _____.

16. **a.** The graph of $f(x) = |x| - 5$ is the same as the graph of $f(x) = |x|$ except that it is shifted __ units ______.

b. The graph of $f(x) = |x - 5|$ is the same as the graph of $f(x) = |x|$ except that it is shifted __ units to the ______.

GUIDED PRACTICE

Refer to the given graph to find each value. See Example 1.

17. **a.** $f(-2)$

b. $f(0)$

c. The value of x for which $f(x) = 4$

d. The value of x for which $f(x) = -2$

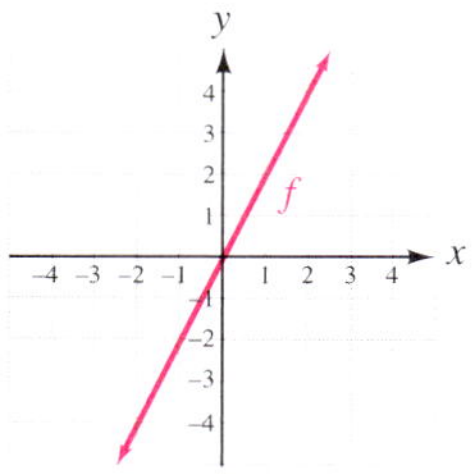

18. **a.** $s(-3)$

b. $s(3)$

c. The values of x for which $s(x) = 0$

d. The values of x for which $s(x) = 3$

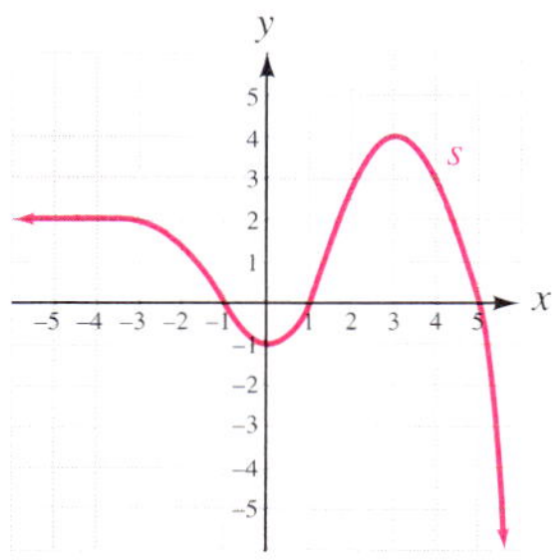

19. **a.** $g(-2)$

b. $g(0)$

c. The value of x for which $g(x) = 3$

d. The values of x for which $g(x) = -1$

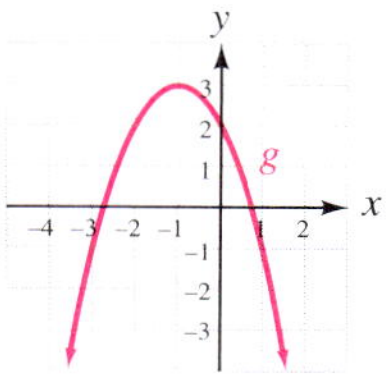

20. **a.** $h(-3)$

b. $h(4)$

c. The values of x for which $h(x) = 1$

d. The value of x for which $h(x) = 0$

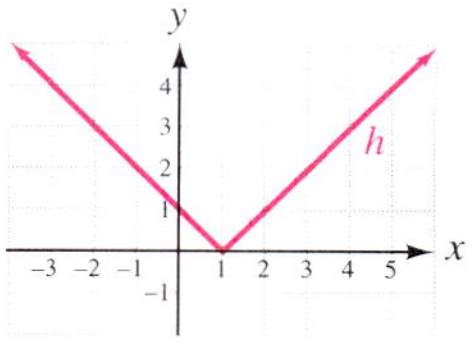

Find the domain and range of each function. See Objective 2 and Example 2.

21.

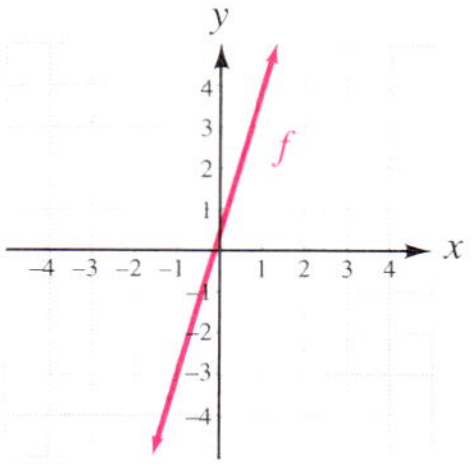

22.

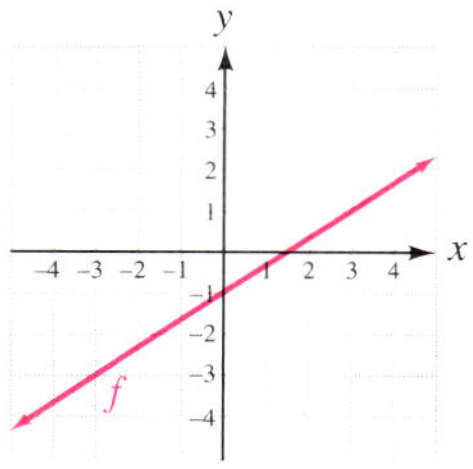

23.

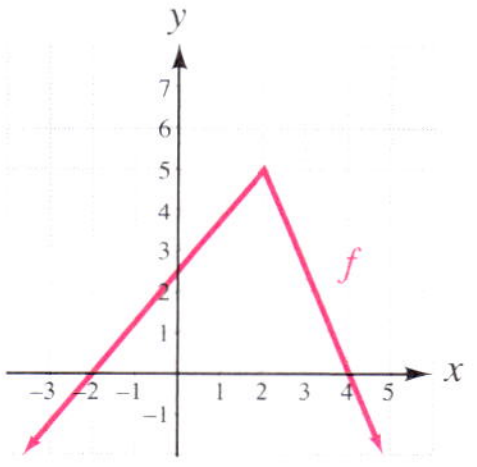

24.

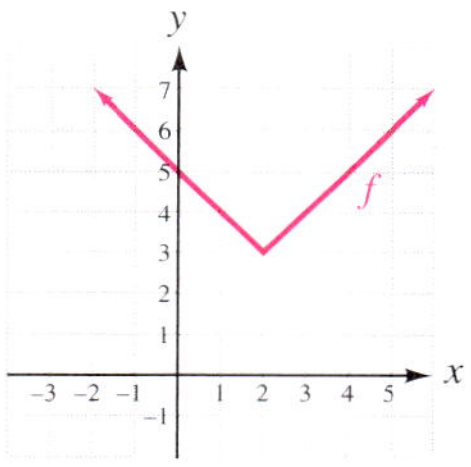

25.

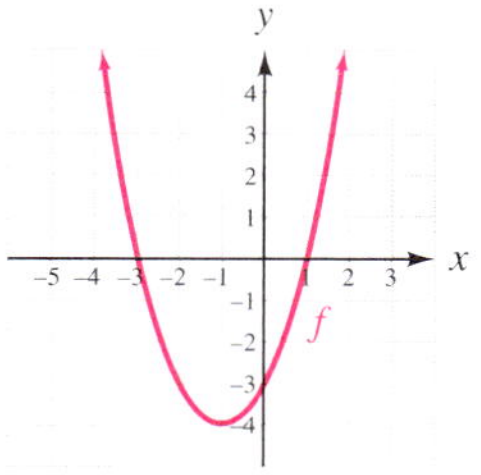

26.

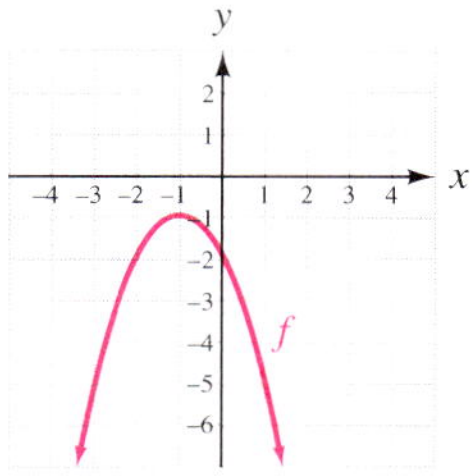

27.

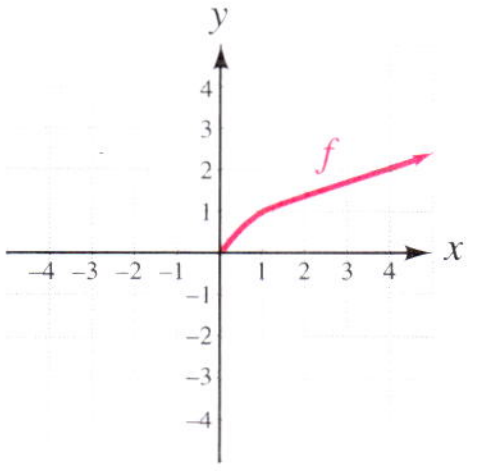

28.

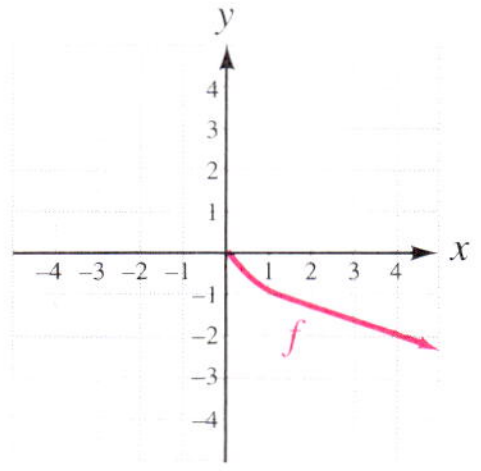

Graph each function by creating a table of function values and plotting points. Give the domain and range of the function. See Examples 2, 3, and 4.

29. $f(x) = x^2 + 2$

30. $f(x) = x^2 - 4$

31. $f(x) = x^3 - 3$

32. $f(x) = x^3 + 2$

33. $f(x) = |x - 1|$

34. $f(x) = |x + 4|$

35. $f(x) = (x + 4)^2$

36. $f(x) = (x - 1)^3$

For each of the following functions, first sketch the graph of its associated function, $f(x) = x^2$, $f(x) = x^3$, or $f(x) = |x|$. Then draw the graph of function g using a translation and give its domain and range. See Examples 5 and 6.

37. $g(x) = |x| - 2$

38. $g(x) = |x + 2|$

39. $g(x) = (x + 1)^3$

40. $g(x) = x^3 + 5$

41. $g(x) = x^2 - 3$

42. $g(x) = (x - 6)^2$

43. $g(x) = (x - 4)^3$

44. $g(x) = |x| + 1$

45. $g(x) = x^3 + 4$

46. $g(x) = x^2 - 5$

47. $g(x) = (x + 4)^2$

48. $g(x) = (x - 1)^3$

For each of the following functions, first sketch the graph of its associated function, $f(x) = x^2$, $f(x) = x^3$, or $f(x) = |x|$. Then draw the graph of function g using translations and/or a reflection. See Examples 7 and 8.

49. $g(x) = |x - 2| - 1$

50. $g(x) = (x + 2)^2 - 1$

51. $g(x) = (x + 1)^3 - 2$

52. $g(x) = |x + 4| + 3$

53. $g(x) = (x - 2)^2 + 4$

54. $g(x) = (x - 4)^2 + 3$

55. $g(x) = |x + 3| + 5$

56. $g(x) = (x - 3)^2 - 2$

57. $g(x) = -x^3$

58. $g(x) = -|x|$

59. $g(x) = -x^2$

60. $g(x) = -(x + 1)^2$

61. $g(x) = -|x + 5|$

62. $g(x) = -(x + 4)^3$

63. $g(x) = -x^2 + 3$

64. $g(x) = -|x| - 4$

Use the graph of the function to find each of the following. See Example 9.

65. Find:
   a. $f(1)$
   b. $f(-3)$
   c. The values of x for which $f(x) = 0$
   d. The domain and range of f

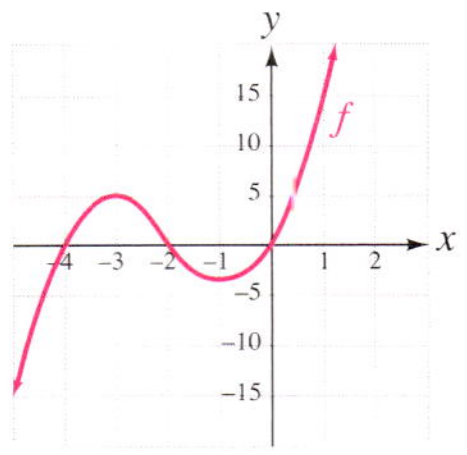

66. Find:
   a. $f(-1)$
   b. $f(0)$
   c. The values of x for which $f(x) = 3$
   d. The domain and range of f

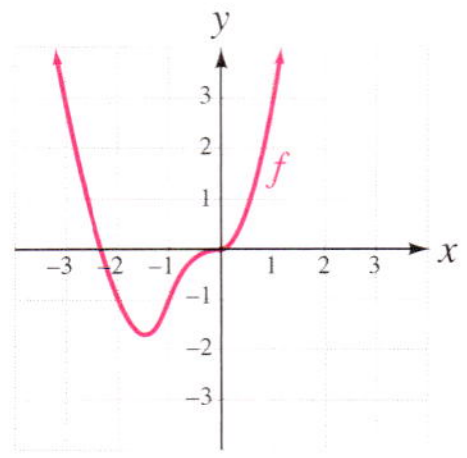

67. Find:
   a. $g(1)$
   b. $g(-4)$
   c. The values of x for which $g(x) = 4$
   d. The domain and range of g

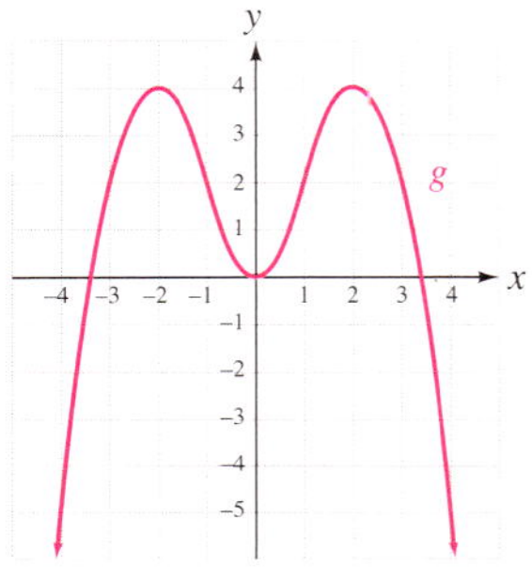

68. Find:
   a. $h(0.5)$
   b. $h(-0.5)$
   c. The values of x for which $h(x) = -1.5$
   d. The domain and range of h

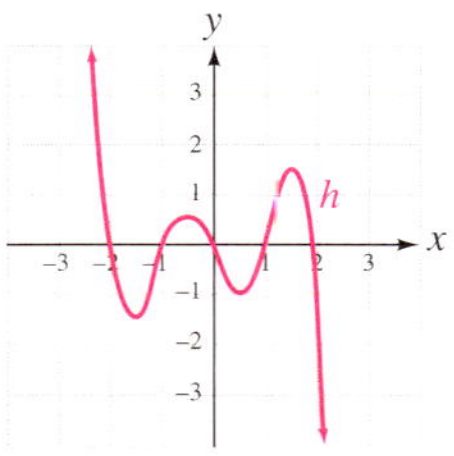

Determine whether each graph is the graph of a function. If it is not, find two ordered pairs where more than one value of y corresponds to a single value of x. See Example 10.

69.

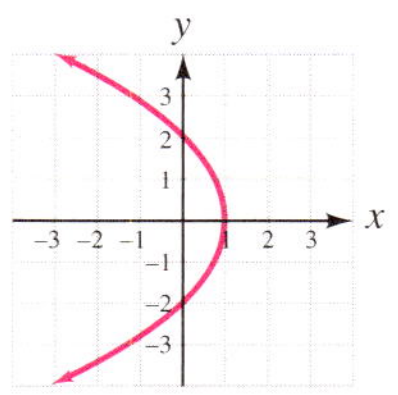

70.

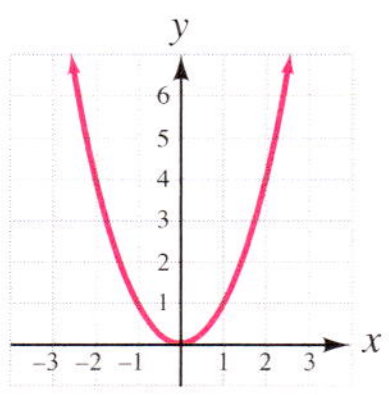

71.
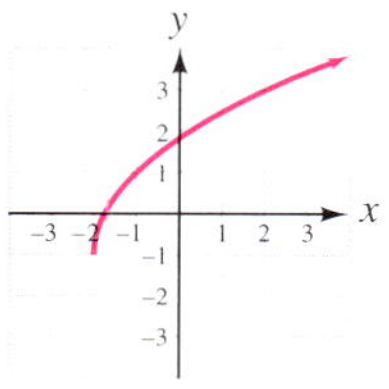

72.
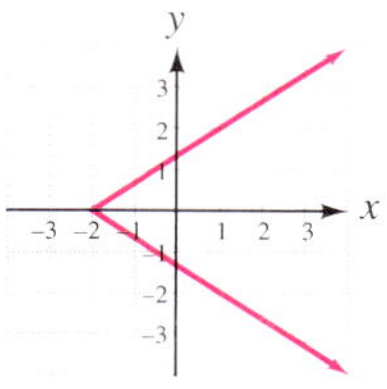

73.
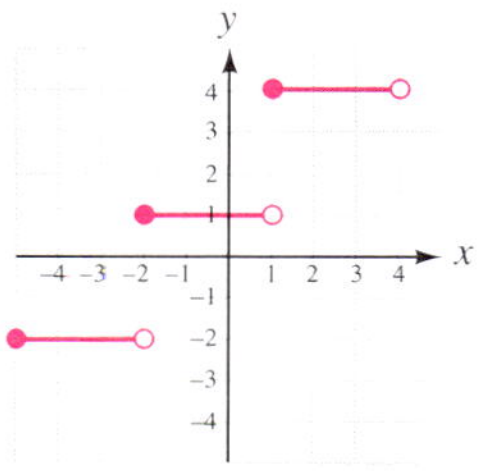

74.
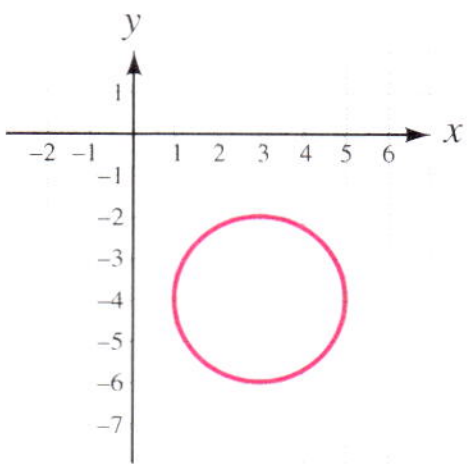

75.
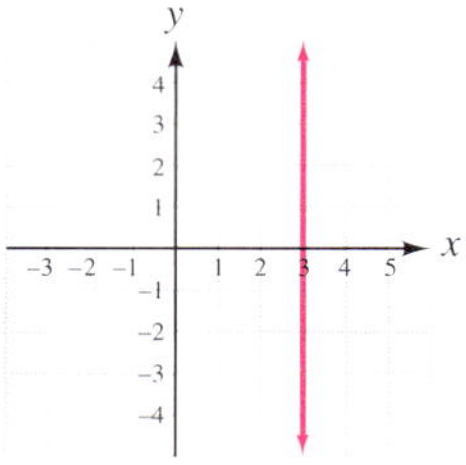

76.
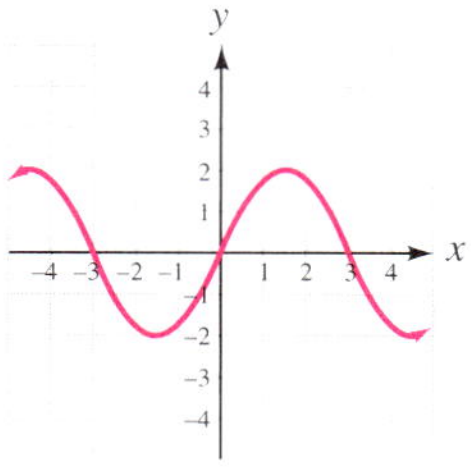

Graph each function using window settings of [−4, 4] for x and [−4, 4] for y. The graph is not what it appears to be. Pick a better viewing window and find a better representation of the true graph. See Using Your Calculator: Graphing Functions.

77. $f(x) = x^2 + 8$

78. $f(x) = x^3 - 8$

79. $f(x) = |x + 5|$

80. $f(x) = |x - 5|$

81. $f(x) = (x - 6)^2$

82. $f(x) = (x + 9)^2$

83. $f(x) = x^3 + 8$

84. $f(x) = x^3 - 12$

APPLICATIONS

85. **Optics.** See the illustration. The **law of reflection** states that the angle of reflection is equal to the angle of incidence. What function studied in this section models the path of the reflected light beam with an angle of incidence measuring 45°?

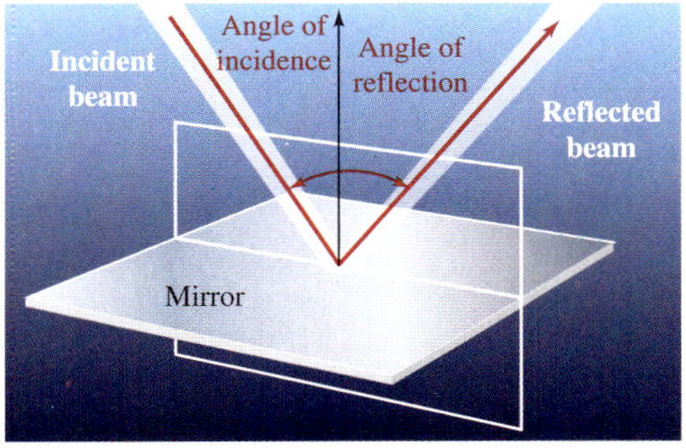

86. **Billiards.** In the illustration, a rectangular coordinate system has been superimposed over a billiard table. Write a function that models the path of the ball that is shown banking off of the cushion.

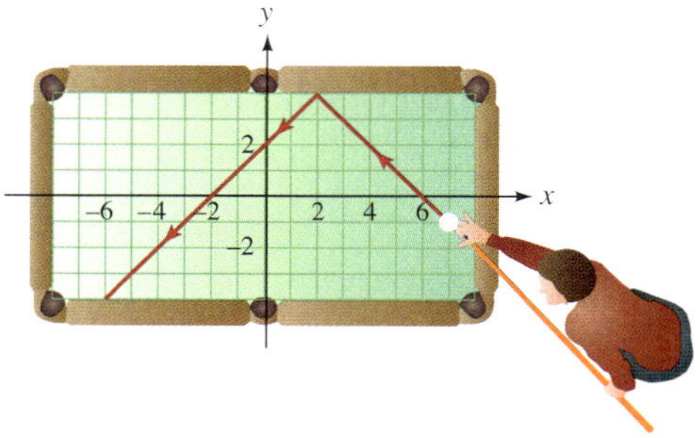

87. **Center of Gravity.** See the illustration. As a diver performs a $1\frac{1}{2}$-somersault in the tuck position, her center of gravity follows a path that can be described by a graph shape studied in this section. What graph shape is that?

88 **Earth's Atmosphere.** The illustration below shows a graph of the temperatures of the atmosphere at various altitudes above Earth's surface. The temperature is expressed in degrees Kelvin, a scale widely used in scientific work.

a. Estimate the coordinates of three points on the graph that have an x-coordinate of 200.

b. Explain why this is not the graph of a function.

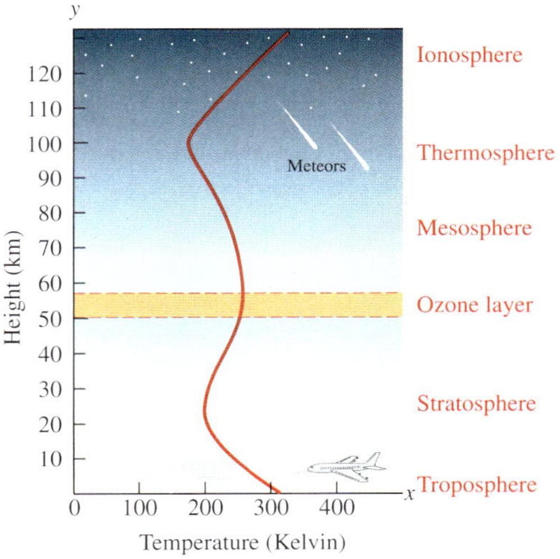

89. Labor Statistics. The polynomial function that is graphed below approximates the number of manufacturing jobs (in millions) in the United States, where x is the number of years after 2000. Use the graph to answer the following questions.

a. Estimate $J(9)$. Explain what the result means.

b. Estimate the value of x for which $J(x) = 14.5$. Explain what the result means.

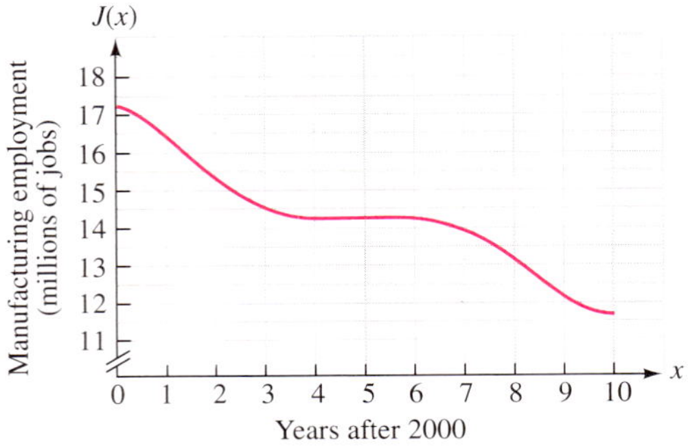

(Source: Bureau of Labor Statistics)

90. Transportation Engineering. The polynomial function A that is graphed below approximates the number of accidents per mile in one year on a 4-lane interstate, where x is the average daily traffic in number of vehicles. Use the graph to answer the following questions.

a. Estimate $A(20{,}000)$. Explain what the result means.

b. Estimate the value of x for which $A(x) = 2$. Explain what the result means.

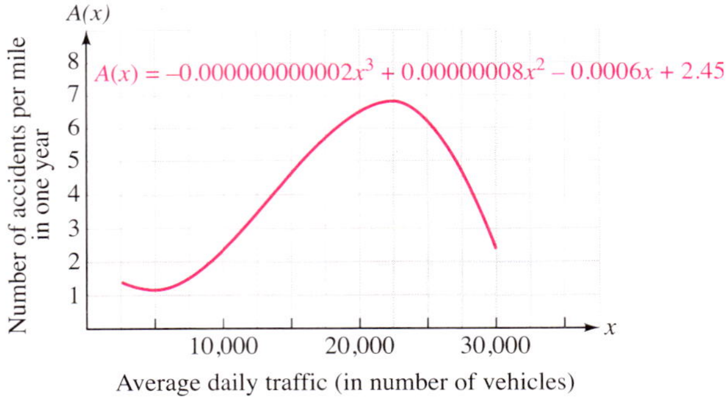

Source: Highway Safety Manual, Colorado Department of Transportation

WRITING

91. Explain how to graph a function by plotting points.

92. Explain how to *project* the graph of a function onto the x-axis. Give an example.

93. a. What does it mean to translate a graph vertically?

b. What does it mean to horizontally translate a graph?

c. What does it mean to reflect the graph of a function about the x-axis?

94. A student was asked to determine whether the graph shown below is the graph of a function. What is wrong with the following reasoning?

When I draw a vertical line through the graph, it intersects the graph only once. By the vertical line test, this is the graph of a function.

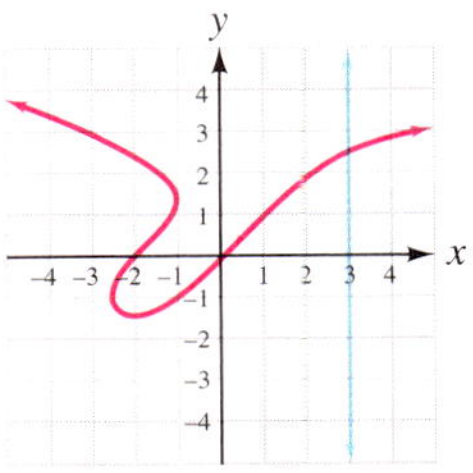

95. Explain why the graph of $g(x) = (x - 2)^2$ is two units to the right of the graph of $f(x) = x^2$.

96. Explain why the range of the polynomial function graphed below is not $(-\infty, \infty)$.

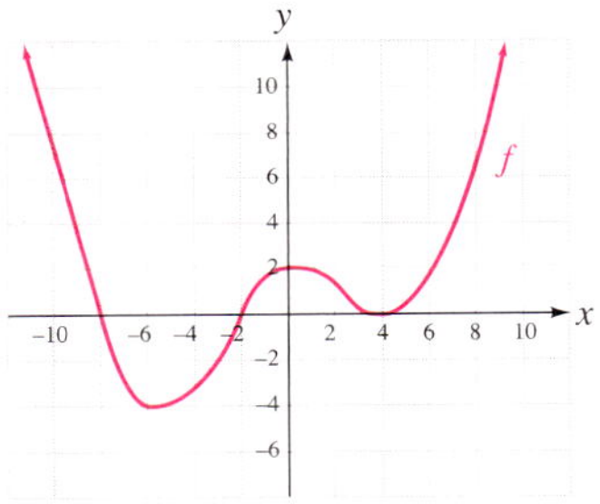

REVIEW

Solve each formula for the indicated variable.

97. $T - W = ma$ for W

98. $a + (n - 1)d = l$ for n

99. $s = \frac{1}{2}gt^2 + vt$ for g

100. $e = mc^2$ for m

CHALLENGE PROBLEMS

Graph each function.

101. $f(x) = \begin{cases} |x| & \text{for } x \geq 0 \\ x^3 & \text{for } x < 0 \end{cases}$

102. $f(x) = \begin{cases} x^2 & \text{for } x \geq 0 \\ |x| & \text{for } x < 0 \end{cases}$

Find the domain and range of each function.

103. a.

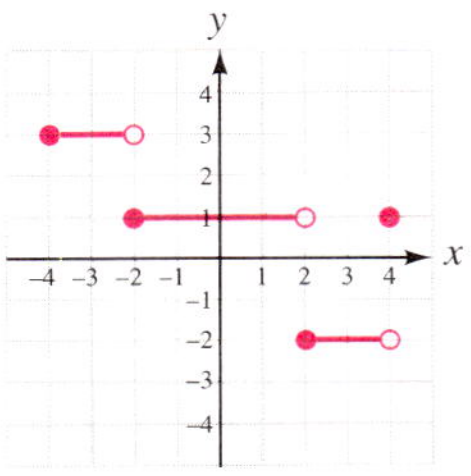

b.

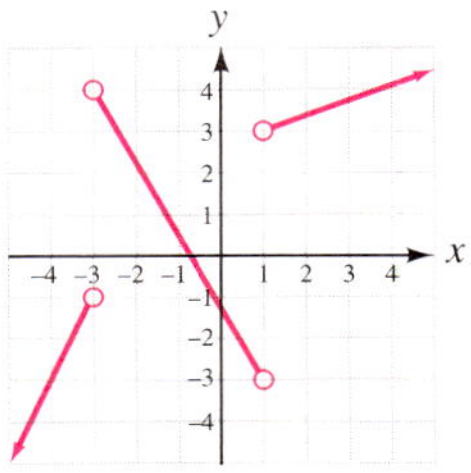

104. Light. Light beams coming from a bulb are reflected outward by a parabolic mirror as parallel rays.

a. The cross-section of a parabolic mirror is given by the function $f(x) = x^2$ for the following values of x: -0.7, -0.6, -0.5, -0.4, -0.3, -0.2, -0.1, 0, 0.1, 0.2, 0.3, 0.4, 0.5, 0.6, 0.7. Sketch the parabolic mirror using the following grid.

b. From the lightbulb filament at (0, 0.25), draw a line segment representing a beam of light that strikes the mirror at (−0.4, 0.16) and then reflects outward, parallel to the y-axis.

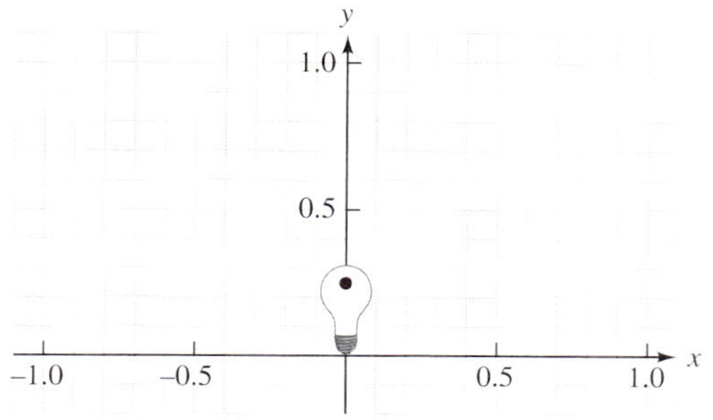

Complete the table and then graph the polynomial function.

105. $f(x) = 2x^3 - 3x^2 - 11x + 6$

x	$f(x)$
−3	
−2	
−1	
0	
1	
2	
3	
4	

106. $f(x) = -x^3 - x^2 + 6x$

x	$f(x)$
−4	
−3	
−2	
−1	
0	
1	
2	
3	

SECTION **8.4**

OBJECTIVES

1 Find the intersection and the union of two sets.
2 Solve compound inequalities containing the word *and*.
3 Solve double linear inequalities.
4 Solve compound inequalities containing the word *or*.

Solving Compound Inequalities

ARE YOU READY?

The following problems review some basic skills that are needed to solve inequalities.

1. Let $A = \{-6, 1, 2, 3, 4\}$ and $B = \{0, 3, 4, 5, 6\}$. What numbers do sets A and B have in common?
2. Solve $6x + 8 + x \geq 5(x - 3) + 9$. Graph the solution set and write it in interval notation.
3. Consider the statements $x > 4$ and $x < 8$. Does the number 7 make both inequalities true?
4. Graph the solution set for $x > 2$ and the solution set for $x \leq -1$ on the same number line.

A label on a first-aid cream warns the user about the temperature at which the medication should be stored. A careful reading reveals that the storage instructions consist of two parts:

The storage temperature should be at least 59°F

and

the storage temperature should be at most 77°F

When the word *and* or the word *or* is used to connect pairs of inequalities, we call the statement a **compound inequality**. To solve compound inequalities, we need to know how to find the *intersection* and *union* of two sets.

1 Find the Intersection and the Union of Two Sets.

Just as operations such as addition and multiplication are performed on real numbers, operations also can be performed on sets. The operation of intersection of two sets produces a new third set that consists of all of the elements that the two given sets have in common.

The Intersection of Two Sets

The **intersection of set A and set B,** written $A \cap B$, is the set of all elements that are common to set A and set B.

The operation of union of two sets produces a third set that is a combination of all of the elements of the two given sets.

The Union of Two Sets

The **union of set A and set B,** written $A \cup B$, is the set of elements that belong to set A or set B or both.

Venn diagrams can be used to illustrate the intersection and union of sets. The area shown in purple in figure (a) represents $A \cap B$ and the area shown in both shades of red in figure (b) represents $A \cup B$.

The Language of Algebra

The **union** of two sets is the collection of elements that belong to either set. The concept is similar to that of a family *reunion,* which brings together the members of several families.

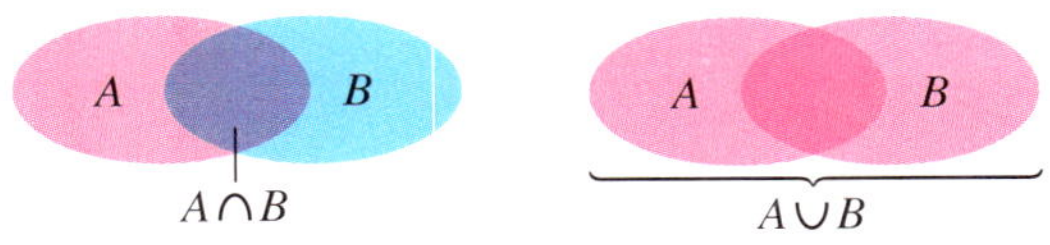

Read as "A intersect B." (a)   Read as "A union B." (b)

EXAMPLE 1 Let $A = \{0, 1, 2, 3, 4, 5, 6\}$ and $B = \{-4, -2, 0, 2, 4\}$.
a. Find $A \cap B$. **b.** Find $A \cup B$.

Strategy In part (a), we will find the elements that sets A and B have in common, and in part (b), we will find the elements that are in one set, or the other set, or both.

Why The symbol $\cap$ means intersection, and the symbol $\cup$ means union.

Solution

a. Since the numbers 0, 2, and 4 are common to both sets A and B, we have

$$A \cap B = \{0, 2, 4\}$$ Since the intersection is, itself, a set, braces are used.

b. Since the numbers in either or both sets are $-4, -2, 0, 1, 2, 3, 4, 5$, and 6, we have

$$A \cup B = \{-4, -2, 0, 1, 2, 3, 4, 5, 6\}$$ 0, 2, and 4 are not listed twice.

The Language of Algebra

The **intersection** of two sets is the collection of elements that they have in common. When two streets cross, we call the area of pavement that they have in common an *intersection.*

Self Check 1 Let $C = \{8, 9, 10, 11\}$ and $D = \{3, 6, 9, 12, 15\}$.
a. Find $C \cap D$. **b.** Find $C \cup D$.

Now Try Problems 17 and 21

2 Solve Compound Inequalities Containing the Word *And.*

When two inequalities are joined with the word *and,* we call the statement a **compound inequality.** Some examples are

$$x \geq -3 \quad \text{and} \quad x \leq 6$$

$$\frac{x}{2} + 1 > 0 \quad \text{and} \quad 2x - 3 < 5$$

$$x + 3 \leq 2x - 1 \quad \text{and} \quad 3x - 2 < 5x - 4$$

The Language of Algebra

Compound means composed of two or more parts, as in *compound* inequalities, chemical *compounds,* and *compound* sentences.

The **solution set of a compound inequality containing the word *and*** includes all numbers that make both of the inequalities true. That is, it is the intersection of their solution sets. We can find the solution set of the compound inequality $x \geq -3$ and $x \leq 6$, for example, by graphing the solution sets of each inequality on the same number line and looking for the numbers common to both graphs.

In the following figure, the graph of the solution set of $x \geq -3$ is shown in red, and the graph of the solution set of $x \leq 6$ is shown in blue.

Preliminary work to determine the graph of the solution set

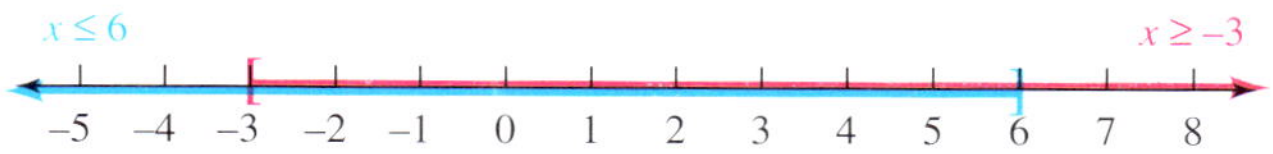

The figure below shows the graph of the solution of the compound inequality $x \geq -3$ and $x \leq 6$. The purple shaded interval, where the red and blue graphs intersect (or overlap), represents the real numbers that are common to the graphs of $x \geq -3$ and $x \leq 6$.

The graph of the solution set

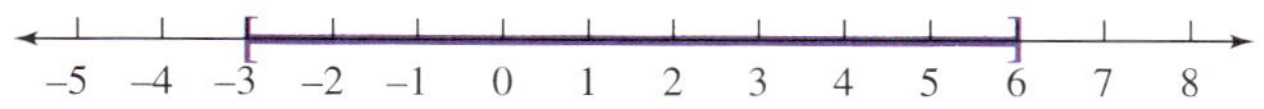

The solution set of $x \geq -3$ and $x \leq 6$ is the **bounded interval** $[-3, 6]$, where the brackets indicate that the endpoints, -3 and 6, are included. It represents all real numbers between -3 and 6, including -3 and 6. Intervals such as this, which contain both endpoints, are called **closed intervals.**

Since the solution set of $x \geq -3$ and $x \leq 6$ is the intersection of the solution sets of the two inequalities, we can write

$$[-3, \infty) \cap (-\infty, 6] = [-3, 6]$$

The solution set of the compound inequality $x \geq -3$ and $x \leq 6$ can be expressed in several ways:

Success Tip

Check with your instructor to see his or her preferred way (or ways) to present solutions of compound inequalities.

1. As a graph: 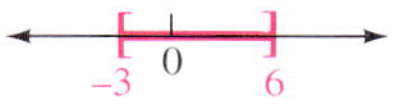

2. In words: all real numbers from -3 to 6
3. In interval notation: $[-3, 6]$
4. Using set-builder notation: $\{x \mid x \geq -3 \text{ and } x \leq 6\}$

EXAMPLE 2 Solve $\frac{x}{2} + 1 > 0$ and $2x - 3 < 5$. Graph the solution set and write it using interval notation and set-builder notation.

Strategy We will solve each inequality separately. Then we will graph the two solution sets on the same number line and determine their intersection.

Why The solution set of a compound inequality containing the word *and* is the intersection of the solution sets of the two inequalities.

Solution In each case, we can use properties of inequality to isolate the variable on one side of the inequality.

$\frac{x}{2} + 1 > 0$	and	$2x - 3 < 5$	This is the compound inequality to solve.
$\frac{x}{2} > -1$		$2x < 8$	
$x > -2$		$x < 4$	

Notation

When graphing on a number line, $(-2, 4)$ represents an *interval.* When graphing on a rectangular coordinate system, $(-2, 4)$ is an *ordered pair* that gives the coordinates of a point.

Next, we graph the solutions of each inequality on the same number line and determine their intersection.

Preliminary work to determine the graph of the solution set

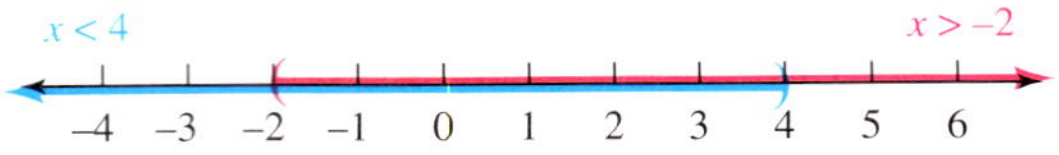

We see that the intersection of the graphs is the set of all real numbers between -2 and 4. The solution set of the compound inequality is the interval $(-2, 4)$, whose graph is shown below. This bounded interval, which does not include either endpoint, is called an **open interval.** Written using set-builder notation, the solution set is $\{x \mid x > -2 \text{ and } x < 4\}$.

The graph of the solution set

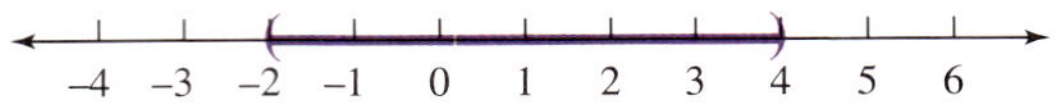

Self Check 2 Solve $3x > -18$ and $\frac{x}{5} - 1 \leq 1$. Graph the solution set and write it using interval notation.

Now Try ▶ Problem 27

The solution of the compound inequality in the Self Check of Example 2 is the interval $(-6, 10]$. A bounded interval such as this, which includes only one endpoint, is called a **half-open interval.** The following chart shows the various types of bounded intervals, along with the inequalities and interval notation that describe them.

Intervals

Open intervals	The interval (a, b) includes all real numbers x such that $a < x < b$.	
Half-open intervals	The interval $[a, b)$ includes all real numbers x such that $a \leq x < b$.	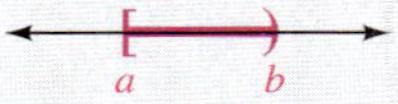
	The interval $(a, b]$ includes all real numbers x such that $a < x \leq b$.	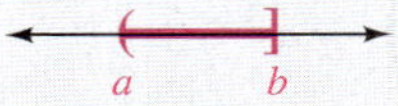
Closed intervals	The interval $[a, b]$ includes all real numbers x such that $a \leq x \leq b$.	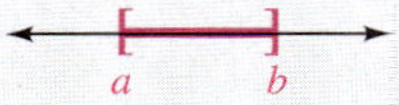

EXAMPLE 3 Solve $x + 3 \leq 2x - 1$ and $3x - 2 < 5x - 4$. Graph the solution set and write it using interval notation and set-builder notation.

Strategy We will solve each inequality separately. Then we will graph the two solution sets on the same number line and determine their intersection.

Why The solution set of a compound inequality containing the word *and* is the intersection of the solution sets of the two inequalities.

Solution In each case, we can use properties of inequality to isolate the variable on one side.

$x + 3 \leq 2x - 1$	and	$3x - 2 < 5x - 4$	This is the compound inequality to solve.
$4 \leq x$		$2 < 2x$	
$x \geq 4$		$1 < x$	
		$x > 1$	

The graph of $x \geq 4$ is shown below in red and the graph of $x > 1$ is shown below in blue.

Preliminary work to determine the graph of the solution set

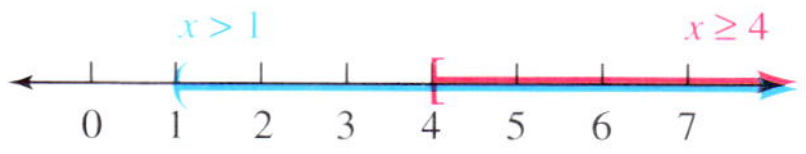

Only those values of x where $x \geq 4$ and $x > 1$ are in the solution set of the compound inequality. Since all numbers greater than or equal to 4 are also greater than 1, the solutions are the numbers x where $x \geq 4$. The solution set is the interval $[4, \infty)$, whose graph is shown below. Written using set-builder notation, the solution set is $\{x \mid x \geq 4\}$.

The graph of the solution set

Self Check 3 Solve $2x + 3 < 4x + 2$ and $3x + 1 < 5x + 3$. Graph the solution set and write it using interval notation.

Now Try Problem 29

EXAMPLE 4 Solve $x - 1 > -3$ and $2x < -8$, if possible.

Strategy We will solve each inequality separately. Then we will graph the two solution sets on the same number line and determine their intersection, if any.

Why The solution set of a compound inequality containing the word *and* is the intersection of the solution sets of the two inequalities.

Solution In each case, we can use properties of inequality to isolate the variable on one side.

$$x - 1 > -3 \quad \text{and} \quad 2x < -8 \qquad \text{This is the compound inequality to solve.}$$
$$x > -2 \qquad\qquad x < -4$$

The graphs of the solution sets shown below do not intersect. Since there are no numbers that make both parts of the original compound inequality true, $x - 1 > -3$ and $2x < -8$ has no solution.

Preliminary work to determine the graph of the solution set

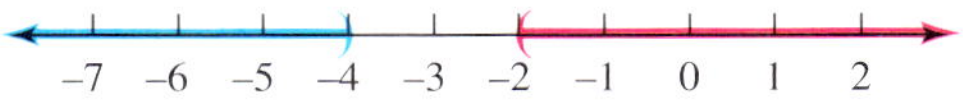

The solution set of the compound inequality is the empty set, which can be written as $\varnothing$. Since there is no solution, a graph is not needed.

Self Check 4 Solve $2x - 3 < x - 2$ and $0 < x - 3.5$, if possible.

Now Try Problem 31

Notation

The graphs of two linear inequalities can intersect at a single point, as shown below. The interval notation used to describe this point of intersection is $[3, 3]$.

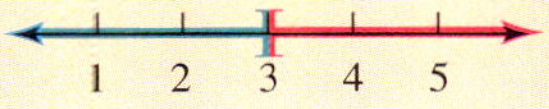

3 Solve Double Linear Inequalities.

Inequalities that contain exactly two inequality symbols are called **double inequalities.** An example is

$$-3 \leq 2x + 5 < 7 \qquad \text{Read as "}-3\text{ is less than or equal to }2x + 5\text{ and }2x + 5\text{ is less than 7."}$$

Any double linear inequality can be written as a compound inequality containing the word *and*. In general, the following is true.

Double Linear Inequalities	The compound inequality $c < x < d$ is equivalent to $c < x$ and $x < d$.

Thus, the double inequality $-3 \le 2x + 5 < 7$ is a shorter form for the compound inequality

$$-3 \le 2x + 5 \quad \text{and} \quad 2x + 5 < 7$$

As you will see in Example 5, it is easier to solve the double inequality because it enables us to solve both inequalities at once.

EXAMPLE 5 Solve $-3 \le 2x + 5 < 7$. Graph the solution set and write it using interval notation and set-builder notation.

Strategy We will solve the double inequality by applying properties of inequality to *all three of its parts* to isolate x in the middle.

Why This double inequality $-3 \le 2x + 5 < 7$ means that $-3 \le 2x + 5$ and $2x + 5 < 7$. We can solve it more easily by leaving it in its original form.

Solution The objective is to isolate x in the middle of the double inequality.

$-3 \le 2x + 5 < 7$ This is the double inequality to solve.

$-3 - 5 \le 2x + 5 - 5 < 7 - 5$ To undo the addition of 5, subtract 5 from all three parts.

$-8 \le 2x < 2$ Perform the subtractions.

$\dfrac{-8}{2} \le \dfrac{2x}{2} < \dfrac{2}{2}$ To isolate x, undo the multiplication by 2 by dividing all three parts by 2.

$-4 \le x < 1$ Perform the divisions.

Notation

Note that the two inequality symbols in $-3 \le 2x + 5 < 7$ point in the same direction and point to the smaller number. And in $-4 \le x < 1$, the numbers -4 and 1 appear in the same order as they do on a number line.

The solution set of the double linear inequality is the half-open interval $[-4, 1)$, whose graph is shown below. Written using set-builder notation, the solution set is $\{x \mid -4 \le x < 1\}$.

Self Check 5 Solve $-5 \le 3x - 8 \le 7$. Graph the solution set and write it using interval notation.

Now Try Problem 35

CAUTION When multiplying or dividing all three parts of a double inequality by a negative number, don't forget to reverse the direction of both inequalities. As an example, we solve $-15 < -5x \le 25$.

$-15 < -5x \le 25$

$\dfrac{-15}{-5} > \dfrac{-5x}{-5} \ge \dfrac{25}{-5}$ Divide all three parts by −5 to isolate x in the middle. Reverse both inequality signs.

$3 > x \ge -5$ Perform the divisions.

$-5 \le x < 3$ Write an equivalent double inequality with the smaller number, −5, on the left.

4 Solve Compound Inequalities Containing the Word *Or.*

A warning on the water temperature gauge of a commercial dishwasher cautions the operator to shut down the unit if

The water temperature goes below 140°

or

The water temperature goes above 160°

When two inequalities are joined with the word *or,* we also call the statement a compound inequality. Some examples are

$$x < 140 \quad \text{or} \quad x > 160$$
$$x \le -3 \quad \text{or} \quad x \ge 2$$
$$\frac{x}{3} > \frac{2}{3} \quad \text{or} \quad -(x-2) > 3$$

Caution

It's incorrect to write the statement $x \le -3$ or $x \ge 2$ as the double inequality $2 \le x \le -3$, because that would imply that $2 \le -3$, which is false.

The **solution set of a compound inequality containing the word *or*** includes all numbers that make one or the other or both inequalities true. That is, it is the union of their solution sets. We can find the solution set of $x \le -3$ or $x \ge 2$, for example, by drawing the graphs of each inequality on the same number line.

In the following figure, the graph of the solution set of $x \le -3$ is shown in red, and the graph of the solution set of $x \ge 2$ is shown in blue.

Preliminary work to determine the graph of the solution set

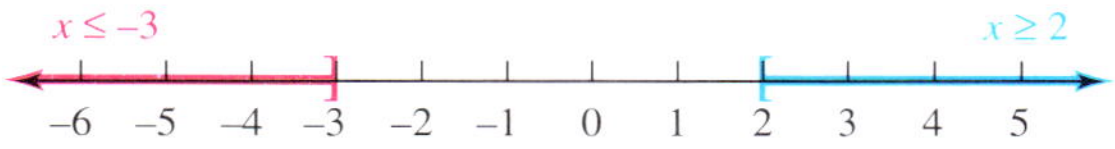

The figure below shows the graph of the solution set of $x \le -3$ or $x \ge 2$. This graph is a union of the graph of $x \le -3$ with the graph of $x \ge 2$.

The graph of the solution set

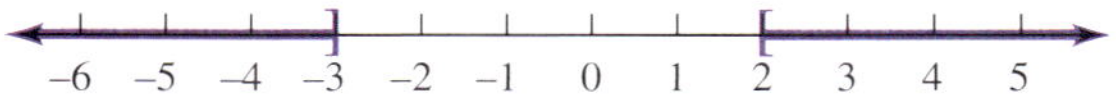

For the compound inequality $x \le -3$ or $x \ge 2$, we can write the solution set as the union of two intervals:

$$(-\infty, -3] \cup [2, \infty)$$

We can express the solution set of the compound inequality $x \le -3$ or $x \ge 2$ in several ways:

1. As a graph: 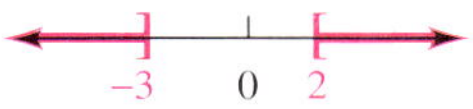

2. In words: all real numbers less than or equal to -3 *or* greater than or equal to 2
3. As the union of two intervals: $(-\infty, -3] \cup [2, \infty)$
4. Using set-builder notation: $\{x \mid x \le -3 \text{ or } x \ge 2\}$

EXAMPLE 6

Solve $\frac{x}{3} > \frac{2}{3}$ or $-(x-2) > 3$. Graph the solution set and write it using interval notation and set-builder notation.

Strategy We will solve each inequality separately. Then we will graph the two solution sets on the same number line to show their union.

Why The solution set of a compound inequality containing the word *or* is the union of the solution sets of the two inequalities.

Solution To solve each inequality, we proceed as follows:

$$\frac{x}{3} > \frac{2}{3} \quad \text{or} \quad -(x-2) > 3 \qquad \text{This is the compound inequality to solve.}$$

$$x > 2 \quad \Big| \quad -x + 2 > 3$$

$$-x > 1$$

$$x < -1$$

The Language of Algebra

The meaning of the word **or** in a compound inequality differs from our everyday use of the word. For example, when we say, "I will go shopping today *or* tomorrow," we mean that we will go one day or the other, but *not* both. With compound inequalities, *or* includes one possibility, or the other, or both.

Next, we graph the solutions of each inequality on the same number line and determine their union.

Preliminary work to determine the graph of the solution set

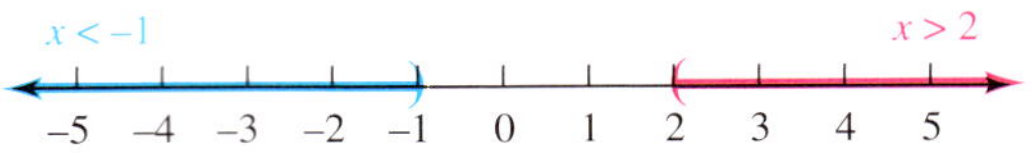

The union of the two solution sets consists of all real numbers less than -1 or greater than 2. The solution set of the compound inequality is the union of two intervals: $(-\infty, -1) \cup (2, \infty)$. Its graph appears below. Written using set-builder notation, the solution set is $\{x \mid x > 2 \text{ or } x < -1\}$.

The graph of the solution set

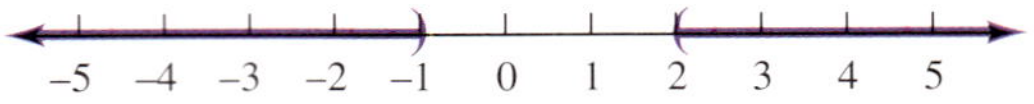

Self Check 6 Solve $\frac{x}{2} > 2$ or $-3(x-2) > 0$. Graph the solution set and write it using interval notation.

Now Try Problem 41

EXAMPLE 7 Solve $x + 3 \geq -3$ or $-x > 0$. Graph the solution set and write it using interval notation and set-builder notation.

Strategy We will solve each inequality separately. Then we will graph the two solution sets on the same number line to show their union.

Why The solution set of a compound inequality containing the word *or* is the union of the solution sets of the two inequalities.

Solution To solve each inequality, we proceed as follows:

$$x + 3 \geq -3 \quad \text{or} \quad -x > 0 \qquad \text{This is the compound inequality to solve.}$$

$$x \geq -6 \quad \Big| \quad x < 0$$

We graph the solution set of each inequality on the same number line and determine their union.

Preliminary work to determine the graph of the solution set

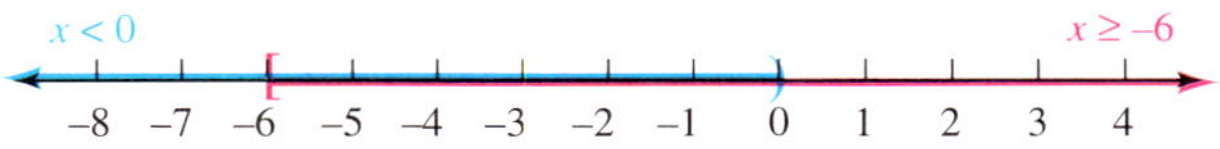

Since the entire number line is shaded, all real numbers satisfy the original compound inequality and the solution set is denoted as $(-\infty, \infty)$ or $\mathbb{R}$. Its graph is shown below. Written using set-builder notation, the solution set is $\{x \mid x \text{ is a real number}\}$.

The graph of the solution set

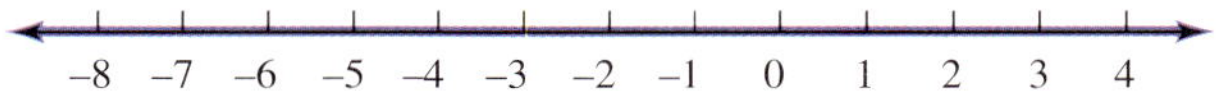

Self Check 7 Solve $x - 1 < 5$ or $-2x \le 10$. Graph the solution set and write it using interval notation.

Now Try Problem 43

Solving Compound Inequalities

1. Solve each inequality separately and graph their solution sets in different colors on the same number line.
2. If the inequalities are connected with the word *and*, find the intersection of the two solution sets. If the inequalities are connected with the word *or*, find the union of the two solution sets.
3. Write the solution set of the compound inequality using interval notation or set-builder notation, and graph it on a new number line.

SECTION 8.4 STUDY SET

VOCABULARY

Fill in the blanks.

1. The ____________ of two sets is the set of elements that are common to both sets and the _______ of two sets is the set of elements that are in one set, or the other, or both.
2. $x \ge 3$ and $x < 4$ is a ___________ inequality.
3. $-6 < x + 1 \le 1$ is a _______ linear inequality.
4. $(2, 3)$ is an example of an open ________, $[-4, 0]$ is an example of a ________ interval, and $(0, 9]$ is an example of a half-______ interval.

CONCEPTS

Fill in the blanks.

5. a. The solution set of a compound inequality containing the word *and* includes all numbers that make _____ inequalities true.
   b. The solution set of a compound inequality containing the word *or* includes all numbers that make _____, or the other, or _____ inequalities true.
6. The double inequality $4 < 3x + 5 \le 15$ is equivalent to $4 < 3x + 5$ _____ $3x + 5 \le 15$.
7. a. When solving a compound inequality containing the word *and*, the solution set is the ___________ of the solution sets of the inequalities.
   b. When solving a compound inequality containing the word *or*, the solution set is the _______ of the solution sets of the inequalities.
8. When multiplying or dividing all three parts of a double inequality by a negative number, the direction of both inequality symbols must be __________.
9. Use a check to determine whether -3 is a solution of the compound inequality.
   a. $\frac{x}{3} + 1 \ge 0$ and $2x - 3 < -10$
   b. $2x \le 0$ or $-3x < -5$
10. Use a check to determine whether -3 is a solution of the double linear inequality.
    a. $-1 < -3x + 4 < 12$
    b. $-1 < -3x + 4 < 14$
11. Use interval notation, if possible, to describe the intersection of each pair of graphs.
    a. 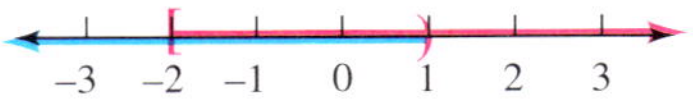

    b. 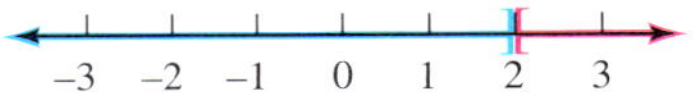

    c. 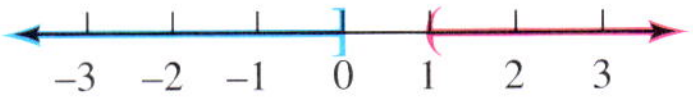

12. Use interval notation to describe the union of each pair of graphs.
    a. 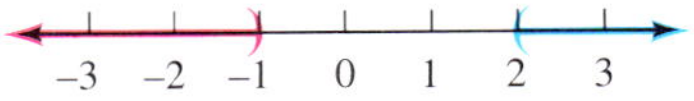

    b. 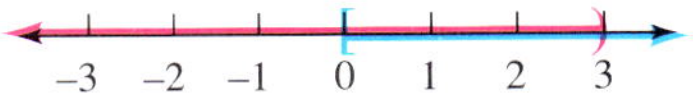

    c. 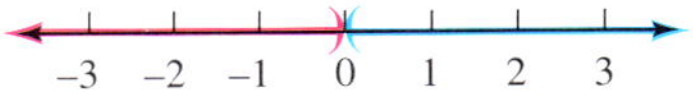

NOTATION

13. Fill in the blanks: We read ∪ as _______ and ∩ as ___________.
14. Match each interval with its corresponding graph.
    a. $[2, 3)$ i. 

    b. $(2, 3)$ ii. 

    c. $[2, 3]$ iii. 

15. What set is represented by the interval notation $(-\infty, \infty)$? Graph it.
16. a. Graph: $(-\infty, 2) \cup [3, \infty)$
    b. Graph: $(-\infty, 3) \cap [-2, \infty)$

GUIDED PRACTICE

Let $A = \{0, 1, 2, 3, 4, 5, 6\}$, $B = \{4, 6, 8, 10\}$, $C = \{-3, -1, 0, 1, 2\}$, *and* $D = \{-3, 1, 2, 5, 8\}$. *Find each set.* See Example 1.

17. $A \cap B$
18. $A \cap D$
19. $C \cap D$
20. $B \cap C$
21. $B \cup C$
22. $A \cup C$
23. $A \cup D$
24. $C \cup D$

Solve each compound inequality, if possible. Graph the solution set (if one exists) and write it using interval notation. See Examples 2–4.

25. $x > -2$ and $x \le 5$
26. $x \le -4$ and $x \ge -7$
27. $2x - 1 > 3$ and $x + 8 \le 11$
28. $5x - 3 \ge 2$ and $6 \ge 4x - 3$
29. $6x + 1 < 5x - 3$ and $\frac{x}{2} + 9 \le 6$
30. $\frac{2}{3}x + 1 > -9$ and $\frac{3}{4}x - 1 > -10$
31. $x + 2 < -\frac{1}{3}x$ and $-6x < 9x$
32. $\frac{3}{2}x + \frac{1}{5} < 5$ and $2x + 1 > 9$

Solve each double inequality. Graph the solution set and write it using interval notation. See Example 5.

33. $4 \le x + 3 \le 7$
34. $-5.3 \le x - 2.3 \le -1.3$
35. $0.9 < 2x - 0.7 < 1.5$
36. $7 < 3x - 2 < 25$

Solve each compound inequality. Graph the solution set and write it using interval notation. See Examples 6 and 7.

37. $x \le -2$ or $x > 6$
38. $x \ge -1$ or $x \le -3$
39. $x - 3 < -4$ or $-x + 2 < 0$
40. $4x < -12$ or $\frac{x}{2} > 4$
41. $3x + 2 < 8$ or $2x - 3 > 11$
42. $3x + 4 < -2$ or $3x + 4 > 10$
43. $2x > x + 3$ or $\frac{x}{8} + 1 < \frac{13}{8}$
44. $2(x + 2) < x - 11$ or $-\frac{x}{5} < 20$

TRY IT YOURSELF

Solve each compound inequality, if possible. Graph the solution set (if one exists) and write it using interval notation.

45. $-4(x + 2) \ge 12$ or $3x + 8 < 11$
46. $4.5x - 1 < -10$ or $6 - 2x \ge 12$
47. $2.2x < -19.8$ and $-4x < 40$
48. $\frac{1}{2}x \le 2$ and $0.75x \ge -6$
49. $-2 < -b + 3 < 5$
50. $2 < -t - 2 < 9$
51. $4.5x - 2 > 2.5$ or $\frac{1}{2}x \le 1$
52. $0 < x$ or $3x - 5 > 4x - 7$
53. $5(x - 2) \ge 0$ and $-3x < 9$
54. $x - 1 \le 2(x + 2)$ and $x \le 2x - 5$
55. $-x < -2x$ and $3x > 2x$
56. $-\frac{x}{4} > -2.5$ and $9x > 2(4x + 5)$
57. $-6 < -3(x - 4) \le 24$
58. $-4 \le -2(x + 8) < 8$
59. $2x + 1 \ge 5$ and $-3(x + 1) \ge -9$
60. $2(-2) \le 3x - 1$ and $3x - 1 \le -1 - 3$
61. $\frac{4.5x - 12}{2} < x$ or $-15.3 > -3(x - 1.4)$
62. $y + 0.52 < 1.05y$ or $9.8 - 15y > -15.7$
63. $\frac{x}{0.7} + 5 > 4$ and $-4.8 \le \frac{3x}{-0.125}$
64. $5(x + 1) \le 4(x + 3)$ and $x + 12 < -3$
65. $-24 < \frac{3}{2}x - 6 \le -15$
66. $-4 > \frac{2}{3}x - 2 > -6$
67. $\frac{x}{3} - \frac{x}{4} > \frac{1}{6}$ or $\frac{x}{2} + \frac{2}{3} \le \frac{3}{4}$
68. $\frac{a}{2} + \frac{7}{4} > 5$ or $\frac{3}{8} + \frac{a}{3} \le \frac{5}{12}$
69. $0 \le \frac{4 - x}{3} \le 2$
70. $-2 \le \frac{5 - 3x}{2} \le 2$
71. $x \le 6 - \frac{1}{2}x$ and $\frac{1}{2}x + 1 \ge 3$
72. $3\left(x + \frac{2}{3}\right) \le -7$ and $2(x + 2) \ge -2$
73. $-6 < f(x) \le 0$ where $f(x) = 3x - 9$
74. $-3 < f(x) < 7$ where $f(x) = \frac{2}{3}x - \frac{1}{3}$
75. Let $f(x) = 5x + 14$ and $g(x) = 2x + 8$. Find all values of x for which $f(x) > 29$ and $g(x) < 20$.
76. Let $f(x) = x - 2$. Find all values of x for which $f(x) > 5$ or $f(x) < -1$.

Look Alikes . . .

Solve the inequality in part a. Graph the solution set and write it in interval notation. Then use your work from part a to determine the solution set for the compound inequality in part b. (No new work is necessary!) Graph the solution set and write it in interval notation.

77. a. $3x - 2 \ge 4$ and $x + 6 \ge 12$
    b. $3x - 2 \ge 4$ or $x + 6 \ge 12$

78. a. $x + 2 \le 10$ or $x - 3 \ge 2$
b. $x + 2 \le 10$ and $x - 3 \ge 2$

79. a. $2x + 1 \le 7$ and $3x + 5 \ge 23$
b. $2x + 1 \le 7$ or $3x + 5 \ge 23$

80. a. $7 \le 4x + 1 < 23$
b. $7 < 4x + 1 \le 23$

APPLICATIONS

81. **Baby Furniture.** Refer to the illustration. A company manufactures various sizes of play yard cribs having perimeters between 128 and 192 inches, inclusive.

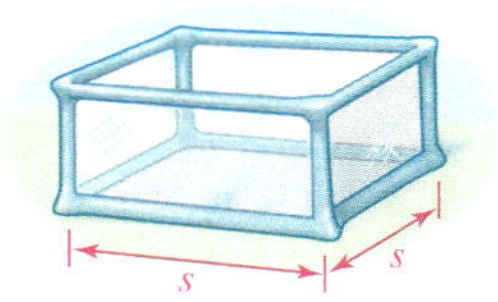

a. Complete the double inequality that describes the range of the perimeters of the play yard shown.

$\le 4s \le$

b. Solve the double inequality to find the range of the side lengths of the play yard.

82. **Trucking.** The distance that a truck can travel in 8 hours, at a constant rate of r mph, is given by $8r$. A trucker wants to travel at least 350 miles, and company regulations don't allow him to exceed 450 miles in one 8-hour shift.

a. Complete the double inequality that describes the mileage range of the truck.

$\le 8r \le$

b. Solve the double inequality to find the range of the average rate (speed) of the truck for the 8-hour trip.

83. **Thermostats.** During business hours, as shown in figure (a), the *Temp range* control in an office is set at 5. This means that the heater comes on when the room temperature gets 5 degrees below the *Temp setting* and the air conditioner comes on when the room temperature gets 5 degrees above the *Temp setting*.

a. Use interval notation to describe the temperature range when neither the heater nor the air conditioner will come on during business hours.

b. After business hours, the *Temp range* setting is changed to save energy. See figure (b). Use interval notation to describe the after-business-hours temperature range when neither the heater nor the air conditioner will come on.

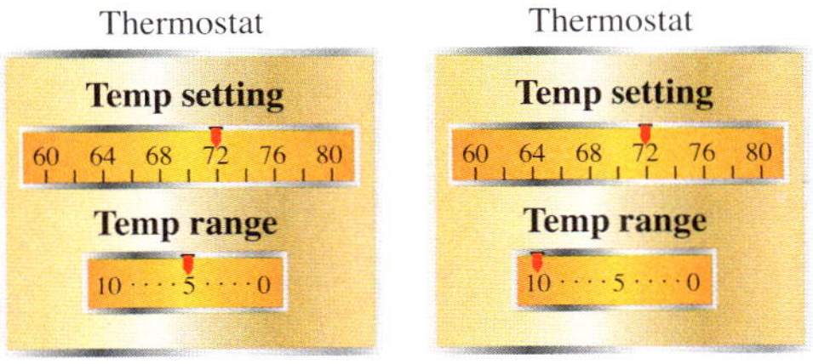

During business hours (a) After business hours (b)

84. **Treating Fevers.** Use the flow chart to determine what action should be taken for a 13-month-old child who has had a 99.8° temperature for 3 days and is not suffering any other symptoms. T represents the child's temperature, A the child's age in months, and S the number of hours the child has experienced the symptoms.

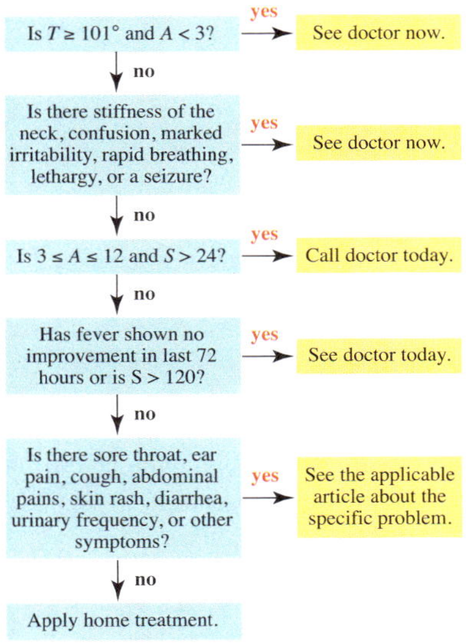

Based on information from *Take Care of Yourself* (Addison-Wesley, 1993)

85. **U.S. Health Care.** Refer to the following graph. Let P represent the percent of children covered by private insurance, M the percent covered by Medicaid, and N the percent not covered. For what years are the following true?

a. $P \ge 63$ and $M \ge 26$
b. $P > 60$ or $M \ge 29$
c. $M \ge 29$ and $N \le 10$
d. $M \ge 30$ or $N < 9.2$

U.S. Health Care Coverage for People Under 18 Years of Age (in percent)

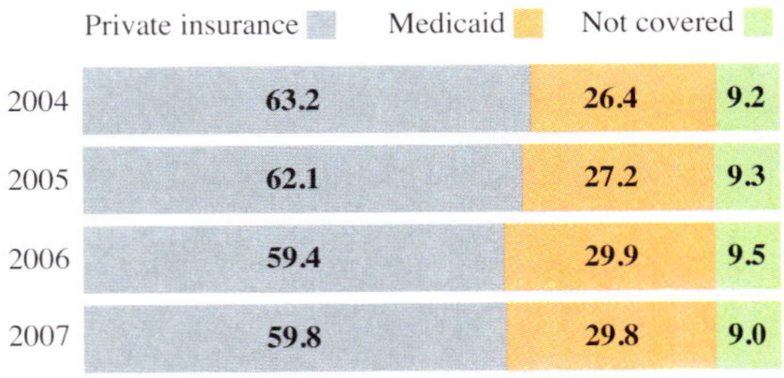

Source: U.S. Department of Health and Human Services

(The small percent of people covered each year by some nonstandard program is not shown.)

86. Polls. For each response to the poll question shown below, the *margin of error* is +/− (read as "plus or minus") 2.8%. This means that for the statistical methods used to do the polling, the actual response could be as much as 2.8 points more or 2.8 points less than shown. Use interval notation to describe the possible interval (in percent) for each response.

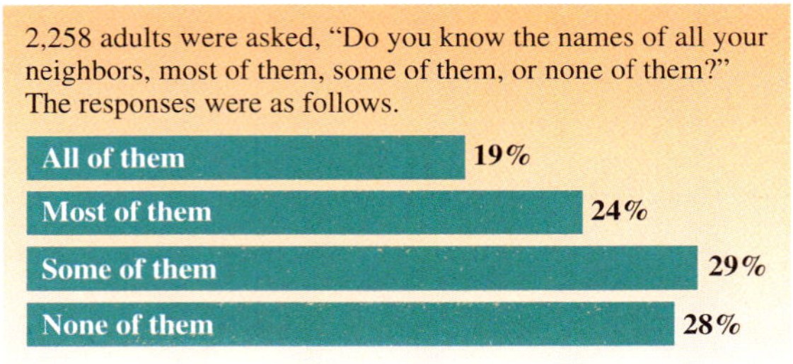

87. Street Intersections. Refer to figure (a) below.

a. Shade the area that represents the intersection of the two streets shown in the illustration.

b. Shade the area that represents the union of the two streets.

(a)

(b)

88. Traffic Signs. The pair of signs shown in figure (b) above are a real-life example of which concept discussed in this section?

WRITING

89. Explain how to find the union and how to find the intersection of $(-\infty, 5)$ and $(-2, \infty)$ graphically.

90. Explain why the double inequality $2 < x < 8$ can be written in the equivalent form $2 < x$ and $x < 8$.

91. Explain the meaning of the notation $(-1, 2)$ for each type of graph.

a. 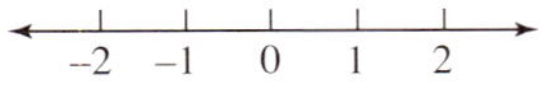

b. 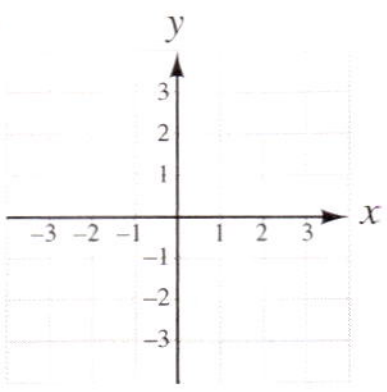

92. The meaning of the word *or* in a compound inequality differs from our everyday use of the word. Explain the difference.

93. Describe each set in words.

a. $(-3, 3)$ **b.** $[7, 12]$

c. $(-\infty, 5] \cup (6, \infty)$

94. What is incorrect about the double inequality $3 < -3x + 4 < -3$?

REVIEW

95. Airplanes. Together, a Delta B747 and a Delta B777 seat 681 passengers. If the B777 seats 125 less people than the B747, how many passengers does each seat? (Source: deltaskymag.com)

96. Denzel. As of October 2010, Denzel Washington's three top domestic grossing films, *American Gangster, Remember the Titans,* and *The Pelican Brief,* had earned a total of \$346.7 million. If *American Gangster* earned \$14.5 million more than *Remember the Titans,* and if *Remember the Titans* earned \$14.9 million more than *The Pelican Brief,* how much did each film earn as of that date? (Source: boxofficemojo.com)

CHALLENGE PROBLEMS

Solve each compound inequality. Graph the solution set and write it in interval notation.

97. $-5 < \frac{x+2}{-2} < 0$ or $2x + 10 \geq 30$

98. $-2 \leq \frac{x-4}{3} \leq 0$ and $\frac{x-5}{2} \geq -3$

99. $x - 12 < 4x < 2x + 16$

100. $6(x - 3) \leq 3(3x + 2) \leq 4(2x + 3)$

SECTION 8.5 Solving Absolute Value Equations and Inequalities

OBJECTIVES

1. Solve equations of the form $|X| = k$.
2. Solve equations with two absolute values.
3. Solve inequalities of the form $|X| < k$.
4. Solve inequalities of the form $|X| > k$.

ARE YOU READY?

The following problems review some basic skills that are needed to solve absolute value equations and inequalities.

1. Find each absolute value.
   a. $|12|$ **b.** $|-7.5|$
2. Tell whether each statement is true or false.
   a. $|-3| \geq 2$ **b.** $|-26| < -27$
3. Solve $3x + 6 \leq -3$ or $3x + 6 \geq 9$. Graph the solution set and write it using interval notation.
4. Solve $-8 < 2x + 8 < 16$. Graph the solution set and write it using interval notation.

Many quantities studied in mathematics, science, and engineering are expressed as positive numbers. To guarantee that a quantity is positive, we often use absolute value. In this section, we will consider equations and inequalities involving the absolute value of an algebraic expression. Some examples are

$$|3x - 2| = 5, \qquad |2x - 3| < 9, \qquad \text{and} \qquad \left|\frac{3 - x}{5}\right| \geq 6$$

To solve these *absolute value equations* and *inequalities,* we write and then solve equivalent compound equations and inequalities.

1 Solve Equations of the Form $|X| = k$.

Recall that the absolute value of a real number is its distance from 0 on a number line. To solve the **absolute value equation** $|x| = 5$, we must find all real numbers x whose distance from 0 on the number line is 5. There are two such numbers: 5 and -5. It follows that the solutions of $|x| = 5$ are 5 and -5 and the solution set is $\{5, -5\}$.

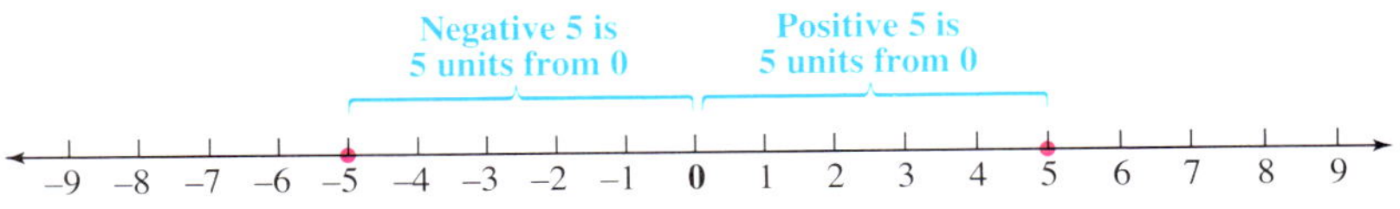

The results from this example suggest the following approach for solving absolute value equations.

Solving Absolute Value Equations

For any positive number k and any algebraic expression X:
To solve $|X| = k$, solve the equivalent compound equation

$$X = k \quad \text{or} \quad X = -k$$

The statement $X = k$ or $X = -k$ is called a **compound equation** because it consists of two equations joined with the word *or.*

EXAMPLE 1 Solve: **a.** $|s| = 0.003$ **b.** $|3x - 2| = 5$ **c.** $|10 - x| = -40$

Strategy To solve the first two equations, we will write and then solve an equivalent compound equation. We will solve the third equation by inspection.

Why All three of the equations are of the form $|X| = k$. However, the standard method for solving absolute value equations cannot be applied to $|10 - x| = -40$ because k is negative.

Solution

The Language of Algebra

When we say that the absolute value equation and a compound equation are **equivalent,** we mean that they have the same solution(s).

Caution

When you see the word *solve,* you probably think of steps such as combining like terms, distributing, or doing something to both sides. However, to solve absolute value equations of the form $|X| = k$, those familiar steps are not used until an equivalent compound equation is written.

a. The absolute value equation $|s| = 0.003$ is equivalent to the compound equation

$$s = 0.003 \quad \text{or} \quad s = -0.003$$

Therefore, the solutions of $|s| = 0.003$ are 0.003 and -0.003, and the solution set is $\{0.003, -0.003\}$.

b. The equation-solving method used in part a can be extended to equations where the expression within absolute value bars is more complicated than a single variable. The absolute value equation $|3x - 2| = 5$ is equivalent to the compound equation

$$3x - 2 = 5 \quad \text{or} \quad 3x - 2 = -5$$

Now we solve each equation for x:

$$\begin{aligned} 3x - 2 &= 5 \\ 3x &= 7 \\ x &= \frac{7}{3} \end{aligned} \quad \text{or} \quad \begin{aligned} 3x - 2 &= -5 \\ 3x &= -3 \\ x &= -1 \end{aligned}$$

The results must be checked separately to see whether each of them produces a true statement. We substitute $\frac{7}{3}$ for x and then -1 for x in the original equation.

Check: ***For*** $x = \frac{7}{3}$

$$\begin{aligned} |3x - 2| &= 5 \\ \left|3\left(\frac{7}{3}\right) - 2\right| &\stackrel{?}{=} 5 \\ |7 - 2| &\stackrel{?}{=} 5 \\ |5| &\stackrel{?}{=} 5 \\ 5 &= 5 \quad \text{True} \end{aligned}$$

For $x = -1$

$$\begin{aligned} |3x - 2| &= 5 \\ |3(-1) - 2| &\stackrel{?}{=} 5 \\ |-3 - 2| &\stackrel{?}{=} 5 \\ |-5| &\stackrel{?}{=} 5 \\ 5 &= 5 \quad \text{True} \end{aligned}$$

The resulting true statements indicate that the equation has two solutions: $\frac{7}{3}$ and -1. The solution set is $\left\{\frac{7}{3}, -1\right\}$.

c. Since an absolute value can never be negative, there are no real numbers x that make $|10 - x| = -40$ true. The equation has no solution and the solution set is $\varnothing$.

Self Check 1 Solve: **a.** $|x| = \frac{1}{2}$ **b.** $|2x - 3| = 7$

c. $\left|\frac{x}{4} - 1\right| = -3$

Now Try ▶ Problems 19, 23, and 29

When solving absolute value equations (or inequalities), we must **isolate the absolute value expression on one side** before writing the equivalent compound statement.

EXAMPLE 2 Solve: $\left|\frac{2}{3}x + 3\right| + 4 = 10$

Strategy We will first isolate $\left|\frac{2}{3}x + 3\right|$ on the left side of the equation and then write and solve an equivalent compound equation.

Why After isolating the absolute value expression on the left, the resulting equation will have the desired form $|X| = k$.

Solution

$$\left|\frac{2}{3}x + 3\right| + 4 = 10$$ This is the equation to solve.

$$\left|\frac{2}{3}x + 3\right| = 6$$ To isolate the absolute value expression, subtract 4 from both sides. The resulting equation is in the form $|X| = k$.

Caution

A common error when solving absolute value equations is to forget to isolate the absolute value expression first. Note:

$$\left|\frac{2}{3}x + 3\right| + 4 = 10$$

does not mean

$$\frac{2}{3}x + 3 + 4 = 10$$

or

$$\frac{2}{3}x + 3 + 4 = -10$$

With the absolute value now isolated, we can solve $\left|\frac{2}{3}x + 3\right| = 6$ by writing and solving an equivalent compound equation:

$$\frac{2}{3}x + 3 = 6 \quad \text{or} \quad \frac{2}{3}x + 3 = -6$$

Now we solve each equation for x:

$$\frac{2}{3}x + 3 = 6 \quad \text{or} \quad \frac{2}{3}x + 3 = -6$$

$$\frac{2}{3}x = 3 \qquad\qquad \frac{2}{3}x = -9$$

$$2x = 9 \qquad\qquad 2x = -27$$

$$x = \frac{9}{2} \qquad\qquad x = -\frac{27}{2}$$

Verify that both $\frac{9}{2}$ and $-\frac{27}{2}$ are solutions by substituting them into the original equation.

Self Check 2 Solve: $|0.4x - 2| - 0.6 = 0.4$

Now Try Problem 33

EXAMPLE 3

Solve: $3\left|\frac{1}{2}x - 5\right| - 4 = -4$

Strategy We will first isolate $\left|\frac{1}{2}x - 5\right|$ on the left side of the equation and then write and solve an equivalent compound equation.

Why After isolating the absolute value expression, the resulting equation will have the desired form $|X| = k$.

Solution

$$3\left|\frac{1}{2}x - 5\right| - 4 = -4$$ This is the equation to solve.

$$3\left|\frac{1}{2}x - 5\right| = 0$$ To isolate the absolute value expression $\left|\frac{1}{2}x - 5\right|$, we first add 4 to both sides.

$$\left|\frac{1}{2}x - 5\right| = 0$$ To complete the process to isolate the absolute value expression, divide both sides by 3. The resulting equation is in the form $|X| = k$.

Success Tip

To solve most absolute value equations, we must consider two cases. However, if an absolute value is equal to 0, we need only consider one: the case when the expression within the absolute value bars is equal to 0.

Since 0 is the only number whose absolute value is 0, the expression $\frac{1}{2}x - 5$ must be 0.

$$\frac{1}{2}x - 5 = 0$$ Set the expression within the absolute value bars equal to 0 and solve for x.

$$\frac{1}{2}x = 5$$ To isolate the variable term $\frac{1}{2}x$, add 5 to both sides.

$$x = 10$$ To isolate x, multiply both sides by 2.

The solution is 10 and the solution set is $\{10\}$. Verify that it satisfies the original equation.

Self Check 3 Solve: $-5\left|\frac{2}{3}x + 4\right| + 1 = 1$

Now Try Problem 39

In Section 8.3 we discussed absolute value functions and their graphs. If we are given an output of an absolute value function, we can work in reverse to find the corresponding input(s).

EXAMPLE 4 Let $f(x) = |x + 4|$. For what value(s) of x is $f(x) = 20$?

Strategy We will substitute 20 for $f(x)$ and solve for x.

Why In the equation, there are two unknowns, x and $f(x)$. If we replace $f(x)$ with 20, we can solve the resulting absolute value equation for x.

Solution

$f(x) = |x + 4|$ This is the given function.

$20 = |x + 4|$ Substitute 20 for f(x).

$|x + 4| = 20$ Rewrite the equation so that the absolute value expression is on the left side.

To solve $|x + 4| = 20$, we write and then solve an equivalent compound equation:

$$x + 4 = 20 \quad \text{or} \quad x + 4 = -20$$

Now we solve each equation for x:

$$x + 4 = 20 \quad \text{or} \quad x + 4 = -20$$
$$x = 16 \quad | \quad x = -24$$

The values of x for which $f(x) = 20$ are 16 and -24. To check, find $f(16)$ and $f(-24)$, and verify that the result is 20 in each case.

Self Check 4 Let $f(x) = |x + 4|$. For what value(s) of x is $f(x) = 11$?

Now Try Problem 71

2 Solve Equations with Two Absolute Values.

Equations can contain two absolute value expressions. To develop a strategy to solve them, consider the following four true statements.

$|3| = |3|$ (The numbers are the same.) or $|-3| = |-3|$ (The numbers are the same.) or $|3| = |-3|$ (The numbers are opposites.) or $|-3| = |3|$ (The numbers are opposites.)

These four possible cases are really just two cases: *Two absolute value expressions are equal when the expressions within the absolute value bars are equal to or opposites of each other.* This observation suggests the following approach for solving equations having two absolute value expressions.

Solving Equations with Two Absolute Values

For any algebraic expressions X and Y:

To solve $|X| = |Y|$, solve the compound equation $X = Y$ or $X = -Y$.

EXAMPLE 5 Solve: $|5x + 3| = |3x + 25|$

Strategy To solve this equation, we will write and then solve an equivalent compound equation.

Why We can use this approach because the equation is of the form $|X| = |Y|$.

Solution The equation $|5x + 3| = |3x + 25|$, with the two absolute value expressions, is equivalent to the following compound equation:

The expressions within the absolute value symbols are equal

$$5x + 3 = 3x + 25$$
$$2x = 22$$
$$x = 11$$

or

The expressions within the absolute value symbols are opposites

$$5x + 3 = -(3x + 25)$$
$$5x + 3 = -3x - 25 \quad \text{Solve each equation.}$$
$$8x = -28$$
$$x = -\frac{28}{8}$$
$$x = -\frac{7}{2} \quad \text{Simplify the fraction.}$$

Verify that both solutions, 11 and $-\frac{7}{2}$, check by substituting them into the original equation. The solution set is $\left\{11, -\frac{7}{2}\right\}$.

Caution

Don't forget to use parentheses to write the opposite of expressions that have more than one term.

Expression	Opposite
$3x + 25$	$-(3x + 25)$

Self Check 5 Solve: $|2x - 3| = |4x + 9|$

Now Try Problem 47

3 Solve Inequalities of the Form $|X| < k$.

To solve the **absolute value inequality** $|x| < 5$, we must find all real numbers x whose distance from 0 on the number line is less than 5. From the graph, we see that there are many such numbers. For example, -4.999, -3, -2.4, $-1\frac{7}{8}$, $-\frac{3}{4}$, 0, 1, 2.8, 3.001, and 4.999 all meet this requirement. We conclude that the solution set is all numbers between -5 and 5, which can be written in interval notation as $(-5, 5)$.

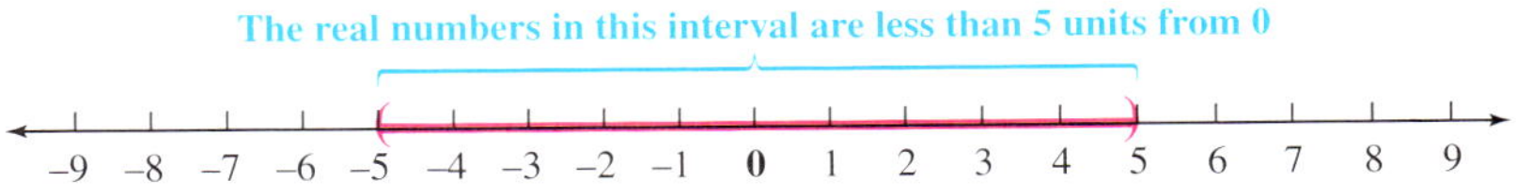

Since x is between -5 and 5, it follows that $|x| < 5$ is equivalent to $-5 < x < 5$. This observation suggests the following approach for solving absolute value inequalities of the form $|X| < k$ and $|X| \leq k$.

Solving $|X| < k$ and $|X| \leq k$

For any positive number k and any algebraic expression X:

To solve $|X| < k$, solve the equivalent double inequality $-k < X < k$.

To solve $|X| \leq k$, solve the equivalent double inequality $-k \leq X \leq k$.

EXAMPLE 6 Solve $|2x - 3| < 9$ and graph the solution set.

Strategy To solve this absolute value inequality, we will write and solve an equivalent double inequality.

Why We can use this approach because the inequality is of the form $|X| < k$, and k is positive.

Solution The absolute value inequality $|2x - 3| < 9$ is equivalent to the double inequality

$$-9 < 2x - 3 < 9$$

which we can solve for x:

$$-9 < 2x - 3 < 9$$

$$-6 < 2x < 12 \quad \text{To isolate the variable term 2x, add 3 to all three parts.}$$

$$-3 < x < 6 \quad \text{To isolate x, divide all parts by 2.}$$

Any number between -3 and 6 is in the solution set, which can be written as $\{x \mid -3 < x < 6\}$. This is the interval $(-3, 6)$; its graph is shown on the right.

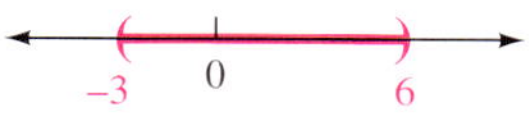

Self Check 6 Solve $|3x + 2| < 4$ and graph the solution set.

Now Try Problems 55 and 59

Because it is related to distance, absolute value can be used to describe the amount of error involved when measurements are taken.

EXAMPLE 7 **Tolerances.** When manufactured parts are inspected by a quality control engineer, they are classified as acceptable if each dimension falls within a given **tolerance range** of the dimensions listed on the blueprint. For the bracket shown in the margin, the distance between the two drilled holes is given as 2.900 inches. Because the tolerance is ± 0.015 inch, this distance can be as much as 0.015 inch longer or 0.015 inch shorter, and the part will be considered acceptable. The acceptable distance d between holes can be represented by the absolute value inequality $|d - 2.900| \leq 0.015$. Solve the inequality and explain the result.

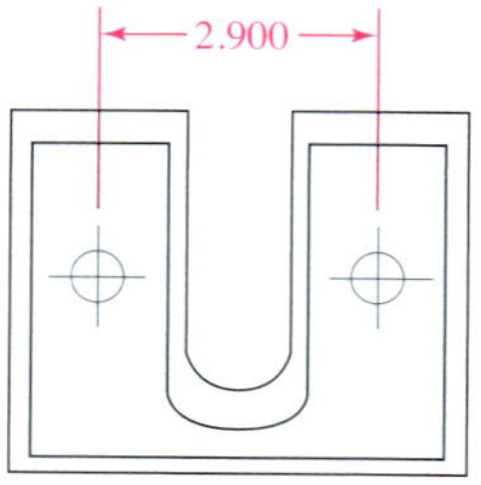

Unless otherwise specified, dimensions are in inches.
Tolerances ±0.015

Strategy To solve $|d - 2.900| \leq 0.015$, we will write and solve an equivalent double inequality.

Why We can use this approach because the inequality is of the form $|X| \leq k$, and k is positive.

Solution The absolute value inequality $|d - 2.900| \leq 0.015$ is equivalent to the double inequality

$$-0.015 \leq d - 2.900 \leq 0.015$$

which we can solve for d:

$$-0.015 \leq d - 2.900 \leq 0.015$$

$$2.885 \leq d \leq 2.915 \quad \text{To isolate d, add 2.900 to all three parts.}$$

The solution set is the interval $[2.885, 2.915]$. This means that the distance between the two holes should be between 2.885 and 2.915 inches, inclusive. If the distance is less than 2.885 inches or more than 2.915 inches, the part should be rejected.

Self Check 7 **Tolerances.** Refer to Example 7. Find the tolerance range if the tolerance is ±0.0015.

Now Try Problem 105

EXAMPLE 8 Solve: $|4x - 5| < -2$

Strategy We will solve this inequality by inspection.

Why The inequality $|4x - 5| < -2$ is of the form $|X| < k$. However, the standard method for solving such inequalities cannot be used because k (in this case, -2) is not positive.

Solution Since $|4x - 5|$ is always greater than or equal to 0 for any real number x, this absolute value inequality has no solution. The solution set is $\varnothing$.

Self Check 8 Solve: $|6x + 24| < -51$

Now Try Problem 61

4 Solve Inequalities of the Form $|X| > k$.

To solve the absolute value inequality $|x| > 5$, we must find all real numbers x whose distance from 0 on the number line is greater than 5. From the following graph, we see that there are many such numbers. For example, -5.001, -6, -7.5, and $-8\frac{3}{8}$, as well as 5.001, 6.2, 7, 8, and $9\frac{1}{2}$ all meet this requirement. We conclude that the solution set is all numbers less than -5 or greater than 5, which can be written as the union of two intervals: $(-\infty, -5) \cup (5, \infty)$.

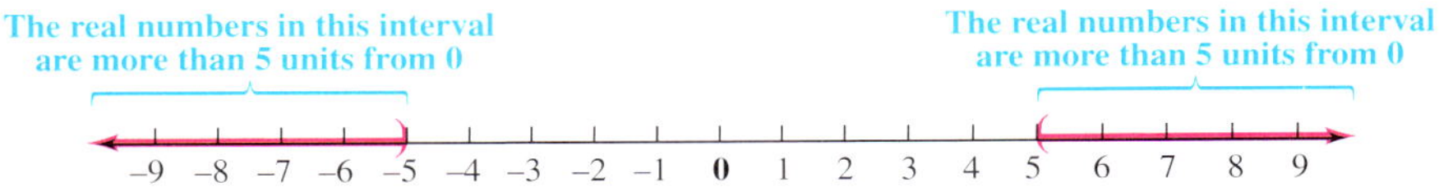

Since x is less than -5 or greater than 5, it follows that $|x| > 5$ is equivalent to $x < -5$ or $x > 5$. This observation suggests the following approach for solving absolute value inequalities of the form $|X| > k$ and $|X| \geq k$.

Solving $|X| > k$ and $|X| \geq k$

For any positive number k and any algebraic expression X:

To solve $|X| > k$, solve the equivalent compound inequality $X < -k$ or $X > k$.

To solve $|X| \geq k$, solve the equivalent compound inequality $X \leq -k$ or $X \geq k$.

EXAMPLE 9 Solve $\left|\frac{3 - x}{5}\right| \geq 6$ and graph the solution set.

Strategy To solve this absolute value inequality, we will write and solve an equivalent compound inequality.

Why We can use this approach because the inequality is of the form $|X| \geq k$, and k is positive.

Solution The absolute value inequality $\left|\frac{3 - x}{5}\right| \geq 6$ is equivalent to the compound inequality

$$\frac{3 - x}{5} \leq -6 \quad \text{or} \quad \frac{3 - x}{5} \geq 6$$

Notation

In this case, the solution set cannot be represented by a single interval. The union symbol $\cup$ must be used to write the solution set in interval notation.

Now we solve each inequality for x:

$$\frac{3-x}{5} \le -6 \quad \text{or} \quad \frac{3-x}{5} \ge 6$$

$$3 - x \le -30 \quad \Big| \quad 3 - x \ge 30 \qquad \text{To clear the fraction, multiply both sides by 5.}$$

$$-x \le -33 \quad \Big| \quad -x \ge 27 \qquad \text{To isolate the variable term } -x, \text{ subtract 3 from both sides.}$$

$$x \ge 33 \quad \Big| \quad x \le -27 \qquad \text{To isolate } x, \text{ divide both sides by } -1 \text{ and reverse the direction of the inequality symbol.}$$

The solution set is the union of two intervals: $(-\infty, -27] \cup [33, \infty)$. Using set-builder notation, the solution set is written as $\{x \mid x \le -27 \text{ or } x \ge 33\}$. Its graph appears on the right.

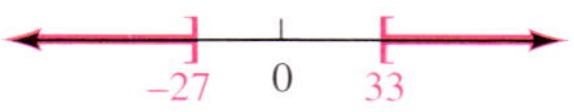

Self Check 9 Solve $\left|\frac{2-x}{4}\right| \ge 1$ and graph the solution set.

Now Try Problems 63 and 65

EXAMPLE 10 Solve $6 < \left|\frac{2}{3}x - 2\right| - 3$ and graph the solution set.

Strategy We will first write the inequality in an equivalent form with the absolute value on the left side.

Why It's usually easier to solve an absolute value inequality if the absolute value appears on the left side of the inequality.

Solution

$$6 < \left|\frac{2}{3}x - 2\right| - 3 \qquad \text{This is the inequality to solve.}$$

$$\left|\frac{2}{3}x - 2\right| - 3 > 6 \qquad \text{Write the inequality with the absolute value on the left side.}$$

$$\left|\frac{2}{3}x - 2\right| > 9 \qquad \text{Add 3 to both sides to isolate the absolute value expression.}$$

Success Tip

To write $6 < \left|\frac{2}{3}x - 2\right| - 3$ with the absolute value on the left, we used the following property:

If $a < b$, then $b > a$.

After isolating the absolute value expression on the left side, the resulting inequality has the form $|X| > k$. To solve this absolute value inequality, we write and solve an equivalent compound inequality:

$$\frac{2}{3}x - 2 < -9 \quad \text{or} \quad \frac{2}{3}x - 2 > 9$$

$$\frac{2}{3}x < -7 \quad \Big| \quad \frac{2}{3}x > 11 \qquad \text{Add 2 to both sides.}$$

$$2x < -21 \quad \Big| \quad 2x > 33 \qquad \text{Multiply both sides by 3.}$$

$$x < -\frac{21}{2} \quad \Big| \quad x > \frac{33}{2} \qquad \text{To isolate } x, \text{ divide both sides by 2.}$$

The solution set is the union of two intervals: $\left(-\infty, -\frac{21}{2}\right) \cup \left(\frac{33}{2}, \infty\right)$. Its graph appears on the right. Using set-builder notation, the solution set can be written as $\left\{x \mid x < -\frac{21}{2} \text{ or } x > \frac{33}{2}\right\}$.

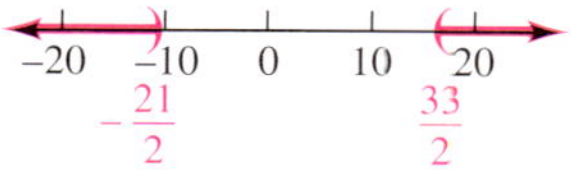

Self Check 10 Solve $3 < \left|\frac{3}{4}x + 2\right| - 1$ and graph the solution set.

Now Try ▶ Problem 67

EXAMPLE 11 Solve $\left|\frac{x}{8} - 1\right| \geq -4$ and graph the solution set.

Strategy We will solve this inequality by inspection.

Why The inequality $\left|\frac{x}{8} - 1\right| \geq -4$ is of the form $|x| \geq k$. However, the standard method for solving such inequalities cannot be used because k is not positive.

Solution Since $\left|\frac{x}{8} - 1\right|$ is always greater than or equal to 0 for any real number x it will also be greater than or equal to -4 for any real number x. Therefore, this absolute value inequality is true for all real numbers. The solution set is the interval $(-\infty, \infty)$ or $\mathbb{R}$. Its graph appears on the right.

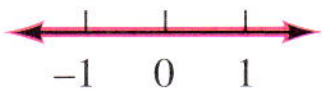

Self Check 11 Solve $|-x - 9| > -0.5$ and graph the solution set.

Now Try ▶ Problem 69

The following summary shows how we can interpret absolute value in three ways. Assume $k > 0$.

Geometric description	*Graphic description*	*Algebraic description*
1. $\lvert x\rvert = k$ means that x is k units from 0 on the number line.	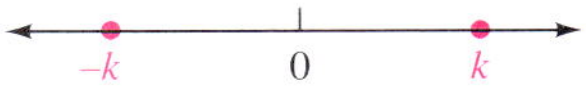	$\lvert x\rvert = k$ is equivalent to $x = k$ or $x = -k$.
2. $\lvert x\rvert < k$ means that x is less than k units from 0 on the number line.	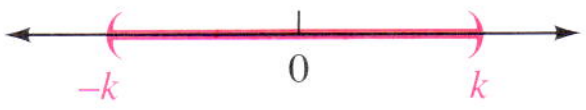	$\lvert x\rvert < k$ is equivalent to $-k < x < k$.
3. $\lvert x\rvert > k$ means that x is more than k units from 0 on the number line.	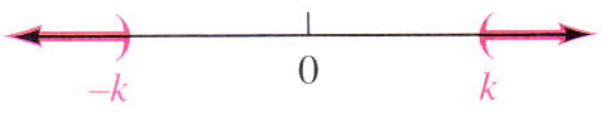	$\lvert x\rvert > k$ is equivalent to $x > k$ or $x < -k$.

Using Your **Calculator** ▶ Solving Absolute Value Equations and Inequalities

We can solve absolute value equations and inequalities with a graphing calculator. For example, to solve $|2x - 3| = 9$, we graph the equations $y = |2x - 3|$ and $y = 9$ on the same coordinate system, as shown in the figure. The equation $|2x - 3| = 9$ will be true for all x-coordinates of points that lie on *both* graphs. Using the TRACE or the INTERSECT feature, we can see that the graphs intersect at the points $(\mathbf{-3}, 9)$ and $(\mathbf{6}, 9)$. Thus, the solutions of the absolute value equation are -3 and 6.

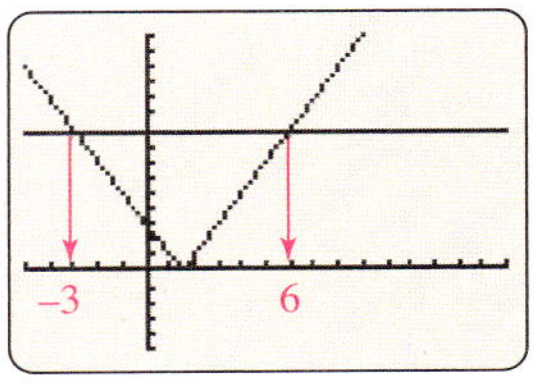

The inequality $|2x - 3| < 9$ will be true for all x-coordinates of points that lie on the graph of $y = |2x - 3|$ and *below* the graph of $y = 9$. We see that these values of x are between -3 and 6. Thus, the solution set is the interval $(-3, 6)$.

The inequality $|2x - 3| > 9$ will be true for all x-coordinates of points that lie on the graph of $y = |2x - 3|$ and *above* the graph of $y = 9$. We see that these values of x are less than -3 or greater than 6. Thus, the solution set is the union of two intervals: $(-\infty, -3) \cup (6, \infty)$.

SECTION 8.5 STUDY SET

VOCABULARY

Fill in the blanks.

1. The ________ ______ of a number is its distance from 0 on a number line.
2. $|2x - 1| = 10$ is an absolute value ________ and $|2x - 1| > 10$ is an absolute value ________.
3. To ________ the absolute value in $|3 - x| - 4 = 5$, we add 4 to both sides.
4. When we say that the absolute value equation and a compound equation are equivalent, we mean that they have the same ________.
5. When two equations are joined by the word *or,* such as $x + 1 = 5$ or $x + 1 = -5$, we call the statement a ________ equation.
6. $f(x) = |6x - 2|$ is called an absolute value ________.

CONCEPTS

Fill in the blanks.

7. To solve absolute value equations and inequalities, we write and solve equivalent ________ equations and inequalities.
8. Two absolute value expressions are equal when the expressions within the absolute value bars are equal to or ________ of each other.
9. Consider the following real numbers:
   $-3, -2.01, -2, -1.99, -1, 0, 1, 1.99, 2, 2.01, 3$
   a. Which of them make $|x| = 2$ true?
   b. Which of them make $|x| < 2$ true?
   c. Which of them make $|x| > 2$ true?
10. Determine whether -3 is a solution of the given equation or inequality.
    a. $|x - 1| = 4$ b. $|x - 1| > 4$
    c. $|x - 1| \leq 4$ d. $|5 - x| = |x + 12|$
11. For each absolute value equation, write an equivalent compound equation.
    a. $|x - 7| = 8$ is equivalent to
       $x - 7 =$ ____ or $x - 7 =$ ____
    b. $|x + 10| = |x - 3|$ is equivalent to
       $x + 10 =$ ____ or $x + 10 =$ ____
12. For each absolute value inequality, write an equivalent compound inequality.
    a. $|x + 5| < 1$ is equivalent to
       ____ $< x + 5 <$ ____
    b. $|x - 6| \geq 3$ is equivalent to
       $x - 6 \leq$ ____ or $x - 6 \geq$ ____
13. For each absolute value equation or inequality, write an equivalent compound equation or inequality.
    a. $|x| = 8$ b. $|x| \geq 8$
    c. $|x| \leq 8$ d. $|5x - 1| = |x + 3|$
14. Perform the necessary steps to isolate the absolute value expression on one side of the equation. *Do not solve.*
    a. $|3x + 2| - 7 = -5$
    b. $6 + 2|5x - 19| \leq 40$
15. Determine the solution set of each absolute value equation or inequality by inspection. (No work is necessary.) Your answer should be either *all real numbers* or *no solution.*
    a. $|7x + 6| = -8$ b. $|7x + 6| \leq -8$
    c. $|7x + 6| \geq -8$
16. Write the inequality $10 > |16x - 3|$ in an equivalent form with the absolute value expression on the left side.

NOTATION

17. Match each equation or inequality with its graph.
    a. $|x| = 1$ i.
    b. $|x| > 1$ ii.
    c. $|x| < 1$ iii.

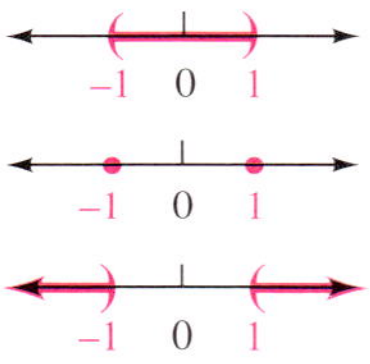

18. Describe the set graphed below using interval notation.

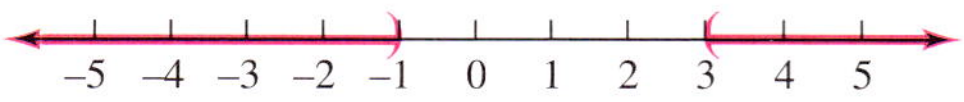

GUIDED PRACTICE

Solve each equation. See Example 1.

19. $|x| = 23$
20. $|x| = 90$
21. $|x - 5| = 8$
22. $|x - 7| = 4$
23. $|3x + 2| = 16$
24. $|5x - 3| = 22$
25. $\left|\frac{x}{5}\right| = 10$
26. $\left|\frac{x}{7}\right| = 2$
27. $|2x + 3.6| = 9.8$
28. $|4x - 24.8| = 32.4$
29. $\left|\frac{7}{2}x + 3\right| = -5$
30. $|x - 2.1| = -16.3$

Solve each equation. See Example 2.

31. $|x - 3| - 19 = 3$
32. $|x - 10| + 30 = 50$
33. $|3x - 7| + 8 = 22$
34. $|6x - 3| + 7 = 28$
35. $|3 - 4x| + 1 = 6$
36. $|8 - 5x| - 8 = 10$
37. $\left|\frac{7}{8}x + 5\right| - 2 = 7$
38. $\left|\frac{3}{4}x + 4\right| - 5 = 11$

Solve each equation. See Example 3.

39. $\left|\frac{1}{5}x + 2\right| - 8 = -8$
40. $\left|\frac{1}{9}x + 4\right| + 25 = 25$
41. $2|3x + 24| = 0$
42. $8\left|\frac{2x}{3} + 10\right| = 0$
43. $-5|2x - 9| + 14 = 14$
44. $-10|16x + 4| - 3 = -3$
45. $6 - 3|10x + 5| = 6$
46. $15 - |12x + 12| = 15$

Solve each equation. See Example 5.

47. $|5x - 12| = |4x - 16|$
48. $|4x - 7| = |3x - 21|$
49. $|10x| = |x - 18|$
50. $|6x| = |x + 45|$
51. $|2 - x| = |3x + 2|$
52. $|4x + 3| = |9 - 2x|$
53. $|5x - 7| = |4(x + 1)|$
54. $|2x + 1| = |3(x + 1)|$

Solve each inequality. Graph the solution set and write it using interval notation. See Examples 6 and 8.

55. $|x| < 4$
56. $|x| < 9$
57. $|x + 9| \le 12$
58. $|x - 8| \le 12$
59. $|3x - 2| < 10$
60. $|4 - 3x| \le 13$
61. $|5x - 12| < -5$
62. $|3x + 2| \le -3$

Solve each inequality. Graph the solution set and write it using interval notation. See Examples 9–11.

63. $|x| > 3$
64. $|x| > 7$
65. $|x - 12| > 24$
66. $|x + 5| \ge 7$
67. $0 \le |5x - 1| - 2$
68. $0 \le |6x - 3| - 5$
69. $|4x + 3| \ge -5$
70. $|7x + 2| \ge -8$

See Examples 4, 6, and 9.

71. Let $f(x) = |x + 3|$. For what value(s) of x is $f(x) = 3$?
72. Let $g(x) = |2 - x|$. For what value(s) of x is $g(x) = 2$?
73. Let $f(x) = |2(x - 1) + 4|$. For what value(s) of x is $f(x) < 4$?
74. Let $h(x) = \left|\frac{x}{5} - \frac{1}{2}\right|$. For what value(s) of x is $h(x) > \frac{9}{10}$?

TRY IT YOURSELF

Solve each equation and inequality. For the inequalities, graph the solution set and write it using interval notation.

75. $|3x + 2| + 1 > 15$
76. $|2x - 5| - 5 > 20$
77. $6\left|\frac{x - 2}{3}\right| \le 24$
78. $8\left|\frac{x - 2}{3}\right| > 32$
79. $-7 = 2 - |0.3x - 3|$
80. $-1 = 1 - |0.1x + 8|$
81. $|2 - 3x| \ge -8$
82. $|-1 - 2x| > 5$
83. $|7x + 12| = |x - 6|$
84. $|8 - x| = |x + 2|$
85. $2 \ge 3|2 - 3x| + 2$
86. $7 \ge |15x - 45| + 7$
87. $-14 = |x - 3|$
88. $-75 = |x + 4|$
89. $\frac{6}{5} = \left|\frac{3x}{5} + \frac{x}{2}\right|$
90. $\frac{11}{12} = \left|\frac{x}{3} - \frac{3x}{4}\right|$
91. $-|2x - 3| < -7$
92. $-|3x + 1| < -8$
93. $|0.5x + 1| < -23$
94. $15 \ge 7 - |1.4x + 9|$

Look Alikes . . .

95. a. $\frac{x}{10} - 1 = 1$ b. $\left|\frac{x}{10} - 1\right| = 1$
c. $\frac{x}{10} - 1 > 1$ d. $\left|\frac{x}{10} - 1\right| > 1$

96. a. $4x - 5 = 15$ b. $|4x - 5| = 15$
c. $4x - 5 \le 15$ d. $|4x - 5| \le 15$

97. a. $0.9 - 0.3x = 8.4$ b. $|0.9 - 0.3x| = 8.4$
c. $0.9 - 0.3x > 8.4$ d. $|0.9 - 0.3x| > 8.4$

98. a. $8(x - 4) = 6x - 44$
b. $|8(x - 4)| = |6x - 44|$

Solve the absolute value inequality in part a. Graph the solution set and write it in interval notation. Then use your work from part a to determine the solution set for the absolute value inequality in part b. (No new work is necessary!) Graph the solution set and write it in interval notation.

99. **a.** $|8x - 40| \le 16$ **b.** $|8x - 40| \ge 16$

100. **a.** $0 \le |14 - 27x|$ **b.** $0 > |14 - 27x|$

101. **a.** $\left|\frac{4x - 4}{3}\right| - 1 > 11$

b. $\left|\frac{4x - 4}{3}\right| - 1 \le 11$

102. **a.** $\left|-\frac{1}{2}x - 3\right| + 2 > 7$

b. $\left|-\frac{1}{2}x - 3\right| + 2 < 7$

APPLICATIONS

103. Temperature Ranges. The temperatures on a sunny summer day satisfied the inequality $|t - 78°| \le 8°$, where t is a temperature in degrees Fahrenheit. Solve this inequality and express the range of temperatures as a double inequality.

104. Operating Temperatures. A car CD player has an operating temperature of $|t - 40°| < 80°$, where t is a temperature in degrees Fahrenheit. Solve the inequality and express this range of temperatures as an interval.

105. Auto Mechanics. On most cars, the bottoms of the front wheels are closer together than the tops, creating a *camber angle.* This lessens road shock to the steering system. (See the illustration.) The specifications for a certain car state that the camber angle c of its wheels should be $0.6° \pm 0.5°$.

a. Express the range with an inequality containing absolute value symbols.

b. Solve the inequality and express this range of camber angles as an interval.

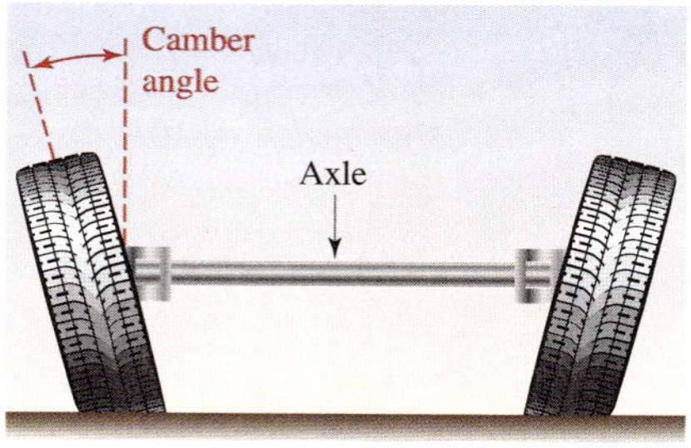

106. Steel Production. A sheet of steel is to be 0.250 inch thick with a tolerance of 0.025 inch.

a. Express this specification with an inequality containing absolute value symbols, using x to represent the thickness of a sheet of steel.

b. Solve the inequality and express the range of thickness as an interval.

107. Error Analysis. In a lab, students measured the percent of copper p in a sample of copper sulfate. The students know that copper sulfate is actually 25.46% copper by mass. They are to compare their results to the actual value and find the amount of *experimental error.* Which measurements shown in the illustration satisfy the absolute value inequality $|p - 25.46| \le 1.00$?

Lab 4 Section A
Title:
"Percent copper (Cu) in copper sulfate ($CuSO_4 \cdot 5H_2O$)"

Results

	% Copper
Trial #1:	22.91%
Trial #2:	26.45%
Trial #3:	26.49%
Trial #4:	24.76%

108. Error Analysis. See Exercise 107. Which measurements satisfy the absolute value inequality $|p - 25.46| > 1.00$?

WRITING

109. Explain the error.

Solve: $|x| + 2 = 6$

$x + 2 = 6$ or $x + 2 = -6$

$x = 4$ | $x = -8$

110. Explain why the equation $|x - 4| = -5$ has no solution.

111. Explain the differences between the solution sets of $|x| = 8$, $|x| < 8$, and $|x| > 8$.

112. Explain how to use the graph in the illustration to solve the following.

a. $|x - 2| = 3$

b. $|x - 2| \le 3$

c. $|x - 2| \ge 3$

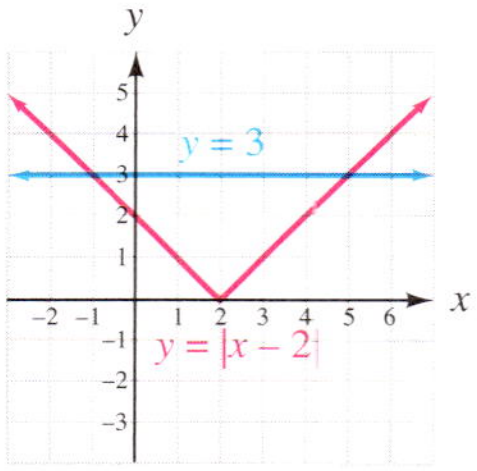

REVIEW

113. Flutes. When it is assembled, a flute is 29 inches long. The middle piece is 4 inches less than twice as long as the first piece. The last piece is two-thirds as long as the first piece. Find the length of each piece of the flute.

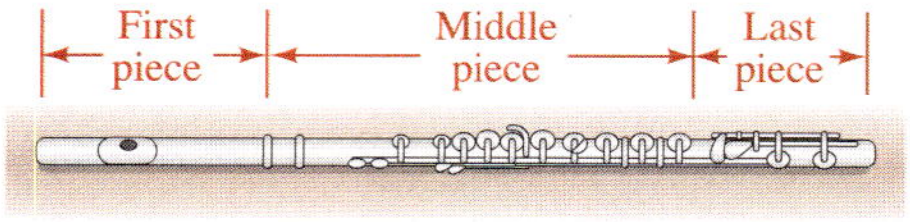

114. Commercials. For the typical "one-hour" prime-time television slot, the number of minutes of commercials is $\frac{3}{7}$ of the number of minutes of the actual program. Determine how many minutes of the program are shown in that one hour.

CHALLENGE PROBLEMS

115. **a.** For what values of k does $|x| + k = 0$ have exactly two solutions?

b. For what values of k does $|x| + k = 0$ have exactly one solution?

116. Solve: $2^{|2x-3|} = 64$

SECTION 8.6

Review of Factoring Methods: GCF, Grouping, Trinomials

OBJECTIVES

1 Factor out the greatest common factor.
2 Factor by grouping.
3 Use factoring to solve formulas for a specified variable.
4 Factor trinomials.
5 Use substitution to factor trinomials.
6 Use the grouping method to factor trinomials.

ARE YOU READY?

The following problems review some basic skills that are needed when factoring expressions.

1. Find the prime factorization of 108.

2. Multiply: $6b(b^3 + 2b + 4)$

3. Multiply: $(x + 8)(x - 6)$

4. Multiply and simplify: $m(m + 2) - 5(m + 2)$

5. Find two integers whose product is 10 and whose sum is 7.

6. Find two integers whose product is -18 and whose sum is 3.

In Chapter 6, we discussed how to factor polynomials. In this section, we will review that material.

1 Factor Out the Greatest Common Factor.

Recall that *when we factor a polynomial, we write a sum of terms as a product of factors.* To perform the most basic type of factoring, we determine whether the terms of the given polynomial have any common factors. This process, called **factoring out the greatest common factor,** is based on the distributive property.

EXAMPLE 1 Factor: $3xy^2z^3 + 6xz^2 - 9xyz^4$

Strategy We will determine the GCF of the terms of the polynomial. Then we will write each term of the polynomial as the product of the GCF and one other factor.

Why We can then use the distributive property to factor out the GCF.

Solution We begin by factoring each term:

$$\left.\begin{aligned} 3xy^2z^3 &= 3 \cdot x \cdot y \cdot y \cdot z \cdot z \cdot z \\ 6xz^2 &= 2 \cdot 3 \cdot x \cdot z \cdot z \\ 9xyz^4 &= 3 \cdot 3 \cdot x \cdot y \cdot z \cdot z \cdot z \cdot z \end{aligned}\right\} \text{GCF} = 3 \cdot x \cdot z \cdot z = 3xz^2$$

Since each term has one factor of 3, one factor of x, and two factors of z, and there are no other common factors, $3xz^2$ is the greatest common factor of the three terms. We write each term as the product of the GCF, $3xz^2$, and one other factor and proceed as follows:

$$3xy^2z^3 + 6xz^2 - 9xyz^4 = 3xz^2 \cdot y^2z + 3xz^2 \cdot 2 - 3xz^2 \cdot 3yz^2$$ This can be done mentally.

$$= 3xz^2(y^2z + 2 - 3yz^2)$$ Factor out the GCF, $3xz^2$.

We can check the factorization using multiplication.

$$3xz^2(y^2z + 2 - 3yz^2) = 3xy^2z^3 + 6xz^2 - 9xyz^4$$ This is the original polynomial.

Self Check 1 Factor: $6a^2b^2 - 4ab^3 + 2ab^2$

Now Try Problem 25

Success Tip

Always verify a factorization by performing the indicated multiplication. The result should be the original polynomial.

When asked to factor a polynomial whose leading coefficient is negative, we factor out the *opposite of the GCF.*

EXAMPLE 2 Factor out the opposite of the GCF from $-6u^2v^3 + 8u^3v^2$.

Strategy We will determine the GCF of the terms of the polynomial. Then we will write each term as the product of the opposite of the GCF and one other factor.

Why We can then use the distributive property to factor out the opposite of the GCF.

Solution Because the greatest common factor of the two terms is $2u^2v^2$, the opposite of the greatest common factor is $-2u^2v^2$. To factor out $-2u^2v^2$, we proceed as follows:

$$-6u^2v^3 + 8u^3v^2 = -2u^2v^2 \cdot 3v - (-2u^2v^2)4u \quad \text{This can be done mentally.}$$
$$= -2u^2v^2(3v - 4u)$$

The leading coefficient of the polynomial within the parentheses is positive.

Success Tip

It is standard practice to factor in such a way that the leading coefficient of the polynomial within the parentheses is positive.

Self Check 2 Factor out the opposite of the GCF: $-3p^3q + 6p^2q^2$

Now Try Problem 35

A polynomial that cannot be factored is called a **prime polynomial** or an **irreducible polynomial.** For example, $9x + 16$ is a prime polynomial because its two terms, $9x$ and 16, have no common factors other than 1.

In the next example, we see that a common factor can have more than one term.

EXAMPLE 3 Factor: $a(x - y + z) - b(x - y + z) + 3(x - y + z)$

Strategy We will factor out the trinomial $x - y + z$ from each term.

Why $x - y + z$ is the GCF of each term of the given expression.

Solution

$$a(x - y + z) - b(x - y + z) + 3(x - y + z) = (x - y + z)(a - b + 3)$$

Self Check 3 Factor: $c^2(y^2 + 1) + d^2(y^2 + 1)$

Now Try Problem 41

2 Factor by Grouping.

Although the terms of many polynomials don't have a common factor, other than 1, it is possible to factor some of them by arranging their terms in convenient groups. This method is called **factoring by grouping.**

EXAMPLE 4 Factor: $2c - 2d + cd - d^2$

Strategy Since the four terms of the polynomial do not have a common factor (other than 1), we will attempt to factor the polynomial by grouping. We will factor out a common factor from the first two terms and from the last two terms.

Why This will produce a common binomial factor that can be factored out.

Solution If we group the terms as highlighted in blue, the first two terms have a common factor, 2, and the last two terms have a common factor, d. When we factor out the common factor from each group, a common binomial factor $c - d$ appears.

Success Tip

You may want to review the steps in the process of factoring by grouping on page 438.

$$2c - 2d + cd - d^2 = 2(c - d) + d(c - d)$$ Factor out 2 from $2c - 2d$ and d from $cd - d^2$. Don't forget the + sign.

$$= (c - d)(2 + d)$$ Factor out the common binomial factor, $c - d$.

We can check by multiplying:

$$(c - d)(2 + d) = 2c + cd - 2d - d^2$$
$$= 2c - 2d + cd - d^2$$ Rearrange the terms to get the original polynomial.

Self Check 4 Factor: $7m - 7n + mn - n^2$

Now Try Problems 43 and 47

The instruction "Factor" means to factor the given expression completely. Each factor of a completely factored expression will be prime.

To factor a polynomial, it is often necessary to factor more than once. When factoring a polynomial, *always look for a common factor first.*

EXAMPLE 5 Factor: $3x^3y - 4x^2y^2 - 6x^2y + 8xy^2$

Strategy Since all four terms have a common factor of xy, we factor it out first. Then we will attempt to factor the resulting polynomial by grouping.

Why Factoring out the GCF first makes factoring by any method easier.

Solution We begin by factoring out the common factor xy.

$$3x^3y - 4x^2y^2 - 6x^2y + 8xy^2 = xy(3x^2 - 4xy - 6x + 8y)$$

We can now factor the resulting four-term polynomial $3x^2 - 4xy - 6x + 8y$ by grouping:

$$3x^3y - 4x^2y^2 - 6x^2y + 8xy^2$$
$$= xy(3x^2 - 4xy - 6x + 8y)$$
$$= xy[x(3x - 4y) - 2(3x - 4y)]$$ Brackets are needed to enclose the factoring by grouping steps. Factor x from $3x^2 - 4xy$ and -2 from $-6x + 8y$.

$$= xy(3x - 4y)(x - 2)$$ Factor out $3x - 4y$. The brackets are no longer needed.

Because xy, $3x - 4y$, and $x - 2$ are prime, no further factoring can be done; the factorization is complete.

Self Check 5 Factor: $3a^3b + 3a^2b - 2a^2b^2 - 2ab^2$

Now Try Problem 51

3 Use Factoring to Solve Formulas for a Specified Variable.

Factoring is often required to solve a formula for one of its variables.

EXAMPLE 6 **Electronics.** The formula $r_1r_2 = rr_2 + rr_1$ is used in electronics to relate the combined resistance, r, of two resistors wired in parallel. The variable r_1 represents the resistance of the first resistor, and the variable r_2 represents the resistance of the second. Solve for r_2.

Strategy To isolate r_2 on one side of the equation, we will get all the terms involving r_2 on the left side and all the terms not involving r_2 on the right side.

Why To *solve a formula for a specified variable* means to isolate that variable on one side of the equation, with all other variables and constants on the opposite side.

Solution

We want to isolate this variable on one side of the equation.

$$r_1r_2 = rr_2 + rr_1$$

$$r_1\mathbf{r_2} - r\mathbf{r_2} = rr_1$$ To eliminate rr_2 on the right side, subtract rr_2 from both sides.

$$\mathbf{r_2}(r_1 - r) = rr_1$$ On the left side, factor out the GCF r_2 from $r_1r_2 - rr_2$.

$$\frac{r_2\overset{1}{\cancel{(r_1 - r)}}}{\underset{1}{\cancel{r_1 - r}}} = \frac{rr_1}{r_1 - r}$$ To isolate r_2 on the left side, divide both sides by $r_1 - r$.

$$r_2 = \frac{rr_1}{r_1 - r}$$ Simplify the left side by removing the common factor $r_1 - r$ from the numerator and denominator.

Self Check 6 Solve $f_1f_2 = ff_1 + ff_2$ for f_1.

Now Try Problem 57

4 Factor Trinomials.

Recall that many trinomials factor as the product of two binomials.

EXAMPLE 7 Factor: $x^2 - 6x + 8$

Strategy We will assume that this trinomial is the product of two binomials. We must find the terms of the binomials.

Why Since the terms of $x^2 - 6x + 8$ do not have a common factor (other than 1), the only option is to try to factor it as the product of two binomials.

Solution We represent the binomials using two sets of parentheses. Since the first term of the trinomial is x^2, we enter x and x as the first terms of the binomial factors.

$$x^2 - 6x + 8 = (x \;\square)(x \;\square)$$ Because $x \cdot x$ will give x^2.

The second terms of the binomials must be two integers whose product is 8 and whose sum is −6. We list all possible integer-pair factors of 8 in the table.

Factors of 8	Sum of the factors of 8
1(8) = 8	1 + 8 = 9
2(4) = 8	2 + 4 = 6
−1(−8) = 8	−1 + (−8) = −9
−2(−4) = 8	**−2 + (−4) = −6**

← This is the pair to choose.

The fourth row of the table contains the correct pair of integers −2 and −4, whose product is 8 and whose sum is −6. To complete the factorization, we enter −2 and −4 as the second terms of the binomial factors.

$$x^2 - 6x + 8 = (x - 2)(x - 4)$$

Success Tip

You may want to review the procedure for factoring trinomials whose leading coefficient is 1 on page 447.

The Language of Algebra

Make sure you understand the following vocabulary: *Many trinomials factor as the product of two binomials.*

$$\underbrace{x^2 - 6x + 8}_{\text{Trinomial}} = \underbrace{(x - 2)(x - 4)}_{\text{Product of two binomials}}$$

Check: We can verify the factorization by multiplication:

$$(x-2)(x-4) = x^2 - 4x - 2x + 8 \quad \text{Use the FOIL method.}$$
$$= x^2 - 6x + 8 \quad \text{This is the original trinomial.}$$

Self Check 7 Factor: $a^2 - 7a + 12$

Now Try Problem 63

EXAMPLE 8 Factor: $2a^2 + 4ab - 30b^2$

Strategy We will factor out the GCF, 2, first. Then we will factor the resulting trinomial.

Why The first step in factoring any polynomial is to factor out the GCF. Factoring out the GCF first makes factoring by any method easier.

Solution Each term in this trinomial has a common factor of 2, which can be factored out.

$$2a^2 + 4ab - 30b^2 = 2(a^2 + 2ab - 15b^2)$$

Next, we factor $a^2 + 2ab - 15b^2$. Since the first term of the trinomial is a^2, the first term of each binomial factor must be a. Since the third term contains b^2, the last term of each binomial factor must contain b. To complete the factorization, we need to determine the coefficient of each b-term.

$$a^2 + 2ab - 15b^2 = (a \,\square\, b)(a \,\square\, b) \quad \text{Because } a \cdot a \text{ will give } a^2 \text{ and } b \cdot b \text{ will give } b^2.$$

The coefficients of b must be two integers whose product is -15 and whose sum is 2. We list the factors of -15 and find the pair whose sum is 2.

This is the pair to choose. ↓

$1(-15)$ $3(-5)$ $5(-3)$ $15(-1)$

Caution

Be sure to include all factors in the final answer. Here, a common error is to forget to write the GCF, which is 2.

The only factorization where the sum of the factors is 2 (which is the coefficient of the middle term of $a^2 + 2ab - 15b^2$) is $5(-3)$. Thus,

$$2a^2 + 4ab - 30b^2 = 2(a^2 + 2ab - 15b^2)$$
$$= 2(a + 5b)(a - 3b)$$

Verify this result by multiplication.

Self Check 8 Factor: $3p^2 + 6pq - 24q^2$

Now Try Problem 67

There are more combinations of coefficients to consider when factoring trinomials with leading coefficients other than 1. Because it is not easy to give specific rules for factoring such trinomials, we will use a method called the **trial-and-check method.**

EXAMPLE 9 Factor: $3p^2 - 4p - 4$

Strategy We will assume that this trinomial is the product of two binomials. To find their terms, we will make educated guesses and then check them using multiplication.

Why Since the terms of the trinomial do not have a common factor (other than 1), the only option is to try to factor it as the product of two binomials.

Solution To factor the trinomial, we note that the first terms of the binomial factors must be $3p$ and p to give the first term of $3p^2$.

$$3p^2 - 4p - 4 = (3p \square)(p \square)$$ Because $3p \cdot p$ will give $3p^2$.

The second terms of the binomials must be two integers whose product is -4. There are three such pairs: $1(-4)$, $-1(4)$, and $-2(2)$. When these pairs are entered, and then reversed, as second terms of the binomials, there are six possibilities to consider.

For 1 and −4: $(3p + 1)(p - 4)$: $-12p + p = -11p$ or $(3p - 4)(p + 1)$: $3p + (-4p) = -p$

For −1 and 4: $(3p - 1)(p + 4)$: $12p + (-p) = 11p$ or $(3p + 4)(p - 1)$: $-3p + 4p = p$

For −2 and 2: $(3p - 2)(p + 2)$: $6p + (-2p) = 4p$ or $(3p + 2)(p - 2)$: $-6p + 2p = -4p$

Of these possibilities, only the one in blue gives the required middle term of $-4p$. Thus,

$$3p^2 - 4p - 4 = (3p + 2)(p - 2)$$

Success Tip

You may want to review the procedure for factoring trinomials using the trial-and-check method on page 458.

Notation

By the commutative property of multiplication, the factors of a trinomial can be written in either order. Thus, we also could write:

$3p^2 - 4p - 4 = (p - 2)(3p + 2)$

Self Check 9 Factor: $4q^2 - 9q - 9$

Now Try Problem 75

EXAMPLE 10 Factor: $6y^3 + 13x^2y^3 + 6x^4y^3$

Strategy We write the expression in descending powers of x.

Why It is easier to factor a trinomial if its terms are written in descending powers of one variable.

Solution We write the expression in descending powers of x and factor out the greatest common factor, y^3.

$$6y^3 + 13x^2y^3 + 6x^4y^3 = 6x^4y^3 + 13x^2y^3 + 6y^3$$
$$= y^3(6x^4 + 13x^2 + 6)$$

To factor $6x^4 + 13x^2 + 6$, we examine its terms.

- Since the first term is $6x^4$, the first terms of the binomial factors must be either $2x^2$ and $3x^2$ or x^2 and $6x^2$.

$$6x^4 + 13x^2 + 6 = (2x^2 \square)(3x^2 \square) \text{ or } (x^2 \square)(6x^2 \square)$$

Success Tip

Always write the terms of a trinomial in descending powers of one variable before attempting to factor it.

- Since the signs of the middle term and the last term of the trinomial are positive, the signs within each binomial factor will be positive.
- Since the product of the last terms of the binomial factors must be 6, we must find two numbers whose product is 6 that will lead to a middle term of $13x^2$.

After trying some combinations, we find the one that works.

$$\begin{aligned} 6x^4y^3 + 13x^2y^3 + 6y^3 &= y^3(6x^4 + 13x^2 + 6) \\ &= y^3(2x^2 + 3)(3x^2 + 2) \end{aligned}$$

Self Check 10 Factor: $4b + 11a^2b + 6a^4b$

Now Try Problem 87

5 Use Substitution to Factor Trinomials.

For more complicated expressions, especially those involving a quantity within parentheses, a substitution sometimes helps to simplify the factoring process.

EXAMPLE 11 Factor: $(x + y)^2 + 7(x + y) + 12$

Strategy We will use a substitution where we will replace each expression $x + y$ with the variable z and factor the resulting trinomial.

Why The resulting trinomial will be easier to factor because it will be in only one variable, z.

Solution If we use the substitution $z = x + y$, we obtain

$$\begin{aligned} (x + y)^2 + 7(x + y) + 12 &= z^2 + 7z + 12 && \text{Replace } x + y \text{ with } z. \\ &= (z + 4)(z + 3) && \text{Factor the trinomial.} \end{aligned}$$

To find the factorization of $(x + y)^2 + 7(x + y) + 12$, we substitute $x + y$ for each z in the expression $(z + 4)(z + 3)$.

$$(z + 4)(z + 3) = (x + y + 4)(x + y + 3)$$

Thus, $(x + y)^2 + 7(x + y) + 12 = (x + y + 4)(x + y + 3)$

Self Check 11 Factor: $(a + b)^2 - 3(a + b) - 10$

Now Try Problem 95

6 Use the Grouping Method to Factor Trinomials.

Another way to factor trinomials is to write them as equivalent four-termed polynomials and factor by grouping.

EXAMPLE 12 Factor by grouping: **a.** $x^2 + 8x + 15$ **b.** $10x^2 + 13xy - 3y^2$

Strategy In each case, we will express the middle term of the trinomial as the sum of two terms.

Why We want to produce an equivalent four-termed polynomial that can be factored by grouping.

Solution

Success Tip

You may want to review the procedure for factoring trinomials using factoring by grouping (the key number method) on page 460.

a. Since $x^2 + 8x + 15 = 1x^2 + 8x + 15$, we identify a as 1, b as 8, and c as 15. The **key number** is $ac = 1(15) = 15$. We must find two integers whose product is the key number 15 and whose sum is $b = 8$. Since the integers must have a positive product and a positive sum, we consider only positive factors of 15.

Key number = 15 $\quad b = 8$

Positive factors of 15	Sum of the factors of 15
$1 \cdot 15 = 15$	$1 + 15 = 16$
$3 \cdot 5 = 15$	$3 + 5 = 8$

The second row of the table contains the correct pair of integers 3 and 5, whose product is 15 and whose sum is 8.

We can express the middle term, $8x$, of the trinomial as the *sum of two terms*, using the integers 3 and 5 as coefficients of the two terms and factor the equivalent four-termed polynomial by grouping:

$$x^2 + 8x + 15 = x^2 + 3x + 5x + 15 \qquad \text{Express } 8x \text{ as } 3x + 5x.$$

$$x^2 + 3x + 5x + 15 = x(x + 3) + 5(x + 3) \qquad \text{Factor } x \text{ out of } x^2 + 3x \text{ and 5 out of } 5x + 15.$$

$$= (x + 3)(x + 5) \qquad \text{Factor out the GCF, } x + 3.$$

Check the factorization by multiplying.

b. In $10x^2 + 13xy - 3y^2$, we have $a = 10$, $b = 13$, and $c = -3$. The key number is $ac = 10(-3) = -30$. We must find a factorization of -30 such that the sum of the factors is $b = 13$. Since the factors must have a negative product, their signs must be different. The possible factor pairs are listed in the table.

Key number = -30 $\quad b = 13$

Factors of -30	Sum of the factors of -30
$1(-30) = -30$	$1 + (-30) = -29$
$2(-15) = -30$	$2 + (-15) = -13$
$3(-10) = -30$	$3 + (-10) = -7$
$5(-6) = -30$	$5 + (-6) = -1$
$6(-5) = -30$	$6 + (-5) = 1$
$10(-3) = -30$	$10 + (-3) = 7$
$15(-2) = -30$	$15 + (-2) = 13$
$30(-1) = -30$	$30 + (-1) = 29$

The seventh row contains the correct pair of numbers 15 and -2, whose product is -30 and whose sum is 13. They serve as the coefficients of two terms, $15xy$ and $-2xy$, that we place between $10x^2$ and $-3y^2$.

Notation

The middle term, $13xy$, may be expressed as $15xy - 2xy$ or as $-2xy + 15xy$ when using factoring by grouping. The resulting factorizations will be equivalent.

$$10x^2 + 13xy - 3y^2 = 10x^2 + 15xy - 2xy - 3y^2 \qquad \text{Express } 13xy \text{ as } 15xy - 2xy.$$

We factor the resulting four-term polynomial by grouping.

$$10x^2 + 15xy - 2xy - 3y^2 = 5x(2x + 3y) - y(2x + 3y) \qquad \text{Factor out } 5x \text{ from } 10x^2 + 15xy. \text{ Factor out } -y \text{ from } -2xy - 3y^2.$$

$$= (2x + 3y)(5x - y) \qquad \text{Factor out the GCF, } 2x + 3y.$$

Thus, $10x^2 + 13xy - 3y^2 = (2x + 3y)(5x - y)$. Check by multiplying.

Self Check 12 Factor by grouping: **a.** $m^2 + 13m + 42$ **b.** $15a^2 + 17ab - 4b^2$

Now Try Problems 63 and 75

SECTION 8.6 STUDY SET

VOCABULARY

Fill in the blanks.

1. When we write $2x + 4$ as $2(x + 2)$, we say that we have ________ $2x + 4$.
2. When we factor a polynomial, we write a sum of terms as a ________ of factors.
3. The abbreviation GCF stands for ________ ________ ________.
4. If a polynomial cannot be factored, it is called a ________ polynomial or an irreducible polynomial.

5. To factor $ab + 6a + 2b + 12$ by ________, we begin by factoring out a from the first two terms and 2 from the last two terms.
6. The trinomial $4a^2 - 5a - 6$ is written in __________ powers of a.
7. The ________ coefficient of $x^2 - 3x + 2$ is 1, the __________ of the middle term is -3, and the last term is ___.
8. The statement $x^2 - x - 12 = (x - 4)(x + 3)$ shows that $x^2 - x - 12$ factors into the ________ of two binomials.

CONCEPTS

9. The prime factorizations of three terms are shown here. Find their GCF.

$2 \cdot 2 \cdot 3 \cdot x \cdot x \cdot y \cdot y \cdot y$

$2 \cdot 3 \cdot 3 \cdot x \cdot y \cdot y \cdot y \cdot y$

$2 \cdot 3 \cdot 3 \cdot 7 \cdot x \cdot x \cdot x \cdot y \cdot y$

10. Use multiplication to determine whether $(3t - 1)(5t - 6)$ is the correct factorization of $15t^2 - 19t + 6$.

11. Complete the table.

Factors of 8	Sum of the factors of 8
$1(8) = 8$	
$2(4) = 8$	
$-1(-8) = 8$	
$-2(-4) = 8$	

12. Find two integers whose
 a. product is 10 and whose sum is 7.
 b. product is 8 and whose sum is -6.
 c. product is -6 and whose sum is 1.
 d. product is -9 and whose sum is -8.
13. Complete the key number table.

Key number = 12 $b = -7$

Negative factors of 12	Sum of the factors of 12
$-1(-12) = 12$	
$-3(-4) = 12$	

14. Use the substitution $x = a + b$ to rewrite the trinomial $6(a + b)^2 - 17(a + b) - 3$.

NOTATION

Complete each factorization.

15. $15c^2d^4 - 25c^2d^4 + 5c^3d^6 = \_\_\_(3c - 5 + cd^2)$
16. $x^3 - x^2 + 2x - 2 = \_\_(x - 1) + \_\_(x - 1)$
 $= (\_\_\_)(x^2 + 2)$
17. $5m^2 + 7m - 3 = (\_\_ - 1)(2m + \_\_)$
18. $2y^2 + 10y + 12 = \_\_(y^2 + 5y + 6)$
 $= 2(y + \_\_)(\_\_ + 2)$

GUIDED PRACTICE

Factor each polynomial. See Example 1.

19. $2x^2 - 6x$
20. $3y^3 + 3y^2$
21. $15x^2y - 10x^2y^2$
22. $63x^3y^2 + 81x^2y^4$
23. $27z^3 + 12z^2 + 3z$
24. $25t^6 - 10t^3 + 5t^2$
25. $24s^3 - 12s^2t + 6st^2$
26. $18y^2z^2 + 12y^2z^3 - 24y^4z^3$
27. $11x^3 - 12y$
28. $14s^3 + 15t^6$
29. $23a^2b^3 + 4x^3y^2$
30. $18p^3q^2 - 5t^5$

Factor each polynomial by factoring out the opposite of the GCF. See Example 2.

31. $-8a - 16$
32. $-6b - 30$
33. $-6x^2 - 3xy$
34. $-15y^3 - 25y^2$
35. $-18a^2b + 12ab^2$
36. $-21t^5 + 28t^3$
37. $-8a^4c^8 + 28a^3c^8 - 20a^2c^9$
38. $-30x^{10}y + 24x^9y^2 - 60x^8y^2$

Factor. See Example 3.

39. $(x + y)u + (x + y)v$
40. $4(x + y) + t(x + y)$
41. $5(a - b + c) - t(a - b + c)$
42. $(a - b - c)r - (a - b - c)s$

Factor by grouping. See Example 4.

43. $ax + bx + ay + by$
44. $ar - br + as - bs$
45. $x^2 + yx - x - y$
46. $d^2 + cd + c + d$
47. $t^3 - 3t^2 - 7t + 21$
48. $b^3 - 4b^2 - 3b + 12$
49. $a^2 - 4b + ab - 4a$
50. $3c - cd + 3d - c^2$

Factor. See Example 5.

51. $6x^3 - 6x^2 + 12x - 12$
52. $3x^3 - 6x^2 + 15x - 30$
53. $28a^3b^3c + 14a^3c - 4b^3c - 2c$
54. $12x^3z + 12xy^2z - 8x^2yz - 8y^3z$

Solve for the specified variable or expression. See Example 6.

55. $2g = ch + dh$ for h
56. $d_1d_2 = fd_2 + fd_1$ for f
57. $r_1r_2 = rr_2 + rr_1$ for r_1
58. $rx - ty = by$ for y
59. $b^2x^2 + a^2y^2 = a^2b^2$ for a^2
60. $b^2x^2 + a^2y^2 = a^2b^2$ for b^2
61. $Sn = (n - 2)180$ for n
62. $S(1 - r) = a - lr$ for r

Factor. See Example 7 or 12.

63. $x^2 - 5x + 6$

64. $y^2 + 7y + 6$

65. $x^2 + x - 30$

66. $c^2 + 3c - 28$

Factor. See Example 8.

67. $3x^2 + 12xy - 63y^2$

68. $2y^2 + 4yz - 48z^2$

69. $6a^2 - 30ab + 24b^2$

70. $4b^2 + 12bc - 16c^2$

71. $n^4 - 28n^3t - 60n^2t^2$

72. $c^4 - 16c^3d - 80c^2d^2$

73. $-3x^2 + 15xy - 18y^2$

74. $-2y^2 - 16yt + 40t^2$

Factor. See Example 9 or 12.

75. $5x^2 + 13x + 6$

76. $5x^2 + 18x + 9$

77. $7a^2 + 12a + 5$

78. $7a^2 + 36a + 5$

79. $11y^2 + 32y - 3$

80. $2y^2 - 9y - 18$

81. $8x^2 - 22x + 5$

82. $4z^2 - 13z + 3$

83. $6y^2 - 13y + 6$

84. $6x^2 - 11x + 3$

85. $15b^2 + 4b - 4$

86. $8a^2 + 6a - 9$

Factor each expression. See Example 10.

87. $30x^4 - 25x^2 - 20$

88. $14x^4 + 77x^2 + 84$

89. $32x^4 - 96x^2 + 72$

90. $20a^4 + 60a^2 + 45$

91. $64h^5 - 4h + 24h^3$

92. $9x^5 - 24x + 30x^3$

93. $-3a^4 - 5a^2b^2 - 2b^4$

94. $-2x^4 + 3x^2y^2 + 5y^4$

Factor by using a substitution. See Example 11.

95. $(a + b)^2 - 2(a + b) - 24$

96. $(x - y)^2 + 3(x - y) - 10$

97. $(x + a)^2 + 2(x + a) + 1$

98. $(a + b)^2 - 2(a + b) + 1$

TRY IT YOURSELF

Factor completely. Factor out all common factors first including −1 if the first term is negative. If an expression is prime, so indicate.

99. $3(m + n + p) + x(m + n + p)$

100. $x(x - y - z) + y(x - y - z)$

101. $-63u^3v^6 + 28u^2v^7 - 21u^3v^3$

102. $-56x^4y^3 - 72x^3y^4 + 80xy^2$

103. $b^4x^2 - 12b^2x^2 + 35x^2$

104. $c^3x^4 + 11c^3x^2 - 42c^3$

105. $1 - n - m + mn$

106. $a^2x^2 - 10 - 2x^2 + 5a^2$

107. $-x^2 + 4xy + 21y^2$

108. $-a^2 - 4ab + 5b^2$

109. $a^2x + bx - a^2 - b$

110. $x^2y - ax - xy + a$

111. $4y^2 + 4y + 1$

112. $9x^2 + 6x + 1$

113. $b^2 + 8b + 18$

114. $x^2 + 4x - 28$

115. $13r + 3r^2 - 10$

116. $-r + 3r^2 - 10$

117. $y^3 - 12 + 3y - 4y^2$

118. $h^3 - 8 + h - 8h^2$

119. $2y^5 - 26y^3 + 60y$

120. $2y^5 - 26y^3 + 84y$

121. $14(q - r)^2 - 17(q - r) - 6$

122. $8(h + s)^2 + 34(h + s) + 35$

APPLICATIONS

123. **Crayons.** The amount of colored wax used to make the crayon shown in the illustration can be found by computing its volume using the formula

$$V = \pi r^2 h_1 + \frac{1}{3}\pi r^2 h_2$$

Factor the expression on the right side of this equation.

124. **Packaging.** The amount of cardboard needed to make the following cereal box can be found by finding the area A, which is given by the formula

$$A = 2wh + 4wl + 2lh$$

where w is the width, h the height, and l the length. Solve the equation for the width.

125. **Ice.** The surface area of the ice cube is represented by the expression $6x^2 + 36x + 54$. Use factoring to find the length of an edge of the cube.

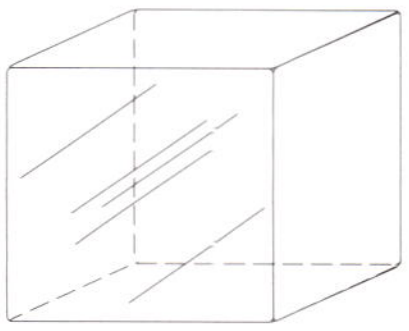

126. **Checkers.** The area of the checkerboard is represented by the expression $25x^2 - 40x + 16$. Use factoring to find the length of each side.

WRITING

127. Explain the error in the following solution.

Solve for r_1: $r_1r_2 = rr_2 + rr_1$

$$\frac{r_1r_2}{r_2} = \frac{rr_2 + rr_1}{r_2}$$

$$r_1 = \frac{rr_2 + rr_1}{r_2}$$

128. Explain the error.

Factor: $2x^2 - 4x - 6 = (2x + 2)(x - 3)$

REVIEW

129. Use the graph to find:

a. $s(-5)$

b. $s(4)$

c. The values of x for which $s(x) = 0$

d. The value of x for which $s(x) = 4$

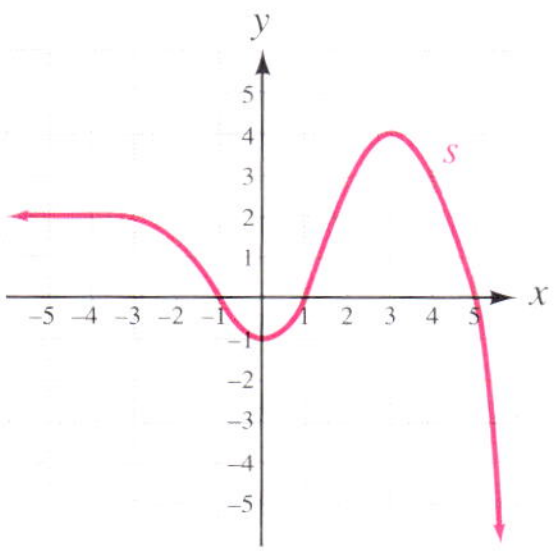

130. Determine whether each equation defines y to be a function of x. If it does not, find two ordered pairs where more than one value of y corresponds to a single value of x.

a. $x = |y|$

b. $y = x^2$

c. $y^2 = x$

CHALLENGE PROBLEMS

Factor out the specified factor.

131. t^{-3} from $t^5 + 4t^{-6}$

132. $7x^{-3n}$ from $21x^{6n} + 7x^{3n} + 14$

Factor. Assume that n is a natural number.

133. $x^{2n} + 2x^n + 1$

134. $2a^{6n} - 3a^{3n} - 2$

135. $x^{4n} + 2x^{2n}y^{2n} + y^{4n}$

136. $6x^{2n} + 7x^n - 3$

SECTION 8.7

OBJECTIVES

1 Factor the difference of two squares.

2 Factor the sum and difference of two cubes.

Review of Factoring Methods: The Difference of Two Squares; the Sum and Difference of Two Cubes

ARE YOU READY?

The following problems review some basic skills that are needed when factoring certain types of binomials.

1. Multiply: $(n + 9)(n - 9)$
2. Simplify: $(8d)^2$
3. Evaluate: **a.** 3^3 **b.** 6^3
4. Simplify: $(4a)^3$
5. Multiply: $(b + 4)(b^2 - 4b + 16)$
6. Multiply: $(2a - 1)(4a^2 + 2a + 1)$

We will now review how to factor the difference of two squares and the sum and difference of two cubes.

The Language of Algebra

The expression $A^2 - B^2$ is a **difference of two squares,** whereas $(A - B)^2$ is the **square of a difference.** They are not equivalent because $(A - B)^2 \neq A^2 - B^2$.

1 Factor the Difference of Two Squares.

Recall the special-product rule for multiplying the sum and difference of the same two terms:

$$(A + B)(A - B) = A^2 - B^2$$

The binomial $A^2 - B^2$ is called a **difference of two squares,** because A^2 is the square of A and B^2 is the square of B. If we reverse this rule, we obtain a method for factoring a difference of two squares.

Factoring a Difference of Two Squares

To factor the square of a First quantity minus the square of a Last quantity, multiply the First plus the Last by the First minus the Last.

$$F^2 - L^2 = (F + L)(F - L)$$

To factor the difference of two squares, it is helpful to know the first twenty **perfect-square integers.** The number 400, for example, is a perfect square, because $400 = 20^2$.

$1 = 1^2$	$25 = 5^2$	$81 = 9^2$	$169 = 13^2$	$289 = 17^2$
$4 = 2^2$	$36 = 6^2$	$100 = 10^2$	$196 = 14^2$	$324 = 18^2$
$9 = 3^2$	$49 = 7^2$	$121 = 11^2$	$225 = 15^2$	$361 = 19^2$
$16 = 4^2$	$64 = 8^2$	$144 = 12^2$	$256 = 16^2$	$400 = 20^2$

EXAMPLE 1 Factor: $49x^2 - 16$

Strategy The terms of this binomial do not have a common factor (other than 1). The only option is to attempt to factor it as a difference of two squares.

Why If a binomial is a difference of two squares, we can factor it using a special product.

Solution $49x^2 - 16$ is the difference of two squares because it can be written as $(7x)^2 - (4)^2$. We can match it to the rule for factoring a difference of two squares to find the factorization.

$$\begin{array}{ccccccc} F^2 & - & L^2 & = & (F + L) & (F - L) \\ \downarrow & & \downarrow & & \downarrow\ \ \downarrow & \downarrow\ \ \downarrow \\ (7x)^2 & - & (4)^2 & = & (7x + 4) & (7x - 4) \end{array}$$

Therefore, $49x^2 - 16 = (7x + 4)(7x - 4)$. We can verify this result using multiplication.

$$(7x + 4)(7x - 4) = 49x^2 - 28x + 28x - 16$$
$$= 49x^2 - 16 \quad \text{This is the original binomial.}$$

Success Tip

Always verify a factorization by doing the indicated multiplication. The result should be the original polynomial.

Self Check 1 Factor: $81p^2 - 25$

Now Try Problems 11 and 17

Expressions such as a^4 and x^6y^8 are also perfect squares, because they can be written as the square of another quantity:

$$a^4 = (a^2)^2, \qquad 81b^4 = (9b^2)^2, \qquad \text{and} \qquad x^6y^8 = (x^3y^4)^2$$

EXAMPLE 2 Factor: $64a^4 - 25b^2$

Strategy The terms of this binomial do not have a common factor (other than 1). The only option is to attempt to factor it as a difference of two squares.

Why If a binomial is a difference of two squares, we can factor it using a special-product rule.

Solution We can write $64a^4 - 25b^2$ in the form $(8a^2)^2 - (5b)^2$ and use the rule for factoring the difference of two squares.

$$\begin{array}{ccccccc} F^2 & - & L^2 & = & (F + L) & (F - L) \\ \downarrow & & \downarrow & & \downarrow\ \ \downarrow & \downarrow\ \ \downarrow \\ (8a^2)^2 & - & (5b)^2 & = & (8a^2 + 5b) & (8a^2 - 5b) \end{array}$$

Therefore, $64a^4 - 25b^2 = (8a^2 + 5b)(8a^2 - 5b)$.

Notation

By the commutative property of multiplication, this factorization could be written

$$(8a^2 - 5b)(8a^2 + 5b)$$

Self Check 2 Factor: $36r^4 - s^2$

Now Try Problems 21 and 25

EXAMPLE 3 Factor: $x^4 - 1$

Strategy The terms of $x^4 - 1$ do not have a common factor (other than 1). To factor this binomial, we will write it in a form that shows it is a difference of two squares.

Why We can then use a special-product rule to factor it.

Solution Because the binomial is the difference of the squares of x^2 and 1, it factors into the sum of x^2 and 1 and the difference of x^2 and 1.

$$x^4 - 1 = (x^2)^2 - (1)^2$$
$$= (x^2 + 1)(x^2 - 1)$$

A prime polynomial (refers to $x^2 + 1$)

Caution

The binomial $x^2 + 1$ is the **sum of two squares.** In general, after removing any common factor, a sum of two squares cannot be factored using real numbers.

The factor $x^2 + 1$ is the sum of two quantities and is prime. However, the factor $x^2 - 1$ is the difference of two squares and can be factored as $(x + 1)(x - 1)$. Thus,

$$x^4 - 1 = (x^2 + 1)(x^2 - 1)$$
$$= (x^2 + 1)(x + 1)(x - 1) \quad \text{Don't forget to write } (x^2 + 1).$$

Self Check 3 Factor: $a^4 - 81$

Now Try Problem 29

At times, a **substitution** can be helpful in simplifying the factorization process.

EXAMPLE 4 Factor: $(x + y)^4 - z^4$

Strategy We will use a substitution to factor this difference of two squares.

Why For more complicated expressions, especially those involving a quantity within parentheses, a substitution often helps simplify the factoring process.

Solution If we use the substitution $a = x + y$, we obtain

$$(x + y)^4 - z^4 = a^4 - z^4 \quad \text{Replace } x + y \text{ with } a.$$
$$= (a^2 + z^2)(a^2 - z^2) \quad \text{Factor the difference of two squares.}$$
$$= (a^2 + z^2)(a + z)(a - z) \quad \text{Factor } a^2 - z^2. \ a^2 + z^2 \text{ is prime.}$$

Caution

A common error after making a substitution is to forget to "undo" or "reverse" it. In this example, if you are to factor an expression in x, y, and z, your answer must involve only x, y, and z. It should not contain the variable a.

To find the factorization of $(x + y)^4 - z^4$, we "reverse" substitute $x + y$ for each a in the expression $(a^2 + z^2)(a + z)(a - z)$.

$$(a^2 + z^2)(a + z)(a - z) = [(x + y)^2 + z^2](x + y + z)(x + y - z)$$

Thus, $(x + y)^4 - z^4 = [(x + y)^2 + z^2](x + y + z)(x + y - z)$.

If we square the binomial within the brackets, we have

$$(x + y)^4 - z^4 = [x^2 + 2xy + y^2 + z^2](x + y + z)(x + y - z)$$

Caution

When factoring a polynomial, be sure to factor it completely. Always check to see whether any of the factors of your result can be factored further.

Self Check 4 Factor: $(a - b)^4 - c^4$

Now Try Problem 35

When possible, we always factor out a common factor before factoring the difference of two squares. The factoring process is easier when all common factors are factored out first.

EXAMPLE 5 Factor: $2x^4y - 32y$

Strategy We will factor out the GCF of $2y$ and factor the resulting difference of two squares.

Why The first step in factoring any polynomial is to factor out the GCF.

Solution

$$2x^4y - 32y = 2y(x^4 - 16) \quad \text{Factor out the GCF, which is } 2y.$$
$$= 2y(x^2 + 4)(x^2 - 4) \quad \text{Factor } x^4 - 16.$$
$$= 2y(x^2 + 4)(x + 2)(x - 2) \quad \text{Factor } x^2 - 4.\ x^2 + 4 \text{ is prime.}$$

Success Tip

Remember that a **difference of two squares** is a binomial. Each term is a square and the terms have different signs. The powers of the variables in the terms must be even.

Self Check 5 Factor: $3a^4 - 3$

Now Try Problems 37 and 43

To factor some expressions, we need to use some creative grouping to begin the process.

EXAMPLE 6 Factor: **a.** $x^2 - y^2 + x - y$ **b.** $x^2 + 6x + 9 - z^2$

Strategy The terms of each expression do not have a common factor (other than 1) and traditional factoring by grouping will not work. Instead, in part a, we will group only the first two terms of the polynomial and in part b we will group the first three terms.

Why Hopefully, those steps will produce equivalent expressions that can be factored.

Solution We group the first two terms, factor them as a difference of two squares, and look for a common factor.

a. $x^2 - y^2 + x - y = (x + y)(x - y) + (x - y)$ Factor $x^2 - y^2$. The terms of the resulting expression have a common binomial factor, $x - y$, that can be factored out later.

$= (x + y)(x - y) + 1(x - y)$ Factor out 1 from $x - y$.

$= (x - y)(x + y + 1)$ Factor out the GCF, $x - y$. Don't forget the 1.

b. We group the first three terms and factor that trinomial to get:

$x^2 + 6x + 9 - z^2 = (x + 3)(x + 3) - z^2$ $x^2 + 6x + 9$ is a perfect-square trinomial.

$= (x + 3)^2 - z^2$ Write $(x + 3)(x + 3)$ as $(x + 3)^2$. The expression that results is a difference of two squares.

$= (x + 3 + z)(x + 3 - z)$ Factor the difference of two squares.

Success Tip

We could use a substitution to factor $(x + 3)^2 - z^2$.

Self Check 6 Factor: **a.** $a^2 - b^2 + a + b$
b. $a^2 + 4a + 4 - b^2$

Now Try Problems 45 and 49

2 Factor the Sum and Difference of Two Cubes.

The number 64 is called a perfect cube, because $4^3 = 64$. To factor the sum or difference of two cubes, it is helpful to know the first ten **perfect-cube integers:**

$\mathbf{1} = 1^3$	$\mathbf{27} = 3^3$	$\mathbf{125} = 5^3$	$\mathbf{343} = 7^3$	$\mathbf{729} = 9^3$
$\mathbf{8} = 2^3$	$\mathbf{64} = 4^3$	$\mathbf{216} = 6^3$	$\mathbf{512} = 8^3$	$\mathbf{1{,}000} = 10^3$

Expressions such as b^6 and $64x^9y^{12}$ are also perfect cubes, because they can be written as the cube of another quantity:

$$b^6 = (b^2)^3 \qquad \text{and} \qquad 64x^9y^{12} = (4x^3y^4)^3$$

To find rules for factoring the sum of two cubes and the difference of two cubes, we need to find the products shown below. Note that each term of the trinomial is multiplied by each term of the binomial.

$$(x + y)(x^2 - xy + y^2) = x^3 - \cancel{x^2y} + \cancel{xy^2} + \cancel{x^2y} - \cancel{xy^2} + y^3$$
$$= x^3 + y^3$$

Combine like terms.

$$(x - y)(x^2 + xy + y^2) = x^3 + \cancel{x^2y} + \cancel{xy^2} - \cancel{x^2y} - \cancel{xy^2} - y^3$$
$$= x^3 - y^3$$

Combine like terms.

We have found that

$$x^3 + y^3 = (x + y)(x^2 - xy + y^2) \qquad \text{and} \qquad x^3 - y^3 = (x - y)(x^2 + xy + y^2)$$

The Language of Algebra

The expression $x^3 + y^3$ is a **sum of two cubes,** whereas $(x + y)^3$ is the **cube of a sum.** If you expand $(x + y)^3$, you will see that they are not equivalent.

The binomial $x^3 + y^3$ is called the **sum of two cubes,** because x^3 represents the cube of x, y^3 represents the cube of y, and $x^3 + y^3$ represents the sum of these cubes. Similarly, $x^3 - y^3$ is called the **difference of two cubes.**

These results justify the rules for factoring the **sum and difference of two cubes.** They are easier to remember if we think of a sum (or a difference) of two cubes as the cube of a **F**irst quantity plus (or minus) the cube of the **L**ast quantity.

Factoring the Sum and Difference of Two Cubes

To factor the cube of a First quantity plus the cube of a Last quantity, multiply the First plus the Last by the First squared, minus the First times the Last, plus the Last squared.

$$F^3 + L^3 = (F + L)(F^2 - FL + L^2)$$

To factor the cube of a First quantity minus the cube of a Last quantity, multiply the First minus the Last by the First squared, plus the First times the Last, plus the Last squared.

$$F^3 - L^3 = (F - L)(F^2 + FL + L^2)$$

EXAMPLE 7 Factor: $a^3 + 8$

Strategy We will write the binomial in a form that shows it is the sum of two cubes.

Why We can then use the rule for factoring the sum of two cubes.

Solution Since $a^3 + 8$ can be written as $a^3 + 2^3$, it is a sum of two cubes, which factors as follows:

$$F^3 + L^3 = (F + L)(F^2 - FL + L^2)$$
$$a^3 + 2^3 = (a + 2)(a^2 - a2 + 2^2)$$
$$= (a + 2)(a^2 - 2a + 4) \quad a^2 - 2a + 4 \text{ does not factor, it is prime.}$$

Therefore, $a^3 + 8 = (a + 2)(a^2 - 2a + 4)$.

We can check by multiplying.

$$(a + 2)(a^2 - 2a + 4) = a^3 - 2a^2 + 4a + 2a^2 - 4a + 8$$
$$= a^3 + 8 \quad \text{This is the original binomial.}$$

Caution

In Example 7, a common error is to try to factor $a^2 - 2a + 4$. It is not a perfect-square trinomial, because the middle term needs to be $-4a$. Furthermore, it cannot be factored by any of the methods of Section 6.2. It is prime.

Self Check 7 Factor: $p^3 + 27$

Now Try Problem 53

You should memorize the rules for factoring the sum and the difference of two cubes. Note that each has the form

(a binomial)(a trinomial)

and that there is a relationship between the signs that appear in these forms.

Success Tip

An easy way to remember the sign patterns within the binomial and trinomial of these factoring forms is with the word "SOAP."

Same, **O**pposite, **A**lways, **P**lus

The Sum of Cubes

The same sign

$$F^3 + L^3 = (F + L)(F^2 - FL + L^2)$$

Opposite signs Always plus

The Difference of Cubes

The same sign

$$F^3 - L^3 = (F - L)(F^2 + FL + L^2)$$

Opposite signs Always plus

EXAMPLE 8 Factor: $27a^3 - 64b^6$

Strategy We will write the binomial in a form that shows it is the difference of two cubes.

Why We can then use the rule for factoring the difference of two cubes.

Solution Since $27a^3 - 64b^6$ can be written as $(3a)^3 - (4b^2)^3$, it is a difference of two cubes, which factors as follows:

$$F^3 - L^3 = (F - L)(F^2 + F\,L + L^2)$$
$$(3a)^3 - (4b^2)^3 = (3a - 4b^2)[(3a)^2 + (3a)(4b^2) + (4b^2)^2]$$
$$= (3a - 4b^2)(9a^2 + 12ab^2 + 16b^4)$$

Thus, $27a^3 - 64b^6 = (3a - 4b^2)(9a^2 + 12ab^2 + 16b^4)$. Multiply to check.

Self Check 8 Factor: $8c^6 - 125d^3$

Now Try Problem 57

EXAMPLE 9 Factor: $a^3 - (c + d)^3$

Strategy To factor this expression, we will use the rule for factoring the difference of two cubes.

Why The terms a^3 and $(c + d)^3$ are perfect cubes.

Solution

$$\begin{array}{ccccccccccc} F^3 & - & L^3 & = (F & - & L &)(F^2 + F & L & + & L^2 &) \\ \downarrow & & \downarrow & & \downarrow & \downarrow & \downarrow \quad \downarrow & \downarrow & & \downarrow & \\ a^3 & - & (c+d)^3 & = [a & - & (c+d)][a^2 & + a(c+d) & & + & (c+d)^2] & \end{array}$$

Now we simplify the expressions within both sets of brackets. Thus,

$$a^3 - (c + d)^3 = (a - c - d)(a^2 + ac + ad + c^2 + 2cd + d^2)$$

Success Tip

We could also use the substitution $x = c + d$, factor $a^3 - x^3$ as $(a - x)(a^2 + ax + x^2)$, and then replace each x with $c + d$.

Self Check 9 Factor: $(p + q)^3 - r^3$

Now Try Problem 61

EXAMPLE 10 Factor: $x^6 - 64$

Strategy This binomial is both the difference of two squares and the difference of two cubes. We will write it in a form that shows it is a difference of two squares to begin the factoring process.

Why It is easier to factor it as the difference of two squares first.

Solution

$$\begin{aligned} x^6 - 64 &= (x^3)^2 - 8^2 \\ &= (x^3 + 8)(x^3 - 8) \end{aligned}$$

Each of these binomial factors can be factored further. The first is the sum of two cubes and the second is the difference of two cubes. Thus,

$$x^6 - 64 = (x + 2)(x^2 - 2x + 4)(x - 2)(x^2 + 2x + 4)$$

Self Check 10 Factor: $1 - x^6$

Now Try Problem 65

EXAMPLE 11 Factor: $2a^5 + 250a^2$

Strategy We will factor out the GCF $2a^2$ and then factor the resulting sum of two cubes.

Why The first step in factoring any polynomial is to factor out the GCF.

Solution We first factor out the common factor $2a^2$ to obtain

$$2a^5 + 250a^2 = 2a^2(a^3 + 125)$$

Then we factor $a^3 + 125$ as the sum of two cubes to obtain

$$2a^5 + 250a^2 = 2a^2(a + 5)(a^2 - 5a + 25) \qquad a^2 - 5a + 25 \text{ is prime.}$$

Self Check 11 Factor: $3x^5 + 24x^2$

Now Try Problem 69

SECTION 8.7 STUDY SET

VOCABULARY

Fill in the blanks.

1. When the polynomial $4x^2 - 25$ is written as $(2x)^2 - (5)^2$, we see that it is the difference of two ________.

2. When the polynomial $8x^3 + 125$ is written as $(2x)^3 + (5)^3$, we see that it is the sum of two ______.

CONCEPTS

3. a. Write the first ten perfect-square integers.
   b. Write the first ten perfect-cube integers.
4. a. Use multiplication to verify that the sum of two squares $x^2 + 25$ does not factor as $(x + 5)(x + 5)$.
   b. Use multiplication to verify that the difference of two squares $x^2 - 25$ factors as $(x + 5)(x - 5)$.
5. Complete each factorization.
   a. $F^2 - L^2 = (F + L)(\quad)$
   b. $F^3 + L^3 = (F + L)(\quad)$
   c. $F^3 - L^3 = (F - L)(\quad)$
6. Factor each binomial.
   a. $5p^2 + 20$
   b. $5p^2 - 20$
   c. $5p^3 + 20$
   d. $5p^3 + 40$

NOTATION

7. Give an example of each.
   a. a difference of two squares
   b. a square of a difference
   c. a sum of two squares
   d. a sum of two cubes
   e. a cube of a sum
8. Fill in the blanks.
   a. $36y^2 - 49m^4 = (\quad)^2 - (\quad)^2$
   b. $125h^3 - 27k^6 = (\quad)^3 - (\quad)^3$

GUIDED PRACTICE

Factor. See Example 1.

9. $x^2 - 16$
10. $y^2 - 49$
11. $9y^2 - 64$
12. $16x^2 - 81$
13. $144 - c^2$
14. $25 - t^2$
15. $100m^2 - 1$
16. $144x^2 - 1$
17. $81a^2 - 49b^2$
18. $64r^2 - 121s^2$
19. $x^2 + 25$
20. $a^2 + 36$

Factor each difference of two squares. See Example 2.

21. $9r^4 - 121s^2$
22. $81a^4 - 16b^2$
23. $16t^2 - 25w^4$
24. $9r^2 - 25s^4$
25. $100r^2s^4 - t^4$
26. $400x^2z^4 - a^4$
27. $36x^4y^2 - 49z^6$
28. $4a^2b^4 - 9d^6$

Factor completely. See Example 3.

29. $x^4 - y^4$
30. $16n^4 - 1$
31. $16a^4 - 81b^4$
32. $81m^4 - 256n^4$

Factor. See Example 4.

33. $(x + y)^2 - z^2$
34. $a^2 - (b - c)^2$
35. $(r - s)^2 - t^4$
36. $(m + n)^2 - p^4$

Factor each expression. Factor out any GCF first. See Example 5.

37. $2x^2 - 288$
38. $8x^2 - 72$
39. $3x^3 - 243x$
40. $2x^3 - 32x$
41. $5ab^4 - 5a$
42. $3ac^4 - 243a$
43. $64b - 4b^5$
44. $1{,}250n - 2n^5$

Factor by first grouping the appropriate terms. See Example 6.

45. $c^2 - d^2 + c + d$
46. $s^2 - t^2 + s - t$
47. $a^2 - b^2 + 2a - 2b$
48. $m^2 - n^2 + 3m + 3n$
49. $x^2 + 12x + 36 - y^2$
50. $x^2 - 6x + 9 - 4y^2$
51. $x^2 - 2x + 1 - 9z^2$
52. $x^2 + 10x + 25 - 16z^2$

Factor each sum of cubes. See Example 7.

53. $a^3 + 125$
54. $b^3 + 64$
55. $8r^3 + s^3$
56. $27t^3 + u^3$

Factor each difference of cubes. See Example 8.

57. $64t^6 - 27v^3$
58. $125m^3 - x^6$
59. $x^3 - 216y^6$
60. $8c^6 - 343w^3$

Factor. See Example 9.

61. $(a - b)^3 + 27$
62. $(b - c)^3 - 1{,}000$
63. $64 - (a + b)^3$
64. $1 - (x + y)^3$

Factor each expression completely. Factor a difference of two squares first. **See Example 10.**

65. $x^6 - 1$
66. $x^6 - y^6$
67. $x^{12} - y^6$
68. $a^{12} - 64$

Factor each sum or difference of cubes. Factor out the GCF first. **See Example 11.**

69. $5x^3 + 625$
70. $2x^3 - 128$
71. $4x^5 - 256x^2$
72. $2x^6 + 54x^3$

TRY IT YOURSELF

Factor each expression.

73. $64a^3 - 125b^6$
74. $8x^6 - 27y^3$
75. $288b^2 - 2b^6$
76. $98x - 2x^5$
77. $x^2 - y^2 + 8x + 8y$
78. $5m - 5n + m^2 - n^2$
79. $x^9 + y^9$
80. $x^6 + y^6$
81. $144a^2t^2 - 169b^6$
82. $25x^6 - 81y^2z^2$
83. $100a^2 + 9b^2$
84. $25s^4 + 16t^2$
85. $81c^4d^4 - 16t^4$
86. $256x^4 - 81y^4$
87. $128u^2v^3 - 2t^3u^2$
88. $56rs^2t^3 + 7rs^2v^6$
89. $y^2 - (2x - t)^2$
90. $(15 - r)^2 - s^2$
91. $x^2 + 20x + 100 - 9z^2$
92. $49a^2 - b^2 - 14b - 49$
93. $(c - d)^3 + 216$
94. $1 - (x + y)^3$
95. $\frac{1}{36} - y^4$
96. $\frac{4}{81} - m^4$
97. $m^6 - 64$
98. $1 - y^6$
99. $(a + b)x^3 + 27(a + b)$
100. $(c - d)r^3 - (c - d)s^3$
101. $x^9 - y^{12}z^{15}$
102. $r^{12} + s^{18}t^{24}$

Look Alikes . . .

103. a. $q^2 - 64$ b. $q^3 - 64$
104. a. $a^2 - b^2$ b. $a^3 - b^3$
105. a. $d^2 - 25$ b. $d^3 - 125$
106. a. $m^2 + 27$ b. $m^3 + 27$

Factor the expression in part a. Then use your answer from part a to give the factorization of the expression in part b. (No new work is necessary!)

107. a. $a^6 - b^3$ b. $a^6 + b^3$
108. a. $c^3 - \frac{1}{8}$ b. $c^3 + \frac{1}{8}$
109. a. $125m^3 + 8n^3$ b. $125m^3 - 8n^3$
110. a. $w^3 + 0.001$ b. $w^3 - 0.001$

APPLICATIONS

111. **Candy.** To find the amount of chocolate used in the outer coating of the malted-milk ball shown, we can find the volume V of the chocolate shell using the formula

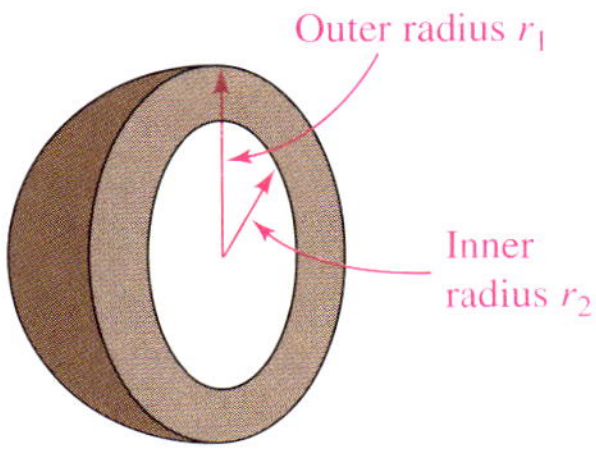

$$V = \frac{4}{3}\pi r_1^3 - \frac{4}{3}\pi r_2^3$$

Factor the expression on the right side of the formula.

112. **Movie Stunts.** The function that gives the distance a stuntwoman is above the ground t seconds after she falls over the side of a 144-foot tall building is $h(t) = 144 - 16t^2$. Factor the right side.

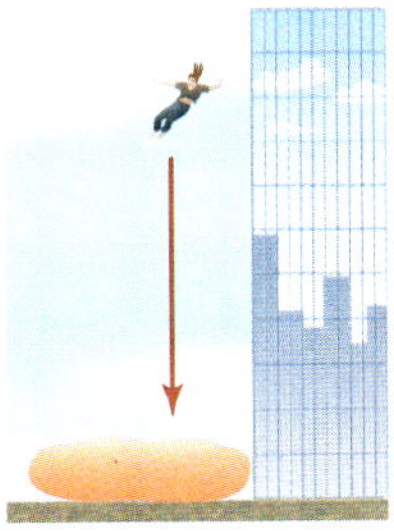

WRITING

113. Explain how the patterns used to factor the sum and difference of two cubes are similar and how they differ.
114. Explain why the factorization is not complete.

Factor: $1 - t^8 = (1 + t^4)(1 - t^4)$

115. Explain the error.

Factor: $4g^2 - 16 = (2g + 4)(2g - 4)$

116. When asked to factor $81t^2 - 16$, one student answered $(9t - 4)(9t + 4)$, and another answered $(9t + 4)(9t - 4)$. Explain why both students are correct.

REVIEW

For each of the following purchases, determine the better buy.

117. Flute lessons: 45 minutes for \$25 or 1 hour for \$35.

118. Tissue paper: 15 sheets for \$1.39 or a dozen sheets for \$1.10.

CHALLENGE PROBLEMS

Factor. Assume all variables represent natural numbers.

119. $4x^{2n} - 9y^{2n}$

120. $25 - x^{6n}$

121. $a^{3b} - c^{3b}$

122. $27x^{3n} + y^{3n}$

123. Factor: $x^{32} - y^{32}$

124. Find the error in this proof that $2 = 1$.

$$x = y$$
$$x^2 = xy$$
$$x^2 - y^2 = xy - y^2$$
$$(x + y)(x - y) = y(x - y)$$
$$\frac{(x + y)\cancel{(x - y)}}{\cancel{(x - y)}} = \frac{y\cancel{(x - y)}}{\cancel{(x - y)}}$$
$$x + y = y$$
$$y + y = y$$
$$2y = y$$
$$\frac{2y}{y} = \frac{y}{y}$$
$$2 = 1$$

SECTION 8.8 Review of Rational Expressions and Rational Equations; Rational Functions

OBJECTIVES

1. Evaluate rational functions.
2. Find the domain of a rational function.
3. Recognize the graphs of rational functions.
4. Simplify rational expressions.
5. Multiply and divide rational expressions.
6. Add and subtract rational expressions.
7. Solve rational equations.

ARE YOU READY?

The following problems review some basic skills that are needed when working with rational expressions.

1. Evaluate: **a.** $\frac{0}{10}$ **b.** $\frac{10}{0}$

2. Simplify: $\frac{36}{28}$

3. Simplify: $\frac{x(x + 5)(x - 7)}{4x(x - 5)(x - 7)}$

4. What is the reciprocal of $\frac{2}{3x}$?

5. **a.** Multiply: $\frac{3}{8} \cdot \frac{1}{7}$

b. Divide: $\frac{21}{25} \div \frac{7}{15}$

6. **a.** Add: $\frac{5}{11} + \frac{3}{11}$

b. Subtract: $\frac{2}{3} - \frac{1}{5}$

Recall that rational expressions are algebraic fractions with polynomial numerators and denominators.

Rational Expressions

A **rational expression** is an expression of the form $\frac{A}{B}$, where A and B are polynomials and B does not equal 0.

Caution

Since division by 0 is undefined, the value of a polynomial in the denominator of a rational expression cannot be 0. For example, x cannot be 7 in the rational expression $\frac{3x}{x - 7}$, because the value of the denominator would be 0.

Some examples of rational expressions are

$$\frac{3x}{x - 7}, \quad \frac{8yz^4}{6y^2z^2}, \quad \frac{5m + n}{m^2 + 4mn + 4n^2}, \quad \text{and} \quad \frac{6a^2 - 13a + 6}{a^3 + 3a^2 + a - 2}$$

The rational expression $\frac{3x}{x - 7}$ is the quotient of the monomial $3x$ and the binomial $x - 7$. It is in one variable, x. The rational expression $\frac{5m + n}{m^2 + 4mn + 4n^2}$ is the quotient of the binomial $5m + n$ and the trinomial $m^2 + 4mn + 4n^2$. It is in two variables, m and n.

Rational expressions in one variable are used to define *rational functions.*

1 Evaluate Rational Functions.

We have previously studied linear and polynomial functions. In this section, we introduce another family of functions known as *rational functions*.

Rational Functions A rational function is a function whose equation is defined by a rational expression in one variable, where the value of the polynomial in the denominator is never zero.

The Language of Algebra

Rational functions get their name from the fact that their defining equation contains a *ratio* (fraction) of two polynomials.

Two examples of rational functions are

$$f(x) = \frac{1}{x + 8}$$

The rational expression $\frac{1}{x+8}$ that defines function f is in one variable, x.

$$s(t) = \frac{5t^2 + t}{t^2 + 6t - 1}$$

The rational expression $\frac{5t^2+t}{t^2+6t-1}$ that defines function s is in one variable, t.

Rational functions can be used to model many types of real-world situations.

EXAMPLE 1

Internet Research. The cost of subscribing to an online research network is $6 a month plus $1.50 per hour of access time. The rational function $c(n) = \frac{1.50n + 6}{n}$ gives the average hourly cost $c(n)$ of using the network for n hours a month. Find the average hourly cost for

a. a student who used the network for 2 hours in a month.

b. an instructor who used the network for 18 hours in a month.

Strategy We will find $c(2)$ and $c(18)$.

Why The notation $c(2)$ represents the average hourly cost for using the network for 2 hours in a month and $c(18)$ represents the average hourly cost for using the network for 18 hours in a month.

Solution

a. To find the average hourly cost the student paid for 2 hours of access time in a month, we find $c(2)$.

$$c(\mathbf{2}) = \frac{1.50(\mathbf{2}) + 6}{\mathbf{2}} = 4.5$$

Substitute 2 for n and evaluate the right side.

The student paid $4.50 per hour to use the online research network for 2 hours.

b. To find the average hourly cost for 18 hours of access time in a month, we find $c(18)$.

$$c(\mathbf{18}) = \frac{1.50(\mathbf{18}) + 6}{\mathbf{18}} = 1.833333333$$

Substitute 18 for n and evaluate the right side.

The instructor paid approximately $1.83 per hour to use the online research network for 18 hours.

Self Check 1 **Internet Research.** Find the average hourly cost when the online research network in Example 1 is used for 100 hours in a month.

Now Try Problem 127

2 Find the Domain of a Rational Function.

Recall that the **domain** of a function is the set of all permissible input values for the variable. Since division by 0 is undefined, any input values that make the denominator 0 in a rational function must be excluded from the domain of the function.

EXAMPLE 2 Find the domain of the function: $f(x) = \dfrac{3x + 2}{x^2 + x - 6}$

Strategy We will set $x^2 + x - 6$ equal to 0 and solve for x.

Why We don't need to examine the numerator of the rational expression; it can be any value, including 0. The domain of the function includes all real numbers, except those that make the *denominator equal to* 0.

Solution

The Language of Algebra

Two other ways that Example 2 could be phrased are:

- **State the restrictions** on the variable.
- Find the values of x for which the function is **undefined.**

$$x^2 + x - 6 = 0 \quad \text{Set the denominator equal to 0.}$$

$$(x + 3)(x - 2) = 0 \quad \text{Factor the trinomial.}$$

$$x + 3 = 0 \quad \text{or} \quad x - 2 = 0 \quad \text{Set each factor equal to 0.}$$

$$x = -3 \quad | \quad x = 2 \quad \text{Solve each linear equation.}$$

Thus, the domain of the function is the set of all real numbers except -3 and 2. Using set-builder notation we can describe the domain as $\{x \mid x \text{ is a real number and } x \neq -3, x \neq 2\}$. In interval notation, the domain is $(-\infty, -3) \cup (-3, 2) \cup (2, \infty)$.

To check the answers, substitute -3 and 2 for x in $x^2 + x - 6$ and verify that the result is 0 for each.

Self Check 2 Find the domain of the function: $f(x) = \dfrac{x^2 + 1}{x^2 - 49}$

Now Try Problem 31

3 Recognize the Graphs of Rational Functions.

The simplest of all rational functions, $f(x) = \frac{1}{x}$, is called the **reciprocal function.** We can use the point-plotting method to graph it. Because of the complex shape of the graph, we must select a large number of x-values, including positive and negative fractions, to determine how it looks. After plotting all of the ordered pairs, we draw a smooth curve through the points that form a "branch" on the left and a smooth curve through the points that form a "branch" on the right. Since 0 is not in the domain of the function, there is no point on the graph for $x = 0$.

$f(x) = \dfrac{1}{x}$

x	$f(x)$	
-4	$-\frac{1}{4}$	$\rightarrow \left(-4, -\frac{1}{4}\right)$
-3	$-\frac{1}{3}$	$\rightarrow \left(-3, -\frac{1}{3}\right)$
-2	$-\frac{1}{2}$	$\rightarrow \left(-2, -\frac{1}{2}\right)$
-1	-1	$\rightarrow (-1, -1)$
$-\frac{1}{2}$	-2	$\rightarrow \left(-\frac{1}{2}, -2\right)$
$-\frac{1}{3}$	-3	$\rightarrow \left(-\frac{1}{3}, -3\right)$
$-\frac{1}{4}$	-4	$\rightarrow \left(-\frac{1}{4}, -4\right)$

↑ Select x. ↑ Find the reciprocal. ↑ Plot the point.

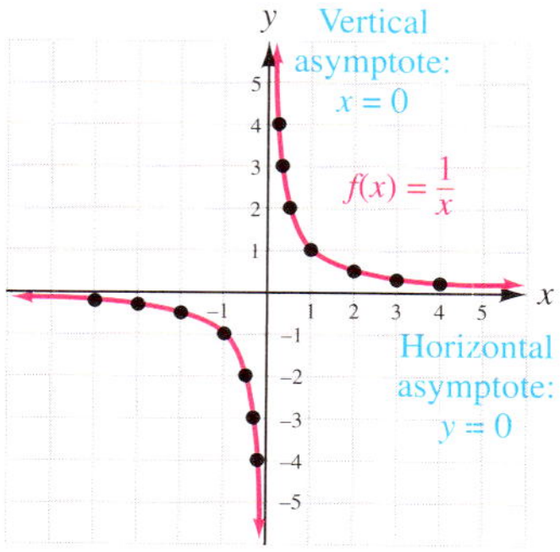

$f(x) = \dfrac{1}{x}$

x	$f(x)$	
4	$\frac{1}{4}$	$\rightarrow \left(4, \frac{1}{4}\right)$
3	$\frac{1}{3}$	$\rightarrow \left(3, \frac{1}{3}\right)$
2	$\frac{1}{2}$	$\rightarrow \left(2, \frac{1}{2}\right)$
1	1	$\rightarrow (1, 1)$
$\frac{1}{2}$	2	$\rightarrow \left(\frac{1}{2}, 2\right)$
$\frac{1}{3}$	3	$\rightarrow \left(\frac{1}{3}, 3\right)$
$\frac{1}{4}$	4	$\rightarrow \left(\frac{1}{4}, 4\right)$

↑ Select x. ↑ Find the reciprocal. ↑ Plot the point.

When a graph approaches a line, we call the line an **asymptote.** In this case, as x gets smaller and approaches 0, the graph of $f(x) = \frac{1}{x}$ approaches the y-axis. We say that the y-axis is a **vertical asymptote** of the graph.

On the far left and far right, the graph gets steadily closer to the x-axis. We say that the x-axis is a **horizontal asymptote** of the graph.

4 Simplify Rational Expressions.

To **simplify a rational expression** means to write it so that the numerator and denominator have no common factors other than 1.

Simplifying Rational Expressions

1. Factor the numerator and denominator completely to determine their common factors.
2. Remove factors equal to 1 by replacing each pair of factors common to the numerator and denominator with the equivalent fraction $\frac{1}{1}$.
3. Multiply the remaining factors in the numerator and in the denominator.

EXAMPLE 3 Simplify: $\dfrac{8yz^4}{6y^2z^2}$

Strategy We will begin by writing the numerator and denominator in factored form. Then we will remove any factors common to the numerator and denominator.

Why The rational expression is simplified when the numerator and denominator have no common factors other than 1.

Solution

$$\frac{8yz^4}{6y^2z^2} = \frac{2 \cdot 2 \cdot 2 \cdot y \cdot z \cdot z \cdot z \cdot z}{2 \cdot 3 \cdot y \cdot y \cdot z \cdot z}$$

To prepare to simplify the rational expression, factor $8yz^4$ and $6y^2z^2$ completely.

$$= \frac{\overset{1}{\cancel{2}} \cdot 2 \cdot 2 \cdot \overset{1}{\cancel{y}} \cdot \overset{1}{\cancel{z}} \cdot \overset{1}{\cancel{z}} \cdot z \cdot z}{\underset{1}{\cancel{2}} \cdot 3 \cdot \underset{1}{\cancel{y}} \cdot y \cdot \underset{1}{\cancel{z}} \cdot \underset{1}{\cancel{z}}}$$

Replace $\frac{2}{2}$, $\frac{y}{y}$, and $\frac{z}{z}$ with $\frac{1}{1}$. This removes the factor $\frac{2 \cdot y \cdot z \cdot z}{2 \cdot y \cdot z \cdot z} = 1$.

$$= \frac{4z^2}{3y}$$

Multiply the remaining factors in the numerator. Multiply the remaining factors in the denominator.

Since $\frac{8yz^4}{6y^2z^2}$ is undefined for $y = 0$ and $z = 0$, the expressions $\frac{8yz^4}{6y^2z^2}$ and $\frac{4z^2}{3y}$ are equal only if $y \neq 0$ and $z \neq 0$. That is,

$$\frac{8yz^4}{6y^2z^2} = \frac{4z^2}{3y} \quad \text{provided } y \neq 0 \text{ and } z \neq 0.$$

An alternate approach is to use the rules for exponents to simplify the rational expressions that are the quotient of two monomials.

$$\frac{8yz^4}{6y^2z^2} = \frac{\overset{1}{\cancel{2}} \cdot 2 \cdot 2 \cdot y^{1-2}z^{4-2}}{\underset{1}{\cancel{2}} \cdot 3} = \frac{4y^{-1}z^2}{3} = \frac{4z^2}{3y}$$

To divide exponential expressions with the same base, keep the base and subtract the exponents.

Self Check 3 Simplify: $\dfrac{12a^4b^2}{20ab^4}$

Now Try Problems 33 and 35

The Language of Algebra

When a rational expression is simplified, the result is an equivalent expression.

Caution

Unless otherwise stated, we will now assume that the variables in rational expressions represent real numbers for which the denominator is not zero. Therefore, as we simplify rational expressions, warnings like the "provided $y \neq 0$ and $z \neq 0$" statement in this example are not necessary.

To simplify rational expressions, we often make use of the factoring methods discussed in Sections 8.6 and 8.7. We also often use the following property about polynomials and their opposites.

The Quotient of Opposites

The quotient of any nonzero polynomials and its opposite is -1.

EXAMPLE 4 Simplify: **a.** $\dfrac{2x^2 + 11x + 12}{3x^2 + 11x - 4}$ **b.** $\dfrac{3x^2 - 10xy - 8y^2}{4y^2 - xy}$

Strategy We will begin by factoring the numerator and denominator completely. Then we will remove any factors common to the numerator and denominator.

Why We need to make sure that the numerator and denominator have no common factors other than 1. If that is the case, then the rational expression is simplified.

Solution

a. $\dfrac{2x^2 + 11x + 12}{3x^2 + 11x - 4} = \dfrac{(2x + 3)\overset{1}{\cancel{(x + 4)}}}{(3x - 1)\underset{1}{\cancel{(x + 4)}}}$ Factor the numerator and denominator. Remove a factor equal to 1: $\frac{x+4}{x+4} = 1$.

$= \dfrac{2x + 3}{3x - 1}$ This expression does not simplify further.

b. $\dfrac{3x^2 - 10xy - 8y^2}{4y^2 - xy} = \dfrac{(3x + 2y)\overset{-1}{\cancel{(x - 4y)}}}{y\underset{1}{\cancel{(4y - x)}}}$ Factor the numerator and denominator. Since $x - 4y$ and $4y - x$ are opposites, simplify by replacing $\frac{x - 4y}{4y - x}$ with the equivalent fraction $\frac{-1}{1} = -1$.

$= \dfrac{-(3x + 2y)}{y}$

This result can also be written as $-\dfrac{3x + 2y}{y}$ or $\dfrac{-3x - 2y}{y}$.

Caution

In Example 4, do not remove the x's in the result $\frac{2x + 3}{3x - 1}$. The x in the numerator is not a factor of the entire numerator. Likewise, the x in the denominator is not a factor of the entire denominator.

$$\frac{\overset{1}{2\cancel{x}} + 3}{\underset{1}{3\cancel{x}} - 1} = \frac{2 + 3}{3 - 1} = \frac{5}{2}$$

Caution

A $-$ symbol in front of a fraction may be placed in the numerator or the denominator, but not in both. For example,

$$-\frac{3x + 2y}{y} \neq \frac{-3x - 2y}{-y}$$

Self Check 4 Simplify: **a.** $\dfrac{2x^2 + 5x + 2}{3x^2 + 5x - 2}$ **b.** $\dfrac{2a^2 - 3ab - 9b^2}{3b^2 - ab}$

Now Try Problems 39 and 43

5 Multiply and Divide Rational Expressions.

Recall that to multiply two fractions, we multiply the numerators and multiply the denominators. We use the same strategy to multiply rational expressions.

Multiplying Rational Expressions

To multiply rational expressions, multiply their numerators and their denominators. Then, if possible, factor and simplify.

For any two rational expressions, $\frac{A}{B}$ and $\frac{C}{D}$,

$$\frac{A}{B} \cdot \frac{C}{D} = \frac{AC}{BD}$$

EXAMPLE 5 Multiply: **a.** $\frac{x^2 - 6x + 9}{20x} \cdot \frac{5x^2}{x^2 - 9}$ **b.** $(2x - x^2) \cdot \frac{x}{x^2 - xb - 2x + 2b}$

Strategy To find the product, we will use the rule for multiplying rational expressions. In the process, we must be prepared to factor the numerators and denominators so that any common factors can be removed.

Why We want to give the result in simplified form.

Solution

a.

$$\frac{x^2 - 6x + 9}{20x} \cdot \frac{5x^2}{x^2 - 9} = \frac{(x^2 - 6x + 9)5x^2}{20x(x^2 - 9)}$$ Multiply the numerators. Multiply the denominators.

$$= \frac{(x - 3)(x - 3)5 \cdot x \cdot x}{4 \cdot 5 \cdot x(x + 3)(x - 3)}$$ To prepare to simplify, factor the numerator and factor the denominator.

$$= \frac{\overset{1}{\cancel{(x - 3)}}(x - 3)\overset{1}{\cancel{5}} \cdot \overset{1}{\cancel{x}} \cdot x}{4 \cdot \underset{1}{\cancel{5}} \cdot \underset{1}{\cancel{x}}(x + 3)\underset{1}{\cancel{(x - 3)}}}$$ Simplify by removing common factors of the numerator and denominator.

$$= \frac{x(x - 3)}{4(x + 3)}$$

Caution

When multiplying rational expressions, always write the result in simplest form by removing any factors common to the numerator and denominator.

We could distribute in the numerator and/or denominator and write the result as $\frac{x^2 - 3x}{4(x + 3)}$ or $\frac{x^2 - 3x}{4x + 12}$. Check with your instructor to see which form of the result he or she prefers.

b. Writing $2x - x^2$ as $\frac{2x - x^2}{1}$ is helpful during the multiplication process when we multiply numerators and multiply denominators.

$$(2x - x^2) \cdot \frac{x}{x^2 - xb - 2x + 2b}$$

$$= \frac{2x - x^2}{1} \cdot \frac{x}{x^2 - xb - 2x + 2b}$$ Write $2x - x^2$ as $\frac{2x - x^2}{1}$.

$$= \frac{(2x - x^2)x}{1(x^2 - xb - 2x + 2b)}$$ Multiply the numerators. Multiply the denominators.

$$= \frac{x(2 - x)x}{1[x(x - b) - 2(x - b)]}$$ To prepare to simplify, factor out x in the numerator. In the denominator, begin factoring by grouping.

$$= \frac{x(2 - x)x}{1(x - b)(x - 2)}$$ In the denominator, complete the factoring by grouping. The brackets [ ] are no longer needed.

$$= \frac{x\overset{-1}{\cancel{(2 - x)}}x}{1(x - b)\underset{1}{\cancel{(x - 2)}}}$$ Simplify: The quotient of any nonzero quantity and its opposite is -1: $\frac{2 - x}{x - 2} = -1$.

$$= -\frac{x^2}{x - b}$$

Success Tip

We would obtain the same answer if we had factored the numerators and denominators first and simplified before we multiplied.

Self Check 5 Multiply: **a.** $\frac{a^2 + 6a + 9}{18a} \cdot \frac{3a^3}{a + 3}$

b. $\frac{x^2 + 5x + 6}{4x + 8 - x^2 - 2x}(x^2 - 4x)$

Now Try Problems 51 and 53

Recall that to divide fractions, we multiply the first fraction by the reciprocal of the second fraction. We use the same strategy to divide rational expressions.

Dividing Rational Expressions

To divide two rational expressions, multiply the first by the reciprocal of the second. Then, if possible, factor and simplify.

For any two rational expressions, $\frac{A}{B}$ and $\frac{C}{D}$, where $\frac{C}{D} \neq 0$,

$$\frac{A}{B} \div \frac{C}{D} = \frac{A}{B} \cdot \frac{D}{C} = \frac{AD}{BC}$$

EXAMPLE 6 Divide: $\frac{x^3 + 8}{4x + 4} \div \frac{x^2 - 2x + 4}{2x^2 - 2}$

Strategy To find the quotient, we will use the rule for dividing rational expressions. After multiplying by the reciprocal, we will factor each polynomial that is not prime and remove any common factors of the numerator and denominator.

Why We want to give the result in simplified form.

Solution

$$\frac{x^3 + 8}{4x + 4} \div \frac{x^2 - 2x + 4}{2x^2 - 2}$$

$$= \frac{x^3 + 8}{4x + 4} \cdot \frac{2x^2 - 2}{x^2 - 2x + 4}$$ Multiply the first rational expression by the reciprocal of the second.

$$= \frac{(x^3 + 8)(2x^2 - 2)}{(4x + 4)(x^2 - 2x + 4)}$$ Multiply the numerators. Multiply the denominators.

$$= \frac{(x + 2)\overset{1}{\cancel{(x^2 - 2x + 4)}}\overset{1}{\cancel{2}}\overset{1}{\cancel{(x + 1)}}(x - 1)}{2 \cdot \underset{1}{\cancel{2}}\underset{1}{\cancel{(x + 1)}}\underset{1}{\cancel{(x^2 - 2x + 4)}}}$$ To prepare to simplify, factor completely. Note that $x^2 - 2x + 4$ is prime. Then simplify.

$$= \frac{(x + 2)(x - 1)}{2}$$ Multiply the remaining monomial factors in the numerator: $1 \cdot 1 \cdot 1 = 1$. Multiply the remaining factors in the denominator: $2 \cdot 1 \cdot 1 \cdot 1 = 2$.

Caution

When dividing rational expressions, always write the result in simplest form by removing any factors common to the numerator and denominator.

Self Check 6 Divide: $\frac{x^3 - 8}{9x - 9} \div \frac{x^2 + 2x + 4}{3x^2 - 3x}$

Now Try Problem 61

6 Add and Subtract Rational Expressions.

To add (or subtract) fractions with like denominators, we add (or subtract) the numerators and keep the same denominator. We use the same strategy to add (or subtract) rational expressions with like denominators.

Adding and Subtracting Rational Expressions That Have the Same Denominator

To add (or subtract) rational expressions that have same denominator, add (or subtract) their numerators and write the sum (or difference) over the common denominator. Then, if possible, factor and simplify.

If $\frac{A}{D}$ and $\frac{B}{D}$ are rational expressions,

$$\frac{A}{D} + \frac{B}{D} = \frac{A + B}{D} \quad \text{and} \quad \frac{A}{D} - \frac{B}{D} = \frac{A - B}{D}$$

EXAMPLE 7 Add $\frac{a^2}{a^2 - 36} + \frac{6a}{a^2 - 36}$ and simplify the result.

Strategy We will add the numerators and write the sum over the common denominator.

Why This is the rule for adding rational expressions that have the *same* denominator.

Solution

$$\frac{a^2}{a^2 - 36} + \frac{6a}{a^2 - 36} = \frac{a^2 + 6a}{a^2 - 36}$$ Add the numerators. Write the sum over the common denominator, $a^2 - 36$.

$$= \frac{a(a + 6)}{(a + 6)(a - 6)}$$ To prepare to simplify, factor the numerator and denominator.

$$= \frac{a\overset{1}{\cancel{(a + 6)}}}{\underset{1}{\cancel{(a + 6)}}(a - 6)}$$ Simplify by removing the common factor of $a + 6$.

$$= \frac{a}{a - 6}$$

Caution

When adding or subtracting rational expressions, always write the result in simplest form by removing any factors common to the numerator and denominator.

Self Check 7 Add: $\frac{2b}{b^2 - 4} + \frac{b^2}{b^2 - 4}$

Now Try Problem 65

To add or subtract rational expressions with unlike denominators, we build them to rational expressions with the same denominator.

Building Rational Expressions

To build a rational expression, multiply it by 1 in the form of $\frac{c}{c}$, where c is any nonzero number or expression.

When adding (or subtracting) rational expressions with unlike denominators, we will write the rational expressions with the smallest common denominator possible, called the **least** (or lowest) **common denominator (LCD).** To find the least common denominator of several rational expressions, we follow these steps.

Finding the LCD

1. Factor each denominator completely.
2. The LCD is a product that uses each different factor obtained in step 1 the greatest number of times it appears in any one factorization.

EXAMPLE 8 Add: $\frac{5a}{24b} + \frac{11a}{18b^2}$

Strategy We will find the LCD of these rational expressions. Then we will build the rational expressions so each one has the LCD as its denominator.

Why Since the denominators are different, we cannot add these rational expressions in their present form.

Solution To find the LCD, we write each denominator as the product of prime numbers and variables.

$$24b = 2 \cdot 2 \cdot 2 \cdot 3 \cdot b = 2^3 \cdot 3 \cdot b$$
$$18b^2 = 2 \cdot 3 \cdot 3 \cdot b \cdot b = 2 \cdot 3^2 \cdot b^2$$

Success Tip

Note that the highest power of each factor is used to form the LCD:

$$24b = 2^{③} \cdot 3 \cdot b$$
$$18b^2 = 2 \cdot 3^{②} \cdot b^{②}$$
$$\text{LCD} = 2^3 \cdot 3^2 \cdot b^2 = 72b^2$$

Then we form a product using each of these factors the greatest number of times it appears in any one factorization.

The greatest number of times the factor 2 appears is three times.
The greatest number of times the factor 3 appears is twice.
The greatest number of times the factor b appears is twice.

$$\text{LCD} = 2 \cdot 2 \cdot 2 \cdot 3 \cdot 3 \cdot b \cdot b$$
$$= 72b^2$$

We now multiply the numerator and denominator of each rational expression by whatever it takes to build their denominators to $72b^2$.

Success Tip

You may want to review the procedure for adding and subtracting rational expressions that have unlike denominators on page 541.

$$\frac{5a}{24b} + \frac{11a}{18b^2} = \frac{5a}{24b} \cdot \frac{3b}{3b} + \frac{11a}{18b^2} \cdot \frac{4}{4}$$ Build each rational expression by multiplying it by a form of 1.

$$= \frac{15ab}{72b^2} + \frac{44a}{72b^2}$$ Multiply the numerators. Multiply the denominators. Note that the expressions now have like denominators.

$$= \frac{15ab + 44a}{72b^2}$$ Add the numerators. Write the sum over the common denominator. The result does not simplify.

Self Check 8 Add: $\frac{3}{28z^3} + \frac{5x}{21z}$

Now Try Problem 73

EXAMPLE 9 Subtract: $\frac{x+1}{x^2 - 2x + 1} - \frac{x-4}{x^2 - 1}$

Strategy We will factor each denominator, find the LCD, and build the rational expressions so each one has the LCD as its denominator.

Why Since the denominators are different, we cannot subtract these rational expressions in their present form.

Solution We factor each denominator to find the LCD:

$$x^2 - 2x + 1 = (x-1)(x-1) = (x-1)^2$$
$$x^2 - 1 = (x+1)(x-1)$$

The greatest number of times $x - 1$ appears is twice. The greatest number of times $x + 1$ appears is once.

The LCD is $(x-1)^2(x+1)$ or $(x-1)(x-1)(x+1)$.

We now write each rational expression with its denominator in factored form. Then we multiply each numerator and denominator by the missing factor, so that each rational expression has a denominator of $(x-1)(x-1)(x+1)$.

$$\frac{x+1}{x^2 - 2x + 1} - \frac{x-4}{x^2 - 1}$$

$$= \frac{x+1}{(x-1)(x-1)} - \frac{x-4}{(x+1)(x-1)}$$ Write each denominator in factored form.

$$= \frac{x+1}{(x-1)(x-1)} \cdot \frac{x+1}{x+1} - \frac{x-4}{(x+1)(x-1)} \cdot \frac{x-1}{x-1}$$ Build each rational expression.

$$= \frac{x^2 + 2x + 1}{(x-1)(x-1)(x+1)} - \frac{x^2 - 5x + 4}{(x-1)(x-1)(x+1)}$$ Multiply the numerators. The denominators are now the same.

Multiply the numerators using the FOIL method to prepare to combine like terms.

Don't multiply the denominators. Leave them in factored form to possibly simplify the result later.

$$= \frac{x^2 + 2x + 1 - (x^2 - 5x + 4)}{(x - 1)(x - 1)(x + 1)}$$ Subtract the numerators. Write the difference over the common denominator. Don't forget the parentheses shown in blue.

$$= \frac{x^2 + 2x + 1 - x^2 + 5x - 4}{(x - 1)(x - 1)(x + 1)}$$ To subtract in the numerator, change the signs of x^2, $-5x$, and 4, and drop the parentheses.

$$= \frac{7x - 3}{(x - 1)(x - 1)(x + 1)}$$ Combine like terms in the numerator. The result does not simplify.

Self Check 9 Subtract: $\frac{a + 2}{a^2 - 4a + 4} - \frac{a - 3}{a^2 - 4}$

Now Try Problem 81

7 Solve Rational Equations.

If an equation contains one or more rational expressions, it is called a **rational equation.** Recall that we use a fraction-clearing strategy to solve rational equations. All possible solutions of a rational equation must be checked. Multiplying both sides of an equation by an expression that contains a variable can lead to **extraneous solutions,** which must be discarded.

EXAMPLE 10 Solve: $1 + \frac{8a}{a^2 + 3a} = \frac{3}{a}$

Strategy We will find the LCD of the rational expressions in the equation and multiply both sides by the LCD.

Why This will clear the equation of fractions.

Solution Since the binomial $a^2 + 3a$ factors as $a(a + 3)$, we can write the given equation as:

$$1 + \frac{8a}{a(a + 3)} = \frac{3}{a}$$ Factor $a^2 + 3a$.

Success Tip

You may want to review the procedure for solving rational equations on page 558.

We see that 0 and -3 cannot be solutions of the equation, because they make rational expressions in the equation undefined.

We can clear the equation of fractions by multiplying both sides by $a(a + 3)$, which is the LCD of the two rational expressions $\frac{8a}{a^2 + 3a}$ and $\frac{3}{a}$.

Caution

After multiplying both sides by the LCD and simplifying, the equation should not contain any fractions. If it does check for an algebraic error, or perhaps your LCD is incorrect

$$a(a + 3)\left[1 + \frac{8a}{a(a + 3)}\right] = a(a + 3)\left(\frac{3}{a}\right)$$ Multiply both sides by the LCD.

$$a(a + 3)1 + a(a + 3)\frac{8a}{a(a + 3)} = a(a + 3)\left(\frac{3}{a}\right)$$ On the left side, distribute the multiplication by $a(a + 3)$.

$$a(a + 3)1 + \frac{\overset{1}{\cancel{a}}\overset{1}{\cancel{(a + 3)}}\,8a}{\underset{1}{\cancel{a}}\underset{1}{\cancel{(a + 3)}}} = \overset{1}{\cancel{a}}(a + 3)\left(\frac{3}{\underset{1}{\cancel{a}}}\right)$$ Remove common factors of the numerator and denominator.

$$a^2 + 3a + 8a = 3a + 9$$ Simplify each side. The resulting quadratic equation does not contain any fractions.

$$a^2 + 8a - 9 = 0$$ To get 0 on the right side, subtract 3a and 9 from both sides.

$$(a + 9)(a - 1) = 0$$ Factor the left side.

$$a + 9 = 0 \quad \text{or} \quad a - 1 = 0$$ Set each factor equal to 0.

$$a = -9 \quad | \quad a = 1$$ Solve each equation.

The solutions are -9 and 1. Verify that both satisfy the original equation.

Self Check 10 Solve: $1 + \frac{2}{2b + 1} = \frac{5}{2b^2 + b}$

Now Try Problem 95

SECTION 8.8 STUDY SET

VOCABULARY

Fill in the blanks.

1. A quotient of two polynomials, such as $\frac{x^2 + x}{x^2 - 3x}$, is called a ________ expression.
2. A ________ function, such as $f(x) = \frac{x - 7}{x^2 - x - 6}$, is a function whose equation is defined by a rational expression in one variable.
3. The ________ of a function is the set of all permissible input values for the variable.
4. The rational function $f(x) = \frac{9x}{x - 10}$ is __________ for $x = 10$. In other words, there is a __________ on the domain of the function: $x \neq 10$.
5. To ________ a rational expression, we remove factors common to the numerator and denominator.
6. The quotient of _________ is -1. For example, $\frac{x - 8}{8 - x} = -1$.
7. In the rational expression $\frac{(x + 2)(3x - 1)}{(x + 2)(4x + 2)}$, the binomial $x + 2$ is a common ______ of the numerator and the denominator.
8. The least _________ _____________ of $\frac{x - 8}{x + 6}$ and $\frac{6 - 5x}{x}$ is $x(x + 6)$.
9. To ______ a rational expression, we multiply it by a form of 1. For example, $\frac{2}{n^2} \cdot \frac{8}{8} = \frac{16}{8n^2}$.
10. Equations that contain one or more rational expressions, such as $\frac{x}{x + 2} = 4 + \frac{10}{x + 1}$, are called ________ equations.

CONCEPTS

11. Let $f(x) = \frac{2x + 1}{x^2 + 3x - 4}$. Find

 a. $f(0)$ b. $f(2)$ c. $f(1)$

12. For what value(s) of x is each function undefined?

 a. $f(x) = \frac{x - 7}{x}$ b. $f(x) = \frac{x + 1}{x - 3}$

 c. $f(x) = \frac{x^2 - 2}{x(x + 8)}$ d. $f(x) = \frac{8x}{(x - 1)(x + 1)}$

13. The graph of rational function f is shown. Find each of the following.

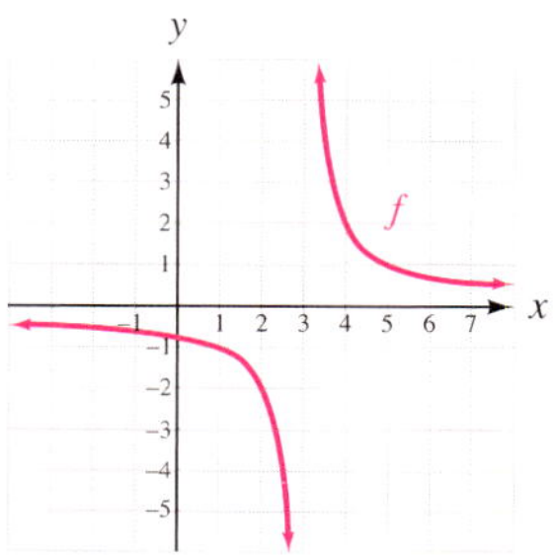

 a. $f(1)$
 b. $f(4)$
 c. The value of x for which $f(x) = -2$
 d. The value of x for which $f(x) = 1$

14. Match each function with the correct graph shown below.

 a. $f(x) = 2$ b. $f(x) = x$ c. $f(x) = x^2$

 d. $f(x) = x^3$ e. $f(x) = |x|$ f. $f(x) = \frac{1}{x}$

i.

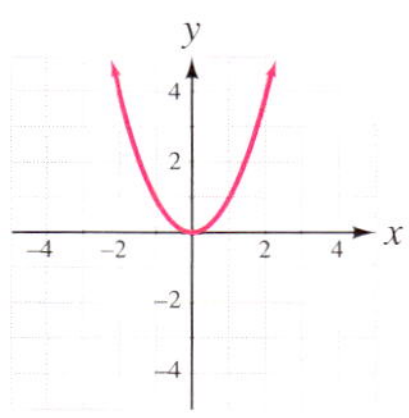

ii.

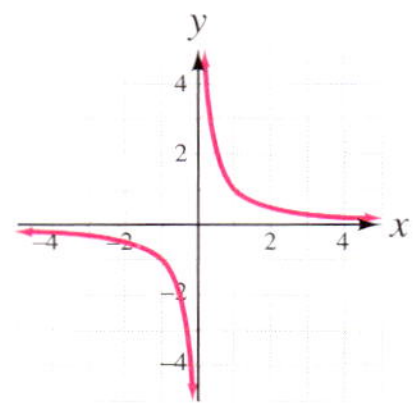

iii.

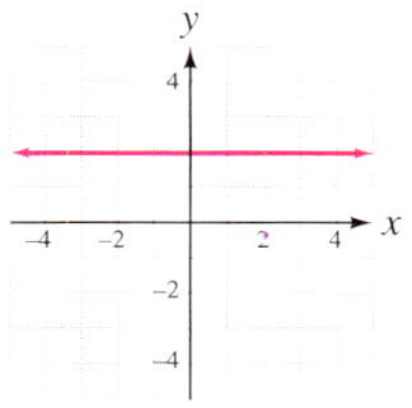

iv.

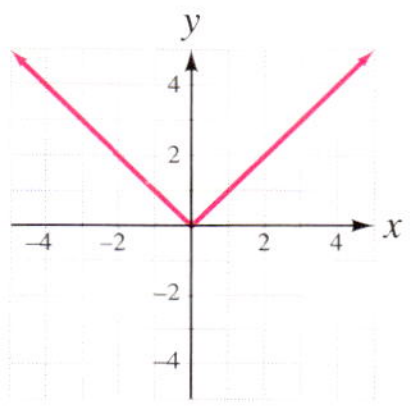

v.

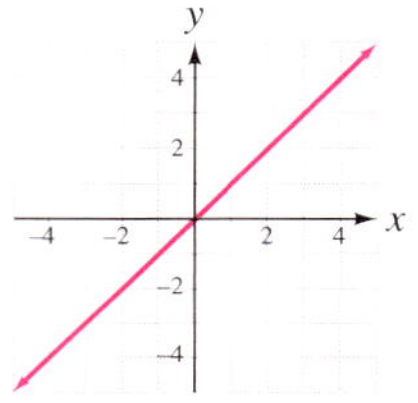

vi.

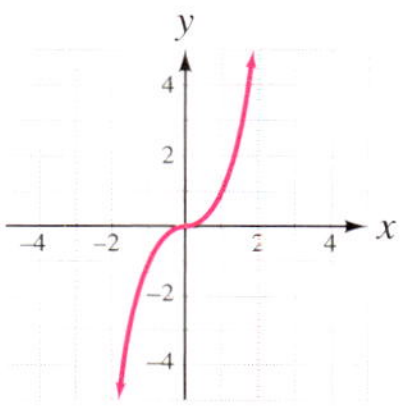

Fill in the blanks.

15. To multiply rational expressions, multiply their __________ and multiply their ______________. To divide two rational expressions, multiply the first by the __________ of the second.
16. To add or subtract rational expressions that have the same denominator, add or subtract the __________ and write the sum or difference over the common __________.
17. To find the least common denominator of several rational expressions, ______ each denominator completely. The LCD is a product that uses each different factor the ________ number of times it appears in any one factorization.
18. The expression $4 - y$ must be multiplied by ______ to obtain $y - 4$.
19. Consider the following factorizations.

$$18x - 36 = 2 \cdot 3 \cdot 3 \cdot (x - 2)$$
$$3x^2 - 3x - 6 = 3(x - 2)(x + 1)$$

 a. What is the greatest number of times the factor 3 appears in any one factorization?
 b. What is the greatest number of times the factor $x - 2$ appears in any one factorization?

20. The LCD for $\frac{2x+1}{x^2+5x+6}$ and $\frac{3x}{x^2-4}$ is

$$\text{LCD} = (x+2)(x+3)(x-2)$$

If we want to subtract these rational expressions, what form of 1 should be used

a. to build $\frac{2x+1}{x^2+5x+6}$?

b. to build $\frac{3x}{x^2-4}$?

21. To clear the following equations of fractions, by what should both sides be multiplied?

a. $\frac{1}{a} = \frac{1}{3} - \frac{2}{3a}$

b. $\frac{2}{x-2} + \frac{10}{x+5} = \frac{2x}{x^2+3x-10}$

22. Perform each multiplication.

a. $4x\left(\frac{3}{4x}\right)$

b. $(x+6)(x-2)\left(\frac{3}{x-2}\right)$

c. $8(x+4)\left[\frac{7x}{2(x+4)}\right]$

d. $6(m-5)\left(\frac{7}{5-m}\right)$

NOTATION

23. a. Write $5x^2 + 35x$ as a fraction.

b. What is the reciprocal of $5x^2 + 35x$?

24. What numbers are not included in each set of real numbers represented using interval notation?

a. $(-\infty, 4) \cup (4, \infty)$

b. $(-\infty, -8) \cup (-8, 0) \cup (0, \infty)$

GUIDED PRACTICE

Find the domain of each rational function. Express your answer in words and using interval notation. See Example 2.

25. $f(x) = \frac{2}{x}$

26. $f(x) = \frac{8}{x-1}$

27. $f(x) = \frac{2x}{x+2}$

28. $f(x) = \frac{2x+1}{x^2-2x}$

29. $f(x) = \frac{3x-1}{x-x^2}$

30. $f(x) = \frac{x^2+36}{x^2-36}$

31. $f(x) = \frac{x^2+3x+2}{x^2-x-56}$

32. $f(x) = \frac{2x^2-3x-2}{x^2+2x-24}$

Simplify each rational expression. See Example 3.

33. $\frac{15a^2}{25a^8}$

34. $\frac{12x}{16x^7}$

35. $\frac{24x^3y^4}{54x^4y^3}$

36. $\frac{15a^5b^4}{21a^2b^5}$

Simplify each rational expression. See Example 4.

37. $\frac{5x^2-10x}{x^2-4x+4}$

38. $\frac{x^2+6x+9}{2x^2+6x}$

39. $\frac{6x^2-7x-5}{2x^2+5x+2}$

40. $\frac{6x^2+x-2}{8x^2+2x-3}$

41. $\frac{4-x^2}{x^2-x-2}$

42. $\frac{x^2-2x-15}{25-x^2}$

43. $\frac{p^3+p^2q-2pq^2}{pq^2+p^2q-2p^3}$

44. $\frac{m^3-mn^2}{mn^2+m^2n-2m^3}$

Perform the operations and simplify, if possible. See Example 5.

45. $\frac{10a^2}{3b^4} \cdot \frac{12b^3}{5a^2}$

46. $\frac{16c^3}{5d^2} \cdot \frac{25d}{12c}$

47. $\frac{3p^2}{6p+24} \cdot \frac{p^2-16}{6p}$

48. $\frac{y^2+6y+9}{15y} \cdot \frac{3y^2}{2y+6}$

49. $\frac{x^2+2x+1}{9x} \cdot \frac{2x^2-2x}{2x^2-2}$

50. $\frac{a+6}{a^2-16} \cdot \frac{3a-12}{3a+18}$

51. $\frac{2x^2-x-3}{x^2-1} \cdot \frac{x^2+x-2}{2x^2+x-6}$

52. $\frac{2p^2-5p-3}{p^2-9} \cdot \frac{2p^2+5p-3}{2p^2+5p+2}$

53. $(6a-a^2) \cdot \frac{a^3}{a^3-6a^2+3a-18}$

54. $(10n-n^2) \cdot \frac{n^6}{n^3-10n^2-2n+20}$

55. $(x^2+x-2cx-2c) \cdot \frac{x^2+3x+2}{4c^2-x^2}$

56. $(2ax-10x-a+5) \cdot \frac{x}{x-2x^2}$

Perform the operations and simplify, if possible. See Example 6.

57. $\frac{m^2n}{4} \div \frac{mn^3}{6}$

58. $\frac{a^4b}{14} \div \frac{a^3b^2}{21}$

59. $\frac{x^2-16}{x^2-25} \div \frac{x+4}{x-5}$

60. $\frac{a^2-9}{a^2-49} \div \frac{a+3}{a+7}$

61. $\frac{5c+1}{6} \div \frac{125c^3+1}{6c+6}$

62. $\frac{6m-8}{9m^3} \div \frac{27m^3-64}{9m+9}$

63. $\frac{3n^2+5n-2}{12n^2-13n+3} \div \frac{n^2+3n+2}{4n^2+5n-6}$

64. $\dfrac{8y^2 - 14y - 15}{6y^2 - 11y - 10} \div \dfrac{4y^2 - 9y - 9}{3y^2 - 7y - 6}$

Perform the operations and simplify, if possible. See Example 7.

65. $\dfrac{3x}{x^2 - 9} - \dfrac{9}{x^2 - 9}$

66. $\dfrac{9x}{x^2 - 1} - \dfrac{9}{x^2 - 1}$

67. $\dfrac{3y - 2}{2y + 6} - \dfrac{2y - 5}{2y + 6}$

68. $\dfrac{5x + 8}{3x + 15} - \dfrac{3x - 2}{3x + 15}$

Perform the operations and simplify, if possible. See Example 8.

69. $\dfrac{3}{4x} + \dfrac{2}{3x}$

70. $\dfrac{2}{5a} + \dfrac{3}{2a}$

71. $\dfrac{8}{9y^2} + \dfrac{1}{6y^4}$

72. $\dfrac{5}{6a^3} + \dfrac{7}{8a^2}$

73. $\dfrac{3}{4ab^2} - \dfrac{5}{2a^2b}$

74. $\dfrac{1}{5xy^3} - \dfrac{2}{15x^2y}$

75. $\dfrac{y - 7}{y^2} - \dfrac{y + 7}{2y}$

76. $\dfrac{x + 5}{xy} - \dfrac{x - 1}{x^2y}$

Perform the operations and simplify, if possible. See Example 9.

77. $\dfrac{3}{x + 2} + \dfrac{5}{x - 4}$

78. $\dfrac{2}{a + 4} - \dfrac{6}{a + 3}$

79. $\dfrac{x + 2}{x + 5} - \dfrac{x - 3}{x + 7}$

80. $\dfrac{7}{x + 3} + \dfrac{4x}{x + 6}$

81. $\dfrac{x}{x^2 + 5x + 6} + \dfrac{x}{x^2 - 4}$

82. $\dfrac{x}{x^2 + 2x + 1} + \dfrac{x}{x^2 - 1}$

83. $\dfrac{4}{x^2 - 2x - 3} - \dfrac{x}{3x^2 - 7x - 6}$

84. $\dfrac{5}{x^2 + 5x - 6} - \dfrac{x}{2x^2 + 7x - 30}$

85. $\dfrac{6}{5d^2 - 5d} - \dfrac{3}{5d - 5}$

86. $\dfrac{9}{2r^2 - 2r} - \dfrac{5}{2r - 2}$

87. $\dfrac{m}{m^2 + 9m + 20} - \dfrac{4}{m^2 + 7m + 12}$

88. $\dfrac{x + 3}{2x^2 - 5x + 2} - \dfrac{3x - 1}{x^2 - x - 2}$

Solve each equation. See Example 10.

89. $\dfrac{3}{y} + \dfrac{7}{2y} = 13$

90. $\dfrac{2}{x} + \dfrac{1}{2} = \dfrac{7}{2x}$

91. $\dfrac{2}{x} + \dfrac{1}{2} = \dfrac{9}{4x} - \dfrac{1}{2x}$

92. $\dfrac{7}{5x} - \dfrac{1}{2} = \dfrac{5}{6x} + \dfrac{1}{3}$

93. $\dfrac{a}{2} = \dfrac{a - 6}{3a - 9} - \dfrac{1}{3}$

94. $\dfrac{b}{5} = \dfrac{b - 14}{2b - 16} - \dfrac{1}{2}$

95. $\dfrac{2}{5x - 5} + \dfrac{x - 2}{15} = \dfrac{4}{5x - 5}$

96. $\dfrac{3}{2x + 4} = \dfrac{x - 2}{2} + \dfrac{x - 5}{2x + 4}$

TRY IT YOURSELF

Simplify each function. List any restrictions on the domain.

97. $f(x) = \dfrac{x^2 + 6x - 16}{x^2 - 4}$

98. $f(x) = \dfrac{5x^2 + 50x}{x^5 + 10x^4}$

99. $g(x) = \dfrac{x^3 + 64}{x^3 + 4x^2 + 3x + 12}$

100. $h(t) = \dfrac{t^3 - 5t^2 - 5t + 25}{t^3 - 125}$

In the following problems, simplify each expression by performing the indicated operations and solve each equation.

101. $\dfrac{p^3 - q^3}{q^2 - p^2} \cdot \dfrac{q^2 + pq}{p^3 + p^2q + pq^2}$

102. $\dfrac{x^3 + y^3}{x^3 - y^3} \div \dfrac{x^2 - xy + y^2}{x^2 + xy + y^2}$

103. $\dfrac{t}{t^2 + 5t + 6} - \dfrac{2}{t^2 + 3t + 2}$

104. $\dfrac{2a}{a^2 - 2a - 8} + \dfrac{3}{a^2 - 5a + 4}$

105. $\dfrac{4}{m^2 - 9} + \dfrac{5}{m^2 - m - 12} = \dfrac{7}{m^2 - 7m + 12}$

106. $\dfrac{34}{x^2} + \dfrac{13}{20x} = \dfrac{3}{2x}$

107. $\dfrac{2}{x - 1} - \dfrac{2x}{x^2 - 1} - \dfrac{x}{x^2 + 2x + 1}$

108. $\dfrac{x - 2}{x^2 - 3x} + \dfrac{2x - 1}{x^2 + 3x} - \dfrac{2}{x^2 - 9}$

109. $(2x^2 - 15x + 25) \div \dfrac{2x^2 - 3x - 5}{x + 1}$

110. $(x^2 - 6x + 9) \div \dfrac{x^2 - 9}{x + 3}$

111. $\dfrac{y^3 - x^3}{2x^2 + 2xy + x + y} \cdot \dfrac{2x^2 - 5x - 3}{yx - 3y - x^2 + 3x}$

112. $\dfrac{ax + ay + bx + by}{x^3 - 27} \cdot \dfrac{x^2 + 3x + 9}{xc + xd + yc + yd}$

113. $\dfrac{24n^4}{16n^4 + 24n^3}$

114. $\dfrac{18m^4}{36m^4 - 9m^3}$

115. $1 + x - \dfrac{x}{x - 5}$

116. $2 - x + \dfrac{3}{x - 9}$

117. $\dfrac{ax + by + ay + bx}{a^2 - b^2}$

118. $\dfrac{3x^2 - 3y^2}{x^2 + 2y + 2x + yx}$

119. $\dfrac{3}{s - 2} + \dfrac{s - 14}{2s^2 - 3s - 2} - \dfrac{4}{2s + 1} = 0$

120. $\dfrac{1}{y^2 - 2y - 3} + \dfrac{1}{y^2 - 4y + 3} - \dfrac{1}{y^2 - 1} = 0$

121. $\dfrac{5}{x + 4} + \dfrac{1}{x + 4} = x - 1$

122. $\dfrac{5}{3x + 12} - \dfrac{1}{9} = \dfrac{x - 1}{3x}$

123. $\dfrac{5x}{x-3} + \dfrac{4x}{3-x}$

124. $\dfrac{8x}{x-4} - \dfrac{10x}{4-x}$

APPLICATIONS

125. Environmental Cleanup. Suppose the cost (in dollars) of removing $p\%$ of the pollution in a river is given by the rational function

$$f(p) = \frac{50{,}000p}{100-p} \quad \text{where } 0 \le p < 100$$

Find the cost of removing each percent of pollution.

a. 50% **b.** 80%

126. Directory Costs. The average (mean) cost for a service club to publish a directory of its members is given by the rational function

$$f(x) = \frac{1.25x + 700}{x}$$

where x is the number of directories printed. Find the average cost per directory if

a. 500 directories are printed.

b. 2,000 directories are printed.

127. Utility Costs. An electric company charges \$7.50 per month plus 9¢ for each kilowatt hour (kwh) of electricity used.

a. Find a linear function that gives the total cost of n kwh of electricity. (*Hint:* See Example 1.)

b. Find a rational function that gives the average cost per kwh when using n kwh.

c. Find the average cost per kwh when 775 kwh are used.

128. Drafting. Among the tools used in drafting are $45° - 45° - 90°$ and $30° - 60° - 90°$ triangles. Find the perimeter of each triangle and express each result as a rational expression.

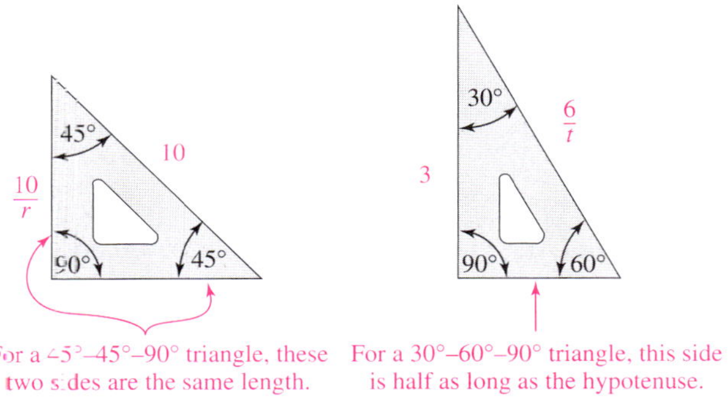

WRITING

129. A student compared his answer, $\frac{a-3b}{2b-a}$, with the answer, $\frac{3b-a}{a-2b}$, in the back of the text. Is the student's work correct? Explain.

130. Explain the error that is made in the following work:

$$\frac{3x^2+1}{3y} = \frac{\cancel{3}x^2+1}{\cancel{3}y} = \frac{x^2+1}{y}$$

131. Write some comments to the student who wrote the following solution, explaining his misunderstanding.

$$\text{Multiply: } \frac{1}{x}\cdot\frac{3}{2} = \frac{1}{x}\cdot\frac{2}{2}\cdot\frac{3}{2}\cdot\frac{x}{x}$$
$$= \frac{2}{2x}\cdot\frac{3x}{2x}$$
$$= \frac{6x}{4x^2}$$

132. Would you use the same approach to answer the following problems? Explain why or why not.

$$\text{Simplify: } \frac{x^2-10}{x^2-1} - \frac{3x}{x-1} - \frac{2x}{x+1}$$

$$\text{Solve: } \frac{x^2-10}{x^2-1} - \frac{3x}{x-1} = -\frac{2x}{x+1}$$

REVIEW

Solve each equation.

133. $-10|16h + 4| - 3 = -3$

134. $-5|2a - 9| + 14 = 14$

CHALLENGE PROBLEMS

135. Simplify: $\dfrac{a^6 - 64}{(a^2 + 2a + 4)(a^2 - 2a + 4)}$

136. Add: $x^{-1} + x^{-2} + x^{-3} + x^{-4} + x^{-5}$

137. Simplify: $[(x^{-1} + 1)^{-1} + 1]^{-1}$

138. Find two rational expressions, each with denominator $x^2 + 5x + 6$, such that their sum is $\frac{1}{x+2}$.

Graph each rational function. Show the vertical asymptote as a dashed line and label it.

139. $f(x) = \dfrac{1}{x-1}$

140. $f(x) = \dfrac{1}{x+4}$

Perform the operations and simplify.

141. $\dfrac{6a^2 - 7a - 3}{2a^2 - 2} \div \dfrac{4a^2 - 12a + 9}{a^2 - 1} \cdot \dfrac{2a^2 - a - 3}{3a^2 - 2a - 1}$

142. $\dfrac{2x^2 - 2x - 4}{x^2 + 2x - 8} \cdot \dfrac{3x^2 + 15x}{x + 1} \div \dfrac{100 - 4x^2}{x^2 - x - 20}$

143. $\dfrac{\dfrac{h}{h^2 + 3h + 2}}{\dfrac{4}{h+2} - \dfrac{4}{h+1}}$

144. $\dfrac{\dfrac{2}{y-1} - \dfrac{2}{y}}{\dfrac{3}{y-1} - \dfrac{1}{1-y}}$

SECTION 8.9 Variation

OBJECTIVES

1. Solve problems involving direct variation.
2. Solve problems involving inverse variation.
3. Solve problems involving joint variation.
4. Solve problems involving combined variation.

ARE YOU READY?

The following problems review some basic skills that are needed when working with variation.

1. Find y if $y = 45x$ and $x = 10$.
2. Find d if $d = \frac{90}{n}$ and $n = 18$.
3. Let $f = 0.0036Av^2$. Find f if $A = 500$ and $v = 30$.
4. Solve each equation:

   a. $26 = 4k$ **b.** $80 = \frac{k}{2.5}$

In this section, we introduce four types of *variation models,* each of which expresses a special relationship between two or more quantities. We will use these models to solve problems involving travel, lighting, geometry, and highway construction.

1 Solve Problems Involving Direct Variation.

To introduce direct variation, we consider the formula for the circumference of a circle

$$C = \pi D$$

where C is the circumference, D is the diameter, and $\pi \approx 3.14159$. If we double the diameter of a circle, we determine another circle with a larger circumference C_1 such that

$$C_1 = \pi(2D) = 2\pi D = 2C$$

Thus, doubling the diameter results in doubling the circumference. Likewise, if we triple the diameter, we will triple the circumference.

In the formula, $C = \pi D$, we say that the variables C and D *vary directly,* or that they are *directly proportional.* This is because C is always found by multiplying D by a constant. In this example, the constant π is called the *constant of variation* or the *constant of proportionality.*

Direct Variation

The words **y varies directly as x** or **y is directly proportional to x** means that $y = kx$ for some nonzero constant k. The constant k is called the **constant of variation** or the **constant of proportionality.**

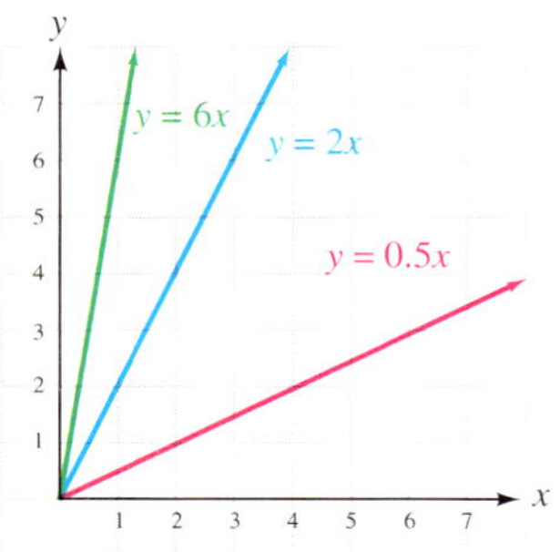

For any **direct variation** equation of the form $y = kx$, where $k > 0$: as x increases, y increases.

Since the formula for direct variation ($y = kx$) defines a linear function, its graph is always a line with a y-intercept at the origin. The graph of $y = kx$ where $x \geq 0$ appears on the left for three positive values of k.

One example of direct variation is Hooke's law from physics. Hooke's law states that the distance a spring will stretch varies directly as the force that is applied to it.

If d represents a distance and f represents a force, this verbal model of Hooke's law can be expressed as

$d = kf$ *The direct variation model can be read as "d is directly proportional to f."*

where k is the constant of variation. Suppose we know that a certain spring stretches 10 inches when a weight of 6 pounds is attached (see the figure). We can find k as follows:

$d = kf$

$10 = k(6)$ *Substitute 10 for d and 6 for f.*

$\frac{10}{6} = k$ *To isolate k, divide both sides by 6.*

$\frac{5}{3} = k$ *Simplify the fraction. This is the constant of variation.*

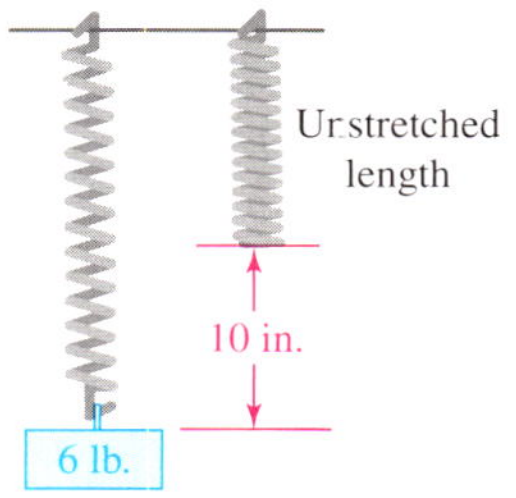

Success Tip

The value that we found for k is for this specific example. Another spring made out of a different type of steel will more than likely have a different value of k.

To find the force required to stretch the spring a distance of 35 inches, we can solve the equation $d = kf$ for f, with $d = 35$ and $k = \frac{5}{3}$.

$$d = kf \quad \text{This is the direct variation model.}$$

$$35 = \frac{5}{3}f \quad \text{Substitute 35 for } d \text{ and } \tfrac{5}{3} \text{ for } k.$$

$$105 = 5f \quad \text{Multiply both sides by 3.}$$

$$21 = f \quad \text{To isolate } f\text{, divide both sides by 5.}$$

Thus, the force required to stretch the spring a distance of 35 inches is 21 pounds.

We can use the following steps to solve variation problems.

Solving Variation Problems

To solve a variation problem:

1. Translate the verbal model into an equation.
2. Substitute the first set of values into the equation from step 1 to determine the value of k.
3. Substitute the value of k into the equation from step 1.
4. Substitute the remaining set of values into the equation from step 3 and solve for the unknown.

EXAMPLE 1

Currency Exchange. The currency calculator shown below converts from U.S. dollars to Russian rubles. When exchanging these currencies, the number of rubles received is directly proportional to the number of dollars to be exchanged. How many rubles will an exchange of $1,200 bring?

Strategy We will use a direct variation model to solve this problem.

Why The words *the number of rubles received is directly proportional to the number of dollars to be exchanged* indicate that this type of model should be used.

Solution

Step 1: The verbal model can be represented by the equation

$$r = kd \quad \text{This is a direct variation model.}$$

where r is the number of rubles, k is the constant of variation, and d is the number of dollars.

Step 2: From the illustration, we see that an exchange of $500 brings 14,900 rubles. To find k, we substitute 500 for d and 14,900 for r, and then we solve for k.

$$r = kd$$

$$14{,}900 = k(500) \quad \text{Substitute for } r \text{ and } d.$$

$$29.8 = k \quad \text{To isolate } k\text{, divide both sides by 500. This is the constant of variation.}$$

Step 3: Now we substitute the value of k, 29.8, into the equation $r = kd$, to get

$$r = 29.8d$$

Step 4: To find how many rubles an exchange of $1,200 will bring, we substitute 1,200 for d in the direct variation model, and then we evaluate the right side.

$$r = 29.8d \quad \text{This is the ruble direct variation model.}$$

$$r = 29.8(1{,}200) \quad \text{Substitute for } d.$$

$$r = 35{,}760 \quad \text{Do the multiplication. The result is in rubles.}$$

An exchange of $1,200 will bring 35,760 rubles.

Self Check 1 **Currency Exchange.** When exchanging currencies, the number of British pounds received is directly proportional to the number of U.S. dollars to be exchanged. If \$800 converts to 392 pounds, how many pounds will be received if \$1,500 is exchanged?

Now Try Problem 35

2 Solve Problems Involving Inverse Variation.

In the formula $w = \frac{12}{l}$, w gets smaller as l gets larger, and w gets larger as l gets smaller. Since these variables vary in opposite directions in a predictable way, we say that the variables *vary inversely*, or that they are *inversely proportional*. The constant 12 is the constant of variation.

Inverse Variation

The words **y varies inversely as x** or **y is inversely proportional to x** mean that $y = \frac{k}{x}$ for some nonzero constant k. The constant k is called the **constant of variation.**

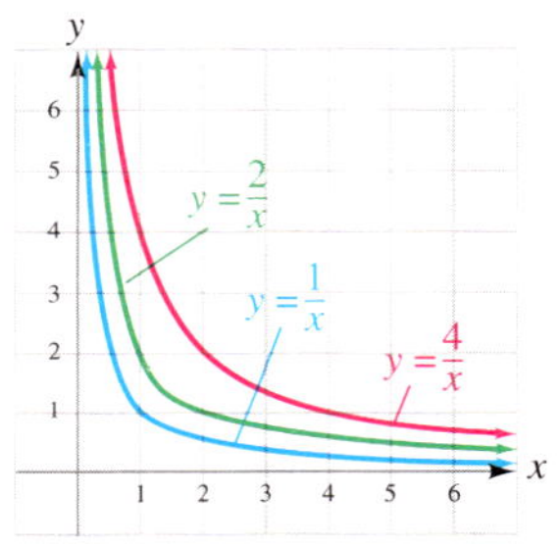

For any **inverse variation** equation of the form $y = \frac{k}{x}$, where $k > 0$: as x increases, y decreases.

The formula for inverse variation, $y = \frac{k}{x}$, defines a rational function whose graph will have the x- and y-axes as asymptotes. The graph of $y = \frac{k}{x}$ where $x > 0$ appears on the left for three positive values of k.

In an elevator, the amount of floor space per person varies inversely as the number of people in the elevator. If f represents the amount of floor space per person and n the number of people in the elevator, the relationship between f and n can be expressed by the equation:

$$f = \frac{k}{n}$$ This inverse variation model also can be read as "f is inversely proportional to n."

The figure on the right shows 6 people in an elevator; each has 8.25 square feet of floor space. To determine how much floor space each person would have if 15 people were in the elevator, we begin by determining k.

$f = \frac{k}{n}$ This is the inverse variation model.

$8.25 = \frac{k}{6}$ Substitute 8.25 for f and 6 for n.

$k = 49.5$ Multiply both sides by 6 to solve for k. This is the constant of variation.

To find the amount of floor space per person if 15 people are in the elevator, we proceed as follows:

$f = \frac{k}{n}$ This is the inverse variation model.

$f = \frac{49.5}{15}$ Substitute 49.5 for k and 15 for n.

$f = 3.3$ Do the division.

If 15 people were in the elevator, each would have 3.3 square feet of floor space.

EXAMPLE 2 **Photography.** The intensity I of light received from a light source varies inversely as the square of the distance from the light source. If a photographer, 16 feet away from his subject, has a light meter reading of 4 foot-candles of luminance, what will the meter read if the photographer moves in for a close-up 4 feet away from the subject?

Strategy We will use the inverse variation model of the form $I = \frac{k}{d^2}$, where I represents the intensity and d^2 represents the square of the distance from the light source.

Why The words *intensity varies inversely as the square of the distance* indicate that this type of model should be used.

Solution

$$I = \frac{k}{d^2}$$ This inverse variation model also can be read as "I is inversely proportional to d^2."

$$4 = \frac{k}{16^2}$$ To find k, we substitute 4 for I and 16 for d and solve for k.

$$4 = \frac{k}{256}$$ Evaluate the denominator: $16^2 = 256$.

$$1{,}024 = k$$ To isolate k, multiply both sides by 256. This is the constant of variation.

To find the intensity when the photographer is 4 feet away from the subject, we substitute 4 for d and 1,024 for k and simplify.

$$I = \frac{k}{d^2}$$ This is the inverse variation model.

$$I = \frac{1{,}024}{4^2}$$ Substitute for k and d.

$$I = 64$$ Evaluate the right side.

The intensity at 4 feet is 64 foot-candles.

Self Check 2 **Photography.** Find the intensity when the photographer is 8 feet away from the subject.

Now Try Problem 39

Success Tip

The constant of variation is usually positive, because most real-life applications involve only positive quantities. However, the definitions of *direct, inverse, joint,* and *combined variation* allow for a negative constant of variation.

3 Solve Problems Involving Joint Variation.

There are times when one variable varies as the product of several variables. For example, the area of a triangle varies directly with the product of its base and height:

$$A = \frac{1}{2}bh$$

Such variation is called *joint variation.*

Joint Variation

If one variable varies directly as the product of two or more variables, the relationship is called **joint variation.** If y varies jointly with x and z, then $y = kxz$. The nonzero constant k is called the **constant of variation.**

EXAMPLE 3 **Force of the Wind.** The force of the wind on a billboard varies jointly as the area of the billboard and the square of the wind velocity. When the wind is blowing at 20 mph, the force on a billboard 30 feet wide and 18 feet high is 972 pounds. Find the force on a billboard having an area of 300 square feet caused by a 40-mph wind.

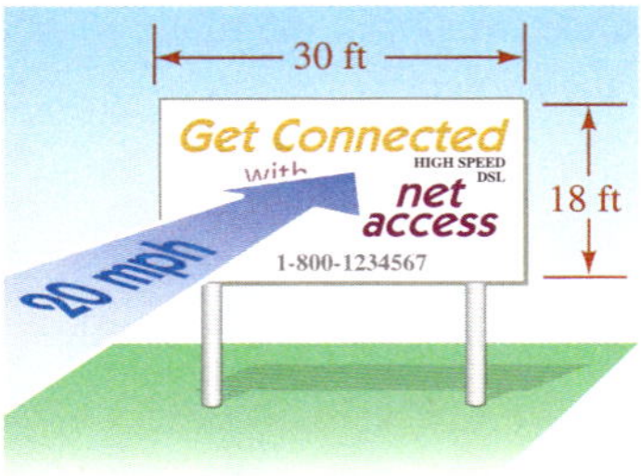

Strategy We will use the joint variation model $f = kAv^2$, where f represents the force of the wind, A represents the area of the billboard, and v^2 represents the square of the velocity of the wind.

Why The words *the force of the wind on a billboard varies jointly as the area of the billboard and the square of the wind velocity* indicate that this type of model should be used.

Solution

$f = kAv^2$ The joint variation model also can be read as "f is directly proportional to the product of A and v^2."

Since the billboard is 30 feet wide and 18 feet high, it has an area of $30 \cdot 18 = 540$ square feet. We can find k by substituting 972 for f, 540 for A, and 20 for v.

$$f = kAv^2$$
$$972 = k(540)(20)^2$$
$$972 = k(216{,}000) \quad \text{Evaluate: } (20)^2 = 400. \text{ Then do the multiplication.}$$
$$0.0045 = k \quad \text{Divide both sides by 216,000 to solve for } k. \text{ This is the constant of variation.}$$

To find the force exerted on a 300-square-foot billboard by a 40-mph wind, we use the formula $f = 0.0045Av^2$ and substitute 300 for A and 40 for v.

$$f = 0.0045Av^2 \quad \text{This is the joint variation model.}$$
$$f = 0.0045(300)(40)^2 \quad \text{Substitute for } A \text{ and } v.$$
$$f = 2{,}160 \quad \text{Evaluate the right side.}$$

The 40-mph wind exerts a force of 2,160 pounds on the billboard.

Self Check 3 **Force of the Wind.** Refer to Example 3. Find the force of a 25-mph wind on a billboard having an area of 375 square feet.

Now Try Problem 43

4 Solve Problems Involving Combined Variation.

Many applied problems involve a combination of direct and inverse variation. Such variation is called **combined variation.**

EXAMPLE 4 **Highway Construction.** The time it takes to build a highway varies directly as the length of the road, and inversely as the number of workers. If it takes 100 workers 4 weeks to build 2 miles of highway, how long will it take 80 workers to build 10 miles of highway?

Strategy We will use the combined variation model $t = \frac{kl}{w}$, where t represents the time in days, l represents the length of road built in miles, and w represents the number of workers.

Why The words *the time it takes to build a highway* ***varies directly*** *as the length of the road, and* ***inversely*** *as the number of workers* indicate that this type of model should be used.

Solution The relationship between these variables can be expressed by the equation

$$t = \frac{kl}{w}$$ This is a combined variation model.

$$4 = \frac{k(2)}{100}$$ Substitute 4 for t, 100 for w, and 2 for l to find k.

$$400 = 2k$$ Multiply both sides by 100.

$$200 = k$$ To isolate k, divide both sides by 2. This is the constant of variation.

We now substitute 80 for w, 10 for l, and 200 for k in the equation $t = \frac{kl}{w}$ and simplify:

$$t = \frac{kl}{w}$$ This is the combined variation model.

$$t = \frac{200(10)}{80}$$ Substitute for k, l, and w.

$$t = 25$$ Evaluate the right side.

It will take 25 weeks for 80 workers to build 10 miles of highway.

Self Check 4 **Highway Construction.** How long will it take 60 workers to build 6 miles of highway?

Now Try Problem 51

SECTION 8.9 STUDY SET

VOCABULARY

Fill in the blanks.

1. The equation $y = kx$ defines ______ variation: As x increases, y ________.
2. The equation $y = \frac{k}{x}$ defines _______ variation: As x increases, y ________.
3. The equation $y = kxz$ defines _____ variation, and $y = \frac{kz}{x}$ defines __________ variation.
4. The equation $y = \frac{kx}{z}$ means that y varies _______ with x and _________ with z.

CONCEPTS

Determine whether direct or inverse variation applies and sketch a possible graph for the situation. (Hint: Refer to the graphs on pages 700 and 702.)

5. a.

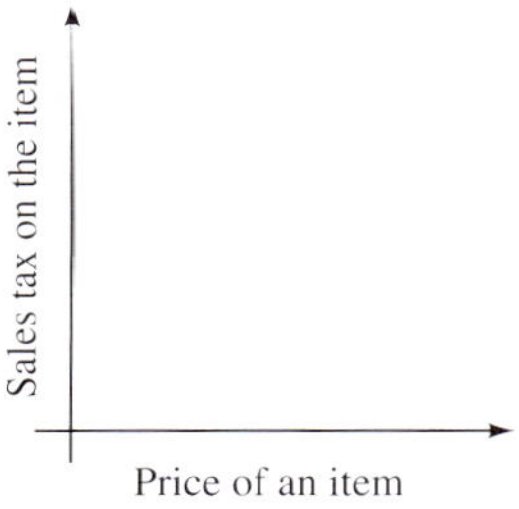

b.

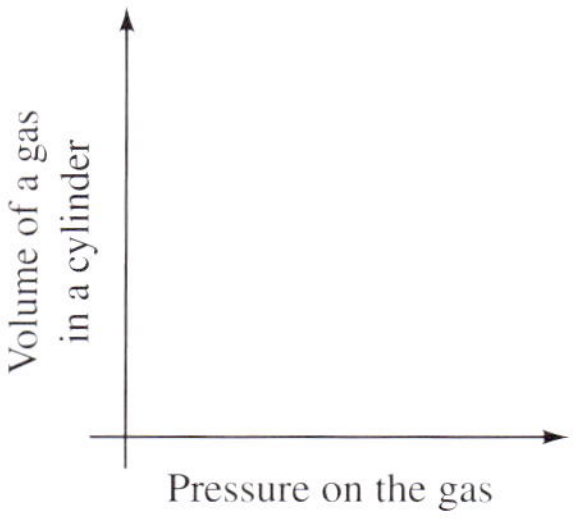

6. a.

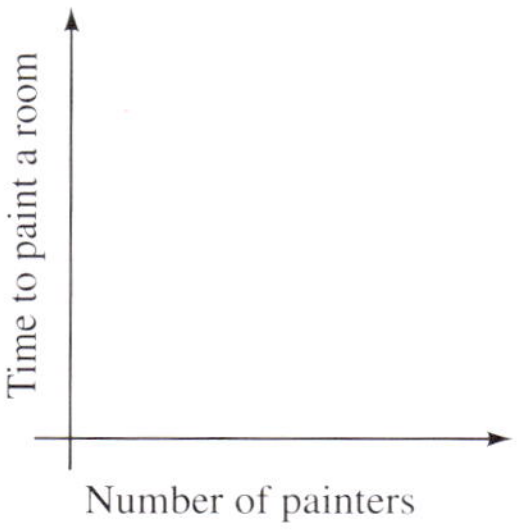

b.

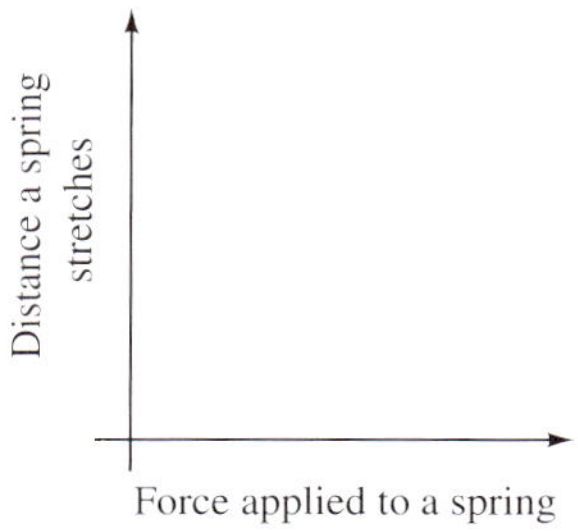

Tell whether each relationship suggests direct or inverse variation.

7. **Recycling.** The amount of money you receive and the number of aluminum cans you return
8. **Karate.** The force needed to break a board and the length of the board
9. **Tools.** The force you must exert on the handle of a wrench to loosen a bolt and the length of the handle
10. **Anatomy.** The volume of blood pumped from your heart each minute and your pulse rate
11. **Swimming Pools.** The amount of chlorine needed in a pool and the amount of water in the pool
12. **Lightning.** The time it takes you to hear the lightning after a strike and your distance from the strike
13. **Desserts.** The number of servings you can get from a wedding cake and the size of the piece that is served
14. **Remodeling.** The cost to remodel a house and the number of square feet to be added

NOTATION

Complete each solution.

15. Determine whether the equation defines direct variation.
    a. $y = kx$
    b. $y = k + x$
    c. $y = \frac{k}{x}$
    d. $m = kc$
16. Determine whether each equation defines inverse variation.
    a. $y = kx$
    b. $y = \frac{k}{x}$
    c. $y = \frac{x}{k}$
    d. $d = \frac{k}{g}$

GUIDED PRACTICE

Express each verbal model in symbols. See Objectives 1 and 2.

17. A varies directly as the square of p.
18. t varies directly as s.
19. z varies inversely as the cube of t.
20. v varies inversely as the square of r.

Express each verbal model in symbols. See Objectives 3 and 4.

21. C varies jointly as x, y, and z.
22. d varies jointly as r and t.
23. P varies directly as the square of a and inversely as the cube of j.
24. M varies inversely as the cube of n and jointly as x and the square of z.

Express each variation model in words. In each equation, k is the constant of variation. See Objectives 1 and 2.

25. $r = kt$
26. $A = kr^3$
27. $b = \frac{k}{h}$
28. $d = \frac{k}{W^4}$

Express each variation model in words. In each equation, k is the constant of variation. See Objectives 3 and 4.

29. $U = krs^2t$
30. $L = kmn$
31. $P = \frac{km}{n}$
32. $R = \frac{kL}{d^2}$

APPLICATIONS

33. from **Campus to Careers**

Webmaster

©Ronen/Shutterstock.com

The language of variation is often used to describe various aspects of the Internet and websites. Determine whether each statement, generally speaking, is true or false.

a. The dollar amount of sales that an Internet website receives is inversely proportional to the amount of Internet traffic that visits the website.
b. The download time of an Internet website varies directly with the bandwidth being used.
c. Search engines like Google place a value on a website that is directly proportional to the number of sites that link to it.

34. **Pendulums.** The time it takes for one complete swing of a pendulum varies directly with the square root of the length L of the pendulum. Write an equation that models this type of variation.

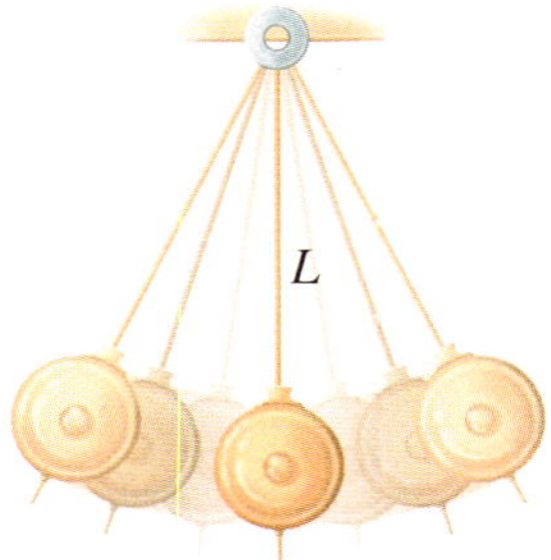

Solve each problem by writing a variation model.

35. **Gravity.** The force of gravity acting on an object varies directly as the mass of the object. The force on a mass of 5 kilograms is 49 newtons. What is the force acting on a mass of 12 kilograms?
36. **Free Fall.** An object in free fall travels a distance s that is directly proportional to the square of the time t. If an object falls 1,024 feet in 8 seconds, how far will it fall in 10 seconds?
37. **Finding Distance.** The distance that a car can go varies directly as the number of gallons of gasoline it consumes. If a car can go 288 miles on 12 gallons of gasoline, how far can it go on a full tank of 18 gallons?
38. **Braking.** Suppose the distance that a vehicle travels after its brakes have been applied varies directly as the square of the speed at which it was traveling. If the stopping distance for such a vehicle going 20 mph is 24 feet, what is the stopping distance for the vehicle traveling at 50 mph?
39. **Farming.** The number of days that a given number of bushels of corn will last when feeding cattle varies inversely as the number of animals. If x bushels will feed 25 cows for 10 days, how long will the feed last for 10 cows?

40. Organ Pipes. The frequency of vibration of air in an organ pipe is inversely proportional to the length of the pipe. If a pipe 2 feet long vibrates 256 times per second, how many times per second will a 6-foot pipe vibrate?

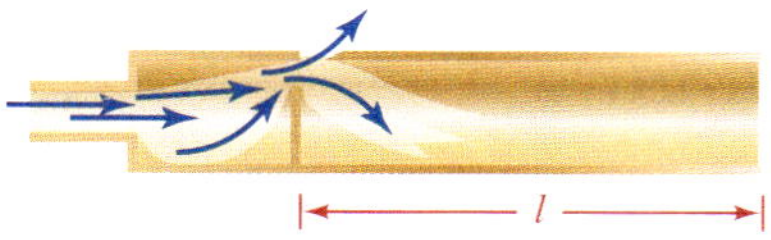

41. Gas Pressure. Under constant temperature, the volume occupied by a gas varies inversely to the pressure applied. If the gas occupies a volume of 20 cubic inches under a pressure of 6 pounds per square inch, find the volume when the gas is subjected to a pressure of 10 pounds per square inch.

42. Real Estate. The following table shows the listing price for three homes in the same general locality. Write the variation model (direct or inverse) that describes the relationship between the listing price and the number of square feet of a house in this area.

Number of square feet	Listing price
1,720	\$180,600
1,205	\$126,525
1,080	\$113,400

43. Trucking Costs. The costs of a trucking company vary jointly as the number of trucks in service and the number of hours they are used. When 4 trucks are used for 6 hours each, the costs are \$1,800. Find the costs of using 10 trucks, each for 12 hours.

44. Oil Storage. The number of gallons of oil that can be stored in a cylindrical tank varies jointly as the height of the tank and the square of the radius of its base. The constant of proportionality is 23.5. Find the number of gallons that can be stored in the cylindrical tank shown.

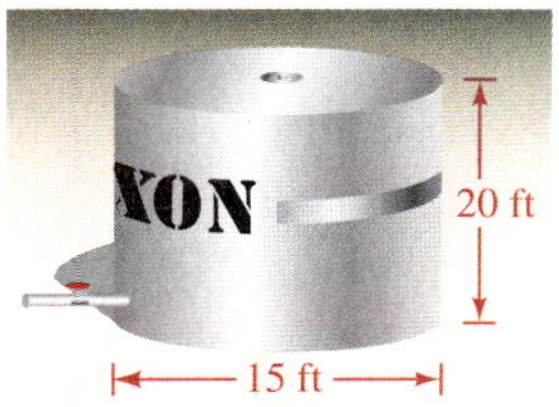

45. Electronics. The voltage (in volts) measured across a resistor is directly proportional to the current (in amperes) flowing through the resistor. The constant of variation is the **resistance** (in ohms). If 6 volts is measured across a resistor carrying a current of 2 amperes, find the resistance.

46. Electronics. The power (in watts) lost in a resistor (in the form of heat) varies directly as the square of the current (in amperes) passing through it. The constant of proportionality is the resistance (in ohms). What power is lost in a 5-ohm resistor carrying a 3-ampere current?

47. Structural Engineering. The deflection of a beam is inversely proportional to its width and the cube of its depth. If the deflection of a 4-inch-wide by 4-inch-deep beam is 1.1 inches, find the deflection of a 2-inch-wide by 8-inch-deep beam positioned as in figure (a) below.

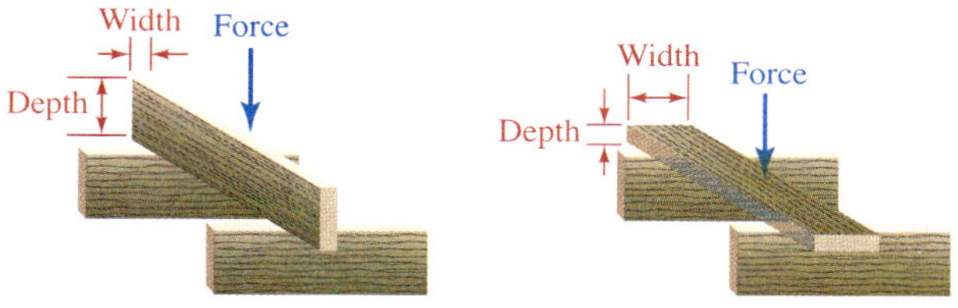

(a) (b)

48. Structural Engineering. Find the deflection of the beam in Exercise 47 when the beam is positioned as in figure (b) above.

49. Electronics. The resistance of a wire is directly proportional to the length of the wire and inversely proportional to the square of the diameter of the wire. If the resistance is 11.2 ohms in a 80-foot-long wire with diameter 0.01inch, what is the resistance in a 160-foot-long wire with diameter 0.04 inch?

50. Business Models. A businessman who sells widgets has found that the revenue from their sale varies directly as the advertising budget and inversely as the price. When \$105,000 was spent on advertising and the widgets were priced at \$19.95, the revenue from their sale was \$200,000. How many widgets would he expect to sell at \$17.50 each if \$700,000 was spent on advertising?

51. Tension in a String. When playing with a Skip It toy, a child swings a weighted ball on the end of a string in a circular motion around one leg while jumping over the revolving string with the other leg. See the illustration. The tension T in the string is directly proportional to the square of the speed s of the ball and inversely proportional to the radius r of the circle. If the tension in the string is 6 pounds when the speed of the ball is 6 feet per second and the radius is 3 feet, find the tension when the speed is 8 feet per second and the radius is 2.5 feet.

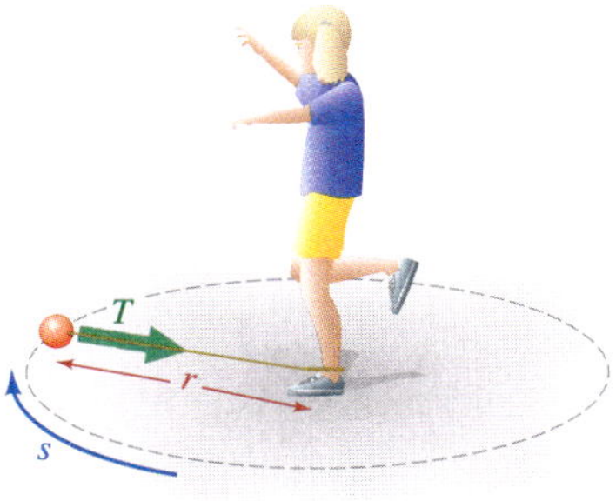

52. Gas Pressure. The pressure of a certain amount of gas is directly proportional to the temperature (measured on the Kelvin scale) and inversely proportional to the volume. A sample of gas at a pressure of 1 atmosphere occupies a volume of 1 cubic meter at a temperature of 273 Kelvin. When heated, the gas expands to twice its volume, but the pressure remains constant. To what temperature is it heated?

WRITING

53. **Computer Printers.** Is there a direct variation or an inverse variation between each pair of quantities? Explain. Draw a graph to support your answer.
    a. The time it takes to print a term paper and the speed of the printer.
    b. The time it takes to print a term paper and the length of the term paper.
54. Give examples of two quantities from everyday life that vary directly and two quantities that vary inversely.

REVIEW

Solve each compound inequality. Graph the solution set and write it using interval notation.

55. $2a + 10 < 7a$ and $5a - 15 < 2a$
56. $-22 \geq 5x - 2$ or $-3x - 4 > 2$
57. $\frac{x}{2} < \frac{x}{6} + 2$ or $0.1x + 0.1 > 1$
58. $\frac{7}{3}x + 2 \leq 16$ and $-8x \geq -48$

CHALLENGE PROBLEMS

59. As the cost of a purchase that is less than \$5 increases, the amount of change received from a five-dollar bill decreases. Is this inverse variation? Explain.

60. You've probably heard of Murphy's first law:

    If anything can go wrong, it will.

    Another of Murphy's laws is:

    The chances of a piece of bread falling with the grape-jelly side down varies directly with the cost of the carpet.

    Write one of your own witty sayings using the phrase *varies directly.*

8 Summary & Review

SECTION 8.1 Review of Solving Linear Equations, Formulas, and Linear Inequalities; Applications

DEFINITIONS AND CONCEPTS

Strategy for Solving Linear Equations in One Variable

1. Clear the equation of fractions or decimals.
2. Simplify each side of the equation by removing all sets of parentheses and combining like terms.
3. Isolate the variable term on one side of the equation.
4. Isolate the variable.
5. Check the result in the original equation.

EXAMPLES

Solve: $\frac{x-1}{6} + x = \frac{2}{3} - \frac{x+2}{6}$

$6\left(\frac{x-1}{6} + x\right) = 6\left(\frac{2}{3} - \frac{x+2}{6}\right)$	Multiply both sides by 6 to clear the fractions.
$6 \cdot \frac{x-1}{6} + 6 \cdot x = 6 \cdot \frac{2}{3} - 6 \cdot \frac{x+2}{6}$	Distribute 6 on both sides.
$x - 1 + 6x = 4 - (x + 2)$	Simplify. Don't forget the parentheses.
$x - 1 + 6x = 4 - x - 2$	Remove parentheses.
$7x - 1 = 2 - x$	Combine like terms on each side.
$7x - 1 + x = 2 - x + x$	To eliminate $-x$ on the right side, add x to both sides.
$8x - 1 = 2$	Combine like terms on each side.
$8x - 1 + 1 = 2 + 1$	To isolate the variable term $8x$, add 1 to both sides.
$8x = 3$	Simplify each side.
$\frac{8x}{8} = \frac{3}{8}$	Isolate the variable x by dividing both sides by 8.
$x = \frac{3}{8}$	

The solution is $\frac{3}{8}$ and the solution set is $\left\{\frac{3}{8}\right\}$. Check this result to verify that it satisfies the *original* equation.

An equation that is satisfied by every number for which both sides are defined is called an **identity**.

When we solve $x + 5 + x = 2x + 5$, the variables drop out and we obtain a true statement $5 = 5$. All real numbers are solutions. The solution set is the set of real numbers written as $\mathbb{R}$.

A **contradiction** is an equation that is never true.

When we solve $y + 2 = y$, the variables drop out and we obtain a false statement $2 = 0$. The equation has no solutions. The solution set contains no elements and can be written as the **empty set** $\{\ \ \}$ or the **null set** $\varnothing$.

To **solve a formula for a specified variable** means to isolate that variable on one side of the equation, with all other variables and constants on the opposite side.

Solve $F = \dfrac{mMg}{r^2}$ for M.

$Fr^2 = mMg$ To clear the fraction, multiply both sides by r^2.

$\dfrac{Fr^2}{mg} = M$ To isolate M, divide both sides by mg.

$M = \dfrac{Fr^2}{mg}$ Write M on the left side.

To **solve a linear inequality** in one variable, we use **properties of inequality** to find the values of its variable that make the inequality true.

If we multiply or divide both sides of an inequality by a negative number, the direction of the **inequality symbol must be reversed** for the inequalities to have the same solutions.

The set of all solutions of an inequality is called its **solution set.**

Solve: $-5x + 7 > 22$

$-5x + 7 - 7 > 22 - 7$ Subtract 7 from both sides.

$-5x > 15$

$\dfrac{-5x}{-5} < \dfrac{15}{-5}$ Divide both sides by -5 and reverse the direction of the inequality symbol.

$x < -3$

The solution set is:

Graph	Interval notation	Set-builder notation
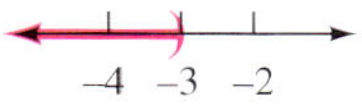	$(-\infty, -3)$	$\{x \mid x < -3\}$

REVIEW EXERCISES

Solve each equation. If an equation is an identity or a contradiction, so indicate.

1. $5x + 12 = 0$
2. $-3x - 7 + x = 6x + 20 - 5x$
3. $4(y - 1) = 28$
4. $2 - 13(x - 1) = 4 - 6x$
5. $\dfrac{8}{3}(x - 5) = \dfrac{2}{5}(x - 4)$
6. $\dfrac{3y}{4} - 14 = -\dfrac{y}{3} - 1$
7. $2x + 4 = 2(x + 3) - 2$
8. $3x - 2 - x = 2(x - 4)$
9. $-\dfrac{5}{4}p = 10$
10. $\dfrac{4t + 1}{3} - \dfrac{t + 5}{6} = \dfrac{t - 3}{6}$

Solve each formula for the indicated variable.

11. $V = \pi r^2 h$ for h
12. $v = \dfrac{1}{6}ab(x + y)$ for x

Solve each inequality. Give each solution set in interval notation and graph it.

13. $0.3x - 0.4 \geq 1.2 - 0.1x$
14. $\dfrac{7}{4}(x + 3) < \dfrac{3}{8}(x - 3)$
15. $-16 < -\dfrac{4}{5}x$
16. $5(2n + 2) - n > 3n - 3(1 - 2n)$

Use the six-step problem-solving strategy for each of the following application problems.

17. **Carpentry.** A carpenter wants to cut a 20-foot rafter so that one piece is 3 times as long as the other. Where should he cut the board?
18. **Geometry.** A rectangle is 4 meters longer than it is wide. If the perimeter of the rectangle is 28 meters, find its length and width.

SECTION 8.2 Functions

DEFINITIONS AND CONCEPTS	EXAMPLES
A **relation** is a set of ordered pairs. The set of first components is called the **domain** of the relation and the set of second components is called the **range.**	The relation $\{(2, 5), (7, -3), (4, 6)\}$ has domain $\{2, 4, 7\}$ and range $\{-3, 5, 6\}$.

DEFINITIONS AND CONCEPTS

A **function** is a set of ordered pairs (a relation) in which to each first component there corresponds exactly one second component.

Since we often work with sets of ordered pairs of the form (x, y), it is helpful to define a function using the variables x and y:

y* is a function of *x: Given a relation in x and y, if to each value of x in the domain there corresponds exactly one value of y in the range, then y is said to be a function of x. Since y depends on x, we call x the **independent variable** and y the **dependent variable.**

EXAMPLES

The relation $\{(2, 5), (7, -3), (4, 6)\}$ defines a function because to each first component there corresponds exactly one second component.

The relation $\{(-9, 1), (3, 8), (0, 0), (3, 24)\}$ *does not* define a function because to the first component 3 there corresponds two second components, 8 and 24.

The arrow diagram *does not* define y as a function of x because to the x-value 4 there corresponds more than one y-value: 2 and 6. The table defines a function and illustrates an important point: *Two different ordered pairs of a function can have the same y-value.*

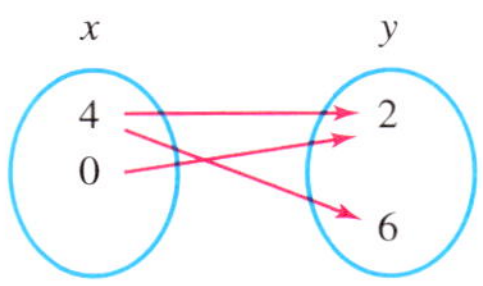

x	y
1	9
2	9
3	9

DEFINITIONS AND CONCEPTS

A function can be defined by an equation, however not all equations in two variables define functions.

EXAMPLES

The equation $y = 2x - 7$ defines y as a function of x because to each value of x there corresponds exactly one value of y.

The equation $x = |y|$ does not define y as a function of x because more than one value of y corresponds to a single value of x. If x is 2, for example, the equation becomes $2 = |y|$ and y can be either 2 or -2.

DEFINITIONS AND CONCEPTS

The **function notation** $y = f(x)$ indicates that the variable y is a function of x. It is read as "f of x."

Think of a function as a machine that takes some **input** x and turns it into some **output** $f(x)$, called a **function value.**

EXAMPLES

If $f(x) = 2x + 1$, find $f(-2)$ and $f(n + 1)$.

$$\begin{aligned} f(x) &= 2x + 1 \\ f(-2) &= 2(-2) + 1 \\ &= -4 + 1 \\ &= -3 \end{aligned}$$

Thus, $f(-2) = -3$.

$$\begin{aligned} f(x) &= 2x + 1 \\ f(n + 1) &= 2(n + 1) + 1 \\ &= 2n + 2 + 1 \\ &= 2n + 3 \end{aligned}$$

Thus, $f(n + 1) = 2n + 3$.

DEFINITIONS AND CONCEPTS

The input-output pairs that a function generates can be written as ordered pairs and plotted on a rectangular coordinate system to give the **graph of the function.**

A **linear function** is a function that can be defined by an equation of the form $f(x) = mx + b$. The graph of a linear function is a straight line.

Linear functions can be graphed using the **point-plotting method,** the **intercept method,** or the **slope–intercept method.**

EXAMPLES

To graph the linear function $f(x) = -4x - 2$, we make a table of values, plot the points, and draw the graph.

x	f(x)	
−1	2	→ (−1, 2)
0	−2	→ (0, −2)
1	−6	→ (1, −6)

↑ Select x. ↑ Find f(x). ↑ Plot the point.

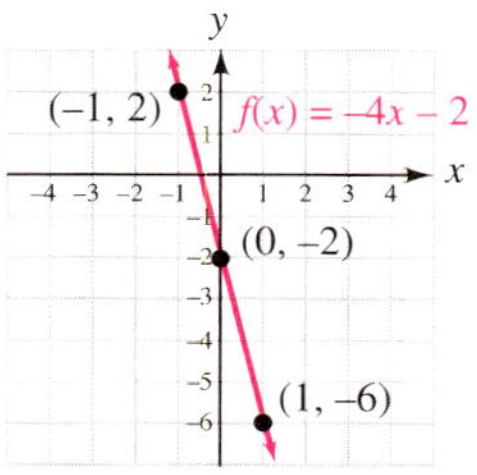

DEFINITIONS AND CONCEPTS

The **domain** of a function is the set of input values. The **range** is the set of output values.

To find the domain of a function defined by an equation, we must identify all the values of the input variable that produce a real-number output.

EXAMPLES

Find the domain of $f(x) = \frac{10}{x + 3}$.

The number -3 cannot be substituted for x, because that would make the denominator equal to 0. Since any real number except -3 can be substituted for x, the domain is *the set of all real numbers except* -3.

To write equations of linear functions, we can use:

slope–intercept form:
$f(x) = mx + b$

point–slope form:
$f(x) - y_1 = m(x - x_1)$

Write an equation for the linear function whose graph has slope -4 and passes through $(1, 8)$.

$f(x) - y_1 = m(x - x_1)$ This is point–slope form.
$f(x) - 8 = -4(x - 1)$ Substitute for m, y_1, and x_1.
$f(x) - 8 = -4x + 4$ Distribute.
$f(x) = -4x + 12$ Solve for $f(x)$.

A **polynomial function** is a function whose equation is defined by a polynomial in one variable.

To **evaluate a polynomial function,** we replace the variable in the defining equation with its value, called the **input.** Then we simplify to find the **output.**

Let $f(x) = x^3 - 3x^2 - 9x + 2$. Find $f(2)$.

$f(x) = x^3 - 3x^2 - 9x + 2$
$f(2) = (2)^3 - 3(2)^2 - 9(2) + 2$ Substitute 2 for x.
$= 8 - 3(4) - 18 + 2$
$= -20$ The output is -20.

REVIEW EXERCISES

19. Find the domain and the range of the relation: $\{(-4, 0), (5, 16), (2, -2), (-1, -2)\}$

20. Fill in the blanks.
   a. A ________ is a set of ordered pairs (a relation) in which to each first component there corresponds exactly one second component.
   b. Given a relation in x and y, if to each value of x in the domain there corresponds exactly one value of y in the range, y is said to be a ________ of x. We call x the independent ________ and y the ________ variable.

Determine whether the relation defines y as a function of x. If it does not, find two ordered pairs where more than one value of y corresponds to a single value of x.

21. a.

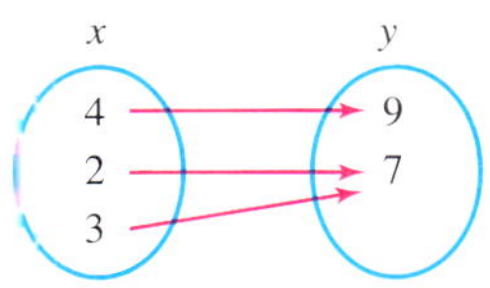

b.

x	y
-1	8
0	5
4	1
-1	9

c. $\{(14, 6), (-1, 14), (6, 0), (-3, 8)\}$

22. For the given input, what value will the function machine output?

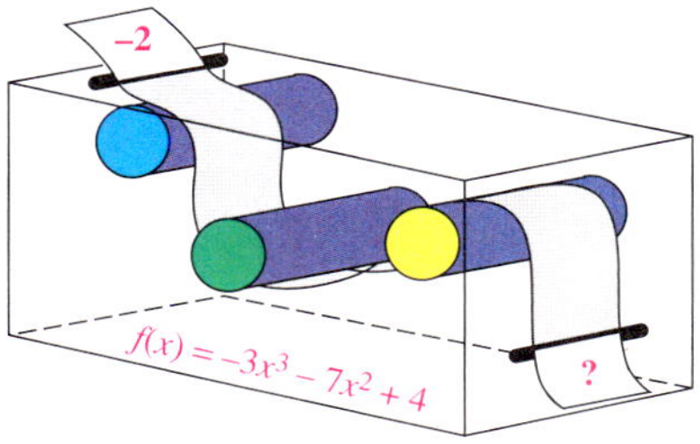

Determine whether each equation defines y as a function of x. If it does not, find two ordered pairs where more than one value of y corresponds to a single value of x.

23. $y = 6x - 4$
24. $y = 4 - x^2$
25. $y^2 = x$
26. $|y| = x + 1$

Let $f(x) = 3x + 2$ and $g(a) = \frac{a^2 - 4a + 4}{2}$. Find each function value.

27. $f(-3)$
28. $g(8)$
29. $g(-2)$
30. $f(t + 2)$

31. Let $f(x) = -5x + 7$. For what value of x is $f(x) = -8$?
32. Let $g(t) = \frac{3}{4}t - 1$. For what value of t is $g(t) = 0$?

Find the domain of each function.

33. $f(x) = 4x - 1$
34. $s(t) = t^2 + 1$
35. $h(x) = \dfrac{4}{2 - x}$
36. $g(b) = \dfrac{12}{5b + 25}$
37. What are the slope and the y-intercept of the graph of $g(x) = -2x - 16$?
38. Graph: $f(x) = \frac{2}{3}x - 2$

Write an equation for a linear function whose graph has the given characteristics.

39. Slope: $\dfrac{9}{10}$, y-intercept: $\left(0, \dfrac{7}{8}\right)$
40. Slope: $\dfrac{1}{5}$, passes through $(10, 1)$
41. Horizontal, passes through $(-8, 12)$
42. Passes through $(2, 5)$, parallel to the graph of $g(x) = 4x - 7$
43. Passes through $(-6, 3)$, perpendicular to the graph of $g(x) = -3x - 12$

44. Electricity. The relationship between the electrical resistance R of a coil of wire and its temperature t can be modeled with a linear function.

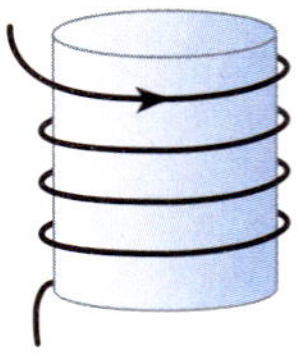

a. Use the following data in the table to write a function that describes this relationship.

b. Use the answer to part (a) to predict the resistance if the temperature of the coil of wire is 100° Celsius.

***t* (in degrees Celsius)**	10	30
***R* (in milliohms)**	5.25	5.65

45. Squirt Guns. The volume of the reservoir on top of the squirt gun is given by the polynomial function $V(r) = 4.19r^3 + 25.13r^2$, where r is the radius in inches. Find $V(2)$ to the nearest cubic inch.

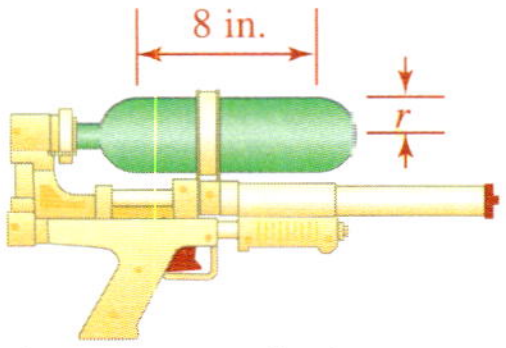

46. Calculus. In the advanced mathematics course called Calculus, an important polynomial function is:

$$f(x) = 1 + x + \frac{x^2}{2} + \frac{x^3}{6} + \frac{x^4}{24}$$

Find $f(1)$.

SECTION 8.3 Graphs of Functions

DEFINITIONS AND CONCEPTS	EXAMPLES
From the graph of a function, we can determine function values. For the example, the value of $f(2)$ is given by the y-coordinate of a point on the graph of f with x-coordinate 2.	To find $f(2)$ from the graph, draw a vertical line through 2 on the x-axis. It intersects the graph at (2, 3). Therefore, 3 corresponds to 2 and it follows that $f(2) = 3$. 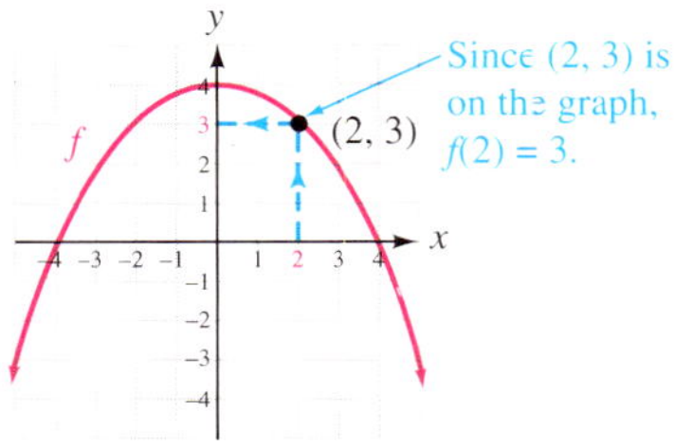
These three basic functions are used so often in algebra that you should memorize their names and their graphs. They are called **nonlinear functions** because their graphs are not lines.	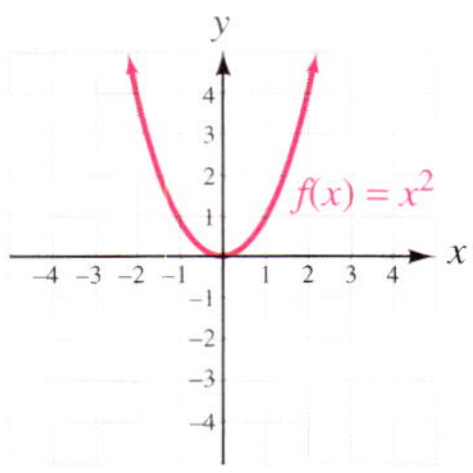 The squaring function 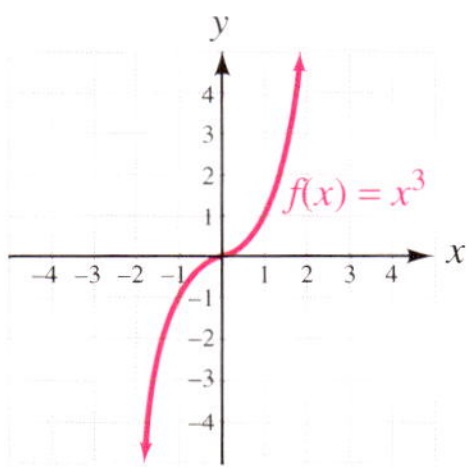 The cubing function 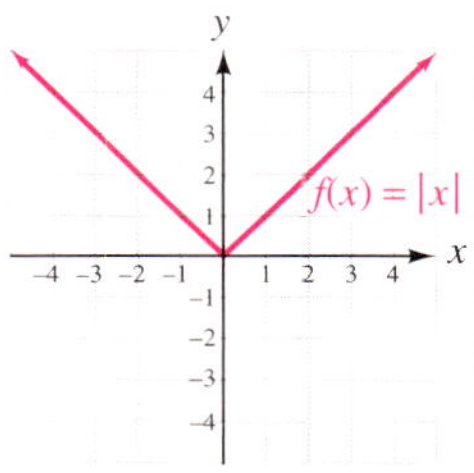 The absolute value function
We can find the domain and range of a function from its graph. The **domain of a function** is the projection of its graph onto the x-axis. **The range of a function** is the projection of its graph onto the y-axis.	Find the domain and range of function f. 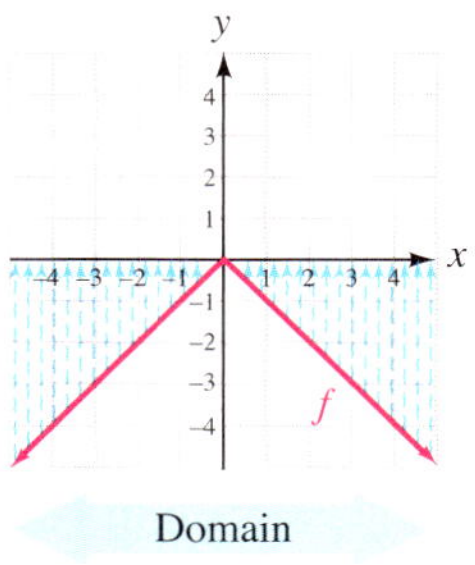 D: The set of real numbers 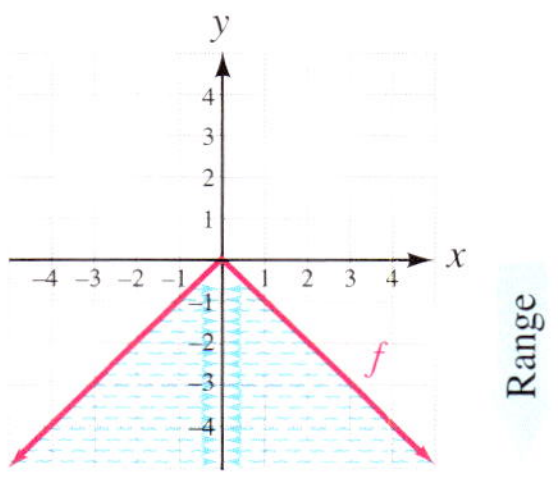 R: The set of nonpositive real numbers

A **vertical translation** shifts a graph upward or downward. A **horizontal translation** shifts a graph left or right. A **reflection** "flips" a graph about the x-axis.

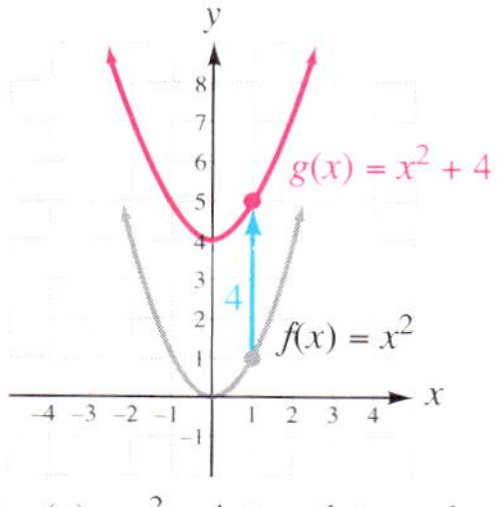

To graph $g(x) = x^2 + 4$, translate each point on the graph of $f(x) = x^2$ up 4 units.

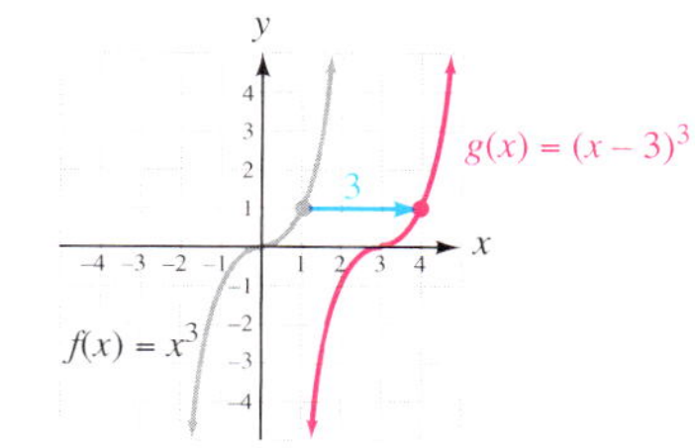

To graph $g(x) = (x - 3)^3$, translate each point on the graph of $f(x) = x^3$ to the right 3 units.

The **vertical line test:** If a vertical line intersects a graph in more than one point, the graph is not the graph of a function.

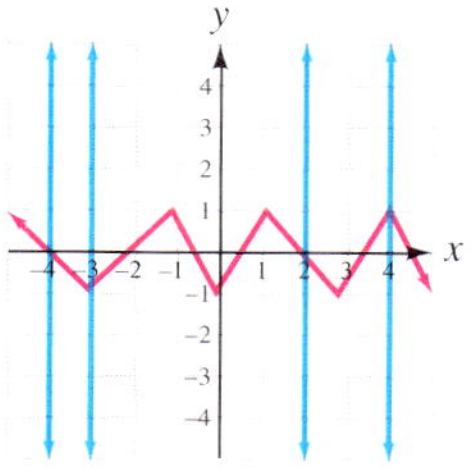

The graph of a function

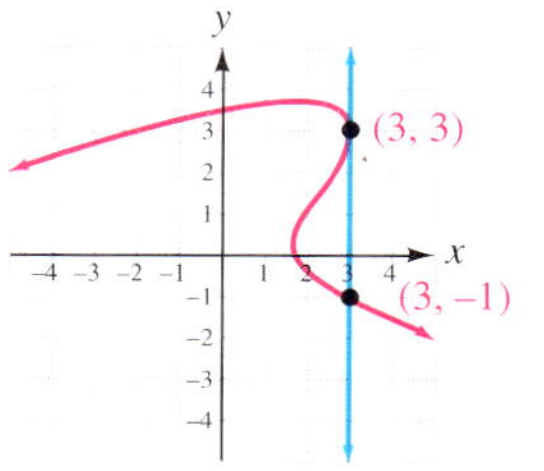

Not the graph of a function

REVIEW EXERCISES

47. Use the graph to find each value.

  a. $f(-2)$

  b. $f(3)$

  c. The value of x for which $f(x) = 0$.

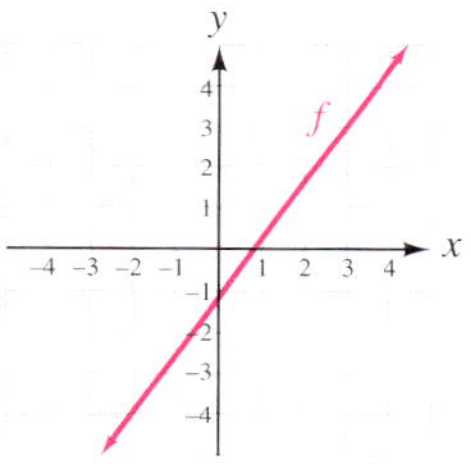

48. Use the graph to find each value.

  a. $g(0)$

  b. $g(-3)$

  c. The value(s) of x for which $g(x) = -4$.

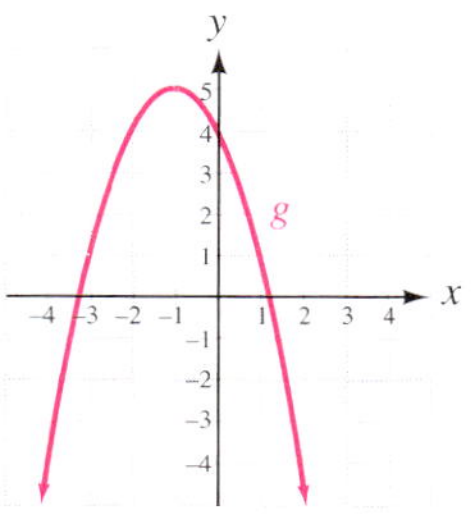

49. Use the graph of function f to find each of the following.

  a. $f(0)$

  b. The values of x for which $f(x) = 0$

  c. Write the domain and range of f in interval notation.

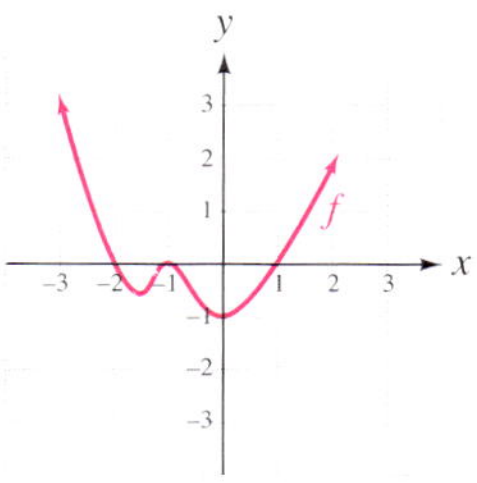

50. What are the nonnegative real numbers?

Give the domain and range of each function graphed below.

51.

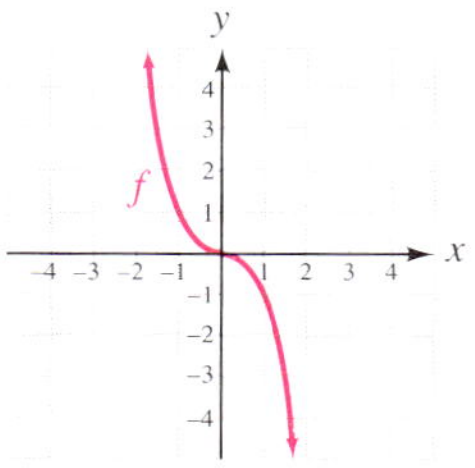

52.

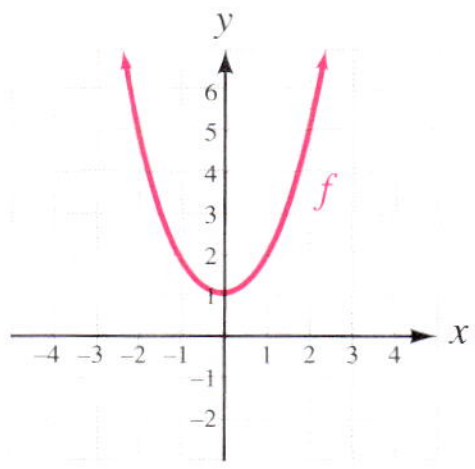

53. Graph $f(x) = |x + 2|$ by creating a table of function values and plotting points. Give the domain and range of the function.

54. Fill in the blanks.

  a. The graph of $f(x) = x^2 + 6$ is the same as the graph of $f(x) = x^2$ except that it is shifted __ units ____.

  b. The graph of $f(x) = (x + 6)^2$ is the same as the graph of $f(x) = x^2$ except that it is shifted __ units to the ____.

For each of the following functions, first sketch the graph of its associated function, $f(x) = x^2$, $f(x) = x^3$, or $f(x) = |x|$. Then draw the graph of function g using a translation and/or a reflection and give its domain and range.

55. $g(x) = x^2 - 3$

56. $g(x) = |x - 4|$

57. $g(x) = (x - 2)^3 + 1$

58. $g(x) = -x^3$

Determine whether each graph is the graph of a function. If it is not, find two ordered pairs where more than one value of y corresponds to a single value of x.

59.

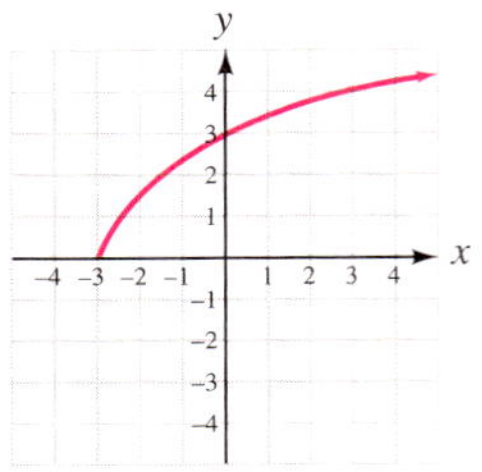

60.

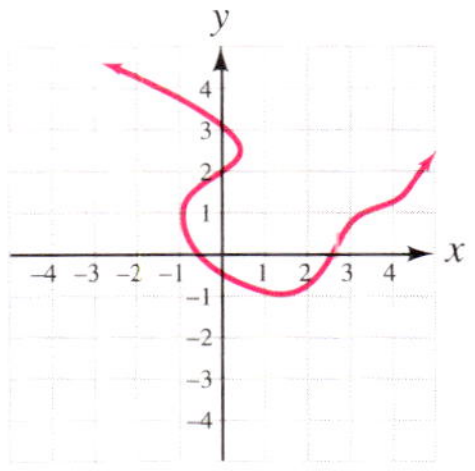

SECTION 8.4 Solving Compound Inequalities

DEFINITIONS AND CONCEPTS	EXAMPLES
The **intersection** of two sets A and B, written $A \cap B$, is the set of all elements that are common to set A and set B. The **union** of two sets A and B, written $A \cup B$, is the set of all elements that are in set A, set B, or both.	Let $A = \{-2, 0, 3, 5\}$ and $B = \{-3, 0, 5, 7\}$. $A \cap B = \{0, 5\}$ The intersection contains the elements that the sets have in common. $A \cup B = \{-3, -2, 0, 3, 5, 7\}$ The union contains the elements that are in one or the other set, or both.

When the word *and* or the word *or* is used to connect pairs of inequalities, we call the statement a **compound inequality.**

The solution set of a **compound inequality containing the word *and*** includes all numbers that make both of the inequalities true. That is, it is the intersection of their solution sets.

Solve: $2x - 1 \le 5$ and $5x + 1 > 4$

We solve each inequality separately. Then we graph the two solution sets (one in red, the other in blue) on the same number line and determine their intersection.

$$2x - 1 \le 5 \quad \text{and} \quad 5x + 1 > 4$$
$$2x \le 6 \qquad\qquad 5x > 3$$
$$x \le 3 \qquad\qquad x > \frac{3}{5}$$

Preliminary Work:

The purple-shaded interval shown below is where the red and blue graphs overlap. Thus, the solution set is:

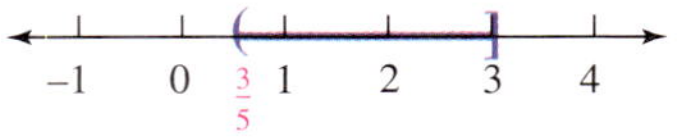

Interval notation: $\left(\frac{3}{5}, 3\right]$

Inequalities that contain exactly two inequality symbols are called **double inequalities.** Any double linear inequality can be written as a compound inequality containing the word *and.* For example:

$c < x < d$ is equivalent to $c < x$ and $x < d$

Solve: $-7 \le 3x - 1 < 5$

We apply properties of inequality to *all three of its parts* to isolate x in the middle.

$$-7 \le 3x - 1 < 5$$
$$-7 + 1 \le 3x - 1 + 1 < 5 + 1 \quad \text{Add 1 to all three parts.}$$
$$-6 \le 3x < 6$$
$$\frac{-6}{3} \le \frac{3x}{3} < \frac{6}{3} \quad \text{Divide each part by 3.}$$
$$-2 \le x < 2$$

The solution set is:

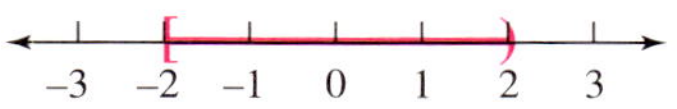

Interval notation: $[-2, 2)$

The solution set of a **compound inequality containing the word *or*** includes all numbers that make one or the other, or both, inequalities true. That is, it is the union of their solution sets.

Solve: $2x - 1 > 5$ or $-(5x - 7) \geq 2$

We solve each inequality separately. Then we graph the two solution sets on the same number line to show their union.

$2x - 1 > 5$	or	$-(5x - 7) \geq 2$
$2x > 6$		$-5x + 7 \geq 2$
$x > 3$		$-5x \geq -5$
		$x \leq 1$

The solution set is:

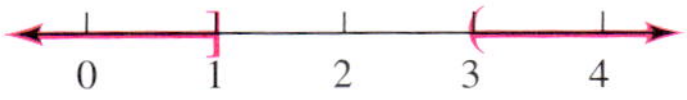

Interval notation: $(-\infty, 1] \cup (3, \infty)$

This is the union of two intervals.

REVIEW EXERCISES

Let $A = \{-6, -3, 0, 3, 6\}$ and $B = \{-5, -3, 3, 8\}$.

61. Find $A \cap B$.

62. Find $A \cup B$.

Check to determine whether -4 is a solution of the compound inequality.

63. $x < 0$ and $x > -5$

64. $x + 3 < -3x - 1$ and $4x - 3 > 3x$

Graph each set.

65. $(-3, 3) \cup [1, 6]$

66. $(-\infty, 2] \cap [1, 4)$

Solve each compound inequality. Graph the solution set and write it using interval notation.

67. $-2x > 8$ and $x + 4 \geq -6$

68. $5(x + 2) \leq 4(x + 1)$ and $11 + x < 0$

69. $\frac{2}{5}x - 2 < -\frac{4}{5}$ and $-\frac{x}{3} < -1$

70. $4\left(x - \frac{1}{4}\right) \leq 3x - 1$ and $x \geq 0$

Solve each double inequality. Graph the solution set and write it using interval notation.

71. $3 < 3x + 4 < 10$

72. $-2 \leq \frac{5 - x}{2} \leq 2$

Check to determine whether -4 is a solution of the compound inequality.

73. $x < 1.6$ or $x > -3.9$

74. $x + 1 < 2x - 1$ or $4x - 3 > 3x$

Solve each compound inequality. Graph the solution set and write it using interval notation.

75. $x + 1 < -4$ or $x - 4 > 0$

76. $\frac{x}{2} + 3 > -2$ or $4 - x > 4$

77. **Rugs.** A manufacturer makes a line of decorator rugs that are 4 feet wide and of varying lengths x (in feet). The floor area covered by the rugs ranges from 17 ft^2 to 25 ft^2. Write and then solve a double linear inequality to find the range of the lengths of the rugs.

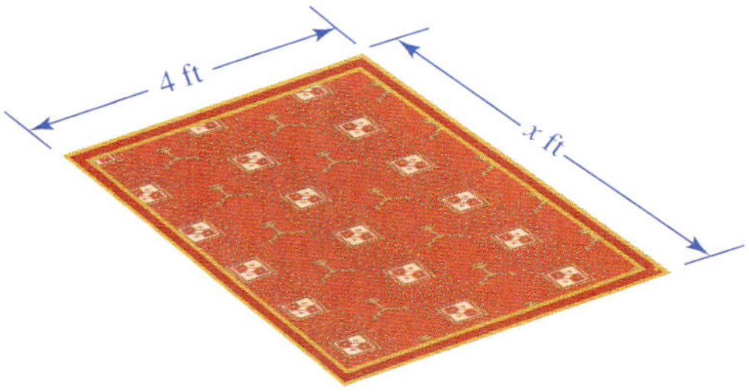

78. Match each word in Column I with *two* associated items in Column II.

Column I	*Column II*
a. or	i. $\cap$
	ii. $\cup$
b. and	iii. intersection
	iv. union

79. Let $f(x) = \frac{5}{4}x - 140$. Find all values of x for which $f(x) < -48$ or $f(x) \geq 32$.

80. Let $f(x) = 3x - 5$. Find all values of x for which $-4 \geq f(x) > -12$.

SECTION 8.5 Solving Absolute Value Equations and Inequalities

DEFINITIONS AND CONCEPTS	EXAMPLES
To **solve absolute value equations** of the form $\|X\| = k$, where $k > 0$, solve the equivalent **compound equation** $X = k \quad \text{or} \quad X = -k$ If k is negative, then $\|X\| = k$ has no solution. Recall that **equivalent equations** have the same solutions.	Solve: $\|2x + 1\| = 7$ This absolute value equation is equivalent to the following compound equation: $2x + 1 = 7 \quad \text{or} \quad 2x + 1 = -7$ $2x = 6 \quad\quad 2x = -8$ $x = 3 \quad\quad x = -4$ This equation has two solutions: 3 and −4. The solution set is $\{-4, 3\}$.
Since absolute value expresses distance, the absolute value of a number is always positive or zero, but never negative.	Solve: $\|4x - 5\| = -3$ Since an absolute value can never be negative, there are no real numbers x that make $\|4x - 5\| = -3$ true. The equation has no solution and the solution set is $\varnothing$.
To **solve absolute value equations** of the form $\|X\| = \|Y\|$, solve the compound equation $X = Y \quad \text{or} \quad X = -Y$ The expressions within the absolute value symbols are equal, or they are opposites.	Solve: $\|3x - 2\| = \|2x + 4\|$ This equation is equivalent to the following compound equation: $3x - 2 = 2x + 4 \quad \text{or} \quad 3x - 2 = -(2x + 4)$ $x - 2 = 4 \quad\quad 3x - 2 = -2x - 4$ $x = 6 \quad\quad 5x - 2 = -4$ $5x = -2$ $x = -\frac{2}{5}$ This equation has two solutions: 6 and $-\frac{2}{5}$. The solution set is $\left\{-\frac{2}{5}, 6\right\}$.
To **solve absolute value inequalities** of the form $\|X\| < k$, where $k > 0$, solve the equivalent double inequality $-k < X < k$. Use a similar approach to solve $\|X\| \le k$. When solving absolute value equations or inequalities, **isolate the absolute value expression** on one side **before** writing the equivalent compound statement.	Solve: $\|4x - 3\| - 2 < 7$ $\|4x - 3\| - 2 + 2 < 7 + 2$ To isolate the absolute value, add 2 to both sides. $\|4x - 3\| < 9$ The resulting inequality is equivalent to the following double inequality: $-9 < 4x - 3 < 9$ $-6 < 4x < 12$ Add 3 to all three parts. $-\frac{3}{2} < x < 3$ Divide each part by 4 and simplify. The solution set is: 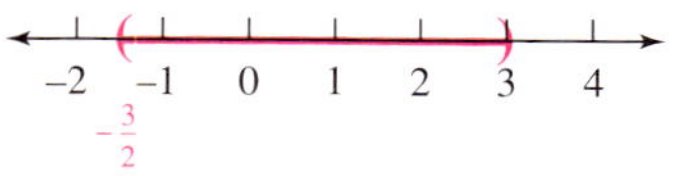 Interval notation: $\left(-\frac{3}{2}, 3\right)$ Set builder: $\left\{x \mid -\frac{3}{2} < x < 3\right\}$

To **solve absolute value inequalities** of the form $\|X\| \geq k$, where $k > 0$, solve the equivalent compound inequality $X \leq -k$ or $X \geq k$. Use a similar approach to solve $\|X\| > k$.	Solve: $\|3x + 1\| \geq 7$ This inequality is equivalent to the following compound inequality: $3x + 1 \leq -7$ or $3x + 1 \geq 7$ $3x \leq -8$ | $3x \geq 6$ $x \leq -\frac{8}{3}$ | $x \geq 2$ 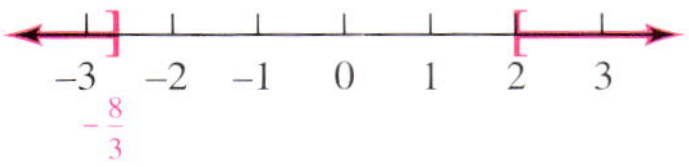 Interval notation: $\left(-\infty, -\frac{8}{3}\right] \cup [2, \infty)$ This is the union of two intervals. Set builder: $\{x \mid x \leq -\frac{8}{3}$ or $x > 2\}$

REVIEW EXERCISES

Solve each absolute value equation.

81. $|4x| = 8$

82. $2|3x + 1| - 1 = 19$

83. $\left|\frac{3}{2}x - 4\right| - 10 = -1$

84. $\left|\frac{2 - x}{3}\right| = -4$

85. $|-4(2x - 6)| = 0$

86. $\left|\frac{3}{8} + \frac{x}{3}\right| = \frac{5}{12}$

87. $|3x + 2| = |2x - 3|$

88. $\left|\frac{2(1 - x) + 1}{2}\right| = \left|\frac{3x - 2}{3}\right|$

Solve each absolute value inequality. Graph the solution set and write it using interval notation.

89. $|x| \leq 3$

90. $|2x + 7| < 3$

91. $2|5 - 3x| \leq 28$

92. $\left|\frac{2}{3}x + 14\right| + 6 < 6$

93. $|x| > 1$

94. $\left|\frac{1 - 5x}{3}\right| \geq 7$

95. $|3x - 8| - 4 > 0$

96. $\left|\frac{3}{2}x - 14\right| \geq 0$

97. **Produce.** Before packing, freshly picked tomatoes are weighed on the scale shown. Tomatoes having a weight w (in ounces) that falls within the highlighted range are sold to grocery stores.

   a. Complete the following absolute value inequality that expresses the acceptable weight range: $|w - \quad| \leq \quad$

   b. Solve the inequality from part (a) and express the acceptable weight range using interval notation.

98. Let $f(x) = \frac{1}{3}|6x| - 1$. For what value(s) of x is $f(x) = 5$?

99. Let $f(x) = 2|3(x + 4)| + 1.5$. Find all values of x for which $f(x) = 25.5$.

100. Let $f(x) = |7 - x|$. Find all values of x for which $f(x) < 5$.

101. Explain why $|0.04x - 8.8| < -2$ has no solution.

102. Explain why the solution set of $\left|\frac{3x}{50} + \frac{1}{45}\right| \geq -\frac{4}{5}$ is the set of all real numbers.

SECTION 8.6 Review of Factoring Methods: GCF, Grouping, Trinomials

DEFINITIONS AND CONCEPTS	EXAMPLES
The first step of factoring a polynomial is to see whether the terms of the polynomial have a common factor. If they do, **factor out the GCF.**	Factor: $14a^4 + 35a^3 - 56a^2 = 7a^2(2a^2 + 5a - 8)$ Factor out the GCF, $7a^2$. Use multiplication to check the factorization: $7a^2(2a^2 + 5a - 8) = 14a^4 + 35a^3 - 56a^2$ This is the original polynomial.

If an expression has four or more terms, try to factor the expression by **grouping.**

Factor: $ax - 2x + 3a - 6$

$$ax - 2x + 3a - 6 = ax - 2x + 3a - 6$$ Group the terms.

$$= x(a - 2) + 3(a - 2)$$ Factor x from $ax - 2x$. Factor 3 from $3a - 6$.

$$= (a - 2)(x + 3)$$ Factor out the GCF, $a - 2$.

Many trinomials factor as the product of two binomials. To **factor a trinomial** of the form $x^2 + bx + c$, whose **leading coefficient is 1,** find two integers whose product is c and whose sum is b.

Factor: $p^2 + 14p + 45$

We must find two integers whose product is 45 and whose sum is 14. Since $5 \cdot 9 = 45$ and $5 + 9 = 14$, two such numbers are **5** and **9**, and we have

$$p^2 + 14p + 45 = (p + 5)(p + 9)$$

Check: $(p + 5)(p + 9) = p^2 + 9p + 5p + 45 = p^2 + 14p + 45$

We can use the **trial-and-check method** to factor trinomials with **leading coefficients other than 1.** Write the trinomial as the product of two binomials and determine four integers.

Factor: $2x^2 - 5x - 12$

Since the first term is $2x^2$, the first terms of the binomial factors must be $2x$ and x.

$$(2x \,\square\,)(x \,\square\,)$$ Because $2x \cdot x$ will give $2x^2$

The second terms of the binomials must be two integers whose product is -12. There are six such pairs:

$$1(-12),\quad 2(-6),\quad \mathbf{3(-4)},\quad 4(-3),\quad 6(-2),\quad \text{and}\quad 12(-1)$$

The pair in blue gives the correct middle term, $-5x$, when placed in the binomial factors shown above. We use the FOIL method to check:

Outer: $-8x$

$$(2x + 3)(x - 4)$$ Combine like terms: $-8x + 3x = -5x$.

Inner: $3x$

Thus, $2x^2 - 5x - 12 = (2x + 3)(x - 4)$.

To factor $ax^2 + bx + c$ by **grouping,** write it as an equivalent four-term polynomial:

$$ax^2 + \square\, x + \square\, x + c$$

The product of these numbers must be ac, and their sum must be b.

Then factor the four-term polynomial by grouping. Use the FOIL method to check.

Factor by grouping: $2x^2 - 5x - 12$

We must find two integers whose product is $ac = 2(-12) = -24$ and whose sum is $b = -5$. Two such numbers are -8 and 3. They serve as the coefficients of $-8x$ and $3x$, the two terms that we use to represent the middle term, $-5x$, of the trinomial.

$$2x^2 - 5x - 12 = 2x^2 - 8x + 3x - 12$$ Express $-5x$ as $-8x + 3x$.

$$= 2x(x - 4) + 3(x - 4)$$

$$= (x - 4)(2x + 3)$$ Factor out the GCF, $(x - 4)$.

REVIEW EXERCISES

Factor, if possible.

103. $z^2 - 11z + 30$

104. $x^4 + 4x^2 + x^2y + 4y$

105. $4a^2 - 5a + 1$

106. $27x^3y^3z^3 + 81x^4y^5z^2 - 90x^2y^3z^7$

107. $15b^2 + 4b - 4$

108. $-x^2 - 3x + 28$

109. $15x^2 - 57xy - 12y^2$

110. $w^8 - w^4 - 90$

111. $r^2y - ar - ry + a + r - 1$

112. $49a^6 + 84a^3b^2 + 36b^4$

113. $3b^2 + 2b + 1$

114. $2a^4 + 4a^3 - 6a^2$

115. Use a substitution to factor: $(s + t)^2 - 2(s + t) + 1$

116. Solve $m_1m_2 = mm_2 + mm_1$ for m_1.

SECTION 8.7 Review of Factoring Methods: The Difference of Two Squares; the Sum and Difference of Two Cubes

DEFINITIONS AND CONCEPTS	EXAMPLES
The **difference of two squares:** To factor the square of a First quantity minus the square of a Last quantity, multiply the First plus the Last by the First minus the Last. $F^2 - L^2 = (F + L)(F - L)$	Factor: $x^2y^2 - 100$ $x^2y^2 - 100 = (xy)^2 - 10^2$ This is a difference of two squares. $= (xy + 10)(xy - 10)$
In general, the **sum of two squares** (with no common factor other than 1) cannot be factored using real numbers.	$x^2 + 100$ and $36y^2 + 49z^4$ are prime polynomials.
The **sum of two cubes:** To factor the cube of a First quantity plus the cube of a Last quantity, multiply the First plus the Last by the First squared, minus the First times the Last, plus the Last squared. $F^3 + L^3 = (F + L)(F^2 - FL + L^2)$	Factor: $y^3 + 27z^6$ $y^3 + 27z^6 = y^3 + (3z^2)^3$ This is a sum of two cubes. $= (y + 3z^2)[y^2 - y \cdot 3z^2 + (3z^2)^2]$ $= (y + 3z^2)(y^2 - 3yz^2 + 9z^4)$
The **difference of two cubes:** To factor the cube of a First quantity minus the cube of a Last quantity, multiply the First minus the Last by the First squared, plus the First times the Last, plus the Last squared. $F^3 - L^3 = (F - L)(F^2 + FL + L^2)$	Factor: $125s^3 - 64$ $125s^3 - 64 = (5s)^3 - 4^3$ This is a difference of two cubes. $= (5s - 4)[(5s)^2 + 5s \cdot 4 + 4^2]$ $= (5s - 4)(25s^2 + 20s + 16)$

REVIEW EXERCISES

Factor, if possible.

117. $z^2 - 16$

118. $x^2y^4 - 64z^6$

119. $a^2b^2 + c^2$

120. $c^2 - (a + b)^2$

121. $32a^4c - 162b^4c$

122. $k^2 + 2k + 1 - 9m^2$

123. $m^2 - n^2 - m - n$

124. $t^3 + 64$

125. $8a^3 - 125b^9$

126. **Spanish Roof Tile.** The amount of clay used to make a roof tile is given by

$$V = \frac{\pi}{2}r_1{}^2h - \frac{\pi}{2}r_2{}^2h$$

Factor the right side of the formula completely.

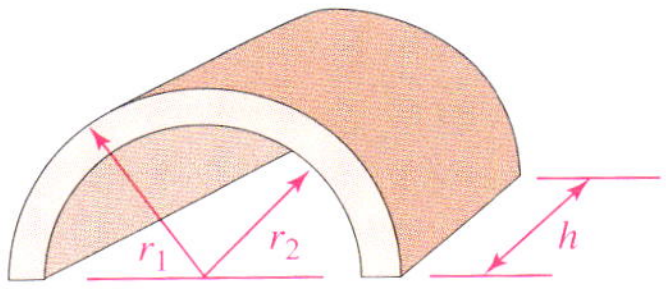

SECTION 8.8 Review of Rational Expressions and Rational Equations; Rational Functions

DEFINITIONS AND CONCEPTS	EXAMPLES
A **rational expression** is an expression of the form $\frac{A}{B}$, where A and B are polynomials and B does not equal 0.	Rational expressions: $\frac{3x^2}{xy}$, $\frac{5b - 15}{b^2 - 25}$, and $\frac{a + 2}{a^2 - 3a - 4}$
A **rational function** is a function whose equation is defined by a rational expression in one variable.	Rational functions: $f(x) = \frac{6x}{x - 2}$ and $f(n) = \frac{n + 3}{n^3 + 2n - 9}$

Since division by 0 is undefined, any values that make the denominator 0 in a rational function must be excluded from the **domain** of the function.

When finding the domain of a rational function, we *don't need to examine the numerator* of the expression; it can be any value, including 0.

Find the domain of the rational function: $f(x) = \dfrac{x+3}{x^2-4}$

$x^2 - 4 = 0$ Set the denominator equal to 0.

$(x+2)(x-2) = 0$ Factor the difference of two squares.

$x + 2 = 0$ or $x - 2 = 0$ Set each factor equal to 0.

$x = -2$ | $x = 2$ Solve each equation.

The domain of the function is the set of all real numbers except -2 and 2. In interval notation, the domain is $(-\infty, -2) \cup (-2, 2) \cup (2, \infty)$.

To simplify a rational expression:

1. Factor the numerator and denominator completely.
2. Remove factors equal to 1 by replacing each pair of factors common to the numerator and denominator with the equivalent fraction $\frac{1}{1}$.
3. Multiply the remaining factors in the numerator and in the denominator.

Simplify: $\dfrac{x^2-4}{2x+4} = \dfrac{\overset{1}{\cancel{(x+2)}}(x-2)}{2\underset{1}{\cancel{(x+2)}}} = \dfrac{x-2}{2}$

Simplify: $\dfrac{2a^3-5a^2-12a}{2a^3-11a^2+12a} = \dfrac{\overset{1}{\cancel{a}}(2a+3)\overset{1}{\cancel{(a-4)}}}{\underset{1}{\cancel{a}}(2a-3)\underset{1}{\cancel{(a-4)}}} = \dfrac{2a+3}{2a-3}$

To **multiply rational expressions,** multiply the numerators and multiply the denominators.

$$\frac{A}{B} \cdot \frac{C}{D} = \frac{AC}{BD}$$

Then simplify, if possible.

Multiply, and then simplify, if possible.

$\dfrac{x^2-4}{x+3} \cdot \dfrac{3x+9}{x+2} = \dfrac{(x^2-4)(3x+9)}{(x+3)(x+2)}$ Multiply the numerators. Multiply the denominators.

$= \dfrac{\overset{1}{\cancel{(x+2)}}(x-2) \cdot 3 \cdot \overset{1}{\cancel{(x+3)}}}{\underset{1}{\cancel{(x+3)}}\underset{1}{\cancel{(x+2)}}}$ Factor completely and then simplify.

$= 3(x-2)$

To **divide rational expressions,** multiply the first by the reciprocal of the second.

$$\frac{A}{B} \div \frac{C}{D} = \frac{A}{B} \cdot \frac{D}{C} = \frac{AD}{BC}$$

Then simplify, if possible.

Divide, and then simplify, if possible.

$\dfrac{x^2+4x+3}{x^2+3x} \div \dfrac{3}{x} = \dfrac{x^2+4x+3}{x^2+3x} \cdot \dfrac{x}{3}$ Multiply the first rational expression by the reciprocal of the second.

$= \dfrac{(x^2+4x+3) \cdot x}{(x^2+3x) \cdot 3}$ Multiply the numerators. Multiply the denominators.

$= \dfrac{(x+1)\overset{1}{\cancel{(x+3)}} \cdot \overset{1}{\cancel{x}}}{\underset{1}{\cancel{x}}\underset{1}{\cancel{(x+3)}} \cdot 3}$ Factor completely and then simplify.

$= \dfrac{x+1}{3}$ Multiply the remaining factors in the numerator. Multiply the remaining factors in the denominator.

To **add (or subtract) two rational expressions with like denominators,** add (or subtract) the numerators and keep the common denominator. Then, if possibe, factor and simplify.

Add:

$\dfrac{x^2-26}{x-5} + \dfrac{1}{x-5} = \dfrac{x^2-26+1}{x-5}$ Add the numerators. Write the sum over the common denominator, $x - 5$.

$= \dfrac{x^2-25}{x-5}$ Combine like terms.

$= \dfrac{(x+5)\overset{1}{\cancel{(x-5)}}}{\underset{1}{\cancel{x-5}}}$ To simplify the result, factor the numerator and remove the factor common to the numerator and denominator.

$= x + 5$

To **add or subtract rational expressions with unlike denominators,** find the LCD and express each rational expression with a denominator that is the LCD. Add (or subtract) the resulting fractions and simplify the result, if possible.

To build a rational expression, multiply it by 1 in the form of $\frac{c}{c}$, where c is any nonzero number or expression.

Subtract:

$$\frac{2x}{x+5} - \frac{1}{x} = \frac{2x}{x+5} \cdot \frac{x}{x} - \frac{1}{x} \cdot \frac{x+5}{x+5}$$ Build each rational expression to have the LCD of $x(x + 5)$.

$$= \frac{2x^2}{x(x+5)} - \frac{x+5}{x(x+5)}$$ Multiply the numerators. Multiply the denominators.

$$= \frac{2x^2 - (x+5)}{x(x+5)}$$ Subtract the numerators. Write the difference over the common denominator.

$$= \frac{2x^2 - x - 5}{x(x+5)}$$ The result does not simplify.

To **solve a rational equation:**

1. Factor all denominators.
2. Determine which numbers cannot be solutions of the equation.
3. Multiply both sides of the equation by the LCD of all rational expressions in the equation.
4. Use the distributive property to remove parentheses, remove any factors equal to 1, and write the result in simplified form.
5. Solve the resulting equation.
6. Check all possible solutions in the original equation.

All possible solutions of a rational equation must be checked. Multiplying both sides of an equation by an expression that contains a variable can lead to **extraneous solutions**, which must be discarded.

Solve: $\frac{3}{2} + \frac{1}{a-4} = \frac{5}{2a-8}$

If we factor the last denominator, the equation can be written as:

$$\frac{3}{2} + \frac{1}{a-4} = \frac{5}{2(a-4)}$$

We see that 4 cannot be a solution of the equation, because it makes at least one of the rational expressions in the equation undefined.

We can clear the equation of fractions by multiplying both sides by $2(a - 4)$, which is the LCD of the three rational expressions.

$$2(a-4)\left(\frac{3}{2} + \frac{1}{a-4}\right) = 2(a-4)\left[\frac{5}{2(a-4)}\right]$$ Multiply both sides by the LCD.

$$2(a-4)\left(\frac{3}{2}\right) + 2(a-4)\left(\frac{1}{a-4}\right) = 2(a-4)\left[\frac{5}{2(a-4)}\right]$$ Distribute.

$$\overset{1}{\cancel{2}}(a-4)\left(\frac{3}{\underset{1}{\cancel{2}}}\right) + 2\overset{1}{\cancel{(a-4)}}\left(\frac{1}{\underset{1}{\cancel{a-4}}}\right) = \overset{1}{\cancel{2}}\overset{1}{\cancel{(a-4)}}\left[\frac{5}{\underset{1}{\cancel{2}}\underset{1}{\cancel{(a-4)}}}\right]$$ Remove common factors.

$$(a-4)3 + 2 = 5$$ Simplify.

$$3a - 12 + 2 = 5$$ Distribute.

$$3a - 10 = 5$$ Combine like terms.

$$3a = 15$$

$$a = 5$$

The solution is 5. Verify that it satisfies the original equation.

REVIEW EXERCISES

127. Use the graph of function f to find each of the following:

a. $f(2)$

b. $f(-1)$

c. The value of x for which $f(x) = -1$

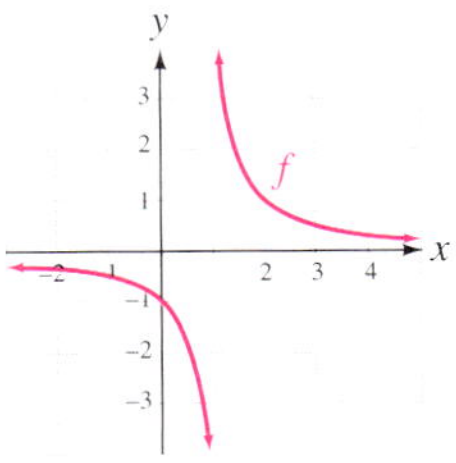

128. Find the domain of the rational function $f(x) = \frac{2x^2 + 8x}{x^2 + 2x - 24}$. Express your answer in words and using interval notation.

129. Injections. A hospital patient was given an injection for pain. The rational function

$$n(t) = \frac{28t}{t^2 + 1}$$

gives the number of milligrams n of pain medication per liter in the patient's bloodstream, where t is the number of hours since the injection. Find $n(3)$ and explain its meaning.

130. Let $f(x) = \frac{4x^2 - 6x + 7}{3x^2 - 27}$. Find $f(3)$.

Simplify each rational expression.

131. $\dfrac{62x^2y}{144xy^2}$

132. $\dfrac{2m - 2n}{n - m}$

Perform the operations and simplify, if possible.

133. $\dfrac{3x^3y^4}{c^2d} \cdot \dfrac{c^3d^2}{21x^5y^4}$

134. $\dfrac{2a^2 - 5a - 3}{a^2 - 9} \div \dfrac{2a^2 + 5a + 2}{2a^2 + 5a - 3}$

135. $\dfrac{m^2 + 3m + 9}{m^2 + mp + mr + pr} \div \dfrac{m^3 - 27}{am + ar + bm + br}$

136. $\dfrac{x^3 + 3x^2 + 2x}{2x^2 - 2x - 12} \cdot \dfrac{3x^2 - 3x}{x^3 - 3x^2 - 4x} \div \dfrac{x^2 + 3x + 2}{2x^2 - 4x - 16}$

137. $\dfrac{d^2}{c^3 - d^3} + \dfrac{c^2 + cd}{c^3 - d^3}$

138. $\dfrac{4}{t - 3} + \dfrac{6}{3 - t}$

139. $\dfrac{5x}{14z^2} + \dfrac{y^2}{16z}$

140. $\dfrac{4}{3xy - 6y} - \dfrac{4}{10 - 5x}$

141. $\dfrac{y + 7}{y + 3} - \dfrac{y - 3}{y + 7}$

142. $\dfrac{2x}{x + 1} + \dfrac{3x}{x + 2} + \dfrac{4x}{x^2 + 3x + 2}$

Solve each equation. If a solution is extraneous, so indicate.

143. $\dfrac{4}{x} - \dfrac{1}{10} = \dfrac{7}{2x}$

144. $\dfrac{3}{y} - \dfrac{2}{y + 1} = \dfrac{1}{2}$

145. $\dfrac{2}{3x + 15} - \dfrac{1}{18} = \dfrac{1}{3x + 12}$

146. $\dfrac{3}{x + 2} = \dfrac{1}{2 - x} + \dfrac{2}{x^2 - 4}$

147. $\dfrac{x + 3}{x - 5} + \dfrac{2x^2 + 6}{x^2 - 7x + 10} = \dfrac{3x}{x - 2}$

148. $\dfrac{5a}{a - 3} - 7 = \dfrac{15}{a - 3}$

SECTION 8.9 Variation

DEFINITIONS AND CONCEPTS	EXAMPLES
The words ***y varies directly as x*** or ***y is directly proportional to x*** mean that $y = kx$ for some nonzero constant k, called the **constant of variation.** The words ***y varies inversely as x*** or ***y is inversely proportional to x*** mean that $y = \frac{k}{x}$ for some nonzero constant k.	The distance d that a spring stretches ***varies directly*** as the force f attached to the spring: $d = kf$. If the voltage in an electric circuit is kept constant, the current I ***varies inversely*** as the resistance R: $I = \dfrac{k}{R}$
Strategy for solving variation problems **1.** Translate the verbal model into an equation. **2.** Substitute the first set of values into the equation from step 1 to determine the value of k. **3.** Substitute the value of k into the equation from step 1. **4.** Substitute the remaining set of values into the equation from step 3 and solve for the unknown.	Suppose d varies inversely as h. If $d = 5$ when $h = 4$, find d when $h = 10$. **1.** The words *d varies inversely as h* translate to $d = \frac{k}{h}$. **2.** If we substitute 5 for d and 4 for h, we have $5 = \dfrac{k}{4}$ $20 = k$ To find k, multiply both sides by 4. This is the constant of variation. **3.** Since $k = 20$, the inverse variation equation is $d = \frac{20}{h}$. **4.** To answer the final question, we substitute 10 for h in the inverse variation model: $d = \dfrac{20}{10} = 2$
Joint variation: One variable varies as the product of several variables. For example, $y = kxz$ (k is a constant).	The number of gallons g of oil that can be stored in a cylindrical tank ***varies jointly*** as the height h of the tank and the square of the radius r of its base: $g = khr^2$.

Combined variation: A combination of direct and inverse variation. For example, $y = \frac{kx}{z}$ (k is a constant)	The gravitational force F between two objects with masses m_1 and m_2 ***varies directly*** as the product of their masses and ***inversely*** as the square of the distance d between them: $F = \frac{km_1m_2}{d^2}$

REVIEW EXERCISES

149. Property Tax. The property tax in a certain county varies directly as assessed valuation. If a tax of $1,575 is charged on a single-family home assessed at $90,000, determine the property tax on an apartment complex assessed at $312,000.

150. Electricity. For a fixed voltage, the current in an electrical circuit varies inversely as the resistance in the circuit. If a certain circuit has a current of $2\frac{1}{2}$ amps when the resistance is 150 ohms, find the current in the circuit when the resistance is doubled.

151. Assume that y varies jointly with x and z. Find the constant of variation if $x = 24$ when $y = 3$ and $z = 4$.

152. Hurricane Winds. The wind force on a vertical surface varies jointly as the area of the surface and the square of the wind's velocity. If a 10-mph wind exerts a force of 1.98 pounds on the sign shown in the next column, find the force on the sign when the wind is blowing at 80 mph.

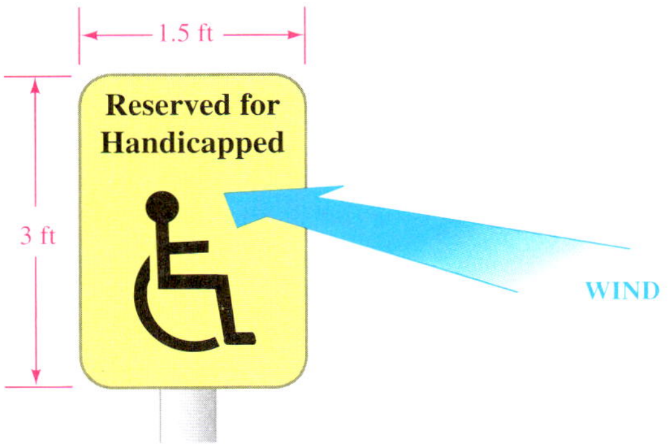

153. Does the graph on the right show direct or inverse variation?

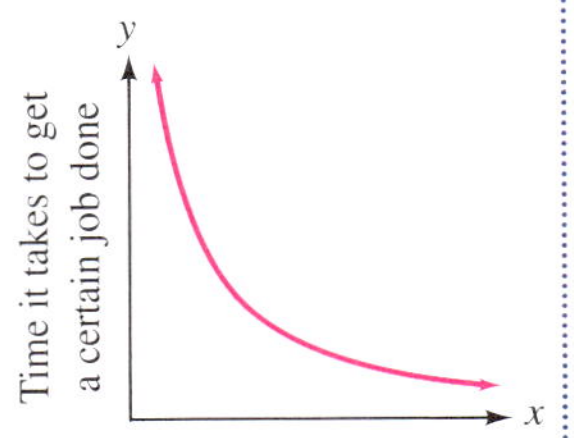

154. Assume that x_1 varies directly with the third power of t and inversely with x_2. Find the constant of variation if $x_1 = 1.6$ when $t = 8$ and $x_2 = 64$.

8 Chapter Test

1. Fill in the blanks.

a. To ______ an equation means to find all of the values of the variable that make the equation true.

b. $<$, $>$, $\le$, and $\ge$ are __________ symbols.

c. The statement $x^2 - x - 12 = (x - 4)(x + 3)$ shows that the trinomial $x^2 - x - 12$ ________ as the product of two binomials.

d. The __________ of $\frac{x+1}{x-7}$ is $\frac{x-7}{x+1}$.

e. Given a relation in x and y, if to each value of x in the domain there corresponds exactly one value of y in the range, y is said to be a ________ of x.

f. For a function, the set of all possible values that can be used for the independent variable is called the ________. The set of all values of the dependent variable is called the ______.

2. Use a check to determine whether 6.7 is a solution of $1.6y + (-3) = y + 1.02$.

Solve each equation.

3. $t + 18 = 5t - 3 + t$

4. $\frac{2}{3}(2s + 2) = \frac{1}{6}(5s + 29) - 4$

5. $6 - (x - 3) - 5x = 3[1 - 2(x + 2)]$

6. Solve $y - y_1 = m(x - x_1)$ for x_1.

7. Hand Tools. With each pass that a craftsman makes with a sander over a piece of fiberglass, he removes 0.03125 inch of thickness. If the fiberglass was originally 0.9375 inch thick, how many passes are needed to obtain the desired thickness of 0.6875 inch?

8. **Averaging Grades.** Use the information from the gradebook to determine what score Karen Nelson-Sims needs on the fifth exam so that her exam average exceeds 80.

Sociology 101 8:00-10:00 pm MW	Exam 1	Exam 2	Exam 3	Exam 4	Exam 5
Nelson-Sims, Karen	70	79	85	88	

9. Determine whether the relation defines y to be a function of x. If it does not, explain why.

a.

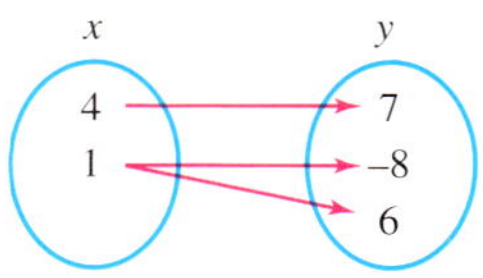

b.

x	y
-3	4
4	-3
1	5
5	1

c. $\{(0, \frac{1}{2}), (-10, \frac{1}{2}), (-20, \frac{1}{2}), (-30, \frac{1}{2})\}$

d. $|y| = x$

10. Find the domain of $f(x) = \frac{15}{12 - 2x}$.

11. Let $f(x) = -\frac{4}{5}x - 12$. For what value of x is $f(x) = 4$?

12. Determine the slope and y-intercept of the graph of $f(x) = 8x - 9$.

13. Write an equation for the linear function whose graph passes through (2, 0) and is perpendicular to the graph of $g(x) = \frac{4}{5}x + \frac{1}{9}$.

14. **Vehicle Performance.** The average fleet-wide performance of light-duty trucks has made constant improvement since 1980. For example, the time for the average 1980 model light-duty truck to accelerate from 0 to 60 mph was about 14.5 seconds. For 2005 models, that time was only 10.5 seconds. (Source: United States EPA)

a. Let m be the model year of a light-duty truck and T be the time in seconds for it to accelerate from 0 to 60 mph. Write a linear function $T(m)$ to model the situation.

b. Use your answer to part a to predict the time it will take the average light-duty truck to accelerate from 0 to 60 mph in 2015, if the trend continues.

Let $f(x) = 3x + 1$ and $g(t) = t^2 - 2t + 1$. Find each value.

15. $f(3)$

16. $g(-6)$

17. $g\left(\frac{1}{4}\right)$

18. $f(r + 8)$

19. **Boating.** The height (in feet) of a warning flare from the surface of the ocean t seconds after being shot into the air is approximated by the polynomial function $h(t) = -16t^2 + 80t + 10$. What is the height of the flare 2.5 seconds after being fired?

20. Use the graph of function f to find each of the following.

a. $f(4)$

b. The values of x for which $f(x) = 2$.

c. The domain and range of f.

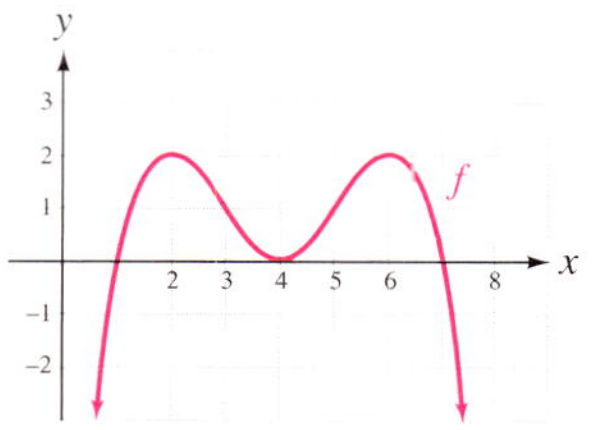

Determine whether each graph is the graph of a function. If it is not, explain why.

21.

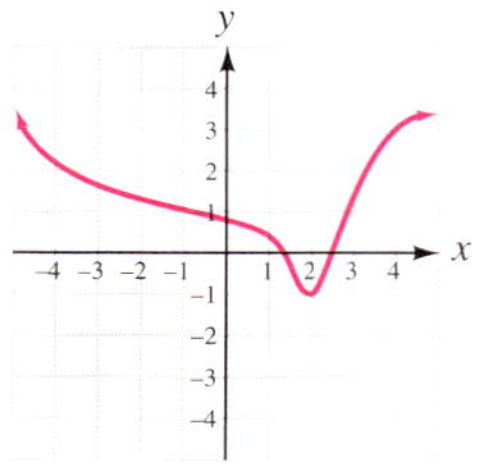

22.

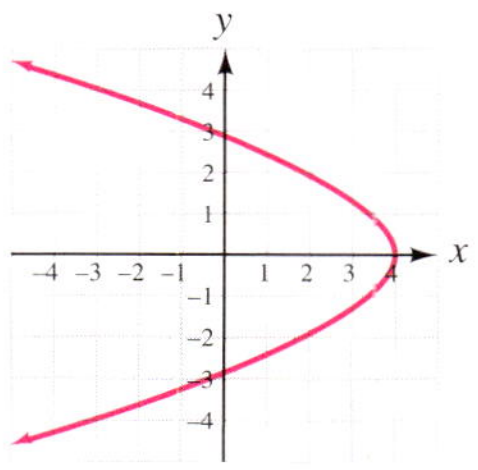

23. Graph $f(x) = |x| + 3$ by creating a table of function values and plotting points. Give the domain and range of the function.

24. Draw the graph of $g(x) = (x - 4)^3 + 1$ using a translation by first sketching the graph of its associated function. Then give the domain and range of function g.

Solve each inequality. Write the solution set in interval notation and graph it.

25. $-2(2x + 3) \geq 14$

26. $-2 < \frac{x - 4}{3} < 4$

27. $3x \geq -2x + 5$ and $7 \geq 4x - 2$

28. $3x < -9$ or $-\frac{x}{4} < -2$

29. $|2x - 4| > 22$

30. $2|3(x - 2)| \leq 4$

Solve each equation.

31. $|2x + 3| - 19 = 0$

32. $|3x + 4| = |x + 12|$

Factor.

33. $12a^3b^2c - 3a^2b^2c^2 + 6abc^3$

34. $4y^4 - 64$

35. $b^3 + 125$

36. $6u^2 + 9u - 6$

37. $ax - xy + ay - y^2$

38. $25m^8 - 60m^4n + 36n^2$

39. $144b^2 + 25$

40. $x^2 + 6x + 9 - y^2$

41. $64a^3 - 125b^6$

42. $(x - y)^2 + 3(x - y) - 10$

43. Refer to the graph of function f on the right. Find

a. $f(5)$

b. $f(-1)$

c. $f(1)$

d. The value of x for which $f(x) = -2$

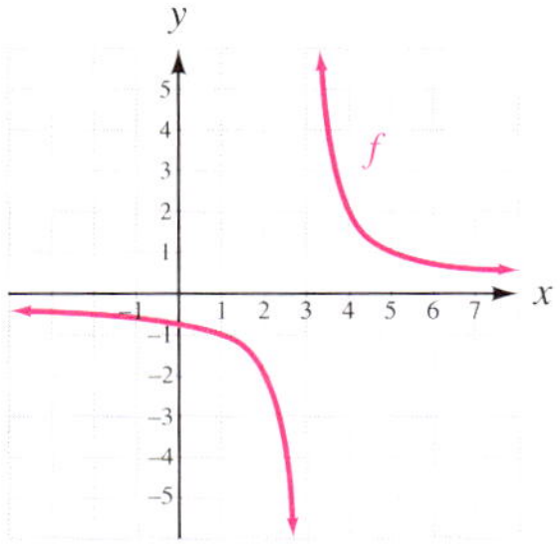

44 Wildlife. The number of prairie dogs p living in a given area of land is approximated by the rational function

$$p(t) = \frac{200t}{t + 1}$$

where t is the time, in months, since the first of the year. Find the number of prairie dogs in that area by the end of July.

45. Are the expressions $\frac{x - 4}{x^2 + x - 18}$ and $-\frac{x + 4}{x^2 + x - 18}$ equivalent?

46. Find the domain of the rational function $f(x) = \frac{x^2 + 6x + 5}{x - x^2}$. Express it using words and interval notation.

Simplify each rational expression.

47. $\frac{3y - 6z}{2z - y}$

48. $\frac{2x^2 + 7xy + 3y^2}{4xy + 12y^2}$

Perform the operations and simplify, if possible.

49. $\frac{x^3 + y^3}{4} \div \frac{x^2 - xy + y^2}{2x + 2y}$

50. $\frac{xu + 2u + 3x + 6}{u^2 - 9} \cdot \frac{13u - 39}{x^2 + 3x + 2}$

51. $\frac{-3t + 4}{t^2 + t - 20} + \frac{5t + 6}{t^2 + t - 20}$

52. $\frac{a + 3}{a^2 - a - 2} - \frac{a - 4}{a^2 - 2a - 3}$

Solve each equation.

53. $\frac{34}{a^2} = \frac{3}{2a} - \frac{13}{20a}$

54. $\frac{u - 2}{u - 3} + 3 = u + \frac{u - 4}{3 - u}$

55. $\frac{3}{x - 2} = \frac{x + 3}{2x}$

56. $\frac{4}{x^2 - 9} = \frac{7}{x^2 - 7x + 12} - \frac{5}{x^2 - x - 12}$

57. Assume that y varies directly with x. If $x = 30$ when $y = 4$, find y when $x = 9$.

58. Sound. Sound intensity (loudness) varies inversely as the square of the distance from the source. If a rock band has a sound intensity of 100 decibels 30 feet away from the amplifier, find the sound intensity 60 feet away from the amplifier.

Group Project

VENN DIAGRAMS

Overview: In this activity, we will discuss several of the fundamental concepts of what is known as *set theory.*

Instructions: *Venn diagrams* are a convenient way to visualize relationships between sets and operations on sets. They were invented by the English mathematician John Venn (1834–1923). To draw a Venn diagram, we begin with a large rectangle, called the *universal set.* Ovals or circles are then drawn in the interior of the rectangle to represent subsets of the universal set.

Form groups of 2 or 3 students. Study the following figures, which illustrate three set operations: union, intersection, and complement.

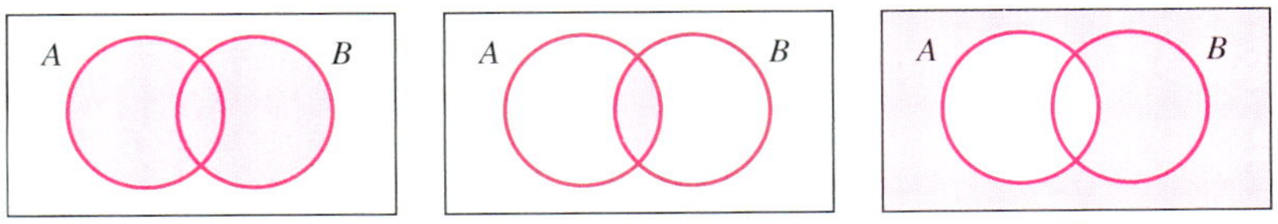

$A \cup B$ The shaded region is the *union* of set A and set B.

$A \cap B$ The shaded region is the *intersection* of set A and set B.

$\overline{A}$ The shaded region is the complement of set A.

For each of the following exercises, sketch the following blank Venn diagram and then shade the indicated region.

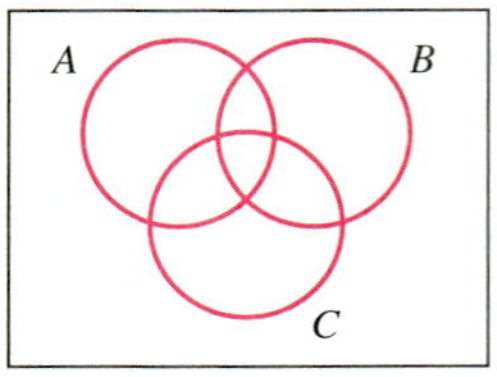

1. $A \cup B$
2. $A \cap B$
3. $A \cap C$
4. $A \cup C$
5. $A \cup B \cup C$
6. $A \cap B \cap C$
7. $(B \cup C) \cap A$
8. $C \cup (A \cap B)$
9. $\overline{A}$
10. $\overline{B} \cup \overline{C}$
11. $\overline{B} \cap \overline{C}$
12. $\overline{A \cup B}$

Radical Expressions and Equations

9

from Campus to Careers

General Contractor

The growing popularity of remodeling has created a boom for general contractors. If it's an additional bedroom you need or a makeover of a dated kitchen or bathroom, they can provide design and construction expertise, as well as knowledge of local building code requirements. From the planning stages of a project through its completion, general contractors use mathematics every step of the way.

Problem 141 in **Study Set 9.2, problem 113** in **Study Set 9.3,** and **problem 109** in **Study Set 9.5** involve situations that a general contractor encounters on the job. The mathematical concepts in this chapter can be used to solve those problems.

Study Skills Workshop ▶ Don't Just Memorize

Many Students attempt to learn algebra by rote memorization. Unfortunately, as the term progresses, they find that does not work. When they encounter problem types slightly different from those that they have memorized, they experience great difficulty. Remember, memorization only provides a superficial grasp of the concepts. When learning a new algebraic procedure, it is most important that you:

UNDERSTAND "WHY": Be able to exaplin the purpose for each stp in the procedure and why they are applied in that order.

UNDERSTAND "WHEN": Be able to explain what types of problems are solved using the procedure and what types are not.

Now Try This ▶

1. Choose five problems in the Guided Practice section of a Study Set to solve. Write a *Strategy* and *Why* statement for each solution in your own words.
2. Select a procedure that is introduced in this chapter and explain when it should be used. Write a *Caution* statement that warns of a possible pitfall when using it. Give an example of an application problem that can be solved using the procedure.

SECTION 9.1 Radical Expressions and Radical Functions

OBJECTIVES

1. Find square roots.
2. Find square roots of expressions containing variables.
3. Graph the square root function.
4. Evaluate radical functions.
5. Find cube roots.
6. Graph the cube root function.
7. Find *n*th roots.

ARE YOU READY?

The following problems review some basic skills that are needed when working with radical expressions and radical functions.

1. Evaluate: **a.** 15^2 **b.** 4^3

2. Evaluate: **a.** $\left(\frac{7}{5}\right)^2$ **b.** $(0.2)^2$

3. Evaluate: **a.** $(-6)^3$ **b.** 3^4

4. Simplify: **a.** $a^4 \cdot a^4$ **b.** $x^3 \cdot x^3 \cdot x^3$

5. Multiply: $(x + 8)(x + 8)$

6. Let $f(x) = |2x - 3|$. Find $f(4)$.

In this section, we will reverse the squaring process and learn how to find *square roots* of numbers. Then we will generalize the concept of root and consider cube roots, fourth roots, and so on. We will also discuss a new family of functions, called *radical functions*.

1 Find Square Roots.

When we raise a number to the second power, we are squaring it, or finding its **square.**

- The square of 5 is 25 because $5^2 = 25$.
- The square of -5 is 25, because $(-5)^2 = 25$.

We can reverse the squaring process to find **square roots** of numbers. For example, to find the square roots of 25, we ask ourselves "What number, when squared, is equal to 25?" There are two possible answers.

- 5 is a square root of 25, because $5^2 = 25$.
- -5 is also a square root of 25, because $(-5)^2 = 25$.

In general, we have the following definition.

Square Root of *a*

The number b is a **square root** of the number a if $b^2 = a$.

The Language of Algebra

Precise vocabulary must be used when discussing square roots. For example, 3 is **a** square root of 9 and -3 is also **a** square root of 9. However, when we speak of **the** square root of a number, we mean only the *nonnegative* square root of that number. Thus, **the** square root of 9 is 3, which is written as $\sqrt{9} = 3$.

Every positive number has two square roots, one positive and one negative. For example, the two square roots of 9 are 3 and -3, and the two square roots of 144 are 12 and -12. The number 0 is the only real number with exactly one square root. In fact, it is its own square root, because $0^2 = 0$.

A **radical symbol** $\sqrt{}$ represents the **positive** or **principal square root** of a number. Since 3 is the positive square root of 9, we can write

$\sqrt{9} = 3$ Read as "the square root of 9."

The symbol $-\sqrt{}$ represents the **negative square root** of a number. It is the opposite of the principal square root. Since -12 is the negative square root of 144, we can write

$-\sqrt{144} = -12$ Read as "the negative square root of 144 is -12" or "the opposite of the square root of 144 is -12."

Square Root Notation

If a is a positive real number,

1. $\sqrt{a}$ represents the **positive** or **principal square root** of a. It is the positive number we square to get a.
2. $-\sqrt{a}$ represents the **negative square root** of a. It is the opposite of the principal square root of a: $-\sqrt{a} = -1 \cdot \sqrt{a}$.
3. The principal square root of 0 is 0: $\sqrt{0} = 0$.

The Language of Algebra

The word **radical** comes from the Latin word *radix*, meaning root. The radical symbol has evolved over the years from ℞ in the 1300s to √ in the 1500s, to the familiar $\sqrt{}$ in the 1600s.

The number or variable expression under a radical symbol is called the **radicand.** Together, the radical symbol and radicand are called a **radical.** An algebraic expression containing a radical is called a **radical expression.**

Radical symbol → $\sqrt{81}$ ← Radicand; Radical (the whole $\sqrt{81}$)

Read $\sqrt{81}$ as "the square root of 81" or as "radical 81."

Some more examples of radical expressions are:

$$\sqrt{100}, \quad \sqrt{2} + 3, \quad \sqrt{x^2}, \quad \text{and} \quad \sqrt{\frac{a-1}{49b^2}}$$

To evaluate (find the value of) square root radical expressions, you need to quickly recognize each of the natural-number **perfect squares** shown below in red.

$\mathbf{1} = 1^2$	$\mathbf{25} = 5^2$	$\mathbf{81} = 9^2$	$\mathbf{169} = 13^2$	$\mathbf{289} = 17^2$
$\mathbf{4} = 2^2$	$\mathbf{36} = 6^2$	$\mathbf{100} = 10^2$	$\mathbf{196} = 14^2$	$\mathbf{324} = 18^2$
$\mathbf{9} = 3^2$	$\mathbf{49} = 7^2$	$\mathbf{121} = 11^2$	$\mathbf{225} = 15^2$	$\mathbf{361} = 19^2$
$\mathbf{16} = 4^2$	$\mathbf{64} = 8^2$	$\mathbf{144} = 12^2$	$\mathbf{256} = 16^2$	$\mathbf{400} = 20^2$

EXAMPLE 1

Evaluate: **a.** $\sqrt{81}$ **b.** $-\sqrt{225}$ **c.** $\sqrt{\frac{49}{4}}$ **d.** $\sqrt{0.36}$

Strategy In each case, we will determine what positive number, when squared, produces the radicand.

Why The symbol $\sqrt{}$ indicates that the positive square root of the number written under it should be found.

Solution **a.** $\sqrt{81} = 9$ Because $9^2 = 81$ **b.** $-\sqrt{225} = -15$ Because $-\sqrt{225} = -1 \cdot \sqrt{225}$

c. $\sqrt{\frac{49}{4}} = \frac{7}{2}$ Because $\left(\frac{7}{2}\right)^2 = \frac{49}{4}$ **d.** $\sqrt{0.36} = 0.6$ Because $(0.6)^2 = 0.36$

Self Check 1 Evaluate: **a.** $\sqrt{64}$ **b.** $-\sqrt{1}$ **c.** $\sqrt{\frac{1}{16}}$
d. $\sqrt{0.09}$

Now Try Problems 23, 27, and 29

Success Tip

Estimation is helpful when approximating square roots. For example, $\sqrt{83}$ must be a number between 9 and 10, because $\sqrt{\mathbf{81}} < \sqrt{83} < \sqrt{\mathbf{100}}$.

A number, such as 81, 225, $\frac{49}{4}$, and 0.36, that is the square of some rational number, is called a **perfect square.** In Example 1, we saw that the square root of a perfect square is a rational number.

If a positive number is not a perfect square, its square root is irrational. For example, $\sqrt{83}$ is an irrational number because 83 is not a perfect square. Since $\sqrt{83}$ is irrational, its decimal representation is nonterminating and nonrepeating. We can find a rational-number approximation of $\sqrt{83}$ using the square root key $\boxed{\sqrt{\ }}$ on a calculator or from the table of square roots found in Appendix I at the back of the book.

$$\sqrt{83} \approx 9.110433579 \quad \text{or} \quad \sqrt{83} \approx 9.11 \quad \text{Round to the nearest hundredth.}$$

CAUTION Square roots of negative numbers are not real numbers. For example, $\sqrt{-9}$ is not a real number, because no real number squared equals -9. Square roots of negative numbers come from a set called the **imaginary numbers,** which we will discuss later in this chapter. If we attempt to evaluate $\sqrt{-9}$ using a calculator, we will get an error message.

Caution

Although they look similar, these radical expressions have very different meanings.

$$-\sqrt{9} = -3$$

$\sqrt{-9}$ is not a real number.

Scientific calculator

Graphing calculator

We summarize three important facts about square roots as follows.

Square Roots

1. If a is a perfect square, then $\sqrt{a}$ is rational.
2. If a is a positive number that is not a perfect square, then $\sqrt{a}$ is irrational.
3. If a is a negative number, then $\sqrt{a}$ is not a real number.

2 Find Square Roots of Expressions Containing Variables.

If $x \neq 0$, the positive number x^2 has x and $-x$ for its two square roots. To denote the positive square root of $\sqrt{x^2}$, we must know whether x is positive or negative.

If x is positive, we can write

$$\sqrt{x^2} = x \quad \sqrt{x^2} \text{ represents the positive square root of } x^2 \text{, which is } x.$$

If x is negative, then $-x$ is positive and we can write

$$\sqrt{x^2} = -x \quad \sqrt{x^2} \text{ represents the positive square root of } x^2 \text{, which is } -x.$$

If we don't know whether x is positive or negative, we can use absolute value symbols to ensure that $\sqrt{x^2}$ is not negative.

Simplifying $\sqrt{x^2}$

For any real number x,

$$\sqrt{x^2} = |x|$$ The principal square root of x is equal to the absolute value of x.

We use this definition to *simplify* square root radical expressions.

EXAMPLE 2 Simplify. Assume all variables are unrestricted. **a.** $\sqrt{16x^2}$ **b.** $\sqrt{t^2 + 2t + 1}$ **c.** $\sqrt{m^6}$ **d.** $\sqrt{49r^8}$

Strategy In each case, we will determine what positive expression, when squared, produces the radicand.

Why The symbol $\sqrt{}$ indicates that the positive square root of the expression written under it should be found.

Solution If x, t, m, and r can be any real number, we have

a. $\sqrt{16x^2} = |4x|$ Because $(4x)^2 = 16x^2$. Since $4x$ could be negative (for example, when $x = -5$, the expression $4x$ is -20), absolute value symbols are needed to ensure that the result is not negative.

$= 4|x|$ Since 4 is a positive constant in the product $4x$, we can write it outside the absolute value symbols.

b. $\sqrt{t^2 + 2t + 1} = \sqrt{(t + 1)^2}$ Factor the radicand: $t^2 + 2t + 1 = (t + 1)^2$.

$= |t + 1|$ Since $t + 1$ can be negative (for example, when $t = -5$, the expression $t + 1$ is -4), absolute value symbols are needed to ensure that the result is not negative.

c. $\sqrt{m^6} = |m^3|$ Because $(m^3)^2 = m^6$. Since m^3 can be negative (for example, when $m = -3$, the expression m^3 is -27), absolute value symbols are needed to ensure that the result is not negative.

d. $\sqrt{49r^8} = 7r^4$ Because $(7r^4)^2 = 49r^8$. Since r^4 is never negative for any value of r, no absolute value symbols are needed.

The Language of Algebra

The phrase **"assume all variables are unrestricted"** in the instructions for Example 2 reminds us that negative numbers are permissible replacement values for the variables. Therefore, absolute value symbols may be necessary to ensure the answer is not negative.

Self Check 2 Simplify. Assume all variables are unrestricted. **a.** $\sqrt{25a^2}$ **b.** $\sqrt{b^{14}}$ **c.** $\sqrt{x^2 - 18x + 81}$ **d.** $\sqrt{100n^8}$

Now Try Problems 39, 41, 45, and 47

If we know that x, t, and m are positive in Example 2, we don't need to use absolute value symbols in the answers. For example,

$\sqrt{16x^2} = 4x$ If x is positive, $4x$ is positive.

$\sqrt{m^6} = m^3$ If m is positive, m^3 is positive.

$\sqrt{t^2 + 2t + 1} = t + 1$ If t is positive, $t + 1$ is positive.

The Language of Algebra

If we are told **"All variables represent positive real numbers,"** then absolute value symbols are not needed when simplifying square root radical expressions.

3 Graph the Square Root Function.

Since there is one principal square root for every nonnegative real number x, the equation $f(x) = \sqrt{x}$ determines a function, called a **square root function.** Square root functions belong to a larger family of functions known as **radical functions.**

EXAMPLE 3 Graph $f(x) = \sqrt{x}$ and find the domain and range of the function.

Strategy We will graph the function by creating a table of function values and plotting the corresponding ordered pairs.

Why After drawing a smooth curve through the plotted points, we will have the graph.

Solution To graph the function, we select several values for x that are perfect squares, such as 0, 1, 4, and 9, and find the corresponding values of $f(x)$. We begin with $x = 0$, since 0 is the smallest input for which $\sqrt{x}$ is defined.

$$f(x) = \sqrt{x} \quad f(0) = \sqrt{0} = 0$$

$$f(x) = \sqrt{x} \quad f(1) = \sqrt{1} = 1$$

$$f(x) = \sqrt{x} \quad f(4) = \sqrt{4} = 2$$

$$f(x) = \sqrt{x} \quad f(9) = \sqrt{9} = 3$$

We also can select values of x that are not perfect squares when creating a table of function values. For example, if $x = 6$, then $f(6) = \sqrt{6}$, and it follows that the point $(2, \sqrt{6})$ is on the graph. To help locate this point on the coordinate system, it is helpful to approximate: $\sqrt{6} \approx 2.45$.

The Language of Algebra

Together, 0 and the positive real numbers are called the **nonnegative real numbers.**

We enter each value of x and its corresponding value of $f(x)$ in the table below. After plotting the ordered pairs, we draw a smooth curve through the points to get the graph shown in figure (a). Since the equation defines a function, its graph passes the vertical line test.

To find the domain of the function graphically, we project the graph onto the x-axis, as shown in figure (b). Because the graph begins at (0, 0) and extends indefinitely to the right, the projection includes 0 and all positive real numbers. Thus, the domain of $f(x) = \sqrt{x}$ is the set of nonnegative real numbers, which can be written in interval notation as $[0, \infty)$.

To find the range of the function graphically, we project the graph onto the y-axis, as shown in figure (b). Because the graph of the function begins at (0, 0) and extends indefinitely upward, the projection includes all nonnegative real numbers. Therefore, the range of the function is $[0, \infty)$.

$f(x) = \sqrt{x}$

x	$f(x)$	
0	0	→ (0, 0)
1	1	→ (1, 1)
4	2	→ (4, 2)
6	$\sqrt{6}$	→ $(6, \sqrt{6})$
9	3	→ (9, 3)
16	4	→ (16, 4)
↑ Select x	↑ Find $f(x)$.	↑ Plot the point.

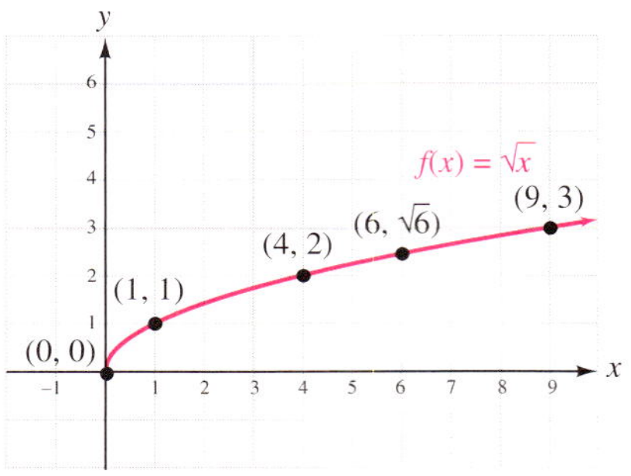

An arrowhead is not drawn on the left end of the graph because it does not extend indefinitely in that direction.

(a)

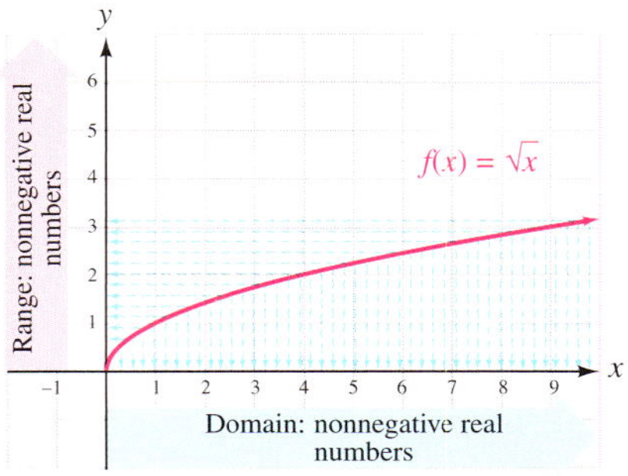

(b)

Self Check 3 Graph: $g(x) = \sqrt{x} + 2$. Then give its domain and range and compare it with the graph of $f(x) = \sqrt{x}$.

Now Try Problem 51

EXAMPLE 4 Let $g(x) = \sqrt{x+3}$. **a.** Find its domain algebraically. **b.** Graph the function. **c.** Find its range.

Strategy We will determine the domain algebraically by finding all the values of x for which $x + 3 \geq 0$.

Why Since the expression $\sqrt{x+3}$ is not a real number when $x + 3$ is negative, we must require that $x + 3 \geq 0$.

Solution

a. To determine the domain of the function, we solve the following inequality:

$$x + 3 \geq 0 \qquad \text{Because we cannot find the square root of a negative number}$$

$$x \geq -3 \qquad \text{To solve for } x \text{, subtract 3 from both sides.}$$

The x-inputs must be real numbers greater than or equal to -3. Thus, the domain of $g(x) = \sqrt{x+3}$ is the interval $[-3, \infty)$.

Success Tip

We solved linear inequalities in one variable, such as $x + 3 \geq 0$, in Section 2.7.

b. To graph the function, we construct a table of function values. We begin by selecting $x = -3$, since -3 is the smallest input for which $\sqrt{x+3}$ is defined. Then we let $x = -2$, 1, and 6 and list each corresponding function value g(x) in the table.

$$g(x) = \sqrt{x+3} \qquad g(-3) = \sqrt{-3+3} = \sqrt{0} = 0$$

$$g(x) = \sqrt{x+3} \qquad g(-2) = \sqrt{-2+3} = \sqrt{1} = 1$$

$$g(x) = \sqrt{x+3} \qquad g(1) = \sqrt{1+3} = \sqrt{4} = 2$$

$$g(x) = \sqrt{x+3} \qquad g(6) = \sqrt{6+3} = \sqrt{9} = 3$$

After plotting the ordered pairs, we draw a smooth curve through the points to get the graph shown in figure (a). In figure (b), we see that the graph of $g(x) = \sqrt{x+3}$ is the graph of $f(x) = \sqrt{x}$, translated 3 units to the left.

$g(x) = \sqrt{x+3}$

x	$g(x)$	
−3	0	→ (−3, 0)
−2	1	→ (−2, 1)
1	2	→ (1, 2)
6	3	→ (6, 3)

↑ Select values of x that make $x + 3$ a perfect square.

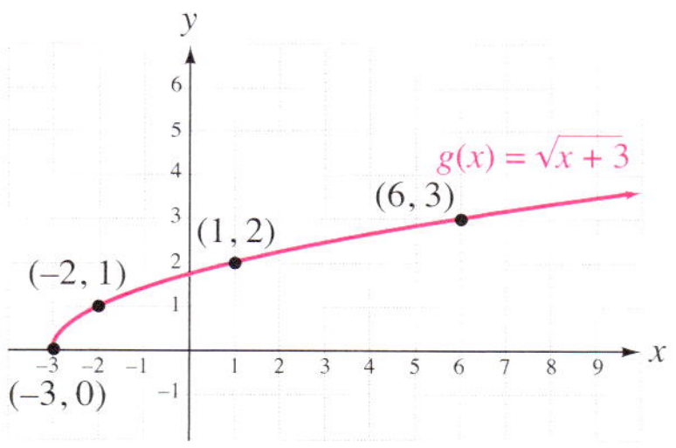

(a)

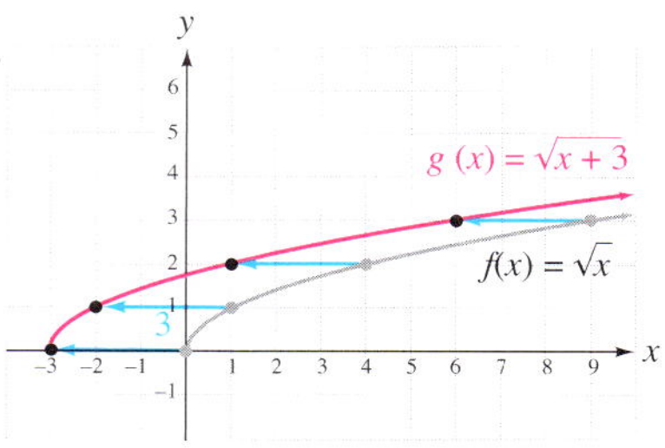

(b)

c. From the graph in figure (a), we see that the range of $g(x) = \sqrt{x+3}$ is $[0, \infty)$.

Self Check 4 Let $h(x) = \sqrt{x-2}$. **a.** Find its domain algebraically. **b.** Graph the function. **c.** Find its range.

Now Try Problems 53 and 55

4 Evaluate Radical Functions.

Radical functions can be used to model certain real-life situations that exhibit growth that eventually levels off. A calculator is often helpful when evaluating a radical function for a given value of the input variable.

EXAMPLE 5 **Pendulums.** The **period of a pendulum** is the time required for the pendulum to swing back and forth to complete one cycle. The period (in seconds) is a function of the pendulum's length L (in feet) and is given by $f(L) = 2\pi\sqrt{\frac{L}{32}}$. Find the period of the 5-foot-long pendulum of a clock. Round the result to the nearest tenth.

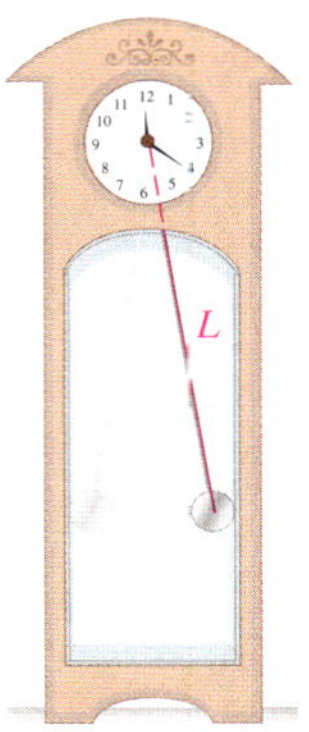

Strategy To find the period of the pendulum we will find $f(5)$.

Why The notation $f(5)$ represents the period (in seconds) of a pendulum whose length L is 5 feet.

Solution

$$f(L) = 2\pi\sqrt{\frac{L}{32}}$$

$$f(5) = 2\pi\sqrt{\frac{5}{32}} \quad \text{Substitute 5 for } L.$$

$$\approx 2.483647066$$ Evaluate the radical expression. Use a calculator to find an approximation. $2\pi\sqrt{\frac{5}{32}}$ means $2 \cdot \pi \cdot \sqrt{\frac{5}{32}}$.

The period is approximately 2.5 seconds.

Self Check 5 **Pendulums.** Find the period of a pendulum that is 3 feet long. Round the result to the nearest hundredth.

Now Try ▶ Problem 123

Using Your Calculator ▶ Graphing and Evaluating a Square Root Function

We can graph radical functions on a graphing calculator. To graph $f(x) = \sqrt{x}$ from Example 3, we press Y = and enter the right side of the equation. If the window settings for x and y are $[-1, 9]$, we get the graph shown in figure (a) when we press GRAPH. The function $g(x) = \sqrt{x + 3}$ from Example 4 is graphed in a similar way with window settings of $[-4, 10]$ for x and $[-1, 9]$ for y. See figure (b).

To answer Example 5 with a graphing calculator, we graph the function $f(x) = 2\pi\sqrt{\frac{x}{32}}$. We then trace and move the cursor toward an x-value of 5 until we see the coordinates shown in figure (c). The pendulum's period is given by the y-value shown on the screen. By zooming in, we can get better results.

After entering $Y_1 = 2\pi\sqrt{\frac{x}{32}}$, we can also use the TABLE mode to find $f(5)$. See figure (d).

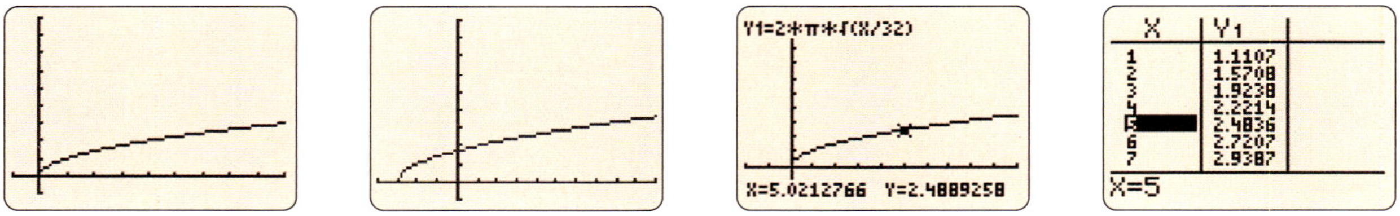

(a) (b) (c) (d)

5 Find Cube Roots.

When we raise a number to the third power, we are cubing it, or finding its **cube.** We can reverse the cubing process to find **cube roots** of numbers. To find the cube root of 8, we ask "What number, when cubed, is equal to 8?" It follows that 2 is a cube root of 8, because $2^3 = 8$.

In general, we have this definition.

Cube Root of a	The number b is a **cube root** of the real number a if $b^3 = a$.

All real numbers have one real cube root. A positive number has a positive cube root, a negative number has a negative cube root, and the cube root of 0 is 0.

Cube Root Notation	The cube root of a is written as $\sqrt[3]{a}$. By definition, $\sqrt[3]{a} = b$ if $b^3 = a$

Notation

For the square root symbol $\sqrt{}$, the unwritten index is understood to be 2.

$$\sqrt{a} = \sqrt[2]{a}$$

Earlier, we determined that the cube root of 8 is 2. In symbols, we can write: $\sqrt[3]{8} = 2$. The number 3 is called the **index,** 8 is called the **radicand,** and the entire expression is called a **radical.**

Index (3) → $\sqrt[3]{8}$ ← Radicand (8); Radical: $\sqrt[3]{8}$

Read as "the cube root of 8."

A number, such as 125, $\frac{1}{64}$, -27, and -8, that is the cube of some rational number is called a **perfect cube.**

To simplify cube root radical expressions, you need to quickly recognize each of the following natural-number perfect cubes shown in blue.

$$\mathbf{1} = 1^3 \qquad \mathbf{27} = 3^3 \qquad \mathbf{125} = 5^3 \qquad \mathbf{343} = 7^3 \qquad \mathbf{729} = 9^3$$
$$\mathbf{8} = 2^3 \qquad \mathbf{64} = 4^3 \qquad \mathbf{216} = 6^3 \qquad \mathbf{512} = 8^3 \qquad \mathbf{1{,}000} = 10^3$$

The following property is also used to simplify cube root radical expressions.

Simplifying $\sqrt[3]{x^3}$	For any real number x, $\sqrt[3]{x^3} = x$

EXAMPLE 6

Simplify: **a.** $\sqrt[3]{125}$ **b.** $\sqrt[3]{-\frac{1}{64}}$ **c.** $\sqrt[3]{27x^3}$ **d.** $\sqrt[3]{-8a^9b^6}$

Strategy In each case, we will determine what number or expression, when cubed, produces the radicand.

Why The symbol $\sqrt[3]{}$ indicates that the cube root of the number written under it should be found.

Solution

a. $\sqrt[3]{125} = 5$ Because $5^3 = 5 \cdot 5 \cdot 5 = 125$

b. $\sqrt[3]{-\frac{1}{64}} = -\frac{1}{4}$ Because $\left(-\frac{1}{4}\right)^3 = \left(-\frac{1}{4}\right)\left(-\frac{1}{4}\right)\left(-\frac{1}{4}\right) = -\frac{1}{64}$

c. $\sqrt[3]{27x^3} = 3x$ Because $(3x)^3 = (3x)(3x)(3x) = 27x^3$. No absolute value symbols are needed.

d. $\sqrt[3]{-8a^9b^6} = -2a^3b^2$ Because $(-2a^3b^2)^3 = (-2a^3b^2)(-2a^3b^2)(-2a^3b^2) = -8a^9b^6$.

Success Tip

Since every real number has exactly one real cube root, absolute value symbols are not to be used when simplifying cube roots.

Self Check 6 Simplify: **a.** $\sqrt[3]{64}$ **b.** $\sqrt[3]{-\frac{1}{1,000}}$

c. $\sqrt[3]{-125a^3}$ **d.** $\sqrt[3]{27m^6n^9}$

Now Try Problems 73, 75, and 81

6 Graph the Cube Root Function.

Since there is one cube root for every real number x, the equation $f(x) = \sqrt[3]{x}$ defines a function, called the **cube root function.** Like square root functions, cube root functions belong to the family of radical functions.

EXAMPLE 7 Consider $f(x) = \sqrt[3]{x}$. **a.** Graph the function. **b.** Find its domain and range. **c.** Graph: $g(x) = \sqrt[3]{x} - 2$

Strategy We will graph the function by creating a table of function values and plotting the corresponding ordered pairs.

Why After drawing a smooth curve through the plotted points, we will have the graph. The answers to parts (b) and (c) can then be determined from the graph.

Solution **a.** To graph the function, we select several values for x, that are perfect cubes, such as $-8, -1, 0, 1,$ and 8, and find the corresponding values of $f(x)$. The results are entered in the table below.

$f(x) = \sqrt[3]{x}$, $f(-8) = \sqrt[3]{-8} = -2$ | $f(x) = \sqrt[3]{x}$, $f(-1) = \sqrt[3]{-1} = -1$ | $f(x) = \sqrt[3]{x}$, $f(0) = \sqrt[3]{0} = 0$ | $f(x) = \sqrt[3]{x}$, $f(1) = \sqrt[3]{1} = 1$ | $f(x) = \sqrt[3]{x}$, $f(8) = \sqrt[3]{8} = 2$

After plotting the ordered pairs, we draw a smooth curve through the points to get the graph shown in figure (a).

$f(x) = \sqrt[3]{x}$

x	$f(x)$	
-8	-2	→ $(-8, -2)$
-1	-1	→ $(-1, -1)$
0	0	→ $(0, 0)$
1	1	→ $(1, 1)$
8	2	→ $(8, 2)$

↑ Select values of x that are perfect cubes.

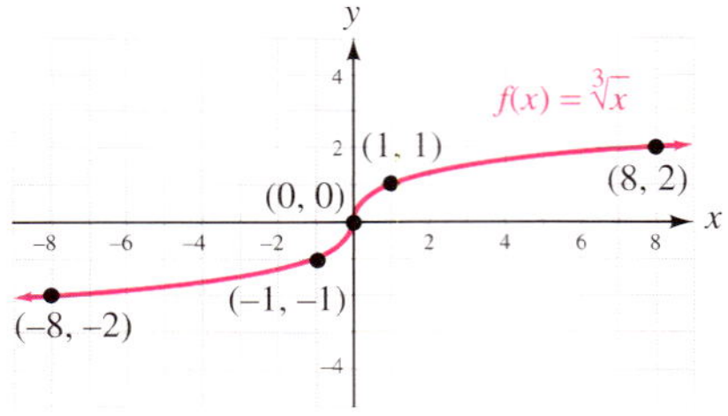

(a)

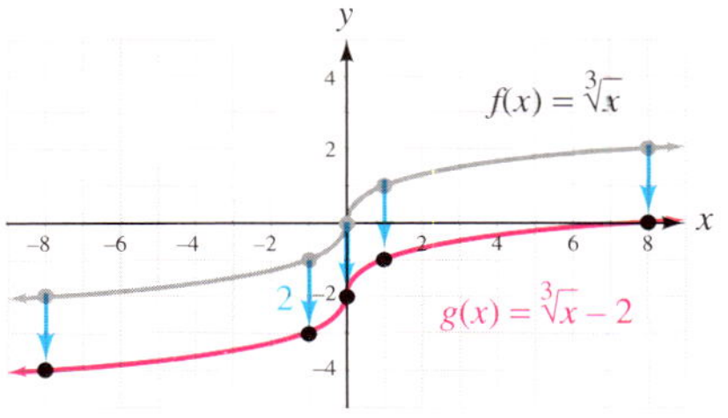

(b)

b. If we project the graph in figure (a) onto the x- and y-axes, we see that the domain and the range of the function are the set of real numbers. Thus, the domain is $(-\infty, \infty)$ and the range is $(-\infty, \infty)$.

c. Refer to figure (b). The graph of $g(x) = \sqrt[3]{x} - 2$ is the graph of $f(x) = \sqrt[3]{x}$, translated 2 units downward.

Self Check 7 Consider $f(x) = \sqrt[3]{x} + 1$. **a.** Graph the function. **b.** Find its domain and range.

Now Try Problem 83

7 Find *n*th Roots.

Just as there are square roots and cube roots, there are fourth roots, fifth roots, sixth roots, and so on. In general, we have the following definition.

nth Roots of a

The nth root of a is written as $\sqrt[n]{a}$, and

$$\sqrt[n]{a} = b \quad \text{if} \quad b^n = a$$

The number n is called the **index** (or **order**) of the radical. If n is an even natural number, a must be positive or zero, and b must be positive.

When n is an odd natural number, the expression $\sqrt[n]{x}$, where $n > 1$, represents an **odd root.** Since every real number has just one real nth root when n is odd, we don't need absolute value symbols when finding odd roots. For example,

$\sqrt[5]{243} = 3$ Because $3^5 = 243$

$\sqrt[5]{-32x^5} = -2x$ Because $(-2x)^5 = -32x^5$

When n is an even natural number, the expression $\sqrt[n]{x}$, where $x > 0$, represents an **even root.** In this case, there will be one positive and one negative real nth root. For example, the real sixth roots of 729 are 3 and -3, because $3^6 = 729$ and $(-3)^6 = 729$. When finding even roots, we can use absolute value symbols to guarantee that the nth root is positive.

$\sqrt[4]{(-3)^4} = |-3| = 3$ We also could simplify this as follows: $\sqrt[4]{(-3)^4} = \sqrt[4]{81} = 3$.

$\sqrt[6]{729x^6} = |3x| = 3|x|$ Because $(3x)^6 = 729x^6$. The absolute value symbols ensure that the result is positive.

In general, we have the following rules.

Success Tip

When simplifying radical expressions, it is helpful to recognize the following perfect fourth and fifth powers:

$16 = 2^4$	$32 = 2^5$
$81 = 3^4$	$243 = 3^5$
$256 = 4^4$	$1,024 = 4^5$
$625 = 5^4$	

Simplifying $\sqrt[n]{x^n}$

If x is a real number and $n > 1$, then

If n is an odd natural number, $\sqrt[n]{x^n} = x$.

If n is an even natural number, $\sqrt[n]{x^n} = |x|$.

EXAMPLE 8 Evaluate: **a.** $\sqrt[4]{625}$ **b.** $\sqrt[4]{-1}$ **c.** $\sqrt[5]{-32}$ **d.** $\sqrt[6]{\frac{1}{64}}$

Strategy In each case, we will determine what number, when raised to the fourth, fifth, or sixth power, produces the radicand.

Why The symbols $\sqrt[4]{\ }$, $\sqrt[5]{\ }$, and $\sqrt[6]{\ }$ indicate that the fourth, fifth, or sixth root of the number written under it should be found.

Solution

a. $\sqrt[4]{625} = 5$, because $5^4 = 625$. Read $\sqrt[4]{625}$ as "the fourth root of 625."

b. $\sqrt[4]{-1}$ is not a real number. An even root of a negative number is not a real number.

c. $\sqrt[5]{-32} = -2$, because $(-2)^5 = -32$. Read $\sqrt[5]{-32}$ as "the fifth root of -32."

d. $\sqrt[6]{\frac{1}{64}} = \frac{1}{2}$, because $\left(\frac{1}{2}\right)^6 = \frac{1}{64}$. Read $\sqrt[6]{\frac{1}{64}}$ as "the sixth root of $\frac{1}{64}$."

Caution

When n is even $(n > 1)$ and $x < 0$, $\sqrt[n]{x}$ is not a real number. For example, $\sqrt[4]{-81}$ is not a real number, because no real number raised to the fourth power is -81.

Self Check 8 Evaluate: **a.** $\sqrt[4]{\frac{1}{81}}$ **b.** $\sqrt[5]{10^5}$ **c.** $\sqrt[6]{-64}$

Now Try Problems 87, 91, and 93

Using Your **Calculator** ▶ Finding Roots

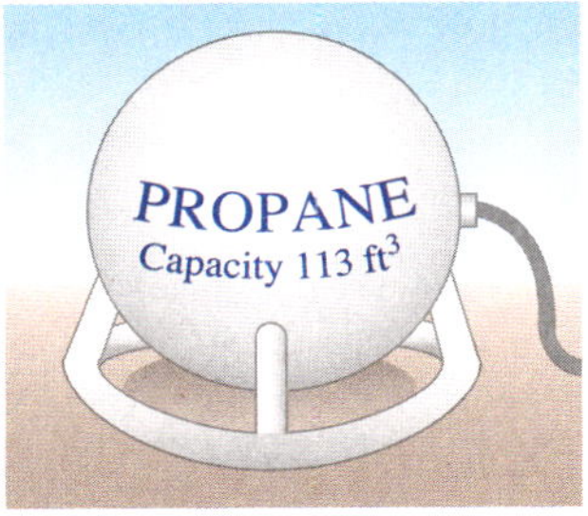

The square root key $\sqrt{\ }$ on a reverse entry scientific calculator can be used to evaluate square roots. To evaluate roots with an index greater than 2, we can use the root key $\sqrt[x]{y}$. For example, the function $r(V) = \sqrt[3]{\frac{3V}{4\pi}}$ gives the radius of a sphere with volume V. To find the radius of the spherical propane tank shown on the left, we substitute 113 for V to get:

$$r(113) = \sqrt[3]{\frac{3(113)}{4\pi}}$$

To evaluate a root, we enter the radicand and press the root key $\sqrt[x]{y}$ followed by the index of the radical, which in this case is 3.

3 [×] 113 [÷] [(] 4 [×] [π] [)] [=] [2nd] [$\sqrt[x]{y}$] 3 [=] `2.999139118`

To evaluate the cube root of $\frac{3(113)}{4\pi}$ using a direct entry calculator, we enter:

3 [2nd] [$\sqrt[x]{y}$] [(] 3 [×] 113 [÷] [(] 4 [×] [π] [)] [)] [ENTER]

To evaluate the cube root of $\frac{3(113)}{4\pi}$ with a graphing calculator, we enter:

[MATH] 4 [(] 3 [×] 113 [)] [÷] [(] 4 [×] [2nd] [π] [)] [)] [ENTER]

```
³√((3*113)/(4*π)
)
        2.999139118
```

If we round the result to the nearest foot, we see that the radius of the propane tank is about 3 feet.

EXAMPLE 9 Simplify each radical expression. Assume that x can be any real number.

a. $\sqrt[5]{x^5}$ **b.** $\sqrt[4]{16x^4}$ **c.** $\sqrt[6]{(x+4)^6}$ **d.** $\sqrt[4]{81x^8}$

Strategy When the index n is odd, we will determine what expression, when raised to the nth power, produces the radicand. When the index n is even, we will determine what positive expression, when raised to the nth power, produces the radicand.

Why This is the definition of nth root.

Solution

a. $\sqrt[5]{x^5} = x$ — Since n is odd, absolute value symbols aren't needed.

b. $\sqrt[4]{16x^4} = |2x| = 2|x|$ — Since n is even and x can be negative, absolute value symbols are needed to ensure that the result is positive.

c. $\sqrt[6]{(x+4)^6} = |x+4|$ — Absolute value symbols are needed to ensure that the result is positive.

d. $\sqrt[4]{81x^8} = 3x^2$ — Because $(3x^2)^4 = 81x^8$. Since $3x^2 \geq 0$ for any value of x, no absolute value symbols are needed.

The Language of Algebra

Another way to say that x can be any real number is to say that **the variable is unrestricted.**

Self Check 9 Simplify. Assume all variables are unrestricted.

a. $\sqrt[6]{x^6}$ **b.** $\sqrt[5]{(a+5)^5}$ **c.** $\sqrt[4]{16a^8}$

Now Try Problems 95, 97, 99, and 101

If we know that x is positive in parts (b) and (c) of Example 9, we don't need to use absolute value symbols. For example, if $x > 0$, then

$\sqrt[4]{16x^4} = 2x$ If x is positive, 2x is positive.

$\sqrt[6]{(x+4)^6} = x + 4$ If x is positive, x + 4 is positive.

We now summarize the definitions concerning nth roots.

Summary of the Definitions of $\sqrt[n]{x}$

If n is a natural number greater than 1 and x is a real number,

If $x > 0$, then $\sqrt[n]{x}$ is the positive number such that $\left(\sqrt[n]{x}\right)^n = x$.

If $x = 0$, then $\sqrt[n]{x} = 0$.

If $x < 0$ $\begin{cases}\text{and } n \text{ is odd, then } \sqrt[n]{x} \text{ is the negative number such that } \left(\sqrt[n]{x}\right)^n = x.\\ \text{and } n \text{ is even, then } \sqrt[n]{x} \text{ is not a real number.}\end{cases}$

SECTION 9.1 STUDY SET

VOCABULARY

Fill in the blanks.

1. $5x^2$ is the ______ root of $25x^4$ because $(5x^2)^2 = 25x^4$. The ______ root of 216 is 6 because $6^3 = 216$.
2. The symbol $\sqrt{}$ is called a ______ symbol or a ______ root symbol.
3. A radical symbol $\sqrt{}$ represents the ______ or principal square root of a number.
4. The number 4 has two square roots, −2 and 2. When we speak of the square root of 4, we mean only the ______ square root of 4, which is 2.
5. The number 100 has two square roots. The positive or ______ square root of 100 is 10.
6. In the expression $\sqrt[3]{27x^6}$, the ______ is 3 and $27x^6$ is the ______.
7. When we write $\sqrt{b^4} = b^2$, we say that we have ______ the radical expression.
8. When n is an odd number, $\sqrt[n]{x}$ represents an ______ root. When n is an ______ number, $\sqrt[n]{x}$ represents an even root.
9. $f(x) = \sqrt{x}$ and $g(x) = \sqrt[3]{x}$ are ______ functions.
10. Together, 0 and the positive real numbers are called the ______ real numbers.

CONCEPTS

Fill in the blanks.

11. b is a square root of a if $b^2 =$ ______.
12. $\sqrt{0} =$ ______ and $\sqrt[3]{0} =$ ______.
13. $\sqrt{-4}$ is not a real number, because no real number ______ equals −4.
14. $\sqrt[3]{x} = y$ if $y^3 =$ ______.
15. $\sqrt{x^2} =$ ______ and $\sqrt[3]{x^3} =$ ______
16. a. The graph of $g(x) = \sqrt{x} + 3$ is the graph of $f(x) = \sqrt{x}$ translated ______ units ______.
    b. The graph of $g(x) = \sqrt{x+5}$ is the graph of $f(x) = \sqrt{x}$ translated ______ units to the ______.
17. The graph of a square root function f is shown below. Find each of the following, if possible.
    a. $f(11)$
    b. $f(2)$
    c. $f(-1)$
    d. The value of x for which $f(x) = 2$
    e. The value of x for which $f(x) = -1$
    f. The domain and range of f (Use interval notation.)

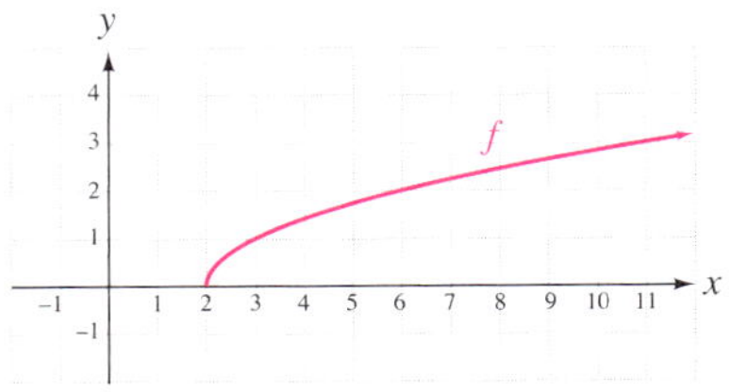

18. Refer to the graph in problem 17. Estimate each of the following function values.
    a. $f(4)$
    b. $f(10)$

19. The graph of a cube root function f is shown below. Find each of the following.

a. $f(-8)$

b. $f(0)$

c. The value of x for which $f(x) = -2$

d. The domain and range of f (Use interval notation.)

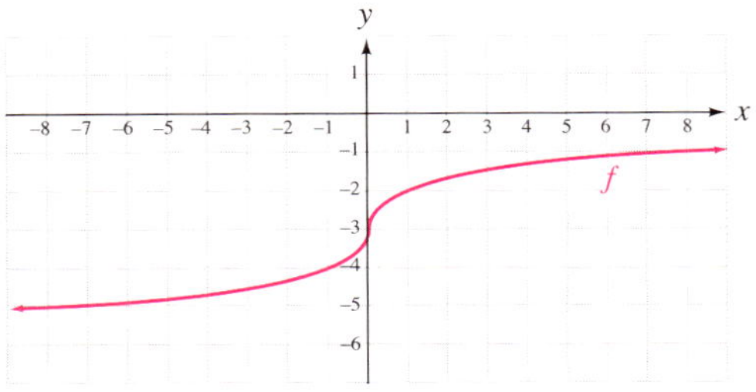

20. Match each function with the correct graph shown below.

a. $f(x) = x - 1$

b. $f(x) = \sqrt{x} - 1$

c. $f(x) = x^2 - 1$

d. $f(x) = \sqrt[3]{x} - 1$

e. $f(x) = |x| - 1$

f. $f(x) = \dfrac{1}{x - 1}$

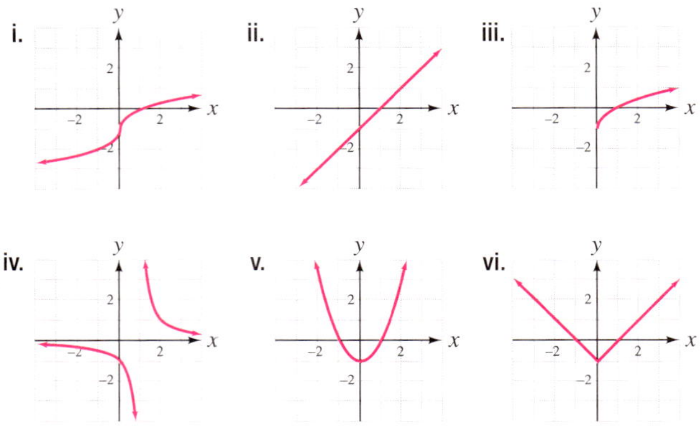

NOTATION

Translate each sentence into mathematical symbols.

21. a. The square root of x squared is the absolute value of x.

b. The cube root of x cubed is x.

c. The fifth root of negative thirty-two is negative two.

22. a. f of x equals the square root of the quantity x minus five.

b. g of x equals the cube root of x squared.

GUIDED PRACTICE

Evaluate each square root without using a calculator. See Objective 1 and Example 1.

23. $\sqrt{100}$

24. $\sqrt{49}$

25. $-\sqrt{64}$

26. $-\sqrt{1}$

27. $\sqrt{\dfrac{1}{9}}$

28. $\sqrt{\dfrac{4}{25}}$

29. $\sqrt{0.25}$

30. $\sqrt{0.16}$

31. $\sqrt{-81}$

32. $-\sqrt{-49}$

33. $\sqrt{121}$

34. $\sqrt{144}$

Use a calculator to find each square root. Give each answer to four decimal places. See Objective 1.

35. $\sqrt{12}$

36. $\sqrt{340}$

37. $\sqrt{679.25}$

38. $\sqrt{0.0063}$

Simplify each expression. Assume that all variables are unrestricted and use absolute value symbols when necessary. See Example 2.

39. $\sqrt{4x^2}$

40. $\sqrt{64t^2}$

41. $\sqrt{81h^4}$

42. $\sqrt{36y^4}$

43. $\sqrt{36s^6}$

44. $\sqrt{9y^6}$

45. $\sqrt{144m^8}$

46. $\sqrt{4n^8}$

47. $\sqrt{y^2 - 2y + 1}$

48. $\sqrt{b^2 - 14b + 49}$

49. $\sqrt{a^4 + 6a^2 + 9}$

50. $\sqrt{x^4 + 10x^2 + 25}$

Complete each table and then graph the function. Give the domain and range. See Examples 3 and 4.

51. $f(x) = -\sqrt{x}$

x	y
0	
1	
4	
9	
16	

52. $f(x) = \sqrt{x} + 2$

x	y
0	
1	
4	
9	
16	

Graph each function. Give the domain and range. See Examples 3 and 4.

53. $f(x) = \sqrt{x + 4}$

54. $f(x) = \sqrt{x} - 1$

Find the domain of each function. See Example 4.

55. $f(x) = \sqrt{x + 6}$

56. $g(x) = \sqrt{x + 12}$

57. $g(x) = \sqrt{8 - 2x}$

58. $h(x) = \sqrt{35 - 5x}$

59. $s(t) = \sqrt{9t - 4}$

60. $T(a) = \sqrt{3a + 17}$

61. $c(x) = \sqrt{0.5x - 20}$

62. $H(b) = \sqrt{0.4b - 36}$

Find each function value, if possible. Do not use a calculator. See Example 5.

63. $f(x) = \sqrt{3x + 1}$

a. $f(8)$

b. $f(-2)$

64. $h(t) = \sqrt{t^2 + t - 3}$

a. $h(-4)$

b. $h(-1)$

65. $g(x) = \sqrt[3]{x - 4}$

a. $g(12)$

b. $g(-23)$

66. $s(a) = -\sqrt[3]{32s}$

a. $s(-2)$

b. $s(2)$

Use a calculator to find each function value. Round to the nearest ten-thousandth. See Example 5 and Using Your Calculator.

67. $f(x) = \sqrt{x^2 + 1}$
    a. $f(4)$ b. $f(2.35)$
68. $g(x) = \sqrt{7 - 4x}$
    a. $g(-\pi)$ b. $g(0.5)$
69. $g(x) = \sqrt[3]{x^2 + 1}$
    a. $g(6)$ b. $g(21.57)$
70. $h(t) = \sqrt[3]{2.1t + 11}$
    a. $h(-0.4)$ b. $h(15)$

Simplify each cube root. See Example 6.

71. $\sqrt[3]{1}$
72. $\sqrt[3]{8}$
73. $\sqrt[3]{-125}$
74. $\sqrt[3]{-27}$
75. $\sqrt[3]{\frac{8}{27}}$
76. $\sqrt[3]{\frac{125}{64}}$
77. $\sqrt[3]{64}$
78. $\sqrt[3]{1{,}000}$
79. $\sqrt[3]{-216a^3}$
80. $\sqrt[3]{-512x^3}$
81. $\sqrt[3]{-1{,}000p^6q^3}$
82. $\sqrt[3]{-343a^6b^3}$

Complete each table and then graph the function. Give the domain and range. See Example 7.

83. $f(x) = \sqrt[3]{x} - 3$

x	y
−8	
−1	
0	
1	
8	

84. $f(x) = -\sqrt[3]{x}$

x	y
−8	
−1	
0	
1	
8	

Graph each function. Give the domain and range. See Example 7.

85. $f(x) = \sqrt[3]{x - 3}$
86. $f(x) = \sqrt[3]{x} + 3$

Evaluate each radical expression, if possible, without using a calculator. See Example 8.

87. $\sqrt[4]{81}$
88. $\sqrt[6]{64}$
89. $-\sqrt[5]{243}$
90. $-\sqrt[4]{625}$
91. $\sqrt[6]{-256}$
92. $\sqrt[6]{-729}$
93. $\sqrt[5]{-\frac{1}{32}}$
94. $\sqrt[5]{-\frac{243}{32}}$

Simplify each radical expression. Assume all variables are unrestricted. See Example 9.

95. $\sqrt[5]{32a^5}$
96. $\sqrt[5]{-32x^5}$
97. $\sqrt[4]{81a^4}$
98. $\sqrt[8]{t^8}$
99. $\sqrt[6]{k^{12}}$
100. $\sqrt[6]{64b^6}$
101. $\sqrt[4]{(m + 4)^8}$
102. $\sqrt[4]{(x - 7)^8}$

TRY IT YOURSELF

Simplify each radical expression, if possible. Assume all variables are unrestricted.

103. $\sqrt[3]{64s^9t^6}$
104. $\sqrt[3]{1{,}000a^6b^6}$
105. $-\sqrt{49b^8}$
106. $-\sqrt{144t^4}$
107. $-\sqrt[5]{-\frac{1}{32}}$
108. $-\sqrt[5]{-243}$
109. $\sqrt[3]{-125m^6}$
110. $\sqrt[3]{-216z^9}$
111. $\sqrt{400m^{16}n^2}$
112. $\sqrt{169p^4q^2}$
113. $\sqrt[6]{64a^6b^6}$
114. $\sqrt[6]{(x + 4)^6}$
115. $\sqrt[4]{-81}$
116. $\sqrt[6]{-1}$
117. $\sqrt{n^2 + 12n + 36}$
118. $\sqrt{s^2 - 20s + 100}$

Look Alikes . . .

119. a. $\sqrt{64}$ b. $\sqrt[3]{64}$
120. a. $\sqrt{-64}$ b. $\sqrt[3]{-64}$
121. a. $\sqrt{81}$ b. $\sqrt[4]{81}$
122. a. $\sqrt{16}$ b. $\sqrt[4]{16}$

APPLICATIONS

Use a calculator to solve each problem. Round answers to the nearest tenth.

123. **Embroidery.** The radius r of a circle is given by the formula $r = \sqrt{\frac{A}{\pi}}$, where A is its area. Find the diameter of the embroidery hoop if there are 38.5 in.^{2} of stretched fabric on which to embroider.

124. **Pendulums.** Find the period of a pendulum with length 1 foot. *See Example 5.*

125. **Shoelaces.** The formula $S = 2\left[H + L + (p - 1)\sqrt{H^2 + V^2}\right]$ can be used to calculate the correct shoelace length for the criss-cross lacing pattern shown in the illustration, where p represents the number of *pairs* of eyelets. Find the correct shoelace length if H(horizontal distance) = 50 mm, L (length of end) = 250 mm, and V(vertical distance) = 20 mm. Round to the nearest tenth. (*Source:* Ian's Shoelace Site at www.fieggen.com)

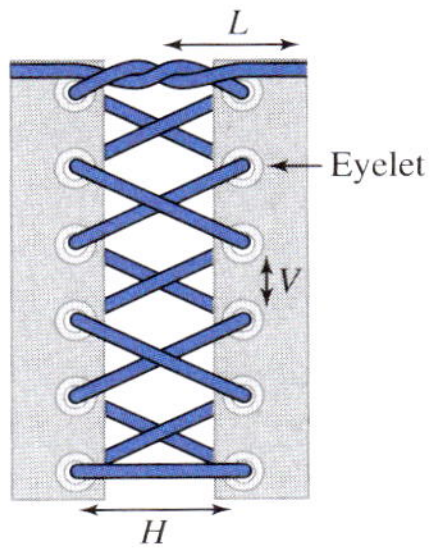

126. Baseball. The length of a diagonal of a square is given by the function $d(s) = \sqrt{2s^2}$, where s is the length of a side of the square. Find the distance from home plate to second base on a softball diamond and on a baseball diamond. Round to the nearest tenth. The illustration gives the dimensions of each type of infield.

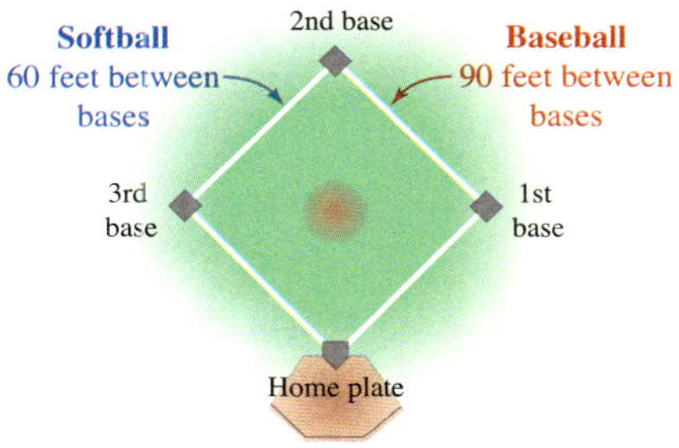

127. Pulse Rates. The approximate pulse rate (in beats per minute) of an adult who is t inches tall is given by the function $p(t) = \frac{590}{\sqrt{t}}$. The *Guinness Book of World Records 2008* lists Leonid Stadnyk of Ukraine as the tallest living man, at 8 feet, 5.5 inches. Find his approximate pulse rate as predicted by the function.

128. The Grand Canyon. The time t (in seconds) that it takes for an object to fall a distance of s feet is given by the formula $t = \frac{\sqrt{s}}{4}$. In some places, the Grand Canyon is one mile (5,280 feet) deep. How long would it take a stone dropped over the edge of the canyon to hit bottom?

129. Biology. Scientists will place five rats inside a clear plastic hemisphere and control the environment to study the rats' behavior. The function $d(V) = \sqrt[3]{12\left(\frac{V}{\pi}\right)}$ gives the diameter of a hemisphere with volume V. Use the function to determine the diameter of the base of the hemisphere, if each rat requires 125 cubic feet of living space.

130. Aquariums. The function $s(g) = \sqrt[3]{\frac{g}{7.5}}$ determines how long (in feet) an edge of a cube-shaped tank must be if it is to hold g gallons of water. What dimensions should a cube-shaped aquarium have if it is to hold 1,250 gallons of water?

131. Collectibles. The *effective rate of interest* r earned by an investment is given by the formula $r = \sqrt[n]{\frac{A}{P}} - 1$, where P is the initial investment that grows to value A after n years. Determine the effective rate of interest earned by a collector on a Lladró porcelain figurine purchased for \$800 and sold for \$950 five years later.

132. Law Enforcement. The graphs of the two radical functions shown in the illustration in the next column can be used to estimate the speed (in mph) of a car involved in an accident. Suppose a police accident report listed skid marks to be 220 feet long but failed to give the road conditions. Estimate the possible speeds the car was traveling prior to the brakes being applied.

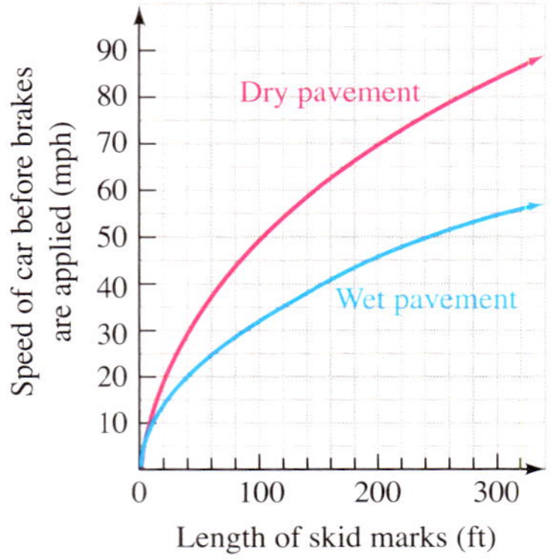

WRITING

133. If x is any real number, that is, if x is unrestricted, then $\sqrt{x^2} = x$ is not correct. Explain why.

134. Explain why $\sqrt{36}$ is just 6, and not also -6.

135. Explain what is wrong with the graph in the illustration if it is supposed to be the graph of $f(x) = \sqrt{x}$.

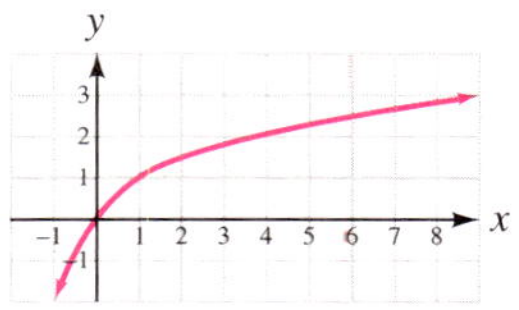

136. Explain how to estimate the domain and range of the radical function whose graph is shown here.

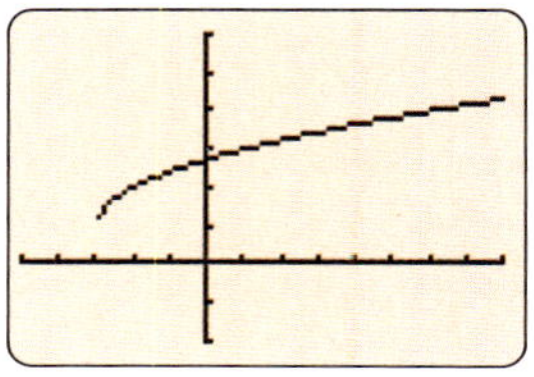

REVIEW

Perform the operations and simplify when possible.

137. $\frac{x^2 - 3xy - 4y^2}{x^2 + cx - 2yx - 2cy} \div \frac{x^2 - 2xy - 3y^2}{x^2 + cx - 4yx - 4cy}$

138. $\frac{2x + 3}{3x - 1} - \frac{x - 4}{2x + 1}$

CHALLENGE PROBLEMS

139. Graph $f(x) = -\sqrt{x - 2} + 3$ and find the domain and range.

140. Simplify $\sqrt{9a^{16} + 12a^8b^{25} + 4b^{50}}$ and assume that $a > 0$ and $b > 0$.

SECTION 9.2

Rational Exponents

OBJECTIVES

1. Simplify expressions of the form $a^{1/n}$.
2. Simplify expressions of the form $a^{m/n}$.
3. Convert between radicals and rational exponents.
4. Simplify expressions with negative rational exponents.
5. Use rules for exponents to simplify expressions.
6. Simplify radical expressions.

ARE YOU READY?

The following problems review some basic skills that are needed when working with rational exponents.

1. Evaluate: **a.** $\sqrt{64}$ **b.** $\sqrt[3]{-64}$
2. Evaluate: **a.** $\sqrt[4]{81}$ **b.** $\sqrt[5]{\frac{1}{32}}$
3. Simplify: **a.** $\sqrt{(x+5)^2}$ **b.** $\sqrt[3]{27a^6}$
4. Evaluate: $\left(\sqrt{36}\right)^3$
5. Simplify: 7^{-2}
6. Simplify: $\frac{x^5x^7}{x^3}$

In this section, we will extend the definition of exponent to include rational (fractional) exponents. We will see how expressions such as $9^{1/2}$, $\left(\frac{1}{16}\right)^{3/4}$, and $(-32x^5)^{-2/5}$ can be simplified by writing them in an equivalent radical form using two new rules for exponents.

1 Simplify Expressions of the Form $a^{1/n}$.

The Language of Algebra

Rational exponents are also called **fractional exponents.**

It is possible to raise numbers to fractional powers. To give meaning to rational exponents, we first consider $\sqrt{7}$. Because $\sqrt{7}$ is the positive number whose square is 7, we have

$$\left(\sqrt{7}\right)^2 = 7$$

We now consider the notation $7^{1/2}$, which is read as "7 to the one-half power." If rational exponents are to follow the same rules as integer exponents, the square of $7^{1/2}$ must be 7, because

$$(7^{1/2})^2 = 7^{1/2 \cdot 2} \quad \text{Keep the base and multiply the exponents.}$$
$$= 7^1 \quad \text{Do the multiplication: } \tfrac{1}{2} \cdot 2 = 1.$$
$$= 7$$

Since the square of $7^{1/2}$ and the square of $\sqrt{7}$ are both equal to 7, we define $7^{1/2}$ to be $\sqrt{7}$. Similarly,

$$7^{1/3} = \sqrt[3]{7}, \quad 7^{1/4} = \sqrt[4]{7}, \quad \text{and} \quad 7^{1/5} = \sqrt[5]{7}$$

In general, we have the following definition.

The Definition of $x^{1/n}$

A rational exponent of $\frac{1}{n}$ indicates the *n*th root of its base.

If n represents a positive integer greater than 1 and $\sqrt[n]{x}$ represents a real number,

$$x^{1/n} = \sqrt[n]{x} \quad \text{Read as "x to the } \tfrac{1}{n} \text{ power equals the nth root of x."}$$

We can use this definition to simplify exponential expressions that have rational exponents with a numerator of 1. For example, to simplify $8^{1/3}$, we write it as an equivalent expression in radical form and proceed as follows:

Index

$$8^{1/3} = \sqrt[3]{8} = 2$$

Radicand

The base of the exponential expression, 8, is the radicand of the radical expression. The denominator of the fractional exponent, 3, is the index of the radical.

Thus, $8^{1/3} = 2$.

EXAMPLE 1 Evaluate: **a.** $9^{1/2}$ **b.** $(-64)^{1/3}$ **c.** $16^{1/4}$ **d.** $-\left(\frac{1}{32}\right)^{1/5}$

Strategy First, we will identify the base and the exponent of the exponential expression. Then we will write the expression in an equivalent radical form using the rule for rational exponents $x^{1/n} = \sqrt[n]{x}$.

Why We can then use the methods from Section 7.1 to evaluate the resulting square root, cube root, fourth root, and fifth root.

Solution

a. $9^{1/2} = \sqrt{9} = 3$ The base is 9 and the exponent is $\frac{1}{2}$. Because the denominator of the exponent $\frac{1}{2}$ is 2, find the square root of the base, 9.

b. $(-64)^{1/3} = \sqrt[3]{-64} = -4$ Read as "-64 to the one-third power." Because the denominator of the exponent $\frac{1}{3}$ is 3, find the cube root of the base, -64.

c. $16^{1/4} = \sqrt[4]{16} = 2$ Because the denominator of the exponent $\frac{1}{4}$ is 4, find the fourth root of the base, 16.

d. $-\left(\frac{1}{32}\right)^{1/5} = -\sqrt[5]{\frac{1}{32}} = -\frac{1}{2}$ Read as "the opposite of the one-fifth power of $\frac{1}{32}$." Because the denominator of the exponent $\frac{1}{5}$ is 5, find the fifth root of the base, $\frac{1}{32}$.

Success Tip

One key to successfully simplifying the exponential expressions in this section is this: When you see an exponent, always ask yourself, "What is the base?"

Self Check 1 Evaluate: **a.** $16^{1/2}$ **b.** $\left(-\frac{27}{8}\right)^{1/3}$ **c.** $-(81)^{1/4}$ **d.** $1^{1/5}$

Now Try Problems 19, 23, 25, and 27

As with radicals, when n is an *odd natural number* in the expression $x^{1/n}$, where $n > 1$, there is exactly one real nth root, and we don't need to use absolute value symbols.

When n is an *even natural number*, there are two nth roots. Since we want the expression $x^{1/n}$ to represent the positive nth root, we often must use absolute value symbols to ensure that the simplified result is positive. Thus, if n is even,

$$(x^n)^{1/n} = |x|$$

When n is even and x is negative, the expression $x^{1/n}$ is not a real number.

EXAMPLE 2 Simplify. Assume that the variables are unrestricted. **a.** $(-27x^3)^{1/3}$ **b.** $(256a^8)^{1/8}$ **c.** $[(y + 4)^2]^{1/2}$ **d.** $(25b^4)^{1/2}$ **e.** $(-256)^{1/4}$

Strategy We will write each exponential expression in an equivalent radical form using the rule for rational exponents $x^{1/n} = \sqrt[n]{x}$.

Why We can then use the methods of Section 7.1 to simplify the resulting radical expression.

Solution

a. $(-27x^3)^{1/3} = \sqrt[3]{-27x^3} = -3x$ Read as "the quantity $-27x^3$ raised to the one-third power." Because $(-3x)^3 = -27x^3$. Since n is odd, no absolute value symbols are needed.

b. $(256a^8)^{1/8} = \sqrt[8]{256a^8} = 2|a|$ Because $(2a)^8 = 256a^8$. Since n is even and a can be any real number, $2a$ can be negative. Thus, absolute value symbols are needed.

c. $\left[(y + 4)^2\right]^{1/2} = \sqrt{(y + 4)^2} = |y + 4|$ Because $|y + 4|^2 = (y + 4)^2$. Since n is even and y can be any real number, $y + 4$ can be negative. Thus, absolute value symbols are needed.

The Language of Algebra

The phrase **"assume that the variables are unrestricted"** in the instructions for Example 2 reminds us that negative numbers are permissible replacement values for the variables. Therefore, absolute value symbols may be necessary to ensure the answer is not negative.

d. $(25b^4)^{1/2} = \sqrt{25b^4} = 5b^2$ Because $(5b^2)^2 = 25b^4$. Since $b^2 \geq 0$, no absolute value symbols are needed.

e. $(-256)^{1/4} = \sqrt[4]{-256}$, which is not a real number. Because no real number raised to the 4th power is -256.

Self Check 2 Simplify. Assume that the variables are unrestricted.
a. $(-8n^3)^{1/3}$ **b.** $(625a^4)^{1/4}$ **c.** $(b^4)^{1/2}$ **d.** $(-64)^{1/6}$

Now Try Problems 29, 35, 37, and 39

If we were told that the variables represent positive real numbers in parts (b) and (c) of Example 2, the absolute value symbols in the answers would not be needed.

$(256a^8)^{1/8} = 2a$ If a represents a positive real number, then $2a$ is positive.

$[(y + 4)^2]^{1/2} = y + 4$ If y represents a positive real number, then $y + 4$ is positive.

We summarize the cases as follows.

Summary of the Definitions of $x^{1/n}$

If n is a natural number greater than 1 and x is a real number,

If $x > 0$, then $x^{1/n}$ is the real number such that $(x^{1/n})^n = x$.

If $x = 0$, then $x^{1/n} = 0$.

If $x < 0$ $\begin{cases} \text{and } n \text{ is odd, then } x^{1/n} \text{ is the negative number such that } (x^{1/n})^n = x. \\ \text{and } n \text{ is even, then } x^{1/n} \text{ is not a real number.} \end{cases}$

2 Simplify Expressions of the Form $a^{m/n}$.

We can extend the definition of $x^{1/n}$ to include fractional exponents with numerators other than 1. For example, since $8^{2/3}$ can be written as $(8^{1/3})^2$, we have

$8^{2/3} = (8^{1/3})^2$ Read $8^{2/3}$ as "8 to the two-thirds power."

$= (\sqrt[3]{8})^2$ Write $8^{1/3}$ in radical form.

$= 2^2$ Find the cube root first: $\sqrt[3]{8} = 2$.

$= 4$ Then find the power.

Thus, we can simplify $8^{2/3}$ by finding the second power of the cube root of 8.

The numerator of the rational exponent is the power.

$$8^{2/3} = (\sqrt[3]{8})^2$$

The base of the exponential expression is the radicand.

The denominator of the rational exponent is the index of the radical.

We also can simplify $8^{2/3}$ by taking the cube root of 8 squared.

$8^{2/3} = (8^2)^{1/3}$

$= 64^{1/3}$ Find the power first: $8^2 = 64$.

$= \sqrt[3]{64}$ Write $64^{1/3}$ in radical form.

$= 4$ Now find the cube root.

In general, we have the following definition.

The Definition of $x^{m/n}$

If m and n represent positive integers ($n \neq 1$) and $\sqrt[n]{x}$ represents a real number,

$$x^{m/n} = \left(\sqrt[n]{x}\right)^m \quad \text{and} \quad x^{m/n} = \sqrt[n]{x^m}$$

We read the first definition given above as "x to the m divided by n power equals the nth root of x, raised to the mth power."

Because of the previous definition, we can interpret $x^{m/n}$ in two ways:

1. $x^{m/n}$ means the nth root of the mth power of x.
2. $x^{m/n}$ means the mth power of the nth root of x.

We can use this definition to evaluate exponential expressions that have rational exponents with a numerator that is not 1. To avoid large numbers, we usually find the root of the base first and then calculate the power using the rule $x^{m/n} = \left(\sqrt[n]{x}\right)^m$.

EXAMPLE 3 Evaluate: **a.** $32^{2/5}$ **b.** $81^{3/4}$ **c.** $(-64)^{2/3}$ **d.** $-\left(\frac{1}{25}\right)^{3/2}$

Strategy First, we will identify the base and the exponent of the exponential expression. Then we will write the expression in an equivalent radical form using the rule for rational exponents $x^{m/n} = \left(\sqrt[n]{x}\right)^m$.

Why We know how to evaluate square roots, cube roots, fourth roots, and fifth roots.

Solution

a. To evaluate $32^{2/5}$, we write it in an equivalent radical form. The denominator of the rational exponent is the same as the index of the corresponding radical. The numerator of the rational exponent indicates the power to which the radical base is raised.

Power / Root

$$32^{2/5} = \left(\sqrt[5]{32}\right)^2 = (2)^2 = 4$$

Read as "32 to the two-fifths power." Because the exponent is 2/5, find the fifth root of the base, 32, to get 2. Then find the second power of 2.

Power / Root

b. $81^{3/4} = \left(\sqrt[4]{81}\right)^3 = (3)^3 = 27$

Read as "81 to the three-fourths power." Because the exponent is 3/4, find the fourth root of the base, 81, to get 3. Then find the third power of 3.

c. For $(-64)^{2/3}$, the base is -64.

Power / Root

$$(-64)^{2/3} = (\sqrt[3]{-64})^2 = (-4)^2 = 16$$

Read as "-64 to the two-thirds power." Because the exponent is 2/3, find the cube root of the base, -64, to get -4. Then find the second power of -4.

d. For $-\left(\frac{1}{25}\right)^{3/2}$, the base is $\frac{1}{25}$, not $-\frac{1}{25}$.

Power / Root

$$-\left(\frac{1}{25}\right)^{3/2} = -\left(\sqrt[2]{\frac{1}{25}}\right)^3 = -\left(\frac{1}{5}\right)^3 = -\frac{1}{125}$$

Read as "the opposite of the three-halves power of $\frac{1}{25}$." Because the exponent is 3/2, find the square root of the base, $\frac{1}{25}$, to get $\frac{1}{5}$. Then find the third power of $\frac{1}{5}$.

Caution

We also can evaluate $x^{m/n}$ using $\sqrt[n]{x^m}$. However, the resulting radicand is often extremely large. For example,

$$\begin{aligned} 81^{3/4} &= \sqrt[4]{81^3} \\ &= \sqrt[4]{531{,}441} \\ &= 27 \end{aligned}$$

Self Check 3 Evaluate: **a.** $16^{3/2}$ **b.** $125^{4/3}$ **c.** $(-216)^{2/3}$
d. $-\left(\frac{1}{32}\right)^{4/5}$

Now Try ▶ Problems 41, 45, and 47

EXAMPLE 4 Simplify. All variables represent positive real numbers. **a.** $(36m^4)^{3/2}$ **b.** $(-8x^3)^{4/3}$ **c.** $-(x^5y^5)^{2/5}$

Strategy We will write each exponential expression in an equivalent radical form using the rule for rational exponents $x^{m/n} = \left(\sqrt[n]{x}\right)^m$.

Why We can then use the methods of Section 9.1 to simplify the resulting radical expression.

Solution

Power
Root

a. $(36m^4)^{3/2} = \left(\sqrt[2]{36m^4}\right)^3 = (6m^2)^3 = 216m^6$

Read as "the quantity of $36m^4$ raised to the three-halves power."
Because the exponent is 3/2, find the square root of the base, $36m^4$, to get $6m^2$. Then find the third power of $6m^2$.

Power
Root

b. $(-8x^3)^{4/3} = \left(\sqrt[3]{-8x^3}\right)^4 = (-2x)^4 = 16x^4$

Read as "the quantity of $-8x^3$ raised to the four-thirds power."
Because the exponent is 4/3, find the cube root of the base, $-8x^3$, to get $-2x$. Then find the fourth power of $-2x$.

Power
Root

c. $-(x^5y^5)^{2/5} = -\left(\sqrt[5]{x^5y^5}\right)^2 = -(xy)^2 = -x^2y^2$

Read as "the opposite of the two-fifths power of the quantity x^5y^5."
Because the exponent is 2/5, find the fifth root of the base, x^5y^5, to get xy. Then find the second power of xy.

Self Check 4 Simplify. All variables represent positive real numbers. **a.** $(4c^4)^{3/2}$
b. $(-27m^3n^3)^{2/3}$ **c.** $-(32a^{10})^{3/5}$

Now Try ▶ Problems 49 and 51

Using Your Calculator ▶ Rational Exponents

We can evaluate expressions containing rational exponents using the exponential key $\boxed{y^x}$ or $\boxed{x^y}$ on a scientific calculator. For example, to evaluate $10^{2/3}$, we enter

10 $\boxed{y^x}$ $\boxed{(}$ 2 $\boxed{\div}$ 3 $\boxed{)}$ $\boxed{=}$ 4.641588834

Note that parentheses were used when entering the power. Without them, the calculator would interpret the entry as $10^2 \div 3$.

To evaluate the exponential expression using a direct entry or graphing calculator, we use the $\boxed{\wedge}$ key, which raises a base to a power. Again, we use parentheses when entering the power.

10 $\boxed{\wedge}$ $\boxed{(}$ 2 $\boxed{\div}$ 3 $\boxed{)}$ $\boxed{\text{ENTER}}$

```
10^(2/3)
     4.641588834
```

To the nearest hundredth, $10^{2/3} \approx 4.64$.

3 Convert Between Radicals and Rational Exponents.

We can use the rules for rational exponents to convert expressions from radical form to exponential form, and vice versa.

EXAMPLE 5 Write $\sqrt{5xyz}$ as an exponential expression with a rational exponent.

Strategy We will use the first rule for rational exponents in reverse: $\sqrt[n]{x} = x^{1/n}$.

Why We are given a radical expression and we want to write an equivalent exponential expression.

Solution The radicand is $5xyz$, so the base of the exponential expression is $5xyz$. The index of the radical is an understood 2, so the denominator of the fractional exponent is 2.

$$\sqrt{5xyz} = (5xyz)^{1/2}$$ Recall: $\sqrt[2]{5xyz} = \sqrt{5xyz}$.

Self Check 5 Write $\sqrt[6]{7ab}$ as an exponential expression with a rational exponent.

Now Try Problem 57

Rational exponents appear in formulas used in many disciplines, such as science and engineering.

EXAMPLE 6 **Satellites.** The formula $r = \left(\frac{GMP^2}{4\pi^2}\right)^{1/3}$ gives the orbital radius (in meters) of a satellite circling Earth, where G and M are constants and P is the time in seconds for the satellite to make one complete revolution. Write the formula using a radical.

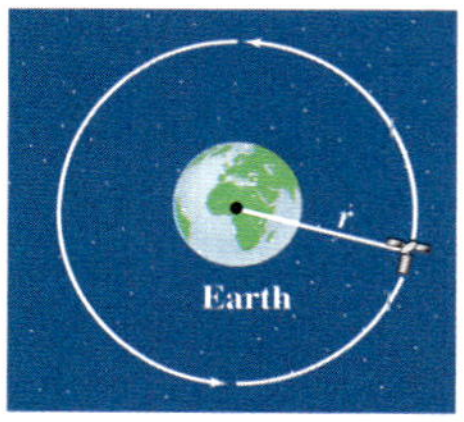

Strategy We will use the first rule for rational exponents: $x^{1/n} = \sqrt[n]{x}$.

Why We are given an exponential expression involving a rational exponent with a numerator of 1 and we want to write an equivalent radical expression.

Solution The fractional exponent $\frac{1}{3}$, with a numerator of 1 and a denominator of 3, indicates that we are to find the cube root of the base of the exponential expression. So we have

$$r = \sqrt[3]{\frac{GMP^2}{4\pi^2}}$$

Self Check 6 **Statistics.** The formula $\sigma = \left(\frac{\Sigma(x - \mu)^2}{N}\right)^{1/2}$ gives the population standard deviation. Write the formula using a radical.

Now Try Problem 61

4 Simplify Expressions with Negative Rational Exponents.

To be consistent with the definition of negative integer exponents, we define $x^{-m/n}$ as follows.

Definition of $x^{-m/n}$

If m and n are positive integers, $\frac{m}{n}$ is in simplified form, and $x^{1/n}$ is a real number, then

$$x^{-m/n} = \frac{1}{x^{m/n}} \quad \text{and} \quad \frac{1}{x^{-m/n}} = x^{m/n} \quad (x \neq 0)$$

From the definition, we see that another way to write $x^{-m/n}$ is to write its reciprocal and change the sign of the exponent.

EXAMPLE 7 Simplify. Assume that x can represent any nonzero real number.

a. $64^{-1/2}$ **b.** $(-16)^{-5/4}$ **c.** $-625^{-3/4}$ **d.** $(-32x^5)^{-2/5}$ **e.** $\frac{1}{25^{-3/2}}$

Strategy We will use one of the rules $x^{-m/n} = \frac{1}{x^{m/n}}$ or $\frac{1}{x^{-m/n}} = x^{m/n}$ to write the reciprocal of each exponential expression and change the exponent's sign to positive.

Why If we can produce an equivalent expression having a positive rational exponent, we can use the methods of this section to simplify it.

Solution

a. $64^{-1/2} = \frac{1}{64^{1/2}} = \frac{1}{\sqrt{64}} = \frac{1}{8}$

Reciprocal

Change sign

Read as "64 to the negative one-half power." Because the exponent is negative, write the reciprocal of $64^{-1/2}$, and change the sign of the exponent.

b. $(-16)^{-5/4}$ is read as "-16 to the negative five-fourths power." It is not a real number because $(-16)^{1/4}$ is not a real number.

c. In $-625^{-3/4}$, the base is 625.

$$-625^{-3/4} = -\frac{1}{625^{3/4}} = -\frac{1}{\left(\sqrt[4]{625}\right)^3} = -\frac{1}{(5)^3} = -\frac{1}{125}$$

d. This is read as "the quantity of $-32x^5$, raised to the negative two-fifths power."

$$(-32x^5)^{-2/5} = \frac{1}{(-32x^5)^{2/5}} = \frac{1}{\left(\sqrt[5]{-32x^5}\right)^2} = \frac{1}{(-2x)^2} = \frac{1}{4x^2}$$

e. This is read as "1 over 25 to the negative three-halves power."

$$\frac{1}{25^{-3/2}} = 25^{3/2} = \left(\sqrt{25}\right)^3 = (5)^3 = 125$$

Because the exponent is negative, write the reciprocal of $\frac{1}{25^{-3/2}}$, and change the sign of the exponent.

Caution

A negative exponent does not, itself, indicate a negative number. For example,

$$64^{-1/2} = \frac{1}{8}$$

Caution

A base of 0 raised to a negative power is undefined. For example, $0^{-2} = \frac{1}{0^2}$ is undefined because we cannot divide by 0.

Self Check 7 Simplify. Assume that a can represent any nonzero real number.
a. $9^{-1/2}$ **b.** $(36)^{-3/2}$ **c.** $(-27a^3)^{-2/3}$ **d.** $-\frac{1}{81^{-3/4}}$

Now Try Problems 65, 69, and 75

5 Use Rules for Exponents to Simplify Expressions.

We can use the rules for exponents to simplify many expressions with fractional exponents. If all variables represent positive real numbers, absolute value symbols are not needed.

EXAMPLE 8 Simplify. All variables represent positive real numbers. Write all answers using positive exponents only. **a.** $5^{2/7} \cdot 5^{3/7}$ **b.** $(11^{2/7})^3$ **c.** $(a^{2/3}b^{1/2})^6$ **d.** $\frac{a^{8/3}a^{1/3}}{a^2}$

Strategy We will use the product, power, and quotient rules for exponents to simplify each expression.

Why The familiar rules for exponents discussed in Chapter 5 are valid for rational exponents.

Solution

a. $5^{2/7} \cdot 5^{3/7} = 5^{2/7+3/7}$ Use the rule $x^m x^n = x^{m+n}$. Do not multiply the bases.

$= 5^{5/7}$ Add the exponents: $\frac{2}{7} + \frac{3}{7} = \frac{5}{7}$.

b. $(11^{2/7})^3 = 11^{(2/7)(3)}$ Use the rule $(x^m)^n = x^{mn}$.

$= 11^{6/7}$ Multiply the exponents: $\frac{2}{7}(3) = \frac{6}{7}$.

c. $(a^{2/3}b^{1/2})^6 = (a^{2/3})^6(b^{1/2})^6$ Use the rule $(xy)^n = x^n y^n$.

$= a^{12/3}b^{6/2}$ Use the rule $(x^m)^n = x^{mn}$ twice. Multiply the exponents.

$= a^4 b^3$ Simplify the exponents.

d. $\dfrac{a^{8/3}a^{1/3}}{a^2} = a^{8/3+1/3-2}$ Use the rules $x^m x^n = x^{m+n}$ and $\frac{x^m}{x^n} = x^{m-n}$.

$= a^{8/3+1/3-6/3}$ To establish an LCD, write -2 as $-\frac{6}{3}$.

$= a^{3/3}$ Simplify: $\frac{8}{3} + \frac{1}{3} - \frac{6}{3} = \frac{3}{3}$.

$= a$ Simplify: $\frac{3}{3} = 1$.

Success Tip

Now is a good time to review the rules for exponents on page 364 in Chapter 5. These rules are not only valid when m and n are integers, but when m and n are rational numbers, as well.

Self Check 8 Simplify. All variables represent positive real numbers.

a. $2^{1/5} \cdot 2^{2/5}$ **b.** $(12^{1/3})^4$ **c.** $(x^{1/3}y^{3/2})^6$ **d.** $\dfrac{x^{5/3}x^{2/3}}{x^{1/3}}$

Now Try Problems 77, 81, and 85

EXAMPLE 9 Perform each multiplication and simplify when possible. Assume all variables represent positive real numbers. Write all answers using positive exponents only.

a. $a^{4/5}(a^{1/5} + a^{3/5})$ **b.** $x^{1/2}(x^{-1/2} - x^{1/2})$

Strategy We will use the distributive property and multiply each term within the parentheses by the term outside the parentheses.

Why The first expression has the form $a(b + c)$ and the second has the form $a(b - c)$.

Solution

a. $a^{4/5}(a^{1/5} + a^{3/5}) = a^{4/5}a^{1/5} + a^{4/5}a^{3/5}$ Use the distributive property.

$= a^{4/5+1/5} + a^{4/5+3/5}$ Use the rule $x^m x^n = x^{m+n}$.

$= a^{5/5} + a^{7/5}$ Add the exponents.

$= a + a^{7/5}$ We cannot add these terms because they are not like terms.

b. $x^{1/2}(x^{-1/2} - x^{1/2}) = x^{1/2}x^{-1/2} - x^{1/2}x^{1/2}$ Use the distributive property.

$= x^{1/2+(-1/2)} - x^{1/2+1/2}$ Use the rule $x^m x^n = x^{m+n}$.

$= x^0 - x^1$ Add the exponents.

$= 1 - x$ Simplify: $x^0 = 1$.

Caution

A common error when distributing $a^{4/5}$ over the terms $a^{1/5}$ and $a^{3/5}$ is to multiply the exponents incorrectly:

$a^{4/5}(a^{1/5} + a^{3/5}) = a^{4/25} + a^{12/25}$

Self Check 9 Simplify: $t^{5/8}(t^{3/8} - t^{-5/8})$. Assume t represents a positive real number.

Now Try Problem 91

6 Simplify Radical Expressions.

We can simplify many radical expressions by using the following steps.

Using Rational Exponents to Simplify Radicals

1. Change the radical expression into an exponential expression.
2. Simplify the rational exponents.
3. Change the exponential expression back into a radical.

EXAMPLE 10 Simplify: **a.** $\sqrt[4]{3^2}$ **b.** $\sqrt[8]{x^6}$ **c.** $\sqrt[9]{27x^6y^3}$ **d.** $\sqrt[5]{\sqrt[3]{t}}$

Strategy We will write each radical expression as an equivalent exponential expression and use rules for exponents to simplify it. Then we will change that result back into a radical.

Why When the given expression is written in an equivalent exponential form, we can use rules for exponents and our arithmetic skills with fractions to simplify the exponents.

Solution

a. $\sqrt[4]{3^2} = 3^{2/4}$ Use the rule $\sqrt[n]{x^m} = x^{m/n}$.

$= 3^{1/2}$ Simplify the fractional exponent: $\frac{2}{4} = \frac{1}{2}$.

$= \sqrt{3}$ Change back to radical form.

b. $\sqrt[8]{x^6} = x^{6/8}$ Use the rule $\sqrt[n]{x^m} = x^{m/n}$.

$= x^{3/4}$ Simplify the fractional exponent: $\frac{6}{8} = \frac{3}{4}$.

$= (x^3)^{1/4}$ Write $\frac{3}{4}$ as $3\left(\frac{1}{4}\right)$.

$= \sqrt[4]{x^3}$ Change back to radical form.

c. $\sqrt[9]{27x^6y^3} = (3^3x^6y^3)^{1/9}$ Write 27 as 3^3 and change the radical to an exponential expression.

$= 3^{3/9}x^{6/9}y^{3/9}$ Raise each factor to the $\frac{1}{9}$ power by multiplying the fractional exponents.

$= 3^{1/3}x^{2/3}y^{1/3}$ Simplify each fractional exponent.

$= (3x^2y)^{1/3}$ Use the rule $(xy)^n = x^ny^n$.

$= \sqrt[3]{3x^2y}$ Change back to radical form.

d. $\sqrt[5]{\sqrt[3]{t}} = \sqrt[5]{t^{1/3}}$ Change the radical $\sqrt[3]{t}$ to exponential notation.

$= (t^{1/3})^{1/5}$ Change the radical $\sqrt[5]{t^{1/3}}$ to exponential notation.

$= t^{1/15}$ Use the rule $(x^m)^n = x^{mn}$. Multiply: $\frac{1}{3} \cdot \frac{1}{5} = \frac{1}{15}$.

$= \sqrt[15]{t}$ Change back to radical form.

Self Check 10 Simplify: **a.** $\sqrt[6]{3^3}$ **b.** $\sqrt[4]{49x^2y^2}$ **c.** $\sqrt[3]{\sqrt[4]{m}}$

Now Try Problems 93, 99, and 101

SECTION 9.2 STUDY SET

VOCABULARY

Fill in the blanks.

1. The expressions $4^{1/2}$ and $(-8)^{-2/3}$ have ________________ exponents.
2. We read $16^{3/2}$ as "16 to the three-_______ power."
3. We read $27^{-1/3}$ as "27 to the ________ one-third power."
4. We read $(-64a^5)^{4/5}$ as "the quantity of $-64a^5$, ______ to the four-fifths power."
5. In the radical expression $\sqrt[4]{16x^8}$, 4 is the ______, and $16x^8$ is the _________.
6. $32^{4/5}$ means the fourth ______ of the fifth _____ of 32.

CONCEPTS

7. Complete the table by writing the given expression in the alternate form. Also give the base and exponent for the exponential form.

Radical form	Exponential form	Base	Exponent
$\sqrt[5]{25}$			
	$(-27)^{2/3}$		
$(\sqrt[4]{16})^{-3}$			
	$81^{3/2}$		
$-\sqrt{\frac{9}{64}}$			

8. In your own words, explain the three rules for rational exponents illustrated in the diagrams below.

a. $(-32)^{1/5} = \sqrt[5]{-32}$

b. $125^{4/3} = (\sqrt[3]{125})^4$

c. $8^{-1/3} = \dfrac{1}{8^{1/3}}$

9. Graph the following real numbers on a number line.

$$\left\{8^{2/3},\ (-125)^{1/3},\ -16^{-1/4},\ 4^{3/2},\ -\left(\frac{9}{100}\right)^{-1/2}\right\}$$

10. a. Evaluate $25^{3/2}$ by writing it in the form $(25^{1/2})^3$.
b. Evaluate $25^{3/2}$ by writing it in the form $(25^3)^{1/2}$.
c. Which way was easier?

Complete each rule for exponents.

11. $x^{1/n} = \underline{\quad}$

12. $x^{m/n} = \underline{\quad} = \sqrt[n]{x^m}$

13. $x^{-m/n} = \underline{\quad}$

14. $\dfrac{1}{x^{-m/n}} = \underline{\quad}$

NOTATION

Complete each solution.

15. Simplify:

$$\begin{aligned}(100a^4)^{3/2} &= \left(\sqrt{\underline{\quad}}\right)^3\\ &= \left(\underline{\quad}\right)^3\\ &= 1{,}000a^6\end{aligned}$$

16. Simplify:

$$\begin{aligned}(m^{1/3}n^{1/2})^6 &= \left(\underline{\quad}\right)^6(n^{1/2})^6\\ &= m^{\underline{\quad}}\,n^{6/2}\\ &= m^2n^3\end{aligned}$$

GUIDED PRACTICE

Evaluate each expression. See Example 1.

17. $125^{1/3}$
18. $8^{1/3}$
19. $81^{1/4}$
20. $625^{1/4}$
21. $32^{1/5}$
22. $0^{1/5}$
23. $(-216)^{1/3}$
24. $(-1{,}000)^{1/3}$
25. $-16^{1/4}$
26. $-125^{1/3}$
27. $\left(\dfrac{1}{4}\right)^{1/2}$
28. $\left(\dfrac{1}{16}\right)^{1/2}$

Simplify each expression. Assume that the variables can be any real number, and use absolute value symbols when necessary. See Example 2.

29. $(4x^4)^{1/2}$
30. $(25a^8)^{1/2}$
31. $(x^2)^{1/2}$
32. $(x^3)^{1/3}$
33. $(-64p^8)^{1/2}$
34. $(-16q^4)^{1/2}$
35. $(-27n^9)^{1/3}$
36. $(-64t^9)^{1/3}$
37. $(-64x^8)^{1/8}$
38. $(243x^{10})^{1/5}$
39. $[(x+1)^6]^{1/6}$
40. $[(x+5)^8]^{1/8}$

Evaluate each expression. See Example 3.

41. $36^{3/2}$
42. $27^{2/3}$
43. $16^{3/4}$
44. $-100^{3/2}$
45. $\left(-\dfrac{1}{216}\right)^{2/3}$
46. $\left(\dfrac{4}{9}\right)^{3/2}$
47. $-4^{5/2}$
48. $(-125)^{4/3}$

Simplify each expression. All variables represent positive real numbers. See Example 4.

49. $(25x^4)^{3/2}$
50. $(27a^3b^3)^{2/3}$
51. $(-8x^6y^3)^{2/3}$
52. $(-32x^{10}y^5)^{4/5}$
53. $(81x^4y^8)^{3/4}$
54. $\left(\dfrac{1}{16}x^8y^4\right)^{3/4}$
55. $-\left(\dfrac{x^5}{32}\right)^{4/5}$
56. $-\left(\dfrac{27}{64y^6}\right)^{2/3}$

Change each radical to an exponential expression. See Example 5.

57. $\sqrt[5]{8abc}$
58. $\sqrt[7]{7p^2q}$
59. $\sqrt[3]{a^2-b^2}$
60. $\sqrt{x^2+y^2}$

Change each exponential expression to a radical. See Example 6.

61. $(6x^3y)^{1/4}$
62. $(7a^2b^2)^{1/5}$
63. $(2s^2-t^2)^{1/2}$
64. $(x^3+y^3)^{1/3}$

Simplify each expression. All variables represent positive real numbers. See Example 7.

65. $4^{-1/2}$
66. $49^{-1/2}$
67. $125^{-1/3}$
68. $8^{-1/3}$
69. $-(1{,}000y^3)^{-2/3}$
70. $-(81c^4)^{-3/2}$
71. $\left(-\dfrac{27}{8}\right)^{-4/3}$
72. $\left(\dfrac{25}{49}\right)^{-3/2}$
73. $\left(\dfrac{16}{81y^4}\right)^{-3/4}$
74. $\left(-\dfrac{8x^3}{27}\right)^{-1/3}$
75. $\dfrac{1}{9^{-5/2}}$
76. $\dfrac{1}{16^{-5/2}}$

Simplify each expression. Write the answers without negative exponents. All variables represent positive real numbers. See Example 8.

77. $9^{3/7}\cdot 9^{2/7}$
78. $4^{2/5}\cdot 4^{2/5}$
79. $6^{-2/3}6^{-4/3}$
80. $5^{1/3}5^{-5/3}$
81. $(m^{2/3}m^{1/3})^6$
82. $(b^{3/5}b^{2/5})^8$
83. $(a^{1/2}b^{1/3})^{3/2}$
84. $(mn^{-2/3})^{-3/5}$
85. $\dfrac{3^{4/3}3^{1/3}}{3^{2/3}}$
86. $\dfrac{2^{5/6}2^{1/3}}{2^{1/2}}$
87. $\dfrac{a^{3/4}a^{3/4}}{a^{1/2}}$
88. $\dfrac{b^{4/5}b^{4/5}}{b^{3/5}}$

Perform the multiplications. All variables represent positive real numbers. See Example 9.

89. $y^{1/3}(y^{2/3}+y^{5/3})$
90. $y^{2/5}(y^{-2/5}+y^{3/5})$
91. $x^{3/5}(x^{7/5}-x^{-3/5}+1)$
92. $x^{4/3}(x^{2/3}+3x^{5/3}-4)$

Use rational exponents to simplify each radical. All variables represent positive real numbers. See Example 10.

93. $\sqrt[4]{5^2}$
94. $\sqrt[6]{7^3}$
95. $\sqrt[9]{11^3}$
96. $\sqrt[12]{13^4}$
97. $\sqrt[6]{p^3}$
98. $\sqrt[8]{q^2}$

99. $\sqrt[10]{x^2y^2}$
100. $\sqrt[6]{x^2y^2}$
101. $\sqrt[9]{\sqrt{c}}$
102. $\sqrt[4]{\sqrt{x}}$
103. $\sqrt[5]{\sqrt[3]{7m}}$
104. $\sqrt[3]{\sqrt[4]{21x}}$

Use a calculator to evaluate each expression. Round to the nearest hundredth. See Using Your Calculator: Rational Exponents.

105. $15^{1/3}$
106. $(50.5)^{1/4}$
107. $(1.045)^{2/5}$
108. $(-1{,}000)^{3/5}$

TRY IT YOURSELF

Simplify each expression. All variables represent positive real numbers.

109. $(25y^2)^{1/2}$
110. $(-27x^3)^{1/3}$
111. $-\left(\frac{a^4}{81}\right)^{3/4}$
112. $-\left(\frac{b^8}{625}\right)^{3/4}$
113. $16^{-3/2}$
114. $(16)^{-5/4}$
115. $\frac{p^{8/5}p^{7/5}}{p^2}$
116. $\frac{c^{2/3}c^{2/3}}{c^{1/3}}$
117. $(-27x^6)^{-1/3}$
118. $(16a^4)^{-1/2}$
119. $\frac{1}{32^{-1/5}}$
120. $\frac{1}{64^{-1/6}}$
121. $n^{1/5}(n^{2/5} - n^{-1/5})$
122. $t^{4/3}(t^{5/3} + t^{-4/3})$
123. $\frac{1}{4^{-5/2}}$
124. $\frac{1}{100^{-5/2}}$
125. $(m^4)^{1/2}$
126. $(a^4)^{1/4}$
127. $\sqrt[4]{25b^2}$
128. $\sqrt[9]{8x^6}$
129. $(16x^4)^{1/4}$
130. $(-x^4)^{1/4}$
131. $-(8a^3b^6)^{-2/3}$
132. $-(25s^4t^6)^{-3/2}$

Look Alikes . . .

133. a. $-125^{2/3}$ b. $(-125)^{2/3}$ c. $-125^{-2/3}$ d. $\frac{1}{(-125)^{-2/3}}$

134. a. $81^{1/4}$ b. $81^{-1/4}$ c. $-81^{1/4}$ d. $\frac{1}{81^{-1/4}}$

135. a. $(64a^4)^{1/2}$ b. $(64a^4)^{-1/2}$ c. $-(64a^4)^{1/2}$ d. $\frac{1}{(64a^4)^{1/2}}$

136. a. $r^{1/3} \cdot r^{1/5}$ b. $(r^{1/3})^{1/5}$ c. $r^{1/3} \cdot r^{-1/5}$ d. $(r^{-1/3})^{-1/5}$

APPLICATIONS

137. **Ballistic Pendulums.** The formula $v = \frac{m + M}{m}(2gh)^{1/2}$ gives the velocity (in ft/sec) of a bullet with weight m fired into a block with weight M, that raises the height of the block h feet after the collision. See the illustration in the next column. The letter g represents a constant, 32. Find the velocity of the bullet to the nearest ft/sec.

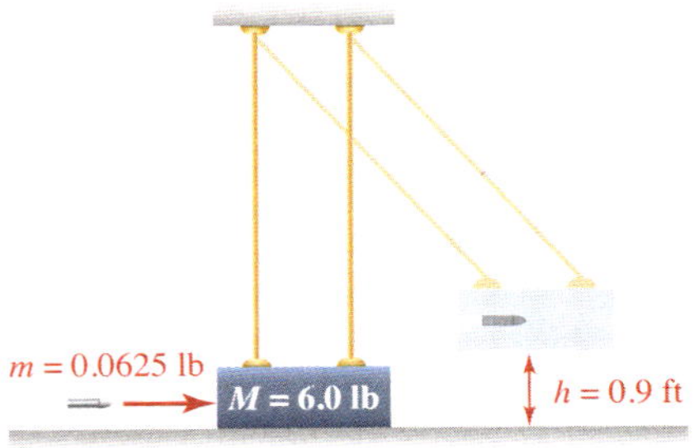

138. **Geography.** The formula $A = [s(s - a)(s - b)(s - c)]^{1/2}$ gives the area of a triangle with sides of length a, b, and c, where s is one-half of the perimeter. Estimate the area of Virginia (to the nearest square mile) using the data given in the illustration.

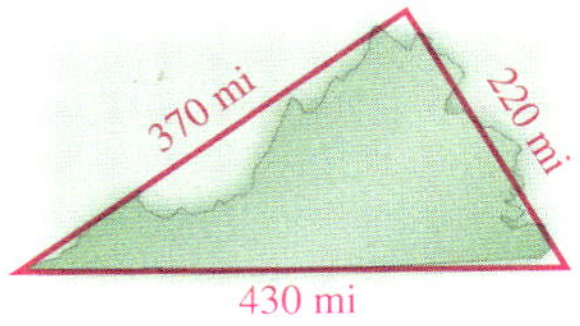

139. **Relativity.** One concept of relativity theory is that an object moving past an observer at a speed near the speed of light appears to have a larger mass because of its motion. If the mass of the object is m_0 when the object is at rest relative to the observer, its mass m will be given by the formula $m = m_0\left(1 - \frac{v^2}{c^2}\right)^{-1/2}$ when it is moving with speed v (in miles per second) past the observer. The variable c is the speed of light, 186,000 mi/sec. If a proton with a rest mass of 1 unit is accelerated by a nuclear accelerator to a speed of 160,000 mi/sec, what mass will the technicians observe it to have? Round to the nearest hundredth.

140. **Logging.** The width w and height h of the strongest rectangular beam that can be cut from a cylindrical log of radius a are given by $w = \frac{2a}{3}\left(3^{1/2}\right)$ and $h = a\left(\frac{8}{3}\right)^{1/2}$. Find the width, height, and cross-sectional area of the strongest beam that can be cut from a log with *diameter* 4 feet. Round to the nearest hundredth.

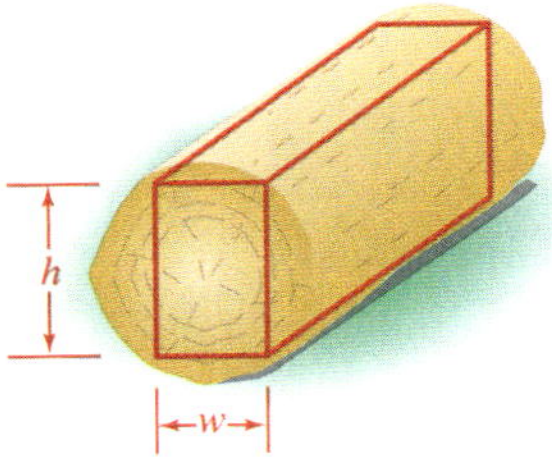

141. from **Campus to Careers**

General Contractor

The length L of the longest board that can be carried horizontally around the right-angle corner of two intersecting hallways is given by the formula $L = (a^{2/3} + b^{2/3})^{3/2}$, where a and b represent the widths of the hallways. Find the longest shelf that a carpenter can carry around the corner if $a = 40$ in. and $b = 64$ in. Give your result in inches and in feet. In each case, round to the nearest tenth.

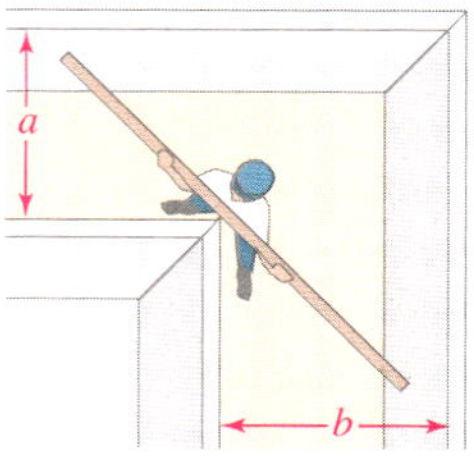

142. Cubicles. The area of the base of a cube is given by the function $A(V) = V^{2/3}$, where V is the volume of the cube. In a preschool room, 18 children's cubicles like the one shown are placed on the floor around the room. Estimate how much floor space is lost to the cubicles. Give your answer in square inches and in square feet.

WRITING

143. What is a rational exponent? Give some examples.

144. Explain how the root key $\boxed{\sqrt[x]{y}}$ on a scientific calculator can be used in combination with other keys to evaluate the expression $16^{3/4}$.

REVIEW

145. Commuting Time. The time it takes a car to travel a certain distance varies inversely with its rate of speed. If a certain trip takes 3 hours at 50 miles per hour, how long will the trip take at 60 miles per hour?

146. Bankruptcy. After filing for bankruptcy, a company was able to pay its creditors only 15 cents on the dollar. If the company owed a lumberyard $9,712, how much could the lumberyard expect to be paid?

CHALLENGE PROBLEMS

147. The fraction $\frac{2}{4}$ is equal to $\frac{1}{2}$. Is $16^{2/4}$ equal to $16^{1/2}$? Explain.

148. Explain how would you evaluate an expression with a mixed-number exponent. For example, what is $8^{1\frac{1}{3}}$? What is $25^{2\frac{1}{2}}$?

SECTION 9.3 Simplifying and Combining Radical Expressions

OBJECTIVES

1. Use the product rule to simplify radical expressions.
2. Use prime factorization to simplify radical expressions.
3. Use the quotient rule to simplify radical expressions.
4. Add and subtract radical expressions.

ARE YOU READY?

The following problems review some basic skills that are needed when adding and subtracting radical expressions.

1. Complete each factorization:

a. $28 = \square \cdot 7$ **b.** $54 = \square \cdot 2$

2. Simplify:

a. $\sqrt[3]{-8}$ **b.** $\sqrt[4]{16}$

3. Multiply: $3 \cdot 3 \cdot 3 \cdot a^4 \cdot b^8$

4. Which of the following are like terms?

$7x$ $3x^2$ $9x$ $2x^3$

5. Combine like terms: $15m^2 + 5m - m^2 - 6m$

6. Simplify: $\dfrac{\sqrt{25}}{\sqrt{36}}$

In algebra, it is often helpful to replace an expression with a simpler equivalent expression. This is certainly true when working with radicals. In most cases, radical expressions should be written in simplified form. We use two rules for radicals to do this.

1 Use the Product Rule to Simplify Radical Expressions.

To introduce the product rule for radicals, we will find $\sqrt{4 \cdot 25}$ and $\sqrt{4}\sqrt{25}$, and compare the results.

Square root of a product	*Product of square roots*
$\sqrt{4 \cdot 25} = \sqrt{100}$	$\sqrt{4}\sqrt{25} = 2 \cdot 5$
$= 10$	$= 10$

In each case, the answer is 10. Thus, $\sqrt{4 \cdot 25} = \sqrt{4}\sqrt{25}$.

Similarly, we will find $\sqrt[3]{8 \cdot 27}$ and $\sqrt[3]{8}\sqrt[3]{27}$ and compare the results.

Cube root of a product	*Product of cube roots*
$\sqrt[3]{8 \cdot 27} = \sqrt[3]{216}$	$\sqrt[3]{8}\sqrt[3]{27} = 2 \cdot 3$
$= 6$	$= 6$

In each case, the answer is 6. Thus, $\sqrt[3]{8 \cdot 27} = \sqrt[3]{8}\sqrt[3]{27}$. These results illustrate the *product rule for radicals.*

Notation

The products $\sqrt{4}\sqrt{25}$ and $\sqrt[3]{8}\sqrt[3]{27}$ also can be written using a raised dot:

$\sqrt{4} \cdot \sqrt{25}$ $\quad$ $\sqrt[3]{8} \cdot \sqrt[3]{27}$

The Product Rule for Radicals

The nth root of the product of two numbers is equal to the product of their nth roots. If $\sqrt[n]{a}$ and $\sqrt[n]{b}$ are real numbers,

$$\sqrt[n]{ab} = \sqrt[n]{a}\sqrt[n]{b}$$

Read as "the nth root of a times b equals the nth root of a times the nth root of b."

CAUTION The product rule for radicals applies to the nth root of a product. There is no such property for sums or differences. For example,

$$\sqrt{9+4} \neq \sqrt{9} + \sqrt{4} \qquad \sqrt{9-4} \neq \sqrt{9} - \sqrt{4}$$
$$\sqrt{13} \neq 3 + 2 \qquad \sqrt{5} \neq 3 - 2$$
$$\sqrt{13} \neq 5 \qquad \sqrt{5} \neq 1$$

Thus, $\sqrt{a+b} \neq \sqrt{a} + \sqrt{b}$ and $\sqrt{a-b} \neq \sqrt{a} - \sqrt{b}$.

The product rule for radicals can be used to simplify radical expressions. When a radical expression is written in **simplified form,** each of the following is true.

Simplified Form of a Radical Expression

1. Each factor in the radicand is to a power that is less than the index of the radical.
2. The radicand contains no fractions or negative numbers.
3. No radicals appear in the denominator of a fraction.

To simplify radical expressions, we must often factor the radicand using two natural-number factors. Since one factor should be a perfect square, perfect cube, perfect-fourth power, and so/on, it is helpful to memorize the following lists.

Perfect squares: **1, 4, 9, 16, 25, 36, 49, 64, 81, 100, 121, 144, 169, 196, 225, . . .**

Perfect cubes: **1, 8, 27, 64, 125, 216, 343, 512, 729, 1,000, . . .**

Perfect-fourth powers: **1, 16, 81, 256, 625, . . .**

Perfect-fifth powers: **1, 32, 243, 1,024, . . .**

EXAMPLE 1 Simplify: **a.** $\sqrt{12}$ **b.** $\sqrt{98}$ **c.** $5\sqrt[3]{54}$ **d.** $-\sqrt[4]{48}$

Strategy We will factor each radicand into two factors, one of which is a perfect square, perfect cube, or perfect-fourth power, depending on the index of the radical. Then we can use the product rule for radicals to simplify the expression.

Why Factoring the radicand in this way leads to a square root, cube root, or fourth root of a perfect square, perfect cube, or perfect-fourth power that we can easily simplify.

Solution

The Language of Algebra

The instructions **simplify** and **approximate** do not mean the same thing.

Simplify: $\sqrt{12} = 2\sqrt{3}$ (exact)

Approximate:

$\sqrt{12} \approx 3.464$ (not exact)

When referrring to a square root, the instruction *simplify* means to remove any perfect-square factors from the radicand.

a. To simplify $\sqrt{12}$, we first factor 12 so that one factor is the largest perfect square that divides 12. Since 4 is the largest perfect-square factor of 12, we write 12 as $4 \cdot 3$, use the product rule for radicals, and simplify.

$\sqrt{12} = \sqrt{4 \cdot 3}$ Write 12 as $12 = 4 \cdot 3$.

Write the perfect-square factor first.

$= \sqrt{4}\sqrt{3}$ The square root of a product is equal to the product of the square roots.

$= 2\sqrt{3}$ Evaluate $\sqrt{4}$. Read as "2 times the square root of 3" or as "2 radical 3."

We say that $2\sqrt{3}$ is the simplified form of $\sqrt{12}$. Note that $\sqrt{12}$ and $2\sqrt{3}$ are just different representations of the same number. When we compare calculator approximations of each, that fact seems reasonable.

$$\sqrt{12} \approx 3.464101615 \qquad 2\sqrt{3} \approx 3.464101615$$

b. The largest perfect-square factor of 98 is 49. Thus,

$\sqrt{98} = \sqrt{49 \cdot 2}$ Write 98 in factored form: $98 = 49 \cdot 2$.

$= \sqrt{49}\sqrt{2}$ The square root of a product is equal to the product of the square roots: $\sqrt{49 \cdot 2} = \sqrt{49}\sqrt{2}$.

$= 7\sqrt{2}$ Evaluate $\sqrt{49}$.

c. Since the largest perfect-cube factor of 54 is 27, we have

The Language of Algebra

In Example 1, **a radical of a product** is written as **a product of radicals:**

$$\sqrt[n]{ab} = \sqrt[n]{a}\sqrt[n]{b}$$

$5\sqrt[3]{54} = 5\sqrt[3]{27 \cdot 2}$ Write 54 as $27 \cdot 2$.

$= 5\sqrt[3]{27}\sqrt[3]{2}$ The cube root of a product is equal to the product of the cube roots: $\sqrt[3]{27 \cdot 2} = \sqrt[3]{27}\sqrt[3]{2}$.

$= 5 \cdot 3\sqrt[3]{2}$ Evaluate $\sqrt[3]{27}$.

$= 15\sqrt[3]{2}$ Multiply: $5 \cdot 3 = 15$.

d. The largest perfect-fourth power factor of 48 is 16. Thus,

$-\sqrt[4]{48} = -\sqrt[4]{16 \cdot 3}$ Write 48 as $16 \cdot 3$.

$= -\sqrt[4]{16}\sqrt[4]{3}$ The fourth root of a product is equal to the product of the fourth roots: $\sqrt[4]{16 \cdot 3} = \sqrt[4]{16} \cdot \sqrt[4]{3}$.

$= -2\sqrt[4]{3}$ Evaluate $\sqrt[4]{16}$.

Self Check 1 Simplify: **a.** $\sqrt{18}$ **b.** $7\sqrt[3]{24}$ **c.** $\sqrt[4]{32}$ **d.** $\sqrt[5]{128}$

Now Try Problems 13, 17, and 19

Variable expressions also can be perfect squares, perfect cubes, perfect-fourth powers, and so on. For example,

Perfect squares: $x^2, x^4, x^6, x^8, x^{10}, \ldots$

Perfect cubes: $x^3, x^6, x^9, x^{12}, x^{15}, \ldots$

Perfect-fourth powers: $x^4, x^8, x^{12}, x^{16}, x^{20}, \ldots$

Perfect-fifth powers: $x^5, x^{10}, x^{15}, x^{20}, x^{25}, \ldots$

EXAMPLE 2

Simplify. All variables represent positive real numbers.

a. $\sqrt{m^9}$ **b.** $10\sqrt{128a^5}$ **c.** $\sqrt[3]{-24x^5}$ **d.** $\sqrt[5]{a^9b^5}$

Strategy We will factor each radicand into two factors, one of which is a perfect nth power.

Why We can then apply the rule *the nth root of a product is the product of the nth roots* to simplify the radical expression.

Solution

a. The largest perfect-square factor of m^9 is m^8.

$$\sqrt{m^9} = \sqrt{m^8 \cdot m} \quad \text{Write } m^9 \text{ in factored form as } m^8 \cdot m.$$
$$= \sqrt{m^8}\sqrt{m} \quad \text{Use the product rule for radicals.}$$
$$= m^4\sqrt{m} \quad \text{Simplify } \sqrt{m^8}.$$

Caution

$\sqrt{m^9} \neq m^3$ because $(m^3)^2 \neq m^9$. By the power rule for exponents, $(m^3)^2 = m^6$. Similarly, $\sqrt{y^{25}} \neq y^5$ because $(y^5)^2 \neq y^{25}$.

b. Since the largest perfect-square factor of 128 is 64 and the largest perfect-square factor of a^5 is a^4, the largest perfect-square factor of $128a^5$ is $64a^4$. We write $128a^5$ as $64a^4 \cdot 2a$ and proceed as follows:

$$10\sqrt{128a^5} = 10\sqrt{64a^4 \cdot 2a} \quad \text{Write } 128a^5 \text{ in factored form as } 64a^4 \cdot 2a.$$
$$= 10\sqrt{64a^4}\sqrt{2a} \quad \text{Use the product rule for radicals.}$$
$$= 10 \cdot 8a^2\sqrt{2a} \quad \text{Simplify } \sqrt{64a^4}.$$
$$= 80a^2\sqrt{2a} \quad \text{Multiply: } 10 \cdot 8 = 80.$$

c. We write $-24x^5$ as $-8x^3 \cdot 3x^2$ and proceed as follows:

$$\sqrt[3]{-24x^5} = \sqrt[3]{-8x^3 \cdot 3x^2} \quad 8x^3 \text{ is the largest perfect-cube factor of } 24x^5. \text{ Since the radicand is negative, we factor it using } -8x^3.$$
$$= \sqrt[3]{-8x^3}\sqrt[3]{3x^2} \quad \text{Use the product rule for radicals.}$$
$$= -2x\sqrt[3]{3x^2} \quad \text{Simplify } \sqrt[3]{-8x^3}.$$

Caution

$\sqrt[3]{-24x^5} = -2x\sqrt[3]{3x^2}$

↑

In simplified form, the radicand should not have any perfect-cube factors.

d. The largest perfect-fifth power factor of a^9 is a^5, and b^5 is a perfect-fifth power.

$$\sqrt[5]{a^9b^5} = \sqrt[5]{a^5b^5 \cdot a^4} \quad a^5b^5 \text{ is the largest perfect-fifth power factor of } a^9b^5.$$
$$= \sqrt[5]{a^5b^5}\sqrt[5]{a^4} \quad \text{Use the product rule for radicals.}$$
$$= ab\sqrt[5]{a^4} \quad \text{Simplify } \sqrt[5]{a^5b^5}.$$

Self Check 2 Simplify. All variables represent positive real numbers.

a. $6\sqrt{98b^3}$ **b.** $\sqrt[3]{-54y^5}$ **c.** $\sqrt[4]{t^8u^{15}}$

Now Try Problems 21, 25, and 27

2 Use Prime Factorization to Simplify Radical Expressions.

When simplifying radical expressions, prime factorization can be helpful in determining how to factor the radicand.

EXAMPLE 3

Simplify. All variables represent positive real numbers.

a. $\sqrt{150}$ **b.** $\sqrt[3]{297b^4}$ **c.** $\sqrt[4]{224s^8t^7}$

Strategy In each case, the way to factor the radicand is not obvious. Another approach is to prime-factor the coefficient of the radicand and look for groups of like factors.

Why Identifying groups of like factors of the radicand leads to a factorization of the radicand that can be easily simplified.

Solution

Caution

To simplify $\sqrt{150}$, we could have factored 150 as $10 \cdot 15$. However, neither of the resulting radical expressions is the square root of a perfect square.

$\sqrt{150} = \sqrt{10}\sqrt{15}$

a. $\sqrt{150} = \sqrt{2 \cdot 3 \cdot 5 \cdot 5}$ — Write 150 in prime-factored form.

$= \sqrt{2 \cdot 3}\sqrt{5 \cdot 5}$ — Group the pair of like factors together and use the product rule for radicals.

$= \sqrt{2 \cdot 3}\sqrt{5^2}$ — Write $5 \cdot 5$ as 5^2.

$= \sqrt{6} \cdot 5$ — Evaluate $\sqrt{5^2}$.

$= 5\sqrt{6}$ — Write the factor 5 first.

$$\begin{array}{r|l} 2 & 150 \\ 3 & 75 \\ 5 & 25 \\ & 5 \end{array}$$

b. $\sqrt[3]{297b^4} = \sqrt[3]{3 \cdot 3 \cdot 3 \cdot 11 \cdot b^3 \cdot b}$ — Write 297 in prime-factored form. The largest perfect-cube factor of b^4 is b^3.

$= \sqrt[3]{3 \cdot 3 \cdot 3 \cdot b^3}\sqrt[3]{11b}$ — Group the three like factors of 3 together and use the product rule for radicals.

$= \sqrt[3]{3^3b^3}\sqrt[3]{11b}$ — Write $3 \cdot 3 \cdot 3$ as 3^3.

$= 3b\sqrt[3]{11b}$ — Simplify $\sqrt[3]{3^3b^3}$.

$$\begin{array}{r|l} 3 & 297 \\ 3 & 99 \\ 3 & 33 \\ & 11 \end{array}$$

c. $\sqrt[4]{224s^8t^7} = \sqrt[4]{2 \cdot 2 \cdot 2 \cdot 2 \cdot 2 \cdot 7 \cdot s^8 \cdot t^4 \cdot t^3}$ — Write 224 in prime-factored form. The largest perfect-fourth power factor of t^7 is t^4.

$= \sqrt[4]{2 \cdot 2 \cdot 2 \cdot 2 \cdot s^8 \cdot t^4}\sqrt[4]{2 \cdot 7 \cdot t^3}$ — Group the four like factors of 2 together and use the product rule for radicals.

$= \sqrt[4]{2^4s^8t^4}\sqrt[4]{2 \cdot 7 \cdot t^3}$ — Write $2 \cdot 2 \cdot 2 \cdot 2$ as 2^4.

$= 2s^2t\sqrt[4]{14t^3}$ — Simplify $\sqrt[4]{2^4s^8t^4}$.

$$\begin{array}{r|l} 2 & 224 \\ 2 & 112 \\ 2 & 56 \\ 2 & 28 \\ 2 & 14 \\ & 7 \end{array}$$

Self Check 3 Simplify. All variables represent positive real numbers.

a. $\sqrt{275}$ **b.** $\sqrt[3]{189c^4d^3}$ **c.** $\sqrt[4]{1{,}250x^8y^6}$

Now Try Problems 29 and 35

3 Use the Quotient Rule to Simplify Radical Expressions.

To introduce the quotient rule for radicals, we will find $\sqrt{\frac{100}{4}}$ and $\frac{\sqrt{100}}{\sqrt{4}}$ and compare the results.

Square root of a quotient	***Quotient of square roots***
$\sqrt{\frac{100}{4}} = \sqrt{25}$	$\frac{\sqrt{100}}{\sqrt{4}} = \frac{10}{2}$
$= 5$	$= 5$

Since the answer is 5 in each case, $\sqrt{\frac{100}{4}} = \frac{\sqrt{100}}{\sqrt{4}}$.

Similarly, we will find $\sqrt[3]{\frac{64}{8}}$ and $\frac{\sqrt[3]{64}}{\sqrt[3]{8}}$, and compare the results.

Cube root of a quotient	***Quotient of cube roots***
$\sqrt[3]{\frac{64}{8}} = \sqrt[3]{8}$	$\frac{\sqrt[3]{64}}{\sqrt[3]{8}} = \frac{4}{2}$
$= 2$	$= 2$

Since the answer is 2 in each case, $\sqrt[3]{\frac{64}{8}} = \frac{\sqrt[3]{64}}{\sqrt[3]{8}}$. These results illustrate the *quotient rule for radicals.*

The Quotient Rule for Radicals

The nth root of the quotient of two numbers is equal to the quotient of their nth roots.

If $\sqrt[n]{a}$ and $\sqrt[n]{b}$ are real numbers, then

$$\sqrt[n]{\frac{a}{b}} = \frac{\sqrt[n]{a}}{\sqrt[n]{b}} \quad (b \neq 0)$$

Read as "the nth root of a divided by b equals the nth root of a divided by the nth root of b."

In the following examples, variables appear in the denominators of radical expressions. To avoid undefined situations, we will assume ***all variables represent positive real numbers.***

EXAMPLE 4 Simplify each expression: **a.** $\sqrt{\frac{7}{64}}$ **b.** $\sqrt{\frac{15}{49x^2}}$ **c.** $\sqrt[3]{\frac{10x^2}{27y^6}}$

Strategy In each case, the radical is not in simplified form because the radicand contains a fraction. To write each of these expressions in simplified form, we will use the quotient rule for radicals.

Why Writing these expressions in $\frac{\sqrt[n]{a}}{\sqrt[n]{b}}$ form leads to square roots of perfect squares and cube roots of perfect cubes that we can easily simplify.

Solution

The Language of Algebra

In Example 4, a **radical of a quotient** is written as a **quotient of radicals**:

$$\sqrt[n]{\frac{a}{b}} = \frac{\sqrt[n]{a}}{\sqrt[n]{b}}$$

a. We can use the quotient rule for radicals to simplify each expression.

$$\sqrt{\frac{7}{64}} = \frac{\sqrt{7}}{\sqrt{64}}$$ The square root of a quotient is equal to the quotient of the square roots.

$$= \frac{\sqrt{7}}{8}$$ Evaluate $\sqrt{64}$.

b. $$\sqrt{\frac{15}{49x^2}} = \frac{\sqrt{15}}{\sqrt{49x^2}}$$ The square root of a quotient is equal to the quotient of the square roots.

$$= \frac{\sqrt{15}}{7x}$$ Simplify the denominator: $\sqrt{49x^2} = 7x$.

c. $$\sqrt[3]{\frac{10x^2}{27y^6}} = \frac{\sqrt[3]{10x^2}}{\sqrt[3]{27y^6}}$$ The cube root of a quotient is equal to the quotient of the cube roots.

$$= \frac{\sqrt[3]{10x^2}}{3y^2}$$ Simplify the denominator.

Self Check 4 Simplify: **a.** $\sqrt[3]{\frac{25}{27}}$ **b.** $\sqrt{\frac{11}{36a^2}}$ **c.** $\sqrt[4]{\frac{a^3}{625y^{12}}}$

Now Try Problems 37, 39, and 43

EXAMPLE 5 Simplify: **a.** $\frac{\sqrt{45xy^2}}{\sqrt{5x}}$ **b.** $\frac{\sqrt[3]{-432x^5}}{\sqrt[3]{8x}}$

Strategy We will use the quotient rule for radicals in reverse: $\frac{\sqrt[n]{a}}{\sqrt[n]{b}} = \sqrt[n]{\frac{a}{b}}$

Why When the radicands are written under a single radical symbol, the result is a rational expression. Our hope is that the rational expression can be simplified.

Solution

a. We can write the quotient of the square roots as the square root of a quotient.

$$\frac{\sqrt{45xy^2}}{\sqrt{5x}} = \sqrt{\frac{45xy^2}{5x}}$$ Use the quotient rule for radicals. Note that the resulting radicand is a rational expression.

$$= \sqrt{9y^2}$$ Simplify the radicand: $\frac{45xy^2}{5x} = \frac{\overset{1}{\cancel{5}} \cdot 9 \cdot \overset{1}{\cancel{x}} \cdot y^2}{\underset{1}{\cancel{5}} \cdot \underset{1}{\cancel{x}}} = 9y^2$.

$$= 3y$$ Simplify the radical.

b. We can write the quotient of the cube roots as the cube root of a quotient.

$$\frac{\sqrt[3]{-432x^5}}{\sqrt[3]{8x}} = \sqrt[3]{-\frac{432x^5}{8x}}$$ Use the quotient rule for radicals. Note that the resulting radicand is a rational expression.

$$= \sqrt[3]{-54x^4}$$ Simplify the radicand: $-\frac{432x^5}{8x} = -54x^4$.

$$= \sqrt[3]{-27x^3 \cdot 2x}$$ $27x^3$ is the largest perfect cube that divides $54x^4$.

$$= \sqrt[3]{-27x^3}\sqrt[3]{2x}$$ Use the product rule for radicals.

$$= -3x\sqrt[3]{2x}$$ Simplify: $\sqrt[3]{-27x^3} = -3x$.

The Language of Algebra

In Example 5, a **quotient of radicals** is written as a **radical of a quotient.**

$$\frac{\sqrt[n]{a}}{\sqrt[n]{b}} = \sqrt[n]{\frac{a}{b}}$$

Success Tip

In parts a and b, we use skills learned in Chapter 7 to simplify the **rational expressions** in each radicand: $\frac{45xy^2}{5x}$ and $-\frac{432x^5}{8x}$.

Self Check 5 Simplify: **a.** $\frac{\sqrt{50ab^2}}{\sqrt{2a}}$ **b.** $\frac{\sqrt[3]{-2{,}000x^5y^3}}{\sqrt[3]{2x}}$

Now Try Problems 47 and 51

4 Add and Subtract Radical Expressions.

Radical expressions with the same index and the same radicand are called **like** or **similar radicals.** For example, $3\sqrt{2}$ and $2\sqrt{2}$ are like radicals. However,

- $3\sqrt{5}$ and $4\sqrt{2}$ are not like radicals, because the radicands are different.
- $3\sqrt[4]{5}$ and $2\sqrt[3]{5}$ are not like radicals, because the indices are different.

For an expression with two or more radical terms, we should attempt to combine like radicals, if possible. For example, to simplify the expression $3\sqrt{2} + 2\sqrt{2}$, we use the distributive property to factor out $\sqrt{2}$ and simplify.

$$3\sqrt{2} + 2\sqrt{2} = (3 + 2)\sqrt{2}$$ Factor out $\sqrt{2}$.

$$= 5\sqrt{2}$$ Do the addition.

Radicals with the same index but different radicands often can be written as like radicals. For example, to simplify the expression $\sqrt{75} - \sqrt{27}$, we simplify both radicals first and then combine the like radicals.

$$\sqrt{75} - \sqrt{27} = \sqrt{25 \cdot 3} - \sqrt{9 \cdot 3}$$ Write 75 and 27 in factored form.

$$= \sqrt{25}\sqrt{3} - \sqrt{9}\sqrt{3}$$ Use the product rule for radicals.

$$= 5\sqrt{3} - 3\sqrt{3}$$ Evaluate $\sqrt{25}$ and $\sqrt{9}$.

$$= (5 - 3)\sqrt{3}$$ Factor out $\sqrt{3}$.

$$= 2\sqrt{3}$$ Do the subtraction.

As the previous examples suggest, we can add or subtract radicals as follows.

Success Tip

Combining like radicals is similar to combining like terms.

$$3\sqrt{2} + 2\sqrt{2} = 5\sqrt{2}$$

$$3x + 2x = 5x$$

Adding and Subtracting Radicals

To add or subtract radicals, simplify each radical, if possible, and combine like radicals.

EXAMPLE 6 Simplify: **a.** $18 + 2\sqrt{12} - 3\sqrt{48} + 9$ **b.** $\sqrt[3]{16} + \sqrt[3]{54} - \sqrt[3]{24}$

Strategy Since the radicals in each part are unlike radicals, we cannot add or subtract them in their current form. However, we will simplify the radicals and hope that like radicals result.

Why Like radicals can be combined.

Solution **a.** We begin by simplifying each radical expression:

$$18 + 2\sqrt{12} - 3\sqrt{48} + 9 = 18 + 2\sqrt{4 \cdot 3} - 3\sqrt{16 \cdot 3} + 9$$ Factor the radicands, 12 and 48.

$$= 18 + 2\sqrt{4}\sqrt{3} - 3\sqrt{16}\sqrt{3} + 9$$ Use the product rule for radicals.

$$= 18 + 2(2)\sqrt{3} - 3(4)\sqrt{3} + 9$$ Evaluate $\sqrt{4}$ and $\sqrt{16}$.

$$= 18 + \mathbf{4}\sqrt{3} - \mathbf{12}\sqrt{3} + 9$$ Both radicals now have the same index, 2, and radicand, 3.

$$= (\mathbf{4} - \mathbf{12})\sqrt{3} + 27$$ Add the constants: $18 + 9 = 27$. Combine like radicals.

$$= -8\sqrt{3} + 27$$ Do the subtraction.

b. We begin by simplifying each radical expression:

$$\sqrt[3]{16} + \sqrt[3]{54} - \sqrt[3]{24} = \sqrt[3]{8 \cdot 2} + \sqrt[3]{27 \cdot 2} - \sqrt[3]{8 \cdot 3}$$ Factor the radicands.

$$= \sqrt[3]{8}\sqrt[3]{2} + \sqrt[3]{27}\sqrt[3]{2} - \sqrt[3]{8}\sqrt[3]{3}$$ Use the product rule.

$$= \mathbf{2}\sqrt[3]{\mathbf{2}} + \mathbf{3}\sqrt[3]{\mathbf{2}} - 2\sqrt[3]{3}$$ Evaluate $\sqrt[3]{8}$ and $\sqrt[3]{27}$.

$$= (\mathbf{2} + \mathbf{3})\sqrt[3]{\mathbf{2}} - 2\sqrt[3]{3}$$ Combine the first two radical expressions because they have the same index and radicand.

$$= 5\sqrt[3]{2} - 2\sqrt[3]{3}$$ Do the addition.

CAUTION Even though the expressions $5\sqrt[3]{2}$ and $2\sqrt[3]{3}$ have the same index, we cannot combine them, because their radicands are different. Neither can we combine radical expressions having the same radicand but a different index. For example, the expression $\sqrt[3]{2} + \sqrt[4]{2}$ cannot be simplified.

Self Check 6 Simplify: **a.** $14 + 3\sqrt{75} - 2\sqrt{12} + 2\sqrt{48} - 8$
b. $\sqrt[3]{24} - \sqrt[3]{16} + \sqrt[3]{54}$

Now Try Problems 57 and 63

EXAMPLE 7 Simplify: $\sqrt[3]{16x^4} + 4\sqrt[3]{54x^4} - x\sqrt[3]{-128x}$

Strategy Since the radicals are unlike radicals, we cannot add or subtract them in their current form. However, we will simplify the radicals and hope that like radicals result.

Why Like radicals can be combined.

Solution We begin by simplifying each radical expression.

$$\sqrt[3]{16x^4} + 4\sqrt[3]{54x^4} - x\sqrt[3]{-128x}$$

$$= \sqrt[3]{8x^3 \cdot 2x} + 4\sqrt[3]{27x^3 \cdot 2x} - x\sqrt[3]{-64 \cdot 2x}$$

$$= \sqrt[3]{\mathbf{8x^3}}\sqrt[3]{2x} + 4\sqrt[3]{\mathbf{27x^3}}\sqrt[3]{2x} - x\sqrt[3]{\mathbf{-64}}\sqrt[3]{2x}$$

$$= \mathbf{2x}\sqrt[3]{2x} + 4 \cdot \mathbf{3x}\sqrt[3]{2x} + \mathbf{x} \cdot \mathbf{4}\sqrt[3]{2x}$$ All three radicals have the same index and radicand.

$$= 2x\sqrt[3]{2x} + 12x\sqrt[3]{2x} + 4x\sqrt[3]{2x}$$ Combine like radicals.

$$= (2x + 12x + 4x)\sqrt[3]{2x}$$ Multiply: $4 \cdot 3x = 12x$ and $x \cdot 4 = 4x$.

$$= 18x\sqrt[3]{2x}$$ Within the parentheses, combine like terms: $2x + 12x + 4x = 18x$.

Self Check 7 Simplify: $\sqrt{32x^3} + 4\sqrt{50x^3} - x\sqrt{18x}$

Now Try Problems 67 and 69

SECTION 9.3 STUDY SET

VOCABULARY

Fill in the blanks.

1. Radical expressions such as $\sqrt[3]{4}$ and $6\sqrt[3]{4}$ with the same index and the same radicand are called _____ radicals.
2. Numbers such as 1, 4, 9, 16, 25, and 36 are called perfect ________. Numbers such as 1, 8, 27, 64, and 125 are called perfect ______. Numbers such as 1, 16, 81, 256, and 625 are called perfect-fourth ________.
3. The largest perfect-square ______ of 27 is 9. The largest ______-cube factor of 16 is 8.
4. To ________ $\sqrt{24}$ means to write it as $2\sqrt{6}$.

CONCEPTS

Fill in the blanks.

5. The product rule for radicals: $\sqrt[n]{ab} =$ ______. In words, the *n*th root of the ________ of two numbers is equal to the product of their *n*th ______.
6. The quotient rule for radicals: $\sqrt[n]{\frac{a}{b}} =$ ______. In words, the *n*th root of the ________ of two numbers is equal to the quotient of their *n*th ______.
7. Consider the expressions $\sqrt{4 \cdot 5}$ and $\sqrt{4}\sqrt{5}$. Which expression is
   a. the square root of a product?
   b. the product of square roots?
   c. How are these two expressions related?
8. Consider $\frac{\sqrt[3]{a}}{\sqrt[3]{x^2}}$ and $\sqrt[3]{\frac{a}{x^2}}$. Which expression is
   a. the cube root of a quotient?
   b. the quotient of cube roots?
   c. How are these two expressions related?
9. a. Write two radical expressions that have the same radicand but a different index. Can the expressions be added?
   b. Write two radical expressions that have the same index but a different radicand. Can the expressions be added?
10. Fill in the blanks.
   a. $5\sqrt{6} + 3\sqrt{6} = (\quad + \quad)\sqrt{6} = \quad\sqrt{6}$
   b. $9\sqrt[3]{n} - 2\sqrt[3]{n} = (\quad - \quad)\sqrt[3]{n} = 7\quad$

NOTATION

Complete each solution.

11. Simplify:
$$\sqrt[3]{32k^4} = \sqrt[3]{\quad \cdot 4k} = \sqrt[3]{\quad}\sqrt[3]{4k} = \quad\sqrt[3]{4k}$$

12. Simplify:
$$\frac{\sqrt{80s^2t^4}}{\sqrt{5s^2}} = \sqrt{\frac{80s^2t^4}{\quad}} = \sqrt{\quad} = \quad$$

GUIDED PRACTICE

Simplify each expression. See Example 1.

13. $\sqrt{50}$
14. $\sqrt{28}$
15. $8\sqrt{45}$
16. $9\sqrt{54}$
17. $\sqrt[3]{32}$
18. $\sqrt[3]{40}$
19. $\sqrt[4]{48}$
20. $\sqrt[4]{32}$

Simplify each radical expression. All variables represent positive real numbers. See Example 2.

21. $\sqrt{75a^2}$
22. $\sqrt{50x^2}$
23. $\sqrt{128a^3b^5}$
24. $\sqrt{75b^8c}$
25. $2\sqrt[3]{-54x^6}$
26. $4\sqrt[3]{-81a^3}$
27. $\sqrt[4]{32x^{12}y^4}$
28. $\sqrt[5]{64x^{10}y^5}$

Simplify each radical expression. All variables represent positive real numbers. See Example 3.

29. $\sqrt{242}$
30. $\sqrt{363}$
31. $\sqrt{112a^3}$
32. $\sqrt{147a^5}$
33. $-\sqrt[5]{96a^4}$
34. $-\sqrt[7]{256t^6}$
35. $\sqrt[3]{405x^{12}y^4}$
36. $\sqrt[3]{280a^5b^6}$

Simplify each radical expression. All variables represent positive real numbers. See Example 4.

37. $\sqrt{\frac{11}{9}}$
38. $\sqrt{\frac{3}{4}}$
39. $\sqrt[4]{\frac{3}{625}}$
40. $\sqrt[5]{\frac{2}{243}}$
41. $\sqrt[5]{\frac{3x^{10}}{32}}$
42. $\sqrt[6]{\frac{5y^{12}}{64}}$
43. $\sqrt{\frac{z^2}{16x^2}}$
44. $\sqrt{\frac{b^4}{64a^8}}$

Simplify each expression. All variables represent positive real numbers. See Example 5.

45. $\dfrac{\sqrt{500}}{\sqrt{5}}$

46. $\dfrac{\sqrt{128}}{\sqrt{2}}$

47. $\dfrac{\sqrt{98x^3}}{\sqrt{2x}}$

48. $\dfrac{\sqrt{75y^5}}{\sqrt{3y}}$

49. $\dfrac{\sqrt[3]{48x^7}}{\sqrt[3]{6x}}$

50. $\dfrac{\sqrt[3]{64y^8}}{\sqrt[3]{8y^2}}$

51. $\dfrac{\sqrt[3]{189a^5}}{\sqrt[3]{7a}}$

52. $\dfrac{\sqrt[3]{243x^8}}{\sqrt[3]{9x}}$

Simplify by combining like radicals. See Objective 4 and Example 6.

53. $5\sqrt{7} + 3\sqrt{7}$

54. $11\sqrt{3} + 2\sqrt{3}$

55. $20\sqrt[3]{4} - 15\sqrt[3]{4}$

56. $30\sqrt[3]{6} - 10\sqrt[3]{6}$

57. $4 + \sqrt{8} + \sqrt{2} + 8$

58. $9 + \sqrt{45} + \sqrt{20} + 16$

59. $\sqrt{98} - \sqrt{50} - \sqrt{72}$

60. $\sqrt{20} + \sqrt{125} - \sqrt{80}$

61. $8 + \sqrt[3]{32} - \sqrt[3]{108} - 7$

62. $12 + \sqrt[3]{80} - \sqrt[3]{10{,}000} + 4$

63. $14\sqrt[4]{32} - 15\sqrt[4]{2}$

64. $23\sqrt[4]{3} + \sqrt[4]{48}$

Simplify by combining like radicals. All variables represent positive real numbers. See Example 7.

65. $4\sqrt{2x} + 6\sqrt{2x}$

66. $6\sqrt[3]{5y} + 3\sqrt[3]{5y}$

67. $\sqrt{18t} + \sqrt{300t} - \sqrt{243t}$

68. $\sqrt{80m} - \sqrt{128m} + \sqrt{288m}$

69. $2\sqrt[3]{16} - \sqrt[3]{54} - 3\sqrt[3]{128}$

70. $\sqrt[3]{250} - 4\sqrt[3]{5} + \sqrt[3]{16}$

71. $\sqrt[4]{32} + 5\sqrt[4]{2} - \sqrt[4]{162}$

72. $\sqrt[4]{48} - \sqrt[4]{243} - \sqrt[4]{768}$

TRY IT YOURSELF

Simplify each expression, if possible. All variables represent positive real numbers.

73. $\sqrt[6]{m^{11}}$

74. $\sqrt[6]{n^{13}}$

75. $2\sqrt[3]{64a} + 2\sqrt[3]{8a}$

76. $3\sqrt[4]{x^4y} - 2\sqrt[4]{x^4y}$

77. $\sqrt{8y^7} + \sqrt{32y^7} - \sqrt{2y^7}$

78. $\sqrt{y^5} - \sqrt{9y^5} - \sqrt{25y^5}$

79. $\sqrt{32b}$

80. $\sqrt{80c}$

81. $\sqrt{\dfrac{125n^5}{64n}}$

82. $\sqrt{\dfrac{72q^7}{25q^3}}$

83. $2\sqrt[3]{125} - 5\sqrt[3]{64}$

84. $3\sqrt[3]{27} + 12\sqrt[3]{216}$

85. $\sqrt{300xy}$

86. $\sqrt{200x^2y}$

87. $\sqrt[4]{\dfrac{5x}{16z^4}}$

88. $\sqrt[3]{\dfrac{11a^2}{125b^6}}$

89. $8\sqrt[5]{7a^2} - 7\sqrt[5]{7a^2}$

90. $10\sqrt[6]{12xy} - \sqrt[6]{12xy}$

91. $\sqrt[5]{x^6y^2} + \sqrt[5]{32x^6y^2} + \sqrt[5]{x^6y^2}$

92. $\sqrt[3]{xy^4} + \sqrt[3]{8xy^4} - \sqrt[3]{27xy^4}$

93. $\sqrt[4]{208m^4n}$

94. $\sqrt[4]{128p^8q^3}$

95. $\sqrt[3]{\dfrac{a^7}{64a}}$

96. $\sqrt[3]{\dfrac{b^3c^8}{125c^5}}$

97. $\sqrt[3]{\dfrac{7}{64}}$

98. $\sqrt[3]{\dfrac{4}{125}}$

99. $\sqrt{80} + \sqrt{45} - \sqrt{27}$

100. $\sqrt{63} + \sqrt{72} - \sqrt{28}$

101. $\sqrt[5]{64t^{11}}$

102. $\sqrt[5]{243r^{22}}$

103. $\sqrt[3]{24x} + \sqrt[3]{3x}$

104. $\sqrt[3]{16y} + \sqrt[3]{128y}$

Look Alikes . . .

105. a. $\sqrt{20} + \sqrt{20}$ b. $\sqrt{21} + \sqrt{21}$

106. a. $\sqrt{2} + \sqrt{18}$ b. $\sqrt{2} + \sqrt{19}$

107. a. $\sqrt{9x^2} - \sqrt{25x^2} + \sqrt{16x^2}$
b. $\sqrt{9x^3} - \sqrt{25x^3} + \sqrt{16x^3}$

108. a. $2\sqrt{5} - \sqrt[3]{5} + 4\sqrt{5} - 6\sqrt[3]{5}$
b. $2\sqrt[3]{5} - \sqrt[3]{5} + 4\sqrt[3]{5} - 6\sqrt[3]{5}$

109. a. $3\sqrt{16} + \sqrt{54}$ b. $3\sqrt[3]{16} + \sqrt[3]{54}$

110. a. $\sqrt[3]{27} - 5\sqrt[3]{8}$ b. $\sqrt{27} - 5\sqrt{8}$

111. a. $24\sqrt[5]{6x} + 16\sqrt[5]{6x}$ b. $24\sqrt[4]{6x} + 16\sqrt[4]{6x}$

112. a. $x\sqrt[3]{64x^6} - x\sqrt[3]{x^6}$ b. $x\sqrt{64x^6} - x\sqrt{x^6}$

APPLICATIONS

First give the exact answer, expressed as a simplified radical expression. Then give an approximation, rounded to the nearest tenth.

113. from Campus to Careers

General Contractor

©Goodluz/Shutterstock.com

Structural engineers have determined that two additional supports (shown in red) need to be added to strengthen the truss shown below. Find the length L of a support using the formula

$$L = \sqrt{\frac{b^2}{2} + \frac{c^2}{2} - \frac{a^2}{4}}$$

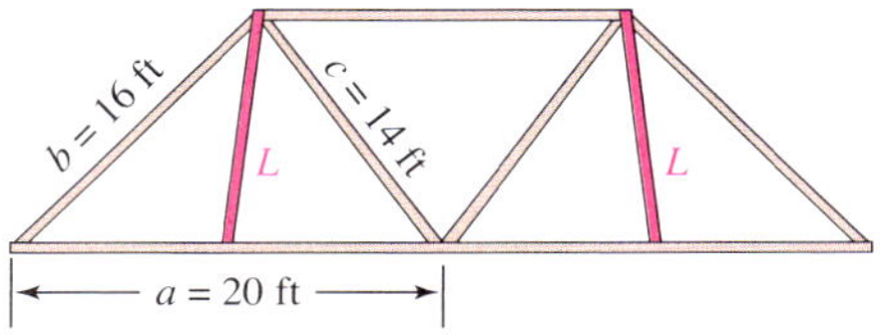

114. Unbrellas. The surface area of a cone is given by the formula $S = \pi r\sqrt{r^2 + h^2}$, where r is the radius of the base and h is its height. Use this formula to find the number of square feet of waterproof cloth used to make the umbrella shown.

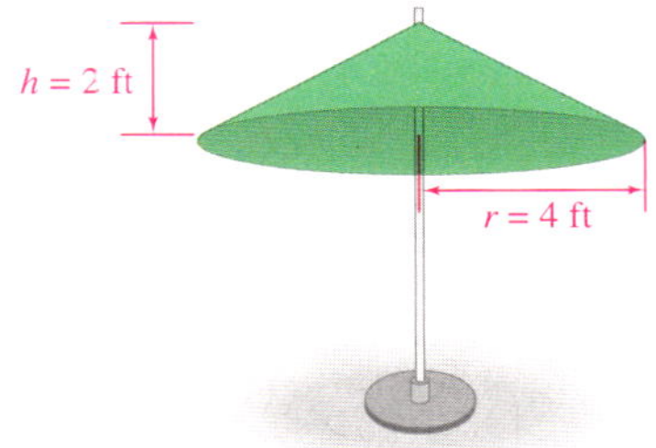

115. Blow Dryers. The current I (in amps), the power P (in watts), and the resistance R (in ohms) are related by the formula $I = \sqrt{\frac{P}{R}}$. What current is needed for a 1,200-watt hair dryer if the resistance is 16 ohms?

116. Communications Satellites. Engineers have determined that a spherical communications satellite needs to have a capacity of 565.2 cubic feet to house all of its operating systems. The volume V of a sphere is related to its radius r by the formula $r = \sqrt[3]{\frac{3V}{4\pi}}$. What radius must the satellite have to meet the engineer's specification? Use 3.14 as an approximation of π.

117. Ductwork. The following pattern is laid out on a sheet of galvanized tin. Then it is cut out and bent on the dashed lines to make an air conditioning duct connection. Find the total length of the cut that must be made. (All measurements are in inches.)

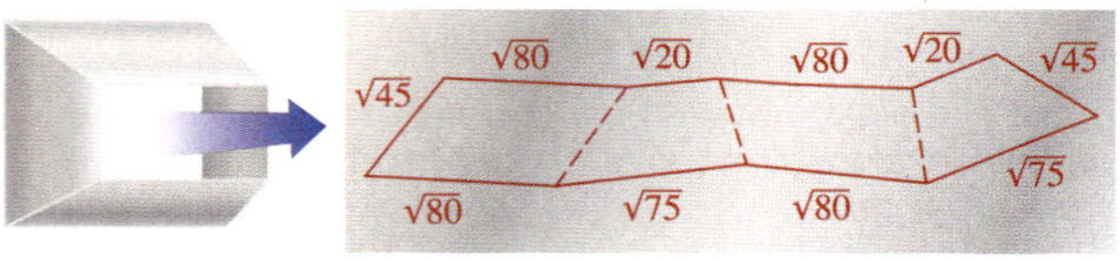

118. Outdoor Cooking. The diameter of a circle is given by the function $d(A) = 2\sqrt{\frac{A}{\pi}}$, where A is the area of the circle. Find the difference between the diameters of the barbecue grills.

WRITING

119. Explain why each expression is not in simplified form.

a. $\sqrt[3]{9x^4}$ **b.** $\sqrt{\frac{24m}{25}}$ **c.** $\frac{\sqrt[4]{c^3}}{\sqrt[4]{16}}$

120. How are the procedures used to simplify $3x + 4x$ and $3\sqrt{x} + 4\sqrt{x}$ similar?

121. Explain the mistake in the student's solution shown below.

Simplify: $\sqrt[3]{54}$

$$\sqrt[3]{54} = \sqrt[3]{27 + 27}$$
$$= \sqrt[3]{27} + \sqrt[3]{27}$$
$$= 3 + 3$$
$$= 6$$

122. Explain how the graphs of $Y_1 = 3\sqrt{24x} + \sqrt{54x}$ (on the left) and $Y_1 = 9\sqrt{6x}$ (on the right) can be used to verify the simplification $3\sqrt{24x} + \sqrt{54x} = 9\sqrt{6x}$. In each graph, settings of $[-5, 20]$ for x and $[-5, 100]$ for y were used.

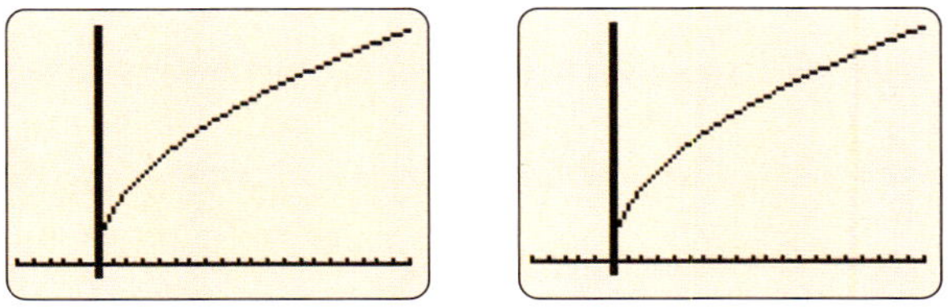

REVIEW

Perform each operation.

123. $3x^2y^3(-5x^3y^{-4})$

124. $(2x^2 - 9x - 5) \cdot \frac{x}{2x^2 + x}$

125. $2p - 5\overline{)6p^2 - 7p - 25}$

126. $\frac{xy}{\frac{1}{x} - \frac{1}{y}}$

CHALLENGE PROBLEMS

Simplify each expression. All variables represent positive real numbers.

127. $\frac{\sqrt{24}}{3} + \frac{\sqrt{6}}{5}$

128. $\sqrt{3} + \sqrt{3^2} + \sqrt{3^3} + \sqrt{3^4} + \sqrt{3^5}$

129. $\sqrt[3]{\frac{3b}{8}} - 9\sqrt[3]{3b}$

130. $\frac{\sqrt[4]{32}}{3y} - \frac{3\sqrt[4]{2}}{7y}$

131. $\sqrt{25x + 25} - \sqrt{x + 1}$

132. $\sqrt[3]{216m^4} + \sqrt[3]{125m}$

SECTION 9.4 Multiplying and Dividing Radical Expressions

OBJECTIVES

1. Multiply radical expressions.
2. Find powers of radical expressions.
3. Rationalize denominators.
4. Rationalize denominators that have two terms.
5. Rationalize numerators.

ARE YOU READY?

The following problems review some basic skills that are needed when multiplying and dividing radical expressions.

Perform the indicated operations and simplify, if possible.

1. $(5a^6)^2$

2. $9t^3(6t^3 + 2t^2)$

3. $(2x + 3)(x - 1)$

4. $7 - 3\sqrt{14} + \sqrt{14} - 6$

5. $(x + 10)(x - 10)$

6. Build an equivalent fraction for $\frac{2}{3a}$ with a denominator of $27a$.

In this section, we will discuss the methods we can use to multiply and divide radical expressions.

1 Multiply Radical Expressions.

We have used the *product rule for radicals* to write radical expressions in simplified form. We also can use this rule to multiply radical expressions that have the same index.

The Product Rule for Radicals

The product of the nth roots of two nonnegative numbers is equal to the nth root of the product of those numbers.

If $\sqrt[n]{a}$ and $\sqrt[n]{b}$ are real numbers,

$$\sqrt[n]{a} \cdot \sqrt[n]{b} = \sqrt[n]{a \cdot b}$$

EXAMPLE 1 Multiply and then simplify: **a.** $\sqrt{5}\sqrt{10}$ **b.** $3\sqrt{6}(2\sqrt{3})$ **c.** $-2\sqrt[3]{7x} \cdot 6\sqrt[3]{49x^2}$

Strategy In each expression, we will use the product rule for radicals to multiply factors of the form $\sqrt[n]{a}$ and $\sqrt[n]{b}$.

Why The product rule for radicals is used to multiply radicals that have the same index.

Solution

a.

$\sqrt{5}\sqrt{10} = \sqrt{5 \cdot 10}$ Use the product rule for radicals.

$= \sqrt{50}$ Multiply under the radical. Note that $\sqrt{50}$ can be simplified.

$= \sqrt{25 \cdot 2}$ Prepare to simplify: factor 50.

$= 5\sqrt{2}$ Simplify: $\sqrt{25 \cdot 2} = \sqrt{25}\sqrt{2} = 5\sqrt{2}$.

Success Tip

Here we use the product rule for radicals in two ways: to *multiply* and to *simplify* the result.

b. We use the commutative and associative properties of multiplication to multiply the integer factors and the radicals separately. Then we simplify any radicals in the product, if possible.

$3\sqrt{6}(2\sqrt{3}) = 3(2)\sqrt{6}\sqrt{3}$ Multiply the integer factors, 3 and 2, and multiply the radicals.

$= 6\sqrt{18}$ Use the product rule for radicals.

$= 6\sqrt{9}\sqrt{2}$ Simplify: $\sqrt{18} = \sqrt{9 \cdot 2} = \sqrt{9}\sqrt{2}$.

$= 6(3)\sqrt{2}$ Evaluate: $\sqrt{9} = 3$.

$= 18\sqrt{2}$ Multiply 6 and 3 to get 18.

Notation

Multiplication of radicals can be shown in several ways:

$\sqrt{5} \cdot \sqrt{10} = \sqrt{5}(\sqrt{10}) = \sqrt{5}\sqrt{10}$

c. $-2\sqrt[3]{7x} \cdot 6\sqrt[3]{49x^2} = -2(6)\sqrt[3]{7x}\sqrt[3]{49x^2}$ Write the integer factors together and the radicals together.

$= -12\sqrt[3]{7x \cdot 49x^2}$ Multiply the integer factors, −2 and 6, and multiply the radicals.

$= -12\sqrt[3]{7x \cdot 7^2x^2}$ Write 49 as 7^2.

$= -12\sqrt[3]{7^3x^3}$ Prepare to simplify: write $7x \cdot 7^2x^2$ as 7^3x^3.

$= -12(7x)$ Simplify: $\sqrt[3]{7^3x^3} = 7x$.

$= -84x$ Multiply.

Caution

Note that to multiply radical expressions, they must have the same index.

$\sqrt[n]{a} \cdot \sqrt[n]{b} = \sqrt[n]{a \cdot b}$

Self Check 1 Multiply and then simplify: **a.** $\sqrt{7}\sqrt{14}$ **b.** $-2\sqrt[3]{2}\left(5\sqrt[3]{12}\right)$ **c.** $\sqrt[4]{4x^3} \cdot 9\sqrt[4]{8x^2}$

Now Try Problems 15, 23, and 25

Recall that to multiply a polynomial by a monomial, we use the distributive property. We use the same technique to multiply a radical expression that has two or more terms by a radical expression that has only one term.

EXAMPLE 2 Multiply and then simplify: $3\sqrt{3}\left(4\sqrt{8} - 5\sqrt{10}\right)$

Strategy We will use the distributive property and multiply each term within the parentheses by the term outside the parentheses.

Why The given expression has the form $a(b - c)$.

Solution

$3\sqrt{3}\left(4\sqrt{8} - 5\sqrt{10}\right)$

$= 3\sqrt{3} \cdot 4\sqrt{8} - 3\sqrt{3} \cdot 5\sqrt{10}$ Distribute the multiplication by $3\sqrt{3}$.

$= 12\sqrt{24} - 15\sqrt{30}$ Multiply the integer factors and use the product rule to multiply the radicals.

$= 12\sqrt{4}\sqrt{6} - 15\sqrt{30}$ Simplify: $\sqrt{24} = \sqrt{4 \cdot 6} = \sqrt{4}\sqrt{6}$.

$= 12(2)\sqrt{6} - 15\sqrt{30}$ Evaluate: $\sqrt{4} = 2$.

$= 24\sqrt{6} - 15\sqrt{30}$ Multiply 12 and 2 to get 24.

Self Check 2 Multiply and then simplify: $4\sqrt{2}\left(3\sqrt{5} - 2\sqrt{8}\right)$

Now Try Problems 31 and 33

Recall that to multiply two binomials, we multiply each term of one binomial by each term of the other binomial and simplify. We multiply two radical expressions, each having two terms, in the same way.

EXAMPLE 3 Multiply and then simplify: **a.** $\left(\sqrt{7} + \sqrt{2}\right)\left(\sqrt{7} - 9\sqrt{2}\right)$
b. $\left(\sqrt[3]{x^2} - 4\sqrt[3]{5}\right)\left(\sqrt[3]{x} + \sqrt[3]{2}\right)$

Strategy As with binomials, we will multiply each term within the first set of parentheses by each term within the second set of parentheses.

Why This is an application of the FOIL method for multiplying binomials.

Solution a. $(\sqrt{7}+\sqrt{2})(\sqrt{7}-9\sqrt{2})$

F O I L

$= \sqrt{7}\sqrt{7} - 9\sqrt{7}\sqrt{2} + \sqrt{2}\sqrt{7} - 9\sqrt{2}\sqrt{2}$ Use the FOIL method.

$= 7 - 9\sqrt{14} + \sqrt{14} - 9(2)$ Perform each multiplication.

$= 7 - 8\sqrt{14} - 18$ Combine like radicals: $-9\sqrt{14} + \sqrt{14} = -8\sqrt{14}$.

$= -11 - 8\sqrt{14}$ Combine like terms: $7 - 18 = -11$.

Caution

A common error is to "simplify" incorrectly by subtracting:

$-11 - 8\sqrt{14} \neq -19\sqrt{14}$

b. $(\sqrt[3]{x^2} - 4\sqrt[3]{5})(\sqrt[3]{x} + \sqrt[3]{2})$

$= \sqrt[3]{x^2}\sqrt[3]{x} + \sqrt[3]{x^2}\sqrt[3]{2} - 4\sqrt[3]{5}\sqrt[3]{x} - 4\sqrt[3]{5}\sqrt[3]{2}$ Use the FOIL method.

$= \sqrt[3]{x^3} + \sqrt[3]{2x^2} - 4\sqrt[3]{5x} - 4\sqrt[3]{10}$ Perform each multiplication.

$= x + \sqrt[3]{2x^2} - 4\sqrt[3]{5x} - 4\sqrt[3]{10}$ Simplify the first term. There are no like radicals or terms to combine.

Self Check 3 Multiply and then simplify:

a. $(\sqrt{5} + 2\sqrt{3})(\sqrt{5} - \sqrt{3})$

b. $(\sqrt[3]{a} + 9\sqrt[3]{2})(\sqrt[3]{a^2} - \sqrt[3]{3})$

Now Try Problems 37 and 41

2 Find Powers of Radical Expressions.

To find the power of a radical expression, such as $(\sqrt{5})^2$ or $(\sqrt[3]{2})^3$, we can use the definition of exponent and the product rule for radicals.

$$(\sqrt{5})^2 = \sqrt{5}\sqrt{5} = \sqrt{25} = 5 \qquad (\sqrt[3]{2})^3 = \sqrt[3]{2}\cdot\sqrt[3]{2}\cdot\sqrt[3]{2} = \sqrt[3]{8} = 2$$

These results illustrate the following property of radicals.

The *n*th Power of the *n*th Root

If $\sqrt[n]{a}$ is a real number,

$$(\sqrt[n]{a})^n = a$$

EXAMPLE 4 Find: a. $(\sqrt{5})^2$ b. $(2\sqrt[3]{7x^2})^3$ c. $(8-\sqrt{3})^2$ d. $(\sqrt{m+1}+2)^2$

Strategy In part (a), we will use the definition of square root. In part (b), we will use a power rule for exponents. In parts (c) and (d), we will use the FOIL method.

Why Part (a) is the square of a square root, part (b) has the form $(xy)^n$, and part (c) has the form $(x + y)^2$.

Solution a. $(\sqrt{5})^2 = 5$ Because the square of the square root of 5 is 5

b. We can use the power of a product rule for exponents to find $(2\sqrt[3]{7x^2})^3$.

$(2\sqrt[3]{7x^2})^3 = 2^3(\sqrt[3]{7x^2})^3$ Raise each factor of $2\sqrt[3]{7x^2}$ to the 3rd power.

$= 8(7x^2)$ Evaluate: $2^3 = 8$. Use $(\sqrt[n]{a})^n = a$.

$= 56x^2$ Multiply: $8 \cdot 7 = 56$.

Success Tip

Since $(8-\sqrt{3})^2$ has the form $(x-y)^2$, we could also use a special-product rule to find this square of a difference quickly.

c. $(8-\sqrt{3})^2 = (8-\sqrt{3})(8-\sqrt{3})$ Write the base $8-\sqrt{3}$ as a factor twice.

$= 64 - 8\sqrt{3} - 8\sqrt{3} + \sqrt{3}\sqrt{3}$ Use the FOIL method.

$= 64 - 16\sqrt{3} + 3$ Combine like radicals: $-8\sqrt{3} - 8\sqrt{3} = -16\sqrt{3}$.

$= 67 - 16\sqrt{3}$ Combine like terms: $64 + 3 = 67$.

Caution

A common error when squaring a two-term expression is to simply square the first term and square the last term, and forget the middle term of the product.

$(\sqrt{m+1}+2)^2 = (\sqrt{m+1})^2 + 2^2 = m + 1 + 4$ (crossed out)

d. We can use the FOIL method to find the product.

$$(\sqrt{m+1}+2)^2 = (\sqrt{m+1}+2)(\sqrt{m+1}+2)$$
$$= (\sqrt{m+1})^2 + 2\sqrt{m+1} + 2\sqrt{m+1} + 2 \cdot 2$$
$$= m + 1 + 2\sqrt{m+1} + 2\sqrt{m+1} + 4 \quad \text{Use } (\sqrt[n]{a})^n = a.$$
$$= m + 4\sqrt{m+1} + 5 \quad \text{Combine like terms.}$$

Self Check 4 Find: **a.** $(\sqrt{11})^2$ **b.** $(3\sqrt[3]{4y})^3$ **c.** $(\sqrt{3}+4)^2$ **d.** $(\sqrt{x-8}-5)^2$

Now Try Problems 43, 49, and 53

3 Rationalize Denominators.

We have seen that when a radical expression is written in simplified form, each of the following statements is true.

Simplified Form of a Radical Expression

1. Each factor in the radicand is to a power that is less than the index of the radical.
2. The radicand contains no fractions or negative numbers.
3. No radicals appear in the denominator of a fraction.

The Language of Algebra

Since $\sqrt{3}$ is an irrational number, $\frac{\sqrt{5}}{\sqrt{3}}$ has an **irrational denominator**. Since 3 is a rational number, $\frac{\sqrt{15}}{3}$ has a **rational denominator**.

Success Tip

As an informal check, we can use a calculator to approximate the value of each expression.

$$\frac{\sqrt{5}}{\sqrt{3}} \approx 1.290994449$$
$$\frac{\sqrt{15}}{3} \approx 1.290994449$$

We now consider radical expressions that do not satisfy requirements 2 or 3. We will introduce an algebraic technique, called **rationalizing the denominator,** that is used to write such expressions in an equivalent simplified form. In this process, we multiply the expression by a form of 1 and use the fact that $\sqrt[n]{a^n} = a$.

As an example, let's consider the following expression:

$\frac{\sqrt{5}}{\sqrt{3}}$ This radical expression is not in simplified form, because a radical appears in the denominator. It doesn't satisfy requirement 3 listed above.

We want to find a fraction equivalent to $\frac{\sqrt{5}}{\sqrt{3}}$ that does not have a radical in its denominator. If we multiply $\frac{\sqrt{5}}{\sqrt{3}}$ by $\frac{\sqrt{3}}{\sqrt{3}}$, the denominator becomes $\sqrt{3}\cdot\sqrt{3} = 3$, a rational number.

$\frac{\sqrt{5}}{\sqrt{3}} = \frac{\sqrt{5}}{\sqrt{3}} \cdot \frac{\sqrt{3}}{\sqrt{3}}$ To build an equivalent fraction, multiply by a form of 1: $\frac{\sqrt{3}}{\sqrt{3}} = 1$.

$= \frac{\sqrt{15}}{3}$ Multiply the numerators: $\sqrt{5}\cdot\sqrt{3} = \sqrt{15}$. Multiply the denominators: $\sqrt{3}\cdot\sqrt{3} = (\sqrt{3})^2 = 3$. The denominator is now a rational number, 3.

Thus, $\frac{\sqrt{5}}{\sqrt{3}} = \frac{\sqrt{15}}{3}$. These equivalent fractions represent the same number, but have different forms. Since there is no radical in the denominator, and $\sqrt{15}$ is in simplest form, the expression $\frac{\sqrt{15}}{3}$ is in simplified form. We say that we have *rationalized the denominator* of $\frac{\sqrt{5}}{\sqrt{3}}$.

EXAMPLE 5 Rationalize the denominator: **a.** $\sqrt{\frac{20}{7}}$ **b.** $\frac{4}{\sqrt[3]{2}}$

Strategy We look at each denominator and ask, "By what must we multiply it to obtain a rational number?" Then we will multiply each expression by a carefully chosen form of 1.

Why We want to produce an equivalent expression that does not have a radical in its denominator.

Solution

a. This radical expression is not in simplified form, because the radicand contains a fraction. (It doesn't satisfy requirement 2.) We begin by writing the square root of the quotient as the quotient of two square roots:

$$\sqrt{\frac{20}{7}} = \frac{\sqrt{20}}{\sqrt{7}}$$ Use the division property of radicals: $\sqrt[n]{\frac{a}{b}} = \frac{\sqrt[n]{a}}{\sqrt[n]{b}}$.

To rationalize the denominator, we proceed as follows:

$$\frac{\sqrt{20}}{\sqrt{7}} = \frac{\sqrt{20}}{\sqrt{7}} \cdot \frac{\sqrt{7}}{\sqrt{7}}$$ To build an equivalent fraction, multiply by $\frac{\sqrt{7}}{\sqrt{7}} = 1$.

$$= \frac{\sqrt{140}}{7}$$ Multiply the numerators. Multiply the denominators: $\sqrt{7} \cdot \sqrt{7} = (\sqrt{7})^2 = 7$. The denominator is now a rational number, 7.

$$= \frac{2\sqrt{35}}{7}$$ Simplify: $\sqrt{140} = \sqrt{4 \cdot 35} = \sqrt{4}\sqrt{35} = 2\sqrt{35}$.

Caution

Do not attempt to remove a common factor of 7 from the numerator and denominator of $\frac{2\sqrt{35}}{7}$. The numerator, $2\sqrt{35}$, does not have a factor of 7.

$$\frac{2\sqrt{35}}{7} = \frac{2 \cdot \sqrt{5 \cdot 7}}{7}$$

b. This expression is not in simplified form because a radical appears in the denominator of a fraction. (It doesn't satisfy requirement 3.) Here, we must rationalize a denominator that is a cube root. We multiply the numerator and the denominator by a number that will give a perfect cube under the radical. Since $2 \cdot 4 = 8$ is a perfect cube, $\sqrt[3]{4}$ is such a number.

$$\frac{4}{\sqrt[3]{2}} = \frac{4}{\sqrt[3]{2}} \cdot \frac{\sqrt[3]{4}}{\sqrt[3]{4}}$$ To build an equivalent fraction, multiply by $\frac{\sqrt[3]{4}}{\sqrt[3]{4}} = 1$.

$$= \frac{4\sqrt[3]{4}}{\sqrt[3]{8}}$$ Multiply the numerators. Multiply the denominators. This radicand is now a perfect cube.

$$= \frac{4\sqrt[3]{4}}{2}$$ Evaluate the denominator: $\sqrt[3]{8} = 2$. The denominator is now a rational number, 2.

$$= 2\sqrt[3]{4}$$ Simplify the fraction: $\frac{4\sqrt[3]{4}}{2} = \frac{\overset{1}{\cancel{2}} \cdot 2\sqrt[3]{4}}{\underset{1}{\cancel{2}}} = 2\sqrt[3]{4}$.

Caution

Multiplying $\frac{4}{\sqrt[3]{2}}$ by $\frac{\sqrt[3]{2}}{\sqrt[3]{2}}$ does not rationalize the denominator.

$$\frac{4}{\sqrt[3]{2}} \cdot \frac{\sqrt[3]{2}}{\sqrt[3]{2}} = \frac{4\sqrt[3]{2}}{\sqrt[3]{4}}$$

Since 4 is not a perfect cube, this radical does not simplify.

An alternate way to rationalize the denominator is to use the equivalent form $\frac{\sqrt[3]{2^2}}{\sqrt[3]{2^2}}$ instead of $\frac{\sqrt[3]{4}}{\sqrt[3]{4}}$ to build an equivalent fraction:

$$\frac{4}{\sqrt[3]{2}} \cdot \frac{\sqrt[3]{2^2}}{\sqrt[3]{2^2}} = \frac{4\sqrt[3]{2^2}}{\sqrt[3]{2^3}}$$ Multiply the numerators. Multiply the denominators. In the denominator, the radicand is now a perfect cube because $2 \cdot 2^2 = 2^3$.

$$= \frac{4\sqrt[3]{4}}{2}$$ In the numerator, $2^2 = 4$. Simplify in the denominator, $\sqrt[3]{2^3} = 2$.

$$= 2\sqrt[3]{4}$$ Simplify the fraction: $\frac{4\sqrt[3]{4}}{2} = \frac{\overset{1}{\cancel{2}} \cdot 2\sqrt[3]{4}}{\underset{1}{\cancel{2}}} = 2\sqrt[3]{4}$.

Self Check 5 Rationalize the denominator: **a.** $\sqrt{\frac{8}{5}}$ **b.** $\frac{5}{\sqrt[3]{9}}$

Now Try Problems 57, 59, and 63

EXAMPLE 6

Rationalize the denominator: $\frac{\sqrt{5xy^2}}{\sqrt{xy^3}}$

Strategy We will begin by using the quotient rule for radicals in reverse: $\frac{\sqrt[n]{a}}{\sqrt[n]{b}} = \sqrt[n]{\frac{a}{b}}$

Why When the radicands are written under a single radical symbol, the result is a rational expression. Our hope is that the rational expression can be simplified, which could possibly make rationalizing the denominator easier.

Solution There are two methods we can use to rationalize the denominator. In each method, we simplify the rational expression $\frac{5xy^2}{xy^3}$ that appears in the radicand first.

Caution

We will assume that all of the variables appearing in the following examples represent positive numbers.

Method 1

$$\frac{\sqrt{5xy^2}}{\sqrt{xy^3}} = \sqrt{\frac{5xy^2}{xy^3}}$$

$$= \sqrt{\frac{5}{y}}$$ Simplify the radicand.

$$= \frac{\sqrt{5}}{\sqrt{y}}$$ Use the quotient rule.

$$= \frac{\sqrt{5}}{\sqrt{y}} \cdot \frac{\sqrt{y}}{\sqrt{y}}$$ Multiply outside the radical.

$$= \frac{\sqrt{5y}}{y}$$ Multiply the numerators and the denominators.

Method 2

$$\frac{\sqrt{5xy^2}}{\sqrt{xy^3}} = \sqrt{\frac{5xy^2}{xy^3}}$$

$$= \sqrt{\frac{5}{y}}$$ Simplify the radicand.

$$= \sqrt{\frac{5}{y} \cdot \frac{y}{y}}$$ Multiply within the radical.

$$= \frac{\sqrt{5y}}{\sqrt{y^2}}$$ Use the quotient rule.

$$= \frac{\sqrt{5y}}{y}$$ Simplify the denominator.

Self Check 6 Rationalize the denominator: $\frac{\sqrt{4ab^3}}{\sqrt{2a^2b^2}}$

Now Try Problems 67 and 73

EXAMPLE 7

Rationalize the denominator: $\frac{11}{\sqrt{20q^5}}$

Strategy We will simplify the radical expression in the denominator before rationalizing the denominator.

Why We could begin by multiplying $\frac{11}{\sqrt{20q^5}}$ by $\frac{\sqrt{20q^5}}{\sqrt{20q^5}}$. However, to work with smaller numbers and simpler radical expressions, it is easier if we simplify $\sqrt{20q^5}$ first, and then rationalize the denominator.

Solution

$$\frac{11}{\sqrt{20q^5}} = \frac{11}{\sqrt{4q^4 \cdot 5q}}$$ To prepare to simplify $\sqrt{20q^5}$, factor $20q^5$ as $4q^4 \cdot 5q$.

$$= \frac{11}{2q^2\sqrt{5q}}$$ Simplify: $\sqrt{4q^4 \cdot 5q} = \sqrt{4q^4}\sqrt{5q} = 2q^2\sqrt{5q}$.

Success Tip

We usually simplify a radical expression before rationalizing the denominator.

$$= \frac{11}{2q^2\sqrt{5q}} \cdot \frac{\sqrt{5q}}{\sqrt{5q}}$$ To rationalize the denominator, multiply by $\frac{\sqrt{5q}}{\sqrt{5q}} = 1$.

$$= \frac{11\sqrt{5q}}{2q^2(5q)}$$ Multiply the numerators. Multiply the denominators: $\sqrt{5q} \cdot \sqrt{5q} = (\sqrt{5q})^2 = 5q$.

$$= \frac{11\sqrt{5q}}{10q^3}$$ Multiply in the denominator: $2 \cdot 5 = 10$ and $q^2 \cdot q = q^3$.

Self Check 7 Rationalize the denominator: $\frac{7}{\sqrt{18c^3}}$

Now Try Problems 75 and 81

EXAMPLE 8 Rationalize each denominator: **a.** $\frac{5}{\sqrt[3]{6n^2}}$ **b.** $\frac{\sqrt[4]{2}}{\sqrt[4]{3a}}$

Strategy In part (a), we will examine the radicand in the denominator and ask, "By what must we multiply it to obtain a perfect cube?" In part (b), we will examine the radicand in the denominator and ask, "By what must we multiply it to obtain a perfect-fourth power?"

Why The answers to those questions will determine what form of 1 we use to rationalize each denominator.

Solution

a. To rationalize the denominator $\sqrt[3]{6n^2}$, we need the radicand to be a perfect cube. Since $6n^2 = 6 \cdot n \cdot n$, the radicand needs two more factors of 6 and one more factor of n. It follows that we should multiply the given expression by $\frac{\sqrt[3]{6^2n}}{\sqrt[3]{6^2n}}$.

$$\frac{5}{\sqrt[3]{6n^2}} = \frac{5}{\sqrt[3]{6n^2}} \cdot \frac{\sqrt[3]{6^2n}}{\sqrt[3]{6^2n}}$$ Multiply by a form of 1 to rationalize the denominator.

$$= \frac{5\sqrt[3]{36n}}{\sqrt[3]{6^3n^3}}$$ Multiply the numerators. Multiply the denominators. This radicand is now a perfect cube.

$$= \frac{5\sqrt[3]{36n}}{6n}$$ Simplify the denominator: $\sqrt[3]{6^3n^3} = 6n$.

Success Tip

An alternate form of 1 that can be used to rationalize the denominator is $\frac{\sqrt[3]{36n}}{\sqrt[3]{36n}}$.

b. To rationalize the denominator $\sqrt[4]{3a}$, we need the radicand to be a perfect-fourth power. Since $3a = 3 \cdot a$, the radicand needs three more factors of 3 and three more factors of a. It follows that we should multiply the given expression by $\frac{\sqrt[4]{3^3a^3}}{\sqrt[4]{3^3a^3}}$.

$$\frac{\sqrt[4]{2}}{\sqrt[4]{3a}} = \frac{\sqrt[4]{2}}{\sqrt[4]{3a}} \cdot \frac{\sqrt[4]{3^3a^3}}{\sqrt[4]{3^3a^3}}$$ Multiply by a form of 1 to rationalize the denominator.

$$= \frac{\sqrt[4]{54a^3}}{\sqrt[4]{3^4a^4}}$$ Multiply the numerators: $2 \cdot 27 = 54$. Multiply the denominators. This radicand is now a perfect-fourth power.

$$= \frac{\sqrt[4]{54a^3}}{3a}$$ Simplify the denominator: $\sqrt[4]{3^4a^4} = 3a$.

Success Tip

An alternate form of 1 that can be used to rationalize the denominator is $\frac{\sqrt[4]{27a^3}}{\sqrt[4]{27a^3}}$.

Self Check 8 Rationalize each denominator: **a.** $\frac{27}{\sqrt[3]{100a}}$ **b.** $\frac{\sqrt[4]{3}}{\sqrt[4]{4y^2}}$

Now Try Problems 83 and 87

4 Rationalize Denominators That Have Two Terms.

So far, we have rationalized denominators that have only one term. We will now discuss a method to rationalize denominators that have two terms.

One-termed denominators			*Two-termed denominators*	
$\frac{\sqrt{5}}{\sqrt{3}}$	$\frac{11}{\sqrt{20q^5}}$	$\frac{4}{\sqrt[3]{2}}$	$\frac{1}{\sqrt{2}+1}$	$\frac{\sqrt{x}+\sqrt{2}}{\sqrt{x}-\sqrt{2}}$

To rationalize the denominator of $\frac{1}{\sqrt{2}+1}$, for example, we multiply the numerator and denominator by $\sqrt{2}-1$, because the product $(\sqrt{2}+1)(\sqrt{2}-1)$ contains no radicals.

$$(\sqrt{2}+1)(\sqrt{2}-1) = (\sqrt{2})^2 - (1)^2 \quad \text{Use a special-product rule.}$$
$$= 2 - 1$$
$$= 1$$

Radical expressions that involve the sum and difference of the same two terms, such as $\sqrt{2}+1$ and $\sqrt{2}-1$, are called **conjugates.**

EXAMPLE 9 Rationalize the denominator: **a.** $\frac{1}{\sqrt{2}+1}$ **b.** $\frac{\sqrt{x}+\sqrt{2}}{\sqrt{x}-\sqrt{2}}$

Strategy In each part, we will rationalize the denominator by multiplying the numerator and the denominator by the conjugate of the denominator.

Why Multiplying each denominator by its conjugate will produce a new denominator that does not contain radicals.

Solution

a. To find a fraction equivalent to $\frac{1}{\sqrt{2}+1}$ that does not have a radical in its denominator, we multiply $\frac{1}{\sqrt{2}+1}$ by a form of 1 that uses the conjugate of $\sqrt{2}+1$.

$$\frac{1}{\sqrt{2}+1} = \frac{1}{\sqrt{2}+1}\cdot\frac{\sqrt{2}-1}{\sqrt{2}-1}$$

$$= \frac{\sqrt{2}-1}{(\sqrt{2})^2-(1)^2} \quad \text{Multiply the numerators. Multiply the denominators using a special-product rule.}$$

$$= \frac{\sqrt{2}-1}{2-1} \quad \text{In the denominator, } (\sqrt{2})^2 = 2.$$

$$= \frac{\sqrt{2}-1}{1}$$

$$= \sqrt{2}-1$$

Success Tip

In the denominator, we can use the special-product rule $(A+B)(A-B) = A^2 - B^2$ to multiply $\sqrt{2}+1$ and $\sqrt{2}-1$ quickly.

b. We multiply the numerator and denominator by $\sqrt{x}+\sqrt{2}$, which is the conjugate of $\sqrt{x}-\sqrt{2}$, and simplify.

$$\frac{\sqrt{x}+\sqrt{2}}{\sqrt{x}-\sqrt{2}} = \frac{\sqrt{x}+\sqrt{2}}{\sqrt{x}-\sqrt{2}}\cdot\frac{\sqrt{x}+\sqrt{2}}{\sqrt{x}+\sqrt{2}}$$

$$= \frac{x+\sqrt{2x}+\sqrt{2x}+2}{(\sqrt{x})^2-(\sqrt{2})^2} \quad \text{Multiply the numerators. Multiply the denominators using a special-product formula.}$$

$$= \frac{x+\sqrt{2x}+\sqrt{2x}+2}{x-2} \quad \text{In the denominator, } (\sqrt{x})^2 = x \text{ and } (\sqrt{2})^2 = 2.$$

$$= \frac{x+2\sqrt{2x}+2}{x-2} \quad \text{In the numerator, combine like radicals: } \sqrt{2x}+\sqrt{2x} = 2\sqrt{2x}.$$

Self Check 9 Rationalize the denominator: $\dfrac{\sqrt{x} - \sqrt{2}}{\sqrt{x} + \sqrt{2}}$

Now Try Problems 91 and 97

5 Rationalize Numerators.

In some advanced mathematics courses, like calculus, we sometimes have to rationalize a numerator by multiplying the numerator and denominator of the fraction by the conjugate of the numerator.

EXAMPLE 10 Rationalize the numerator: $\dfrac{\sqrt{x} - 3}{\sqrt{x}}$

Strategy To rationalize the numerator, we will multiply the numerator and the denominator by the conjugate of the numerator.

Why After rationalizing the numerator, we can simplify the expression. Although the result will not be in simplified form, this nonsimplified form is often desirable in calculus.

Solution We multiply the numerator and denominator by $\sqrt{x} + 3$, which is the conjugate of the numerator.

$$\frac{\sqrt{x} - 3}{\sqrt{x}} = \frac{\sqrt{x} - 3}{\sqrt{x}} \cdot \frac{\sqrt{x} + 3}{\sqrt{x} + 3}$$ Multiply by a form of 1 to rationalize the numerator.

$$= \frac{(\sqrt{x})^2 - 3^2}{x + 3\sqrt{x}}$$ Multiply the numerators using a special-product rule. Multiply the denominators using the distributive property.

$$= \frac{x - 9}{x + 3\sqrt{x}}$$ In the numerator, $(\sqrt{x})^2 = x$ and $3^2 = 9$.

Self Check 10 Rationalize the numerator: $\dfrac{\sqrt{x} + 3}{\sqrt{x}}$

Now Try Problem 101

SECTION 9.4 STUDY SET

VOCABULARY

Fill in the blanks.

1. In this section, we used the ________ rule for radicals in reverse: $\sqrt[n]{a} \cdot \sqrt[n]{b} = \sqrt[n]{ab}$.
2. To multiply $2\sqrt{5}(3\sqrt{8} + \sqrt{3})$, use the __________ property.
3. To __________ the denominator of $\frac{4}{\sqrt{5}}$, we multiply the fraction by $\frac{\sqrt{5}}{\sqrt{5}}$.
4. The denominator of the fraction $\frac{4}{\sqrt{5}}$ is an __________ number.
5. To obtain a ________-cube radicand in the denominator of $\frac{\sqrt[3]{7}}{\sqrt[3]{5n}}$, we multiply the fraction by $\frac{\sqrt[3]{25n^2}}{\sqrt[3]{25n^2}}$.
6. The __________ of $\sqrt{x} + 1$ is $\sqrt{x} - 1$.

CONCEPTS

7. Tell why each of the following expressions is not in simplified radical form. Then simplify it. Finally, use a calculator to approximate its value.

	Why isn't it in simplified form?	Simplified form	Approximation
$\frac{3}{\sqrt{2}}$			
$\frac{\sqrt{18}}{2}$			
$\sqrt{\frac{9}{2}}$			

8. Fill in the blank: To rationalize the denominator of $\frac{3}{\sqrt{2}}$, we multiply it by $\frac{\sqrt{2}}{\sqrt{2}}$, which is a form of ___.

9. Fill in the blanks to complete this special product:

$$(5 - \sqrt{x})^2 = (\ \ )^2 - \ \ (5)(\sqrt{x}) + (\ \ )^2$$
$$= \ \ - 10\sqrt{x} + \ \ $$

10. Fill in the blanks to complete this special product:

$$(\sqrt{7} + 2)(\sqrt{7} - 2) = (\sqrt{7})^{\ } - (2)^{\ }$$
$$= \ \ - 4$$
$$= \ \ $$

11. Perform each operation, if possible.

a. $4\sqrt{6} + 2\sqrt{6}$
b. $4\sqrt{6}(2\sqrt{6})$
c. $3\sqrt{2} - 2\sqrt{3}$
d. $3\sqrt{2}(-2\sqrt{3})$

12. Perform each operation, if possible.

a. $5 + 6\sqrt[3]{6}$
b. $5(6\sqrt[3]{6})$
c. $\frac{30\sqrt[3]{15}}{5}$
d. $\frac{\sqrt[3]{15}}{5}$

NOTATION

Fill in the blanks.

13. Multiply:

$$5\sqrt{8} \cdot 7\sqrt{6} = 5(7)\sqrt{8}\ \ $$
$$= 35\sqrt{\ \ }$$
$$= 35\sqrt{\ \ \cdot 3}$$
$$= 35(\ \ )\sqrt{3}$$
$$= 140\sqrt{3}$$

14. Rationalize the denominator:

$$\frac{9}{\sqrt[3]{4a^2}} = \frac{9}{\sqrt[3]{4a^2}} \cdot \frac{\sqrt[3]{2a}}{\ \ }$$
$$= \frac{9\sqrt[3]{2a}}{\sqrt[3]{\ \ }}$$
$$= \frac{9\sqrt[3]{2a}}{\ \ }$$

GUIDED PRACTICE

Multiply and simplify. All variables represent positive real numbers. See Example 1.

15. $\sqrt{3}\sqrt{15}$
16. $\sqrt{5}\sqrt{15}$
17. $2\sqrt{3}\sqrt{6}$
18. $-3\sqrt{11}\sqrt{33}$
19. $(3\sqrt[3]{9})(2\sqrt[3]{3})$
20. $(2\sqrt[3]{16})(-\sqrt[3]{4})$
21. $\sqrt[3]{2} \cdot \sqrt[3]{12}$
22. $\sqrt[3]{3} \cdot \sqrt[3]{18}$
23. $6\sqrt{ab^3}(8\sqrt{ab})$
24. $3\sqrt{8x}(2\sqrt{2x^3y})$
25. $\sqrt[4]{5a^3}\sqrt[4]{125a^2}$
26. $\sqrt[4]{2r^3}\sqrt[4]{8r^2}$

Multiply and simplify. All variables represent positive real numbers. See Example 2.

27. $3\sqrt{5}(4 - \sqrt{5})$
28. $2\sqrt{7}(3 - \sqrt{7})$
29. $\sqrt{2}(4\sqrt{6} + 2\sqrt{7})$
30. $-\sqrt{3}(\sqrt{7} - \sqrt{15})$
31. $-2\sqrt{5x}(4\sqrt{2x} - 3\sqrt{3})$
32. $3\sqrt{7t}(2\sqrt{7t} + 3\sqrt{3t^2})$
33. $\sqrt[3]{2}(4\sqrt[3]{4} + \sqrt[3]{12})$
34. $\sqrt[3]{3}(2\sqrt[3]{9} + \sqrt[3]{18})$

Multiply and simplify. All variables represent positive real numbers. See Example 3.

35. $(\sqrt{2} + 1)(\sqrt{2} - 3)$
36. $(2\sqrt{3} + 1)(\sqrt{3} - 1)$
37. $(\sqrt{3x} - \sqrt{2y})(\sqrt{3x} + \sqrt{2y})$
38. $(\sqrt{3m} + \sqrt{2n})(\sqrt{3m} - \sqrt{2n})$
39. $(2\sqrt[3]{4} - 3\sqrt[3]{2})(3\sqrt[3]{4} + 2\sqrt[3]{10})$
40. $(4\sqrt[3]{9} - 3\sqrt[3]{3})(4\sqrt[3]{3} + 2\sqrt[3]{6})$
41. $(\sqrt[3]{5z} + \sqrt[3]{3})(\sqrt[3]{5z} + 2\sqrt[3]{3})$
42. $(\sqrt[3]{3p} - 2\sqrt[3]{2})(\sqrt[3]{3p} + \sqrt[3]{2})$

Square or cube each quantity and simplify the result. See Example 4.

43. $(\sqrt{7})^2$
44. $(\sqrt{11})^2$
45. $(\sqrt[3]{12})^3$
46. $(\sqrt[3]{9})^3$
47. $(3\sqrt{2})^2$
48. $(2\sqrt{5})^2$

49. $\left(-2\sqrt[3]{2x^2}\right)^3$
50. $\left(-3\sqrt[3]{10y^3}\right)^3$
51. $\left(6-\sqrt{3}\right)^2$
52. $\left(9-\sqrt{11}\right)^2$
53. $\left(\sqrt{3x}+\sqrt{3}\right)^2$
54. $\left(\sqrt{5x}-\sqrt{3}\right)^2$

Rationalize each denominator. See Example 5.

55. $\sqrt{\frac{2}{7}}$
56. $\sqrt{\frac{5}{3}}$
57. $\sqrt{\frac{8}{3}}$
58. $\sqrt{\frac{8}{7}}$
59. $\frac{4}{\sqrt{6}}$
60. $\frac{8}{\sqrt{10}}$
61. $\frac{1}{\sqrt[3]{2}}$
62. $\frac{2}{\sqrt[3]{6}}$
63. $\frac{3}{\sqrt[3]{9}}$
64. $\frac{2}{\sqrt[3]{a}}$
65. $\frac{1}{\sqrt[4]{8}}$
66. $\frac{1}{\sqrt[5]{2}}$

Rationalize each denominator. All variables represent positive real numbers. See Example 6.

67. $\frac{\sqrt{10y^2}}{\sqrt{2y^3}}$
68. $\frac{\sqrt{15b^2}}{\sqrt{5b^3}}$
69. $\frac{\sqrt{48x^2}}{\sqrt{8x^2y}}$
70. $\frac{\sqrt{9xy}}{\sqrt{3x^2y}}$
71. $\frac{\sqrt[3]{12t^3}}{\sqrt[3]{54t^2}}$
72. $\frac{\sqrt[3]{15m^4}}{\sqrt[3]{12m^3}}$
73. $\frac{\sqrt[3]{4a^6}}{\sqrt[3]{2a^5b}}$
74. $\frac{\sqrt[3]{9x^5y^4}}{\sqrt[3]{3x^5y^5}}$

Rationalize each denominator. All variables represent positive real numbers. See Example 7.

75. $\frac{23}{\sqrt{50p^5}}$
76. $\frac{11}{\sqrt{75s^5}}$
77. $\frac{7}{\sqrt{24b^3}}$
78. $\frac{13}{\sqrt{32n^3}}$

Rationalize each denominator. All variables represent positive real numbers. See Example 8.

79. $\sqrt[3]{\frac{5}{16}}$
80. $\sqrt[3]{\frac{2}{81}}$
81. $\sqrt[3]{\frac{4}{81}}$
82. $\sqrt[3]{\frac{7}{16}}$
83. $\frac{19}{\sqrt[3]{5c^2}}$
84. $\frac{1}{\sqrt[3]{4m^2}}$
85. $\frac{\sqrt[3]{3}}{\sqrt[3]{2r}}$
86. $\frac{\sqrt[3]{7}}{\sqrt[3]{100s}}$
87. $\frac{\sqrt[4]{2}}{\sqrt[4]{3t^2}}$
88. $\frac{\sqrt[4]{3}}{\sqrt[4]{5b^3}}$
89. $\frac{25}{\sqrt[4]{8a}}$
90. $\frac{4}{\sqrt[4]{9t}}$

Rationalize each denominator. All variables represent positive real numbers. See Example 9.

91. $\frac{\sqrt{2}}{\sqrt{5}+3}$
92. $\frac{\sqrt{3}}{\sqrt{3}-2}$
93. $\frac{2}{\sqrt{x}+1}$
94. $\frac{3}{\sqrt{x}-2}$
95. $\frac{\sqrt{7}-\sqrt{2}}{\sqrt{2}+\sqrt{7}}$
96. $\frac{\sqrt{3}+\sqrt{2}}{\sqrt{3}-\sqrt{2}}$
97. $\frac{\sqrt{x}-\sqrt{y}}{\sqrt{x}+\sqrt{y}}$
98. $\frac{\sqrt{x}+\sqrt{y}}{\sqrt{x}-\sqrt{y}}$

Rationalize each numerator. All variables represent positive real numbers. See Example 10.

99. $\frac{\sqrt{x}+3}{x}$
100. $\frac{2+\sqrt{x}}{5x}$
101. $\frac{\sqrt{x}+\sqrt{y}}{\sqrt{x}}$
102. $\frac{\sqrt{x}-\sqrt{y}}{\sqrt{x}+\sqrt{y}}$

TRY IT YOURSELF

The following problems involve addition, subtraction, and multiplication of radical expressions, as well as rationalizing the denominator. Perform the operations and simplify, if possible. All variables represent positive real numbers.

103. $\sqrt{x}\left(\sqrt{14x}+\sqrt{2}\right)$
104. $2\sqrt[3]{16}-3\sqrt[3]{128}-\sqrt[3]{54}$
105. $\frac{3\sqrt{2}-5\sqrt{3}}{2\sqrt{3}-3\sqrt{2}}$
106. $\frac{3\sqrt{6}+5\sqrt{5}}{2\sqrt{5}-3\sqrt{6}}$
107. $\left(10\sqrt[3]{2x}\right)^3$
108. $\frac{\sqrt{3}}{\sqrt{98x^2}}$
109. $-4\sqrt[3]{5r^2s}\left(5\sqrt[3]{2r}\right)$
110. $-\sqrt[3]{3xy^2}\left(-\sqrt[3]{9x^3}\right)$
111. $\left(3p+\sqrt{5}\right)^2$
112. $\sqrt{288t}+\sqrt{80t}-\sqrt{128t}$
113. $\sqrt{\frac{72m^8}{25m^3}}$
114. $\left(\sqrt{14x}+\sqrt{3}\right)\left(\sqrt{14x}-\sqrt{3}\right)$
115. $\sqrt[4]{3n^2}\sqrt[4]{27n^3}$
116. $\frac{\sqrt{y}-2}{\sqrt{y}+3}$
117. $\frac{\sqrt[3]{x}}{\sqrt[3]{9}}$
118. $\sqrt[5]{\frac{2}{243}}$

119. $\left(3\sqrt{2r} - 2\right)^2$

120. $\left(2\sqrt{3t} + 5\right)^2$

121. $\sqrt{x(x+3)}\sqrt{x^3(x+3)}$

122. $\sqrt{y^2(x+y)}\sqrt{(x+y)^3}$

123. $\dfrac{2z - 1}{\sqrt{2z} - 1}$
(*Hint:* Do not perform the multiplication of the numerators.)

124. $\dfrac{3t - 1}{\sqrt{3t} + 1}$
(*Hint:* Do not perform the multiplication of the numerators.)

Look Alikes . . .

125. a. $\left(3\sqrt{a}\right)^2$ b. $\left(3 + \sqrt{a}\right)^2$

126. a. $\left(9\sqrt{x - 5}\right)^2$ b. $\left(9 + \sqrt{x - 5}\right)^2$

127. a. $\left(\sqrt{m - 6}\right)^2$ b. $\left(\sqrt{m} - 6\right)^2$

128. a. $\dfrac{1}{\sqrt{xy}}$ b. $\dfrac{1}{\sqrt{x} + \sqrt{y}}$

APPLICATIONS

129. Statistics. An example of a normal distribution curve, or *bell-shaped* curve, is shown. A fraction that is part of the equation that models this curve is $\dfrac{1}{\sigma\sqrt{2\pi}}$, where σ is a letter from the Greek alphabet. Rationalize the denominator of the fraction.

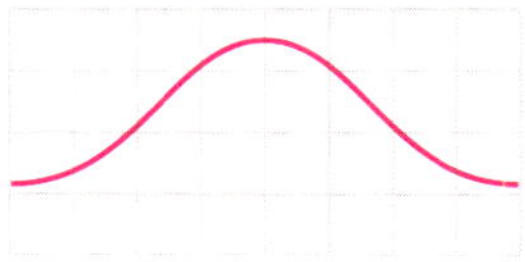

130. Analytical Geometry. The length of the perpendicular segment drawn from $(-2, 2)$ to the line with equation $2x - 4y = 4$ is given by

$$L = \frac{|2(-2) + (-4)(2) + (-4)|}{\sqrt{(2)^2 + (-4)^2}}$$

Find L. Express the result in simplified radical form. Then give an approximation to the nearest tenth.

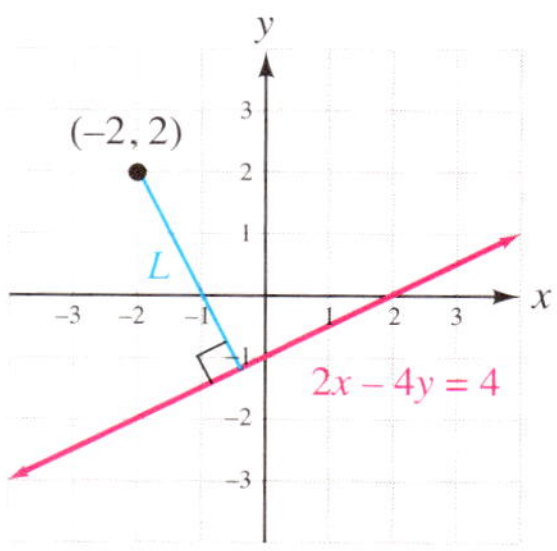

131. Trigonometry. In trigonometry, we often must find the ratio of the lengths of two sides of right triangles. Use the information in the illustration to find the ratio

$$\frac{\text{length of side } AC}{\text{length of side } AB}$$

Write the result in simplified radical form.

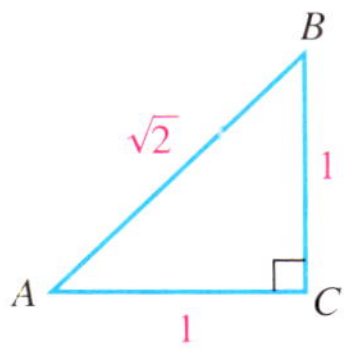

132. Engineering. Refer to the illustration below that shows a block connected to two walls by springs. A measure of how fast the block will oscillate when the spring system is set in motion is given by the formula $\omega = \sqrt{\dfrac{k_1 + k_2}{m}}$ where k_1 and k_2 indicate the stiffness of the springs and m is the mass of the block. Rationalize the right side and restate the formula.

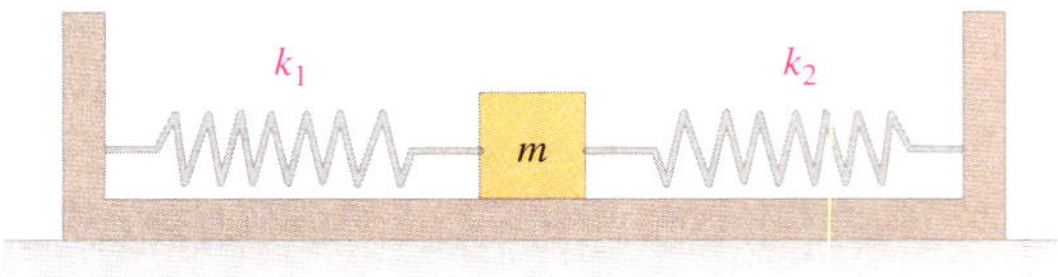

WRITING

133. Consider $\dfrac{\sqrt{3}}{\sqrt{7}} = \dfrac{\sqrt{3}}{\sqrt{7}} \cdot \dfrac{\sqrt{7}}{\sqrt{7}}$. Explain why the expressions on the left side and the right side of the equation are equal.

134. To rationalize the denominator of $\dfrac{\sqrt[4]{12}}{\sqrt[4]{3}}$, why wouldn't we multiply the numerator and denominator by $\dfrac{\sqrt[4]{3}}{\sqrt[4]{3}}$?

135. Explain why $\dfrac{\sqrt[3]{12}}{\sqrt[3]{5}}$ is not in simplified form.

136. Explain why $\sqrt{\dfrac{3a}{11k}}$ is not in simplified form.

137. Explain why $\sqrt{m} \cdot \sqrt{m} = m$ but $\sqrt[3]{m} \cdot \sqrt[3]{m} \neq m$. Assume that m represents a positive number.

138. Explain why the product of $\sqrt{m} + 3$ and $\sqrt{m} - 3$ does not contain a radical.

REVIEW

Solve each equation.

139. $\dfrac{8}{b - 2} + \dfrac{3}{2 - b} = -\dfrac{1}{b}$

140. $\dfrac{2}{x - 2} + \dfrac{1}{x + 1} = \dfrac{1}{(x + 1)(x - 2)}$

CHALLENGE PROBLEMS

141. Multiply: $\sqrt{2} \cdot \sqrt[3]{2}$. (*Hint:* Keep in mind two things. The indices (plural for *index*) must be the same to use the product rule for radicals, and radical expressions can be written using rational exponents.)

142. Show that $\dfrac{\sqrt[3]{a^2} + \sqrt[3]{a}\sqrt[3]{b} + \sqrt[3]{b^2}}{\sqrt[3]{a^2} + \sqrt[3]{a}\sqrt[3]{b} + \sqrt[3]{b^2}}$ can be used to rationalize the denominator of $\dfrac{1}{\sqrt[3]{a} - \sqrt[3]{b}}$.

SECTION 9.5 Solving Radical Equations

OBJECTIVES

1. Solve equations containing one radical.
2. Solve equations containing two radicals.
3. Solve formulas containing radicals.

ARE YOU READY?

The following problems review some basic skills that are needed when solving radical equations.

1. Simplify: **a.** $\left(\sqrt{x-1}\right)^2$ **b.** $\left(\sqrt[3]{x^3+7}\right)^3$
2. Simplify: $\left(3\sqrt{2x+5}\right)^2$
3. Expand: $(x-4)^2$
4. Solve: $x^2 - 6x - 27 = 0$
5. Expand: $(x+2)^3$
6. Multiply: $\left(2-\sqrt{x}\right)^2$

When we solve equations containing fractions, we clear them of the fractions by multiplying both sides by the LCD. To solve equations containing radical expressions, we take a similar approach. The first step is to clear them of the radicals by raising both sides to a power.

1 Solve Equations Containing One Radical.

Radical equations contain at least one radical expression with a variable in the radicand. Some examples are

$$\sqrt{x+3} = 4 \qquad \sqrt[3]{x^3+7} = x+1 \qquad \sqrt{x} + \sqrt{x+2} = 2$$

To **solve a radical equation,** we find all the values of the variable that make the equation true. The goal when solving a radical equation is to use the following **power rule** to find an equivalent equation that we already know how to solve, such as a linear equation in one variable or a quadratic equation.

The Power Rule for Solving Radical Equations

If we raise two equal quantities to the same power, the results are equal quantities. If x, y, and n are real numbers and $x = y$, then

$$x^n = y^n \quad \text{for any exponent } n.$$

Caution

A similar warning about checking proposed solutions was given in Chapter 7 when we solved rational equations, such as:

$$\frac{11x}{x-5} = 6 + \frac{55}{x-5}$$

If both sides of an equation are raised to the same power, all solutions of the original equation are also solutions of the new equation. However, the resulting equation might not be equivalent to the original equation. For example, if we square both sides of the equation $x = 3$ with a solution set of $\{3\}$, we obtain the equation $x^2 = 9$ with a solution set of $\{3, -3\}$.

The equations $x = 3$ and $x^2 = 9$ are not equivalent, because they have different solution sets. The solution -3 of $x^2 = 9$ does not satisfy the equation $x = 3$. Since raising both sides of an equation to the same power can produce an equation with proposed solutions that don't satisfy the original equation, **we must always check each proposed solution in the original equation** and discard any **extraneous solutions.**

When we use the power rule to solve square root radical equations, it produces expressions of the form $\left(\sqrt{a}\right)^2$. We have seen that when this expression is simplified, the radical symbol is removed.

The Square of a Square Root

For any nonnegative real number a, $\left(\sqrt{a}\right)^2 = a$.

Here are some examples of the square of a square root. Notice how squaring such an expression *removes the square root symbol.*

$$\left(\sqrt{4n}\right)^2 = 4n, \qquad \left(\sqrt{x-3}\right)^2 = x - 3, \qquad \text{and} \qquad \left(\sqrt{5a+8}\right)^2 = 5a + 8$$

EXAMPLE 1 Solve: $\sqrt{x+3} = 4$

Strategy We will use the power rule and square both sides of the equation.

Why Squaring both sides will produce, on the left side, the expression $\left(\sqrt{x+3}\right)^2$ that simplifies to $x + 3$. This step clears the equation of the radical.

Solution

$\sqrt{x+3} = 4$	This is the equation to solve.
$\left(\sqrt{x+3}\right)^2 = (4)^2$	To clear the equation of the square root, square both sides.
$x + 3 = 16$	Perform the operations on each side.
$x = 13$	Solve the resulting linear equation by subtracting 3 from both sides.

The Language of Algebra

When we square both sides of an equation, we are **raising both sides to the second power.**

We must check the proposed solution 13 to see whether it satisfies the original equation.

The Language of Algebra

Proposed solutions are also called **potential** or **possible** solutions.

Evaluate the left side. Do not square both sides when checking!

Check: $\sqrt{x+3} = 4$	This is the original equation.
$\sqrt{13+3} \stackrel{?}{=} 4$	Substitute 13 for x.
$\sqrt{16} \stackrel{?}{=} 4$	
$4 = 4$	True

Since 13 satisfies the original equation, it is the solution. The solution set is $\{13\}$.

Self Check 1 Solve: $\sqrt{a-2} = 3$

Now Try Problems 15 and 19

The method used in Example 1 to solve a radical equation containing a square root can be generalized, as follows.

Solving an Equation Containing Radicals

1. Isolate a radical term on one side of the equation.
2. Raise both sides of the equation to the power that is the same as the index of the radical.
3. If it still contains a radical, go back to step 1. If it does not contain a radical, solve the resulting equation.
4. Check the proposed solutions in the original equation.

EXAMPLE 2 **Free Fall.** The distance d in feet that an object will fall in t seconds is given by the formula $t = \sqrt{\frac{d}{16}}$. If the designers of the amusement park attraction want the riders to experience 3 seconds of vertical free fall, what length of vertical drop is needed?

Strategy We will begin by substituting 3 for the time t in the formula.

Why We can then solve the resulting radical equation in one variable to find the unknown distance d.

Solution

$$t = \sqrt{\frac{d}{16}}$$ This is the given formula.

$$3 = \sqrt{\frac{d}{16}}$$ Substitute 3 for t. Here the radical is isolated on the right side.

$$(3)^2 = \left(\sqrt{\frac{d}{16}}\right)^2$$ To clear the equation of the square root, square both sides.

$$9 = \frac{d}{16}$$ Perform the operations on each side.

$$144 = d$$ Solve the resulting equation by multiplying both sides by 16.

The amount of vertical drop needs to be 144 feet.

Caution

When using the power rule, don't forget to raise both sides to the same power. For this example, a common error would be to write

$$3 = \left(\sqrt{\frac{d}{16}}\right)^2$$

Self Check 2 **Free Fall.** How long a vertical drop is needed if the riders are to free fall for 3.5 seconds?

Now Try Problem 105

When solving radical equations, always isolate a radical term before using the power rule.

EXAMPLE 3 Solve: $\sqrt{3x + 1} + 1 = x$

Strategy Since 1 is outside the square root symbol, there are two terms on the left side of the equation. To isolate the radical term, we will subtract 1 from both sides.

Why This will put the equation in a form in which we can square both sides to clear the radical.

Solution

$$\sqrt{3x + 1} + 1 = x$$ This is the equation to solve.

$$\sqrt{3x + 1} = x - 1$$ To isolate the radical on the left side, subtract 1 from both sides.

$$\left(\sqrt{3x + 1}\right)^2 = (x - 1)^2$$ Square both sides to eliminate the square root. Don't forget the parentheses.

$$3x + 1 = x^2 - 2x + 1$$ On the right side, use a special-product formula to square the binomial: $(x - 1)^2 = x^2 - 2x + 1$. The resulting equation is quadratic.

$$0 = x^2 - 5x$$ To get 0 on the left side, subtract $3x$ and 1 from both sides.

$$0 = x(x - 5)$$ Factor out the GCF, x.

$$x = 0 \quad \text{or} \quad x - 5 = 0$$ Set each factor equal to 0.

$$x = 0 \quad | \quad x = 5$$

Caution

If both sides of the equation in Example 3 are squared, as is, the radical is not cleared. In fact, the equation becomes more complicated.

$$\left(\sqrt{3x + 1} + 1\right)^2 = (x)^2$$

$$3x + 1 + 2\sqrt{3x + 1} + 1 = x^2$$

We must check each proposed solution to see whether it satisfies the original equation.

This is the check for **0**:

$$\sqrt{3x + 1} + 1 = x$$

$$\sqrt{3(0) + 1} + 1 \stackrel{?}{=} 0$$

$$\sqrt{1} + 1 \stackrel{?}{=} 0$$

$$2 = 0$$ False

This is the check for **5**:

$$\sqrt{3x + 1} + 1 = x$$ This is the original equation.

$$\sqrt{3(5) + 1} + 1 \stackrel{?}{=} 5$$

$$\sqrt{16} + 1 \stackrel{?}{=} 5$$

$$5 = 5$$ True

The proposed solution 0 does not check; it must be discarded. Since the only solution is 5, the solution set is $\{5\}$.

Success Tip

Even if you are certain that no algebraic mistakes were made when solving a radical equation, you must still check your solutions. Raising both sides to an even power can introduce extraneous solutions that must be discarded.

Self Check 3 Solve: $\sqrt{4x+1}+1=x$

Now Try ▶ Problems 23 and 27

Using Your Calculator ▶ Solving Radical Equations

To find solutions for $\sqrt{3x+1}+1=x$ with a graphing calculator, we graph the functions $f(x)=\sqrt{3x+1}+1$ and $g(x)=x$, as in figure (a). We then trace to find the approximate x-coordinate of their intersection point, as in figure (b). After repeated zooms, we will see that $x=5$.

We also can use the INTERSECT feature to approximate the point of intersection of the graphs. See figure (c). The intersection point of (5, 5), with x-coordinate 5, implies that 5 is a solution of the radical equation.

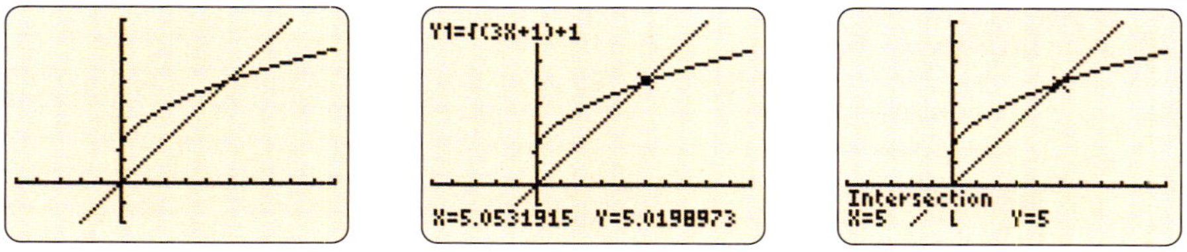

(a) (b) (c)

EXAMPLE 4 Solve: $\sqrt{3x}+8=2$

Strategy Since 8 is outside the square root symbol, there are two terms on the left side of the equation. To isolate the radical, we will subtract 8 from both sides.

Why This will put the equation in a form in which we can square both sides to clear the radical.

Solution

$$\sqrt{3x}+8=2 \quad \text{This is the equation to solve.}$$
$$\sqrt{3x}=-6 \quad \text{To isolate the radical on the left side, subtract 8 from both sides.}$$
$$\left(\sqrt{3x}\right)^2=(-6)^2 \quad \text{Square both sides to eliminate the square root.}$$
$$3x=36 \quad \text{Perform the operations on each side.}$$
$$x=12 \quad \text{To solve the resulting linear equation, divide both sides by 3.}$$

We check the proposed solution 12 in the original equation.

$$\sqrt{3x}+8=2$$
$$\sqrt{3(12)}+8\stackrel{?}{=}2 \quad \text{Substitute 12 for x.}$$
$$\sqrt{36}+8\stackrel{?}{=}2$$
$$6+8\stackrel{?}{=}2$$
$$14=2 \quad \text{False}$$

Since 12 does not satisfy the original equation, it is extraneous. The equation $\sqrt{3x}+8=2$ has no solution. The solution set is $\varnothing$.

Success Tip

After isolating the radical, we obtained the equation $\sqrt{3x}=-6$. Since the principal square root of a number cannot be negative, we immediately know that $\sqrt{3x}+8=2$ has no solution.

Self Check 4 Solve: $\sqrt{a-9}+3=0$

Now Try ▶ Problem 31

Recall the following property of radicals from Section 9.4.

The *n*th Power of the *n*th Root

If $\sqrt[n]{a}$ is a real number,

$$\left(\sqrt[n]{a}\right)^n = a$$

Here are some examples of the cube of a cube root, and the fourth power of a fourth root. Notice how raising each expression to the appropriate power *removes the radical symbol.*

$$\left(\sqrt[3]{5x}\right)^3 = 5x \qquad \left(\sqrt[3]{2a-11}\right)^3 = 2a - 11 \qquad \left(\sqrt[4]{6-y}\right)^4 = 6 - y$$

The power rule and the *n*th power of the *n*th root property can be used in combination to solve radical equations that involve cube roots, fourth roots, fifth roots, and so on.

EXAMPLE 5 Solve: $(x^3 + 7)^{1/3} = x + 1$

Strategy We will use the definition of a fractional exponent with numerator 1 to rewrite the given equation as an equivalent radical equation.

Why Then we can use the equation-solving strategy of this section to find the solution(s).

Solution Recall from Section 9.2 that $x^{1/n} = \sqrt[n]{x}$. Thus, we can replace $(x^3 + 7)^{1/3}$ on the left side of the given equation with $\sqrt[3]{x^3 + 7}$.

$(x^3 + 7)^{1/3} = x + 1$ This is the equation to solve.

$\sqrt[3]{x^3 + 7} = x + 1$ Write $(x^3 + 7)^{1/3}$ using radical notation.

$\left(\sqrt[3]{x^3 + 7}\right)^3 = (x + 1)^3$ Cube both sides to eliminate the cube root. Don't forget the parentheses.

$x^3 + 7 = x^3 + 3x^2 + 3x + 1$ Perform the operations on each side. On the right: $(x + 1)^3 = (x + 1)(x + 1)^2$.

$0 = 3x^2 + 3x - 6$ To get 0 on the left side, subtract x^3 and 7 from both sides. This is a quadratic equation.

$0 = x^2 + x - 2$ Divide both sides by 3.

$0 = (x + 2)(x - 1)$ Factor the trinomial.

$x + 2 = 0$ or $x - 1 = 0$ Set each factor equal to 0.

$x = -2$ | $x = 1$ Solve each linear equation.

We check each proposed solution, −2 and 1, to see whether they satisfy the original equation.

Check:

$(x^3 + 7)^{1/3} = x + 1$	$(x^3 + 7)^{1/3} = x + 1$
$\sqrt[3]{x^3 + 7} = x + 1$	$\sqrt[3]{x^3 + 7} = x + 1$
$\sqrt[3]{(-2)^3 + 7} \stackrel{?}{=} -2 + 1$	$\sqrt[3]{1^3 + 7} \stackrel{?}{=} 1 + 1$
$\sqrt[3]{-8 + 7} \stackrel{?}{=} -1$	$\sqrt[3]{1 + 7} \stackrel{?}{=} 2$
$\sqrt[3]{-1} \stackrel{?}{=} -1$	$\sqrt[3]{8} \stackrel{?}{=} 2$
$-1 = -1$ True	$2 = 2$ True

Both −2 and 1 satisfy the original equation. Thus, the solution set is $\{-2, 1\}$.

Self Check 5 Solve: $(x^3 + 8)^{1/3} = x + 2$

Now Try Problems 35 and 39

Success Tip

Be careful cubing the binomial $x + 1$.

$(x + 1)^3 \neq x^3 + 1^3$

Success Tip

Raising both sides of an equation to an even power may not produce an equivalent equation. Raising both sides to an odd power does produce an equivalent equation and, in such cases, there will be no extraneous solutions. However, it is still recommended that all proposed solutions be checked regardless of the type of root involved.

EXAMPLE 6 Let $f(x) = \sqrt[4]{2x + 1}$. For what value(s) of x is $f(x) = 5$?

Strategy We will substitute 5 for $f(x)$ and solve the equation $5 = \sqrt[4]{2x + 1}$. To do so, we will raise both sides of the equation to the fourth power.

Why Raising both sides to the fourth power will produce, on the right side, the expression $\left(\sqrt[4]{2x + 1}\right)^4$ that simplifies to $2x + 1$. This step clears the equation of the radical.

Solution To find the value(s) for which $f(x) = 5$, we substitute 5 for $f(x)$ and solve for x.

$$f(x) = \sqrt[4]{2x + 1}$$

$$5 = \sqrt[4]{2x + 1} \quad \text{This is the equation to solve.}$$

Since the equation contains a fourth root, we raise both sides to the fourth power to solve for x.

$$(5)^4 = \left(\sqrt[4]{2x + 1}\right)^4 \quad \text{Use the power rule to clear the radical.}$$

$$625 = 2x + 1 \quad \text{Perform the operations on each side.}$$

$$624 = 2x \quad \text{To solve the resulting equation, subtract 1 from both sides.}$$

$$312 = x \quad \text{Divide both sides by 2.}$$

If $x = 312$, then $f(x) = 5$. Verify this by evaluating $f(312)$ using a calculator, if necessary.

Self Check 6 Let $g(x) = \sqrt[5]{10x + 1}$. For what value(s) of x is $g(x) = 1$?

Now Try Problem 43

2 Solve Equations Containing Two Radicals.

To solve an equation containing two radicals, we want to have one radical on the left side and one radical on the right side.

EXAMPLE 7 Solve: $\sqrt{5x + 9} = 2\sqrt{3x + 4}$

Strategy We will square both sides to clear the equation of both radicals.

Why We can square both sides immediately since each radical is isolated on one side of the equation.

Solution

$$\sqrt{5x + 9} = 2\sqrt{3x + 4} \quad \text{This is the equation to solve.}$$

$$\left(\sqrt{5x + 9}\right)^2 = \left(2\sqrt{3x + 4}\right)^2 \quad \text{Square both sides to eliminate the radicals.}$$

$$5x + 9 = 2^2\left(\sqrt{3x + 4}\right)^2 \quad \text{Simplify on the left. On the right, raise each factor of the product } 2\sqrt{3x+4} \text{ to the second power.}$$

$$5x + 9 = 4(3x + 4) \quad \text{Perform the operations on the right.}$$

$$5x + 9 = 12x + 16 \quad \text{To solve the resulting linear equation, distribute the multiplication by 4.}$$

$$-7 = 7x \quad \text{Subtract 5x and 16 from both sides.}$$

$$-1 = x \quad \text{Divide both sides by 7.}$$

Caution

When finding $\left(2\sqrt{3x + 4}\right)^2$, remember to square both of the factors, 2 and $\sqrt{3x + 4}$, to get:

$$2^2\left(\sqrt{3x + 4}\right)^2$$

This is an application of the power of a product rule for exponents:

$$(xy)^n = x^n y^n$$

We check the solution by substituting -1 for x in the original equation.

$$\sqrt{5x + 9} = 2\sqrt{3x + 4}$$

$$\sqrt{5(-1) + 9} \stackrel{?}{=} 2\sqrt{3(-1) + 4} \quad \text{Substitute } -1 \text{ for } x.$$

$$\sqrt{4} \stackrel{?}{=} 2\sqrt{1}$$

$$2 = 2 \quad \text{True}$$

The solution is -1 and the solution set is $\{-1\}$.

Self Check 7 Solve: $\sqrt{x-4} = 2\sqrt{x-16}$

Now Try Problem 49

When more than one radical appears in an equation, we often must use the power rule more than once.

EXAMPLE 8 Solve: $\sqrt{x} + \sqrt{x+2} = 2$

Strategy We will isolate $\sqrt{x+2}$ on the left side of the equation and square both sides to eliminate it. After simplifying the resulting equation, we will isolate the remaining radical term and square both sides a second time to eliminate it.

Why Each time that we square both sides, we are able to clear the equation of one radical.

Solution

$$\sqrt{x} + \sqrt{x+2} = 2 \quad \text{This is the equation to solve.}$$

$$\sqrt{x+2} = 2 - \sqrt{x} \quad \text{To isolate } \sqrt{x+2}\text{, subtract } \sqrt{x} \text{ from both sides.}$$

$$\left(\sqrt{x+2}\right)^2 = \left(2 - \sqrt{x}\right)^2 \quad \text{Square both sides to eliminate the square root on the left side.}$$

$$x + 2 = \left(2 - \sqrt{x}\right)^2 \quad \text{Perform the operation on the left side.}$$

To square the expression $2 - \sqrt{x}$ on the right side, we can use a special-product rule:

$$x + 2 = \underbrace{2^2}_{\text{Square the first term, 2.}} - \underbrace{2(2)\left(\sqrt{x}\right)}_{\text{Twice the product of both terms}} + \underbrace{\left(\sqrt{x}\right)^2}_{\text{Square the last term, } \sqrt{x}.}$$

$$x + 2 = \mathbf{4 - 4\sqrt{x} + x}$$

Since the equation still contains a radical, we need to square both sides again. Before doing that, we must isolate the radical on one side.

$$2 = 4 - 4\sqrt{x} \quad \text{Subtract } x \text{ from both sides.}$$

$$-2 = -4\sqrt{x} \quad \text{To isolate the radical term } -4\sqrt{x}\text{, subtract 4 from both sides.}$$

$$\frac{1}{2} = \sqrt{x} \quad \text{To isolate the radical, divide both sides by } -4.$$

$$\left(\frac{1}{2}\right)^2 = \left(\sqrt{x}\right)^2 \quad \text{To eliminate the radical, square both sides again.}$$

$$\frac{1}{4} = x \quad \text{Perform the operations on each side.}$$

Check:

$$\sqrt{x} + \sqrt{x+2} = 2 \quad \text{This is the original equation.}$$

$$\sqrt{\frac{1}{4}} + \sqrt{\frac{1}{4} + 2} \stackrel{?}{=} 2 \quad \text{Substitute } \tfrac{1}{4} \text{ for } x.$$

$$\frac{1}{2} + \sqrt{\frac{9}{4}} \stackrel{?}{=} 2 \quad \text{Think of 2 as } \tfrac{8}{4} \text{ and add: } \tfrac{1}{4} + \tfrac{8}{4} = \tfrac{9}{4}.$$

$$\frac{1}{2} + \frac{3}{2} \stackrel{?}{=} 2 \quad \text{Evaluate } \sqrt{\tfrac{9}{4}}.$$

$$2 = 2 \quad \text{True}$$

The result $\frac{1}{4}$ checks. The solution set is $\left\{\frac{1}{4}\right\}$.

Success Tip

Note in this example that we isolate the more complicated radical term, $\sqrt{x+2}$. As a result, we work with the less complicated radical term $\sqrt{x}$ when applying the special-product rule.

Caution

When finding $\left(2 - \sqrt{x}\right)^2$, it is like squaring a binomial. Remember to use a special-product rule (or FOIL). **Do not just square the first term and the last term.**

$$\left(2 - \sqrt{x}\right)^2 \neq 4 + x$$

Self Check 8 Solve: $\sqrt{a} + \sqrt{a + 3} = 3$

Now Try Problems 55 and 59

Using Your Calculator ▶ Solving Radical Equations

To find solutions for $\sqrt{x} + \sqrt{x + 2} = 4$ (an equation similar to Example 8) with a graphing calculator, we graph the functions $f(x) = \sqrt{x} + \sqrt{x + 2}$ and $g(x) = 4$. We then trace to find an approximation of the x-coordinate of their intersection point, as in figure (a). From the figure, we can see that $x \approx 2.98$. We can zoom to get better results.

Figure (b) shows that the INTERSECT feature gives the approximate coordinates of the point of intersection of the two graphs as (3.06, 4). Therefore, an approximate solution of the radical equation is 3.06. Check its reasonableness.

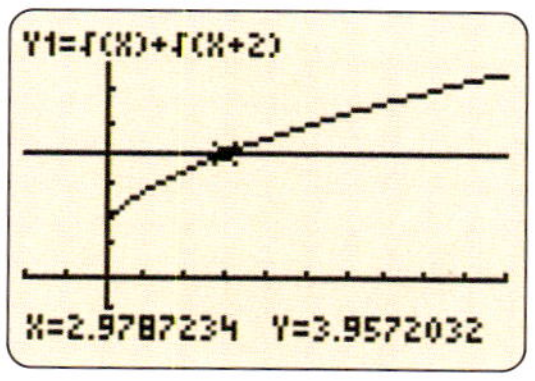

(a)

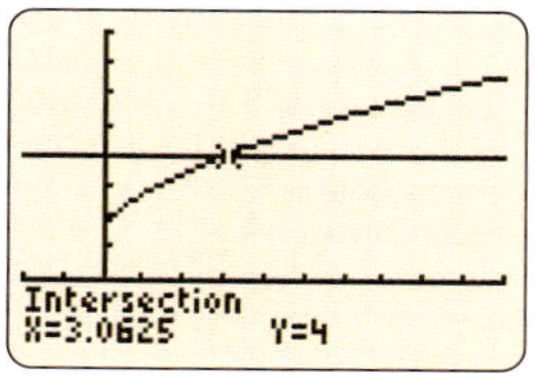

(b)

3 Solve Formulas Containing Radicals.

To *solve a formula for a variable* means to isolate that variable on one side of the equation, with all other quantities on the other side.

EXAMPLE 9 **Depreciation Rates.** Some office equipment that is now worth V dollars originally cost C dollars 3 years ago. The rate r at which it has depreciated is given by $r = 1 - \sqrt[3]{\frac{V}{C}}$. Solve the formula for C.

Strategy To isolate the radical, we will subtract 1 from both sides. We can then eliminate the radical by cubing both sides.

Why Cubing both sides will produce, on the right, the expression $\left(\sqrt[3]{\frac{V}{C}}\right)^3$ that simplifies to $\frac{V}{C}$. This step clears the equation of the radical.

Solution We begin by isolating the cube root on the right side of the equation.

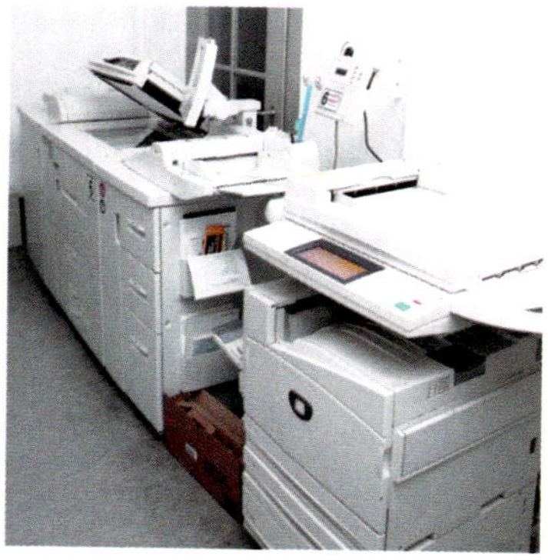
iStockphoto.com/wakila

$$r = 1 - \sqrt[3]{\frac{V}{C}}$$ This is the depreciation model.

$$r - 1 = -\sqrt[3]{\frac{V}{C}}$$ Subtract 1 from both sides to isolate the radical.

$$(r - 1)^3 = \left(-\sqrt[3]{\frac{V}{C}}\right)^3$$ To eliminate the radical, cube both sides.

$$(r - 1)^3 = -\frac{V}{C}$$ Simplify the right side.

$$C(r - 1)^3 = -V$$ To clear the equation of the fraction, multiply both sides by C.

$$C = -\frac{V}{(r - 1)^3}$$ To isolate C, divide both sides by $(r - 1)^3$.

Self Check 9 **Statistics.** A formula used in statistics to determine the size of a sample to obtain a desired degree of accuracy is $E = z_0\sqrt{\frac{pq}{n}}$. Solve the formula for n.

Now Try Problem 67

SECTION 9.5 STUDY SET

VOCABULARY

Fill in the blanks.

1. Equations such as $\sqrt{x+4} - 4 = 5$ and $\sqrt[3]{x+1} = 12$ are called ________ equations.
2. To solve a radical equation, we find all the values of the variable that make the equation ______
3. When we square both sides of a radical equation, we say we are ________ both sides to the second power.
4. When solving equations containing radicals, first we ________ one radical expression on one side of the equation.
5. Proposed solutions of a radical equation that don't satisfy it are called ____________ solutions.
6. To ________ a proposed solution means to substitute it into the original equation and see whether a true statement results.

CONCEPTS

7. Fill in the blanks.
   a. The power rule for solving radical equations states that if x, y, and n are real numbers and $x = y$, then

      $x^{\quad} = y^{\quad}$

   b. If $\sqrt[n]{a}$ is a real number, then $\left(\sqrt[n]{a}\right)^n = \quad$.
8. Determine whether 6 is a solution of each radical equation.
   a. $\sqrt{x+3} = x - 3$
   b. $\sqrt[3]{5x-3} + 9 = x$
9. What is the first step in solving each equation?
   a. $\sqrt{x} + 11 = 5$
   b. $\sqrt[3]{5x+4} + 3 = 30$
   c. $\sqrt{x+8} - \sqrt{2x} = 1$
10. Simplify each expression.
    a. $\left(\sqrt{x}\right)^2$
    b. $\left(\sqrt{x-5}\right)^2$
    c. $\left(\sqrt[3]{4x-8}\right)^3$
    d. $\left(\sqrt[4]{8x}\right)^4$
    e. $\left(4\sqrt{2x}\right)^2$
    f. $\left(3\sqrt[3]{x+1}\right)^3$
11. Find: $\left(\sqrt{x} - 3\right)^2$
12. Find: $\left(\sqrt{5x+2} - 4\right)^2$

NOTATION

Complete each solution.

13. Solve: $\sqrt{3x+3} - 1 = 5$

$$\sqrt{3x+3} = \quad$$
$$\left(\sqrt{3x+3}\right)^{\quad} = (6)^{\quad}$$
$$\quad = 36$$
$$3x = \quad$$
$$x = \quad$$

    Does the proposed solution check?
14. Fill in the blanks. Write each radical equation using a rational exponent.
    a. $\sqrt{x+10} + 5 = 15$ can be written $(x+10)^{\quad} + 5 = 15$
    b. $\sqrt[3]{2t+4} = t - 1$ can be written $(2t+4)^{\quad} = t - 1$

GUIDED PRACTICE

Solve each equation. See Example 1.

15. $\sqrt{a-3} = 1$
16. $\sqrt{x-10} = 1$
17. $\sqrt{4x+5} = 5$
18. $\sqrt{5x-6} = 2$
19. $\sqrt{6x+13} = 7$
20. $\sqrt{6x+1} = 5$
21. $\sqrt{\frac{1}{3}x - 2} = 8$
22. $\sqrt{\frac{1}{2}x + 3} = 6$

Solve each equation. Write all proposed solutions. Cross out those that are extraneous. See Example 3.

23. $\sqrt{2x+11} + 2 = x$
24. $\sqrt{2a-3} + 3 = a$
25. $\sqrt{2r-3} + 9 = r$
26. $\sqrt{-x+2} + 2 = x$
27. $\sqrt{3t+7} - t = 1$
28. $\sqrt{t+3} - t = 1$
29. $\sqrt{9-a} - a = 3$
30. $\sqrt{4-a} - a = 2$

Solve each equation. Write all proposed solutions. Cross out those that are extraneous. See Example 4.

31. $\sqrt{5x} + 10 = 8$
32. $\sqrt{3x} + 5 = 2$
33. $\sqrt{5-x} + 10 = 9$
34. $1 = 2 + \sqrt{4x+75}$

Solve each equation. See Example 5.

35. $\sqrt[3]{7n-1} = 3$
36. $\sqrt[3]{12m+4} = 4$
37. $\sqrt[3]{x^3-7} = x - 1$
38. $\sqrt[3]{b^3-63} = b - 3$

39. $(m^3 + 26)^{1/3} = m + 2$

40. $(x^3 + 56)^{1/3} = x + 2$

41. $(5r + 14)^{1/3} = 4$

42. $(2b + 29)^{1/3} = 3$

See Example 6.

43. Let $f(x) = \sqrt[4]{3x + 1}$. For what value(s) of x is $f(x) = 4$?

44. Let $f(x) = \sqrt{2x^2 - 7x}$. For what value(s) of x is $f(x) = 2$?

45. Let $f(x) = \sqrt[3]{3x - 6}$. For what value(s) of x is $f(x) = -3$?

46. Let $f(x) = \sqrt[5]{4x - 4}$. For what value(s) of x is $f(x) = -2$?

Solve each equation. See Example 7.

47. $\sqrt{3x + 12} = \sqrt{5x - 12}$

48. $\sqrt{m + 4} = \sqrt{2m - 5}$

49. $2\sqrt{4x + 1} = \sqrt{x + 4}$

50. $\sqrt{6 - 2x} = 4\sqrt{x - 3}$

51. $\sqrt{6t + 9} = 3\sqrt{t}$

52. $\sqrt{12x + 24} = 6\sqrt{x}$

53. $(34x + 26)^{1/3} = 4(x - 1)^{1/3}$

54. $(a^2 + 2a)^{1/3} = 2(a - 1)^{1/3}$

Solve each equation. Write all proposed solutions. Cross out those that are extraneous. See Example 8.

55. $\sqrt{x - 5} + \sqrt{x} = 5$

56. $\sqrt{x - 7} + \sqrt{x} = 7$

57. $\sqrt{z + 3} - \sqrt{z} = 1$

58. $\sqrt{x + 12} + \sqrt{x} = 6$

59. $3 = \sqrt{y + 4} - \sqrt{y + 7}$

60. $3 = \sqrt{u - 3} - \sqrt{u}$

61. $2 = \sqrt{2u + 7} - \sqrt{u}$

62. $1 = \sqrt{4s + 5} - \sqrt{2s + 2}$

Solve each equation for the specified variable or expression. See Example 9.

63. $v = \sqrt{2gh}$ for h

64. $d = 1.4\sqrt{h}$ for h

65. $T = 2\pi\sqrt{\dfrac{l}{32}}$ for l

66. $d = \sqrt[3]{\dfrac{12V}{\pi}}$ for V

67. $r = \sqrt[3]{\dfrac{A}{P}} - 1$ for A

68. $r = \sqrt[3]{\dfrac{A}{P}} - 1$ for P

69. $L_A = L_B\sqrt{1 - \dfrac{v^2}{c^2}}$ for v^2

70. $R_1 = \sqrt{\dfrac{A}{\pi} - R_2{}^2}$ for A

TRY IT YOURSELF

Solve each equation. Write all proposed solutions. Cross out those that are extraneous.

71. $2\sqrt{x} = \sqrt{5x - 16}$

72. $3\sqrt{x} = \sqrt{3x + 54}$

73. $\sqrt{x + 5} + \sqrt{x - 3} = 4$

74. $\sqrt{b + 7} - \sqrt{b - 5} = 2$

75. $n = (n^3 + n^2 - 1)^{1/3}$

76. $(m^4 + m^2 - 25)^{1/4} = m$

77. $\sqrt{y + 2} + y = 4$

78. $\sqrt{22y + 86} - y = 9$

79. $\sqrt[3]{x + 8} = -2$

80. $\sqrt[3]{x + 4} = -1$

81. $2 = \sqrt{x + 5} - \sqrt{x} + 1$

82. $4 = \sqrt{x + 8} - \sqrt{x} + 2$

83. $x = \dfrac{\sqrt{12x - 5}}{2}$

84. $x = \dfrac{\sqrt{16x - 12}}{2}$

85. $(n^2 + 6n + 3)^{1/2} = (n^2 - 6n - 3)^{1/2}$

86. $(m^2 - 12m - 3)^{1/2} = (m^2 + 12m + 3)^{1/2}$

87. $\sqrt{x - 5} - \sqrt{x + 3} = 4$

88. $\sqrt{x + 8} - \sqrt{x - 4} = -2$

89. $\sqrt[4]{10y + 6} = 2\sqrt[4]{y}$

90. $\sqrt[4]{21a + 39} = 3\sqrt[4]{a - 1}$

91. $\sqrt{-5x + 24} = 6 - x$

92. $-s - 3 = 2\sqrt{5 - s}$

93. $\sqrt{2x} + 5 = 1$

94. $\sqrt{3x} + 10 = 1$

95. $\sqrt{6x + 2} - \sqrt{5x + 3} = 0$

96. $\sqrt{5x + 2} - \sqrt{x + 10} = 0$

97. Let $f(x) = \sqrt{x + 16}$ and $g(x) = 7 - \sqrt{x + 9}$. Find all values of x for which $f(x) = g(x)$.

98. Let $s(t) = \sqrt{t + 8}$ and $h(t) = 6 - \sqrt{t - 4}$. Find all values of t for which $s(t) = h(t)$.

99. Let $f(x) = \sqrt[4]{x + 8} - \sqrt[4]{2x}$. Find all values of x for which $f(x) = 0$.

100. Let $h(a) = \sqrt[4]{a + 11} - \sqrt[4]{2a + 6}$. Find all values of a for which $h(a) = 0$.

Look Alikes . . .

101. a. $3\sqrt{5n - 9} = \sqrt{5n}$ b. $3 + \sqrt{5n - 9} = \sqrt{5n}$

102. a. $\sqrt{8 + a} = 2\sqrt{a}$ b. $\sqrt{8 + a} = 2 + \sqrt{a}$

103. a. $\sqrt{2x} - 10 = 0$ b. $\sqrt{2x} + 10 = 0$

104. a. $x^{1/2} + 6 = 8$ b. $x^{1/4} + 6 = 8$

APPLICATIONS

105. **Highway Design.** A curved road will accommodate traffic traveling s mph if the radius of the curve is r feet, according to the formula $s = 3\sqrt{r}$. If engineers expect 40-mph traffic, what radius should they specify? Give the result to the nearest foot.

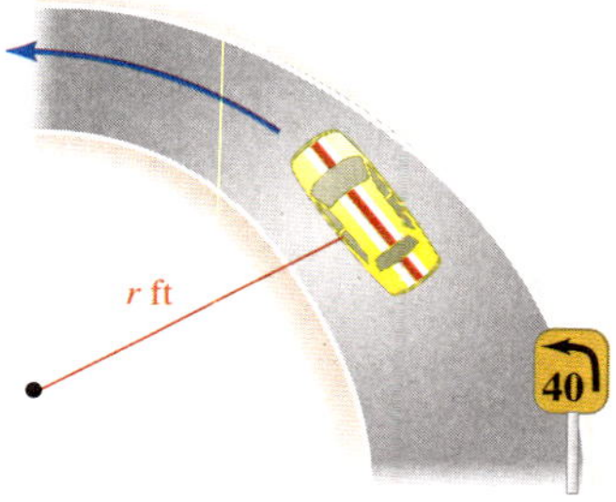

106. Forestry. The taller a lookout tower, the farther an observer can see. That distance d (called the *horizon distance,* measured in miles) is related to the height h of the observer (measured in feet) by the formula $d = 1.22\sqrt{h}$. How tall must a lookout tower be to see the edge of the forest, 25 miles away? (Round to the nearest foot.)

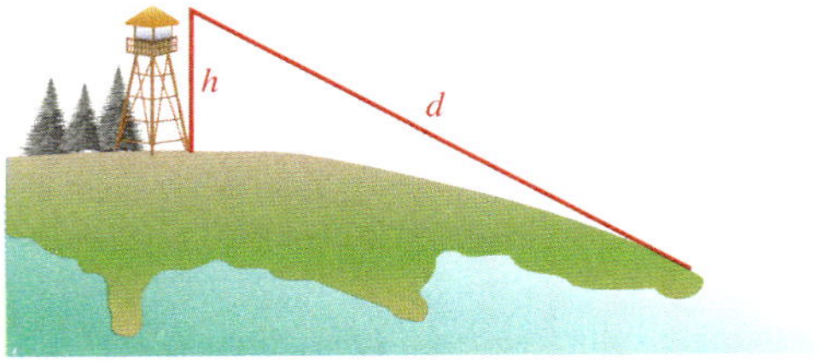

107. Wind Power. The power generated by a windmill is related to the velocity of the wind by the formula $v = \sqrt[3]{\frac{P}{0.02}}$ where P is the power (in watts) and v is the velocity of the wind (in mph). Find how much power the windmill is generating when the wind is 29 mph.

108. Diamonds. The *effective rate of interest* r earned by an investment is given by the formula $r = \sqrt[n]{\frac{A}{P}} - 1$ where P is the initial investment that grows to value A after n years. If a diamond buyer got \$4,000 for a 1.73-carat diamond that he had purchased 4 years earlier, and earned an annual rate of return of 6.5% on the investment, what did he originally pay for the diamond?

109. **from Campus to Careers**

General Contractor

©Goodluz/Shutterstock.com

During construction, carpenters often brace walls as shown in the illustration, where the length L of the brace is given by the formula $L = \sqrt{f^2 + h^2}$. If a carpenter nails a 10-ft brace to the wall 6 feet above the floor, how far from the base of the wall should he nail the brace to the floor?

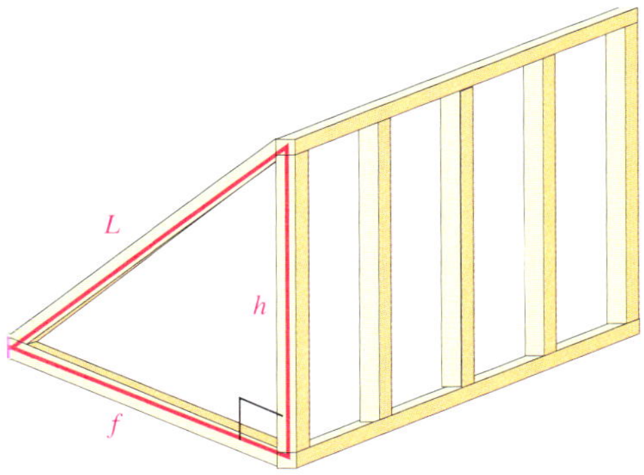

110. Theater Productions. The ropes, pulleys, and sandbags shown in the illustration are part of a mechanical system used to raise and lower scenery for a stage play. For the scenery to be in the proper position, the following formula must apply: $w_2 = \sqrt{{w_1}^2 + {w_3}^2}$. If $w_2 = 12.5$ lb and $w_3 = 7.5$ lb, find w_1.

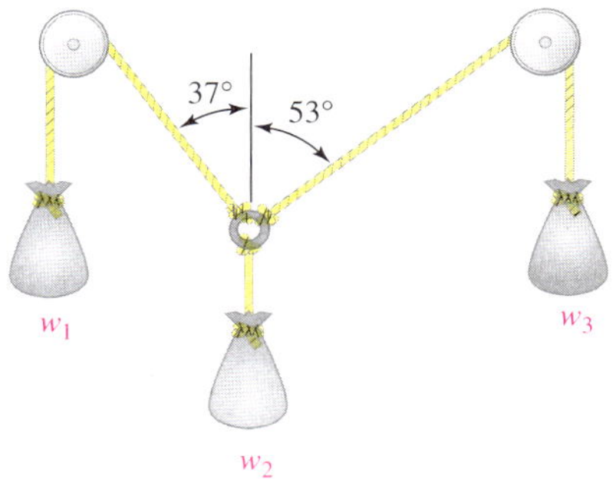

111. Supply and Demand. The number of wrenches that will be produced at a given price can be predicted by the formula $s = \sqrt{5x}$, where s is the supply (in thousands) and x is the price (in dollars). The demand d for wrenches can be predicted by the formula $d = \sqrt{100 - 3x^2}$. Find the equilibrium price—that is, find the price at which supply will equal demand.

112. Supply and Demand. The number of mirrors that will be produced at a given price can be predicted by the formula $s = \sqrt{23x}$, where s is the supply (in thousands) and x is the price (in dollars). The demand d for mirrors can be predicted by the formula $d = \sqrt{312 - 2x^2}$. Find the equilibrium price—that is, find the price at which supply will equal demand.

WRITING

113. What is wrong with the work shown below?

Solve:
$$\sqrt[3]{x+1} - 3 = 8$$
$$\sqrt[3]{x+1} = 11$$
$$\left(\sqrt[3]{x+1}\right)^3 = 11$$
$$x + 1 = 11$$
$$x = 10$$

114. The first step of a student's solution is shown below. What is a better way to begin the solution?

Solve:
$$\sqrt{x} + \sqrt{x+22} = 12$$
$$\left(\sqrt{x} + \sqrt{x+22}\right)^2 = 12^2$$

115. Explain the error in the following work.

Solve:
$$\sqrt{2y+1} = \sqrt{y+7} + 3$$
$$\left(\sqrt{2y+1}\right)^2 = \left(\sqrt{y+7} + 3\right)^2$$
$$2y + 1 = y + 7 + 9$$

116. Explain why it is immediately apparent that $\sqrt{8x - 7} = -2$ has no solution.

117. To solve the equation $\sqrt{2x + 7} = \sqrt{x}$ we need only square both sides once. To solve the equation $\sqrt{2x + 7} = \sqrt{x} + 2$ we have to square both sides twice. Why does the second equation require more work?

118. Explain how to solve $\sqrt{x - 2} + 2 = 4$ using the graphs below.

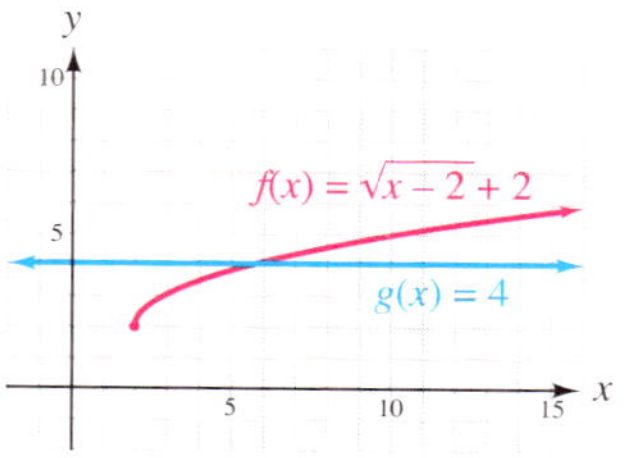

119. Explain how the table can be used to solve $\sqrt{4x - 3} - 2 = \sqrt{2x - 5}$ if $Y_1 = \sqrt{4x - 3} - 2$ and $Y_2 = \sqrt{2x - 5}$.

X	Y1	Y2
2	.23607	ERROR
3	1	1
4	1.6056	1.7321
5	2.1231	2.2361
6	2.5826	2.6458
7	3	3
8	3.3852	3.3166

X=2

120. Explain how to use the graph of $f(x) = \sqrt[3]{x - 0.5} - 1$, shown in the illustration, to approximate the solution of $\sqrt[3]{x - 0.5} = 1$.

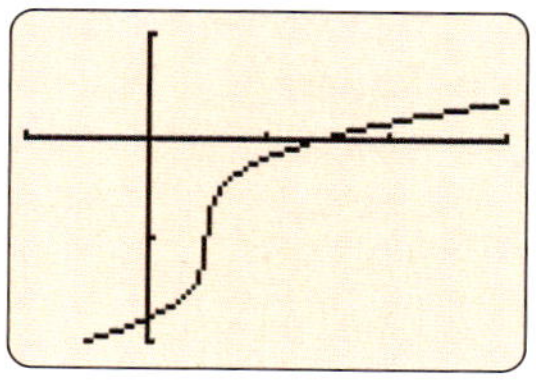

REVIEW

121. Lighting. The intensity of light from a lightbulb varies inversely as the square of the distance from the bulb. If you are 5 feet away from a bulb and the intensity is 40 foot-candles, what will the intensity be if you move 20 feet away from the bulb?

122. Property Tax. The property tax in a certain county varies directly as assessed valuation. If a tax of \$1,575 is charged on a single-family home assessed at \$90,000, determine the property tax on an apartment complex assessed at \$312,000.

123. Typesetting. If 12-point type is 0.166044 inch tall, how tall is 30-point type?

124. Guitar Strings. The frequency of vibration of a string varies directly as the square root of the tension and inversely as the length of the string. Suppose a string 2.5 feet long, under a tension of 16 pounds, vibrates 25 times per second. Find k, the constant of proportionality.

CHALLENGE PROBLEMS

Solve each equation. Write all proposed solutions. Cross out those that are extraneous.

125. $\sqrt[4]{x} = \sqrt{\dfrac{x}{4}}$

126. $\sqrt[3]{2x} = \sqrt{x}$

127. $\sqrt{x + 2} + \sqrt{2x} = \sqrt{18 - x}$

128. $\sqrt{8 - x} - \sqrt{3x - 8} = \sqrt{x - 4}$

129. $\sqrt{2\sqrt{x + 1}} = \sqrt{16 - 4x}$

130. $(2x - 1)^{2/3} = x^{1/3}$

SECTION 9.6 Geometric Applications of Radicals

OBJECTIVES

1. Use the Pythagorean theorem to solve problems.
2. Solve problems involving 45°–45°–90° triangles.
3. Solve problems involving 30°–60°–90° triangles.
4. Use the distance formula to solve problems.
5. Find the midpoint of a line segment.

ARE YOU READY?

The following problems review some basic skills that are needed when working with special triangles.

1. What is the sum of the measures of the angles of any triangle?

2. Simplify: $\sqrt{3a^2}$

3. Rationalize the denominator: $\dfrac{25}{\sqrt{3}}$

4. Evaluate: $\sqrt{(9 - 4)^2 + (-1 - 11)^2}$

5. Approximate to the nearest hundredth: $2\sqrt{3}$

6. Multiply: $2 \cdot \dfrac{7\sqrt{3}}{3}$

We will now consider applications of square roots in geometry. Then we will find the distance between two points on a rectangular coordinate system, using a formula that contains a square root. We begin by considering an important theorem about right triangles.

1 Use the Pythagorean Theorem to Solve Problems.

If we know the lengths of two legs of a right triangle, we can find the length of the **hypotenuse** (the side opposite the 90° angle) by using the **Pythagorean theorem.**

The Pythagorean Theorem

If a and b are the lengths of two legs of a right triangle and c is the length of the hypotenuse,

$$a^2 + b^2 = c^2$$

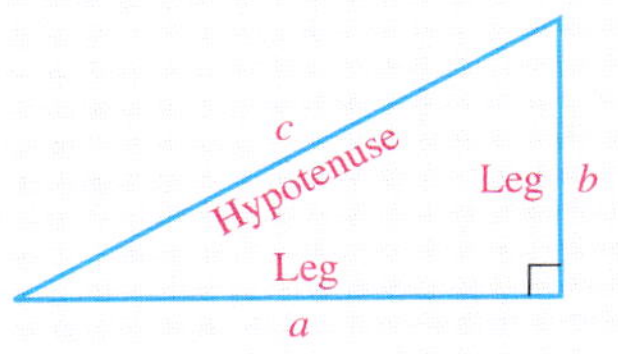

In words, the Pythagorean theorem is expressed as follows:

In a right triangle, the sum of the squares of the lengths of two legs is equal to the square of the length of the hypotenuse.

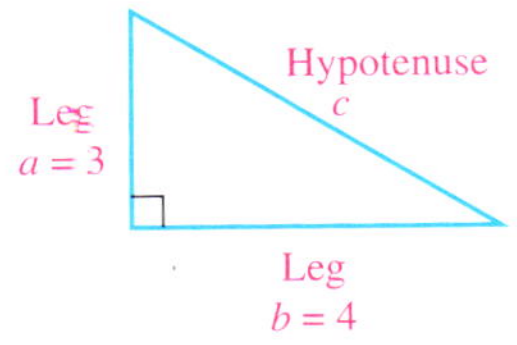

Suppose the right triangle shown in the margin has legs of length 3 and 4 units. To find the length of the hypotenuse, we use the **Pythagorean equation.**

$a^2 + b^2 = c^2$

$3^2 + 4^2 = c^2$ Substitute 3 for a and 4 for b.

$9 + 16 = c^2$

$25 = c^2$

To find c, we ask "What number, when squared, is equal to 25?" There are two such numbers: the positive square root of 25 and the negative square root of 25. Since c represents the length of the hypotenuse, and it cannot be negative, it follows that c is the positive square root of 25.

$\sqrt{25} = c$ Recall that a radical symbol $\sqrt{\ }$ is used to represent the positive, or principal, square root of a number.

$5 = c$

The length of the hypotenuse is 5 units.

The Language of Algebra

A **theorem** is a mathematical statement that can be proved. The Pythagorean theorem is named after Pythagoras, a Greek mathematician who lived about 2,500 years ago. He is thought to have been the first to prove the theorem.

EXAMPLE 1

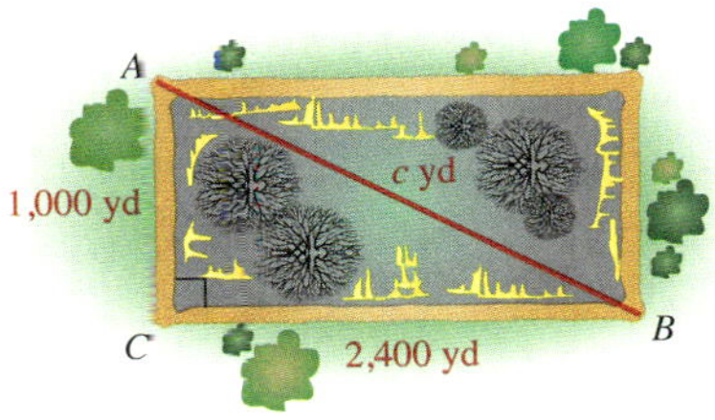

Firefighting. To fight a fire, the forestry department plans to clear a rectangular firebreak around the fire, as shown in the illustration on the left. Crews are equipped with mobile communications that have a 3,000-yard range. Can crews at points A and B remain in radio contact?

Strategy We will use the Pythagorean theorem to find the distance between points A and B.

Why If this distance is less than 3,000 yards, they can communicate. If it is greater than 3,000 yards, they cannot communicate.

Solution The line segments connecting points A, B, and C form a right triangle. To find the distance c from point A to point B, we can use the Pythagorean theorem, substituting 2,400 for a and 1,000 for b and solving for c.

$a^2 + b^2 = c^2$ This is the Pythagorean equation.

$2{,}400^2 + 1{,}000^2 = c^2$ Substitute for a and b.

$5{,}760{,}000 + 1{,}000{,}000 = c^2$ Evaluate each exponential expression.

$6{,}760{,}000 = c^2$ Do the addition.

$\sqrt{6{,}760{,}000} = c$ If $c^2 = 6{,}760{,}000$, then c must be a square root of 6,760,000. Because c represents a length, it must be the positive square root of 6,760,000.

$2{,}600 = c$ Use a calculator to find the square root.

The two crews are 2,600 yards apart. Because this distance is less than the 3,000-yard range of the radios, they can communicate by radio.

Caution

When using the Pythagorean equation $a^2 + b^2 = c^2$, we can let a represent the length of either leg of the right triangle. We then let b represent the length of the other leg. The variable c must always represent the length of the hypotenuse.

Self Check 1 **Firefighting.** In Example 1, can the crews communicate if $b = 1{,}500$ yards?

Now Try Problems 15 and 55

2 Solve Problems Involving 45°–45°–90° Triangles.

An **isosceles right triangle** is a right triangle with two legs of equal length. Isosceles right triangles have angle measures of 45°, 45°, and 90°. If we know the length of one leg of an isosceles right triangle, we can use the Pythagorean theorem to find the length of the hypotenuse. Since the triangle shown in the margin is a right triangle, we have

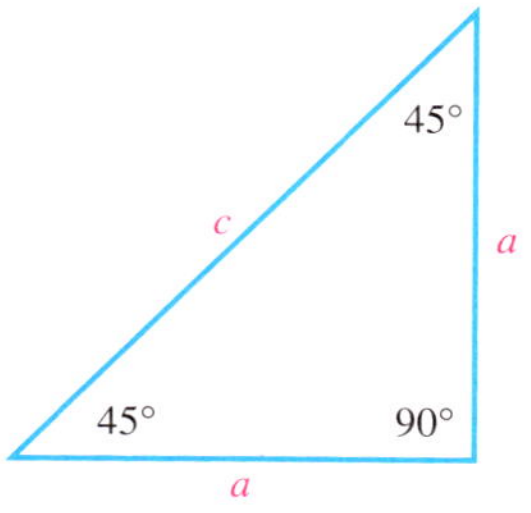

$c^2 = a^2 + b^2$ This is the Pythagorean equation.

$c^2 = a^2 + a^2$ Both legs are a units long, so replace b with a.

$c^2 = 2a^2$ Combine like terms.

$c = \sqrt{2a^2}$ If $c^2 = 2a^2$, then c must be a square root of $2a^2$. Because c represents a length, it must be the positive square root of $2a^2$.

$c = a\sqrt{2}$ Simplify the radical: $\sqrt{2a^2} = \sqrt{2}\sqrt{a^2} = \sqrt{2}a = a\sqrt{2}$.

Thus, **in an isosceles right triangle, the length of the hypotenuse is $\sqrt{2}$ times the length of one leg.**

EXAMPLE 2 If one leg of an isosceles right triangle is 10 feet long, find the exact length of the hypotenuse. Then approximate the length to two decimal places.

Strategy We will multiply the length of the known leg by $\sqrt{2}$.

Why The length of the hypotenuse of an isosceles right triangle is $\sqrt{2}$ times the length of one leg.

Solution Since the length of the hypotenuse is the length of a leg times $\sqrt{2}$, we have

$$c = 10\sqrt{2}$$

The exact length of the hypotenuse is $10\sqrt{2}$ feet. If we approximate to two decimal places, the length is 14.14 feet.

Self Check 2 Find the exact length of the hypotenuse of an isosceles right triangle if one leg is 12 meters long. Then approximate the length to two decimal places.

Now Try Problem 23

EXAMPLE 3 Find the exact length of each leg of the isosceles right triangle shown in the margin. Then approximate the lengths to two decimal places.

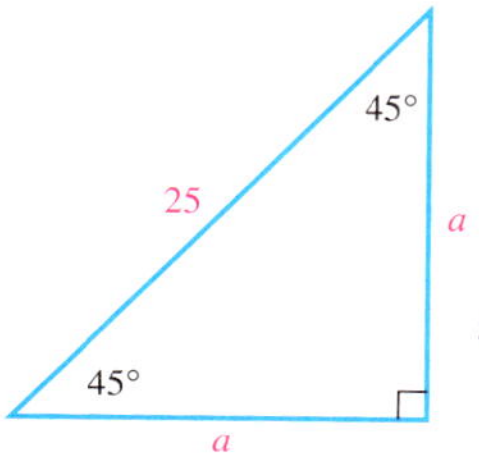

Strategy We will find the length of each leg of the triangle by substituting 25 for c in the formula $c = a\sqrt{2}$ and solving for a.

Why The variable a represents the unknown length of each leg of the triangle.

Solution

$c = a\sqrt{2}$ This is the formula for the length of the hypotenuse of an isosceles right triangle.

$25 = a\sqrt{2}$ Substitute 25 for c, the length of the hypotenuse.

$\dfrac{25}{\sqrt{2}} = \dfrac{a\sqrt{2}}{\sqrt{2}}$ To isolate a, undo the multiplication by $\sqrt{2}$ by dividing both sides by $\sqrt{2}$.

$$\frac{25}{\sqrt{2}} = a$$ Simplify the right side.

$$a = \frac{25}{\sqrt{2}}$$ Reverse the sides of the equation so that *a* is on the left.

$$a = \frac{25}{\sqrt{2}} \cdot \frac{\sqrt{2}}{\sqrt{2}}$$ Rationalize the denominator.

$$a = \frac{25\sqrt{2}}{2}$$ Simplify the denominator: $\sqrt{2} \cdot \sqrt{2} = 2$.

The exact length of each leg is $\frac{25\sqrt{2}}{2}$ units. If we approximate to two decimal places, the length is 17.68 units.

Self Check 3 Find the exact length of each leg of an isosceles right triangle if the length of the hypotenuse is 9 inches. Then approximate the lengths to two decimal places.

Now Try Problem 27

3 Solve Problems Involving 30°–60°–90° Triangles.

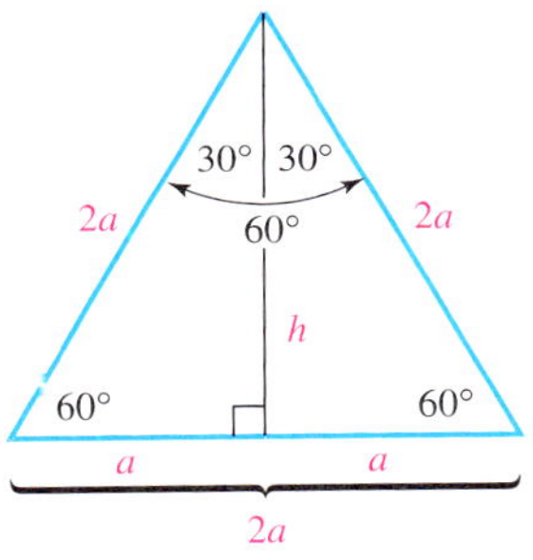

From geometry, we know that an **equilateral triangle** is a triangle with three sides of equal length and three 60° angles. Each side of the equilateral triangle shown in the margin is $2a$ units long. If an **altitude** (height) is drawn to its base, the altitude divides the base into two segments of equal length and divides the equilateral triangle into two 30°–60°–90° triangles. From the figure, we can see that the shorter leg of each 30°–60°–90° triangle (the side *opposite* the 30° angle) is a units long. Thus,

The length of the hypotenuse of a 30°–60°–90° triangle is twice as long as the shorter leg.

We can discover another relationship between the legs of a 30°–60°–90° triangle if we find the length of the altitude h in the figure. We begin by applying the Pythagorean theorem to one of the 30°–60°–90° triangles.

$$a^2 + b^2 = c^2$$ This is the Pythagorean equation.

$$a^2 + h^2 = (2a)^2$$ The altitude is h units long, so replace b with h. The hypotenuse is $2a$ units long, so replace c with $2a$.

$$a^2 + h^2 = 4a^2$$ $(2a)^2 = (2a)(2a) = 4a^2$.

$$h^2 = 3a^2$$ Subtract a^2 from both sides.

$$h = \sqrt{3a^2}$$ If $h^2 = 3a^2$, then h must be the positive square root of $3a^2$.

$$h = a\sqrt{3}$$ Simplify the radical: $\sqrt{3a^2} = \sqrt{3}\sqrt{a^2} = a\sqrt{3}$.

We see that the altitude—the longer leg of the 30°–60°–90° triangle—is $\sqrt{3}$ times as long as the shorter leg. Thus,

The length of the longer leg of a 30°–60°–90° triangle is $\sqrt{3}$ times the length of the shorter leg.

EXAMPLE 4 Find the length of the hypotenuse and the length of the longer leg of the 30°–60°–90° triangle shown in the margin.

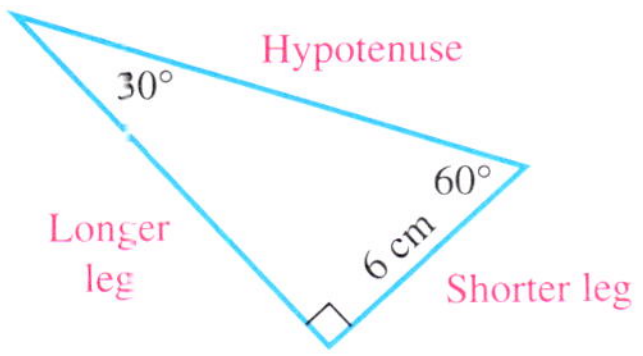

Strategy To find the length of the hypotenuse, we will multiply the length of the shorter leg by 2. To find the length of the longer leg, we will multiply the length of the shorter leg by $\sqrt{3}$.

Why These side-length relationships are true for any 30°–60°–90° triangle.

Solution Since the length of the hypotenuse of a $30°–60°–90°$ triangle is twice as long as the shorter leg, and the length of the shorter leg is 6 cm, the hypotenuse is $2 \cdot 6 = 12$ cm.

Since the length of the longer leg is $\sqrt{3}$ times the length of the shorter leg, and the length of the shorter leg is 6 cm, the longer leg is $6\sqrt{3}$ cm (about 10.39 cm).

Self Check 4 Find the length of the hypotenuse and the longer leg of a $30°–60°–90°$ triangle if the shorter leg is 8 centimeters long.

Now Try Problem 31

EXAMPLE 5 Find the length of the hypotenuse and the length of the shorter leg of the $30°–60°–90°$ triangle shown in the margin. Then approximate the lengths to two decimal places.

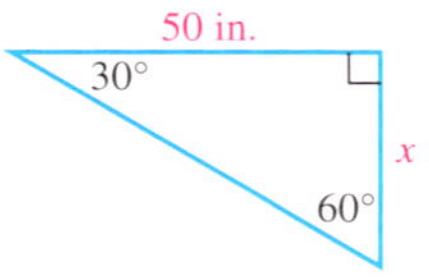

Strategy We will find the length of the shorter leg first.

Why Once we know the length of the shorter leg, we can multiply it by 2 to find the length of the hypotenuse.

Solution If we let x = the length in inches of the shorter leg of the triangle, we can form an equation by translating the following statement:

Success Tip

In a $30°–60°–90°$ triangle, the side opposite the $30°$ angle is the shorter leg, the side opposite the $60°$ angle is the longer leg, and the hypotenuse is opposite the $90°$ angle.

The length of the longer leg of a $30°–60°–90°$ triangle	is	$\sqrt{3}$	times	the length of the shorter leg.
50	=	$\sqrt{3}$	$\cdot$	x

To find the length of the shorter leg, we solve the equation for x.

$$50 = \sqrt{3}x$$

$$\frac{50}{\sqrt{3}} = \frac{\sqrt{3}x}{\sqrt{3}} \quad \text{To isolate } x\text{, divide both sides by } \sqrt{3}.$$

$$\frac{50}{\sqrt{3}} = x$$

The length of the shorter leg is exactly $\frac{50}{\sqrt{3}}$ inches. To write this number in simplified radical form, we rationalize the denominator.

$$\frac{50}{\sqrt{3}} \cdot \frac{\sqrt{3}}{\sqrt{3}} = \frac{50\sqrt{3}}{3}$$

Thus, the length of the shorter leg is exactly $\frac{50\sqrt{3}}{3}$ inches (about 28.87 inches).

Since the length of the hypotenuse of a $30°–60°–90°$ triangle is twice as long as the shorter leg, the hypotenuse is $2 \cdot \frac{50\sqrt{3}}{3} = \frac{100\sqrt{3}}{3}$ inches (about 57.74 inches).

Self Check 5 Find the length of the hypotenuse and the shorter leg of a $30°–60°–90°$ triangle if the longer leg is 15 feet long. Then approximate the lengths to two decimal places.

Now Try Problem 35

EXAMPLE 6 **Stretching Exercises.** A doctor prescribed the exercise shown in figure (a) on the next page. The patient was instructed to raise his leg to an angle of $60°$ and hold the position for 10 seconds. If the patient's leg is 36 inches long, how high off the floor will his foot be when his leg is held at the proper angle?

Strategy This situation is modeled by a $30°–60°–90°$ triangle. We will begin by finding the length of the shorter leg.

Why Once we know the length of the shorter leg, we can easily find the length of the longer leg, which represents the distance the patient's foot is off the ground.

Solution In figure (b), we see right triangle ABC, which models the situation. Since the length of the hypotenuse is twice as long as the side opposite the 30° angle, side AC is half as long as the hypotenuse. Since the hypotenuse is given to be 36 inches long, side AC must be 18 inches long.

Since the length of the longer leg (the leg opposite the 60° angle) is $\sqrt{3}$ times the length of the shorter leg (side AC), side BC is $18\sqrt{3}$, or about 31 inches long. So the patient's foot will be about 31 inches from the floor when his leg is in the proper position.

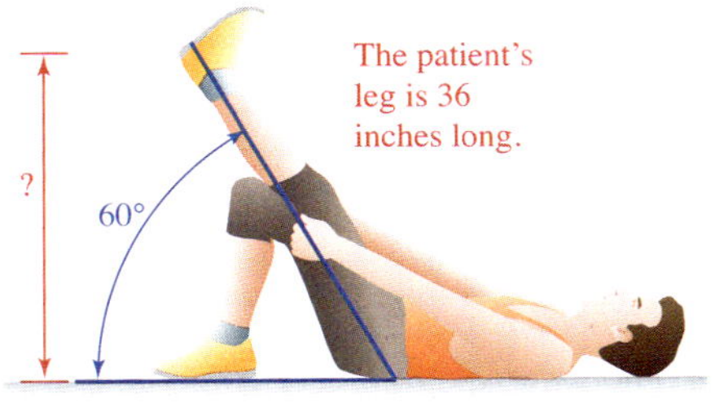

(a)

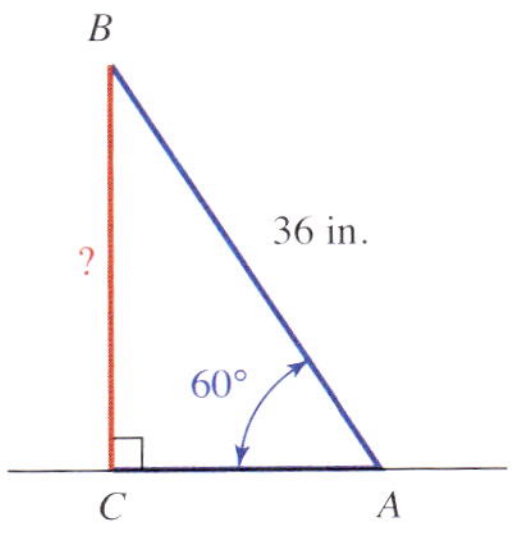

(b)

Self Check 6 **Stretching Exercises.** Refer to Example 6. The doctor prescribed the same exercise for a patient whose leg is 29 inches long. Find how high off the floor this patient's foot is when her leg is held at the proper angle.

Now Try Problems 39 and 61

4 Use the Distance Formula to Solve Problems.

With the **distance formula,** we can find the distance between any two points graphed on a rectangular coordinate system. To find the distance d between points $P(x_1, y_1)$ and $Q(x_2, y_2)$ shown in the figure on the right, we construct the right triangle PRQ. The distance between P and R is $|x_2 - x_1|$, and the distance between R and Q is $|y_2 - y_1|$. We apply the Pythagorean theorem to the right triangle PRQ to get

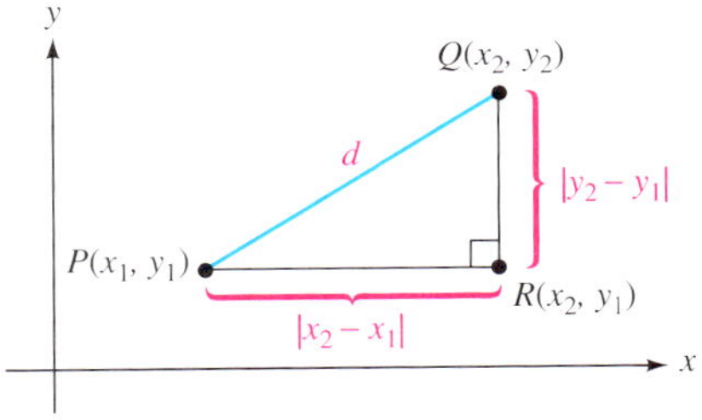

$$d^2 = |x_2 - x_1|^2 + |y_2 - y_1|^2$$
$$= (x_2 - x_1)^2 + (y_2 - y_1)^2 \quad \text{Because } |x_2 - x_1|^2 = (x_2 - x_1)^2 \text{ and } |y_2 - y_1|^2 = (y_2 - y_1)^2$$

Because d represents the distance between two points, it must be equal to the positive square root of $(x_2 - x_1)^2 + (y_2 - y_1)^2$.

$$d = \sqrt{(x_2 - x_1)^2 + (y_2 - y_1)^2}$$

We call this result the *distance formula.*

Distance Formula

The distance d between two points with coordinates (x_1, y_1) and (x_2, y_2) is given by

$$d = \sqrt{(x_2 - x_1)^2 + (y_2 - y_1)^2}$$

EXAMPLE 7 Find the distance between the points: **a.** $(-2, 3)$ and $(4, -5)$ **b.** $(27, 46)$ and $(33, 50)$

Strategy We will use the distance formula.

Why We know the x- and y-coordinates of both points.

Solution **a.** To find the distance, we can use the distance formula by substituting 4 for x_2, -2 for x_1, -5 for y_2, and 3 for y_1.

$$d = \sqrt{(x_2 - x_1)^2 + (y_2 - y_1)^2}$$ This is the distance formula.

$$= \sqrt{[4 - (-2)]^2 + (-5 - 3)^2}$$ Substitute for x_1, x_2, y_1, and y_2.

$$= \sqrt{(4 + 2)^2 + (-5 - 3)^2}$$ Simplify within the parentheses.

$$= \sqrt{6^2 + (-8)^2}$$ Do the addition and subtraction.

$$= \sqrt{36 + 64}$$ Evaluate each exponential expression.

$$= \sqrt{100}$$ Do the addition.

$$= 10$$ Evaluate the square root.

The distance between the points is 10 units.

b.

$$d = \sqrt{(x_2 - x_1)^2 + (y_2 - y_1)^2}$$ This is the distance formula.

$$d = \sqrt{(33 - 27)^2 + (50 - 46)^2}$$ Substitute 33 for x_2, 27 for x_1, 50 for y_2, and 46 for y_1.

$$= \sqrt{6^2 + 4^2}$$ Do the subtraction.

$$= \sqrt{52}$$ Evaluate: $6^2 + 4^2 = 36 + 16 = 52$.

$$= 2\sqrt{13}$$ Simplify: $\sqrt{52} = \sqrt{4 \cdot 13} = 2\sqrt{13}$.

The distance between the points is exactly $2\sqrt{13}$ units, which is about 7.21 units.

Self Check 7 Find the distance between the points: **a.** $(-2, -2)$ and $(3, 10)$ **b.** $(55, 29)$ and $(61, 32)$

Now Try Problems 47 and 49

5 Find the Midpoint of a Line Segment.

If point M in the figure on the right lies midway between point P and point Q, it is called the **midpoint** of line segment PQ. We call the points P and Q, the **endpoints** of the segment.

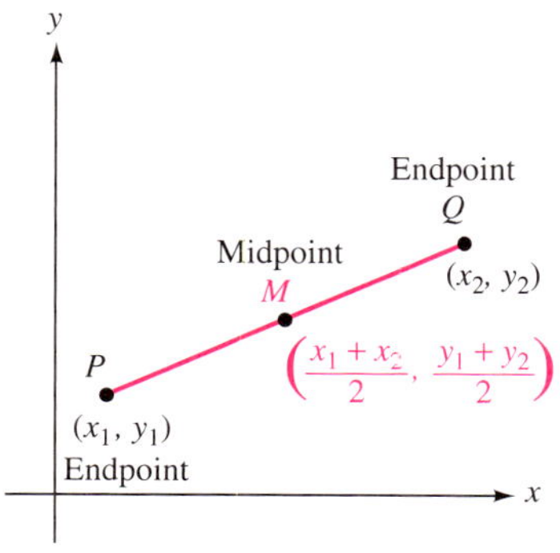

To distinguish between the coordinates of the endpoints of a line segment, we can use *subscript notation.* In the figure, the point P with coordinates (x_1, y_1) is read as "point P with coordinates x sub 1 and y sub 1," and the point Q with coordinates (x_2, y_2) is read as "point Q with coordinates x sub 2 and y sub 2."

To find the coordinates of point M, we find the average of the x-coordinates and the average of the y-coordinates of points P and Q. Using subscript notation, we can write the midpoint formula in the following way.

The Midpoint Formula

The **midpoint** of a line segment with endpoints (x_1, y_1) and (x_2, y_2) is the point with coordinates

$$\left(\frac{x_1 + x_2}{2}, \frac{y_1 + y_2}{2}\right)$$

EXAMPLE 8 Find the midpoint of the line segment with endpoints $(-2, 5)$ and $(4, -2)$.

Strategy To find the coordinates of the midpoint, we find the average of the x-coordinates and the average of the y-coordinates of the endpoints.

Why This is what is called for by the expressions $\frac{x_1 + x_2}{2}$ and $\frac{y_1 + y_2}{2}$ of the midpoint formula.

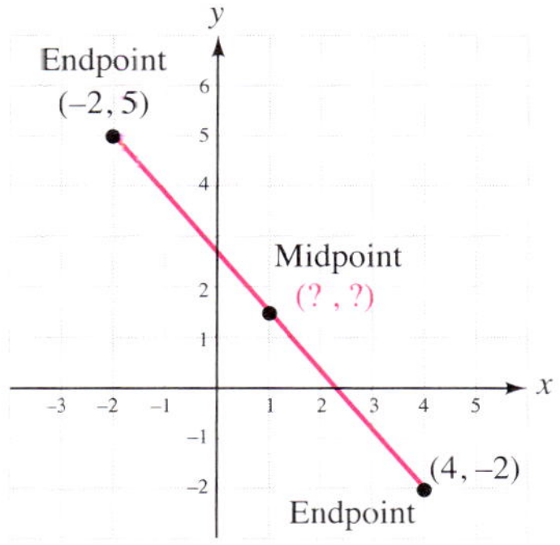

Solution We can let $(x_1, y_1) = (-2, 5)$ and $(x_2, y_2) = (4, -2)$. After substituting these values into the expressions for the x- and y-coordinates in the midpoint formula, we evaluate each expression to find the coordinates of the midpoint.

Find the x-coordinate of the midpoint:

$$\frac{x_1 + x_2}{2} = \frac{-2 + 4}{2} = \frac{2}{2} = 1$$

Find the y-coordinate of the midpoint:

$$\frac{y_1 + y_2}{2} = \frac{5 + (-2)}{2} = \frac{3}{2}$$

Thus, the midpoint is $\left(1, \frac{3}{2}\right)$.

Self Check 8 Find the midpoint of the line segment with endpoints $(-1, 8)$ and $(5, 2)$.

Now Try Problem 57

EXAMPLE 9 The midpoint of the line segment joining $(-5, -3)$ and a point Q is the point $(-1, 2)$. Find the coordinates of point Q.

Strategy As in Example 8, we will use the midpoint formula to find the unknown coordinates. However, this time, we need to find x_2 and y_2.

Why We want to find the coordinates of one of the endpoints.

Solution We can let $(x_1, y_1) = (-5, -3)$ and $(x_M, y_M) = (-1, 2)$, where x_M represents the x-coordinate and y_M represents the y-coordinate of the midpoint. To find the coordinates of point Q, we substitute for x_1, x_M, y_1, and y_M in the expressions for the coordinates in the midpoint formula and solve the resulting equations for x_2 and y_2.

$$x_M = \frac{x_1 + x_2}{2}$$

$$-1 = \frac{-5 + x_2}{2} \quad \text{Substitute.}$$

$$-2 = -5 + x_2 \quad \text{Multiply both sides by 2.}$$

$$3 = x_2 \quad \text{Add 5 to both sides.}$$

$$y_M = \frac{y_1 + y_2}{2}$$

Read x_M as "x sub M" and y_M as "y sub M."

$$2 = \frac{-3 + y_2}{2} \quad \text{Substitute.}$$

$$4 = -3 + y_2 \quad \text{Multiply both sides by 2.}$$

$$7 = y_2 \quad \text{Add 3 to both sides.}$$

Since $x_2 = 3$ and $y_2 = 7$, the coordinates of point Q are $(3, 7)$.

Self Check 9 The midpoint of the line segment joining $(-5, -3)$ and a point P is the point $(-2, 5)$. Find the coordinates of point P.

Now Try Problem 63

SECTION 9.6 STUDY SET

VOCABULARY

Fill in the blanks.

1. In a right triangle, the side opposite the 90° angle is called the ___________.
2. An _________ right triangle is a right triangle with two legs of equal length.
3. The ___________ theorem states that in a right triangle, the sum of the squares of the lengths of the two legs is equal to the square of the hypotenuse.
4. An __________ triangle has three sides of equal length and three 60° angles.

CONCEPTS

Fill in the blanks.

5. If a and b are the lengths of the legs of a right triangle and c is the length of the hypotenuse, then $\quad + \quad = \quad$. This is called the Pythagorean _________.
6. In any right triangle, the square of the hypotenuse is equal to the _____ of the squares of the two _____.
7. In an isosceles right triangle, the length of the hypotenuse is $\quad$ times the length of one leg.
8. The shorter leg of a 30°–60°–90° triangle is _____ as long as the hypotenuse.
9. The length of the longer leg of a 30°–60°–90° triangle is $\quad$ times the length of the shorter leg.
10. In a 30°–60°–90° triangle, the shorter leg is opposite the $\quad$ angle, and the longer leg is opposite the $\quad$ angle.
11. The formula to find the distance between points (x_1, y_1) and (x_2, y_2) is $d = \sqrt{\quad + \quad}$.
12. Solve for c, where c represents the length of the hypotenuse of a right triangle. Simplify the result, if possible.
    a. $c^2 = 64$
    b. $c^2 = 15$
    c. $c^2 = 24$

NOTATION

Complete each solution.

13. Evaluate. Approximate to two decimal places.

$$\begin{aligned}\sqrt{(-1-3)^2 + [2-(-4)]^2} &= \sqrt{(-4)^2 + [\quad]^2}\\ &= \sqrt{\quad}\\ &= \sqrt{\quad \cdot 13}\\ &= \quad\sqrt{13}\\ &\approx \quad\end{aligned}$$

14. Solve $8^2 + 4^2 = c^2$ and assume $c > 0$. Approximate to two decimal places.

$$\begin{aligned}\quad + 16 &= c^2\\ \quad &= c^2\\ \sqrt{\quad} &= \quad\\ \sqrt{\quad \cdot 5} &= c\\ \quad\sqrt{5} &= c\\ c \quad & 8.94\end{aligned}$$

GUIDED PRACTICE

The lengths of two sides of the right triangle ABC are given. Find the length of the missing side. See Example 1.

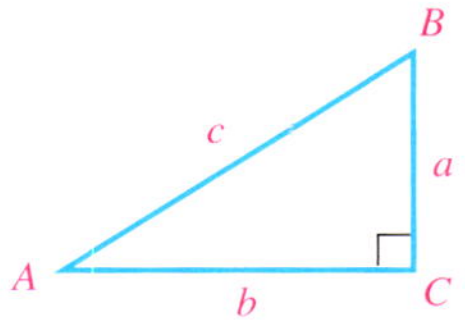

15. $a = 6$ ft and $b = 8$ ft
16. $a = 5$ in. and $b = 12$ in.
17. $a = 8$ ft and $b = 15$ ft
18. $a = 24$ yd and $b = 7$ yd
19. $b = 9$ ft and $c = 41$ ft
20. $b = 18$ m and $c = 82$ m
21. $a = 10$ cm and $c = 26$ cm
22. $a = 14$ in. and $c = 50$ in.

Find the missing side lengths in each triangle. Give the exact answer and then an approximation to two decimal places when appropriate. See Example 2.

23.

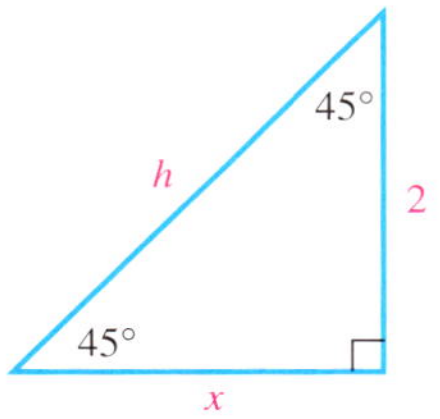

24.

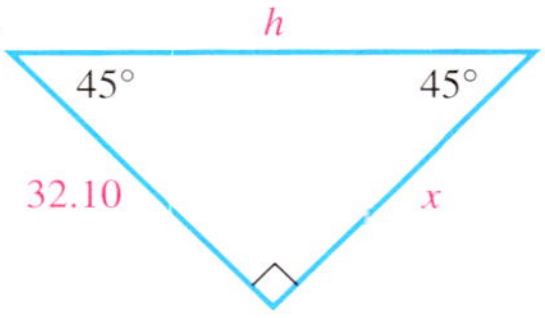

25. One leg of an isosceles right triangle is 3.2 feet long. Find the length of its hypotenuse. Give the exact answer and then an approximation to two decimal places.
26. One side of a square is $5\frac{1}{2}$ in. long. Find the length of its diagonal. Give the exact answer and then an approximation to two decimal places.

Find the missing side lengths in each triangle. Give the exact answer and then an approximation to two decimal places. See Example 3.

27.

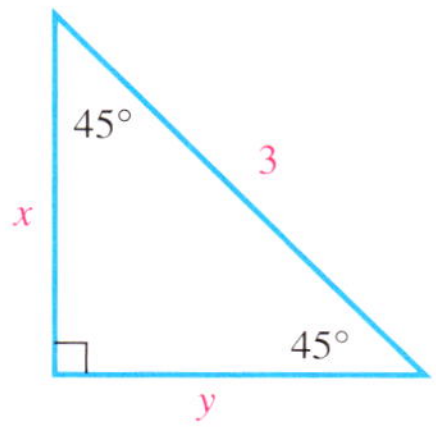

28.

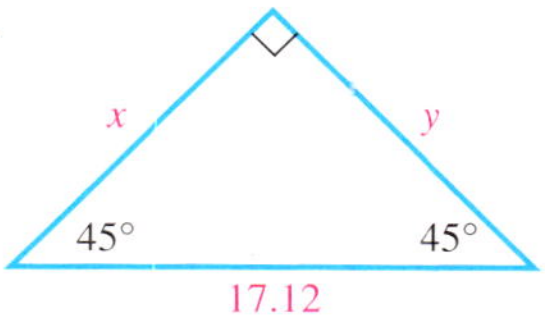

29. **Photographs.** The diagonal of a square photograph measures 10 inches. Find the length of one of its sides. Give the exact answer and then an approximation to two decimal places.

30. **Parking Lots.** The diagonal of a square parking lot is approximately 1,414 feet long.

a. Find the length of one side of the parking lot. Round to the nearest foot.

b. Find the approximate area of the parking lot.

Find the missing side lengths in each triangle. Give the exact answer and then an approximation to two decimal places, when appropriate. See Example 4.

31.
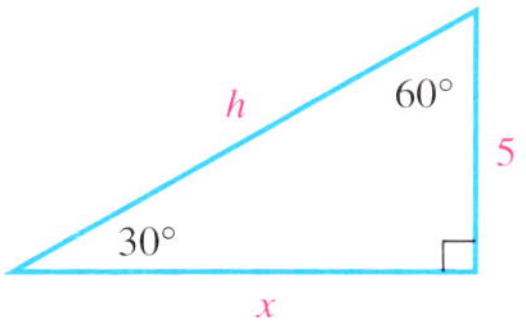

32.
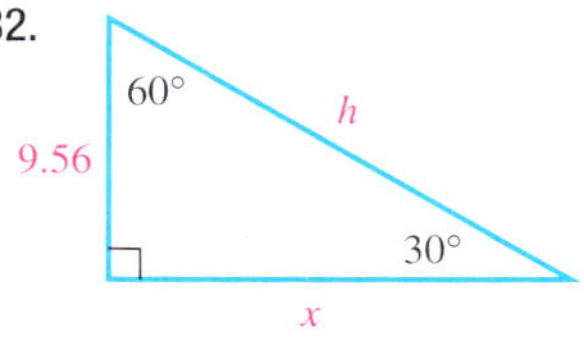

33. In a 30°–60°–90° triangle, the length of the leg opposite the 30° angle is 75 cm. Find the length of the leg opposite the 60° angle and the length of the hypotenuse. Give the exact answer and then an approximation to two decimal places, when appropriate.

34. In a 30°–60°–90° triangle, the length of the shorter leg is $5\sqrt{2}$ inches. Find the length of the hypotenuse and the length of the longer leg. Give the exact answer and then an approximation to two decimal places.

Find the missing lengths in each triangle. Give the exact answer and then an approximation to two decimal places. See Example 5.

35.
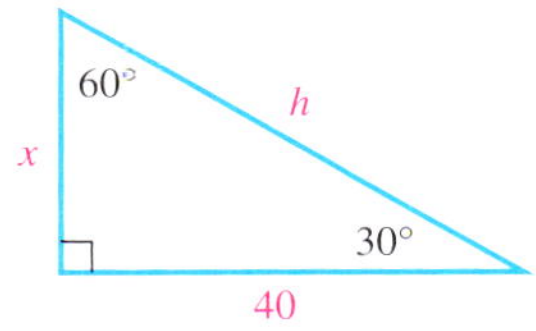

36.
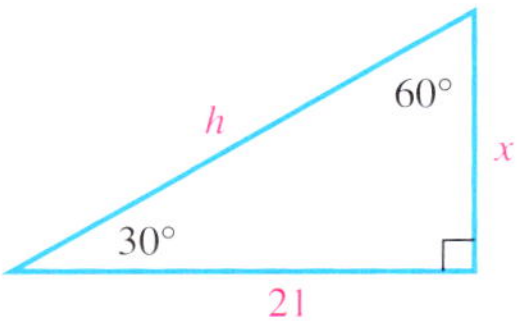

37. In a 30°–60°–90° right triangle, the length of the leg opposite the 60° angle is 55 millimeters. Find the length of the leg opposite the 30° angle and the length of the hypotenuse. Give the exact answer and then an approximation to two decimal places.

38. In a 30°–60°–90° right triangle, the length of the longer leg is 24 yards. Find the length of the hypotenuse and the length of the shorter leg. Give the exact answer and then an approximation to two decimal places.

Find the missing lengths in each triangle. Give the exact answer and then an approximation to two decimal places, when appropriate. See Example 6.

39.
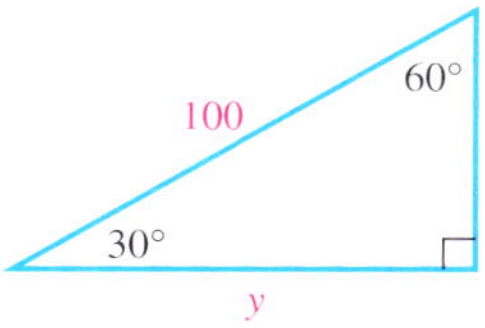

40.
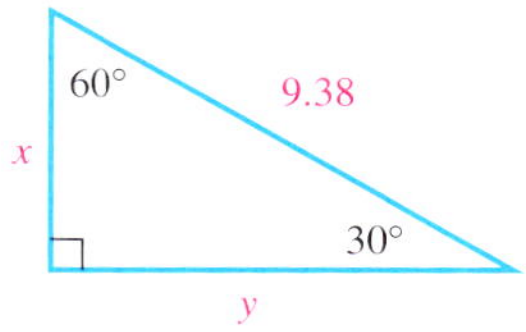

41. In a 30°–60°–90° right triangle, the length of the hypotenuse is 1.5 feet. To the nearest hundredth, find the length of the shorter leg and the length of the longer leg. Give the exact answer and then an approximation to two decimal places, when appropriate.

42. In a 30°–60°–90° right triangle, the length of the hypotenuse is $12\sqrt{3}$ inches. Find the length of the leg opposite the 30° angle and the length of the leg opposite the 60° angle. Give the exact answer and then an approximation to two decimal places, when appropriate.

Find the exact distance between each pair of points. See Example 7.

43. (0, 0), (3, −4)
44. (0, 0), (−12, 16)
45. (−2, −8), (3, 4)
46. (−5, −2), (7, 3)
47. (6, 8), (12, 16)
48. (10, 4), (2, −2)
49. (−2, 1), (3, 4)
50. (2, −3), (4, −8)
51. (−1, −6), (3, −4)
52. (−3, 5), (−5, −5)
53. (−2, −1), (−5, 8)
54. (4, 7), (−4, −5)

Find the midpoint of the line segment with the given endpoints. See Example 8.

55. (0, 0), (6, 8)
56. (10, 12), (0, 0)
57. (6, 8), (12, 16)
58. (10, 4), (2, −2)
59. (−2, −8), (3, −8)
60. (−5, −2), (7, 3)
61. (7, 1), (−10, 4)
62. (−4, −3), (4, −8)

Solve each problem. See Example 9.

63. If (−2, 3) is the midpoint of segment PQ and the coordinates of P are (−8, 5), find the coordinates of Q.

64. If (6, −5) is the midpoint of segment PQ and the coordinates of Q are (−5, −8), find the coordinates of P.

65. If (−7, −3) is the midpoint of segment QP and the coordinates of Q are (6, −3), find the coordinates of P.

66. If $\left(\frac{1}{2}, -2\right)$ is the midpoint of segment QP and the coordinates of P are $\left(-\frac{5}{2}, 5\right)$, find the coordinates of Q.

APPLICATIONS

67. Soccer. The allowable length of a rectangular soccer field used for international adult matches can be from 100 to 110 meters and the width can be from 64 to 75 meters.

a. Find the length of the diagonal of the field that has the minimum allowable length and minimum allowable width. Give an approximation to two decimal places.

b. Find the length of the diagonal of the field that has the maximum allowable length and maximum allowable width. Give the exact answer and an approximation to two decimal places.

68. Cubes. Find the exact length of the diagonal (in blue) of one of the *faces* of the cube shown here.

69. Cubes. Find the exact length of the diagonal (in green) of the cube shown here.

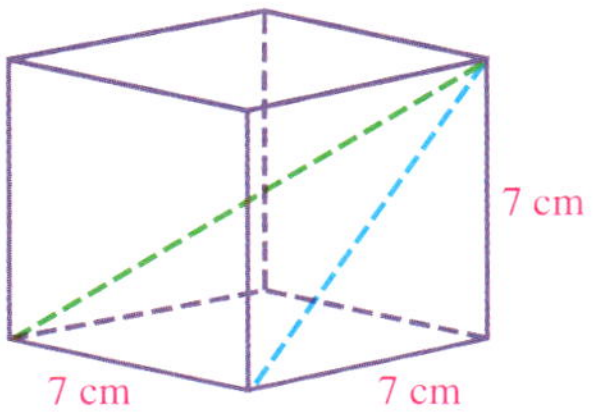

70. Geometry. Use the distance formula to show that a triangle with vertices $(-2, 4)$, $(2, 8)$, and $(6, 4)$ is isosceles.

71. Washington, D.C. The square in the map shows the 100-square-mile site selected by George Washington in 1790 to serve as a permanent capital for the United States. In 1847, the part of the district lying on the west bank of the Potomac was returned to Virginia. Find the exact coordinates of each corner of the original square that outlined the District of Columbia and approximations to two decimal places.

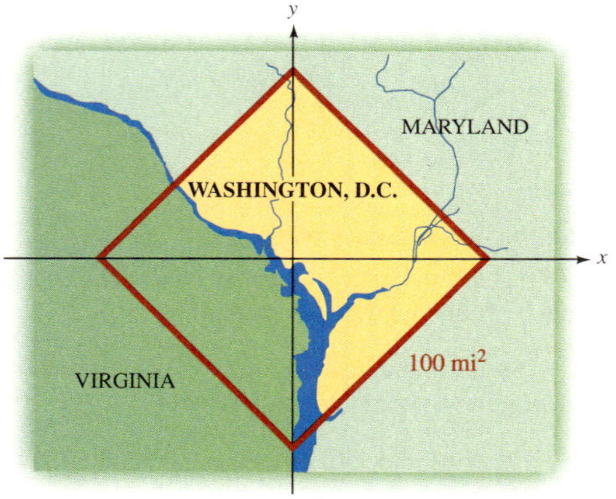

72. Paper Airplanes. The illustration gives the directions for making a paper airplane from a square piece of paper with sides 8 inches long. Find the length l of the plane when it is completed. Give the exact answer and an approximation to two decimal places.

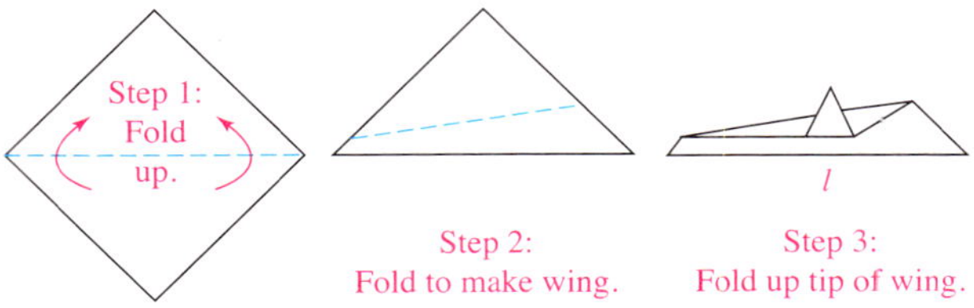

73. Hardware. The sides of a regular hexagonal nut are 10 millimeters long. Find the height h of the nut. Give the exact answer and an approximation to two decimal places.

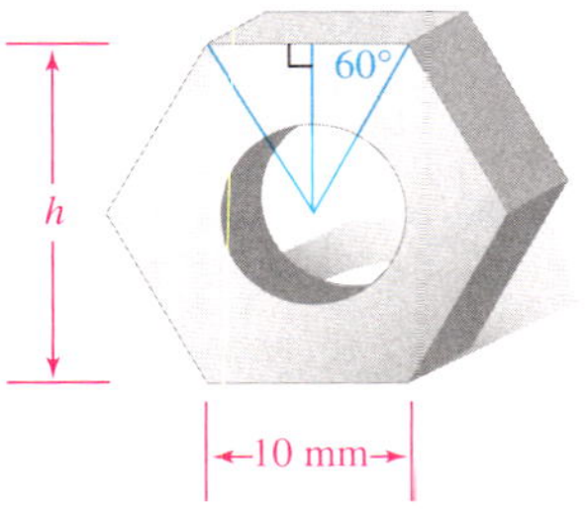

74. Ironing Boards. Find the height h of the ironing board shown in the illustration. Give the exact answer and an approximation to two decimal places.

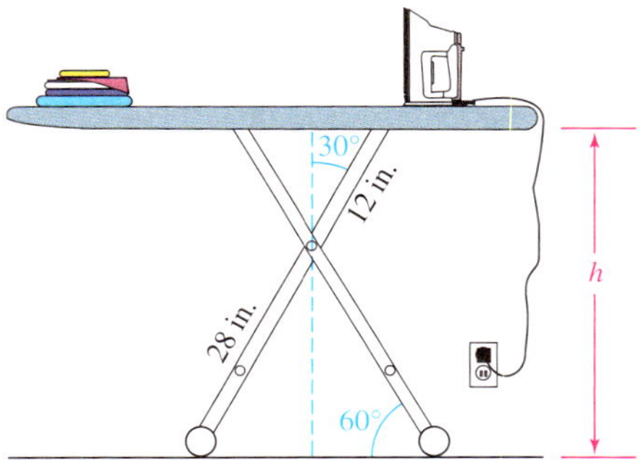

75. Baseball. A baseball diamond is a square, 90 feet on a side. If the third baseman fields a ground ball 10 feet directly behind third base, how far must he throw the ball to throw a runner out at first base? Give the exact answer and an approximation to two decimal places.

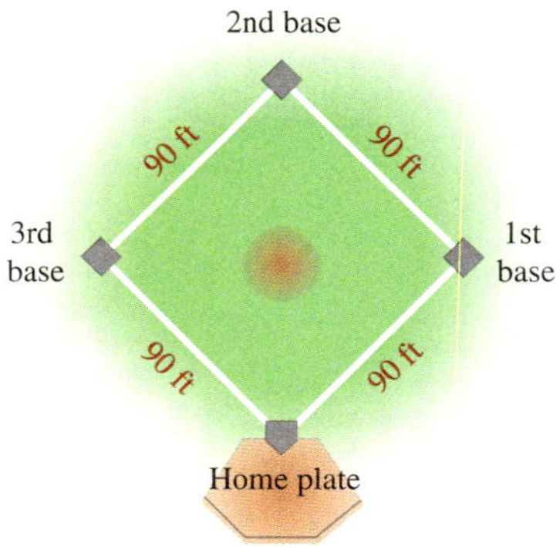

76. Baseball. A shortstop fields a grounder at a point one-third of the way from second base to third base. How far will he have to throw the ball to make an out at first base? Give the exact answer and an approximation to two decimal places.

77. Clotheslines. A pair of damp jeans are hung on a clothesline to dry. They pull the center down 1 foot. By how much is the line stretched? Give an approximation to two decimal places.

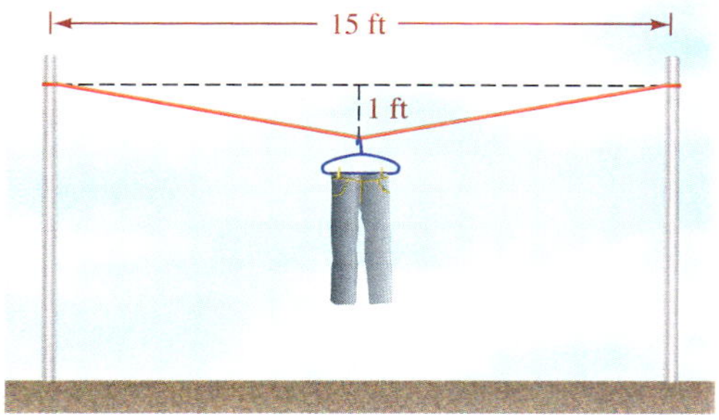

78. Firefighting. The base of the 37-foot ladder is 9 feet from the wall. Will the top reach a window ledge that is 35 feet above the ground? Verify your result.

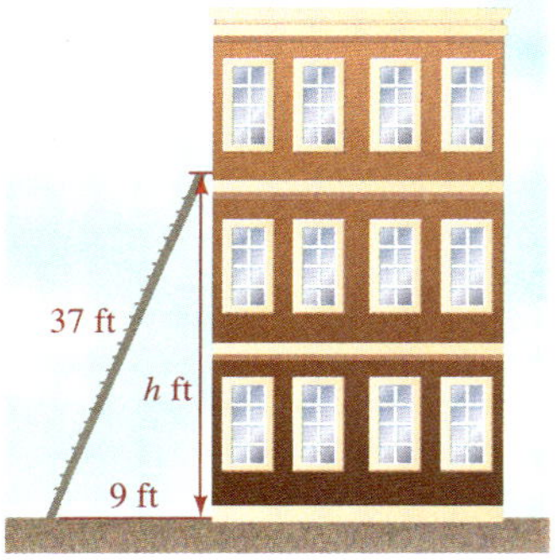

79. Art History. A figure displaying some of the characteristics of Egyptian art is shown in the illustration. Use the distance formula to find the following dimensions of the drawing. Round your answers to two decimal places.

a. From the foot to the eye

b. From the belt to the hand holding the staff

c. From the shoulder to the symbol held in the hand

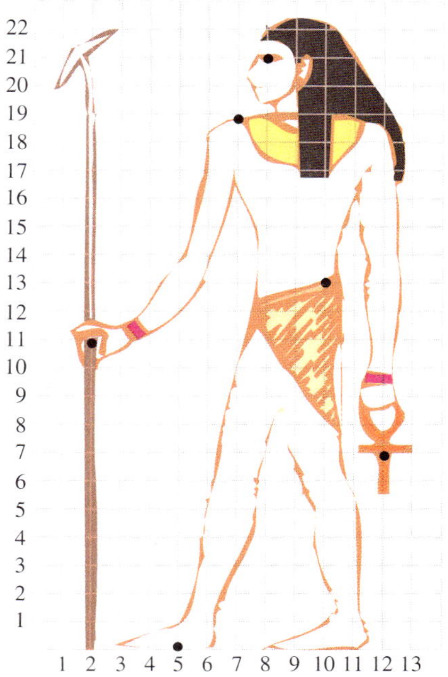

80. Packaging. The diagonal d of a rectangular box with dimensions $a \times b \times c$ is given by $d = \sqrt{a^2 + b^2 + c^2}$. Will the umbrella fit in the shipping carton in the illustration? Verify your result.

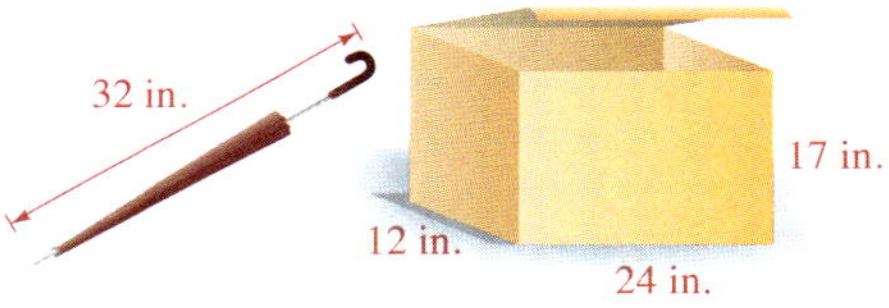

81. Packaging. An archaeologist wants to ship a 34-inch femur bone. Will it fit in a 4-inch-tall box that has a square base with sides 24 inches long? (See Exercise 68.) Verify your result.

82. Telephone. The telephone cable in the illustration runs from A to B to C to D. How much cable is required to run from A to D directly?

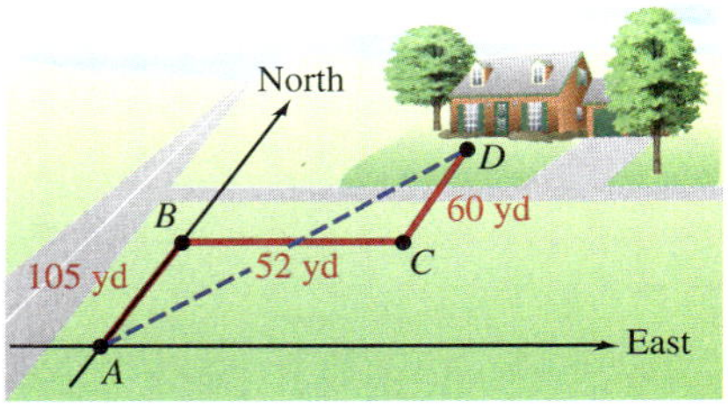

WRITING

83. State the Pythagorean theorem in words.

84. List the facts that you learned about special right triangles in this section.

85. When the lengths of the sides of a certain triangle are substituted into the equation of the Pythagorean theorem, the result is a false statement. Explain why.

$$a^2 + b^2 = c^2$$
$$2^2 + 4^2 = 5^2$$
$$4 + 16 = 25$$
$$20 = 25 \quad \text{False}$$

86. Explain how the distance formula and the Pythagorean theorem can be used to show that a triangle with vertices $(2, 3)$, $(-3, 4)$, and $(1, -2)$ is a right triangle.

REVIEW

87. **Discount Buying.** A repairman purchased some washing-machine motors for a total of \$224. When the unit cost decreased by \$4, he was able to buy one extra motor for the same total price. How many motors did he buy originally?

88. **Aviation.** An airplane can fly 650 miles with the wind in the same amount of time as it can fly 475 miles against the wind. If the wind speed is 40 mph, find the speed of the plane in still air.

CHALLENGE PROBLEMS

89. Find the length of the diagonal of the cube shown in figure (a) below.

90. Show that the length of the diagonal of the rectangular solid shown in figure (b) below is $\sqrt{a^2 + b^2 + c^2}$ cm.

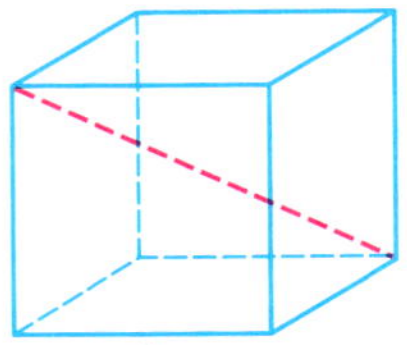

(a)

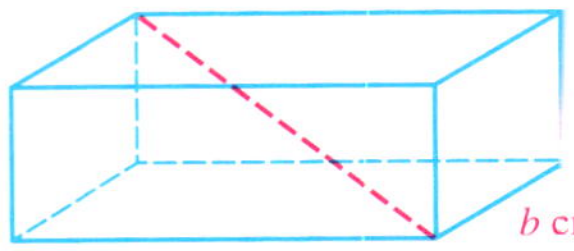

(b)

Find the distance between each pair of points.

91. $\left(\sqrt{48}, \sqrt{150}\right)$ and $\left(\sqrt{12}, \sqrt{24}\right)$

92. $\left(\sqrt{8}, -\sqrt{20}\right)$ and $\left(\sqrt{50}, -\sqrt{45}\right)$

SECTION 9.7 Complex Numbers

OBJECTIVES

1. Express square roots of negative numbers in terms of i.
2. Write complex numbers in the form $a + bi$.
3. Add and subtract complex numbers.
4. Multiply complex numbers.
5. Divide complex numbers.
6. Perform operations involving powers of i.

ARE YOU READY?

The following problems review some basic skills that are needed when working with complex numbers.

1. Explain why $\sqrt{-16}$ is not a real number.
2. Simplify: $(x^2 + 8x) + (5x^2 - 10x)$
3. Multiply: $(2n + 3)(7n - 1)$
4. Simplify: $\sqrt{63}$
5. Rationalize the denominator: $\dfrac{7}{\sqrt{x} + 4}$
6. Divide and give the remainder: $4\overline{)87}$

Recall that the square root of a negative number is not a real number. However, an expanded number system, called the *complex number system,* gives meaning to square roots of negative numbers, such as $\sqrt{-9}$ and $\sqrt{-25}$. To define complex numbers, we use a number that is denoted by the letter i.

1 Express Square Roots of Negative Numbers in Terms of *i*.

Some equations do not have real-number solutions. For example, $x^2 = -1$ has no real-number solutions because the square of a real number is never negative. To provide a solution to this equation, mathematicians have defined the number i so that $i^2 = -1$.

The Number *i*

The **imaginary number *i*** is defined as

$$i = \sqrt{-1}$$

From the definition, it follows that $i^2 = -1$.

This definition enables us to write the square root of any negative number in terms of i.

We can use extensions of the product and quotient rules for radicals to write the square root of a negative number as the product of a real number and i.

EXAMPLE 1 Write each expression in terms of i: **a.** $\sqrt{-9}$ **b.** $\sqrt{-7}$ **c.** $-\sqrt{-18}$

d. $\sqrt{-\frac{24}{49}}$

Strategy We will write each radicand as the product of -1 and a positive number. Then we will apply the appropriate rules for radicals.

Why We want our work to produce a factor of $\sqrt{-1}$ so that we can replace it with i.

Solution After factoring the radicand, we use an extension of the product rule for radicals.

a. $\sqrt{-9} = \sqrt{-1 \cdot 9} = \sqrt{-1}\sqrt{9} = i \cdot 3 = 3i$ Replace $\sqrt{-1}$ with i.

b. $\sqrt{-7} = \sqrt{-1 \cdot 7} = \sqrt{-1}\sqrt{7} = i\sqrt{7}$ or $\sqrt{7}i$ Replace $\sqrt{-1}$ with i.

c. $-\sqrt{-18} = -\sqrt{-1 \cdot 9 \cdot 2} = -\sqrt{-1}\sqrt{9}\sqrt{2} = -i \cdot 3 \cdot \sqrt{2} = -3i\sqrt{2}$ or $-3\sqrt{2}i$

d. After factoring the radicand, use an extension of the product and quotient rules for radicals.

$$\sqrt{-\frac{24}{49}} = \sqrt{-1 \cdot \frac{24}{49}} = \frac{\sqrt{-1 \cdot 24}}{\sqrt{49}} = \frac{\sqrt{-1}\sqrt{4}\sqrt{6}}{\sqrt{49}} = \frac{2i\sqrt{6}}{7} \quad \text{or} \quad \frac{2\sqrt{6}}{7}i$$

Notation

Since it is easy to confuse $\sqrt{bi}$ with $\sqrt{b}i$, we usually write i first so that it is clear that the i is not under the radical symbol. However, both $i\sqrt{b}$ and $\sqrt{b}i$ are correct. For example: $i\sqrt{7} = \sqrt{7}i$.

Self Check 1 Write each expression in terms of i: **a.** $\sqrt{-25}$ **b.** $-\sqrt{-19}$ **c.** $\sqrt{-45}$ **d.** $\sqrt{-\frac{50}{81}}$

Now Try Problems 19, 21, and 27

The results from Example 1 illustrate a rule for simplifying square roots of negative numbers.

Square Root of a Negative Number

For any positive real number b,

$$\sqrt{-b} = i\sqrt{b}$$

To justify this rule, we use the fact that $\sqrt{-1} = i$.

$$\sqrt{-b} = \sqrt{-1 \cdot b}$$
$$= \sqrt{-1}\sqrt{b}$$
$$= i\sqrt{b}$$

2 Write Complex Numbers in the Form $a + bi$.

The imaginary number i is used to define *complex numbers.*

Complex Numbers

A **complex number** is any number that can be written in the form $a + bi$, where a and b are real numbers and $i = \sqrt{-1}$.

Complex numbers of the form $a + bi$, where $b \neq 0$, are also called **imaginary numbers.***

*Some textbooks define imaginary numbers as complex numbers with $a = 0$ and $b \neq 0$.

Notation

It is acceptable to use $a - bi$ as a substitute for the form $a + bi$. For example:

$$6 - 9i = 6 + (-9)i$$

For a complex number written in the **standard form** $a + bi$, we call a the **real part** and b the **imaginary part.** Some examples of complex numbers written in standard form are

$2 + 11i$	$6 - 9i$	$-\frac{1}{2} + 0i$	$0 + i\sqrt{3}$
$a = 2, b = 11$	$a = 6, b = -9$	$a = -\frac{1}{2}, b = 0$	$a = 0, b = \sqrt{3}$

Two complex numbers $a + bi$ and $c + di$ are equal if and only if $a = c$ and $b = d$. Thus, $0.5 + 0.9i = \frac{1}{2} + \frac{9}{10}i$ because $0.5 = \frac{1}{2}$ and $0.9 = \frac{9}{10}$.

EXAMPLE 2 Write each number in the form $a + bi$: **a.** 6 **b.** $\sqrt{-64}$ **c.** $-2 + \sqrt{-63}$

Strategy We will determine a, the real part, and we will simplify the radical (if necessary) to determine the bi part.

Why We can put the two parts together to produce the desired $a + bi$ form.

Solution

a. $6 = 6 + 0i$ The real part a is 6. The imaginary part b is 0.

b. $\sqrt{-64} = 0 + 8i$ The real part a is 0. Simplify: $\sqrt{-64} = \sqrt{-1}\sqrt{64} = 8i$. Thus, b is 8.

c. $-2 + \sqrt{-63} = -2 + 3i\sqrt{7}$ The real part a is -2. Simplify: $\sqrt{-63} = \sqrt{-1}\sqrt{63} = \sqrt{-1}\sqrt{9}\sqrt{7} = 3i\sqrt{7}$. Thus, $b = 3\sqrt{7}$.

Caution

Use i only when working with the *square root* of a negative number. It does not apply to cube roots.

$\sqrt{-64} = 8i$ and $\sqrt[3]{-64} = -4$

Self Check 2 Write each number in the form $a + bi$: **a.** -18 **b.** $\sqrt{-36}$ **c.** $1 + \sqrt{-24}$

Now Try Problems 29 and 33

The following illustration shows the relationship between the real numbers, the imaginary numbers, and the complex numbers.

The complex numbers: $a + bi$

The real numbers: $a + bi$ where $b = 0$				The imaginary numbers: $a + bi$ where $b \neq 0$		
-6	$\frac{5}{16}$	-1.75	π	$9 + 7i$	$-2i$	$\frac{1}{4} - \frac{3}{4}i$
$48 + 0i$	0	$-\sqrt{10}$	$-\frac{7}{2}$	$0.56i$	$\sqrt{-10}$	$6 + i\sqrt{3}$
$\sqrt[3]{-10}$	$4 + \pi$	$15\frac{5}{8}$	$\sqrt[4]{99}$	$25i\sqrt{5}$	$\frac{7}{8}i$	$\sqrt{3} + 3i\sqrt{2}$

Success Tip

Just as real numbers are either rational or irrational, but not both, complex numbers are either real or imaginary, but not both.

3 Add and Subtract Complex Numbers.

Adding and subtracting complex numbers is similar to adding and subtracting polynomials.

Addition and Subtraction of Complex Numbers

1. To add complex numbers, add their real parts and add their imaginary parts.
2. To subtract complex numbers, add the opposite of the complex number being subtracted.

EXAMPLE 3 Perform each operation. Write the answers in the form $a + bi$. **a.** $(8 + 4i) + (12 + 8i)$ **b.** $(-6 + 4i) - (3 + 2i)$ **c.** $\left(7 - \sqrt{-16}\right) + \left(9 + \sqrt{-4}\right)$

Strategy To add the complex numbers, we will add their real parts and add their imaginary parts. To subtract the complex numbers, we will add the opposite of the complex number to be subtracted.

Why We perform the indicated operations as if the complex numbers were polynomials with i as a variable.

Solution

a. $(8 + 4i) + (12 + 8i) = (8 + 12) + (4 + 8)i$ — The sum of the imaginary parts; The sum of the real parts

$= 20 + 12i$

b. $(-6 + 4i) - (3 + 2i) = (-6 + 4i) + (-3 - 2i)$ — Add; the opposite. To find the opposite, change the sign of each term of $3 + 2i$.

$= [-6 + (-3)] + [4 + (-2)]i$ — Add the real parts. Add the imaginary parts.

$= -9 + 2i$

c. $\left(7 - \sqrt{-16}\right) + \left(9 + \sqrt{-4}\right)$

$= (7 - 4i) + (9 + 2i)$ — Write $\sqrt{-16}$ and $\sqrt{-4}$ in terms of i.

$= (7 + 9) + (-4 + 2)i$ — Add the real parts. Add the imaginary parts.

$= 16 - 2i$ — Write $16 + (-2i)$ in the form $16 - 2i$.

The Language of Algebra

For years, mathematicians thought numbers like $\sqrt{-4}$ and $\sqrt{-16}$ were useless. In the 17th century, French mathematician René Descartes (1596–1650) called them **imaginary numbers.** Today they have important uses such as describing alternating electric current.

Success Tip

Always express square roots of negative numbers in terms of i before performing any operations.

Self Check 3 Perform the operations. Write the answers in the form $a + bi$.

a. $(3 - 5i) + (-2 + 7i)$

b. $\left(3 - \sqrt{-25}\right) - \left(-2 + \sqrt{-49}\right)$

Now Try Problems 37 and 43

4 Multiply Complex Numbers.

Since imaginary numbers are not real numbers, some properties of real numbers do not apply to imaginary numbers. For example, we cannot use the product rule for radicals to multiply two imaginary numbers.

CAUTION If a and b are both negative, then $\sqrt{a}\sqrt{b} \neq \sqrt{ab}$. For example, if $a = -4$ and $b = -9$,

$$\sout{\sqrt{-4} \cdot \sqrt{-9} = \sqrt{-4(-9)} = \sqrt{36} = 6} \qquad \sqrt{-4}\sqrt{-9} = 2i(3i) = 6i^2 = 6(-1) = -6$$

EXAMPLE 4 Multiply: $\sqrt{-2} \cdot \sqrt{-20}$

Strategy To multiply the imaginary numbers, first we will write $\sqrt{-2}$ and $\sqrt{-20}$ in $i\sqrt{b}$ form. Then we will use the product rule for radicals.

Why We cannot use the product rule for radicals immediately because it does not apply when both radicands are negative.

Solution

$$\sqrt{-2} \cdot \sqrt{-20} = (i\sqrt{2})(2i\sqrt{5}) \quad \text{Simplify: } \sqrt{-20} = i\sqrt{20} = 2i\sqrt{5}.$$
$$= 2i^2\sqrt{2 \cdot 5} \quad \text{Multiply: } i \cdot 2i = 2i^2\text{. Use the product rule for radicals.}$$
$$= 2i^2\sqrt{10}$$
$$= 2(-1)\sqrt{10} \quad \text{Replace } i^2 \text{ with } -1.$$
$$= -2\sqrt{10} \quad \text{Multiply.}$$

Success Tip

The algebraic methods we have used with polynomials and rational expressions (such as combining like terms, distributive property, FOIL, and building fractions) are also used with complex numbers. However, when you encounter i^2 in your work, remember to replace it with -1.

Self Check 4 Multiply: $\sqrt{-3} \cdot \sqrt{-32}$

Now Try Problem 47

Multiplying complex numbers is similar to multiplying polynomials.

EXAMPLE 5 Multiply. Write the answers in the form $a + bi$. **a.** $6(2 + 9i)$ **b.** $-5i(4 - 8i)$

Strategy We will use the distributive property to find the products.

Why We perform the indicated operations as if the complex numbers were polynomials with i as a variable.

Solution

a. $6(2 + 9i) = 6(2) + 6(9i)$ Use the distributive property.

$= 12 + 54i$ Perform each multiplication.

b. $-5i(4 - 8i) = -5i(4) - (-5i)8i$ Use the distributive property.

$= -20i + 40i^2$ Perform each multiplication.

$= -20i + 40(-1)$ Replace i^2 with -1.

$= -20i - 40$ Multiply.

$= -40 - 20i$ Write the real part, -40, as the first term.

Caution

A common mistake is to replace i with -1. Remember, $i \neq -1$. By definition, $i = \sqrt{-1}$ and $i^2 = -1$.

Self Check 5 Multiply. Write the answers in the form $a + bi$.
a. $-2(-9 - i)$ **b.** $10i(7 + 4i)$

Now Try Problems 49 and 55

EXAMPLE 6 Multiply. Write the answers in the form $a + bi$. **a.** $(2 + 3i)(3 - 2i)$
b. $(-4 + 2i)(2 + i)$

Strategy We will use the FOIL method to multiply the two complex numbers.

Why We perform the indicated operations as if the complex numbers were binomials with i as a variable.

Solution

F O I L

a. $(2 + 3i)(3 - 2i) = 6 - 4i + 9i - 6i^2$ Use the FOIL method.
$= 6 + 5i - 6(-1)$ Combine the imaginary terms: $-4i + 9i = 5i$. Replace i^2 with -1.
$= 6 + 5i + 6$ Simplify the last term.
$= 12 + 5i$ Combine like terms.

b. $(-4 + 2i)(2 + i) = -8 - 4i + 4i + 2i^2$ Use the FOIL method.
$= -8 + 0i + 2(-1)$ Combine like terms: $-4i + 4i = 0i$. Replace i^2 with -1.
$= -8 + 0i - 2$ Multiply.
$= -10 + 0i$ Combine like terms.

Success Tip

i is not a variable, but you can think of it as one when adding, subtracting, and multiplying. For example:

$$-4i + 9i = 5i$$
$$6i - 2i = 4i$$
$$i \cdot i = i^2$$

Remember that the expression i^2 simplifies to -1.

Self Check 6 Multiply. Write the answers in the form $a + bi$. $(-2 + 3i)(3 - 2i)$

Now Try Problem 59

5 Divide Complex Numbers.

Before we can discuss division of complex numbers, we must introduce an important fact about *complex conjugates*.

Complex Conjugates

The complex numbers $a + bi$ and $a - bi$ are called **complex conjugates**.

For example,

- $7 + 4i$ and $7 - 4i$ are complex conjugates.
- $5 - i$ and $5 + i$ are complex conjugates.
- $-6i$ and $6i$ are complex conjugates, because $-6i = 0 - 6i$ and $6i = 0 + 6i$.

In general, the product of the complex number $a + bi$ and its complex conjugate $a - bi$ is the real number $a^2 + b^2$, as the following work shows:

$(a + bi)(a - bi) = a^2 - abi + abi - b^2i^2$ Use the FOIL method.
$= a^2 - b^2(-1)$ $-abi + abi = 0$. Replace i^2 with -1.
$= a^2 + b^2$

The Language of Algebra

Recall that the word **conjugate** was used earlier when we rationalized the denominators of radical expressions such as

$$\frac{5}{\sqrt{6} - 1}$$

EXAMPLE 7 Find the product of $3 + 5i$ and its complex conjugate.

Strategy The complex conjugate of $3 + 5i$ is $3 - 5i$. We will find their product by using the FOIL method.

Why We perform the indicated operations as if the complex numbers were binomials with i as a variable.

Solution We can find the product as follows:

$$\begin{aligned}(3 + 5i)(3 - 5i) &= 9 - 15i + 15i - 25i^2 && \text{Use the FOIL method.}\\ &= 9 - 25i^2 && \text{Combine like terms: } -15i + 15i = 0.\\ &= 9 - 25(-1) && \text{Replace } i^2 \text{ with } -1.\\ &= 9 + 25\\ &= 34\end{aligned}$$

The product of $3 + 5i$ and its complex conjugate $3 - 5i$ is the real number 34.

Self Check 7 Multiply: $(2 + 3i)(2 - 3i)$

Now Try Problem 65

Recall that to divide *radical expressions,* we rationalized the denominator. We will use a similar approach to divide complex numbers. To divide two complex numbers when the divisor has two terms, we use the following strategy.

Division of Complex Numbers

To divide complex numbers, multiply the numerator and denominator by the complex conjugate of the denominator.

EXAMPLE 8 Divide. Write the answers in the form $a + bi$. **a.** $\frac{3}{6 + i}$ **b.** $\frac{1 + 2i}{3 - 4i}$

Strategy We will build each fraction by multiplying it by a form of 1 that uses the conjugate of the denominator.

Why This step produces a *real number* in the denominator so that the result then can be written in the form $a + bi$.

Solution **a.** We want to build a fraction equivalent to $\frac{3}{6 + i}$ that does not have i in the denominator. To make the denominator, $6 + i$, a real number, we need to multiply it by its complex conjugate, $6 - i$. It follows that $\frac{6 - i}{6 - i}$ should be the form of 1 that is used to build $\frac{3}{6 + i}$.

$$\begin{aligned}\frac{3}{6 + i} &= \frac{3}{6 + i} \cdot \frac{6 - i}{6 - i} && \text{To build an equivalent fraction, multiply by } \tfrac{6 - i}{6 - i} = 1.\\ &= \frac{18 - 3i}{36 - 6i + 6i - i^2} && \text{To multiply the numerators, distribute the multiplication by 3. Use the FOIL method to multiply the denominators.}\\ &= \frac{18 - 3i}{36 - (-1)} && \text{Combine like terms: } -6i + 6i = 0. \text{ Replace } i^2 \text{ with } -1. \text{ Note that the denominator no longer contains } i\\ &= \frac{18 - 3i}{37} && \text{Simplify the denominator. This notation represents the difference of two fractions that have the common denominator 37: } \tfrac{18}{37} \text{ and } \tfrac{3i}{37}.\\ &= \frac{18}{37} - \frac{3}{37}i && \text{Write the complex number in the form } a + bi.\end{aligned}$$

Success Tip

When multiplying the denominator and its complex conjugate, we can use the special-product rule for the sum and difference of the same two terms to simplify the calculations:

$$(3 - 4i)(3 + 4i) = 3^2 - (4i)^2 = 9 - 16i^2$$

Caution

Remember, we can remove only factors common to the entire numerator and denominator. Don't make the mistake of removing a common term, as shown below:

$$\frac{\overset{1}{\cancel{-5}} + 10i}{\underset{5}{\cancel{25}}}$$

b. We can make the denominator of $\frac{1 + 2i}{3 - 4i}$ a real number by multiplying it by the complex conjugate of $3 - 4i$, which is $3 + 4i$. It follows that $\frac{3 + 4i}{3 + 4i}$ should be the form of 1 that is used to build $\frac{1 + 2i}{3 - 4i}$.

$$\frac{1 + 2i}{3 - 4i} = \frac{1 + 2i}{3 - 4i} \cdot \frac{3 + 4i}{3 + 4i}$$ To build an equivalent fraction, multiply by $\frac{3 + 4i}{3 + 4i} = 1$.

$$= \frac{3 + 4i + 6i + 8i^2}{9 + 12i - 12i - 16i^2}$$ Use the FOIL method to multiply the numerators and the denominators.

$$= \frac{3 + 10i + 8(-1)}{9 - 16(-1)}$$ Combine like terms in the numerator and denominator. Replace i^2 with -1. The denominator is now a real number.

$$= \frac{3 + 10i - 8}{9 + 16}$$ Simplify the numerator and denominator.

$$= \frac{-5 + 10i}{25}$$ Combine like terms in the numerator and denominator.

$$= \frac{\overset{1}{\cancel{5}}(-1 + 2i)}{\underset{1}{\cancel{5}} \cdot 5}$$ Factor out 5 in the numerator and remove the common factor of 5.

$$= \frac{-1 + 2i}{5}$$ Simplify. This notation represents the sum of two fractions that have the common denominator 5.

$$= -\frac{1}{5} + \frac{2}{5}i$$ Write the complex number in the form $a + bi$.

Self Check 8 Divide. Write the answers in the form $a + bi$. **a.** $\frac{6}{5 + 2i}$ **b.** $\frac{2 - 4i}{5 - 3i}$

Now Try Problems 71 and 77

EXAMPLE 9 Divide and write the answer in the form $a + bi$: $\frac{4 + \sqrt{-16}}{2 + \sqrt{-4}}$

Strategy We will begin by writing $\sqrt{-16}$ and $\sqrt{-4}$ in $i\sqrt{b}$ form.

Why To perform any operations, the numerator and denominator should be written in the form $a + bi$.

Solution

$$\frac{4 + \sqrt{-16}}{2 + \sqrt{-4}} = \frac{4 + 4i}{2 + 2i}$$ Simplify: $\sqrt{-16} = \sqrt{-1}\sqrt{16} = 4i$ and $\sqrt{-4} = \sqrt{-1}\sqrt{4} = 2i$.

$$= \frac{2\overset{1}{\cancel{(2 + 2i)}}}{\underset{1}{\cancel{2 + 2i}}}$$ Factor out 2 in the numerator and remove the common factor of $2 + 2i$.

$$= 2$$

$$= 2 + 0i$$ Write 2 in the form $a + bi$.

Self Check 9 Divide and write the answer in the form $a + bi$: $\frac{21 - \sqrt{-49}}{3 - \sqrt{-1}}$

Now Try Problem 81

EXAMPLE 10 Divide and write the result in the form $a + bi$: $\frac{7}{2i}$

Strategy We will use $\frac{-2i}{-2i}$ as the form of 1 to build $\frac{7}{2i}$.

Why Since the denominator $2i$ can be expressed as $0 + 2i$, its complex conjugate is $0 - 2i$. However, instead of building with $\frac{0 - 2i}{0 - 2i}$, we will drop the zeros and just use $\frac{-2i}{-2i}$.

Solution

$$\frac{7}{2i} = \frac{7}{2i} \cdot \frac{-2i}{-2i}$$ To build an equivalent fraction, multiply by $\frac{-2i}{-2i} = 1$.

$$= \frac{-14i}{-4i^2}$$ Multiply the numerators and multiply the denominators.

$$= \frac{-14i}{-4(-1)}$$ Replace i^2 with -1. The denominator is now a real number.

$$= \frac{-14i}{4}$$ Simplify the denominator.

$$= -\frac{7i}{2}$$ Simplify the fraction: $-\frac{\overset{1}{\cancel{2}} \cdot 7i}{\underset{1}{\cancel{2}} \cdot 2}$.

$$= 0 - \frac{7}{2}i$$ Write in the form $a + bi$.

Success Tip

In this example, the denominator of $\frac{7}{2i}$ is of the form bi. In such cases, we can eliminate i in the denominator by simply multiplying by $\frac{i}{i}$.

$$\frac{7}{2i} = \frac{7}{2i} \cdot \frac{i}{i} = \frac{7i}{2i^2} = -\frac{7}{2}i$$

Self Check 10 Divide and write the result in the form $a + bi$: $\frac{3}{4i}$

Now Try Problem 85

6 Perform Operations Involving Powers of *i*.

The powers of i produce an interesting pattern:

$$i = \sqrt{-1} = i \qquad i^5 = i^4 i = 1i = i$$
$$i^2 = \left(\sqrt{-1}\right)^2 = -1 \qquad i^6 = i^4 i^2 = 1(-1) = -1$$
$$i^3 = i^2 i = -1i = -i \qquad i^7 = i^4 i^3 = 1(-i) = -i$$
$$i^4 = i^2 i^2 = (-1)(-1) = 1 \qquad i^8 = i^4 i^4 = (1)(1) = 1$$

The pattern continues: $i, -1, -i, 1, \ldots$.

Success Tip

Note that the powers of i cycle through four possible outcomes:

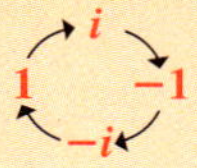

Larger powers of i can be simplified by using the fact that $i^4 = 1$. For example, to simplify i^{29}, we note that 29 divided by 4 gives a quotient of 7 and a remainder of 1. Thus, $29 = 4 \cdot 7 + 1$ and

$$i^{29} = i^{4 \cdot 7 + 1}$$ $4 \cdot 7 = 28$.

$$= (i^4)^7 \cdot i^1$$ Use the rules for exponents $x^{m \cdot n} = (x^m)^n$ and $x^{m+n} = x^m \cdot x^n$.

$$= 1^7 \cdot i$$ Simplify: $i^4 = 1$.

$$= i$$ Simplify: $1 \cdot i = i$.

The result of this example illustrates the following fact.

Powers of *i* If n is a natural number that has a remainder of R when divided by 4, then

$$i^n = i^R$$

EXAMPLE 11 Simplify: **a.** i^{55} **b.** i^{98}

Strategy We will examine the remainder when we divide the exponents 55 and 98 by 4.

Why The remainder determines the power to which i is raised in the simplified form.

Solution

a. We divide 55 by 4 and get a remainder of 3. Therefore,

$$i^{55} = i^3 = -i \qquad \begin{array}{r} 13\text{R } 3 \\ 4\overline{)55} \\ \underline{-4} \\ 15 \\ \underline{-12} \\ 3 \end{array}$$

b. We divide 98 by 4 and get a remainder of 2. Therefore,

$$i^{98} = i^2 = -1 \qquad \begin{array}{r} 24\text{R } 2 \\ 4\overline{)98} \\ \underline{-8} \\ 18 \\ \underline{-16} \\ 2 \end{array}$$

Success Tip

If we divide the natural number exponent n of a power of i by 4, the remainder indicates the simplified form of i^n:

R = 1: i
R = 2: -1
R = 3: $-i$
R = 0: 1

Self Check 11 Simplify: **a.** i^{62} **b.** i^{105}

Now Try Problem 91

SECTION 9.7 STUDY SET

VOCABULARY

Fill in the blanks.

1. The __________ number i is defined as $i = \sqrt{-1}$. We call i^{25} a ________ of i.
2. A _________ number is any number that can be written in the form $a + bi$, where a and b are real numbers and $i = \sqrt{-1}$.
3. For the complex number $2 + 5i$, we call 2 the _____ part and 5 the __________ part.
4. $6 + 3i$ and $6 - 3i$ are called complex __________.

CONCEPTS

Fill in the blanks.

5. a. $i =$ b. $i^2 =$
   c. $i^3 =$ d. $i^4 =$
   e. In general, the powers of i cycle through _____ possible outcomes.
6. Simplify:
   $$\sqrt{-36} = \sqrt{ \cdot 36} = \sqrt{}\sqrt{36} = 6$$
7. a. To add (or subtract) complex numbers, add (or subtract) their _____ parts and add (or subtract) their __________ parts.
   b. To multiply two complex numbers, such as $(2 + 3i)(3 + 5i)$, we can use the _______ method.
8. To divide $6 + 7i$ by $1 - 8i$, we multiply $\frac{6 + 7i}{1 - 8i}$ by 1 in the form of $\frac{}{}$.
9. Give the complex conjugate of each number.
   a. $2 - 3i$ b. 2 c. $-3i$
10. Factor each numerator. Then remove the factor common to the numerator and denominator. Write the result in the form $a + bi$.
    a. $\frac{3 + 6i}{3}$ b. $\frac{15 + 25i}{10}$
11. Complete the illustration. Label the *real numbers,* the *imaginary numbers,* the *complex numbers,* the *rational numbers,* and the *irrational numbers.*

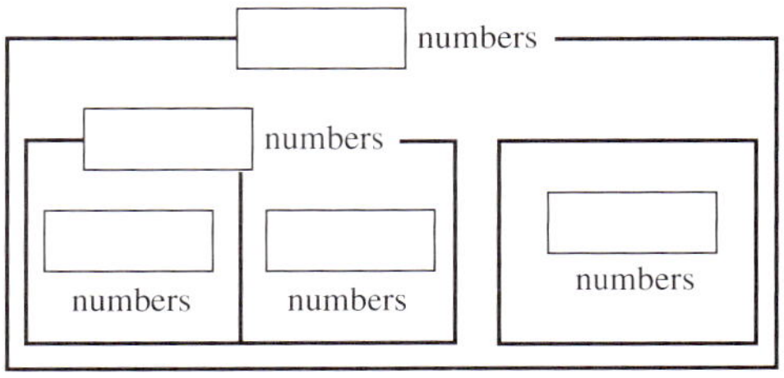

12. Determine whether each statement is true or false.
    a. Every complex number is a real number.
    b. Every real number is a complex number.
    c. i is a real number.
    d. The square root of a negative number is an imaginary number.
    e. The product of a complex number and its complex conjugate is always a real number.

NOTATION

Complete each solution.

13. $(3 + 2i)(3 - i) = \underline{\qquad} - 3i + \underline{\qquad} - 2i^2$
$= 9 + 3i + \underline{\qquad}$
$= \underline{\qquad} + 3i$

14. $\dfrac{3}{2 - i} = \dfrac{3}{2 - i} \cdot \dfrac{\underline{\qquad}}{\underline{\qquad}}$
$= \dfrac{6 + \underline{\qquad}}{4 - \underline{\qquad}}$
$= \dfrac{6 + 3i}{\underline{\qquad}}$
$= \dfrac{\underline{\qquad}}{\underline{\qquad}} + \dfrac{3}{5}i$

15. Determine whether each statement is true or false.
a. $\sqrt{6i} = i\sqrt{6}$ b. $\sqrt{8i} = \sqrt{8}i$
c. $\sqrt{-25} = -\sqrt{25}$ d. $-i = i$

16. Write each number in the form $a + bi$.
a. $\dfrac{9 + 11i}{4}$ b. $\dfrac{1 - i}{18}$

GUIDED PRACTICE

Express each number in terms of i. See Example 1.

17. $\sqrt{-9}$
18. $\sqrt{-4}$
19. $\sqrt{-7}$
20. $\sqrt{-11}$
21. $\sqrt{-24}$
22. $\sqrt{-28}$
23. $-\sqrt{-72}$
24. $-\sqrt{-24}$
25. $5\sqrt{-81}$
26. $6\sqrt{-49}$
27. $\sqrt{-\dfrac{25}{9}}$
28. $-\sqrt{-\dfrac{121}{144}}$

Write each number in the form a + bi. See Example 2.

29. a. 5
b. $\sqrt{-49}$
30. a. -43
b. $\sqrt{-169}$
31. a. $1 + \sqrt{-25}$
b. $-3 + \sqrt{-8}$
32. a. $21 + \sqrt{-16}$
b. $-9 + \sqrt{-12}$
33. a. $76 - \sqrt{-54}$
b. $-7 + \sqrt{-19}$
34. a. $88 - \sqrt{-98}$
b. $-2 + \sqrt{-35}$
35. a. $-6 - \sqrt{-9}$
b. $3 + \sqrt{-6}$
36. a. $-45 - \sqrt{-81}$
b. $8 + \sqrt{-7}$

Perform the operations. Write all answers in the form a + bi. See Example 3.

37. $(3 + 4i) + (5 - 6i)$
38. $(8 + 3i) + (-7 - 2i)$
39. $(6 - i) + (9 + 3i)$
40. $(5 + 3i) - (6 - 9i)$
41. $(7 - 3i) - (4 + 2i)$
42. $(5 - 4i) - (3 + 2i)$
43. $\left(8 + \sqrt{-25}\right) - \left(7 + \sqrt{-4}\right)$
44. $\left(-7 + \sqrt{-81}\right) - \left(-2 - \sqrt{-64}\right)$

Multiply. See Example 4.

45. $\sqrt{-1} \cdot \sqrt{-36}$
46. $\sqrt{-9} \cdot \sqrt{-100}$
47. $\sqrt{-2}\sqrt{-12}$
48. $\sqrt{-3}\sqrt{-45}$

Multiply. Write all answers in the form a + bi. See Example 5.

49. $3(2 - 9i)$
50. $-4(3 + 4i)$
51. $7(5 - 4i)$
52. $-5(3 + 2i)$
53. $2i(7 - 3i)$
54. $i(8 + 2i)$
55. $-5i(5 - 5i)$
56. $2i(7 + 2i)$

Multiply. Write all answers in the form a + bi. See Example 6.

57. $(2 + i)(3 - i)$
58. $(4 - i)(2 + i)$
59. $(3 - 2i)(2 + 3i)$
60. $(3 - i)(2 + 3i)$
61. $\left(4 + \sqrt{-1}\right)\left(3 - \sqrt{-1}\right)$
62. $\left(1 - \sqrt{-25}\right)\left(1 - \sqrt{-16}\right)$
63. $(2 + i)^2$
64. $(3 - 2i)^2$

Find the product of the given complex number and its conjugate. See Example 7.

65. $2 + 6i$
66. $5 + 2i$
67. $-4 - 7i$
68. $-10 - 9i$

Divide. Write all answers in the form a + bi. See Example 8.

69. $\dfrac{9}{5 + i}$
70. $\dfrac{4}{2 - i}$
71. $\dfrac{11i}{4 - 7i}$
72. $\dfrac{2i}{3 + 8i}$
73. $\dfrac{3 - 2i}{4 - i}$
74. $\dfrac{6 - i}{2 + i}$
75. $\dfrac{7 + 4i}{2 - 5i}$
76. $\dfrac{2 + 3i}{2 - 3i}$
77. $\dfrac{7 + 3i}{4 - 2i}$
78. $\dfrac{5 - 3i}{4 + 2i}$
79. $\dfrac{1 - 3i}{3 + i}$
80. $\dfrac{3 + 5i}{1 - i}$

Divide. Write all answers in the form a + bi. See Example 9. (*Hint:* Factor the numerator.)

81. $\dfrac{8 + \sqrt{-144}}{2 + \sqrt{-9}}$
82. $\dfrac{3 + \sqrt{-36}}{1 + \sqrt{-4}}$
83. $\dfrac{-4 - \sqrt{-4}}{2 + \sqrt{-1}}$
84. $\dfrac{-5 - \sqrt{-25}}{1 + \sqrt{-1}}$

Divide. Write all answers in the form a + bi. See Example 10.

85. $\dfrac{5}{3i}$
86. $\dfrac{3}{8i}$
87. $-\dfrac{2}{7i}$
88. $-\dfrac{8}{5i}$

Simplify each expression. See Example 11.

89. i^{21}
90. i^{19}
91. i^{27}
92. i^{22}
93. i^{100}
94. i^{97}
95. i^{42}
96. i^{200}

TRY IT YOURSELF

Perform the operations. Write all answers in the form a + bi.

97. $(3 - i) - (-1 + 10i)$
98. $(14 + 4i) - (-9 - i)$
99. $(2 - \sqrt{-16})(3 + \sqrt{-4})$
100. $(3 - \sqrt{-4})(4 - \sqrt{-9})$
101. $(-6 - 9i) + (4 + 3i)$
102. $(-3 + 11i) + (-1 - 6i)$
103. $\frac{-2i}{3 + 2i}$
104. $\frac{-4i}{2 - 6i}$
105. $6i(2 - 3i)$
106. $-9i(4 - 6i)$
107. $\frac{4}{5i^{35}}$
108. $\frac{3}{2i^{17}}$
109. $(2 + i\sqrt{2})(3 - i\sqrt{2})$
110. $(5 + i\sqrt{3})(2 - i\sqrt{3})$
111. $\frac{5 + 9i}{1 - i}$
112. $\frac{5 - i}{3 + 2i}$
113. $(4 - 8i)^2$
114. $(7 - 3i)^2$
115. $\frac{\sqrt{5} - i\sqrt{3}}{\sqrt{5} + i\sqrt{3}}$
116. $\frac{\sqrt{3} + i\sqrt{2}}{\sqrt{3} - i\sqrt{2}}$

Look Alikes . . .

117. a. $\sqrt{-8}$ b. $\sqrt[3]{-8}$
118. a. $(3 + i) + (2 + 4i)$ b. $(3 + i)(2 + 4i)$
119. a. $(2i)^2$ b. $(2 + i)^2$
120. a. $\sqrt{-9}\sqrt{-16}$ b. $\sqrt{9}\sqrt{16}$

APPLICATIONS

121. **Fractals.** Complex numbers are fundamental in the creation of the intricate geometric shape shown below, called a *fractal*. The process of creating this image is based on the following sequence of steps, which begins by picking any complex number, which we will call z.

1. Square z, and then add that result to z.
2. Square the result from step 1, and then add it to z.
3. Square the result from step 2, and then add it to z.

If we begin with the complex number i, what is the result after performing steps 1, 2, and 3?

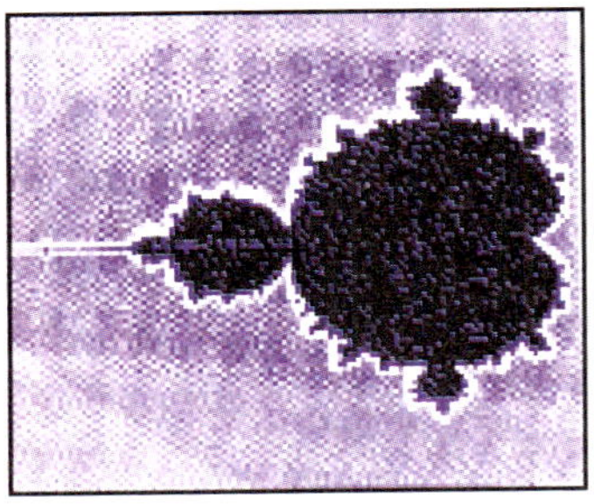

122. **Electronics.** The impedance Z in an AC (alternating current) circuit is a measure of how much the circuit impedes (hinders) the flow of current through it. The impedance is related to the voltage V and the current I by the formula $V = IZ$. If a circuit has a current of $(0.5 + 2.0i)$ amps and an impedance of $(0.4 - 3.0i)$ ohms, find the voltage.

WRITING

123. What is an imaginary number? What is a complex number?
124. The method used to divide complex numbers is similar to the method used to divide radical expressions. Explain why. Give an example.
125. Explain the error. Then find the correct result.
   a. Add: $\sqrt{-16} + \sqrt{-9} = \sqrt{-25}$
   b. Multiply: $\sqrt{-2}\sqrt{-3} = \sqrt{-2(-3)} = \sqrt{6}$
126. Determine whether the pair of complex numbers are equal. Explain your reasoning.
   a. $4 - \frac{2}{5}i, \frac{8}{2} - 0.4i$
   b. $0.25 + 0.7i, \frac{1}{4} + \frac{7}{10}i$

REVIEW

127. **Wind Speeds.** A plane that can fly 200 mph in still air makes a 330-mile flight with a tail wind and returns, flying into the same wind. Find the speed of the wind if the total flying time is $3\frac{1}{3}$ hours.
128. **Finding Rates.** A student drove a distance of 135 miles at an average speed of 50 mph. How much faster would she have to drive on the return trip to save 30 minutes of driving time?

CHALLENGE PROBLEMS

129. Simplify: $(i^{349})^{-i^{456}}$
130. Simplify $(2 + 3i)^{-2}$ and write the result in the form $a + bi$.

9 Summary & Review

SECTION 9.1 Radical Expressions and Radical Functions

DEFINITIONS AND CONCEPTS	EXAMPLES
The number b is a **square root of a** if $b^2 = a$. Every positive real number has two square roots.	7 is a square root of 49 because $7^2 = 49$. -7 is also a square root of 49 because $(-7)^2 = 49$.
A **radical symbol** $\sqrt{\ }$ represents the **positive** or **principal square root** of a number. For any real number x, $\sqrt{x^2} = \lvert x\rvert$ The symbol $-\sqrt{\ }$ represents the **negative square root** of a number. Review the list of perfect squares on page 729.	$\sqrt{25} = 5$ because $5^2 = 25$. $\sqrt{36x^2} = \lvert 6x\rvert = 6\lvert x\rvert$ because $(6x)^2 = 36x^2$. Since x could be negative, absolute value symbols ensure that the result is not negative. $\sqrt{\frac{r^8}{100}} = \frac{r^4}{10}$ because $\left(\frac{r^4}{10}\right)^2 = \frac{r^8}{100}$. Since $\frac{r^4}{10}$ cannot be negative, no absolute value symbols are needed. $-\sqrt{81} = -9$ and $\sqrt{-81}$ is not a real number.
A function of the form $f(x) = \sqrt{x}$ is called a **square root function.** Review its graph on page 732.	Let $f(x) = \sqrt{x - 2}$. Find $f(38)$. $f(38) = \sqrt{38 - 2} = \sqrt{36} = 6$
The **domain of a square root function** is the set of all real numbers for which the radicand is nonnegative. To find the domain, set the radicand greater than or equal to 0 and solve for the variable.	Find the domain of $g(x) = \sqrt{2x - 4}$. Since $\sqrt{2x - 4}$ is not a real number when $2x - 4$ is negative, we must require that $2x - 4 \geq 0$ Because we cannot find the square root of a negative number $2x \geq 4$ Solve for x. $x \geq 2$ The domain of g is $[2, \infty)$.
The **cube root** of x is denoted as $\sqrt[3]{x}$ and is defined as $\sqrt[3]{x} = y$ if $y^3 = x$ Review the list of perfect cubes on page 735.	$\sqrt[3]{8} = 2$ because $2^3 = 8$. $\sqrt[3]{-64} = -4$ because $(-4)^3 = -64$. $\sqrt[3]{-27a^6} = -3a^2$ because $(-3a^2)^3 = -27a^6$.
A function of the form $f(x) = \sqrt[3]{x}$ is called a **cube root function.** Review its graph on page 736.	Let $f(x) = \sqrt[3]{x + 4}$. Find $f(23)$. $f(23) = \sqrt[3]{23 + 4} = \sqrt[3]{27} = 3$
The ***n*th root of *x*** is denoted as $\sqrt[n]{x}$. Index → $\sqrt[4]{16a^8}$ ← Radicand If x is a real number and $n > 1$, then: If n is an odd natural number, $\sqrt[n]{x^n} = x$. If n is an even natural number, $\sqrt[n]{x^n} = \lvert x\rvert$.	$\sqrt[4]{81x^4} = \lvert 3x\rvert = 3\lvert x\rvert$ because $(3x)^4 = 81x^4$. $\sqrt[4]{-16}$ Not a real number $\sqrt[5]{32a^{10}} = 2a^2$ because $(2a^2)^5 = 32a^{10}$. No absolute value symbols are needed. $\sqrt[6]{(m - 4)^6} = \lvert m - 4\rvert$ because $(m - 4)^6 = (m - 4)^6$.

REVIEW EXERCISES

1. The graph of a square root function f is shown here. Find each of the following:
   a. $f(1)$ b. $f(-3)$
   c. The value of x for which $f(x) = 3$
   c. The domain and range of f

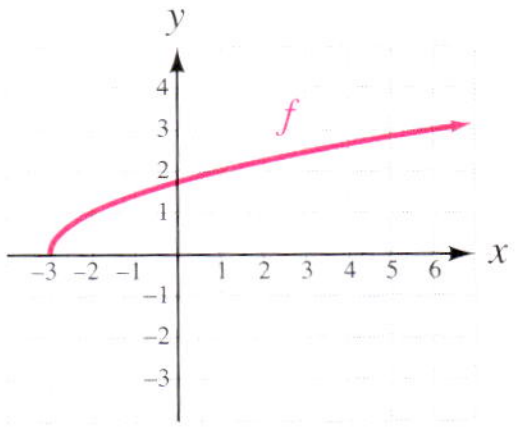

2. a. Simplify $\sqrt{100a^2}$. Assume that the variable a is unrestricted.
   b. Simplify $\sqrt{100a^2}$. Assume that the variable a is a positive real number.

Simplify each expression. Assume that all variables are unrestricted and use absolute value symbols when necessary.

3. $\sqrt{49}$
4. $-\sqrt{121}$
5. $\sqrt{\frac{225}{49}}$
6. $\sqrt{-4}$
7. $\sqrt{100a^{12}}$
8. $\sqrt{25x^2}$
9. $\sqrt{x^8}$
10. $\sqrt{x^2 + 4x + 4}$
11. $\sqrt[3]{-27}$
12. $-\sqrt[3]{216}$
13. $\sqrt[3]{64x^6y^3}$
14. $\sqrt[3]{\frac{x^9}{125}}$
15. $\sqrt[6]{64}$
16. $\sqrt[5]{-32}$
17. $\sqrt[4]{256x^8y^4}$
18. $\sqrt[5]{(x+1)^5}$
19. $-\sqrt[4]{\frac{1}{16}}$
20. $\sqrt[4]{-81}$
21. $\sqrt[6]{-1}$
22. $\sqrt[3]{0}$
23. Find the domain of $f(x) = \sqrt{3x + 15}$.
24. **Geometry.** The side of a square with area A square feet is given by the function $s(A) = \sqrt{A}$. Find the length of one side of a square that has an area of 169 ft^2.
25. **Cubes.** The total surface area of a cube is related to its volume V by the function $A(V) = 6\sqrt[3]{V^2}$. Find the surface area of a cube with a volume of 8 cm^3.
26. Let $g(x) = \sqrt[3]{x^2 + 9}$. Use a calculator fo find $g(-1.9)$ to four decimal places.

Graph each function. Find the domain and range.

27. $f(x) = \sqrt{x}$
28. $f(x) = \sqrt[3]{x}$
29. $f(x) = \sqrt{x + 2}$
30. $f(x) = -\sqrt[3]{x} + 3$

SECTION 9.2 Rational Exponents

DEFINITIONS AND CONCEPTS	EXAMPLES
To simplify exponential expressions involving **rational (fractional) exponents,** use the following rules to write the expressions in an equivalent radical form. $x^{1/n} = \sqrt[n]{x}$ $x^{m/n} = (\sqrt[n]{x})^m = \sqrt[n]{x^m}$	Simplify. All variables represent positive real numbers. $25^{1/2} = \sqrt{25} = 5$ $\left(\frac{256}{d^4}\right)^{1/4} = \sqrt[4]{\frac{256}{d^4}} = \frac{4}{d}$ $8^{2/3} = (\sqrt[3]{8})^2 = (2)^2 = 4$ $(t^{10})^{6/5} = (\sqrt[5]{t^{10}})^6 = (t^2)^6 = t^{12}$
To be consistent with the definition of negative integer exponents, we define $x^{-m/n}$ as follows. $x^{-m/n} = \frac{1}{x^{m/n}}$ $\frac{1}{x^{-m/n}} = x^{m/n}$	Simplify. All variables represent positive real numbers. $(125)^{-2/3} = \frac{1}{(125)^{2/3}} = \frac{1}{(\sqrt[3]{125})^2} = \frac{1}{25}$ $\frac{1}{(-32x^5)^{-3/5}} = (-32x^5)^{3/5} = (\sqrt[5]{-32x^5})^3 = -8x^3$
The **rules for exponents** can be used to simplify expressions with rational (fractional) exponents.	Simplify: $\frac{p^{5/3}p^{8/3}}{p^4} = p^{5/3+8/3-4} = p^{5/3+8/3-12/3} = p^{1/3}$
We can write certain radical expressions as an equivalent exponential expression and use rules for exponents to simplify it. Then we can change that result back into a radical.	Simplify: $\sqrt[4]{9} = \sqrt[4]{3^2} = 3^{2/4} = 3^{1/2} = \sqrt{3}$

REVIEW EXERCISES

Write each expression in radical form.

31. $t^{1/2}$
32. $(5xy^3)^{1/4}$

Simplify each expression, if possible. Assume that all variables represent positive real numbers.

33. $25^{1/2}$
34. $-36^{1/2}$
35. $(-36)^{1/2}$
36. $1^{1/5}$
37. $\left(\frac{9}{x^2}\right)^{1/2}$
38. $(-8)^{1/3}$
39. $625^{1/4}$
40. $(81c^4d^4)^{1/4}$
41. $9^{3/2}$
42. $8^{-2/3}$
43. $-49^{5/2}$
44. $\frac{1}{100^{-1/2}}$
45. $\left(\frac{4}{9}\right)^{-3/2}$
46. $\frac{1}{25^{5/2}}$
47. $(25x^2y^4)^{3/2}$
48. $(8u^6v^3)^{-2/3}$

Perform the operations. Write answers without negative exponents. Assume that all variables represent positive real numbers.

49. $5^{1/4} \cdot 5^{1/2}$
50. $a^{3/7}a^{-2/7}$
51. $(k^{4/5})^{10}$
52. $\frac{3^{5/6}3^{1/3}}{3^{1/2}}$

Perform the multiplications. Assume all variables represent positive real numbers.

53. $u^{1/2}(u^{1/2} - u^{-1/2})$
54. $v^{2/3}(v^{1/3} + v^{4/3})$

Use rational exponents to simplify each radical. All variables represent positive real numbers.

55. $\sqrt[4]{a^2}$
56. $\sqrt[3]{\sqrt{c}}$

57. **Visibility.** The distance d in miles a person in an airplane can see to the horizon on a clear day is given by the formula $d = 1.22a^{1/2}$, where a is the altitude of the plane in feet. Find d.

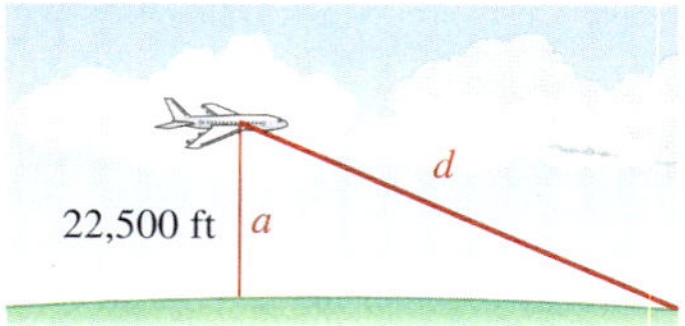

58. Substitute the x- and y-coordinates of each point labeled in the graph into the equation

$$x^{2/3} + y^{2/3} = 32$$

Show that each one satisfies the equation.

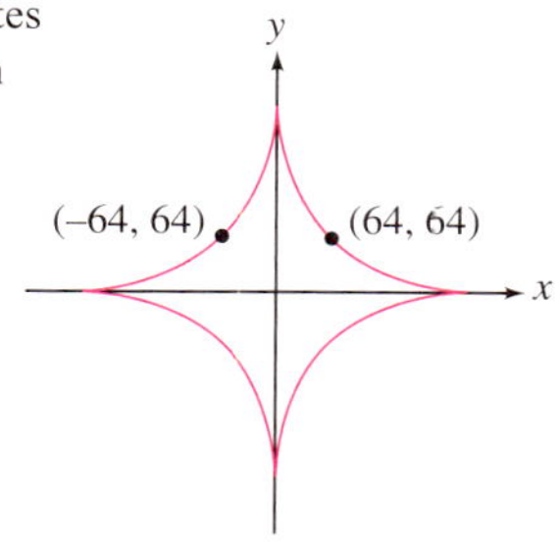

SECTION 9.3 Simplifying and Combining Radical Expressions

DEFINITIONS AND CONCEPTS	EXAMPLES
Product rule for radicals: $\sqrt[n]{ab} = \sqrt[n]{a}\sqrt[n]{b}$ The product rule for radicals can be used to simplify radical expressions.	Simplify: $\sqrt{98} = \sqrt{49 \cdot 2}$ Write 98 as the product of its greatest perfect-square factor and one other factor. $= \sqrt{49}\sqrt{2}$ The square root of a product is equal to the product of the square roots. $= 7\sqrt{2}$ Evaluate $\sqrt{49}$.
Simplified form of a radical: **1.** Except for 1, the radicand has no perfect-square factors. **2.** No fraction appears in the radicand. **3.** No radical appears in the denominator. Review the lists of perfect squares, cubes, fourth powers, and fifth powers on pages 755 and 756.	Simplify: $\sqrt[3]{16x^4} = \sqrt[3]{8x^3 \cdot 2x}$ Write $16x^4$ as the product of its greatest perfect-cube factor and one other factor. $= \sqrt[3]{8x^3}\sqrt[3]{2x}$ The cube root of a product is equal to the product of the cube roots. $= 2x\sqrt[3]{2x}$ Simplify $\sqrt[3]{8x^3}$.

Quotient rule for radicals: $\sqrt[n]{\dfrac{a}{b}} = \dfrac{\sqrt[n]{a}}{\sqrt[n]{b}}$	Simplify: $\sqrt{\dfrac{10}{25x^4}} = \dfrac{\sqrt{10}}{\sqrt{25x^4}} = \dfrac{\sqrt{10}}{5x^2}$ Simplify: $\sqrt[3]{\dfrac{16y^3}{125a^3}} = \dfrac{\sqrt[3]{16y^3}}{\sqrt[3]{125a^3}} = \dfrac{\sqrt[3]{8y^3}\sqrt[3]{2}}{5a} = \dfrac{2y\sqrt[3]{2}}{5a}$
Radical expressions with the same index and radicand are called **like radicals.** Like radicals can be combined by addition and subtraction. To **combine like radicals,** we use the distributive property in reverse.	Add: $3\sqrt{6} + 5\sqrt{6} = (3 + 5)\sqrt{6} = 8\sqrt{6}$ Subtract: $8\sqrt[4]{2y} - 9\sqrt[4]{2y} = (8 - 9)\sqrt[4]{2y} = -\sqrt[4]{2y}$
If a sum or difference involves unlike radicals, make sure that each one is written in simplified form. After doing so, like radicals may result that can be combined.	Simplify: $\sqrt[3]{54a^4} - \sqrt[3]{16a^4} = \sqrt[3]{27a^3 \cdot 2a} - \sqrt[3]{8a^3 \cdot 2a}$ $= \sqrt[3]{27a^3}\sqrt[3]{2a} - \sqrt[3]{8a^3}\sqrt[3]{2a}$ $= 3a\sqrt[3]{2a} - 2a\sqrt[3]{2a}$ $= (3a - 2a)\sqrt[3]{2a}$ $= a\sqrt[3]{2a}$

REVIEW EXERCISES

Simplify each expression. All variables represent positive real numbers.

59. $\sqrt{80}$
60. $\sqrt[3]{54}$
61. $\sqrt[4]{160}$
62. $\sqrt[5]{-96}$
63. $\sqrt{8x^5}$
64. $\sqrt[4]{r^{17}}$
65. $\sqrt[3]{-27j^7k}$
66. $\sqrt[3]{-16x^5y^4}$
67. $\sqrt{\dfrac{m}{144n^{12}}}$
68. $\sqrt{\dfrac{17xy}{64a^4}}$
69. $\dfrac{\sqrt[5]{64x^8}}{\sqrt[5]{2x^3}}$
70. $\dfrac{\sqrt[5]{243x^{16}}}{\sqrt[5]{x}}$

Simplify and combine like radicals. All variables represent positive real numbers.

71. $\sqrt{2} + 2\sqrt{2}$
72. $6\sqrt{20} - \sqrt{5}$
73. $2\sqrt[3]{3} - \sqrt[3]{24}$
74. $-\sqrt[4]{32a^5} - 2\sqrt[4]{162a^5}$
75. $2x\sqrt{8} + 2\sqrt{200x^2} + \sqrt{50x^2}$
76. $\sqrt[3]{54x^3} - 3\sqrt[3]{16x^3} + 4\sqrt[3]{128x^3}$
77. $2\sqrt[4]{32t^3} - 8\sqrt[4]{6t^3} + 5\sqrt[4]{2t^3}$
78. $10\sqrt[4]{16x^9} - 8x^2\sqrt[4]{x} + 5\sqrt[4]{x^5}$

79. Explain the error in each simplification.
    a. $2\sqrt{5x} + 3\sqrt{5x} = 5\sqrt{10x}$
    b. $30 + 30\sqrt[4]{2} = 60\sqrt[4]{2}$
    c. $7\sqrt[3]{y^2} - 5\sqrt[3]{y^2} = 2$
    d. $6\sqrt{11ab} - 3\sqrt{5ab} = 3\sqrt{6ab}$

80. **Sewing.** A corner of fabric is folded over to form a collar and stitched down as shown. From the dimensions given in the figure, determine the exact number of inches of stitching that must be made. Then give an approximation to one decimal place. (All measurements are in inches.)

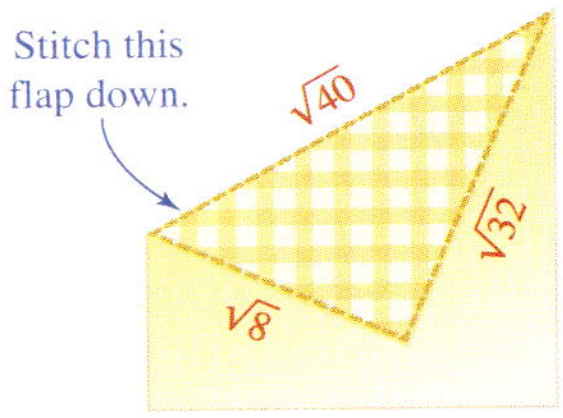

SECTION 9.4 Multiplying and Dividing Radical Expressions

DEFINITIONS AND CONCEPTS	EXAMPLES
We can use the product rule for radicals to **multiply radical expressions** that have the same index: $\sqrt[n]{a}\sqrt[n]{b} = \sqrt[n]{ab}$ provided $\sqrt[n]{a}$ and $\sqrt[n]{b}$ are real numbers.	Multiply and then simplify, if possible: $\sqrt{6}\sqrt{8} = \sqrt{6 \cdot 8} = \sqrt{48} = \sqrt{16 \cdot 3} = \sqrt{16}\sqrt{3} = 4\sqrt{3}$ $\sqrt[3]{9x^4}\sqrt[3]{3x^2} = \sqrt[3]{9x^4 \cdot 3x^2} = \sqrt[3]{27x^6} = 3x^2$
We can use the **distributive property** to multiply a radical expression with two or more terms by a radical expression with one term.	Multiply and then simplify, if possible: $2\sqrt{3}(4\sqrt{5} - 5\sqrt{2}) = 2\sqrt{3} \cdot 4\sqrt{5} - 2\sqrt{3} \cdot 5\sqrt{2}$ $= 2 \cdot 4\sqrt{3 \cdot 5} - 2 \cdot 5\sqrt{3 \cdot 2}$ $= 8\sqrt{15} - 10\sqrt{6}$
We can use the **FOIL method** to multiply a radical expression with two terms by another radical expression with two terms.	Multiply and then simplify, if possible: F O I L $(\sqrt[3]{x} - \sqrt[3]{3})(\sqrt[3]{x} + \sqrt[3]{9}) = \sqrt[3]{x}\sqrt[3]{x} + \sqrt[3]{x}\sqrt[3]{9} - \sqrt[3]{3}\sqrt[3]{x} - \sqrt[3]{3}\sqrt[3]{9}$ $= \sqrt[3]{x^2} + \sqrt[3]{9x} - \sqrt[3]{3x} - \sqrt[3]{27}$ $= \sqrt[3]{x^2} + \sqrt[3]{9x} - \sqrt[3]{3x} - 3$
If a radical appears in a denominator of a fraction, or if a radicand contains a fraction, we can write the radical in simplest form by **rationalizing the denominator.** To rationalize a denominator, we multiply the given expression by a carefully chosen form of 1.	Rationalize the denominator: $\sqrt{\frac{10}{3}} = \frac{\sqrt{10}}{\sqrt{3}} \cdot \frac{\sqrt{3}}{\sqrt{3}} = \frac{\sqrt{30}}{3}$ $\sqrt[3]{\frac{5}{2p^2}} = \frac{\sqrt[3]{5}}{\sqrt[3]{2p^2}} \cdot \frac{\sqrt[3]{4p}}{\sqrt[3]{4p}} = \frac{\sqrt[3]{20p}}{\sqrt[3]{8p^3}} = \frac{\sqrt[3]{20p}}{2p}$
Radical expressions that involve the sum and difference of the same two terms are called **conjugates.**	Conjugates: $\sqrt{2x} + 3$ and $\sqrt{2x} - 3$
To **rationalize a two-termed denominator** of a fraction, multiply the numerator and the denominator by the **conjugate** of the denominator.	Rationalize the denominator: $\frac{\sqrt{x} - 2}{\sqrt{x} + 2} = \frac{\sqrt{x} - 2}{\sqrt{x} + 2} \cdot \frac{\sqrt{x} - 2}{\sqrt{x} - 2}$ $= \frac{\sqrt{x}\sqrt{x} - 2\sqrt{x} - 2\sqrt{x} + 4}{\sqrt{x}\sqrt{x} - 2\sqrt{x} + 2\sqrt{x} - 4}$ $= \frac{x - 4\sqrt{x} + 4}{x - 4}$

REVIEW EXERCISES

Simplify each expression. All variables represent positive real numbers.

81. $\sqrt{7}\sqrt{7}$
82. $(2\sqrt{5})(3\sqrt{2})$
83. $(-2\sqrt{8})^2$
84. $2\sqrt{6}\sqrt{15}$
85. $\sqrt{9x}\sqrt{x}$
86. $(\sqrt[3]{x+1})^3$
87. $-\sqrt[3]{2x^2}\sqrt[3]{4x^8}$
88. $\sqrt[5]{9}\cdot\sqrt[5]{27}$
89. $3\sqrt{7t}(2\sqrt{7t}+3\sqrt{3t^2})$
90. $-\sqrt[4]{4x^5y^{11}}\sqrt[4]{8x^9y^3}$
91. $(\sqrt{3b}+\sqrt{3})^2$
92. $(\sqrt[3]{3p}-2\sqrt[3]{2})(\sqrt[3]{3p}+\sqrt[3]{2})$

Rationalize each denominator. All variables represent positive real numbers.

93. $\dfrac{10}{\sqrt{3}}$
94. $\sqrt{\dfrac{3}{5xy}}$
95. $\dfrac{\sqrt[3]{6u}}{\sqrt[3]{u^5}}$
96. $\dfrac{\sqrt[4]{a}}{\sqrt[4]{3b^2}}$
97. $\dfrac{2}{\sqrt{2}-1}$
98. $\dfrac{4\sqrt{x}-2\sqrt{z}}{\sqrt{z}+4\sqrt{x}}$
99. Rationalize the numerator: $\dfrac{\sqrt{a}-\sqrt{b}}{\sqrt{a}}$
100. **Volume.** The formula relating the radius r of a sphere and its volume V is $r=\sqrt[3]{\frac{3V}{4\pi}}$. Write the radical in simplest form.

SECTION 9.5 Solving Radical Equations

DEFINITIONS AND CONCEPTS	EXAMPLES
We can use the **power rule** to solve radical equations. If $x=y$, then $x^n=y^n$. To **solve equations containing radicals:** **1.** Isolate one radical expression on one side of the equation. **2.** Raise both sides of the equation to the power that is the same as the index. **3.** If it still contains a radical, go back to step 1. If it does not contain a radical, solve the resulting equation. **4.** Check the proposed solutions in the original equation to eliminate **extraneous** solutions.	Solve each equation. $\sqrt{2x-2}+1=x$ $\sqrt{2x-2}=x-1$ $(\sqrt{2x-2})^2=(x-1)^2$ $2x-2=x^2-2x+1$ $0=x^2-4x+3$ $0=(x-3)(x-1)$ $x-3=0$ or $x-1=0$ $x=3$ | $x=1$ The solutions are 3 and 1. Verify that each satisfies the original equation. $\sqrt[3]{x+2}=3$ $(\sqrt[3]{x+2})^3=3^3$ $x+2=27$ $x=25$ The solution is 25. Verify that it satisfies the original equation.
When **more than one radical** appears in an equation, we often must use the power rule more than once to solve the equation. The **special-product rules** are often helpful when squaring expressions containing two terms, such as $(5-\sqrt{x})^2$. Warning: The result is not $25+x$.	Solve: $\sqrt{x}+\sqrt{x+5}=5$ $\sqrt{x+5}=5-\sqrt{x}$ — To isolate one radical, subtract $\sqrt{x}$ from both sides. $(\sqrt{x+5})^2=(5-\sqrt{x})^2$ — To eliminate one radical, square both sides. $x+5=25-10\sqrt{x}+x$ — Perform the operations on each side. $-20=-10\sqrt{x}$ — To isolate the radical term, subtract 25 and x from both sides. $2=\sqrt{x}$ — To isolate the radical, divide both sides by −10. $(2)^2=(\sqrt{x})^2$ — To eliminate the radical, square both sides again. $4=x$ The solution is 4. Verify that it satisfies the original equation.

REVIEW EXERCISES

Solve each equation. Write all proposed solutions. Cross out those that are extraneous.

101. $\sqrt{7x - 10} - 1 = 11$

102. $u = \sqrt{25u - 144}$

103. $2\sqrt{y - 3} = \sqrt{2y + 1}$

104. $\sqrt{z + 1} + \sqrt{z} = 2$

105. $\sqrt[3]{x^3 + 56} - 2 = x$

106. $a = \sqrt{a^2 + 5a - 35}$

107. $(x + 2)^{1/2} - (4 - x)^{1/2} = 0$

108. $\sqrt{b^2 + b} = \sqrt{3 - b^2}$

109. $\sqrt[4]{8x - 8} + 2 = 0$

110. $\sqrt{2m + 4} - \sqrt{m + 3} = 1$

111. Let $f(x) = \sqrt{5x + 1}$ and $g(x) = x + 1$. For what values of x is $f(x) = g(x)$?

112. Let $f(x) = \sqrt{2x^2 - 7x}$. For what value(s) of x is $f(x) = 2$?

Solve each equation for the specified variable.

113. $r = \sqrt{\dfrac{A}{P}} - 1$ for P

114. $h = \sqrt[3]{\dfrac{12I}{b}}$ for I

SECTION 9.6 Geometric Applications of Radicals

DEFINITIONS AND CONCEPTS	EXAMPLES

DEFINITIONS AND CONCEPTS

The Pythagorean theorem:

If a and b are the lengths of the **legs** of a right triangle and c is the length of the **hypotenuse,** then $a^2 + b^2 = c^2$.

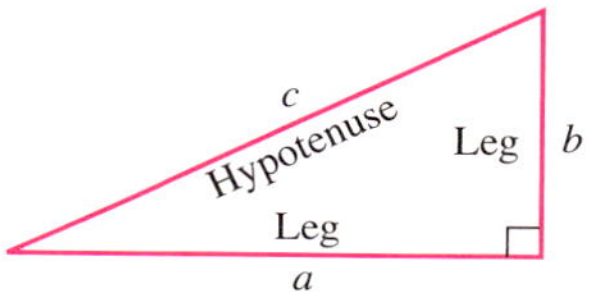

EXAMPLES

Find the length of the third side of the right triangle.

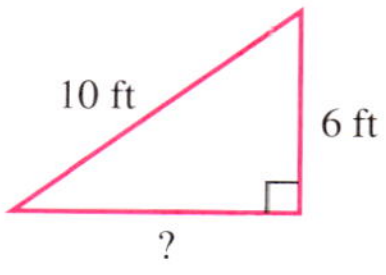

$a^2 + b^2 = c^2$ — This is the Pythagorean equation.

$6^2 + b^2 = 10^2$ — Substitute 6 for a and 10 for c.

$36 + b^2 = 100$

$b^2 = 64$ — To isolate b^2, subtract 36 from both sides.

$b = \sqrt{64}$ — Since b must be positive, find the positive square root of 64.

$b = 8$

The length of the third side of the triangle is 8 ft.

DEFINITIONS AND CONCEPTS

In an **isosceles right triangle,** the length of the hypotenuse is $\sqrt{2}$ times the length of one leg.

EXAMPLES

If the length of one leg of an isosceles right triangle is 7 feet, the length of the hypotenuse is $7\sqrt{2}$ feet.

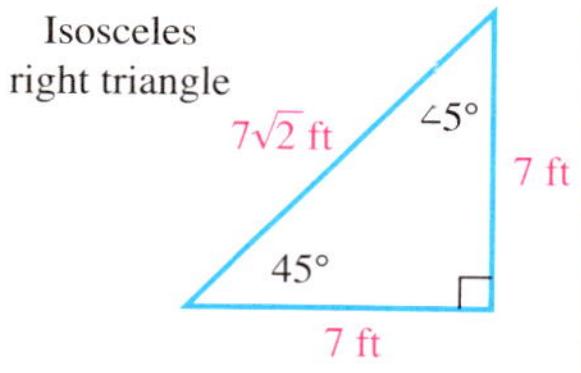

DEFINITIONS AND CONCEPTS

The hypotenuse of a **30°–60°–90° triangle** is twice as long as the shorter leg (the leg opposite the 30° angle.) The length of the longer leg (the leg opposite the 60° angle) is $\sqrt{3}$ times the length of the shorter leg.

EXAMPLES

If the shorter leg of a 30°–60°–90° triangle is 9 inches long:

- The length of the hypotenuse is $2 \cdot 9 = 18$ inches.
- The length of the longer leg is $\sqrt{3} \cdot 9 = 9\sqrt{3}$ inches.

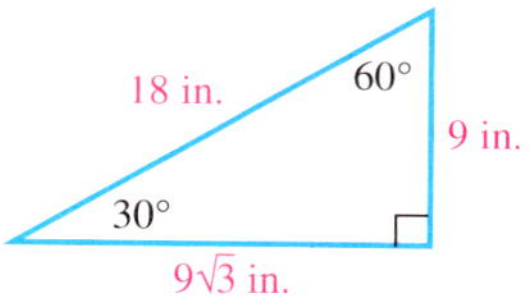

30°–60°–90° Triangle

DEFINITIONS AND CONCEPTS

The distance d between two points with coordinates (x_1, y_1) and (x_2, y_2) is given by **the distance formula:**

$$d = \sqrt{(x_2 - x_1)^2 + (y_2 - y_1)^2}$$

EXAMPLES

The distance between points $(-2, 3)$ and $(1, 7)$ is:

$$\begin{aligned} d &= \sqrt{(x_2 - x_1)^2 + (y_2 - y_1)^2} \\ &= \sqrt{[1 - (-2)]^2 + (7 - 3)^2} \\ &= \sqrt{3^2 + 4^2} \\ &= \sqrt{9 + 16} \\ &= \sqrt{25} \\ &= 5 \end{aligned}$$

The **midpoint** of a line segment with endpoints (x_1, y_1) and (x_2, y_2) is the point with coordinates

$$\left(\frac{x_1 + x_2}{2}, \frac{y_1 + y_2}{2}\right)$$

Find the midpoint of the segment joining $(-3, 7)$ and $(5, -8)$.

We let $(x_1, y_1) = (-3, 7)$ and $(x_2, y_2) = (5, -8)$ and substitute the coordinates into the midpoint formula.

$$\left(\frac{x_1 + x_2}{2}, \frac{y_1 + y_2}{2}\right) = \left(\frac{-3 + 5}{2}, \frac{7 + (-8)}{2}\right)$$

$$= \left(1, -\frac{1}{2}\right)$$ This is the midpoint.

REVIEW EXERCISES

115. Carpentry. The gable end of the roof shown below is divided in half by a vertical brace, 8 feet in height. Find the length of the roof line.

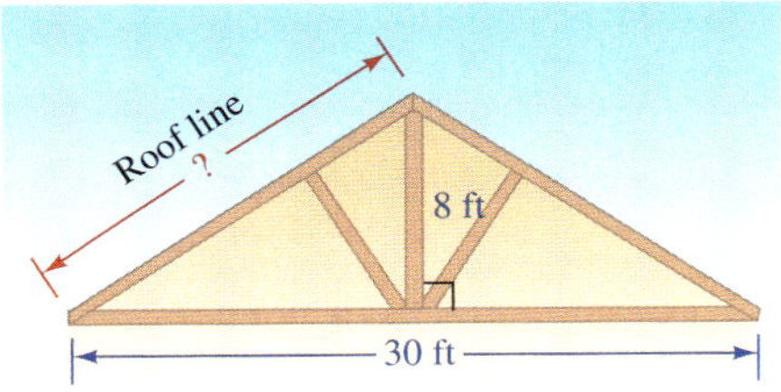

116. Sailing. A technique called *tacking* allows a sailboat to make progress into the wind. A sailboat follows the course shown below. Find d, the distance the boat advances into the wind after tacking.

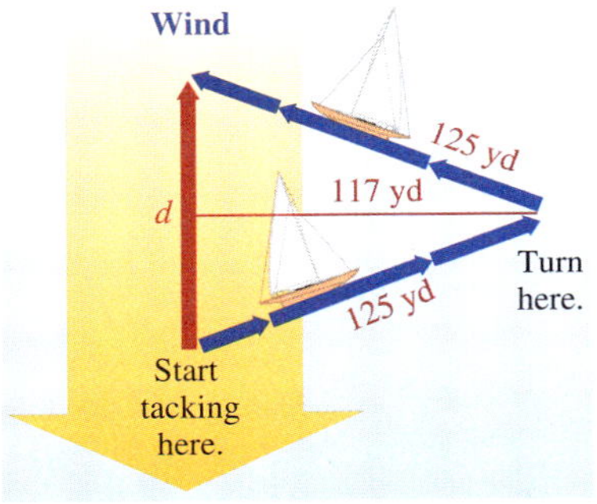

For problems 117–122, give the exact answer and then an approximation to two decimal places, when appropriate.

117. Find the length of the hypotenuse of an isosceles right triangle if the length of one leg is 7 meters.

118. The length of the hypotenuse of an isosceles right triangle is 15 yards. Find the length of one leg of the triangle.

119. The length of the hypotenuse of a 30°–60°–90° triangle is 12 centimeters. Find the length of each leg.

120. In a 30°–60°–90° triangle, the length of the longer leg is 60 feet. Find the length of the hypotenuse and the length of the shorter leg.

121. Find x and y.

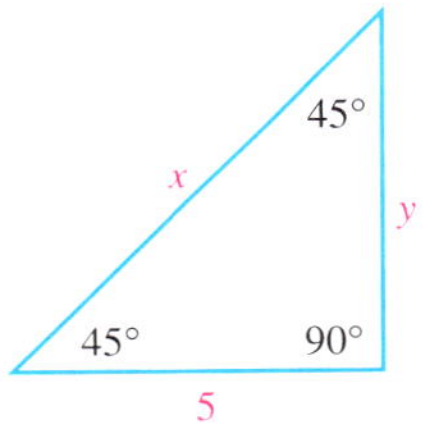

122. Find x and y.

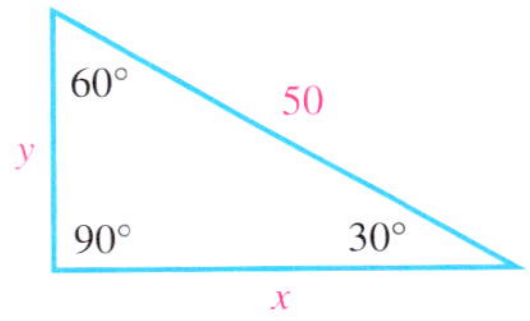

Find the distance between the points.

123. $(1, 3)$ and $(6, -9)$

124. $(-4, 6)$ and $(-2, 8)$

125. Find the midpoint of the segment joining $(8, -2)$ and $(6, -4)$.

126. If $(6, 1)$ is the midpoint of segment PQ and the coordinates of P are $(10, 4)$, find the coordinates of Q.

SECTION 9.7 Complex Numbers

DEFINITIONS AND CONCEPTS

The **imaginary number** i is defined as

$$i = \sqrt{-1}$$

From the definition, it follows that $i^2 = -1$.

EXAMPLES

Write each expression in terms of i:

$$\begin{aligned}\sqrt{-81} &= \sqrt{-1 \cdot 81} \\ &= \sqrt{-1}\sqrt{81} \\ &= i \cdot 9 \\ &= 9i\end{aligned}$$

$$\begin{aligned}\sqrt{-24} &= \sqrt{-1 \cdot 24} \\ &= \sqrt{-1}\sqrt{24} \\ &= i\sqrt{4}\sqrt{6} \\ &= 2i\sqrt{6} \quad \text{or} \quad 2\sqrt{6}i\end{aligned}$$

A **complex number** is any number that can be written in the form $a + bi$, where a and b are real numbers and $i = \sqrt{-1}$. We call a the **real part** and b the **imaginary part.**

Complex numbers:

$5 + 3i$ — 5 is the real part a and 3 is the imaginary part b.

$16 = 16 + 0i$ — 16 is the real part a and 0 is the imaginary part b.

$9i = 0 + 9i$ — 0 is the real part a and 9 is the imaginary part b.

Adding and subtracting complex numbers is similar to adding and subtracting polynomials. To **add two complex numbers,** add their real parts and add their imaginary parts.

To **subtract two complex numbers,** add the opposite of the complex number being subtracted.

Add. Write the answer in the form $a + bi$.

$$(7 - 5i) + (3 + 9i) = (7 + 3) + (-5 + 9)i$$
$$= 10 + 4i$$

Add the real parts. Add the imaginary parts.

Subtract. Write the answer in the form $a + bi$.

$$(8 - i) - (-1 + 6i) = (8 - i) + (1 - 6i)$$

Add the opposite of $-1 + 6i$.

$$= (8 + 1) + [-1 + (-6)]i$$

Add the real parts. Add the imaginary parts.

$$= 9 - 7i$$

Multiplying complex numbers is similar to multiplying polynomials.

Multiply. Write the answers in the form $a + bi$.

$$3i(6 - 4i) = 18i - 12i^2$$
$$= 18i - 12(-1)$$
$$= 18i + 12$$
$$= 12 + 18i$$

$$(4 + 7i)(2 - i) = 8 - 4i + 14i - 7i^2$$
$$= 8 + 10i - 7(-1)$$
$$= 8 + 10i + 7$$
$$= 15 + 10i$$

The complex numbers $a + bi$ and $a - bi$ are called **complex conjugates.**

The complex numbers $7 - 2i$ and $7 + 2i$ are complex conjugates.

To **divide complex numbers,** multiply the numerator and denominator by the complex conjugate of the denominator. The process is similar to rationalizing denominators.

Divide. Write the answers in the form $a + bi$.

$$\frac{3}{1 + i} \cdot \frac{1 - i}{1 - i} = \frac{3(1 - i)}{1 - i + i - i^2}$$
$$= \frac{3(1 - i)}{1 - (-1)}$$
$$= \frac{3 - 3i}{2}$$
$$= \frac{3}{2} - \frac{3}{2}i$$

$$\frac{6 + i}{2 - i} \cdot \frac{2 + i}{2 + i} = \frac{12 + 6i + 2i + i^2}{4 + 2i - 2i - i^2}$$
$$= \frac{12 + 8i + (-1)}{4 - (-1)}$$
$$= \frac{11 + 8i}{5}$$
$$= \frac{11}{5} + \frac{8}{5}i$$

The **powers of i** cycle through four possible outcomes: i, -1, $-i$, and 1.

$i^1 = i = i^5 = i^9 = \ldots$ R = 1

$i^2 = -1 = i^6 = i^{10} = \ldots$ R = 2

$i^3 = -i = i^7 = i^{11} = \ldots$ R = 3

$i^4 = 1 = i^8 = i^{12} = \ldots$ R = 0

Simplify: i^{66}

We divide 66 by 4 to get a remainder of 2. The remainder determines the power to which i is raised in the simplified form.

Thus, $i^{66} = i^2 = -1$.

```
  16 R2
4)66
 -4
  26
 -24
   2
```

REVIEW EXERCISES

Write each expression in terms of i.

127. $\sqrt{-25}$

128. $\sqrt{-18}$

129. $-\sqrt{-6}$

130. $\sqrt{-\frac{9}{64}}$

131. Complete the diagram.

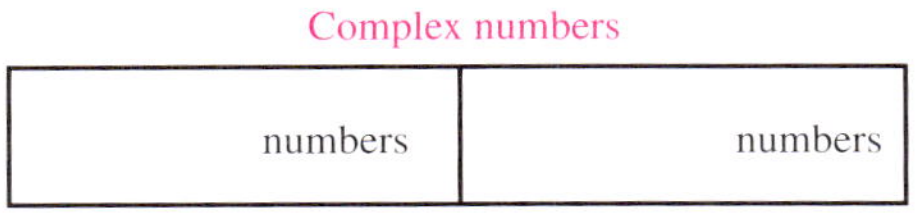

132. Determine whether each statement is true or false.

a. Every real number is a complex number.

b. $3 - 4i$ is a complex number.

c. $\sqrt{-4}$ is a real number.

d. i is a real number.

Give the complex conjugate of each number.

133. a. $3 + 6i$
b. $19i$

134. a. $-1 - 7i$
b. $-i$

Perform the operations. Write all answers in the form a + bi.

135. $(3 + 4i) + (5 - 6i)$

136. $\left(7 - \sqrt{-9}\right) - \left(4 + \sqrt{-4}\right)$

137. $3i(2 - i)$

138. $(2 - 7i)(-3 + 4i)$

139. $\sqrt{-3} \cdot \sqrt{-9}$

140. $(9i)^2$

141. $\frac{5 + 14i}{2 + 3i}$

142. $\frac{3}{11i}$

Simplify each expression.

143. i^{42}

144. i^{97}

9 Chapter Test

1. Fill in the blanks.

a. The symbol $\sqrt{\ }$ is called a _______ symbol.

b. The _________ number i is defined as $i = \sqrt{-1}$.

c. Squaring both sides of an equation can introduce _________ solutions.

d. An _________ right triangle is a right triangle with two legs of equal length.

e. To _________ the denominator of $\frac{4}{\sqrt{5}}$, we multiply the fraction by $\frac{\sqrt{5}}{\sqrt{5}}$.

f. A _________ number is any number that can be written in the form $a + bi$, where a and b are real numbers and $i = \sqrt{-1}$.

2. a. State the product rule for radicals.

b. State the quotient rule for radicals.

c. Explain why $\sqrt[4]{-16}$ is not a real number.

3. Graph $f(x) = \sqrt{x - 1}$. Find the domain and range of the function.

4. Diving. Refer to the illustration in the next column. The velocity v of an object in feet per second after it has fallen a distance of d feet is approximated by the function $v(d) = \sqrt{64.4d}$. Olympic diving platforms are 10 meters tall (approximately 32.8 feet). Estimate the velocity at which a diver hits the water from this height. Round to the nearest foot per second.

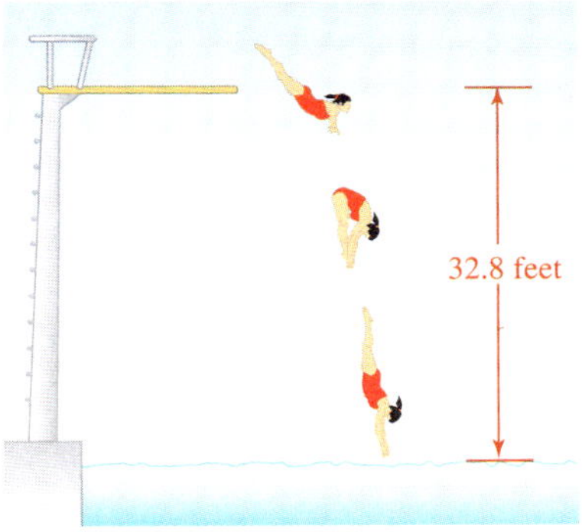

5. Use the graph of function f below to find each of the following.

a. $f(-1)$

b. $f(8)$

c. The value of x for which $f(x) = 1$

d. The domain and range of f

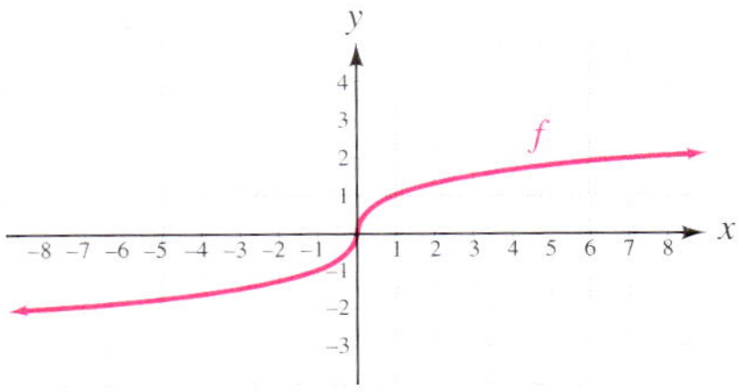

6. Find the domain of $f(x) = \sqrt{10x + 50}$.

Simplify each expression. All variables represent positive real numbers. Write answers without using negative exponents.

7. $(49x^4)^{1/2}$
8. $-27^{2/3}$
9. $36^{-3/2}$
10. $\left(-\dfrac{8}{125n^6}\right)^{-2/3}$
11. $\dfrac{2^{5/3}2^{1/6}}{2^{1/2}}$
12. $(a^{2/3})^{1/6}$

Simplify each expression. The variables are unrestricted.

13. $\sqrt{x^2}$
14. $\sqrt{y^2 - 10y + 25}$

Simplify each expression. All variables represent positive real numbers.

15. $\sqrt[3]{-64x^3y^6}$
16. $\sqrt{\dfrac{4a^2}{9}}$
17. $\sqrt[5]{(t+8)^5}$
18. $\sqrt{540x^3y^5}$
19. $\dfrac{\sqrt[3]{24x^{15}y^4}}{\sqrt[3]{y}}$
20. $\sqrt[4]{32}$

Perform the operations and simplify. All variables represent positive real numbers.

21. $2\sqrt{48y^5} - 3y\sqrt{12y^3}$
22. $2\sqrt[3]{40} - \sqrt[3]{5{,}000} + 4\sqrt[3]{625}$
23. $\sqrt[4]{243z^{13}} + z\sqrt[4]{48z^9}$
24. $-2\sqrt{xy}\left(3\sqrt{x} + \sqrt{xy^3}\right)$
25. $\left(3\sqrt{2} + \sqrt{3}\right)\left(2\sqrt{2} - 3\sqrt{3}\right)$
26. $\left(\sqrt[3]{2a} + 9\right)^2$
27. $\dfrac{8}{\sqrt{10}}$
28. $\dfrac{\sqrt{x} + \sqrt{y}}{\sqrt{x} - \sqrt{y}}$
29. $\sqrt[3]{\dfrac{9}{4a}}$
30. Rationalize the numerator: $\dfrac{\sqrt{5} + 3}{-4\sqrt{2}}$

Solve each equation. Write all proposed solutions. Cross out those that are extraneous.

31. $4\sqrt{x} = \sqrt{x+1}$
32. $\sqrt[3]{6n+4} - 4 = 0$
33. $1 = \sqrt{u-3} + \sqrt{u}$
34. $(2m^2 - 9)^{1/2} = m$
35. $\sqrt{t-2} - t + 2 = 0$
36. $\sqrt{x-8} + 10 = 0$
37. Let $f(x) = \sqrt[4]{15 - x}$ and $g(x) = \sqrt[4]{13 - 2x}$. Find all values of x for which $f(x) = g(x)$.
38. Solve $r = \sqrt[3]{\dfrac{GMt^2}{4\pi^2}}$ for G.

Find the missing side lengths in each triangle. Give the exact answer and then an approximation to two decimal places, when appropriate.

39.
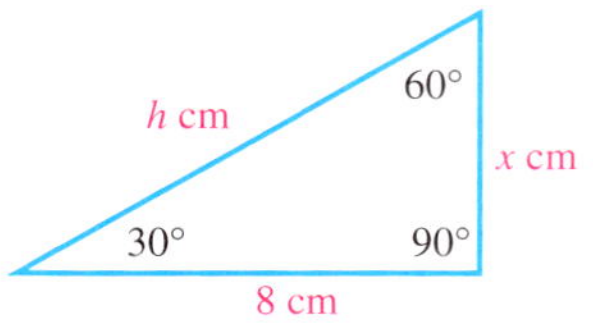

40.
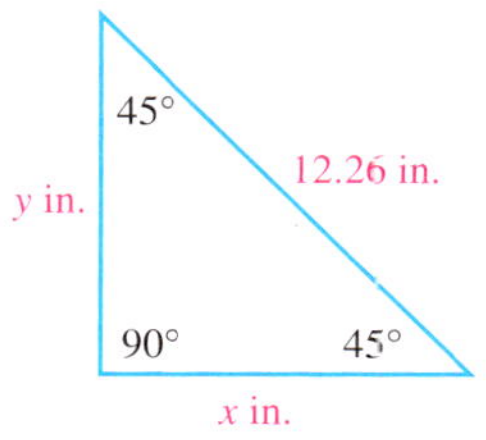

41. Find the distance between $(-2, 5)$ and $(22, 12)$.
42. Find the coordinates of the midpoint of the line segment joining $(-2, -5)$ and $(7, -11)$.
43. If $(3, 0)$ is the midpoint of segment PQ and the coordinates of P are $(-1, -3)$, find the coordinates of Q.
44. **Shipping Crates.** The diagonal brace on the shipping crate in the illustration is 53 inches. Find the height h of the crate.

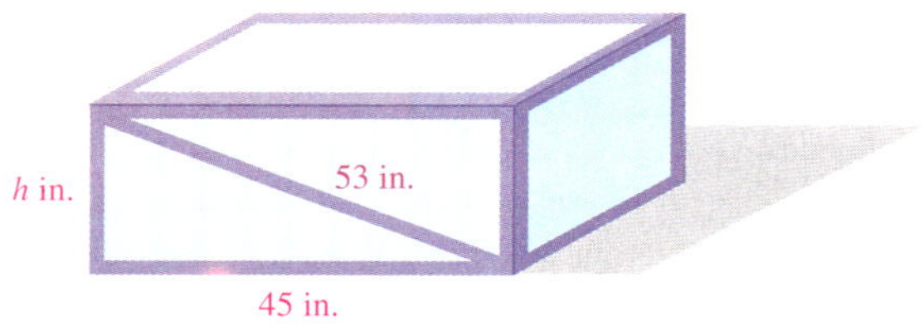

45. Express $\sqrt{-45}$ in terms of i.
46. Simplify: i^{106}

Perform the operations. Write all answers in the form $a + bi$.

47. $(9 + 4i) + (-13 + 7i)$
48. $\left(3 - \sqrt{-9}\right) - \left(-1 + \sqrt{-16}\right)$
49. $15i(3 - 5i)$
50. $(8 + 10i)(-7 - i)$
51. $\dfrac{1}{i\sqrt{2}}$
52. $\dfrac{2 + i}{3 - i}$

Group Project

A Spiral of Roots

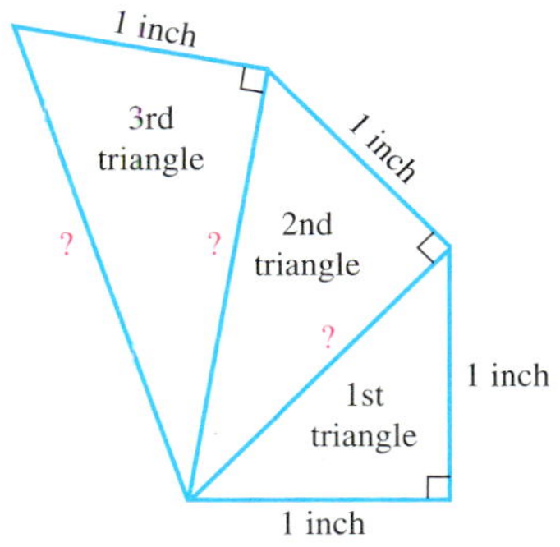

Overview: In this activity, you will create a visual representation of a collection of square roots.

Instructions: Form groups of 2 or 3 students. You will need a piece of unlined paper, a protractor, a ruler, and a pencil. Begin by drawing an isosceles right triangle with legs of length 1 inch in the middle of the paper. (See the illustration.) Use the Pythagorean theorem to determine the length of the hypotenuse. Draw a second right triangle using the hypotenuse of the first right triangle as one leg. Draw its second leg with length 1 inch. Find the length of the hypotenuse of the second triangle.

Continue creating right triangles, using the previous hypotenuse as one leg and drawing a new second leg of length 1 inch each time. Calculate the length of each resulting hypotenuse. When the figure begins to spiral onto itself, you may stop the process. Make a list of the lengths of each hypotenuse. What pattern do you see?

Graphing in Three Dimensions

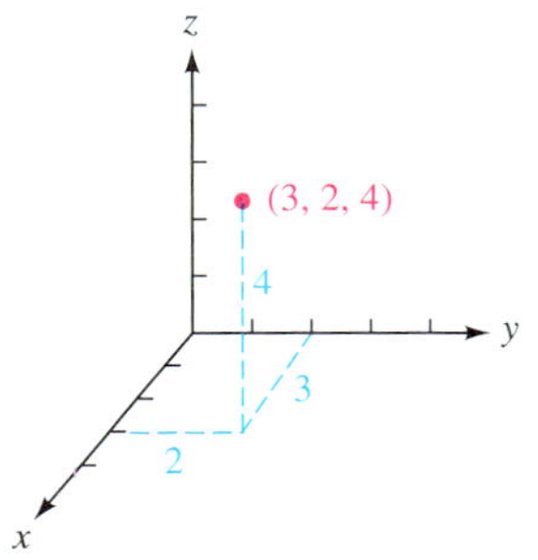

Overview: In this activity, you will find the distance between two points that lie in three-dimensional space.

Instructions: Form groups of 2 or 3 students. In a three-dimensional Cartesian coordinate system, the positive x-axis is horizontal and pointing toward the viewer (out of the page), the positive y-axis is also horizontal and pointing to the right, and the positive z-axis is vertical, pointing up. A point is located by plotting an ordered triple of numbers (x, y, z). In the illustration, the point $(3, 2, 4)$ is plotted.

In three dimensions, the distance formula is

$$d = \sqrt{(x_2 - x_1)^2 + (y_2 - y_1)^2 + (z_2 - z_1)^2}$$

1. Copy the illustration shown. Then plot the point $(1, 4, 3)$. Use the distance formula to find the distance between these two points.
2. Draw another three-dimensional coordinate system and plot the points $(-3, 3, -4)$ and $(2, -3, 2)$. Use the distance formula to find the distance between these two points.

Quadratic Equations, Functions, and Inequalities

10

from Campus to Careers

Police Patrol Officer

The responsibilities of a police patrol officer are extremely broad. Quite often, he or she must make a split-second decision while under enormous pressure. One internet vocational website cautions anyone considering such a career to "pay attention in your mathematics and science classes. Those classes help sharpen your ability to think things through and solve problems—an important part of police work."

Problem 93 in **Study Set 10.2, problem 87** in **Study Set 10.3,** and **problem 89** in **Study Set 10.4** involve situations that a police patrol officer encounters on the job. The mathematical concepts discussed in this chapter can be used to solve those problems.

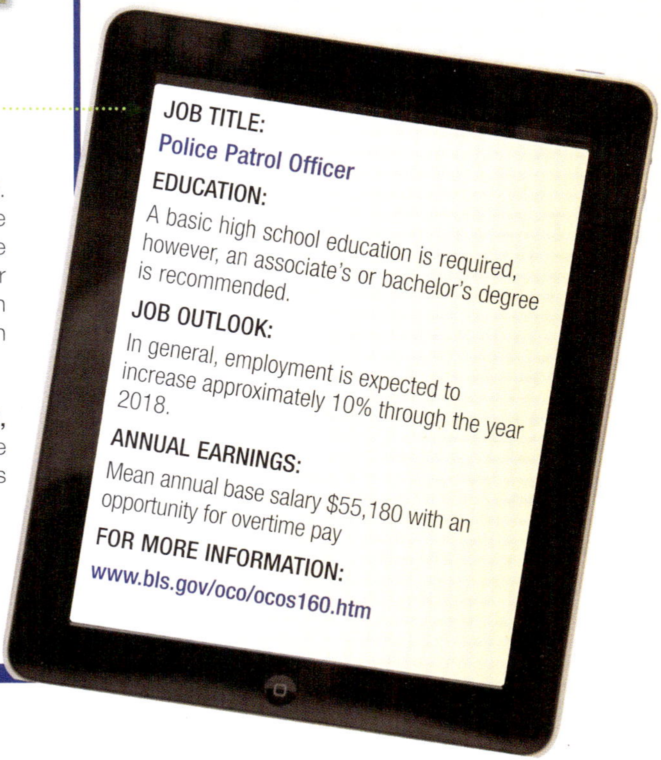

Study Skills Workshop ▶ Organizing Your Notebook

If you're like most students, your algebra notebook could probably use some attention at this stage of the course. You will definitely appreciate a well-organized notebook when it comes time to study for the final exam. Here are some suggestions to put it in tip-top shape.

ORGANIZE YOUR NOTEBOOK INTO SECTIONS: Create a separate section in the notebook for each chapter (or unit of study) that your class has covered this term.

ORGANIZE THE PAPERS WITHIN EACH SECTION: One recommended order is to begin each section with your class notes, followed by your completed homework assignments, then any study sheets or handouts, and, finally, all graded quizzes and tests.

Now Try This ▶

1. Organize your algebra notebook using the guidelines given above.
2. Write a Table of Contents to place at the beginning of your notebook. List each chapter (or unit of study) and include the dates over which the material was covered.
3. Compare your completed notebook with those of other students in your class. Have you overlooked any important items that would be useful when studying for the final exam?

SECTION 10.1 The Square Root Property and Completing the Square

OBJECTIVES

1. Use the square root property to solve quadratic equations.
2. Solve quadratic equations by completing the square.

ARE YOU READY?

The following problems review some basic skills that are needed when using the square root property and completing the square.

1. Simplify: **a.** $\sqrt{28}$ **b.** $\sqrt{-36}$
2. Rationalize: $\sqrt{\frac{5}{6}}$
3. What is one-half of 9?
4. Find the square of $\frac{7}{2}$.
5. Fill in the blank: $1 + \frac{25}{144} = \frac{\quad}{144}$
6. Factor: $x^2 - 8x + 16$

Recall that a *quadratic equation* is an equation of the form $ax^2 + bx + c = 0$, where a, b, and c are real numbers and $a \neq 0$. We have solved quadratic equations, such as $6x^2 - 7x - 3 = 0$, using factoring and the zero-factor property as shown below.

$$6x^2 - 7x - 3 = 0$$

$$(2x - 3)(3x + 1) = 0 \qquad \text{Factor the trinomial.}$$

$$2x - 3 = 0 \quad \text{or} \quad 3x + 1 = 0 \qquad \text{Set each factor equal to 0.}$$

$$x = \frac{3}{2} \quad \Big| \quad x = -\frac{1}{3} \qquad \text{Solve each linear equation.}$$

The solutions are $-\frac{1}{3}$ and $\frac{3}{2}$.

Many expressions do not factor as easily as $6x^2 - 7x - 3$. For example, it would be difficult to solve $2x^2 + 4x + 1 = 0$ by factoring, because $2x^2 + 4x + 1$ cannot be factored by using only integers. In this section and the next, we will develop more general methods that enable us to solve any quadratic equation. Those methods are based on the *square root property*.

1 Use the Square Root Property to Solve Quadratic Equations.

To develop general methods for solving quadratic equations, we first consider the equation $x^2 = c$. If $c \geq 0$, we can find the real solutions of $x^2 = c$ as follows:

$$x^2 = c$$

$$x^2 - c = 0 \quad \text{Subtract } c \text{ from both sides.}$$

$$x^2 - (\sqrt{c})^2 = 0 \quad \text{Replace } c \text{ with } (\sqrt{c})^2\text{, since } c = (\sqrt{c})^2.$$

$$(x + \sqrt{c})(x - \sqrt{c}) = 0 \quad \text{Factor the difference of two squares.}$$

$$x + \sqrt{c} = 0 \quad \text{or} \quad x - \sqrt{c} = 0 \quad \text{Set each factor equal to 0.}$$

$$x = -\sqrt{c} \quad | \quad x = \sqrt{c} \quad \text{Solve each linear equation.}$$

The solutions of $x^2 = c$ are $\sqrt{c}$ and $-\sqrt{c}$.

The Square Root Property

For any nonnegative real number c, if $x^2 = c$, then

$$x = \sqrt{c} \quad \text{or} \quad x = -\sqrt{c}$$

The Language of Algebra

The ± symbol is often seen in political polls. A candidate with 48% (±4%) support could be between 48 + 4 = 52% and 48 − 4 = 44%

We can write the conclusion of the square root property in a more compact form, called **double-sign notation.**

$$x = \pm\sqrt{c}$$

This is formally read as "x equals the positive or negative square root of c." However, it is often read more informally as "x equals plus or minus the square root of c."

EXAMPLE 1 Solve: $x^2 - 12 = 0$

Strategy Since $x^2 - 12$ does not factor as a difference of two integer squares, we must take an alternate approach. We will add 12 to both sides of the equation and use the square root property to solve for x.

Why After adding 12 to both sides, the resulting equivalent equation will have the desired form $x^2 = c$.

Solution

$$x^2 - 12 = 0 \quad \text{This is the equation to solve. It is a quadratic equation that is missing an } x\text{-term.}$$

$$x^2 = 12 \quad \text{To isolate } x^2 \text{ on the left side, add 12 to both sides.}$$

$$x = \sqrt{12} \quad \text{or} \quad x = -\sqrt{12} \quad \text{Use the square root property.}$$

$$x = 2\sqrt{3} \quad | \quad x = -2\sqrt{3} \quad \text{Simplify: } \sqrt{12} = \sqrt{4}\sqrt{3} = 2\sqrt{3}.$$

Check:

$$x^2 - 12 = 0 \qquad\qquad x^2 - 12 = 0$$

$$(2\sqrt{3})^2 - 12 \stackrel{?}{=} 0 \qquad\qquad (-2\sqrt{3})^2 - 12 \stackrel{?}{=} 0$$

$$12 - 12 \stackrel{?}{=} 0 \qquad\qquad 12 - 12 \stackrel{?}{=} 0$$

$$0 = 0 \quad \text{True} \qquad\qquad 0 = 0 \quad \text{True}$$

The exact solutions are $2\sqrt{3}$ and $-2\sqrt{3}$, which can be written using double-sign notation as $\pm 2\sqrt{3}$. The solution set is $\{-2\sqrt{3}, 2\sqrt{3}\}$. We can use a calculator to approximate the solutions. To the nearest hundredth, they are ± 3.46.

Self Check 1 Solve: $x^2 - 18 = 0$

Now Try Problems 15 and 19

EXAMPLE 2 **Phonograph Records.** Before compact discs, music was recorded on thin vinyl discs. The discs used for long-playing records had a surface area of about 111 square inches per side. Find the radius of a long-playing record to the nearst tenth of an inch.

Strategy The area A of a circle with radius r is given by the formula $A = \pi r^2$. We will find the radius of a record by substituting 111 for A and dividing both sides by π. Then we will use the square root property to solve for r.

Why After substituting 111 for A and dividing both sides by π, the resulting equivalent equation will have the desired form $r^2 = c$.

Solution

$$A = \pi r^2 \qquad \text{This is the formula for the area of a circle.}$$

$$111 = \pi r^2 \qquad \text{Substitute 111 for A.}$$

$$\frac{111}{\pi} = r^2 \qquad \text{To undo the multiplication by } \pi\text{, divide both sides by } \pi.$$

$$r = \sqrt{\frac{111}{\pi}} \quad \text{or} \quad r = -\sqrt{\frac{111}{\pi}} \qquad \text{Use the square root property. Since the radius of the record cannot be negative, discard the second solution.}$$

The radius of a long-playing record is $\sqrt{\frac{111}{\pi}}$ inches—to the nearest tenth, 5.9 inches.

Self Check 2 **Restaurant Seating.** A restaurant requires 14 square feet of floor area for one of their round tables. Find the radius of a table. Round to the nearest hundredth.

Now Try Problems 35 and 103

Some quadratic equations have solutions that are not real numbers.

EXAMPLE 3 Solve: $4x^2 + 25 = 0$

Strategy We will subtract 25 from both sides of the equation and divide both sides by 4. Then we will use the square root property to solve for x.

Why After subtracting 25 from both sides and dividing both sides by 4, the resulting equivalent equation will have the desired form $x^2 =$ a constant.

Solution

$$4x^2 + 25 = 0 \qquad \text{This is the equation to solve.}$$

$$x^2 = -\frac{25}{4} \qquad \text{To isolate } x^2\text{, subtract 25 from both sides and divide both sides by 4.}$$

$$x = \pm\sqrt{-\frac{25}{4}} \qquad \text{Use the square root property and write the result using double-sign notation.}$$

$$x = \pm\frac{5}{2}i \qquad \text{Simplify the radical: } \sqrt{-\frac{25}{4}} = \sqrt{-1 \cdot \frac{25}{4}} = \sqrt{-1}\frac{\sqrt{25}}{\sqrt{4}} = \frac{5}{2}i$$

Since the solutions are $\frac{5}{2}i$ and $-\frac{5}{2}i$, the solution set is $\left\{\frac{5}{2}i, -\frac{5}{2}i\right\}$.

Check:

$$4x^2 + 25 = 0 \qquad\qquad 4x^2 + 25 = 0$$
$$4\left(\frac{5}{2}i\right)^2 + 25 \stackrel{?}{=} 0 \qquad\qquad 4\left(-\frac{5}{2}i\right)^2 + 25 \stackrel{?}{=} 0$$
$$4\left(\frac{25}{4}\right)i^2 + 25 \stackrel{?}{=} 0 \qquad\qquad 4\left(\frac{25}{4}\right)i^2 + 25 \stackrel{?}{=} 0$$
$$25(-1) + 25 \stackrel{?}{=} 0 \qquad\qquad 25(-1) + 25 \stackrel{?}{=} 0$$
$$0 = 0 \quad \text{True} \qquad\qquad 0 = 0 \quad \text{True}$$

Self Check 3 Solve: $16x^2 + 49 = 0$

Now Try Problem 27

We can extend the square root property to solve equations that involve the square of a binomial and a constant.

EXAMPLE 4 Use the square root property to solve $(x - 1)^2 = 16$.

Strategy Instead of a variable squared on the left side, we have a quantity squared. We still use the square root property to solve the equation.

Why We want to eliminate the square on the binomial, so that we can eventually isolate the variable on one side of the equation.

Solution

$(x - 1)^2 = 16$	This is the equation to solve.
$x - 1 = \pm\sqrt{16}$	Use the square root property and write the result using double-sign notation.
$x - 1 = \pm 4$	Simplify: $\sqrt{16} = 4$.
$x = 1 \pm 4$	To isolate x, add 1 to both sides. It is standard practice to write the addition of the 1 in front of the ± symbol. Read as "one plus or minus four."
$x = 1 + 4$ or $x = 1 - 4$	To find one solution, use +. To find the other, use −.
$x = 5$ | $x = -3$	Add (subtract).

Verify that 5 and −3 satisfy the original equation.

Caution

It might be tempting to square the binomial on the left side of $(x - 1)^2 = 16$. However, that causes unnecessary, additional steps to solve the equation another way.

$$(x - 1)^2 = 16$$
$$\cancel{x^2 - 2x + 1 = 16}$$

Self Check 4 Use the square root property to solve $(x + 2)^2 = 9$.

Now Try Problems 31 and 33

EXAMPLE 5 Let $f(x) = (3x + 8)^2$. For what value(s) of x is $f(x) = 6$?

Strategy We will substitute 6 for $f(x)$ and solve for x.

Why In the equation, there are two unknowns, x and $f(x)$. If we replace $f(x)$ with 6, we can use equation-solving techniques to find x.

Solution

$f(x) = (3x + 8)^2$ — This is the given function.

$6 = (3x + 8)^2$ — Substitute 6 for $f(x)$.

$\pm\sqrt{6} = 3x + 8$ — Use the square root property and write the result using double-sign notation.

$-8 \pm \sqrt{6} = 3x$ — To isolate the variable term $3x$, subtract 8 from both sides. It is standard practice to write the subtraction of -8 in front of the $\pm$.

$\frac{-8 \pm \sqrt{6}}{3} = x$ — To isolate x, undo the multiplication by 3 by dividing both sides by 3.

We have found that $f(x) = 6$ when $x = \frac{-8 - \sqrt{6}}{3}$ or $x = \frac{-8 + \sqrt{6}}{3}$. To check, verify that $f\left(\frac{-8 - \sqrt{6}}{3}\right) = 6$ and $f\left(\frac{-8 + \sqrt{6}}{3}\right) = 6$.

Self Check 5 Let $f(x) = (4x + 11)^2$. For what values of x is $f(x) = 3$?

Now Try Problem 35

2 Solve Quadratic Equations by Completing the Square.

When the polynomial in a quadratic equation doesn't factor easily, we can solve the equation by **completing the square.** This method is based on the following perfect-square trinomials (with leading coefficients of 1) and their factored forms:

$$x^2 + 2bx + b^2 = (x + b)^2 \quad \text{and} \quad x^2 - 2bx + b^2 = (x - b)^2$$

The Language of Algebra

Recall that trinomials that are the square of a binomial are called **perfect-square trinomials.**

In each of these perfect-square trinomials, the third term is the square of one-half of the coefficient of x.

- In $x^2 + 2bx + b^2$, the coefficient of x is $2b$. If we find $\frac{1}{2} \cdot 2b$, which is b, and square it, we get the third term, b^2.
- In $x^2 - 2bx + b^2$, the coefficient of x is $-2b$. If we find $\frac{1}{2}(-2b)$, which is $-b$, and square it, we get the third term: $(-b)^2 = b^2$.

We can use these observations to change certain binomials into perfect-square trinomials using a 3-step process. For example, to change $x^2 + 12x$ into a perfect-square trinomial, we find one-half of the coefficient of x, square the result, and add the square to $x^2 + 12x$.

$$x^2 + 12x + \square$$

Step 1: Find one-half of the coefficient of x.

$$\frac{1}{2} \cdot 12 = 6 \qquad 6^2 = 36$$

Step 2: Square the result.

Step 3: Add the square to the binomial.

We obtain the perfect-square trinomial $x^2 + 12x + 36$, which factors as $(x + 6)^2$. By adding 36 to $x^2 + 12x$, we say that we have *completed the square on* $x^2 + 12x$.

Completing the Square

To complete the square on $x^2 + bx$, add the square of one-half of the coefficient of x:

$$x^2 + bx + \left(\frac{1}{2}b\right)^2$$

EXAMPLE 6 Complete the square and factor the resulting perfect-square trinomial: **a.** $x^2 + 10x$ **b.** $x^2 - 11x$ **c.** $x^2 + \frac{7}{3}x$

Strategy We will add the square of one-half of the coefficient of x to the given binomial.

Why Adding such a term will change the binomial into a perfect-square trinomial that will factor.

Solution

a. To make $x^2 + 10x$ a perfect-square trinomial, we find one-half of 10, square it, and add the result to $x^2 + 10x$.

$$x^2 + 10x + 25 \qquad \tfrac{1}{2} \cdot 10 = 5 \text{ and } 5^2 = 25. \text{ Add 25 to the given binomial.}$$

This trinomial factors as $(x + 5)^2$. To check, we square $x + 5$ and verify that the result is $x^2 + 10x + 25$.

b. To make $x^2 - 11x$ a perfect-square trinomial, we find one-half of -11, square it, and add the result to $x^2 - 11x$.

$$x^2 - 11x + \frac{121}{4} \qquad \tfrac{1}{2}(-11) = -\tfrac{11}{2} \text{ and } \left(-\tfrac{11}{2}\right)^2 = \tfrac{121}{4}. \text{ Add } \tfrac{121}{4} \text{ to the given binomial.}$$

This trinomial factors as $\left(x - \frac{11}{2}\right)^2$. Check using multiplication or using a special-product rule.

c. To make $x^2 + \frac{7}{3}x$ a perfect-square trinomial, we find one-half of $\frac{7}{3}$, square it, and add the result to $x^2 + \frac{7}{3}x$.

$$x^2 + \frac{7}{3}x + \frac{49}{36} \qquad \text{To complete the square: } \tfrac{1}{2}\left(\tfrac{7}{3}\right) = \tfrac{7}{6} \text{ and } \left(\tfrac{7}{6}\right)^2 = \tfrac{49}{36}. \text{ Add } \tfrac{49}{36} \text{ to the given binomial.}$$

The trinomial factors as $\left(x + \frac{7}{6}\right)^2$.

Self Check 6 Complete the square on $a^2 - 5a$ and factor the resulting trinomial.

Now Try Problems 39 and 41

Caution

Realize that when we complete the square on a binomial, we are not writing an equivalent trinomial expression. Since the result is a completely different polynomial, it would be incorrect to use an = symbol between the two.

$$x^2 + 10x = x^2 + 10x + 25$$ (crossed out)

The Language of Algebra

When we add $\frac{121}{4}$ to $x^2 - 11x$, we say we have **completed the square** on $x^2 - 11x$. For that reason, binomials such as $x^2 - 11x$ are often called **incomplete squares.**

To solve an equation of the form $ax^2 + bx + c = 0$ by completing the square, we use the following steps.

Completing the Square to Solve a Quadratic Equation in x

1. If the coefficient of x^2 is 1, go to step 2. If it is not, make it 1 by dividing both sides of the equation by the coefficient of x^2.
2. Get all variable terms on one side of the equation and constants on the other side.
3. Complete the square by finding one-half of the coefficient of x, squaring the result, and adding the square to both sides of the equation.
4. Factor the perfect-square trinomial as the square of a binomial.
5. Solve the resulting equation using the square root property.
6. Check your answers in the original equation.

EXAMPLE 7 Solve by completing the square: $x^2 - 8x - 5 = 0$. Approximate the solutions to the nearest hundredth.

Strategy We will use the addition property of equality and add 5 to both sides. Then we will complete the square to solve for x.

Why To prepare to complete the square, we need to isolate the variable terms, x^2 and $-8x$, on the left side of the equation and the constant term on the right side.

The Language of Algebra

For $x^2 - 8x - 5$, we call x^2 the **leading term,** the **first term,** or the **quadratic term.** We call $-8x$ the **second term** or the **linear term.** Together, x^2 and $-8x$ are called **variable terms.** We call -5 the **constant term** or the **third term.**

Solution

$x^2 - 8x - 5 = 0$ This is the equation to solve.

$x^2 - 8x = 5$ Add 5 to both sides so that the constant term is on the right side.

In $x^2 - 8x$, the coefficient of x is -8. One-half of -8 is -4, and $(-4)^2 = 16$. If we add 16 to $x^2 - 8x$, it becomes a perfect-square trinomial.

$x^2 - 8x + 16 = 5 + 16$ To complete the square on the left side, add 16 to both sides.

$(x - 4)^2 = 21$ Factor the perfect-square trinomial. Add on the right side.

$x - 4 = \pm\sqrt{21}$ By the square root property, $x - 4 = \sqrt{21}$ or $x - 4 = -\sqrt{21}$. Use double-sign notation to show this.

$x = 4 \pm \sqrt{21}$ To isolate x, add 4 to both sides. It is common practice to write the 4 in front of the $\pm$ symbol.

The exact solutions of $x^2 - 8x - 5 = 0$ are two irrational numbers, $4 + \sqrt{21}$ and $4 - \sqrt{21}$.

We also can approximate each solution by using the decimal approximation of $\sqrt{21}$, which is 4.582575695:

$$4 + \sqrt{21} \approx 4 + 4.582575695 \approx 8.58 \qquad 4 - \sqrt{21} \approx 4 - 4.582575695 \approx -0.58$$

To the nearest hundredth, the solutions are 8.58 and -0.58. These approximations can be used to check the exact solutions informally by substituting each of them into the original equation.

Self Check 7 Solve by completing the square: $x^2 - 10x - 4 = 0$. Approximate the solutions to the nearest hundredth.

Now Try Problem 47

If the coefficient of the squared variable (called the **leading coefficient**) of a quadratic equation is not 1, we must make it 1 before we can complete the square.

EXAMPLE 8 Solve $6x^2 + 5x - 6 = 0$ by completing the square.

Strategy We will begin by dividing both sides of the equation by 6.

Why This will create a leading coefficient that is 1 so that we can proceed to complete the square to solve the equation.

Solution *Step 1:* To make the coefficient of x^2 equal to 1, we divide both sides of the equation by 6.

$6x^2 + 5x - 6 = 0$ This is the equation to solve.

$\frac{6x^2}{6} + \frac{5x}{6} - \frac{6}{6} = \frac{0}{6}$ Divide both sides by 6, term-by-term.

$x^2 + \frac{5}{6}x - 1 = 0$ Simplify. The coefficient of x^2 is now 1.

Success Tip

When solving a quadratic equation using the factoring method, one side of the equation must be 0. When we complete the square, we are not concerned with that requirement.

Step 2: To have the constant term on one side of the equation and the variable terms on the other, add 1 to both sides.

$$x^2 + \frac{5}{6}x \quad\quad = 1$$

Some students find it helpful to leave additional space here to prepare for completing the square.

Step 3: The coefficient of x is $\frac{5}{6}$, one-half of $\frac{5}{6}$ is $\frac{5}{12}$, and $\left(\frac{5}{12}\right)^2 = \frac{25}{144}$. To complete the square, we add $\frac{25}{144}$ to both sides.

$$x^2 + \frac{5}{6}x + \mathbf{\frac{25}{144}} = 1 + \mathbf{\frac{25}{144}}$$

$$x^2 + \frac{5}{6}x + \frac{25}{144} = \frac{169}{144}$$ On the right side, add: $1 + \frac{25}{144} = \frac{144}{144} + \frac{25}{144} = \frac{169}{144}$.

Step 4: Factor the left side of the equation.

$$\left(x + \frac{5}{12}\right)^2 = \frac{169}{144}$$ $x^2 + \frac{5}{6}x + \frac{25}{144}$ is a perfect-square trinomial.

Caution

When using the square root property to solve an equation, always write the ± symbol. If you forget, you will lose one of the solutions.

Step 5: We can solve the resulting equation by using the square root property.

$$x + \frac{5}{12} = \pm\sqrt{\frac{169}{144}}$$ Don't forget to write the ± symbol.

$$x + \frac{5}{12} = \pm\frac{13}{12}$$ Simplify: $\pm\sqrt{\frac{169}{144}} = \pm\frac{13}{12}$.

$$x = -\frac{5}{12} \pm \frac{13}{12}$$ To isolate x, subtract $\frac{5}{12}$ from both sides. These are like terms that can be added and subtracted.

Success Tip

Since the solutions of $6x^2 + 5x - 6 = 0$ are rational numbers, that indicates it can also be solved by factoring:

$$(3x - 2)(2x + 3) = 0$$

$$x = \frac{2}{3} \quad \text{or} \quad x = -\frac{3}{2}$$

To find the first solution, we evaluate the expression using the + symbol. To find the second solution, we evaluate the expression using the − symbol.

$$x = -\frac{5}{12} + \frac{13}{12} \quad \text{or} \quad x = -\frac{5}{12} - \frac{13}{12}$$

$$x = \frac{8}{12} \qquad\qquad x = -\frac{18}{12}$$ Add (subtract) the fractions.

$$x = \frac{2}{3} \qquad\qquad x = -\frac{3}{2}$$ Simplify each fraction. The solutions are two rational numbers.

Step 6: Check each solution, $\frac{2}{3}$ and $-\frac{3}{2}$, in the original equation.

Self Check 8 Solve by completing the square: $3x^2 + 2x - 8 = 0$

Now Try Problem 55

EXAMPLE 9 Solve: $2x^2 + 4x + 1 = 0$. Approximate the solutions to the nearest hundredth.

Strategy We will follow the steps for solving a quadratic equation by completing the square.

Why Since the trinomial $2x^2 + 4x + 1$ cannot be factored using only integers, solving the equation by completing the square is our only option at this time.

Caution

A common error is to add a constant to one side of an equation to complete the square and forget to add it to the other side.

Solution

$$2x^2 + 4x + 1 = 0 \quad \text{This is the equation to solve.}$$

$$x^2 + 2x + \frac{1}{2} = 0 \quad \text{Divide both sides by 2 to make the coefficient of } x^2 \text{ equal to 1.}$$

$$x^2 + 2x = -\frac{1}{2} \quad \text{Subtract } \tfrac{1}{2} \text{ from both sides so that the constant term is on the right side.}$$

$$x^2 + 2x + \mathbf{1} = -\frac{1}{2} + \mathbf{1} \quad \text{To complete the square, square one-half of the coefficient of } x \text{ and add it to both sides.}$$

$$(x + 1)^2 = \frac{1}{2} \quad \text{Factor on the left side and do the addition on the right side.}$$

$$x + 1 = \pm\sqrt{\frac{1}{2}} \quad \text{Use the square root property. Don't forget the } \pm \text{ symbol.}$$

$$x = -1 \pm \sqrt{\frac{1}{2}} \quad \text{To isolate } x \text{, subtract 1 from both sides.}$$

To write $\sqrt{\frac{1}{2}}$ in simplified radical form, we use the quotient rule for radicals and then rationalize the denominator.

$$x = -1 + \frac{\sqrt{2}}{2} \quad \text{or} \quad x = -1 - \frac{\sqrt{2}}{2} \qquad \text{Rationalize: } \sqrt{\tfrac{1}{2}} = \frac{\sqrt{1}}{\sqrt{2}} = \frac{1 \cdot \sqrt{2}}{\sqrt{2}\sqrt{2}} = \frac{\sqrt{2}}{2}.$$

We can express each solution in an alternate form if we write -1 as a fraction with a denominator of 2.

Caution

Recall that to simplify a fraction, we remove common factors of the numerator and denominator. Since -2 is a term of the numerator of $\frac{-2 + \sqrt{2}}{2}$, no further simplification of this expression can be made.

$$x = -\frac{2}{2} + \frac{\sqrt{2}}{2} \quad \text{or} \quad x = -\frac{2}{2} - \frac{\sqrt{2}}{2} \qquad \text{Write } -1 \text{ as } -\tfrac{2}{2}.$$

$$x = \frac{-2 + \sqrt{2}}{2} \quad \Big| \quad x = \frac{-2 - \sqrt{2}}{2} \qquad \text{Add (subtract) the numerators and keep the common denominator 2.}$$

The exact solutions are $\frac{-2 + \sqrt{2}}{2}$ and $\frac{-2 - \sqrt{2}}{2}$, or simply, $\frac{-2 \pm \sqrt{2}}{2}$. We can use a calculator to approximate them. To the nearest hundredth, they are -0.29 and -1.71.

Self Check 9 Solve: $3x^2 + 6x + 1 = 0$. Approximate the solutions to the nearest hundredth.

Now Try ▶ Problem 59

Using Your Calculator ▶ Checking Solutions of Quadratic Equations

We can use a graphing calculator to check the solutions of the quadratic equation $2x^2 + 4x + 1 = 0$ found in Example 9. After entering $Y_1 = 2x^2 + 4x + 1$, we call up the home screen by pressing 2nd QUIT. Then we press VARS, arrow to Y-VARS, press ENTER, and enter 1 to get the display shown in figure (a). We evaluate $2x^2 + 4x + 1$ for $x = \frac{-2 + \sqrt{2}}{2}$ by entering the solution using function notation, as shown in figure (b). When ENTER is pressed, the result of 0 is confirmation that $x = \frac{-2 + \sqrt{2}}{2}$ is a solution of the equation.

(a)

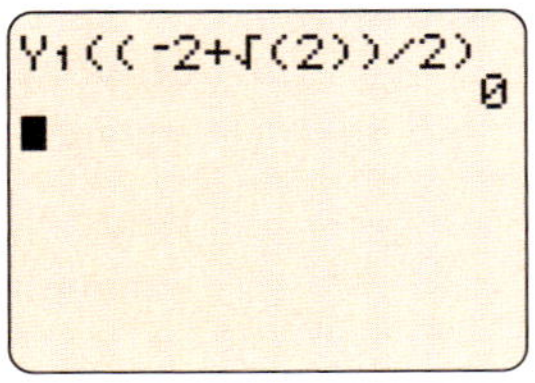

(b)

The solutions of some quadratic equations are complex numbers that contain i.

EXAMPLE 10 Solve: **a.** $x^2 - 4x + 16 = 0$ **b.** $x^2 + \frac{2}{3}x + 6 = 0$

Strategy We will follow the steps for solving a quadratic equation by completing the square.

Why Since the trinomials $x^2 - 4x + 16$ and $x^2 + \frac{2}{3}x + 6 = 0$ cannot be factored using only integers, solving the equation by completing the square is our only option.

Solution

a.

$x^2 - 4x + 16 = 0$	This is the equation to solve.
$x^2 - 4x = -16$	Subtract 16 from both sides so that the constant term is on the right side.
$x^2 - 4x + 4 = -16 + 4$	Complete the square: $\frac{1}{2}(-4) = -2$ and $(-2)^2 = 4$. Add 4 to both sides.
$(x - 2)^2 = -12$	Factor the left side. On the right side, add.
$x - 2 = \pm\sqrt{-12}$	Use the square root property. Don't forget the $\pm$ symbol.
$x = 2 \pm \sqrt{-12}$	To isolate x, add 2 to both sides.
$x = 2 \pm 2i\sqrt{3}$	Simplify the radical: $\sqrt{-12} = \sqrt{-1 \cdot 4 \cdot 3} = \sqrt{-1}\sqrt{4}\sqrt{3} = 2i\sqrt{3}$.

There are two complex number solutions involving i: $2 + 2i\sqrt{3}$ and $2 - 2i\sqrt{3}$. Verify this by checking them in the original equation.

b.

$x^2 + \frac{2}{3}x + 6 = 0$	This is the equation to solve.
$x^2 + \frac{2}{3}x = -6$	Subtract 6 from both sides so that the constant term is on the right side.
$x^2 + \frac{2}{3}x + \frac{1}{9} = \frac{1}{9} - 6$	Complete the square: $\frac{1}{2} \cdot \frac{2}{3} = \frac{1}{3}$ and $\left(\frac{1}{3}\right)^2 = \frac{1}{9}$. Add $\frac{1}{9}$ to both sides.
$\left(x + \frac{1}{3}\right)^2 = \frac{1}{9} - \frac{6}{1} \cdot \frac{9}{9}$	Factor the left side. On the right side, prepare to subtract. Write 6 as $\frac{6}{1}$ and multiply by $\frac{9}{9}$ to build an equivalent fraction with denominator 9.
$\left(x + \frac{1}{3}\right)^2 = \frac{1}{9} - \frac{54}{9}$	On the right side, multiply the numerators. Multiply the denominators.
$\left(x + \frac{1}{3}\right)^2 = -\frac{53}{9}$	On the right side, subtract the fractions.
$x + \frac{1}{3} = \pm\sqrt{-\frac{53}{9}}$	Use the square root property. Don't forget the $\pm$ symbol.
$x = -\frac{1}{3} \pm \sqrt{-\frac{53}{9}}$	To isolate x, subtract $\frac{1}{3}$ from both sides.
$x = -\frac{1}{3} \pm \frac{\sqrt{53}}{3}i$	Simplify the radical: $\sqrt{-\frac{53}{9}} = \sqrt{-1}\frac{\sqrt{53}}{\sqrt{9}} = \frac{\sqrt{53}}{3}i$.

The solutions are $-\frac{1}{3} + \frac{\sqrt{53}}{3}i$ and $-\frac{1}{3} - \frac{\sqrt{53}}{3}i$.

Self Check 10 Solve: **a.** $x^2 - 6x + 27 = 0$ **b.** $x^2 + \frac{2}{5}x + 3 = 0$

Now Try Problems 63 and 67

Success Tip

Example 8 can be solved by factoring. However, Examples 7, 9, and 10 cannot. This illustrates an important fact: completing the square can be used to solve any quadratic equation.

Success Tip

Notice how we multiply $\frac{6}{1}$ by a form of 1 to build an equivalent fraction with a denominator 9:

$$\frac{6}{1} \cdot \frac{9}{9}$$

Notation

The solutions are written in complex number form $a + bi$. They also could be written as

$$\frac{-1 \pm i\sqrt{53}}{3}$$

SECTION 10.1 STUDY SET

VOCABULARY

Fill in the blanks.

1. An equation of the form $ax^2 + bx + c = 0$, where $a \neq 0$, is called a __________ equation.
2. $x^2 + 6x + 9$ is called a ________-square trinomial because it factors as $(x + 3)^2$.
3. When we add 16 to $x^2 + 8x$, we say that we have completed the ________ on $x^2 + 8x$.
4. The ________ coefficient of $5x^2 - 2x + 7$ is 5 and the ________ term is 7.

CONCEPTS

Fill in the blanks.

5. For any nonnegative number c, if $x^2 = c$, then $x =$ or $x =$.
6. To complete the square on $x^2 + 10x$, add the square of ________ of the coefficient of x.
7. Find one-half of the given number and square the result.
   a. 12
   b. -5
8. Fill in the blanks to factor the perfect-square trinomial.
   a. $x^2 + 8x + 16 = (x + \quad)^2$
   b. $x^2 - 9x + \frac{81}{4} = \left(x - \quad\right)^2$
9. What is the first step to solve each equation by completing the square? **Do not solve.**
   a. $x^2 + 9x + 7 = 0$
   b. $4x^2 + 5x - 16 = 0$
10. Divide both sides of the equation by the proper number to make the coefficient of x^2 equal to 1. **Do not solve.**
    a. $2x^2 - 3x + 6 = 0$
    b. $3x^2 + 6x - 4 = 0$
11. Use a check to determine whether $-2 + \sqrt{2}$ is a solution of $x^2 + 4x + 2 = 0$.
12. Determine whether each statement is true or false.
    a. Any quadratic equation can be solved by the factoring method.
    b. Any quadratic equation can be solved by completing the square.

NOTATION

13. We read $8 \pm \sqrt{3}$ as "eight _____ ___ ______ the square root of 3."
14. When solving a quadratic equation, a student obtains $x = \frac{-5 \pm \sqrt{7}}{3}$.
    a. How many solutions are represented by this notation? List them.
    b. Approximate the solutions to the nearest hundredth.

GUIDED PRACTICE

Use the square root property to solve each equation. See Example 1.

15. $t^2 - 11 = 0$
16. $w^2 - 47 = 0$
17. $x^2 - 35 = 0$
18. $x^2 - 101 = 0$
19. $z^2 - 50 = 0$
20. $u^2 - 24 = 0$
21. $3x^2 - 16 = 0$
22. $5x^2 - 49 = 0$

Use the square root property to solve each equation. See Example 3.

23. $p^2 = -16$
24. $q^2 = -25$
25. $a^2 + 8 = 0$
26. $m^2 + 18 = 0$
27. $4m^2 + 81 = 0$
28. $9n^2 + 121 = 0$
29. $6b^2 + 144 = 0$
30. $9n^2 + 288 = 0$

Use the square root property to solve each equation. See Example 4.

31. $(x + 5)^2 = 9$
32. $(x - 1)^2 = 4$
33. $(t + 4)^2 = 16$
34. $(s - 7)^2 = 9$

See Example 5.

35. Let $f(x) = (x + 5)^2$. For what value(s) of x is $f(x) = 3$?
36. Let $f(x) = (x + 3)^2$. For what value(s) of x is $f(x) = 7$?
37. Let $g(a) = (7a - 2)^2$. For what value(s) of a is $g(a) = 8$?
38. Let $h(c) = (11c - 2)^2$. For what value(s) of c is $h(c) = 12$?

Complete the square and factor the resulting perfect-square trinomial. See Example 6.

39. $x^2 + 24x$
40. $y^2 - 18y$
41. $a^2 - 7a$
42. $b^2 + 11b$
43. $x^2 + \frac{2}{3}x$
44. $x^2 + \frac{2}{5}x$
45. $m^2 - \frac{5}{6}m$
46. $n^2 - \frac{5}{3}n$

Use completing the square to solve each equation. Approximate each solution to the nearest hundredth. See Example 7.

47. $x^2 - 4x - 2 = 0$
48. $x^2 - 6x - 4 = 0$
49. $x^2 - 12x + 1 = 0$
50. $x^2 - 18x + 2 = 0$
51. $t^2 + 20t + 25 = 0$
52. $t^2 + 22t + 93 = 0$

53. $t^2 + 16t - 16 = 0$

54. $t^2 + 14t - 7 = 0$

Use completing the square to solve each equation. See Example 8.

55. $2x^2 - x - 1 = 0$

56. $2x^2 - 5x + 2 = 0$

57. $12t^2 - 5t - 3 = 0$

58. $5m^2 + 13m - 6 = 0$

Use completing the square to solve each equation. Approximate each solution to the nearest hundredth. See Example 9.

59. $3x^2 - 12x + 1 = 0$

60. $6x^2 - 12x + 1 = 0$

61. $2x^2 + 5x - 2 = 0$

62. $2x^2 - 8x + 5 = 0$

Use completing the square to solve each equation. See Example 10.

63. $p^2 + 2p + 2 = 0$

64. $x^2 - 6x + 10 = 0$

65. $y^2 + 8y + 18 = 0$

66. $n^2 + 10n + 28 = 0$

67. $x^2 + \frac{2}{3}x + 7 = 0$

68. $x^2 + \frac{2}{5}x + 6 = 0$

69. $a^2 - \frac{1}{2}a + 1 = 0$

70. $b^2 - \frac{1}{4}b + 1 = 0$

TRY IT YOURSELF

Solve each equation. Approximate the solutions to the nearest hundredth when appropriate.

71. $(3x - 1)^2 = 25$

72. $(5x - 2)^2 = 64$

73. $3x^2 - 6x = 1$

74. $2x^2 - 6x = -3$

75. $x^2 + 8x + 6 = 0$

76. $x^2 + 6x + 4 = 0$

77. $6x^2 + 72 = 0$

78. $5x^2 + 40 = 0$

79. $x^2 - 2x = 17$

80. $x^2 + 10x = 7$

81. $m^2 - 7m + 3 = 0$

82. $m^2 - 5m + 3 = 0$

83. $7h^2 = 35$

84. $9n^2 = 99$

85. $\frac{7x + 1}{5} = -x^2$

86. $\frac{3}{8}x^2 = \frac{1}{8} - x$

87. $t^2 + t + 3 = 0$

88. $b^2 - b + 5 = 0$

89. $(8x + 5)^2 = 24$

90. $(3y - 2)^2 = 18$

91. $r^2 - 6r - 27 = 0$

92. $s^2 - 6s - 40 = 0$

93. $4p^2 + 2p + 3 = 0$

94. $3m^2 - 2m + 3 = 0$

Look Alikes . . .

95. a. $x^2 - 24 = 0$ b. $x^2 + 24 = 0$

96. a. $x^2 + 7x - 8 = 0$ b. $x^2 + 7x - 9 = 0$

97. a. $2m^2 - 8m = 0$ b. $2m^2 - 8m = 1$

98. a. $a^2 + a - 7 = 0$ b. $a^2 - a - 7 = 0$

99. a. $x^2 - 4x + 20 = 0$ b. $x^2 - 4x - 20 = 0$

100. a. $a^2 = 6a - 3$ b. $n^2 = 6n + 3$

101. a. $2r^2 - 4r + 3 = 0$ b. $2r^2 - 4r - 3 = 0$

102. a. $5y^2 + 15y + 12 = 0$ b. $5y^2 + 15y - 12 = 0$

APPLICATIONS

103. **Movie Stunts.** According to the *Guinness Book of World Records,* stuntman Dan Koko fell a distance of 312 feet into an airbag after jumping from the Vegas World Hotel and Casino. The distance d in feet traveled by a free-falling object in t seconds is given by the formula $d = 16t^2$. To the nearest tenth of a second, how long did the fall last?

104. **Geography.** The surface area S of a sphere is given by the formula $S = 4\pi r^2$, where r is the radius of the sphere. An almanac lists the surface area of the Earth as 196,938,800 square miles. Assuming the Earth to be spherical, what is its radius to the nearest mile?

105. **Accidents.** The height h (in feet) of an object that is dropped from a height of s feet is given by the formula $h = s - 16t^2$, where t is the time the object has been falling. A 5-foot-tall woman on a sidewalk looks directly overhead and sees a window washer drop a bottle from four stories up. How long does she have to get out of the way? Round to the nearest tenth. (A story is 12 feet.)

106. **Flags.** In 1912, an order by President Taft fixed the width and length of the U.S. flag in the ratio 1 to 1.9. If 100 square feet of cloth are to be used to make a U.S. flag, estimate its dimensions to the nearest $\frac{1}{4}$ foot.

107. Automobile Engines. As the piston shown moves upward, it pushes a cylinder of a gasoline/air mixture that is ignited by the spark plug. The formula that gives the volume of a cylinder is $V = \pi r^2 h$, where r is the radius and h the height. Find the radius of the piston (to the nearest hundredth of an inch) if it displaces 47.75 cubic inches of gasoline/air mixture as it moves from its lowest to its highest point.

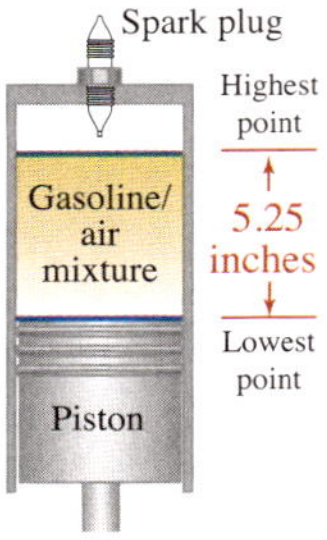

108. Investments. If P dollars are deposited in an account that pays an annual rate of interest r, then in n years, the amount of money A in the account is given by the formula $A = P(1 + r)^n$. A savings account was opened on January 3, 2006, with a deposit of \$10,000 and closed on January 2, 2008, with an ending balance of \$11,772.25. Find the rate of interest.

109. Physics. Albert Einstein discovered a connection between energy and mass. This relationship (*energy equals mass times the velocity of light squared*) is expressed in the equation $E = mc^2$. Solve for c.

110. Right Triangles. The Pythagorean theorem relates the lengths of the sides in a right triangle: $a^2 + b^2 = c^2$, where a and b represent the lengths of the legs and c represents the length of the hypotenuse. Solve for b.

WRITING

111. Give an example of a perfect-square trinomial. Why do you think the word "perfect" is used to describe it?

112. Explain why completing the square on $x^2 + 5x$ is more difficult than completing the square on $x^2 + 4x$.

113. Explain the error in the work shown below.

a. $\dfrac{4 \pm \sqrt{3}}{8} = \dfrac{\overset{1}{\cancel{4}} \pm \sqrt{3}}{\underset{1}{\cancel{4}} \cdot 2} = \dfrac{1 \pm \sqrt{3}}{2}$

b. $\dfrac{1 \pm \sqrt{5}}{5} = \dfrac{1 \pm \sqrt{\overset{1}{\cancel{5}}}}{\underset{1}{\cancel{5}}} = \dfrac{1 \pm 1}{1}$

114. Explain the steps involved in expressing $8 \pm \frac{\sqrt{15}}{2}$ as a single fraction with denominator 2.

REVIEW

Simplify each expression. All variables represent positive real numbers.

115. $\sqrt[3]{40a^3b^6}$

116. $\sqrt[8]{x^{24}}$

117. $\sqrt[4]{\dfrac{16}{625}}$

118. $\sqrt{175a^2b^3}$

CHALLENGE PROBLEMS

119. What number must be added to $x^2 + \sqrt{3}x$ to make a perfect-square trinomial?

120. Solve $x^2 + \sqrt{3}x - \frac{1}{4} = 0$ by completing the square.

SECTION 10.2 The Quadratic Formula

OBJECTIVES

1. Derive the quadratic formula.
2. Solve quadratic equations using the quadratic formula.
3. Write equivalent equations to make quadratic formula calculations easier.
4. Use the quadratic formula to solve application problems.

ARE YOU READY?

The following problems review some basic skills that are needed when solving quadratic equations using the quadratic formula.

1. Evaluate: $\sqrt{5^2 - 4(4)(-6)}$

2. Simplify: $\sqrt{45}$

3. How many terms does $2x^2 - x + 7$ have? What is the coefficient of each term?

4. Evaluate: $\dfrac{-5 \pm 11}{8}$

5. Classify $\sqrt{-28}$ as a rational, irrational, or not a real number.

6. Approximate $\dfrac{6 + \sqrt{3}}{2}$ to the nearest hundredth.

We can solve quadratic equations by completing the square, but that method is often lengthy. In this section, we will develop a formula, called the *quadratic formula,* that enables us to solve quadratic equations with less effort.

1 Derive the Quadratic Formula.

To develop a formula that will produce the solutions of any given quadratic equation, we start with a quadratic equation in **standard form**, $ax^2 + bx + c = 0$, where $a > 0$. We can solve for x by completing the square.

$$ax^2 + bx + c = 0$$

$$\frac{ax^2}{a} + \frac{bx}{a} + \frac{c}{a} = \frac{0}{a}$$ Divide both sides by a so that the coefficient of x^2 is 1.

$$x^2 + \frac{b}{a}x + \frac{c}{a} = 0$$ Simplify: $\frac{ax^2}{a} = x^2$. Write $\frac{bx}{a}$ as $\frac{b}{a}x$

$$x^2 + \frac{b}{a}x = -\frac{c}{a}$$ Subtract $\frac{c}{a}$ from both sides so that only the variable terms are on the left side of the equation and the constant is on the right side.

We can complete the square on $x^2 + \frac{b}{a}x$ by adding the square of one-half of the coefficient of x. Since the coefficient of x is $\frac{b}{a}$, we have $\frac{1}{2} \cdot \frac{b}{a} = \frac{b}{2a}$ and $\left(\frac{b}{2a}\right)^2 = \frac{b^2}{4a^2}$.

$$x^2 + \frac{b}{a}x + \frac{b^2}{4a^2} = -\frac{c}{a} + \frac{b^2}{4a^2}$$ To complete the square, add $\frac{b^2}{4a^2}$ to both sides.

$$x^2 + \frac{b}{a}x + \frac{b^2}{4a^2} = -\frac{4ac}{4aa} + \frac{b^2}{4a^2}$$ On the right side, build $-\frac{c}{a}$ by multiplying it by $\frac{4a}{4a}$. Now the fractions on that side have the common denominator $4a^2$.

$$\left(x + \frac{b}{2a}\right)^2 = \frac{b^2 - 4ac}{4a^2}$$ On the left side, factor the perfect-square trinomial. On the right side, add the fractions. In the numerator, write $-4ac + b^2$ as $b^2 - 4ac$.

$$x + \frac{b}{2a} = \pm\sqrt{\frac{b^2 - 4ac}{4a^2}}$$ Use the square root property.

$$x + \frac{b}{2a} = \pm\frac{\sqrt{b^2 - 4ac}}{\sqrt{4a^2}}$$ On the right side, the square root of a quotient is the quotient of square roots.

$$x + \frac{b}{2a} = \pm\frac{\sqrt{b^2 - 4ac}}{2a}$$ On the right side, simplify the denominator. Since $a > 0$, $\sqrt{4a^2} = 2a$.

$$x = -\frac{b}{2a} \pm \frac{\sqrt{b^2 - 4ac}}{2a}$$ To isolate x, subtract $\frac{b}{2a}$ from both sides.

$$x = \frac{-b \pm \sqrt{b^2 - 4ac}}{2a}$$ Combine the fractions. Write the sum (and difference) over the common denominator $2a$.

The Language of Algebra

To **derive** means to obtain by reasoning. To *derive* the quadratic formula means to solve $ax^2 + bx + c = 0$ for x, using the series of steps shown here, to obtain

$$x = \frac{-b \pm \sqrt{b^2 - 4ac}}{2a}$$

This result is called the **quadratic formula.** To develop this formula, we assumed that a was positive. If a is negative, similar steps are used, and we obtain the same result. This formula is very useful and should be memorized.

The Quadratic Formula

The solutions of $ax^2 + bx + c = 0$, with $a \neq 0$, are given by

$$x = \frac{-b \pm \sqrt{b^2 - 4ac}}{2a}$$ Read as "x equals the opposite of b plus or minus the square root of b squared minus $4ac$, all over $2a$."

The quadratic formula is a compact way of representing two solutions:

$$x = \frac{-b + \sqrt{b^2 - 4ac}}{2a} \quad \text{or} \quad x = \frac{-b - \sqrt{b^2 - 4ac}}{2a}$$

2 Solve Quadratic Equations Using the Quadratic Formula.

In the next example, we will use the quadratic formula to solve a quadratic equation.

EXAMPLE 1 Solve $2x^2 - 5x - 3 = 0$ by using the quadratic formula.

Strategy We will begin by comparing $2x^2 - 5x - 3 = 0$ to the standard form $ax^2 + bx + c = 0$.

Why To use the quadratic formula, we need to identify the values of a, b, and c.

Solution

$$2x^2 - 5x - 3 = 0 \quad \text{This is the equation to solve.}$$
$$\uparrow \quad \uparrow \quad \uparrow$$
$$ax^2 + bx + c = 0$$

Caution

Make sure to include the correct sign when determining a, b, and c. In this example, $b = -5$ and $c = -3$.

We see that $a = 2$, $b = -5$, and $c = -3$. To find the solutions of the equation, we substitute these values into the quadratic formula and evaluate the right side.

$$x = \frac{-b \pm \sqrt{b^2 - 4ac}}{2a} \quad \text{This is the quadratic formula.}$$

$$x = \frac{-(-5) \pm \sqrt{(-5)^2 - 4(2)(-3)}}{2(2)} \quad \text{Substitute 2 for } a, -5 \text{ for } b, \text{ and } -3 \text{ for } c.$$

$$x = \frac{5 \pm \sqrt{25 - (-24)}}{4} \quad \text{Simplify: } -(-5) = 5. \text{ Evaluate the power and multiply within the radical. Multiply in the denominator.}$$

$$x = \frac{5 \pm \sqrt{49}}{4} \quad \text{Simplify within the radical.}$$

$$x = \frac{5 \pm 7}{4} \quad \text{Evaluate the radical: } \sqrt{49} = 7.$$

Caution

When writing the quadratic formula, be careful to draw the fraction bar so that it includes the entire numerator. Do not write

$$x = -b \pm \frac{\sqrt{b^2 - 4ac}}{2a}$$

or

$$x = -b \pm \sqrt{\frac{b^2 - 4ac}{2a}}$$

To find the first solution, we evaluate the expression using the + symbol. To find the second solution, we evaluate the expression using the − symbol.

$$x = \frac{5 + 7}{4} \qquad \text{or} \qquad x = \frac{5 - 7}{4}$$
$$x = \frac{12}{4} \qquad\qquad x = \frac{-2}{4}$$
$$x = 3 \qquad\qquad x = -\frac{1}{2}$$

The solutions are 3 and $-\frac{1}{2}$ and the solution set is $\{3, -\frac{1}{2}\}$. Check each solution in the original equation.

Self Check 1 Solve $4x^2 - 7x - 2 = 0$ by using the quadratic formula.

Now Try Problem 13

To solve a quadratic equation in x using the quadratic formula, we follow these steps.

Solving a Quadratic Equation in x Using the Quadratic Formula

1. Write the equation in standard form: $ax^2 + bx + c = 0$.
2. Identify a, b, and c.
3. Substitute the values for a, b, and c in the quadratic formula and evaluate the right side to obtain the solutions.

$$x = \frac{-b \pm \sqrt{b^2 - 4ac}}{2a}$$

EXAMPLE 2 Solve: $2x^2 = -4x - 1$

Strategy We will write the equation in standard form $ax^2 + bx + c = 0$. Then we will identify the values of a, b, and c, and substitute these values into the quadratic formula.

Why The quadratic equation must be in standard form to identify the values of a, b, and c.

Solution To write the equation in standard form, we need to have all nonzero terms on the left side and 0 on the right side.

$$2x^2 = -4x - 1 \quad \text{This is the equation to solve.}$$

$$2x^2 + 4x + 1 = 0 \quad \text{To get 0 on the right side, add 4x and 1 to both sides.}$$

In the resulting equivalent equation, $a = 2$, $b = 4$, and $c = 1$.

$$x = \frac{-b \pm \sqrt{b^2 - 4ac}}{2a} \quad \text{This is the quadratic formula.}$$

$$x = \frac{-4 \pm \sqrt{4^2 - 4(2)(1)}}{2(2)} \quad \text{Substitute 2 for } a\text{, 4 for } b\text{, and 1 for } c.$$

$$x = \frac{-4 \pm \sqrt{16 - 8}}{4} \quad \text{Evaluate the expression within the radical. Multiply in the denominator.}$$

$$x = \frac{-4 \pm \sqrt{8}}{4} \quad \text{Do the subtraction within the radical.}$$

$$x = \frac{-4 \pm 2\sqrt{2}}{4} \quad \text{Simplify the radical: } \sqrt{8} = \sqrt{4 \cdot 2} = 2\sqrt{2}.$$

We can write the solutions in simpler form by factoring out 2 from the two terms in the numerator and removing the common factor of 2 in the numerator and denominator.

$$x = \frac{-4 \pm 2\sqrt{2}}{4} = \frac{2(-2 \pm \sqrt{2})}{4} = \frac{\overset{1}{\cancel{2}}(-2 \pm \sqrt{2})}{\underset{1}{\cancel{2}} \cdot 2} = \frac{-2 \pm \sqrt{2}}{2} \quad \text{Factor first.}$$

The two irrational solutions are $\frac{-2 + \sqrt{2}}{2}$ and $\frac{-2 - \sqrt{2}}{2}$ and the solution set is $\left\{\frac{-2 + \sqrt{2}}{2}, \frac{-2 - \sqrt{2}}{2}\right\}$. We can approximate the solutions using a calculator. To the nearest hundredth, they are -0.29 and -1.71.

Self Check 2 Solve $3x^2 = 2x + 3$. Approximate the solutions to the nearest hundredth.

Now Try Problem 25

Success Tip

Perhaps you noticed that Example 1 could be solved by factoring. However, that is not the case for Example 2. These observations illustrate an important fact: The quadratic formula can be used to solve *any* quadratic equation.

Notation

The solutions also can be written:

$$-\frac{2}{2} \pm \frac{\sqrt{2}}{2} = -1 \pm \frac{\sqrt{2}}{2}$$

EXAMPLE 3 Solve: $m^2 + m = -1$

Strategy We will write the equation in standard form $am^2 + bm + c = 0$. Then we will identify the values of a, b, and c, and substitute these values into the quadratic formula.

Why The quadratic equation must be in standard form to identify the values of a, b, and c.

Solution To write $m^2 + m = -1$ in standard form, we add 1 to both sides, to get

$m^2 + m + 1 = 0$ This is the equation to solve.

In the resulting equivalent equation, $a = 1$, $b = 1$, and $c = 1$:

$m = \frac{-b \pm \sqrt{b^2 - 4ac}}{2a}$ This is the quadratic formula.

$m = \frac{-1 \pm \sqrt{1^2 - 4(1)(1)}}{2(1)}$ Substitute 1 for a, 1 for b, and 1 for c.

$m = \frac{-1 \pm \sqrt{1 - 4}}{2}$ Evaluate the expression within the radical. Multiply in the denominator.

$m = \frac{-1 \pm \sqrt{-3}}{2}$ Do the subtraction within the radical.

$m = \frac{-1 \pm i\sqrt{3}}{2}$ Simplify the radical: $\sqrt{-3} = \sqrt{-1 \cdot 3} = \sqrt{-1}\sqrt{3} = i\sqrt{3}$.

The solutions are two complex numbers involving i: $-\frac{1}{2} + \frac{\sqrt{3}}{2}i$ and $-\frac{1}{2} - \frac{\sqrt{3}}{2}i$ and the solution set is $\left\{-\frac{1}{2} + \frac{\sqrt{3}}{2}i, -\frac{1}{2} - \frac{\sqrt{3}}{2}i\right\}$.

Caution

Since the variable in the equation is m, not x, we must change the variable in the quadratic formula to reflect this:

$$m = \frac{-b \pm \sqrt{b^2 - 4ac}}{2a}$$

Notation

The solutions are written in complex number form $a + bi$. They also could be written as

$$\frac{-1 \pm i\sqrt{3}}{2}$$

Self Check 3 Solve: $a^2 + 3a = -5$

Now Try Problem 29

3 Write Equivalent Equations to Make Quadratic Formula Calculations Easier.

When solving a quadratic equation by the quadratic formula, we often can simplify the calculations by solving a simpler, but equivalent equation.

EXAMPLE 4 For each equation below, write an equivalent equation so that the quadratic formula calculations will be simpler.

a. $-2x^2 + 4x - 1 = 0$ **b.** $x^2 + \frac{4}{5}x - \frac{1}{3} = 0$

c. $20x^2 - 60x - 40 = 0$ **d.** $0.03x^2 - 0.04x - 0.01 = 0$

Strategy We will multiply both sides of each equation by a carefully chosen number.

Why In each case, the objective is to find an equivalent equation whose values of a, b, and c are easier to work with than those of the given equation.

Solution **a.** It is often easier to solve a quadratic equation using the quadratic formula if a is positive. If we multiply (or divide) both sides of $-2x^2 + 4x - 1 = 0$ by -1, we obtain an equivalent equation with $a > 0$.

$-2x^2 + 4x - 1 = 0$ Here, $a = -2$.

$-1(-2x^2 + 4x - 1) = -1(0)$ Don't forget to multiply each term by -1.

$2x^2 - 4x + 1 = 0$ Now $a = 2$.

Success Tip

Unlike completing the square, the quadratic formula does not require the leading coefficient to be 1.

b. For $x^2 + \frac{4}{5}x - \frac{1}{3} = 0$, two coefficients are fractions: $b = \frac{4}{5}$ and $c = -\frac{1}{3}$. We can multiply both sides of the equation by their least common denominator, 15, to obtain an equivalent equation having coefficients that are integers.

$$x^2 + \frac{4}{5}x - \frac{1}{3} = 0 \quad \text{Here, } a = 1,\ b = \tfrac{4}{5}, \text{ and } c = -\tfrac{1}{3}.$$

$$15\left(x^2 + \frac{4}{5}x - \frac{1}{3}\right) = 15(0)$$

$$15x^2 + 12x - 5 = 0 \quad \text{On the left side, distribute the multiplication by 15.}$$

Now $a = 15$, $b = 12$, and $c = -5$.

The Language of Algebra

Recall that **equivalent equations** have the same solutions.

c. For $20x^2 - 60x - 40 = 0$, the coefficients 20, −60, and −40 have a common factor of 20. If we divide both sides of the equation by their GCF, we obtain an equivalent equation having smaller coefficients.

$$20x^2 - 60x - 40 = 0 \quad \text{Here, } a = 20,\ b = -60, \text{ and } c = -40.$$

$$\frac{20x^2}{20} - \frac{60x}{20} - \frac{40}{20} = \frac{0}{20} \quad \text{The division by 20 is done term-by-term.}$$

$$x^2 - 3x - 2 = 0 \quad \text{Now } a = 1,\ b = -3, \text{ and } c = -2.$$

d. For $0.03x^2 - 0.04x - 0.01 = 0$, all three coefficients are decimals. We can multiply both sides of the equation by 100 to obtain an equivalent equation having coefficients that are integers.

$$0.03x^2 - 0.04x - 0.01 = 0 \quad \text{Here, } a = 0.03,\ b = -0.04, \text{ and } c = -0.01.$$

$$100(0.03x^2 - 0.04x - 0.01) = 100(0)$$

$$3x^2 - 4x - 1 = 0 \quad \text{On the left side, distribute 100.}$$

Now $a = 3$, $b = -4$, and $c = -1$.

Self Check 4 For each equation, write an equivalent equation so that the quadratic formula calculations will be simpler.

a. $-6x^2 + 7x - 9 = 0$

b. $\frac{1}{3}x^2 - \frac{2}{3}x - \frac{5}{6} = 0$

c. $44x^2 + 66x - 99 = 0$

d. $0.08x^2 - 0.07x - 0.02 = 0$

Now Try Problems 37 and 39

4 Use the Quadratic Formula to Solve Application Problems.

A variety of real-world applications can be modeled by quadratic equations. However, such equations are often difficult or even impossible to solve using the factoring method. In those cases, we can use the quadratic formula to solve the equation.

EXAMPLE 5 **Shortcuts.** Instead of using the hallways, students are wearing a path through a planted quad area to walk 195 feet directly from the classrooms to the cafeteria. If the length of the hallway from the office to the cafeteria is 105 feet longer than the hallway from the office to the classrooms, how much walking are the students saving by taking the shortcut?

Analyze The two hallways and the shortcut form a right triangle with a hypotenuse 195 feet long. We will use the Pythagorean theorem to solve this problem.

Assign If we let $x =$ the length (in feet) of the hallway from the classrooms to the office, then the length of the hallway from the office to the cafeteria is $(x + 105)$ feet.

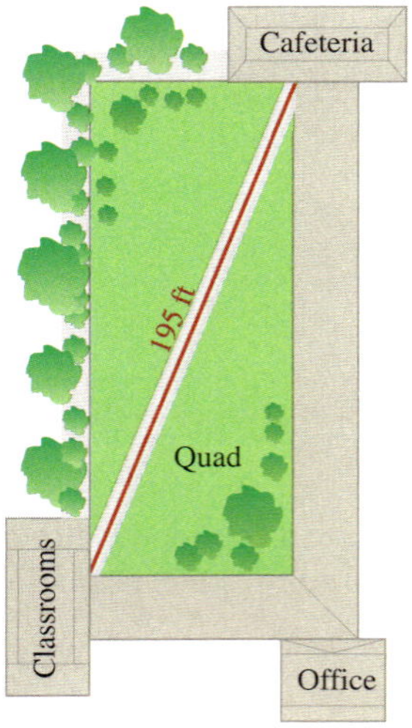

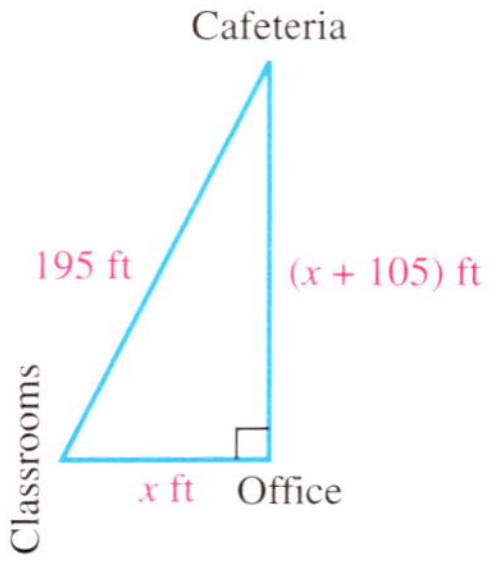

Form Substituting the lengths into the Pythagorean equation, we have

$$a^2 + b^2 = c^2 \quad \text{This is the Pythagorean equation.}$$

$$x^2 + (x + 105)^2 = 195^2 \quad \text{Substitute } x \text{ for } a, (x + 105) \text{ for } b, \text{ and 195 for } c.$$

$$x^2 + x^2 + 105x + 105x + 11{,}025 = 38{,}025 \quad \text{Find } (x + 105)^2.$$

$$2x^2 + 210x + 11{,}025 = 38{,}025 \quad \text{Combine like terms.}$$

$$2x^2 + 210x - 27{,}000 = 0 \quad \text{To get 0 on the right side, subtract 38,025 from both sides. This is a quadratic equation.}$$

$$x^2 + 105x - 13{,}500 = 0 \quad \text{The coefficients have a common factor of 2: Divide both sides by 2.}$$

Solve To solve $x^2 + 105x - 13{,}500 = 0$, we will use the quadratic formula with $a = 1$, $b = 105$, and $c = -13{,}500$.

$$x = \frac{-b \pm \sqrt{b^2 - 4ac}}{2a} \quad \text{This is the quadratic formula.}$$

$$x = \frac{-105 \pm \sqrt{105^2 - 4(1)(-13{,}500)}}{2(1)} \quad \text{Substitute for } a, b, \text{ and } c.$$

$$x = \frac{-105 \pm \sqrt{65{,}025}}{2}$$

Evaluate within the radical: $105^2 - 4(1)(-13{,}500) = 11{,}025 + 54{,}000 = 65{,}025$. Multiply in the denominator.

$$x = \frac{-105 \pm 255}{2} \quad \text{Use a calculator: } \sqrt{65{,}025} = 255.$$

$$x = \frac{150}{2} \quad \text{or} \quad x = \frac{-360}{2} \quad \text{Add: } -105 + 255 = 150. \text{ Subtract: } -105 - 255 = -360.$$

$$x = 75 \quad \bigg| \quad \cancel{x = -180}$$

Do the division. Since the length of the hallway can't be negative, discard the solution −180.

State The length of the hallway from the classrooms to the office is 75 feet. The length of the hallway from the office to the cafeteria is $75 + 105 = 180$ feet. Instead of using the hallways, a distance of $75 + 180 = 255$ feet, the students are taking the 195-foot shortcut to the cafeteria, a savings of $(255 - 195)$, or 60 feet.

Check The length of the 180-foot hallway is 105 feet longer than the length of the 75-foot hallway. The sum of the squares of the lengths of the hallways is $75^2 + 180^2 = 38{,}025$. This equals the square of the length of the 195-foot shortcut: $195^2 = 38{,}025$. The result checks.

Self Check 5 **Right Triangles.** The hypotenuse of a right triangle is 41 in. long. The longer leg is 31 inches longer than the shorter leg. Find the lengths of the legs of the triangle.

Now Try Problem 81

EXAMPLE 6 **Mass Transit.** A bus company has 4,000 passengers daily, each currently paying a 75¢ fare. For each 15¢ fare increase, the company estimates that it will lose 50 passengers. If the company needs to bring in $6,570 per day to stay in business, what fare must be charged to produce this amount of revenue?

Analyze To understand how a fare increase affects the number of passengers, let's consider what happens if there are two fare increases. We organize the data in a table. The fares are expressed in terms of dollars.

Number of increases	New fare	Number of passengers
One \$0.15 increase	\$0.75 + \$0.15(1) = \$0.90	4,000 − 50(1) = 3,950
Two \$0.15 increases	\$0.75 + \$0.15(2) = \$1.05	4,000 − 50(2) = 3,900

In general, the new fare will be the old fare (\$0.75) plus the number of fare increases times \$0.15. The number of passengers who will pay the new fare is 4,000 minus 50 times the number of \$0.15 fare increases.

Assign If we let $x =$ the number of \$0.15 fare increases necessary to bring in \$6,570 daily, then $\$(0.75 + 0.15x)$ is the fare that must be charged. The number of passengers who will pay this fare is $4,000 - 50x$.

Form We can now form an equation, using the words of the problem.

The bus fare	times	the number of passengers who will pay that fare	equals	\$6,570.
$(0.75 + 0.15x)$	$\cdot$	$(4,000 - 50x)$	$=$	$6,570$

Solve

$$(0.75 + 0.15x)(4,000 - 50x) = 6,570$$

$$3,000 - 37.5x + 600x - 7.5x^2 = 6,570$$ Multiply the binomials on the left side.

$$-7.5x^2 + 562.5x + 3,000 = 6,570$$ Combine like terms: $-37.5x + 600x = 562.5x$.

$$-7.5x^2 + 562.5x - 3,570 = 0$$ To get 0 on the right side, subtract 6,570 from both sides. This is a quadratic equation.

$$7.5x^2 - 562.5x + 3,570 = 0$$ Multiply both sides by −1 so that the value of *a*, 7.5, is positive.

To solve this equation, we will use the quadratic formula.

$$x = \frac{-b \pm \sqrt{b^2 - 4ac}}{2a}$$ This is the quadratic formula.

$$x = \frac{-(-562.5) \pm \sqrt{(-562.5)^2 - 4(7.5)(3,570)}}{2(7.5)}$$ Substitute 7.5 for *a*, −562.5 for *b*, and 3,570 for *c*.

$$x = \frac{562.5 \pm \sqrt{209,306.25}}{15}$$ Evaluate within the radical: $(-562.5)^2 - 4(7.5)(3,570) = 316,406.25 - 107.100 = 209,306.25$. Multiply in the denominator.

$$x = \frac{562.5 \pm 457.5}{15}$$ Use a calculator: $\sqrt{209,306.25} = 457.5$.

$$x = \frac{1,020}{15} \quad \text{or} \quad x = \frac{105}{15}$$ Add: $562.5 + 457.5 = 1,020$. Subtract: $562.5 - 457.5 = 105$.

$$x = 68 \quad | \quad x = 7$$ Do the division.

State If there are 7 fifteen-cent increases in the fare, the new fare will be \$0.75 + \$0.15(7) = \$1.80. If there are 68 fifteen-cent increases in the fare, the new fare will be \$0.75 + \$0.15(68) = \$10.95. Although this fare would bring in the necessary revenue, a \$10.95 bus fare is unreasonable, so we discard it.

Check A fare of \$1.80 will be paid by $[4{,}000 - 50(7)] = 3{,}650$ bus riders. The amount of revenue brought in would be $\$1.80(3{,}650) = \$6{,}570$. The result checks.

Self Check 6 **Airport Shuttles.** A bus company shuttles 1,120 passengers daily between Rockford, Illinois, and O'Hare Airport. The current one-way fare is \$10. For each 25¢ increase in the fare, the company predicts that it will lose 48 passengers. What increase in fare will produce daily revenue of \$10, 208?

Now Try Problem 91

EXAMPLE 7 **Lawyers.** The number of lawyers in the United States is approximated by the function $N(x) = 45x^2 + 21{,}000x + 560{,}000$, where $N(x)$ is the number of lawyers and x is the number of years after 1980. In what year does this model indicate that the United States had one million lawyers? (Based on data from the American Bar Association)

Strategy We will substitute 1,000,000 for $N(x)$ in the equation and solve for x.

Why The value of x will give the number of years after 1980 that the United States had 1,000,000 lawyers.

Solution

$$N(x) = 45x^2 + 21{,}000x + 560{,}000$$ This is the quadratic function model.

$$1{,}000{,}000 = 45x^2 + 21{,}000x + 560{,}000$$ Replace N(x) with 1,000,000.

$$0 = 45x^2 + 21{,}000x - 440{,}000$$ To get 0 on the left side, subtract 1,000,000 from both sides.

We can simplify the calculations by dividing both sides of the equation by 5, which is the greatest common factor of 45, 21,000, and 440,000.

$$9x^2 + 4{,}200x - 88{,}000 = 0$$ Divide both sides by 5.

We solve this equation using the quadratic formula.

$$x = \frac{-b \pm \sqrt{b^2 - 4ac}}{2a}$$

$$x = \frac{-4{,}200 \pm \sqrt{(4{,}200)^2 - 4(9)(-88{,}000)}}{2(9)}$$ Substitute 9 for a, 4,200 for b, and −88,000 for c.

$$x = \frac{-4{,}200 \pm \sqrt{20{,}808{,}000}}{18}$$ Evaluate the expression within the radical. Multiply in the denominator.

$$x \approx \frac{362}{18} \quad \text{or} \quad x \approx \frac{-8{,}762}{18}$$ Use a calculator to evaluate each numerator.

$$x \approx 20.1 \qquad\qquad x \approx -486.8$$ Do the division. Since the model is defined for only positive values of x, we discard the second solution.

In 20.1 years after 1980, or in early 2000, the model predicts that the United States had approximately 1,000,000 lawyers.

Self Check 7 **Lawyers.** See Example 7. In what year does the model indicate that the United States had three-quarters of a million lawyers?

Now Try Problem 95

EXAMPLE 8 **Graduation Announcements.** To create the announcement shown, a graphic artist must follow two design requirements.

- A border of uniform width should surround the text.
- Equal areas should be devoted to the text and to the border.

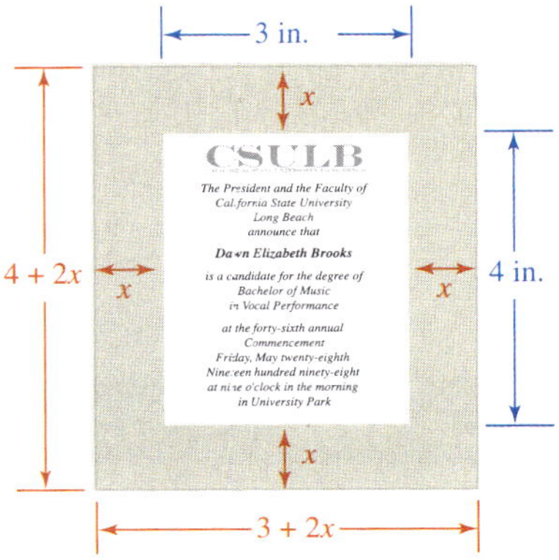

To meet these requirements, how wide should the border be?

Analyze The text occupies $4 \cdot 3 = 12$ in.^{2} of space. The border must also have an area of 12 in.^{2}.

Assign If we let x = the width of the border in inches, the length of the announcement is $(4 + 2x)$ inches and the width is $(3 + 2x)$ inches.

Form We can now form an equation. Recall that the area of a rectangle is the product of its length and width.

The area of the announcement	minus	the area of the text	equals	the area of the border.
$(4 + 2x)(3 + 2x)$	$-$	12	$=$	12

Solve

$$(4 + 2x)(3 + 2x) - 12 = 12$$

$$12 + 8x + 6x + 4x^2 - 12 = 12 \quad \text{On the left side, multiply the binomials.}$$

$$4x^2 + 14x = 12 \quad \text{Combine like terms. This is a quadratic equation.}$$

$$4x^2 + 14x - 12 = 0 \quad \text{To get 0 on the right side, subtract 12 from both sides.}$$

$$2x^2 + 7x - 6 = 0 \quad \text{The coefficients have a common factor of 2. Divide both sides by 2.}$$

To solve this equation, we will use the quadratic formula with $a = 2$, $b = 7$, and $c = -6$.

$$x = \frac{-b \pm \sqrt{b^2 - 4ac}}{2a}$$

$$x = \frac{-7 \pm \sqrt{7^2 - 4(2)(-6)}}{2(2)} \quad \text{Substitute for } a, b, \text{ and } c.$$

$$x = \frac{-7 \pm \sqrt{97}}{4} \quad \text{Evaluate within the radical: } 7^2 - 4(2)(-6) = 49 + 48 = 97. \text{ Multiply in the denominator.}$$

$$x \approx \frac{-7 + \sqrt{97}}{4} \quad \text{or} \quad x \approx \frac{-7 - \sqrt{97}}{4} \quad \text{These are the two exact solutions.}$$

$$x \approx 0.71 \qquad\qquad x \approx -4.21$$

State The width of the border should be about 0.71 inch. (We discard the solution $\frac{-7 - \sqrt{97}}{4}$ since it is negative.)

Check If the border is 0.71 inch wide, the announcement has an area of about $5.42 \cdot 4.42 \approx 23.96$ in.^{2}. If we subtract the area of the text from the area of the announcement, we get $23.96 - 12 = 11.96$ in.^{2}. This represents the area of the border, which was to be 12 in.^{2}. The result seems reasonable.

Self Check 8 **Graduation Announcements.** See Example 8. Find the width of the border if the text occupies an area 3 inches by 5 inches.

Now Try Problem 97

SECTION 10.2 STUDY SET

VOCABULARY

Fill in the blanks.

1. The standard form of a ________ equation is $ax^2 + bx + c = 0$.
2. $x = \dfrac{-b \pm \sqrt{b^2 - 4ac}}{2a}$ is called the ________ formula.

CONCEPTS

3. Write each quadratic equation in standard form.
   a. $x^2 + 2x = -5$ b. $3x^2 = -2x + 1$
4. For each quadratic equation, find the values of a, b, and c.
   a. $x^2 + 5x + 6 = 0$ b. $8x^2 - x = 10$
5. Determine whether each statement is true or false.
   a. Any quadratic equation can be solved by using the quadratic formula.
   b. Any quadratic equation can be solved by completing the square.
   c. Any quadratic equation can be solved by factoring using integers.
6. What is wrong with the beginning of the solution shown below?

   Solve: $x^2 - 3x = 2$

   $a = 1 \quad b = -3 \quad c = 2$
7. Evaluate each expression.
   a. $\dfrac{-2 \pm \sqrt{2^2 - 4(1)(-8)}}{2(1)}$
   b. $\dfrac{-(-1) \pm \sqrt{(-1)^2 - 4(2)(-4)}}{2(2)}$
8. A student used the quadratic formula to solve a quadratic equation and obtained $x = \frac{-2 \pm \sqrt{3}}{2}$.
   a. How many solutions does the equation have? What are they exactly?
   b. Graph the solutions on a number line.
9. Simplify each of the following.
   a. $\dfrac{3 \pm 6\sqrt{2}}{3}$ b. $\dfrac{-12 \pm 4\sqrt{7}}{8}$
10. a. Write an expression that represents the width of the larger rectangle shown in red.
    b. Write an expression that represents the length of the larger rectangle shown in red.

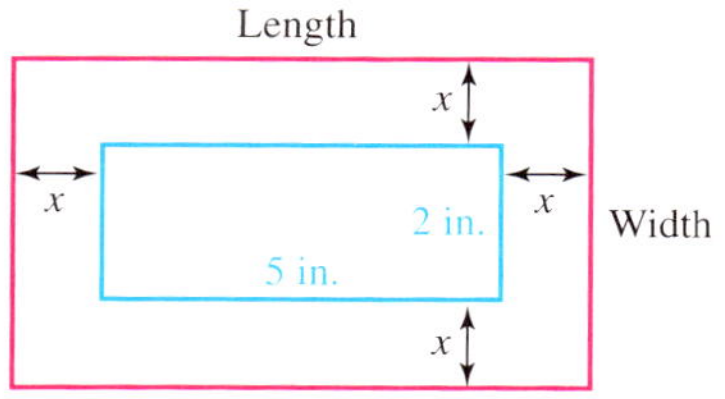

NOTATION

11. On a quiz, students were asked to write the quadratic formula. What is wrong with each answer shown below?
    a. $x = -b \pm \dfrac{\sqrt{b^2 - 4ac}}{2a}$
    b. $x = \dfrac{-b\sqrt{b^2 - 4ac}}{2a}$
12. When reading $\frac{-b \pm \sqrt{b^2 - 4ac}}{2a}$, we say, "The ________ of b, plus or ______ the square ______ of b ________ minus ___ times a times c, all ______________ $2a$."

GUIDED PRACTICE

Use the quadratic formula to solve each equation. See Example 1.

13. $x^2 - 3x + 2 = 0$
14. $x^2 + 3x + 2 = 0$
15. $x^2 + 12x = -36$
16. $x^2 - 18x + 81 = 0$
17. $2x^2 + x - 3 = 0$
18. $6x^2 - x - 1 = 0$
19. $12t^2 - 5t - 2 = 0$
20. $12z^2 + 5z - 3 = 0$

Solve each equation. Approximate the solutions to the nearest hundredth. See Example 2.

21. $x^2 = x + 7$
22. $t^2 = t + 4$
23. $5x^2 + 5x = -1$
24. $2x^2 + 7x = -1$

25. $3y^2 + 1 = -6y$

26. $4w^2 + 1 = -6w$

27. $4m^2 = 4m + 19$

28. $3y^2 = 12y - 4$

Solve each equation. See Example 3.

29. $2x^2 + x + 1 = 0$

30. $2x^2 + 3x + 5 = 0$

31. $3x^2 - 2x + 1 = 0$

32. $3x^2 - 2x + 5 = 0$

33. $x^2 - 2x + 2 = 0$

34. $x^2 - 4x + 8 = 0$

35. $4a^2 + 4a + 5 = 0$

36. $4b^2 + 4b + 17 = 0$

For each equation, write an equivalent quadratic equation that will be easier to solve. Do not solve the equation. See Example 4.

37. a. $-5x^2 + 9x - 2 = 0$
    b. $1.6t^2 + 2.4t - 0.9 = 0$

38. a. $\frac{1}{8}x^2 + \frac{1}{2}x - \frac{3}{4} = 0$
    b. $33y^2 + 99y - 66 = 0$

39. a. $45x^2 + 30x - 15 = 0$
    b. $\frac{1}{3}m^2 - \frac{1}{2}m - \frac{1}{3} = 0$

40. a. $0.6t^2 - 0.1t - 0.2 = 0$
    b. $-a^2 - 15a + 12 = 0$

TRY IT YOURSELF

Solve each equation. Approximate the solutions to the nearest hundredth when appropriate.

41. $x^2 - \frac{14}{15}x = \frac{8}{15}$

42. $x^2 = -\frac{5}{4}x + \frac{3}{2}$

43. $3x^2 - 4x = -2$

44. $2x^2 + 3x = -3$

45. $-16y^2 - 8y + 3 = 0$

46. $-16x^2 - 16x - 3 = 0$

47. $2x^2 - 3x - 1 = 0$

48. $3x^2 - 9x - 2 = 0$

49. $-x^2 + 10x = 18$

50. $-x^2 - 6x - 2 = 0$

51. $x(x - 6) = 391$

52. $x(x - 27) = 280$

53. $x^2 + 5x - 5 = 0$

54. $x^2 - 3x - 27 = 0$

55. $9h^2 - 6h + 7 = 0$

56. $5x^2 = 2x - 1$

57. $50x^2 + 30x - 10 = 0$

58. $120b^2 + 120b - 40 = 0$

59. $0.6x^2 + 0.03 - 0.4x = 0$

60. $2x^2 + 0.1x = 0.04$

61. $\frac{1}{8}x^2 - \frac{1}{2}x + 1 = 0$

62. $\frac{1}{2}x^2 + 3x + \frac{13}{2} = 0$

63. $\frac{a^2}{10} - \frac{3a}{5} + \frac{7}{5} = 0$

64. $\frac{c^2}{4} + c + \frac{11}{4} = 0$

65. $\frac{x^2}{2} + \frac{5}{2}x = -1$

66. $\frac{x^2}{8} - \frac{x}{4} = \frac{1}{2}$

67. $900x^2 - 8{,}100x = 1{,}800$

68. $14x^2 - 21x = 49$

69. $\frac{1}{4}x^2 - \frac{1}{6}x - \frac{1}{6} = 0$

70. $81x^2 + 12x - 80 = 0$

71. Let $f(x) = 0.7x^2 - 3.5x$ For what value(s) of x is $f(x) = 25$?

72. Let $g(x) = 4.5x^2 + 0.2x$. For what value(s) of x is $g(x) = 3.75$?

Look Alikes . . .

73. a. $a^2 + 4a - 7 = 0$
    b. $a^2 - 4a - 7 = 0$

74. a. $n^2 + 6n - 2 = 0$
    b. $n^2 + 6n + 2 = 0$

75. a. $(x + 2)(x - 4) = 16$
    b. $(x + 2)(x - 4) = -16$

76. a. $5n^2 - 14n - 3 = 0$
    b. $-5n^2 - 14n - 3 = 0$

77. a. $x^2 - 42x + 441 = 0$
    b. $x^2 + 42x + 441 = 0$

78. a. $0.3y^2 + 0.6y + 0.5 = 0$
    b. $0.003y^2 + 0.006y + 0.005 = 0$

APPLICATIONS

79. **Crosswalks.** Refer to the illustration below. Instead of using the Main Street and First Avenue crosswalks to get from Nordstrom to Best Buy, a shopper uses the diagonal crosswalk to walk 97 feet directly from one corner to the other. If the length of the Main Street crosswalk is 7 feet longer than the First Avenue crosswalk, how much walking does the shopper save by using the diagonal crosswalk?

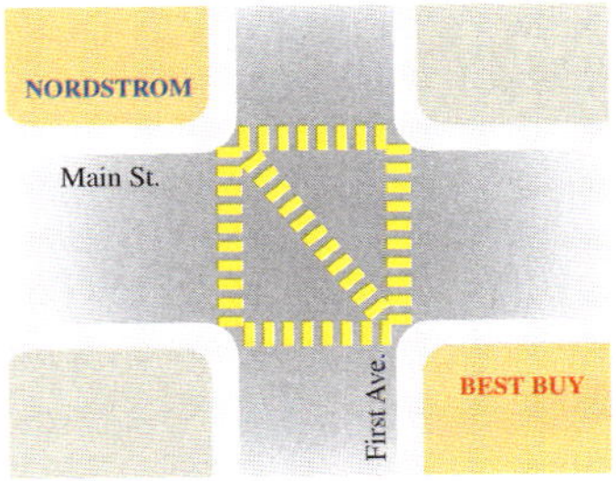

80. Badminton. The person who wrote the instructions for setting up the badminton net shown below forgot to give the specific dimensions for securing the pole. How long is the support string?

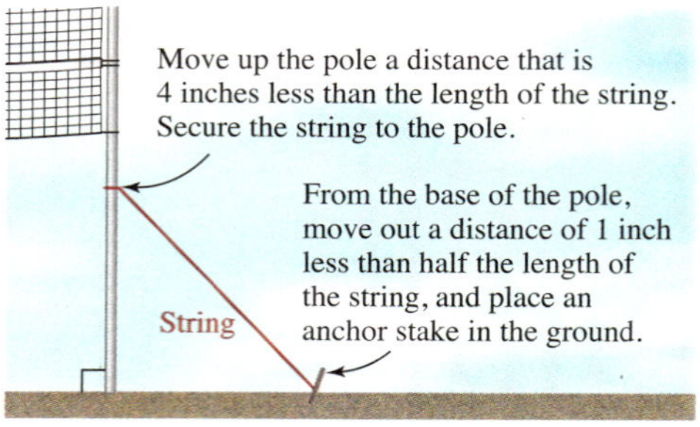

81. Right Triangles. The hypotenuse of a right triangle is 2.5 units long. The longer leg is 1.7 units longer than the shorter leg. Find the lengths of the legs of the triangle.

82. Televisions. The screen size of a television is measured diagonally from one corner to the opposite corner. In 2007, Sharp developed the world's largest TV screen to date—a 108-inch flat-panel liquid crystal display (LCD). Find the width and the height of the rectangular screen if the width were 41 inches greater than the height. Round to the nearest inch.

83. IMAX Screens. The largest permanent movie screen is in the Panasonic Imax theater at Darling Harbor, Sydney, Australia. The rectangular screen has an area of 11,349 square feet. Find the dimensions of the screen if it is 20 feet longer than it is wide.

84. World's Largest LED Screen. A huge suspended LED screen is the centerpiece of The Place, a popular mall in Beijing, China. Find the length and width of the rectangular screen if the length is 10 meters more than 8 times its width, and the viewable area is 7,500 square meters.

©China Photos/Alamy

85. Parks. Central Park is one of New York's best-known landmarks. Rectangular in shape, its length is 5 times its width. When measured in miles, its perimeter numerically exceeds its area by 4.75. Find the dimensions of Central Park if we know that its width is less than 1 mile.

86. History. One of the important cities of the ancient world was Babylon. Greek historians wrote that the city was square. Measured in miles, its area numerically exceeded its perimeter by about 124. Find its dimensions. (Round to the nearest tenth.)

87. Polygons. A five-sided polygon, called a *pentagon,* has 5 diagonals. The number of diagonals d of a polygon of n sides is given by the formula $d = \frac{n(n-3)}{2}$. Find the number of sides of a polygon if it has 275 diagonals.

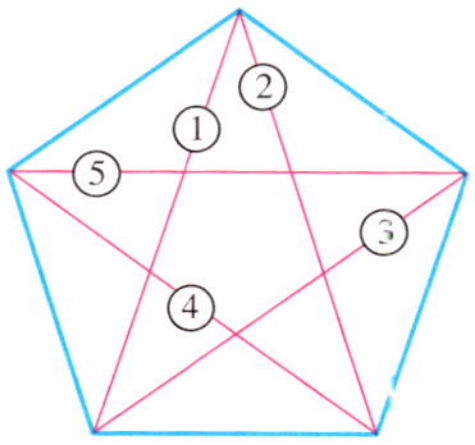

88. Metal Fabrication. A box with no top is to be made by cutting a 2-inch square from each corner of a square sheet of metal. After bending up the sides, the volume of the box is to be 220 cubic inches. Find the length of a side of the square sheet of metal that should be used in the construction of the box. Round to the nearest hundredth.

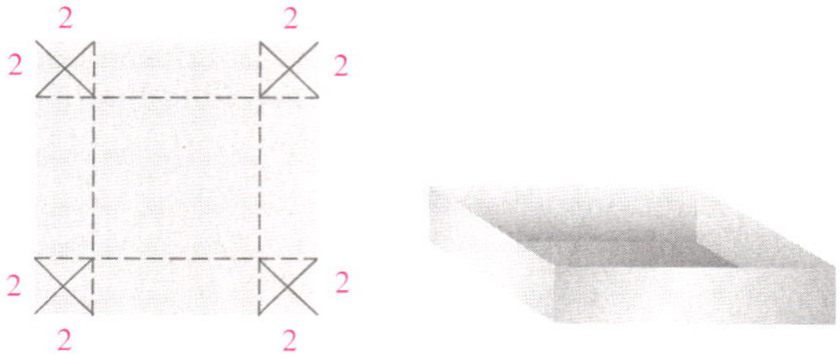

89. Dances. Tickets to a school dance cost \$4 and the projected attendance is 300 people. It is further projected that for every 10¢ increase in ticket price, the average attendance will decrease by 5. At what ticket price will the receipts from the dance be \$1,248?

90. Ticket Sales. A carnival usually sells three thousand 75¢ ride tickets on a Saturday. For each 15¢ increase in price, management estimates that 80 fewer tickets will be sold. What increase in ticket price will produce \$2,982 of revenue on Saturday?

91. Magazine Sales. The *Gazette's* profit is \$20 per year for each of its 3,000 subscribers. Management estimates that the profit per subscriber will increase by 1¢ for each additional subscriber over the current 3,000. How many subscribers will bring a total profit of \$120,000?

92. Investment Rates. A woman invests \$1,000 in a fund for which interest is compounded annually at a rate r. After one year, she deposits an additional \$2,000. After two years, the balance in the account is $\$1,000(1 + r)^2 + \$2,000(1 + r)$. If this amount is \$3,368.10, find r.

93. from **Campus to Careers**

Police Patrol Officer

© Robert E Daemmrich/Getty Images

The number of female police officers (sworn) in the United States is approximated by the function $f(t) = -24t^2 + 1{,}534t + 72{,}065$, where t is the number of years after 2000. In what year does the model indicate that the number of female officers reached 75,000? (Source: *Crime in the United States, 2000–2010*)

94. Shopping Centers. The number of shopping centers in the United States is approximated by the function $s(t) = 48t^2 + 581t + 77{,}383$, where t is the number of years after 1990. In what year does the model indicate that the number of shopping centers reached 100,000? (Source: U.S. Census Bureau)

95. Picture Framing. The matting around the picture has a uniform width. How wide is the matting if its area equals the area of the picture? Round to the nearest hundredth of an inch.

96. Swimming Pools. In the advertisement shown, how wide will the free concrete decking be if a uniform width is constructed around the perimeter of the pool? Round to the nearest hundredth of a yard. (*Hint:* Note the difference in units. Convert the dimensions of the pool to yards.)

97. Dimensions of Rectangle. A rectangle is 4 feet longer than it is wide, and its area is 20 square feet. Find its dimensions to the nearest tenth of a foot.

98. Dimensions of a Triangle. The height of a triangle is 4 meters longer than twice its base. Find the base and height if the area of the triangle is 10 square meters. Round to the nearest hundredth of a meter.

WRITING

99. Explain why the quadratic formula, in most cases, is easier to use to solve a quadratic equation than is the method of completing the square.

100. On an exam, a student was asked to solve the equation $-4w^2 - 6w - 1 = 0$. Her first step was to multiply both sides of the equation by -1. She then used the quadratic formula to solve $4w^2 + 6w + 1 = 0$ instead. Is this a valid approach? Explain.

REVIEW

Change each radical to an exponential expression.

101. $\sqrt{n}$

102. $\sqrt[7]{8r^2s}$

103. $\sqrt[4]{3b}$

104. $3\sqrt[3]{c^2 - d^2}$

Write each expression in radical form.

105. $t^{1/3}$

106. $(3m^2n^2)^{1/5}$

107. $(3t)^{1/4}$

108. $(c^2 + d^2)^{1/2}$

CHALLENGE PROBLEMS

All of the equations we have solved so far have had rational-number coefficients. However, the quadratic formula can be used to solve quadratic equations with irrational or even imaginary coefficients. Solve each equation.

109. $x^2 + 2\sqrt{2}x - 6 = 0$

110. $\sqrt{2}x^2 + x - \sqrt{2} = 0$

111. $x^2 - 3ix - 2 = 0$

112. $100ix^2 + 300x - 200i = 0$

SECTION 10.3 The Discriminant and Equations That Can Be Written in Quadratic Form

OBJECTIVES

1. Use the discriminant to determine number and type of solutions.
2. Solve equations that are quadratic in form.
3. Solve application problems using quadratic equations.

ARE YOU READY?

The following problems review some basic skills that are needed when working with discriminants and equations that can be written in quadratic form.

1. Find the value of the expression within the radical symbol only: $\dfrac{6 \pm \sqrt{(-6)^2 - 4(3)(-4)}}{6}$

2. Fill in the blanks: **a.** $x^4 = (x^{\square})^{\square}$ **b.** $(x^{1/3})^{\square} = x$

3. Fill in the blanks: **a.** $(\sqrt{x})^2 = \square$ **b.** $15a^{-2} = \dfrac{15}{\square}$

4. Solve: $x^2 = -4$

5. Solve: **a.** $\sqrt{x} = 3$ **b.** $\sqrt[3]{x} = -\dfrac{1}{2}$

6. Solve: $\dfrac{1}{a} = \dfrac{1}{5}$

We have seen that solutions of the quadratic equation $ax^2 + bx + c = 0$ with $a \neq 0$ are given by the formula

$$x = \frac{-b \pm \sqrt{b^2 - 4ac}}{2a}$$

In this section, we will examine the radicand within the quadratic formula to distinguish or "*discriminate*" among the three types of solutions—rational, irrational, or imaginary.

1 Use the Discriminant to Determine Number and Type of Solutions.

The expression $b^2 - 4ac$ that appears under the radical symbol in the quadratic formula is called the **discriminant.** The discriminant can be used to predict what kind of solutions a quadratic equation has without solving it.

The Discriminant

For a quadratic equation of the form $ax^2 + bx + c = 0$ with rational-number coefficients and $a \neq 0$, the expression $b^2 - 4ac$ is called the **discriminant** and can be used to determine the number and type of the solutions of the equation.

Discriminant: $b^2 - 4ac$	*Number and type of solutions*
Positive	Two different real numbers
0	One repeated solution, a rational number
Negative	Two different imaginary numbers that are complex conjugates

Discriminant: $b^2 - 4ac$	*Number and type of solutions*
A perfect square	Two different rational numbers
Positive and not a perfect square	Two different irrational numbers

EXAMPLE 1 Determine the number and type of solutions for each equation:

a. $x^2 + x + 1 = 0$ **b.** $3x^2 + 5x + 2 = 0$

Strategy We will identify the values of a, b, and c in each equation. Then we will use those values to compute $b^2 - 4ac$, the discriminant.

Why Once we know whether the discriminant is positive, 0, or negative, and whether it is a perfect square, we can determine the number and type of the solutions of the equation.

Solution

a. For $x^2 + x + 1 = 0$, the discriminant is:

$$b^2 - 4ac = 1^2 - 4(1)(1) \quad \text{Substitute: } a = 1, b = 1, \text{ and } c = 1.$$
$$= -3 \quad \text{The result is a negative number.}$$

Since $b^2 - 4ac < 0$, the solutions of $x^2 + x + 1 = 0$ are two different imaginary numbers that are complex conjugates.

b. For $3x^2 + 5x + 2 = 0$, the discriminant is:

$$b^2 - 4ac = 5^2 - 4(3)(2) \quad \text{Substitute: } a = 3, b = 5, \text{ and } c = 2.$$
$$= 25 - 24$$
$$= 1 \quad \text{The result is a positive number.}$$

Since $b^2 - 4ac > 0$ and $b^2 - 4ac$ is a perfect square, the solutions of $3x^2 + 5x + 2 = 0$ are two different rational numbers.

Success Tip

The discriminant also can be used to determine factorability. The trinomial $ax^2 + bx + c$ with integer coefficients and $a \neq 0$ will factor as the product of two binomials with integer coefficients if the value of $b^2 - 4ac$ is a perfect square. If $b^2 - 4ac = 0$, the factors will be the same.

Self Check 1 Determine the number and type of solutions for:

a. $x^2 + x - 1 = 0$

b. $3x^2 + 4x + 2 = 0$

Now Try Problems 11, 13, and 15

2 Solve Equations That Are Quadratic in Form.

We have discussed four methods that are used to solve quadratic equations. The table below shows some advantages and disadvantages of each method.

Method	Advantages	Disadvantages	Examples
Factoring and the zero-factor property	It can be very fast. When each factor is set equal to 0, the resulting equations are usually easy to solve.	Some polynomials may be difficult to factor and others impossible.	$x^2 - 2x - 24 = 0$ $4a^2 - a = 0$
Square root property	It is the fastest way to solve equations of the form $ax^2 = n$ (n = a number) or $(ax + b)^2 = n$.	It applies only to equations that are in these forms.	$x^2 = 27$ $(2y + 3)^2 = 25$
Completing the square*	It can be used to solve any quadratic equation. It works well with equations of the form $x^2 + bx = n$, where b is even.	It involves more steps than the other methods. The algebra can be cumbersome if the leading coefficient is not 1 or if b is odd or a fraction.	$x^2 + 4x = -1$ $t^2 - 14t - 9 = 0$
Quadratic formula	It can be used to solve any quadratic equation.	It involves several calculations in which sign errors can be made. Often the result must be simplified.	$x^2 + 3x - 33 = 0$ $4s^2 - 10s + 5 = 0$

*The quadratic formula is just a condensed version of completing the square and is usually easier to use. However, you need to know how to complete the square because it is used in more advanced mathematics courses.

Success Tip

The solutions of a quadratic equation can always be found using the quadratic formula. The solutions cannot always be found using factoring.

To determine the most efficient method for a given equation, we can use the following strategy.

Strategy for Solving Quadratic Equations

1. See whether the equation is in a form such that the **square root method** is easily applied.
2. If step 1 does not apply, write the equation in $ax^2 + bx + c = 0$ form.
3. See whether the equation can be solved using the **factoring method.**
4. If you can't factor, solve the equation by the **quadratic formula. Completing the square** can also be used.

Many nonquadratic equations can be written in quadratic form ($ax^2 + bx + c = 0$) and solved using the techniques discussed in previous sections. For example, a careful inspection of the equation $x^4 - 5x^2 + 4 = 0$ leads to the following observations:

$$x^4 - 5x^2 + 4 = 0$$

The leading term, x^4, is the square of the expression x^2 in the middle term: $x^4 = (x^2)^2$.

The last term is a constant.

Equations that contain an expression, the same expression squared, and a constant term are said to be **quadratic in form**. One method used to solve such equations is to make a substitution.

EXAMPLE 2 Solve: $x^4 - 3x^2 - 4 = 0$

Strategy Since the leading term, x^4, is the square of the expression x^2 in the middle term, we will substitute y for x^2.

Why Our hope is that such a substitution will produce an equation that we can solve using one of the methods previously discussed.

Solution If we write x^4 as $(x^2)^2$, the equation takes the form

$$(x^2)^2 - 3x^2 - 4 = 0$$

and it is said to be **quadratic in x^2**. We can solve this equation by letting $y = x^2$.

$$y^2 - 3y - 4 = 0 \quad \text{Replace each } x^2 \text{ with } y.$$

We can solve the resulting quadratic equation by factoring.

$(y - 4)(y + 1) = 0$ — Factor $y^2 - 3y - 4$.

$y - 4 = 0$ or $y + 1 = 0$ — Set each factor equal to 0.

$y = 4$ | $y = -1$ — Solve for y.

These are not the solutions for x. To find x, we reverse the substitution by replacing each y with x^2 and proceed as follows:

$x^2 = 4$ or $x^2 = -1$ — Undo the substitution. Substitute x^2 for y.

$x = \pm\sqrt{4}$ | $x = \pm\sqrt{-1}$ — Use the square root property. Don't forget $\pm$.

$x = \pm 2$ | $x = \pm i$ — Simplify each radical.

This equation has four solutions: 2, -2, i, and $-i$. Check each one in the original equation. In general, a fourth-degree polynomial equation, such as this, can have up to four distinct solutions.

Self Check 2 Solve: $x^4 - 5x^2 - 36 = 0$

Now Try Problem 23

Notation

The choice of the letter y for the substitution is arbitrary. We could just as well let $b = x^2$.

Caution

If you are solving an equation in x, you can't answer with values of y. Remember to reverse (undo) any substitutions, and solve for the variable in the original equation.

EXAMPLE 3 Solve: $x - 7\sqrt{x} + 12 = 0$

Strategy Since the leading term, x, is the square of the expression $\sqrt{x}$ in the middle term, we will substitute y for $\sqrt{x}$.

Why Our hope is that such a substitution will produce an equation that we can solve using one of the methods previously discussed.

Solution We examine the leading term and the middle term.

The leading term, x, is the square of the expression $\sqrt{x}$ in middle term: $x = (\sqrt{x})^2$.

$$x - 7\sqrt{x} + 12 = 0$$

The last term is a constant.

If we write x as $(\sqrt{x})^2$, the equation takes the form

$$(\sqrt{x})^2 - 7\sqrt{x} + 12 = 0$$

and it is said to be **quadratic in $\sqrt{x}$**. We can solve this equation by letting $y = \sqrt{x}$ and factoring.

$y^2 - 7y + 12 = 0$ — Replace each $\sqrt{x}$ with y. This is a quadratic equation.

$(y - 3)(y - 4) = 0$ — Factor the trinomial.

$y - 3 = 0$ or $y - 4 = 0$ — Set each factor equal to 0.

$y = 3$ | $y = 4$ — Solve for y.

The Language of Algebra

Equations such as $x - 7\sqrt{x} + 12 = 0$ that are quadratic in form are also said to be **reducible** to a quadratic.

Success Tip

Recall from Chapter 9 that we must check each proposed solution of a radical equation and eliminate any extraneous solutions.

To find x, we reverse the substitution and replace each y with $\sqrt{x}$. Then we solve the resulting radical equations by squaring both sides.

$$\sqrt{x} = 3 \quad \text{or} \quad \sqrt{x} = 4 \qquad \text{Undo the substitution.}$$
$$x = 9 \qquad\qquad x = 16$$

The solutions are 9 and 16. Check each solution in the original equation.

Self Check 3 Solve: $x + \sqrt{x} - 6 = 0$

Now Try Problem 27

EXAMPLE 4 Solve: $2m^{2/3} - 2 = 3m^{1/3}$

Strategy We will write the equation in descending powers of m and look for a possible substitution to make.

Why Our hope is that a substitution will produce an equation that we can solve using one of the methods previously discussed.

Solution After writing the equation in descending powers of m, we see that

$$2m^{2/3} - 3m^{1/3} - 2 = 0$$

The Language of Algebra

Recall that in $2m^{2/3}$, we call 2/3 a **rational** or **fractional exponent.**

is **quadratic in $m^{1/3}$**, because $m^{2/3} = (m^{1/3})^2$. We will use the substitution $y = m^{1/3}$ to write this equation in quadratic form.

$$2m^{2/3} - 3m^{1/3} - 2 = 0 \qquad \text{This is the equation to solve.}$$
$$2(m^{1/3})^2 - 3m^{1/3} - 2 = 0 \qquad \text{Write } m^{2/3} \text{ as } (m^{1/3})^2.$$
$$2y^2 - 3y - 2 = 0 \qquad \text{Substitute: Replace each } m^{1/3} \text{ with } y. \text{ This is a quadratic equation.}$$
$$(2y + 1)(y - 2) = 0 \qquad \text{Factor } 2y^2 - 3y - 2.$$
$$2y + 1 = 0 \quad \text{or} \quad y - 2 = 0 \qquad \text{Set each factor equal to 0.}$$
$$y = -\frac{1}{2} \qquad\qquad y = 2 \qquad \text{Solve for } y.$$

To find m, we reverse the substitution and replace each y with $m^{1/3}$. Then we solve the resulting equations by cubing both sides.

Success Tip

We could write $m^{1/3} = -\frac{1}{2}$ in the equivalent radical form $\sqrt[3]{m} = -\frac{1}{2}$ and solve in the same way—by cubing both sides.

$$m^{1/3} = -\frac{1}{2} \quad \text{or} \quad m^{1/3} = 2 \qquad \text{Undo the substitution.}$$
$$(m^{1/3})^3 = \left(-\frac{1}{2}\right)^3 \qquad (m^{1/3})^3 = (2)^3 \qquad \text{Recall that } m^{1/3} = \sqrt[3]{m}. \text{ To solve for } m, \text{ cube both sides.}$$
$$m = -\frac{1}{8} \qquad\qquad m = 8 \qquad \text{Find each power.}$$

The solutions are $-\frac{1}{8}$ and 8. Check each solution in the original equation.

Self Check 4 Solve: $a^{2/3} = -3a^{1/3} + 10$

Now Try Problem 31

EXAMPLE 5 Solve: $(4t + 2)^2 - 30(4t + 2) + 224 = 0$

Strategy Since the leading term, $(4t + 2)^2$, is the square of the expression $4t + 2$ in the middle term, we will substitute y for $4t + 2$.

Why Our hope is that such a substitution will produce an equation that we can solve using one of the methods previously discussed.

Solution This equation is **quadratic in $4t + 2$**. If we make the substitution $y = 4t + 2$, we obtain

$$y^2 - 30y + 224 = 0$$

which can be solved by using the quadratic formula.

$$y = \frac{-b \pm \sqrt{b^2 - 4ac}}{2a}$$

$$y = \frac{-(-30) \pm \sqrt{(-30)^2 - 4(1)(224)}}{2(1)}$$ Substitute 1 for *a*, −30 for *b*, and 224 for *c*.

$$y = \frac{30 \pm \sqrt{4}}{2}$$ Evaluate within the radical.

$$y = \frac{30 \pm 2}{2}$$ Evaluate the radical: $\sqrt{4} = 2$.

$y = 16$ or $y = 14$ Evaluate: $\frac{30+2}{2} = 16$ and $\frac{30-2}{2} = 14$

To find t, we reverse the substitution and replace y with $4t + 2$. Then we solve for t.

$4t + 2 = 16$ or $4t + 2 = 14$ Undo the substitution.

$4t = 14$ | $4t = 12$ Isolate the variable term, $4t$.

$t = 3.5$ | $t = 3$ Solve for *t*.

Verify that 3.5 and 3 satisfy the original equation.

Self Check 5 Solve: $(n + 3)^2 - 6(n + 3) = -8$

Now Try Problem 35

EXAMPLE 6 Solve: $15a^{-2} - 8a^{-1} + 1 = 0$

Strategy We will write the equation with positive exponents and look for a possible substitution to make.

Why Our hope is that a substitution will produce an equation that we can solve using one of the methods previously discussed.

Solution When we write the terms $15a^{-2}$ and $-8a^{-1}$ using positive exponents, we see that this equation is **quadratic in $\frac{1}{a}$**.

Success Tip

We could also solve this equation by multiplying both sides by the LCD, a^2.

$$\frac{15}{a^2} - \frac{8}{a} + 1 = 0$$ Think of this equation as $15 \cdot \left(\frac{1}{a}\right)^2 - 8 \cdot \frac{1}{a} + 1 = 0$. Note that 0 is not a possible solution.

If we let $y = \frac{1}{a}$, the resulting quadratic equation can be solved by factoring.

$15y^2 - 8y + 1 = 0$ Substitute y^2 for $\frac{1}{a^2}$ and y for $\frac{1}{a}$. This is a quadratic equation.

$(5y - 1)(3y - 1) = 0$ Factor $15y^2 - 8y + 1$.

$5y - 1 = 0$ or $3y - 1 = 0$ Set each factor equal to 0.

$y = \frac{1}{5}$ | $y = \frac{1}{3}$ Solve for *y*.

Success Tip

Recall that we use cross products to solve proportions.

$$\frac{1}{a} \times \frac{1}{5}$$

$$5 = a$$

To find a, we reverse the substitution and replace each y with $\frac{1}{a}$. Then we proceed as follows:

$$\frac{1}{a} = \frac{1}{5} \quad \text{or} \quad \frac{1}{a} = \frac{1}{3} \qquad \text{Undo the substitution.}$$

$$5 = a \qquad\qquad 3 = a \qquad \text{Solve the proportions by finding the cross products.}$$

The solutions are 5 and 3. Check each solution in the original equation.

Self Check 6 Solve: $28c^{-2} - 3c^{-1} - 1 = 0$

Now Try ▶ Problem 41

3 Solve Application Problems Using Quadratic Equations.

EXAMPLE 7

Household Appliances. The illustration shows a temperature control on a washing machine. When the *warm* setting is selected, both the hot and cold water pipes open to fill the tub in 2 minutes 15 seconds. When the *cold* setting is chosen, the tub fills 45 seconds faster than when the *hot* setting is used. How long does it take to fill the washing machine if the *hot* setting is used?

Electronic Temperature Control

Water Temp

Analyze It is helpful to organize the facts of this shared-work problem in a table. We note that the hot and cold water inlets will be open for the same time: 2 minutes 15 seconds, or 135 seconds.

Assign Let x = the number of seconds it takes to fill the tub with hot water. Since the cold water inlet fills the tub in 45 seconds less time, $x - 45$ = the number of seconds it takes to fill the tub with cold water.

Form To determine the work completed by each inlet, multiply the rate by the time.

	Rate	· Time	= Work completed
Hot water	$\frac{1}{x}$	135	$\frac{135}{x}$
Cold water	$\frac{1}{x-45}$	135	$\frac{135}{x-45}$

Enter this information first. Multiply to get each of these entries: $W = rt$.

In shared-work problems, 1 represents one whole job completed. So we have

The fraction of tub filled with hot water	plus	the fraction of the tub filled with cold water	equals	1 tub filled.
$\frac{135}{x}$	$+$	$\frac{135}{x-45}$	$=$	1

Success Tip

An alternate way to form an equation is to note that what the hot water inlet can do in 1 second plus what the cold water inlet can do in 1 second equals what they can do together in 1 second:

$$\frac{1}{x} + \frac{1}{x-45} = \frac{1}{135}$$

Solve

$$\frac{135}{x} + \frac{135}{x - 45} = 1$$ This is a rational equation.

$$x(x - 45)\left(\frac{135}{x} + \frac{135}{x - 45}\right) = x(x - 45)(1)$$ Multiply both sides by the LCD $x(x - 45)$ to clear the equation of fractions.

$$x(x - 45)\frac{135}{x} + x(x - 45)\frac{135}{x - 45} = x(x - 45)(1)$$ Distribute $x(x - 45)$.

$$135(x - 45) + 135x = x(x - 45)$$ Simplify each side.

$$135x - 6{,}075 + 135x = x^2 - 45x$$ Distribute 135 and x.

$$270x - 6{,}075 = x^2 - 45x$$ Combine like terms.

$$0 = x^2 - 315x + 6{,}075$$ Get 0 on the left side.

To solve for x, we will use the quadratic formula: $a = 1$, $b = -315$, and $c = 6{,}075$.

$$x = \frac{-b \pm \sqrt{b^2 - 4ac}}{2a}$$

$$x = \frac{-(-315) \pm \sqrt{(-315)^2 - 4(1)(6{,}075)}}{2(1)}$$ Substitute 1 for a, -315 for b, and 6,075 for c.

$$x = \frac{315 \pm \sqrt{74{,}925}}{2}$$ Evaluate the expression within the radical.

$x \approx 294$ or $x \approx 21$ (crossed out) Use a calculator to find each solution.

State We can discard the solution of 21 seconds, because this would imply that the cold water inlet fills the tub in a negative number of seconds $(21 - 45 = -24)$. Therefore, the hot water inlet fills the washing machine tub in about 294 seconds, which is 4 minutes 54 seconds.

Check Use estimation to check the result. The work completed by the hot water inlet is $\frac{135}{294} \approx 0.46$ and the work completed by the cold water inlet is $\frac{135}{294 - 45} \approx 0.54$. Since $0.46 + 0.54 \approx 1$, the result seems reasonable.

Self Check 7 **Mowing Lawns.** Carly can mow a lawn in 1 hour less time than her friend Lindsey. Together they can finish the job in 5 hours. How long would it take Carly if she worked alone?

Now Try Problem 87

SECTION 10.3 STUDY SET

VOCABULARY

Fill in the blanks.

1. For the quadratic equation $ax^2 + bx + c = 0$, the ____________ is $b^2 - 4ac$.
2. We can solve $x - 2\sqrt{x} - 8 = 0$ by making the ____________ $y = \sqrt{x}$.

CONCEPTS

Consider the quadratic equation $ax^2 - bx + c = 0$, where a, b, and c represent rational numbers, and fill in the blanks.

3. If $b^2 - 4ac < 0$, the solutions of the equation are two different imaginary numbers that are complex ____________.
4. If $b^2 - 4ac =$, the equation has one repeated rational-number solution.

5. If $b^2 - 4ac$ is a perfect square, the solutions of the equation are two different ________ numbers.
6. If $b^2 - 4ac$ is positive and not a perfect square, the solutions of the equation are two different ________ numbers.
7. For each equation, determine the substitution that should be made to write the equation in quadratic form.
   a. $x^4 - 12x^2 + 27 = 0$ Let $y =$
   b. $x - 13\sqrt{x} + 40 = 0$ Let $y =$
   c. $x^{2/3} + 2x^{1/3} - 3 = 0$ Let $y =$
   d. $x^{-2} - x^{-1} - 30 = 0$ Let $y =$
   e. $(x + 1)^2 - (x + 1) - 6 = 0$ Let $y =$
8. Fill in the blanks.
   a. $x^4 = (\quad)^2$ b. $x = (\quad)^2$
   c. $x^{2/3} = (\quad)^2$ d. $\frac{1}{x^2} = (\quad)^2$

NOTATION

Complete the solution.

9. To find the type of solutions for the equation $x^2 + 5x + 6 = 0$, we calculate the discriminant.

$$b^2 - \quad = \quad ^2 - 4(1)(\quad)$$
$$= 25 - $$
$$= 1$$

Since a, b, and c are rational numbers and the value of the discriminant is a perfect square, the solutions are two different ________ numbers.

10. Fill in the blanks to write each equation in quadratic form.
   a. $x^4 - 2x^2 - 15 = 0 \rightarrow (\quad)^2 - 2 \quad - 15 = 0$
   b. $x - 2\sqrt{x} + 3 = 0 \rightarrow (\quad)^2 - 2 \quad + 3 = 0$
   c. $8m^{2/3} - 10m^{1/3} - 3 = 0 \rightarrow 8(\quad)^2 - 10 \quad - 3 = 0$

GUIDED PRACTICE

Use the discriminant to determine the number and type of solutions for each equation. Do not solve. See Example 1.

11. $4x^2 - 4x + 1 = 0$
12. $6x^2 - 5x - 6 = 0$
13. $5x^2 + x + 2 = 0$
14. $3x^2 + 10x - 2 = 0$
15. $2x^2 = 4x - 1$
16. $9x^2 = 12x - 4$
17. $x(2x - 3) = 20$
18. $x(x - 3) = -10$
19. $3x^2 - 10 = 0$
20. $5x^2 - 24 = 0$
21. $x^2 - \frac{14}{15}x = \frac{8}{15}$
22. $x^2 = -\frac{5}{4}x + \frac{3}{2}$

Solve each equation. See Example 2.

23. $x^4 - 17x^2 + 16 = 0$
24. $x^4 - 10x^2 + 9 = 0$
25. $x^4 + 5x^2 - 36 = 0$
26. $x^4 - 15x^2 - 16 = 0$

Solve each equation. See Example 3.

27. $x - 13\sqrt{x} + 40 = 0$
28. $x - 9\sqrt{x} + 18 = 0$
29. $2x + \sqrt{x} - 3 = 0$
30. $2x - \sqrt{x} - 1 = 0$

Solve each equation. See Example 4.

31. $a^{2/3} - 2a^{1/3} = 3$
32. $r^{2/3} + 4r^{1/3} = 5$
33. $x^{2/3} + 2x^{1/3} - 8 = 0$
34. $x^{2/3} - 7x^{1/3} + 12 = 0$

Solve each equation. See Example 5.

35. $(c + 1)^2 - 4(c + 1) + 3 = 0$
36. $(a - 5)^2 - 4(a - 5) - 21 = 0$
37. $2(2x + 1)^2 - 7(2x + 1) + 6 = 0$
38. $3(2 - x)^2 + 10(2 - x) - 8 = 0$

Solve each equation. See Example 6.

39. $m^{-2} + m^{-1} - 6 = 0$
40. $t^{-2} + t^{-1} - 42 = 0$
41. $8x^{-2} - 10x^{-1} - 3 = 0$
42. $2x^{-2} - 5x^{-1} - 3 = 0$

Solve each equation. See Example 7.

43. $1 - \frac{5}{x} = \frac{10}{x^2}$
44. $1 - \frac{3}{x} = \frac{5}{x^2}$
45. $\frac{1}{2} + \frac{1}{b} = \frac{1}{b - 7}$
46. $\frac{1}{4} - \frac{1}{n} = \frac{1}{n + 3}$

TRY IT YOURSELF

Solve each equation.

47. $2x - \sqrt{x} = 3$
48. $3x + 4\sqrt{x} = 4$
49. $x^{-2} + 2x^{-1} - 3 = 0$
50. $x^{-2} + 2x^{-1} - 8 = 0$
51. $x^4 + 19x^2 + 18 = 0$
52. $t^4 + 4t^2 - 5 = 0$
53. $(k - 7)^2 + 6(k - 7) + 10 = 0$
54. $(d + 9)^2 - 4(d + 9) + 8 = 0$
55. $\frac{2}{x - 1} + \frac{1}{x + 1} = 3$
56. $\frac{3}{x - 2} - \frac{1}{x + 2} = 5$
57. $x - 6x^{1/2} = -8$
58. $x - 5x^{1/2} + 4 = 0$
59. $(y^2 - 9)^2 + 2(y^2 - 9) - 99 = 0$
60. $(a^2 - 4)^2 - 4(a^2 - 4) - 32 = 0$

61. $x^{-4} - 2x^{-2} + 1 = 0$

62. $4x^{-4} + 1 = 5x^{-2}$

63. $t^4 + 3t^2 = 28$

64. $3h^4 + h^2 - 2 = 0$

65. $2x^{2/5} - 5x^{1/5} = -3$

66. $2x^{2/5} + 3x^{1/5} = -1$

67. $9\left(\frac{3m + 2}{m}\right)^2 - 30\left(\frac{3m + 2}{m}\right) + 25 = 0$

68. $4\left(\frac{c - 7}{c}\right)^2 - 12\left(\frac{c - 7}{c}\right) + 9 = 0$

69. $\frac{3}{a - 1} = 1 - \frac{2}{a}$

70. $1 + \frac{4}{x} = \frac{3}{x^2}$

71. $\left(8 - \sqrt{a}\right)^2 + 6\left(8 - \sqrt{a}\right) - 7 = 0$

72. $\left(10 - \sqrt{t}\right)^2 - 4\left(10 - \sqrt{t}\right) - 45 = 0$

73. $x + \frac{2}{x - 2} = 0$

74. $x + \frac{x + 5}{x - 3} = 0$

75. $3x + 5\sqrt{x} + 2 = 0$

76. $3x - 4\sqrt{x} + 1 = 0$

77. $x^4 - 6x^2 + 5 = 0$

78. $2x^4 - 26x^2 + 24 = 0$

79. $8(t + 1)^{-2} - 30(t + 1)^{-1} + 7 = 0$

80. $2(s - 2)^{-2} + 3(s - 2)^{-1} - 5 = 0$

81. $\frac{1}{x + 2} + \frac{24}{x + 3} = 13$

82. $\frac{3}{x} + \frac{4}{x + 1} = 2$

Look Alikes . . .

83. a. $y^{2/3} + y^{1/3} - 20 = 0$ b. $y^{-2} + y^{-1} - 20 = 0$

84. a. $x + 6\sqrt{x} - 16 = 0$ b. $x^{2/3} + 6x^{1/3} - 16 = 0$

85. a. $\frac{1}{x} = \frac{2x}{x + 1}$ b. $\frac{1}{x} + \frac{2x}{x + 1} = 1$

86. a. $(m^2 - 1)^2 - (m^2 - 1) = 2$ b. $m^4 - m^2 = 2$

APPLICATIONS

87. from **Campus to Careers**

Police Patrol Officer

Crowd Control. After a sporting event at a stadium, police have found that a public parking lot can be emptied in 60 minutes if both the east and west exits are opened. If just the east exit is used, it takes 40 minutes longer to clear the lot than it does if just the west exit is opened. How long does it take to clear the parking lot if every car must use the west exit? Round to the nearest minute.

88. **Paper Routes.** When a father, in a car, and his son, on a bicycle, work together to distribute the morning newspaper, it takes them 35 minutes to complete the route. Working alone, it takes the son 25 minutes longer than the father. To the nearest minute, how long does it take the son to cover the route on his bicycle?

89. **Assembly Lines.** A newly manufactured product traveled 300 feet on a high-speed conveyor belt at a rate of r feet per second. It could have traveled the 300 feet in 3 seconds less time if the speed of the conveyor belt was increased by 5 feet per second. Find r.

90. **Bicycling.** Tina bicycles 160 miles at the rate of r mph. The same trip would have taken 2 hours longer if she had decreased her speed by 4 mph. Find r.

91. **Architecture.** A **golden rectangle** is one of the most visually appealing of all geometric forms. The Parthenon, built by the Greeks in the 5th century B.C., fits into a golden rectangle if its ruined triangular pediment is included. See the illustration. In a golden rectangle, the length l and width w must satisfy the equation $\frac{l}{w} = \frac{w}{l - w}$. If a rectangular billboard is to have a width of 20 feet, what should its length be so that it is a golden rectangle? Round to the nearest tenth.

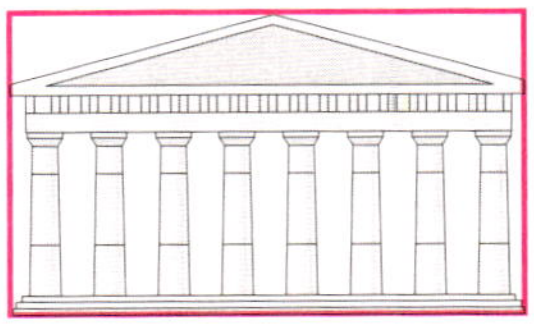

92. **Door Designs.** An architect needs to determine the height h of the window shown in the illustration. The radius r, the width w, and the height h of the circular-shaped window are related by the formula $r = \frac{4h^2 + w^2}{8h}$. If w is to be 34 inches and r is to be 18 inches, find h to the nearest tenth of an inch.

WRITING

93. Describe how to predict what type of solutions the equation $3x^2 - 4x + 5 = 0$ will have.

94. What **error** is made in the following solution?

Solve: $x^4 - 12x^2 + 27 = 0$

$y^2 - 12y + 27 = 0$ Let $y = x^2$.

$(y - 9)(y - 3) = 0$

$y - 9 = 0$ or $y - 3 = 0$

$y = 9$ | $y = 3$

The solutions of $x^4 - 12x^2 + 27 = 0$ are 9 and 3.

REVIEW

95. Write an equation of the vertical line that passes through (3, 4).

96. Find an equation for a linear function whose graph passes through (−1, −6) and (−2, −1).

97. Write an equation of the line with slope $\frac{2}{3}$ that passes through the origin.

98. Find an equation of the line that passes through $(2, -3)$ and is perpendicular to the line whose equation is $y = \frac{x}{5} + 6$. Write the equation in slope–intercept form.

CHALLENGE PROBLEMS

99. Find the real-number solutions of $x^4 - 3x^2 - 2 = 0$. Rationalize the denominators of the solutions.

100. Solve: $x^6 + 17x^3 + 16 = 0$

101. Solve: $x^3 - x^2 + 16x - 16 = 0$

102. Solve: $\sqrt{x^2 + 10} = 4\sqrt{x}$

SECTION 10.4 Quadratic Functions and Their Graphs

OBJECTIVES

1. Graph functions of the form $f(x) = ax^2$ and $f(x) = ax^2 + k$.
2. Graph functions of the form $f(x) = a(x - h)^2$ and $f(x) = a(x - h)^2 + k$.
3. Graph functions of the form $f(x) = ax^2 + bx + c$ by completing the square.
4. Find the vertex using $-\frac{b}{2a}$.
5. Determine minimum and maximum values.
6. Solve quadratic equations graphically.

ARE YOU READY?

The following problems review some basic skills that are needed when graphing quadratic functions.

1. Graph: $f(x) = x^2$
2. Let $f(x) = 2(x - 3)^2 - 4$. Find $f(4)$.
3. Complete the square on $x^2 + 8x$. Then factor the resulting trinomial.
4. Complete the square on $x^2 + x$. Then factor the resulting trinomial.
5. Solve: $x^2 + 6x + 9 = 0$
6. Solve: $-2x^2 - 8x - 8 = 0$
7. Evaluate $-\frac{b}{2a}$ for $a = -2$ and $b = -20$.
8. Is the graph of $x = -1$ a horizontal or a vertical line?

In this section, we will discuss methods for graphing *quadratic functions.*

Quadratic Functions

A **quadratic function** is a second-degree polynomial function that can be written in the form

$$f(x) = ax^2 + bx + c$$

where a, b, and c are real numbers and $a \neq 0$.

Quadratic functions often are written in another form, called **standard form,**

$$f(x) = a(x - h)^2 + k$$

where a, h, and k are real numbers and $a \neq 0$. This form is useful because a, h, and k give us important information about the graph of the function. To develop a strategy for graphing quadratic functions written in standard form, we will begin by considering the simplest case, $f(x) = ax^2$.

Notation

Since $y = f(x)$, quadratic functions can also be written as $y = ax^2 + bx + c$ and $y = a(x - h)^2 + k$.

1 Graph Functions of the Form $f(x) = ax^2$ and $f(x) = ax^2 + k$.

One way to graph quadratic functions is to plot points.

EXAMPLE 1 Graph: **a.** $f(x) = x^2$ **b.** $g(x) = 3x^2$ **c.** $s(x) = \frac{1}{3}x^2$

Strategy We can make a table of values for each function, plot each point, and connect them with a smooth curve.

Why At this time, this method is our only option.

Solution After graphing each curve, we see that the graph of $g(x) = 3x^2$ is narrower than the graph of $f(x) = x^2$, and the graph of $s(x) = \frac{1}{3}x^2$ is wider than the graph of $f(x) = x^2$. For $f(x) = ax^2$, the smaller the value of $|a|$, the wider the graph.

$f(x) = x^2$

x	$f(x)$
-2	4
-1	1
0	0
1	1
2	4

$g(x) = 3x^2$

x	$g(x)$
-2	12
-1	3
0	0
1	3
2	12

$s(x) = \frac{1}{3}x^2$

x	$s(x)$
-2	$\frac{4}{3}$
-1	$\frac{1}{3}$
0	0
1	$\frac{1}{3}$
2	$\frac{4}{3}$

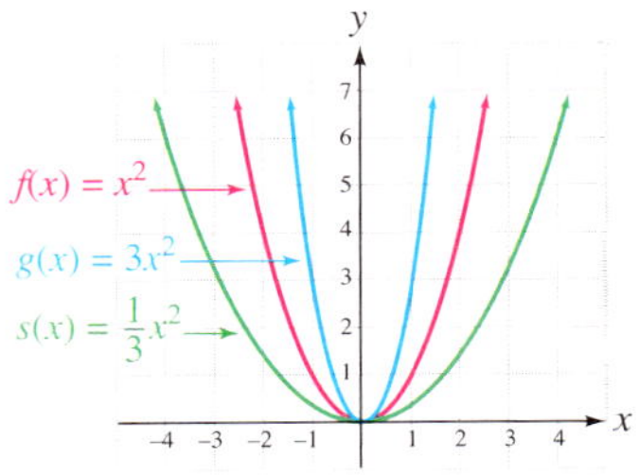

The values of $g(x)$ increase faster than the values of $f(x)$, making its graph steeper.

The values of $s(x)$ increase more slowly than the values of $f(x)$, making its graph flatter.

Self Check 1 Graph: $g(x) = \frac{2}{3}x^2$.

Now Try Problem 15

EXAMPLE 2 Graph: $f(x) = -3x^2$

Strategy We make a table of values for the function, plot each point, and connect them with a smooth curve.

Why At this time, this method is our only option.

Solution After graphing the curve, we see that it opens downward and has the same shape as the graph of $g(x) = 3x^2$ that was graphed in Example 1.

$f(x) = -3x^2$

x	$f(x)$	
-2	-12	→ $(-2, -12)$
-1	-3	→ $(-1, -3)$
0	0	→ $(0, 0)$
1	-3	→ $(1, -3)$
2	-12	→ $(2, -12)$

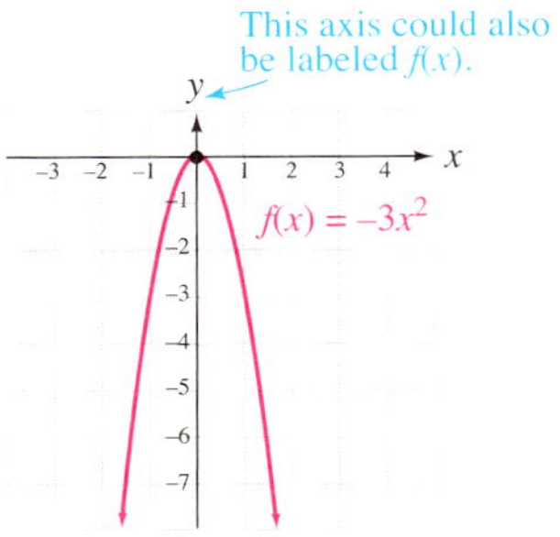

Self Check 2 Graph: $f(x) = -\frac{1}{3}x^2$

Now Try Problem 17

The graphs of functions of the form $f(x) = ax^2$ are **parabolas.** The lowest point on a parabola that opens upward, or the highest point on a parabola that opens downward, is called the **vertex** of the parabola. The vertical line, called an **axis of symmetry,** that passes through the vertex divides the parabola into two congruent halves. If we fold the paper along the axis of symmetry, the two sides of the parabola will match.

The Language of Algebra

An axis of symmetry divides a parabola into two matching sides. The sides are said to be **mirror images** of each other.

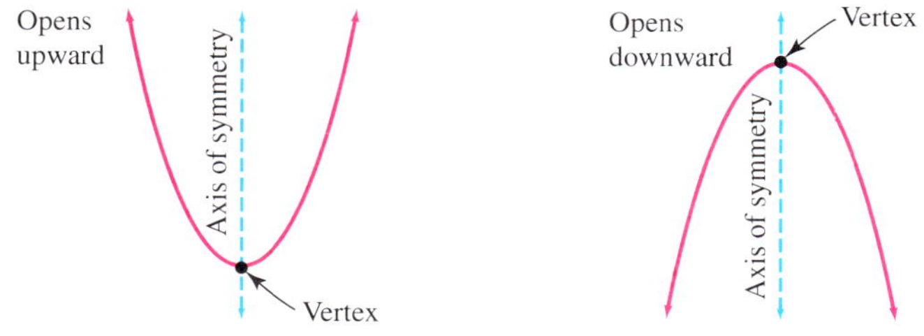

The results from Examples 1 and 2 confirm the following facts.

The Graph of $f(x) = ax^2$

The graph of $f(x) = ax^2$ is a parabola opening upward when $a > 0$ and downward when $a < 0$, with vertex at the point $(0, 0)$ and axis of symmetry the line $x = 0$.

EXAMPLE 3 Graph: **a.** $f(x) = 2x^2$ **b.** $g(x) = 2x^2 + 3$ **c.** $s(x) = 2x^2 - 3$

Strategy We make a table of values for each function, plot each point, and connect them with a smooth curve.

Why At this time, this method is our only option.

Solution After graphing the curves, we see that the graph of $g(x) = 2x^2 + 3$ is identical to the graph of $f(x) = 2x^2$, except that it has been translated 3 units upward. The graph of $s(x) = 2x^2 - 3$ is identical to the graph of $f(x) = 2x^2$, except that it has been translated 3 units downward. In each case, the axis of symmetry is the line $x = 0$.

$f(x) = 2x^2$

x	$f(x)$
−2	8
−1	2
0	0
1	2
2	8

$g(x) = 2x^2 + 3$

x	$g(x)$
−2	11
−1	5
0	3
1	5
2	11

$s(x) = 2x^2 - 3$

x	$s(x)$
−2	5
−1	−1
0	−3
1	−1
2	5

For each x-value, g(x) is 3 more than f(x).

For each x-value, s(x) is 3 less than f(x).

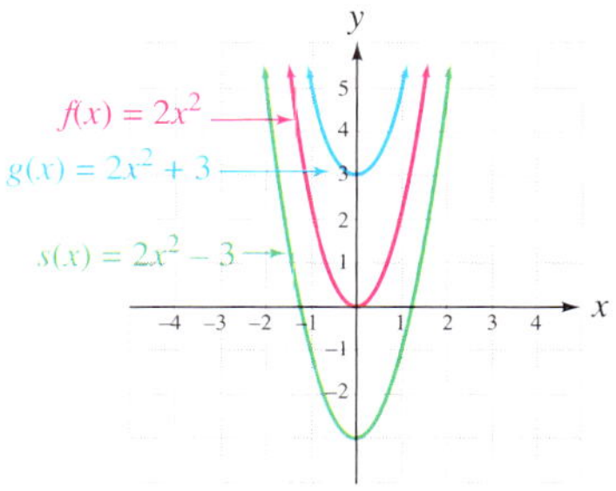

Self Check 3 Graph: $f(x) = 2x^2 + 1$

Now Try Problem 19

The results of Example 3 confirm the following facts.

The Graph of $f(x) = ax^2 + k$

The graph of $f(x) = ax^2 + k$ is a parabola having the same shape as $f(x) = ax^2$ but translated k units upward if k is positive and $|k|$ units downward if k is negative. The vertex is at the point $(0, k)$, and the axis of symmetry is the line $x = 0$.

2 Graph Functions of the Form $f(x) = a(x - h)^2$ and $f(x) = a(x - h)^2 + k$.

EXAMPLE 4 Graph: **a.** $f(x) = 2x^2$ **b.** $g(x) = 2(x - 3)^2$ **c.** $s(x) = 2(x + 3)^2$

Strategy We make a table of values for each function, plot each point, and connect them with a smooth curve.

Why At this time, this method is our only option.

Solution We note that the graph of $g(x) = 2(x - 3)^2$ below is identical to the graph of $f(x) = 2x^2$, except that it has been translated 3 units to the right. The graph of $s(x) = 2(x + 3)^2$ is identical to the graph of $f(x) = 2x^2$, except that it has been translated 3 units to the left.

$f(x) = 2x^2$

x	$f(x)$
−2	8
−1	2
0	0
1	2
2	8

$g(x) = 2(x - 3)^2$

x	$g(x)$
1	8
2	2
3	0
4	2
5	8

$s(x) = 2(x + 3)^2$

x	$s(x)$
−5	8
−4	2
−3	0
−2	2
−1	8

When an x-value is increased by 3, the function's outputs are the same.

When an x-value is decreased by 3, the function's outputs are the same.

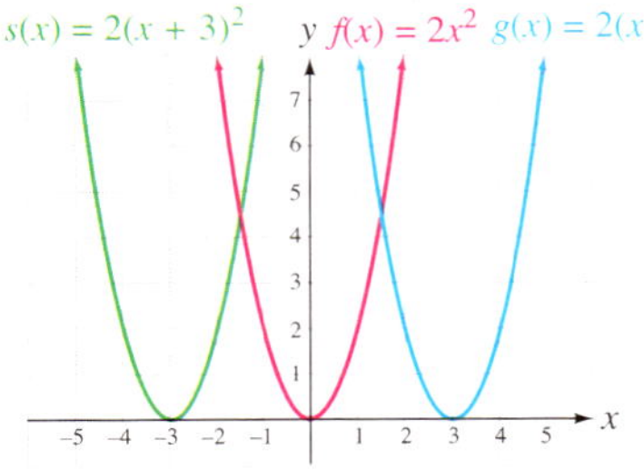

Self Check 4 Graph: $g(x) = 2(x + 1)^2$

Now Try Problem 21

The results of Example 4 confirm the following facts.

The Graph of $f(x) = a(x - h)^2$

The graph of $f(x) = a(x - h)^2$ is a parabola having the same shape as $f(x) = ax^2$ but translated h units to the right if h is positive and $|h|$ units to the left if h is negative. The vertex is at the point $(h, 0)$, and the axis of symmetry is the line $x = h$.

The results of Examples 1–4 suggest a general strategy for graphing quadratic functions that are written in the form $f(x) = a(x - h)^2 + k$.

Graphing a Quadratic Function in Standard Form

The graph of the quadratic function $f(x) = a(x - h)^2 + k$, where $a \neq 0$, is a parabola with vertex at (h, k). The axis of symmetry is the line $x = h$. The parabola opens upward when $a > 0$ and downward when $a < 0$.

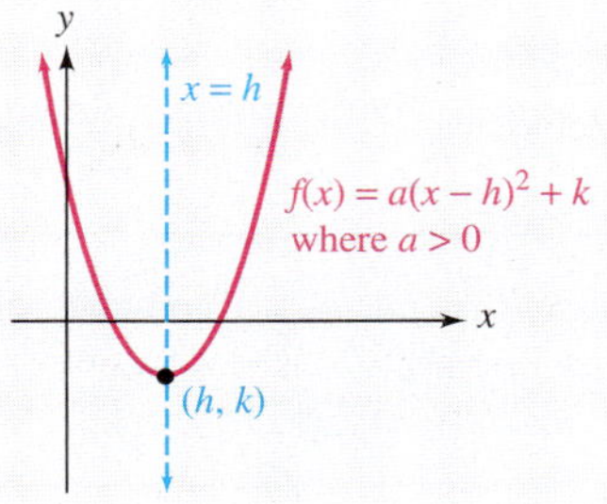

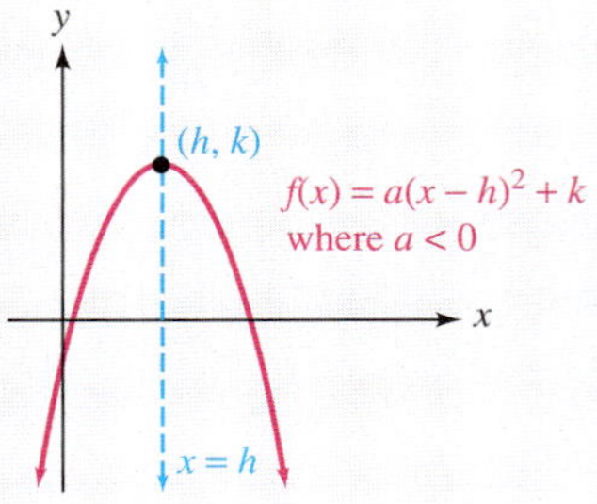

EXAMPLE 5 Graph: $f(x) = 2(x - 3)^2 - 4$. Label the vertex and draw the axis of symmetry.

Strategy We will determine whether the graph opens upward or downward and find its vertex and axis of symmetry. Then we will plot some points and complete the graph.

Why This method will be more efficient than plotting many points.

Solution The graph of $f(x) = 2(x - 3)^2 - 4$ is identical to the graph of $g(x) = 2(x - 3)^2$, except that it has been translated 4 units downward. The graph of $g(x) = 2(x - 3)^2$ is identical to the graph of $s(x) = 2x^2$, except that it has been translated 3 units to the right.

We can learn more about the graph of $f(x) = 2(x - 3)^2 - 4$ by determining a, h, and k.

Success Tip

The most important point to find when graphing a quadratic function is the vertex.

$$\left.\begin{array}{l} f(x) = 2(x - 3)^2 - 4 \\ \quad\uparrow \quad \uparrow \quad \uparrow \\ f(x) = a(x - h)^2 + k \end{array}\right\} \quad a = 2,\ h = 3, \text{ and } k = -4$$

Upward/downward: Since $a = 2$ and $2 > 0$, the parabola opens upward.

Vertex: The vertex of the parabola is $(h, k) = (3, -4)$, as shown below.

Axis of symmetry: Since $h = 3$, the axis of symmetry is the line $x = 3$, as shown below.

Plotting points: We can construct a table of values to determine several points on the parabola. Since the x-coordinate of the vertex is 3, we choose the x-values of 4 and 5, find $f(4)$ and $f(5)$, and record the results in a table. Then we plot $(4, -2)$ and $(5, 4)$, and use symmetry to locate two other points on the parabola: $(2, -2)$ and $(1, 4)$. Finally, we draw a smooth curve through the points to get the graph.

Success Tip

When graphing, remember that any point on a parabola to the right of the axis of symmetry yields a second point to the left of the axis of symmetry, and vice versa. Think of it as two-for-one.

$f(x) = 2(x - 3)^2 - 4$

x	$f(x)$	
4	−2	→ (4, −2)
5	4	→ (5, 4)

↑ The x-coordinate of the vertex is 3. Choose values for x close to 3 and on the same side of the axis of symmetry.

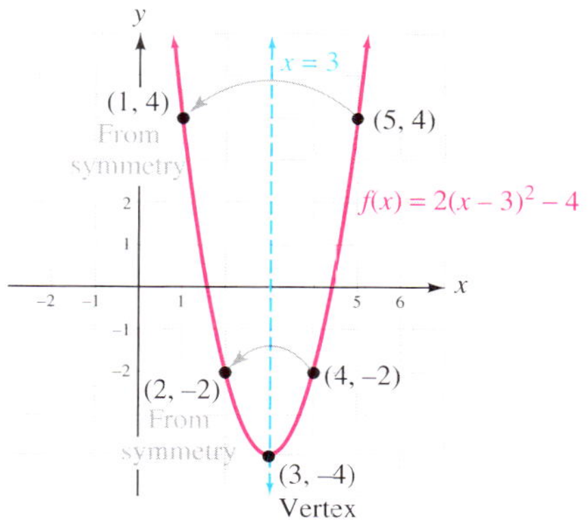

Self Check 5 Graph: $f(x) = 2(x - 1)^2 - 2$. Label the vertex and draw the axis of symmetry.

Now Try Problems 25 and 31

3 Graph Functions of the Form $f(x) = ax^2 + bx + c$ by Completing the Square.

To graph functions of the form $f(x) = ax^2 + bx + c$, we can complete the square to write the function in standard form $f(x) = a(x - h)^2 + k$.

EXAMPLE 6 Determine the vertex and the axis of symmetry of the graph of $f(x) = x^2 + 8x + 21$. Will the graph open upward or downward?

Strategy To find the vertex and the axis of symmetry, we will complete the square on x and write the equation of the function in standard form.

Why Once the equation is written in standard form, we can determine the values of a, h, and k. The coordinates of the vertex will be (h, k) and the equation of the axis of symmetry will be $x = h$. The graph will open upward if $a > 0$ or downward if $a < 0$.

Solution To determine the vertex and the axis of symmetry of the graph, we complete the square on the right side so we can write the function in $f(x) = a(x - h)^2 + k$ form.

$$f(x) = x^2 + 8x + 21$$
$$f(x) = (x^2 + 8x \qquad) + 21$$ Prepare to complete the square on x by writing parentheses around $x^2 + 8x$.

To complete the square on $x^2 + 8x$, we note that one-half of the coefficient of x is $\frac{1}{2} \cdot 8 = 4$, and $4^2 = 16$. If we add 16 to $x^2 + 8x$, we obtain a perfect-square trinomial within the parentheses. Since this step adds 16 to the right side, we must also subtract 16 from the right side so that it remains in an equivalent form.

Success Tip

When a number is added to and that same number is subtracted from one side of an equation, the value of that side of the equation remains the same.

Add 16 to the right side. Subtract 16 from the right side to counteract the addition of 16.

$$f(x) = (x^2 + 8x + 16) + 21 - 16$$
$$f(x) = (x + 4)^2 + 5$$ Factor $x^2 + 8x + 16$ and combine like terms.

The function is now written in standard form, and we can determine a, h, and k.

The standard form requires a minus symbol here.

$$f(x) = \left[(x - (-4)\right]^2 + 5$$ Write $x + 4$ as $x - (-4)$ to determine h.

$a = 1$ $h = -4$ $k = 5$

The vertex is $(h, k) = (-4, 5)$ and the axis of symmetry is the line $x = -4$. Since $a = 1$ and $1 > 0$, the parabola opens upward.

Self Check 6 Determine the vertex and the axis of symmetry of the graph of $f(x) = x^2 + 4x + 10$. Will the graph open upward or downward?

Now Try Problem 39

EXAMPLE 7 Graph: $f(x) = 2x^2 - 4x - 1$

Strategy We will complete the square on x and write the equation of the function in standard form, $f(x) = a(x - h)^2 + k$.

Why When the equation is in standard form, we can identify the values of a, h, and k from the equation. This information will help us sketch the graph.

Solution Recall that to complete the square on $2x^2 - 4x$, the coefficient of x^2 must be equal to 1. Therefore, we factor 2 from $2x^2 - 4x$.

$$f(x) = 2x^2 - 4x - 1$$
$$f(x) = 2(x^2 - 2x \qquad) - 1$$

To complete the square on $x^2 - 2x$, we note that one-half of the coefficient of x is $\frac{1}{2}(-2) = -1$, and $(-1)^2 = 1$. If we add 1 to $x^2 - 2x$, we obtain a perfect-square trinomial within the parentheses. Since this step adds 2 to the right side, we must also subtract 2 from the right side so that it remains in an equivalent form.

By the distributive property, when 1 is added to the expression within the parentheses, $2 \cdot 1 = 2$ is added to the right side.

Subtract 2 to counteract the addition of 2 shown in red.

$$f(x) = 2(x^2 - 2x + 1) - 1 - 2$$
$$f(x) = 2(x - 1)^2 - 3 \qquad \text{Factor } x^2 - 2x + 1 \text{ and combine like terms.}$$

We see that $a = 2$, $h = 1$, and $k = -3$. Thus, the vertex is at the point $(1, -3)$, and the axis of symmetry is $x = 1$. Since $a = 2$ and $2 > 0$, the parabola opens upward. We plot the vertex and axis of symmetry as shown below.

Finally, we construct a table of values, plot the points, use symmetry to plot the corresponding points, and then draw the graph.

$$f(x) = 2x^2 - 4x - 1$$
or
$$f(x) = 2(x - 1)^2 - 3$$

x	$f(x)$	
2	−1	→ (2, −1)
3	5	→ (3, 5)

The x-coordinate of the vertex is 1. Choose values for x close to 1 and on the same side of the axis of symmetry.

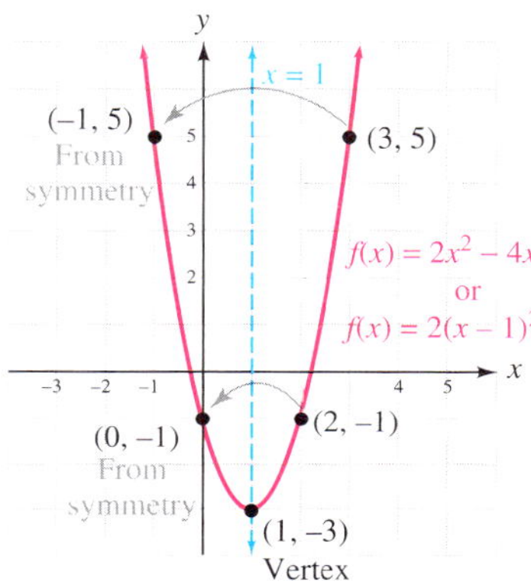

Success Tip

To find additional points on the graph, select values of x that are close to the x-coordinate of the vertex.

Self Check 7 Graph: $f(x) = 3x^2 - 12x + 8$.

Now Try Problem 45

4 Find the Vertex Using $-\frac{b}{2a}$.

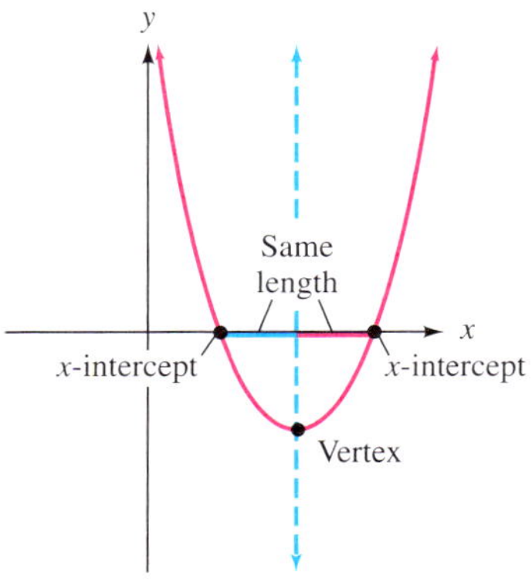

Because of symmetry, if a parabola has two x-intercepts, the x-coordinate of the vertex is exactly midway between them. We can use this fact to derive a formula to find the vertex of a parabola.

In general, if a parabola has two x-intercepts, they can be found by solving $0 = ax^2 + bx + c$ for x. We can use the quadratic formula to find the solutions. They are

$$x = \frac{-b - \sqrt{b^2 - 4ac}}{2a} \quad \text{and} \quad x = \frac{-b + \sqrt{b^2 - 4ac}}{2a}$$

Thus, the parabola's x-intercepts are $\left(\frac{-b - \sqrt{b^2 - 4ac}}{2a}, 0\right)$ and $\left(\frac{-b + \sqrt{b^2 - 4ac}}{2a}, 0\right)$.

Since the x-value of the vertex of a parabola is halfway between the two x-intercepts, we can find this value by finding the average, or $\frac{1}{2}$ of the sum of the x-coordinates of the x-intercepts.

$$x = \frac{1}{2}\left(\frac{-b - \sqrt{b^2 - 4ac}}{2a} + \frac{-b + \sqrt{b^2 - 4ac}}{2a}\right)$$

$$x = \frac{1}{2}\left(\frac{-b - \sqrt{b^2 - 4ac} + (-b) + \sqrt{b^2 - 4ac}}{2a}\right)$$ Add the numerators and keep the common denominator.

$$x = \frac{1}{2}\left(\frac{-2b}{2a}\right)$$ Combine like terms: $-b + (-b) = -2b$ and $-\sqrt{b^2 - 4ac} + \sqrt{b^2 - 4ac} = 0$.

$$x = -\frac{b}{2a}$$ Remove the common factor of 2 in the numerator and denominator and simplify.

This result is true even if the graph has no x-intercepts.

Formula for the Vertex of a Parabola

The vertex of the graph of the quadratic function $f(x) = ax^2 + bx + c$ is

$$\left(-\frac{b}{2a}, f\left(-\frac{b}{2a}\right)\right)$$

and the axis of symmetry of the parabola is the line $x = -\frac{b}{2a}$.

EXAMPLE 8 Find the vertex of the graph of $f(x) = 2x^2 - 4x - 1$.

Strategy We will determine the values of a and b and substitute into the formula for the vertex of a parabola.

Why It is easier to find the coordinates of the vertex using the formula than it is to complete the square on $2x^2 - 4x - 1$.

Solution The function is written in $f(x) = ax^2 + bx + c$ form, where $a = 2$ and $b = -4$. To find the vertex of its graph, we calculate

Success Tip

We can find the vertex of the graph of a quadratic function by completing the square or by using the formula.

$$-\frac{b}{2a} = -\frac{-4}{2(2)} = -\frac{-4}{4} = 1$$ This is the x-coordinate of the vertex.

$$f\left(-\frac{b}{2a}\right) = f(1) = 2(1)^2 - 4(1) - 1 = -3$$ This is the y-coordinate of the vertex.

The vertex is the point $(1, -3)$. This agrees with the result we obtained in Example 7 by completing the square.

Self Check 8 Find the vertex of the graph of $f(x) = 3x^2 - 12x + 8$.

Now Try Problem 55

Using Your Calculator ▶ Finding the Vertex

We can use a graphing calculator to graph the function $f(x) = 2x^2 + 6x - 3$ and find the coordinates of the vertex and the axis of symmetry of the parabola. If we enter the function, we will obtain the graph shown in figure (a) on the next page.

We then trace to move the cursor to the lowest point on the graph, as shown in figure (b). By zooming in, we can see that the vertex is the point $(-1.5, -7.5)$, or $\left(-\frac{3}{2}, -\frac{15}{2}\right)$, and that the line $x = -\frac{3}{2}$ is the axis of symmetry.

Some calculators have an fmin or fmax feature that also can be used to find the vertex.

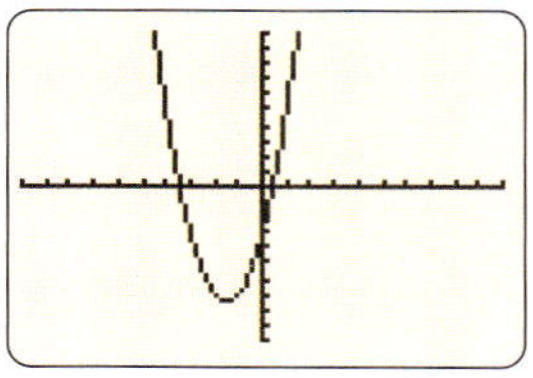

(a)

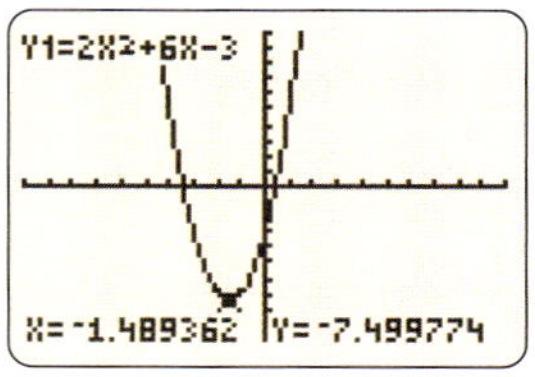

(b)

We can determine much about the graph of $f(x) = ax^2 + bx + c$ from the coefficients a, b, and c. This information is summarized as follows:

Graphing a Quadratic Function $f(x) = ax^2 + bx + c$

- Determine whether the parabola opens upward or downward by finding the value of a.
- The x-coordinate of the vertex of the parabola is $x = -\frac{b}{2a}$.
- To find the y-coordinate of the vertex, substitute $-\frac{b}{2a}$ for x and find $f\left(-\frac{b}{2a}\right)$.
- The axis of symmetry is the vertical line passing through the vertex.
- The y-intercept is determined by the value of $f(x)$ when $x = 0$: the y-intercept is $(0, c)$.
- The x-intercepts (if any) are determined by the values of x that make $f(x) = 0$. To find them, solve the quadratic equation $ax^2 + bx + c = 0$.

EXAMPLE 9 Graph: $f(x) = -2x^2 - 8x - 8$

Strategy We will follow the steps for graphing a quadratic function.

Why This is the most efficient way to graph a general quadratic function.

Solution ***Step 1: Determine whether the parabola opens upward or downward.*** The function is in the form $f(x) = ax^2 + bx + c$, with $a = -2$, $b = -8$, and $c = -8$. Since $a < 0$, the parabola opens downward.

Step 2: Find the vertex and draw the axis of symmetry. To find the coordinates of the vertex, we calculate

$$x = -\frac{b}{2a}$$

$$x = -\frac{-8}{2(-2)}$$ Substitute -2 for a and -8 for b.

$$= -2$$ This is the x-coordinate of the vertex.

$$f\left(-\frac{b}{2a}\right) = f(-2)$$

$$= -2(-2)^2 - 8(-2) - 8$$

$$= -8 + 16 - 8$$

$$= 0$$ This is the y-coordinate of the vertex.

The vertex of the parabola is the point $(-2, 0)$. This point is in blue on the graph. The axis of symmetry is the line $x = -2$.

Success Tip

An easy way to remember the vertex formula is to note that $x = \frac{-b}{2a}$ is part of the quadratic formula:

$$x = \frac{-b \pm \sqrt{b^2 - 4ac}}{2a}$$

Step 3: ***Find the x- and y-intercepts.*** Since $c = -8$, the y-intercept of the parabola is $(0, -8)$. The point $(-4, -8)$, two units to the left of the axis of symmetry, must also be on the graph. We plot both points in black on the graph.

To find the x-intercepts, we set $f(x)$ equal to 0 and solve the resulting quadratic equation.

$$f(x) = -2x^2 - 8x - 8 \quad \text{This is the function to graph.}$$
$$0 = -2x^2 - 8x - 8 \quad \text{Set } f(x) = 0.$$
$$0 = x^2 + 4x + 4 \quad \text{Divide both sides by } -2.$$
$$0 = (x + 2)(x + 2) \quad \text{Factor the trinomial.}$$
$$x + 2 = 0 \quad \text{or} \quad x + 2 = 0 \quad \text{Set each factor equal to 0.}$$
$$x = -2 \quad | \quad x = -2$$

Since the solutions are the same, the graph has only one x-intercept: $(-2, 0)$. This point is the vertex of the parabola and has already been plotted.

Step 4: ***Plot another point.*** Finally, we find another point on the parabola. If $x = -3$, then $f(-3) = -2$. We plot $(-3, -2)$ and use symmetry to determine that $(-1, -2)$ is also on the graph. Both points are in green.

Step 5: ***Draw a smooth curve through the points.***

$f(x) = -2x^2 - 8x - 8$

x	$f(x)$	
-3	-2	→ $(-3, -2)$

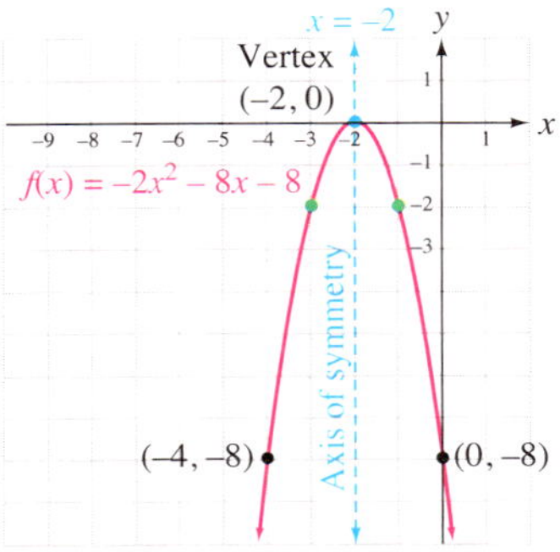

Self Check 9 Graph: $f(x) = -2x^2 + 12x - 16$

Now Try ▶ Problems 59 and 69

5 Determine Minimum and Maximum Values.

It is often useful to know the smallest or largest possible value a quantity can assume. For example, companies try to minimize their costs and maximize their profits. If the quantity is expressed by a quadratic function, the y-coordinate of the vertex of the graph of the function gives its minimum or maximum value.

EXAMPLE 10

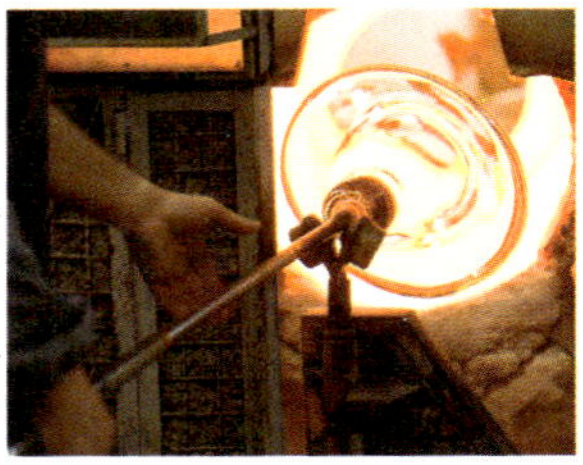

Minimizing Costs. A glassworks that makes lead crystal vases has daily production costs given by the function $C(x) = 0.2x^2 - 10x + 650$, where x is the number of vases made each day. How many vases should be produced to minimize the per-day costs? What will the costs be?

Strategy We will find the vertex of the graph of the quadratic function.

Why The x-coordinate of the vertex indicates the number of vases to make to keep costs at a minimum, and the y-coordinate indicates the minimum cost.

Solution The graph of $C(x) = 0.2x^2 - 10x + 650$ is a parabola opening upward. The vertex is the lowest point on the graph. To find the vertex, we calculate

The Language of Algebra

We say that 25 is the value for which the function $C(x) = 0.2x^2 - 10x + 650$ is a minimum.

$$-\frac{b}{2a} = -\frac{-10}{2(0.2)} \quad b = -10 \text{ and } a = 0.2.$$
$$= -\frac{-10}{0.4}$$
$$= 25$$

$$f\left(-\frac{b}{2a}\right) = f(25)$$
$$= 0.2(25)^2 - 10(25) + 650$$
$$= 525$$

The vertex is (25, 525), and it indicates that the costs are a minimum of $525 when 25 vases are made daily.

To solve this problem with a graphing calculator, we graph the function $C(x) = 0.2x^2 - 10x + 650$. By using TRACE and ZOOM, we can locate the vertex of the graph. The coordinates of the vertex indicate that the minimum cost is $525 when the number of vases produced is 25.

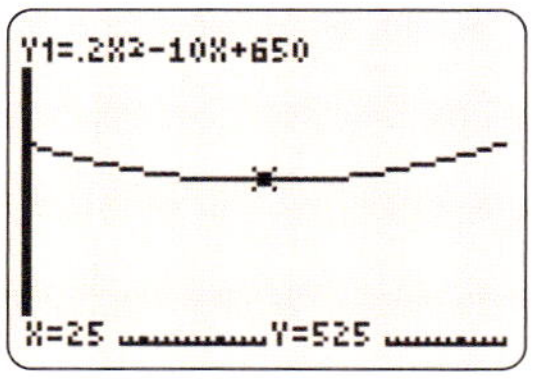

Self Check 10 **Minimizing Costs.** A manufacturing company has a daily production cost of $c(x) = 0.25x^2 - 10x + 800$, where x is the number of items produced and $c(x)$ is the cost. How many items should be produced to minimize the per-day cost? What is the minimum cost?

Now Try Problem 85

6 Solve Quadratic Equations Graphically.

When solving quadratic equations graphically, we must consider three possibilities. If the graph of the associated quadratic function has two x-intercepts, the quadratic equation has two real-number solutions. Figure (a) shows an example of this. If the graph has one x-intercept, as shown in figure (b), the equation has one repeated real-number solution. Finally, if the graph does not have an x-intercept, as shown in figure (c), the equation does not have any real-number solutions.

Success Tip

Note that the solutions of each quadratic equation are given by the x-coordinate of the x-intercept of each respective graph.

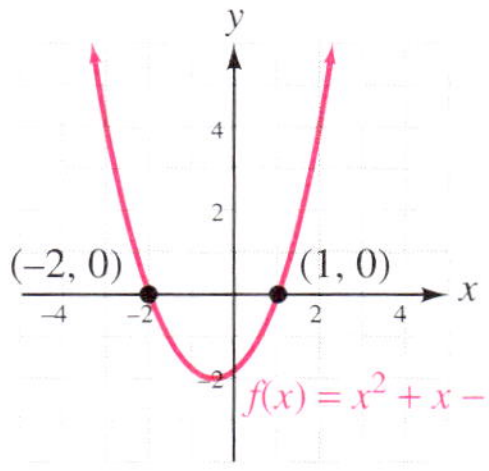

$x^2 + x - 2 = 0$ has two solutions, −2 and 1.

(a)

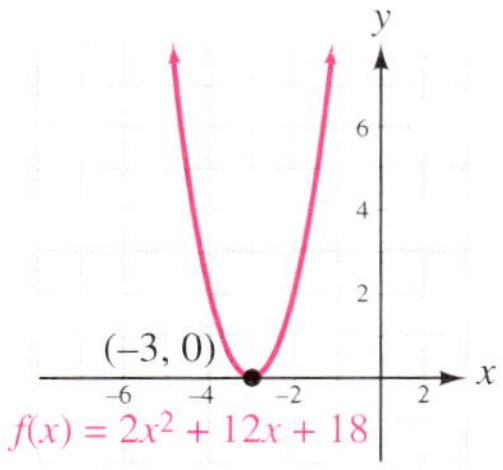

$2x^2 + 12x + 18 = 0$ has one repeated solution, −3.

(b)

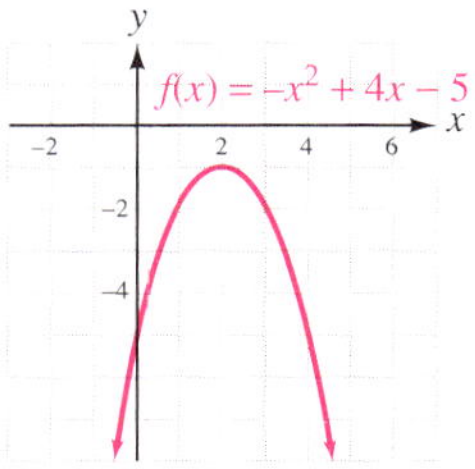

$-x^2 + 4x - 5 = 0$ has no real-number solutions.

(c)

Using Your Calculator ▶ **Solving Quadratic Equations Graphically**

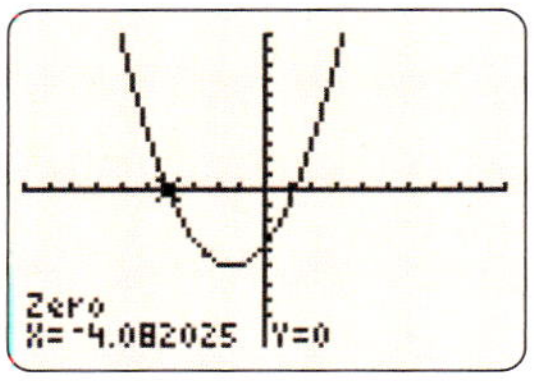

We can use a graphing calculator to find approximate solutions of quadratic equations. For example, the solutions of $0.7x^2 + 2x - 3.5 = 0$ are the numbers x that will make $y = 0$ in the quadratic function $f(x) = 0.7x^2 + 2x - 3.5$. To approximate these numbers, we graph the quadratic function and read the x-intercepts from the graph using the ZERO feature. (The ZERO feature can be found by pressing 2nd, CALC, and then 2.) In the figure, we see that the x-coordinate of the left-most x-intercept of the graph is given as −4.082025. This means that an approximate solution of the equation is −4.08. To find the positive x-intercept, we use similar steps.

SECTION 10.4 STUDY SET

VOCABULARY

Refer to the graph. Fill in the blanks.

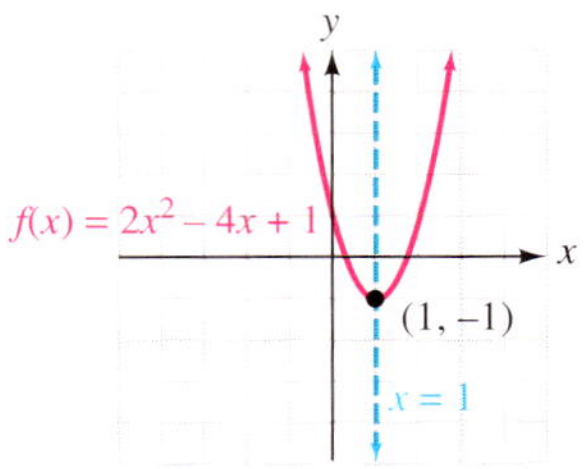

1. $f(x) = 2x^2 - 4x + 1$ is called a ________ function. Its graph is a cup-shaped figure called a ________.
2. The lowest point on the graph is $(1, -1)$. This is called the ______ of the parabola.
3. The vertical line $x = 1$ divides the parabola into two halves. This line is called the _____ ___ ________.
4. $f(x) = a(x - h)^2 + k$ is called the ________ form of the equation of a quadratic function.

CONCEPTS

5. Refer to the graph.

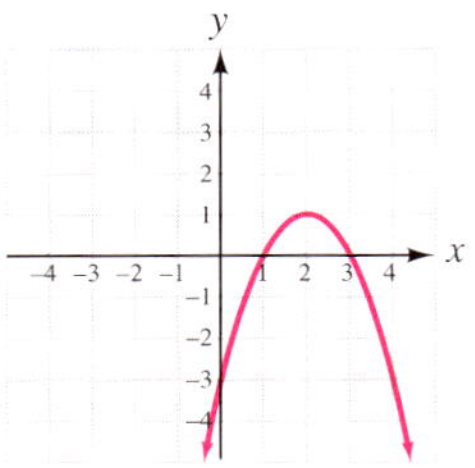

   a. What are the x-intercepts of the graph?
   b. What is the y-intercept of the graph?
   c. What is the vertex?
   d. What is the axis of symmetry?
   e. What are the domain and the range of the function?

6. The vertex of a parabola is at $(1, -3)$, its y-intercept is $(0, -2)$, and it passes through the point $(3, 1)$, as shown in the illustration. Use the axis of symmetry shown in blue to help determine two other points on the parabola.

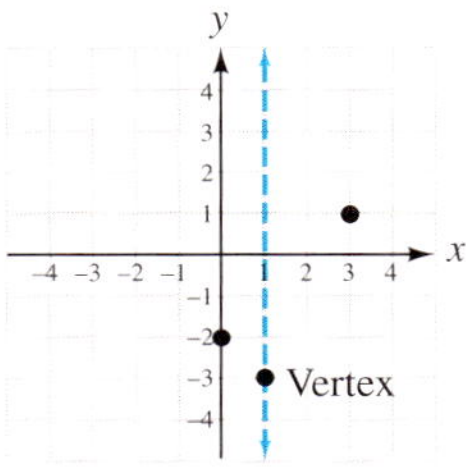

7. Draw the graph of a quadratic function using the given facts about its graph.
   - Opens upward
   - y-intercept: $(0, -3)$
   - Vertex: $(-1, -4)$
   - x-intercepts: $(-3, 0)$, $(1, 0)$
   -

x	$f(x)$
2	5

8. For $f(x) = -x^2 + 6x - 7$, the value of $-\frac{b}{2a}$ is 3. Find the y-coordinate of the vertex of the graph of this function.
9. Fill in the blanks.
   a. To complete the square on the right side of $f(x) = 2x^2 + 12x + 11$, what should be factored from the first two terms?

   $$f(x) = \quad (x^2 + 6x) + 11$$

   b. To complete the square on $x^2 + 6x$ shown below, what should be added within the parentheses and what should be subtracted outside the parentheses?

   $$f(x) = 2(x^2 + 6x + \quad) + 11 - $$

10. Fill in the blanks. To complete the square on $x^2 + 4x$ shown below, what should be added within the parentheses and what should be added outside the parentheses?

   $$f(x) = -5(x^2 + 4x + \quad) + 7 + $$

11. Use the graph of $f(x) = \frac{1}{10}x^2 - \frac{1}{5}x - \frac{3}{2}$, shown here, to estimate the solutions of the equation $\frac{1}{10}x^2 - \frac{1}{5}x - \frac{3}{2} = 0$.

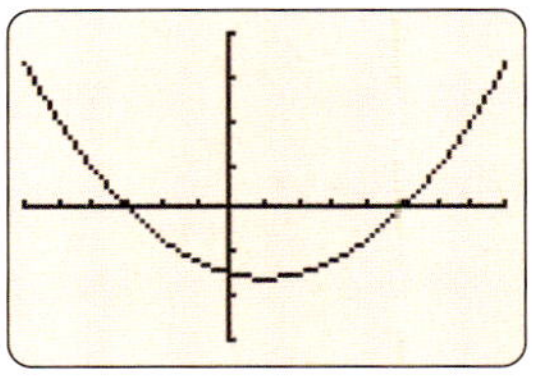

12. Three quadratic equations are to be solved graphically. The graphs of their associated quadratic functions are shown here. Determine which graph indicates that the equation has
   a. two real solutions.
   b. one repeated real solution.
   c. no real solutions.

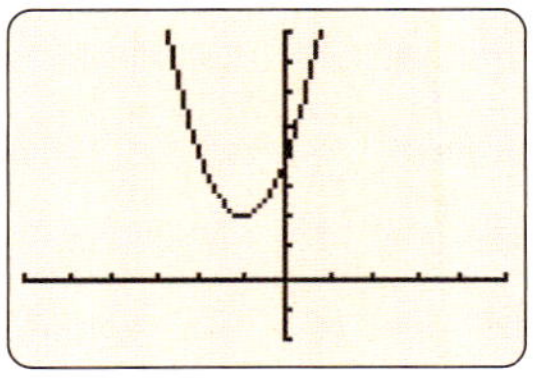

(i)

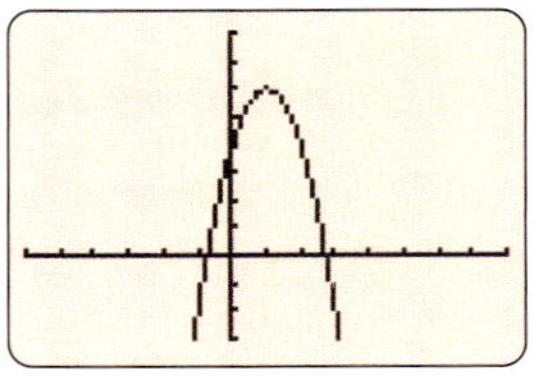

(ii)

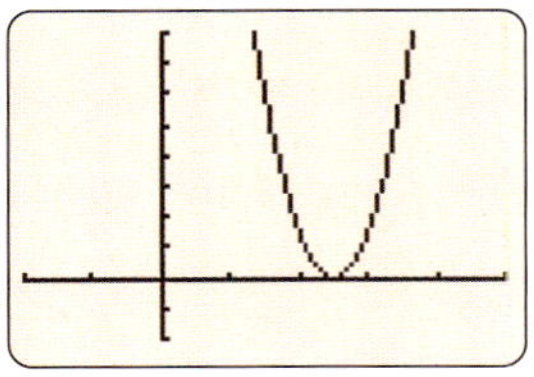

(iii)

NOTATION

13. The function $f(x) = 2(x + 1)^2 + 6$ is written in the form $f(x) = a(x - h)^2 + k$. Is $h = -1$ or is $h = 1$? Explain.
14. Consider the function $f(x) = 2x^2 + 4x - 8$.
   a. What are a, b, and c?
   b. Find $-\dfrac{b}{2a}$.

GUIDED PRACTICE

***Graph each group of functions on the same coordinate system.* See Example 1.**

15. $f(x) = x^2$, $g(x) = 2x^2$, $s(x) = \frac{1}{2}x^2$
16. $f(x) = x^2$, $g(x) = 4x^2$, $s(x) = \frac{1}{4}x^2$

***Graph each pair of functions on the same coordinate system.* See Example 2.**

17. $f(x) = 2x^2$, $g(x) = -2x^2$
18. $f(x) = \frac{1}{2}x^2$, $g(x) = -\frac{1}{2}x^2$

Graph each group of functions on the same coordinate system. See Example 3.

19. $f(x) = 4x^2$, $g(x) = 4x^2 + 3$, $s(x) = 4x^2 - 2$
20. $f(x) = \frac{1}{3}x^2$, $g(x) = \frac{1}{3}x^2 + 4$, $s(x) = \frac{1}{3}x^2 - 3$

Graph each group of functions on the same coordinate system and describe how the graphs are similar and how they are different. See Example 4.

21. $f(x) = 3x^2$, $g(x) = 3(x + 2)^2$, $s(x) = 3(x - 3)^2$
22. $f(x) = \frac{1}{2}x^2$, $g(x) = \frac{1}{2}(x + 3)^2$, $s(x) = \frac{1}{2}(x - 2)^2$

Find the vertex and the axis of symmetry of the graph of each function. Do not graph the function, but determine whether the graph will open upward or downward. See Example 5.

23. $f(x) = (x - 1)^2 + 2$
24. $f(x) = 2(x - 2)^2 - 1$
25. $f(x) = -2(x + 3)^2 - 4$
26. $f(x) = -3(x + 1)^2 + 3$
27. $f(x) = -0.5(x - 7.5)^2 + 8.5$
28. $f(x) = -\frac{3}{2}\left(x + \frac{1}{4}\right)^2 + \frac{7}{8}$
29. $f(x) = 2x^2 - 4$
30. $f(x) = 3x^2 - 3$

Determine the vertex and the axis of symmetry of the graph of each function. Then plot several points and complete the graph. See Example 5.

31. $f(x) = (x - 3)^2 + 2$
32. $f(x) = (x + 1)^2 - 2$
33. $f(x) = -(x - 2)^2$
34. $f(x) = -(x + 2)^2$
35. $f(x) = -2(x + 3)^2 + 4$
36. $f(x) = -2(x - 2)^2 - 4$
37. $f(x) = \frac{1}{2}(x + 1)^2 - 3$
38. $f(x) = \frac{1}{3}(x - 1)^2 + 2$

Determine the vertex and the axis of symmetry of the graph of each function. Will the graph open upward or downward? See Example 6.

39. $f(x) = x^2 + 4x + 5$
40. $f(x) = x^2 - 4x - 1$
41. $f(x) = -x^2 + 6x - 15$
42. $f(x) = -x^2 - 6x + 3$

Complete the square to write each function in $f(x) = a(x - h)^2 + k$ form. Determine the vertex and the axis of symmetry of the graph of the function. Then plot several points and complete the graph. See Examples 6 and 7.

43. $f(x) = x^2 + 2x - 3$
44. $f(x) = x^2 + 6x + 5$
45. $f(x) = 4x^2 + 24x + 37$
46. $f(x) = 3x^2 - 12x + 10$
47. $f(x) = x^2 + x - 6$
48. $f(x) = x^2 - x - 6$
49. $f(x) = -4x^2 + 16x - 10$
50. $f(x) = -2x^2 + 4x + 3$
51. $f(x) = 2x^2 + 8x + 6$
52. $f(x) = 3x^2 - 12x + 9$
53. $f(x) = -x^2 - 8x - 17$
54. $f(x) = -x^2 + 6x - 8$

Use the vertex formula to find the vertex of the graph of each function. See Example 8.

55. $f(x) = x^2 + 2x - 5$
56. $f(x) = -x^2 + 4x - 5$
57. $f(x) = 2x^2 - 3x + 4$
58. $f(x) = 2x^2 - 7x - 4$

Find the x- and y-intercepts of the graph of the quadratic function. See Example 9.

59. $f(x) = x^2 - 2x - 35$
60. $f(x) = -x^2 - 10x - 21$
61. $f(x) = -2x^2 + 4x$
62. $f(x) = 3x^2 + 6x - 9$

Determine the coordinates of the vertex of the graph of each function using the vertex formula. Then determine the x- and y-intercepts of the graph. Finally, plot several points and complete the graph. See Example 9.

63. $f(x) = x^2 + 4x + 4$
64. $f(x) = x^2 - 6x + 9$
65. $f(x) = -x^2 + 2x - 1$
66. $f(x) = -x^2 - 2x - 1$
67. $f(x) = x^2 - 2x$
68. $f(x) = x^2 + x$
69. $f(x) = 2x^2 - 8x + 6$
70. $f(x) = 3x^2 - 12x + 12$
71. $f(x) = -6x^2 - 12x - 8$
72. $f(x) = -2x^2 + 8x - 10$
73. $f(x) = 4x^2 - 12x + 9$
74. $f(x) = 4x^2 + 4x - 3$

Use a graphing calculator to find the coordinates of the vertex of the graph of each quadratic function. Round to the nearest hundredth. See Using Your Calculator: Finding the Vertex.

75. $f(x) = 2x^2 - x + 1$
76. $f(x) = x^2 + 5x - 6$

77. $f(x) = -x^2 + x + 7$

78. $f(x) = 2x^2 - 3x + 2$

Use a graphing calculator to solve each equation. If an answer is not exact, round to the nearest hundredth. **See Using Your Calculator: Solving Quadratic Equations Graphically.**

79. $x^2 + x - 6 = 0$

80. $2x^2 - 5x - 3 = 0$

81. $0.5x^2 - 0.7x - 3 = 0$

82. $2x^2 - 0.5x - 2 = 0$

APPLICATIONS

83. **Crossword Puzzles.** Darken the appropriate squares to the right of the dashed red line so that the puzzle has symmetry with respect to that line.

84. **Graphic Arts.** Draw an axis of symmetry over the letter shown here.

85. **Operating Costs.** The cost C in dollars of operating a certain concrete-cutting machine is related to the number of minutes n the machine is run by the function $C(n) = 2.2n^2 - 66n + 655$. For what number of minutes is the cost of running the machine a minimum? What is the minimum cost?

86. **Water Usage.** The height (in feet) of the water level in a reservoir over a 1-year period is modeled by the function $H(t) = 3.3(t - 9)^2 + 14$ where $t = 1$ represents January, $t = 2$ represents February, and so on. How low did the water level get that year, and when did it reach the low mark?

87. **Fireworks.** A fireworks shell is shot straight up with an initial velocity of 120 feet per second. Its height s in feet after t seconds is approximated by the equation $s = 120t - 16t^2$. If the shell is designed to explode when it reaches its maximum height, how long after being fired, and at what height, will the fireworks appear in the sky?

88. **Projectiles.** A ball is thrown straight upward from the top of a building with an initial velocity of 32 feet per second. The equation $s = -16t^2 + 32t + 48$ gives the height s of the ball in feet t seconds after it is thrown. Find the maximum height reached by the ball and the time it takes for the ball to hit the ground.

89. **from Campus to Careers**

Police Patrol Officer

©Robert E Daemmrich/Getty Images

Suppose you are a police patrol officer and you have a 300-foot-long roll of yellow "DO NOT CROSS" barricade tape to seal off an automobile accident, as shown in the illustration. What dimensions should you use to seal off the maximum rectangular area around the collision? What is the maximum area?

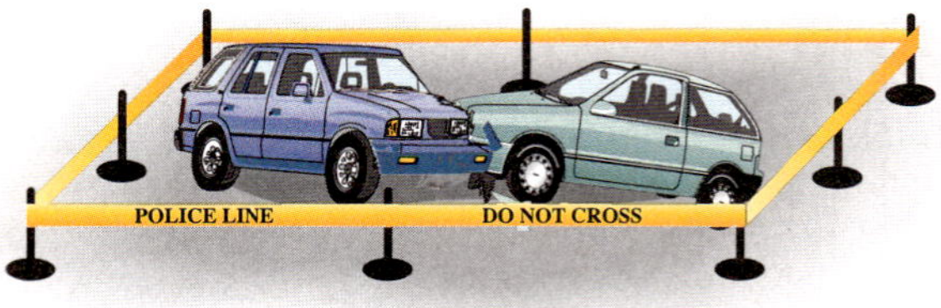

90. **Ranching.** See the illustration. A farmer wants to fence in three sides of a rectangular field with 1,000 feet of fencing. The other side of the rectangle will be a river. If the enclosed area is to be maximum, find the dimensions of the field.

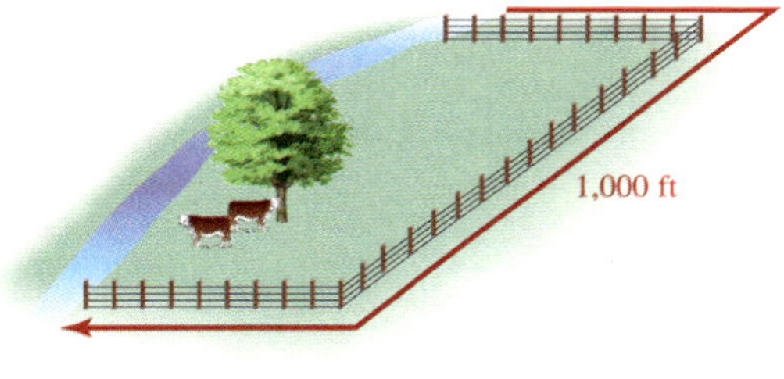

91. **Military History.** The function $N(x) = -0.0534x^2 + 0.337x + 0.97$ gives the number of active-duty military personnel in the United States Army (in millions) for the years 1965–1972, where $x = 0$ corresponds to 1965, $x = 1$ corresponds to 1966, and so on. For this period, when was the army's personnel strength level at its highest, and what was it? Historically, can you explain why?

92. **School Enrollment.** After peaking in 1970, school enrollment in the United States fell during the 1970s and 1980s. The total annual enrollment (in millions) in U.S. elementary and secondary schools for the years 1975–1996 is given by the model $E(x) = 0.058x^2 - 1.162x + 50.604$ where $x = 0$ corresponds to 1975, $x = 1$ corresponds to 1976, and so on. For this period, when was enrollment the lowest? What was the enrollment?

93. **Maximizing Revenue.** The revenue R received for selling x stereos is given by the formula $R = -\frac{x^2}{5} + 80x - 1{,}000$. How many stereos must be sold to obtain the maximum revenue? Find the maximum revenue.

94. Maximizing Revenue. When priced at \$30 each, a toy has annual sales of 4,000 units. The manufacturer estimates that each \$1 increase in price will decrease sales by 100 units. Find the unit price that will maximize total revenue. (*Hint:* Total revenue = price · the number of units sold.)

WRITING

95. Use the example of a stream of water from a drinking fountain to explain the concepts of the vertex and the axis of symmetry of a parabola. Draw a picture.

96. What are some quantities that are good to maximize? What are some quantities that are good to minimize?

97. A mirror is held against the y-axis of the graph of a quadratic function. What fact about parabolas does this illustrate?

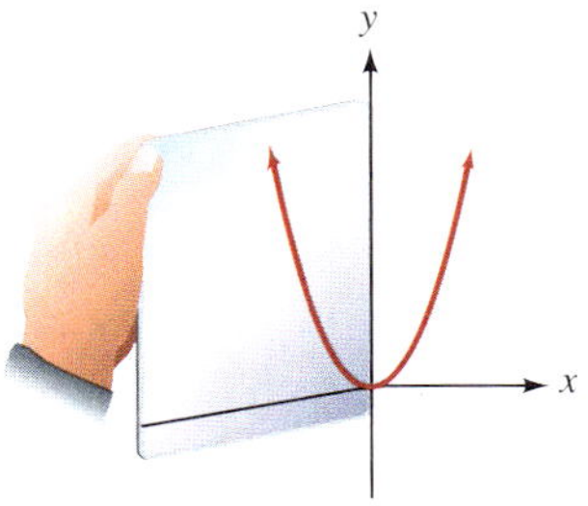

98. The vertex of a quadratic function $f(x) = ax^2 + bx + c$ is given by the formula $\left(-\frac{b}{2a}, f\left(-\frac{b}{2a}\right)\right)$. Explain what is meant by the notation $f\left(-\frac{b}{2a}\right)$.

99. A table of values for $f(x) = 2x^2 - 4x + 3$ is shown. Explain why it appears that the vertex of the graph of f is the point $(1, 1)$.

X	Y1
.25	2.125
.5	1.5
.75	1.125
1	1
1.25	1.125
1.5	1.5
1.75	2.125

X=.25

100. The illustration shows the graph of the quadratic function $f(x) = -4x^2 + 12x$ with domain $[0, 3]$. Explain how the value of $f(x)$ changes as the value of x increases from 0 to 3.

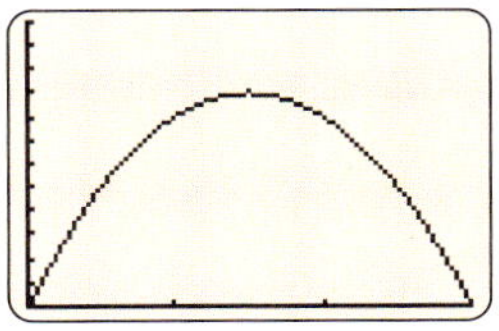

REVIEW

Simplify each expression. Assume all variables represent positive numbers.

101. $\dfrac{\sqrt{3}}{\sqrt{50}}$

102. $\dfrac{3}{\sqrt[3]{9}}$

103. $3\left(\sqrt{5b} - \sqrt{3}\right)^2$

104. $-2\sqrt{5b}\left(4\sqrt{2b} - 3\sqrt{3}\right)$

CHALLENGE PROBLEMS

105. Find a number between 0 and 1 such that the difference of the number and its square is a maximum.

106. Determine a quadratic function whose graph has x-intercepts of $(2, 0)$ and $(-4, 0)$.

SECTION 10.5 Quadratic and Other Nonlinear Inequalities

OBJECTIVES

1. Solve quadratic inequalities.
2. Solve rational inequalities.
3. Graph nonlinear inequalities in two variables.

ARE YOU READY?

The following problems review some basic skills that are needed when working with quadratic and nonlinear inequalities?

1. Does $x = -2$ satisfy $x^2 + x - 6 < 0$?

2. Solve: $x^2 - 5x - 50 = 0$

3. Graph the set of real numbers between -3 and 2 on a number line.

4. Graph the set of real numbers greater than 1or less than or equal to -2 on a number line.

5. What values of x make the denominator of $\dfrac{x + 1}{x^2 - 9}$ equal to 0?

6. Graph: $y = -x^2 + 4$

We have previously solved *linear* inequalities in one variable such as $2x + 3 > 8$ and $6x - 7 < 4x - 9$. To find their solution sets, we used properties of inequalities to isolate the variable on one side of the inequality.

In this section, we will solve *quadratic* inequalities in one variable such as $x^2 + x - 6 < 0$ and $x^2 + 4x \geq 5$. We will use an interval testing method on the number line to determine their solution sets.

1 Solve Quadratic Inequalities.

Recall that a quadratic equation can be written in the form $ax^2 + bx + c = 0$. If we replace the $=$ symbol with an inequality symbol, we have a quadratic inequality.

Quadratic Inequalities

A **quadratic inequality** can be written in one of the standard forms

$$ax^2 + bx + c < 0 \qquad ax^2 + bx + c > 0 \qquad ax^2 + bx + c \le 0 \qquad ax^2 + bx + c \ge 0$$

where a, b, and c are real numbers and $a \ne 0$.

To solve a quadratic inequality in one variable, we will use the following steps to find the values of the variable that make the inequality true.

Solving Quadratic Inequalities

1. Write the inequality in standard form and solve its related quadratic equation.
2. Locate the solutions (called **critical numbers**) of the related quadratic equation on a number line.
3. Test each interval on the number line created in step 2 by choosing a test value from the interval and determining whether it satisfies the inequality. The solution set includes the interval(s) whose test value makes the inequality true.
4. Determine whether the endpoints of the intervals are included in the solution set.

EXAMPLE 1 Solve: $x^2 + x - 6 < 0$

Strategy We will solve the related quadratic equation $x^2 + x - 6 = 0$ by factoring to determine the critical numbers. These critical numbers will separate the number line into intervals.

Why We can test each interval to see whether numbers in the interval are in the solution set of the inequality.

Solution The expression $x^2 + x - 6$ can be positive, negative, or 0, depending on what value is substituted for x. Solutions of the inequality are x-values that make $x^2 + x - 6$ less than 0. To find them, we will follow the steps for solving quadratic inequalities.

Step 1: Solve the related quadratic equation. For the quadratic inequality $x^2 + x - 6 < 0$, the related quadratic *equation* is $x^2 + x - 6 = 0$.

$$x^2 + x - 6 = 0$$

$$(x + 3)(x - 2) = 0 \qquad \text{Factor the trinomial.}$$

$$x + 3 = 0 \quad \text{or} \quad x - 2 = 0 \qquad \text{Set each factor equal to 0.}$$

$$x = -3 \quad | \quad x = 2 \qquad \text{Solve each equation.}$$

The solutions of $x^2 + x - 6 = 0$ are -3 and 2. These solutions are the critical numbers.

Step 2: Locate the critical numbers on a number line. When we highlight -3 and 2 on a number line, they separate the number line into three intervals:

The Language of Algebra

We say that the critical numbers **partition** the real-number line into test intervals. Interior decorators use freestanding screens to *partition* off parts of a room.

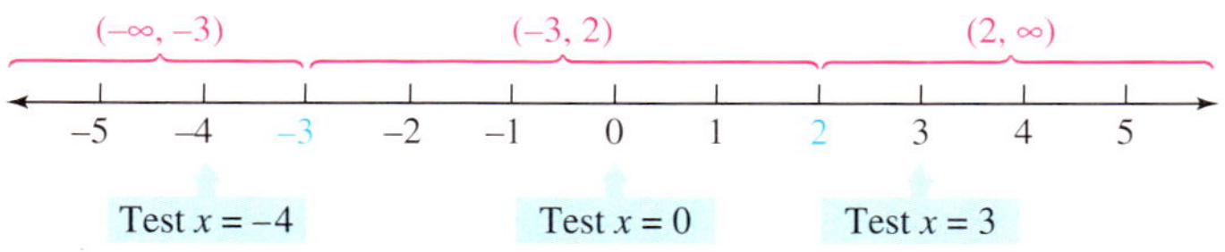

Step 3:* *Test each interval. To determine whether the numbers in $(-\infty, -3)$ are solutions of the inequality, we choose a number from that interval, substitute it for x, and see whether it satisfies $x^2 + x - 6 < 0$. *If one number in that interval satisfies the inequality, all numbers in that interval will satisfy the inequality.*

If we choose -4 from $(-\infty, -3)$, we have:

$$x^2 + x - 6 < 0 \quad \text{This is the original inequality.}$$
$$(-4)^2 + (-4) - 6 \stackrel{?}{<} 0 \quad \text{Substitute } -4 \text{ for } x.$$
$$16 + (-4) - 6 \stackrel{?}{<} 0$$
$$6 < 0 \quad \text{False}$$

Since -4 does not satisfy the inequality, none of the numbers in $(-\infty, -3)$ are solutions.

To test the second interval, $(-3, 2)$, we choose $x = 0$.

$$x^2 + x - 6 < 0 \quad \text{This is the original inequality.}$$
$$0^2 + 0 - 6 \stackrel{?}{<} 0 \quad \text{Substitute 0 for } x.$$
$$-6 < 0 \quad \text{True}$$

Since 0 satisfies the inequality, all of the numbers in $(-3, 2)$ are solutions.

To test the third interval, $(2, \infty)$, we choose $x = 3$.

$$x^2 + x - 6 < 0 \quad \text{This is the original inequality.}$$
$$3^2 + 3 - 6 \stackrel{?}{<} 0 \quad \text{Substitute 3 for } x.$$
$$9 + 3 - 6 \stackrel{?}{<} 0$$
$$6 < 0 \quad \text{False}$$

Since 3 does not satisfy the inequality, none of the numbers in $(2, \infty)$ are solutions.

Success Tip

If a quadratic inequality contains $\le$ or $\ge$, the endpoints of the intervals are included in the solution set. If the inequality contains $<$ or $>$, they are not.

Step 4:* *Are the endpoints included? From the interval testing, we see that only numbers from $(-3, 2)$ satisfy $x^2 + x - 6 < 0$. The endpoints -3 and 2 are not included in the solution set because they do not satisfy the inequality. (Recall that -3 and 2 make $x^2 + x - 6$ equal to 0.) The solution set is the interval $(-3, 2)$ as graphed on the right.

Self Check 1 Solve: $x^2 + x - 12 < 0$

Now Try Problem 15

EXAMPLE 2 Solve: $x^2 + 4x \ge 5$

Strategy This inequality is not in standard form because it does not have 0 on the right side. We will write it in standard form and solve its related quadratic equation to find any critical numbers. These critical numbers will separate the number line into intervals.

Why We can then test each interval to see whether numbers in the interval are in the solution set of the inequality.

Solution To get 0 on the right side, we subtract 5 from both sides.

$$x^2 + 4x \ge 5 \quad \text{This is the inequality to solve.}$$
$$x^2 + 4x - 5 \ge 0 \quad \text{Write the inequality in the equivalent form } ax^2 + bx + c \ge 0.$$

We can solve the related quadratic *equation* $x^2 + 4x - 5 = 0$ by factoring.

$$x^2 + 4x - 5 = 0$$
$$(x + 5)(x - 1) = 0 \quad \text{Factor the trinomial.}$$
$$x + 5 = 0 \quad \text{or} \quad x - 1 = 0 \quad \text{Set each factor equal to 0.}$$
$$x = -5 \quad\quad x = 1$$

The critical numbers -5 and 1 separate the number line into three intervals. We pick a test value from each interval to see whether it satisfies $x^2 + 4x - 5 \geq 0$.

Success Tip

When choosing a test value from an interval, pick a convenient number that makes the calculations easy. When applicable, 0 is an obvious choice.

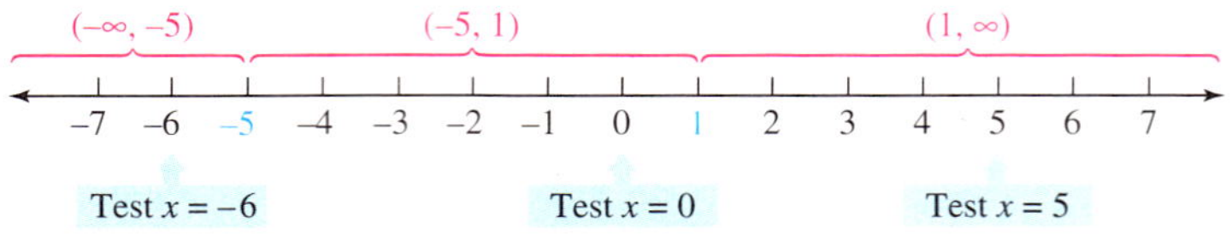

$$x^2 + 4x - 5 \geq 0 \qquad x^2 + 4x - 5 \geq 0 \qquad x^2 + 4x - 5 \geq 0$$

$$(-6)^2 + 4(-6) - 5 \overset{?}{\geq} 0 \qquad 0^2 + 4(0) - 5 \overset{?}{\geq} 0 \qquad 5^2 + 4(5) - 5 \overset{?}{\geq} 0$$

$$7 \geq 0 \quad \text{True} \qquad -5 \geq 0 \quad \text{False} \qquad 40 \geq 0 \quad \text{True}$$

The numbers in the intervals $(-\infty, -5)$ and $(1, \infty)$ satisfy the inequality. Since the endpoints -5 and 1 also satisfy $x^2 + 4x - 5 \geq 0$, they are included in the solution set. (Recall that -5 and 1 make $x^2 + 4x - 5$ equal to 0.) Thus, the solution set is the union of two intervals: $(-\infty, -5] \cup [1, \infty)$. The graph of the solution set is shown on the right.

Self Check 2 Solve: $x^2 + 3x \geq 40$

Now Try Problem 19

2 Solve Rational Inequalities.

Rational inequalities in one variable such as $\frac{9}{x} < 8$ and $\frac{x^2 + x - 2}{x - 4} \geq 0$ can also be solved using the interval testing method.

Solving Rational Inequalities

1. Write the inequality in standard form with a single quotient on the left side and 0 on the right side. Then solve its related rational equation.
2. Set the denominator equal to zero and solve that equation.
3. Locate the solutions (called **critical numbers**) found in steps 1 and 2 on a number line.
4. Test each interval on the number line created in step 3 by choosing a test value from the interval and determining whether it satisfies the inequality. The solution set includes the interval(s) whose test value makes the inequality true.
5. Determine whether the endpoints of the intervals are included in the solution set. Exclude any values that make the denominator 0.

EXAMPLE 3 Solve: $\frac{9}{x} < 8$

Strategy This rational inequality is not in standard form because it does not have 0 on the right side. We will write it in standard form and solve its related rational equation to find any critical numbers. These critical numbers will separate the number line into intervals.

Why We can test each interval to see whether numbers in the interval are in the solution set of the inequality.

Caution

When solving rational inequalities such as $\frac{9}{x} < 8$, a common error is to multiply both sides by x to clear it of the fraction. However, we don't know whether x is positive or negative, so we don't know whether or not to reverse the inequality symbol.

Solution To get 0 on the right side, we subtract 8 from both sides. We then find a common denominator to write the left side as a single quotient.

$$\frac{9}{x} < 8 \quad \text{This is the inequality to solve.}$$

$$\frac{9}{x} - 8 < 0 \quad \text{Subtract 8 from both sides.}$$

$$\frac{9}{x} - 8 \cdot \frac{x}{x} < 0 \quad \text{To write the left side as a single quotient, build 8 to a fraction with denominator } x.$$

$$\frac{9}{x} - \frac{8x}{x} < 0$$

$$\frac{9 - 8x}{x} < 0 \quad \text{Subtract the numerators and keep the common denominator, } x.$$

Now we solve the related rational equation.

Success Tip

Recall that if a fraction is equal to 0, its numerator must be 0.

$$\frac{9 - 8x}{x} = 0$$

$$9 - 8x = 0 \quad \text{If } x \neq 0\text{, we can clear the equation of the fraction by multiplying both sides by } x.$$

$$-8x = -9 \quad \text{To isolate the variable term } -8x\text{, subtract 9 from both sides.}$$

$$x = \frac{9}{8} \quad \text{Solve for } x\text{. This is a critical number.}$$

If we set the denominator of $\frac{9 - 8x}{x}$ equal to 0, we obtain a second critical number, $x = 0$. When graphed, the critical numbers 0 and $\frac{9}{8}$ separate the number line into three intervals. We pick a test value from each interval to see whether it satisfies $\frac{9 - 8x}{x} < 0$.

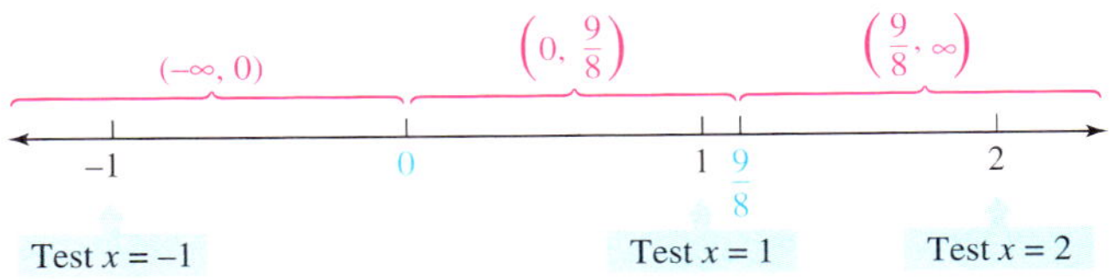

$\frac{9 - 8x}{x} < 0$	$\frac{9 - 8x}{x} < 0$	$\frac{9 - 8x}{x} < 0$
$\frac{9 - 8(-1)}{-1} \stackrel{?}{<} 0$	$\frac{9 - 8(1)}{1} \stackrel{?}{<} 0$	$\frac{9 - 8(2)}{2} \stackrel{?}{<} 0$
$-17 < 0$ True	$1 < 0$ False	$-\frac{7}{2} < 0$ True

The numbers in the intervals $(-\infty, 0)$ and $\left(\frac{9}{8}, \infty\right)$ satisfy the inequality. We do not include the endpoint 0 in the solution set, because it makes the denominator of the original inequality 0. Neither do we include $\frac{9}{8}$, because it does not satisfy $\frac{9 - 8x}{x} < 0$. (Recall that $\frac{9}{8}$ makes $\frac{9 - 8x}{x}$ equal to 0.) Thus, the solution set is the union of two intervals: $(-\infty, 0) \cup \left(\frac{9}{8}, \infty\right)$. Its graph is shown on the right.

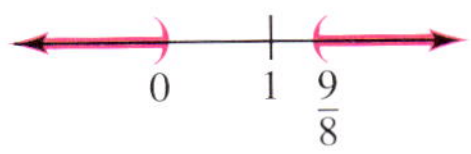

Self Check 3 Solve: $\frac{3}{x} < 5$

Now Try ▶ Problem 23

EXAMPLE 4 Solve: $\frac{x^2 + x - 2}{x - 4} \geq 0$

Strategy This inequality is in standard form. We will solve its related rational equation to find any critical numbers. These critical numbers will separate the number line into intervals.

Why We can test each interval to see whether numbers in the interval are in the solution set of the inequality.

Solution To solve the related rational equation, we proceed as follows:

$$\frac{x^2 + x - 2}{x - 4} = 0$$

$$x^2 + x - 2 = 0 \quad \text{If } x \neq 4 \text{, we can clear the equation of the fraction by multiplying both sides by } x - 4.$$

$$(x + 2)(x - 1) = 0 \quad \text{Factor the trinomial.}$$

$$x + 2 = 0 \quad \text{or} \quad x - 1 = 0 \quad \text{Set each factor equal to 0.}$$

$$x = -2 \quad | \quad x = 1 \quad \text{These are critical numbers.}$$

Caution

We cannot multiply both sides of the inequality by $x - 4$ to clear it of fractions. We don't know if $x - 4$ is positive or negative, so we don't know whether or not to reverse the inequality symbol.

If we set the denominator of $\frac{x^2 + x - 2}{x - 4}$ equal to 0, we see that $x = 4$ is also a critical number. When graphed, the critical numbers, -2, 1, and 4, separate the number line into four intervals. We pick a test value from each interval to see whether it satisfies $\frac{x^2 + x - 2}{x - 4} \geq 0$.

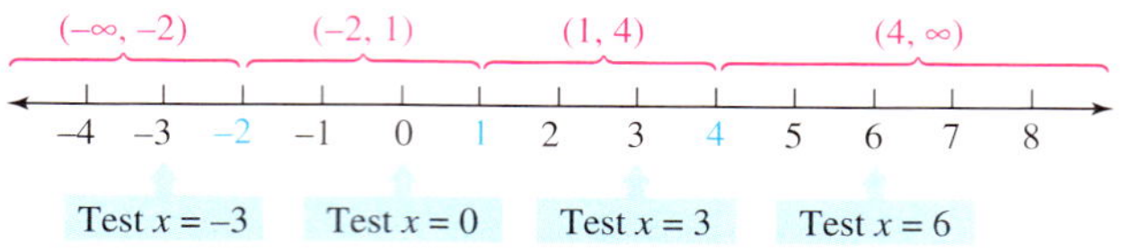

$$\frac{(-3)^2 + (-3) - 2}{-3 - 4} \overset{?}{\geq} 0 \qquad \frac{0^2 + 0 - 2}{0 - 4} \overset{?}{\geq} 0 \qquad \frac{3^2 + 3 - 2}{3 - 4} \overset{?}{\geq} 0 \qquad \frac{6^2 + 6 - 2}{6 - 4} \overset{?}{\geq} 0$$

$$-\frac{4}{7} \geq 0 \qquad \frac{1}{2} \geq 0 \qquad -10 \geq 0 \qquad 20 \geq 0$$

False True False True

The numbers in the intervals $(-2, 1)$ and $(4, \infty)$ satisfy the inequality. We include the endpoints -2 and 1 in the solution set because they satisfy the inequality. We do not include 4 because it makes the denominator of the fraction 0. Thus, the solution set is the union of two intervals $[-2, 1] \cup (4, \infty)$, as graphed on the right.

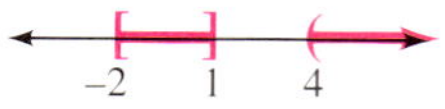

Self Check 4 Solve: $\frac{x + 2}{x^2 - 2x - 3} \geq 0$

Now Try Problem 27

EXAMPLE 5 Solve: $\frac{3}{x - 1} < \frac{2}{x}$

Strategy We will subtract $\frac{2}{x}$ from both sides to get 0 on the right side and solve the resulting related rational equation to find any critical numbers. These critical numbers will separate the number line into intervals.

Why We can test each interval to see whether numbers in the interval are in the solution set of the inequality.

Solution

$$\frac{3}{x-1} < \frac{2}{x}$$ This is the inequality to solve.

$$\frac{3}{x-1} - \frac{2}{x} < 0$$ Subtract $\frac{2}{x}$ from both sides.

$$\frac{3}{x-1} \cdot \frac{x}{x} - \frac{2}{x} \cdot \frac{x-1}{x-1} < 0$$ To get a single quotient on the left side, build each rational expression to have the common denominator $x(x - 1)$.

$$\frac{3x - 2x + 2}{x(x-1)} < 0$$ Subtract the numerators and keep the common denominator.

$$\frac{x+2}{x(x-1)} < 0$$ Combine like terms.

The only solution of the related rational equation $\frac{x+2}{x(x-1)} = 0$ is -2. Thus, -2 is a critical number. When we set the denominator equal to 0 and solve $x(x - 1) = 0$, we find two more critical numbers, 0 and 1. These three critical numbers create four intervals to test.

Success Tip

When the endpoints of an interval are consecutive integers, such as with the third interval $(0, 1)$, we cannot choose an integer as a test value. For these cases, choose a fraction or decimal that lies within the interval.

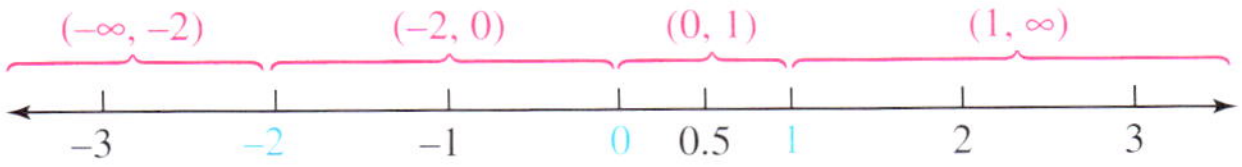

Test $x = -3$ | Test $x = -1$ | Test $x = 0.5$ | Test $x = 3$

$$\frac{-3+2}{-3(-3-1)} \overset{?}{<} 0 \qquad \frac{-1}{-3(-4)} \overset{?}{<} 0 \qquad -\frac{1}{12} < 0$$

True

$$\frac{-1+2}{-1(-1-1)} \overset{?}{<} 0 \qquad \frac{1}{-1(-2)} \overset{?}{<} 0 \qquad \frac{1}{2} < 0$$

False

$$\frac{0.5+2}{0.5(0.5-1)} \overset{?}{<} 0 \qquad \frac{2.5}{0.5(-0.5)} \overset{?}{<} 0 \qquad -10 < 0$$

True

$$\frac{3+2}{3(3-1)} \overset{?}{<} 0 \qquad \frac{5}{3(2)} \overset{?}{<} 0 \qquad \frac{5}{6} < 0$$

False

The numbers 0 and 1 are not included in the solution set because they make the denominator 0, and the number -2 is not included because it does not satisfy the inequality. The solution set is the union of two intervals $(-\infty, -2) \cup (0, 1)$, as graphed on the right.

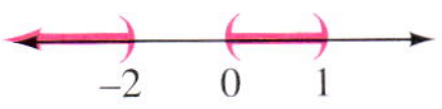

Self Check 5 Solve: $\frac{2}{x+1} > \frac{1}{x}$

Now Try ▶ Problem 31

Using Your Calculator ▶ Solving Inequalities Graphically

We can solve $x^2 + 4x \geq 5$ (Example 2) graphically by writing the inequality as $x^2 + 4x - 5 \geq 0$ and graphing the quadratic function $f(x) = x^2 + 4x - 5$, as shown in figure (a) on the next page. The solution set of the inequality will be those values of x for which the graph lies on or above the x-axis. We can trace to determine that this is the union of two intervals: $(-\infty, -5] \cup [1, \infty)$.

To solve $\frac{3}{x-1} < \frac{2}{x}$ (Example 5) graphically, we first write the inequality in the form $\frac{x+2}{x(x-1)} < 0$ and then graph the rational function $f(x) = \frac{x+2}{x(x-1)}$, as shown in figure (b) on the next page. The solution of the inequality will be those values of x for which the graph lies below the axis.

We can trace to see that the graph is below the x-axis when x is less than -2. Since we cannot see the graph in the interval $0 < x < 1$, we redraw the graph using window settings of $[-1, 2]$ for x and $[-25, 10]$ for y, as shown in figure (c).

Now we see that the graph is below the x-axis in the interval $(0, 1)$. Thus, the solution set of the inequality is the union of the two intervals: $(-\infty, -2) \cup (0, 1)$.

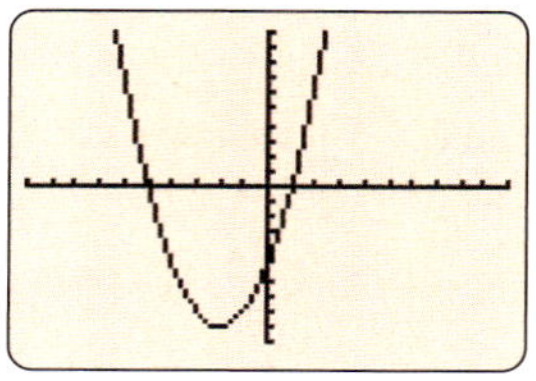
(a)

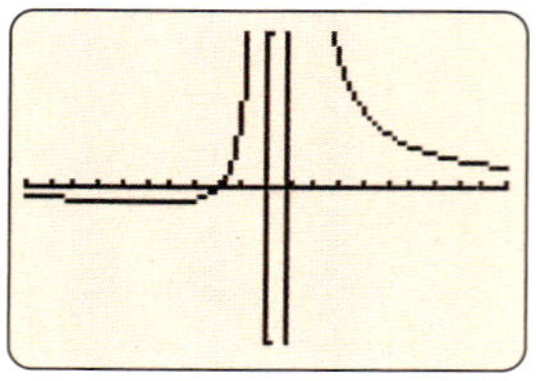
(b)

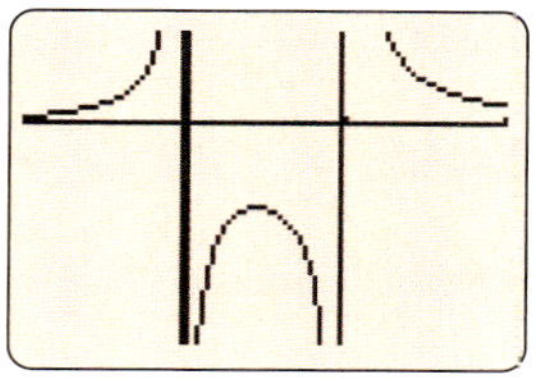
(c)

3 Graph Nonlinear Inequalities in Two Variables.

We have previously graphed linear inequalities in two variables such as $y > 3x + 2$ and $2x - 3y \leq 6$ using the following steps.

Graphing Inequalities in Two Variables

1. Graph the related equation to find the boundary line of the region. If the inequality allows equality (the symbol is either $\leq$ or $\geq$), draw the boundary as a solid line. If equality is not allowed ($<$ or $>$), draw the boundary as a dashed line.
2. Pick a test point that is on one side of the boundary line. (Use the origin if possible.) Replace x and y in the original inequality with the coordinates of that point. If the inequality is satisfied, shade the side that contains that point. If the inequality is not satisfied, shade the other side of the boundary.

We use the same procedure to graph **nonlinear inequalities** in two variables.

EXAMPLE 6 Graph: $y < -x^2 + 4$

Strategy We will graph the related equation $y = -x^2 + 4$ to establish a boundary parabola. Then we will determine which side of the boundary parabola represents the solution set of the inequality.

Why To *graph a nonlinear inequality* in two variables means to draw a "picture" of the ordered pairs (x, y) that make the inequality true.

Solution The graph of the boundary $y = -x^2 + 4$ is a parabola opening downward, with vertex at $(0, 4)$ and axis of symmetry $x = 0$ (the y-axis). Since the inequality contains an $<$ symbol and equality is not allowed, we draw the parabola using a dashed curve.

To determine which region to shade, we pick the test point $(0, 0)$ and substitute its coordinates into the inequality. We shade the region containing $(0, 0)$ because its coordinates satisfy $y < -x^2 + 4$.

Solve each inequality. Write the solution set in interval notation and graph it. See Example 5.

31. $\dfrac{3}{x-2} < \dfrac{4}{x}$

32. $\dfrac{-6}{x+1} \geq \dfrac{1}{x}$

33. $\dfrac{7}{x-3} \geq \dfrac{2}{x+4}$

34. $\dfrac{-5}{x-4} < \dfrac{3}{x+1}$

Graph each inequality. See Example 6.

35. $y < x^2 + 1$

36. $y > x^2 - 3$

37. $y \leq x^2 + 5x + 6$

38. $y \geq x^2 + 5x + 4$

Graph each inequality. See Example 7.

39. $y < |x + 4|$

40. $y \leq |x - 3|$

41. $y \geq -|x| + 2$

42. $y > |x| - 2$

Use a graphing calculator to solve each inequality. Write the solution set in interval notation. See Using Your Calculator: Solving Inequalities Graphically.

43. $x^2 - 2x - 3 < 0$

44. $x^2 + x - 6 > 0$

45. $\dfrac{x+3}{x-2} > 0$

46. $\dfrac{3}{x} < 2$

TRY IT YOURSELF

Solve each inequality. Write the solution set in interval notation and graph it.

47. $\dfrac{x}{x+4} \leq \dfrac{1}{x+1}$

48. $\dfrac{x}{x+9} \geq \dfrac{1}{x+1}$

49. $x^2 \geq 9$

50. $x^2 \geq 16$

51. $x^2 + 6x \geq -9$

52. $x^2 + 8x < -16$

53. $\dfrac{x^2 + x - 2}{x - 3} > 0$

54. $\dfrac{x-2}{x^2-1} > 0$

55. $2x^2 - 50 < 0$

56. $3x^2 - 243 < 0$

57. $\dfrac{2x-3}{3x+1} < 0$

58. $\dfrac{x-5}{x+1} < 0$

59. $x^2 - 6x + 9 < 0$

60. $x^2 + 4x + 4 > 0$

61. $\dfrac{5}{x+1} > \dfrac{3}{x-4}$

62. $\dfrac{3}{x-2} \leq -\dfrac{2}{x+3}$

APPLICATIONS

63. **Bridges.** If an x-axis is superimposed over the roadway of the Golden Gate Bridge, with the origin at the center of the bridge, the length L in feet of a vertical support cable can be approximated by the formula

$$L = \frac{1}{9{,}000}x^2 + 5$$

For the Golden Gate Bridge, $-2{,}100 < x < 2{,}100$. For what intervals along the x-axis are the vertical cables more than 95 feet long?

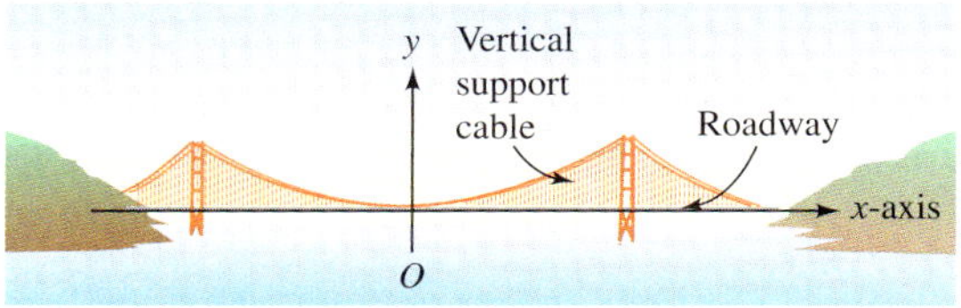

64. **Malls.** The number of people n in a mall is modeled by the formula

$$n = -100x^2 + 1{,}200x$$

where x is the number of hours since the mall opened. If the mall opened at 9 A.M., when were there 2,000 or more people in it?

WRITING

65. How are critical numbers used when solving a quadratic inequality in one variable?

66. Explain how to graph $y \geq x^2$.

67. The graph of $f(x) = x^2 - 3x + 4$ is shown here. Explain why the quadratic inequality $x^2 - 3x + 4 < 0$ has no solution.

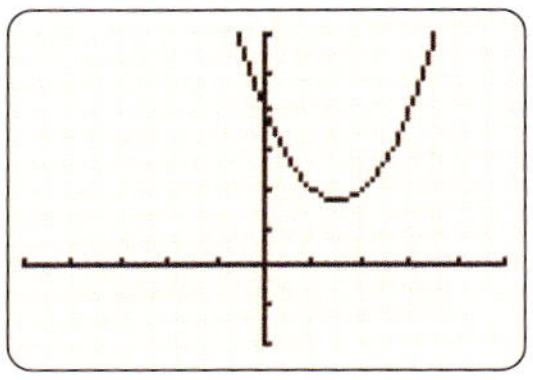

68. Describe the following solution set of a rational inequality in words: $(-\infty, 4] \cup (6, 7)$.

REVIEW

Translate each statement into an equation.

69. x varies directly as y.

70. y varies inversely as t.

71. t varies jointly as x and y.

72. d varies directly as t and inversely as u^2.

CHALLENGE PROBLEMS

73. **a.** Solve: $x^2 - x - 12 > 0$

b. Find a rational inequality in one variable that has the same solution set as the quadratic inequality in part (a).

74. **a.** Solve: $\dfrac{1}{x} < 1$

b. Now incorrectly "solve" $\frac{1}{x} < 1$ by multiplying both sides by x to clear it of the fraction. What part of the solution set is not obtained with this incorrect approach?

10 Summary & Review

SECTION 10.1 The Square Root Property and Completing the Square

DEFINITIONS AND CONCEPTS	EXAMPLES
We can use the **square root property** to solve equations of the form $x^2 = c$, where $c > 0$. The two solutions are $x = \sqrt{c}$ or $x = -\sqrt{c}$ We can write $x = \sqrt{c}$ or $x = -\sqrt{c}$ in more compact form using **double-sign notation:** $x = \pm\sqrt{c}$	Solve: $x^2 = 24$ $x = \sqrt{24}$ or $x = -\sqrt{24}$ Use the square root property. $x = \pm\sqrt{24}$ Use double-sign notation. $x = \pm 2\sqrt{6}$ Simplify $\sqrt{24}$. The solutions are $2\sqrt{6}$ and $-2\sqrt{6}$. Verify this using a check. Solve: $(x - 3)^2 = -81$ $x - 3 = \pm\sqrt{-81}$ Use the square root property and double-sign notation. $x = 3 \pm \sqrt{-81}$ To isolate x, add 3 to both sides. $x = 3 \pm 9i$ Simplify the radical expression. The solutions are $3 + 9i$ and $3 - 9i$. Verify this using a check.
To **complete the square** on $x^2 + bx$, add the square of one-half of the coefficient of x. $x^2 + bx + \left(\frac{1}{2}b\right)^2$	Complete the square on $x^2 + 8x$ and factor the resulting perfect-square trinomial. $x^2 + 8x + 16$ The coefficient of x is 8. To complete the square: $\frac{1}{2} \cdot 8 = 4$ and $4^2 = 16$. Add 16 to the binomial. Now we factor: $x^2 + 8x + 16 = (x + 4)^2$
To **solve a quadratic equation in x by completing the square:** **1.** If necessary, divide both sides of the equation by the coefficient of x^2 to make its coefficient 1. **2.** Get all variable terms on one side of the equation and all constants on the other side. **3.** Complete the square. **4.** Factor the perfect-square trinomial. **5.** Solve the resulting equation by using the square root property. **6.** Check your answers in the original equation.	Solve: $3x^2 - 12x + 6 = 0$ $\frac{3x^2}{3} - \frac{12x}{3} + \frac{6}{3} = \frac{0}{3}$ To make the leading coefficient 1, divide both sides by 3, term-by-term. $x^2 - 4x + 2 = 0$ Do the division. $x^2 - 4x = -2$ Subtract 2 from both sides so that the constant term, −2, is on the right side. $x^2 - 4x + 4 = -2 + 4$ The coefficient of x is −4. To complete the square: $\frac{1}{2}(-4) = -2$ and $(-2)^2 = 4$. Add 4 to both sides. $(x - 2)^2 = 2$ Factor the perfect-square trinomial on the left side. Add on the right side. $x - 2 = \pm\sqrt{2}$ Use the square root property. $x = 2 \pm \sqrt{2}$ To isolate x, add 2 to both sides. The solutions are $2 + \sqrt{2}$ and $2 - \sqrt{2}$. Verify this using a check.

REVIEW EXERCISES

Solve each equation by factoring.

1. $x^2 + 9x + 20 = 0$
2. $6x^2 + 17x + 5 = 0$

Solve each equation using the square root property. Approximate the solutions to the nearest hundredth when appropriate.

3. $x^2 = 28$
4. $(t + 2)^2 = 36$
5. $a^2 + 25 = 0$
6. $5x^2 - 49 = 0$
7. Solve $A = \pi r^2$ for r. Assume all variables represent positive numbers. Express the result in simplified radical form.
8. Complete the square on $x^2 - x$ and then factor the resulting perfect-square trinomial.

Solve each equation by completing the square. Approximate the solutions to the nearest hundredth when appropriate.

9. $x^2 + 6x + 8 = 0$
10. $2x^2 - 6x + 3 = 0$

11. $6a^2 - 12a = -1$

12. $x^2 - 2x = -13$

13. Let $f(x) = x^2$. Find all values of x for which $f(x) = 32$.

14. Let $g(x) = (7x - 51)^2$. Find all values of x for which $g(x) = 11$.

15. Explain why completing the square on $x^2 + 7x$ is more difficult than completing the square on $x^2 + 6x$.

16. **Happy New Year.** As part of a New Year's Eve celebration, a huge ball is to be dropped from the top of a 605-foot-tall building at the proper moment so that it strikes the ground at exactly 12:00 midnight. The distance d in feet traveled by a free-falling object in t seconds is given by the formula $d = 16t^2$. To the nearest second, when should the ball be dropped from the building?

SECTION 10.2 The Quadratic Formula

DEFINITIONS AND CONCEPTS

To **solve a quadratic equation in x using the quadratic formula:**

1. Write the equation in standard form:

$$ax^2 + bx + c = 0$$

2. Identify a, b, and c.
3. Substitute the values for a, b, and c in the quadratic formula

$$x = \frac{-b \pm \sqrt{b^2 - 4ac}}{2a}$$

and evaluate the right side to obtain the solutions.

EXAMPLES

Solve: $3x^2 - 2x - 2 = 0$

Here, $a = 3$, $b = -2$, and $c = -2$.

$$x = \frac{-b \pm \sqrt{b^2 - 4ac}}{2a}$$ This is the quadratic formula.

$$x = \frac{-(-2) \pm \sqrt{(-2)^2 - 4(3)(-2)}}{2(3)}$$ Substitute 3 for *a*, −2 for *b*, and −2 for *c*.

$$x = \frac{2 \pm \sqrt{28}}{6}$$ Evaluate the expression within the radical.

$$x = \frac{2 \pm 2\sqrt{7}}{6}$$ Simplify the radical: $\sqrt{28} = \sqrt{4}\sqrt{7} = 2\sqrt{7}$.

$$x = \frac{\overset{1}{\cancel{2}}(1 \pm \sqrt{7})}{\underset{1}{\cancel{2}} \cdot 3}$$ Factor out the GCF, 2, from the two terms in the numerator. In the denominator, factor 6 as $2 \cdot 3$. Then remove the common factor, 2.

$$x = \frac{1 \pm \sqrt{7}}{3}$$

The exact solutions are $\frac{1 + \sqrt{7}}{3}$ and $\frac{1 - \sqrt{7}}{3}$. We can use a calculator to approximate them. To the nearest hundredth, they are 1.22 and −0.55.

When solving a quadratic equation using the quadratic formula, we can often **simplify the calculations** by solving an equivalent equation that does not involve fractions or decimals, and whose leading coefficient is positive.

Before solving . . .	**do this . . .**	**to get this**
$-3x^2 + 5x - 1 = 0$	Multiply both sides by −1.	$3x^2 - 5x + 1 = 0$
$x^2 + \frac{7}{8}x - \frac{1}{2} = 0$	Multiply both sides by 8.	$8x^2 + 7x - 4 = 0$
$60x^2 - 40x + 90 = 0$	Divide both sides by 10.	$6x^2 - 4x + 9 = 0$
$0.05x^2 + 0.16x + 0.71 = 0$	Multiply both sides by 100.	$5x^2 + 16x + 71 = 0$

REVIEW EXERCISES

Solve each equation using the quadratic formula. Approximate the solutions to the nearest hundredth when appropriate.

17. $2x^2 + 13x = 7$

18. $-x^2 + 10x - 18 = 0$

19. $x^2 - 10x = 0$

20. $3y^2 = 26y - 2$

21. $\frac{1}{3}p^2 + \frac{1}{2}p + \frac{1}{2} = 0$

22. $3{,}000t^2 - 4{,}000t = -2{,}000$

23. $0.5x^2 + 0.3x - 0.1 = 0$

24. $x^2 - 3x - 27 = 0$

25. Let $h(x) = x^2 + 3x - 7$. Find all values of x for which $h(x) = 1$.

26. Let $T(x) = -4x^2 - 2x$. Find all values of x for which $T(x) = -3$.

27. Explain the error: $\frac{2 \pm \sqrt{7}}{2} = \frac{\overset{1}{\cancel{2}} \pm \sqrt{7}}{\underset{1}{\cancel{2}}}$

28. a. Write an expression that represents the width of the larger rectangle shown in red.

b. Write an expression that represents the length of the larger rectangle shown in red.

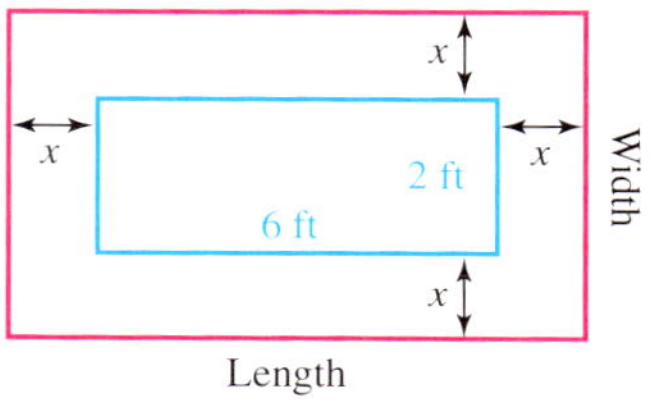

29. **Posters.** The specifications for a poster of Cesar Chavez call for a 615-square-inch photograph to be surrounded by a green border. The borders on the top and bottom of the poster are to be twice as wide as those on the sides. Find the width of each border.

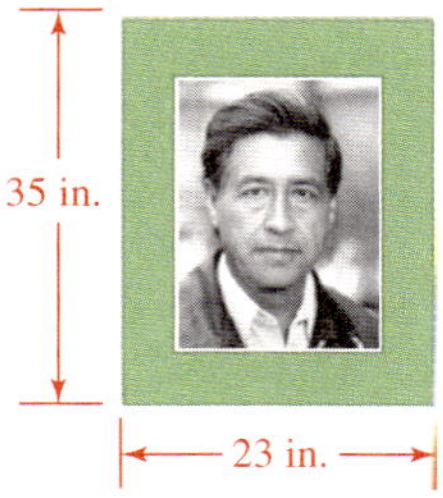

30. **Tutoring.** A private tutoring company charges \$20 for a 1-hour session. Currently, 300 students are tutored each week. Since the company is losing money, the owner has decided to increase the price. For each 50¢ increase, she estimates that 5 fewer students will participate. If the company needs to bring in \$6,240 per week to stay in business, what price must be charged for a 1-hour tutoring session to produce this amount of revenue?

31. **Acrobats.** To begin his routine on a trapeze, an acrobat is catapulted upward as shown in the illustration. His distance d (in feet) from the arena floor during this maneuver is given by the function $d(t) = -16t^2 + 40t + 5$, where t is the time in seconds since being launched. If the trapeze bar is 25 feet in the air, at what two times will he be able to grab it? Round to the nearest tenth.

32. **Triangles.** The length of the longer leg of a right triangle exceeds the length of the shorter leg by 23 inches and the length of the hypotenuse is 65 inches. Find the length of each leg of the triangle.

SECTION 10.3 The Discriminant and Equations That Can Be Written in Quadratic Form

DEFINITIONS AND CONCEPTS	EXAMPLES
The **discriminant** predicts the type of solutions of $ax^2 + bx + c = 0$, where a, b, and c are rational numbers and $a \neq 0$. Review the table on page 852.	In the quadratic equation $2x^2 - 5x - 3 = 0$, we have $a = 2$, $b = -5$, and $c = -3$. So the value of the discriminant is $b^2 - 4ac = (-5)^2 - 4(2)(-3) = 25 + 24 = 49$ Since the value of the discriminant is positive and a perfect square, the equation $2x^2 - 5x - 3 = 0$ has two different rational-number solutions.

Solution

$$\frac{3}{x-1} < \frac{2}{x}$$ This is the inequality to solve.

$$\frac{3}{x-1} - \frac{2}{x} < 0$$ Subtract $\frac{2}{x}$ from both sides.

$$\frac{3}{x-1} \cdot \frac{x}{x} - \frac{2}{x} \cdot \frac{x-1}{x-1} < 0$$ To get a single quotient on the left side, build each rational expression to have the common denominator $x(x-1)$.

$$\frac{3x - 2x + 2}{x(x-1)} < 0$$ Subtract the numerators and keep the common denominator.

$$\frac{x+2}{x(x-1)} < 0$$ Combine like terms.

The only solution of the related rational equation $\frac{x+2}{x(x-1)} = 0$ is -2. Thus, -2 is a critical number. When we set the denominator equal to 0 and solve $x(x-1) = 0$, we find two more critical numbers, 0 and 1. These three critical numbers create four intervals to test.

Success Tip

When the endpoints of an interval are consecutive integers, such as with the third interval $(0, 1)$, we cannot choose an integer as a test value. For these cases, choose a fraction or decimal that lies within the interval.

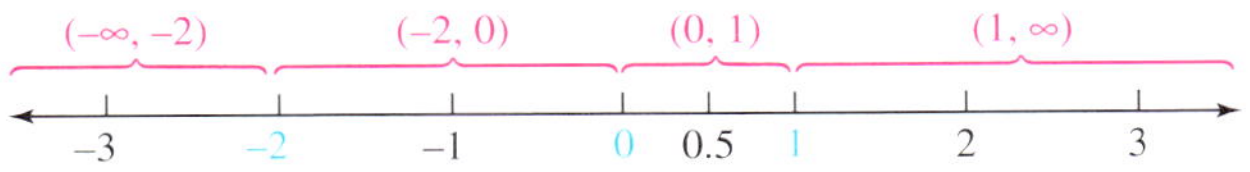

Test $x = -3$	Test $x = -1$	Test $x = 0.5$	Test $x = 3$
$\frac{-3+2}{-3(-3-1)} \stackrel{?}{<} 0$	$\frac{-1+2}{-1(-1-1)} \stackrel{?}{<} 0$	$\frac{0.5+2}{0.5(0.5-1)} \stackrel{?}{<} 0$	$\frac{3+2}{3(3-1)} \stackrel{?}{<} 0$
$\frac{-1}{-3(-4)} \stackrel{?}{<} 0$	$\frac{1}{-1(-2)} \stackrel{?}{<} 0$	$\frac{2.5}{0.5(-0.5)} \stackrel{?}{<} 0$	$\frac{5}{3(2)} \stackrel{?}{<} 0$
$-\frac{1}{12} < 0$	$\frac{1}{2} < 0$	$-10 < 0$	$\frac{5}{6} < 0$
True	False	True	False

The numbers 0 and 1 are not included in the solution set because they make the denominator 0, and the number -2 is not included because it does not satisfy the inequality. The solution set is the union of two intervals $(-\infty, -2) \cup (0, 1)$, as graphed on the right.

Self Check 5 Solve: $\frac{2}{x+1} > \frac{1}{x}$

Now Try ▶ Problem 31

Using Your Calculator ▶ Solving Inequalities Graphically

We can solve $x^2 + 4x \geq 5$ (Example 2) graphically by writing the inequality as $x^2 + 4x - 5 \geq 0$ and graphing the quadratic function $f(x) = x^2 + 4x - 5$, as shown in figure (a) on the next page. The solution set of the inequality will be those values of x for which the graph lies on or above the x-axis. We can trace to determine that this is the union of two intervals: $(-\infty, -5] \cup [1, \infty)$.

To solve $\frac{3}{x-1} < \frac{2}{x}$ (Example 5) graphically, we first write the inequality in the form $\frac{x+2}{x(x-1)} < 0$ and then graph the rational function $f(x) = \frac{x+2}{x(x-1)}$, as shown in figure (b) on the next page. The solution of the inequality will be those values of x for which the graph lies below the axis.

We can trace to see that the graph is below the x-axis when x is less than -2. Since we cannot see the graph in the interval $0 < x < 1$, we redraw the graph using window settings of $[-1, 2]$ for x and $[-25, 10]$ for y, as shown in figure (c).

Now we see that the graph is below the x-axis in the interval $(0, 1)$. Thus, the solution set of the inequality is the union of the two intervals: $(-\infty, -2) \cup (0, 1)$.

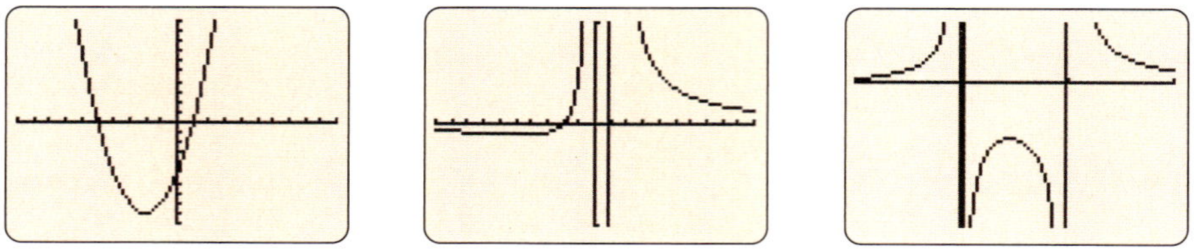

(a) (b) (c)

3 Graph Nonlinear Inequalities in Two Variables.

We have previously graphed linear inequalities in two variables such as $y > 3x + 2$ and $2x - 3y \le 6$ using the following steps.

Graphing Inequalities in Two Variables

1. Graph the related equation to find the boundary line of the region. If the inequality allows equality (the symbol is either $\le$ or $\ge$), draw the boundary as a solid line. If equality is not allowed ($<$ or $>$), draw the boundary as a dashed line.
2. Pick a test point that is on one side of the boundary line. (Use the origin if possible.) Replace x and y in the original inequality with the coordinates of that point. If the inequality is satisfied, shade the side that contains that point. If the inequality is not satisfied, shade the other side of the boundary.

We use the same procedure to graph **nonlinear inequalities** in two variables.

EXAMPLE 6 Graph: $y < -x^2 + 4$

Strategy We will graph the related equation $y = -x^2 + 4$ to establish a boundary parabola. Then we will determine which side of the boundary parabola represents the solution set of the inequality.

Why To *graph a nonlinear inequality* in two variables means to draw a "picture" of the ordered pairs (x, y) that make the inequality true.

Solution The graph of the boundary $y = -x^2 + 4$ is a parabola opening downward, with vertex at $(0, 4)$ and axis of symmetry $x = 0$ (the y-axis). Since the inequality contains an $<$ symbol and equality is not allowed, we draw the parabola using a dashed curve.

To determine which region to shade, we pick the test point $(0, 0)$ and substitute its coordinates into the inequality. We shade the region containing $(0, 0)$ because its coordinates satisfy $y < -x^2 + 4$.

Graph the boundary

$y = -x^2 + 4$

Compare to $y = a(x - h)^2 + k$

$a = -1$: Opens downward

$h = 0$ and $k = 4$: Vertex $(0, 4)$

Axis of symmetry $x = 0$

x	y
1	3
2	0

***Shading: Use the test point* (0, 0)**

$y < -x^2 + 4$

$0 \overset{?}{<} -0^2 + 4$

$0 < 4$ True

Since $0 < 4$ is true, $(0, 0)$ is a solution of $y < -x^2 + 4$.

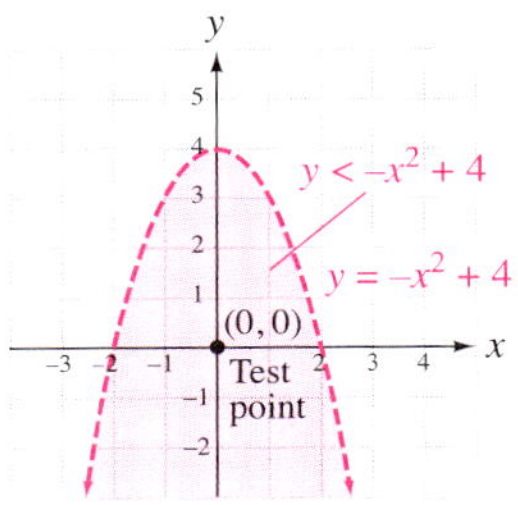

Self Check 6 Graph: $y \geq -x^2 + 4$

Now Try Problem 35

EXAMPLE 7 Graph: $x \leq |y|$

Strategy We will graph the related equation $x = |y|$ to establish a boundary. Then we will determine which side of the boundary represents the solution set of the inequality.

Why To *graph a nonlinear inequality* in two variables means to draw a "picture" of the ordered pairs (x, y) that make the inequality true.

Solution To graph the boundary, $x = |y|$, we construct a table of solutions, as shown in figure (a). In figure (b), the boundary is a solid line because the inequality contains a $\leq$ symbol and equality is permitted. Since the origin is on the graph, we cannot use it as a test point. However, any other point, such as $(1, 0)$, will do. We substitute 1 for x and 0 for y into the inequality to get

$x \leq |y|$ This is the inequality to graph.

$1 \overset{?}{\leq} |0|$ Substitute.

$1 \leq 0$ False

Since $1 \leq 0$ is a false statement, the point $(1, 0)$ does not satisfy the inequality and is not part of the graph. Thus, the graph of $x \leq |y|$ is to the left of the boundary.

The complete graph is shown in figure (c).

$x = |y|$

x	y
0	0
1	1
1	−1
2	2
2	−2

(a)

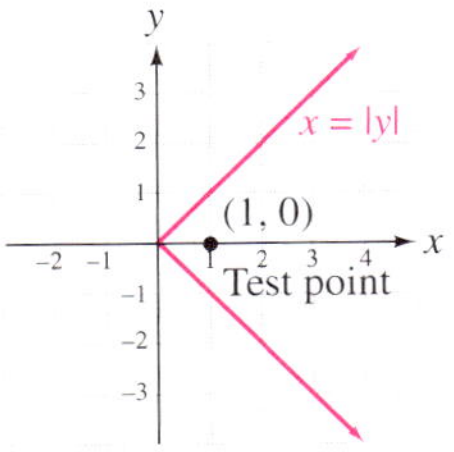

(b)

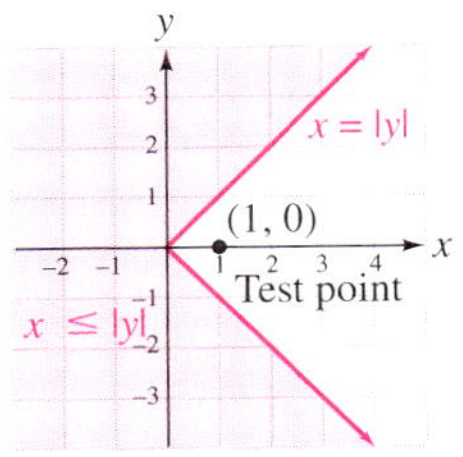

(c)

Self Check 7 Graph: $x \geq -|y|$

Now Try Problem 39

SECTION 10.5 STUDY SET

VOCABULARY

Fill in the blanks.

1. $x^2 + 3x - 18 < 0$ is an example of a ________ inequality in one variable.
2. $\frac{x-1}{x^2-x-20} \le 0$ is an example of a ________ inequality in one variable.
3. $y \le x^2 - 4x + 3$ is an example of a nonlinear inequality in _____ variables.
4. The set of real numbers greater than 3 can be represented using the ________ notation $(3, \infty)$.

CONCEPTS

5. The critical numbers of a quadratic inequality are highlighted in red on the number line shown below. Use interval notation to represent each interval that must be tested to solve the inequality.

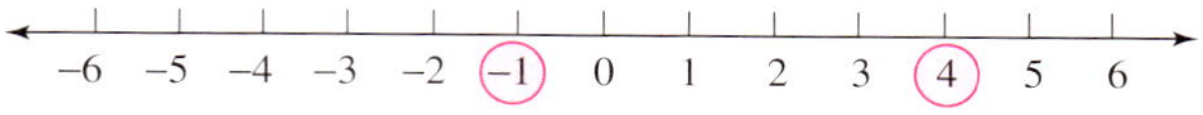

6. Graph each of the following solution sets.
   a. $(-2, 4)$ b. $(-\infty, -2) \cup (3, 5]$
7. The graph of the solution set of a rational inequality in one variable is shown. Determine whether each of the following numbers is a solution of the inequality.

   a. -10 b. -5
   c. 0 d. 4
8. What are the critical numbers for each inequality?
   a. $x^2 - 2x - 48 \ge 0$ b. $\frac{x-3}{x(x+4)} > 0$
9. a. The results after interval testing for a quadratic inequality containing a $>$ symbol are shown below. (The critical numbers are highlighted in red.) What is the solution set?

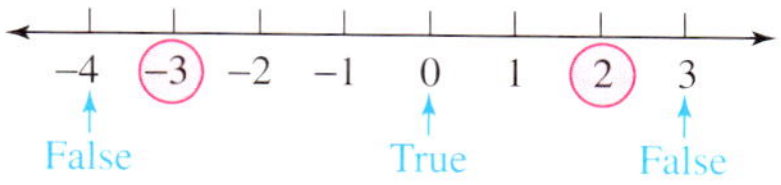

   b. The results after interval testing for a quadratic inequality containing a $\le$ symbol are shown below. (The critical numbers are highlighted in red.) What is the solution set?

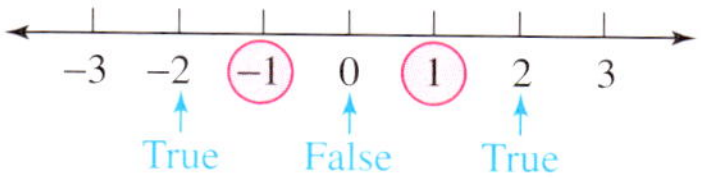

10. Fill in the blank to complete this important fact about the interval testing method discussed in this section: *If one number in an interval satisfies the inequality,* ___ *numbers in that interval will satisfy the inequality.*

11. a. When graphing the solution of $y \le x^2 + 2x + 1$, should the boundary be solid or dashed?
    b. Does the test point $(0, 0)$ satisfy the inequality?
12. a. Estimate the solution of $x^2 - x - 6 > 0$ using the graph of $y = x^2 - x - 6$ shown in figure (a) below.
    b. Estimate the solution of $\frac{x-3}{x} \le 0$ using the graph of $y = \frac{x-3}{x}$ shown in figure (b) below.

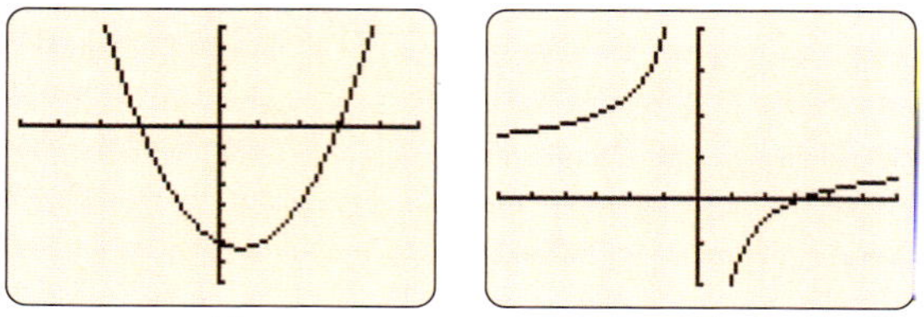

(a) (b)

NOTATION

13. Write the quadratic inequality $x^2 - 6x \ge 7$ in standard form.
14. The solution set of a rational inequality consists of the intervals $(-1, 4]$ and $(7, \infty)$. When writing the solution set, what symbol is used between the two intervals?

GUIDED PRACTICE

Solve each inequality. Write the solution set in interval notation and graph it. See Example 1.

15. $x^2 - 5x + 4 < 0$
16. $x^2 + 2x - 8 < 0$
17. $x^2 - 8x + 15 > 0$
18. $x^2 - 3x - 4 > 0$

Solve each inequality. Write the solution set in interval notation and graph it. See Example 2.

19. $x^2 - x \ge 42$
20. $x^2 - x \ge 72$
21. $x^2 + x \le 12$
22. $x^2 - 8x \le -15$

Solve each inequality. Write the solution set in interval notation and graph it. See Example 3.

23. $\frac{1}{x} < 2$
24. $\frac{1}{x} < 3$
25. $\frac{5}{x} \ge -3$
26. $\frac{4}{x} \ge 8$

Solve each inequality. Write the solution set in interval notation and graph it. See Example 4.

27. $\frac{x^2 - x - 12}{x - 1} < 0$
28. $\frac{x^2 + x - 6}{x - 4} \ge 0$
29. $\frac{6x^2 - 5x + 1}{2x + 1} \ge 0$
30. $\frac{6x^2 + 11x + 3}{3x - 1} < 0$

Definitions and Concepts	Examples
Equations that contain an expression, the same expression squared, and a constant term are said to be **quadratic in form.** One method used to solve such equations is to make a **substitution.** The leading term, x, is the square of the expression $\sqrt{x}$ in the middle term: $x = (\sqrt{x})^2$. $x - 9\sqrt{x} + 20 = 0$ The last term is a constant.	Solve: $x^{2/3} - 6x^{1/3} + 5 = 0$ The equation can be written in quadratic form: $(x^{1/3})^2 - 6x^{1/3} + 5 = 0$ Because $x^{2/3} = (x^{1/3})^2$ We substitute y for $x^{1/3}$ and use factoring to solve the resulting quadratic equation $y^2 - 6y + 5 = 0$. $(y - 1)(y - 5) = 0$ Let $y = x^{1/3}$. $y = 1$ or $y = 5$ Now we reverse (undo) the substitution $y = x^{1/3}$ and solve for x. $x^{1/3} = 1$ or $x^{1/3} = 5$ $(x^{1/3})^3 = (1)^3$ $(x^{1/3})^3 = (5)^3$ Recall: $x^{1/3} = \sqrt[3]{x}$. $x = 1$ $x = 125$ The solutions are 1 and 125. Check both in the original equation.

REVIEW EXERCISES

Use the discriminant to determine the number and type of solutions for each equation.

33. $3x^2 + 4x - 3 = 0$
34. $4x^2 - 5x + 7 = 0$
35. $3x^2 - 4x + \frac{4}{3} = 0$
36. $m(2m - 3) = 20$

Solve each equation.

37. $x - 13\sqrt{x} + 12 = 0$
38. $a^{2/3} + a^{1/3} - 6 = 0$
39. $3x^4 + x^2 - 2 = 0$
40. $\frac{6}{x + 2} + \frac{6}{x + 1} = 5$
41. $(x - 7)^2 + 6(x - 7) + 10 = 0$
42. $m^{-4} - 2m^{-2} + 1 = 0$
43. $4\left(\frac{x + 1}{x}\right)^2 + 12\left(\frac{x + 1}{x}\right) + 9 = 0$
44. $2m^{2/5} - 5m^{1/5} + 2 = 0$
45. **Weekly Chores.** Working together, two sisters can do the yard work at their house in 45 minutes. When the older girl does it all herself, she can complete the job in 20 minutes less time than it takes the younger girl working alone. How long does it take the older girl to do the yard work?
46. **Road Trips.** A woman drives her automobile 150 miles at a rate of r mph. She could have gone the same distance in 2 hours less time if she had increased her speed by 20 mph. Find r.

SECTION 10.4 Quadratic Functions and Their Graphs

DEFINITIONS AND CONCEPTS	EXAMPLES
A **quadratic function** is a second-degree polynomial function of the form $f(x) = ax^2 + bx + c$	Quadratic functions: $f(x) = 2x^2 - 3x + 5,$ $g(x) = -x^2 + 4x,$ and $s(x) = \frac{1}{4}x^2 - 10$

The graph of the quadratic function $f(x) = a(x - h)^2 + k$ where $a \neq 0$ is a **parabola** with **vertex** at (h, k). The **axis of symmetry** is the line $x = h$. The parabola opens upward when $a > 0$ and downward when $a < 0$.

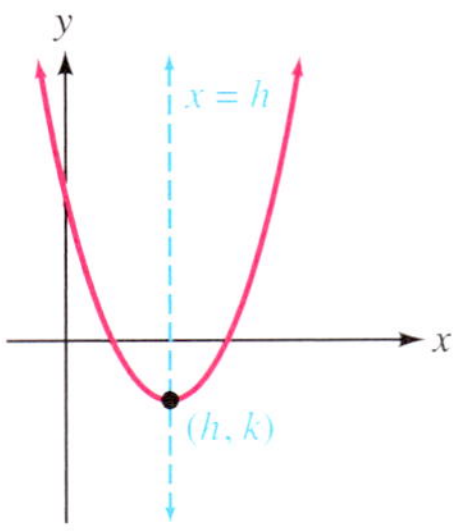

$f(x) = a(x - h)^2 + k$ where $a > 0$

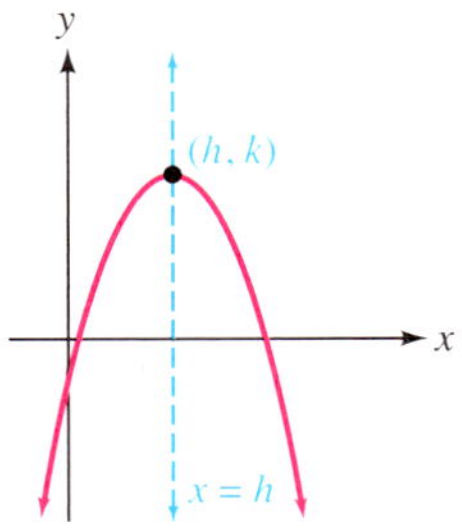

$f(x) = a(x - h)^2 + k$ where $a < 0$

Graph: $f(x) = 2(x + 1)^2 - 8$

$$f(x) = 2[x - (-1)]^2 - 8$$

$$f(x) = a(x - h)^2 + k$$

We see that $a = 2$, $h = -1$, and $k = -8$. The graph is a parabola with vertex $(h, k) = (-1, -8)$ and axis of symmetry $x = -1$. Since a is positive, the parabola opens upward.

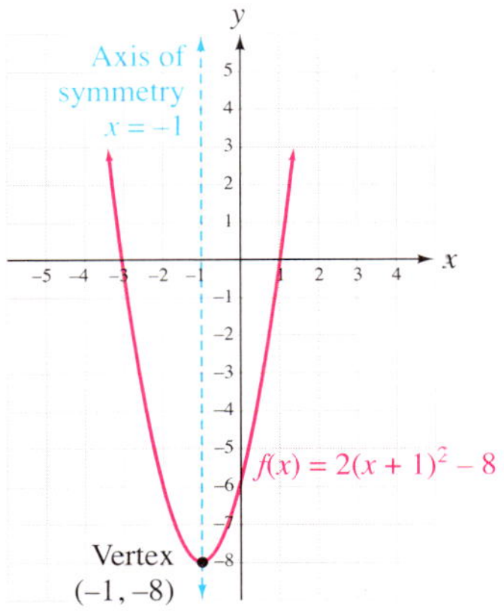

The **vertex** of the graph of $f(x) = ax^2 + bx + c$ is

$$\left(-\frac{b}{2a}, f\left(-\frac{b}{2a}\right)\right)$$

and the axis of symmetry is the line

$$x = -\frac{b}{2a}$$

The y-coordinate of the vertex of the graph of a quadratic function gives the **minimum or maximum value** of the function.

Graph: $f(x) = -x^2 + 3x + 4$

Here, $a = -1$, $b = 3$, and $c = 4$.

- Since $a < 0$, the graph opens downward.
- The x-coordinate of the vertex of the graph is:

$$-\frac{b}{2a} = -\frac{3}{2(-1)} = \frac{3}{2}$$

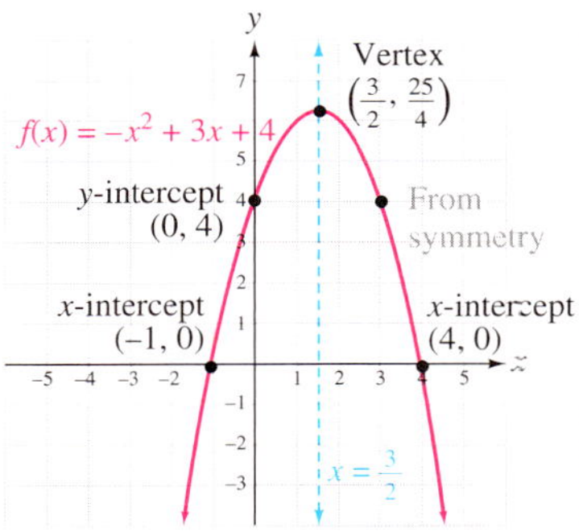

To find the y-coordinate of the vertex, we substitute $\frac{3}{2}$ for x in the function.

$$f\left(\frac{3}{2}\right) = -\left(\frac{3}{2}\right)^2 + 3\left(\frac{3}{2}\right) + 4 = \frac{25}{4}$$

The vertex of the parabola is $\left(\frac{3}{2}, \frac{25}{4}\right)$.

The **y-intercept** is determined by the value of $f(x)$ when $x = 0$: the y-intercept is $(0, c)$.

To find the **x-intercepts**, let $f(x) = 0$ and solve $ax^2 + bx + c = 0$.

- The y-intercept is the value of the function when $x = 0$. Since $f(0) = 4$, the y-intercept is $(0, 4)$.
- To find the x-intercepts, we solve:

$$-x^2 + 3x + 4 = 0 \quad \text{Let } f(x) = 0.$$

$$x^2 - 3x - 4 = 0 \quad \text{Multiply both sides by } -1.$$

$$(x + 1)(x - 4) = 0 \quad \text{Factor.}$$

$$x = -1 \quad \text{or} \quad x = 4$$

The x-intercepts are $(-1, 0)$ and $(4, 0)$.

REVIEW EXERCISES

47. Hospitals. The annual number of in-patient admissions to U.S. community hospitals for the years 1980–2008 can be modeled by the quadratic function $A(x) = 0.03x^2 - 0.88x + 37.3$, where $A(x)$ is the number of admissions in millions and x is the number of years after 1980. Use the function to estimate the number of in-patient admissions for the year 2005. Round to the nearest tenth of one million. (Source: American Hospital Association)

48. Fill in the blanks. The graph of the quadratic function $f(x) = a(x - h)^2 + k$ is a parabola with vertex at (,). The axis of symmetry is the line $\quad = h$. The parabola opens upward when $a > 0$ and downward when $a < 0$.

Graph each pair of functions on the same coordinate system.

49. $f(x) = 2x^2$, $g(x) = 2x^2 - 3$

50. $f(x) = -\frac{1}{4}x^2$, $g(x) = -\frac{1}{4}(x + 2)^2$

51. Find the vertex and the axis of symmetry of the graph of $f(x) = -2(x - 1)^2 + 4$. Then plot several points and complete the graph.

52. Complete the square to write $f(x) = 4x^2 + 16x + 9$ in the form $f(x) = a(x - h)^2 + k$. Determine the vertex and the axis of symmetry of the graph. Then plot several points and complete the graph.

53. Find the vertex of the graph of $f(x) = -2x^2 + 4x - 8$ using the vertex formula.

54. First determine the coordinates of the vertex and the axis of symmetry of the graph of $f(x) = x^2 + x - 2$ using the vertex formula. Then determine the x- and y-intercepts of the graph. Finally, plot several points and complete the graph.

55. Farming. The number of farms in the United States for the years 1870–1970 is approximated by

$$N(x) = -1{,}526x^2 + 155{,}652x + 2{,}500{,}200$$

where $x = 0$ represents 1870, $x = 1$ represents 1871, and so on. For this period, when was the number of U.S. farms a maximum? How many farms were there?

56. Estimate the solutions of $-3x^2 - 5x + 2 = 0$ from the graph of $f(x) = -3x^2 - 5x + 2$, shown here.

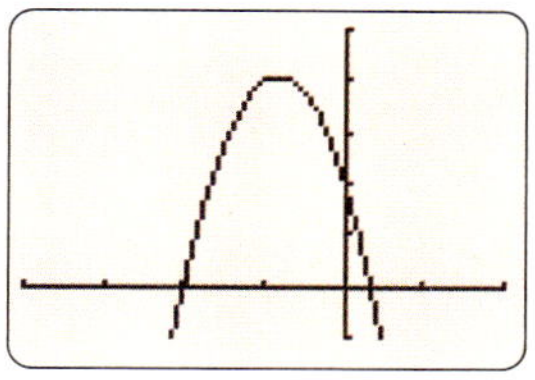

SECTION 10.5 Quadratic and Other Nonlinear Inequalities

DEFINITIONS AND CONCEPTS	EXAMPLES
To solve a quadratic inequality, get 0 on the right side and solve the related quadratic equation. Then locate the **critical numbers** on a number line, test each interval, and check the endpoints.	To solve $x^2 - x - 6 \geq 0$, we solve the related quadratic equation:

$$x^2 - x - 6 = 0$$

$$(x - 3)(x + 2) = 0 \qquad \text{Factor.}$$

$$x = 3 \quad \text{or} \quad x = -2 \qquad \text{These are the critical numbers that divide the number line into three intervals.}$$

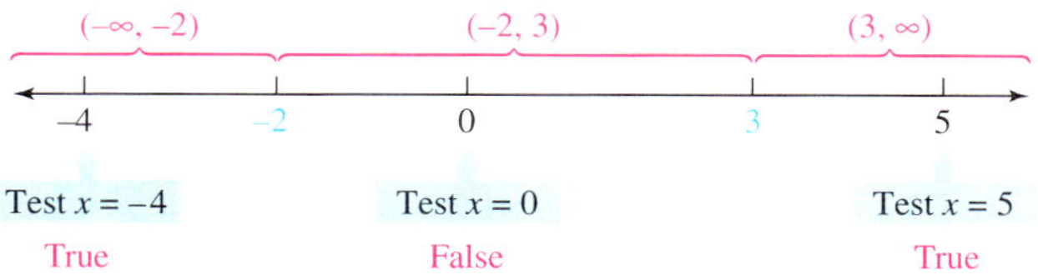

After testing each interval and noting that 3 and −2 satisfy the inequality, we see that the solution set is $(-\infty, -2] \cup [3, \infty)$.

To solve a rational inequality, get 0 on the right side and solve the related rational equation. Then locate the **critical numbers** (including any values that make the denominator 0) on a number line, test each interval, and check the endpoints.

To solve $\frac{x+1}{x-4} < 0$, we solve the related rational equation $\frac{x+1}{x-4} = 0$ to obtain the solution $x = -1$, which is a critical number. Another critical number is $x = 4$, the value that makes the denominator 0. These critical numbers divide the number line into three intervals.

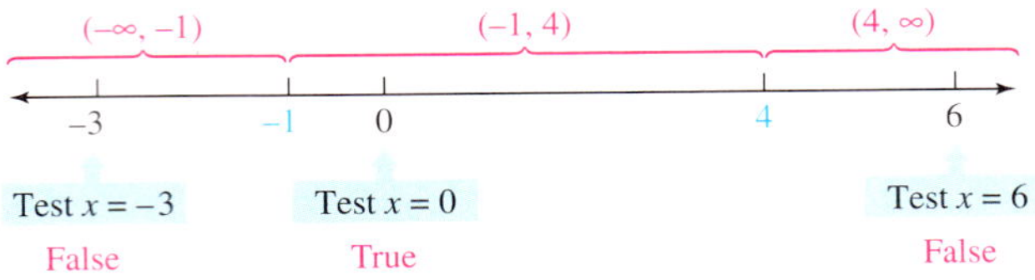

After testing each interval and noting that -1 and 4 do not satisfy the inequality, we see that the solution set is the interval $(-1, 4)$.

To graph a nonlinear inequality in two variables, first graph the boundary. Then use a test point to determine which side of the boundary to shade.

This is the graph of $y \leq x^2 + 5x + 4$.

Since the inequality contains the symbol $\leq$, and equality is allowed, we draw the parabola determined by $y = x^2 + 5x + 4$ using a **solid curve.**

We shade the region containing the test point $(0, 0)$ because its coordinates satisfy $y \leq x^2 + 5x + 4$.

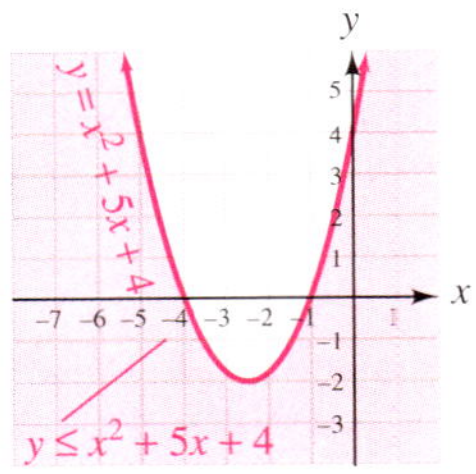

REVIEW EXERCISES

Solve each inequality. Write the solution set in interval notation and graph it.

57. $x^2 + 2x - 35 > 0$

58. $x^2 \leq 81$

59. $\frac{3}{x} \leq 5$

60. $\frac{2x^2 - x - 28}{x - 1} > 0$

61. Estimate the solution set of $3x^2 + 10x - 8 \leq 0$ from the graph of $f(x) = 3x^2 + 10x - 8$ shown in figure (a).

62. Estimate the solution set of $\frac{x-1}{x} > 0$ from the graph of $f(x) = \frac{x-1}{x}$ shown in figure (b).

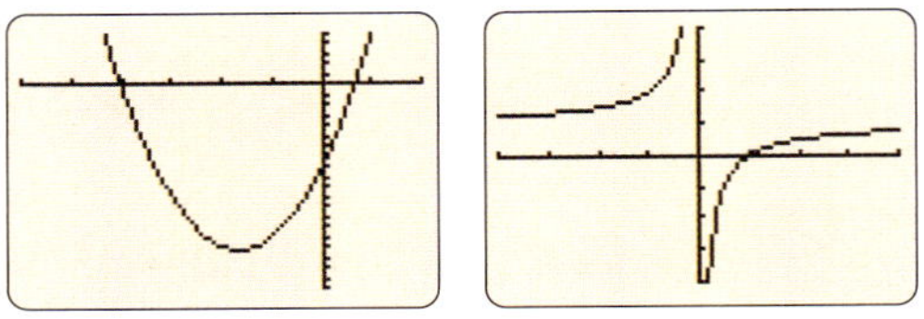

(a) (b)

Graph each inequality.

63. $y < \frac{1}{2}x^2 - 1$

64. $y \geq -|x|$

10 Chapter Test

1. Fill in the blanks.
   a. An equation of the form $ax^2 + bx + c = 0$, where $a \neq 0$, is called a ________ equation.
   b. When we add 81 to $x^2 + 18x$, we say that we have ________ the ________ on $x^2 + 18x$.
   c. The lowest point on a parabola that opens upward, or the highest point on a parabola that opens downward, is called the ________ of the parabola.
   d. $\frac{x-5}{x^2 - x - 56} > 0$ is an example of a ________ inequality in one variable.
   e. $y \leq x^2 - 4x + 3$ is an example of a ________ inequality in two variables.

2. Solve $x^2 - 63 = 0$ using the square root property. Approximate the solutions to the nearest hundredth.

Solve each equation using the square root property.

3. $(a + 7)^2 = 50$

4. $m^2 + 4 = 0$

5. Add a number to make $x^2 + 11x$ a perfect-square trinomial. Then factor the result.

6. Solve $x^2 + 3x - 2 = 0$ by completing the square. Approximate the solutions to the nearest hundredth.

7. Solve $2x^2 + 8x + 12 = 0$ by completing the square.

8. Let $f(x) = (3x - 2)^2$. Find all values of x for which $f(x) = 18$.

Use the quadratic formula to solve each equation. Approximate each solution to the nearest hundredth, when appropriate.

9. $4x^2 + 4x - 1 = 0$

10. $\frac{1}{8}t^2 - \frac{1}{4}t = \frac{1}{2}$

11. $-t^2 + 4t - 13 = 0$

12. $0.01x^2 = -0.08x - 0.15$

13. $m^2 - 94m = -2{,}209$

14. Let $g(x) = 3x^2 - 20$. Find all values of x for which $g(x) = 10$.

Solve each equation by any method. Approximate each solution to the nearest hundredth, when appropriate.

15. $2y - 3\sqrt{y} + 1 = 0$

16. $3 = m^{-2} - 2m^{-1}$

17. $x^4 - x^2 - 12 = 0$

18. $4\left(\frac{x+2}{3x}\right)^2 - 4\left(\frac{x+2}{3x}\right) - 3 = 0$

19. $\frac{1}{n+2} = \frac{1}{3} - \frac{1}{n}$

20. $5a^{2/3} + 11a^{1/3} = -2$

21. $10x(x + 1) = -3$

22. $a^3 - 3a^2 + 8a - 24 = 0$

23. Solve $E = mc^2$ for c. Assume that all variables represent positive numbers. Express any radical in simplified form.

24. Use the discriminant to determine the number and type of solutions for each equation.
   a. $3x^2 + 5x + 17 = 0$
   b. $9m^2 - 12m = -4$

25. **Tablecloths.** In 1990, Sportex of Highland, Illinois, made what was at the time the world's longest tablecloth. Find the dimensions of the rectangular tablecloth if it covered an area of 6,759 square feet and its length was 8 feet more than 332 times its width.

26. **Cooking.** Working together, a chef and his assistant can make a pastry dessert in 25 minutes. When the chef makes it himself, it takes him 8 minutes less time than it takes his assistant working alone. How long does it take the chef to make the dessert?

27. **Drawing.** An artist uses four equal-sized right triangles to block out a perspective drawing of an old hotel. See the illustration in the next column. For each triangle, the leg on the horizontal line is 14 inches longer than the leg on the center line. The length of each hypotenuse is 26 inches. On the center line of the drawing, what is the length of the segment extending from the ground to the top of the building?

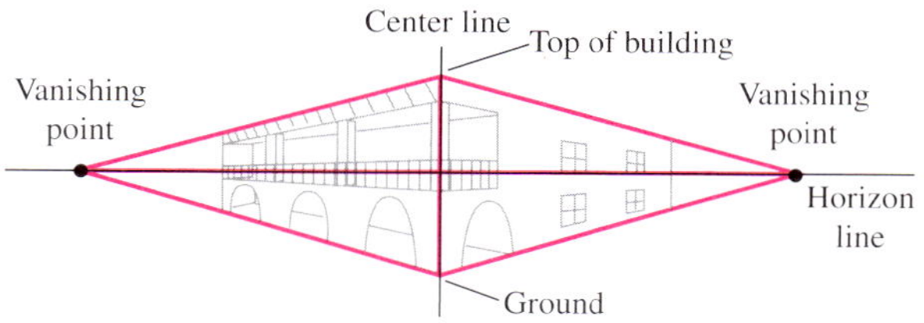

28. **Home Decorating.** A woman is going to put a cork border of uniform width around an 18-inch by 24-inch wall picture, as shown here. If the area covered by the cork is to be the same as the surface area of the picture, how wide should the border be? Round to the nearest tenth of an inch.

29. **ER Rooms.** The number of hospital emergency departments in the United States is approximated by the function $E(t) = 2.29t^2 - 75.72t + 5{,}206.95$, where t is the number of years after 1990. In what year does the model indicate that the number of hospital emergency departments was 5,000? (Source: American Hospital Association)

30. **Anthropology.** Anthropologists refer to the shape of the human jaw as a *parabolic dental arcade.* Which function is the best mathematical model of the parabola shown in the illustration?

   i. $f(x) = -\frac{3}{8}(x - 4)^2 + 6$
   ii. $f(x) = -\frac{3}{8}(x - 6)^2 + 4$
   iii. $f(x) = -\frac{3}{8}x^2 + 6$
   iv. $f(x) = \frac{3}{8}x^2 + 6$

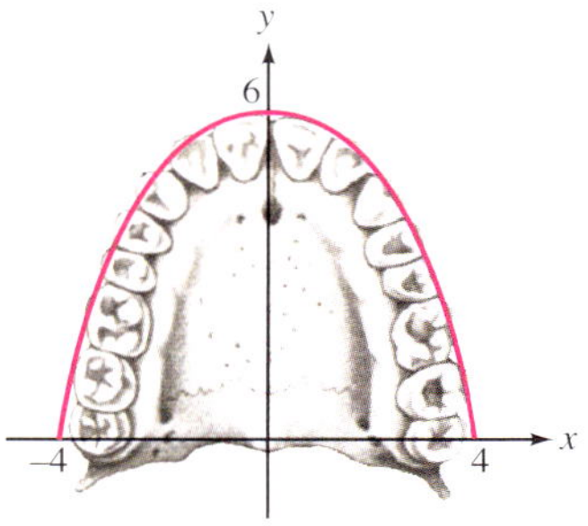

31. Find the vertex and the axis of symmetry of the graph of $f(x) = -3(x - 1)^2 + 2$. Then plot several points and complete the graph.

32. Complete the square to write the function $f(x) = 5x^2 + 10x - 1$ in the form $f(x) = a(x - h)^2 + k$. Determine the vertex and the axis of symmetry of the graph. Then plot several points and complete the graph.

33. First determine the coordinates of the vertex and the axis of symmetry of the graph of $f(x) = 2x^2 + x - 1$ using the vertex formula. Then determine the x- and y-intercepts of the graph. Finally, plot several points and complete the graph.

34. **Distress Signals.** A flare is fired directly upward into the air from a boat that is experiencing engine problems. The height of the flare (in feet) above the water, t seconds after being fired, is given by the function $h(t) = -16t^2 + 112t + 15$. If the flare is designed to explode when it reaches its highest point, at what height will this occur?

Solve each inequality. Write the solution set in interval notation and then graph it.

35. $x^2 - 2x > 8$

36. $\dfrac{x - 2}{x + 3} \le 0$

37. **Climate.** The average monthly temperature in San Antonio, Texas, is approximated by the function $T(m) = -1.1m^2 + 15.3m + 29.5$, where m is the number of the month (January = 1, February = 2, and so on). Use the function to approximate the average monthly temperature in July. (Source: cityrating.com)

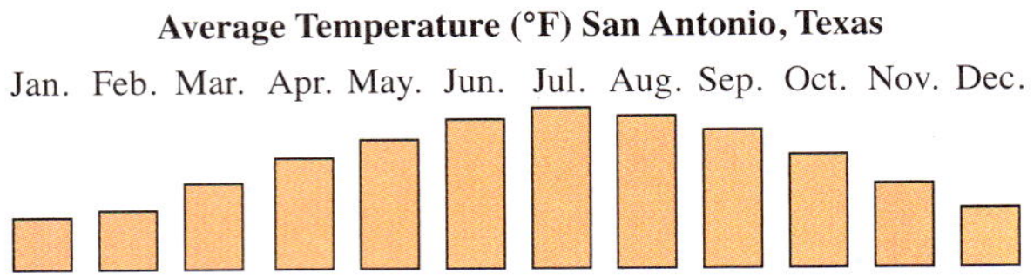

38. Graph: $y \le -x^2 + 3$

39. The graph of a quadratic function of the form $f(x) = ax^2 + bx + c$ is shown. Estimate the solutions of the corresponding quadratic equation $ax^2 + bx + c = 0$.

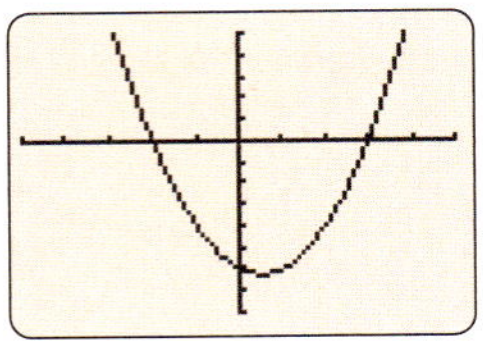

40. See Exercise 39. Estimate the solution of the quadratic inequality $ax^2 + bx + c \le 0$.

Group Project

Picture Framing

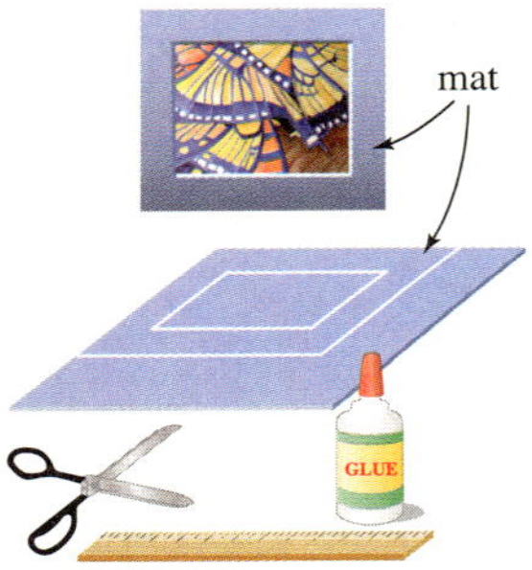

Overview: When framing pictures, mats are often used to enhance the images and give them a sense of depth. In this activity, you will use the quadratic formula to design the matting for several pictures.

Instructions: Form groups of 3 students. Each person in your group is to bring a picture to class. You can use a picture from a magazine or newspaper, a picture postcard, or a photograph that is no larger than 5 in. × 7 in. You will also need a pair of scissors, a ruler, glue, and three pieces of construction paper (12 in. × 18 in.).

Select one of the pictures and find its area. A mat of *uniform* width is to be placed around the picture. The area of the mat should equal the area of the picture. To determine the proper width of the matting, follow the steps of Example 8 in Section 10.2. Once you have determined the proper width, cut out the mat from the construction paper and glue it to the picture.

Then, choose another picture and find its area. Determine the uniform width that a matting should have so that its area is double that of the picture. Cut out the proper-size matting from the construction paper and glue it to the second picture.

Finally, find the area of the third picture and determine the uniform width that a matting should have so that its area is one-half that of the picture. Cut out the proper-size matting from the construction paper and glue it to the third picture.

Is one size matting more visually appealing than another? Discuss this among the members of your group.

CUMULATIVE REVIEW Chapters 1–10

1. Determine whether each statement is true or false. [Section 1.3]
   a. All whole numbers are integers.
   b. π is a rational number.
   c. A real number is either rational or irrational.
2. Evaluate: $\dfrac{-3(3+2)^2-(-5)}{17-|-22|}$ [Section 1.7]
3. Simplify: $-9(3a-9)-7(2a-7)$ [Section 1.9]
4. Use a check to determine whether -11 is a solution of
   $$\frac{3(b+2)}{2}=\frac{4b-10}{4}$$ [Section 2.2]
5. Solve: $2-(4x+7)=3+2(x+2)$ [Section 2.2]
6. Solve: $2(6n+5)=4(3n+2)+2$ [Section 2.2]
7. **Backpacks.** Pediatricians advise that children should not carry more than 20% of their own body weight in a backpack. According to this warning, how much weight can a fifth-grade girl who weighs 85 pounds safely carry in her backpack? [Section 2.3]
8. **Surface Area.** The total surface area A of a box with dimensions l, w, and h is given by the formula $A = 2lw + 2wh + 2lh$. If $A = 202$ square inches, $l = 9$ inches, and $w = 5$ inches, find h. [Section 2.4]

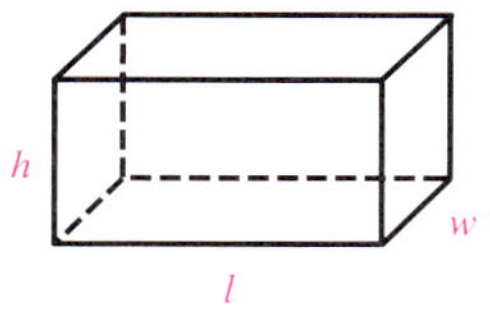

9. **Geometry.** If the vertex angle of an isosceles triangle is 53°, find the measure of each base angle. [Section 2.5]
10. **Search and Rescue.** Two search and rescue teams leave base at the same time, looking for a lost boy. The first team, on foot, heads north at 2 mph and the other, on horseback, south at 4 mph. How long will it take them to search a distance of 21 miles between them? [Section 2.6]
11. **Pharmacist.** How many liters of a 1% glucose solution should a pharmacist mix with 2 liters of a 5% glucose solution to obtain a 2% glucose solution? [Section 2.6]
12. **Blending Coffee.** A store sells regular coffee for \$8 a pound and gourmet coffee for \$14 a pound. Using 40 pounds of the gourmet coffee, the owner makes a blend to put on sale for \$10 a pound. How many pounds of regular coffee should he use? [Section 2.6]
13. Solve: $3-3x \ge 6+x$. Graph the solution set. Then describe the graph using interval notation. [Section 2.7]
14. Check to determine whether $(-6,-7)$ is a solution of $4x-3y=-4$. [Section 3.1]

Graph each equation.

15. $y=\frac{1}{2}x$ [Section 3.2]
16. $3x-4y=12$ [Section 3.3]
17. $x=5$ [Section 3.3]
18. $y=2x^2-3$ [Section 5.4]
19. Find the x-intercept and the y-intercept of the graph of the linear equation $5x-3y=6$. [Section 3.3]
20. **Skype.** The line graph below shows the approximate growth in the number of subscribers to Skype, a software application that allows users to make video calls over the Internet. Find the rate of change in the number of subscribers to Skype during the years 2007–2010 by finding the slope of the line. [Section 3.4]

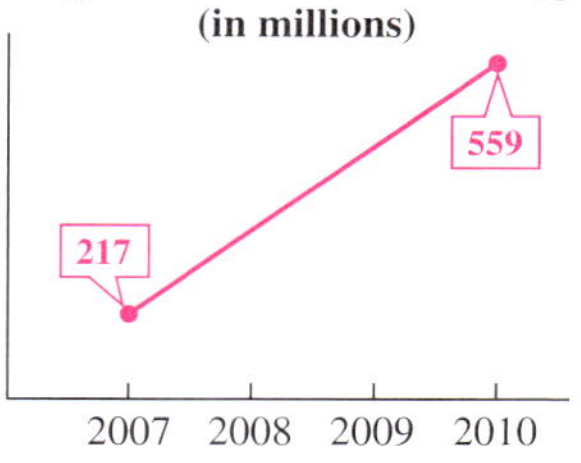

Source: businessinsider.com

21. What is the slope of the line defined by each equation? [Section 3.5]
    a. $y=3x-7$
    b. $5x-6y=13$
22. What is the slope of the line whose equation is $y=-3$? [Section 3.5]

Find an equation of the line with the given properties. Write the equation in slope–intercept form.

23. Slope 3, passes through $(-2,-4)$ [Section 3.5]
24. Parallel to the graph of $2x+3y=6$ and passes through $(0,-2)$ [Section 3.6]
25. If $f(x)=2x^2-3x+1$, find $f(-3)$. [Section 3.8]
26. **Boating.** The graph shows the vertical distance from a point on the tip of a propeller to the centerline as the propeller spins. Is this the graph of a function? [Section 3.8]

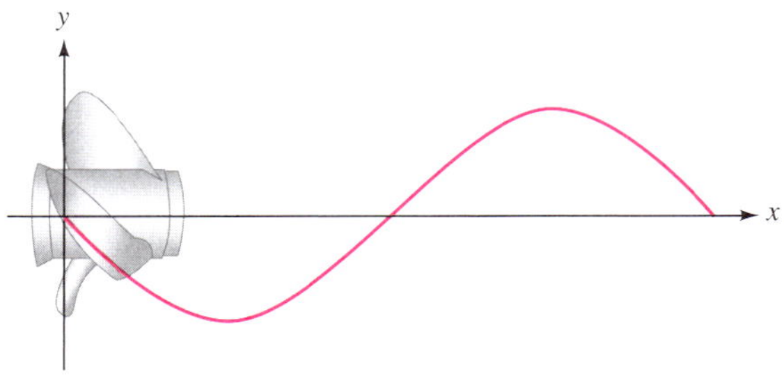

27. Use a check to determine whether $\left(2,\frac{1}{3}\right)$ is a solution of
    $$\begin{cases} x-3y=1 \\ -2x+6=-6y \end{cases}$$ [Section 4.1]
28. Solve the system $\begin{cases} x+y=4 \\ y=x+6 \end{cases}$ by graphing. [Section 4.1]

Solve each system of equations.

29. $\begin{cases} x = y + 4 \\ 2x + y = 5 \end{cases}$ [Section 4.2]

30. $\begin{cases} 3s + 4t = 5 \\ 2s - 3t = -8 \end{cases}$ [Section 4.3]

31. **Financial Planning.** In investing \$6,000 of a couple's money, a financial planner put some of it into a savings account paying 6% annual interest. The rest was invested in a riskier mini-mall development plan paying 12% annually. The combined interest earned for the first year was \$540. How much money was invested at each rate? Use two variables to solve this problem. [Section 4.4]

32. Graph: $\begin{cases} 3x + 2y \geq 6 \\ x + 3y \leq 6 \end{cases}$ [Section 4.5]

Simplify. Use only positive exponents in your answers.

33. $(x^5)^2(x^7)^3$ [Section 5.1]

34. $\left(\frac{a^3b}{c^4}\right)^5$ [Section 5.1]

35. $4^{-3} \cdot 4^{-2} \cdot 4^5$ [Section 5.2]

36. $\left(\frac{2a^2b^3c^{-4}}{5a^{-2}b^{-1}c^3}\right)^{-3}$ [Section 5.2]

37. $5n^0$

38. $\frac{4b^{-4}}{(a^{-2})^{-5}}$

39. Write each number in scientific notation and perform the operations. Give the answer in scientific notation and in standard notation:

$$\frac{(1{,}280{,}000{,}000)(2{,}700{,}000)}{240{,}000}$$ [Section 5.3]

40. Simplify: $\frac{3}{5}s^2 - \frac{2}{5}t^2 - \frac{1}{2}s^2 - \frac{7}{10}st - \frac{3}{10}st$ [Section 5.5]

Perform the indicated operations.

41. $(-8.9t^3 - 2.4t) - (2.1t^3 + 0.8t^2 - t)$ [Section 5.5]

42. $(2a - b)(4a^2 + 2ab + b^2)$ [Section 5.6]

43. $(-3t + 2s)(2t - 3s)$ [Section 5.6]

44. $(4b - 8)^2$ [Section 5.7]

45. $\left(6b + \frac{1}{2}\right)\left(6b - \frac{1}{2}\right)$ [Section 5.7]

46. $x + 2\overline{)2x^2 + 3x - 2}$ [Section 5.8]

Factor each expression.

47. $12uvw^3 - 18uv^2w^2$ [Section 6.1]

48. $x^2 + 4y - xy - 4x$ [Section 6.1]

49. $x^2 + 7x + 10$ [Section 6.2]

50. $6 + 3x^2 + x$ [Section 6.3]

51. $6a^2 - 7a - 20$ [Section 6.3]

52. $30a^4 - 4a^3 - 16a^2$ [Section 6.3]

53. $49s^6 - 84s^3n^2 + 36n^4$ [Section 6.4]

54. $x^4 - 16y^4$ [Section 6.4]

55. $x^3 - 64$ [Section 6.4]

56. $8x^6 + 125y^3$ [Section 6.5]

Solve each equation.

57. $x^2 + 3x + 2 = 0$ [Section 6.7]

58. $5x^2 = 10x$ [Section 6.7]

59. $6x^2 - x = 2$ [Section 6.7]

60. $a^2 - 25 = 0$ [Section 6.7]

61. $(m + 4)(2m + 3) - 22 = 10m$ [Section 6.7]

62. $6a^3 - 2a = a^2$ [Section 6.7]

63. **Children's Stickers.** A rectangular-shaped sticker has an area of 20 cm^2. The width is 1 cm shorter than the length. Find the length of the sticker. [Section 6.8]

64. Simplify: $\frac{x^2 + 2x + 1}{x^2 - 1}$ [Section 7.1]

Perform the operations. Simplify, if possible.

65. $\frac{p^2 - p - 6}{3p - 9} \div \frac{p^2 + 6p + 9}{p^2 - 9}$ [Section 7.2]

66. $\frac{12x^2}{7 - x} \cdot \frac{x - 7}{20x^3}$ [Section 7.2]

67. $\frac{13}{15a} - \frac{8}{15a}$ [Section 7.3]

68. $\frac{x + 2}{x + 5} - \frac{x - 3}{x + 7}$ [Section 7.4]

69. $\frac{1}{6b^4} - \frac{8}{9b^2}$ [Section 7.4]

70. $\frac{\frac{1}{x} + \frac{1}{y}}{\frac{1}{x} - \frac{1}{y}}$ [Section 7.5]

71. Solve: $\frac{7}{a^2 - a - 2} + \frac{1}{a + 1} = \frac{3}{a - 2}$ [Section 7.6]

72. Solve: $\frac{x - 4}{x - 3} + \frac{x - 2}{x - 3} = x - 3$ [Section 7.6]

73. Solve for R: $\frac{1}{R} = \frac{1}{R_1} + \frac{1}{R_2} + \frac{1}{R_3}$ [Section 7.6]

74. **Filling a Pool.** An inlet pipe can fill an empty swimming pool in 5 hours, and another inlet pipe can fill the pool in 4 hours. How long will it take both pipes to fill the pool? [Section 7.7]

75. **Online Sales.** A company found that, on average, it made 9 online sales transactions for every 500 hits on its Internet Web site. If the company's Web site had 360,000 hits in one year, how many sales transactions did it have that year? [Section 7.8]

76. The triangles shown below are similar. Find x. [Section 7.8]

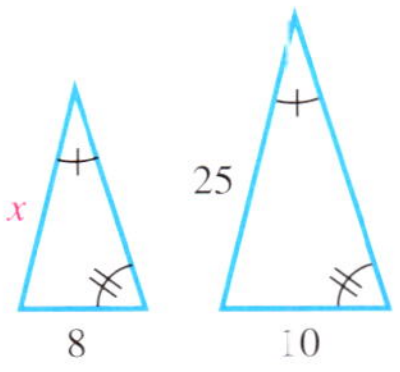

77. Solve: $\frac{1}{6}(a + 12) + 1 = \frac{a}{3}$ [Section 8.1]

78. Solve: $2x + 1 = 5(0.2x + 1) - (4 - x)$ [Section 8.1]

79. **Grades.** A student has scores of 75, 83, 91, and 68 on four algebra tests. What score does she need on a fifth test to give her an average of 80 or better? [Section 8.1]

80. Fill in the blank: Given a relation in x and y, if to each value of x in the domain there corresponds exactly one value of y in the range, then y is said to be a ________ of x. [Section 8.2]

Refer to the following graph of function f.

81. a. Find $f(1)$. [Section 8.2]

b. Find the value of x for which $f(x) = 1$. [Section 8.2]

82. Find the domain and range of the function. [Section 8.3]

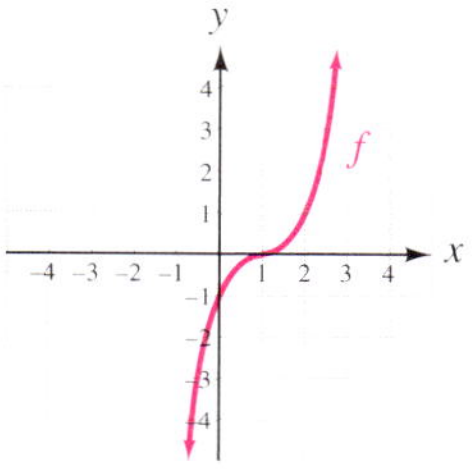

83. Find the domain and range of the relation $\{(1, -12), (-6, 7), (5, 8), (0, 7), (0, 4)\}$. Is it a function? [Section 8.2]

84. Does $y^2 = x$ define y as a function of x? If it does not, find two ordered pairs where more than one value of y corresponds to a single value of x. [Section 8.2]

85. Let $h(x) = -\frac{1}{5}x - 12$. For what value of x is $h(x) = 0$? [Section 8.2]

86. Let $f(x) = 3x^2 + 2$ and $g(x) = -2x - 1$. Find each function value. [Section 8.2]

a. $g(-2)$ b. $f(-r)$

87. Find the domain of the function $f(x) = \frac{5}{x + 1}$. [Section 8.2]

88. What are the slope and y-intercept of the graph of the linear function $f(x) = 6x + 15$? [Section 8.2]

89. Write an equation for the linear function whose graph is horizontal and passes through $\left(\frac{3}{2}, -\frac{7}{8}\right)$. [Section 8.2]

90. **Driving.** The total number of miles driven each day in the United States on all roads by all vehicles since the year 1980 can be modeled by a linear function. In 1990, approximately 5.76 billion vehicle miles were driven daily. In 2005, approximately 8.31 billion vehicle miles were driven daily. (Source: FHWA Highway Statistics) [Section 8.2]

a. Let t be the number of years after 1980 and M be the number of billion vehicle miles driven daily. Write a linear function $M(t)$.

b. Use your answer to part a to estimate how many billion vehicle miles will be driven daily in 2020.

91. Graph $f(x) = -x^2 + 1$ by plotting points. Then give the domain and range of the function. [Section 8.3]

92. First sketch the graph of the basic function associated with $g(x) = |x - 3| - 4$. Then draw the graph of function g using a translation. Give the domain and range of function g. [Section 8.3]

Solve each equation.

93. $2|4x - 3| + 1 = 19$ [Section 8.5]

94. $|2x - 1| = |3x + 4|$ [Section 8.5]

Solve each inequality. Write the solution set in interval notation and then graph it.

95. $5(-2x + 2) > 20 - x$ [Section 8.1]

96. $3x + 4 < -2$ or $3x + 4 > 10$ [Section 8.4]

97. $5x - 3 \geq 2$ and $6 \geq 4x - 3$ [Section 8.4]

98. $|2x - 5| \geq 25$ [Section 8.5]

99. $|3x - 2| \leq 4$ [Section 8.5]

100. $5|4 - x| + 6 < 1$ [Section 8.5]

101. Use the substitution to factor $(x + y)^2 + 7(x + y) + 12$. [Section 8.6]

102. Solve $d_1d_2 - fd_2 = fd_1$ for f. [Section 8.6]

103. Factor $x^6 - 1$ by factoring it as a difference of two squares first. [Section 8.7]

104. Find the domain of the function $f(x) = \frac{2x + 1}{x^2 - 2x}$. Express your answer in words and using interval notation. [Section 8.8]

105. Match each function with the correct graph shown below.

a. $f(x) = x$ [Section 8.2]

b. $f(x) = x^2$ [Section 8.3]

c. $f(x) = x^3$ [Section 8.3]

d. $f(x) = \frac{1}{x}$ [Section 8.8]

i.

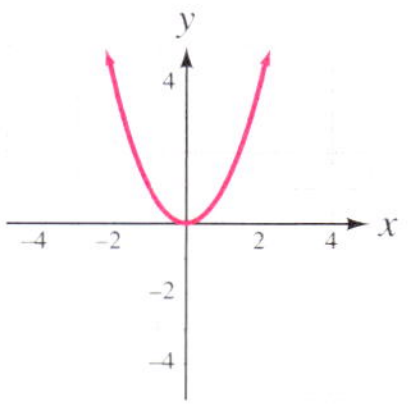

ii.

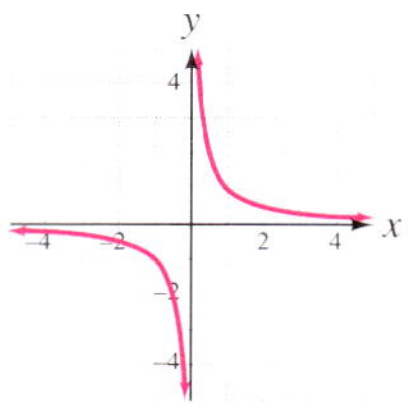

iii.

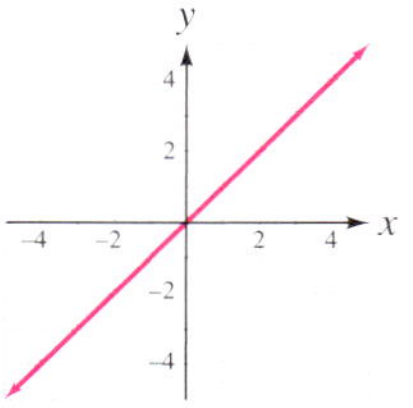

iv.

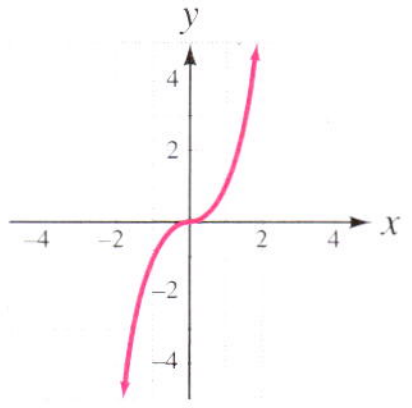

106. **Deliveries.** The costs of a delivery company vary jointly with the number of trucks in service and the number of hours they are used. When 8 trucks are used for 12 hours each, the costs are \$3,600. Find the costs of using 20 trucks, each for 12 hours. [Section 8.9]

107. Graph the function $f(x) = \sqrt{x - 2}$ and give its domain and range. [Section 9.1]

108. Find the domain of the function $f(x) = \sqrt{0.4x - 36}$. [Section 9.1]

Simplify each expression. All variables represent real numbers.

109. $\sqrt[3]{-27x^3}$ [Section 9.1]

110. $\sqrt{48t^3}$ [Section 9.1]

111. $64^{-2/3}$ [Section 9.2]

112. $\dfrac{x^{5/3}x^{1/2}}{x^{3/4}}$ [Section 9.2]

113. $-3\sqrt[4]{32} - 2\sqrt[4]{162} + 5\sqrt[4]{48}$ [Section 9.3]

114. $3\sqrt{2}\left(2\sqrt{3} - 4\sqrt{12}\right)$ [Section 9.4]

115. $\dfrac{\sqrt{x} + 2}{\sqrt{x} - 1}$ [Section 9.4]

116. $\dfrac{5}{\sqrt[3]{x}}$ [Section 9.4]

Solve each equation.

117. $5\sqrt{x + 2} = x + 8$ [Section 9.5]

118. $\sqrt{x} + \sqrt{x + 2} = 2$ [Section 9.5]

119. Let $f(x) = \sqrt[3]{x^2 + 2x}$ and $g(x) = 2\sqrt[3]{x - 1}$. Find all values of x for which $f(x) = g(x)$. [Section 9.5]

120. a. Find the length of the hypotenuse of the right triangle in figure (a). [Section 9.6]

b. Find the length of the hypotenuse of the right triangle in figure (b). [Section 9.6]

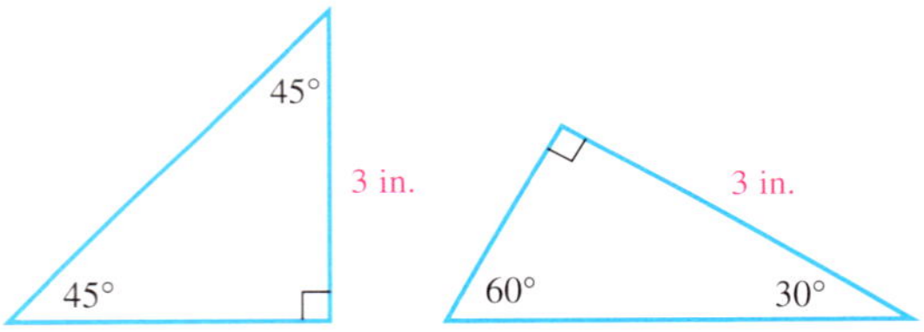

(a) (b)

121. a. Find the distance between $(-2, 6)$ and $(4, 14)$. [Section 9.6]

b. Find the midpoint of the line segment with endpoints $(7, 1)$ and $(-10, 4)$. [Section 9.6]

122. Simplify: i^{43} [Section 9.7]

Perform the indicated operations. Write each result in $a + bi$ form.

123. $\left(-7 + \sqrt{-81}\right) - \left(-2 - \sqrt{-64}\right)$ [Section 9.7]

124. $\dfrac{5}{3 - i}$ [Section 9.7]

125. $(2 + i)^2$ [Section 9.7]

126. $\dfrac{-4}{6i^7}$ [Section 9.7]

Solve each equation.

127. $x^2 = 28$ [Section 10.1]

128. $(x - 19)^2 = -5$ [Section 10.1]

129. Use the method of completing the square to solve $2x^2 - 6x + 3 = 0$. [Section 10.1]

130. Use the quadratic formula to solve $a^2 - \frac{2}{5}a = -\frac{1}{5}$. [Section 10.2]

131. **Community Gardens.** Residents of a community can work their own 16-ft × 24-ft plot of city-owned land if they agree to the following conditions:

- The area of the garden cannot exceed 180 square feet.
- A path of uniform width must be maintained around the garden.

Find the dimensions of the largest possible garden. [Section 10.2]

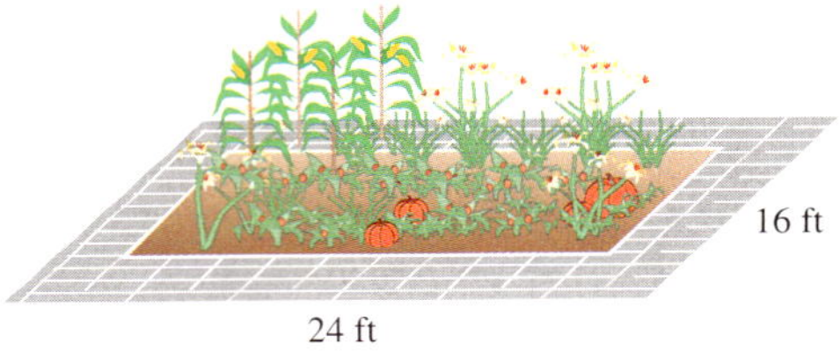

132. **Sidewalks.** A 170-meter-long sidewalk from the mathematics building M to the student center C is shown in red in the illustration. However, students prefer to walk directly from M to C, across a lawn. How long are the two segments of the existing sidewalk? [Section 10.2]

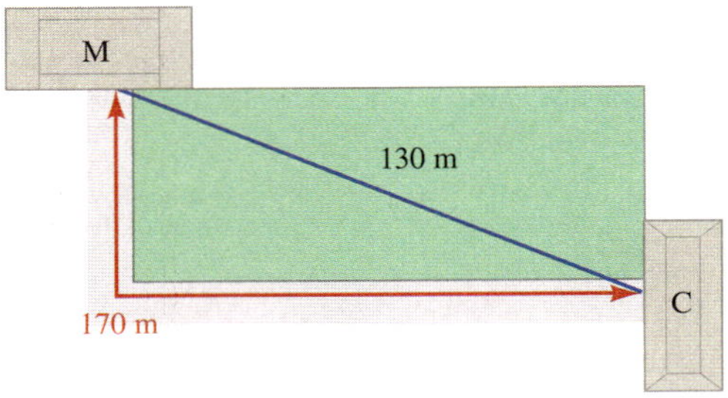

Solve each equation.

133. $t^{2/3} - t^{1/3} = 6$ [Section 10.3]

134. $x^{-4} - 2x^{-2} + 1 = 0$ [Section 10.3]

135. First determine the vertex and the axis of symmetry of the graph of $f(x) = -x^2 - 4x$ using the vertex formula. Then determine the x- and y-intercepts of the graph. Finally, plot several points and complete the graph. [Section 10.4]

Solve each inequality. Write the solution set in interval notation and then graph it.

136. $x^2 - 81 < 0$ [Section 10.5]

137. $\dfrac{1}{x + 1} \geq \dfrac{x}{x + 4}$ [Section 10.5]

138. a. The graph of $f(x) = 16x^2 + 24x + 9$ is shown below. Estimate the solution(s) of $16x^2 + 24x + 9 = 0$. [Section 10.5]

b. Use the graph to determine the solution of $16x^2 + 24x + 9 < 0$. [Section 10.5]

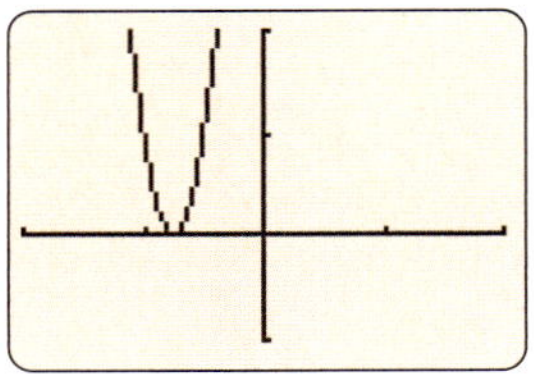

Exponential and Logarithmic Functions

11

from Campus to Careers

Social Worker

For those with a desire to help improve other people's lives, social work is one career option to consider. Social workers offer guidance and counseling to people in crisis. They must be critical thinkers—able to use their logic and reasoning to brainstorm alternate solutions to problems faced by their clients. Social workers use their mathematical skills to construct family budgets, plan personnel schedules, gather and interpret data, and comprehend the statistical methods used in research studies.

Problem 49 in **Study Set 11.3, problem 115** in **Study Set 11.4,** and **problem 87** in **Study Set 11.5** involve situations that a social worker encounters on the job. The mathematical concepts discussed in this chapter can be used to solve those problems.

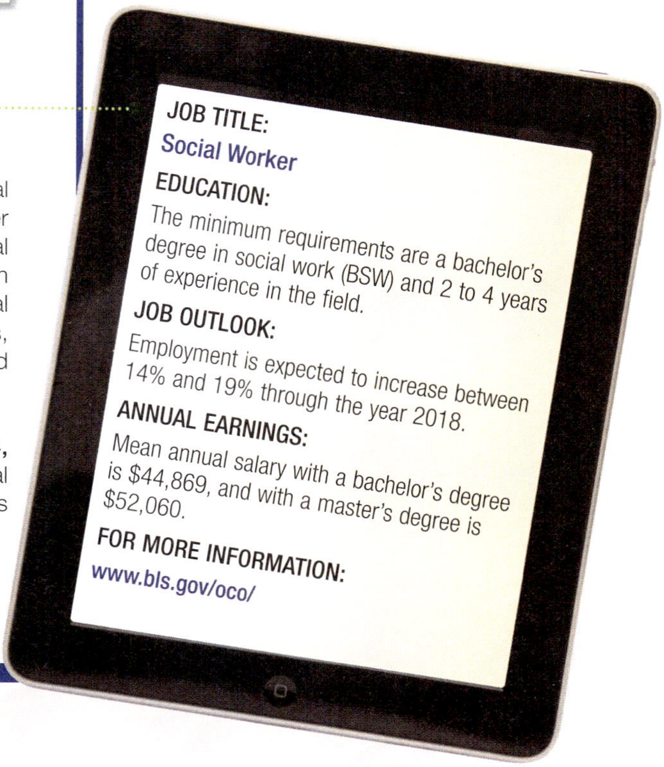

Study Skills Workshop ▶ Participating in Class

One of the keys to success in algebra is to learn as much as you can in class. To get the most out of class meetings, and to make them more enjoyable, you should participate in the following ways:

ASK QUESTIONS: During class, clear up any questions that may arise from your homework assignments or from your instructor's lectures. Also, pay close attention when other students ask questions. You never know when you might face the same difficulty.

ANSWER QUESTIONS: Many instructors direct questions to the class while lecturing. Take advantage of this opportunity to increase your knowledge by attempting to answer all such questions from your instructor.

INTERACT WITH CLASSMATES: Before class begins and after class ends, regularly discuss the material that you are studying with fellow classmates.

Now Try This

1. List the reasons why you do not feel comfortable asking questions in class.
2. While working on your next homework assignment, write down any questions that occur to you so that you will not forget to ask them in class.
3. Exchange a written question about a homework problem with a classmate. See if you can answer each other's question.

SECTION 11.1 Algebra and Composition of Functions

OBJECTIVES

1. Add, subtract, multiply, and divide functions.
2. Find the composition of functions.
3. Use graphs to evaluate functions.
4. Use composite functions to solve problems.

ARE YOU READY?

The following problems review some basic skills that are needed when performing the algebra of functions.

1. Add: $(7x^2 + 5x - 12) + (2x^2 - 3x + 4)$
2. Subtract: $(8x^3 - 4x^2) - (6x^3 - 3x^2)$
3. Multiply: $(8x - 5)(9x^2 - 4)$
4. Divide: $(x^2 - x - 6) \div (x - 3)$

Just as it is possible to perform arithmetic operations on real numbers, it is possible to perform those operations on functions. We call the process of adding, subtracting, multiplying, and dividing functions the **algebra of functions.**

1 Add, Subtract, Multiply, and Divide Functions.

We have seen that the sum, difference, product, and quotient of two functions are themselves functions. The new functions that result from such operations can be represented using the following notation.

Operations on Functions

Sum: $(f + g)(x) = f(x) + g(x)$ Read as "*f* plus *g* of *x* equals *f* of *x* plus *g* of *x*."

Difference: $(f - g)(x) = f(x) - g(x)$ Read as "*f* minus *g* of *x* equals *f* of *x* minus *g* of *x*."

Product: $(f \cdot g)(x) = f(x)g(x)$ Read as "*f* times *g* of *x* equals *f* of *x* times *g* of *x*."

Quotient: $(f/g)(x) = \dfrac{f(x)}{g(x)}$ where $g(x) \neq 0$ Read as "*f* divided by *g* of *x* equals *f* of *x* divided by *g* of *x*."

The domain of each sum, difference, and product function shown above is the set of real numbers x that are in the domains of both f and g. The domain of the quotient function is the set of real numbers x that are the domains of f and g, excluding any values of x where $g(x) = 0$.

EXAMPLE 1 Let $f(x) = 2x^2 + 1$ and $g(x) = 5x - 3$. Find each function and give its domain:
a. $f + g$ **b.** $f - g$ **c.** $f \cdot g$ **d.** f/g

Strategy We will add, subtract, multiply, and divide the functions as if they were binomials.

Why We add because of the plus symbol in $f + g$, we subtract because of the minus symbol in $f - g$, we multiply because of the raised dot in $f \cdot g$, and we divide because of the fraction bar in f/g.

Solution

a.
$$\begin{aligned}(f + g)(x) &= f(x) + g(x) && \text{This is the definition of a sum function.}\\ &= (2x^2 + 1) + (5x - 3) && \text{Replace } f(x) \text{ with } 2x^2 + 1 \text{ and } g(x) \text{ with } 5x - 3.\\ &= 2x^2 + 5x - 2 && \text{Drop the parentheses and combine like terms.}\end{aligned}$$

The domain of $f + g$ is the set of real numbers that are in the domain of both f and g. Since the domain of both f and g is $(-\infty, \infty)$, the domain of $f + g$ is the $(-\infty, \infty)$.

b.
$$\begin{aligned}(f - g)(x) &= f(x) - g(x) && \text{This is the definition of a difference function.}\\ &= (2x^2 + 1) - (5x - 3) && \text{Replace } f(x) \text{ with } 2x^2 + 1 \text{ and } g(x) \text{ with } 5x - 3.\\ &= 2x^2 + 1 - 5x + 3 && \text{Change the sign of each term of } 5x - 3 \text{ and drop the parentheses.}\\ &= 2x^2 - 5x + 4 && \text{Combine like terms.}\end{aligned}$$

Since the domain of both f and g is $(-\infty, \infty)$, the domain of $f - g$ is $(-\infty, \infty)$.

Notation

The parentheses around $\frac{3}{5}$ in the notation $\left(-\infty, \frac{3}{5}\right) \cup \left(\frac{3}{5}, \infty\right)$ indicate that $\frac{3}{5}$ is not included in the domain of f/g. The domain could also be shown graphically:

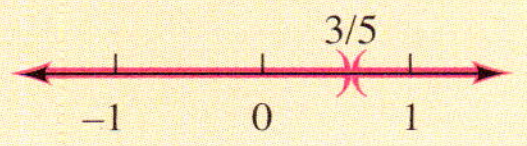

c.
$$\begin{aligned}(f \cdot g)(x) &= f(x) \cdot g(x) && \text{This is the definition of a product function.}\\ &= (2x^2 + 1)(5x - 3) && \text{Replace } f(x) \text{ with } 2x^2 + 1 \text{ and } g(x) \text{ with } 5x - 3.\\ &= 10x^3 - 6x^2 + 5x - 3 && \text{Multiply the binomials. There are no like terms.}\end{aligned}$$

The domain of $f \cdot g$ is the set of real numbers that are in the domain of both f and g. Since the domain of both f and g is $(-\infty, \infty)$, the domain of $f \cdot g$ is $(-\infty, \infty)$.

d.
$$\begin{aligned}(f/g)(x) &= \frac{f(x)}{g(x)} && \text{This is the definition of a quotient function.}\\ &= \frac{2x^2 + 1}{5x - 3} && \text{Replace } f(x) \text{ with } 2x^2 + 1 \text{ and } g(x) \text{ with } 5x - 3. \text{ The result does not simplify.}\end{aligned}$$

Since the denominator of the fraction cannot be 0, it follows that this function is undefined if $5x - 3 = 0$. If we solve for x, we see that x cannot be $\frac{3}{5}$. Thus, the domain of f/g is the union of two intervals: $\left(-\infty, \frac{3}{5}\right) \cup \left(\frac{3}{5}, \infty\right)$.

Self Check 1 Let $f(x) = 3x - 2$ and $g(x) = 2x^2 + 3x$. Find each function and give its domain:
a. $f + g$
b. $f - g$
c. $f \cdot g$
d. f/g

Now Try Problems 13, 15, 17, and 19

EXAMPLE 2 Use the results from Example 1 to find: **a.** $(f + g)(-3)$ **b.** $(f - g)(6)$
c. $(f \cdot g)(0)$ **d.** $(f/g)(10)$

Strategy We will substitute the given values within the second set of parentheses for each x in the sum, difference, product, and quotient functions found in Example 1. Then we will evaluate the right side of each equation.

Why The number that is within the second set of parentheses is the input of the function.

Solution

a. $(f+g)(x) = 2x^2 + 5x - 2$ This is the sum function found in Example 1, part a.

$(f+g)(-3) = 2(-3)^2 + 5(-3) - 2$ Substitute −3 for each x.

$= 18 + (-15) - 2$ Evaluate the right side.

$= 1$

b. $(f-g)(x) = 2x^2 - 5x + 4$ This is the difference function found in Example 1, part b.

$(f-g)(6) = 2(6)^2 - 5(6) + 4$ Substitute 6 for each x.

$= 72 - 30 + 4$ Evaluate the right side.

$= 46$

c. $(f \cdot g)(x) = 10x^3 - 6x^2 + 5x - 3$ This is the product function found in Example 1, part c.

$(f \cdot g)(0) = 10(0)^3 - 6(0)^2 + 5(0) - 3$ Substitute 0 for each x.

$= -3$ Evaluate the right side.

d. $(f/g)(x) = \dfrac{2x^2 + 1}{5x - 3}$ This is the quotient function found in Example 1, part d.

$(f/g)(10) = \dfrac{2(10)^2 + 1}{5(10) - 3}$ Substitute 10 for each x.

$= \dfrac{2(100) + 1}{50 - 3}$ Evaluate the right side.

$= \dfrac{201}{47}$

Self Check 2 Use the results from Self Check 1 to find: **a.** $(f+g)(5)$ **b.** $(f-g)(-9)$ **c.** $(f \cdot g)(2)$ **d.** $(f/g)(0)$

Now Try Problems 29 and 31

There is a relationship that can be seen between the graphs of two functions and the graph of their sum (or difference) function. For example, in the following illustration, the graph of $f + g$, which gives the total number of elementary and secondary students in the United States, can be found by adding the graph of f, which gives the number of secondary students, to the graph of g, which gives the number of elementary students. For any given x-value, we simply add the two corresponding y-values to get the graph of the sum function $f + g$.

Actual and Projected Enrollment in U. S. Public and Private Elementary and Secondary Schools

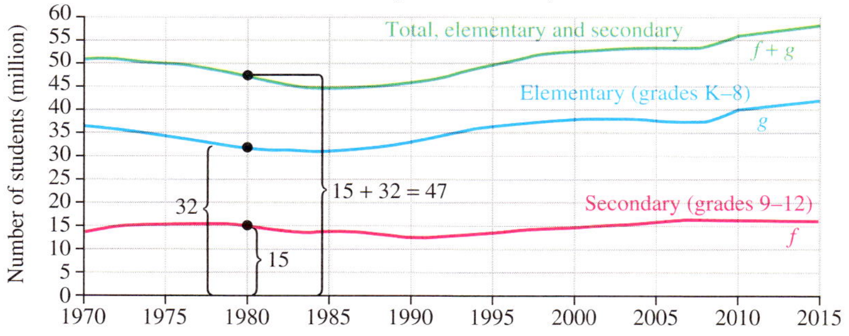

Source: U.S. Department of Education

To find the total enrollment in 1980, add the elementary enrollment of 32 million to the secondary enrollment of 15 million to get 47 million.

2 Find the Composition of Functions.

We have seen that a function can be represented by a machine: We put in a number from the domain, and a number from the range comes out. For example, if we put the number 2 into the machine shown on the right, the number $f(2) = 8$ comes out. In general, if we put x into the machine, the value $f(x)$ comes out.

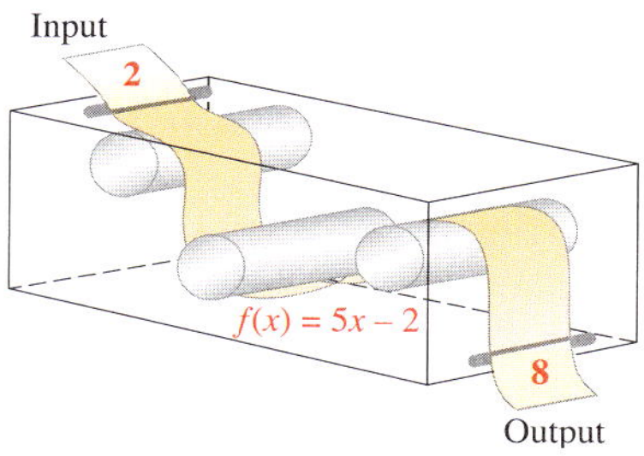

Often one quantity is a function of a second quantity that depends, in turn, on a third quantity. For example, the cost of a car trip is a function of the gasoline consumed. The amount of gasoline consumed, in turn, is a function of the number of miles driven. Such chains of dependence can be analyzed mathematically as **compositions of functions.**

The Language of Algebra

A **composite** is something that is formed by bringing together distinct parts. For example, a police artist creates a *composite* sketch from eyewitness descriptions of the suspect.

Suppose that $y = f(x)$ and $y = g(x)$ define two functions. Any number x in the domain of g will produce the corresponding value $g(x)$ in the range of g. If $g(x)$ is in the domain of function f, then $g(x)$ can be substituted into f, and a corresponding value $f(g(x))$ will be determined. Because of the **nested parentheses** in $f(g(x))$, we read it as "f of g of x." This two-step process defines a new function, called a **composite function,** denoted by $f \circ g$. (This is read as "f composed with g" or "the composition of f and g" or "f circle g.")

The Language of Algebra

$f(g(x))$

We call these **nested parentheses.**

The function machines shown below illustrate the composition $f \circ g$. When we put a number into the function g, a value $g(x)$ comes out. The value $g(x)$ then goes into function f, which transforms $g(x)$ into $f(g(x))$. If the function machines for g and f were connected to make a single machine, that machine would be named $f \circ g$.

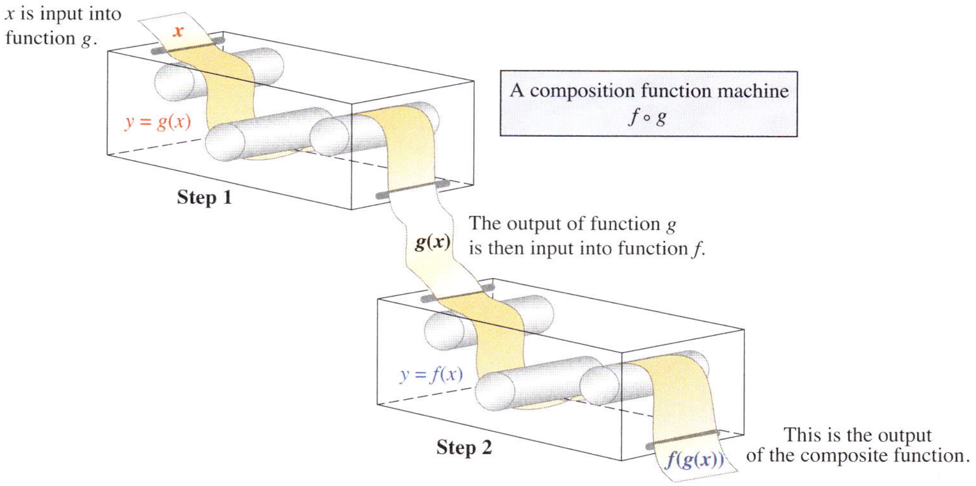

To be in the domain of the composite function $f \circ g$, a number x has to be in the domain of g and the output of g must be in the domain of f. Thus, the domain of $f \circ g$ consists of those numbers x that are in the domain of g, and for which $g(x)$ is in the domain of f.

Composite Functions

The composite function $f \circ g$ is defined by

$$(f \circ g)(x) = f(g(x))$$

If $f(x) = 4x$ and $g(x) = 3x + 2$, to find $f \circ g$ and $g \circ f$, we proceed as follows.

$$\begin{aligned}(f \circ g)(x) &= f(g(x)) \\ &= f(3x + 2) \\ &= 4(3x + 2) \\ &= 12x + 8\end{aligned} \qquad \begin{aligned}(g \circ f)(x) &= g(f(x)) \\ &= g(4x) \\ &= 3(4x) + 2 \\ &= 12x + 2\end{aligned}$$

Different results

Caution

$(f \circ g)(x)$ — Composition

does not mean

$(f \cdot g)(x)$ — Multiplication

The different results illustrate that the composition of functions is not commutative. Usually, we will find that $(f \circ g)(x) \neq (g \circ f)(x)$.

EXAMPLE 3 Let $f(x) = 2x + 1$ and $g(x) = x - 4$. Find: **a.** $(f \circ g)(9)$ **b.** $(f \circ g)(x)$ **c.** $(g \circ f)(-2)$

Strategy In part (a), we will find $f(g(9))$. In part (b), we will find $f(g(x))$. In part (c), we will find $g(f(-2))$.

Why To evaluate a composite function written with the circle $\circ$ notation, we rewrite it using nested parentheses.

Solution **a.** $(f \circ g)(9)$ means $f(g(9))$, where the function g is applied first and function f is applied second.

$$(f \circ g)(9) = f(g(9))$$

Apply first

In figure (a) on the next page, we see that function g receives the number 9, subtracts 4, and releases the number $g(9) = 5$. Then 5 goes into the f function, which doubles 5 and adds 1. The final result, 11, is the output of the composite function $f \circ g$.

Read as "f composed with g of 9." Read as "f of g of 9."

$(f \circ g)(9) = f(g(9))$ Change from $\circ$ notation to nested parentheses notation.

$= f(5)$ Evaluate: $g(9) = 9 - 4 = 5$.

$= 2(5) + 1$ Evaluate $f(5)$ using $f(x) = 2x + 1$.

$= 11$

Thus, $(f \circ g)(9) = 11$.

b. $(f \circ g)(x)$ means $f(g(x))$. In figure (a) on the next page, function g receives the number x, subtracts 4, and releases the number $x - 4$. Then $x - 4$ goes into the f function, which doubles $x - 4$ and adds 1. The final result, $2x - 7$, is the output of the composite function $f \circ g$.

Read as "f composed with g of x." Read as "f of g of x."

$(f \circ g)(x) = f(g(x))$ Change from $\circ$ notation to nested parentheses notation.

$= f(x - 4)$ We are given $g(x) = x - 4$. Replace $g(x)$ with $x - 4$.

$= 2(x - 4) + 1$ Find $f(x - 4)$ using $f(x) = 2x + 1$.

$= 2x - 8 + 1$ Distribute the multiplication by 2.

$= 2x - 7$ Combine like terms.

Thus, $(f \circ g)(x) = 2x - 7$.

c. $(g \circ f)(-2)$ means $g(f(-2))$. In figure (b) on the next page, function f receives the number -2, doubles it and adds 1, and releases -3 into the g function. Function g subtracts 4 from -3 and outputs a final result of -7. Thus,

Read as "g composed with f of −2." Read as "g of f of −2."

$(g \circ f)(-2) = g(f(-2))$ Change from $\circ$ notation to nested parentheses notation.

$= g(-3)$ Evaluate $f(-2)$ using $f(x) = 2x + 1$.

$= -3 - 4$ Evaluate $g(-3)$ using $g(x) = x - 4$.

$= -7$ Do the subtraction.

Thus, $(g \circ f)(-2) = -7$.

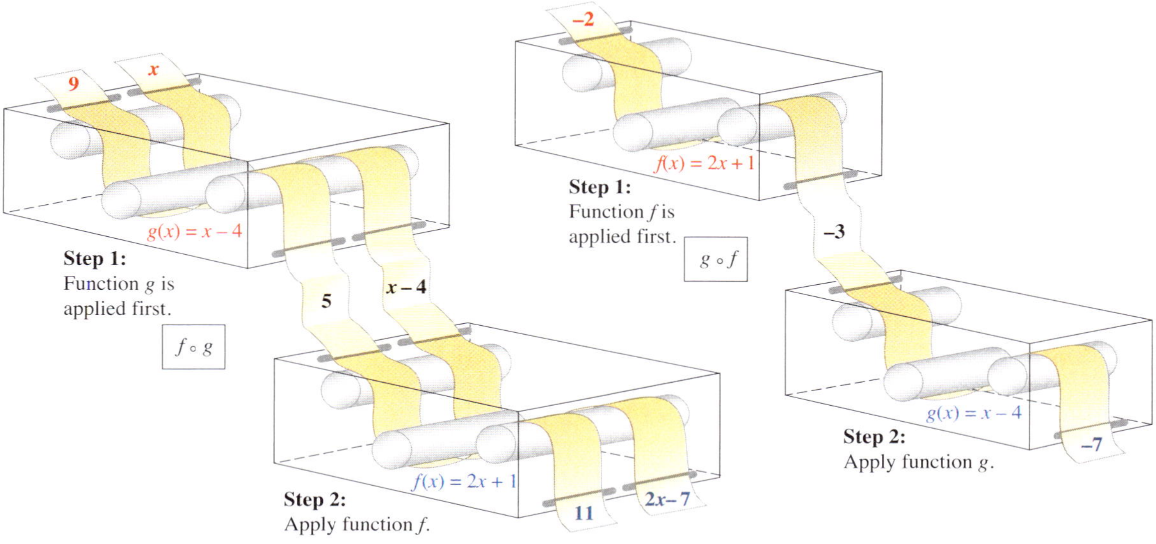

(a) (b)

Self Check 3 Let $f(x) = x^3$ and $g(x) = 6 - x$. Find: **a.** $(f \circ g)(8)$ **b.** $(g \circ f)(1)$ **c.** $(g \circ f)(x)$

Now Try Problems 37, 39, and 43

3 Use Graphs to Evaluate Functions.

EXAMPLE 4 Refer to the graphs of functions f and g on the left to find each of the following.
a. $(f + g)(-4)$ **b.** $(f \cdot g)(2)$ **c.** $(f \circ g)(-3)$

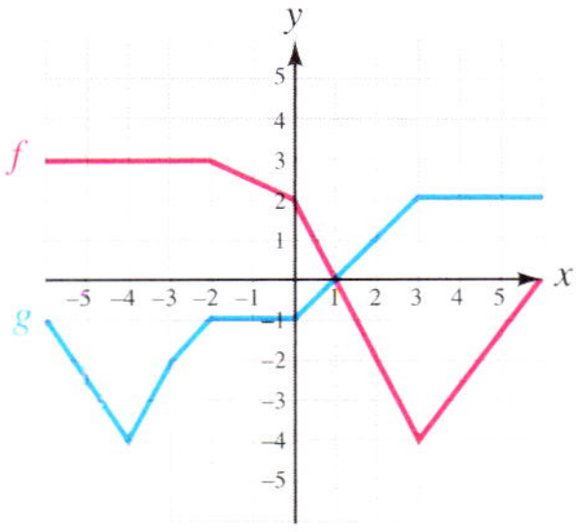

Strategy We will express the sum, product, and composite functions using the functions from which they are formed.

Why We can evaluate sum, product, and composite functions at a given x-value by evaluating each function from which they are formed at that x-value.

Solution

a. The value of $f(-4)$ is found by looking in quadrant II of the graph and the value of $g(-4)$ by looking in quadrant III.

$$
\begin{aligned}
(f + g)(-4) &= f(-4) + g(-4) \\
&= 3 + (-4) \\
&= -1
\end{aligned}
$$

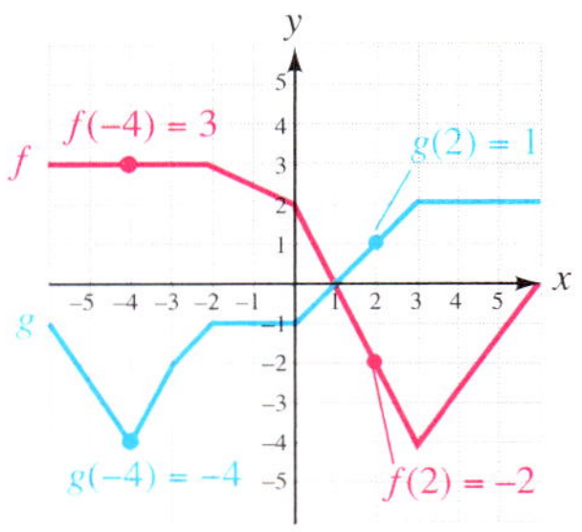

b. The value of $f(2)$ is found by looking in quadrant IV of the graph and the value of $g(2)$ by looking in quadrant I.

$$
\begin{aligned}
(f \cdot g)(2) &= f(2) \cdot g(2) \\
&= -2 \cdot 1 \\
&= -2
\end{aligned}
$$

c. The value of $g(-3)$ is found by looking in quadrant III of the graph and the value of $f(-2)$ by looking in quadrant II.

$$
\begin{aligned}
(f \circ g)(-3) &= f(g(-3)) \quad \text{Change from } \circ \text{ notation to nested parentheses notation.} \\
&= f(-2) \\
&= 3
\end{aligned}
$$

Self Check 4 Refer to the graph in Example 4 to find each of the following.

a. $(f - g)(3)$ **b.** $\left(\frac{f}{g}\right)(-2)$ **c.** $(g \circ f)(3)$

Now Try Problems 45 and 47

EXAMPLE 5 If $h(x) = (4x - 6)^2$, find f and g such that $h(x) = (f \circ g)(x)$.

Strategy We will determine the order in which we would evaluate function h for a given value of x.

Why If we can see the evaluation as a two-step process, it will help us determine the unknown functions f and g.

Solution If we were to evaluate function h for a given value of x, we would first find $4x - 6$, and then we would square that result. This suggests that the first function that operates in the composition, function g, should receive an input x and produce $4x - 6$. Thus, $g(x) = 4x - 6$. The second function that operates in the composition, function f, should receive an input and square it. Thus, $f(x) = x^2$.

We can check these results by forming the composition function.

$$(f \circ g)(x) = f(g(x)) \quad \text{Change from } \circ \text{ notation to nested parentheses.}$$
$$= f(4x - 6) \quad \text{Replace } g(x) \text{ with } 4x - 6.$$
$$= (4x - 6)^2 \quad \text{Find } f(4x - 6) \text{ using } f(x) = x^2.$$

The composition of functions $f(x) = x^2$ and $g(x) = 4x - 6$ does indeed produce $h(x) = (4x - 6)^2$. It is important to note that there are many other possibilities for f and g, but the ones that we found here are the most obvious.

Self Check 5 If $h(x) = \sqrt{x + 15}$, find f and g such that $h(x) = (f \circ g)(x)$.

Now Try Problems 49 and 51

4 Use Composite Functions to Solve Problems.

EXAMPLE 6 **Biological Research.** A specimen is stored in refrigeration at a temperature of 15° Fahrenheit. Biologists remove the specimen and warm it at a controlled rate of 3°F per hour. Express its Celsius temperature as a function of the time t since it was removed from refrigeration.

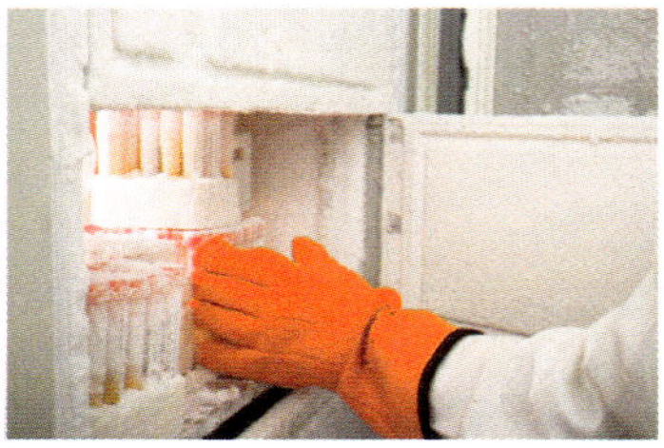

Strategy We will express the Fahrenheit temperature of the specimen as a function of the time t since it was removed from refrigeration. Then we will express the Celsius temperature of the specimen as a function of its Fahrenheit temperature and find the composition of the two functions.

Why The Celsius temperature of the specimen is a function of its Fahrenheit temperature. Its Fahrenheit temperature is a function of the time since it was removed from refrigeration. This chain of dependence suggests that we write a composition of functions.

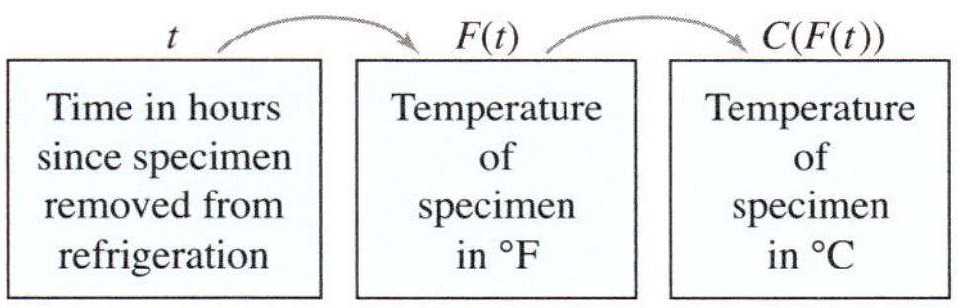

Solution The temperature of the specimen is 15°F when the time $t = 0$. Because it warms at a rate of 3°F per hour, its initial temperature of 15°F increases by $3t$°F in t hours. The Fahrenheit temperature at time t of the specimen is given by the function

$$F(t) = 3t + 15$$

The Celsius temperature C is a function of this Fahrenheit temperature F, given by the function

$$C(F) = \frac{5}{9}(F - 32)$$

To express the specimen's Celsius temperature as a function of *time,* we find the composite function $(C \circ F)(t)$.

$$
\begin{aligned}
(C \circ F)(t) &= C(F(t)) && \text{Change from } \circ \text{ notation to nested parentheses notation.}\\
&= C(3t + 15) && \text{Substitute } 3t + 15 \text{ for } F(t).\\
&= \frac{5}{9}\left(3t + 15 - 32\right) && \text{Find } C(3t + 15) \text{ by substituting } 3t + 15 \text{ for } F \text{ in } C(F) = \tfrac{5}{9}(F - 32).\\
&= \frac{5}{9}(3t - 17) && \text{Subtract within the parentheses: } 15 - 32 = -17.\\
&= \frac{15}{9}t - \frac{85}{9} && \text{Distribute the multiplication by } \tfrac{5}{9}.\\
&= \frac{5}{3}t - \frac{85}{9} && \text{Simplify } \tfrac{15}{9}\text{: } \tfrac{\overset{1}{\cancel{3}} \cdot 5}{\underset{1}{\cancel{3}} \cdot 3} = \tfrac{5}{3}.
\end{aligned}
$$

The composite function, $C(t) = \frac{5}{3}t - \frac{85}{9}$, gives the temperature of the specimen in degrees Celsius t hours after it is removed from refrigeration.

Self Check 6 **Weather Forecasting.** A low-pressure area is bringing in colder weather for the next 12 hours. The temperature is now 86º Fahrenheit and is expected to fall 3º every 2 hours. Write a composite function that expresses the Celsius temperature as a function of the number of hours from now.

Now Try Problem 87

SECTION 11.1 STUDY SET

VOCABULARY

Fill in the blanks.

1. The ____ of f and g, denoted as $f + g$, is defined by $(f + g)(x) =$ ________ and the __________ of f and g, denoted as $f - g$, is defined by $(f - g)(x) =$ ________.
2. The _________ of f and g, denoted as $f \cdot g$, is defined by $(f \cdot g)(x) =$ ________ and the _________ of f and g, denoted as f/g, is defined by $(f/g)(x) =$ ________.
3. The ________ of the function $f + g$ is the set of real numbers x that are in the domain of both f and g.
4. The __________ function $f \circ g$ is defined by $(f \circ g)(x) =$ ________.
5. When we write $(f \circ g)(x)$ as $f(g(x))$, we have changed from $\circ$ notation to ______ parentheses notation.
6. When reading the notation $f(g(x))$, we say "f ___ g ___ x."

CONCEPTS

7. Fill in the blanks.
   a. $(f \circ g)(3) = f(\quad)$
   b. To find $f(g(3))$, we first find ______ and then substitute that value for x in $f(x)$.
8. a. If $f(x) = 3x + 1$ and $g(x) = 1 - 2x$, find $f(g(3))$ and $g(f(3))$.
   b. Is the composition of functions commutative?

9. Fill in the three blanks in the drawing of the function machines that show how to compute $g(f(-2))$.

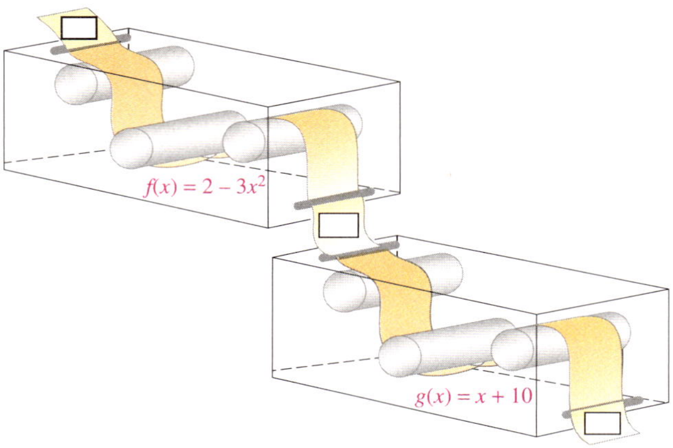

10. If $f(x) = x^2 + 3$ and $g(x) = x - 4$, then $(f/g)(x) = \dfrac{x^2 + 3}{x - 4}$.

a. What value of x makes $g(x) = 0$?

b. Fill in the blank: The domain of f/g is $(-\infty, 4)$ ___ $(4, \infty)$.

NOTATION

Complete each solution.

11. Let $f(x) = 3x - 1$ and $g(x) = 2x + 3$. Find $f \cdot g$.

$(f \cdot g)(x) = f(x) \cdot$ ___

$= $ ___ $(2x + 3)$

$= 6x^2 +$ ___ $-$ ___ $- 3$

$(f \cdot g)(x) = 6x^2 + 7x - 3$

12. Let $f(x) = 3x - 1$ and $g(x) = 2x + 3$. Find $f \circ g$.

$(f \circ g)(x) = f($ ___ $)$

$= f($ ___ $)$

$= 3($ ___ $) - 1$

$=$ ___ $+$ ___ $- 1$

$(f \circ g)(x) = 6x + 8$

GUIDED PRACTICE

Let f(x) = 2x + 1 and g(x) = x − 3. Find each function and give its domain. See Example 1.

13. $f + g$
14. $f - g$
15. $g - f$
16. $g + f$
17. $f \cdot g$
18. f/g
19. g/f
20. $g \cdot f$

Let f(x) = 3x and g(x) = 4x. Find each function and give its domain. See Example 1.

21. $f + g$
22. $f - g$
23. $g - f$
24. $g + f$
25. $f \cdot g$
26. f/g
27. g/f
28. $g \cdot f$

Let f(x) = 2x − 5 and g(x) = x + 1. Find each of the following function values. See Example 2.

29. $(f + g)(8)$
30. $(f - g)(-4)$
31. $(f \cdot g)(0)$
32. $(f/g)(2)$

Let s(x) = 3 − x and t(x) = x² − x − 6. Find each function value. See Example 2.

33. $(s \cdot t)(-2)$
34. $(s + t)(3)$
35. $(s/t)(1)$
36. $(s - t)(12)$

Let f(x) = 2x + 1 and g(x) = x² − 1. Find each of the following. See Example 3.

37. $(f \circ g)(2)$
38. $(g \circ f)(2)$
39. $(g \circ f)(-3)$
40. $(f \circ g)(-3)$
41. $(f \circ g)\left(\frac{1}{2}\right)$
42. $(g \circ f)\left(\frac{1}{3}\right)$
43. $(g \circ f)(2x)$
44. $(f \circ g)(2x)$

Refer to graphs at the right. Find each function value. See Example 4.

45. a. $(f + g)(-5)$
b. $(f - g)(3)$
c. $(f \cdot g)(-3)$

46. a. $(f/g)(0)$
b. $(f \circ g)(3)$
c. $(g \circ f)(2)$

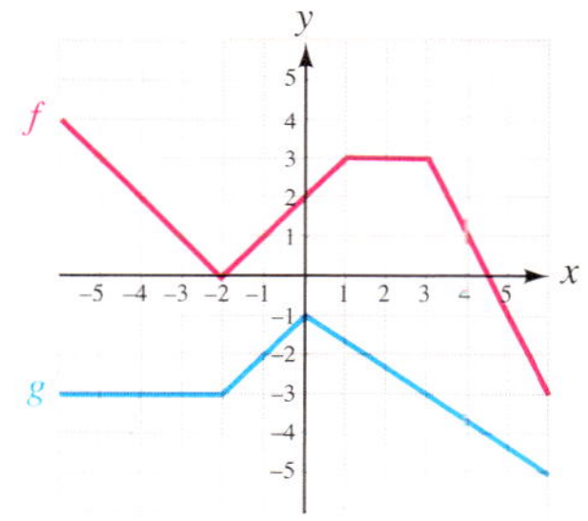

Refer to graphs at the right. Find each function value. See Example 4.

47. a. $(g + f)(2)$
b. $(g - f)(-5)$
c. $(g \cdot f)(1)$

48. a. $(g/f)(-6)$
b. $(g \circ f)(4)$
c. $(f \circ g)(6)$

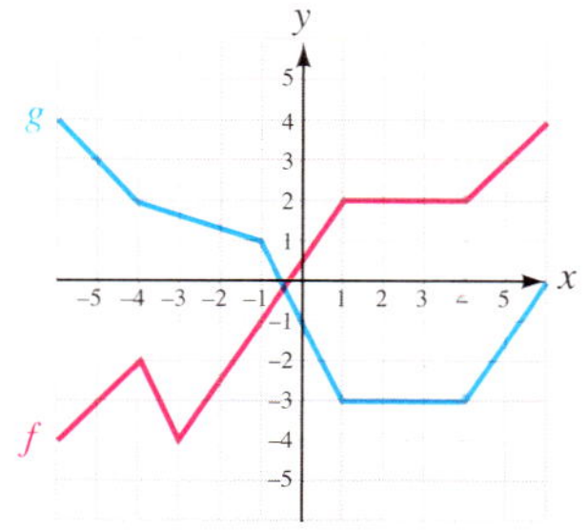

Find f(x) and g(x) such that h(x) = (f ∘ g)(x). Answers may vary. See Example 5.

49. $h(x) = (x + 15)^2$
50. $h(x) = (x - 9)^3$
51. $h(x) = x^5 + 9$
52. $h(x) = x^6 - 100$
53. $h(x) = \sqrt{16x - 1}$
54. $h(x) = \sqrt[3]{10 - x}$
55. $h(x) = \dfrac{1}{x - 4}$
56. $h(x) = \dfrac{1}{3x - 16}$

TRY IT YOURSELF

Let $f(x) = 3x - 2$ and $g(x) = x^2 + x$. Find each of the following.

57. $(f \circ g)(4)$
58. $(g \circ f)(4)$
59. $(g \circ f)(-3)$
60. $(f \circ g)(-3)$
61. $(g \circ f)(0)$
62. $(f \circ g)(0)$
63. $(g \circ f)(x)$
64. $(f \circ g)(x)$

Let $f(x) = 3x - 2$ and $g(x) = 2x^2 + 1$. Find each function and give its domain.

65. $f - g$
66. $f + g$
67. f/g
68. $f \cdot g$

Let $f(x) = \frac{1}{x}$ and $g(x) = \frac{1}{x^2}$. Find each of the following.

69. $(g \circ f)\left(\frac{1}{3}\right)$
70. $(g \circ f)\left(\frac{1}{10}\right)$
71. $(g \circ f)(8x)$
72. $(f \circ g)(5x)$

Let $f(x) = x^2 - 1$ and $g(x) = x^2 - 4$. Find each function and give its domain.

73. $f - g$
74. $f + g$
75. g/f
76. $g \cdot f$

Let $h(t) = \sqrt{t + 3}$ and $k(t) = t - 5$. Find each of the following.

77. $(h \circ k)(18)$
78. $(h \circ k)(11)$
79. $(k \circ h)(22)$
80. $(k \circ h)(-2)$

81. Use the tables of values for functions f and g to find each of the following.
   a. $(f + g)(1)$ b. $(f - g)(5)$
   c. $(f \cdot g)(1)$ d. $(g/f)(5)$

x	$f(x)$
1	3
5	8

x	$g(x)$
1	4
5	0

82. Use the table of values for functions f and g to find each of the following.
   a. $(f \circ g)(1)$ b. $(g \circ f)(2)$

x	$f(x)$
2	5
4	7

x	$g(x)$
1	2
5	−3

83. If $f(x) = x + 1$ and $g(x) = 2x - 5$, show that $(f \circ g)(x) \neq (g \circ f)(x)$.
84. If $f(x) = x^2 + 1$ and $g(x) = 3x^2 - 2$, show that $(f \circ g)(x) \neq (g \circ f)(x)$.

APPLICATIONS

85. **SAT Scores.** The graph of function m in the next column gives the average score on the mathematics portion of the SAT college entrance exam, the graph of function r gives the average score on the critical reading portion, and x represents the number of years since 2000.
   a. Find $(m + r)(4)$ and explain what information about SAT scores it gives.
   b. Find $(m - r)(4)$ and explain what information about SAT scores it gives.
   c. Find: $(m + r)(9)$
   d. Find: $(m - r)(9)$

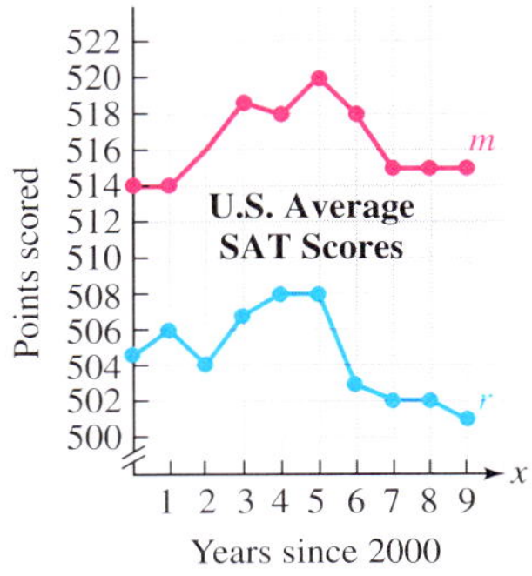

Source: National Center for Education Statistics

86. **Bachelor's Degrees.** The graph of function m below gives the number of bachelor's degrees awarded to men, and the graph of function w gives the number of bachelor's degrees awarded to women in the U. S. for the years 1990 through 2008.
   a. Estimate $(w + m)(2004)$ and explain what information about bachelor's degrees it gives.
   b. Estimate $(w - m)(2004)$ and explain what information about bachelor's degrees it gives.
   c. Estimate: $(w + m)(1994)$
   d. Estimate: $(w - m)(2000)$

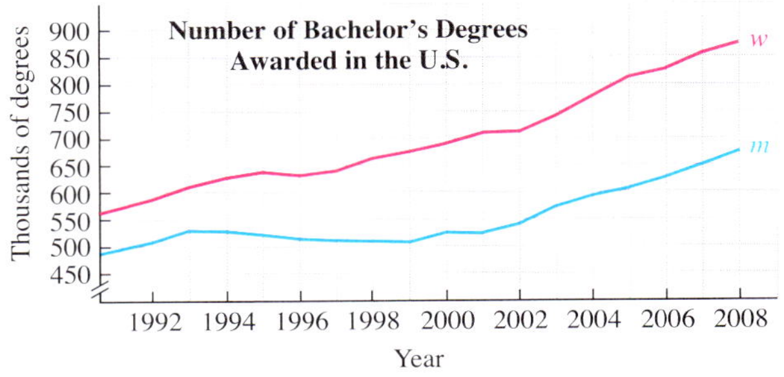

Source: National Center for Education Statistics

87. **Metallurgy.** A molten alloy must be cooled slowly to control crystallization. When removed from the furnace, its temperature is 2,700°F, and it will be cooled at 200° per hour. Write a composition function that expresses the Celsius temperature as a function of the number of hours t since cooling began. (*Hint*: $C(F) = \frac{5}{9}(F - 32)$.)
88. **Weather Forecasting.** A high-pressure area promises increasingly warmer weather for the next 48 hours. The temperature is now 34° Celsius and is expected to rise 1° every 6 hours. Write a composition function that expresses the Fahrenheit temperature as a function of the number of hours from now. (*Hint*: $F(C) = \frac{9}{5}C + 32$.)

89. Vacation Mileage Costs.

a. Use the following graphs to determine the cost of the gasoline consumed if a family drove 500 miles on a summer vacation.

b. Write a composition function that expresses the cost of the gasoline consumed on the vacation as a function of the miles driven.

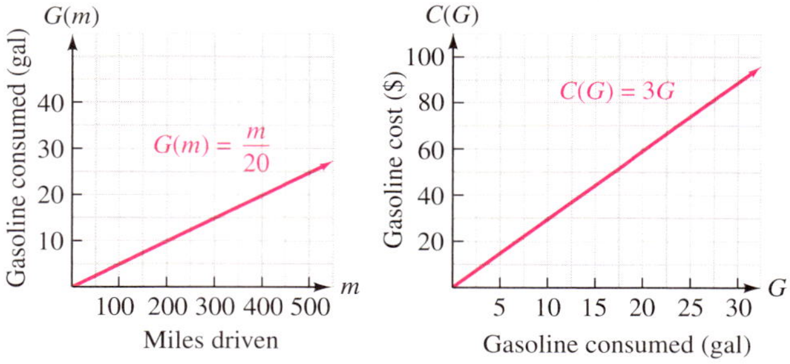

90. Halloween Costumes. The tables on the back of a pattern package can be used to determine the number of yards of material needed to make a rabbit costume for a child.

a. How many yards of material are needed if the child's chest measures 29 inches?

b. In this exercise, one quantity is a function of a second quantity that depends, in turn, on a third quantity. Explain this dependence.

PATTERN 9810 **Simplicity**

BODY MEASUREMENTS

Chest (in.)	21	22	23	25	26	27	$28\frac{1}{2}$	29	30
Pattern Size	2	3	4	6	7	8	10	11	12

YARDAGE NEEDED

Pattern Size	2-4	6-8	10-12
Yards	$2\frac{5}{8}$	$3\frac{3}{8}$	$3\frac{3}{4}$

WRITING

91. Exercise 89 illustrates a chain of dependence between the cost of the gasoline, the gasoline consumed, and the miles driven. Describe another chain of dependence that could be represented by a composition function.

92. In this section, what operations are performed on functions? Give an example of each.

93. Write out in words how to say each of the following:

$$(f \circ g)(2) \qquad g(f(-8))$$

94. If $Y_1 = f(x)$ and $Y_2 = g(x)$, explain how to use the following tables to find $g(f(2))$.

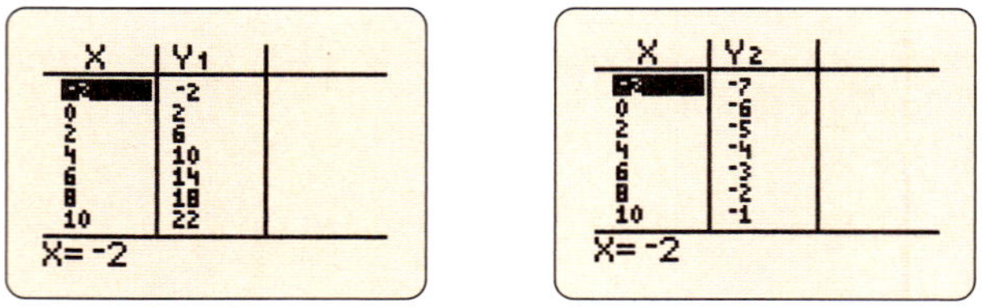

REVIEW

Simplify each complex fraction.

95. $\dfrac{\dfrac{ac - ad - c + d}{a^3 - 1}}{\dfrac{c^2 - 2cd + d^2}{a^2 + a + 1}}$

96. $\dfrac{2 + \dfrac{1}{x^2 - 1}}{1 + \dfrac{1}{x - 1}}$

CHALLENGE PROBLEMS

Fill in the blanks.

97. If $f(x) = x^2$ and $g(x) =$ ______, then $(f \circ g)(x) = 4x^2 + 20x + 25$.

98. If $f(x) = \sqrt{3x}$ and $g(x) =$ ______, then $(g \circ f)(x) = 9x^2 + 7$.

Refer to the following graphs of functions f and g.

99. Graph the sum function $f + g$ on the given coordinate system.

100. Graph the difference function $f - g$ on the given coordinate system.

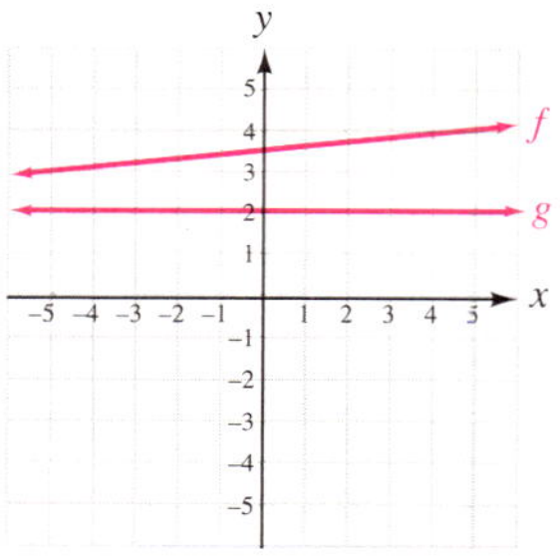

SECTION 11.2

OBJECTIVES

1 Determine whether a function is a one-to-one function.

2 Use the horizontal line test to determine whether a function is one-to-one.

3 Find the equation of the inverse of a function.

4 Find the composition of a function and its inverse.

5 Graph a function and its inverse.

Inverse Functions

ARE YOU READY?

The following problems review some basic skills that are needed when working with inverse functions.

1. What are the domain and the range of the function $\{(-2, 8), (3, -3), (5, 10), (9, 1)\}$?
2. Is a parabola that opens upward the graph of a function?
3. Fill in the blank: If y is a ________ of x, the symbols y and $f(x)$ are interchangeable.
4. Let $f(x) = 2x + 6$. Find $f(8x)$.

In the previous section, we created new functions from given functions by using the operations of arithmetic and composition. Another way to create new functions is to find the *inverse of a function.*

1 Determine Whether a Function Is a One-to-One Function.

In figure (a) below, the arrow diagram defines a function f. If we reverse the arrows as shown in figure (b), we obtain a new correspondence where the range of f becomes the domain of the new correspondence, and the domain of f becomes the range. The new correspondence is a function because to each member of the domain, there corresponds exactly one member of the range. We call this new correspondence the **inverse** of f, or f inverse.

(a) (b)

This reversing process does not always produce a function. For example, if we reverse the arrows in function g defined by the diagram in figure (a) below, the resulting correspondence shown in figure (b) is not a function. This is because to the number 2 in the domain, there corresponds two members of the range: 8 and 4.

A function

Not a function

(a) (b)

The question that arises is, "What must be true of an original function to guarantee that the reversing process produces a function?" The answer is: *The original function must be one-to-one.*

We have seen that in a function, each input determines exactly one output. For some functions, different inputs determine different outputs, as in figure (a) below. For other functions, different inputs might determine the *same* output, as in figure (b). When a function has the property that different inputs determine different outputs, as in figure (a), we say the function is *one-to-one.*

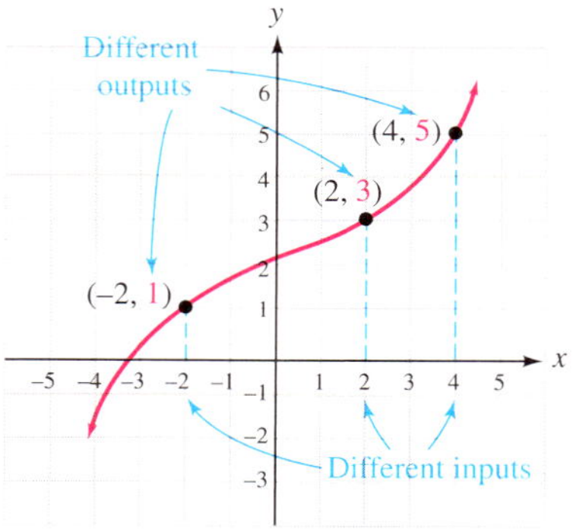

A one-to-one function

(a)

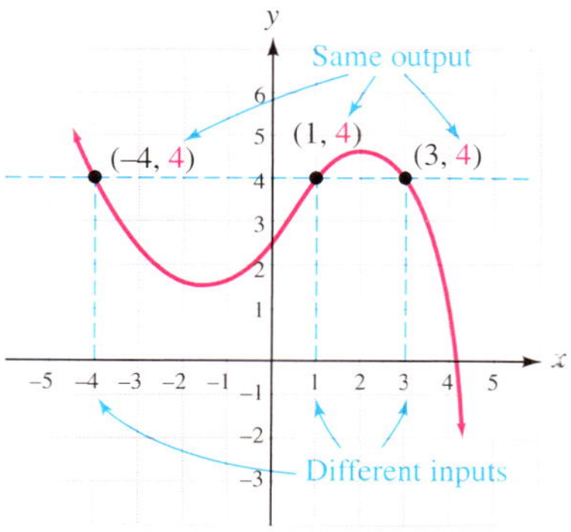

Not a one-to-one function

(b)

One-to-One Functions A function is called a **one-to-one function** if different inputs determine different outputs.

EXAMPLE 1 Determine whether each function is one-to-one. **a.** $f(x) = x^2$ **b.** $f(x) = x^3$

Strategy We will determine whether different inputs have different outputs.

Why If different inputs have different outputs, the function is one-to-one. If different inputs have the same output, the function is not one-to-one.

Solution

Success Tip

Example 1 illustrates that not every function is one-to-one.

a. Since two different inputs, -3 and 3, have the same output 9, $f(x) = x^2$ is not a one-to-one function.

$f(-3) = (-3)^2 = 9$ and $f(3) = 3^2 = 9$

x	$f(x)$
-3	9
3	9

The output 9 does not correspond to exactly one input.

b. Since different numbers have different cubes, each input of $f(x) = x^3$ determines a different output. This function is one-to-one.

Self Check 1 Determine whether each function is one-to-one. If not, find an output that corresponds to more than one input. **a.** $f(x) = 2x + 3$ **b.** $f(x) = x^4$

Now Try Problems 19 and 21

2 Use the Horizontal Line Test to Determine Whether a Function Is One-to-One.

To determine whether a function is one-to-one, it is often easier to view its graph rather than its defining equation. If two (or more) points on the graph of a function have the same y-coordinate, the function is not one-to-one. This observation suggests the following **horizontal line test.**

The Horizontal Line Test A function is one-to-one if each horizontal line that intersects its graph does so exactly once.

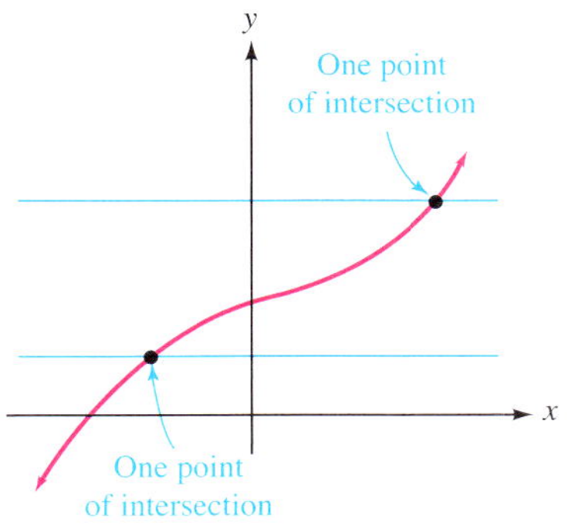

A one-to-one function

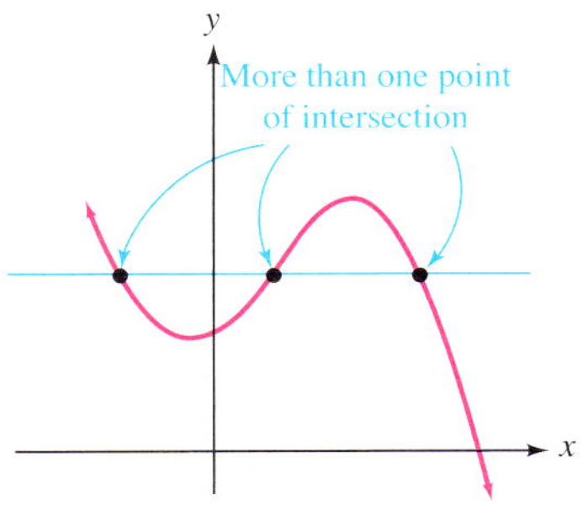

Not a one-to-one function

EXAMPLE 2 Use the horizontal line test to determine whether the following graphs of functions represent one-to-one functions.

Strategy We will draw horizontal lines through the graph of the function and see how many times each line intersects the graph.

Why If each horizontal line intersects the graph of the function exactly once, the graph represents a one-to-one function. If any horizontal line intersects the graph of the function more than once, the graph does not represent a one-to-one function.

Solution

a. Because every horizontal line that intersects the graph of $f(x) = -\frac{3}{4}x - 2$ in figure (a) does so exactly once, the graph represents a one-to-one function. We simply say, the function $f(x) = -\frac{3}{4}x - 2$ is one-to-one.

Success Tip

Recall that we use the **vertical line test** to determine whether a graph represents a function. We use the **horizontal line test** to determine whether the function that is graphed is one-to-one.

b. Refer to figure (b). Because we can draw a horizontal line that intersects the graph of $f(x) = x^2 - 4$ twice, the graph does not represent a one-to-one function. We simply say, the function $f(x) = x^2 - 4$ is not one-to-one.

c. Because every horizontal line that intersects the graph of $f(x) = \sqrt{x}$ in figure (c) does so exactly once, the graph represents a one-to-one function. We simply say, the function $f(x) = \sqrt{x}$ is one-to-one.

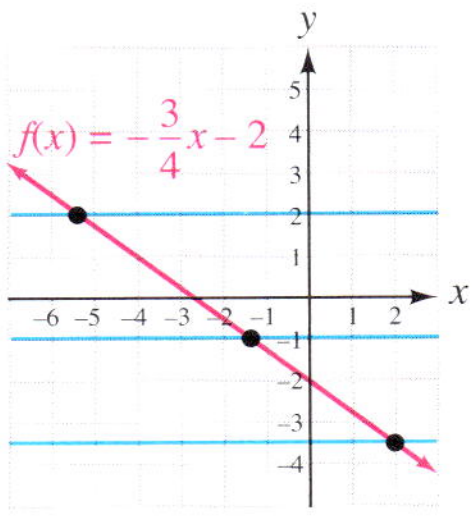

(a)

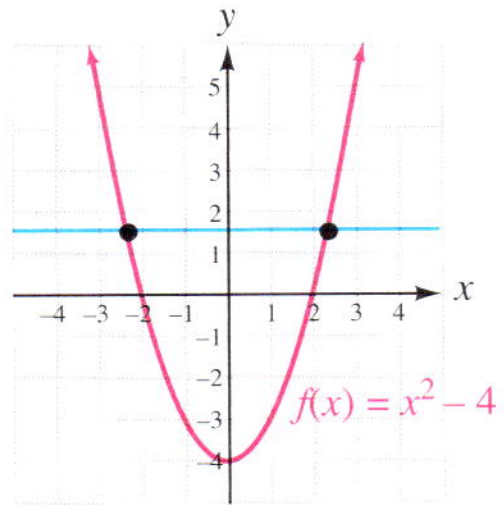

(b)

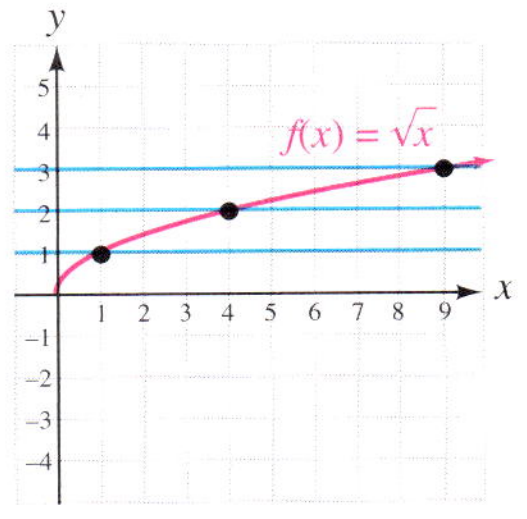

(c)

Self Check 2 Use the horizontal line test to determine whether the graph represents a one-to-one function.

a.

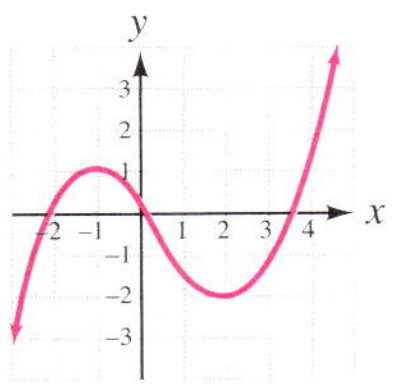

b.

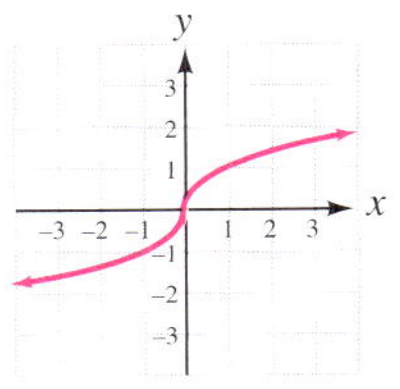

Now Try Problems 27 and 29

3 Find the Equation of the Inverse of a Function.

If f is the one-to-one function defined by the arrow diagram in figure (a), it turns the number 1 into 10, 2 into 20, and 3 into 30. The ordered pairs that define f can be listed in a table. Since the inverse of f must turn 10 back into 1, 20 back into 2, and 30 back into 3, it consists of the ordered pairs shown in the table in figure (b).

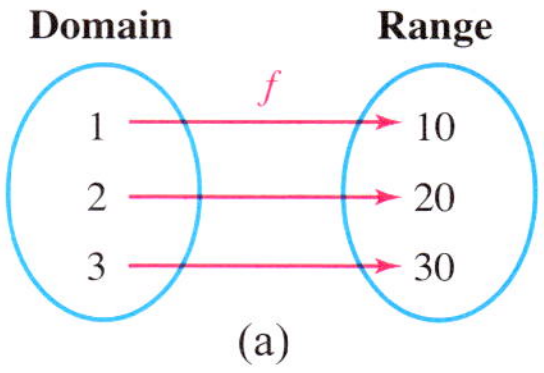

(a)

Function f

x	y	(x, y)
1	10	(1, 10)
2	20	(2, 20)
3	30	(3, 30)

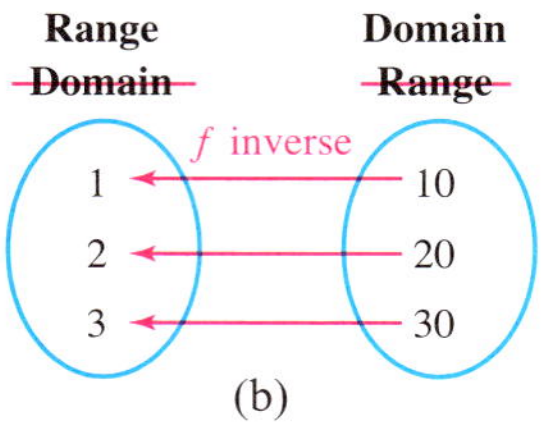

(b)

f inverse

x	y	(x, y)
10	1	(10, 1)
20	2	(20, 2)
30	3	(30, 3)

The x- and y-coordinates are interchanged.

Caution

The -1 in the notation $f^{-1}(x)$ is not an exponent:

$$f^{-1}(x) \neq \frac{1}{f(x)}$$

We note that the domain of f and the range of its inverse is $\{1, 2, 3\}$. The range of f and the domain of its inverse is $\{10, 20, 30\}$.

This example suggests that to form the inverse of a function f, we simply interchange the coordinates of each ordered pair that determines f. When the inverse of a function is also a function, we call it ***f* inverse** and denote it with the symbol f^{-1}. The symbol $f^{-1}(x)$ is read as "the inverse of $f(x)$" or "f inverse of x."

The Inverse of a Function

If f is a one-to-one function consisting of ordered pairs of the form (x, y), **the inverse of** f, denoted f^{-1}, is the one-to-one function consisting of all ordered pairs of the form (y, x).

When a one-to-one function is defined by an equation, we use the following method to find the equation of its inverse.

Finding the Equation of the Inverse of a Function

If a function is one-to-one, we find its inverse as follows:

1. If the function is written using function notation, replace $f(x)$ with y.
2. Interchange the variables x and y.
3. Solve the resulting equation for y.
4. Substitute $f^{-1}(x)$ for y.

EXAMPLE 3 Determine whether each function is one-to-one. If so, find the equation of its inverse.
a. $f(x) = 4x + 2$ **b.** $f(x) = x^3$

Strategy We will determine whether each function is one-to-one. If it is, we can find the equation of its inverse by replacing $f(x)$ with y, interchanging x and y, and solving for y.

Why The reason for interchanging the variables is this: If a one-to-one function takes an input x into an output y, by definition, its inverse function has the reverse effect.

Solution

a. We recognize $f(x) = 4x + 2$ as a linear function whose graph is a straight line with slope 4 and y-intercept (0, 2). Since such a graph would pass the horizontal line test, we conclude that f is one-to-one.

Success Tip

Recall that $f(x) = 4x + 2$ is a linear function. Every linear function, except those of the form $f(x) = c$, where c is a constant, is one-to-one.

To find the inverse of function f, we proceed as follows:

$f(x) = 4x + 2$ Function f multiplies all inputs by 4 and then adds 2.

$y = 4x + 2$ Replace f(x) with y.

$x = 4y + 2$ Interchange the variables x and y.

$x - 2 = 4y$ To isolate the term 4y, subtract 2 from both sides.

$\frac{x-2}{4} = y$ To solve for y, divide both sides by 4.

$y = \frac{x-2}{4}$ Write the equation with y on the left side.

To denote that this equation is the inverse of function f, we replace y with $f^{-1}(x)$.

$f^{-1}(x) = \frac{x-2}{4}$ Function f^{-1} subtracts 2 from each input and then divides by 4.

As an informal check, we see below that if $x = 1$, function f produces an output of 6. And if $x = 6$, function f^{-1} produces an output of 1.

Success Tip

Notice that f^{-1} operates on an input using the inverse operations of f in the reverse order.

$f(1) = 4(1) + 2$	$f^{-1}(6) = \frac{6-2}{4}$
$= 4 + 2$	$= \frac{4}{4}$
$= 6$	$= 1$
Ordered pair: (1, 6)	Ordered pair: (6, 1)

The coordinates are interchanged.

b. The graph of $f(x) = x^3$ is shown on the right. Since such a graph would pass the horizontal line test, we conclude that f is a one-to-one function.

To find its inverse, we proceed as follows:

$f(x) = x^3$ Function f cubes all inputs.

$y = x^3$ Replace f(x) with y.

$x = y^3$ Interchange the variables x and y.

$\sqrt[3]{x} = y$ To solve for y, take the cube root of both sides.

$y = \sqrt[3]{x}$ Write the equation with y on the left side.

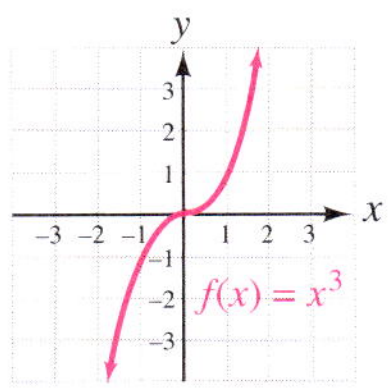

Caution

Only one-to-one functions have inverse functions.

Replacing y with $f^{-1}(x)$, we have

$f^{-1}(x) = \sqrt[3]{x}$ Function f^{-1} finds the cube root of each input.

As an informal check, let $x = 4$ and determine whether f and f^{-1} produce ordered pairs whose coordinates are reversed.

Self Check 3 Determine whether each function is one-to-one. If it is, find the equation of its inverse. **a.** $f(x) = -5x - 3$
b. $f(x) = x^5$

Now Try Problems 35 and 45

4 Find the Composition of a Function and Its Inverse.

To emphasize a relationship between a function and its inverse, we substitute some number x, such as $x = 3$, into the function $f(x) = 4x + 2$ of Example 3(a). The corresponding value of y that is produced is

$f(3) = 4(3) + 2 = 14$ f determines the ordered pair (3, 14).

If we substitute 14 into the inverse function, $f^{-1}(x) = \frac{x-2}{4}$, the corresponding value of y that is produced is

$$f^{-1}(\mathbf{14}) = \frac{\mathbf{14} - 2}{4} = \mathbf{3} \quad f^{-1} \text{ determines the ordered pair } (14, 3).$$

Thus, the function f turns 3 into 14, and the inverse function f^{-1} turns 14 back into 3.

In general, the composition of a function and its inverse function is the identity function, $f(x) = x$, such that any input x has the output x. This fact can be stated symbolically as follows.

The Composition of Inverse Functions

For any one-to-one function f and its inverse, f^{-1},

$$(f \circ f^{-1})(x) = x \quad \text{and} \quad (f^{-1} \circ f)(x) = x$$

We can use this property to determine whether two functions are inverses.

EXAMPLE 4 Show that $f(x) = 4x + 2$ and $f^{-1}(x) = \frac{x-2}{4}$ are inverses.

Strategy We will find the composition of $f(x)$ and $f^{-1}(x)$ in both directions and show that the result is x.

Why Only when the result of the composition is x in both directions are the functions inverses.

Solution To show that $f(x) = 4x + 2$ and $f^{-1}(x) = \frac{x-2}{4}$ are inverses, we must show that for each composition, an input of x gives an output of x.

$$\begin{aligned}(f \circ f^{-1})(x) &= f(\mathbf{f^{-1}(x)}) \\ &= f\left(\frac{\mathbf{x-2}}{\mathbf{4}}\right) \\ &= 4\left(\frac{\mathbf{x-2}}{\mathbf{4}}\right) + 2 \\ &= x - 2 + 2 \\ &= x\end{aligned}$$

$$\begin{aligned}(f^{-1} \circ f)(x) &= f^{-1}(\mathbf{f(x)}) \\ &= f^{-1}(\mathbf{4x + 2}) \\ &= \frac{\mathbf{4x + 2} - 2}{4} \\ &= \frac{4x}{4} \\ &= x\end{aligned}$$

Because $(f \circ f^{-1})(x) = x$ and $(f^{-1} \circ f)(x) = x$, the functions are inverses.

Self Check 4 Show that $f(x) = x - 4$ and $g(x) = x + 4$ are inverses.

Now Try Problem 55

Success Tip

Recall that the line $y = x$ passes through points whose x- and y-coordinates are equal: $(-1, -1)$, $(0, 0)$, $(1, 1)$, $(2, 2)$, and so on.

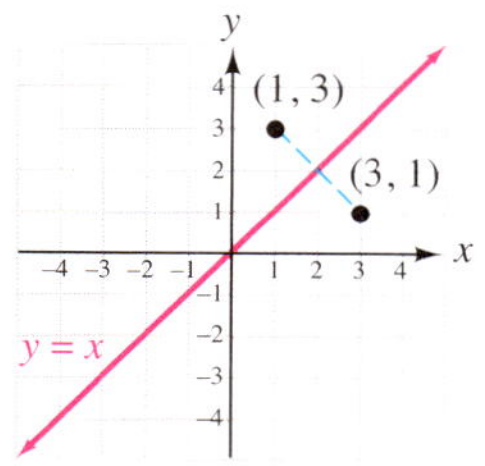

5 Graph a Function and Its Inverse.

If a point (a, b) is on the graph of function f, it follows that the point (b, a) is on the graph of f^{-1}, and vice versa. There is a geometric relationship between a pair of points whose coordinates are interchanged. For example, in the graph, we see that the line segment between $(1, 3)$ and $(3, 1)$ is perpendicular to and cut in half by the line $y = x$. We say that $(1, 3)$ and $(3, 1)$ are mirror images of each other with respect to $y = x$.

Since each point on the graph of f^{-1} is a mirror image of a point on the graph of f, and vice versa, the graphs of f and f^{-1} must be mirror images of each other with respect to $y = x$.

EXAMPLE 5 Find the equation of the inverse of $f(x) = -\frac{3}{2}x + 3$. Then graph f and its inverse on one coordinate system.

Strategy We will determine whether the function has an inverse. If so, we will replace $f(x)$ with y, interchange x and y, and solve for y to obtain the equation of the inverse.

Why The reason for interchanging the variables is this: If a one-to-one function takes an input x into an output y, by definition, its inverse function has the reverse effect.

Solution Since $f(x) = -\frac{3}{2}x + 3$ is a linear function, it is one-to-one and has an inverse. To find the inverse function, we replace $f(x)$ with y, and interchange x and y to obtain

$$x = -\frac{3}{2}y + 3$$

Then we solve for y to get

$$x - 3 = -\frac{3}{2}y \quad \text{Subtract 3 from both sides.}$$

$$-\frac{2}{3}x + 2 = y \quad \text{To isolate } y \text{, multiply both sides by } -\tfrac{2}{3}.$$

When we replace y with $f^{-1}(x)$, we have $f^{-1}(x) = -\frac{2}{3}x + 2$.

To graph f, we construct a table of values, plot points, and draw the graph in red as shown below. To graph f^{-1}, we don't need to do any calculations to construct a table of values. We can simply interchange the coordinates of the ordered pairs in the table for f and use them to graph f^{-1}. The result is the graph in blue shown below. Because the functions are inverses of each other, their graphs are **mirror images** about the line $y = x$.

Success Tip

If the graphs of f and f^{-1} intersect, it will always be on the line $y = x$.

$f(x) = -\frac{3}{2}x + 3$

x	$f(x)$	
0	3	$\rightarrow (0, 3)$
2	0	$\rightarrow (2, 0)$
4	-3	$\rightarrow (4, -3)$

$f^{-1}(x) = -\frac{2}{3}x + 2$

x	$f^{-1}(x)$	
3	0	$\rightarrow (3, 0)$
0	2	$\rightarrow (0, 2)$
-3	4	$\rightarrow (-3, 4)$

Simply interchange the coordinates to graph f^{-1}.

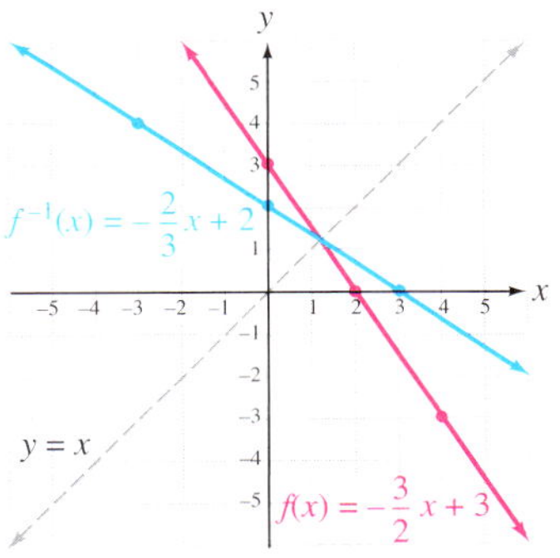

Self Check 5 Find the inverse of $f(x) = \frac{2}{3}x - 2$. Then graph the function and its inverse on one coordinate system.

Now Try Problem 59

Using Your Calculator ▶ Graphing the Inverse of a Function

We can use a graphing calculator to check the result found in Example 5. First, we enter $f(x) = -\frac{3}{2}x + 3$ and then enter what we believe to be the inverse function, $f^{-1}(x) = -\frac{2}{3}x + 2$, as well as the equation $y = x$. See figure (a) on the next page. Before graphing, we adjust the display so that the graphing grid will be composed of squares. The line of symmetry $y = x$ is then at a 45° angle to the positive x-axis.

In figure (b), it appears that the two graphs are symmetric about the line $y = x$. Although it is not definitive, this visual check does help to validate the result of Example 5.

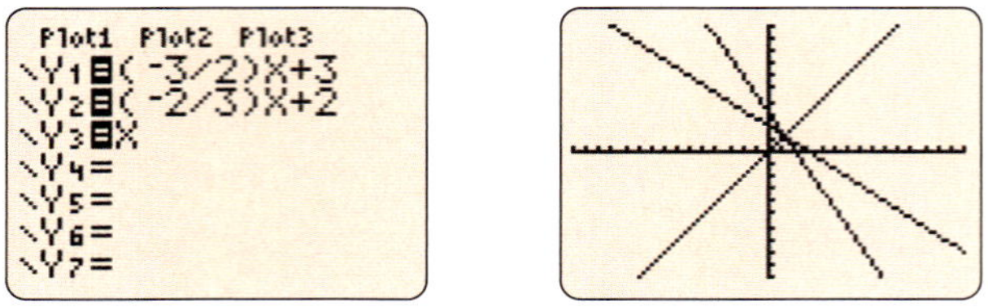

(a) (b)

EXAMPLE 6 Graph the inverse of function f shown in figure (a).

Strategy We will find the coordinates of several points on the graph of f in figure (a). After interchanging the coordinates of these points, we will plot them as shown in figure (b).

Why The reason for interchanging the coordinates is this: If (a, b) is a point on the graph of a one-to-one function, then the point (b, a) is on the graph of its inverse.

Solution In figure (a), we see that the points $(-5, -3)$, $(-2, -1)$, $(0, 2)$, $(3, 3)$, $(5, 4)$, and $(7, 5)$ lie on the graph of function f. To graph the inverse, we interchange their coordinates, and plot them in blue, as shown in figure (b). Then we graph the line $y = x$ and use symmetry to draw a smooth curve through those points to get the graph of f^{-1}.

The Language of Algebra

We also can say that the graphs of f and f^{-1} are **reflections** of each other about the line $y = x$, or they are **symmetric about** $y = x$.

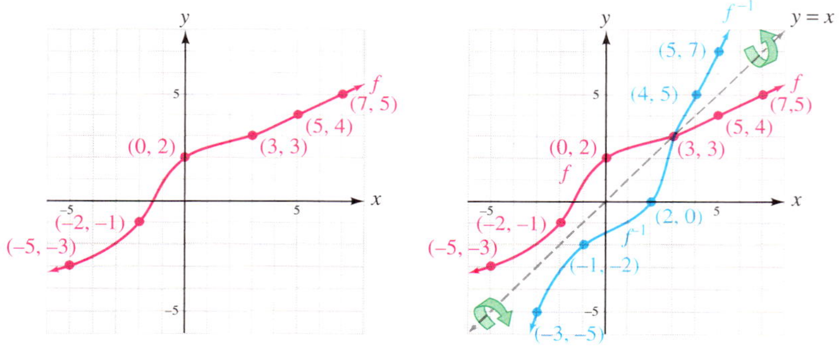

(a) (b)

Now Try Problem 65

SECTION 11.2 STUDY SET

VOCABULARY

Fill in the blanks.

1. A function is called a __________ function if different inputs determine different outputs.
2. The __________ line test can be used to determine whether the graph of a function represents a one-to-one function.
3. The functions f and f^{-1} are __________.
4. The graphs of a function and its inverse are __________ images of each other with respect to $y = x$. We also say that their graphs are __________ with respect to the line $y = x$.

CONCEPTS

Fill in the blanks.

5. If any horizontal line that intersects the graph of a function does so more than once, the function is not __________.
6. To find the inverse of the function $f(x) = 2x - 3$, we begin by replacing $f(x)$ with y, and then we __________ x and y.

7. If f is a one-to-one function, the domain of f is the ______ of f^{-1}, and the range of f is the ________ of f^{-1}.

8. If a function turns an input of 2 into an output of 5, the inverse function will turn an input of 5 into the output .

9. If f is a one-to-one function, and if $f(1) = 6$, then $f^{-1}(6) =$.

10. If the point $(9, -4)$ is on the graph of the one-to-one function f, then the point (,) is on the graph of f^{-1}.

11. a. Is the correspondence defined by the arrow diagram in figure (a) below a one-to-one function?

b. Is the correspondence defined by the table in figure (b) below a one-to-one function?

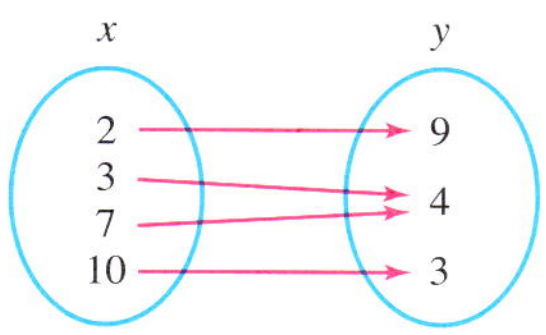

(a)

x	$f(x)$
−2	4
−1	1
0	0
2	4
3	9

(b)

12. Is the inverse of a one-to-one function always a function?

13. Use the table of values of the one-to-one function f to complete a table of values for f^{-1}.

x	$f(x)$
−4	−2
0	0
8	4

x	$f^{-1}(x)$
−2	
0	
4	

14. Redraw the graph of function f. Then graph f^{-1} and the axis of symmetry on the same coordinate system.

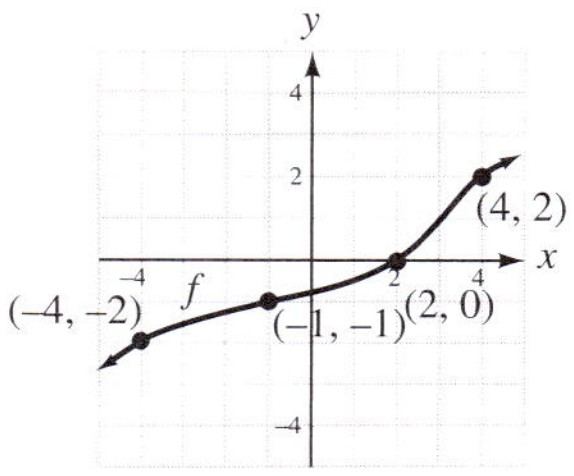

NOTATION

Complete each solution.

15. Find the inverse of $f(x) = 2x - 3$.

$$\quad = 2x - 3$$
$$x = \quad - 3$$
$$x + \quad = 2y$$
$$\frac{x+3}{2} = \quad$$

The inverse of $f(x) = 2x - 3$ is $\quad (x) = \frac{x+3}{2}$.

16. Find the inverse of $f(x) = \sqrt[3]{x} + 2$.

$$\quad = \sqrt[3]{x} + 2$$
$$x = \sqrt[3]{\quad} + 2$$
$$x - \quad = \sqrt[3]{y}$$
$$(x-2)^3 = \quad$$

The inverse of $f(x) = \sqrt[3]{x} + 2$ is $\quad (x) = (x-2)^3$.

17. The symbol f^{-1} is read as "the ________ of f" or "f ________."

18. Explain the difference in the meaning of the -1 in the notation $f^{-1}(x)$ as compared with a^{-1}.

GUIDED PRACTICE

Determine whether each function is one-to-one. See Example 1.

19. $f(x) = 2x$

20. $f(x) = |x|$

21. $f(x) = x^4$

22. $f(x) = x^3 + 1$

23. $f(x) = -x^2 + 3x$

24. $f(x) = \frac{2}{3}x + 8$

25. $\{(1, 1), (2, 1), (3, 1), (4, 1)\}$

26. $\{(3, 2), (2, 1), (1,0)\}$

Each graph represents a function. Use the horizontal line test to determine whether the function is one-to-one. See Example 2.

27.

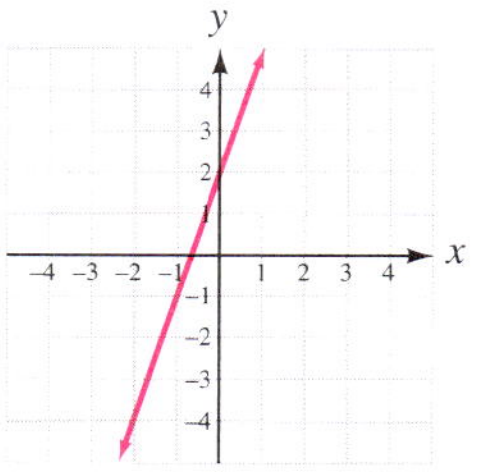

28.

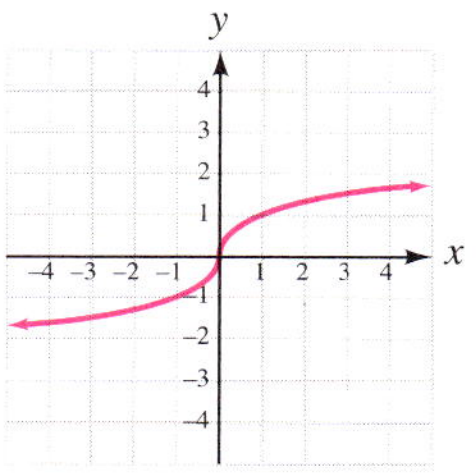

29.

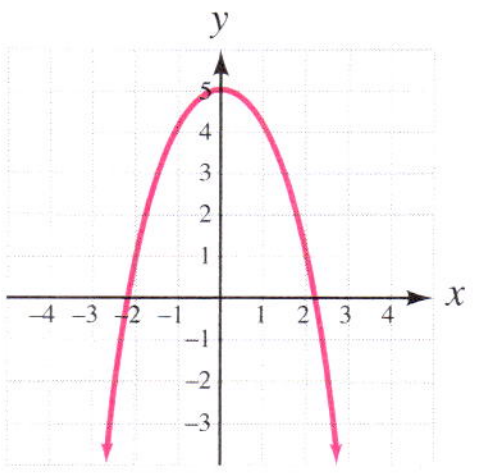

30.

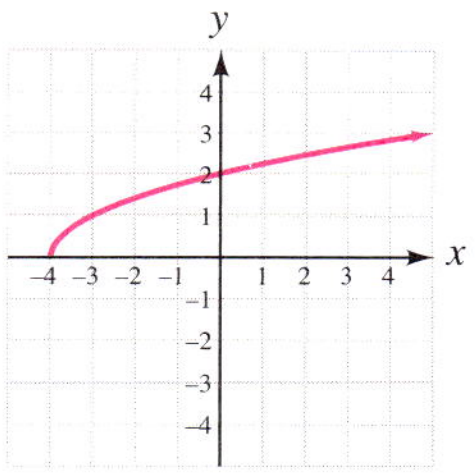

31.

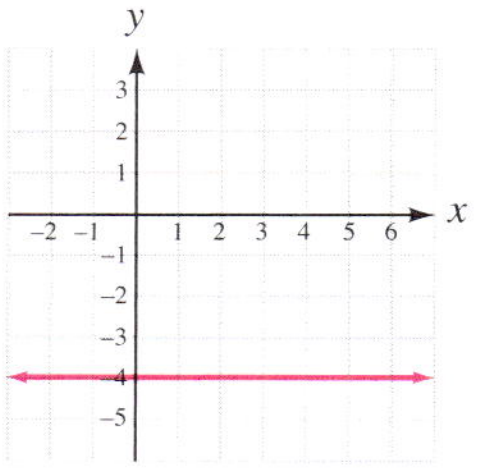

32.

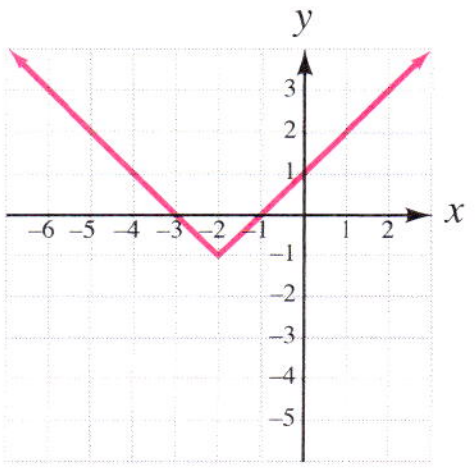

33. 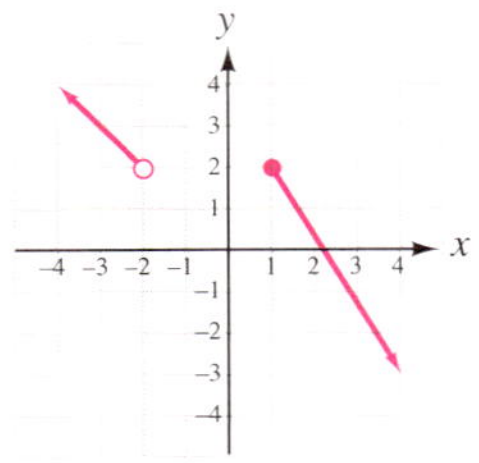

34. 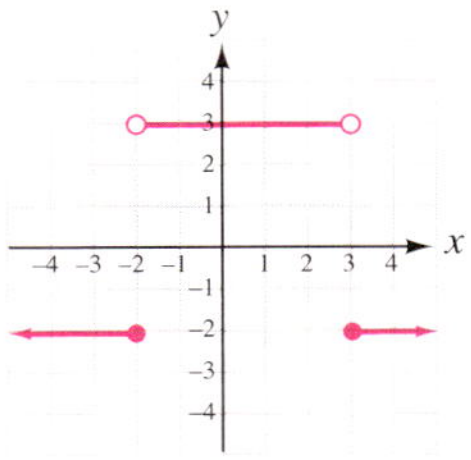

Each of the following functions is one-to-one. Find the inverse of each function and express it using $f^{-1}(x)$ notation. **See Example 3.**

35. $f(x) = 2x + 4$

36. $f(x) = 5x - 1$

37. $f(x) = \dfrac{x}{5} + \dfrac{4}{5}$

38. $f(x) = \dfrac{x}{3} - \dfrac{1}{3}$

39. $f(x) = \dfrac{x - 4}{5}$

40. $f(x) = \dfrac{2x + 6}{3}$

41. $f(x) = \dfrac{2}{x - 3}$

42. $f(x) = \dfrac{3}{x + 1}$

43. $f(x) = \dfrac{4}{x}$

44. $f(x) = \dfrac{1}{x}$

45. $f(x) = x^3 + 8$

46. $f(x) = x^3 - 4$

47. $f(x) = \sqrt[3]{x}$

48. $f(x) = \sqrt[3]{x - 5}$

49. $f(x) = (x + 10)^3$

50. $f(x) = (x - 9)^3$

51. $f(x) = 2x^3 - 3$

52. $f(x) = \dfrac{3}{x^3} - 1$

53. $f(x) = \dfrac{x^7}{2}$

54. $f(x) = \dfrac{x^9}{4}$

Show that each pair of functions are inverses. **See Example 4.**

55. $f(x) = 2x + 9,\ f^{-1}(x) = \dfrac{x - 9}{2}$

56. $f(x) = 5x - 1,\ f^{-1}(x) = \dfrac{x + 1}{5}$

57. $f(x) = \dfrac{2}{x - 3},\ f^{-1}(x) = \dfrac{2}{x} + 3$

58. $f(x) = \sqrt[3]{x - 6},\ f^{-1}(x) = x^3 + 6$

Find the inverse of each function. Then graph the function and its inverse on one coordinate system. Show the line of symmetry on the graph. **See Examples 5 and 6.**

59. $f(x) = 2x$

60. $f(x) = -3x$

61. $f(x) = 4x + 3$

62. $f(x) = \dfrac{x}{3} + \dfrac{1}{3}$

63. $f(x) = -\dfrac{2}{3}x + 3$

64. $f(x) = -\dfrac{1}{3}x + \dfrac{4}{3}$

65. $f(x) = x^3$

66. $f(x) = x^3 + 1$

67. $f(x) = x^2 - 1\ (x \geq 0)$

68. $f(x) = x^2 + 1\ (x \geq 0)$

APPLICATIONS

69. **Interpersonal Relationships.** Feelings of anxiety in a relationship can increase or decrease, depending on what is going on in the relationship. The graph shows how a person's anxiety might vary as a relationship develops over time.

a. Is this the graph of a function? Is its inverse a function?

b. Does each anxiety level correspond to exactly one point in time? Use the dashed lined labeled *Maximum threshold* to explain.

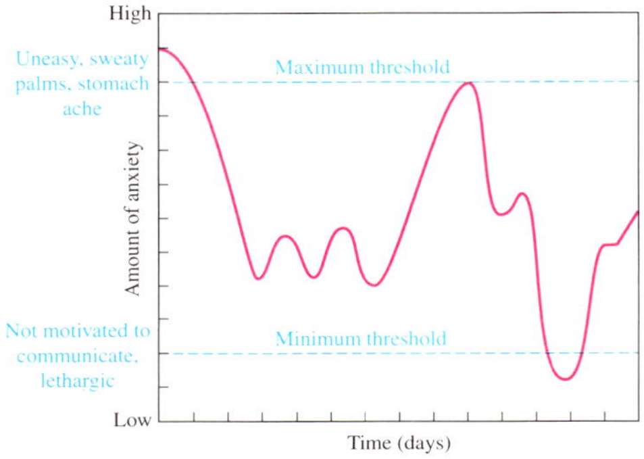

Source: Gudykunst, *Building Bridges: Interpersonal Skills for a Changing World* (Houghton Mifflin, 1994)

70. **Lighting Levels.** The ability of the eye to see detail increases as the level of illumination increases. This relationship can be modeled by a function E, whose graph is shown here.

a. From the graph, determine $E(240)$.

b. Is function E one-to-one? Does E have an inverse?

c. If the effectiveness of seeing in an office is 7, what is the illumination in the office? How can this question be asked using inverse function notation?

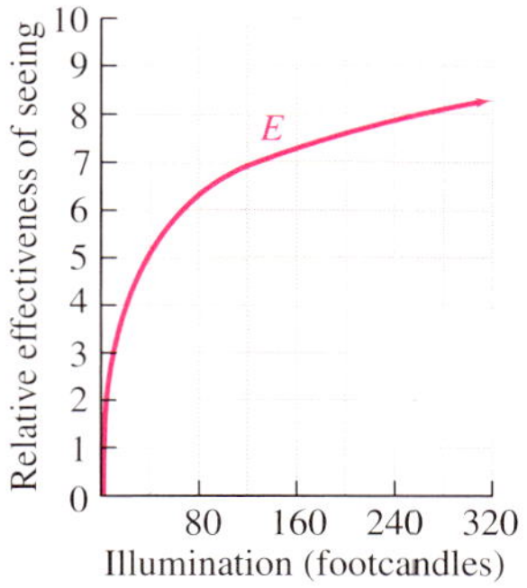

Based on information from *World Book Encyclopedia*

WRITING

71. In your own words, what is a one-to-one function?

72. Two functions are graphed on the square grid on the right along with the line $y = x$. Explain why the functions cannot be inverses of each other.

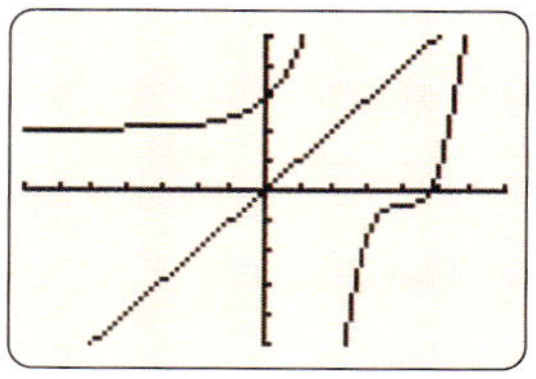

73. Explain how the graph of a one-to-one function can be used to draw the graph of its inverse function.

74. **a.** Explain the purpose of the vertical line test.
    b. Explain the purpose of the horizontal line test.

75. In the illustration, a function f and its inverse f^{-1} have been graphed on the same coordinate system. Explain what concept can be demonstrated by folding the graph paper on the dashed line.

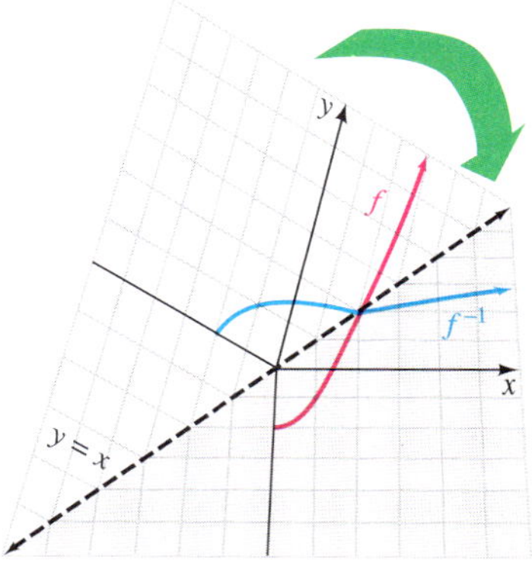

76. Write in words how to read the notation.
    a. $f^{-1}(x) = \frac{1}{2}x - 3$
    b. $(f \circ f^{-1})(x) = x$

REVIEW

Simplify. Write the result in the form a + bi.

77. $3 - \sqrt{-64}$

78. $(2 - 3i) + (4 + 5i)$

79. $(3 + 4i)(2 - 3i)$

80. $\dfrac{6 + 7i}{3 - 4i}$

81. $(6 - 8i)^2$

82. i^{100}

CHALLENGE PROBLEMS

83. Find the inverse of $f(x) = \dfrac{x + 1}{x - 1}$.

84. Using the functions of Exercise 83, show that $(f \circ f^{-1})(x) = x$ and $(f^{-1} \circ f)(x) = x$.

85. A table of values for a function f is shown in figure (a). A table of values for f^{-1} is shown in figure (b). Use the tables to find $f^{-1}(f(4))$ and $f(f^{-1}(2))$.

X	Y1
-2	-2
0	2
2	6
4	10
6	14
8	18
10	22

X=-2

(a)

X	Y1
-2	-2
0	-1
2	0
4	1
6	2
8	3
10	4

X=-2

(b)

86. **a.** The graph of a one-to-one function lies entirely in quadrant I. In what quadrant does the graph of its inverse lie?
    b. The graph of a one-to-one function lies entirely in quadrant II. In what quadrant does the graph of its inverse lie?
    c. The graph of a one-to-one function lies entirely in quadrant III. In what quadrant does the graph of its inverse lie?
    d. The graph of a one-to-one function lies entirely in quadrant IV. In what quadrant does the graph of its inverse lie?

SECTION 11.3 Exponential Functions

OBJECTIVES

1. Define exponential functions.
2. Graph exponential functions.
3. Use exponential functions in applications involving growth or decay.

ARE YOU READY?

The following problems review some basic skills that are needed when working with exponential functions.

1. Simplify: **a.** $3^4 \cdot 3^8$ **b.** $(3^4)^8$
2. Evaluate: **a.** 2^3 **b.** 2^0 **c.** 2^{-2}
3. Evaluate: **a.** $\left(\frac{1}{3}\right)^2$ **b.** $\left(\frac{1}{3}\right)^0$ **c.** $\left(\frac{1}{3}\right)^{-3}$
4. Fill in the blanks: The graph of $f(x) = x$ is a ____ and the graph of $f(x) = x^2$ is a ________.

In previous chapters, we have discussed linear functions, polynomial functions, rational functions, and radical functions. We now begin a study of a new family of functions known as *exponential functions*.

As an example, consider the graph in figure (a) below, which models the soaring popularity of the social network website Twitter in recent years. The rapidly rising red curve is the graph of an exponential function.

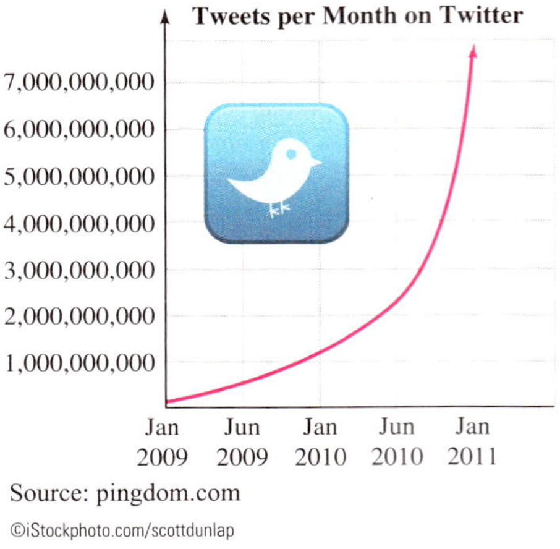

(a)

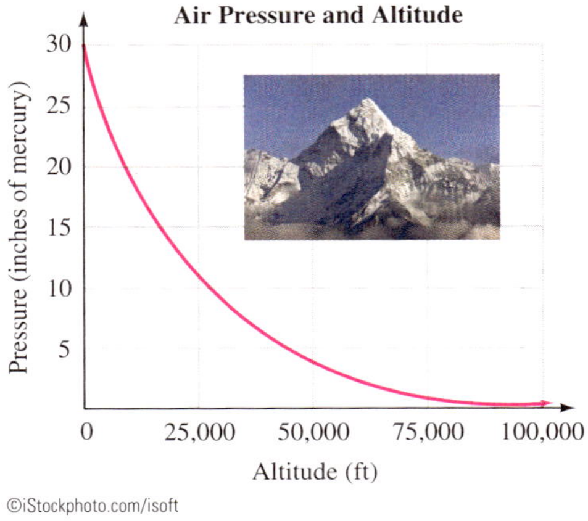

(b)

If you have ever climbed a high mountain or gone up in an airplane that does not have a pressurized cabin, you have probably felt the effects of low air pressure. The graph in figure (b) above shows how the atmospheric pressure decreases with increasing altitude. The rapidly falling red curve is also the graph of an exponential function.

Exponential functions are used to model many other situations, such as population growth, the spread of an epidemic, the temperature of a heated object as it cools, and radioactive decay.

1 Define Exponential Functions.

In this course, we have evaluated exponential expressions with integer exponents, such as 2^5, 7^0, and 5^{-1}, and rational number exponents, such as $9^{1/2}$, $64^{3/4}$, and $8^{-2/3}$. To define exponential functions, we also must be able to evaluate exponential expressions that have irrational exponents. For example, let's consider the expression

$5^{\sqrt{2}}$ where $\sqrt{2}$ is the irrational number 1.414213562 . . .

We can successively approximate $5^{\sqrt{2}}$ using the following *rational* powers:

$$5^{1.4}, \quad 5^{1.41}, \quad 5^{1.414}, \quad 5^{1.4142}, \quad 5^{1.41421}, \ldots$$

Using concepts from advanced mathematics, it can be shown that there is exactly one number that these powers approach. We define $5^{\sqrt{2}}$ to be that number. This process can be used to approximate $5^{\sqrt{2}}$ to as many decimal places as desired. Any other positive irrational exponent can be defined in the same manner, and negative irrational exponents can be defined using reciprocals.

This discussion leads us to the following conclusion: If b is positive, the exponential expression b^x has meaning and can be evaluated for any real number exponent x. Furthermore, it can be shown that all of the familiar rules of exponents are also true for irrational exponents.

Using Your Calculator ▶ Evaluating Exponential Expressions

We can use a calculator to obtain a very good approximation of an exponential expression with an irrational exponent. To find the value of $5^{\sqrt{2}}$ with a reverse entry scientific calculator, we enter:

5 y^x 2 $\sqrt{\ }$ = 9.738517742

With a direct entry graphing calculator, we enter:

5 ^ 2nd √ 2) ENTER

```
5^√(2)
        9.738517742
```

If $b > 0$ and $b \neq 1$, the function $f(x) = b^x$ is called an **exponential function.** Since x can be any real number, its domain is the set of real numbers, which can be written as $(-\infty, \infty)$.

Because b is positive, the value of $f(x)$ is positive, and the range is the set of positive numbers, which can be written as $(0, \infty)$.

Since $b \neq 1$, an exponential function cannot be the constant function $f(x) = 1^x$, in which $f(x) = 1$ for every real number x.

Exponential Functions

An exponential function with base b is defined by the equations

$$f(x) = b^x \quad \text{or} \quad y = b^x$$

where $b > 0$, $b \neq 1$, and x is a real number. The domain of $f(x) = b^x$ is the interval $(-\infty, \infty)$, and the range is the interval $(0, \infty)$.

Exponential functions have a *constant base* and a *variable exponent.* Some examples of exponential functions are:

$f(x) = 2^x$ — The base is 2.

$g(x) = \left(\frac{1}{3}\right)^x$ — The base is $\frac{1}{3}$.

$s(x) = 4^{x+1}$ — The base is 4.

$P(t) = (1.45)^{-3.04t}$ — The base is 1.45.

In the third example, we see that the exponent of an exponential function doesn't have to be just x. It can be a variable expression, such as $x + 1$. In the fourth example, we see that the base can be a positive decimal, and the input variable can be a letter other than x.

The following functions are not exponential functions.

$f(x) = x^2$ $\quad g(x) = x^{1/3}$ — These have a variable base and a constant exponent.

$h(x) = (-4)^x$ — The base cannot be negative.

$f(x) = 1^x$ — The base cannot be 1.

2 Graph Exponential Functions.

Since the domain and range of $f(x) = b^x$ are sets of real numbers, we can graph exponential functions on a rectangular coordinate system. To do this, we will use the familiar **point-plotting method.**

EXAMPLE 1 Graph: $f(x) = 2^x$

Strategy We will graph the function by creating a table of function values and plotting the corresponding ordered pairs.

Why After drawing a smooth curve through the plotted points, we will have the graph.

Solution To graph $f(x) = 2^x$, we select several values for x and find the corresponding values of $f(x)$. If x is -3, and if x is -2, we have:

$$f(x) = 2^x$$
$$f(-3) = 2^{-3} \quad \text{Substitute } -3 \text{ for } x.$$
$$= \frac{1}{2^3}$$
$$= \frac{1}{8}$$

$$f(x) = 2^x$$
$$f(-2) = 2^{-2} \quad \text{Substitute } -2 \text{ for } x.$$
$$= \frac{1}{2^2}$$
$$= \frac{1}{4}$$

Caution

We have previously graphed the linear function $f(x) = 2x$ and the squaring function $f(x) = x^2$. For the exponential function $f(x) = 2^x$, note that the variable is in the exponent.

The points $\left(-3, \frac{1}{8}\right)$ and $\left(-2, \frac{1}{4}\right)$ are on the graph of $f(x) = 2^x$. In a similar way, we find the corresponding values of $f(x)$ for x values of -1, 0, 1, 2, 3, and 4 and list them in a table. Then we plot the ordered pairs and draw a smooth curve through them, as shown below. Notice with the ordered pairs, as the value of x increases, the value of y also increases, but very rapidly as compared with x. This graph is an example of **exponential growth.**

$f(x) = 2^x$

x	$f(x)$	
-3	$\frac{1}{8}$	$\rightarrow \left(-3, \frac{1}{8}\right)$
-2	$\frac{1}{4}$	$\rightarrow \left(-2, \frac{1}{4}\right)$
-1	$\frac{1}{2}$	$\rightarrow \left(-1, \frac{1}{2}\right)$
0	1	$\rightarrow (0, 1)$
1	2	$\rightarrow (1, 2)$
2	4	$\rightarrow (2, 4)$
3	8	$\rightarrow (3, 8)$
4	16	$\rightarrow (4, 16)$

Because of the variable exponent in their equations, the graphs of exponential functions rise or fall sharply. When graphing them, make sure you plot enough points to show this.

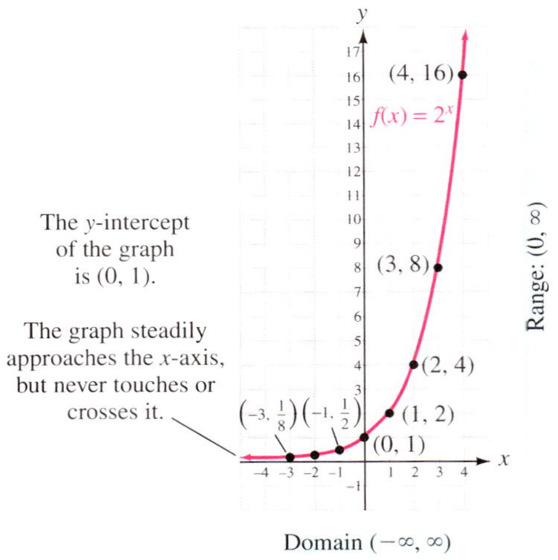

By the vertical line test, we see that this is indeed the graph of a function. Because the graph extends indefinitely to the left and right, the projection of the graph onto the x-axis includes all real numbers. Thus, the domain of $f(x) = 2^x$ is $(-\infty, \infty)$.

Because the projection of the graph onto the y-axis covers only the positive portion of that axis, the range of the function is $(0, \infty)$. Since the graph passes the horizontal line test, the function is one-to-one.

Note that as x decreases, the values of $f(x)$ decrease and approach 0. Thus, the x-axis is a **horizontal asymptote** of the graph. The graph does not have an x-intercept, the y-intercept is $(0, 1)$, and the graph passes through the point $(1, 2)$.

Self Check 1 Graph: $g(x) = 4^x$

Now Try Problem 19

Exponential functions can have a base that is a real number between 0 and 1.

EXAMPLE 2 Graph: $f(x) = \left(\frac{1}{3}\right)^x$

Strategy We will graph the function by creating a table of function values and plotting the corresponding ordered pairs.

Why After drawing a smooth curve through the plotted points, we will have the graph.

Solution If $x = -2$ and if $x = -1$, we have

$$f(x) = \left(\frac{1}{3}\right)^x$$
$$f(-2) = \left(\frac{1}{3}\right)^{-2}$$
$$= \left(\frac{3}{1}\right)^2 \quad \text{Recall: } \left(\frac{x}{y}\right)^{-n} = \left(\frac{y}{x}\right)^n.$$
$$= 9$$

$$f(x) = \left(\frac{1}{3}\right)^x$$
$$f(-1) = \left(\frac{1}{3}\right)^{-1}$$
$$= \left(\frac{3}{1}\right)^1$$
$$= 3$$

Success Tip

Since $\frac{1}{3} = 3^{-1}$, and since the rules for exponents hold for any real number, we can rewrite this function as:

$$f(x) = \left(\frac{1}{3}\right)^x = (3^{-1})^x = 3^{-x}$$

The points $(-2, 9)$ and $(-1, 3)$ are on the graph of $f(x) = \left(\frac{1}{3}\right)^x$. In a similar way, we find the corresponding values of $f(x)$ for $x = 0$, 1, and 2 and list them in a table. Then we plot the ordered pairs and draw a smooth curve through them, as shown below. Notice with the ordered pairs, as the value of x increases, the value of y decreases very rapidly as compared with x. This graph is an example of **exponential decay.**

$f(x) = \left(\frac{1}{3}\right)^x$

x	$f(x)$	
-2	9	$\to (-2, 9)$
-1	3	$\to (-1, 3)$
0	1	$\to (0, 1)$
1	$\frac{1}{3}$	$\to \left(1, \frac{1}{3}\right)$
2	$\frac{1}{9}$	$\to \left(2, \frac{1}{9}\right)$

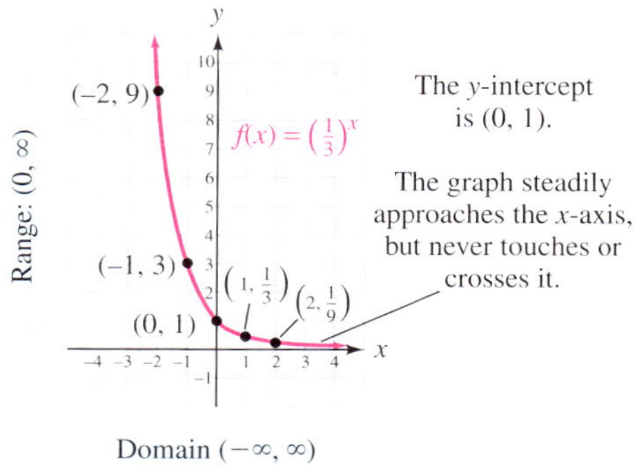

The graph passes the vertical line test, and so it is indeed the graph of a function. Because the graph extends indefinitely to the left and right, the projection of the graph onto the x-axis includes all real numbers. Thus, the domain of $f(x) = \left(\frac{1}{3}\right)^x$ is $(-\infty, \infty)$. Because the projection of the graph onto the y-axis covers only the positive portion of that axis, the range of the function is $(0, \infty)$.

Note that as x increases, the values of $f(x)$ decrease and approach 0. Thus, the x-axis is a horizontal asymptote of the graph. The graph does not have an x-intercept, the y-intercept is $(0, 1)$, and the graph passes through the point $\left(1, \frac{1}{3}\right)$.

Self Check 2 Graph: $g(x) = \left(\frac{1}{2}\right)^x$

Now Try Problem 23

In Example 1 (where $b = 2$), the values of y increase as the values of x increase. Since the graph rises as we move to the right, we call the function an *increasing function.* When $b > 1$, the larger the value of b, the steeper the curve, as shown in figure (a) below.

In Example 2 $\left(\text{where } b = \frac{1}{3}\right)$, the values of y decrease as the values of x increase. Since the graph drops as we move to the right, we call the function a *decreasing function.* When $0 < b < 1$, the smaller the value of b, the steeper the curve, as shown in figure (b) below.

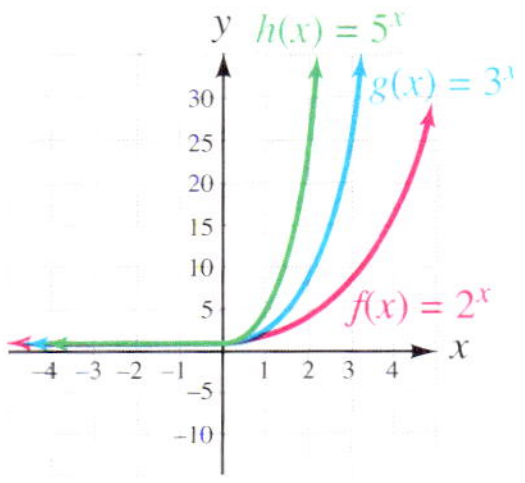

The bases of these exponential functions are 2, 3, and 5. Notice that the largest base, 5, has the steepest graph.

(a)

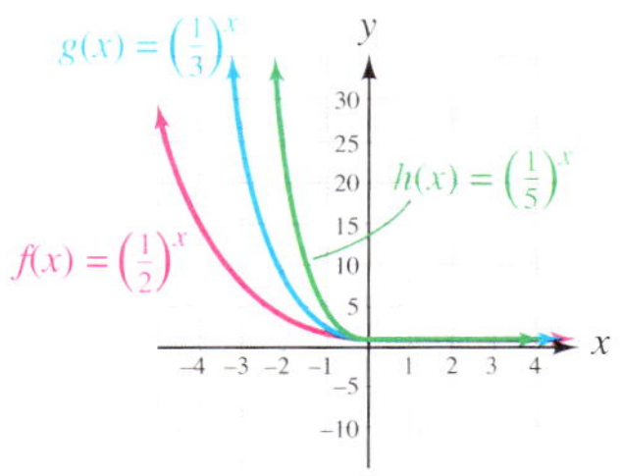

The bases of these exponential functions are $\frac{1}{2}$, $\frac{1}{3}$, and $\frac{1}{5}$, which are numbers between 0 and 1. Notice that the smallest base, $\frac{1}{5}$, has the steepest graph.

(b)

Examples 1 and 2 illustrate the following properties of exponential functions.

Properties of Exponential Functions

1. The domain of the exponential function $f(x) = b^x$ is the interval $(-\infty, \infty)$ and the range is the interval $(0, \infty)$.
2. The graph has a y-intercept of $(0, 1)$.
3. The x-axis is an asymptote of the graph.
4. The graph of $f(x) = b^x$ passes through the point $(1, b)$.
5. Exponential functions are one-to-one.
6. If $b > 1$, then $f(x) = b^x$ is an **increasing function.**

   If $0 < b < 1$, then $f(x) = b^x$ is a **decreasing function.**

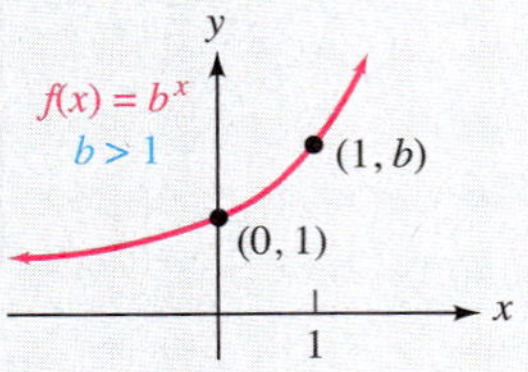

Increasing function

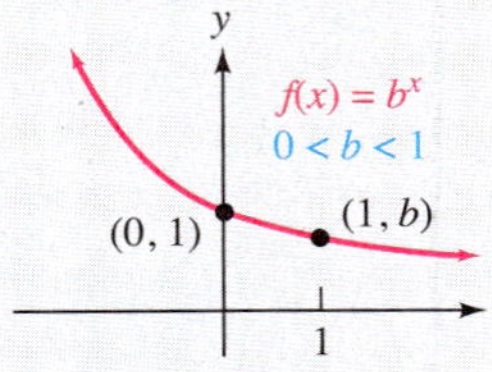

Decreasing function

Using Your Calculator ▶ Graphing Exponential Functions

To use a graphing calculator to graph $f(x) = \left(\frac{2}{3}\right)^x$ and $g(x) = \left(\frac{3}{2}\right)^x$, we enter the right sides of the equations after the symbols $Y_1 =$ and $Y_2 =$. The screen will show the following equations.

$Y_1 = (2/3)$^X
$Y_2 = (3/2)$^X

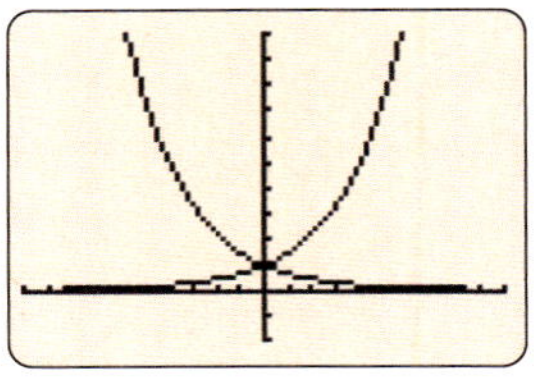

If we press GRAPH, we will obtain the display shown.

We note that the graph of $f(x) = \left(\frac{2}{3}\right)^x$ passes through $(0, 1)$. Since $\frac{2}{3} < 1$, the function is decreasing. The graph of $g(x) = \left(\frac{3}{2}\right)^x$ also passes through $(0, 1)$. Since $\frac{3}{2} > 1$, the function is increasing. Since both graphs pass the horizontal line test, each function is one-to-one.

The graphs of many exponential functions are horizontal and vertical translations of basic graphs.

EXAMPLE 3 Graph each function by using a translation:

a. $g(x) = 2^x - 4$ **b.** $g(x) = \left(\frac{1}{3}\right)^{x+3}$

Strategy We will graph $g(x) = 2^x - 4$ by translating the graph of $f(x) = 2^x$ downward 4 units. We will graph $g(x) = \left(\frac{1}{3}\right)^{x+3}$ by translating the graph of $f(x) = \left(\frac{1}{3}\right)^x$ to the left 3 units.

Why The subtraction of 4 in $g(x) = 2^x - 4$ causes a vertical shift of the graph of the base-2 exponential function 4 units downward. The addition of 3 to x in $g(x) = \left(\frac{1}{3}\right)^{x+3}$ causes a horizontal shift of the graph of the base-$\frac{1}{3}$ exponential function 3 units to the left.

Solution **a.** The graph of $g(x) = 2^x - 4$ will be the same shape as the graph of $f(x) = 2^x$. We call this a **vertical translation.** To graph $g(x) = 2^x - 4$, simply translate each point on the graph of $f(x) = 2^x$ down 4 units. See figure (a) below.

b. The graph of $g(x) = \left(\frac{1}{3}\right)^{x+3}$ will be the same shape as the graph of $f(x) = \left(\frac{1}{3}\right)^x$. We call this a **horizontal translation**. To graph $g(x) = \left(\frac{1}{3}\right)^{x+3}$, simply translate each point on the graph of $f(x) = \left(\frac{1}{3}\right)^x$ to the left 3 units. See figure (b) below.

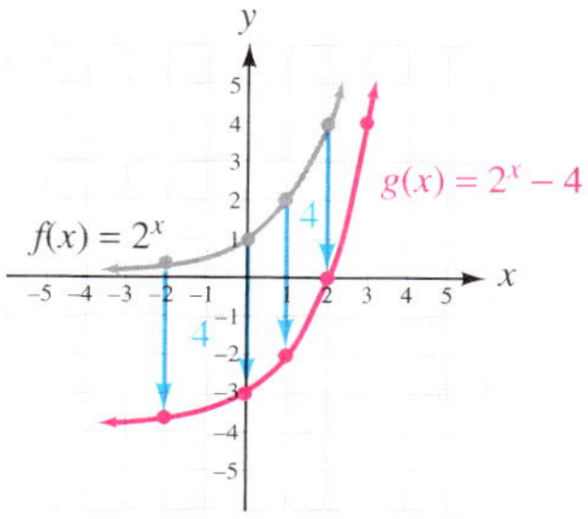

(a)

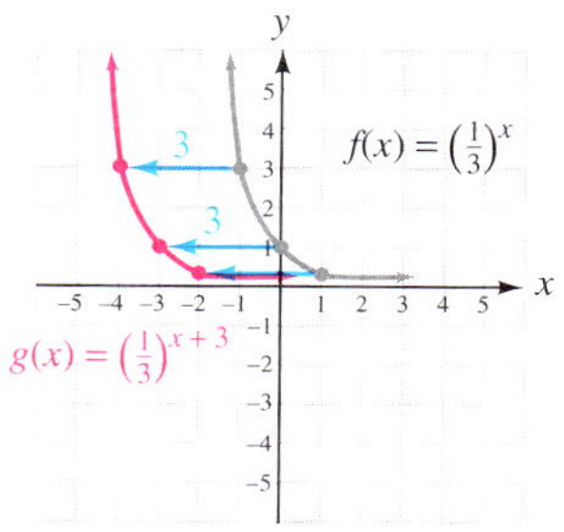

(b)

Self Check 3 Graph each function by using a translation:

a. $g(x) = \left(\frac{1}{4}\right)^x + 2$ **b.** $g(x) = 4^{x-3}$

Now Try Problems 27 and 29

3 Use Exponential Functions in Applications Involving Growth or Decay.

Many real-world situations can be modeled by exponential functions that describe how a quantity grows or decays over time. Some examples of this include the studies of populations, bacteria, heat transfer, radioactive substances, drug concentrations, and financial accounts. Two examples of such functions are:

Exponential growth:

$c(t) = 5(1.034)^t$

A constant — A base greater than 1

Exponential decay:

$f(n) = 650(0.94)^n$

A constant — A base between 0 and 1

EXAMPLE 4 **Professional Baseball Salaries.** The exponential function $s(t) = 650{,}000(1.09)^t$ approximates the average annual salary of a major league baseball player, where t is the number of years after 1990. (Source: Baseball Almanac) **a.** Graph the function. **b.** Use the function to determine the average annual salary in 2020, if the current trend continues.

©jason aron/Shutterstock.com

Strategy For part a, we will graph the function by creating a table of function values and plotting the resulting ordered pairs. For part b, we will find $s(30)$.

Why After drawing a smooth curve through the plotted points, we will have the graph. Since the year 2020 is 30 years after 1990, $t = 30$.

Solution

The Language of Algebra

The word **exponential** is used in many settings to describe rapid growth. For example, we hear that the processing power of computers is growing *exponentially*.

a. The function values for $t = 0$ and $t = 5$ are calculated as follows:

$t = 0$ (the year 1990)

$$s(t) = 650{,}000(1.09)^t$$
$$s(0) = 650{,}000(1.09)^0$$
$$= 650{,}000(1)$$
$$= 650{,}000$$

$t = 5$ (the year 1995)

$$s(t) = 650{,}000(1.09)^t$$
$$s(5) = 650{,}000(1.09)^5$$
$$\approx 1{,}000{,}106 \quad \text{Use a calculator.}$$

To approximate $s(5)$, use the keystrokes 650000 [×] 1.09 [y^x] 5 [=] on a scientific calculator and 650000 [×] 1.09 [^] 5 [ENTER] on a graphing calculator.

In a similar way, we find the corresponding values of $s(t)$ for t-values of 10, 15, and 20 and list them in a table. Then we plot the ordered pairs and draw a smooth curve through them to get the graph shown here.

t	$s(t)$
0	650,000
5	1,000,106
10	1,538,786
15	2,367,614
20	3,642,867

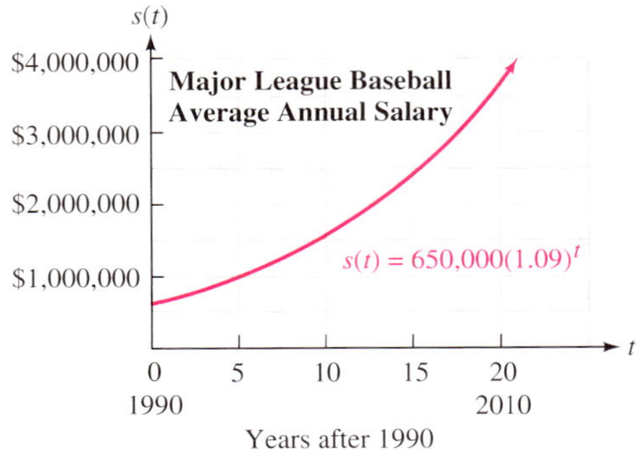

b. To estimate the average annual major league baseball salary in 2020, which is 30 years after 1990, we find $s(30)$.

$$s(t) = 650{,}000(1.09)^t \quad \text{This is the exponential growth model.}$$
$$s(30) = 650{,}000(1.09)^{30} \quad \text{Substitute 30 for } t.$$
$$\approx 8{,}623{,}991 \quad \text{Use a calculator.}$$

If the trend continues, in 2020, the average annual salary will be approximately $8,623,991.

Self Check 4 **Salaries.** Use the function in Example 4 to determine the average annual salary in 2018, if the current trend continues.

Now Try Problem 43

Using Your Calculator ▶ Graphing Exponential Functions

To use a graphing calculator to graph the exponential function $s(t) = 170{,}000(1.12)^t$, we enter the right side of the equation after the symbol $Y_1 =$ and replace the variable t with x. The display will show the equation

$$Y_1 = 170000(1.12\text{^}X)$$

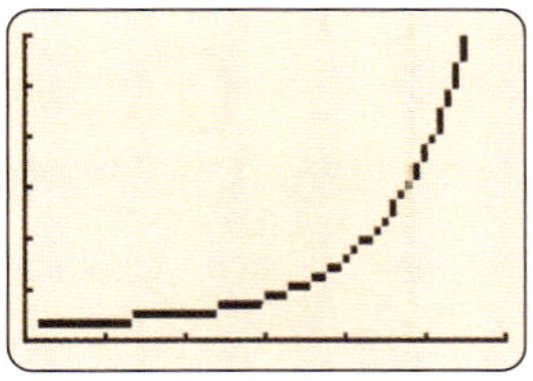

With window settings [0, 30] for x and Xscale = 5 and [0, 3000000] for y and Yscale = 500000, we obtain the display shown when we press [GRAPH].

Up to this point, the financial application problems that we have solved involved **simple interest,** which is calculated using the formula $I = Prt$. However, most savings accounts and investments pay compound interest rather than simple interest. **Compound interest** is paid more than once a year on the principal and *previously earned interest.* The following compound interest formula is a useful application of exponential functions.

Formula for Compound Interest

If $\$P$ is deposited in an account and interest is paid k times a year at an annual rate r, the amount A in the account after t years is given by

$$A = P\left(1 + \frac{r}{k}\right)^{kt}$$

EXAMPLE 5 **Educational Savings Plan.** To save for college, parents of a newborn child invest \$12,000 in a mutual fund at 10% interest, compounded quarterly.

©iStockphoto.com/sjlocke

a. Find a function for the amount in the account after t years.

b. If the quarterly interest paid is continually reinvested, how much money will be in the account when the child is 18 years old?

Strategy To write a function for the amount in the account after t years, we will substitute the given values for P, r, and k into the compound interest formula.

Why The resulting equation will involve only two variables, A and t. Then we can write that equation using function notation.

Solution

a. When we substitute 12,000 for P, 0.10 for r, and 4 for k in the formula for compound interest, the resulting formula involves only two variables, A and t.

$$A = P\left(1 + \frac{r}{k}\right)^{kt}$$ This is the compound interest model.

$$A = \mathbf{12{,}000}\left(1 + \frac{\mathbf{0.10}}{\mathbf{4}}\right)^{4t}$$ Since the interest is compounded quarterly, $k = 4$. Express $r = 10\%$ as a decimal.

Since the value of A depends on the value of t, we can express this relationship using function notation.

$$A(t) = 12{,}000\left(1 + \frac{0.10}{4}\right)^{4t}$$

$$A(t) = 12{,}000(1 + 0.025)^{4t}$$ Evaluate within the parentheses: $\frac{0.10}{4} = 0.025$.

$$A(t) = 12{,}000(1.025)^{4t}$$ The base of this exponential function is 1.025.

b. To find how much money will be in the account when the child is 18 years old, we need to find $A(18)$.

$$A(t) = 12{,}000(1.025)^{4(t)}$$ This is the exponential growth model.

$$A(\mathbf{18}) = 12{,}000(1.025)^{4(\mathbf{18})}$$ Substitute 18 for t.

$$= 12{,}000(1.025)^{72}$$ Evaluate the exponent: $4(18) = 72$.

$$\approx 71{,}006.74$$ Use a scientific calculator and press these keys: 12000 × 1.025 y^x 72 = .

When the child is 18 years old, the account will contain \$71,006.74.

The Language of Algebra

The following words indicate the number of times that interest is paid by a financial institution in one year.

Annually: 1 time
Semiannually: 2 times
Quarterly: 4 times
Monthly: 12 times
Daily: 365 times

Self Check 5 **Savings Plans.** In Example 5, how much money would be in the account after 18 years if the parents initially invested \$20,000?

Now Try Problem 53

Using Your Calculator ▶ Solving Investment Problems

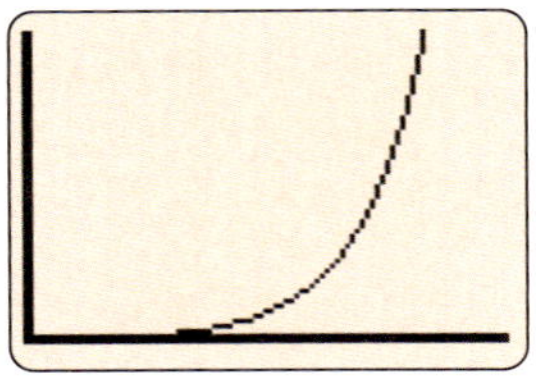

Suppose \$1 is deposited in an account earning 6% annual interest, compounded monthly. To use a graphing calculator to estimate how much will be in the account in 100 years, we can substitute 1 for P, 0.06 for r, and 12 for k in the formula and simplify.

$$A = P\left(1 + \frac{r}{k}\right)^{kt} = 1\left(1 + \frac{0.06}{12}\right)^{12t} = (1.005)^{12t}$$

We now graph the function $A(t) = (1.005)^{12t}$ using window settings of [0, 120] and [0, 400] with Xscale = 1 and Yscale = 1 to obtain the graph shown. We can then trace and zoom to estimate that \$1 grows to be approximately \$397 in 100 years. From the graph, we can see that the money grows slowly in the early years and rapidly in the later years.

Examples 4 and 5 are applications illustrating exponential growth. In the next example, we see an application of exponential decay.

EXAMPLE 6 **Medications.** The most common way people take medications is orally (by mouth). Medications, when swallowed, travel from the stomach or small intestine into the bloodstream, but they are eventually eliminated from the body by the kidneys and the liver. If a patient takes a 250-milligram dose of an antibiotic, the function $A(t) = 250(0.58)^t$ approximates the amount of medication (in milligrams) left in the patient's bloodstream t hours after it is taken.

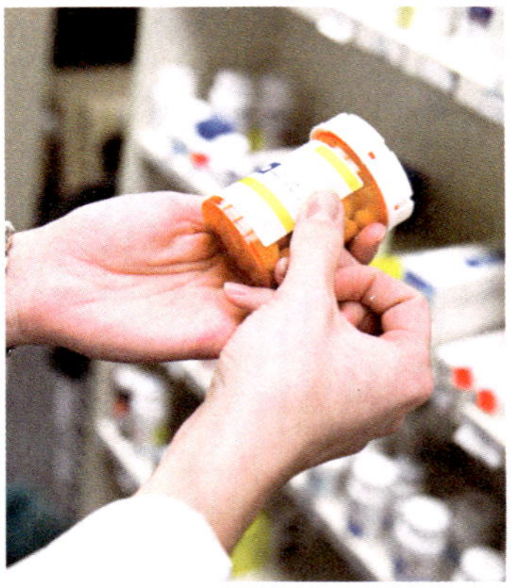

a. Graph the function.

b. Use the function to determine the amount of medication in the patient's bloodstream 10 hours after taking the dose.

Strategy For part a, we will graph the function by creating a table of function values and plotting the resulting ordered pairs. For part b, we will find $A(10)$.

Why After drawing a smooth curve through the plotted points, we will have the graph. Since the variable t represents the time since taking the dose, $t = 10$.

Solution **a.** The function values for $t = 0$ and $t = 2$ are calculated as follows:

$t = 0$:

$$\begin{aligned} A(t) &= 250(0.58)^t \\ A(0) &= 250(0.58)^0 \\ &= 250(1) \\ &= 250 \end{aligned}$$

$t = 2$:

$$\begin{aligned} A(t) &= 250(0.58)^t \\ A(2) &= 250(0.58)^2 \\ &\approx 84.1 \end{aligned}$$

Use a calculator.

Success Tip

$A(t) = 250(0.58)^t$ is a decreasing function because the base, 0.58, is such that $0 < 0.58 < 1$.

In a similar way, we find the corresponding values of $A(t)$ for t-values of 4 and 6, and list them in a table. Then we plot the ordered pairs and draw a smooth curve through them to get the graph shown below.

t	$A(t)$
0	250
2	84.1
4	28.3
6	9.5

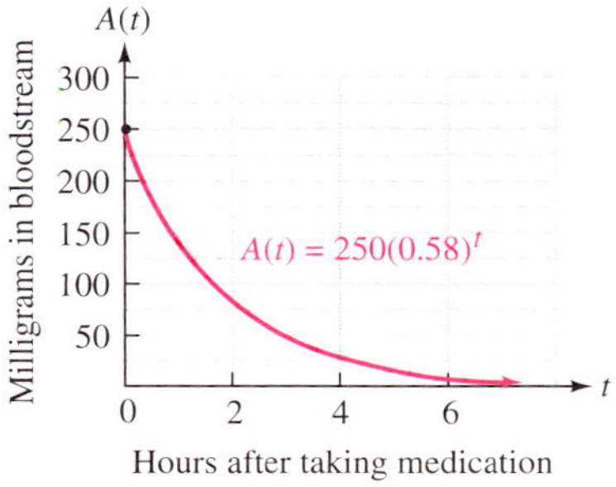

b. To estimate the amount of medication in the patient's bloodstream 10 hours after taking the dose, we find $A(10)$.

$$A(t) = 250(0.58)^t \quad \text{This is the exponential decay model.}$$
$$A(10) = 250(0.58)^{10} \quad \text{Substitute 10 for } t.$$
$$\approx 1.1 \quad \text{Use a calculator.}$$

In 10 hours, there will be approximately 1.1 milligrams of medication in the patient's bloodstream.

Self Check 6 **Medications.** Use the function in Example 6 to determine the amount of medication in the patient's bloodstream $8\frac{1}{2}$ hours after taking the dose.

Now Try Problem 44

SECTION 11.3 STUDY SET

VOCABULARY

Fill in the blanks.

1. $f(x) = 2^x$ and $f(x) = \left(\frac{1}{4}\right)^x$ are examples of __________ functions.
2. Exponential functions have a constant base and a variable __________ .
3. The graph of $f(x) = 3^x$ approaches, but never touches, the negative portion of the x-axis. Thus, the x-axis is an __________ of the graph.
4. __________ interest is paid on the principal and previously earned interest.

CONCEPTS

5. Refer to the graph shown at the right.
   a. What type of function is $f(x) = 3^x$?
   b. What is the domain of the function?
   c. What is the range of the function?
   d. What is the y-intercept of the graph? What is the x-intercept of the graph?
   e. Is the function one-to-one?
   f. What is an asymptote of the graph?
   g. Is f an increasing or a decreasing function?
   h. The graph passes through the point $(1, y)$. What is y?

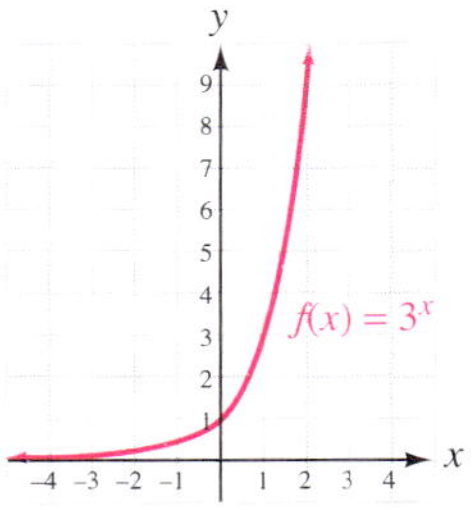

6. Which of the following functions are exponential functions?

   a. $f(x) = x^2$ b. $g(x) = 4x$ c. $h(x) = 8^x$

   d. $s(x) = \dfrac{1}{x}$ e. $T(x) = (0.92)^{x+1}$ f. $r(x) = x^3$

   g. $P(x) = \sqrt{x}$ h. $d(x) = |x|$

7. Evaluate each expression without a calculator.

   a. 3^{-2} b. $\left(\dfrac{1}{2}\right)^4$ c. $\left(\dfrac{1}{5}\right)^{-2}$

8. Evaluate each expression using a calculator. Round to the nearest tenth.

   a. $20{,}000(1.036)^{52}$ b. $92(0.88)^6$

9. Match each function with its graph shown below.

   a. $f(x) = x^2$ b. $f(x) = 2^x$

   c. $f(x) = 2$ d. $f(x) = 2x$

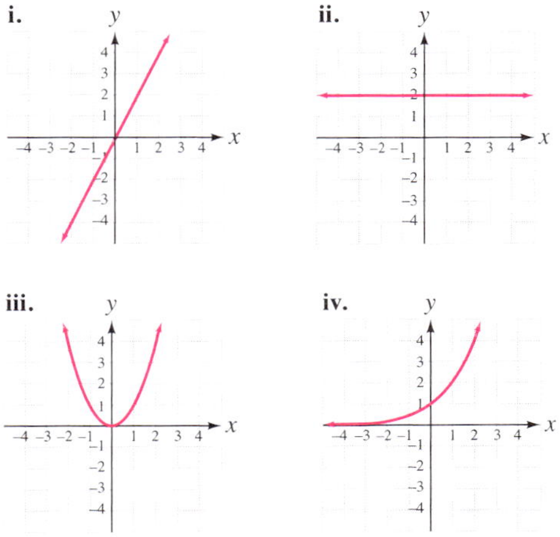

10. a. Two exponential functions of the form $f(x) = b^x$ are graphed in figure (a) below. Which function has the larger base b, the one graphed in red or the one graphed in blue?

    b. Two exponential functions of the form $f(x) = b^x$ are graphed in figure (b) below. Which function has the smaller base b?

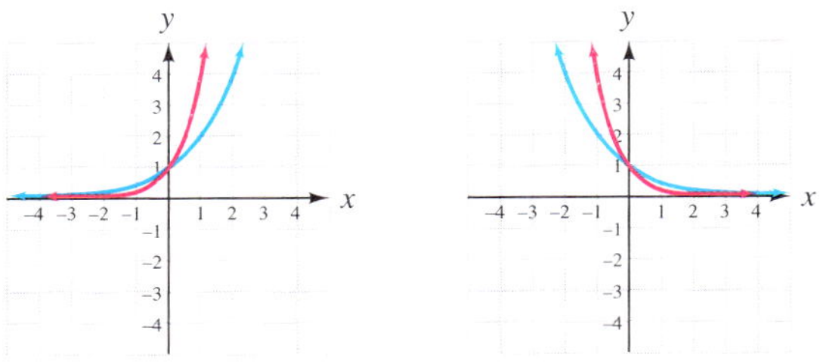

(a) (b)

11. Determine the domain and range of each exponential function graphed below.

a.

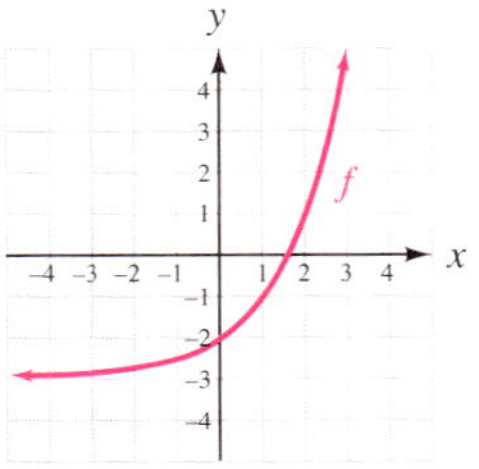

b.

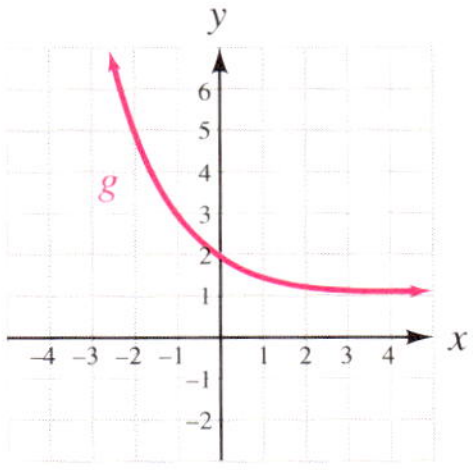

12. Fill in the blanks.

a. The graph of $g(x) = 4^x + 3$ is similar to the graph of $f(x) = 4^x$, but it is translated 3 units ________.

b. The graph of $g(x) = 4^{x-3}$ is similar to the graph of $f(x) = 4^x$, but it is translated 3 units to the _____.

13. Complete the table of function values shown here.

$f(x) = 5^x$

x	$f(x)$
−3	
−2	
−1	
0	
1	
2	
3	

14. Determine whether each of the following functions model exponential growth or exponential decay.

a. $D(t) = 150(0.44)^t$ **b.** $H(t) = 15{,}000(1.03)^t$

15. Match each situation to the exponential graph that best models it.

a. The number of cell phone subscribers in the world over the past 5 years

b. The level of caffeine in the bloodstream after drinking a cup of coffee

c. The amount of money in a bank account earning interest compounded quarterly

d. The number of rabbits in a population with a high birth rate

e. The amount of water in a shirt that was just washed and hung on a clothesline to dry

i.

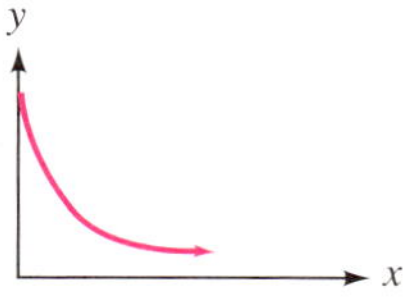

ii.

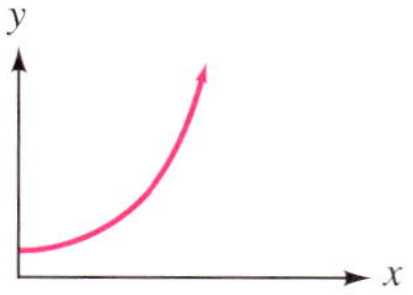

16. What formula is used to determine the amount of money in a savings account earning compound interest?

NOTATION

17. For an exponential function of the form $f(x) = b^x$, what are the restrictions on b?

18. In $A(t) = 16{,}000\left(1 + \frac{0.05}{365}\right)^{365t}$, what is the base and what is the exponent?

GUIDED PRACTICE

Graph each function. See Examples 1 and 2.

19. $f(x) = 3^x$

20. $f(x) = 6^x$

21. $f(x) = 5^x$

22. $f(x) = 7^x$

23. $f(x) = \left(\frac{1}{4}\right)^x$

24. $f(x) = \left(\frac{1}{5}\right)^x$

25. $f(x) = \left(\frac{1}{6}\right)^x$

26. $f(x) = \left(\frac{1}{8}\right)^x$

Graph each function by plotting points or using a translation. See Example 4.

27. $g(x) = 3^x - 2$

28. $g(x) = 2^x + 1$

29. $g(x) = 2^{x+1}$

30. $g(x) = 3^{x-1}$

31. $g(x) = 4^{x-1} + 2$

32. $g(x) = 4^{x+1} - 2$

33. $g(x) = -2^x$

34. $g(x) = -3^x$

Use a graphing calculator to graph each function. Determine whether the function is an increasing or a decreasing function. See Using Your Calculator: Graphing Exponential Functions.

35. $f(x) = \frac{1}{2}(3^{x/2})$

36. $f(x) = -3(2^{x/3})$

37. $f(x) = 2(3^{-x/2})$

38. $f(x) = -\frac{1}{4}(2^{-x/2})$

APPLICATIONS

39. **CO_2 Concentration.** The exponential growth model below illustrates the rise in atmospheric carbon dioxide from 1744 to 2006. The historical data (shown with blue points) comes from ice cores, and modern data (shown with red points) was collected from the Mauna Loa Observatory in Hawaii.

a. Estimate the atmospheric carbon dioxide concentrations in 1800, 1900, and 2000.

b. In approximately what year did the concentration surpass 325 parts per million?

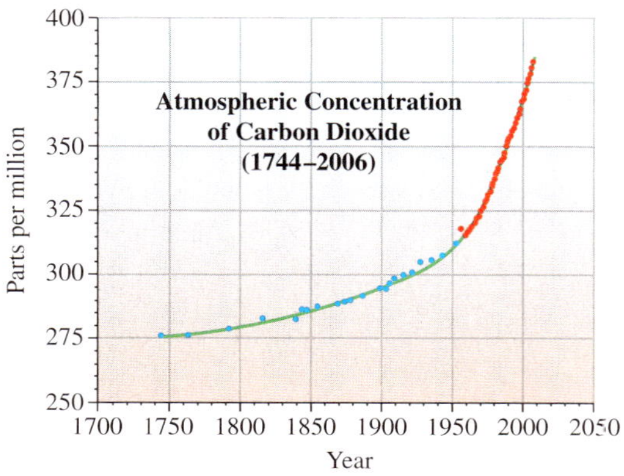

Source: www.eoearth.org

40. Global Warming. The following graph from the United States Environmental Protection Agency shows the projected sea level changes due to anticipated global warming.

a. What type of function does it appear could be used to model the sea level change?

b. When were the earliest instrumental records of sea level change made?

c. For the year 2100, what is the upper-end projection for sea level change? What is the lower-end projection?

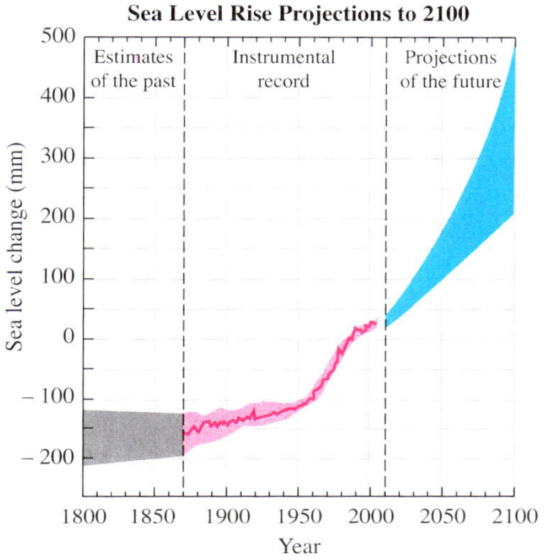

Source: United States Environmental Protection Agency

41. Value of a Car. The graph shows how the value of the average car depreciates as a percent of its original value over a 10-year period. It also shows the yearly maintenance costs as a percent of the car's value.

a. When is the car worth half of its purchase price?

b. When is the car worth a quarter of its purchase price?

c. When do the average yearly maintenance costs surpass the value of the car?

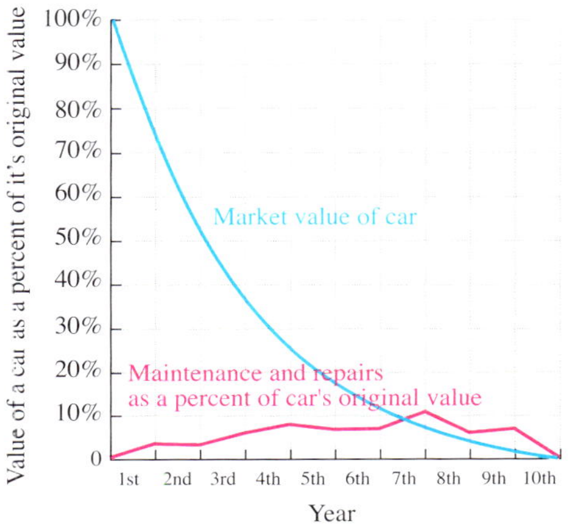

Source: U.S. Department of Transportation

42. Diving. *Bottom time* is the time a scuba diver spends descending plus the actual time spent at a certain depth. Graph the bottom time limits given in the table as ordered pairs of the form (depth, bottom time).Then draw a smooth curve through the points.

Depth (ft)	Bottom time limit (min)
30	no limit
35	310
40	200
50	100
60	60
70	50

Depth (ft)	Bottom time limit (min)
80	40
90	30
100	25
110	20
120	15
130	10

43. Computer Viruses. Suppose the number of computers infected by the spread of a virus through an e-mail is described by the exponential function $c(t) = 5(1.034)^t$, where t is the number of minutes since the first infected e-mail was opened.

a. Graph the function. Scale the t-axis from 0 to 400, in units of 50. Scale the $c(t)$-axis from 0 to 800,000 in units of 100,000.

b. Use the function to determine the number of infected computers in 8 hours, which is 480 minutes.

44. Salvage Value. A small business purchased a computer for \$5,000. The value (in dollars) of the computer, t years after its purchase, is given by the exponential function $v(t) = 5{,}000(0.75)^t$.

a. Graph the function. Scale the t-axis from 0 to 10 in units of 2. Scale the $v(t)$-axis from 0 to 6,000 in units of 1,000.

b. Use the function to determine the value of the computer 12 years after it is purchased.

45. Guitars. The frets on the neck of a guitar are placed so that pressing a string against them determines the strings' vibrating length. The exponential function $f(n) = 650(0.94)^n$ gives the vibrating length (in millimeters) of a string on a certain guitar for the fret number n. Find the length of the vibrating string when a guitarist holds down a string at the 7th fret.

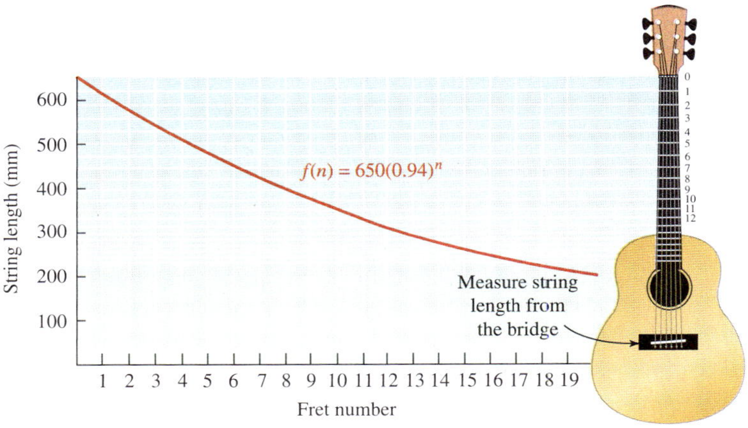

46. Bacterial Cultures. A colony of 6 million bacteria was determined to be growing in the culture medium shown in illustration (a). If the population P of bacteria after t hours is given by the function $P(t) = 6{,}000{,}000(2.3)^t$, find the population in the culture later in the day using the information given in illustration (b).

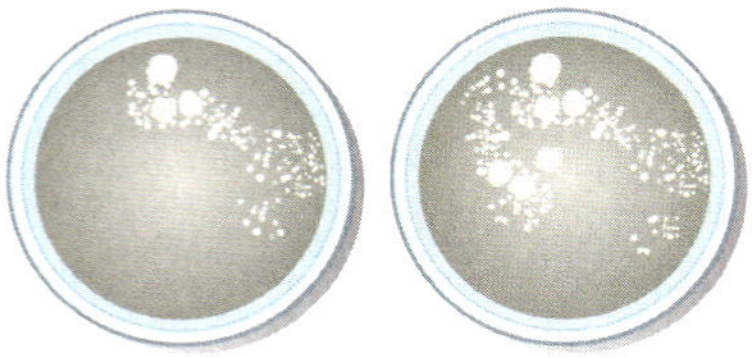

Date: 8-16-08 Time: 6:00 AM (a) | Date: 8-16-08 Time: 10:00 AM (b)

47. Radioactive Decay. Five hundred grams of a radioactive material decays according to the formula $A = 500\left(\frac{2}{3}\right)^t$, where t is measured in years. Find the amount present in 10 years. Round to the nearest one-tenth of a gram.

48. Discharging a Battery. The charge remaining in a battery decreases as the battery discharges. The charge C (in coulombs) after t days is given by the function $C(t) = 0.0003(0.7)^t$. Find the charge after 5 days.

49. from **Campus to Careers**

Social Worker

©iStockphoto/Chris Schmidt

The function $P(t) = 35.8(1.06)^t$ approximates the number of people (in millions) in the United States living in poverty, where t is the number of years after 2006. Use the function to complete the table below. Round to the nearest tenth. (Source: U.S. Census Bureau)

Year	2006	2007	2008	2009
Number in poverty (in millions)				

50. Publishing Books. The function $W(f) = 4{,}066(0.8753)^f$ approximates the number of words that can be typeset on a standard page using the Times Roman font size f. Find the number of words that can be typeset on a page using the font size 12. (Source: writersservices.com)

51. Population Growth. The population of North Rivers is decreasing exponentially according to the formula $P = 3{,}745(0.93)^t$, where t is measured in years from the present date. Find the population in 6 years, 9 months.

52. The Louisiana Purchase. In 1803, the United States negotiated the Louisiana Purchase with France. The country doubled its territory by adding 827,000 square miles of land for $15 million. If the land appreciated at the rate of 6% each year, what would one square mile of land be worth in 2005?

In Exercises 53–58, assume that there are no deposits or withdrawals.

53. Compound Interest. An initial deposit of $10,000 earns 8% interest, compounded quarterly. How much will be in the account after 10 years?

54. Compound Interest. An initial deposit of $10,000 earns 8% interest, compounded monthly. How much will be in the account after 10 years?

55. Comparing Interest Rates. How much more interest could $1,000 earn in 5 years, compounded quarterly, if the annual interest rate were $5\frac{1}{2}\%$ instead of 5%?

56. Comparing Savings Plans. Which institution in the ads provides the better investment?

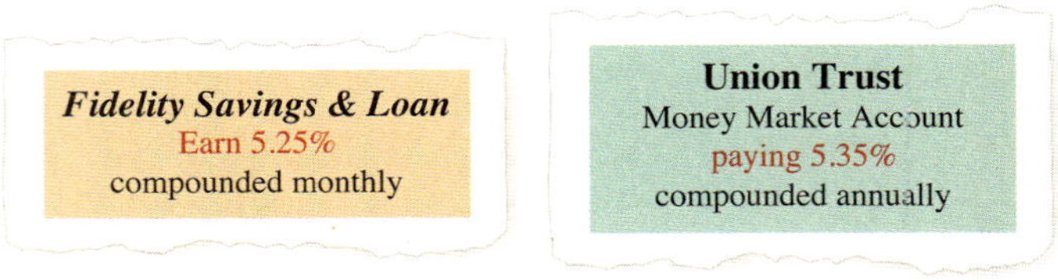

57. Compound Interest. If $1 had been invested on July 4, 1776, at 5% interest, compounded annually, what would it be worth on July 4, 2076?

58. Frequency of Compounding. $10,000 is invested in each of two accounts, both paying 6% annual interest. In the first account, interest compounds quarterly, and in the second account, interest compounds daily. Find the difference between the accounts after 20 years.

WRITING

59. If world population is increasing exponentially, why is there cause for concern?

60. How do the graphs of $f(x) = 3^x$ and $g(x) = \left(\frac{1}{3}\right)^x$ differ? How are they similar?

61. A snowball rolling downhill grows *exponentially* with time. Explain what this means. Sketch a simple graph that models the situation.

62. Explain why the change in temperature of a cup of hot coffee left unattended on a kitchen table is an example of exponential decay.

63. Let $f(x) = \left(\frac{1}{5}\right)^x$. Explain why we can rewrite the function equation as $f(x) = 5^{-x}$.

64. Explain why the graph of $f(x) = 3^x$ gets closer and closer to the x-axis as the values of x decrease. Does the graph ever cross the x-axis? Explain why or why not.

65. Describe the graphs of $f(x) = x^2$ and $g(x) = 2^x$ in words.

66. Write a paragraph explaining the concept that is illustrated in the graph.

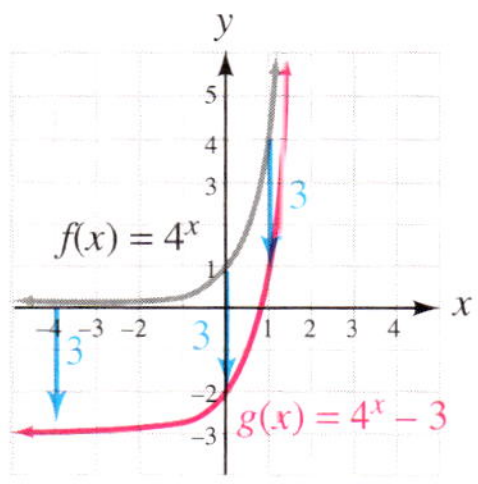

67. In the definition of the exponential function, b could not be negative. Why?
68. How does exponential growth differ from linear growth? Give an example.

REVIEW

In Exercises 69–72, refer to the illustration below in which lines r and s are parallel.

69. Find x.
70. Find the measure of $\angle 1$.
71. Find the measure of $\angle 2$.
72. Find the measure of $\angle 3$.

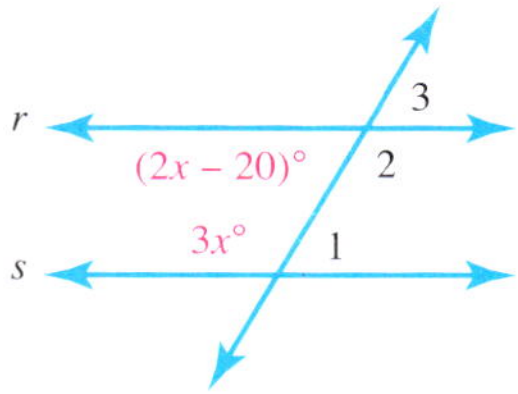

CHALLENGE PROBLEMS

73. Graph $f(x) = 3^x$. Then use the graph to estimate the value of $3^{1.5}$.
74. Graph $y = x^{1/2}$ and $y = \left(\frac{1}{2}\right)^x$ on the same set of coordinate axes. Estimate the coordinates of any point(s) that the graphs have in common.
75. Find the value of b that would cause the graph of $f(x) = b^x$ to look like the graph on the right.

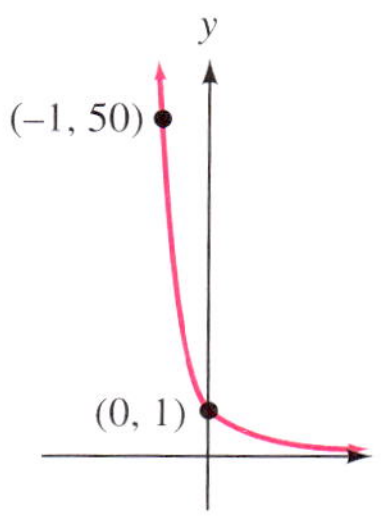

Simplify each expression. Write answers using positive exponents.

76. **a.** $\left(2^{\sqrt{3}}\right)^{\sqrt{3}}$ **b.** $7^{\sqrt{3}}7^{\sqrt{12}}$
c. $\dfrac{5^{6\sqrt{2}}}{5^{4\sqrt{2}}}$ **d.** $5^{-\sqrt{5}}$
77. Graph: $f(x) = 2^{|x|}$
78. Graph the inverse of $f(x) = 3^x$.

SECTION 11.4 Logarithmic Functions

OBJECTIVES

1. Define logarithm.
2. Write logarithmic equations as exponential equations.
3. Write exponential equations as logarithmic equations.
4. Evaluate logarithmic expressions.
5. Graph logarithmic functions.
6. Use logarithmic formulas and functions in applications.

ARE YOU READY?

The following problems review some basic skills that are needed when working with logarithmic functions.

1. A table of values for a one-to-one function f is shown below. Complete the table of values for f^{-1}.

x	$f(x)$
0	1
1	2
2	4

x	$f^{-1}(x)$

2. Fill in the blanks: **a.** $5^{-3} = \dfrac{1}{}$ **b.** $3^{} = 27$
3. Fill in the blanks: **a.** $4^{} = 4$ **b.** $7^{} = \sqrt{7}$
4. Evaluate: **a.** 10^3 **b.** 10^{-2}

In this section, we will discuss inverses of exponential functions. These functions are called *logarithmic functions,* and they can be used to solve problems from fields such as electronics, seismology (the study of earthquakes), and business.

1 Define Logarithm.

The graph of the exponential function $f(x) = 2^x$ is shown in red below. Since it passes the horizontal line test, it is a one-to-one function and has an inverse. To graph f^{-1}, we interchange the coordinates of the ordered pairs in the table, plot those points, and draw a smooth curve through them, as shown in blue. As expected, the graphs of f and f^{-1} are symmetric with respect to the line $y = x$.

$f(x) = 2^x$

To graph f^{-1}, interchange each pair of coordinates.

x	$f(x)$		
-3	$\frac{1}{8}$	$\rightarrow \left(-3, \frac{1}{8}\right)$	$\left(\frac{1}{8}, -3\right)$
-2	$\frac{1}{4}$	$\rightarrow \left(-2, \frac{1}{4}\right)$	$\left(\frac{1}{4}, -2\right)$
-1	$\frac{1}{2}$	$\rightarrow \left(-1, \frac{1}{2}\right)$	$\left(\frac{1}{2}, -1\right)$
0	1	$\rightarrow (0, 1)$	$(1, 0)$
1	2	$\rightarrow (1, 2)$	$(2, 1)$
2	4	$\rightarrow (2, 4)$	$(4, 2)$
3	8	$\rightarrow (3, 8)$	$(8, 3)$

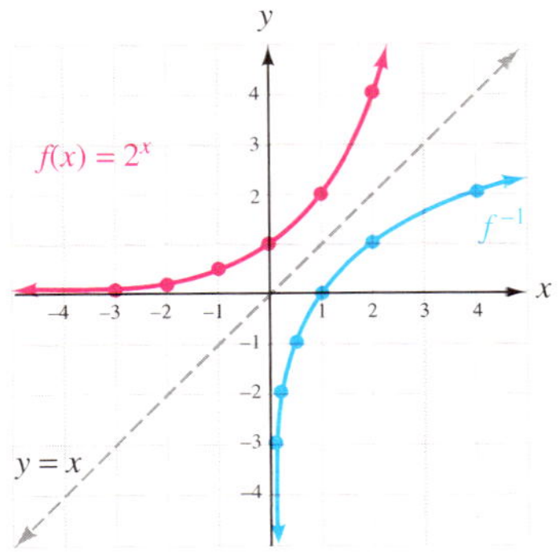

To write an equation for the inverse of $f(x) = 2^x$, we proceed as follows:

$$f(x) = 2^x$$

$$y = 2^x \quad \text{Replace } f(x) \text{ with } y.$$

$$x = 2^y \quad \text{Interchange the variables } x \text{ and } y.$$

We cannot solve the equation for y because we have not discussed methods for solving equations with a variable exponent. However, we can translate the relationship $x = 2^y$ into words:

$$y = \text{ the power to which we raise 2 to get } x$$

If we substitute the notation $f^{-1}(x)$ for y, we see that

$$f^{-1}(x) = \text{ the power to which we raise 2 to get } x$$

The Language of Algebra

The abbreviation log is used for the word **logarithm.** A *logarithm* is an exponent.

If we define the symbol $\log_2 x$ to mean *the power to which we raise* 2 *to get* x, we can write the equation for the inverse as

$$f^{-1}(x) = \log_2 x$$ Read $\log_2 x$ as "the logarithm, base 2, of x" or "log, base 2, of x."

We have found that the inverse of the exponential function $f(x) = 2^x$ is $f^{-1}(x) = \log_2 x$. To find the inverse of exponential functions with other bases, such as $f(x) = 3^x$ and $f(x) = 10^x$, we define *logarithm* in the following way.

Definition of Logarithm

For all positive numbers b, where $b \neq 1$, and all positive numbers x,

$$y = \log_b x \quad \text{is equivalent to} \quad x = b^y$$

Success Tip

Here are two examples of inverses of other exponential functions:

$f(x) = 3^x \quad f^{-1}(x) = \log_3 x$

$g(x) = 10^x \quad g^{-1}(x) = \log_{10} x$

This definition guarantees that any pair (x, y) that satisfies the logarithmic equation $y = \log_b x$ also satisfies the exponential equation $x = b^y$. Because of this relationship, a statement written in logarithmic form can be written in an equivalent exponential form, and vice versa. They are just different ways of expressing the same thing. The following diagram will help you remember the respective positions of the exponent and base in each form.

Exponent

$$y = \log_b x \quad \text{means} \quad x = b^y$$

Base

2 Write Logarithmic Equations as Exponential Equations.

Success Tip

Study the relationship between exponential and logarithmic statements carefully. Your success with the material in the rest of this chapter depends greatly on your understanding of this definition.

The following table shows the relationship between logarithmic and exponential notation. We need to be able to work in both directions.

Logarithmic equation		*Exponential equation*
$\log_2 8 = 3$	means	$2^3 = 8$
$\log_3 81 = 4$		$3^4 = 81$
$\log_4 4 = 1$		$4^1 = 4$
$\log_5 \frac{1}{125} = -3$		$5^{-3} = \frac{1}{125}$

EXAMPLE 1 Write each logarithmic equation as an exponential equation:

a. $\log_4 64 = 3$ **b.** $\log_7 \sqrt{7} = \frac{1}{2}$ **c.** $\log_6 \frac{1}{36} = -2$

Strategy To write an equivalent exponential equation, we will determine which number will serve as the base and which will serve as the exponent.

Why We can then use the definition of logarithm to move from one form to the other: $\log_b x = y$ is equivalent to $x = b^y$.

Solution

a. $\log_4 64 = 3$ means $4^3 = 64$. Read as "log, base 4, of 64 equals 3."

(Exponent; Base) The base in each form is the same: 4.

Caution

Recall that $f(x)$ is not read as $f \cdot x$. Similarly, $\log_4 64$ is not read as $\log_4 \cdot 64$.

b. $\log_7 \sqrt{7} = \frac{1}{2}$ means $7^{1/2} = \sqrt{7}$. Read as "log, base 7, of $\sqrt{7}$ equals $\frac{1}{2}$."

(Exponent; Base) The base in each form is the same: 7.

c. $\log_6 \frac{1}{36} = -2$ means $6^{-2} = \frac{1}{36}$. Read as "log, base 6, of $\frac{1}{36}$ equals -2."

(Exponent; Base) The base in each form in the same: 6.

Self Check 1 Write $\log_2 128 = 7$ as an exponential equation.

Now Try Problems 23, 27, and 35

3 Write Exponential Equations as Logarithmic Equations.

EXAMPLE 2 Write each exponential equation as a logarithmic equation:

a. $8^0 = 1$ **b.** $6^{1/3} = \sqrt[3]{6}$ **c.** $\left(\frac{1}{4}\right)^2 = \frac{1}{16}$

Strategy To write an equivalent logarithmic equation, we will determine which number will serve as the base and where we will place the exponent.

Why We can then use the definition of logarithm to move from one form to the other: $x = b^y$ is equivalent to $\log_b x = y$.

Solution

a. $8^0 = 1$ means $\log_8 1 = 0$ *In each form, the base is 8.*

(Exponent: 0; Base: 8)

The Language of Algebra

Just as 8 is called the **base** in the exponential form $8^0 = 1$, we call 8 the **base** in the logarithmic form $\log_8 1 = 0$.

b. $6^{1/3} = \sqrt[3]{6}$ means $\log_6 \sqrt[3]{6} = \frac{1}{3}$ *In each form, the base is 6.*

(Exponent: $\frac{1}{3}$; Base: 6)

c. $\left(\frac{1}{4}\right)^2 = \frac{1}{16}$ means $\log_{1/4} \frac{1}{16} = 2$ *In each form, the base is $\frac{1}{4}$.*

(Exponent: 2; Base: $\frac{1}{4}$)

Self Check 2 Write $9^{-1} = \frac{1}{9}$ as a logarithmic equation.

Now Try Problems 39, 43, and 49

Certain logarithmic equations can be solved by writing them as exponential equations.

EXAMPLE 3 Solve each equation for x: **a.** $\log_x 25 = 2$ **b.** $\log_3 x = -3$ **c.** $\log_{1/2} \frac{1}{16} = x$

Strategy To solve each logarithmic equation, we will instead write and solve an equivalent exponential equation.

Why The resulting exponential equation is easier to solve because the variable term is often isolated on one side.

Solution

a. Since $\log_x 25 = 2$ is equivalent to $x^2 = 25$, we can solve $x^2 = 25$ to find x.

(Exponent: 2; Base: x)

The Language of Algebra

Recall that **equivalent equations** have the same solutions.

$$x^2 = 25$$
$$x = \pm\sqrt{25} \quad \text{Use the square root property.}$$
$$x = \pm 5$$

In the expression $\log_x 25$, the base of the logarithm is x. Because the base must be positive, we discard -5 and we have

$$x = 5$$

To check the solution of 5, verify that $\log_5 25 = 2$. The solution set is written as $\{5\}$.

b. Since $\log_3 x = -3$ is equivalent to $3^{-3} = x$, we can instead solve $3^{-3} = x$ to find x.

(Exponent: -3; Base: 3)

$$3^{-3} = x$$
$$\frac{1}{3^3} = x$$
$$x = \frac{1}{27}$$

To check the solution of $\frac{1}{27}$, verify that $\log_3 \frac{1}{27} = -3$. The solution set is $\left\{\frac{1}{27}\right\}$.

Success Tip

To solve this equation, we note that if the bases are equal, the exponents must be equal.

$$\left(\frac{1}{2}\right)^x = \left(\frac{1}{2}\right)^4$$

c. Since $\log_{1/2} \frac{1}{16} = x$ is equivalent to $\left(\frac{1}{2}\right)^x = \frac{1}{16}$, we can instead solve $\left(\frac{1}{2}\right)^x = \frac{1}{16}$ to find x.

(Exponent: x; Base: $\frac{1}{2}$)

$$\left(\frac{1}{2}\right)^x = \frac{1}{16}$$

$$\left(\frac{1}{2}\right)^x = \left(\frac{1}{2}\right)^4 \quad \text{Write } \tfrac{1}{16} \text{ as a power of } \tfrac{1}{2} \text{ to match the bases: } \tfrac{1}{2}\cdot\tfrac{1}{2}\cdot\tfrac{1}{2}\cdot\tfrac{1}{2} = \tfrac{1}{16}.$$

$$x = 4 \quad \text{Since the bases are the same, and since exponential functions are one-to-one, the exponents must be equal.}$$

To check the solution of 4, verify that $\log_{1/2} \frac{1}{16} = 4$. The solution set is $\{4\}$.

Self Check 3 Solve each equation for x: **a.** $\log_x 49 = 2$ **b.** $\log_{1/3} x = 2$ **c.** $\log_6 216 = x$

Now Try Problems 51, 53, and 55

4 Evaluate Logarithmic Expressions.

In the previous examples, we have seen that the logarithm of a number is an exponent. In fact,

$\log_b x$ is the exponent to which b is raised to get x.

Translating this statement into symbols, we have

$$b^{\log_b x} = x$$

EXAMPLE 4 Evaluate each expression: **a.** $\log_8 64$ **b.** $\log_3 \frac{1}{3}$ **c.** $\log_4 2$

Strategy After identifying the base, we will ask "To what power must the base be raised to get the other number?"

Why That power is the value of the logarithmic expression.

Solution

a. $\log_8 64 = 2$ Ask: "To what power must we raise 8 to get 64?" Since $8^2 = 64$, the answer is the 2nd power.

An alternate approach for this evaluation problem is to let $\log_8 64 = x$. When we write the equivalent exponential equation $8^x = 64$, it is easy to see that $x = 2$.

b. $\log_3 \frac{1}{3} = -1$ Ask: "To what power must we raise 3 to get $\frac{1}{3}$?" Since $3^{-1} = \frac{1}{3}$, the answer is the -1 power.

We could also let $\log_3 \frac{1}{3} = x$. The equivalent exponential equation is $3^x = \frac{1}{3}$. Thus, x must be -1.

c. $\log_4 2 = \frac{1}{2}$ Ask: "To what power must we raise 4 to get 2?" Since $\sqrt{4} = 4^{1/2} = 2$, the answer is the $\frac{1}{2}$ power.

We could also let $\log_4 2 = x$. Then the equivalent exponential equation is $4^x = 2$. Thus, x must be $\frac{1}{2}$.

Self Check 4 Evaluate each expression: **a.** $\log_9 81$ **b.** $\log_4 \frac{1}{16}$ **c.** $\log_9 3$

Now Try Problems 75 and 77

The Language of Algebra

London professor Henry Briggs (1561–1630) and Scottish lord John Napier (1550–1617) are credited with developing the concept of **common logarithms.** Their tables of logarithms were useful tools at that time for those performing large calculations.

In many applications, base-10 logarithms (also called **common logarithms**) are used. When the base b is not indicated in the notation $\log x$, we assume that $b = 10$:

$$\log x \quad \text{means} \quad \log_{10} x$$

The table below shows the relationship between base-10 logarithmic notation and exponential notation.

Logarithmic form		*Exponential form*	
$\log 100 = 2$	means	$10^2 = 100$	Read log 100 as "log of 100."
$\log \frac{1}{10} = -1$	means	$10^{-1} = \frac{1}{10}$	
$\log 1 = 0$	means	$10^0 = 1$	

In general, we have

$$\log 10^x = x$$

EXAMPLE 5 Evaluate each expression: **a.** $\log 1{,}000$ **b.** $\log \frac{1}{100}$ **c.** $\log 10$ **d.** $\log(-10)$

Strategy After identifying the base, we will ask "To what power must 10 be raised to get the other number?"

Why That power is the value of the logarithmic expression.

Solution

a. $\log 1{,}000 = 3$ Ask: "To what power must we raise 10 to get 1,000?" Since $10^3 = 1{,}000$, the answer is: the 3rd power.

b. $\log \frac{1}{100} = -2$ Ask: "To what power must we raise 10 to get $\frac{1}{100}$?" Since $10^{-2} = \frac{1}{100}$, the answer is: the -2 power.

c. $\log 10 = 1$ Ask: "To what power must we raise 10 to get 10?" Since $10^1 = 10$, the answer is: the 1st power.

d. To find $\log(-10)$, we must find a power of 10 such that $10^? = -10$. There is no such number. Thus, $\log(-10)$ is undefined.

Self Check 5 Evaluate each expression: **a.** $\log 10{,}000$ **b.** $\log \frac{1}{1{,}000}$ **c.** $\log 0$

Now Try Problems 79 and 81

Many logarithmic expressions cannot be evaluated by inspection. For example, to find $\log 2.34$, we ask, "To what power must we raise 10 to get 2.34?" This answer isn't obvious. In such cases, we use a calculator.

Using Your Calculator ▶ Evaluating Logarithms

To find $\log 2.34$ with a scientific calculator we enter

2.34 [LOG] .369215857

On some calculators, the [10^x] key also serves as the [LOG] key when [2nd] or [SHIFT] is pressed. This is because $f(x) = 10^x$ and $f(x) = \log x$ are inverses.

To use a graphing calculator, we enter

[LOG] 2.34 [)] [ENTER]

```
log(2.34)
          .369215857
```

To four decimal places, $\log 2.34 = 0.3692$. This means, $10^{0.3692} \approx 2.34$.

If we attempt to evaluate logarithmic expressions such as log 0, or the logarithm of a negative number, such as $\log(-5)$, an error message like the following will be displayed.

Error

```
ERR:DOMAIN
1:QUIT
2:Go to
```

```
ERR:NONREAL ANS
1:QUIT
2:Go to
```

EXAMPLE 6 Solve $\log x = 0.3568$ and round to four decimal places.

Strategy To solve this logarithmic equation, we will instead write and solve an equivalent exponential equation.

Why The resulting exponential equation is easier to solve because the variable term is isolated on one side.

Solution The equation $\log x = 0.3568$ means $\log_{10} x = 0.3568$, which is equivalent to $10^{0.3568} = x$. Since we cannot determine $10^{0.3568}$ by inspection, we will use a calculator to find an approximate solution. We enter

10 y^x .3568 =

The display reads 2.274049951. To four decimal places,

$$x = 2.2740$$

If your calculator has a 10^x key, enter .3568 and press it to get the same result. The solution is 2.2740. To check, use your calculator to verify that $\log 2.2740 \approx 0.3568$.

Self Check 6 Solve $\log x = 1.87737$ and round to four decimal places.

Now Try Problem 91

5 Graph Logarithmic Functions.

Because an exponential function defined by $f(x) = b^x$ is one-to-one, it has an inverse function that is defined by $x = b^y$. When we write $x = b^y$ in the equivalent form $y = \log_b x$, the result is called a *logarithmic function*.

Logarithmic Functions

If $b > 0$ and $b \neq 1$, the **logarithmic function with base *b*** is defined by the equations

$$f(x) = \log_b x \quad \text{or} \quad y = \log_b x$$

The domain of $f(x) = \log_b x$ is the interval $(0, \infty)$ and the range is the interval $(-\infty, \infty)$.

Caution

Since the domain of the logarithmic function is the set of positive real numbers, it is impossible to find the logarithm of 0 or the logarithm of a negative number. For example, $\log_2(-4)$ and $\log_2 0$ are undefined.

Since every logarithmic function is the inverse of a one-to-one exponential function, logarithmic functions are one-to-one.

We can plot points to graph logarithmic functions. For example, to graph $f(x) = \log_2 x$, we construct a table of function values, plot the resulting ordered pairs, and draw a smooth curve through the points to get the graph, as shown in figure (a) on the next page. To graph $f(x) = \log_{1/2} x$, we use the same method, as shown in figure (b).

By the vertical line test, we see that each graph is indeed the graph of a function. Because in each case the projection of the graph onto the x-axis covers only the positive portion of that axis, the domain of each function is $(0, \infty)$. Because the graphs extend indefinitely upward and downward, the projection of the graphs onto the y-axis includes all real numbers. Thus, the range of each function is $(-\infty, \infty)$.

$f(x) = \log_2 x$

x	$f(x)$	
$\frac{1}{4}$	-2	$\to \left(\frac{1}{4}, -2\right)$
$\frac{1}{2}$	-1	$\to \left(\frac{1}{2}, -1\right)$
1	0	$\to (1, 0)$
2	1	$\to (2, 1)$
4	2	$\to (4, 2)$
8	3	$\to (8, 3)$

↑ Because the base of the function is 2, choose values for x that are integer powers of 2.

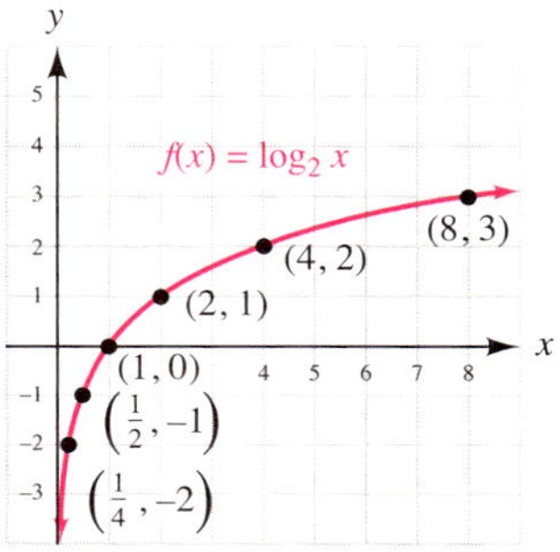

(a)

$f(x) = \log_{1/2} x$

x	$f(x)$	
$\frac{1}{4}$	2	$\to \left(\frac{1}{4}, 2\right)$
$\frac{1}{2}$	1	$\to \left(\frac{1}{2}, 1\right)$
1	0	$\to (1, 0)$
2	-1	$\to (2, -1)$
4	-2	$\to (4, -2)$
8	-3	$\to (8, -3)$

↑ Because the base of the function is $\frac{1}{2}$, choose values for x that are integer powers of $\frac{1}{2}$.

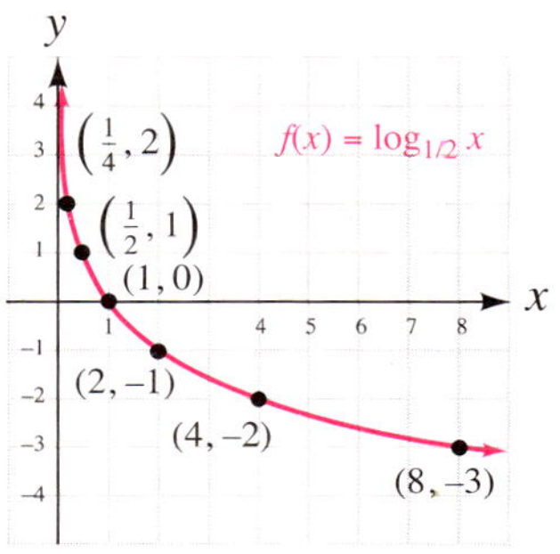

(b)

The graphs of all logarithmic functions are similar to those shown below. If $b > 1$, the logarithmic function is increasing, as in figure (a). If $0 < b < 1$, the logarithmic function is decreasing, as in figure (b).

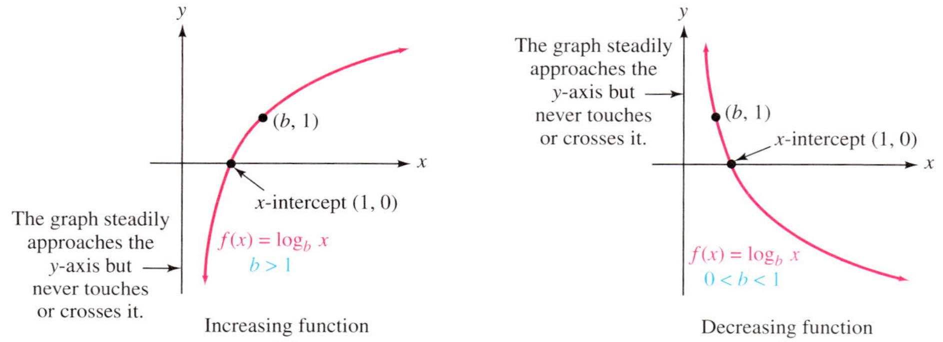

Increasing function
(a)

Decreasing function
(b)

Properties of Logarithmic Functions	The graph of $f(x) = \log_b x$ (or $y = \log_b x$) has the following properties. 1. It passes through the point $(1, 0)$. 2. It passes through the point $(b, 1)$. 3. The y-axis (the line $x = 0$) is an asymptote. 4. The domain is the interval $(0, \infty)$ and the range is the interval $(-\infty, \infty)$.

The exponential and logarithmic functions are inverses of each other, so their graphs have symmetry about the line $y = x$. The graphs of $f(x) = \log_b x$ and $g(x) = b^x$ are shown in figure (a) when $b > 1$ and in figure (b) when $0 < b < 1$.

Notation

If it is helpful, imagine parentheses around the input value in the logarithmic function notation.

$f(8) = 3$

Function name — Input — Output

$\log_2 (8) = 3$

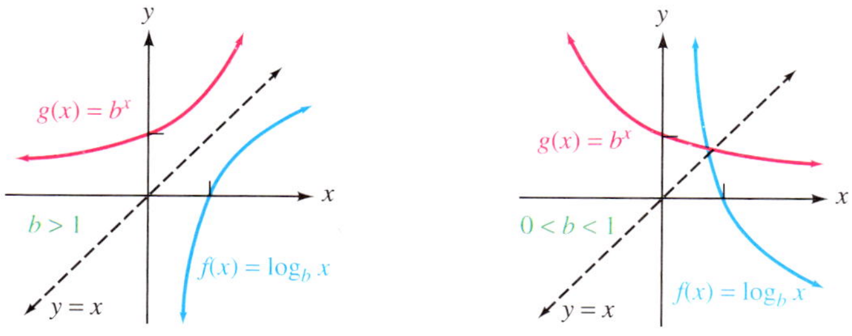

(a) (b)

The graphs of many functions involving logarithms are translations of the basic logarithmic graphs.

EXAMPLE 7 Graph each function by using a translation: **a.** $g(x) = 3 + \log_2 x$ **b.** $g(x) = \log_{1/2}(x - 1)$

Notation

Since $y = f(x)$, we can write

$$f(x) = \log_2 x$$
as
$$y = \log_2 x$$

Strategy We will graph $g(x) = 3 + \log_2 x$ by translating the graph of $f(x) = \log_2 x$ upward 3 units. We will graph $g(x) = \log_{1/2}(x - 1)$ by translating the graph of $f(x) = \log_{1/2} x$ to the right 1 unit.

Why The addition of 3 in $g(x) = \mathbf{3 +} \log_2 x$ causes a vertical shift of the graph of the base-2 logarithmic function 3 units upward. The subtraction of 1 from x in $g(x) = \log_{1/2}(x \mathbf{- 1})$ causes a horizontal shift of the graph of the base-$\frac{1}{2}$ logarithmic function 1 unit to the right.

Solution

a. The graph of $g(x) = 3 + \log_2 x$ will be the same shape as the graph of $f(x) = \log_2 x$. We simply translate each point on the graph of $f(x) = \log_2 x$ up 3 units. See figure (a) below.

b. The graph of $g(x) = \log_{1/2}(x - 1)$ will be the same shape as the graph of $f(x) = \log_{1/2} x$. We simply translate each point on the graph of $f(x) = \log_{1/2} x$ to the right 1 unit. See figure (b) below.

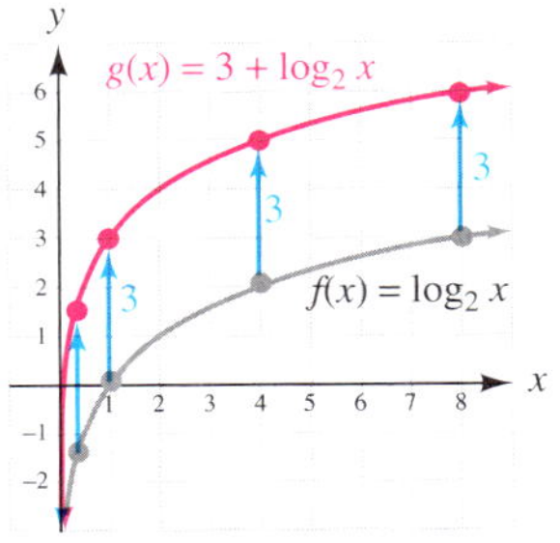

(a)

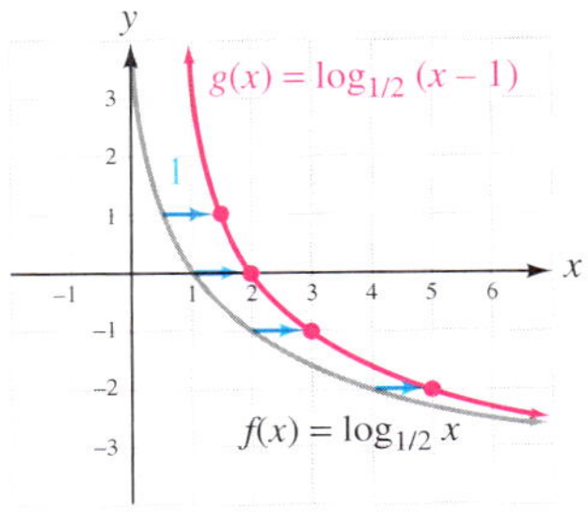

(b)

Self Check 7 Graph each function by using a translation: **a.** $g(x) = (\log_3 x) - 2$ **b.** $g(x) = \log_{1/3}(x + 2)$

Now Try ▶ Problems 103 and 105

To graph more complicated logarithmic functions, a graphing calculator is a useful tool.

Using Your Calculator ▶ Graphing Logarithmic Functions

To use a calculator to graph the logarithmic function $f(x) = -2 + \log_{10}\frac{x}{2}$, we enter the right side of the equation after the symbol $Y_1 =$. The display will show the equation

$$Y_1 = -2 + \log(X/2)$$

If we use window settings of $[-1, 5]$ for x and $[-4, 1]$ for y and press the GRAPH key, we will obtain the graph shown.

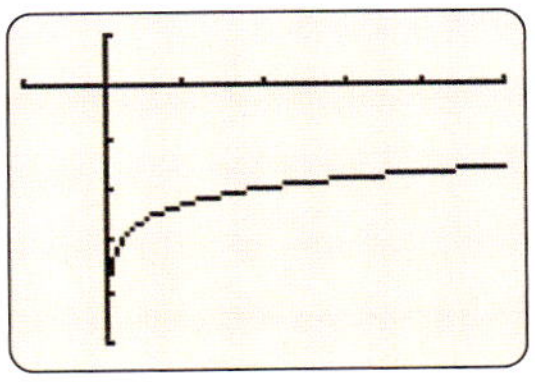

6 Use Logarithmic Formulas and Functions in Applications.

Logarithmic functions, like exponential functions, can be used to model certain types of growth and decay. Logarithms are especially useful when measuring a quantity that varies over a very large range of values, such as the intensity of earthquakes, the loudness of sounds, and the brightness of stars.

The following common logarithm formula is used in electrical engineering to express the gain (or loss) of an electronic device such as an amplifier as it takes an input signal and produces an output signal. The unit of gain (or loss) in such cases is called the **decibel,** which is abbreviated **dB.**

Decibel Voltage Gain

If E_O is the output voltage of a device and E_I is the input voltage, the decibel voltage gain of the device (dB gain) is given by

$$\text{dB gain} = 20 \log \frac{E_O}{E_I}$$

EXAMPLE 8

dB Gain. If the input to an amplifier is 0.5 volt and the output is 40 volts, find the decibel voltage gain of the amplifier.

Strategy We will substitute into the formula for dB gain and evaluate the right side using a calculator.

Why We can use this formula to find the dB gain because we are given the input voltage E_I and the output voltage E_O.

Solution We can find the decibel voltage gain by substituting 0.5 for E_I and 40 for E_O into the formula for dB gain:

Notation

$20 \log \frac{E_O}{E_I}$ means $20 \cdot \log \frac{E_O}{E_I}$

$$\text{dB gain} = 20 \log \frac{E_O}{E_I}$$ Read as "20 times the log of E sub O divided by E sub I."

$$\text{dB gain} = 20 \log \frac{40}{0.5}$$ Substitute the input voltage 0.5 for E_I and the output voltage 40 for E_O.

$$= 20 \log 80$$ Divide: $\frac{40}{0.5} = 80$.

$$\approx 38$$ Use a scientific calculator and press: 20 × 80 LOG =.

The amplifier provides a 38-decibel voltage gain.

Self Check 8 **dB Gain.** If the input to an amplifier is 0.6 volt and the output is 40 volts, find the decibel voltage gain of the amplifier.

Now Try Problem 111

EXAMPLE 9

Stocking Lakes. To create the proper environmental balance, 250 hybrid bluegill were introduced into a lake by the local Fish and Game Department. Department biologists found that the number of bluegill in the lake could be approximated by the logarithmic function $f(t) = 250 + 400 \log (t + 1)$, where t is the number of years since the lake was stocked. Find the bluegill population in the lake after 5 years.

Strategy We will find $f(5)$.

Why Since the variable t represents the time since the lake was stocked, $t = 5$.

Solution

$$f(t) = 250 + 400 \log (t + 1)$$ This is the logarithmic growth model.

$$f(5) = 250 + 400 \log (5 + 1)$$ Substitute 5 for t.

$$= 250 + 400 \log 6$$ Do the addition within the parentheses.

$$\approx 561$$ Use a scientific calculator and press: 250 + 400 × 6 LOG =.

There were approximately 561 bluegill in the lake 5 years after it was stocked.

Self Check 9 **Stocking Ponds.** A rancher stocked a pond on his property with 50 catfish. He was told that with the proper care, the catfish population could be approximated by the logarithmic function $f(t) = 50 + 22 \log (t + 1)$, where t is the number of years since the pond was stocked. Find the number of catfish he can expect in the pond in 3 years.

Now Try Problem 117

SECTION 11.4 STUDY SET

VOCABULARY

Fill in the blanks.

1. $f(x) = \log_2 x$ and $g(x) = \log x$ are examples of __________ functions.
2. Base-10 logarithms are called __________ logarithms.
3. The graph of $f(x) = \log_2 x$ approaches, but never touches, the negative portion of the y-axis. Thus the y- axis is an __________ of the graph.
4. $\log_x 81 = 4$ is __________ to $x^4 = 81$.

CONCEPTS

5. Refer to the graph on the right.

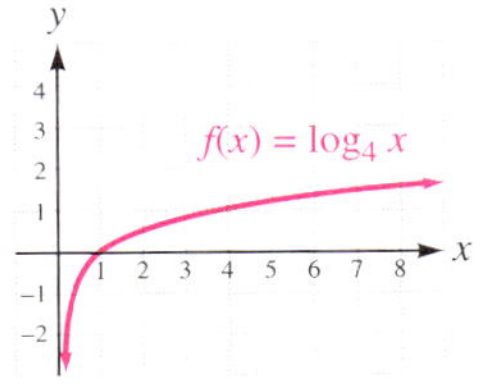

   a. What type of function is $f(x) = \log_4 x$?
   b. What is the domain of the function? What is the range of the function?
   c. What is the y-intercept of the graph? What is the x-intercept of the graph?
   d. Is f a one-to-one function?
   e. What is an asymptote of the graph?
   f. Is f an increasing or a decreasing function?
   g. The graph passes through the point $(4, y)$. What is y?
6. Determine the domain and range of each logarithmic function graphed below.

a.

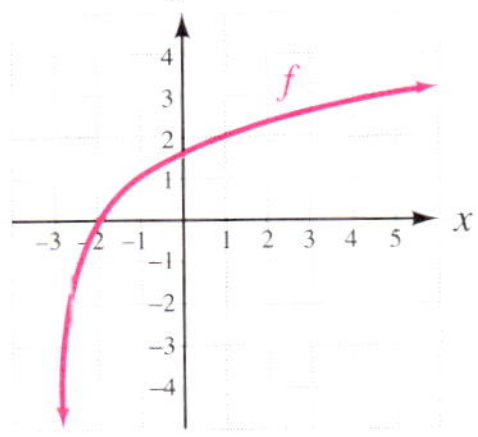

b.

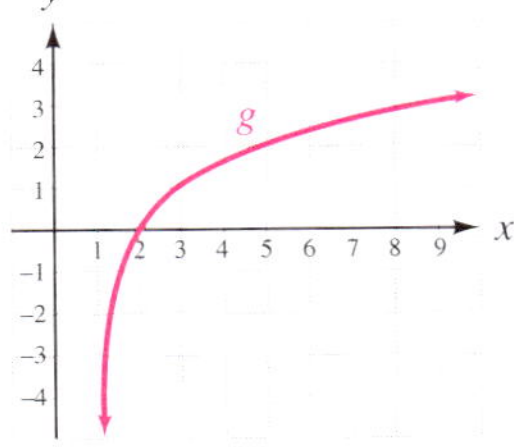

7. Match each function with its graph shown below.

   a. $f(x) = x^3$ b. $f(x) = \log_3 x$

   c. $f(x) = \sqrt[3]{x}$ d. $f(x) = 3x$

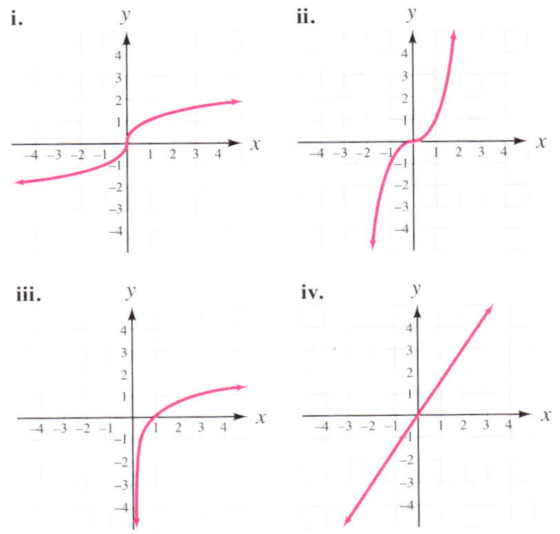

8. a. An exponential function is graphed below. Graph its inverse and the axis of symmetry on the same coordinate system.
   b. What type of function is the inverse function?

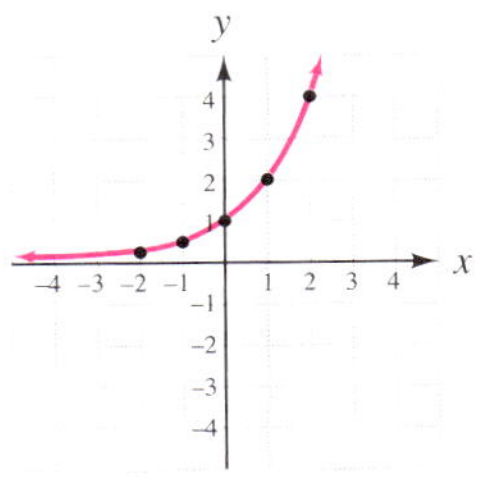

Fill in the blanks.

9. $\log_6 36 = 2$ means ____ = ____.
10. $\log x = -2$ is equivalent to ____ = ____.
11. $\log_b x$ is the __________ to which b is raised to get x.
12. The functions $f(x) = \log_{10} x$ and $f(x) = 10^x$ are __________ functions.
13. The inverse of an exponential function is called a __________ function.

14. Fill in the blanks.

a. The graph of $g(x) = 4 + \log x$ is similar to the graph of $f(x) = \log x$, but it is translated 4 units ________.

b. The graph of $g(x) = \log_4 (x + 2)$ is similar to the graph of $f(x) = \log_4 x$, but it is translated 2 units to the _____.

Complete the table of values.

15. $f(x) = \log x$

x	$f(x)$
100	
$\frac{1}{100}$	

16. $f(x) = \log_5 x$

x	$f(x)$
25	
$\frac{1}{25}$	

17. $f(x) = \log_6 x$

Input	Output
6	
−6	
0	

18. $f(x) = \log_8 x$

Input	Output
8	
−8	
0	

19. a. Use a calculator to complete the table of values for $f(x) = \log x$. Round to the nearest hundredth.

b. Graph $f(x) = \log x$. Note that the units on the x- and y-axes are different.

x	$f(x)$
0.5	
1	
2	
4	
6	
8	
10	

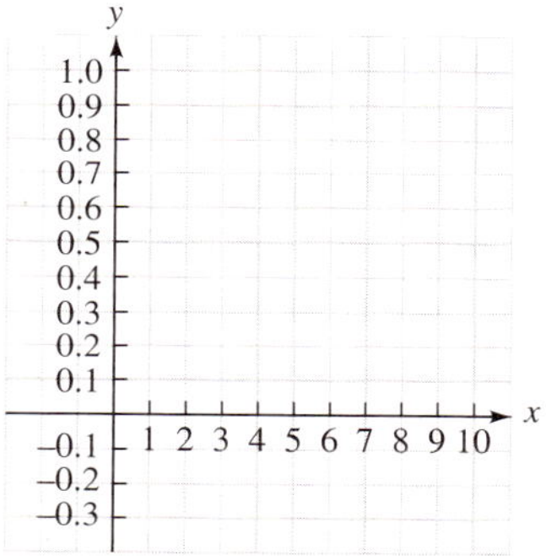

20. For each function, determine its inverse, $f^{-1}(x)$.

a. $f(x) = 10^x$

b. $f(x) = 3^x$

c. $f(x) = \log x$

d. $f(x) = \log_2 x$

NOTATION

Fill in the blanks.

21. a. $\log x = \log \quad x$

b. $\log_{10} 10^x =$

22. a. We read $\log_5 25$ as "log, 5, 25."

b. We read $\log x$ as " of x."

GUIDED PRACTICE

Write each logarithmic equation as an exponential equation. See Example 1. Do not solve.

23. $\log_3 81 = 4$

24. $\log_7 7 = 1$

25. $\log_{10} 10 = 1$

26. $\log_{10} 100 = 2$

27. $\log_4 \frac{1}{64} = -3$

28. $\log_6 \frac{1}{36} = -2$

29. $\log_5 \sqrt{5} = \frac{1}{2}$

30. $\log_8 \sqrt[3]{8} = \frac{1}{3}$

31. $\log 0.1 = -1$

32. $\log 0.01 = -2$

33. $x = \log_8 64$

34. $x = \log_9 81$

35. $t = \log_b T_1$

36. $n = \log_b R_1$

37. $\log_n C = -42$

38. $\log_m P = 101$

Write each exponential equation as a logarithmic equation. See Example 2.

39. $8^2 = 64$

40. $10^3 = 1{,}000$

41. $4^{-2} = \frac{1}{16}$

42. $3^{-4} = \frac{1}{81}$

43. $\left(\frac{1}{2}\right)^{-5} = 32$

44. $\left(\frac{1}{3}\right)^{-3} = 27$

45. $x^y = z$

46. $m^n = p$

47. $y^t = 8.6$

48. $b^r = \frac{2}{3}$

49. $7^{4.3} = B + 1$

50. $12^{-2.6} = N + 1$

Solve for x. See Example 3.

51. $\log_x 81 = 2$

52. $\log_x 9 = 2$

53. $\log_8 x = 2$

54. $\log_7 x = 0$

55. $\log_5 125 = x$

56. $\log_4 16 = x$

57. $\log_5 x = -2$

58. $\log_3 x = -4$

59. $\log_{36} x = -\frac{1}{2}$

60 $\log_{27} x = -\frac{1}{3}$

61. $\log_x 0.01 = -2$

62. $\log_x 0.001 = -3$

63. $\log_{27} 9 = x$

64. $\log_{12} x = 0$

65. $\log_x 5^3 = 3$

66. $\log_x 5 = 1$

67. $\log_{100} x = \frac{3}{2}$

68. $\log_x \frac{1}{1{,}000} = -\frac{3}{2}$

69. $\log_x \frac{1}{64} = -3$

70. $\log_x \frac{1}{100} = -2$

71. $\log_8 x = 0$

72. $\log_4 8 = x$

73. $\log_x \frac{\sqrt{3}}{3} = \frac{1}{2}$

74. $\log_x \frac{9}{4} = 2$

Evaluate each logarithmic expression. See Examples 4 and 5.

75. $\log_2 8$

76. $\log_3 9$

77. $\log_4 16$

78. $\log_6 216$

79. $\log 1{,}000{,}000$

80 $\log 100{,}000$

81. $\log \frac{1}{10}$

82. $\log \frac{1}{10{,}000}$

83. $\log_{1/2} \frac{1}{32}$

84. $\log_{1/3} \frac{1}{81}$

85. $\log_9 3$

86. $\log_{125} 5$

Use a calculator to find each value. Give answers to four decimal places. See Using Your Calculator: Evaluating Logarithms.

87. $\log 3.25$

88. $\log 0.57$

89. $\log 0.00467$

90. $\log 375.876$

Use a calculator to solve each equation. Round answers to four decimal places. See Example 6.

91. $\log x = 3.7813$

92. $\log x = 2.8945$

93. $\log x = -0.7630$

94. $\log x = -1.3587$

95. $\log x = -0.5$

96. $\log x = -0.926$

97. $\log x = -1.71$

98. $\log x = 1.4023$

Graph each function. Determine whether each function is an increasing or a decreasing function. See Objective 5.

99. $f(x) = \log_3 x$

100. $f(x) = \log_{1/3} x$

101. $y = \log_{1/2} x$

102. $y = \log_4 x$

Graph each function by plotting points or by using a translation. (The basic logarithmic functions graphed in Exercises 99–102 will be helpful.) See Example 7.

103. $f(x) = 3 + \log_3 x$

104. $f(x) = (\log_{1/3} x) - 1$

105. $y = \log_{1/2} (x - 2)$

106. $y = \log_4 (x + 2)$

Graph each pair of inverse functions on the same coordinate system. Draw the axis of symmetry. See Objective 1.

107. $f(x) = 6^x$
$f^{-1}(x) = \log_6 x$

108. $f(x) = 3^x$
$f^{-1}(x) = \log_3 x$

109. $f(x) = 5^x$
$f^{-1}(x) = \log_5 x$

110. $f(x) = 8^x$
$f^{-1}(x) = \log_8 x$

APPLICATIONS

111. **dB Gain.** Find the dB gain of the amplifier shown below. Round to the nearest tenth.

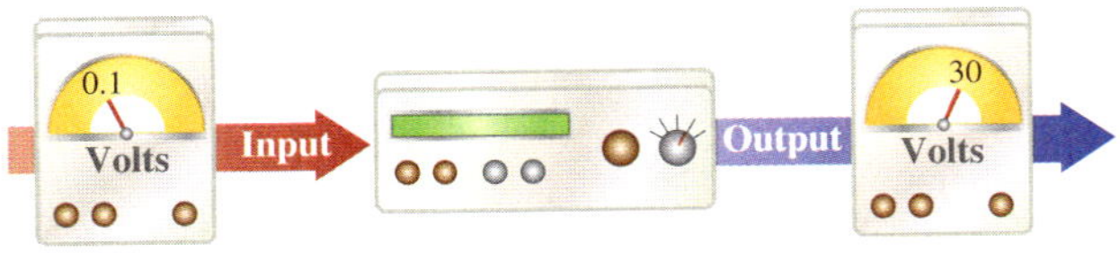

112. **Output Voltage.** Find the dB gain of an amplifier if the output voltage is 2.8 volts when the input voltage is 0.05 volt. Round to the nearest dB.

113. **Earthquakes.** Refer to the illustration in the next column. Common logarithms are used to measure the intensity of earthquakes. If R is the intensity of an earthquake on the **Richter scale,** A is the amplitude (measured in micrometers) of the ground motion and P is the period (the time of one oscillation of the Earth's surface measured in seconds), then $R = \log \frac{A}{P}$. If an earthquake has amplitude of 5,000 micrometers and a period of 0.2 second, what is its measure on the Richter scale? Round to the nearest tenth.

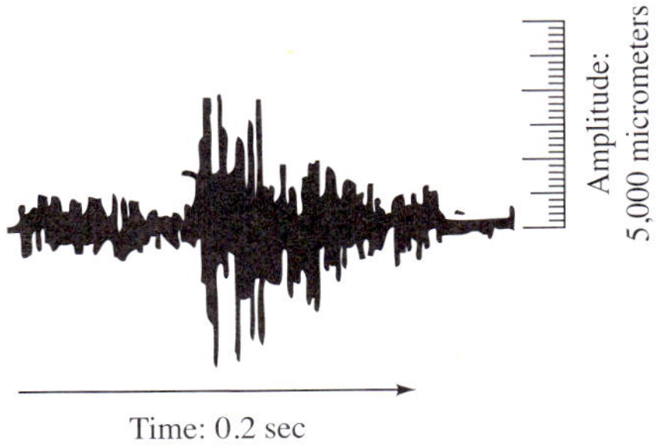

114. **Earthquakes.** If an earthquake has amplitude of 95,000 micrometers and a period of $\frac{1}{4}$ second, what is its measure on the Richter scale? See problem 113. Round to the nearest tenth.

115. **from Campus to Careers**

Social Worker

The logarithmic function $c(m) = 500 \log (m + 1)$ approximates the total number of cases handled one year by the staff at a county department of social services. The variable m represents the month of the year, where January is 1, February is 2, and so on. Find the total number of cases handled by the department that year by the end of June.

116. **Advertising.** The dollar amount of sales of a certain new product is approximated by the logarithmic function $D(m) = 20{,}000 + 100{,}000 \log (15m + 1)$, where m is the number of minutes of advertising of the product that is shown on cable television. Find the total sales if 30 minutes of cable television advertising time is purchased.

117. **Stocking Lakes.** A farmer stocked a lake on her property with 75 sunfish. She was told that with the proper oversight, the sunfish population could be approximated by the logarithmic function $f(t) = 75 + 45 \log (t + 1)$, where t is the number of years since the lake was stocked. Find the number of sunfish she can expect in the lake in $2\frac{1}{2}$ years.

118. **Zoology.** A trap-and-release program run by zoologists found that the ground squirrel population in a wilderness area could be estimated by the logarithmic function $s(t) = 800 + 600 \log (50t + 1)$, where t is the number of months after the program started. Find the ground squirrel population 3 years after the program began. (*Hint*: Be careful to use the correct units in your solution.)

119. **Children's Height.** The logarithmic function $h(A) = 29 + 48.8 \log(A + 1)$ gives the percent of the adult height a male child A years old has attained. If a boy is 9 years old, what percent of his adult height will he have reached?

120. Depreciation. In business, equipment is often depreciated using the double declining-balance method. In this method, a piece of equipment with a life expectancy of N years, costing $\$C$, will depreciate to a value of $\$V$ in n years, where n is given by the formula

$$n = \frac{\log V - \log C}{\log\left(1 - \frac{2}{N}\right)}$$

A computer that cost \$37,000 has a life expectancy of 5 years. If it has depreciated to a value of \$8,000, how old is it?

121. Investing. If $\$P$ is invested at the end of each year in an annuity earning annual interest at a rate r, the amount in the account will be $\$A$ after n years, where

$$n = \frac{\log\left(\frac{Ar}{P} + 1\right)}{\log(1 + r)}$$

If \$1,000 is invested each year in an annuity earning 12% annual interest, how long will it take for the account to be worth \$20,000? Round to the nearest tenth of a year.

122. Growth of Money. If \$5,000 is invested each year in an annuity earning 8% annual interest, how long will it take for the account to be worth \$50,000? (See Exercise 121.) Round to the nearest tenth of a year.

WRITING

123. Explain the mathematical relationship between $f(x) = \log x$ and $g(x) = 10^x$.

124. Explain why it is impossible to find the logarithm of a negative number.

125. A table of solutions for $f(x) = \log x$ is shown here. As x decreases and gets close to 0, what happens to the values of $f(x)$?

X	Y1
1	0
.9	-.0458
.8	-.0969
.7	-.1549
.6	-.2218
.5	-.301
.4	-.3979

X=1

126. What question should be asked when evaluating the expression $\log_4 16$?

REVIEW

Solve each equation.

127. $\sqrt[3]{6x + 4} = 4$

128. $\sqrt{3x + 4} = \sqrt{7x + 2}$

129. $\sqrt{a + 1} - 1 = 3a$

130. $3 - \sqrt{t - 3} = \sqrt{t}$

CHALLENGE PROBLEMS

131. Without graphing, determine the domain of the function $f(x) = \log_5 (x^2 - 1)$. Express the result in interval notation.

132. Evaluate: $\log_6(\log_5(\log_4 1{,}024))$

133. Earthquakes. In 1985, Mexico City experienced an earthquake of magnitude 8.1 on the Richter scale. In 1989, the San Francisco Bay area was rocked by an earthquake measuring 7.1. By what factor must the amplitude of an earthquake change to increase its severity by 1 point on the Richter scale? (Assume that the period remains constant.)

134. Graph: $f(x) = \log_2|x|$

SECTION 11.5 Base-*e* Exponential and Logarithmic Functions

OBJECTIVES

1. Define the natural exponential function.
2. Graph the natural exponential function.
3. Use base-e exponential formulas and functions in applications.
4. Define base-e logarithms.
5. Evaluate natural logarithmic expressions.
6. Graph the natural logarithmic function.
7. Use base-e logarithmic formulas and functions in applications.

ARE YOU READY?

The following problems review some basic skills that are needed when working with base-e exponential and logarithmic functions.

1. Evaluate $\left(1 + \frac{1}{n}\right)^n$ for $n = 2$.

2. Evaluate: $(2.718)^0$

3. Round 2.718281828459 to the nearest tenth.

4. Use a calculator to evaluate $55(2.718)^{0.4}$. Round to the nearest hundredth.

5. Fill in the blank: $\log_3 x = 6$ is equivalent to ___ = ___.

6. Fill in the blank: If (3, 8) is on the graph of a one-to-one function f, then the point (___, ___) is on the graph of f^{-1}.

Any positive real number not equal to 1 can be used as a base of an exponential or a logarithmic function. However, some bases are used more often than others. Exponential and logarithmic functions that have many applications are ones whose base is an irrational number represented by the letter e.

n	$\left(1 + \frac{1}{n}\right)^n$
1	2
2	2.25
4	2.44140625 ...
12	2.61303529 ...
365	2.71456748 ...
1,000	2.71692393 ...
100,000	2.71826830 ...
1,000,000	2.71828137 ...

A scientific calculator was used to evaluate the expression.

1 Define the Natural Exponential Function.

The number called e is defined to be the value that $\left(1 + \frac{1}{n}\right)^n$ approaches as n gets larger and larger. The table in the margin shows the value of that expression as n increases from 1 to 1,000,000. It can be shown that as n approaches infinity, the value of $\left(1 + \frac{1}{n}\right)^n$ approaches:

$$e = 2.718281828459\ldots$$

We give this number a letter-name because it makes communication easier. Reciting its first thirteen digits every time we refer to it would be overwhelming. So, we simply call it by the name "e."

Like π, the number e is an irrational number. That means its decimal representation is nonterminating and nonrepeating. Rounded to four decimal places, $e \approx 2.7183$.

Of all possible bases for an exponential function, e is the most convenient for problems involving growth or decay. Since these situations occur often in natural settings, we call $f(x) = e^x$ the *natural exponential function.*

The Natural Exponential Function

The function defined by $f(x) = e^x$ is the **natural exponential function** (or the **base-*e* exponential function**) where $e = 2.71828\ldots$ The domain of $f(x) = e^x$ is the interval $(-\infty, \infty)$. The range is the interval $(0, \infty)$.

The e^x key on a calculator is used to find powers of e.

Using Your Calculator ▶ Finding powers of *e*

To find the value of e^5 with a reverse entry scientific calculator, we press:

5 e^x `148.4131591`

On some calculators, the e^x key also serves as the LN key. (Later in this section we will see why.) To activate the e^x key, we must begin by pressing 2nd or SHIFT.

With a direct entry graphing calculator, we press:

2nd e^x 5) ENTER   `e^(5)` `148.4131591`

Since powers of e are irrational numbers (nonterminating, nonrepeating decimals), we often round such answers. For example, to the nearest hundredth, $e^5 \approx 148.41$

2 Graph the Natural Exponential Function.

To graph $f(x) = e^x$, we construct a table of function values by choosing several values for x and finding the corresponding values of $f(x)$. For example, if $x = -2$, we have

$$f(x) = e^x$$

$$f(-2) = e^{-2}$$ Substitute -2 for each x.

$$= 0.135335283\ldots$$ Use a calculator. On a scientific calculator, press: 2 +/− 2nd e^x.

$$\approx 0.1$$ Round to the nearest tenth.

We enter $(-2, 0.1)$ in the table on the next page. Similarly, we find $f(-1)$, $f(0)$, $f(1)$, and $f(2)$, enter each result in the table, and plot the ordered pairs. We draw a smooth curve through the points to get the graph.

From the graph, we can verify that the domain of $f(x) = e^x$ is the interval $(-\infty, \infty)$ and the range is the interval $(0, \infty)$. Since the graph passes the horizontal line test, the function is one-to-one.

Notation

Swiss born Leonhard Euler (1707–1783) is said to have published more than any mathematician in history. Through his work, the symbol e came into common use.

Note that as x decreases, the values of $f(x)$ decrease and approach 0. Thus, the x-axis is an asymptote of the graph. The graph does not have an x-intercept, the y-intercept is $(0, 1)$, and the graph passes through the point $(1, e)$.

$f(x) = e^x$

x	$f(x)$	
-2	$\frac{1}{e^2} \approx 0.1$	$\rightarrow (-2, 0.1)$
-1	$\frac{1}{e^1} \approx 0.4$	$\rightarrow (-1, 0.4)$
0	$e^0 = 1$	$\rightarrow (0, 1)$
1	$e^1 \approx 2.7$	$\rightarrow (1, 2.7)$
2	$e^2 \approx 7.4$	$\rightarrow (2, 7.4)$

The outputs can be found using the e^x key on a calculator. In such cases, round to the nearest tenth to make point-plotting easier.

The graph steadily approaches the x-axis, but never touches or crosses it.

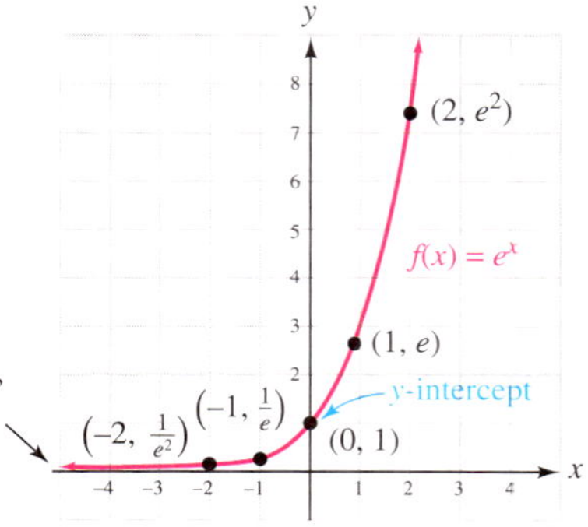

The graph of the natural exponential function can be translated horizontally and vertically, as shown below.

Success Tip

Recall that exponential functions have a constant base and a variable exponent. For the natural exponential function $f(x) = e^x$, the constant base is e.

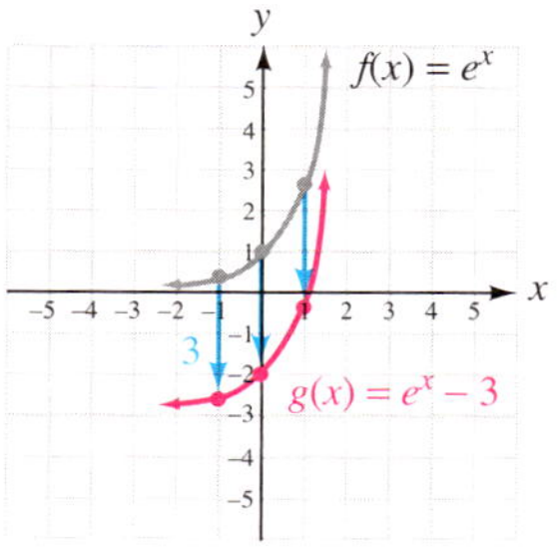

To graph $g(x) = e^x - 3$, translate each point on the graph of $f(x) = e^x$ down 3 units.

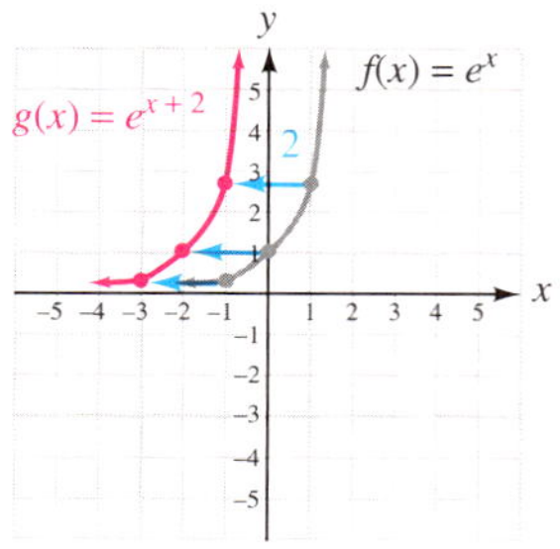

To graph $g(x) = e^{x+2}$, translate each point on the graph of $f(x) = e^x$ to the left 2 units.

We can illustrate the effects of vertical and horizontal translations of the natural exponential function by using a graphing calculator.

Using Your Calculator ▶ Graphing Base e (Natural) Exponential Functions

Figure (a) shows the calculator graphs of $f(x) = e^x$, $g(x) = e^x + 5$, and $h(x) = e^x - 3$. To graph them, we enter the right sides of the equations after $Y_1 =$, $Y_2 =$, and $Y_3 =$. The display will show:

$$Y_1 = e\text{^}(X) \qquad Y_2 = e\text{^}(X) + 5 \qquad Y_3 = e\text{^}(X) - 3$$

The graph of $g(x) = e^x + 5$ is 5 units above the graph of $f(x) = e^x$ and the graph of $h(x) = e^x - 3$ is 3 units below the graph of $f(x) = e^x$.

Figure (b) shows the calculator graphs of $f(x) = e^x$, $g(x) = e^{x+5}$, and $h(x) = e^{x-3}$. The graph of $g(x) = e^{x+5}$ is 5 units to the left of the graph of $f(x) = e^x$ and the graph of $h(x) = e^{x-3}$ is 3 units to the right of the graph of $f(x) = e^x$.

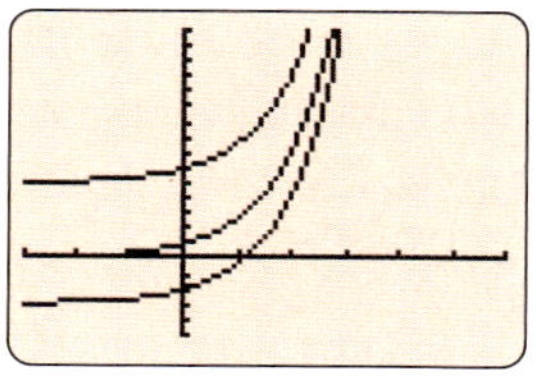

(a)

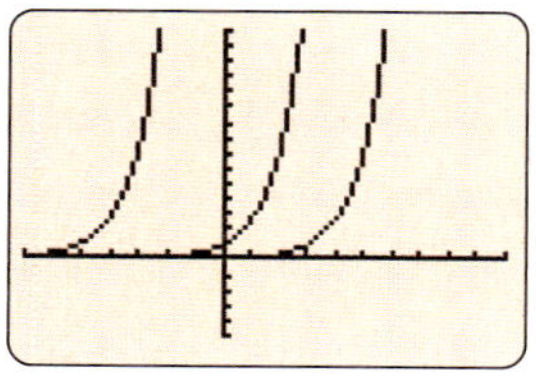

(b)

To graph complicated natural exponential functions, point-plotting can be tedious. In such cases, we will use a graphing calculator. For example, the figure shows the calculator graph of $f(x) = 3e^{-x/2}$. To graph this function, we enter the right side of the equation after the symbol $Y_1 =$. The display will show the equation

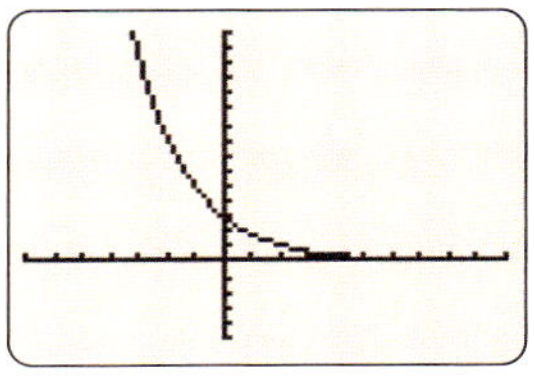

$$Y_1 = 3(e^{\wedge}(-X/2))$$

3 Use Base-*e* Exponential Formulas and Functions in Applications.

The following formula involving e provides a mathematical model for growth and decay applications. It is used in many areas, such as continuous compound interest, population trends, heat transfer, radioactivity, and learning retention.

Formula for Exponential Growth/Decay

If a quantity P increases or decreases at an annual rate r, compounded continuously, the amount A after t years is given by

$A = Pe^{rt}$ Read as "A equals P times e to the rt power."

For the formula above, if the time is measured in years, r is called the **annual growth rate.** If r is negative, the growth represents a decrease.

For a given quantity P, say 10,000, and a given rate r, say 5%, we can write the formula for exponential growth using function notation:

$$A(t) = 10{,}000e^{0.05t}$$

Recall that if a bank pays interest twice a year, we say the interest is compounded *semiannually.* If it pays interest four times a year, we say the interest is paid *quarterly.* If it pays interest continuously (infinitely many times a year), we say that the interest is **compounded continuously.**

EXAMPLE 1 **Investing.** If \$25,000 accumulates interest at an annual rate of 8%, compounded continuously, find the balance in the account in 50 years.

Strategy We will substitute 25,000 for P, 0.08 for r, and 50 for t in the formula $A = Pe^{rt}$ and calculate the value of A.

Why The words *compounded continuously* indicate that we should use the base-e exponential growth/decay formula.

Solution

$A = Pe^{rt}$ This is the formula for continuous compound interest.

$A = 25{,}000e^{0.08(50)}$ Write 8% as 0.08. Substitute for P, r, and t.

$= 25{,}000e^{4}$ Evaluate the exponent: $0.08(50) = 4$.

$\approx 1{,}364{,}953.751$ Use a calculator. On a scientific calculator, press: 25000 × 4 2nd e^x = .

$\approx 1{,}364{,}953.75$ Round to the nearest hundredth.

In 50 years, the balance will be \$1,364,953.75—more than a million dollars.

Self Check 1 **Investing.** In Example 1, find the balance in 60 years.

Now Try Problems 33 and 73

EXAMPLE 2 **City Planning.** The population of a city is currently 15,000, but economic conditions are causing the population to decrease 3% each year. If this trend continues, find the population in 30 years.

Strategy We will substitute 15,000 for P, -0.03 for r, and 30 for t in the formula $A = Pe^{rt}$ and calculate the value of A.

Why Since the population is decreasing 3% each year, the annual growth rate is -3%, or -0.03.

Solution

$A = Pe^{rt}$ This is the model for population growth/decay.

$A = 15{,}000e^{-0.03(30)}$ Substitute for P, r, and t.

$= 15{,}000e^{-0.9}$ Evaluate the exponent: $-0.03(30) = -0.9$.

$\approx 6{,}098.544896$ Use a calculator. On a scientific calculator, press: 15000 × .9 +/− 2nd e^x =.

$\approx 6{,}099$ Round to the nearest whole number.

Success Tip

For quantities that are decreasing, remember to enter a negative value for r, the annual rate, in the formula $A = Pe^{rt}$.

In 30 years, the expected population will be 6,099.

Self Check 2 **City Planning.** In Example 2, find the population in 50 years.

Now Try Problems 37 and 81

EXAMPLE 3 **Baking.** A mother takes a cake out of the oven and sets it on a rack to cool. The function $T(t) = 68 + 220e^{-0.18t}$ gives the cake's temperature in degrees Fahrenheit after it has cooled for t minutes. If her children will be home from school in 20 minutes, will the cake have cooled enough for the children to eat it? (Assume that 80°F, or cooler, would be a comfortable eating temperature.)

Strategy We will substitute 20 for t in the function $T(t) = 68 + 220e^{-0.18t}$.

Why The variable t represents the number of minutes the cake has cooled.

Solution When the children arrive home, the cake will have cooled for 20 minutes. To find the temperature of the cake at that time, we need to find $T(20)$.

$T(t) = 68 + 220e^{-0.18t}$ This is the cooling model.

$T(20) = 68 + 220e^{-0.18(20)}$ Substitute 20 for t.

$= 68 + 220e^{-3.6}$ Evaluate the exponent: $-0.18(20) = -3.6$.

≈ 74.01121894 Use a calculator. On a scientific calculator, press: 68 + 220 × 3.6 +/− 2nd e^x =.

≈ 74.0 Round to the nearest tenth.

When the children return home, the temperature of the cake will be about 74°, and it can be eaten.

Self Check 3 **Baking.** In Example 3, find the temperature of the cake 10 minutes after it is removed from the oven. Round to the nearest tenth.

Now Try Problem 91

4 Define Base-e Logarithms.

Of all possible bases for a logarithmic function, e is the most convenient for problems involving growth or decay. Since these situations occur often in natural settings, base-e

logarithms are called **natural logarithms** or **Napierian logarithms** after John Napier (1550–1617). They are usually written as $\ln x$ rather than $\log_e x$:

$\ln x$ means $\log_e x$ Read ln x letter-by-letter as "ℓ . . . n . . . of x."

In general, the logarithm of a number is an exponent. For natural logarithms,

ln x is the exponent to which e is raised to get x.

Translating this statement into symbols, we have

$$e^{\ln x} = x$$

Caution

Because of the font used to print the natural log of x, some students initially misread the notation as In x. In handwriting, ln x should look like *ℓn x*.

5 Evaluate Natural Logarithmic Expressions.

EXAMPLE 4 Evaluate each natural logarithmic expression:

a. $\ln e$ **b.** $\ln \frac{1}{e^2}$ **c.** $\ln 1$ **d.** $\ln \sqrt{e}$

Strategy Since the base is e in each case, we will ask "To what power must e be raised to get the given number?"

Why That power is the value of the logarithmic expression.

Solution

a. $\ln e = 1$ Ask: "To what power must we raise e to get e?" Since $e^1 = e$, the answer is: the 1st power.

b. $\ln \frac{1}{e^2} = -2$ Ask: "To what power must we raise e to get $\frac{1}{e^2}$?" Since $e^{-2} = \frac{1}{e^2}$, the answer is: the -2 power.

c. $\ln 1 = 0$ Ask: "To what power must we raise e to get 1?" Since $e^0 = 1$, the answer is: the 0 power.

d. $\ln \sqrt{e} = \frac{1}{2}$ Ask: "To what power must we raise e to get $\sqrt{e}$?" Since $e^{1/2} = \sqrt{e}$, the answer is: the $\frac{1}{2}$ power.

Success Tip

Every natural logarithmic equation has a corresponding natural exponential equation. For example:

$\ln \frac{1}{e^2} = -2$ means $e^{-2} = \frac{1}{e^2}$

and

$\ln \sqrt{e} = \frac{1}{2}$ means $e^{1/2} = \sqrt{e}$

Self Check 4 Evaluate each expression: **a.** $\ln e^3$ **b.** $\ln \frac{1}{e}$ **c.** $\ln \sqrt[3]{e}$

Now Try Problems 41, 45, and 47

Many natural logarithmic expressions are not as easy to evaluate as those in the previous example. For example, to find ln 2.34, we ask, "To what power must we raise e to get 2.34?" The answer isn't obvious. In such cases, we use a calculator.

Using Your Calculator ▶ Evaluating Base-*e* (Natural) Logarithms

To find ln 2.34 with a reverse entry scientific calculator, we press:

2.34 [LN] .850150929

On some calculators, the [e^x] key also serves as the [LN] key when [2nd] or [SHIFT] is pressed. We will see why this is so later in this section.

To use a direct entry graphing calculator, we press:

[LN] 2.34 [)] [ENTER]

```
ln(2.34)
    .8501509294
```

To four decimal places, $\ln 2.34 \approx 0.8502$. This means that $e^{0.8502} \approx 2.34$.

If we attempt to evaluate logarithmic expressions such as ln 0, or the logarithm of a negative number, such as ln (−5), then one of the following error statements will be displayed.

Error

```
ERR:DOMAIN
1:QUIT
2:Go to
```

```
ERR:NONREAL ANS
1:QUIT
2:Go to
```

Certain natural logarithmic equations can be solved by writing them as equivalent natural exponential equations.

EXAMPLE 5 Solve each equation: **a.** $\ln x = 1.335$ and **b.** $\ln x = -5.5$. Give each result to four decimal places.

Strategy To solve these logarithmic equations, we will instead write and solve equivalent exponential equations.

Why The resulting exponential equations are easier to solve because the variable term is isolated on one side.

Solution

a. Since the base of the natural logarithmic function is e:

Exponent

$\ln x = 1.335$ is equivalent to $e^{1.335} = x$

Base

To use a reverse entry scientific calculator to find x, press:

1.335 [2nd] [e^x] 3.799995946

To four decimal places, $x = 3.8000$. Thus, the solution is 3.8000. To check, use your calculator to verify that $\ln 3.8000 \approx 1.335$.

b. The logarithmic equation $\ln x = -5.5$ is equivalent to the exponential equation $e^{-5.5} = x$. To use a reverse entry scientific calculator to find x, press:

5.5 [+/−] [2nd] [e^x] 0.004086771

To four decimal places, $x = 0.0041$. Thus, the solution is 0.0041. To check, use your calculator to verify that $\ln 0.0041 \approx -5.5$.

Success Tip

Recall that equivalent equations have the same solutions.

Self Check 5 Solve each equation. Give each result to four decimal places.

a. $\ln x = 1.9344$ **b.** $-3 = \ln x$

Now Try Problems 61 and 65

6 Graph the Natural Logarithmic Function.

Because the natural exponential function defined by $f(x) = e^x$ is one-to-one, it has an inverse function that is defined by $x = e^y$. When we write $x = e^y$ in the equivalent form $y = \ln x$, the result is called the *natural logarithmic function.*

The Natural Logarithmic Function

The **natural logarithmic function** with base e is defined by the equations

$$f(x) = \ln x \text{ or } y = \ln x, \text{ where } \ln x = \log_e x$$

The domain of $f(x) = \ln x$ is the interval $(0, \infty)$, and the range is the interval $(-\infty, \infty)$.

Since the natural logarithmic function is the inverse of the one-to-one natural exponential function, the natural logarithmic function is one-to-one.

To graph $f(x) = \ln x$, we can construct a table of function values, plot the resulting ordered pairs, and draw a smooth curve through the points to get the graph shown in figure (a). Figure (b) shows the calculator graph of $f(x) = \ln x$. We see that the domain of the natural logarithmic function is $(0, \infty)$ and the range is $(-\infty, \infty)$.

$f(x) = \ln x$

To plot these ordered pairs, use a calculator to approximate *x*.

x	$f(x)$		
$\frac{1}{e}$	-1	$\to \left(\frac{1}{e}, -1\right)$	$\to (0.4, -1)$
1	0	$\to (1, 0)$	$\to (1, 0)$
e	1	$\to (e, 1)$	$\to (2.7, 1)$
e^2	2	$\to (e^2, 2)$	$\to (7.4, 2)$

Since the base of the natural logarithmic function is *e*, choose *x*-values that are integer powers of *e*.

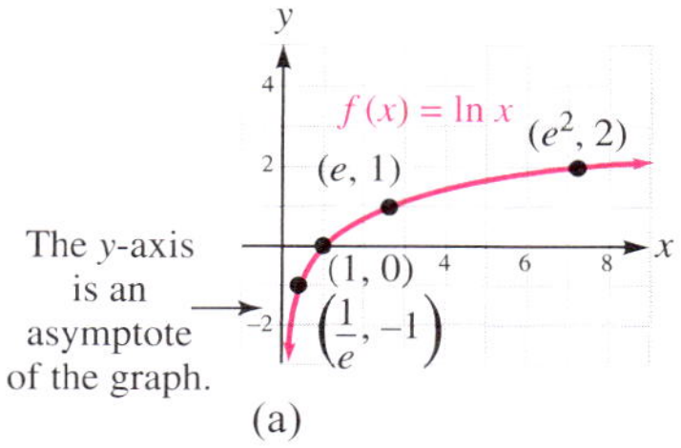

(a)

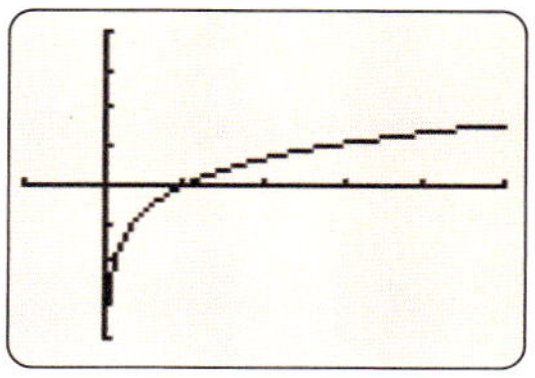
(b)

Notation

If it is helpful, imagine parentheses around the input value in the natural logarithmic function notation.

$f(e) = 1$ — Function name, Input, Output

$\ln (e) = 1$

The natural exponential function and the natural logarithm function are inverse functions. The figure shows that their graphs are symmetric to the line $y = x$.

You may want to confirm that these functions are indeed inverses using the composition of inverses test in Section 11.2.

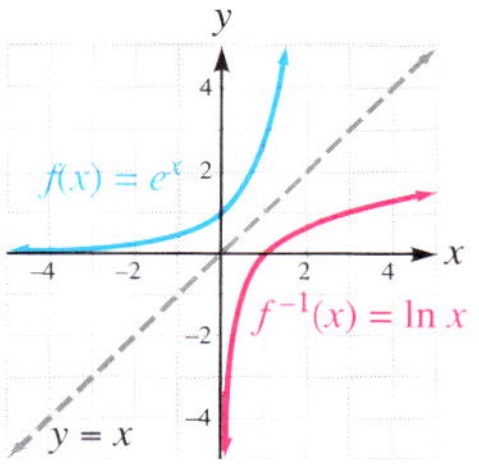

Using Your Calculator ▶ Graphing Base-*e* Logarithmic Functions

Many graphs of logarithmic functions involve translations of the graph of $f(x) = \ln x$. For example, the figure below shows calculator graphs of the functions $f(x) = \ln x$, $g(x) = (\ln x) + 2$, and $h(x) = (\ln x) - 3$.

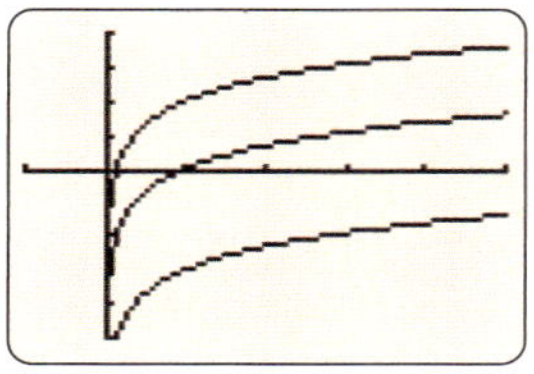

The graph of $g(x) = (\ln x) + 2$ is 2 units above the graph of $f(x) = \ln x$.

The graph of $h(x) = (\ln x) - 3$ is 3 units below the graph of $f(x) = \ln x$.

The next figure shows the calculator graph of the functions $f(x) = \ln x$, $g(x) = \ln (x - 2)$, and $h(x) = \ln (x + 3)$.

The graph of $h(x) = \ln (x + 3)$ is 3 units to the left of the graph of $f(x) = \ln x$.

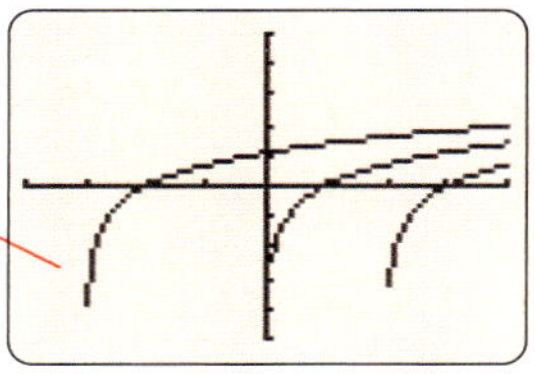

The graph of $g(x) = \ln (x - 2)$ is 2 units to the right of the graph of $f(x) = \ln x$.

7 Use Base-e Logarithmic Formulas and Functions in Applications.

If a population grows exponentially at a certain annual rate, the time required for the population to double is called the **doubling time.** It is given by the following formula.

Formula for Doubling Time

If r is the annual rate, compounded continuously, and t is the time required for a population to double, then

$$t = \frac{\ln 2}{r}$$

EXAMPLE 6

©iStockphoto.com/adisa

Doubling Time. The population of the Earth is growing at the approximate rate of 1.133% per year. If this rate continues, how long will it take for the population to double? (Source: CIA World Fact Book, 2009 data)

Strategy We will substitute 1.133% expressed as a decimal for r in the formula for doubling time and evaluate the right side using a calculator.

Why We can use this formula because we are given the annual rate of continuous compounding.

Solution Since the population is growing at the rate of 1.133% per year, we substitute 0.0133 for r in the formula for doubling time and simplify.

$$t = \frac{\ln 2}{r}$$ Don't forget to substitute the decimal form of 1.133%, which is 0.0133, for r.

$$t = \frac{\ln 2}{0.0133}$$

$$\approx 52.11632937$$ Use a calculator. On a scientific calculator, press: 2 LN ÷ .0033 =.

$$\approx 52$$ Round to the nearest year.

At the current growth rate, the population of the Earth will double in about 52 years.

Self Check 6 **Doubling Time.** If the annual growth rate of the Earth's population could be reduced to 1.1% per year, what would be the doubling time?

Now Try Problem 93

EXAMPLE 7

©Elena Elisseeva/Shutterstock.com

The Pace of Life. A study by psychologists M. H. Bornstein and H. G. Bornstein found that the average walking speed s, in feet per second, of a pedestrian in a city of population p, is approximated by the natural logarithmic function $s(p) = 0.05 + 0.37 \ln p$. According to this study, what is the average walking speed of a pedestrian in Detroit if its population is 910,000? (Source: *Nature*, Volume 259, Feb. 19, 1976 and infoplease.com)

Strategy We will find $s(910{,}000)$.

Why Since the variable p represents the population of the city, $p = 910{,}000$.

Solution

$$s(p) = 0.05 + 0.37 \ln p$$ This is the natural logarithmic model.

$$s(910{,}000) = 0.05 + 0.37 \ln 910{,}000$$ Substitute 910,000 for p.

$$\approx 5.126843955$$ Use a calculator. On a scientific calculator, press: 0.05 + 0.37 × 910000 LN =.

$$\approx 5.1$$ Round to the nearest tenth.

The average walking speed of a pedestrian in Detroit is about 5.1 feet per second.

Self Check 7 **The Pace of Life.** Use the natural logarithmic function from Example 7 to find the average walking speed of a pedestrian in Dallas if its population is 1,300,000.

Now Try ▶ Problem 101

SECTION 11.5 STUDY SET

VOCABULARY

Fill in the blanks.

1. $f(x) = e^x$ is called the natural __________ function. The base is .
2. $f(x) = \ln x$ is called the _______ logarithmic function. The base is .
3. If a bank pays interest infinitely many times a year, we say that the interest is compounded ___________
4. Like π, the number e is an ________ number. Its decimal representation is nonterminating and ___________.

CONCEPTS

5. Refer to the graph on the right.

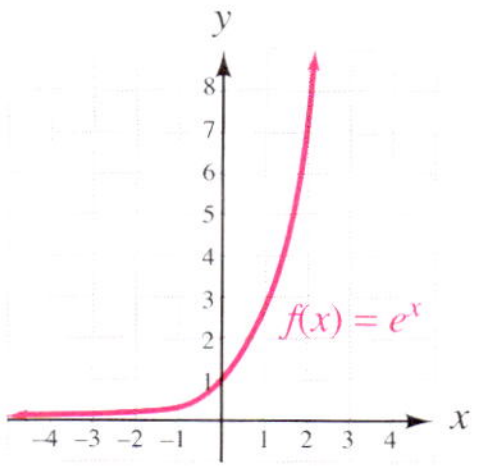

   a. What is the name of the function $f(x) = e^x$?
   b. What is the domain of the function? What is the range of the function?
   c. What is the y-intercept of the graph? What is the x-intercept of the graph?
   d. Is the function one-to-one?
   e. What is an asymptote of the graph?
   f. Is f an increasing or a decreasing function?
   g. The graph passes through the point $(1, y)$. What is y?

6. a. Use a calculator to complete the table of values in the next column for $f(x) = \ln x$. Round to the nearest hundredth.
   b. Graph $f(x) = \ln x$. Note that the units on the x- and y-axes are different.
   c. What are the domain and range of the function?
   d. What is the x-intercept of the graph? What is the y-intercept?
   e. What is an asymptote of the graph?
   f. Is f increasing or decreasing?
   g. Is the function one-to-one?

x	$f(x)$
0.5	
1	
2	
4	
6	
8	
10	

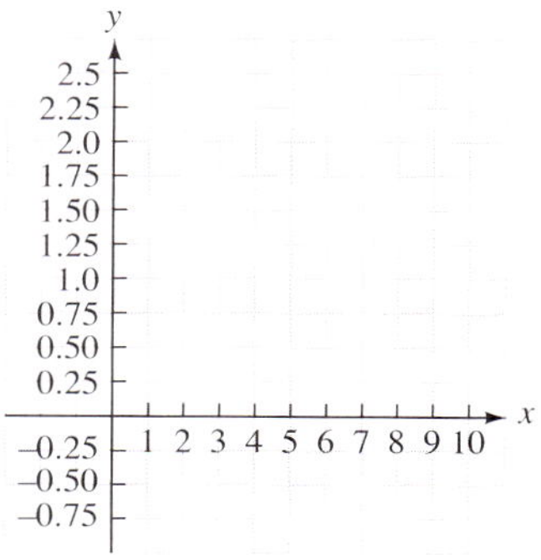

Fill in the blanks.

7. $e =$ ___ . ___ ___ ___ ___ ___ ___ ___ ___ ___ ___ ___ ___ . . .
8. To two decimal places, the value of e is .
9. If n gets larger and larger, the value of $\left(1 + \frac{1}{n}\right)^n$ approaches the value of .
10. The formula for exponential growth/decay is $A =$ e .
11. To find $\ln e^2$, we ask, "To what power must we raise to get e^2?" Since the answer is the 2nd power, $\ln e^2 =$.
12. The logarithmic equation $\ln x = 1.5318$ is equivalent to the exponential equation $=$.
13. Graph each irrational number on the number line: $\{\pi, e, \sqrt{2}\}$.

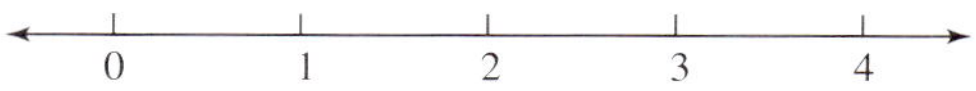

14. Complete the table of values. Use a calculator when necessary and round to the nearest hundredth.

x	−2	−1	0	1	2
e^x					

15. a. The function $f(x) = e^x$ is graphed in figure (a) below and the TRACE feature is used. What is the y-coordinate of the point on the graph having an x-coordinate of 1? What is the symbol that represents this number?

b. Figure (b) below shows a table of values for $f(x) = e^x$. As x decreases, what happens to the values of $f(x)$ listed in the Y_1 column? Will the value of $f(x)$ ever be 0 or negative?

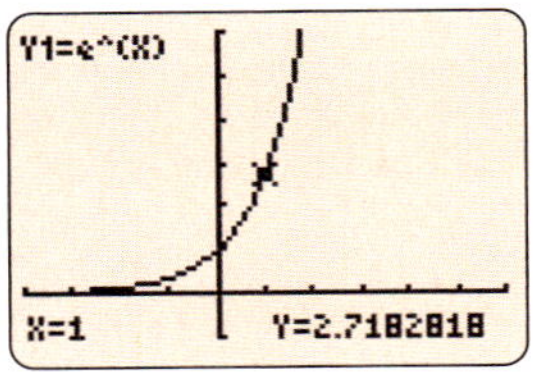

(a)

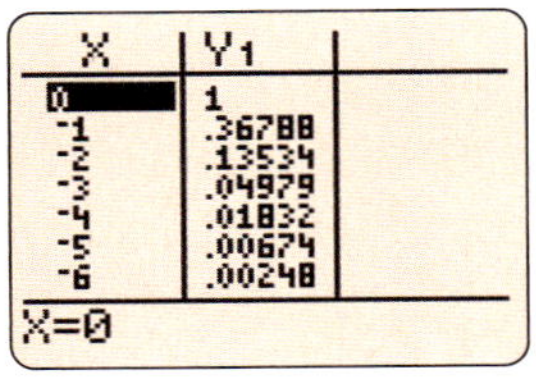

(b)

16. a. The illustration shows the graph of $f(x) = \ln x$, as well as a vertical translation of that graph. Using the notation $g(x)$ for the translation, write the defining equation for that function.

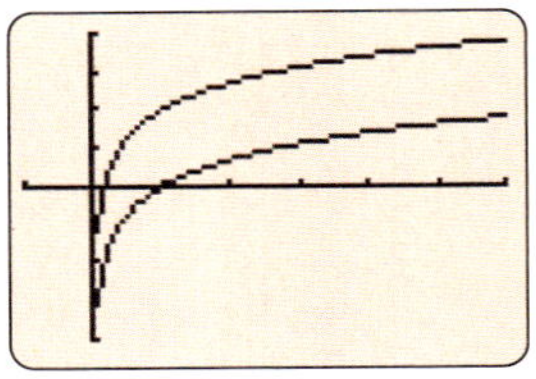

b. In the illustration, $f(x) = \ln x$ was graphed, and the TRACE feature was used. What is the x-coordinate of the point on the graph having a y-coordinate of 1? What is the name given this number?

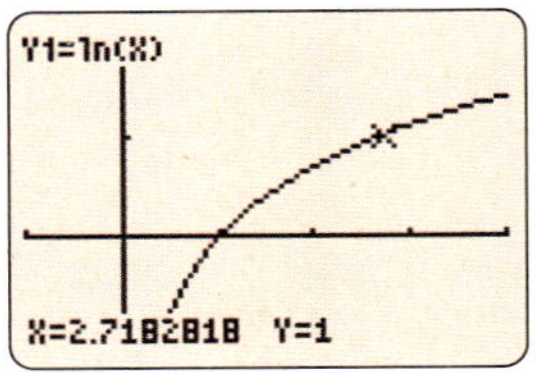

17. What is the inverse of the natural logarithmic function $f(x) = \ln x$?

18. Let $f(x) = 75 + 3{,}570 \ln x$. Find $f(28.1)$. Round to the nearest tenth.

NOTATION

Find A using the formula $A = Pe^{rt}$ given the following values of P, r, and t. Round to the nearest tenth.

19. $P = 1{,}000$, $r = 0.09$, and $t = 10$

$A = \quad e^{(0.09)(\quad)}$

$= 1{,}000e$

$\approx$ Use a calculator.

20. $P = 50{,}000$, $r = -0.12$, and $t = 50$

$A = 50{,}000e^{(\quad)(50)}$

$= 50{,}000e$

$\approx$ Use a calculator.

Fill in the blanks.

21. We read $\ln x$ letter-by-letter as " ____ . . . ____ . . . ____ of x."

22. a. $\ln 2$ means $\log$ ____ 2.

b. $\log 2$ means $\log$ ____ 2.

23. To evaluate a base-10 logarithm with a calculator, use the ____ key. To evaluate the base-e logarithm, use the ____ key.

24. If a population grows exponentially at a rate r, the time it will take the population to double is given by the formula $t =$ ____.

GUIDED PRACTICE

Graph each function. See Objective 2.

25. $f(x) = e^x$

26. $f(x) = -e^x$

27. $f(x) = e^x + 1$

28. $f(x) = e^x - 2$

29. $y = e^{x+3}$

30. $y = e^{x-5}$

31. $f(x) = 2e^x$

32. $f(x) = \frac{1}{2}e^x$

Find A using the formula $A = Pe^{rt}$ given the following values of P, r, and t. Round to the nearest hundredth. See Example 1.

33. $P = 5{,}000$, $r = 8\%$, $t = 20$ years

34. $P = 15{,}000$, $r = 6\%$, $t = 40$ years

35. $P = 20{,}000$, $r = 10.5\%$, $t = 50$ years

36. $P = 25{,}000$, $r = 6.5\%$, $t = 100$ years

Find A using the formula $A = Pe^{rt}$ given the following values of P, r, and t. Round to the nearest hundredth. See Example 2.

37. $P = 15{,}895$, $r = -2\%$, $t = 16$ years

38. $P = 33{,}999$, $r = -4\%$, $t = 21$ years

39. $P = 565$, $r = -0.5\%$, $t = 8$ years

40. $P = 110$, $r = -0.25\%$, $t = 9$ years

Evaluate each expression without using a calculator. See Example 4.

41. $\ln e^5$

42. $\ln e^2$

43. $\ln e^6$

44. $\ln e^4$

45. $\ln \frac{1}{e}$

46. $\ln \frac{1}{e^3}$

47. $\ln \sqrt[4]{e}$

48. $\ln \sqrt[5]{e}$

49. $\ln \sqrt[3]{e^2}$

50. $\ln \sqrt[4]{e^3}$

51. $\ln e^{-7}$

52. $\ln e^{-10}$

Use a calculator to evaluate each expression, if possible. Express all answers to four decimal places. See Using Your Calculator: Evaluating Base-e (Natural) Logarithms.

53. $\ln 35.15$

54. $\ln 0.675$

55. $\ln 0.00465$

56. $\ln 378.96$

57. $\ln 1.72$

58. $\ln 2.7$

59. $\ln(-0.1)$

60. $\ln(-10)$

Solve each equation. Express all answers to four decimal places. See Example 5.

61. $\ln x = 1.4023$

62. $\ln x = 2.6490$

63. $\ln x = 4.24$

64. $\ln x = 0.926$

65. $\ln x = -3.71$

66. $\ln x = -0.28$

67. $\ln x = 1.001$

68. $\ln x = -0.001$

Use a graphing calculator to graph each function. See Objective 2. See Using Your Calculator: Graph Base-e Logarithmic Functions.

69. $f(x) = \ln\left(\frac{1}{2}x\right)$

70. $f(x) = \ln x^2$

71. $f(x) = \ln(-x)$

72. $f(x) = \ln(3x)$

APPLICATIONS

In Exercises 73–78, assume that there are no deposits or withdrawals.

73. Continuous Compound Interest. An initial investment of \$5,000 earns 8.2% interest, compounded continuously. What will the investment be worth in 12 years?

74. Continuous Compound Interest. An initial investment of \$2,000 earns 8% interest, compounded continuously. What will the investment be worth in 15 years?

75. Comparison of Compounding Methods. An initial deposit of \$5,000 grows at an annual rate of 8.5% for 5 years. Compare the final balances resulting from annual compounding and continuous compounding.

76. Comparison of Compounding Methods. An initial deposit of \$30,000 grows at an annual rate of 8% for 20 years. Compare the final balances resulting from annual compounding and continuous compounding.

77. Determining the Initial Deposit. An account now contains \$11,180 and has been accumulating interest at 7% annual interest, compounded continuously, for 7 years. Find the initial deposit.

78. Determining the Previous Balance. An account now contains \$3,610 and has been accumulating interest at 8% annual interest, compounded continuously. How much was in the account 4 years ago?

79. The 20th Century. The exponential function $A(t) = 123e^{0.0117t}$ approximates the population of the United States (in millions), where t is the number of years after 1930. Use the function to estimate the U.S. population for these important dates:

- 1937 The Golden Gate Bridge is completed
- 1941 The United States enters World War II
- 1955 Rosa Parks refuses to give up her seat on a Montgomery, Alabama, bus
- 1969 Astronaut Neil Armstrong walks on the moon
- 1974 President Nixon resigns
- 1986 The *Challenger* space shuttle explodes
- 1997 *The Simpsons* becomes the longest running cartoon television series in history

80. World Population Growth. The population of Earth is approximately 6.8 billion people and is growing at an annual rate of 1.133%. Use the exponential growth model to predict the world population in 30 years.

81. Highs and Lows. Kuwait, located at the head of the Persian Gulf, has one of the greatest population growth rates in the world. Bulgaria, in southeastern Europe, has one of the smallest. Use an exponential growth/decay model to complete the table.

Country	Population 2010	Annual growth rate	Estimated population 2025
Kuwait	2,789,132	3.501%	
Bulgaria	7,148,785	−0.768%	

Source: CIA World Factbook

82. Bed Bugs. If not checked, the population of a colony of bed bugs will grow exponentially at a rate of 65% per week. If a colony currently has 50 bed bugs, how many will there be in 6 weeks?

83. Epidemics. The spread of hoof-and-mouth disease through a herd of cattle can be modeled by the function $P(t) = 2e^{0.27t}$ (t is in days). If a rancher does not quickly treat the two cows that now have the disease, how many cattle will have the disease in 12 days?

84. Oceanography. The width w (in millimeters) of successive growth spirals of the sea shell *Catapulus voluto,* shown below, is given by the exponential function $w(n) = 1.54e^{0.503n}$ where n is the spiral number. Find the width, to the nearest tenth of a millimeter, of the sixth spiral.

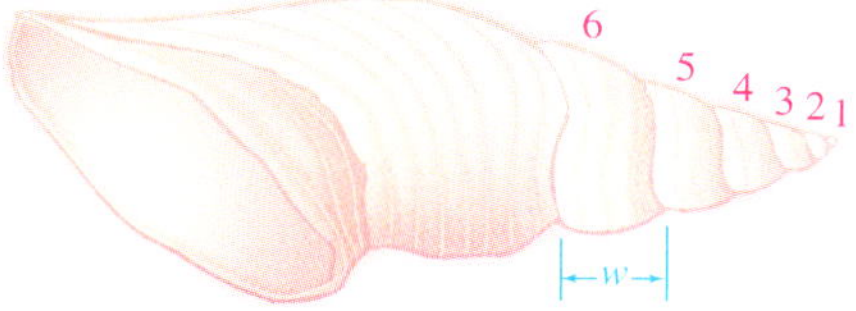

85. Ants. Shortly after an explorer ant discovers a food source, a recruitment process begins in which numerous additional ants travel to the source. The number of ants at the source grows exponentially according to the function $a(t) = 1.36\left(\frac{e}{2.5}\right)^t$, where t is the number of minutes since the explorer discovered the food. How many ants will be at the source in 40 minutes?

86. Half-Life of a Drug. The quantity of a prescription drug in the bloodstream of a patient t hours after it is administered can be modeled by an exponential function. (See the graph.) From the graph determine the time it takes to eliminate half of the initial dose from the body.

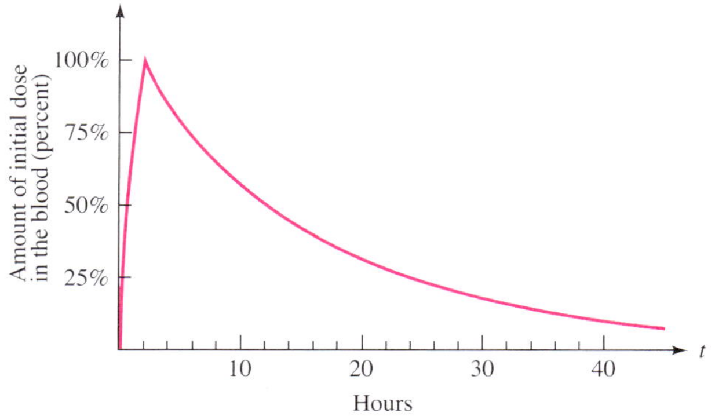

87. from **Campus to Careers**

Social Worker

Social workers often use occupational test results when counseling their clients about employment options. The "learning curve" below shows that as a factory trainee assembled more chairs, the assembly time per chair generally decreased. If company standards required an average assembly time of 10 minutes or less, how many chairs did the trainee have to assemble before meeting company standards? (Notice that the graph is a model of exponential decay.)

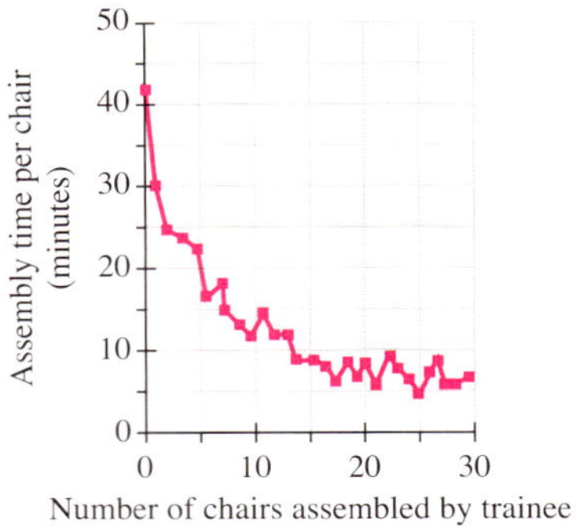

88. Ozone Concentrations. A ***Dobson*** unit is the most basic measure used in ozone research. Roughly 300 Dobson units are equivalent to the height of 2 pennies stacked on top of each other. Suppose the ozone layer thickness (in Dobsons) over a certain city is modeled by the function $A(t) = 300e^{-0.0011t}$, where t is the number of years after 1990. Estimate how thick the ozone layer will be in 2015.

89. Disinfectants. The exponential function $A(t) = 2{,}000{,}000e^{-0.588t}$ approximates the number of germs on a table top, t minutes after disinfectant was sprayed on it. Estimate the germ count on the table 5 minutes after it is sprayed.

90. Medicine. The concentration of a certain prescription drug in an organ after t minutes is modeled by the function $f(t) = 0.08\left(1 - e^{-0.1t}\right)$ where $f(t)$ is the concentration at time t. Find the concentration of the drug at 30 minutes.

91. Sky Diving. Before the parachute opens, a skydiver's velocity in meters per second is modeled by the function $f(t) = 50\left(1 - e^{-0.2t}\right)$ where $f(t)$ is the velocity at time t. Find the velocity after 20 seconds of free fall.

92. Free Fall. After t seconds a certain falling object has a velocity in meters per second given by the function $f(t) = 50\left(1 - e^{-0.3t}\right)$. Which is falling faster after 2 seconds—the object or the skydiver in Exercise 91?

93. The Tarheel State. The 4.3% annual population growth rate for the Raleigh-Cary metropolitan area in North Carolina is one of the largest of any metropolitan area in the United States. If its growth rate remains constant, how long will it take for its population to double? (Source: U.S. Bureau of the Census)

94. The Big Easy. New Orleans has steadily won back some of the population it lost in the wake of Hurricane Katrina in 2005. If the current 8.2% annual increase in population remains constant, how long will it take for its population to double? (Source: money.cnn.com)

95. The Equality State. In 2009, the state with the fastest annual population growth rate was Wyoming. If the 2.13% annual increase in population remains constant, what is the first full year that the population of Wyoming will be double what it was in 2009? (Source: U.S. Bureau of the Census)

96. Doubling Money. How long will it take $1,000 to double if it is invested at an annual rate of 5% compounded continuously?

97. Population Growth. A population growing continuously at an annual rate r will triple in a time t given by the formula $t = \dfrac{\ln 3}{r}$. How long will it take the population of a town to triple if it is growing at the rate of 12% per year?

98. Tripling Money. Find the length of time for $25,000 to triple when it is invested at 6% annual interest, compounded continuously. See Exercise 97.

99. Forensic Medicine. To estimate the number of hours t that a murder victim had been dead, a coroner used the formula $t = \dfrac{1}{0.25} \ln \dfrac{98.6 - T_s}{82 - T_s}$ where T_s is the temperature of the surroundings where the body was found. If the crime took place in an apartment where the thermostat was set at 70°F, approximately how long ago did the murder occur?

100. Making Jello. After the contents of a package of JELL-O are combined with boiling water, the mixture is placed in a refrigerator whose temperature remains a constant 42°F. Estimate the number of hours t that it will take for the JELL-O to cool to 50°F using the formula $t = -\dfrac{1}{0.9} \ln \dfrac{50 - T_r}{200 - T_r}$ where T_r is the temperature of the refrigerator.

101. Cross Country Skiing. The function $H(s) = -47.73 + 107.38 \ln s$ approximates the heart rate (in beats/minute) for an Olympic-class cross country skier traveling at s miles per hour, where $s > 5$ mph. Find the heart rate of a skier traveling at a rate of 7.5 miles per hour. (Source: btc.ontana.edu/Olympics/physiology)

102. Strength Loss. After participating in an eight-week weight training program, a college student was selected to be part of a study to see how much strength he would lose if he discontinued working out. The function $M_B(w) = 225 - 14 \ln (4w + 1)$ approximates his maximum bench press (in pounds) w weeks after stopping weight training. What was his maximum bench press:

a. At the end of the eight-week training course?

b. 6 weeks after stopping the weight training?

103. The Pace of Life. According to the study discussed in Example 7, how much faster does the average pedestrian in New York City (population 8,392,000) walk than the average pedestrian in Atlanta (population 541,000)? (Source: infoplease.com)

104. **Maturity Levels.** The function $P(a) = 41.0 + 20.4 \ln a$ approximates the percent of adult height attained by an early-maturing girl of age a years, for $1 \le a \le 18$. The function $P(a) = 37.5 + 20.2 \ln a$ does the same for a late-maturing girl. Find the difference in percent of their adult height for both maturity types on their 10th birthday. (Source: Growth, Maturation, and Physical Activity, Human Kinetic Books, Robert Malina)

WRITING

105. Explain why the graph of $y = e^x - 5$ is five units below the graph of $y = e^x$.

106. A feature article in a newspaper stated that the sport of snowboarding was growing *exponentially*. Explain what the author of the article meant by that.

107. As of 2007, the population growth rate for Russia was -0.37% annually. What are some of the consequences for a country that has a negative population growth?

108. What is e?

109. Explain the difference between the functions $f(x) = \log x$ and $g(x) = \ln x$.

110. How are the functions $f(x) = \ln x$ and $g(x) = e^x$ related?

111. Explain why $\ln e = 1$.

112. Why is $f(x) = \ln x$ called the natural logarithmic function?

113. A table of values for $f(x) = \ln x$ is shown in figure (a) below. Explain why ERROR appears in the Y_1 column for the first three entries.

114. The graphs of $f(x) = \ln x$, $g(x) = e^x$, and $y = x$ are shown in figure (b) below. Describe the relationship between the graphs in words.

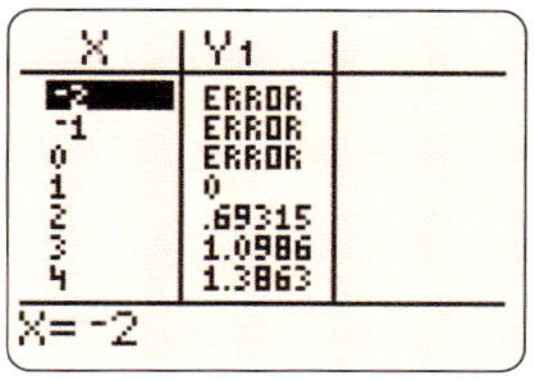

(a)

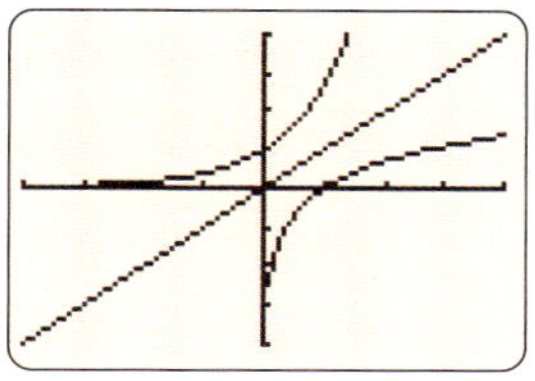

(b)

REVIEW

Simplify each expression. Assume that all variables represent positive numbers.

115. $\sqrt{240x^5}$

116. $\sqrt[3]{-125x^5y^4}$

117. $4\sqrt{48y^3} - 3y\sqrt{12y}$

118. $\sqrt[4]{48z^5} + \sqrt[4]{768z^5}$

CHALLENGE PROBLEMS

119. Without using a calculator, determine whether the statement $e^e > e^3$ is true or false. Explain your reasoning.

120. Graph the function defined by the equation $f(x) = \dfrac{e^x + e^{-x}}{2}$ from $x = -2$ to $x = 2$. The graph will look like a parabola, but it is not. The graph, called a **catenary,** is important in the design of power distribution networks, because it represents the shape of a uniform flexible cable whose ends are suspended from the same height.

121. If $e^{t+5} = ke^t$, find k.

122. If $e^{5t} = k^t$, find k.

123. Use the formula $P = P_0e^{rt}$ to verify that P will be twice P_0 when $t = \frac{\ln 2}{r}$.

124. Use the formula $P = P_0e^{rt}$ to verify that P will be three times as large as P_0 when $t = \frac{\ln 3}{r}$.

125. Use a graphing calculator to graph the function $f(x) = \frac{1}{1 + e^{-2x}}$. Describe its graph in words.

126. **Food Shortages.** Suppose that a country with a population of 1,000 people is growing according to the formula $P = 1{,}000e^{0.02t}$ where t is in years. Furthermore, assume that the food supply F, measured in adequate food per day per person, is growing linearly according to the formula $F = 30.625t + 2{,}000$ (t is time in years). Use a graphing calculator to determine in how many years the population will outstrip the food supply.

SECTION 11.6

Properties of Logarithms

OBJECTIVES

1. Use the four basic properties of logarithms.
2. Use the product rule for logarithms.
3. Use the quotient rule for logarithms.
4. Use the power rule for logarithms.
5. Write logarithmic expressions as a single logarithm.
6. Use the change-of-base formula.
7. Use properties of logarithms to solve application problems.

ARE YOU READY?

The following problems review some basic skills that are needed when working with properties of logarithms.

1. Evaluate: $\log_2 8 + \log_2 1$
2. Evaluate: $\log 10{,}000 - \log 10$
3. Evaluate: $9 \log_3 \frac{1}{3}$
4. Evaluate: $\dfrac{\log_7 49}{\log_7 7}$
5. **a.** Write $\sqrt{x}$ using a fractional exponent.
   b. Write $(x - 2)^{1/2}$ using radical notation.
6. Use a calculator to find ln 5. Round to four decimal places.

Since a logarithm is an exponent, we would expect there to be properties of logarithms just as there are properties of exponents. In this section, we will introduce seven properties of logarithms and use them to simplify and expand logarithmic expressions.

1 Use the Four Basic Properties of Logarithms.

The first four properties of logarithms follow directly from the definition of logarithm.

Properties of Logarithms

For all positive numbers b, where $b \neq 1$,

1. $\log_b 1 = 0$ 2. $\log_b b = 1$ 3. $\log_b b^x = x$ 4. $b^{\log_b x} = x \quad (x > 0)$

We can use the definition of logarithm to prove that these properties are true.

1. $\log_b 1 = 0$, because $b^0 = 1$. Read as "the log base b of 1 equals 0."
2. $\log_b b = 1$, because $b^1 = b$. Read as "the log base b of b equals 1."
3. $\log_b b^x = x$, because $b^x = b^x$. Read as "the log base b of b to the x power equals x."
4. $b^{\log_b x} = x$, because $\log_b x$ is the exponent to which b is raised to get x.
   Read as "b raised to the log base b of x power equals x."

Properties 3 and 4 also indicate that the composition of the exponential and logarithmic functions (in both directions) is the identity function. This is expected, because the exponential and logarithmic functions are inverse functions.

EXAMPLE 1 Simplify: **a.** $\log_5 1$ **b.** $\log_3 3$ **c.** $\ln e^3$ **d.** $6^{\log_6 7}$

Strategy We will compare each logarithmic expression to the left side of the previous four properties of logarithms.

Why When we get a match, the property will provide the answer.

Solution

a. By property 1, $\log_5 1 = 0$, because $5^0 = 1$.

b. By property 2, $\log_3 3 = 1$, because $3^1 = 3$.

c. By property 3, $\ln e^3 = 3$, because $e^3 = e^3$.

d. By property 4, $6^{\log_6 7} = 7$, because $\log_6 7$ is the power to which 6 is raised to get 7.

Self Check 1 Simplify: **a.** $\log_4 1$ **b.** $\log_4 4$ **c.** $\log_2 2^4$ **d.** $5^{\log_5 2}$

Now Try Problems 19, 21, 23, and 27

2 Use the Product Rule for Logarithms.

The next property of logarithms is related to the product rule for exponents: $x^m \cdot x^n = x^{m+n}$.

The Product Rule for Logarithms

The logarithm of a product equals the sum of the logarithms of the factors. For all positive real numbers M, N, and b, where $b \neq 1$,

$$\log_b MN = \log_b M + \log_b N$$

Read as "the log base *b* of *M* times *N* equals the log base *b* of *M* plus the log base *b* of *N*."

As we apply properties of logarithms to rewrite expressions, we assume that all variables represent positive numbers.

EXAMPLE 2 Write each expression as a sum of logarithms. Then simplify, if possible. **a.** $\log_2(2 \cdot 7)$ **b.** $\log 100x$ **c.** $\log_5 125yz$

Strategy In each case, we will use the product rule for logarithms.

Why We use the product rule because each of the logarithmic expressions has the form $\log_b MN$.

Solution

a. To avoid any confusion, the product $2 \cdot 7$ is written within parentheses.

$\log_2(2 \cdot 7) = \log_2 2 + \log_2 7$ Read as "the log base 2 of 2 times 7." The log of the product $2 \cdot 7$ is the sum of the logs of its two factors.

$= 1 + \log_2 7$ Simplify: By property 2, $\log_2 2 = 1$.

b. Recall that $100x$ means $100 \cdot x$.

$\log 100x = \log 100 + \log x$ The log of the product $100x$ is the sum of the logs of its two factors.

$= 2 + \log x$ Simplify: By property 3, $\log 100 = \log 10^2 = 2$.

c. The product $125yz$ has three factors: $125 \cdot y \cdot z$.

$\log_5 125yz = \log_5 125 + \log_5 y + \log_5 z$ The log of the product $125yz$ is the sum of the logs of its three factors.

$= 3 + \log_5 y + \log_5 z$ Simplify: By property 3, $\log_5 125 = \log_5 5^3 = 3$.

Success Tip

Your eyes should go immediately to the "input" of each log function, and then ask:

$\log 100x$

"What are the factors?"

$\log_5 125yz$

Self Check 2 Write each expression as the sum of logarithms. Then simplify, if possible. **a.** $\log_3(3 \cdot 4)$ **b.** $\log 1{,}000y$ **c.** $\log_5 25cd$

Now Try Problems 31 and 35

PROOF To prove the product rule for logarithms, we let $x = \log_b M$, $y = \log_b N$, and use the definition of logarithm to write each equation in exponential form.

$$M = b^x \text{ and } N = b^y$$

Then $MN = b^x b^y$, and a property of exponents gives

$MN = b^{x+y}$ Keep the base and add the exponents: $b^x b^y = b^{x+y}$.

We write this exponential equation in logarithmic form as

$$\log_b MN = x + y$$

Substituting the values of x and y completes the proof.

$\log_b MN = \log_b M + \log_b N$ This is the product rule for logarithms.

Caution

The log of a sum **does not equal** the sum of the logs. The log of a difference **does not equal** the difference of the logs.

CAUTION By the product rule, the logarithm of a *product* is equal to the *sum* of the logarithms. The logarithm of a sum or a difference usually does not simplify. **Do not incorrectly apply the distributive property in such cases.** In general,

$$\log_b (M + N) \neq \log_b M + \log_b N \quad \text{and} \quad \log_b (M - N) \neq \log_b M - \log_b N$$

For example,

$$\log_2 (2 + 7) \neq \log_2 2 + \log_2 7 \quad \text{and} \quad \log (100 - y) \neq \log 100 - \log y$$

Using Your Calculator ▶ Verifying Properties of Logarithms

We can use a calculator to illustrate the product rule for logarithms by showing that

$$\log (3.7 \cdot 15.9) = \log 3.7 + \log 15.9$$

We calculate the left and right sides of the equation separately and compare the results. To use a scientific calculator to find $\log (3.7 \cdot 15.9)$, we enter

3.7 [×] 15.9 [=] [LOG]   `1.769598848`

To find $\log 3.7 + \log 15.9$, we enter

3.7 [LOG] [+] 15.9 [LOG] [=]   `1.769598848`

Since the screen displays for the left and right sides are equal, that would suggest that the equation $\log (3.7 \cdot 15.9) = \log 3.7 + \log 15.9$ is true.

3 Use the Quotient Rule for Logarithms.

The next property of logarithms is related to the quotient rule for exponents: $\frac{x^m}{x^n} = x^{m-n}$.

The Quotient Rule for Logarithms

The logarithm of a quotient equals the difference of the logarithms of the numerator and denominator. For all positive real numbers M, N, and b, where $b \neq 1$,

$$\log_b \frac{M}{N} = \log_b M - \log_b N$$

Read as "the log base b of M divided by N equals the log base b of M minus the log base b of N."

The proof of the quotient rule for logarithms is similar to the proof for the product rule for logarithms.

EXAMPLE 3 Write each expression as a difference of logarithms. Then simplify, if possible.

a. $\ln \frac{10}{7}$ **b.** $\log_4 \frac{x}{64}$

Strategy In both cases, we will apply the quotient rule for logarithms.

Why We use the quotient rule because each of the logarithmic expressions has the form $\log_b \frac{M}{N}$.

Solution

a. $\ln \frac{10}{7} = \ln 10 - \ln 7$ Recall that $\ln \frac{10}{7}$ is a natural logarithm and means $\log_e \frac{10}{7}$. The log of the quotient $\frac{10}{7}$ is the difference of the logs of its numerator and denominator.

b. $\log_4 \frac{x}{64} = \log_4 x - \log_4 64$ The log of the quotient $\frac{x}{64}$ is the difference of the logs of its numerator and denominator.

$= \log_4 x - 3$ Simplify: $\log_4 64 = \log_4 4^3 = 3$.

Self Check 3 Write each expression as a difference of logarithms. Then simplify, if possible. **a.** $\log_6 \frac{6}{5}$ **b.** $\ln \frac{y}{100}$

Now Try Problem 39

CAUTION By the quotient rule, the logarithm of a *quotient* is equal to the *difference* of the logarithms. The logarithm of a quotient is not the quotient of the logarithms:

$$\log_b \frac{M}{N} \neq \frac{\log_b M}{\log_b N}$$

For example,

$$\ln \frac{10}{7} \neq \frac{\ln 10}{\ln 7} \quad \text{and} \quad \log_4 \frac{x}{64} \neq \frac{\log_4 x}{\log_4 64}$$

Using a calculator, we have:
$\ln \frac{10}{7} \approx 0.356674944$
$\frac{\ln 10}{\ln 7} \approx 1.183294662$

In the next example, the product and quotient rules for logarithms are used in combination to rewrite an expression.

EXAMPLE 4 Write $\log \frac{xy}{10z}$ as the sum and/or difference of logarithms of a single quantity. Then simplify, if possible.

Strategy We will use the quotient rule for logarithms and then the product rule.

Why We use the quotient rule because $\log \frac{xy}{10z}$ has the form $\log_b \frac{M}{N}$. We later use the product rule because the numerator and denominator of $\frac{xy}{10z}$ contain products.

Solution We begin by applying the quotient rule for logarithms.

$\log \frac{xy}{10z} = \log xy - \log 10z$ The log of a quotient is the difference of the logs.

$= \log x + \log y - (\log 10 + \log z)$ The log of a product is the sum of the logs.
Write parentheses here so that both terms of the sum $\log 10 + \log z$ are subtracted.

$= \log x + \log y - \log 10 - \log z$ Change the sign of each term of $\log 10 + \log z$ and drop the parentheses.

$= \log x + \log y - 1 - \log z$ Simplify: $\log 10 = 1$.

Success Tip

Your eyes should go immediately to the "input" of the log function, and identify any quotients and/or products.

$$\log \frac{xy}{10z}$$

Self Check 4 Write $\log_b \frac{x}{yz}$ as the sum and/or difference of logarithms of a single quantity. Then simplify, if possible.

Now Try Problem 45

4 Use the Power Rule for Logarithms.

The next property of logarithms is related to the power rule for exponents: $(x^m)^n = x^{mn}$.

The Power Rule for Logarithms

The logarithm of a number raised to a power equals the power times the logarithm of the number.

For all positive real numbers M and b, where $b \neq 1$, and any real number p,

$$\log_b M^p = p \log_b M$$

Read as "the log base b of M to the p power equals p times the log base b of M."

EXAMPLE 5 Write each logarithm without an exponent or a square root: **a.** $\log_5 6^2$ **b.** $\log \sqrt{10}$

Strategy In each case, we will use the power rule for logarithms.

Why We use the power rule because $\log_5 6^2$ has the form $\log_b M^p$, as will $\log \sqrt{10}$ if we write $\sqrt{10}$ as $10^{1/2}$.

Solution

a. $\log_5 6^2 = 2 \log_5 6$ The log of a power is equal to the power times the log. Write the exponent 2 in front of $\log_5 6$.

b. $\log \sqrt{10} = \log 10^{1/2}$ Write $\sqrt{10}$ using a fractional exponent: $\sqrt{10} = (10)^{1/2}$.

$= \frac{1}{2} \log 10$ The log of a power is equal to the power times the log. Write the exponent $\frac{1}{2}$ in front of log 10.

$= \frac{1}{2}$ Simplify: $\log 10 = 1$.

Self Check 5 Write each logarithm without an exponent or a cube root:

a. $\ln x^4$ **b.** $\log_2 \sqrt[3]{3}$

Now Try Problems 51 and 53

PROOF

To prove the power rule, we let $x = \log_b M$, write the expression in exponential form, and raise both sides to the pth power:

$M = b^x$

$(M)^p = (b^x)^p$ Raise both sides to the pth power.

$M^p = b^{px}$ Keep the base and multiply the exponents.

Using the definition of logarithms gives

$\log_b M^p = px$

Substituting $\log_b M$ for x completes the proof.

$\log_b M^p = p \log_b M$ This is the power rule.

It is often necessary to use more than one rule for logarithms to **expand a logarithmic expression.**

EXAMPLE 6 Write each expression as the sum and/or difference of logarithms of a single quantity:

a. $\log_b x^2 y^3 z$ **b.** $\ln \frac{y^3 \sqrt[4]{x}}{z}$

Strategy In part (a), we will use the product rule and the power rules for logarithms. In part (b), we will use the quotient rule, the product rule, and the power rule for logarithms.

Why In part (a), we first use the product rule because the expression has the form $\log_b MN$. In part (b), we first use the quotient rule because the expression has the form $\log_b \frac{M}{N}$.

Solution

The Language of Algebra

In Examples 2, 3, 4, and 6, we use properties of logarithms to **expand** single logarithmic expressions. For example:

Expand →

$\log_b MN = \log_b M + \log_b N$

a. The expression $\log_b x^2y^3z$ is the logarithm of a product.

$$\log_b x^2y^3z = \log_b x^2 + \log_b y^3 + \log_b z \quad \text{The log of a product is the sum of the logs.}$$
$$= 2\log_b x + 3\log_b y + \log_b z \quad \text{The log of a power is the power times the log.}$$

Success Tip

Your eyes should go immediately to the "input" of the log function, and identify any quotients, products, powers, or roots.

$\ln \frac{y^3\sqrt[4]{x}}{z}$ ↑

b. The expression $\ln \frac{y^3\sqrt[4]{x}}{z}$ is the logarithm of a quotient.

$$\ln \frac{y^3\sqrt[4]{x}}{z} = \ln y^3\sqrt[4]{x} - \ln z \quad \text{This is a natural log expression. The log of a quotient is the difference of the logs.}$$
$$= \ln y^3 + \ln \sqrt[4]{x} - \ln z \quad \text{The log of a product is the sum of the logs.}$$
$$= \ln y^3 + \ln x^{1/4} - \ln z \quad \text{Write } \sqrt[4]{x} \text{ as } x^{1/4}.$$
$$= 3\ln y + \frac{1}{4}\ln x - \ln z \quad \text{The log of a power is the power times the log.}$$

Self Check 6 Expand: $\log \sqrt[4]{\frac{x^3y}{z}}$

Now Try Problems 59 and 61

5 Write Logarithmic Expressions as a Single Logarithm.

We also can use properties of logarithms in reverse to **condense logarithmic expressions** having two or more terms.

EXAMPLE 7 Write each logarithmic expression as one logarithm:

a. $16\log_8 x + \frac{1}{3}\log_8 y$ **b.** $\frac{1}{2}\log_b(x-2) - \log_b y + 3\log_b z$

Strategy In part (a), we will use the power rule and product rule for logarithms in reverse. In part (b), we will use the power rule, the quotient rule, and the product rule for logarithms in reverse.

Why We use the power rule because we see expressions of the form $p\log_b M$. The $+$ symbol between logarithmic terms suggests that we use the product rule and the $-$ symbol between such terms suggests that we use the quotient rule.

Solution

Success Tip

When condensing logarithmic expressions, you should first identify the coefficients of any logarithmic terms, and then apply the power rule in reverse.

$16\log_8 x + \frac{1}{3}\log_8 y$ (arrows ↑ under 16 and $\frac{1}{3}$)

a. We begin by using the power rule on both terms of the expression.

$$16\log_8 x + \frac{1}{3}\log_8 y = \log_8 x^{16} + \log_8 y^{1/3} \quad \text{A power times a log is the log of the power.}$$
$$= \log_8 (x^{16} \cdot y^{1/3}) \quad \text{The sum of two logs is the log of the product.}$$
$$= \log_8 x^{16}y^{1/3}$$
$$= \log_8 x^{16}\sqrt[3]{y} \quad \text{Write } y^{1/3} \text{ using radical notation: } \sqrt[3]{y}.$$

The Language of Algebra

In these examples, we use properties of logarithms in "reverse" to **condense** the given expression into a single logarithmic expression. For example:

Condense

$\log_b MN = \log_b M + \log_b N$

b. The first and third terms of this expression can be rewritten using the power rule of logarithms. Note that the base of each logarithm is b. We do not need to know the value of b to apply properties of logarithms.

$$\frac{1}{2}\log_b (x-2) - \log_b y + 3\log_b z$$

$$= \log_b (x-2)^{1/2} - \log_b y + \log_b z^3$$ A power times a log is the log of the power.

$$= \log_b \frac{(x-2)^{1/2}}{y} + \log_b z^3$$ The difference of two logs is the log of the quotient.

$$= \log_b \frac{\sqrt{x-2}}{y} + \log_b z^3$$ Write $(x-2)^{1/2}$ using a radical notation: $\sqrt{x-2}$.

$$= \log_b \left(\frac{\sqrt{x-2}}{y} \cdot z^3\right)$$ The sum of two logs is the log of the product.

$$= \log_b \frac{z^3\sqrt{x-2}}{y}$$

Self Check 7 Write the expression as one logarithm:

$$2\log_a x + \frac{1}{2}\log_a y - 2\log_a (x-y)$$

Now Try Problems 75 and 79

The properties of logarithms can be used when working with numerical values.

EXAMPLE 8 If $\log 2 \approx 0.3010$ and $\log 3 \approx 0.4771$, use properties of logarithms to find approximations for: **a.** $\log 6$ **b.** $\log 18$

Strategy We will express 6 and 18 using factors of 2 and 3 and then use properties of logarithms to simplify each resulting expression.

Why We express 6 and 18 using factors of 2 and 3 because we are given values of log 2 and log 3.

Solution

a. $\log 6 = \log (2 \cdot 3)$ Write 6 using the factors 2 and 3.

$= \log 2 + \log 3$ The log of a product is the sum of the logs.

$\approx 0.3010 + 0.4771$ Substitute the given values for log 2 and log 3. We must now use an $\approx$ symbol. Do the addition.

≈ 0.7781

b. $\log 18 = \log (2 \cdot 3^2)$ Write 18 using the factors 2 and 3.

$= \log 2 + \log 3^2$ The log of a product is the sum of the logs.

$= \log 2 + 2\log 3$ The log of a power is the power times the log. Write the exponent 2 in front of log 3.

$\approx 0.3010 + 2(0.4771)$ Substitute the given value for log 2 and log 3. We must now use an $\approx$ symbol.

≈ 1.2552 Evaluate the expression.

Self Check 8 See Example 8. Find approximations for each logarithm.

a. $\log 1.5$ **b.** $\log 0.75$

Now Try Problems 87 and 89

We summarize the properties of logarithms as follows.

Properties of Logarithms

If b, M, and N are positive real numbers, $b \neq 1$, and p is any real number,

1. $\log_b 1 = 0$
2. $\log_b b = 1$
3. $\log_b b^x = x$
4. $b^{\log_b x} = x$
5. $\log_b MN = \log_b M + \log_b N$
6. $\log_b \frac{M}{N} = \log_b M - \log_b N$
7. $\log_b M^p = p \log_b M$

6 Use the Change-of-Base Formula.

Most calculators can find common logarithms (base 10) and natural logarithms (base e). If we need to find a logarithm with some other base, we can use the following **change-of-base formula.**

Change-of-Base Formula

For any logarithmic bases a and b, and any positive real number x,

$$\log_b x = \frac{\log_a x}{\log_a b}$$

This formula converts a logarithm of one base to a ratio of logarithms of a different base.

We can use any positive number other than 1 for base a in the change-of-base formula. However, we usually use 10 or e because of the capabilities of a standard calculator.

EXAMPLE 9 Find: $\log_3 5$

Strategy To evaluate this base-3 logarithm, we will substitute into the change-of-base formula.

Why We assume that the reader does not have a calculator that evaluates base-3 logarithms (at least not directly). Thus, the only alternative is to change the base.

Solution To find $\log_3 5$, we substitute 3 for b, 10 for a, and 5 for x in the change-of-base formula and simplify:

$$\log_b x = \frac{\log_a x}{\log_a b}$$ This is the change-of-base formula.

$$\log_3 5 = \frac{\log_{10} 5}{\log_{10} 3}$$ The old base is 3. The new base we want to introduce is 10. Substitute: $b = 3$, $x = 5$, and $a = 10$.

$$\approx 1.464973521$$ Approximate. On a scientific calculator, enter: 5 LOG ÷ 3 LOG = .

To four decimal places, $\log_3 5 = 1.4650$. To check this result, use a calculator to verify that $3^{1.4650} \approx 5$.

Caution

Don't misapply the quotient rule: $\frac{\log_{10} 5}{\log_{10} 3}$ means $\log_{10} 5 \div \log_{10} 3$. It is the expression $\log_{10} \frac{5}{3}$ that means $\log_{10} 5 - \log_{10} 3$.

Caution

Wait until the final calculation has been made to round. Don't round any values when performing intermediate calculations. That could make the final result incorrect because of a build-up of rounding errors.

We also can use the natural logarithm function (base e) in the change-of-base formula to find a base-3 logarithm.

$$\log_b x = \frac{\log_a x}{\log_a b}$$ This is the change-of-base formula.

$$\log_3 5 = \frac{\log_e 5}{\log_e 3}$$ The old base is 3. The new base we want to introduce is e. Substitute: $b = 3$, $x = 5$, and $a = e$.

$$\log_3 5 = \frac{\ln 5}{\ln 3}$$ Write $\log_e 5$ as $\ln 5$ and $\log_e 3$ as $\ln 3$.

$$\approx 1.464973521$$ Approximate. On a scientific calculator, enter: 5 LN ÷ 3 LN = .

We obtain the same result.

Self Check 9 Find $\log_5 3$ to four decimal places.

Now Try Problem 95

PROOF To prove the change-of-base formula, we begin with the equation $\log_b x = y$.

$$y = \log_b x$$

$$x = b^y$$ Change the equation from logarithmic to exponential form.

$$\log_a x = \log_a b^y$$ Take the base-a logarithm of both sides.

$$\log_a x = y \log_a b$$ The log of a power is the power times the log.

$$y = \frac{\log_a x}{\log_a b}$$ Divide both sides by $\log_a b$.

$$\log_b x = \frac{\log_a x}{\log_a b}$$ Refer to the first equation, $y = \log_b x$ and substitute $\log_b x$ for y. This is the change-of-base formula.

7 Use Properties of Logarithms to Solve Application Problems.

In chemistry, common logarithms are used to express how basic or acidic a solution is. The more acidic a solution, the greater the concentration of hydrogen ions. (A **hydrogen ion** is a positively charged hydrogen atom missing its electron.) The concentration of hydrogen ions in a solution is commonly measured using the **pH scale.** The pH of a solution is defined as follows.

pH of a Solution

If $[H^+]$ is the hydrogen ion concentration in gram-ions per liter, then

$$\text{pH} = -\log[H^+]$$

EXAMPLE 10 **pH Meters.** One of the most accurate ways to measure pH is with a probe and meter. What reading should the meter give for lemon juice if it has a hydrogen ion concentration $[H^+]$ of approximately 6.2×10^{-3} gram-ions per liter?

Strategy We will substitute into the formula for pH and use the power rule for logarithms to simplify the right side.

Why After substituting 6.2×10^{-3} for $[H^+]$ in $-\log[H^+]$, the resulting expression will have the form $\log_b M^p$.

Solution Since lemon juice has approximately 6.2×10^{-3} gram-ions per liter, its pH is

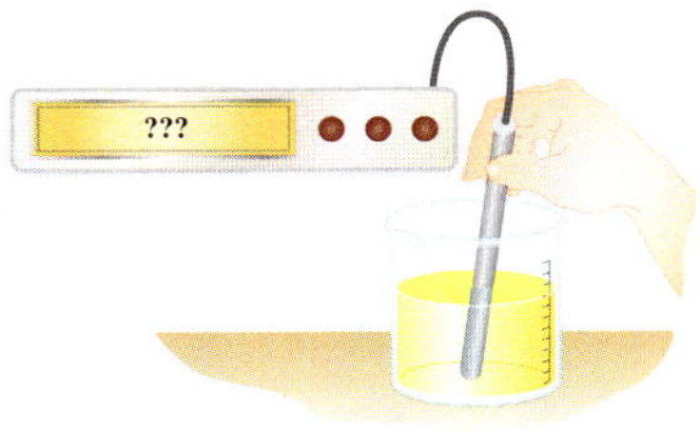

$\text{pH} = -\log[\mathbf{H^+}]$	This is the formula for pH. Read as "the opposite of the log of the hydrogen ion concentration."
$\text{pH} = -\log(\mathbf{6.2 \times 10^{-3}})$	Substitute 6.2×10^{-3} for $[H^+]$.
$= -(\log 6.2 + \log 10^{-3})$	The log of a product is the sum of the logs.
$= -[\log 6.2 + (-3)\,\mathbf{\log 10}]$	The log of a power is the power times the log.
$= -[\log 6.2 + (-3) \cdot \mathbf{1}]$	Evaluate: $\log 10 = 1$.
≈ 2.207608311	Use a calculator.
≈ 2.2	Round to the nearest tenth.

The meter should give a reading of approximately 2.2.

Self Check 10 **pH Meters.** The hydrogen ion concentration range for a freshwater aquarium has a low-end value of 2.5×10^{-8}. Find the pH level that corresponds to this value.

Now Try ▶ Problem 107

SECTION 11.6 STUDY SET

VOCABULARY

Fill in the blanks.

1. The logarithm of a ________, such as $\log_3 4x$, equals the sum of the logarithms of the factors.
2. The logarithm of a ________, such as $\log_2 \frac{5}{x}$, equals the difference of the logarithms of the numerator and denominator.
3. The logarithm of a number to a ______, such as $\log_4 5^3$, equals the power times the logarithm of the number.
4. The ________-of-base formula converts a logarithm of one base to a ratio of logarithms of a different base.

CONCEPTS

Fill in the blanks. In problem 6, also give the name of each rule.

5. a. $\log_b 1 =$ b. $\log_b b =$
   c. $\log_b b^x =$ d. $b^{\log_b x} =$
6. a. $\log_b MN = \log_b \quad + \log_b \quad$ ________ rule
   b. $\log_b \frac{M}{N} = \log_b M \quad \log_b N$ ________ rule
   c. $\log_b M^p = p \log_b \quad$ ________ rule
   d. $\log_b x = \frac{\log_a \quad}{\log \quad b}$ ________-of-______ rule

Use a calculator to verify that each equation is true. See Using Your Calculator: Verifying Properties of Logarithms.

7. $\log(2.5 \cdot 3.7) = \log 2.5 + \log 3.7$
8. $\ln(2.25)^4 = 4 \ln 2.25$
9. $\ln \frac{11.3}{6.1} = \ln 11.3 - \ln 6.1$
10. $\log \sqrt{24.3} = \frac{1}{2} \log 24.3$

Match each expression with an equivalent expression from the list on the right.

11. $\log_3 10$ a. $\frac{\log 11}{\log 3}$
12. $\log_3 \frac{10}{11}$ b. $11 \log_3 10$
13. $\log_3 10^{11}$ c. $\log_3 5 + \log_3 2$
14. $\log_3 11$ d. $\log_3 10 - \log_3 11$

NOTATION

Complete each solution.

15. $\log_8 8a^3 = \log_8 \quad + \log_8 \quad$
    $= \log_8 8 + \quad \log_8 a$
    $= \quad + 3 \log_8 a$

16. $\log \frac{r}{st} = \log r - \log(\quad)$
    $= \log r - (\log \quad + \log t)$
    $= \log r - \log s$

17. True or False?

$$\log 10{,}000x = \log(10{,}000x)$$

18. Fill in the blanks:

a. $y^{1/3} = \sqrt{}$ b. $\sqrt[5]{x} = x$

GUIDED PRACTICE

In this Study Set, assume that all variables represent positive numbers and b ≠ 1.

Evaluate each expression. See Example 1.

19. $\log_6 1$
20. $\log_9 9$
21. $\log_4 4^7$
22. $\ln e^8$
23. $5^{\log_5 10}$
24. $8^{\log_8 10}$
25. $\log_5 5^2$
26. $\log_4 4^2$
27. $\ln e$
28. $\log_7 1$
29. $\log_3 3^7$
30. $5^{\log_5 8}$

Write each logarithm as a sum. Then simplify, if possible. See Example 2.

31. $\log_2 (4 \cdot 5)$
32. $\log_3 (27 \cdot 5)$
33. $\log 25y$
34. $\log xy$
35. $\log 100pq$
36. $\log 1{,}000rs$
37. $\log 5xyz$
38. $\log 10abc$

Write each logarithm as a difference. Then simplify, if possible. See Example 3.

39. $\log \dfrac{100}{9}$
40. $\ln \dfrac{27}{e}$
41. $\log_6 \dfrac{x}{36}$
42. $\log_8 \dfrac{y}{8}$

Write each logarithm as the sum and/or difference of logarithms of a single quantity. Then simplify, if possible. See Example 4.

43. $\log \dfrac{7c}{2}$
44. $\log \dfrac{9t}{4}$
45. $\log \dfrac{10x}{y}$
46. $\log_2 \dfrac{ab}{4}$
47. $\ln \dfrac{exy}{z}$
48. $\ln \dfrac{5p}{e}$
49. $\log_8 \dfrac{1}{8m}$
50. $\log_6 \dfrac{1}{36r}$

Write each logarithm without an exponent or a radical symbol. Then simplify, if possible. See Example 5.

51. $\ln y^7$
52. $\ln z^9$
53. $\log \sqrt{5}$
54. $\log \sqrt[3]{7}$
55. $\log e^{-3}$
56. $\log e^{-1}$
57. $\log_7 \left(\sqrt[5]{100}\right)^3$
58. $\log_3 \left(\sqrt{10}\right)^5$

Write each logarithm as the sum and/or difference of logarithms of a single quantity. Then simplify, if possible. See Example 6.

59. $\log xyz^2$
60. $\log 4xz^2$
61. $\log_2 \dfrac{2\sqrt[3]{x}}{y}$
62. $\log_3 \dfrac{\sqrt[4]{x}}{yz}$
63. $\log x^3y^2$
64. $\log xy^2z^3$
65. $\log_b \sqrt{xy}$
66. $\log_b x^3\sqrt{y}$
67. $\log_a \dfrac{\sqrt[3]{x}}{\sqrt[4]{yz}}$
68. $\log_b \sqrt[4]{\dfrac{x^3y^2}{z^4}}$
69. $\ln x^{20}\sqrt{z}$
70. $\ln \sqrt{xy}$
71. $\log_5 \left(\dfrac{1}{t^3}\right)^d$
72. $\log_6 \left(\dfrac{1}{x^4}\right)^t$
73. $\ln \sqrt{ex}$
74. $\ln \sqrt[3]{e^2x}$

Write each logarithmic expression as one logarithm. See Example 7.

75. $\log_2 (x + 1) + 9 \log_2 x$
76. $2 \log x + \dfrac{1}{2} \log y$
77. $\log_3 x + \log_3 (x + 2) - \log_3 8$
78. $-2 \log x - 3 \log y + \log z$
79. $-3 \log_b x - 2 \log_b y + \dfrac{1}{2} \log_b z$
80. $3 \log_b (x + 1) - 2 \log_b (x + 2) + \log_b x$
81. $\dfrac{1}{3}\left[\log_b (M^2 - 9) - \log_b (M + 3)\right]$
82. $\dfrac{1}{4}\left[\log_r (n^2 - 16) - \log_r (n - 4)\right]$
83. $\ln\left(\dfrac{x}{z} + x\right) - \ln\left(\dfrac{y}{z} + y\right)$
84. $\ln (xy + y^2) - \ln (xz + yz) + \ln z$
85. $\dfrac{1}{2} \log_6 (x^2 + 1) - \log_6 (x^2 + 2)$
86. $\dfrac{1}{2} \log_8 (x^2 + 5) - \log_8 (x^2 + 5)$

Assume that $\log 4 \approx 0.6021$, $\log 7 \approx 0.8451$, and $\log 9 \approx 0.9542$. Use these values to evaluate each logarithm. See Example 8.

87. $\log_b 28$
88. $\log_b \dfrac{7}{4}$
89. $\log_b \dfrac{4}{63}$
90. $\log_b 36$

91. $\log_b \frac{63}{4}$

92. $\log_b 2.25$

93. $\log_b 64$

94. $\log_b 49$

Use the change-of-base formula to find each logarithm to four decimal places. See Example 9.

95. $\log_3 7$

96. $\log_7 3$

97. $\log_{1/3} 3$

98. $\log_{1/2} 6$

99. $\log_3 8$

100. $\log_5 10$

101. $\log_{\sqrt{2}} \sqrt{5}$

102. $\log_\pi e$

Look Alikes . . .

Which pair of expressions in each list are equivalent?

103. a. $\log (9 \cdot 3)$ b. $\log 9 \cdot \log 3$ c. $\log 9 + \log 3$

104. a. $\log_6 \frac{7}{9}$ b. $\frac{\log_6 7}{\log_6 9}$ c. $\log_6 7 - \log_6 9$

105. a. $\log_2 11^4$ b. $4 \log_2 11$ c. $(\log_2 11)^4$

106. a. $\ln \sqrt{t}$ b. $\sqrt{\ln t}$ c. $\frac{1}{2} \ln t$

APPLICATIONS

107. **pH of a Solution.** Find the pH of a solution with a hydrogen ion concentration of 1.7×10^{-5} gram-ions per liter.

108. **pH of Pickles.** The hydrogen ion concentration of sour pickles is 6.31×10^{-4}. Find the pH.

109. **Formulas.** Use properties of logarithms to write the right side of each formula in an equivalent condensed form.

a. From sound engineering: $B = 10(\log I - \log I_0)$

b. From medicine: $T = \frac{1}{k}(\ln C_2 - \ln C_1)$

110. **Doubling Time.** The formula $t = \frac{\ln 2}{r}$ gives the time t for a population to double, where r is the annual rate of continuous compounding. Write the formula in an equivalent form so that it involves a common logarithm, not a natural logarithm.

WRITING

111. Explain the difference between a logarithm of a product and the product of logarithms.

112. How can the LOG key on a calculator be used to find $\log_2 7$?

Explain why each statement is false.

113. $\log AB = (\log A)(\log B)$

114. $\log (A + B) = \log A + \log B$

115. $\log_b (A - B) = \frac{\log_b A}{\log_b B}$

116. $\frac{\log_b A}{\log_b B} = \log_b A - \log_b B$

117. Explain the meaning of the arrow:

$\log 7^{15}$ (arrow from the exponent 15 to the front of log)

118. When is the change-of-base formula helpful?

REVIEW

Consider the line that passes through P(−2, 3) and Q(4, −4).

119. Find the slope of line PQ.

120. Find the distance between P and Q.

121. Find the midpoint of line segment PQ.

122. Write the equation in slope–intercept form of line PQ.

CHALLENGE PROBLEMS

123. Explain why $e^{\ln x} = x$.

124. If $\log_b 3x = 1 + \log_b x$, find b.

125. Show that $\log_{b^2} x = \frac{1}{2} \log_b x$.

126. Show that $e^{x \ln a} = a^x$.

SECTION 11.7 Exponential and Logarithmic Equations

OBJECTIVES

1 Solve exponential equations.

2 Solve logarithmic equations.

3 Use exponential and logarithmic equations to solve application problems.

ARE YOU READY?

The following problems review some basic skills that are needed when solving exponential and logarithmic equations.

1. Fill in the blanks: **a.** $16 = 2$ **b.** $\frac{1}{125} = 5$

2. Find $\frac{\log 12}{\log 9}$. Round to four decimal places.

3. Use the power rule for logarithms: $\log_3 8^x$

4. Evaluate: $\ln e$

5. Use a property of logarithms to write each expression as a single logarithm:

a. $\log_2 5 + \log_2 x$

b. $\ln 10 - \ln (2t + 1)$

6. Evaluate: $\log (-6)$

In earlier chapters, we have solved linear equations, absolute value equations, quadratic equations, rational equations, and radical equations. In this section, we will solve two new types of equations: exponential equations and logarithmic equations.

1 Solve Exponential Equations.

An **exponential equation** contains a variable in one of its exponents. Some examples of exponential equations are

$$3^{x+1} = 81, \qquad 6^{x-3} = 2^x, \qquad \text{and} \qquad e^{0.9t} = 8$$

If both sides of an exponential equation can be expressed as a power of the same base, we can use the following property to solve it.

Exponent Property of Equality

If two exponential expressions with the same base are equal, their exponents are equal. For any real number b, where $b \neq -1, 0,$ or 1,

$$b^x = b^y \quad \text{is equivalent to} \quad x = y$$

EXAMPLE 1 Solve: $3^{x+1} = 81$

Strategy We will express the right side of the equation as a power of 3.

Why If each side of the equation is expressed as a power of the *same* base (in this case, 3), we can use the exponent property of equality to set the exponents equal and solve for x.

Solution

$3^{x+1} = 81$ This is the equation to solve.

$3^{x+1} = 3^4$ Write 81 as a power of 3: $81 = 3^4$.

$x + 1 = 4$ If two exponential expressions with the same base are equal, their exponents are equal.

$x = 3$ Solve for x by subtracting 1 from both sides.

The solution is 3 and the solution set is $\{3\}$. To check this result, we substitute 3 for x in the original equation.

Check:

$$3^{x+1} = 81$$
$$3^{3+1} \stackrel{?}{=} 81$$
$$3^4 \stackrel{?}{=} 81$$
$$81 = 81 \quad \text{True}$$

Self Check 1 Solve: $5^{3x-4} = 25$

Now Try Problem 21

EXAMPLE 2 Solve: $2^{x^2+2x} = \frac{1}{2}$

Strategy We will express the right side of the equation as a power of 2.

Why If each side of the equation is expressed as a power of the same base (in this case, 2), we can use the exponent property of equality to set the exponents equal and solve for x.

Solution

$2^{x^2+2x} = \frac{1}{2}$ This is the equation to solve.

$2^{x^2+2x} = 2^{-1}$ Write $\frac{1}{2}$ as a power of 2: $\frac{1}{2} = 2^{-1}$.

$x^2 + 2x = -1$ If two exponential expressions with the same base are equal, their exponents are equal.

$x^2 + 2x + 1 = 0$ Add 1 to both sides.

$(x + 1)(x + 1) = 0$ Factor the trinomial.

$x + 1 = 0$ or $x + 1 = 0$ Set each factor equal to 0.

$x = -1$ | $x = -1$ Solve each linear equation.

We see that the two solutions are the same. Thus, -1 is a repeated solution and the solution set is $\{-1\}$. Verify that -1 satisfies the original equation.

Self Check 2 Solve: $3^{x^2-2x} = \frac{1}{3}$

Now Try Problem 25

Using Your Calculator ▶ Solving Exponential Equations Graphically

To use a graphing calculator to approximate the solutions of $2^{x^2+2x} = \frac{1}{2}$ (see Example 2), we can subtract $\frac{1}{2}$ from both sides of the equation to get $2^{x^2+2x} - \frac{1}{2} = 0$ and graph the corresponding function $f(x) = 2^{x^2+2x} - \frac{1}{2}$ as shown in figure (a).

The solutions of $2^{x^2+2x} - \frac{1}{2} = 0$ are the x-coordinates of the x-intercepts of the graph of $f(x) = 2^{x^2+2x} - \frac{1}{2}$. Using the ZERO feature, we see in figure (a) that the graph has only one x-intercept, $(-1, 0)$. Therefore, -1 is the only solution of $2^{x^2+2x} - \frac{1}{2} = 0$.

We also can solve $2^{x^2+2x} = \frac{1}{2}$ using the INTERSECT feature found on most graphing calculators. After graphing $Y_1 = 2^{x^2+2x}$ and $Y_2 = \frac{1}{2}$, we select INTERSECT, which approximates the coordinates of the point of intersection of the two graphs. From the display shown in figure (b), we can conclude that the solution is -1. Verify this by checking.

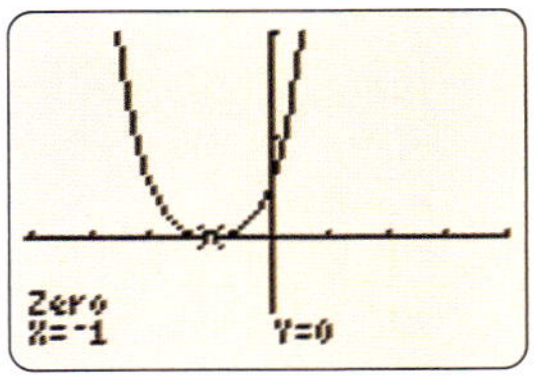

(a)

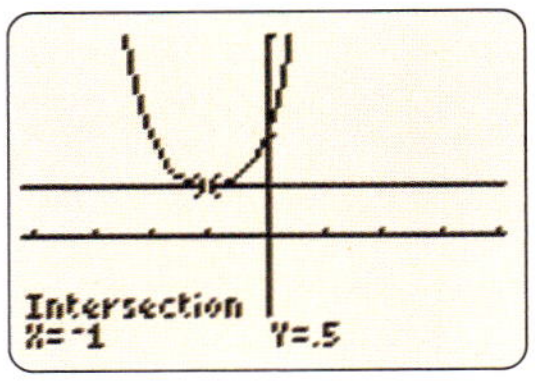

(b)

When it is difficult or impossible to write each side of an exponential equation as a power of the same base, we often can use the following property of logarithms to solve the equation.

Logarithm Property of Equality

If two positive numbers are equal, the logarithms base-b of the numbers are equal. For any positive number b, where $b \neq 1$, and positive numbers x and y,

$\log_b x = \log_b y$ is equivalent to $x = y$

EXAMPLE 3 Solve: $3^x = 5$

Strategy We will take the base-10 logarithm on both sides of the equation.

Why We can then use the power rule of logarithms to move the variable x from its current position as an exponent to a position as a factor.

Solution Unlike Example 1, where we solved $3^{x+1} = 81$, it is not possible to write each side of $3^x = 5$ as an integer power of the same base 3. Instead, we use the logarithm property of equality and *take the logarithm on each side* to solve the equation. Although any base logarithm can be chosen, the calculations with a calculator are usually simplest if we use a common or natural logarithm.

The Language of Algebra

Some instructors phrase the first step of the solution process as "taking the logarithm *of* both sides."

$3^x = 5$ — This is the equation to solve.

$\log 3^x = \log 5$ — Take the common logarithm on each side.

$x \log 3 = \log 5$ — The log of a power is the power times the log: $\log 3^x = x \log 3$. The variable x is now a factor of $x \log 3$ and not an exponent.

$\frac{x \log 3}{\log 3} = \frac{\log 5}{\log 3}$ — $x \log 3$ means $x \cdot \log 3$. To isolate x, undo the multiplication by log 3 by dividing both sides by log 3.

$\frac{x \cancel{\log 3}^{1}}{\cancel{\log 3}_{1}} = \frac{\log 5}{\log 3}$ — Simplify the left side by removing the common factor of log 3 from the numerator and denominator.

$x = \frac{\log 5}{\log 3}$ — This is the exact solution.

$x \approx 1.464973521$ — Approximate. On a reverse-entry scientific calculator, press: 5 LOG ÷ 3 LOG =.

Success Tip

The power rule of logarithms provides a way of **moving the variable** x **from its position in an exponent to a position as a factor** of $x \log 3$.

$$\log 3^x = \log 5$$
$$x \log 3 = \log 5$$

The exact solution is $\frac{\log 5}{\log 3}$ and the solution set is $\left\{\frac{\log 5}{\log 3}\right\}$. Rounded to four decimal places, an approximate solution is 1.4650.

We also can take the natural logarithm on each side of the equation to solve for x.

$3^x = 5$

$\ln 3^x = \ln 5$ — Take the natural logarithm on each side.

$x \ln 3 = \ln 5$ — Use the power rule of logarithms: $\ln 3^x = x \ln 3$.

$\frac{x \ln 3}{\ln 3} = \frac{\ln 5}{\ln 3}$ — $x \ln 3$ means $x \cdot \ln 3$. To isolate x, undo the multiplication by ln 3 by dividing both sides by ln 3.

$\frac{x \cancel{\ln 3}^{1}}{\cancel{\ln 3}_{1}} = \frac{\ln 5}{\ln 3}$ — Simplify the left side by removing the common factor of ln 3 from the numerator and denominator.

$x = \frac{\ln 5}{\ln 3}$ — This is the exact solution.

$x \approx 1.464973521$ — Approximate. On a scientific calculator, press: 5 LN ÷ 3 LN =.

Caution

Don't misuse the quotient rule:

$\frac{\ln 5}{\ln 3}$ means $\ln 5 \div \ln 3$

It is the expression $\ln \frac{5}{3}$ that means $\ln 5 - \ln 3$.

The result is the same using the natural logarithm. To check the approximate solution, we substitute 1.4650 for x in 3^x and see if $3^{1.4650}$ is approximately 5.

Check:

$$3^x = 5$$
$$3^{1.4650} \stackrel{?}{=} 5$$
$$5.000145454 \approx 5$$

On a scientific calculator, press: 3 y^x 1.4650 =.

Self Check 3 Solve: $5^x = 4$

Now Try Problem 29

EXAMPLE 4 Solve: $6^{x-3} = 2^x$

Strategy We will take the common logarithm on both sides of the equation.

Why We can then use the power rule of logarithms to move the expressions $x - 3$ and x from their current positions as exponents to positions as factors.

Solution

$6^{x-3} = 2^x$ — This is the equation to solve.

$\log 6^{x-3} = \log 2^x$ — Take the common logarithm on each side.

$(x - 3)\log 6 = x \log 2$ — The log of a power is the power times the log. The expression $x - 3$ is now a factor of $(x - 3)\log 6$ and not an exponent.

$x \log 6 - 3\log 6 = x \log 2$ — Distribute the multiplication by $\log 6$.

$x \log 6 - x \log 2 = 3 \log 6$ — To get the terms involving x on the left side, add $3 \log 6$ and subtract $x \log 2$ on both sides.

$x(\log 6 - \log 2) = 3 \log 6$ — Factor out x on the left side.

$\frac{x(\log 6 - \log 2)}{\log 6 - \log 2} = \frac{3 \log 6}{\log 6 - \log 2}$ — To isolate x, undo the multiplication by $\log 6 - \log 2$ by dividing both side by $\log 6 - \log 2$. Then simplify the left side.

$x = \frac{3 \log 6}{\log 6 - \log 2}$ — This is the exact solution.

$x \approx 4.892789261$ — Approximate. On a reverse-entry scientific calculator, press: 3 × 6 LOG ÷ (6 LOG − 2 LOG) =.

The Language of Algebra

$\frac{3 \log 6}{\log 6 - \log 2}$ is the **exact** solution of $6^{x-3} = 2^x$. An **approximate** solution is 4.8928.

The solution is $\frac{3 \log 6}{\log 6 - \log 2}$ and the solution set is $\left\{\frac{3 \log 6}{\log 6 - \log 2}\right\}$. To four decimal places, an approximate solution is 4.8928. To check, we substitute 4.8928 for each x in $6^{x-3} = 2^x$. The resulting values on the left and right sides of the equation should be approximately equal.

Self Check 4 Solve: $5^{x-2} = 3^x$

Now Try Problem 33

When an exponential equation involves an exponential expression with base e, it is easiest to take the natural logarithm on both sides to solve the equation.

EXAMPLE 5 Solve: $e^{0.9t} = 10$

Strategy We will take the natural (base-e) logarithm on both sides of the equation.

Why We can then use the power rule of logarithms to move the expression $0.9t$ from its current position as an exponent to a position as a factor.

Solution The exponential expression on the left side has base e. In such cases, the calculations are easier when we take the natural logarithm of each side.

Success Tip

When we take the natural logarithm of both sides, that conveniently produces the factor $\ln e$, which is equal to 1, on the left side of the equation.

$e^{0.9t} = 10$ — This is the equation to solve.

$\ln e^{0.9t} = \ln 10$ — Take the natural logarithm on each side.

$0.9t \ln e = \ln 10$ — Use the power rule of logarithms: $\ln e^{0.9t} = 0.9t \ln e$. The expression $0.9t$ is now a factor of $0.9t \ln e$ and not an exponent.

$0.9t \cdot \mathbf{1} = \ln 10$ Simplify: $\ln e = 1$.

$0.9t = \ln 10$ Simplify the left side.

$t = \frac{\ln 10}{0.9}$ To isolate t, undo the multiplication by 0.9 by dividing both sides by 0.9.

$t \approx 2.558427881$ Approximate. On a reverse-entry scientific calculator, press: 10 LN ÷ .9 =.

The exact solution is $\frac{\ln 10}{0.9}$. To four decimal places, an approximate solution is 2.5584. Verify this by using a calculator to show that $e^{0.9(\mathbf{2.5584})} \approx 10$.

Self Check 5 Solve: $e^{2.1t} = 35$

Now Try Problem 37

Strategy for Solving Exponential Equations

1. Isolate one of the exponential expressions in the equation.
2. If both sides of the equation can be written as exponential expressions with the same base, do so. Then set the exponents equal and solve the resulting equation.
3. If step 2 is difficult or impossible, take the common or natural logarithm on both sides. Use the power rule of logarithms to write the variable exponent as a factor, and then solve the resulting equation.
4. Check the results in the original equation.

2 Solve Logarithmic Equations.

A **logarithmic equation** is an equation with a logarithmic expression that contains a variable. Some examples of logarithmic equations are

$$\log 5x = 3, \qquad \log(3x + 2) = \log(2x - 3), \qquad \text{and} \qquad \log_2 7 - \log_2 x = 5$$

Some logarithmic equations can be solved by rewriting them in equivalent exponential form.

EXAMPLE 6 Solve: $\log 5x = 3$

Strategy Recall that $\log 5x = \log_{10} 5x$. To solve $\log 5x = 3$, we will instead write and solve an equivalent base-10 exponential equation.

Why The resulting exponential equation is easier to solve because the variable term is isolated on one side.

Solution

$\log 5x = 3$ This is the equation to solve.

$\log_{\mathbf{10}} 5x = 3$ The base of the logarithm is 10.

$10^3 = 5x$ Write the equivalent base-10 exponential equation.

$1{,}000 = 5x$ Simplify: $10^3 = 1{,}000$.

$200 = x$ To isolate x, divide both sides by 5.

The solution is 200 and the solution set is $\{200\}$.

Check:

$\log 5\mathbf{x} = 3$ This is the original equation.

$\log 5(\mathbf{200}) \stackrel{?}{=} 3$ Substitute 200 for x.

$\log 1{,}000 \stackrel{?}{=} 3$ Multiply $5(200) = 1{,}000$.

$3 = 3$ Evaluate: $\log 1{,}000 = \log 10^3 = 3$.

Caution

Always check your solutions to a logarithmic equation to identify any extraneous solutions.

Self Check 6 Solve: $\log_2 (x - 3) = -1$

Now Try ▶ Problem 41

There is a possibility of obtaining **extraneous solutions** when solving logarithmic equations. Always discard any possible solutions that produce the logarithm of a negative number or the logarithm of 0 in the original equation.

EXAMPLE 7

Solve: $\log(3x + 2) = \log(2x - 3)$

Strategy We will use the logarithmic property of equality to see that $3x + 2 = 2x - 3$.

Why We can use the logarithm property of equality because the given equation, $\log(3x + 2) = \log(2x - 3)$, has the form $\log_b x = \log_b y$.

Solution

$\log(3x + 2) = \log(2x - 3)$	This is the equation to solve.
$3x + 2 = 2x - 3$	If the logarithms of two numbers are equal, the numbers are equal.
$x + 2 = -3$	Subtract 2x from both sides.
$x = -5$	To isolate x, subtract 2 from both sides.

Check:

$\log(3x + 2) = \log(2x - 3)$	This is the original equation.
$\log[3(-5) + 2] \stackrel{?}{=} \log[2(-5) - 3]$	Substitute −5 for x.
~~$\log(-13) \stackrel{?}{=} \log(-13)$~~	Evaluate within brackets. Recall that $\log(-13)$ is undefined.

Since the logarithm of a negative number does not exist, the proposed solution of -5 must be discarded. This equation has no solution. Its solution set is $\varnothing$.

Caution

Don't make this error of trying to "distribute" log:

$\log(3x + 2)$

The notation log is not a number, it is the name of a function and cannot be distributed.

Self Check 7 Solve: $\log(5x + 14) = \log(7x - 2)$

Now Try ▶ Problem 49

In Examples 8 and 9, we will use the product and quotient rules of logarithms to "condense" one side of the equation first, before solving for the variable.

EXAMPLE 8

Solve: $\log x + \log(x - 3) = 1$

Strategy We will use the product rule for logarithms in reverse: The sum of two logarithms is equal to the logarithm of a product. Then we will write and solve an equivalent exponential equation.

Why We use the product rule of logarithms because the left side of the equation, $\log x + \log(x - 3)$, has the form $\log_b M + \log_b N$.

Solution

$\log x + \log(x - 3) = 1$	This is the equation to solve.
$\log x(x - 3) = 1$	On the left side, use the product rule of logarithms.
$\log_{10} x(x - 3) = 1$	The base of the logarithm is 10.
$x(x - 3) = 10^1$	Write the equivalent base-10 exponential equation.
$x^2 - 3x - 10 = 0$	Distribute the multiplication by x, and then subtract 10 from both sides.
$(x + 2)(x - 5) = 0$	Factor the trinomial.
$x + 2 = 0$ or $x - 5 = 0$	Set each factor equal to 0.
$x = -2$	$x = 5$	

Success Tip

The objective is to use the product rule to "condense" the left side of the equation. We want to write an equivalent equation in which the variable x appears in only a single logarithmic expression.

Caution

Don't automatically discard proposed solutions that are negative. Only discard proposed solutions that produce undefined logarithms in the original equation.

Check: The number -2 is not a solution because it does not satisfy the equation (a negative number does not have a logarithm). We will check the other result, 5.

$$\log x + \log (x - 3) = 1$$ This is the original equation.

$$\log 5 + \log (5 - 3) \stackrel{?}{=} 1$$ Substitute 5 for *x*.

$$\log 5 + \log 2 \stackrel{?}{=} 1$$ Do the subtraction within the parentheses.

$$\log 10 \stackrel{?}{=} 1$$ Use the product rule of logarithms: $\log 5 + \log 2 = \log (5 \cdot 2) = \log 10$.

$$1 = 1$$ Evaluate: $\log 10 = 1$.

Since 5 satisfies the equation, it is the solution.

Self Check 8 Solve: $\log x + \log (x + 3) = 1$

Now Try ▶ Problem 53

Using Your Calculator ▶ Solving Logarithmic Equations Graphically

To use a graphing calculator to approximate the solutions of the logarithmic equation $\log x + \log (x - 3) = 1$ (see Example 8), we can subtract 1 from both sides of the equation to get $\log x + \log (x - 3) - 1 = 0$ and graph the corresponding function $f(x) = \log x + \log (x - 3) - 1$ as shown in figure (a). Since the solution of the equation is the x-value that makes $f(x) = 0$, the solution is the x-coordinate of the x-intercept of the graph. We can use the ZERO feature to find that this x-value is 5.

We also can solve $\log x + \log (x - 3) = 1$ using the INTERSECT feature. After graphing $Y_1 = \log x + \log (x - 3)$ and $Y_2 = 1$, we select INTERSECT, which approximates the coordinates of the point of intersection of the two graphs. From the display shown in figure (b), we can conclude that the solution is 5.

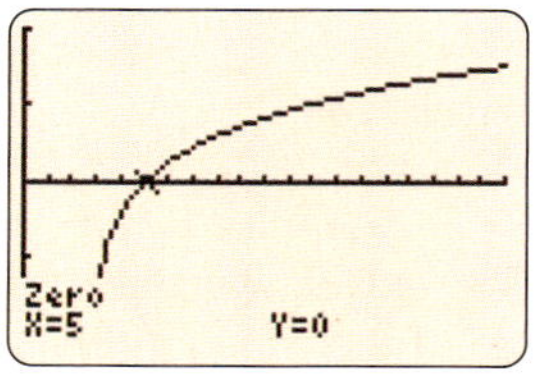

(a)

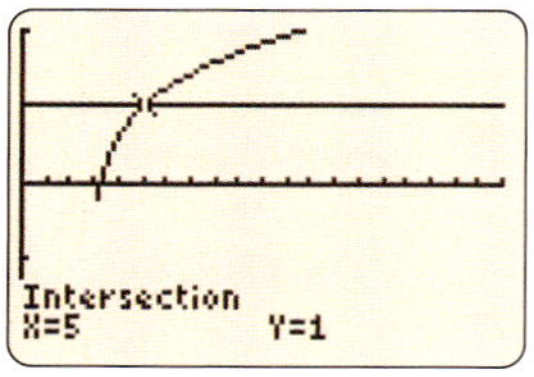

(b)

EXAMPLE 9 Solve: $\log_2 7 - \log_2 x = 5$

Strategy We will use the quotient rule for logarithms in reverse: The difference of two logarithms is equal to the logarithm of a quotient. Then we will write and solve an equivalent exponential equation.

Why We use the quotient rule for logarithms because the left side of the equation, $\log_2 7 - \log_2 x$, has the form $\log_b M - \log_b N$.

Solution

$$\log_2 7 - \log_2 x = 5$$ This is the equation to solve.

$$\log_2 \frac{7}{x} = 5$$ On the left side, use the quotient rule for logarithms.

$$\frac{7}{x} = 2^5$$ Write the equivalent base-2 exponential equation.

$$\frac{7}{x} = 32$$ Evaluate: $2^5 = 32$.

Success Tip

The objective is to use the quotient rule to "condense" the left side of the equation. We want to write an equivalent equation in which the variable x appears in only a single logarithmic expression.

$$7 = 32x$$ To clear the equation of the fraction, multiply both sides by x.

$$\frac{7}{32} = x$$ To isolate x, divide both sides by 32.

The solution is $\frac{7}{32}$. Verify that it satisfies the original equation.

Self Check 9 Solve: $\log_2 9 - \log_2 x = 4$

Now Try Problem 57

3 Use Exponential and Logarithmic Equations to Solve Application Problems.

Recall from Section 11.6 that a **hydrogen ion** $[H^+]$ is the positively charged nucleus of a hydrogen atom, without its electron. The concentration of hydrogen ions in a solution is commonly measured using the **pH scale.** The pH of a solution is defined as follows.

pH of a Solution

If $[H^+]$ is the hydrogen ion concentration in gram-ions per liter, then

$$\text{pH} = -\log[H^+]$$

EXAMPLE 10

Hydrogen Ion Concentration. Find the hydrogen ion concentration of seawater if its pH is 8.5.

Strategy To find the hydrogen ion concentration, we will substitute 8.5 for pH in the formula $\text{pH} = -\log[H^+]$ and solve the resulting equation for $[H^+]$.

Why After substituting for pH, the resulting logarithmic equation can be solved by solving an equivalent exponential equation.

Solution

Notation

In this logarithmic equation, the variable is the symbol $[H^+]$.

$$\text{pH} = -\log[H^+]$$ This is the formula for pH.

$$8.5 = -\log[H^+]$$ Substitute 8.5 for pH.

$$-8.5 = \log[H^+]$$ Multiply both sides by −1.

$$-8.5 = \log_{10}[H^+]$$ The base of the logarithm is 10.

$$[H^+] = 10^{-8.5}$$ Write the equivalent base-10 exponential equation.

$$[H^+] \approx 0.000000003$$ Approximate. On a reverse-entry scientific calculator, press: 10 y^x 8.5 +/− = .

We can write the result using scientific notation:

$$[H^+] \approx 3.0 \times 10^{-9} \text{ gram-ions per liter}$$

Self Check 10 **Hydrogen Ion Concentration.** Find the hydrogen ion concentration of a solution with a pH value of 4.8.

Now Try Problem 101

Experiments have determined the time it takes for half of a sample of a radioactive material to decompose. This time is a constant, called the material's **half-life.**

When living organisms die, the oxygen–carbon dioxide cycle common to all living things ceases, and carbon-14, a radioactive isotope with a half-life of 5,700 years, is no longer absorbed. By measuring the amount of carbon-14 present in an ancient object, archaeologists can estimate the object's age by using the radioactive decay formula.

Radioactive Decay Formula

If A is the amount of radioactive material present at time t, A_0 was the amount present at $t = 0$, and h is the material's half-life, then

$$A = A_0 2^{-t/h}$$

EXAMPLE 11 **Carbon-14 Dating.** How old is a piece of wood that retains only one-third of its original carbon-14 content?

Strategy If A_0 is the original carbon-14 content, then today's content $A = \frac{1}{3}A_0$. We will substitute $\frac{A_0}{3}$ for A and 5,700 for h in the radioactive decay formula and solve for t.

Why The value of t is the estimated age of the piece of wood.

Solution

To find the time t when $A = \frac{1}{3}A_0$, we substitute $\frac{A_0}{3}$ for A and 5,700 for h in the radioactive decay formula and solve for t:

$$A = A_0 2^{-t/h}$$ This is the radioactive decay model.

$$\frac{A_0}{3} = A_0 2^{-t/5,700}$$ The half-life of carbon-14 is 5,700 years.

$$1 = 3(2^{-t/5,700})$$ Divide both sides by A_0 and multiply both sides by 3.

$$\log 1 = \log 3(2^{-t/5,700})$$ Take the common logarithm on both sides.

$$0 = \log 3 + \log 2^{-t/5,700}$$ $\log 1 = 0$, and use the product rule for logarithms.

$$-\log 3 = -\frac{t}{5,700}\log 2$$ Subtract log 3 from both sides and use the power rule of logarithms.

$$5,700\left(\frac{\log 3}{\log 2}\right) = t$$ Multiply both sides by $-\frac{5,700}{\log 2}$.

$$t \approx 9,034.286254$$ Approximate. On a reverse-entry scientific calculator, press: 5700 × 3 LOG ÷ 2 LOG =.

The piece of wood is approximately 9,000 years old.

Notation

The initial amount of radioactive material is represented by A_0, and it is read as "A sub 0."

Self Check 11 **Carbon-14 Dating.** How old is a piece of wood that retains 25% of its original carbon-14 content?

Now Try Problem 103

When there is sufficient food and space available, populations of living organisms tend to increase exponentially according to the following growth model.

Exponential Growth Model

If P is the population at some time t, P_0 is the initial population at $t = 0$, and k depends on the rate of growth, then

$$P = P_0 e^{kt}$$

EXAMPLE 12 **Population Growth.** The bacteria in a laboratory culture increased from an initial population of 500 to 1,500 in 3 hours. How long will it take for the population to reach 10,000?

Strategy We will substitute 500 for P_0, 1,500 for P, and 3 for t in the exponential growth model and solve for k.

Why Once we know the value of k, we can substitute 10,000 for P, 500 for P_0, and the value of k in the exponential growth model and solve for the time t.

Solution

$P = P_0e^{kt}$ This is the population growth formula.

$1{,}500 = 500(e^{k3})$ Substitute 1,500 for P, 500 for P_0, and 3 for t.

$3 = e^{3k}$ Divide both sides by 500.

$3k = \ln 3$ Write the equivalent base-e logarithmic equation.

$k = \frac{\ln 3}{3}$ Divide both sides by 3.

Notation

The initial population of bacteria is represented by P_0, and it is read as "P sub 0."

To find when the population will reach 10,000, we substitute 10,000 for P, 500 for P_0, and $\frac{\ln 3}{3}$ for k in the growth model and solve for t:

$P = P_0e^{kt}$

$10{,}000 = 500e^{[(\ln 3)/3]t}$

$20 = e^{[(\ln 3)/3]t}$ Divide both sides by 500.

$\left(\frac{\ln 3}{3}\right)t = \ln 20$ Write the equivalent base-e logarithmic equation.

$t = \frac{3 \ln 20}{\ln 3}$ To isolate t, multiply both sides by the reciprocal of $\frac{\ln 3}{3}$, which is $\frac{3}{\ln 3}$.

≈ 8.180499084 Approximate. On a reverse-entry scientific calculator, press: 3 × 20 LN ÷ 3 LN = .

The culture will reach 10,000 bacteria in about 8 hours.

Self Check 12 **Population Growth.** In Example 12, how long will it take the population to reach 20,000?

Now Try Problem 115

SECTION 11.7 STUDY SET

VOCABULARY

Fill in the blanks.

1. An equation with a positive constant base and a variable in its exponent, such as $3^{2x} = 8$, is called an __________ equation.
2. An equation with a logarithmic expression that contains a variable, such as $\log_5 (2x - 3) = \log_5 (x + 4)$, is a __________ equation.

CONCEPTS

Fill in the blanks.

3. a. If two exponential expressions with the same base are equal, their exponents are ______.

   $b^x = b^y$ is equivalent to $\quad = \quad$.

   b. If the logarithms base-b of two numbers are equal, the numbers are ______.

   $\log_b x = \log_b y$ is equivalent to $\quad = \quad$.
4. The right side of the exponential equation $5^{x-3} = 125$ can be written as a power of $\quad$.
5. If $6^{4x} = 6^{-2}$, then $4x = \quad$.
6. a. Write the equivalent base-10 exponential equation for $\log (x + 1) = 2$.

   b. Write the equivalent base-e exponential equation for $\ln (x + 1) = 2$.

Fill in the blanks.

7. To solve $5^x = 2$, we can take the __________ of both sides of the equation to get $\log 5^x = \log 2$.
8. a. For $5^x = 2$, the power rule for logarithms provides a way of moving the variable x from its position as an __________ to a position as a factor.

   b. If the power rule for logarithms is used on the left side of the equation $\log 5^x = 2$, the resulting equation is $\quad \log 5 = 2$.
9. If $e^{x+2} = 4$, then $\ln e^{x+2} = \quad$.
10. Perform a check to determine whether -2 is a solution of $5^{2x+3} = \frac{1}{5}$.
11. Perform a check to determine whether 4 is a solution of $\log_5 (x + 1) = 2$.

12. Use a calculator to determine whether 2.5646 is an approximate solution of $2^{2x+1} = 70$.

13. a. How do we solve $x \ln 3 = \ln 5$ for x?

b. What is the exact solution?

c. What is an approximate solution to four decimal places?

14. Use a property of logarithms to condense the left side of each equation to a single logarithm. **Do not solve.**

a. $\log_5 x + \log_5 \cdot (4x - 1) = 1$

b. $\log_3 4x - \log_3 7 = 2$

15. a. Find $\frac{\log 8}{\log 5}$. Round to four decimal places.

b. Find $\frac{3 \ln 12}{\ln 4 - \ln 2}$. Round to four decimal places.

16. Does $\frac{\log 7}{\log 3} = \log 7 - \log 3$?

17. Complete each formula.

a. $\text{pH} = -\quad [\text{H}^+]$

b. Radioactive decay: $A =$

c. Population growth: $P =$

18. Use the graphs below to estimate the solution of each equation.

a. $2^x = 3^{-x+3}$

b. $3 \log (x - 1) = 2 \log x$

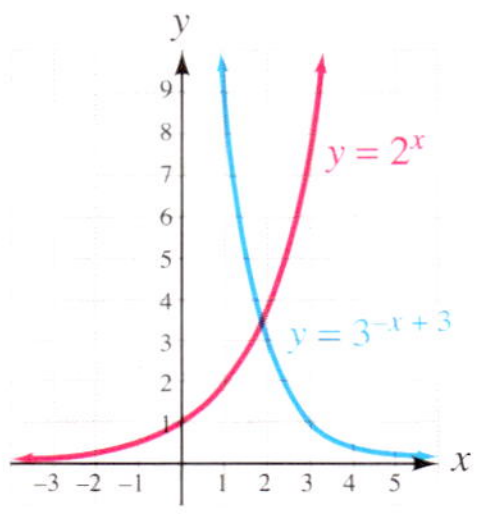

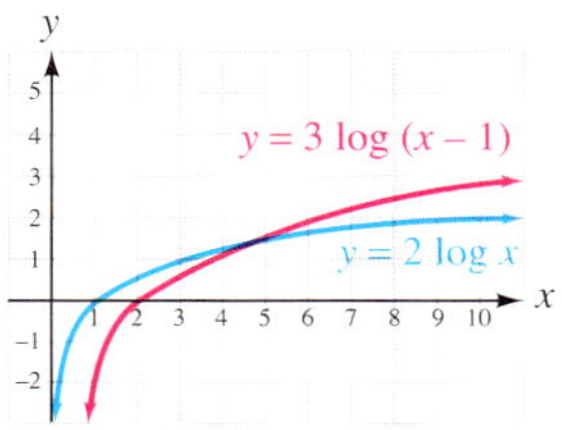

NOTATION

Complete each solution.

19. Solve: $2^x = 7$

$2^x = \log 7$

$x \qquad = \log 7$

$x = \frac{\log 7}{\log 2}$

$x \approx$

20. Solve: $\log_2 (2x - 3) = \log_2 (x + 4)$

$= x + 4$

$x =$

GUIDED PRACTICE

Solve each equation. See Example 1.

21. $6^{x-2} = 36$

22. $3^{x+1} = 27$

23. $5^{4x} = \frac{1}{125}$

24. $8^{-2x+1} = \frac{1}{64}$

Solve each equation. See Example 2.

25. $2^{x^2-2x} = 8$

26. $3^{x^2-3x} = 81$

27. $3^{x^2+4x} = \frac{1}{81}$

28. $7^{x^2+3x} = \frac{1}{49}$

Solve each equation. Give the exact solution and an approximation to four decimal places. See Example 3.

29. $4^x = 5$

30. $7^x = 12$

31. $13^{x-1} = 2$

32. $5^{x+1} = 3$

Solve each equation. Give the exact solution and an approximation to four decimal places. See Example 4.

33. $2^{x+1} = 3^x$

34. $6^x = 7^{x-4}$

35. $5^{x-3} = 3^{2x}$

36. $8^{3x} = 9^{x+1}$

Solve each equation. Give the exact solution and an approximation to four decimal places. See Example 5.

37. $e^{2.9x} = 4.5$

38. $e^{3.3t} = 9.1$

39. $e^{-0.2t} = 14.2$

40. $e^{-0.7x} = 6.2$

Solve each equation. See Example 6.

41. $\log 2x = 4$

42. $\log 5x = 4$

43. $\log_3 (x - 3) = 2$

44. $\log_4 (2x - 1) = 3$

45. $\log (7 - x) = 2$

46. $\log (2 - x) = 3$

47. $\log \frac{1}{8}x = -2$

48. $\log \frac{1}{5}x = -3$

Solve each equation. See Example 7.

49. $\log (3 - 2x) = \log (x + 24)$

50. $\log (3x + 5) = \log (2x + 6)$

51. $\ln (3x + 1) = \ln (x + 7)$

52. $\ln (x^2 + 4x) = \ln (x^2 + 16)$

Solve each equation. See Example 8.

53. $\log x + \log (x - 48) = 2$

54. $\log x + \log (x + 9) = 1$

55. $\log_5 (4x - 1) + \log_5 x = 1$

56. $\log_2 (x - 7) + \log_2 x = 3$

Solve each equation. See Example 9.

57. $\log 5 - \log x = 1$

58. $\log 11 - \log x = 2$

59. $\log_3 4x - \log_3 7 = 2$

60. $\log_2 5x - \log_2 3 = 4$

TRY IT YOURSELF

Solve each equation. Give the exact solution and, when appropriate, an approximation to four decimal places.

61. $\log 2x = \log 4$

62. $\log 3x = \log 9$

63. $\ln x = 1$

64. $\ln x = 5$

65. $7^{x^2} = 10$

66. $8^{x^2} = 11$

67. $\log (x + 90) + \log x = 3$

68. $\log (x - 90) + \log x = 3$

69. $3^{x-6} = 81$

70. $5^{x+4} = 125$

71. $\log \frac{4x + 1}{2x + 9} = 0$

72. $\log \frac{2 - 5x}{2(x + 8)} = 0$

73. $15 = 9^{x+2}$

74. $29 = 5^{x-6}$

75. $\log x^2 = 2$

76. $\log x^3 = 3$

77. $\log (x - 6) - \log (x - 2) = \log \frac{5}{x}$

78. $\log (3 - 2x) - \log (x + 9) = 0$

79. $\log_3 x = \log_3 \left(\frac{1}{x}\right) + 4$

80. $\log_5 (7 + x) + \log_5 (8 - x) - \log_5 2 = 2$

81. $2 \log_2 x = 3 + \log_2 (x - 2)$

82. $2 \log_3 x - \log_3 (x - 4) = 2 + \log_3 2$

83. $\log (7y + 1) = 2 \log (y + 3) - \log 2$

84. $2 \log (y + 2) = \log (y + 2) - \log 12$

85. $e^{3x} = 9$

86. $e^{4x} = 60$

87. $\frac{\log (5x + 6)}{2} = \log x$

88. $\frac{1}{2} \log (4x + 5) = \log x$

Look Alikes . . .

89. a. $\log 5x = 1.7$ b. $\ln 5x = 1.7$

90. a. $\log_2 (x^2 - x) = 1$ b. $\log_6 (x^2 - x) = 1$

91. a. $4^{3x-5} = 90$ b. $e^{3x-5} = 90$

92. a. $\log x + 2 \log x = \log 8$ b. $\log x - 2 \log x = \log 8$

93. a. $\log_2 (x + 5) - \log_2 4x = \log_2 x$
b. $\ln (x + 5) - \ln 4x = \ln x$

94. a. $5^{9x-1} = 125$ b. $5^{9x-1} = 124$

95. a. $\left(\frac{2}{3}\right)^{6-x} = \frac{8}{27}$ b. $\left(\frac{2}{3}\right)^{6-x} = \frac{16}{81}$

96. a. $\log x - \log (x + 7) = -1$ b. $\log x - \log (x + 7) = 1$

Use a graphing calculator to solve each equation. If an answer is not exact, round to the nearest tenth. See Using Your Calculator: Solving Exponential Equations Graphically or Solving Logarithmic Equations Graphically.

97. $2^{x+1} = 7$

98. $3^x - 10 = 3^{-x}$

99. $\log x + \log (x - 15) = 2$

100. $\ln (2x + 5) - \ln 3 = \ln (x - 1)$

APPLICATIONS

101. **Hydrogen Ion Concentration.** Find the hydrogen ion concentration of a saturated solution of calcium hydroxide whose pH is 13.2.

102. **Aquariums.** The safe pH range for a freshwater aquarium is shown on the scale in the next column. Find the corresponding hydrogen ion concentration.

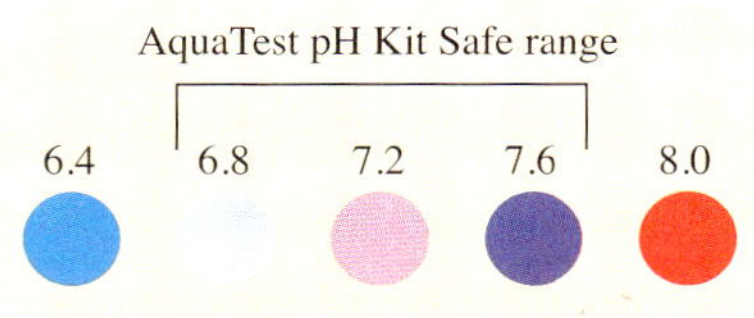

103. **Tritium Decay.** The half-life of tritium is 12.4 years. How long will it take for 25% of a sample of tritium to decompose?

104. **Radioactive Decay.** In 2 years, 20% of a radioactive element decays. Find its half-life.

105. **Thorium Decay.** An isotope of thorium, written as ^{227}Th, has a half-life of 18.4 days. How long will it take for 80% of the sample to decompose?

106. **Lead Decay.** An isotope of lead, written as ^{201}Pb, has a half-life of 8.4 hours. How many hours ago was there 30% more of the substance?

107. **Carbon-14 Dating.** A bone fragment analyzed by archaeologists contains 60% of the carbon-14 that it is assumed to have had initially. How old is it?

108. **Carbon-14 Dating.** Only 10% of the carbon-14 in a small wooden bowl remains. How old is the bowl?

109. **Compound Interest.** If \$500 is deposited in an account paying 8.5% annual interest, compounded semiannually, how long will it take for the account to increase to \$800?

110. **Continuous Compund Interest.** In Exercise 109, how long will it take if the interest is compounded continuously?

111. **Compound Interest.** If \$1,300 is deposited in a savings account paying 9% interest, compounded quarterly, how long will it take the account to increase to \$2,100?

112. **Compound Interest.** A sum of \$5,000 deposited in an account grows to \$7,000 in 5 years. Assuming annual compounding, what interest rate is being paid?

113. **Rule of Seventy.** A rule of thumb for finding how long it takes an investment to double is called the **rule of seventy.** To apply the rule, divide 70 by the interest rate written as a percent. At 5%, an investment takes $\frac{70}{5} = 14$ years to double. At 7%, it takes $\frac{70}{7} = 10$ years. Explain why this formula works.

114. **Bacterial Growth.** A bacterial culture grows according to the function $P(t) = P_0 a^t$. If it takes 5 days for the culture to triple in size, how long will it take to double in size?

115. **Rodent Control.** The rodent population in a city is currently estimated at 30,000. If it is expected to double every 5 years, when will the population reach 1 million?

116. **Population Growth.** The population of a city is expected to triple every 15 years. When can the city planners expect the present population of 140 persons to double?

117. **Bacterial Culture.** A bacteria culture doubles in size every 24 hours. By how much will it have increased in 36 hours?

118. Oceanography. The intensity I of a light a distance x meters beneath the surface of a lake decreases exponentially. Use the data in the illustration to find the depth at which the intensity will be 20%.

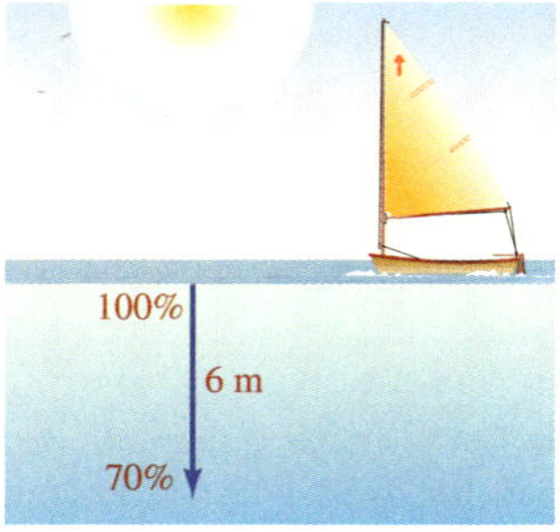

119. Newton's Law of Cooling. Water initially at 100°C is left to cool in a room at temperature 60°C. After 3 minutes, the water temperature is 90°. The water temperature T is a function of time t given by the formula $T = 60 + 40e^{kt}$. Find k.

120. Newton's Law of Cooling. Refer to Exercise 119 and find the time for the water temperature to reach 70°C.

WRITING

121. Explain how to solve the equation $2^{x+1} = 31$.

122. Explain how to solve the equation $2^{x+1} = 32$.

123. Write a justification for each step of the solution.

$15^x = 9$ This is the equation to solve.

$\log 15^x = \log 9$ ________________.

$x \log 15 = \log 9$ ________________.

$x = \dfrac{\log 9}{\log 15}$ ________________.

124. What is meant by the term *half-life*?

REVIEW

125. Find the length of leg AC.

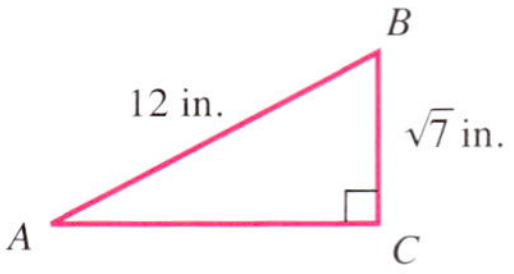

126. Dosages. The amount of medicine a patient should take is often proportional to his or her weight. If a patient weighing 83 kilograms needs 150 milligrams of medicine, how much will be needed by a person weighing 99.6 kilograms?

CHALLENGE PROBLEMS

Solve each equation.

127. $\log_3 x + \log_3 (x + 2) = 2$

128. $x^{\log x} = 10{,}000$

129. $\dfrac{\log_2 (6x - 8)}{\log_2 x} = 2$

130. $\dfrac{\log (3x - 4)}{\log x} = 2$

11 Summary & Review

SECTION 11.1 Algebra and Composition of Functions

DEFINITIONS AND CONCEPTS	EXAMPLES
Just as it is possible to perform arithmetic operations on real numbers, it is possible to perform those operations on functions. The **sum, difference, product,** and **quotient functions** are defined as: $(f + g)(x) = f(x) + g(x)$ $(f - g)(x) = f(x) - g(x)$ $(f \cdot g)(x) = f(x)g(x)$ $(f/g)(x) = \dfrac{f(x)}{g(x)}$, with $g(x) \neq 0$	Let $f(x) = 2x + 1$ and $g(x) = x^2$. $(f + g)(x) = f(x) + g(x)$ $= 2x + 1 + x^2$ $= x^2 + 2x + 1$ $(f - g)(x) = f(x) - g(x)$ $= 2x + 1 - x^2$ $= -x^2 + 2x + 1$ $(f \cdot g)(x) = f(x) \cdot g(x)$ $= (2x + 1)x^2$ $= 2x^3 + x^2$ $(f/g)(x) = \dfrac{f(x)}{g(x)}$ $= \dfrac{2x + 1}{x^2}$

Often one quantity is a function of a second quantity that depends, in turn, on a third quantity. Such chains of dependence can be modeled by a **composition of functions.** Read as "f of g of x." $(f \circ g)(x) = f(g(x))$ Read as "f composed with g of x."	Let $f(x) = 4x - 9$ and $g(x) = x^3$. Find $(f \circ g)(2)$ and $(f \circ g)(x)$. $(f \circ g)(2) = f(g(2))$ Change to nested parentheses notation. $= f(8)$ Evaluate: $g(2) = 2^3 = 8$. $= 4(8) - 9$ Evaluate $f(8)$ using $f(x) = 4x - 9$. $= 23$ $(f \circ g)(x) = f(g(x)) = f(x^3) = 4x^3 - 9$

REVIEW EXERCISES

Let $f(x) = 2x$ and $g(x) = x + 1$. Find each function and give its domain.

1. $f + g$
2. $f - g$
3. $f \cdot g$
4. f/g

Let $f(x) = x^2 + 2$ and $g(x) = 2x + 1$. Find each of the following.

5. $(f \circ g)(-1)$
6. $(g \circ f)(0)$
7. $(f \circ g)(x)$
8. $(g \circ f)(x)$

9. Use the graphs of functions f and g to find each of the following.
   a. $(f + g)(2)$
   b. $(f \cdot g)(-4)$
   c. $(f \circ g)(4)$
   d. $(g \circ f)(6)$

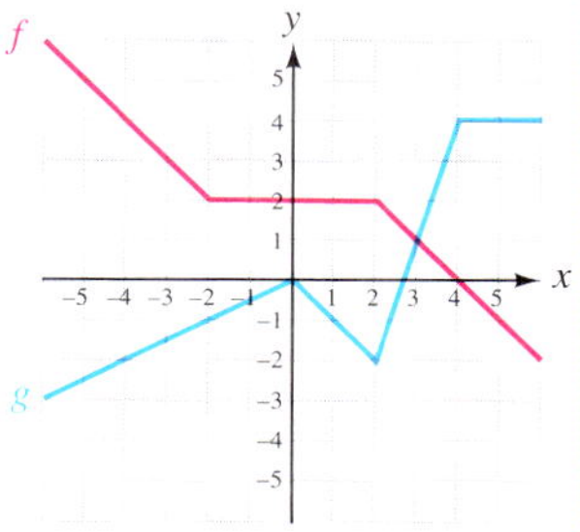

10. **Mileage Costs.** The function $f(m) = \frac{m}{8}$ gives the number of gallons of fuel consumed if a bus travels m miles. The function $C(f) = 3.25f$ gives the cost (in dollars) of f gallons of fuel. Write a composition function that expresses the cost of the fuel consumed as a function of the number of miles driven.

SECTION 11.2 Inverse Functions

DEFINITIONS AND CONCEPTS	EXAMPLES
A function is called a **one-to-one function** if different inputs determine different outputs.	The function $f(x) = 3x - 5$ is a one-to-one function because different inputs have different outputs. Since two different inputs, -2 and 2, have the same output 16, the function $f(x) = x^4$ is not one-to-one.
Horizontal line test: A function is one-to-one if every horizontal line intersects the graph of the function at most once.	The function $f(x) = \lvert x \rvert - 2$ is not a one-to-one function because we can draw a horizontal line that intersects its graph twice. 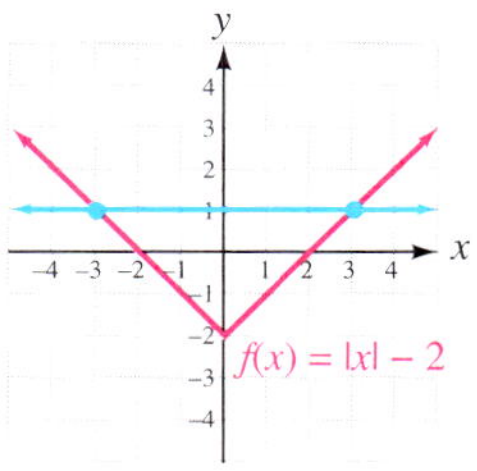
To find the inverse of a function, replace $f(x)$ with y, interchange the variables x and y, solve for y, and then replace y with $f^{-1}(x)$.	Find the inverse of the one-to-one function $f(x) = 2x + 1$. $f(x) = 2x + 1$ $y = 2x + 1$ Replace $f(x)$ with y. $x = 2y + 1$ Interchange the variables x and y. $\frac{x - 1}{2} = y$ Solve for y. $f^{-1}(x) = \frac{x - 1}{2}$ Replace y with $f^{-1}(x)$.

If a point (a, b) is on the graph of function f, it follows that the point (b, a) is on the graph of f^{-1}, and vice versa.

The graph of a function and its inverse are **symmetric about the line $y = x$.**

The graphs of $f(x) = 2x + 1$ and $f^{-1}(x) = \frac{x-1}{2}$ are symmetric about the line $y = x$ as shown in the illustration.

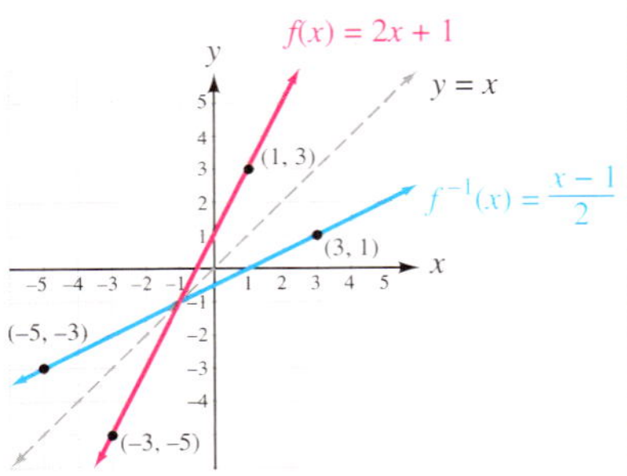

For any one-to-one function f and its inverse, f^{-1},

$$(f \circ f^{-1})(x) = x \quad \text{and} \quad (f^{-1} \circ f)(x) = x$$

The composition of $f(x) = 2x + 1$ and its inverse $f^{-1}(x) = \frac{x-1}{2}$ is the identity function $f(x) = x$.

$$(f \circ f^{-1})(x) = f(f^{-1}(x)) = f\left(\frac{x-1}{2}\right) = 2\left(\frac{x-1}{2}\right) + 1 = x - 1 + 1 = x$$

$$(f^{-1} \circ f)(x) = f^{-1}(f(x)) = f^{-1}(2x + 1) = \frac{2x + 1 - 1}{2} = \frac{2x}{2} = x$$

REVIEW EXERCISES

In Exercises 11–16, determine whether the function is one-to-one.

11. $f(x) = x^2 + 3$

12. $f(x) = \frac{1}{3}x - 8$

13. $\{(3, 4), (5, 10), (10, -1), (6, 6)\}$

14.

x	$f(x)$
0	−5
2	10
4	−5
6	15

15.

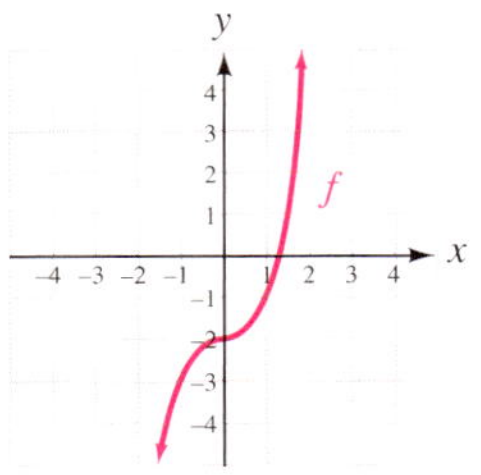

16.

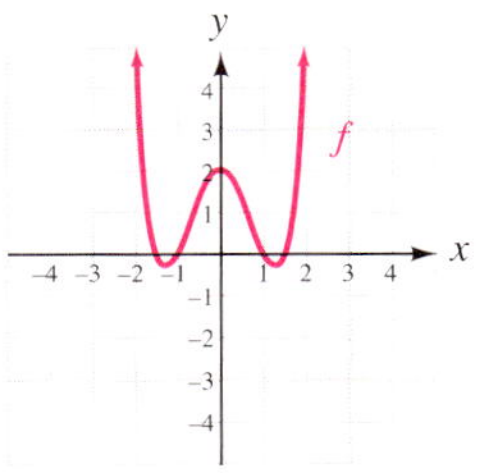

17. Use the table of values of the one-to-one function f to complete a table of values for f^{-1}.

x	$f(x)$
−6	−6
−1	−3
7	12
20	3

x	$f^{-1}(x)$
−6	
−3	
12	
3	

18. Given the graph of function f, graph f^{-1} on the same coordinate axes. Label the axis of symmetry.

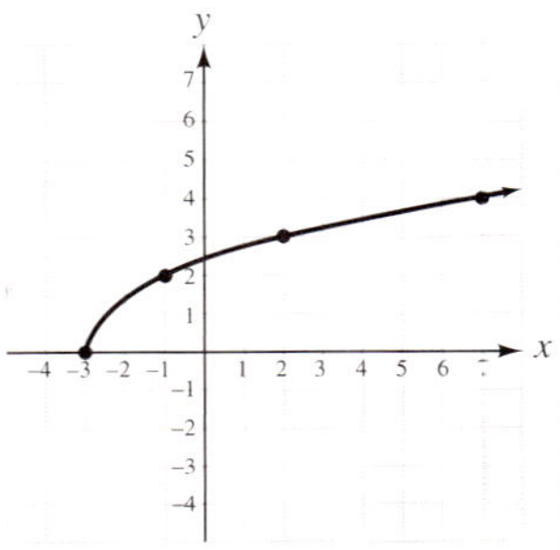

Find the inverse of each function.

19. $f(x) = 6x - 3$

20. $f(x) = \frac{4}{x - 1}$

21. $f(x) = (x + 2)^3$

22. $f(x) = \frac{x}{6} - \frac{1}{6}$

23. Find the inverse of $f(x) = \sqrt[3]{x - 1}$. Then graph the function and its inverse on one coordinate system. Show the axis of symmetry on the graph.

24. Use composition to show that $f(x) = 5 - 4x$ and $f^{-1}(x) = -\frac{x-5}{4}$ are inverse functions.

SECTION 11.3 Exponential Functions

DEFINITIONS AND CONCEPTS	EXAMPLES
Exponential functions have a *constant base* and a *variable exponent* and are defined by the equation $f(x) = b^x$, with $b > 0$, $b \neq 1$ Properties of an exponential function $f(x) = b^x$: The **domain** is the interval $(-\infty, \infty)$. The **range** is the interval $(0, \infty)$. Its graph has a **y-intercept** of $(0, 1)$. The x-axis is an **asymptote** of its graph. The graph **passes through** the point $(1, b)$. If $b > 1$, then $f(x) = b^x$ is an **increasing function.** If $0 < b < 1$, then $f(x) = b^x$ is a **decreasing function.**	The graphs of $f(x) = 2^x$ and $g(x) = \left(\frac{1}{2}\right)^x$ are shown below. (see tables and graph below) Since the base 2 is greater than 1, the function $f(x) = 2^x$ is an increasing function. Since the base $\frac{1}{2}$ is such that $0 < \frac{1}{2} < 1$, the function $g(x) = \left(\frac{1}{2}\right)^x$ is a decreasing function.
The graphs of exponential functions can be **translated** horizontally and vertically.	$h(x) = 2^{x+3}$ The graph of $f(x) = 2^x$ is moved 3 units to the left. $s(x) = \left(\frac{1}{2}\right)^x - 4$ The graph of $g(x) = \left(\frac{1}{2}\right)^x$ is moved 4 units downward.
Exponential functions are used to model many situations, such as population **growth,** the spread of an epidemic, the temperature of a heated object as it cools, and radioactive **decay.** Exponential functions are suitable models for describing **compound interest.** If $\$P$ is the deposit, and interest is paid k times a year at an annual rate r, the amount A in the account after t years is given by $A = P\left(1 + \frac{r}{k}\right)^{kt}$	If \$15,000 is deposited in an account paying an annual interest rate of 7.5%, compounded monthly, how much will be in the account in 60 years? (see work below)

$g(x) = \left(\frac{1}{2}\right)^x$

x	$g(x)$
-3	8
-2	4
-1	2
0	1
1	$\frac{1}{2}$
2	$\frac{1}{4}$

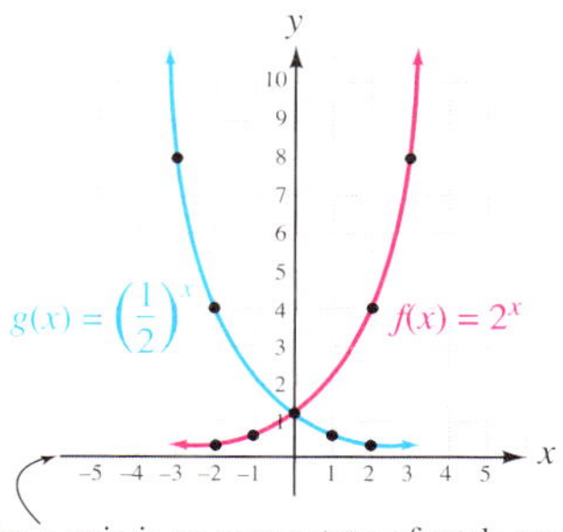

The x-axis is an asymptote of each graph.

$f(x) = 2^x$

x	$f(x)$
-2	$\frac{1}{4}$
-1	$\frac{1}{2}$
0	1
1	2
2	4
3	8

$$A(t) = 15{,}000\left(1 + \frac{0.075}{12}\right)^{12t}$$ To write the formula in function notation, substitute for *P*, *r*, and *k*.

$$A(60) = 15{,}000\left(1 + \frac{0.075}{12}\right)^{12(60)}$$ Substitute 60 for the time *t*.

$$= 15{,}000\left(1 + \frac{0.075}{12}\right)^{720}$$ Evaluate the exponent: 12(60) = 720.

$$\approx 1{,}331{,}479.52$$ Use a calculator with an exponential key: y^x or ^.

In 60 years, the account will contain about \$1,331,479.52.

REVIEW EXERCISES

25. a. Which of the following are exponential functions?

$f(x) = 2x$ $\quad g(x) = x^2$ $\quad h(a) = \sqrt{a}$

$n(x) = 2^x$ $\quad t(x) = \frac{1}{x}$ $\quad s(t) = 1.08^t$

b. Use a calculator to find $0.9(1.42)^{14}$. Round to four decimal places.

26. Determine whether each application is an example of exponential growth or decay.

a.

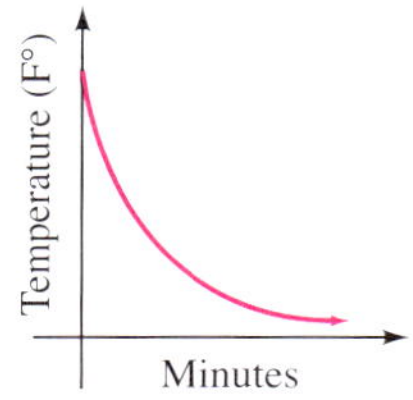

b.

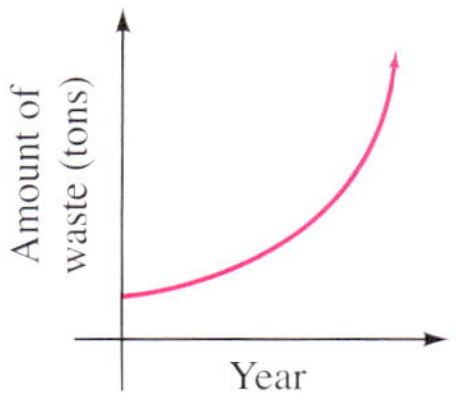

Graph each function and give the domain and the range. Label the y-intercept.

27. $f(x) = 3^x$

28. $f(x) = \left(\frac{1}{3}\right)^x$

29. $f(x) = \left(\frac{1}{2}\right)^x - 2$

30. $f(x) = 3^{x-1}$

31. In Exercise 30, what is the asymptote of the graph of $f(x) = 3^{x-1}$?

32. **Coal Production.** The function $c(t) = 128{,}000\,(1.08)^t$ approximates the number of tons of coal produced in the United States for the years 1800–1910, where t is the number of years after 1800. How many tons of coal did the U.S. produce in:

a. 1800? b. 1900?

33. Compound Interest. How much will \$10,500 become if it earns 9% annual interest, compounded quarterly, for 60 years?

34. Depreciation. The value (in dollars) of a certain model car is given by the function $V(t) = 12{,}000(10^{-0.155t})$, where t is the number of years from the present. Find the value of the car in 5 years.

SECTION 11.4 Logarithmic Functions

DEFINITIONS AND CONCEPTS	EXAMPLES
Definition of logarithm: If $b > 0$, $b \neq 1$, and x is positive, then $y = \log_b x$ is equivalent to $x = b^y$ (Exponent: y; Base: b)	***Logarithmic form*** / ***Exponential form*** $\log_5 125 = 3$ means $5^3 = 125$ (Exponent: 3; Base: 5) $\log_2 \frac{1}{16} = -4$ means $2^{-4} = \frac{1}{16}$ (Exponent: −4; Base: 2)
$\log_b x$ is the exponent to which b is raised to get x. $b^{\log_b x} = x$	To evaluate $\log_4 16$ we ask: "To what power must we raise 4 to get 16?" Since $4^2 = 16$, the answer is: the 2nd power. Thus, $\log_4 16 = 2$.
We cannot find the logarithm of 0 or a negative number.	$\log_4 0$ is undefined. $\log_8 (-64)$ is undefined.
For calculation purposes and in many applications, we use base-10 logarithms, called **common logarithms.** $\log x$ means $\log_{10} x$	$\log 1{,}000{,}000 = 6$ because $10^6 = 1{,}000{,}000$. $\log \frac{1}{1{,}000} = -3$ because $10^{-3} = \frac{1}{1{,}000}$.
If $b > 0$ and $b \neq 1$, the **logarithmic function with base b** is defined by $f(x) = \log_b x$. The domain is $(0, \infty)$ and the range is $(-\infty, \infty)$. If $b > 1$, then $f(x) = \log_b x$ is an increasing function. If $0 < b < 1$, then $f(x) = \log_b x$ is a decreasing function. The graphs of logarithmic functions can be **translated** horizontally and vertically.	The graph of the logarithmic function $f(x) = \log_2 x$ is shown at the right. From the graph, we see that $f(x) = \log_2 x$ is an increasing function. To graph $g(x) = \log_2 (x - 3)$, move the graph of $f(x) = \log_2 x$ to the right 3 units. 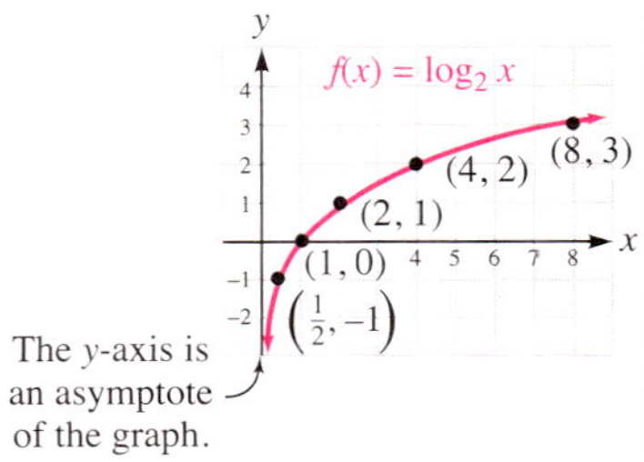
The exponential function $f(x) = b^x$ and the logarithmic function $f(x) = \log_b x$ are inverses of each other.	$f(x) = 3^x$ and $f^{-1}(x) = \log_3 x$ are inverses of each other. Their graphs are symmetric about the line $y = x$. Similarly, $f(x) = 10^x$ and $f^{-1}(x) = \log x$ are inverses. 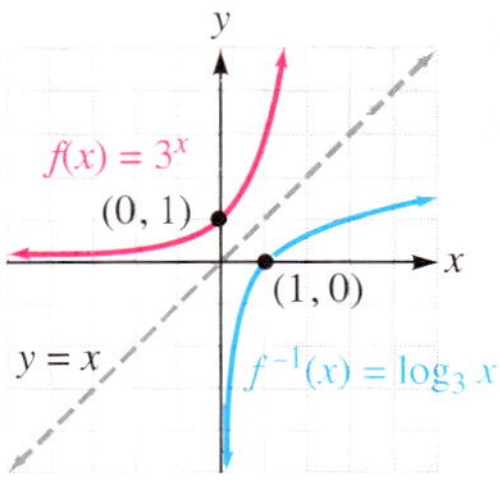

Logarithmic functions, like exponential functions, can be used to **model** certain types of growth and decay. **Decibel voltage gain:** dB gain $= 20 \log \frac{E_O}{E_I}$ **The Richter scale:** $R = \log \frac{A}{P}$	The input to an amplifier is 0.4 volt and the output is 30 volts. Find the dB gain. $\text{dB gain} = 20 \log \frac{E_O}{E_I}$ $= 20 \log \frac{30}{0.4}$ Substitute 30 for E_O and 0.4 for E_I. ≈ 37.50122527 Use a calculator with a LOG key. The dB gain is about 38 decibels.

REVIEW EXERCISES

35. Give the domain and range of $f(x) = \log x$.

36. Explain why a student got an error message when she used a calculator to evaluate log 0.

37. Write the statement $\log_4 64 = 3$ in exponential form.

38. Write the statement $7^{-1} = \frac{1}{7}$ in logarithmic form.

Evaluate.

39. $\log_3 9$

40 $\log_9 \frac{1}{81}$

41. $\log_{1/2} 1$

42. $\log_5 (-25)$

43. $\log_6 \sqrt{6}$

44. $\log 1{,}000$

Solve for x.

45. $\log_2 x = 5$

46. $\log_3 x = -4$

47. $\log_x 16 = 2$

48. $\log_x \frac{1}{100} = -2$

49. $\log_9 3 = x$

50. $\log_{27} 3 = x$

Use a calculator to find the value of x to four decimal places.

51. $\log 4.51 = x$

52. $\log x = 1.43$

Graph each function and its inverse on the same coordinate system. Draw the axis of symmetry.

53. $f(x) = \log_4 x$ and $g(x) = 4^x$

54. $f(x) = \log_{1/3} x$ and $g(x) = \left(\frac{1}{3}\right)^x$

Graph each function. Label the x-intercept.

55. $f(x) = \log (x - 2)$

56. $f(x) = 3 + \log x$

57. Electrical Engineering. Find the dB gain of an amplifier with an output of 18 volts and an input of 0.04 volt.

58. Earthquakes. An earthquake had a period of 0.3 second and an amplitude of 7,500 micrometers. Use the formula $R = \log \frac{A}{P}$ to find its measure on the Richter scale.

59. Organ Pipes. The design for a set of brass pipes for a church organ is shown below. The function $h(n) = 52 + 25 \log n$ approximates the height (in centimeters) of the pipe with number n. Find the height of

a. Pipe 1

b. Pipe 8

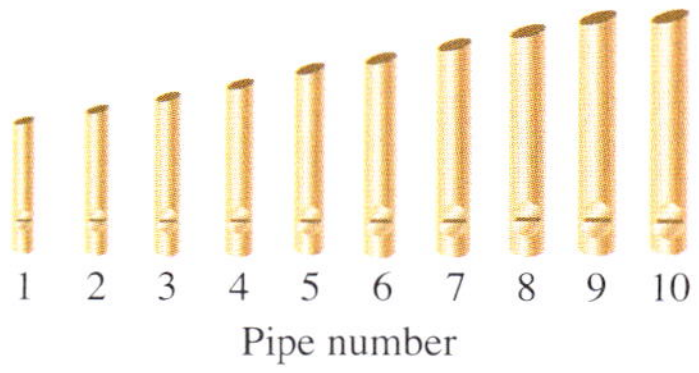

60. Girls' Heights. The function $P(A) = 61.8 + 34.9 \log(A - 4)$ approximates the percent of the adult height a female child A years old has attained, where $5 \le A \le 15$. What percent of her adult height will a girl have reached the day she first becomes a teenager?

SECTION 11.5 Base-*e* Exponential and Logarithmic Functions

DEFINITIONS AND CONCEPTS	EXAMPLES
Of all possible bases for an exponential function, ***e*** is the most convenient for problems involving growth or decay. $e = 2.718281828459 \ldots$ As n approaches infinity, the value of $\left(1 + \frac{1}{n}\right)^n$ approaches e. The function defined by $\mathbf{f(x) = e^x}$ is called the **natural exponential function.** It is called this because it can be applied to many natural settings.	From the graph, we see that the domain of the natural exponential function is $(-\infty, \infty)$ and the range is $(0, \infty)$. The graph of $f(x) = e^x$ can be **translated** horizontally and vertically. 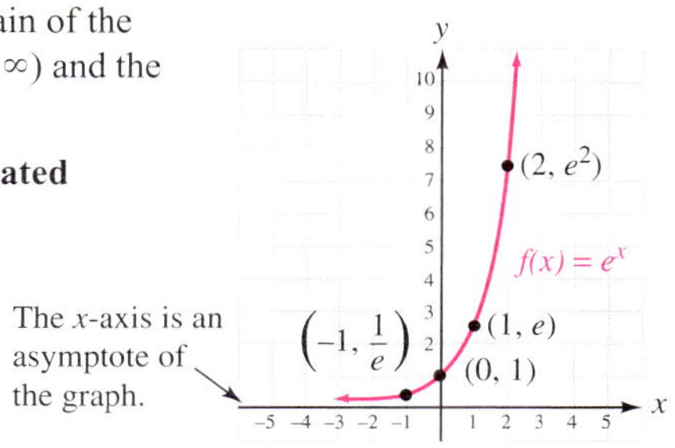

Exponential growth/decay: If a quantity increases or decreases at an annual rate r, **compounded continuously,** the amount A after t years is given by

$$A = Pe^{rt}$$

If r is negative, the amount decreases.

If interest is paid coutinuously (infinitely many times a year), we say that the interest is **compounded continuously.**

If \$30,000 accumulates interest at an annual rate of 9%, compounded continuously, find the amount in the account after 25 years.

$A = Pe^{rt}$	This is the formula for continuous compound interest.
$= 30{,}000e^{0.09 \cdot 25}$	Substitute 30,000 for P, 0.09 for r, and 25 for t.
$= 30{,}000e^{2.25}$	Evaluate the exponent: $0.09 \cdot 25 = 2.25$
$\approx 284{,}632.08$	Use a calculator with an e^x key.

In 30 years, the account will contain \$284,632.08.

Suppose the population of a city of 50,000 people is decreasing exponentially according to the function $P(t) = 50{,}000e^{-0.003t}$, where t is measured in years from the present date. Find the expected population of the city in 20 years.

$P(t) = 50{,}000e^{-0.003t}$	Since r is negative, this is the exponential decay model.
$P(20) = 50{,}000e^{-0.003(20)}$	Substitute 20 for t.
$= 50{,}000e^{-0.06}$	Evaluate the exponent: $-0.003(20) = -0.06$.
$\approx 47{,}088$	Use a calculator with an e^x key.

After 20 years, the expected population will be about 47,088 people.

Of all possible bases for a logarithmic function, e is the most convenient for problems involving growth or decay. Since these situations occur often in natural settings, base-e logarithms are called **natural logarithms:**

$\ln x$ means $\log_e x$

To evaluate $\ln \frac{1}{e^4}$ we ask: "To what power must we raise e to get $\frac{1}{e^4}$?"

Since $e^{-4} = \frac{1}{e^4}$, the answer is: the -4th power. Thus,

$$\ln \frac{1}{e^4} = -4$$

$\ln \mathbf{0}$ and $\ln(\mathbf{-5})$ are undefined.

$\ln x$ is the exponent to which e is raised to get x.

$\ln 9 \approx \mathbf{2.1972}$ means $e^{\mathbf{2.1972}} \approx 9$

The **natural logarithmic function** with base e is defined by

$$f(x) = \ln x$$

The domain is the interval $(0, \infty)$ and the range is the interval $(-\infty, \infty)$.

The graph of the natural logarithmic function $f(x) = \ln x$ is shown on the right.

From the graph, we see that $f(x) = \ln x$ is an increasing function.

The graph of $f(x) = \ln x$ can be translated horizontally and vertically.

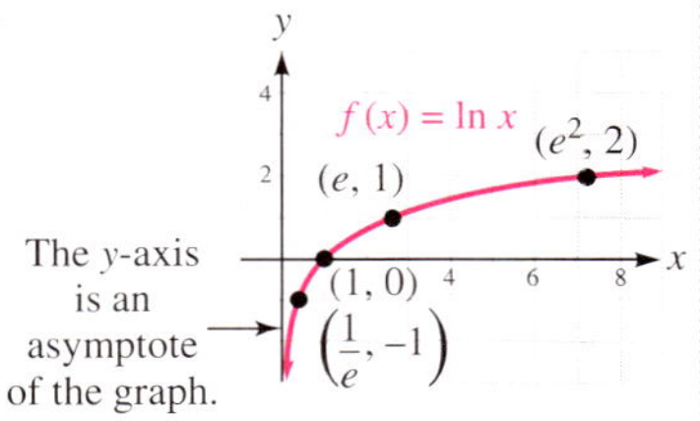

The natural exponential function $f(x) = e^x$ and the natural logarithmic function $f^{-1}(x) = \ln x$ are **inverses** of each other.

The graphs are symmetric about the line $y = x$.

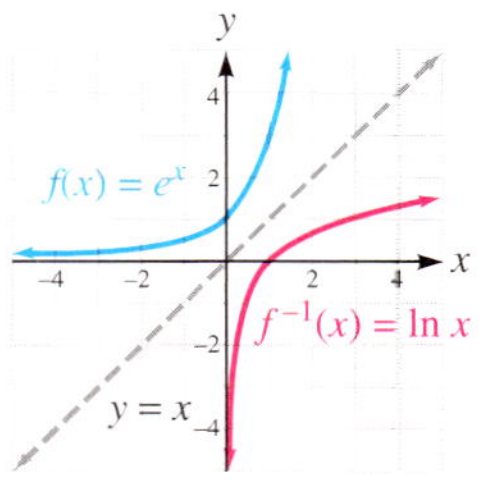

If a population grows exponentially at a certain annual rate r, the time required for the population to double is called the **doubling time.** It is given by the formula:

$$t = \frac{\ln 2}{r}$$

The population of a town is growing at a rate of 3% per year. If this rate continues, how long will it take the population to double?

We substitute 0.03 for r and use a calculator to perform the calculation.

$$t = \frac{\ln 2}{r} = \frac{\ln 2}{0.03} \approx 23.10490602$$ Use a calculator with an LN key.

The population will double in about 23.1 years.

REVIEW EXERCISES

61. a. Approximate e to the nearest hundredth.
 b. Fill in the blanks:
 $\ln 15 \approx 2.7081$ means $e^{\_\_\_\_} \approx \_\_\_\_$

62. Use a calculator to find $16.4 + 252.7e^{-0.76(9)}$. Round to four decimal places.

Graph each function, and give the domain and the range.

63. $f(x) = e^x + 1$

64. $f(x) = e^{x-3}$

65. **Interest Compounded Continuously.** If \$10,500 accumulates interest at an annual rate of 9%, compounded continuously, how much will be in the account in 60 years?

66. **The Sooner State.** In 2009, Oklahoma had the largest *gross domestic product* growth rate of all fifty states: 6.6%. (The GDP, as it is called, is the value of all goods and services produced within a state.) If the 2009 GDP totaled \$142.5 billion, predict Oklahoma's GDP in 2015, assuming the growth rate remains the same. (Source: huffingtonpost.com)

67. **Mortgage Rates.** There was the housing boom in the 1980s as the baby boomers (those born from 1946 through 1964) bought their homes. The average annual interest rate in percent on a 30-year fixed-rate home mortgage for the years 1980–1996 can be approximated by the function $r(t) = 13.9e^{-0.035t}$, where t is the number of years since 1980. To the nearest hundredth of a percent, what does this model predict was the 30-year fixed rate in 1980? In 1985? In 1990?

68. **Medical Tests.** A radioactive dye is injected into a patient as part of a test to detect heart disease. The amount of dye remaining in his bloodstream t hours after the injection is given by the function $f(t) = 10e^{-0.27t}$. How can you determine from the function that the amount of dye in the bloodstream is decreasing?

Evaluate each expression. Do not use a calculator.

69. $\ln e$

70. $\ln e^2$

71. $\ln \frac{1}{e^5}$

72. $\ln \sqrt{e}$

73. $\ln(-e)$

74. $\ln 0$

75. $\ln 1$

76. $\ln e^{-7}$

Use a calculator to evaluate each expression. Express all answers to four decimal places.

77. $\ln 452$

78. $\ln 0.85$

Solve each equation. Express all answers to four decimal places.

79. $\ln x = 2.336$

80. $\ln x = -8.8$

81. Explain the difference between the functions $f(x) = \log x$ and $g(x) = \ln x$.

82. What function is the inverse of $f(x) = \ln x$?

Graph each function.

83. $f(x) = 1 + \ln x$

84. $f(x) = \ln(x + 1)$

85. **Population Growth.** How long will it take the population of Mexico to double if the growth rate is currently about 1.118%? (Source: CIA World Fact Book)

86. **Botany.** The height (in inches) of a certain plant is approximated by the function $H(a) = 13 + 20.03 \ln a$, where a is its age in years. How tall will it be when it is 19 years old?

SECTION 11.6 Properties of Logarithms

DEFINITIONS AND CONCEPTS	EXAMPLES
Properties of logarithms: If M, N, and b are positive real numbers, $b \neq 1$	Apply a property of logarithms and then simplify, if possible.
1. $\log_b 1 = 0$ 2. $\log_b b = 1$ 3. $\log_b b^x = x$ 4. $b^{\log_b x} = x$	1. $\log_3 1 = 0$ 2. $\log_7 7 = 1$ 3. $\log_5 5^3 = 3$ 4. $9^{\log_9 10} = 10$
5. ***Product rule for logarithms:*** $\log_b MN = \log_b M + \log_b N$	5. $\log_2(6 \cdot 8) = \log_2 6 + \log_2 8$ $= \log_2 6 + 3$
6. ***Quotient rule for logarithms:*** $\log_b \frac{M}{N} = \log_b M - \log_b N$	6. $\log_3 \frac{81}{x} = \log_3 81 - \log_3 x$ $= 4 - \log_3 x$
7. ***Power rule for logarithms:*** $\log_b M^p = p \log_b M$	7. $\log_8 7^3 = 3 \log_8 7$ and $\log \sqrt[4]{x} = \log x^{\frac{1}{4}} = \frac{1}{4} \log x$

Properties of logarithms can be used to **expand** logarithmic expressions.	Write $\log_3 (x^2y^3)$ as the sum and/or difference of logarithms of a single quantity. $\log_3 (x^2y^3) = \log_3 x^2 + \log_3 y^3$ The log of a product is the sum of the logs. $= 2 \log_3 x + 3 \log_3 y$ The log of a power is the power times the log.
Properties of logarithms can be used to **condense** certain logarithmic expressions. To condense, apply the power rule first to make the coefficients of logarithms 1. Then use the product and quotient rules.	Write $3 \ln x - \frac{1}{2} \ln y$ as a single logarithm. $3 \ln x - \frac{1}{2} \ln y = \ln x^3 - \ln y^{1/2}$ A power times a log is the log of the power. $= \ln \frac{x^3}{y^{1/2}}$ The difference of two logs is the log of the quotient. $= \ln \frac{x^3}{\sqrt{y}}$ Write $y^{1/2}$ as $\sqrt{y}$.
If we need to find a logarithm with some base other than 10 or *e*, we can use a conversion formula. **Change-of-base formula:** $\log_b x = \frac{\log_a x}{\log_a b}$	Find $\log_7 6$ to four decimal places. $\log_7 6 = \frac{\log 6}{\log 7} \approx 0.920782221$ Change to the ratio of base-10 logarithms. To four decimal places, $\log_7 6 = 0.9208$. To check, verify that $7^{0.9208}$ is approximately 6.
In chemistry, common logarithms are used to express the acidity of solutions using pH. **pH scale:** $\text{pH} = -\log[\text{H}^+]$ where the symbol $[\text{H}^+]$ represents the hydrogen ion concentration in gram-ions per liter.	Find the pH of a liquid with a hydrogen ion concentration of 10^{-8} gram-ions per liter. $\text{pH} = -\log [\text{H}^+]$ This is the pH formula. $\text{pH} = -\log 10^{-8}$ Substitute 10^{-8} for $[\text{H}^+]$. $= -(-8) \log 10$ The log of a power is the power times the log. $= 8$ Simplify: $\log 10 = 1$.

REVIEW EXERCISES

Simplify each expression.

87. $\log_2 1$
88. $\log_9 9$
89. $\log 10^3$
90. $7^{\log_7 4}$

Write each logarithm as the sum and/or difference of logarithms of a single quantity. Then simplify, if possible.

91. $\log_3 27x$
92. $\log \frac{100}{x}$
93. $\log_5 \sqrt{27}$
94. $\log_b 10ab$

Write each logarithm as the sum and/or difference of logarithms of a single quantity.

95. $\log_b \frac{x^2y^3}{z}$
96. $\ln \sqrt{\frac{x}{yz^2}}$

Write each logarithmic expression as one logarithm.

97. $3 \log_2 x - 5 \log_2 y + 7 \log_2 z$
98. $-3 \log_b y - 7 \log_b z + \frac{1}{2} \log_b (x + 2)$
99. $\log_b (a^2 - 25) - \log_b (a + 5)$
100. $3 \log_8 x + 4 \log_8 x$

Assume that $\log_b 5 = 1.1609$ and $\log_b 8 = 1.5000$ and find each value to four decimal places.

101. $\log_b 40$
102. $\log_b 64$
103. Find $\log_5 17$ to four decimal places.
104. **pH of Grapefruit.** Find the pH of grapefruit juice if its hydrogen ion concentration is 7.9×10^{-4} gram-ions per liter. Round to the nearest tenth.

SECTION 11.7 Exponential and Logarithmic Equations

DEFINITIONS AND CONCEPTS	EXAMPLES
An **exponential equation** contains a variable in one of its exponents. Two examples are: $6^{x-3} = 9$ and $e^{-2.5t} = 56$ If both sides of an exponential equation can be expressed as a power of the same base, we can use the following property to solve it: $b^x = b^y$ is equivalent to $x = y$	Solve: $3^{x+2} = 27$ $3^{x+2} = 3^3$ Express the right side of the equation as a power of 3: $27 = 3^3$. $x + 2 = 3$ If two exponential expressions with the same base are equal, their exponents are equal. $x = 1$ The solution is 1. Check it in the original equation.
When it is difficult to write each side of an exponential equation as a power of the same base, **take the logarithm on each side.** With this method, we often obtain **exact solutions** involving logarithms that we can **approximate.**	Solve: $4^x = 7$. Give an approximate solution to four decimal places. $\log 4^x = \log 7$ Take the base-10 logarithm on both sides of the equation. $x \log 4 = \log 7$ The log of a power is the power times the log. $x = \frac{\log 7}{\log 4}$ To isolate x, divide both sides by log 4. This is the exact solution. $x \approx 1.4037$ Use a calculator with a LOG key. To four decimal places, the approximate solution is 1.4037. To check it, we substitute 1.4037 for x in $4^x = 7$ and use a calculator to evaluate the left side: $4^{1.4037} \approx 7$
To solve an exponential equation that contains a base-e exponential expression, take the natural logarithm on both sides of the equation.	Solve: $e^{5t} = 11$ $\ln e^{5t} = \ln 11$ Take the base-e logarithm on both sides. $5t \ln e = \ln 11$ The log of a power is the power times the log. $5t \cdot 1 = \ln 11$ Simplify: $\ln e = 1$. $5t = \ln 11$ Simplify the left side. $t = \frac{\ln 11}{5}$ This is the exact solution. An approximate solution to four decimal places is 0.4796.
A **logarithmic equation** is an equation containing a variable in a logarithmic expression. Two examples are: $2 \log_3 x + \log_3 5x = 4$ and $\ln(x + 1) = \ln(3x - 4)$ Certain logarithmic equations can be solved using the following property: $\log_b x = \log_b y$ is equivalent to $x = y$	Solve: $\log(4x - 3) = \log(2x + 7)$ $4x - 3 = 2x + 7$ If the logarithms of two numbers are equal, the numbers are equal. $2x = 10$ $x = 5$ The solution is 5. Check it in the original equation.
To solve some logarithmic equations, we write and solve an equivalent exponential equation.	Solve: $\log_4(x + 1) = 2$ $x + 1 = 4^2$ Write the equivalent base-4 exponential equation. $x + 1 = 16$ $x = 15$ The solution is 15. Check it in the original equation.

To solve some logarithmic equations, we first apply properties of logarithms, such as:

- The product rule
- The quotient rule
- The power rule

Solve: $\log_2 (x + 14) - \log_2 x = 3$

$$\log_2 \frac{x + 14}{x} = 3$$ On the left side, use the quotient rule for logarithms.

$$\frac{x + 14}{x} = 2^3$$ Write the equivalent base-2 exponential equation.

$$\frac{x + 14}{x} = 8$$ Evaluate: $2^3 = 8$.

$$x + 14 = 8x$$ Multiply both sides by x.

$$14 = 7x$$ Subtract x from both sides.

$$2 = x$$ Divide both sides by 7.

The solution is 2. Check by substituting it into the original equation.

When there is sufficient food and space available, populations of living organisms tend to increase exponentially according to the following **growth model.**

Population growth: $P = P_0e^{kt}$

Find the number of bacteria in a culture of 1,000 bacteria if they are allowed to reproduce for 5 hours. Assume $k = \frac{\ln 3}{3}$.

$$P = P_0e^{kt}$$ This is the population growth model.

$$= 1{,}000e^{\frac{\ln 3}{3} \cdot 5}$$ Substitute for P_0, k, and t.

$$\approx 6{,}240$$ Use a calculator with LN and e^x keys.

In 5 hours, there will be approximately 6,240 bacteria.

REVIEW EXERCISES

Solve each equation. Give the exact solution and an approximate solution to four decimal places, when appropriate.

105. $5^{x+6} = 25$

106. $2^{x^2+4x} = \frac{1}{8}$

107. $3^x = 7$

108. $2^x = 3^{x-4}$

109. $e^x = 7$

110. $e^{-0.4t} = 25$

Solve each equation.

111. $\left(\frac{2}{5}\right)^{3x-4} = \frac{8}{125}$

112. $9^{x^2} = 33$

113. $\log (x - 4) = 2$

114. $\ln (2x - 3) = \ln 15$

115. $\log x + \log (29 - x) = 2$

116. $\log_2 x + \log_2 (x - 2) = 3$

117. $\frac{\log (7x - 12)}{\log x} = 2$

118. $\log_2 (x + 2) + \log_2 (x - 1) = 2$

119. $\log x + \log (x - 5) = \log 6$

120. $\log 3 - \log (x - 1) = -1$

121. Evaluate both sides of the statement $\frac{\log 8}{\log 15} \neq \log 8 - \log 15$ to show that the sides are indeed not equal.

122. **Carbon-14 Dating.** A wooden statue found in Egypt has a carbon-14 content that is two-thirds of that found in living wood. If the half-life of carbon-14 is 5,700 years, how old is the statue?

123. **Ants.** The number of ants in a colony is estimated to be 800. If the ant population is expected to triple every 14 days, how long will it take for the population to reach one million?

124. The approximate coordinates of the points of intersection of the graphs of $f(x) = \log x$ and $g(x) = 1 - \log (7 - x)$ are shown in parts (a) and (b) of the illustration. Use the graphs to estimate the solutions of the logarithmic equation $\log x = 1 - \log (7 - x)$. Then check your answers.

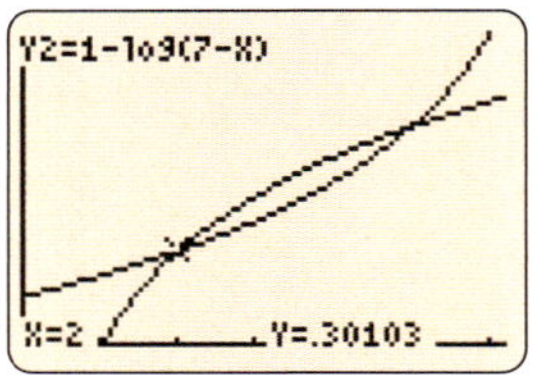

(a)

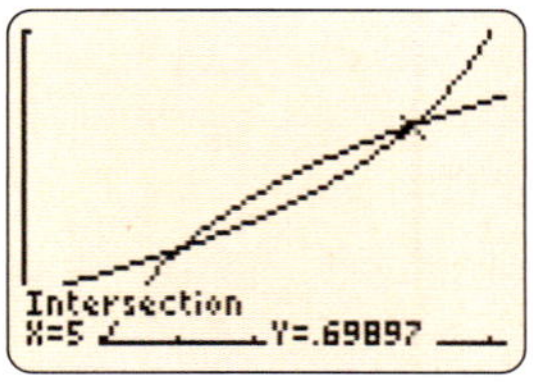

(b)

11 Chapter Test

1. Fill in the blanks.
   a. A __________ function is denoted by $f \circ g$.
   b. $f(x) = e^x$ is the _______ exponential function.
   c. In __________ compound interest, the number of compoundings is infinitely large.
   d. The functions $f(x) = \log_{10} x$ and $f(x) = 10^x$ are _______ functions.
   e. $f(x) = \log_4 x$ is a __________ function.
2. Write out in words how to say each of the following:
   a. $(f \circ g)(x)$
   b. $g(f(8))$
   c. $f^{-1}(x)$

Let $f(x) = x + 9$ and $g(x) = 4x^2 - 3x + 2$. Find each function and give its domain.

3. $f + g$
4. g/f

Let $f(x) = 2x^2 + 3$ and $g(x) = 4x - 8$. Find each composition.

5. $(g \circ f)(-3)$
6. $(f \circ g)(x)$

Use the tables of values for functions f and g to find each of the following.

7. a. $(f \cdot g)(9)$
   b. $(f \circ g)(-3)$

x	$f(x)$
9	−1
10	17

x	$g(x)$
−3	10
9	16

8. Refer to the graphs of functions f and g below to find each of the following.
   a. $(g/f)(-4)$
   b. $(f \circ g)(1)$
   c. $(f + g)(2)$
   d. $(f \cdot g)(0)$
   e. $(g - f)(1)$

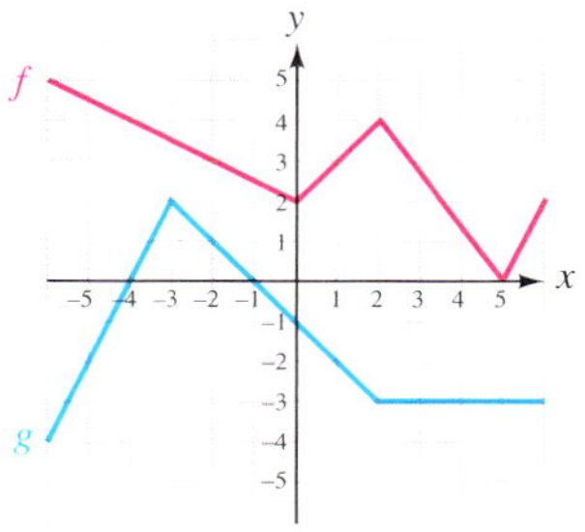

9. Determine whether each function is one-to-one.
   a. $f(x) = |x|$
   b. $\{(1,7), (7,1), (2,8), (8,2)\}$
   c.

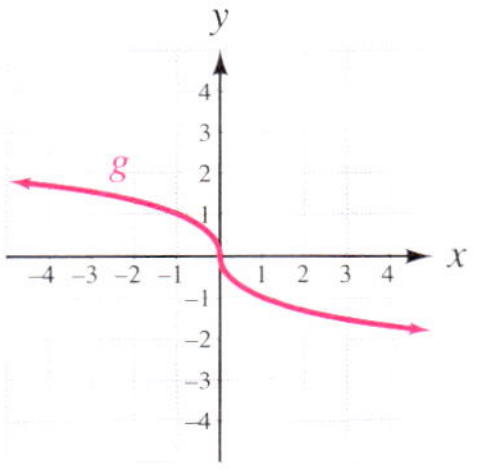

   d.

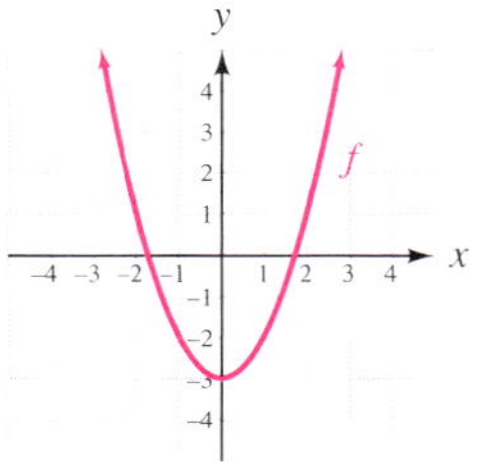

10. Find the inverse of $f(x) = -\frac{1}{3}x$ and then graph f and its inverse on the same coordinate axes. Label the axis of symmetry.
11. Determine whether $f(x) = \frac{1}{3}x + 2$ is a one-to-one function. If it is, find its inverse.
12. Find the inverse of $f(x) = (x - 15)^3$.
13. Use composition to show that $f(x) = 4x + 4$ and $f^{-1}(x) = \frac{x - 4}{4}$ are inverse functions.
14. Consider the following graph of the function f.
    a. Is f a one-to-one function?
    b. Is its inverse a function?
    c. What is $f^{-1}(260)$? What information does it give?

Relationship Between Car Speed and Tire Temperature

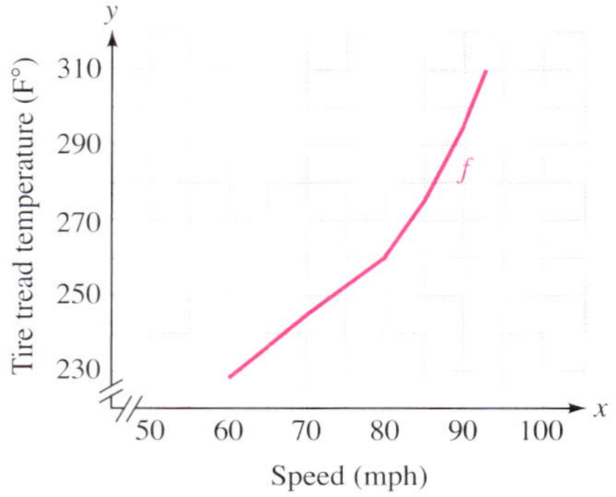

Graph each function and give the domain and the range.

15. $f(x) = 2^x + 1$
16. $f(x) = 3^{-x}$
17. **Radioactive Decay.** A radioactive material decays according to the formula $A = A_0(2)^{-t}$. How much of a 3-gram sample will be left in 6 years?
18. **Compound Interest.** An initial deposit of \$1,000 earns 6% interest, compounded twice a year. How much will be in the account in one year?
19. a. Graph $f(x) = e^x$. Label the y-intercept and the asymptote of the graph.
    b. Give the domain and range.
    c. What is the inverse of $f(x) = e^x$?
20. **Population Growth.** As of July 2010, the population of India was estimated to be 1,173,108,018, with an annual growth rate of 1.376%. If the growth rate remains the same, how large will the population be in July, 2020? Round to the nearest thousand. (Source: CIA World Fact Book)

21. Biology. Human growth hormone, known as HGH, is produced by the pituitary gland in the brain and released into the blood stream. It stimulates growth and cell production. After the age of 20, levels of HGH in the body decrease dramatically, as shown in the graph. Use the given function to approximate the amount of HGH produced per day by a person 55 years old.

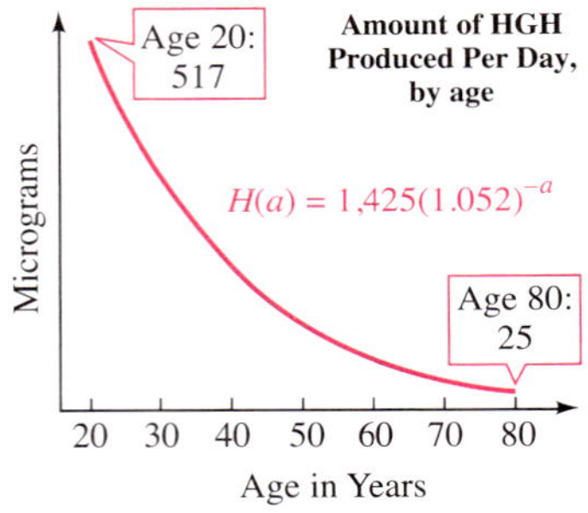

Source: Aging Well, James L.Holly, MD

22. Write the statement $\log_6 \frac{1}{36} = -2$ in exponential form.

23. a. What are the domain and range of the function $f(x) = \log x$?

b. What is the inverse of $f(x) = \log x$?

24. Botany. Which phrase best describes the relationship between the rate of photosynthesis in plants and light intensity that is graphed below: linear growth, exponential growth, or logarithmic growth?

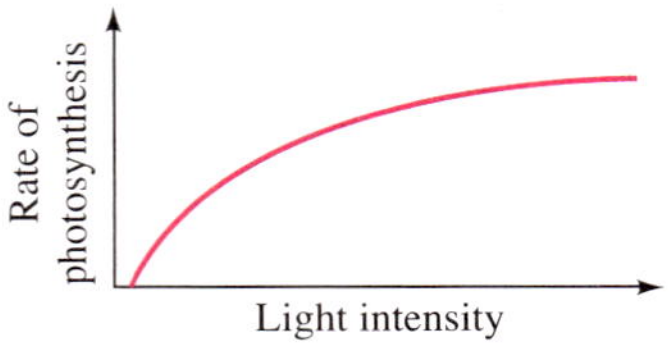

Evaluate each logarithmic expression, if possible.

25. $\log_5 25$

26. $\log_9 \frac{1}{81}$

27. $\log(-100)$

28. $\ln \frac{1}{e^6}$

29. $\log_4 2$

30. $\log_{1/3} 1$

Solve for x.

31. $\log_x 32 = 5$

32. $\log_8 x = \frac{4}{3}$

33. $\log_3 x = -3$

34. $\ln x = 1$

Graph each function.

35. $f(x) = -\log_3 x$

36. $f(x) = \ln x$

37. Chemistry. Find the pH of a solution with a hydrogen ion concentration of 3.7×10^{-7} gram-ions per liter. (*Hint:* $\text{pH} = -\log[\text{H}^+]$.)

38. Electronics. Find the dB gain of an amplifier when $E_O = 60$ volts and $E_I = 0.3$ volt. *Hint:* dB gain $= 20 \log \frac{E_O}{E_I}$.

39. Use a calculator to find x to four decimal places: $\log x = -1.06$

40. Use the change-of-base formula to find $\log_7 3$ to four decimal places.

41. Write the expression $\log_b a^2bc^3$ as the sum and/or difference of logarithms of a single quantity. Then simplify, if possible.

42. Write the expression $\frac{1}{2}\ln(a + 2) + \ln b - 3\ln c$ as a logarithm of a single quantity.

Solve each equation. Give the exact solution and an approximate solution to four decimal places, when appropriate.

43. $5^x = 3$

44. $3^{x-1} = 27$

45. $\left(\frac{3}{2}\right)^{6x+2} = \frac{27}{8}$

46. $e^{0.08t} = 4$

47. $2\log x = \log 25$

48. $\log_2(x + 2) - \log_2(x - 5) = 3$

49. $\ln(5x + 2) = \ln(2x + 5)$

50. $\log x + \log(x - 9) = 1$

51. The illustration shows the graphs of $y = \frac{1}{2}\ln(x - 1)$ and $y = \ln 2$ and the approximate coordinates of their point of intersection. Estimate the solution of the logarithmic equation $\frac{1}{2}\ln(x - 1) = \ln 2$. Then check the result.

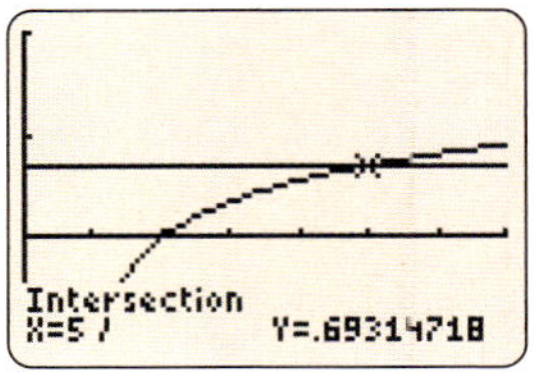

52. Insects. The number of insects attracted to a bright light is currently 5. If the number is expected to quadruple every 6 minutes, how long will it take for the number to reach 500?

Group Project

The Number *e*

Overview: In this activity, you will use a calculator to find progressively more accurate approximations of *e*.

Instructions: Form groups of two students. Each student will need a scientific calculator.

Begin by finding an approximation of *e* using the e^x key on your calculator. Copy the table shown below, and write the number displayed on the calculator screen at the top of the table.

The value of *e* can be calculated to any degree of accuracy by adding the terms of the following pattern:

$$e = 1 + 1 + \frac{1}{2} + \frac{1}{2 \cdot 3} + \frac{1}{2 \cdot 3 \cdot 4} + \frac{1}{2 \cdot 3 \cdot 4 \cdot 5} + \cdots$$

The more terms that are added, the closer the sum will be to *e*.

You are to add as many terms as necessary until you obtain a sum that matches the value of *e*, given by the e^x key on your calculator. Work together as a team. One member of the group should compute the fractional form of the term to be added. (See the middle column of the table.) The other member should take that information and calculate the cumulative sum. (See the right column of the table.)

How many terms must be added so that the cumulative sum approximation and the e^x key approximation match in each decimal place?

Approximation of *e* found using the e^x key: $e \approx$ ____________________

Number of terms in the sum	Term (expressed as a fraction)	Cumulative sum (an approximation of *e*)
1	1	1
2	1	2
3	$\frac{1}{2}$	2.5
4	$\frac{1}{2 \cdot 3} = \frac{1}{6}$	2.666666667
⋮	⋮	⋮

APPENDIX 1 Roots and Powers

n	n^2	$\sqrt{n}$	n^3	$\sqrt[3]{n}$	n	n^2	$\sqrt{n}$	n^3	$\sqrt[3]{n}$
1	1	1.000	1	1.000	51	2,601	7.141	132,651	3.708
2	4	1.414	8	1.260	52	2,704	7.211	140,608	3.733
3	9	1.732	27	1.442	53	2,809	7.280	148,877	3.756
4	16	2.000	64	1.587	54	2,916	7.348	157,464	3.780
5	25	2.236	125	1.710	55	3,025	7.416	166,375	3.803
6	36	2.449	216	1.817	56	3,136	7.483	175,616	3.826
7	49	2.646	343	1.913	57	3,249	7.550	185,193	3.849
8	64	2.828	512	2.000	58	3,364	7.616	195,112	3.871
9	81	3.000	729	2.080	59	3,481	7.681	205,379	3.893
10	100	3.162	1,000	2.154	60	3,600	7.746	216,000	3.915
11	121	3.317	1,331	2.224	61	3,721	7.810	226,981	3.936
12	144	3.464	1,728	2.289	62	3,844	7.874	238,328	3.958
13	169	3.606	2,197	2.351	63	3,969	7.937	250,047	3.979
14	196	3.742	2,744	2.410	64	4,096	8.000	262,144	4.000
15	225	3.873	3,375	2.466	65	4,225	8.062	274,625	4.021
16	256	4.000	4,096	2.520	66	4,356	8.124	287,496	4.041
17	289	4.123	4,913	2.571	67	4,489	8.185	300,763	4.062
18	324	4.243	5,832	2.621	68	4,624	8.246	314,432	4.082
19	361	4.359	6,859	2.668	69	4,761	8.307	328,509	4.102
20	400	4.472	8,000	2.714	70	4,900	8.367	343,000	4.121
21	441	4.583	9,261	2.759	71	5,041	8.426	357,911	4.141
22	484	4.690	10,648	2.802	72	5,184	8.485	373,248	4.160
23	529	4.796	12,167	2.844	73	5,329	8.544	389,017	4.179
24	576	4.899	13,824	2.884	74	5,476	8.602	405,224	4.198
25	625	5.000	15,625	2.924	75	5,625	8.660	421,875	4.217
26	676	5.099	17,576	2.962	76	5,776	8.718	438,976	4.236
27	729	5.196	19,683	3.000	77	5,929	8.775	456,533	4.254
28	784	5.292	21,952	3.037	78	6,084	8.832	474,552	4.273
29	841	5.385	24,389	3.072	79	6,241	8.888	493,039	4.291
30	900	5.477	27,000	3.107	80	6,400	8.944	512,000	4.309
31	961	5.568	29,791	3.141	81	6,561	9.000	531,441	4.327
32	1,024	5.657	32,768	3.175	82	6,724	9.055	551,368	4.344
33	1,089	5.745	35,937	3.208	83	6,889	9.110	571,787	4.362
34	1,156	5.831	39,304	3.240	84	7,056	9.165	592,704	4.380
35	1,225	5.916	42,875	3.271	85	7,225	9.220	614,125	4.397
36	1,296	6.000	46,656	3.302	86	7,396	9.274	636,056	4.414
37	1,369	6.083	50,653	3.332	87	7,569	9.327	658,503	4.431
38	1,444	6.164	54,872	3.362	88	7,744	9.381	681,472	4.448
39	1,521	6.245	59,319	3.391	89	7,921	9.434	704,969	4.465
40	1,600	6.325	64,000	3.420	90	8,100	9.487	729,000	4.481
41	1,681	6.403	68,921	3.448	91	8,281	9.539	753,571	4.498
42	1,764	6.481	74,088	3.476	92	8,464	9.592	778,688	4.514
43	1,849	6.557	79,507	3.503	93	8,649	9.644	804,357	4.531
44	1,936	6.633	85,184	3.530	94	8,836	9.695	830,584	4.547
45	2,025	6.708	91,125	3.557	95	9,025	9.747	857,375	4.563
46	2,116	6.782	97,336	3.583	96	9,216	9.798	884,736	4.579
47	2,209	6.856	103,823	3.609	97	9,409	9.849	912,673	4.595
48	2,304	6.928	110,592	3.634	98	9,604	9.899	941,192	4.610
49	2,401	7.000	117,649	3.659	99	9,801	9.950	970,299	4.626
50	2,500	7.071	125,000	3.684	100	10,000	10.000	1,000,000	4.642

APPENDIX 2 Synthetic Division

We have discussed how to divide polynomials by polynomials using a long division process. We will now discuss a shortcut method, called **synthetic division,** that we can use to divide a polynomial by a binomial of the form $x - k$.

1 Synthetic Division.

To see how synthetic division works, we consider $(4x^3 - 5x^2 - 11x + 20) \div (x - 2)$. On the left below is the long division, and on the right is the same division with the variables and their exponents removed. The various powers of x can be remembered without actually writing them, because the exponents of the terms in the divisor, dividend, and quotient were written in descending order.

$$
\begin{array}{r|l}
 & 4x^2 + 3x - 5 \\
\hline
x - 2 & 4x^3 - 5x^2 - 11x + 20 \\
 & \underline{4x^3 - 8x^2} \\
 & \quad 3x^2 - 11x \\
 & \quad \underline{3x^2 - 6x} \\
 & \qquad -5x + 20 \\
 & \qquad \underline{-5x + 10} \\
 & \qquad\qquad 10 \quad \text{(remainder)}
\end{array}
\qquad
\begin{array}{r|l}
 & 4 \quad 3 - 5 \\
\hline
1 - 2 & 4 - 5 - 11 \quad 20 \\
 & \underline{\mathbf{4} - 8} \\
 & \quad 3 - \mathbf{11} \\
 & \quad \underline{\mathbf{3} - 6} \\
 & \qquad -5 \quad \mathbf{20} \\
 & \qquad \underline{\mathbf{-5} \quad 10} \\
 & \qquad\qquad 10 \quad \text{(remainder)}
\end{array}
$$

The numbers printed in color need not be written, because they are duplicates of the numbers above them. Thus, we can write the division in the form shown below on the left. We can shorten the process further by compressing the work vertically and eliminating the 1 (the coefficient of x in the divisor) as shown below on the right.

$$
\begin{array}{r|l}
 & 4 \quad 3 - 5 \\
\hline
1 - 2 & 4 - 5 - 11 \quad 20 \\
 & \underline{\quad - 8} \\
 & \quad 3 \\
 & \quad \underline{- 6} \\
 & \qquad - 5 \\
 & \qquad \underline{\qquad 10} \\
 & \qquad\qquad 10
\end{array}
\qquad
\begin{array}{r|rrrr}
 & 4 & 3 & -5 & \\
\hline
-2 & 4 & -5 & -11 & 20 \\
 & & -8 & -6 & 10 \\
\hline
 & & 3 & -5 & 10
\end{array}
$$

The Language of Algebra

Synthetic means devised to imitate something natural. You've probably heard of *synthetic* fuels or *synthetic* fibers. *Synthetic* division imitates the long division process.

If we write the 4 in the quotient on the bottom line, that line gives the coefficients of the quotient and the remainder. If we eliminate the top line, the division appears as follows:

$$\begin{array}{r|rrrr} -2 & 4 & -5 & -11 & 20 \\ & & -8 & -6 & 10 \\ \hline & 4 & 3 & -5 & 10 \end{array}$$

The Language of Algebra

Synthetic division is used to divide a polynomial by a binomial of the form $x - k$. We call k the synthetic divisor. In this example, we are dividing by $x - 2$, so k is 2.

The bottom line was obtained by subtracting the middle line from the top line. If we replace the -2 in the divisor by 2, the division process will reverse the signs of every entry in the middle line, and then the bottom line can be obtained by addition. This gives the final form of the synthetic division.

$$\begin{array}{r|rrrr} 2 & 4 & -5 & -11 & 20 \\ & & 8 & 6 & -10 \\ \hline & 4 & 3 & -5 & 10 \\ & \downarrow & \downarrow & \downarrow & \downarrow \\ & 4x^2 & +\ 3x & -\ 5 & +\ \dfrac{10}{x-2} \end{array}$$

These are the coefficients of the dividend.

These are the coefficients of the quotient and the remainder.

Read the result from the bottom row.

Thus,

$$\frac{4x^3 - 5x^2 - 11x + 20}{x - 2} = 4x^2 + 3x - 5 + \frac{10}{x - 2}$$

EXAMPLE 1 Use synthetic division to find $(6x^2 + 5x - 2) \div (x - 5)$.

Solution We write the coefficients in the dividend and the 5 in the divisor in the following form:

Since we are dividing the polynomial by $x - 5$, the synthetic divisor is 5.

$$\begin{array}{r|rrr} 5 & 6 & 5 & -2 \\ \hline & & & \end{array}$$ ← This represents the dividend $6x^2 + 5x - 2$.

Then we follow these steps:

$$\begin{array}{r|rrr} 5 & 6 & 5 & -2 \\ & \downarrow & & \\ \hline & 6 & & \end{array}$$ Begin by bringing down the 6.

$$\begin{array}{r|rrr} 5 & 6 & 5 & -2 \\ & & 30 & \\ \hline & 6 & & \end{array}$$ Multiply 5 by 6 to get 30.

$$\begin{array}{r|rrr} 5 & 6 & 5 & -2 \\ & & 30 & \\ \hline & 6 & 35 & \end{array}$$ Add 5 and 30 to get 35.

$$\begin{array}{r|rrr} 5 & 6 & 5 & -2 \\ & & 30 & 175 \\ \hline & 6 & 35 & \end{array}$$ Multiply 35 by 5 to get 175.

$$\begin{array}{r|rrr} 5 & 6 & 5 & -2 \\ & & 30 & 175 \\ \hline & 6 & 35 & 173 \end{array}$$ Add -2 and 175 to get 173.

Success Tip

In his process, numbers below the line are multiplied by the synthetic divisor and that product is carried above the line to the next column. Numbers above the horizontal line are added.

The numbers 6 and 35 represent the quotient $6x + 35$, and 173 is the remainder. Thus,

$$\frac{6x^2 + 5x - 2}{x - 5} = 6x + 35 + \frac{173}{x - 5}$$

Self Check 1 Divide $5x^2 - 4x + 2$ by $x - 3$.

EXAMPLE 2

Use synthetic division to find $\dfrac{x^3 + x^2 - 1}{x - 3}$.

Solution We begin by writing

$$\begin{array}{r|rrrr} 3 & 1 & 1 & 0 & -1 \\ & & & & \\ \hline \end{array}$$

Write 0 for the coefficient of x, the missing term.

and complete the division as follows.

$$\begin{array}{r|rrrr} 3 & 1 & 1 & 0 & -1 \\ & & 3 & & \\ \hline & 1 & 4 & & \end{array} \qquad \begin{array}{r|rrrr} 3 & 1 & 1 & 0 & -1 \\ & & 3 & 12 & \\ \hline & 1 & 4 & 12 & \end{array} \qquad \begin{array}{r|rrrr} 3 & 1 & 1 & 0 & -1 \\ & & 3 & 12 & 36 \\ \hline & 1 & 4 & 12 & 35 \end{array}$$

Multiply, then add. Multiply, then add. Multiply, then add.

Thus,

$$\frac{x^3 + x^2 - 1}{x - 3} = x^2 + 4x + 12 + \frac{35}{x - 3}$$

Self Check 2 Use synthetic division to find $\dfrac{x^3 + 3x - 90}{x - 4}$.

EXAMPLE 3

Use synthetic division to divide $5x^2 + 6x^3 + 2 - 4x$ by $x + 2$.

Solution First, we write the dividend with the exponents in descending order.

$$6x^3 + 5x^2 - 4x + 2$$

Then we write the divisor in $x - k$ form: $x - (\mathbf{-2})$. Thus, $k = -2$. Using synthetic division, we begin by writing

This represents division by x + 2.

$$\begin{array}{r|rrrr} -2 & 6 & 5 & -4 & 2 \\ & & & & \\ \hline \end{array}$$

and complete the division.

$$\begin{array}{r|rrrr} -2 & 6 & 5 & -4 & 2 \\ & & -12 & 14 & -20 \\ \hline & 6 & -7 & 10 & -18 \end{array}$$

The remainder is negative.

Thus,

$$\frac{5x^2 + 6x^3 + 2 - 4x}{x + 2} = 6x^2 - 7x + 10 - \frac{18}{x + 2}$$

Notation

Because the remainder is negative, we can also write the result as

$$6x^2 - 7x + 10 + \frac{-18}{x + 2}$$

Self Check 3 Divide $2x - 4x^2 + 3x^3 - 3$ by $x + 1$.

2 The Remainder Theorem.

Synthetic division is important because of the **remainder theorem.**

Remainder Theorem If a polynomial $P(x)$ is divided by $x - k$, the remainder is $P(k)$.

It follows from the remainder theorem that we can evaluate polynomials using synthetic division. We illustrate this in the following example.

EXAMPLE 4 Let $P(x) = 2x^3 - 3x^2 - 2x + 1$. Find **a.** $P(3)$ and **b.** the remainder when $P(x)$ is divided by $x - 3$.

Notation

Naming the function with the letter P, instead of f, stresses that we are working with a polynomial function.

Solution **a.** To find $P(3)$ we evaluate the function for $x = 3$.

$$\begin{aligned} P(3) &= 2(3)^3 - 3(3)^2 - 2(3) + 1 \quad \text{Substitute 3 for } x. \\ &= 2(27) - 3(9) - 6 + 1 \\ &= 54 - 27 - 6 + 1 \\ &= 22 \end{aligned}$$

Thus, $P(3) = 22$.

b. We can use synthetic division to find the remainder when $P(x)$ is divided by $x - 3$.

$$\begin{array}{r|rrrr} 3 & 2 & -3 & -2 & 1 \\ & & 6 & 9 & 21 \\ \hline & 2 & 3 & 7 & 22 \end{array} \quad P(x) = 2x^3 - 3x^2 - 2x + 1$$

Thus, the remainder is 22.

Success Tip

It is often easier to find $P(k)$ by using synthetic division than by substituting k for x in $P(x)$. This is especially true if k is a decimal.

The same results in parts a and b show that rather than substituting 3 for x in $P(x) = 2x^3 - 3x^2 - 2x + 1$, we can divide $2x^3 - 3x^2 - 2x + 1$ by $x - 3$ to find $P(3)$.

Self Check 4 Let $P(x) = 5x^3 - 3x^2 + x + 6$. Find **a.** $P(1)$ and **b.** use synthetic division to find the remainder when $P(x)$ is divided by $x - 1$.

3 The Factor Theorem.

If two quantities are multiplied, each is called a **factor** of the product. Thus, $x - 2$ is a factor of $6x - 12$, because $6(x - 2) = 6x - 12$. A theorem, called the **factor theorem,** tells us how to find one factor of a polynomial if the remainder of a certain division is 0.

Factor Theorem If $P(x)$ is a polynomial in x, then

$P(k) = 0$ if and only if $x - k$ is a factor of $P(x)$

If $P(x)$ is a polynomial in x and if $P(k) = 0$, k is called a **zero of the polynomial function.**

EXAMPLE 5 Let $P(x) = 3x^3 - 5x^2 + 3x - 10$. Show that **a.** $P(2) = 0$ and **b.** $x - 2$ is a factor of $P(x)$.

Solution

a. We can use the remainder theorem to evaluate $P(2)$ by dividing $P(x)$ by $x - 2$.

$$\begin{array}{r|rrrr} 2 & 3 & -5 & 3 & -10 \\ & & 6 & 2 & 10 \\ \hline & 3 & 1 & 5 & 0 \end{array} \qquad P(x) = 3x^2 - 5x^2 + 3x - 10$$

The remainder in this division is 0. By the remainder theorem, the remainder is $P(2)$. Thus, $P(2) = 0$, and 2 is a zero of the polynomial.

b. Because the remainder is 0, the numbers 3, 1, and 5 in the synthetic division in part a represent the quotient $3x^2 + x + 5$. Thus,

$$\underbrace{(x-2)}_{\text{Divisor}} \cdot \underbrace{(3x^2 + x + 5)}_{\text{quotient}} + \underbrace{0}_{\text{remainder}} = \underbrace{3x^3 - 5x^2 + 3x - 10}_{\text{the dividend, } P(x)}$$

or

$$(x - 2)(3x^2 + x + 5) = 3x^3 - 5x^2 + 3x - 10$$

Thus, $x - 2$ is a factor of $3x^3 - 5x^2 + 3x - 10$.

Self Check 5 Let $P(x) = x^3 - 4x^2 + x + 6$. Show that $x + 1$ is a factor of $P(x)$ using synthetic division.

The Language of Algebra

The phrase *if and only if* in the factor theorem means:

If $P(2) = 0$, then $x - 2$ is a factor of $P(x)$

and

If $x - 2$ is a factor of $P(x)$, then $P(2) = 0$.

The result in Example 5 is true, because the remainder, $P(2)$, is 0. If the remainder had not been 0, then $x - 2$ would not have been a factor of $P(x)$.

APPENDIX 2 STUDY SET

VOCABULARY

Fill in the blanks.

1. The method of dividing $x^2 + 2x - 9$ by $x - 4$ shown below is called ________ division.

$$\begin{array}{rrrr} 4 & 1 & 2 & -9 \\ & & 4 & 24 \\ \hline & 1 & 6 & 15 \end{array}$$

2. Synthetic division is used to divide a polynomial by a ________ of the form $x - k$.
3. In Exercise 1, the synthetic ________ is 4.
4. By the ________ theorem, if a polynomial $P(x)$ is divided by $x - k$, the remainder is $P(k)$.
5. The factor ________ tells us how to find one factor of a polynomial if the remainder of a certain division is 0.
6. If $P(x)$ is a polynomial and if $P(k) = 0$, then k is called a _____ of the polynomial.

CONCEPTS

7. **a.** What division is represented below?

   b. What is the answer?

$$\begin{array}{r|rrrr} -2 & 5 & 0 & 1 & -3 \\ & & -10 & 20 & -42 \\ \hline & 5 & -10 & 21 & -45 \end{array}$$

Fill in the blanks.

8. In the synthetic division process, numbers below the line are ________ by the synthetic divisor and that product is carried above the line to the next column. Numbers above the horizontal line are ______.
9. Rather than substituting 8 for x in $P(x) = 6x^3 - x^2 - 17x + 9$, we can divide the polynomial ____________ by _____ to find $P(8)$.
10. For $P(x) = x^3 - 4x^2 + x + 6$, suppose we know that $P(3) = 0$. Then _____ is a factor of $x^3 - 4x^2 + x + 6$.

NOTATION

Complete each synthetic division.

11. Divide $6x^3 + x^2 - 23x + 2$ by $x - 2$.

⌋				
	6		−23	2
				6
		13	3	

12. Divide $2x^3 - 4x^2 - 25x + 15$ by $x + 3$.

⌋				
	2	−4		15
			30	
				0

GUIDED PRACTICE

Use synthetic division to perform each division.

13. $\dfrac{x^2 + x - 2}{x - 1}$

14. $\dfrac{x^2 + x - 6}{x - 2}$

15. $\dfrac{x^2 - 7x + 12}{x - 4}$

16. $\dfrac{x^2 - 6x + 5}{x - 5}$

17. $\dfrac{x^2 + 8 + 6x}{x + 4}$

18. $\dfrac{x^2 - 15 - 2x}{x + 3}$

19. $\dfrac{x^2 - 5x + 14}{x + 2}$

20. $\dfrac{x^2 + 13x + 42}{x + 6}$

21. $\dfrac{3x^3 - 10x^2 + 5x - 6}{x - 3}$

22. $\dfrac{2x^3 - 9x^2 + 10x - 3}{x - 3}$

23. $\dfrac{2x^3 - 6 - 5x}{x - 2}$

24. $\dfrac{4x^3 - 1 + 5x^2}{x + 2}$

25. $\dfrac{5x^2 + 4 + 6x^3}{x + 1}$

26. $\dfrac{4 - 3x^2 + x}{x - 4}$

27. $\dfrac{t^3 + t^2 + t + 2}{t + 1}$

28. $\dfrac{m^3 - m^2 - m - 1}{m - 1}$

29. $\dfrac{a^5 - 1}{a - 1}$

30. $\dfrac{b^4 - 81}{b - 3}$

31. $\dfrac{-5x^5 + 4x^4 + 30x^3 + 2x^2 + 20x + 3}{x - 3}$

32. $\dfrac{-6c^5 + 14c^4 + 38c^3 + 4c^2 + 25c - 36}{c - 4}$

33. $\dfrac{8t^3 - 4t^2 + 2t - 1}{t - \frac{1}{2}}$

34. $\dfrac{9a^3 + 3a^2 - 21a - 7}{a + \frac{1}{3}}$

35. $\dfrac{x^4 - x^3 - 56x^2 - 2x + 16}{x - 8}$

36. $\dfrac{x^4 - 9x^3 + x^2 - 7x - 20}{x - 9}$

Use a calculator and synthetic division to perform each division.

37. $\dfrac{7.2x^2 - 2.1x + 0.5}{x - 0.2}$

38. $\dfrac{2.7x^2 + x - 5.2}{x + 1.7}$

39. $\dfrac{9x^3 - 25}{x + 57}$

40. $\dfrac{0.5x^3 + x}{x - 2.3}$

Let $P(x) = 2x^3 - 4x^2 + 2x - 1$. Evaluate $P(x)$ by substituting the given value of x into the polynomial and simplifying. Then evaluate the polynomial by using the remainder theorem and synthetic division.

41. $P(1)$
42. $P(2)$
43. $P(-2)$
44. $P(-1)$
45. $P(3)$
46. $P(-4)$
47. $P(0)$
48. $P(4)$

Let $Q(x) = x^4 - 3x^3 + 2x^2 + x - 3$. Evaluate $Q(x)$ by substituting the given value of x into the polynomial and simplifying. Then evaluate the polynomial by using the remainder theorem and synthetic division.

49. $Q(-1)$
50. $Q(1)$
51. $Q(2)$
52. $Q(-2)$
53. $Q(3)$
54. $Q(0)$
55. $Q(-3)$
56. $Q(-4)$

Use the remainder theorem and synthetic division to find $P(k)$.

57. $P(x) = x^3 - 4x^2 + x - 2;\ k = 2$
58. $P(x) = x^3 - 3x^2 + x + 1;\ k = 1$
59. $P(x) = 2x^3 + x + 2;\ k = 3$
60. $P(x) = x^3 + x^2 + 1;\ k = -2$
61. $P(x) = x^4 - 2x^3 + x^2 - 3x + 2;\ k = -2$
62. $P(x) = x^5 + 3x^4 - x^2 + 1;\ k = -1$
63. $P(x) = 3x^5 + 1;\ k = -\dfrac{1}{2}$
64. $P(x) = 5x^7 - 7x^4 + x^2 + 1;\ k = 2$

Use the factor theorem and determine whether the first expression is a factor of $P(x)$.

65. $x - 3;\ P(x) = x^3 - 3x^2 + 5x - 15$
66. $x + 1;\ P(x) = x^3 + 2x^2 - 2x - 3$
(*Hint:* Write $x + 1$ as $x - (-1)$.)
67. $x + 2;\ P(x) = 3x^2 - 7x + 4$
(*Hint:* Write $x + 2$ as $x - (-2)$.)
68. $x;\ P(x) = 7x^3 - 5x^2 - 8x$
(*Hint:* $x = x - 0$.)

WRITING

69. When dividing a polynomial by a binomial of the form $x - k$, synthetic division is considered to be faster than long division. Explain why.
70. Let $P(x) = x^3 - 6x^2 - 9x + 4$. You now know two ways to find $P(6)$. What are they? Which method do you prefer?
71. Explain the factor theorem.
72. What is a *zero* of a polynomial function?

REVIEW

Evaluate each expression for $x = -3$, $y = -5$, and $z = 0$.

73. $x^2z(y^3 - z)$

74. $|y^3 - z|$

75. $\dfrac{x - y^2}{2y - 1 + x}$

76. $\dfrac{2y + 1}{x} - x$

CHALLENGE PROBLEMS

Suppose that
$P(x) = x^{100} - x^{99} + x^{98} - x^{97} + \cdots + x^2 - x + 1.$

77. Find the remainder when $P(x)$ is divided by $x - 1$.

78. Find the remainder when $P(x)$ is divided by $x + 1$.

APPENDIX 3 Answers to Selected Exercises

ARE YOU READY? 1.1 (page 2)

1. 210 **2.** 1,750 **3.** 1,092 **4.** 9

SELF CHECKS 1.1

1. 350 calories **2.** The product of 22 and 11 equals 242. **3.** $u = 500 - p$ (Answers may vary depending on the variables used.) **4.** 480 calories **5.**

m	c
8	80
75	750

STUDY SET SECTION 1.1 (page 7)

1. sum, difference, product, quotient **3.** constant **5.** equation; expression **7.** horizontal **9. a.** Equation **b.** Algebraic Expression **11. a.** Algebraic Expression **b.** Equation **13.** Addition, multiplication, division; t

15.

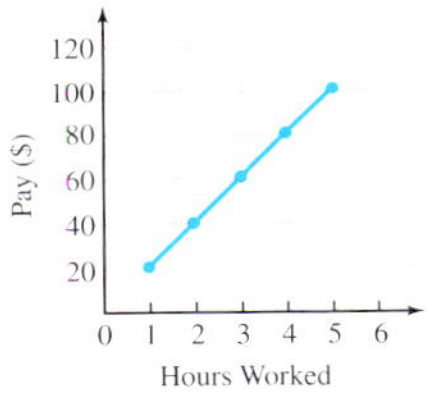

17. is not equal to **19.** $5 \cdot 6$, $5(6)$ **21.** $4x$ **23.** $2w$ **25.** $\frac{32}{x}$ **27.** $\frac{55}{5}$ **29.** 15-year-old machinery is worth \$35,000. **31.** \$250 **33.** The product of 8 and 2 equals 16. **35.** The difference of 11 and 9 equals 2. **37.** The sum of x and 2 equals 10. **39.** The quotient of 66 and 11 equals 6. **41.** $p = 100 - d$ **43.** $7d = h$ **45.** $s = 3c$ **47.** $w = e + 1{,}200$ **49.** $p = r - 600$ **51.** $\frac{l}{4} = m$ **53.** 390, 400, 405 **55.** 1,300; 1,200; 1,100 **57.** 12 **59.** 2

61. a. The number of calories burned is the product of 3 and the number of minutes cleaning. **b.** $c = 3m$ **c.** 30, 60, 90, 120, 150, 180

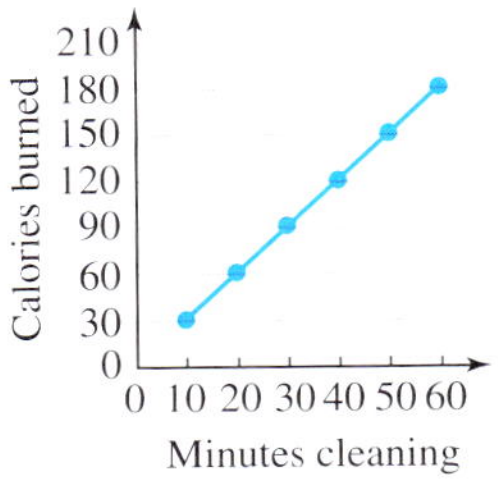

ARE YOU READY? 1.2 (page 9)

1. 1 **2.** 150 **3.** Yes **4.** four-fifths

SELF CHECKS 1.2

1. $189 = 3 \cdot 3 \cdot 3 \cdot 7$ **2.** $\frac{10}{27}$ **3.** $\frac{12}{25}$ **4.** $\frac{15}{24}$ **5. a.** $\frac{3}{7}$ **b.** In simplest form **6.** $\frac{9}{5}$ **7.** $\frac{13}{240}$ **8. a.** 28 **b.** $\frac{4}{5}$ **9.** $\frac{81}{8} = 10\frac{1}{8}$ **10.** $6\frac{11}{24}$

STUDY SET SECTION 1.2 (page 20)

1. multiplied **3.** prime-factored **5.** equivalent **7.** least or lowest **9. a.** 1 **b.** a **c.** $\frac{a \cdot c}{b \cdot d}$ **d.** $\frac{a \cdot d}{b \cdot c}$ **e.** $\frac{a + b}{d}$ **f.** $\frac{a - b}{d}$ **11. a.** 1 **b.** 1 **13. a.** $\frac{5}{5}, \frac{25}{30}$ **b.** $2, 7, \frac{2}{7}$ **15.** $3 \cdot 5 \cdot 5$ **17.** $2 \cdot 2 \cdot 7$ **19.** $3 \cdot 3 \cdot 3 \cdot 3$ **21.** $3 \cdot 3 \cdot 13$ **23.** $2 \cdot 2 \cdot 5 \cdot 11$ **25.** $2 \cdot 3 \cdot 11 \cdot 19$ **27.** $\frac{5}{48}$ **29.** $\frac{21}{55}$ **31.** $\frac{15}{8}$ **33.** $\frac{42}{25}$ **35.** $\frac{3}{9}$ **37.** $\frac{24}{54}$ **39.** $\frac{35}{5}$ **41.** $\frac{35}{7}$ **43.** $\frac{1}{3}$ **45.** $\frac{6}{7}$ **47.** $\frac{3}{8}$ **49.** Simplest form **51.** $\frac{2}{3}$ **53.** $\frac{4}{25}$ **55.** $\frac{6}{5}$ **57.** $\frac{4}{7}$ **59.** $\frac{5}{24}$ **61.** $\frac{22}{35}$ **63.** $\frac{41}{45}$ **65.** $\frac{5}{12}$ **67.** 24 **69.** 4 **71.** $\frac{7}{9}$ **73.** $\frac{7}{20}$ **75.** $32\frac{2}{3}$ **77.** $2\frac{1}{2}$ **79.** $\frac{5}{9}$ **81.** $5\frac{19}{48}$ **83.** $\frac{19}{15}$ **85.** 70 **87.** $13\frac{3}{4}$ **89.** $\frac{1}{2}$ **91.** $\frac{14}{5}$ **93.** $\frac{8}{5}$ **95.** $\frac{3}{35}$ **97.** $\frac{9}{4}$ **99.** $\frac{3}{25}$ **101.** $1\frac{9}{11}$ **103.** $\frac{1}{7}$ **105.** 0 **107.** 14 **109. a.** $\frac{55}{63}$ **b.** $\frac{1}{63}$ **c.** $\frac{4}{21}$ **d.** $\frac{28}{27}$ **111. a.** $\frac{7}{32}$ in. **b.** $\frac{3}{32}$ in. **113.** $63\frac{1}{8}$ in. **115.** $40\frac{1}{2}$ in. **121.** Variables

ARE YOU READY? 1.3 (page 22)

1. 28 **2.** -10 **3.** a decimal (3.5 GPA) **4.** a fraction ($\frac{3}{4}$ cup of sugar)

SELF CHECKS 1.3

1. Natural numbers: 45; whole numbers: 45; integers: 45, -2; rational numbers: 0.1, $-\frac{2}{7}$, 45, -2, $\frac{13}{4}$, $-6\frac{7}{8}$; irrational numbers: none; real numbers: all **2. a.** $>$ **b.** $<$ **c.** $>$ **d.** $<$

3.

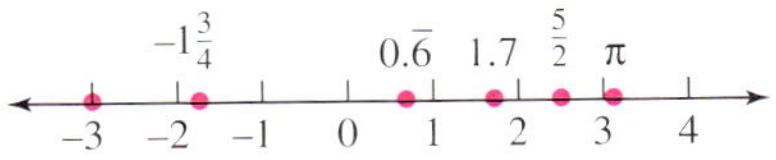

4. a. 100 **b.** 4.7 **c.** $\sqrt{2}$

STUDY SET SECTION 1.3 (page 29)

1. whole **3.** number line **5.** signed **7.** inequality **9.** terminating; repeating **11.** decimal **13. a.** $-\$15$ million **b.** $\frac{5}{16}$ in. or $+\frac{5}{16}$ in. **15. a.** -20 **b.** $\frac{2}{3}$ **17.** -14 and -4 **19.** square, root **21.** is approximately equal to **23.** Greek **25.** $-4, -5$

27.

	5	0	−3	$\frac{7}{8}$	0.17	$-9\frac{1}{4}$	$\sqrt{2}$	π
Real	✓	✓	✓	✓	✓	✓	✓	✓
Irrational							✓	✓
Rational	✓	✓	✓	✓	✓	✓		
Integer	✓	✓	✓					
Whole	✓	✓						
Natural	✓							

29. True **31.** False **33.** True **35.** True **37.** > **39.** <
41. > **43.** < **45.** > **47.** < **49.** 0.625 **51.** $0.0\overline{3}$
53. $0.01\overline{6}$ **55.** 0.42
57.
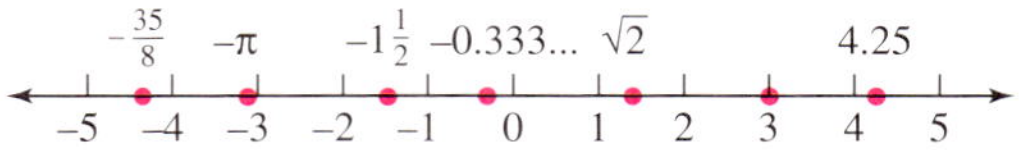

59.
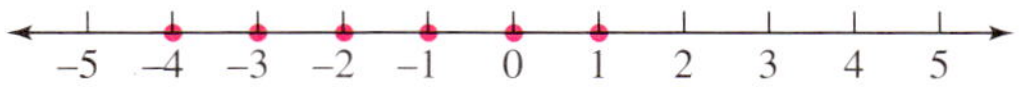

61. 83 **63.** $\frac{4}{3}$ **65.** 11 **67.** 6.1 **69.** >
71. < **73.** = **75.** = **77.** < **79.** >
81. Natural, whole, integers: 9; rational: 9, $\frac{15}{16}$, $3\frac{1}{8}$, 1.765; irrational: 2π, 3π, $\sqrt{89}$; real: all **83.** iii
85. a. 2006; −\$90 billion **b.** 2009; −\$45 billion **93.** $\frac{4}{9}$
95. $2\frac{5}{23}$

ARE YOU READY? 1.4 (page 32)

1. 3, 5; −5 has the larger absolute value. **2.** 7.17 **3.** 621 **4.** $\frac{2}{15}$

SELF CHECKS 1.4

1. a. −60 **b.** −13.18 **c.** $-\frac{11}{12}$ **2. a.** −24 **b.** 1.73 **c.** $\frac{2}{5}$
3. 569 **4.** 139 **5.** −5 **6.** 11

STUDY SET SECTION 1.4 (page 37)

1. sum **3.** commutative, associative **5. a.** 6 **b.** −9.2
7. a. Negative **b.** Positive **9. a.** 1 + (−5) **b.** −80.5 + 15
c. 20 + 4 **d.** 3 + 2.1 **11.** Step 1: Commutative Property of Addition; Step 2: Associative Property of Addition
13. a. $x + y = y + x$ **b.** $(x + y) + z = x + (y + z)$ **15.** −9
17. −17 **19.** −74 **21.** −10.3 **23.** $-\frac{17}{12}$ **25.** $-\frac{7}{20}$ **27.** −3
29. 39 **31.** 0 **33.** 2.25 **35.** $-\frac{4}{15}$ **37.** $\frac{3}{8}$ **39.** 16 **41.** −15
43. −21 **45.** 195 **47.** 215 **49.** −112 **51.** $1\frac{2}{3}$ **53.** 15.4
55. 9 **57.** 1 **59.** 70 **61.** −6.6 **63.** $-\frac{1}{8}$ **65.** −14 **67.** 0
69. −26 **71.** 0.67 **73.** −5 **75.** −22.1 **77.** 2,167 **79.** −1.7
81. 68 **83.** $-\frac{15}{28}$ **85.** −0.9 **87. a.** 27 **b.** 3 **c.** −27 **d.** −3
89. a. $\frac{13}{18}$ **b.** $-\frac{5}{18}$ **c.** $-\frac{13}{18}$ **d.** $\frac{5}{18}$ **91.** 2,150 m
93. Yang: −8, Woods: −5, Westwood: −3, McIlroy: −3
95. 1,242.86 **97.** −\$99,650,000 **99.** −1, 3
101. 79 feet above sea level **103.** \$6,276 million **107.** True
109. −9 and 3

ARE YOU READY? 1.5 (page 40)

1. −6, 15 **2.** 22 − 6 **3.** 12 **4.** −9

SELF CHECKS 1.5

1. a. 1 **b.** y **c.** −500 **2. a.** −57 **b.** 2.9 **c.** $\frac{5}{12}$ **3. a.** 2.7
b. −2.7 **4.** −8 **5.** 41°F **6.** −11 ft

STUDY SET SECTION 1.5 (page 44)

1. Subtraction **3.** range **5. a.** −12 **b.** $\frac{1}{5}$ **c.** −2.71 **d.** 0
7. +, 9, 10 **9.** −10 + (−8) + (−23) + 5 + 34 **11. a.** −6 ⊖ (−4)
b. 7 + (−3) ⊖ 5 ⊖ (−2) **13. a.** 1 − (−7) = 8 **b.** −(−2) = 2
c. −|−3| = −3 **d.** 2 − 6 = −4 **15.** 55 **17.** x **19.** −25
21. $-\frac{3}{16}$ **23.** −3 **25.** −10 **27.** 11 **29.** −6 **31.** 2
33. 40 **35.** 5 **37.** 12 **39.** −6.9 **41.** −2.31 **43.** $-\frac{1}{2}$ **45.** $-\frac{5}{12}$
47. 22 **49.** −25 **51.** −7 **53.** −1 **55.** 256 **57.** 0 **59.** −2.1
61. $\frac{47}{56}$ **63.** 3 **65.** −47.5 **67.** 149 **69.** −171 **71.** 4.63
73. $-\frac{19}{12}$ **75.** −11 **77.** −88 **79.** −1.1 **81.** −50 **83.** −3.5
85. $-\frac{5}{16}$ **87.** 1 **89.** −13 **91. a.** −53 **b.** −47 **93. a.** $-\frac{13}{13}$
b. $-\frac{7}{18}$ **95.** 160°F **97.** 21 points **99.** Orlando: −115,000 passengers; Ft Lauderdale: −50,000 passengers **101.** 1,030 ft
103. an increase of 315°F **105.** −179, −206 **111.** 2 · 3 · 5
113. True

ARE YOU READY? 1.6 (page 47)

1. 14, 6.75 **2.** Different signs **3.** $\frac{5}{16}$ **4.** 3.7

SELF CHECKS 1.6

1. a. −60 **b.** −15 **c.** −11.18 **d.** $-\frac{2}{5}$ **2. a.** 120 **b.** $\frac{1}{12}$
3. a. −300 **b.** 18 **4. a.** $-\frac{16}{15}$ **b.** $\frac{16}{15}$ **c.** $-\frac{1}{27}$ **5. a.** 7 **b.** −3
c. −0.2 **d.** $-\frac{6}{5}$ **6.** The house depreciated \$2,250 per year.
7. a. Undefined **b.** 0

STUDY SET SECTION 1.6 (page 55)

1. product, quotient **3.** associative **5. a.** positive **b.** negative
7. a. −3, 3, −9 **b.** 0, 8, 0 **9. a.** a **b.** 1 **c.** 0
d. Undefined **11. a.** Positive **b.** Negative **c.** Negative
13. a. 8 · 5 **b.** (−2 · 6)9 **c.** $\frac{1}{5}$ **d.** 1 **15. a.** NEG
b. Not possible to tell **c.** POS **d.** NEG **17.** −4(−5) = 20
19. −4 **21.** −16 **23.** −60 **25.** −66 **27.** −0.48 **29.** $-\frac{1}{4}$
31. 7 **33.** 54 **35.** 9 **37.** −441 **39.** 2.4 **41.** $\frac{1}{12}$
43. 66 **45.** −720 **47.** −861 **49.** −216 **51.** $\frac{9}{7}$; 1
53. $-\frac{1}{13}$; 1 **55.** 10 **57.** 3 **59.** −17 **61.** 1 **63.** −9
65. −0.005 **67.** $-\frac{5}{12}$ **69.** $\frac{15}{4}$ **71.** 0 **73.** Undefined
75. −4 **77.** −1 **79.** 16 **81.** 0 **83.** −4.7 **85.** −520
87. $\frac{1}{24}$ **89.** −11 **91.** $1\frac{1}{2}$ **93.** 30.24 **95.** $-\frac{3}{8}$
97. $-\frac{3}{20}$ **99.** 30.3 **101.** $\frac{15}{16}$ **103. a.** 1.8 **b.** 3.6 **c.** −2.43
d. −3 **105.** −67 **107.** 6 **109.** −\$8,000 per year **111.** −72°
113. a. ii **b.** 36 lb **115.** −51°F **117.** (160), (250), (800)
119. a. 5, −10 **b.** 2.5, −5 **c.** 7.5, −15 **d.** 10, −20
125. −5 **127.** $1.08\overline{3}$

ARE YOU READY? 1.7 (page 58)

1. Subtraction, multiplication **2.** Addition, multiplication, division **3.** 81 **4.** −125

SELF CHECKS 1.7

1. a. 12^3 **b.** $2 \cdot 9^4$ **c.** $(-30)^2$ **d.** y^6 **e.** $8b^3c$ **2. a.** 32
b. $-\frac{27}{64}$ **c.** 9 **d.** 0.09 **e.** 36 **f.** −125 **3.** −625 **4. a.** 35
b. −66 **c.** 27 **d.** 256 **5. a.** 216 **b.** −35 **6.** −130 **7.** −1
8. 1,003 **9.** 2

STUDY SET SECTION 1.7 (page 65)

1. base, exponent, power **3.** exponent **5.** order
7. a. Subtraction **b.** Division **c.** Addition **d.** Power
9. a. Parentheses, brackets, braces, absolute value symbols, fraction bar **b.** Innermost: parentheses; outermost: brackets **11. a.** −5
b. 5 **13.** 3, 9, 27, 54, −73 **15.** 8^3 **17.** $7^3 12^2$ **19.** x^3 **21.** r^4s^2
23. 49 **25.** 216 **27.** 625 **29.** 0.01 **31.** $-\frac{1}{64}$ **33.** $\frac{8}{27}$
35. 36, −36 **37.** 64, −64 **39.** −17 **41.** 30 **43.** 43
45. 8 **47.** −34 **49.** −118 **51.** −44 **53.** 0 **55.** −148

57. 100 **59.** 53 **61.** −86 **63.** −392 **65.** 3 **67.** $\frac{1}{2}$ **69.** 0 **71.** $-\frac{8}{9}$ **73.** 13 **75.** 2 **77.** −31 **79.** 11 **81.** 1 **83.** −32 **85.** 86 **87.** −8 **89.** −19 **91.** −500 **93.** −376 **95.** 12 **97.** 39 **99.** Undefined **101.** −275 **103.** −54 **105.** $\frac{1}{8}$ **107.** 10 **109.** −1 **111. a.** 22 **b.** −13 **113. a.** −40 **b.** −10 **115.** 2^2 square units, 3^2 square units, 4^2 square units **117.** 12 min **119. a.** \$11,875 **b.** \$95 **121.** 81 in. **127.** −17, −5

ARE YOU READY? 1.8 (page 68)

1. $9x$ **2.** m **3.** $-t$ **4.** $\frac{7}{8}y$ **5.** Quotient **6.** Difference **7.** Product **8.** Sum

SELF CHECKS 1.8

1. 1, −12, 3, −4 **2. a.** Factor **b.** Term **3. a.** $t - 80$ **b.** $\frac{2}{3}T$ **c.** $(2a - 15)^2$ **4.** $(m + 15)$ min **5.** s = amount donated to scholarship fund in dollars; $900 - s$ = amount donated to building fund in dollars **6.** x = number of votes received by the challenger; $3x - 55$ = number of votes received by the incumbent **7.** Daughter's: x; Kayla's: $x + 5$; Son's: $x + 2$ **8.** $\frac{h}{24}$ days **9. a.** \$300 **b.** \$100t **c.** \$1,000$(x - 4)$ **10. a.** 17 **b.** −18 **c.** 0 **11.** 96 ft

STUDY SET SECTION 1.8 (page 75)

1. expressions **3.** terms **5.** coefficient **7.** 7, 14, 21, $7w$ **9.** $12 - h$ **11.** $x + 20$ **13. a.** $b - 15$ **b.** $p + 15$ **15.** 5, 25, 45 **17. a.** $8y$ **b.** $2cd$ **c.** Commutative **19. a.** 4 **b.** 3, 11, −1, 9 **21.** Term **23.** Factor **25.** $l + 15$ **27.** $50x$ **29.** $\frac{w}{l}$ **31.** $P + \frac{2}{3}p$ **33.** $k^2 - 2{,}005$ **35.** $2a - 1$ **37.** $\frac{1{,}000}{n}$ **39.** $2p + 90$ **41.** $3(35 + h + 300)$ **43.** $p - 680$ **45.** $4d - 15$ **47.** $2(200 + t)$ **49.** $|a - 2|$ **51.** $0.1d$ or $\frac{1}{10}d$ **53.** Three-fourths of r **55.** 50 less than t **57.** The product of x, y, and z **59.** Twice m, increased by 5 **61.** $x + 2$ **63.** $36 - x$ **65.** $60h$ **67.** $\frac{i}{12}$ **69.** \$8x **71.** $49xt$ **73.** \$2t **75.** \$25$(x + 2)$ **77.** 2 **79.** 13 **81.** 20 **83.** −12 **85.** −5 **87.** $-\frac{1}{5}$ **89.** 17 **91.** 36 **93.** 255 **95.** 8 **97.** −1, −2, −28 **99.** 41, 11, 2 **101.** 150, −450 **103.** 0, 0, 5 **105. a.** $x + 7^2$ **b.** $(x + 7)^2$ **107. a.** $4(x + 2)$ **b.** $4x + 2$ **109. a.** Let x = weight of the Element, $2x - 340$ = weight of the Hummer **b.** 6,400 lb **111. a.** Let x = age of Apple; $x + 80$ = age of IBM; $x - 9$ = age of Dell **b.** IBM: 112 yr; Dell: 23 yr **117.** 60 **119.** $\frac{8}{27}$

ARE YOU READY? 1.9 (page 78)

1. The terms are in a different order. **2.** The position of the parentheses is different. **3.** 30, 30; same result **4.** Different variable factors (x and y); the same coefficient (4)

SELF CHECKS 1.9

1. a. $(5x + 3x) + 1$ **b.** $-15a$ **2. a.** $54s$ **b.** $48u$ **c.** m **d.** $8y$ **3. a.** $7m + 14$ **b.** $-640x - 240$ **c.** $4y + 9$ **4. a.** $10x - 5$ **b.** $9y + 36$ **c.** $-c + 22$ **5. a.** $-2x - 8$ **b.** $54c - 108d$ **c.** $-1.4r - 3.5s + 5.6$ **6.** $5x - 18$ **7. a.** $-2y$ and $7y$ **b.** $5p^2$ and $17p^2$; −12 and 2 **8. a.** $8x$ **b.** $-3y$ **c.** $0.5s^4$ **d.** Doesn't simplify **e.** $\frac{6}{7}c$ **9. a.** $8h$ **b.** $10h$ **c.** h **d.** $-h$ **10.** $7y^2 + 19y - 6$ **11.** $21y - 8$

STUDY SET SECTION 1.9 (page 86)

1. simplify **3.** distributive **5.** like **7. a.** 4, 9, 36 **b.** Associative Property of Multiplication **9. a.** + **b.** − **c.** − **d.** + **11. a.** $10x$ **b.** Can't be simplified **c.** $-42x$ **d.** Can't be simplified **e.** $18x$ **f.** $3x + 5$ **13. a.** $6(h - 4)$ **b.** $-(z + 16)$ **15.** $(8 + 7) + a$ **17.** $11y$ **19.** $8d(2 \cdot 6)$ **21.** $t + 4$ **23.** $12t$ **25.** $-35q$ **27.** $11.2x$ **29.** $60c$ **31.** g **33.** $5x$ **35.** $5x + 15$ **37.** $-12x - 27$ **39.** $9x + 10$ **41.** $0.4x + 1.6$ **43.** $36c - 42$ **45.** $-78c + 18$ **47.** $30t + 90$ **49.** $4a - 1$ **51.** $24t + 16$ **53.** $2w - 4$ **55.** $56y + 32$ **57.** $5a - 7.5b + 2.5$ **59.** $-x + 7$ **61.** $5.6y - 7$ **63.** $3x$ and $-2x$ **65.** $-3m^3$ and $-m^3$ **67.** $10x$ **69.** $20b^2$ **71.** $28y$ **73.** $\frac{4}{5}t$ **75.** r **77.** $-s^3$ **79.** $-6y - 10$ **81.** $-2x + 5$ **83.** $9m^2 + 6m - 4$ **85.** $4x^2 - 3x + 9$ **87.** $7z - 25$ **89.** $s^2 - 12$ **91.** $-\frac{5}{8}x$ **93.** $-3.6c$ **95.** $-96m$ **97.** 0 **99.** $0.4r$ **101.** $6y$ **103.** $-41r + 130$ **105.** $63m$ **107.** $12c + 34$ **109.** $300t$ **111.** $8x - 9$ **113.** $-10r$ **115.** $-20r$ **117.** $3a$ **119.** $9r - 16$ **121.** Doesn't simplify **123.** $c - 13$ **125.** $a^3 - 8$ **127. a.** $70x$ **b.** $14x + 10$ **129.** $12x$ in. **133.** 2

CHAPTER 1 REVIEW (page 89)

1. 1 hr; 100 cars **2.** 100 **3.** 7 P.M. **4.** 12 A.M. (midnight) **5.** The difference of 15 and 3 equals 12. **6.** The sum of 15 and 3 equals 18. **7.** The quotient of 15 and 3 equals 5. **8.** The product of 15 and 3 equals 45. **9. a.** $4 \cdot 9$; 4(9) **b.** $\frac{9}{3}$ **10. a.** $8b$ **b.** Prt **11. a.** Equation **b.** Expression **12.** 10, 15, 25 **13. a.** $2 \cdot 12$, $3 \cdot 8$ (Answers may vary) **b.** $2 \cdot 2 \cdot 6$ (Answers may vary) **c.** 1, 2, 3, 4, 6, 8, 12, 24 **14.** Equivalent **15.** $2 \cdot 3^3$ **16.** $3 \cdot 7^2$ **17.** $5 \cdot 7 \cdot 11$ **18.** Prime **19.** $\frac{4}{7}$ **20.** $\frac{4}{3}$ **21.** $\frac{40}{64}$ **22.** $\frac{36}{3}$ **23.** 90 **24.** 210 **25.** $\frac{7}{64}$ **26.** $\frac{5}{21}$ **27.** $\frac{16}{45}$ **28.** $3\frac{1}{4}$ **29.** $\frac{2}{5}$ **30.** $\frac{5}{22}$ **31.** $\frac{59}{60}$ **32.** $\frac{5}{18}$ **33.** $52\frac{1}{2}$ million **34.** $\frac{17}{96}$ in. **35. a.** 0 **b.** {. . . , −2, −1, 0, 1, 2, . . .} **36.** −206 ft **37. a.** < **b.** > **38. a.** $\frac{7}{10}$ **b.** $\frac{14}{3}$ **39.** 0.004 **40.** $0.\overline{772}$ **41.**

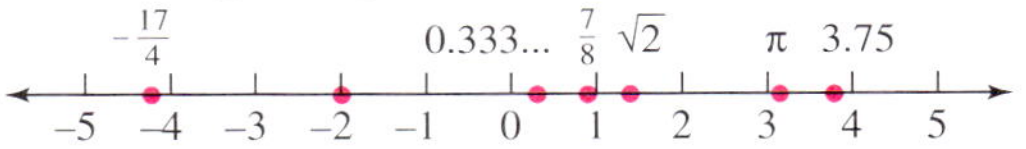

42. Natural: 8; whole: 0, 8; integers: 0, −12, 8; rational: $-\frac{4}{5}$, 99.99, 0, −12, $4\frac{1}{2}$, 0.666 . . . , 8; irrational: $\sqrt{2}$; real: all **43.** False **44.** False **45.** True **46.** True **47.** > **48.** < **49.** −82 **50.** 12 **51.** −7 **52.** 0 **53.** −11 **54.** −12.3 **55.** $-\frac{3}{16}$ **56.** 11 **57. a.** Commutative Property of Addition **b.** Associative Property of Addition **c.** Addition Property of Opposites (Inverse Property of Addition) **d.** Addition Property of 0 (Identity Property of Addition) **58.** 118°F **59. a.** −10 **b.** 3 **60. a.** $\frac{9}{16}$ **b.** −4 **61.** −19 **62.** $-\frac{14}{15}$ **63.** 5 **64.** 5.7 **65.** −10 **66.** −29 **67.** 65,233 ft; 65,233 + (−36,205) = 29,028 **68.** 287 B.C.; (−287); −287 + 75 = −212 **69.** −56 **70.** 1 **71.** 12 **72.** −12 **73.** 6.36 **74.** −2 **75.** $-\frac{2}{15}$ **76.** 0 **77.** High: 3, low: −4.5 **78. a.** Associative Property of Multiplication **b.** Commutative Property of Multiplication **c.** Multiplication Property of 1 (Identity Property of Multiplication) **d.** Inverse Property of Multiplication **79.** −1 **80.** −17 **81.** 3 **82.** $-\frac{6}{5}$ **83.** Undefined **84.** −4.5 **85.** 0, 18, 0 **86.** −\$360 **87. a.** 8^5 **b.** $9\pi r^2$ **88. a.** 81 **b.** $-\frac{8}{27}$ **c.** 32 **d.** 50 **89.** 17 **90.** −36 **91.** −169 **92.** 23 **93.** −420 **94.** $-\frac{7}{19}$ **95.** 113 **96.** Undefined **97. a.** $(-9)^2 = 81$ **b.** $-9^2 = -81$

98. \$20 **99. a.** 3 **b.** 1 **100. a.** 16, −5, 25 **b.** $\frac{1}{2}$, 1
101. $h + 25$ **102.** $3s - 15$ **103.** $\frac{1}{2}t - 6$
104. $|2 - a^2|$ **105.** $n + 4$ **106.** $b - 4$ **107.** $10d$
108. $x - 5$ **109.** 30, $10d$ **110.** 0, 19, −16 **111.** 40
112. −36 **113.** $150a$ **114.** $(9 + 1) + 7y$ **115.** $(2.7 \cdot 10)b$
116. $2x^2 + x$ **117.** $-28w$ **118.** $24x$ **119.** $2.08f$ **120.** r
121. $5x + 15$ **122.** $-2x - 3 + y$ **123.** $3c - 6$
124. $12.6c + 29.4$ **125.** $9p$ **126.** $-7m$ **127.** $4n$
128. $-p - 18$ **129.** $0.1k^2$ **130.** $8a^3 - 1$ **131.** w
132. $4h - 15$ **133.** $4.6t - 7.7$ **134. a.** x **b.** $-x$
c. $4x + 1$ **d.** $4x - 1$

CHAPTER 1 TEST (page 98)

1. a. equivalent **b.** product **c.** reciprocal **d.** like terms
e. undefined **2. a.** \$24 **b.** 5 hr **3.** 3, 20, 70
4. $2 \cdot 2 \cdot 3 \cdot 3 \cdot 5 = 2^2 \cdot 3^2 \cdot 5$ **5.** $\frac{2}{5}$ **6.** $\frac{3}{2} = 1\frac{1}{2}$ **7.** $\frac{27}{35}$ **8.** $6\frac{11}{15}$
9. a. $4\frac{1}{4}$ lb **b.** \$3.57 **10.** $0.8\overline{3}$
11. a.

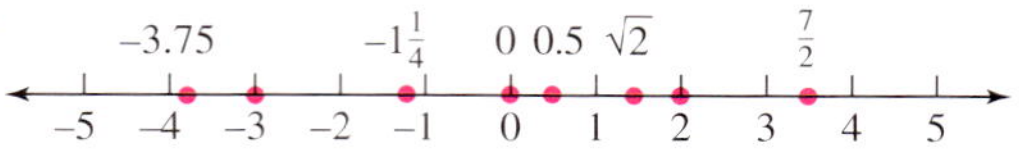

b. Natural numbers: 2; whole numbers: 0, 2; integers: 0, 2, −3; rational numbers: $-1\frac{1}{4}$, 0, −3.75, 2, $\frac{7}{2}$, 0.5, −3; irrational numbers: $\sqrt{2}$; real numbers: all **12. a.** True **b.** False **c.** True **d.** True
e. True **13. a.** > **b.** < **c.** < **d.** >
14. A gain of 0.6 of a rating point **15.** −7.6 **16.** −2 **17.** $\frac{3}{8}$
18. a. −6 **b.** $-6 + (-4) = -10$ **19. a.** 14
b. $14(-0.9) = -12.6$ **20.** −30 **21.** −2.44 **22.** 0 **23.** $-\frac{27}{125}$
24. 0 **25.** −3 **26.** 50 **27.** 14 **28.** a net loss of 78 in.
29. −\$1,275 **30. a.** $-12 + (97 + 3)$ **b.** $2x + 14$ **c.** $-2(5)m$
d. 1 **e.** $15x$ **31. a.** 9^5 **b.** $3x^2z^3$ **32.** 4, 17, −59 **33.** 170
34. −12 **35.** −99 **36.** −351 **37.** 20 **38.** −60 **39.** $2w - 7$
40. a. $x - 2$ **b.** $25q$¢ **41.** 3 **42.** 1, −6, −1, 10 **43.** $-20x$
44. $224t$ **45.** $-4a + 4$ **46.** $-5.9d^3$ **47.** $14x + 3$
48. $3m^2 + 2m - 4$ **49. a.** True **b.** False **c.** False **d.** False
e. True **f.** False

ARE YOU READY? 2.1 (page 102)

1. − **2.** + **3.** 7 **4.** · **5.** $\frac{2}{5}$ **6.** $-\frac{8}{9}$

SELF CHECKS 2.1

1. Yes **2.** 49 **3. a.** 33 **b.** 49 **4. a.** $\frac{29}{15}$ **b.** −0.5 **5.** 72
6. a. 6 **b.** $-\frac{11}{21}$ **7. a.** 11 **b.** −25.1 **8.** 12

STUDY SET SECTION 2.1 (page 110)

1. equation **3.** solve **5.** equivalent **7. a.** $x + 6$ **b.** Neither
c. No **d.** Yes **9. a.** c, c **b.** c, c **11. a.** x **b.** y **c.** t
d. h **13.** 5, 5, 50, 50, $\stackrel{?}{=}$, 45, 50 **15. a.** Is possibly equal to
b. Yes **17.** No **19.** No **21.** No **23.** No **25.** Yes **27.** No
29. No **31.** Yes **33.** Yes **35.** Yes **37.** 71 **39.** 18 **41.** −12
43. 3 **45.** $\frac{11}{10}$ **47.** −2.3 **49.** 45 **51.** −48 **53.** 20 **55.** $-\frac{5}{42}$
57. 4 **59.** 0.5 **61.** −18 **63.** $-\frac{4}{21}$ **65.** 7 **67.** $\frac{8}{9}$ **69.** 0
71. −0.9 **73.** 15 **75.** 20 **77.** $-\frac{1}{25}$ **79.** 21 **81.** 0 **83.** −2.64
85. 1 **87.** 4 **89.** 13 **91.** 2.5 **93.** $-\frac{8}{3}$ **95.** $-\frac{21}{16}$ **97.** −5
99. −200 **101.** $\frac{4}{3}$ **103.** 4 **105. a.** $\frac{13}{20}$ **b.** $\frac{17}{20}$ **c.** $\frac{15}{2}$ **d.** $\frac{3}{40}$
107. 65° **109.** \$6,000,000 **115.** 0 **117.** $45 - x$

ARE YOU READY? 2.2 (page 112)

1. −12 **2.** $2a$ **3.** $-7m + 18$ **4.** $3x$ **5.** $24n$ **6.** 8

SELF CHECKS 2.2

1. 7 **2.** −36 **3.** −3.9 **4. a.** 1 **b.** −11 **5.** −16 **6.** $-\frac{5}{2}$
7. 6,000 **8.** 0 **9.** All real numbers; the equation is an identity.
10. No solution; the equation is a contradiction.

STUDY SET SECTION 2.2 (page 120)

1. equation **3.** identity **5. a.** subtraction, multiplication
b. addition, division **7.** No **9. a.** 6 **b.** 10
11. 7, 7, 2, 2, 14, $\stackrel{?}{=}$, 28, 21, 14 **13.** −9 **15.** −5 **17.** 18 **19.** 16
21. −4 **23.** 2.9 **25.** $\frac{11}{5}$ **27.** −6 **29.** −21 **31.** 1 **33.** $\frac{2}{15}$
35. 6 **37.** 5 **39.** 200 **41.** −4 **43.** −1 **45.** All real numbers
47. No solution **49.** $\frac{1}{4}$ **51.** 12 **53.** −1 **55.** 6 **57.** 1 **59.** 30
61. No solution **63.** −11 **65.** 7 **67.** −11 **69.** $\frac{9}{2}$ **71.** $-\frac{12}{5}$
73. −0.25 **75.** −7 **77.** $\frac{10}{3}$ **79.** 5 **81.** 1,000 **83.** −3 **85.** $\frac{27}{5}$
87. 200 **89.** −11 **91.** 3 **93.** $\frac{52}{9}$ **95.** 0.04 **97.** 80 **99.** $-\frac{5}{2}$
101. −20 **103.** No solution **105.** All real numbers **107.** −6
109. −6 **111. a.** $x - 20$ **b.** −5 **113. a.** $0.4 - 0.2x$ **b.** 0
119. Commutative property of multiplication **121.** Associative property of addition

ARE YOU READY? 2.3 (page 122)

1. 61 **2.** 2 **3.** 0.5, 0.75 **4.** 0.27 **5.** 3.72 **6.** 49.5

SELF CHECKS 2.3

1. 2.24 **2.** 570 million **3.** 1.45% **4.** 21% **5.** \$120 **6.** \$600

STUDY SET SECTION 2.3 (page 127)

1. Percent **3.** multiplication, is **5.** $\frac{51}{100}$, 0.51, 51%
7. amount, percent, base **9. a.** 639 **b.** 639, what, 3,618
11. a. 0.35 **b.** 0.085 **c.** 1.5 **d.** 0.0275 **e.** 0.0925 **f.** 0.015
13. 312 **15.** 46.2 **17.** 300 **19.** 1,464 **21.** 26% **23.** 2.5%
25. 0.48 oz **27. a.** \$1,012.6 billion **b.** \$715.2 billion **29.** \$10.45
31. \$24.20 **33.** 60%, 40% **35.** 19% **37.** No (66%)
39. 120 children **41. a.** 5 g; 25% **b.** 20 g **43. a.** 11% **b.** 46%
45. 12% **47. a.** 6% **b.** 23.5 mpg **49.** \$75 **51.** \$300
53. \$95,000 **55.** \$25,600 **61.** $\frac{12}{5} = 2\frac{2}{5}$ **63.** No

ARE YOU READY? 2.4 (page 130)

1. a. one **b.** three **2.** d **3.** x **4.** $8c - cx$ **5.** Yes **6.** 1

SELF CHECKS 2.4

1. \$1,510.50 **2.** 2.5 yr **3.** 1.25 min **4.** −283°F **5.** 230 ft
6. 62.83 in. **7.** 1,357.2 m^3 **8.** $c = r - p$ **9.** $a = \frac{2A}{r^2}$
10. $c = \frac{B - 4d}{3}$ **11.** $y = 4 - \frac{1}{3}x$ or $y = -\frac{1}{3}x + 4$
12. $s = \frac{A + 3xy}{xy}$

STUDY SET SECTION 2.4 (page 137)

1. formula **3.** volume **5. a.** $d = rt$ **b.** $r = c + m$
c. $p = r - c$ **d.** $I = Prt$ **7.** 11,176,920 mi, 65,280 ft
9. Ax, Ax, B, B, B **11. a.** $\pi r^2 h$ **b.** The radius of a cylinder; the height of a cylinder **13.** \$240 million **15.** \$931 **17.** 3.5%
19. \$6,000 **21.** 2.5 mph **23.** 4.5 hours **25.** 185°C

27. $-454°F$ **29.** 20 in. **31.** 1,885 mm^3 **33.** $c = r - m$
35. $b = P - a - c$ **37.** $h = \frac{3V}{B}$ **39.** $R = \frac{E}{I}$ **41.** $r = \frac{T - 2t}{2}$
43. $x = \frac{C - By}{A}$ **45.** $y = -\frac{2}{7}x + 3$ **47.** $y = \frac{9}{2}x + 4$
49. $m = \frac{T - 4ab}{4ab}$ **51.** $r = \frac{G + g}{4g}$ **53.** $c = 3A - a - b$
55. $y = -3x + 9$ **57.** $m = \frac{2K}{v^2}$ **59.** $r = \frac{C}{2\pi}$ **61.** $M = 4.2B + 19.8$
63. $f = \frac{s}{w}$ **65.** $y = \frac{1}{3}x + 3$ **67.** $b = \frac{2A}{h} - d$ or $b = \frac{2A - hd}{h}$
69. $a^2 = c^2 - b^2$ **71.** $c = \frac{72 - 8w}{7}$ **73.** $b = \frac{m - 70 - at}{t}$
75. $l = \frac{V}{wh}$ **77.** $t = T - 18E$ **79.** $r^2 = \frac{s}{4\pi}$ **81. a.** $R = A - ab$
b. $a = \frac{A - R}{b}$ **83. a.** $h = \frac{S - 4lw}{2w}$ **b.** $l = \frac{S - 2wh}{4w}$
85. Horsepower $= \frac{\text{RPM} \cdot \text{Torque}}{5{,}252}$ **87.** 168.4, 192.8 **89.** 14 in.
91. 50 in. **93.** 25 in., 2.5 in. **95.** 18.1 $in.^2$ **97.** 2,463 ft^2
99. 3,150 cm^2 **101.** 6 in. **103.** 8 ft **105.** 348 ft^3 **107.** 254 $in.^2$
109. $R = \frac{L - 2D - 3.25r}{3.25}$ or $R = \frac{L - 2D}{3.25} - r$ **115.** 137.76
117. 15%

ARE YOU READY? 2.5 (page 142)

1. $4x + 4$ **2.** 49 **3.** $39.15 **4.** $0.28x$ **5.** $P = 2l + 2w$
6. 180° **7.** $2x - 8$ **8.** 0.06

SELF CHECKS 2.5

1. 16 mi, 18 mi, 20 mi, and 22 mi **2.** 47 shirts **3.** $2,762.77
4. Page 579 **5.** 5 ft by 11 ft **6.** 12 cm, 12 cm

STUDY SET SECTION 2.5 (page 147)

1. consecutive **3.** vertex, base **5.** 17, $x + 2$, $3x$ **7.** $$0.03x$
9. 180° **11. a.** $x + 1$ **b.** $x + 2$ **c.** $x + 2$, $x + 4$ **13.** 4 ft, 8 ft
15. Day 1: 102 mi; day 2: 108 mi; day 3: 114 mi; day 4: 120 mi
17. 7.3 ft, 10.7 ft **19.** *Guitar Hero:* $2.99; *Call of Duty: World at War: Zombies II:* $9.99; *Tom Tom USA:* $39.99
21. 250 calories in ice cream, 600 calories in pie **23.** 7 hr
25. 580 mi **27.** 20 hr **29.** $50,000 **31.** $240 **33.** $5,250
35. Ronaldo: 15 goals; Mueller: 14 goals
37. *Friends:* 236 episodes; *Leave It to Beaver:* 234 episodes
39. July 22, 24, 26 **41.** Width: 27 ft; length: 78 ft
43. 21 in. by 30.25 in. **45.** 7 ft, 7 ft; 11 ft **47.** 20°
49. At steering column: 42.5°; at seat support: 70°; at pedal gear: 67.5° **51.** $x = 11°$; 22°, 68°
53. Maximum stride angle: 106° **59.** -24 **61.** $-\frac{40}{37}$

ARE YOU READY? 2.6 (page 151)

1. $400 **2.** 135 mi **3.** 3.6 gal of antifreeze **4.** $19.60
5. $14,000 **6.** $3x$

SELF CHECKS 2.6

1. $2,400 at 2%, $1,800 at 3% **2.** 3.5 hr **3.** 1.5 hr **4.** 10 gal
5. 40 lb **6.** 12 iPods, 4 skins, 24 cards

STUDY SET SECTION 2.6 (page 158)

1. investment, motion **3.** $30{,}000 - x$ **5.** $r - 150$
7. $35t$; t; $45t$; 80; $35t + 45t = 80$
9. a. $0.25x$, $0.50(6)$, $6 + x$, $0.30(6 + x)$, $0.25x + 0.50(6) = 0.30(6 + x)$
b. $10 - x$, $0.06x$, $0.03(10 - x)$, $0.05(10)$, $0.03(10 - x) + 0.06x = 0.05(10)$ **11.** 0.06, 0.152 **13.** 4
15. 6,000 **17.** $15,000 at 4%; $10,000 at 7% **19.** Silver: $1,500; gold: $2,000 **21.** $26,000 **23.** 822: $9,000; 721: $6,000 **25.** $4,900
27. Credit union: $13,500; stocks: $4,500 **29.** 2 hr
31. $\frac{1}{4}$ hr = 15 min **33.** 1 hr **35.** 4 hr **37.** 55 mph
39. 50 gal **41.** 4%: 5 gal; 1%: 10 gal
43. 32 ounces of 8%; 32 ounces of 22% **45.** 6 gal **47.** 50 lb
49. 20 scoops **51.** 15 **53.** $4.25 **55.** 17 **57.** 90
59. 40 pennies, 20 dimes, 60 nickels
61. 2-pointers: 50; 3-pointers: 4 **67.** $-36a - 48b + 384$
69. $30t + 6$

ARE YOU READY? 2.7 (page 162)

1. is less than **2.** True **3.**
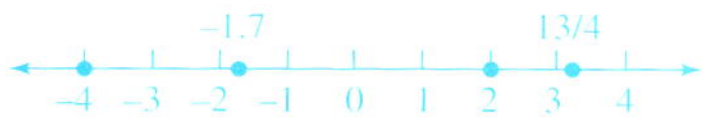

4. $0 < 10$

SELF CHECKS 2.7

1. Yes **2.** $[0, \infty)$

3. $(-\infty, 1)$

4. a. $[-200, \infty)$
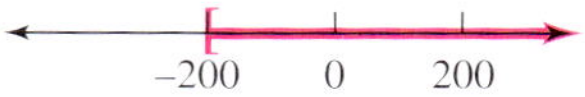

b. $(-\infty, 12)$
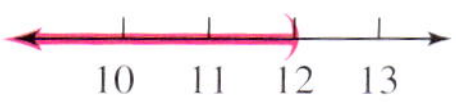

5. $(-3, \infty)$
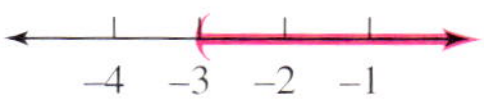

6. $(-\infty, 2.1)$
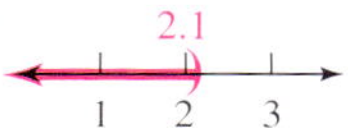

7. $\left[\frac{13}{4}, \infty\right)$
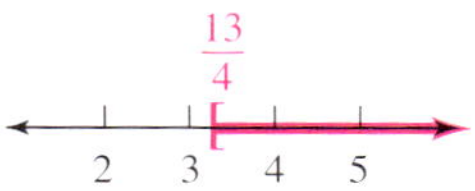

8. $\left(-\infty, \frac{21}{2}\right)$
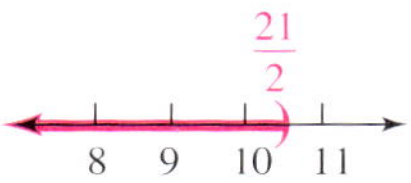

9. $[-2, 1)$
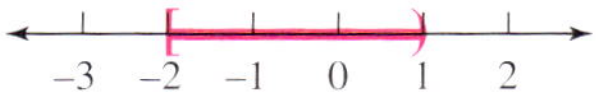

10. $[-4, 0]$
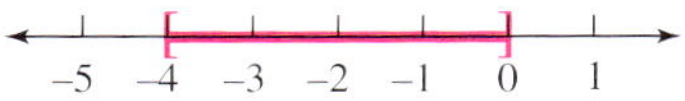

11. 84% or better

STUDY SET SECTION 2.7 (page 171)

1. inequality **3.** interval **5. a.** both **b.** positive
c. negative **7.** $x > 32$ **9. a.** -1, $-\infty$ **b.** 2, ∞ **11. a.** $\le$
b. ∞ **c.** [or] **d.** > **13.** 5, 5, 12, 4, 4, 3 **15. a.** Yes **b.** No
17. a. No **b.** Yes **19.** $(-\infty, 5)$

21. $(-3, 1]$

23. $(3, \infty)$

25. $[10, \infty)$

27. $(-\infty, 48]$

29. $[3, \infty)$

31. $[2, \infty)$

33. $(-\infty, 6)$

35. $(7, \infty)$

37. $(-\infty, -5]$

39. $(-\infty, 0.4]$

41. $(2.4, \infty)$

43. $(-5, \infty)$

45. $\left[-\frac{5}{3}, \infty\right)$

47. $\left(\frac{5}{4}, \infty\right)$

49. $(-\infty, 15]$ **51.** $[-2, 3)$ **53.** $\left(-\frac{7}{4}, 2\right)$ **55.** $(7, 10)$ **57.** $[-10, 0]$ **59.** $[2, 3)$ **61.** $(-3, 6]$ **63.** $\left(-\infty, \frac{3}{2}\right]$ **65.** $(-\infty, 12]$ **67.** $(-2, 1]$ **69.** $(-\infty, 1.5]$ **71.** $\left(-\infty, \frac{1}{8}\right]$ **73.** $(-\infty, 0)$ **75.** $\left[\frac{9}{4}, \infty\right)$ **77.** $(-5, -2)$ **79.** $(-\infty, 2]$ **81.** $(-\infty, -27)$ **83.** $\left(-\infty, \frac{17}{21}\right]$ **85.** $[-13, \infty)$ **87.** $(6, \infty)$ **89.** $[-32, 48]$ **91.** $(-\infty, 1.5]$ **93.** $\left[-\frac{3}{8}, \infty\right)$ **95.** $\left(-\infty, \frac{1}{2}\right)$ **97.** $(-\infty, -1]$ **99. a.** $\left(\frac{1}{8}, \infty\right)$ **b.** $\frac{1}{8}$ **101. a.** $[5, \infty)$ **b.** $[5, 12)$

103. 98% or better **105.** More than 27 mpg **107.** 19 ft or less **109.** More than 5 ft **111.** 40 or less **113.** 12.5 in. or less **115.** 26, 27, 28, 29, 30 **119.** 1, −3, 6

CHAPTER 2 REVIEW (page 175)

1. Yes **2.** No **3.** No **4.** No **5.** Yes **6.** Yes **7.** equation **8.** True **9.** 21 **10.** 32 **11.** −20.6 **12.** 107 **13.** 24 **14.** $\frac{16}{21}$ **15.** −9 **16.** −7.8 **17.** 0 **18.** $-\frac{16}{5}$ **19.** 2 **20.** −30.6 **21.** 30 **22.** −19 **23.** 4 **24.** 1 **25.** $\frac{5}{4}$ **26.** $\frac{47}{13}$ **27.** 6 **28.** $-\frac{22}{75}$ **29.** 5 **30.** 1 **31.** Identity; all real numbers **32.** Contradiction; no solution **33. a.** Percent **b.** discount **c.** commission **34.** 192.4 **35.** 142.5 **36.** 12% **37. a.** 28.8% **b.** 221 million **38.** \$26.74 **39.** No **40.** \$450 **41.** \$150 **42.** 1,567% **43.** \$176 **44.** \$11,800 **45.** 8 min **46.** 4.5% **47.** 1,949°F **48. a.** 168 in. **b.** 1,440 in.^{2} **c.** 4,320 in.^{3} **49.** 76.5 m^2 **50.** 144 in.^{2} **51. a.** 50.27 cm **b.** 201 cm^2 **52.** 9.4 ft^3 **53.** 381.70 in.^{3} **54.** 120 ft^3 **55.** $h = \frac{A}{2\pi r}$ **56.** $G = 3A - 3BC + K$ **57.** $t = \frac{4C}{s} + d$ **58.** $y = \frac{3}{4}x + 4$ **59.** 8 ft **60.** 200 signatures **61.** \$2,500,000 **62.** Labonte: 43; Petty: 45 **63.** 24.875 in. × 29.875 in. $\left(24\frac{7}{8}\text{ in.} \times 29\frac{7}{8}\text{ in.}\right)$ **64.** 76.5°, 76.5° **65.** \$16,000 at 7%, \$11,000 at 9% **66.** 20 min **67.** $1\frac{2}{3}$ hr = 1 hr 40 min **68.** TV celebrities: 12 autographs; movie stars: 4 autographs **69.** 10 lb of each **70.** 2 gal **71.** $(-\infty, 1)$ **72.** $(-\infty, 12]$ **73.** $\left(\frac{5}{4}, \infty\right)$ **74.** $[3, \infty)$ **75.** $(-\infty, 40]$ **76.** $(7, \infty)$ **77.** $(6, 11)$ **78.** $\left(-\frac{7}{2}, \frac{3}{2}\right]$ **79.** $2.40\text{ g} \le w \le 2.53\text{ g}$ **80.** $0\text{ in.} < l \le 48\text{ in.}$; 48 inches or less

CHAPTER 2 TEST (page 182)

1. a. solve **b.** Percent **c.** circumference **d.** inequality **e.** multiplication, equality **2.** No **3.** 2 **4.** 22 **5.** −5 **6.** $\frac{12}{7}$ **7.** 1,336 **8.** All real numbers (an identity) **9.** $\frac{7}{4}$ **10.** 55 **11.** 0 **12.** 0 **13.** −4 **14.** No solution (a contradiction) **15.** 12.16 **16.** \$76,000 **17.** 6% **18.** \$30 **19.** \$295 **20.** −10°C **21.** 66,480 ft **22.** 393 in.^{3} **23.** $h = \frac{V}{\pi r^2}$ **24.** $r = \frac{A - P}{Pt}$ **25.** $c = 4A - a - b - d$ **26.** $y = \frac{2}{3}x - 3$ **27.** 20 in.^{2} **28.** Programming: 22 min; commercials: 8 min **29.** \$40.55 **30.** 80 balcony seats, 800 floor seats **31.** \$120,000 **32.** 380 mi, 280 mi **33.** Green: 16 lb; herbal: 4 lb **34.** 412, 413 **35.** $\frac{3}{5}$ hr **36.** 10 liters **37.** $\frac{1}{3}$ hr = 20 min **38.** 68° **39.** \$5,250 **40.** No **41.** $[-3, \infty)$ **42.** $(-\infty, 6.4)$ **43.** $[-7, 4)$ **44.** $(-\infty, -13)$ **45.** $(-\infty, 5]$ **46.** 180 words

ARE YOU READY? 3.1 (page 186)

1.

2. a. 8 **b.** −3.5 **3.** I, II, III, IV **4.** $4\frac{1}{2}, -3\frac{2}{3}$

SELF CHECK 3.1

1.

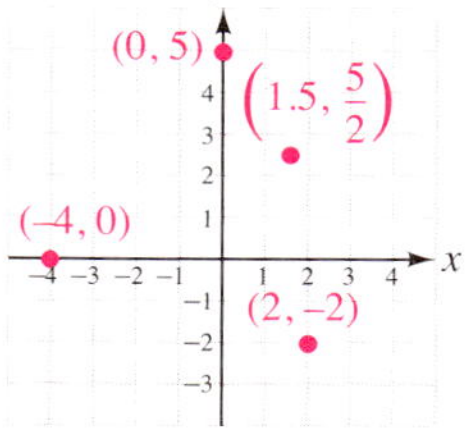

2. $A(4, 0)$; $B(0, 1)$; $C(-3.5, -2.5)$; $D(2, -4)$ **3. a.** 30 min before and 85 min after taping began **b.** 200 **c.** 40 min before taping began

STUDY SET SECTION 3.1 (page 191)

1. ordered **3.** axis, axis, origin **5.** rectangular
7. a. origin, left, up **b.** origin, right, down **9. a.** I and II
b. II and III **c.** IV **d.** The y-axis
11. (3, 5) is an ordered pair, 3(5) = 3 · 5
13. Yes **15.** Horizontal
17.

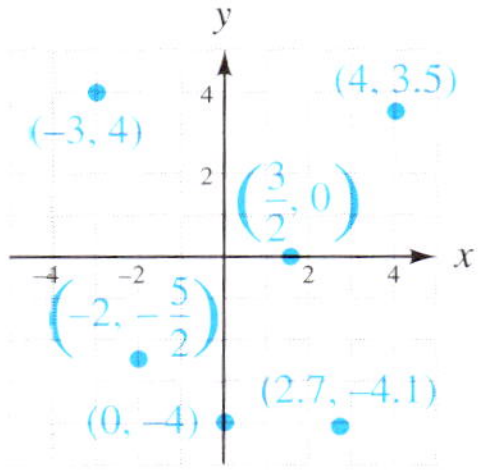

19. (4, 3), (0, 4), (−5, 0), (−4, −5), (3, −3)
21. a. 60 beats/min **b.** 10 min
23. a. 5 min and 50 min after starting **b.** 20 min
25. a. 2 hr **b.** −1,000 ft **27. a.** It ascends (rises) 500 ft
b. −500 ft **29.** Rivets: (−60, 0), (−20, 0), (20, 0), (60, 0); welds: (−40, 30), (0, 30), (40, 30); anchors: (−60, −30), (60, −30)
31. (G, 2), (G, 3), (G, 4) **33. a.** 8 teeth **b.** It represents the patient's left side. **35. a.** 60°; 4 ft **b.** 30°; 4 ft **37.** 10 square units
39.

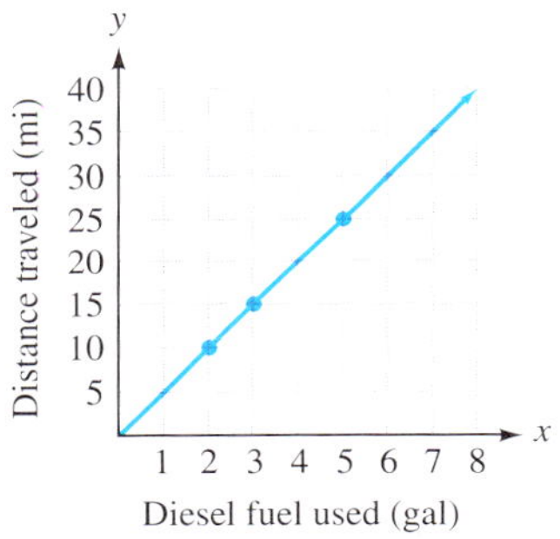

a. 20 mi. **b.** 6 gal **c.** 35 mi

41.

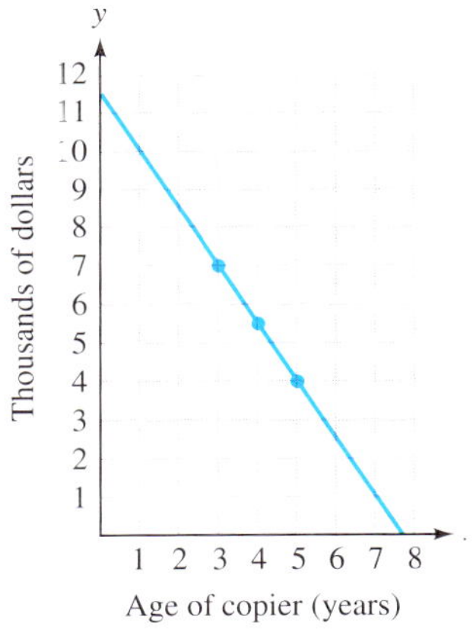

a. A 3-yr-old copier is worth \$7,000. **b.** \$1,000
c. 6 yr
47. $h = \frac{3(AC + T)}{2}$ or $h = \frac{3AC + 3T}{2}$
49. −1

ARE YOU READY? 3.2 (page 195)

1. False **2.** 5 **3.** 2 **4.** $y = -\frac{4}{5}x - 3$ **5.** 3 **6.** −5

SELF CHECK 3.2

1. Yes **2.** (−2, −10) **3.**

x	y	(x, y)
3	−2	(3, −2)
5	−5	(5, −5)

4.

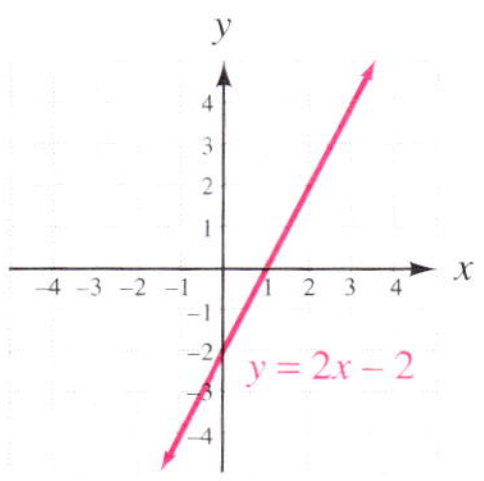

5.

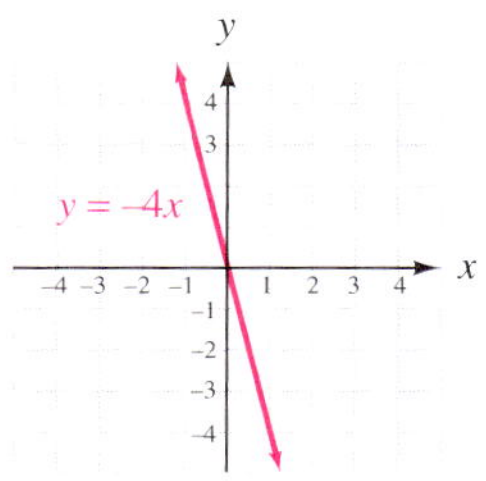

6.

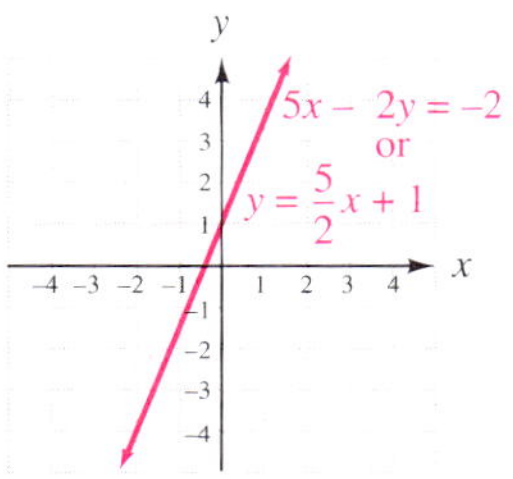

7. $c = 15 + 10n$

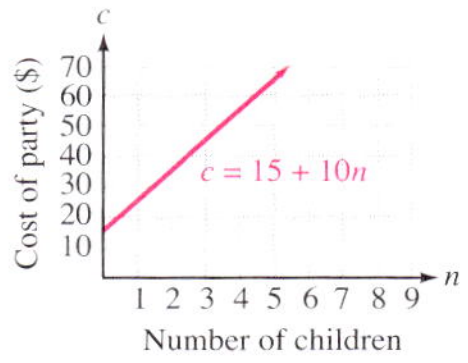

STUDY SET SECTION 3.2 (page 203)

1. two **3.** table **5.** linear **7. a.** 2 **b.** Yes **c.** No
d. Infinitely many **9.** solution, point
11. a. −5, 0, 5 (Answers may vary)
b. −10, 0, 10 (Answers may vary) **13.** 6, −2, 2, 6
15. a. 1's **b.** The exponent on x is not 1. **17.** Yes
19. No **21.** Yes **23.** Yes **25.** No **27.** No **29.** 11
31. 4 **33.** 13 **35.** $-\frac{8}{7}$
37.

x	y	(x, y)
8	12	(8, 12)
6	8	(6, 8)

39.

x	y	(x, y)
−5	−13	(−5, −13)
−1	−1	(−1, −1)

41.

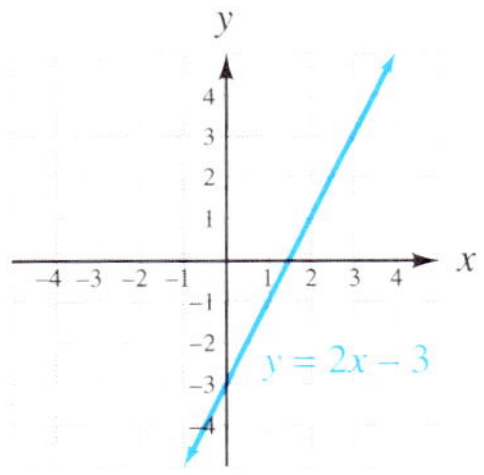

43.

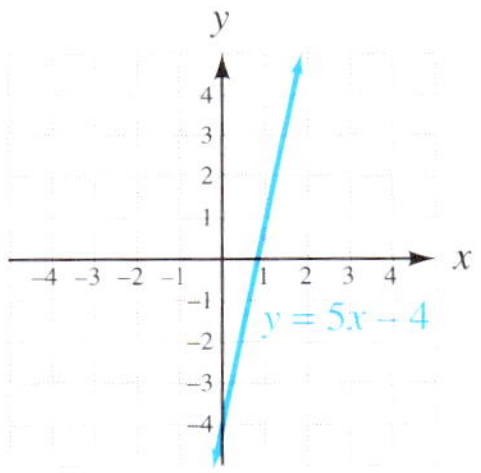

45.

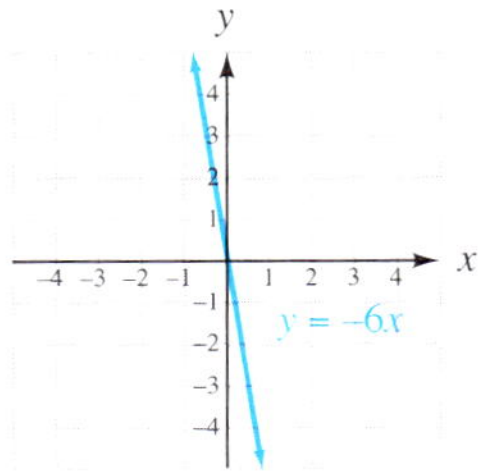

47.

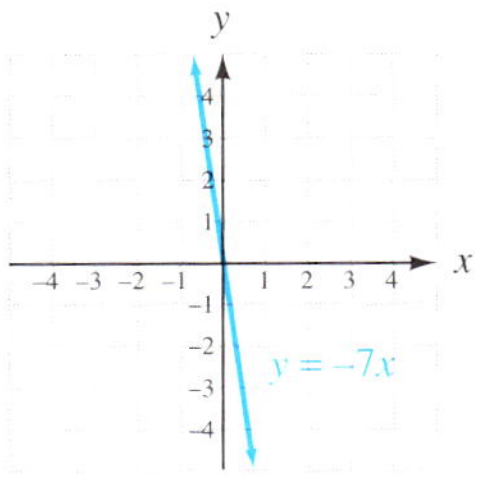

49.

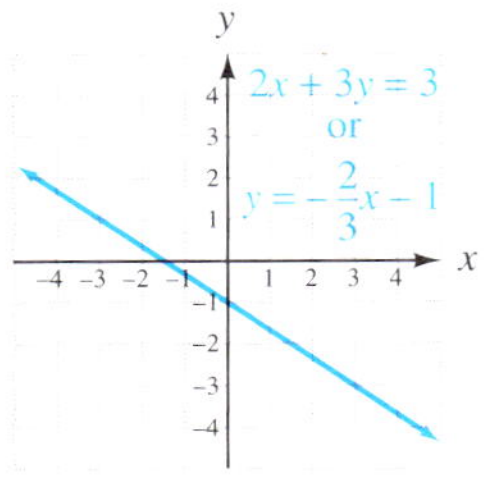

51.

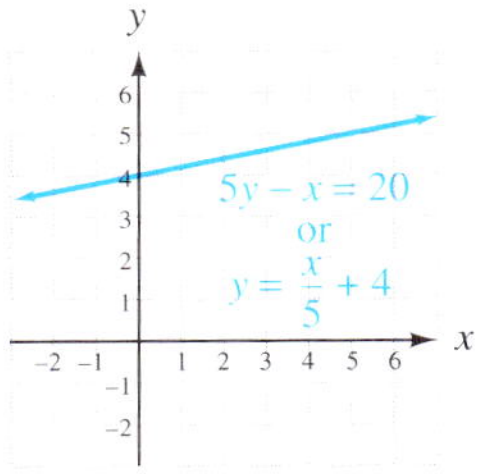

53.

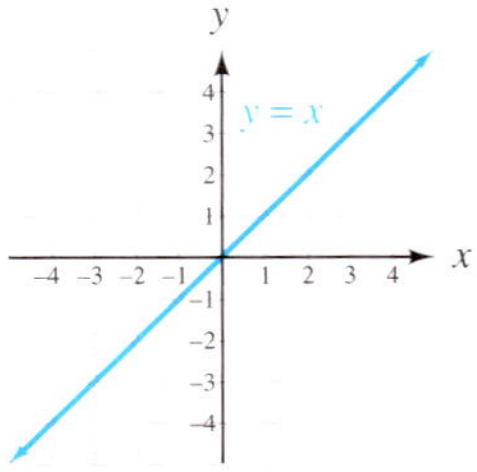

55.

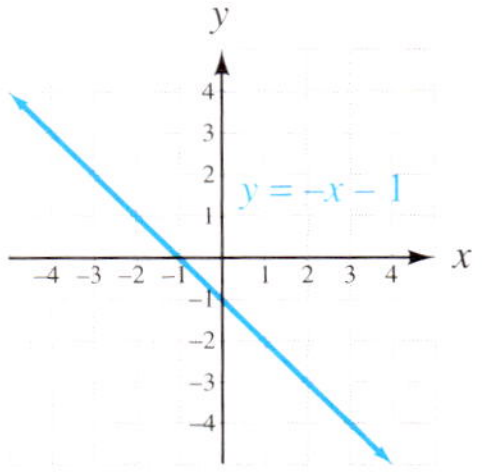

57.

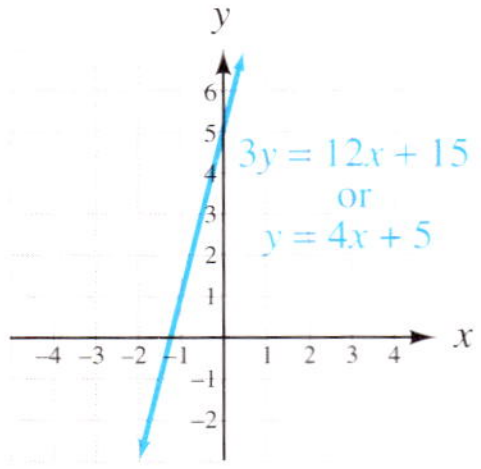

59.

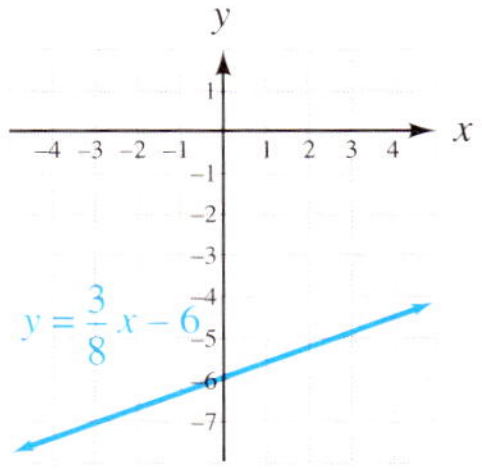

61.

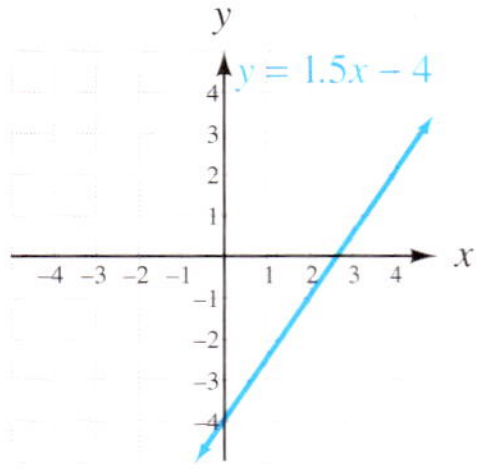

63.

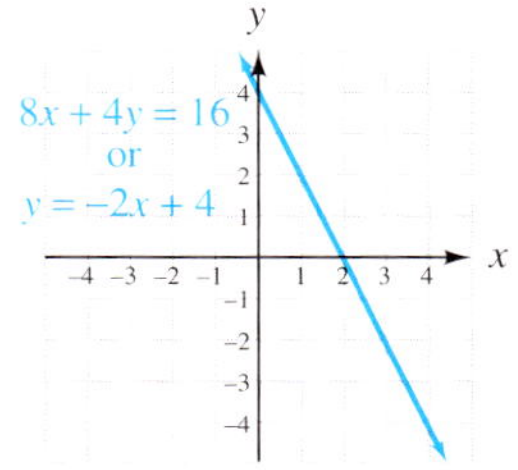

65.

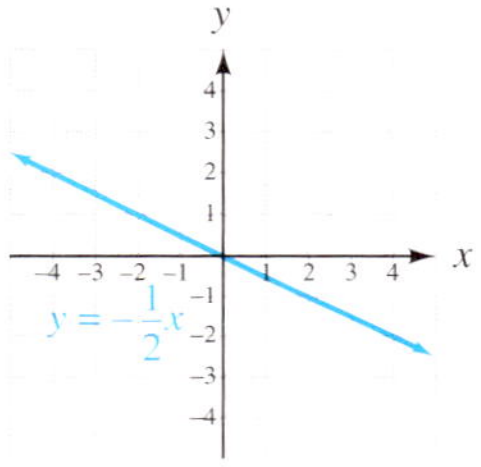

67.

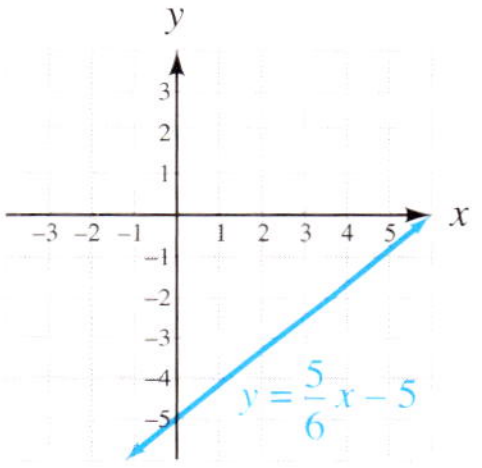

69.

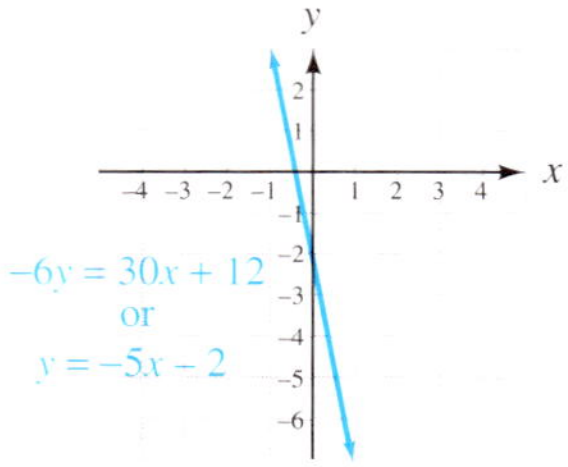

71.

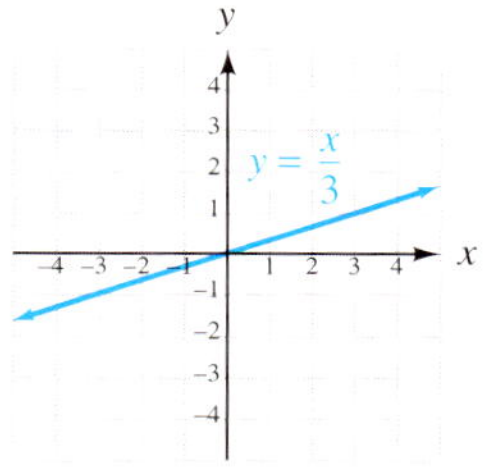

73.

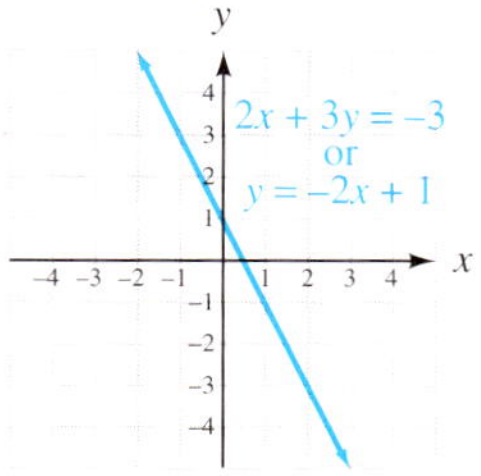

75.

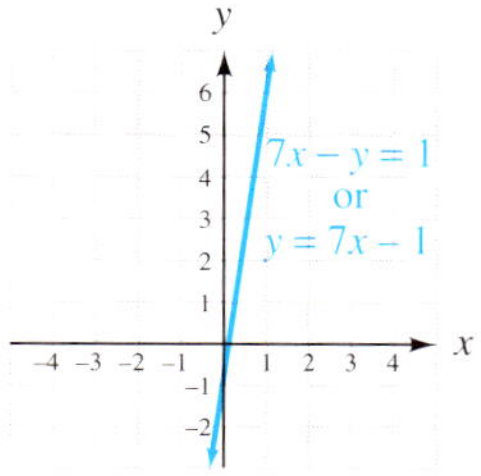

77.

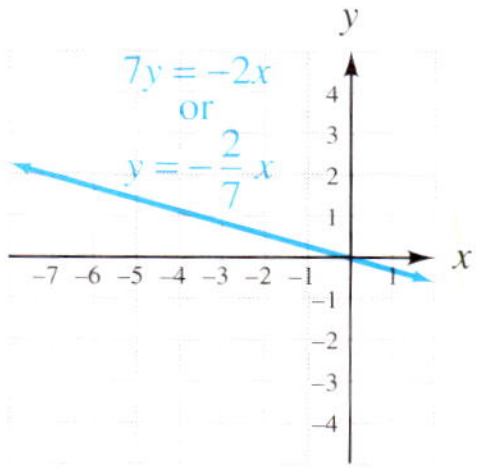

79.

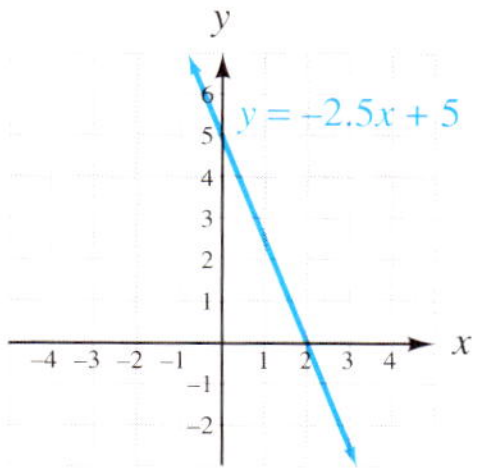

81.

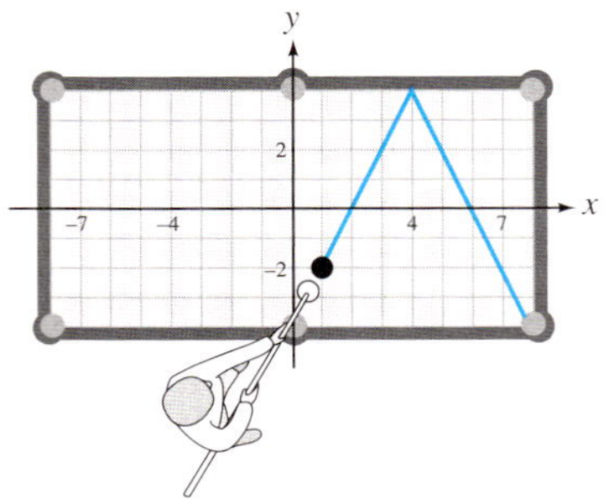

83. About 125 hr

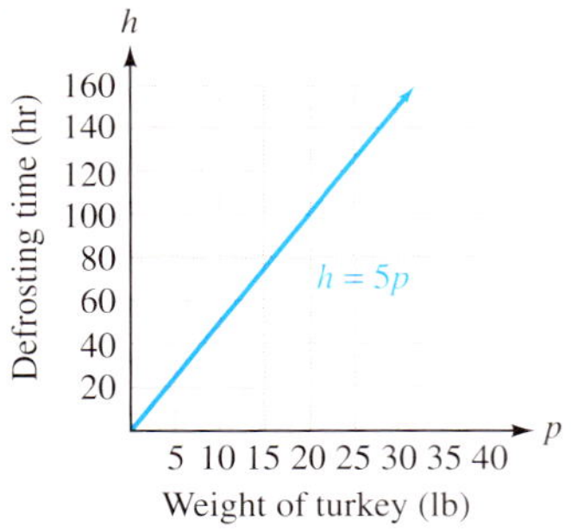

85. About 3 oz

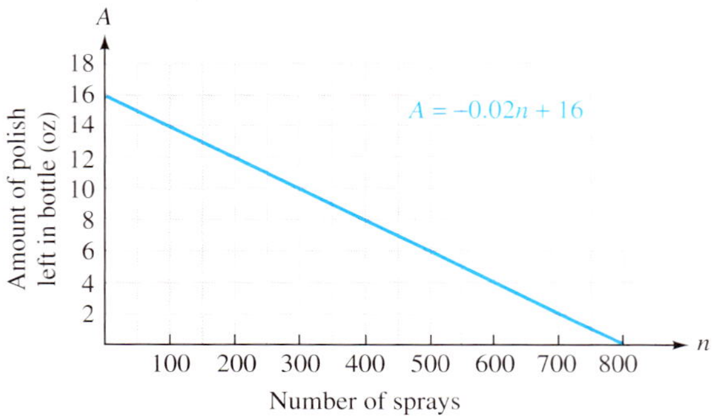

87. About $100

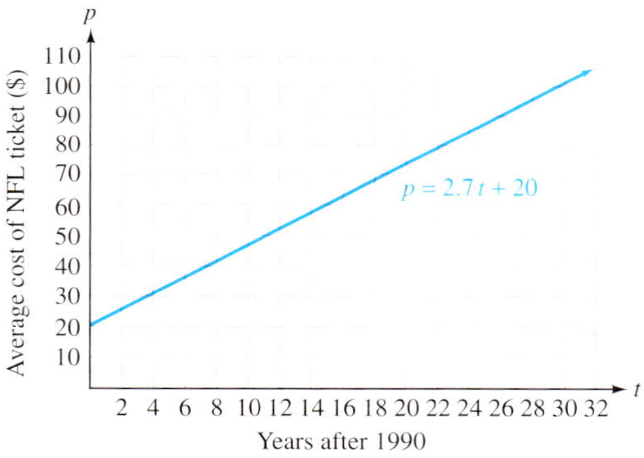

89. About 180 tickets

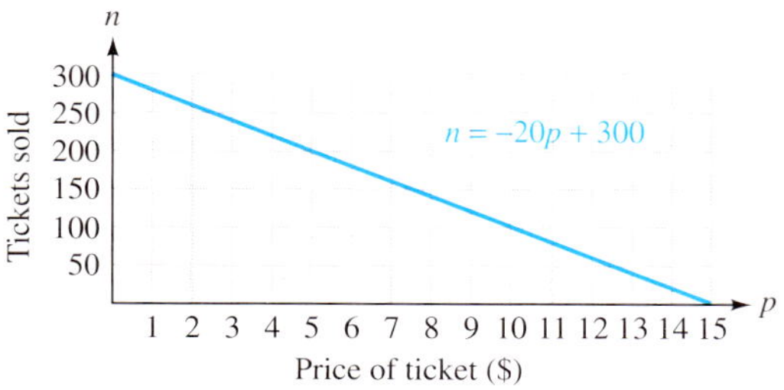

91. 2030

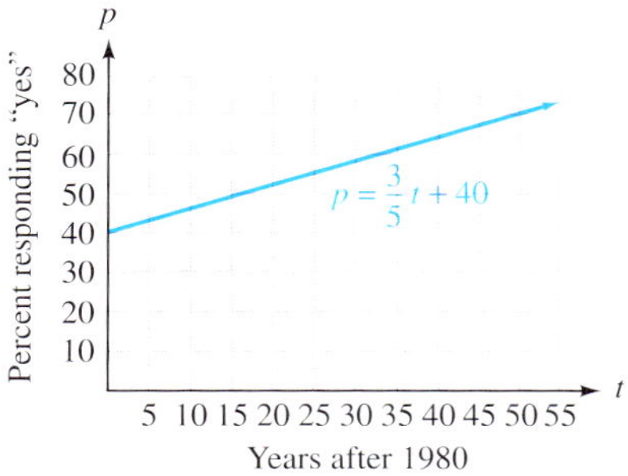

101. $5 + 4c$ or $4c + 5$ **103.** 904.8 ft^3

ARE YOU READY? 3.3 (page 206)

1.

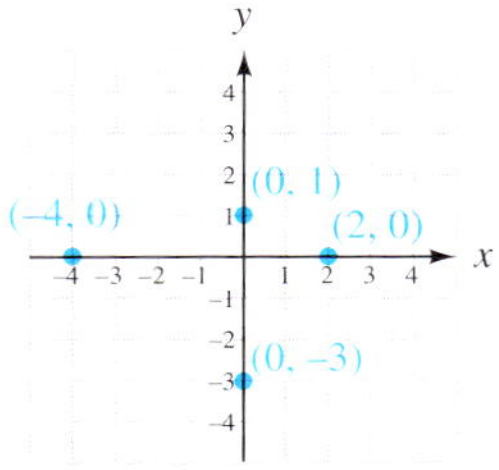

2. 0 **3.** The origin (0, 0) **4.** 5

SELF CHECK 3.3

1. (−1, 0); (0, −3)

2.

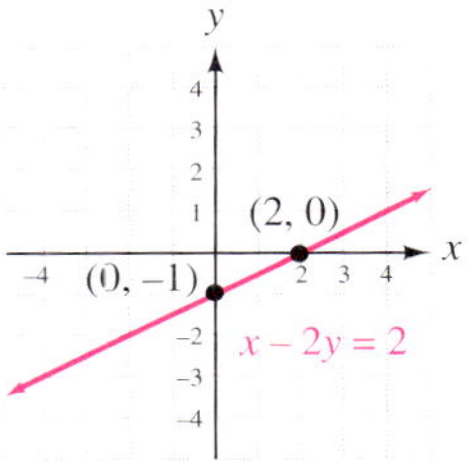

3.

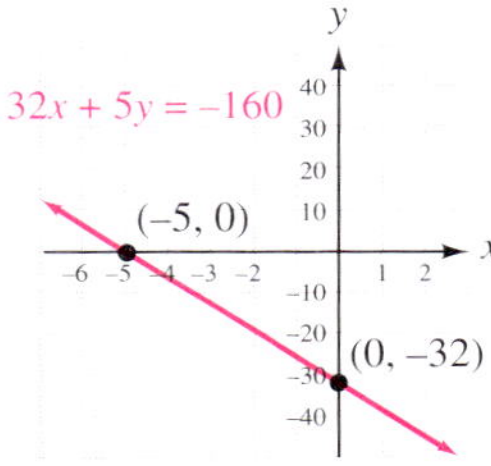

4.

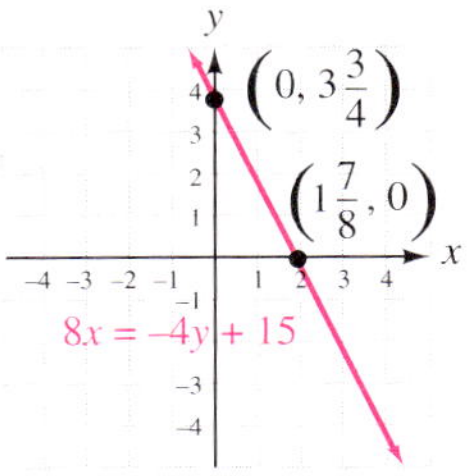

5.

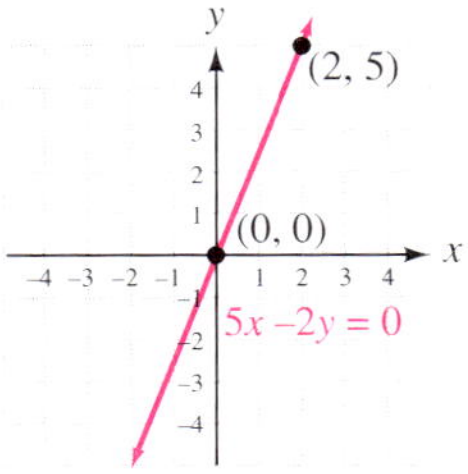

6.

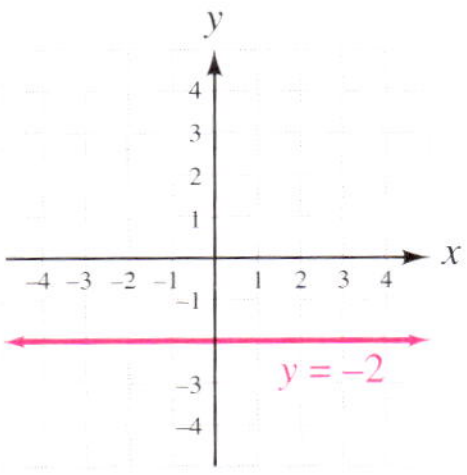

7.

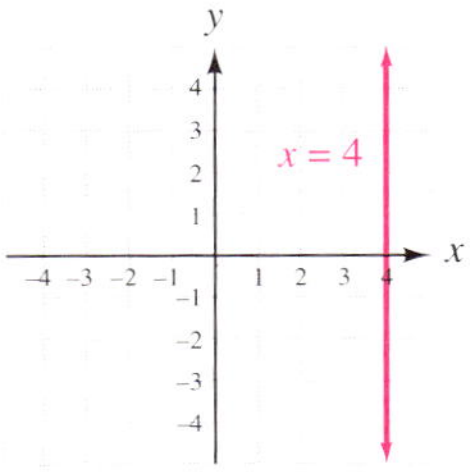

STUDY SET SECTION 3.3 (page 214)

1. x-intercept **3.** horizontal, vertical **5. a.** 0, y **b.** 0, x

7. a. y-intercept: (0, 80,000); \$80,000

b. x-intercept: (30, 0); 30 years after purchase **9.** $y = 0$; $x = 0$

11. x-intercept: (4, 0), y-intercept: (0, 3)

13. x-intercept: (−5, 0), y-intercept: (0, −4)

15. No x-intercept, y-intercept: (0, 2)

17. x-intercept: $\left(-2\frac{1}{2}, 0\right)$; y-intercept: $\left(0, \frac{2}{3}\right)$ (Answers may vary)

19. (3, 0); (0, 8) **21.** (4, 0); (0, −14) **23.** (−2, 0); $\left(0, -\frac{10}{3}\right)$

25. $\left(\frac{3}{2}, 0\right)$; (0, 9)

27.

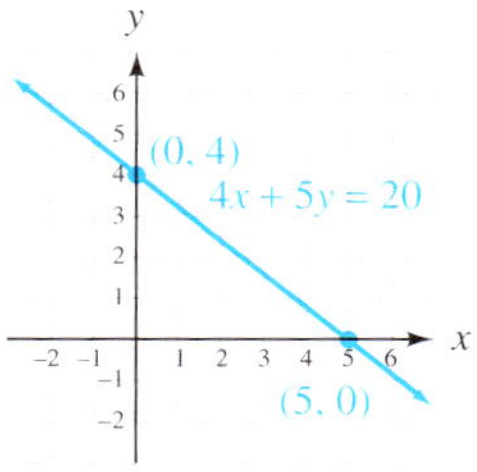

29.

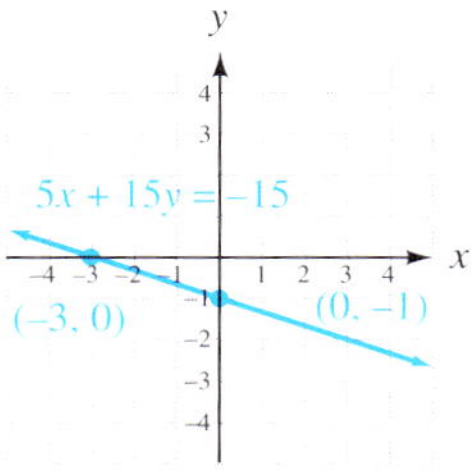

31.

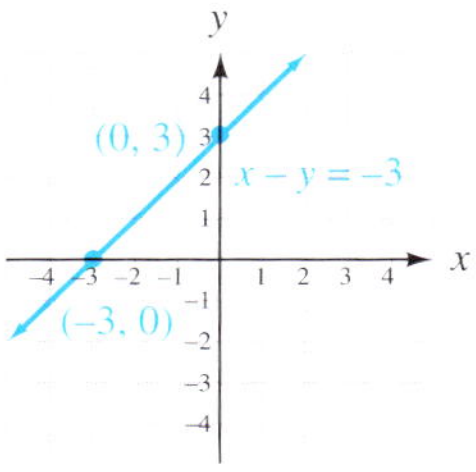

33.

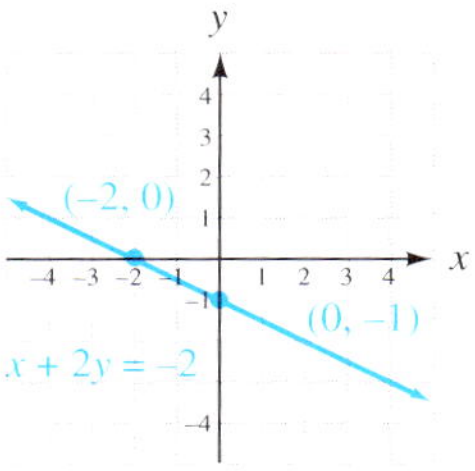

35.

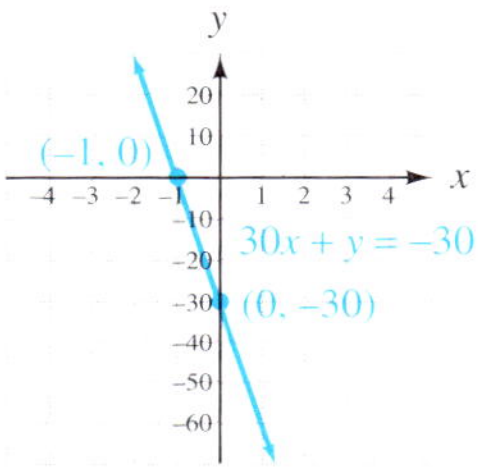

37.

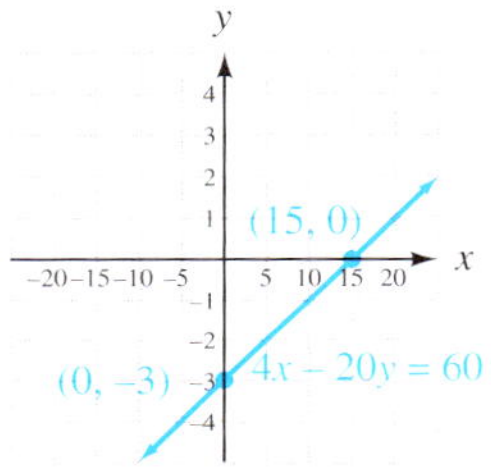

39.

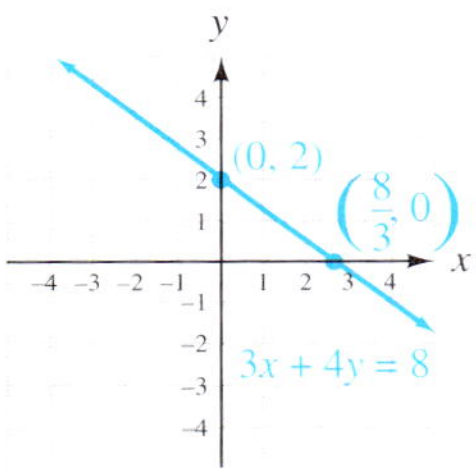

41.

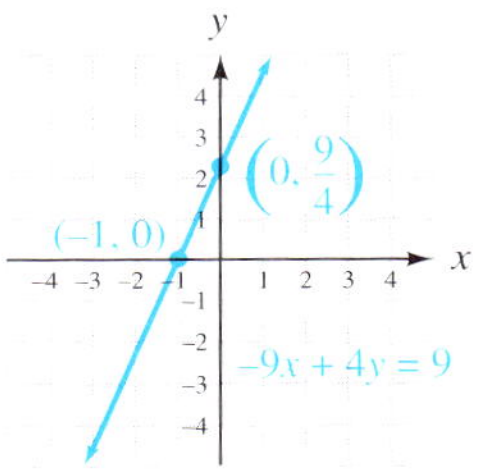

43.

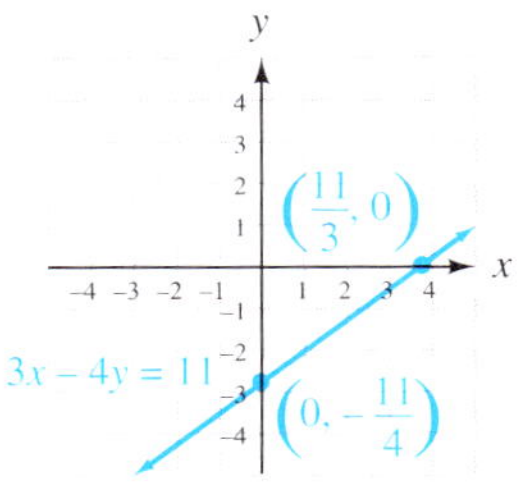

45.

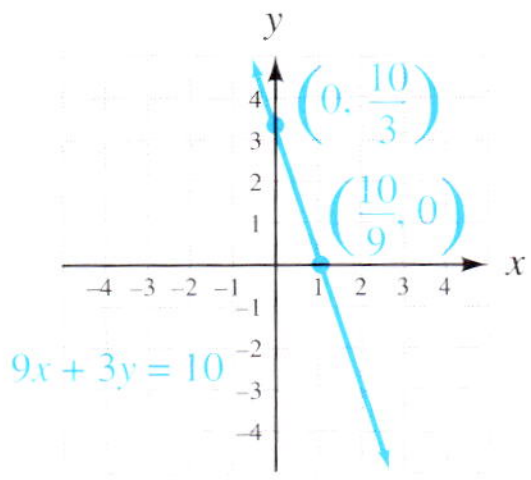

47.

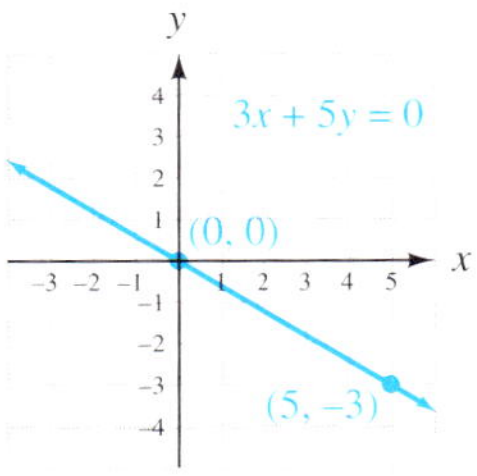

49.

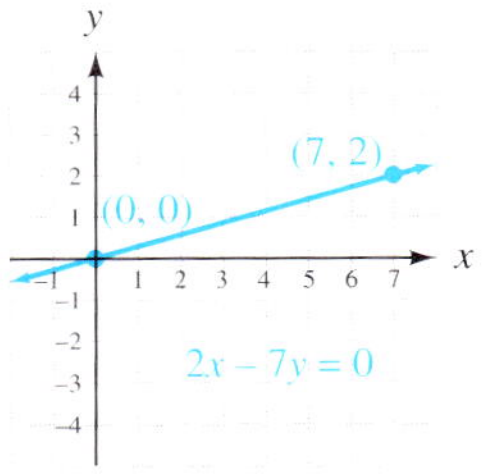

51.

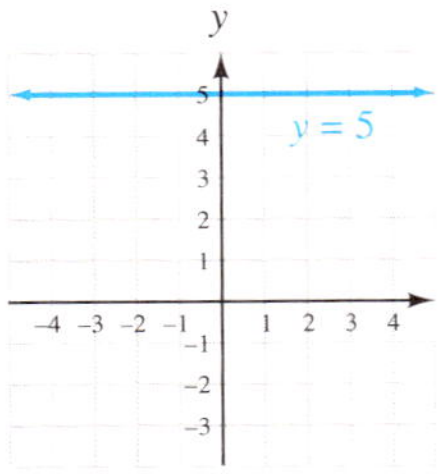

53.

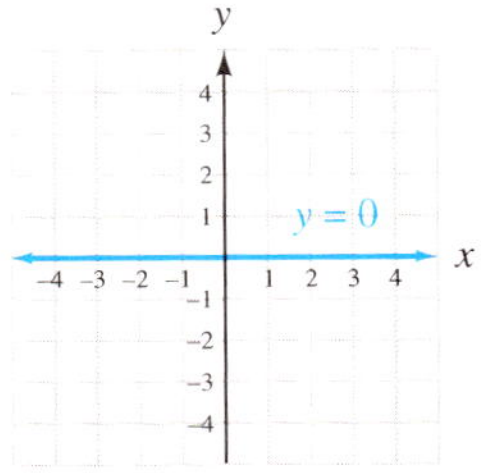

55.

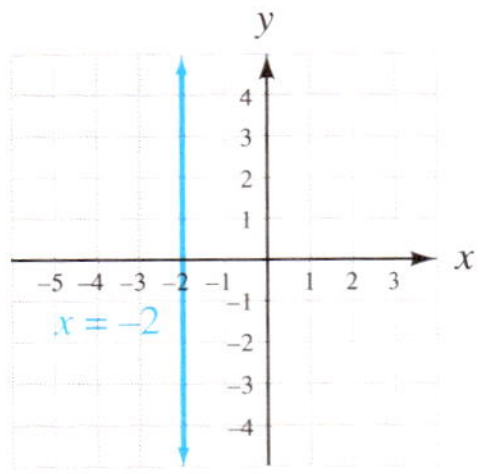

57.

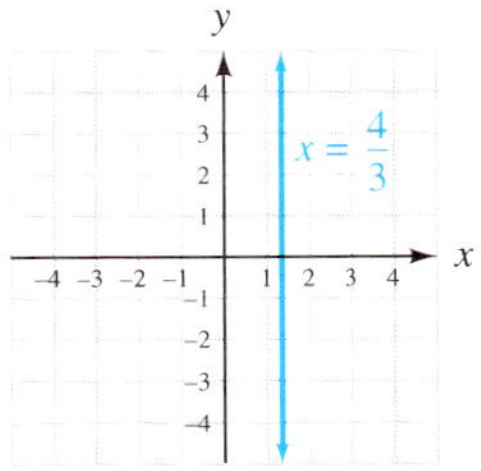

59. $y = 2$

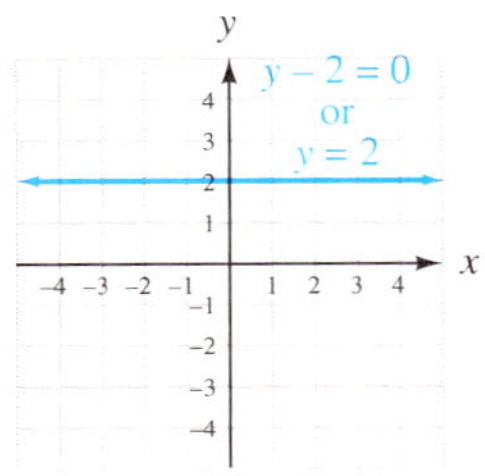

61. $x = 1.5$

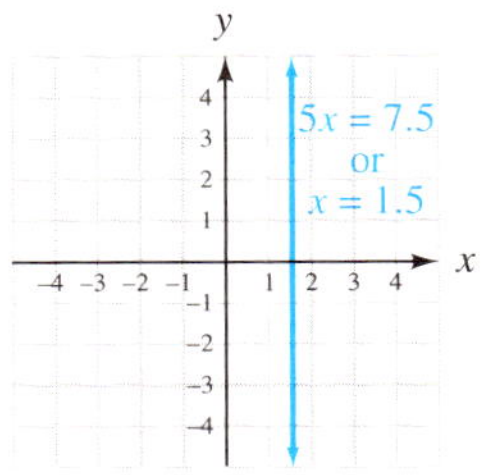

63.

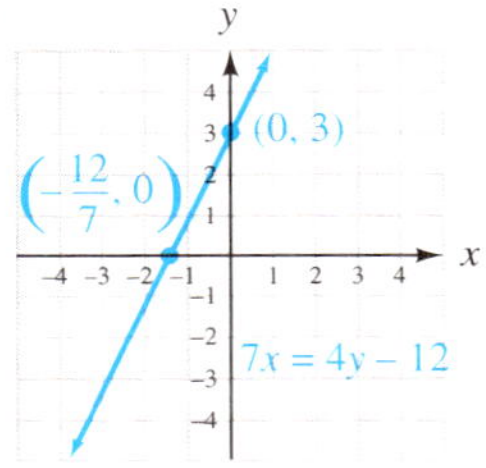

65.

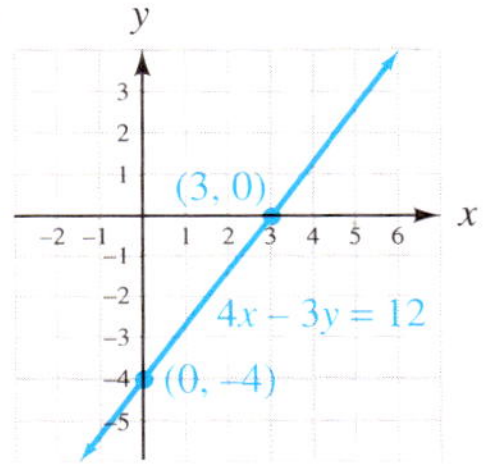

67.

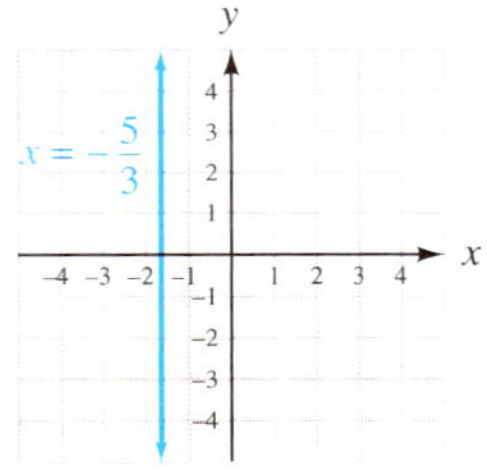

69.

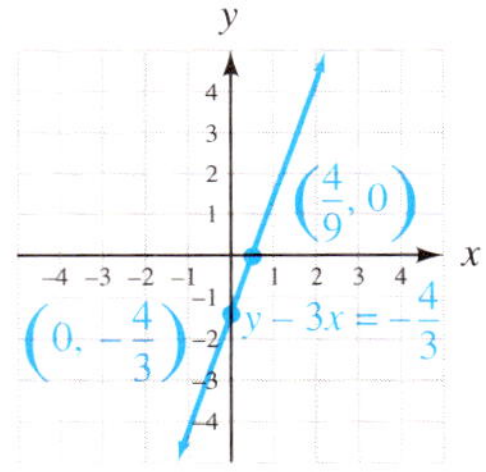

71.

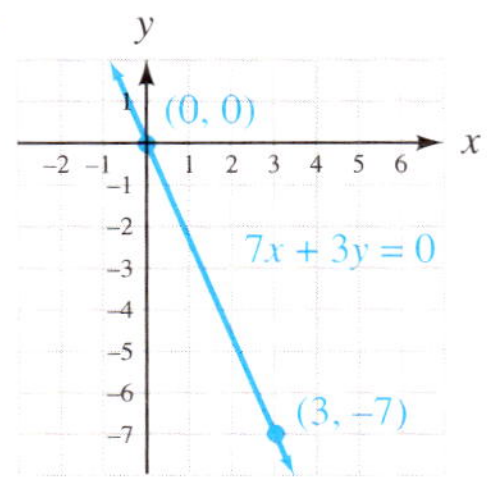

73.

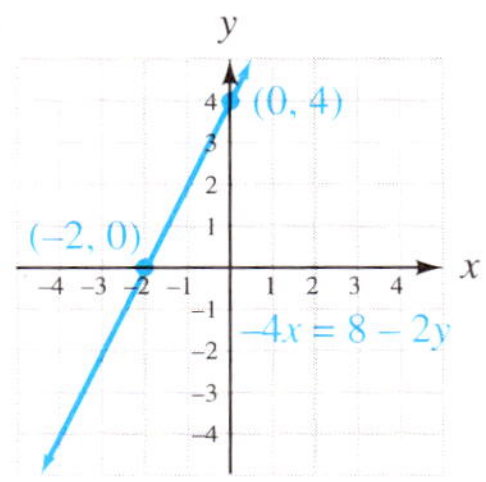

75.

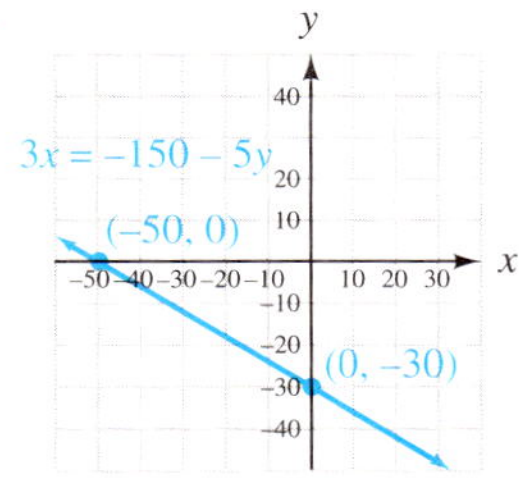

77.

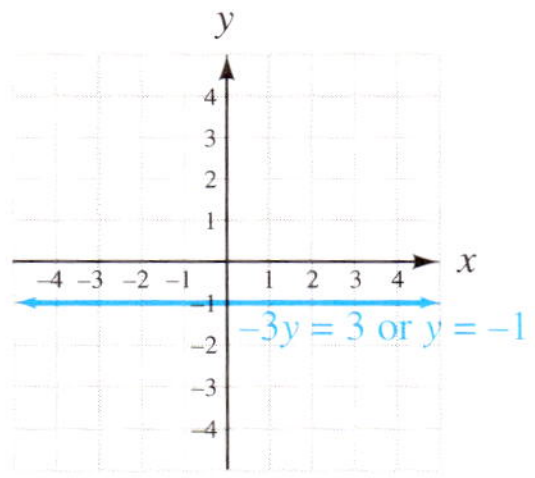

79. a. About $-270°C$ **b.** 0 milliliters

81. The g-intercept is (0, 5): Before any cups of water have been served from the bottle, it contains 5 gallons of water. The c-intercept is $\left(106\frac{2}{3}, 0\right)$: The bottle will be empty after $106\frac{2}{3}$ six-ounce cups of water have been served from it.

83.

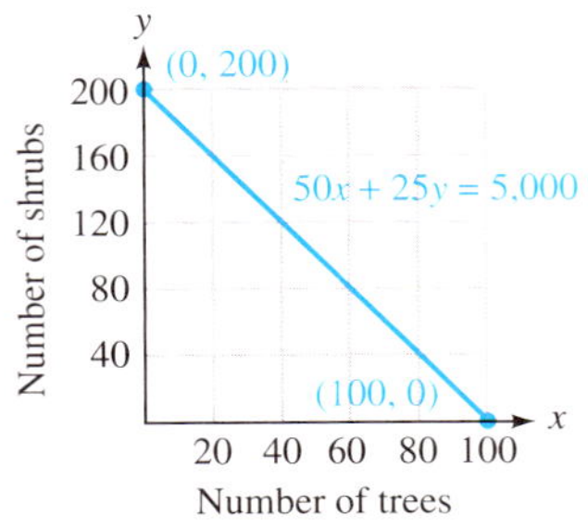

a. If only shrubs are purchased, he can buy 200.
b. If only trees are purchased, he can buy 100.

89. $\frac{1}{5}$ **91.** $2x - 6$

ARE YOU READY? 3.4 (page 217)

1. $\frac{3}{5}$ **2.** undefined **3.** -1 **4.** $\frac{5}{6}$

SELF CHECK 3.4

1. $-\frac{1}{2}$ **2.** 5 **3.** $-\frac{5}{2}$ **4.** 0 **5.** Undefined slope **6.** $\frac{5}{12}$
7. An increase of 650 million bushels per year **8.** Neither
9. $-\frac{13}{4}$

STUDY SET SECTION 3.4 (page 225)

1. slope, ratio **3.** change **5. a.** Line 2 **b.** Line 1 **c.** Line 4 **d.** Line 3 **7. a.** Line 1 **b.** Line 1 **c.** Line 2 **9. a.** $\frac{1}{3}$ **b.** $\frac{4}{12} = \frac{1}{3}$ **c.** same **11. a.** 0 **b.** Undefined **c.** $\frac{1}{4}$ **d.** 2 **13.** 40% **15. a.** $-\frac{1}{6}$ **b.** $\frac{8}{7}$ **c.** 1 **17. a.** $m = \frac{y_2 - y_1}{x_2 - x_1}$ **b.** sub, sub, over (divided by), two, one **19.** 1 **21.** $\frac{2}{3}$ **23.** $\frac{4}{3}$ **25.** -2 **27.** 0 **29.** $-\frac{1}{5}$ **31.** $\frac{1}{2}$ **33.** 1 **35.** -3 **37.** $\frac{5}{4}$ **39.** $-\frac{1}{2}$ **41.** $\frac{3}{5}$ **43.** 0 **45.** Undefined **47.** $-\frac{2}{3}$ **49.** -4.75 **51.** 0 **53.** $\frac{7}{5}$ **55.** $-\frac{2}{5}$ **57.** $\frac{3}{4}$ **59.** 0 **61.** 0 **63.** 0 **65.** Undefined **67.** Undefined **69.** 0 **71.** Undefined **73.** Parallel **75.** Perpendicular **77.** Neither **79.** Perpendicular **81.** Parallel **83.** Neither **85.** $\frac{5}{9}$ **87.** $-\frac{2}{3}$ **89.** -1 **91.** $\frac{1}{2}$ **93.** $-\frac{2}{5}$ **95.** $\frac{1}{20}$; 5% **97.** $\frac{3}{25}$; 12% **99.** Front: $\frac{3}{2}$; side: $\frac{3}{5}$
101. a decrease of 875 gal per hour (-875 gal per hr)
103. 319 lb per yr **105.** An increase of 325 students per year
111. 40 lb licorice; 20 lb gumdrops

ARE YOU READY? 3.5 (page 230)

1. a. $3x, -6$ **b.** 3 **2.** $y = -\frac{2}{5}x + 3$ **3.** True **4.** $\frac{3}{1}$
5. the y-axis **6.** True

SELF CHECK 3.5

1. a. $m = -9; (0, 4)$ **b.** $m = \frac{1}{11}; (0, -2)$ **c.** $m = -5; \left(0, -\frac{7}{2}\right)$
2. $y = x - 12$ **3.** $y = -\frac{3}{2}x + 2$

4.

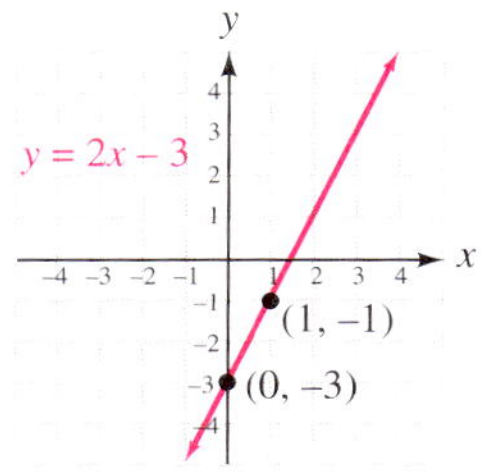

5.

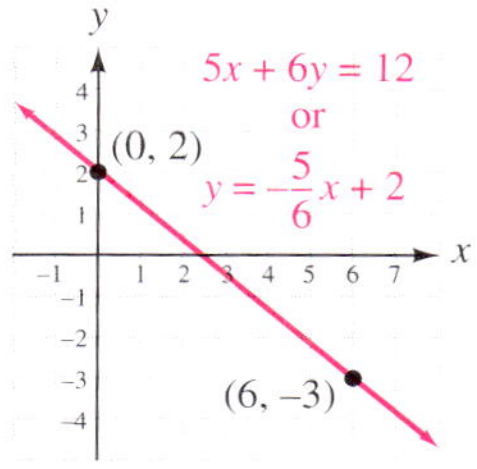

6. Neither **7.** $c = -10.50p + 4{,}500$

STUDY SET SECTION 3.5 (page 237)

1. slope–intercept **3. a.** No **b.** No **c.** Yes **d.** No
5. a. $y = 2x + 8$ **b.** $y = -5x - 3$ **c.** $y = \frac{1}{3}x - 1$
d. $y = \frac{9}{5}x + 4$ **7.** $-2x, 5y, 5, 5, 5, -\frac{2}{5}, 3, -\frac{2}{5}, 0, 3$ **9.** $-2, -3$
11. $4, (0, 2)$ **13.** $-5, (0, -8)$ **15.** $25, (0, -9)$ **17.** $-1, (0, 11)$
19. $\frac{1}{2}, (0, 6)$ **21.** $\frac{1}{4}, \left(0, -\frac{1}{2}\right)$ **23.** $-5, (0, 0)$ **25.** $1, (0, 0)$
27. $0, (0, -2)$ **29.** $0, \left(0, -\frac{2}{5}\right)$ **31.** $-1, (0, 8)$ **33.** $\frac{1}{6}, (0, -1)$
35. $-\frac{3}{2}, (0, 1)$ **37.** $-\frac{2}{3}, (0, 2)$ **39.** $\frac{3}{5}, (0, -3)$ **41.** $1, \left(0, -\frac{11}{6}\right)$

43. $y = 5x - 3$

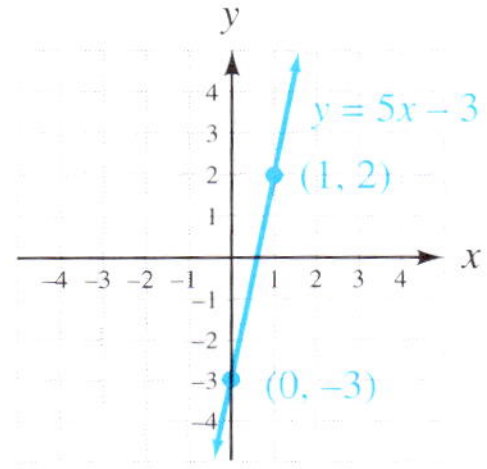

45. $y = -3x + 6$

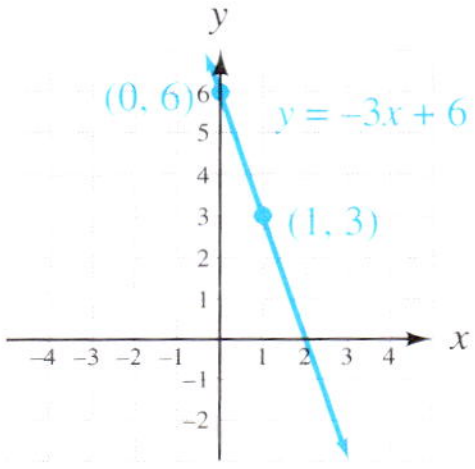

47. $y = \frac{1}{4}x - 2$

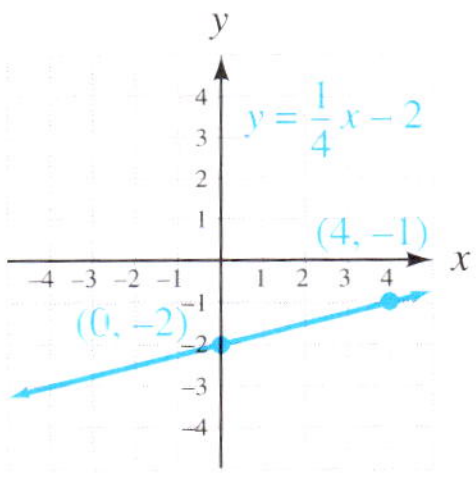

49. $y = -\frac{8}{3}x + 5$

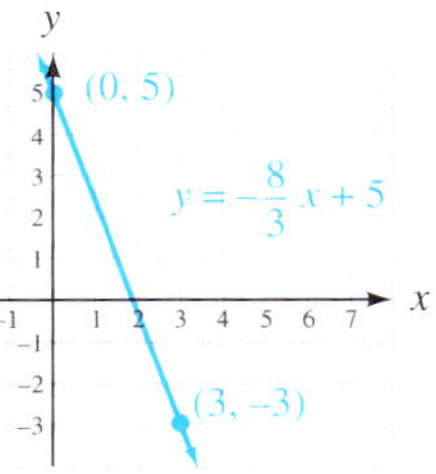

51. $y = \frac{6}{5}x$

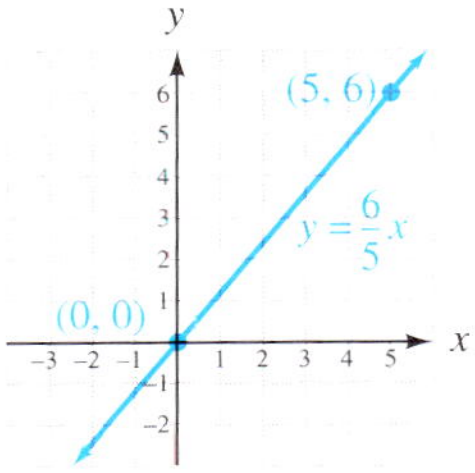

53. $y = -2x + \frac{1}{2}$

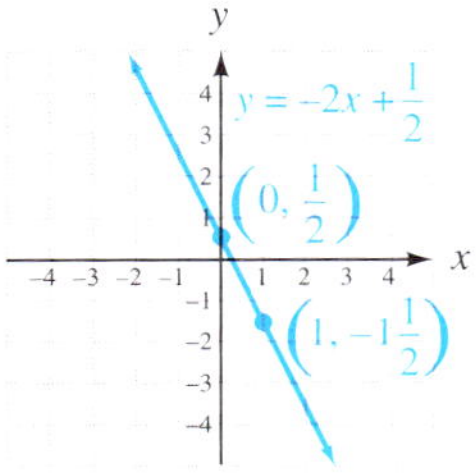

55. $y = 5x - 1$ **57.** $y = -2x + 3$ **59.** $y = \frac{4}{5}x - 2$
61. $y = -\frac{5}{3}x + 2$

63. $3, (0, 3)$

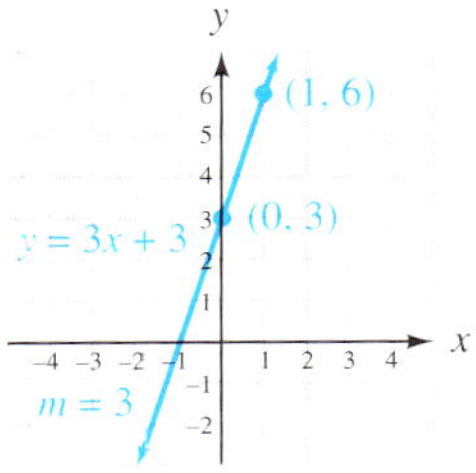

65. $\frac{1}{2}, (0, 2)$

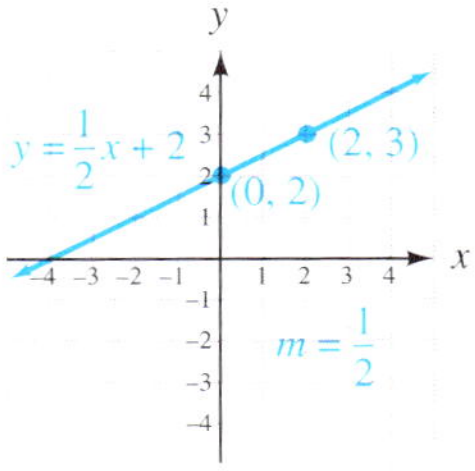

67. $-3, (0, 0)$

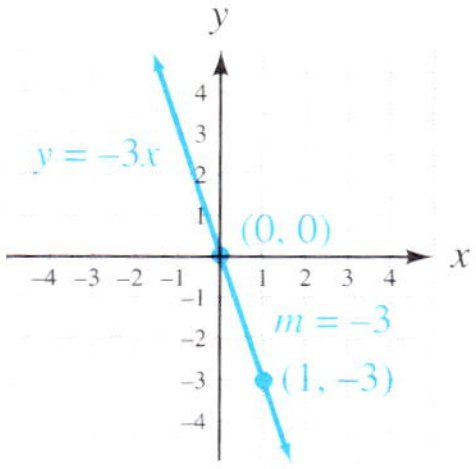

69. $-4, (0, -4)$

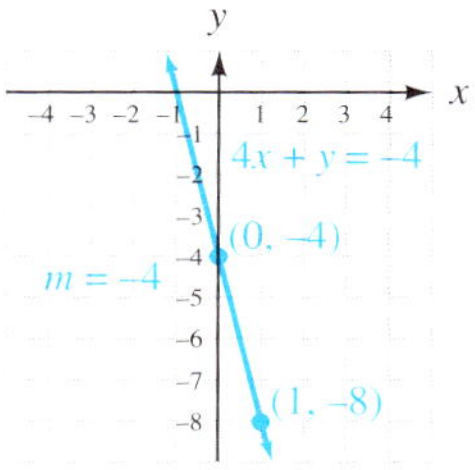

71. $-\frac{3}{4}, (0, 4)$

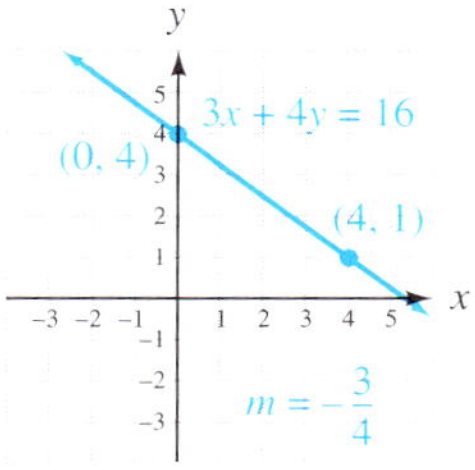

73. $2, (0, -1)$

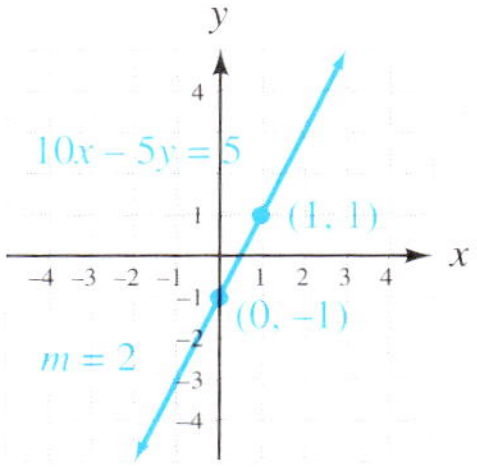

75. Parallel **77.** Perpendicular **79.** Parallel **81.** Neither
83. Perpendicular **85.** Parallel **87.** Perpendicular **89.** Neither
91. a. $c = 2{,}000h + 5{,}000$ **b.** \$21,000 **93.** $F = 5t - 10$
95. $c = -20m + 500$ **97.** $c = 5x + 20$
99. a. $c = 14.95m + 629.99$ **b.** \$988.79
101. a. When there are no head waves, the ship can travel at 18 knots. **b.** $-\frac{1}{2}$ knot/ft **c.** $y = -\frac{1}{2}x + 18$
105. 42 ft, 45 ft, 48 ft, 51 ft

ARE YOU READY? 3.6 (page 240)

1. $\frac{4}{5}$ **2.** $x + 5$ **3.** $y = 6x - 44$ **4.** $\frac{29}{4}$

SELF CHECK 3.6

1. $y = -2x + 5$ **2.** $y = -\frac{10}{13}x + \frac{2}{13}$ **3. a.** $y = 2$ **b.** $x = -1$

4.

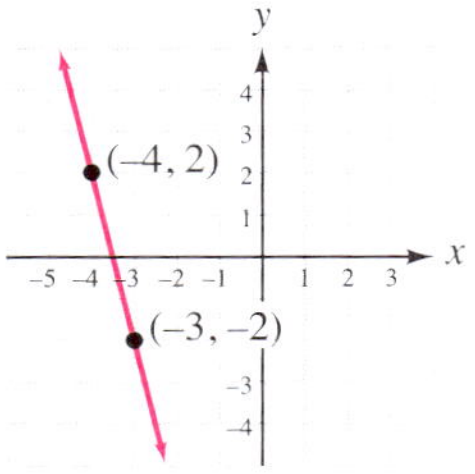

5. $F = \frac{9}{5}C + 32$ **6.** $c = 5n + 12$

STUDY SET SECTION 3.6 (page 246)

1. point–slope, sub, times, minus, one **3. a.** point–slope
b. slope–intercept **5. a.** $(-2, -3)$ **b.** $\frac{5}{6}$ **c.** $y + 3 = \frac{5}{6}(x + 2)$
7. (67, 170), (79, 220) **9.** 5, −1, +, 2, 3
11. point–slope, slope–intercept **13.** $y - 1 = 3(x - 2)$
15. $y + 1 = \frac{4}{5}(x + 5)$ **17.** $y = 2x - 1$ **19.** $y = -5x - 37$
21. $y = -3x$ **23.** $y = \frac{1}{5}x - 1$ **25.** $y = -\frac{4}{3}x + 4$
27. $y = -\frac{11}{6}x - \frac{7}{3}$ **29.** $y = 2x + 5$ **31.** $y = -\frac{1}{2}x + 1$
33. $y = 5$ **35.** $y = \frac{1}{10}x + \frac{1}{2}$ **37.** $x = -8$ **39.** $y = \frac{1}{2}x$
41. $x = 4$ **43.** $y = 5$

45.

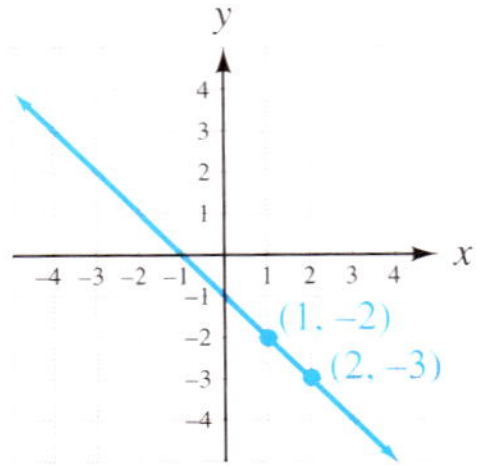

47.

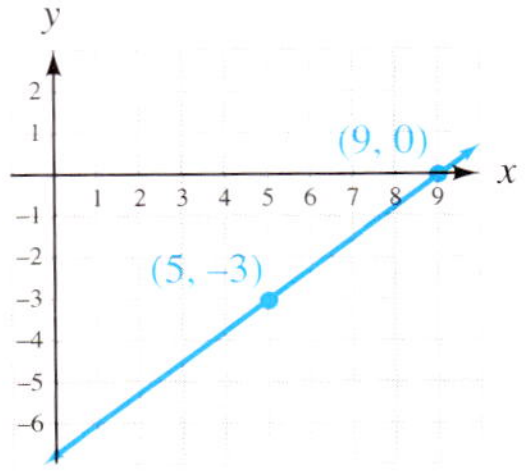

49.

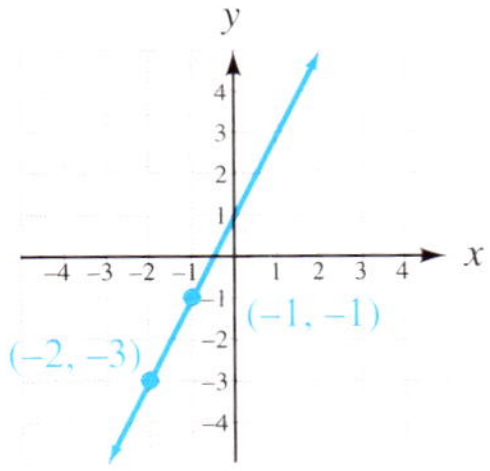

51.

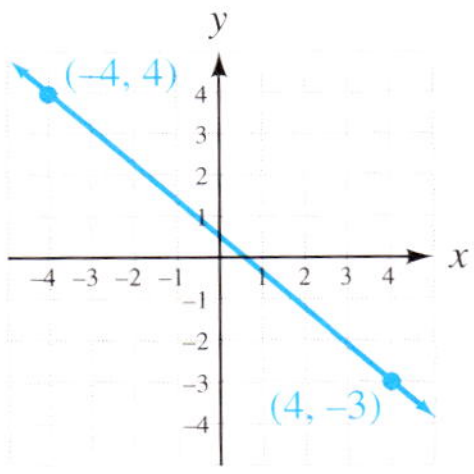

53. $y = \frac{1}{4}x - \frac{5}{4}$ **55.** $y = 12$ **57.** $y = -\frac{1}{4}x + \frac{7}{8}$
59. $y = -\frac{2}{3}x + 2$ **61.** $y = 8x + 4$ **63.** $x = -3$ **65.** $y = 7x - 11$
67. $y = -4x - 9$ **69.** $y = \frac{2}{7}x - 2$ **71.** $y = \frac{1}{10}x$ **73.** $x = -\frac{1}{8}$
75. $y = 1.7x - 2.8$ **77.** $h = 3.9r + 28.9$
79. $y = -\frac{2}{5}x + 4, y = -7x + 70, x = 10$
81. a. $y = -40m + 920$ **b.** 440 yd³ **83.** $l = \frac{25}{4}r + \frac{1}{4}$
85. a. $y = -\frac{3}{10}x + \frac{283}{10}$ or $y = -0.3x + 28.3$ **b.** 16.3 gal
91. 17 in. by 39 in.

ARE YOU READY? 3.7 (page 249)

1. False **2.** True **3.**

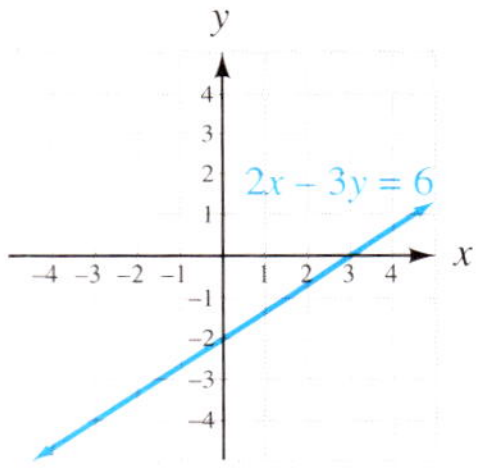

4. a. Below **b.** On **c.** Above

SELF CHECK 3.7

1. a. Not a solution **b.** Solution **c.** Solution **d.** Solution

2.

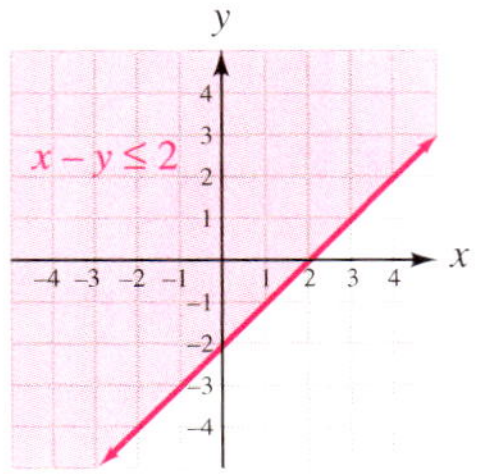

3.

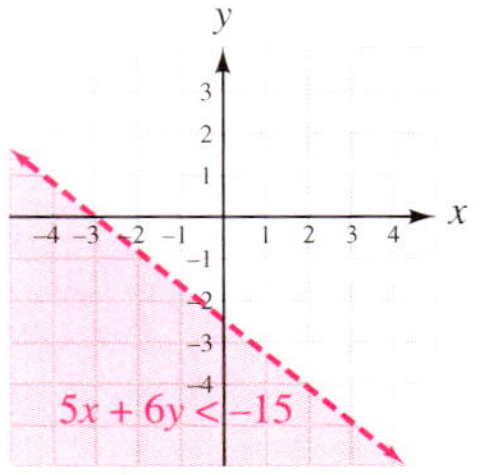

4.

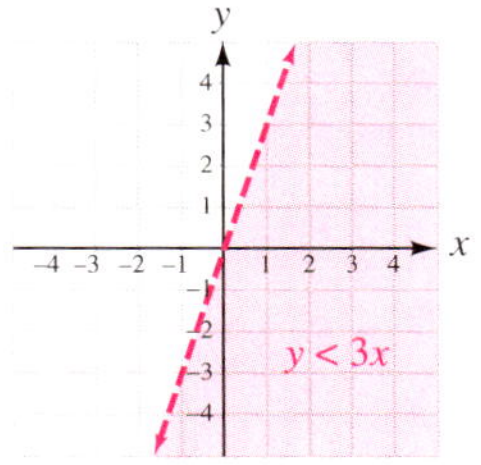

5.

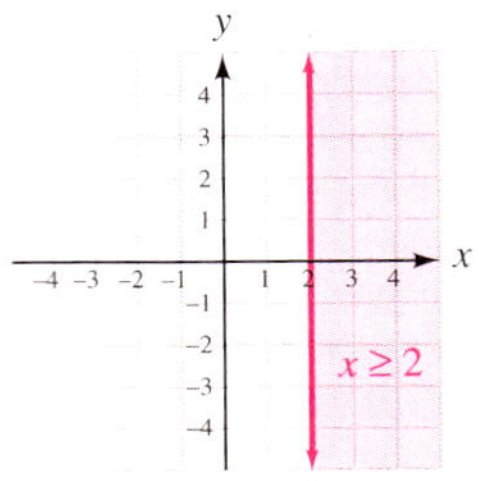

(a)

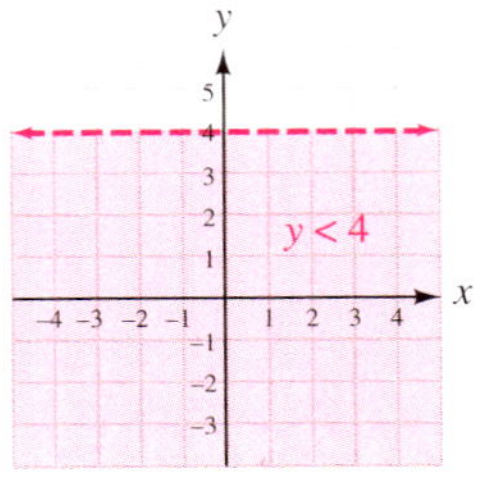

(b)

6. $x + 15y \le 150$

STUDY SET SECTION 3.7 (page 255)

1. inequality **3.** satisfies **5.** half-planes **7.** Yes
9. dashed, solid **11.** The half-plane opposite that in which the test point lies **13. a.** Yes **b.** No **c.** No **d.** Yes
15. a. Is less than **b.** Is greater than or equal to
c. Is less than or equal to **d.** Is possibly greater than **17.** $=, <$
19. Yes **21.** No **23.** No **25.** Yes

27.

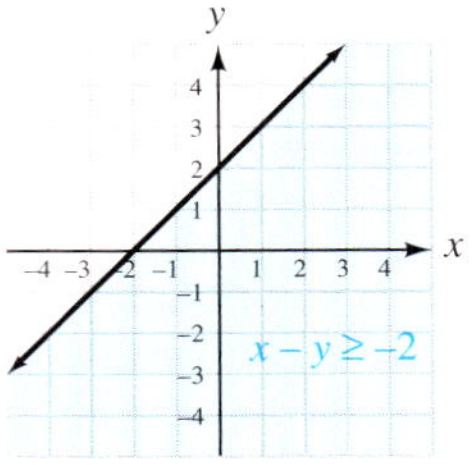

29.

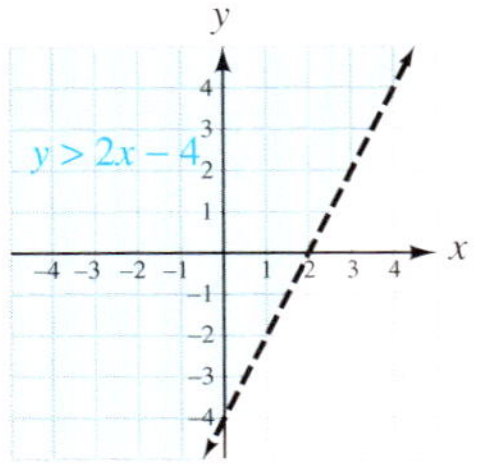

31.

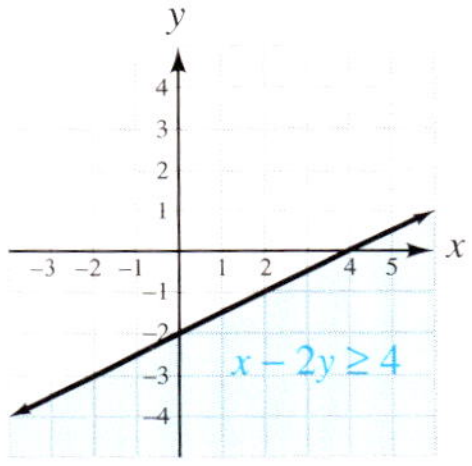

33.

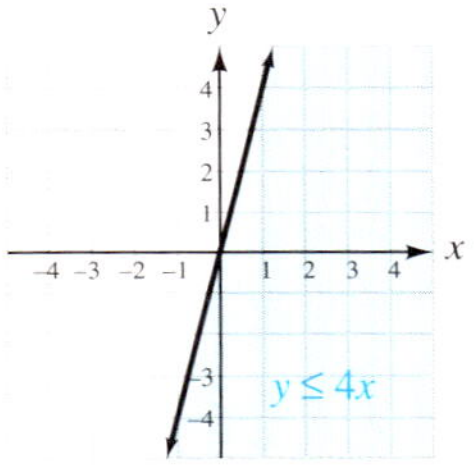

35.

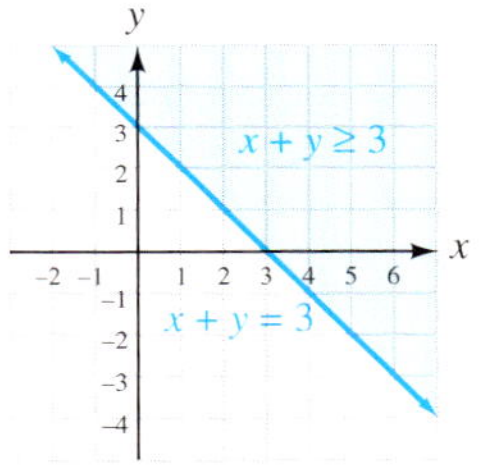

37.

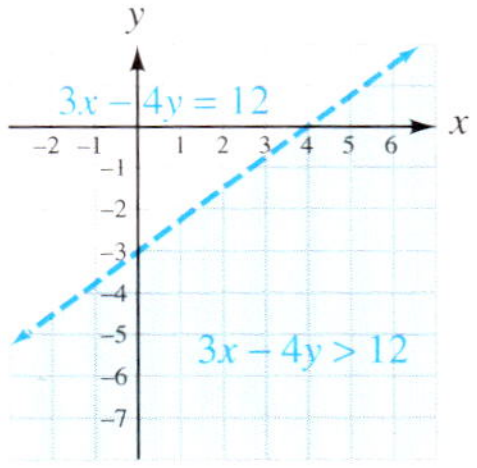

39.

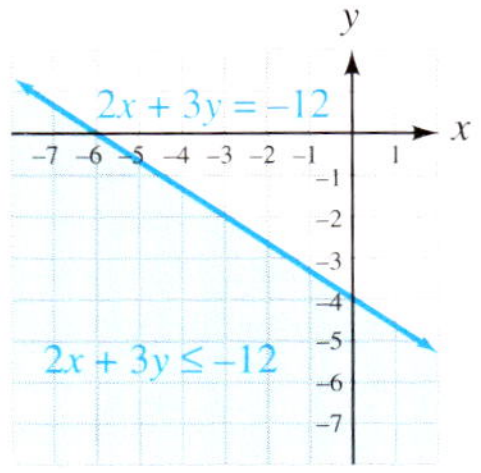

41.

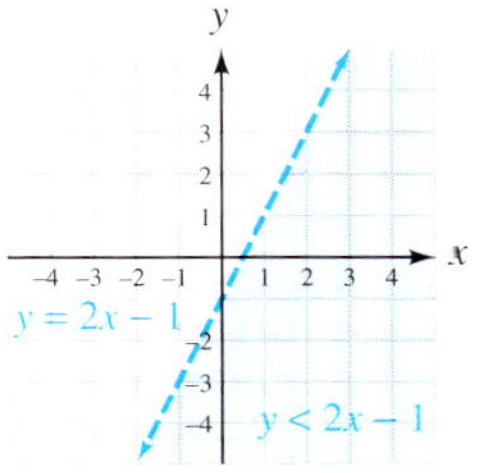

43.

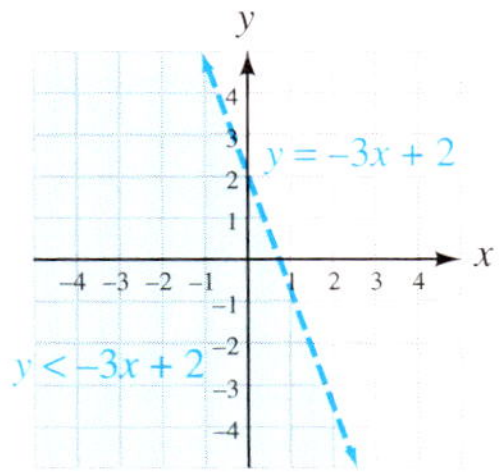

45.

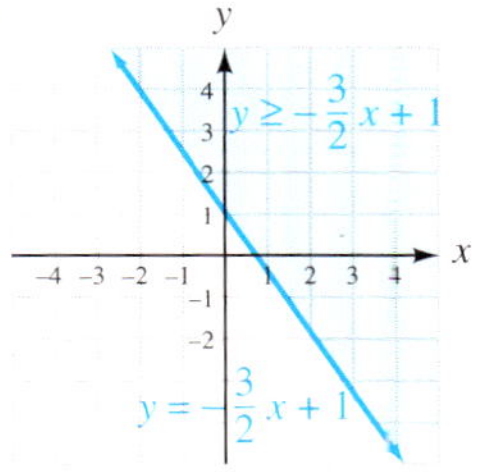

47.

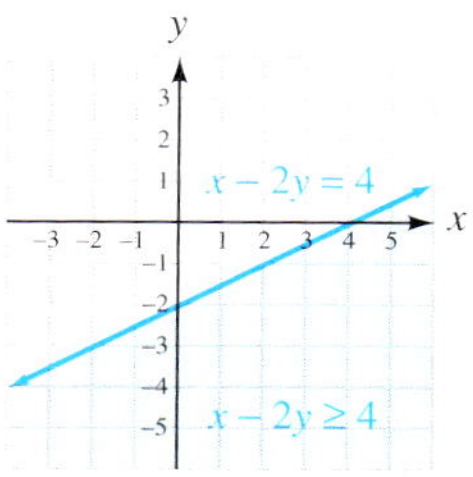

49.

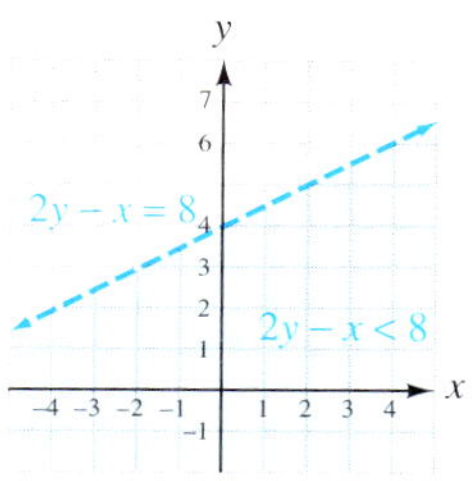

51.

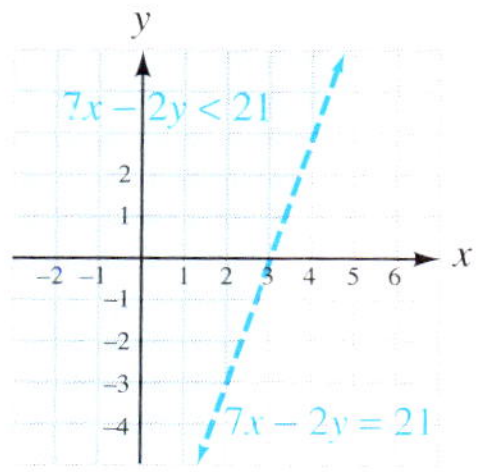

53.

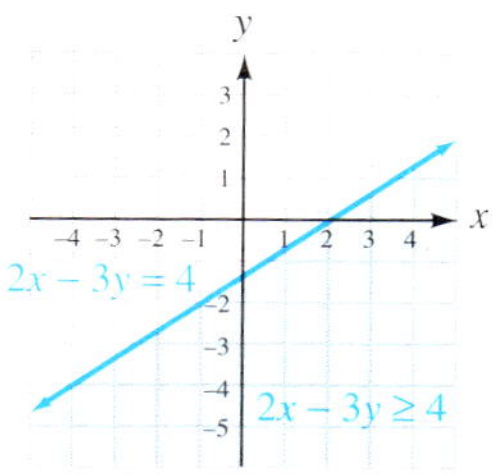

55.

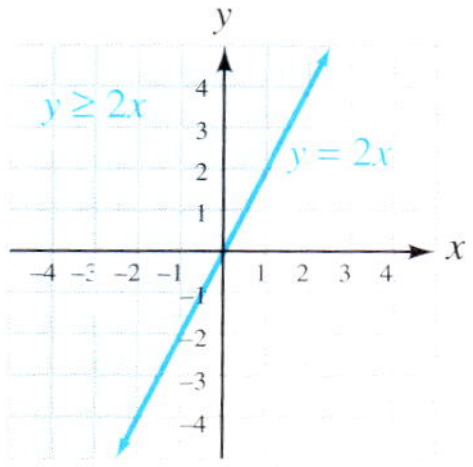

57.

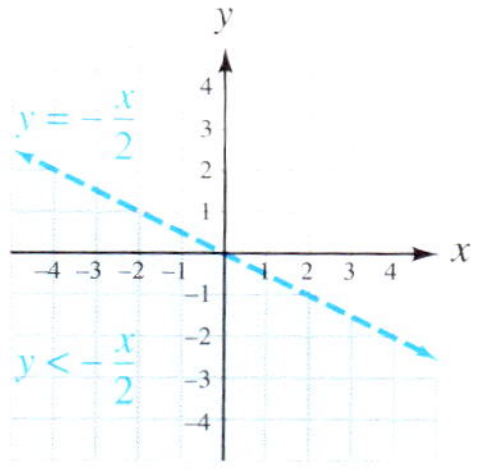

59.

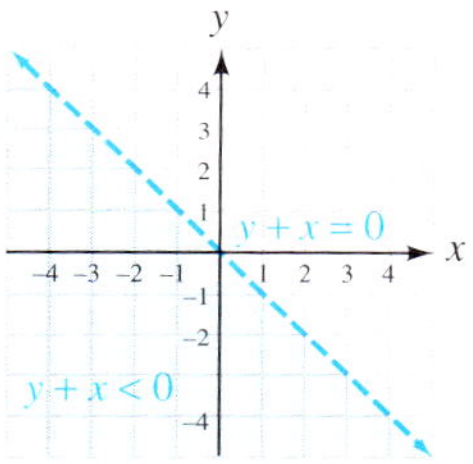

61.

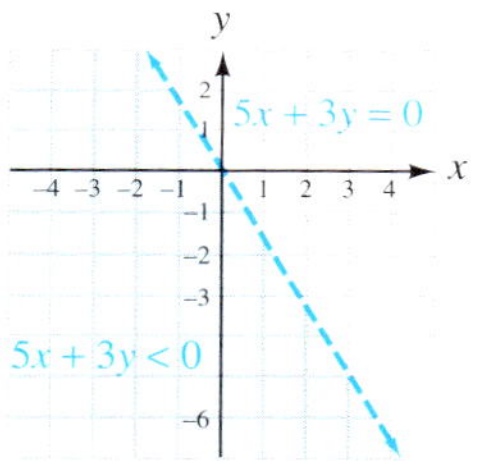

63.

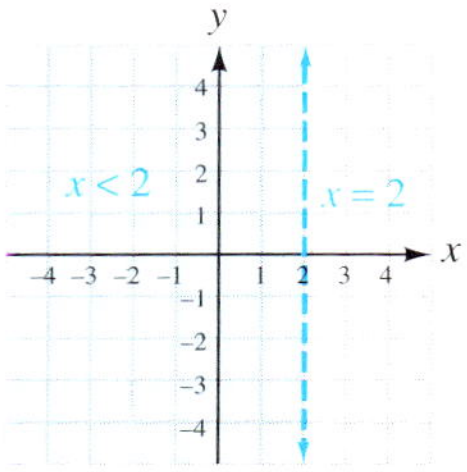

65.

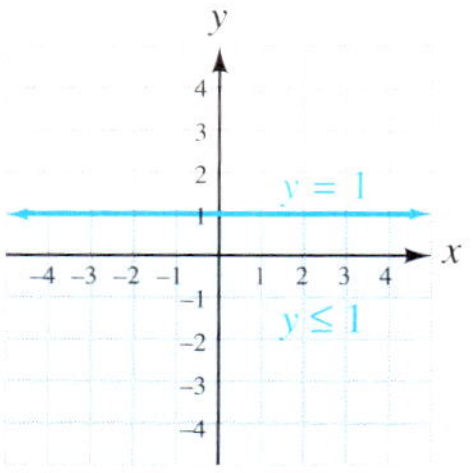

67.

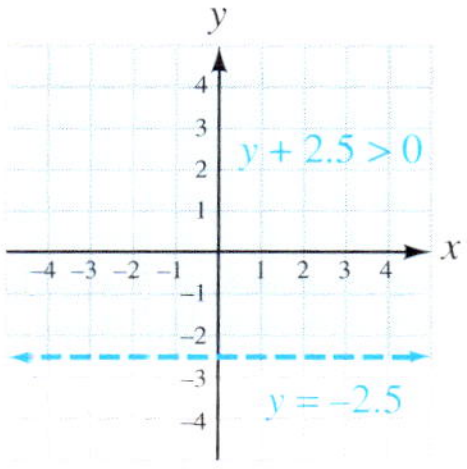

69.

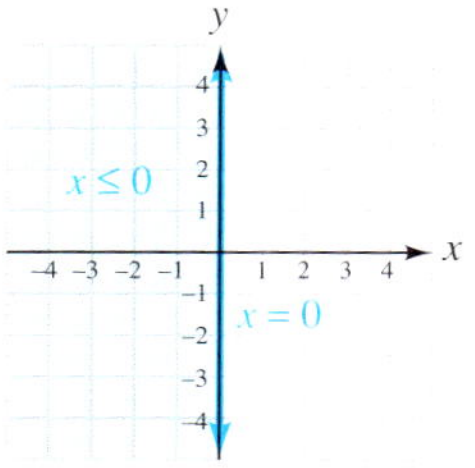

71. a.

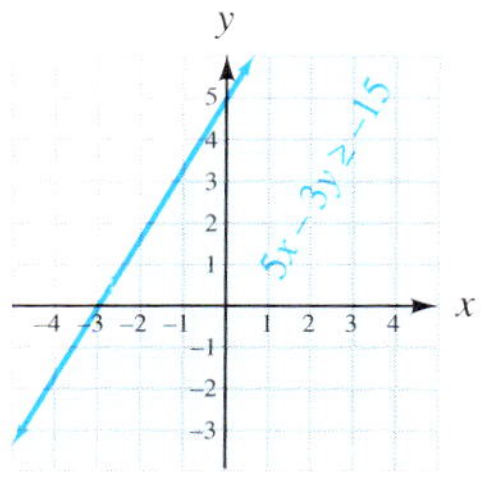

b.

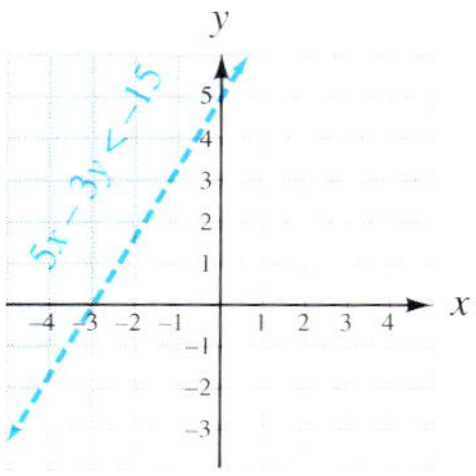

73. a.

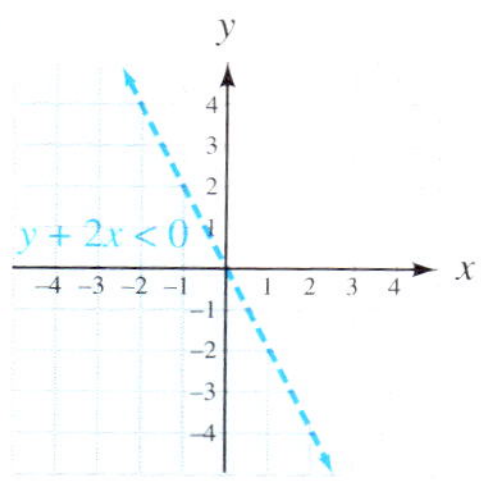

b.

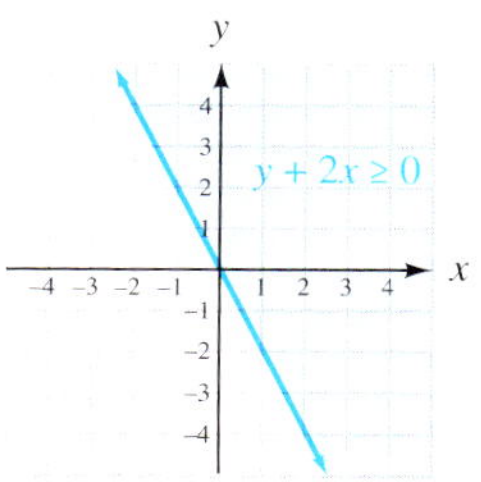

75. No **77.** (2, 7), (4, 6), (9, 2); answers may vary.

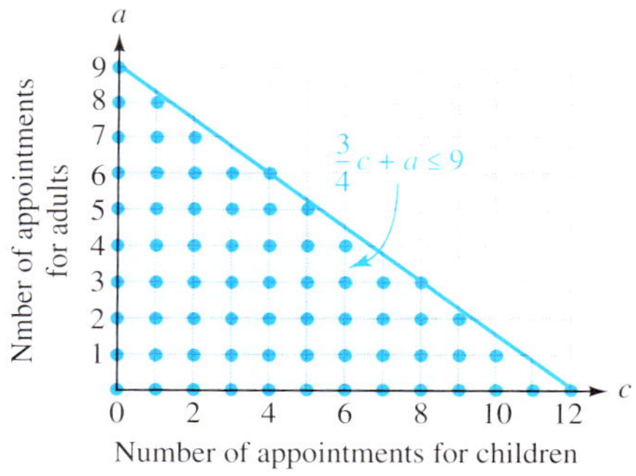

79. (10, 10), (20, 10), (10, 20); Answers may vary

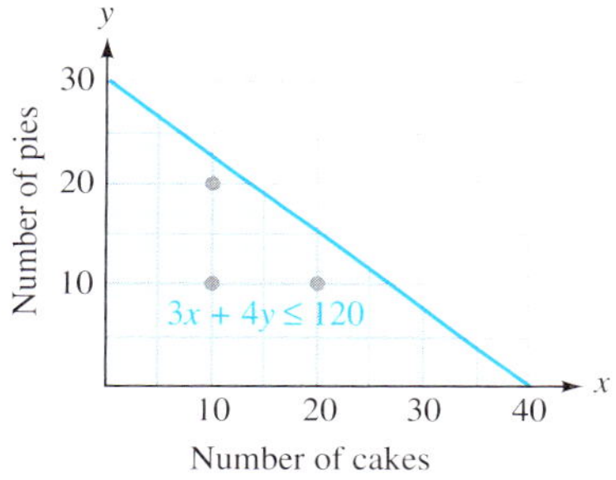

81. (40, 30), (30, 40), (40, 20); Answers may vary

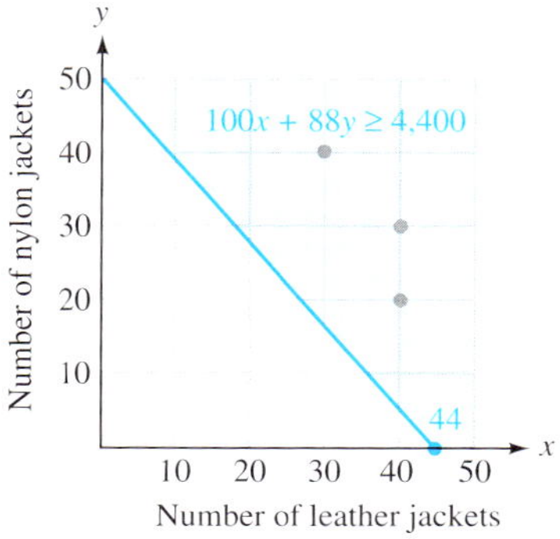

87. $t = \frac{A - P}{Pr}$
89. $15x + 22$

ARE YOU READY? 3.8 (page 258)

1. 17 **2.** (3, 5), (3, 0) **3.** (8, 4), (−3, 4) **4. a.** 5 **b.** 2

SELF CHECK 3.8

1. Domain: {−5, −1, 6, 8}; range: {−5, 2, 10} **2. a.** Not a function, −1 → 4 and −1 → 5 **b.** Function; domain: {−6, 4, 5}; range: {−6, 5, 8} **c.** Function; domain: {1, 3, 4, 9}; range: {4, 9} **3. a.** −5 **b.** 5 **c.** 3 **4.** 3, 79

5.

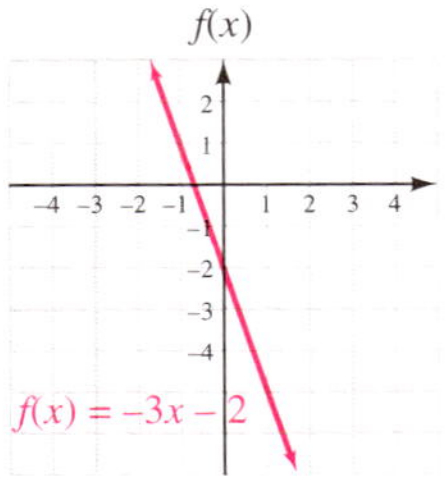

6.

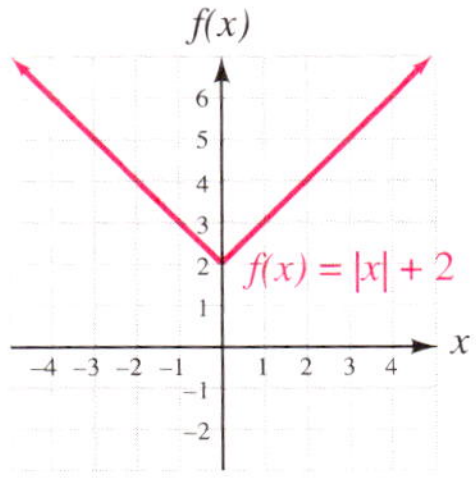

7. a. Function **b.** Not a function **8.** $140

STUDY SET SECTION 3.8 (page 266)

1. relation **3.** domain, range **5.** value

7. Domain Range

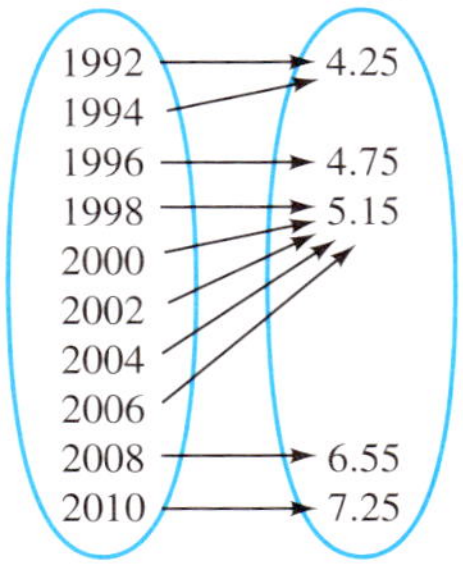

9. 33 **11.** of **13.** 4, 5, 4, 5 **15.** Domain: {−6, −1, 6, 8}; range: {−10, −5, −1, 2} **17.** Domain: {−8, 0, 6}; range: {9, 50} **19.** Yes; domain: {10, 20, 30}; range: {20, 40, 60} **21.** No; (4, 2), (4, 4), (4, 6) (Answers may vary) **23.** Yes; domain: {1, 2, 3, 4, 5}; range: {7, 8, 15, 16, 23} **25.** No; (−1, 0), (−1, 2) **27.** No; (3, 4), (3, −4) or (4, 3), (4, −3) **29.** Yes; domain: {−3, 1, 5, 6}; range: {−8, 0, 4, 9} **31.** No, (3, 4), (3, −4) or (4, 3), (4, −3) **33.** Yes; domain: {−2, −1, 0, 1}; range: {7, 10, 13, 16} **35. a.** 3 **b.** −9 **c.** 0 **d.** 199 **37. a.** 0.32 **b.** 18 **c.** 2,000,000 **d.** $\frac{1}{32}$ **39. a.** 7 **b.** 14 **c.** 0 **d.** 1 **41. a.** 0 **b.** 990 **c.** −24 **d.** 210 **43. a.** 36 **b.** 0 **c.** 9 **d.** 4 **45.** 1.166

47.

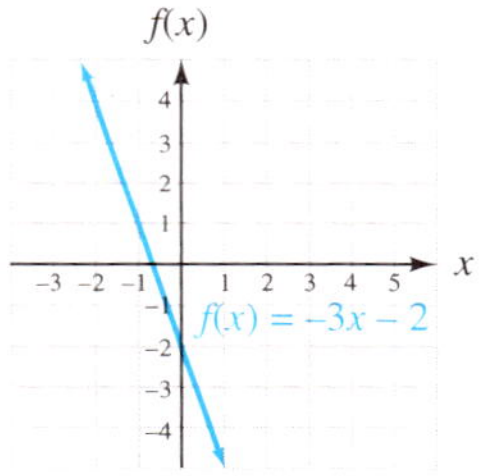

x	$f(x)$
−2	4
−1	1
0	−2
1	−5

49.

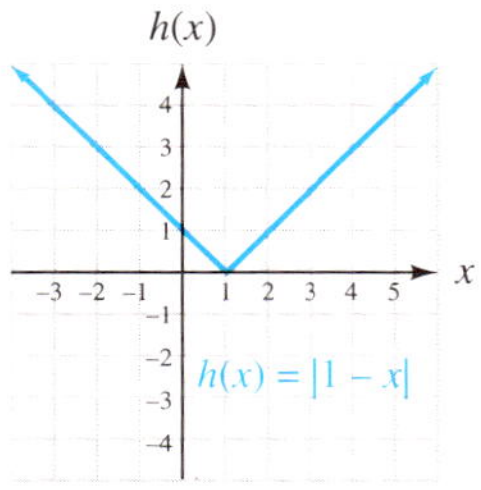

x	$h(x)$
−2	3
−1	2
0	1
1	0
2	1
3	2
4	3

51.

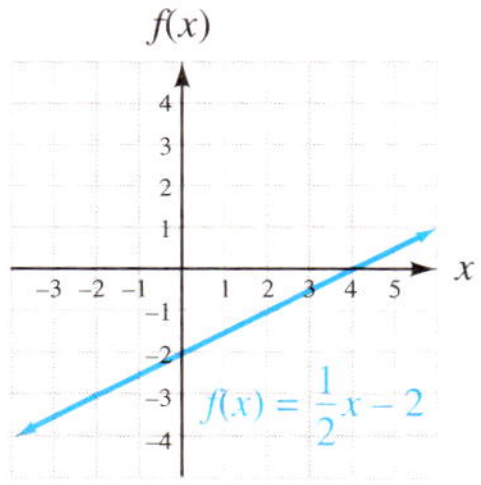

53.

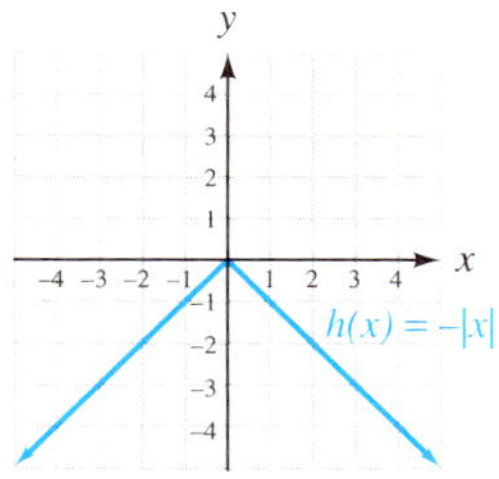

55. Yes **57.** No; (3, 4), (3, −1) (Answers may vary) **59.** No; (0, 2), (0, −4) (Answers may vary) **61.** No; (3, 0), (3, 1) (Answers may vary) **63.** $f(x) = |x|$ **65.** $900 **67.** 78.5 ft², 1,256.6 ft² **69.** Yes **77.** 80 lb of regular coffee

CHAPTER 3 REVIEW (page 270)

1.

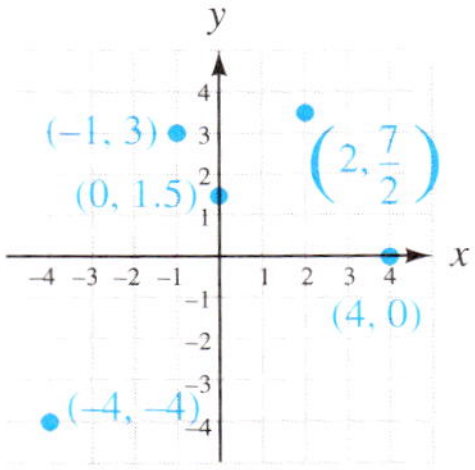

2. (158, 21.5) **3.** Quadrant III **4.** (0, 0) **5.** (1, 4); 36 square units **6. a.** 2,500; week 2 **b.** 1,000 **c.** 1st week and 5th week **7.** Yes

8.

x	y	(x, y)
−2	−6	(−2, −6)
−8	3	(−8, 3)

9. $y = x^2 + 1$ and $y - x^3 = 0$ **10. a.** True **b.** False

11.

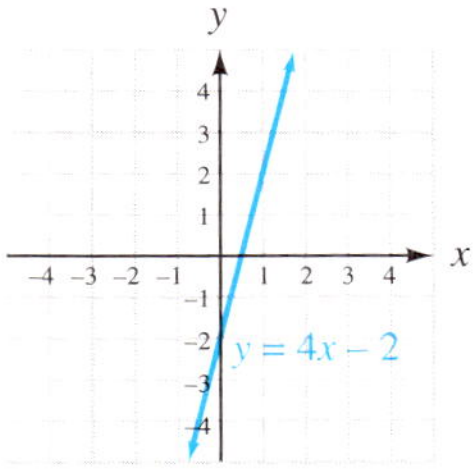

12.

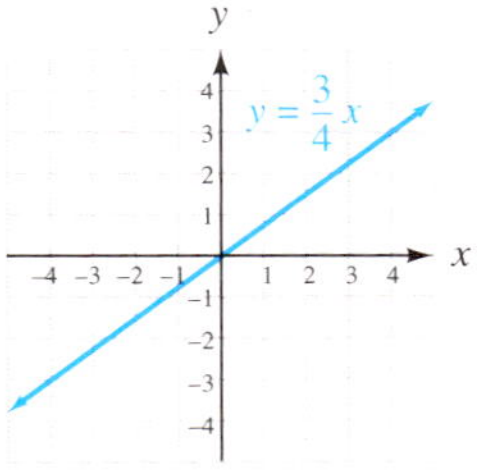

13.

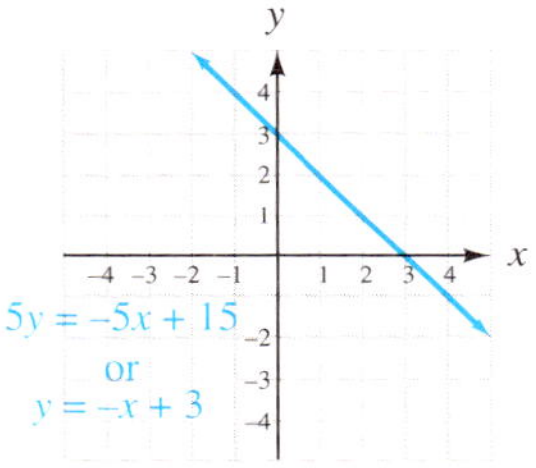

14.

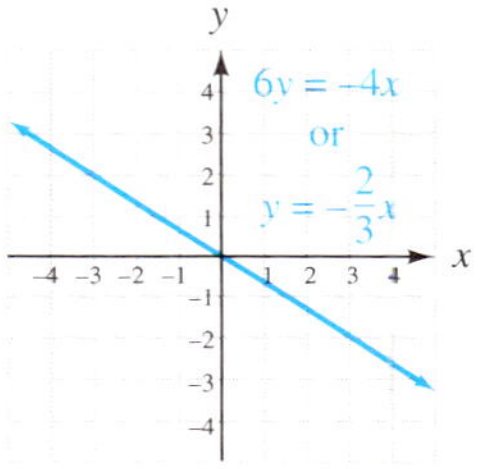

15. About $190

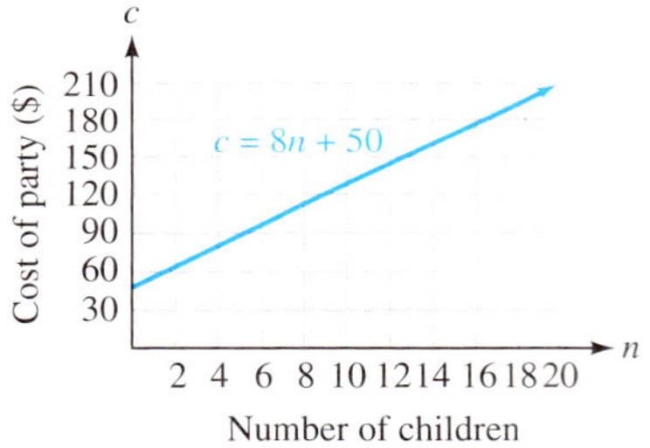

16. a. False **b.** True **17.** (−3, 0), (0, 2.5) **18.** (0, 25,000); the equipment was originally valued at $25,000. (10, 0); in 10 years, the sound equipment had no value.

19. x-intercept: (−2, 0); y-intercept: (0, 4)

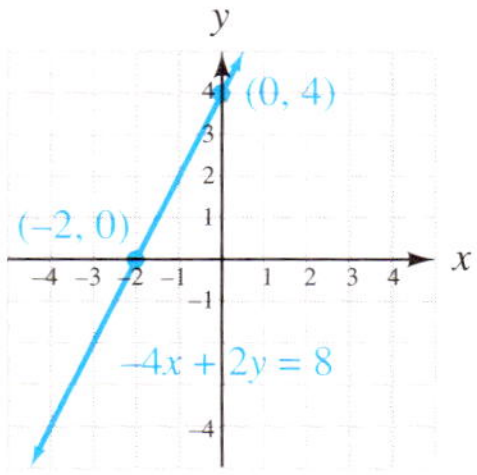

20. x-intercept: $\left(\frac{13}{5}, 0\right)$; y-intercept: $\left(0, -\frac{13}{4}\right)$

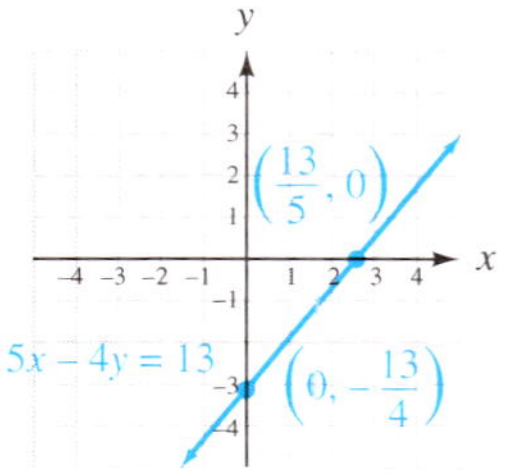

21.

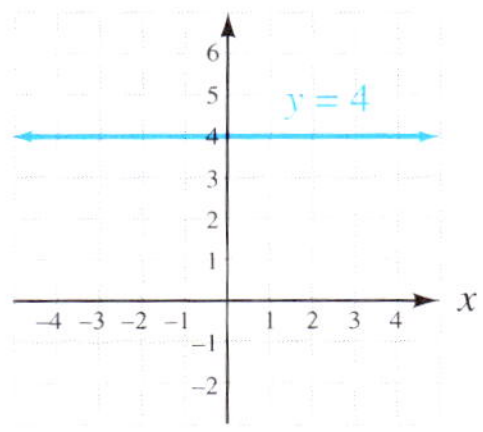

22.

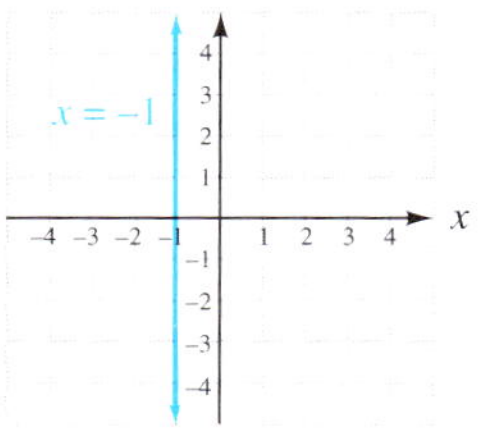

23. $\frac{1}{4}$ **24.** $-\frac{7}{8}$ **25.** -7 **26.** $-\frac{3}{2}$

27.

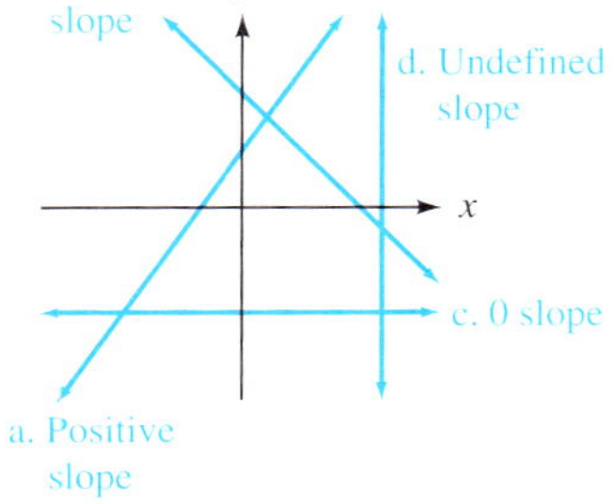

28. $\frac{3}{4}$ **29.** 8.3% **30.** 1.5 gal/yr **31.** They are neither.

32. $-\frac{7}{5}$ **33.** $m = \frac{3}{4}$; y-intercept: $(0, -2)$

34. $m = -4$; y-intercept: $(0, 0)$ **35.** $m = \frac{1}{8}$; y-intercept: $(0, 10)$

36. $m = -\frac{7}{5}$; y-intercept: $\left(0, -\frac{21}{5}\right)$

37. $y = -4x - 1$

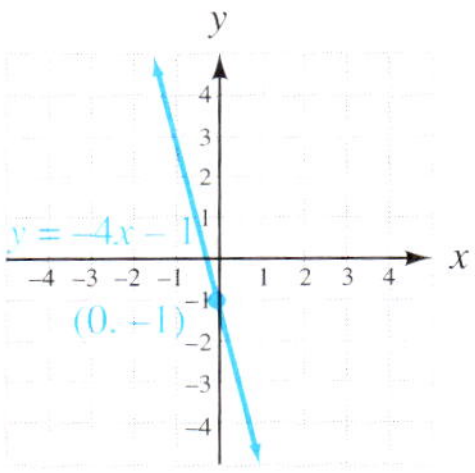

38. $y = \frac{3}{2}x - 3$

39. $m = 3$; y-intercept: $(0, -5)$

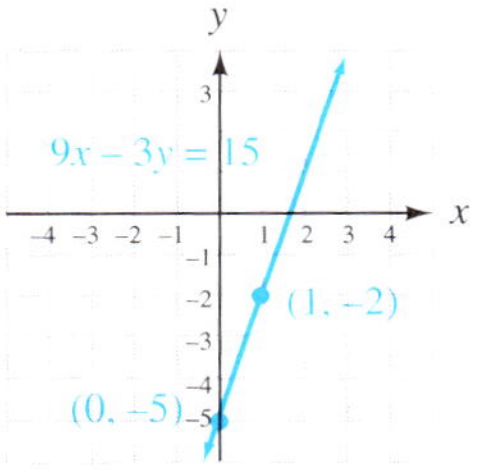

40. **a.** $c = 300w + 75{,}000$ **b.** 90,600 copies **41.** Parallel

42. Perpendicular

43. $y = 3x + 2$

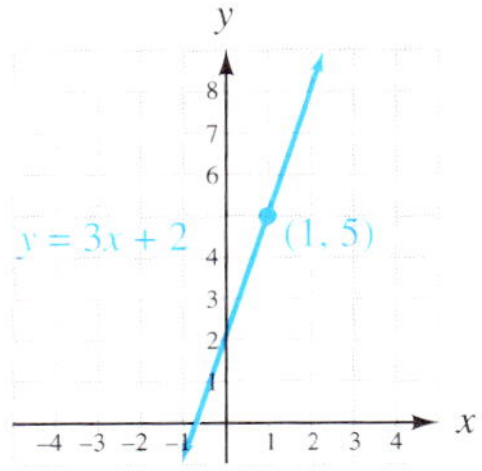

44. $y = -\frac{1}{2}x - 3$

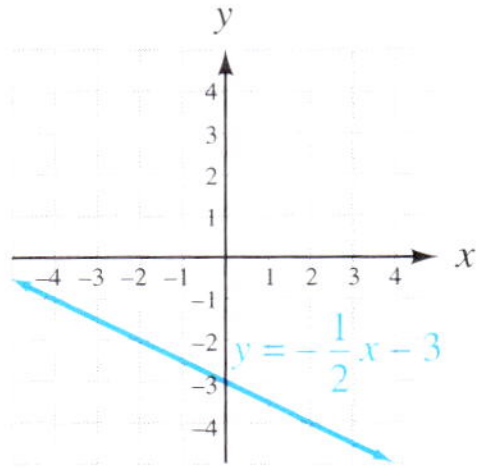

45. $y = \frac{2}{3}x + 5$ **46.** $y = -8$ **47.** $f = -35x + 450$

48. **a.** $P = \frac{3}{2}t + 310$ **b.** 400 parts per million

49. **a.** Yes **b.** Yes **c.** Yes **d.** No **50.** $=, >$

51.

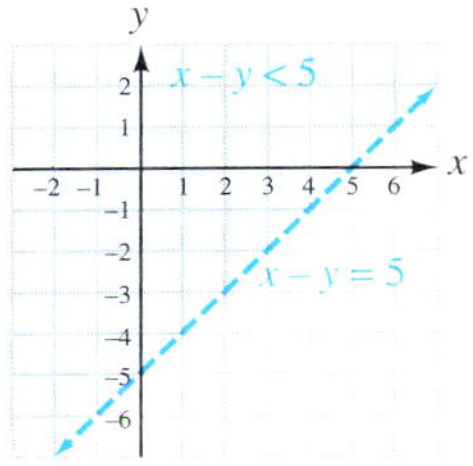

52.

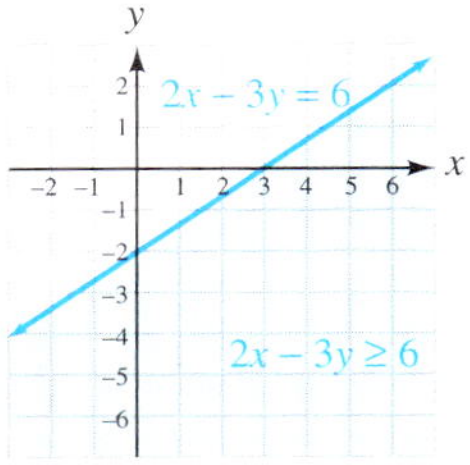

53.

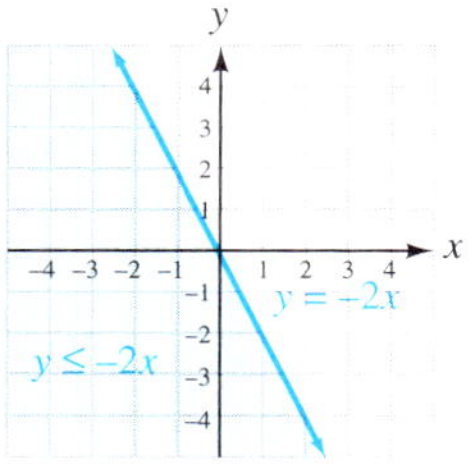

54.

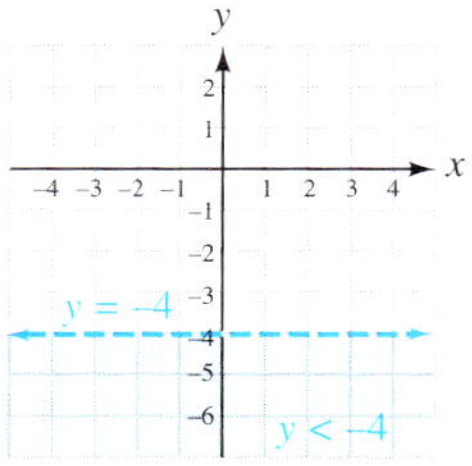

55. **a.** True **b.** False **c.** False

56. (2, 4), (5, 3), (6, 2); answers may vary

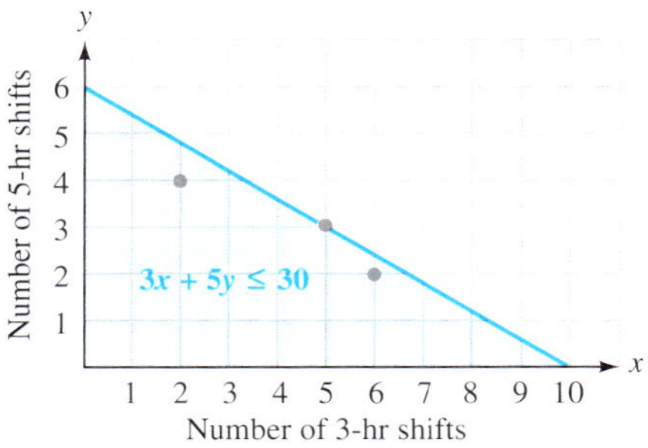

57. Domain: $\{-5, 0, 4, 7\}$; range: $\{-11, -3, 4, 9\}$

58. Domain: $\{-6, 1, 2, 15\}$; range: $\{-8, -2, 9\}$

59. Yes; domain: $\{1, 4, 8\}$; range: $\{0, 6, 9\}$

60. Yes; domain: $\{2, 3, 5, 6\}$; range: $\{1, 4\}$

61. Yes; domain: $\{3, 5, 7, 9\}$; range: $\{9, 25, 49, 81\}$

62. No; $(-1, 2), (-1, 4)$ **63.** Yes; domain: $\{-1, 0, 1, 2\}$; range: $\{6\}$

64. No; (4, 4), (4, 6) **65.** domain, range **66.** $f(x)$ **67.** -3

68. 0 **69.** 21 **70.** $-\frac{7}{4}$ **71.** -5 **72.** 37 **73.** -2

74. -8 **75.** No; (1, 0.5), (1, 4), (answers may vary) **76.** Yes

77.

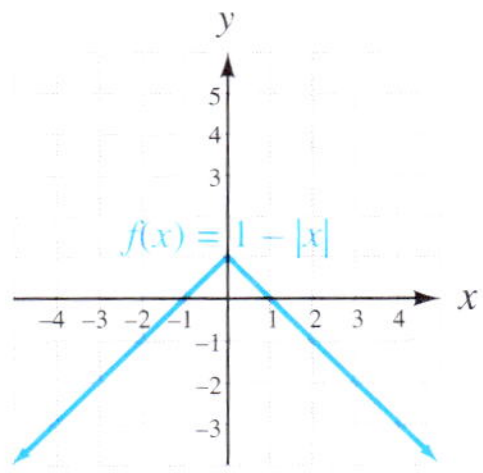

x	$f(x)$
0	1
1	0
3	-2
-1	0
-3	-2

78. 1,004.8 in.^3

CHAPTER 3 TEST (page 279)

1. **a.** axis, axis **b.** solution **c.** linear **d.** slope **e.** function **2.** 10 dogs **3.** 60 dogs **4.** 1 day before and the 3rd day of the holiday **5.** 50 dogs were in the kennel when the holiday began.

6.

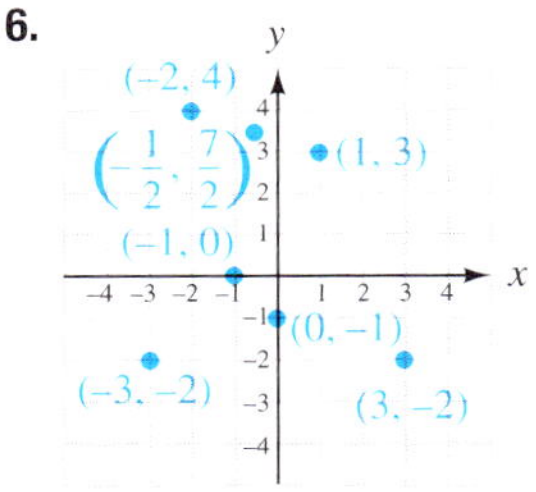

7. $A(2, 4), B(-3, 3), C(-2, -3), D(4, -3), E(-4, 0), F(3.5, 1.5)$

8. **a.** III **b.** IV

9. Yes

10.

x	y	(x, y)
2	1	(2, 1)
-6	3	$(-6, 3)$

11. **a.** False **b.** True **12.**

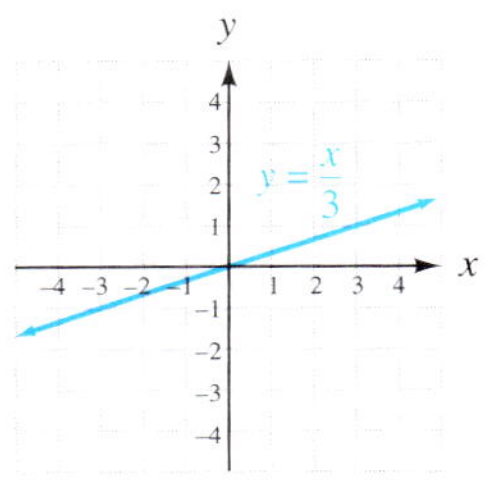

13. x-intercept: (3, 0); y-intercept: $(0, -2)$

14.
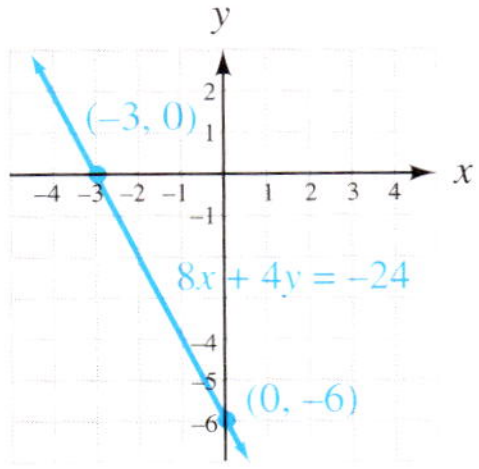

15. $\frac{8}{7}$ **16.** −1 **17.** 0 **18.** 10% **19.** Perpendicular **20.** Parallel **21.** −15 ft per mi **22.** 25 ft per mi

23.
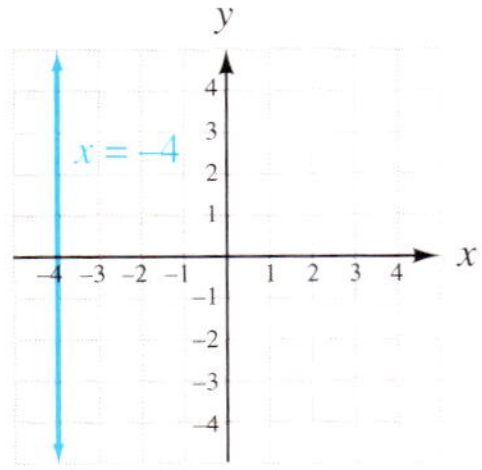

24.
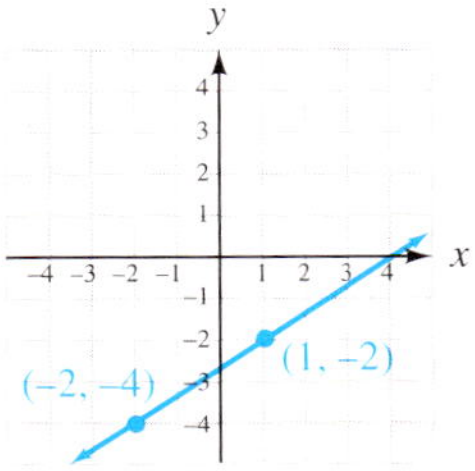

25. $m = -\frac{1}{2}$; (0, 4) **26.** $y = 7x + 19$ **27.** $y = -2x - 5$ **28. a.** $v = -1{,}500x + 15{,}000$ **b.** \$3,000 **29.** Yes **30.** $y = -\frac{1}{5}T + 41$ **31. a.** Yes **b.** No **c.** No **32.** Yes

33.
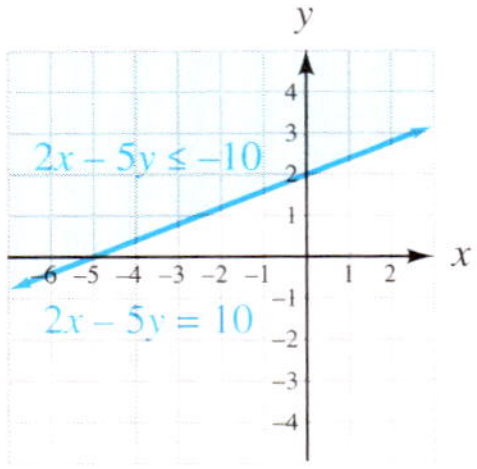

34. Domain: {−4, 0, 1, 5}; range: {−8, 3, 12} **35.** Yes; domain: {1, 2, 3, 4}; range: {1, 2, 3, 4} **36.** No; (−3, 9), (−3, −7); **37.** Yes; domain: {6, 7, 8, 9, 10}; range: {5} **38.** No; (2, 6), (2, 2) **39.** No; (2, 3.5), (2, −3.5); (answers may vary) **40.** No; (−2, 2), (−2, −1); (answers may vary) **41.** −13 **42.** 756 **43.** $C(45) = 28.50$; it costs \$28.50 to make 45 calls.

44.
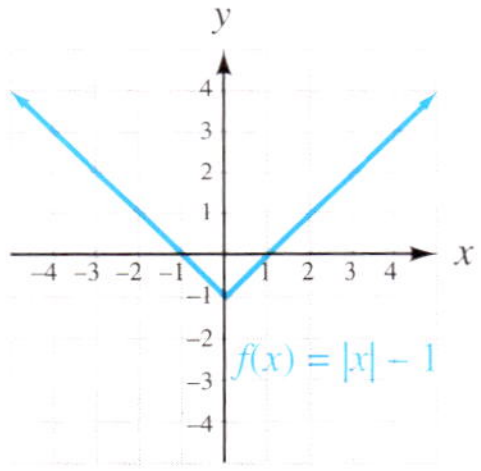

CUMULATIVE REVIEW CHAPTERS 1–3 (page 281)

1. $2^2 \cdot 3^3$ **2.** 0.004 **3. a.** True **b.** True **c.** True **4.** −15 **5.** −0.77 **6.** −945 **7.** 30 **8.** 2 **9.** 32 **10.** $500 - x$ **11.** 3, −2 **12. a.** $2x + 8$ **b.** $-2x + 8$ **13.** $4a + 10$ **14.** $-63t$ **15.** $4b^2$ **16.** 0 **17.** 4 **18.** $-160a$ **19.** $-3y$ **20.** $7x - 12$ **21.** 6 **22.** 2.9 **23.** 9 **24.** −19 **25.** $\frac{1}{7}$ **26.** 1 **27.** $-\frac{55}{6}$ **28.** No solution, contradiction **29.** −99 **30.** $-\frac{1}{4}$ **31.** 1,100 **32.** $h = \frac{S - 2\pi r^2}{2\pi r}$ **33.** $3\frac{1}{8}$ in., $\frac{39}{64}$ in.^{2} **34.** 45°

35.

	% acid	Liters	Amount of acid
50% solution	0.50	x	$0.50x$
25% solution	0.25	$13 - x$	$0.25(13 - x)$
30% mixture	0.30	13	0.30(13)

36. 7.5 hr **37.** 80 lb candy corn, 120 lb gumdrops **38.** $(-\infty, 48]$

39. $(0, \infty)$

40. I and II **41.** No

42.
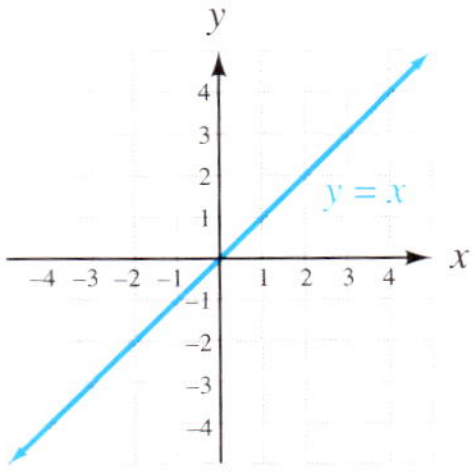

43.
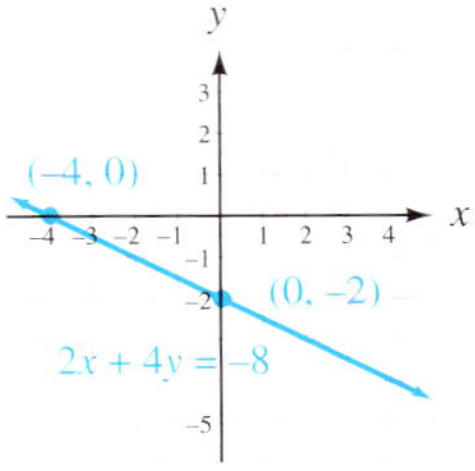

44. 0 **45.** $-\frac{10}{7}$ **46.** $\frac{7}{12}$ **47.** $\frac{2}{3}$, (0, 2) **48.** $y = -2x + 1$ **49.** $y + 9 = -\frac{7}{8}(x - 2)$; $y = -\frac{7}{8}x - \frac{29}{4}$ **50.** Yes

51.
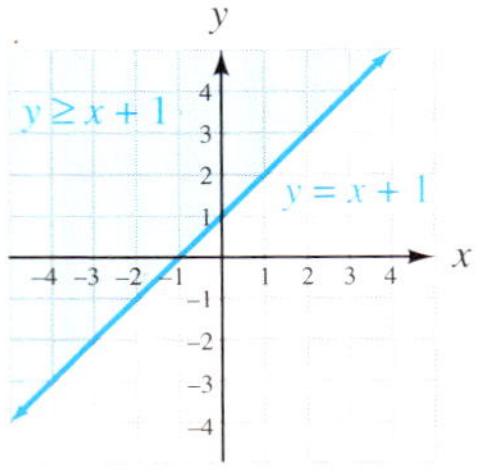

52.
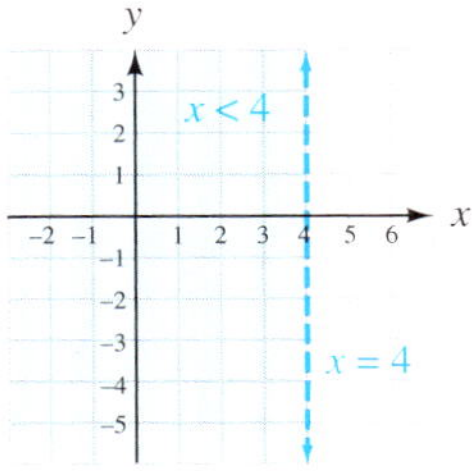

53. 78 **54.** No

ARE YOU READY? 4.1 (page 284)

1. Not a solution

2.
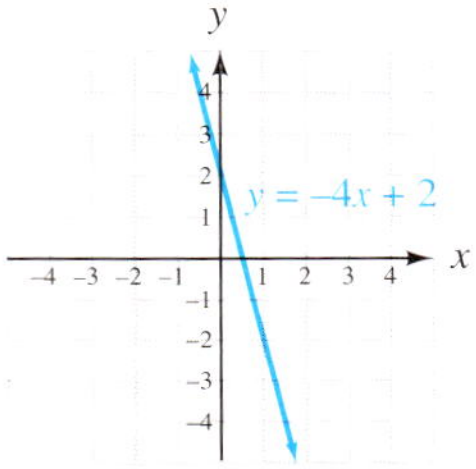

3.
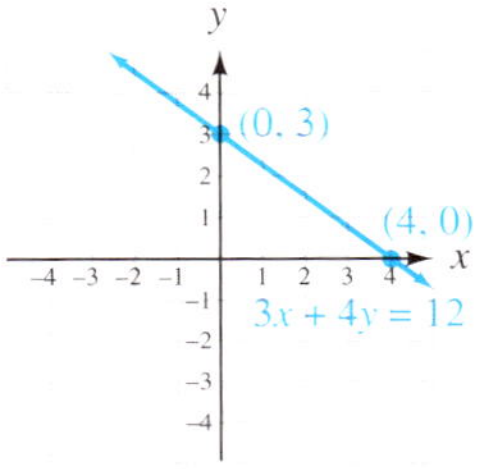

4. Parallel

SELF CHECKS 4.1

1. Not a solution

2. (−2, 1)
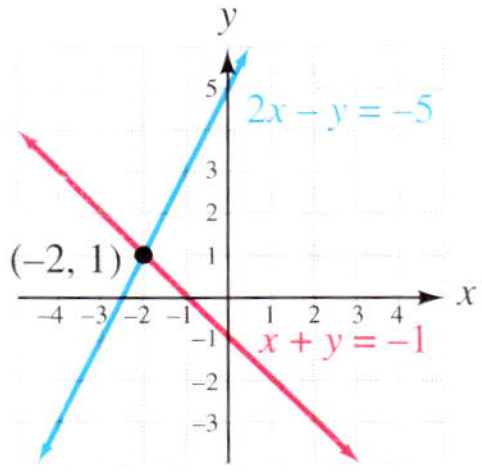

3. No solution
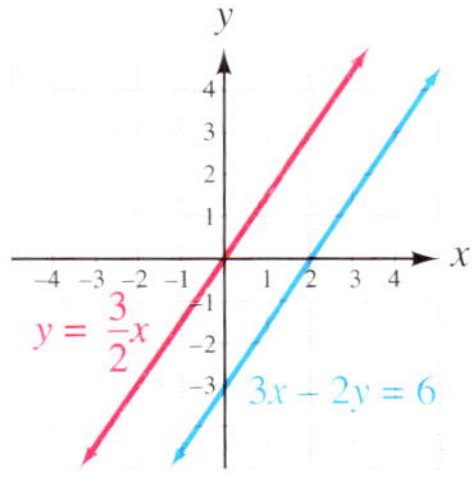

4. Infinitely many solutions
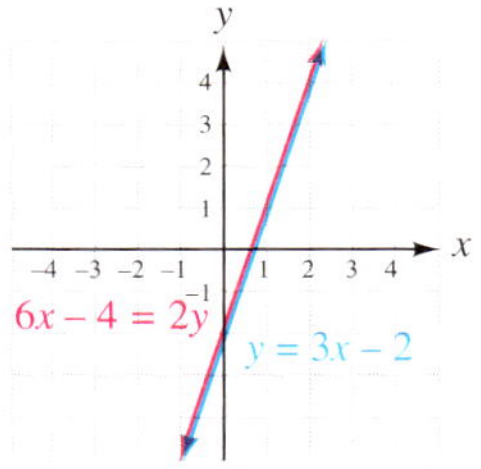

5. No solution

STUDY SET SECTION 4.1 (page 290)

1. system **3.** intersection **5.** consistent, inconsistent **7. a.** True **b.** True **9. a.** −5, 2 **b.** 3, 3, (0, −2) **11.** No solution; independent **13.** A solution **15.** A solution

17. Not a solution **19.** Not a solution **21.** Not a solution **23.** A solution **25.** (3, 2)

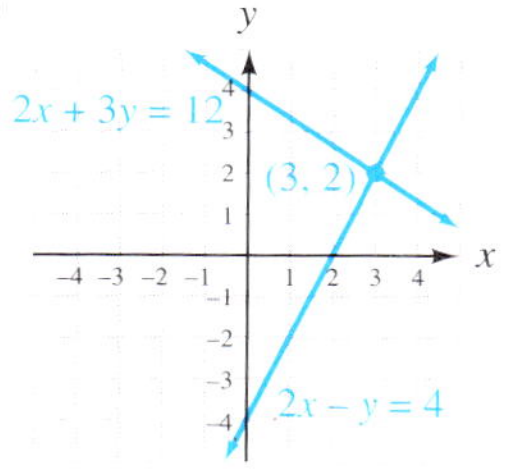

27. (−1, 5)

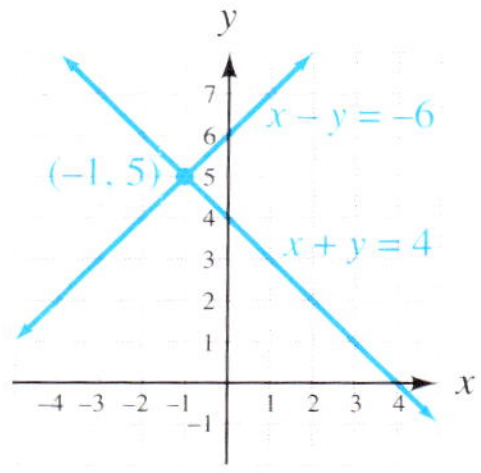

29. No solution

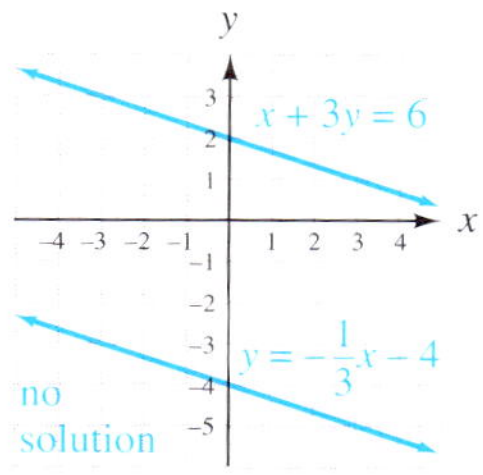

31. No solution

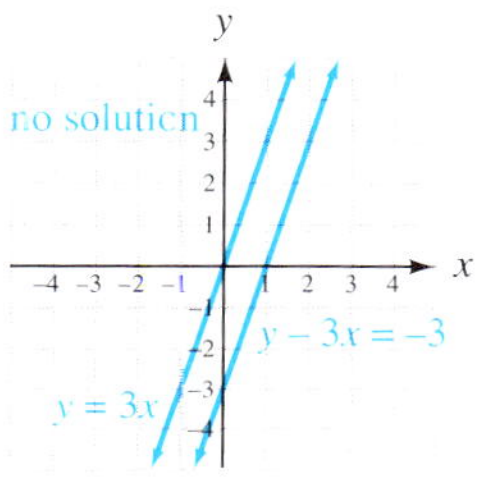

33. Infinitely many solutions

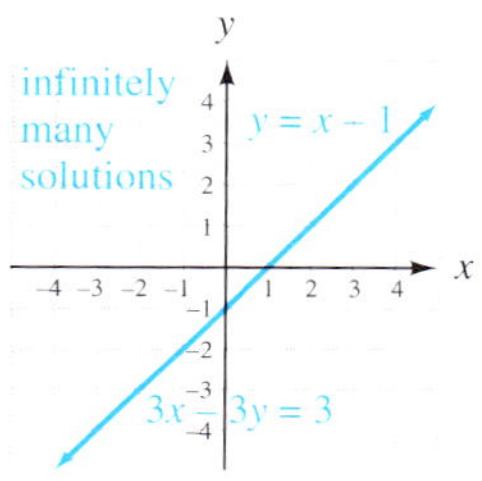

35. Infinitely many solutions

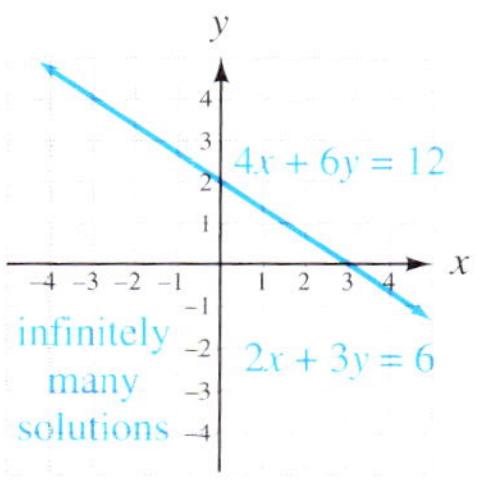

37. 1 solution **39.** Infinitely many solutions **41.** No solution **43.** 1 solution **45.** (1, 3) **47.** No solution

49. (−2, 0)

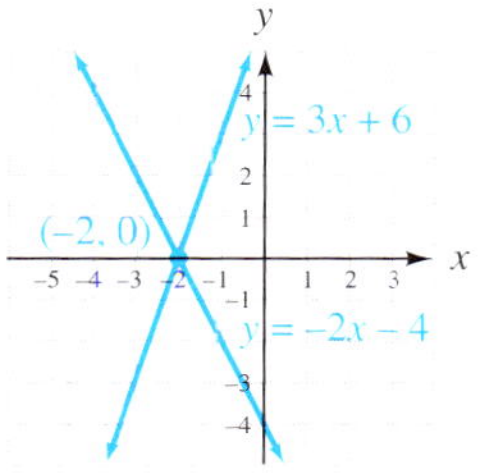

51. No solution

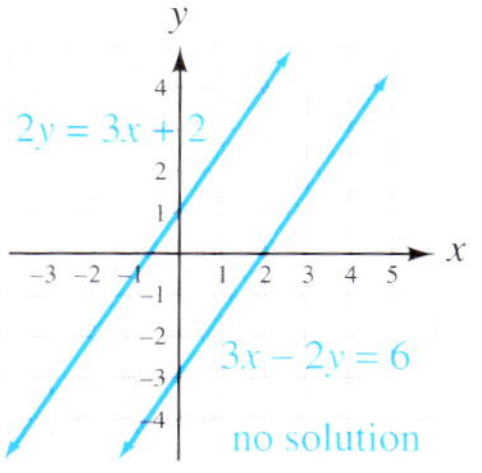

53. (3, −1)

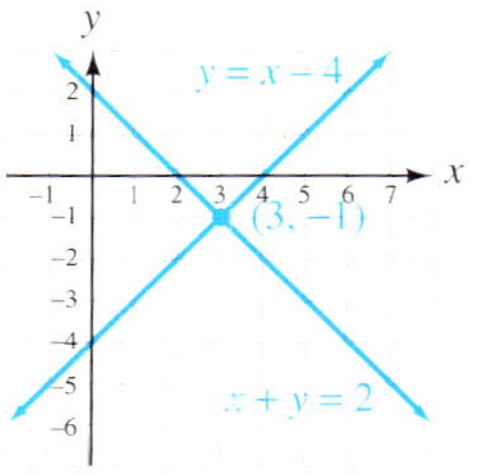

55. (3, 0)

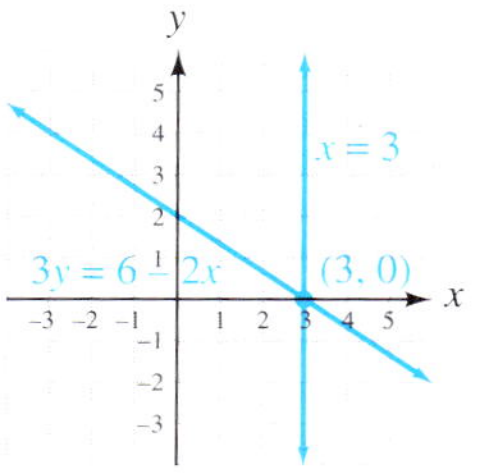

57. (−4, 0)

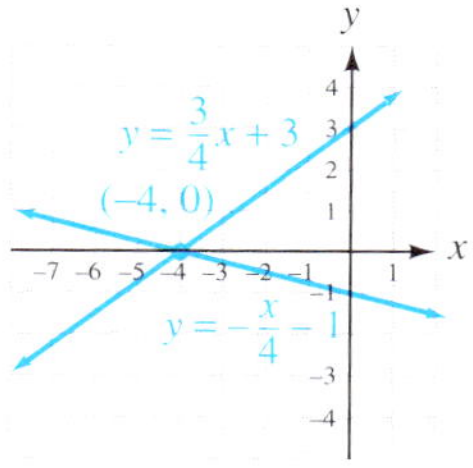

59. (4, −6)

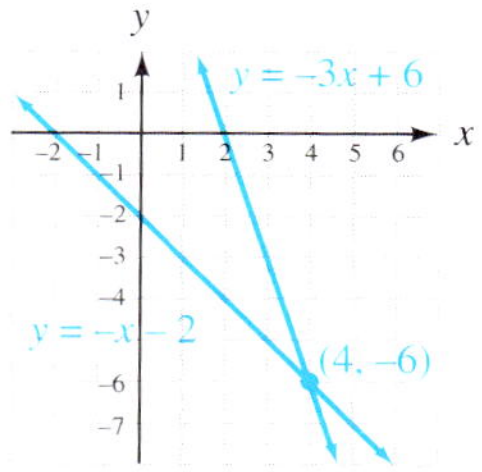

61. (5, −2)

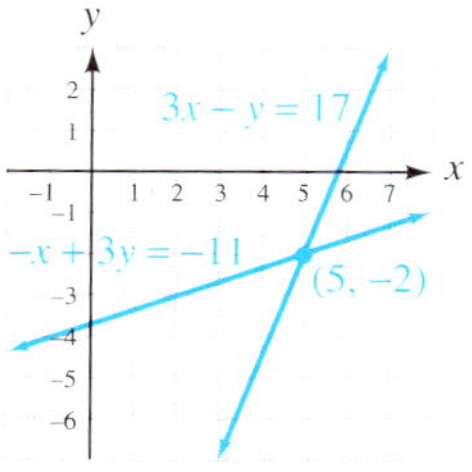

63. (1, 1)

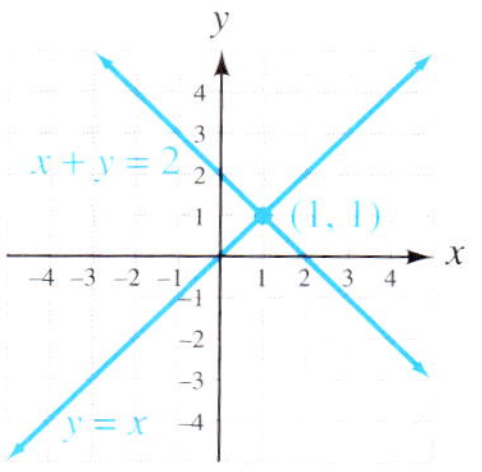

65. Infinitely many solutions

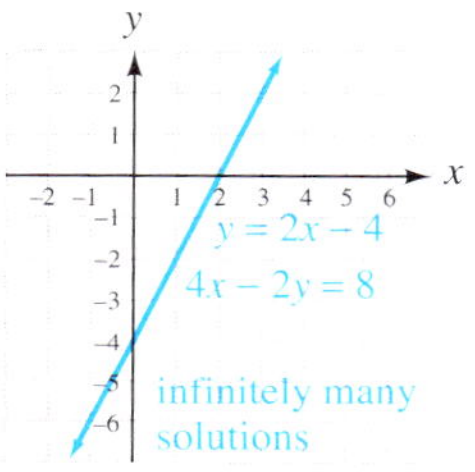

67. (−6, 1)

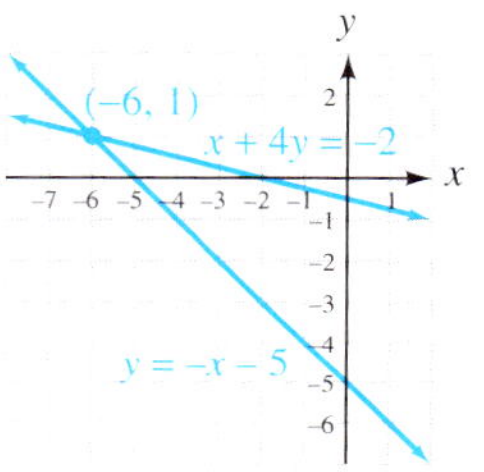

69. (−2, −3)

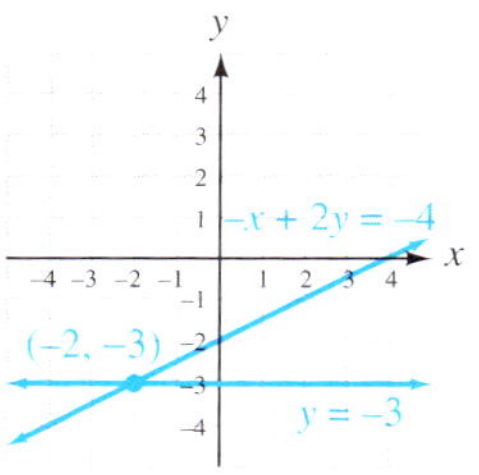

71. (4, −4)

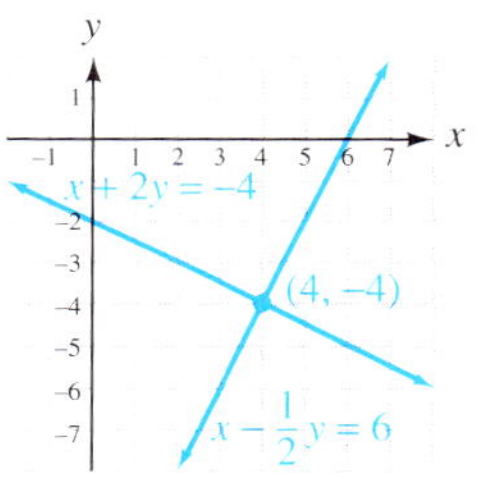

73. May ’09; about 70 million **75. a.** Houston, New Orleans, St. Augustine **b.** St. Louis, Memphis, New Orleans **c.** New Orleans **77. a.** The incumbent; 7% **b.** November 2 **c.** The challenger; 3 **79.** (6, 4), (6, 8), (12, 4), (12, 8); yes

87. $[-3, \infty)$

89. $(-\infty, 2]$

ARE YOU READY? 4.2 (page 294)

1. 1 **2.** $y = 2x + 3$ **3.** 18 **4.** 17 **5.** $3x$

SELF CHECKS 4.2

1. (3, 1) **2.** (−8, 2) **3.** (5, 0) **4.** $\left(\frac{18}{7}, \frac{8}{7}\right)$ **5.** (−3, −3) **6.** No solution **7.** Infinitely many solutions

STUDY SET SECTION 4.2 (page 301)

1. substituting **3.** $y = -3x$ **5.** $x + 3(x - 4) = 8$ **7.** Substitute 3 for a in the second equation. **9. a.** No **b.** ii **11.** $3x$, 4, −2, −2, −6, −2, −6 **13.** (2, 4) **15.** (3, 0)

17. $(-10, 2)$ **19.** $(2, -10)$ **21.** $(3, 2)$ **23.** $(-5, 5)$ **25.** $(-4, -6)$ **27.** $(3, -2)$ **29.** $(-6, 4)$ **31.** $\left(10, \frac{15}{2}\right)$ **33.** No solution **35.** No solution **37.** Infinitely many solutions **39.** Infinitely many solutions **41.** $(3, -2)$ **43.** $(1, 1)$ **45.** $(9, 11)$ **47.** No solution **49.** $\left(\frac{1}{3}, \frac{2}{3}\right)$ **51.** $\left(\frac{1}{5}, 4\right)$ **53.** $(-2, -2)$ **55.** $\left(\frac{1}{2}, \frac{1}{3}\right)$ **57.** $\left(\frac{2}{3}, -\frac{1}{3}\right)$ **59.** $\left(-4, \frac{5}{4}\right)$ **61.** $(4, -2)$ **63.** $(-5, -1)$ **65.** $(4, 2)$ **67.** $(-4, -9)$ **69.** $(-2, 3)$ **71.** Infinitely many solutions **73.** $(-10, -24)$ **75.** $\left(\frac{1}{2}, 2\right)$ **77.** $\left(-1, \frac{2}{3}\right)$ **79.** Infinitely many solutions **81.** $(-4, -1)$ **83.** Angle of approach: 40°; angle of departure: 37° **85.** 22.5°, 67.5° **93.** $3^3 \cdot 7$ **95.** $\frac{2}{3}$

ARE YOU READY? 4.3 (page 304)

1. 6 **2.** 0 **3.** -1 **4.** $-12x - 36y = -4$

SELF CHECKS 4.3

1. $(2, 4)$ **2.** $(-3, -3)$ **3.** $(0, 5)$ **4.** $(-5, 6)$ **5.** $\left(-\frac{22}{9}, \frac{10}{9}\right)$ **6.** No solution **7.** Infinitely many solutions

STUDY SET SECTION 4.3 (page 311)

1. opposites **3.** $7y$ and $-7y$ **5. a.** $5a = -4$ **b.** $-4y = 1$ **7. a.** -2 **b.** 3 **9. a.** Multiply both sides by 15. **b.** Multiply both sides by 10. **11.** $2x$, 1, 1, 4, 1, 4 **13.** $(3, 2)$ **15.** $(-2, -3)$ **17.** $(0, 8)$ **19.** $(-3, 4)$ **21.** $(-12, 1)$ **23.** $(3, 11)$ **25.** $(-2, 7)$ **27.** $(1, 1)$ **29.** $(-2, 5)$ **31.** $(2, -1)$ **33.** $\left(\frac{5}{2}, -\frac{1}{2}\right)$ **35.** $\left(-\frac{1}{2}, -\frac{2}{3}\right)$ **37.** No solution **39.** No solution **41.** Infinitely many solutions **43.** Infinitely many solutions **45.** $(2, 3)$ **47.** $\left(\frac{3}{4}, \frac{1}{3}\right)$ **49.** $\left(\frac{1}{3}, 3\right)$ **51.** $(3, -2)$ **53.** Infinitely many solutions **55.** $\left(\frac{10}{3}, \frac{10}{3}\right)$ **57.** $(6, -2)$ **59.** $(3, 0)$ **61.** $(10, 9)$ **63.** $(1, -1)$ **65.** $\left(\frac{7}{25}, -\frac{1}{25}\right)$ **67.** $(6, 8)$ **69.** $(4, -2)$ **71.** $\left(\frac{13}{75}, \frac{14}{75}\right)$ **73.** $(2.6, -4.5)$ **75.** $(0, -2)$ **77.** $(12, -9)$ **79.** $\left(1, -\frac{5}{2}\right)$ **81.** $(-4, -1)$ **83.** $(-4, 5)$ **85.** $(-4, -5)$ **87.** $(-5, 10)$ **89.** No solution **91.** 1991 **93.** After 24 days **99.** $y = -\frac{11}{6}x - \frac{7}{3}$ **101.** -80

ARE YOU READY? 4.4 (page 314)

1. $P = 2l + 2w$ **2.** \$38.85 **3.** \$2,200 **4.** 420 mi **5.** 0.96 oz alcohol

SELF CHECKS 4.4

1. 4 ft, 8 ft **2.** 101°, 79° **3.** Length: 55 ft; width: 20 ft **4.** Practice cone: \$0.50, portable goal: \$20 **5.** \$2,500 at 9%, \$7,500 at 10% **6.** Boat: 10 mph; current: 2 mph **7.** 1%: 40 L; 4%: 20 L **8.** Planting mix: 120 yd^3; topsoil: 80 yd^3

STUDY SET SECTION 4.4 (page 323)

1. complementary, supplementary **3.** $x + y = 20, y = 2x - 1$ **5.** $x + y = 180, y = x - 25$ **7.** $5x + 2y = 15$ **9.** Downstream: $(x + c)$, upstream: $(x - c)$ mph **11. a.** $(x + y)$ mL **b.** 33% **13.** 20°, 70° **15.** 50°, 130° **17.** Upper: 22 ft, lower: 29 ft **19.** President: \$400,000; vice president: \$227,300 **21.** ∠1: 115°; ∠2: 65° **23.** Length: 96 ft, width: 70 ft **25.** Length: 15 m, width: 10 m **27.** Printer: \$2; copier: \$15 **29.** \$29.50 for a 10 × 14; \$21.00 for an 8 × 10 **31.** Elvis: 29¢; Liberty: 34¢ **33.** Cones: 85, sundaes: 63 **35.** Nursing: \$2,000; business: \$3,000 **37.** International fund: \$18,500; offshore bank: \$21,500 **39.** 4% account: \$11,000; biotech: \$11,000 **41.** Still water: 25 mph, current: 5 mph **43.** Still air: 180 mph, wind: 20 mph **45.** 8 gal 6% salt water, 24 gal 2% salt water **47.** 4% solution: 48 oz; 12% solution: 80 oz **49.** 52 lb of the \$8.75, 48 lb of the \$3.75 **51.** $\frac{20}{3} = 6\frac{2}{3}$ pints of mushrooms; $\frac{40}{3} = 13\frac{1}{3}$ pints of olives

55. $(-\infty, 4)$ 

57. $(-1, 2]$ 

ARE YOU READY? 4.5 (page 328)

1. True **2.**

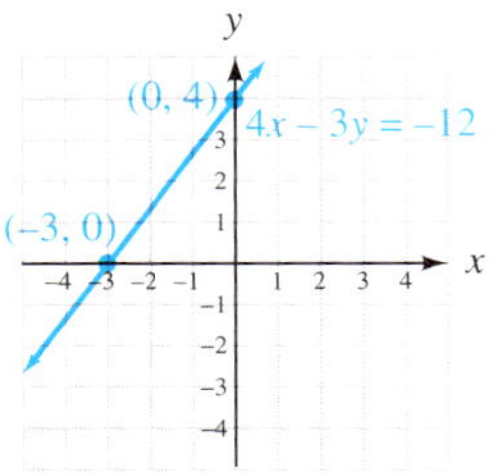

3. a. Above **b.** Below **c.** On **4.** Dashed

SELF CHECKS 4.5

1.

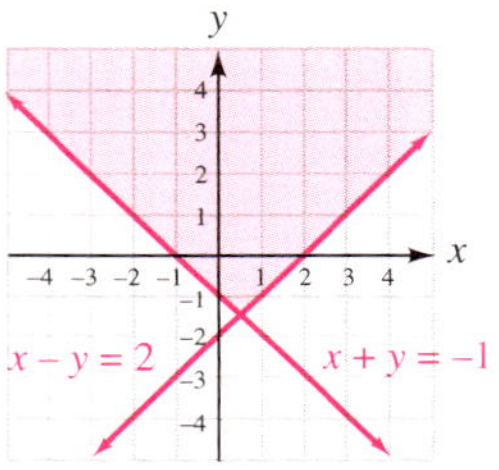

2.

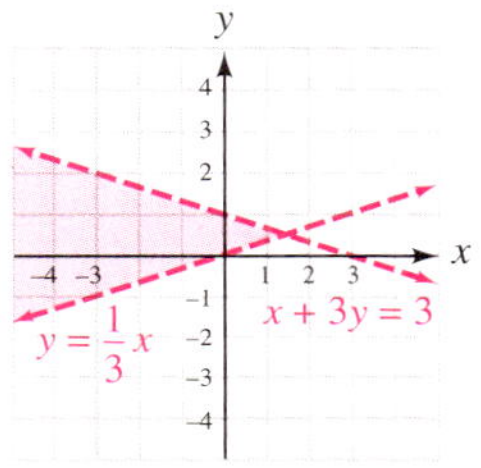

3.

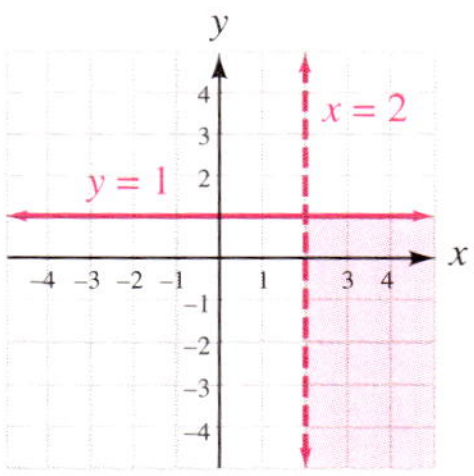

4.

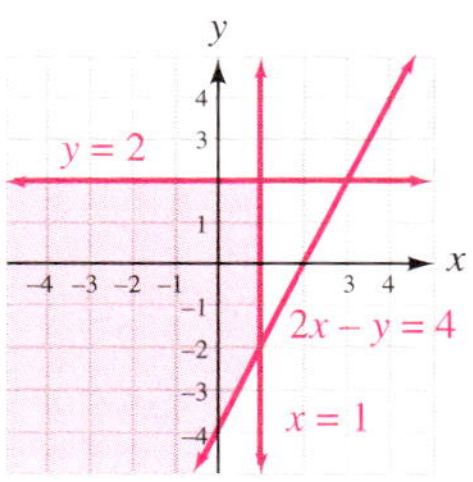

5.

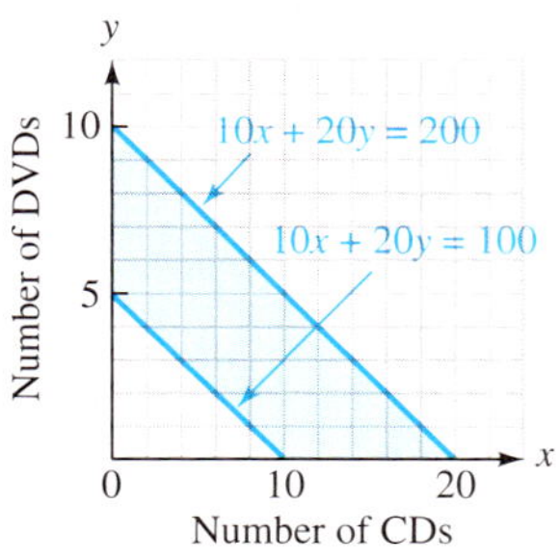

Any ordered pair in the shaded region with whole number coordinates is a possible combination.

STUDY SET SECTION 4.5 (page 334)

1. inequalities **3.** intersection **5. a.** $3x - y = 5$ **b.** Dashed **7.** Slope: $4 = \frac{4}{1}$, y-intercept: $(0, -3)$ **9. a.** No **b.** Above **11. a.** Yes **b.** No **c.** No **13. a.** ii **b.** iii **c.** iv **d.** i

15.
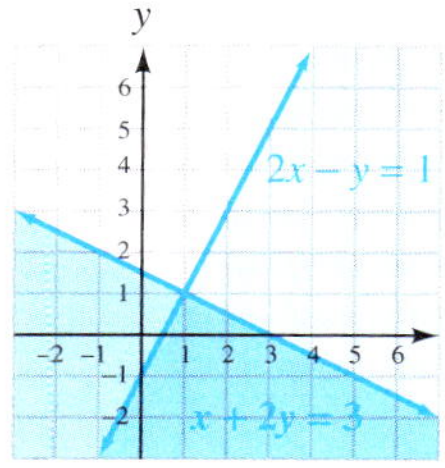

17.
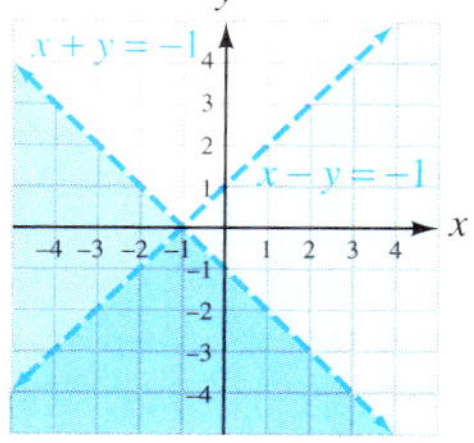

19.
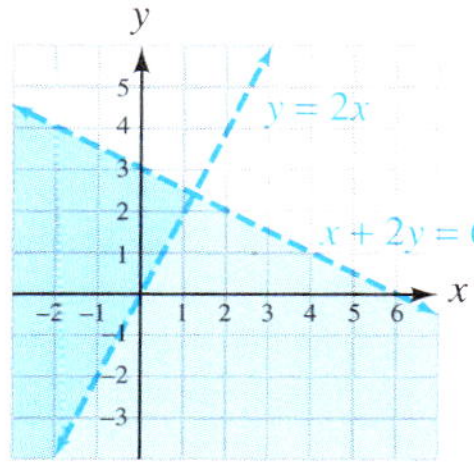

21.
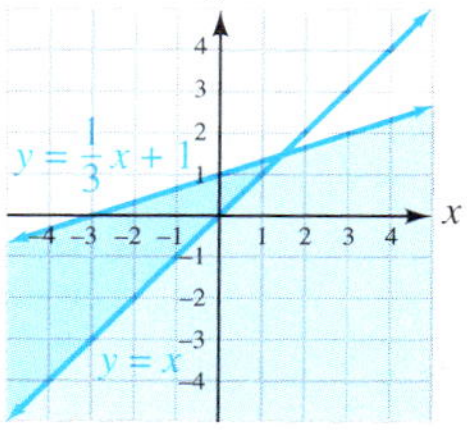

23.
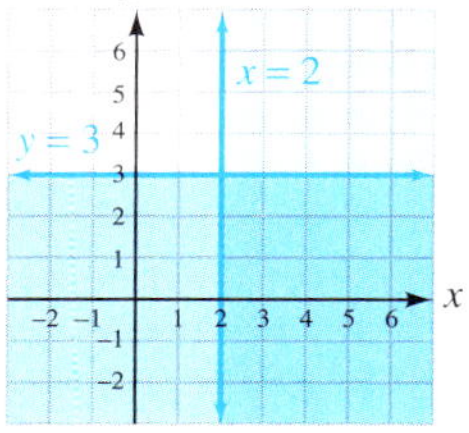

25.
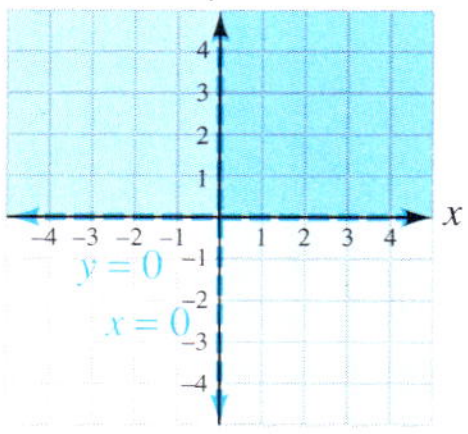

27.
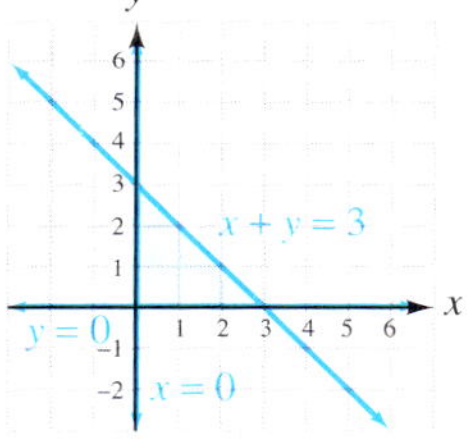

29.
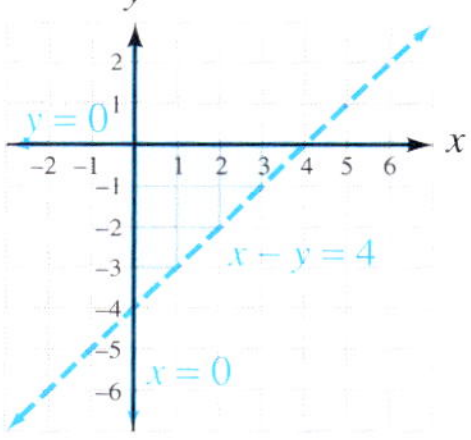

31.
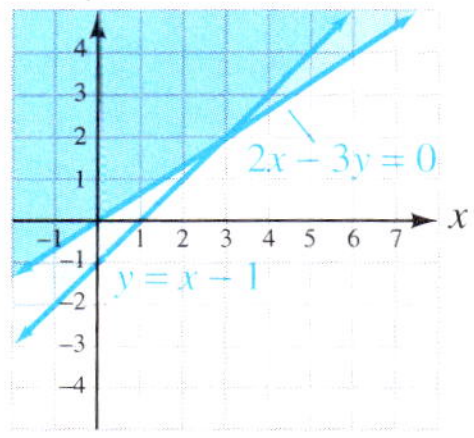

33.
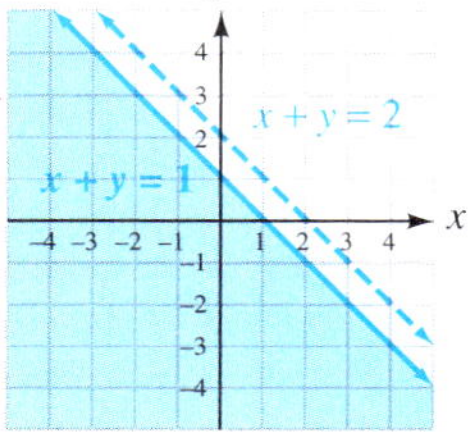

35.
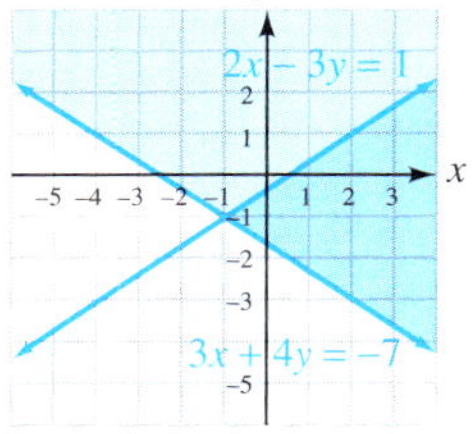

37.
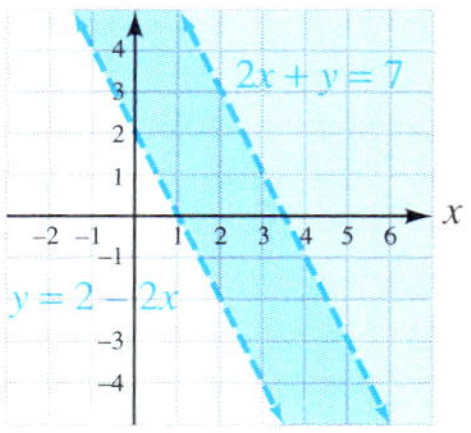

39.
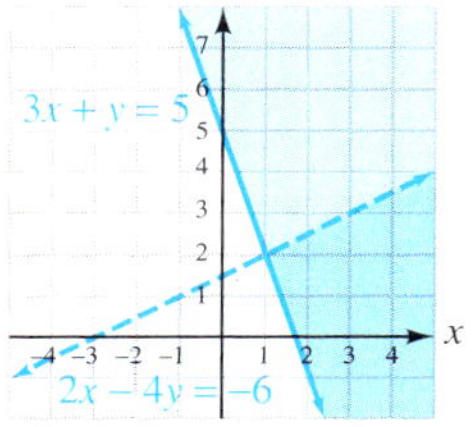

41.
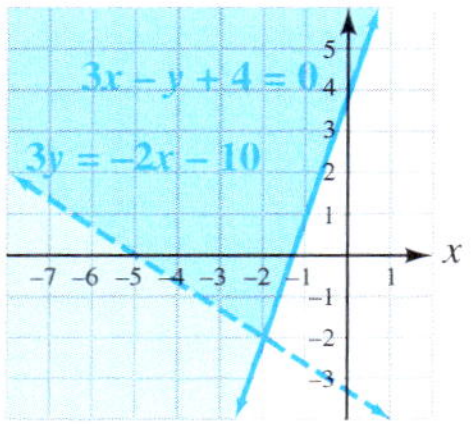

43.
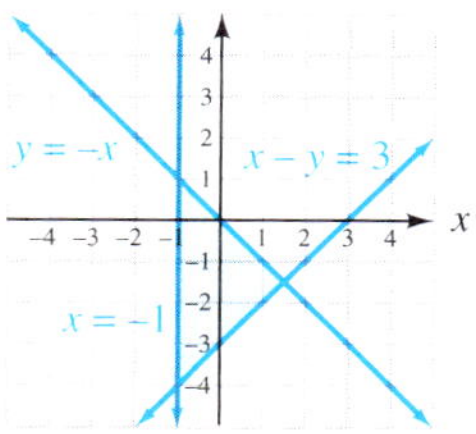

45.
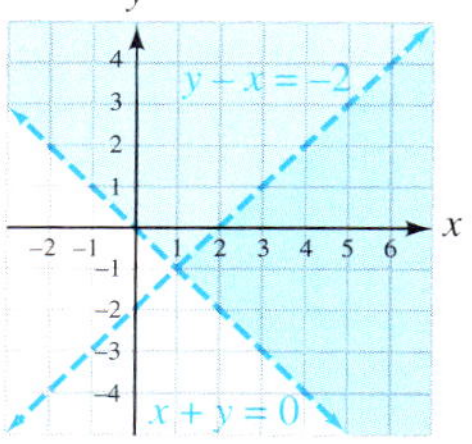

47. a.
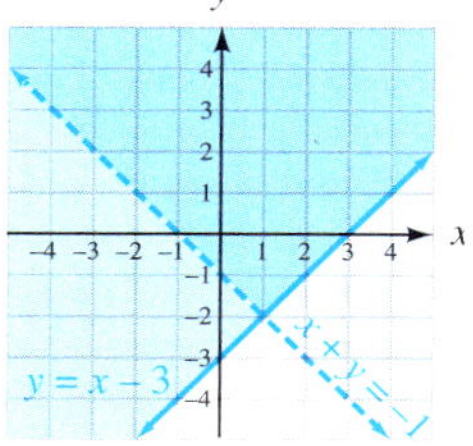

b. $(1, -2)$

49. c.
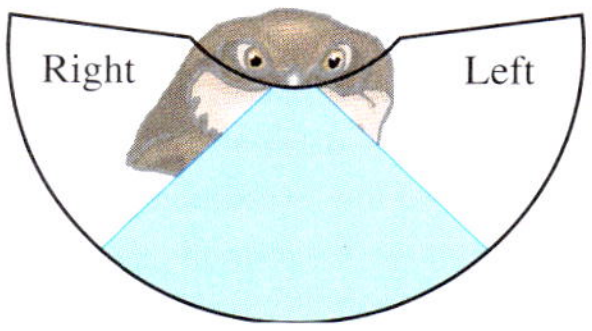

51. 1 \$10 CD and 2 \$15 CDs; 4 \$10 CDs and 1 \$15 CD (answers may vary)

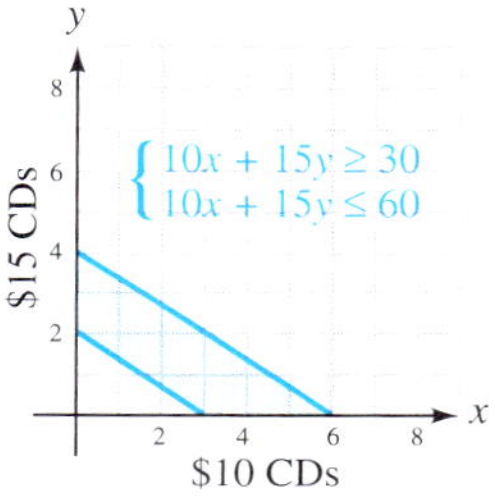

53. 2 desk chairs and 4 side chairs; 1 desk chair and 5 side chairs (answers may vary)

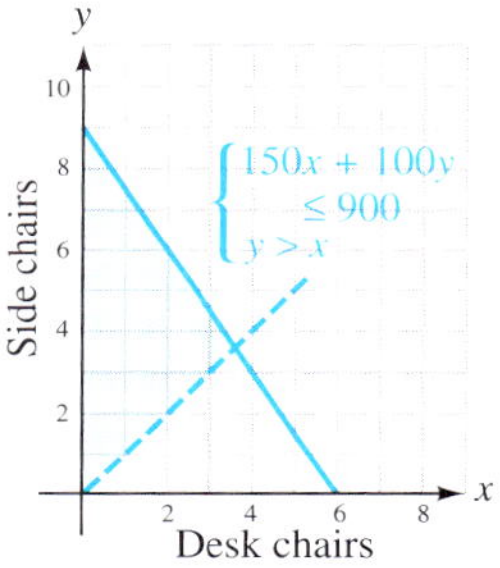

55.
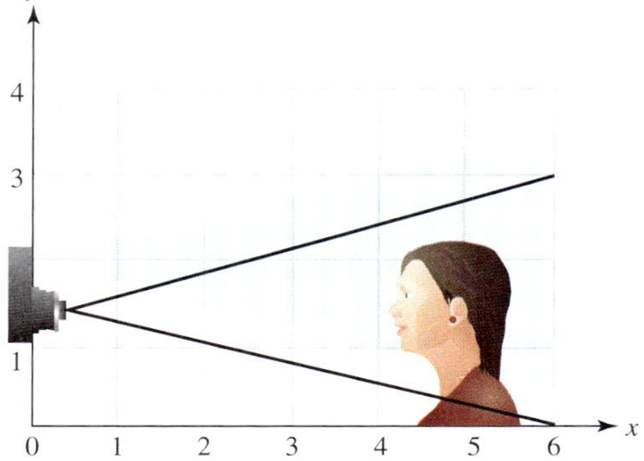

61. $6t$

63. $-\frac{5}{8}x$

65.
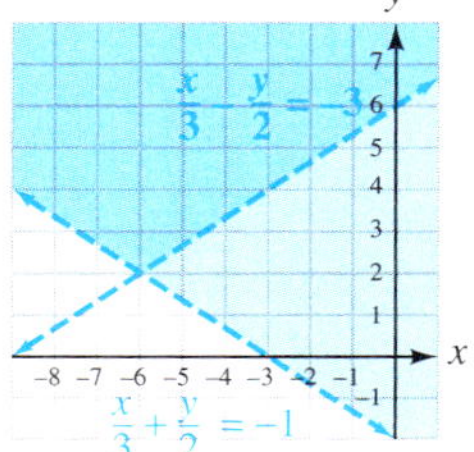

67.
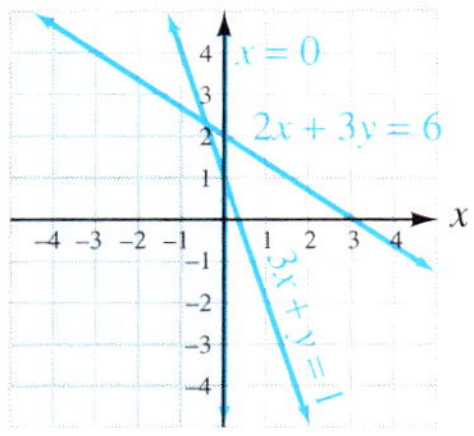

CHAPTER 4 REVIEW (page 337)

1. Yes **2.** No

3. (4, 3) **4.** (3, −1)

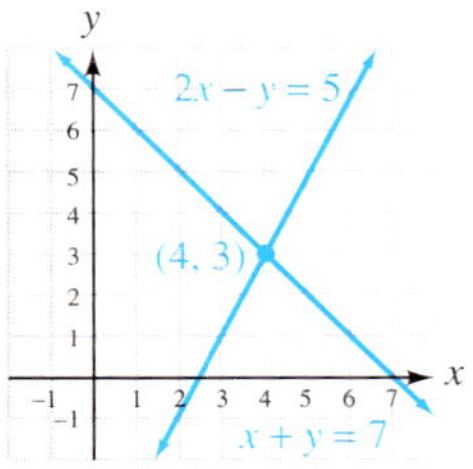

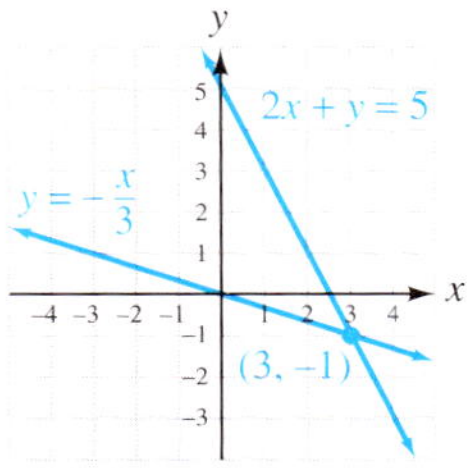

5. Infinitely many solutions **6.** No solution

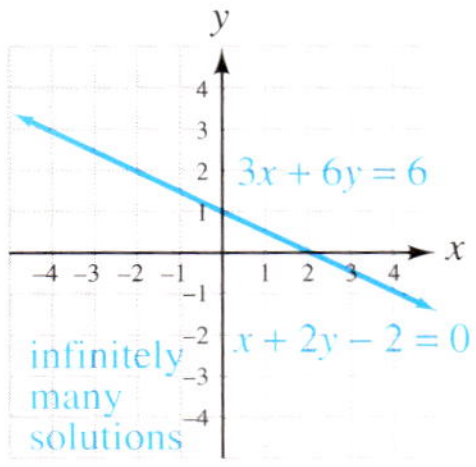

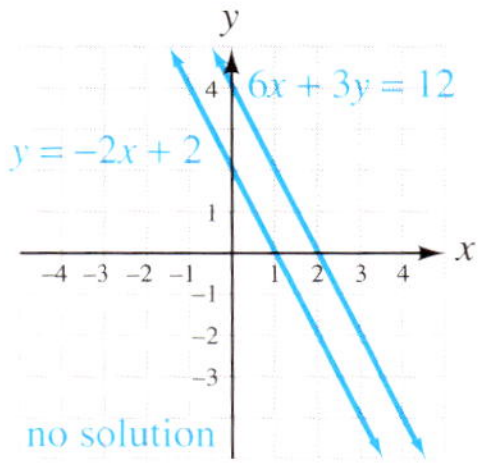

7. No solution **8.** (1980: 480,000); In 1980, the number of men and women awarded Bachelor's degrees in the U.S. was the same, 480,000. **9.** (5, 0) **10.** (3, 3) **11.** $\left(-\frac{1}{2}, \frac{7}{2}\right)$ **12.** (1, −2) **13.** Infinitely many solutions **14.** (12, 10) **15. a.** No solution **b.** Two parallel lines **c.** Inconsistent system **16.** one

17. $\begin{cases} 4x + 2y = 7 \\ 5x - 3y = -6 \end{cases}$ **18.** one **19.** (3, −5) **20.** $\left(3, \frac{1}{2}\right)$

21. (−1, 7) **22.** (0, 9) **23.** Infinitely many solutions **24.** (−5, 2) **25.** (1, −1) **26.** No solution **27.** Elimination; no variables have a coefficient of 1 or −1. **28.** Substitution; equation 1 is solved for x. **29.** Las Vegas: 2,000 ft; Baltimore: 100 ft **30.** Base: 21 ft; extension: 14 ft **31.** 65°, 25° **32.** 10,800 yd^2 **33. a.** $0.02x$, $0.09y$, $0.08(100)$ **b.** $5(s + w)$, $7(s - w)$ **c.** $0.11x$, $0.06y$ **d.** $4x$, $8y$, $10(5)$ **34.** 12 lb worms, 18 lb bears **35.** 3 mph **36.** \$16.40, \$10.20 **37.** \$750

38. $13\frac{1}{3}$ gal 40%, $6\frac{2}{3}$ gal 70%

39.

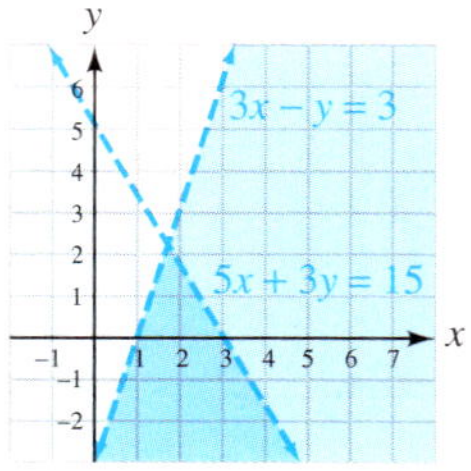

40.

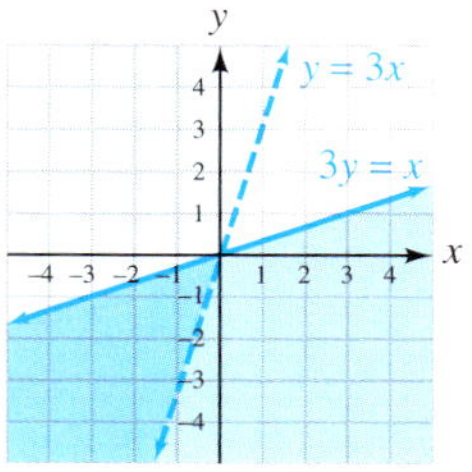

41.

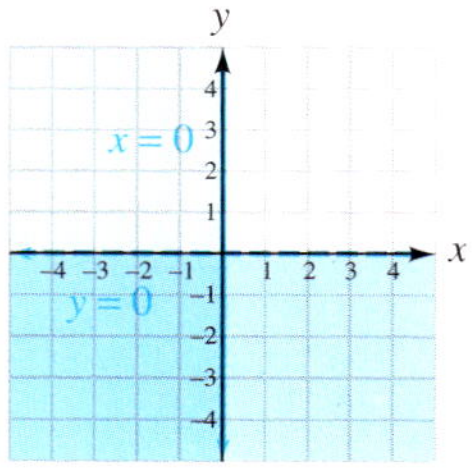

42.

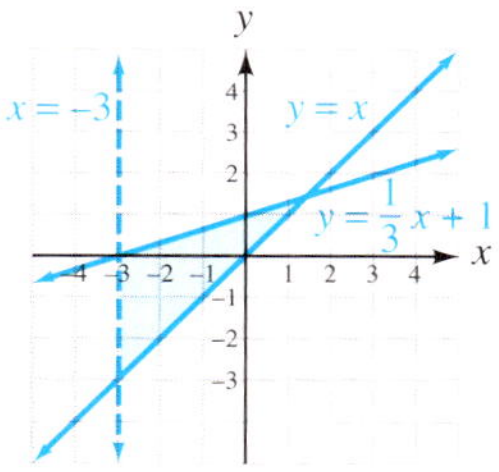

43. a. Yes **b.** No **44.** $10x + 20y \geq 40$, $10x + 20y \leq 60$; (3, 1): 3 shirts and 1 pair of pants; (1, 2): 1 shirt and 2 pairs of pants (Answers may vary)

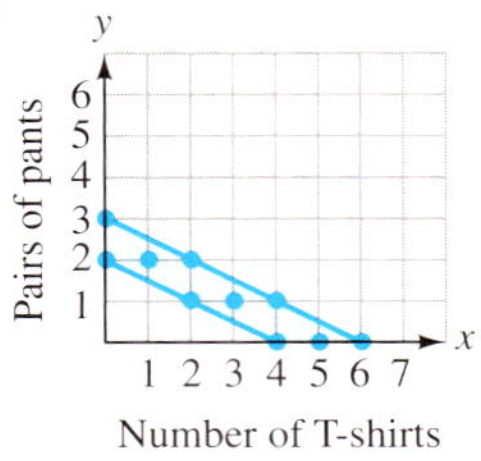

CHAPTER 4 TEST (page 342)

1. Yes **2.** No **3. a.** solution **b.** consistent **c.** inconsistent **d.** independent **e.** dependent **f.** supplementary **4.** (2005, 770); In 2005, the amount of electricity generated by natural gas and nuclear sources was the same, about 770 billion kilowatt hours each.

5. (2, 3) **6.** No solution

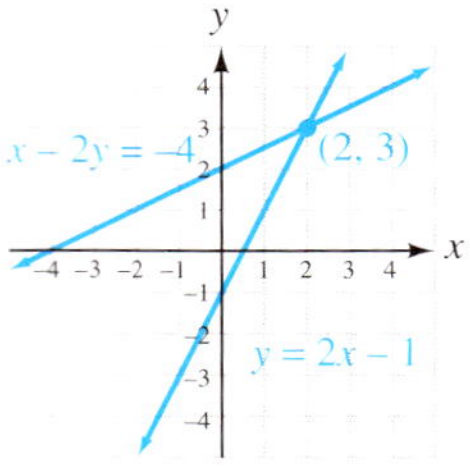

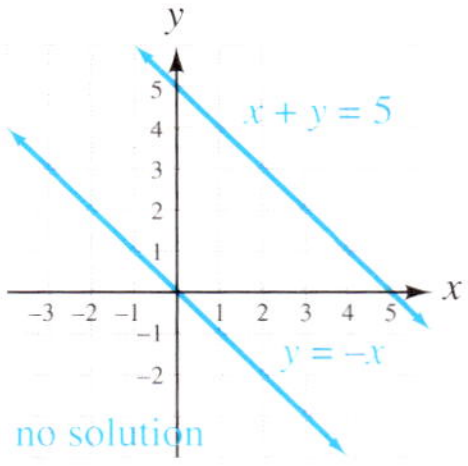

7. Since the lines have different slopes, they will intersect at one point. The system has 1 solution. **8.** Infinitely many solutions; (0,2), (3, 1), (6,0) (answers may vary) **9.** (−2, −3) **10.** Infinitely many solutions **11.** (2, 4) **12.** (−3, 3) **13.** No solution **14.** (−1, −1) **15.** (5, 14) **16.** (0, 0)

17. $\left(\frac{1}{2}, -\frac{1}{4}\right)$ **18.** (10, 5) **19.** (3, 3) **20.** $\left(3, -\frac{1}{2}\right)$

21. 1st part: 8 mi, 2nd part: 14 mi **22.** 3 adult tickets; 4 child tickets **23.** \$6,000 at 8%, \$4,000 at 9% **24.** Speed of calm air: 165 mph, speed of wind: 15 mph **25.** Larger: 70°, smaller: 20° **26.** 5%: 4 pints; 20%: 8 pints **27.** \$1.50 sunscreen: 3 oz; \$0.80 sunscreen: 7 oz **28.** No

29.

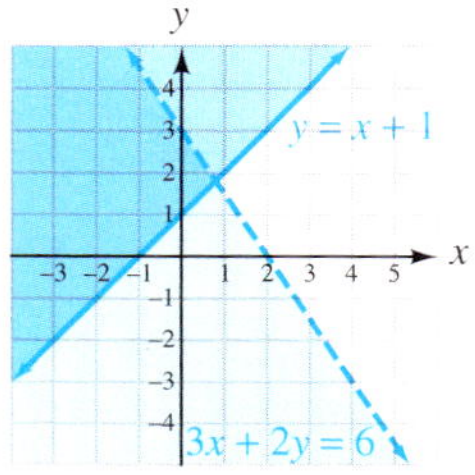

30.

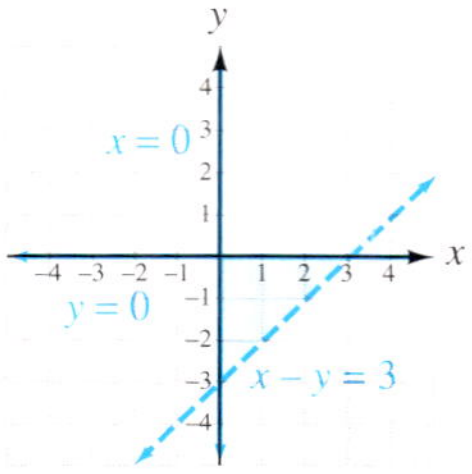

31. (1, 2), (2, 2), (3, 1) (Answers may vary)

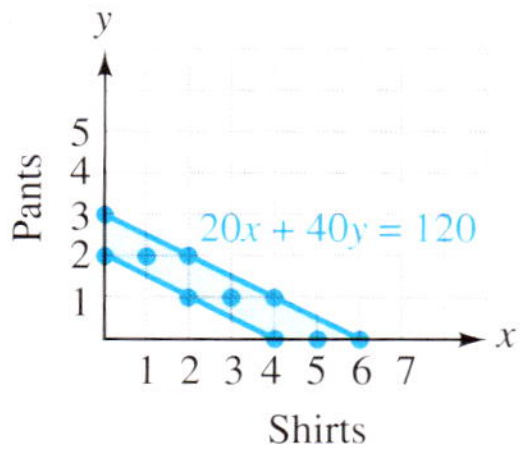

32. a. iii **b.** ii **c.** i **d.** iv

CUMULATIVE REVIEW CHAPTERS 1–4 (page 344)

1. sum, difference, product, quotient **2.** $2^2 \cdot 5^2$ **3.** $\frac{5}{8}$ **4.** $\frac{22}{35}$
5. Irrational **6.**

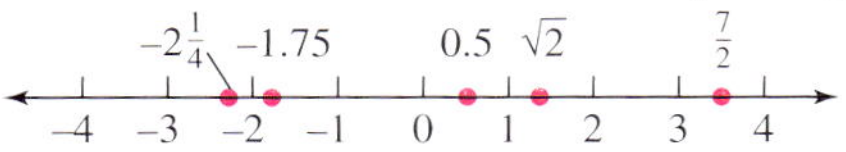

7. $0.\overline{6}$ **8.** Associative property of multiplication **9.** 0 **10.** 1
11. 4 **12.** −2 **13.** $250 - x$ **14.** $10d$ cents **15.** r **16.** $18x$
17. $3d - 11$ **18.** $-78c + 18$ **19.** 13 **20.** 41 **21.** $\frac{10}{9}$
22. −24 **23.** 140 children **24.** 1,941 teragrams **25.** $h = \frac{2A}{b + B}$
26. Men: 116,000; women: 103,000 **27.** 20 lb of \$3.80 candy; 10 lb of \$4.40 candy **28.** $(-\infty, -14)$

29. (Ninth Avenue, 44th Street) **30.** 38 units **31.** II **32.** No
33.

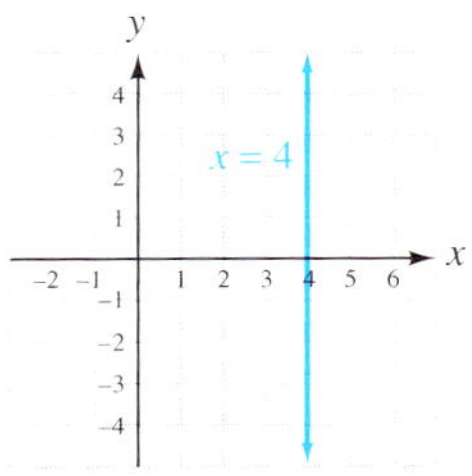

34.

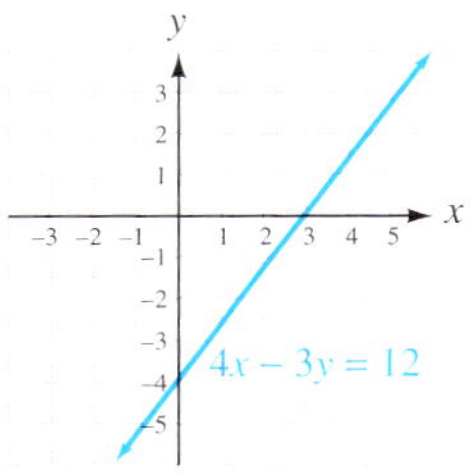

35. $\frac{1}{2}$ **36.** 0 **37.** $\frac{2}{3}$ **38.** Perpendicular **39.** Slope: 2; $(-1, 0)$, $(0, 2)$ **40.** A decrease of 1.6% per year **41.** $y = \frac{2}{3}x + 5$
42. $y = \frac{3}{4}x + \frac{11}{2}$ **43.** $y = 4$ **44.**

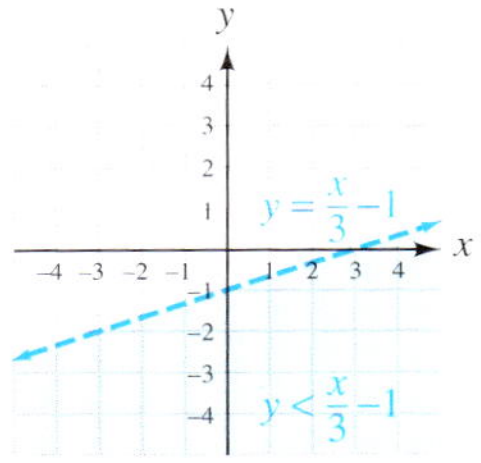

45. 1 **46.** No: (1, 2), (1, −2) (answers may vary)
47. (−6, 1)

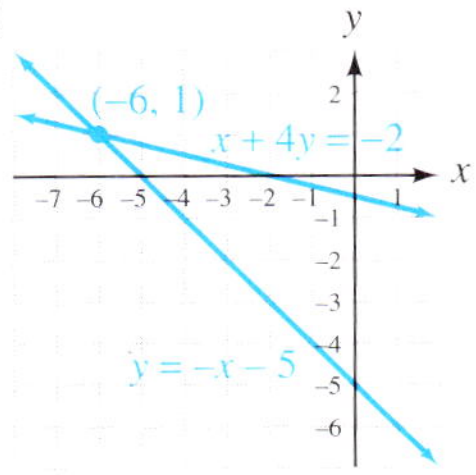

48.

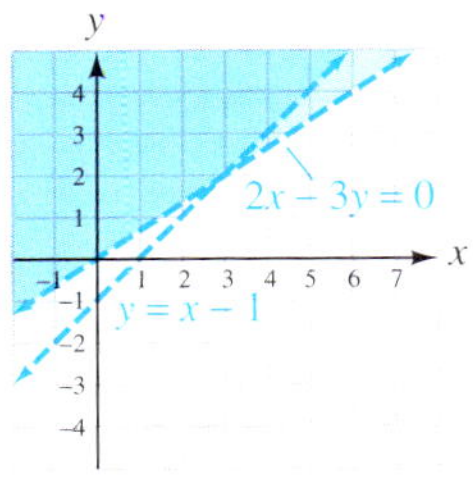

49. (4, 1) **50.** $\left(\frac{2}{3}, \frac{3}{2}\right)$ **51.** Noodles: 2 servings, rice: 3 servings
52. \$3,000 at 10%, \$5,000 at 12%

ARE YOU READY? 5.1 (page 348)

1. a. 15 **b.** 125 **2. a.** 64 **b.** 12 **3.** x **4.** 4

SELF CHECKS 5.1

1. a. Base: 16; exponent: 2 **b.** Base: y; exponent: 4 **c.** Base: $3y$; exponent: 4 **d.** Base: m; exponent: 15 **2. a.** $9a^2b^4$ **b.** $(9a)^2$
c. $(x + y)^5$ **3. a.** 7^{15} **b.** x^6 **c.** $(y - 1)^{10}$ **d.** s^8t^7 **4.** a^{12} in.^{2}
5. a. 55^{25} **b.** a^2 **c.** $8.9t$ **d.** $b^{11}c^3$ **6.** b **7. a.** 4^{30} **b.** y^{10}
8. a. a^{21} **b.** a^{17} **9. a.** $16t^4$ **b.** $c^{18}d^{24}$ **c.** $-\frac{1}{27}a^3b^{15}$
10. $c^{10}d^{12}$ **11.** $64h^6$ **12. a.** $\frac{x^3}{343}$ **b.** $\frac{16x^{12}}{81y^8}$

STUDY SET SECTION 5.1 (page 356)

1. exponential **3. a.** $3x, 3x, 3x, 3x$ **b.** $(-5y)^3$
5. a. Subtract **b.** Add **c.** Multiply **d.** Multiply
7. a. $2x^2$ **b.** x^4 **c.** Doesn't simplify **d.** x^3 **9.** x^6, 18
11. a. power **b.** quantity, quantity **c.** sixth, ninth
13. a. Base: 4, exponent: 3 **b.** Base: 4, exponent: 3
c. Base: −4, exponent: 3 **15. a.** Base: $-3x$, exponent: 2
b. Base: x, exponent: 2 **c.** Base: $-3x$, exponent: 2
17. a. Base: m, exponent: 12 **b.** Base: $9m$, exponent: 12
c. Base: m, exponent: 12 **19.** $(4t)^4$ **21.** $-4t^5$ **23.** $\left(\frac{t}{2}\right)^3$
25. $(x - y)^2$ **27.** 5^7 **29.** b^6 **31.** $(y - 2)^7$ **33.** a^5b^6
35. a^{10} mi^2 **37.** x^9 ft^3 **39.** 8^8 **41.** x^{12} **43.** $(3.7p)^5$
45. c^2d^6 **47.** y^4 **49.** a **51.** 3^8 **53.** $(-4.3)^{24}$ **55.** m^{500}
57. y^{15} **59.** x^{25} **61.** p^{25} **63.** t^{18} **65.** u^{14} **67.** $36a^2$
69. $625y^4$ **71.** $-8r^6s^9$ **73.** $-\frac{1}{243}y^{10}z^{20}$ **75.** ab^4 **77.** $r^{13}s^3$
79. $216k^3$ **81.** $9q^2$ **83.** $\frac{a^3}{b^3}$ **85.** $\frac{64a^4}{121b^{10}}$ **87.** $\frac{x^{10}}{y^{15}}$ **89.** y^9
91. 15^3 **93.** t^7 **95.** $(k - 2)^{14}$ **97.** c^2d^5 **99.** y^{15} **101.** s^2
103. $-216a^9b^6$ **105.** $\frac{243m^{20}}{32n^{25}}$ **107.** $a^{21}b^{21}$ **109.** n^{33}
111. $36h^2$ **113.** x^3y^4 **115.** $\frac{m^4}{81}$ **117. a.** a^6 **b.** a^9 **c.** $2a^3$
119. a. b^9 **b.** b^{20} **c.** b **121. a.** $25x^2$ ft^2 **b.** $9a^2\pi$ ft^2
123. $\frac{1}{8,192}$ **127.** c **129.** d

ARE YOU READY? 5.2 (page 359)

1. $\frac{1}{2}$ **2.** $\frac{1}{x^2}$ **3. a.** 5 **b.** $\frac{1}{w}$ **4. a.** m, 0 **b.** 7, −2

SELF CHECKS 5.2

1. a. 1 **b.** $-5d$ **c.** 1 **2. a.** $\frac{1}{64}$ **b.** $\frac{1}{x^5}$ **c.** $-\frac{1}{27}$ **d.** $\frac{1}{16}$
3. a. $\frac{12}{h^9}$ **b.** $-\frac{1}{16}$ **4. a.** w^5 **b.** $\frac{64}{25}$ **c.** $-\frac{8a^7}{h^6}$ **5.** $\frac{81}{c^2}$ **6. a.** t^4
b. $\frac{1}{a^5}$ **c.** $\frac{1}{n^{20}}$ **d.** $\frac{64c^6}{d^3}$ **e.** $\frac{8}{c^{12}}$ **7. a.** $\frac{1}{a^6}$ **b.** $\frac{81x^2}{y^3}$ **c.** $\frac{d^{15}}{c^{18}}$

STUDY SET SECTION 5.2 (page 367)

1. negative **3.** zero
5.

Expression	Base	Exponent
4^{-2}	4	−2
$6x^{-5}$	x	−5
$\left(\frac{3}{y}\right)^{-8}$	$\frac{3}{y}$	−8
-7^{-1}	7	−1
$(-2)^{-3}$	−2	−3
$10a^0$	a	0

7.

x	3^x
2	9
1	3
0	1
−1	$\frac{1}{3}$
−2	$\frac{1}{9}$

9. a. 3 **b.** 6 **c.** sign, 3, 2 **d.** reciprocal, 2 **11.** y^8, −40, 40
13. 1 **15.** 1 **17.** 2 **19.** $\frac{5}{2}$ **21.** $\frac{1}{4}$ **23.** $\frac{1}{6}$ **25.** $\frac{1}{b^5}$ **27.** $-\frac{1}{5}$ **29.** $\frac{1}{2}$
31. $\frac{8}{9}$ **33.** $\frac{15}{g^6}$ **35.** $\frac{5}{x^3}$ **37.** $-\frac{1}{27}$ **39.** $-\frac{1}{64}$ **41.** 125 **43.** $8s$
45. $\frac{3}{16}$ **47.** $-\frac{4p^{10}}{d}$ **49.** 36 **51.** 8 **53.** $\frac{d^8}{c^8}$ **55.** $\frac{m^4}{81}$ **57.** y^6
59. b^7 **61.** $\frac{1}{y}$ **63.** $\frac{1}{h^7}$ **65.** $\frac{1}{x^{12}}$ **67.** $\frac{1}{b^8}$ **69.** $\frac{36s^8}{t^{14}}$ **71.** $\frac{x^9}{64}$ **73.** y^3
75. $\frac{1}{a^6}$ **77.** $\frac{9a^2}{2b^2}$ **79.** $\frac{y^{14}}{z^{10}}$ **81.** $\frac{8b^3}{a^{12}}$ **83.** r^{20} **85.** $\frac{b^2}{a^5}$ **87.** $-\frac{1}{1,000}$

89. 1 **91.** $\frac{64s^2}{81t^4}$ **93.** $\frac{32v^{25}}{u^{10}}$ **95.** $\frac{9}{y^8}$ **97.** $-15y$ **99.** $\frac{h^{20}}{16}$ **101.** $\frac{1}{x^6}$
103. $\frac{c^{12}}{d^{27}}$ **105.** 15 **107.** $\frac{9}{4g^2}$ **109.** $\frac{125}{d^6}$ **111.** $\frac{x^{28}}{y^{20}}$ **113.** $\frac{32x^{15}}{y^{10}}$
115. t^{10} **117.** $-\frac{4t^2}{s^5}$ **119.** $\frac{1}{x^3}$ **121. a.** $\frac{1}{8}$ **b.** $-\frac{1}{8}$ **c.** $-\frac{1}{8}$ **d.** $\frac{1}{8}$
123. a. $\frac{4x}{y^2}$ **b.** $\frac{1}{16x^2y^2}$ **c.** $\frac{4y}{x^2}$ **d.** $\frac{xy}{16}$

125.

Type of Sound	Intensity
Front row rock concert	10^{-1}
Normal conversation	10^{-6}
Vacuum cleaner	10^{-4}
Military jet takeoff	10^2
Whisper	10^{-10}

129. $-\frac{3}{2}$ **131.** $y = \frac{3}{4}x - 5$

ARE YOU READY? 5.3 (page 369)

1. 100 **2.** 4,528 **3.** $\frac{1}{10}$ **4.** 0.0622

SELF CHECKS 5.3

1. a. 4,880,000 **b.** 0.0098 **2. a.** 9.3×10^7 **b.** 9.055×10^{-5} **c.** 8.5×10^{-2} **3.** 1.0414×10^9 **4.** 3.3×10^{-22} g

STUDY SET SECTION 5.3 (page 374)

1. scientific, standard **3.** right, left **5. a.** positive **b.** negative **7. a.** 7.7 **b.** 5.0 **c.** 8 **9. a.** $(5.1 \times 1.5)(10^9 \times 10^{22})$ **b.** $\frac{8.8}{2.2} \times \frac{10^{30}}{10^{19}}$ **11.** 1, 10, integer **13.** 230 **15.** 812,000 **17.** 0.00115 **19.** 0.000976 **21.** 6,001,000 **23.** 2.718 **25.** 0.06789 **27.** 0.00002 **29.** 2.3×10^4 **31.** 1.7×10^6 **33.** 6.2×10^{-2} **35.** 5.1×10^{-6} **37.** 5.0×10^9 **39.** 3.0×10^{-7} **41.** 9.09×10^8 **43.** 3.45×10^{-2} **45.** 9.0×10^0 **47.** 1.1×10^1 **49.** 1.718×10^{18} **51.** 1.23×10^{-14} **53.** 7.3×10^5 **55.** 2.018×10^{17} **57.** 7.3×10^{-5} **59.** 3.602×10^{-19} **61.** 7.14×10^5; 714,000 **63.** 4.032×10^{-3}; 0.004032 **65.** 4.0×10^{-4}; 0.0004 **67.** 3.0×10^4; 30,000 **69.** 4.3×10^{-3}; 0.0043 **71.** 3.08×10^{-2}; 0.0308 **73.** 2.0×10^5; 200,000 **75.** 7.5×10^{-11}; 0.000000000075 **77.** $9.038030748 \times 10^{15}$ **79.** $1.734152992 \times 10^{-12}$ **81.** 2.57×10^{13} mi **83.** 197,000,000 mi²; 109,000,000,000,000,000 mi²; 14,600,000 mi² **85.** 4.5×10^{-10} oz **87.** g, x, u, v, i, m, r **89.** 3.3×10^{-1} km/sec **91.** 3.09936×10^{16} ft **93.** 3.04×10^{11} dollars **95.** 1.0×10^6, 1.0×10^9, 1.0×10^{12}, 1.0×10^{15}, 1.0×10^{18} **101.** 5 **103.** $c = 30t + 45$

ARE YOU READY? 5.4 (page 376)

1. 3 terms **2.** 6 **3.** $5b^4$ **4.** -26

SELF CHECKS 5.4

1. a. A trinomial in one variable of degree 2 written in descending powers of x; terms: x^2, $4x$, -16; coefficients: 1, 4, -16; degree of terms: 2, 1, 0 **b.** A binomial in two variables of degree 7 written in descending powers of s and ascending powers of t; terms: $-14s^5t$, s^4t^3; coefficients: -14, 1; degree of terms: 6, 7 **2.** 33 **3.** 55 cans **4.** 16

5.

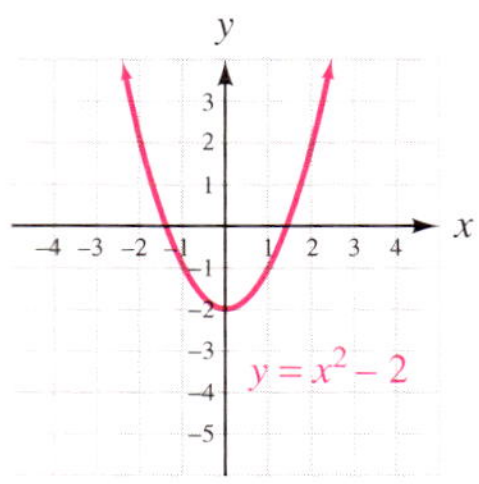

6.

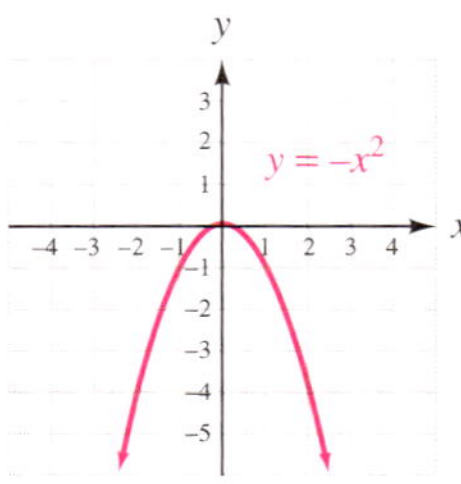

7.

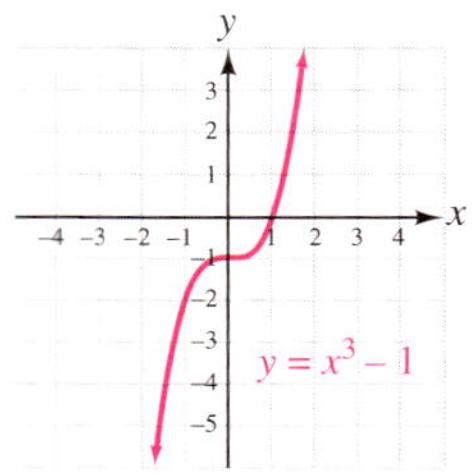

STUDY SET SECTION 5.4 (page 382)

1. polynomial **3.** one, descending, two, ascending **5.** monomial, binomial, trinomial **7.** evaluate **9. a.** Yes **b.** No **c.** No **d.** Yes **e.** Yes **f.** Yes **11.**

Term	Coefficient	Degree
$8x^2$	8	2
x	1	1
-7	-7	0

Degree of the polynomial: 2

13.

Term	Coefficient	Degree
$8a^6b^3$	8	9
$-27ab$	-27	2

Degree of the polynomial: 9

15. a. $5x^3 + 3x^2 + x - 9$ **b.** $x^2 - 2xy + y^2$ **17.** Binomial **19.** Trinomial **21.** Monomial **23.** Binomial **25.** Trinomial **27.** None of these **29.** None of these **31.** Trinomial **33.** 4th **35.** 2nd **37.** 1st **39.** 4th **41.** 12th **43.** 0th **45.** 18th **47.** 3rd **49. a.** 3 **b.** 13 **51. a.** -6 **b.** -8 **53. a.** 7 **b.** 34 **55. a.** -11.6 **b.** -40.2 **57. a.** 28 **b.** 4 **59. a.** 2 **b.** 0 **61.** 72 **63.** 19 **65.** -35 **67.** -257

69.

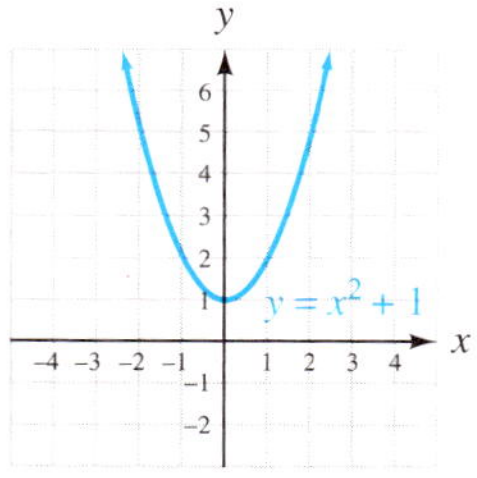

71.

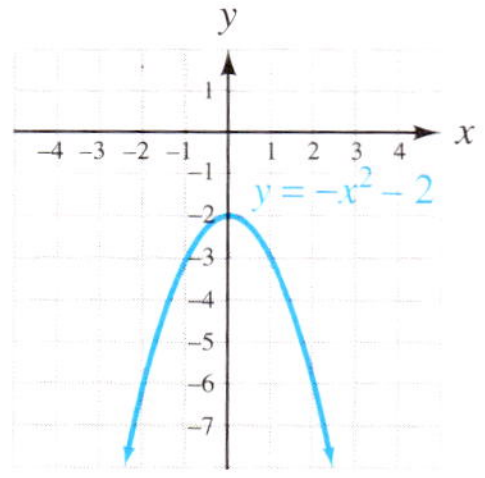

73.

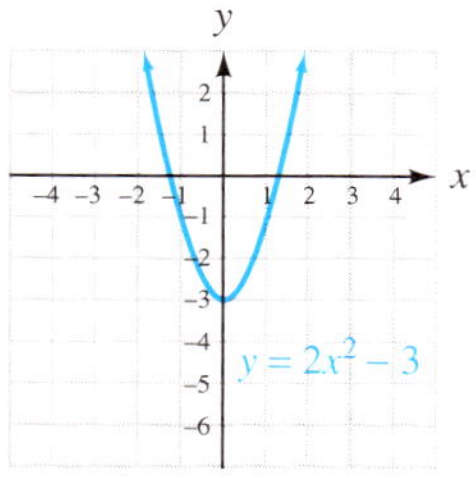

75.

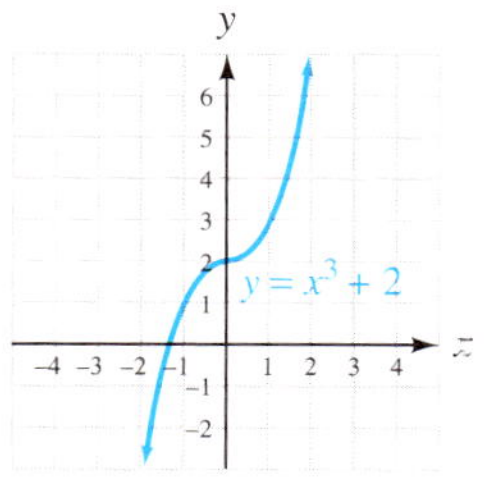

77.

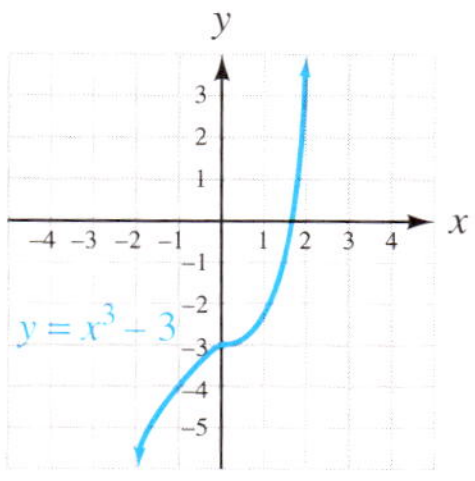

79.

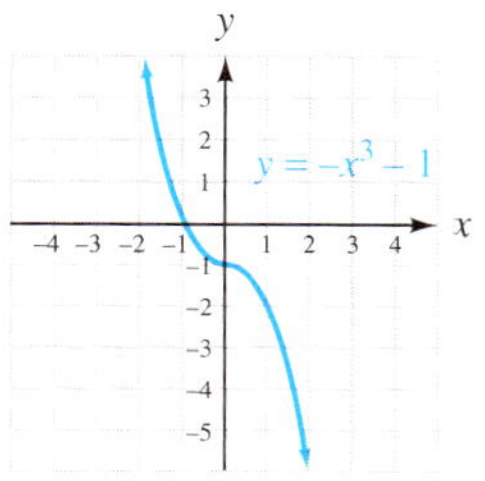

81. 91 cantaloupes **83.** 63 ft **85.** 28.6 billion downloads
87.

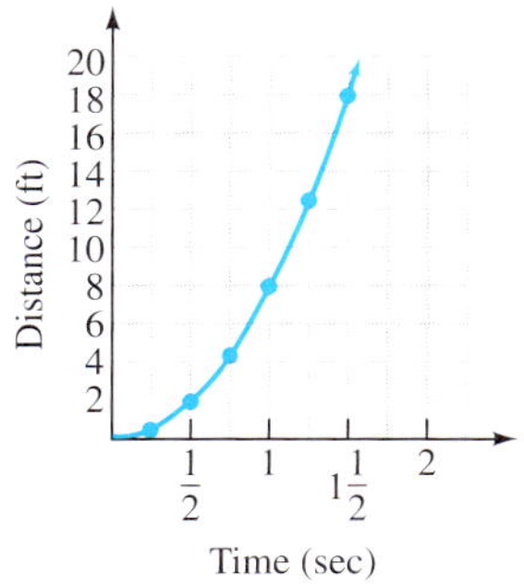

93. $[-3, \infty)$

95. x^{18} **97.** y^9

ARE YOU READY? 5.5 (page 385)

1. $17x$ **2.** $1.2a^3$ **3.** $5t^2u$ **4.** Does not simplify **5.** $-4b^2 + 9b - 1$ **6.** -8

SELF CHECKS 5.5

1. a. $9m^4$ **b.** x **c.** $1.1s^3t + 0.3s^2t$ **d.** $\frac{11}{8}c^5 - \frac{4}{15}d^5 - 8$ **2. a.** $7a^2 + 5a - 1$ **b.** $\frac{9}{4}b^3 - \frac{3}{10}b - 3$ **c.** $11x^2 - 2xy - 2y^2$ **3.** $(29h^2 - 15h)$ in. **4.** $6q^2 - 8q + 2$ **5. a.** $7a^3 - 4a^2 + 12$ **b.** $x^2y - 8x - 8y$ **6. a.** $3p^2 - 8p + 15$ **b.** $-12m^3 + 6m^2 + 9m$ **7.** $0.6q^2 - 0.3q$ **8.** $(7{,}000x + 220{,}000)$

STUDY SET SECTION 5.5 (page 391)

1. polynomials **3.** Like **5.** combine **7. a.** $5x^2$ **b.** $14m^3$ **c.** $7a^3b$ **d.** $6cd + 4c^2d$ **9. a.** $-5x^2 + 8x - 23$ **b.** $5y^4 - 3y^2 + 7$ **11.** $4x^2, 2x, 1, 10x^2, 4$ **13.** $12t^2$ **15.** $20x^2 - 19x$ **17.** $x^2 + x$ **19.** $\frac{13}{15}x^2 - \frac{1}{8}x$ **21.** $1.3x^3$ **23.** $2st$ **25.** $-ab$ **27.** $-4x^3y + x^2y + 5$ **29.** $5q^2 - 4q - 5$ **31.** $y^3 + \frac{19}{20}y^2 + \frac{1}{3}$ **33.** $0.7p - 0.9q$ **35.** $7x^2 + xy + 2y^2$ **37.** $(3x^2 + 6x - 2)$ yd **39.** $(7x^2 + 5x + 6)$ mi **41.** $5x^2 + x + 11$ **43.** $-3a^2 + 7a + 7$ **45.** $10z^3 + z - 2$ **47.** $-x^3y^2 + 4x^2y + 5x + 6$ **49.** $2a^2 + a - 3$ **51.** $-5h^3 + 5h^2 + 30$ **53.** $\frac{1}{24}s^8 - \frac{19}{20}s^7$ **55.** $b^2 + 4ab - 2$ **57.** $x^2 + 6x + 2$ **59.** $4s^2 - 4s + 7$ **61.** $9a^3 - 8a^2 + 28a - 11$ **63.** $0.6x^3 + 1.2x^2 + 1.3x - 0.3$ **65.** $-9x^2 - 3x + 1$ **67.** $t^3 + 3t^2 + 6t - 5$ **69.** $13a^2 + a$ **71.** $3y^5 - 6y^4 + 1.2$ **73.** $10r^4 - 4r$ **75.** $-0.14f^2 + 0.25f + 2.09$ **77.** $\frac{5}{4}r^4 + \frac{11}{9}r^2 - 2$ **79.** $9c^2 - 6c - 14$ **81.** $19.4h^3 + 11h^2$ **83.** $-5r^2t - 9rt + 30$ **85.** $6x^2 + x - 5$ **87.** $\frac{7}{12}c^2 - \frac{1}{2}cd + d^2$ **89.** $7x + 4$ **91.** $-1.3t^2 + 0.7t + 0.6$ **93.** $-48u^3$ **95.** $5d^2 + 14d$ **97.** $7x^3y^2 - 2x^2y - 2x + 15$ **99.** $3x + 1$ **101.** $9x^3 + 2x^2 - 7x + 6$ **103. a.** $-19x^2 + 3x + 10$ **b.** $3x^2 - 9x - 10$ **105. a.** $(x^2 - 8x + 12)$ ft **b.** $(x^2 + 2x - 8)$ ft **107.** $(2a^2 + 6a + 5)$ in. **109. a.** $(22t + 20)$ ft **b.** 108 ft **117.** 180° **119.**

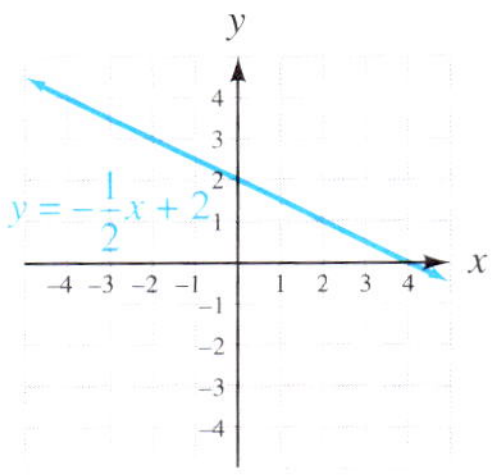

ARE YOU READY? 5.6 (page 394)

1. $50a$ **2.** $10x - 15$ **3.** $28y + 56$ **4.** x^{14} **5.** -30 **6.** $9x^2 + 3x + 2$

SELF CHECKS 5.6

1. a. $18t^2$ **b.** $60d^{11}$ **c.** $4y^{14}$ **d.** $-30a^7b^8$ **2. a.** $220x^6 + 22x^5$ **b.** $20c^5 - 45c^3 - 40c^2$ **c.** $s^6t^4 - s^5t^5 + s^4t^6 - 7s^3t^2$ **d.** $6w^{12} - 12w^6$ **3.** $(3n^5 + 2n^4 - 9n^3)\text{ m}^2$ **4.** $9y^2 - 33y - 12$ **5. a.** $y^2 + 4y + 3$ **b.** $6a^2 + a - 2$ **c.** $16x^2 + x - \frac{3}{8}$ **d.** $10y^6 - 39by^3 + 14b^2$ **6.** $6a^6 - 5a^4 - 3a^3 + a^2 + a$ **7. a.** $6x^3 - 8x^2 + 7x + 10$ **b.** $-4x^4 + 8x^3 + 8x^2 - 12x - 3$ **8.** $-6y^3 - 14y^2 + 12y$

STUDY SET SECTION 5.6 (page 400)

1. monomials, binomials **3.** first, outer, inner, last **5. a.** each, each **b.** any, third **7. a.** $6x^2 + x - 12$ **b.** $5x^4 + 8ax^2 + 3a^2$ **9.** $8, n^3, 72n^5$ **11.** $2x, 5, 5, 4x, 15x, 11x$ **13.** $5m^2$ **15.** $12x^5$ **17.** $6c^6$ **19.** $-24b^6$ **21.** $8x^5y^5$ **23.** $-2a^{11}$ **25.** $3x^2 + 12x$ **27.** $-4t^3 + 28t$ **29.** $-6x^5 + 2x^4 - 2x^3$ **31.** $\frac{5}{8}t^8 + 5t^4$ **33.** $-12x^4z - 4x^2z^3 - 4x^3z^2 + 4x^2z$ **35.** $6x^{14} - 72x^{13}$ **37.** $(7h^2 + 3h)\text{ in.}^2$ **39.** $(4w^2 - 2w)\text{ ft}^2$ **41.** $y^2 + 8y + 15$ **43.** $m^2 - 3m - 54$ **45.** $4y^2 + 23y - 35$ **47.** $12x^2 - 28x + 15$ **49.** $7.6y^2 - 5.8y + 1$ **51.** $18m^2 - 10m + \frac{8}{9}$ **53.** $t^4 - 7t^2 + 12$ **55.** $12a^2 - 5ab - 2b^2$ **57.** $x^3 - x + 6$ **59.** $4t^3 + 11t^2 + 18t + 9$ **61.** $2x^3 + 7x^2 - 16x - 35$ **63.** $r^4 - 5r^3 + 2r^2 - 7r - 15$ **65.** $x^3 - 3x + 2$ **67.** $12x^3 + 17x^2 - 6x - 8$ **69.** $8x^3 - 12x^2 - 8x$ **71.** $-3a^3 + 3ab^2$ **73.** $18a^6 - 12a^5$ **75.** $x^3 - 6x^2 + 5x + 12$ **77.** $30x^2 - 17x + 2$ **79.** $6x^4 + 8x^3 - 14x^2$ **81.** $2t^2 + 2t - 24$ **83.** $6a^4 + 5a^3 + 5a^2 + 10a + 4$ **85.** $9t^2 + 15st - 6s^2$ **87.** $2a^{10}$ **89.** $16a^2 - 2ar - \frac{15}{16}r^2$ **91.** $a^2 + 2ab + b^2$ **93.** $x^4 + 11x^3 + 26x^2 - 28x - 24$ **95.** $9x^4 - 18x^3 + 54x^2$ **97.** $4y^3 + 40y^2 + 84y$ **99.** $0.12p^9 - 1.8p^7$ **101.** $16.4p^2q^2 - 24.6p^2q + 41pq^2$ **103.** $-3x^3 + 25x^2y - 56xy^2 + 16y^3$ **105. a.** $x^2 + 3x + 2$ **b.** $x^3 - 8$ **107. a.** Does not simplify **b.** $-18x^3z^8$ **109. a.** $-x^2 + 2x$ **b.** $6x^4 - 9x^3 + 3x^2$ **111. a.** $13a + 1$ **b.** $72a^3 + 6a^2 - 6a$ **113.** $(6x^2 + x - 1)\text{ cm}^2$ **115.** $(0.785x^2 - 0.785)\text{ in.}^2$ **117.** $(2x^3 - 4x^2 - 6x)\text{ in.}^3$ **125. a.** 1 **b.** Undefined **c.** $-\frac{2}{3}$ **d.** 0

ARE YOU READY? 5.7 (page 403)

1. $x^2 + 4x + 4$ **2. a.** Base: $x + 4$; exponent: 2 **b.** Base: $x - 1$; exponent: 3 **3.** $a^2 + 10a + 25$ **4.** $a^2 - 25$

SELF CHECKS 5.7

1. a. $r^2 + 12r + 36$ **b.** $49g^2 - 28g + 4$ **c.** $v^2 + 1.6v + 0.64$ **d.** $w^8 - 3w^4y + \frac{9}{4}y^2$ **2. a.** $b^2 - 16$ **b.** $25m^2 - 81$ **c.** $s^2 - \frac{9}{16}$ **d.** $c^6 - 4d^2$ **3.** $n^3 - 9n^2 + 27n - 27$ **4. a.** $7a^2 + 4a$ **b.** $11x^2 - 3x - 24$ **c.** $-3a^2 + 16a - 97$ **5.** $(72a^2 - 2)\text{ ft}^2$

STUDY SET SECTION 5.7 (page 409)

1. products **3. a.** square, Twice, first **b.** second, square **5.** $x, 4, 4, 8x$ **7.** $s, 5, 25$ **9.** $x^2 + 2x + 1$ **11.** $m^2 - 12m + 36$ **13.** $16x^2 + 40x + 25$ **15.** $49m^2 - 28m + 4$ **17.** $1 - 6y + 9y^2$ **19.** $y^2 + 1.8y + 0.81$ **21.** $a^4 + 2a^2b^2 + b^4$ **23.** $s^2 + \frac{3}{2}s + \frac{9}{16}$ **25.** $x^2 - 9$ **27.** $4p^2 - 49$ **29.** $9n^2 - 1$ **31.** $c^2 - \frac{9}{16}$

33. $0.16 - 81m^4$ **35.** $25 - 36g^2$ **37.** $x^3 + 12x^2 + 48x + 64$ **39.** $n^3 - 18n^2 + 108n - 216$ **41.** $8g^3 - 36g^2 + 54g - 27$ **43.** $a^3 + 3a^2b + 3ab^2 + b^3$ **45.** $-x^2 + 20x - 8$ **47.** $4x^2 - 5x - 11$ **49.** $-80d^3 + 40d^2 - 5d$ **51.** $4d^5 - 4dg^6$ **53.** $(2x^2 - 2)$ yd² **55.** $(9x^2 + 6x + 1)$ ft² **57.** $4v^6 - 32v^3 + 64$ **59.** $12x^3 + 36x^2 + 27x$ **61.** $16f^2 - 0.16$ **63.** $r^4 + 20r^2s + 100s^2$ **65.** $6x - 2$ **67.** $d^8 + \frac{1}{2}d^4 + \frac{1}{16}$ **69.** $d^2 - 49$ **71.** $4a^2 - 12ab + 9b^2$ **73.** $n^2 - 36$ **75.** $36m + 36$ **77.** $8m^3 + 12m^2n + 6mn^2 + n^3$ **79.** $25m^2 - 12m + \frac{36}{25}$ **81.** $r^4 - 2r^2s^2 + s^4$ **83.** $x^2 - 4x + 4$ **85.** $r^2 + 4r + 4$ **87.** $n^4 - 8n^3 + 24n^2 - 32n + 16$ **89.** $17y^2 + 2y - 60$ **91.** $13x^2 - 8x + 5$ **93.** $f^2 - 16f + 64$ **95.** $36b^2 - \frac{1}{4}$ **97.** $4y^2 + 6y - 1$ **99.** $36 - 24d^3 + 4d^6$ **101.** $8e^3 + 12e^2 + 6e + 1$ **103.** $64x^2 + 48x + 9$ **105. a.** x^2y^2 **b.** $x^2 + 2xy + y^2$ **107. a.** $4b^4d^2$ **b.** $4b^4 + 4b^2d + d^2$ **109.** $(x^2 + 12x + 36)$ in.² **111.** $\pi hR^2 - \pi hr^2$ **117.** $\frac{5}{6}$ **119.** $\frac{21}{40}$

ARE YOU READY? 5.8 (page 411)

1. a. $\frac{2}{5}$ **b.** a^2 **2.** $\frac{a+b}{d}$ **3.** 36 **4.** $2x^2$

SELF CHECKS 5.8

1. a. $6y^2$ **b.** $\frac{d^4}{4c^3}$ **2. a.** $10h + 1$ **b.** $2s^3 - \frac{s^2t}{11} + 4t^2$ **3.** $x + 4$ **4.** $4x - 3 + \frac{6}{2x+3}$ **5.** $3x^2 + 2x - 4$ **6.** $x + 3$ **7.** $(9x + 3)$ in.

STUDY SET SECTION 5.8 (page 417)

1. monomial **3.** binomial **5.** Divide, multiply, subtract, bring down **7.** quotient, dividend **9.** $7x^2, x^3, 7x^2, 5, 2, 7, 2, 2, 4x^3, 5, 7$ **11.** $5x^4 + 0x^3 + 2x^2 + 0x - 1$ **13.** x^3 **15.** $\frac{4h^2}{3}$ **17.** $-\frac{1}{5d^4}$ **19.** $\frac{10}{s}$ **21.** $\frac{x^2}{5y^4}$ **23.** $\frac{4r}{y^5}$ **25.** $2x + 1$ **27.** $\frac{1}{a^3} - \frac{1}{a} + 1$ **29.** $\frac{h^2}{4} + \frac{2}{h}$ **31.** $3s^5 - 6s^2 + 4s$ **33.** $c^3 + 3c^2 - 2c - \frac{5}{c}$ **35.** $5y - \frac{6}{x} + \frac{1}{xy}$ **37.** $x + 6$ **39.** $x - 2$ **41.** $x + 1 + \frac{-1}{2x+3}$ **43.** $2x - 3 + \frac{-1}{3x-1}$ **45.** $2x - 1$ **47.** $2x + 1$ **49.** $a - 5$ **51.** $x + 1$ **53.** $2x - 3$ **55.** $9b + 7$ **57.** $y + 12 + \frac{1}{y+1}$ **59.** $3a^5 - \frac{2b^3}{a}$ **61.** $2x^2 + 2x + 1$ **63.** $2x^5 - 8x^2$ **65.** $3a - 2$ **67.** $5m^5$ **69.** $b + 3$ **71.** $x + 3$ **73.** $x^2 - x + 1$ **75.** $-\frac{13}{3rs^3}$ **77.** $-2w^2 - \frac{1}{w^4}$ **79.** $9 - \frac{6}{m}$ **81.** $y^2 + 2y + 5 + \frac{10}{y-2}$ **83.** $\frac{x}{5} - \frac{2}{5x^2}$ **85.** $x^2 - 2x + 1$ **87.** $x + 1 + \frac{10}{x+5}$ **89.** $3x^2y - 2x - \frac{1}{y}$ **91.** $a - 12 + \frac{4}{a-5}$ **93.** $a^2 + a + 1$ **95.** $2x^2 + x + 1 + \frac{2}{3x-1}$ **97.** $\frac{x^2}{2y^{10}}$ **99.** $3m - 8$ **101. a.** $4x - 4 - \frac{5}{4x}$ **b.** $4x - 5$ **103.** $(x - 6)$ in. **105.** $(2x^2 - x + 3)$ in. **111.** $y = -\frac{11}{6}x - \frac{7}{3}$

CHAPTER 5 REVIEW (page 420)

1. a. Base n, exponent 12 **b.** Base $2x$, exponent 6 **c.** Base r, exponent 4 **d.** Base $y - 7$, exponent 3 **2. a.** m^5 **b.** $-3x^4$ **c.** $(x + 8)^2$ **d.** $\left(\frac{1}{2}pq\right)^3$ **3.** 7^{12} **4.** m^2n^2 **5.** y^{21} **6.** $81x^4$ **7.** b^9 **8.** $-b^{12}$ **9.** $256s^{10}$ **10.** $4.41x^4y^2$ **11.** $(-9)^{15}$ **12.** a^{23} **13.** $\frac{1}{8}x^{15}$ **14.** $\frac{x^{12}}{9y^2}$ **15.** $(m - 25)^{12}$ **16.** $125yz^4$ **17.** a^{11} **18.** c^5d^5 **19.** $64x^{12}$ in.³ **20.** y^4 ft² **21.** 1 **22.** 1 **23.** 3 **24.** $\frac{1}{1{,}000}$ **25.** $-\frac{1}{25}$ **26.** $\frac{1}{t^6}$ **27.** $8x^5$ **28.** $-\frac{6}{y}$ **29.** $\frac{8}{49}$ **30.** x^{14} **31.** $-\frac{27}{r^9}$ **32.** $\frac{1}{16z^2}$ **33.** $\frac{8c}{9d^5}$ **34.** t^{30} **35.** w^{22} **36.** $\frac{f^{40}}{4^{10}}$ **37.** 7.2×10^8 **38.** 9.37×10^{15} **39.** 9.42×10^{-9} **40.** 1.3×10^{-4} **41.** 1.8×10^{-4} **42.** 8.53×10^5 **43.** 126,000 **44.** 0.00000003919 **45.** 2.68 **46.** 57.6 **47.** 3.0×10^{-4}; 0.0003 **48.** 1.6×10^8; 160,000,000 **49.** 6,570,000,000; 6.57×10^9 **50.** $1.0 \times 10^5 = 100{,}000$ **51. a.** 4 **b.** $3x^3$ **c.** 3, −1, 1, 10 **d.** 10 **52. a.** 7th, monomial **b.** 3rd, monomial **c.** 2nd, binomial **d.** 5th, trinomial **e.** 6th, binomial **f.** 4th, none of these **53.** 3, −13 **54.** 8 in.

55.

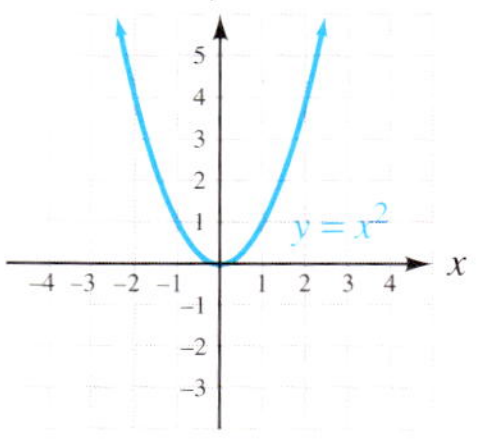

x	−2	−1	0	1	2
y	4	1	0	1	4

56.

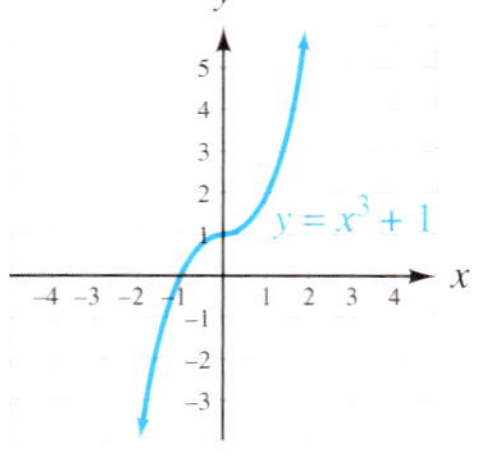

x	−2	−1	0	1	2
y	−7	0	1	2	9

57. $13y^3$ **58.** $-4a^3b + a^2b + 6$ **59.** $\frac{7}{12}x^2 - \frac{3}{4}xy + y^2$ **60.** $6.7c^5 + 8.1c^4 - 2.1c^3$ **61.** $25r^6 + 9r^3 + 5r$ **62.** $3.7a^2 + 6.1a - 17.6$ **63.** $4r^3s - 7r^2s^2 - 7rs^3 - 2s^4$ **64.** $\frac{5}{8}m^4 - m^3$ **65.** $-z^3 + 2z^2 + 5z - 17$ **66.** $(x^2 + x + 3)$ in. **67.** $4x^2 + 2x + 8$ **68.** $8x^3 - 7x^2 + 19x$ **69.** $10x^3$ **70.** $-6x^{10}z^5$ **71.** $120b^{11}$ **72.** $2h^{14} + 8h^{11}$ **73.** $9n^4 - 15n^3 + 6n^2$ **74.** $x^2y^3 - x^3y^2$ **75.** $6x^6 + 12x^5$ **76.** $a^6b^4 - a^5b^5 + a^3b^6 - 7a^3b^2$ **77.** $x^2 + 5x + 6$ **78.** $2x^2 - x - 1$ **79.** $6t^2 - 6$ **80.** $6n^8 - 13n^6 + 5n^4$ **81.** $-5a^9 + 4a^7b + a^5b^2$ **82.** $6.6a^2 - 6.6$ **83.** $18t^2 + 3t - \frac{5}{9}$ **84.** $24b^2 - 34b + 11$ **85.** $8a^3 - 27$ **86.** $56x^4 + 15x^3 - 21x^2 - 3x + 2$ **87.** $8x^3 + 1$ **88. a.** $(6x + 10)$ in. **b.** $(2x^2 + 11x - 6)$ in.² **c.** $(6x^3 + 33x^2 - 18x)$ in.³ **89.** $a^2 - 6a + 9$ **90.** $m^3 + 6m^2 + 12m + 8$ **91.** $x^2 - 49$ **92.** $4x^2 - 0.81$ **93.** $4y^2 + 4y + 1$ **94.** $y^4 - 1$ **95.** $36r^4 + 120r^2s + 100s^2$ **96.** $-64a^2 + 48ac - 9c^2$ **97.** $80r^4s - 80s^5$ **98.** $36b^3 - 96b^2 + 64b$ **99.** $t^2 - \frac{3}{2}t + \frac{9}{16}$ **100.** $x^2 + \frac{8}{3}x + \frac{16}{9}$ **101.** $5x^2 + 19x + 3$ **102.** $24c^2 - 10c + 37$ **103.** $(x^2 - 4)$ in.² **104.** $(50x^2 - 8)$ in.² **105.** $2n^3$ **106.** $-\frac{2x}{3y^2}$ **107.** $\frac{a^3}{6} - \frac{4}{a^4}$ **108.** $3a^3 + \frac{b}{5a} - \frac{5}{a^2}$ **109.** $x - 5$ **110.** $2x + 1$ **111.** $5x - 6 + \frac{4}{3x+2}$ **112.** $5y - 3$ **113.** $3x^2 - x - 4$ **114.** $3x^2 + 2x + 1 + \frac{2}{2x-1}$ **115.** $(y + 3)(3y + 2) = 3y^2 + 11y + 6$ **116.** $(2x^2 + 3x - 4)$ in.

CHAPTER 5 TEST (page 427)

1. a. base, exponent **b.** monomial, binomial, trinomial **c.** degree **d.** special **2.** $2x^3y^4$ **3.** y^6 **4.** $\frac{1}{32}x^{21}$ **5.** 3.5 **6.** $\frac{2}{y^3}$ **7.** $\frac{1}{125}$ **8.** $(x + 1)^9$ **9.** y^{21} **10.** $\frac{b^3}{64a^3}$ **11.** $\frac{m^{12}}{64}$ **12.** $-6ab^9$ **13.** $1{,}000y^{12}$ in.³ **14.** 6.25×10^{18} **15.** 0.000093 **16.** 9.2×10^3; 9,200 **17.** Trinomial

Term	Coefficient	Degree
x^4	1	4
$8x^2$	8	2
−12	−12	0

Degree of the polynomial: 4

18. 5th degree **19.**

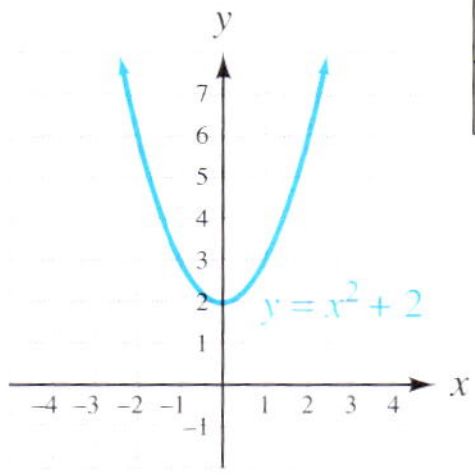

x	−2	−1	0	1	2
y	6	3	2	3	6

20. 0 ft; the rock hits the canyon floor 18 seconds after being dropped.
21. $\frac{1}{10}x^2 + \frac{7}{12}x - 2$ **22.** $-4a^3b + a^2b + 5$
23. $19.4h^3 - 11.1h^2 - 0.6$ **24.** $6b^3c - 2bc - 12$
25. $-3y^3 + 18y^2 - 17y + 35$ **26.** $(10a^2 + 8a - 20)$ in.
27. $10x^5y^{11}$ **28.** $-72b^8$ **29.** $3y^4 - 6y^3 + 9y^2$
30. $0.24p^{11} - 0.54p^8$ **31.** $\frac{3}{4}s^7t^{17} + 12s^4t^{10}$
32. $3x^2 - 11x - 20$ **33.** $12t^2 - 8t - \frac{3}{4}$
34. $7.6m^2 - 5.8m + 1$ **35.** $a^6 + a^3 - 42$
36. $2x^3 - 7x^2 + 14x - 12$ **37.** $1 - 100c^2$
38. $49b^6 - 42b^3t + 9t^2$ **39.** $2.2a^3 + 4.4a^2 - 33a$
40. $2x^2 + 2xy$ **41.** $\frac{a}{4b} - \frac{b}{2a}$ **42.** $x - 2$
43. $3x^2 + 2x + 1 + \frac{2}{2x - 1}$ **44.** $(x - 5)$ ft
45. Yes; $(5m + 1)(m - 6) = 5m^2 - 29m - 6$
46. No; $(a + b)^2 = a^2 + 2ab + b^2$

CUMULATIVE REVIEW CHAPTERS 1–5 (page 429)

1. $2 \cdot 3^3 \cdot 5$ **2. a.** $a + b = b + a$ **b.** $(xy)z = x(yz)$ **3.** −37
4. 28 **5.** $18x$ **6.** 0 **7.** −2 **8.** 15 **9.** $2.21 billion
10. 1.2 ft³ **11.** 30° **12.** $6,250 **13.** Mutual fund: $25,000; bonds: $20,000 **14.** $\left(-\infty, -\frac{11}{4}\right)$

15.

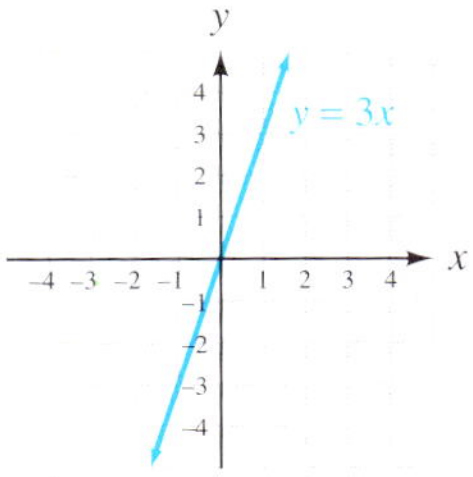

16.

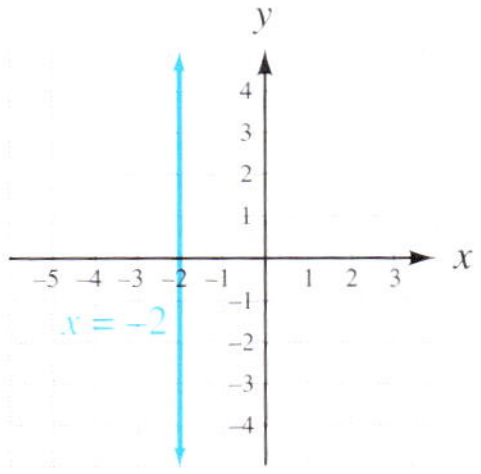

17. $-\frac{4}{9}$ **18.** 0.75 million people per year or $\frac{3}{4}$ million people per year **19.** $m = 3$, $(0, -2)$; $y = 3x - 2$ **20.** Perpendicular
21. $y = -4x + 2$ **22.** No **23.** 26 **24.** Not a function; (1, 2), (1, −2); answers may vary. **25.** No
26. (4, 1)

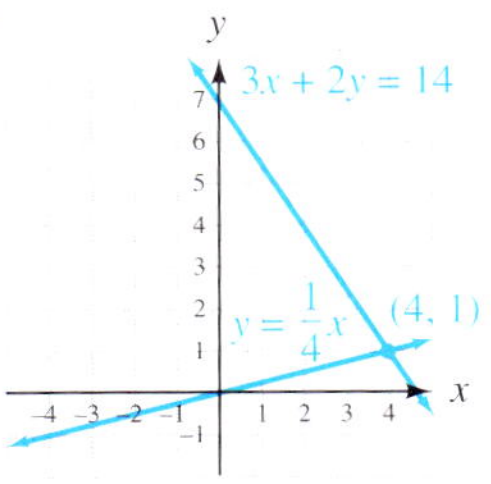

27. (−4, 3) **28.** (−2, 4)
29. Adult: $61; child: $51

30.

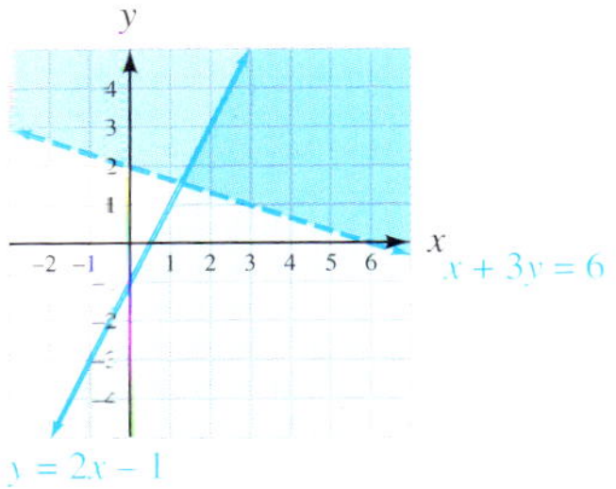

31. $9x^4y^8$ **32.** y^{22}
33. $a^2b^7c^6$ **34.** $\frac{64t^{12}}{27}$
35. $\frac{1}{16y^4}$ **36.** a^7 **37.** $-\frac{1}{25}$
38. $\frac{x^{10}}{a^{10}}$ **39.** 6.15×10^5
40. 1.3×10^{-6}

41.

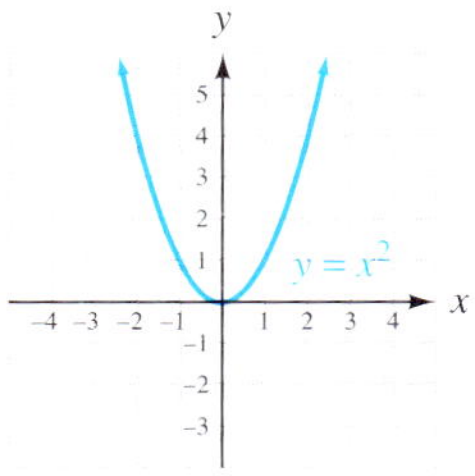

42. 1.5 in. **43.** $7c^2 + 7c$
44. $-6x^4 - 17x^2 - 68x + 11$
45. $6t^2 + 7st - 3s^2$
46. $12x^3 + 36x^2 + 27x$
47. $2x + 1$ **48.** $\frac{1}{8} - \frac{2}{x}$

ARE YOU READY? 6.1 (page 432)

1. $2 \cdot 3 \cdot 3 \cdot 3 = 2 \cdot 3^3$ **2.** a^6 **3.** $-x + 7$ **4.** 4
5. $4d^4 + 8d^2$ **6.** $x^3 + x^2 - 3x - 3$

SELF CHECKS 6.1

1. a. 2 **b.** 1 **c.** 15 **2. a.** $11c$ **b.** $21s^2t^2$ **3. a.** $6(f + 6)$ **b.** $6s^2t(4t - 7s)$ **c.** $y^3(y^3 - 10y - 1)$ **4. a.** $-(b^4 + 3b^2 - 2)$ **b.** $-(t - 9)$ **5.** $-11(4c - 5)$ **6.** $(y - 1)(2y + 7)$
7. a. $(3n + 2)(n^2 + 3)$ **b.** $(x - y)(7 + y)$
8. a. $(a + 11)(a^4 + 1)$ **b.** $(b - c)(b - 1)$ **9.** $(y + 3)(y^2 - 2)$
10. $4(t + s)(1 + z)$

STUDY SET SECTION 6.1 (page 440)

1. factor **3.** grouping **5. a.** 3 **b.** 7, h **c.** 3, y, y
7. a. $2x + 4$ (Answers may vary) **b.** $x^3 + x^2 + x$ (Answers may vary) **9.** $2x$, $x + 3$ **11. a.** 4 **b.** No **c.** 2; h **13.** 8
15. b^2, 2, $b - 6$ **17.** 2 **19.** 6 **21.** 7 **23.** 8 **25.** m^3
27. 5 **29.** $4c$ **31.** $9a^3$ **33.** $8a$ **35.** $3m^3n$ **37.** $x + 7$
39. $p - t$ **41.** $3(x + 2)$ **43.** $9(2m - 1)$ **45.** $d(d - 7)$
47. $5(3c^3 + 5)$ **49.** $8a(3 - 2a)$ **51.** $7(2x^2 - x - 1)$
53. $t^2(t^2 + t + 2)$ **55.** $3xy^2(7xy + 1)$ **57.** $-(a + b)$
59. $-(x^2 + x - 16)$ **61.** $-(-5 + x)$ or $-(x - 5)$
63. $-(-9 + 4a)$ or $-(4a - 9)$ **65.** $-3x(x + 2)$
67. $-4a^2(b - 3a)$ **69.** $-12x^2(2x^2 + 4x - 3)$
71. $-2ab^2(2a^2 - 7a + 5)$ **73.** $(x + 2)(y + 3)$
75. $(p - q)(m - 5)$ **77.** $(x + y)(2 + a)$ **79.** $(s - u)(r + 8w)$
81. $(7m - 2)(m^2 + 2)$ **83.** $(5x - 1)(x^2 + 2)$ **85.** $(b + c)(a + 1)$
87. $(r + 4s)(s - 1)$ **89.** $(2x - 3)(a + b)$ **91.** $(m - n)(p - q)$
93. $(m + 1)(5m^2 + 6)$ **95.** $(y^2 + 3)(y - 4)$
97. $a(x - 2)(x^2 + 5)$ **99.** $6(x^2 + 2)(x - 1)$
101. $(14 + r)(h^2 + 5)$ **103.** $11a^2(2a - 3)$ **105.** $(a + b)(x - 1)$
107. $3r^5(5r^3 - 6r - 10)$ **109.** $3(3p + q)(3m - n)$
111. $-20pt^2(3p + 4t)$ **113.** $-(2x - 5)$ **115.** $(3x - y)(2x - 5)$
117. $2z(x - 2)(x^2 + 16)$ **119.** $6uvw^2(2w - 9v)$
121. $(x + 1)(x^2 + 1)$ **123.** $-(3r - 2s + 3)$
125. a. $(5t + 6)(t^2 + 3)$ **b.** $3(t^3 + 2t^2 + 5t + 6)$
127. $(x^2 + 5)$ ft; $(x + 4)$ ft **129. a.** $V = \pi r^2\left(h_1 + \frac{1}{3}h_2\right)$
b. A 40 in.³ block of wax is needed. **135.** 12%

ARE YOU READY? 6.2 (page 443)

1. 1 **2.** $x^2 + 7x - 8$ **3.** 2 **4.** 2 **5.** 2 and 4 **6.** −3 and 5

SELF CHECKS 6.2

1. $(y + 2)(y + 5)$ **2.** $(p - 2)(p - 4)$ **3.** $(m + 7)(m - 6)$
4. $(q + 4)(q - 6)$ **5.** $-(x - 4)(x - 7)$ **6.** $(s + 7t)(s - t)$
7. $4m^3(m + 4)(m - 2)$ **8.** $t(t - 2)(t + 6)$ **9.** Prime trinomial
10. $(m + 7)(m - 6)$ **11.** $(q + 4t)(q - 6t)$
12. $3m(m - 8)(m - 1)$

STUDY SET SECTION 6.2 (page 452)

1. factors **3.** leading **5. a.** descending **b.** common **7.** 3, 5 **9. a.** No **b.** Yes **11. a.** They are both positive or both negative. **b.** One will be positive, the other negative. **13.** +3, −2 **15.** $(x + 2)(x + 1)$ **17.** $(z + 4)(z + 3)$ **19.** $(m - 3)(m - 2)$ **21.** $(t - 7)(t - 4)$ **23.** $(x + 8)(x - 3)$ **25.** $(t - 3)(t + 16)$ **27.** $(a - 8)(a + 2)$ **29.** $(b - 12)(b + 3)$ **31.** $-(x + 5)(x + 2)$ **33.** $-(t + 6)(t - 5)$ **35.** $-(r + 9)(r - 6)$ **37.** $-(m - 7)(m - 11)$ **39.** $(a + 3b)(a + b)$ **41.** $(x - 7y)(x + y)$ **43.** $(r + 2s)(r - s)$ **45.** $(a - 3b)(a - 2b)$ **47.** $2(x + 3)(x + 2)$ **49.** $6(a - 4)(a - 1)$ **51.** $5(a - 3)(a - 2)$ **53.** $-z(z - 4)(z - 25)$ **55.** $(x - 4)(x - 20)$ **57.** $(y + 9)(y + 1)$ **59.** $r(r - 2)(r + 8)$ **61.** $r(r + 3x)(r + x)$ **63.** Prime **65.** Prime **67.** $(x + 3)(5 + y)$ **69.** $2n(13n - 4)$ **71.** $(a - 5)(a + 1)$ **73.** $-(x - 22)(x + 1)$ **75.** $4(y - 1)(x + 7)$ **77.** $12b^2(b^2 - 4b - 3)$ **79.** $(r - 3)(r - 6)$ **81.** $-n^2(n - 30)(n + 2)$ **83.** $(x + 2y)(x + 2y) = (x + 2y)^2$ **85.** $(a - 6b)(a + 2b)$ **87.** $4x^2(x + 2)(x + 2) = 4x^2(x + 2)^2$ **89.** $(a - 45)(a - 1)$ **91.** Prime **93.** $(x + 2)(t + 7)$ **95.** $s^2(s + 13)(s - 2)$ **97.** $15(s^3 + 5)$ **99.** $(y - 14)(y + 1)$ **101.** $2(x - 2)(x - 4)$ **103. a.** $(x - 4)(x - 6)$ **b.** $(x - 12)(x + 2)$ **105.** $(x + 9)$ in., x in., $(x + 3)$ in. **113.** $\frac{1}{x^2}$ **115.** $\frac{1}{x^{10}}$

ARE YOU READY? 6.3 (page 454)

1. 3 **2.** $5y^2$ **3.** $6x^2 - 13x - 5$ **4.** $1(-3)$ and $-1(3)$ **5.** 6, −3, 7 **6.** $(2b + 3)(5b - 1)$

SELF CHECKS 6.3

1. $(2x + 1)(x + 2)$ **2.** $(6b - 1)(b - 3)$ **3.** $(5t + 2)(t - 5)$ **4.** $(2x + 3y)(2x - y)$ **5.** $-2y(7y + 3)(y - 2)$ **6.** $(3a + 4)(5a - 1)$ **7.** $a^2(7a - 2)(3a - 1)$

STUDY SET SECTION 6.3 (page 462)

1. leading **3.** $5y, y, 1, 3$ **5.** $10x$ and x, $5x$ and $2x$ **7. a.** descending, GCF, coefficient **b.** $3s^2$ **c.** $-(2d^2 - 19d + 8)$ **9.** negative **11.** different **13.** −13; −6, −8; −3, −7 **15. a.** 12, 20, −9 **b.** −108 **17.** $3t, 2, 4t + 3$ **19.** $(2x + 1)(x + 1)$ **21.** $(3a + 1)(a + 3)$ **23.** $(5x + 2)(x + 1)$ **25.** $(7x + 11)(x + 1)$ **27.** $(2x - 3)(2x - 1)$ **29.** $(4x - 1)(2x - 5)$ **31.** $(5t - 7)(3t - 1)$ **33.** $(6y - 1)(y - 2)$ **35.** $(3x + 7)(x - 3)$ **37.** $(5m + 3)(m - 2)$ **39.** $(7y - 1)(y + 8)$ **41.** $(11y - 4)(y + 1)$ **43.** $(3r + 2s)(2r - s)$ **45.** $(2x + 3y)(2x + y)$ **47.** $(8m + 3n)(m + 11n)$ **49.** $(5x + 3y)(3x - 2y)$ **51.** $2(3x + 2)(x - 5)$ **53.** $a(2a - 5)(4a - 3)$ **55.** $(2u + 3v)(u - 2v)$ **57.** $4(9y - 4)(y - 2)$ **59.** $(2t - 5)(3t + 4)$ **61.** $(3p - q)(5p + q)$ **63.** $(2t - 1)(2t - 7)$ **65.** $10(13r - 11)(r + 1)$ **67.** $(4y + 1)(2y - 1)$ **69.** $(18x - 5)(x + 2)$ **71.** $-y(y + 12)(y + 1)$ **73.** Prime **75.** $-3x^2(2x + 1)(x - 3)$ **77.** $(3p - q)(2p + q)$ **79.** $3r^3(5r - 2)(2r + 5)$ **81.** $2mn(4m + 3n)(2m + n)$ **83.** Prime **85.** $-(4y - 3)(3y - 4)$ **87.** $(m + 7)(m - 4)$ **89.** $3a^2(2a + 5)$ **91.** $(x - 2)(x^2 + 5)$ **93.** $(5y - 3)(y - 1)$ **95.** $-2(x + 2)(x + 3)$ **97.** $3x^2y^2(4xy - 6y + 5)$ **99.** $(a - 5b)(a - 2b)$ **101.** $u^4(9u + 1)(u - 8)$ **103.** $(2x + 11)$ in., $(2x - 1)$ in. **109.** −49 **111.** 1 **113.** 49

ARE YOU READY? 6.4 (page 464)

1. $9y^2 + 12y + 4$ **2.** $m^2 - 81$ **3.** $64d^2$ **4.** $x^2 + 25$

SELF CHECKS 6.4

1. a. Yes **b.** No **c.** No **d.** No **2. a.** $(x + 9)^2$ **b.** $(4x - y)^2$ **3.** $x(7x - 1)^2$ **4. a.** $(c + 2)(c - 2)$ **b.** $(11 + t)(11 - t)$ **c.** Prime **d.** Prime **5. a.** $(4y + 3)(4y - 3)$ **b.** $(3m + 8n^2)(3m - 8n^2)$ **c.** $(10 + a^2)(10 - a^2)$ or $-1(a^2 + 10)(a^2 - 10)$ **6.** $2(p + 10)(p - 10)$ **7.** $(a^2 + 9)(a + 3)(a - 3)$

STUDY SET SECTION 6.4 (page 470)

1. perfect **3. a.** $5x$ **b.** 3 **c.** $5x, 3$ **5. a.** x, y **b.** − **c.** $+, x, y$ **7.** 1, 4, 9, 16, 25, 36, 49, 64, 81, 100, 121, 144, 169, 196, 225, 256, 289, 324, 361, 400 **9.** 2 **11.** +, − **13.** Yes **15.** No **17.** No **19.** Yes **21.** $(x + 3)^2$ **23.** $(b + 1)^2$ **25.** $(c - 6)^2$ **27.** $(2x + 3)^2$ **29.** $(6m + 5n)^2$ **31.** $(9x - 4y)^2$ **33.** $3(u - 3)^2$ **35.** $x(6x + 1)^2$ **37.** $(x + 2)(x - 2)$ **39.** $(x + 4)(x - 4)$ **41.** $(6 + y)(6 - y)$ **43.** $(t + 5)(t - 5)$ **45.** Prime **47.** Prime **49.** $(5t + 8)(5t - 8)$ **51.** $(9y + 1)(9y - 1)$ **53.** $(3x^2 + y)(3x^2 - y)$ **55.** $(4c + 7d^2)(4c - 7d^2)$ or $-(7d^2 + 4c)(7d^2 - 4c)$ **57.** $8(x + 2y)(x - 2y)$ **59.** $7(3a + 1)(3a - 1)$ **61.** $(9 + s^2)(3 + s)(3 - s)$ **63.** $(b^2 + 16)(b + 4)(b - 4)$ **65.** $(a^2 + 12b)(a^2 - 12b)$ **67.** $(3xy + 5)^2$ **69.** $16(t^2 + s^2)(t + s)(t - s)$ **71.** $(t - 10)^2$ **73.** $(3y - 4)^2$ **75.** $(z + 8)(z - 8)$ **77.** $25(m^2 + 1)(m + 1)(m - 1)$ **79.** $2a^3(3a + 7b)^2$ **81.** $x(x + 12)(x - 12)$ **83.** $(7t - 2s)^2$ **85.** $3(m^2 + n^2)(m + n)(m - n)$ **87.** $(5m + 7)^2$ **89.** $-(10t - 1)^2$ **91.** $6x^2(x + y)(x - y)$ **93.** Prime **95.** $(5x + 13)(5x - 13)$ or $-(13 + 5x)(13 - 5x)$ **97.** $(x + 7)(x - 6)$ **99.** $(x + 3)(x - 3)$ **101.** $8a^2b(3a - 2)$ **103.** $-2(r - 10)(r - 4)$ **105.** $(x + 3)(x^2 + 4)$ **107.** $(2b - 5)^2$ **109.** $(p + q)^2$ **111.** $0.5g(t_1 + t_2)(t_1 - t_2)$ **117.** $3cd + \frac{3c}{2} + d$

ARE YOU READY? 6.5 (page 472)

1. $x^3 + 64$ **2.** $8h^3 - 1$ **3. a.** 27 **b.** 125 **4.** There are no two integers whose product is 4 and whose sum is −2.

SELF CHECKS 6.5

1. $(h + 3)(h^2 - 3h + 9)$ **2.** $(2c - 1)(4c^2 + 2c + 1)$ **3.** $4(c + d)(c^2 - cd + d^2)$

STUDY SET SECTION 6.5 (page 475)

1. sum, cubes **3. a.** F, L **b.** $-, F^2, L^2$ **5.** $6n, 5$ **7.** 1, 8, 27, 64, 125, 216, 343, 512, 729, 1,000 **9.** No **11.** $2a$ **13.** $b + 3$ **15. a.** $x^3 + 8$ (Answers may vary.) **b.** $(x + 8)^3$ **17.** $(y + 5)(y^2 - 5y + 25)$ **19.** $(a + 4)(a^2 - 4a + 16)$ **21.** $(n + 8)(n^2 - 8n + 64)$ **23.** $(2 + t)(4 - 2t + t^2)$ **25.** $(a + 10b)(a^2 - 10ab + 100b^2)$ **27.** $(5c + 3d)(25c^2 - 15cd + 9d^2)$ **29.** $(a - 3)(a^2 + 3a + 9)$ **31.** $(m - 7)(m^2 + 7m + 49)$ **33.** $(6 - v)(36 + 6v + v^2)$ **35.** $(2s - t)(4s^2 + 2st + t^2)$ **37.** $(10a - w)(100a^2 + 10aw + w^2)$ **39.** $(4x - 3y)(16x^2 + 12xy + 9y^2)$ **41.** $2(x + 1)(x^2 - x + 1)$ **43.** $3(d + 3)(d^2 - 3d + 9)$ **45.** $x(x - 6)(x^2 + 6x + 36)$ **47.** $8x(2m - n)(4m^2 + 2mn + n^2)$ **49.** $(x + 4)^2$ **51.** $(3r + 4s)(3r - 4s)$ **53.** $(x - t)(y + s)$ **55.** $4(p + 2q)(p^2 - 2pq + 4q^2)$ **57.** $2ct^2(4c + 3t)(2c + t)$ **59.** $36(e^2 + 1)(e + 1)(e - 1)$ **61.** $7a^2b^2(5a - 2b + 2ab)$ **63.** $(6r + 5s)^2$ **65. a.** $(x + 1)(x - 1)$ **b.** $(x - 1)(x^2 + x + 1)$ **67. a.** $x(x + 2)$ **b.** $(x + 1)^2$ **69.** $(1{,}000 - x^3)$ in.^{3}; $(10 - x)(100 + 10x + x^2)$ **73.** Repeating **75.** −3

ARE YOU READY? 6.6 (page 477)

1. a. 4 **b.** 3 **2.** No **3.** $3n^3 - 27n^2 - 210n$ **4.** $5cd^2$

SELF CHECKS 6.6

1. $11a^2(a^2 + 1)(a + 1)(a - 1)$ **2.** $-2h^2(4h + 5)^2$ **3.** $(b - 1)(b + 2)(b^2 - 2b + 4)$ **4.** $6m(m^2 + m - 9)$ **5.** $3y(2y - 1)(y + 4)$

STUDY SET SECTION 6.6 (page 481)

1. product **3.** Factor out the GCF **5.** Perfect-square trinomial **7.** Sum of two cubes **9.** Trinomial factoring **11.** Is there a common factor? **13.** $14m, m$ **15.** $2(b + 6)(b - 2)$ **17.** $4p^2q^3(2pq^4 + 1)$ **19.** $2(2y + 1)(10y + 1)$ **21.** $8(x^2 + 1)(x + 1)(x - 1)$ **23.** $(c + 21)(c - 7)$ **25.** Prime **27.** $-2x^2(x - 4)(x^2 + 4x + 16)$ **29.** $(c + d^2)(a^2 + b)$ **31.** $-(3x - 1)^2$ **33.** $-5m(2m + 5)^2$ **35.** $(2c + d)(c - 3d)$ **37.** $(p - 2)^2(p^2 + 2p + 4)$ **39.** $(x - a)(a + b)(a - b)$ **41.** $(ab + 12)(ab - 12)$ **43.** $(x + 5)(2x^2 + 1)$ **45.** $v^2(v^2 - 14v + 8)$ **47.** $2(3a - b)(3a + 7c)$ **49.** $2x(2ax + b)(2ax - b)$ **51.** $2(3x - 4)(x - 1)$ **53.** $y^2(2x + 1)^2$ **55.** $4m^2(m + 5)(m^2 - 5m + 25)$ **57.** $(a - 6)(a + 2)(a - 2)$ **59.** Prime **61.** $2a^2(2a - 3)(4a^2 + 6a + 9)$ **63.** $27(x - y - z)$ **65.** $(x - t)(y + s)$ **67.** $x^6(7x + 1)(5x - 1)$ **69.** $5(x - 2)(1 + 2y)$ **71.** $(7p + 2q)^2$ **73.** $4(t^2 + 9)$ **75.** $(n + 3)(n - 3)(m^2 + 3)$

81.

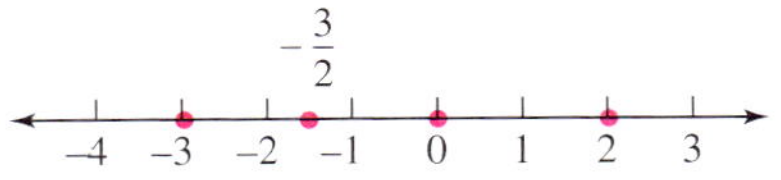

83.

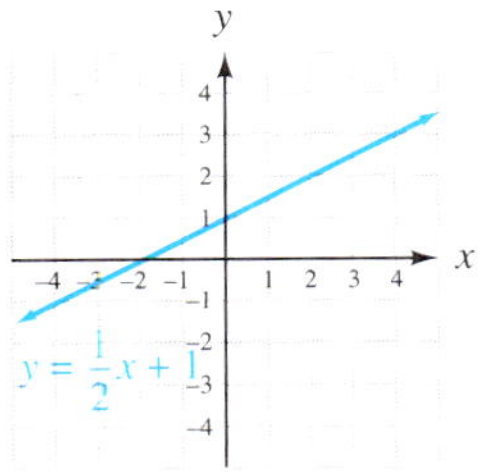

ARE YOU READY? 6.7 (page 482)

1. 0 **2.** 0 **3.** −4 **4.** 0 **5.** $(x - 3)(x + 2)$ **6.** $(3n + 2)(n - 1)$

SELF CHECKS 6.7

1. $12, -\frac{6}{5}$ **2.** $-2, -3$ **3.** $-7, 7$ **4.** $0, 5$ **5.** $\frac{2}{3}, -4$ **6.** $-\frac{3}{2}$ repeated **7.** $0, \frac{2}{5}, -\frac{1}{2}$

STUDY SET SECTION 6.7 (page 487)

1. quadratic **3.** zero-factor, 0, 0 **5. a.** Yes **b.** No **c.** Yes **d.** No **7.** $-\frac{4}{5}$ **9. a.** Add 6 to both sides. **b.** Distribute the multiplication by x and subtract 3 from both sides. **11.** $0, x + 7, -7$ **13.** $p, p - 3, 0, 3, -2$ **15.** $3, 2$ **17.** $-7, 7$ **19.** $0, \frac{5}{2}$ **21.** $0, -\frac{10}{3}$ **23.** $0, 6, -8$ **25.** $1, -2, 3$ **27.** $12, 1$ **29.** $-3, 7$ **31.** $8, 1$ **33.** $-3, -5$ **35.** $-9, 9$ **37.** $-5, 5$ **39.** $-\frac{1}{2}, \frac{1}{2}$ **41.** $-\frac{7}{3}, \frac{7}{3}$ **43.** $0, 7$ **45.** $0, 16$ **47.** $0, 3$ **49.** $0, -\frac{8}{3}$ **51.** $-2, \frac{1}{3}$ **53.** $-\frac{3}{2}, 1$ **55.** $\frac{1}{5}, 1$ **57.** $-\frac{5}{2}, 4$ **59.** $-\frac{7}{2}$ repeated **61.** $\frac{5}{3}$ repeated **63.** $0, -1, -2$ **65.** $0, 9, -3$ **67.** $-\frac{9}{2}, \frac{9}{2}$ **69.** 8 repeated **71.** $\frac{5}{2}, -6$ **73.** $2, 10$ **75.** $0, -5, 4$ **77.** $-10, 10$ **79.** $3, 4$ **81.** $-\frac{1}{3}, 5$ **83.** $2, 7, 1$ **85.** $-3, -2$ **87.** 0, 3 repeated **89.** $0, 2$ **91.** $-\frac{2}{3}, -\frac{3}{2}$ **93.** $0, -1, -\frac{1}{3}$ **95.** $0, -1, 2$ **97.** $\frac{2}{3}, -\frac{1}{5}$ **99.** $-\frac{11}{2}, \frac{11}{2}$ **101.** $\frac{1}{8}, 1$ **103. a.** $(x + 7)(x - 3)$ **b.** $-7, 3$ **105. a.** $(4n + 1)(3n - 2)$ **b.** $-\frac{1}{4}, \frac{2}{3}$ **113.** $15 \text{ min} \le t < 30 \text{ min}$

ARE YOU READY? 6.8 (page 490)

1. $A = lw$ **2.** $A = \frac{1}{2}bh$ **3.** $x + 1$ **4.** 3

SELF CHECKS 6.8

1. Width: 5 m; length: 11 m **2.** Base: 3 yd; height: 10 yd **3.** 23 and 24 **4.** 8 in., 15 in., and 17 in. **5.** 5 sec

STUDY SET SECTION 6.8 (page 495)

1. consecutive **3.** hypotenuse, legs **5.** ii. **7.** $20 = b(b + 5)$ **9. a.** A right triangle **b.** x ft; $(x + 1)$ ft **c.** 9 ft **11.** $-16, 1, 0, 0, 3, -1$ **13.** Width: 3 ft; length: 6 ft **15.** 8 in., 10 in. **17.** 3 ft by 9 ft **19.** Base: 6 cm; height: 5 cm **21.** Foot: 4 ft; luff: 12 feet **23.** Kahne: 9; Riggs: 10 **25.** 12 **27.** (11, 13) **29.** 10 yd **31.** 8 ft **33.** 5 m, 12 m, 13 m **35.** 5 sec **37.** 4 sec **39.** 1 sec **41.** 8 **47.** $25b^2 - 20b + 4$ **49.** $s^4 + 8s^2 + 16$ **51.** $81x^2 - 36$

CHAPTER 6 REVIEW (page 499)

1. $5 \cdot 7$ **2.** $2^5 \cdot 3$ **3.** 7 **4.** $18a^3$ **5.** $3(x + 3y)$ **6.** $5a(x^2 + 3)$ **7.** $7s^3(s^2 + 2)$ **8.** $\pi a(b - c)$ **9.** $12x(2x^2 + 5x - 4)$ **10.** $xy^3z^2(x^4 + y^2z - 1)$ **11.** $-5ab(b - 2a + 3)$ **12.** $(x - 2)(4 - x)$ **13.** $-(a + 7)$ **14.** $-(4t^2 - 3t + 1)$ **15.** $(c + d)(2 + a)$ **16.** $(y + 6)(3x - 5)$ **17.** $(a + 1)(2a^2 - 1)$ **18.** $4m(n + 3)(m - 2)$ **19.** 1

20.

Factors of 6	Sum of the factors of 6
1(6)	7
2(3)	5
−1(−6)	−7
−2(−3)	−5

21. $(x + 6)(x - 4)$ **22.** $(x - 20)(x + 2)$ **23.** $(x - 5)(x - 9)$ **24.** Prime **25.** $-(y - 8)(y - 7)$ **26.** $(y + 9)(y + 1)$ **27.** $(c + 5d)(c - 2d)$ **28.** $(m - 2n)(m - n)$ **29.** Multiply **30.** There are no two integers whose product is 11 and whose sum is 7. **31.** $5a^3(a + 10)(a - 1)$ **32.** $-4x(x + 3y)(x - 2y)$ **33.** $(2x + 1)(x - 3)$ **34.** $(7y + 5)(5y - 2)$ **35.** $-(3x + 5)(x - 6)$ **36.** $3p(6p + 1)(p - 2)$ **37.** $(4b - c)(b - 4c)$ **38.** Prime **39.** $(4x + 1)$ in., $(3x - 1)$ in. **40.** The signs of the second terms must be negative. **41.** $(x + 5)^2$ **42.** $(3y - 4)^2$ **43.** $-(z - 1)^2$ **44.** $(5a + 2b)^2$ **45.** $(x + 3)(x - 3)$ **46.** $(7t + 11y)(7t - 11y)$ **47.** $(xy + 20)(xy - 20)$ **48.** $8a(t + 2)(t - 2)$ **49.** $(c^2 + 16)(c + 4)(c - 4)$ **50.** Prime **51.** $(b + 1)(b^2 - b + 1)$ **52.** $(x - 6)(x^2 + 6x + 36)$ **53.** $(p + 5q)(p^2 - 5pq + 25q^2)$ **54.** $2x^2(2x - 3y)(4x^2 + 6xy + 9y^2)$ **55.** $2y^2(3y - 5)(y + 4)$ **56.** $5(t + u^2)(s^2 + v)$ **57.** $(j^2 + 4)(j + 2)(j - 2)$ **58.** $-3(j + 2)(j^2 - 2j + 4)$ **59.** $(x + 1)(20 + m)(20 - m)$ **60.** $3w^2(2w - 3)^2$ **61.** $2(t^3 + 5)$ **62.** Prime **63.** $z(x + 8y)^2$ **64.** $6c^2d(3cd - 2c - 4)$ **65.** $0, 6$ **66.** $\frac{7}{4}, -1$ **67.** $0, -2$ **68.** $-3, 3$ **69.** $-\frac{5}{12}, \frac{5}{12}$ **70.** $3, 4$ **71.** -7 repeated **72.** $6, -4$ **73.** $1, \frac{1}{5}$ **74.** $0, -1, 2$ **75.** Width: 9 in.; length: 11 in. **76.** 15 m **77.** $x + 1$; $x + 2$; $x + 2$

78. Jackson: 12 nominations; West: 10 nominations **79.** 5 m **80.** 10 sec

CHAPTER 6 TEST (page 505)

1. a. greatest, common, factor **b.** product **c.** Pythagorean **d.** difference **e.** binomials **2. a.** $45 = 3^2 \cdot 5$; $30 = 2 \cdot 3 \cdot 5$ **b.** $15x^3$ **3.** $4(x + 4)$ **4.** $(q + 9)(q - 9)$ **5.** $5ab(6ab^2 - 4a^2b + 1)$ **6.** Prime **7.** $(x + 1)(2x + 3)$ **8.** $(x + 3)(x + 1)$ **9.** $-(x - 11)(x + 2)$ **10.** $x^2(x - 30)(x - 2)$ **11.** $(a - b)(9 + x)$ **12.** $(2a - 3)(a + 4)$ **13.** $2(3x + 5y)^2$ **14.** $(x + 2)(x^2 - 2x + 4)$ **15.** $15m^6(4m^2 - 3)$ **16.** $3(a - 3)(a^2 + 3a + 9)$ **17.** $(4x^2 + 9)(2x + 3)(2x - 3)$ **18.** $(a + 5)(a^2 + 1)$ **19.** $(a + 6)(a^3 - 4)$ **20.** $(5d + 4)(2d - 1)$ **21.** $8(m + 10)(m - 10)$ **22.** $(6n - 7)^2$ **23.** $(4r - 1)(2r - 3)$ **24.** Prime **25.** $(5x - 4)$ in. **26.** $(x - 9)(x + 6)$; $x^2 + 6x - 9x - 54 = x^2 - 3x - 54$ **27.** $-3, 2$ **28.** $-5, 5$ **29.** $0, \frac{1}{6}$ **30.** -3 repeated **31.** $\frac{1}{3}, -\frac{1}{2}$ **32.** $9, -2$ **33.** $0, -1, -6$ **34.** 6 ft by 9 ft **35.** 5 sec **36.** Base: 6 in.; height: 11 in. **37.** 12, 13 **38.** 10 **39.** A quadratic equation is an equation that can be written in the form $ax^2 + bx + c = 0$; $x^2 - 2x + 1 = 0$. (Answers may vary.) **40.** At least one of them is 0.

CUMULATIVE REVIEW CHAPTERS 1–6 (page 508)

1. About 35 beats/min difference **2.** $2 \cdot 5^3$ **3.** $\frac{24}{25}$ **4.** 0.992 **5. a.** False **b.** True **c.** True **6.** $\frac{5}{0}$ **7.** -39 **8.** -5 **9.** -27 **10.** $\$20x$ **11.** 3 **12.** $8, -1, 9$ **13.** $-13y^2 + 6$ **14.** $3y + z$ **15.** $-\frac{3}{2}$ **16.** -2 **17.** 9 **18.** $-\frac{55}{6}$ **19.** 248 lb **20.** 330 mi **21.** $I = Prt$ **22.** 12.6 in.^{2} **23.** $t = \frac{A - P}{Pr}$ **24.** 22nd president, 24th president **25.** Los Angeles: 6 wk; Las Vegas: 4 wk; Dallas: 7 wk **26.** 4 L **27.** 5 lb apple slices, 5 lb banana chips **28.** $(-\infty, -2)$

29. Yes

30.

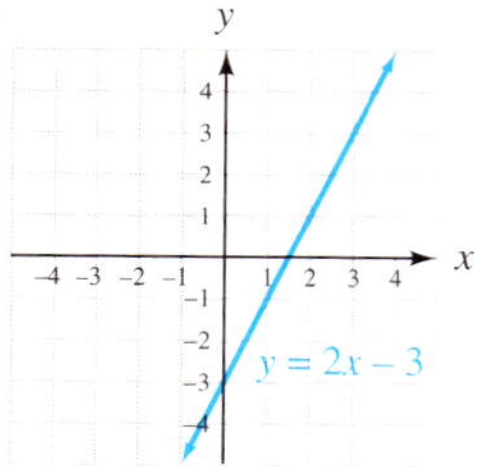

31. Vertical line **32.** They are the same. **33.** An increase of 536,000 articles per year **34.** 1; $(0, -2)$ **35.** $y = 7x + 19$

36.

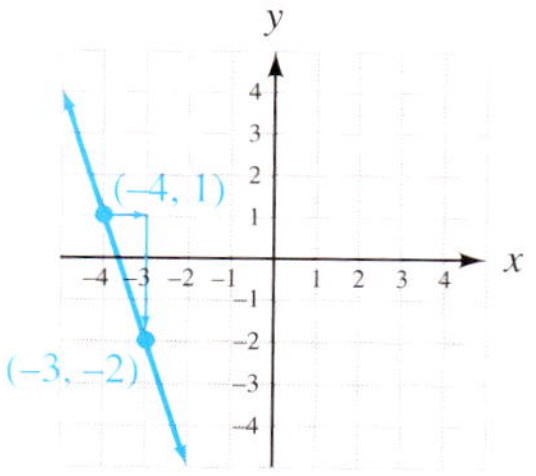

37.

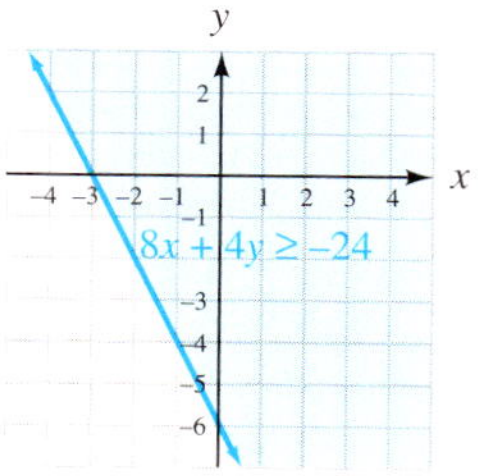

38. 17 **39.** Yes **40.**

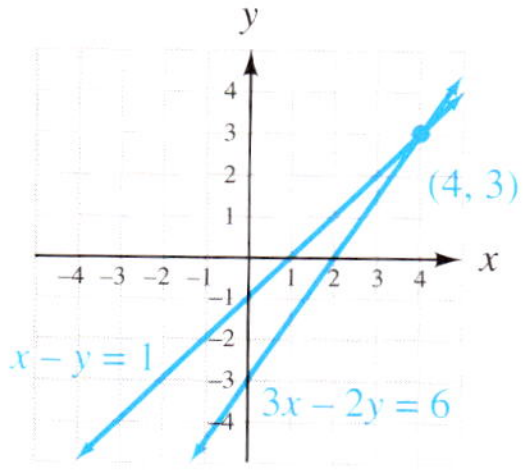

41. $\left(\frac{3}{4}, -2\right)$ **42.** $\left(-2, \frac{2}{3}\right)$ **43.** Newspaper: 8 tons; cardboard: 6 tons **44.**

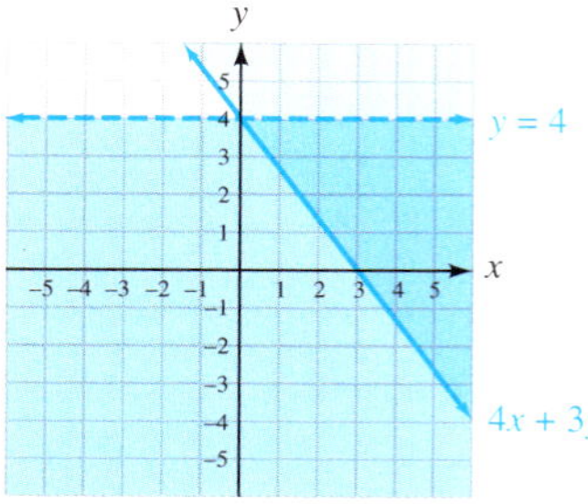

45. $-4y^5$

46. x^4y^{23} **47.** b^7 **48.** 2 **49.** 9.011×10^{-5} **50.** 1.7×10^6 **51.** 3 **52.**

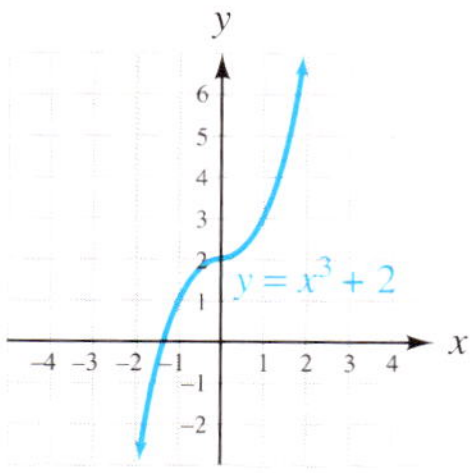

53. $-2x^2 - 4x + 5$ **54.** $8b^5 - 8b^4$ **55.** $3x^2 + 10x - 8$ **56.** $y^2 - 12y + 36$ **57.** $3ab - 2a - 1$ **58.** $2x + 1$ **59. a.** $(4x + 8)$ in. **b.** $(x^2 + 4x + 3)$ in.^{2} **c.** $(x^3 + 4x^2 + 3x)$ in.^{3} **60.** $6x^5y$ **61.** $9b^2(b - 3)$ **62.** $(x + y)(a + b)$ **63.** $(u + 3)(u - 1)$ **64.** $(2x + 1)(5x - 2)$ **65.** $(2a - 3)^2$ **66.** $(3z + 1)(3z - 1)$ **67.** $(t - 2)(t^2 + 2t + 4)$ **68.** $3(b^2 - 2)(a + 1)(a - 1)$ **69.** $0, \frac{4}{3}$ **70.** $\frac{1}{2}, 2$ **71.** $0, -1, -2$ **72.** 9 in. by 12 in.

ARE YOU READY? 7.1 (page 512)

1. a. 0 **b.** Undefined **2. a.** $\frac{3}{4}$ **b.** $\frac{1}{9}$ **3.** $4(3x - 2)$ **4.** $(a + 4)(a - 4)$ **5.** $(5x + 2)(x - 5)$ **6.** -1

SELF CHECKS 7.1

1. $\frac{13}{50}$, -1 **2. a.** -9 **b.** $-5, 5$ **c.** None **d.** None **3.** $\frac{4a^2}{3}$ **4. a.** $\frac{t - 5}{3}$ **b.** $\frac{x + 2}{x}$ **c.** $2x^3$ **5. a.** $\frac{2x + 3}{2x(2x + 5)}$ **b.** $(a + 3b)^3$ **6.** $\frac{4}{3}$ **7.** -1 **8. a.** $-\frac{m + 10}{m}$ **b.** Does not simplify

STUDY SET SECTION 7.1 (page 519)

1. rational **3.** undefined **5.** $\frac{6}{24} = \frac{1}{4}$ **7. a.** 1 **b.** -1 **c.** 1 **d.** Does not simplify **9.** $x, 1, \cancel{x + 1}, x + 3$ **11.** 4 **13.** 0 **15.** Undefined **17.** $-\frac{2}{11}$ **19.** $\frac{1}{6}$ **21.** Undefined **23.** 2 **25.** 0 **27.** None **29.** $\frac{1}{2}$ **31.** None **33.** $-6, 6$ **35.** $-2, 1$ **37.** 20 **39.** $\frac{5}{a}$ **41.** $\frac{3x^2}{2}$ **43.** $\frac{2x + 1}{3}$ **45.** $\frac{1}{3}$ **47.** $\frac{x + 2}{x - 4}$ **49.** $\frac{x + 1}{x}$ **51.** $\frac{m - n}{7(m + n)}$ or $\frac{m - n}{7m + 7n}$ **53.** $\frac{1}{2b + 1}$ **55.** $\frac{10}{3}$ **57.** $\frac{2x}{x - 6}$ **59.** -1 **61.** $-\frac{1}{2}$ **63.** $-\frac{1}{a + 1}$ **65.** $-\frac{5}{m + 5}$ **67.** $\frac{1}{a}$ **69.** $-\frac{x + 2}{x + 1}$ **71.** -6 **73.** $\frac{x + 1}{x - 1}$ **75.** $\frac{3x}{y}$ **77.** x^7 **79.** $\frac{4 - x}{4 + x}$ or $-\frac{x - 4}{x + 4}$ **81.** 4 **83.** $\frac{3(x + 3)}{2x + 1}$ or $\frac{3x + 9}{2x + 1}$ **85.** $-\frac{3x + 11}{x + 3}$ **87.** $\frac{7c^2}{3d^2}$ **89.** Does not simplify **91.** $\frac{2u - 3}{u^3}$ **93.** $(2x + 3)^2$ **95.** 9 **97.** $\frac{3x}{5y}$ **99.** Does not simplify **101.** $\frac{2}{x - 9}$ **103.** $85\frac{1}{3}$ **105.** 2, 1.6, and 1.2 milligrams per liter **113. a.** $(a + b) + c = a + (b + c)$ **b.** $ab = ba$

ARE YOU READY? 7.2 (page 522)

1. $\frac{3}{20}$ **2.** $\frac{9}{8}$ **3.** $\frac{11}{14}$ **4.** $x(1 - x)$

SELF CHECKS 7.2

1. a. $\frac{6(a+7)}{5a^4}$ **b.** $\frac{3a}{11}$ **2. a.** $\frac{(n-3)(3n-2)}{2}$ **b.** $-\frac{2(m+1)}{3m+1}$
3. a. 6 **b.** $4x + 12$ **4.** $\frac{3b}{2a}$ **5.** $\frac{z+1}{z}$ **6.** $-a$ **7.** 600 yd^2
8. 36,000 beats per minute

STUDY SET SECTION 7.2 (page 528)

1. reciprocal **3. a.** numerators, denominators, reciprocal **b.** AC, BD, D, C **5.** $-\frac{y^2}{y+1}$ **7.** 1 **9.** ft **11.** $\frac{3y}{14}$ **13.** $\frac{3(y+2)}{y^3}$ or $\frac{3y+6}{y^3}$ **15.** $\frac{20}{3n}$ **17.** x^2y^2 **19.** $\frac{x}{5}$ **21.** $\frac{3}{2x}$ **23.** $x + 1$ **25.** $-(x-2)$ or $-x+2$ **27.** $\frac{(x-2)^2}{x}$ **29.** $\frac{(m-2)(m-3)}{2(m+2)}$ **31.** 35 **33.** $3x + 3$ **35.** $10y - 16$ **37.** $\frac{36a-60}{a}$ **39.** $\frac{3}{2y}$ **41.** $\frac{3a}{5}$ **43.** $\frac{x^2}{3}$ **45.** $\frac{9p^3}{5}$ **47.** $\frac{5(a-2)}{4a^3}$ **49.** $-\frac{x+2}{3}$ **51.** $\frac{3x}{x+7}$ **53.** $-(m+5)$ or $-m-5$ **55.** $t + 7$ **57.** $\frac{2(x-7)}{x+9}$ **59.** $\frac{1}{3}$ **61.** 1 **63.** $\frac{1}{12(2r-3s)}$ **65.** $\frac{4(n-1)}{3n}$ **67.** 450 ft **69.** $\frac{3}{4}$ gal **71.** $\frac{1}{2}$ mi per min **73.** 1,800 m per min **75.** $\frac{b-3}{b}$ **77.** $\frac{1}{(x+1)^2}$ **79.** $\frac{3(a+3)^2}{2a^3}$ **81.** $\frac{d(6c-7d)}{6}$ **83.** $25h - 15$ **85.** $n - 1$ **87.** $\frac{5r^3}{2s^2}$ **89.** $\frac{7(p+2)}{3p^4}$ or $\frac{7p+14}{3p^4}$ **91.** $\frac{x-2}{x-3}$ **93.** $\frac{2x-3y}{y(2x+3y)}$ **95.** -2 **97. a.** $(x+2)^2$ **b.** $\frac{9}{16}$ **99. a.** $\frac{(x-3)^2}{x-2}$ **b.** $\frac{x-2}{4}$ **101.** $\frac{x^2}{10}$ ft^2 **103.** 4,380,000 **105.** 8 yd^2 **107.** $\frac{1}{2}$ mi per min **109.** $\frac{1}{4}$ mi^2 **115.** $w = 6$ in., $l = 10$ in.

ARE YOU READY? 7.3 (page 531)

1. $\frac{8}{11}$ **2.** $\frac{1}{3}$ **3.** $x^2 + 5x + 1$ **4.** $\frac{1}{x-7}$ **5.** $2 \cdot 2 \cdot 2 \cdot 3 \cdot x \cdot x$ **6.** $5(x+1)$

SELF CHECKS 7.3

1. a. $\frac{2x}{5}$ **b.** $\frac{3m-6}{23n}$ or $\frac{3(m-2)}{23n}$ **2. a.** $\frac{2}{3}$ **b.** $\frac{2c^2-11c}{(c-1)(c+2)}$ or $\frac{c(2c-11)}{(c-1)(c+2)}$ **3.** $\frac{1}{n+4}$ **4. a.** $x - 1$ **b.** $\frac{6y^2-y}{(y+3)(y-3)}$ **5. a.** $150y^3$ **b.** $a(a+3)$ **6. a.** $2x(x-6)$ **b.** $(m+3)(m-3)^2$ **7. a.** $\frac{21m}{60m^3}$ **b.** $\frac{2c^2+6c}{(c+1)(c+3)}$ **8.** $\frac{x^2+5x-24}{x(x-4)(x+8)}$

STUDY SET SECTION 7.3 (page 538)

1. denominator **3.** build **5.** numerators, denominator, $A + B, D, A - B, D$ **7.** $\frac{4}{5}$ **9. a.** Twice **b.** Once **11.** $3, 4a + 1, 2, 4a + 1, 2$ **13.** $\frac{11}{x}$ **15.** $\frac{x+5}{18}$ **17.** $\frac{1}{3a^2}$ **19.** $\frac{x+4}{y}$ **21.** $\frac{1}{r-5}$ **23.** 9 **25.** $\frac{x}{25}$ **27.** $\frac{m-6}{6m^2}$ **29.** $\frac{1}{t+2}$ **31.** $\frac{2}{w(w-9)}$ **33.** $\frac{1}{2}$ **35.** $2x - 5$ **37.** $\frac{1}{y}$ **39.** 0 **41.** $6x$ **43.** $30a^3$ **45.** $3a^2b^3$ **47.** $c(c+2)$ **49.** $12(x-1)$ **51.** $12(b+2)$ **53.** $8k(k+2)$ **55.** $(x+1)(x-1)$ **57.** $(x+1)(x+5)(x-5)$ **59.** $(2n+5)(n+4)^2$ **61.** $\frac{50}{10r}$ **63.** $\frac{8xy}{x^2y}$ **65.** $\frac{27b}{12b^2}$ **67.** $\frac{3x^2+3x}{(x+1)^2}$ **69.** $\frac{x^2+9x}{x^2(x+5)}$ **71.** $\frac{t^2+14t+45}{4(t+2)(t+9)}$ **73.** $\frac{4y^2+12y}{4y(y-2)(y-3)}$ **75.** $\frac{36-3h}{3(h+9)(h-9)}$ **77.** $\frac{3}{t-7}$ **79.** $\frac{1}{c+d}$ **81.** $\frac{3}{4}$ **83.** $\frac{5a}{a+2}$ **85.** $\frac{2x}{3-x^2}$ **87.** $\frac{7n+1}{(n+4)(n-2)}$ **89.** $\frac{3}{r}$ **91.** $\frac{5}{9y}$ **93.** $-\frac{1}{3x-1}$ **95.** $3x - 2$ **97. a.** $\frac{t}{2}$ **b.** $\frac{5t^2}{144}$ **c.** $\frac{1}{5}$ **99. a.** $\frac{4}{5}$ **b.** $\frac{m^2+8m+12}{25}$ **c.** $\frac{m+6}{m+2}$ **101.** $\frac{2x+6}{x+2}$ ft **109. a.** $I = Prt$ **b.** $A = \frac{1}{2}bh$ **c.** $P = 2l + 2w$

ARE YOU READY? 7.4 (page 540)

1. $\frac{20a}{16a^2}$ **2.** $2 \cdot 2 \cdot 3 \cdot 3 \cdot x \cdot x \cdot x$ **3.** $\frac{11x}{9}$ **4.** $t - 8$

SELF CHECKS 7.4

1. $\frac{19y}{14}$ **2.** $\frac{20-9z}{84z^2}$ **3.** $\frac{2}{5(x-5)}$ **4.** $\frac{6x+63}{(x+9)(x+8)}$ **5.** $\frac{b+3}{(b+2)(b+5)}$ **6.** $\frac{14y+ny}{n+4}$ **7.** $\frac{n-12}{n-8}$

STUDY SET SECTION 7.4 (page 546)

1. unlike **3. a.** $2 \cdot 2 \cdot 5 \cdot x \cdot x$ **b.** $(x-2)(x+6)$ **5.** $(x+6)(x+3)$ **7.** $\frac{5}{5}$ **9.** $3x, 5, 15x, 15x, 35$ **11.** $\frac{13x}{21}$ **13.** $\frac{67a}{40}$ **15.** $\frac{7-2m}{m^2}$ **17.** $\frac{6-5p}{10p^2}$ **19.** $\frac{4t^2-33}{24t^3}$ **21.** $\frac{3-16c^2}{18c^4}$ **23.** $\frac{a+8}{2(a+2)(a-2)}$ **25.** $\frac{6a+9}{(3a+2)(3a-2)}$ **27.** $\frac{4a+1}{(a+2)^2}$ **29.** $\frac{6-3m}{5m(m-1)}$ **31.** $\frac{17t+42}{(t+3)(t+2)}$ **33.** $\frac{2x^2+11x}{(2x-1)(2x+3)}$ **35.** $\frac{14s+58}{(s+3)(s+7)}$ **37.** $\frac{2m^2+20m-6}{(m-2)(m+5)}$ **39.** $\frac{s^2+8s+4}{(s+4)(s+1)(s+1)}$ **41.** $\frac{2x+13}{(x-8)(x-1)(x+2)}$ **43.** $\frac{1}{a+1}$ **45.** $\frac{1}{(y+3)(y+4)}$ **47.** $\frac{5y+xy}{x-4}$ **49.** $\frac{8+xz}{x}$ **51.** $\frac{2}{a-4}$ **53.** $\frac{c+d}{7c-d}$ **55.** $\frac{1}{(x-1)(x-2)}$ **57.** $\frac{6d-3}{d-9}$ **59.** $\frac{xy-y+10}{x-1}$ **61.** $\frac{2b+1}{(b+1)(b+2)}$ **63.** $\frac{1}{g+2}$ **65.** $\frac{5y}{2}$ **67.** $\frac{35x^2+x+5}{5x(x+5)}$ **69.** $\frac{41}{30x}$ **71.** $\frac{2x^2-1}{x(x+1)}$ **73.** $\frac{y+4}{y-1}$ **75.** $\frac{2n+2}{15}$ **77.** $\frac{y^2+7y+6}{15y^2}$ **79.** $\frac{x+2}{x-2}$ **81.** $\frac{a^2b-3}{a^2}$ **83.** $\frac{10a-14}{3a(a-2)}$ **85.** $\frac{17x+3}{x^2}$ **87.** $\frac{-4x-3}{x+1}$ or $-\frac{4x+3}{x+1}$ **89.** $\frac{11b}{12}$ **91. a.** $\frac{75+8x^2}{30x}$ **b.** $\frac{2}{3}$ **93. a.** $\frac{t^2+4t}{(t-5)(t+5)}$ **b.** $t + 5$ **95.** $\frac{20x+9}{6x^2}$ cm **101.** 8; (0, 2) **103.** 0

ARE YOU READY? 7.5 (page 549)

1. Division **2.** $\frac{3}{2}$ **3.** $2x$ **4.** $8a - 10$

SELF CHECKS 7.5

1. $\frac{5y}{6}$ **2.** $\frac{10x+30}{6x^2-15x}$ **3.** $\frac{b}{a}$ **4.** $\frac{15x-60}{12x^2+20x}$ **5.** $-\frac{2}{n}$ **6.** $\frac{2x+4}{2x+5}$

STUDY SET SECTION 7.5 (page 555)

1. complex, complex **3.** single, division, reciprocal **5. a.** $\frac{x-3}{4}$, yes **b.** $\frac{1}{12} - \frac{x}{6}$; no **7.** $\div$ **9.** $\frac{8}{9}$ **11.** $\frac{5x}{12}$ **13.** $\frac{x^2}{y}$ **15.** $\frac{n^3}{8}$ **17.** $\frac{10}{3a^3}$ **19.** $-\frac{x^2}{14}$ **21.** $\frac{5}{7}$ **23.** $\frac{3y+12}{4y^2-6y}$ **25.** $\frac{2-5y}{6}$ **27.** $\frac{24-c^2}{12}$ **29.** $\frac{5}{4}$ **31.** $\frac{1-3x}{5+2x}$ **33.** $\frac{1+x}{2+x}$ **35.** $\frac{3-x}{x-1}$ **37.** $\frac{x-12}{x+6}$ **39.** $a - 7$ **41.** $\frac{5d^2+16d}{10d+10}$ **43.** $\frac{y}{x-2y}$ **45.** $\frac{1}{x+2}$ **47.** $\frac{1}{x+3}$ **49.** $\frac{q+p}{q}$ **51.** $18x$ **53.** $\frac{2c}{2-c}$ **55.** $\frac{r-1}{r+1}$ **57.** $\frac{b+9}{8a}$ **59.** $\frac{5x}{3}$ **61.** $\frac{32h-1}{96h+6}$ **63.** $\frac{m^2+n^2}{m^2-n^2}$ **65.** $\frac{8}{4c+5c^2}$ **67.** $\frac{t^3}{2}$ **69.** $\frac{s^2-s}{2+2s}$ **71.** $\frac{t+2}{t-3}$ **73.** $\frac{xy}{y+x}$ **75.** $-\frac{10x^3}{3}$ **77.** $\frac{x}{x-2}$

79. $\frac{d-3}{2d}$ **81.** $\frac{b-5ab}{3ab-7a}$ **83.** $2m-1$ **85.** $\frac{7}{6}$ **87.** $\frac{R_1R_2}{R_2+R_1}$
93. 1 **95.** $\frac{25}{16x^{12}}$

ARE YOU READY? 7.6 (page 557)

1. $18x$ **2.** $5x$ **3.** $-7, 8$ **4.** -8

SELF CHECKS 7.6

1. 1 **2.** 22 **3.** $3, -8$ **4.** No solution, 6 is extraneous **5.** $\frac{5}{2}$
6. $S = \frac{RT-10R}{e}$ **7.** $p = \frac{fq}{q-f}$

STUDY SET SECTION 7.6 (page 564)

1. rational **3.** clear **5. a.** Yes **b.** No **7. a.** 3, 0 **b.** 3, 0 **c.** 3, 0 **9. a.** y **b.** $(x+2)(x-2)$ **11. a.** 3 **b.** $3x+18$ **13.** $2a, 2a, 2a, 2a, 2a, 4, 7, 4, 3$ **15.** 1 **17.** 0 **19.** $\frac{3}{5}$ **21.** 9 **23.** $-\frac{4}{3}$ **25.** $-\frac{12}{7}$ **27.** -48 **29.** $\frac{3}{10}$ **31.** 2, 4 **33.** $-5, 2$ **35.** $-4, -5$ **37.** $3, -\frac{5}{3}$ **39.** No solution; 5 is extraneous **41.** No solution; -2 is extraneous **43.** $\frac{5}{2}$ **45.** 0, 3 **47.** $A = \frac{h(b+d)}{2}$ **49.** $r = \frac{E-IR}{I}$ **51.** $x = \frac{5yz}{5y+4z}$ **53.** $r = \frac{st}{s-t}$ **55.** $P = nrt$ **57.** $d = \frac{bc}{a}$ **59.** $a = \frac{b}{b-1}$ **61.** $L^2 = 6dF - 3d^2$ **63.** 6 **65.** 1 **67.** $-1, 6$ **69.** -40 **71.** No solution; -1 is extraneous **73.** 7 **75.** 1 **77.** $-4, 3$ **79.** 1 **81.** 3 **83.** No solution; 2 is extraneous **85.** $\frac{9}{40}$ **87.** 0 **89.** -3 **91.** 1, 2 **93.** 7 **95. a.** $\frac{2a+3}{5}$ **b.** $-\frac{9}{4}$ **97. a.** $\frac{x^2-4x+2}{(x-2)(x-3)}$ **b.** 4 **99.** $R = \frac{HB}{B-H}$ **101.** $r = \frac{r_1r_2}{r_2+r_1}$ **107.** 20

ARE YOU READY? 7.7 (page 566)

1. $t = \frac{d}{r}$ **2.** $\frac{x}{5}$ **3.** $I = Prt$ **4.** 63

SELF CHECKS 7.7

1. 9 **2.** 4 mph **3.** $\frac{3}{4}$ of the job **4.** $3\frac{3}{7}$ hr **5.** 1%

STUDY SET SECTION 7.7 (page 573)

1. motion, investment, work **3.** iii **5. a.** $\frac{1}{45}$ of the job per minute **b.** $\frac{x}{4}$ **7. a.** $t = \frac{d}{r}$ **b.** $P = \frac{I}{rt}$ **9.** $\frac{x}{15}, \frac{x}{8}$ **11.** $6\frac{1}{9}$ days **13.** 4 **15.** 2 **17.** 5 **19.** $\frac{2}{3}$ or $\frac{3}{2}$ **21.** 8 **23.** Garin: 16 mph; Armstrong: 26 mph **25.** 1st: $1\frac{1}{2}$ ft per sec; 2nd: $\frac{1}{2}$ ft per sec **27.** Canada goose: 30 mph; great blue heron: 20 mph **29.** $\frac{300}{255+x}, \frac{210}{255-x}$; 45 mph **31.** $2\frac{6}{11}$ days **33.** No, after the pipes are opened, the swimming is scheduled to take place in 6 hours. It takes 7.2 hr (7 hr 12 min) to fill the pool. **35.** 8 hr **37.** 20 min **39.** $1\frac{4}{5}$ hr = 1.8 hr **41.** Credit union: 4%; bonds: 6% **43.** 7% and 8% **47.** (1, 3) **49.** Yes

ARE YOU READY? 7.8 (page 576)

1. $\frac{7}{9}$ **2. a.** ab **b.** bc **3.** 8.5 **4.** $-2, 9$

SELF CHECKS 7.8

1. a. $\frac{15}{2}$ **b.** $\frac{1}{4}$ **2.** No **3.** 24 **4.** $-5, 6$ **5.** \$187.50 **6.** 30 ft **7.** 3 lb for \$6.89 **8.** 66 ft

STUDY SET SECTION 7.8 (page 583)

1. ratio, rate **3.** extremes, means **5.** unit **7.** equal, ad, bc **9.** 2, 1,000, x **11.** 2.19, 1 **13.** x, 288, 18, 18, 16 **15.** as, to **17.** $\frac{4}{15}$ **19.** $\frac{3}{4}$ **21.** $\frac{5}{4}$ **23.** $\frac{1}{2}$ **25.** $\frac{1}{2}$ **27.** $\frac{25}{22}$ **29.** Yes **31.** No **33.** 4 **35.** 14 **37.** 0 **39.** $-\frac{3}{2}$ **41.** -2 **43.** -27 **45.** $2, -2$ **47.** $6, -1$ **49.** $-1, 16$ **51.** $-\frac{5}{2}, -1$ **53.** 15 **55.** 8 **57.** $-\frac{1}{3}, 2$ **59.** 2 **61.** $-\frac{27}{2}$ **63.** $-10, 10$ **65.** $-4, 3$ **67.** $\frac{26}{9}$ **69. a.** $-\frac{15}{8}$ **b.** $\frac{65}{8}$ **71. a.** $\frac{8}{5}$ **b.** $\frac{2}{3}, 4$ **73.** \$62.50 **75.** 14 breaths **77. a.** 462 **b.** $\frac{11}{12}$; 11:12 **79.** \$309 **81.** 45,000 tweets **83.** 568, 13, 14 **85.** Not exactly, but close **87.** 140 **89.** 522 in.; 43.5 ft **91.** 10 ft **93.** 45 min for \$25 **95.** 150 for \$12.99 **97.** 6-pack for \$1.50 **99.** Four 4-oz cartons **101.** 39 ft **103.** $46\frac{7}{8}$ ft **105.** 8 **111.** 90% **113.** 480

CHAPTER 7 REVIEW (page 587)

1. $4, -4$ **2.** $-\frac{3}{7}$ **3.** $\frac{1}{2x}$ **4.** $\frac{5}{2x}$ **5.** $\frac{x}{x+1}$ **6.** $a-2$ **7.** -1 **8.** $-\frac{1}{x+3}$ **9.** $\frac{x}{x-1}$ **10.** Does not simplify **11.** $\frac{1}{x-y}$ **12.** $\frac{4}{3}$ **13.** x is not a common factor of the numerator and the denominator; x is a term of the numerator. **14.** 150 mg **15.** $\frac{3x}{y}$ **16.** 96 **17.** $\frac{x-1}{x+2}$ **18.** $\frac{2x}{x+1}$ **19.** $\frac{3y}{2}$ **20.** $-x-2$ **21. a.** Yes **b.** No **c.** Yes **d.** Yes **22.** $\frac{1}{3}$ mi per min **23.** $\frac{1}{3d}$ **24.** 1 **25.** $\frac{2x+2}{x-7}$ **26.** $\frac{1}{a-4}$ **27.** $9x$ **28.** $8x^3$ **29.** $m(m-8)$ **30.** $(5x+1)(5x-1)$ **31.** $(a+5)(a-5)$ **32.** $(2t+7)(t+5)^2$ **33.** $\frac{63}{7a}$ **34.** $\frac{2xy+x}{x(x-9)}$ **35.** $\frac{2b+14}{6(b-5)}$ **36.** $\frac{9r^2-36r}{(r+1)(r-4)(r+5)}$ **37.** $\frac{a-7}{7a}$ **38.** $\frac{x^2+x-1}{x(x-1)}$ **39.** $\frac{1}{t+1}$ **40.** $\frac{x^2+4x-4}{2x^2}$ **41.** $\frac{b+6}{b-1}$ **42.** $\frac{6c+8}{c}$ **43.** $\frac{14n+58}{(n+3)(n+7)}$ **44.** $\frac{4t+1}{(t+2)^2}$ **45.** $\frac{1}{(a+3)(a+2)}$ **46.** $\frac{17y-2}{12(y-2)(y+2)}$ **47.** Yes **48.** $\frac{14x+28}{(x+6)(x-1)}$ units, $\frac{12}{(x+6)(x-1)}$ square units **49.** $\frac{n^3}{14}$ **50.** $\frac{r+9}{8s}$ **51.** $\frac{1+y}{1-y}$ **52.** $\frac{21}{3a+10a^2}$ **53.** x^2+3 **54.** $\frac{y-5xy}{3xy-7x}$ **55.** 3 **56.** No solution; 5 is extraneous **57.** 3 **58.** 2, 4 **59.** 0 **60.** $-4, 3$ **61.** $T_1 = \frac{T_2}{1-E}$ **62.** $y = \frac{xz}{z-x}$ **63.** 3 **64.** 5 mph **65.** $\frac{1}{4}$ of the job per hr **66.** $5\frac{5}{6}$ days **67.** 5% **68.** 40 mph **69.** No **70.** Yes **71.** $\frac{9}{2}$ **72.** 0 **73.** 7 **74.** $4, -\frac{3}{2}$ **75.** 255 **76.** 20 ft **77.** 5 ft 6 in. **78.** 250 for \$98

CHAPTER 7 TEST (page 595)

1. a. rational **b.** similar **c.** proportion **d.** build **e.** factors **2.** 10 words **3.** 0 **4.** $-3, 2$ **5.** 3,360,000 or 3,360K bits per minute **6.** 5 is not a common factor of the numerator, and therefore cannot be removed. 5 is a term of the numerator. **7.** $\frac{8x}{9y}$ **8.** -7 **9.** $\frac{x+1}{2x+3}$ **10.** 1 **11.** $3c^2d^3$ **12.** $(n+1)(n+5)(n-5)$ **13.** $\frac{5y^2}{4}$ **14.** $\frac{x+1}{3(x-2)}$ **15.** $-\frac{x^2}{3}$ **16.** $\frac{1}{6}$ **17.** 3 **18.** $\frac{4n-5mn}{10m}$ **19.** $\frac{2x^2+x+1}{x(x+1)}$ **20.** $\frac{2a+7}{a-1}$ **21.** $\frac{c^2-4c+9}{c-4}$ **22.** $\frac{1}{(t+3)(t+2)}$ **23.** $\frac{12}{5m}$ **24.** $\frac{a+2s}{2as^2-3a^2}$ **25.** 11 **26.** No solution; 6 is extraneous **27.** 1 **28.** 1, 2 **29.** $\frac{2}{3}$ **30.** -10; 3 is extraneous **31.** $B = \frac{HR}{R-H}$ **32.** $s = \frac{rt}{r-t}$ **33.** Yes **34.** 1,785

35. 171 ft **36.** 80 sheets for \$3.89 **37.** $3\frac{15}{16}$ hr **38.** 4 mph **39.** 2 **40.** To simplify $\frac{1}{x} + \frac{1}{4}$, we build each fraction to have the LCD of $4x$. To solve $\frac{1}{x} + \frac{1}{4} = \frac{1}{2}$, we multiply both sides by the LCD $4x$ to eliminate the denominators and clear the equation of fractions.

CUMULATIVE REVIEW CHAPTERS 1–7 (page 597)

1. a. False **b.** False **c.** True **d.** True **2.** $<$ **3.** 36 **4.** 77 **5.** $6c + 62$ **6.** -5 **7.** 3 **8.** About 26% **9.** $B = \frac{A - c - r}{2}$ **10.** 104°F **11.** 240 ft³ **12.** 12 lb of the \$6.40 tea and 8 lb of the \$4 tea **13.** 500 mph **14.** $[-1, \infty)$ 

15.
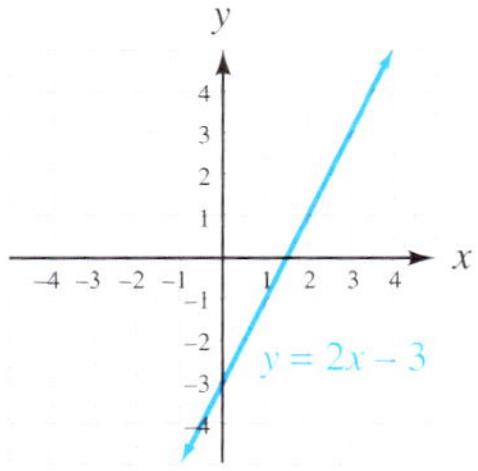

16. -1 **17.** 0.008 mm/m **18.** $\frac{8}{7}$ **19.** $y = 3x + 2$ **20.**
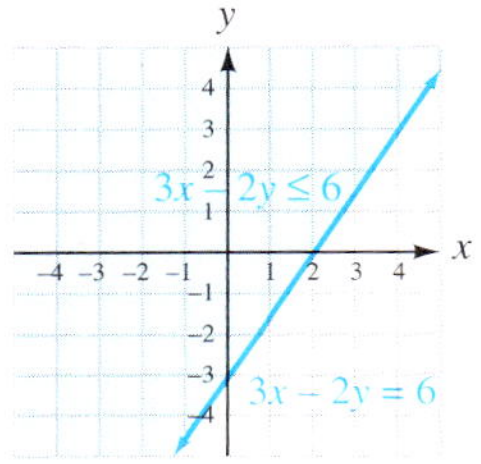

21. 0 **22.** domain, range **23.**
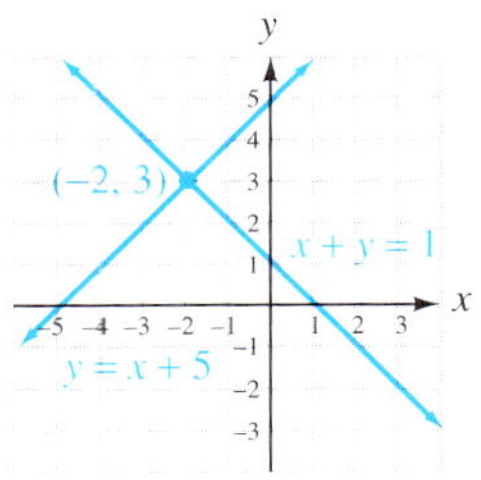

24. (5, 2) **25.** $(1, -1)$ **26.** Red: 8, blue: 15 **27.** x^7 **28.** x^{25} **29.** $\frac{y^3}{8}$ **30.** $-\frac{32a^5}{b^5}$ **31.** $\frac{a^8}{b^{12}}$ **32.** $3b^2$ **33.** 2.9×10^5 **34.** 3 **35.**
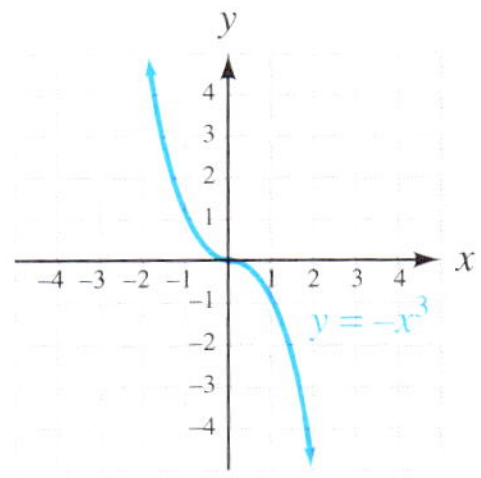

36. $A = \pi R^2 - \pi r^2$ **37.** $6x^2 + x - 5$ **38.** $-\frac{1}{2}t^3 - \frac{7}{4}t^2 - \frac{1}{12}$ **39.** $6x^4y^5$ **40.** $6y^2 - y - 35$ **41.** $-12x^4z + 4x^2z^2$ **42.** $9a^2 - 24a + 16$ **43.** $2x + 3$ **44.** $x^2 + 2x - 1$ **45.** $k^2t(k - 3)$ **46.** $(b + c)(2a + 3)$ **47.** $(u - 9)^2$ **48.** $-(r - 2)(r + 1)$ **49.** Prime **50.** $(2x - 9)(3x + 7)$ **51.** $2(a + 10b)(a - 10b)$ **52.** $(b + 5)(b^2 - 5b + 25)$ **53.** $0, -\frac{1}{5}$ **54.** $\frac{1}{3}, \frac{1}{2}$ **55.** 16 in., 10 in. **56.** $5, -5$ **57.** $\frac{2x}{x - 2}$ **58.** $-\frac{x^3}{3}$ **59.** $2m - 5$ **60.** $-\frac{1}{x - 3}$ **61.** $\frac{m - 3}{(m + 3)(m + 1)}$ **62.** $\frac{x^2}{(x + 1)^2}$ **63.** $\frac{17}{25}$ **64.** 2 **65.** $14\frac{2}{5}$ hr **66.** 34.8 ft

ARE YOU READY? 8.1 (page 600)

1. $-11m + 8$ **2.** $18n$ **3.** 27 **4.** True **5.** 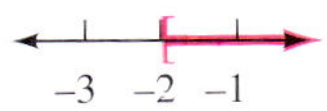

6. $x < 6$

SELF CHECKS 8.1

1. -3 **2.** $-\frac{53}{37}$ **3. a.** All real numbers, ℝ; identity **b.** No solution, ∅; contradiction **4.** $r = \frac{A - P}{Pt}$ **5.** $t = \frac{7S + 360}{180}$ **6.** $(-8, \infty)$ 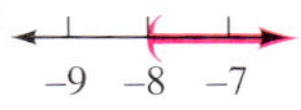

7. $[1, \infty)$ 

8. $(-\infty, \infty)$ 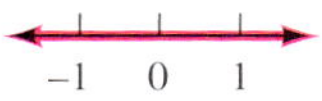

9. 30 people **10.** Be Thrifty's plan is better if the car is going to be driven more than 100 miles that day.

STUDY SET SECTION 8.1 (page 609)

1. equation **3.** solution **5.** identity **7.** inequality **9. a.** same **b.** same **11. a.** No **b.** Yes **13. a.** iii **b.** i **c.** ii **15.** 3 **17.** 4 **19.** -2 **21.** 1 **23.** -2 **25.** 24 **27.** 18 **29.** $-\frac{1}{6}$ **31.** No solution, ∅; contradiction **33.** All real numbers, ℝ; identity **35.** All real numbers, ℝ; identity **37.** No solution, ∅; contradiction **39.** $w = \frac{P - 2l}{2}$ **41.** $B = \frac{3V}{h}$ **43.** $W = T - ma$ **45.** $x = z\sigma + \mu$ **47.** $l = \frac{2S - na}{n}$ or $l = \frac{2S}{n} - a$ **49.** $d = \frac{l - a}{n - 1}$ **51.** $(2, \infty)$ 

53. $(-\infty, 2]$ 

55. $(-\infty, 20]$ 

57. $(0, \infty)$ 

59. $[-2, \infty)$ 

61. $\left[-\frac{2}{5}, \infty\right)$ 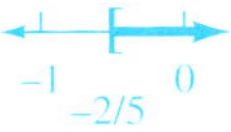

63. $(-\infty, 3)$ 

65. $(-\infty, 7]$ 

67. $(-\infty, \infty)$; ℝ 

69. No solution; ∅ **71.** 2 **73.** -11 **75.** $-\frac{5}{2}$ **77.** $-\frac{1}{2}$ **79.** All real numbers, ℝ; identity **81.** 2 **83.** $\frac{21}{19}$ **85.** No solution, ∅; contradiction **87.** 1,000 **89.** $\frac{9}{13}$ **91.** $\left(-\infty, -\frac{8}{5}\right)$ 

93. $(-\infty, 6]$ 

95. $(-\infty, \infty)$; ℝ 

97. $(-\infty, 1.5]$ 

99. $[-36, \infty)$ 

101. $(6, \infty)$ 

103. a. $-x$ **b.** $\frac{12}{7}$ **105. a.** $(-\infty, 3.25)$ **b.** $(3.25, \infty)$ **107.** 20 **109.** 310 mi **111.** 20 ft by 45 ft **113.** 9,431,000 views **115.** 8 hr **117.** 7 hr **121.** $\frac{1}{t^{12}}$

ARE YOU READY? 8.2 (page 613)

1. (3, 5), (3, 0) **2.** 7 **3.** 3, $(0, -8)$ **4.** $-\frac{3}{2}$

SELF CHECKS 8.2

1. D: $\{-12, -6, 5, 8\}$; R: $\{-6, 4, 6\}$ **2. a.** No; (0, 2), (0, 3) **b.** Yes **c.** No; $(4, -1)$, (4, 4) **d.** Yes **3. a.** Yes **b.** No; (1, 2), $(1, -2)$ **4. a.** -5 **b.** 5 **c.** $2t - 1$ **d.** 0.42 **5.** -27 **6. a.** The set of real numbers **b.** The set of all real numbers except -3 **7. a.** $f(x) = -\frac{1}{5}x + 9$ **b.** $f(x) = 2x - 13$

c. $f(x) = \frac{1}{10}x - 2$ **8. a.** $E(t) = 8t + 398$ **b.** 638 quadrillion Btu **9.** 96 in.^3

STUDY SET SECTION 8.2 (page 624)

1. relation, domain, range **3.** function, variable, dependent **5.** linear, one **7. a.** {(2000, 63), (2001, 56), (2002, 54), (2003, 50), (2004, 52), (2005, 51), (2006, 51)} **b.** D: {2000, 2001, 2002, 2003, 2004, 2005, 2006}; R: {50, 51, 52, 54, 56, 63}

c.

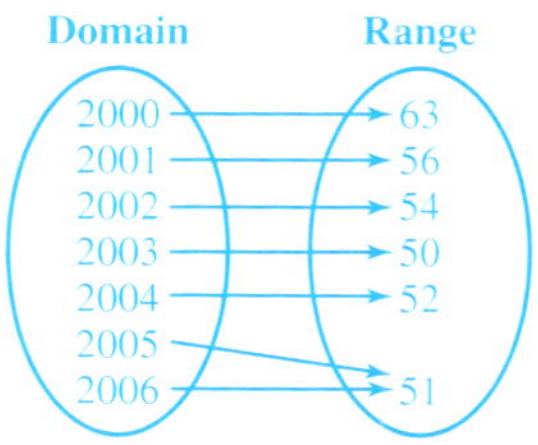

9. If $x = -4$, the denominator of $\frac{1}{x+4}$ is 0 and the fraction is undefined. **11. a.** $(0, -4)$ **b.** $\left(-\frac{2}{3}, 0\right)$ **13. a.** of **b.** of **15.** $f(8)$ **17.** $f(x), y$ **19.** D: {−5, −1, 7, 8}; R: {−11, −6, −1, 3} **21.** D: {−23, 0, 7}; R: {1, 35} **23.** Yes **25.** No; (4, 2), (4, 4), (4, 6) **27.** No; (3, 4), (3, −4) or (4, 3), (4, −3) **29.** Yes **31.** Yes **33.** No; (−1, 0), (−1, 2) **35.** Yes **37.** Yes **39.** No; (1, 1), (1, −1) **41.** Yes **43.** Yes **45.** No; (1, 1), (1, −1) **47.** 9, −3 **49.** 3, −5 **51.** −6, −1 **53.** −6, −24 **55.** 9, 16 **57.** 7, 16 **59.** 7, 4 **61.** $\frac{1}{8}$, 1 **63.** $\frac{5}{2}, \frac{2}{5}$ **65.** 0, undefined **67.** 3.7, 1.1, 3.4 **69.** $-\frac{27}{64}, \frac{1}{216}, \frac{125}{8}$ **71.** $2w, 2w + 2$ **73.** $3w - 5, 3w - 2$ **75.** 0 **77.** 1 **79. a.** The set of real numbers **b.** The set of all real numbers except 4 **81. a.** The set of real numbers **b.** The set of all real numbers except $-\frac{1}{2}$

83.

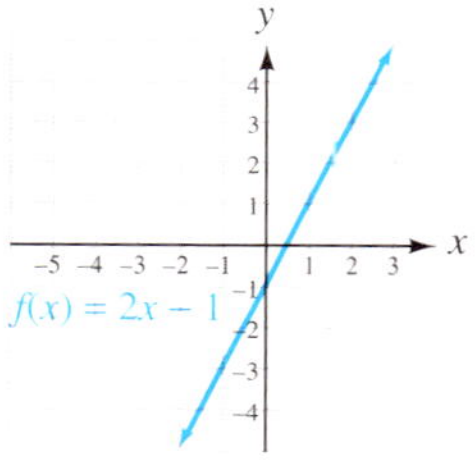

85.

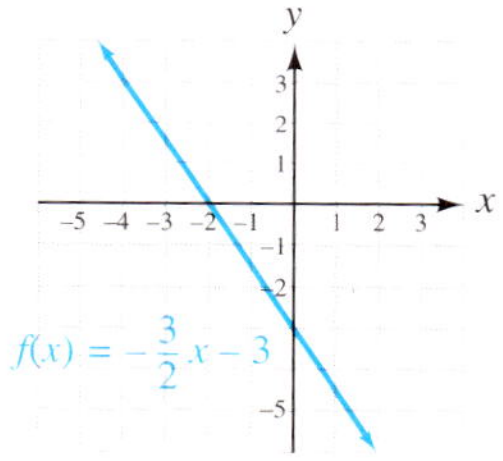

87.

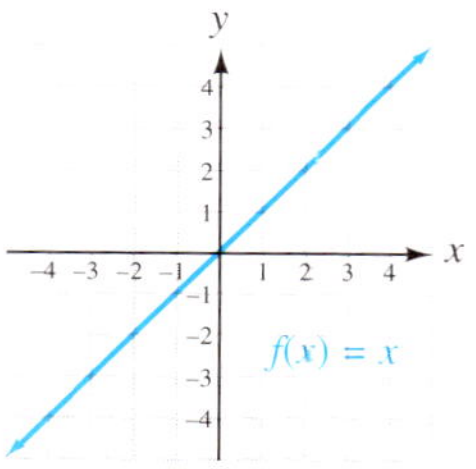

89.

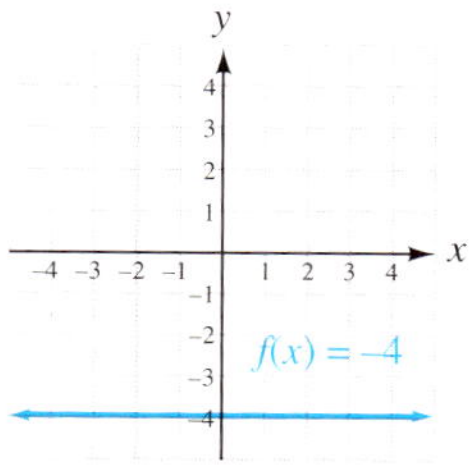

91.

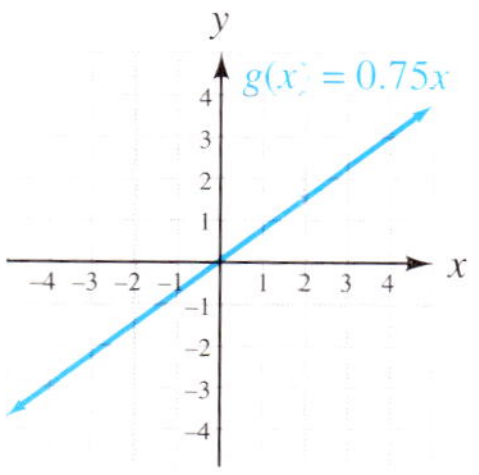

93.

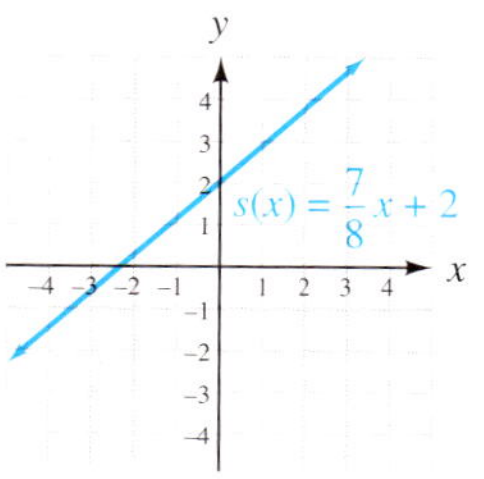

95. $f(x) = 5x - 3$ **97.** $f(x) = \frac{1}{5}x - 1$ **99.** $f(x) = 2x + 5$ **101.** $f(x) = -\frac{2}{3}x + 2$ **103.** $f(x) = 6x - 4$ **105.** $f(x) = 12$ **107.** 105.17 million websites **109. a.** $p(b) = 4.75b - 125$ **b.** \$397.50 **111. a.** $N(t) = 41{,}100t + 1{,}970{,}000$ **b.** 2,997,500 registered nurses **113. a.** $L(a) = -1.2a + 132$ **b.** 36% **115. a.** \$3,331.25; the tax on an adjusted gross income of \$25,000 is \$ 3,331.25. **b.** $T(a) = 4{,}681.25 + 0.25(a - 34{,}000)$ **117.** 10 m, 42 m, 26 m, 10 m **119.** 2,160 in.^3 **125.** No solution, ∅; contradiction

ARE YOU READY? 8.3 (page 628)

1. 3, 9 **2.** 1, 1 **3.** −8, 8 **4.** 4, 4

SELF CHECKS 8.3

1. a. −2 **b.** 2

2. D: the set of real numbers, R: the set of all real numbers greater than or equal to −2; the graph has the same shape, but is 2 units lower.

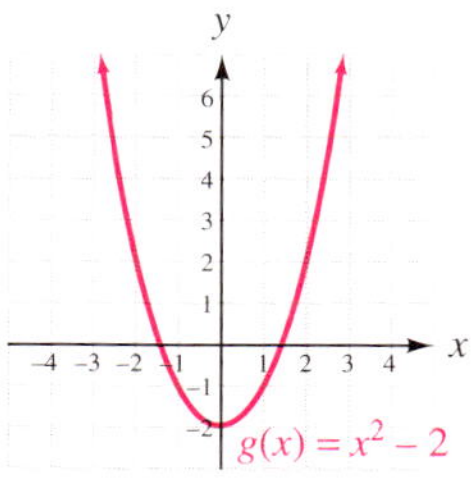

3. D: the set of real numbers, R: the set of all real numbers; the graph has the same shape, but is 1 unit higher.

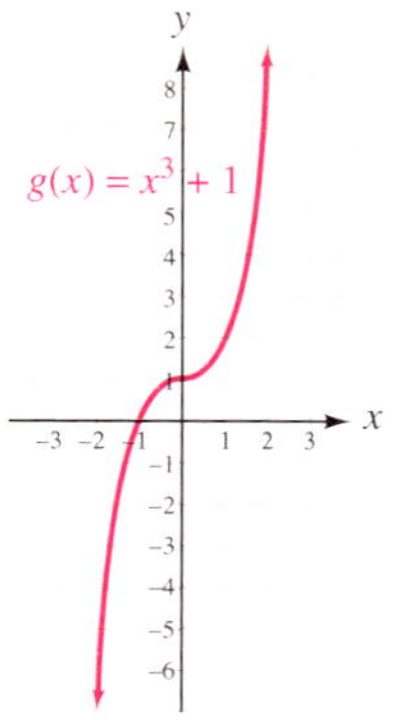

4. D: the set of real numbers, R: the set of nonnegative real numbers; the graph has the same shape, but is 2 units to the right.

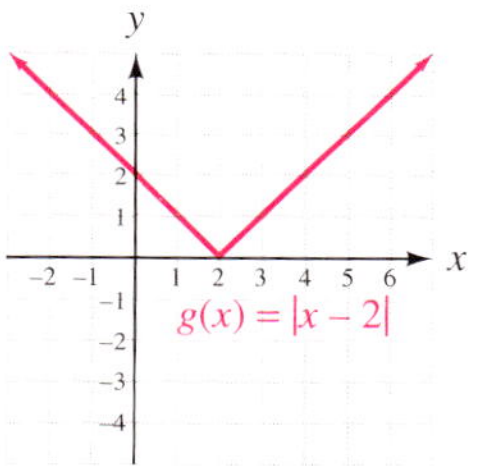

5.

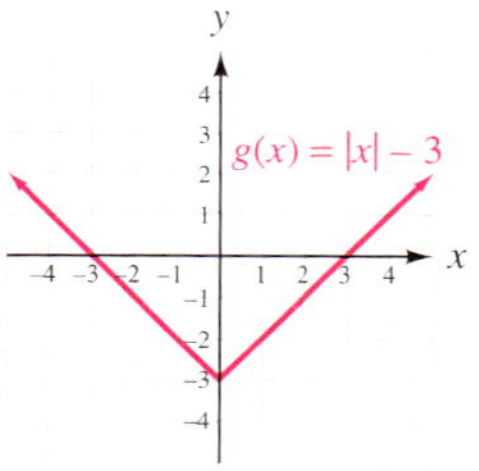

6.

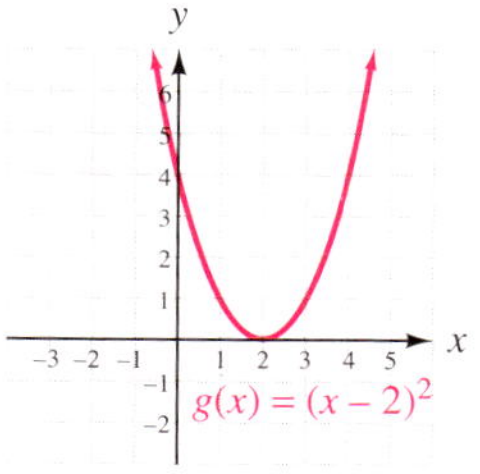

7.

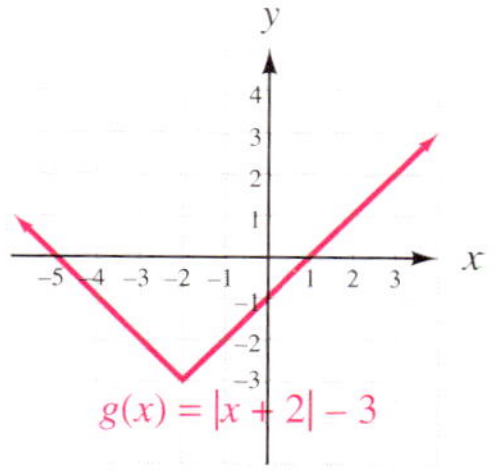

8.

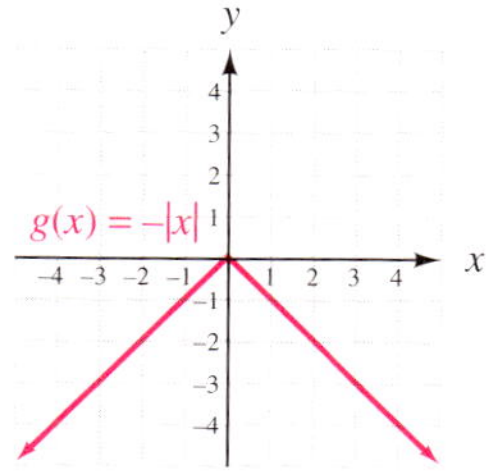

9. a. −10 **b.** −2, 0.2, 4.8 **10.** Not a function

STUDY SET SECTION 8.3 (page 638)

1. nonlinear **3.** nonnegative

5. a. The squaring function

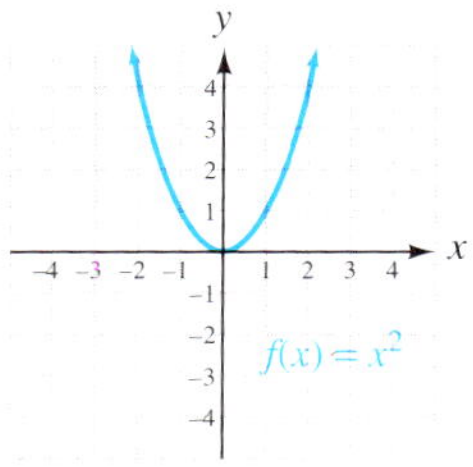

b. The cubing function

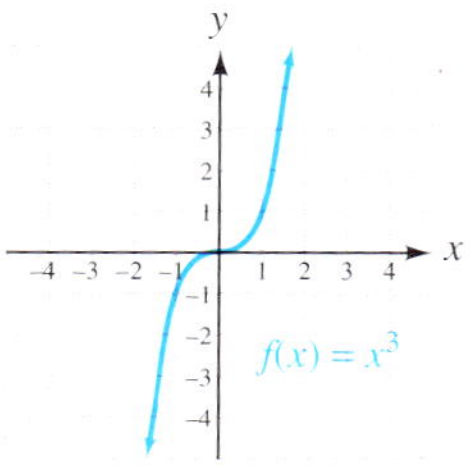

c. The absolute value function

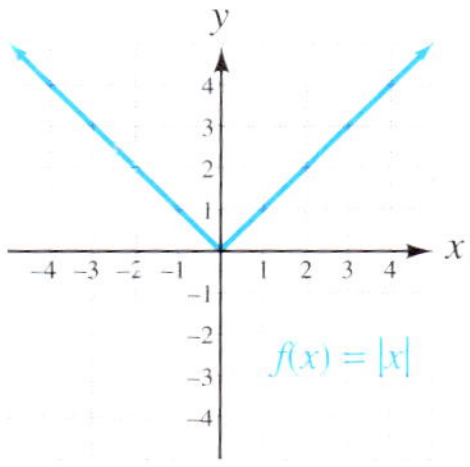

7. reflection

9. a.

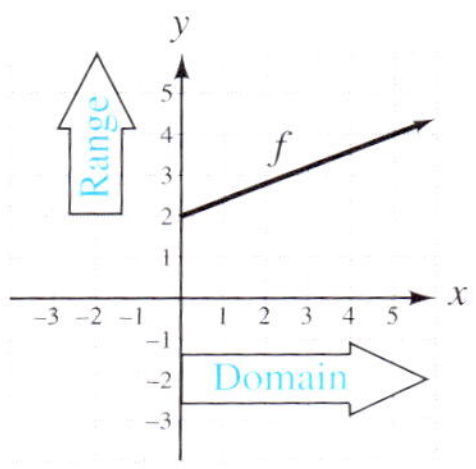

b. D: the set of nonnegative real numbers, R: the set of real numbers greater than or equal to 2 **11. a.** 4 **b.** 0 **c.** -2 **13. a.** $(-2, 4), (-2, -4)$ **b.** No; to the x-value -2, there corresponds more than one y-value (4 and -4). **15. a.** 4, left **b.** 4, up **17. a.** -4 **b.** 0 **c.** 2 **d.** -1 **19. a.** 2 **b.** 2 **c.** -1 **d.** -3, 1 **21.** D: the set of real numbers, R: the set of real numbers **23.** D: the set of real numbers, R: the set of real numbers less than or equal to 5 **25.** D: the set of real numbers, R: the set of real numbers greater than or equal to -4 **27.** D: the set of nonnegative real numbers, R: the set of nonnegative real numbers

29. D: the set of real numbers, R: the set of real numbers greater than or equal to 2

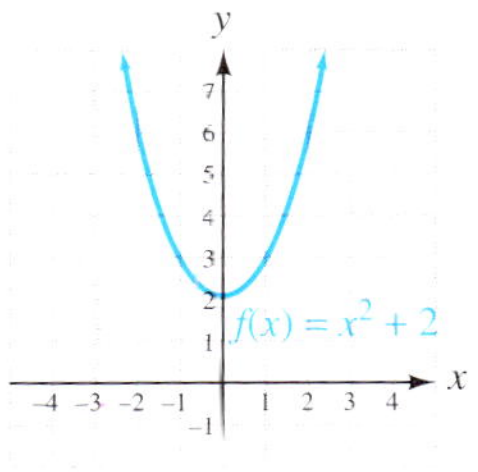

31. D: the set of real numbers, R: the set of real numbers

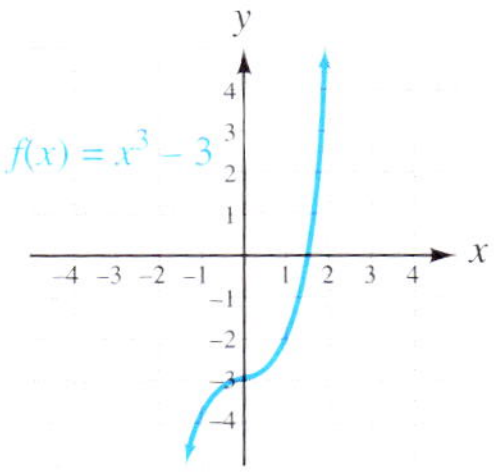

33. D: the set of real numbers, R: the set of nonnegative real numbers

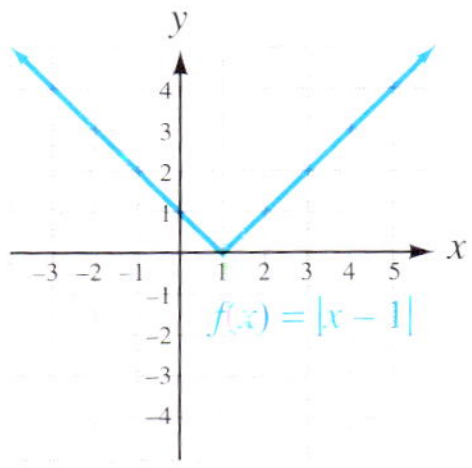

35. D: the set of real numbers, R: the set of nonnegative real numbers

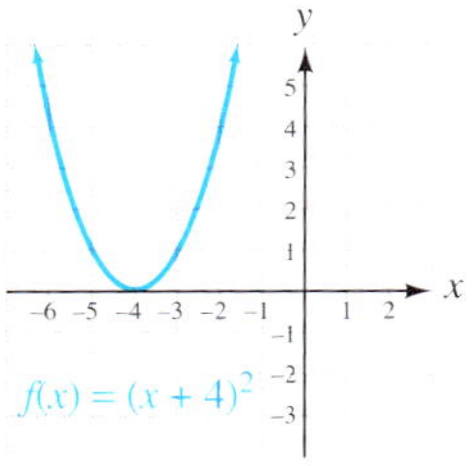

37. D: the set of real numbers, R: the set of real numbers greater than or equal to -2

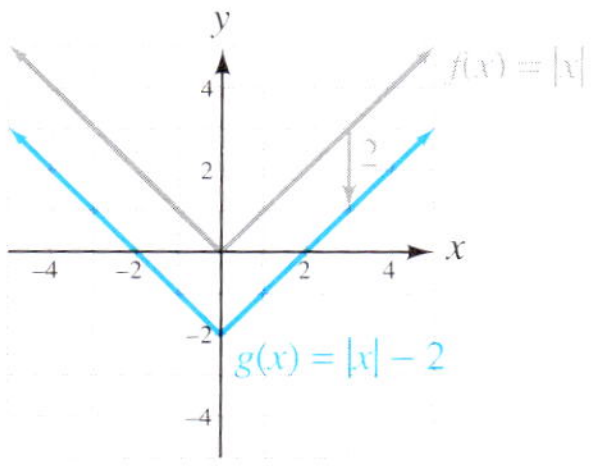

39. D: the set of real numbers, R: the set of real numbers

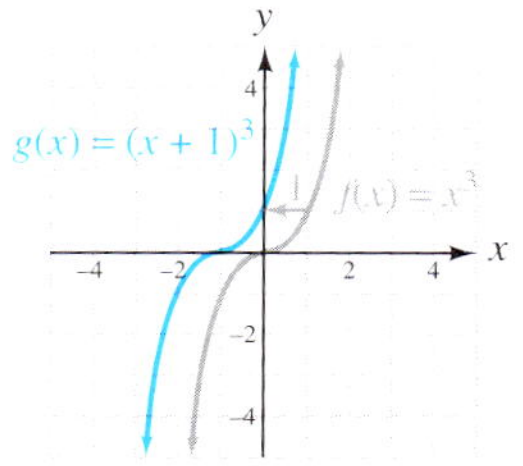

41. D: the set of real numbers, R: the set of all real numbers greater than or equal to -3

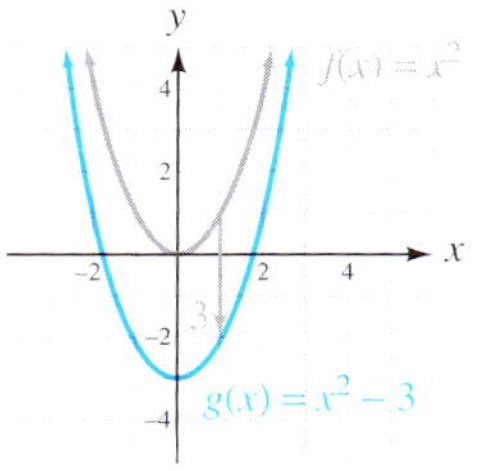

43. D: the set of real numbers, R: the set of real numbers

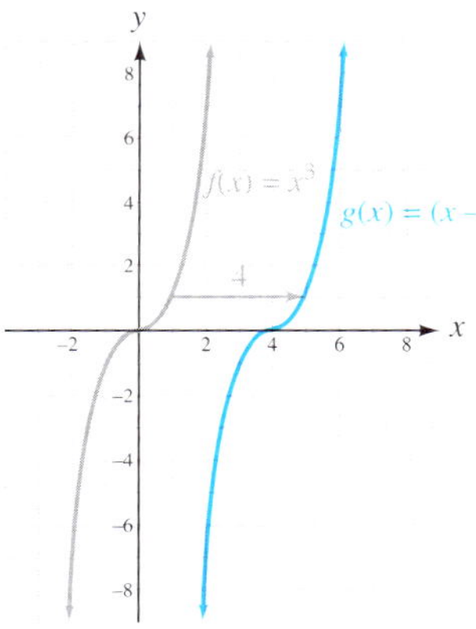

45. D: the set of real numbers, R: the set of real numbers

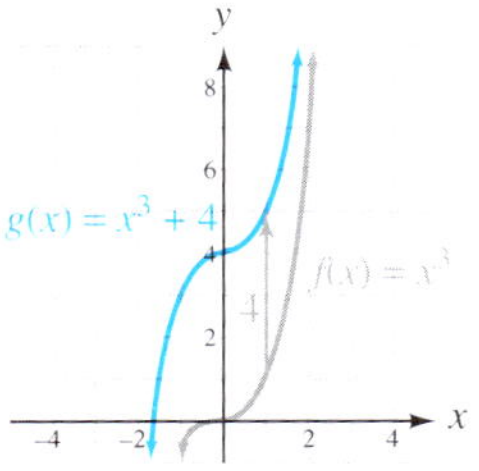

47. D: the set of real numbers, R: the set of nonnegative real numbers

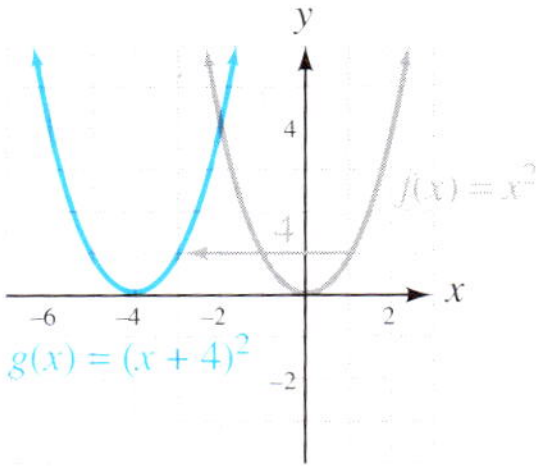

49.

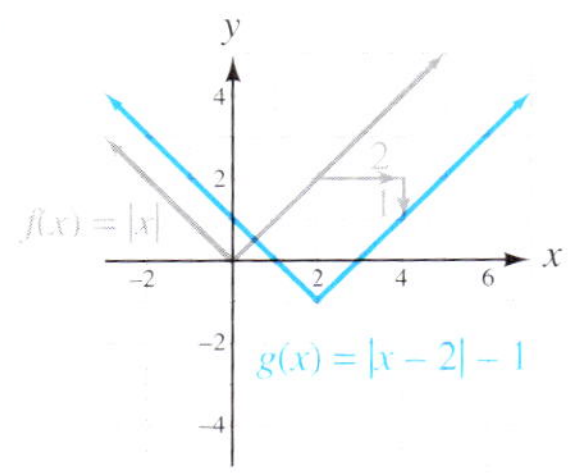

51.

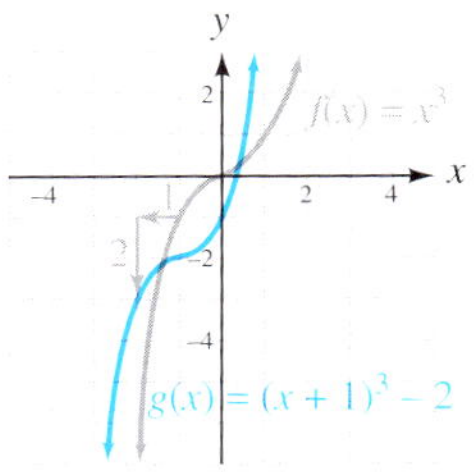

53.

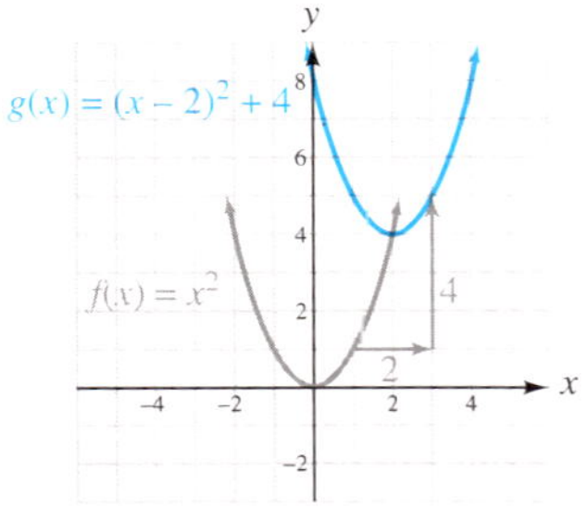

55.

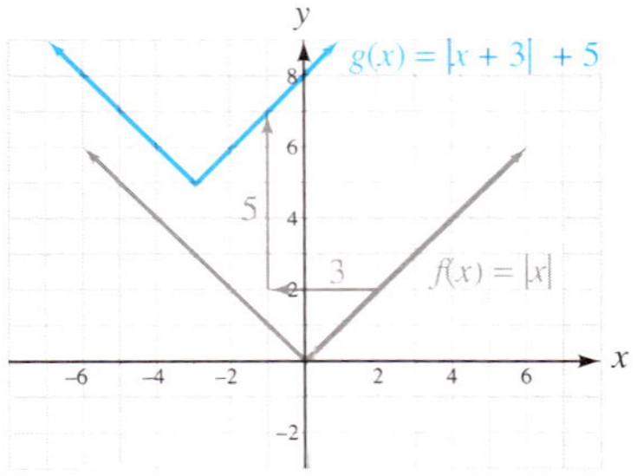

57.

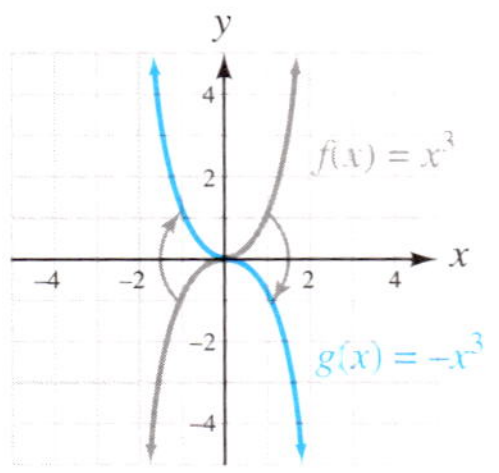

59.

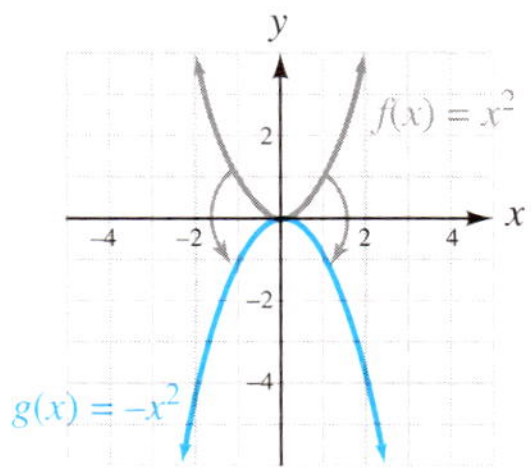

61.

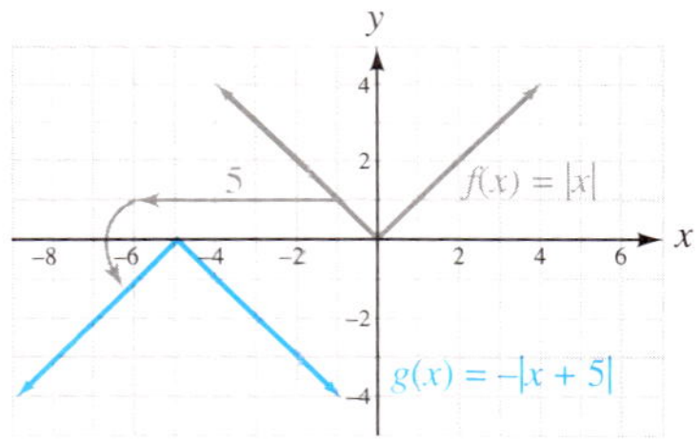

63.

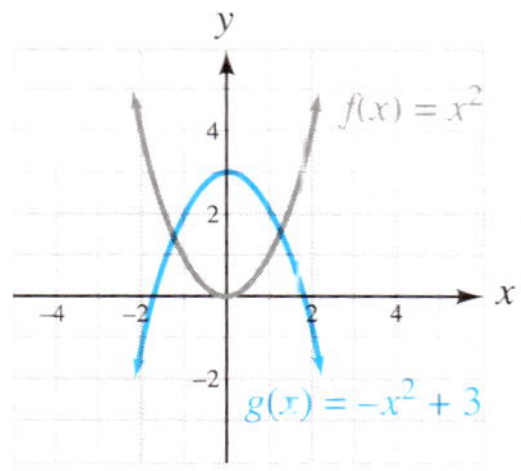

65. a. 15 **b.** 5 **c.** 0, −2, −4 **d.** D: $(-\infty, \infty)$; R: $(-\infty, \infty)$
67. a. 2 **b.** −5 **c.** −2, 2 **d.** D: $(-\infty, \infty)$; R: $(-\infty, 4]$
69. No, (0, 2), (0, −2) **71.** Yes **73.** Yes **75.** No, (3, 0), (3, 1)
77. 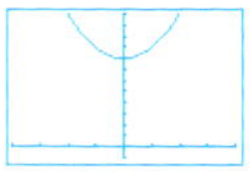 **79.** 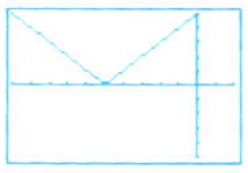 **81.** 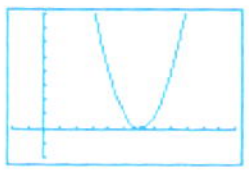
83. 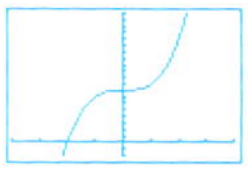 **85.** $f(x) = |x|$ **87.** A parabola
89. a. $J(9) \approx 12.2$; in 2009, there were about 12.2 million manufacturing jobs in the U.S.
b. $x \approx 3$; in 2003, there were about 14.5 million manufacturing jobs in the U.S. **97.** $W = T - ma$ **99.** $g = \frac{2(s - vt)}{t^2}$

ARE YOU READY? 8.4 (page 643)

1. 3 and 4 **2.** $[-7, \infty)$ 

3. Yes

4. 

SELF CHECKS 8.4

1. a. {9} **b.** {3, 6, 8, 9, 10, 11, 12, 15}
2. $(-6, 10]$ 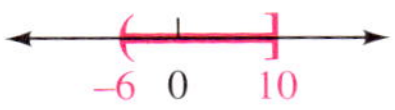

3. $\left(\frac{1}{2}, \infty\right)$ 

4. No solution; ∅ **5.** [1, 5] 

6. $(-\infty, 2) \cup (4, \infty)$ 

7. $(-\infty, \infty)$ 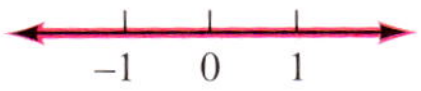

STUDY SET SECTION 8.4 (page 651)

1. intersection, union **3.** double **5. a.** both **b.** one, both
7. a. intersection **b.** union **9. a.** No **b.** Yes
11. a. $[-2, 1)$ **b.** [2, 2] **c.** ∅ **13.** union, intersection
15. All real numbers 

17. {4, 6} **19.** {−3, 1, 2}
21. {−3, −1, 0, 1, 2, 4, 6, 8, 10} **23.** {−3, 0, 1, 2, 3, 4, 5, 6, 8}
25. $(-2, 5]$ 

27. (2, 3] 

29. $(-\infty, -6]$ 

31. No solution; ∅
33. [1, 4] 

35. (0.8, 1.1) 

37. $(-\infty, -2] \cup (6, \infty)$ 

39. $(-\infty, -1) \cup (2, \infty)$ 

41. $(-\infty, 2) \cup (7, \infty)$ 

43. $(-\infty, \infty)$ 

45. $(-\infty, 1)$ 

47. (−10, −9) 

49. (−2, 5) 

51. $(-\infty, \infty)$ 

53. $[2, \infty)$ 

55. No solution; ∅
57. $[-4, 6)$ 

59. [2, 2] 

61. $(-\infty, 4.8) \cup (6.5, \infty)$ 

63. (−0.7, 0.2] 

65. (−12, −6] 

67. $\left(-\infty, \frac{1}{6}\right] \cup (2, \infty)$ **69.** $[-2, 4]$ **71.** $[4, 4]$ **73.** $(1, 3]$ **75.** $(3, 6)$ **77. a.** $[6, \infty)$ **b.** $[2, \infty)$ **79. a.** No solution; $\varnothing$ **b.** $(-\infty, 3] \cup [6, \infty)$

81. a. 128, 192 **b.** $32 \le s \le 48$, [32, 48] **83. a.** (67, 77) **b.** (62, 82) **85. a.** 2004 **b.** 2004, 2005, 2006, 2007 **c.** 2006, 2007 **d.** 2007 **87. a.**  **b.**  **95.** B747: 403 passengers; B777: 278 passengers

ARE YOU READY? 8.5 (page 655)

1. a. 12 **b.** 7.5 **2. a.** True **b.** False **3.** $(-\infty, -3] \cup [1, \infty)$ **4.** $(-8, 4)$

SELF CHECKS 8.5

1. a. $\frac{1}{2}, -\frac{1}{2}$ **b.** $5, -2$ **c.** No solution, $\varnothing$ **2.** 7.5, 2.5 **3.** -6 **4.** $7, -15$ **5.** $-1, -6$ **6.** $\left(-2, \frac{2}{3}\right)$ **7.** [2.8985, 2.9015] **8.** No solution, $\varnothing$ **9.** $(-\infty, -2] \cup [6, \infty)$ **10.** $(-\infty, -8) \cup \left(\frac{8}{3}, \infty\right)$ **11.** $(-\infty, \infty)$

STUDY SET SECTION 8.5 (page 664)

1. absolute value **3.** isolate **5.** compound **7.** compound **9. a.** $-2, 2$ **b.** $-1.99, -1, 0, 1, 1.99$ **c.** $-3, -2.01, 2.01, 3$ **11. a.** $8, -8$ **b.** $x - 3, -(x - 3)$ **13. a.** $x = 8$ or $x = -8$ **b.** $x \le -8$ or $x \ge 8$ **c.** $-8 \le x \le 8$ **d.** $5x - 1 = x + 3$ or $5x - 1 = -(x + 3)$ **15. a.** No solution **b.** No solution **c.** All real numbers **17. a.** ii **b.** iii **c.** i **19.** $23, -23$ **21.** $13, -3$ **23.** $\frac{14}{3}, -6$ **25.** $50, -50$ **27.** $3.1, -6.7$ **29.** No solution; $\varnothing$ **31.** $25, -19$ **33.** $7, -\frac{7}{3}$ **35.** $2, -\frac{1}{2}$ **37.** $\frac{32}{7}, -16$ **39.** -10 **41.** -8 **43.** $\frac{9}{2}$ **45.** $-\frac{1}{2}$ **47.** $-4, \frac{28}{9}$ **49.** $-2, \frac{18}{11}$ **51.** $0, -2$ **53.** $11, \frac{1}{3}$ **55.** $(-4, 4)$ **57.** $[-21, 3]$ **59.** $\left(-\frac{8}{3}, 4\right)$ **61.** No solution; $\varnothing$ **63.** $(-\infty, -3) \cup (3, \infty)$ **65.** $(-\infty, -12) \cup (36, \infty)$ **67.** $\left(-\infty, -\frac{1}{5}\right] \cup \left[\frac{3}{5}, \infty\right)$ **69.** $(-\infty, \infty)$ **71.** $0, -6$ **73.** $(-3, 1)$ **75.** $\left(-\infty, -\frac{16}{3}\right) \cup (4, \infty)$ **77.** $[-10, 14]$ **79.** $40, -20$ **81.** $(-\infty, \infty)$ **83.** $-3, -\frac{3}{4}$ **85.** $\frac{2}{3}$ **87.** No solution; $\varnothing$ **89.** $\frac{12}{11}, -\frac{12}{11}$ **91.** $(-\infty, -2) \cup (5, \infty)$ **93.** No solution; $\varnothing$ **95. a.** 20 **b.** 20, 0 **c.** $(20, \infty)$ **d.** $(-\infty, 0) \cup (20, \infty)$ **97. a.** -25 **b.** $-25, 31$ **c.** $(-\infty, -25)$ **d.** $(-\infty, -25) \cup (31, \infty)$ **99. a.** $[3, 7]$ **b.** $(-\infty, 3] \cup [7, \infty)$ **101. a.** $(-\infty, -8) \cup (10, \infty)$ **b.** $[-8, 10]$ **103.** $70° \le t \le 86°$ **105. a.** $|c - 0.6°| \le 0.5°$ **b.** [0.1°, 1.1°] **107.** 26.45%, 24.76% **113.** First piece: 9 in.; middle piece: 14 in.; last piece: 6 in.

ARE YOU READY? 8.6 (page 667)

1. $2 \cdot 2 \cdot 3 \cdot 3 \cdot 3 = 2^2 \cdot 3^3$ **2.** $6b^4 + 12b^2 + 24b$ **3.** $x^2 + 2x - 48$ **4.** $m^2 - 3m - 10$ **5.** 2 and 5 **6.** -3 and 6

SELF CHECKS 8.6

1. $2ab^2(3a - 2b + 1)$ **2.** $-3p^2q(p - 2q)$ **3.** $(y^2 + 1)(c^2 + d^2)$ **4.** $(m - n)(7 + n)$ **5.** $ab(3a - 2b)(a + 1)$ **6.** $f_1 = \frac{ff_2}{f_2 - f}$ **7** $(a - 4)(a - 3)$ **8.** $3(p + 4q)(p - 2q)$ **9.** $(4q + 3)(q - 3)$ **10.** $b(2a^2 + 1)(3a^2 + 4)$ **11.** $(a + b + 2)(a + b - 5)$ **12. a.** $(m + 7)(m + 6)$ **b.** $(3a + 4b)(5a - b)$

STUDY SET SECTION 8.6 (page 674)

1. factored **3.** greatest common factor **5.** grouping **7.** leading, coefficient, 2 **9.** $6xy^2$ **11.** $9, 6, -9, -6$ **13.** $-1 + (-12) = -13$, $-2(-6) = 12$, $-2 + (-6) = -8$, $-3 + (-4) = -7$ **15.** $5c^2d^4$ **17.** $3m, 3$ **19.** $2x(x - 3)$ **21.** $5x^2y(3 - 2y)$ **23.** $3z(9z^2 + 4z + 1)$

25. $6s(4s^2 - 2st + t^2)$ **27.** Prime **29.** Prime **31.** $-8(a + 2)$
33. $-3x(2x + y)$ **35.** $-6ab(3a - 2b)$
37. $-4a^2c^8(2a^2 - 7a + 5c)$ **39.** $(x + y)(u + v)$
41. $(a - b + c)(5 - t)$ **43.** $(x + y)(a + b)$ **45.** $(x + y)(x - 1)$
47. $(t - 3)(t^2 - 7)$ **49.** $(a + b)(a - 4)$ **51.** $6(x - 1)(x^2 + 2)$
53. $2c(2b^3 + 1)(7a^3 - 1)$ **55.** $h = \frac{2g}{c + d}$ **57.** $r_1 = \frac{rr_2}{r_2 - r}$
59. $a^2 = \frac{b^2x^2}{b^2 - y^2}$ **61.** $n = \frac{360}{180 - S}$ **63.** $(x - 3)(x - 2)$
65. $(x + 6)(x - 5)$ **67.** $3(x + 7y)(x - 3y)$
69. $6(a - 4b)(a - b)$ **71.** $n^2(n - 30t)(n + 2t)$
73. $-3(x - 3y)(x - 2y)$ **75.** $(5x + 3)(x + 2)$
77. $(7a + 5)(a + 1)$ **79.** $(11y - 1)(y + 3)$
81. $(4x - 1)(2x - 5)$ **83.** $(3y - 2)(2y - 3)$
85. $(5b - 2)(3b + 2)$ **87.** $5(3x^2 - 4)(2x^2 + 1)$
89. $8(2x^2 - 3)^2$ **91.** $4h(8h^2 - 1)(2h^2 + 1)$
93. $-(3a^2 + 2b^2)(a^2 + b^2)$ **95.** $(a + b + 4)(a + b - 6)$
97. $(x + a + 1)^2$ **99.** $(m + n + p)(3 + x)$
101. $-7u^2v^3(9uv^3 - 4v^4 + 3u)$ **103.** $x^2(b^2 - 7)(b^2 - 5)$
105. $(1 - n)(1 - m)$ **107.** $-(x + 3y)(x - 7y)$
109. $(a^2 + b)(x - 1)$ **111.** $(2y + 1)^2$ **113.** Prime
115. $(3r - 2)(r + 5)$ **117.** $(y^2 + 3)(y - 4)$
119. $2y(y^2 - 10)(y^2 - 3)$ **121.** $(7q - 7r + 2)(2q - 2r - 3)$
123. $\pi r^2\left(h_1 + \frac{1}{3}h_2\right)$ **125.** $x + 3$ **129. a.** 2 **b.** 3 **c.** $-1, 1, 5$
d. 3

ARE YOU READY? 8.7 (page 677)

1. $n^2 - 81$ **2.** $64d^2$ **3. a.** 27 **b.** 216 **4.** $64a^3$ **5.** $b^3 + 64$
6. $8a^3 - 1$

SELF CHECKS 8.7

1. $(9p + 5)(9p - 5)$ **2.** $(6r^2 + s)(6r^2 - s)$
3. $(a^2 + 9)(a + 3)(a - 3)$
4. $(a^2 - 2ab + b^2 + c^2)(a - b + c)(a - b - c)$
5. $3(a^2 + 1)(a + 1)(a - 1)$ **6. a.** $(a + b)(a - b + 1)$
b. $(a + 2 + b)(a + 2 - b)$ **7.** $(p + 3)(p^2 - 3p + 9)$
8. $(2c^2 - 5d)(4c^4 + 10c^2d + 25d^2)$
9. $(p + q - r)(p^2 + 2pq + q^2 + pr + qr + r^2)$
10. $(1 + x)(1 - x + x^2)(1 - x)(1 + x + x^2)$
11. $3x^2(x + 2)(x^2 - 2x + 4)$

STUDY SET SECTION 8.7 (page 683)

1. squares **3. a.** 1, 4, 9, 16, 25, 36, 49, 64, 81, 100
b. 1, 8, 27, 64, 125, 216, 343, 512, 729, 1,000 **5. a.** F − L
b. $F^2 - FL + L^2$ **c.** $F^2 + FL + L^2$ **7. a.** $x^2 - 4$ (Answers may vary) **b.** $(x - 4)^2$ (Answers may vary) **c.** $x^2 + 4$ (Answers may vary) **d.** $x^3 + 8$ (Answers may vary) **e.** $(x + 8)^3$ (Answers may vary) **9.** $(x + 4)(x - 4)$ **11.** $(3y + 8)(3y - 8)$
13. $(12 + c)(12 - c)$ **15.** $(10m + 1)(10m - 1)$
17. $(9a + 7b)(9a - 7b)$ **19.** Prime **21.** $(3r^2 + 11s)(3r^2 - 11s)$
23. $(4t + 5w^2)(4t - 5w^2)$ **25.** $(10rs^2 + t^2)(10rs^2 - t^2)$
27. $(6x^2y + 7z^3)(6x^2y - 7z^3)$ **29.** $(x^2 + y^2)(x + y)(x - y)$
31. $(4a^2 + 9b^2)(2a + 3b)(2a - 3b)$ **33.** $(x + y + z)(x + y - z)$
35. $(r - s + t^2)(r - s - t^2)$ **37.** $2(x + 12)(x - 12)$
39. $3x(x + 9)(x - 9)$ **41.** $5a(b^2 + 1)(b + 1)(b - 1)$
43. $4b(4 + b^2)(2 + b)(2 - b)$ **45.** $(c + d)(c - d + 1)$
47. $(a - b)(a + b + 2)$ **49.** $(x + 6 + y)(x + 6 - y)$
51. $(x - 1 + 3z)(x - 1 - 3z)$ **53.** $(a + 5)(a^2 - 5a + 25)$
55. $(2r + s)(4r^2 - 2rs + s^2)$
57. $(4t^2 - 3v)(16t^4 + 12t^2v + 9v^2)$
59. $(x - 6y^2)(x^2 + 6xy^2 + 36y^4)$
61. $(a - b + 3)(a^2 - 2ab + b^2 - 3a + 3b + 9)$
63. $(4 - a - b)(16 + 4a + 4b + a^2 + 2ab + b^2)$
65. $(x + 1)(x^2 - x + 1)(x - 1)(x^2 + x + 1)$
67. $(x^2 + y)(x^4 - x^2y + y^2)(x^2 - y)(x^4 + x^2y + y^2)$
69. $5(x + 5)(x^2 - 5x + 25)$ **71.** $4x^2(x - 4)(x^2 + 4x + 16)$
73. $(4a - 5b^2)(16a^2 + 20ab^2 + 25b^4)$
75. $2b^2(12 + b^2)(12 - b^2)$ **77.** $(x + y)(x - y + 8)$
79. $(x + y)(x^2 - xy + y^2)(x^6 - x^3y^3 + y^6)$
81. $(12at + 13b^3)(12at - 13b^3)$ **83.** Prime
85. $(9c^2d^2 + 4t^2)(3cd + 2t)(3cd - 2t)$
87. $2u^2(4v - t)(16v^2 + 4tv + t^2)$ **89.** $(y + 2x - t)(y - 2x + t)$
91. $(x + 10 + 3z)(x + 10 - 3z)$
93. $(c - d + 6)(c^2 - 2cd + d^2 - 6c + 6d + 36)$
95. $\left(\frac{1}{6} + y^2\right)\left(\frac{1}{6} - y^2\right)$
97. $(m + 2)(m^2 - 2m + 4)(m - 2)(m^2 + 2m + 4)$
99. $(a + b)(x + 3)(x^2 - 3x + 9)$
101. $(x^3 - y^4z^5)(x^6 + x^3y^4z^5 + y^8z^{10})$ **103. a.** $(q + 8)(q - 8)$
b. $(q - 4)(q^2 + 4q + 16)$ **105. a.** $(d + 5)(d - 5)$
b. $(d - 5)(d^2 + 5d + 25)$ **107. a.** $(a^2 - b)(a^4 + a^2b + b^2)$
b. $(a^2 + b)(a^4 - a^2b + b^2)$
109. a. $(5m + 2n)(25m^2 - 10mn + 4n^2)$
b. $(5m - 2n)(25m^2 + 10mn + 4n^2)$
111. $\frac{4}{3}\pi(r_1 - r_2)(r_1{}^2 + r_1r_2 + r_2{}^2)$
117. 45 minutes for \$25

ARE YOU READY? 8.8 (page 686)

1. a. 0 **b.** Undefined **2.** $\frac{9}{7}$ **3.** $\frac{x + 5}{4(x - 5)}$ **4.** $\frac{3x}{2}$ **5. a.** $\frac{3}{56}$ **b.** $\frac{9}{5}$
6. a. $\frac{8}{11}$ **b.** $\frac{7}{15}$

SELF CHECKS 8.8

1. \$1.56 per hour **2.** The domain is set of all real numbers except -7 and 7: $(-\infty, -7) \cup (-7, 7) \cup (7, \infty)$ **3.** $\frac{3a^3}{5b^2}$ **4. a.** $\frac{2x + 1}{3x - 1}$
b. $-\frac{2a + 3b}{b}$ or $\frac{-2a - 3b}{b}$ **5. a.** $\frac{a^2(a + 3)}{6}$ **b.** $-x(x + 3)$
6. $\frac{x(x - 2)}{3}$ **7.** $\frac{b}{b - 2}$ **8.** $\frac{9 + 20xz^2}{84z^3}$ **9.** $\frac{9a - 2}{(a - 2)(a - 2)(a + 2)}$
10. $-\frac{5}{2}, 1$

STUDY SET SECTION 8.8 (page 696)

1. rational **3.** domain **5.** simplify **7.** factor **9.** build
11. a. $-\frac{1}{4}$ **b.** $\frac{5}{6}$ **c.** Undefined **13. a.** -1 **b.** 2 **c.** 2 **d.** 5
15. numerators, denominators, reciprocal **17.** factor, greatest
19. a. Twice **b.** Once **21. a.** 3a **b.** $(x - 2)(x + 5)$
23. a. $\frac{5x^2 + 35x}{1}$ **b.** $\frac{1}{5x^2 + 35x}$
25. All real numbers except 0; $(-\infty, 0) \cup (0, \infty)$
27. All real numbers except -2; $(-\infty, -2) \cup (-2, \infty)$
29. All real numbers except 0 and 1; $(-\infty, 0) \cup (0, 1) \cup (1, \infty)$
31. All real numbers except -7 and 8; $(-\infty, -7) \cup (-7, 8) \cup (8, \infty)$
33. $\frac{3}{5a^6}$ **35.** $\frac{4y}{9x}$ **37.** $\frac{5x}{x - 2}$ **39.** $\frac{3x - 5}{x + 2}$ **41.** $-\frac{x + 2}{x + 1}$
43. $-\frac{p + 2q}{q + 2p}$ **45.** $\frac{8}{b}$ **47.** $\frac{p(p - 4)}{12}$ **49.** $\frac{x + 1}{9}$ **51.** 1
53. $-\frac{a^4}{a^2 + 3}$ **55.** $-\frac{(x + 1)^2(x + 2)}{2c + x}$ **57.** $\frac{3m}{2n^2}$ **59.** $\frac{x - 4}{x + 5}$
61. $\frac{c + 1}{25c^2 - 5c + 1}$ **63.** $\frac{n + 2}{n + 1}$ **65.** $\frac{3}{x + 3}$ **67.** $\frac{1}{2}$ **69.** $\frac{17}{12x}$
71. $\frac{16y^2 + 3}{18y^4}$ **73.** $\frac{3a - 10b}{4a^2b^2}$ **75.** $-\frac{y^2 + 5y + 14}{2y^2}$ **77.** $\frac{8x - 2}{(x + 2)(x - 4)}$
79. $\frac{7x + 29}{(x + 5)(x + 7)}$ **81.** $\frac{2x^2 + x}{(x + 3)(x + 2)(x - 2)}$
83. $\frac{-x^2 + 11x + 8}{(3x + 2)(x + 1)(x - 3)}$ **85.** $\frac{6 - 3d}{5d(d - 1)}$ **87.** $\frac{m - 5}{(m + 3)(m + 5)}$

89. $\frac{1}{2}$ **91.** $-\frac{1}{2}$ **93.** 1, 2 **95.** 4, −1
97. $f(x) = \frac{x+8}{x+2}$, provided $x \neq -2, x \neq 2$ **99.** $g(x) = \frac{x^2-4x+16}{x^2+3}$, provided $x \neq -4$ **101.** $-\frac{q}{p}$ **103.** $\frac{t-3}{(t+3)(t+1)}$ **105.** 26
107. $\frac{-x^2+3x+2}{(x-1)(x+1)^2}$ **109.** $x - 5$ **111.** $\frac{y^2+xy+x^2}{x+y}$ **113.** $\frac{3n}{2n+3}$
115. $\frac{x^2-5x-5}{x-5}$ **117.** $\frac{x+y}{a-b}$ **119.** 1 **121.** 2, −5 **123.** $\frac{x}{x-3}$
125. a. \$50,000 **b.** \$200,000 **127. a.** $c(n) = 0.09n + 7.50$
b. $c(n) = \frac{0.09n + 7.50}{n}$ **c.** About 10¢ **133.** $-\frac{1}{4}$

ARE YOU READY? 8.9 (page 700)

1. 450 **2.** 5 **3.** 1,620 **4. a.** $\frac{2}{13}$ **b.** 200

SELF CHECKS 8.9

1. 735 British pounds **2.** 16 foot-candles
3. Approximately 1,055 lb **4.** 20 weeks

STUDY SET SECTION 8.9 (page 705)

1. direct, increases **3.** joint, combined **5. a.** Direct **b.** Inverse
7. Direct **9.** Inverse **11.** Direct **13.** Inverse **15. a.** Yes
b. No **c.** No **d.** Yes **17.** $A = kp^2$ **19.** $z = \frac{k}{t^3}$
21. $C = kxyz$ **23.** $P = \frac{ka^2}{j^3}$ **25.** r varies directly as t.
27. b varies inversely as h. **29.** U varies jointly as r, the square of s, and t. **31.** P varies directly as m and inversely as n.
33. a. False **b.** False **c.** True **35.** 117.6 newtons
37. 432 mi **39.** 25 days **41.** 12 in.^{3} **43.** \$9,000 **45.** 3 ohms
47. 0.275 in. **49.** 1.4 ohms **51.** 12.8 lb
55. (2, 5)

57. $(-\infty, 6) \cup (9, \infty)$

CHAPTER 8 REVIEW (page 708)

1. $-\frac{12}{5}$ **2.** −9 **3.** 8 **4.** $\frac{11}{7}$ **5.** $\frac{88}{17}$ **6.** 12 **7.** All real numbers, ℝ; identity **8.** No solution, ∅; contradiction **9.** −8
10. 0 **11.** $h = \frac{V}{\pi r^2}$ **12.** $x = \frac{6v - aby}{ab}$ or $x = \frac{6v}{ab} - y$
13. $[4, \infty)$

14. $\left(-\infty, -\frac{51}{11}\right)$

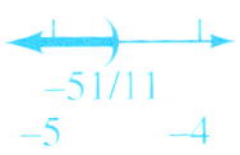

15. $(-\infty, 20)$

16. $(-\infty, \infty)$, ℝ

17. 5 ft from one end **18.** Length: 9 m, width: 5 m
19. D: {−4, −1, 2, 5}, R: {−2, 0, 16} **20. a.** function
b. function, variable, dependent **21. a.** Yes
b. No; (−1, 8), (−1, 9) **c.** Yes **22.** 0 **23.** Yes **24.** Yes
25. No; (25, 5), (25, −5) **26.** No; (3, 4), (3, −4)
27. −7 **28.** 18 **29.** 8 **30.** $3t + 8$ **31.** 3 **32.** $\frac{4}{3}$
33. The set of real numbers **34.** The set of real numbers
35. The set of all real numbers except 2 **36.** The set of all real numbers except −5 **37.** −2, (0, −16)

38.

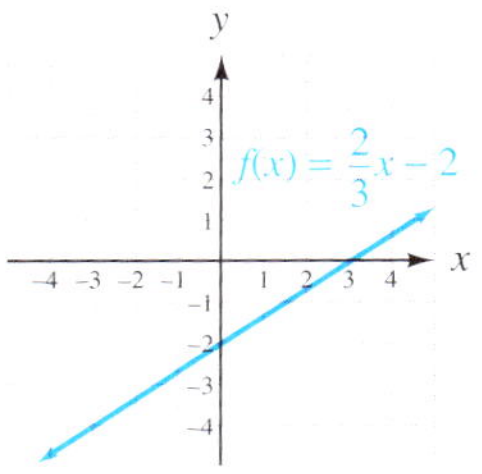

39. $f(x) = \frac{9}{10}x + \frac{7}{8}$
40. $f(x) = \frac{1}{5}x - 1$ **41.** $f(x) = 12$
42. $f(x) = 4x - 3$
43. $f(x) = \frac{1}{3}x + 5$
44. a. $R(t) = 0.02t + 5.05$
b. 7.05 milliohms
45. 134 in.^{3} **46.** $\frac{65}{24} = 2\frac{17}{24}$
47. a. −4 **b.** 3 **c.** 1
48. a. 4 **b.** 1 **c.** −4, 2
49. a. −1 **b.** −2, −1, 1 **c.** D: $(-\infty, \infty)$; R: $[-1, \infty)$
50. The real numbers greater than or equal to 0
51. D: the set of real numbers, R: the set of real numbers
52. D: the set of real numbers, R: the set of real numbers greater than or equal to 1
53. D: the set of real numbers, R: the set of nonnegative real numbers **54. a.** 6, up **b.** 6, left

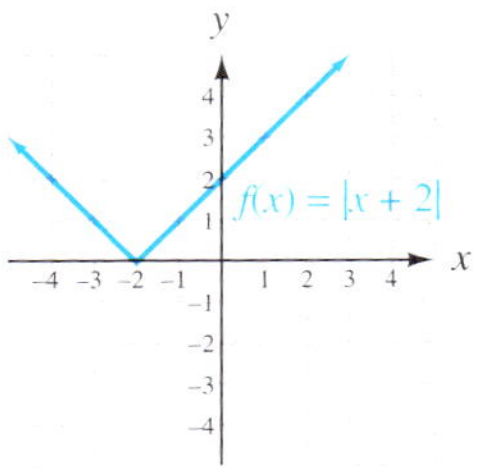

55. D: the set of real numbers, R: the set of real numbers greater than or equal to −3

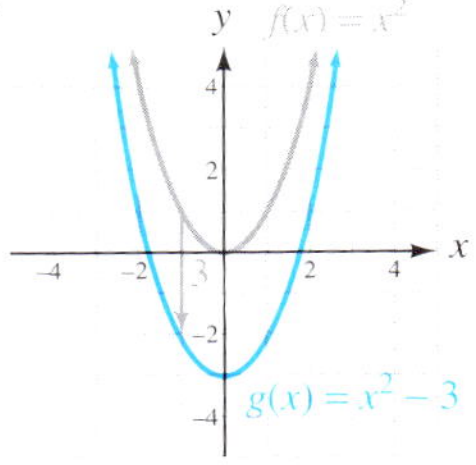

56. D: the set of real numbers, R: the set of nonnegative real numbers

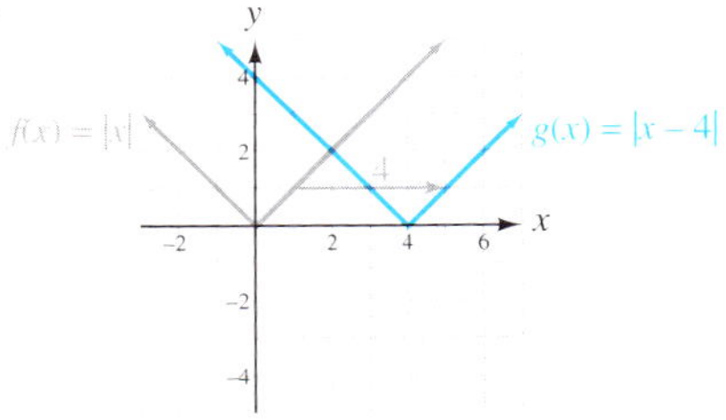

57. D: the set of real numbers, R: the set of real numbers

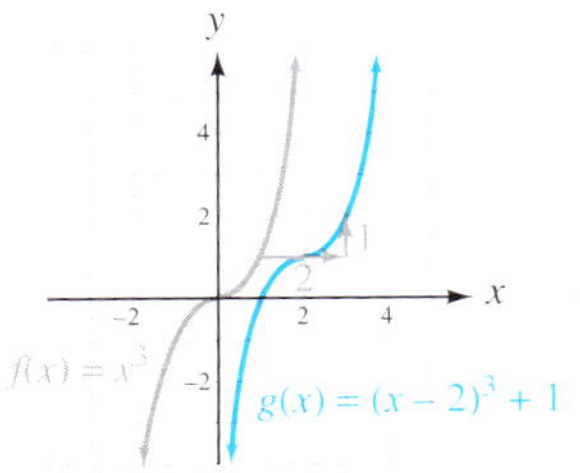

58. D: the set of real numbers, R: the set of real numbers

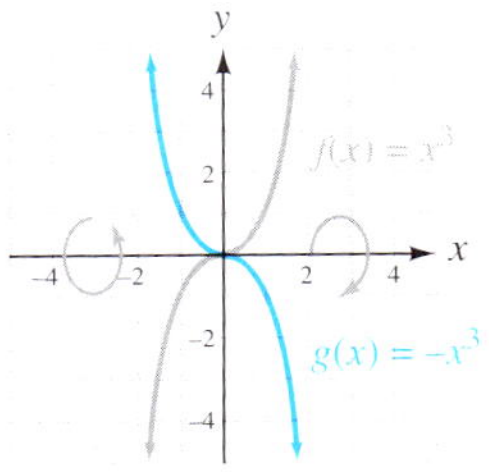

59. Function **60.** Not a function; (0, 2), (0, 3) **61.** {−3, 3}
62. {−6, −5, −3, 0, 3, 6, 8} **63.** Yes **64.** No
65.

66.

67. [−10, −4)

68. $(-\infty, -11)$ 

69. No solution; $\varnothing$

70. $[0, 0]$ 

71. $\left(-\frac{1}{3}, 2\right)$ 

72. $[1, 9]$ 

73. Yes **74.** No

75. $(-\infty, -5) \cup (4, \infty)$ 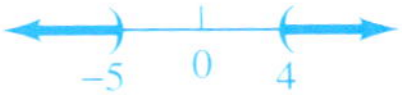

76. $(-\infty, \infty)$ 

77. $17 \le 4x \le 25$, $4.25 \text{ ft} \le x \le 6.25 \text{ ft}$, $[4.25, 6.25]$

78. a. ii, iv **b.** i, iii **79.** $(-\infty, 73.6) \cup [137.6, \infty)$

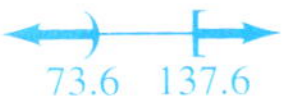

80. $\left(-\frac{7}{3}, \frac{1}{3}\right]$ 

81. $2, -2$ **82.** $3, -\frac{11}{3}$ **83.** $\frac{26}{3}, -\frac{10}{3}$

84. No solution; $\varnothing$ **85.** 3 **86.** $\frac{1}{8}, -\frac{19}{8}$ **87.** $\frac{1}{5}, -5$ **88.** $\frac{13}{12}$

89. $[-3, 3]$ 

90. $(-5, -2)$ 

91. $\left[-3, \frac{19}{3}\right]$ 

92. No solution; $\varnothing$

93. $(-\infty, -1) \cup (1, \infty)$ 

94. $(-\infty, -4] \cup \left[\frac{22}{5}, \infty\right)$ 

95. $\left(-\infty, \frac{4}{3}\right) \cup (4, \infty)$ 

96. $(-\infty, \infty)$, $\mathbb{R}$ 

97. a. 8, 2 **b.** $[6, 10]$ **98.** $3, -3$ **99.** $0, -8$

100. $(2, 12)$ 

101. Since $|0.04x - 8.8|$ is always greater than or equal to 0 for any real number x, it can never be less than -2. This absolute value inequality has no solution.

102. Since $\left|\frac{3x}{50} + \frac{1}{45}\right|$ is always greater than or equal to 0 for any real number x, it will always be greater than or equal to $-\frac{4}{5}$. This absolute value inequality is true for all real numbers.

103. $(z - 5)(z - 6)$ **104.** $(x^2 + 4)(x^2 + y)$

105. $(4a - 1)(a - 1)$ **106.** $9x^2y^3z^2(3xz + 9x^2y^2 - 10z^5)$

107. $(5b - 2)(3b + 2)$ **108.** $-(x + 7)(x - 4)$

109. $3(5x + y)(x - 4y)$ **110.** $(w^4 - 10)(w^4 + 9)$

111. $(ry - a + 1)(r - 1)$ **112.** $(7a^3 + 6b^2)^2$ **113.** Prime

114. $2a^2(a + 3)(a - 1)$ **115.** $(s + t - 1)^2$ **116.** $m_1 = \frac{mm_2}{m_2 - m}$

117. $(z + 4)(z - 4)$ **118.** $(xy^2 + 8z^3)(xy^2 - 8z^3)$ **119.** Prime

120. $(c + a + b)(c - a - b)$

121. $2c(4a^2 + 9b^2)(2a + 3b)(2a - 3b)$

122. $(k + 1 + 3m)(k + 1 - 3m)$ **123.** $(m + n)(m - n - 1)$

124. $(t + 4)(t^2 - 4t + 16)$

125. $(2a - 5b^3)(4a^2 + 10ab^3 + 25b^6)$

126. $\frac{\pi}{2}h(r_1 + r_2)(r_1 - r_2)$ **127. a.** 1 **b.** $-\frac{1}{2}$ **c.** 0 **128.** The domain is the set of all real numbers except -6 and 4: $(-\infty, -6) \cup (-6, 4) \cup (4, \infty)$.

129. $n(3) = 8.4$; Three hours after the injection, the concentration of pain medication in the patient's bloodstream was 8.4 milligrams per liter. **130.** Undefined **131.** $\frac{31x}{72y}$ **132.** -2 **133.** $\frac{cd}{7x^2}$

134. $\frac{2a - 1}{a + 2}$ **135.** $\frac{a + b}{(m + p)(m - 3)}$ **136.** $\frac{3x(x - 1)}{(x - 3)(x + 1)}$ **137.** $\frac{-}{c - d}$

138. $-\frac{2}{t - 3}$ **139.** $\frac{40x + 7y^2z}{112z^2}$ **140.** $\frac{12y + 20}{15y(x - 2)}$ **141.** $\frac{14y + 58}{(y + 3)(y + 7)}$

142. $\frac{5x^2 + 11x}{(x + 1)(x + 2)}$ **143.** 5 **144.** $-2, 3$ **145.** $-1, -2$ **146.** $\frac{3}{2}$

147. 0 **148.** No solution; 3 is extraneous **149.** \$5,460

150. 1.25 amps **151.** $\frac{1}{32}$ **152.** 126.72 lb **153.** Inverse variation

154. 0.2

CHAPTER 8 TEST (page 723)

1. a. solve **b.** inequality **c.** factors **d.** reciprocal **e.** function **f.** domain, range **2.** Yes **3.** $\frac{21}{5}$ **4.** -1

5. No solution, $\varnothing$; contradiction **6.** $x_1 = \frac{y_1 + mx - y}{m}$ **7.** 8

8. More than 78 **9. a.** No; $(1, -8), (1, 6)$ **b.** Yes **c.** Yes **d.** No; $(4, -4), (4, 4)$ **10.** The set of all real numbers except 6

11. -20 **12.** $8, (0, -9)$ **13.** $f(x) = -\frac{5}{4}x + \frac{5}{2}$

14. a. $T(m) = -0.16m + 331.3$ **b.** 8.9 sec **15.** 10 **16.** 49

17. $\frac{9}{16}$ **18.** $3r + 25$ **19.** 110 ft **20. a.** 0 **b.** 2, 6 **c.** D: $(-\infty, \infty)$; R: $(-\infty, 2]$ **21.** Function **22.** Not a function; $(2, 2), (2, -2)$ **23.** D: the set of real numbers, R: the set of real numbers greater than or equal to 3

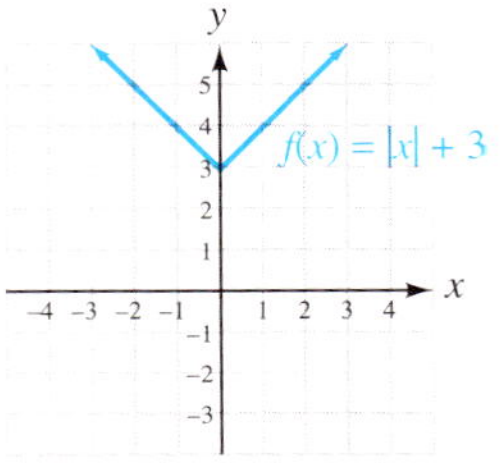

24.

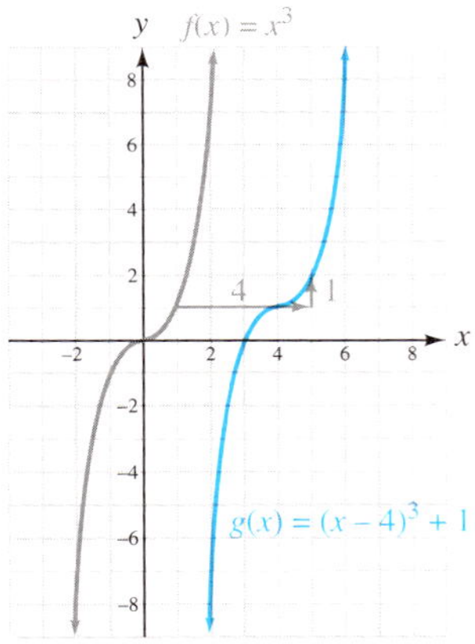

D: the set of real numbers,
R: the set of real numbers

25. $(-\infty, -5]$ 

26. $(-2, 16)$ 

27. $\left[1, \frac{9}{4}\right]$ 

28. $(-\infty, -3) \cup (8, \infty)$ 

29. $(-\infty, -9) \cup (13, \infty)$ 

30. $\left[\frac{4}{3}, \frac{8}{3}\right]$ 

31. $8, -11$ **32.** $4, -4$ **33.** $3abc(4a^2b - abc + 2c^2)$

34. $4(y^2 + 4)(y + 2)(y - 2)$
35. $(b + 5)(b^2 - 5b + 25)$ **36.** $3(u + 2)(2u - 1)$
37. $(a - y)(x + y)$ **38.** $(5m^4 - 6n)^2$ **39.** Prime
40. $(x + 3 + y)(x + 3 - y)$
41. $(4a - 5b^2)(16a^2 + 20ab^2 + 25b^4)$
42. $(x - y + 5)(x - y - 2)$ **43. a.** 1 **b.** $-\frac{1}{2}$ **c.** -1 **d.** 2
44. $p(7) = 175$ **45.** No **46.** The set of all real numbers except 0 and 1; $(-\infty, 0) \cup (0, 1) \cup (1, \infty)$ **47.** -3 **48.** $\frac{2x + y}{4y}$
49. $\frac{(x + y)^2}{2}$ **50.** $\frac{13}{x + 1}$ **51.** $\frac{2}{t - 4}$ **52.** $\frac{6a - 17}{(a + 1)(a - 2)(a - 3)}$
53. 40 **54.** 5; 3 is extraneous **55.** 6, -1 **56.** 26 **57.** $\frac{6}{5}$
58. 25 decibels

ARE YOU READY? 9.1 (page 728)

1. a. 225 **b.** 64 **2. a.** $\frac{49}{25}$ **b.** 0.04 **3. a.** -216 **b.** 81
4. a. a^8 **b.** x^9 **5.** $x^2 + 16x + 64$ **6.** 5

SELF-CHECKS 9.1

1. a. 8 **b.** -1 **c.** $\frac{1}{4}$ **d.** 0.3 **2. a.** $5|a|$ **b.** $|b^7|$ **c.** $|x - 9|$
d. $10n^4$ **3.**

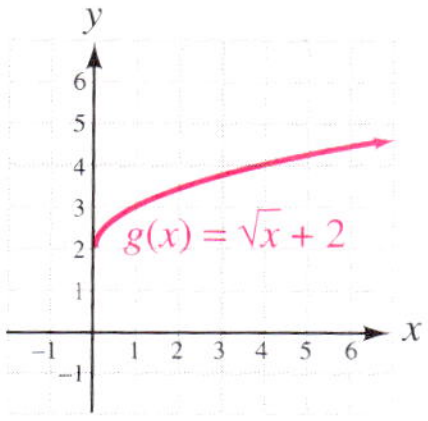

D: $[0, \infty)$; R: $[2, \infty)$; the graph is 2 units higher

4. a. $[2, \infty)$ **b.**

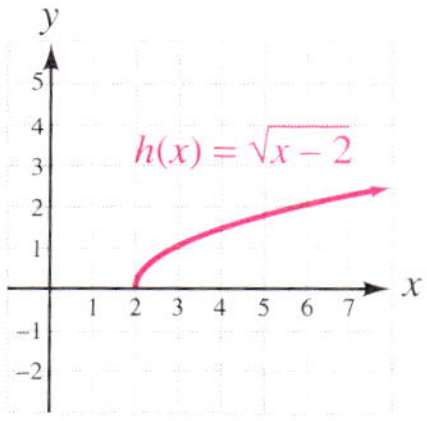

c. $[0, \infty)$

5. 1.92 sec **6. a.** 4 **b.** $-\frac{1}{10}$ **c.** $-5a$ **d.** $3m^2n^3$
7. a.

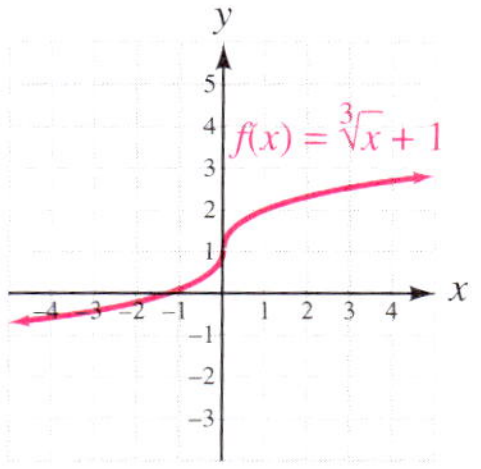

b. D: $(-\infty, \infty)$; R: $(-\infty, \infty)$

8. a. $\frac{1}{3}$ **b.** 10 **c.** Not a real number **9. a.** $|x|$ **b.** $a + 5$
c. $2a^2$

STUDY SET SECTION 9.1 (page 739)

1. square, cube **3.** positive **5.** principal **7.** simplified
9. radical **11.** a **13.** squared **15.** $|x|, x$ **17. a.** 3 **b.** 0
c. Undefined **d.** 6 **e.** None **f.** D: $[2, \infty)$; R: $[0, \infty)$ **19. a.** -5
b. -3 **c.** 1 **d.** D: $(-\infty, \infty)$, R: $(-\infty, \infty)$ **21. a.** $\sqrt{x^2} = |x|$
b. $\sqrt[3]{x^3} = x$ **c.** $\sqrt[5]{-32} = -2$ **23.** 10 **25.** -8 **27.** $\frac{1}{3}$
29. 0.5 **31.** Not a real number **33.** 11 **35.** 3.4641
37. 26.0624 **39.** $2|x|$ **41.** $9h^2$ **43.** $6|s^3|$ **45.** $12m^4$
47. $|y - 1|$ **49.** $a^2 + 3$

51. 0, -1, -2, -3, -4; D: $[0, \infty)$; R: $(-\infty, 0]$

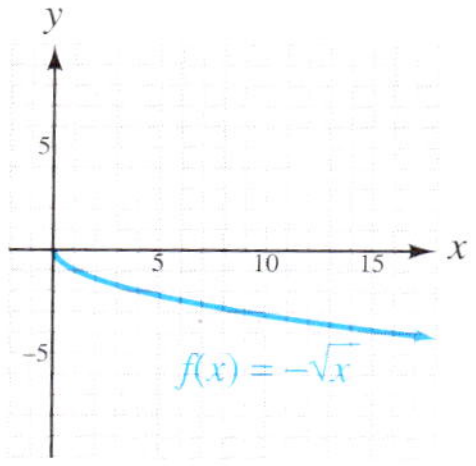

53. D: $[-4, \infty)$; R: $[0, \infty)$

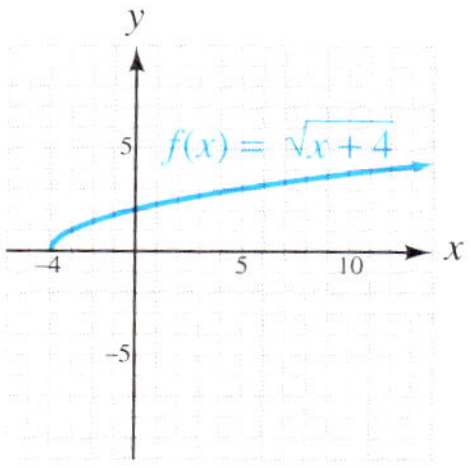

55. D: $[-6, \infty)$ **57.** D: $(-\infty, 4]$ **59.** D: $\left[\frac{4}{9}, \infty\right)$ **61.** D: $[40, \infty)$
63. a. 5 **b.** Undefined **65. a.** 2 **b.** -3 **67. a.** 4.1231
b. 2.5539 **69. a.** 3.3322 **b.** 7.7543 **71.** 1 **73.** -5 **75.** $\frac{2}{3}$
77. 4 **79.** $-6a$ **81.** $-10p^2q$
83. $-5, -4, -3, -2, -1$; D: $(-\infty, \infty)$; R: $(-\infty, \infty)$

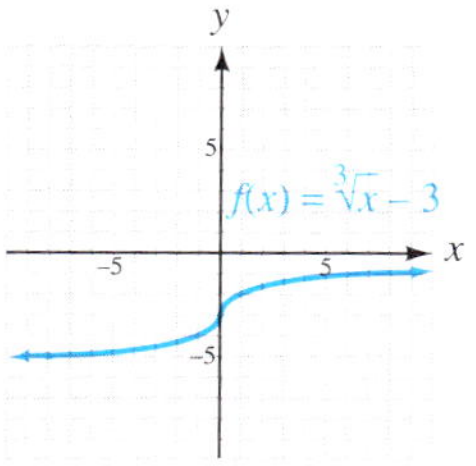

85. D: $(-\infty, \infty)$; R: $(-\infty, \infty)$

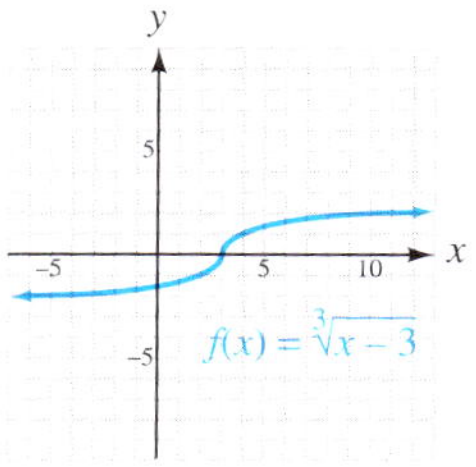

87. 3 **89.** -3 **91.** Not a real number **93.** $-\frac{1}{2}$ **95.** $2a$
97. $3|a|$ **99.** k^2 **101.** $(m + 4)^2$ **103.** $4s^3t^2$ **105.** $-7b^4$
107. $\frac{1}{2}$ **109.** $-5m^2$ **111.** $20m^8|n|$ **113.** $2|ab|$ **115.** Not a real number **117.** $|n + 6|$ **119. a.** 8 **b.** 4 **121. a.** 9 **b.** 3
123. 7.0 in. **125.** 1,138.5 mm **127.** About 58.6 beats/min
129. 13.4 ft **131.** 3.5% **137.** $\frac{(x - 4y)^2}{(x - 2y)(x - 3y)}$

ARE YOU READY? 9.2 (page 743)

1. a. 8 **b.** -4 **2. a.** 3 **b.** $\frac{1}{2}$ **3. a.** $|x + 5|$ **b.** $3a^2$ **4.** 216
5. $\frac{1}{49}$ **6.** x^9

SELF-CHECKS 9.2

1. a. 4 **b.** $-\frac{3}{2}$ **c.** -3 **d.** 1 **2. a.** $-2n$ **b.** $5|a|$ **c.** b^2
d. Not a real number **3. a.** 64 **b.** 625 **c.** 36 **d.** $-\frac{1}{16}$
4. a. $8c^6$ **b.** $9m^2n^2$ **c.** $-8a^6$ **5.** $(7ab)^{1/6}$ **6.** $\sigma = \sqrt{\frac{\Sigma(x - \mu)^2}{N}}$
7. a. $\frac{1}{3}$ **b.** $\frac{1}{216}$ **c.** $\frac{1}{9a^2}$ **d.** -27 **8. a.** $2^{3/5}$ **b.** $12^{4/3}$ **c.** x^2y^9
d. x^2 **9.** $t - 1$ **10. a.** $\sqrt{3}$ **b.** $\sqrt{7xy}$ **c.** $\sqrt[12]{m}$

STUDY SET SECTION 9.2 (page 751)

1. rational (or fractional) **3.** negative **5.** index, radicand
7.

Radical form	Exponential form	Base	Exponent
$\sqrt[5]{25}$	$25^{1/5}$	25	$\frac{1}{5}$
$(\sqrt[3]{-27})^2$	$(-27)^{2/3}$	-27	$\frac{2}{3}$
$(\sqrt[4]{16})^{-3}$	$16^{-3/4}$	16	$-\frac{3}{4}$
$(\sqrt{81})^3$	$81^{3/2}$	81	$\frac{3}{2}$
$-\sqrt{\frac{9}{64}}$	$-\left(\frac{9}{64}\right)^{1/2}$	$\frac{9}{64}$	$\frac{1}{2}$

9.

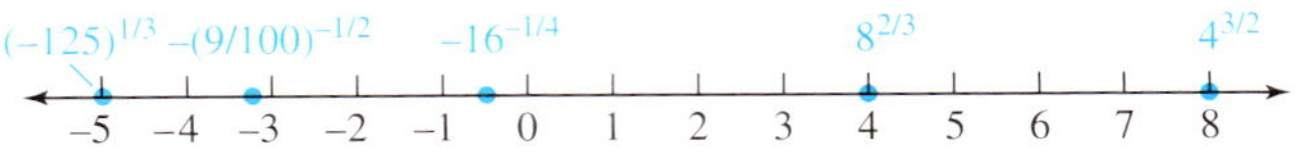

11. $\sqrt[n]{x}$ **13.** $\frac{1}{x^{m/n}}$ **15.** $100a^4, 10a^2$ **17.** 5 **19.** 3 **21.** 2 **23.** −6 **25.** −2 **27.** $\frac{1}{2}$ **29.** $2x^2$ **31.** $|x|$ **33.** Not a real number **35.** $-3n^3$ **37.** Not a real number **39.** $|x+1|$ **41.** 216 **43.** 8 **45.** $\frac{1}{36}$ **47.** −32 **49.** $125x^6$ **51.** $4x^4y^2$ **53.** $27x^3y^6$ **55.** $-\frac{x^4}{16}$ **57.** $(8abc)^{1/5}$ **59.** $(a^2-b^2)^{1/3}$ **61.** $\sqrt[4]{6x^3y}$ **63.** $\sqrt{2s^2-t^2}$ **65.** $\frac{1}{2}$ **67.** $\frac{1}{5}$ **69.** $-\frac{1}{100y^2}$ **71.** $\frac{16}{81}$ **73.** $\frac{27y^3}{8}$ **75.** 243 **77.** $9^{5/7}$ **79.** $\frac{1}{36}$ **81.** m^6 **83.** $a^{3/4}b^{1/2}$ **85.** 3 **87.** a **89.** $y+y^2$ **91.** $x^2-1+x^{3/5}$ **93.** $\sqrt{5}$ **95.** $\sqrt[3]{11}$ **97.** $\sqrt{p}$ **99.** $\sqrt[5]{xy}$ **101.** $\sqrt[18]{c}$ **103.** $\sqrt[15]{7m}$ **105.** 2.47 **107.** 1.02 **109.** $5y$ **111.** $-\frac{a^3}{27}$ **113.** $\frac{1}{64}$ **115.** p **117.** $-\frac{1}{3x^2}$ **119.** 2 **121.** $n^{3/5}-1$ **123.** 32 **125.** m^2 **127.** $\sqrt{5b}$ **129.** $2x$ **131.** $-\frac{1}{4a^2b^4}$ **133. a.** −25 **b.** 25 **c.** $-\frac{1}{25}$ **d.** 25 **135. a.** $8a^2$ **b.** $\frac{1}{8a^2}$ **c.** $-8a^2$ **d.** $\frac{1}{8a^2}$ **137.** 736 ft/sec **139.** 1.96 units **141.** 145.8 in. or 12.1 ft **145.** $2\frac{1}{2}$ hr

ARE YOU READY? 9.3 (page 754)

1. a. 4 **b.** 27 **2. a.** −2 **b.** 2 **3.** $27a^4b^8$ **4.** $7x, 9x$ **5.** $14m^2-m$ **6.** $\frac{5}{6}$

SELF-CHECKS 9.3

1. a. $3\sqrt{2}$ **b.** $14\sqrt[3]{3}$ **c.** $2\sqrt[4]{2}$ **d.** $2\sqrt[5]{4}$ **2. a.** $42b\sqrt{2b}$ **b.** $-3y\sqrt[3]{2y^2}$ **c.** $t^2u^3\sqrt[4]{u^3}$ **3. a.** $5\sqrt{11}$ **b.** $3cd\sqrt[3]{7c}$ **c.** $5x^2y\sqrt[4]{2y^2}$ **4. a.** $\frac{\sqrt[3]{25}}{3}$ **b.** $\frac{\sqrt{11}}{6a}$ **c.** $\frac{\sqrt[4]{a^3}}{5y^3}$ **5. a.** $5b$ **b.** $-10xy\sqrt[3]{x}$ **6. a.** $19\sqrt{3}+6$ **b.** $2\sqrt[3]{3}+\sqrt[4]{2}$ **7.** $21x\sqrt{2x}$

STUDY SET SECTION 9.3 (page 762)

1. like **3.** factor, perfect **5.** $\sqrt[n]{a}\sqrt[n]{b}$, product, roots **7. a.** $\sqrt{4\cdot 5}$ **b.** $\sqrt{4}\sqrt{5}$ **c.** $\sqrt{4\cdot 5}=\sqrt{4}\sqrt{5}$ **9. a.** $\sqrt{5}$, $\sqrt[3]{5}$ (Answers may vary); no **b.** $\sqrt{5}, \sqrt{6}$ (Answers may vary); no **11.** $8k^3, 8k^3, 2k$ **13.** $5\sqrt{2}$ **15.** $24\sqrt{5}$ **17.** $2\sqrt[3]{4}$ **19.** $2\sqrt[4]{3}$ **21.** $5a\sqrt{3}$ **23.** $8ab^2\sqrt{2ab}$ **25.** $-6x^2\sqrt[3]{2}$ **27.** $2x^3y\sqrt[4]{2}$ **29.** $11\sqrt{2}$ **31.** $4a\sqrt{7a}$ **33.** $-2\sqrt[5]{3a^4}$ **35.** $3x^4y\sqrt[3]{15y}$ **37.** $\frac{\sqrt{11}}{3}$ **39.** $\frac{\sqrt[4]{3}}{5}$ **41.** $\frac{x^2\sqrt[5]{3}}{2}$ **43.** $\frac{z}{4x}$ **45.** 10 **47.** $7x$ **49.** $2x^2$ **51.** $3a\sqrt[3]{a}$ **53.** $8\sqrt{7}$ **55.** $5\sqrt[3]{4}$ **57.** $12+3\sqrt{2}$ **59.** $-4\sqrt{2}$ **61.** $1-\sqrt[3]{4}$ **63.** $13\sqrt[3]{2}$ **65** $10\sqrt{2x}$ **67.** $3\sqrt{2t}+\sqrt{3t}$ **69.** $-11\sqrt[3]{2}$ **71.** $4\sqrt[4]{2}$ **73.** $m\sqrt[6]{m^5}$ **75.** $12\sqrt[3]{a}$ **77.** $5y^3\sqrt{2y}$ **79.** $4\sqrt{2b}$ **81.** $\frac{5n^2\sqrt{5}}{8}$ **83.** −10 **85.** $10\sqrt{3xy}$ **87.** $\frac{\sqrt[4]{5x}}{2z}$ **89.** $\sqrt[5]{7a^2}$ **91.** $4x\sqrt[5]{xy^2}$ **93.** $2m\sqrt[4]{13n}$ **95.** $\frac{a^2}{4}$ **97.** $\frac{\sqrt[3]{7}}{4}$ **99.** $7\sqrt{5}-3\sqrt{3}$ **101.** $2t^2\sqrt[5]{2t}$ **103.** $3\sqrt[3]{3x}$ **105. a.** $4\sqrt{5}$ **b.** $2\sqrt{21}$ **107. a.** $2x$ **b.** $2x\sqrt{x}$ **109. a.** $12+3\sqrt{6}$ **b.** $9\sqrt[3]{2}$ **111. a.** $40\sqrt[6]{6x}$ **b.** $40\sqrt[4]{6x}$ **113.** $3\sqrt{14}$ ft, 11.2 ft **115.** $5\sqrt{3}$ amps; 8.7 amps **117.** $\left(26\sqrt{5}+10\sqrt{3}\right)$ in.; 75.5 in. **123.** $-\frac{15x^5}{y}$ **125.** $3p+4-\frac{5}{2p-5}$

ARE YOU READY? 9.4 (page 765)

1. a. $25a^{12}$ **2.** $54t^6+18t^5$ **3.** $2x^2+x-3$ **4.** $1-2\sqrt{14}$ **5.** x^2-100 **6.** $\frac{18}{27a}$

SELF-CHECKS 9.4

1. a. $7\sqrt{2}$ **b.** $-20\sqrt[3]{3}$ **c.** $18x\sqrt[4]{2x}$ **2.** $12\sqrt{10}-32$ **3. a.** $-1+\sqrt{15}$ **b.** $a-\sqrt[3]{3a}+9\sqrt[3]{2a^2}-9\sqrt[3]{6}$ **4. a.** 11 **b.** $108y$ **c.** $19+8\sqrt{3}$ **d.** $x-10\sqrt{x-8}+17$ **5. a.** $\frac{2\sqrt{10}}{5}$ **b.** $\frac{5\sqrt[3]{3}}{3}$ **6.** $\frac{\sqrt{2ab}}{a}$ **7.** $\frac{7\sqrt{2c}}{6c^2}$ **8. a.** $\frac{27\sqrt[3]{10a^2}}{10a}$ **b.** $\frac{\sqrt[3]{12y^2}}{2y}$ **9.** $\frac{x-2\sqrt{2x}+2}{x-2}$ **10.** $\frac{x-9}{x-3\sqrt{x}}$

STUDY SET SECTION 9.4 (page 773)

1. product **3.** rationalize **5.** perfect

7.

	Why isn't it in simplified form?	Simplified form	Approximation
$\frac{3}{\sqrt{2}}$	A radical appears in the denominator.	$\frac{3\sqrt{2}}{2}$	2.121320344
$\frac{\sqrt{18}}{2}$	There is a perfect-square factor in the radicand: 9	$\frac{3\sqrt{2}}{2}$	2.121320344
$\sqrt{\frac{9}{2}}$	The radicand contains a fraction.	$\frac{3\sqrt{2}}{2}$	2.121320344

9. $5, 2, \sqrt{x}, 25, x$ **11. a.** $6\sqrt{6}$ **b.** 48 **c.** Can't be simplified **d.** $-6\sqrt{6}$ **13.** $\sqrt{6}, 48, 16, 4$ **15.** $3\sqrt{5}$ **17.** $6\sqrt{2}$ **19.** 18 **21.** $2\sqrt[3]{3}$ **23.** $48ab^2$ **25.** $5a\sqrt[4]{a}$ **27.** $12\sqrt{5}-15$ **29.** $8\sqrt{3}+2\sqrt{14}$ **31.** $-8x\sqrt{10}+6\sqrt{15x}$ **33.** $8+2\sqrt[3]{3}$ **35.** $-1-2\sqrt{2}$ **37.** $3x-2y$ **39.** $12\sqrt[3]{2}+8\sqrt[3]{5}-18-6\sqrt[3]{20}$ **41.** $\sqrt[3]{25z^2}+3\sqrt[3]{15z}+2\sqrt[3]{9}$ **43.** 7 **45.** 12 **47.** 18 **49.** $-16x^2$ **51.** $39-12\sqrt{3}$ **53.** $3x+6\sqrt{x}+3$ **55.** $\frac{\sqrt{14}}{7}$ **57.** $\frac{2\sqrt{6}}{3}$ **59.** $\frac{2\sqrt{6}}{3}$ **61.** $\frac{\sqrt[3]{4}}{2}$ **63.** $\sqrt[3]{3}$ **65.** $\frac{\sqrt[4]{2}}{2}$ **67.** $\frac{\sqrt{5y}}{y}$ **69.** $\frac{\sqrt{6y}}{y}$ **71.** $\frac{\sqrt[3]{6t}}{3}$ **73.** $\frac{\sqrt[3]{2ab^2}}{b}$ **75.** $\frac{23\sqrt{2p}}{10p^3}$ **77.** $\frac{7\sqrt{6b}}{12b^2}$ **79.** $\frac{\sqrt[3]{20}}{4}$ **81.** $\frac{\sqrt[3]{36}}{9}$ **83.** $\frac{19\sqrt[3]{25c}}{5c}$ **85.** $\frac{\sqrt[3]{12r^2}}{2r}$ **87.** $\frac{\sqrt[4]{54t^2}}{3t}$ **89.** $\frac{25\sqrt[3]{2a^3}}{2a}$ **91.** $\frac{3\sqrt{2}-\sqrt{10}}{4}$ **93.** $\frac{2(\sqrt{x}-1)}{x-1}$ or $\frac{2\sqrt{x}-2}{x-1}$ **95.** $\frac{9-2\sqrt{14}}{5}$ **97.** $\frac{x-2\sqrt{xy}+y}{x-y}$ **99.** $\frac{x-9}{x(\sqrt{x}-3)}$ **101.** $\frac{x-y}{\sqrt{x}(\sqrt{x}-\sqrt{y})}$ **103.** $x\sqrt{14}+\sqrt{2x}$ **105.** $\frac{3\sqrt{6}+4}{2}$ **107.** $2{,}000x$ **109.** $-20r\sqrt[3]{10s}$ **111.** $9p^2+6p\sqrt{5}+5$ **113.** $\frac{6m^2\sqrt{2m}}{5}$ **115.** $3n\sqrt[4]{n}$ **117.** $\frac{\sqrt[3]{3x}}{3}$ **119.** $18r-12\sqrt{2r}+4$ **121.** $x^2(x+3)$ **123.** $\sqrt{2z}+1$ **125. a.** $9a$ **b.** $9+6\sqrt{a}+a$ **127. a.** $m-6$ **b.** $m-12\sqrt{m}+36$ **129.** $\frac{\sqrt{2\pi}}{2\pi\sigma}$ **131.** $\frac{\sqrt{2}}{2}$ **139.** $\frac{1}{3}$

ARE YOU READY? 9.5 (page 777)

1. a. $x-1$ **b.** x^3+7 **2.** $18x+45$ **3.** $x^2-8x+16$ **4.** −3, 9 **5.** $x^3+6x^2+12x+8$ **6.** $4-4\sqrt{x}+x$

SELF-CHECKS 9.5

1. 11 **2.** 196 ft **3.** 6, 0 is extraneous **4.** 18 is extraneous, no solution, $\varnothing$ **5.** 0, -2 **6.** 0 **7.** 20 **8.** 1 **9.** $n = \frac{z_0{}^2 pq}{E^2}$

STUDY SET SECTION 9.5 (page 785)

1. radical **3.** raising **5.** extraneous **7. a.** n, n **b.** a **9. a.** Square both sides. **b.** Subtract 3 from both sides. **c.** Add $\sqrt{2x}$ to both sides. **11.** $x - 6\sqrt{x} + 9$ **13.** 6, 2, 2, $3x + 3$, 33, 11, Yes **15.** 4 **17.** 5 **19.** 6 **21.** 198 **23.** 7, ~~-1~~ **25.** 14, ~~6~~ **27.** 3, ~~-2~~ **29.** 0, ~~-7~~ **31.** ~~$\frac{4}{5}$~~, no solution **33.** ~~4~~, no solution **35.** 4 **37.** 2, -1 **39.** 1, -3 **41.** 10 **43.** 85 **45.** -7 **47.** 12 **49.** 0 **51.** 3 **53.** 3 **55.** 9 **57.** 1 **59.** ~~-3~~, no solution **61.** 1, 9 **63.** $h = \frac{v^2}{2g}$ **65.** $l = \frac{8T^2}{\pi^2}$ **67.** $A = P(r + 1)^3$ **69.** $v^2 = c^2\left(1 - \frac{L_A{}^2}{L_B{}^2}\right)$ **71.** 16 **73.** 4 **75.** -1, 1 **77.** 2, ~~7~~ **79.** -16 **81.** 4 **83.** $\frac{5}{2}$, $\frac{1}{2}$ **85.** $-\frac{1}{2}$ **87.** ~~6~~, no solution **89.** 1 **91.** 4, 3 **93.** ~~8~~, no solution **95.** 1 **97.** 0 **99.** 8 **101. a.** $\frac{81}{40}$ **b.** $\frac{9}{5}$ **103. a.** 50 **b.** ~~50~~, no solution **105.** 178 ft **107.** About 488 watts **109.** 8 ft **111.** \$5 **121.** 2.5 foot-candles **123.** 0.41511 in.

ARE YOU READY? 9.6 (page 788)

1. 180° **2.** $a\sqrt{3}$ **3.** $\frac{25\sqrt{3}}{3}$ **4.** 13 **5.** 3.46 **6.** $\frac{14\sqrt{3}}{3}$

SELF-CHECKS 9.6

1. Yes **2.** $12\sqrt{2}$ m $\approx$ 16.97 m **3.** $\frac{9\sqrt{2}}{2}$ in. $\approx$ 6.36 in. **4.** 16 cm, $8\sqrt{3}$ cm **5.** $10\sqrt{3}$ in. $\approx$ 17.32 in., $5\sqrt{3}$ in. $\approx$ 8.66 in. **6.** $\frac{29\sqrt{3}}{2}$ in. $\approx$ 25 in. **7. a.** 13 units **b.** $3\sqrt{5} \approx 6.71$ units **8.** (2, 5) **9.** (1, 13)

STUDY SET SECTION 9.6 (page 796)

1. hypotenuse **3.** Pythagorean **5.** a^2, b^2, c^2, equation **7.** $\sqrt{2}$ **9.** $\sqrt{3}$ **11.** $(x_2 - x_1)^2$, $(y_2 - y_1)^2$ **13.** 6, 52, 4, 2, 7.21 **15.** 10 ft **17.** 17 ft **19.** 40 ft **21.** 24 cm **23.** $x = 2$, $h = 2\sqrt{2} \approx 2.83$ **25.** $3.2\sqrt{2}$ ft $\approx$ 4.53 ft **27.** $x = \frac{3\sqrt{2}}{2} \approx 2.12$, $y = \frac{3\sqrt{2}}{2} \approx 2.12$ **29.** $5\sqrt{2}$ in. $\approx$ 7.07 in. **31.** $x = 5\sqrt{3} \approx 8.66$, $h = 10$ **33.** $75\sqrt{3}$ cm $\approx$ 129.90 cm, 150 cm **35.** $x = \frac{40\sqrt{3}}{3} \approx 23.09$, $h = \frac{80\sqrt{3}}{3} \approx 46.19$ **37.** $\frac{55\sqrt{3}}{3}$ mm $\approx$ 31.75 mm, $\frac{110\sqrt{3}}{3}$ mm $\approx$ 63.51 mm **39.** $x = 50$, $y = 50\sqrt{3} \approx 86.60$ **41.** 0.75 ft, $0.75\sqrt{3}$ ft $\approx$ 1.30 ft **43.** 5 **45.** 13 **47.** 10 **49.** $\sqrt{34}$ **51.** $2\sqrt{5}$ **53.** $3\sqrt{10}$ **55.** (3, 4) **57.** (9, 12) **59.** $\left(\frac{1}{2}, -8\right)$ **61.** $\left(-\frac{3}{2}, \frac{5}{2}\right)$ **63.** (4, 1) **65.** $(-20, -3)$ **67. a.** 118.73 m **b.** 133.14 m **69.** $7\sqrt{3}$ cm **71.** $\left(5\sqrt{2}, 0\right)$, $\left(0, 5\sqrt{2}\right)$, $\left(-5\sqrt{2}, 0\right)$, $\left(0, -5\sqrt{2}\right)$; (7.07, 0), (0, 7.07), $(-7.07, 0)$, $(0, -7.07)$ **73.** $10\sqrt{3}$ mm $\approx$ 17.32 mm **75.** $10\sqrt{181}$ ft $\approx$ 134.54 ft **77.** About 0.13 ft **79. a.** 21.21 units **b.** 8.25 units **c.** 13.00 units **81.** Yes **87.** 7

ARE YOU READY? 9.7 (page 800)

1. No real number squared is equal to -16. **2.** $6x^2 - 2x$ **3.** $14n^2 + 19n - 3$ **4.** $3\sqrt{7}$ **5.** $\frac{7(\sqrt{x} - 4)}{x - 16}$ **6.** 21R3

SELF-CHECKS 9.7

1. a. $5i$ **b.** $-i\sqrt{19}$ **c.** $3i\sqrt{5}$ **d.** $\frac{5\sqrt{2}}{9}i$ **2. a.** $-18 + 0i$ **b.** $0 + 6i$ **c.** $1 + 2i\sqrt{6}$ **3. a.** $1 + 2i$ **b.** $5 - 12i$ **4.** $-4\sqrt{6}$ **5. a.** $18 + 2i$ **b.** $-40 + 70i$ **6.** $0 + 13i$ **7.** 13 **8. a.** $\frac{30}{29} - \frac{12}{29}i$ **b.** $\frac{11}{17} - \frac{7}{17}i$ **9.** $7 + 0i$ **10.** $0 - \frac{3}{4}i$ **11. a.** -1 **b.** i

STUDY SET SECTION 9.7 (page 809)

1. imaginary, power **3.** real, imaginary **5. a.** $\sqrt{-1}$ **b.** -1 **c.** $-i$ **d.** 1 **e.** four **7. a.** real, imaginary **b.** FOIL **9. a.** $2 + 3i$ **b.** $2 - 0i$ **c.** $0 + 3i$ **11.**

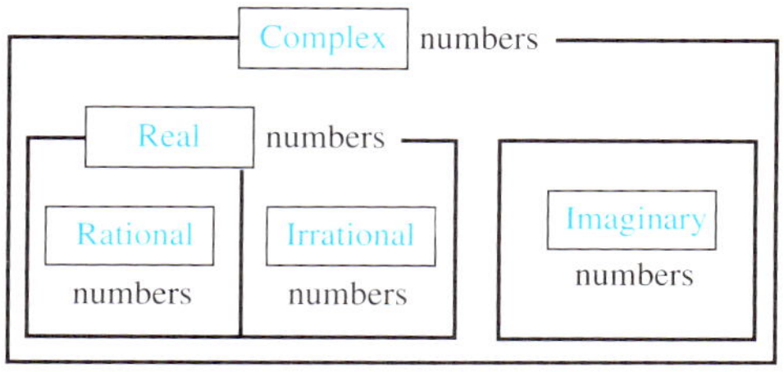

13. 9, $6i$, 2, 11 **15. a.** True **b.** False **c.** False **d.** False **17.** $3i$ **19.** $\sqrt{7}i$ or $i\sqrt{7}$ **21.** $2\sqrt{6}i$ or $2i\sqrt{6}$ **23.** $-6\sqrt{2}i$ or $-6i\sqrt{2}$ **25.** $45i$ **27.** $\frac{5}{3}i$ **29. a.** $5 + 0i$ **b.** $0 + 7i$ **31. a.** $1 + 5i$ **b.** $-3 + 2i\sqrt{2}$ **33. a.** $76 - 3i\sqrt{6}$ **b.** $-7 + i\sqrt{19}$ **35. a.** $-6 - 3i$ **b.** $3 + i\sqrt{6}$ **37.** $8 - 2i$ **39.** $15 + 2i$ **41.** $3 - 5i$ **43.** $1 + 3i$ **45.** -6 **47.** $-2\sqrt{6}$ **49.** $6 - 27i$ **51.** $35 - 28i$ **53.** $6 + 14i$ **55.** $-25 - 25i$ **57.** $7 + i$ **59.** $12 + 5i$ **61.** $13 - i$ **63.** $3 + 4i$ **65.** 40 **67.** 65 **69.** $\frac{45}{26} - \frac{9}{26}i$ **71.** $-\frac{77}{65} + \frac{44}{65}i$ **73.** $\frac{14}{17} - \frac{5}{17}i$ **75.** $-\frac{6}{29} + \frac{43}{29}i$ **77.** $\frac{11}{10} + \frac{13}{10}i$ **79.** $0 - i$ **81.** $4 + 0i$ **83.** $-2 + 0i$ **85.** $0 - \frac{5}{3}i$ **87.** $0 + \frac{2}{7}i$ **89.** i **91.** $-i$ **93.** 1 **95.** -1 **97.** $4 - 11i$ **99.** $14 - 8i$ **101.** $-2 - 6i$ **103.** $-\frac{4}{13} - \frac{6}{13}i$ **105.** $18 + 12i$ **107.** $0 + \frac{4}{5}i$ **109.** $8 + \sqrt{2}i$ or $8 + i\sqrt{2}$ **111.** $-2 + 7i$ **113.** $-48 - 64i$ **115.** $\frac{1}{4} - \frac{\sqrt{15}}{4}i$ **117. a.** $2i\sqrt{2}$ **b.** -2 **119. a.** $-4 + 0i$ **b.** $3 + 4i$ **121.** $-1 + i$ **127.** 20 mph

CHAPTER 9 REVIEW (page 812)

1. a. 2 **b.** 0 **c.** 6 **d.** D: $[-3, \infty)$, R: $[0, \infty)$ **2. a.** $10|a|$ **b.** $10a$ **3.** 7 **4.** -11 **5.** $\frac{15}{7}$ **6.** Not a real number **7.** $10a^6$ **8.** $5|x|$ **9.** x^4 **10.** $|x + 2|$ **11.** -3 **12.** -6 **13.** $4x^2y$ **14.** $\frac{x^3}{5}$ **15.** 2 **16.** -2 **17.** $4x^2|y|$ **18.** $x + 1$ **19.** $-\frac{1}{2}$ **20.** Not a real number **21.** Not a real number **22.** 0 **23.** $[-5, \infty)$ **24.** 13 ft **25.** 24 cm^2 **26.** 2.3276

27. D: $[0, \infty)$; R: $[0, \infty)$

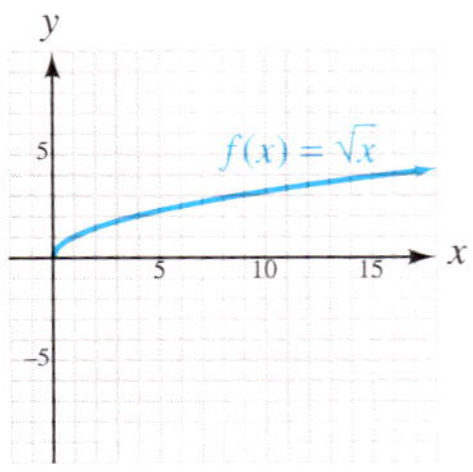

28. D: $(-\infty, \infty)$; R: $(-\infty, \infty)$

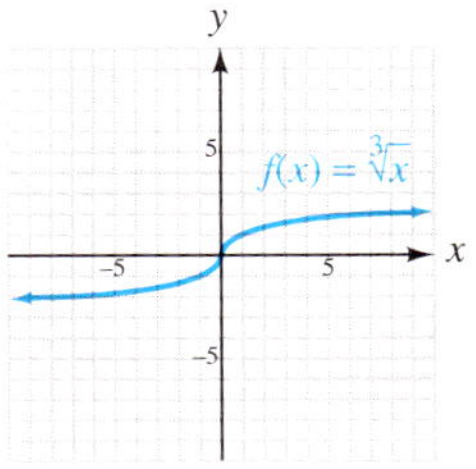

29. D: $[-2, \infty)$; R: $[0, \infty)$

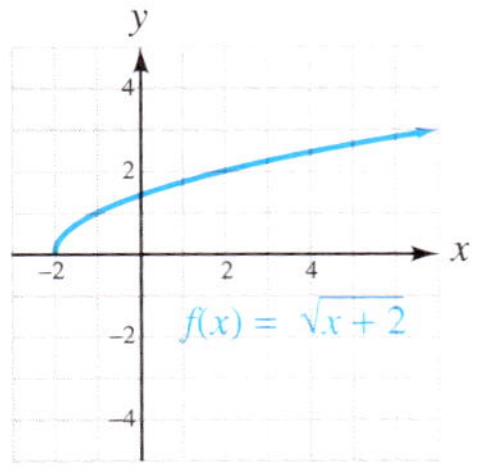

30. D: $(-\infty, \infty)$; R: $(-\infty, \infty)$

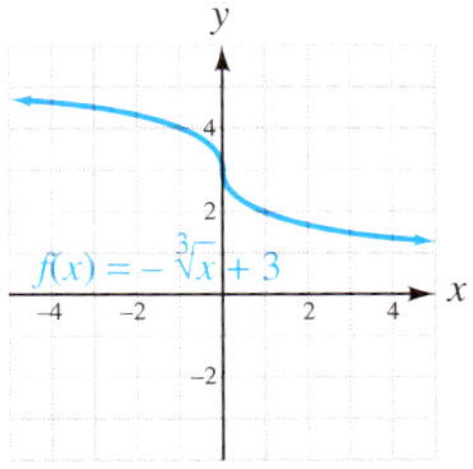

31. $\sqrt{t}$ **32.** $\sqrt[4]{5xy^3}$ **33.** 5 **34.** -6 **35.** Not a real number **36.** 1 **37.** $\frac{3}{x}$ **38.** -2 **39.** 5 **40.** $3cd$ **41.** 27 **42.** $\frac{1}{4}$ **43.** $-16{,}807$ **44.** 10 **45.** $\frac{27}{8}$ **46.** $\frac{1}{3{,}125}$ **47.** $125x^3y^6$ **48.** $\frac{1}{4u^4v^2}$ **49.** $5^{3/4}$ **50.** $a^{1/7}$ **51.** k^8 **52.** $3^{2/3}$ **53.** $u - 1$ **54.** $v + v^2$ **55.** $\sqrt{a}$ **56.** $\sqrt[6]{c}$ **57.** 183 mi **58.** Two true statements result: $32 = 32$. **59.** $4\sqrt{5}$ **60.** $3\sqrt[3]{2}$ **61.** $2\sqrt[4]{10}$ **62.** $-2\sqrt[5]{3}$ **63.** $2x^2\sqrt{2x}$ **64.** $r^4\sqrt[4]{r}$ **65.** $-3j^2\sqrt[3]{jk}$ **66.** $-2xy\sqrt[3]{2x^2y}$ **67.** $\frac{\sqrt{m}}{12n^6}$ **68.** $\frac{\sqrt{17xy}}{8a^2}$ **69.** $2x$ **70.** $3x^3$ **71.** $3\sqrt{2}$ **72.** $11\sqrt{5}$ **73.** 0 **74.** $-8a\sqrt[4]{2a}$ **75.** $29x\sqrt{2}$ **76.** $13x\sqrt[3]{2}$ **77.** $9\sqrt[4]{2t^3} - 8\sqrt[4]{6t^3}$ **78.** $12x^2\sqrt[4]{x} + 5x\sqrt[4]{x}$ **80.** $\left(6\sqrt{2} + 2\sqrt{10}\right)$ in., 14.8 in. **81.** 7 **82.** $6\sqrt{10}$ **83.** 32 **84.** $6\sqrt{10}$ **85.** $3x$ **86.** $x + 1$ **87.** $-2x^3\sqrt[3]{x}$ **88.** 3 **89.** $42t + 9t\sqrt{21t}$ **90.** $-2x^3y^3\sqrt[3]{2x^2y^2}$ **91.** $3b + 6\sqrt{b} + 3$ **92.** $\sqrt[3]{9p^2} - \sqrt[3]{6p} - 2\sqrt[3]{4}$ **93.** $\frac{10\sqrt{3}}{3}$ **94.** $\frac{\sqrt{15xy}}{5xy}$ **95.** $\frac{\sqrt[3]{6u^2}}{u^2}$ **96.** $\frac{\sqrt[4]{27ab^2}}{3b}$ **97.** $2\left(\sqrt{2} + 1\right)$ or $2\sqrt{2} + 2$ **98.** $\frac{12\sqrt{xz} - 16x - 2z}{z - 16x}$ **99.** $\frac{a - b}{a + \sqrt{ab}}$ **100.** $r = \frac{\sqrt[3]{6\pi^2V}}{2\pi}$ **101.** 22 **102.** 16, 9 **103.** $\frac{13}{2}$ **104.** $\frac{9}{16}$ **105.** 2, -4 **106.** 7 **107.** 1 **108.** $-\frac{3}{2}$, 1 **109.** ~~3~~, no solution **110.** 6, ~~-2~~ **111.** 0, 3 **112.** $-\frac{1}{2}$, 4 **113.** $P = \frac{A}{(r+1)^2}$ **114.** $I = \frac{h^3b}{12}$ **115.** 17 ft **116.** 88 yd **117.** $7\sqrt{2}$ m ≈ 9.90 m **118.** $\frac{15\sqrt{2}}{2}$ yd ≈ 10.61 yd **119.** Shorter leg: 6 cm, longer leg: $6\sqrt{3}$ cm ≈ 10.39 cm **120.** $40\sqrt{3}$ ft ≈ 69.28 ft, $20\sqrt{3}$ ft ≈ 34.64 ft **121.** $x = 5\sqrt{2} \approx 7.07$, $y = 5$ **122.** $x = 25\sqrt{3} \approx 43.30$, $y = 25$ **123.** 13 **124.** $2\sqrt{2}$ **125.** $(7, -3)$ **126.** $(2, -2)$ **127.** $5i$ **128.** $3i\sqrt{2}$ **129.** $-i\sqrt{6}$ **130.** $\frac{3}{8}i$ **131.** Real, Imaginary **132. a.** True **b.** True **c.** False **d.** False **133. a.** $3 - 6i$ **b.** $0 - 19i$ **134. a.** $-1 + 7i$ **b.** $0 + i$ **135.** $8 - 2i$ **136.** $3 - 5i$ **137.** $3 + 6i$ **138.** $22 + 29i$ **139.** $-3\sqrt{3} + 0i$ **140.** $-81 + 0i$ **141.** $4 + i$ **142.** $0 - \frac{3}{11}i$ **143.** -1 **144.** i

CHAPTER 9 TEST (page 821)

1. a. radical **b.** imaginary **c.** extraneous **d.** isosceles **e.** rationalize **f.** complex **2. a.** If $\sqrt[n]{a}$ and $\sqrt[n]{b}$ are real numbers, then $\sqrt[n]{ab} = \sqrt[n]{a}\sqrt[n]{b}$. **b.** If $\sqrt[n]{a}$ and $\sqrt[n]{b}$ are real numbers, then $\sqrt[n]{\frac{a}{b}} = \frac{\sqrt[n]{a}}{\sqrt[n]{b}}$, $(b \neq 0)$. **c.** No real number raised to the fourth power is -16.

3. D: $[1, \infty)$; R: $[0, \infty)$

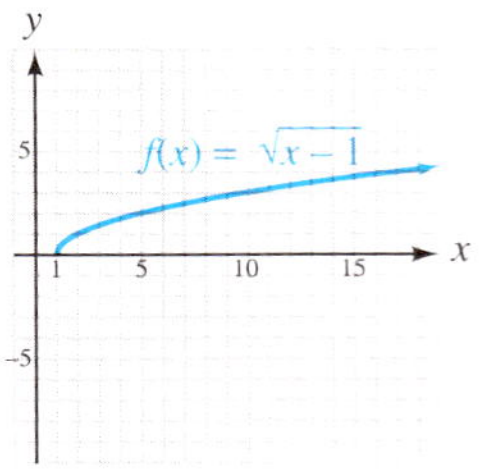

4. 46 ft/sec **5. a.** -1 **b.** 2 **c.** 1 **d.** D: $(-\infty, \infty)$; R: $(-\infty, \infty)$ **6.** $[-5, \infty)$ **7.** $7x^2$ **8.** -9 **9.** $\frac{1}{216}$ **10.** $\frac{25n^4}{4}$ **11.** $2^{4/3}$ **12.** $a^{1/9}$ **13.** $|x|$ **14.** $|y - 5|$ **15.** $-4xy^2$ **16.** $\frac{2}{3}a$ **17.** $t + 8$ **18.** $6xy^2\sqrt{15xy}$ **19.** $2x^5y\sqrt[3]{3}$ **20.** $2\sqrt[4]{2}$ **21.** $2y^2\sqrt{3y}$ **22.** $14\sqrt[3]{5}$ **23.** $5z^3\sqrt[3]{3z}$ **24.** $-6x\sqrt{y} - 2xy^2$ **25.** $3 - 7\sqrt{6}$ **26.** $\sqrt[3]{4a^2} + 18\sqrt[3]{2a} + 81$ **27.** $\frac{4\sqrt{10}}{5}$ **28.** $\frac{x + 2\sqrt{xy} + y}{x - y}$ **29.** $\frac{\sqrt[3]{18a^2}}{2a}$ **30.** $\frac{1}{\sqrt{2}(\sqrt{5} - 3)} = \frac{1}{\sqrt{10} - 3\sqrt{2}}$ **31.** $\frac{1}{15}$ **32.** 10 **33.** ~~4~~, no solution **34.** 3, ~~-3~~ **35.** 2, 3 **36.** ~~108~~, no solution **37.** -2 **38.** $G = \frac{4\pi^2r^3}{Mt^2}$ **39.** $x = \frac{8\sqrt{3}}{3}$ cm ≈ 4.62 cm, $h = \frac{16\sqrt{3}}{3}$ cm ≈ 9.24 cm **40.** $x = 6.13\sqrt{2}$ in. ≈ 8.67 in., $y = 6.13\sqrt{2}$ in. ≈ 8.67 in. **41.** 25 **42.** $\left(\frac{5}{2}, -8\right)$ **43.** $(7, 3)$ **44.** 28 in. **45.** $3i\sqrt{5}$ **46.** -1 **47.** $-4 + 11i$ **48.** $4 - 7i$ **49.** $75 + 45i$ **50.** $-46 - 78i$ **51.** $0 - \frac{\sqrt{2}}{2}i$ **52.** $\frac{1}{2} + \frac{1}{2}i$

ARE YOU READY? 10.1 (page 826)

1. a. $2\sqrt{7}$ **b.** $6i$ **2.** $\frac{\sqrt{30}}{6}$ **3.** $\frac{9}{2}$ **4.** $\frac{49}{4}$ **5.** 169 **6.** $(x - 4)^2$

SELF CHECKS 10.1

1. $\pm 3\sqrt{2}$ **2.** $\sqrt{\frac{14}{\pi}}$ ft ≈ 2.11 ft **3.** $\pm\frac{7}{4}i$ **4.** 1, -5 **5.** $\frac{-11 \pm \sqrt{3}}{4}$ **6.** $a^2 - 5a + \frac{25}{4} = \left(a - \frac{5}{2}\right)^2$ **7.** $5 \pm \sqrt{29}$; 10.39, -0.39 **8.** $\frac{4}{3}$, -2 **9.** $\frac{-3 \pm \sqrt{6}}{3}$; -0.18, -1.82 **10. a.** $3 \pm 3i\sqrt{2}$ **b.** $-\frac{3}{5} \pm \frac{i\sqrt{74}}{5}$

STUDY SET SECTION 10.1 (page 836)

1. quadratic **3.** square **5.** $\sqrt{c}, -\sqrt{c}$ **7. a.** 36 **b.** $\frac{25}{4}$ **9. a.** Subtract 7 from both sides. **b.** Divide both sides by 4. **11.** It is a solution. **13.** plus or minus **15.** $\pm\sqrt{11}$ **17.** $\pm\sqrt{35}$ **19.** $\pm 5\sqrt{2}$ **21.** $\pm\frac{4\sqrt{3}}{3}$ **23.** $\pm 4i$ **25.** $\pm 2i\sqrt{2}$ **27.** $\pm\frac{9}{2}i$ **29.** $\pm 2i\sqrt{6}$ **31.** -8, -2 **33.** 0, -8 **35.** $-5 \pm \sqrt{3}$ **37.** $\frac{2 \pm 2\sqrt{2}}{7}$ **39.** $x^2 + 24x + 144 = (x + 12)^2$ **41.** $a^2 - 7a + \frac{49}{4} = \left(a - \frac{7}{2}\right)^2$ **43.** $x^2 + \frac{2}{3}x + \frac{1}{9} = \left(x + \frac{1}{3}\right)^2$ **45.** $m^2 - \frac{5}{6}m + \frac{25}{144} = \left(m - \frac{5}{12}\right)^2$ **47.** $2 \pm \sqrt{6}$; 4.45, -0.45 **49.** $6 \pm \sqrt{35}$; 11.92, 0.08 **51.** $-10 \pm 5\sqrt{3}$; -1.34, -18.66

53. $-8 \pm 4\sqrt{5}$; 0.94, −16.94 **55.** $-\frac{1}{2}, 1$ **57.** $\frac{3}{4}, -\frac{1}{3}$
59. $\frac{6 \pm \sqrt{33}}{3}$; 3.91, 0.09 **61.** $\frac{-5 \pm \sqrt{41}}{4}$; 0.35, −2.85
63. $-1 \pm i$ **65.** $-4 \pm i\sqrt{2}$ **67.** $-\frac{1}{3} \pm \frac{i\sqrt{62}}{3}$ **69.** $\frac{1}{4} \pm \frac{i\sqrt{15}}{4}$
71. $2, -\frac{4}{3}$ **73.** $\frac{3 \pm 2\sqrt{3}}{3}$; 2.15, −0.15 **75.** $-4 \pm \sqrt{10}$; −7.16, −0.84 **77.** $\pm 2i\sqrt{3}$ **79.** $1 \pm 3\sqrt{2}$; 5.24, −3.24
81. $\frac{7 \pm \sqrt{37}}{2}$; 6.54, 0.46 **83.** $\pm\sqrt{5}$; ± 2.24
85. $\frac{-7 \pm \sqrt{29}}{10}$; −0.16, −1.24 **87.** $-\frac{1}{2} \pm \frac{\sqrt{11}}{2}i$
89. $\frac{-5 \pm 2\sqrt{6}}{8}$; −1.24, −0.01 **91.** −3, 9 **93.** $-\frac{1}{4} \pm \frac{\sqrt{11}}{4}i$
95. a. $\pm 2\sqrt{6}$; ± 4.90 **b.** $\pm 2i\sqrt{6}$ **97. a.** 0, 4 **b.** $\frac{4 \pm 3\sqrt{2}}{2}$; 4.12, −0.12 **99. a.** $2 \pm 4i$ **b.** $2 \pm 2\sqrt{6}$; 6.90, −2.90
101. a. $\frac{2 \pm i\sqrt{2}}{2}$ **b.** $\frac{2 \pm \sqrt{10}}{2}$; 2.58, −0.58 **103.** 4.4 sec
105. 1.6 sec **107.** 1.70 in. **109.** $c = \frac{\sqrt{Em}}{m}$ **115.** $2ab^2\sqrt[3]{5}$
117. $\frac{2}{5}$

ARE YOU READY? 10.2 (page 838)

1. 11 **2.** $3\sqrt{5}$ **3.** 3; 2, −1, 7 **4.** $\frac{3}{4}$, −2 **5.** Not a real number **6.** 3.87

SELF CHECKS 10.2

1. $2, -\frac{1}{4}$ **2.** $\frac{1 \pm \sqrt{10}}{3}$; −0.72, 1.39 **3.** $-\frac{3}{2} \pm \frac{\sqrt{11}}{2}i$
4. a. $6x^2 - 7x + 9 = 0$ **b.** $2x^2 - 4x - 5 = 0$ **c.** $4x^2 + 6x - 9 = 0$ **d.** $8x^2 - 7x - 2 = 0$ **5.** 9 in., 40 in.
6. \$1 **7.** Late 1988 **8.** $\frac{4 + \sqrt{31}}{2}$ in. ≈ 0.78 in.

STUDY SET SECTION 10.2 (page 848)

1. quadratic **3. a.** $x^2 + 2x + 5 = 0$ **b.** $3x^2 + 2x - 1 = 0$
5. a. True **b.** True **c.** False **7. a.** 2, −4 **b.** $\frac{1 \pm \sqrt{33}}{4}$
9. a. $1 \pm 2\sqrt{2}$ **b.** $\frac{-3 \pm \sqrt{7}}{2}$ **11. a.** The fraction bar wasn't drawn under both terms of the numerator. **b.** A ± sign wasn't written between $-b$ and the radical. **13.** 1, 2
15. A repeated solution of −6 **17.** $-\frac{3}{2}, 1$ **19.** $\frac{2}{3}, -\frac{1}{4}$
21. $\frac{1 \pm \sqrt{29}}{2}$; 3.19, −2.19 **23.** $\frac{-5 \pm \sqrt{5}}{10}$; −0.28, −0.72
25. $\frac{-3 \pm \sqrt{6}}{3}$; −0.18, −1.82 **27.** $\frac{1 \pm 2\sqrt{5}}{2}$; 2.74, −1.74
29. $-\frac{1}{4} \pm \frac{\sqrt{7}}{4}i$ **31.** $\frac{1}{3} \pm \frac{\sqrt{2}}{3}i$ **33.** $1 \pm i$ **35.** $-\frac{1}{2} \pm i$
37. a. $5x^2 - 9x + 2 = 0$ **b.** $16t^2 + 24t - 9 = 0$
39. a. $3x^2 + 2x - 1 = 0$ **b.** $2m^2 - 3m - 2 = 0$ **41.** $\frac{4}{3}, -\frac{2}{5}$
43. $\frac{2}{3} \pm \frac{\sqrt{2}}{3}i$ **45.** $\frac{1}{4}, -\frac{3}{4}$ **47.** $\frac{3 \pm \sqrt{17}}{4}$; 1.78; −0.28
49. $5 \pm \sqrt{7}$; 2.35, 7.65 **51.** 23, −17
53. $\frac{-5 \pm 3\sqrt{5}}{2}$; 0.85, −5.85 **55.** $\frac{1}{3} \pm \frac{\sqrt{6}}{3}i$
57. $\frac{-3 \pm \sqrt{29}}{10}$; 0.24, −0.84 **59.** $\frac{10 \pm \sqrt{55}}{30}$; 0.58, 0.09
61. $2 \pm 2i$ **63.** $3 \pm i\sqrt{5}$ **65.** $\frac{-5 \pm \sqrt{17}}{2}$; −0.44, −4.56
67. $\frac{9 \pm \sqrt{89}}{2}$; 9.22, −0.22 **69.** $\frac{1 \pm \sqrt{7}}{3}$; 1.22, −0.55
71. $\frac{35 \pm 5\sqrt{329}}{14}$; 8.98, −3.98 **73. a.** $-2 \pm \sqrt{11}$; 1.32, −5.32
b. $2 \pm \sqrt{11}$; 5.32, −1.32 **75. a.** 6, −4 **b.** $1 \pm i\sqrt{7}$
77. a. A repeated solution of 21 **b.** A repeated solution of −21
79. 40 ft **81.** 0.7 units, 2.4 units **83.** 97 ft by 117 ft
85. About 0.5 mi by 2.5 mi **87.** 25 sides **89.** \$4.80 or \$5.20
91. 4,000 subscribers **93.** 2002 **95.** 0.92 in. **97.** 2.9 ft, 6.9 ft
101. $n^{1/2}$ **103.** $(3b)^{1/4}$ **105.** $\sqrt[3]{t}$ **107.** $\sqrt[4]{3t}$

ARE YOU READY? 10.3 (page 851)

1. 84 **2. a.** 2 **b.** 2/3 **3. a.** x **b.** a^2 **4.** $\pm 2i$ **5. a.** 9 **b.** $-\frac{1}{8}$ **6.** 5

SELF CHECKS 10.3

1. a. Two different irrational numbers **b.** Two different imaginary numbers that are complex conjugates
2. 3, −3, $2i$, $-2i$ **3.** 4, −9 does not check **4.** −125, 8
5. −1, 1 **6.** −7, 4 **7.** About 9.5 hours

STUDY SET SECTION 10.3 (page 858)

1. discriminant **3.** conjugates **5.** rational **7. a.** x^2 **b.** $\sqrt{x}$ **c.** $x^{1/3}$ **d.** $\frac{1}{x}$ **e.** $x + 1$ **9.** $4ac$, 5, 6, 24, rational
11. One repeated rational-number solution **13.** Two imaginary-number solutions (complex conjugates) **15.** Two different irrational-number solutions **17.** Two different rational-number solutions **19.** Two different irrational-number solutions
21. Two different rational-number solutions **23.** −1, 1, −4, 4
25. 2, −2, $3i$, $-3i$ **27.** 25, 64 **29.** 1, $\frac{9}{4}$ is extraneous
31. −1, 27 **33.** −64, 8 **35.** 0, 2 **37.** $\frac{1}{4}, \frac{1}{2}$ **39.** $-\frac{1}{3}, \frac{1}{2}$
41. $-4, \frac{2}{3}$ **43.** $\frac{5 \pm \sqrt{65}}{2}$ **45.** $\frac{7 \pm \sqrt{105}}{2}$ **47.** $\frac{9}{4}$, 1 is extraneous
49. $-\frac{1}{3}, 1$ **51.** $-i, i, -3i\sqrt{2}, 3i\sqrt{2}$ **53.** $4 \pm i$
55. $\frac{3 \pm \sqrt{57}}{6}$ **57.** 16, 4 **59.** $\pm i\sqrt{2}, \pm 3\sqrt{2}$
61. Repeated solutions of 1 and −1 **63.** $2, -2, i\sqrt{7}, -i\sqrt{7}$
65. $\frac{243}{32}$, 1 **67.** A repeated solution of $-\frac{3}{2}$ **69.** $3 \pm \sqrt{7}$
71. 49, 225 **73.** $1 \pm i$ **75.** No solution **77.** $1, -1, \sqrt{5}, -\sqrt{5}$
79. $-\frac{5}{7}, 3$ **81.** $-1, -\frac{27}{13}$ **83. a.** 64, −125 **b.** $\frac{1}{4}, -\frac{1}{5}$
85. a. $1, -\frac{1}{2}$ **b.** $\pm i$ **87.** 103 min **89.** 20 feet per second
91. 32.4 ft **95.** $x = 3$ **97.** $y = \frac{2}{3}x$

ARE YOU READY? 10.4 (page 861)

1.

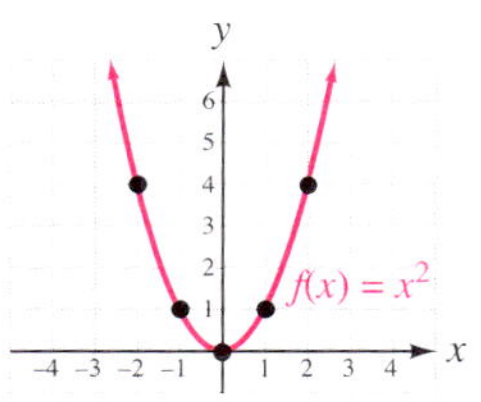

2. −2
3. $x^2 + 8x + 16 = (x + 4)^2$
4. $x^2 + x + \frac{1}{4} = \left(x + \frac{1}{2}\right)^2$
5. A repeated solution of −3
6. A repeated solution of −2
7. −5 **8.** A vertical line

SELF CHECKS 10.4

1.

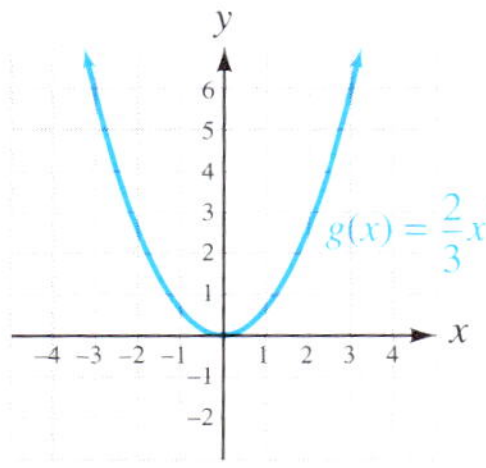

2.

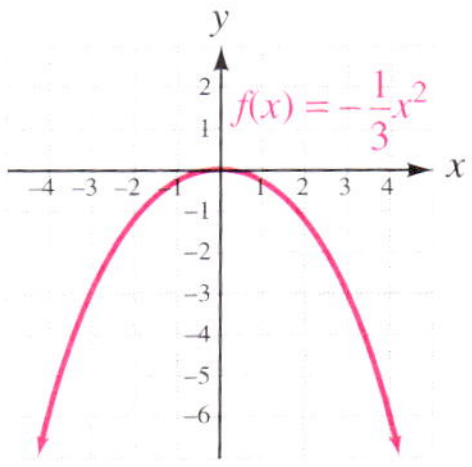

3.

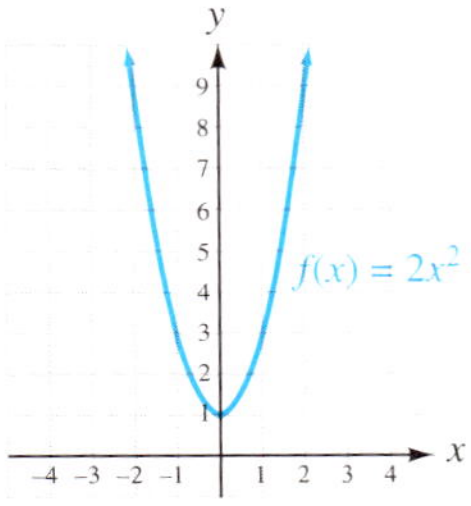

4.

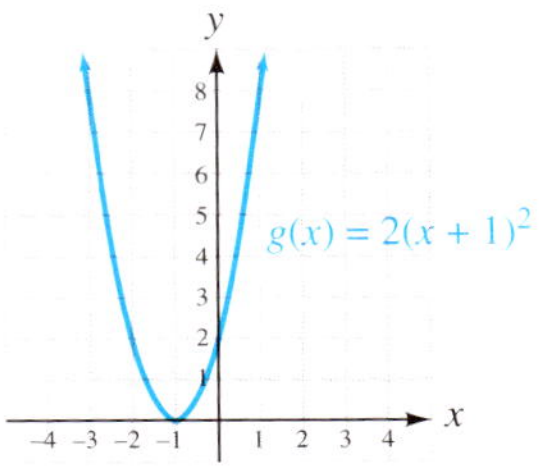

5.

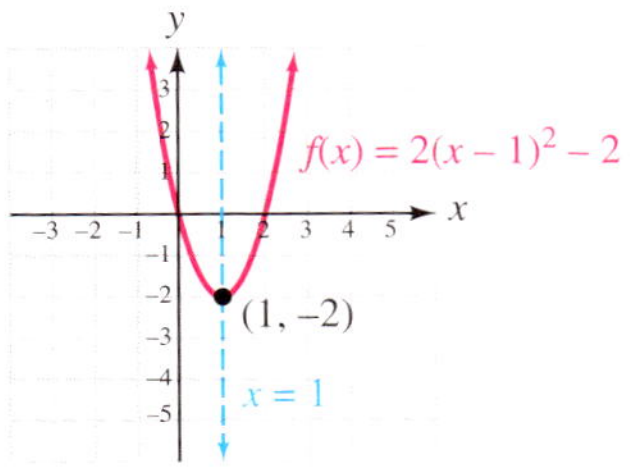

6. $(-2, 6)$; $x = -2$; opens upward

7.

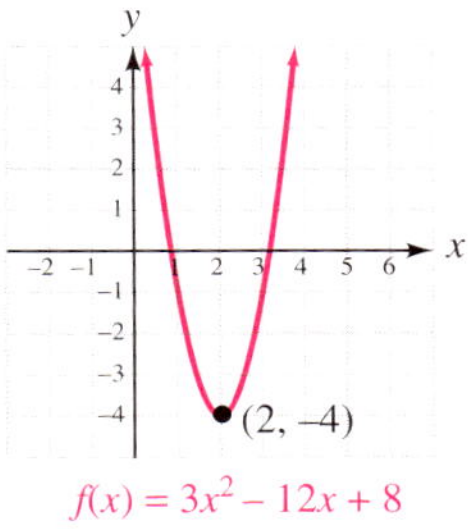

8. $(2, -4)$

9.

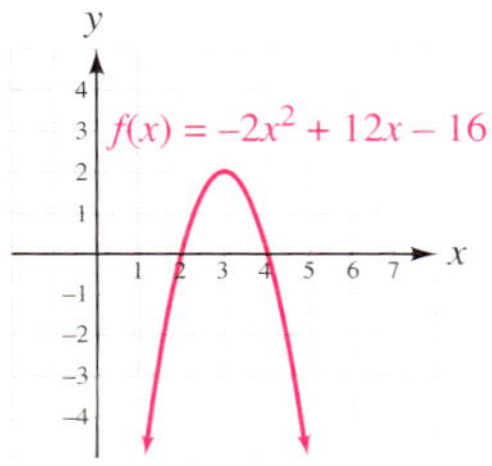

10. 20, $700

17.

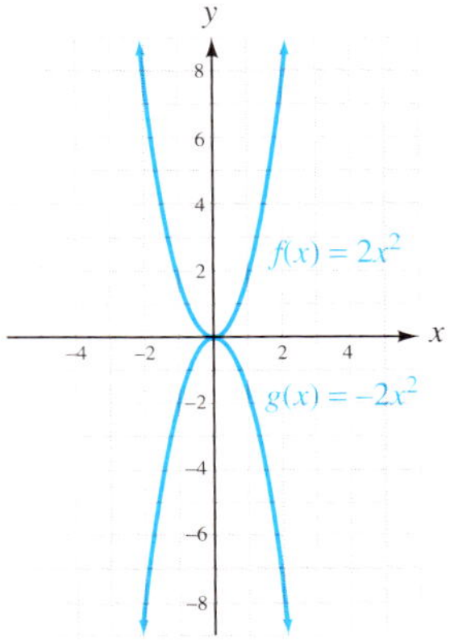

19.

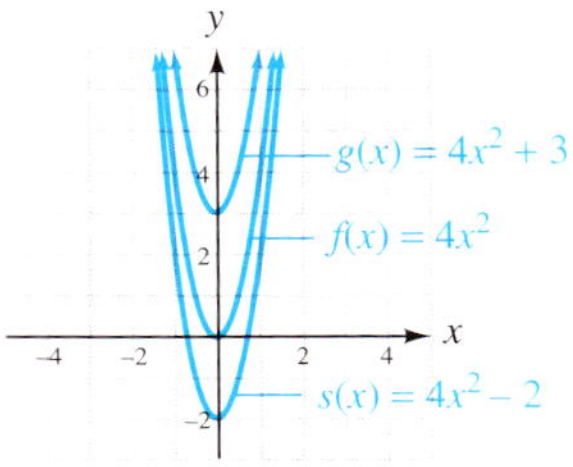

21.

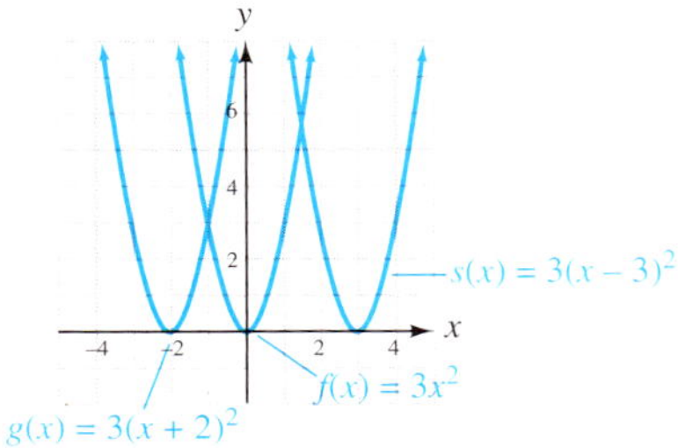

23. $(1, 2)$; $x = 1$; upward **25.** $(-3, -4)$; $x = -3$; downward
27. $(7.5, 8.5)$; $x = 7.5$; downward **29.** $(0, -4)$; $x = 0$; upward
31. $(3, 2)$, $x = 3$ **33.** $(2, 0)$, $x = 2$

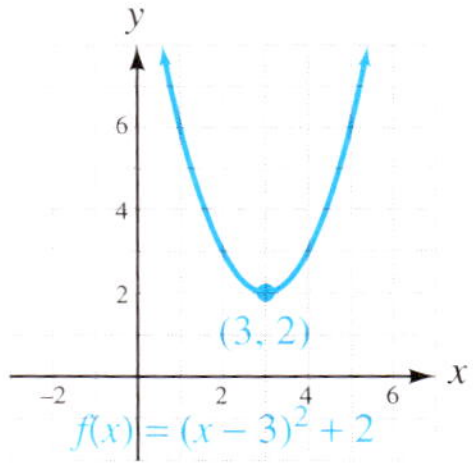

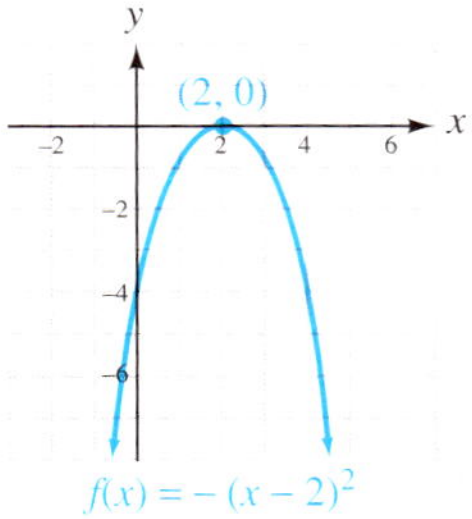

35. $(-3, 4)$, $x = -3$ **37.** $(-1, -3)$, $x = 1$

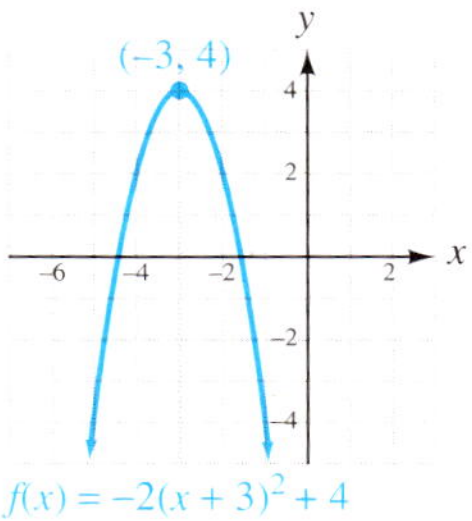

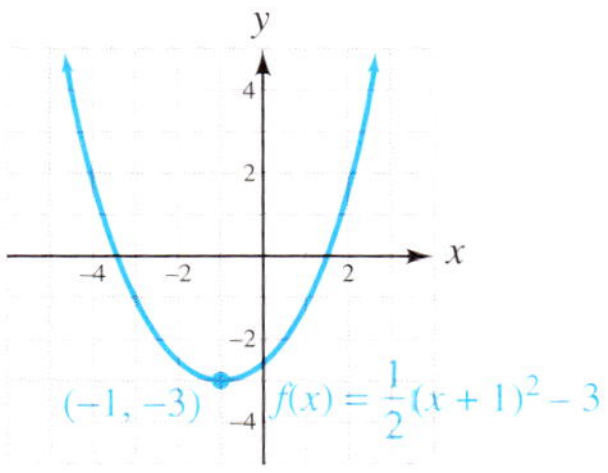

39. $(-2, 1)$; $x = -2$; upward **41.** $(3, -6)$; $x = 3$; downward

STUDY SET SECTION 10.4 (page 872)

1. quadratic, parabola **3.** axis of symmetry
5. a. $(1, 0), (3, 0)$ **b.** $(0, -3)$ **c.** $(2, 1)$ **d.** $x = 2$
e. Domain: $(-\infty, \infty)$; range: $(-\infty, 1]$

7.

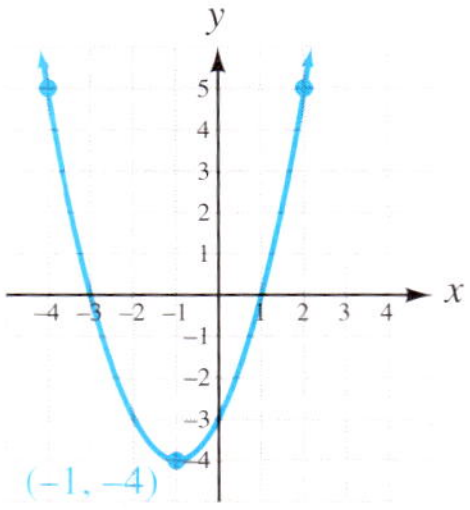

9. a. 2 **b.** 9, 18 **11.** $-3, 5$
13.
$h = -1$; $f(x) = 2[x - (-1)]^2 + 6$

15.

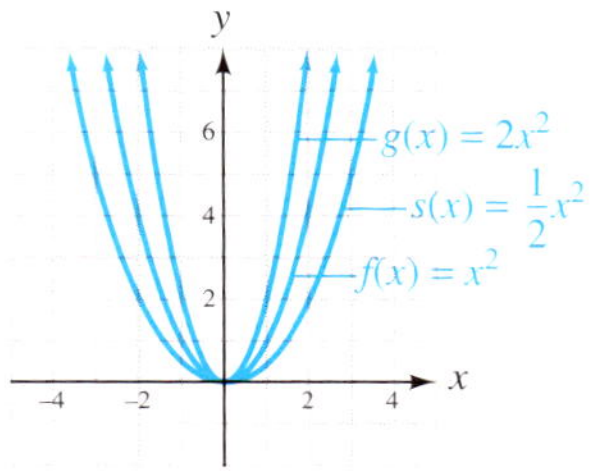

43. $f(x) = (x + 1)^2 - 4$; $(-1, -4)$, $x = -1$

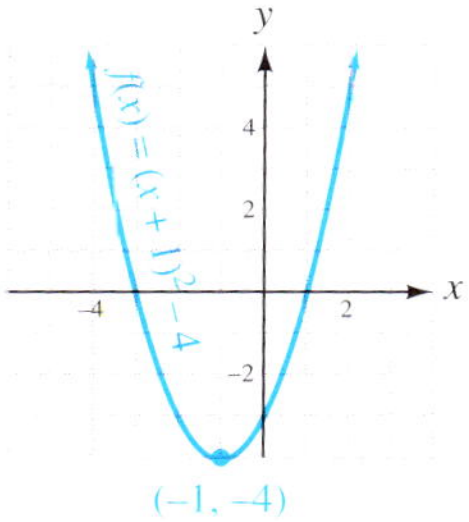

45. $f(x) = 4(x + 3)^2 + 1$; $(-3, 1)$, $x = -3$

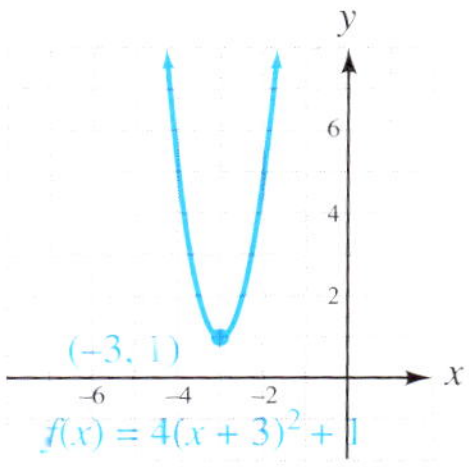

47. $f(x) = \left(x + \frac{1}{2}\right)^2 - \frac{25}{4}$; $\left(-\frac{1}{2}, -\frac{25}{4}\right)$; $x = -\frac{1}{2}$

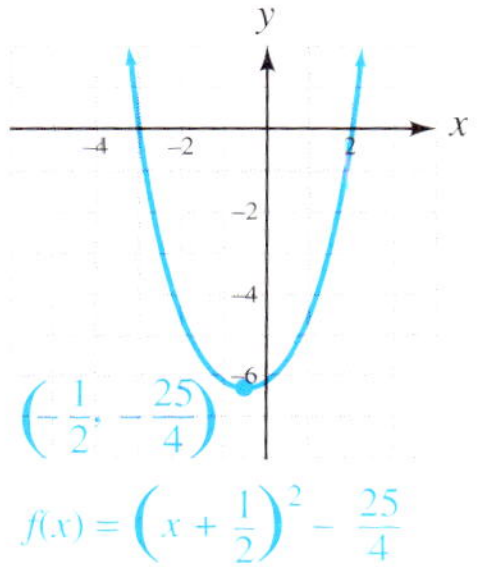

49. $f(x) = -4(x - 2)^2 + 6$; $(2, 6)$, $x = 2$

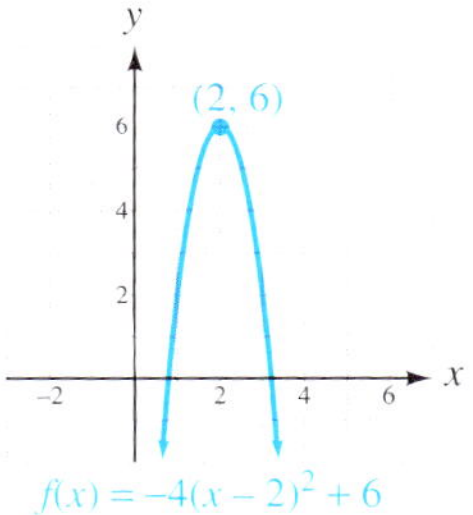

51. $f(x) = 2(x + 2)^2 - 2$; $(-2, -2)$, $x = -2$

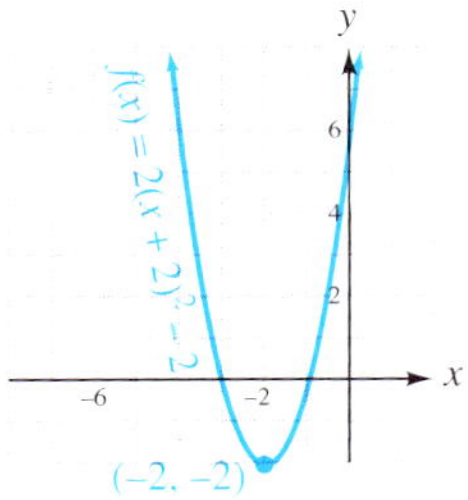

53. $f(x) = -(x + 4)^2 - 1$; $(-4, -1)$, $x = -4$

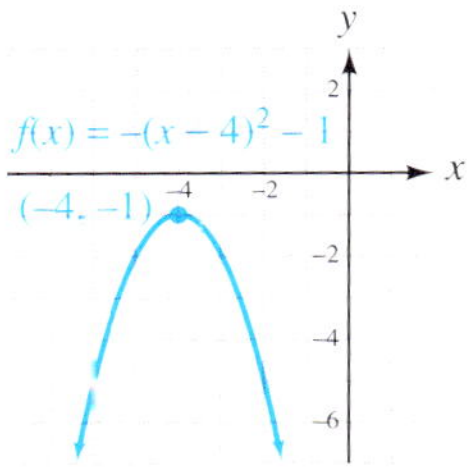

55. $(-1, -6)$ **57.** $\left(\frac{3}{4}, \frac{23}{8}\right)$ **59.** $(-5, 0), (7, 0); (0, -35)$

61. $(0, 0), (2, 0); (0, 0)$

63. $(-2, 0); (-2, 0); (0, 4)$

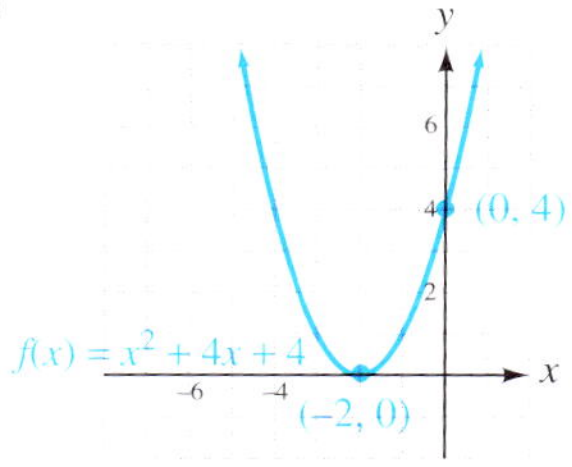

65. $(1, 0); (1, 0); (0, -1)$

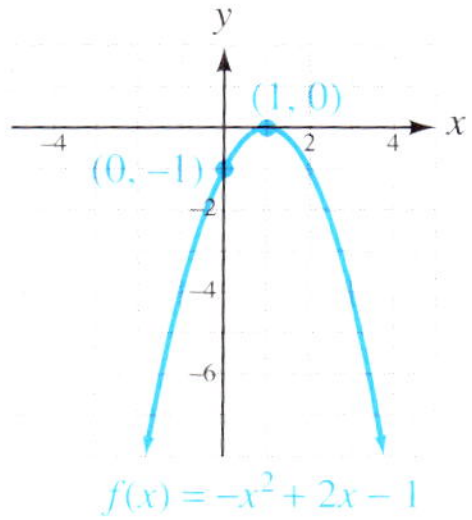

67. $(1, -1); (0, 0), (2, 0); (0, 0)$

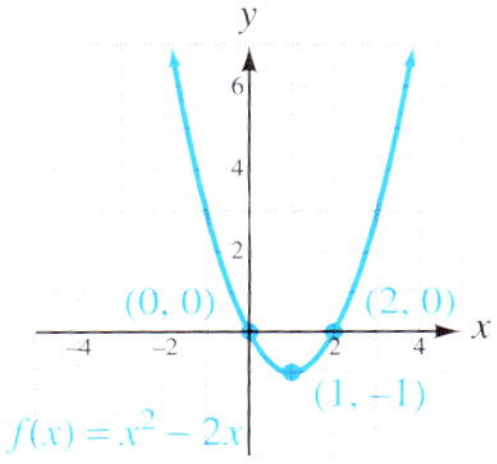

69. $(2, -2); (1, 0), (3, 0); (0, 6)$

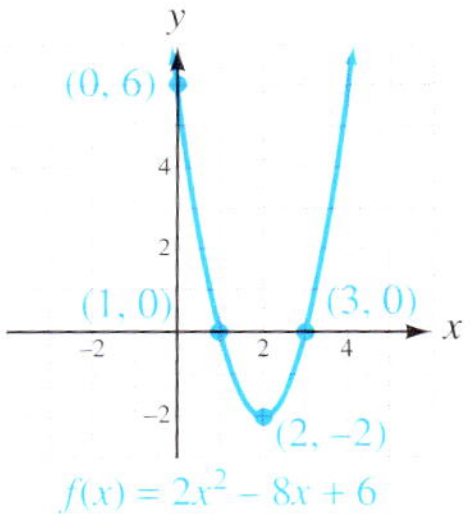

71. $(-1, -2)$; no x-intercept; $(0, -8)$

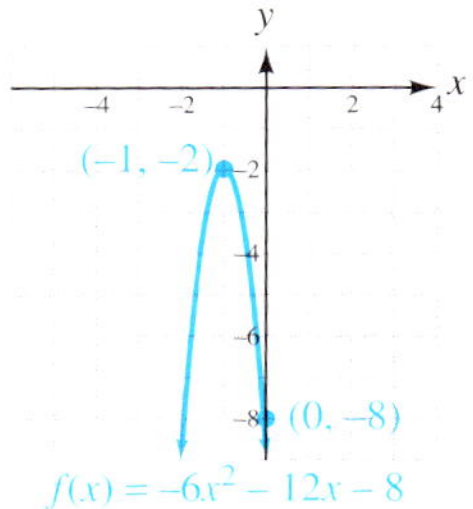

73. $\left(\frac{3}{2}, 0\right); \left(\frac{3}{2}, 0\right); (0, 9)$

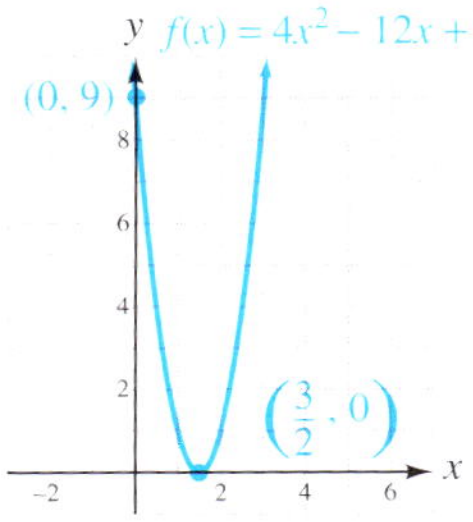

75. $(0.25, 0.88)$ **77.** $(0.50, 7.25)$ **79.** $2, -3$ **81.** $-1.85, 3.25$

83.

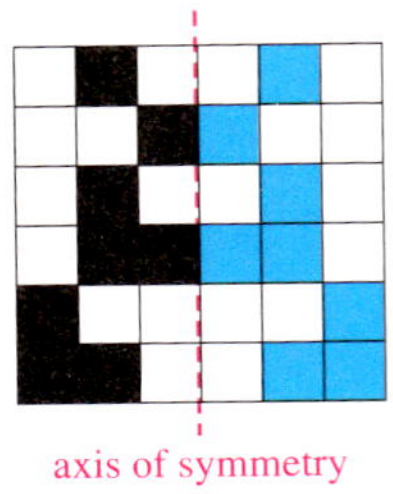

85. 15 min, \$160 **87.** 3.75 sec, 225 ft **89.** 75 ft by 75 ft, 5,625 ft^2 **91.** 1968, about 1.5 million; the U.S. involvement in the war in Vietnam was at its peak **93.** 200, \$7,000 **101.** $\frac{\sqrt{6}}{10}$ **103.** $15b - 6\sqrt{15b} + 9$

ARE YOU READY? 10.5 (page 875)

1. Yes **2.** $-5, 10$ **3.**

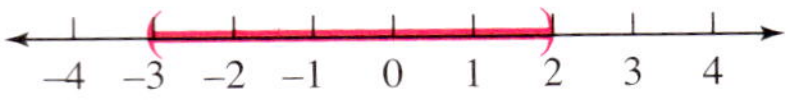

4.

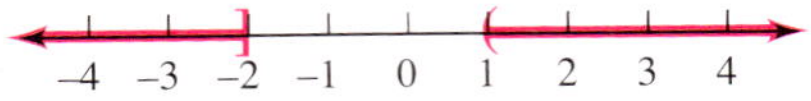

5. $-3, 3$ **6.**

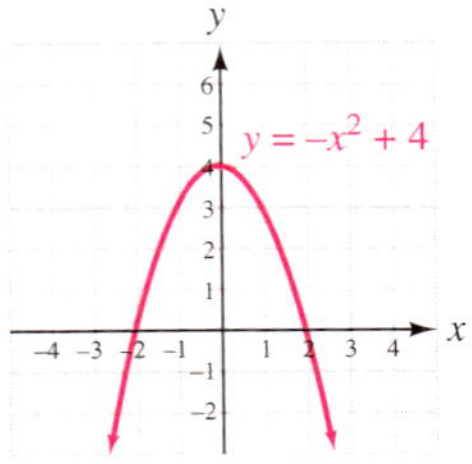

SELF CHECKS 10.5

1. $(-4, 3)$

2. $(-\infty, -8] \cup [5, \infty)$

3. $(-\infty, 0) \cup \left(\frac{3}{5}, \infty\right)$

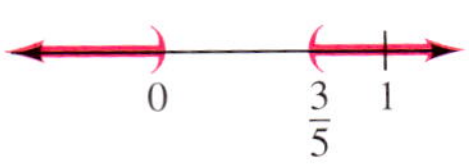

4. $[-2, -1) \cup (3, \infty)$

5. $(-1, 0) \cup (1, \infty)$

6.

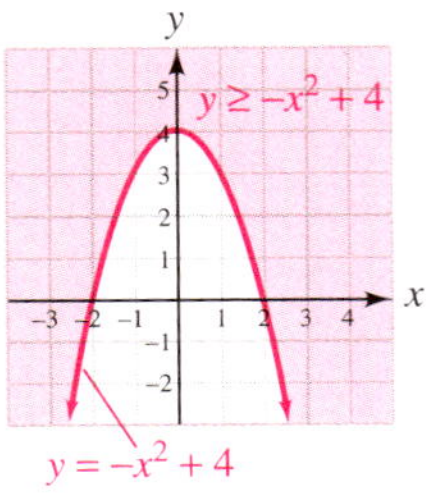

7.

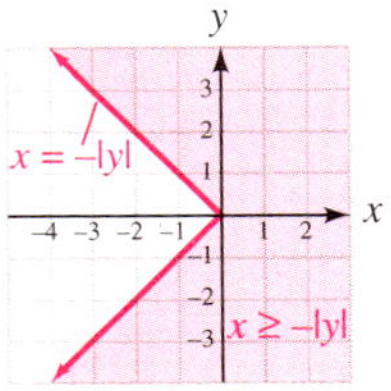

STUDY SET SECTION 10.5 (page 884)

1. quadratic **3.** two **5.** $(-\infty, -1), (-1, 4), (4, \infty)$ **7. a.** Yes **b.** No **c.** Yes **d.** No **9. a.** $(-3, 2)$ **b.** $(-\infty, -1] \cup [1, \infty)$ **11. a.** Solid **b.** Yes **13.** $x^2 - 6x - 7 \geq 0$ **15.** $(1, 4)$

17. $(-\infty, 3) \cup (5, \infty)$

19. $(-\infty, -6] \cup [7, \infty)$

21. $[-4, 3]$

23. $(-\infty, 0) \cup \left(\frac{1}{2}, \infty\right)$

25. $\left(-\infty, -\frac{5}{3}\right] \cup (0, \infty)$

27. $(-\infty, -3) \cup (1, 4)$

29. $\left(-\frac{1}{2}, \frac{1}{3}\right] \cup \left[\frac{1}{2}, \infty\right)$

31. $(0, 2) \cup (8, \infty)$

33. $\left[-\frac{34}{5}, -4\right) \cup (3, \infty)$

35.

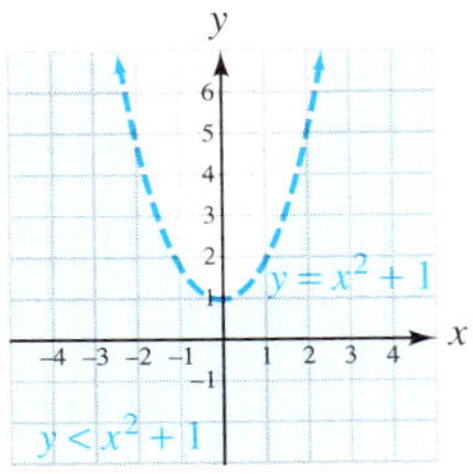

37.

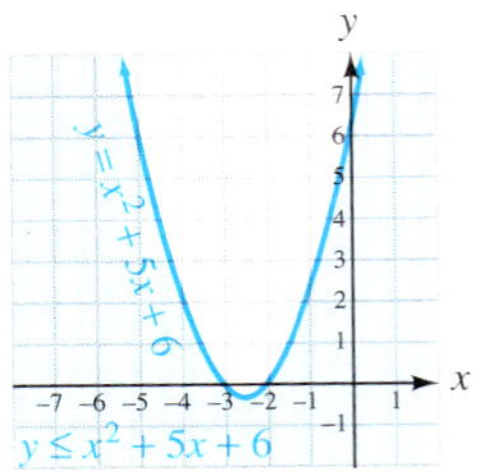

39.

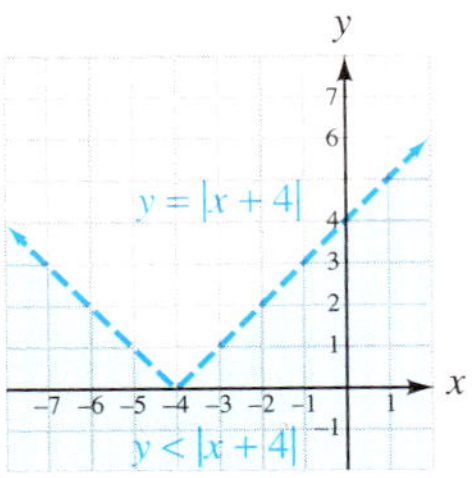

41.

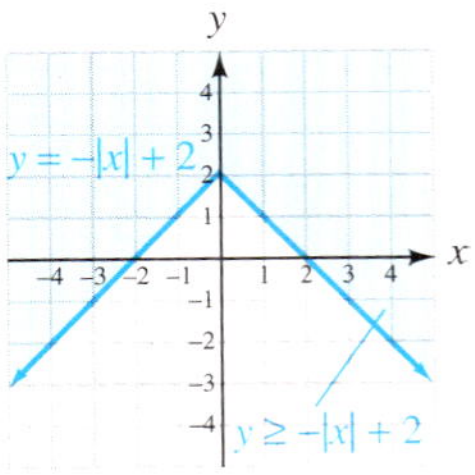

43. $(-1, 3)$ **45.** $(-\infty, -3) \cup (2, \infty)$ **47.** $(-4, -2] \cup (-1, 2]$

49. $(-\infty, -3] \cup [3, \infty)$

51. $(-\infty, \infty)$

53. $(-2, 1) \cup (3, \infty)$

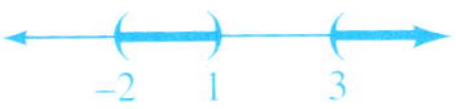

55. $(-5, 5)$

57. $\left(-\frac{1}{3}, \frac{3}{2}\right)$

59. No solutions

61. $(-1, 4) \cup \left(\frac{23}{2}, \infty\right)$

63. $(-2{,}100, -900) \cup (900, 2{,}100)$ **69.** $x = ky$ **71.** $t = kxy$

CHAPTER 10 REVIEW (page 886)

1. $-5, -4$ **2.** $-\frac{1}{3}, -\frac{5}{2}$ **3.** $\pm 2\sqrt{7}; \pm 5.29$ **4.** $4, -8$ **5.** $\pm 5i$ **6.** $\pm\frac{7\sqrt{5}}{5}; \pm 3.13$ **7.** $r = \frac{\sqrt{\pi A}}{\pi}$ **8.** $x^2 - x + \frac{1}{4} = \left(x - \frac{1}{2}\right)^2$ **9.** $-4, -2$ **10.** $\frac{3 \pm \sqrt{3}}{2}$, 2.37, 0.63 **11.** $\frac{6 \pm \sqrt{30}}{6}$; 1.91, 0.09 **12.** $1 \pm 2i\sqrt{3}$ **13.** $\pm 4\sqrt{2}$ **14.** $\frac{51 \pm \sqrt{11}}{7}$ **15.** Because 7 is an odd number and not divisible by 2, the calculations involved in completing the square on $x^2 + 7x$ involve fractions. The calculations involved in completing the square on $x^2 + 6x$ do not. **16.** 6 seconds before midnight

17. $\frac{1}{2}, -7$ **18.** $5 \pm \sqrt{7}$; 7.65, 2.35 **19.** 0, 10
20. $\frac{13 \pm \sqrt{163}}{3}$; 8.59, 0.08 **21.** $-\frac{3}{4} \pm \frac{\sqrt{15}}{4}i$ **22.** $\frac{2}{3} \pm \frac{\sqrt{2}}{3}i$
23. $\frac{-3 \pm \sqrt{29}}{10}$; 0.24, −0.84 **24.** $\frac{3 \pm 3\sqrt{13}}{2}$; 6.91, −3.91
25. $\frac{-3 \pm \sqrt{41}}{2}$ **26.** $\frac{-1 \pm \sqrt{13}}{4}$ **27.** 2 is not a factor of the numerator—it is a term. Only common factors of the numerator and denominator can be removed. **28. a.** $(2 + 2x)$ ft **b.** $(6 + 2x)$ ft
29. Sides: 1.25 in. wide; top/bottom: 2.5 in. wide
30. \$24 or \$26 **31.** 0.7 sec, 1.8 sec **32.** 33 in., 56 in.
33. Two different irrational-number solutions
34. Two imaginary-number solutions that are complex conjugates
35. One repeated solution, a rational number **36.** Two different rational-number solutions **37.** 1, 144 **38.** 8, −27
39. $i, -i, \frac{\sqrt{6}}{3}, -\frac{\sqrt{6}}{3}$ **40.** $1, -\frac{8}{5}$ **41.** $4 \pm i$ **42.** Repeated solutions of −1 and 1 **43.** A repeated solution of $-\frac{2}{5}$ **44.** $\frac{1}{32}$, 32
45. About 81 min **46.** 30 mph **47.** 34.1 million **48.** h, k, x
49.

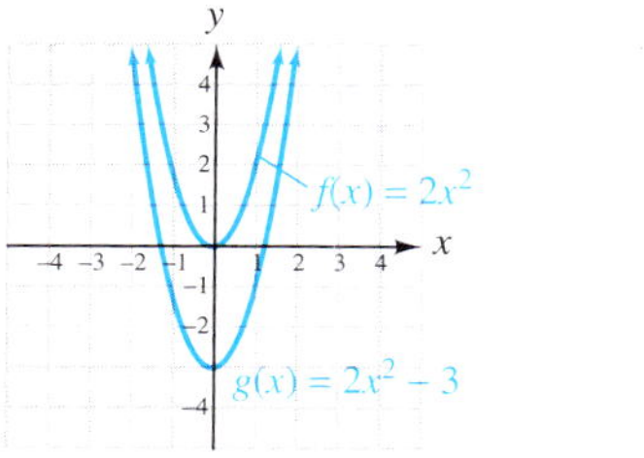

50.

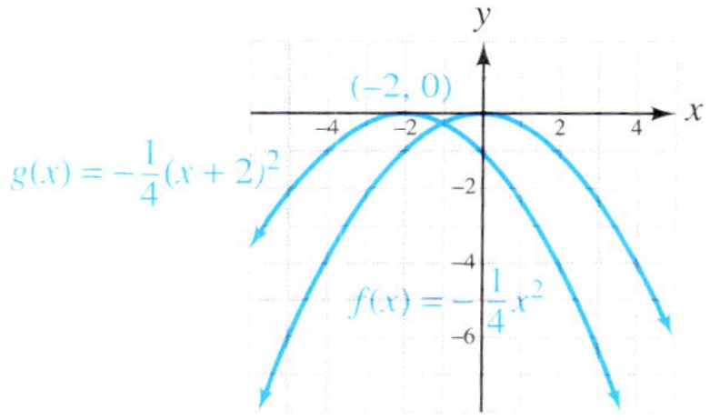

51. (1, 4), $x = 1$

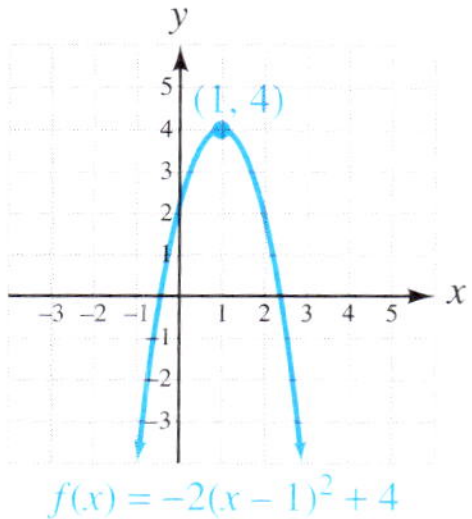

52. $f(x) = 4(x + 2)^2 - 7$; $(-2, -7)$, $x = -2$

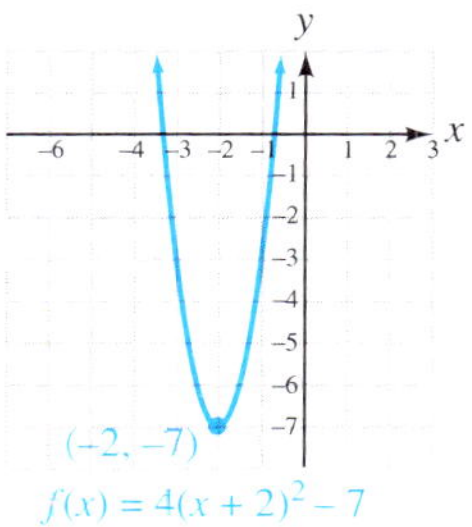

53. (1, −6)

54. $\left(-\frac{1}{2}, -\frac{9}{4}\right)$; $x = -\frac{1}{2}$; (−2, 0), (1, 0); (0, −2)

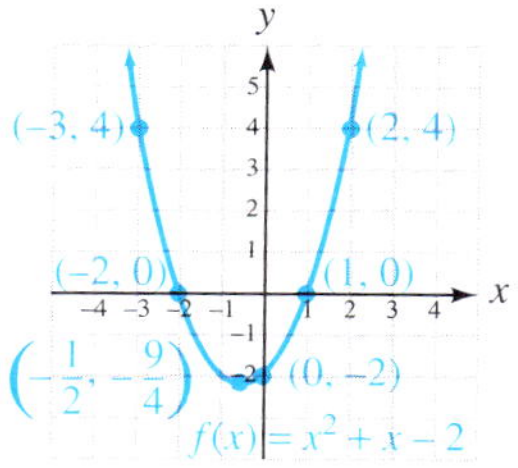

55. 1921; 6,469,326 **56.** $-2, \frac{1}{3}$
57. $(-\infty, -7) \cup (5, \infty)$

58. [−9, 9]

59. $(-\infty, 0) \cup \left[\frac{3}{5}, \infty\right)$

60. $\left(-\frac{7}{2}, 1\right) \cup (4, \infty)$

61. $\left[-4, \frac{2}{3}\right]$ **62.** $(-\infty, 0) \cup (1, \infty)$
63.

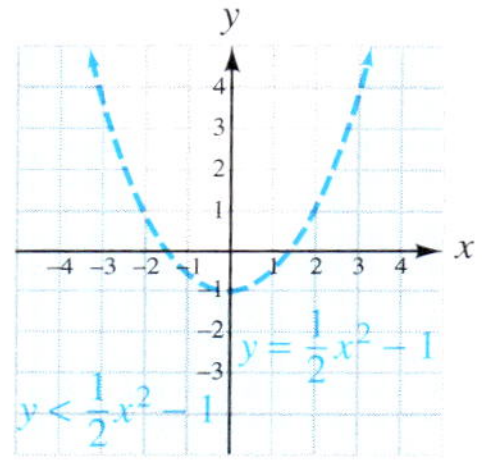

64.

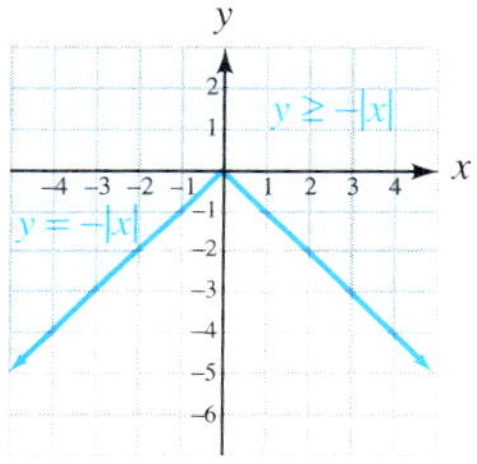

CHAPTER 10 TEST (page 892)

1. a. quadratic **b.** completed, square **c.** vertex **d.** rational **e.** nonlinear **2.** $\pm 3\sqrt{7} \approx \pm 7.94$ **3.** $-7 \pm 5\sqrt{2}$ **4.** $\pm 2i$
5. $x^2 + 11x + \frac{121}{4} = \left(x + \frac{11}{2}\right)^2$ **6.** $\frac{-3 \pm \sqrt{17}}{2}$; −3.56; 0.56
7. $-2 \pm i\sqrt{2}$ **8.** $\frac{2 \pm 3\sqrt{2}}{3}$ **9.** $\frac{-1 \pm \sqrt{2}}{2}$; −1.21; 0.21
10. $1 \pm \sqrt{5}$; −1.24; 3.24 **11.** $2 \pm 3i$ **12.** −5, −3
13. A repeated solution of 47 **14.** $\pm\sqrt{10}$ **15.** $1, \frac{1}{4}$ **16.** $-1, \frac{1}{3}$
17. $2, -2, i\sqrt{3}, -i\sqrt{3}$ **18.** $-\frac{4}{5}, \frac{4}{7}$ **19.** $2 \pm \sqrt{10}$; 5.16, −1.16
20. $-8, -\frac{1}{125}$ **21.** $-\frac{1}{2} \pm \frac{\sqrt{5}}{10}i$ **22.** $3, \pm 2i\sqrt{2}$ **23.** $c = \frac{\sqrt{Em}}{m}$
24. a. Two different imaginary-number solutions that are complex conjugates **b.** One repeated solution, a rational number
25. 4.5 ft by 1,502 ft **26.** About 46 min **27.** 20 in. **28.** 4.3 in.
29. 1993 **30.** iii
31. (1, 2), $x = 1$

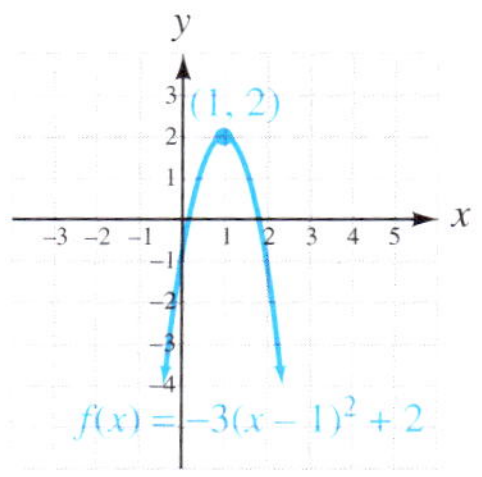

32. $f(x) = 5(x+1)^2 - 6$; $(-1, -6)$, $x = -1$

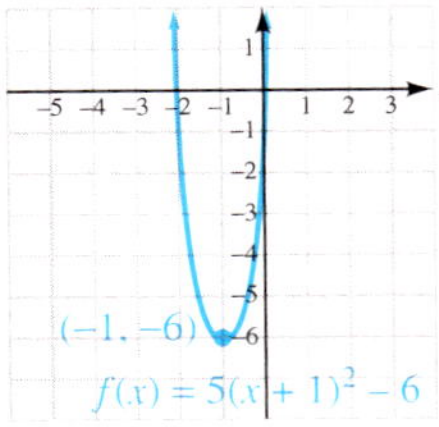

33. $\left(-\frac{1}{4}, -\frac{9}{8}\right)$, $x = -\frac{1}{4}$, $(-1, 0)$, $\left(\frac{1}{2}, 0\right)$; $(0, -1)$

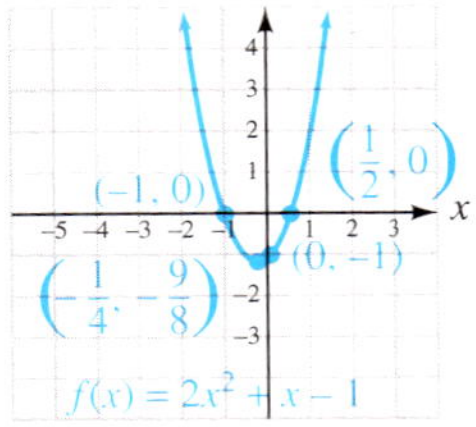

34. 211 ft **35.** $(-\infty, -2) \cup (4, \infty)$

36. $(-3, 2]$

37. 82.7° F

38.

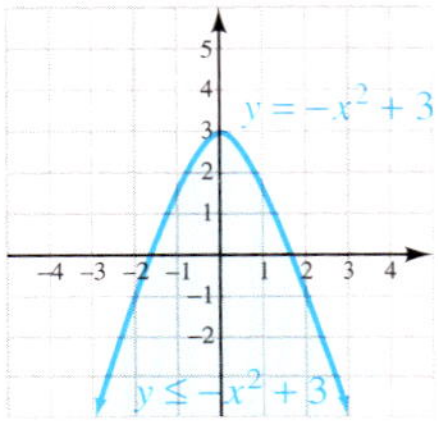

39. $-2, 3$ **40.** $[-2, 3]$

CUMULATIVE REVIEW CHAPTERS 1–10 (page 895)

1. a. True **b.** False **c.** True **2.** 14 **3.** $-41a + 130$ **4.** It is a solution. **5.** -2 **6.** All real numbers **7.** 17 lb **8.** 4 in. **9.** 63.5° **10.** 3.5 hr **11.** 6 L **12.** 80 lb **13.** $\left(-\infty, -\frac{3}{4}\right]$

14. No

15.

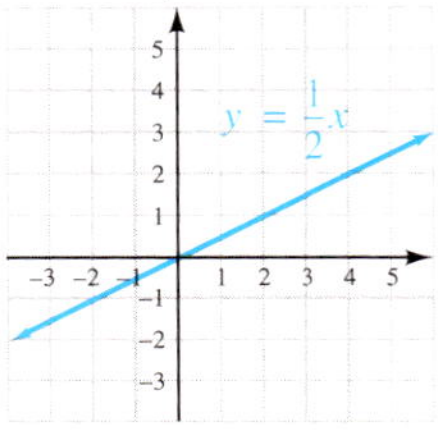

16.

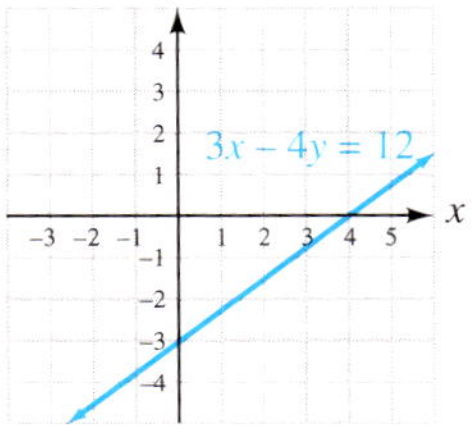

17.

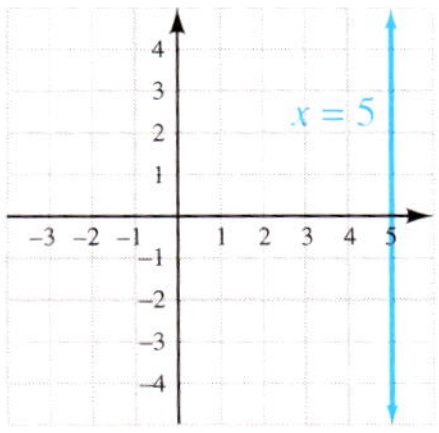

18.

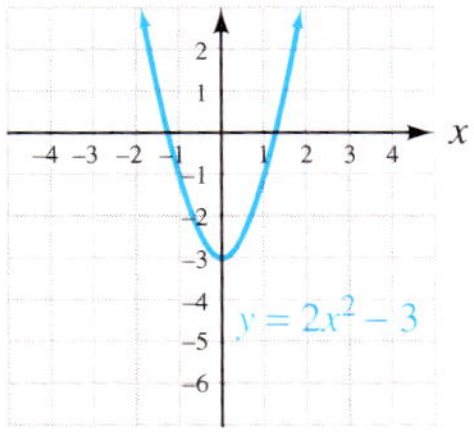

19. x-intercept: $\left(\frac{6}{5}, 0\right)$; y-intercept: $(0, -2)$ **20.** An increase of 114 million subscribers per year **21. a.** 3 **b.** $\frac{5}{6}$ **22.** 0 **23.** $y = 3x + 2$ **24.** $y = -\frac{2}{3}x - 2$

25. 28 **26.** Yes **27.** It is not a solution. **28.** $(-1, 5)$

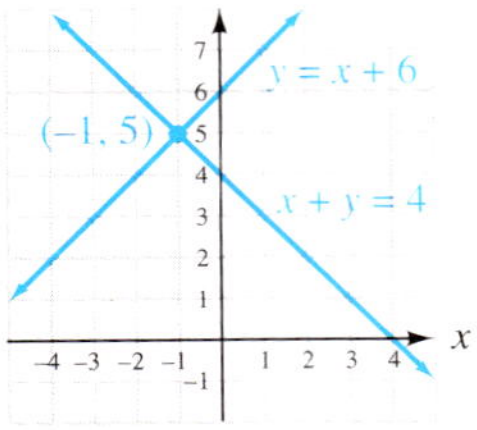

29. $(3, -1)$ **30.** $(-1, 2)$ **31.** 6%: \$3,000; 12%: \$3,000

32.

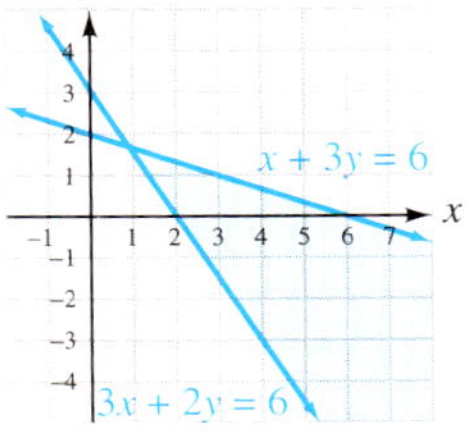

33. x^{31} **34.** $\frac{a^{15}b^5}{c^{20}}$ **35.** 1 **36.** $\frac{125c^{21}}{8a^{12}b^{12}}$ **37.** 5 **38.** $\frac{4}{a^{10}b^4}$ **39.** 1.44×10^{10}; 14,400,000,000 **40.** $\frac{1}{10}s^2 - st - \frac{2}{5}t^2$ **41.** $-11t^3 + 0.8t^2 - 1.4t$ **42.** $8a^3 - b^3$ **43.** $-6t^2 + 13st - 6s^2$ **44.** $16b^2 - 64b + 64$ **45.** $36b^2 - \frac{1}{4}$ **46.** $2x - 1$ **47.** $6uvw^2(2w - 3v)$ **48.** $(x - y)(x - 4)$ **49.** $(x + 5)(x + 2)$ **50.** Prime **51.** $(3a + 4)(2a - 5)$ **52.** $2a^2(3a + 2)(5a - 4)$ **53.** $(7s^3 - 6n^2)^2$ **54.** $(x^2 + 4y^2)(x + 2y)(x - 2y)$ **55.** $(x - 4)(x^2 + 4x + 16)$ **56.** $(2x^2 + 5y)(4x^4 - 10x^2y + 25y^2)$ **57.** $-1, -2$ **58.** 0, 2 **59.** $\frac{2}{3}, -\frac{1}{2}$ **60.** $-5, 5$ **61.** $2, -\frac{5}{2}$ **62.** $0, \frac{2}{3}, -\frac{1}{2}$ **63.** 5 cm **64.** $\frac{x+1}{x-1}$ **65.** $\frac{(p-3)(p+2)}{3(p+3)}$ **66.** $-\frac{3}{5x}$ **67.** $\frac{1}{3a}$ **68.** $\frac{7x+29}{(x+5)(x+7)}$ **69.** $\frac{3-16b^2}{18b^4}$ **70.** $\frac{y+x}{y-x}$ **71.** 1 **72.** 5; 3 is extraneous **73.** $R = \frac{R_1R_2R_3}{R_2R_3 + R_1R_3 + R_1R_2}$ **74.** $2\frac{2}{9}$ hr **75.** 6,480 **76.** 20 **77.** 18 **78.** All real numbers, $\mathbb{R}$; identity **79.** 83 or better **80.** function **81. a.** 0 **b.** 2 **82.** D: The set of real numbers, R: The set of real numbers **83.** D: $\{-6, 0, 1, 5\}$, R: $\{-12, 4, 7, 8\}$; no **84.** No; $(4, 2), (4, -2)$ **85.** -60 **86. a.** 3 **b.** $3r^2 + 2$ **87.** The set of all real numbers except -1 **88.** 6, (0, 15) **89.** $f(x) = -\frac{7}{8}$ **90. a.** $M(t) = 0.17t + 4.06$ **b.** 10.86 billion vehicle miles

91. D: the set of real numbers, R: the set of real numbers less than or equal to 1

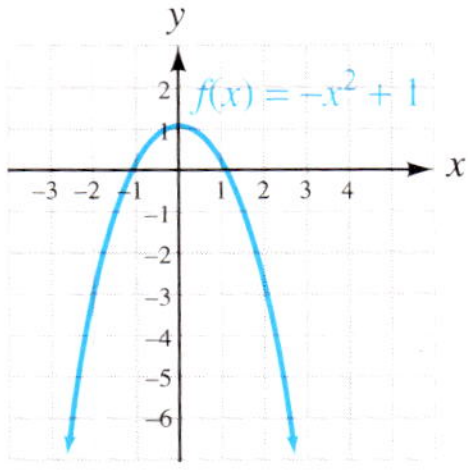

92. D: the set of real numbers, R: the set of real numbers greater than or equal to -4

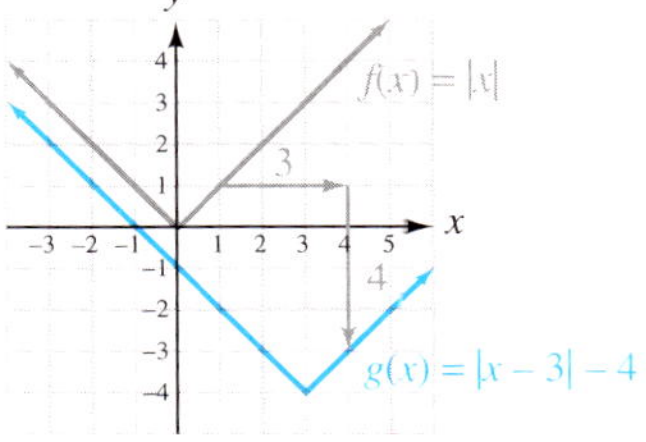

93. $3, -\frac{3}{2}$ **94.** $-5, -\frac{3}{5}$ **95.** $\left(-\infty, -\frac{10}{9}\right)$

96. $(-\infty, -2) \cup (2, \infty)$

97. $\left[1, \frac{9}{4}\right]$

98. $(-\infty, -10] \cup [15, \infty)$

99. $\left[-\frac{2}{3}, 2\right]$ 

100. No solution

101. $(x + y + 4)(x + y + 3)$ **102.** $f = \frac{d_1 d_2}{d_2 + d_1}$

103. $(x + 1)(x^2 - x + 1)(x - 1)(x^2 + x + 1)$

104. All real numbers except 0 and 2; $(-\infty, 0) \cup (0, 2) \cup (2, \infty)$ **105. a.** iii. **b.** i. **c.** iv. **d.** ii.

106. \$9,000

107. D: $[2, \infty)$; R: $[0, \infty)$

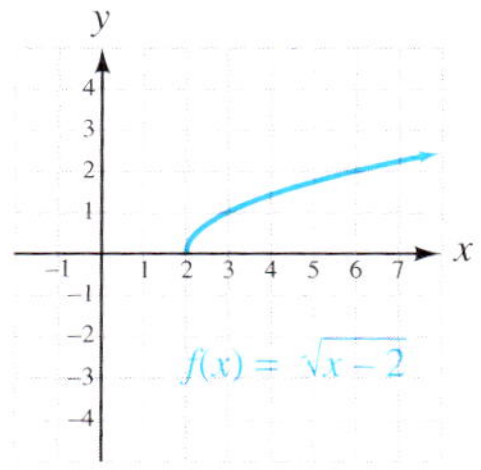

108. D: $[90, \infty)$ **109.** $-3x$ **110.** $4t\sqrt{3t}$ **111.** $\frac{1}{16}$ **112.** $x^{17/12}$

113. $-12\sqrt[4]{2} + 10\sqrt[4]{3}$ **114.** $-18\sqrt{6}$ **115.** $\frac{x + 3\sqrt{x} + 2}{x - 1}$

116. $\frac{5\sqrt[3]{x^2}}{x}$ **117.** 2, 7 **118.** $\frac{1}{4}$ **119.** 2, 4 **120. a.** $3\sqrt{2}$ in. **b.** $2\sqrt{3}$ in. **121. a.** 10 **b.** $\left(-\frac{3}{2}, \frac{5}{2}\right)$ **122.** $-i$ **123.** $-5 + 17i$

124. $\frac{3}{2} + \frac{1}{2}i$ **125.** $3 + 4i$ **126.** $0 - \frac{2}{3}i$ **127.** $\pm 2\sqrt{7}$

128. $19 \pm i\sqrt{5}$ **129.** $\frac{3 \pm \sqrt{3}}{2}$ **130.** $\frac{1}{5} \pm \frac{2}{5}i$ **131.** 10 ft by 18 ft

132. 50 m and 120 m **133.** $-8, 27$

134. Repeated solutions of -1 and 1

135. $(-2, 4), x = -2; (-4, 0), (0, 0); (0, 0)$

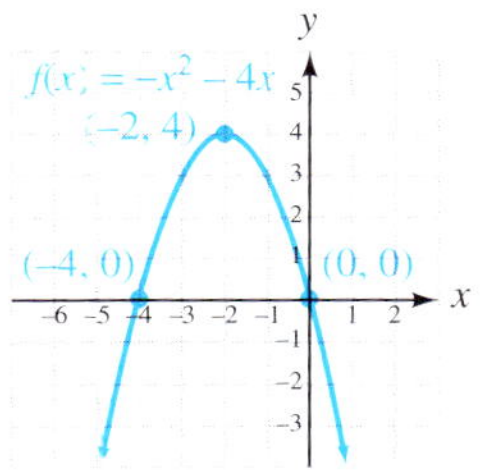

136. $(-9, 9)$ 

137. $(-4, -2] \cup (-1, 2]$ 

138. a. $-\frac{3}{4}$ **b.** No solution

ARE YOU READY? 11.1 (page 900)

1. $9x^2 + 2x - 8$ **2.** $2x^3 - x^2$ **3.** $72x^3 - 45x^2 - 32x + 20$ **4.** $x + 2$

SELF CHECKS 11.1

1. a. $(f + g)(x) = 2x^2 + 6x - 2$; D: $(-\infty, \infty)$ **b.** $(f - g)(x) = -2x^2 - 2$; D: $(-\infty, \infty)$ **c.** $(f \cdot g)(x) = 6x^3 + 5x^2 - 6x$; D: $(-\infty, \infty)$ **d.** $(f/g)(x) = \frac{3x - 2}{2x^2 + 3x}$; $\left(-\infty, -\frac{3}{2}\right) \cup \left(-\frac{3}{2}, 0\right) \cup (0, \infty)$ **2. a.** 78 **b.** -164 **c.** 56 **d.** Undefined **3. a.** -8 **b.** 5 **c.** $(g \circ f)(x) = 6 - x^3$ **4. a.** -6 **b.** -3 **c.** -4 **5.** $f(x) = \sqrt{x}, g(x) = x + 15$ **6.** $C(t) = -\frac{5}{6}t + 30$

STUDY SET SECTION 11.1 (page 907)

1. sum, $f(x) + g(x)$, difference, $f(x) - g(x)$ **3.** domain **5.** nested **7. a.** $g(3)$ **b.** $g(3)$

9.

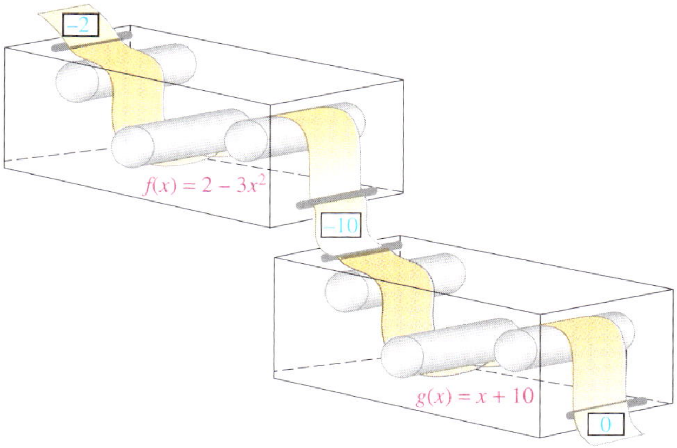

11. $g(x)$, $(3x - 1)$, $9x$, $2x$ **13.** $(f + g)(x) = 3x - 2, (-\infty, \infty)$ **15.** $(g - f)(x) = -x - 4, (-\infty, \infty)$ **17.** $(f \cdot g)(x) = 2x^2 - 5x - 3, (-\infty, \infty)$ **19.** $(g/f)(x) = \frac{x - 3}{2x + 1}, \left(-\infty, -\frac{1}{2}\right) \cup \left(-\frac{1}{2}, \infty\right)$ **21.** $(f + g)(x) = 7x, (-\infty, \infty)$ **23.** $(g - f)(x) = x, (-\infty, \infty)$ **25.** $(f \cdot g)(x) = 12x^2, (-\infty, \infty)$ **27.** $(g/f)(x) = \frac{4}{3}, (-\infty, 0) \cup (0, \infty)$ **29.** 20 **31.** -5 **33.** 0 **35.** $-\frac{1}{3}$ **37.** 7 **39.** 24 **41.** $-\frac{1}{2}$ **43.** $(g \circ f)(2x) = 16x^2 + 8x$ **45. a.** 0 **b.** 6 **c.** -3 **47. a.** -1 **b.** 6 **c.** -6 **49.** $f(x) = x^2$; $g(x) = x + 15$ **51.** $f(x) = x + 9$; $g(x) = x^5$ **53.** $f(x) = \sqrt{x}$; $g(x) = 16x - 1$ **55.** $f(x) = \frac{1}{x}$; $g(x) = x - 4$ **57.** 58 **59.** 110 **61.** 2 **63.** $(g \circ f)(x) = 9x^2 - 9x + 2$ **65.** $(f - g)(x) = -2x^2 + 3x - 3, (-\infty, \infty)$ **67.** $(f/g)(x) = \frac{3x - 2}{2x^2 + 1}, (-\infty, \infty)$ **69.** $\frac{1}{9}$ **71.** $(g \circ f)(8x) = 64x^2$ **73.** $(f - g)(x) = 3, (-\infty, \infty)$ **75.** $(g/f)(x) = \frac{x^2 - 4}{x^2 - 1}, (-\infty, -1) \cup (-1, 1) \cup (1, \infty)$ **77.** 4 **79.** 0 **81. a.** 7 **b.** 8 **c.** 12 **d.** 0 **85. a.** 1,026; in 2004, the average combined math and reading score was 1,026. **b.** 10; in 2004, the average difference in the math and reading scores was 10. **c.** 1,016 **d.** 14 **87.** $C(t) = \frac{5}{9}(2{,}668 - 200t)$ **89. a.** About \$75 **b.** $C(m) = \frac{3m}{20} = 0.15m$ **95.** $\frac{1}{c - d}$

ARE YOU READY? 11.2 (page 911)

1. D: $\{-2, 3, 5, 9\}$, R: $\{-3, 1, 8, 10\}$ **2.** Yes **3.** function **4.** $16x + 6$

SELF CHECKS 11.2

1. a. Yes **b.** No, $(-1, 1), (1, 1)$ **2. a.** No **b.** Yes **3. a.** $f^{-1}(x) = \frac{-x - 3}{5}$ **b.** $f^{-1}(x) = \sqrt[5]{x}$

5.

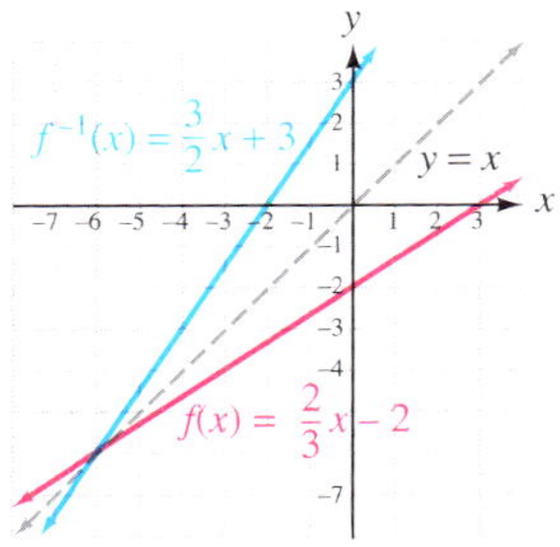

STUDY SET SECTION 11.2 (page 918)

1. one-to-one **3.** inverses **5.** one-to-one **7.** range, domain **9.** 1 **11. a.** No **b.** No **13.** $-4, 0, 8$ **15.** $y, 2y, 3, y, f^{-1}$ **17.** inverse, inverse **19.** Yes **21.** No **23.** No **25.** No **27.** One-to-one **29.** Not one-to-one **31.** Not one-to-one **33.** One-to-one **35.** $f^{-1}(x) = \frac{x-4}{2}$ **37.** $f^{-1}(x) = 5x - 4$ **39.** $f^{-1}(x) = 5x + 4$ **41.** $f^{-1}(x) = \frac{2}{x} + 3$ **43.** $f^{-1}(x) = \frac{4}{x}$ **45.** $f^{-1}(x) = \sqrt[3]{x-8}$ **47.** $f^{-1}(x) = x^3$ **49.** $f^{-1}(x) = \sqrt[3]{x} - 10$ **51.** $f^{-1}(x) = \sqrt[3]{\frac{x+3}{2}}$ **53.** $f^{-1}(x) = \sqrt[7]{2x}$

59. $f^{-1}(x) = \frac{1}{2}x$

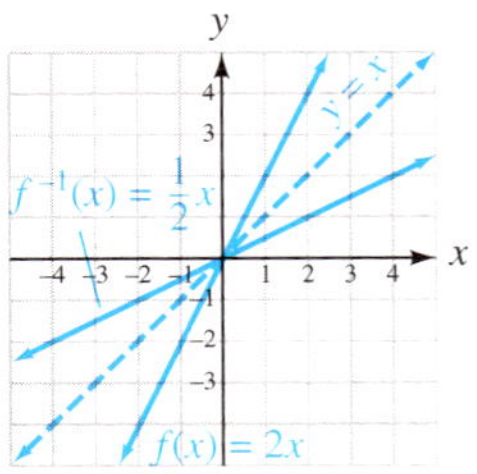

61. $f^{-1}(x) = \frac{x-3}{4}$

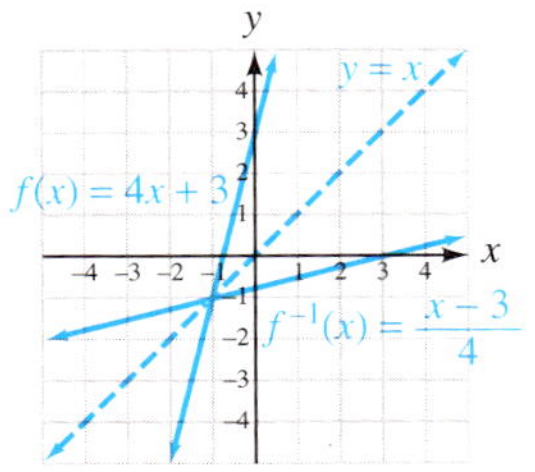

63. $f^{-1}(x) = -\frac{3}{2}x + \frac{9}{2}$

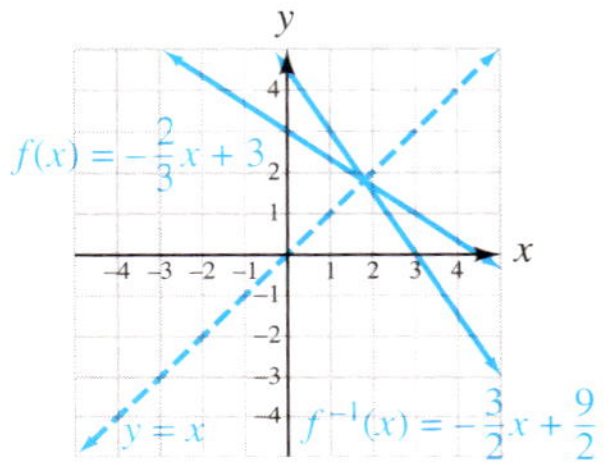

65. $f^{-1}(x) = \sqrt[3]{x}$

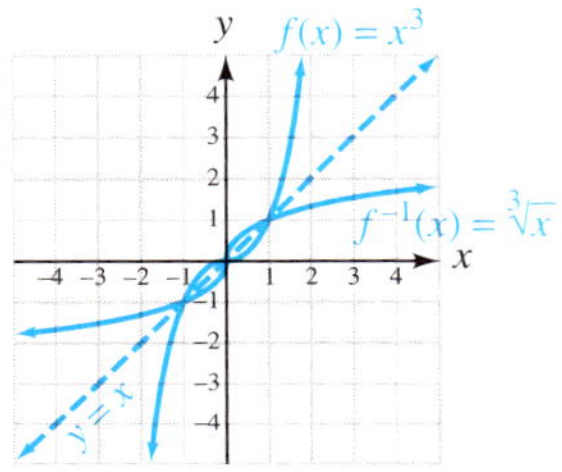

67. $f^{-1}(x) = \sqrt{x+1}$

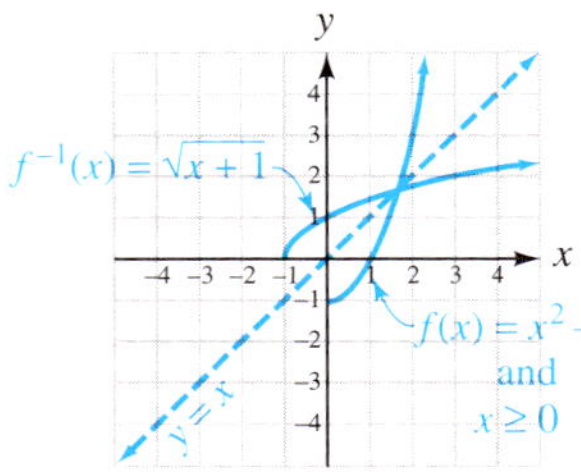

69. a. Yes; no **b.** No. Twice during this period, the person's anxiety level was at the maximum threshold value. **77.** $3 - 8i$ **79.** $18 - i$ **81.** $-28 - 96i$

ARE YOU READY? 11.3 (page 921)

1. a. 3^{12} **b.** 3^{32} **2. a.** 8 **b.** 1 **c.** $\frac{1}{4}$ **3. a.** $\frac{1}{9}$ **b.** 1 **c.** 27 **4.** line, parabola

SELF CHECKS 11.3

1.

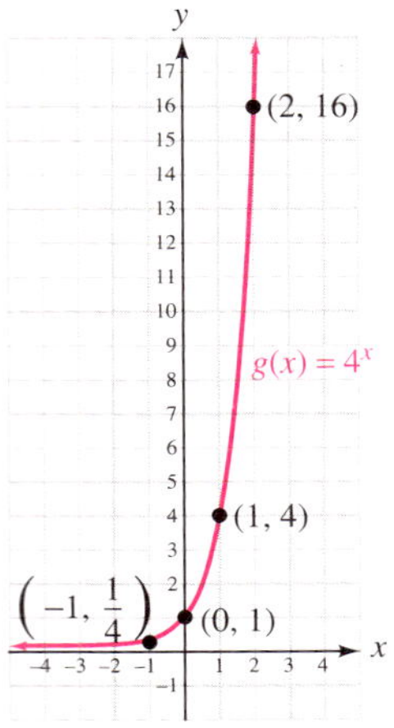

2.

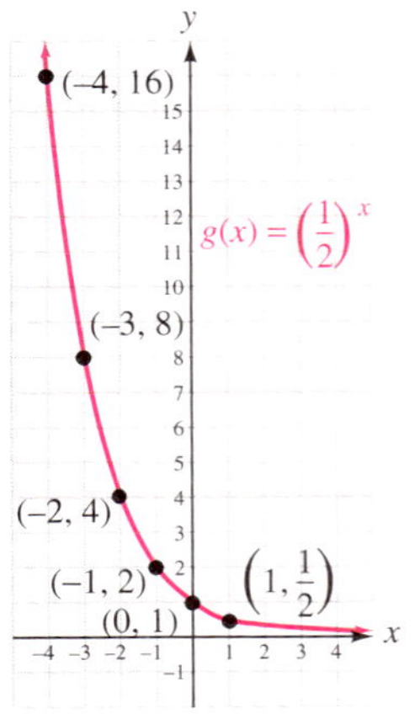

3. a.

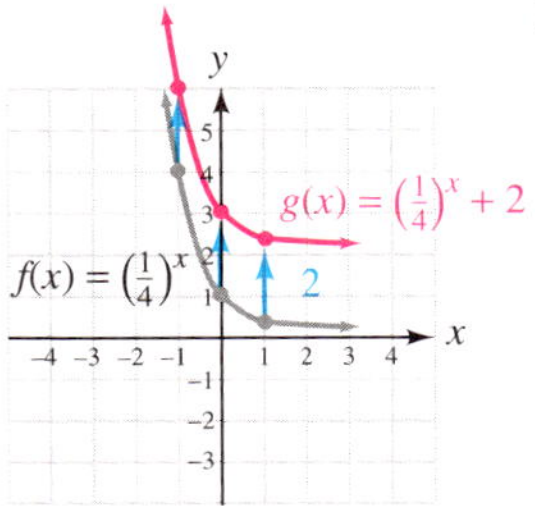

b.

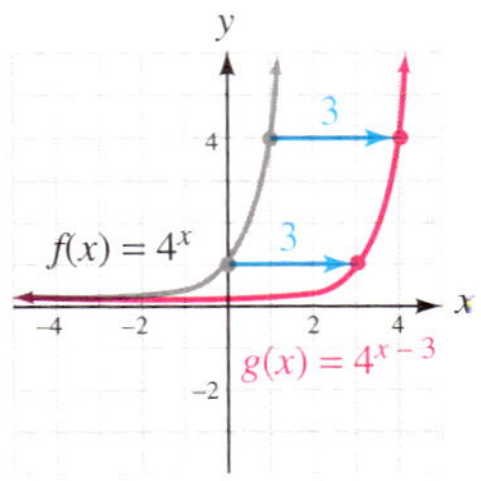

4. \$7,258,641 **5.** \$118,344.56 **6.** About 2.4 milligrams

STUDY SET SECTION 11.3 (page 931)

1. exponential **3.** asymptote **5. a.** Exponential **b.** $(-\infty, \infty)$ **c.** $(0, \infty)$ **d.** $(0, 1)$, none **e.** Yes **f.** The x-axis $(y = 0)$ **g.** Increasing **h.** 3 **7. a.** $\frac{1}{9}$ **b.** $\frac{1}{16}$ **c.** 25 **9. a.** iii **b.** iv **c.** ii **d.** i **11. a.** D: $(-\infty, \infty)$, R: $(-3, \infty)$ **b.** D: $(-\infty, \infty)$, R: $(1, \infty)$ **13.** $f(x) = 5^x$

x	$f(x)$
-3	$\frac{1}{125}$
-2	$\frac{1}{25}$
-1	$\frac{1}{5}$
0	1
1	5
2	25
3	125

15. a. ii **b.** i **c.** ii **d.** ii **e.** i **17.** $b > 0, b \neq 1$

19.

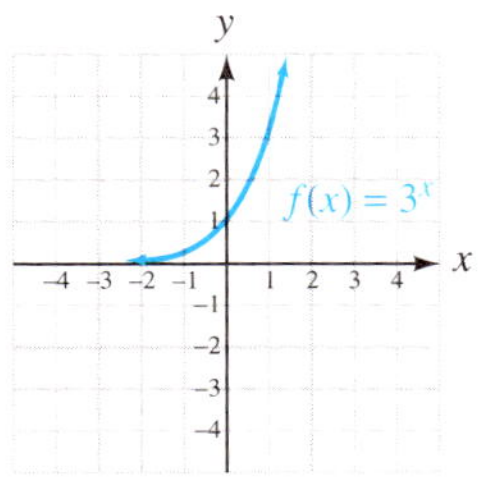

21.

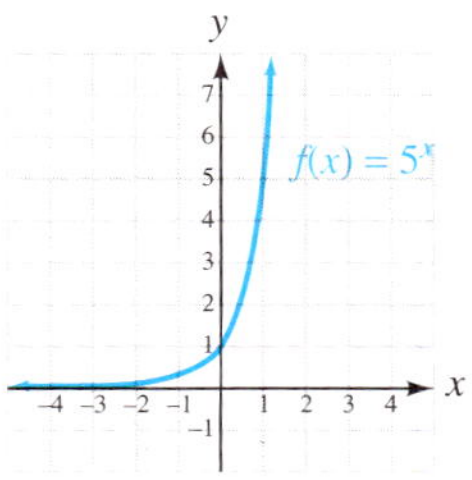

23.

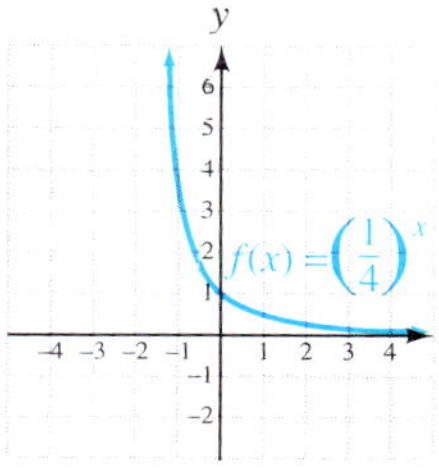

25.

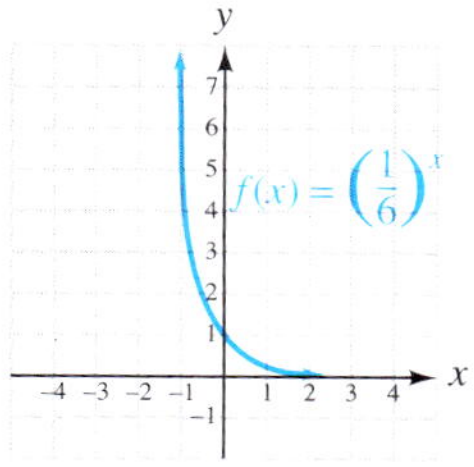

27.

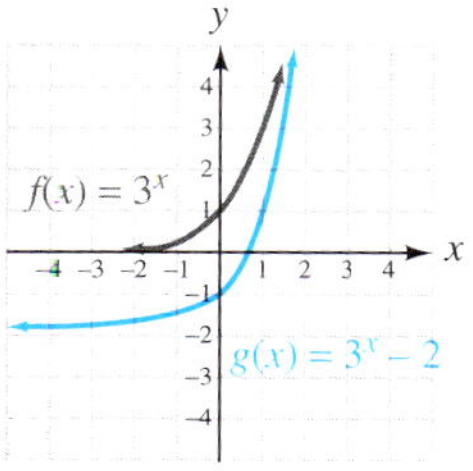

29.

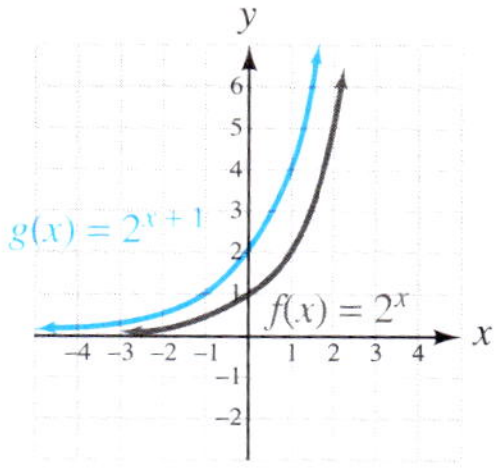

31.

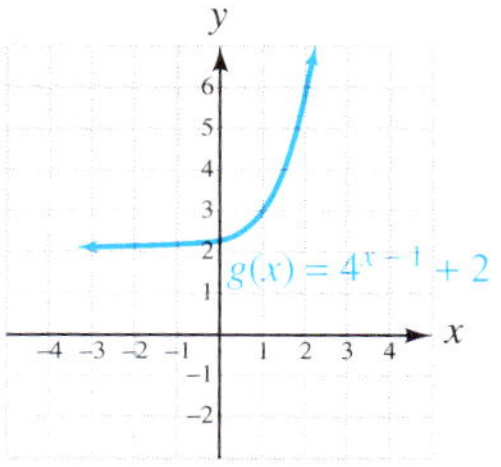

33.

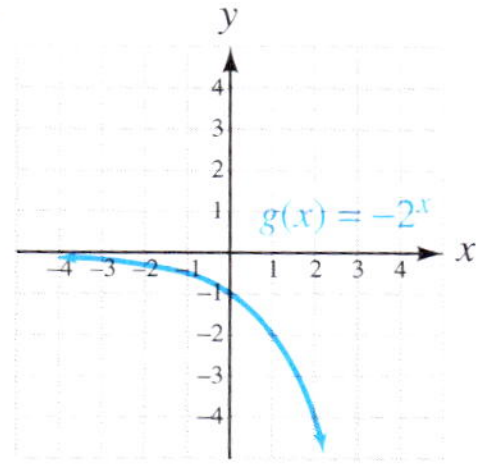

35. Increasing

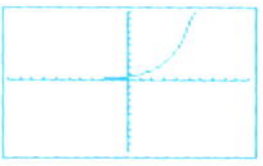

37. Decreasing

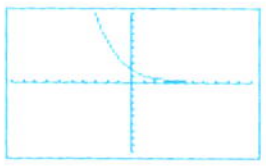

39. a. 280 ppm, 295 ppm, 370 ppm **b.** About 1970
41. a. At the end of the 2nd year **b.** At the end of the 4th year
c. During the 7th year
43. a.

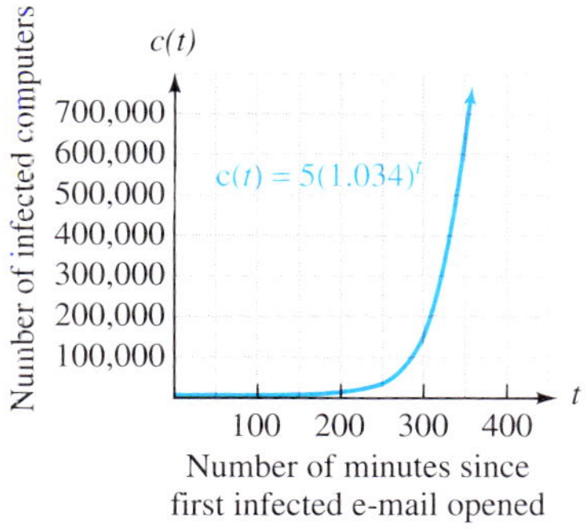

b. About 46,650,000 computers **45.** About 422 mm **47.** 8.7 gm

49.

Year	2006	2007	2008	2009
Number in poverty (in millions)	35.8	37.9	40.2	42.6

51. 2,295 **53.** \$22,080.40 **55.** \$32.03 **57.** \$2,273,996.13
69. 40 **71.** 120°

ARE YOU READY? 11.4 (page 935)

1.

x	$f^{-1}(x)$
1	0
2	1
4	2

2. a. 125 **b.** 3 **3. a.** 1 **b.** 1/2
4. a. 1,000 **b.** $\frac{1}{100}$

SELF CHECKS 11.4

1. $2^7 = 128$ **2.** $\log_9 \frac{1}{9} = -1$ **3. a.** 7 **b.** $\frac{1}{9}$ **c.** 3 **4. a.** 2 **b.** -2 **c.** $\frac{1}{2}$ **5. a.** 4 **b.** -3 **c.** Undefined **6.** 75.3998
7. a.

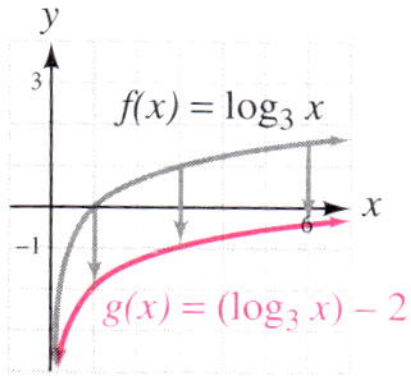

b.

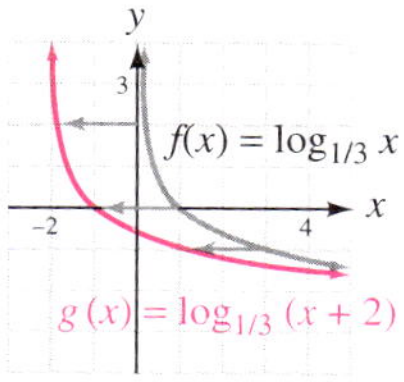

8. About 36 dB **9.** About 63 catfish

STUDY SET SECTION 11.4 (page 945)

1. logarithmic **3.** asymptote **5. a.** Logarithmic
b. D: $(0, \infty)$; R: $(-\infty, \infty)$ **c.** None, (1, 0) **d.** Yes
e. The y-axis ($x = 0$) **f.** Increasing **g.** 1 **7. a.** ii **b.** iii **c.** i
d. iv **9.** 6^2, 36 **11.** exponent **13.** logarithmic
15. 2, -2 **17.** 1, undefined, undefined
19. a. -0.30, 0, 0.30, 0.60, 0.78, 0.90, 1
b.

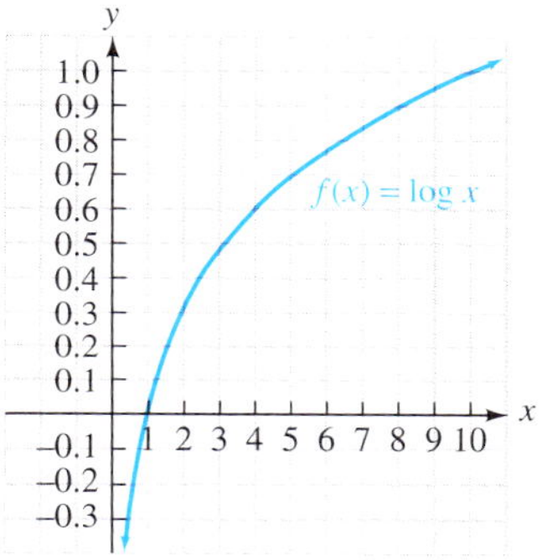

21. a. 10 **b.** x
23. $3^4 = 81$
25. $10^1 = 10$
27. $4^{-3} = \frac{1}{64}$
29. $5^{1/2} = \sqrt{5}$
31. $10^{-1} = 0.1$
33. $8^x = 64$
35. $b^t = T_1$
37. $n^{-42} = C$
39. $\log_8 64 = 2$
41. $\log_4 \frac{1}{16} = -2$
43. $\log_{1/2} 32 = -5$
45. $\log_x z = y$ **47.** $\log_y 8.6 = t$ **49.** $\log_7 (B + 1) = 4.3$
51. 9 **53.** 64 **55.** 3 **57.** $\frac{1}{25}$ **59.** $\frac{1}{6}$ **61.** 10 **63.** $\frac{2}{3}$
65. 5 **67.** 1,000 **69.** 4 **71.** 1 **73.** $\frac{1}{3}$ **75.** 3 **77.** 2
79. 6 **81.** -1 **83.** 5 **85.** $\frac{1}{2}$ **87.** 0.5119 **89.** -2.3307
91. 6,043.6597 **93.** 0.1726 **95.** 0.3162 **97.** 0.0195
99. Increasing

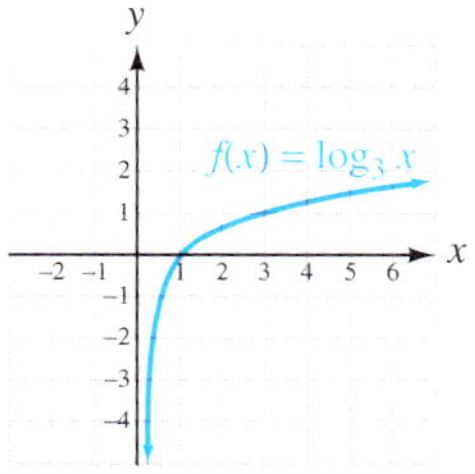

101. Decreasing

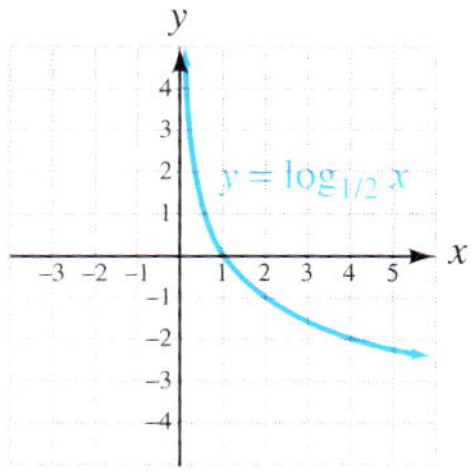

103.

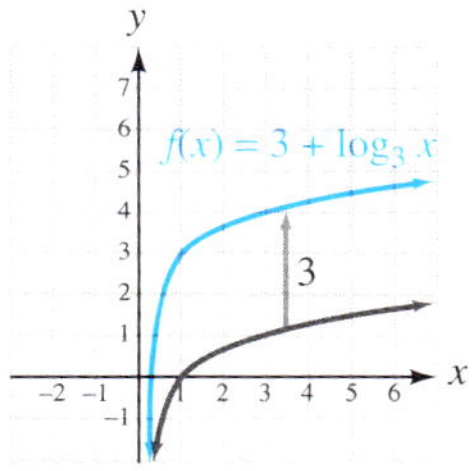

105.

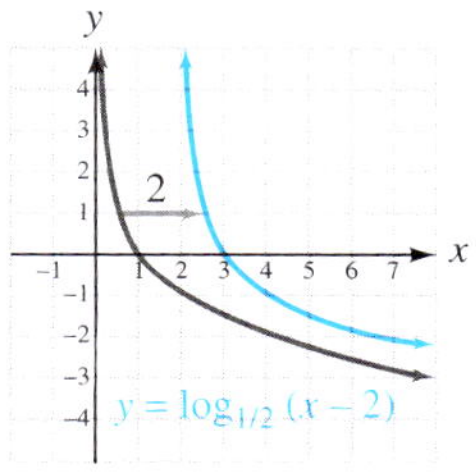

107.

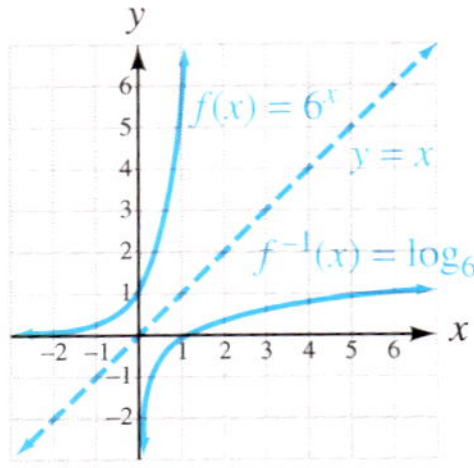

109.

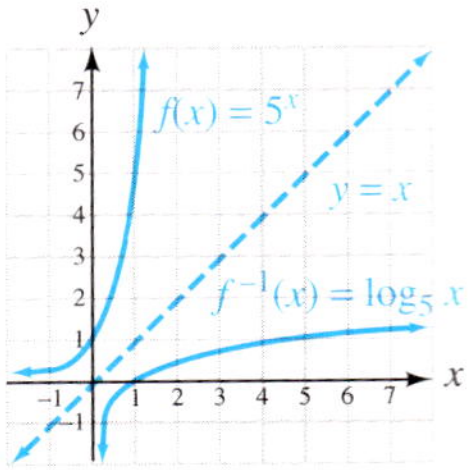

111. 49.5 dB **113.** 4.4 **115.** About 423 cases **117.** About 99 sunfish **119.** 77.8% **121.** 10.8 yr **127.** 10 **129.** 0; $-\frac{5}{9}$ is extraneous

ARE YOU READY? 11.5 (page 948)

1. 2.25 **2.** 1 **3.** 2.7 **4.** 82.05 **5.** 3^6, x **6.** 8, 3

SELF CHECKS 11.5

1. \$3,037,760.44 **2.** 3,347 **3.** 104.4° **4. a.** 3 **b.** -1 **c.** $\frac{1}{3}$ **5. a.** 6.9199 **b.** 0.0498 **6.** About 63 years **7.** About 5.3 feet per second

STUDY SET SECTION 11.5 (page 957)

1. exponential; e **3.** continuously **5. a.** The natural exponential function **b.** D: $(-\infty, \infty)$; R: $(0, \infty)$ **c.** $(0, 1)$, none **d.** Yes **e.** The x-axis $(y = 0)$ **f.** Increasing **g.** e **7.** 2.718281828459 **9.** e **11.** e, 2

13.

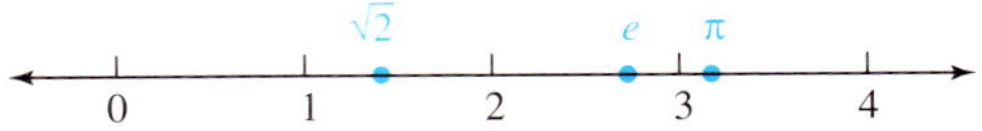

15. a. 2.7182818 . . .; e **b.** They decrease, no **17.** $f^{-1}(x) = e^x$ **19.** 1,000, 10, 0.9, 2,459.6 **21.** l, n **23.** LOG, LN

25.

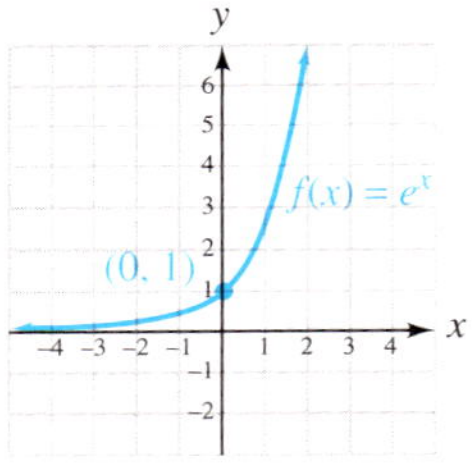

27.

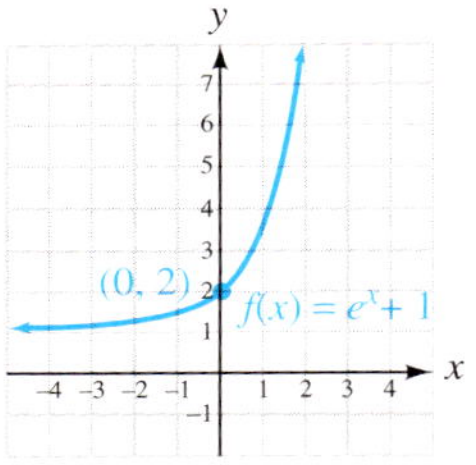

29.

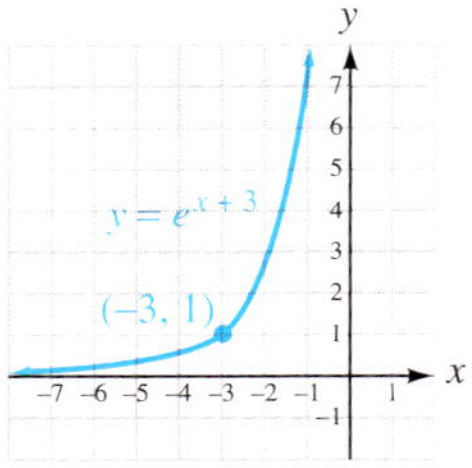

31.

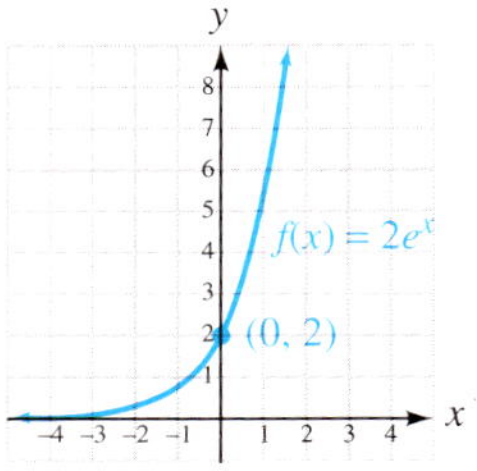

33. 24,765.16 **35.** 3,811,325.37 **37.** 11,542.14 **39.** 542.85 **41.** 5 **43.** 6 **45.** -1 **47.** $\frac{1}{4}$ **49.** $\frac{2}{3}$ **51.** -7 **53.** 3.5596 **55.** -5.3709 **57.** 0.5423 **59.** Undefined **61.** 4.0645 **63.** 69.4079 **65.** 0.0245 **67.** 2.7210

69.

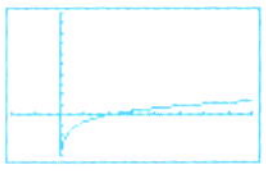

71.

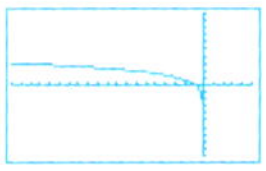

73. \$13,375.68 **75.** \$7,518.28 from annual compounding, \$7,647.95 from continuous compounding **77.** \$6,849.16 **79.** 133 million, 140 million, 165 million, 194 million, 206 million, 237 million, 269 million **81.** 4,715,620; 6,370,911 **83.** About 51 **85.** About 39 **87.** 14 chairs **89.** 105,731 germs **91.** About 49 meters per second **93.** About 16 years **95.** The year 2042 **97.** About 9.2 yr **99.** About 3.5 hr **101.** About 169 beats per minute **103.** About 1 foot per second faster (5.9 ft/sec − 4.9 ft/sec) **115.** $4x^2\sqrt{15x}$ **117.** $10y\sqrt{3y}$

ARE YOU READY? 11.6 (page 962)

1. 3 **2.** 3 **3.** -9 **4.** 2 **5. a.** $x^{1/2}$ **b.** $\sqrt{x-2}$ **6.** 1.6094

SELF CHECKS 11.6

1. a. 0 **b.** 1 **c.** 4 **d.** 2 **2. a.** $1 + \log_3 4$ **b.** $3 + \log y$ **c.** $2 + \log_5 c + \log_5 d$ **3. a.** $1 - \log_6 5$ **b.** $\ln y - \ln 100$ **4.** $\log_b x - \log_b y - \log_b z$ **5. a.** $4 \ln x$ **b.** $\frac{1}{3}\log_2 3$ **6.** $\frac{3}{4}\log x + \frac{1}{4}\log y - \frac{1}{4}\log z$ **7.** $\log_a \frac{x^2\sqrt{y}}{(x-y)^2}$ **8. a.** 0.1761 **b.** -0.1249 **9.** 0.6826 **10.** 7.6

STUDY SET SECTION 11.6 (page 971)

1. product **3.** power **5. a.** 0 **b.** 1 **c.** x **d.** x **11.** c **13.** b **15.** 8, a^3, 3, 1 **17.** True **19.** 0 **21.** 7 **23.** 10 **25.** 2 **27.** 1 **29.** 7 **31.** $2 + \log_2 5$ **33.** $\log 25 + \log y$ **35.** $2 + \log p + \log q$ **37.** $\log 5 + \log x + \log y + \log z$ **39.** $2 - \log 9$ **41.** $\log_6 x - 2$ **43.** $\log 7 + \log c - \log 2$ **45.** $1 + \log x - \log y$ **47.** $1 + \ln x + \ln y - \ln z$ **49.** $-1 - \log_8 m$ **51.** $7 \ln y$ **53.** $\frac{1}{2}\log 5$ **55.** $-3\log e$ **57.** $\frac{3}{5}\log_7 100$ **59.** $\log x + \log y + 2\log z$ **61.** $1 + \frac{1}{3}\log_2 x - \log_2 y$ **63.** $3\log x + 2\log y$ **65.** $\frac{1}{2}\log_b x + \frac{1}{2}\log_b y$ **67.** $\frac{1}{3}\log_a x - \frac{1}{4}\log_a y - \frac{1}{4}\log_a z$ **69.** $20\ln x + \frac{1}{2}\ln z$ **71.** $-3d\log_5 t$ **73.** $\frac{1}{2} + \frac{1}{2}\ln x$ **75.** $\log_2 x^9(x+1)$ **77.** $\log_3 \frac{x(x+2)}{8}$ **79.** $\log_b \frac{\sqrt{z}}{x^3y^2}$ **81.** $\log_b \sqrt[3]{M-3}$ **83.** $\ln \frac{\frac{x}{z}+x}{\frac{y}{z}+y} = \ln \frac{x}{y}$ **85.** $\log_6 \frac{\sqrt{x^2+1}}{x^2+2}$ **87.** 1.4472 **89.** -1.1972 **91.** 1.1972 **93.** 1.8063 **95.** 1.7712 **97.** -1.0000 **99.** 1.8928 **101.** 2.3219 **103.** a, c **105.** a, b **107.** About 4.8 **109. a.** $B = \log\left(\frac{I}{I_0}\right)^{10}$ **b.** $T = \ln\sqrt[k]{\frac{C_2}{C_1}}$ **119.** $-\frac{7}{6}$ **121.** $\left(1, -\frac{1}{2}\right)$

ARE YOU READY? 11.7 (page 973)

1. a. 4 **b.** -3 **2.** 1.1309 **3.** $x\log_3 8$ **4.** 1 **5. a.** $\log_2 5x$ **b.** $\ln \frac{10}{2t+1}$ **6.** Undefined

SELF CHECKS 11.7

1. 2 **2.** A repeated solution of 1 **3.** $\frac{\log 4}{\log 5} \approx 0.8614$ **4.** $\frac{2\log 5}{\log 5 - \log 3} \approx 6.3013$ **5.** $\frac{\ln 35}{2.1} \approx 1.6930$ **6.** $\frac{7}{2}$ **7.** 8 **8.** 2 **9.** $\frac{9}{16}$ **10.** About 1.58×10^{-5} gram-ions per liter **11.** About 11,400 years **12.** About 10 hours

STUDY SET SECTION 11.7 (page 983)

1. exponential **3. a.** equal; x, y **b.** equal; x, y **5.** -2 **7.** logarithm **9.** ln 4 **11.** Not a solution **13. a.** Divide both sides by ln 3. **b.** $\frac{\ln 5}{\ln 3}$ **c.** 1.4650 **15. a.** 1.2920 **b.** 10.7549 **17. a.** log **b.** $A_0 2^{-t/h}$ **c.** $P_0 e^{kt}$ **19.** log, log 2, 2.8074 **21.** 4 **23.** $-\frac{3}{4} = -0.75$ **25.** 3, -1 **27.** A repeated solution of -2 **29.** $\frac{\log 5}{\log 4} \approx 1.1610$ **31.** $\frac{\log 2 + \log 13}{\log 13} \approx 1.2702$ **33.** $\frac{\log 2}{\log 3 - \log 2} \approx 1.7095$ **35.** $\frac{3\log 5}{\log 5 - 2\log 3} \approx -8.2144$ **37.** $\frac{\ln 4.5}{2.9} \approx 0.5186$

39. $-\frac{\ln 14.2}{0.2} \approx -13.2662$ **41.** 5,000 **43.** 12 **45.** -93
47. 0.08 **49.** -7 **51.** 3 **53.** 50 **55.** $\frac{5}{4} = 1.25$ **57.** 0.5
59. 15.75 **61.** 2 **63.** $e \approx 2.7183$ **65.** $\pm\sqrt{\frac{\log 10}{\log 7}} \approx \pm 1.0878$
67. 10 **69.** 10 **71.** 4 **73.** $\frac{\log 15 - 2\log 9}{\log 9} \approx -0.7675$
75. $10, -10$ **77.** 10 **79.** 9 **81.** A repeated solution of 4
83. 1, 7 **85.** $\frac{\ln 9}{3} \approx 0.7324$ **87.** 6 **89. a.** $\frac{10^{1.7}}{5} \approx 10.0237$
b. $\frac{e^{1.7}}{5} \approx 1.0948$ **91. a.** $\frac{\log 90 + 5\log 4}{3\log 4} \approx 2.7486$
b. $\frac{5 + \ln 90}{3} \approx 3.1666$ **93. a.** $\frac{5}{4}$ **b.** $\frac{5}{4}$ **95. a.** 3 **b.** 2 **97.** 1.8
99. 20 **101.** About 6.3×10^{-14} gram-ions per liter
103. About 5.1 yr **105.** About 42.7 days **107.** About 4,200 yr
109. About 5.6 yr **111.** About 5.4 yr **113.** Because $\ln 2 \approx 0.7$
115. About 25.3 yr **117.** 2.828 times larger
119. $\frac{1}{3}\ln 0.75 \approx -0.0959$ **125.** $\sqrt{137}$ in.

CHAPTER 11 REVIEW (page 986)

1. $(f - g)(x) = 3x + 1, (-\infty, \infty)$
2. $(f - g)(x) = x - 1, (-\infty, \infty)$
3. $(f \cdot g)(x) = 2x^2 + 2x, (-\infty, \infty)$
4. $(f/g)(x) = \frac{2x}{x+1}, (-\infty, -1) \cup (-1, \infty)$ **5.** 3 **6.** 5
7. $(f \circ g)(x) = 4x^2 + 4x + 3$ **8.** $(g \circ f)(x) = 2x^2 + 5$
9. a. 0 **b.** -8 **c.** 0 **d.** -1 **10.** $C(m) = \frac{3.25m}{8}$
11. No **12.** Yes **13.** Yes **14.** No **15.** Yes **16.** No
17. $-6, -1, 7, 20$ **18.**

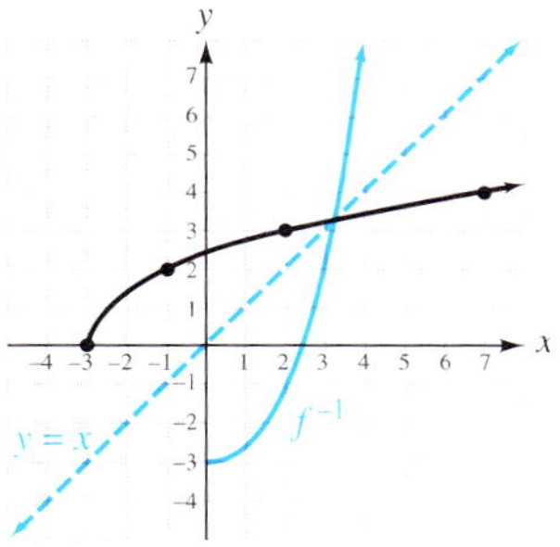

19. $f^{-1}(x) = \frac{x+3}{6}$ **20.** $f^{-1}(x) = \frac{4}{x} + 1$ **21.** $f^{-1}(x) = \sqrt[3]{x} - 2$
22. $f^{-1}(x) = 6x + 1$
23. $f^{-1}(x) = x^3 + 1$

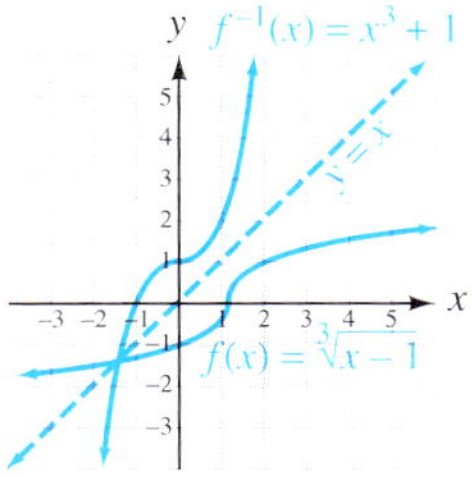

25. a. $n(x) = 2^x, s(t) = 1.08^t$ **b.** 121.9774
26. a. Exponential decay **b.** Exponential growth
27. D: $(-\infty, \infty)$; R: $(0, \infty)$ **28.** D: $(-\infty, \infty)$; R: $(0, \infty)$

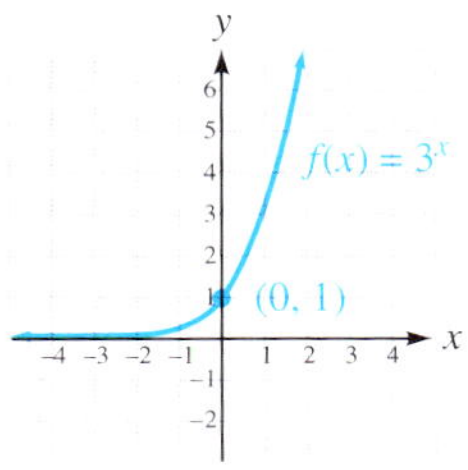

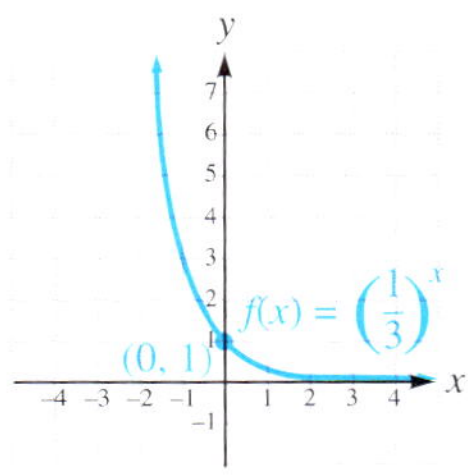

29. D: $(-\infty, \infty)$; R: $(-2, \infty)$ **30.** D: $(-\infty, \infty)$; R: $(0, \infty)$

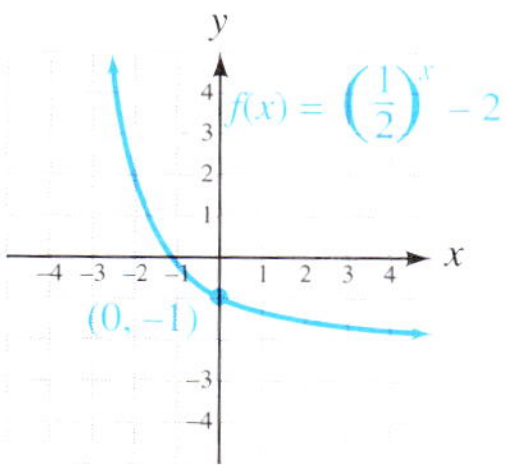

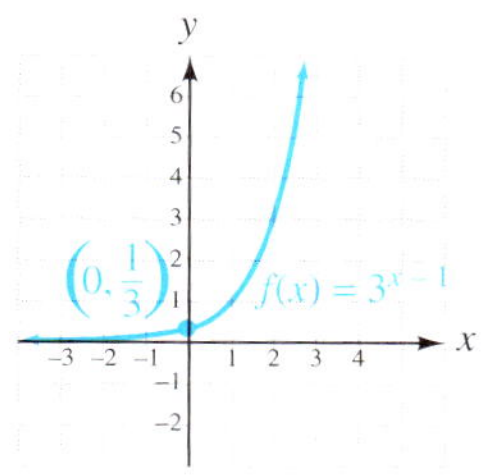

31. The x-axis ($y = 0$) **32. a.** 128,000 tons
b. About 281,569,441 tons **33.** About \$2,189,703.45
34. About \$2,015 **35.** D: $(0, \infty)$; R: $(-\infty, \infty)$ **36.** Since there is no real number such that $10^? = 0$, log 0 is undefined. **37.** $4^3 = 64$
38. $\log_7 \frac{1}{7} = -1$ **39.** 2 **40.** -2 **41.** 0 **42.** Undefined
43. $\frac{1}{2}$ **44.** 3 **45.** 32 **46.** $\frac{1}{81}$ **47.** 4 **48.** 10 **49.** $\frac{1}{2}$
50. $\frac{1}{3}$ **51.** 0.6542 **52.** 26.9153
53.

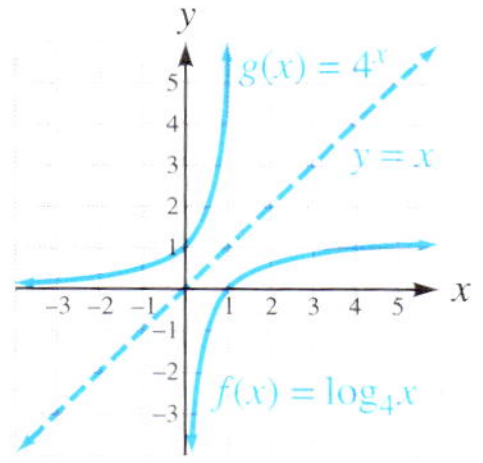

54.

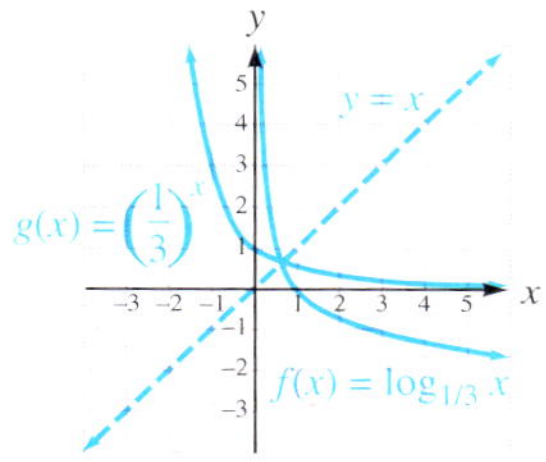

55.

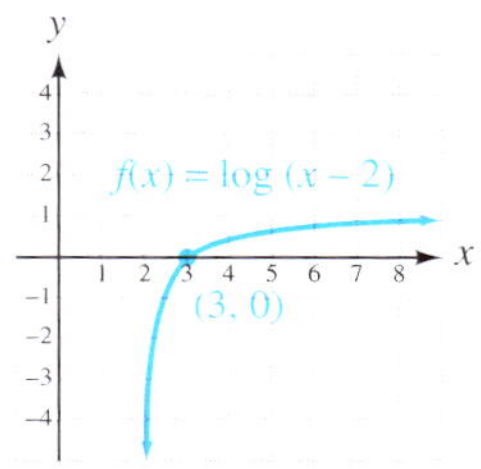

56.

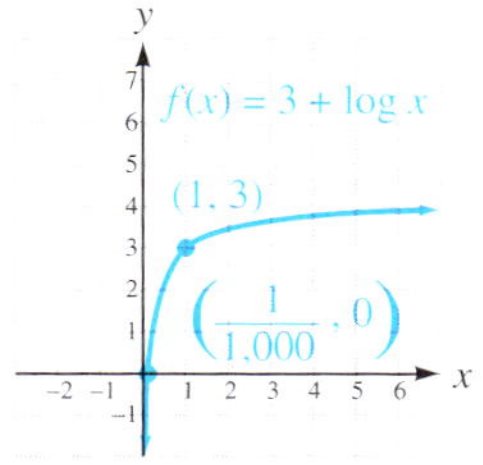

57. About 53 dB **58.** About 4.4 **59. a.** 52 cm **b.** About 74.6 cm
60. About 95% **61. a.** $e \approx 2.72$ **b.** 2.7081, 15 **62.** 16.6704
63. D: $(-\infty, \infty)$; R: $(1, \infty)$ **64.** D: $(-\infty, \infty)$; R: $(0, \infty)$

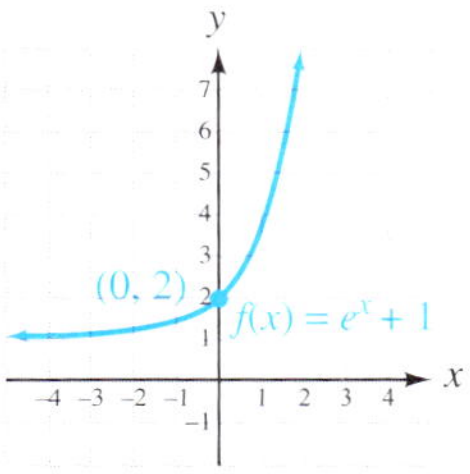

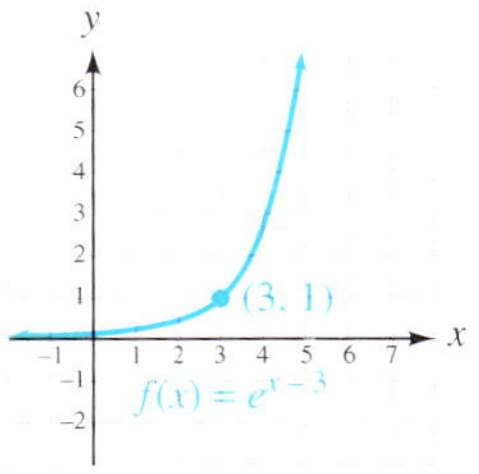

65. \$2,324,767.37 **66.** About \$211.7 billion **67.** 13.9%, 11.67%, 9.80% **68.** The exponent on the base e is negative.
69. 1 **70.** 2 **71.** -5 **72.** $\frac{1}{2}$ **73.** Undefined **74.** Undefined
75. 0 **76.** -7 **77.** 6.1137 **78.** -0.1625 **79.** 10.3398
80. 0.0002
81. They have different bases: $\log x = \log_{10} x$ and $\ln x = \log_e x$.

82. $f^{-1}(x) = e^x$

83.

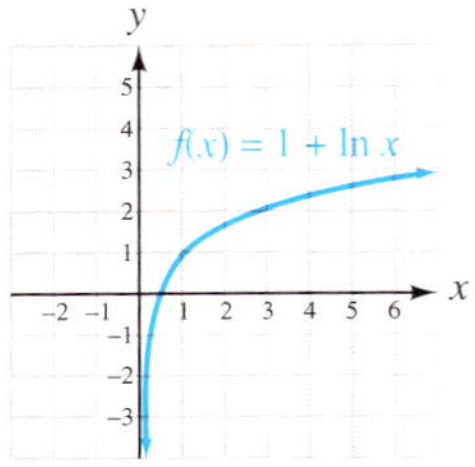

84.

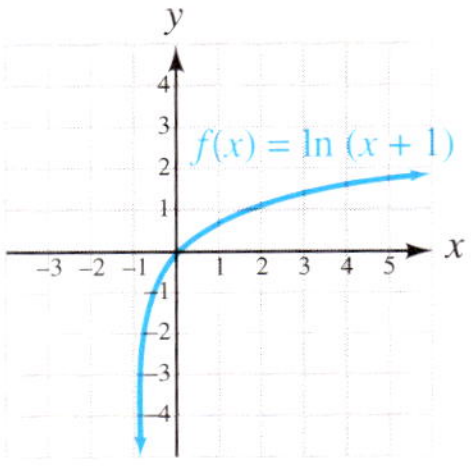

85. About 62 yr **86.** About 72 in. (6 ft) **87.** 0 **88.** 1 **89.** 3 **90.** 4 **91.** $3 + \log_3 x$ **92.** $2 - \log x$ **93.** $\frac{1}{2}\log_5 27$ **94.** $\log_b 10 + \log_b a + 1$ **95.** $2\log_b x + 3\log_b y - \log_b z$ **96.** $\frac{1}{2}\ln x - \frac{1}{2}\ln y - \ln z$ **97.** $\log_2 \frac{x^3z^7}{y^5}$ **98.** $\log_b \frac{\sqrt{x+2}}{y^3z^7}$ **99.** $\log_b(a - 5)$ **100.** $\log_8 x^7$ **101.** 2.6609 **102.** 3.0000 **103.** 1.7604 **104.** 3.1 **105.** −4 **106.** −3, −1 **107.** $\frac{\log 7}{\log 3} \approx 1.7712$ **108.** $\frac{4\log 3}{\log 3 - \log 2} \approx 10.8380$ **109.** $\ln 7 \approx 1.9459$ **110.** $-\frac{\ln 25}{0.4} \approx -8.0472$ **111.** $\frac{7}{3}$ **112.** $\pm\sqrt{\frac{\log 33}{\log 9}} \approx \pm 1.2615$ **113.** 104 **114.** 9 **115.** 25, 4 **116.** 4 **117.** 4, 3 **118.** 2 **119.** 6 **120.** 31 **121.** $0.76787 \neq -0.27300$ **122.** About 3,300 yr **123.** About 91 days **124.** 2, 5

CHAPTER 11 TEST (page 997)

1. a. composite **b.** natural **c.** continuous **d.** inverse **e.** logarithmic **2. a.** f composed with g of x **b.** g of f of eight **c.** f inverse of x **3.** $(f + g)(x) = 4x^2 - 2x + 11, (-\infty, \infty)$ **4.** $(g/f)(x) = \frac{4x^2 - 3x + 2}{x + 9}, (-\infty, -9) \cup (-9, \infty)$ **5.** 76 **6.** $32x^2 - 128x + 131$ **7. a.** −16 **b.** 17 **8. a.** 0 **b.** 3 **c.** 1 **d.** −2 **e.** −5 **9. a.** No **b.** Yes **c.** Yes **d.** No **10.** $f^{-1}(x) = -3x$

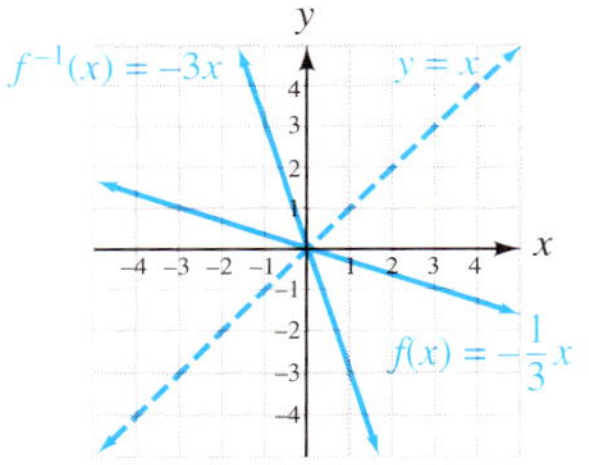

11. One-to-one function; $f^{-1}(x) = 3x - 6$ **12.** $f^{-1}(x) = \sqrt[3]{x} + 15$ **14. a.** Yes **b.** Yes **c.** 80; when the temperature of the tire tread is 260°, the vehicle is traveling 80 mph **15.** D: $(-\infty, \infty)$; R: $(1, \infty)$

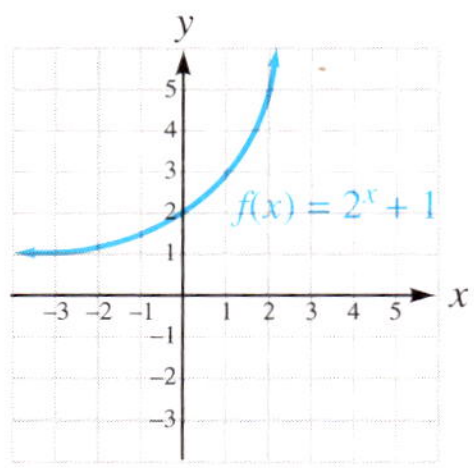

16. D: $(-\infty, \infty)$; R: $(0, \infty)$

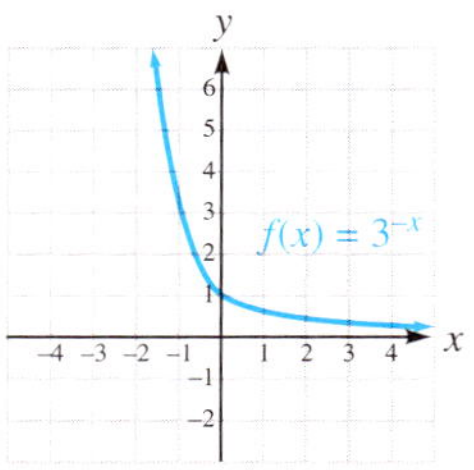

17. $\frac{3}{64}$ g = 0.046875 g **18.** About $1,060.90

19. a.

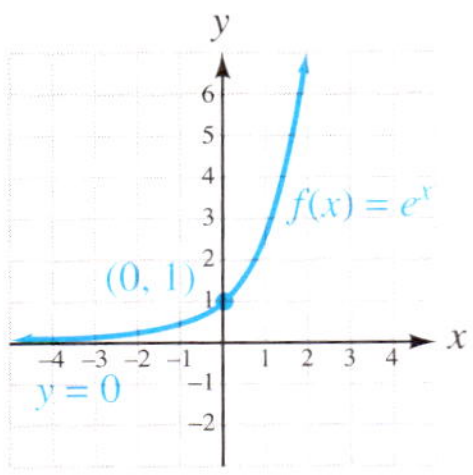

b. D: $(-\infty, \infty)$; R: $(0, \infty)$ **c.** $f^{-1}(x) = \ln x$ **20.** 1,346,161,000 **21.** About 88 micrograms per day **22.** $6^{-2} = \frac{1}{36}$ **23. a.** D: $(0, \infty)$; R: $(-\infty, \infty)$ **b.** $f^{-1}(x) = 10^x$ **24.** Logarithmic growth **25.** 2 **26.** −2 **27.** Undefined **28.** −6 **29.** $\frac{1}{2}$ **30.** 0 **31.** 2 **32.** 16 **33.** $\frac{1}{27}$ **34.** e

35.

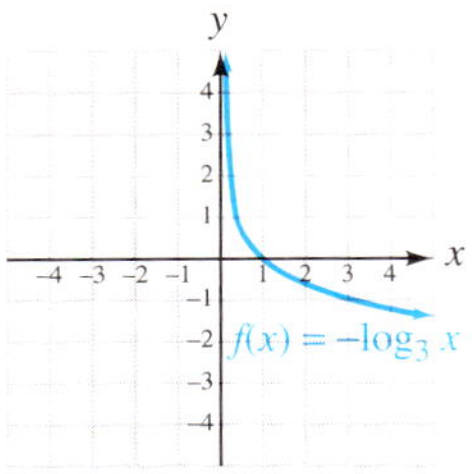

36.

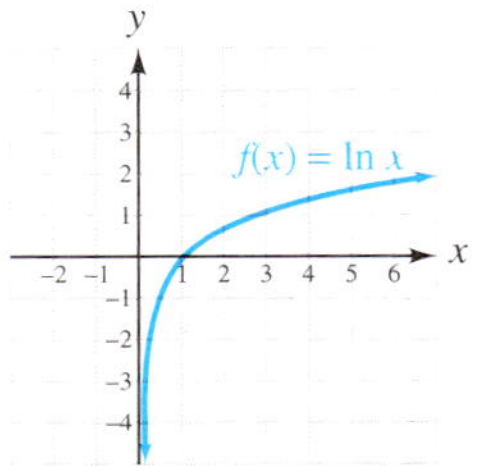

37. About 6.4 **38.** About 46 dB **39.** 0.0871 **40.** 0.5646 **41.** $2\log_b a + 1 + 3\log_b c$ **42.** $\ln \frac{b\sqrt{a+2}}{c^3}$ **43.** $\frac{\log 3}{\log 5} \approx 0.6826$ **44.** 4 **45.** $\frac{1}{6}$ **46.** $\frac{\ln 4}{0.08} \approx 17.3287$ **47.** 5 **48.** 6 **49.** 1 **50.** 10 **51.** 5 **52.** About 20 min

ARE YOU READY? 12.1 (page 1002)

1. It is not a solution.

2.

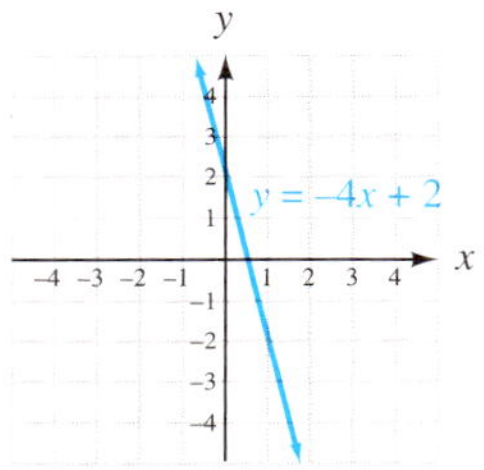

3.

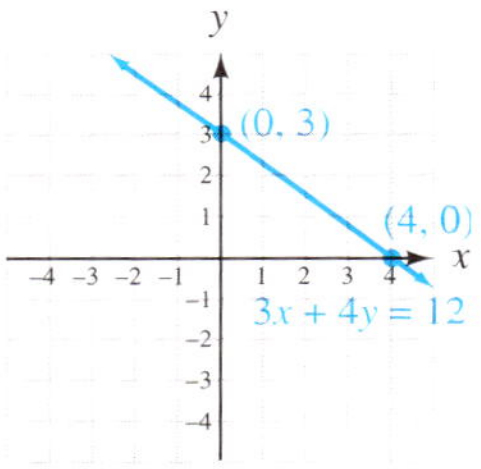

4. $y = 5x + 4$ **5.** −9 **6.** $-28x + 4y = -36$

SELF CHECKS 12.1

1. It is a solution.

2. (1, 2)

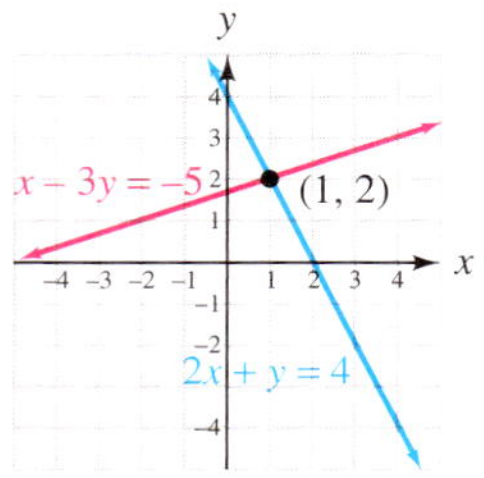

3. No solution; ∅

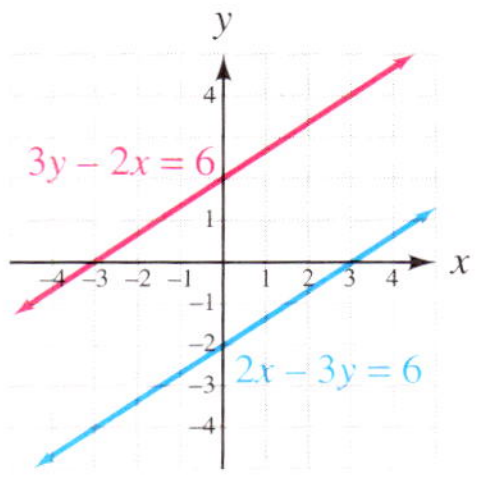

4. $\{(x, y) \mid 2x - y = 4\}$; infinitely many solutions

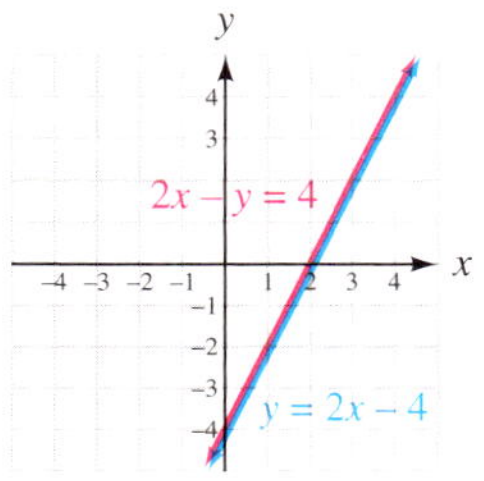

INDEX

T

U

V

How to Reduce Math Test Anxiety

CHAPTER 3

In this chapter you will first learn about math anxiety and then explore test anxiety. They are different issues that can affect each other. It is best to learn about them as separate entities first.

Math anxiety is a common problem for many high school, college, and university students. It is especially difficult for students in developmental courses, who normally have more math anxiety than other students. However, there are students in higher-level math courses who also struggle with this problem. It is very common for students to have anxiety only about math and not in their other subjects.

Most students think that math anxiety affects them only when they are taking tests, but it also affects other areas. It can affect the way you do your homework, the way you learn in the classroom or through distance learning courses, and the way you choose a career. Students who have math anxiety may procrastinate in doing their homework or put off sitting down and completing an online lesson. This can lead to math failure. Students also select majors based on the amount of math that is required, which could lead to lower-paying or dissatisfying careers. However, most students with math anxiety meet it face to face during tests, experiencing test anxiety as well.

CHAPTER GOALS

What You Will Learn

3.1 Understanding Math Anxiety
3.2 How to Recognize Test Anxiety
3.3 The Causes of Test Anxiety
3.4 The Different Types of Test Anxiety
3.5 How to Reduce Test Anxiety

Mild test anxiety can be a motivational factor in that it can make students properly prepare for a test. However, high test anxiety can cause major problems in both learning and test taking, as students avoid studying for the test when anxiety begins to afflict their thought processes. Reducing test anxiety is the key for many students to become successful in math. Such students need to learn the causes of test anxiety and how to reduce the test anxiety that affects their learning and grades.

Several techniques have proven helpful in reducing both math anxiety and math test anxiety. However, reducing them does not guarantee good math grades. It must be coupled with effective study skills and a desire to do well in math.

DAN'S TAKE

In 10th grade, I achieved the lowest test score of my life. The class: trigonometry. The grade: well, you could count my score on one hand.

As if this weren't embarrassing enough, when my teacher handed the test back to me, the entire group sitting at my table took notice before I could manage to stuff the exam underneath my textbook. But then, how could they miss it? The paper was so thoroughly covered with thick, radiating red-pen marks that it may well have been visible to the astronauts on the International Space Station.

But before you judge me, let's follow my thought process before the exam:

1. I am bad at math. I do not get it.
2. This test is the third in the semester and I have already failed the previous two.
3. If I failed the other tests, what are the chances that I'll pass the third?
4. Conclusion: there is absolutely no point in studying. I can't possibly do any worse right?

I was wrong. Horribly wrong. I *could* do worse . . . and I did. For one great night of videogames and television reruns, I had backed myself into a corner that nearly sunk my academic future. What's more, I nearly lost my spot on the basketball team as my midterm GPA was within a 10th of a point of making me ineligible.

When I chose not to study for this particular exam, I was attempting to avoid the shame of trying *and still* failing. I somehow justified the act of failing without studying as a smaller blow to my pride than knowing that I had put in the proper amount of time and effort and still come up flat.

What I did not realize at the time was that I was simply looking for an excuse to cover up the real issue behind my poor math grades: math anxiety. I convinced myself that I was just bad at math. Even a simple thought about that subject was enough to send me into a spiral of self-doubt and discomfort. What I did not know was that by merely thinking these thoughts, I was setting myself up to fail from the very beginning. My fear made me "draw blanks" during my exams and I often wound up failing.

It was not until I went to a tutor that I realized I could get over these fears. I had to stop using the excuse "I am just bad at math." With a little help, I pulled my final grade up to a C. It turned out I was not bad at math at all—I was merely afraid of it!

3.1 Understanding Math Anxiety

Definitions of Math Anxiety

Math anxiety is a relatively new concept in education. During the 1970s, certain educators began referring to *mathophobia* or *mathemaphobia* as a possible cause for children's unwillingness to learn math. Math anxiety is an extreme emotional and/or physical reaction to a very negative attitude toward math. There is a strong relationship between low math confidence and high math test anxiety (Fox, 1977).

Math anxiety is the feeling of tension and anxiety that interferes with the manipulation of numbers and the solving of math problems during tests (Richardson and Suinn, 1973). Math anxiety is a state of panic, helplessness, paralysis, and mental disorganization that occurs in some students when they are required to solve math problems. This discomfort varies in intensity and is the outcome of numerous experiences students have had in their past learning situations (Tobias, 1978).

Today, math anxiety is accepted as one of the major problems students have in completing their math courses. It is real, but it can be overcome.

Types of Math Anxiety

Math anxiety can be divided into three separate anxieties: math test anxiety, numerical anxiety, and abstraction anxiety. Math test anxiety involves anticipation of, completion of, and feedback on math tests. Numerical anxiety refers to everyday situations requiring people to work with numbers and perform arithmetic calculations. Numerical anxiety can also include students who are trying to figure out the amount of a tip, thinking about mathematics, doing math homework, or listening to or seeing math instruction. Abstraction anxiety involves working with variables and mathematical concepts used to solve equations. Students can have all three math anxieties or only one anxiety. Most often, the students I have worked with have had math test anxiety and abstraction anxiety. These students don't have any anxiety working with numbers, but once they start learning algebra, they develop both conditions. This may have happened in high school or college.

REMEMBER

When people try to overcome anxiety of any kind, including math anxiety, they must seek advice and support. It is much easier with support!

The Causes of Math Anxiety

Since math anxiety is a learned condition, its causes are unique to each student, but they are all rooted in individuals' past experiences. Bad experiences in elementary school are one of the most common sources of students' math anxiety: coming in last in math races at the blackboard, watching classmates finish problems twice as fast as they do, teachers saying, "That's okay, you just aren't good in math, you are better in English," and classmates and teachers calling them stupid. These words and experiences remain with people; they can still hear the words and eventually begin telling themselves the same thing. When these students walk into a classroom or open a math book, or take a test, these "mental tapes" play in their minds. When asked, many students indicate that they were made fun of when trying to solve math problems at the chalkboard. When they could not solve the problems, the teacher and/or students would call them "stupid."

Teacher and peer embarrassment and humiliation become the conditioning experiences that cause some students' math anxiety. Over the years, this math anxiety is reinforced and even increases in magnitude. In fact, many math anxious students—now 30, 40, and 50 years old—still have extreme fear about working math problems on the board. One 56-year-old indicated that he had a great deal of fear that the instructor would call him to the board. Even if he knew how to do the problem, displaying that knowledge to his peers was very difficult. Some students have said that they absolutely refused to go to the board.

Being embarrassed by family members can also cause math anxiety. According to students who have been interviewed on the matter, their parents tried to help them with math and this sometimes led to serious trauma. These students claim that the tutoring from their guardians, mainly their fathers, often resulted in scolding when they were not able to complete the problems. One student reported that his father hit him every time he got a multiplication problem wrong. Brothers and sisters can also tease one another about being "dumb" in math. This is particularly true of boys telling girls that they cannot do math. When people hear these statements often enough, they may start to believe them and associate these bad feelings with the word *math*. So, for students many years later, just hearing the word "math" triggers a response of anxiety, consciously or unconsciously recalling the bad feelings, and becoming uneasy.

A good example of this is a student I worked with who had completed her B.S. degree 15 years ago at a college that did not require much math. She was returning to college to be an elementary school teacher, which required her to take math and a math placement test. As soon as I mentioned that she had to take math, she said, "I can't do math and I will have to wait a few days to get psychologically ready to take the math placement test." She indicated that her old feeling of not being able to do math rushed through her and she almost had an anxiety attack. This is an extreme case, but a true example, of math anxiety. In most cases, math anxiety is not this bad, but it is disruptive enough to cause learning and testing problems.

In order to overcome math anxiety, it is necessary to identify when it first started and what caused it. Math anxiety usually starts from one or a series of events in a student's past that were negative. Negative feelings were imbedded in memories and then anxiety started whenever the student was placed in similar situations.

Answer the questions on the Exploring Past Math Experiences Worksheet.

If some of the memories are tough ones, just make a short note of them right now. Don't dwell on them, because they are in the past. This can be frustrating for some students, but it is worth working through it. You will realize that math anxiety is usually a result of events, and does not occur because you are "dumb" or have other personal flaws.

If you have had a very positive experience in the past when you were studying math, brainstorm answers to the appropriate questions. You may want to keep in mind friends you know who struggle with math and think about the other questions.

This brainstorming will help you write your math autobiography assignment at the end of this chapter.

SOME FACTS TO REMEMBER

- Math anxiety is usually a result of past experiences that were negative. As a result of these experiences, any time a person is put in a similar situation, he or she gets anxious.
- Math anxiety is not related to how smart a person is.
- Math anxiety is a learned condition; therefore, in most cases, it can be unlearned or at least managed.
- A person must be willing to change and find strategies to practice continually that will help manage math anxiety.

EXPLORING PAST MATH EXPERIENCES WORKSHEET

At the end of this chapter you will write a "Math Autobiography" to help you explore why you might have math anxiety. This knowledge is the first step in learning how to manage or overcome the anxiety. This worksheet will help you prepare for writing the autobiography. Answer the following questions. For some students, these questions will be easy to answer; for others, they will be more difficult because their past experiences are quite negative. The more honest you can be, the more you will be able to confront and work on any math anxiety.

1. When was the first time you were nervous or anxious about learning math? (Do not include taking tests.) Was it in elementary, middle, or high school? Can you recall a specific incident? If so, briefly describe it. It could have taken place in school or at home.

   Examples: Getting scolded by a teacher or parent for a poor homework grade. Losing a math race on the board. Going to recess late because you did not get your math work done in time. Older sibling always comparing math grades with you.

   ____

   ____

2. When was the last time you told yourself that you couldn't do math or were just not a math type?

   ____

   ____

3. Even though you might have math anxiety, there were probably some positive experiences associated with math in the past. Try to think of at least one of them.

   Examples: A good grade. Completing an assignment in class. A parent or teacher praising you for your math work.

   ____

   ____

4. If you have math anxiety, take heart. Most students can learn to minimize it and control it. Write a statement declaring that you will set a goal to learn to manage your math anxiety this semester.

   ____

   ____

How Math Anxiety Affects Learning

Math anxiety can cause learning problems in several ways. It can affect how you do your homework and your participation in the classroom and in study groups. Let's first start by looking at how math anxiety could affect your homework. Students with high math anxiety may have difficulty starting or completing their homework.

Doing homework reminds some students of their learning problems in math. More specifically, it reminds them of their previous math failures, which causes further anxiety. This anxiety can lead to total avoidance of homework or "approach-avoidance" behavior.

Total homework avoidance is called *procrastination*. The very thought of doing their homework causes these students anxiety, which causes them to put off tackling their homework. This makes them feel better for a short amount of time—*until test day*.

Math anxiety can also affect your classroom participation and learning. Usually students with math anxiety are afraid to speak out in class and ask questions. They remember that in the past they were made fun of when they gave incorrect answers. They are also afraid of asking a question that others, including the teacher, will consider "dumb." So they sit in class fearful of being asked a question, looking like they understand the lecture so they will not be called on. They also take a lot of notes even though they don't understand them, in order to give the illusion of comprehension. If you are one of these students, these are hard habits to break. However, these habits may cause you to be unsuccessful in your math class. Here are some suggestions that may help you break these habits:

1. Make an appointment to talk to your math instructor. Math instructors want you to talk to them. When I do my consulting around the country, one of the major complaints I get from math instructors is that the students don't come and see them. Make an appointment to see your math instructor before the first major test to discuss your math history and to ask for learning suggestions. In most cases it is easier to talk to the instructor before you get your first grade.
2. Before class, ask the instructor to work one homework problem. You might want to write the problem on the board before the instructor arrives. This is less stressful because you are not asking the question in front of the whole class. In fact, one of my good friends, Dr. Mike Hamm, suggests that his students put the problems they do not know how to solve on the board before class. Other students go to the board and solve the problems. Dr. Hamm then solves the ones the students cannot solve.
3. Prepare one question from your homework and ask it within the first 15 minutes of class. Instructors are more likely to answer questions in the first part of class when they have

time instead of the end of class when time is running out.

4. Ask a question to which you already know the answer. The question will begin discussion with the instructor and other students when you can contribute since you understand the problem. It will set you up for a positive experience.
5. Use email to send questions to your instructor. This way, you can still get the answer with very little anxiety.

By using these suggestions, you can reduce your math anxiety and learn more mathematics. A question unanswered could be a missed test question.

> **REMEMBER**
> The instructor's job is to answer your questions, and you are paying for the course.

Math-anxious students sometimes avoid doing additional math outside of the classroom. They avoid study groups and supplemental instruction. It is like asking a person with hydrophobia (fear of water) to take a vacation at the beach. However, a person with hydrophobia can go to the beach and enjoy himself or herself and not get wet. In other words, students can still attend study groups and supplemental instruction and just listen. When they are ready to get their feet wet, they can ask a few questions. Don't let these great opportunities go by.

Math anxiety can affect how you learn mathematics. It can be overcome with your effort. You don't have to live in the past with your math fears. Today is a new day, and you can change how math affects you. The next step is to understand how test anxiety can affect your demonstration of math knowledge.

SECTION 3.1 REVIEW

1. List and explain the three types of math anxiety.

   First Type: ____________________

   Second Type: ____________________

   Third Type: ____________________

2. List and explain two causes of math anxiety.

   First Cause: ____________________

   Second Cause: ____________________

3. List three ways math anxiety can cause learning problems.

   First Way: ____________________

   Second Way: ____________________

   Third Way: ____________________

4. List and explain three ways students can manage or avoid the effects of math anxiety.

   First Way: ____________________

   Second Way: ____________________

   Third Way: ____________________

3.2 How to Recognize Test Anxiety

Test anxiety has existed for as long as tests have been used to evaluate student performance. Because it is so common and because it has survived the test of time, test anxiety has been carefully studied over the last 50 years. Pioneering studies indicate that test anxiety generally leads to low test scores. At the University of South Florida (Tampa), Dr. Charles Spielberger investigated the relationship between test anxiety and intellectual ability. The study results suggested that anxiety coupled with high ability can improve academic performance; but anxiety coupled with low or average ability can interfere with academic performance. That is:

anxiety + high ability = improvement

anxiety + low or average ability = no improvement

EXAMPLE

Students with average ability and low test anxiety had better performance and higher grades than did students with average ability and high test anxiety. However, there are students who make good grades, take calculus, and still have test anxiety.

Test anxiety is a *learned* response; a person is not born with it. An environmental situation brings about test anxiety. The good news is that because it is a learned response, it can be *unlearned.* Test anxiety is a special kind of general stress. General stress is considered to be "strained exertion," which can lead to physical and psychological problems.

Math Test Anxiety

There are several definitions of test anxiety. One definition states, "Test anxiety is a conditioned emotional habit to either a single terrifying experience, recurring experience of high anxiety, or a continuous condition of anxiety" (Wolpe, 1958).

Another definition of test anxiety relates to the educational system. The educational system develops evaluations that measure one's mental performance, and this creates test anxiety. This definition suggests that test anxiety is the *anticipation* of some realistic or nonrealistic situational threat (Cattell, 1966). The "test" can be a research paper, an oral report, work at the chalkboard, a multiple-choice exam, a written essay, or a math test.

Math test anxiety is a relatively new concept in education. *Ms.* magazine published "Math Anxiety: Why Is a Smart Girl Like You Counting on Your Fingers?" (Tobias, 1976) and coined the phrase *math anxiety.* During the 1970s, other educators began referring to *mathophobia* or *mathemaphobia* as a possible cause of children's unwillingness to learn math. Additional studies on the graduate level discovered that math anxiety was common among adults as well as children.

One of my students once described math test anxiety as "being in a burning house with no way out." No matter how you define it, math test anxiety is real, and it affects millions of students.

Why Math Tests Create Anxiety

It has been shown that math anxiety exists among many students who usually do not suffer from other tensions. Counselors at a major university reported that one-third of the students who enrolled in behavior therapy programs offered through counseling centers had problems with math anxiety (Suinn, 1988).

Educators know that math anxiety is common among college students and is more prevalent in women than in men. They also know that math anxiety frequently occurs in students with poor high school math backgrounds. These students were found to have the greatest amount of anxiety.

Approximately half of the students in college prep math courses (designed for students with inadequate high school math backgrounds or low placement scores) could be considered to have math anxiety. However, math anxiety also occurs in students in high-level math courses, such as college algebra and calculus.

Educators investigating the relationship between anxiety and math have indicated that anxiety contributes to poor grades in math. They have also found that simply *reducing* math test anxiety does not guarantee higher math grades. Students often have other problems that affect their math grades, such as poor study skills, poor test-taking skills, or poor class attendance.

SECTION 3.2 REVIEW

1. What is your personal definition of test anxiety?

____

2. What type of student does math anxiety affect most?

____

3.3 The Causes of Test Anxiety

The causes of test anxiety can be different for each student, but they can be explained by seven basic concepts. See Figure 3.

The causes of math test anxiety can be different for each student. It could possibly have first occurred in middle school or high school. However, for many students it first occurs in college when passing tests is the only way to pass the course. Homework and extra credit in most college courses don't count toward your grade. Now students must have a passing average and in some cases must pass the departmental final exam. Additional pressure also exists because not passing algebra means you won't graduate and you might not get the job you want. As you can see, there are more reasons to have math test anxiety in college than in high school.

If you have math test anxiety, I want you to try to remember the first time it surfaced. Was it in middle school, high school, or college? Can you recall a specific incident? Was it your first algebra test? Was it your first math test after being out of school for a long time? Was it after you decided to get serious about college? Was it a college algebra course that was required for your major? Was it an instructor who told you that if you didn't pass the next math test you would fail the course? Was it when you needed to pass the next math test to pass the course so you could keep your financial aid? Was it your children asking you why you failed your last math test? If you cannot remember a specific incident when you had test anxiety, do you expect to have any major anxiety on the next test you take?

FIGURE 3 Causes of Test Anxiety

- Test anxiety can be a learned behavior resulting from the expectations of parents, teachers, or other significant people in the student's life.
- Test anxiety can be caused by the association between grades and a student's personal worth.
- Test anxiety develops from fear of alienating parents, family, or friends due to poor grades.
- Test anxiety can stem from a feeling of lack of control and an inability to change one's life situation.
- Test anxiety can be caused by a student's being embarrassed by the teacher or other students when trying to do math problems.
- Test anxiety can be caused by timed tests and the fear of not finishing the test, even if one can solve all the problems.
- Test anxiety can be caused by being put in math courses above one's level of competence.
- What is (are) the cause(s) of your *test* anxiety?

Students with math test anxiety also have had positive experiences with taking math tests. Now try to remember your first positive experience with taking a math test. Was it in middle school, high school, or college? Was it after studying many hours for the test? Now think back to your last positive experience with a math test. How did it make you feel?

Since we have already explored your experiences with taking math tests, let's look at some of the direct causes of your math test anxiety. If you do have test anxiety, what is the main cause? If you don't know, then review the seven causes of test anxiety in Figure 3. Does one of these reasons fit you? If you don't have test anxiety, what would be a reason that could cause test anxiety?

Like you did before, brainstorm answers to the questions listed in the previous three paragraphs. You will also use this brainstorming to help you write your autobiography.

REMEMBER
For some students, just writing about their previous math history helps them.

SECTION 3.3 REVIEW

1. List and describe five causes of test anxiety.

   First Cause: ______________________________

   ____

   ____

   Second Cause: ______________________________

   ____

   ____

   Third Cause: ______________________________

   ____

   ____

   Fourth Cause: ______________________________

   ____

   ____

   Fifth Cause: ______________________________

   ____

   ____

2. Put into your own words your cause(s) of test anxiety. If you don't have test anxiety, list what you think is the major cause of test anxiety.

   ____

   ____

3.4 The Different Types of Test Anxiety

The two basic types of test anxiety are emotional anxiety (educators call this *somatic* anxiety) and worry anxiety (educators call this *cognitive* anxiety). Students with high test anxiety have *both* emotional and worry anxiety.

Signs of emotional anxiety are upset stomach; nausea; sweaty palms; pain in the neck; stiff shoulders; high blood pressure; rapid, shallow breathing; rapid heartbeat; and general feelings of nervousness. As anxiety increases, these feelings intensify. Some students even run to the bathroom to throw up or have diarrhea.

Even though these *feelings* are caused by anxiety, the physical response is real. These feelings and physical inconveniences can affect your concentration and your testing speed, and they can cause you to "draw a blank."

Worry anxiety causes the student to think about failing the test. These negative thoughts can happen either before or during the test. This negative "self-talk" causes students to focus on their anxiety instead of recalling math concepts.

The effects of test anxiety range from a "mental block" on a test to avoiding homework. One of the most common side effects of test anxiety is getting the test and immediately forgetting information that you know. Some students describe this event as having a mental block, going blank, or finding that the test looks like a foreign language.

After 5 or 10 minutes into the test, some of these students can refocus on the test and start working the problems. They have, however, lost valuable time. For other students, anxiety persists throughout the test and they cannot recall the needed math information. It is only after they walk out the door that they can remember how to work the problems.

Sometimes math anxiety does not cause students to go blank but slows down their mental processing speed. This means it takes longer to recall formulas and concepts and to work problems. The result is frustration and loss of time, leading to *more* anxiety. Since, in most cases, math tests are *speed* tests (those in which you have a certain amount of time to complete the test), you may not have enough time to work all the problems or to check the answers if you have mentally slowed down. The result is a lower test score, because even though you know the material, you do not complete all of the questions before test time runs out.

Not using all of the time allotted for the test is another problem caused by test anxiety. Students know that they should use all the test time to check their answers. In fact, math is one of the few subjects in which you can check test problems to find out if your work is correct. However, most students do not use all of the test time, and this results in lower test scores. Why does this happen?

Students with high test anxiety do not want to stay in the classroom. This is especially true of students whose test anxiety increases as the test progresses. The test anxiety gets so bad that they would rather leave early and receive a lower grade than stay in that "burning house."

Students have another reason for leaving the test early: the fear of what the instructor and other students will think about them for being the last one to hand in the test. These students refuse to be the last ones to finish the test because they think that the instructor or other students will think they are "dumb." This is middle school thinking, but the feelings are still real—no matter what the age of the student. These students do not realize that some students who turn in their tests first fail, whereas many students who turn in their tests last make As and Bs.

Another effect of test anxiety relates to completing homework assignments. Students who have high test anxiety may have difficulty starting or completing their math homework. Doing homework reminds some students of their learning problems in math. More specifically, it reminds them of their previous math failures, which causes further anxiety. This anxiety can lead to total homework avoidance or "approach-avoidance" behavior.

Total homework avoidance is called *procrastination*. The very thought of doing their homework causes these students anxiety, which causes them to put off tackling their homework. This makes them feel better for a short amount of time—*until test day.*

EXAMPLE

Some students begin their homework and work some problems successfully. They then get stuck on a problem that causes them anxiety, so they take a break. During their break the anxiety disappears until they start doing their homework again. Continuing their homework causes more anxiety, which leads to another break. The breaks become more frequent. Finally, the student ends up taking one long break and not doing the homework. Quitting, to them, means *no more anxiety* until the next homework assignment.

The effects of math test anxiety can be different for each student. Students can have several of the mentioned characteristics that can interfere with math learning and test taking. However, there are certain myths about math that each student needs to know. Review Figure 4, The 12 Myths About Test Anxiety, to see which ones you believe. If you have test anxiety, which of the items listed is (are) true for you?

FIGURE 4 The 12 Myths About Test Anxiety

1. Students are born with test anxiety.
2. Test anxiety is a mental illness.
3. Test anxiety cannot be reduced.
4. Any level of test anxiety is bad.
5. All students who are not prepared have test anxiety.
6. Students with test anxiety cannot learn math.
7. Students who are well prepared will not have test anxiety.
8. Very intelligent students and students taking high-level courses, such as calculus, do not have test anxiety.
9. Attending class and doing homework should reduce all test anxiety.
10. Being told to relax during a test will make a student relaxed.
11. Doing nothing about test anxiety will make it go away.
12. Reducing test anxiety will guarantee better grades.

SECTION 3.4 REVIEW

1. List and describe the two basic types of test anxiety.

   First Type: ____________________

   Second Type: ____________________

2. List two reasons students leave the test room early instead of checking their answers.

   First Reason: ____________________

   Second Reason: ____________________

3. List 6 of the 12 myths about test anxiety that you most believed.

   First Myth: ____________________

   Second Myth: ____________________

   Third Myth: ____________________

   Fourth Myth: ____________________

   Fifth Myth: ____________________

   Sixth Myth: ____________________

3.5 How to Reduce Test Anxiety

To reduce math test anxiety, you need to understand both the relaxation response and how negative self-talk undermines your abilities.

Relaxation Techniques

The relaxation response is any technique or procedure that helps you to become relaxed. It will take the place of an anxiety response. Someone simply telling you to relax or even telling yourself to relax, however, does little to reduce your test anxiety. There are both short-term and long-term relaxation response techniques that help control emotional (somatic) math test anxiety. These techniques will also help reduce worry (cognitive) anxiety. Effective *short-term* techniques include the tensing and differential relaxation method, the palming method, and deep breathing.

Short-Term Relaxation Techniques

The Tensing and Differential Relaxation Method The tensing and differential relaxation method helps you relax by tensing and relaxing your muscles all at once. Follow these procedures while you are sitting at your desk before taking a test:

1. Put your feet flat on the floor.
2. With your hands, grab underneath the chair.
3. Push down with your feet and pull up on your chair at the same time for about 5 seconds.
4. Relax for 5 to 10 seconds.
5. Repeat the procedure two or three times.
6. Relax all your muscles except the ones that are actually used to take the test.

The Palming Method The palming method is a visualization procedure used to reduce test anxiety. While you are at your desk before or during a test, follow these procedures:

1. Close and cover your eyes using the center of the palms of your hands.
2. Prevent your hands from touching your eyes by resting the lower parts of your palms on your cheekbones and placing your fingers on your forehead. Your eyeballs must not be touched, rubbed, or handled in any way.
3. Think of some real or imaginary relaxing scene. Mentally visualize this scene. Picture the scene as if you were actually there, looking through your own eyes.
4. Visualize this relaxing scene for 1 to 2 minutes.
5. Open your eyes and wait about 2 minutes and visualize the scene again. This time also imagine any sounds or smells that can enhance your scene. For example, if you are imagining a beach scene, imagine the sound of the waves on the beach and the smell of the salty air. This technique can also be completed without having your hands over your eyes by just closing your eyes.

Practice visualizing this scene for several days before taking a test and the effectiveness of this relaxation procedure will improve.

Deep Breathing Deep breathing is another short-term relaxation technique that can help you relax. Proper breathing is a way to reduce stress and decrease test anxiety. When you are breathing properly, enough oxygen gets into your bloodstream to nourish your body and mind. A lack of oxygen in your blood contributes to an anxiety state that makes it more difficult to react to stress. Proper deep breathing can help you control your test anxiety.

Deep breathing can replace the rapid, shallow breathing that sometimes accompanies test anxiety, or it can prevent test anxiety. Here are the steps in the deep-breathing technique:

1. Sit straight up in your chair in a good posture position.
2. Slowly inhale through your nose.
3. As you inhale, first fill the lower section of your lungs and work your way up to the upper part of your lungs.
4. Hold your breath for a few seconds.

5. Exhale slowly through your mouth.
6. Wait a few seconds and repeat the cycle.
7. Keep doing this exercise for 4 or 5 minutes. This should involve going through about 10 breathing cycles. Remember to take two normal breaths between cycles. If you start to feel lightheaded during this exercise, stop for 30 to 45 seconds and then start again.
8. Throughout the entire exercise, make sure you keep breathing smoothly and in a regular rhythm without gulping air or suddenly exhaling.
9. As an extra way to improve your relaxation, say "relax" or "be calm" to yourself as you exhale. This can start a conditioned response that can trigger relaxation when you repeat the words during anxious situations. As you keep practicing, this conditioned response will strengthen. Practice is the key to success.

You need to practice this breathing exercise for several weeks before using the technique during tests. If you don't practice this technique, it will not work. After taking your first test, keep doing the exercise several times a week to strengthen the relaxation response.

The CD *How to Reduce Test Anxiety* further explains test anxiety and discusses these and other short-term relaxation response techniques. Short-term relaxation techniques can be learned quickly but are not as successful as the long-term relaxation technique. Short-term techniques are intended to be used while learning the long-term technique.

Long-Term Relaxation Technique

The cue-controlled relaxation response technique is the best long-term relaxation technique. It is presented on the CD *How To Reduce Test Anxiety*. Cue-controlled relaxation means you can induce your own relaxation based on repeating certain cue words to yourself. In essence, you are taught to relax and then silently repeat cue words, such as, "I am relaxed." After enough practice, you can relax during math tests. The cue-controlled relaxation technique has worked with thousands of students. For a better understanding of test anxiety and how to reduce it, listen to *How to Reduce Test Anxiety*.

Negative Self-Talk

According to cognitive psychologists, self-talk is what we say to ourselves as a response to an event or situation. This self-talk determines our feelings about that event or situation. Sometimes we say it so quickly and automatically that we don't even hear ourselves. We then think it is the situation that causes the feeling, but in reality it is our interactions or thoughts about the experience that are controlling our emotions. This sequence of events can be represented by the following time line:

External Events
(Math test)

⇩

Interpretation of Events and Self-Talk
(How you feel about the test
and what you are telling yourself)

⇩

Feelings and Emotions
(Happy, glad, angry, mad, upset)

In most cases this means that you are responsible for how and what you feel. You can engage in positive self-talk or negative self-talk about taking a math test. Yes, some students practice positive self-talk about math tests and see each test as a challenge and something to accomplish that makes them feel good, while others see it as an upsetting event that leads to anger and anxiety. In other words, you are what you tell yourself.

Negative self-talk is a form of worry (cognitive) anxiety. This type of worrying can interfere with your test preparation and can keep you from concentrating on the test. Worrying can motivate you to study, but too much worrying may prevent you from studying at all.

Negative self-talk is defined as the negative statements you tell yourself before and during tests. Negative self-talk causes students to lose confidence and to give up on tests. Furthermore, it can give you an inappropriate excuse for failing math and cause you to give up on learning math.

Examples of Negative Self-Talk

"It doesn't matter how hard I work. I am going to fail the class no matter what."

"I have already failed this class before. There is no way I can pass it this time, either."

"I do not get these problems. I am going to fail this test and flunk the course."

"I knew how to do these problems not 15 minutes ago, and now . . . nothing! What is up with me, I always do this when I take math tests."

"If I can't even pass this easy test, how can I expect to pass the math classes I need to graduate?"

Students who have test anxiety are prone to negative self-talk. Test anxiety can be generated or heightened by repeatedly making statements to yourself that usually begin with What if? For example, What if I fail the test? or What if I fail this class again? These what-if statements generate more anxiety, which can cause students to feel sick. These statements tell them to be anxious. Some other aspects of self-talk are as follows:

- Self-talk can be in telegraphic form with short words or images.
- Self-talk can be illogical, but at the time, the person believes it.
- Negative self-talk can lead to avoidance such as not taking a test or skipping classes.
- Negative self-talk can cause more anxiety.
- Negative self-talk can lead to depression and a feeling of helplessness.
- Negative self-talk is a bad habit that can be changed.

There are different types of negative self-talk. If you have negative self-talk, then review the following types and see which one fits you best. You may use a combination of them.

1. *The Critic* is the person inside us who is always trying to put us down. It constantly judges our behaviors and finds fault even if there is none. It jumps on any mistake and exaggerates it to cause more anxiety. The Critic puts us down for mistakes on tests and blames us for not controlling our anxiety. The Critic reminds us of previous comments from real people who have criticized us. It compares us with other students who are doing better in the class. It loves to say, "That was a stupid mistake!" or "You are a total disappointment. You can't pass this math class like everyone else can!" The Critic's goal is to promote low self-esteem.
2. *The Worrier* is the person inside us who looks at the worst-case scenario. It wants to scare us with ideas of disaster and complete failure. When it sees the first sign of anxiety, it "blows it out of proportion" to the extent that we will not remember anything and totally fail the test. The Worrier creates more anxiety than normal. The Worrier anticipates the worst, underestimates our ability, and sees us not only failing the test but also "failing life." The Worrier loves to ask What if? For example, What if I fail the math test and don't graduate? or What if I can't control my anxiety and throw up in math class? The goal of the Worrier is to cause more anxiety so we will quit.
3. *The Victim* is the person inside us who wants us to feel helpless or hopeless. It wants us to believe that no matter what we do, we will not be successful in math. The Victim does not blame other events (poor schooling) or people (bad teachers) for our math failures. It blames us. It dooms us and puts us into a learned helpless mode, meaning that if we try to learn math, we will fail, or if we don't try to learn math, we will fail. So why try? The Victim likes to say, "I can't learn math." The goal of the Victim is to cause depression and make us stop trying.
4. *The Perfectionist* is similar to the Critic but is the opposite of the Victim. It wants us to do the best we can and will guide us into doing better. It tells us that we are not studying enough for math tests and that a B is not good enough and that we must make an A. In fact, sometimes an A is not good enough unless it is a score of 100%. So, if we make a B on a test, the Perfectionist says, "An A or a B is just like making an F." The Perfectionist is the hard-driving part of us that wants the best but cannot stand mistakes or poor grades. It can drive us to mental and physical exhaustion to make that perfect grade. It is not interested in

self-worth, just perfect grades. Students driven by the Perfectionist often drop a math course because they only have a B average. The Perfectionist loves to repeat, "I should have . . ." or "I must. . . ." The goal of the Perfectionist is to cause chronic anxiety that leads to burnout.

Review these types of personalities to see which one may fit you best. We may have a little of each one in us, but what is the dominant one for you in math? Now we can look at how to stop these negative thoughts.

Managing Negative Self-Talk

Students need to change their negative self-talk to positive self-talk without making unrealistic statements.

Positive self-statements can improve your studying and test preparation. During tests, positive self-talk can build confidence and decrease your test anxiety. These positive statements (see examples) can help reduce your test anxiety and improve your grades. Some more examples of positive self-statements are on the CD *How to Reduce Test Anxiety*. Before each test, make up some positive statements to tell yourself.

There are several ways to counter and control negative self-talk. Students can replace negative self-talk with positive statements and questions that make them think in a realistic way about the situation. Another way is to develop thought-stopping techniques to reduce or eliminate the negative thoughts. Try each way or a combination to see what works best for you.

Countering self-talk involves writing down and practicing positive statements that can replace negative statements. Students can develop their own positive statements. Some rules for writing positive statements are

1. Use the first-person present tense. For example, "I can control my anxiety and relax myself."
2. Don't use negatives in the statement. For example, don't say, "I am not going to have anxiety." Instead, say, "I will be calm during the test."
3. Have a positive and realistic belief in the statement. For example, say, "I am going to pass this test," instead of saying, "I am going to make an A on this test" when you have not made an A on any of the tests.

The statements used to counter negative thoughts can be based on the type of negative self-talk we tend to engage in. The Critic, who puts us down by saying, "Your test anxiety is going to cause you to fail the test," can be countered with, "I have test anxiety, but I am learning to control it." The Worrier, who asks, "What if I fail the test?" can be countered with "So what? I will do better on the next test." The Victim, who thinks things are hopeless and says, "I will never be able to pass math," can be countered with, "I have studied differently for this test, and I can pass the math course." The Perfectionist, who says, "I need to make an A on the test or I will drop out of school," can be countered with, " I don't need to make an A to please anyone. All I need is to pass the math course to get the career I want." These are some of the examples of positive statements that can control anxiety.

Examples of Positive Self-Talk

"I might have failed the course last semester, but things are going to be different this time now that I have learned how to study math."

"Now that I know how to fight off test anxiety, there is no way I will panic during an exam like I did before."

"Just because I haven't had perfect results in the past doesn't mean that I am bad at math. This time, I know how to study math, so this course should not be a problem."

"If I put in the work, then I will pass with flying colors."

"I have done everything I needed to do to prepare for this test. If I can keep my composure during the test, and not get frustrated if I don't understand one or two problems, I should do all right."

"I will not let this one problem bother me. I have answered all the other problems correctly, so I cannot let this one make me miss the rest of the questions on the test."

"I will not hurry in order to be the first person to finish this test. I can go ahead and take my time to make sure I don't make any careless mistakes."

Thought-Stopping Technique

Some students have difficulty stopping their negative self-talk. These students cannot just tell themselves to eliminate those thoughts. These students need to use a thought-stopping technique to overcome their worry and become relaxed.

Thought stopping involves focusing on the unwanted thoughts and, after a few seconds, suddenly stopping those thoughts by emptying your mind. Using the command Stop! or a loud noise like clapping your hands can effectively interrupt the negative self-talk. In a homework situation, you may be able to use a loud noise to stop your thoughts, but don't use it during a test.

To stop your thoughts in the classroom or during a test, silently shout to yourself, "Stop!" or "Stop thinking about that!" After your silent shout, either relax yourself or repeat one of your positive self-talk statements. You may have to shout to yourself several times during a test or while doing homework to control negative self-talk. After every shout, use a different relaxation technique/scene or positive self-talk statement.

Thought stopping works because it interrupts the worry response before it can cause high negative emotions. During that interruption, you can replace the negative self-talk with positive statements or relaxation. However, students with high worry anxiety should practice this technique for three days to one week before taking a test. Contact your counselor if you have additional questions about the thought-stopping technique.

Writing Your Math Autobiography

An autobiography relates how you remember and feel about your past experiences. In addition, many autobiographies explore how these past feelings and experiences shape current life. While some people write autobiographies so that others can learn about their lives, many people write private autobiographies in order to understand what is going on in their lives.

Your final product will be a typed (or handwritten if your instructor approves) paper that summarizes your experiences learning math, how you feel about these experiences, and how they shape your current perspective on learning math.

SECTION 3.5 REVIEW

1. Describe your best short-term relaxation technique.

____

2. You can use the palming method by closing your eyes and visualizing a scene without putting your hands to your face. Describe a very relaxing scene that you could visualize. Make sure to include some sounds and visual images in your scene.

____

____

____

____

3. Practice your relaxation scene for 3 to 5 minutes for the next 5 days. List the times and dates you practiced your scene.

Date: ____________________ Time: ____________________

Date: ____________________ Time: ____________________

Date: ____________________ Time: ____________________

Date: ______________________ Time: ______________________

Date: ______________________ Time: ______________________

4. How does negative self-talk cause you to have text anxiety?

____

____

____

____

5. Make up three positive self-talk statements that are not listed in this text.

   Statement 1: ______________________

   Statement 2: ______________________

   Statement 3: ______________________

6. Describe the thought-stopping technique in your own words.

____

7. What word or words will you use as your silent shout?

____

8. What will you do after your silent shout?

____

9. How does the thought-stopping technique work?

____

Name: ______________________ Date: ____________

CHAPTER 3 REVIEW

1. Reducing math test anxiety ____________ ____________ guarantee good math grades. (Answer on page 38)
2. Test anxiety is a ____________ response; a person is not ____________ with it. (Answers on page 37)
3. Math test anxiety involves ____________, ____________, and ____________ of math tests.
4. One cause of test anxiety is that a student goes to the board to work a problem but is called ____________ when he or she cannot work the problem.
5. The two basic types of test anxiety are ____________ and ____________.
6. The effects of test anxiety range from a "____________ ____________" on a test to ____________ homework.
7. The ____________ and ____________ ____________ method helps you relax by tensing and relaxing your muscles all at once.
8. The palming method is a ____________ procedure used to reduce text anxiety.
9. Negative self-talk is a form of ____________ anxiety.
10. Thought stopping works because it interrupts the ____________ response before it can cause ____________ emotions.

What is the most important information you learned from this chapter?

____

____

____

How can you immediately use it?

____

____

____

DAN'S TAKE REVIEW

1. What was Dan's mistake in preparing for his math test?

____

____

____

Name: ______________________ Date: ____________

2. What were the potential consequences of his decision-making process?

____

____

____

3. How could he have avoided the issue, and how can you avoid this issue in the future?

____

____

____

LECTURE NOTES

How to Improve Your Listening and Note-Taking Skills

CHAPTER 4

Listening and note-taking skills in a math class are very important, since most students either do not read the math text or have difficulty understanding it. In most of your other classes, if you do not understand the lecture, you can read the text and get almost all the information. In math class, however, the instructor can usually explain the textbook better than the students can read and understand it.

Students who do not have good listening skills or note-taking skills will be at a disadvantage in learning math. Most math understanding takes place in the classroom. Students must learn how to take advantage of learning math in the classroom by becoming effective listeners, calculator users, and note takers.

CHAPTER GOALS

What You Will Learn:

4.1 How to Become an Effective Listener

4.2 How to Become a Good Note Taker

4.3 The Seven Steps to Math Note Taking

4.4 Classroom Participation: Becoming an Active Questioner

4.5 How to Rework Your Notes

DAN'S TAKE

My university became an "e-campus" my freshman year. Gone were the days of pencil and paper. Now they gave us laptops. The tapping of computer keys became a common soundtrack for class, as students tried frantically to type the last words of a PowerPoint slide into their notes before the professor hit the "next slide" button—or perhaps more accurately, as they instant messaged their friends across campus.

There is no question that students today have more distractions in the classroom than ever before. More campuses are using computers in the classroom, making paying attention nearly impossible. How can professors compete in a world with instant access to celebrity gossip rags and up-to-the-second BCS standings?

Yet math teachers can rejoice. It is very difficult to write math problems using a computer. Even when using online math courses, all the work has to be done on paper. This means there is no time for distraction.

On my particular campus, this made for a rather bizarre phenomenon. To many us, our computers were an extension of our bodies. They might well have been attached to our fingers. So math classes at my school presented an odd and old-school set of problems. The lack of distraction actually became a distraction itself. If one of your hands is missing, well, that's the kind of thing a person takes notice of.

So to fill my diversionary void, I fell into a pattern of behavior that can only be described as idiotic. I cannot tell you how many times I would find myself doing something utterly pointless instead of taking notes. Things like flipping a pencil or staring at the cute sophomore sitting on the other side of the room.

Here is the problem with this scenario. Math is entirely based upon sequential learning. Two flips of a pencil and one love-stricken gaze might be enough to derail an entire lecture's worth of understanding. In my case, by the time I would refocus I'd be too far behind. This would require me to borrow someone else's notebook or figure out the correct processes through the homework itself.

The key to staying focused in class is to enter the room with a plan. The note-taking methods described in this chapter are invaluable. They really help you keep an organized notebook. It is important to be able to reference your notes easily during homework and remember which information is needed to answer each type of problem.

4.1 How to Become an Effective Listener

Becoming an effective listener is the foundation for good note taking. You can become an effective listener using a set of skills that you can learn and practice. To become an effective listener, you must prepare yourself both physically and mentally.

Sitting in the Golden Triangle

The physical preparation for becoming an effective listener involves *where you sit* in the classroom. Sit in the best area to obtain high grades, that is, in "the golden triangle of success." The golden triangle of success begins with seats in the front row facing the instructor's desk (see Figure 5).

Students seated in this area (especially in the front row) directly face the teacher and are most likely to pay attention to the lecture. This is a great seating location for visual learners. There is also less tendency for them to be distracted by activities outside the classroom or by students making noise within the classroom.

The middle seat in the back row is another point in the golden triangle for students to sit, especially those who are auditory (hearing) learners. You can hear the instructor better because the instructor's voice is projected to that point. This means that there is less chance of misunderstanding the instructor, and you can hear well enough to ask appropriate questions.

By sitting in the golden triangle of success, you can force yourself to pay more attention during class and be less distracted by other students. This is very important for math students, because math instructors usually go over a point once and continue on to the next point. If you miss that point in the lesson, then you could be lost for the remainder of the class.

FIGURE 5 The Golden Triangle of Success

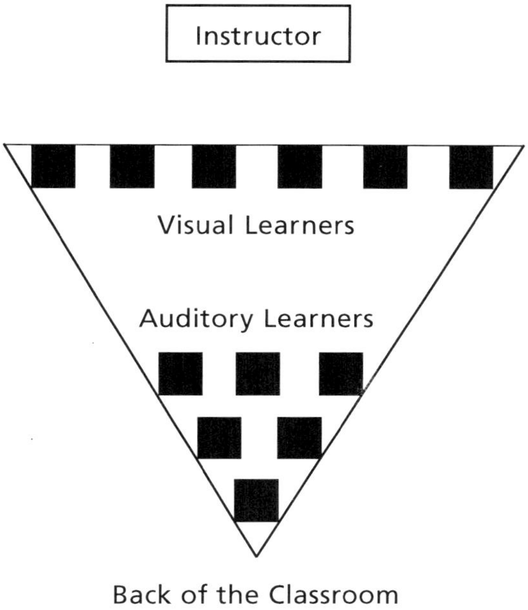

This may seem obvious, but I know too many students who do the following if they have early morning classes. They roll out of bed, put on the closest clothes to them, brush their hair, maybe their teeth, and race to class. Don't do this. You won't wake up until the end of your class! Train yourself to be somewhere on campus to review your notes at least 15 minutes before class starts and follow some of the other strategies suggested in the next section. So, another physical way to be ready is to be awake and nourished!

Warming Up for Math Class

The mental preparation for note taking involves "warming up" before class begins and becoming an active listener. Just as an athlete must warm up before a game begins, you must warm up before taking notes. Warm up by

- reviewing the previous day's notes
- reviewing the reading material
- reviewing the homework
- preparing questions
- working one or two unassigned homework problems

This mental warm-up before the lecture allows you to refresh your memory and prepare pertinent questions, making it easier to learn the new lecture material.

How to Become an Active Listener

Becoming an active listener is the second part of the mental preparation for note taking.

Do not anticipate what the instructor is going to say or immediately judge the instructor's information before the point is made. This will distract you from learning the information.

EXAMPLE

Watch the speaker, listen for main ideas, and nod your head or say to yourself, "I understand" when agreeing with the instructor.

Expend energy looking for interesting topics in the lecture. When the instructor discusses information that you need to know, immediately repeat it to yourself to begin the learning process.

You can practice this exercise by viewing math DVD's and repeating important information. This is an especially good learning technique for auditory learners.

Ask appropriate questions when you have them. Sometimes if you are so confused you do not know what question to ask, still let the instructor know you are confused. Also, mark your notes where you got lost so that if the instructor doesn't have time to answer your question, you can visit during office hours to get an answer.

REMEMBER

Class time is an intense study period that should not be wasted.

Listening and Learning

Some students think that listening to the instructor and taking notes are a waste of valuable time. Students too often sit in class and use only a fraction of their learning ability. Class time should be considered a valuable study period where you can listen, take notes, and learn at the same time. One way to do this is by memorizing important facts when the instructor is talking about material you already know. Another technique is to repeat to yourself the important concepts right after the instructor discusses them in class. Using class time to learn math is an efficient learning system.

SECTION 4.1 REVIEW

1. Explain how sitting in the golden triangle of success makes you a better listener.

____

____

____

2. List five ways you can warm up before math class begins.

   First Way: _______________

   Second Way: _______________

   Third Way: _______________

   Fourth Way: _______________

   Fifth Way: _______________

3. Select your best warm-up process and try it before your next math class. What was your warm-up process, and how did it help you?

____

____

____

____

4. How can you listen and learn at the same time?

____

____

____

____

4.2 How to Become a Good Note Taker

Becoming a good note taker requires two basic strategies. One strategy is to be specific in detail. In other words, *copy* the problems down, step by step. The second strategy is to *understand* the general principles, general concepts, and general ideas.

Copying from the Board

While taking math notes, you need to copy each and every step of each problem even though you may already know every step. While in the classroom, you might understand each step, but a week later you might not remember how to do the problem unless you have written down all the steps. In addition, as you write down each step, you are memorizing it. Make sure to copy every step of each problem written on the board.

There will be times when you will get lost while listening to a lecture. Nevertheless, you should keep taking notes, even though you do not understand the problem. This will provide you with a reference point for further study. Put a question mark by each step that you do not understand; then, after class, review the steps you did not understand with the instructor, your tutor, or another student.

If you seem to think that many other students are lost, too, ask questions or let the instructor know.

Taking Notes

The goal of note taking is to take the least amount of notes and get the greatest amount of information on your paper. This could be the opposite of what most instructors have told you. Some instructors tell you to take down everything. This is not necessarily a good note-taking system, since it is very difficult to take precise, specific notes while trying to understand the instructor.

What you need to develop is a note-taking system in which you write the least amount possible and get the most information down while still understanding what the instructor is saying.

Developing an Abbreviation List

To reduce the amount of written notes, an abbreviation system is needed. An abbreviation system is your own system of reducing long words to shorter versions that you still can understand. By writing less, you can listen more and gain a better understanding of the material.

Figure 6 provides a list of abbreviations. Add your own abbreviations to this list. By using abbreviations as much as possible, you can obtain the same meaning from your notes and have more time to listen to the instructor.

EXAMPLE

When the instructor starts explaining the commutative property, you need to write "commutative property" out the first time. After that, use "com." You should develop abbreviations for all the most commonly used words in math.

FIGURE 6 Abbreviations

e.g.	(for example)
cf	(compare, remember in context)
N.B.	(note well, this is important)
\	(therefore)
∴	(because)
É	(implies, it follows from this)
>	(greater than)
<	(less than)
=	(equals, is the same)
≠	(does not equal, is not the same)
( )	(parentheses in the margin, around a sentence or group of sentences, indicates an important idea)
?	(used to indicate that you do not understand the material)
O	(a circle around a word may indicate that you are not familiar with it; look it up)
TQ	(test questions)
1, 2, 3	(used to indicate a series of facts)
D	(shows disagreement with a statement or passage)
REF	(reference)
et al.	(and others)
bk	(book)
p	(page)
etc.	(and so forth)
V	(see)
VS	(see above)
SC	(namely)
SQ	(the following)
com.	(Commutative)
dis.	(Distributive)
APA	(associative property of addition)
AI	(additive inverse)
IPM	(identity property of multiplication)

When to Take Notes

To become a better note taker, you must know when to take notes and when not to take notes. The instructor will give cues that indicate which material is important. These cues include

- presenting facts or ideas
- writing on the board
- summarizing
- pausing
- repeating statements
- enumerating, such as "1, 2, 3" or "A, B, C"
- working several examples of the same type of problem on the chalkboard
- saying This is a tricky problem
- saying This is the most difficult step in the problem
- indicating that certain types of problems will be on the test, such as coin or age problems
- explaining boldface words.

You must learn the cues your instructor gives to indicate important material. If you are in doubt about the importance of the class material, do not hesitate to ask the instructor about its importance.

While taking notes, you may become confused about math material. At that point, take as many notes as possible, and do not give up on note taking. One thing you shouldn't do—bother the student next to you for an explanation. This can be annoying, and both of you will miss important information while you are talking.

As you take notes on confusing problem steps, leave extra space; then go back and fill in information that clarifies your misunderstanding of the steps in question. Ask your tutor or instructor for help with the uncompleted problem steps, and write down the reasons for each step in the space provided.

Another procedure for saving time while taking notes is to stop writing complete sentences. Write your main thoughts in phrases. Phrases are easier to jot down and easier to memorize.

SECTION 4.2 REVIEW

1. Note taking requires two basic strategies: ______________________________ and ______________________________.

2. List five abbreviations (and their meanings) that you use in your math class (one abbreviation and meaning per line).

____

____

____

____

____

3. List five cues that your instructor gives that indicate which material is important (one cue per line).

____

____

____

____

____

4.3 The Seven Steps in Math Note Taking

No matter what format you use to take notes, a good set of notes contains

- examples of problems
- explanations of how to complete the problems
- appropriate math vocabulary and rules that are used for each problem

In addition, it is wise to make your notes as organized as possible. It may get on your nerves for a while, but when you see that organized notes save time when you are studying, you will be more willing to take the time to organize them.

Many successful math students, whether in developmental math courses or calculus, use a form of the three-column note-taking system. It organizes the information, making it easy to review. While creating the three-column system, students are also learning the math, so it is not busywork. It takes a few weeks to feel comfortable with this system, but it is worth sticking with it. Many students have adopted this system and improved the efficiency of their study time.

The seven steps lead you through several important stages in the learning process. Steps 1 to 3 help you *record the information* accurately, completely, and in an organized fashion. Steps 4 to 7 help you *rehearse and understand the information*. These steps also help you learn to *recall the information*, thus helping you prepare for tests.

To prepare for taking notes this way, get a good chunk of paper ready. Here are some choices:

- Use regular paper, portrait or landscape style, and draw two lines to create three columns.
- See Figure 7, for a completed example.
- Some students purchase graph paper because it helps them keep numbers lined up correctly.
- Other students design the page on a computer and print out copies.
- Choose two columns next to each other to be the "Examples" and the "Explanations" columns.

FIGURE 7 Three-Column Note Taking System

Key Words / Rules	Example	Explanations
Solve a linear equation	$4(x + 4) + 3(x - 4) = 2(x - 1)$	Have to get x on one side of the = and numbers on the other side of the =.
Distributive Property	$4x + 16 + 3x - 12 = 2x - 2$	Multiply numbers to the left of the ( ) by each variable and number in the ( ).
Commutative Property	$4x + 3x + 16 - 12 = 2x - 2$	Regroup numbers and variables.
Combine like terms	$7x + 4 = 2x - 2$	
Additive Inverse Property	$7x - 2x + 4 = 2x - 2x - 2$ $5x + 4 = -2$	Subtract 2x from both sides in order to get variables all on the right side of the =.
Additive Inverse Property	$5x + 4 - 4 = -2 - 4$ $5x = -6$	Subtract 4 from both sides to get numbers all on the right side of the =.
Multiplicative Inverse Property	$\frac{5x}{5} = \frac{-6}{5}$	Divide both sides by 5 to get x by itself on the left side of the =.
Simplify	$x = -1\frac{1}{5}$	Write as a whole number and fraction.
	Insert new problem.	

- Label them as such. These are the two columns you will use all the time during class.
- Label the other column "Key Words/Rules." You might fill this column in during class but will probably complete it outside of class.

If you do this ahead of time, you do not have to draw lines frantically while the instructor is talking. If you forget or run out of paper, it might be quicker to fold the paper into three columns.

Draw a line dividing the end of one problem from the beginning of the next. Or, do one problem per page to leave plenty of room for notes. Graph paper can help you keep numbers and words lined up. Some students like to turn the note books or regular paper sideways and then organize their notes in three columns.

During class you will use two columns most of the time—the "Examples" and the "Explanations" columns. Sometimes, when instructors really want their students to use this method, they will organize their board work to reflect what the notes should look like. Most of the time, however, this does not happen.

The "Examples" column is where you will place the examples the instructor uses during class.

The "Explanations" column is where you will put the instructor's verbal explanations of the examples. If the instructor tends to explain by just doing the problem without talking about it, ask questions that will force the instructor to explain the problem verbally.

When the instructor simply explains a vocabulary word or rule, write the vocabulary word or a phrase naming the rule in the "Key Words/Rules" column. Write the explanation in the "Explanations" column. If the instructor provides an example, obviously it goes in the "Examples" column. If no example is provided, ask for one. Then if your wish is not granted, look one up in the textbook after class.

Here are the seven steps in math note taking. Remember that these are the basic steps. You can adapt them to work for you.

Step 1 *Record the instructor's example problems in the "Examples" column.*

Step 2 *Record the reasons for each step in the "Explanations" column.* After completing a problem, draw a horizontal line, separating it from the next problem.

If you get lost during the lecture, place a question mark in this column and move on.

Step 3 *If possible, record the key words and rules in the first column.* Here is an opportunity for you to ask what rules are being used if the instructor is assuming that the class knows them and you do not. Otherwise, you will need to refer to the textbook, work with a tutor, or use online resources after class to complete this column.

It is important to understand the vocabulary words, rules, and concepts when learning how to complete a math problem. You will be able to figure out other problems that may be slightly different.

Step 4 *In order to keep everything in your head, use the notes to review.* Cover up the "Examples" and "Explanations" columns and quiz yourself on the vocabulary and rules. Or, fold back the "Key Words/Rules" column, look at the examples and explanations, and choose the correct key words and rules used to complete the examples.

Step 5 *Place a check mark by the key words and concepts that you did not know.*

Step 6 *Review the information that you checked until you understand and remember it.*

Step 7 *Develop a glossary of the math terms and rules that are difficult for you to remember.*

For a while you might spend more time organizing this system outside of class. Some students still have to depend on their current process, concentrating on getting the information down. Then they transfer the information into the three-column format. However, it is best to learn how to do most of the process during math class.

Remember the learning modalities and learning styles? This process can use all of them!

- *Kinesthetic:* Writing out the notes
- *Auditory:* Saying information out loud
- *Visual:* Looking at the notes and thinking; color-code each column with highlighters, making sure you do not smear the pencil writing

- *Group:* Working on the notes with one or two other people to make sure they are complete and accurate
- *Individual:* Reading the notes to yourself while waiting to meet someone

Current research is discovering that using language (vocabulary, discussion) to learn math helps students understand the concepts more deeply. Then completing many problems helps them master the skills and increase the speed that is important in preparing for tests. These notes will help you understand the math so that it will stick with you. You will remember more information. Homework will help you master the skills and increase the speed at which you can complete problems.

A Math Glossary

A math glossary is created to define math vocabulary in your own words. Since math is considered a foreign language, understanding the math vocabulary becomes the key to comprehending math. Creating a glossary for each chapter of your textbook will help you understand math.

Your glossary should include all words printed in boldface in the text and any words you do not understand form the lecture. If you cannot explain the math vocabulary in your own words, ask your instructor or tutor for help. You may want to use the last pages in your notebook to develop a math glossary for each chapter in your textbook. Some students use note cards. Review your math glossary before each test.

SECTION 4.3 REVIEW

1. List and describe the seven steps in math note taking.

   First Step: ____________________

   Second Step: ____________________

   Third Step: ____________________

   Fourth Step: ____________________

   Fifth Step: ____________________

   Sixth Step: ____________________

   Seventh Step: ____________________

2. Make several copies of the modified three-column note-taking system, and take notes using this system in your next math class. Or, develop the modified three-column note-taking system in your own notebook.

3. List two benefits of creating a math glossary.

   Benefit 1: ____________________

   Benefit 2: ____________________

4. From your current math chapter, list and define five words that you can put in your math glossary.

   Word 1: ____________________

   Word 2: ____________________

   Word 3: ____________________

Word 4: ______________________________

Word 5: ______________________________

4.4 Classroom Participation: Becoming an Active Questioner

Not that anyone should have to tell you this, but college isn't free. In fact, it is quite expensive. While you probably received a free education in high school, you are paying top dollar for your higher education, so you might as well take advantage of the tools at your disposal. If you have a question for your professor, you should never be afraid to ask it.

To begin with, there is no such thing as a stupid question. If you need something cleared up, even a basic idea, it is much better to have it explained by an expert than to have to figure it out on your own while doing your homework. Also, if you are confused about something, there is almost always another student in the class that will be stumped by the same information.

It is also true that professors generally prefer active questioners as these students help them stay on focus and to double-check that their message is being properly understood. It slows the professors down and gives them the peace of mind that they are doing a good job. Questions help them to further develop subtle points about complex processes, and are therefore an absolute must for a proper and successful lecture.

If you are afraid to ask questions on the spot, you should try reading parts of the text before class and walking in with a few questions written down in your notebook. Doing this should endear you to your professor, as he or she will see that you have an active interest in the subject matter and are therefore taking the course seriously.

In the case that you are absolutely too scared to ask a question in class, you can always email your professors—but this should be the absolute last case scenario, and should never become an ongoing habit.

We've touched on this a little bit in other areas of the text, but the importance of fully understanding the processes behind math problems cannot be understated. If you are confused about something, don't be afraid to ask the questions required to get you on the right track. Put that money of yours to work!

SECTION 4.4 REVIEW

1. Is there such thing as a stupid question? Why or why not?

____

____

____

2. Can asking questions actually help your professor teach? How so?

____

____

____

3. Name one of the methods mentioned that can help you walk in to class and ask questions with confidence.

____

____

____

4.5 How to Rework Your Notes

The note-taking system does not stop when you leave the classroom. As soon as possible after class, rework your notes. You can rework the notes between classes or as soon as you get home. By reworking your notes as soon as possible, you can decrease the amount of information you forget. This is an excellent procedure for transferring math information from short-term memory to long-term memory.

REMEMBER
Most forgetting occurs right after the material is learned. You need to rework your notes as soon as possible. Waiting means that you probably will not understand what was written.

Here are six important steps in reworking your notes:

Step 1 *Rewrite the material you cannot read or will not be able to understand a few weeks later.* If you do not rework your notes, you will be frustrated when studying for a test if you come across notes you cannot read or completing your homework. Another benefit of rewriting the notes is that you immediately learn the new material. Waiting means it will take more time to learn the material.

Step 2 *Fill in the gaps.* Most of the time, when you are listening to a lecture, you cannot write down everything. It is almost impossible to write down everything, even if you know shorthand. Locate the portions of your notes that are incomplete. Fill in the concepts that were left out. In the future, skip two or three lines in your notebook page for anticipated lecture gaps.

Step 3 *Add additional key words and rules in the "Key Words/Rules" column.* These key words and rules are the ones that were not recorded during the lecture.

EXAMPLE
You did not know that you should add the *opposite* of 18 to solve a particular problem, and you incorrectly added 18. Put additional important key words and ideas (such as "opposite" and "negative of") in the notes; these are the words that will improve your understanding of math.

Step 4 *Make a problem log of those problems that the teacher worked in class.* The problem log is a separate section of your notebook that contains a listing of the problems (without explanations—just the problems) that your teacher worked in class. If your teacher chose those problems to work in class, you can bet that they are considered important. The problems in this log can be used as a practice test for the next exam. Your regular class notes will contain not only the solutions but also all the steps involved in arriving at those solutions and can be used as a reference when you take your practice test.

Write the problem on the left half of the page. Then save some room on the right side for writing out what the test instructions might be for the problem. Sometimes we forget what the test instructions are, and we miss points on the test by following the directions incorrectly. Skip several lines to complete the problem when you use the problem log as a practice test.

Step 5 *Make a calculator handbook and put in it your keystroke sequences.* The calculator handbook can be a spiral-bound set of note cards or a separate section of your notebook that holds only calculator-related information. Your handbook should also include an explanation of when that particular set of keystrokes is to be used.

Step 6 *Reflection and synthesis.* Once you have finished going over your notes, review the major points in your mind. Combine your new notes with your previous knowledge to gain a better understanding of what you have learned today.

If you are a group learner and even if you are not, it helps to review and reflect with other students.

Use a Tape Recorder

If you have a math class during which you cannot get all the information down while listening to the lecture, ask your instructor about using a tape recorder. To ensure success, the tape recorder must have a tape counter and must be voice activated.

The tape counter displays a number indicating the amount of tape to which you have listened. When you find you are in an area of confusing information, write the beginning and ending tape counter numbers in the left margin of your notes. When reviewing your notes, the tape count number will be a reference point for obtaining information to work the problem. You can also reduce the time it takes to listen to the tape by using the pause button to stop the recording of unnecessary material.

Ask Questions

To obtain the most from a lecture, you must ask questions in class. By asking questions, you improve your understanding of the material and decrease your homework time. By not asking questions, you create for yourself unnecessary confusion during the remainder of the class period. Also, it is much easier to ask questions in class about potential homework problems than it is to spend hours trying to figure out the problems on your own at a later time.

If you are shy about asking questions in class, write down the questions and read them to your instructor. If the instructor seems confused about the questions, tell him or her that you will discuss the problem after class. To encourage yourself to ask questions, remember the following:

- You have paid for the instructor's help.
- Five other students probably have the same question.
- The instructor needs feedback on his or her teaching to help the class learn the material.
- There is no such thing as a stupid question
- If you have trouble asking questions, then email your instructor the question.

Record Each Problem Step

The final suggestion on note taking is to record each step of every problem written or verbally explained. By recording each problem step, you begin overlearning how to work the problems. This will increase your problem-solving speed during future tests. If you get stuck on the homework, you will also have complete examples to review.

The major reason for recording every step of a problem is to understand how to do the problem while the instructor is explaining it instead of trying to remember unwritten steps later. Recording each step will help when completing your homework because each step of a similar problem can be reviewed to help you solve the homework problem. Although it may seem time consuming, it pays off during homework and test time.

SECTION 4.5 REVIEW

1. List and describe the six steps in reworking your notes.

   First Step: ______________________________

   Second Step: ______________________________

   Third Step: ______________________________

Fourth Step: ______________________________

Fifth Step: ______________________________

Sixth Step: ______________________________

2. How can using a tape recorder in class improve your learning?

____

3. List three reasons for asking questions in class.

First Reason: ______________________________

Second Reason: ______________________________

Third Reason: ______________________________

4. What is the major reason for recording each problem step?

____

Name: ________________ Date: ________________

CHAPTER 4 REVIEW

1. Sitting in the golden triangle of success can help you pay more ________________ during class. (Answer on page 52)
2. Class time is an intense ________________ that should not be wasted. (Answer on page 53)
3. Auditory learners should review the math videotape and ________________ back the important information.
4. The goal of note taking is to take the ________________ amount of notes and get the ________________ amount of information.
5. Using abbreviations can help you write ________________ and listen ________________.
6. Your ________________ should include any words you don't know and their explanations.
7. Rework your notes to transfer information from ________________-________________ memory to ________________-________________ memory.
8. A ________________ ________________ is a separate section in your notebook that contains problems that your teacher worked in class.
9. Asking ________________ in class will improve your understanding and decrease your homework time.
10. ________________ each step of the problem will increase your problem-solving speed and memory.

What is the most important information you learned from this chapter?

____

____

____

____

How can you immediately use it?

____

____

____

____

DAN'S TAKE REVIEW

1. What were the consequences of Dan not paying attention while he was trying to take notes?

____

____

____

Name: ______________________ Date: ______________

2. Were his distractions limited to new technologies, or has there always been a certain level of distraction that might vary depending upon a given situation?

___

___

___

3. Having read what you have in this chapter, what would you tell him to do to avoid distractions in the future?

___

___

___

LECTURE NOTES

How to Improve Your Reading, Homework, and Study Techniques

5

CHAPTER

Reading a math textbook is more difficult than reading other textbooks. Math textbooks are written differently than English and social science textbooks. Math textbooks contain condensed material and, therefore, take longer to read.

Mathematicians can reduce a page of writing to one paragraph using math formulas and symbols. To make sure you understood that same information, an English instructor would take that original page of writing and expand it into two pages. Mathematicians pride themselves on how little they can write and still cover the concept. This is one reason it may take you two to three times as long to read your math text as it would any other text.

CHAPTER GOALS

What You Will Learn:

5.1 How to Read a Math Textbook
5.2 How to Do Your Homework
5.3 How to Solve Word Problems
5.4 How to Work with a Study Buddy
5.5 The Benefit of Study Breaks
5.6 Using Online/Computer Resources to Support Learning
5.7 The Benefits of Time Management and Study Breaks

REMEMBER
Reading your math text will take longer than reading your other texts.

Math students are expected to know how to do their homework; however, most math students do not have a homework system. Most students begin their homework by going directly to the problems and trying to work them. When they get stuck, they usually quit. This is not a good homework system.

A good homework system will improve your homework success and math learning at the same time.

DAN'S TAKE

College students are busy people. With hours of work and multiple deadlines, procrastination seems very appealing. Why suffer every night when you can put things off?

This type of logic stayed with me most of the way through college, and to be honest, it worked fairly well. Then I took my first college math course.

In this class, homework was going to be collected only four times during the semester. I thought I would need to spend only four nights the entire semester doing math homework. Granted, I knew that these nights would be rough on me and require hours of uninterrupted study, but to limit myself to just a few of them over the course of the semester seemed smarter to me than stressing out every single day.

I went to class every day and took very good notes. Then I went home at night and let them gather dust. By the time the first test rolled around, I had not done a single math problem. The night before the exam I sat down, opened my book, and looked at the homework assignment list. There were *ten chapters* due! I worked the whole night, and got a low C on my first test.

When I actually thought about it, I got really mad at myself. The homework was the easiest way to earn points. All I had to do was complete the work and I would have been given full credit. I also didn't realize that taking good notes was not enough to succeed in a college-level math course. By putting the homework off, I forgot how to do the math and had to relearn it all in one night.

Math is totally different from other courses such as science, history, and literature. Putting off homework can cause you to fall far behind, and catching up is very difficult. It's much easier to do a little bit every day than to try to cram an entire semester's worth of material into one night!

5.1 How to Read a Math Textbook

The way you read a math textbook is different from the traditional way students are taught to read textbooks in high school and college. Students are taught to read quickly or skim the material. If you do not understand a word, you are supposed to keep on reading. Instructors of other courses want students to continue to read so they can pick up the unknown words and their meanings from context.

This reading technique may work with your other classes, but using it in your math course will be totally confusing. If you skip some major concept words or boldface words, you will not understand the math textbook or be able to do the homework. Reading a math textbook takes more time and concentration than reading your other textbooks.

If you have a reading problem, it would be wise to take a developmental reading course before taking math. This is especially true with reform math courses, where reading and writing are emphasized.

Reform math classes deal more with word problems than traditional math courses do. If you cannot take the developmental reading course before taking math, then take it during the same semester as the math course.

Ten Steps in Understanding Reading Materials

There are several appropriate steps in reading a math textbook:

Step 1 *Skim the assigned reading material.* Skim the material to get a general idea about the major topics. Read the chapter introduction and each section summary. You do not want to learn the material at this time; you simply want to get an overview of the assignment. Then think about similar math topics that you already know.

EXAMPLE

Skimming will allow you to see if problems presented in one chapter section are further explained in later chapter sections.

Step 2 *As you skim the chapter, circle (using a pencil) the new words that you do not understand.* If you do not understand these new words after reading the assignment, ask the instructor for help. Skimming the reading assignments should take only 5 to 10 minutes.

Step 3 *Put all your concentration into reading.* While reading the textbook, highlight the material that is important to you. However, do not highlight more than 50% of any page, because doing so will not sufficiently narrow the material for future study. Especially highlight the material that is also discussed in the lecture. Material discussed in both the textbook and the lecture will usually appear on the test. The purpose of highlighting is to emphasize the important material for future study. Do not skip reading assignments.

REMEMBER

Reading a math textbook is very difficult. It might take you half an hour to read and understand just one page.

Step 4 *When you get to the examples, go through each step.* If the example skips any steps, make sure you write down each one of those skipped steps in the textbook for better understanding. Later on, when you go back and review, the steps will already be filled in. You will understand how each step was completed. Also, by filling in the extra steps, you will start to overlearn the material for better recall on future tests.

Step 5 *Mark the concepts and words that you do not know.* Maybe you marked them the first time while skimming. If you understand them now, erase the marks. If you do not understand the words or concepts, then reread the page or look them up in the glossary. Try not to read any further until you understand all the words and concepts.

Step 6 If you do not clearly understand some words or concepts, add these words to the note-taking glossary in the back of your notebook. Your glossary will contain the boldface words that you do not understand. If you have difficulty understanding the boldface words, ask the instructor for a better explanation. You should know all the words and concepts in your notebook's glossary before taking the test.

Step 7 *If you do not understand the material, follow these eight points, one after the other, until you do understand the material:*

Point 1—Go back to the previous page and reread the information to maintain a train of thought.

Point 2—Read ahead to the next page to discover if any additional information better explains the misunderstood material.

Point 3—Locate and review any diagrams, examples, or rules that explain the misunderstood material.

Point 4—Read the misunderstood paragraph(s) several times aloud to better understand their meaning.

Point 5—Refer to your math notes for a better explanation of the misunderstood material.

Point 6—Refer to another math textbook, computer software program, or videotape that expands the explanation of the misunderstood material.

Point 7—Define exactly what you do not understand and call your study buddy for help.

Point 8—Contact your math tutor or math instructor for help in understanding the material.

Step 8 *Recall important material.* Try to recall some of the important concepts and rules. If you have difficulty recalling this information, look at the highlighted areas of the text, boldface words, or formulas. If you are a visual learner, write down these words several times on a separate sheet of paper. If you are an auditory learner, repeat these words several times out loud. If you are a kinesthetic (hands-on) learner, write out the formulas twice on two separate cards. Cut one of them up into puzzle pieces. Mix up the parts of the puzzle. Turn over the other formula card so you cannot see it. Put the puzzle back together. Do this several times. For boldface words, write the word on one card and its explanation twice on two other separate cards. Make another puzzle. Mix up the pieces of the explanation and put them back together. Then add these words and formulas to your note-taking glossary. Recalling important material right after you read it improves your ability to remember that information.

Step 9 *Reflect on what you have read.* Combine what you already know with the new information that you just read. Think about how this new information enhances your math knowledge. Prepare questions for your instructor on the confusing information. Ask those questions at the next class meeting.

Step 10 *Write anticipated test questions.* Research has noted that students have about 80% accuracy in predicting test questions. Think about what is the most important concept you just read and what problems the instructor could give you that would test the knowledge of that concept. Make up three to five problems and add them to your problem log (this was part of your note-taking system). Indicate in the log that these are your questions, not the instructor's.

By using this reading technique, you narrow down the important material to be learned. You skim the textbook to get an overview of the assignment. You carefully read the material and highlight the important parts. You then add unfamiliar words or concepts to your note-taking glossary.

REMEMBER

You should review the highlighted material before doing the homework problems, and you should learn 100% of the glossary before taking the test.

How Reading Ahead Can Help

Reading ahead is another way to improve learning. If you read ahead, do not expect to understand everything. Read ahead two or three sections and put question marks (in pencil) by the material you do not understand.

When the instructor starts discussing that material, have your questions prepared and take good notes. Also, if the lecture is about to end, ask the instructor to explain the confusing material in the textbook. Reading ahead will take more time and effort, but it will better prepare you for the lectures.

How to Establish Study Period Goals

Before beginning your homework, establish goals for the study period. Do not just do the homework problems.

Ask yourself this question: "What am I going to do tonight to become more successful in math?"

By setting up short-term homework goals and reaching them, you will feel more confident about math. This also improves your self-esteem and helps you become a more internally motivated student. Set up homework tasks that you can complete. Be realistic.

Study period goals are set up either on a time-line basis or an item-line basis. Studying on a time-line basis is studying math for a certain amount of time.

EXAMPLE

You may want to study math for an hour and then switch to another subject. You will be studying on a time-line basis.

Studying on an item-line basis means you will study your math until you have completed a certain number of homework problems.

EXAMPLE

You might set a goal to study math until you have completed all the odd-numbered problems in the chapter review. The odd-numbered problems are the most important problems to work, because, in most texts, they are answered in the answer section in the back of the book. Such problems provide the opportunity to recheck your work if you do not get the answer correct. Once you have completed these problems, do the even-numbered problems.

No matter what homework system you use, remember this important rule: Always finish a homework session by understanding a concept or doing a homework problem correctly.

Do not end a homework session with a problem you cannot complete. You will lose confidence, since all you will think about is the last problem you could not solve instead of the 50 problems you solved correctly. If you do quit on a problem you cannot solve, rework some of the problems you have done correctly.

REMEMBER

Do not end your study period with a problem you cannot complete.

SECTION 5.1 REVIEW

1. List and describe the 10 steps in reading a math textbook.

Step 1: ____________________

Step 2: ____________________

Step 3: ____________________

Step 4: ____________________

Step 5: ____________________

Step 6: ____________________

Step 7: ____________________

Step 8: ____________________

Step 9: ____________________

Step 10: ____________________

2. After trying each of the 10 steps, make up your own condensed version of these steps.
   Step 1: ____________
   Step 2: ____________
   Step 3: ____________
   Step 4: ____________
   Step 5: ____________
   Step 6: ____________
   Step 7: ____________
   Step 8: ____________
   Step 9: ____________
   Step 10: ____________
3. List two reasons why reading ahead can improve your learning.
   First Reason: ____________
   Second Reason: ____________
4. Describe the two types of study period goals. Which one do you use?
   First Type: ____________
   ____
   Second Type: ____________
   ____

5.2 How to Do Your Homework

Doing your homework can be frustrating or rewarding. Most students jump right into their homework, become frustrated, and stop studying. These students usually go directly to the math problems and start working them without any preparation. When they get stuck on one problem, they flip to the back of the text for the answer. Then they either try to work the problem backward to understand the problem steps or they just copy down the answer.

Other students go to the solution guide or the textbook website and copy the steps. After getting stuck several times, these students will inevitably quit doing their homework. Their homework becomes a frustrating experience and they may even quit doing their math homework altogether. Motivation decreases every time a student does not feel that they learned anything from his or her homework to the point that they may begin skipping classes and subsequently fail. Students may not have to successfully complete each and every problem, but they do need to feel successful. In addition, they must also develop a plan on how to get help solving the problems they cannot complete.

The Importance of Doing Your Homework

Do you know the reasons for doing math homework? I have asked this question to math instructors and to tutors and was surprised by their answers.

Students have told me that homework is assigned to waste their time because the instructor did not want to teach them or because the instructor did not like them and wanted them to suffer. Reinforcing this theory, the instructors did not even grade the homework or return it to them in time to prepare for the test. Even in this instance, the student's perception is not necessarily true. Instructors almost always want you to learn and will assign you homework only in order to improve your learning and success.

Now let's look at the importance of doing your homework. In sports or in learning to play a musical instrument, you need to practice in order to have a good game or perform well in a recital. In the case of math, homework is that "practice" so you can take the information taught in the classroom and put it into abstract and long-term memory (you will learn more about this concept in Chapter 6). You need to be able to practice the math problems often enough to understand the concepts and then automatically recognize the problems and know how to solve them. In other words, you do your homework to remember how to solve the problems during the test. However, just memorizing how to do specific problems is not enough. Most instructors will not put homework problems on the test, so it becomes important to have a full grasp on the concepts behind them in order for the problems to become interchangeable.

Doing your homework provides an excellent opportunity to practice the problems as they might appear on a test and thereby reinforce the mathematical concepts used to solve them. This is perhaps the best way to figure out what problems you cannot solve *before* taking the test. Not doing your homework or not completing your homework means that the information never reaches long-term memory. This leads to poor math learning and potential math failure. Successfully completing homework leads to a positive learning experience.

Now that you know its importance, has anyone every taught you the best way to complete your math homework? Probably not. The next section will teach you a math homework system that has helped thousands of students better learn math and thereby improve their grades.

Ten Steps in Doing Your Homework

To improve your homework success and learning, refer to these 10 steps:

Step 1 *Review the textbook material that relates to the homework.* A proper review will increase your chances of successfully completing your homework. If you get stuck on a problem, you will have a better chance of finding similar problems. If you do not review prior to doing your homework, you could get stuck and not know where to find help in the textbook.

REMEMBER
To be successful in learning the material and in completing homework assignments, you must first review your textbook.

Step 2 *Review your lecture notes that relate to the homework.* If you cannot understand the explanation in the textbook on how to complete the homework assignment, then review your notes.

REMEMBER
Reviewing your notes will give you a better idea about how to complete your homework assignment.

Step 3 *Do your homework as neatly as possible.* Doing your homework neatly has several benefits. When approaching your instructor about problems with your homework, he or she will be able to understand your previous attempts to solve the problem. The instructor will easily locate the mistakes and show you how to correct the steps without having to decipher your handwriting. Another benefit is that when you review for midterm or final exams, you can quickly relearn the homework material without having to decipher your own writing.

REMEMBER
Neatly prepared homework can help you now and in the future.

Step 4 *When doing your homework, write down every step of the problem.* Even if you can do the step in your head, write it down anyway. This will increase the amount of homework time, but you will be over-learning how to solve problems, which improves your memory. Writing down every step is an easy way to memorize and understand the material. Another advantage is that when you rework the problems you did wrong, it is easy to review each step and to find your mistake.

REMEMBER
In the long run, writing down every step of your homework problems will save you time and frustration.

Step 5 *Understand the reasons for each problem step and check your answers.* Do not get into the bad habit of memorizing how to do problems without knowing the reasons for each step. Many students are smart enough to memorize the procedures required to complete a set of homework problems. However, when similar homework problems are presented on a test, some of these students cannot solve them. To avoid this dilemma, keep reminding yourself about the rules, laws, or properties used to solve problems.

EXAMPLE
Problem: $2(a + 5) = 0$. What property allows you to change this equation to $2a + 10 = 0$? *Answer:* the distributive property.

Once you know the correct reason for going from one step to another in solving a math problem, you can solve any problem requiring that property. Students who simply memorize how to do problems instead of understanding the reasons for correctly working the steps are more likely to fail their math course.

Checking your homework answers should be part of your homework duties. Checking your answers can improve your learning and help you prepare for tests.

Check the answers of the problems for which solutions are not given. These may be the even-numbered or odd-numbered problems or the problems not answered in the solutions manual.

First, check each answer by estimating the correct answer.

EXAMPLE
If you are multiplying 2.234 by 5.102, remember that 2 times 5 is 10. The answer should be a little over 10.

You can also check your answers by substituting the answer back into the equation or doing the opposite function required to answer the question. The more answers you check, the faster you will become. This is very important because increasing your answer-checking speed can help you catch more careless errors on future tests.

Step 6 If you do not understand how to do a problem, refer to the following points:

Point 1—Review the textbook material that relates to the problem.

Point 2—Review the lecture notes that relate to the problem.

Point 3—Review any similar problems, diagrams, examples, or rules that explain the misunderstood material.

Point 4—Refer to another math textbook, solutions guide, math computer program software, or videotape to obtain a better understanding of the material.

Point 5—Call your study buddy.

Point 6—Skip the problem and ask your tutor or math instructor for help as soon as possible.

Step 7 *Always finish your homework by successfully completing problems.* Even if you get stuck, go back and successfully complete previous problems before quitting. You want to end your homework assignment with feelings of success.

Step 8 *After finishing your homework assignment, recall to yourself or write down the most important concepts you have learned.* Recalling this information will increase your ability to learn these new concepts. Additional information about Step 8 will be presented later in this chapter.

Step 9 *Make up note cards containing hard-to-remember problems or concepts.* Note cards are an excellent way to review material for a test. More information on the use of note cards as learning tools is presented later in this chapter.

Step 10 *Do not fall behind.* As mentioned in Chapter 1, learning math is a sequential process. If you get behind, it is difficult to catch up because each topic builds on the previous topic. Falling behind in math is like going to Spanish class without learning the last set of vocabulary words. The teacher will be talking to you using the new vocabulary, but you will not understand what is being said.

Do Not Fall Behind

To keep up with your homework, it is recommended that you work on it every school day and even on weekends. Working on your homework for one-half hour each day for two days in a row is better than working on it for one hour every other day.

If you have to get behind in one of your courses, *make sure it is not math.* Fall behind in a course that does not require a sequential learning process, such as psychology or history. After using the ten steps in doing your homework, you may be able to combine two steps into one. Find your best combination of homework steps and use them.

REMEMBER
Getting behind in math homework is the fastest way to fail the course.

SECTION 5.2 REVIEW

1. List the 10 steps you should follow in doing your math homework.

Step 1: ______________________

Step 2: ______________________

Step 3: ______________________

Step 4: ______________________

Step 5: ______________________

Step 6: ______________________

Step 7: ______________________

Step 8: ______________________

Step 9: ______________________

Step 10: ______________________

2. After trying each of the 10 steps, make up your own condensed version of these steps.

Step 1: ______________________

Step 2: ______________________

Step 3: ____________________

Step 4: ____________________

Step 5: ____________________

Step 6: ____________________

Step 7: ____________________

Step 8: ____________________

Step 9: ____________________

Step 10: ____________________

3. How will reviewing your notes or your math text help you complete your homework more successfully?

____

____

____

4. List two ways you can check your homework answers.

First Way: ____________________

Second Way: ____________________

5. What happens to students who fall behind in their math homework?

____

____

____

5.3 How to Solve Word Problems

The most difficult homework assignment for most math students is working story/word problems. Solving word problems requires excellent reading comprehension and translating skills.

Students often have difficulty substituting algebraic symbols and equations for English terms. But once an equation is written, it is usually easily solved. To help you solve word problems, follow these ten steps:

Step 1 *Read the problem three times.* Read the problem quickly the first time as a scanning procedure. As you are reading the problem the second time, answer these three questions:

1. *What is this problem asking me?* (This is usually at the end of the problem.)
2. *What is this problem telling me that is useful?* (Cross out unneeded information.)
3. *What is this problem implying?* (This is usually something you have been told to remember.)

Now read the problem a third time to make sure that you fully understand its meaning.

Step 2 *Draw a simple picture of the problem to make it more real to you* (e.g., a circle with an arrow can represent travel in any form—by train, by boat, by plane, by car, or by foot).

Step 3 *Make a table of information and leave a blank space for the information you are not told.*

Step 4 *Use as few unknowns in your table as possible.* If you can represent all the unknown information in terms of a single letter, do so! When using more than one unknown, use letters that remind you of each unknown. Then write down what each unknown represents. This eliminates the problem of assigning the right answer to the wrong unknown. Remember that you have to create as many separate equations as you have unknowns.

Step 5 Translate the word problem into an algebraic equation using the lists of terms in Figures 8 and 9. Remember, English terms are sometimes stated in a different order than the corresponding algebraic terms.

Step 6 *Immediately retranslate the equation, as you now have it, back into words.* The translation may not sound like a normal word problem, but the meaning should be the same as the original problem. If the meaning is not the same, the equation is incorrect and needs to be rewritten. Rewrite the equation until it means the same as the word problem.

Step 7 *Review the equation to see if it is similar to equations from your homework and if it makes sense.* Some formulas dealing with specific word problems may need to be rewritten. Distance problems, for example, may need to be written solving for each of the other variables in the formula. Given distance = rate × time, time = distance/rate, and rate = distance/time. Usually, a distance problem will identify the specific variable for which you are to solve.

Step 8 *Solve the equation using the rules of algebra.*

REMEMBER

Whatever is done to one side of the equation must be done to the other side of the equation. The unknown must end up on one side of the equation, by itself. If you have more than one unknown, then use the substitution method or the elimination method to solve the equations.

Step 9 *Look at your answer to see if it makes sense.*

EXAMPLES

If tax is added to an item, it should cost more; if a discount is applied to an item, it should cost less. Is there more than one answer? Does your answer match the original question? Does your answer have the correct units?

Step 10 *Substitute your answer back into the original equation to see if it is correct.* If one side of the equation equals the other side of the equation, then you have the correct answer. If you do not have the correct answer, go back to Step 5.

The most difficult part of solving word problems is translating part of a sentence into algebraic symbols and then into algebraic expressions. Review Figure 8 (Translating English Words and Phrases into Algebraic Symbols) and Figure 9 (Translating English Phrases into Algebraic Expressions).

How to Recall What You Have Learned

After completing your homework problems, a good visual learning technique is to make note cards. Note cards are 3" × 5" index cards on which you place information that is difficult to learn or material you think will be on a test.

On the front of each note card, write a math problem or information that you need to know. Color-code the important information in red or blue. On the back of the note card, write how to work the problem or explain important information.

EXAMPLE

If you are having difficulty remembering the rules for multiplying positive and negative numbers, write some examples on the front of a note card and write the answers on the back.

Make note cards on important information you might forget. Every time you have 5 spare minutes, pull out your note cards and review them. You can glance at the front of the card, repeat the answer to yourself, and check it by looking at the back of the card. If you are correct and know the information

FIGURE 8 Translating English Words and Phrases into Algebraic Symbols

English Word or Phrase	Algebraic Symbol
Sum	+
Add	+
In addition	+
More than	+
Increased	+
In excess	+
Greater	+
Decreased by	−
Less than	−
Subtract	−
Difference	−
Diminished	−
Reduced	−
Remainder	+
Times as much	×
Percent of	×
Product	×
Interest on	×
Per	÷
Divide	÷
Quotient	÷
Quantity	( )
Is	=
Was	=
Equal	=
Will be	=
Results	=
Greater than	>
Greater than or equal to	≥
Less than	<
Less than or equal to	≤

on the card, do not put it back in the deck. Mix up the cards you do not know and pick another card. Keep doing this until there are no cards left in the deck.

If you are an auditory learner, use a tape recorder instead of note cards. Record the important information as you would on the front of a note card. Then leave a blank space on the recording before recording the answer. Play the tape back. When you hear the silence, put the tape on pause. Then say the answer out loud to yourself. Take the tape player off pause and see if you were correct.

FIGURE 9 Translating English Phrases into Algebraic Expressions

English Phrases	Algebraic Expression
Ten more than x	$x + 10$
A number added to 5	$5 + x$
A number increased by 13	$x + 13$
Five less than 10	$10 - 5$
A number decreased by 7	$x - 7$
Difference between x and 3	$x - 3$
Difference between 3 and x	$3 - x$
Twice a number	$2x$
Ten percent of x	$0.10x$
Ten times x	$10x$
Quotient of x and 3	$x/3$
Quotient of 3 and x	$3/x$
Five is 3 more than a number	$5 = x + 3$
The product of 2 and a number is 10	$2x = 10$
One-half of a number is 10	$x/2 = 10$
Five times the sum of x and 2	$5(x + 2)$
Seven is greater than x	$7 > x$
Five times the difference of a number and 4	$5(x - 4)$
Ten subtracted from 10 times a number is that number plus 5	$10x - 10 = x + 5$
The sum of $5x$ and 10 is equal to the product of x and 15	$5x + 10 = 15x$
The sum of two consecutive integers	$(x) + (x + 1)$
The sum of two consecutive even integers	$(x) + (x + 2)$
The sum of two consecutive odd integers	$(x) + (x + 2)$

Review What You Have Learned

After finishing your homework, close the textbook and try to remember what you have learned. Ask yourself these questions: What major concepts did I learn tonight? What test questions might the instructor ask on this material?

For about 3 to 4 minutes, recall the major points of the assignment, especially the areas you had difficulty understanding. Write down questions for the instructor or tutor. Since most forgetting occurs right after the material is learned, this short review will help you retain the new material.

SECTION 5.3 REVIEW

1. List the 10 steps in solving word problems.

   Step 1: ______

   Step 2: ______

   Step 3: ______

   Step 4: ______

   Step 5: ______

   Step 6: ______

   Step 7: ______

   Step 8: ______

   Step 9: ______

   Step 10: ______

2. What is your best way to recall what you have learned?

   ______

   ______

   ______

3. How does reviewing right after doing your homework help you remember?

   ______

   ______

   ______

5.4 How to Work with a Study Buddy

You need to have a study buddy when you miss class or when you do your homework. A study buddy is a friend or classmate who is taking the same course. You can find a study buddy by talking to your classmates or making friends in the math lab. Try to find a study buddy who knows more about math than you do. Tell the class instructor that you are trying to find a study buddy and ask which students make the best grades. Meet with your study buddy several times a week to work on problems and to discuss math. If you miss class, get the notes from your study buddy so you will not fall behind.

Call your study buddy when you get stuck on your homework. You can solve math problems over the phone. Do not sit for half an hour or an hour trying to work one problem; that will destroy your confidence, waste valuable time, and possibly alienate your study buddy. Think how much you could have learned by trying the problem for 15 minutes and then calling your study buddy for help. Spend, at the maximum, 15 minutes on one problem before going on to the next problem or calling your study buddy.

> **REMEMBER**
> A study buddy can improve your learning while helping you complete your homework assignments. Just do not overuse your study buddy or expect your study buddy to do your homework for you.

SECTION 5.4 REVIEW

1. How do you select a study buddy? __________
2. How can a study buddy help you learn math? __________
3. Who is your study buddy? __________

5.5 The Benefits of Study Breaks

Psychologists have discovered that learning decreases if you do not take study breaks. Therefore, take a break after studying math for 45 minutes to 1 hour.

If you have studied for only 15 or 20 minutes and feel you are not retaining the information or your mind is wandering, take a break. If you continue to force yourself to study, you will not learn the material. After taking a break, return to studying.

If you still cannot study after taking a break, review your purpose for studying and your educational goals. Think about what is required to graduate. It will probably come down to the fact that you will have to pass math. Think about how studying math today will help you pass the next test; this will increase your chances of passing the course and of graduating.

On an index card write three positive statements about yourself and three positive statements about studying. Look at this index card every time you have a study problem. Use every opportunity available to reinforce your study habits.

SECTION 5.5 REVIEW

1. How can study breaks improve your learning?

____

____

____

____

2. What do you do when you don't want to study? How do you get started?

____

____

____

____

5.6 Using Online/Computer Resources to Support Learning

Many math textbooks include CDs and access to online websites that provide further practice and learning. There are even online homework programs that offer instant feedback and help. While these resources may seem unfamiliar or difficult to use at first, online support and CDs can improve some student learning and are worth a try.

What Are Online/Computer Resources?

Computer and web resources come in a variety of formats. Some math texts have CDs that can provide additional practice exercises and suggestions to improve learning. Other books come with passwords to websites that contain online tutorials and extra practice problems. Other publishers provide live tutorial help, such as *Enhanced WebAssign®*, which is offered by Cengage and provides 24-hour-a-day support for students. There are even math tutorial sites that are not meant for use with your textbook—just Google "math" and see what comes up! Finally, your own math department, math lab, or learning resource center may have other CDs, computer programs, or websites that can help you. Find out what online/computer resources are available to supplement your math learning.

Computer Technology Requirements

It is important to find out if the computer you use on a regular basis will support the online/computer program resources for further math learning. Check the operating or installation instructions for any program or website you want to use to make sure that your computer meets the minimum hardware and Internet connection requirements.

If your own computer is inadequate, do not despair and decide not to try the online or CD resources. Most campuses provide excellent computer and online access for students, but remember that most computers in the lab may not let you save your work on their hard drives. Be sure to bring your own flash drive or recordable CD so you can save your work.

After finding a computer that is adequate for the program's technical requirements, do a trial run. It is helpful if you can do this in a lab or library that has technical assistants in case you run into problems during installation or setup. Once the math program or website comes up, explore the different aspects of that program. If you have difficulty running the programs or need input on which program may be best for you, don't hesitate to ask your instructor for help. If your instructor is not available, then ask the lab learning assistants or a tutor. If they are not available, then ask your fellow students for help.

During the Semester

Using online/computer resources can help in various ways. First, try each online/computer resource available to you and see which ones work best for you. Some students like the drill and practice that is provided by an online quiz, while other students want more in-depth knowledge on the reason for doing each step and prefer interactive tutorials. Second, try using the online/computer resources both before and after your class or tutoring session to see which schedule helps you most. Some students like to use these resources as a preview to a math lesson or as a review after the lesson. Sometimes it might help to use the resources before *and* after the lesson.

Treat the time you spend using the resources as if it were class or lab time. First, set up a time and place to use the online/computer resources. Try to use the resources at the same place and same time. For example, if you have an hour between classes, go to the lab and practice, or practice right when you get home. This is a good way to develop excellent study habits. Second, keep a log for several weeks on which resources you used and how much time you spent on them. If it is obvious that certain times, places, or programs work best for you, keep the same study system. If your system is not working for you, adjust and revise it. Ask a lab assistant or instructor for advice.

OVERCOMING TECHNOPHOBIA

If you are experiencing technophobia (fear of technology, including computers), this is a wonderful time to overcome it. You can start with these strategies:

- Get in touch with the benefits of technology. Being comfortable with computers can give you an edge in almost every aspect of being a student, from doing library research to planning your career. In the eyes of many employers, experience with computers is sometimes a necessity and almost always a plus.
- Sign up for a computer class for beginners.
- Ask questions. When it comes to computers, there truly aren't any "dumb" questions.
- Find a competent teacher—someone who remembers what it was like to know nothing about computers.
- Just experiment. Sit down, do something, and see what happens. Short of dropping a computer or hitting it with a hammer, you can't hurt the thing.
- Remember that computers are not always user friendly—at least not yet. Learning how to use them takes patience and time. Knowing this up front can put you at ease and prepare you for the cyberspace adventures ahead.
- Also remind yourself of past successes in making transitions. As far as technology is concerned, this includes everything from writing with a pen to driving a car. You're already mastering a major life change—the transition to higher education. In doing so, you've shown that you have what it takes to tame any technology that enters your life.

Learning Styles and Online/Computer Learning

Your learning modality style may determine how helpful online/computer learning may be for you. Visual learners who like to learn by themselves usually work well with these types of resources, while auditory learners who need to hear the information may not like the resources unless they are also auditorily enhanced. Also, hands-on learners may not benefit from these programs unless there is some type of movement with the learning. Nevertheless, these programs are worth a try no matter what the learning style. Why? We all learn visually to some extent, and auditory and hands-on learners can adjust these resources to be more auditory or hands-on. For instance, students can read the screens out loud as they are using them, or can work with partners. Learning styles are just preferred ways to learn. Productive students learn how to use all available modalities when learning math.

Evaluate the Experience

After trying these programs for several weeks, evaluate whether they are helping you improve your learning or your grades. Evaluate each program. Take into consideration how much time you spent with each program. Look through your log to remind yourself of these details. You may have to wait until the next major test to see if your grades improve. Ask yourself, Did the resources improve my learning? and Did the resources improve my test scores? If the answer to either of these questions is yes, then keep on using them!

SECTION 5.6 REVIEW

1. Look in the preface of your math textbook and list some of the online/computer resources that are included with your book.

____

____

____

2. Which resource listed sounds as if it would be the most helpful to you? Why?

____

____

____

3. Remember your learning modality style from Chapter 2. How could you use one of the online/computer resources so that it worked with your personal learning modality?

____

____

5.7 The Benefits of Time Management and Study Breaks

Now that you have learned some of the best ways to read the textbook, complete your homework, solve word problems, work with a study buddy, and use online resources, it is important to figure out when will you have time to complete these activities. Many students have told me that they don't even have enough time to do their homework, let alone all these other activities that **will** improve their grades. **I do not believe them.** If you are truly an adult, then you can find ways to control your time. The truth is, many students simply choose not to control their time and, therefore, allow time to control them. For many students it comes down to priorities and making good choices.

If you want to obtain a better career for yourself or your family, then making time to study math should be your top priority. Most math students need to study between 8 and 10 hours a week to get an A or a B. Notice that I did not mention Cs because students who make a C in an algebra-based course will most likely fail their next algebra course.

Another way to improve math grades is to become an effective math learner. The truth is, throughout your academic careers, most of you have probably never been taught how to learn math. This, of course, is what this particular text is all about. Some general learning efficiency suggestions are to

- Set up a study time to review your notes or do your homework as soon as possible after math class.
- Leave a 1-hour opening in your schedule right after math class.
- Go to the math lab or Learning Resource Center right after math class and/or right before math class.
- Study math right when you get home, when you are alert.
- Review your problems right before you go to bed the night before your test (without watching TV or playing video games).
- Find the best time to study math according to your personal biological clock (morning, noon, early afternoon, night).
- Use the other suggestion in this text to improve your learning.

The next step is to list your priorities, as well as how much time you need to spend on your top priorities. Examples of top priorities may include work, attending class, family, sleep, eating, church, or others. You need to rank these priorities from most important to the least important. Then, next to each activity, put the number of hours you can allow for that priority. Remember you need to study math about 10 hours a week to get an A or a B. Also, it is important to remember that there are 168 hours in a week.

Next, you need to write down the times you have decided to study math. You may want to put the times in your notebook or on a 3" × 5" card and then put the card on the refrigerator. You will then need to remind yourself about the study times and keep a weekly log of the hours you have actually studied.

Next you need to set a goal for your math grade. After studying for several weeks and receiving your first major math test grade, you may have to adjust your study hours based upon the results.

Of all the students with whom I have worked who have flunked out of college or high school, most pointed to time management as one of the major culprits. Proper time management can help you become successful in college, just as bad time management can cause you to fail a math course and eventually drop out. The choice is yours!

SECTION 5.7 REVIEW

1. How can time management help you or hurt you while trying to become successful in math and college?

____

____

____

____

2. How can study breaks improve your learning?

____

____

____

____

3. What do you do when you don't want to study? How do you get started?

____

____

____

____

Name: ______________________ Date: ______________

CHAPTER 5 REVIEW

1. Reading one page of a math text might take you ________ ________ ________.

2. After reading your textbooks, ________ important material can improve your ability to remember that information.

3. Write anticipated test questions in your ________ and indicate that they are your questions.

4. Checking your homework answers can improve your ________ and help you prepare for ________.

5. Matching:

   (For students in a prealgebra class)

______ Product	A. +
______ Increased	B. −
______ Sum	C. ×
______ Per	D. ÷
______ Is	E. =
______ Reduced	
______ Remainder	

   (For students taking elementary algebra or a higher-level math course)

______ Difference between x and 3	A. $10x$
______ Quotient of x and 3	B. $x/3$
______ Five times the sum of x and 2	C. $5(x - 4)$
______ A number added to 5	D. $5(x + 2)$
______ Five less than 10	E. $5 = x + 3$
______ Ten times x	F. $5 + x$
______ Five is 3 more than a number	G. $10 - 5$
______ Five times the difference of a number and 4	H. $x - 3$

6. Make sure you finish your homework by working problems you ________________ do.

7. Using ________________ ________________ is a visual way to recall what you learned in math.

Name: ______________________ Date: ______________

8. Using a ____________ ____________ is an auditory way to recall what you learned in math.

9. Most ____________ occurs right after learning; a short review will help you ____________ the new material.

10. A ____________ ____________ can improve your learning while helping you complete your homework assignments.

What is the most important information you learned from this chapter?

____

____

____

____

How can you immediately use it?

____

____

____

____

DAN'S TAKE REVIEW

1. What mistake did Dan make while scheduling his time for his first college math class?

____

____

____

2. What was the problem with his logic that it would be better to cram four times during the semester rather than doing the homework on a daily basis?

____

____

____

3. How could Dan avoid such problems in the future?

____

____

____

How to Improve Your Math Test-Taking Skills

CHAPTER 7

Taking a math test is different from taking tests in other subjects. Math tests require you not only to recall the information, but also to apply the information. Multiple-choice tests, for example, usually test you on recall, and if you do not know the answers, you can guess.

Math tests build on each other, whereas history tests often do not test you on previous material. Most math tests are conceptional tests, where you cannot memorize the answers; most social science tests are designed to test long-term memory and for everyone to finish.

Math test preparation and test-taking skills are different from preparation and skills needed for other tests. You need to have a test-taking plan and a test analysis plan to demonstrate your total knowledge on math tests. Students with these plans make better grades compared with students without them. Math instructors want to measure your math knowledge, not your test-taking skills.

CHAPTER GOALS

What You Will Learn:

DAN'S TAKE

The moment the professor hands out a math test, everyone dives in, frantically jotting down answers as fast as they can. The first few questions are easy, giving the test takers a false sense of security heading into the second page. At this point, those who know what they are doing begin to surge ahead, panicking their neighbors and pressuring them to hurry through their tests to keep up.

Then the first person finishes, and one by one, those taking their time watch their classmates leave. The few students that remain are left with a sense of failure before they even turn in their tests.

I don't understand why college students believe that a test is a race. Everyone knows that those who finish first don't always score the highest. But for some reason, we still feel the need to avoid the dreaded "last person to finish" label.

Math and speed do not mix well together. In fact, trying to finish a test quickly often causes mistakes. Speed "kills" math grades. Even if you know the steps, a small error can make you get a question wrong. It might be that you just carried a number incorrectly. These mistakes happen when you rush through a test.

In high school I often fell into this trap. I would study and prepare for a test and then be shocked when I got a C. During the test, as I watched my peers pick up their things and leave the class, I would panic. I, like so many others that age, was terrified that I might appear stupid if I were the last person to leave the room.

Everything changed when I realized that I didn't need to be the first person out the door after a test. By taking my time, I was able to catch all the little mistakes that were making me lose points. I could leave with the confidence that I honestly represented my knowledge of the material at hand.

It is very important that you not feel pressured during a math test. When people finish before you, ignore them. What's the big deal? It's not fair to measure yourself by how quickly those around you work. Instead of worrying about getting things done quickly, worry about getting them done correctly and making a good grade and thus passing the course.

7.1 Why Attending Class and Doing Your Homework May Not Be Enough to Pass

Most students and some instructors believe that attending class and doing all the homework ensures an A or a B on tests. This is far from true. Doing all the homework and getting the correct answers is very different in many ways from taking tests:

1. While doing homework, there is little anxiety. A test situation is just the opposite.
2. You are not under a time constraint while doing your homework; you may have to complete a test in 55 minutes or less.
3. If you get stuck on a homework problem, your textbook, notes, and Websites are there to assist you. This is not true for most tests.
4. Once you learn how to do several problems in a homework assignment, the rest are similar. In a test, the problems may be in random order and measure different concepts.
5. In doing homework, you have the answers to at least half the problems in the back of the text and answers to all the problems in the solutions guide. This is not true for tests.
6. While doing homework, you have time to figure out how to use your calculator. During the test, you can waste valuable time figuring out how to use your calculator.
7. When doing homework, you can call your study buddy or ask the tutor for help, something you cannot do during the test.
8. While doing your homework, you can go to Websites for online tutoring to help you solve problems you do not understand.

9. While doing online homework, you have the hints if you get stuck and other resources. These resources are not available during tests.
10. While doing your homework in the math lab or Learning Resource Center, when students leave it does not bother you. In a test situation when students get up and leave, it may cause anxiety and worry and make you want to leave before checking your problems.

Do not develop a false sense of security by believing you can make an A or a B by just doing your homework. Tests measure more than just your math knowledge.

The General Pretest Rules

General rules are important when taking any type of test:

1. *Get a good night's sleep before taking a test.* This is true for the ACT, the SAT, and your math tests. If you imagine you are going to cram all night and perform well on your test with 3 to 4 hours of sleep, you are wrong. It would be better to get 7 or 8 hours of sleep and be fresh enough to use your memory to recall information needed to answer the questions.
2. *Start studying for the test at least 3 days ahead of time.* Make sure you take a practice test to find out, before the test, what you do not know. Review your problem log and work the problems. Review the concept errors you made on the last test. (How to identify and correct your concept errors will be discussed later in this chapter.) Meet with your instructor or tutor for help on those questions you cannot solve.
3. *Review only the material you have already learned the night before a test.*
4. *Make sure you know all the information on your mental cheat sheet.* Review your notebook and glossary to make sure you understand the concepts. Work a few problems and recall the information on your mental cheat sheet right before you go to bed. Go directly to bed; do not watch television, listen to the radio, or party. While you are asleep, your mind will work on and remember the last thing you did before you went to sleep.
5. *Get up in the morning at your usual time and review your notes and problem log.* Do not do any new problems, but make sure your calculator is working.

SECTION 7.1 REVIEW

1. List three reasons why only attending class and doing your homework may not be enough to pass your math course.

   First Reason: ____________________

   ____

   ____

   ____

   ____

   Second Reason: ____________________

   ____

   ____

   ____

   ____

Third Reason: ____________________

____

____

____

____

2. List and explain three general pretest rules that best apply to you.

First Rule: ____________________

____

____

____

____

Second Rule: ____________________

____

____

____

____

Third Rule: ____________________

____

____

____

____

PRETEST CHECKOFF LIST

Just like a pilot who conducts a preflight check list, you can do a pretest check list in order take off safely and not "crash" on your test. Complete as many of the items on the checkoff list as you can 2 days or more *before* the test.

Reducing Test Anxiety

Students who have test anxiety need to practice the relaxation techniques that they will use before and during the test. Reducing test anxiety can help students recall information for the test.

I have practiced my best short-term relaxation technique. *Yes* ____ *No* ____ *Don't have test anxiety* ____

Students with test anxiety usually say very negative statements to themselves, which intensifies their test anxiety further.

What I usually say to myself during a test is not positive. *Yes* ____ *No* ____

What are these negative statements?

1. ____________________
2. ____________________
3. ____________________

I have written and practiced saying positive statements to replace the negative ones. Yes _____ No _____

What are these positive statements?

1. ____________________

2. ____________________

3. ____________________

Reviewing Notes and Problem Log

Understanding math vocabulary is key to learning math. Talking about the meaning and uses for vocabulary words helps to review the key concepts you must remember for a test. To review math vocabulary, use your math glossary and/or the key words in your note-taking system.

I have reviewed my math glossary or key words at least three or four times a week. Yes _____ No _____
Did not develop a glossary or set of key words ______

I reviewed my math glossary or key words at least once a day starting 2 days before the test. Yes _____ No _____
Did not develop a glossary or set of key words ______

Reviewing all your notes before the test is a good strategy to help you remember math concepts and how to do math problems.

I have reviewed all my notes 2 days before the test. Yes _____ No _____
Did not take notes ______

Reviewing your problem log of instructor examples is another way to understand math concepts and problem-solving strategies.

I have reviewed my problem log examples by 2 days before the test. Yes _____ No _____
Did not develop a problem log ______

Reviewing the Textbook

Reviewing the examples in the textbook is another way to refresh your math skills.

I have reviewed the examples in the textbook 2 days before the test. Yes _____ No _____

One of the strategies of reading a textbook is to develop anticipated test questions.

I reviewed my anticipated test questions. Yes _____ No _____
Did not develop anticipated test questions _____

Reviewing Homework

Reviewing your homework problems is a good way to help remember how to do the problems.

I reviewed homework problems 2 days before the test. Yes _____ No _____
Did not do the homework problems _____

To study productively, it is important to identify the problems and concepts you do not know well. Instead of repeatedly reviewing what you already know well, focus your energy on improving your comprehension and ability to complete the more challenging problems.

I have identified what I do not know well enough for the test at least a week before the test. Yes _____ No _____

Memory Techniques

Memory techniques can be used for hard-to-remember rules, concepts, or problems. Memory techniques, which include repeating concepts or problems out loud, making associations, using mnemonic devices, and using acronyms, will improve your test-taking skills.

I have developed memory techniques for my test. Yes _____ No _____

Reviewing Concept Errors

One of the six types of test-taking errors, which are discussed later in this chapter, is concept error. Concept errors from previous tests need to be reviewed in order not to make the same mistakes again. Reviewing your concept errors can improve your understanding of mathematics.

Did you review your concept errors from previous tests? *Yes ____ No ____*
Did not record concept errors ____

Using Practice Tests

Practice tests help you find out what you don't know before the test instead of during the test. The best way to develop a practice test is with other students, because you can exchange questions. Developing practice tests can improve your ability to solve math problems on the real test.

I worked with other students to develop and take a practice test at lease 2 days before the test. *Yes ____ No ____*
Did not develop practice tests ____

Taking instructor-made practice tests can also help you to find out what problems you need to work on before, instead of during, the test. Using practice tests can help you find out what you need to study before the test.

I have worked the problems on the instructor's practice test. *Yes ____ No ____*
There was no instructor practice test ____

Memory Data Dump

The information from a memory data dump can be used to help you during the test. Using a memory data dump on your test can improve your test scores. You can also make up memory techniques based on the problems that you missed on the practice tests.

I have developed memory devices to use during a data dump. *Yes ____ No ____*

Test-Taking Strategies

Some students use the "Ten Steps to Better Test-Taking" as a test-taking guide while others develop their own test-taking strategies. Either technique can improve your test-taking strategy.

I have developed and practiced my test-taking strategy. *Yes ____ No ____*

Are you wondering why we are asking you to complete the preceding items 2 days or more before the test instead of the night before the test? If you get stuck on a problem, find you don't understand a concept, or cannot develop one of the strategies, then you will still have at least 1 day to get help before the test. Remember, the night before the test you need to be reviewing material you already have learned—not learning new material. The most productive way to complete this checklist is to include these steps in your study time every day.

7.2 Ten Steps to Better Test Taking

You need to have a game plan to take a math test. This plan is different from plans for taking history, English, humanities, and some science tests. The game plan is to get the most points in the least amount of time. Many students lose test points because they use the wrong test-taking strategies for math. By following these ten steps you can demonstrate more knowledge on the test and get more problems right.

Step 1 *Use a memory data dump.* When you get your test, turn it over and write down the information that you might forget. Remember, this is your mental cheat sheet that you should have already made while preparing for the test. Your mental cheat sheet has now turned into a mental list, and writing down this information is not cheating. Do not put your name on the test, do not skim it, just turn it over and write down those facts, figures, and formulas from your mental cheat sheet or other information you might not remember during the test. This is called your *first memory data dump.* The data dump provides memory cues for test questions.

EXAMPLE

It might take you a while to remember how to do a coin problem. However, if you had immediately turned your test over and written down different ways of solving coin problems, it would be easier to solve the coin problem.

Step 2 *Preview the test.* Put your name on the test and start previewing. Previewing the test requires you to look through the entire test to find different types of problems and their point values. Put a mark by each question that you can do without thinking. These are the questions that you will solve first.

Step 3 *Do a second memory data dump.* The second data dump is for writing down material that was jarred from your memory while previewing the test. Write this information on the back of the test.

Step 4 *Develop a test progress schedule.* When you begin setting up a test schedule, determine the point value for each question. Some test questions might be worth more points than others.

In some tests, word problems are worth 5 points and other questions might be worth 2 or 3 points. You must decide the best way to get the most points in the least amount of time. This might mean working the questions worth 2 or 3 points first and leaving the more difficult word problems for last.

Decide how many problems should be completed halfway through the test. You should have more than half the problems completed by that time.

Step 5 *Answer the easiest problems first.* Solve, in order, the problems you marked while previewing the test. Then review the answers to see if they make sense. Start working through the test as fast as you can while being accurate. Answers should be reasonable.

EXAMPLE

The answer to a problem asking you to find the area of a rectangle cannot be negative, and the answer to a land-rate-distance problem cannot be 1000 miles per hour.

Clearly write down each step to get partial credit. Even if you wind up with an incorrect answer, you might get some credit. In most math tests, the easier problems are near the beginning of the first page; you need to answer them efficiently and quickly. This will give you both more time for the harder problems and time to review.

Step 6 *Skip difficult problems.* If you find a problem that you do not know how to work, read it twice and automatically skip it. Reading it twice will help you understand the problem and put it into your working memory. While you are solving other problems, your mind will still be working on that problem. A difficult problem could be a type of problem you have never seen before or a problem in which you get stuck on the second or third step. In either case, skip the problem and go on to the next one.

Step 7 *Review the skipped problems.* When working the skipped problems, think how you have solved other similar problems. Also try to remember how the instructor solved that type of problem on the board.

While reviewing skipped problems, or at any other time, you may have the "Aha!" response. The "Aha!" response is your remembering how to do a skipped problem. Do not wait to finish your current problem. Go to the problem on which you had the "Aha!" and finish that problem. If you wait to finish your current problem, your "Aha!" response could turn into an "Oh, no!" response.

Step 8 *Guess at the remaining problems.* Do as much work as you can on each problem, even if it is just writing down the first step. If you cannot write down the first step, rewrite the problem. Sometimes rewriting the problem can jar your memory enough to do the first step or the entire problem. This is particularly true for tactile/concrete learners, for whom writing can trigger the memory process of solving the problem. Remember, you did not learn how to solve math problems with your hands tied behind your back. If you leave the problem blank, you will get a zero. Do not waste too much time on guessing or trying to work the problems you cannot do.

Step 9 *Review the test.* Look for careless errors or other errors you may have made. Students usually lose 2 to 5 test points on errors that could have been caught in review. Do not talk yourself out of an answer just because it may not look right. This often happens when an answer does not come out even. It is possible for the answer to be a fraction or a decimal.

REMEMBER
Research reveals that the odds of changing a right answer to a wrong answer are greater than the odds of changing a wrong answer to a right one.

Step 10 *Use all the allowed test time.* Review each problem by substituting the answer back into the equation or doing the opposite function required to answer the question. If you cannot check the problem in either of these two ways, rework the problem on a separate sheet of paper and compare the answers. Do not leave the test room until you have reviewed each problem two times or until the bell rings.

Even though we encourage students to work until the end of the test period, most students leave the classroom before the end of the period. These students state that even though they know they should use all the test time, they cannot stay in the room until the end of the test time. These students also know that their grades would probably improve if they kept checking their answers or kept working the problems they were having difficulty with. After talking to hundreds of these students, I discovered two different themes for leaving the classroom early. First, test anxiety gets so overwhelming that they cannot stay in the room. The relief from the test anxiety (leaving the room) is worth more than getting a better grade. If you are one of these students, you must learn how to decrease your test anxiety by following the suggestions in Chapter 3 or by using the *How to Reduce Test Anxiety* CD. Don't let test anxiety control your grades!

The other reason for leaving the test early is that they do not want to be the last student or one of the last few students to turn in their tests. They still believe that students who turn their tests in last are "dumb and stupid." These students also believe that students who turn their tests in first make As and Bs and students who turn their tests in last make Ds and Fs. If you are one of these students, you don't need to care about what other students think about you (it's usually wrong anyway). YOU need to fight the urge to leave early and use all the test time. Remember, passing mathematics is the best way to get a high-paying job and support yourself or your family.

REMEMBER
There is no prize for handing your test in first, and some students who turn their papers in last make As.

Handing in your scratch paper with your math test has several advantages:

- If you miscopied an answer from the scratch paper, you will probably get credit for the problem.
- If you get the answer incorrect due to a careless error, your work on the scratch paper could give you a few points.
- If you do get the problem wrong, it will be easier to locate the errors when the instructor reviews the test. This will prevent you from making the same mistakes on the next math test.

REMEMBER
Handing in your scratch paper may get you extra points or improve your next test score.

SECTION 7.2 REVIEW

1. List and explain the ten steps to better test taking.

Step 1: ______________________________

____

Step 2: ______________________________

____

Step 3: ______________________________

____

Step 4: ______________________________

____

Step 5: ______________________________

____

Step 6: ______________________________

____

Step 7: ______________________________

____

Step 8: ______________________________

____

Step 9: ______________________________

____

Step 10: ______________________________

____

2. After trying the ten steps to better test taking, develop your own personalized test-taking steps.

   Step 1: ____________________

   Step 2: ____________________

   Step 3: ____________________

   Step 4: ____________________

   Step 5: ____________________

   Step 6: ____________________

   Step 7: ____________________

   Step 8: ____________________

   Step 9: ____________________

   Step 10: ____________________

   Additional steps: ____________________

   ____

   ____

   ____

   ____

   ____

7.3 Six Types of Test-Taking Errors

To improve future test scores, you must conduct a test analysis of previous tests. In analyzing your tests, you should look for the following kinds of errors:

1. Misread-directions errors
2. Careless errors
3. Concept errors
4. Application errors
5. Test-taking errors
6. Study errors

Students who conduct math test analyses will improve their total test scores.

Misread-directions errors occur when you skip directions or misunderstand directions but do the problem anyway.

EXAMPLES

Suppose you have this type of problem to solve:

$$(x + 1)(x + 1)$$

Some students will try to solve for x, but the problem calls only for multiplication. You would solve for x only if you have an equation such as $(x + 1)(x + 1) = 0$.

Another common mistake is not reading the directions before doing several word problems or statistical problems. All too often, when a test is returned, you find only three out of the five problems had to be completed. Even if you did get all five of them correct, it cost you valuable time that could have been used obtaining additional test points.

To avoid misread-directions errors, carefully read and interpret all the directions. Look for anything that is unusual, or notice if the directions have two parts. If you do not understand the directions, ask the instructor for clarification. If you feel uneasy about asking the instructor for interpretation of the question, remember the instructor in most cases does not want to test you on your interpretation of the question but how you answer it. Also, you don't want to make the mistake of assuming that the instructor will not interpret the question. Let the instructor make the decision to interpret the question, not you.

Careless errors are mistakes that you can catch automatically upon reviewing the test. Both good and poor math students make careless errors. Such errors can cost a student a higher letter grade on a test.

EXAMPLES

Dropping the sign: $-3(2x) = 6x$ instead of $-6x$, which is the correct answer.

Not simplifying your answer: Leaving $(3x - 12)/3$ as your answer instead of simplifying it to $x - 4$.

Adding fractions: $1/2 + 1/3 = 2/5$ instead of $5/6$, which is the correct answer.

Word problems: $x = 15$ instead of "The student had 15 tickets."

However, many students want all their errors to be careless errors. This means that the students did know the math but simply made silly mistakes. In such cases, I ask the student to solve the problem immediately, while I watch.

If the student can solve the problem or point out his or her mistake in a few seconds, it is a careless error. If the student cannot find an error immediately, it is not a careless error and is probably a concept error.

When working with students who make careless errors, I ask them two questions. First, I ask, "How many points did you lose due to careless errors?" Then I follow with, "How much time was left in the class period when you handed in your test?" Students who lose test points due to careless errors are giving away points if they hand in their test papers before the test period ends.

To reduce careless errors, you must realize the types of careless errors that students make and recognize them when reviewing your test. If you cannot solve a missed problem immediately, it is not a careless error. If your major error is not simplifying answers, review each answer as if it were a new problem and try to reduce it.

Concept errors are mistakes made when you do not understand the properties or principles required to work math problems. Concept errors, if not corrected, will follow you from test to test, causing a loss of test points.

EXAMPLES

Some common concept errors are not knowing

- $(-)(-)x = x$, *not* $-x$
- $1(2) > x(-1)$ gives $2 < x$, *not* $2 > x$
- 5/0 is undefined, *not* 0
- $(a + x)/x$ is *not* reduced to a
- the order of operations

Concept errors must be corrected to improve your next math test score. Students who make numerous concept errors will fail the next test and the course if concepts are not understood. Just going back to rework a concept error problem is not good enough. You must go back to your textbook or notes and learn why you missed that type of problem, not just the one problem itself.

The best way to learn how to work the types of problems on which you tend to make concept errors is to set up a concept error problem page in the back of your notebook. Label the first page "Test One Concept Errors." Write down all your concept errors and how to solve the problems. Then, work five more problems that involve the same concept. Now, in your own words, write the concepts that you are using to solve these problems.

If you cannot write a concept in your own words, you do not understand it. Get assistance from your instructor if you need help finding similar problems using the same concept or if you cannot understand the concept. Do this for every test.

Application errors occur when you know a concept but cannot apply it to a problem. Application errors usually are found in word problems, deducing formulas (such as the quadratic formula), and graphing. Even some better students become frustrated with application errors; they understand the material but cannot apply it to problems.

To reduce application errors, you must predict the types of application problems that will be on the test. You must then think through and practice solving those types of problems using the appropriate concepts.

EXAMPLE

If you must derive the quadratic formula, you should practice doing it backward and forward while telling yourself the concept used to move from one step to the next.

Application errors are common with word problems. After completing a word problem, reread the question to make sure you have properly applied the answer to the intended question. Application errors can be avoided by appropriate practice and insight.

Test-taking errors apply to the specific way you take tests. Some students consistently make the same types of test-taking errors. Through recognition, these bad test-taking habits can be replaced by good test-taking habits. The result will be higher test scores. The list that follows includes the test-taking errors that can cause you to lose many points on an exam:

1. Missing more questions in the first third, second third, or last third of a test is considered a test-taking error.

   Missing more questions in the first third of a test can be due to carelessness when doing easy problems or due to test anxiety.

   Missing questions in the last third of the test can result from the fact that the last problems are more difficult than the earlier questions or from increasing your test speed to finish the test.

   If you consistently miss more questions in a certain part of the test, use your remaining test time to review that section of the test first. This means you may review the last part of your test first.
2. *Not completing a problem to its last step* is another test-taking error. If you have this bad habit, review the last step of each test problem first, before doing an in-depth test review.
3. *Changing test answers from correct ones to incorrect ones* is a problem for some students. Find out if you are a good or bad answer changer by comparing the numbers of answers you change to correct and to incorrect answers. If you are a bad answer changer, write on your test, "Don't change answers." Change answers only if you can prove to yourself or the instructor that the changed answer is correct.
4. *Getting stuck on one problem and spending too much time on it* is another test-taking error. You need to set a time limit on each problem before moving to the next problem. Working too long on a problem without success will increase your test anxiety and waste valuable time that could be used in solving other problems or in reviewing your test.
5. *Rushing through the easiest part of the test and making careless errors* is a common test-taking error for the better student. If you have the bad habit of getting more points taken off for the easy problems than for the hard problems, review the easy problems first, and the hard problems later.
6. *Miscopying an answer from your scratch paper to the test* is an uncommon test-taking error, but it does cost some students points. To avoid this kind of error, systematically compare your last problem step on scratch paper with the answer written on the test. In addition, always hand in your scratch paper with your test.
7. *Leaving answers blank* will get you zero points. If you look at a problem and cannot figure out how to solve it, do not leave it blank. Write down some information about the problem, rewrite the problem, or try to do at least the first step.

REMEMBER
Writing down the first step of a problem is the key to solving the problem and obtaining partial credit.

8. *Solving only the first step of a two-step problem* causes some students to lose points. These students get so excited when answering the first step of the problem that they forget about the second step. This is especially true on two-step word problems. To correct this test-taking error, write "two" in the margin of the problem. That will remind you that there are two steps or two answers to this problem.
9. *Not understanding all the functions of your calculator* can cause major testing problems. Some students only barely learn how to use the critical calculator functions. They then forget or have to relearn how to use their calculators, which costs test points and test time. Do not wait to learn how to use your calculator during the test. Overlearn the use of your calculator before the test.
10. *Leaving the test early without checking all your answers* is a costly habit. Do not worry about the first person who finishes the test and leaves. Many students start to get nervous when students start to leave after finishing the test. This can lead to test anxiety, mental blocks, and loss of recall.

According to research, the first students finishing the test do not always get the best grades. It sometimes is the exact opposite. Ignore the exiting students, and always use the full time allowed.

Make sure you follow the ten steps to better test taking. Review your test-taking procedures for discrepancies in following the ten steps. Deviating from these proven ten steps will cost you points.

Study errors, the last type of mistake to look for in test analysis, occur when you study the wrong type of material or do not spend enough time on pertinent material. Review your test to find out if you missed problems because you did not practice those types of problems or because you did practice them but forgot how to do them during the test. Study errors will take some time to track down, but correcting study errors will help you on future tests.

If you are like most students, after analyzing one or several tests, you will recognize at least one major, common test-taking error. Understanding the effects of this test-taking error should change your study techniques or test-taking strategy.

EXAMPLE

If there are 7 minutes left in the test, should you review for careless errors or try to answer those two problems you could not totally solve? This is a trick question. The real question is, do you miss more points due to careless errors or concept errors, or are the missed points about even? The answer to this question should determine how you will spend the final minutes of the test. If you miss more points due to careless errors or miss about the same number of points due to careless errors and concept errors, review for careless errors.

Careless errors are easier to correct than concept errors. However, if you make very few or no careless errors, you should be working on those last two problems to get the greatest number of test points. Knowing your test-taking errors can add more points to your test by changing your test-taking procedure.

Now that the Six Types of Test-Taking Errors have been reviewed a real life example may be helpful in applying this analysis to tests. Question four in the Section Review pertains to reviewing Figure 17 and recording the error types, points lost for each error, and examples of errors. Since this is a real test not all the test-taking errors are used and some errors are used twice. The answer key is at the end of the Section Review.

FIGURE 17 Math Test for Prealgebra

The answers are in boldface. The correct answers to missed questions are shaded. Identify the type of error based on the Six Types of Test-Taking Errors. The student's test score is 70.

1. Write in words: 32.685

   Thirty-two and six hundred eighty-five thousandths

2. Write as a fraction and simplify: 0.078

   $\frac{\mathbf{78}}{\mathbf{1000}}$ $\frac{\mathbf{39}}{\mathbf{500}}$ (-2)

3. Round to the nearest hundredth: 64.8653

   64.865 **64.87** (-4)

4. Combine like terms: $6.78x - 3.21 + 7.23x - 6.19$

   $\mathbf{= 6.78x + 7.23x + (-3.21) + (-6.19)}$

   $\mathbf{= 14.01x - 9.4}$

5. Divide and round to the nearest hundredth: $68.1357 \div 2.1$

   32.4454 → 32.45

6. Write as a decimal: $\frac{5}{16}$

   0.3125

7. Insert < or > to make a true statement:

   $\frac{3}{8} < \frac{6}{13}$

8. Solve: $\frac{3}{x} = \frac{9}{12}$

   $9x = 3(12)$

   $\frac{x}{9} = \frac{36}{9}$

   $x = 5$ **x = 4** (-2)

9. What number is 35% of 60?

   2100 **21.00** (-4)

10. 20.8 is 40% of what number?

    52 (-8)

11. 567 is what percent of 756?

    $\frac{756}{567} = \frac{9}{100}$

    = 133.3% **75%** (-8)

12. Multiply:

    $(-6.03)\ (-2.31) = 13.9$ **13.9293** (-2)

Answer Key for Prealgebra Test

1. Correct
2. *Misread-directions error*—Forgot to simplify by reducing the fraction.
3. *Concept error*—Did not know that hundredths is two places to the right of the decimal.
4. Correct
5. Correct
6. Correct
7. Correct
8. *Careless error*—Divided incorrectly in the last step.
9. *Test-taking error*—Did not follow Step 5 in test taking steps: reviewing answers to see if they make sense. The number that equals 35% of 60 can't be larger than 60.
10. *Test-taking error*—Did not follow Steps 7 and 8 in test-taking steps: don't leave an answer blank.
11. *Application error*—Solved the equation correctly but the equation setup was wrong.
12. *Concept error*—Did not know that when you multiply with one number in the hundredths, the answer must include the hundredths column.

SECTION 7.3 REVIEW

1. List the six types of test-taking errors.

   Type 1: ____________________

   Type 2: ____________________

   Type 3: ____________________

   Type 4: ____________________

   Type 5: ____________________

   Type 6: ____________________

2. Which of these types of errors have you made in the past?

   ____

   ____

3. How can you avoid these types of errors on the next test?

   ____

   ____

4. Review the Math Test for Prealgebra (Figure 17) if you are in a prealgebra course. For each error, list the type of error, the points lost for the error, and a different example of that type of error.

   Error Type: __________ Points Lost for Error: __________

   Example of Error: ____________________

   Error Type: __________ Points Lost for Error: __________

   Example of Error: ____________________

   Error Type: __________ Points Lost for Error: __________

   Example of Error: ____________________

   Error Type: __________ Points Lost for Error: __________

   Example of Error: ____________________

   Error Type: __________ Points Lost for Error: __________

   Example of Error: ____________________

   Error Type: __________ Points Lost for Error: __________

   Example of Error: ____________________

   Error Type: __________ Points Lost for Error: __________

   Example of Error: ____________________

5. Review your last test and list the types of errors you made, the points lost for each error, and examples of the errors. You may have more errors than allowed for in the list below. Use a separate sheet of paper if you need to list more. (Remember: careless errors are problems you can solve immediately.)

   Error Type: ____________________ Points Lost for Error: ____________________

   Problem with the Error in It: ____

   Error Type: ____________________ Points Lost for Error: ____________________

   Problem with the Error in It: ____

   Error Type: ____________________ Points Lost for Error: ____________________

   Problem with the Error in It: ____

   Error Type: ____________________ Points Lost for Error: ____________________

   Problem with the Error in It: ____

   Using the pages in the back of your math notebook, rework the problems on which you made concept errors and do five math problems just like each one. For each concept error, write why you could not solve the problem on the test and what new information you learned that allowed you to solve the problem. Do this for every major test. Give an example of one of your concept errors and the reasons why you can now solve the problem.

   Concept Error: ____

   Reasons Why I Can Now Solve the Problem: ____

   ____

   ____

7.4 How to Prepare for the Final Exam

The first day of class is when you start preparing for the final exam. Look at the syllabus or ask the instructor if the final exam is cumulative. A cumulative exam covers everything from the first chapter to the last chapter. Most math final exams are cumulative.

The second question you should ask is if the final exam is a departmental exam or if it is made up by your instructor. In most cases, departmental exams are more difficult and need a little different preparation. If you have a departmental final, you need to ask for last year's test and ask other students what their instructors say will be on the test.

The third question is how much the final exam will count. Does it carry the same weight as a regular test or, as in some cases, will it count as a third of your grade? If the latter is true, the final exam will usually make a letter grade difference in your final grade. The final exam could also determine if you pass or fail the course. Knowing this information before the final exam will help you prepare for it.

Preparing for the final exam is similar to preparing for each chapter test. You must create a pretest to discover what you have forgotten. You can use questions from the textbook chapter tests or questions from your study group.

Review the concept errors that you recorded in the back of your notebook labeled "Test 1," "Test 2," and so on. Review your problem log for questions you consistently miss. Even review material that you knew for the first and second tests but that was not used on any other tests. Students forget how to work some of these problems.

If you do not have a concept-error page for each chapter and did not keep a problem log, you need to develop a pretest before taking the final exam. If you are preparing for the final individually, then copy each chapter test and do every fourth problem to see what errors you may make. However, it is better to have a study group where each of the four members brings in ten problems with the answers worked out on a separate page. Then each group member can take a 30-question test made up from the problems brought in by the other group members to find out what they need to study. You can refer to the answers to help you solve the problems that you miss. Remember, that you want to find out what you don't know before the test, not during the test. Developing and using these methods before each test will improve your grades.

Make sure to use the ten steps to better test taking and the information gained from your test analysis. Use all the time on the final exam, because you could improve your final grade by a full letter if you make an A on the final.

SECTION 7.4 REVIEW

What are the three questions you need to ask about your final exam?

Question 1: ______________________________

Question 2: ______________________________

Question 3: ______________________________

7.5 Motivating Yourself to Learn Math

This text has taught you the skills necessary to improve your math learning and success. Now is the time to apply these new math study skills and test-taking behaviors and to eliminate the old behaviors that may have caused previous math failure. To accomplish this task you need to motivate yourself to use these skills, which can involve self-esteem, locus of control, and self-efficacy. No longer can you use poor math study skills and test-taking skills as an excuse for not being successful in math. Now is the time to take more responsibility for your success. Thousands of students have taken up this responsibility, passed math and graduated. In the process, they have also changed their lives.

First, let's take a look at self-esteem. Self-esteem is the part of the personality that allows us to feel good about ourselves. Many students have positive self-esteem in their personal lives but do not in regards to their abilities to succeed in math or college. Other students have positive self-esteem in college but not in their personal lives. Having positive self-esteem means that you can accept yourself and then develop a "can-do" attitude that is geared toward accomplishing your goals. Students with poor self-esteem may not put forth as much effort in accomplishing their goals. They give up easily. The first step in recovering from a low self-esteem is to replace negative thoughts with positive thoughts—thoughts that express a willingness to succeed in life instead of worrying about failure. This can be accomplished by setting short-term goals and taking pride in any subsequent successes.

The next area of taking responsibility involves the locus of control. A locus of control is the degree to which a student claims control over his or her own life. Do students feel that they are in control of their own lives, or is it their environment that dictates their choices and actions? Students who feel that they are not in control of their lives are considered "external" students, while students who feel they are in control of their lives are referred to as "internal." Internal students accept the responsibility of their behaviors and adjust their behaviors in order to become more successful. For example, the internal person does not blame instructors for his/her poor grades. This student also realizes that he or she can change his/her study behaviors instead of simply complaining about things that are out of his or her control. Are you internal or external?

Finally, there is self-efficacy. Self-efficacy is a new concept that is related to locus of control but is more specific to the belief that you can accomplish a specific goal. You need to ask yourself if you have the skills to be successful at _______. You can fill in the blank with tasks such as running a marathon, getting a better job, passing English or passing math. You may have self-efficacy in one or two of these areas but not all of them. Self-efficacy is an important success predictor as to how likely you are to succeed in math or any other of these other areas. In other words, students who believe they can be successful in math have a higher chance of passing a math course. With these new math study skills and test-taking skills you've developed through this workbook, your self-efficacy should have improved in the area of succeeding in math. This means that you should become more persistent and that if certain learning and test-taking behaviors do not work out, you should change them and try different techniques from this text. Having self-efficacy in math means that you will keep trying to succeed even if it means getting help from others.

Motivation is internal but you may need some help in getting it out. If you need extra help in improving motivation, talk to your instructor or contact your advisor or counselor to help you start the process. This does not mean that you have to like math, you just have to pass it. Now is the time to tell yourself that **I CAN BECOME SUCCESSFUL IN MATH.**

SECTION 7.5 REVIEW

1. Define self-esteem and explain how it affects math learning.

____

____

____

2. Define locus of control and explain how it affects math learning.

____

____

____

3. Define self-efficacy and how it affects math learning.

____

____

____

4. Who should you contact to help you improve your motivation to learn math?

____

____

____

Name: ______________________ Date: ____________

CHAPTER 7 REVIEW

1. You should start studying for a test at least __________ days ahead of time.
2. You should be reviewing only the material you have __________ the night before the test.
3. When previewing a test, put a __________ by each question you can solve without thinking, and do those problems __________.
4. When taking a test, you must decide the best way to get the __________ points in the __________ amount of time.
5. Halfway through the test, you should have __________ than half of the problems completed.
6. If you cannot write down the first step of a problem, you should __________ the problem to help you remember how to solve it.
7. Do not leave the test room until you have reviewed each problem __________ times or until the bell rings.
8. In reviewing your test, if you cannot find an error immediately (within a few seconds), it is not a __________ error and is probably a __________ error.
9. Students who make numerous __________ errors will fail the next test and the course if the __________ are not understood.
10. The __________ day of class is when you start preparing for the final exam.

What is the most important information you learned from this chapter?

____

____

____

How can you immediately use it?

____

____

____

DAN'S TAKE REVIEW

1. Why did Dan feel the need to rush through the test?

____

____

____

Name: ______________________________ Date: ______________

2. What often resulted from his fear of being the last to complete his exam?

____

____

____

3. How did he eventually overcome this problem?

____

____

____

WebAssign®

- When you click on some WebAssign chemistry or math questions an input palette will open. These palettes, called chemPad and mathPad, will help you enter your answer in proper notation.

- Some questions may require the use of an Active Figure simulation. Active Figures require the free Macromedia Flash Player plug-in, downloadable from www.macromedia.com.

- If your instructor allows it, you can save your work without grading by selecting the Save Work button at the end of the question. After you save your work, it will be available to you the next time you click the assignment.

- Please note that WebAssign will **not** automatically submit your answers for scoring if you only **Save** your work. Your teacher will not be able to see your submissions. Please be sure to **Submit** prior to the due date and time.

- If your instructor allows it, you can submit answers by question part or for the entire assignment. To submit an individual question answer for grading, click the **Submit New Answers to Question __** button at the bottom of each question. To submit the entire assignment for grading, click the **Submit All New Answers** button at the end of the assignment.

Technical Support

If you are having difficulty logging in, please be sure to check with your teacher and verify whether an account has been created for you or whether you need to self-enroll. In either case your teacher needs to provide the appropriate information (username, institution code and password OR Class Key).

To email WebAssign Support go to http://www.webassign.net/info/support/report.html. This page also lists answers to **Common Problems**, and provides links to the **Student Guide**.

August 7, 2008

WebAssign

WebAssign Student Quick Start Guide

You can use WebAssign to access your homework, quizzes, and tests — whatever your instructor chooses — at any time of day or night, from any computer with a connection to the Internet and a Web browser. Your instructor creates your assignments, schedules them, and decides how many submissions you get. Your instructor also determines if you can have an extension, if you can save your work without submitting it at the time, and how much feedback you get after you submit an assignment.

The WebAssign support staff cannot change your username or password, give extensions, change your score, give you extra submissions, or help you with the content of your assignments.

Logging In

You can log in to WebAssign using any Web browser connected to the Internet. There are two different ways to log in to WebAssign. Each requires information from your teacher. If you are unsure about how to log in, please check with your teacher or another student in your class.

Go to the login page at http://webassign.net/login.html or the web address provided by your teacher. The way you log in depends on how your instructor set up the class:

- If your teacher created a WebAssign account for you, they will provide you with a **Username**, an **Institution** code and a **Password**. Simply enter this information in the boxes provided and click the **Log In** button.

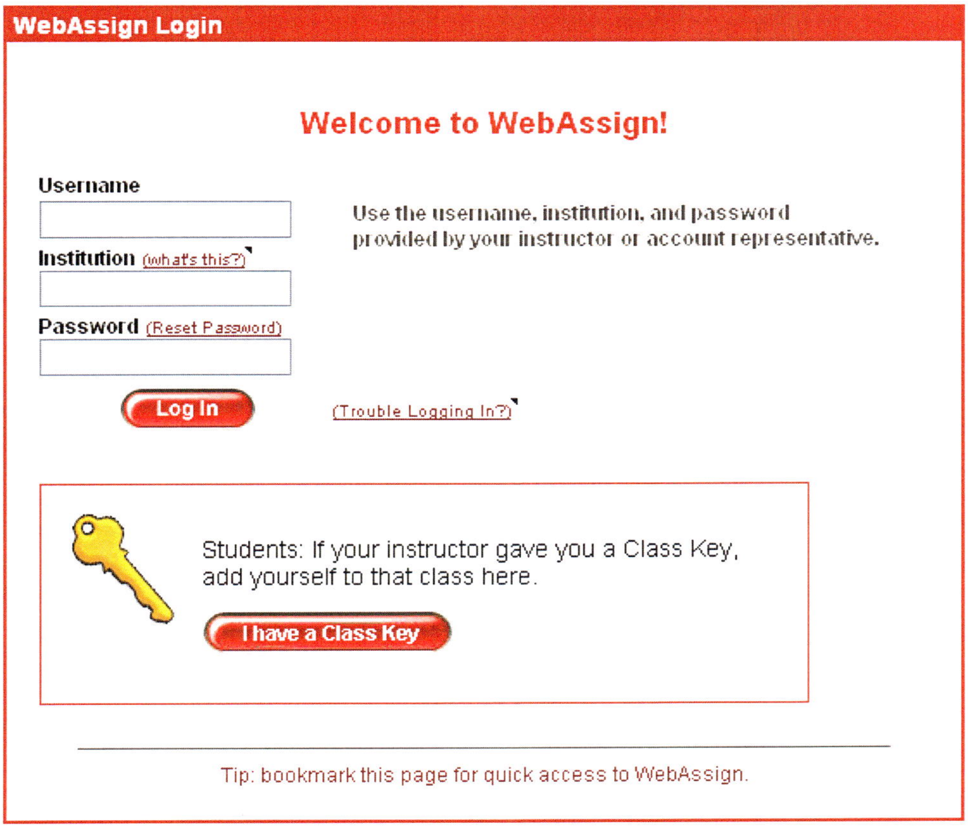

WebAssign

- If your teacher wants you to **Self-Enroll** in the WebAssign course they will provide you with a **Class Key**. You will create your own username and password. It is important that you remember this information so you can log in for the remainder of the class. In this case, just click the **I have a Class Key** button. You don't need to enter any other information on this page.

  Then, enter the **Class Key** your instructor provided and click **Submit.** Verify you are enrolling in the correct class on the next page.

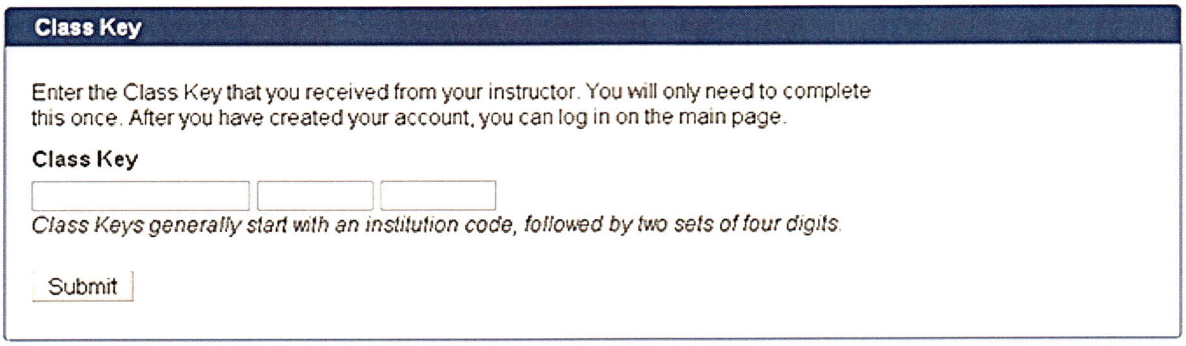

- Enter your preferred Login and Student information.
- Click the **Create My Account** button to complete the enrollment process.
- A review screen will display, showing your username, institution code, and password. **Retain a copy of this information.** You will need it to log into WebAssign.

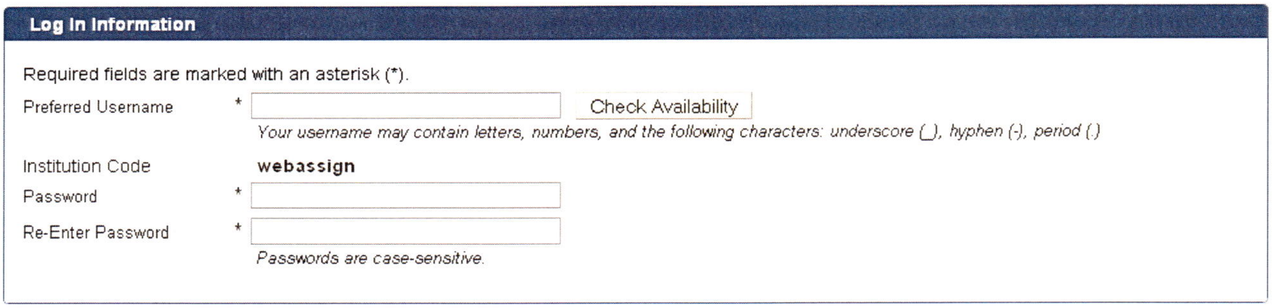

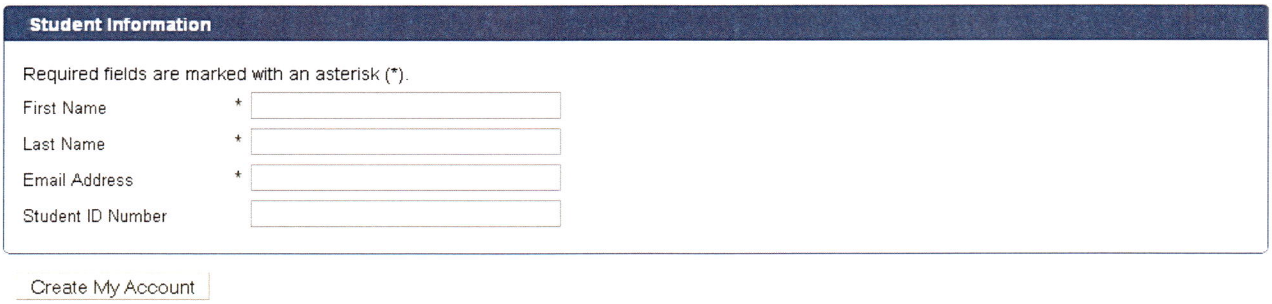

WebAssign

Access Codes

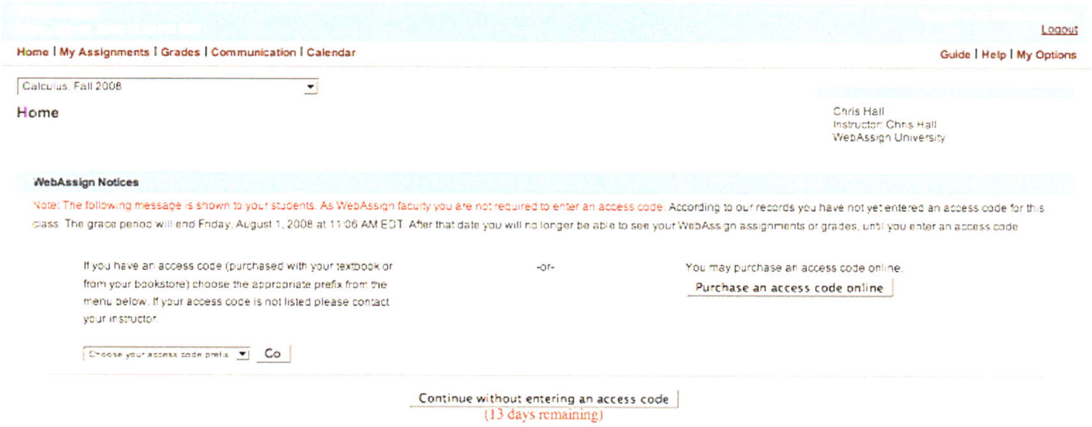

Once you log in, you may see a WebAssign Notice about entering an access code for your class. You can get an Access Code from any of the following places if you need to use one:

- A new textbook you purchased for the class.
- Your bookstore, which may sell Access Code cards.
- Online, where you can purchase an access code with a credit card.

You have a 14 day grace period to use WebAssign, starting with the WebAssign class start date. During this time you can work on and view your WebAssign assignments without registering a code.

After the grace period is over you will only see the code registration message until you submit or purchase a code.

There are two types of WebAssign access code cards. The small card requires you to scratch off the silver surface in order to reveal the complete access code.

WebAssign

The larger security envelope card requires you to open the card to reveal the access code number.

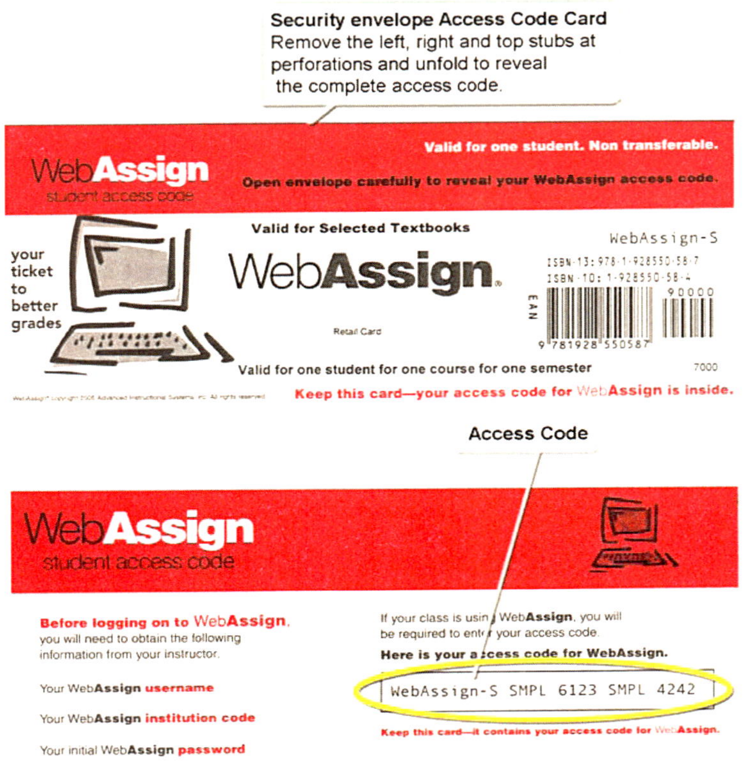

If you would like to purchase an access code directly from WebAssign online, you may do so with a credit card. Your code will be automatically registered to your WebAssign account as soon as the transaction is complete. You will receive an email confirmation. Please keep a copy for your records.

Your WebAssign Home Page

Once you have successfully logged in you will see your WebAssign homepage. If you are taking more than one WebAssign class, you will need to select which class you wish to view first.

The upper right corner features links to a complete Student **Guide**, as well as a link to WebAssign Technical Support under **Help**. If you want to change your password or add or update your email address, simply click **My Options** in the upper right hand corner.

You will see your assignments and due dates listed, as well as any Communications, Grades, and Announcements posted by your teacher.

WebAssign

Answering Questions

WebAssign has a variety of different question types, ranging from multiple choice to fill-in-the-blank to symbolic questions. Here are some things to keep in mind as you work through your assignments:

- Some questions may include numbers or words that appear in red. This signifies that the number or word has been randomized, so that you receive a different version of the same basic question from your classmates.

- Some WebAssign questions check the number of significant figures in your answer. If you enter the correct value with the wrong number of significant figures, you will not receive credit, but you will receive a hint that your number does not have the correct number of significant figures.

- Some questions require entering symbolic notation. Answer symbolic questions by using calculator notation. You must use the exact variables specified in the questions. The order is not important as long as it is mathematically correct. Clicking on the eye button previews the expression you enter in proper mathematical notation. Clicking on the symbolic formatting help button provides tips for using the correct keystrokes.

CALCULATOR INDEX